Public Transportation 2018: Parts 1-6

Transportation Research Record: Journal of the Transportation Research Board Volume 2672, Issue 8

ISBN: 978-1-5108-8708-4

Printed from e-media with permission by:

Curran Associates, Inc.
57 Morehouse Lane
Red Hook, NY 12571

Some format issues inherent in the e-media version may also appear in this print version.

Published by Sage Publications USA

ISBN (Print) 978-1-5108-8708-4
ISBN 2018 Printed Set (All Issues) 978-1-5108-7735-1

Printed by Curran Associates, Inc. (2019)

Additional copies of this publication are available from:

Curran Associates, Inc.
57 Morehouse Lane
Red Hook, NY 12571 USA
Phone: 845-758-0400
Fax: 845-758-2633
Email: curran@proceedings.com
Web: www.proceedings.com

Volume 2672 Number 8 December 2018

TRR
JOURNAL OF THE TRANSPORTATION RESEARCH BOARD

Contents

Public Transportation 2018: Parts 1–6

Part 1: Urban and Rural Bus Services

Articles

Part 3: Management and Performance, Quality of Service, and Marketing Fare Policy

Articles

2018 WILLIAM W. MILLAR AWARD: Outstanding Paper in Public Transportation

Article

2018 FRED BURGGRAF AWARD: Public Transportation

Articles

Part 4: Paratransit, Accessibility, and Mobility

Articles

Part 5: Planning & Parking

Articles

Part 6: Transportation Demand Management & the Sharing Economy

Articles

---

* Membership as of February 2019.

TRB

TRANSPORTATION RESEARCH BOARD

CORE SPONSORS

Transportation Departments of the 50 States and the District of Columbia

Federal Government

U.S. Department of Transportation
- Federal Aviation Administration (FAA)
- Federal Highway Administration
- Federal Motor Carrier Safety Administration (FMCSA)
- Federal Railroad Administration (FRA)
- Federal Transit Administration
- National Highway Traffic Safety Administration (NHTSA)
- Office of the Assistant Secretary of Transportation for Research and Technology (OST-R)

U.S. Air Force Civil Engineer Center (AFCEC)
U.S. Coast Guard (USCG)
U.S. Department of Energy
U.S. Department of the Interior (DOI)
U.S. Environmental Protection Agency (EPA)

Nongovernmental Organizations

American Public Transportation Association (APTA)
Association of American Railroads (AAR)
California Air Resources Board

The National Academies of

SCIENCES · ENGINEERING · MEDICINE

The **National Academy of Sciences** was established in 1863 by an Act of Congress, signed by President Lincoln, as a private, nongovernmental institution to advise the nation on issues related to science and technology. Members are elected by their peers for outstanding contributions to research. Dr. Marcia McNutt is president.

The **National Academy of Engineering** was established in 1964 under the charter of the National Academy of Sciences to bring the practices of engineering to advising the nation. Members are elected by their peers for extraordinary contributions to engineering. Dr. C. D. Mote, Jr., is president.

The **National Academy of Medicine** (formerly the Institute of Medicine) was established in 1970 under the charter of the National Academy of Sciences to advise the nation on medical and health issues. Members are elected by their peers for distinguished contributions to medicine and health. Dr. Victor J. Dzau is president.

The three Academies work together as the National Academies of Sciences, Engineering, and Medicine to provide independent, objective analysis and advice to the nation and conduct other activities to solve complex problems and inform public policy decisions. The Academies also encourage education and research, recognize outstanding contributions to knowledge, and increase public understanding in matters of science, engineering, and medicine.

Learn more about the National Academies of Sciences, Engineering, and Medicine at **www.national-academies.org.**

The **Transportation Research Board** is one of seven major programs of the National Academies of Sciences, Engineering, and Medicine. The mission of the Transportation Research Board is to increase the benefits that transportation contributes to society by providing leadership in transportation innovation and progress through research and information exchange, conducted within a setting that is objective, interdisciplinary, and multimodal. The Board's varied activities annually engage about 7,000 engineers, scientists, and other transportation researchers and practitioners from the public and private sectors and academia, all of whom contribute their expertise in the public interest. The program is supported by state transportation departments, federal agencies including the component administrations of the U.S. Department of Transportation, and other organizations and individuals interested in the development of transportation.

Learn more about the Transportation Research Board at **www.TRB.org**.

TRR
JOURNAL OF THE TRANSPORTATION RESEARCH BOARD

Article

Transportation Research Record
2018, Vol. 2672(8) 1–9

DOI: 10.1177/0361198118758039
journals.sagepub.com/home/trr
SAGE

Successful Practices and Training Initiatives to Reduce Bus Accidents and Incidents at Transit Agencies: Abridged Version of TCRP Synthesis 126

Jodi Godfrey[1], Lisa Staes[1], Jenifer Flynn[1], and Roberta Yegidis[1]

Abstract

The purpose of this study is to document successful practices and training initiatives that have been effective in reducing transit accidents and incidents. Recognizing that safety does not occur in a vacuum, this study also focuses on other system approaches that have been implemented to address safety hazards. These approaches include various technology applications, infrastructure modifications, and other programs and initiatives, such as driver incentive programs and close call/near miss reporting. The research team conducted a literature review; performed a survey of selected transit agencies; and identified and conducted detailed case studies of 11 public transit agencies, selected from the survey respondents based on their responses to the survey. Effective safety management is comprehensive and multifocused in nature. At the case study agencies, safety management methods are often undertaken concurrently or within a period that does not provide agencies the ability to establish, with certainty, that one method or strategy contributed more to improved safety versus another. Successful programs examined in the case studies have been effective because of multifaceted, coordinated efforts to address transit safety. Strategies such as increased or modified operator training, technology applications, infrastructure modifications, and safety campaigns and promotions were often implemented simultaneously. Thus, judging the relative success of one strategy in reducing transit incidents when multiple strategies were implemented concurrently was difficult. Case study participants agreed that the progress toward a mature safety management system (SMS) framework within their agencies would contribute to an overall reduction in transit safety risks.

For this study, the definition of safety was adopted from the *Transit Safety Management and Performance Measurement* guidebook (p. 17) as, "the state in which the risk of injury to persons or damage to property is reduced to, and maintained at or below, an acceptable level through a continuing process of hazard identification and risk management" (*1, 2*).

Along with the definition of safety used in this study, the institution of safety management systems (SMSs) was central to the safety cultures that were examined. Mature SMS structures and associated activities do not occur in a vacuum. Risk and associated mitigation measures include myriad strategies; risks associated with transit collisions, as an example, are not strictly confined to the training of bus operators. SMSs establish that system approaches must be implemented to address safety hazards, including those that result in, or can result in, transit incidents or other systemic safety failures. In the transit industry, some of the measures undertaken to address system risks include, for example, various technology applications, infrastructure modifications, and other programs and initiatives, such as operator incentive programs and close call/near miss reporting. These strategies are reflected in the case studies and other report elements.

The primary objective of this study was to document public transit agency bus operator training programs and other strategies that have been successful in reducing accidents and incidents, resulting in safer transit systems for the communities they serve. Technology applications and other internal operational and personnel related policies and procedures were also examined. After a comprehensive literature review was conducted to gain an understanding of the impacts of training programs and incentives, technology solutions, and other safety rewards, the initial research objective was completed by means of an online survey that was distributed via email to 42 public transit agencies in the U.S. The transit agencies were chosen based on previous recognition of successful training programs and other collision mitigation methods, in addition to their size and geographical distributions. A total of 37 agencies responded, yielding an 88% response rate. From those 37 respondent agencies, 11 case

[1]Center for Urban Transportation Research, University of South Florida's College of Engineering, Tampa, FL

Corresponding Author:
Address correspondence to Jodi Godfrey Jodis@cutr.usf.edu

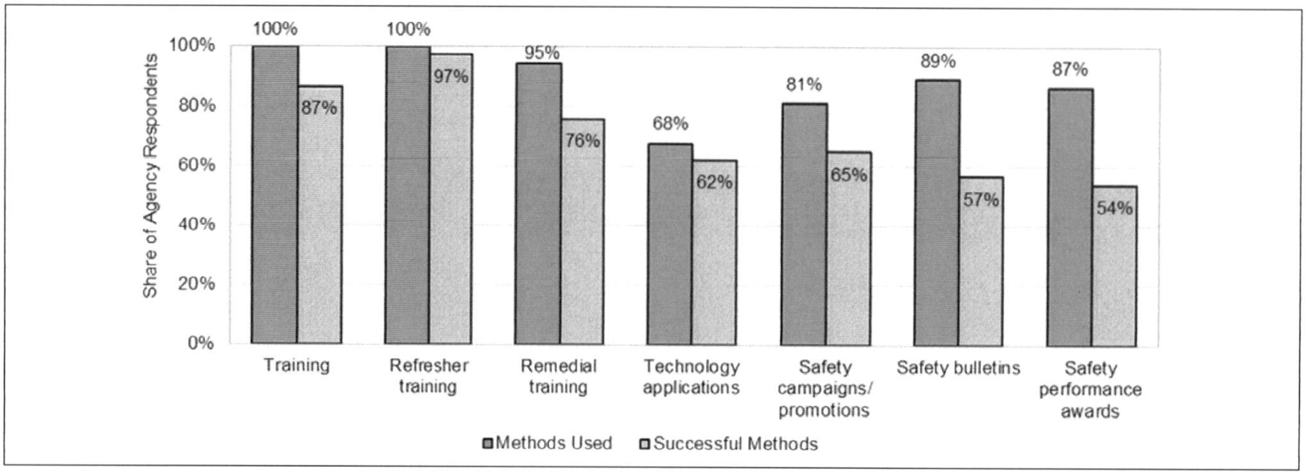

Figure 1. Methods used to mitigate identified safety-related issues and success rates.

study agencies were selected to highlight specific successful safety improvement strategies.

Survey Results

The 28-question survey collected information from 37 transit agencies that represented 23 states. The questions addressed system characteristics, accident review and risk assessment processes, bus operator training, use of technologies, and other agency policies or model practices.

Accident Review and Risk Assessment

Of the 37 survey respondents, 87% have an accident review board or use a similar process. Most of responding agencies (94%) task their accident review board with preventability determinations. Most agencies, 92% of survey respondents, indicated that their agency examines and tracks causal or contributing factors in safety event investigations to determine preventability. Human error is the most prevalent contributing factor reported by survey respondents (78%), followed closely by distractions (73%). Most survey respondents (83%) indicated that they use a database or electronic reporting system to track safety incidents.

Bus Operator Training

Once the safety-related issues are identified, the transit agencies use combinations of several methods to mitigate those issues. Given that these agencies were chosen as participants due, at least partially, to their known successful training initiatives and technology applications, it is no surprise that mitigation methods related to training have such a high indication of use among the respondents, as shown in Figure 1. Refresher training was reported as being the most successful mitigating method used to reduce prevalent safety issues within the transit agencies, with 97% of agencies reporting it as successful.

The subjects included in bus operator refresher safety training vary by agency, with over 90% of reporting agencies indicating they cover safety policies/procedures and defensive driving. Security is the least reported topic that agencies cover in their operator refresher training. In addition, 61% of the agencies indicated that new training opportunities have been offered, specifically targeting the reduction of incidents prevalent in their particular system. The agencies that utilize training simulators indicated that this method is primarily used to target common areas of risk based on agency safety trends. Telemetric video-based event monitoring systems were also identified as valuable tools for addressing not only particular agency incidents, but also risky operator behaviors and near miss incidents. Agencies indicated that this technology provides opportunities to counsel bus operators with the goal of adjusting their driving techniques and behaviors to prevent or reduce the probability of an incident occurring. This proactive approach to addressing safety-related incidents was noted by survey respondents as effective and successful.

The training delivery methods utilized by transit agencies vary depending on the agency. On-site, instructor-led (classroom) safety training is the most common refresher training delivery method reported, with 87% of the agencies indicating a preference for on-site classroom training. Behind-the-wheel training closely followed, with 77% of agencies preferring that delivery method.

Use of Technologies

The number of safety enhancing technologies used by transit agencies varied from 2 to 12 of the 13 technological enhancement options listed in this survey question. Of the listed technologies, all responding agencies indicated the use of a vehicle tracking system, whereas only 5.6% indicated the

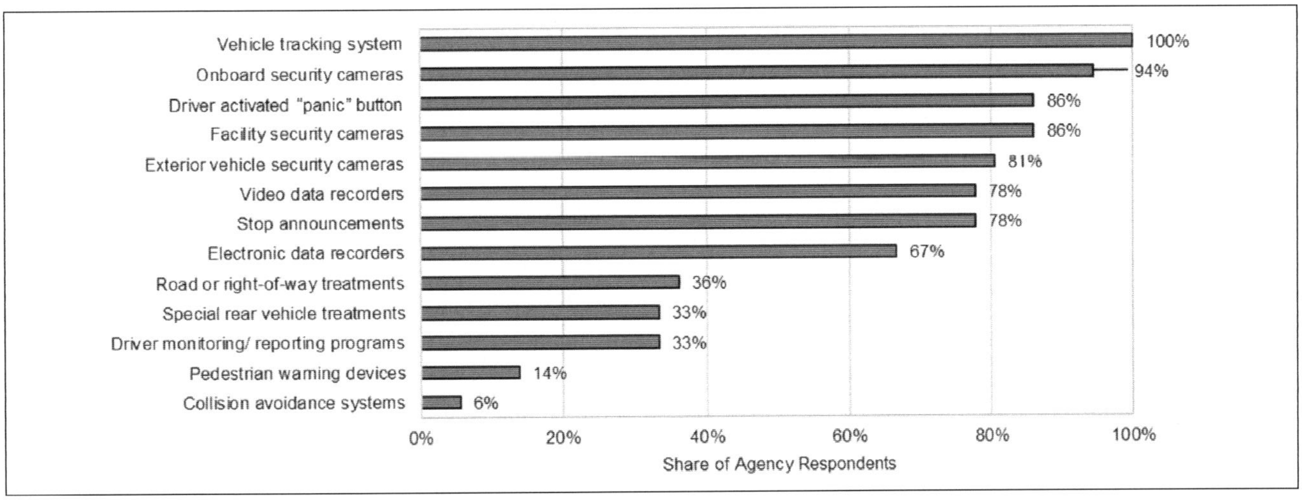

Figure 2. Share of technology use by type.

use of collision avoidance systems. Figure 2 displays the share of agencies that report the use of each safety technology.

The technological safety improvements have resulted in documented success in 58% of the agency responses. Of all the listed safety-enhancing technology applications, 42% of respondents indicated that the use of video and/or operator monitoring programs have been the most successful, adding that the use of onboard video attributed to documented success in the reduction of transit incidents.

Other Agency Policies and Model Practices

Most (71%) of the reporting transit agencies require their bus operators to report outside employment. Seven out of 10 agencies require their operators to report the use of over-the-counter medications. Nearly all respondents (92%) indicated that their agencies have a distracted driving policy for their bus operators. Most of the responding agencies (92%) follow a standard, rule, policy, or regulation related to operator hours of service.

Key trends that were echoed across agencies that reported model practices were the importance of regular refresher training and agencywide accountability. The safety culture of the agency is dependent on support from everyone from the general manager (GM) to operations, maintenance, administration, and safety. Respondents noted that it is critical that in agencies with labor unions, members of those unions are at the table and that safety is an area of focus in collective bargaining agreements. One of the central pillars to SMS is the ability to not only evaluate and react to existing trends, but also to identify areas of risk for which events have not yet occurred. This is when the telemetry-based driver monitoring and video surveillance systems demonstrate their usefulness and successes, as reported by the survey respondents.

Case Studies

A total of 11 case study sites were selected from the survey respondents based on their successes in instituting programs that demonstrated improved safety results, including the reduction of transit collisions or other incidents, as reported in the responses to the survey questionnaire. Details regarding each of the 11 case study agencies are expanded on in alphabetical order in this section of the paper.

Charlotte Area Transit System (Charlotte, North Carolina)

The mission of Charlotte Area Transit System (CATS) is "to improve the quality of life for everyone in the greater Charlotte region by providing outstanding communitywide public transit services while proactively contributing to focused growth and sustainable regional development" (*3*). CATS' training program was overhauled in 2014 to improve the consistency and structure of bus operator training. The new bus operator training at CATS is a seven-week program consisting of 90 hours of platform time. In addition, 8 hours of refresher training is provided to all operators bi-annually.

CATS buses are equipped with onboard video recording with audio, which was indicated as being the most effective technology in use by the agency because of the ability to substantiate complaints, identify poorly performing operators, and address issues before an incident occurring. The videos are also used for training purposes, as CATS finds it important that the operators observe and recognize practices performed correctly and those performed incorrectly.

CATS launched the "CATS See Say" app to enable riders to alert transit police if they notice something that seems concerning. This app was launched for the system because, as CATS Chief Executive Officer (CEO) Mr Lewis said, "Safety is and always will be a primary concern of CATS.

Through the use of this mobile app, riders can have a bigger hand in helping our public transit system operate as safely as possible" (*4*).

CATS' hours of service regulation are based on Federal Motor Carrier Safety Regulations (FMCSRs), and on-duty time is closely monitored to reduce incidents involving operator fatigue.

The GM is supportive of the training program, reviewed the entire overhaul of the training curriculum, and provided feedback throughout. Having a safety-focused accountable executive shapes the safety culture for the entire agency. Whereas CATS has had varying rates of collisions, probably at least partly attributable to increased population and economic growth, the monthly trend of preventable incidents has been declining over the past 2 years. This trend is especially significant when ridership, population, and construction are also increasing, meaning the chance of collisions is increasing.

City of Madison Metro Transit (Madison, Wisconsin)

The mission of City of Madison Metro Transit (Metro) is to provide safe, reliable, convenient, and efficient public transit to the citizens and visitors of the Metro service area through the efforts of dedicated, well-trained employees. In 2014, Metro officials began a bus stop consolidation project as an approach to address schedule issues caused by increased ridership. Following an analysis of bus stop spacing, a strategic reduction of stops improved on-time performance, while also improving safety and customer satisfaction. Metro has adopted the SMS framework, and includes initiatives such as increased data collection, risk identification and management, and the assurance of a safety manager who reports directly to the accountable executive: Metro's GM.

In 2008, Metro began the process of revamping its training curriculum. The agency reviewed all content used in training programs and modernized the curriculum, focusing more on the local needs specific to Madison. Metro indicated benefits to having as few as two trainees per instructor per bus, and providing more drive time for each trainee. More resources were allocated to training and were used to hire more trainers, reducing the training class sizes. Metro has also allocated an additional bus used in the training program to ensure that each trainee is allotted 4–5 h of behind-the-wheel time per day of training. These training improvements have resulted in noticeable positive feedback from the operators. Metro's annual refresher training follows a roundtable discussion structure. This training structure was designed to gather information and data directly from bus operators to recognize safety concerns and identify problem areas. The purpose of this approach to refresher training is to develop a safety culture within the agency that improves both accident prevention and company morale.

Madison Metro buses are equipped with video and audio surveillance equipment. The system records speed, braking, and turn-signal use. The use of this equipment was reported as beneficial, allowing for event specific one-on-one coaching and counseling of bus operators. In addition, the surveillance equipment is also an effective tool for investigating customer service complaints.

In 2014, Metro installed audible turn signals on all fixed-route buses that make a chirping noise when signaling a turn in traffic or pulling into a bus stop. In response to negative public feedback coupled with questionable effectiveness, Metro turned the audible turn signals off in 2015.

Metro launched the Safe Streets Campaign in November of 2014, which included the release of videos of safety-related close call incidents captured by Metro bus surveillance cameras. The goal of the campaign was to increase safety for everyone and prevent accidents by increasing community education and awareness.

Whereas there are no hours of service restrictions in Wisconsin, Metro operators have the right, through union restrictions, to refuse any work over 10 h behind the wheel, 12 h scheduled, or 13.5 h of spread time.

Metro's GM has been supportive in the development of modernized training manuals, and across-the-agency communication and collaboration on safety-related issues and mitigation strategies. Metro holds biweekly customer service and safety and security meetings in which the GM, deputy GM, marketing manager, customer service manager, and the operations and safety manager discuss issues and trends.

Whereas Metro has taken a multifaceted approach to improve safety, interviewing their general operations supervisor of safety and security revealed that open communication throughout the entire agency is highlighted as the most successful approach to increased safety. Renewed training curriculum along with several other campaigns, promotions, and technology applications, resulted in a 24% reduction in preventable/chargeable collisions from 2008 to 2015.

Greater Bridgeport Transit (Bridgeport, Connecticut)

Greater Bridgeport Transit (GBT) is committed to providing a safe, convenient, reliable bus service to ensure that the travel needs of visitors and residents are met. In late 2012, GBT began restructuring the agency, which included the introduction of two new positions related to safety training, the manager of safety training and the manager of transportation operations. Before the restructuring, safety training focused predominately on remedial areas such as post-accident training, or safety training for repetitive problems or repeat offenses. Since the restructuring, resources dedicated to training have increased, and in addition to new-hire training and remedial training, all GBT operators attend quarterly training sessions.

The new-hire training program is between 6 and 8 weeks in duration depending on the size of the class; 1 week of this is classroom training and the remainder of the time is spent on the road. On the final day of new-hire training, the trainee is in a room with a trainer and is quizzed on every run, every turn, every street, and every stop. On answering all questions correctly, the trainee graduates from the training program.

Two technology solutions have provided major assistance to GBT staff. These include an onboard surveillance system which allows for visual monitoring of activities in and around the bus, and an automated vehicle location (AVL) system, installed in 2009, that provides real-time information for dispatchers, road supervisors, and administrative staff. The combination of these tools allows GBT managers to have a full understanding of safety events and to address them properly.

The FMCSR hours of service regulations restrict operators from driving more than 10 h at a time and no more than 15 h per day.

GBT's CEO and Chief Operating Officer are heavily involved in the agency's safety program from policy development to day-to-day training. GBT credits their entire leadership team with the safety success of the agency. When asked what successful measure attributed to the increased safety culture at GBT, the interviewees indicated teamwork as being most successful, citing that working together to solve common issues has proven to be the most beneficial. The rolling 3-month preventable incident rate at GBT improved by an astounding 57% between October of 2012 and February of 2016.

Greater Cleveland Regional Transit Authority (Cleveland, Ohio)

The mission of the Greater Cleveland Regional Transit Authority (RTA) is "to provide safe, reliable, clean and courteous public transportation," and protecting the safety of employees, riders, and the public is the agency's top priority. RTA's safety program is guided by a culture of teamwork between departments, and accountability throughout the entire chain of command, from the CEO to frontline employees.

RTA installed telemetry-based driver monitoring system (DMS) equipment fleetwide in 2014 to help protect its operators, reduce costs associated with risky driving behaviors, and create a safer experience for riders. RTA makes use of feedback from the DMS, which records events and behaviors such as hard braking or following too closely, to determine when refresher or remedial training is needed for specific operators. Video footage will be shown to the operator and is sometimes shown to other operators for training purposes. Safety rounds are a part of the executive safety committee meeting in which near misses and other incidents are discussed to identify causal factors and ensure that employees are made aware of specific behaviors and actions to avoid. Most topics discussed during the safety rounds originate from video footage.

Exceptionally safe employees at RTA are nominated each month for the "Champions of Safety" award. Awardees are recognized at the executive safety committee's monthly meeting and receive a clock as a reminder that safety should be the top priority "around the clock." The following year, a calendar is distributed throughout RTA, with each calendar month including the Champion(s) of Safety recognized for that month in the year before. RTA also has an annual awards night, which is usually celebrated in April.

RTA follows the 10-hour rest rule, with shifts no longer than 14 hours.

When asked to comment on the keys to the success of RTA's safety program, staff cited the DMS, open and honest dialog with employees, a relentless pursuit of improvement, recognizing and rewarding employees for exceptional safety performance, and using teamwork between the safety and operations departments to create a true atmosphere of safety. Staff also strongly emphasized the importance of having a consistent, agencywide definition of safety, and commented that an agency can adjust service, but not an injury, because when safety is compromised the consequences can be permanent. The efforts at RTA have led to reduced collision rates in the agency's busiest districts.

Jacksonville Transportation Authority (Jacksonville, Florida)

The mission statement of Jacksonville Transportation Authority (JTA) is "to improve Northeast Florida's economy, environment, and quality of life by providing safe, reliable, and efficient multimodal transportation services and facilities" (*5*). With the arrival of JTA's new CEO in 2014, the agency undertook an aggressive restructuring and re-design initiative "Blueprint 2020," the primary catalyst for the evolution of JTA. One initiative of many that were implemented to support Blueprint 2020 was JTA's Route Optimization Initiative (ROI). The ROI was described by JTA as a "massive overhaul" of the agency's system, the first time in 30 years a comprehensive analysis and restructuring had been performed. Central to the ROI was a concerted effort to not only evaluate and modify JTA's legacy fixed-route system from operational and planning perspectives, but also from a position of mitigating safety risks inherent in the system.

In 2013, JTA instituted mandatory quarterly safety training meetings in which refresher training is conducted. During these meetings, representatives from JTA's safety, operations, and training departments combine to provide focused training on various topics, including those related to current areas of concern for the agency. Many of these topics are drawn from registering events from the telemetry-based video monitoring system and video footage. JTA installed DMS technology on 50 fixed-route buses in 2015, and at the time of data collection for this study, they were in the process of fitting all buses with this technology. JTA indicated that the system has been a valuable tool for individual coaching with bus operators, and has led to a decrease in liability claims and insurance premiums.

JTA also utilizes a training simulator, which reportedly allows drivers the opportunity to associate muscle memory, reaction time, problem solving, decision making, and judgment all at the same time in a repetitive manner with none of the repercussions that would be associated with the same scenarios in the real world.

In November 2015, JTA kicked off the communitywide "Keep it in Your Pocket" distracted driving campaign, specifically emphasizing social media marketing. The success of the campaign was attributed to cooperation between JTA, the Jacksonville Sheriff's Office, Jacksonville Fire Department, and the Northeast Florida Safety Council.

The CEO at JTA, Nathaniel Ford, was identified as the driving force behind the agency's improved safety culture, particularly his agencywide challenge to do things differently. The combination of many efforts has evolved as JTA's SMS approach to improving bus safety, which has also led to improvements in preventable and non-preventable incidents and collisions.

Kansas City Area Transportation Authority (Kansas City, Missouri)

The mission statement of Kansas City Area Transit Authority (KCATA) is "We connect people to opportunities" (*6*). In response to some alarming safety trends, KCATA developed a comprehensive corrective action plan to reduce bus accidents. This five-point plan began in 2013 and focused on the five "E"s: engineering, education, enforcement, encouragement, and evaluation. In the area of engineering, KCATA spent $350,000 retrofitting its bus fleet with accelerometers to monitor operator driving behaviors. In the area of education, refresher and remedial training focused on pedestrian awareness, along with understanding and recognizing blind spots and how to mitigate them while driving. The agency's accident discipline policies and associated enforcement were strengthened, as well as the penalties associated with noncompliance with those policies. Recognizing the importance of encouragement, the agency created a safety suggestion program with recognition and rewards for employees who promote safe driving behavior. Finally, they established an evaluation program with metrics that are monitored and tracked.

KCATA utilizes bus simulators in all segments of its training program, including new operator, refresher, and remedial training. While the agency does not have data to establish the success of the simulator in training bus operators, KCATA indicated that the simulators were "absolutely effective." Onboard audio and video recordings of bus collisions and other events are saved in a video library and utilized in both refresher and remedial bus operator training.

The KCATA Passenger Code of Conduct was developed to inform passengers of rules and guidelines to promote safe behavior. KCATA celebrates Pedestrian Safety Week annually in May and pedestrian safety tips are communicated through KCATA's Transit Talk articles. They also adopted FTA's "Transit Watch," "Be Alert," and "Is This Yours?" public awareness outreach campaigns. KCATA representatives suggested that the combination of these passenger and pedestrian campaigns and programs be considered as a model for agencies across the country.

KCATA utilizes an incident occurrence tracking system to record transit incidents, both those meeting the National Transit Database (NTD) major incident reporting thresholds and minor incidents. KCATA tracks and trends incidents for routine discussion at KCATA's Safety Committee meetings in which mitigation measures are discussed and put forward for implementation.

The GM was very supportive of the safety programs instituted and he was central to the decision to purchase (and obtain funding) for the agency's simulators. The combination of many efforts along with GM support have led to reductions in pedestrian collisions, a 10% reduction in total bus collisions, and a 92% reduction in total claims paid by KCATA between 2012 and 2015.

King County Metro (Seattle, Washington)

Protecting the safety and security of customers, employees, and facilities is King County Metro's (Metro) top priority. In response to a spike in pedestrian-related events, Metro added several key components to its safety program that resulted in significant safety improvements throughout the agency. Metro requires annual refresher training (previously required every 3 years) for all bus operators, with an emphasis on pedestrian awareness. Metro also uses agency onboard video recordings in its training curriculum.

Metro upgraded its ride-check program to ensure that practical drift was not present in the pre-trip inspection process. The agency uses data analyses to determine and educate drivers of the places and times that are statistically of greatest risk. There has also been examination and piloting of several technologies at Metro such as audible bus turning warnings, strobe lights, automatic braking, and warning decals on the mirror frames to remind drivers to look for passengers.

Metro hosts an annual Safety Awareness Day with a barbeque, dunk tank, and other activities to emphasize the agency's safety message. In addition, Metro has adopted national safety campaigns such as "Be Seen, Be Safe, Be Smart," and even added "Look Up" to the campaign to remind pedestrians to be cognizant of their surroundings.

Metro has contracted with King County Police for an Operator Assaults Reduction Initiative. As part of this program, police trainers visit all Metro venues to explain the initiative and to educate operators on techniques they can use, such as "verbal judo" and other common-sense tools, to de-escalate tense situations. This initiative was reported as successful by the agency with a 6.1% decrease in operator assaults from 2014 to 2015.

Metro started a nonpunitive near miss pilot program to make sure operators have an opportunity to communicate hazards and other things that affect their ability to be safe in the workplace. Feedback is also received through green cards, a hazardous reporting program, safety committee meetings, and safety awareness days.

Metro observes FMCSRs, allowing operators to drive for 14 hours, with 10 hours of rest.

Metro saw a 35% reduction in pedestrian events compared with 2013, and the reduction has continued into 2015. Overall, liability claims of all types against Metro were down 8% in 2014 versus 2013, and down 6.8% from the agency's 2009–2013 average. In 2014, claim payouts were down 39% from those in the 2009–2013 period. Moreover, reducing the cost of these claims has allowed Metro to put the savings back into its core operations.

Lane Transit District (Springfield, Oregon)

The mission of the Lane Transit District (LTD) is "to provide and improve services to all our diverse communities in ways that are sensitive and responsive to cultural differences, including accessibility for persons with disabilities" (*7*). LTD's overall approach to safety has been guided by an agencywide focus on customer service. The emphasis on customer service is so strong, in fact, that LTD places its advertisements for new operators in the customer service category rather than the driving category.

LTD has two types of refresher training. There is an annual "back to basics" refresher training that is mandatory for all operators. Mandatory refresher training also includes video footage of actual LTD accidents, a practice that has been found to be highly effective for keeping trainees fully engaged. The second type of refresher training is provided by a proprietary driver training system and is tailored specifically to operators who have had an accident.

LTD staff also distribute a monthly written "Review of Accidents" to operators discussing all accidents, preventable and non-preventable, that occurred during the previous month. According to LTD, the Review of Accidents is a benefit to all operators because it allows them to see what went wrong from a wider perspective, one that includes the entire bus, not just the view from behind the windshield. One thing that is heavily stressed in the discussion is the impact major collisions and hard braking incidents have on LTD passengers. This has made a strong impression on operators by allowing them to witness shocking real-life examples that they would be far less likely to notice while focusing on driving.

All LTD buses are equipped with video surveillance equipment, with up to 16 cameras per bus and at least 10 cameras on smaller vehicles. LTD reported using a telemetry-based DMS on a trial basis on a select number of its buses. Ultimately, it was determined that the system was not a good match for LTD's needs, due in part to the cost burden given the small size of the agency.

At LTD, hours of service regulations require that operators have a break of at least 9.5 h before beginning a new shift and drive no more than 14 h consecutively.

Route timing is reviewed on a regular basis by LTD's planning department, and the safety department is included in the process of planning for a new route or when a different style of vehicle is placed on a route. LTD has an open-door policy as well as a "blue card" reporting system. Operators document unsafe conditions or other concerns on specially designated blue cards, which are then submitted to an assigned employee and routed internally to the appropriate department. LTD maintains a database on all accidents, and the database includes information on causal and contributing factors for preventable accidents. LTD's management has been supportive of safety programs, which has helped lead to improvements in both total and preventable accidents.

Solano County Transit: Operated by National Express (Vallejo, California)

The mission of Solano County Transit (SolTrans) is to "deliver safe, reliable and efficient transportation services that link people, jobs, and our communities" (*8*). SolTrans representatives suggested that a minimum duration of 20 h of behind-the-wheel driving be a nationwide standard for new bus operators. Currently SolTrans requires new bus operators to receive 25 h of actual behind-the-wheel training before they are allowed to transition into revenue service. Annual recertification training is eight hours in length and is required for all employees. The recertification training includes both behind-the-wheel and classroom training.

SolTrans has continuous video feeds on its bus fleet, which includes audio capabilities. Videos are used for remedial training and addressing customer complaints. Whereas the videos can be used for coaching opportunities or punitive measures, the focus of the video usage is to recognize exceptional operators. SolTrans uses an asset management and tracking company with a technology-based software platform that has GPS-enabled advanced fleet monitoring, real-time vehicle location, and an asset management system, which they describe as beneficial to the agency. At the time of the case study, SolTrans had been utilizing the technology for 1.5 years for ride checks, road observations, accident reporting, and evaluations. This technology collects, stores, and transforms raw data into usable data that facilitates data-driven analysis and decision making (*9*).

SolTrans representatives indicated that "positivity and safety culture are interrelated. At SolTrans employees are treated as internal customers, with the understanding that an agency's employees are the most important people in an organization. It is necessary to thank them for a job well done and provide them with the opportunity to provide good customer service" (*8*).

SolTrans follows the company policy for hours of service regulations, which restricts operators from being on duty for more than 10 h a day.

SolTrans' GM has championed the safety culture at the agency. Quarterly town hall meetings are held as an opportunity to reveal any issues to corporate regional managers.

SolTrans is working toward SMS implementation with its close call reporting pocket-sized safety culture guidebook, which is a small book with carbon copies that the operators carry with them at all times. The book is used to document any near misses or unsafe conditions, which are then turned in. The operator keeps a copy as a way to track them, and management is required to respond within 48 hours.

Initial results indicated a 71% decrease in total incidents in the last 6 months of 2013 compared with the first 6 months of 2013. During that same time, passenger falls decreased 75% and employee injuries decreased 66%, indicating great initial success of the restructuring of SolTrans' safety practices. These successes have been followed by continued success through 2015. Preventable collision rates decreased 52% from 2013 through 2015, during which time the total injury rate decreased 65%.

TriMet (Portland, Oregon)

TriMet's mission is "to provide valued transit service that is safe, dependable and easy to use" (*10*). TriMet's GM created the Safety & Service Excellence Task Force in July 2010. According to TriMet representatives, the Task Force was charged with addressing "how to migrate TriMet to the highest levels of safety performance, and thereby improve performance in all areas of its business" (*10*). The Task Force made several recommendations which resulted in the following actions: required annual recertification training for all bus operators, realigned organizational structure with the safety and security division as direct reports to the GM, appointment of an executive director of safety with a seat at the leadership table, and review of reports and operational data to track close calls to eliminate potential hazards in the system.

TriMet buses include GPS-based AVL systems, a seat alarm which sounds if the operator leaves the operator's seat without setting the parking break, front and rear wheel turn lights that provide illumination of crosswalks during turns, turn signals on exterior mirrors, rear-facing cameras on both the operator's side and the curb side, and a message display unit called the vehicle control head (VCH). The VCH has an emergency alarm button that alerts dispatch when police, fire, or rescue services are needed; messages may also be sent to all buses or an individual bus.

TriMet recently acquired a training simulator that had only been used for new operator training for approximately 1 month before the interview. The agency plans to use the simulator for all training, including recertification and remedial training.

There is also a major emphasis on public education and outreach. Every fall TriMet rolls out its "Be Seen, Be Safe" campaign, which stresses the importance of wearing bright clothing, reflective materials, and personal lights to be visible, especially during the fall season when daylight savings time comes to an end. TriMet runs a safety awareness campaign every spring as well, and the 2016 theme was "Stay Alert, Stay Alive."

In 2013, the hours of rest for operators was increased from 7 h to 10 h. Per TriMet policy, safety sensitive employees cannot work more than 70 h in any 7-day period and cannot work more than 13 days consecutively.

Operators can report close calls of various types with the push of a control head button. These events can be instantly recorded, mapped with GPS, and aggregated with other data to reveal potential "hot spots" in need of attention. This type of proactive hazard identification is paramount for an SMS-style safety culture. TriMet has made safety its core value and the lens through which the organization makes all of its operational, planning, and strategic decisions. This includes everything from hiring and training employees to operating and maintaining vehicles. In addition, every TriMet employee is charged with embracing safety as a value.

Utah Transit Authority (Salt Lake City, Utah)

The mission of Utah Transit Authority (UTA) is to "strengthen and connect communities enabling individuals to pursue a fuller life with greater ease and convenience by leading through partnering, planning, and wise investment of physical, economic, and human resources" (*11*). The agency stated that the hiring of UTA's chief safety officer (CSO) 5 years ago has improved the overall safety culture of the entire organization.

Refresher training at UTA is conducted at the business unit level and is focused on the current prevalent incidents and training needs determined at that business unit. As of 2016, each business unit received approval to conduct up to 16 h of refresher training for all bus operators. This training is based on predetermined training needs of the business unit, but the agency works to ensure a consistent training message between units.

In May of 2014, UTA installed a telemetry-based DMS on its entire fleet of buses. The DMS is not only used for incident investigation, but the system also provides bus operators with an extra level of protection via capabilities to manually activate recording in the case of circumstances such as robbery, unruly passengers, and road rage. In addition, the DMS gives UTA the opportunity to track incidents by type, allowing for focused coaching on current prevalent incidents. The system can also produce reports over a specified time period, which is used to identify the top three driving concerns that are addressed via safety bulletins and coaching.

In 2015, an illuminated yield sign was added to the rear lighting configuration of UTA's buses to increase visibility in an attempt to reduce rear-ended collisions. Although there are no data available to prove the effectiveness yet, bus operators feel favorably about it.

UTA has a distracted driving policy, with discipline dependent on the type of distraction. The cell phone and audio policy at UTA is zero tolerance, meaning cell phone use while driving is terms for immediate termination. There is a civil ordinance that allows UTA transit police to ticket members of the general public for distracted behavior exhibited near its rail lines.

UTA hours of service restrictions include no more than 10 h driving in a 15 h maximum spread following 8 h consecutively off duty.

Improvements in the safety culture at UTA are largely due to the safety championing of authoritative personnel, with the CSO reporting directly to the GM/CEO. The appointment of the CSO has led to UTA's improved safety culture, as proven by a reduction in avoidable bus collisions. Between 2012 and 2015, UTA's avoidable bus collisions were decreased by 36% whereas the number of claims payments were decreased by 9% during the same period.

Conclusions

The purpose of this study was to document successful practices and training initiatives, including bus operator training and retraining programs, that have been effective in reducing transit bus accidents and incidents. The study also focused on other system approaches that have been implemented to address safety hazards, including various technology applications, infrastructure modifications, and other programs and initiatives, such as close call/near miss reporting.

Training is an effective and necessary method that must be used to ensure and improve transit bus system safety. Agencies with well-designed and consistent bus operator refresher training and post-event or event avoidance technology-directed refresher training and counseling were very effective in reducing the riskiest behaviors among bus operators.

The use of telemetry-based DMSs was found to be more prevalent across the industry. These systems automatically collect data for analysis of many safety-related events such as incidents, near misses, speeding, and distracted driving. The systems used by four case study agencies, GBT, Greater Cleveland RTA, JTA, and UTA, capture, identify, prioritize, and analyze causes of poor or risky driving before an incident occurring, allowing for corrective actions to be undertaken by the transit agency (*12*).

Successful programs utilize multipronged or multifaceted coordinated efforts to address transit safety. In all case study agencies, there was no single method identified that led to improved safety. Judging the relative success of one strategy in reducing transit incidents when multiple strategies were implemented concurrently is difficult. However, other agencies that are considering the institution of various corrective actions to improve safety issue areas should embrace what is reflected in this study: strategies that include multiple approaches to improve bus safety do find success.

Several topics that would benefit from further research based on the survey results include longitudinal research on the effectiveness of simulator training over time, and an independent assessment of telemetry-based DMSs to evaluate the effectiveness of the technology. In addition, the actions of other drivers that result in collisions with transit buses are a critical point that could be addressed through research to evaluate public outreach programs and other strategies designed to increase the public's awareness of the dangers of distraction and other collision-related contributing factors.

Acknowledgments

This research was sponsored by the Transportation Research Board's Transit Cooperative Research Program. We also thank Mariela Garcia-Colberg, the National Academies' staff contact for the study.

Author Contributions

The authors confirm contribution to the paper as follows: study conception and design: Jodi Godfrey, Lisa Staes, Jennifer Flynn, Roberta Yegidis; data collection: Jodi Godfrey, Lisa Staes, Jennifer Flynn, Roberta Yegidis; analysis and interpretation of results: Jodi Godfrey, Lisa Staes, Jennifer Flynn, Roberta Yegidis; draft manuscript preparation: Jodi Godfrey, Lisa Staes, Jennifer Flynn, Roberta Yegidis. All authors reviewed the results and approved the final version of the manuscript.

References

1. Staes, L., J. Godfrey, J. Flynn and R. Yegidis *TCRP Synthesis 93: Successful Practices and Training Initiatives to Reduce Bus Accidents and Incidents at Transit Agencies*. Transportation Research Board of the National Academies, Washington, D.C., 2017.
2. Ahmed, S.A. *Transit Safety Management and Performance Measurement*. Report FTA-OK-26-7007.2011.1. Federal Transit Administration, 2011.
3. City of Charlotte, North Carolina. *Charlotte Area Transit System, Mission, Visions & Goals*. 2018. http://charlottenc.gov/cats/about/Pages/mission.aspx.
4. Greater Cleveland Regional Transit Authority. *Mission, Vision, Values*. 2018. http://www.riderta.com/mission-save.
5. Jacksonville Transportation Authority. *About JTA*. 2018. https://www.jtafla.com/about-jta/.
6. Kansas City Area Transportation Authority. *Mission*. 2018. http://www.kcata.org/about_kcata/entries/mission.
7. Lane Transit District. *2016–2025 Capital Improvements Plan*. 2015. https://www.ltd.org/file_viewer.php?id=1356
8. *Solano County Transit (SolTrans) Performance Goals and Measures*. 2015. http://www.soltransride.com/wp-content/uploads/2012/05/FINAL-FY-15-16-SolTrans-Performance-Measure.pdf.
9. TrackIt, 2016. http://www.trackit.co.nz/.
10. TriMet. *Our Vision, Mission, and Values*. 2018. https://trimet.org/about/mission.htm.
11. Utah Transit Authority. *Mission and Vision*. 2005. https://www.rideuta.com/About-UTA/Mission-and-Vision.
12. Lytx. Fleet Safety Management Solutions 2016. http://www.lytx.com/our-solutions/drivecam-programs.

The Task Force on Transit Safety and Security (AP018T) peer-reviewed this paper (18-01040).

Article

TRR
JOURNAL OF THE TRANSPORTATION RESEARCH BOARD

Transportation Research Record
2018, Vol. 2672(8) 10–18

DOI: 10.1177/0361198118770168
journals.sagepub.com/home/trr
SAGE

Assessing Effects of Transit Signal Priority on Traffic Safety: Empirical Bayes Before–After Study using King County, Washington, Data

Yu Song[1] and David Noyce[1]

Abstract

Transit signal priority (TSP) has become popular among cities in the United States as a transit preferential treatment that improves transit operational performance through adjusting signal timing when transit vehicles are approaching signalized intersections. The majority of attention has been given to the operational effects of TSP. A small number of prior studies have discussed such a system's effects on traffic safety, with few studies investigating North American implementations. Prior assessment results were mixed, leaving the understanding of TSP's safety effects unclear. The purpose of this study was to add to the existing research on this topic, and to get a better understanding of TSP's safety effects. An empirical Bayes (EB) before–after analysis was carried out using TSP implementation data from King County, Washington. Eleven transit corridors having had TSP activated in 2010–2014 were assessed, and 75 street sections without TSP treatment were selected to form a reference group. The EB analysis results showed a 13% reduction in total crashes with the implementation of TSP, a 16% reduction in property-damage-only crashes, and a 5% reduction in fatal and injury crashes. These results are consistent with most existing TSP safety studies. The study implies that TSP brings road safety benefits to transit corridors in a U.S. metropolitan setting, and suggests that transportation and transit agencies add safety assessment into future TSP practices.

Transit signal priority (TSP) is a type of transit preferential treatment implemented at signalized intersections to reduce mass transit delays by providing extended or early green signal to transit vehicles in operation, when they are approaching signal-controlled intersections (*1, 2*). Prior TSP deployments in the United States generally yielded average bus travel time savings between 2% and 20%, with 8% to 12% representing the most typical range (*3*). Although improving transit operational performance along its route, TSP does bring increases in delays to cross-street traffic, which can be kept to a minimum by using GPS-based or adaptive TSP systems (*4*). However, TSP involves signal timing adjustments, which directly and indirectly affect traffic movements and road user behavior at and around intersections (*5–9*). Given the high density of signalized intersections in U.S. cities, the effects of TSP can cover an entire urban corridor.

Incorporating safety assessments into project development processes is becoming standard practice for transportation agencies' decision-making. Furthermore, there are available analytical tools like the *Highway Safety Manual*, which provides rigorous and rational procedures for traffic safety assessments (*10*). However, in the case of TSP, there has not been many safety assessments carried out by transportation agencies or research entities. Although the purpose of implementing TSP is to improve transit operational performance, its potential effects on traffic safety should not be overlooked. If TSP benefits traffic safety, an assessment could help quantify these benefits and further justify the wider implementation of TSP. If TSP correlates with a worsening of road safety, transportation agencies will need to make a detailed assessment and consider carefully before making any decisions to implement TSP. There are several existing studies on the traffic safety effects of TSP, reporting mixed results. Therefore, a comprehensive understanding of TSP's safety effectiveness has not been reached. Moreover, few studies on TSP safety effects have discussed specifically North American practices.

In this paper, an empirical Bayes (EB) before–after study was carried out to assess the effects of TSP on traffic crashes along 11 transit corridors in King County, Washington, which had had TSP installed and activated in 2010, 2011, 2012, and 2014. The EB before–after study method is included in the *Highway Safety Manual*, and is considered an advanced

[1]Department of Civil and Environmental Engineering, Traffic Operations and Safety Laboratory, University of Wisconsin-Madison, Madison, WI

Corresponding Author:
Address correspondence to Yu Song: yu.song@wisc.edu

method of safety evaluation (*11*). Corridor-level assessments were carried out considering all traffic crashes, property-damage-only (PDO) crashes, and fatal and injury (FI) crashes. This study will add to the existing studies on TSP safety assessments, and help enhance understanding of the potential benefits or costs of TSP, from a traffic safety perspective. The results will be of interest to decision-makers in city transportation and transit agencies, especially in U.S. cities, as this is one of the first safety assessments of TSP implemented in the United States.

Following this section, this article has five sections: a literature review synthesizing previous published studies on safety assessments of TSP is provided in the next section. The analytical method used for this study is then elaborated, followed by a summary of results with interpretation and discussion. The conclusions and recommendations for future work are presented in the final section of this paper.

Literature Review

There are small number of prior studies on the safety effects of TSP, few of which assessed North American implementations. In the existing TSP-related studies, more attention has been given to the operational effectiveness of TSP, which is the outcome of interest when such a treatment is designed and implemented. However, about a decade ago, safety researchers started to identify the need to take transit safety (road safety issues involving transit vehicles or facilities) into consideration (*12*). In recent years, with more transit preferential treatments (including TSP) being applied to improve transit operational performance, the effects of these treatments on road safety started to receive more attention. Four studies on TSP's road safety effects, referenced in this paper, were carried out by two research teams from Canada and Australia (*13–16*).

To study the relationship between transit facilities and road traffic crashes at a sophisticated level, Shahla et al. analyzed transit facilities' safety effects at signalized intersections (*13*). Three types of generalized linear models, with a negative binomial error structure, were applied in a cross-sectional study of a 5-year (1999–2003) data set from Toronto, Canada with fields including intersection crashes, turning movements, annual average daily traffic (AADT), bus stop locations (near-side or far-side) and the appearance of TSP. Model I was for transit-involved crashes at signalized intersections with both regular traffic and transit service operations; Model II was for all crashes at the same signalized intersections; and Model III was for all crashes at all signalized intersections, including those without transit service. Modeling coefficients from Shahla et al.'s study indicated a positive relationship between TSP existence and number of traffic crashes at signalized intersections, and a negative relationship between the existence of far-side stops and number of crashes at signalized intersections.

Researchers from the Public Transport Research Group at Monash University, Australia, have carried out a series of studies on transit priority treatments' road safety effects. Goh et al. explored the road safety impacts of bus priority treatments including bus lanes and TSP (*14*). An EB before–after study was applied for an aggregate-level analysis to explore changes in expected crash frequency at locations (intersections and roadway sections) where transit priority measures were implemented through the "smartBus Program" (a bus rapid transit system) in Melbourne, Australia in 2006. Other multiple types of intersections and roadway sections, with corresponding crash counts, were used as a reference group. Safety reviews and audits were also applied for a disaggregate-level analysis to identify safety impacts of changes in road configurations associated with the implementation of bus priority treatments. Results showed a lower expected crash frequency during the "after" period with TSP implemented (effectiveness = 0.889, or an 11.1% reduction) or non-TSP bus priority treatments (bus lanes and road-side parking restrictions, effectiveness = 0.818, or an 18.2% reduction), compared with the "after" period crash frequency estimates had the treatments not been implemented.

Goh et al. carried out another study on bus-involved accidents under situations with and without bus priority treatments including signal priority and right-of-way priority (*15*). Mixed-effects negative binomial (MENB) modeling and back-propagation neural network (BPNN) modeling methods were used to analyze a Melbourne data set of segment-level crashes, bus route lengths, AADT, bus service frequency, stop density, and presence of bus priority. Modeling results showed that with bus priority treatments, the transit-involved crash frequency significantly reduced (including crashes with buses hitting stationary objects and vehicles, as well as those involving collisions into/out of bus stops). Furthermore, the reductions were likely due to the effect of bus priority facilitating bus movements. The BPNN model results showed that bus priority had the effect of reducing route-section bus crash frequency by 53.4%, and the MENB model showed a reduction of 53.5%.

Naznin et al., also from the Public Transport Research Group at Monash University, studied the safety effects of streetcar priority (29 intersections with signal priority and 23 arterials with tram lane priority) in Melbourne (*16*). An EB before–after study method was used. Study results showed a statistically significant adjusted crash reduction rate of 16.4% after implementation of tram priority measures, with signal priority measures found to reduce crashes by 13.9% and lane priority measures by 19.4%. A disaggregate-level, simple before–after analysis indicated reductions in total and serious crashes as well as vehicle, pedestrian, and motorcycle-involved crashes.

Table 1 summarizes the TSP safety assessment results from the four cited prior studies. Effectiveness is a factor to be multiplied to the predicted crash frequency without

Table 1. Estimated TSP Effectiveness from Prior Studies

Study	Crash type	TSP effectiveness	95% CI
Shahla et al. (*13*)	Streetcar/bus-involved crashes[a]	1.5153	[1.1689, 1.9642]
		1.2597	[1.0041, 1.5804]
	All crashes, streetcar TSP	1.3184	[1.2252, 1.4186]
	All crashes, bus TSP	1.2839	[1.1672, 1.4122]
Goh et al. (*14*)	Fatal + injury crashes	0.8890	[0.6734, 1.1046]
Goh et al. (*15*)	Bus-involved crashes	0.4652[b]	NA
Naznin et al. (*16*)	Streetcar-involved crashes	0.8610	[0.6990, 1.0220]

Note: CI = confidence interval.
[a]Results from two different models.
[b]Combined effectiveness of TSP and non-TSP bus priority treatments.

Table 2. Selected TSP Corridors (Treatment Group)

ID	City	Corridor	Limit	Activation year
1	Kent/Des Moines	SR-99 Pacific Hwy	S 216th St. to S 272nd St.	2010
2	SeaTac	SR-99 Pacific Hwy	S 170th St. to S 208th St.	2010
3	Redmond	148th Ave.NE	NE 40th St. to Willow Rd	2011
4	Redmond	NE 40th St.	148th Ave. NE to 156th Ave. NE	2011
5	Seattle	Queen Anne Ave. N	Denny Way to Mercer St.	2012
6	Seattle	Elliott Ave. W – 15th Ave. W	Mercer Pl. to NW 85th St.	2012
7	Seattle	California Ave. SW	SW Alaska St. to Fauntleroy Way SW	2012
8	Renton	Rainier Ave. S	S 2nd St. to S 7th St.	2014
9	Seattle	SR-99 Aurora Ave. N	N 80th St. to N 105th St.	2014
10	Seattle	SR-99 Aurora Ave. N	N 125th St. to N 192nd St.	2014
11	Renton	Logan Ave. N	Park Ave. N to Airport Way	2014

Note: SR = State Route; Hwy = Highway; St. = Street; Ave. = Avenue; Pl. = Place.

TSP, to obtain the crash frequency with TSP. Thus, by comparing situations with and without TSP, an effectiveness index >1 indicates an increase in the number of crashes with TSP implemented, while an effectiveness index <1 indicates a decrease in the number of crashes with TSP implemented.

Data

King County, Washington, has been implementing TSP on its Metro Transit System since the early 1990s, with the first demonstration initiated in 1991 (*1*). The corridors selected as study sites for this paper had TSP installed with the introduction of King County Metro's "RapidRide" express bus system. Information including the TSP installation locations and activation dates were obtained from King County Metro Transit. Based on the TSP data set, 11 TSP corridors were selected as the treatment group, which are listed in Table 2. Locations of selected TSP corridors are shown in Figure 1. To form a reference group for EB analysis, 75 street sections without TSP, also in King County, were selected. To ensure a good similarity between the reference group and the treatment group in geometric design features and traffic patterns, the selection of reference sections considered only urban streets with the same functional classifications, the same posted speed limits, and the same number of lanes as the treated sections.

The crash data of the treatment group and reference group from 2002–2015, were obtained from the Washington State Department of Transportation and the Highway Safety Information System (HSIS) of the Federal Highway Administration. Traffic volume data and roadway information were obtained from multiple sources including open-source city traffic volume maps and HSIS. Descriptive statistics of variables used in the analysis are presented in Table 3. Traffic volumes were provided as AADT). Crash data were aggregated to each street section by year as annual frequencies. Apart from the total crash frequency, which counted for crashes of all severity levels, categories of PDO and FI crash frequencies were also used for separate analyses in this study. Roadway information including section length, number of lanes, and posted speed limit were used in estimating the safety performance function (SPF).

Methodology

Based on Hauer's implementation of an EB before–after study in road safety, to describe the safety effectiveness of a treatment (*17*), we define:

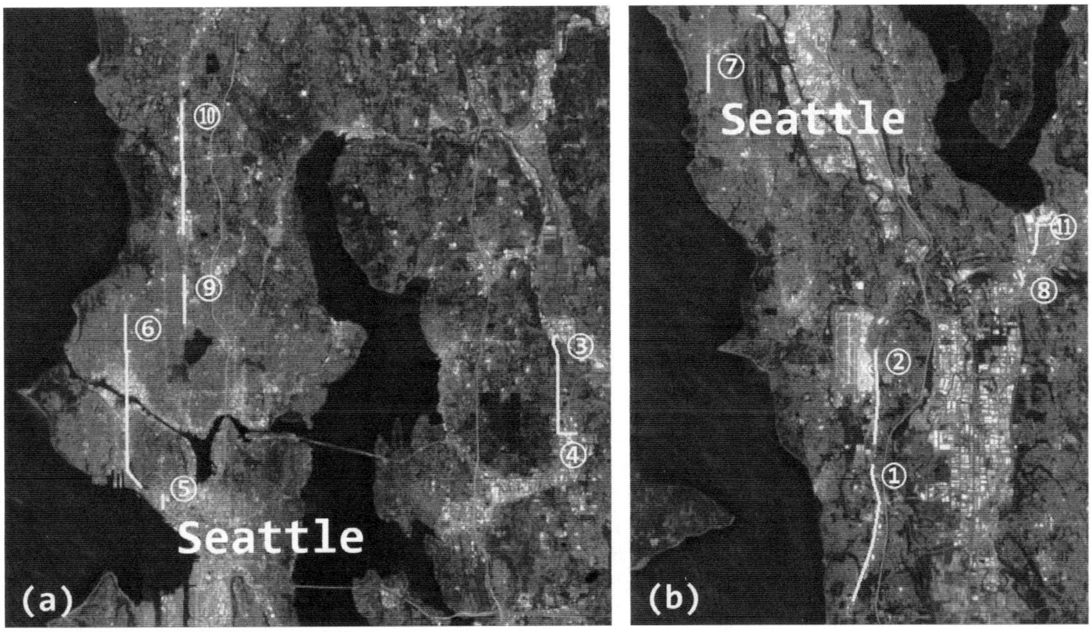

Figure 1. Locations of selected TSP corridors: (*a*) north section; (*b*) south section.

- π = the expected number of crashes (of the type of interest, severity, etc.) of a specific entity in the "after" period had the treatment not been implemented. This value needs to be predicted.
- λ = the expected number of crashes (of the type of interest, severity, etc.) of the entity in the "after" period. This value needs to be estimated.

The treatment effects can then be expressed in two ways:

- $\delta = \pi - \lambda$, the reduction in expected number of crashes of interest in the "after" period.
- $\theta = \lambda / \pi$, the ratio of expected number of crashes of interest in the "after" period to the expected number of crashes of interest in the "after" period without the treatment.

According to Hauer, there are two existing issues in the naïve and comparison-group before–after study approaches that lead to bias in estimating the "before" period crash frequency of the treatment entities (*17*). One issue is that the selection of sites for a treatment may be based on the sites' high crash frequencies, and this link causes a regression-to-the-mean bias. In the context of this study, TSP was not implemented based on road safety performance, however, it is unclear if there were other indirect links between the selection of TSP's implementation locations and the associated corridor's crash history. The second issue is that the "before" period varies from 1 year, to 2 years, to multiple years, but estimating the "before" period crash frequency solely based on the arithmetic average of counted crash frequencies also leads to bias. The EB method provides more precise estimations by addressing the above two issues.

The EB approach employs an SPF developed with reference group data, which is then used in combination with the observed "before" crash frequency of the treatment group to predict the "would have" expected crash frequency, π, of each treated entity in the "after" period, had the treatment not been implemented. Negative binomial regression models are often used to estimate SPFs in road safety studies, and are also used with the EB method provided by the *Highway Safety Manual* (*11, 18*). In this paper, a negative binomial model was developed based on available traffic and roadway data, and possible issues of temporal change or crash report inconsistency were addressed by using yearly calibration multipliers (*19, 20*). The equation expression for the negative binomial model used in this study is:

$$E(Y_i) = L_i T \times AADT_i^{\beta_1} \times \exp(\beta_0 + \beta_2 nl_i + \beta_3 sl_i) \times \gamma_y$$

where

$E(Y_i)$ = expected crash frequency of reference street section i,
L_i = length (mile [mi]) of reference street section i,
T = time period (year) over which crashes were counted,
$AADT_i$ = average AADT (vehicles per day [vpd]) through time period T of reference street section i,
nl_i = number of lanes of reference street section i,

Table 3. Descriptive Statistics

		Treatment group				Reference group			
Variable	Year	Min.	Max.	Mean	SD	Min.	Max.	Mean	SD
AADT	2002	8,151	43,979	27,392.9	11,556.0	3,400	82,237	28,550.7	15,946.1
	2003	7,964	45,033	26,508.2	12,077.6	3,427	118,977	28,971.7	18,257.6
	2004	7,715	44,833	27,002.6	12,037.1	3,461	123,617	27,857.8	17,109.6
	2005	7,653	44,833	27,199.5	11,468.4	3,682	121,952	27,253.4	16,986.2
	2006	7,404	45,733	27,340.2	12,114.9	2,665	121,232	27,759.2	17,191.2
	2007	7,311	44,883	26,923.8	11,884.3	2,673	119,995	27,673.5	17,363.6
	2008	7,218	44,033	26,006.1	11,003.0	2,503	95,430	26,451.6	15,449.0
	2009	7,591	42,233	25,987.3	10,643.4	3,419	93,933	26,274.3	14,962.4
	2010	6,906	43,286	26,297.7	11,237.9	3,391	97,277	26,334.8	14,866.5
	2011	6,346	48,567	26,621.0	12,471.5	3,672	98,408	26,027.5	14,693.1
	2012	7,031	44,138	26,449.5	11,505.5	2,785	101,976	26,292.3	15,004.0
	2013	6,160	44,138	26,174.8	11,589.6	2,785	101,976	26,434.7	15,241.5
	2014	7,529	44,204	26,936.7	11,136.4	2,668	102,130	26,975.1	15,377.4
	2015	6,766	44,167	27,034.1	11,669.5	2,738	105,725	27,831.2	15,889.6
Total crashes	2002	7	337	120.5	118.5	0	281	38.6	47.3
	2003	7	273	112.9	102.6	0	316	38.4	50.4
	2004	8	294	109.2	104.1	0	294	39.4	48.4
	2005	2	345	123.3	120.3	1	288	42.2	47.8
	2006	7	310	114.0	103.2	3	318	43.3	47.5
	2007	4	260	100.8	91.8	4	276	39.9	42.7
	2008	14	361	110.5	109.5	2	225	36.3	38.1
	2009	7	235	88.5	77.3	3	199	33.3	34.8
	2010	9	202	83.4	70.0	2	182	33.6	33.9
	2011	11	231	87.4	78.4	3	191	33.7	34.5
	2012	13	274	95.5	89.3	1	248	38.5	40.5
	2013	14	280	88.2	88.8	1	240	38.0	39.9
	2014	7	297	105.1	104.3	1	268	41.3	44.4
	2015	5	386	111.4	121.7	2	249	44.4	43.3
PDO crashes	2002	4	214	72.5	72.6	0	175	23.4	28.7
	2003	5	158	68.3	59.9	0	186	23.9	30.5
	2004	2	170	67.2	62.4	0	171	24.7	28.7
	2005	2	209	76.2	71.8	0	171	25.9	29.7
	2006	4	194	70.1	62.4	1	203	27.7	31.0
	2007	4	176	63.5	57.7	2	185	25.9	28.0
	2008	9	238	70.5	70.3	0	141	23.0	24.6
	2009	6	142	53.5	45.8	1	128	21.1	22.1
	2010	8	130	52.9	44.4	1	123	21.8	22.5
	2011	9	149	53.2	48.2	1	131	21.4	22.2
	2012	10	181	58.8	54.9	1	162	24.5	25.9
	2013	9	176	55.8	55.6	1	160	25.4	26.8
	2014	4	218	67.6	70.2	0	183	27.1	29.2
	2015	4	250	68.2	76.2	1	160	29.1	28.3
FI crashes	2002	2	130	48.0	47.9	0	108	15.3	19.3
	2003	1	116	44.6	43.8	0	130	14.5	20.6
	2004	1	129	42.0	42.7	0	123	14.8	20.5
	2005	0	136	47.1	48.8	0	117	16.3	18.8
	2006	2	116	43.9	41.5	0	115	15.5	17.3
	2007	0	84	37.4	35.2	0	91	14.0	15.3
	2008	3	123	40.0	39.7	0	84	13.4	14.0
	2009	1	93	35.0	31.7	0	71	12.2	13.1
	2010	1	72	30.5	26.4	0	59	11.8	12.2

(continued)

Table 3. (continued)

Variable	Year	Treatment group				Reference group			
		Min.	Max.	Mean	SD	Min.	Max.	Mean	SD
	2011	2	82	34.2	30.7	0	61	12.3	13.1
	2012	3	93	36.7	35.4	0	86	14.0	15.3
	2013	3	104	32.4	33.5	0	80	12.6	13.5
	2014	3	105	37.5	36.2	0	85	14.2	16.0
	2015	1	136	43.2	46.4	0	89	15.3	16.1
Length	All	0.41	6.10	2.13	1.94	0.12	24.49	1.80	3.39
No. of lanes	All	2	6	4.18	1.33	2	6	3.49	1.18
Speed limit	All	25	45	36.36	6.36	25	45	38.20	5.30

Note: SD = standard deviation.

sl_i = posted speed limit (miles per hour [mph]) of reference street section *i*,

$\gamma_y = \Sigma_{iy}$ (observed crash frequency of reference section *i* for year *y*)/Σ_{i2002} (observed crash frequency of reference section *i* for year 2002), yearly calibration multiplier, adjustment factor for year-to-year changes in possible crash reporting inconsistency.

The SPF estimates are listed in Table 4. This SPF was then brought to the treatment group data set to estimate the "before" period crash frequency of treated street sections:

$$E\left(K_{jy}\right) = a_y \times [L_j \times AADT_j^{\beta_1} \times \exp\left(\beta_0 + \beta_2 nl_j + \beta_3 sl_j\right) \times \gamma_y]$$

where

$E(K_{jy})$ = expected crash frequency of treated street section *j* for year *y*,

$a_y = \Sigma_{iy}$ (observed crash frequency of reference section *i* for year *y*)/Σ_{iy} (SPF predicted crash frequency of reference section *i* for year *y*), adjustment factor for year-to-year changes in traffic volume compared with the average value used in estimating SPF, and year-to-year changes in factors that are "not measured" or "not understood",

L_j = length (mile [mi]) of treated street section *j*,

$AADT_j$ = average AADT (vehicles per day [vpd]) through time period *T* of treated street section *j*,

nl_j = number of lanes of treated street section *j*,

sl_j = posted speed limit (miles per hour [mph]) of treated street section *j*.

The expected crash frequency during the "before" period at treated section *j* is then:

$$M_j = w_j \times P_j + \left(1 - w_j\right) \times K_j$$

where

M_j = expected crash frequency during the "before" period at treated section *j*,

$w_j = 1/(1 + kP_j)$; *k* is the dispersion parameter of the negative binomial SPF,

P_j = sum of annual SPF predictions of treated segment *j* during the "before" period,

K_j = counted crash frequency of treated segment *j* during the "before" period.

Then, the expected crash frequency during the "after" period that would have occurred at treated street section *j* without implementing the treatment is predicted, and its variance:

$$\hat{\pi}_j = C_j \times M_j$$

$$V\hat{A}R\left(\hat{\pi}_j\right) = C_j^2\, V\hat{A}R\left(M_j\right) = C_j^2\left(1 - w_j\right)M_j$$

where $\hat{\pi}_j$ = expected crash frequency during the "after" period at treated section *j* without the treatment;

where $C_j = \Sigma_{jy}$ (SPF predicted crash frequency of treated section *j* for "after" period year *y*)/Σ_{jx} (SPF predicted crash frequency of treated section *j* for "before" period year *x*).

The next steps of the EB method are to sum up the predicted "after" period crashes had the treatment not been implemented, and the observed actual "after" period crashes with treatment. Then, calculate the effectiveness of the treatment. The sum of the predicted crashes and its variance are denoted by:

$$\hat{\pi} = \sum_{j=1}^{J} \hat{\pi}_j$$

$$V\hat{A}R\left(\hat{\pi}\right) = \sum_{j=1}^{J} V\hat{A}R\left(\hat{\pi}_j\right)$$

The sum of the observed crashes and its variance are expressed by:

Table 4. SPF Estimates[a]

Variable	Total crashes	PDO crashes	FI crashes
Length × Time (Exposure)	1	1	1
Ln (AADT)[b]	0.9352	0.9615	0.8903
Posted speed limit	−0.0484	−0.0513	−0.0432
Number of lanes	0.1936	0.1916	0.1987
Intercept	−4.8377	−5.4220	−5.6448
Yearly multiplier γ			
2002	1	1	1
2003	0.9931	1.0223	0.9485
2004	1.0204	1.0554	0.9668
2005	1.0918	1.1090	1.0655
2006	1.1198	1.1872	1.0166
2007	1.0342	1.1096	0.9188
2008	0.9403	0.9829	0.8751
2009	0.8633	0.9047	0.8000
2010	0.8692	0.9321	0.7729
2011	0.8726	0.9167	0.8052
2012	0.9955	1.0468	0.9170
2013	0.9831	1.0873	0.8236
2014	1.0704	1.1615	0.9310
2015	1.1502	1.2471	1.0017
Dispersion parameter k	0.3787	0.4131	0.3393

[a]Coefficients significant at the 95% confidence level.
[b]Natural log of AADT.

$$\hat{\lambda} = L = \sum_{j=1}^{J} L_j$$

$$V\hat{A}R\left(\hat{\lambda}\right) = \hat{\lambda}$$

Then the estimations of two measures of treatment effectiveness, $\hat{\delta}$ and $\hat{\theta}$, and their variances are calculated.

$$\hat{\delta} = \hat{\pi} - \hat{\lambda}$$

$$V\hat{A}R\left(\hat{\delta}\right) = V\hat{A}R\left(\hat{\pi}\right) - V\hat{A}R\left(\hat{\lambda}\right)$$

$$\hat{\theta} = \frac{\hat{\lambda} / \hat{\pi}}{1 + \frac{V\hat{A}R\left(\hat{\pi}\right)}{\hat{\pi}^2}}$$

$$V\hat{A}R\left(\hat{\theta}\right) = \frac{\hat{\theta}^2 \left(\frac{V\hat{A}R\left(\hat{\lambda}\right)}{\hat{\lambda}^2} + \frac{V\hat{A}R\left(\hat{\pi}\right)}{\hat{\pi}^2} \right)}{\left(1 + \frac{V\hat{A}R\left(\hat{\pi}\right)}{\hat{\pi}^2} \right)^2}$$

Of these two measures of effectiveness, $\hat{\delta}$ is an estimation of the reduction in the expected number of crashes, whereas $\hat{\theta}$ is an estimation of the index of effectiveness. With the index of effectiveness, the percentage reduction in expected number of crashes is calculated as 100% × (1 - $\hat{\theta}$).

Results

Following the steps of the EB before–after study, the safety assessment of 11 King County, Washington, TSP corridors was carried out. Results are summarized in Table 5. The difference between predicted and observed "after" period crashes ($\hat{\delta}$) indicate that in general, with TSP implementation, the corridor-level number of crashes reduced. When comparing each corridor's observed number of total crashes (with TSP) with the predicted number of total crashes (without TSP), 9 out of 11 corridors showed a significant reduction, whereas 2 out of 11 corridors showed a slight increase (less than 10%). In terms of PDO crashes, 9 out of 11 corridors showed a significant reduction. In terms of FI crashes, 7 out of 11 corridors showed a significant reduction. The effects of TSP on PDO crashes and FI crashes were significantly different at a 95% confidence level, with an average reduction in PDO crashes of 21%, larger than the 4% average reduction in FI crashes for each treated section.

The index of effectiveness ($\hat{\delta}$) values give a direct sense of the percentage of crash reduction correlated with the implementation of TSP. With the implementation of TSP, there was a 13% reduction in total crashes. The effectiveness was 0.8732, 95% CI [0.8387, 0.9077], which indicates a 10% to 16% reduction. A 16% reduction in PDO crashes, and a 5% reduction in FI crashes were found, with effectiveness indices of 0.8408, 95% CI [0.7990, 0.8826], and 0.9489, 95% CI [0.8872, 1.0107].

Comparing the results from this study with the results from the four prior studies cited in the literature review, this study is consistent with three out of the four studies. The Australian studies carried out by Goh et al. and Naznin et al. all reported a reduction in crashes with the implementation of TSP (*14–16*), whereas the study by Shahla et al. reported an increase in crashes with the implementation of TSP (*13*).

Conclusions

TSP has become popular among U.S. cities as a transit preferential treatment that improves transit operational performance through adjusting signal timing when transit vehicles are approaching signalized intersections. The majority of attention has been given to the operational effects of TSP. Only a handful of prior studies discussed such a system's effects on traffic safety, with even fewer studies investigating North American cases. Prior assessment results have been mixed, leaving the understanding of TSP's safety effects unclear. The purpose of this study was to add to the existing research on this topic, by carrying out an EB before–after analysis of TSP implementation cases from U.S. cities.

Table 5. Assessment Results

Corridor ID	Observed "After" period crashes			Predicted "After" period crashes (SE)		
	Total	PDO	FI	Total	PDO	FI
1	1,098	656	442	1,149.3 (28.4)	776.9 (23.9)	370.8 (15.4)
2	1,468	974	494	1,628.1 (33.3)	1,065.3 (27.5)	556.9 (18.6)
3	171	135	36	159.6 (8.5)	117.1 (7.4)	42.8 (4.1)
4	39	27	12	47.2 (4.4)	38.0 (4.0)	10.0 (1.9)
5	46	32	14	67.8 (4.5)	55.2 (4.1)	11.7 (1.7)
6	381	219	162	508.1 (12.9)	307.7 (10.3)	194.6 (7.6)
7	71	51	20	105.5 (5.8)	73.1 (4.9)	31.5 (2.9)
8	57	34	23	74.2 (2.7)	47.4 (2.2)	26.0 (1.5)
9	98	50	48	136.1 (3.7)	82.4 (3.0)	51.2 (2.1)
10	187	115	72	268.9 (5.3)	165.7 (4.3)	99.7 (3.0)
11	18	11	7	16.7 (1.4)	10.8 (1.2)	5.9 (0.8)
Sum	3,634	2,304	1,330	4,161.3 (47.8)	2,739.6 (39.8)	1,401.1 (26.3)
Reduction in expected crashes (SE)				527.3 (47.8)	435.6 (39.8)	71.1 (26.3)
95% confidence interval				[433.7, 620.9]	[357.7, 513.6]	[19.6, 122.6]
Index of effectiveness (SE)				0.8732 (0.0176)	0.8408 (0.0213)	0.9489 (0.0315)
95% confidence interval				[0.8387, 0.9077]	[0.7990, 0.8826]	[0.8872, 1.0107]

Note: SE = standard error; PDO = property-damage-only crashes; FI = fatal and injury crashes.

The EB before–after study was carried out using data of 11 transit corridors in King County, Washington, which had TSP activated in 2010–2014. Another 75 street sections without TSP treatment, also from King County, were selected to form a reference group. The findings from the EB before–after analysis are as follows.

- The implementation of TSP correlates with the reduction of total corridor-level crashes, with an index of effectiveness of 0.8732 (about a 13% reduction).
- The implementation of TSP correlates with the reduction of PDO crashes and FI crashes, at a corridor level, with indices of effectiveness of 0.8387 and 0.9077, respectively.
- The assessment results from this study are consistent with most prior studies assessing the safety effects of TSP.

The findings of this study indicate that TSP, apart from improving operational performance of transit systems, also brings road safety benefits. Safety assessment of operational-performance-targeted transportation treatments should be encouraged, as the study findings suggest that the safety benefits from TSP are significant, but current TSP decision-making and design practices lack such a process. In addition to the operational benefit evaluation, when carrying out cost–benefit analyses for TSP, the traffic safety benefits should also be considered.

This study was accomplished using a widely recognized, robust analytical method; therefore, the results from this study, together with the results from prior studies and future studies on this topic will provide legitimate crash modification factors for TSP assessments. In terms of future studies, TSP's effects on different crash types could be assessed. Furthermore, crashes involving different groups of road users (e.g., pedestrians, cyclists, motorcyclists) could be analyzed separately in further studies when relevant traffic exposure data (i.e., pedestrian, and bike traffic volumes) are available. With more studies carried out on different types of TSP practices in different places, the understanding of TSP's effects will be enhanced, which will benefit future practices involving TSP.

Acknowledgments

The authors thank the Federal Highway Administration, Washington State Department of Transportation, and King County Metro Transit for the data they provided for this research.

References

1. Smith, H. R., B. Hemily, and M. Ivanovic. *Transit Signal Priority (TSP): A Planning and Implementation Handbook*, ITS America, May, 2005.
2. Danaher, A. R. *TCRP Synthesis 83: Bus and Rail Transit Preferential Treatments in Mixed Traffic*. Transportation Research Board of the National Academies, Washington, D.C., 2010.
3. Danaher, A. R., K. Blume, H. S. Levinson, and S. Zimmerman. *TCRP Report 118: – Bus Rapid Transit Practitioner's Guide*. Transportation Research Board of the National Academies, Washington, D.C., 2007.
4. Song, Y., M. Zlatkovic, and R. J. Porter. Evaluation of GPS-Based Transit Signal Priority for Mixed-Traffic Bus Rapid

Transit. *Transportation Research Record: Journal of the Transportation Research Board*, 2016. 2539: 30–39.
5. Federal Highway Administration (FHWA). *Manual on Uniform Traffic Control Devices*. 2009.
6. Federal Highway Administration (FHWA). *Signalized Intersection Informational Guide*, 2nd ed. July, 2013.
7. Retting, R., and M. Greene. Influence of Traffic Signal Timing on Red-Light Running and Potential Vehicle Conflicts at Urban Intersections. *Transportation Research Record: Journal of the Transportation Research Board*, 1997. 1595: 1–7.
8. Retting, R. A., J. F. Chapline, and A. F. Williams. Changes in Crash Risk Following Re-Timing of Traffic Signal Change Intervals. *Accident Analysis & Prevention*, Vol. 34, No. 2, 2002, pp. 215–220.
9. Papaioannou, P. Driver Behaviour, Dilemma Zone and Safety Effects at Urban Signalised Intersections in Greece. *Accident Analysis & Prevention*, Vol. 39, No. 1, 2007, pp. 147–158.
10. Federal Highway Administration (FHWA). *Scale and Scope of Safety Assessment Methods in the Project Development Process*. November, 2016.
11. American Association of State Highway and Transportation Officials (AASHTO). *Highway Safety Manual*. 2009.
12. Cheung, C., A. Shalaby, B. Persaud, and A. Hadayeghi. Models for Safety Analysis of Road Surface Transit. *Transportation Research Record: Journal of the Transportation Research Board*, 2008. 2063: 168–175.
13. Shahla, F., A. Shalaby, B. Persaud, and A. Hadayeghi. Analysis of Transit Safety at Signalized Intersections in Toronto, Ontario, Canada. *Transportation Research Record: Journal of the Transportation Research Board*, 2009. 2102: 108–114.
14. Goh, K.C.K., G. Currie, M. Sarvi, and D. Logan. Road Safety Benefits from Bus Priority: An Empirical Study. *Transportation Research Record: Journal of the Transportation Research Board*, 2013. 2352: 41–49.
15. Goh, K.C.K., G. Currie, M. Sarvi, and D. Logan. Bus Accident Analysis of Routes with/without Bus Priority. *Accident Analysis & Prevention*, Vol. 65, 2014, pp. 18–27.
16. Naznin, F., G. Currie, M. Sarvi, and D. Logan. An Empirical Bayes Safety Evaluation of Tram/Streetcar Signal and Lane Priority Measures in Melbourne. *Traffic Injury Prevention*, Vol. 17, No. 1, 2016, pp. 91–97.
17. Hauer, E. *Observational Before–After Studies in Road Safety: Estimating the Effects of Highway and Traffic Engineering Measures on Road Safety*. Emerald Group Publishing Limited, United Kingdom, 1997.
18. Wood, J. S., E. T. Donnell, and R. J. Porter. Comparison of Safety Effect Estimates Obtained from Empirical Bayes Before–After Study, Propensity Scores-Potential Outcomes Framework, and Regression Model with Cross-Sectional Data. *Accident Analysis & Prevention*, Vol. 75, 2015, pp. 144–154.
19. Wu, H., Z. Han, M. R. Murphy, and Z. Zhang. Empirical Bayes Before–After Study on Safety Effect of Narrow Pavement Widening Projects in Texas. *Transportation Research Record: Journal of the Transportation Research Board*, 2015. 2515: 63–69.
20. Harkey, D. L., R. Srinivasan, J. Baek, F. M. Council, K. Eccles, N. Lefler, F. Gross, B. Persaud, C. Lyon, E. Hauer, and J. A. Bonneson. *National Cooperative Highway Research Program (NCHRP) Report 617: – Accident Modification Factors for Traffic Engineering and ITS Improvements*. Transportation Research Board of the National Academies, Washington, D.C., 2008.

The Task Force on Transit Safety and Security (AP018T) peer-reviewed this paper (18-03340).

The work presented in this paper remains the sole responsibility of the authors.

Article

TRR
JOURNAL OF THE TRANSPORTATION RESEARCH BOARD

Transportation Research Record
2018, Vol. 2672(8) 19–28

DOI: 10.1177/0361198118758677
journals.sagepub.com/home/trr
SAGE

Individual, Travel, and Bus Stop Characteristics Influencing Travelers' Safety Perceptions

Roberto F. Abenoza[1], Vania Ceccato[1], Yusak O. Susilo[1], and Oded Cats[2,3]

Abstract

Ensuring safety during door-to-door public transport trips is a fundamental challenge to service providers, as safety influences individuals' mobility. Using reported safety perceptions of travelers waiting at six bus stops with different characteristics in Stockholm, this study investigates factors that have an impact on determining travelers' perceived safety and crime perceptions. This is done by assessing the importance of real-time information provision and the environmental characteristics of bus stops during the day and at night for different types of crime, after controlling for travelers' individual and trip characteristics, and their previous experiences of victimization. Interaction effects of age, gender, and travel frequency are also tested. The results suggest that bus shelter characteristics, natural surveillance, and trustworthy real-time information are the most important factors influencing safety and crime perceptions. Additionally, safety perceptions are strongly influenced by previous experiences of victimization. The effect of perceived feelings about crime and safety are found to be nuanced by age and gender. Unlike some common beliefs, travelers: (1) feel less worried about becoming a victim of crime at bus stops associated with high crime rates; (2) prefer opaque shelters at night; and (3) have higher safety perceptions when the stop is located in an area of mixed land use. The impact of a bus stop's number of passers-by is found to be insignificant. No direct or indirect effects can be attributed to frequency of travel by bus, indicating that familiar places and routine behavior have no effect on declared crime and safety perceptions.

Safety is one of many factors influencing individuals' mobility. Ensuring safety along door-to-door public transport (PT) trips is a fundamental challenge to those responsible for providing the service. The experience of PT consists of moving times (on board and getting to and from the boarding and alighting points) and waiting and transferring times. Waiting times are paramount for the entire trip as they are usually despised by travelers and have been found to be a key component when forming an overall satisfaction evaluation of a door-to-door PT trip (*1*). It is known that places where PT travelers spend their waiting time, including bus stops and bus stations, tend to be more criminogenic than other locations (*2, 3*).

Travelers' perceived safety is more important than actual safety in influencing perceived travel and waiting times (*4*) and influences travel behavior, the decision to use PT (*5*), travel time, and travel mode choice (*6*). Moreover, there is a strong link between travelers' safety perceptions and overall travel satisfaction, which in turn affects PT customer retention and recommendation. Travelers' safety perceptions have consistently been found to be amongst the most important determinants of travel satisfaction (*7*), and this includes the results of a particular analysis of a 13-year period (*8*). Therefore, it is essential to investigate how safety is perceived by travelers in waiting environments.

Previous studies (*5, 3*) have demonstrated that fear of crime, and safety perceptions are correlated with actual crime levels, the surrounding environment, and the overall design quality and features of the transport facilities. However, hitherto, there have been no scientific findings concerning the impact of bus stop factors [real-time information (RTI), the bus stops themselves, and their surrounding characteristics] on safety perceptions, in particular in relation

[1]Department of Urban Planning and Built Environment, Royal Institute of Technology (KTH), Stockholm, Sweden
[2]Department of Transport and Planning, Delft University of Technology, Delft, the Netherlands
[3]Department of Civil and Architectural Engineering, Royal Institute of Technology (KTH), Stockholm, Sweden

Corresponding Author:
Address correspondence to Roberto F. Abenoza: rfa@kth.se

to variations according to the time of day or night and the type of crime (property or person). It also remains unknown whether travel frequency moderates the effect of the factors.

To explore these questions, the objectives of this study are twofold. First, the paper takes a new look at the factors that might influence travelers' self-reported safety perceptions. We focus on travelers' experiences while waiting at bus stops because buses constitute the primary and most heavily used PT travel mode in many urban environments. This is done by assessing the importance of RTI provision and the environmental characteristics of bus stops during the day and at night and for different types of crime (property or person), after controlling for travelers' individual socio-demographic and travel characteristics, and their previous experiences of victimization. Second, the study aims to extend the current knowledge of how age, gender, and travel frequency moderate the effect of bus stop design, RTI, surrounding characteristics, and the impact of previous experiences of crime on safety perceptions. Obtaining the key design and information factors that minimize travelers' perceptions of being unsafe will allow stakeholders (urban and transport planners) and those responsible for the design and maintenance of bus stops to provide environments that are perceived to be safer when waiting for a bus.

This study begins by reviewing the relevant literature on crime, looking at variable aspects that influence travelers' safety perceptions of PT, such as socio-demographic and travel characteristics, the bus stops themselves and the type of environment that surrounds them, previous experiences of crime, and RTI. Section 3 presents the dataset and the bus stops investigated and Section 4 continues with descriptive statistics and exhibiting and discussing model estimation results. Section 5 completes this study with a discussion of policy implications, study limitations, and directions for potential future research.

Literature Review

There are several factors that affect travelers' perceived safety at bus stops. Some of them are related to the characteristics of those who fear (e.g., gender, age, disability, previous experiences of victimization), whereas others are triggered by the environment (e.g., the characteristics of the bus stop, the kind of neighborhood, the type of transportation system) or by other, less tangible factors that affect individuals' overall level of anxiety (e.g., fears about terrorism and the future). In this study we focus on three dimensions, the bus stops' design, the surrounding environment of the bus stops, and the travelers' individual characteristics.

Travelers' Socio-Demographic and Travel Characteristics

There is a considerable amount of evidence in the criminology and transport literature indicating that socio-demographic characteristics, such as ethnicity, income level, gender, and age, affect travelers' safety perceptions (*8–10*). Safety is gendered and age dependent and interacts with the environment surrounding bus stops in a variety of ways. For example, feeling safe while waiting for PT was more important for women than men (*11*) and young women feel less safe than men when traveling by PT (*12*). Furthermore, Tucker (*13*) concluded that women and the elderly are especially fearful of crime and are more apprehensive while waiting for the bus. Women's fear was attributed to their feeling more exposed to affective crimes and also being responsible for children. Elderly people, who use bus services frequently, perceive that they are less safe (*10*). Yavuz and Welch (*9*) studied interaction effects based on gender for train trips. At peak hour, as would be expected with overcrowded railway stations, men feel safer than women. Male travelers' security perceptions are highly affected by the reliability of the service, followed by the presence of police and their previous experiences of crime. By contrast, for female travelers, it is mainly their previous experiences of crime that affects their security perceptions, followed by the presence of police, reliability of the service, and the presence of CCTV cameras.

However, some studies did not find that age and gender had any effect on security perception with regard to PT. For example, Delbosc and Currie's (*14*) structural equation model showed indirect negative effects for women and elderly people through feeling safe in the home (age and gender) and in the neighborhood (age). In addition, Currie et al., (*15*) found that for young travelers, their safety perceptions were mainly influenced by feeling comfortable when traveling alongside people who were unknown to them. In their study, gender and actual experiences of unsafe events were found to exert a moderate effect. The relationship between gender and safety goes beyond the female–male dichotomy. Ceccato and Paz (*16*) indicate the need to consider safety from the perspective of those who are potentially more targeted on PT, such as the lesbian, gay, bisexual, trans, and queer community, as recent research shows that gay and transgendered persons are more often targets of harassment and violence on PT.

Very few studies have been published with a focus on the influence of travel characteristics, such as trip purpose or travel frequency, on crime and safety perceptions. Nothing has been found on trip purpose and the very little found on travel frequency has been inconclusive. For example, Currie et al. (*15*), with regard to young travelers, and Yavuz and Welch (*9*), with regard to train passengers, found that travel frequency did not affect influence travelers' safety perceptions. In contrast, the Department for Transport in the UK indicates that familiarity with an area and the transport mode are important factors that positively influence perceived safety. In other words, those who frequently use PT feel safer than infrequent users (*17*). In addition, Wallace et al. (*18*) demonstrated that travelers who used PT more frequently were more likely to notice transport security measures, thus, indirectly feeling safer.

Bus Stop Design: Environmental Characteristics

Travelers' declared perceived safety at bus stops depends on multiscale environmental and temporal features that take effect when they leave their starting point of their trip. Yet, there are particular sites (such as bus stops and bus stations) that shape travelers' safety perceptions because people spend a considerable amount of time at or in them when traveling. The international literature has long shown that the environmental features of these sites are bound to affect travelers' perceived safety (*13*). This implies that environments can be planned in a way that reduces the possibility of crime occurring and improves overall perceived safety. In this line of thought, Crime Prevention Through Environmental Design (CPTED) suggests that "the proper design and effective use of the built environment can lead to a reduction in the fear of crime and the incidence of crime, and to an improvement in the quality of life" (*19*). CPTED points out certain environmental principles that when used in design can stimulate natural surveillance, foster territoriality, and reduce areas of conflict by controlling access. Corroborating CPTED principles, Tucker's review (*13*) indicates that there are several features of bus stops that contribute to increasing travelers' security perceptions. These include: shelters, benches, lighting, location, surrounding environment, design, maintenance and cleanliness of the bus stop, number of people waiting and passing by, the amount of time waiting, monitoring of the bus stop (CCTV), and the access provided to and from it.

Travelers' declared perceived safety is also affected by what happens at bus stops and in surrounding areas. Many people can be concentrated at bus stops, which makes it easier for offenders to commit crime, and bus stops can potentially pull motivated offenders toward them by the types of environments they offer (*2, 3, 20*). Liggett et al. (*3*) found that, in the United States, opaque shelters, litter, and bus stops located in or near empty areas and close to liquor stores, pubs, and establishments for adults led to a higher incidence of crime. With regard to metro stations in Stockholm, Ceccato et al. (*11*) found that around half of the variation in travelers' perceived safety concerns was explained by the environmental conditions at underground stations and conditions in the surrounding areas. Unsafe underground stations were associated with visible social disturbance in lobbies, poor surveillance, higher rates of violence, and more public disorder. In turn, safe underground stations had effective formal social control—CCTV cameras or presence of guards—and had a high potential for promoting natural surveillance.

Yet, crime does not happen at random at bus stops. Crime tends to follow individuals' daily rhythmic patterns of activities, and crime may just occur in a particular area and at a time when a potential offender is aware of an opening. According to routine activity theory (*21*), for a crime to happen at a bus stop, some conditions must be in place: a motivated offender, a potential victim, and a lack of controllers (handlers, guardians, and place managers). A motivated offender could be controlled by a handler. If the target is an individual, then guardians can be other travelers who are at the same bus stop. Place managers can be, for example, shop managers whose premises happen to be just in front of the bus stop and who are able to watch what happens there. In general, crimes can be categorized into property and person (*22*). The main difference between them is that the former involves violence aiming to acquire anything tangible, such as belongings or money (theft), whereas the latter has an emotional component and includes assault, murder, disorder, and rape.

Risk with regard to crime and travelers' perceived safety varies temporarily, hourly, daily, weekly, and seasonally (*20*). On top of light conditions and day or night time hours, peak hour periods, specifically, are characterized by larger flows of travelers (targets) with more potential guardians than off-peak hours. Therefore, declared safety at bus stops may reflect several other conditions experienced along the trip in a diverse array of transit environments during daily activities (*23*). These conditions are bound to have an effect on travelers' perceived safety.

RTI and Previous Experiences of Victimization

Some authors have partially explored how RTI variables affect safety and crime perceptions. Dziekan and Kottenhoff (*24*) synthesized a series of benefits attributed to at-stop RTI. Amongst these, the most relevant for this study were: perceived reduced waiting time; travel behavior adjustment; travel mode choice; and positive psychological effects that led to an increased perception of personal safety. Zhang et al. (*25*) concluded that travelers' safety perceptions and overall travel satisfaction increased when RTI was introduced at bus stops, and RTI was made available through handheld devices. For example, the results of a trial implementation of mobile RTI in Seattle showed that travelers' feel safer, experience reduced waiting times, are more satisfied with their trip, and increase their weekly ridership (*26*). In turn, Brakewood et al. (*27*) determined that in addition to the improvement in travelers' perceptions of their waiting time, stress and anxiety were reduced and their sense of personal safety was increased. Moreover, thanks to RTI, travelers' safety perceptions increased during the day with regard to the control group. However, no significant differences with regard to travelers' security perceptions were found at night.

Fear of crime has been studied widely. In an in-depth review (*28*), some evidence was found that previous experience of direct crime does not have a strong relationship with fear of crime (crime and safety perceptions). In contrast, rumors, media and crime experienced by acquaintances exert an impact on these perceptions (*28*). However, Teseloni and Zarafonitou (*29*) demonstrated that previous experience of direct and indirect crime is associated with feeling unsafe when walking alone at night. In addition, they found that people who were more exposed to crime by living either in criminogenic areas or by being more active in their day-to-day life (commuters) were more prone to feeling victimized. Furthermore, Quann and Hung (*30*) showed that the relationship between victimization, safety, and perceptions of crime varies according to crime type (person or property).

Hypotheses of Study

Taken together, the existing literature suggests that individual and environmental factors have an impact on travelers' declared perceived safety at bus stops. For the purposes of this study, we follow the recent strand of Western research on perceived safety in PT environments and hypothesize that perceived safety at bus stops is related not only to the environmental conditions at the bus stop itself but also to its surroundings, such as land use, socio-economic conditions, and the particular city context. Travelers' individual characteristics are also expected to affect perceived safety. Therefore, this study will test the following set of hypotheses:

(1) Assuming CPTED principles, travelers' perceived safety is reduced by bus stops with poor capacity to promote natural surveillance (opaque surface, low traffic density, few passers-by, poor PT service). Their perceived safety is affected negatively at bus stops that are crime attractors or generators (more criminogenic). Equally important are the bus stop surroundings. Travelers' perceived safety is affected by the surroundings at bus stops and mixed land use is more criminogenic than other land use types.
(2) Travelers' individual characteristics also matter in determining declared perceived safety levels. It is expected that women will declare feeling less safe than men. Travelers who have been victimized by crime tend to declare feeling less safe at bus stops than those who have not been a victim of crime.
(3) Frequency of travel from the bus stop should have an impact on safety and crime perceptions. Those who are frequent bus users and who are familiar with the schedules, the security measures of the stop, and the characteristics of the environment will be more satisfied with their safety (familiarity).

Methodology and Survey Description

A hardcopy survey was designed to evaluate travelers' safety perceptions at different bus stops. The survey was carried out in autumn 2016 with a random sample of 123 travelers who waited at six different bus stops in Stockholm, Sweden. No significant events were reported during the data collection that could affect the survey results. After the dataset was cleaned and verified for completeness, 108 samples were kept. Out of the 108, almost 75% were collected at three of the bus stops (Arkitektur-Moderna museet, Barnängen, and Mariatorget). The remaining 29 respondents were waiting at the three remaining stops (Hötorget, Erstagatan, and Slussen). The bus stops were selected on the basis of all being located in inner city areas but varying in their environmental characteristics such as land use, number of passers-by, and crime counts. The stops also differ as far as their service characteristics are concerned, characteristics such as design, frequency of service, and passenger volumes.

The survey included questions related to the travelers' safety perceptions (general, during the day and at night) for different types of crime (person and property), previous experiences of crime, socio-demographic and travel characteristics, and travel information and planning-related issues variables. The questionnaire was designed so that it could be completed in approximately 5 min.

With regard to studying the factors that influence safety perceptions at bus stops, the following variables were employed:

- *Socio-demographic and travel characteristics*: Gender; age; children in the household; marital status; frequency of travel from the given bus stop; and travel purpose.
- *Safety perceptions*: Feeling safe in the vicinity of or at the bus stop (safe day); feeling safe at night in the vicinity of or at the bus stop (safe night); and worried about becoming a victim of violence (crime person) or theft (crime property) in the vicinity of or at the bus stop.
- *Previous victimization*: Ever having been a victim of a crime (victim crime); having been a victim of theft during the past two years (victim theft); having been a victim of violence during the past two years (victim violence); and knowing anyone, either family or friends, who has been victim of a crime during the last two years (family).
- *Travel information and planning-related issues*: Planning the trip so as to spend as little time as possible at the bus stop (plan stop); trusting RTI displayed on the panels (trust info); feeling safer when RTI shows that the bus will arrive soon (real time soon); and feeling safer because of RTI (real time presence).

In addition to the aforementioned variables that were collected in the survey, the following variables were also included in the survey and were specified based on site visits, geographical information system tools, and PT service information:

- *Characteristics of the immediate bus stop surroundings*: Crime rate and number of passers-by—high or low; land use—mixed or other (commercial or residential).
- *Bus stop characteristics*: Presence of shelter; natural surveillance and presence of CCTV; frequency of service at bus stop; number of passers-by and road traffic level (these have the same response pattern (low/low/low or vice versa) for the same stops and, thus, are interchangeable).

The scale of measurement of travel information, planning-related issues, and safety perceptions is a Likert scale from 1

Table 1. Summary Statistics of Sample Profile

Variable	Category		Share (%)
Socio-demographic and travel characteristics (%)	Gender	Female (Male)	42.6
	Age	<30	32.4
		31–50	52.8
		>50	14.8
	Children in the household	Yes (No)	41.7
	Marital status	Single*	52.8
	Frequency trip	Frequent (weekly)	73.2
	Travel purpose	Commuting*	53.7
Safety perceptions (mean)	Safe day		4.15
	Safe night		3.83
	Crime person		2.71
	Crime property		2.81
Previous experiences of crime (%)	Victim crime	Yes (No)	25
	Victim theft	Yes (No)	9.3
	Victim violence	Yes (No)	13
	Family	Yes (No)	23.1
Travel information available and planning-related issues (mean)	Plan stop		4
	Trust info		3.96
	Real time soon		4
	Real time presence		3.91
Characteristics of the surrounding area (%)	Crime rate	High (Low)	44.4
	Passers-by	High (Low)	48.1
	Land use	Mixed*	56.5
Characteristics of the bus stop (%)	Natural surveillance	Good (Bad)	51.9
	Shelter	Opaque (Transparent)	21.3
	Service frequency	High <10 min in peak time (Low)	48.1
	CCTV	Yes (No)	14.8

Note: *Other.

(completely disagree) to 5 (completely agree). Safety perceptions, previous experiences of crime, and bus stop characteristics variables are defined as dummy variables. Natural surveillance was assessed by considering aspects such as direct view, the view from the outside toward the inside of the bus shelter, lighting at the location, and objects obstructing a direct view. Following Loukaitou-Sideris et al. (*31*), the crime rate at bus stops is calculated by normalizing crime counts by the yearly ridership per bus stop. The threshold between high and low crime rate was based on quantile classification and set to 250 boarding passengers per crime.

Analysis and Results

Descriptive Analysis

Table 1 presents the summary statistics of the socio-demographic and travel characteristics, safety perceptions, prior experiences of crime, relevance of available travel information and planning-related issues, characteristics of the bus stop, and the characteristics of the surrounding area. In some cases, the data is shown as the percentage of respondents by category (%), whereas for the Likert scale variables the mean is shown. Variables shown with an asterisk have "other" as the alternative response category. For the remaining nominal variables the alternative response category is shown in brackets.

About half of the respondents are middle-aged and single and have one or more children in the household. Almost three-quarters of people in the sample travel frequently (weekly or more often) and for most of them it was a commuting trip. Not surprisingly, and in line with previous research (*9*), safety perceptions are higher during the day than at night. Bus stops are characterized by: transparent shelters; no CCTV surveillance; good lighting and being located along the street; no litter; no signs of vandalism; no barriers; and windows in multiple sides. Furthermore, about half of the bus stops are very well served by PT and are subject to natural surveillance. Around half of the stops are located in areas with mixed land use and have high crime rates and number of passers-by.

Every fourth respondent has either an acquaintance who has been a victim of crime or they have been a victim themselves in the past. However, amongst these, only about 10%

have been subject to theft or violence in the last two years. Overall, travelers trust the RTI displayed at bus stops and in planning their trips so they wait for as little time as possible. In addition, travelers report feeling quite safe when RTI is displayed at bus stops and when it shows the bus is coming soon.

A *t*-test identified a significant average difference between being a victim of a specific type of offense – crime person and crime property (sig. = 0.063). In addition, significant average differences were also found when comparing safe day and safe night (sig. = 0.000). The results of the *t*-tests provide evidence that different types of offenses and safety perceptions at different types of the day should be examined individually.

Safety Perception Models

To systematically investigate both the factors influencing safety perceptions at different times of the day, and crime perceptions for different type of offenses [person (violence) and property (theft)] four regression models were estimated. The first two models specify dependent variables "safe day" for Model 1 and "safe night" for Model 2. In turn, "crime person" and "crime property" are the dependent variables of the remaining two models, M3 and M4, respectively. The model specification of all four models is composed of the same set of explanatory variables. This set of variables includes all the factors listed in Table 1: socio-demographic and travel characteristics; prior experiences of crime; travel information available and planning-related issues; characteristics of the bus stop; and characteristics of the surrounding area. Models 1 to 4 control for many factors that influence crime and safety (socio-demographic and travel characteristics) and, therefore, will allow for better identification of the most influential ones amongst them. The independent variables were tested for multi-collinearity issues, which highlighted CCTV, resulting in this being dropped from the models. As explained in Section 3, bus service frequency and road traffic levels are equivalent to passers-by and, therefore, neither were included in the models.

Models 1 to 4 are also enriched with the inclusion of and testing for interaction effects. Two-way interaction terms were included with the aim of examining whether gender, age, and travel frequency exert a differential impact on safety and crime perceptions. These base variables were selected based on the findings of previous research (*8, 13*) and tested against travel information, bus stop characteristics and characteristics of the surrounding area variables. A manual stepwise backward method was employed to keep the significant interaction effects in the model. This method consists of including, at first, main effect variables and all two-way interaction effects. Then, after checking the model output, all main effects are retained but only significant interaction effects. Several iterations are run until the models converge into a model specification that includes all main effects but only the significant interaction effects. The total number of iterations per model was five for Model 1 (crime person) and three for Models 2, 3, and 4 (crime property, safe day, and safe night).

As the dependent variables are ordinal, ranging from 1 (completely disagree) to 5 (completely agree), an ordered logit model is the most appropriate. This can be expressed as:

$$y_k^* = X_k\beta + \varepsilon_k \tag{1}$$

where y_k^* is the latent dependent variable of individual k and X_k is the explanatory variable set of individual k, which includes all the aforementioned main and interaction effects for individual k. Note that the intercept is dropped for identification issues. Here β is the corresponding vector of parameters to be estimated and ε_k is the error term that is assumed to be an identically distributed logistic error term. The latent dependent variables are then associated with the observed dependent variables, y_k (5-point Likert scale), with m = 1–5, defined as follows:

$$y_k = \begin{cases} 1, \text{ if } -\infty < y_k^* < \mu_1 \\ 2, \text{ if } \mu_1 < y_k^* < \mu_2 \\ \quad \cdots \\ m, \text{ iff } \mu_{m-1} < y_k^* < +\infty \end{cases} \tag{2}$$

For each of the four models, Table 2 displays the estimated coefficients in one column (Estim.) and the significance values (Sig.) in another. Most of the significant values are at a 99% significance level. However, the table also shows significant values at 95% and 90% confidence intervals represented by one or two asterisks respectively. The insignificant variables (<90%) are marked with "ns". Not applicable (na) refers to the interaction terms that were found insignificant and which, thus, were not included in the models.

Table 2 shows the widely used Nagelkerke pseudo R^2 index. It is evident that all safety models have a high degree of fit, explaining between 36% (M1 crime person) and 45% (M4 safe night) of the variation in safety and crime perceptions. All models are superior to the intercept-only models according to the log-likelihood ratio test.

As expected, and in accordance with previous research (*6, 9*), female travelers feel less safe than male travelers. In contrast to some previous findings (*14*), travelers younger than 50 feel safer compared with older ones. Travelers with no children in the household and, thus, conceivably traveling without being in charge of anybody, are found to be less worried with regard to becoming a victim of crime in M1 (crime person). This may indicate that when traveling with children, travelers are more concerned about personal offenses because of their sense of responsibility. However, the models do not allow for determining whether this finding is gender-specific as in some previous results (*13*) or if it is cross-gender.

Variables related to travel characteristics, trip purpose, and travel frequency are found to be insignificant. This indicates that there are no safety benefits arising from being familiar with the stop and schedule and is in disagreement

Table 2. Safety Perception Models

	M1 Crime Person		M2 Crime Property		M3 Safe Day		M4 Safe Night	
	Estim.	Sig.	Estim.	Sig.	Estim.	Sig.	Estim.	Sig.
Gender (female)	1.218	.004	1.609	0.006	ns	ns	−1.086*	.011
Children in household (no)	−.866**	.099	ns	ns	ns	ns	ns	ns
Low travel frequency	ns	ns	ns	ns	ns	ns	ns	ns
Purpose (commuting)	ns	ns	ns	ns	ns	ns	ns	ns
Purpose (other)	Ref. value		Ref. value		Ref. value		Ref. value	
Married or living together	ns	ns	ns	ns	ns	ns	ns	ns
Single	Ref. value		Ref. value		Ref. value		Ref. value	
Age (<30)	−20.574	.000	−19.948	.000	2.347*	.029	2.180*	.018
Age (30–50)	−21.155	.000	−21.009	.000	1.730**	.051	2.013	.010
Age (>50)	Ref. value		Ref. value		Ref. value		Ref. value	
Plan stop	−.403*	.048	−.376**	.072	−1.170	.005	ns	ns
Trust info	ns	ns	ns	ns	.838*	.033	.730*	.040
Real time soon	ns	ns	ns	ns	ns	ns	−3.098	.000
Real time presence	ns	ns	ns	ns	ns	ns	ns	ns
Victim crime (yes)	ns	ns	ns	ns	−22.378	.000	−21.814	.000
Theft (yes)	ns	ns	ns	ns	−2.477	.003	ns	ns
Violence (yes)	2.318	.010	ns	ns	ns	ns	−1.180**	.068
Family (yes)	1.481*	.040	ns	ns	23.331	.000	21.021	.000
Crime rate (high)	−19.498	.000	−18.554	.000	ns	ns	−3.999*	.024
Land use (mixed)	ns	ns	ns	ns	ns	ns	3.226	.004
Land use (other)	Ref. value		Ref. value		Ref. value		Ref. value	
Natural surveillance (yes)	−16.974	.000	−16.498	.000	ns.	ns.	−4.377	.000
Shelter (opaque)	19.447	.000	18.339	.000	−1.726**	.082	1.958	.032
Passers-by (high)	ns	ns	ns	ns	ns	ns	ns	ns
<30* (Crime rate = high)	20.304	.000	19.467	.000	na	na	na	na
30–50* (Crime rate = high)	20.783	.000	20.203	.000	na	na	na	na
<30* (Natural surveillance = good)	16.989	.000	16.843	.000	na	na	na	na
30-50* Natural surveillance = good	18.429	.000	19.542	.000	na	na	na	na
<30* (Shelter = opaque)	−20.928	.000	−21.157	.000	na	na	na	na
30-50* (Theft = yes)	3.509*	.044	na	na	na	na	na	na
Female* (Victim crime = yes)	na	na	na	na	−1.945**	.076	na	na
<30* (Family = yes)	na	na	na	na	−22.595	.000	−21.520	.000
30-50* (Family = yes)	na	na	na	na	−24.152	.000	−21.474	.000
<30* (Victim crime = yes)	na	na	na	na	21.680	.000	22.224	.000
30-50* Plan stop	na	na	na	na	1.534	.004	na	na
<30* Real time soon	na	na	na	na	na	na	2.734	.001
30–50* Real time soon	na	na	na	na	na	na	2.564	.002
Female* Natural surveillance = good	na	na	−2.235*	.015	na	na	na	na
Female* (Shelter = opaque)	na	na	3.499	.006	na	na	na	na

	M1 Crime Person	M2 Crime Property	M3 Safe Day	M4 Safe Night
Log-LL zero	339.182	340.592	264.474	304.064
Log-LL final	293.069	286.496	210.211	244.026
Nagelkerke R^2	.363	.411	.432	.454
N	108	108	108	108

Note: Significance levels: ns = not significant; **90%; *95%; otherwise 99%. na = not applicable. Estim. = estimated; Sig. = significant; Ref. value = Reference or base category.

with findings showing that perceived waiting times are longer for utilitarian trips (commuters) (*32*). Moreover, marital status has no influence on safety and crime perceptions.

Overall, RTI-related variables are insignificant, thus contradicting previous evidence (*25, 26*). The exception is the positive impact that trusting RTI displayed at bus stops has on safety perceptions during the day and at night (M3 and M4). This proves that showing trustworthy and accurate RTI to travelers is worthwhile for the effect it has on them. Interaction effects show that travelers between the ages of 30 and 50 who spend as little time as possible at the location (plan stop) tend to have higher safety perceptions.

Surprisingly, model estimation results indicate that RTI showing that the bus is coming soon negatively influences safety perceptions at night. This finding substantiates Brakewood's (*27*) insignificant results of RTI at night. Interaction effects unveil that feeling unsafe when RTI shows that the bus is coming soon applies only to the older traveler segment (over 50 years old). This counter-intuitive finding might be attributed to different causes—perhaps older people do not pay the same amount of attention to RTI. Alternatively, their perception of waiting time at night might be perceived as longer, considering that this age group does not use their smartphone as intensively as others and there are not many other people around to pose a distraction.

In accordance with previous research (*9, 29*), travelers who have been subject to crime in the past feel less safe at and around bus stops both during the day and at night (M3 and M4) when compared with those who have not. However, interaction effects show that this only holds true for travelers over the age of 30 and for female travelers (M3). The age effect might be explained by people feeling more vulnerable over time (*33*). This gender moderating effect of direct victimization on safety perceptions has already been noted by Yavuz and Welch (*9*). Furthermore, travelers who have been victims of theft [M3 (safe day)] and of violence (M1 and M4) have moderate higher crime and lower safety perceptions. Curiously, travelers with a family member or friend who has been a victim of crime feel safer during the day (M3) as well as at night (M4). Interaction effects, however, indicate that this only applies to those over 50. This may suggest that only direct experiences and not third-person ones influence safety perceptions of the elderly. Moreover, interaction effects show that travelers between the ages of 30 and 50 who have been victims of theft are more worried than others about becoming a victim of violence (M1). All in all, unlike Quann and Hung's (*30*) results, no evidence of crime-specific differences in relation to previous victimization was found.

Model estimation results suggest that not only do safety perceptions not correspond to reported crime rates, they even exercise a negative correlation. In other words, the higher the crime rate, the lower travelers' crime perceptions become [crime person (M1) and crime property (M2)]. This negative correlation concurs with several earlier research findings (*20*). Interaction effects, however, reveal that this is only true for travelers older than 50. At night (M4), and as expected, travelers feel less safe waiting at bus stops with high crime rates.

In line with Ceccato et al. (*11*), the results indicate that for crime models M1 and M2, a good natural surveillance may decrease the perception of crime. Interaction effects, however, show that this is only the case for travelers over 50. Additionally, there is a weak negative impact on property crime perception for women (M2), which indicates that the positive influence of having good natural surveillance is greater for men. Natural surveillance also has a negative influence on safety perceptions at night (M4).

In general, opaque shelters reduce travelers' safety perceptions. This is in line with previous results (*13*) showing that glass or transparent shelters were preferred by travelers. However, at night, (M4) travelers feel the opposite. This might be explained by a feeling that a person is less visible. Alternatively, it might also be caused by the influence of a fishbowl effect (*34*), which would make people waiting at brightly lit bus stops with dark surroundings feel less safe. Interaction effects indicate that women especially are more worried about becoming a victim of theft when waiting at bus stops with opaque shelters. In contrast, travelers under 30 are much less worried about becoming a victim of person and property crimes (M1 and M2) when the bus shelter is opaque.

In contrast to earlier findings (*13*), many passers-by and, by extension, a frequent bus service that is close to a road with high levels of traffic, are found to be insignificant in all models. It is reasonable to think that a frequent bus service would be associated with higher perceptions of safety as travelers' exposure time to crime is reduced. At the same time, bus stops located in areas that have many passers-by and heavy traffic should also be associated with higher safety perception levels because there are more informal guardians who can prevent crime from occurring. The insignificance of these results may indicate that the differences between bus stops are too small and, therefore, the subjective thresholds between low and high are not different enough. The results may also indicate that some other important elements might have been left out of the model. For example, whereas it is true that the Arkitektur-Moderna museet stop has a infrequent bus service (and light traffic and few passers-by), there might be some other features of the surrounding built environment (in this case, a small island that is host to a several museums and hotels and linked to the mainland by a single bridge) not captured by land use, that may somehow attenuate the, in theory, negative effect on perceived safety. A mixed land use urban environment positively affects safety perceptions at night (M4), signifying that having a balanced and adequate amount of activity and of passers-by in the dark is preferred when compared with commercial and residential only land uses.

No differences across the crime models were found and this indicates that crime associated with higher levels of crowding [crime property (petty crime)] and crime associated with lower levels of crowding (crime person) do not diverge.

All in all, the interaction effects tested were mainly found to be significant with age variables, but also with gender variables. No significant interaction effects were found with frequency of travel, indicating that familiarity with PT does not play a mediating effect. In general, the estimated coefficients for the crime (M1 and M2) and safe (M3 and M4) models are consistent with regard to sign and magnitude. In addition, although main effects were very similar across crime and safe models they were significantly different with regard to the interaction terms.

Conclusions

In this study we have investigated the factors that have the greatest effect on determining travelers' perceptions of safety and crime. The analysis is based on reported safety perceptions of travelers waiting at six bus stops with different characteristics in Stockholm, at the same time controlling for a range of factors that affect these perceptions. The analysis was performed using a modest sample size and the generalization of the results should, therefore, be done with caution. Notwithstanding its limitations, the results of a series of Ordered Logit Model estimations suggest that, across the board, shelter, natural surveillance, and trusting RTI are the most important factors influencing safety and crime perceptions. In addition, safety perceptions (M3 and M4) are highly influenced by prior direct and indirect experiences of victimization.

With regard to characteristics of bus stops (Hypothesis 1) there is a mix of anticipated and unexpected results. As hypothesized, bus stops providing good natural surveillance and transparent shelters are found to increase safety perceptions (*3, 10*). In contrast, surprisingly, travelers feel less worried about becoming a victim of any type of crime at bus stops associated with high crime rates, and no significant impact is found for the number of passers-by (traffic and bus service frequency). Mixed land use increases safety perceptions at night and the reason for this may be that bus stops located in areas where there are commercial and retail premises (stores, bars, restaurants, cinemas) have people around them at all times, promoting more guardianship.

With regard to individual attributes (Hypothesis 2), women feel less safe than men across all models. As a wealth of literature has already determined (*8, 9*), the effects of perceived crime and safety feelings are found to be nuanced by age and gender. As expected, for the safety models (M3 and M4), travelers who have had previous experiences of crime feel less safe. Contrary to expectations, the results show that no direct or indirect effects can be attributed to frequency of travel by bus (Hypothesis 3). This indicates that familiar places and routine behavior have no effect on declared crime and safety perceptions.

The finding indicating preference for natural surveillance in the form of opaque shelters at night might be addressed by introducing smart bus stops, at which the shelter color can be dimmed according to the time of day or night. The insignificant effect of travel characteristics suggests that the inclusion of these variables in perceived safety surveys can be disregarded in favor of some others that have greater effect. In general, the results of the crime models (M1 and M2) show that travelers' perceptions do not seem to be affected by crime type (person or property).

This paper provides an important understanding of all the factors that affect travelers' declared safety and crime perceptions with regard to bus stops and PT. However, there are important issues to be addressed in future studies. It would be particularly insightful to repeat this study at bus stops with even more contrasting characteristics, including a very different land use, a much different location, CCTV, RTI, lighting, litter, vandalism, windows, or existing barriers to the stop. This would allow for identifying the impact of more tangible elements that stakeholders could directly implement within the bus network to boost safety perceptions. Furthermore, different urban development and cultural contexts could be considered.

Acknowledgments

We thank Michael Boulis for his kind assistance with collecting the survey data. We also thank Oscar Willems for helping transfer data from the hard copy survey to datasheets and Nerma Muhovic for collecting data on land use and the areas surrounding bus stops. We are also indebted to the Stockholm police officer who provided the data for the analysis and to Stockholm County for providing the SLL Innovation funding that supported the authors' research activities.

References

1. Abenoza, R. F., O. Cats, and Y. O. Susilo. Travel Satisfaction With Public Transport: Determinants, User Classes, Regional Disparities and their Evolution. *Transportation Research Part A*, Vol. 95, 2017, pp. 64–84.
2. Smith, M. J., and R. V. Clarke. Crime and Public Transport. *Crime and Justice*, Vol. 27, 2000, pp. 169–233.
3. Liggett, R., A. Loukaitou-Sideris, and H. Iseki. Bus Stop–Environment Connection: Do Characteristics of the Built Environment Correlate With Bus Stop Crime? *Transportation Research Record: Journal of the Transportation Research Board*, 2001. 1760: 20–27.
4. Iseki, H., and B. D. Taylor. Not All Transfers Are Created Equal: Towards a Framework Relating Transfer Connectivity to Travel Behaviour. *Transport Reviews*, Vol. 29, No. 6, 2009, pp. 777–800.
5. Ingalls, G. L., D. T. Hartgen, and T. W. Owens. Public Fear of Crime and Its Role in Bus Transit Use. *Transportation Research Record*, 1994. 1433: 201–211.
6. Loukaitou-Sideris, A., A. Bornstein, C. Fink, S. Gerami, and L. Samuels. *How to Ease Women's Fear of Transportation Environments: Case Studies and Best Practices*. MTI Report 09-01. U.S. Department of Transportation, 2009.
7. Abenoza, R. F., O. Cats, and Y. O. Susilo. How Does Travel Satisfaction Sum Up? Decomposing the Door-to-Door Experience for Multimodal Trips. *Transportation*, 2017, pp. 1-28.
8. Cats, O., Abenoza, R. F., Liu, C. and Y. O. Susilo. Identifying Priority Areas Based on a Thirteen Years Evolution of Satisfaction With Public Transport and Its Determinants. *Transportation Research Record: Journal of the Transportation Research Board*, 2015. 2323:99–109.
9. Yavuz, N., and E. W. Welch. Addressing Fear of Crime in Public Space: Gender Differences in Reaction to Safety Measures in Train Transit. *Urban Studies*, Vol. 47, No. 12, 2010, pp. 2491–2515.
10. Levine, N., and M. Wachs. Bus Crime in Los Angeles: Victims and Public Impact. *Transportation Research Part A*, Vol. 20, No.4, 1986, pp. 285–293.
11. Ceccato, V., A. Uittenbogaard, and R. Bamzar. Security in Stockholm's Underground Stations: The Importance of

Environmental Attributes and Context. *Security Journal*, Vol. 26, No.1, 2013, pp. 33–59.

12. Susilo, Y. O., and O. Cats. Exploring Key Determinants of Travel Satisfaction for Multi-Modal Trips by Different Travellers' Groups. *Transportation Research Part A*, Vol. 67, 2014.
13. Tucker, L. E. *Safer Stops for Vulnerable Customers*. Final Report. NCTR-473-13. Center for Urban Transportation Research at the University of South Florida, Tampa, Fla., 2003.
14. Delbosc, A., and G. Currie. Modelling the Causes and Impacts of Personal Safety Perceptions on Public Transport Ridership. *Transport Policy*, Vol. 24, 2012, pp. 302–309.
15. Currie, G., A. Delbosc, and S. Mahmoud. Factors Influencing Young Peoples' Perceptions of Personal Safety on Public Transport. *Journal of Public Transportation*, Vol. 16, No. 1, 2013, pp 1–19.
16. Ceccato, V., and Y. Paz. Crime in Sao Paulo's Metro System: Sexual Crimes Against Women. *Crime Prevention and Community Safety*, Vol. 19, No. 3–4, 2017, pp. 211–226.
17. Department of the Environment, Transport and the Regions. Personal Security on Public Transport for Operators. Why Personal Security? TSO, London, 1996.
18. Wallace, R. R., D. A. Rodriguez, C. White, and J. Levine. Who Noticed, Who Cares? Passenger Reactions to Transit Safety Measures. *Transportation Research Record: Journal of the Transportation Research Board*, 1999. 1666: 133–138.
19. Crowe, T. D. *Crime Prevention Through Environmental Design*. 2nd ed. Oxford: Butterworth Heinemann, 2000.
20. Ceccato, V. *Moving Safely: Crime and Perceived Safety in Stockholm's Subway Stations*. Plymouth, UK: Lexington Books. 2013.
21. Cohen, L. E., and M. Felson. Social Change and Crime Rate Trends: A Routine Activity Approach. *American Sociological Review*, Vol. 44, No. 4, 1979, pp. 588–608.
22. Cohn, E. G., and J. Rotton. Even Criminals Take a Holyday: Instrumental and Expressive Crimes on Major and Minor Holidays. *Journal of Criminal Justice*, Vol. 31, No. 4, 2003, pp. 351–360.
23. Wiebe, D., T. Richmond, J. Poster, W. Guo, P. Allison, and C. Branas. Adolescents' Fears of Violence in Transit Environments During Daily Activities. In *Safety and Security in Transit Environments* (V. Ceccato and A. Newton, eds.), Palgrave Macmillan, London, 2015, pp. 137–155.
24. Dziekan, K., and K. Kottenhoff. Dynamic At-Stop Real-Time Information Displays for Public Transport: Effects on Customers. *Transportation Research Part A*, Vol. 41, No. 6, 2007, pp. 489–501.
25. Zhang, F., Q. Shen, and K. Clifton. Examination of Traveler Responses to Real-Time Information About Bus Arrivals Using Panel Data. *Transportation Research Record: Journal of the Transportation Research Board*, 2008. 2082: 107–115.
26. Ferris, B., K. Watkings, and A. Borning. OneBusAway: Results from Providing Real-Time Arrival Information for Public Transit. *Proc., SIGCHI Conference on Human Factors in Computing Systems*. Atlanta, Ga., ACM, New York, 2010, pp. 1807–1816.
27. Brakewood, C., S. Barbeau, and K. Watkins. An Experiment Evaluating the Impacts of Real-Time Transit Information on Bus Riders in Tampa, Florida. *Transportation Research Part A*, Vol. 69, 2014, pp. 409–422.
28. Hale, C. Fear of Crime: A Review of the Literature. *International Review of Victimology*, Vol. 4, No. 2, 1996, pp. 79–150.
29. Tseloni, A., and C. Zarafonitou. Fear of Crime and Victimization. *European Journal of Criminology*, Vol. 5, No. 4, 2008, pp. 387–409.
30. Quann, N., and K. Hung. Victimization Experience and the Fear of Crime. A Cross-National study. In *Crime Victimization in Comparative Perspective. Results from the International Crime Victims Survey 1989–2000* (P. Nieuwbeerta, ed.), Netherlands Institute for the Study of Crime and Law Enforcement, Amsterdam, the Netherlands; Boom Juridische Uitgevers, The Hague, the Netherlands, 2002, pp. 301–316.
31. Loukaitou-Sideris, A., R. Liggett, and H. Iseki. The Geography of Transit Crime: Documentation and Evaluation of Crime Incidence on and Around the Green Line Stations in Los Angeles. *Journal of Planning Education and Research*, Vol. 22, No. 2, 2002, pp. 131–151.
32. Psarros, I., K. Kepaptsoglou, and M. G. Karlaftis. An Empirical Investigation of Passenger Wait Time Perceptions Using Hazard-Based Duration Models. *Journal of Public Transportation*, Vol. 14, No. 3, pp. 109–122.
33. Ceccato, V., and R. Bamzar. Elderly Victimization and Fear of Crime in Public Spaces. *International Criminal Justice Review*, Vol. 26, No. 2, 2016, pp. 115–133.
34. Loukaitou-Sideris, A., and C. Fink. Addressing Women's Fear of Victimization in Transportation Settings. A Survey of U.S. Transit Agencies. *Urban Affairs Review*, Vol. 44, No. 4, 2009, pp. 554–587.

The Task Force on Transit Safety and Security (AP018T) peer-reviewed this paper (18-05252).

TRR
JOURNAL OF THE TRANSPORTATION RESEARCH BOARD

Article

Transportation Research Record
2018, Vol. 2672(8) 29–40

DOI: 10.1177/0361198118791914
journals.sagepub.com/home/trr
SAGE

Strategies and Barriers in Effective Bus Lane Implementation and Management: Best Practices for Use in the Greater Washington, D.C. Region

Burak Cesme[1], Richard Roisman[2,3], Ramona Burns[4], Katie List[5], Arrianna Koudounas[2], Jesus Cuellar[1], Meredyth Sanders[1], Kevin Lee[1], and David Miller[5]

Abstract

Bus lanes have the potential to significantly improve bus speeds and reliability. These benefits are typically not solely achieved through the design and installation of a bus lane, however. A successful bus lane must also have sufficient public support for regulating bus lane use, as well as enforcing those regulations. Effective design, enforcement, legislation, and education strategies are critical during both the planning and post-implementation phases to limit bus lane violations and increase compliance, thereby improving bus lane efficiency. Despite the importance of developing a robust management plan to limit violations and maintain bus lane efficiency, there is little documented guidance for bus lane implementation and management in the literature, particularly related to enforcement and education strategies. This paper develops a best practices review to respond to that gap in guidance by conducting a comprehensive literature review of best practice in bus lane management along with interviews with agency staff. The objective of this paper is to research, identify, and develop strategies to improve bus lane observance and enforcement, and to create an implementation plan for agencies with specific recommendations and strategies to ensure the success of bus lane initiatives.

Urban areas of all sizes in the U.S. are experiencing increased vehicle congestion; in 95 of America's 100 largest metro areas congestion increased from 2013 to 2014 (*1*, *2*). While all road users experience the negative impacts of congested conditions, the effects on public transit users riding buses operating in mixed traffic tend to be more significant as transit routes are typically fixed, not allowing buses to change course to avoid congestion. Furthermore, due to frequent stops, buses generally travel in the right-hand lane, which tends to have the most "friction" with parking and loading activities, taxis, and right-turning vehicles. A study conducted by the Washington Metropolitan Area Transit Authority (WMATA) in 2010 showed average bus operating speeds of less than 10 miles per hour (mph) on most corridors in the Washington, D.C. region and less than 5 mph in downtown D.C. (*3*). Two recent mobility studies conducted by the New York City Department of Transportation (NYCDOT) and the District Department of Transportation (DDOT) confirmed this finding and indicated that bus speeds are often less than 5 mph in central business districts (*4*, *5*). As bus speeds continue to drop and bus performance suffers from congestion, jurisdictions have recognized and responded to the need for preferential treatments for public transit, including bus lanes. As a result, new bus lanes are operating in several U.S. cities, and many more are being planned.

Bus lanes have the potential to significantly improve bus speed and reliability by limiting the negative impacts of traffic congestion (*6*). For transit agencies, bus lanes can result in shorter running times, as well as decreased schedule recovery times, due to improved reliability. For

[1]Kittelson & Associates, Inc., Washington, D.C.
[2]Metropolitan Washington Council of Governments, Department of Transportation Planning, Washington, D.C.
[3]Regional Transportation Planning Program Coordinator, Arlington County, VA
[4]Washington Metropolitan Area Transit Authority, Washington, D.C.
[5]Foursquare Integrated Transportation Planning, MD

Corresponding Author:
Address correspondence to Burak Cesme: bcesme@kittelson.com

passengers, bus lanes can shorten in-vehicle travel time, reduce average waiting time at stops, and decrease vehicle crowding as a result of improved reliability. This increases the relative attractiveness of transit and potentially increases ridership. However, these benefits cannot be achieved solely through bus lane design and installation. A successful bus lane must also have sufficient public support for regulating bus lane use, as well as enforcing those regulations. Effective design, enforcement, and education strategies are critical during both the planning and post-implementation phases to limit bus lane violations and increase compliance, thereby improving bus lane efficiency (Figure 1). A study from San Francisco showed that over 60% of vehicles along a transit lane in a busy corridor violated bus lane regulations in the peak direction during the evening peak hour, limiting the lanes' benefits (*7*). Similar findings were reported from New York City's Select Bus Service (SBS), where 30% to 50% of buses face some sort of obstruction in bus lanes, with the obstruction rate increasing in sections with more pedestrians and taxis (*8*).

Despite the importance of developing a robust management plan to limit violations and maintain bus lane efficiency, there is little documented bus lane implementation and management guidance in the literature, particularly related to enforcement and education strategies. This best practices review responds to that gap in guidance by conducting a comprehensive literature review of bus lane management best practices and interviews with agency staff to research, identify, and develop strategies to improve bus lane observance and enforcement, which are critical to the success of bus lane initiatives. This best practices review focuses primarily on bus lane implementation and management. The assessment and feasibility of bus lanes, which generally occurs earlier on in the planning process, was beyond the scope of this review.

Effective Bus Lane Management

A comprehensive bus lane management literature review, along with interviews with local and national transportation agencies, was conducted to identify successful enforcement, legislative, and educational techniques in

Figure 1. Key elements of effective bus lane implementation.

the U.S. and abroad (*9*). This section discusses key findings, identifies barriers to implementation, and presents strategies for agencies to manage bus lanes effectively, with a focus on:

- Stakeholder coordination
- Enforcement
- Legislation
- Education
- Monitoring

Stakeholder Coordination

Interagency coordination plays a critical role in the overall success of any bus lane implementation project. Transit operating agencies are rarely exclusively responsible for maintaining lane markings, setting traffic signal timings, and performing other essential activities of a bus lane. Case studies from across the U.S. reiterated the need for cooperation at all phases of implementation among state, regional, and local agencies, as well as between traffic engineering and transit service planning officials. Interagency coordination is essential not just in the planning, design, and construction of bus lanes, but also in their operation.

As an example, transit-only lane implementation in New York City is a "joint venture" of two different agencies, NYCDOT and Metropolitan Transportation Authority (MTA) New York City Transit (an entity of the State of New York). Throughout New York City's ongoing transit-only lane implementation process, there has been a high level of interagency cooperation to implement both the initial pilot transit-only lanes and subsequent transit-only lane projects (*9*).

Given that projects of this nature often require collaboration from multiple agencies and community stakeholders, getting these groups on the same page can greatly improve the project's chance of success. Reviewing past projects and identifying best practices can also help future projects succeed, but agencies must be willing to reevaluate their own past practices—and those of others—to manage effective bus lanes successfully.

Enforcement

Several studies from the literature indicated that the perception of limited bus lane enforcement increases violation rates, diminishing the effectiveness of bus lanes and resulting in reduced bus speeds (*8*, *10*). Although practices vary, police enforcement and automated enforcement (such as on-board cameras) are the two most common enforcement tools used to reduce violations. Figure 2 summarizes the enforcement strategies used by the agencies surveyed for this study, including fines for bus lane violation. Detailed information and further findings from the locations surveyed can be found in the final report of the project (*9*).

Police Enforcement. When automated camera enforcement is not practical, some level of police presence is needed to discourage potential violators from entering the bus lanes. Agency interviews indicated that typically jurisdictions place more emphasis on police enforcement when bus lanes first open (e.g., the first month after opening). However, targeted enforcement tends to diminish afterwards due to several challenges associated with police enforcement:

- **Resources:** Police enforcement requires considerable financial and human resources. Budget limitations and conflicting priorities can make it difficult to sustain a continuous police enforcement program.
- **Authorization:** For most agencies, transit agency staff (including transit police) are rarely authorized to enforce bus lane restrictions or moving violations. This increases reliance on police enforcement, which compounds budget and resource allocation issues.
- **Physical infrastructure:** Low-cost, low-resource bus lane concepts, such as curbside lanes with no paint, are the easiest to implement but also the most difficult to enforce consistently (*8*). It is necessary to find a balance between building a "self-enforcing" lane (e.g., offset bus lanes with red paint) and paying to enforce restrictions.
- **Compliance impact on operations:** Pulling over non-compliant vehicles in the bus lanes can block buses, negatively affecting bus operations. To address this, on recently implemented bus lanes in Baltimore City, Baltimore police pull violators over on side streets.
- **Other permitted users:** Curbside bus lanes often allow other vehicles such as taxis, shuttles, and right-turning vehicles to use bus lanes. While allowing other vehicles in bus lanes increases utilization of roadway space, it creates enforcement challenges.

Automated Enforcement. Generally, transit and law enforcement agencies use two types of camera enforcement to automate the enforcement process: stationary cameras installed at selected locations/corridors, and cameras on buses. Both types can generate citations for both moving and parking violations. Compared with active police enforcement, which is resource-intensive, automated enforcement can have significant fiscal and enforcement benefits.

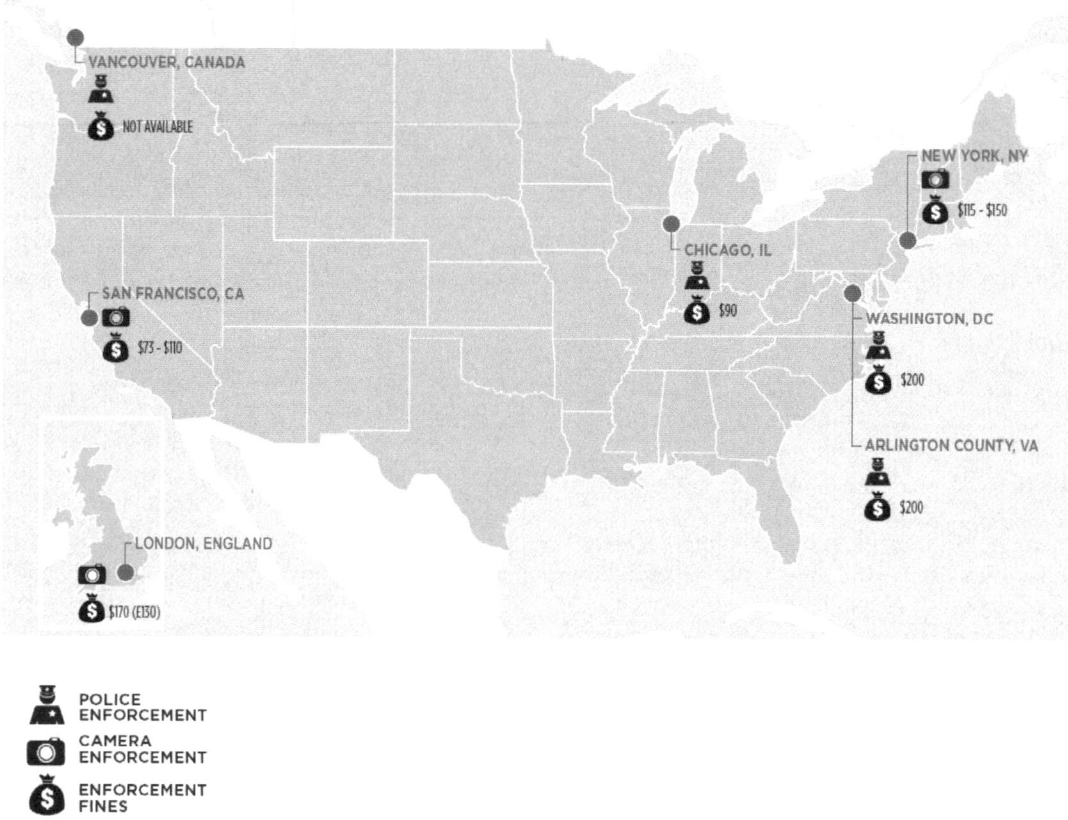

Figure 2. Enforcement strategies used and associated fines for bus lane violation in the jurisdictions studied.

However, transit agencies are rarely authorized to enforce restrictions in the bus lanes in which they operate, presenting challenges in bus lane enforcement (*11*). Automated camera enforcement in bus lanes usually requires new enabling legislation, as most states do not currently have laws permitting it (see below for legislation details and recommendations). New York and California are the only states in the U.S. which have specifically enabled bus lane camera enforcement. Targeted legislation in each state enabled specific cities to begin camera-based bus lane enforcement as pilot or demonstration programs. Each state then used an iterative legislative process to extend and expand the pilot or demonstration programs to new corridors.

Lessons Learned from New York and California. New York's initial automated camera enforcement legislation (2010) granted NYCDOT and MTA New York City Transit the authority to install bus lane enforcement cameras on five specified SBS routes. In 2015, the New York State Legislature and Governor extended the law for ten years, allowing the city to use bus lane cameras on up to 15 additional routes. New York's enabling legislation includes a maximum fine amount, requirements for camera-related signage along corridors, and a time span for enforcement (bus lane cameras may only be operated on designated bus lanes during weekdays from 7:00 a.m. to 7:00 p.m.) (*12*).

Two types of camera enforcement have been used in New York City to date: stationary cameras and on-bus cameras. On-bus cameras record standing violations, such as parking, while stationary cameras primarily record driving violations in the bus lane. Stationary cameras, installed along SBS corridors, are operated by NYCDOT; a pilot program with on-bus cameras was administered by MTA New York City Transit. Each enforcement method was designed to capture multiple photos to ensure that a violation was being committed, and to allow MTA New York City Transit staff (on-bus cameras) or NYCDOT staff (stationary cameras) to determine if there was a legitimate reason for a private vehicle to enter the bus lane. An adjudication process, managed by the New York City Department of Finance, was also established to allow drivers who felt they were wrongly cited to appeal the fine. As of 2012, only 2% of all citations were overturned (*13*).

San Francisco uses forward-facing cameras on buses for its Transit-Only Lane Enforcement (TOLE) program. If a camera detects a vehicle using the lane illegally, the camera photographs the vehicle's license plate and a citation is issued to the vehicle's owner. The process does not require a bus driver to initiate it (*14*). San Francisco's legal ability to install cameras on city-owned public transit vehicles is enabled by changes made to the California Vehicles Code, as well as municipal regulations (*15*). The City and County of San Francisco can issue citations (civil penalties) for violations captured by cameras in bus lanes during the posted hours of operation for a transit-only traffic lane; the video image is confidential, and destroyed after six months, or 60 days after the final disposition of the citation.

Legislation

Although states and municipalities have varying regulations, there are typically two ways in which bus lane violations are processed:

- Infractions, in which a police officer files charges directly against a vehicle operator, resulting in a court hearing, fines, driver's license penalties, or possibly jail time; and
- Administrative or civil violations (such as parking tickets), which are issued to the registered owner of a vehicle (not necessarily the person who was operating the vehicle), resulting in fines, but not necessarily a court hearing. Administrative violations can be issued by government agents other than police officers, and typically require less evidence (and result in less paperwork) than infractions.

In New York City, bus lane moving violations issued by police officers are infractions, and may result in both fines and points against a driver's license. In contrast, a bus lane violation captured on camera may result in a fine, but will not be included in a driver's operating record, or used for insurance purposes (*16*). It can be difficult for camera-based systems to meet evidence standards required for infractions, such as proof of the driver's identity (*11*).

There are various types of camera-based enforcement of parking or moving violations, but New York City and San Francisco have the most robust, most explicit, on-board camera enforcement of violations in bus lanes in the U.S. Each city required enabling legislation from its respective state to develop its bus lane camera enforcement program, and each city used an iterative policy and legislation development process that began with pilot or demonstration projects and developed into broader programs. Key elements of their respective enabling legislation are summarized in Figure 3. The four main branches under state enabling legislation illustrate the primary factors and sub factors to consider for legislation. Likewise, the branches under Municipal Code illustrate three primary areas where adjustments/additions to the municipal code may be required. It should be noted that Figure 3 provides a general overview of the process and reflects bus lane camera enforcement elements found in New York and California. Other states and municipalities may have different circumstances and requirements.

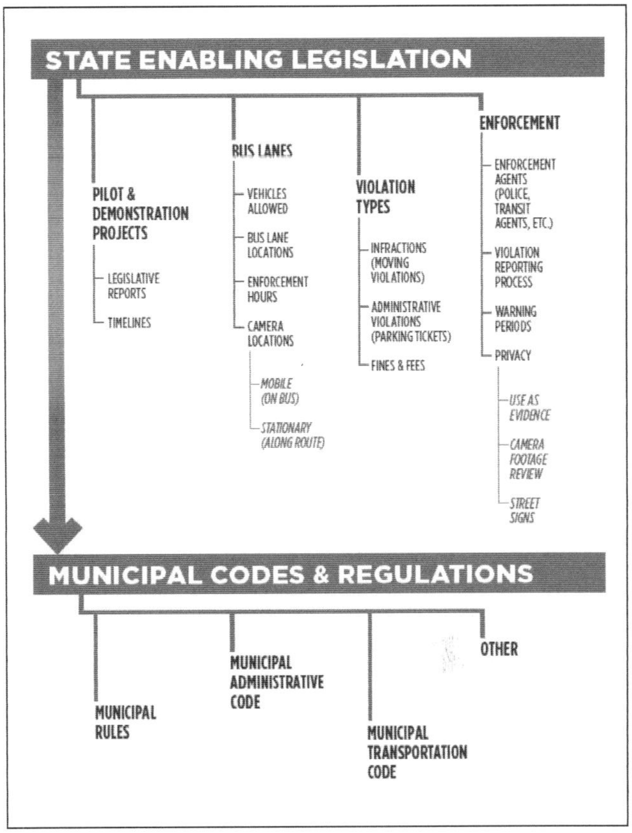

Figure 3. Bus lane camera enforcement: general legislative factors to consider.

Education

As noted in the section on Stakeholder Coordination, above, effective bus lane management often requires collaboration from multiple agencies and community stakeholders, and getting these groups on the same page can greatly improve the project's chance of success. Best practices for bus lane public outreach—which helps communicate the intent and use of the bus lane to community stakeholders and the general public—include bus lane infrastructure markings and targeted outreach tailored and scaled to the needs of the project.

Signal the Exclusivity of a Bus Lane. As demonstrated in San Francisco and many other locations, installing lane markings, colored lanes (e.g., red paint), and signs to indicate the existence of a bus lane is the simplest, most practical, and perhaps the most necessary form of public education during bus lane projects. This intervention effectively educates all road users simultaneously, including pedestrians, cyclists, taxi drivers, private vehicle drivers, and transit operators.

Provide Clear and Information Project Details through Website/Social Media. When promoting a bus lane project, the presence of easy-to-read, sufficiently detailed information on project details, frequently asked questions, upcoming meetings, and discussion forums on websites, blogs, and social media is crucial to the processes of educating the public, thereby improving compliance and bus lane efficiency. The use of digital information allows for real-time updates on information that may shift as a plan progresses.

Earned, Paid, and Produced Media All Have a Role. Media coverage can increase exposure, expanding ways to reach a larger audience and amplify key messages. Agencies can attract extra attention to a project by purchasing advertisements or working with reporters to spread information. Press releases can also be an effective tool in garnering media and public attention. In short, transit agencies can use media as another tool to provide answers to the public on such questions as:

- How will dedicated lanes change my commute?
- Will travel times by car or bus be shorter or longer?
- When are the lanes scheduled to open?

Print Materials Are Important, Too. Not everyone has access to a computer. Sometimes the best way to reach potentially affected travelers in the corridor is via print materials, which can be distributed in person, to vehicles entering parking garages, on a transit vehicle, or via direct mail. Figure 4 provides two examples of brochures from the Chicago Transit Authority (CTA) and the Jacksonville Transportation Authority that help increase public awareness of bus lane rules and locations.

Tailor Engagement Methods to Fit the Project. Outreach efforts should be tailored and scaled to the needs of the project. With a dedicated bus lane, all road users—including pedestrians, cyclists, drivers, and transit operators—will be affected. Agencies should target outreach toward residents, homeowner associations, community centers, major community organizations, educational or religious institutions, store owners, and jurisdictional leaders within close proximity of the proposed or in-place right of way and bus stop station areas.

Prior to implementing public outreach, agencies should perform an identification assessment of populations likely to be affected using geographic information system (GIS) and other research methods. As noted in the Transit Cooperative Research Program's *Public Participation Strategies for Transit*, agencies can use a variety of data sources and consultation methods to accomplish this goal (*17*).

Depending on the project, agencies may wish to perform targeted outreach toward certain demographic groups, including seniors, persons with disabilities, transit-dependent populations, low-income residents, minorities, students, choice riders, and non-English speakers.

Educate Transit Vehicle Operators. Wherever bus lanes are implemented, transit vehicle operators will also require education. When implementing bus lane projects, agencies should update operator manuals and offer training prior to and during implementation to help transit vehicle operators avoid conflicts with other road users, take advantage of time-saving techniques such as off-board fare collection or all-door boarding (if applicable), and generally know what to expect when a new project opens.

Monitoring

Stakeholder coordination, enforcement, legislation, and outreach activities are all critical elements of implementing effective bus lanes. However, a successful bus lane also requires continuous monitoring after bus lanes are installed. Monitoring should include performance measures that are meaningful and measurable for evaluating the effectiveness of bus lanes, as well as compliance and violation rates. Key measures to assess the efficacy of enforcement tactics on bus lanes include:

- **Compliance**: Post-implementation evaluation should track the number of vehicles complying with the bus lane relative to the number of vehicles driving illegally in the bus lane, as well as the number of stationary vehicles in the bus lane. Changes in the type of enforcement (e.g., from police to camera enforcement) should be monitored to document the effect of enforcement strategies on adherence to bus lane rules. Figure 5 shows an example from a recent study in San Francisco displaying the total number of violations before and after the red paint treatment on 3rd Street (*18*).
- **Repeat offenders**: Initial non-compliance with bus lanes may be attributed to a lack of understanding regarding the purpose and/or function of the

Figure 4. Examples of brochures to educate public and raise awareness about bus lanes.

facilities. Post-implementation monitoring should assess the frequency of repeat offending to determine the effectiveness of painted bus lanes, enforcement, educational campaigns, etc.

- **Bus travel time comparison**: Post-implementation monitoring should focus on change in bus travel time to help assess the effectiveness of bus lanes. This measure can also help agencies identify

Figure 5. San Francisco 3rd Street bus lane before and after red treatment and corresponding violations per hour along the corridor (*18*).

segments that require more targeted enforcement strategies to improve bus operations.

Implementation Plan

As noted previously, this best practices review focuses primarily on bus lane implementation and management. While the assessment and feasibility of bus lanes, which occurs earlier on in the planning process, is beyond the scope of this review, a general framework for the planning process is prescribed in this section. Agencies looking for detailed guidance in earlier stages of bus priority treatment planning should review the following documents:

- Shared-Use Bus Priority Lanes on City Streets: Case Studies in Design and Management (*11*)
- TCRP Report 183: A Guidebook on Transit-Supportive Roadway Strategies (*19*)

The phases and associated recommendations for successful implementation of bus lane projects are summarized in Figure 6.

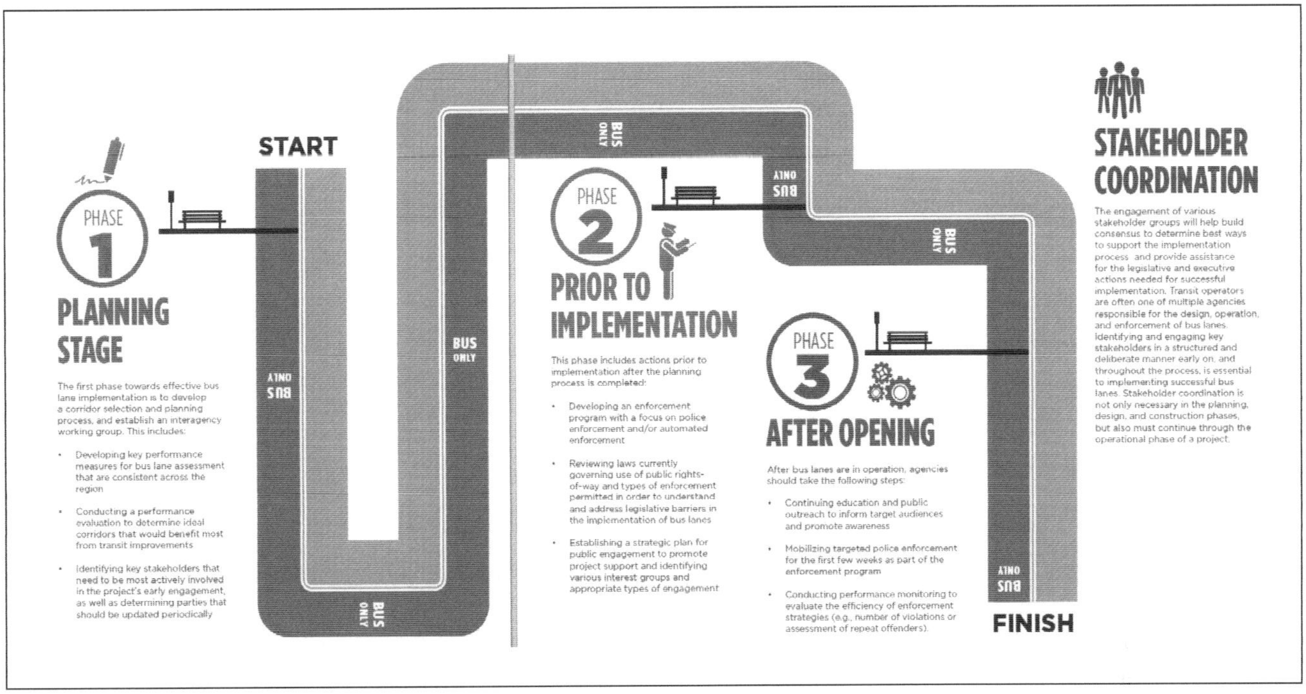

Figure 6. Implementation plan for effective installation and management of bus lane.

Table 1. Benefit-Cost Analysis (BCA): Cost Elements and Units

Cost element	Cost	Unit
Standard bus lane—white pavement striping (capital cost)	\$100,000 (*20*)	Per mile
Standard bus lane—white pavement striping (maintenance cost)	\$10,000 (*20*)	Per mile per year
Red paint bus lane (capital cost)	\$5 (*21*)	Per square feet
	\$308,000 (*21*)	Per mile per every five years[a]
Red paint bus lane (maintenance cost)[b]	\$10,000 (*20*)	Per mile per year
Manual enforcement (police enforcement)	\$49.50 (22)[c]	Per hour per staff
Bus-mounted camera enforcement (capital cost)	\$9,500 (*23*)	Per bus
Bus-mounted camera enforcement (maintenance cost)	\$15 (*23*)	Per bus per week
Stationary camera enforcement (capital cost)	\$64,945 (*13*)	Per camera
Stationary camera enforcement (maintenance cost)	\$414 (*13*)	Per camera per week

[a]Red paint needs to be re-applied every five years. The cost associated with re-application is considered as capital cost instead of maintenance cost.
[b]Maintenance cost for bus lanes with red paint is assumed to be equal to those for white pavement striping, as agencies still need to provide all the MUTCD-required elements including the white stripe between the bus lane and the adjacent general purpose lane, "Bus Lane" word markings, and bus lane signage when red paint is used.
[c]Hourly wage is inflated by 40% to incorporate overhead costs.

Benefit-Cost Analysis

Agencies may use a benefit-cost analysis (BCA) to help identify efficient bus lane enforcement strategies for their particular regions. BCAs look at the net present value of the benefits, and divide them by the net present value of costs. A benefit-cost ratio (BCR) greater than one indicates that benefits exceed costs and that the investment is promising. A BCR below one indicates that costs outweigh benefits, and that the project will need further study or innovative strategies to identify benefits that may not have been adequately quantified to justify the project.

Cost Elements

Table 1 summarizes the cost elements included in the BCA; detailed information on BCA methodology is provided in a separate document (*9*). Within this section, "manual enforcement" refers to police enforcement of bus lanes.

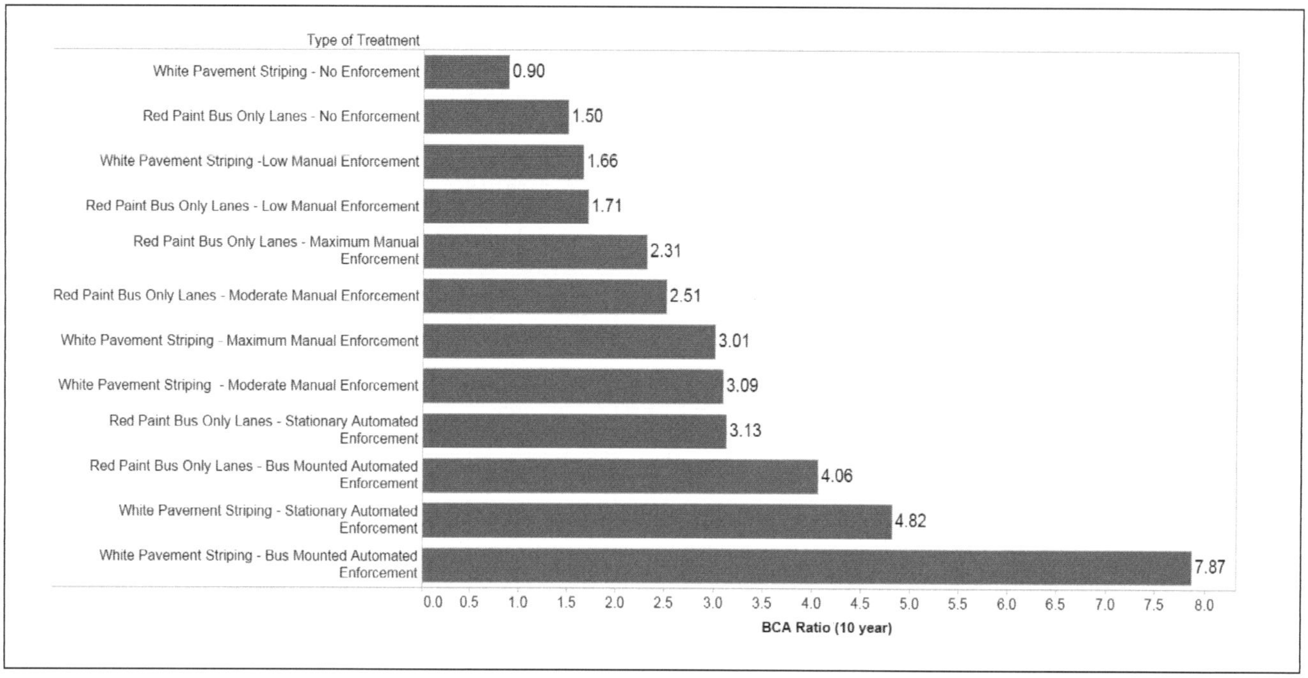

Figure 7. Benefit-cost ratios of bus lane enforcement strategies.

The capital and maintenance costs associated with standard bus lane white pavement striping were obtained from TCRP Synthesis 83 (*20*). The synthesis indicates that, where existing lanes are converted to bus lanes, capital costs may range from \$50,000 to \$100,000 per mile. Since the report was published in 2010, the higher end of the range, that is \$100,000, was assumed to account roughly for inflation.

Benefit Elements

For the benefit calculation, the analysis considers passenger travel time savings and fleet savings. Due to limitations in the data held by agencies about the effects of enforcement, the travel time savings and fleet saving benefits associated with 12 potential implementation strategies identified in Figure 7 were quantified using methods outlined in the Transit Capacity and Quality of Service Manual (TCQSM) (*24*). The methodology here estimates bus running time rates (i.e., minutes per mile) under various enforcement levels to calculate passenger travel time savings. Table 2 summarizes the methodology and shows travel time savings under each enforcement scenario. In addition to the travel time savings in Table 2, it is assumed that red paint treatment provides additional 0.25 minutes per mile travel time savings with no, low, or moderate enforcement and 0.10 minutes per mile savings under strong enforcement. This is because it is anticipated that additional benefits resulting from red paint would be lower in high enforcement scenarios, compared with no, low, or moderate enforcement scenarios. It should be noted that 0.10 and 0.25 minutes per mile values were assumed based on the authors' judgment. Finally, while there is limited data based on agencies' experience, the percent saving results in Table 2 were compared with the results from San Francisco and New York as a "reality check".

For fleet savings calculation, the analysis is based on average travel time savings and an assumed fleet saving threshold (fleet savings are achieved when total travel time savings are greater than bus headway). As noted above, detailed information on fleet savings assumptions and findings is provided in a separate document (*9*).

Benefit-Cost Ratio Results

To capture the effects of multi-year costs and benefits, including the cost of re-applying red paint to bus lanes, a ten-year benefit-cost ratio (BCR) was calculated.

Figure 7 summarizes the BCR calculated for each enforcement scenario. The results are calculated based on the assumptions that each bus lane would operate for five days a week during peak periods (6 hours per day) at a frequency of 15 buses per hour. Each one mile bus lane is assumed to operate for 50 weeks (approximately one year, excluding major holidays).

Results indicate that the strategies with no enforcement scenarios have the lowest benefit-cost ratios (with a

Table 2. Estimated Travel Time Savings Under Various Enforcement Scenarios

Enforcement scenario	Unimpeded bus running time (min/mile)	Additional travel time due to congestion (min/mile)	Total travel time (min/mile)	Percent saving (%)
Mixed traffic operation	7.0	3.0	10.0 (6.0 mph[a])	N/A
Bus lane—no police enforcement	7.0	2.95	9.95 (6.0 mph[a])	0.5%
Bus lane—low police enforcement	7.0	2.85	9.85 (6.1 mph[a])	1.5%
Bus lane—moderate police enforcement	7.0	2.4	9.4 (6.4 mph[a])	6.0%
Bus lane—maximum police or automated enforcement	7.0	2.0[b]	9.0 (6.7 mph[a])	10.0%

[a]Indicates associated speed in miles per hour (mph).
[b]Additional travel time in the maximum enforcement scenario is due to the congestion resulting from right-turning vehicles using the bus lane.

BCR of 0.90), while the strategies with standard lane treatments and automated enforcement scenarios have the highest benefit-cost ratios (BCR of 7.87 and 4.82). Red paint bus lanes fall in the middle range due to the high cost of installing and maintaining red paint bus lanes. However, it is important to note that the BCA assumes agencies have adequate resources to provide a moderate to maximum level of manual enforcement. For agencies with limited resources, red paint treatment yields a higher BCR compared with the standard lane treatment under the no enforcement (1.50 versus 0.90) and low manual enforcement scenarios (1.71 versus 1.66) as red paint serves as both an educational and enforcement tool. In addition, the authors acknowledge that some of the results presented here are order-of-magnitude estimates and further sensitivity analysis needs to be conducted to make conclusions.

Another interesting finding is that white pavement striping with moderate manual (police) enforcement yields a slightly higher benefit-cost ratio than the white pavement striping with a maximum manual enforcement due to the high cost of manual enforcement (3.09 versus 3.01). Finally, eight of the 12 strategies evaluated have benefit-cost ratios that exceed 2.0. These promising ratios indicate that a moderate to strong enforcement program can ensure the success of bus lanes with a return on investment in terms of travel time and fleet savings. However, as noted previously, future research is needed with more field data to have a better understanding of benefit-cost effects associated with various bus lane enforcement strategies and to draw strong conclusions from the analyses.

Summary and Future Research

The initial objective of this research was to identify best practices in bus lane management strategies related to enforcement, legislation, and education. However, agency interviews highlighted the importance of stakeholder coordination at all phases of bus lane implementation. The interviews also revealed that agencies need to establish effective and lasting stakeholder engagement processes, as the management of bus lanes requires coordination and input from many constituents.

The state of the practice indicated that some level of enforcement, either through police or automated enforcement, is required to limit bus lane violations and improve the effectiveness of bus lanes. While police enforcement of bus lanes may be generally effective and feasible for small corridors, the expansion of bus lanes can make continuous police enforcement of lanes impractical due to limitations of budget and human resources. Financial barriers associated with enforcement can be overcome by automating the enforcement process through use of cameras. However, examples from California and New York show that automated enforcement requires new enabling legislation and administrative processes, and that final authorization may take several years. Therefore, agencies interested in developing camera-based enforcement should begin the legislative process early, and conduct a robust education and outreach program to address potential public concerns over privacy issues.

Furthermore, education is a crucial piece of an effective bus lane management process. Identifying project partners early and targeting constituents with relevant messages, both during and after implementation, are found to be the most effective educational strategies. Furthermore, installing strong visual cues (e.g., lane striping, red paint, and/or signs) is recommended as both a form of education and enforcement.

Finally, this paper provides a comprehensive review of current best practices for bus lanes with a focus on enforcement, legislation, and education. The findings presented here lay a general groundwork for future research for establishing nationwide strategies for effective bus lane management and major barriers to implementation.

Acknowledgments

This paper builds on an earlier work performed by the National Capital Region Transportation Planning Board (TPB), in coordination with WMATA. The authors would like to thank TPB and WMATA for funding this study and for their valuable suggestions and comments.

Author Contributions

The authors confirm contribution to the paper as follows: study conception and design Ramona Burns, Richard Roisman, Burak Cesme, Kevin Lee, and David Miller; data collection: Jesus Cuellar, Katie List, Meredyth Sander; analysis and interpretation of results: Meredyth Sanders, Jesus Cuellar, Burak Cesme and Katie List; draft manuscript preparation: Burak Cesme, Katie List, David Miller, Kevin Lee, and Arianna Koudounas. All authors reviewed the results and approved the final version of the manuscript.

References

1. Schrank, D., B. Eisele, T. Lomax, and J. Bak. *Urban Mobility Scorecard*. Texas Transportation Institute, 2015. https://static.tti.tamu.edu/tti.tamu.edu/documents/mobility-scorecard-2015.pdf
2. Federal Highway Administration. *Status of the Nation's Highways, Bridges, and Transit: Conditions and Performance*. Federal Highway Administration, 2015. https://www.fhwa.dot.gov/policy/2015cpr/
3. [Online]. https://www.wmata.com/initiatives/strategic-plans/upload/3Metro-2025-Metrobus-Priority-Corridor-Network-Secure.pdf. Accessed July 25, 2017.
4. NYC Department of Transportation. *New York City Mobility Report*. NYC Department of Transportation, October 2016. http://www.nyc.gov/html/dot/html/about/mobilityreport.shtml
5. District Department of Transportation. *District Mobility Report*. District Department of Transportation, 2016. https://districtmobility.org/documents/district-mobility-project-summary-report.pdf
6. Beaton, E., J. Barr, J. Chiarmonte, T. Orosz, D. Paukowits, and A. Sugiura. Select Bus Service on M15 in New York City. *Transportation Research Record: Journal of the Transportation Research Board*, 2012. 2277: 1–10.
7. Kiesling, M., and M. Ridgway. Effective Bus-Only Lanes. *Institute of Transportation Engineers Journal*, 2006. 76(7): 1–12.
8. Safran, J., E. Beaton, and R. Thompson. Factors Contributing to Bus Lane Obstruction and Usage in New York City: Does Design Matter? *Transportation Research Record: Journal of the Transportation Research Board*, 2014: 2418, 58–65.
9. *MWCOG Bus Lane Enforcement Study*. [Online]. https://www.mwcog.org/assets/1/28/10062017_-_Item_12_-_DO_NOT_PRINT_-_Bus_Lane_Enforcement_Study_Final_Report.pdf. Accessed November 13, 2017.
10. Gavanas, N., A. Tsakalidis, A. Aggelakakis, and M. Pitsiava-Latinopoulou. Assessment of Bus Lane Violations in Relation to Road Infrastructure, Traffic and Land-Use Features: The Case of Thessaloniki, Greece. *European Transport*, Vol. 55, 2013, pp. 1–20.
11. Agrawal, A. W., T. Goldman, and N. Hannaford. *Shared-Use Bus Priority Lanes on City Streets: Case Studies in Design and Management*. Mineta Transportation Institute, 2012. https://transweb.sjsu.edu/sites/default/files/2606-shared-use-bus-priority-lanes-city-streets.pdf
12. *Laws of New York, Vehicle and Traffic Law, § 1111-c.*
13. *2012 Bus Lane Camera Enforcement Update Report*. New York City Department of Transportation, 2012.
14. Red Light Camera and Other Automated Enforcement. [Online]. https://www.sfmta.com/services/permits-citations/camera-enforcement. Accessed July 25, 2017.
15. California *Assembly Bill No. 1041*. [Online]. http://www.leginfo.ca.gov/pub/11-12/bill/asm/ab_1001-1050/ab_1041_bill_20110926_chaptered.pdf. Accessed July 25, 2017.
16. *New York State Assembly Bill No. A10882*, 2018.
17. Giering, S. *TCRP Synthesis 89: Public Participation Strategies for Transit*. Transportation Research Board, 2011. http://www.trb.org/Publications/Blurbs/165652.aspx
18. San Francisco Municipal Transportation Agency. *Red Transit Lanes Final Evaluation Report*. San Francisco Municipal Transportation Agency, 2017. https://www.sfmta.com/sites/default/files/reports/2017/Red%20Transit%20Lanes%20Final%20Evaluation%20Report%202-10-2017.pdf
19. Ryus, P., K. Laustsen, K. Blume, S. Beaird, and S. Langdon. *TCRP Report 183: A Guidebook on Transit-Supportive Roadway Strategies*. Transportation Research Board of the National Academies, Washington, D.C., 2016.
20. Danaher, A. R. *TCRP Synthesis 83: Bus and Rail Transit Preferential Treatments in Mixed Traffic*. Transportation Research Board of the National Academies, Washington, D.C., 2010.
21. San Francisco Municipal Transportation Agency. *TIGER Benefit Cost Analysis: 22 Filmore Transit Priority Project*. San Francisco Municipal Transportation Agency, 2015 https://www.sfmta.com/sites/default/files/projects/2015/SFMTA%2022%20Fillmore%20Benefit-Cost%20Analysis.pdf
22. [Online]. Occupational Employment Statistics https://www.bls.gov/oes/current/oes_47900.htm. Accessed July 26, 2017.
23. San Francisco Municipal Transportation Agency. *Transit-Only Lane Enforcement (TOLE) Pilot Program Evaluation*. San Francisco Municipal Transportation Agency, 2011.
24. Transportation Research Board. *TCRP Report 165: Transit Capacity and Quality of Service Manual*, 3rd ed. Transportation Research Board of the National Academies, Washington, D.C., 2013.

The Standing Committee on Bus Transit Systems (AP050) peer-reviewed this paper (18-03098).

TRR
JOURNAL OF THE TRANSPORTATION RESEARCH BOARD

Transportation Research Record
2018, Vol. 2672(8) 41–51

DOI: 10.1177/0361198118782268
journals.sagepub.com/home/trr
SAGE

Characteristics of Bus Transit Vehicles in the United States: A 30-Year National Trend Analysis

Li Tang[1], Albert Gan[2], Fabian Cevallos[2], and Priyanka Alluri[2]

Abstract

In 2015, transit agencies in the United States spent over 60% of their bus transit capital funds on their revenue vehicles. Using 30 years of data from the National Transit Database (NTD), this paper examines the national trends of seven major characteristics associated specifically with bus revenue vehicles. These trends can provide important information on where the market might be heading and aid in planning decisions on transit investments. The characteristics examined include the number of vehicles, spare ratio, average age, average capacity, Americans with Disabilities Act (ADA) accessibility, vehicle reliability, and vehicle operations and maintenance expenses. Some findings from the trend data include: (1) a steady increase in the privatization of bus services; (2) the average spare ratios have consistently exceeded the maximum of 20%, as suggested by the Federal Transit Administration (FTA) for systems operating with more than 50 vehicles; (3) vehicles operated by contractors tended to be significantly newer than those operated directly by transit agencies, although the gap has narrowed in recent years; (4) vehicles operated directly by transit agencies tended to have higher seating and standing capacities than those operated by contractors; (5) there was a tendency among transit agencies to trade seats for more standing room; (6) by 2015, nearly all bus vehicles were ADA-compliant, and increasingly vehicles with lifts were converted to vehicles with ramps/low-floor; (7) reliability of vehicles considering the number of mechanical failures per million revenue vehicle miles has significantly improved over the years; and (8) contractors spent significantly less on average than transit agencies in operating and maintaining their vehicles.

Bus transit systems have continued to be the dominant mode of public transportation in the United States (U.S.) Figure 1 shows that in 2015, urban bus transit systems in the U.S. served a total of 4.96 billion unlinked passenger trips. This number represents an increase from the 4.51 billion trips in 1996 but is essentially unchanged from the 4.95 billion trips achieved in 1986. Figure 1 also shows a steady decline in the transit market share for buses over the years, decreasing from 62.3% in 1986 to 47.8% in 2015, or about a 15% reduction. During the same period, the market share for heavy rail systems increased from 29.3% to 37.2%. Another major increase was for light rail systems, which increased from 1.6% to 4.6%. Together, they indicate a gradual but significant transit mode shift from bus to rail systems over the last 30 years. Other transit modes, including commuter rail, automated guideway, monorail, demand response, jitney, and so forth, collectively increased only slightly, from 6.7% to 10.3% over the same period. If the current trends hold, the market share of bus transit systems is likely to continue to shrink in the coming years, while their total ridership remains stable.

For most bus transit agencies, investments in revenue vehicles usually represent a major use of their capital funds. Figure 2 shows that in 2015, capital expenses for the acquisition and rehabilitation of revenue vehicles accounted for 60.9% of the total capital funds of bus transit agencies, compared to 15.1% for building maintenance, 6.8% for passenger stations, 6.6% for information and communication systems, and 10.7% for miscellaneous other capital investments. This paper will examine the national trends of seven major characteristics associated specifically with bus transit vehicles. These trends provide important information on where the market might be heading and can aid in the planning of transit investments. The characteristics examined include number of vehicles, vehicle spare ratio, average vehicle age, average vehicle capacity, ADA accessibility, vehicle reliability, and vehicle operations and maintenance expenses.

Data Sources

The data used in this study came from the National Transit Database (NTD). They are, thus, limited to only those transit

[1]Amazon, Inc., Seattle, WA
[2]Florida International University, Miami, FL

Corresponding Author:
Address correspondence to Albert Gan: gana@fiu.edu

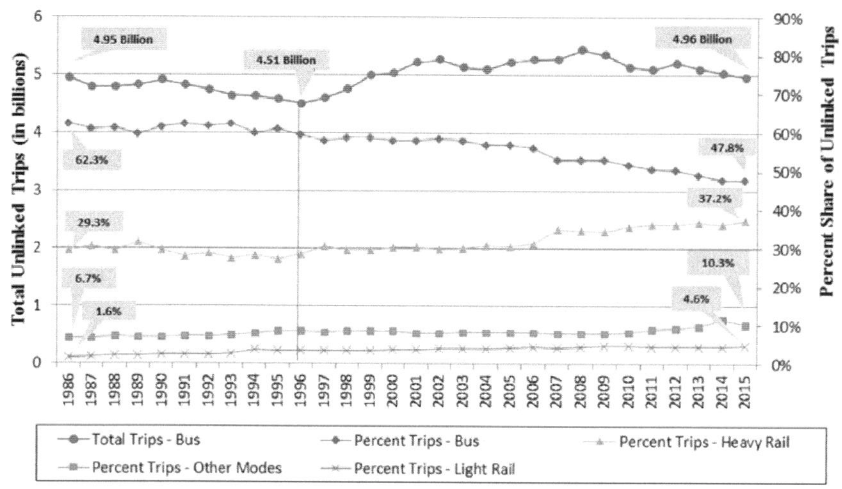

Figure 1. Total and percent share of unlinked bus passenger trips.

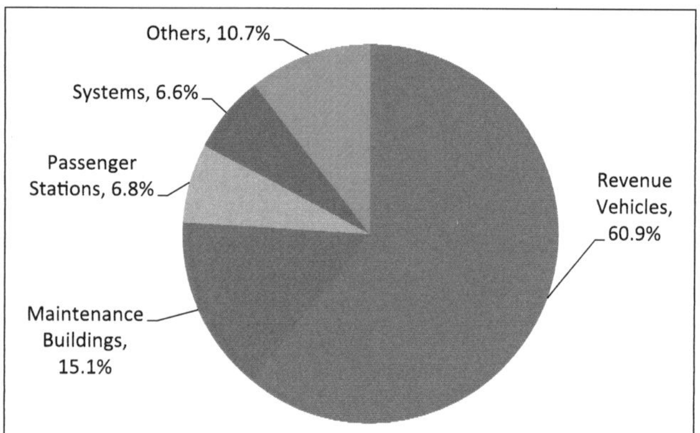

Figure 2. Distribution of capital fund uses for bus transit agencies.

service providers that reported to the NTD under the Urbanized Area Formula Program. NTD is a uniform data set collected as part of the requirements for transit agencies to receive federal grant funds. First collected in 1978 and distributed by the then Urban Mass Transportation Administration (UMTA), NTD has become the sole source of standardized and comprehensive data for use by all constituencies of the U.S. transit industry (*1*). NTD includes data on transit organization characteristics, vehicle fleet characteristics, revenues and subsidies, operating and maintenance costs, vehicle fleet reliability and inventory, services consumed and supplied, and safety and security. These data not only provide direct information on the urban transit systems in the U.S., but can also be used to derive transit performance measures.

The latest available NTD data are from 2015, which are reported on a set of 17 standard forms covering seven different reporting areas. Among these forms, some are reported for the entire transit agencies, while others are for individual transit modes and/or service types. A service type can be either directly operated (DO), if the service is provided in-house, or purchased transportation (PT), if the service is contracted to an outside service provider. When a service is contracted out, the NTD data may be reported either by the transit agency or by the subcontractor.

While the majority of the NTD data collected are made available to the public through the Federal Transit Administration (FTA), access to these data has not been easy. One major reason is because NTD data have been collected and distributed annually on separate files. As a result, trend data must be manually extracted from individual files and then combined for analysis. The process is both time consuming and often error-prone, especially for novice users (*2*).

To facilitate the access and analysis of multiple years of NTD data, a web-based system called the Integrated National Transit Database (iNTD) system (*3*) was developed by the authors. The system integrates multiple years of NTD data in a single standardized database and provides a user-friendly interface for quick data retrieval and analysis. The system offers the single most complete and comprehensive NTD database that goes back to 1984. Based on the 1986–2015 NTD data extracted from the system, the following sections examine the national trends from over the past 30 years, from 1986 to 2015, for each of the aforementioned seven characteristics associated with bus transit vehicles.

Average Fleet Size

The NTD includes two variables for the reporting of fleet size, that is, vehicles available for maximum service and vehicles operated in maximum service. Vehicles available for maximum service include the number of vehicles available for use by a transit agency to meet the annual

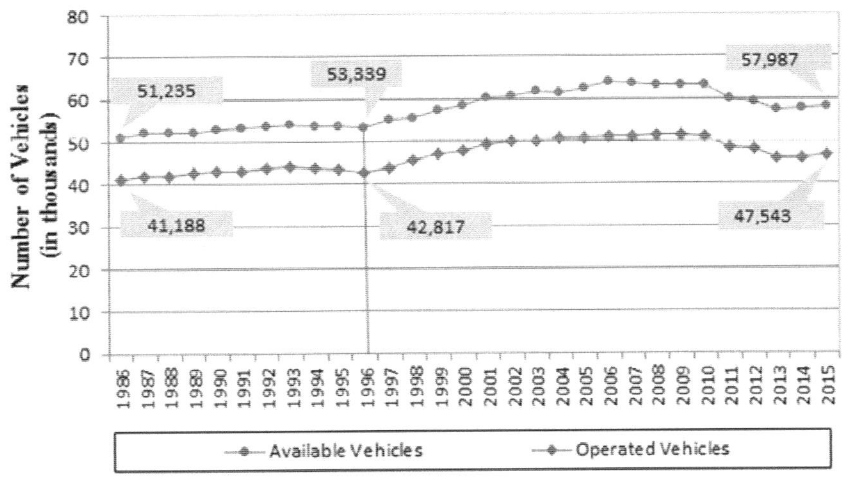

Figure 3. Total available and total operated vehicles.

maximum service requirement. Specifically, they include spare vehicles, out-of-service vehicles, and vehicles in or awaiting maintenance, but exclude vehicles awaiting sale, and emergency and contingency vehicles. Vehicles operated in maximum service, on the other hand, include the number of revenue vehicles operated to meet the annual maximum service requirement, that is, the revenue vehicle count during the peak hours of the peak days/weeks of the peak season (*4*).

Figure 3 gives the nationwide totals from 1986 to 2015 for bus vehicles available and operated for maximum service. The figure shows that, from 1986 to 1996, the number of vehicles nationwide remained relatively stable. However, starting from 1996 through to 2006, the number of vehicles increased steadily, at an average rate of about 1,000 additional buses a year. However, the number of vehicles remained flat from 2006 to 2010. There was a significant drop in the number of vehicles in 2011, likely as a result of the economic downturn that started in 2009. However, the trend has remained stable between 2011 and 2015.

Considering the vehicles per transit system, Figure 4 shows that the average number of buses operated directly by transit agencies has decreased between 1986 and 2009. Despite the recent drop in the total number of vehicles, the figure shows that, on average, agencies actually operated more vehicles. Figure 5 plots the contracted vehicles as a percent of the total vehicles. The figure clearly shows an increasing trend in privatizing bus services, from 3.3% in 1986 to 18.5% in 2015.

Spare Ratios

Spare vehicles are usually maintained as backups for interruptions. Two equations have been used to calculate the spare ratio (SR):

$$SR_1 = \frac{N_a - N_o}{N_o} \times 100 \qquad (1)$$

$$SR_2 = \frac{N_a - N_o}{N_a} \times 100 \qquad (2)$$

In which N_a is the number of vehicles available for maximum service, and N_o is the number of vehicles operated for maximum service. The main difference between the two equations is their divisors. Equation 1 normalizes the spare vehicles by the number of operated vehicles while Equation 2 does so by the number of available vehicles. As by definition N_a is greater than N_o, the range for SR_1 is [0, ∞] and the range for SR_2 is [0, 1]. It can easily be shown that SR_1 is always greater than SR_2, with the only exception being when both the spare ratios are zero. It can also be shown that the difference between the two equations tends to increase with the increase in spare vehicles. For example, if one out of ten vehicles is used as a spare vehicle, SR_1 = 11% and SR_2 = 10%, with a difference of 1%. If four out of ten vehicles are used as spare vehicles, SR_1 = 150% and SR_2 = 60%, with a difference of 90%. It is thus highly critical that one finds out the equation(s) used when comparing spare ratios from different sources.

Figure 6 shows the average spare ratios based on both equations for both DO and PT service types. The figure shows that the spare ratios calculated using the two different equations, even at the aggregated national level, differ by as much as 9%. The same figure also shows no significant difference in spare ratio between the DO and PT services, especially for those after 1990. Overall, the spare ratios have remained relatively constant over the years.

Although the [0, 1] range of Equation 2 is easier to understand and compare, Equation 1 with an open range of [0, ∞]

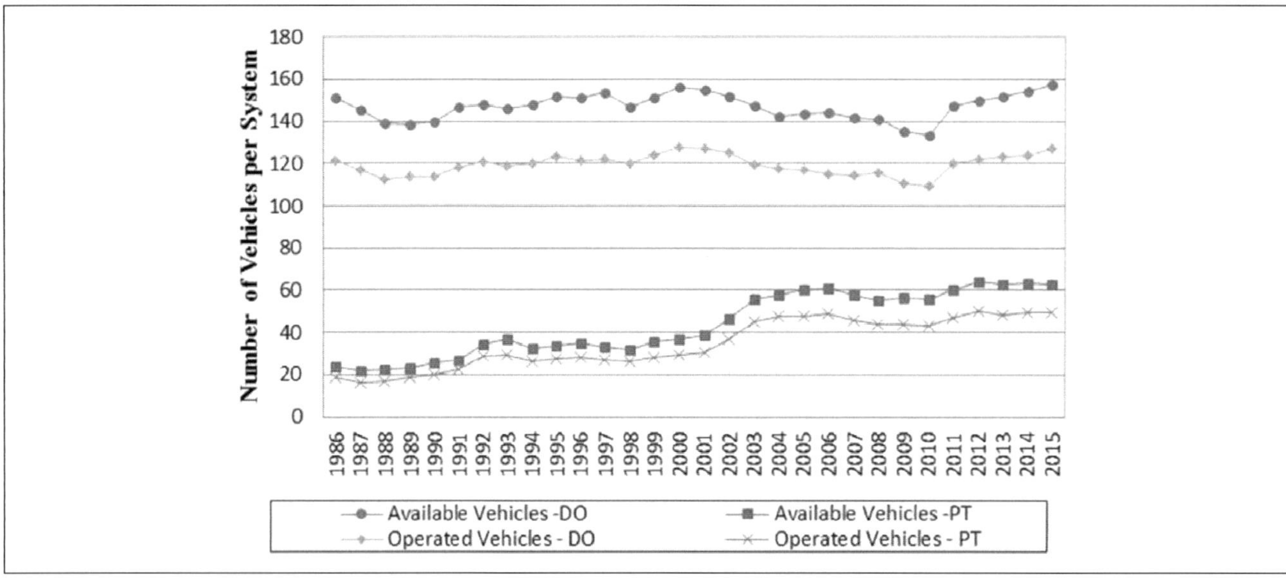

Figure 4. Available and operated vehicles per system by service type.

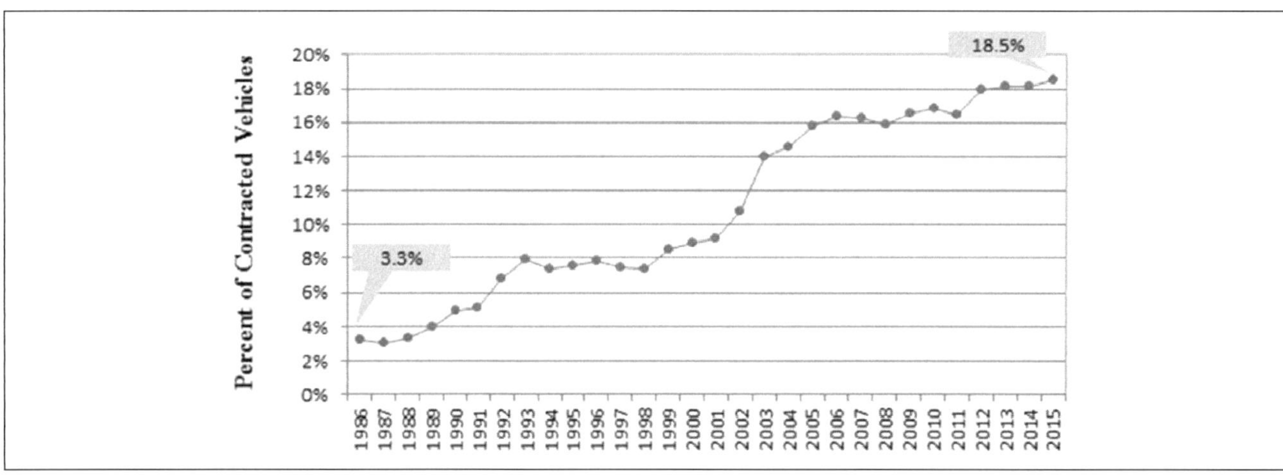

Figure 5. Percent share of contracted vehicles in total fleet.

is the equation officially adopted by the FTA (*5*). The FTA further suggests that the spare ratio for agencies operating 50 or more revenue vehicles should not exceed 20% (*5*). However, using Equation 1, Figure 7 shows that, nationwide, the spare ratio had consistently exceeded 20% for systems operating with more than 50 vehicles. For smaller systems with less than 50 vehicles, the spare ratios are significantly higher.

Average Age of Bus Vehicles

Average fleet age is an important measure as older buses tend to require more maintenance and result in poorer condition ratings. A study by Karlaftis shows that one additional year of age contributes to a 30% increase in expected maintenance time (*6*). As part of its Revenue Vehicle Inventory form, NTD provides the following two variables that can be used to calculate the average age of vehicle fleets: number of active vehicles and years of manufacture (*4*).

Figure 8 shows that overall, the average fleet age during the period from 1991 through 2015 ranged from 7.0 and 8.8 years and the numbers exhibited an interesting cyclic pattern. The same figure also shows that contracted bus vehicles tend to be newer. However, the gap has narrowed from a high of about two years in 1992, to less than one year in recent years. The shapes of the curves show that the narrowing of the gap was not caused by the lowering of the average fleet age of contracted vehicles, but to an increase in the number of new

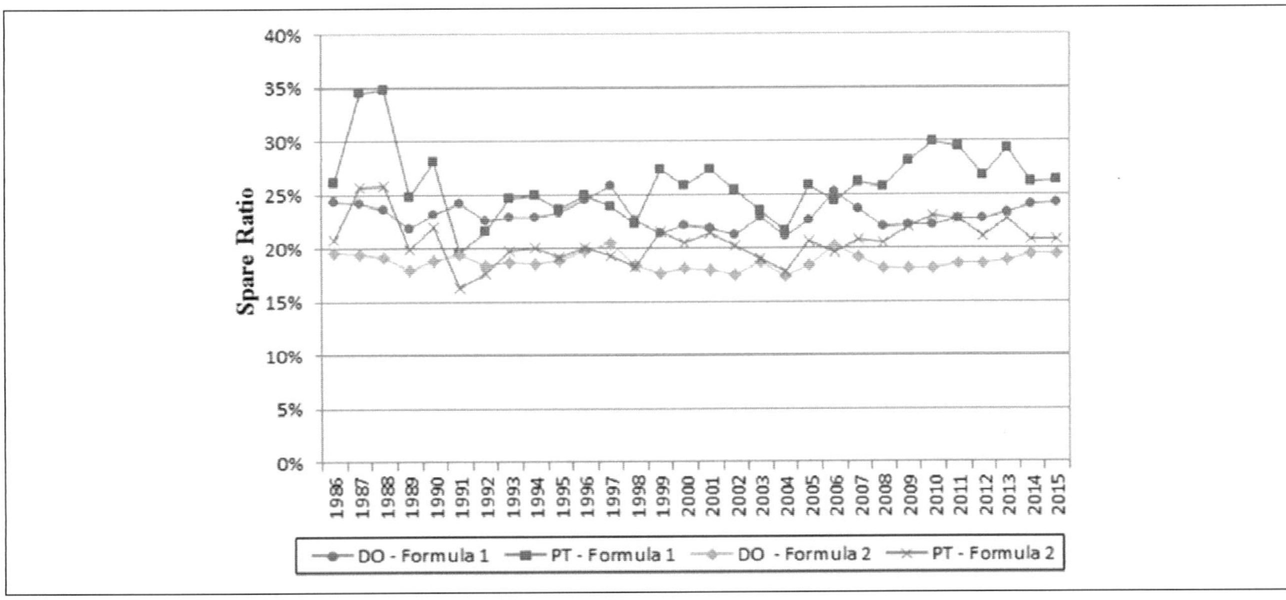

Figure 6. Spare ratios by different service types and equations.

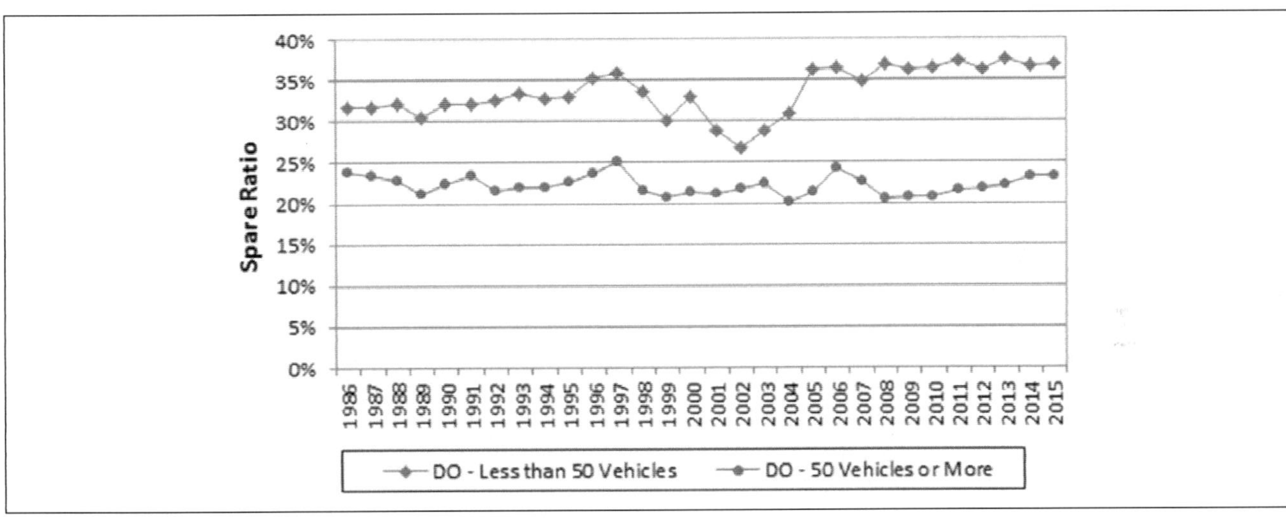

Figure 7. Spare ratios by system fleet size.

buses purchased by transit agencies. Note that the data before 1991 are not plotted because the vehicle inventory data in NTD before 1991 did not identify the specific mode that the vehicles were used for, making it impossible to determine if a vehicle was used for fixed-route bus service or some other services such as demand response.

Average Passenger Capacities

Vehicle seating capacity and standing capacity were added to the NTD in 1991. The seating capacity is the actual number of seats on-board the vehicle and is usually cited in the specifications used to manufacture the vehicle. Standing capacity, on the other hand, is the number of standees allowed by transit agency policy. Zero is reported if local policy prohibits standing. In the absence of local policy on the maximum number of standees, the manufacturer-rated standing capacity is reported (*4*).

Figure 9 plots the seating, standing, and total capacities for both the DO and PT service types. The figure clearly shows that vehicles operated directly by transit agencies maintained higher seating and standing capacities than those provided by outside contractors.

The American Public Transportation Association (APTA) (*7*) suggests for procurement purposes a minimum seating capacity of 43 for a 40-foot bus. Figure 9 shows that, for

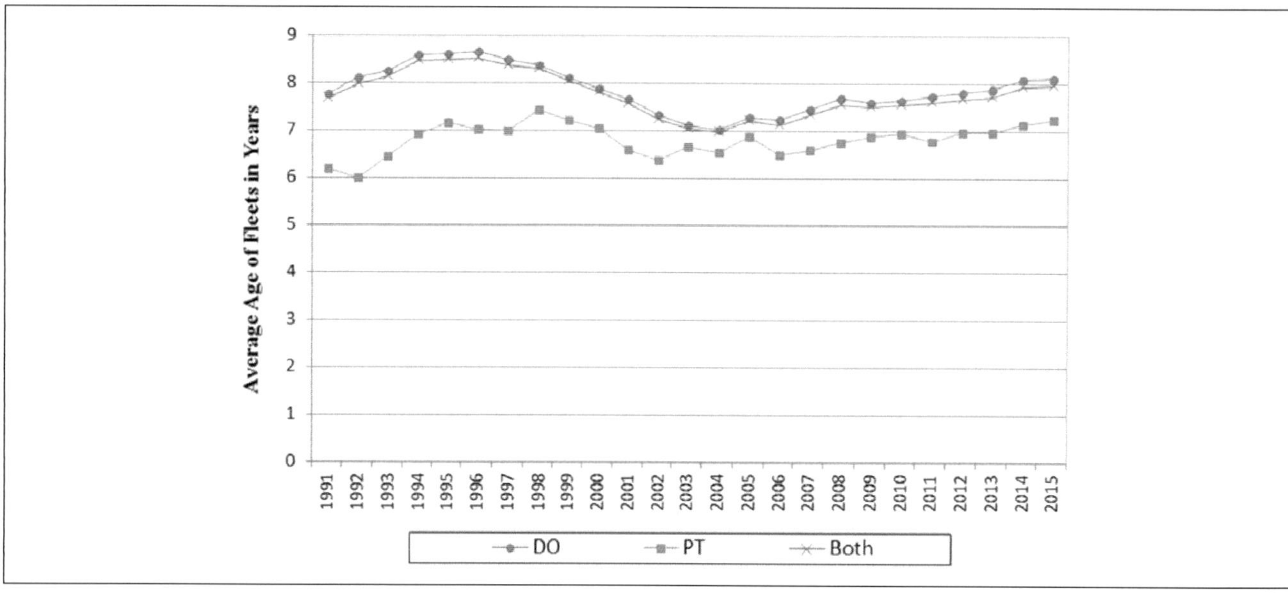

Figure 8. Average ages of fleets.

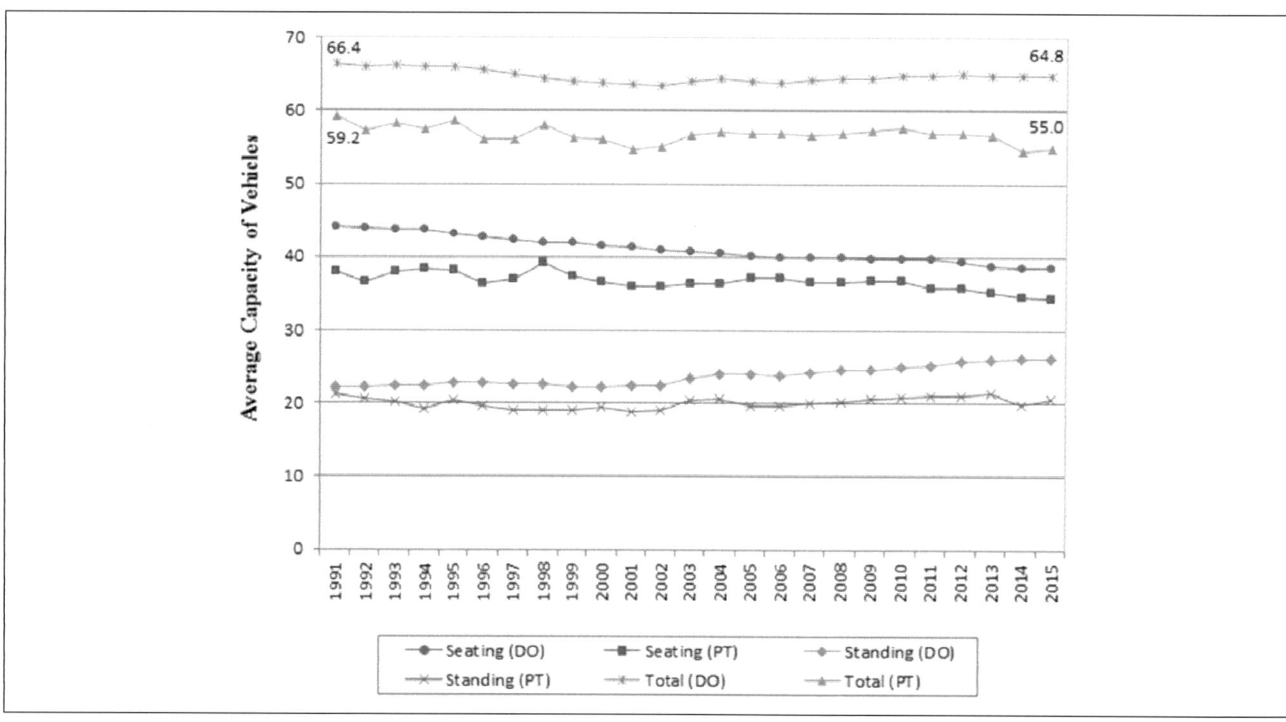

Figure 9. Average passenger capacities by service types.

directly operated vehicles, this suggested minimum was met in the early 1990's. However, the average seating capacity has since fallen to below 40. The same figure also shows a tendency to trade seating for more standing room in directly operated vehicles. This is evidenced from the decreasing seating capacity and increasing standing capacity over the years. One explanation for this compromise could be because of the enactment of the Americans with Disabilities Act (ADA) in 1990, which required vehicles to provide for wheelchairs access, thus reducing the number of seats.

Reporting of vehicle length was added to NTD in 2002. Consistent with the lower capacity of contracted vehicles,

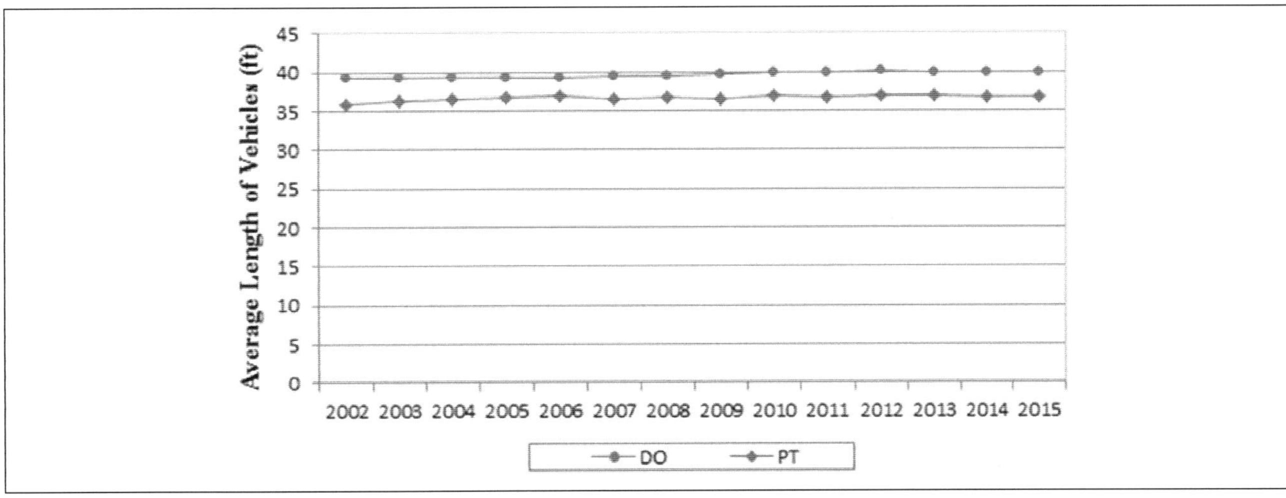

Figure 10. Average vehicle lengths by service types.

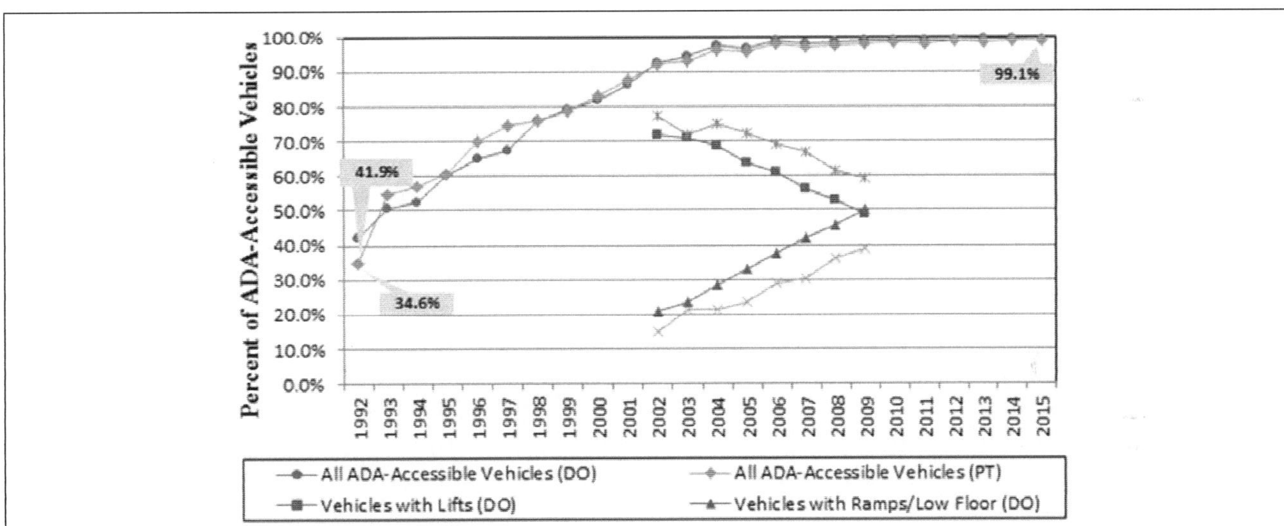

Figure 11. Percentage of ADA-accessible vehicles.

Figure 10 shows that the average vehicle lengths have remained steady over the years, and that contracted vehicles have consistently been about three feet shorter than those of the directly operated vehicles, which has averaged around 40 feet.

ADA Accessibility

The status of ADA accessibility of transit vehicles was first reported in the NTD in 1992. From 1992 to 2001, agencies were only required to report the number of ADA-accessible vehicles. Additional details were added between 2002 and 2009 when agencies were required to also report the number of vehicles equipped with lifts and the number of vehicles equipped with ramps/low floor. Lift-equipped vehicles use wheelchair lifts. Low floor vehicles do not have stairs inside the front or rear doors but are usually equipped with a front-door accessible ramp and a kneeling feature that allows easier access for persons with mobility aids, such as those who have difficulty climbing steps or who may use a wheelchair (*4*). A vehicle is considered ADA-accessible if it is equipped with a lift or with ramps/low floor.

Figure 11 plots the percentages of ADA-accessible bus vehicles for both the DO and PT services. The figure shows that when the data were first collected in 1992, the percentages of vehicles that meet the ADA requirements were 42% and 35% for DO and PT, respectively. Overall, there was no significant difference between DO and PT services. By 2015, nearly all of the vehicles were considered ADA-accessible, with a small number of about 1% remaining non-compliant.

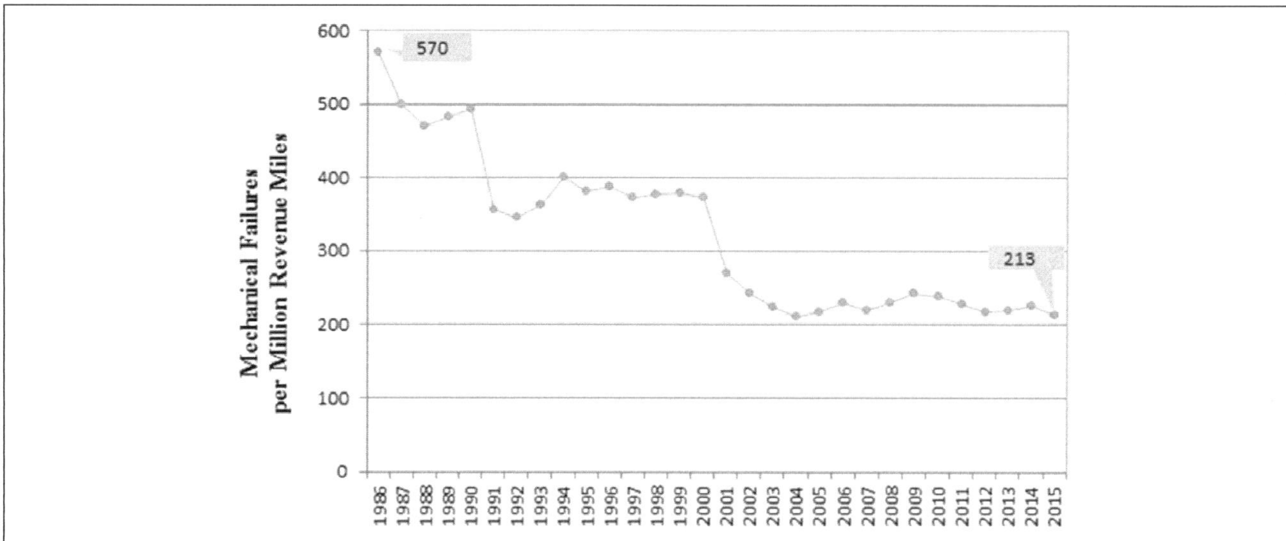

Figure 12. Average number of mechanical failures per million revenue miles.

Figure 11 also plots the percentages of vehicles with lifts and the percentages of buses with ramps/low floor for the period from 2002 to 2009. The figure clearly shows a trend to convert vehicles with lifts to vehicles with ramps/low-floor. It can also be seen that, compared to directly operated vehicles, contracted vehicles were more likely to be equipped with lifts than with ramps/low floor.

Vehicle Reliability

NTD includes two variables that can be used to assess the mechanical reliability of revenue vehicles. The first is the number of major mechanical system failures. These include the failure of the mechanical components of major systems that prevent the specific revenue vehicle from completing a scheduled revenue trip, or from starting the next scheduled revenue trip, because actual movement is limited or because of safety concerns. The second variable reports some other mechanical element of the revenue vehicle that, because of local agency policy, prevents the revenue vehicle from completing a scheduled revenue trip or from starting the next scheduled revenue trip, even though the vehicle is physically able to continue in revenue service. The third variable is simply the sum of the first two variables (*4*).

Figure 12 plots the average total number of mechanical failures per million revenue vehicle miles (RVM) as a measure of vehicle reliability. Data for the first two variables are not plotted in this figure to avoid skewed totals because some agencies reported only the total mechanical failures. It can be seen that vehicle reliability has improved significantly over the years. Specifically, from 1986 to 2004, the vehicles as a whole experienced a remarkable reduction of over 60% in the number of mechanical failures per million RVM, before it started to taper off at slightly above 200 mechanical failures per million RVM. It is interesting to note that the improvements over the 30-year period did not occur in a continuous manner, but with intermittent periods of improvement and stagnation. Figure 13 shows that the period of improvement in vehicle reliability occurred despite a general increase in vehicle usage between 1986 and 2009, as measured by the average revenue vehicle miles served per operated vehicle.

Vehicle Operations and Maintenance Expenses

Expenses reported in NTD are either for operations or capital investments. Operating expenses are reported in four categories: vehicle operations, vehicle maintenance, non-vehicle maintenance, and general administration. The first two categories are expended on vehicles and are the focus of this section. Vehicle operations include all activities associated with dispatching and running vehicles in revenue service to carry passengers, including administrative and clerical support. Vehicle maintenance includes all the activities associated with ensuring revenue vehicles and service vehicles are operable, cleaned, fueled, inspected, and repaired. Non-vehicle maintenance includes all the activities associated with ensuring buildings, grounds, and equipment (garages, passenger stations and shelters, administration buildings); fare collection equipment; and communications systems, track, structures, tunnels, and power systems are operable. Finally, general administration includes all the activities associated with supporting the provision of the transit service (*4*).

Figure 14 shows the percentages of funds expended on each of the four categories over the last 30 years. The figure shows that the percentages of funds expended on vehicle operations had been inching toward 60%. Together with

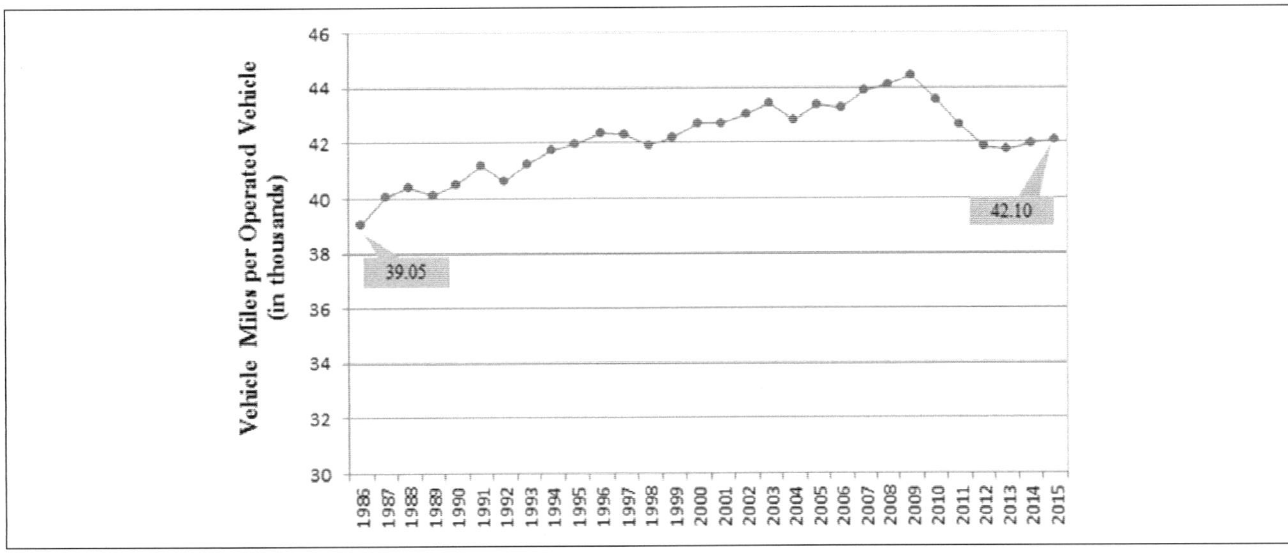

Figure 13. Vehicle usage measured by revenue miles served per operated vehicle.

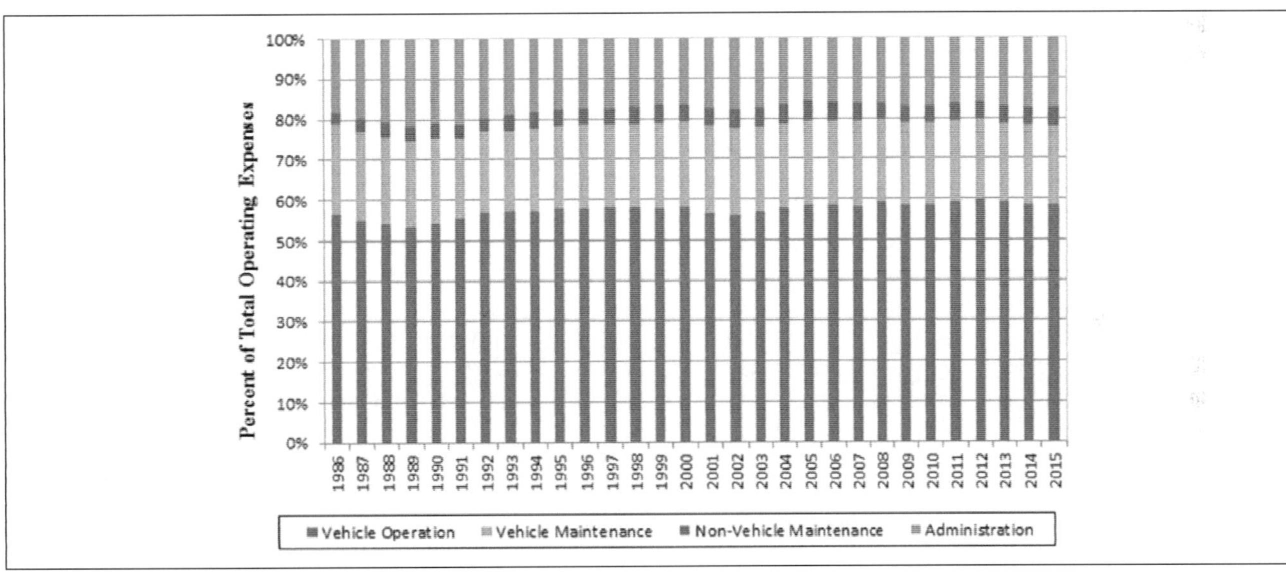

Figure 14. Percentage distribution of total operating expenses.

vehicle maintenance, they accounted for nearly 80% of the total operating expenses. Figure 15 shows that the expenses (in 2015 dollars) per operated vehicle have clearly increased from 1995 to 2015, with expenses on vehicle operations increasing at the fastest rate. In contrast, expenses on vehicle maintenance have increased much slower. It can also be seen from the same figure that transit agencies spent significantly more on vehicle operation and maintenance per vehicle than outside contractors. In 2015, vehicle operation expenses for contracted vehicles are about 35% lower. For vehicle maintenance expenses, the difference was even lower, it was 42% lower. These results clearly show that, as far as the vehicle operation and maintenance cost are concerned, privatization could be a more financially efficient choice, especially for cash-strapped local governments. Note that in Figure 15, data are plotted for 1995 onwards because data before 1995 were not stratified by service types.

Summary and Conclusion

Investments in bus revenue vehicles are a major use of capital funds for transit agencies. Their importance is further emphasized under Section 20019 of the Moving Ahead for Progress in the 21st Century Act (or MAP-21), which requires that transit providers develop and implement transit asset management (TAM) plans that not only keep an inventory of their

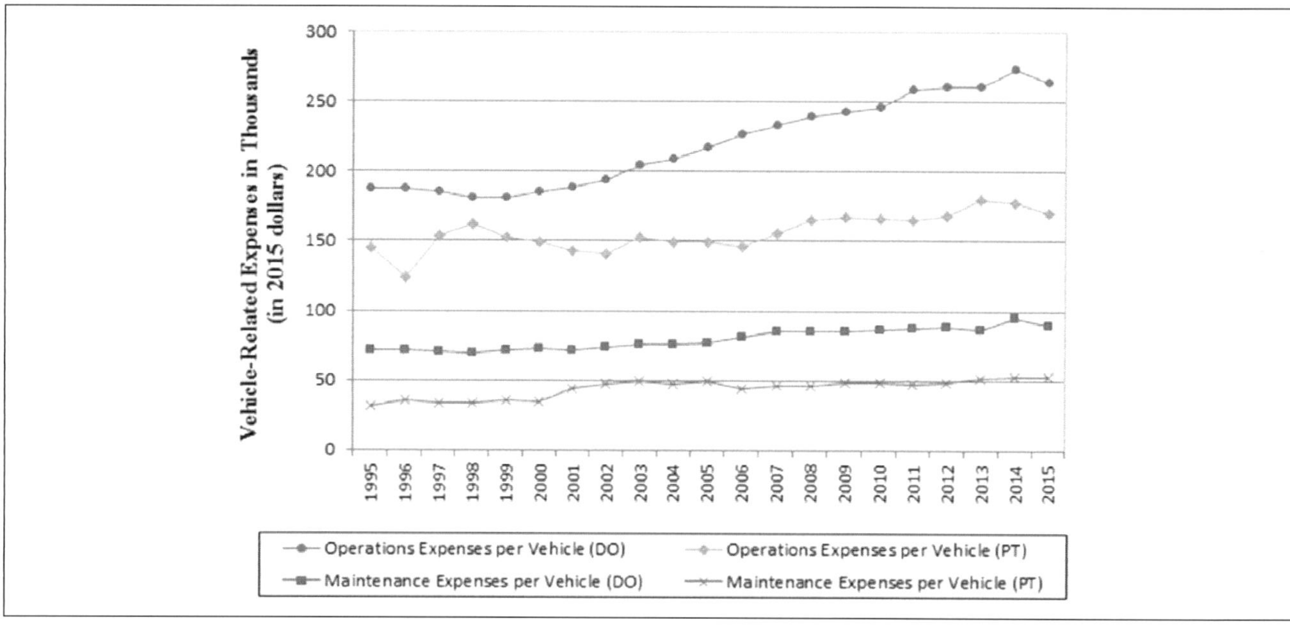

Figure 15. Operations and maintenance expenses per operated bus vehicle.

vehicles, but also monitor and report the conditions of those vehicles to improve their state of good repair.

This paper examined the national trends for seven characteristics specifically associated with bus revenue vehicles. They included vehicle fleet sizes, vehicle spare ratio, average vehicle age, average vehicle capacity, ADA-compliance, vehicle reliability, and vehicle operations and maintenance expenses. Using comprehensive data from an integrated system that provides access to the 1986 through 2015 National Transit Database (NTD), this paper has revealed several interesting national trends for these characteristics.

When considering vehicle fleet sizes, it was found that nationally there was a significant increase starting from 1996 through 2009, but this has decreased over the recent years. However, the average number of vehicles operated by each system has continued to increase over the same years. The trend in fleet sizes also showed that, nationally, there was a move over the years to privatize bus transit services. This could be considered a positive trend by some. As the data show, private service providers had spent significantly less on operation and maintenance per vehicle.

Two equations have been used to calculate the average spare ratios which, individually, may yield very different numbers. As such, it is highly critical that one be aware of the specific equations(s) used when comparing spare ratios from different sources. The FTA has suggested that the spare ratio should not exceed 20% for agencies operating 50 or more revenue vehicles. However, the results show that, nationally, this limit has been consistently exceeded.

The national trend in average fleet age showed that contracted bus vehicles tended to be newer than those operated directly by transit agencies, although the gap has narrowed in recent years. This narrowing of the gap appears to be largely a result of new vehicles purchased by transit agencies. Considering vehicle design, the national trends showed a tendency to trade seats for more standing room, and a clear trend to convert vehicles with lifts to vehicles with ramps/low-floor for ADA accessibility. By 2006, nearly all bus vehicles were ADA-compliant. Another major improvement found was in the reliability of vehicles. This was evidenced from a reduction of over 60% in the number of mechanical failures per million revenue vehicle miles served between 1986 and 2015.

Acknowledgment

The authors are grateful to the Florida Department of Transportation (FDOT) Public Transit Office for its financial support for development of the Integrated National Transit Database (iNTD) system, which makes this paper possible.

Author Contributions

The authors confirm contribution to the paper as follows: study conception and design: Albert Gan, Li Tang; data collection: Li Tang; analysis and interpretation of results: Albert Gan, Li Tang, Fabian Cevallos, Priyanka Alluri; draft manuscript preparation: Albert Gan, Li Tang, Fabian Cevallos, Priyanka Alluri. All authors reviewed the results and approved the final version of the manuscript.

References

1. Lyons, W. M., and E. R. Fleischman. New Future for the Federal Transit Administration Section 15 Program. *Transportation Research Record: Journal of the Transportation Research Board*, 1992. 1349: 18–27.

2. Gan, A., and F. Gui. Development of a Web-Based Information System for Integration and Analysis of Multiple-Year National Transit Database (NTD). Presented at 88th Annual Meeting of the Transportation Research Board, Washington, D.C., 2009.
3. Florida Department of Transportation. *Urban Integrated National Transit Database (iNTD)*. 2017. http://ftis.org/urban_iNTD.aspx. Accessed Novmber 1, 2017.
4. Federal Transit Administration. *NTD Annual Reporting Manual, National Transit Database*. Federal Transit Administration, Washington, D.C., 2008.
5. Federal Transit Administration, Circular 5010.1D. Grant Management Requirements, 2008.
6. Karlaftis, M. G. Modeling Transit Vehicle Repair Duration and Active Service Time. *Transportation Planning and Technology*, Vol. 34, No. 5, 2011, pp. 433-442.
7. American Public Transit Association, Standard Bus Procurement Guidelines, 1997. http://www.apta.com/resources/reportsandpublications/Documents/comterms.pdf

The Standing Committee on Bus Transit Systems (AP050) peer-reviewed this paper (18-06344).

Article

TRR
JOURNAL OF THE TRANSPORTATION RESEARCH BOARD

Transportation Research Record
2018, Vol. 2672(8) 52–63

DOI: 10.1177/0361198118797827
journals.sagepub.com/home/trr
SAGE

Developing Dedicated Bus Lane Screening Criteria in Baltimore, Maryland

Adam Vest[1], Patrick J. McMahon[2], and Jesus Cuellar[1]

Abstract

Despite the increasing number of dedicated bus lanes in various cities across the United States, no clear screening criteria or thresholds have emerged for determining appropriate locations for these dedicated facilities. The Maryland Department of Transportation Maryland Transit Administration (MDOT MTA), in coordination with the Baltimore City Department of Transportation, conducted a detailed review of literature, selection of performance measures, and screening evaluation to identify candidate corridors for dedicated bus lanes and criteria to be used for future assessments. A national-level literature review and review of best practices was used to identify performance measures by which to assess the benefits and tradeoffs of installing dedicated bus lanes. As a result of this literature review and evaluation, design was undertaken on the selected corridor segments and implementation has started on the selected corridor segments. By the end of 2017 nearly 5.5 lane miles of dedicated bus lanes were in place, with 4.9 mi of full-time lanes with red paint (methyl methacrylate) and appropriate signing and markings, and 0.5 mi of peak-only lanes with signage and pavement markings but no paint. This effort was conducted in support of BaltimoreLink, MDOT MTA's overhaul and rebranding of its core bus service, which included redesigning the entire local and express bus system throughout the Baltimore region in 2017.

The Maryland Department of Transportation Maryland Transit Administration's (MDOT MTA) updated surface transit system, BaltimoreLink, included redesigning the entire local and express bus system throughout the Baltimore region in 2017. The redesign also included the introduction of 12 high-frequency, color-coded bus routes to improve connections to jobs, places of interest, and other transit modes. The new high-frequency routes, CityLink, operate at 10 min or lower headways during peak periods and 15 min or lower headways during off-peak periods. Transit priority treatments were implemented in selected locations to improve the reliability and speed of CityLink routes: transit signal priority (TSP) on two corridors (Loch Raven Boulevard and Greenmount Avenue/York Road), and dedicated bus lanes on nine streets in downtown Baltimore.

MDOT MTA, in close coordination with the Baltimore City Department of Transportation (BCDOT), completed a comprehensive evaluation to assess the applicability of dedicated bus lanes on Baltimore's city streets. This effort began with a detailed literature review of national research publications and case studies from various cities to better understand performance measures most appropriate for screening study corridors. Using a selected list of performance measures, the evaluation began with nearly 25 candidate streets across the city that were selected based on planned bus service, existing high ridership bus service, and locations where bus service encounters significant delays. Nine streets totaling 6 mi were eventually chosen for dedicated bus lanes. By the end of 2017 nearly 5.5 lane miles of dedicated bus lanes were in place, with 4.9 mi of full-time lanes with red paint (methyl methacrylate) and appropriate signing and markings, and 0.5 mi of peak-only lanes with signage and pavement markings but no paint.

Background and Literature Review

Although the operation of urban dedicated bus lanes is not as well studied as general traffic operations, a significant body of work exists that both catalogs the results of dedicated bus lane projects in case studies, and

[1]Kittelson & Associates, Inc., Philadelphia, PA
[2]Maryland Transit Administration, Baltimore, MD

Corresponding Author:
Address correspondence to Adam Vest: avest@kittelson.com

Table 1. Performance Measures: Mobility (Literature Review)

Measure	Mode	Dedicated bus lanes thresholds from case studies and publications
Person throughput	Bus	A bus-only lane is only cost effective if the daily person throughput of the corridor is over 20,000 people (Ang-Olson and Mahendra [1]).
		A bus-only lane may be warranted if the buses serving a particular corridor currently carry (or are expected to carry) as many people as people traveling in automobiles (Vuchic [2]).
	Auto	The bus-only lane should carry as many people as in the adjacent general traffic lane (Ryus et al. [3]). Various case studies have shown however, that there is some flexibility in this comparison and that an approximate range above and below the number of existing users may also be appropriate.
Person delay	Person	A strategy that results in a net reduction in person delay would be considered acceptable, even if automobile delay increases somewhat, as long as the intersection or roadway as a whole operates below capacity (Ryus et al. [4]).
Volume (peak hour)/ frequency	Auto	A general rule of thumb is to assume a single travel lane could accommodate approximately 10,000 vehicles, and as such could be used to assess reasonableness of replacing a travel lane with a dedicated bus-only lane.
	Bus	60 buses per hour (for bus in curb lane) to 180 buses per hour (for two exclusive lanes) (Levinson et al. [5]).
		Minimum one-way peak-hour volume: 30 to 40 buses for curb bus-only lane with normal (QJ)(?h2)flow; 50 to 80 buses for central business district (CBD) curb bus-only lanes, main street (QJ)(?h2)(Ryus et al. [3]).
		20 to 60 buses per hour (Washington Metropolitan Area Transit Authority [6]).
		Introducing a lane for fewer than 40 buses per hour can result in the loss of vehicular (QJ)(?h2)capacity of up to 30% to 50% (St. Jacques and Levinson [7]).
		NYC Select Bus Service on various corridors carries approximately 14 to 22 buses per (QJ)(?h2)hour in bus-only lanes.
	Bus/ bike	10 to 80 buses per hour. Buses and cyclists may share the wide curb lanes when there are: a) 0 to 100 cyclists per hour and bus headways ranging from 15–30 min; and b) separate facilities if there are >100 cyclists per hour and bus headways are <15 min (Hillsman et al. [8]).
Ridership (passengers per hour)	Bus	2,000–3,200 for curb lanes in CBD with commercial frontage; 1,200–1,600 for curb lanes with normal flow and at least two lanes available for other traffic; and 2,4003,600 for median bus-only lanes and at least two lanes available for other traffic (Ryus et al. [3]).
Average speed	Bus	If average bus speeds are: a) ~8 mph or lower, then exclusive bus lane can result in greater mode shift and travel time benefits; and b) ~12 mph or greater, then exclusive bus lane is unlikely to significantly improve speeds (Institute for Transportation and Development Policy [9]).
Level of service (LOS)[a]/ delay and ratio of volume to capacity (v/c)[b]	Auto	Numerous case studies from across the country were reviewed and most did not use a specific threshold for LOS or Delay. However, the Kansas City Area Transit Authority will not approve a lane conversion that results in auto LOS worse than "D" (i.e., 35-55 seconds of average vehicle control delay).
		Bus-only lanes are probably appropriate with v/c < 0.8; and possibly appropriate when v/c of "new" less than v/c of "old," or change in v/c < 0.1 (Washington Metropolitan Area Transit Authority [6]).

[a]Function of the average vehicle control delay, measured using grades A through F where A represents the least average vehicle delay and F representing the highest average vehicle delay (*10*).
[b]Volume to capacity ratio, also referred to as degree of saturation, represents the sufficiency of an intersection to accommodate the vehicular demand (*10*).

analyzes the aggregated results of dedicated bus lane installation in the United States and other countries. In order to assess whether a dedicated bus lane would deliver enough benefit to justify setting street space aside from general traffic use in Baltimore, Maryland, 10 dedicated bus lane case studies and 16 meta-analyses and reports on the subject of dedicated bus lane operations were reviewed. Relevant performance measures from these studies have been summarized and condensed in Tables 1, 2, and 3 under three key areas: mobility, access, and design adequacy. Additionally, MDOT MTA and BCDOT staff conducted interviews with four cities (New York City, Seattle, San Francisco, and Chicago) that had recently installed red dedicated bus lanes. These interviews focused on materials and cost, approval process, bicycle compatibility, enforcement, design, and bus operations. It should also be noted that these aforementioned cities may or may not rely solely on red bus lanes to promote improved transit operations and may have coupled them with other priority treatments.

Though there is no clear consensus on specific performance measures that should be used for selecting streets

Table 2. Performance Measures: Access (Literature Review)

Measure	Mode	Dedicated bus lanes thresholds from case studies and publications
Parking and loading impacts	Auto	For curbside bus lanes, parking should be removed when there are 500 to 600 personal vehicles or more; the level of service (LOS) for the street is an "E" or "F" (i.e., 55-80 and >80 seconds of average vehicle control delay, respectively); and travel speeds fall below 20 mph (Levinson et al. [*11*]).
Population near routes	Bus	Population within close proximity to transit corridors.
Transit-dependent population near routes	Bus	Determine potential conflicts with existing emergency routes.
Access to jobs	Bus	Determine potential conflicts with existing freight routes.
Connectivity/transfers	Bus	Transit-dependent population within close proximity to transit corridors.
Emergency routes	Auto/bus	Determine potential conflicts with existing emergency routes.
Freight routes	Auto/bus	Determine potential conflicts with existing freight routes.

Table 3. Performance Measures: Design Adequacy (Literature Review)

Measure	Mode	Dedicated bus lanes thresholds from case studies and publications
Lane width	Bus/bike	10–18 ft (Hillsman et al. [*8*]). Lane width thresholds based on speed: <45 mph, SBBL 13.5 ft; >45 mph < 50 mph, SBBL 15–16 ft; and >50 mph, no SBBL (Queensland Transport [*12*]). San Francisco and Albuquerque recommend separate but adjacent bike and bus lanes for lane widths between 14 and 17 ft (Hillsman et al. [*8*]). When more than 16 ft of width is available, consider providing separate bus and bicycle facilities (Ryus et al. [*4*]).
Right turns at intersections	Auto	At intersections with <100 right-turning vehicles per hour, vehicles can use the bus lane; and >100 right-turning vehicles per hour, an exclusive alternative should be constructed (Florida Department of Transportation [*13*]). Bus lanes lose nearly half of their travel time savings benefit in central business district (CBD) areas when right turns are allowed at the intersections (Ryus et al. [*4*]).

where dedicated bus lanes may work best, there are some clear considerations that must be considered.

Frequency of bus service on a study street should be a key factor in determining whether a dedicated bus lane is warranted on a street. If bus service is relatively infrequent, the lane may sit empty for large periods of time, resulting in underutilized street space and a public perception that motorists in non-bus modes of transportation are being penalized unjustly. Although some research, particularly that focusing more on traditional bus-rapid transit suggest frequencies of 60 + buses per hour, cities across the United States have installed dedicated bus lanes on streets with as few as 14 buses per hour (i.e., approximate headways of 4 min). In some cases, such as New York City's Select Bus Service, a variety of transit priority treatments, including dedicated bus lanes, have been implemented to create key east–west transit options to improve transit speeds and reliability, even if bus headways are not considered "high" by bus-rapid transit standards. As such, subjective observations across the country suggest cities that have relied on lower frequency thresholds have also encountered higher numbers of infractions related to loading, unloading, deliveries, and parking because the bus lanes are less frequently in use by transit vehicles. These lower thresholds can be offset to some degree by increased enforcement and through additional priority treatments such as off-board fare payment, low-boarding buses, TSP, and other potential enhancements.

Person throughput is being used more regularly to compare the number of people being moved by private vehicles and bus. This moves the discussion away from average automobile delay, a measure that the industry has been using for decades to assess mobility, and focuses on overall mobility, regardless of mode. Some literature also likens this approach to a more equitable assessment of mobility, which does not penalize those who cannot afford a private vehicle or those who choose not to own a private vehicle. Most sources are in agreement that a dedicated bus lane should be carrying at least as many people as adjacent general travel lanes during congested periods.

Table 4. Selected Performance Measures: Mobility

Performance measure	Auto	Bus
Person throughput	Bus lane should carry approximately 80% to 120% of the adjacent auto lane. The flexibility/range allows for consideration of exclusive bus lanes, business access and transit lanes, peak period bus lanes, and use of consecutive turn lanes and on-street parking conversion.	
Person delay	Change in person delay (passengers/riders/operators of autos and buses) with conversion to bus lane.	
Volume (peak hour, peak direction)/ Frequency	Peak hour: >1,000 vehs requires more than 1 auto lane; Daily: >10,000 vehs requires more than 1 auto lane.	Curb lane: ⩾24 buses (1 bus every 2.5 min); offset lane (i.e., adjacent to parking): ⩾18 buses (1 bus every 3.3 min).
Passengers per hour	Not applicable (1.15 passengers per vehicle assumed system-wide per Baltimore Metropolitan Council).	Curb lane (CBD): 2,000–3,000; curb (normal flow): 1,200–1,600; offset lane (i.e., adjacent to parking): >800 (all expressed as peak hour).
Travel time	Projected impacts to be assessed on case-by-case basis, balancing need to move the greatest number of people.	
Average speed	>10 mph below speed limit: bus lane detrimental to corridor mobility; 0–10 mph below speed limit: bus lane may have limited mobility impacts; additional case-by-case consideration given to intersection impacts.	<8 mph: substantial benefits to bus lane; 8–12 mph: potential benefits to bus lane; >12 mph: limited benefits to bus lane; additional case-by-case consideration given to intersection impacts and potential for transit preferential treatments (e.g., transit signal priority, queue jumps, etc.).
Level of service (LOS)/ delay and v/c	Expected change in LOS/delay and v/c (LOS/ delay may be appropriate at LOS "E" [55-80 seconds of average vehicle control delay] if benefits to bus travel are substantial); v/c < 1.0).	Expected change in delay.

Average speed and *reliability* both provide a snapshot of how transit service may be operating on a street, and depending on data availability, can provide a detailed synopsis on problem areas such as congested intersections, issues with re-entry, and bus stops with high boardings and alightings. Some research suggests that candidate streets for dedicated lanes should have average transit speeds less than 8 mph, whereas streets with average transit speeds greater than 12 mph may not benefit from dedicated bus lanes.

Automobile level of service (LOS) (i.e., average vehicle delay) must be considered to understand modal tradeoffs, traffic impacts, and potential congestion hot spots. Though it most likely should not be relied upon in isolation, it can be used to identify locations where dedicated bus lanes may create more problems than they solve. Some reviewed city case studies suggest that resulting impacts from the installation of dedicated bus lanes that result in traffic LOS to be worse than LOS "D" are not tenable (LOS "D" is equivalent to 35-55 seconds of average vehicle control delay); however, several cities with more congested conditions do not use a LOS threshold, as many urban streets are already congested.

Selected Screening Criteria

After reviewing various criteria from a number of sources, MDOT MTA and BCDOT identified a list of performance measures considered most relevant for both agencies. The relative importance of each measure varies, and not every performance measure would be analyzed for every candidate corridor. Of these performance measures, some address how many people would benefit from the bus travel time savings that could be expected from a dedicated bus lane, some address whether roadway conditions would interfere with how much bus travel time savings could be achieved, and some address the feasibility of installing a dedicated bus lane. The measures will also be key criteria of post-construction evaluations to assess effectiveness of the dedicated bus lanes. In particular, some of the listed measures may serve minimal function for assessing corridors before installation of the dedicated bus lanes, although they will be key measures of effectiveness after construction is complete. The selected corridor screening criteria are summarized in Tables 4, 5, and 6 under three key areas: mobility, access, and design adequacy.

Preliminary Screening

The initial round of screening for dedicated bus lanes was performed on 20 study streets, each of which is shown in Figure 1. These initial study streets were identified as corridors with a combination of relatively high frequency and ridership, and some level of travel time delay and reliability issues. Of the study streets, 14 operate as one-

Table 5. Selected Performance Measures: Access

Performance measure	Auto	Bus
Parking and loading/unloading impacts	Case-by-case basis to determine potential impacts; likely only applicable for curbside bus lanes, but consideration will also be given to any potential parking and loading/unloading impacts.	
Population near routes	na	% population accessed within 5-min walk of corridor, bus routes on corridor, or both.
Transit-dependent population near routes	na	% transit-dependent population accessed within 5-min walk of corridor, bus routes on corridor, or both.
Access to jobs	na	# of jobs accessed by bus routes on corridor.
Connectivity/Transfers	na	# of direct connections/transfers to high-capacity transit (i.e., Metro, light rail, CityLink).
Emergency routes	Yes/No	
Freight routes	Yes/No	
Lane width	10–12 ft: bus lane appropriate; 12–14 ft: consider painted buffer or consider bus/bike lane; >=14 ft: consider separate adjacent bike (green) and bus (red) lane.	
Right turns at intersections	<100 right turns per hour: motorists can use bus lane; >100 right turns per hour: exclusive alternative should be considered (i.e., bus bypass lane, queue jump).	na

Note: na = not applicable.

Table 6. Selected Performance Measures: Design Adequacy

Performance measure	Auto	Bus
Lane width	10–12 ft: bus lane appropriate; 12–14 ft: consider painted buffer or consider bus/bike lane; >=14 ft: consider separate adjacent bike (green) and bus (red) lane.	
Right turns at intersections	<100 right turns per hour: motorists can use bus lane; >100 right turns per hour: exclusive alternative should be considered (i.e., bus bypass lane, queue jump).	na

way couplets in downtown Baltimore, effectively operating as seven corridors. The primary measure used at this level of analysis was hourly frequency of buses (i.e., number of buses per hour) that would be expected to utilize a proposed dedicated bus lane during the AM and PM peak periods. Though there are other important measures when weighing different candidate corridors against one another, a threshold of 18 buses per hour was used as the initial screening criteria for which a dedicated bus lane may constitute a wise investment. The proposed peak-hour bus frequency is shown in Figure 2. As detailed in Figure 2, the highest bus frequencies are mostly concentrated in the downtown core, where some streets had more than 40 buses per hour. It should be noted that the analysis was undertaken in the midst of a redesign of the bus network and as such some of the bus volumes on streets have changed since the analysis was undertaken, but this graphic represents the volumes at the time when corridors were selected.

Detailed Screening

The initial bus frequency (i.e., buses per hour) performance measure assesses how many vehicles are likely to benefit from a dedicated bus lane. Another important measure for a dedicated bus lane is person throughput, an estimate of how many people the lane will move, when compared with an adjacent general purpose lane. Based on research on other cities' dedicated bus lane projects, a successful dedicated bus lane should carry more than 80% of the number of people as an adjacent travel lane. Analysis of the study corridors was tiered, meaning that a preliminary screening of whether enough bus service was planned on a corridor to justify a dedicated bus lane eliminated a number of the corridors from consideration. Then, person throughput by mode and spatial feasibility were analyzed for the remaining corridors, further eliminating some candidate streets. A full operations analysis was only conducted for the remaining corridors.

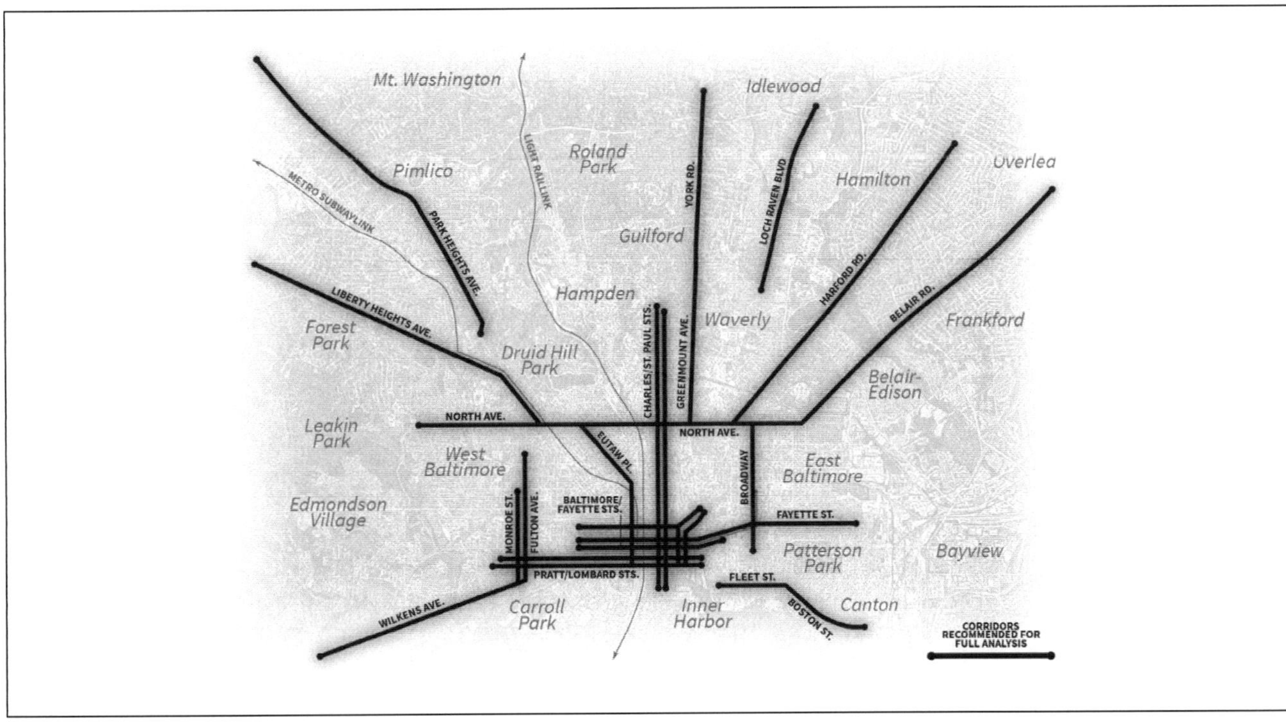

Figure 1. Streets considered for dedicated bus lanes.

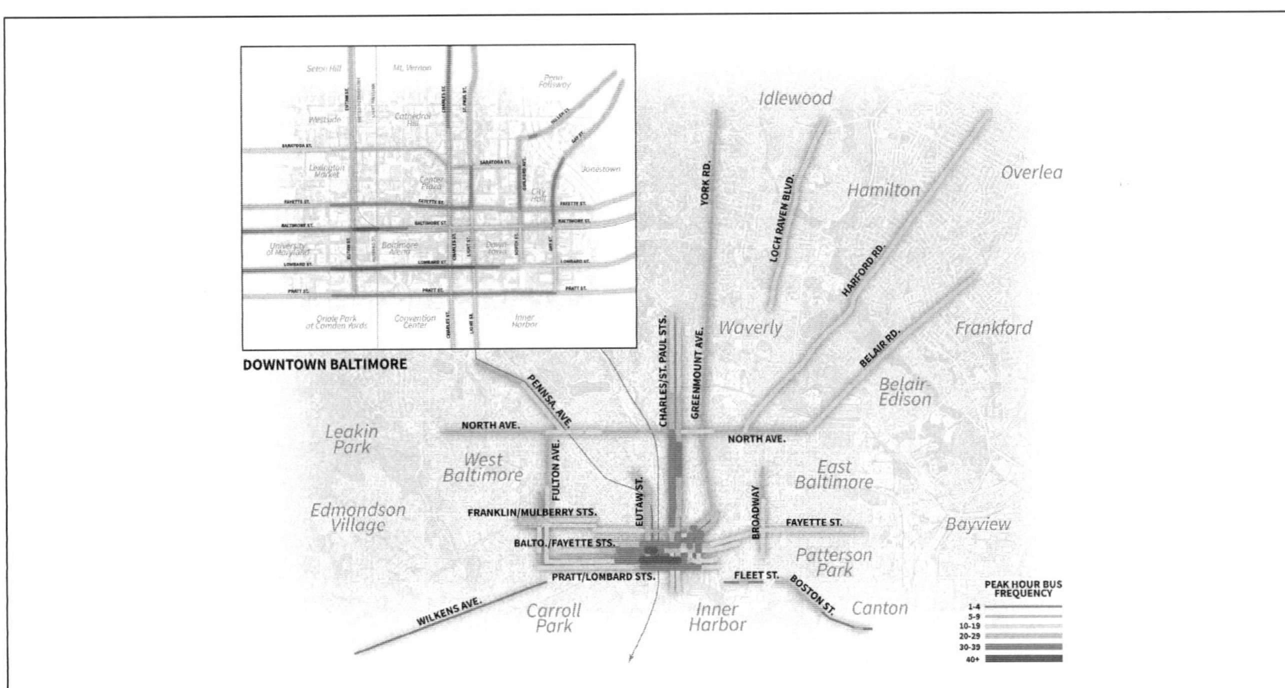

Figure 2. Bus frequency.

In order to assess person throughput, average peak period ridership (from MDOT MTA's Bus Network Improvement Project existing conditions data from 2014) was used to approximate the number of passengers that could be expected to ride each CityLink, LocalLink, and non-MTA bus expected to use the dedicated bus lane

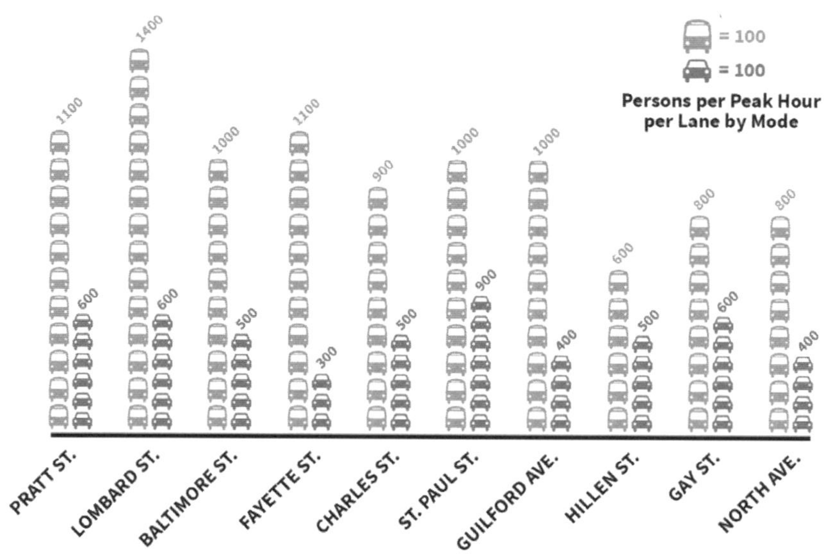

Figure 3. Person throughput per lane (peak hour).

during the peak travel period. The number of passengers was then assigned to the dedicated bus lane, whereas the number of vehicle passengers based on traffic counts and car occupancy assumptions was divided evenly between the remaining travel lanes (the exception being right-turning vehicles that would also use a dedicated bus lane). The occupancy assumption used for cars was 1.15, taken from *A Profile of Travel Trends: A Statistical Abstract for 1980, 1990, 2000, and 2006* compiled by the Baltimore Metropolitan Council. A unique consideration for calculating person throughput for this project is that the dedicated bus lanes would be rolled out with a completely reconstructed bus network with new bus routes and headways. As a result, bus ridership was based on existing routes that operate similarly to future routes, and this existing ridership was then applied to future routes and headways.

Person throughput was perhaps the most useful performance measure for assessing how streets are currently being utilized, moving the conversation toward equitable transportation solutions instead of transportation by private vehicle. Even though a transportation professional may be aware of high bus activity on a particular street, that subjective understanding takes on new meaning once person throughput is used to frame person mobility in quantifiable terms. As summarized in Figure 3, a number of streets in downtown Baltimore were carrying more than twice as many people in buses than private vehicles on a per lane basis.

Based on a combination of the strength of this person throughput measure, the other measures detailed in Tables 4, 5, and 6, and feasibility based on road geometry, 10 streets were removed from further consideration for dedicated bus lanes, but would still likely benefit from some transit priority treatments. A total of 10 streets emerged from this detailed screening as strong candidates for dedicated bus lanes and were the subjects of full analysis, including traffic operations analysis, parking impacts, and concept-level design recommendations. These corridors and their extents are shown in Figure 4 and listed below. Although the results are not detailed in this document, screening criteria identified in Tables 4, 5, and 6 were also considered in the detailed screening evaluation though not shown here for simplicity and brevity:

- Pratt Street, from Gilmor Street to President Street; and Lombard Street, from Martin Luther King Junior Boulevard to South Street.
- Baltimore Street, from Arlington Avenue to Central Avenue; and Fayette Street, from Martin Luther King Junior Boulevard to Broadway.
- Hillen Street, from Forrest Street to Guilford Avenue; Guilford Avenue/South Street, from Pleasant Street to Lombard Street; and Gay Street/Ensor Street, from Forrest Street to Pratt Street.
- Charles Street and St. Paul Street, from Lee Street to University Parkway.
- North Avenue, from Hilton Street to Milton Avenue. (This was ultimately awarded a TIGER

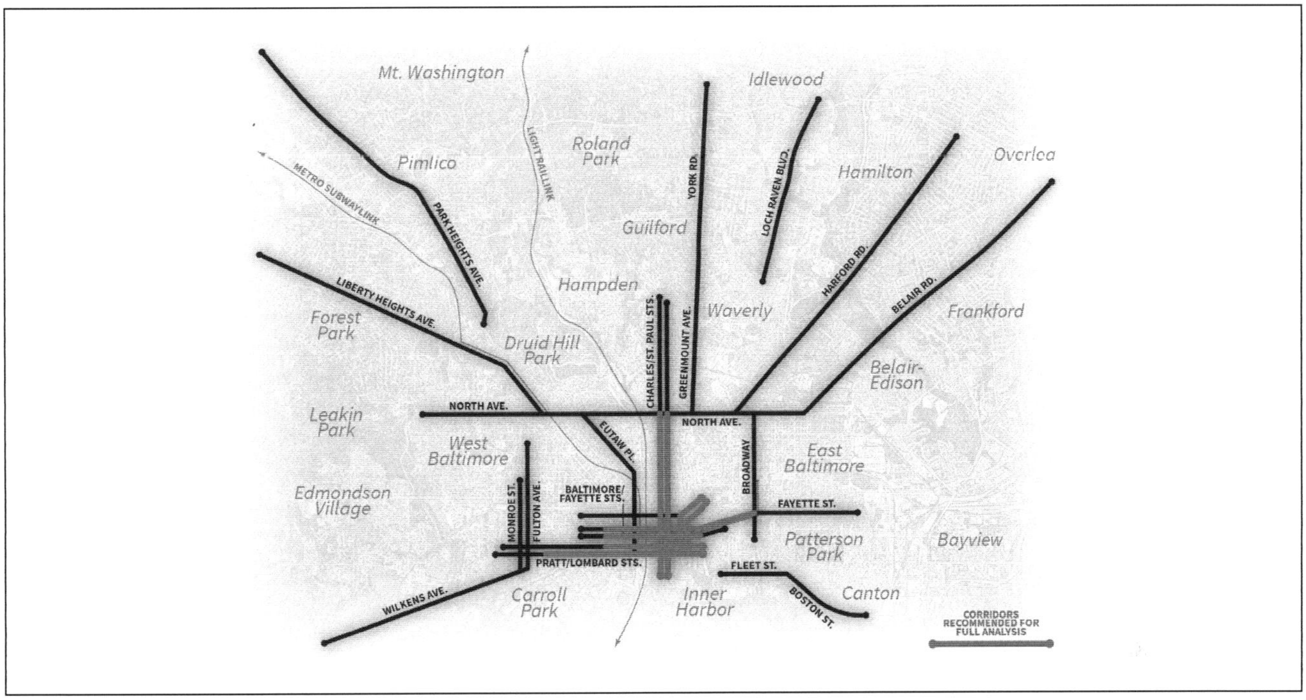

Figure 4. Corridors recommended for full analysis.

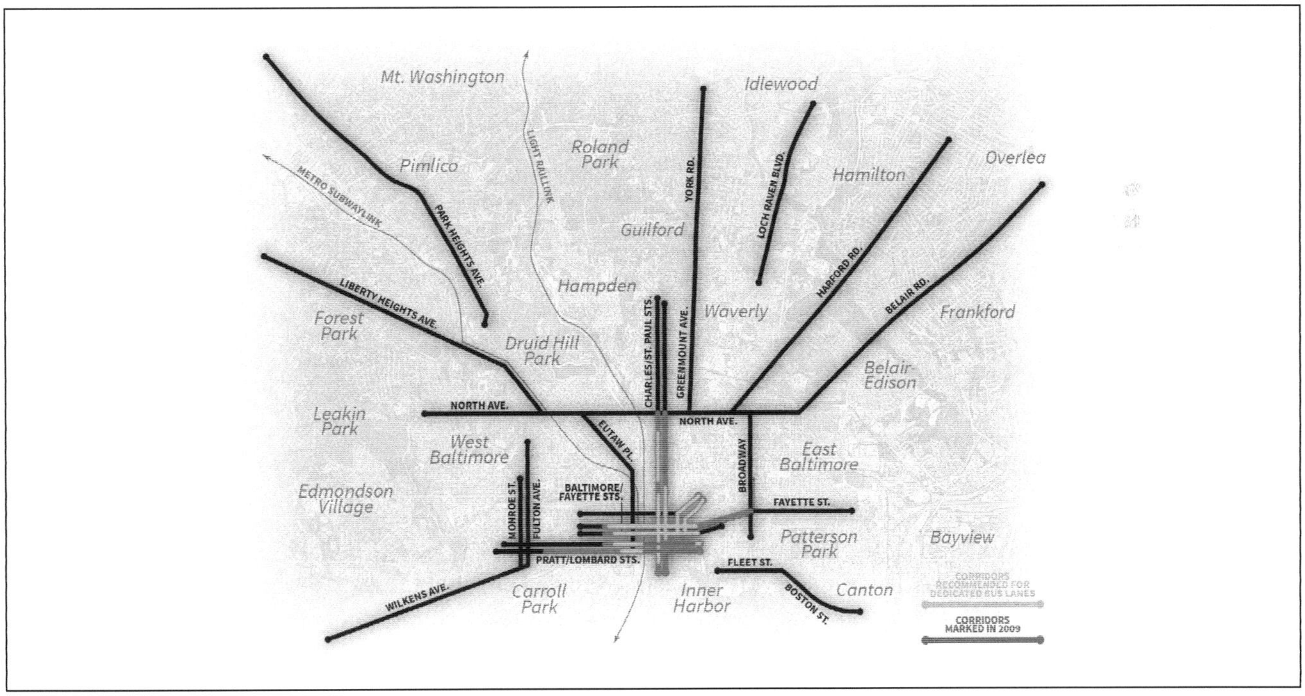

Figure 5. Recommended sections for dedicated bus lanes.

grant for a variety of investments including dedicated bus lanes, and as such is no longer detailed in this evaluation.)

Full Analysis

The remaining nine streets (i.e., four corridors) were analyzed to predict the impact that dedicated bus lane installation would have on parking and on traffic operations. Where necessary, recommendations were made at intersections to mitigate the added traffic delay from dedicated bus lane installation.

Existing Curbside Parking Restrictions

A variety of parking restrictions are present on the corridors recommended for full analysis, including no parking, and combinations of AM, PM, and AM and PM peak period parking restrictions. Parking restrictions increase vehicular capacity during the morning or evening peak periods, or both, and accommodate parking demand during the off-peak periods. For the analysis of the recommended corridors, parking restrictions informed the type and location of the dedicated bus lane that could be proposed. In most cases, on streets with peak period parking restrictions, parking was removed on one side of the street to accommodate a dedicated bus lane while the other side of the street accommodates full-time parking. There are also a few exceptions where full-time parking is maintained and a dedicated bus lane runs offset to the on-street parking. In some other locations a decision was made to include a peak-only bus lane during times of the pre-existing parking restrictions, with a variety of factors contributing to this decision, including ongoing construction projects and pre-existing permits for use of the space.

Traffic Operations

Synchro Models. The models used to evaluate the corridors recommended for full analysis came from a variety of sources (e.g., St. Paul Street and Calvert Street Two-Way Study and BCDOT). The files were reviewed to ensure that they accurately reflected existing conditions (lane configuration, signal timing, peak period restrictions, etc.) observed in the field. The intersection volumes were updated to reflect the most recent counts available and balanced where counts were not recent (i.e., older than 2010) or readily available.

Traffic Operations Thresholds. The quality of service of a roadway can be assessed in a variety ways. For the corridors recommended for full analysis, delay, LOS, and volume-to-capacity (v/c) ratio were evaluated to determine the effect of a dedicated bus lane on traffic operations. Specifically, the performance measures were evaluated to highlight where the proposed changes would result in the traffic volumes exceeding the roadway capacity. The threshold at which traffic exceeds capacity for delay, LOS, and v/c ratio are described further below. The models were analyzed during the AM and PM peak periods using the *Highway Capacity Manual 2010* (*10*) methodology and Synchro 9 software.

Delay. Delay refers to the amount of additional automobile travel time incurred at an intersection due to changes in traffic volumes, roadway geometry, or operations. Control delay is an important measure in determining LOS. HCM 2010 defines six categories for average control delay. Signalized intersections with control delays over 55 s per vehicle are experiencing severe delays. When control delay is greater than 80 s per vehicle, the available demand is determined to exceed the capacity.

Level of Service. LOS is a quantitative performance measure used to describe operational conditions within a traffic stream. HCM 2010 defines six LOS ranging from A to F. Intersections with LOS A to D are generally considered acceptable, whereas intersections with LOS E and F have demand that is determined to exceed capacity.

Volume-to-Capacity Ratio. The v/c ratio is a measure of the existing vehicle demand on a roadway segment or intersection as a proportion of the maximum number of vehicles that can be processed over a period of time (typically an hour). A v/c ratio closer to 1.0 represents a facility nearing capacity, whereas a lower value represents a facility that has excess capacity. A v/c ratio over 1.0 represents a facility operating over its capacity.

Design Recommendations

The streets (and their extents) that were recommended for dedicated bus lanes are shown in Figure 5 and the segments that were ultimately implemented are detailed below and shown in Figure 6.

Fayette Street

The 0.87-mi (1.40-km) dedicated bus lane on Fayette Street is a permanent, curbside running facility between Arch Street and President Street, except between Holliday Street and Gay Street where the dedicated bus lane is a permanent, offset running facility to maintain parking. Construction was completed in summer 2017.

Baltimore Street

The 0.87-mi (1.40-km) dedicated bus lane on Baltimore Street is a permanent, curbside running facility between Arch Street and Gay Street. To accommodate a curbside dedicated bus lane, parking was removed between

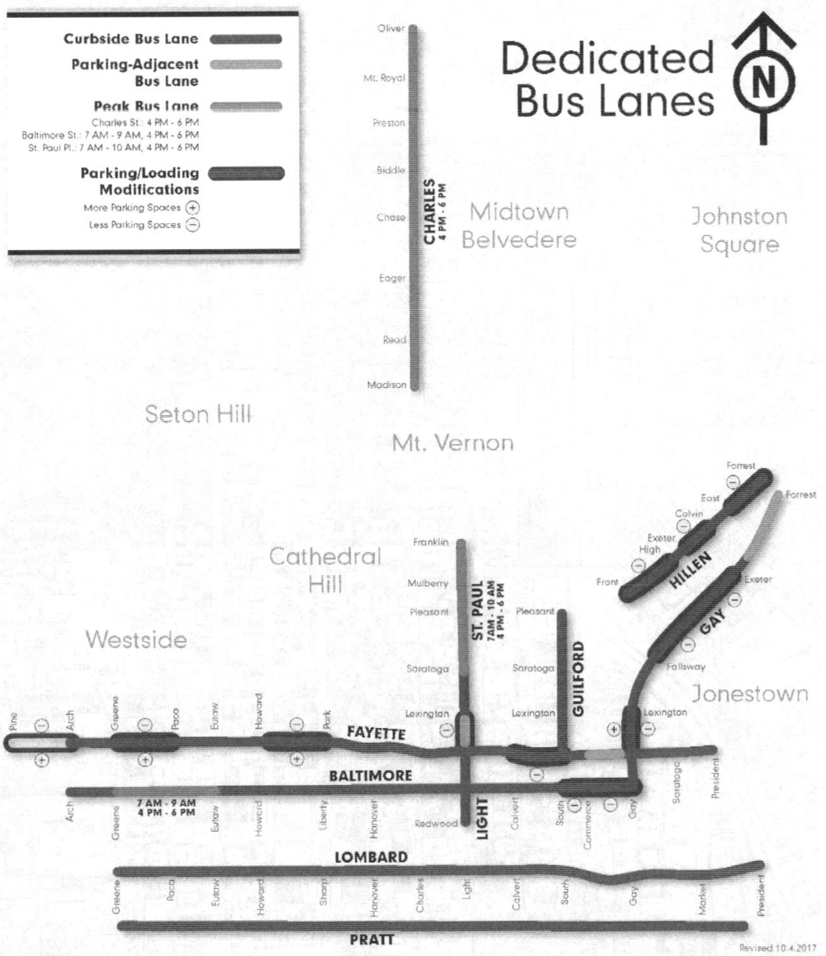

Figure 6. Final details for recommended dedicated bus lanes.

Calvert Street and Gay Street. Parking regulations currently prohibit on-street parking during AM and PM peak periods between Greene Street and Eutaw Street and this segment was preserved as an unpainted AM and PM peak period-only bus lane for the time being to minimize impacts on existing permits for food trucks and for loading of the Hippodrome Theater.

Charles Street

The 0.55-mi (0.88-km) dedicated bus lane on Charles Street is a part-time PM peak period curbside running facility along Charles Street between Madison Street and the I-83 ramp at Penn Station. No existing parking was modified to accommodate the dedicated bus lane because parking is currently prohibited on the east side of the street during the PM peak period. Likewise, red paint was not used and the dedicated bus lane is only in effect for the PM peak period.

St. Paul Street

The 0.42-mi (0.67-km) dedicated bus lane on St. Paul Street is a combination of peak-only, permanent curbside, and offset running facilities. An ongoing construction project along St. Paul Place prevented implementation of a full-time bus lane between Franklin Street and Saratoga Street and so that segment is being implemented by the contractor for that project as an AM and PM peak period curbside running facility following the pre-existing parking restrictions on that street. The bus lane will be curbside most of St. Paul Street between Saratoga Street and Redwood Street, except between Lexington Street and Franklin Street, where the dedicated bus lane is a permanent, offset running facility.

Hillen Street/Guilford Avenue/South Street

The 0.52-mi (0.83-km) dedicated bus lane is a permanent, curbside running facility on Hillen Street between Forrest Street and Front Street and on Guilford Avenue between Pleasant Street and Fayette Street, with a short offset segment on Guilford between Saratoga Street and Lexington Street. To accommodate the dedicated bus lane, parking was removed between East Street and Front Street. For most of the segment on Guilford Avenue a pre-existing bicycle lane is between the curb and the bus lane.

Gay Street/Ensor Street

The 0.53-mi (0.85-kim) proposed dedicated bus lane is a permanent, curbside running facility along Gay Street between Baltimore Street and Forrest Street, except on Ensor Street between Colvin Street and Forrest Street where the dedicated bus lane is a permanent, offset running facility. A dedicated bus lane was not proposed between Baltimore Street and Pratt Street.

Lombard Street and Pratt Street

Dedicated bus/bike lanes were introduced on Lombard Street and Pratt Street between Greene Street and Market Place by BCDOT in 2009. At that point the dedicated lanes consisted of a limited amount of signing and roadway striping and there was little enforcement of the lanes.

At the time the evaluation was underway, BCDOT had applied red paint on Lombard Street between Howard Street and President Street and Pratt Street between Howard Street and Constellation Way to reinforce the existing curbside running dedicated bus lane. This project extended the red-painted permanent curbside running facility on Lombard Street between Greene Street and Howard Street and Pratt Street between Greene Street and Howard Street.

Conclusion

Although the number of dedicated bus lanes is growing in the United States, there have been few evaluations that have developed strategic performance measures and thresholds for assessing impacts and applicability of these facilities on our city streets. This study includes a detailed literature review that ultimately led to the development of a range of performance measures that were then used to assess operational impacts of dedicated bus lanes. This detailed evaluation led to the approval of more than 5 mi of dedicated bus lanes that were implemented as part of BaltimoreLink, a major redesign of the Baltimore region's entire bus system. The selected performance measures were used to determine the most appropriate streets for dedicated bus lanes; however, they will also be used during ongoing post-construction evaluations to assess the benefits and drawbacks of the bus lanes, particularly impacts on transit operations such as speed, reliability, travel time, and on-time performance. The final summary of dedicated bus lanes, including the bus lane placement and parking modifications, is shown in Figure 6.

Lessons Learned

Though removing general travel lanes and parking lanes to prioritize other modes will always be challenging, being able to clearly describe the analysis undertaken during this process and explaining that the person throughput for the bus lanes would be greater than the adjacent travel lanes has been incredibly helpful. The decision to change how modes are prioritized is clearly a policy decision and one that is inherently political, but having a data-driven process makes that effort more effective.

Opportunities for Future Research

Effectiveness of Red Dedicated Bus Lanes. Given the ongoing implementation of the dedicated bus lanes in Baltimore, it is too soon to have a thorough analysis of their effectiveness; however MDOT MTA and BCDOT are in the beginning stages of a thorough before–after study that will evaluate the effectiveness of the dedicated bus lanes. The evaluation will incorporate a variety of key performance indicators, including bus on-time performance, bus travel time, illegal motor vehicle travel within the dedicated bus lanes, and collisions.

Bicycle Utilization of Red Dedicated Bus Lanes and the Impact on Bus Operations. The dedicated bus lanes on Lombard Street and Pratt Street have been legally shared with bicycles since their implementation in 2009 and all of the new dedicated bus lanes are signed and striped as shared lanes, but little analysis has been done about the extent to which they are being utilized for bicycling, the types of bicyclists utilizing the lanes, and how the use for bicycling affects their effectiveness as bus lanes. Although the potential for a single dedicated lane to benefit both bicycling and transit is an attractive concept, the volume of buses in these downtown corridors may limit the number of bicyclists utilizing the lanes and the presence of bicycling in the lanes likely has some impact on the speed of buses in the lanes. It will be helpful for both transit and bicycle planning professionals to have a better understanding of the utilization and impacts of these two modes on each other in order to plan for future facilities.

Acknowledgments

The authors acknowledge the contribution of the staff of the Maryland Department of Transportation Maryland Transit Administration, Maryland Environmental Service, Baltimore City Department of Transportation, Parking Authority of Baltimore City, and the Mayor's Office of Baltimore that made development and implementation of this project possible. Additionally we thank the staff of Kittelson & Associates that worked on the analysis and the staff at RK&K who worked on the design of the dedicated bus lanes.

Author Contributions

The authors confirm contribution to the paper as follows: background and literature review: PM, AV, JC; data collection: JC; selection of screening criteria: PM, AV; multimodal analysis: AV, JC; draft and final manuscript preparation and editing: AV, JC, PM. All authors reviewed the results and approved the final version of the manuscript.

References

1. Ang-Olson, J., and A. Mahendra. *NCHRP Research Results Digest 352: Cost/Benefit Analysis of Converting a Lane for Bus Rapid Transit*. Transportation Research Board of the National Academies, Washington, D.C., 2011.
2. Vuchic, V. *Urban Transit Systems and Technology*. John Wiley & Sons, Inc., Hoboken, NJ, 2007.
3. Ryus, P., A. Danaher, M. Walker, F. Nichols, W. Carter, E. Ellis, L. Cherrington, and A. Bruzzone. *TCRP Report 165: Transit Capacity and Quality of Service Manual*. Transportation Research Board of the National Academies, Washington, D.C., 2013.
4. Ryus, P., K. Laustsen, K. Blume, S. Beaird, and S. Langdon. *TCRP Report 183: A Guidebook on Transit-Supportive Roadway Strategies*. Transportation Research Board of the National Academies, Washington, D.C., 2016.
5. Levinson, H., L. Lennon, and J. Cherry. *Public Transit Research: Rail, Bus, and New Technology*. TRB, National Research Council, Washington, D.C., 1991.
6. Washington Metropolitan Area Transit Authority. *State of the Practice Memorandum*. Washington, D.C., 2013.
7. St. Jacques, K., and H. Levinson. *TCRP Report 26: Operational Analysis of Bus Lanes on Arterials*. Transportation Research Board of the National Academies, Washington, D.C., 1997.
8. Hillsman, E. L., S. J. Hendricks, and J. Fiebe. *A Summary of Design, Policies and Operational Characteristics for Shared Bicycle/Bus Lanes*. Florida Department of Transportation Research Center, Tallahassee, FL, 2012.
9. Institute for Transportation and Development Policy. *Bus Rapid Transit Planning Guide*. Institute for Transportation and Development Policy, New York, NY, 2007. https://www.itdp.org/wp-content/uploads/2014/07/Bus-Rapid-Transit-Guide-Complete-Guide.pdf.
10. *Highway Capacity Manual 2010*. Transportation Research Board of the National Academies, Washington, D.C., 2010.
11. Levinson, H., S. Zimmerman, J. Clinger, J. Gast, S. Rutherford, and E. Bruhn. *TCRP Report 90: Bus Rapid Transit: Implementation Guidelines*. Transportation Research Board of the National Academies, Washington, D.C., 2003.
12. Queensland Transport. *Bicycles and Bus Lanes*. Queensland Transport, Fortitude Valley, Queensland, Australia, 2006. https://www.tmr.qld.gov.au/-/media/Travelandtransport/Cycling/Bike-user-guide/Technical-information/Pdf_b9_bicycles_and_bus_lanes.pdf?la=en.
13. Florida Department of Transportation. *Typical Sections for Exclusive Transit Running Ways*. Florida Department of Transportation, Tallahassee, FL, 2013.

The Standing Committee on Bus Transit Systems (AP050) peer-reviewed this paper (18-05464).

Article

TRR
JOURNAL OF THE TRANSPORTATION RESEARCH BOARD

Transportation Research Record
2018, Vol. 2672(8) 64–74

DOI: 10.1177/0361198118792330
journals.sagepub.com/home/trr
SAGE

Large-Scale Transit Signal Priority Implementation: District of Columbia's Path to Success

Bailey Lozner[1], Kevin Lee[2], A. Wasim Raja[3], Mohammed G. Habib[3], and Burak Cesme[1]

Abstract

In 2016, the District Department of Transportation (DDOT) deployed transit signal priority (TSP) at 195 intersections in highly urbanized areas of Washington, D.C. Years before deployment, DDOT outlined a multi-year process that would aid in the realization of their ambitious objectives. In collaboration with a broader regional implementation and in partnership with the Washington Metropolitan Area Transit Authority (WMATA), DDOT set out to apply a systems engineering driven process to identify, design, test, and accept a large-scale TSP system. Application of this systems engineering process led to the successful, widespread deployment of TSP in an urbanized operating environment. Although numerous research efforts to date focus on smaller scale efforts, this paper seeks to close gaps in the literature with a process-focused case study, highlighting intersection selection, system verification testing, and validation for acceptance as applied by DDOT for establishment of an operational, large-scale system. The paper concludes with a discussion of the lessons learned.

A decade ago, the vison for bus transit in the National Capital area was a fully integrated, regional system that would bring a quality, reliable service to riders across the region. The vision was ambitious, calling for large-scale bus service improvements across state lines. With these improvements as motivation, regional agencies embarked on a collaborative journey to implement priority treatments for Washington Metropolitan Area Transit Authority (WMATA) bus routes. Among these improvements was the implementation of transit signal priority (TSP) on nearly a dozen routes operating throughout the District of Columbia, northern Virginia, and areas of Maryland. One agency, the District Department of Transportation (DDOT), worked in concert with the regional stakeholders to spearhead the deployment of TSP in the District. Through rigorous planning and application of a systems engineering process, DDOT brought the vision of a large-scale TSP system to fruition, deploying the functionality at nearly 200 traffic signals. The process the agency utilized for successful implementation of a large-scale TSP system is the subject matter of this paper.

TSP is a tool used to improve transit performance and reliability. The operational strategy facilitates the movement of transit vehicles through traffic signals, resulting in reduced transit travel times, improved schedule adherence, improved transit efficiency, and increased road network efficiency as measured by person mobility (*1*). The most common TSP strategies either extend a phase to allow a transit vehicle to pass (i.e., green extension) or terminate conflicting phases to allow early service and reduce red time (i.e., red truncation, or early return to green) (*2*). These signal timing adjustments can result in reduced transit delay when applied at signalized intersections. TSP has been utilized for decades to improve transit service in urbanized and suburban areas across the U.S.. Research shows that overall route travel time reductions of up to 10% with the implementation of TSP are common. Priority systems also have shown to reduce variability in bus delays, thereby increasing reliability (*3*).

Common U.S. implementations of TSP are at the corridor level; large-scale TSP deployments in urban operating environments are much less common, and research

[1]Kittelson & Associates, Inc, Washington, D.C.
[2]Kittelson & Associates, Inc., Cincinnati, OH
[3]District Department of Transportation, Washington, D.C.

Corresponding Author:
Address correspondence to Bailey Lozner: blozner@kittelson.com

documentation is limited, resulting in gaps in the literature. The objectives of this research paper are to:

- Outline a practice-ready systems engineering process for large-scale deployment of TSP
- Document design considerations in deployment of TSP
- Demonstrate that technology-specific conditions influence testing and acceptance

The paper places a particular focus on the portions of the systems engineering process for TSP implementation completed to date under the primary direction of the DDOT. It is this process as applied by DDOT and regional partners that delivered on the arduous undertaking of a regional TSP system. Thus, as applied by DDOT, four systems engineering elements are examined in the sections that follow: design, implementation, verification, and validation. Owing to the complexity of the implementation as it relates to a broader regional effort, the authors review contextual background and detail technology functionality in sections that precede the main elements of the report.

Background

In 2010, five jurisdictions in the National Capital Region set out to implement a series of bus priority treatments in the greater Washington, D.C. area. The effort, funded by a grant award under the Transportation Investments Generating Economic Recovery (TIGER) program (*4*), aimed to improve the quality and reliability of bus transit service operated by WMATA. Under the administrative oversight of the Metropolitan Washington Council of Governments (MWCOG), regional agencies—including WMATA, DDOT, the Maryland Department of Transportation (MDOT), the City of Alexandria, and the Potomac and Rappahannock Transportation Commission (PRTC)—worked in partnership to see the transit treatments through to implementation. A multitude of projects resulted from the grant award, including implementation of TSP at intersections spanning several traffic agency jurisdictions. Although TSP in the District of Columbia was one of several TSP implementation efforts resulting from the award, this implementation was the largest, resulting in active TSP at 195 DDOT intersections.

To aid in multi-jurisdictional collaboration, WMATA developed a regional Concept of Operations in 2009, outlining that a single TSP technology would function on all traffic signals in the region. Following the grant award, the regional agencies each documented needs and requirements for the TSP technology, incorporating the many conditions and constraints specific to each jurisdiction. These conditions amalgamated in a system requirements document that drove the technology selection process. The system requirements document later served as the basis for each of the traffic agencies to move forward with procurement, design, and verification testing.

In 2013, WMATA accepted a proposal consistent with the completed system requirements document to customize and implement a TSP technology that met traffic agency needs and capitalized on the transit agency's existing onboard system hardware.

TSP System Components and Functionality

System requirements both define the minimum technical needs of a system and serve as the baseline for test plan development (*5*). The components and functionality of the TSP system deployed in the National Capital Region were verified against these system requirements to meet the needs of the regional stakeholders. The following sections describe the architecture, components, and functionality as deployed in the District of Columbia in 2016.

System Architecture

The TSP architecture implemented in the National Capital Region is distributed; that is, priority decisions are made at the intersection level. The TSP System was formed under the high-level guidance of the National Transportation Communications for ITS Protocol Object Definitions for Signal Control and Prioritization (NTCIP 1211), though the components of the system do not exclusively adhere to the 1211 communications protocol. The architecture of the regional TSP system most closely resembles the Physical Architecture Example 4 of the 2008 1211 Standard, in which the priority request generator (PRG) resides within the fleet vehicle and uses predefined criteria to generate a priority request. The PRG sends the priority request directly to the priority request server (PRS) housed in the traffic signal cabinet. The PRS receives the priority request, processes the request, and passes the request on to the traffic signal controller to be processed by the coordinator (*6*).

It is within a version of this distributed architecture that the main system elements function. These elements include the on-board vehicle equipment, the wayside equipment, the central management software (CMS), and the mechanisms by which the elements communicate. Collectively, these elements and their sub-systems enable TSP messages to be transmitted via cellular communication from bus to intersection and for conditional priority (i.e., priority requested only when certain requirements are met (*7*)) to function. A brief description of components is summarized below:

- *Vehicle equipment*: On-board vehicle equipment (OBE) consists of bus systems vendor Clever Devices' Intelligent Vehicle Network (IVN) Unit and cellular communications hardware. The IVN unit runs a TSP-enabling software to transmit real-time transit vehicle information to the wayside equipment through WMATA's dedicated cellular communications network.
- *Fleet management software*: The fleet management software refers to the Clever Devices bus firmware and/or software that permits configuration of global, intersection-specific, and route-specific TSP-related parameters to enable TSP operation and determine qualifications for priority.
- *Wayside equipment*: The wayside equipment (WSE) processes transit vehicle information and transmits TSP requests to the traffic signal controllers. WSE includes a cabinet-mounted cellular antenna and in-cabinet cellular modem, phase selector card, and auxiliary interface panel (AIP) located within the traffic signal cabinet at the TSP-enabled intersection.
- *Phase selector*: The phase selector synthesizes the received TSP requests and communicates the calls to the traffic signal controller to provide priority. The phase selector card is installed in the input file of a rack-mount cabinet and is doublewide, requiring two adjacent slots in the input file rack for installation. For National Electrical Manufacturers Association (NEMA) standard cabinets, the phase selector may be housed in a dedicated harness and wired as a stand-alone input.
- *AIP*: The AIP detects green phases from the traffic signal controller and provides additional outputs for the phase selector card to interconnect with the traffic signal cabinet.
- *Traffic signal controller firmware*: The traffic signal controller firmware installed on the traffic signal controller at a TSP-enabled intersection processes and responds to TSP requests.
- *TSP CMS*: The CMS operates on a dedicated server machine located at traffic agency (i.e., DDOT) offices. The CMS facilitates configuration of the phase selectors from a central location and logs the TSP events for operational reports.
- *Cellular network*: The cellular communications network provides a dedicated communications medium for the TSP System to communicate reliably between the OBE and the WSE and between the WSE and the CMS.
  - *Cellular antenna*: Detects electronic signals over a wireless cellular network and facilitates data transmission and reception for the modem over a wireless cellular network. Cellular antenna and modem allow TSP-related information such as signal priority requests, TSP parameters and logs from the phase selector card to be transmitted wirelessly to and received from transit vehicles and the CMS.
  - *Cellular modem*: provides communications connectivity to the phase selector card to communicate with the CMS and transit vehicles.

Figure 1 illustrates the integration of each of these components. When the OBE determines all requirements (e.g., TSP-enabled route, time-of-day, passenger load, schedule adherence, etc.) for priority to be met, the system sends a signal from the transit vehicle to the phase selector, which places a low priority call into the traffic signal controller. This signal is transmitted through a cellular communications network to a cellular modem within the traffic signal cabinet. The cellular modem sends the signal to the phase selector, which provides a TSP input to the traffic signal controller.

The Role of the Traffic Signal Controller. The traffic signal controller utilized for the District of Columbia deployment—Type 170E hardware running McCain MC1 firmware—is capable of running green extension and early return to green (i.e., early green) priority routines. If green extension is enabled and a priority call is made near the end of the transit vehicle approach phase's green interval, the traffic signal controller will extend the green interval, extending phase service and allowing the transit vehicle to proceed through the intersection. Green extension terminates either when the low priority (i.e., TSP) call is dropped—no longer active in the traffic signal controller—or when the extension reaches the DDOT-determined and programmed maximum green extension time. If early green is enabled and a priority call is placed during the transit vehicle approach phase's red interval, the traffic signal controller will truncate (up to the maximum DDOT-determined and programmed maximum early green time) the variable green of the non-sync phases within the coordinated cycle to return to phase service earlier. Early green terminates after service of the requested transit vehicle approach phase. For this TSP deployment, 10-second thresholds were programmed for both green extension and early green traffic signal controller parameters at a majority of TSP intersections.

The MC1 firmware and its predecessor BiTrans RV1 were bench tested extensively as part of Concept Exploration early on in the effort. Various TSP scenarios were simulated in the traffic signal controller via a suitcase tester under a multitude of phasing and timing configurations. Testing was conducted to verify vendor

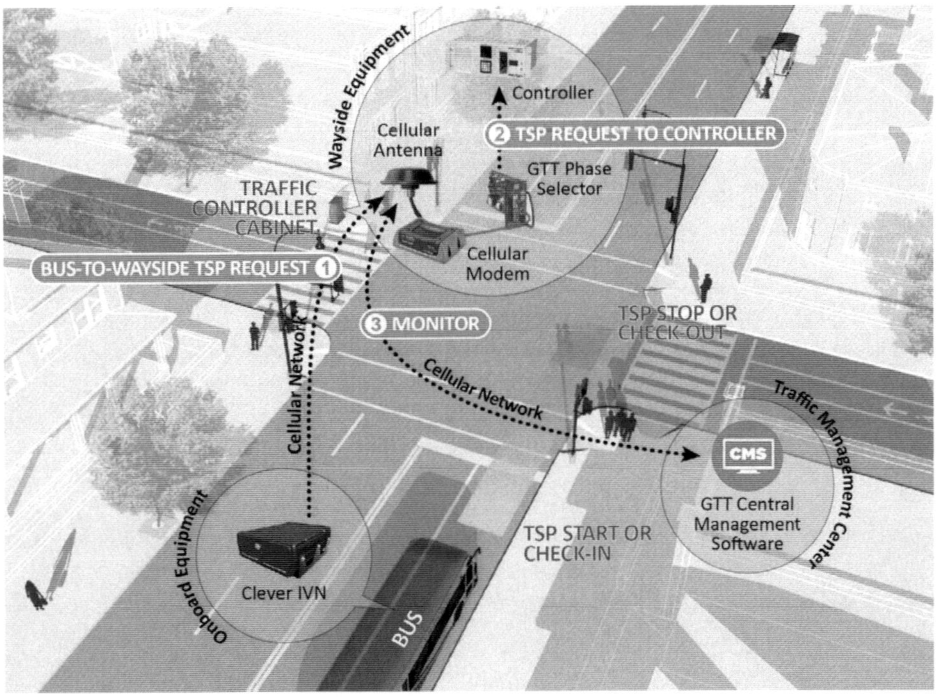

Figure 1. Integration of TSP system components.

documentation of TSP algorithms, and reduce the risk of unanticipated issues with widespread, active TSP.

Conditions for TSP Service

The regional TSP System is able to be constrained by both conditions for priority and by time of day operation.

Conditions for TSP Request. As stated above, conditional priority describes a priority treatment that is applied only if a TSP-enabled bus meets a set of predefined criteria. The TSP system implemented in the National Capital Region is a conditional priority system that uses both a lateness threshold (i.e., schedule adherence) and a minimum passenger loading criterion as a qualifying basis for the transmission of a TSP request. The system also requires a "doors closed" status to be active for the generation of a TSP request.

For schedule-based routes, WMATA considers a bus "on-time" when buses arrive at a stop two minutes early and up to seven minutes behind schedule (*8*). The initial deployment of the TSP system was configured such that buses can only make priority requests when the buses are seven minutes behind schedule (consistent with the WMATA's on-time threshold) and have a minimum of 20 passengers on board. However, the authors recognize that conditional priority requires an "optimal" service design (i.e., running time schedules) that is purposely altered for TSP in conjunction with the appropriate lateness threshold to take full advantage of conditional priority (*7*, *9*). As a result, an ongoing study is currently analyzing schedules for TSP routes, as well as various lateness thresholds to improve the effectiveness of conditional priority by refining the seven minutes lateness threshold.

In addition to the schedule deviation and passenger loading conditions, the transit vehicle must be within a certain proximity to the intersection to request priority. The distance to the intersection when a bus is able to initiate communication is known as the TSP start distance (also known as "check-in" distance). The TSP start distance was chosen so that the bus is able to clear the intersection during the green time extension. The TSP start distance is based on the minimum of the three limiting distance criteria:

1. Distance to an upstream intersection
2. Distance to a near-side bus stop (if present)
3. Distance corresponding to a 10 second travel time assuming an average bus speed

Time-of-Day Restrictions. The TSP system in the District operates on nine priority routes during peak periods only. For TSP-enabled commuter corridors providing

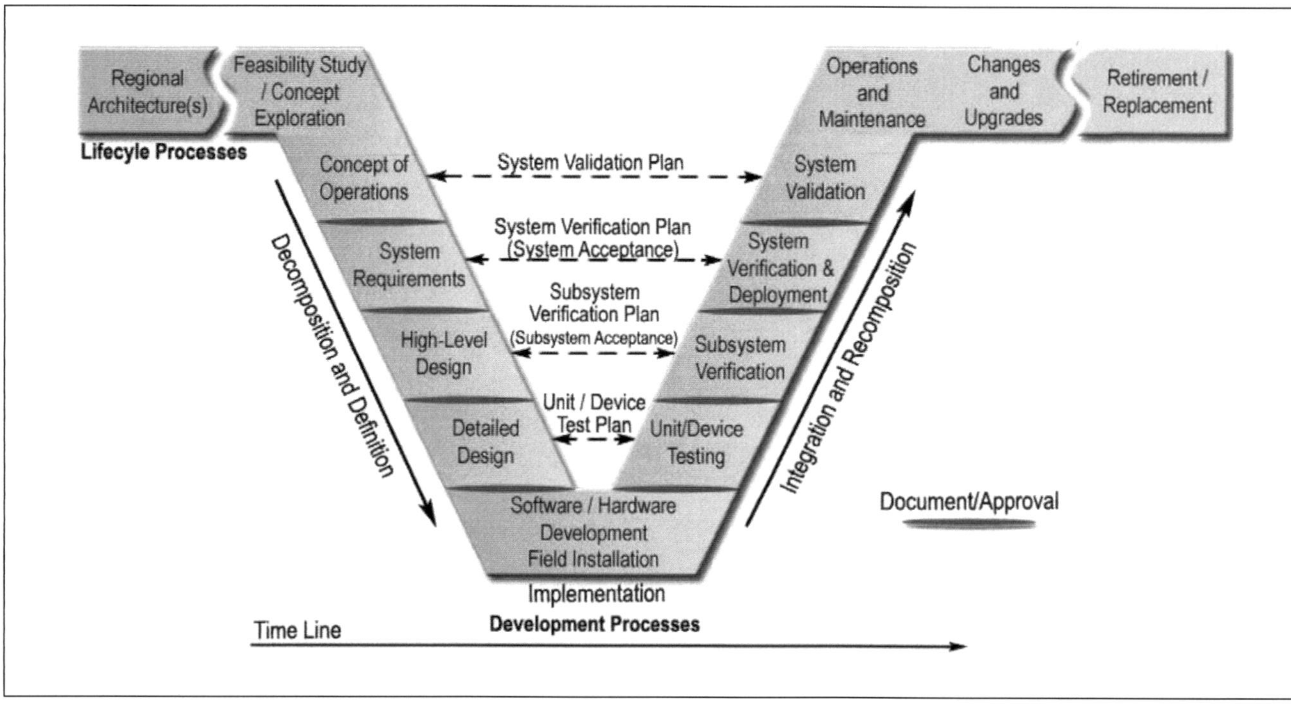

Figure 2. Systems engineering "V" diagram.

access to and from D.C.'s urban core, TSP operates in the peak direction: inbound during the AM peak (6–9AM) and outbound during the PM peak (4–6PM).

Systems Engineering Applied

Following the model set forth by WMATA, DDOT took action in the continued application of the systems engineering process (see Figure 2). The traffic agency outlined an action plan that would carry the project through to functional operation and meet defined needs. Through the process, a screening methodology was developed for intersection and approach selection. A detailed design process was additionally applied that met the needs of the project while fast-tracking the design schedule. As part of the design bid package, technical specifications were detailed that built off the system requirements and included a four-tiered approach to testing for system acceptance.

Each step of the definition and integration processes (illustrated in Figure 2) was iterative and required not only adherence to a structured approach, but a high degree of inter-agency coordination as the individual TSP efforts within the region moved forward in parallel. Although the systems engineering model framework was fundamental to the success of this deployment, it should be noted that the framework was followed as a construct and not a standard to be strictly adhered to.

Design

Design was an integral phase in the deployment that progressed the project from specific system requirements to an implementable framework specific to the traffic agency and the D.C. operating environment. The systems engineering model applied to the DDOT's approach drove high-level design with intersection selection and parameters development. Detailed design advanced the development of final plans, technical specifications, and traffic signal controller parameters.

High-Level Design: Intersection Selection Process and Parameters Development

As part of the intersection selection process, route alignment, schedules, and bus stop locations for the identified priority routes were reviewed. Field reviews were conducted at hundreds of candidate intersections within the areas of consideration to survey intersection conditions, verify bus stop locations and amenities, and document in-cabinet equipment and capacity. The data was then synthesized to determine the most viable candidates for TSP operation. The following factors influence the selection and the subsequent TSP parameters developed.

Bus Stop Location. The location of a bus stop plays a critical role in the effectiveness of TSP. Far-side stops are generally preferred over near-side stops for TSP implementation as there is less uncertainty in the prediction of arrival time of a bus at an intersection (*10*). TSP technology types also affect the degree of influence that near-side stop placement may have on TSP functionality. For TSP in the National Capital Region, the system utilizes cell-based communication. As such, communication zones at a given TSP-enabled intersection are defined by communication "TSP start" and "TSP stop" distances, which are referenced to the route waypoint nearest the intersection. Route waypoints are geocoded location references spaced every 50–100 feet on the centerline of the route alignment, and location of the waypoints to which the TSP start and stop distances are referenced has a low degree of accuracy in alignment with the center of the intersection. Owing to the low reference point accuracy, designing communication zones to exclude near-side stops is only intermittently successful.

Within the design process, intersections with nearside bus stops were screened as candidates for bus stop relocation to support TSP operation. Relocation efforts proved challenging, however, as competing priorities and initiatives in highly urbanized areas—where public right of way is constrained—often surface in coordination efforts, and a modification to support one initiative may deteriorate another. Bus stops within the District of Columbia can serve multiple routes and multiple transit service providers; relocation to benefit one route or service provider may not benefit another. Adequate right of way for shelter relocation is also challenging, particularly where intersections are closely spaced, such as in many areas of downtown Washington. As such, out of nearly 20 bus stop relocation candidates, only four relocations were ultimately implemented.

Variable Green Time. To provide TSP, sufficient variable green time (also defined as available green time, or slack time) is required. The total variable green time is determined by several factors, including cycle length, number of phases, clearance times, minimum green intervals, and pedestrian intervals.

At each intersection, variable green times were calculated for peak-period coordination timing plans. Cycle lengths, minimum green time, clearance time, and pedestrian interval timing were reviewed against phase splits to confirm the presence of sufficient variable green. Where present, pedestrian walk and "flashing don't walk" (FDW) intervals were taken into account as a minimum fixed time. Although TSP operation does not violate minimum fixed time in the MC1 firmware, including pedestrian intervals, variable green is required for effective TSP operation. Calculations assumed a minimum variable green time of 10 seconds to be necessary for TSP.

Spatial Distance. The methodology applied to calculate TSP start distances—the distance upstream of an intersection at which a bus enters communication with the wayside equipment for priority request—considered three constraints: upstream block length, upstream distance to mid-block bus stop, and distance traveled at average speed for a 10-second window. Should the distance to an upstream stop be determined insufficient, the intersection was classified as an inadequate candidate for TSP.

Once an intersection was selected as a viable candidate, the minimum of the three measured distances was determined to be the point at which a TSP-enabled bus would begin communication with an intersection for a given approach.

Traffic Operations. Traffic operations were reviewed for intersections on the priority transit routes. Intersections were screened for excessively high volume-to-capacity ratios, as well as for critical movements that may have been affected by TSP operation. A qualitative approach was taken to assess the effects of TSP in a coordinated environment. Transit routes in the urban core were logically segmented, and time-space diagrams were generated in traffic analysis software for the corresponding stretches of roadway. The design team walked through the time-space diagrams, running a series of scenarios to qualitatively simulate the service of priority routines at various points in the segments. Approaches were screened to prevent excessive queuing, side street delay, and disruption of coordination on the corridor.

Intersection Screening. Screening results were compiled to determine the most viable intersection candidates for TSP operation. With consideration of bus stop location, available variable green, distance constraints, and traffic operations, the process resulted in the selection of 195 TSP intersections across four key areas: three radial arterials and the downtown core, as shown in Figure 3.

OBE Parameter Development. The on-board equipment requires programming of parameters that are documented in the intersection screening process. These include:

- Intersection approach
- Bus minimum and maximum heading
- Approach phase
- TSP channel
- Intersection location (latitude, longitude)
- TSP start ("check-in") distance
- TSP stop ("check-out") distance

The compilation of this information is input as guidance for the OBE programming and sets the stage for detailed design.

Figure 3. Map of TSP in the District of Columbia.

Detailed Design: Plans and Technical Specifications

For the District of Columbia's deployment, final design plans indicated the positions of installation for proposed equipment within the traffic signal cabinet at each intersection. The plans called out locations where detector inputs were to be moved as they conflicted with the proposed standard position of installation for the phase selectors. The plan set additionally included a communications diagram and a general installation detail that depicts the assemblies and assembly interfaces. The technical specifications detailed the minimum functional requirements of these assemblies. The plan diagrams and technical specifications incorporated the system requirements as defined by WMATA, and layered requirements for needs specific to DDOT.

System Testing

For full system acceptance, a series of acceptance tests were defined to verify and validate end-to-end functionality of the TSP system. In accordance with the contract specifications and systems engineering process, a four-stage acceptance testing regimen was executed following equipment installation and system configuration:

- Prototype testing
- On-site stand alone testing (SAT)
- Final system acceptance testing (FSAT)
- Operational acceptance testing (OAT)

Testing stage objectives and sample scenarios for each are detailed in Table 1. Testing was to be executed in

Table 1. Testing Stage Objectives and Sample Testing Scenarios

Type of acceptance test	Objective	Example testing scenarios
Prototype testing	Verification of end-to-end functionality of the TSP system	• Bus running on time • Bus running late • Red light stop simulation • Cancel request simulation • Clear message simulation • Bus running on time, then late
Stand-alone testing (SAT)	Demonstration of all functions of on-site intersection TSP system at each traffic signal cabinet following TSP equipment installation	• Hardware installation check • Functionality check for cellular modem and phase selector • Cellular modem connection check • Signal controller TSP parameters check • Phase selector configuration check • Consistency of TOD restrictions with configuration pack dial sheets
Final system acceptance testing (FSAT)	Verification of end-to-end functionality for each intersection and each bus	• Communication initiation check on bus arrivals • Bus action check in accordance with conditional priority parameters • Bus message transmittal check • Message received confirmation by the wayside equipment • Bus checked out confirmation downstream of the intersection
Operational acceptance testing (OAT)	Full operation validation of the TSP system for a pre-determined span of time	• Intersection and bus activity monitoring for the entire system for a span of 30 days

Note: TSP = transit signal priority.

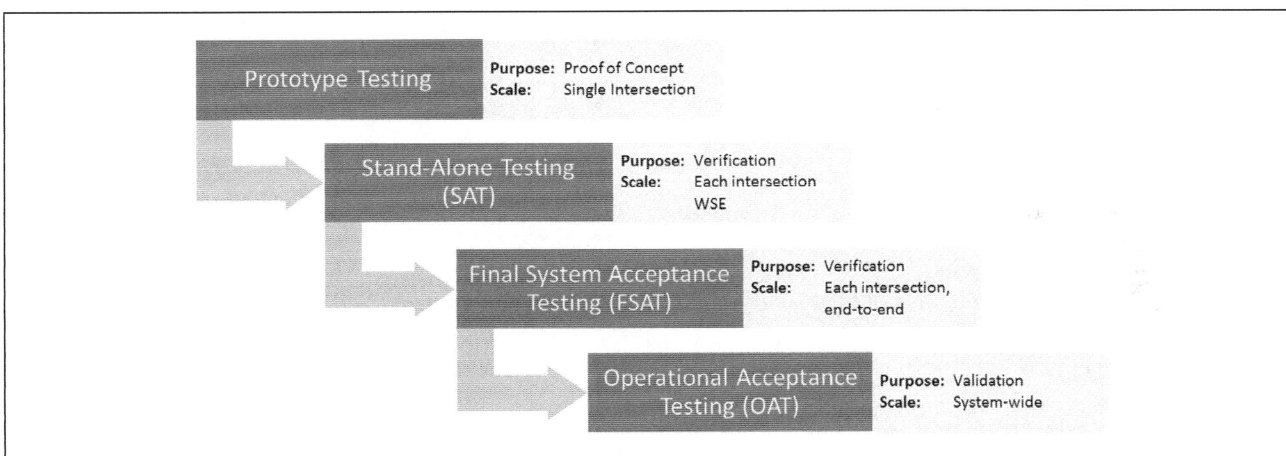

Figure 4. Systems engineering driven testing process.

series (Figure 4); system components could not advance to the next stage of testing without passing one or more prior stages.

Test plan development incorporated the system requirements as defined before the technology development. These requirements were considered to be the necessary minimum operation of the system.

Prototype Testing

The objective of prototype testing was to verify the end-to-end functionality of the TSP design. WMATA provided a TSP-enabled bus and was in attendance to observe demonstration of the technology.

Prototype testing demonstrated the successful end-to-end communication of a TSP request from bus to intersection in a DC operating environment. The test required a single TSP-enabled bus to maneuver through a TSP-enabled intersection under conditions determined by a series of pre-defined scenarios with expected outcomes. Scenarios included:

- Bus running on time
  - Scenario description: Bus approaches intersection, running on time. Bus continues through intersection on green with no TSP.
  - Expected outcome: Verify log only call is sent to intersection but no priority request is sent.

- Bus running late
  - Scenario description: Bus approaches intersection running late and requests priority. Bus continues through the intersection on green extension with TSP served.
  - Expected outcome: Verify priority request is sent and acknowledged, and clear messaged is sent and received.
- Red light stop simulation
  - Scenario description: Bus approaches intersection running late and requests priority. Bus stops for a red signal, then continues through on green signal with TSP served.
  - Expected outcome: Verify priority request is sent and acknowledged, and clear messaged is sent and received.
- Cancel request simulation
  - Scenario description: Bus approaches intersection running late and requests priority. Bus stops for a near-side bus stop. Bus sends priority cancel message with doors open.
  - Expected outcome: Verify priority request is send and acknowledged, and cancel message is sent and received.
- Clear message simulation
  - Scenario description: Bus approaches intersection running late and requests priority. Communication is disabled. Request for priority is cleared automatically.
  - Expected outcome: Verify priority request is sent and acknowledged, and request is cleared automatically.
- Bus running on time, then late
  - Scenario description: Bus approaches intersection running on time. Bus slows as it approaches the intersection and is then running late. Once late, bus sends request for priority message and continues through on TSP green extension.
  - Expected outcome: Verify "log only" call is sent, followed by a priority request. Verify priority request is acknowledged and clear message is sent and received.

The bus OBE and intersection WSE each logged the communication events for each scenario. The TSP events (i.e., OBE logs) and phase selector logs (i.e., WSE logs) collected during the prototype test were pulled and analyzed for six test scenarios to verify test results. Collectively, the data sets were reviewed to verify that the bus transmits conditional priority requests to the wayside phase selector; the phase selector then receives the priority request and places a low priority (i.e., TSP) call to the traffic signal controller; and the traffic signal controller responds to serve priority for the approaching bus. Visual confirmation of traffic signal controller priority service was necessary because of the logging limitations of the traffic signal controller firmware.

On-Site Stand-Alone Testing

SAT requires demonstration of all functions of the on-site intersection TSP system. SAT testing was performed at each traffic signal cabinet following TSP equipment installation. A laptop running Opticom On-site and Opticom AVL simulator was used to perform diagnostics on the equipment. The SAT verified the following for each intersection:

- Hardware installed properly
- Cellular modem, phase selector and traffic signal controller power up and are functional
- Cellular modem is connected
- Traffic signal controller is configured with TSP timing parameters
- The phase selector configuration of low priority channels and pin assignments is consistent with traffic signal controller configuration
- TOD restrictions are consistent with configuration pack dial sheet
- Correct call shows up on the traffic signal controller
- Verify the correct TSP request call shows up on the right channel on the phase selector card

SAT closely aligned with the installation schedule and was conducted immediately following equipment installation at a majority of the locations. Installation and SAT status was documented in detail for each of the 195 intersections on a standardized form. Although the prototype testing verified the functionality of the technology in a DC operating environment, the SAT verified WSE functionality at each intersection. Equivalent verification testing for the OBE was beyond the purview of DDOT.

Acceptance

Following the verification of the TSP technology concept through prototype testing and of WSE equipment functionality through stand-alone testing, DDOT's action plan progressed the effort to acceptance through verification and validation of the full scope of TSP within the District. The final stages of acceptance testing were designed to both verify the functional operation of the whole DC TSP system against the system requirements via FSAT, and to validate the system against pre-defined needs via OAT.

Final System Acceptance Testing

FSAT verifies that the end-to-end TSP system is operating as stated in the project documentation. FSAT was conducted after the successful completion of the SAT. FSAT tested for TSP request generation, transmission of the request from the bus to the phase selector via cellular communication, and the successful placement of the TSP call from the phase selector to the local traffic signal controller. The following pass/fail tests were conducted at each intersection:

1. Does the bus initiate communication on approach?
2. Does the bus take action in accordance with conditional priority parameters?
3. Is a message transmitted by the bus?
4. Is a message received by the wayside equipment?
5. Does the bus check out downstream of the intersection?

The testing oversight team executed FSAT through the review of the TSP events log data (documented via onboard equipment) and CMS log data (documented via wayside equipment) to verify end-to-end communication. The oversight team identified TSP activity in TSP events log data and cross-referenced the events in the CMS log data to confirm that the transmitted TSP events were received by the intersection equipment.

In addition to the cross-reference of activity, the oversight team reviewed logs to confirm passenger load and schedule deviation thresholds. The team also reviewed the data to confirm that priority requests were only sent when the bus doors were closed, and that a priority cancel message was sent any time the doors were opened.

Operational Acceptance Testing: System Validation

OAT validates full operation of the TSP System; the testing confirms that the entire system can function as specified for a pre-determined span of time under typical operating conditions. OAT was performed after the completion of the FSAT at all locations, and after training is conducted to prepare DDOT to operate the system as described in the specifications. OAT was performed across the entire system at the same time for a span of 30 days.

Intersection and bus activity were monitored at regular intervals. Activity that violated expected outcomes was documented and steps were taken to correct the erred activity. In accordance with the testing plan, erroneous activity reset the observation period, and the 30-day monitoring began again with issue resolution. The system was considered to be fully accepted with the completion of OAT. DDOT established full acceptance of the TSP system in 2016.

Lessons Learned

Deployment of a large-scale system is not without lessons learned. The following are those realized by the implementation team.

- *Establish formal agency roles and responsibilities before the commencement of the project.* A noted challenge with this regional implementation was the distribution of roles across agencies. Projects were slow to move forward without substantial regional coordination, and processes for decision-making where regional traffic agencies, municipalities, and community representatives required engagement were ill defined. This was limited, in part, by the terms of the TIGER Grant, agencies could have benefited from a mutually agreed and documented governance of responsibilities and roles for increased accountability.
- *Foster open discussion of system capabilities and limitations.* Throughout the TSP technology concept development, a large focus was placed on schedule and contractual matters, although open discussions on functionality were limited to intermittent milestones. This was inhibiting from a traffic agency perspective, as these agencies were tasked with moving forward with high-level design without clear definition of shortcomings. Care should be taken to facilitate forums for open discussion to establish a thorough understanding of capabilities.
- *Identify contingencies for cost overruns.* As with many product development and pilot efforts, total cost is difficult to estimate with a high degree of certainty. With large-scale efforts, as in this TSP deployment, small relative variabilities in cost can be quite substantial in absolute terms. When planning for such efforts, conservative estimations and identification of contingency funding is recommended.
- *Highlight the importance of oversight and review.* As the system design was distributed across agencies, quality control and review was paramount. A high degree of importance was placed on verification of testing results and parameter review across agencies. DDOT in particular sought to confirm parameter exports and reported functionality against log outputs and independent data sets. DDOT summarized these reviews and provided them to the regional agencies for review and comment resolution. These cross-examinations as part

of the system verification and acceptance greatly contributed to issue resolution.

- *Plan for evaluation and refinement.* Although trailblazing implementations are momentous, many can be greatly improved on with refinement. A full-scale system evaluation is planned for TSP in the District of Columbia to assess the overall benefit as compared to defined measures of effectiveness. The results of the evaluation will guide recommendations for refinement of TSP parameters to optimize performance.

Conclusions

Almost 10 years ago, the National Capital region embarked on a journey to improve bus transit for its ridership with the widespread deployment of TSP. Today, the realization of that vision can be attributed to the systems engineering driven process set in place by regional stakeholders. This practice-ready process documented in the sections above highlights design considerations for successful implementation and demonstrates that conditions specific to the selected TSP technology influence testing and acceptance. The outline of a systems engineering process and diligent execution of that process charted the course for effective operation of the deployed system.

Through extensive collaboration, detailed design, and the application of multi-tiered acceptance testing, TSP is now operational at nearly 200 intersections across multiple priority routes in the District of Columbia. As the District looks to maximize the lifecycle of the TSP system, plans for evaluation and refinement are in progress.

Acknowledgments

The authors express tremendous gratitude to the District Department of Transportation for sponsorship of this research endeavor. In addition, the authors thank the Washington Metropolitan Area Transit Authority for providing access to data and resources throughout the implementation. Finally, the authors thank United States Federal Transit Administration and the Metropolitan Washington Council of Governments for their roles in the TIGER I grant award and administration.

Author Contributions

The authors confirm contribution to the paper as follows: study conception and design: Kevin Lee, Bailey Lozner, and A. Wasim Raja; data collection: Kevin Lee and Bailey Lozner; analysis and interpretation of results: Burak Cesme, Kevin Lee, Bailey Lozner, and A. Wasim Raja; draft manuscript preparation: Burak Cesme, M.G. Habib, Kevin Lee, Bailey Lozner, and A. Wasim Raja. All authors reviewed the results and approved the final version of the manuscript.

References

1. Li, Y., Koonce, P., Li M., Zhou, K., Li, Y., Beaird, S., Zhang, W., Hegen, L., Hu, K., Skabardonis, A., and Z. Sun. *Transit Signal Priority Research Tools.* Publication FTA-CA-26-7049-2008.1. U.S. Department of Transportation, 2008.
2. Kittelson & Associates, Inc. *NCHRP Report 812: Signal Timing Manual, Second Edition.* Transportation Research Board of the National Academies, Washington, D.C., 2014.
3. Levinson, H., S. Zimmerman, J. Clinger, J. Gast, S. Rutherford, and E. Bruhn. *TCRP Report 90: Bus Rapid Transit, Volume 2: Implementation Guidelines.* Transportation Research Board of the National Academies, Washington, D.C., 2003.
4. U.S. Department of Transportation. *Transportation Investment Generating Economic Recovery (TIGER) Grants.* February 17, 2010. https://www.transportation.gov/sites/dot.gov/files/docs/Tiger_I_Awards.pdf. Accessed July 20, 2017.
5. U.S. Department of Transportation. *Systems Engineering for Intelligent Transportation Systems.* Publication FHWA-HOP-07-069. FHWA and FTA, U.S. Department of Transportation. January 2007.
6. *NCTIP 1211: National Transportation Communications for ITS Protocol Object Definitions for Signal Control and Prioritization (SCP), Version 01.* American Association of State Highway and Transportation Officials (AASHTO), Institute of Transportation Engineers (ITE), National Electrical Manufacturers Association (NEMA), May 2008.
7. Albright, E., and M. Figliozzi. Factors Influencing Effectiveness of Transit Signal Priority and Late-Bus Recovery at Signalized-Intersection Level. *Transportation Research Record: Journal of the Transportation Research Board,* 2014. 2311: 186–194.
8. Washington Metropolitan Area Transit Authority. *Vital Signs January - December 2016 Annual Report.* February 2017. https://www.wmata.com/about/records/scorecard/upload/Vital-Signs_Report_2016-Q4_030117.pdf. Accessed July 20, 2017.
9. Altun, Z. S., and P. G. Furth. Scheduling Buses to Take Advantage of Transit Signal Priority. *Transportation Research Record: Journal of the Transportation Research Board,* 2014. 2111: 50–59.
10. Chada, S. and R. Newland. *Effectiveness of Bus Signal Priority: Final Report.* Publication NCTR-416-04.Office of Research and Special Programs, U.S. Department of Transportation, Washington, D.C., 2002.

The Standing Committee on Bus Transit Systems (AP050) peer-reviewed this paper (18-04233).

TRR
JOURNAL OF THE TRANSPORTATION RESEARCH BOARD

Article

Transportation Research Record
2018, Vol. 2672(8) 75–85

DOI: 10.1177/0361198118797218
journals.sagepub.com/home/trr
SAGE

Applying Spatial Aggregation Methods to Identify Opportunities for New Bus Services in London

Cecilia Viggiano[1,2], Haris N. Koutsopoulos[3], Nigel H.M. Wilson[1], and John Attanucci[1]

Abstract

Innovative analyses of origin–destination (OD) data derived from automatic fare collection and automatic vehicle location systems in public transport networks enable planners to gain new insights into how passengers travel in the network and the quality of service provided, and can even inform decisions about network improvements. Particularly in large, complex networks, systematic, data-driven approaches to network evaluation and planning are essential. New methodologies are needed to transform OD data into informative metrics and planning recommendations. This paper proposes a framework for this process and applies it to London's public transport network. Though there are many ways to improve public transport networks, this paper focuses on the addition of new bus routes to reduce circuity. The proposed framework includes three steps that combine OD-level analysis with spatial aggregation methodologies for the identification of corridors for new bus services. First, bus stops and rail stations were clustered into geographic zones. Second, a subset of zonal OD pairs with circuitous service were identified as candidates for improvement through new bus routes, based on performance standards established with user-defined parameters. Third, an algorithm that clusters OD pairs into corridors was applied to identify promising corridors for new bus services. This paper discusses corridors identified for new services in the London case study.

Innovative approaches, including quantitative and spatial aggregation techniques, can produce new insights for transportation evaluation and planning. In tandem with these techniques, increased availability of origin–destination (OD) data from new sources presents opportunities for improved understanding of behavior and better evaluation of mobility and accessibility, which can lead to better planning.

The framework for data-driven sketch planning applied in this paper represents an alternative to existing approaches to public transport planning. One challenge, adding new bus routes to reduce circuity and transfers, is explored here, with an application to London's public transport network. In this paper, bus routes are defined as circuitous when the distance traveled by bus significantly exceeds the shortest path distance. Using geographic zones that reflect the structure of the existing public transport network, OD-level analysis is conducted to determine which OD pairs can be improved by new bus services that reduce circuity or eliminate transfers. Improvable OD pairs are spatially aggregated into corridors that are appropriate for new bus services. This paper includes a review of related literature, an overview of the framework applied, and a discussion of the implementation and results of the framework for the London network.

Opportunity for a Data-Driven Approach

The increasing availability of travel information at the OD level presents an opportunity for improved bus planning. Automatic fare collection (AFC) and automatic vehicle location (AVL) data are routinely collected in many public transport networks. Recently, methodologies have been developed to infer complete journey information from this data (*1*, *2*). These methodologies infer destination and interchange information for a large share of boardings that use fare cards, often upwards of 70%

[1]Massachusetts Institute of Technology, Cambridge, MA
[2]Economic Development Research Group, Boston, MA
[3]Northeastern University, Boston, MA

Corresponding Author:
Address correspondence to Cecilia Viggiano: cviggiano@edrgroup.com

(*2*). Compared with OD data collected from surveys, which is generally collected for a small number of passengers on a few selected days, this vastly increases the sample size, improving the accuracy and precision of estimates of OD-level demand and travel time. Making use of emerging information for planning requires new metrics and methods for evaluation and aggregation.

Traditional Network Planning Approaches

A methodology that could be applied to an existing public transport network to identify opportunities for improvement would fill a gap in the public transport planning literature. Traditional approaches to public transport network design use partial optimization methods to identify the set of routes or lines that best serve a specified demand matrix. However, these approaches may produce a network that bears little resemblance to the existing network. As a result, within established networks, planners generally have little use for such results. Instead, planners typically make incremental changes to the existing network over time. Because changes to bus services are significantly less costly (and less permanent) than in rail networks, planners may use a trial-and-error approach to planning, piloting routes to assess the impacts. Though this can be effective in small networks, in large, complex networks, it is inefficient. This leaves an unaddressed need to systematically identify opportunities for incremental improvement.

The Role of Circuity

This paper focuses on the introduction of new routes to reduce circuity and transfers in a public transport network. Huang and Levinson found evidence that commuters live in places where their commute trip will have low circuity—circuity is measured as the ratio of public transport distance to straight-line distance (*3*). They also found that public transport accessibility measures for 36 metropolitan areas were negatively correlated with circuity. This suggests that circuity is not only undesirable for specific trips, it may also degrade the overall quality of the public transport system. In support of this, many graph theoretic metrics, designed to assess public transportation networks, characterize them in terms of circuity or directness (*4*, *5*).

Transferring is widely regarded as detracting from public transport attractiveness and quality of service, with a large body of research dedicated to understanding the transfer penalty. Although the penalty can be reduced by coordination of service or high frequency of service, eliminating transfers can be beneficial.

Introducing new services to eliminate circuity and transfers requires investment of money and resources. A methodology that systematically assesses the benefits of reduced circuity could help planners make resource allocation decisions.

London's Public Transport Network

The London example contextualizes the need for a bus planning framework that can be applied to an existing network to identify opportunities for new services. Like many urban areas, London has experienced growth in recent years. From 2001 to 2011, the population grew at an average rate of 87,000 people per year. The number of jobs in London is projected to increase from 4.9 million in 2011 to 5.8 million in 2036 (*6*).

London has an extensive public transport system that plays a critical role in mobility. Almost half of all Londoners use the bus at least 2 days per week (*7*). Demand for bus services grew by 64% from 1999 to 2013. However, from 2014 to 2017 bus journeys decreased by 5.4%, correlating with decreases in bus speeds (*8*).

Given the vast size of the London bus network, trial-and-error approaches are inefficient. Instead, a data-driven approach could guide investment. The Oyster card, a smart payment card that was first issued in 2003, was used by approximately 80% of rail passengers and 90% of bus passengers in 2012, generating large quantities of data (*2*).

Literature Review

This paper demonstrates one application using OD data. Other applications are summarized in the following section. Next, a review of the use spatial aggregation in transport analysis is provided, and a summary of traditional network planning and sketch planning techniques.

Use of Automatically Collected Origin–Destination Data

Automated data collection systems in public transport networks have been lauded as a source for better estimates of existing measures and the development of new measures (*9*). Trépanier et al. demonstrated the estimation of a variety of supply and demand-oriented measures from AFC systems (*10*). Uniman et al. developed new metrics for service reliability based on AFC data (*11*). Zhu used AFC data and train tracking data to assign passengers to specific trains (*12*).

Individual itineraries can be inferred from AFC and AVL data (*1*, *2*). In networks where smart card usage is prevalent, this provides a comprehensive picture of public transport travel. Vanderwaart et al. developed a framework for using complete OD-level journey data for service planning (*13*). The framework includes the selection of target locations, evaluation of network and demand characteristics, and proposed service changes. Compared with Vanderwaart's research, the proposed framework generates service alternatives directly from data, rather than relying on interpretation of statistics and maps. This makes the proposed framework more appropriate for large and complex networks. To do so, it focuses on the addition of new bus routes, rather than the diverse service changes considered in Vanderwaart's framework.

Spatial Aggregation

Much of the early work on network structure modeled transport networks as graphs composed of nodes and edges. Garrison and Marble defined summary metrics to assess connectivity of transport networks (*4*). More recently, Derrible and Kennedy used concepts from graph theory to develop indicators of the maturity of a network, the relationship of the network to the built environment, and network directness (*5*).

For spatial clustering of OD data, methods have been developed based on trajectory clustering. Lee et al. developed a method that clusters similar parts of trajectories (*14*). This methodology was adapted for desire line clustering by Bahbouh and Morency (*15*), and further by Bahbouh et al. (*16*). Their methodology, applied to identify opportunities for walkability improvements, decomposes desire lines into line segments, and imposes maximum width and maximum angle requirements to ensure that clusters form a reasonable travel shed.

The research presented here builds on the methodology developed by Bahbouh et al. However, corridors must be appropriate for bus service, and are prioritized by expected benefits. This requires several adaptations, including minimum and maximum length constraints, the requirement that full desire lines are clustered, and a new prioritization method.

Network Planning and Sketch Planning

This paper demonstrates a network planning approach that differs significantly from traditional approaches. The transit network design problem is usually formulated as an optimization problem, with methodologies tailored for bus (*17*, *18*), rail (*19*), and for general application to public transport networks (*20*). Most methodologies can be applied to existing or new networks, but are intended for full network redesign and usually recommend drastic changes. As an example, when Cipriani et al. applied their framework to Rome, only 20% of existing routes were retained (*21*).

In existing networks, planners are more likely to employ sketch planning methods, such as the one presented in this paper. Other examples of sketch planning methodologies in the literature include the use of EU standards to evaluate public transport proposals (*22*), and a methodology for demand and cost estimation for proposed extensions of the Bay Area Rapid Transit system (*23*).

The approach in this paper uses a network evaluation process to identify opportunities for new bus routes. One example of the use of performance indicators to locate improvement opportunities comes from highway infrastructure planning in Minnesota, where corridors were prioritized based on demand, growth trends, and connections to regional trade centers, and evaluated based on how closely actual speeds matched target speeds along the corridor (*24*).

Other approaches make recommendations about network design more generally. Daganzo proposed a methodology to determine the appropriate network shape (grid or hub-and-spoke) and modes, given network size and demographics (*25*). Badia et al. compared the performance and cost of different network structures (*26*). Compared with the proposed framework, the recommendations from these methodologies require more interpretation to be translated into planning decisions, particularly in existing networks.

Methodology

This paper applies a method developed and described in detail by Viggiano (*27*). Here, an overview of the approach is presented. The framework consists of three steps: zone definition, OD analysis, and corridor identification.

Step 1

The first step in the framework is the definition of geographic zones, which create a set of zonal pairs to which all public transport journeys are assigned, based on their origin and destination bus stop or rail station. This serves to group journeys on multiple paths (including parallel and non-parallel bus and rail lines) that connect similar origins and destinations.

Though many cities and regions have existing zonal schemes, such as postcodes or census tracts, these schemes usually use roads as boundaries, resulting in a large share of bus stops along zonal boundaries. Spatially clustering bus stops and rail stations based on their locations results in zones that reflect the structure of the existing public transport system and avoids the location of bus stops and rail stations close to the boundaries of zones. Viggiano et al. explain the zone-definition methodology in detail (*28*).

Step 2

In an existing network, not all OD pairs can be improved by new service. Some OD pairs are already well-served by the existing network or have no anticipated travel time benefits from new service. The first part of the OD-level analysis filters out these pairs. Planners may also manually eliminate OD pairs they know are not good candidates for bus service due to other constraints, such as road network limitations.

The framework uses specific definitions of well-served and improvable OD pairs. OD pairs are defined as well-served if they are served by direct, single-stage service. In this framework, direct and circuitous are used to refer to how closely a path through the road network resembles the shortest possible path. Single stage always refers to a journey requiring a single vehicle. It can refer to one bus stage (one boarding, one alighting) or one rail stage (one entry and one exit without any in-station or out-of-station interchange). Of those OD pairs that are not well-served, improvable OD pairs are defined as those that are expected to see travel time improvements from new service.

Following the filtering step, the potential benefits of new service are quantified for improvable OD pairs. The benefits of new service are twofold. There are travel time savings accumulated by existing public transport passengers who shift to the new service, and there are new passengers attracted from outside the public transport system. This includes passengers who shift from car or other modes and new journeys that previously were not made. Individuals may choose to make more trips due to improved service and, over longer time periods, individuals may change their home and work locations in response to new services. Expected new demand depends on many factors including the change in bus travel time, the demand and performance of alternative modes, and socio-demographic and land use factors. Predicting demand changes is challenging, both because reliable data on these factors is not always available and because of complexity and variation in how these factors influence individual behavior. The best estimates available for the planning time horizon desired will produce the most insightful results.

Step 3

The final step in the framework is to spatially aggregate improvable OD pairs into corridors. New services will not influence OD pairs in isolation. Rather, they will provide benefits along a linear (or pseudo-linear) corridor. Step 3 clusters OD pairs into corridors that can be served by a single bus route and have demand characteristics that can support new service. For each corridor, expected benefits are defined as the weighted sum of travel time savings for existing public transport passengers and the total journey-minutes of new journeys attracted from outside the public transport network. Depending on the planer perspective, different definitions of benefits could be applied. For example, economic benefit analysis often uses the "rule of half" to quantify the benefits from behavioral responses to cost or travel time changes, and a similar approach could be adopted. Here, benefits are used only to prioritize corridors within the corridor identification algorithm. Therefore, the travel time savings for existing public transport passengers and the total journey-minutes of new journeys attracted are simply summed.

Within the framework, corridor definitions can incorporate varying degrees of spatial specificity and complexity. For instance, corridors can be represented simply as straight lines, they can bend to account for barriers such as rivers and parks, or they can reflect actual paths through the road network. Similarly, the algorithm used to identify corridors can incorporate road network distance and barriers when assessing the similarity of OD pairs, or can use Euclidean distance as a simple proxy.

London Implementation and Results

This section summarizes the process and results of applying this three-step framework to the London public transport network.

Data

The inputs to the framework were AFC and AVL data, which provided OD-level demand, distance, and travel time estimates within the public transport network. Also required were OD-level shortest path distance and travel time estimates through the road network, which were used to estimate potential travel time and distance for new bus routes. The fare card data used was from 10 weekday AM peak periods from October 2012 to February 2013. The AFC and AVL data were combined to infer origin bus stops, alighting bus stops, and link stages of multi-stage journeys using a methodology developed by Gordon et al. (*2*).

Estimates of shortest path distances and travel times for OD pairs were queried using the Google Maps Distance Matrix API. The API provides estimates of point-to-point car distance and travel time for a specified time of day, based on historical data, thus accounting for expected traffic conditions. A study comparing Google Maps estimates of travel times to estimates from GIS networks found the estimates to be reasonably consistent (*29*).

Step 1: Zones

The process of developing appropriately sized zones that reflect the structure of the existing public transport

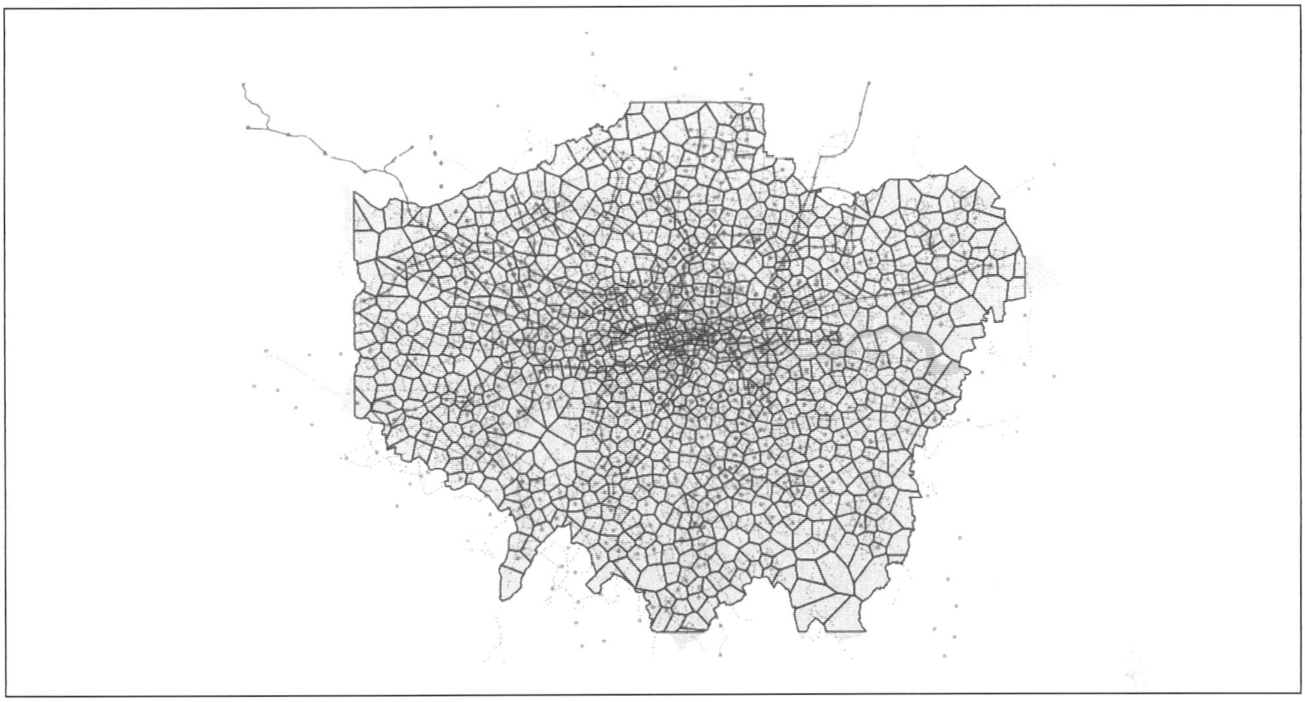

Figure 1. London zones.

system in London is described in detail in Viggiano et al. (*28*). The result is 1,000 zones (shown in Figure 1) with an average radius of approximately 0.7 km, which corresponds to the average access distance to public transport (bus and rail) stops and stations, and to the 75th percentile of access distances specifically to bus stops in London (*30*). By spatially clustering bus stop and rail station locations, the resulting zones have fewer bus stops and rail stations along zonal boundaries, compared with existing London zonal schemes.

Step 2: Improvable OD Pairs

In London, the peripheral region is more likely to have opportunities for new services. OD pairs that originate or end in Fare Zone 1 (Central London) were excluded from the analysis. This left 950 zones and 902,500 zonal OD pairs. To ensure travel times could be reliably estimated, a minimum of 23 journeys over the 10 AM peaks analyzed was required. Only 45,916 OD pairs met this threshold, demonstrating the sparsity of the OD matrix. This constraint removed just 16% of journeys.

As described in the methodology section, OD pairs are considered well-served by the existing network if they are served by direct, single-stage service. For OD pairs served by bus, this requires comparing the bus distance traveled in the current network with the shortest path distance through the road network.

Shortest paths were estimated using the full road network, excluding highways. The actual path of a potential new bus route is not expected to follow the shortest path exactly. Paths that pass important locations such as schools or hospitals may be desirable, and some roads and junctions cannot accommodate buses. Therefore, OD pairs are considered to have direct service if the current bus distance is less than or equal to the shortest path distance, increased by a multiplicative factor, m, and an additive factor, a. m reflects the additional distance that is proportional to the shortest path distance for the OD pair. a reflects the fixed additional distance that is independent of the shortest path distance. Planners could use these parameters to specify the standard for directness they wish new bus routes to meet.

For the London application, relatively large values of m and a were used: 1.1 and 1.4 km (0.9 mi), respectively. The choice of m and a was based on analysis of the directness of existing bus routes, with larger values selected to focus the analysis on OD pairs with significant potential for improvement in terms of directness. Of those OD pairs served by single-stage bus, 91% were considered well-served, given these parameters. The remaining 9% (1,737 OD pairs) were not well-served in terms of directness. In addition, there were 17,942 OD pairs that were served only by multi-stage service.

The next filtering step determines which of these OD pairs are improvable in terms of travel time. Here, the median travel time in the existing public transport network is compared with the expected travel time. The expected travel time is based on the shortest path travel time, the expected number of stops, and the estimated time required to make a stop. Again, this estimate can be modified with multiplicative or additive factors to reflect acceptable deviations from the shortest path.

Transport for London planners aim for a stop spacing of approximately 400 m (0.25 mi) (*31*). Dwell time varies based on the number of boardings and alightings as well as characteristics of the bus and stop. Using the middle of the ranges estimated by York (*32*), and the average acceleration/deceleration time found by Robinson (*33*), 33 s were added for each stop. In addition, a multiplicative factor of 1.1 was applied.

Given these parameters, of the 1,737 OD pairs served by single-stage bus and identified as not well-served, 1,621 had travel times greater than the expected travel time for the OD pair. This is reasonable; most OD pairs served by circuitous bus service are improvable in terms of travel time. For OD pairs served by multi-stage service, the travel time filter has a much greater impact. Of 17,942 OD pairs served only by multi-stage service, 5,856 OD pairs had travel times that exceed the pair's expected travel time. This is also unsurprising; many of these OD pairs are served by multi-stage rail, which is often faster than bus.

For the improvable OD pairs, the final part of the OD analysis estimated the demand a new bus service could be expected to attract. There are many methods for predicting demand, including four-step models, direct demand models, discrete choice models, and elasticity-based methods (*34*). Elasticity methods require only two inputs, which are already used in the analysis: existing demand and anticipated travel time savings. Therefore, the elasticity method was used, eliminating additional data requirements. Based on a review of studies in the UK, elasticities of demand with respect to bus in-vehicle time of -0.4, -0.5, and -0.6 were applied for short trips (less than 20 min), medium trips (20 to 40 min) and long trips (more than 40 min), respectively (*35*). The varied elasticities account for the fact that for shorter trips, in-vehicle time generally makes up a smaller percentage of the total journey time. This method does not account for demand changes due to contemporaneous population or land use changes, and it may underestimate demand on OD pairs where current public transport demand is low due to poor existing service.

Step 3: Corridors

The corridor identification methodology applied was developed by Viggiano (*27*). It uses simplified spatial representations. Corridors are defined as straight lines with surrounding buffers. Consideration of barriers and constraints on termini location are left to post-analysis. The effectiveness of this approach is discussed with respect to an example corridor.

The methodology requires the specification of a minimum and maximum route length, a maximum distance, a maximum angle, and a minimum flow. The length parameters restrict the length of the corridors identified to ensure they are reasonable for bus service. For this application, based on the lengths of existing bus routes in London, corridors 5 to 9 mi in (straight line) length are considered. Actual bus routes are expected to be significantly longer than this straight-line distance.

The maximum distance parameter constrains the Euclidean distance from zonal OD pair centroids to the straight line that represents the center of the corridor. Because zone size was set based on public transport access distances, ideal corridors should be one zone wide, that is, with all centroids of assigned zonal OD pairs aligned along the linear corridor. In fact, zone centroids do not always align, so choosing a maximum distance that is too small eliminates many potential corridors. Through trial-and-error, it was determined that specifying a maximum distance of 0.6 km (0.4 mi) ensures that corridors are mostly only one zone in width, without eliminating too many potential corridors.

Some short OD pairs may have origins and destination zone centroids within the maximum distance, but if the direction of travel on the OD pair and the route is not similar, the potential route would not serve the OD pair. To eliminate these OD pairs, angular similarity is defined as in Figure 2, by computing the minimum angle formed by the connected centroids of the OD pair and the potential linear corridor. If this angle is within the

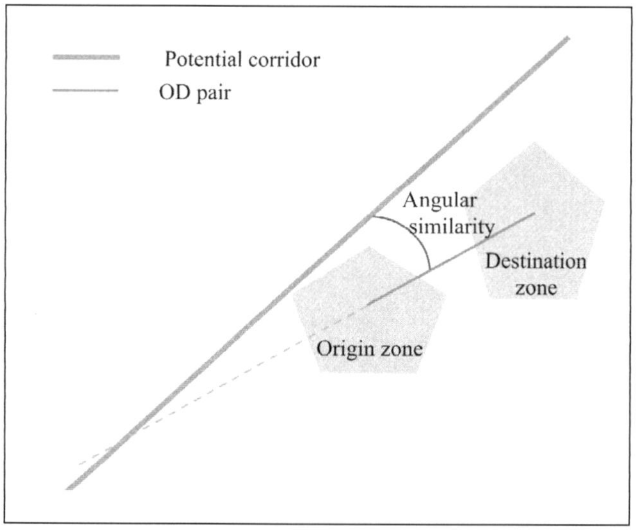

Figure 2. Angular similarity.

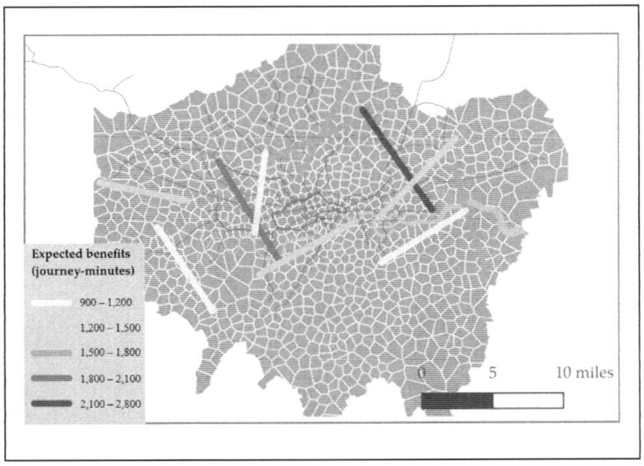

Figure 3. Corridors identified.

maximum angle specified, OD pairs can be assigned to a potential corridor. A maximum angle of 22.5° was applied.

Corridors identified should have sufficient expected demand to justify adding bus service. This is enforced through a minimum flow constraint. Flow on a potential corridor can be estimated as the expected hourly flow averaged along the corridor. This measure does not fully capture the shape of demand, such as spatial peaking, but it does account for flow relative to the length of the route, and it is easy to understand. For this implementation, a moderate minimum flow of 50 passengers per hour is used. This accounts for the fact that the methodology likely underestimates demand. More detailed demand estimation could be conducted at the corridor level in post-analysis.

Using these parameters, 11 corridors were identified (See Figure 3). Each corridor was prioritized by the sum of potential journeys-minutes saved (for existing passengers) and new journey-minutes induced; both benefits were estimated per AM peak hour. Only 5% of improvable OD pairs (9% of journeys on improvable OD pairs) were assigned to a corridor. This reflects the fact that the improvable OD pairs are spatially distributed across the network. The parameters selected also influence the number of corridors identified and the percent of OD pairs and journeys assigned to corridors. Reducing the minimum flow identifies more corridors and assigns more OD pairs and journeys to corridors. However, these corridors may not have sufficient demand to support bus services. Increasing the maximum distance or angle also results in more corridors being identified, as loosening these shape requirements allows more corridors to meet the minimum flow requirement. However, with less stringent shape requirements, it may be difficult to serve the entire corridor with one bus route. Overall, sensitivity analysis of the resulting corridors for varied shape and flow parameters revealed that the highest priority corridors were identified consistently across parameter choices.

The results were also influenced by parameter choices in Step 2 of the framework. Reducing the multiplicative and additive factors, and thus decreasing expected distance and travel times, identifies more improvable OD pairs. This increases the spatial density of OD pairs to be clustered into corridors and results in more corridors being identified.

Example Corridor

The corridor discussed here is the second highest priority corridor, extending from Wembley Park to Fulham. In addition to having significant expected benefits, this corridor has the greatest expected demand of the corridors identified. Almost 60% of the current journeys on the improvable OD pairs assigned to the corridor are multistage bus journeys. Approximately 7% of the journeys are multi-stage rail journeys (defined as rail journeys in which the origin and destination stations are not served by the same line or rail journeys with out-of-station transfers) and 15% are combined bus and rail journeys. The remaining journeys are served by single-stage bus that did not meet the directness threshold.

The expected benefits on this corridor, consisting of potential travel time savings for current passengers and anticipated new journey-minutes, were distributed over 41 OD pairs. Approximately 75% of the benefits, however, were concentrated on 15 OD pairs, shown in Figure 4. In the direction of Fulham Broadway, demand was distributed along the corridor and several long OD pairs were expected to receive significant benefits. In the direction of Wembley Park, demand was concentrated in the northern part of the corridor. The three highest demand OD pairs combined had 59 expected journeys per AM peak hour, and expected travel time savings of 6 to 10 min.

Figure 5 shows the main bus routes serving the corridor, which serve 80% of the demand on the OD pairs shown in Figure 4. Route 206 traverses the northern part of the corridor, but is circuitous. Route 220 serves the middle part of the corridor. However, several of the OD pairs included in the corridor begin north of the terminus of Route 220 and extend south, requiring either multistage journeys or circuitous travel on Route 266 or Route 260. Extending a single route the length of the corridor would serve these journeys better.

The southern half of the corridor is two zones wide in many places. As described in Step 3: Corridors, this is because corridors are linear, but zone centroids are not always aligned. The corridor width poses challenges to designing a route that serves all OD pairs. Figure 6 shows

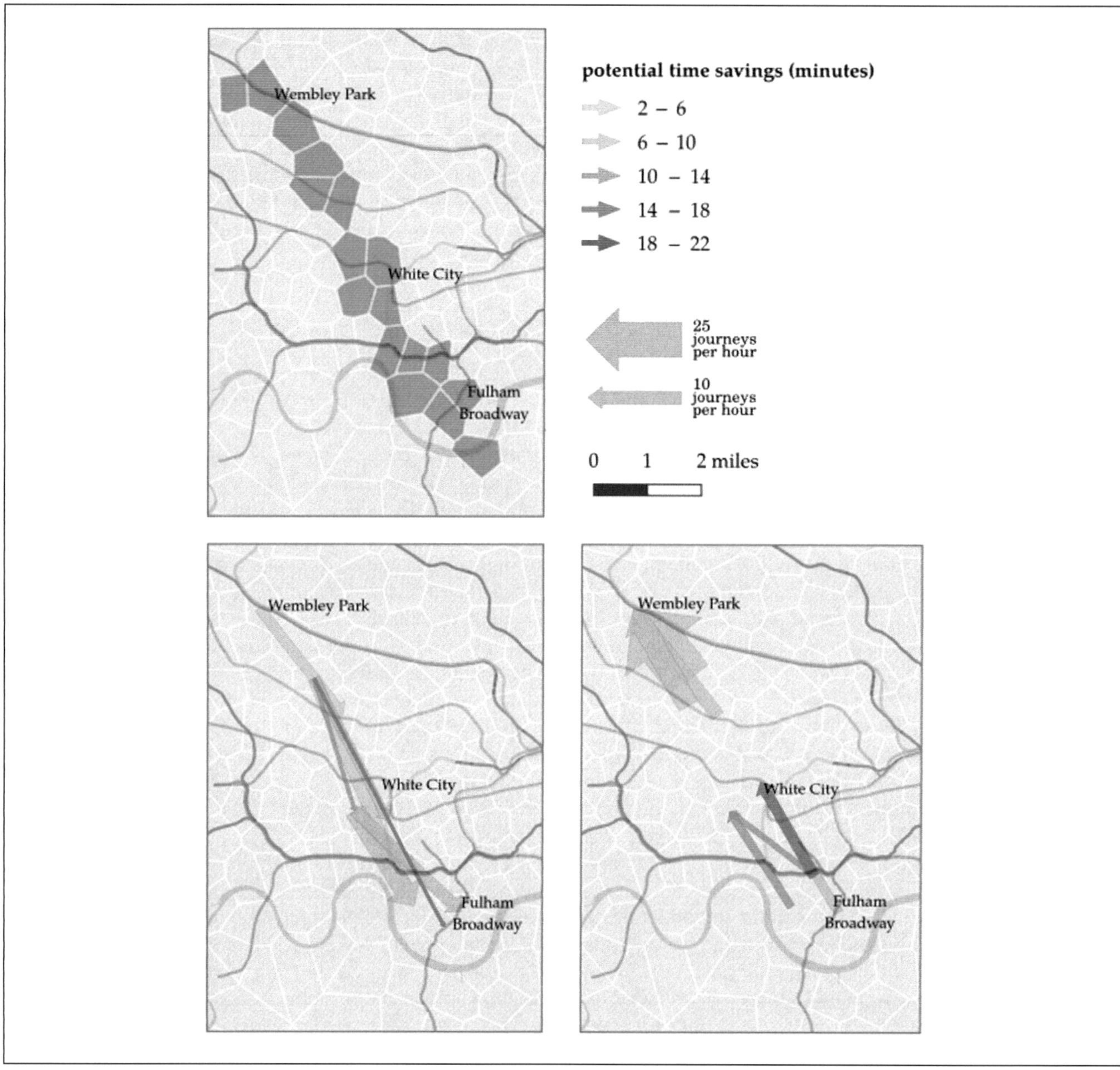

Figure 4. Demand and potential travel time savings in example corridor.

a potential design for a route. In the southern part of the corridor it serves the western side of the corridor more so than the eastern, as the eastern side has more poorly served demand.

Discussion

The implementation of the framework revealed two possible areas for improvement. The first is the choice of spatial representation. More complex representation of corridors as actual paths through the road network, and the use of road network distance in place of Euclidean distance in the corridor identification algorithm could limit cases as in the southern part of the example corridor, where the corridor width and characteristics of the road network make it difficult to serve all OD pairs with a single route. The second area for improvement is the incorporation of a more sophisticated demand model to better account for new demand on OD pairs that are currently not well-served.

A large percentage of improvable OD pairs were not assigned to any corridor. These OD pairs were not good

Figure 5. Existing bus routes and rail services in example corridor.

candidates for improvement through new bus services because they were not located within a corridor with sufficient demand. In networks where a small percentage of OD pairs identified for improvement are assigned to corridors, planners might consider services other than a fixed route bus for these OD pairs.

Two extensions to the framework might also be useful. The first is to allow for new routes to provide improved service while still allowing a transfer, which may be appropriate in areas of sparser demand. This would require changes to the corridor identification algorithm such that demand could be served by a combination of two routes.

The second extension would be to provide recommendations for existing routes that should be removed. Given good estimates of how journeys shift from existing to new services, an iterative approach could be taken. In each iteration, routes with insufficient demand would be eliminated and the analysis would be repeated for the new network.

Conclusion

This paper presents an innovative application of OD data for bus network evaluation and planning. Spatial aggregation approaches, such as those applied in this paper, could help planners understand characteristics of transport networks and the quality of service they provide. In this case, the analysis evaluates OD-level circuity and helps planners identify opportunities for new bus services.

The success of this approach suggests that the application of spatial and quantitative methods to OD data could be adapted to plan other modes, such as on-demand service. In addition, similar approaches could be applied to other planning challenges, such as decisions about where to implement bus priority measures, and where to add express bus services. For bus priority measures, metrics and estimates to assess potential benefits must be developed, and spatial methods are needed to aggregate potential benefits to the corridor level. For express service, the demand patterns that are generally associated with this type of service, such as long journeys and concentrated OD-level demand would need to be identified through OD analysis and spatial aggregation. The framework applied here could form the groundwork for the development of new metrics assessing transportation system performance, mobility, and accessibility, at the OD and corridor level.

More broadly, with the development of appropriate methods, OD data is likely to play a growing role in transportation policy and planning. OD data is becoming more plentiful and is already supporting an improved understanding of travel behavior and system performance. The strategic definition of zonal OD pairs, development of appropriate metrics, and spatial aggregation of disaggregate data, as demonstrated here, allow

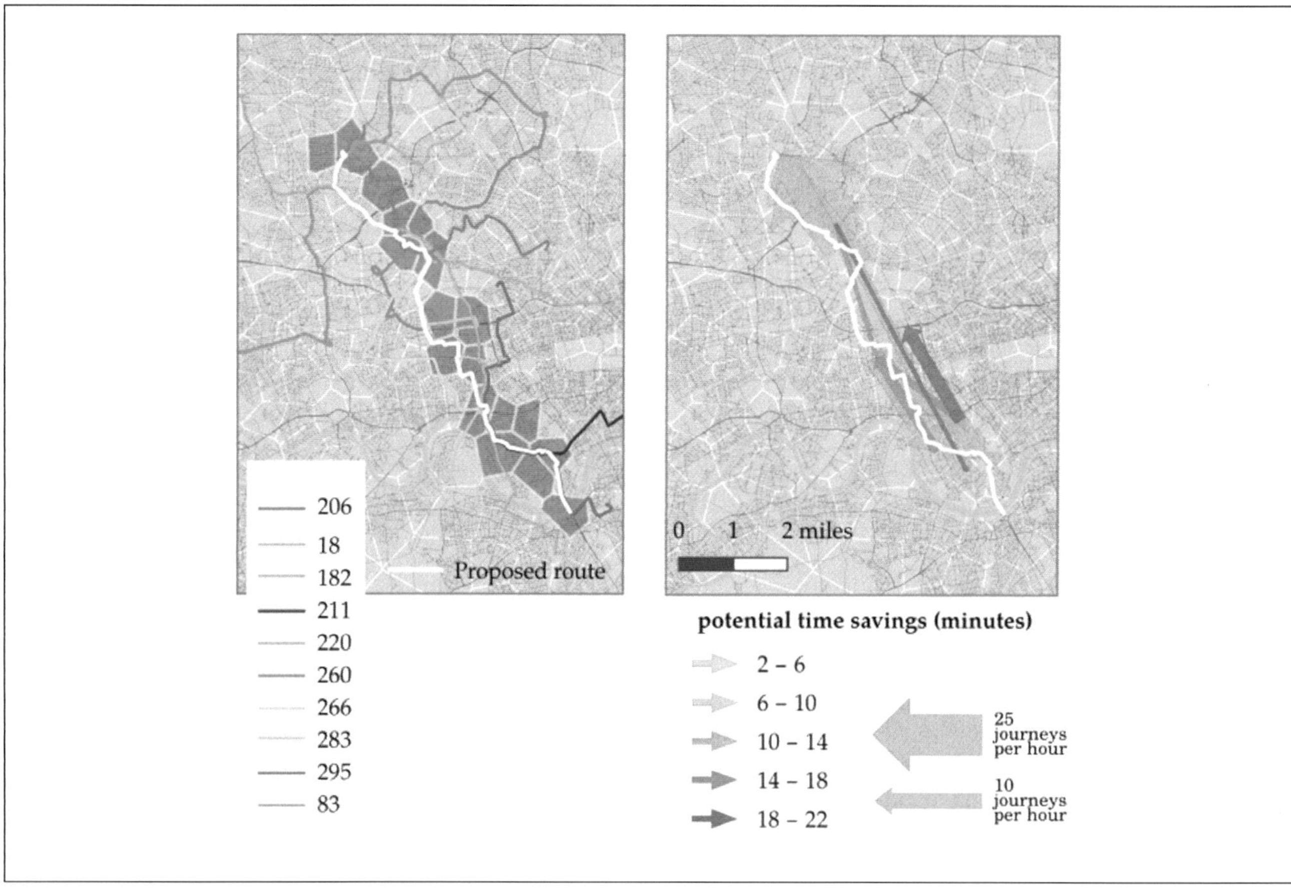

Figure 6. Potential new route design for example corridor.

planners to use the improved understanding for better planning.

Acknowledgments

This research received funding from Transport for London and support from the staff of London Buses and TfL Analytics.

Author Contributions

The authors confirm contribution to the paper as follows: study conception and design: CV, HK, NW, JA; data collection: CV; analysis and interpretation of results: CV, HK, NW, JA; draft manuscript preparation: CV, HK, NW, JA. All authors reviewed the results and approved the final version of the manuscript.

References

1. Chu, A. K. K., and R. Chapleau. Enriching Archived Smart Card Transaction Data for Transit Demand Modeling. *Transportation Research Record: Journal of the Transportation Research Board*, 2008. 2063: 63–72.
2. Gordon, J., H. Koutsopoulos, N. Wilson, and J. Attanucci. Automated Inference of Linked Transit Journeys in London Using Fare-Transaction and Vehicle Location Data. *Transportation Research Record: Journal of the Transportation Research Board*, 2013. 2343: 17–24.
3. Huang, J., and D. M. Levinson. Circuity in Urban Transit Networks. *Journal of Transport Geography*, Vol. 48, 2015, pp. 145–153.
4. Garrison, W. L., and D. F. Marble. Factor-Analytic Study of the Connectivity of a Transportation Network. *Papers in Regional Science*, Vol. 12, No. 1, 1964, pp. 231–238.
5. Derrible, S., and C. Kennedy. Characterizing Metro Networks: State, Form, and Structure. *Transportation*, Vol. 37, No. 2, 2010, pp. 275–297.
6. Greater London Authority. *The London Plan: The Spatial Development Strategy for London Consolidated with Alterations Since 2011*. London: Transport Committee of the London Assembly, 2016.
7. Greater London Authority. *Bus Services in London*. London: Transport Committee of the London Assembly, 2013.
8. Transport for London. *Transport for London Budget 2017/18*. London: Transport for London, 2017.
9. Wilson, N. H. M., J. Zhao, and A. Rahbee. The Potential Impact of Automated Data Collection Systems on Urban Public Transport Planning. In *Schedule-Based Modeling of Transportation Networks*, Springer, Boston, MA, 2009, pp. 1–25. https://doi.org/10.1007/978-0-387-84812-9

10. Trépanier, M., C. Morency, and B. Agard. Calculation of Transit Performance Measures Using Smartcard Data. *Journal of Public Transportation*, Vol. 12, No.1, 2009, pp. 79–96.
11. Uniman, D., J. Attanucci, R. Mishalani, and N. Wilson. Service Reliability Measurement Using Automated Fare Card Data: Application to the London Underground. *Transportation Research Record: Journal of the Transportation Research Board*, 2010. 2143: 92–99.
12. Zhu, Y., H. N. Koutsopoulos, and N. H. M. Wilson. A Probabilistic Passenger-to-Train Assignment Model Based on Automated Data. *Transportation Research Part B: Methodological*, Vol. 104, 2017, pp. 522–542.
13. Vanderwaart, C., J. P. Attanucci, and F. P. Salvucci. Applications of Inferred Origins, Destinations, and Interchanges in Bus Service Planning. *Transportation Research Record: Journal of the Transportation Research Board*, 2017. 2652: 70–77.
14. Lee, J., J. Han, and K. Whang. Trajectory Clustering: A Partition-and-Group Framework. *Proceedings of the 2007 ACM SIGMOD International Conference on Management of Data*, Beijing, China, 2007, pp. 593–604.
15. Bahbouh, K., and C. Morency. Encapsulating and Visualizing Disaggregated Origin–Destination Desire Lines to Identify Demand Corridors. *Transportation Research Record: Journal of the Transportation Research Board*, 2014. 2430: 162–169.
16. Bahbouh, K., J. R. Wagner, C. Morency, and C. Berdier. Traclus-dl: A Desire Line Clustering Framework to Identify Demand Corridors. Presented at 94th Annual Meeting of the Transportation Research Board, Washington, D.C., 2015.
17. Ciaffi, F., E. Cipriani, and M. Petrelli. Feeder Bus Network Design Problem: A New Metaheuristic Procedure and Real Size Applications. *Procedia-Social and Behavioral Sciences*, Vol. 54, 2012, pp. 798–807.
18. Bielli, M., M. Caramia, and P. Carotenuto. Genetic Algorithms in Bus Network Optimization. *Transportation Research Part C: Emerging Technologies*, Vol. 10, No. 1, 2002, pp. 19–34.
19. Guan, J. F., H. Yang, and S. C. Wirasinghe. Simultaneous Optimization of Transit Line Configuration and Passenger Line Assignment. *Transportation Research Part B: Methodological*, Vol. 40, No. 10, 2006, pp. 885–902.
20. Bagloee, S. A., and A. Ceder. Transit-Network Design Methodology for Actual-Size Road Networks. *Transportation Research Part B: Methodological*, Vol. 45, No. 10, 2011, pp. 1787–1804.
21. Cipriani, E., S. Gori, and M. Petrelli. Transit Network Design: A Procedure and an Application to a Large Urban Area. *Transportation Research Part C: Emerging Technologies*, Vol. 20, No. 1, 2012, pp. 3–14.
22. Cascetta, E., and A. Carteni. A Quality-Based Approach to Public Transportation Planning: Theory and a Case Study. *International Journal of Sustainable Transportation*, Vol. 8, No. 1, 2014, pp. 84–106.
23. Deakin, E., M. Payne, and V. Menotti. Development of Bay Area Rapid Transit System Expansion Criteria and Process. *Transportation Research Record: Journal of the Transportation Research Board*, 2004. 1887: 18–25.
24. Zemotel, L, and D. Montebello. Interregional Corridors: Prioritizing and Managing Critical Connections between Minnesota's Economic Centers. *Transportation Research Record: Journal of the Transportation Research Board*, 2002. 1817: 79–87.
25. Daganzo, C. F. Structure of Competitive Transit Networks. *Transportation Research Part B: Methodological*, Vol. 44, No. 4, 2010, pp. 434–446.
26. Badia, H., M. Estrada, and F. Robusté. Bus Network Structure and Mobility Pattern: A Monocentric Analytical Approach on a Grid Street Layout. *Transportation Research Part B: Methodological*, Vol. 93, Part A, 2016, pp. 37–56.
27. Viggiano, C. *Bus Network Sketch Planning with Origin–Destination Travel Data*. PhD Thesis, Massachusetts Institute of Technology, Cambridge, MA, 2017.
28. Viggiano, C., H. N. Koutsopoulos, N. H. M. Wilson, and J. Attanucci. Journey-Based Characterization of Multi-Modal Public Transportation Networks. *Public Transport*, Vol. 9, No. 1–2, 2017, pp. 437–461.
29. Wang, F., and Y. Xu. Estimating O–D Travel Time Matrix by Google Maps API: Implementation, Advantages, and Implications. *Annals of GIS*, Vol. 17, No. 4, 2011, pp. 199–209.
30. Viggiano, C., H. N. Koutsopoulos, J. Attanucci, and N. H. M. Wilson. Inferring Public Transport Access Distance from Smart Card Registration and Transaction Data. *Transportation Research Record: Journal of the Transportation Research Board*, 2016. 2544: 55–62.
31. Bus Priority Team. *Accessible Bus Stop Design Guidance*. Transport for London, London, United Kingdom, 2006.
32. York, I. O. Factors Affecting Bus-Stop Times. *TRRL Project Report 2*. Technical Report, Transport and Road Research Laboratory, Wokingham, United Kingdom, 1993.
33. Robinson, S. Measuring Bus Stop Dwell Time and Time Lost Serving Stop with London Ibus Automatic Vehicle Location Data. *Transportation Research Record: Journal of the Transportation Research Board*, 2013. 2352: 68–75.
34. Ortuzar, J., and L. G. Willumsen. *Modelling Transport*. John Wiley & Sons, Hoboken, NJ, 2011.
35. Wardman, M. Review and Meta-Analysis of UK Time Elasticities of Travel Demand. *Transportation*, Vol. 39, No. 3, 2012, pp. 465–490.

The Standing Committee on Bus Transit Systems (AP050) peer-reviewed this paper (18-04936).

Article

TRR
JOURNAL OF THE TRANSPORTATION RESEARCH BOARD

Transportation Research Record
2018, Vol. 2672(8) 86–95

Article reuse guidelines:
sagepub.com/journals-permissions
DOI: 10.1177/0361198118796717
journals.sagepub.com/home/trr
SAGE

A Data-Driven Approach to Prioritizing Bus Schedule Revisions at New York City Transit

Malik Coleman[1], Lauren Tarte[1], Steve Chau[1], Brian Levine[2], and Alla Reddy[1]

Abstract

Over two million trips are taken every weekday across the New York City Transit (NYCT) bus network. Revising the schedules for each of these routes is a labor-intensive process, and because of limited resources, fewer than half of all routes are examined each year. Traditionally, schedules have been revised on a first-in, first-out basis, with most schedules rewritten once every 2 years. This approach leaves no room for reviewing routes that need more frequent changes, meaning service may not catch up to changes in demand or traffic patterns for several years. It also requires staff to spend valuable staff time analyzing routes that may be inactive. To better address rapidly evolving bus corridors, NYCT developed a methodology to pinpoint the routes most in need of schedule revisions. This data-driven approach uses automatic vehicle location data and a ridership algorithm that combines automated fare collection data with other sources to infer stop-by-stop boardings and alightings for individual trips. Each route's schedule is evaluated on service capacity versus actual ridership, and scheduled versus actual running times. Routes that show the greatest discrepancies are designated for later in-depth review. This methodology was applied to develop the 2018 list of schedule revisions. As this process identifies routes with too much capacity or running time, as well as those with too little, resource-costly schedule adjustments can be offset with resource-saving ones. Using this approach allows scheduling staff to react more quickly to changes in customer demand and new development, thereby providing better service to passengers.

The Metropolitan Transportation Agency (MTA) operates the largest bus fleet in the U.S. (*1*) and one of the most extensive bus networks in the world, with 315 routes moving over 2.4 million passengers each weekday (*2*, *3*). New York City Transit (NYCT) operates the majority of New York City bus routes, and is the focus of this paper. MTA Bus Company, a separate agency under the MTA, is responsible for a smaller set of routes. The NYCT bus network functions as both an important feeder to the subway system and as a primary transportation service in the many areas lacking proximate rail transit, particularly outside of Manhattan. In recent years, new development and shifts in neighborhood demographics, as well as changes to other transportation modes have affected both ridership patterns and bus speeds on many routes (*4*). Adjusting the bus service to keep up with these changes has been facilitated the move by NYCT from manually collected data to automatically collected data sources.

Historically, bus planning and scheduling processes have relied on manually collected data by traffic surveyors; a source that is both time-consuming and costly to collect and process, and which typically yields small sample sizes. The implementation of bus time, the automatic vehicle location (AVL) system used by MTA across the full bus network provided a complete source of bus running time and performance data (*5*). By integrating bus time data with automated fare collection (AFC) data, NYCT developed an inference model that determines passenger boarding and alighting locations at the individual stop and trip level (*6*). Using this algorithm, a significantly larger (and therefore more representative) ridership dataset can be generated than manual traffic checks can provide. Data from this model has since been used for service planning (*7*) and for revising

[1]Data, Research, Development (DRD), Operations Planning, New York City Transit Authority, New York, NY
[2]Strategic Analysis & Forecasting, Aviation Department, Port Authority of New York & New Jersey, New York, NY

Corresponding Author:
Address correspondence to Brian Levine: Blevine@panynj.gov

individual bus schedules. The efforts described in this paper extend the applications of the ridership model data to prioritizing the selection of bus schedules that are revised each year.

Background

Bus Scheduling at New York City Transit

NYCT operate 235 bus routes which constitute over 650 distinct schedules. Each route has different schedules for weekdays, Saturdays, and Sundays, and for a minority of routes, separate schedules are developed for the summer months to accommodate the increase in weekend leisure trips and decrease in student ridership. Some routes also have different schedules on weekdays when schools are closed. Before the implementation of a prioritization methodology for revising bus schedules, a set of standards dictated how frequently routes were analyzed for potential schedule revisions. For weekday schedules, the minimum review period was two years, while weekend schedules needed to be revised only once every four years and express routes were checked yearly. These standards were developed well before the advent of automated data sources, when schedule revisions relied wholly on manually collected data. A major constraint that led to these specific requirements was the amount of staff time needed to capture the necessary ridership and performance data. Manual data collection aboard buses was an extremely time-intensive effort, particularly given the long travel times to reach survey locations for routes in the outer boroughs. Using automatically collected data sources removes this limitation, and has opened the door to re-examining the timeline by which schedules are reviewed.

The annual schedule revision process begins by developing a list of schedules to be revised the following year. In the past, this list was almost entirely comprised of routes due for review by the fixed two and four-year cycles, although scheduling staff had some leeway to include routes based on anecdotal knowledge. The revisions on this list are then distributed throughout the year, to be completed in time for regular changes in bus operator assignments. Operator work assignments are reapportioned quarterly in a procedure by which bus drivers, in order of seniority, pick their run for the next three months. Each of these four "picks" is also an opportunity on the planning side to change route paths, stops, and schedules. Routes selected for revision may be allocated into any one of these four picks. Ideally, schedule revisions are distributed within the year so that bus additions and reductions balance out, and to spread each season's work as evenly as possible across staff members.

The schedule-making process entails analyzing the most recent available ridership and running time data, and making modifications to the current schedule as necessary. Loading guidelines set minimum trip frequencies and maximum bus loads for given passenger volumes and vary by bus type, route type, and time of day. During peak hours, trips are scheduled to 100% of a bus's physical capacity, ensuring that all seats and standing room are used. By contrast, schedules for off-peak time periods are written to ensure most passengers get a seat. Thus, when discussing off-peak hours, the term "capacity" does not refer to the maximum number of passengers a bus can carry, but rather the desired maximum load as set by the guideline. When ridership increases above guideline capacity, service frequencies are increased, even if the existing trips could accommodate more standing passengers. Running times are examined for both end-to-end changes and intra-trip adjustments.

Writing schedules is now a process highly dependent on NYCT's suite of automatically collected data sources. For running time calculations, AVL data from the NYCT bus time system provides actual bus locations at a 30-second frequency. An algorithm then matches this data with locations of timepoints from the schedule to calculate running times. Moving from manually collected to AVL data was the first of many improvements to the schedule-making process facilitated by new, automatically collected data streams.

Further enhancements followed with the development of the NYCT ridership model. This algorithm, built entirely in-house, combines AVL data, AFC data from the MetroCard system, and general transit feed specification (GTFS) schedule data to provide detailed, stop-level boarding and alighting data for every trip on a given day. Further inputs account for fare evasion and cash payments, to ensure 100% of passengers are represented. Like most bus systems, NYCT lacks exit data, and uses next-swipe inference to determine alighting locations. In its current form, several manual inputs and adjustments are required to run the model for a given dataset (typically one month). Owing to the time required to prepare the input data, and because schedule-making and service planning primarily use data from high-ridership months only, the model is used to generate ridership data for a limited range of dates; typically one to two months in the spring and fall, when passenger volumes are highest, and a similar timeframe in summer, which is subject to distinctly different ridership patterns.

Automated ridership data is also combined with manually collected data gathered by traffic surveyors. While manual traffic checks have decreased in number with the advent of the ridership model, they are still used to supplement model data in cases when AFC data may have large errors, such as routes with large fractions of passengers using cash, or with higher fare evasion rates. Traffic checks were also used while developing the model to prove its accuracy, and continue to be used periodically to ensure the validity of the automated ridership data.

These automatically collected data sources have had wide-ranging applications across the agency. In recent years, several intensive bus studies have leveraged the highly detailed ridership data provided by the model to improve bus service in a focused geographic area (*7*). The first of these covered ten distinct routes in Co-Op City, a neighborhood in the Bronx, in 2013. Detailed ridership data showed that the most underserved origin-destination pairs were on trips taken to other neighborhoods. This finding contradicted anecdotal reports of insufficient intra-neighborhood service. Minor changes were implemented to several routes as a result, including added stops and one re-route. Following positive feedback from the Co-Op City community, a second study was conducted in Northeast Queens in 2015. Larger in scale, this effort revealed the most important bus corridors in the area, and confirmed that these routes primarily serve as feeders to the subway system. Several corridors were identified as possibilities for future Select Bus Service upgrades. NYCT is currently conducting its most extensive bus study to date in Staten Island. The Staten Island Bus Study involves analyzing 51 local, limited, and express bus routes that serve a borough of over 450,000 people. The first phase of results proposes a major restructuring of express routes serving the island (*8*). The express route network will be streamlined and simplified, and end-to-end running times will be reduced by an average of 20 minutes. Consequently, most passengers will experience more frequent service, faster trips, and improved reliability. These studies allow bus planners and schedulers to focus on changing routings and headways for multiple routes at the same time, and using detailed ridership data illuminates the finer nuances of passenger demand.

While the previous process of devising schedules has seen many refinements with the introduction of automatically generated ridership and running time data, by contrast the preliminary step of selecting routes for review was not data-driven at all. The fixed, first-in, first-out cycle was not sufficiently reactive to change. Traffic speeds may decline, and development may occur in between review periods. Yet, a route experiencing these types of changes was only revised once the two to four years has elapsed since the prior review. With the exception of major service changes (such as route splits or upgrading a route to Select Bus Service), routes were not analyzed other than when scheduled for review.

This rigid timeline also caused scheduling staff to split their time equally between routes, rather than allocating more time to routes that have significant issues. Even if a route had little or no change in ridership and running times, schedule managers were required to spend time analyzing this data, only to conclude that no changes were needed. This limited the time they could spend on routes that needed major changes. The methodology in this paper was developed to improve the first step of the annual schedule revision process, by ensuring that the routes placed on each year's review list are those most in need of changes.

Bus Scheduling at Other Agencies

According to a 2009 USDOT-sponsored report (*9*), 53% of agencies monitor bus route performance on a monthly basis, although the survey did not distinguish between monitoring for the purpose of performance reporting and monitoring for service evaluation. Five percent of respondents stated that performance measures were not used for service evaluation at all. Out of the 59 agencies, only NYCT and St. John's County (FL) evaluated on a less than annual basis. As for ranking and prioritizing routes, AC Transit (CA) analyzes routes that perform at the 25th percentile or below on annual subsidy or passengers per hour. Madison Metro puts routes that perform below 60% of the average for three consecutive quarters on a review list. Most agencies have a more frequent review process than NYCT, but also have fewer routes and trips to review. At a fleet size of over 4,000 vehicles, NYCT was by far the largest agency included in the study, with the next largest having fewer than 1,100 vehicles.

Case Studies: Routes B32, M20, and S93

Three routes in the NYCT system with notable demand changes over the past five years are the B32, M20, and S93. The B32 and S93 had major increases in ridership, while the M20 had a sizeable decrease. All three routes also had their service revised after a periodic review under the previous process. These routes were used to demonstrate how the new methodology reduces the lag between changes in a route's performance and consequential schedule updates.

The B32 serves Williamsburg and Greenpoint in Brooklyn, and Long Island City in Queens. This route was introduced in September 2013 with the intent of meeting the increasing demand in these waterfront neighborhoods, which are growing both in commercial and residential demand, partially owing to increased development (*10*). An existing route, the B62, provides some service to this area, but travels three blocks further east from the waterfront (Figure 1*a*).

As shown in Figure 2, ridership on the B32 increased 10% year over year in both 2015 and 2016. One northbound trip in the AM peak was added in July 2017,

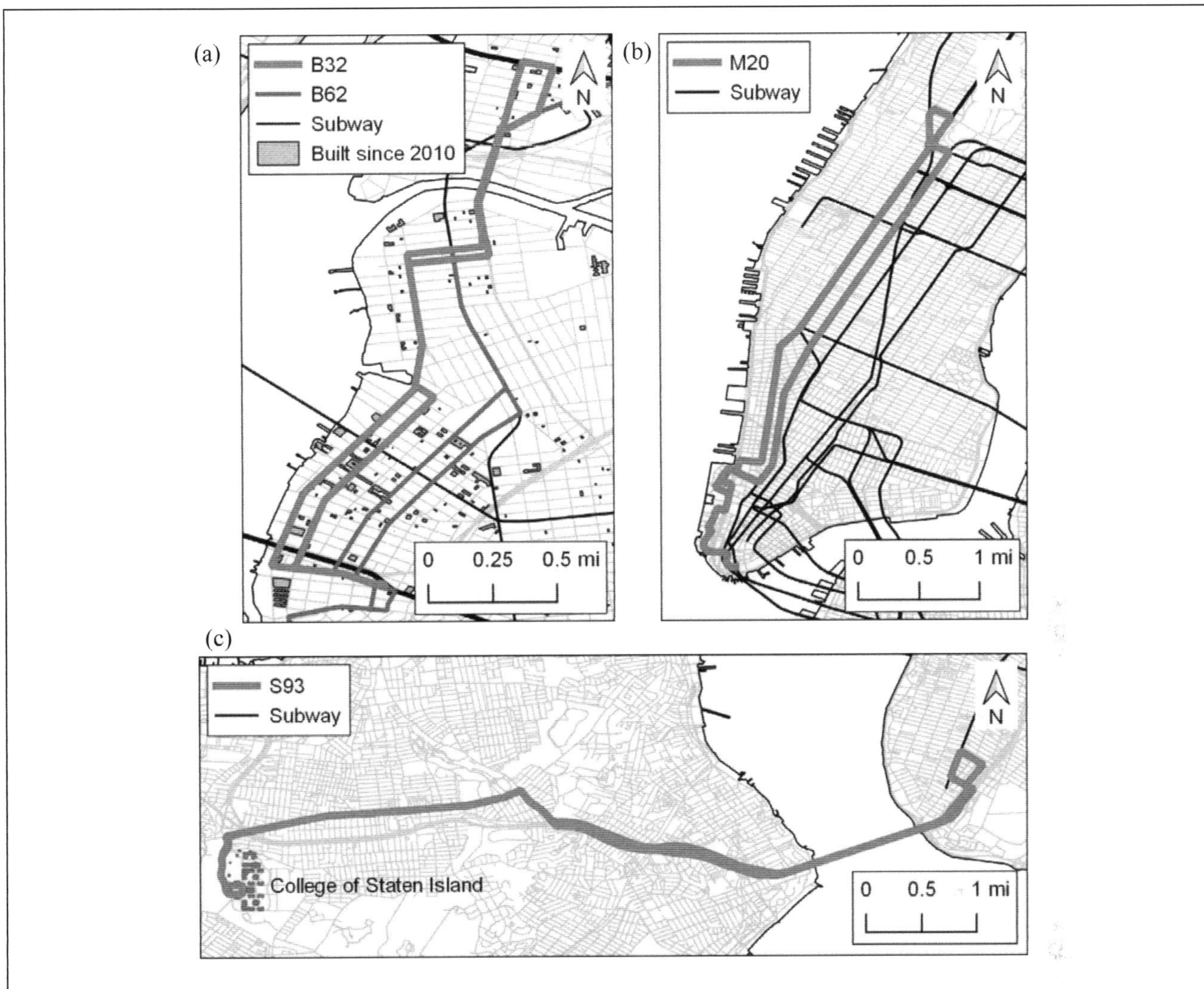

Figure 1. Maps of the (*a*) B32, (*b*) M20, and (*c*) S93.

although this change trailed the ridership increases. In this same period, ridership decreased 7% per year on the B62.

In contrast to the growing B32 is the M20, which runs in Manhattan from 66th Street down to South Ferry at the southern tip of Manhattan (Figure 1*b*). The M20 is largely parallel to the subway service on the 1 line. As the 1 line is a local service, its stations are only five to eight blocks apart, and its higher speed and frequency makes it more attractive for most riders. However, few of the nearby 1 line stations have elevators, sending passengers with disabilities or those carrying large items to the M20 instead. In contrast, all NYCT buses are fully accessible.

Ridership on the M20 has declined severely in the past five years (Figure 2). Annual ridership in 2016 was down 40% from 2011, with double-digit decreases in each of the past three years. Service frequencies were cut from four to three buses per hour during peak periods only recently, despite this steady decline. This reduction in service interacts in a negative feedback loop with passenger volumes, as longer headways discourage potential passengers, and dropping passenger volumes incurs further service cuts.

A final example of a route with growing passenger volumes is the S93. This route runs between Staten Island and Brooklyn across the Verrazano-Narrows Bridge (Figure 1*c*), with limited stops. It is one of only four transit connections between the island and the rest of New York City. Two other bus routes (the S53 Local and S79 Select Bus Service) connect to the NYCT subway system in Brooklyn, and the Staten Island Ferry, a free passenger ferry, runs to Manhattan every 15–30 minutes.

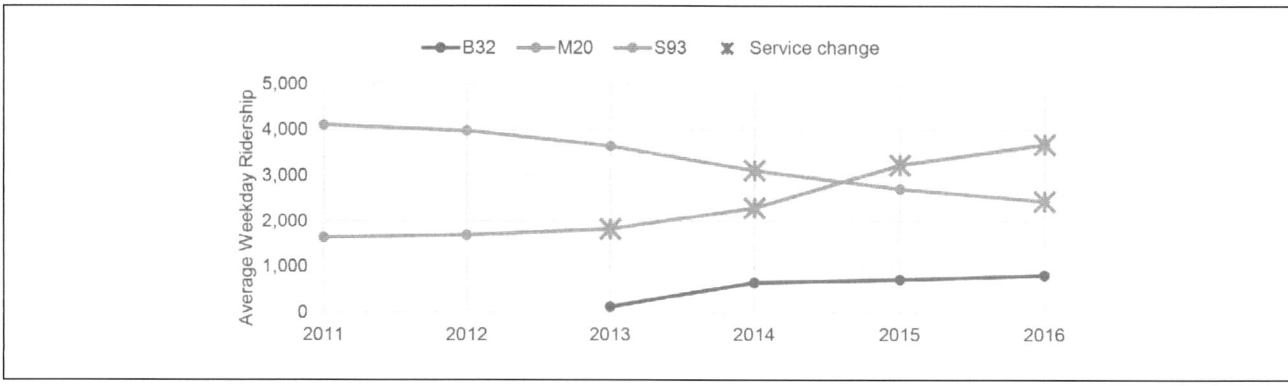

Figure 2. Average weekday ridership per year for the B32, M20, and S93.

The S93's western terminal is at the College of Staten Island (CSI), a campus of 13,775 students. Whereas CSI runs shuttle buses from the ferry to campus, the S93 is a major transportation link for students and others to buses and subway lines in Brooklyn. In January 2013, the S93 was extended to the inside of CSI, from its previous terminal a half-mile from the main building. In September 2014, a midday service was added. These two changes helped ridership more than double between 2012 and 2016 (Figure 2).

The S93 is a unique exception in the context of the previous revision process, in that it has been privileged with consecutive schedule changes over the past few years to keep pace with its growth in passenger volumes (Figure 2). This is partially because of extra efforts made by scheduling staff when a route is known to have above-average growth. Schedule managers often conduct supplementary passenger counts themselves, going out in person to record bus performance and crowding. These observations may be prompted by unofficial reports of overcrowding from staff on the ground in the Road Operations department. Although this level of involvement does produce more accurate bus schedules, it cannot be relied on to identify all routes needing frequent revisions. It is heavily dependent on anecdotal knowledge of a route's ridership by either the scheduling department themselves or Road Operations. In addition, attention is rarely called to routes with declining ridership, such as the M20. Furthermore, this approach is a notable drain on scheduling staff's time. A comprehensive, objective means of determining high-need routes as presented here allows staff to spend their time more efficiently, and review a greater number of routes each year.

Methodology

The approach described here is designed to be applied to every route and for the weekday, Saturday, and Sunday schedules (each schedule is evaluated independently, though this paper uses the term "route" for simplicity). The final output is a list of the approximately 650 schedules, ranked from most in need of revision to least. In practice, this process is carried out once per year as part of the annual budgeting exercise, though it may be run more often if required.

Design Objectives

An additional aim of the project, further to simply creating a list of routes recommended for revision, was to ensure such recommendations were feasible, given the resource constraints at NYCT. The size of the bus fleet is a hard constraint in schedule revisions, and for each revision cycle, ensuring that all changes are net resource-neutral is a principal objective. This meant that a set of recommendations should not consist solely of adding services, as NYCT is currently bus constrained in peak hours. Instead, recommendations for additional trips must be balanced wherever possible with reductions elsewhere. In addition, peak hour service is more important than off-peak service, both because the full bus fleet operates during peak hours, and also because issues during this time naturally affect the greatest number of riders.

These aims are accomplished by judicious filtering after the calculations are complete. If more routes require bus additions than require bus reductions, some additions are delayed to a future pick. Service reductions, however, do not need to be balanced with corresponding additions. Where feasible, bus changes should also be balanced by division or depot, although this is a lower priority. NYCT has 20 depots, in six divisions, in the five boroughs of New York City, and changing routes and buses between depots is sometimes necessary, but not simple.

Performance Measures

The methodology evaluates schedules on two metrics: the ratio of passenger volumes to capacity (V/C), and the difference between actual and scheduled running times. Volumes are calculated using data from the bus ridership model, which provides the number of passengers boarding and alighting at every stop along each bus trip on a given day. The dates selected for the volume versus capacity analysis include one month in spring (typically May) and one month in fall (typically October). Summer months are not included owing to notably different ridership patterns in summer. For different routes, the spring pick may have a different schedule than the fall period, so the ridership and trip counts per route, trip type, direction, and time period may not match. The ridership is calculated separately for each pick and then averaged across the two picks.

Ridership is aggregated to yield average stop-level ridership by route, trip type, and direction, in a manner consistent with the schedule-making process. The passenger ridership at the peak load point for every 30 minute time period is then calculated.

This peak load point ridership is then compared to the capacities that are established by the NYCT loading guidelines. Using the actual trip counts during each time period, the provided capacity from the guidelines is compared to the actual ridership volume. For each time period, the ratio of average daily ridership at the maximum load point for each route per direction to its corresponding capacity is then calculated. The ratios are calculated separately for weekdays, Saturday, and Sunday. For weekday data, Monday and Friday are not used because of lower ridership. These ratios are the metrics used for the volume capacity portion of the new bus schedule prioritization analysis.

Secondly, running time metrics are calculated by subtracting the actual end-to-end running time in minutes from the scheduled end-to-end running time. Running time data is taken from the most recently completed season (typically winter). The scheduled running time can then be classified as insufficient (positive running time difference), surplus (negative running time difference), or neither (when the difference is zero). Similar to volume-to-capacity ratios, running time metrics are aggregated by route, trip type, direction and time period, then averaged across the day.

It should be noted that the two metrics use data from different temporal scopes. Generating ridership data is a time-intensive procedure that requires seasonal tweaking and validation. Thus, data is only available after weeks of processing and only for select months. AVL data, on the other hand, needs much less computation and is processed daily on a 72-hour lag. This means that much more recent data is available for running times than for ridership. Using the most current running time data, rather than data from the same time frame as for capacity calculations, was a deliberate choice to ensure that recent schedule changes are incorporated.

Time Periods

Capacity and running times are evaluated over each route's full span of service, although different times of day are evaluated at different levels of detail. As bus performance can fluctuate drastically in a short time span, standard multi-hour time periods are too broad to illuminate specific time periods that will benefit from schedule tweaks. Therefore, performance measures are evaluated in narrow time "slices" of 15–60 minutes. Peak-hour time slices are half the duration of off-peak time slices, which allows for finer-grained analysis of peak trips. Capacity is measured using 30 minute time slices during peak hours, and 60 minute slices during the off-peak. The length of the capacity time slices mirror those used in the schedule-making process. For running times, peak and off-peak time slices are 15 and 30 minutes long, respectively. This higher level of detail is because of the greater precision of AVL data, and the high variance of traffic speeds in New York City.

Performance Thresholds

For both the capacity and running time measures, routes may "fail" for two possible reasons: having insufficient capacity or scheduled running time, or having surplus capacity or running time. Therefore, all routes are examined in four distinct categories. Identifying routes with too much capacity or running time allows buses to be reallocated to routes with too little of either, to satisfy the resource-neutral objective. The thresholds for failure are given in Table 1.

It is possible for a route to have both too much and too little running time. This may happen when the current schedule has too much running time in off-peak periods, but has too little running time in peak periods. It is

Table 1. Failure Thresholds for Performance Measures

	Capacity	Running time
Insufficient	Volume > 100% of capacity	Actual running time > 2 min. above schedule
Surplus	Volume < 50% of capacity	Actual running time < 2 min. below schedule

also possible if the total running time is not split between the directions correctly, with too much running time in one direction and not enough in the reverse. Similarly, some routes may have volumes higher than scheduled capacities in the peak hours, but are well under capacity in off-peak periods.

Scoring Methodology

Every route is scored on a scale of zero to four in each of the above categories. Four indicates high priority, and is used when a route has significant problems. Minor issues warrant fewer points, and no or negligible issues receive zero points. Zero points does not mean that a route is perfect in that category, only that any discrepancies are extremely minor. The scoring criteria are given in Table 2. In all categories, issues in the peak period are weighted heavier than issues in the off peak period, consistent with the previously stated objective of prioritizing peak service. These criteria may be adjusted in the future, as needs change or the prevalence of severe problems falls.

The categories are weighted evenly, then combined to give a summary number of the route's current condition. The categories are combined multiplicatively to produce each route's final ranking:

$$Rank = (C_i + 1)(C_s + 1)(R_i + 1)(R_s + 1)$$

In which C_i, C_s, R_i, and R_s are as given in Table 2.

Multiplying, rather than adding, the category scores ensures that routes with a smaller number of major issues are given higher priority than routes with many small issues. For example, a route with four points in both the excess running time and too little capacity categories is higher ranked than a route that has two points in every category. This helps prioritize routes in which changes by bus schedules will have the highest impact.

Other Considerations

Although this approach provides a purely data-driven means of selecting routes for review, it is not the only avenue by which schedules are updated. The bus planning department also adds routes to the final list each year at their discretion. Such routes are the subject of in-depth studies that include field work and public outreach, and that typically result in service changes along with schedule revisions. These include route splits or combinations, Select Bus Service introductions, or bus lane additions.

In other cases, routes may merit service reductions according to the prioritization process, but already operate at the NYCT minimum frequency guidelines. Frequencies on these routes cannot be reduced regardless of their surplus capacity score, so they are removed before providing the list to the bus scheduling group. Finally, there is a ceiling on the number of years a route can go without being analyzed. Once a route hits the ceiling, it must be looked at in detail, even if it does not get a high priority ranking, and may need no changes. This is to avoid routes remaining unrevised for multiple years.

Application

This section describes the application of the above methodology to the three case study routes. A complete implementation of the prioritization process involves analyzing every route in the bus network across the full span of the

Table 2. Thresholds for Prioritization Point Categories

Points	Insufficient capacity C_i	Surplus capacity C_s	Insufficient running time R_i	Surplus running time R_s
4	More than 10% of peak time slices have V/C ratio >1	More than 33% of peak time slices have V/C ratio <0.5	More than 33% of time slices have median RT > 2 min. late	More than 33% of time slices have median RT > 2 min. early
3	One or more peak time slice(s) have V/C ratio >1	Between 20–33% of peak time slices have V/C ratio <0.5	Both peak time periods have median RT > 2 min. late (sum of both directions)	Both peak time periods have median RT > 2 min. early (sum of both directions)
2	More than 10% of off-peak time slices have V/C ratio >1	More than 33% of off-peak time slices have V/C ratio <0.5	One peak time period has median RT > 2 min. late (sum of both directions)	One peak time period has median RT > 2 min. early (sum of both directions)
1	One or more off-peak time slice(s) have V/C ratio >1	Between 20–33% of off-peak time slices have V/C ratio <0.5	More than 25% of time slices have median RT > 2 min. late	More than 25% of time slices have median RT > 2 min. early

Note: V/C = ratio of passenger volumes to capacity; RT = running time.

ROUTE (Peak Dir)	7:00 - 7:29	7:30 - 7:59	8:00 - 8:29	8:30 - 8:59	16:00 - 16:29	16:30 - 16:59	17:00 - 17:29	17:30 - 17:59	18:00 - 18:29	18:30 - 18:59
B32 (NB)	0.8	1.1	1.4	1.3			0.4			
B32 (SB)					0.3	0.4		0.4	0.6	0.5
M20 (NB)	0.9	0.6	0.4	0.4	0.4	0.5	0.5	0.4	0.3	0.4
M20 (SB)										
S93 (EB)			0.6		0.5	0.8			0.6	0.7
S93 (WB)	0.6	0.6		0.3			0.6	0.6		

Figure 3. Excerpt from volume-to-capacity ratios by time slice report (data from May and October 2016).

ROUTE (Dir)	7:00 - 7:14	7:15 - 7:29	7:30 - 7:44	7:45 - 7:59	8:00 - 8:14	8:15 - 8:29	8:30 - 8:44	8:45 - 8:59
B32 (NB)	-4.0		-0.5		0.9		-1.0	
B32 (SB)	-5.8		-2.5		-0.9		0.5	
M20 (NB)	-1.0	-9.9		-11.3	-6.4	-5.4		-6.2
M20 (SB)	2.2		-2.0	0.8	5.2	8.8	-1.8	
S93 (EB)	10.5	6.5	9.8	3.9	6.4		3.1	
S93 (WB)	1.1	0.4	1.1	-1.7	-2.6	-0.9	0.3	-1.0

Figure 4. Excerpt from end-to-end running time deviation report (data from January–March 2017).

service. The full prioritization process was executed in spring 2017 to develop the bus schedule department's workload and budget for 2018. The data and figures below are excerpts from that exercise.

As described in the Performance Measures section above, running time differentials can be produced using much more current data than is available for developing volume-to-capacity ratios. This application is based on ridership data from May and October 2016, and running time data from winter 2017. Figure 3 shows an intermediate output of the prioritization report for the B32, M20, and S93. The full report shows volume-to-capacity ratios for all time slices in the day. For each time slice, only the peak direction is shown, as capacity is not analyzed in the off-peak direction. Time periods highlighted in grey, such as the B32 southbound from 16:00–16:29, are already scheduled at minimum frequencies.

The B32 shows a ridership over the guidelines between 7:30 and 8:59, highlighted in red. Using older data, this report shows overcrowding as far back as 2014, the first full year of the route. The new prioritization process would have highlighted the need for extra service sooner.

Despite the 2016 service cuts, the M20 still has surplus capacity (Figure 3). In fact, as far back as 2014, ridership was low enough that cutting services was feasible in 40% of peak time slices. By 2015, this fraction rose to 80%, and remained that high in 2016.

Meanwhile on the S93, ridership growth has outpaced service increases, leading to heavier average loads. The percentage of time slices with average loads at or above 75% of guidelines has increased from 14% to 41%, and the percentage of time slices with surplus capacity (as defined in Table 1) has decreased from 23% to 9%.

Figure 4 shows an extract from the running time report, highlighting some of the AM peak issues with these three routes. For example, the B32's AM peak actual and scheduled running times match closely, with only a moderate reduction needed between 7:00 and 7:14. Overall running times on the M20 deteriorated between 2014 and 2016. In 2014, only 12% of time slices averaged more than two minutes late, that percentage increased in 2015 and again in 2016, growing to 22% of time slices.

While the S93 has only small discrepancies in the westbound direction, it lacks sufficient running time

Table 3. Final Rankings

Route	Insufficient capacity C_i	Surplus capacity C_s	Insufficient running time R_i	Surplus running time R_s	Rank
B32	4	0	0	4	25
M20	0	4	0	4	25
S93	2	0	0	2	9

Table 4. Weekday Headways in Minutes for Selected Hours

(a) B32	7:00–8:00	8:00–9:00	9:00–10:00
2016 Fall	30	30	30
2017 Fall	30	20 ↓	30

(b) M20	7:00–8:00	8:00–9:00	16:00–17:00	17:00–18:00
2015 Fall	20	15	15	15
2017 Fall	20	20 ↑	20 ↑	20 ↑

(c) S93	7:00–8:00	8:00–9:00	9:00–10:00	10:00–11:00
2015 Fall	10	8	10	30
2017 Fall	9 ↓	7 ↓	12 ↑	30

eastbound. Eastbound is the peak direction in the morning, and traffic congestion over the Verrazano-Narrows Bridge is common. A high-occupancy vehicle (HOV) lane was installed on the bridge in June 2017, which should reduce running times. However, running times may still need to be adjusted if the HOV lane does not provide sufficient improvement in running times.

Aggregate rankings for each route are calculated as shown in Table 3. The B32's high AM peak crowding results in an insufficient capacity score of 4. The B32 would also merit 4 points in the surplus capacity category, but as scheduled frequencies during these time slices are already at minimum guidelines, no points are awarded. The M20 had ridership much lower than guidelines in all periods, and therefore receives 4 points in the surplus capacity category. The B32 and the M20 each net 4 points for surplus running time, as both had running times faster than scheduled in multiple time slices. The S93 had high ridership in some off peak periods, and ran faster than scheduled in a few time periods, resulting in scores of 2 points in both the insufficient capacity and surplus running time categories. The overall rankings, shown in the last column, are calculated according to the previously given equation. Both the B32 and the M20 had a final rank of 25 points, and were recommended for revision.

Table 4 shows the changes in average headways for the three case study routes. The dates of some of these changes were previously shown in Figure 2. Selected time periods are included to show changes, but schedules were adjusted in all time periods where needed. The bulk of changes resulting from the prioritization process, for all routes selected for review based on their final rankings, will be implemented in 2018.

Conclusions and Future Bus Scheduling Work

By developing a prioritization process for bus schedule review, NYCT has taken an important step in making the schedule revision process faster and more responsive to changing conditions. Although it is a less aggressive approach than some peer agencies, this change is a necessary step in moving a large agency to better serve its customers. By using the new methodology, growing ridership on expanded services, as well as lower ridership owing to service cuts, can both be closely monitored and responded to in a more dynamic manner.

Future enhancements include determining routes that need changes in service frequency and running times on a quarterly basis instead of making a list only once per year. This would allow more flexibility and more timely revisions to routes in which conditions are changing rapidly. Additionally, with better and more frequent data on ridership and running times, schedules can be more readily aligned with differences in ridership patterns that occur by days of the week.

Author Contributions

The authors confirm contribution to the paper as follows: study conception and design: Malik Coleman, Lauren Tarte, Brian Levine, Steve Chau, Alla Reddy; data collection: Malik Coleman, Lauren Tarte, Brian Levine, Steve Chau; analysis and interpretation of results: Lauren Tarte, Malik Coleman, Brian Levine; draft manuscript preparation: Lauren Tarte and Malik Coleman. All authors reviewed the results and approved the final version of the manuscript.

References

1. National Transit Database Office of Budget and Policy. *Transit Profiles: 2016 Top 50 Agencies Summary*. Federal Transit Administration, U.S. Department of Transportation, Washington, D.C., 2017. https://cms.fta.dot.gov/ntd/2016-top-50-summary-and-complete-profile-set-top-50-agencies. Accessed November 13, 2017.
2. Metropolitan Transportation Authority. *The MTA Network*. December 31, 2015. http://web.mta.info/mta/network.htm. Accessed June 27, 2017.
3. Metropolitan Transportation Authority. *Transit & Bus Committee Meeting*. March 20, 2017. http://web.mta.info/mta/news/books/pdf/170320_1030_Transit.pdf. Accessed June 28, 2017.
4. Meyer, D. Decline in NYC Bus Ridership Concentrated in Manhattan and Brooklyn. *Streetsblog NYC*. February 21, 2017. https://nyc.streetsblog.org/2017/02/21/decline-in-nyc-bus-ridership-concentrated-in-manhattan-and-brooklyn/. Accessed November 13, 2017.
5. Levine, B., S. Iyer, and A. Reddy. Development of an Automated Vehicle Location (AVL) Data System to Improve Bus Service at New York City Transit. Presented at 93rd Annual Meeting of the Transportation Research Board, Washington, D.C., 2014.
6. Zeng, Q., A. Reddy, A. Lu, and B. Levine. Development of Application for Estimating Daily Boarding and Alighting Counts on New York City Buses. *Transportation Research Record: Journal of the Transportation Research Board*, 2015. 2535: 1–14.
7. Hanft, J., S. Iyer, B. Levine, and A. Reddy. Transforming Bus Service Planning Using Integrated Electronic Data Sources at NYC Transit. *Journal of Public Transportation*, Vol. 19, No. 2, 2016, pp. 89–108. http://dx.doi.org/10.5038/2375-0901.19.2.6.
8. Metropolitan Transportation Authority. *Reimagining Staten Island Bus Service*. May 2017. http://nymta.civicconnect.com/sites/default/files/MTA%20Presentation%205-31-07%20Final.pdf. Accessed August 1, 2017.
9. Mistretta, M., J. A. Goodwill, R. Gregg, and C. DeAnnutis. *Best Practices in Transit Service Planning, National Center for Transit Research*. University of South Florida, Tampa, FL, 2009.
10. Metropolitan Transportation Authority. *New Bus Service Coming to Brooklyn and Queens Starting*. September 8. http://web.mta.info/nyct/service/NewBusServiceComing_BrooklynQueens_September2013.htm. Accessed May 16, 2017.

The Standing Committee on Bus Transit Systems (AP050) peer-reviewed this paper (18-04069).

Article

TRR
JOURNAL OF THE TRANSPORTATION RESEARCH BOARD

Transportation Research Record
2018, Vol. 2672(8) 96–107

DOI: 10.1177/0361198118772953
journals.sagepub.com/home/trr
SAGE

Planning of Fast-Charging Stations for a Battery Electric Bus System under Energy Consumption Uncertainty

Zhaocai Liu[1], Ziqi Song[1], and Yi He[1]

Abstract

Battery-powered electric buses are gaining popularity as an energy-efficient and emission-free alternative for bus fleets. However, battery electric buses continue to struggle with concerns related to their limited driving range and time-consuming recharging processes. Fast-charging technology, which utilizes dwelling time at bus stops or terminals to recharge buses in operation employing high power, can raise battery electric buses to the same level of capability as their diesel counterparts in terms of driving range and operating time. To develop an economical and effective battery electric bus system using fast-charging technology, fast-charging stations must be strategically deployed. Moreover, due to the instability of traffic conditions and travel demands, the energy consumption uncertainty of buses should also be considered. This study addresses the planning problem of fast-charging stations that is inherent in a battery electric bus system in light of the energy consumption uncertainty of buses. A robust optimization model that represents a mixed integer linear program is developed with the objective of minimizing the total implementation cost. The model is then demonstrated using a real-world bus system. The performances of deterministic solutions and robust solutions are compared under a worst-case scenario. The results demonstrate that the proposed robust model can provide an optimal plan for a fast-charging battery electric bus system that is robust against the energy consumption uncertainty of buses. The trade-off between system cost and system robustness is also addressed.

The electrification of bus fleets has been widely considered as an effective method of abating local emissions, improving fuel economy, and reducing oil dependence in conventional diesel-powered bus systems (*1–3*). Battery electric buses (BEBs) are among the most popular electric buses today. Compared to trolleybuses, which are attached to continuous overhead wires along the entire bus line, BEBs are more flexible in their operation. However, due to the limitations of battery technology, the energy density of batteries for BEBs is rather low when compared to diesel. BEBs therefore suffer from the disadvantages of cumbersome and costly on-board batteries and limited driving ranges. Moreover, the refueling of BEBs using either standard or slow-charging methods takes much longer when compared to diesel buses. The emerging technology of fast-charging BEBs provides promising potential in helping to offset these drawbacks. This technique follows the opportunity charging concept, which utilizes dwelling times at bus stops or terminals to recharge buses in operation using high charging power (*3*). By fast-charging en route, BEBs are as capable as their diesel counterparts in terms of range and operating time. Promising results have been reported by many fast-charging BEB demonstration projects (*4–6*).

A fast-charging BEB system requires comprehensive planning and design to ensure effective and economical implementation. Two key design factors of the system are the deployment of fast-charging stations and on-board battery size. The planning problem is twofold. First, the combination of deployed fast-charging facilities and the design of battery sizes should ensure the normal operation of the bus system. Second, one must consider the trade-off between on-board battery sizes and the number of fast-charging facilities.

Although several studies have investigated the planning problem of a fast-charging electric bus system (e.g., *1, 3, 7*), none have considered the energy consumption uncertainty of buses. For a fast-charging electric bus

[1]Department of Civil and Environmental Engineering, Utah State University, Logan, UT

Corresponding Author:
Address correspondence to Ziqi Song: ziqi.song@usu.edu

system, the energy consumption of BEBs is a key parameter in determining the locations of charging stations and the battery sizes of BEBs. In the above studies, energy consumption is assumed to be predefined and is estimated either based on collected sample data or through simulation. In reality, however, the energy consumption of BEBs may be uncertain. For instance, the energy consumption of a BEB between two stops will change in conjunction with traffic conditions and travel demands. Ignoring the uncertainty of energy consumption of BEBs could lead to a suboptimal or even an infeasible plan for a fast-charging electric bus system. Therefore, this study explicitly addresses energy consumption uncertainty in the planning of a fast-charging electric bus system.

To model the energy consumption uncertainty of BEBs, this study applies the technique of set-based robust optimization (RO). This technique has been developed and improved by many researchers (e.g., *8–10*). It allows the parameters with uncertainty to vary within a given set and requires solutions to be feasible for any realizations of the parameters in the set. Compared with other modeling methods regarding data uncertainty, such as stochastic programming and scenario-based RO, the set-based RO has the virtues of not requiring probability distribution of uncertain data and providing computationally tractable formulations. The set-based RO approach has been applied in many study areas, including humanitarian relief supply chain problems with demand uncertainty (*11*), multi-period fleet allocation problems in a bike-sharing system with demand uncertainty (*12*), robust dynamic network design problems with demand uncertainty (*13*), and robust planning of dynamic wireless charging infrastructure for BEBs (*14*).

In this study, we first provide a deterministic optimization model for the planning problem of a fast-charging electric bus system. We then develop its robust counterpart model to address the energy consumption uncertainty of BEBs. The approach developed by Ben-Tal et al. (*10*) is adopted to derive the robust counterpart for a given uncertainty set.

Deterministic Optimization Model

In this section, we introduce the deterministic optimization model for the planning of a fast-charging electric bus system. To develop a feasible and cost-effective system, the battery sizes of BEBs and the deployment of fast-charging stations need to be simultaneously determined. Because there are multiple types of fast-charging stations with different power levels and costs, we also need to determine the most suitable type for each charging station. The combination of designed battery sizes and deployed charging stations should first meet the energy requirements for normal operation of the bus system. Under this premise, we then seek the optimal combination with minimum total cost. Moreover, our model is based on a fast-charging electric bus system with the following assumptions:

1. Each bus line in the bus system operates on a fixed route.
2. Each bus line has a base station, where all buses begin and end each of their service loops. Different bus lines may share a base station if their service routes converge at the base station.
3. After finishing a service loop, an electric bus will be fully charged at the base station before it starts another service loop.

Assumption 1. is not restrictive, because we can always treat a bus line with multiple routes as multiple lines that share the same bus fleet and require the bus fleet to be feasible for all lines. Assumption 2. is accordant with the common practice of bus systems. The base station is usually the location where drivers have a rest and buses are checked. Assumption 3. means that a fast-charging station must be installed at the base station, and that an electric bus will utilize its long dwell time between two service loops to get fully charged. Assumption 3. is consistent with the operational rules in the demonstration project of a fast-charging electric bus line, as reported by Eudy et al. (*4*). We note that some other studies, such as that of Qin et al. (*15*), allow electric buses to operate more than one service loop before they are fully charged. However, this may result in a queue at the charging station and require additional scheduling efforts to ensure normal operations. Because the primary focus of this study is to develop a modeling framework for the robust optimal planning of a fast-charging electric bus system, we leave the work of capturing more flexible operational rules to future studies.

Notation

For the convenience of readers, we list all notations of the deterministic model in Table 1.

Model Formulation

$$(\mathrm{S1}): \min_{(x,y,e^{\max},s,e^{\mathrm{arriving}},e^{\mathrm{leaving}})} \sum_{k\in K} a^{\mathrm{battery}}\zeta_k e_k^{\max} + \sum_{i\in\tilde{N}}\sum_{t\in T} a_t^{\mathrm{charger}} x_{i,t} + \sum_{k\in K}\sum_{i\in N\backslash\tilde{N}}\sum_{t\in T} a_t^{\mathrm{charger}} y_{k,i,t}$$

s.t.

$$\sum_{t\in T} x_{i,t} \leq 1 \quad \forall i\in\tilde{N} \tag{1}$$

$$\sum_{t\in T} y_{k,i,t} \leq 1 \quad \forall k\in K, i\in N_k\backslash\tilde{N} \tag{2}$$

Table 1. Notation of the Deterministic Model

	Description
Sets	
N	Set of bus stations (including base stations and terminals), indexed by i and j
L	Set of directed road segments connecting sequential stations, indexed by (i,j)
T	Set of fast-charging station types, indexed by t
K	Set of bus lines, indexed by k
$\tilde{N}$	Set of bus stations (excluding base stations and terminals), indexed by i and j
L_k	Set of all the links that form bus line $k \in K$
N_k	Set of all the nodes that form bus line $k \in K$
Parameters	
o_k	A node representing the base station of bus line $k \in K$
a^{battery}	Amortized battery cost per kWh, considering replacement and discount rate
a_t^{charger}	Amortized cost of installing a type $t \in T$ fast-charging station
ζ_k	Total number of buses on line $k \in K$
$c_{k,i,j}$	Energy consumption of an electric bus from line $k \in K$ on link $(i,j) \in L_k$
$\tau_{k,i}$	Dwell time of an electric bus from line $k \in K$ at station $i \in N$
p_t	Power of a type $t \in T$ fast-charging station
ϵ^{upper}	Upper bound of remaining battery power (in percentage of battery capacity)
ϵ^{lower}	Lower bound of remaining battery power (in percentage of battery capacity)
Variables	
$x_{i,t}$	A binary variable, representing whether to build a type $t \in T$ fast-charging station at bus station $i \in \tilde{N}$. If yes, $x_{i,t} = 1$; otherwise, $x_{i,t} = 0$
$y_{k,i,t}$	A binary variable, representing whether to build a type $t \in T$ fast-charging station at base station or terminal station $i \in N \backslash \tilde{N}$. If yes, $y_{k,i,t} = 1$; otherwise, $y_{k,i,t} = 0$
$e_k^{\max}$	Battery size for electric buses on line $k \in K$
$e_{k,i}^{\text{arriving}}$	Remaining battery power of an electric bus on line $k \in K$ when it arrives at station $i \in N$
$e_{k,i}^{\text{leaving}}$	Remaining battery power of an electric bus on line $k \in K$ when it leaves station $i \in N$
$s_{k,i}$	Energy supply at station $i \in N$ for an electric bus from line $k \in K$

$$x_{i,t} \in \{0,1\} \quad \forall i \in N, t \in T \tag{3}$$

$$y_{k,i,t} \in \{0,1\} \quad \forall i \in N, t \in T \tag{4}$$

$$e_{k,i}^{\text{leaving}} = \epsilon^{\text{upper}} e_k^{\max} \quad \forall k \in K, i = o_k \tag{5}$$

$$e_{k,j}^{\text{arriving}} = e_{k,i}^{\text{leaving}} - c_{k,i,j} \quad \forall k \in K, (i,j) \in L_k \tag{6}$$

$$e_{k,i}^{\text{leaving}} = e_{k,i}^{\text{arriving}} + s_{k,i} \quad \forall k \in K, i \in N_k \tag{7}$$

$$s_{k,i} \leq \sum_{t \in T} p_t x_{i,t} \tau_{k,i} \quad \forall k \in K, i \in N_k \cap \tilde{N} \tag{8}$$

$$s_{k,i} \leq \sum_{t \in T} p_t y_{k,i,t} \tau_{k,i} \quad \forall k \in K, i \in N_k \backslash \tilde{N} \tag{9}$$

$$e_{k,i}^{\text{arriving}} \geq \epsilon^{\text{lower}} e_k^{\max} \quad \forall k \in K, i \in N_k \tag{10}$$

$$e_{k,i}^{\text{leaving}} \leq \epsilon^{\text{upper}} e_k^{\max} \quad \forall k \in K, i \in N_k \tag{11}$$

$$e_k^{\max} \geq 0 \quad \forall k \in K \tag{12}$$

The objective function minimizes the total system cost, including the cost of batteries and the cost of installing fast-charging stations. Note that because the dwelling time or the potential charging time at a base station or terminal station is long, to avoid a queue, we assume that different bus lines cannot share charging stations at a shared base station or terminal station. While at a bus station that is neither a base station nor a terminal station, we assume that different bus lines can share one charging station. We note that to realize charging station sharing, a bus system must be systematically and comprehensively scheduled. We leave the work of developing a schedule-based modeling framework to future studies.

Constraint (1) ensures that only one type of fast-charging station can be installed at bus station $i \in \tilde{N}$. Similarly, constraint (2) ensures that for bus line $k \in K$, only one type of charging station can be deployed at the base station or terminal station $i \in N_k \backslash \tilde{N}$. Constraints (3) and (4) require variables $x_{i,t}$ and $y_{k,i,t}$ to be binary. Constraint (5) specifies that a BEB is fully charged when it begins its service from the base station. Constraints (6) and (7) describe the change in battery power on a bus due to energy consumption and potential fast-charging. Constraints (8) and (9) ensure that an electric bus can be charged at a bus station only when it has an available charging station, and that the potential energy supply cannot exceed the product of charging power and dwell time. Constraints (10) and (11) specify the lower bound and upper bound of the battery power of an electric bus. Constraint (12) ensures that battery capacity cannot be negative.

By eliminating variables, $e_{k,i}^{\text{arriving}}$ and $s_{k,i}$, model S1 can be equivalently reformulated as a more concise program as

$$(\text{S2}): \min_{(\boldsymbol{x},\boldsymbol{y},\boldsymbol{e}^{\max},\boldsymbol{e}^{\text{leaving}})} \sum_{k\in K} a^{\text{battery}}\zeta_k e_k^{\max} + \sum_{j\in\tilde{N}}\sum_{t\in T} a_t^{\text{charger}} x_{j,t} + \sum_{k\in K}\sum_{j\in N\setminus\tilde{N}}\sum_{t\in T} a_t^{\text{charger}} y_{k,j,t}$$

s.t. (1)-(5), (11)-(12)

$$e_{k,j}^{\text{leaving}} \leq e_{k,i}^{\text{leaving}} - c_{k,i,j} + \sum_{t\in T} p_t x_{j,t}\tau_{k,j} \quad \forall k\in K, j\in N_k\cap\tilde{N}, (i,j)\in L_k \tag{13}$$

$$e_{k,j}^{\text{leaving}} \leq e_{k,i}^{\text{leaving}} - c_{k,i,j} + \sum_{t\in T} p_t y_{k,j,t}\tau_{k,j} \quad \forall k\in K, j\in N_k\setminus\tilde{N}, (i,j)\in L_k \tag{14}$$

$$e_{k,i}^{\text{leaving}} - c_{k,i,j} \geq \epsilon^{\text{lower}} e_k^{\max} \quad \forall k\in K, (i,j)\in L_k \tag{15}$$

Robust Formulation

As formulated, the above deterministic optimization model is a mixed integer linear programming problem. In this section, we apply the set-based RO methodology to address the uncertainty in energy consumption of BEBs and to identify a robust solution for the planning problem of a fast-charging BEB system.

Energy Consumption Uncertainty

Under the set-based RO approach, it is assumed that energy consumption $c_{k,i,j}$ is unknown and belongs to a specified uncertainty set. Specifically, the following box uncertainty set is commonly used:

$$c_{k,i,j} \in U_k^b \equiv \left[\bar{c}_{k,i,j}, \bar{c}_{k,i,j} + \hat{c}_{k,i,j}\sigma_{k,i,j}\right]$$

where $\sigma_{k,i,j}$ is the uncertainty level coefficient; $\bar{c}_{k,i,j}$ and $\hat{c}_{k,i,j}$ are the nominal value and maximum deviation of energy consumption of a BEB from link $k\in K$ on link $(i,j)\in L_k$, respectively. In practice, both $\bar{c}_{k,i,j}$ and $\hat{c}_{k,i,j}$ can be calculated based on collected bus driving profile and ridership data. Here, possible negative deviations in energy consumption are neglected, because they are not adverse factors for the bus system.

The box uncertainty set only restricts the maximum deviation of each individual parameter $c_{k,i,j}$. Under the box uncertainty set, all energy consumption parameters can reach their worst-case value (i.e., the maximum value) simultaneously. In practice, however, it is very unlikely that there are heavy passenger demands and extremely bad traffic conditions in all parts of the bus route. Therefore, to find a less conservative solution, we impose Γ_k as a budget for uncertainty to set an upper limit for the cumulative deviation in energy consumption along a route. Thus, the following polyhedral uncertainty set is used to replace the box uncertainty set as

$$\boldsymbol{c}_k \in U_k^p \equiv \left\{ \boldsymbol{c}_k \in \mathbb{R}^{|L_k|} \middle| \bar{c}_{k,i,j} \leq c_{k,i,j} \leq \bar{c}_{k,i,j} + \hat{c}_{k,i,j}\sigma_{k,i,j}, \sum_{(i,j)\in L_k} \left(c_{k,i,j} - \bar{c}_{k,i,j}\right) \leq \Gamma_k \right\}$$

where $\boldsymbol{c}_k$ denotes the vector of $\{\cdots, c_{k,i,j}, \cdots\}$.

Affinely Adjustable Robust Counterpart

In the deterministic model S2, if constraints (13)–(15) must be satisfied for all realizations of $\boldsymbol{c}_k \in U_k^p$, then the corresponding solution is deemed robust against the uncertainty set U_k^p. A traditional set-based RO approach focuses on static problems and has an underlying assumption that all decision variables represent "here and now" decisions and therefore should be assigned specific numerical values as a result of solving the problem before the actual data "reveals itself" (*10*). In our problem, however, the change in battery power for a BEB depends on the actual realizations of uncertain energy consumption and should be a dynamic process. To consider such a dynamic nature, we adopt a more advanced approach of an affinely adjustable robust counterpart (AARC), as developed by Ben-Tal et al. (*9*). AARC allows some of the decision variables, including auxiliary variables (e.g., slack or surplus variables) and variables representing "wait and see" decisions (i.e., decisions that can be made when part of the uncertain data becomes known) (*9*), to be adjustable based on different realizations of uncertain data by introducing functional relationships between these decision variables and uncertain data. Moreover, by restricting the functional relationships to the affine, the AARC approach ensures tractable robust formulations.

For our problem, the decision variables include $x_{j,t}, y_{k,j,t}, e_k^{\max}$, and $e_{k,i}^{\text{leaving}}$, where $x_{j,t}$ and $y_{k,j,t}$ determine the deployment of fast-charging stations, $e_k^{\max}$ determines on-board battery sizes, and $e_{k,i}^{\text{leaving}}$ represents the remaining battery power of an electric bus at a certain station. From the perspective of planning, the deployment of fast-charging stations and the battery sizes of electric buses should be determined before construction of a fast-charging electric bus system. Thus, decision variables $x_{i,t}, y_{k,i,t}$, and $e_k^{\max}$ should represent "here and now" decisions and should not be adjustable variables. However, because the remaining battery power of an electric bus at a certain station is directly affected by actual energy consumption on those links that have been passed, variable $e_{k,i}^{\text{leaving}}$ should represent a "wait and see" decision (its value can be determined only when the actual energy consumption data on the links that have been passed "reveals itself"). Therefore, variable $e_{k,i}^{\text{leaving}}$ should be an adjustable variable.

To derive the AARC of our problem, we should first define a functional relationship between the adjustable variable $e_{k,i}^{\text{leaving}}$ and uncertain energy consumption data. As discussed above, the value of $e_{k,i}^{\text{leaving}}$ depends on uncertain energy consumption data on those links that have been passed. Let L_k^i denote the set of links that an electric bus needs to pass from the base station to station $i \in N_k$ along the route of bus line $k \in K$. Note that for the base station of a bus line $i = o_k$, we set $L_k^i = L_k$. Let $\boldsymbol{c}_k^i$ denote the vector of $\{\cdots, c_{k,m,n}, \cdots\}$, where $(m,n) \in L_k^i$. The adjustable variable $e_{k,i}^{\text{leaving}}$ can then be represented as the affine function

$$e_{k,i}^{\text{leaving}}\left(\boldsymbol{c}_k^i\right) = \theta_k^i + \sum_{(m,n)\in L_k^i} \lambda_{k,m,n}^i c_{k,m,n}$$

where θ_k^i and $\lambda_{k,m,n}^i$ are new decision variables that are not adjustable.

Substituting the adjustable variable, we arrive at the AARC formulation

$$(\text{S2}-\text{AARC}): \min_{(\boldsymbol{x},\boldsymbol{y},\boldsymbol{e}^{\max},\boldsymbol{\theta},\boldsymbol{\lambda})} \sum_{k\in K} a^{\text{battery}} \zeta_k e_k^{\max} + \sum_{j\in \tilde{N}} \sum_{t\in T} a_t^{\text{charger}} x_{j,t} + \sum_{k\in K} \sum_{j\in N\backslash \tilde{N}} \sum_{t\in T} a_t^{\text{charger}} y_{k,j,t}$$

s.t. (1)–(4), (12)

$$\theta_k^i + \sum_{(m,n)\in L_k^i} \lambda_{k,m,n}^i c_{k,m,n} = \epsilon^{\text{upper}} e_k^{\max} \quad \forall k \in K, i = o_k, \boldsymbol{c}_k \in U_k^p \tag{16}$$

$$\theta_k^j + \sum_{(m,n)\in L_k^j} \lambda_{k,m,n}^j c_{k,m,n} \le \theta_k^i + \sum_{(m,n)\in L_k^i} \lambda_{k,m,n}^i c_{k,m,n} - c_{k,i,j} + \sum_{t\in T} p_t x_{j,t} \tau_{k,j} \quad \forall k \in K, j \in N_k \cap \tilde{N}, (i,j) \in L_k, \boldsymbol{c}_k \in U_k^p \tag{17}$$

$$\theta_k^j + \sum_{(m,n)\in L_k^j} \lambda_{k,m,n}^j c_{k,m,n} \le \theta_k^i + \sum_{(m,n)\in L_k^i} \lambda_{k,m,n}^i c_{k,m,n} - c_{k,i,j} + \sum_{t\in T} p_t y_{k,j,t} \tau_{k,j} \quad \forall k \in K, j \in N_k \backslash \tilde{N}, (i,j) \in L_k, \boldsymbol{c}_k \in U_k^p \tag{18}$$

$$\theta_k^i + \sum_{(m,n)\in L_k^i} \lambda_{k,m,n}^i c_{k,m,n} - c_{k,i,j} \ge \epsilon^{\text{lower}} e_k^{\max} \quad \forall k \in K, (i,j) \in L_k, \boldsymbol{c}_k \in U_k^p \tag{19}$$

$$\theta_k^i + \sum_{(m,n)\in L_k^i} \lambda_{k,m,n}^i c_{k,m,n} \le \epsilon^{\text{upper}} e_k^{\max} \quad \forall k \in K, i \in N_k, \boldsymbol{c}_k \in U_k^p \tag{20}$$

In S2-AARC, the adjustable variable $e_{k,i}^{\text{leaving}}$ is no longer a decision variable, but the coefficients of the affine function of $e_{k,i}^{\text{leaving}}$, including θ_k^i and $\lambda_{k,m,n}^i$, become new decision variables. This means that the solution of S2-AARC only determines the relationship between $e_{k,i}^{\text{leaving}}$ and uncertain energy consumption. Specific values of $e_{k,i}^{\text{leaving}}$ can be calculated following the realization of energy consumption.

Tractable Reformulation of S2-AARC

In the above formulation, we observe that each piece of constraint in constraints (16)–(20) (i.e., for a certain combination of superscripts and subscripts) actually consists of a continuum, or an infinite number, of constraints, because the formulation must be satisfied for all realizations of uncertain energy consumption. Therefore, S2-AARC is a semi-infinite programming problem (the number of variables is finite) that is computationally intractable. As shown in Proposition 1, we can further reformulate S2-AARC as a tractable optimization problem.

Proposition 1. Given the polyhedral uncertainty set, U_k^p, S2-AARC can be equivalently reformulated as the mixed integer linear programming problem

$$(\text{S2}-\text{AARC}-\text{T}): \min_{(\boldsymbol{x},\boldsymbol{y},\boldsymbol{e}^{\max},\boldsymbol{\theta},\boldsymbol{\lambda},\boldsymbol{\omega})} \sum_{k\in K} a^{\text{battery}} \zeta_k e_k^{\max} + \sum_{j\in \tilde{N}} \sum_{t\in T} a_t^{\text{charger}} x_{j,t} + \sum_{k\in K} \sum_{j\in N\backslash \tilde{N}} \sum_{t\in T} a_t^{\text{charger}} y_{k,j,t}$$

s.t. (1)–(4), (12)

$$\theta_k^i = \epsilon^{\text{upper}} e_k^{\max} \quad \forall k \in K, i = o_k \tag{21}$$

$$\lambda_{k,m,n}^i = 0 \quad \forall k \in K, i = o_k, (m,n) \in L_k^i \tag{22}$$

$$\sum_{(m,n)\in L_k} \left(\bar{c}_{k,i,j} + \hat{c}_{k,i,j}\sigma_{k,i,j}\right)\omega_{kmn}^{i1} - \sum_{(m,n)\in L_k} \bar{c}_{k,m,n}\omega_{kmn}^{i2} + \left(\Gamma_k + \sum_{(m,n)\in L_k} \bar{c}_{k,m,n}\right)\omega_k^{i3} \le \theta_k^i - \theta_k^j + \sum_{t\in T} p_t x_{j,t}\tau_{k,j} \quad \forall k \in K, j \in N_k \cap \tilde{N}, (i,j) \in L_k \tag{23}$$

$$\omega_{kmn}^{i1} - \omega_{kmn}^{i2} + \omega_k^{i3} = \lambda_{k,m,n}^j - \lambda_{k,m,n}^i \quad \forall k \in K, j \in N_k \cap \tilde{N}, (i,j) \in L_k, (m,n) \in L_k^i \tag{24}$$

$$\omega_{kij}^{i1} - \omega_{kij}^{i2} + \omega_k^{i3} = \lambda_{k,i,j}^j + 1 \quad \forall k \in K, j \in N_k \cap \tilde{N}, (i,j) \in L_k \tag{25}$$

$$\omega_{kmn}^{i1} - \omega_{kmn}^{i2} + \omega_k^{i3} = 0 \quad \forall k \in K, j \in N_k \cap \tilde{N}, (i,j) \in L_k, (m,n) \in L_k \backslash L_k^j \tag{26}$$

$$\omega_{kmn}^{i1}, \omega_{kmn}^{i2}, \omega_{k}^{i3} \geq 0$$
$$\forall k \in K, j \in N_k \cap \tilde{N}, (i,j) \in L_k, (m,n) \in L_k \quad (27)$$

$$\sum_{(m,n)\in L_k} \left(\bar{c}_{k,i,j} + \hat{c}_{k,i,j}\sigma_{k,i,j}\right)\omega_{kmn}^{i4} - \sum_{(m,n)\in L_k} \bar{c}_{k,m,n}\omega_{kmn}^{i5} + \left(\Gamma_k + \sum_{(m,n)\in L_k} \bar{c}_{k,m,n}\right)\omega_k^{i6} \leq \theta_k^i - \theta_k^j + \sum_{t\in T} p_t y_{k,j,t}\tau_{k,j} \quad \forall k \in K, j \in N_k \backslash \tilde{N}, (i,j) \in L_k \quad (28)$$

$$\omega_{kmn}^{i4} - \omega_{kmn}^{i5} + \omega_k^{i6} = \lambda_{k,m,n}^{j} - \lambda_{k,m,n}^{i}$$
$$\forall k \in K, j \in N_k \backslash \tilde{N}, (i,j) \in L_k, (m,n) \in L_k^i \quad (29)$$

$$\omega_{kmn}^{i4} - \omega_{kmn}^{i5} + \omega_k^{i6} = \lambda_{k,i,j}^{j} + 1$$
$$\forall k \in K, j \in N_k \backslash \tilde{N}, (i,j) \in L_k \quad (30)$$

$$\omega_{kmn}^{i4} - \omega_{kmn}^{i5} + \omega_k^{i6} = 0$$
$$\forall k \in K, j \in N_k \backslash \tilde{N}, (i,j) \in L_k, (m,n) \in L_k \backslash L_k^j \quad (31)$$

$$\omega_{kmn}^{i4}, \omega_{kmn}^{i5}, \omega_k^{i6} \geq 0$$
$$\forall k \in K, j \in N_k \backslash \tilde{N}, (i,j) \in L_k, (m,n) \in L_k \quad (32)$$

$$\sum_{(m,n)\in L_k} \left(\bar{c}_{k,i,j} + \hat{c}_{k,i,j}\sigma_{k,i,j}\right)\omega_{kmn}^{i7} - \sum_{(m,n)\in L_k} \bar{c}_{k,m,n}\omega_{kmn}^{i8} + \left(\Gamma_k + \sum_{(m,n)\in L_k} \bar{c}_{k,m,n}\right)\omega_k^{i9} \leq \theta_k^i - \epsilon^{\text{lower}} e_k^{\max} \quad \forall k \in K, (i,j) \in L_k \quad (33)$$

$$\omega_{kmn}^{i7} - \omega_{kmn}^{i8} + \omega_k^{i9} = -\lambda_{k,m,n}^{i}$$
$$\forall k \in K, i \in N_k, (i,j) \in L_k, (m,n) \in L_k^i \quad (34)$$

$$\omega_{kmn}^{i7} - \omega_{kmn}^{i8} + \omega_k^{i9} = 1$$
$$\forall k \in K, i \in N_k, (i,j) \in L_k \quad (35)$$

$$\omega_{kmn}^{i7} - \omega_{kmn}^{i8} + \omega_k^{i9} = 0$$
$$\forall k \in K, i \in N_k, (i,j) \in L_k, (m,n) \in L_k \backslash L_k^j \quad (36)$$

$$\omega_{kmn}^{i7}, \omega_{kmn}^{i8}, \omega_k^{i9} \geq 0 \quad \forall k \in K, i \in N_k, (i,j) \in L_k, (m,n) \in L_k \quad (37)$$

$$\sum_{(m,n)\in L_k} \left(\bar{c}_{k,i,j} + \hat{c}_{k,i,j}\sigma_{k,i,j}\right)\omega_{kmn}^{i10} - \sum_{(m,n)\in L_k} \bar{c}_{k,m,n}\omega_{kmn}^{i11} + \left(\Gamma_k + \sum_{(m,n)\in L_k} \bar{c}_{k,m,n}\right)\omega_k^{i12} \leq \epsilon^{\text{upper}} e_k^{\max} - \theta_k^i \quad \forall k \in K, i \in N_k \quad (38)$$

$$\omega_{kmn}^{i10} - \omega_{kmn}^{i11} + \omega_k^{i12} = \lambda_{k,m,n}^{i} \quad \forall k \in K, i \in N_k, (m,n) \in L_k^i \quad (39)$$

$$\omega_{kmn}^{i10} - \omega_{kmn}^{i11} + \omega_k^{i12} = 0 \quad \forall k \in K, i \in N_k, (m,n) \in L_k \backslash L_k^i \quad (40)$$

$$\omega_{kmn}^{i10}, \omega_{kmn}^{i11}, \omega_k^{i12} \geq 0 \quad \forall k \in K, i \in N_k, (m,n) \in L_k \backslash L_k^i \quad (41)$$

where ω is a set of dual variables. Note that the numerical superscripts are used for simplicity of notation.

Proof. For constraint (16) in S2-AARC, because it must be satisfied for all realizations of uncertain energy consumption $\boldsymbol{c}_k \in U_k^p$, it is straightforward to prove its equivalency with constraints (21) and (22). Except for constraint (16), each constraint affected by energy consumption uncertainty in S2-AARC (i.e., each of constraints (17)–(20)) can be generalized by

$$\sum_{(m,n)\in L_k} \alpha_{k,m,n}^{i} c_{k,m,n} \leq h^i \quad \forall \boldsymbol{c}_k \in U_k^p \quad (42)$$

For instance, for constraint (17), $\alpha_{k,m,n}^{i} = \lambda_{k,m,n}^{j} - \lambda_{k,m,n}^{i}$ for $\forall (m,n) \in L_k^i$, $\alpha_{k,i,j}^{i} = \lambda_{k,i,j}^{j} + 1$, $\alpha_{k,m,n}^{i} = 0$ for $\forall (m,n) \in L_k \backslash L_k^j$, and $h^i = \theta_k^i - \theta_k^j + \sum_{t\in T} p_t x_{j,t}\tau_{k,j}$. Note that constraints (17)–(19) are link-based, and superscript i in constraint (42) corresponds to the starting node of link $(i,j) \in L_k$.

Constraint (42) can be equivalently given as

$$\max_{c_k\in U_k^p} \sum_{(m,n)\in L_k} \alpha_{k,m,n}^{i} c_{k,m,n} \leq h^i \quad (43)$$

Without loss of generality, $U_k^p = \left\{ \boldsymbol{c}_k \in \mathbb{R}^{|L_k|} \mid \bar{c}_{k,i,j} \leq c_{k,i,j} \leq \bar{c}_{k,i,j} + \hat{c}_{k,i,j}\sigma_{k,i,j}, \sum_{(i,j)\in L_k} \left(c_{k,i,j} - \bar{c}_{k,i,j}\right) \leq \Gamma_k \right\}$ can be represented as $A\boldsymbol{c}_k \leq \boldsymbol{b}$, and $\max_{c_k\in U_k^p} \sum_{(m,n)\in L_k} \alpha_{k,m,n}^{i} c_{k,m,n}$ can be written as the optimization problem

$$(P): \max_{c_k} \boldsymbol{\alpha}_k^i \boldsymbol{c}_k$$

s.t.

$$A\boldsymbol{c}_k \leq \boldsymbol{b}$$
$$\boldsymbol{c}_k \geq 0$$

According to the property of strong duality (*10, 16*), the equivalent dual problem (D) of (P) is given as

$$(\mathrm{D}) : \min_{\omega} \boldsymbol{b\omega}$$

s.t.

$$A^T \boldsymbol{\omega} = \boldsymbol{\alpha}_k^i$$

$$\boldsymbol{\omega} \geq 0$$

where $\boldsymbol{\omega}$ is a vector of dual variables.

Because the programming of (D) is equivalent to that of (P), they will have the same optimal value. Therefore, constraint (43), which requires that the optimal value of (P), $\leq h^i$, is equivalent to the optimal value of (D), is achieved and is $\leq h^i$. Because (D) is a minimization problem, the requirement that the optimal value of (D) is $\leq h^i$ is equivalent to the requirement that (D) has a feasible solution $\boldsymbol{\omega}$ with $\boldsymbol{b\omega} \leq h^i$. Therefore, constraint (43) can be equivalently replaced by $\boldsymbol{b\omega} \leq h^i$, $A^T\boldsymbol{\omega} = \boldsymbol{\alpha}_k^i$ and $\boldsymbol{\omega} \geq 0$, and the reformulation of S2-AARC-T becomes a mixed integer linear programming problem, which is tractable.

Numerical Study

To demonstrate the effectiveness of the proposed model, a numerical study is presented. The case study is based on the bus system in downtown Salt Lake City, Utah, in the United States.

Bus System and Model Parameters

A subnetwork of the Salt Lake City bus system, as shown in Figure 1, is considered in the case study. The subnetwork covers 91.4 kilometers of road segments and includes eight bus lines, with a total of 34 buses serving 315 bus stations. Assuming that the transit agency wants to transform this bus system into a fast-charging electric bus system, the locations and types of charging stations, as well as the battery capacity of the electric buses, must be determined.

To evaluate the energy consumption of an electric bus on each link for each bus line, an energy consumption model proposed by Wang et al. (*17*) has been adopted. The model is derived from the analysis of forces exerted on a running electric bus, which considers air resistance, rolling resistance, climbing resistance, acceleration resistance and regenerative braking. Table 2*a* reveals the simulation parameters applied to calculate energy consumption. For simplicity, it is assumed that all roads in the network have the same friction factor, and that the eight bus lines use the same type of electric buses. The slope profile is calculated based on Digital Elevation Model (DEM) data from the Utah Automated Geographic Reference Center (AGRC). The driving profiles are extracted from live Service Interface for Real Time Information (SIRI) vehicle monitoring data, which is collected using the "VehicleMonitoring by Route" service implemented by the Utah Transit Authority (UTA) (*18*). The data was collected from October 10, 2017, to November 5, 2017. In total, 1,086, 216, 510, 608, 324, 486, 590 and 414 driving loop profiles for bus lines 2, 2X, 3, 6, 11, 500, 519 and 520, respectively were obtained. Figure 2*a* lists some sample data for bus line 2, which corresponds to two service loops of a vehicle on October 12. The dwelling time data can be extracted

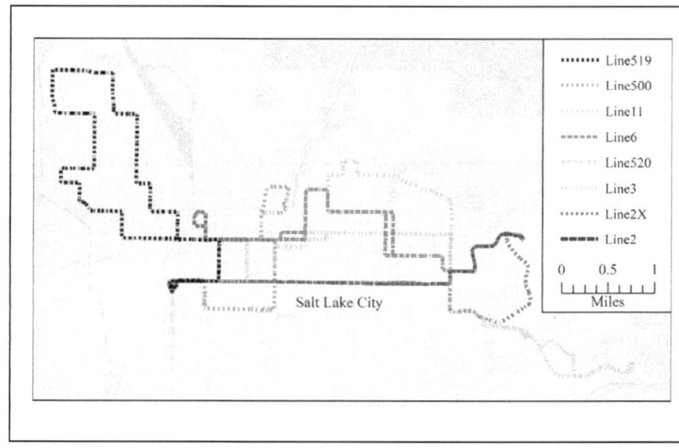

Figure 1. Salt Lake City bus system.

Table 2. Model Parameters

	Value
Parameters of Energy Consumption Model	
Parameter	
Friction factor	0.02
Total mass of a bus (Kg)	20,400
Gravity acceleration (m/s^2)	9.81
Air density (kg/m^3)	1.2
Air resistance coefficient	0.7
Bus frontal area (m^2)	7.5
Energy output efficiency	60%
Energy input efficiency	50%
Parameters of Batteries and Charging Stations	
Parameter	
Charging power of a type 1 charging station (kW)	90
Charging power of a type 2 charging station (kW)	250
Amortized cost of installing and maintaining a type 1 charging station ($)	4,637
Amortized cost of installing and maintaining a type 2 charging station ($)	11,589
Amortized cost of battery, including replacement ($/kWh)	148
Lower bound of remaining battery power (in percentage of battery capacity)	0.3
Upper bound of remaining battery power (in percentage of battery capacity)	0.8
Parameters of Uncertainty Set	
Uncertainty level of box uncertainty set	1.0, 2.0, 3.0
Uncertainty level of budget uncertainty set	0-1.0

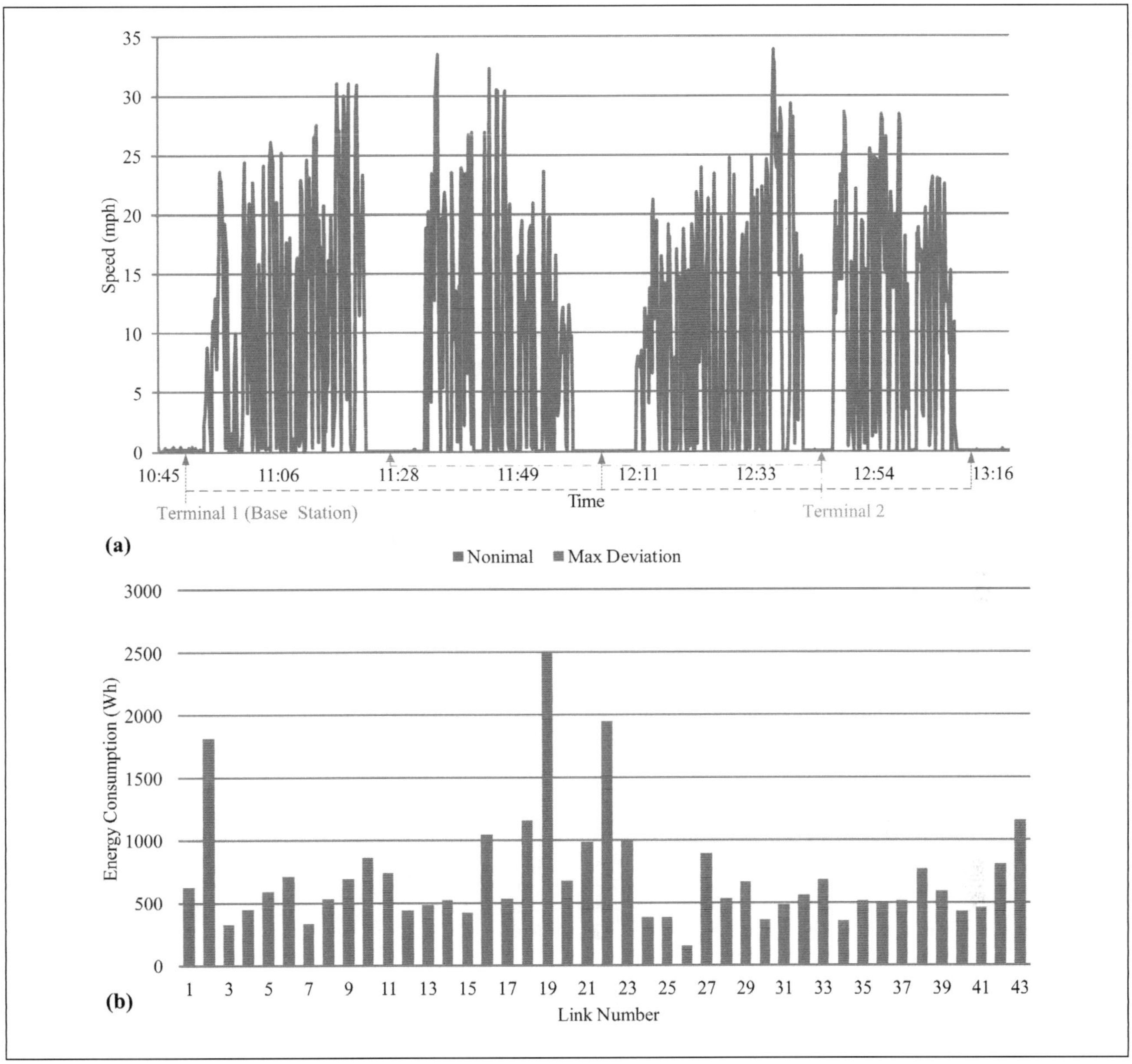

Figure 2. (*a*) Sample speed profile data for line 2, and (*b*) energy consumption data for line 2.

from the driving profiles. The ridership data was requested from the UTA. The average weight of a passenger is assumed to be 70 kg. Based on the collected data, the energy consumption data can then be calculated. Figure 2*b* reveals the energy consumption data for bus line 2.

The parameters relating to batteries and fast-charging stations are given in Table 2*b*. For simplicity, we only consider two types of fast-charging stations, whose charging power and costs are 90 kW, $50,000 (*19*) and 250 kW, $125,000 (*20*), respectively. We further assume that the service life of charging stations is 12 years, and that the maintenance cost is 1% of the installation cost per year. For the on-board batteries, we consider the lithium-titanate battery (referred to as LTO in the battery industry) utilized in Proterra's battery electric CATALYST Fast Charge transit bus. The battery has a service life of six years with a cost of around $1000/kWh (*21*). Considering a 0.01 discount rate, we can then calculate the amortized cost of charging stations and batteries.

The uncertainty set in our robust model is given by $\boldsymbol{c_k} \in U_k^p \equiv \left\{ \boldsymbol{c_k} \in \mathbb{R}^{|L_k|} | \bar{c}_{k,i,j} \leq c_{k,i,j} \leq \bar{c}_{k,i,j} + \hat{c}_{k,i,j}\sigma_{k,i,j}, \right.$

$\left.\sum_{(i,j)\in L_k} (c_{k,i,j} - \bar{c}_{k,i,j}) \leq \Gamma_k \right\}$. $\bar{c}_{k,i,j}$ and $\hat{c}_{k,i,j}$ are calculated based on real-world data. Parameters $\sigma_{k,i,j}$ and Γ_k represent the conservation level of the robust model and are determined in practice by decision makers or planners. In this numerical study, we assume that $\sigma_{k,i,j}$ (representing the conservation level of the box uncertainty set) is uniform for all bus lines and all links and considers three conservation levels: $\sigma_{k,i,j} \in \{1.0, 2.0, 3.0\}$. For the budget of uncertainty, we use the ratio of the uncertainty budget to the total possible energy consumption deviation along a route (i.e., $\Gamma^k / \sum_{(i,j)\in L_k} \hat{c}_{k,i,j}\sigma_{k,i,j}$), denoted as γ^k, to represent the conservation level. We further assume that γ^k is identical for all bus lines and consider 11 levels of uncertainty with γ^k ranging between 0 and 1, with a step size of 0.1.

Results and Analysis

Based on the network of the Salt Lake City bus system, we obtain a robust model with 171,951 variables (644 binary variables) and 86,217 constraints. We solve this model using GAMS (*22*) with a CPLEX (*23*) solver on a 3.40 GHz Dell Computer with 16 GB of RAM. With a 0.1% relative optimality gap, the computation time of robust models ranges from 10 to 60 minutes, depending on the uncertainty level. Note that when γ^k is zero, the robust model will have the same solution as the deterministic model. Table 3 lists the comparison between the results of one robust model with an uncertainty level of $\sigma_{k,i,j} = 2.0$ and $\gamma^k = 1.0$ and the results of the corresponding deterministic model. Compared to the deterministic model, the robust model requires larger batteries for all eight bus lines. Moreover, as shown in Figure 3*a*, *b*, the deployment and the number of fast-charging stations in the robust model is also different from that of the deterministic model. We observe that when compared to the deterministic solution, the robust solution deploys more 250 kW fast-charging stations and less 90 kW fast-charging stations. In addition, bus lines 6, 11, and 500 use 90 kW fast-charging stations at the base station in the deterministic solution, whereas they require 250 kW fast-charging stations in the robust solution. When considering the uncertainty of energy consumption at the levels of $\sigma_{k,i,j} = 2.0$ and $\gamma^k = 1.0$, the amortized cost of the electric bus system will increase from \$347,835 to \$481,277 under the robust model.

Although the robust optimal solution requires greater investments, the corresponding electric bus system can operate uninterrupted when energy consumption experiences deviations within the uncertainty set. Consider a worst-case scenario in which all of the energy consumption $c_{k,i,j}$ has a deviation from the expected value $\bar{c}_{k,i,j}$ which is double that of the calculated maximum deviation $\hat{c}_{k,i,j}$. Under this scenario, the solutions of robust models with conservation levels of $\sigma_{k,i,j} \geq 2.0$ and $\gamma^k \geq 1.0$ will still be feasible, and the corresponding electric bus system can operate normally, whereas the solution of the deterministic model will become infeasible, and the corresponding electric bus system cannot satisfy the energy requirements for normal operations. Under the worst-case scenario, Figure 4 lists comparisons of the

Table 3. Comparison between the Deterministic Model and A Robust Model ($\sigma_{k,i,j} = 2.0$, $\gamma^k = 1.0$)

Battery size comparison			
	Battery capacity (kWh)		
Shuttle Line	Deterministic model	Robust model	Battery size increase
519	55.9	75.5	35.1%
520	55.2	75.8	37.3%
500	32.0	60.8	90.0%
11	33.2	65.3	96.7%
6	32.4	69.2	113.6%
3	59.2	82.3	39.0%
2	15.8	25.7	62.7%
2X	25.0	28.5	14.0%
Comparison of fast-charging stations			
	Number		–
Items	Deterministic model	Robust model	
90 kW station	9	6	
250 kW station	11	15	
Cost comparison			
	Cost		
Items	Deterministic model	Robust model	Cost increase
Battery	\$178,623	\$279,620	56.5%
Fast-charging stations	\$169,212	\$201,657	19.2%
Total	\$347,835	\$481,277	38.4%

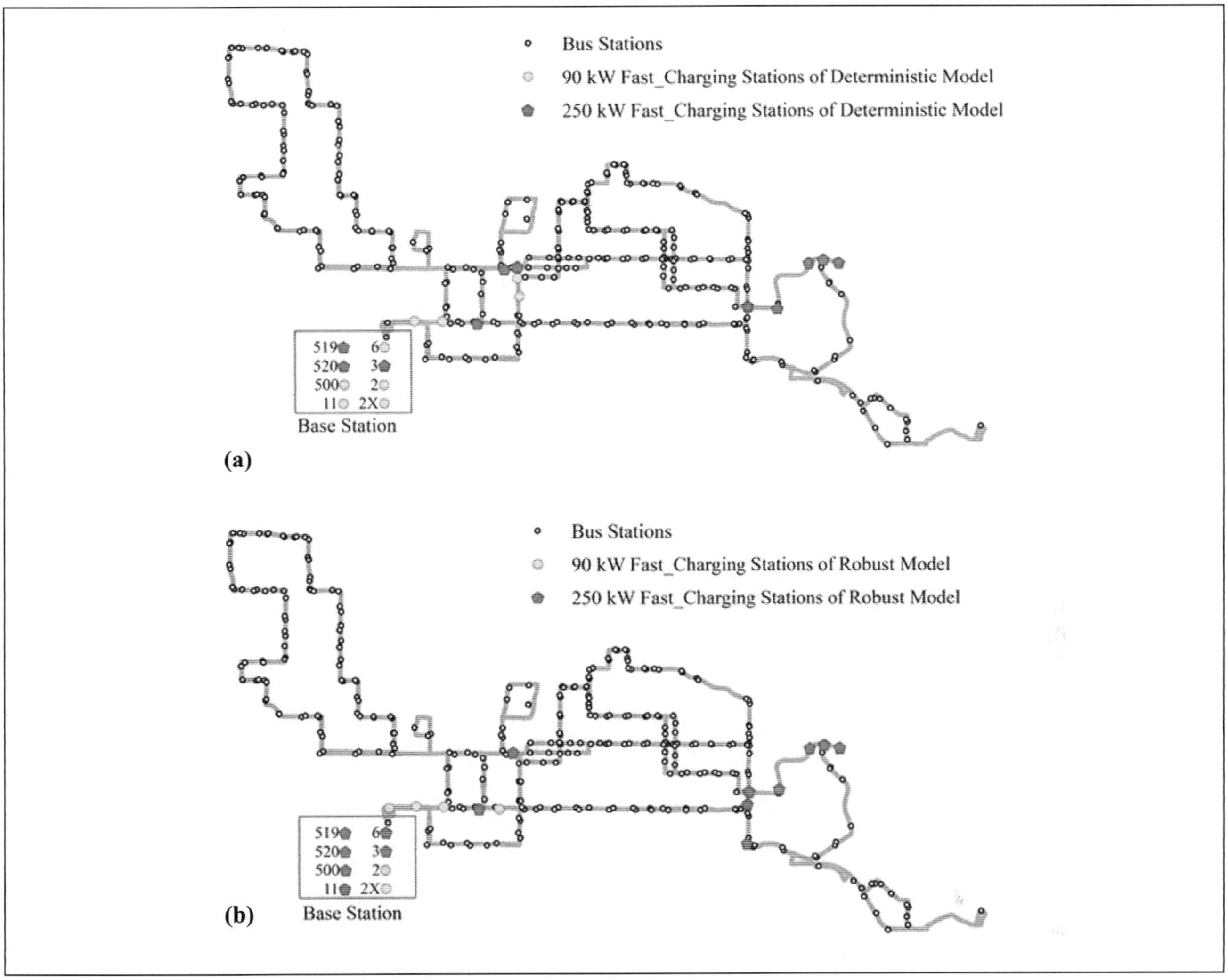

Figure 3. (*a*) Deployment of fast-charging stations in the deterministic model, and (*b*) deployment of fast-charging stations in a robust model.

battery level profiles of bus line 6 in its first service loop between the deterministic solution and the robust solution, with $\sigma_{k,i,j} = 2.0$ and $\gamma^k = 1.0$. We can observe that bus line 6 under the robust solution can operate normally within the given range of the battery level, whereas bus line 6 under the deterministic solution will run out of energy at the 63rd station and cannot return to the base station. Therefore, the performance of the robust solution is superior to that of the deterministic solution under the worst-case scenario.

With our robust model, we can obtain the optimal design for a fast-charging electric bus system that is robust against the uncertainty of energy consumption. However, additional investments will be required when an optimal robust design is sought. In practice, decision makers must determine the trade-off between robustness and the total cost of the bus system. For the bus system under study, Figure 5 compares the total amortized cost under different levels of conservation (or robustness). We can observe that under a certain box uncertainty level $\sigma_{k,i,j}$, the total cost increases as the budget uncertainty level γ^k increases. Moreover, under a certain budget uncertainty level $\gamma^k > 0$, a higher box uncertainty level leads to a higher total cost. This is expected under the robust model, because as the size of the uncertainty set for energy consumption increases, the electric bus system will require larger batteries, or a greater number of higher power fast-charging stations, or both, thus leading to a higher total cost. Under a certain box uncertainty level $\sigma_{k,i,j}$, the size of the polyhedral uncertainty set will increase as the budget uncertainty level γ^k increases. Under a certain budget uncertainty level $\gamma^k > 0$, the size of the polyhedral uncertainty set will increase as the box uncertainty level $\sigma_{k,i,j}$ increases.

Conclusion

In this study, we address the robust optimal planning problem of a fast-charging electric bus system considering

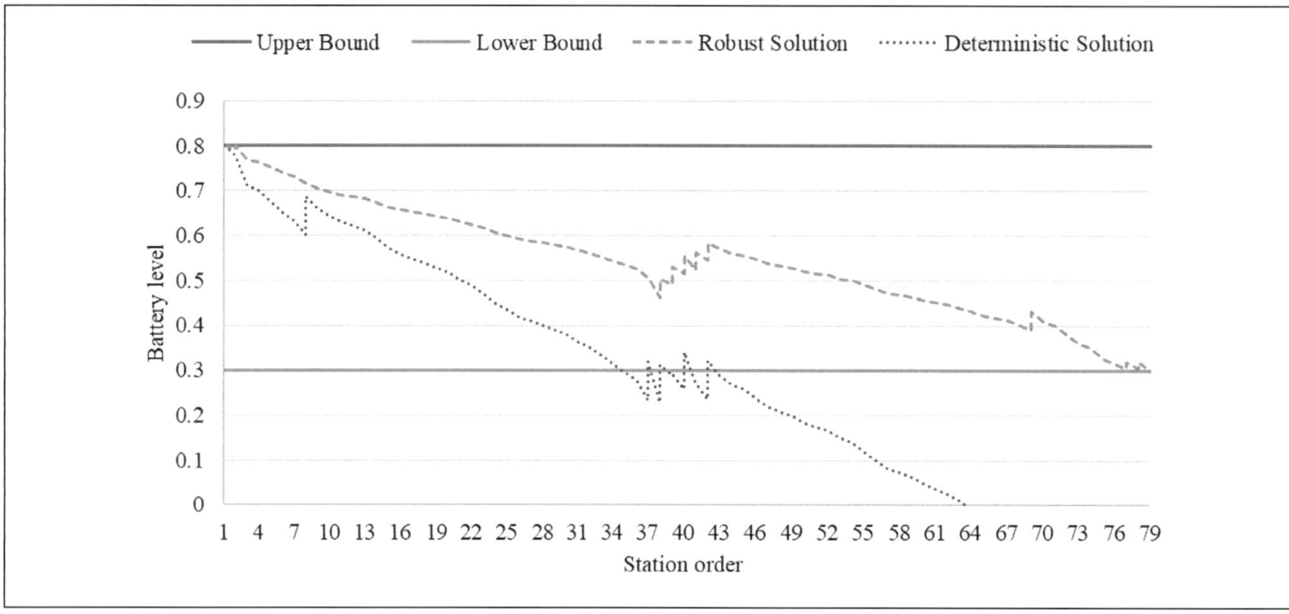

Figure 4. Comparison of battery level profiles of bus line 6.

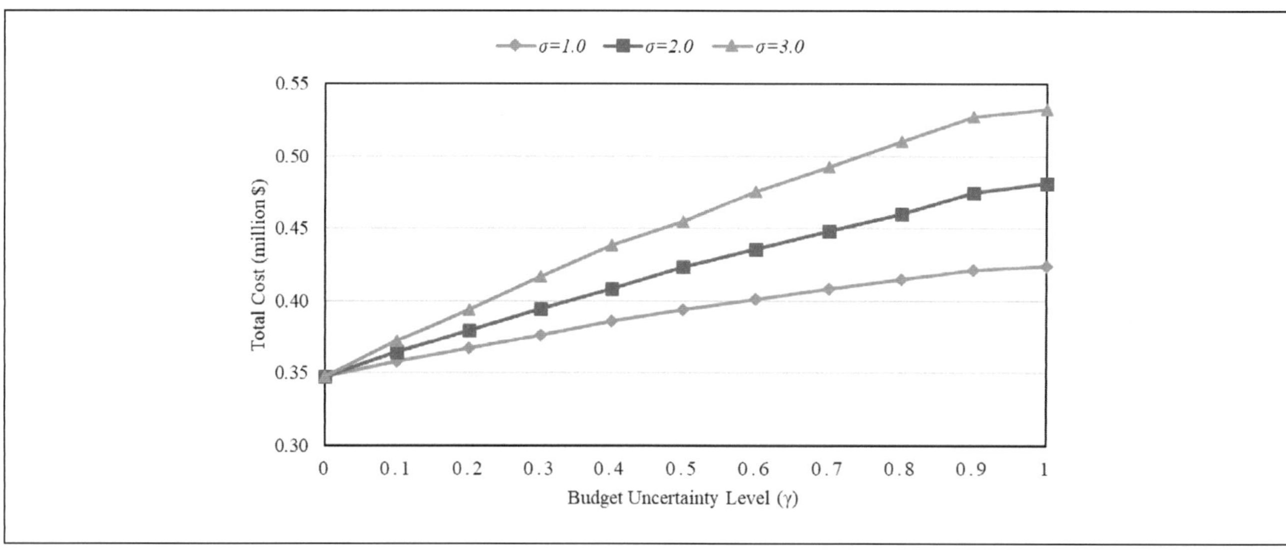

Figure 5. System cost under different uncertainty levels.

the uncertain energy consumption of buses. We first formulate a deterministic optimization model to determine the types and the deployment of fast-charging stations and the sizes of on-board batteries. Using the RO technique, we then derive the robust counterpart of the deterministic model. The uncertainty of energy consumption is bound by the intersection of the box uncertainty set and the budget uncertainty set. We apply the AARC approach to derive a less conservative and computationally tractable robust model. The model is tested with a real-world, large-scale numerical example. The results demonstrate that our model is able to solve the robust planning problem of a fast-charging electric bus system, and the optimal design is robust against uncertain energy consumption data. The comparison between the solutions of the deterministic model and those of the robust model under a worst-case scenario demonstrate that the robust model provides solutions that are robust against energy consumption uncertainty. Using different uncertainty levels, we investigate the relationship between the total cost and the level of robustness in a fast-charging electric bus system. The results may help decision-makers to determine the best trade-off between investments and the level of robustness in a fast-charging electric bus system.

The fast-charging electric bus system, as an emerging clean and sustainable alternative to a traditional diesel bus system, may be widely adopted in the near future. The proposed modeling framework in this study provides practitioners with an effective tool in the design of a robust optimal fast-charging electric bus system.

Acknowledgments

The study was partially sponsored by the U.S. Department of Energy (DE-EE0007997).

Author Contributions

The authors confirm contribution to the paper as follows: study conception and design: Zhaocai Liu, Ziqi Song, Yi He; data collection: Yi He; analysis and interpretation of results: Zhaocai Liu, Ziqi Song, Yi He; draft manuscript preparation: Zhaocai Liu. All authors reviewed the results and approved the final version of the manuscript.

References

1. Chen, J., B. Atasoy, T. Robenek, M. Bierlaire, and M. Thémans. Planning of Feeding Station Installment for Electric Urban Public Mass-transportation System. *Proc., 13th Swiss Transportation Research Conference*, 2013, No. EPFL-CONF-195881.
2. Liu, Z., Z. Song, and Y. He. Optimal Deployment of Dynamic Wireless Charging Facilities for an Electric Bus System. *Transportation Research Record: Journal of the Transportation Research Board*, 2017. 2647: 100–108.
3. Kunith, A., R. Mendelevitch, A. Kuschmierz, and D. Goehlich. Optimization of Fast Charging Infrastructure for Electric Bus Systems–Electrification of a City Bus Network. *Proc., 29th International Electric Vehicle Symposium (EVS29), EDTA*, Washington, D.C., 2016, pp. 1–12.
4. Eudy, L., R. Prohaska, K. Kelly, and M. Post. *Foothill Transit Battery Electric Bus Demonstration Results*. NREL/TP-5400-65274. National Renewable Energy Laboratory, Golden, CO, November 2015.
5. Miles, J., and S. Potter. Developing a Viable Electric Bus Service: The Milton Keynes Demonstration Project. *Research in Transportation Economics*, Vol. 48, 2014, pp. 357–363.
6. Li, J. Q. Battery-electric Transit Bus Developments and Operations: A Review. *International Journal of Sustainable Transportation*, Vol. 10, No. 3, 2016, pp. 157–169.
7. Kunith, A., R. Mendelevitch, and D. Goehlich. Electrification of a City Bus Network–An Optimization Model for Cost-effective Placing of Charging Infrastructure and Battery Sizing of Fast-charging Electric Bus Systems *International Journal of Sustainable Transportation*, Vol. 11, No. 10, 2017, pp. 707–720.
8. Ben-Tal, A., and A. Nemirovski. Robust Solutions of Uncertain Linear Programs. *Operations Research Letters*, Vol. 25, No. 1, 1999, pp. 1–13.
9. Ben-Tal, A., A. Goryashko, E. Guslitzer, and A. Nemirovski. Adjustable Robust Solutions of Uncertain Linear Programs. *Mathematical Programming*, Vol. 99, No. 2, 2004, pp. 351–376.
10. Ben-Tal, A., L. El Ghaoui, and A. Nemirovski. *Robust Optimization*. Princeton University Press, 2009.
11. Ben-Tal, A., B. Do Chung, S. R. Mandala, and T. Yao. Robust Optimization for Emergency Logistics Planning: Risk Mitigation in Humanitarian Relief Supply Chains. *Transportation Research Part B: Methodological*, Vol. 45, No. 8, 2011, pp. 1177–1189.
12. Lu, C. C. Robust Multi-period Fleet Allocation Models for Bike-sharing Systems. *Networks and Spatial Economics*, Vol. 16, No. 1, 2016, pp. 61–82.
13. Chung, D. B., T. Yao, C. Xie, and A. Thorsen. Robust Optimization Model for a Dynamic Network Design Problem Under Demand Uncertainty. *Networks and Spatial Economics*, Vol. 11, No. 2, 2011, pp. 371–389.
14. Liu, Z., and Z. Song. Robust Planning of Dynamic Wireless Charging Infrastructure for Battery Electric Buses. *Transportation Research Part C: Emerging Technologies*, Vol. 83, 2017, pp. 77–103.
15. Qin, N., A. Gusrialdi, R. P. Brooker, and T. Ali. Numerical Analysis of Electric Bus Fast-charging Strategies for Demand Charge Reduction. *Transportation Research Part A: Policy and Practice*, Vol. 94, 2016, pp. 386–396.
16. Glineur, F. *Topics in Convex Optimization: Interior-point Methods, Conic Duality and approximations*. U Mons, Belgium, 2001, Chapter 3, pp. 65–66.
17. Wang, Y., J. Jiang, and T. Mu. Context-aware and Energy-driven Route Optimization for Fully Electric Vehicles via Crowd Sourcing. *Intelligent Transportation Systems*, Vol. 14, No. 3, 2013, pp. 1331–1345.
18. SIRI VehicleMonitoring Call. *SIRI VehicleMonitoring Request*. http://developer.rideuta.com/DeveloperResources/VehicleByVehcileInstructions.aspx (accessed April 19, 2018)
19. National Research Council. *Transitions to Alternative Vehicles and Fuels*. National Academies Press, Washington, D.C., 2013.
20. Schroeder, A., and T. Traber. The Economics of Fast-charging Infrastructure for Electric Vehicles. *Energy Policy*, Vol. 43, 2012, pp. 136–144.
21. California Environmental Protection Agency. *Advanced Clean Transit Battery Cost for Heavy-Duty Electric Vehicles (Discussion Draft)*. Air Resources Board, Califonia, 2016.
22. Rosenthal, R. E. *GAMS—A User's Guide*. GAMS Development Corporation, Washington, D.C., 2012.
23. CPLEX, I. I. V12. 1: User's Manual for CPLEX. *International Business Machines Corporation*, Vol. 46, No. 53, 2009, p. 157.

The Standing Committee on Bus Transit Systems (AP050) peer-reviewed this paper (18-03360).

The views expressed are those of the authors and do not reflect the official policy or position of the project's sponsors.

Article

TRR
JOURNAL OF THE TRANSPORTATION RESEARCH BOARD

Transportation Research Record
2018, Vol. 2672(8) 108–120

DOI: 10.1177/0361198118777358
journals.sagepub.com/home/trr
SAGE

Bus Capacity Estimation using Stochastic Queuing Models for Isolated Bus Stops in China

Chao Wang[1,2,3], Zhirui Ye[1,2], Jon D. Fricker[3], Yunlong Zhang[4], and Satish V. Ukkusuri[3]

Abstract

When serving passengers at a bus stop, a dwelling bus may cause a bottleneck that constrains traffic flow near the bus stop. A major element in bus system capacity is bus stop capacity (veh/h). Therefore, this study proposes a method estimating the capacity of bus stops isolated from the influences of traffic signals and other bus stops. Data collected for the seven most common bus stop configurations in China were used to validate the proposed model. The results indicated that the arrival process obeys a Poisson distribution and the service times fit a lognormal distribution. Stochastic queuing models for both single-berth and multi-berth stops were developed to estimate bus stop capacity and bus delay time. To enable comparison, the Highway Capacity Manual (HCM) model for bus stop capacity estimation was also used. The results showed that the proposed method was more accurate and reliable, with a mean absolute percentage error (MAPE) of 8.44% compared with a MAPE of 17.42% from the HCM method. In addition, the proposed method reduced root mean squared error (RMSE) by 58.8%, mean absolute error (MAE) by 57.9%, and variance of absolute percentage error (VAPE) by 52.6% compared with the values from the HCM model. Sensitivity analyses were also conducted to investigate the effects of bus arrival rate, bus service time, and the number of bus berths on the capacity of bus stops. The results indicated that the proposed method can be used as a guide for designing bus stop facilities under specified conditions.

With the advance of modern technologies and the requirements to improve levels of service for public transportation, there has been a transition from "vehicle-oriented" (private transportation) to "people-oriented" (public transportation) in the development of urban transportation networks (*1–3*). Urban public transportation systems perform an essential function in the mobility of citizens in metropolitan areas around the world (*4–6*). Recently, more and more researchers have focused on studying the role of bus capacity, which is relevant to optimizing resources on the basis of ridership. Many studies have found that the bus stop, which serves as a crucial part in the bus system, has a great effect on traffic flow (*7*, *8*). When serving passengers at a busy stop, buses can interact in ways that limit their discharge flows (i.e., bus capacity at stops) (*9*). Thus, a dwelling bus may become a bottleneck that constrains traffic flow near the bus stop, and may degrade the bus system's overall service quality (*10–12*).

In general, any bus stop consists of a stop area, where vehicles stop, and an adjacent platform, where passengers wait for vehicles and boarding/alighting operations take place. Under light ridership conditions, bus stops seem to be a rather simple mechanism to model. Passenger inter-arrival times and bus headways are large, so the number of bus berths is enough for dwelling buses. In high demand situations, however, this is not the case. For example, at peak periods at a bus stop, if there is no unoccupied berth for an arriving bus, the bus must wait on the road, blocking traffic and impeding the progress of passengers wishing to board and alight (*10*). Thus, transit system capacity is controled by stop capacity in public transportation systems (*13*). The Highway

[1]Jiangsu Key Laboratory of Urban ITS, Southeast University, Jiangsu Province, China
[2]Jiangsu Province Collaborative Innovation Center of Modern Urban Traffic Technologies, School of Transportation, Southeast University, Jiangsu Province, China
[3]Lyles School of Civil Engineering, Purdue University, West Lafayette, IN
[4]Texas A&M University, Zachry Department of Civil Engineering, College Station, TX

Corresponding Author:
Address correspondence to Satish V. Ukkusuri: sukkusur@purdue.edu

Capacity Manual (HCM) reports that the capacity of a bus stop is inversely proportional to the average bus service time and a second term that accounts for the variation in this service time such as delay time at bus stop (*14*). Delay at stop is the sum of two average delays: one in the entry queue, and the other in the berth after the bus has finished serving its passengers (*15*). Gu et al. (*16*) found that bus delays could be increased by placing the bus stop upstream of its neighboring intersection, rather than downstream of it. In contrast, car delays were increased by placing the stop some distance downstream of the intersection, rather than upstream.

Two categories of methods for modeling stop capacity have been reported in the literature. The first is the use of analytical models to estimate capacity with a steady state formula (*17*). The best known is the HCM formula (*14*). The parameters in this model include green time ratio, clearance time, average dwell time, standard normal variable corresponding to a desired failure rate, and coefficient of variation of dwell time. Fernández et al. (*18*) introduced a concept called capacity of divided bus stops. A divided bus stop contains berths that are separated to reduce bus interference and increase bus capacity. Jaiswal et al. (*19*) introduced bus lost time as an additional component of the minimum headway to calculate bus platform capacity. Accordingly, a busway loading bus capacity model was developed with lost time variables. The second category is the use of simulation models to calculate capacity as well as other performance measures of stop operations (*10*). Two specially-designed simulation models of stop operations can be found in the literature: IRENE (*12*) and PASSION (*11, 20, 21*). PASSION is developed as part of broader studies on the interactions between buses, passengers, and traffic at bus stops.

Although a variety of methods have been developed to estimate bus stop capacity, the performance of these models varies. For instance, Fernández and Planzer (*11*) reported that the HCM formulas tended to under-predict the stop capacity because the stochastic processes at bus stops may limit the usefulness of HCM formulas. In previous studies, the arrival rate of a bus stream and the service time of a bus stop were considered to be constant, neglecting the variation of bus arrival headway and the differences in bus stop designs (*17, 20, 22*). For a multi-berth stop, the HCM formula calculates the capacity as the product of the single-berth capacity and the number of effective berths, ignoring the influence of the bus arrival process on stop capacity (*9*).

In addition, most of the existing methods have relied (to greater or lesser degrees) on computer simulation. The dearth of analytical models for evaluating capacity has thus far hampered the understanding of cause-and-effect relations at bus stops. In light of this, an estimation method that takes into account the statistical test for bus arrival process and service time and an analytical model for single-berth and multi-berth stops is developed. In this study, the bus stops are isolated from the influences of traffic signals and other bus stops, where sufficient space exists for accommodating the bus queues.

The remaining sections of the paper are organized as follows. The second and third sections describe the data collection and proposed methodology to model bus stop capacity. The fourth section validates the proposed method and evaluates the performance by analysis and comparison. The findings and conclusions are provided in the final section.

Data Collection

Based on the right-of-way, bus lanes at bus stops can be divided into three categories: grade-separated bus lanes, at-grade bus lanes, and non-exclusive bus lanes (*23*). The bus dwell time will be affected by the layout of the bus stop (*24*). In general, the more exclusive the design, that is, the less interaction that a transit vehicle has with other traffic, the fewer impacts there will be on bus dwell time. In terms of the form, bus stops can be classified into two categories: on-line and off-line (*23*). Compared with an on-line bus stop, additional time is required for buses at an off-line stop to find an acceptable gap between consecutive vehicles to rejoin the traffic stream. It can thus be concluded that the form of a bus stop has an impact on the bus dwell time. According to the location of the cross-section, bus stops can be divided into the two categories of median and curbside. Moreover, based on the station design, bus stops can be classified into two categories: enclosed bus stop (having gate controled access to bus doors) and non-enclosed bus stop. According to the above classifications, seven types of bus stop are most commonly observed in China (*25*), as shown in Figure 1.

(a) Type 1: The at-grade bus lanes are separated from motor vehicle lanes by traffic markings. Bus stops are on-line and on the curbside.
(b) Type 2: There is no exclusive bus lane. Bus stops are on-line and on the curbside.
(c) Type 3: The at-grade bus lanes are separated from motor vehicle lanes by traffic markings. Bus stops are off-line (bay-style) and on the curbside.
(d) Type 4: There is no exclusive bus lane. Bus stops are off-line (bay-style) and on the curbside.
(e) Type 5: The grade-separated bus lanes are separated from motor vehicle lanes by separation strips. Bus stops are on-line and in the median of the cross-section.

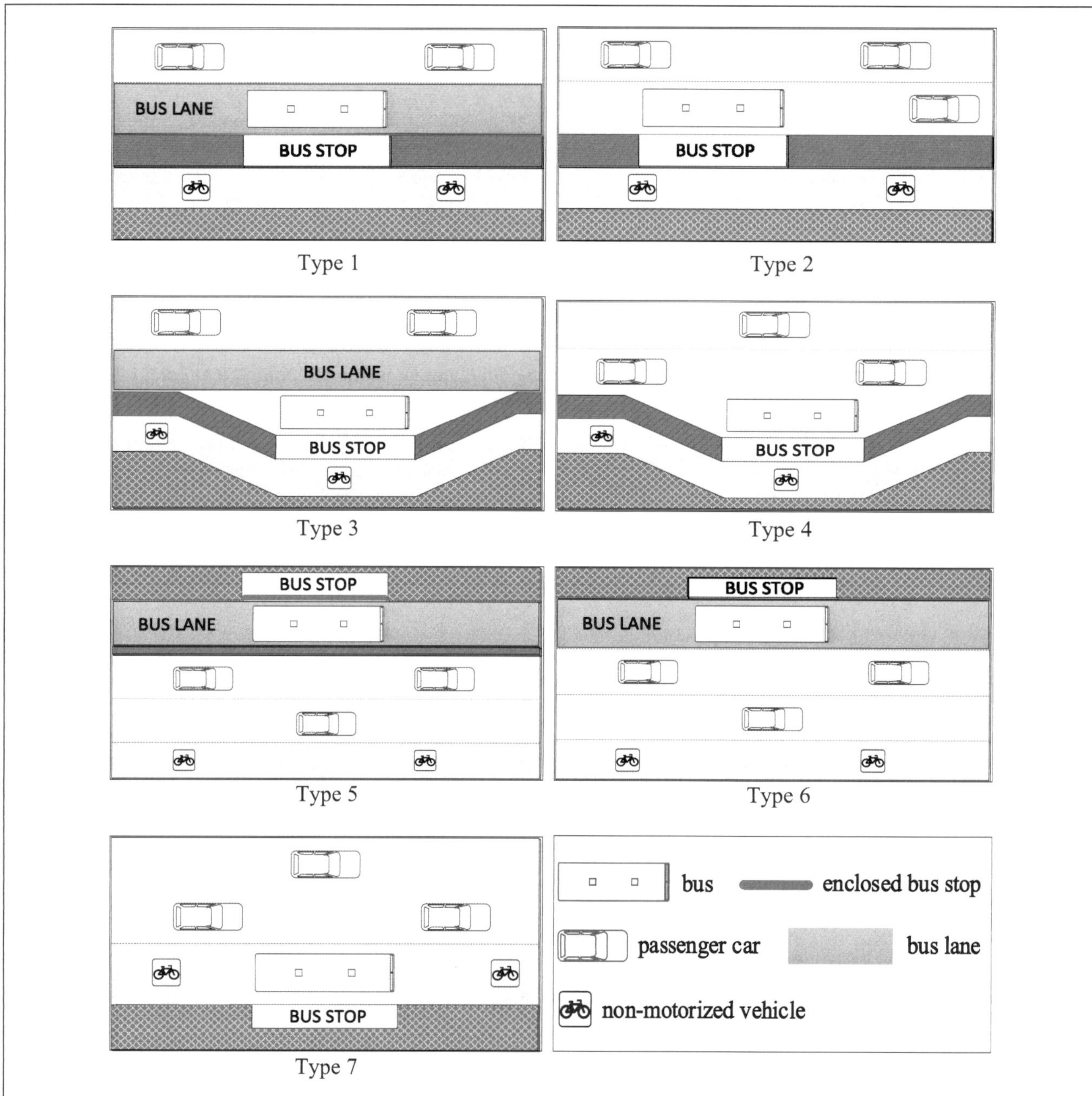

Figure 1. Schematic diagram of seven most common bus stops.

(f) Type 6: The at-grade bus lanes are separated from motor vehicle lanes by traffic markings. Bus stops are on-line and in the median of the cross-section.
(g) Type 7: There is no exclusive bus lane. Bus stops are on-line and on the curbside. Buses pull over to the curbside and occupy bicycle lanes to dwell.

In this study, data were collected for these seven different types of bus stops in the cities of Nanjing, Changzhou, and Guangzhou, China. The data were collected under good weather conditions between May 19, 2014 and June 15, 2014 to exclude potential influences of adverse weather. In addition, there was no curb parking around the stops.

Table 1. Site and Traffic Characteristics of the Bus Stops and Service Times

No.	Bus stop	Type	BN	TC	SS	V	C	AST	STV	SVC
1	Gulou North	Type 1	4	Peak	67	2677	4500	30.13	83.69	0.30
2				Non-peak	47	1723	4500	32.38	79.37	0.27
3	Beiji Huitang	Type 2	2	Peak	113	3017	3900	23.90	82.98	0.38
4				Non-peak	108	2155	3900	19.01	44.53	0.35
5	Public Transport Corporation	Type 3	5	Peak	96	3378	4500	33.64	97.10	0.29
6				Non-peak	101	2286	4500	30.80	71.32	0.27
7	Xuanwuhu Park	Type 4	5	Peak	40	3850	4500	35.83	145.43	0.33
8				Non-peak	56	3054	4500	40.96	156.25	0.30
9	Renmin Park	Type 5	3	Peak	34	2398	4000	34.37	71.35	0.25
10				Non-peak	40	1755	4000	32.86	40.27	0.20
11	Gangding	Type 6	4	Peak	51	2848	4500	31.06	56.26	0.24
12				Non-peak	52	2205	4500	26.77	61.36	0.29
13	Danfeng Street	Type 7	1	Peak	39	2078	3000	26.33	63.49	0.30
14				Non-peak	41	1386	3000	24.95	42.65	0.26

Note: BN = number of bus berths at the bus stop; TC = traffic condition (peak or non-peak period); SS = sample size (buses); V = traffic flow rate (veh/h); C = capacity (veh/h); AST = average service time of buses (s), denoted by μ; STV = variations in bus service time (s^2), denoted by σ^2; SVC = coefficient of variation in bus service time, defined as σ divided by μ.

Three video cameras were used at each stop to record traffic data. One camera was set up at a high location and two other video cameras were in front of and behind the bus stop, respectively. The recorded videos were reviewed by several trained graduate students to obtain traffic volume, bus average service time, variances in bus service time and coefficient of variation in bus service time (defined as standard deviation in bus service time divided by average service time of buses). The site and traffic characteristics of the bus stops and service times are shown in Table 1. It should be noted that the service time at several stops, such as Gulou North and Xuanwuhu Park, is higher during non-peak than peak time. This is because longer departure intervals result in more boarding and alighting passengers during non-peak.

Bus Stop Capacity Model

In operational terms, a bus stop can be considered as a concurrent queuing system between buses and passengers (*10*). It is assumed that bus stops operate in a steady state, such that the long-run average bus arrival rate never exceeds the stop's capacity (*9*). Bus stop capacity is closely related to the bus average service time at bus stops (*14*). In addition, a bus halts at the stop area for a certain duration, which depends on the number of waiting passengers. If there is no unoccupied berth for an arriving bus, the bus may have to wait on the road, blocking the traffic and impeding the progress of boarding and alighting (*10*). Consequently, the arrival rate of buses can affect the bus service time, and thus affect the bus stop capacity.

It has been assumed that bus arrival at stops follows a Poisson process (*26*, *27*). The interarrival time is an independent, exponentially distributed random variable with arrival rate of buses λ. Some studies also show that bus service time at stops follows some general distribution, and the distribution of an individual bus service time is independent of the number of berths (*9*, *28*, *29*). To examine and explore further the distributions of bus arrivals and service time, we collected data on 885 buses from the 14 sites, as shown in Table 1.

Bus Arrival Process

Different bus routes have different departure intervals (i.e., pre-set headways). The distribution of bus arrivals with multiple routes would also be influenced by buses on other bus routes. In addition, many stochastic variables have effects on bus arrivals, for example, traffic conditions (*30*). Thus, to validate the Poisson process for bus arrivals, the Kolmogorov-Smirnov test (K-S test) is introduced as a criterion. The K-S test can be used to compare a sample with a reference probability distribution, or to compare two samples (*31*). The K-S statistic quantifies a distance between the empirical distribution function of the sample and the cumulative distribution function of the reference distribution, or between the empirical distribution functions of two samples. The null distribution of this statistic is calculated under the null hypothesis that the samples are drawn from the same distribution (in the two-sample case) or that the sample is drawn from the reference distribution (in the one-sample case). In this section, the Poisson process for bus arrivals is a one-sample K-S test. The results of K-S tests show that all of the P values for the 14 sites are much greater than 0.05 (varying between 0.58 and 1.00). Hence, the tests show that the results are consistent with previous studies that bus arrivals follow Poisson distribution for isolated bus stops.

Bus Service Time

In this section, three common distributions including normal, lognormal, and Weibull distribution are selected to fit the bus service time. In addition, the exponential distribution is also introduced as a comparison. Similar to the way of validating bus arrival process, distribution for bus service time is examined using the K-S test. In general, the smaller the K-S value and the larger the P value, the better the fitting performance. As shown in Table 2, among the four distributions, the lognormal distribution (boldface in Table 2) has the best performance for all bus stops during peak and non-peak periods.

In order to provide visual comparison of the fitting results for bus service time, the probability density function of normal distribution, lognormal distribution, Weibull distribution, exponential distribution, and sample data for Beiji Huitang bus stop (Type 2) and Public Transport Corporation bus stop (Type 3) in peak and non-peak periods are chosen to be presented in Figure 2. The results show that the sample data follow the probability density function of lognormal distribution better. The distribution parameters of the lognormal distribution for each bus stop in peak and non-peak periods are also displayed in the figure.

Based on the above testing results, an M/G-L/1 queuing model for single-berth stops and an M/G-L/S queuing model for multi-berth stops are developed to estimate bus stop capacity, where M represents Poisson arrivals, G-L represents lognormal bus service time, and 1(S) represents the number of bus berths. Because of the stochastic nature of bus arrivals at bus stops, the proposed models are called stochastic queuing models (SQMs).

Model for Single-Berth Stops

The Highway Capacity Manual (*14*) reports that the capacity of a bus stop is inversely proportional to the average bus service time and a second term that accounts for the variation in this service time. In this study, the expected time spent in the queuing system W_s is used to represent the sum of average bus service time and variation in this service time. Thus, the capacity for single-berth stops can be expressed as follows:

$$C_s = 3600/W_s \tag{1}$$

where C_s is the capacity for single-berth stops (veh/h); W_s is the expected time spent in the system (s).

Table 2. K-S Test for Bus Service Time

Type	Bus stop	State	Sample size	K-S value (the smaller, the better)				P value ($P > .05$)			
				Normal distribution	Lognormal distribution	Weibull distribution	Exponential distribution	Normal distribution	Lognormal distribution	Weibull distribution	Exponential distribution
Type 1	Gulou North	Peak	67	0.148	0.113	0.163	0.457	0.108	0.362	0.051	0.000
		Non-peak	47	0.113	0.061	0.113	0.444	0.587	0.994	0.552	0.000
Type 2	Beiji Huitang	Peak	113	0.149	0.087	0.146	0.393	0.013	0.363	0.016	0.000
		Non-peak	108	0.126	0.082	0.114	0.384	0.064	0.461	0.119	0.000
Type 3	Public Transport Corporation	Peak	96	0.131	0.078	0.127	0.433	0.074	0.607	0.082	0.000
		Non-peak	101	0.119	0.097	0.141	0.484	0.114	0.297	0.036	0.000
Type 4	Xuanwuhu Park	Peak	40	0.127	0.081	0.137	0.438	0.536	0.955	0.405	0.000
		Non-peak	56	0.112	0.061	0.113	0.430	0.488	0.985	0.456	0.000
Type 5	Renmin Park	Peak	34	0.102	0.082	0.109	0.470	0.964	0.997	0.909	0.000
		Non-peak	40	0.093	0.080	0.099	0.487	0.879	0.960	0.786	0.000
Type 6	Gangding	Peak	51	0.148	0.117	0.137	0.472	0.211	0.491	0.269	0.000
		Non-peak	52	0.109	0.070	0.115	0.431	0.572	0.960	0.461	0.000
Type 7	Danfeng Street	Peak	39	0.161	0.139	0.158	0.470	0.266	0.439	0.258	0.000
		Non-peak	41	0.125	0.094	0.134	0.421	0.542	0.864	0.421	0.000

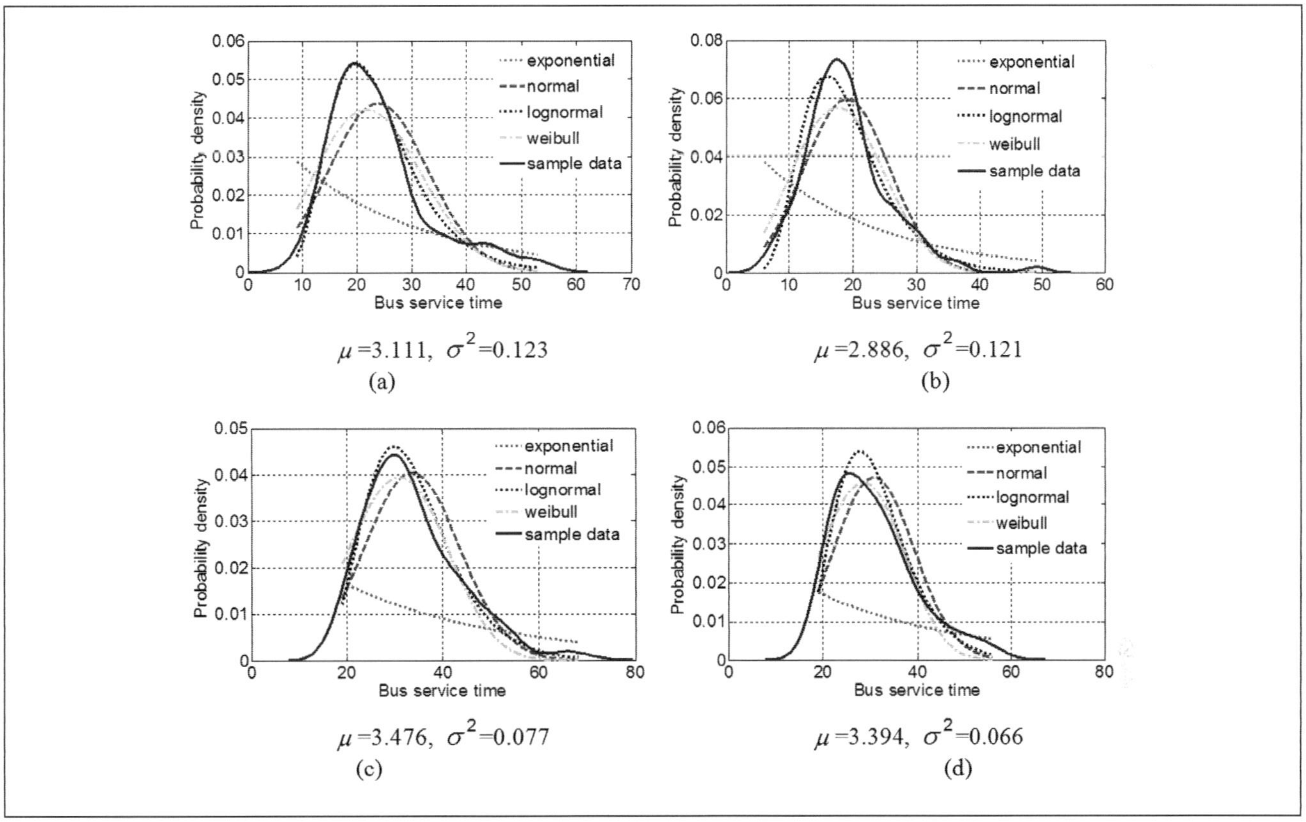

Figure 2. Probability density function of four distributions and sample data at selected bus stops: (*a*) Beiji Huitang (peak); (*b*) Beiji Huitang (non-peak); (*c*) Public Transport Corporation (peak); (*d*) Public Transport Corporation (non-peak).

The input into the system is formed by the buses approaching from upstream, and there are Poisson arrivals with a mean headway of $1/\lambda$ s. The service time at a bus stop is expressed by generally distributed variables with mean $1/u$ s. The service counter is the bus berths. According to Little's Law (*32*, *33*), W_s in the M/G-L/1 system is given by:

$$W_s = L_s/\lambda \tag{2}$$

where L_s is the expected number of buses in the system (veh); λ is the arrival rate of buses (veh/s). The expected number of buses in the system L_s can be represented as:

$$L_s = \left(\rho_s^2 + \lambda^2 \times D(T_n)\right)/(2 \times (1 - \rho_s)) + \rho_s \tag{3}$$

$$\rho_s = \lambda/u = \lambda \times E(T_n) \tag{4}$$

where ρ_s is the traffic intensity for single-berth stops; $D(T_n)$ represents the variance of service time for lognormal distribution (s^2); u is the service rate (veh/s); $E(T_n)$ represents the mean service time for lognormal distribution (s). Substituting Equations 2 to 4 into Equation 1, the capacity for single-berth stops can be expressed by:

$$C_s = (3600 \times 2 \times (1 - \lambda \times E(T_n)))/\left(2 \times E(T_n) - \lambda \times E^2(T_n) + \lambda \times D(T_n)\right) \tag{5}$$

Model for Multi-Berth Stops

In real-world bus stops, a bus's entering and exiting maneuvers will be affected by other buses dwelling at neighboring berths. Thus, the concept of the number of effective berths is introduced for multi-berth stops. The capacity for multi-berth stops can be expressed by:

$$C_m = (3600 \times N_{el})/W_s \tag{6}$$

where C_m is the capacity for multi-berth stops (veh/h) and N_{el} represents the number of effective berths. According to HCM (*14*), the first berth is 100% efficient in terms of stop capacity. If a second berth is added, it increases the capacity by 75% (N_{el} = 1.75) and a third berth may add 70% extra capacity (N_{el} = 2.45). However, the fourth and fifth berths only increase capacity by 17% (N_{el} = 2.62) and 13% (N_{el} = 2.75), respectively.

As for the multi-berth stops, similar to single-berth stops, there are Poisson arrival headways with mean $1/\lambda$

s and generally distributed service time with mean $1/u$ s. Explicit exact solutions for the M/G-L/S system have not been obtained except for some special cases, for example, the M/M/S system. It is, however, possible to obtain approximate solutions for a relatively wide class of service time distributions (*34*, *35*). Thus, the expected time W_s in the M/G-L/S system can be approximately given by:

$$W_s = W_q + E(T_n) \tag{7}$$

$$W_q = \frac{\rho_m^S \times (D(T_n) + E^2(T_n))}{(S-1)! \times 2 \times E(T_n) \times (S-\rho_m)^2} \times \left(\sum_{j=1}^{S-1} \frac{\rho_m^j}{j!} + \frac{\rho_m^S}{(S-1)! \times (S-\rho_m)} \right)^{-1} \tag{8}$$

$$\rho_m = \lambda/(S \times u) = (\lambda \times E(T_n))/S \tag{9}$$

where W_q is the expected time spent in the queue (s); ρ_m is the traffic intensity for multi-berth stops; and S denotes the number of bus berths. Substituting Equations 7 to 9 into Equation 6, the capacity for multi-berth stops can be expressed as follows:

$$C_m = \frac{3600 \times N_{el} \times (S-1)! \times 2 \times E(T_n) \times \left(S - \frac{\lambda \times E(T_n)}{S}\right)^2}{\left(\frac{\lambda \times E(T_n)}{S}\right)^S \times (D(T_n) + E^2(T_n)) \times \Pi + (S-1)! \times 2 \times E(T_n) \times \left(S - \frac{\lambda \times E(T_n)}{S}\right)^2 \times E(T_n)} \tag{10}$$

where

$$\Pi = \left(\sum_{j=1}^{S-1} \left(\frac{\lambda \times E(T_n)}{S}\right)^j /j! + \left(\frac{\lambda \times E(T_n)}{S}\right)^S / \left((S-1)! \times \left(S - \frac{\lambda \times E(T_n)}{S}\right)\right)\right)^{-1} \tag{11}$$

In addition, the mean service time $E(T_n)$ and the variance of service time $D(T_n)$ for lognormal distribution can be expressed as:

$$E(T_n) = exp\left(\mu + \frac{\sigma^2}{2}\right) \tag{12}$$

$$D(T_n) = exp(2\mu + \sigma^2)(exp(\sigma^2) - 1) \tag{13}$$

Bus Delays at Stops

In this study, bus delays at stops consist of average delay for re-entering the traffic stream and other additional delays such as boarding lost time and bus stop failure time (*25*). Average delay for re-entering the traffic stream is a function of the capacity and the degree of saturation in the vicinity of a bus stop (*14*, *17*). The analytical model used to estimate average delay assumes that the demand is less than capacity for the period of analysis. According to the HCM (*14*), in most cases, the recommended analysis period is 15 minutes.

$$t_{ad} = \frac{3600}{C} + 900 \times T \times \left(\frac{V}{C} - 1 + \sqrt{\left(\frac{V}{C} - 1\right)^2 + \frac{(3600/C) \times (V/C)}{(450 \times T)}} \right) \tag{14}$$

where t_{ad} is average delay for re-entering the car stream, and T represents the analysis time period, $T = 0.25$ for a 15-minute period.

In addition to average delay for re-entering the traffic stream, there are two additional causes of delay. One is boarding lost time, the time spent waiting for passengers to walk to the bus door(s) from their waiting position at the stop. When passengers wait at bus stops with multiple loading areas, such as high-volume stations served by multiple routes, they do not know in advance at which loading area the bus will stop when it arrives. When a bus arrives, there is typically a delay from when the bus doors open to when the first passenger arrives to board the bus (*25*). According to the field investigations, the amount of boarding lost time was found to vary by different types of bus stop design, with median values ranging from 2.9 s to 4.1 s and interquartile range values ranging from 0.4 s to 0.9 s. The second cause of delay is bus stop failure time. If bus arrives at a stop to find all the loading areas are occupied, it is forced to wait until other buses leave the stop. In addition, when a bus is ready to depart from a stop, it also has to wait if it is blocked by other buses at the stop. In this case, the bus will be delayed (i.e., bus stop failure time) while waiting for all of the buses at the stop to finish serving their passengers (*25*). According to the field investigations, the amount of bus stop failure time was also found to vary by different types of bus stops, with median values ranging from 4.9 s to 6.4 s and interquartile range values ranging from 0.5 s to 0.9 s.

Model Validation

Comparison of Results

In addition to the SQM method, the model based on the HCM formula (*14*), which is one of the most well-known analytical models, is used for bus stop capacity estimation and comparison. According to the HCM, the capacity is computed as follows:

$$C_s = \frac{3600 \times (g/c)}{t_c + (g/c) \times t_d + Z \times C_v \times t_d} \quad (15)$$

$$C_m = N_{el} \times \frac{3600 \times (g/c)}{t_c + (g/c) \times t_d + Z \times C_v \times t_d} \quad (16)$$

where g/c is the effective green time of a downstream traffic signal (as for the isolated bus stops in this study, $g/c = 1$); t_c is the clearance time between successive buses (s); and t_d is the average dwell time (s). In this study, the bus service time includes t_c and t_d; C_v is the coefficient of variation in bus service time; and Z is the standard normal variable corresponding to a desired failure rate. The above values are provided in Table 1.

The SQM method and HCM method are compared using field data at the seven bus stops. The bus flow that can be served by the stop is equivalent to the flow that can arrive at and depart from it. The literature commonly refers to this maximum bus flow as the bus stop capacity (*15*). According to relevant references (*10*, *15*, *36*), actual capacity (i.e., the number of buses that could be served by a stop) is defined as the maximum hourly equivalent of a period under saturated conditions, such as 5-minute, 10-minute, or 15-minute. Instruments and equipment used for the data collection included digital video cameras, tripods, stopwatch, and computer. Between May and June 2014, around 4 hours of data collection was conducted to observe the number of buses at each bus stop for both peak and non-peak periods. Taking 5 minutes, 10 minutes, and 15 minutes as analysis intervals respectively, the maximum number of buses for these periods from all the three types of intervals was selected. The number of buses was then derived from an hourly volume, which is the actual capacity of a bus stop.

Several measures of effectiveness (MOEs), including mean absolute percentage error (MAPE), root mean squared error (RMSE), mean absolute error (MAE), and variance of absolute percentage error (VAPE), were chosen to evaluate the performance of the SQM and HCM methods (*37*). These measures are calculated by:

$$MAPE = \frac{1}{n}\sum_{i=1}^{N}\left|\frac{\hat{y}_l - y_i}{y_i}\right| \quad (17)$$

$$RMSE = \sqrt{\frac{\sum_{i=1}^{N}(\hat{y}_l - y_i)^2}{N}} \quad (18)$$

$$MAE = \frac{\sum_{i=1}^{N}|\hat{y}_l - y_i|}{N} \quad (19)$$

$$VAPE = \sqrt{\frac{N\sum_{i=1}^{N}\left(\frac{|\hat{y}_l - y_i|}{y_i}\right)^2 - \left(\sum_{i=1}^{N}\frac{|\hat{y}_l - y_i|}{y_i}\right)^2}{N(N-1)}} \quad (20)$$

where y_i and $\hat{y}_i$ are actual and estimated capacity, respectively.

To evaluate the performance of the two methods, the comparison data of MAPE at seven bus stops in peak and non-peak periods are presented in Table 3. According to the results, the SQM method has more accurate and reliable estimation (with 8.09% of MAPE on average) than the HCM method (with 16.53% of MAPE on average). In addition, the MAPE values of the SQM method at all bus stops are below 15%, with the lowest value of 0.53% at Gangding bus stop in non-peak period. The performance of the HCM method, however, is inferior to the SQM method, with almost half of the MAPE values over 20%.

Table 4 summarizes all of the values of the MOEs for the two methods. Overall, the results clearly show that the SQM method has more accurate and reliable estimation. Compared with the values of the HCM model, the SQM method reduced MAPE by 51.1%, RMSE by 58.8%, MAE by 57.9%, and VAPE by 52.6%.

In addition, the results of bus delays for different types of bus stops are shown in Figure 3. From Figure 3*a* and *b*, cumulative curves for bus delays at Type 1, Type 3, Type 5, and Type 6 bus stops were invariably to the left of the curves for other types of bus stops during peak and non-peak periods. This is because the buses in grade-separated busways (Type 5) and at-grade busways (Types 1, 3, and 6) cannot be disturbed by other non-bus vehicles. Overall, bus delay times at stops during peak period were longer than the values during non-peak period.

Sensitivity Analyses

In order to analyze fully the proposed SQM method, we use sensitivity analyses to investigate the effects of bus arrival rate, service time, and the number of bus berths on the capacity of bus stops. Differences in the arrival rate of the bus stream affect bus capacity at single-berth stops, as shown in Figure 4. At the same bus service time, the probability of no stopped bus at the stop

Table 3. Comparison of MAPE at Bus Stops in Peak and Non-Peak Periods

				SQM method		HCM method	
No.	Bus stop/ number of berths	State	Field data (veh/h)	Capacity (veh/h)	MAPE (%)	Capacity (veh/h)	MAPE (%)
1	Gulou North/ 4	Peak	316	314	0.63	242	23.42
2		Non-peak	270	290	7.41	225	16.67
3	Beiji Huitang/ 2	Peak	282	245	13.12	201	28.72
4		Non-peak	274	315	14.96	247	9.85
5	Public Transport Corporation/ 5	Peak	318	296	6.92	240	24.53
6		Non-peak	362	319	11.88	259	28.45
7	Xuanwuhu Park/ 5	Peak	264	275	4.17	218	17.42
8		Non-peak	223	241	8.07	187	16.14
9	Renmin Park/ 3	Peak	231	256	10.82	220	4.76
10		Non-peak	234	267	14.10	238	1.71
11	Gangding/ 4	Peak	335	356	6.27	254	24.18
12		Non-peak	379	377	0.53	287	24.27
13	Danfeng Street/ 1	Peak	103	111	7.77	107	3.88
14		Non-peak	122	114	6.56	113	7.38
Average				MAPE	8.09	MAPE	16.53

Table 4. Comparison of MOEs

	Measures of effectiveness			
Method	MAPE (%)	RMSE	MAE	VAPE
SQM method	8.086	24.668	20.786	0.045
HCM model	16.527	59.918	49.357	0.095

decreases with the increasing arrival rates of the bus stream, and further reduces the bus stop capacity. In addition, the service time of buses at the stop affects bus stop capacity. At the same arrival rate, the increasing bus service time increases the probability of stopped bus at the bus stop. The stopped bus blocks bus streams near the stop and impedes the progress of boarding and alighting, and further reduces bus stop capacity from 217 veh/h to 67 veh/h at the arrival rate 0.01 veh/s, from 204 veh/h to 51 veh/h at the arrival rate 0.015 veh/s, and from 190 veh/h to 29 veh/h at the arrival rate 0.02 veh/s, respectively. In this study, when the arrival rate is 0.02 veh/s and bus service time is more than 35 s, the bus stop capacity is reduced to fewer than 50 veh/h. This is in accordance with the actual bus operations at bus stops, which becomes unsteady when bus arrival rate is near to or exceeds the service rate, that is, when the traffic intensity is close to or exceeds 1.0.

The results of capacity estimation for multi-berth stops are shown in Figure 5. Similar to single-berth stops, at a same arrival rate, the bus stop capacity reduces as the bus service time increases. For example, as for the three-berth stops, bus stop capacity is reduced from 439 veh/h to 170 veh/h at the arrival rate 0.06 veh/s, as bus service time increases from 20 s to 50 s. In addition, as shown in Figures 5*a* and *b*, at a same bus service time, the increase in arrival rates of the bus stream significantly reduces the bus stop capacity at two-berth and three-berth stops. However, the trend does not occur at the stops whose number of bus berths is more than three. This is reasonable because the probability of unoccupied berth(s) at the stop increases with the increase of the number of bus berths. Hence, as for the four-berth (Figure 5*c*) and five-berth (Figure 5*d*) stops, the bus stop capacity is not sensitive to arrival rates. For instance, when the number of bus berths is equal to four, the maximum reduction of bus stop capacity is merely 7 veh/h when the arrival rate of buses changes from 0.01 veh/s to 0.15 veh/s.

To understand the degree to which bus berths and bus service time affect capacity at bus stops, the SQM method is further used to explore the change of bus stop capacity when arrival rate of buses is 0.01 veh/s (36 veh/h) and bus service time is between 15 s and 45s. The curves in Figure 6*a* display the differences in how the number of bus berths affects bus stop capacity. When the bus service time is short, additional berths could produce increasing returns in capacity. For example, Figure 6*a* shows that when bus service time is equal to 20 s, two-berth stops increase bus stop capacity by 141 veh/h over one-berth stops. This favorable trend does not continue, however, as bus service time increases. In addition, at the same bus service time, the incremental changes in bus stop capacity diminish dramatically as the number of bus berths increases (Figure 6*b*). When bus service time is 25 s, the increasing returns in capacity are reduced from 134

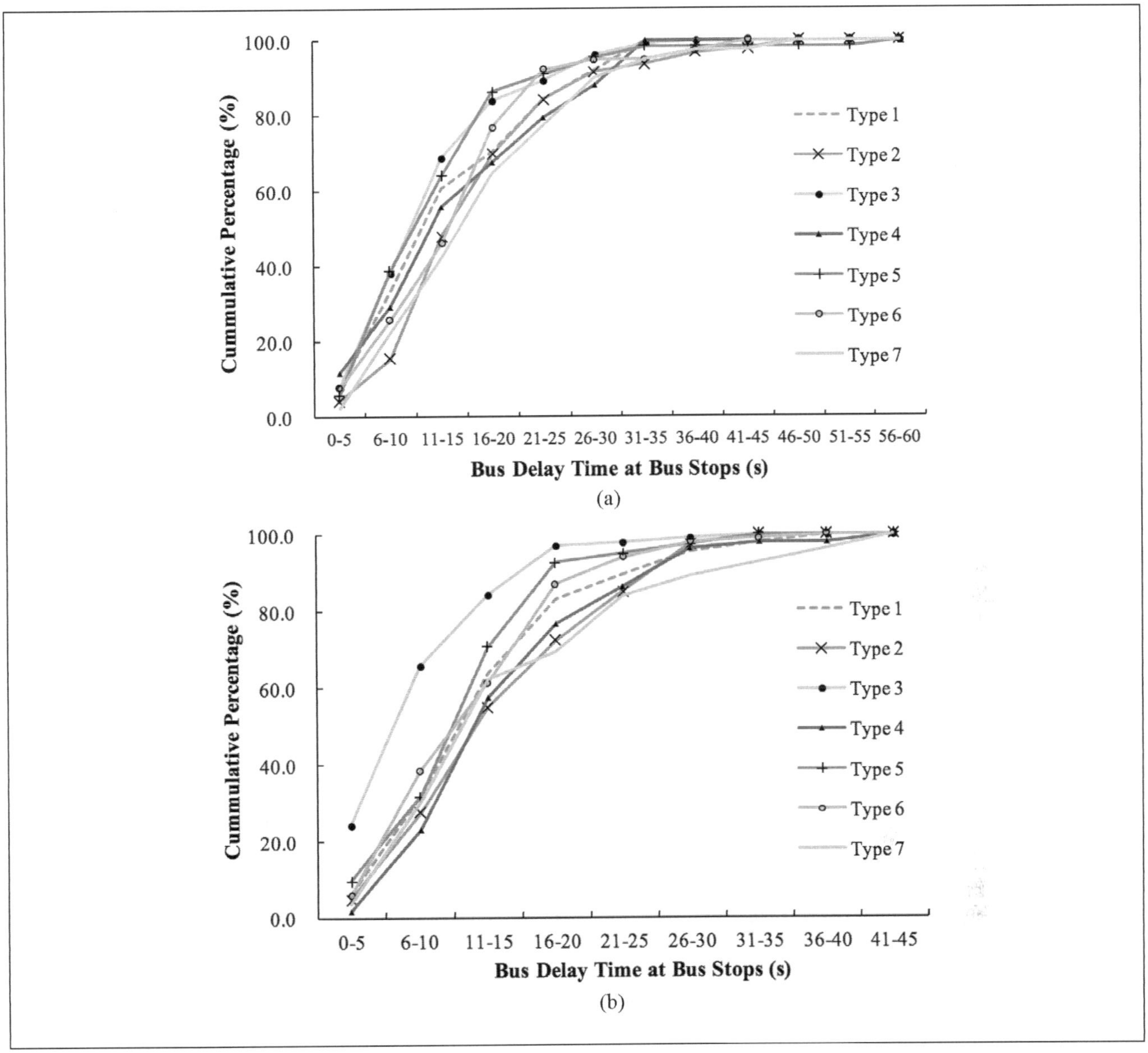

Figure 3. Bus delay time at different types of bus stops for (a) peak period and (b) non-peak period.

veh/h to 19 veh/h for the first berth through the fifth berth, especially for the fourth and fifth berths, with increases of 25 veh/h and 19 veh/h, respectively. This finding is in accordance with the actual bus operations at bus stops; current practice states that it is not efficient to build bus stops with more than three adjacent berths (9). If more than three berths are necessary, the stop should be split in two multi-berth stops (e.g., a four-berth stop can be split into two two-berth stops). The results further indicate that the method is well validated and could be used in practice for the analysis and estimation of bus stop capacity in China.

Conclusions

This study investigated bus arrival process and bus service time distribution, and applied a stochastic queuing method to estimate the capacity for isolated bus stops. The method not only takes into account the statistical test for bus arrival process and bus service time, but also considers the analytical model for single-berth and multi-berth stops.

In this study, data collected from seven bus stop configurations in China were used to validate the SQM. A K-S test was performed to explore the bus arrival process and bus service time distribution. According to the P

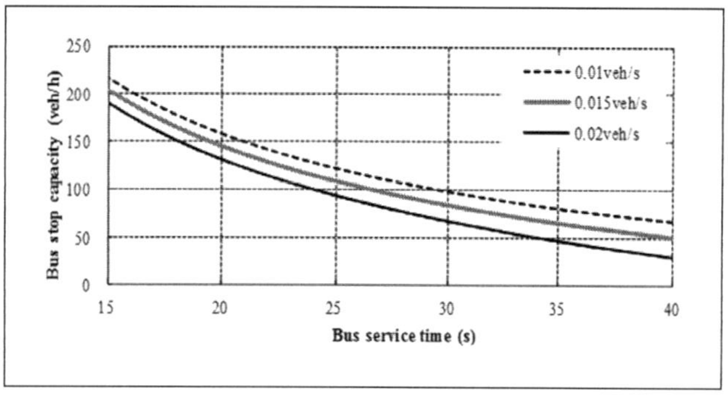

Figure 4. Bus stop capacity with different arrival rates of bus stream and bus service time at single-berth stops.

values and K-S values, bus service times followed the log-normal distribution, and bus arrival process obeyed the Poisson distribution. An M/G-L/1 queuing model for single-berth stops and an M/G-L/S model for multi-berth stops were developed to estimate bus stop capacity and bus delay time. To evaluate the SQM's performance, a model based on the HCM formula was selected for comparison. According to the analysis results, the five main findings were as follows:

(1) The SQM method had more accurate and reliable estimation, with an average MAPE of 8.44% compared with an average MAPE of 17.42% from the HCM model. Compared with the values of the HCM model, the SQM method reduced MAPE by 51.1%, RMSE by 58.8%, MAE by 57.9%, and VAPE by 52.6%.

(2) Bus delay times at Type 1, Type 3, Type 5, and Type 6 bus stops were shorter than for other types of bus stops during peak and non-peak periods. This was because the buses in grade-separated and at-grade busways cannot be disturbed by other non-bus vehicles. Moreover, bus delay times at stops during peak period were longer than the values during non-peak period.

(3) Sensitivity analyses were conducted to investigate the effects of bus arrival rate, number of bus berths, and bus service time on the capacity of bus stops. For single-berth stops, at the same bus service time, the probability of no stopped bus at the stop decreases with the increasing arrival rates of the buses and further reduces the bus stop capacity. At the same arrival rate, increasing bus service time increases the probability of stopped bus at the bus stop, and further reduces bus stop capacity from 217 veh/h to 67 veh/h at the arrival rate 0.01 veh/s, from 204 veh/h to 51 veh/h at the arrival rate 0.015 veh/s, and from

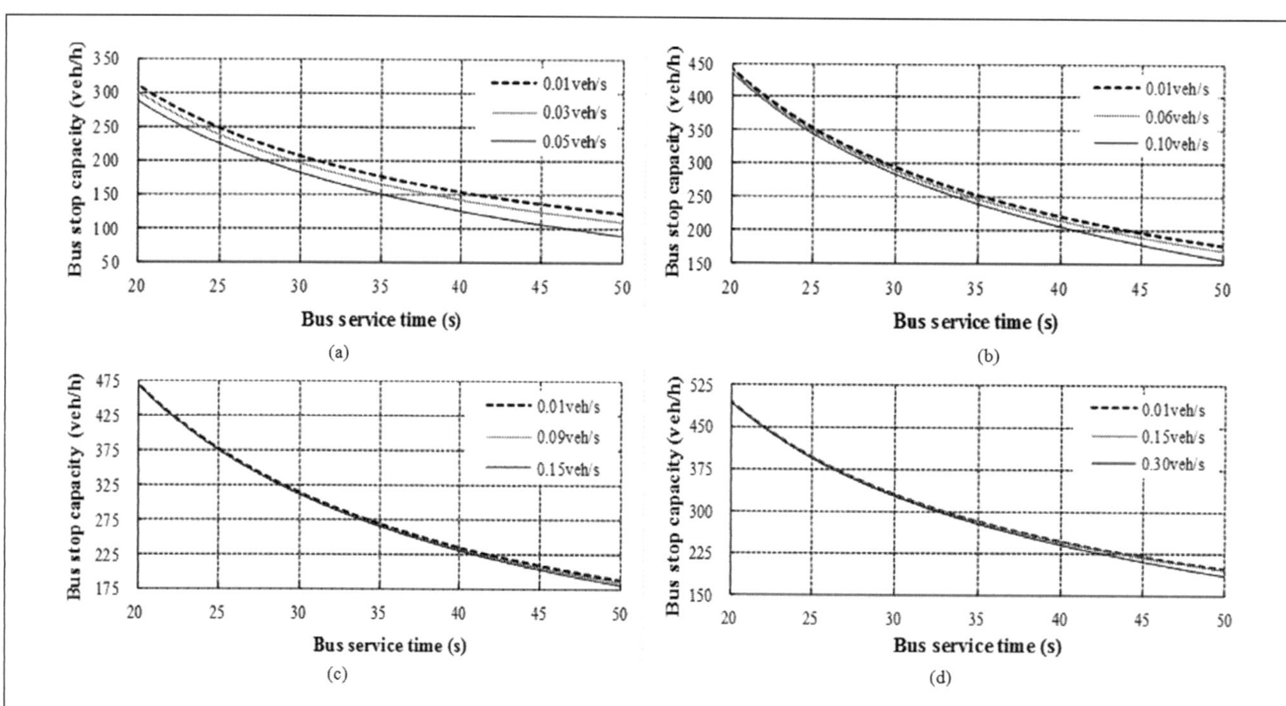

Figure 5. Bus stop capacity with different arrival rates of bus stream and bus service times at multi-berth stops: (*a*) two berths; (*b*) three berths; (*c*) four berths; (*d*) five berths.

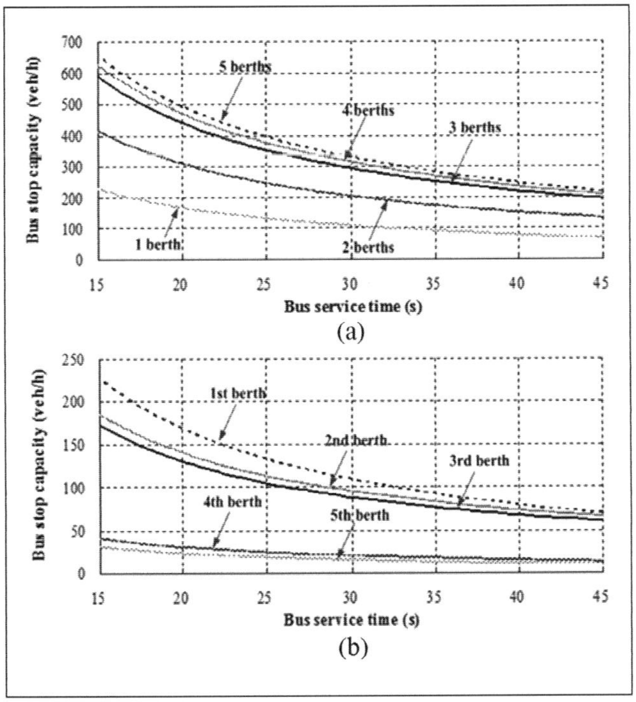

Figure 6. Bus stop capacity and incremental changes in capacity versus bus service time for different berths: (*a*) bus stop capacity versus bus service time; (*b*) incremental changes in capacity versus bus service time.

190 veh/h to 29 veh/h at the arrival rate 0.02 veh/s, respectively.

(4) Similar to single-berth stops, at a same arrival rate, the bus stop capacity for multi-berth stops reduces as the bus service time increases. At the same bus service time, the increase in arrival rates of the bus stream significantly reduced the bus stop capacity at two-berth and three-berth stops. However, this trend does not occur at stops where the number of bus berths is more than three. This showed that the probability of unoccupied berth(s) at the stop increased with the increase of the number of bus berths.

(5) In addition, the results show that it is not efficient to build bus stops with more than three adjacent berths. If more than three berths are necessary, the stop should be split in two multi-berth stops (e.g., a four-berth stop can be split into two two-berth stops).

This study explores the capacity of isolated bus stops in general weather conditions. In different weather conditions, such as inclement weather, the bus average service time and delay time at bus stop may be different. Thus, the proposed method can be applied in different weather conditions. To adapt the method, the research team will need to collect traffic data in other weather conditions to obtain the corresponding characteristics, such as bus service time, and conduct further research to investigate their impacts.

Acknowledgments

This study was sponsored by the Scientific Research Foundation of the Graduate School of Southeast University (No. YBJJ1633), the Fundamental Research Funds for the Central Universities, the Scientific Research Program of the Graduate School of Jiangsu Province (No. KYLX16-0280), and the China Scholarship Council.

Author Contributions

The authors confirm contribution to the paper as follows: study conception and design: Chao Wang, Zhirui Ye, Satish V. Ukkusuri; data collection: Chao Wang, Zhirui Ye; analysis and interpretation of results: Chao Wang, Jon D. Fricker, Satish V. Ukkusuri; draft manuscript preparation: Chao Wang, Yunlong Zhang, Jon D. Fricker, Satish V. Ukkusuri, Zhirui Ye. All authors reviewed the results and approved the final version of the manuscript.

References

1. Yu, B., Z. Yang, P. Jin, S. Wu, and B. Yao. Transit Route Network Design-Maximizing Direct and Transfer Demand Density. *Transportation Research Part C*, Vol. 22, 2012, pp. 58–75.
2. Bagloee, S. A., and A. Ceder. Transit-Network Design Methodology for Actual-Size Road Networks. *Transportation Research Part B*, Vol. 45, 2011, pp. 1787–1804.
3. Das, S., and D. Pandit. Determination of Level-of-Service Scale Values for Quantitative Bus Transit Service Attributes Based on User Perception. *Transportmetrica A: Transport Science*, Vol. 11, No. 1, 2015, pp. 1–21.
4. Gonzalez, E. M., M. G. Romana, and O. M. Alvaro. Bus Dwell-Time Models of Main Urban Route Stops. *Transportation Research Record: Journal of the Transportation Research Board*, 2012. 2274: 126–134.
5. Bordagaray, M., L. dell'Olio, A. Ibeas, and P. Cecin. Modeling User Perception of Bus Transit Quality Considering User and Service Heterogeneity. *Transportmetrica A: Transport Science*, Vol. 10, No. 8, 2014, pp. 705–721.
6. Chang, H., D. Park, S. Lee, H. Lee, and S. Back. Dynamic Multi-Interval Bus Travel Time Prediction Using Bus Transit Data. *Transportmetrica*, Vol. 6, No. 1, 2010, pp. 19–38.
7. Ding, H., W. Wang, T. Luo, Z. Yang, Y. Li, and Z. Li. Cellular Automata Based Modeling for Evaluating Different Bus Stop Designs In China. *Discrete Dynamics in Nature and Society*, Vol. 2015, 2015, pp. 1–10.
8. Tirachini, A. Bus Dwell Time: The Effect of Different Fare Collection Systems, Bus Floor Level and Age of Passengers. *Transportmetrica A: Transport Science*, Vol. 9, No. 1, 2013, pp. 28–49.

9. Gu, W., Y. Li, M. J. Cassidy, and J. B. Griswold. On the Capacity of Isolated, Curbside Bus Stops. *Transportation Research Part B*, Vol. 45, No. 4, 2011, pp. 714–723.
10. Fernández, R. Modeling Public Transport Stops by Microscopic Simulation. *Transportation Research Part C*, Vol. 18, No. 6, 2010, pp. 856–868.
11. Fernández, R., and R. Planzer. On the Capacity of Bus Transit Systems. *Transport Reviews*, Vol. 22, No. 3, 2002, pp. 267–293.
12. Gibson, J., I. Baeza, and L. G. Willumsen. Bus Stops, Congestion, and Congested Bus Stops. *Traffic Engineering and Control*, Vol. 30, No. 6, 1989, pp. 291–302.
13. Vuchic, V. R. *Urban Transit Operations Planning and Economics*. Wiley, Hoboken, NJ, 2005.
14. *Highway Capacity Manual*. Transportation Research Board, Washington, D.C., 2000.
15. Gu, W., M. J. Cassidy, and Y. Li. Models of Bus Queueing at Curbside Stops. *Transportation Science*, Vol. 49, No. 2, 2015, pp. 204–212.
16. Gu, W., V. V. Gayah, M. J. Cassidy, and N. Saade. On the Impacts of Bus Stops near Signalized Intersections: Models of Car and Bus Delays. *Transportation Research Part B*, Vol. 68, 2014, pp. 123–140.
17. Yang, X., Z. Gao, X. Zhao, and B. Si. Road Capacity at Bus Stops with Mixed Traffic Flow in China. *Transportation Research Record: Journal of the Transportation Research Board*, 2009. 2111: 18–23.
18. Fernández, R., V. Burgos, and C. E. Cortés. Results of the Microscopic Modeling of Traffic Interactions at Stops, Junctions and Roads for Designing of Bus Rapid Transit Facilities. In: *European Transport Conference 2007*, Association for European Transport and Contributors, Noordwijkerhout, Netherlands, 2007, pp. 17–19.
19. Jaiswal, S., J. Bunker, and L. Ferreira. Modeling the Relationships between Passenger Demand and Bus Delays at Busway Stations. Presented at 88th Annual Meeting of the Transportation Research Board, Washington, D.C., 2009.
20. Fernández, R. A New Approach to Bus Stop Modeling. *Traffic Engineering and Control*, Vol. 42, No. 7, 2001, pp. 240–246.
21. Fernández, R., and N. Tyler. Effect of Passenger–Bus-Traffic Interactions on Bus Stop Operation. *Transportation Planning and Technology*, Vol. 28, No. 4, 2005, pp. 273–292.
22. Koshy, R. Z., and V. A. Arasan. Influence of Bus Stops on Flow Characteristics of Mixed Traffic. *Journal of Transportation Engineering*, Vol. 131, No. 8, 2005, pp. 640–643.
23. KFH Group. *Transit Capacity and Quality of Service Manual*, 3rd ed. Transportation Research Board, Washington, D.C., 2013.
24. TCRP Report 19. Guidelines for the Location and Design of Bus Stops. http://www.gulliver.trb.org/publications/tcrp/tcrp_rpt_19-a.pdf. Accessed January 16, 1996.
25. Wang, C., Z. Ye, Y. Wang, Y. Xu, and W. Wang. Modeling Bus Dwell Time and Time Lost Serving Stop in China. *Journal of Public Transportation*, Vol. 19, No. 3, 2016, pp. 55–77.
26. Danas, A. Arrivals of Passengers and Buses at Two London Bus Stops. *Traffic Engineering and Control*, Vol. 21, No. 10, 1980, pp. 472–475.
27. Kohler, U. Capacity of Transit Lanes. *Proc., International Symposium on Highway Capacity*, Karlsruhe, Germany, 1991.
28. Gross, D., J. F. Shortle, J. M. Thompson, and C. M. Harris. *Fundamentals of Queueing Theory*, 4th ed. John Wiley & Sons, Inc., Hoboken, N.J., 2008.
29. Gu, W., and M. J. Cassidy. Maximizing Bus Discharge Flows from Multi-Berth Stops by Regulating Exit Maneuvers. *Transportation Research Part B*, Vol. 56, 2013, pp. 254–264.
30. Yu, B., W. H. K. Lam, and M. L. Tam. Bus Arrival Time Prediction at Bus Stop with Multiple Routes. *Transportation Research Part C*, Vol. 19, 2011, pp. 1157–1170.
31. Corder, G. W., and D. I. Foreman. *Nonparametric Statistics: A Step-By-Step Approach*. Wiley, Hoboken, N.J., 2014.
32. Little, J. D. C. A Proof for the Queuing Formula: $L = \lambda W$. *Operations Research*, Vol. 9, No. 3, 1961, pp. 383–387.
33. Sztrik, J. *Basic Queueing Theory*. University of Debrecen: Faculty of Informatics, 2011.
34. Kimura, T. Diffusion Approximation for an M/G/m Queue. *Operations Research*, Vol. 31, No. 2, 1983, pp. 304–321.
35. Zhang, W., L. Chen, Y. Huang, L. Niu, M. Huang, D. Zhang, and W. Shi. Application of M/G/K Queuing Model in Queuing System of Electric Taxi Charging Station. *Power System Technology*, Vol. 39, No. 3, 2015, pp. 724–729.
36. Sun, F., D. Wang, D. Ma, and X. Ma. Method for Calculating the Capacity of Bus Bay. *Journal of Southeast University (English Edition)*, Vol. 28, No. 4, 2012, pp. 485–489.
37. Wang, C., and Z. Ye. Traffic Flow Forecasting Based on a Hybrid Model. *Journal of Intelligent Transportation Systems*, Vol. 20, No. 5, 2016, pp. 428–437.

The Standing Committee on Bus Transit Systems (AP050) peer-reviewed this paper (18-03627).

Article

TRR
JOURNAL OF THE TRANSPORTATION RESEARCH BOARD

Transportation Research Record
2018, Vol. 2672(8) 121–134

DOI: 10.1177/0361198118790132
journals.sagepub.com/home/trr
SAGE

Does Roadway Performance Affect Transit Headway Unreliability? Evidence from Mixed-Traffic Transit Corridors in Toronto, Canada

Christopher Yuen[1] and Matthias N. Sweet[2]

Abstract

This research assesses the links between roadway performance and transit headway unreliability. Frequent and reliable transit services are valued by transit users and feature as critical components affecting generalized travel costs. But retaining high quality and reliable transit services in mixed-traffic urban contexts is challenging because more and more users are competing for limited physical road space, making an understanding of service unreliability critical for effective transportation system management in growing urban contexts. This study focuses transit headway unreliability on three major downtown Toronto arterial transit corridors—King Street, Queen Street, and Dundas Street. These corridors are primarily served by streetcars but are also occasionally served by buses in cases of diversions or street closures. This research uses Inrix, Inc. probe data to measure traffic congestion and roadway service unreliability and to explore its links with transit headway reliability. Both inferential models and descriptive statistics are presented to identify patterns in transit headway reliability and estimate the strength of association with traffic congestion. Results indicate that roadway performance levels and terminus departure headway unreliability are linked with stop-level transit headway unreliability. Nevertheless, much of the variation in transit headway unreliability remains unexplained by inferential models. Results indicate that insofar as traffic management actions can affect congestion levels, this may improve transit headway reliability. However, one should temper expectations of congestion management affecting mixed-traffic transit services, particularly in a highly urbanized, mixed-traffic environment such as downtown Toronto. Instead, findings suggest meaningful improvements in stop-level headway reliability from better initial scheduling and headways at the beginning of these transit routes.

Improving public transit has long been viewed as an antidote to or alternative to urban traffic congestion, but the relationship between urban growth and transportation services among competing users is particularly challenging in downtown cores (*1–3*). In Toronto, Ontario, Canada's largest city, several downtown transit corridors illustrate the challenges of balancing service levels and needs of different users while background congestion grows. Despite new transit services on grade-separated (ROW A) or exclusive right-of-way corridors (ROW B) in the Greater Toronto and Hamilton Area (the region around the City of Toronto), much of the surface transit network in the downtown (including both streetcars and buses) continues to operate in mixed traffic (ROW C). Mixed-traffic transit is uniquely challenged by the competing needs of other users (*3*), including the very congestion associated with urban growth. Potential implications include congestion-induced degradations to travel times, schedules, and headway reliability, making an understanding of the links between mixed-traffic transit, such as streetcars, and roadway transportation services of critical importance.

But the evidence on the links between transit performance and other competing transportation needs in a mixed-use environment—focusing on road congestion and traffic reliability—are largely untested despite their potential importance. Some studies indicate that road traffic slows public transit travel times (*4–6*), but impacts on headways and schedules (also of significant

[1]Jarrett Walker & Associates, Portland, OR
[2]School of Urban & Regional Planning, Ryerson University, Toronto, ON, Canada

Corresponding Author:
Address correspondence to Matthias N. Sweet: matthiassweet@ryerson.ca

importance to users) are unclear. The link between transit and traffic congestion is often central to the political economy of justifying transportation investments (*7*). Yet, the very traffic congestion to which transit provides an alternative may likewise affect transit services in mixed-traffic urban contexts—suggesting a need to re-examine how user and mode priority are allocated in highly congested urban mixed-use rights-of-way.

Focusing on improving transit headway unreliability, it is unclear under what circumstances transportation planners should prioritize transit operational improvements or roadway service improvements and congestion alleviation measures (e.g., right-of-way priority). If reduced transit performance is chiefly a function of traffic congestion and roadway performance, this would suggest policy recommendations such as increased transportation demand management strategies or signal timing changes may be more effective. However, if transit reliability is fundamentally determined by initial transit operating conditions, this suggests a focus on improving transit operating procedures would yield the best transit service improvements. From a practical perspective, the decision which city operating divisions (in this case, the Toronto Transit Commission or the Transportation Services Division) should act most decisively (respectively by focusing on transit operations or alleviating road congestion) to improve transit services hinges on the answer to this question.

The purpose of this study is to test the links between traffic congestion and transit unreliability on three urban streetcar corridors in urban Toronto, focusing on predictors of transit headway unreliability. This study is limited to the King, Queen, and Dundas Street transit corridors in Toronto during the period of January 1, 2014, to June 30, 2016, but it provides a framework which could be adapted in other contexts.

Literature Review

Transit service improvements are often justified based on multiple potential objectives, including improving transit customer experiences, shifting travel behavior, and alleviating traffic congestion. Research on transit service provision has identified several causes and consequences of transit services elements which shape service quality and customer experiences in different contexts. In particular, high-growth regions with increasing traffic congestion and limited roadway space in mixed-traffic environments, such as in Toronto (*8*), establish particularly challenging conditions to balance needs of transit customers and transit service provision options.

Transit customers value frequent, regular transit services at station stops and place high importance on transit headways and headway reliability (*1*, *9–11*). The links between travel unreliability, traffic, and travel behavior have been explored long before the ubiquity of global positioning system devices and smartphones (*12*, *13*), leading to conclusions regarding the importance of travel unreliability but identifying measurement challenges in better understanding the causes and consequences of unreliability. Significant research on unreliability has further formalized its theoretical role in travel behavior and has been empirically tested with newer data and information technology sources (*13*, *14*). More widely available information technology, global positioning systems, and electronic databases have enabled new research on travel unreliability and broader proliferation of such research. Nevertheless, despite the history of research in this domain, there remains less evidence to inform policymaking on the complex links between roadway service levels and transit services, particularly in mixed-traffic contexts.

Transit services are influenced by many factors that are generally subdivided into "internal" and "external" factors (*15*). Internal factors are under a transit agency's control, while external factors are shaped by other forces, such as weather, topography, roadway conditions, urban form, and roadway design (*15*, *16*). Roadway levels of service and traffic congestion are clearly outside the purview of transit operators. It is broadly unclear what the relative roles of external or internal factors may be in shaping transit ridership and performance (*16*), but other studies suggest that, on balance, transit ridership (admittedly only one indicator of service quality) may be most chiefly influenced by external factors, but that internal factors still play an important role (*17*, *18*).

Transit travel time variability for on-road public transport has been explored using empirical data within the utility theory framework (*19*), providing and testing theoretical expectations of the links between transit services, performance, and user responses. There are multiple transit service performance metrics, each of which focus on different elements of the service delivery or user experience. One type of measure, reliability, reflects not only the speed and/or convenience of service, but also the degree to which one can expect a given service experience (*20*). Based on a generalized cost framework, Casello et al. (*19*) provide a theoretical framework through which simulations based on available data illustrate how ignoring transit service unreliability is likely to underestimate generalized costs by more than 20% and that unreliability itself either directly or indirectly accounts for most of the additional travel cost. The impact of travel unreliability has been estimated by several authors, providing evidence that unreliability affects travel behavior (*13*, *14*) and incurs a significant cost to users (*21*). Some have identified impacts of transit unreliability on patronage (*22*) on stop-level ridership (*23*), travel satisfaction (*24*), and on service quality (*25*).

Transit agencies use several reliability metrics, one of which is schedule adherence. Schedule adherence may be particularly important to the customer experience when headways exceed 15 minutes (*26*). But for transit services operating at short headways, passengers value headway adherence (*27*). In such cases, whether a given transit vehicle is late or early may not be of interest. Rather, in cases of high-frequency headways where transit users arrive at random without first consulting schedules, headway reliability may be a more meaningful transit performance metric.

Likewise, while congestion alleviation is a common goal of most transportation programs, estimates of how congestion affects transit services is less clear. Existing research suggests that traffic congestion can influence the relative attractiveness of non-auto services (*28*) and mode choice (*29*). Some have used road performance measures as a means of estimating transit run times (*5, 6, 30*). For example, Mazloumi et al. (*4*) found links between transit departure delay and travel time variability in Melbourne. McKnight et al. (*6*) similarly identify links between congestion and transit operations when comparing observed and free-flow operating conditions. Likewise, Cooper and Gould (*31*) find that a 15% increase in scheduled travel time is associated with a 160% increase in deviation from scheduled headway.

The links between congestion and slower transit services in a mixed-traffic environment appear to be intuitive and have been documented: slow cars are logically expected to impede and slow transit vehicles (*30, 5*). McKnight et al. (*6*), explore both the impacts of roadway traffic congestion on both bus transit performance and bus transit costs, finding significant impacts. Other studies focus on the congestion–transit run time link simply as a means of estimating transit run times in the absence of better transit data (*5, 30*). Nevertheless, despite the slowing effects of traffic on public transit, the links between roadway performance and transit headway unreliability (the focus of this study) are less clear. The potential for congestion to degrade headway unreliability is of particular note given the potential for traffic congestion on transit lines with short headways to exacerbate transit vehicle bunching (*32, 33*). Bunching occurs when a delayed transit vehicle is further delayed by increasing dwell times as the number of waiting customers builds up during abnormally long headways. A transit vehicle scheduled immediately following the delayed vehicle may see very few customers because of the short time that has elapsed since the previous departure and will therefore catch up to the previous delayed vehicle. Phenomena such as bunching are expected to be one mechanism whereby traffic congestion could more rapidly deteriorate transit headway unreliability, but the effect of bunching from other sources of unreliability are not distinguished in this work.

Research Design

This study measures transit reliability and traffic congestion on three major downtown Toronto transit corridors (Queen Street, King Street, and Dundas Street) and explores the links between roadway performance and transit headway unreliability using both descriptive statistics and a regression modeling framework. Descriptive analyses and modeling are conducted to respectively explore general correlations between roadway performance and transit headway unreliability and to estimate short-term operational factors (including roadway congestion and transit operations) which shape transit headway unreliability.

This study combines transit probe data (from the Toronto Transit Commission) with roadway probe data (from Inrix, Inc.) to explore predictors of transit unreliability. Data processing and modeling was conducted using R and each data source is discussed below.

Performance Measures

First, measures of traffic congestion and transit unreliability are developed covering the period from January 1, 2014, through June 30, 2016. Transit data for the Queen Street (Route 501), King Street (Route 504), and Dundas Street (Route 505) corridors are obtained from the Toronto Transit Commission's Communications and Information System (CIS). These routes were chosen as high-frequency surface routes operating in mixed traffic in the growing urban Toronto environment. The 504 (King Street) route is the busiest surface route in Toronto, with over 65,000 boardings per weekday (*34*).

Streetcars are the primary transit vehicles serving each route, but both the King Street and Dundas Street routes are supplemented with buses. These differences are documented in the CIS data. Buses operate on the same street with the same operational characteristics and they serve the same customers as streetcars, so this study does not distinguish between streetcar and bus modes in estimating the links between congestion and transit unreliability. For this study, the analysis of the Queen Street corridor was truncated to eliminate the western portion of the route, some of which operates on dedicated right-of-way (ROW B).

Transit Performance. Transit service data used in this study was obtained from the Toronto Transit Commission (TTC), the city's public transit operator, and is derived from their CIS. Such probe data are increasingly employed for assessment of transit performance (*35*). Transit stop locations were drawn from the TTC's General Transit Feed Specification (GTFS) dataset published on the City of Toronto's Open Data Portal (*36*).

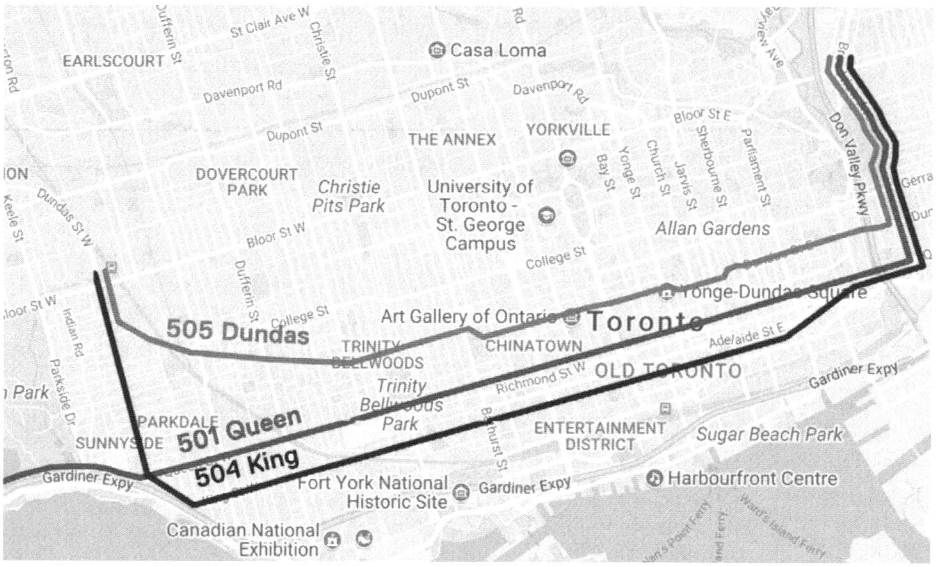

Figure 1. Transit routes locations analyzed for Dundas Street (Route 505, top), Queen Street (Route 501, middle), and King Street (Route 504, bottom).

The TTC provided archived location data for each transit vehicle on the study corridors. This CIS system collects location-at-time data every 20 seconds to create a full record of bus and streetcar daily movements. Every record includes a time stamp, a route identifier, a run number, a vehicle identifier, and geographic coordinates.

Directionally specific stop locations are mapped below in Figure 1 and illustrate that not all routes are strictly east–west; instead, Queen Street (east and west) and Dundas Street (eastern portion) proceed in a northerly direction for part of their routes. Two directionally specific stop locations on the Queen Street route are omitted based on data quality.

Transit performance measures were estimated by first identifying stop locations at which metrics would be calculated and secondly by processing TTC data to measure headway unreliability. Using the GTFS to Shapefile conversion tool (*37*), stops were mapped and time points on the study corridors were isolated. The terminus stations were removed and replaced with the second stop and the penultimate stop on each route to facilitate the analysis (e.g., identifying directionality and the beginning of runs). Because this study focuses on TTC time points at which operational adjustments to improve headway adherence may be most likely, observed headway deviations may be higher at other stops. As such, the estimates of headway variability in this study may be interpreted as lower bounds.

Since the TTC's CIS data is in "location-at-time" format, this study processes the data to estimate "time-and-location" at the transit stops of interest. The initial CIS dataset included more than 13 gigabytes of "location-at-time" records for each route, some of which included rogue data points (e.g., GPS coordinates nowhere near the transit routes). As many transit systems are installing computer-aided dispatch/automated vehicle location systems that directly record time-at-location data, similar studies in the future may be able to omit this analysis step in this study.

To subset the CIS data by the geographic location of each station, a 250 by 250 meter bounding box around each station was used to identify candidate transit vehicle locations from among which the closest point to the stop would be selected (see Figure 2). All data points outside of the bounding box of each stop were omitted to ensure that software computations are tractable. Arrival times for each stopping event were estimated using the time stamp of the CIS data point closest to each stop for each individual run on each individual day for each evening peak period (defined as 4 to 6 p.m.).

Transit operating schedules consist of "blocks," the daily assignment for an individual vehicle, "runs," a driver's daily work assignment, and "trips," a vehicle or driver's travel from one terminus to the next (*38*). Of these, the CIS data only provide data at the "run" level—meaning that multiple trips are grouped together. To avoid double-counting trips in this study, the analysis was limited to the two-hour evening peak period (4 to 6 p.m.) since no driver's run would be expected to include consecutive trips crossing the same location within a

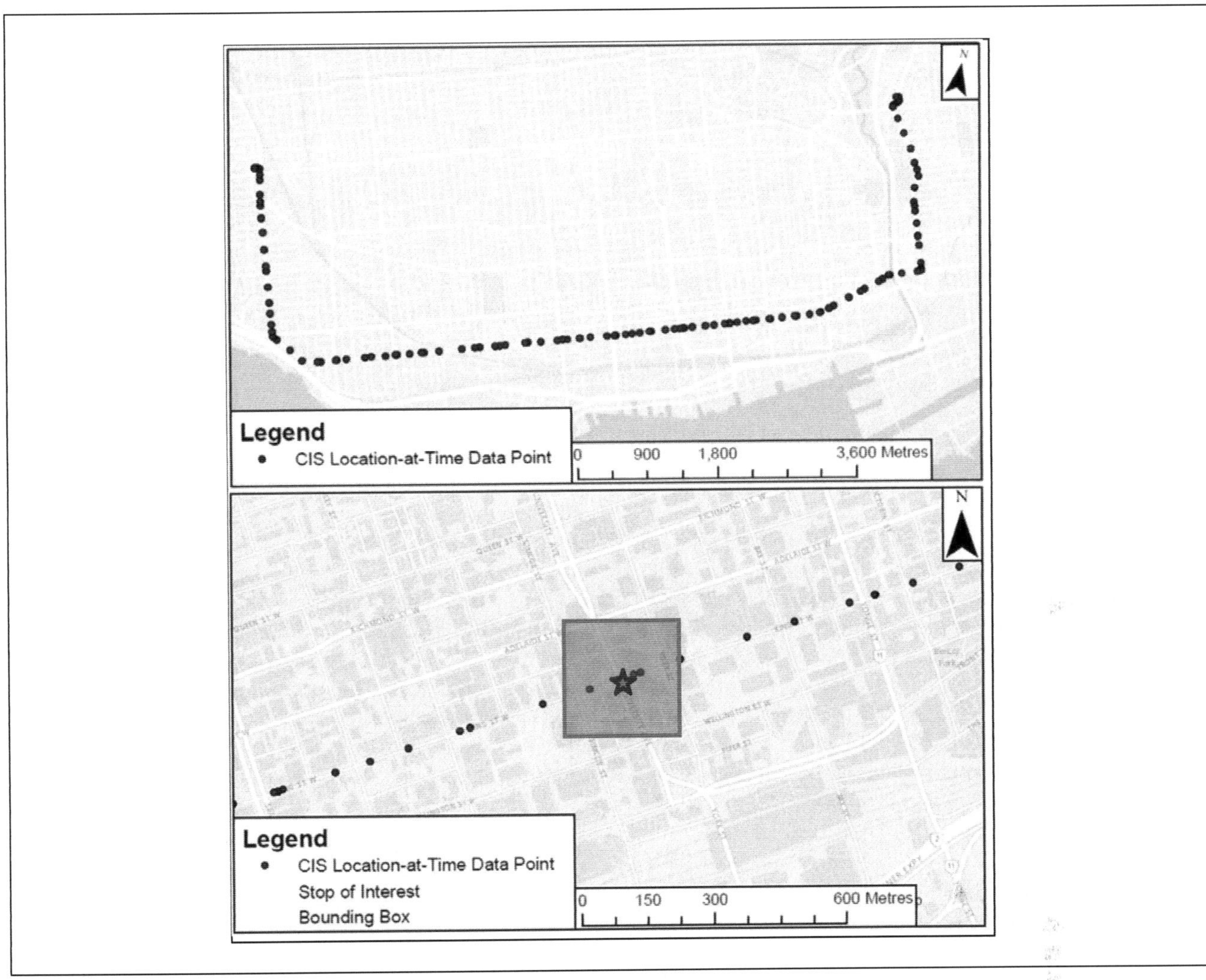

Figure 2. TTC CIS data (one day, one direction) and bounding box.

two-hour period based on transit schedules. The scheduled travel time for the shortest route, the Dundas Street Route 505, was 59 minutes in one direction with some layover time at each terminus. Stop-specific metrics were aggregated to reflect typical weekday conditions during each specific month between January 2014 and June 2016.

The direction of travel is estimated for each CIS data point based on the relative change in location of the specific vehicle from 2 minutes and 20 seconds earlier (seven 20-second lags). This "lag" value was chosen by modeling the observed mean headway with different "lag" values on the basis of its stability. While one might expect overly short lags to be insufficient to estimate movement in stop–start conditions, overly long lags would be expected to incorrectly code data points near each turn and terminus.

Congestion Metrics. Data used to estimate traffic congestion levels was purchased from Inrix, Inc. based on a network of directional-specific road links. The City of Toronto data for January 1, 2014, to June 30, 2016, consist of billions of anonymous, crowdsourced "GPS probe" speed data points collected from private and commercial vehicles, as well as from some GPS-enabled smartphones. Because smartphones (and not only commercial vehicles) make up a portion of the Inrix, Inc. speed data, some measures of traffic may represent individuals' smartphones riding in public transit services rather than automobile travel. The Inrix, Inc. algorithms for extracting and cleaning this data are proprietary and not transparent. As such, comparisons between streets with or without transit services should be made very carefully. As this study focuses on three transit corridors, only those corridors which all have transit services running at regular intervals are included in the analysis. Speed observations are recorded at the link level and are coded by directionally specific "traffic message center" (TMC) identifiers. TMCs are approximately 2 kilometers in length and are not as granular as the transit vehicle location data. To estimate roadway transportation services, congestion (travel time index) and unreliability (buffer time index) metrics are derived for typical evening peak periods (4 to 6 p.m.) on weekdays during an average month on a specific directionally specific link. To estimate central tendencies and distributions, the speeds for individual 15-minute intervals (e.g., the 5:15 to 5:30 p.m. time on Tuesday, March 1, 2016) on specific directionally specific links are used as the individual component observations.

Modeling Framework

Regression models of transit headway unreliability were estimated using explanatory variables, focusing on transportation service measures including congestion and traffic unreliability. Other control variables included annual differences (2014, 2015, and 2016) and fixed differences between each of the two directions for each of the two corridors (Queen East, Queen West, King East, etc.). Each of these model variables are discussed in turn. Transit stops and congestion metrics are linked on the basis of a one-to-many relationship—according to which each directionally specific roadway link (from the Inrix, Inc. data) is linked with no, one, or several directionally specific transit stops. All measures are for the evening peak period (defined as 4 to 6 p.m.) and are calculated for all weekdays in a specific month between January 2014 and June 2016 (e.g., all weekdays for January 2015).

Transit Services. Three principle transit service measures are estimated for the regression framework. First, transit unreliability (the outcome of interest) is measured using a metric reflecting headway buffer time index for the 85th percentile worst conditions (henceforth called Transit Headway BTI). This measure is the difference between 85th percentile headways and 50th percentile headways, divided by the mean headway. For example, based on a Transit Headway BTI of 2.0, a route with a scheduled headway of every 5 minutes would see actual headways of shorter than 10 minutes, 85% of the time. Each measure reflects the difference between the 85th percentile longest headways and the 50th percentile or mean headways during all evening peak periods (4 to 6 p.m.) during weekdays in a specific month.

Second, models account for transit base operating conditions using simplified measures to control for this impact on downstream station headway unreliability. Initial Transit Headway BTI (henceforth called Initial Transit BTI) was estimated at the first stop beyond the terminus as a proxy to estimate the regularity of departures from the starting terminus. It is expected that unreliability at the first stop would be positively linked with unreliability further down the line. Given that this measure captures transit service conditions immediately on deploying transit vehicles, it is designed to capture operational characteristics and scheduling by the TTC with little to no downstream interference from competing users in mixed-traffic conditions. This control does not account for potential real-time adjustments which can be made by the TTC to change transit vehicle deployments and improve services and reduce bunching on the fly—particularly at the very time-point stations used in this analysis. As such, any expected links between covariates and transit headway unreliability likely represent a lower-bound for the impact because some of the TTC's real-time operational adjustments could simultaneously be reducing congestion-induced degradation of transit services and headway unreliability.

Third, to account for the ordered distance from the start of each route, a variable is defined (henceforth called Stop Number) as the order of a stop from among those analyzed from the terminus in the direction of travel (beginning with Stop Number 1). Distances between stops were not equal, so this variable reflects a simplified approach to measuring distance from the starting terminus. Stop Order is expected to be positively linked with transit unreliability—for example, based on findings from Feng and Figliozzi (*32*).

Roadway Service Levels. Two chief measures of roadway service levels are estimated using Inrix, Inc. data—one of which captures traffic unreliability while the other reflects recurring traffic congestion. First, traffic unreliability is estimated using the Buffer Time Index (Traffic BTI). This measure reflects the amount of extra "buffer" time a person must budget to be on time 95% of the time. Since BTI captures extreme values, this measure is indicative of seasonal variation in mean and extreme travel conditions. Especially in mixed-traffic environments like King Street, Queen Street, and Dundas Street, it was expected that Traffic BTI would be positively linked with transit unreliability.

Second, traffic congestion is measured using the Traffic Travel Time Index (henceforth referred to as Traffic TTI). The Travel Time Index (TTI) is the ratio of mean travel time during the peak period and travel time under free-flow conditions. Measures range from 1.1 to 2.9 for any given typical weekday evening peak period during a given month. This measure more closely reflects recurring congestion rather than extreme experiences. But like Traffic BTI, Traffic TTI is expected to be associated with less reliable transit headways.

Table 1. King, Queen, and Dundas Street Corridor Descriptive Statistics (January 2014–June 2016)[a]

Measure	N	Mean	St. dev	Minimum	Maximum
Queen Street					
Transit headway mean	1,320	5.1	5.5	1.2	50.3
Transit headway (85th percentile)	1,320	9.9	10.3	2.0	92.5
Transit headway (50th percentile)	1,320	3.4	4.7	0.0	48.0
Transit headway buffer time index (BTI)	1,320	2.0	0.2	1.2	2.5
Traffic mean speed (mph)	1,320	13.2	3.0	6.4	19.8
Traffic speed (5th percentile)	1,320	10.3	2.9	3.9	18.0
Traffic speed (50th percentile)	1,320	13.0	3.1	6.0	20.0
Traffic buffer time index (BTI)	1,320	1.3	0.1	1.1	2.0
Traffic free-flow speed (mph)	1,320	20.2	2.4	16.5	24.0
Traffic travel time index (TTI)	1,320	1.6	0.3	1.1	2.9
King Street					
Transit headway mean	838	3	2.6	0.9	22.5
Transit headway (85th percentile)	838	5.9	5	2	48
Transit headway (50th percentile)	838	1.8	1.7	0	18
Transit headway buffer time index (BTI)	838	2	0.2	1.6	2.7
Traffic mean speed (mph)	838	12.9	2.9	7.7	20.8
Traffic speed (5th percentile)	838	10.1	2.7	5.4	18.7
Traffic speed (50th percentile)	838	12.7	2.9	7	20.5
Traffic buffer time index (BTI)	838	1.3	0.1	1.1	1.8
Traffic free-flow speed (mph)	838	19.9	2.5	16	23
Traffic travel time index (TTI)	838	1.6	0.3	1.1	2.2
Dundas Street					
Transit headway mean	868	6.1	3.8	2.8	31.8
Transit headway (85th percentile)	868	11.9	7.1	6	53.6
Transit headway (50th percentile)	868	4.1	2.9	1	32
Transit headway buffer time index (BTI)	868	2	0.2	1.1	2.5
Traffic mean speed (mph)	868	12.5	2.9	6.3	20.3
Traffic speed (5th percentile)	868	10	2.7	4.4	16.5
Traffic speed (50th percentile)	868	12.4	3	6.2	20
Traffic buffer time index (BTI)	868	1.3	0.1	1.1	1.7
Traffic free-flow speed (mph)	868	19.2	2.2	16.4	24
Traffic travel time index (TTI)	868	1.6	0.3	1.1	2.6

[a]Transit performance measures calculated using tTC CIS data and roadway performance measures calculated using inrix, inc. data.

Results

This study explores the potential links between traffic congestion, transit base operating conditions, and transit headway unreliability. Results are presented first by focusing on general descriptive statistics and then by highlighting inferential model results.

Descriptive Statistics

Overall descriptive statistics are presented in Table 1 and show that mean headways on Queen, King, and Dundas Streets are respectively, 5.1, 3.0, and 6.1 minutes during typical evening peak periods (4 to 6 p.m.) on weekdays during months between January 2014 and June 2016. These headways reflect scheduled headways. While King Street, which has the highest frequency transit services, has the highest maximum and minimum values of Transit Headway BTI, there is very little variation in typical transit headway unreliability between the three corridors.

Transit headway BTI metrics are graphed (see Figures 3 and 4), focusing on both the typical metrics of unreliability and the distribution of unreliability metrics using a box-plot to show the range of monthly BTI metrics for each stop between January 2014 and June 2016. Transit Headway BTI metrics suggest that even when a specific station typically has very low transit headway unreliability, there is significant variability in transit headway unreliability measures during the January 2014 to June 2016 study time frame. These variations underscore one of the sources of difference between observing general trends between typical transit headway unreliability and traffic congestion levels which indicate longer-term trends, on the one hand, and estimated inferential model results (discussed in the next section) which are designed to be more sensitive to operational differences, on the other.

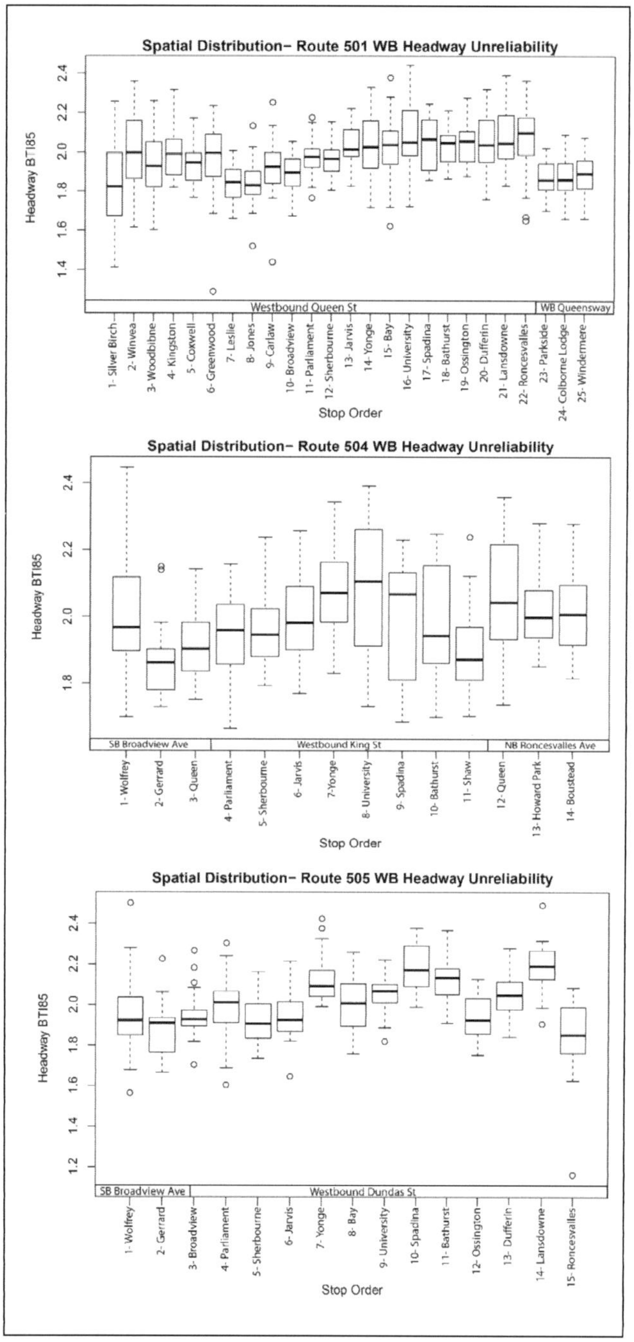

Figure 3. Westbound transit headway unreliability on Queen Street (501), King Street (504), and Dundas Street (505).

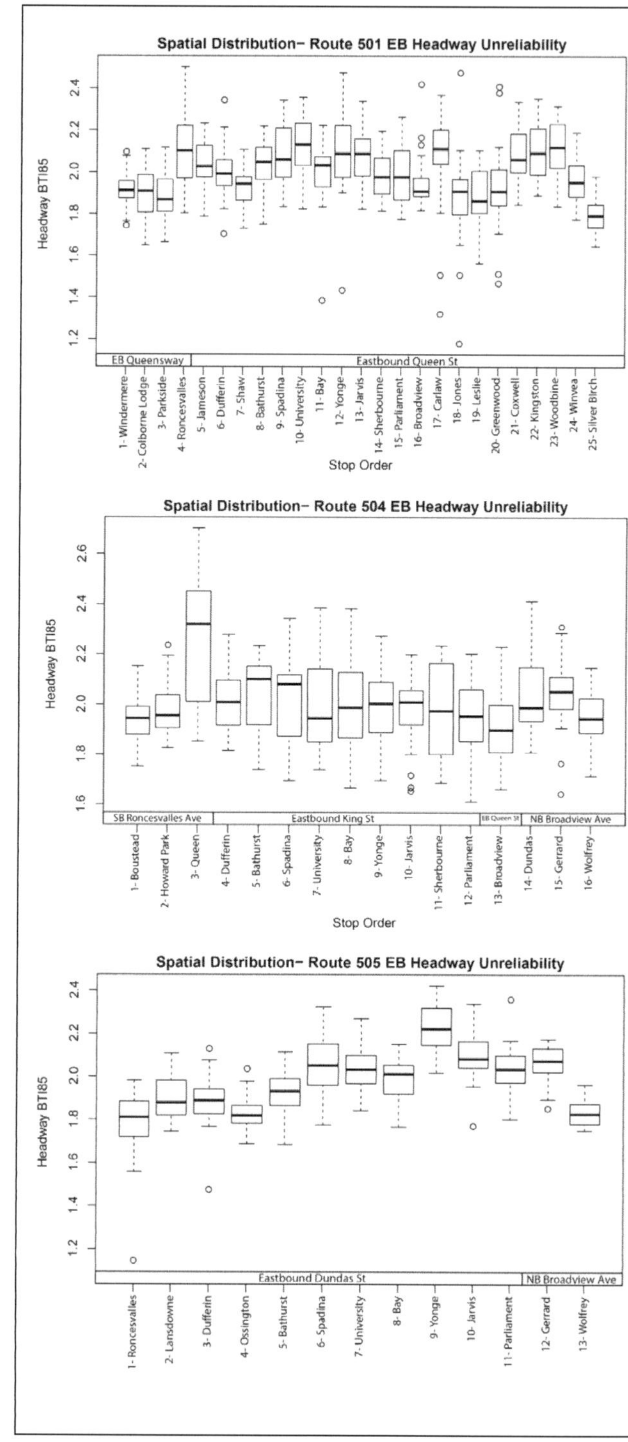

Figure 4. Eastbound transit headway unreliability on Queen Street (501), King Street (504), and Dundas Street (505).

Congestion metrics (using Traffic TTI) and transit headway unreliability metrics are next graphed for the three corridors (see Figures 5 and 6). Plots comparing traffic congestion and transit unreliability along each corridor indicate that congestion and transit unreliability appear to correlate closely in the case of most direction-specific routes. The same congestion data point (observed at a road segment level) may appear several times because there are typically several transit stops analyzed within each Inrix TMC. Only sections along King, Queen, and Dundas Streets are included. Headway unreliability appears to become more severe as transit vehicles continue further along the route (as Stop

Figure 5. Westbound transit headway unreliability and congestion on Queen Street, King Street, and Dundas Street.

Number increases). However, this trend is not clear, especially as the route approaches its terminus. The strength and cause of this attenuation in the trend is not clear, but it could be explained by real-time route dispatch adjustments by the TTC to improve headways at time points.

Each corridor has a unique spatial profile for transit unreliability and traffic congestion, with some observable linkages between the two, as well as some unexplained variation. These spatial distributions are explored in detail below.

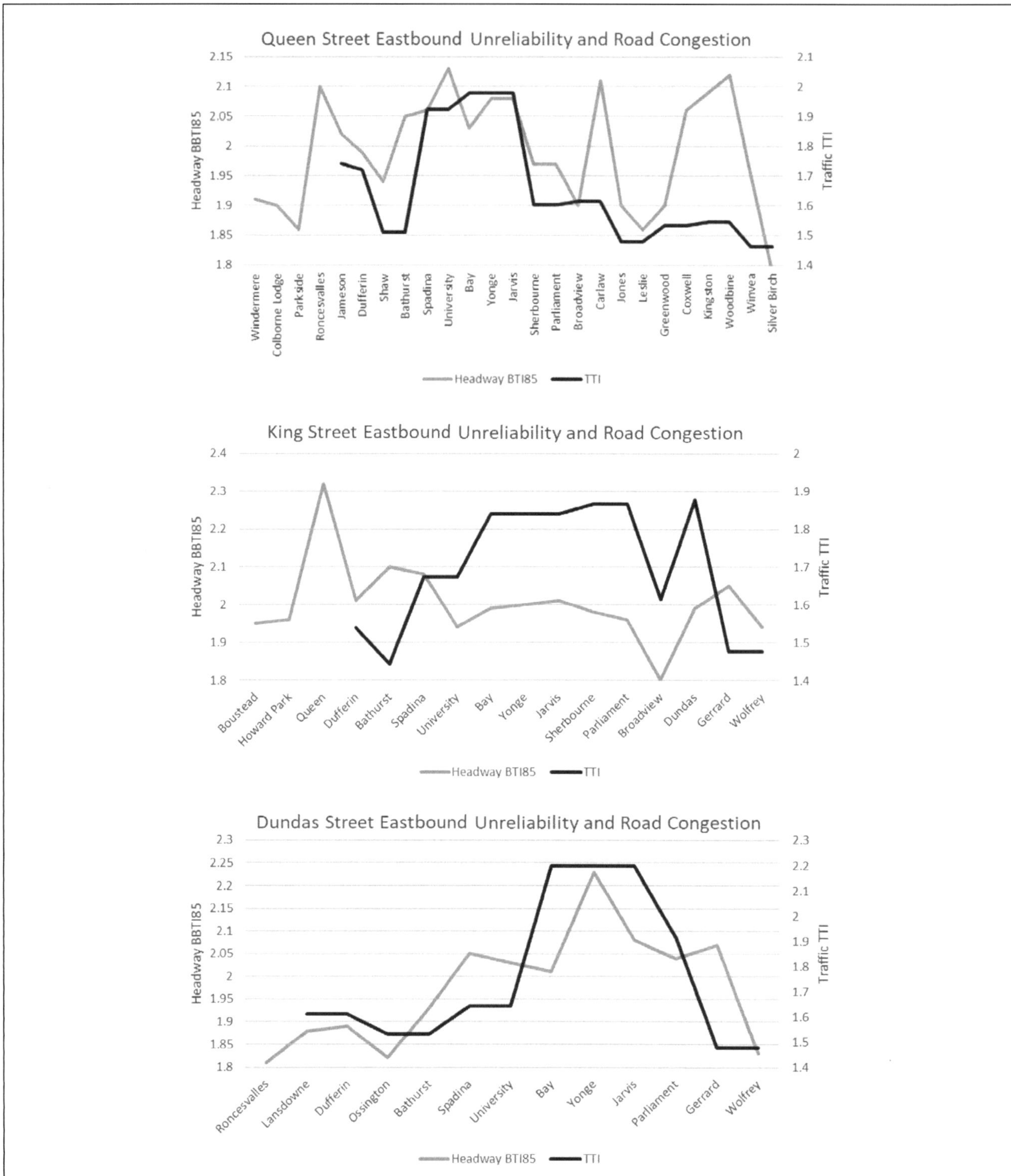

Figure 6. Eastbound transit headway unreliability and congestion on Queen Street, King Street, and Dundas Street.

Throughout all three routes in both the eastbound and westbound direction, transit unreliability appears to roughly mirror the spatial trends in traffic congestion. Along Dundas Street (505), both congestion and transit unreliability are highest near Yonge Street (a major arterial in the heart of downtown Toronto). However, in

Table 2. Model Results: Predictors of Transit Headway Unreliability (January 2014–June 2016)

Dependent variable = Transit headway unreliability (buffer time index)	Model 1	Model 2	Model 3
Variables	Coefficient estimate	Coefficient estimate	Coefficient estimate
(Intercept)	0.781**	0.709***	0.702***
Headway mean (ln)	−0.055**	−0.051***	−0.050***
Transit operations			
Stop order*	−0.001*		
Initial transit headway BTI (ln) at start of route		0.091***	0.095***
Traffic congestion			
Buffer time index (ln)	0.072**	0.065***	
Travel time index (ln)			0.051***
Dummy variables			
Year 2014	Reference	Reference	Reference
Year 2015	0.004	0.003	−0.006
Year 2016	−0.010	−0.007	−0.015**
Route 501 eastbound	−0.030**	−0.040***	−0.038***
Route 501 westbound	−0.021**	−0.018***	−0.017***
Route 504 eastbound	−0.055**	−0.053***	−0.051***
Route 504 westbound	−0.052**	−0.051***	−0.050***
Route 505 eastbound	−0.021**	−0.011	−0.012*
Route 505 westbound	Reference	Reference	Reference
Adjusted R-squared:	0.118	0.125	0.127
Degrees of freedom	3,015	2,990	2,990

*p = 0.10. **p = 0.05. ***p = 0.01.

the case of the Queen Street eastbound direction after Sherbourne Street, traffic congestion and transit unreliability appear to be less closely aligned. While traffic congestion appears to be stable, transit unreliability fluctuates significantly.

Models of Headway Unreliability

This study produced three inferential models using ordinary least squares regression, all of which predict Transit Headway BTI using variables related to transit operations and roadway performance. The differences between the three models reflect how transit operation variables (Stop Order or Initial Transit BTI) or roadway performance (Traffic BTI or Traffic TTI) are modeled. All three models control for year, route, and direction. All continuous variables, except Stop Order, are natural logged. Notably, seasonal factors are omitted from models—as much variation in congestion is associated with seasonality. As such, the estimated effects in these models also reflect how the seasonal variation in congestion influences transit headway unreliability. These models are presented in Table 2.

Model 1 estimates predictors of transit headway unreliability while measuring transit operations using Stop Order and measuring roadway performance using Traffic BTI—a metric of traffic unreliability. Both variables are statistically significant predictors, but the impact of Stop Order is extremely small and opposite of what was expected, implying that unreliability becomes less severe as transit vehicles continue further on any given route. This model suggests a 7.2% decrease in headway unreliability can be expected for every unit decrease in Traffic BTI. The adjusted R-squared for this model is 0.118, indicating that over 88% of variation in headway unreliability remains unexplained.

In Model 2, Stop Order is removed due to its low explanatory power and instead, Initial Transit BTI from at the start of every route is introduced as a new variable to measure transit operations. Like in Model 1, Traffic BTI is used to measure roadway performance. All Stop 1 records are removed from Model 2, resulting in a smaller sample size, because Initial Transit BTI is measured as the first timing station beyond the route turning point—leading those observations to be removed from the model. Both transit service and roadway performance variables are statistically significant predictors of headway unreliability. This model suggests a 9.1% increase in Transit Headway BTI throughout each route can be expected by doubling Initial Transit BTI at the start of the route—in short, improving initial headway unreliability would be expected to be associated with unreliability improvements downstream. The unreliability of traffic conditions is likewise a statistically significant predictor of headway unreliability. A 6.5% increase in headway unreliability can be expected for every doubling in

Traffic BTI. The adjusted R-squared for this model is 0.125 higher than Model 1 (perhaps due to the change in the number of observations), but still leaving more than 87% of variation in headway unreliability unexplained.

In Model 3, Initial Transit BTI is used to measure transit performance while roadway performance is measured using Traffic TTI (a metric of roadway congestion) rather than using Traffic BTI (a measure of roadway unreliability). This metric therefore captures average peak hour levels of traffic congestion instead of extreme values that may be experienced only several days a month. This model suggests that a 9.5% increase in headway unreliability throughout each route can be expected by doubling unreliability at the start of the route (an elasticity of 0.095). A 5.1% increase in headway unreliability can be expected by doubling traffic congestion as measured by Traffic TTI (an elasticity of 0.051). In fact, given that this model does not include seasonality controls and when those are included the congestion variables are not expected to be significant, this model suggests that the broad seasonality of congestion, rather than day-to-day variations in congestion, are most closely linked with changes in transit headway unreliability. The adjusted R-squared for this model is 0.127, still leaving most of variation in headway unreliability unexplained.

Overall, none of the model explained more than 12.7% of the variation in transit headway unreliability. However, while the expected impact of congestion using a measure of recurring congestion (travel time index) is associated with changes in transit headway unreliability, the unreliability of roadway travel times (the buffer time index) is more strongly linked with transit headway unreliability.

Conclusion

This study focuses on the Queen Street, King Street, and Dundas Street transit corridors in Toronto, Canada to explore the links between roadway performance and transit headway unreliability in urban mixed-traffic transit corridors. Focusing on the links between transit headway unreliability and roadway congestion, this study uses both general descriptive statistics and inferential models to explore opportunities to identify means to improve transit system performance.

Using month-level data aggregated at two-hour peak periods (4 to 6 p.m.), inferential models indicate that roadway performance (congestion and unreliability) is positively associated with transit headway unreliability, with elasticities between 0.051 and 0.072. These findings should be interpreted carefully because they reflect seasonal differences in congestion levels rather than those of day-to-day. Likewise, the irregularity of transit departures from the originating terminus is also positively linked to transit headway unreliability at downstream stations, with elasticities between 0.091 and 0.095. These findings imply that halving congestion or halving the irregularity of departures from the terminus would be associated with station-level headway reliability improvements of between approximately 3% and 5%. But when considering the feasibility of alleviating congestion, it is unclear whether consequent transit improvements would be meaningful from achievable improvements in roadway congestion.

Findings indicate that the underlying causes of transit headway unreliability are complex and that there is no single panacea for improving reliability. In fact, almost 88% of variation in station-level transit unreliability remains unexplained. Descriptive statistics and graphs (see Figures 5 and 6) suggest that transit headway unreliability and roadway performance mirror one another on aggregate at the station level—perhaps related to site context and design.

This study indicates some limited opportunities for transit reliability improvements through traffic congestion management in mixed-traffic contexts—insofar that either broad patterns in road congestion or traffic reliability can be improved. But the longer-term links between traffic and transit unreliability should be better explored and monitored in response to alternate policy actions. Notably, improving transit scheduling practices was more significantly linked with station headway unreliability than traffic congestion. Nevertheless, findings indicated that the vast majority of transit station headway unreliability in these highly urbanized downtown Toronto mixed-traffic transit corridors remains unexplained—suggesting that policymakers should temper expectations.

Acknowledgments

This project was funded by the City of Toronto's Transportation Services Division and benefitted from data and information provided by the Transportation Services Big Data and Innovation Group, led by Jesse Coleman. Thank you to the Toronto Transit Commission (TTC) and to Inrix, Inc.—both of whom sold us data used in this project.

Author Contributions

The authors confirm contributing to the paper as follows: study conception and design: Chris Yuen and Matthias Sweet; data collection: Chris Yuen and Matthias Sweet; analysis and interpretation of results: Chris Yuen and Matthias Sweet; draft manuscript preparation: Chris Yuen; manuscript finalization and editing: Chris Yuen and Matthias Sweet. All authors reviewed the results and approved the final version of the manuscript.

References

1. Iseki, H., and B. D. Taylor. Style versus Service? An Analysis of User Perceptions of Transit Stops and Stations. *Journal of Public Transportation*, Vol. 13, No. 3, 2010, pp. 39–63.
2. Currie, G., and A. Shalaby. Success and Challenges in Modernizing Streetcar Systems: Experiences in Melbourne, Australia, and Toronto, Canada. *Transportation Research Record: Journal of the Transportation Research Board*, 2007. 2006: 31–39.
3. Currie, G., M. Sarvi, and B. Young. A New Approach to Evaluating On-Road Public Transport Priority Projects: Balancing the Demand for Limited Road-Space. *Transportation*, Vol. 34, No. 4, 2007, pp. 413–428.
4. Mazloumi, E., G. Currie, and G. Rose. Causes of Travel Time Unreliability - A Melbourne Case Study. *Proc., 31st Australasian Transport Research Forum (ATRF2008)* (Morris, J., G. Rose, and E. Ramsay, eds.), Department of Transport Victoria, Melbourne VIC Australia, 2008, pp. 195–202.
5. Mazloumi, E., G. Rose, G. Currie, and M. Sarvi. An Integrated Framework to Predict Bus Travel Time and Its Variability Using Traffic Flow Data. *Journal of Intelligent Transportation Systems*, Vol. 15, No. 2, 2011, pp. 75–90.
6. McKnight, C. E., H. S. Levinson, K. Ozbay, C. Kamga, and R. E. Paaswell. *Impact of Congestion on Bus Operations and Costs*. Publication FHWA-NJ-2003-008. New Jersey Department of Transportation Division of Research and Technology and the U.S. Department of Transportation, 2003.
7. Taylor, B. D. The Politics of Congestion Mitigation. *Transport Policy*, Vol. 11, No. 3, 2004, pp. 299–302.
8. Sweet, M., C. Harrison, and P. Kanaroglou. *Congestion Trends in the City of Toronto: 2011-2014*. McMaster Institute for Transportation and Logistics, Hamilton, Ontario, Canada, 2015.
9. Graham, C.. The Demand Performance of Bus Rapid Transit. *Journal of Public Transportation*, Vol. 8, No. 1, 2005, pp. 41–55.
10. Thompson, G., J. Brown, and T. Bhattacharya. What Really Matters for Increasing Transit Ridership: Understanding the Determinants of Transit Ridership Demand in Broward County, Florida. *Urban Studies*, Vol. 49, No. 15, 2012, pp. 3327–3345.
11. Yoh, A., H. Iseki, M. Smart, and B. D. Taylor. Hate to Wait: Effects of Wait Time on Public Transit Travelers' Perceptions. *Transportation Research Record: Journal of the Transportation Research Board*, 2011. 2216: 116–124.
12. Abkowitz, M. D. *The Impact of Service Reliability on Work Travel Behavior*. Submitted in partial fulfillment of the requirements for the Degree of Doctor of Philosophy in Civil Engineering. Massachusetts Institute of Technology, Boston, 1980.
13. Bowman, L. A., and M. A. Turnquist. Service Frequency, Schedule Reliability and Passenger Wait Times at Transit Stops. *Transportation Research Part A: General*, Vol. 15A, No. 6, 1981, pp. 465–471.
14. Noland, R., and P. John. Travel Time Variability: A Review of Theoretical and Empirical Issues. *Transport Reviews*, Vol. 22, No. 1, 2002, pp. 39–54.
15. Noland, R. Commuter Responses to Travel Time Uncertainty under Congested Conditions: Expected Cost and the Provision of Information. *Journal of Urban Economics*, Vol. 41, No. 3, 1997, pp. 377–406.
16. Transportation Research Board. *Transit Capacity and Quality of Service Manual*, 3rd ed. Transportation Research Board, Washington, D.C., 2013.
17. Currie, G., and J. Rose. Growing Patronage - Challenges and what has been found to Work. *Research in Transportation Economics*, Vol. 22, No. 1, 2008, pp. 5–11.
18. Guerra, E., and C. Robert. Cost of a Ride: The Effects of Densities on Fixed-Guideway Transit Ridership and Costs. *Journal of the American Planning Association*, Vol. 77, No. 3, 2011, pp. 267–290.
19. Taylor, B. D., D. Miller, H. Iseki, and C. Fink. Nature and/or Nurture? Analyzing the Determinants of Transit Ridership across US Urbanized Areas. *Transportation Research Part A: Policy and Practice*, Vol. 43, No. 1, 2008, pp. 60–77.
20. Casello, J. M., A. Nour, and B. Hellinga. Quantifying the Impacts of Transit Reliability on User Costs. *Transportation Research Record: Journal of the Transportation Research Board*, 2009. 2112: 136–141.
21. Mesbah, M., J. Lin, and G. Currie. "Weather" Transit is Reliable? Using AVL Data to Explore Tram Performance in Melbourne, Australia. *Journal of Traffic and Transportation Engineering*, Vol. 2, No. 3, 2015, pp. 125–135.
22. Bates, J., J. Polak, P. Jones, and A. Cook. The Valuation of Reliability for Personal Travel. *Transportation Research Part E: Logistics and Transportation Review*, Vol. 37, No. 2–3, 2001, pp. 191–229.
23. Chakrabarti, S., and G. Giuliano. Does Service Reliability Determine Transit Patronage? Insights from the Los Angeles Metro Bus System. *Transport Policy*, Vol. 42, 2015, pp. 12–20.
24. Chakrabarti, S. The Demand for Reliable Transit Service: New Evidence Using Stop Level Data from the Los Angeles Metro Bus System. *Journal of Transport Geography*, Vol. 48, 2015, pp. 154–164.
25. Nagy, E., and C. Csiszar. Analysis of Delay Causes in Railway Passenger Transportation. *Periodica Polytechnica Transportation Engineering*, Vol. 43, No. 2, 2015, pp. 73–80.
26. Duran-Hormazabal, E., and T. Alejandro. Estimation of Travel Time Variability for Cars, Buses, Metro and Door-to-Door Public Transport Trips in Santiago, Chile. *Research in Transportation Economics*, Vol. 59, 2016, pp. 26–39.
27. Saberi Kalaee, M. *Investigating Freeway Speed-Flow Relationships for Traffic Assignment Applications*. Civil and Environmental Engineering. Portland State University, 2010.
28. Saberi, M., and A. K. Zockaie. Definition and Properties of Alternative Bus Service Reliability Measures at the Stop Level. *Journal of Public Transportation*, Vol. 16, No. 1, 2013, pp. 97–122.

29. Chatman, D. G. Deconstructing Development Density: Quality, Quantity and Price Effects on Household Travel. *Transportation Research A: Policy and Practice*, Vol. 42, No. 7, 2008, pp. 1008–1030.
30. Sweet, M., and C. Mengke. Does Regional Travel Time Unreliability influence Mode Choice? *Transportation*, Vol. 38, No. 4, 2011, pp. 625–642.
31. El-Geneidy, A. M., J. Hourdos, and J. Horning. Bus Transit Service Planning and Operations in a Competitive Environment. *Journal of Public Transportation*, Vol. 12, No. 3, 2009, pp. 39–59.
32. Cooper, S., and L. Gould. *Faster than Walking? Street Congestion and New York City Transit Buses*. New York City Transit Authority, Operation Planning, 1994.
33. Feng, W., and M. Figliozzi. *Using Archived AVL/APC Bus Data to Identify Spatial-Temporal*. Portland State University, 2010.
34. Pilachowski, J. M. *An Approach to Reducing Bus Bunching*. University of California, Berkley, 2009.
35. Toronto Transit Commission. *Ridership numbers and Revenues Summary- Data Catalogue- Open Data and City of Toronto*. City of Toronto [Online]. January 15, 2016. http://www1.toronto.ca/wps/portal/contentonly?vgnextoid=a3b7c87477438310VgnVCM1000003dd60f89RCRD. Accessed August 1, 2017.
36. Mesbah, M., G. Currie, C. Lennon, and T. Northcott. Spatial and Temporal Visualization of Transit Operations Performance Data at a Network Level. *Journal of Transport Geography*, Vol. 25, 2012, pp. 15–26.
37. Toronto Transit Commission. *TTC Routes and Schedules-Data Catalogue- Open Data and City of Toronto*. City of Toronto [Online]. November 15, 2016. http://www1.toronto.ca/wps/portal/contentonly?vgnextoid=96f236899e02b210VgnVCM1000003dd60f89RCRD. Accessed August 1, 2017.
38. Morang, M. transit.melindamorang.com [Online]. transit.melindamorang.com/overview_DisplayGTFSInArcGIS.html. Accessed June 23, 2017.

The Standing Committee on Bus Transit Systems (AP050) peer-reviewed this paper (18-01060).

This work was completed as part of Chris Yuen's graduate research at Ryerson University's School of Urban and Regional Planning. The analyses and conclusions reflect the views of the study authors and in no way represent the position of the authors' current or past employers or the City of Toronto.

Article

TRR
JOURNAL OF THE TRANSPORTATION RESEARCH BOARD

Transportation Research Record
2018, Vol. 2672(8) 135–142

DOI: 10.1177/0361198118784134
journals.sagepub.com/home/trr
SAGE

Environmental Effect of Bus Priority Measures Applied on a Road Network in Santiago, Chile

Xenia Karekla[1], Rodrigo Fernandez[2], and Nick Tyler[1]

Abstract
Air pollution is at the highest levels ever and there is currently a worldwide initiative by transport engineers and urban planners to redesign public transport modes and cities to become more sustainable and environmentally friendly. The environmental impact of everyday activities is more apparent in developing cities which take longer to adapt to advanced methods of running public transport modes. This study aims to investigate the reduction of bus energy consumption and carbon emissions through bus priority measures in a bus route in the city of Santiago, Chile. Two bus priority schemes are tested in this study: Bus Only Lanes and Bus Signal Priority. The microscopic traffic simulator TSIS-CORSIM is used to quantify the environmental impact of these schemes. The results have shown that both schemes lead to lower fuel consumption and emissions, especially for the bus service. The environmental improvements are mostly apparent at traffic flows below 1000 veh/h, with clear benefits for both the bus service and passenger cars when dedicated bus lanes are included in the road infrastructure.

Air pollution is an environmental phenomenon which puts a burden on people's lives and affects everyday activities. Especially in urban metropolises, where more than half of the global population is accommodated (*1*), and more than 70% of the world's carbon emissions are recorded (*2*). The world's 10% richest people alone emit 50% of the worldwide greenhouse gas (GHG) emissions (*3*). These statistics show that the activities undertaken in cities are harming the environment and there is a clear need for changing our habits in the way we attempt everyday activities. It is essential to find ways to reduce air pollution and climate change whilst preserving people's health and improving their wellbeing.

One of the greatest GHG emitters in urban environments, with documented effects on air pollution, is transport systems (*4*, *5*). Transporting people and goods is a vital activity in cities, however energy related emissions from transportation amounts to 25% of GHG emissions worldwide (*6*). For reference, in 2015 the transportation sector in Europe emitted 26% of the world's GHG (*7*) and the USA emitted 27% (*8*). Transportation related emissions in China doubled between 2000 and 2010 and an extra 54% increase is expected up to 2020 (*9*). Road transportation, in particular, is considered to be responsible for 70% of all GHG emissions, making it the largest polluter of all transport modes (*10*). As an example from the Latin American context, in Chile, 28% of the GHG emissions are generated by the transport sector, with private cars gaining ground over public transportation year after year, the levels of GHG emissions are constantly increasing. Indicated by 45% of GHG emissions from transportation in Chile being produced by cars and taxis, 10% by buses, and 1% by trains (*11*).

In transport economics, road pollution is considered to be an abuse of the road that has been provided, whilst other users on the street are passive victims (*12*). Therefore, it is necessary to provide alternative means of transport, such as public transport modes, as well as to improve the existing ones so that people stop the abuse of roads and start favoring environmentally friendly means over their private cars. Enabling people to choose more sustainable means of transport for their everyday mobility could significantly contribute to the reduction of GHG emissions worldwide.

In regard to energy consumption, urban areas consume up to 75% of global energy, the majority of which is absorbed by the transportation sector (*13*). In 2010, transportation systems around the world consumed 19% of global energy supplies, 69% of which was consumed by road transportation (*14*). This number is expected to grow rapidly, and by 2050 transport energy consumption is expected to reach an 80% increase. In other words, by 2050 30% of the global energy

[1]Department of Civil, Environmental and Geomatic Engineering, Centre of Transport Studies, University College London, London, UK
[2]Faculty of Engineering and Applied Sciences, Universidad de los Andes, Santiago, Chile

Corresponding Author:
Address correspondence to Xenia Karekla: x.karekla@ucl.ac.uk

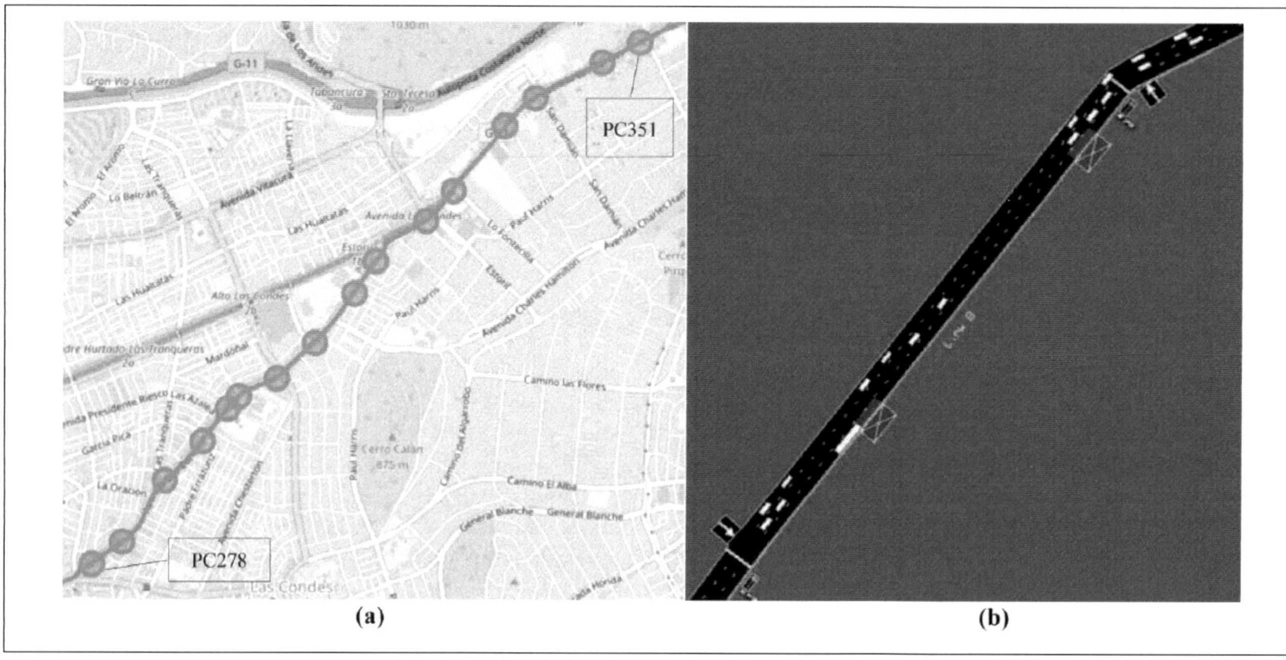

Figure 1. (*a*) Bus stops of bus route C01 along Avenida Las Condes and (*b*) simulation model of part of the network when testing the effect of bus lanes.

will be channeled toward transportation, with the highest demand arising from developing countries and areas undergoing strong economic and population growth, such as Latin America and China. Statistics concerning car ownership rates highlight the global need to increase the attractiveness of public transport modes and to invest in energy efficient transport systems; car ownership in China has been growing by 12% per year, by 1.6% per year in the US, whereas only a slight increase in vehicle ownership is expected for European countries that are members of the Organization for Economic Cooperation and Development (OECD-Europe) by 2050 (*14*).

Bus priority schemes have been widely tested in various environments, such as in Japan, the UK, and Canada, with the aim to reduce traffic congestion, to improve bus journey times and road safety (examples are included but not limited to *15–17*), but their environmental effect is still being explored. To tackle the environmental challenges discussed above, the main objective of this study is to examine how bus priority schemes may reduce energy consumption and carbon emissions. It focuses on the bus system in Santiago, Chile, and with the use of the microscopic traffic simulator TSIS-CORSIM, the current road network and bus system is replicated. Two bus priority schemes—Bus Only Lanes and Bus Signal Priority—were tested during the morning peak period.

The collection of data, the model parameters and the examined bus priority schemes are described in the methodology section. The results obtained for energy consumption and emissions for each of the schemes are presented in the results section for both the cars and the bus system. The results are then discussed and the traffic flow and bus priority scheme for which the highest environmental improvement is achieved is proposed.

Methodology

Bus route C01 travels across the district of Las Condes, one of the most affluent districts in Santiago, Chile, and presents one of the highest boarding and alighting rates of bus routes in that area. In addition, the highest car volumes of Santiago are concentrated in the area of Las Condes (*18*), which is reflected in an increase in road congestion and air pollution. Therefore, these characteristics make the area of Las Condes and the road segment served by bus route C01 an excellent candidate for evaluation.

The operator of bus route C01, Redbus Urbano S.A of the Transdev Group, was contacted to provide the necessary data about the bus system. Bus frequency, number of bus stops and the distance between them, passenger demand during boarding and alighting, bus dwell times, vehicle length, capacity and velocity, were collected. Bus route C01 is in fact a combination of two bus routes—C01 and a shorter one, C01c, that serves only the middle stations of the general route to increase the service frequency during peak hours—and therefore it was decided to study 16 bus stops that are served by both services (Figure 1*a*). Only the direction away from the city center (north-east) was considered, with the first bus station being Juan Montalvo (PC278) and the last being San Francisco De Asís (PC351). The length of the studied network is around 6 km.

On-site observations were also used to gather data about the traffic flow on the examined road network. This data involved the number of cars at each section of the network, location of traffic signals (nodes), signal times and cycle times, location of pedestrian crossings and pedestrian flow and road parking. Considering the road infrastructure, data such as road gradient, length of road between traffic signals (links), number of lanes on the street and the coordinates of each node were also collected.

The collected data were organized in a database and were then entered in the simulation package (Figure 1*b*). The answer to the research question was attempted through the alteration of three variables: reduction of the current traffic flow and introduction of two bus priority schemes—Bus Only Lanes and Bus Signal Priority (BSP).

Traffic Flow

The current traffic flow along bus route C01 was measured to be 1500 veh/h. Given the frequency of buses on the C01 bus route (26 buses per hour), the maximum flow of cars on this road should not exceed 1000 veh/h to implement a bus lane (*19*). Therefore, the environmental effect of lower traffic flows, such as 1000 and 800 veh/h was examined. Although these traffic flows are lower than the current, they do not reflect a free flow condition, but rather a 'medium' level of traffic congestion on the road. Furthermore, a higher traffic flow of 2000 veh/h was also tested to reveal the trends in a congested environment as well as to have a clearer understanding of the effect of traffic flow on energy consumption and emissions.

Bus Lanes

In regard to road infrastructure, the examined network does not provide dedicated lanes for the bus service and therefore a mixed-traffic usage of the road is observed. Having buses travelling on separate lanes to private cars would reveal whether this measure benefits the bus service as well as the environment.

Bus Signal Priority

Traffic signals in Avenida Las Condes allow a cycle of 110 sec. To test the particular scheme, traffic signal cycles were reduced to 60 sec to enable buses to spend less time at each traffic signal, and therefore reduce their journey time (*20*). The effect of bus signal priority (BSP) on energy consumption and emissions was also tested.

The vehicle fleet considered in all three cases consisted of high performance passenger cars (75% of the examined traffic flow), low performance passenger cars (25% of the examined traffic flow) and buses. An average occupancy of 1.30 passengers per vehicle was considered for both the high and low performance passenger cars, whereas it was assumed that each bus travelling along the simulated network is carrying 25 passengers on average. The length of each vehicle (high and low passenger car and bus) is 5, 4.30 and 12 m respectively and the maximum non-emergency acceleration/deceleration of both cars and buses was set at 2.5 m/s^2.

Vehicle energy consumption (L) and emissions (grams/km) are calculated internally based on tabulated data embedded in the TSIS-CORSIM software. Information about the vehicle's performance and speed is used to access the software tables and define the rate of fuel consumption (scaled by 0.0001 gallons/sec, later converted to liters) and emission (scaled by 0.001 grams/sec) as a function of the vehicle's acceleration. The methods used for the development of these tables can be found in (*21*). The effect of each bus priority scheme on a vehicles' speed (km/h) and fuel efficiency (km/L) was also derived. Enhanced understanding of the way a scheme affects a combination of vehicle performance aspects is important to reach a conclusion as to which combination of traffic flow and scheme would be best to be applied to improve the environment.

Results

This section presents the effect of each of the proposed schemes on fuel consumption and emissions, and reference is also made to vehicles' speed and fuel efficiency. The impact of the proposed schemes on GHG emissions specifically is examined through three common diesel pollutants which contribute to global warming and include unburned hydrocarbons (HC), carbon monoxide (CO) and nitrogen oxides (NOx). The results derived from each scheme are discussed in comparison to the current traffic situation at the examined network (mixed-traffic) and the percentage change of each environmental factor between the two schemes is presented. Traffic flow is not presented separately, but in conjunction with each of the bus priority schemes. The numerical results of each case, especially for vehicle speed and fuel efficiency for which graphs are not presented, can be provided upon request.

Scheme 1: Bus Only Lanes

Having dedicated bus lanes so that buses and cars do not mix on the road has proven beneficial for the environment as well as for its users. First, having buses using a dedicated lane increases the speed of buses by up to 11% compared to the mixed-traffic case. One effect is that the average speed of cars reduces dramatically, especially at high traffic flows (2000 veh/h) in which a reduction of 30% is observed. At lower traffic flows of 1000 or 800 veh/h, which imply a 'medium' level of traffic congestion, the average speed of buses still increases by 4% and 6% respectively, whereas the

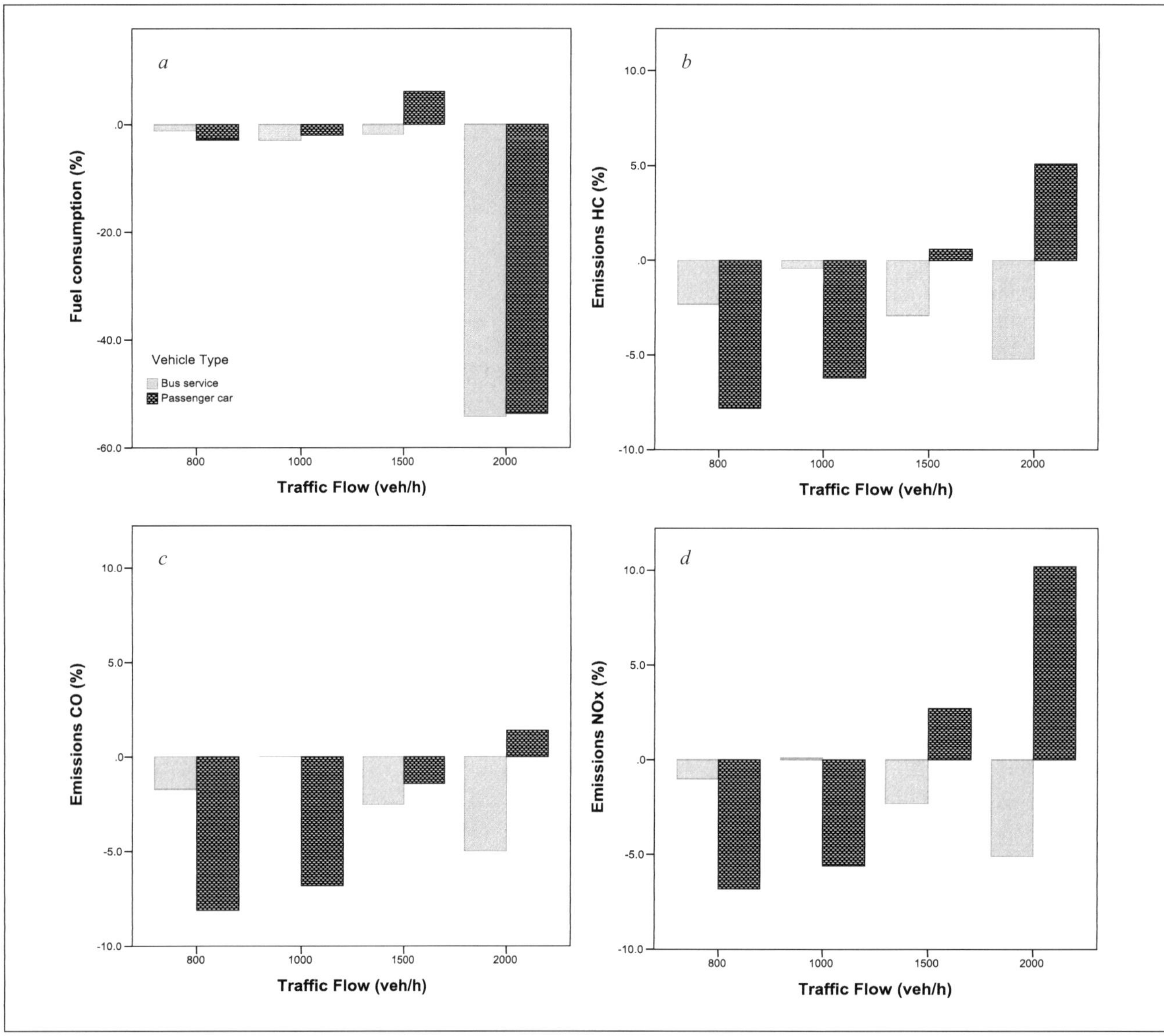

Figure 2. Percentage change of fuel consumption and emissions for cars and buses when bus lanes are introduced.

average speed of cars presents the smallest change (reduction of 3% and 2% respectively). Thus, it is observed that the gain in speed for the buses is almost equal to the loss in speed for the cars at the lowest traffic flows.

Focusing on fuel consumption (Figure 2*a*), enabling buses to travel in dedicated bus lanes whilst traffic flow remains unchanged (1500 veh/h) reduces the amount of fuel buses need to travel along the examined road network by 2%. However, cars appear to use 6% more fuel because of the road congestion (their average speed drops by 7%). At a higher traffic flow (2000 veh/h) congestion becomes so high that it is inevitable that cars appear to use 53.6% less fuel (their average speed drops by 30% and they are almost idle). On the other hand, at a traffic flow of 2000 veh/h, buses move faster but spend less time accelerating and decelerating, therefore a 54.2% decrease in their fuel consumption is observed. At lower traffic flows (1000 and 800 veh/h), fuel consumption for both the buses and cars reduces by 3% at the most.

Fuel efficiency follows the opposite trend of fuel consumption. When buses travel on dedicated bus lanes they use less fuel in all traffic flow conditions, and therefore their fuel efficiency increases. Disregarding the cases at which congestion is very high (1500 and 2000 veh/h) and therefore fuel efficiency presents the highest increase, buses benefit the most with regard to fuel efficiency when traffic flow is at 800 veh/h. Cars, on the other hand, present a decrease in fuel efficiency in all traffic flow cases apart from when 800 veh/h use the road. Although it is a small increase (+1%), it is the only case for which cars can travel longer distances with the same amount of fuel.

When considering pollution (Figure 2, *b–d*), the implementation of a bus lane reduces emissions for both buses and cars, especially at a 'medium' congestion level when traffic flow is less than 1000 veh/h (the flow for which a bus lane is required). The observed reduction in emissions follows the same tendency for HC, CO and NOx gases. The highest decrease in GHG emissions for passenger cars is observed at 800 veh/h (8% for HC and CO and 7% for NOx). Similarly, the bus service emits the least amount of GHG at 800 veh/h, however the benefit is less than that observed for the cars (decreases of 2% for HC and CO and 1% for NOx). In the cases of actual flow and high congestion (1500 and 2000 veh/h), bus lanes also appear to benefit the environment when buses are taken into account (reduction of 2.5% for GHG emissions at 1500 veh/h and of 5% at 2000 veh/h). However, cars appear to be high emitters of GHG when traffic flow is 1500 veh/h and higher. Especially at 1500 veh/h, passenger cars emit more HC (+0.6%) and NOx (+2.7%) gases, but less CO (−1.4%). This occurs when, during combustion, the quantity of oxygen in the air is higher than normal and creates a lean mixture of air and fuel (air/fuel ratio greater than 1) because of this a lower concentration of CO is produced, and if combustion is not complete and occurs at high temperatures, a higher concentration of HC and NOx gases is produced (*22*).

Scheme 2: Bus Signal Priority

In the case of the BSP scheme, the average speed of both cars and buses increases at all examined traffic flows between 6% and 8% in comparison to the current signal timing. This shows that both buses and cars benefit from this scheme regardless of the congestion on the road. The highest increase of average speed for cars (8.3%) is observed at a traffic flow of 1000 veh/h, whereas buses can travel 7.6% faster at the current traffic flow (1500 veh/h) when they are given BSP.

When fuel consumption is examined (Figure 3*a*), both cars and buses benefit from a BSP scheme as less energy is required to travel along the studied network compared to the case when no BSP is given. This effect is seen at all examined traffic flows, however the highest reduction in fuel consumption for cars (22%) is observed when 1000 veh/h are using the road. At this traffic flow, cars can travel up to 4% longer distances with the same amount of fuel, compared to the current situation (1500 veh/h). Reduction in fuel consumption for buses is not as significant as it is for cars. At a traffic flow of 1000 veh/hour, for which cars present the highest benefit, buses appear to have the lowest reduction of fuel consumption (1.2%). This might be because of the high average speed they develop (26.2 km/h) as well as the number of stops (bus stops and traffic signals) they make compared to cars. At this level of traffic flow, cars also reach high average speed (37.3 km/h) but only stop at the traffic signals. The highest reduction in fuel consumption for buses is observed at 2000 veh/h and at 800 veh/h (1.8%).

Subsequently, fuel efficiency for cars and buses increases, with the highest change in fuel efficiency for buses observed at the most congested cases of 1500 and 2000 veh/h (+2%). In regard to fuel efficiency for the cars, the highest gain is seen at 1000 veh/h (+4%), as well as at the most congested cases of 1500 and 2000 veh/h (around +2%).

When the BSP scheme is implemented on a road network, it can be seen that cars emit more greenhouse gases, whereas buses are more environmentally friendly, as the amount of GHG they emit reduces (Figure 3, *b–d*). More specifically, cars emit more HC, CO and NOx (7% more HC and CO and 5.4% more NOx) when BSP is applied to the current traffic flow (1500 veh/h). The lowest change in emissions for cars between BSP and the current signal timing, and therefore the most environmentally friendly, is observed at 1000 veh/h (+1.2% for HC, +2.3% for CO and +0.2% for NOx). The difference between the two schemes for GHG emissions is less when buses are examined. However, buses emit the least amount of greenhouse gases when traffic flow is at 1500 veh/h (−2.8% for HC, −2.6% for CO and −2.3% for NOx). At the traffic flow for which cars are the most environmentally friendly (1000 veh/h), buses emit less HC (−1.6%), less CO (−1.6%) and less NOx (−1.2%) compared to the no BSP scheme.

Discussion and Conclusions

Microscopic traffic simulation has been used to examine the effect of two bus priority measures on the environment, and more specifically on reducing fuel consumption and GHG emissions. The tested bus priority measures included bus lanes and bus signal priority.

The selected road network presents high congestion, especially during peak hours, and is located in a district of Santiago, Chile that is served by several bus routes. Bus lanes are not included in the current road infrastructure; therefore a mixed-traffic lane usage is experienced by road users. The selected bus route, C01, serves the 16 bus stops of the examined road network with 26 buses/h, whilst the current traffic flow is 1500 veh/h. According to Vuchic (*19*), traffic flow should be less or equal to 1000 veh/h for the bus lane measure to be implemented. Therefore, the influence of traffic flows lower than the current flows, such as 1000 and 800 veh/h, was also investigated.

The results have shown that both bus priority measures contribute toward the improvement of the environment by reducing the amount of consumed energy and emitted GHG, especially by the bus service. In particular, for the bus lane scheme, limiting the traffic flow of this road network, or any other road network of similar characteristics, to 800 veh/h would result in the highest reduction of energy consumption and GHG emissions by both the cars and the bus service. However, the current traffic flow of the examined network is

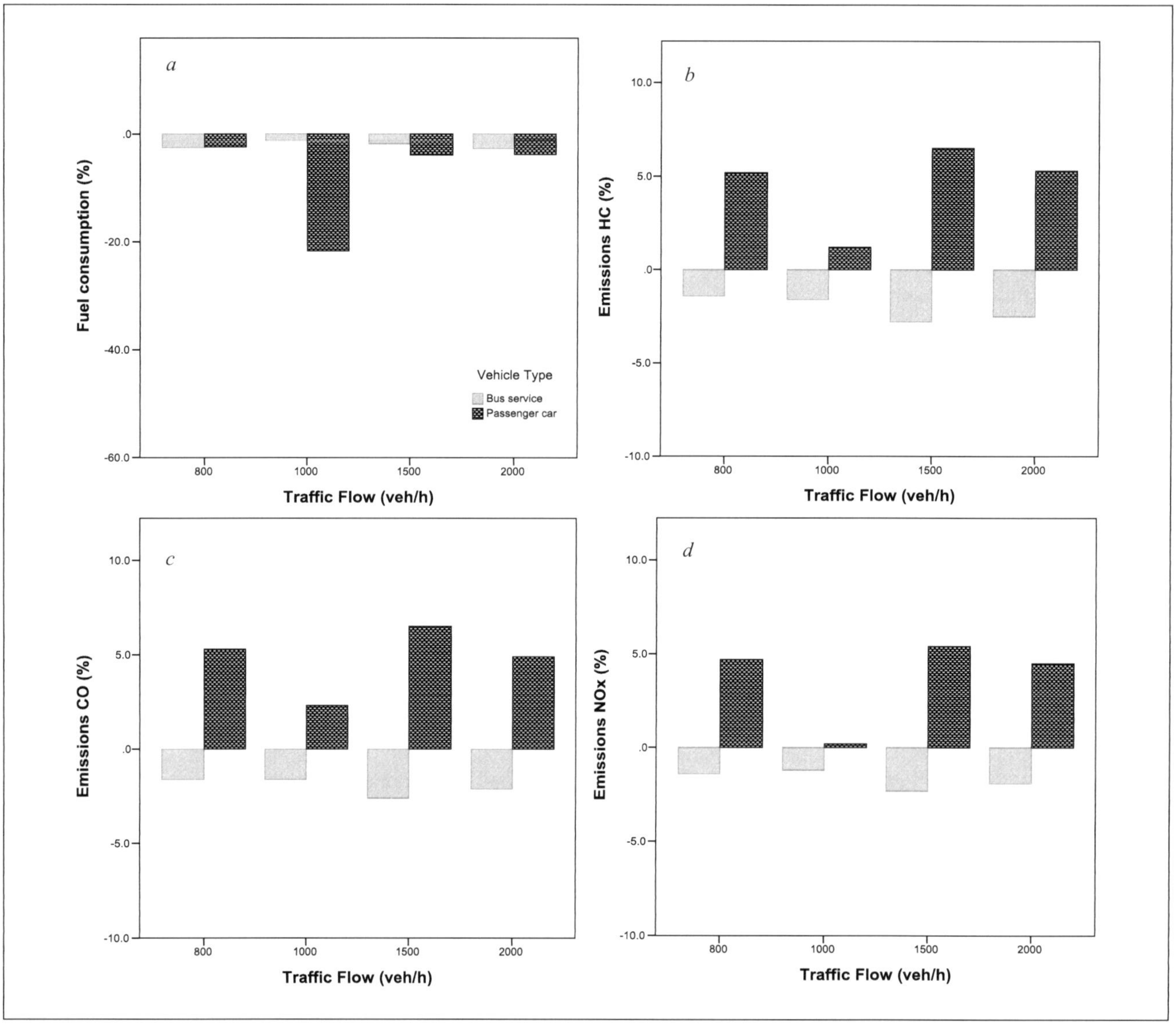

Figure 3. Percentage change of fuel consumption and emissions for cars and buses when BSP is applied.

much higher than 800 veh/h, therefore transport authorities will need to think of ways to reduce traffic on this street, as well as to direct road users toward more sustainable means of transport, such as public transport modes and cycling. This may not seem to be too hard to achieve, as, with the implementation of bus lanes, car users will experience lower average speeds which will then have an effect on their journeys by making them longer. As a result, car users will be dissatisfied by this experience and they might start searching for alternative means for their everyday commute, turning away from private cars. At the same time, bus services will be reaching higher average speeds and bus passengers will be reaching their destination much quicker than car users. Bus passenger satisfaction of the provided service will increase and passenger demand would be expected to follow the same trend.

The environmental impact of the BSP scheme is not as high as that of the bus lane scheme for cars and buses, but it also reduces energy consumption and emissions for the bus service. This reduction can be seen at all traffic flows for the bus service, nevertheless the BSP scheme has the biggest effect when applied to the current traffic flow (1500 veh/h). Therefore, if transport authorities decide to apply this measure without making any changes to the road infrastructure or finding ways to redirect traffic, the environmental improvement will be particularly apparent. When the results are considered in the case of cars, energy consumption and GHG emissions increase, however for a traffic flow of 1000 veh/h this increase is minimal.

At all traffic flows and when BSP is applied, road users of the main artery enjoy higher average speeds and therefore shorter journeys. In the short run, this has the advantage of high satisfaction for both the car users and the bus passengers. However, considering the consequences in the long run, such a measure has the implication of not promoting active transportation. Car users will continue using their cars, as the applied bus priority scheme works to their advantage, and more and more people will be choosing the comfort of their private car compared to the provided public transport service or other more sustainable modes of transport. To achieve the highest possible environmental benefit and accomplish the 2050 goals in regards to the global environmental challenges, transport authorities should be proposing solutions that turn road users away from their private cars. The additional benefits of such implementations would reflect on people's health and wellbeing and would dramatically reduce obesity (*23*).

Subsequently, it is evident that the 'Bus Only Lane' priority scheme has the highest and most enduring environmental impact and contributes the most to the reduction of fuel consumption and emissions of cars and buses when applied at traffic flows up to 1000 veh/h. To implement this measure, transport authorities as well as policy makers will need to enforce the creation of bus lanes dedicated solely to the bus service and to find ways to reduce the current traffic flow to at least 1000 veh/h. Changing societal habits can be the biggest challenge, therefore it is essential that road monitoring is taking place to ensure that bus lanes are respected by car users. For the purposes of reducing the traffic flow, directing traffic to adjacent arteries will only shift the problem to other areas. The most effective way of achieving traffic flow reduction is by attracting people to public transport modes, and this should be taken on board by transport authorities. In the case that an immediate environmental improvement is required, transport authorities can implement the BSP scheme to the current traffic flow (1500 veh/h) and road infrastructure (mixed-traffic), but this measure does not ensure long-term reduction in fuel consumption and emissions.

Transport policy worldwide is encouraging people to use more active modes of transport in preference to the car, as it provides the means for non-car users to travel distances that are too great to walk without difficulty. Public transport, especially buses, plays an important part in the implementation of this policy. Organizations such as the Active Living Research program, the Department of Transportation in the USA, and the WHO Regional Office in Europe provide information on how transport and mobility infrastructure have a positive impact on health (e.g., reduce obesity and improve respiratory diseases) and employability, reduce societal inequalities and increase connectivity (*21*). The critical role of urban planners and policy makers in this is greatly highlighted. The findings of this study show that priorities for buses, such as bus lanes and bus signal priority, have effects not only in reducing bus travel times, but also in fuel consumption and GHG emissions. These outcomes can be used in mass media campaigns to inform people of the benefits of active transportation, for the introduction of schemes aimed to reduce congestion—such as the congestion charge zone scheme in London, UK, for the reduction of street parking and the promotion of multi-modal transportation. The authors also believe that the results presented in this study can be taken as an example by other developing or rapidly growing countries with the aim to reduce their local energy consumption and GHG emissions challenges. Such countries could be other Latin American countries with similar road networks and China. It is undoubtable that advanced cities in the US and Europe can also benefit from these findings. Road networks in cities like Los Angeles, New York, Munich and many others, that present high congestion and mixed-traffic lane usage can reduce their energy consumption and emissions by following the recommendations of this paper.

Despite the significance of the outcomes of this study, the employed methods are subject to limitations and weaknesses that need to be recognized. TSIS-CORSIM is a software developed in the early 2000s, and therefore the values considered in the environmental tables and used to calculate fuel consumption and emission are outdated and reflect only a part of the vehicle fleet (mainly diesel cars) using the roads nowadays. Re-running the simulation model using updated environmental tables will increase the accuracy of the environmental impact of the proposed measures. Furthermore, considering a more diverse vehicle fleet—such as trucks, articulated and double decker buses—as well as the application to a road network, will provide a more complete overall picture of the environmental problems that need to be addressed.

As part of their future work, the authors consider testing the effect of bus acceleration on the environment. Buses are modelled in the examined network to accelerate and decelerate at 2.5 m/s^2 (absolute value). It has been shown by previous studies that the reduction of a vehicle's acceleration can greatly reduce fuel consumption and GHG emissions and increase passenger safety (*24–27*). A bus acceleration of 1.5 m/s^2 or lower has been proven to improve comfort during bus journeys and to enable passengers to walk naturally inside the bus when searching for a seat (*28*). This, however, cannot be achieved with the TSIS-CORSIM software as it is using a model to calculate fuel consumption and emissions which is not sensitive to changes in acceleration. An alternative approach will have to be followed.

Acknowledgments

This is a preliminary work done for the "City-Wide Analysis to Propel Cities towards Resource Efficiency and Better Wellbeing" project, which is part of the Chinese Low Carbon Cities Development and is funded by EPSRC (EP/N010779/1). The authors would like to thank Redbus Urbano for their help in providing the necessary data for this study.

Author Contributions

The authors confirm contribution to the paper as follows: study conception and design: Xenia Karekla, Rodrigo Fernandez, Nick

Tyler; data collection: Xenia Karekla, Rodrigo Fernandez; analysis and interpretation of results: Xenia Karekla, Rodrigo Fernandez, Nick Tyler; draft manuscript preparation: Xenia Karekla. All authors reviewed the results and approved the final version of the manuscript.

References

1. WHO. *Urban Population Growth* [Internet]. WHO, 2017. http://www.who.int/gho/urban_health/situation_trends/urban_population_growth_text/en/. Accessed July 2017
2. Dena Levitz. *How Do Greenhouse Gas Emissions Compare in Cities Around the World?* [Internet]. Hopes & Fears, 2015. http://www.hopesandfears.com/hopes/city/city_index/216917-city-index-carbon-emissions. Accessed July 2017
3. Oxfam International. *Extreme Carbon Inequality* [Internet], 2017. https://www.oxfam.org/en/research/extreme-carbon-inequality. Accessed July 2017
4. WHO. *Protecting Health from Climate Change* [Internet], 2008. http://www.who.int/world-health-day/toolkit/report_web.pdf. Accessed July 2017
5. US Department of Transportation. *Livability and Sustainability* [Internet], 2012. https://www.fhwa.dot.gov/publications/research/general/utc/2012webinar/sustainability/index.cfm. Accessed July 2017
6. International Energy Agency. *Transport, Energy and CO_2* [Internet], 2009, p. 418. https://www.iea.org/publications/freepublications/publication/transport2009.pdf. Accessed July 2017
7. European Environment Agency. *Greenhouse Gas Emissions from Transport* [Internet]. European Environment Agency. https://www.eea.europa.eu/data-and-maps/indicators/transport-emissions-of-greenhouse-gases/transport-emissions-of-greenhouse-gases-10. Accessed July 2017
8. US EPA. *Global Greenhouse Gas Emissions Data* [Internet]. US EPA, 2016. https://www.epa.gov/ghgemissions/global-greenhouse-gas-emissions-data. Accessed July 2017
9. ICCT. *International Council on Clean Transportation: China* [Internet], 2017. http://www.theicct.org/china. Accessed July 2017
10. European Commission. *Reducing Emissions from Transport* [Internet]. Climate Action: European Commission, 2016. https://ec.europa.eu/clima/policies/transport_en. Accessed July 2017
11. O'Ryan, R., D. Sperling, M. Delucchi, and T. Turrentine. *Transportation in Developing Countries: Greenhouse Gas Scenarios for Chile* [Internet]. Center for Climate and Energy Solutions, 2002. https://www.c2es.org/docUploads/transportation_chile.pdf. Accessed July 2017
12. Button, K. *Transport Economics*, 3rd ed. Edward Elgar Publishing Ltd, Aldershot, Hants; Northampton, 2010, p. 528.
13. Creutzig, F., G. Baiocchi, R. Bierkandt, P.-P. Pichler, and K. C. Seto. Global Typology of Urban Energy Use and Potentials for an Urbanization Mitigation Wedge. *Proceedings of the National Academy of Sciences*, Vol. 112, No. 20, 2015, pp. 6283–6288.
14. World Energy Council. *Global Transport Scenarios 2050* [Internet], 2011. https://www.worldenergy.org/wp-content/uploads/2012/09/wec_transport_scenarios_2050.pdf. Accessed July 2017
15. Cueto, J. L., A. M. Petrovici, R. Hernández, and F. Fernández]. Analysis of the Impact of Bus Signal Priority on Urban Noise. *Acta Acustica United with Acustica*, Vol. 103, No. 4, 2017, pp. 561–573.
16. Li, L., B. Persaud, and A. Shalaby. Using Micro-Simulation to Investigate the Safety Impacts of Transit Design Alternatives at Signalized Intersections. *Accident Analysis & Prevention*, Vol. 100, 2017, pp. 123–132.
17. Oguchi, T., A. Mitsuyasu, D. Oshima, and T. Imagawa. An Evaluation Study on Advanced Public Transport Priority System Using Traffic Simulation. *International Journal of Urban Sciences*, Vol. 21, 2017, pp. 43–53.
18. Instituto Nacional de Estadistica (Spanish Statistical Office). *Transport of Passengers Statistic* [Internet], 2017. http://www.ine.es/en/welcome.shtml. Accessed July 2017
19. Vuchic, V. R. *Urban Transit Systems and Technology*, 1st ed. John Wiley & Sons, Hoboken, NJ, 2007, p. 624.
20. Fernandez, R. A Study of Bus Operations on Arterial Roads by Simulation. *ITE*, Vol. 73, No. 4, 2003, pp. 77–81.
21. McGill, R. *Fuel Consumption and Emission Values for Traffic Models*. Report no. FHWA-RD-85-053. FHWA, Washington, D.C., 1985.
22. Schirmer, W. N., L. Z. Olanyk, C. L. B. Guedes, T. P. Quessada, C. B. Ribeiro, and M. A. Capanema. Effects of Air/Fuel Ratio on Gas Emissions in a Small Spark-Ignited Non-Road Engine Operating with Different Gasoline/Ethanol Blends. *Environmental Science and Pollution Research*, Vol. 24, No. 25, 2017, pp. 20354–20359.
23. ARUP, BRE, University College London, AREA Research. *Health + Mobility: A Design Protocol for Mobilising Healthy Living* [Internet], 2016. http://publications.arup.com/publications/h/health_and_mobility\. Accessed October 20, 2016.
24. Pampel, S. M., S. L. Jamson, D. L. Hibberd, and Y. Barnard. How I Reduce Fuel Consumption: An Experimental Study on Mental Models of Eco-Driving. *Transportation Research Part C: Emerging Technologies*, Vol. 58, 2015, pp. 669–680.
25. Lai, W.-T. The Effects of Eco-Driving Motivation, Knowledge and Reward Intervention on Fuel Efficiency. *Transportation Research Part D: Transport and Environment*, Vol. 34, 2015, pp. 155–160.
26. Martin, N. P. D., J. D. K. Bishop, R. Choudhary, and A. M. Boies. Can UK Passenger Vehicles Be Designed to Meet 2020 Emissions Targets? A Novel Methodology to Forecast Fuel Consumption with Uncertainty Analysis. *Applied Energy*, Vol. 157, 2015, pp. 929–939.
27. Walnum, H. J., and M. Simonsen. Does Driving Behavior Matter? An Analysis of Fuel Consumption Data from Heavy-Duty Trucks. *Transportation Research Part D: Transport and Environment*, Vol. 36, 2015, pp. 107–120.
28. Karekla, X. *Improving Accessibility of Public Transport Systems: The Influence of Double-Decker Bus Acceleration on Passenger Movement*. PhD thesis. University College, London, 2016.

The Standing Committee on Bus Transit Systems (AP050) peer-reviewed this paper (18-01703).

Article

TRR
JOURNAL OF THE TRANSPORTATION RESEARCH BOARD

Transportation Research Record
2018, Vol. 2672(8) 143–151

DOI: 10.1177/0361198118798722
journals.sagepub.com/home/trr
SAGE

Provision of Bus Real-Time Information: Turning Passengers from Being Contributors of Headway Irregularity to Controllers

Pengfei Wang[1,2,3], Xuewu Chen[1,3], Wendong Chen[1,3], Long Cheng[1,4], and Da Lei[1,3]

Abstract
In frequently serviced bus routes passengers are more concerned about bus headway regularity than actual punctuality to the schedule. The buses tend to bunch naturally partly because of the participation of passengers in the system. Thus, the irregularity of headway of routes with high passenger demand is extremely serious. With the development of technique, bus real-time information containing arrival times and passenger-loads of following buses can be provided for passengers. A special "control way" which is particularly adaptive to the high-frequency routes is proposed based on this background. It is thought that passengers would control the headway regularity themselves when provided with bus real-time information. To test the effects of this special "control way," bus-focused stochastic models are used for the simulation. The results show that providing real-time information does improve headway regularity as expected. The impacts are significant in morning-peak simulation. There is a reduction of 30.1% of headway irregularity after providing real-time information in morning peak, as well as, 18.5% in off-peak. Besides, through the comparison of line without passenger and normal line, it is confirmed that passengers are a vital factor contributing to the headway irregularity. By providing real-time information, the moderating effects of passengers approximately offset the negative effects from them both in morning peak and in off-peak. Lastly, as expected, passenger loads among buses became more even both in morning peak and in off-peak.

In frequently serviced bus routes passengers are more concerned about bus headway regularity than actual punctuality of bus arrival to the schedule. Buses arriving within very large or very small headways (bus bunching) are of particular concern and much less desirable for both operators and passengers. There are many factors contributing to headway irregularity including unpredictable traffic congestion on links between stations, delays caused by traffic signals, and dwell time at stations. Bus systems are considered to be naturally unstable. The buses tend to bunch when the system is uncontrolled; first discovered and explained by Newell and Potts (*1*). The time passengers take to board the bus at a station increases with the time between two consecutive bus arrivals. Consequently, the first bus experiences an increasing delay, whereas the next bus becomes early on its schedule. These two buses finally bunch. The instability increases with the passenger demand intensity. In the metropolises of China, passenger demands on popular routes are so huge that even when departing frequently the buses are often too crowded to carry more people. Thus, the irregularity of headway of routes with high passenger demand is extremely serious.

Strategies to maintain the regularity of bus systems have received an increasing amount of interest over the last decades. Existing control strategies are based on the following ideas: allowing buses to skip forthcoming stations (*2*), holding buses at specific locations (*3–6*), and implementing a transit signal priority (*7*). To analyze the bus system dynamics and to evaluate the control strategies, bus traveling models have received concern as well. Bus-focused deterministic models simplify the traveling process, focusing on the theory mainly (*8*). Bus-focused stochastic models reproduce the variability of traffic

[1]Jiangsu Key Laboratory of Urban ITS, Southeast University, Nanjing, China
[2]Jiangsu Province Collaborative Innovation Center of Modern Urban Traffic Technologies, Southeast University, Nanjing, China
[3]School of Transportation, Southeast University, Nanjing, China
[4]Department of Geography, Ghent University, Ghent, Belgium

Corresponding Author:
Address correspondence to Xuewu Chen: chenxuewu@seu.edu.cn

conditions and passenger demand to generate travel times of buses (9, 10). Static models are also based on probabilistic distributions, but only to express the mean and the variance of headways of buses at different positions (11). In addition to the control strategies mentioned above, this paper tries to present a special "control way" which is particularly adaptive to high-frequency routes with high passenger demand.

In recent years, with the development of technique, AVL (automatic vehicle location) and APC (automatic passenger count) systems have been used widely in bus systems. Information about arrival times of the following buses is provided for passengers. The information is shown on LED display at stations or in smartphones in passengers' hands. However, the real-time information about in-vehicle crowding of the following buses has not commonly appeared yet. When the real-time arrival information and in-vehicle crowding information are both provided, passengers would have comprehensive understanding about the conditions of following buses which may help them make better decisions.

Under such circumstances, if bunching or short headway happens, passengers would be likely to choose the second bus if a more comfortable in-vehicle environment is known to them. Irregular headway or bunching is usually accompanied by uneven passenger loads; with more passengers on the first bus and fewer on the next in a bunch. So, it is reasonable to assume a situation like this. Then, if some passengers chose to wait for the second bus, their boarding time would make the second bus dwell longer and enlarge its headway from the previous bus. Theoretically, the closer two consecutive buses are, the more passengers choose the latter. As a result, passengers themselves would not let bunching happen. So, passengers here are used as the special holding tool. All these effects are based on the provision of both arrival time and in-vehicle crowding information. If there is no in-vehicle crowding information, passengers would probably not wait for another bus with unknown in-vehicle condition. In this paper the bus real-time information, which will appear frequently later, represents the total of arrival time information and in-vehicle crowding information. The paper aims at testing the effects of the proposed special "control way." Bus-focused stochastic models are used for the simulation to realize it.

The paper is organized as follows. The next section prepares the methodology used for simulation, including the models and the passenger boarding choice with real-time information. Then the simulation and results are described, including the calibration of the simulation models and two levels of analyses on headway regularity. The final section shows the conclusions.

Methodology

Bus Route Settings and Simulation Models

Consider $n = 1...N$ bus trips departing within a certain time interval on a fixed bus route equipped with $s = 1...S$ bus stations. Let $T_{n,s}$ be the arrival time of bus n at station s. If $d_{n,s}$ denotes the dwell time of bus n at station s and $t_{n,s}$ denotes the travel time of bus n on link s between station s and station $s+1$, then:

$$T_{n,s+1} = T_{n,s} + d_{n,s} + t_{n,s} \tag{1}$$

The travel time $t_{n,s}$ of a bus n on a link s is considered to relate to the length of link s, the traffic condition, and setting of traffic signals along link s. In order to simplify the models, the distances between two consecutive stations are assumed to be equal along the route. In addition, the traffic condition and setting of traffic signals in each link are assumed to be the same as each other. Therefore, $t_{n,s}$ of any bus n on any link s could be considered to follow a same probability distribution, for example, a normal distribution in this paper, here:

$$t_{n,s} \sim N(\mu_t, \sigma_t^2) \tag{2}$$

where μ_t and σ_t are the mean and standard deviation of travel time of a bus on a link.

The dwell time $d_{n,s}$ of a bus n at a station s consists of two parts of time. One is the time the bus opens and closes the doors, noted as constant c_0, which is usually about 10 s. The other is the actual loading time, depending on the number of boarding passengers $B_{n,s}$ or alighting passengers $A_{n,s}$. According to previous studies, the average boarding time per passenger c_B is within a range between 2 and 5 s and the average alighting time per passenger c_A is generally less than c_B by 0.5 s. For buses on high-frequency routes equipped with two doors, boarding and alighting happen simultaneously. Actual loading time is the bigger out of the boarding time and the alighting time. Then, dwell time $d_{n,s}$ is expressed as:

$$d_{n,s} = c_0 + \mathrm{Max}(c_B \cdot B_{n,s}, c_A \cdot A_{n,s}) \tag{3}$$

Passengers arrive at station s randomly, and arrival of passengers is considered to follow Poisson distribution. Supposing passenger arrival rate at station s is λ_s, the mean of number of passengers who wait for bus n is $\lambda_s(T_{n,s} - T_{n-1,s})$. Here, the difference between dwell time of bus $n-1$ and bus n is negligible compared with the headway between them. Thus:

$$B_{n,s} \sim P(\lambda_s(T_{n,s} - T_{n-1,s})) \tag{4}$$

For each passenger boarding at station s, the riding distance (express as number of passing stations) is assumed to follow Poisson distribution. The parameter in Poisson distribution λ_L equals the average riding distance L. Notice that the average riding distance of passengers who board near the end of the route cannot be L anymore, λ_L is assumed to be $S-s$ when $s>S-L$. Altogether, the riding distance L_s of a passenger boarding at station s is shown as:

$$L_s \sim P(\lambda_L) \text{ where } \lambda_L = \begin{cases} L & s \leq S-L \\ S-s & s>S-L \end{cases} \quad (5)$$

Similarly, passenger arrival rate λ_s should be adjusted when the station is near the end. It is assumed that λ_s keeps unchanged when $s \leq S-L$ and decreases linearly when $s>S-L$. Let λ indicate the average passenger arrival rate at a station, then:

$$\lambda_s = \begin{cases} \lambda & s \leq S-L \\ \lambda \cdot (S-s)/L & s>S-L \end{cases} \quad (6)$$

Using Equation 4–6, passenger activities including both boarding and alighting could be simulated approximately and then dwell time $d_{n,s}$ in Equation 3 is determined. Let $\text{Load}_{n,s}$ be the passenger load on bus n at station s (consider alighting but not boarding at station s), then:

$$\text{Load}_{n,s} = \text{Load}_{n,s-1} + B_{n,s-1} - A_{n,s} \quad (7)$$

$\text{Load}_{n,s}$ will be the judgement basis when passengers make boarding choices in the next section. $\text{Load}_{n,s} + B_{n,s}$ is the passenger load on the link s.

At last, departure time at originating station should be discussed. If the departure interval is h_0, then the departure time of each bus is got easily. However, the departure interval of high-frequency buses can hardly remain stable in reality. Therefore, a uniform deviation between $-\Delta$ and Δ is added to the departure time. If the departure time of the first bus is 0s, then departure time of bus n is expressed as:

$$T_{n,1} = h_0 \cdot (n-1) + \delta \cdot \Delta \quad (n = 2, 3, ..., N) \quad (8)$$

with uniformly distributed $\delta \in [-1, 1]$.

Having necessary initial values, arrival time $T_{n,s}$ and dwell time $d_{n,s}$ of any bus n at any station s can be calculated step by step through Equation 1–3.

Passenger Boarding Choice with Real-Time Information

As we know, bus real-time information containing both arrival times and passenger loads of subsequent buses has not been realized in China or other countries yet. Passenger boarding choices with adequate real-time information are obtained by asking passengers their choices in case of certain situations, namely by stated-preference (SP) survey. Yu et al. (*12*) has conducted detailed research on passenger boarding choice prediction with real-time information. A series of surveys about passenger boarding choice under different situations have been done. Among them, the survey about passengers' choices between the first and second buses on the same route is what this paper focuses on.

This survey was conducted between 6:30:00 and 8:00:00 in weekdays in November 2012, with 705 passengers participating. The route discussed in the survey is a high-frequency line (12 buses per hour) in Dalian City, China. According to cloud model theory, the survey classified in-vehicle crowding into 4 categories: "seat available," "not crowded," "slightly crowded," and "very crowded." Similarly, more waiting time for the second bus was classified into 3 categories: "approximately 1–3 min," "approximately 5 min," and "approximately 10 min." Passengers were asked their choice between the first and second buses, such as in the following situation: *If the first bus to arrive is very crowded and the second bus is not crowded, and you only need to wait approximately 3 more minutes for the second bus to arrive, would you wait for the second bus?* Other questions were designed similarly. Then, a probability of choosing the second bus was obtained for each combination of passenger load and more waiting time for the second bus.

For simplification, passenger-load categories and waiting-time categories correspond to certain intervals in this paper. According to Yu (*12*), for a bus with 27 seats, "seat available" means passenger load less than 27 and "not crowded" is considered to be equal to or greater than 27 but less than 40. Similarly, "slightly crowded" is equal to or greater than 40 but less than 55, and "very crowded" is equal to or greater than 55. As for the more waiting time for the second bus, since the research object in this paper is a high-frequency line whose headways are rarely longer than 10 min, the category "approximately 10 min" is dropped. In this paper, "approximately 1–3 min" and "approximately 5 min" are considered to be less than 3 min, and greater than 3 min but less than 7 min, respectively. Thus, inferences to the choice regarding in-vehicle crowding and more waiting time are shown as Table 1.

From the survey results shown in Table 1, when the first bus is "very crowded" or "slightly crowded" and the second bus is "seat available," no matter how much more waiting time is needed (less than 7 min), passengers are very likely to choose the second bus to avoid the crowd. When more waiting time is less than 3 min, most passengers are willing to choose the second bus for a more comfortable in-vehicle environment.

With provision of real-time in-vehicle crowding information, as well as bus-arrival information, passengers have the choice to board the current bus or the next.

Table 1. Inferences to the Choice Regarding In-vehicle Crowding and More Waiting Time

In-vehicle crowding (passenger load)		More waiting time for the second bus (min)	Probability of choosing the second bus (%)
First bus	Second bus		
≥55 (very crowded)	0~27(seat available)	3~7	94
		0~3	98
40~54 (slightly crowded)	0~27(seat available)	3~7	91
		0~3	95
27~39 (not crowded)	0~27(seat available)	3~7	44
		0~3	61
≥55 (very crowded)	27~39(not crowded)	3~7	65
		0~3	82
40~54 (slightly crowded)	27~39(not crowded)	3~7	39
		0~3	77
≥55 (very crowded)	40~54(slightly crowded)	3~7	51
		0~3	73

Every passenger's choice is assumed to be independent. The number of passengers who choose to wait for the next bus $\chi_{n,s}$ at station s is thought to follow a binomial distribution:

$$\chi_{n,s} \sim B(B_{n,s}, p_{n,s}) \tag{9}$$

where $B_{n,s}$ is the number of passengers waiting for bus n initially at station s, and $p_{n,s}$ is the probability of choosing the next bus according to the inferences in Table 1. Here, passengers get on either the current bus or the next. It is assumed no one chooses the third bus.

Then actual number of passengers getting on bus n at station s $ACTB_{n,s}$ is expressed as:

$$ACTB_{n,s} = (B_{n,s} - \chi_{n,s}) + \chi_{n-1,s} \tag{10}$$

When computing dwell time $d_{n,s}$ of a bus in Equation 3, $B_{n,s}$ should be replaced by $ACTB_{n,s}$ under the consideration of passenger boarding choice with real-time information. So far, the process of passenger boarding choice has been discussed completely. Simulation models could be run combining the bus running models at the start of this section and passenger boarding choice described here.

Measure Evaluating Bus Headway Regularity

Coefficient of variation of headway (c_{vh}) is the indicator to characterize bus headway regularity in this paper. c_{vh} is one of the most common indicators to evaluate headway regularity for high-frequency buses. TCQSM (*13*) suggests level of service corresponding to value of c_{vh}.

In this paper, the arrival time of bus n at station s, $T_{n,s}$ can be generated by the simulation models mentioned above. Let $h_{n,s}$ be the headway between bus $n-1$ and bus n at station s, then:

$$h_{n,s} = T_{n,s} - T_{n-1,s} \tag{11}$$

For each station s, c_{vh} is calculated among all headways at station s. Let μ_s^h and σ_s^h be the mean and standard deviation of observed headways at station s, respectively, then:

$$c_{vh,s} = \frac{\sigma_s^h}{\mu_s^h} \quad s = 1 \ldots S \tag{12}$$

To evaluate headway regularity of a route, the mean coefficient of variance of headway of all stations C_{vh} is proposed, as:

$$C_{vh} = \frac{\sum_{s=1}^{S} c_{vh,s}}{S} \tag{13}$$

To summarize, c_{vh} is used to evaluate headway regularity at a station and C_{vh} is used to evaluate headway regularity of a whole line.

Simulation and Analysis

Calibration of the Simulation Models

The parameters in the simulation models are shown in Table 2. A route in Changzhou, China is used to calibrate the parameters. Changzhou, a prefecture-level city with 4.7 million population, is located in eastern China. Changzhou is famous for its bus rapid transit (BRT) and it also has more than 200 conventional bus lines. The route studied in this paper is the busiest conventional route in Changzhou, namely line 2, with a volume of 26,000 passengers per day. The vehicles serviced on line 2 are also large-sized buses, with 27 seats for passengers.

There are 36 stations on line 2. Passenger arrival rate is considered to be constant in the simulating time period, so the number of bus trips should not be so large. Then, 11 bus trips are adopted in the models so that there are 10 headways at each station. The links are assumed to be

Table 2. Calibration of Parameters in the Simulation Models

Notation	Explanation	Unit	Value	
			In morning peak	In off-peak
N	Bus trips in the model	/	11	11
S	Station number in the route	/	36	36
μ_t	Average travel time on a link	s	90	80
σ_t	Standard deviation of travel time on a link	s	25	16
c_0	Time opening and closing doors	s	10	10
c_B	Average boarding time per passenger	s	2.5	2.5
c_A	Average alighting time per passenger	s	2.0	2.0
λ	Average passenger arrival rate at a station	per/min	0.75	0.45
L	Average passenger riding distance	stops	12	12
h_0	Departure interval	min	6	8
Δ	Uniform deviations in departure time	min	1.5	1
$d_{1,s}$	First bus dwell time at any station s	s	25	20

uniform in the models. Average travel time μ_t and standard deviation of travel time σ_t on a link are calibrated by using AVL data from line 2. μ_t is different for different time periods depending on traffic condition at that time, as well as σ_t. For example, μ_t and σ_t are 90 s and 25 s in morning peak (a.m. 7:00–8:00), 80 s and 16 s in off-peak (p.m. 14:00–15:00), respectively.

Time opening and closing doors c_0 is set as 10 s. Actually, c_0 is not influencing when calculating headways. Average boarding time per passenger c_B approximates 2.5 s according to a study by Lei et al. (*14*), which focuses on passenger boarding time in Changzhou. And average alighting time per passenger c_A is the empirical value which is less than c_B by 0.5 s, namely 2.0 s.

As shown in IC smart card data from line 2, there are about 1350 passengers boarding line 2 between a.m. 7:00–8:00 in one direction. Then average passenger arrival rate λ at a station is $1350/30/60 = 0.75$ per/min. The value of λ in off-peak is calculated in the same way. Also, according to the analysis results from integrated circuit (IC) smart card data, average passenger riding distance L is set as 12 (stops).

To run the simulation models, necessary initial values are needed. Departure intervals h_0 are 6 min in morning peak and 8 min in off-peak. Uniform deviations in departure time Δ are 1.5 min and 1 min for morning peak and off-peak, respectively, referring to data from line 2. In addition, since the first bus has no boarding passenger according to the models, its dwell time $d_{1,s}$ should be set to fixed values, which are 25 s in morning peak and 20 s in off-peak. The complete calibration results are shown in Table 2.

Simulation

A program is designed to realize the simulation models. The algorithm below outlines the general framework of the bus models.

Algorithm. Simulation of bus models

```
Generate the departure time T_{n,1} of all buses
for each station s = 2 : S do
  for each bus n = 1 : N do
    Compute the travel time t_{n,s-1} of bus n between stations
    s - 1 and s.
    Compute the arrival time T_{n,s} = T_{n,s-1} + d_{n,s-1} + t_{n,s-1}.
    Compute the dwell time d_{n,s} of bus n at station s preparing
    for next loop.
  end
end
```

There is a test after computing arrival time of all buses at a station s to avoid overtaking. The output of the program is a series of arrival time $T_{n,s}[n \times s]$ and a series of passenger loads on links $\text{Load}_{n,s}[n \times (s-1)]$. Situations with and without bus real-time information are both tested.

Figure 1 is the trajectories of 10 bus trips without and with bus real-time information in morning peak, respectively. The color of a line represents its passenger load. Green means low passenger load and red means high passenger load.

As shown in Figure 1(*a*), bunching happens frequently in conventional line. Meanwhile, passenger loads are uneven in a pair of bunching buses, usually crowded in the former one and fewer passengers in the latter one, which is consistent with the reality. When providing real-time information as shown in Figure 1(*b*), the headway regularity is better although there are bunches near the end of the route. Passenger loads are also more even than that of conventional line. Quantitative analyses regarding headway regularity and passenger load evenness are described in the next section.

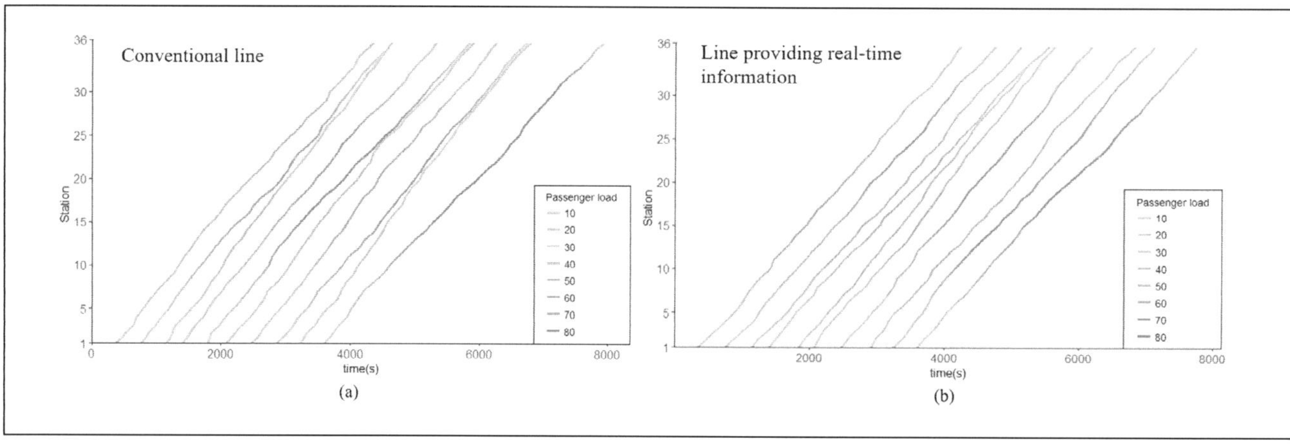

Figure 1. Trajectories of 10 bus trips of (*a*) conventional line and (*b*) line providing real-time information.

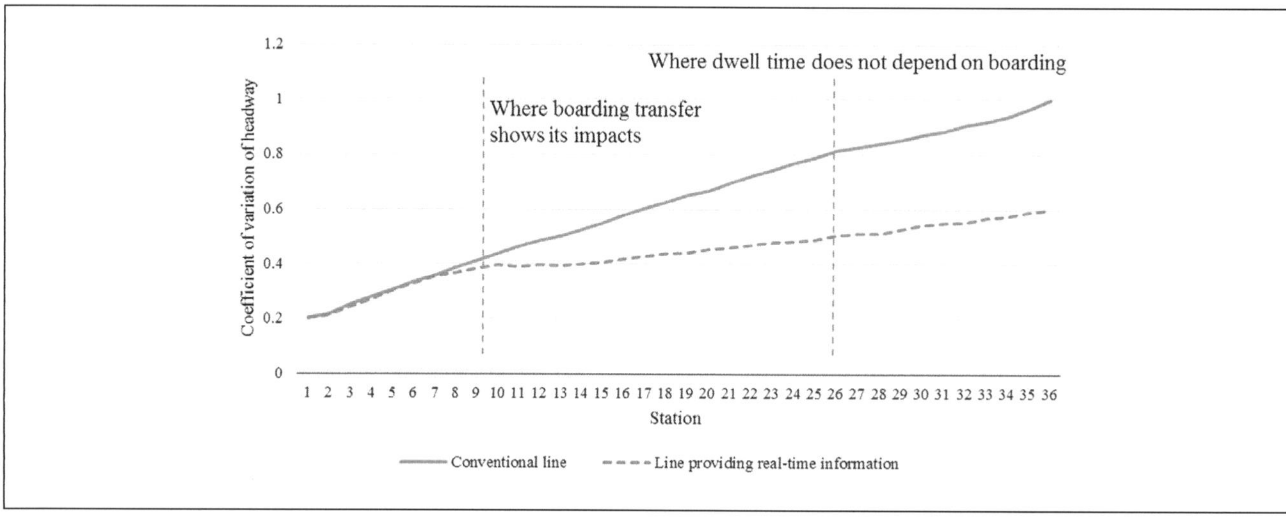

Figure 2. Coefficient of variation of headway at stations in morning peak.

Results and Analyses

To address the randomness in the models, 50 repeated trials are conducted for situations without or with bus real-time information.

Headway Regularity at Stations. The average coefficient of variation of headway c_{vh} at stations among 50 replications in morning peak is shown in Figure 2.

Overall, with the provision of real-time information, c_{vh} has a significant decline, which means much better headway regularity, as expected. c_{vh} of conventional line goes up to 1.0 at the end of the route, whereas c_{vh} of the line providing real-time information only reaches 0.6 at the end.

In detail, the variation of c_{vh} with stations implicates more information. The uptrends of two lines are almost the same until station 9. From station 9, in-vehicle passengers accumulate to a certain amount and some buses get crowded. Then a portion of passengers choose to transfer to the next bus, resulting in variation in the headways. c_{vh} of the line providing real-time information goes up slowly from station 9 compared with conventional line. After station 26, tiny changes exist in the trend of c_{vh} of the two lines. c_{vh} of conventional line increases a bit more slowly than before, whereas c_{vh} of line providing real-time information is on the contrary. That is because the dwell times do not depend on boarding gradually after station 26. Without the effects from boarding passengers, c_{vh} of the conventional line increases a bit more slowly than before.

Similarly, the average coefficient of variation of headway c_{vh} at stations among 50 replications in off-peak is shown in Figure 3.

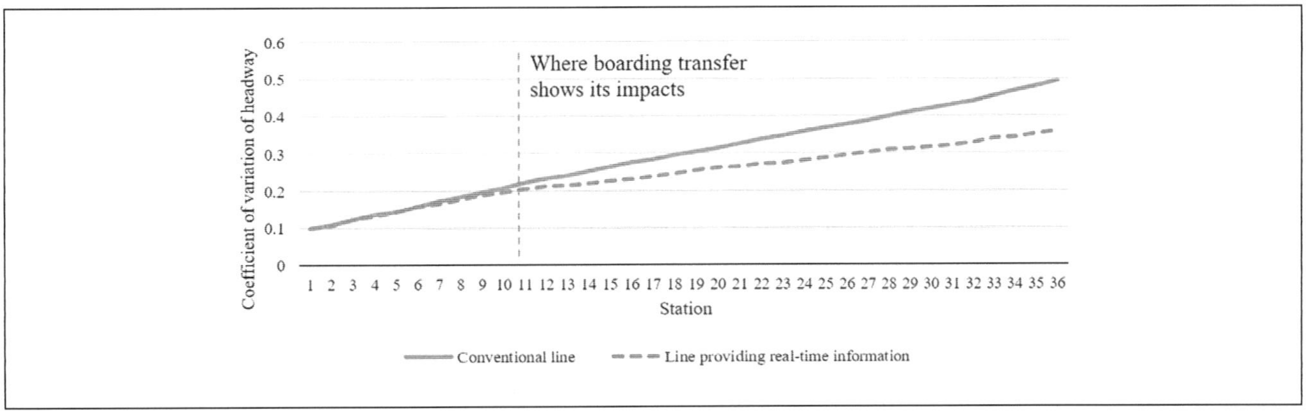

Figure 3. Coefficient of variation of headway at stations in off-peak.

Table 3. Headway Regularity of Route in Morning Peak and in Off-Peak

C_{vh}		Line without passenger	(a) Conventional line	(b) Line providing real-time information	Reduction comparing (b) with (a)
In morning peak	Mean	0.401	0.622	0.435	30.1%
	SD	0.082	0.129	0.082	36.4%
In off-peak	Mean	0.215	0.298	0.243	18.5%
	SD	0.049	0.065	0.050	23.1%

Note: SD = standard deviation

In general, c_{vh} in off-peak is much smaller than that in morning peak. It shows that bus headway irregularity is not that serious in off-peak for the passenger demands are smaller and traffic conditions are better in off-peak. Also, the effects of boarding transfers when providing real-time information are not as obvious as in morning peak.

The two lines show their difference from station 11, which is a bit larger than that in morning peak. c_{vh} of the line providing real-time information increases a bit more slowly after station 11, whereas c_{vh} of the conventional line keeps growing as before until the end.

In summary, providing bus real-time information makes certain passengers transfer to the next bus and thus makes better headway regularity as expected. The decrease of c_{vh} is significant in morning peak. The boarding transfers show impacts on c_{vh} usually from about station 10 where some buses get crowded. The decrease of c_{vh} in off-peak is not as obvious as in morning peak.

Headway Regularity of Route and Passenger Load Evenness. The mean of coefficient of variation of headway among all stations C_{vh} is to evaluate the headway regularity of the whole line. Results of 50 replications mentioned above are used for analysis. For interest, the results of a line without passengers ($\lambda = 0$) are used to figure out how the headway regularity is if there were only traffic condition counted.

In addition, passenger boarding transfers when providing real-time information would make passenger loads more even among buses theoretically. To study passenger load along a bus trip, the number of passengers serviced by a bus trip (N_p) is calculated by adding all boarding passengers of the bus together. Then there are 10 N_p within one sample. The standard deviation of N_p indicates the evenness of passenger loads among buses. 50 samples are included here.

The headway regularity of route in morning peak and in off-peak are shown in Table 3. The passenger load evenness in morning peak and in off-peak are shown in Table 4.

In morning peak, C_{vh} of line without passengers is 0.401 whereas that of conventional line is 0.622, which confirms passengers are a vital factor contributing to headway irregularity. When providing real-time information, the reduction of C_{vh} is 30.1% compared with conventional line. The value 0.435 approaches that of line without passengers. It indicates that after providing real-time information the moderating effects of passengers approximately offset the negative effects from them. Besides, the standard deviation of C_{vh} of line providing

Table 4. Passenger Load Evenness in Morning Peak and in Off-Peak

	Conventional line		Line providing real-time information		
Mean and SD of number of passengers serviced by a bus trip	μ_N	σ_N	μ_N	σ_N	Reduction of σ_N
In morning peak	130.6	68.6	132.9	32.8	52.2%
In off-peak	106.7	29.2	105.8	15.9	45.5%

real-time information is much smaller than the one of conventional line. The system stability of line providing real-time information is better, which means with the moderating effects of passengers extremely irregular headways would be less likely to appear. The reduction of C_{vh} in off-peak is not as obvious as that in morning peak, as illustrated in Figure 3. Also, the value 0.243 approaches that of line without passengers.

As shown in Table 4, it is clearly seen that after providing real-time information the standard deviation of number of passengers serviced by one bus trip becomes a half of that in conventional line. Passenger loads among buses are much more even than before. An obvious reduction of σ_N also exists in off-peak.

Conclusion

In recent years, bus real-time information, containing arrival times and passenger loads of subsequent buses, could be provided for passengers. When arrive time information and passenger load information are both provided, it may help to control the headway regularity of high-frequency buses. To evaluate the effects of this special control way, bus simulation models are established. The models consider the randomness of departure headways, travel time on links, passenger arrivals at stations and passenger alighting positions. To determine the passenger boarding choice, the results from a survey conducted in Dalian City, China are used, and the inferences to the choice regarding in-vehicle crowding and more waiting times are given. The parameters in the simulation models are calibrated by the data from line 2 in Changzhou, China. Parameters of morning peak and of off-peak are both given.

Coefficient of variation of headway is adopted to evaluate headway regularity. There are two levels: coefficient of variation of headway at a station (c_{vh}) and the mean of it (C_{vh}) for a route. Conventional line and line providing real-time information are tested in the simulation models. The results show that providing real-time information does make better headway regularity as expected. The decrease of c_{vh} is significant in morning peak. The boarding transfers show its impacts on c_{vh} usually from about station 10 where some buses get crowded. In off-peak, although c_{vh} of the conventional line itself is not large, the decline still exists. As for the route level, there is a reduction of 30.1% of C_{vh} after providing real-time information in morning peak, as well as 18.5% in off-peak. Besides, through a comparison of line without passengers and normal line, it is confirmed that passengers are a vital factor contributing to headway irregularity. And after providing real-time information the moderating effects of passengers approximately offset the negative effects from them both in morning peak and off-peak. In addition, decline in standard deviation of C_{vh} shows the special control way is extremely effective faced with high headway irregularity. Lastly, the standard deviation of number of passengers serviced by a bus trip is calculated to find out passenger load evenness. As expected, passenger loads among buses become more even both in morning peak and in off-peak for a reduction of 52.2% and 45.5% of standard deviation, respectively.

Our research still includes some limitations. More efforts are needed to figure out the influencing factors of passenger boarding choice, like passengers' sociodemographic characteristics. Furthermore, some external factors could affect the boarding choices as well, like weather or waiting environment at stations. So, the results in this paper represent a normal situation without special weather. The conditions tested in the paper are simple. More complex conditions including varying passenger-arrival rate, varying departure interval and varying standard deviation of travel time on links should be tested in future research.

Acknowledgments

This research is sponsored by the National Natural Science Foundation of China (51338003). The author appreciates Changzhou Public Transit Company for providing the data used in this study.

Author Contributions

The authors confirm contribution to the paper as follows: study conception and design: Pengfei Wang, Xuewu Chen; data collection: Wendong Chen; analysis and interpretation of results:

Pengfei Wang, Da Lei; draft manuscript preparation: Pengfei Wang, Long Cheng. All authors reviewed the results and approved the final version of the manuscript.

References

1. Newell, G. F., and R. B. Potts. Maintaining a Bus Schedule. *Australian Road Research Board (ARRB) Conference*, Melbourne, Australia, 2nd ed. Vol. 2, No. 1, 1964.
2. Sun, A., and M. Hickman. The Real–Time Stop–Skipping Problem. *Journal of Intelligent Transportation Systems*, Vol. 9, No. 2, 2005, pp. 91–109.
3. Eberlein, X. J., N. H. Wilson, and D. Bernstein. The Holding Problem with Real–time Information Available. *Transportation Science*, Vol. 35, No. 1, 2001, pp. 1–18.
4. Van Oort, N., N. Wilson, and R. Van Nes. Reliability Improvement in Short Headway Transit Services: Schedule-and Headway-Based Holding Strategies. *Transportation Research Record: Journal of the Transportation Research Board*, 2010. 2143: 67–76.
5. Daganzo, C. F. A Headway-Based Approach to Eliminate Bus Bunching: Systematic Analysis and Comparisons. *Transportation Research Part B: Methodological*, Vol. 43, No. 10, 2009, pp. 913–921.
6. Daganzo, C. F., and J. Pilachowski. Reducing Bunching with Bus-to-bus Cooperation. *Transportation Research Part B: Methodological*, Vol. 45, No. 1, 2011, pp. 267–277.
7. Hongchao, L., J. Zhang, and D. Cheng. Analytical Approach to Evaluating Transit Signal Priority. *Journal of Transportation Systems Engineering and Information Technology*, Vol. 8, No. 2, 2008, pp. 48–57.
8. Ruan, M. Probability-Based Bus Headway Regularity Measure. *IET Intelligent Transport Systems*, Vol. 3, No. 4, 2009, pp. 400–408.
9. Hans, E., N. Chiabaut, and L. Leclercq. Investigating the Irregularity of Bus Routes: Highlighting How Underlying Assumptions of Bus Models Impact the Regularity Results. *Journal of Advanced Transportation*, Vol. 49, No. 3, 2015, pp. 358–370.
10. Cats, O., A. Larijani, H. Koutsopoulos, and W. Burghout. Impacts of Holding Control Strategies on Transit Performance: Bus Simulation Model Analysis. *Transportation Research Record: Journal of the Transportation Research Board*, 2011. 2216: 51–58.
11. Adebisi, O. A Mathematical Model for Headway Variance of Fixed-route Buses. *Transportation Research Part B: Methodological*, Vol. 20, No. 1, 1986, pp. 59–70.
12. Yu, B., T. Li, L. Kong, K. Wang, and Y. Wu. Passenger Boarding Choice Prediction at a Bus Hub with Real-time Information. *Transportmetrica B: Transport Dynamics*, Vol. 3, No. 3, 2015, pp. 192–221.
13. Capacity, T. *Quality of Service Manual (TCQSM), (2013)*. Part–7, 2nd ed. Kittleson and Associates, 2014.
14. Lei, D., X. Chen, L. Cheng, H. Li, P. Wang, and Z. Xie. Exploring Passenger Boarding Time Differences of Adults and Seniors: A Smart Card Data Based-Study of Changzhou, China. Presented at 96th Annual Meeting of the Transportation Research Board, Washington, D.C., 2017.

The Standing Committee on Bus Transit Systems (AP050) peer-reviewed this paper (18-06043).

Article

TRR
JOURNAL OF THE TRANSPORTATION RESEARCH BOARD

Transportation Research Record
2018, Vol. 2672(8) 152–164

DOI: 10.1177/0361198118797778
journals.sagepub.com/home/trr
SAGE

Evaluation of Transit Ecodriving in Rural, Suburban, and Urban Environments

Xiaodan Xu[1], Hanyan "Ann" Li[1], Haobing Liu[1], Michael O. Rodgers[1], and Randall Guensler[1]

Abstract

Ecodriving (also known as eco-driving) is a widely recognized strategy to reduce transit vehicle fuel consumption and emissions. Previous simulation modeling and field studies have demonstrated that ecodriving can reduce fuel consumption by 2%–27%. However, prior studies have typically focused on urban transit operations. More than 1700 transit agencies operate in rural areas within the United States, operating under very different conditions than urban transit systems. Assessing the variability of fuel savings across various transit operating conditions and physical terrain conditions will help transit agencies predict the potential benefits of deploying ecodriving strategies for their systems. The objective of this study is to assess the potential benefits and limitations of deploying ecodriving strategies for different transit services, service areas, fleet composition, and road topographies. There are three preliminary tasks involved in this study. The first element collects and processes the operation data across different transit services. Second-by-second operation data are collected from one urban, suburban, and rural transit agency. The second element evaluates the operation characteristics of each type of transit service. The third element assesses the potential benefits of introducing the transit ecodriving strategy to the operations of these fleets. The fuel consumption is evaluated by matching the second-by-second MOVES model operating mode bin with corresponding emission rates. Overall, the ecodriving strategy can help reduce 1%–5% fuel use across these agencies, and save $ 0.011–$ 0.045 operational cost per mile due to the fuel saving. The actual benefits vary by service, mileage, and road grade impact.

In 2014, about 43.5% of total transit expenses in the United States were for operating costs (*1*). Fuel cost is often a significant portion of operating costs and even 1% or 2% of fuel cost saving would result in notable savings for operating costs. For example, in 2017 fiscal year, Metropolitan Atlanta Rapid Transit Authority (MARTA) would spend $7.7 million on diesel and $6.6 million on compressed natural gas (CNG) fuels in their 2017's budget, which is $14.3 million in total and 7.5% of total bus operating cost (24.6% of non-labor operating cost) (*2*). Transit agencies are always seeking opportunities to reduce fuel consumption to lower operating costs, which also provides emission reduction at the same time (*3*). Transit fleet fuel consumption is primarily affected by: vehicle fleet composition, including vehicle and engine characteristics; transit service parameters, including route length and frequency and location of stops; and on-road operating conditions, including speed and acceleration and interactions with road grade (*3–6*). Therefore, the return on investment (ROI) of potential fuel-saving strategies should also be investigated within a local operations context.

Ecodriving (also known as eco-driving) is receiving a great deal of attention due to the significant potential for fuel saving and emission reductions in transit fleets that have been demonstrated through simulation modeling and field measurement (*3, 4, 7–12*). The ecodriving strategy changes the on-road driving cycle by encouraging the driver to implement specific driving behaviors (*13*): anticipating driving conditions, reducing high-speed operations, avoiding hard accelerations, shifting to the highest available gear, driving smoothly, and limiting idling.

With respect to field measurement studies, Zarkadoula, et al. (*12*) provided ecodriving training for three bus drivers in Athens, Greece, and recorded fuel economy before and after training. They found the overall fuel saving was about 4.35% after ecodriving training.

[1]School of Civil and Environmental Engineering, Georgia Institute of Technology, Atlanta, GA

Corresponding Author:
Address correspondence to Xiaodan Xu: xxu312@gatech.edu

Wåhlberg (*11*) and Strömberg and Karlsson (*10*) measured the effectiveness of ecodriving classes and real-time driving guidance in Sweden. Wåhlberg found that the real-time driving feedback can provide additional fuel saving benefits, while Strömberg and Karlsson did not find a significant difference. Carrese (*9*) observed the fuel economy of drivers with different driving styles in Rome, Italy, where the researchers concluded that the difference in driving behavior could lead to a 27% reduction in fuel use. Sullman, et al. (*7*) assessed the long-term fuel-saving benefits after ecodriving training in Finland and found that the observed improvement on fuel economy six months after the ecodriving training was 16.9%.

With respect to simulation modeling, Zheng and Zhang (*8*) proposed a simulated ecodriving scenario in Beijing, China with embedded traffic lights timing model to assess the effectiveness of ecodriving strategy along with signal timing. This study also incorporated a two-degree grade for road grade consideration. The simulation results suggested that ecodriving could help to reduce engine load (expressed as vehicle-specific power or scaled tractive power in the U.S. Environmental Protection Agency's MOVES model), which ultimately leads to fuel savings. Xu, et al. (*3*) compared the system-wide energy consumption with real-world operation data and simulated ecodriving operation data with a case study in Atlanta, GA. Ecodriving was expected to reduce fuel consumption by 5% in local transit service, and 7% in express bus service (*3*). The aforementioned studies found that by adopting an ecodriving strategy, fuel savings for transit fleet could vary from 2%–27% across different fleet compositions, training strategies, and on-road operating conditions.

However, many of the aforementioned research projects strengthened the relation between fuel savings and driver characteristics, with less effort spent on examining the relation between fuel reduction and service parameters, such as fleet composition, service conditions, and road grade. The bus routes were typically selected from urbanized areas, with small urban areas, suburban areas, and rural areas often being ignored. Also, as many of the studies were performed in European countries with their own fuel standards and emission models, the fuel-saving benefits might also be very different for United States vehicles operating under local conditions. Hence, it seems appropriate to investigate the potential benefits and limitations of ecodriving strategies within urban, suburban, and rural contexts.

In this study, the potential effectiveness of ecodriving was examined across various transit routes that serve different areas, use different fuels, and undertake different service conditions (mileage and on-road operating conditions). The first element collects and processes the operation data across different transit services. The second element of this study evaluated the on-road operating characteristics of each of these transit services. The third element of the study assessed the potential benefits of introducing the ecodriving strategy previously developed by the team (*3*) to all three types of transit fleets. The potential reductions in fuel consumption were derived from the operational improvements. The research applied the MOVES-Matrix framework developed by the Georgia Tech research team for energy and emission analysis (*14*). Fuel consumption was evaluated by matching the second-by-second MOVES operating mode bin with corresponding emission rates. The potential benefits and limitations of deploying the ecodriving strategy were identified corresponding to various service parameters.

Methodology

In this paper, the second-by-second vehicle operation data, including speed, acceleration, and road grade, were used to analyze transit routes, on-road operating condition, and energy consumption. The overall methodology was illustrated in Figure 1 below. Second-by-second speed, acceleration, latitude, and longitude data for transit operations were collected from three types of transit service: 1) transit routes in the center of Atlanta, Georgia, a large metropolitan area (referred to as "urban" data in the analyses); 2) bus routes in a suburb of metropolitan Atlanta (referred to as "suburban" in the analyses); and 3) from a small town/rural area in Henderson County, North Carolina (referred to as "rural" in the analyses). After collecting second-by-second GPS data, the data were post-processed to minimize GPS errors, remove duplicate records, infill missing values, and match data with road grade information. The speed–acceleration frequency distributions, idling time fraction, operating mode distribution, and road grade for these different services were quantified and compared during post-processing. Finally, a transit bus ecodriving algorithm that utilizes second-by-second operation data to provide real-time feedback (*3*) was applied and results were compared. The simulated cycles reflected the driver behavior modifications, predominantly limiting vehicle acceleration rates and maximum speeds.

Energy consumption was assessed using MOVES-Matrix (*3*, *14*), and fuel savings from ecodriving were assessed at the system level. Finally, to quantify the impact of terrain roughness on the effectiveness of ecodriving, a flat terrain scenario (road grade equals zero) was included in every step of analysis in contrast with raw driving cycles with grade. The operation patterns and energy use with and without grade were compared across different transit services.

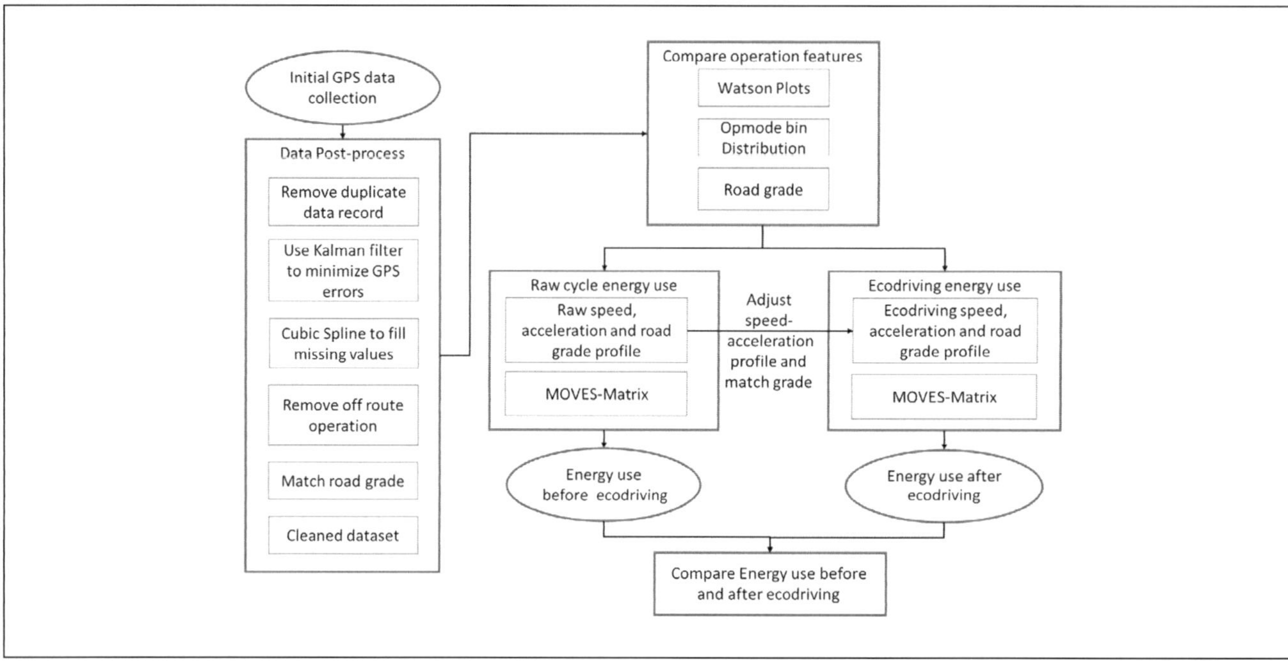

Figure 1. Methodology flow chart.

Data Collection and Post-Processing

Second-by-second transit data were collected from: Apple Country Transit rural transit operations in Henderson County, NC; MARTA suburban bus transit operations in metropolitan Atlanta; and MARTA urban transit operations in downtown Atlanta, GA. Geographic location of measured bus routes is shown in Figure 2 below. In the metropolitan Atlanta area, MARTA operates 149 diesel buses and 426 CNG buses with body length ranging from 35–40 ft. The urban and suburban operation data were collected on 13 buses for 381 days (June 28, 2004, to Oct 24, 2005) using the GT Trip Data Collector (*3*). The urban operation data were collected from bus routes #4, #16, and # 32. Suburban bus operation data were collected from MARTA routes #86, #103, and #189. Rural data were collected from Apple Country Transit, which operates six Ford E450 Cutaways (about 22 ft long) with CNG fuel. Android phones equipped with the *Commute Warrior* app (*15*) were installed on the six buses operating on three bus routes. The rural transit operation data were collected during a two-week period (September 28 to October 10, 2016) from routes #1, #2, and #3.

A series of post-processing steps were performed to clean and infill the raw data. The post-processing steps include:

1. Remove duplicated data records: First, the collected data were screened to remove some duplicate data records.
2. Kalman filter data smoothing: A data smoothing technique, known as modified Kalman filter (*16*), was applied to modify erroneous GPS points to ensure a realistic vehicle trace.
3. Cubic spline to infill missing values: Whenever a few seconds of missing values were encountered, data were interpolated with the cubic spline method.
4. Remove off-route operations: After data cleaning, all the data were plotted within a GIS system along with the route maps. Operations that did not appear on revenue routes were removed from the dataset. Because the system did not monitor engine on/off conditions, data reflecting inactivity (data with zero motion) after the vehicles reached transit stations and maintenance facilities were also trimmed from the analyses. If these vehicles were indeed running their engines while stationary (engine idling), the ecodriving strategy would result in drivers turning off their engines. Hence, the potential ecodriving benefits estimated in this paper likely represented an underestimation of the actual benefits that would accrue.
5. Match road grade: The study utilized the digital elevation model (DEM), a digital cartographic/geographic dataset of elevations, maintained by U.S. Geological Survey (USGS) to generate road grade for energy and emission modeling (*17*). In this study, the 3 × 3 m DEM dataset was used because it is the highest resolution available for all of the routes. A streamlined method extracted

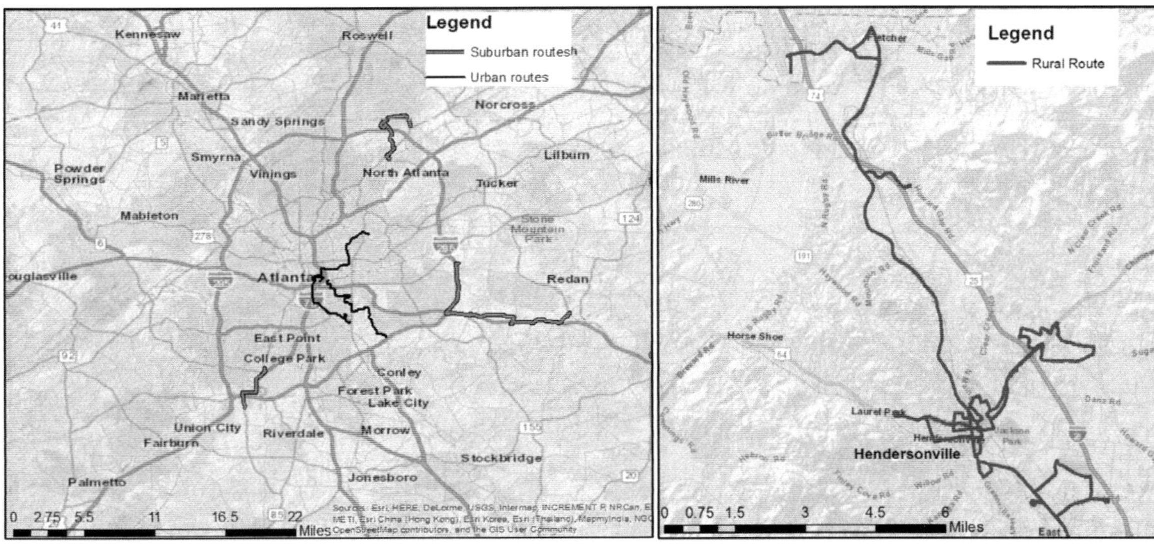

Figure 2. Data collection from three types of transit service operations.

roadway elevation profile from DEM database, cleaned and refilled erroneous elevation data, and applied a cubic smoothing spline to generate road grade at high resolution (*17*). Based upon field verification of road grade data, the grade generation method generated highly accurate road grade results, with root-mean-square error (RMSE) of estimated road grades 0.2%–0.23% for highways and 0.5%–0.6% for local roads (*17*). The terrain elevations and grades for ground positions were sampled at 20–30 ft horizontal spacing along the roads. The road grade of each second of operations was generated by looking at the closest points on road grade profile with GPS data. The sign (+ /−) of the grade value was verified with elevation data to ensure the uphill/downhill data were correctly assigned.

After data post-processing, the cleaned datasets were analyzed for each of the three transit service fleets. The summary statistics for on-road operation data were tabulated in Table 1 below. Based on the information from the table, the ranges of speed, acceleration, and road grade varied significantly across these three sources. For example, the rural transit service has the highest average speed among the three. All three transit services exhibited different operation patterns in terms of route characteristics and speed and acceleration rates. Hence, each fleet might achieve very different fuel savings when adopting the proposed ecodriving strategy.

Operation Pattern Analysis

In this study, modal-based emission modeling framework developed by U.S. Environmental Protection Agency (EPA) in their MOtor Vehicle Emission Simulator (MOVES) was used for energy and emissions analysis (*18*, *19*). MOVES is the state-of-practice tool for emission analysis and is required for use in all regulatory analyses. The MOVES model uses scaled tractive power (STP) as a surrogate for engine load, and STP is a function of vehicle speed, acceleration, road grade, and vehicle mass (*3*). The STP values, together with speed and acceleration, are grouped into operating mode (OpMode) bins, where each OpMode bin is associated with a specific MOVES emission rate. By calculating the STP bin for each second in each driving cycle, users can quantify the MOVES energy consumption for different transit services. The second-by-second observed speed, acceleration, and estimated grade were used first to analyze the transit operation characteristics. Then, the following formula was used to calculate second-by-second STP and estimate the energy and emissions for all driving traces collected:

$$STP = \left(\frac{A}{M}\right)V + \left(\frac{B}{M}\right)V^2 + \left(\frac{C}{M}\right)V^3 + \left(\frac{m}{M}\right)(a + g^{*}sin\theta)V \quad (1)$$

where

A = rolling resistance coefficient [kW • sec/m];
B = rotational resistance coefficient [kW • sec^2/m^2];

Table 1. Summary of Operation Data from Three Transit Systems

Service	Urban	Suburban	Rural
Agency	MARTA	MARTA	Apple Country Transit
Number of routes	3	3	3
Total distance (mile)	407.9	129.4	178.8
Total duration (h)	4774.9	2190.8	3575.0
Average speed (mph)	11.7	16.9	20.0
2.5th percentile speed (mph)	0.0	0.0	0.0
50th percentile speed (mph)	7.2	14.5	19.0
97.5th percentile speed (mph)	38.3	52.9	48.5
2.5th percentile acceleration (mph/s)	−3.4	−3.4	−3.4
50th percentile acceleration (mph/s)	0.0	0.0	0.0
97.5th percentile acceleration (mph/s)	2.7	2.4	2.8
2.5th percentile grade (%)	−5.0	−4.3	−6.0
50th percentile grade (%)	0.4	0.0	−0.2
97.5th percentile grade (%)	5.0	7.7	6.0

C = aerodynamic drag coefficient [$kW \cdot sec^3/m^3$];
m = mass of individual test vehicle [metric tonnes];
M = fixed mass factor;
v = instantaneous vehicle velocity at time t [m/s];
a = instantaneous vehicle acceleration [m/s^2];
g = gravitational acceleration with the value 9.8 m/s^2; and
θ = fractional road grade in percent grade angle.

The gross vehicle weight rating (GVWR) of all vehicles in this study were greater than 8500 lb.; hence, all vehicles fell into the MOVES transit bus category (*20*). Parameters for the STP formula were based on Transit Bus (source type = 42) in the MOVES default database.

The same process outlined above was applied to assess the energy and emissions from the second-by-second vehicle traces collected in the field, but with zero grade assumed for all operations. This allowed the research team to compare the impacts on energy analysis that arise from the assumption of flat terrain, which most planning agencies applied in their development of emissions inventories and microscale air quality impact assessments. The differences in modeled fuel consumption for operations on observed grade and an assumed zero grade were quite different, indicating that ecodriving strategies do need to incorporate local road grades.

Applying the Ecodriving Strategy and Estimating Energy Savings

The ecodriving strategy applied in this study generated simulated ecodriving cycles that assume that drivers respond to real-time driving guidance based on second-by-second GPS data (*3*). By applying this ecodriving strategy, the drivers were expected to limit vehicle engine load by reducing hard acceleration activity and reducing top speeds. Because the MOVES modeling approach was adopted in this study, the engine load was estimated by STP and calculated using Equation 1 above. The higher level of STP generally leads to higher energy consumption (within specific speed ranges). In this case, the optimization goal of the ecodriving strategy was to maintain STP values below certain thresholds in various speed bins by limiting maximum speed and acceleration. The first step in the ecodriving process was to set an STP value limit (STP_L). For each speed V and road grade θ, the maximum acceleration acc_L was determined to prevent STP from exceeding STP_L. The equation of calculating acc_L is:

$$acc_L = \frac{STP_L * M}{mV} - g*sin\theta - \left(\frac{A}{m}\right) - \left(\frac{B}{m}\right)V - \left(\frac{C}{m}\right)V^2 \quad (2)$$

After setting the STP limit value, the second-by-second operation data were screened to identify all occurrences where instantaneous STP exceeded the STP limit. If the STP was less than the STP limit, the current operation (status quo) was maintained without adjustment to speed and acceleration. If the STP exceeded the limit, the maximum acceleration from Equation 2 was assigned to current speed, and following speeds, to smooth the acceleration curve, until the acceleration curves reached the speed limit. Under the ecodriving routine, the lower speeds and acceleration rates of ecodriving usually prevented the vehicle from reaching its destination in the travel time that was observed during field data collection. Hence, a few minutes of additional driving was often required to reach the destination. To conserve the same travel distance, the ecodriving cycle was extended by adding additional travel time under cruise until the travel

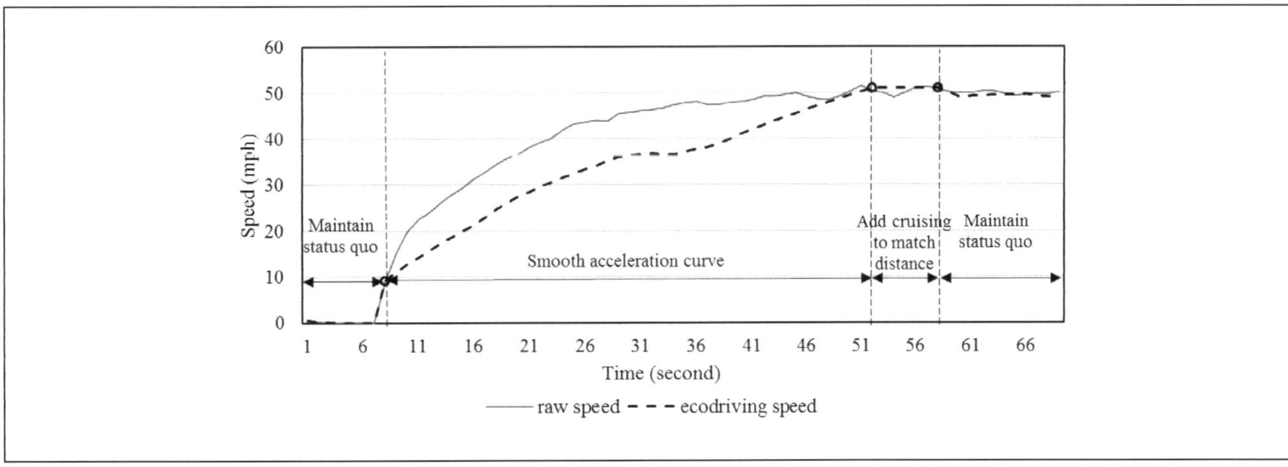

Figure 3. Illustration of the ecodriving strategy.

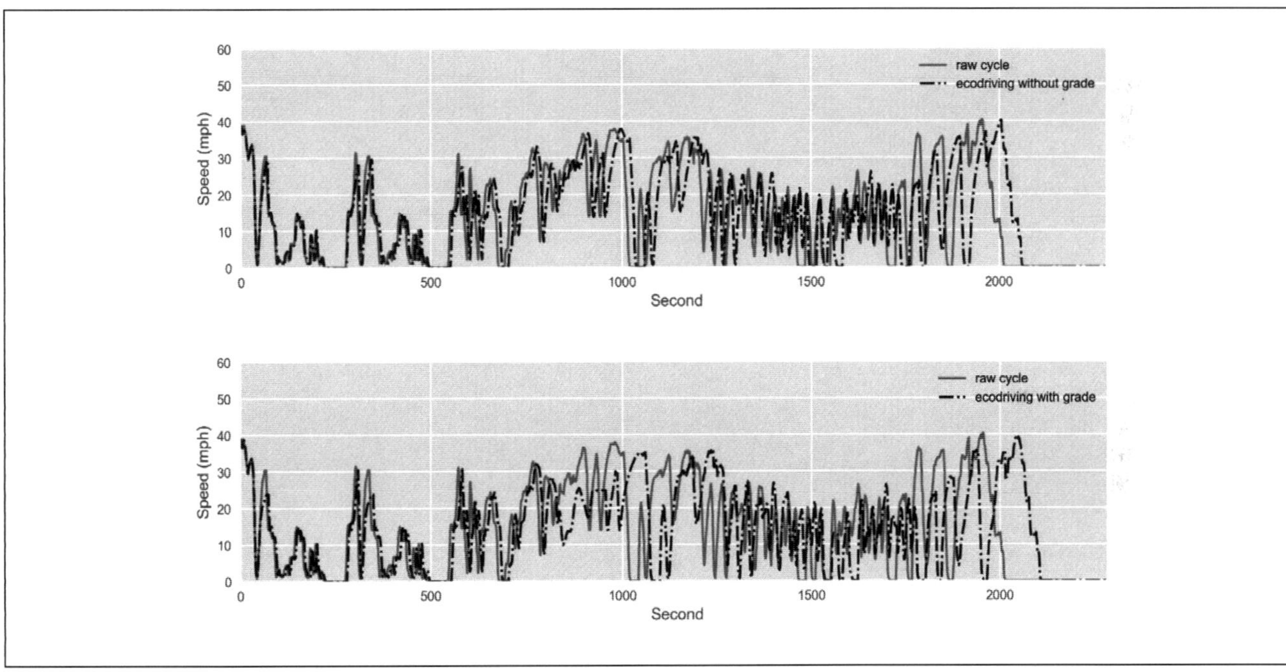

Figure 4. Simulated ecodriving cycles with and without grade.

distance matched the original cycle. For the time being, additional potential complex interaction of ecodriving with traffic signal timing was ignored in these analyses. The ecodriving strategy can be illustrated using the following diagram.

By deploying above ecodriving algorithm, ecodriving cycles were generated for cleaned dataset with grade and for flat terrain. The sample ecodriving cycles ($STP_L = 6$) are shown in Figure 4 below. For the flat terrain analysis, the bus required about 1 min longer to reach its destination. Because grade constitutes an acceleration against gravity, the linear acceleration rates and top speeds need to be lower in the ecodriving cycle than for the ecodriving under flat terrain. Hence, ecodriving on rough terrain generally requires longer travel times to cover the same route distance. The benefits and limitations of deploying ecodriving strategy for different transit services will be further discussed in the following section.

Results

In this section, the on-road operating patterns and energy use for each transit service were compared, with

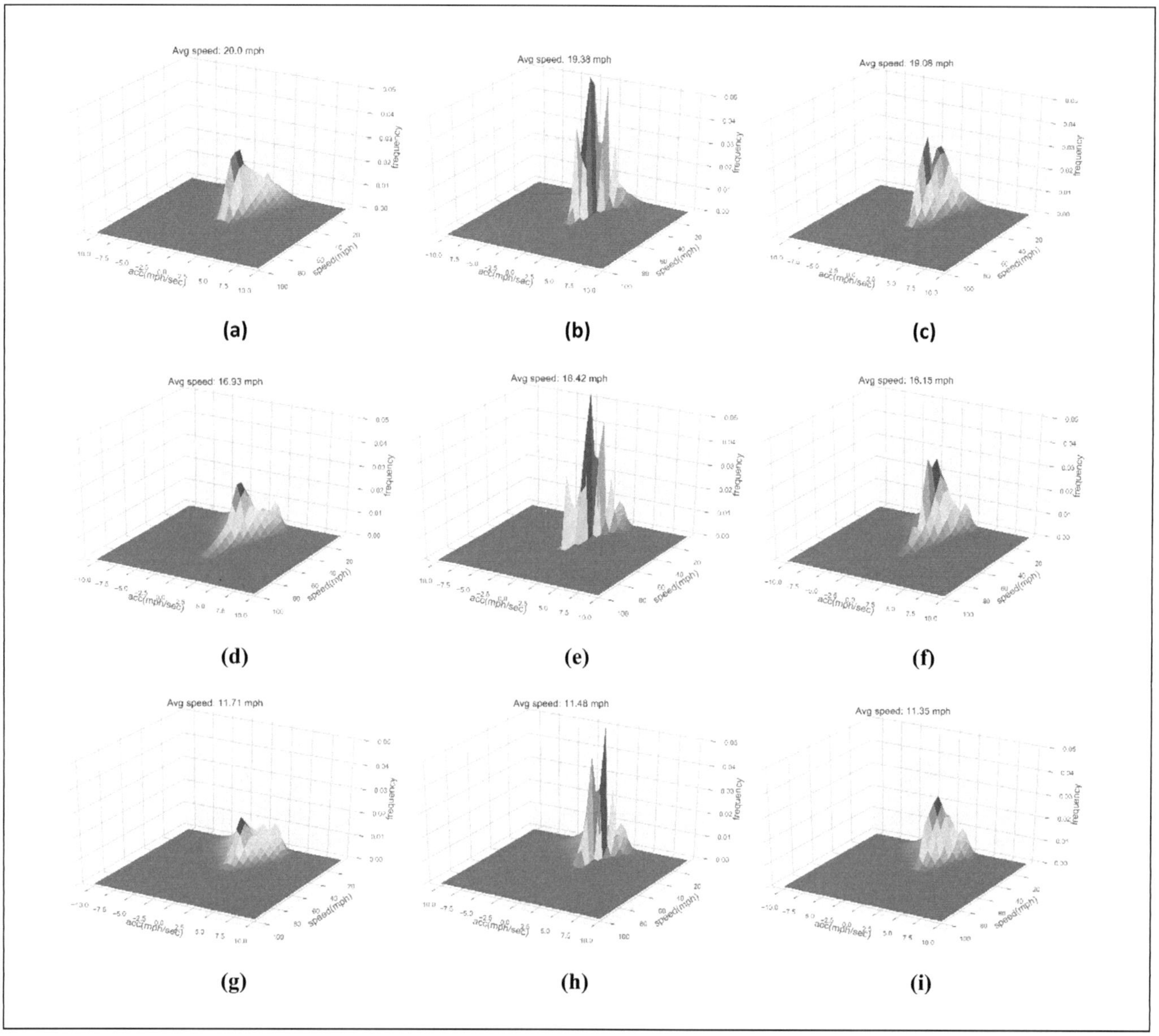

Figure 5. Speed–acceleration distributions for different service operations: (*a*) rural, raw cycle; (*b*) rural, ecodriving cycle, no grade; (*c*) rural, ecodriving cycle, with grade; (*d*) suburban, raw cycle; (e) suburban, ecodriving cycle, no grade; (*f*) suburban, ecodriving cycle, with grade; (*g*) urban, raw cycle; (*h*) urban, ecodriving cycle, no grade; (*i*) urban, ecodriving cycle, with grade.

and without road grade. The speed–acceleration frequency distributions were compared first. Then, the MOVES OpMode bins estimated using Equation 1 and distributions were compared before and after deploying ecodriving. For energy use, the system-level fuel use (for these vehicles operating on their routes) was generated before and after adopting an ecodriving strategy. The upper limit of STP, $STP_L = 6$, was applied for generating all of the cycles after adopting the ecodriving strategy. Real-world fuel economy was used for the vehicles in the fleet to scale the fuel economy estimates generated by MOVES and to compare relative fuel savings.

Comparison of On-Road Operating Conditions (Speed–Acceleration Profiles)

The compilation of second-by-second speed and acceleration distributions for each scenario were plotted. The idling operations (speed < 1 mph) were not displayed on the graph (the large amount of idle in each graph makes the rest of the operations difficult to discern when plotted in the same figure). In the original collected data, the idling operations accounted for 20% of time in rural transit, 32% in suburban, and 39% in urban transit. Figure 5 illustrates acceleration (mph/s) on the x-axis,

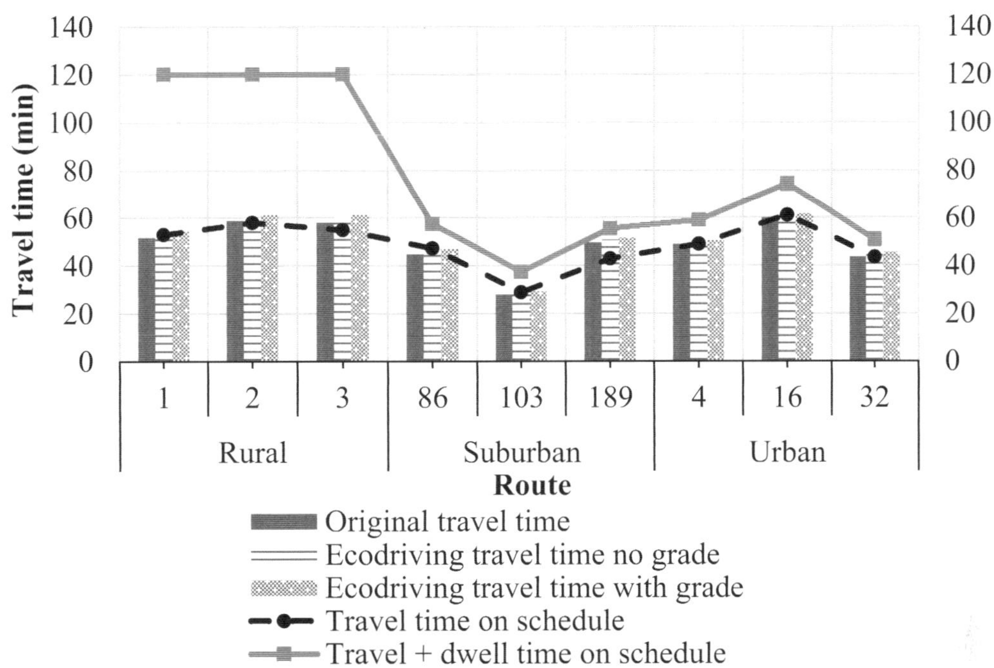

Figure 6. Travel time before and after implementing ecodriving.

speed (mph) on the y-axis, and relative frequency on the z-axis.

The distributions in Figure 5 indicated that the frequency of high-speed and high-acceleration activity declined after adopting ecodriving. The ecodriving strategy without grade had high levels of activity in medium-speed low-acceleration operations, while the ecodriving with grade exhibited more dispersed speed–acceleration distributions because speeds and accelerations were more limited on uphill operations and allowed to attain higher levels during downhill operations.

Rural transit service exhibited the highest average speed, while the transit service in downtown Atlanta (urban) had the lowest average speed. After deploying the ecodriving strategy, the average speed of each scenario slightly decreased, which lengthened the travel time in the real-world operation. However, during real-world operation, the agencies typically allow for dwell time at the end of each trip. Hence, even if ecodriving adds additional travel time to a trip, schedules will not be impacted if dwell time is sufficient. In this study, the travel time from selected routes were checked against the route departure times on the published agency schedules. As presented in Figure 6, even though additional time was required under the ecodriving cycle, the dwell time was sufficient for each route so that the buses would remain on schedule.

The analyses indicated that ecodriving should result in a maximum of 3% decrease in average speed without considering grade, and about 5% decrease in average speed when the grade effect is taken into account. On average, applying the ecodriving strategy would result in extra travel time (normally 2–3 min per trip), but the delay would not significantly affect the schedule given the current scheduled dwell time between trips. In addition, the ecodriving strategy in this study only included controlling top speed and acceleration, while the full package of ecodriving training also includes idling reduction, anticipating traffic, and anticipating signal. The team is collecting traffic, signal, and stop information to add the full spectrum of ecodriving into simulation, which is likely to reduce the bus delay.

Next, the STP bin distributions of original cycles and ecodriving cycles were analyzed both with and without grade. The fractions of time under different operating mode were plotted by different speed ranges (in mph) and by STP level. The distributions in Figure 7 indicated that the time fractions that fall into each STP level were affected by road grade, and therefore affected the operating mode distributions after applying ecodriving. Assuming flat terrain, the ecodriving strategy tended to strictly limit the STP level under designated STP level (STP_L = 6). However, with road grade, a small portion of high STP operations failed to be removed after

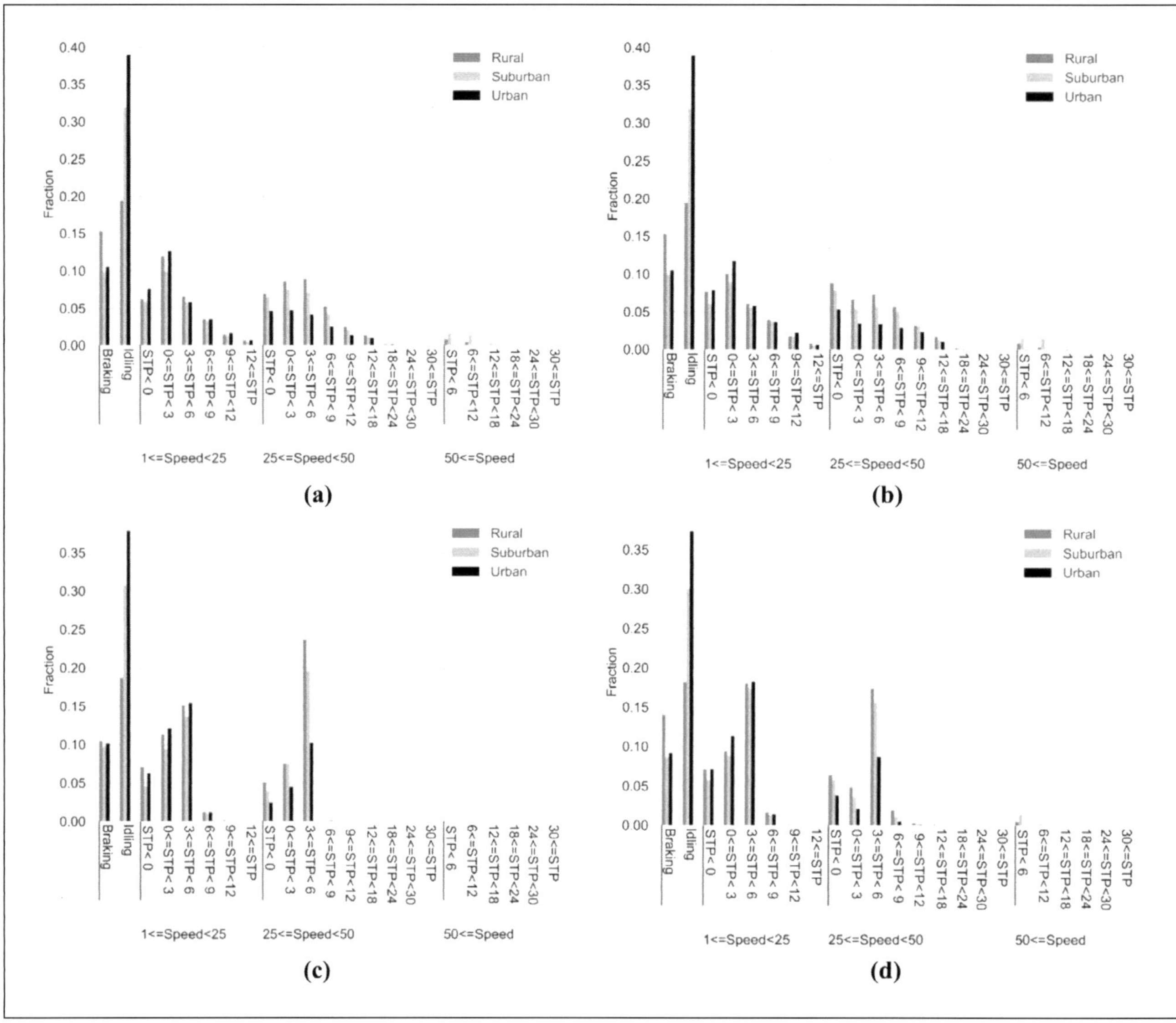

Figure 7. MOVES OpMode bin distribution before and after ecodriving: (*a*) raw cycle, no grade; (*b*) raw cycle, with grade; (*c*) ecodriving cycle, no grade; (*d*) ecodriving cycle, with grade.

adopting ecodriving due to the impact of road grade. The high level of engine power was still sometimes needed to overcome some severe slopes along the trips. That is, traditional ecodriving strategies should not be implemented on uphill grades without considering the impact on uphill vehicle performance. It would not be wise to limit vehicle acceleration so severely that resulting transit vehicle speeds on the uphill grade impede normal roadway operations. There are times when high energy demand was required.

Based on discussions above, the operation patterns of different transit services were quite different both before and after applying the ecodriving strategy. Road grade also significantly affected the operating patterns under ecodriving condition. The impact of ecodriving on fuel savings will be compared in the following section, with the ecodriving operating conditions serving as the analytical input.

Energy Use and Cost Comparison

In this section, on-road fuel consumption was calculated by matching the OpMode bin distribution from the previous section with the appropriate MOVES-Matrix emission rates (*14*). MOVES-Matrix rates were generated by running MOVES across all combinations of vehicle source type, fuel, environmental, operating conditions, and other parameters affecting MOVES outputs, and saving the data in a 2 Terabyte emission rate array so that the emission rates can be used in other analyses

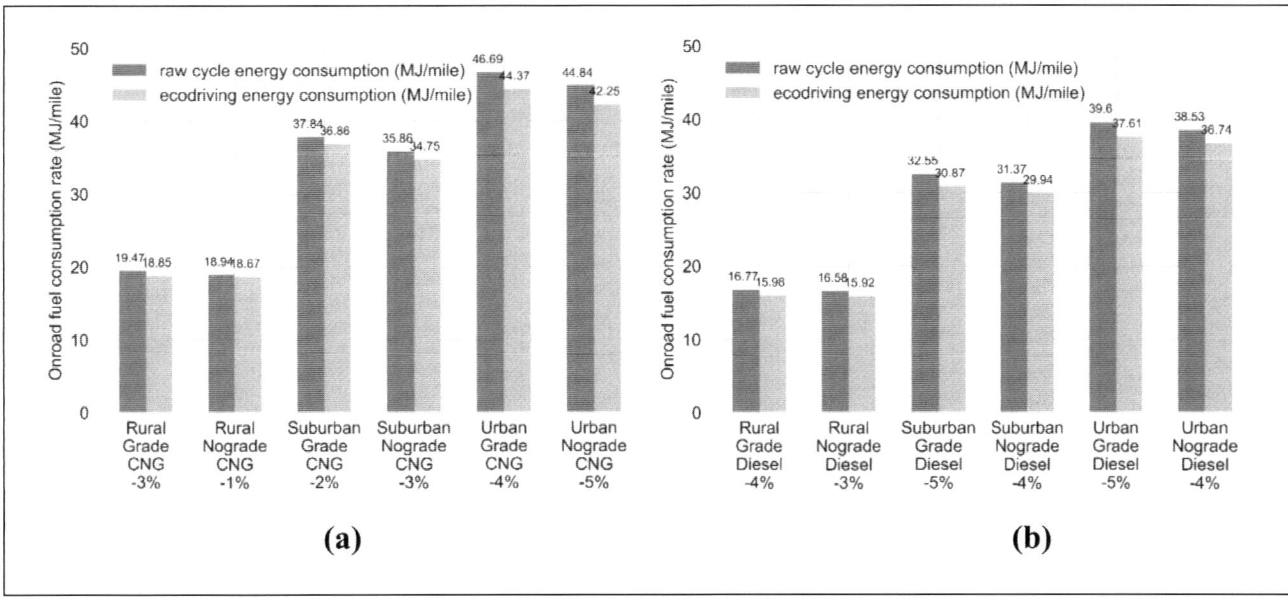

Figure 8. Energy consumption rate before and after ecodriving with (*a*) CNG and (*b*) diesel.

without re-running MOVES. Utilizing MOVES-Matrix generates exactly the same results as running MOVES, but drastically reduces (about 99.5% reduction) run time. In this study, the emission rates were retrieved from MOVES-Matrix using the following MOVES parameters for a hot summer morning peak analysis:

- Calendar Year: 2017
- Region: Fulton County, Atlanta for MARTA; Henderson County, NC for Apple Country Transit
- Month: July
- Day and time: Weekday, 7:00–8:00 AM
- Temperature: 85°F
- Humidity: 65%
- Fuel: Default fuel supply and fuel share from MOVES
- I/M strategy: Default 2015 I/M strategy from MOVES
- Vehicle type: Transit bus
- Model year: 2011

After calculating the energy consumption for all cycles, the fuel economy was estimated using the conversion factor 116,090 Btu/gasoline gallon equivalent (GGE) (122.5 MJ/GGE). The real-world fuel economy data was applied to adjust the MOVES fuel economy (which is a fleet average) to better match the actual fleet. The energy consumption rates for each scenario are shown in Figure 8.

The ecodriving strategy could reduce fuel consumption for urban, suburban, and rural transit service, regardless of grade condition and fuel type. Predicted energy savings ranged from 1%–5%, and the energy savings with and without grade were quite different (1% or 2%). For rural transit service, the fuel savings were actually larger with grade than for the flat terrain scenario, while for the other two scenarios the fuel savings decreased with the presence of grade. It is likely that rural transit had a higher portion of high engine load compared with urban transit (see Figure 7), due to a combination of high operating speed and higher grades. Whereas considering road grade, the STP level of rural transit was much higher than assuming flat terrain, and ecodriving strategy would gain higher potential fuel savings by reducing these loads.

Next, the fuel and cost savings for each transit service were assessed (with road grade included). The team applied \$2.30/gallon as the diesel fuel cost for Georgia and \$2.10/gallon of diesel for North Carolina. For CNG fuel, \$2.4/GGE was used for Georgia and \$2.00/GGE for North Carolina. The results are summarized in Table 2 below. The total cost savings varied greatly due to large differences in annual fleet mileage. The unit savings varied from \$ 0.011–\$ 0.045 per mile, due to the difference in operation patterns and fuel prices. The urban routes seemed to provide the largest unit cost savings, while the rural transit agency only achieved modest cost savings. The cost saving was also affected by fuel type for different services due to different fuel economy improvements under different fuels. The direct benefit in fuel savings per bus per year in urban and suburban areas was sufficient to warrant the cost of the implementation of ecodriving strategies (*3*). However, the modest savings in the

Table 2. Annual Energy and Cost Saving for Selected Routes

Diesel			
Service	Rural	Suburban	Urban
Annual mileage	163,373	981,856	730,005
Number of buses	6	11	11
Before fuel rate (miles/GGE)	7.3	3.8	3.1
Before fuel usage (DGE)	19,686	229,601	207,688
After fuel rate (miles/GGE)	7.7	4.0	3.3
After fuel usage (DGE)	18,759	217,702	197,237
Fuel saving (DGE)	927	11,899	10,451
Assumed fuel price ($/DGE)	$2.10	$2.30	$2.30
Cost saving per bus ($/bus/year)	$324.40	$2488.00	$2185.00
Unit saving ($/mile)	$ 0.012	$ 0.028	$ 0.033

CNG			
Service	Rural	Suburban	Urban
Annual mileage	163,373	981,856	730,005
Number of buses	6	11	11
Before fuel rate (miles/GGE)	6.3	3.2	2.6
Before fuel usage (GGE)	25,971	303,298	278,241
After fuel rate (miles/GGE)	6.5	3.3	2.8
After fuel usage (GGE)	25,140	295,434	264,401
Fuel saving (GGE)	829	7,864	13,840
Unit price ($/GGE)	$2.10	$2.40	$2.40
Cost saving per bus ($/bus/year)	$290.00	$1716.00	$3020.00
Unit saving ($/mile)	$ 0.011	$ 0.019	$ 0.045

Note: GGE = gasoline gallon equivalent; DGE = diesel gallon equivalent; CNG = compressed natural gas.

rural transit service may, or may not, cover the costs associated with implementing ecodriving.

Summary and Conclusion

In this study, the benefits of applying ecodriving strategy for transit were assessed for urban, suburban, and rural transit services. The second-by-second on-road operation data were first collected from three different transit fleets. The team found that the three transit services exhibited significantly different on-road operating patterns (speed–acceleration profiles). The data were collected from actual on-road operations, so the team integrated road grade with the second-by-second data streams for use in energy analysis. After data cleaning, MOVES STP bin assignment was used to assess second-by-second energy consumption. The energy analysis was also conducted under the assumption that the vehicles were operating on flat terrain (which was typically assumed in emissions inventory work for air quality planning and microscale air quality impact assessment for simplicity). Then, an ecodriving strategy was introduced to the three services, to simulate the driving cycles expected to result after introducing the real-world driver feedback system designed to limit maximum acceleration rates and speeds. The results indicated that the simulated ecodriving cycle was different with and without the impact of road grade. The energy-saving and cost-saving results were also distinctly different by service type and road grade conditions. Overall, the ecodriving strategy could help reduce 1%–5% fuel use for transit agencies and save $ 0.011–$ 0.045 operation cost per mile due to the fuel saving. This study suggested that although ecodriving could generally help agencies save fuel, the scale of benefits is subject to local conditions. Nevertheless, the economic benefits from fuel savings on the urban and suburban routes was quite large, and likely warranted the costs of ecodriving implementation (*see also, 3*). Based on the research team's prior experience in monitoring fleet in the Cobb County (Georgia) School District (*21*), preliminary cost estimates for adopting ecodriving would be about $650/bus/year. The cost for following years would be much lower since no cost spent on equipment. The fuel savings are mostly sufficient to pay for such monitoring.

Several study limitations need to be addressed through further research. First, the routes selected in this study only covered the operation on local routes and did not include highway network operations (such as express buses). The conclusions from this study were limited by the selected area and might change under other scenarios. Second, the benefits of ecodriving were analyzed on an average basis; the variance associated with ecodriving benefits has not yet been fully investigated. For some periods of the day, the ecodriving strategy will, by definition, result in slightly longer travel times between stations. It is unclear whether these buses will negatively impact surrounding traffic or improve traffic flow through smoother operations (acting like a pace car). The net will likely be dependent on the complex interactions with local traffic signal timing. That said, it may be possible to synchronize corridor traffic signal timing to directly support ecodriving operations. Finally, the current ecodriving simulations assumed that drivers would respond to ecodriving training and in-vehicle signals. Real-world implementation is needed to assess how actual (measured) fuel savings compare with simulation results to verify cost effectiveness.

Acknowledgments

The research team would like to thank MARTA for allowing us to collect second-by-second data from their urban routes. The research team especially acknowledges and appreciates the participation of the Henderson County Planning Department (Apple Country Transit) for assisting in data collection and providing technical support for this project. This work was supported by the National Center for Sustainable Transportation, a National University Transportation Center sponsored by the U.S. Department of Transportation.

Author Contributions

The authors confirm contribution to the paper as follows: study conception and design: Xiaodan Xu, Hanyan “Ann” Li, Haobing Liu, Michael O. Rodgers, Randall Guensler; data collection: Xiaodan Xu, Haobing Liu; analysis and interpretation of results: Xiaodan Xu; draft manuscript preparation: Xiaodan Xu, Hanyan “Ann” Li, Haobing Liu, Michael O. Rodgers, Randall Guensler. All authors reviewed the results and approved the final version of the manuscript.

References

1. American Public Transportation Association. *Public Transportation Fact Book*. American Public Transportation Association, Washington, D.C., 2017. http://www.apta.com/resources/statistics/Documents/FactBook/2016-APTA-Fact-Book.pdf. Accessed July 25, 2017.
2. Metropolitan Atlanta Rapid Transit Authority. *FY2017 Adopted Budget Book*. Department of Finance, Office of Management & Budget, Metropolitan Atlanta Rapid Transit Authority, Atlanta, GA, 2016. http://www.itsmarta.com/uploadedfiles/FY17AdoptedBudgetBook.pdf. Accessed July 25, 2017.
3. Xu, Y., H. Li, H. Liu, M. O. Rodgers, and R. L. Guensler. Eco-Driving for Transit: An Effective Strategy to Conserve Fuel and Emissions. *Applied Energy*, Vol. 194, 2017, pp. 784–797. https://doi.org/10.1016/j.apenergy.2016.09.101.
4. Xu, A., A. Li, H. Liu, M. O. Rodgers, and R. Guensler. *Ecodriving for Transit (Ecotransit)*. A National Center for Sustainable Transportation Report. Prepared for the National Center for Sustainable Transportation, University Transportation Center, Georgia Institute of Technology. Atlanta, GA, 2016. https://ncst.ucdavis.edu/wp-content/uploads/2014/08/04-15-2016-NCST-EcoDriving-White-Paper-FINAL.pdf. Accessed July 25, 2017.
5. Frey, H. C., N. M. Rouphail, H. Zhai, T. L. Farias, and G. A. Gonçalves. Comparing Real-World Fuel Consumption for Diesel-and Hydrogen-Fueled Transit Buses and Implication for Emissions. *Transportation Research Part D: Transport and Environment*, Vol. 12, No. 4, 2007, pp. 281–291. https://doi.org/10.1016/j.trd.2007.03.003.
6. Ang, B. W., and T. F. Fwa. A Study on the Fuel-Consumption Characteristics of Public Buses. *Energy*, Vol. 14, No. 12, 1989, pp. 797–803. https://doi.org/10.1016/0360-5442(89)90033-9.
7. Sullman, M. J., L. Dorn, and P. Niemi. Eco-Driving Training of Professional Bus Drivers–Does it Work? *Transportation Research Part C: Emerging Technologies*, Vol. 58, 2015, pp. 749–759. https://doi.org/10.1016/j.trc.2015.04.010.
8. Zheng, X., and L. Zhang. Ecodriving for Reduction of Bus Transit Emission with Vehicle's Hybrid Dynamic Model. *Mathematical Problems in Engineering*, Vol. 2015, 2015, pp. 1–8. http://dx.doi.org/10.1155/2015/543429.
9. Carrese, S., A. Gemma, and S. La Spada. Impacts of Driving Behaviours, Slope and Vehicle Load Factor on Bus Fuel Consumption and Emissions: A Real Case Study in the City of Rome. *Procedia-Social and Behavioral Sciences*, Vol. 87, 2013, pp. 211–221. https://doi.org/10.1016/j.sbspro.2013.10.605.
10. Strömberg, H. K., and I. M. Karlsson. Comparative Effects of Eco-Driving Initiatives Aimed at Urban Bus Drivers–Results from a Field Trial. *Transportation Research Part D: Transport and Environment*, Vol. 22, 2013, pp. 28–33. https://doi.org/10.1016/j.trd.2013.02.011.
11. af Wåhlberg, A. E. Long-Term Effects of Training in Economical Driving: Fuel Consumption, Accidents, Driver Acceleration Behavior and Technical Feedback. *International Journal of Industrial Ergonomics*, Vol. 37, No. 4, 2007, pp. 333–343. https://doi.org/10.1016/j.ergon.2006.12.003.
12. Zarkadoula, M., G. Zoidis, and E. Tritopoulou. Training Urban Bus Drivers to Promote Smart Driving: A Note on a Greek Eco-Driving Pilot Program. *Transportation Research Part D: Transport and Environment*, Vol. 12, No. 6, 2007, pp. 449–451. https://doi.org/10.1016/j.trd.2007.05.002.
13. Saynor, B. *ECOWILL Deliverable 6.1: Guidelines for Delivering Short-Duration Eco-Driving Training Courses*. Intelligent Energy Europe, 2011. https://ec.europa.eu/energy/intelligent/projects/sites/iee-projects/files/projects/documents/ecowill_guidelines_short_duration_trainings_en.pdf. Accessed July 25, 2017.
14. Guensler, R., H. Liu, X. Xu, Y. Xu, and M. Rodgers. MOVES-Matrix: Setup, Implementation, and Application

(No. 16-6362). Presented at 95th Annual Meeting of the Transportation Research Board, Washington, D.C., 2016.
15. Akanser, A., A. Grossman, Y. Xu, and R. Guensler. Commute Warrior: An Android Application to Collect Longitudinal Travel Survey Data. Presented at European Transport Conference, Barcelona, Spain, 2016. http://abstracts.aetransport.org/paper/index/id/5081/confid/21. Accessed July 25, 2017.
16. Jun, J., R. Guensler, and J. Ogle. Smoothing Methods Designed to Minimize the Impact of GPS Random Error on Travel Distance, Speed, and Acceleration Profile Estimates. *Transportation Research Record: Journal of the Transportation Research Board*, 2006. 1972: 141–150.
17. Liu, H., H. Li, M. Rodgers, and R. Guensler. Development of Road Grade Data using the United States Geological Survey Digital Elevation Model. *Transportation Research Part C: Emerging Technologies*, Vol. 92, 2018, pp. 243–257.
18. U.S. Environmental Protection Agency. *Population and Activity of On-road Vehicles in MOVES2014*. Publication EPA-420-R-16-003a. U.S. Environmental Protection Agency, U.S. Government Printing Office, Washington, D.C., 2016.
19. U.S. Environmental Protection Agency. *User Guide for MOVES2014*. Publication EPA-420-B-14-055. Office of Transportation and Air Quality, U.S. Environmental Protection Agency, Ann Arbor, MI, 2014.
20. Liu, H., Y. Xu, M. O. Rodgers, and R. Guensler. Developing Vehicle Classification Inputs for Project-Level MOVES Analysis. *Transportation Research Record: Journal of the Transportation Research Board*, 2015. 2503: 81–90.
21. Xu, Y., V. Elango, R. Guensler, and S. Khoeini. Idle Monitoring, Real-Time Intervention, and Emission Reductions from Cobb County, Georgia, School Buses. *Transportation Research Record: Journal of the Transportation Research Board*, 2013. 2340: 59–65.

The Standing Committee on Rural Public and Intercity Bus Transportation (AP055) peer-reviewed this paper (18-05181).

TRR
JOURNAL OF THE TRANSPORTATION RESEARCH BOARD

Article

Transportation Research Record
2018, Vol. 2672(8) 165–174

DOI: 10.1177/0361198118801338
journals.sagepub.com/home/trr
SAGE

Getting Business People on the Coach: A Stated Preference Experiment for Intercity Long Distance Coach Travel

Steven Lannoo[1], Veronique Van Acker[2,5], Roselinde Kessels[3,4], Daniel Palhazi Cuervo[4], and Frank Witlox[5,6,7]

Abstract

From the middle of the 1990s, the traditional coach industry in Western Europe has been in decline. However, recent regulatory changes have created new opportunities in the sector of scheduled intercity services, resulting in fast growth of both coach lines and passengers traveling on these lines. Until now, operators have persuaded a public consisting mainly of students and people traveling for leisure purposes. In this paper we analyze whether business travelers could also be an interesting target group and what service characteristics are most convincing for them. For this purpose, we organized a stated preference experiment in which we gathered data from 63 Belgian business travelers. Analysis of the data revealed that for business travelers, price is the dominant factor in seducing customers. However, journey length, higher commercial travel speeds, ample leg space, on-board Wi-Fi and the entertainment system also play a role. Moreover, business travelers are prepared to pay for extra services. We conclude that when an adjusted service is offered, business travelers form an interesting (additional) target group for the intercity coach business. Our findings could be used by coach operators for product development and help to understand travel market segmentation, and eventually also have impact on developing a more sustainable travel policy.

For over two decades, the traditional coach market has been waning in most Western European countries (*1*). For instance, in the Belgian tourism market the number of long holidays (i.e., 4 nights or more) for which the coach was the main transport mode has declined from 1,250,000 in 1994 to less than 500,000 in 2014. Market share for coaches in the tourism market has also declined from 12% in the early 1990s to only 4.5% in 2014. The decline in volumes and market shares can be observed in all segments of the market and for all target groups (*2*). Long-distance shuttle services for instance, have lost many clients to low-cost airlines and the classic closed-doors tour-holiday by coach appeals to an ever more restricted group of travelers. Those customers that have remained loyal to the coach are rather traditional and conservative: they book their journeys through travel agencies instead of booking online, they prefer traditional destinations, and they are less interested in modern technologies (*2*). Consequently, coach companies risk getting trapped in a vicious circle: a traditional market discourages investments in modern marketing, innovative technologies and new concepts of travel. In turn, a conservative industry does not attract many new customers. To break this vicious circle, coach companies should invest more effort in developing new products and exploring new markets.

An example of such a new market has recently appeared through the liberalization of long distance scheduled coach services. In most countries in Western Europe, creating intercity connections by coach was for a long time seriously hampered (or even prohibited) by legislation aimed to protect national railway companies

[1]Institute for Coach and Bus, Brussels, Belgium
[2]Luxembourg Institute of Socio-Economic Research (LISER), Urban Development & Mobility, Esch-sur-Alzette, Luxembourg
[3]School of Economics, University of Amsterdam, Amsterdam, The Netherlands
[4]Faculty of Business and Economics, University of Antwerp, Antwerp, Belgium
[5]Department of Geography, Ghent University, Ghent, Belgium
[6]College of Civil Aviation, Nanjing University of Aeronautics and Astronautics, Nanjing, China
[7]Department of Geography, University of Tartu, Tartu, Estonia

Corresponding Author:
Address correspondence to Steven Lannoo: Steven.Lannoo@ICB-institute.be

from competition. With the Transport Act of 1980, the UK was the first country to allow private companies to set up long distance bus connections. Sweden, Norway and Italy gradually liberalized their markets in the 1990s and early 2000s (*3*). However, the big break-through came with the opening up of the German market in 2013, and the French market in 2015.

Liberalized markets have seen a rapid development, both in demand and supply. In Germany, in 2016, 24 million travelers made use of the so-called *Fernbuslinien* (*4*), representing a quarter of the total coach market in Germany. In France, in the first year of liberalization, over 5 million passengers traveled by intercity coach, representing an equally fast growth rate as in Germany. A few years' later, a dense network of scheduled coach services emerged that connects all mayor cities in Western Europe (*5*).

Research on the profile of travelers shows that in particular young people, students and people traveling for leisure purposes have made use of the intercity coach, with people traveling for business only making up a small amount of travelers (*1*). In this paper we want to investigate if and how more business travelers can be convinced to make use of the intercity coach. The remainder of this paper is structured as follows. The next section describes the evolution of the intercity coach market in Western Europe, especially after the liberalization of this market. The third section summarizes why people choose to ride a coach compared with other transport modes, with a particular focus on modal choices for business travel. To get better insights into this modal choice, we designed a stated preference experiment in which 105 Belgian business travelers participated. Details of this experiment are described in the fourth section and the results are presented in the subsequent section. The final section summarizes the major conclusions and discusses implications for the coach sector.

The Intercity Coach Market in Western Europe

At the turn of the twentieth century, (collective) road transport started to challenge the then dominant mode of railway transportation. Motor coaches had some advantages over rail with regards to flexibility and cost, and were therefore particularly successful on less frequently used connections. However, companies operating these coaches soon started services that ran parallel to existing rail connections. Authorities reacted by creating legislation that tempered competition between two scheduled services. The legislation took different forms in different countries but nearly always favored the railway over the road. Consequently, private bus and coach companies focused on occasional services and subcontracting to large state-owned monopolists. Although, in the second half of the twentieth century these monopolists replaced some loss-making railway sections by road services, in most countries a network of long-distance scheduled coach services did not develop (*6*).

This situation first began to change in the UK with the Transport Act of 1980 that allowed private companies to establish freely scheduled bus or coach services. In the late nineties, Norway and Sweden gradually entered the path of liberalization, as did Italy in 2007. However, it was the liberalization of the German (2013) and the French (2015) markets that was the game changer for the European intercity coach market. Before reform in the different member states, international scheduled services had already been liberalized. However, their development was seriously hampered by the prohibition to selling national tickets in large European countries. With liberalization in France and Germany, the creation of a dense pan-European network of bus connections suddenly became much more evident. This dynamic even affected small European countries such as Belgium that have not changed their legislation, but saw numerous new international coach connections develop on their territory.

During the first years of liberalization in the UK, fierce competition between operators led to a considerable increase in the number of services offered, some improvements in comfort level and a decrease in price. National Express was the only company that could maintain a nationwide network and survive. Consequently, the British market turned into a quasi-monopoly, which put growth on hold both for ridership and service innovation. Analysts argued that new companies had a hard time competing because they lacked access to the bus terminals that served as the central points for information and sale (*7*). In the early 21st century, the democratization of the internet opened the way for the curbside bus model. The possibility of online sales and promotion made the presence in bus terminals less important, which brought down entrance costs resulting in more competition and a new market dynamism (*8*).

Although times had changed between 1980 and 2013, the story of the liberalization of the German market is remarkably similar. Once the new legislation was in place, a period of fierce competition began, leading to a fast growth in ridership and rapid consolidation on the supply side (*9*). Currently the dominant player has a market share of over 90% and ridership has hit a ceiling (*4*).

To summarize, market deregulation in several European countries has resulted in the creation of a dense pan-European network of intercity coach connections. The fast growth in ridership shows that intercity bus connections satisfy a latent demand that was present in the population (*3, 10, 11*). Therefore, the question arises what kind of people are using these new transportation services. Studies on the demographic profile of coach travelers show that their average income is lower than for the general population, and many of the clients do not own their own car. Passengers tend to be younger on average and students are heavily overrepresented. In most cases, coach users travel alone to meet friends or family or for leisure purposes. People traveling for business reasons only make up between 3% (UK) and 18% (Spain) of the clients (*1*, *8*, *12–14*). In the next section, we review the literature, considering why and under which circumstances travelers in general, and business travels in particular, prefer to use a coach over any other transport mode. As such, we are looking first to answer the questions if and how business travelers can be convinced to travel by intercity coach.

Determinants of the Preference for Intercity Coach Transport

For a long time, rail systems were believed to attract significantly more passengers compared with bus/coach systems, even if both systems offer similar level-of-service (LOS) characteristics such as travel time, fares, frequency and the number of transfers. Given the choice between seemingly equivalent systems, travelers are believed to prefer rail over bus/coach. This preference for rail may be explained by "soft" or rather intangible factors such as reliability, comfort and safety, for which rail is often perceived to be superior. Vuchic and Stanger, for example, confirmed that most attributes favor rail services (*15*). Their study is based on a comparison of two very different types of services in different metropolitan areas: the Lindenwold "Hi-Speed Line" (between Philadelphia, Pennsylvania, and Lindenwold, New Jersey) and the Shirley Highway Busway (between Washington, D.C., and Richmond, Virginia). This situation does not, however, really represent a true choice context between rail and bus. It would be better to have information from corridors where both transit modes coexist and compete. Kottenhoff and Lindh, for example, present such evidence from Blekinge, in the south of Sweden (*16*). In January 1992, a new high-standard train service replaced a combination of existing old rail cars and standard coaches. Furthermore, 18 months later a new high comfort coach service was introduced as a complement to the train timetable and to increase the number of departures. In 1994, 500 passengers were interviewed of which almost half were carried out on board trains and buses respectively. Respondents who used the coach service indicated that this is mainly because the coach departure time was more convenient. The majority however prefer a train service above an express coach service for journeys longer than one hour. Also, the modern express coach is valued lower compared with the train service, and this is even more evident among train riders. However, investing in additional leg-room in coaches might convince at least some of the train passengers. Coach passengers' value 10 cm (4 inch) more leg-room in coaches at about 5% of the fare, whereas train passengers value this at about 13% of the fare.

Another alternative source of data is from stated preference surveys (SPS) in which travelers are asked to respond to hypothetical scenarios of coach and rail services. It is interesting to note that such SPS often do not find a particular preference for rail over coach services (*17*, *18*). SPS are not only useful in determining the overall preference for a specific transport mode, but also to value the specific attributes of this transport mode (*19*). With respect to coach services, almost all studies found traditional LOS attributes such as travel time and cost to be highly significant (*17*, *18*, *20*). Findings related to rather intangible attributes such as reliability, comfort and safety are more mixed. For example, Ahern and Tapley (*18*) found that comfort defined by having a toilet on-board a coach was significant, but it obtained an unexpected negative sign. This might indicate that the presence of on-board services only is not sufficient; the quality of these services should also be guaranteed. Toilets on-board coaches or trains are indeed often poorly maintained. Furthermore, reliability is often found to be insignificant, but several studies point out that the existing coach services already perform well and therefore respondents might rate reliability as less important (*18*, *20*).

However, these previously mentioned studies do not specifically focus on business trips. It is nevertheless important to control the travel purpose in the evaluation of a coach alternative as people on business-trips tend to assign different levels of importance to their attributes of choice. In their practical guide to the demand for public transport, Balcombe et al. point out that an employer is likely to be less sensitive to a fare increase than a passenger paying his or her own fare if a local business journey needs to be made (*21*). Consequently, very low price elasticities are generally found for business trips with public transport. Moreover, business travelers value time much more compared with people traveling for other purposes, particularly if the chosen mode is public transport. However, most existing evidence on modal choices for business trips focus on the choice between airplane, car, or train (*22–24*). For example, Fowkes et al. found that a short journey time and convenient departure time are important factors for long-distance business trips by airplane, car and train (*25*). In addition, the ability to work while traveling was found to be an important reason to choose the train above the other modes. Many business travelers did not perceive the coach as a viable option

Table 1. Variables Used in the Study

Coach Journey Attributes and Levels		Respondent-Level Variables	
Price		Age	
Cheap		Mean	45
Average		SD	11
Expensive		Travel frequency	
Speed		Less than once a year	8%
Fast		About once a year	17%
Average		Multiple times a year	68%
Slow		Monthly or more	6%
Wi-fi		Educational level	
Present		Bachelor	33%
Not present		Higher secondary	14%
Leg space		Lower secondary	3%
Standard		Master	46%
Luxurious		No degree	3%
Catering		Profession	
No catering		Employee	48%
Snacks		Manual worker	8%
Hot meal		Other	5%
Entertainment system		Retired	5%
Collective		Self-employed	14%
Individual		Staff member	21%
Power plug		Pickup location	
Present		Antwerp	16%
Not present		Arlon	2%
Journey length (hours:minutes)		Bruges	5%
Mean	2:24	Brussels	44%
SD	2:06	Charleroi	5%
		Ghent	13%
		Hasselt	5%
		Liège	10%
		Namur	2%
		Destination	
		Amsterdam	21%
		Cologne	22%
		Frankfurt	22%
		Lille	19%
		Paris	16%
		Time to pickup location (min.)	
		Mean	26
		SD	17
		Sex	
		Men	71%
		Women	29%

because it is considered to be very slow compared with other transport modes. In addition to the journey length and departure times, other quality of service factors might determine modal choices for business trips. Balcombe et al. summarized the values for such quality of service factors and, even though these values apply to rail transport, it might be useful to consider these for coaches too (*21*). For instance, business travelers paying for themselves particularly value a quiet environment, whereas business travelers whose employers pay particularly value free drinks and a quality newspaper or magazine. Having breakfast or a light snack on-board is valued lower by business travelers compared with leisure travelers. In addition, business travelers value the availability of free car parking or taxis.

To summarize, coaches are generally valued lower than competing modes of transport by all travelers. Investing in travel times, comfort and on-board facilities might increase the attractiveness for travelers in general and business travelers in particular. However, evidence is missing on the modal choices for business trips using coaches compared with other modes of transport. The remainder of this paper therefore describes the details of a stated choice experiment conducted among business travelers that was constructed to answer these questions.

Method

In our stated preference experiment, we confronted business travelers with hypothetical offerings and asked them whether, given the price and level of service offered, they would buy a ticket for the coach or whether they would prefer to travel by another mode. The design used was a Bayesian D-optimal partial profile design that considers the no-choice option (*26*).

We selected the participants from the consumer panel of GFK-Belgium, a firm that maintains a demographically representative sample of consumers from the different regions of Belgium that are willing to participate in surveys. As we wanted to focus our research on business travelers, we only selected those people that indicated they sometimes traveled for business to cities such as Paris, Amsterdam or Cologne. Potential participants were pre-screened by asking them whether they would, under certain conditions of price and luxury on board, ever consider traveling to such places by coach. Only those giving a positive answer (44%) were invited to participate in the experiment. In total 105 respondents completed the experiment, but only 63 questionnaires satisfied all quality criteria and were included in the analysis. The experiment was conducted online and participants were asked to participate through mailed invitations. During the experiment participants had to choose 16 times between two coach trip alternatives or the option to not take the coach and take any other transport mode such as the car, train or airplane instead. Each participant received questions for one of the five destinations under study: Paris, Lille, Frankfurt, Cologne and Amsterdam. The participants first had to choose from nine pick-up locations in Belgium: Bruges, Ghent, Antwerp, Charleroi, Brussels, Hasselt, Namur, Liège and Arlon. The pick-up location remained the same throughout the experiment. Next, a series of coach trip alternatives were presented at random. The trips differed in the level of service including Wi-Fi (yes/no), an individual power plug (yes/no), catering (none/light

Imagine you have to make a journey from Ghent (St-Pieters Station) to Frankfurt? Which one of the options mentioned beneath would you choose?

	Coach journey A	Coach journey B	Other
Travel TIme	5:42	4:59	Another mode (e.g. train, plane, car)
Wi-Fi	Not present	Not present	
Leg space	Luxurious	Standard	
Catering	Snacks	Hot meal	
Entertainment	Collective	Collective	
Individual power plug	Not present	Not present	
Ticket price	€ 24.10	€ 36.20	

Figure 1. Example of a stated preference choice set.

snack/hot meal), leg space (standard/luxurious), and entertainment system (collective/individual).

The trip alternatives also differed in price (between 6 and 37 euro) and travel time (between 1 and 8.5 hours). To determine the ticket price we made an analysis of the market situation in Belgium in 2015. A regression analysis revealed a negative exponential relationship between distance and price per kilometer. This functional relationship was then used to determine an average price level for every combination of pickup location and destination. Additionally, for every combination, a cheap and an expensive price alternative were calculated. The cheap version offered a 50% discount, whereas the costly version was 50% more expensive. To determine the travel time, the travel times mentioned in offerings from different intercity coach companies based in Belgium in 2015 were analyzed. Commercial speeds varied between 60 and 80 kilometres per hour. Therefore, 70 km/h was used as the average speed version, 80 km/h as the fast speed version and 60 km/hour as the slow speed version. Participants were only shown specific prices and travel times. As a consequence, they were not aware whether they were offered a discount, an average or a more expensive ticket. In addition, they did not know whether they were confronted with a fast, average or slow speed version.

We also collected information on the participants in the experiment, that is their age, travel frequency, educational level, profession, gender and the time they needed to reach their pickup location (see Table 1). Information and descriptive statistics on the different variables in the study can be found in Figure 1.

In our stated preference experiment, we also introduced a so-called no-choice alternative (i.e., the option to travel by "any other mode"). For such an experimental design, statistical analysis can be performed using the nested logit model or using the multinomial logit (MNL) model with a no-choice constant. Following Haaijer, we preferred the latter option because it is easier to estimate and because it outperforms the MNL model on predictive fit (*27*, *28*). The models were estimated in JMP 12.0.

Results

Table 2 lists the results of the no-choice multinomial logit model on the stated preference data we gathered from our sample of business travelers. As can be seen in the table, not all respondent-level variables mentioned in Table 1 are entered in the model for which the results are presented. As respondent-level variables have to be entered as interactions, the full model is too large to be estimated. The model presented is a more parsimonious model in which only significant interaction terms are retained.

The parameters estimated show the effect of the journey feature on the utility of the coach alternatives for the respondents. The parameter for Wi-Fi, indicates that having Wi-Fi on board increases the utility of the coach by 0.54 (= 0.27 + 0.27). Offering luxurious leg space also has a positive effect. However, the magnitude of the effect depends on how frequently the respondent travels, as is shown by the significant interaction effect of leg space and travel frequency. Respondents traveling monthly or more value leg space most, whereas the effect for the other groups is smaller. Offering catering also has a positive effect, as offering snacks increases the utility and offering a hot meal increases it even further.

As was the case for luxurious leg space, the effect of an individual entertainment system on board depends on the respondent's frequency of travel. Although the effect is positive for the most frequent travelers (traveling monthly or more), such a system does not make the coach more attractive for less frequent travelers.

Table 2. Results of the No-Choice Multinomial Logit Model

	Estimate	Std. error	L-R chisquare	DF	Prob>ChiSq
Wifi	0.27	0.0612	19.6810	1	0.000
Ample legspace	0.43	0.0859	26.7480	1	0.000
Ample legspace*Trip frequency			10.3850	3	0.016
Less than once a year	–0.20	0.1565			
About once a year	–0.12	0.1225			
Multiple times a year, but not monthly	–0.21	0.0950			
(Monthly or more = reference category)					
Catering			9.0800	2	0.011
None	–0.18	0.0662			
Snacks	0.02	0.0604			
(Hot meal = reference category)					
Individual entertainment	0.24	0.0849	8.5150	1	0.004
Individual entertainment*Trip frequency			15.6090	3	0.001
Less than once a year	–0.43	0.1597			
About once a year	–0.03	0.1220			
Multiple times a year, but not monthly	–0.17	0.0954			
(Monthly or more=reference category)					
Power plug	0.05	0.0550	0.6920	1	0.405
Journey length	0.01	0.0038	8.6740	1	0.003
Journey length^2	0.00	0.0000	8.8330	1	0.003
Speed			28.7470	2	0.000
60 km/h	–0.42	0.0823			
70 km/h	0.08	0.0594			
(80 km/h=reference category)					
Price	–0.07	0.0065	119.9750	1	0.000
No-choice parameter	0.35	0.2257	2.5780	1	0.108

Note: Std. = standard; DF = degrees of freedom.

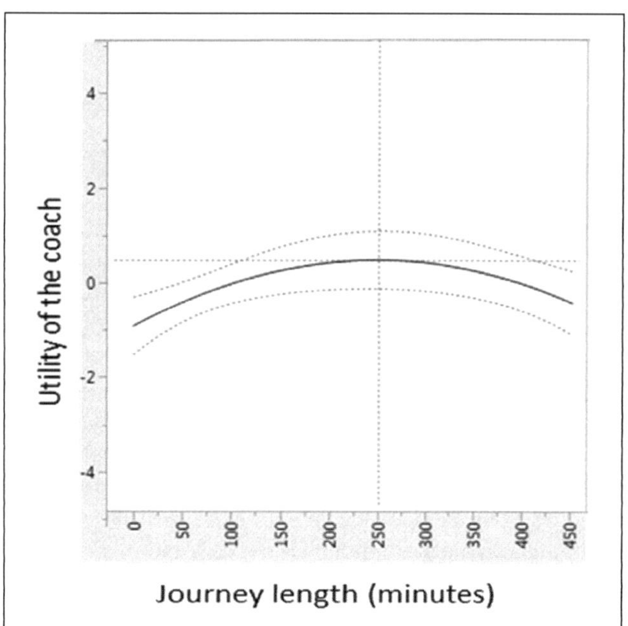

Figure 2. Effect of journey length.

The presence of a power plug on board is the only travel feature for which there is no evidence that it exerts any effect on the utility of the coach.

The relationship between the journey length and utility was best represented by a second order polynomial term. As is evident from the graphic representation of the relationship in Figure 2, the utility of the coach journey increases until a journey length of about 4 hours and then starts to decrease again. The coach seems to be most attractive for journeys between 2 and 6 hours. For shorter journeys, the effort of coming to the pickup location probably decreases the attractiveness relative to the car. On the other hand, for very long journeys planes and high-speed rail start to take over.

As we presented different commercial speeds for the same trajectories, we are able to model the effect of speed regardless of journey length. Speed logically has a positive effect, resulting in a decreased utility for the slowest journeys. Also, in line with expectations, the effect of price on the utility of a bus alternative is negative and highly significant.

The no-choice alternative was not significant, so we can say there is no evidence that the participants in our survey preferred either the coach or another mode of transport (*27*). As only people that would consider traveling by coach were included in the survey, this result is also in line with expectations.

Figure 3 shows us the relative importance of the different parameters in the model. The values shown are based

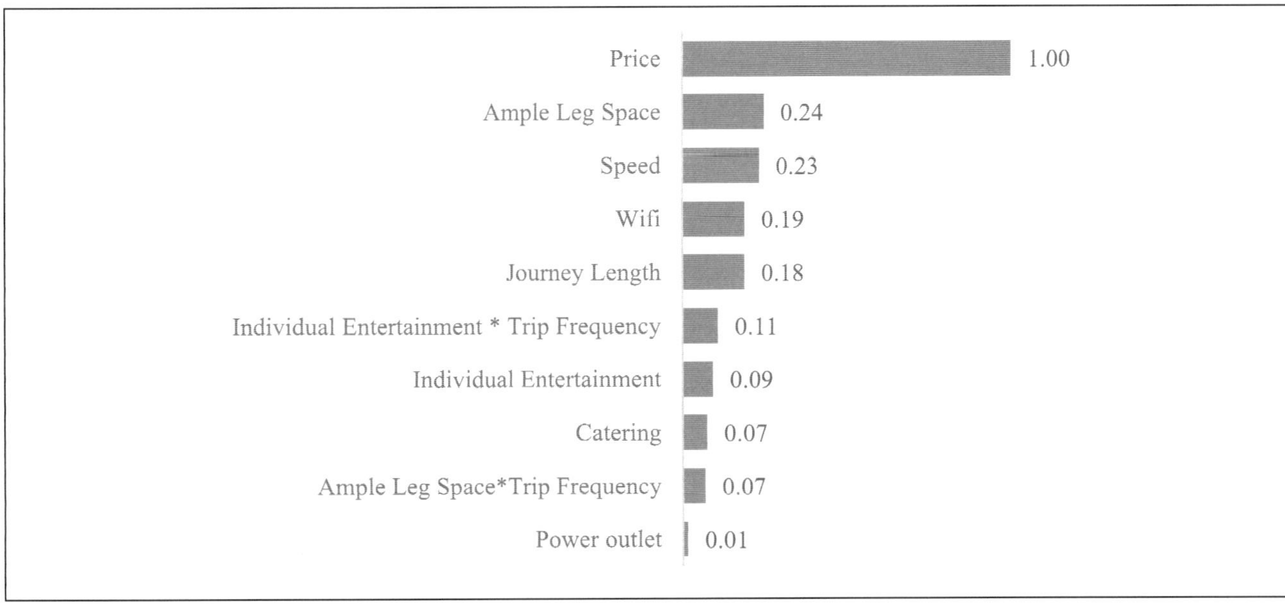

Figure 3. Relative importance of parameters.

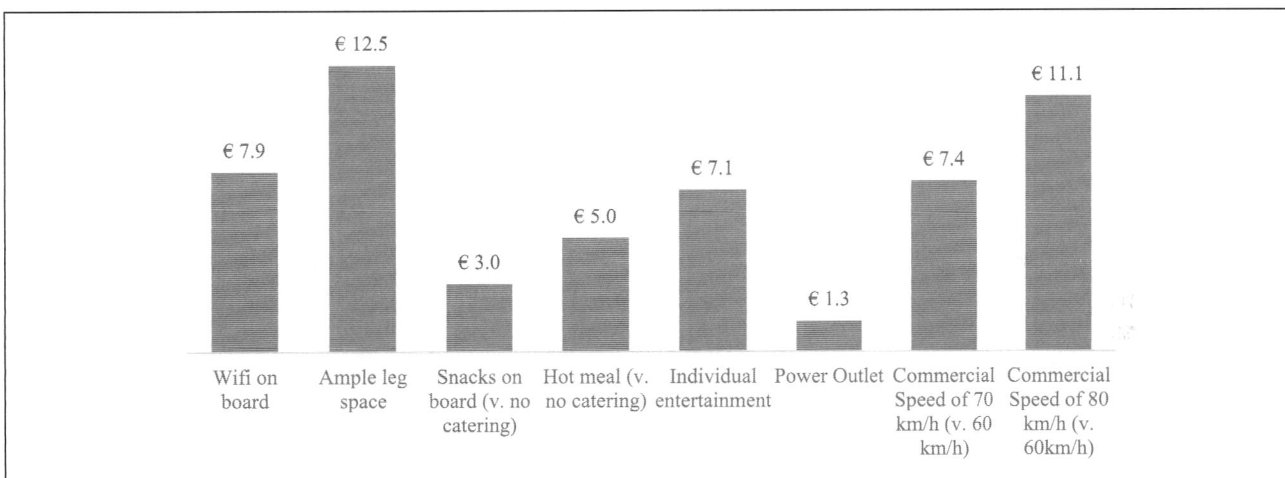

Figure 4. Estimated willingness to pay (average over travel frequencies).

on the negative logarithm of the p-values expressed as a percentage of the strongest real effect (i.e., as a percentage of the value for price). We can see that price is the dominant factor in predicting coach utility. Ample leg space is the second most important factor, followed by speed and Wi-fi. The difference in importance between these last three characteristics is rather small, but the difference from price is very large. This finding underscores the importance of price for the attractiveness of the coach, even for this specific group of people traveling for business purposes.

To further illustrate the effect of the service characteristics, we estimated the willingness to pay (WTP) for the attributes studied. Given the significant interaction of travel frequency with individual entertainment and ample leg space, the WTP for the service characteristics depends on the frequency of travel of the respondent. In Figure 4 we show the results of averaging WTP over the different travel frequency categories, whereas Figure 5 shows the results split up by travel frequency.

The results show that ample leg space allows coach operators to drive up their ticket prices by 12.5 euros. The most frequent travelers in particular are willing to pay extra for this feature (WTP estimated at 28 euros), less frequent travelers are only willing to pay between 6.4 and 9 euros. Increasing speed from 60 to 70 km/h allows the operator to increase the ticket price by 7.4 euros.

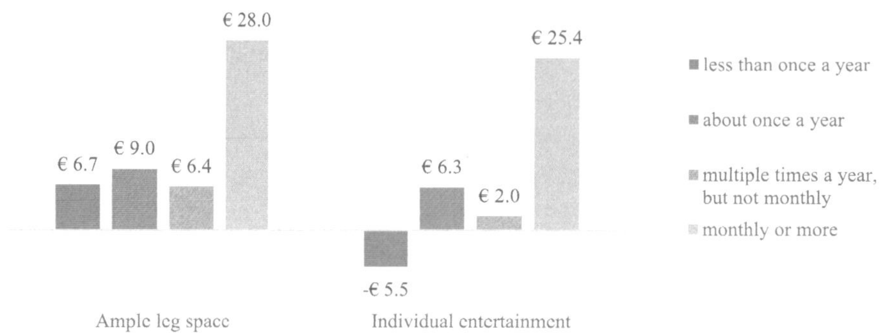

Figure 5. Estimated willingness to pay by travel frequency.

Driving commercial speed up to 80 km/h allows a further increase of 11.1 euros. On-board Wi-Fi is worth 7.9 euros and individual entertainment 7.1 euros. As we have seen previously, the attractiveness of individual entertainment is also dependent on the travel frequency. The most frequent travelers are willing to pay a premium of 25 euros for the service, while the less frequent travelers prefer a collective service.

Finally, the WTP for snacks is 3 euros and for a hot meal 5 euros. This is rather low, certainly when one considers the additional costs associated with this service. The WTP for the power plug can be ignored as the coefficient wasn't significantly different from 0.

Conclusions and Further Research

From the mid 1990s the traditional coach industry in most Western European countries has been in decline. However, recent regulatory changes have created new opportunities. In Germany and France deregulation of intercity coach services was followed by a rapid growth in both supply and demand. Nevertheless, the majority of travelers embracing the coach are students and people traveling for leisure purposes. The question arises as to if and how intercity coach services can appeal to business travelers. A review of the literature showed that coaches are generally considered less appealing, but that investments in luxury, on-board facilities and travel time can stimulate attractiveness. In this study we conducted a stated preference experiment to investigate if and how such investments can convince business travelers to get on board.

The results of our experiment show that price is the dominant factor in attracting passengers. This might be surprising as one could expect business travelers to be less price sensitive. After all, it is generally observed that business travelers place a high value on time, and are very concerned about maximizing the productivity of their travel time. Additionally, many business travelers do not pay for travel themselves, which also decreases price sensitivity in many cases (*21*, *29*). However, other studies have already pointed to the importance of the price-factor in modal choice in general, and for the choice of buses in particular (*18*, *30*). Intercity buses are a very price competitive mode of transport and even though their level of service and comfort might be as good as that of other modes, price remains a predominant factor determining their relative attractiveness.

This said, other factors also play a role. Speed, ample leg space and on-board Wi-Fi can stimulate business travelers to prefer using intercity buses. In addition, they are willing to pay for the extra services. Clearly, these services are important for increasing the productiveness of their journey, a key factor for business travelers as mentioned previously. Entertainment did affect attractiveness, but mostly so for frequent travelers. Catering also had a negative effect, but the WTP can be considered to be low relative to the cost of the service. On the contrary, we did not find any evidence of the effect of an individual power plug. Journey time was probably too short to make the presence of a power plug a necessity. Additionally, such services are increasingly considered standard services that should be offered on the journey, and therefore do not present an added value. Finally, a second-order polynomial relationship between the utility of the coach trip and journey length was revealed. Journeys between 2 and 4 hours were found to be most attractive.

To summarize, the intercity coach operators could potentially attract more people traveling for business purposes than they do currently, if an adjusted service was offered. Furthermore, although price remains a key factor for these customers, offering additional services that can increase the productivity of their journey, results in a higher willingness to pay for coach tickets.

These empirical findings open up several avenues for further research. Firstly, this research project looked at the attractiveness of intercity coach travel in general, but did not consider the competitive position of the coach vis-à-vis specific other travel modes (train, plane, etc.). It would be interesting to investigate under which conditions the coach competes with these specific modes. Secondly, as our experiment has been conducted on a Belgian public only, it is unclear whether and how the specific national context of the country influences our conclusions. As such, repeating the experiment in countries with a recently liberated bus market (Germany, France, etc.) or with a strong bus culture (Spain, Turkey, etc.) could increase our insights on the matter. Also, it could be interesting to see how this young market in Belgium evolves in the near future and whether this affects the attitude of the public toward the service. Finally, on a more abstract level, research should address the role that the intercity coach can play in the sustainable transport systems of tomorrow.

Author Contributions

The authors confirm contribution to the paper as follows: study conception and design: SL, VVA, FW, RK, DPC; data collection: SL; analysis and interpretation of results: SL, RK, VVA, FW; draft manuscript preparation: SL, VVA, FW. All authors reviewed the results and approved the final version of the manuscript.

References

1. Gleave, S. D. *Study of Passenger Transport by Coach*. Final Report. European Commission, Brussels, Belgium, 2009.
2. Verhaege, A. *De positie van de autocar op de Belgische reismarkt*. WES, Bruges, Belgium, 2015.
3. Van de Velde, D. Long-Distance Coach Services in Europe. In *Regulating Transport in Europe* (Finger, M., and T. Holvad, eds.), Edward Elgar, Cheltenham, United Kingdom, 2013, pp. 115–139.
4. IGES-Institute. *Fernbusmarkt: weniger Fahrten bei weiter steigenden Fahrgastzahlen*. http://www.iges.com/presse/2017/fernbusmarkt-quartal-i/index_ger.html. Accessed April 10, 2017.
5. Guihéry, L. New Interurban Coach Services in Germany and France: Which European Perspective? The Key Role of Bus Stations...but Where to Locate Them? In *Mobility in a Globalised World 2015* (Sucky, E., J. Werner, R. Kolke, and N. Biethahn, eds.), University of Bamberg Press, Bamberg, Germany, 2016, pp. 83–90.
6. Schiefelbusch, M. Past and Future Regulation of Interurban Coach Services in Germany. *Journal of Transport Economics and Policy*, Vol. 47, 2013, pp. 299–305.
7. Robbin, D. K., and P. R. White. The Experience of Express Coach Deregulation in Great Britain. *Transportation*, Vol. 13, 1986, pp. 359–384.
8. Klein, N. J. Get on the (Curbside) Bus: The New Intercity Bus. *Journal of Transport and Land Use*, Vol. 8, 2015, pp. 155–169.
9. Dürr, N. S., S. Heim, and K. Hüschelrath. Deregulation, Competition, and Consolidation: The Case of the German Interurban Bus Industry. *Journal of Transport Economics and Policy*, Vol. 50, 2016, pp.164–188.
10. Beria, P., R. Grimaldi, and A. Laurino. Long Distance Coach Transport in Italy: State of the Art and Perspectives. Presented at the 13th World Conference on Transport Research, Rio de Janeiro, Brasil, 2013.
11. Aarhaug, J., and N. Fearnley. Deregulation of the Norwegian Long Distance Express Coach Market. *Transport Policy*. Vol. 46, 2015, pp. 1–6.
12. Harrington, A., and B. Parolin. Factors Affecting the Use of Bus and Rail Services in a Deregulated Environment. *Transportation*, Vol. 18, 1991, pp. 175–193.
13. Krämer, A., and J. Hercher. *Mobilitätstrends 2016. Reisen mit dem Fernlinienbus (FLB): Vergleichende Analyse für die D-A-CH-Region*. Rogator, Bonn, Germany, 2016.
14. ARAFER. *Les pratiques de mobilité de longue-distance des voyageurs sur les lignes régulières d'autocar librement organisées. Enquêtes de terrain réalisées entre octobre et décembre 2016*. Autorité de Régulation des Activités Ferroviaires et Routières, Paris, France, 2017.
15. Vuchic, V. R., and R. M. Stanger. Lindenwold Rail Line and Shirley Busway: A Comparison. *Highway Research Record*, Vol. 459, 1973, pp. 13–28.
16. Kottenhoff, K., and C. Lindh. The Value and Effects of Introducing High Standard Train and Bus Concepts in Blekinge, Sweden. *Transport Policy*, Vol. 2, 1996, pp. 235–241.
17. Ben-Akiva, M., and T. Morikawa. Comparing Ridership Attraction of Rail and Bus. *Transport Policy*, Vol. 9, 2002, pp. 107–116.
18. Ahern, A. A., and N. Tapley. The Use of Stated Preference Techniques to Model Modal Choices on Interurban Trips in Ireland. *Transportation Research Part A*, Vol. 42, 2008, pp. 15–27.
19. Witlox, F., and E. Vandaele. Determining the Monetary Value of Quality Attributes in Freight Transportation Using a Stated Preference Approach. *Transportation Planning and Technology*, Vol. 28, No. 2, 2005, pp. 77–92.
20. Ortúzar, J., de Dios, and A. Iacobelli. Mixed Modelling of Interurban Trips by Coach and Train. *Transportation Research Part A*, Vol. 32, 1998, pp. 345–357.
21. Balcombe, R., R. Mackett, N. Paulley, J. Preston, J. Shires, H. Titheridge, M. Wardman, and P. White. *The Demand for Public Transport: A Practical Guide*. TRL Report 593. Transport and Road Research Laboratory, London, UK, 2004.
22. Bhat, C. An Endogenous Segmentation Mode Choice Model with an Application to Intercity Travel. *Transportation Science*, Vol. 31, 1997, pp. 34–48. https://doi.org/10.1287/trsc.31.1.34
23. Limtanakool, N., M. Dijst, and T. Schwanen. The Influence of Socioeconomic Characteristics, Land Use and Travel Time Considerations on Mode Choice for Medium- and Longer-Distance Trips. *Journal of Transport Geography*, Vol. 14, 2006, pp. 327–341.
24. Derudder, B., J. V. Beaverstock, J. R. Faulconbridge, T. Storme, and F. Witlox. You are the Way You Fly: On the

Association between Business Travel and Business Class Travel. *Journal of Transport Geography*, Vol. 19, 2011, pp. 997–1000.
25. Fowkes, A. S., I. Johnson, and P. Marks. *Long Distance Business Travel and Mode Choice: The Results of Two Surveys of Business Travellers*. Working Paper of the Leeds Institute of Transport Studies, Leeds, United Kingdom, 1985.
26. Kessels, R., D. Palhazi Duervo, and K. Sörensen. *Optimal Design of Discrete Choice Experiments with Partial Profiles Involving a No-Choice Option*. Paper Presented at the International Choice Modelling Conference, Cape Town, South Africa, 2017.
27. Haaijer, M. E., W. A. Kamakura, and M. Wedel. The 'No-Choice' Alternative in Conjoint Choice Experiments. *International Journal of Market Research*, Vol. 43, 2001, pp. 93–106.
28. Haaijer, M. E. *Modeling Conjoint Choice Experiments with the Probit Model*. Labyrint Publications, Capelle a/d Ijssel, The Netherlands, 1999.
29. Brons, M., E. Pels, P. Nijkamp, and P. Rietveld. Price Elasticities of Demand for Passenger Air Travel: A Meta-Analysis. *Journal of Air Transport Management*, Vol. 8, 2002, pp. 165–175.
30. Creemers, L., M. Cools, H. Tormans, P.-J. Lateur, D. Janssens, and G. Wets. Identifying the Determinants of Light Rail Mode Choice for Medium/Long Distance Trips. Results from a Stated Preference Study. *Transportation Research Record: Journal of the Transportation Research Board*, 2012. 2275: 30–38.

The Standing Committee on Rural Public and Intercity Bus Transportation (AP055) peer-reviewed this paper (18-00361).

TRR
JOURNAL OF THE TRANSPORTATION RESEARCH BOARD

Article

Transportation Research Record
2018, Vol. 2672(8) 175–186

Article reuse guidelines:
sagepub.com/journalsPermissions.nav
DOI: 10.1177/0361198118781166
journals.sagepub.com/home/trr
SAGE

Constructing Spatiotemporal Load Profiles of Transit Vehicles with Multiple Data Sources

Ding Luo[1], Loïc Bonnetain[2], Oded Cats[1], and Hans van Lint[1]

Abstract

Obtaining load profiles of transit vehicles has remained as a difficult task for transit operators because of technical and financial constraints. Although a significant advance in transit demand and supply data collection has been achieved over the past decade, information related to load profiles at the vehicular level is either impossible or very difficult to retrieve from them. It is not even uncommon to see that these data are underutilized by transit operators owing to considerable deficiencies and shortcomings in the data themselves, and/or the processing algorithms needed to process them. This study is therefore dedicated to addressing this challenge that has largely been overlooked by both researchers and practitioners. First, the issues which hinder the construction of load profiles based on three prevailing transit data sources are identified, including automatic fare collection (AFC), automatic vehicle location (AVL), and general transit feed specification (GTFS) data. Second, a methodology is developed for sequentially addressing all the issues and generating desirable vehicle load profiles. The methodology consists of four steps, including (1) data pre-processing, (2) matching trips in GTFS and AVL, (3) matching passenger rides to vehicle trajectories, and (4) improving vehicle trajectories. The resulting spatiotemporal load profiles of transit vehicles enable detailed investigation into vehicle movements and demand patterns over time and space, including service utilization and the propagation of delays and crowding. Data collected from the urban transit network in The Hague, The Netherlands are used to demonstrate the proposed methodology. The visualization of spatiotemporal load profiles through space-time seat occupancy graphs provides operators with a compact and powerful reference for the improvement of their services.

Knowing the on-board load of transit vehicles is key to improving transit services from both planning and operational perspectives. However, obtaining such information for transit operators has remained as a difficult task for a long time because of technical and financial constraints. Although manual surveys have often been used to estimate on-board passenger loads, such surveys are too costly to be conducted daily over all offered services, and are also subject to error and bias. Opportunities to change this situation, however, have emerged in recent years with the fast-growing data richness in transit research and practice, including automatic vehicle location (AVL) (*1*); automatic fare collection (AFC) (*2*, *3*); and general transit feed specification (GTFS) data (*4*, *5*). In many cities and regions around the world, transit demand and supply data have been continuously collected and managed with fine granularity, accuracy, and spatiotemporal scale. Notwithstanding, it is not uncommon to see transit operators still struggle to obtain some fundamentally important information, such as service utilization (i.e., passenger load). Data are often underutilized as a result of considerable deficiencies and shortcomings that can be frequently overlooked. To unlock their potential, it becomes necessary to develop sound techniques to overcome these issues and achieve valuable information, such as the transit vehicle load, by processing and integrating different data sources, including AFC, AVL, and GTFS. The current study is therefore dedicated to this specific challenge.

To the best of the authors' knowledge, few scientific studies and practical reports have attempted to address a similar problem. One of the main causes for this scarcity could be the rather limited access to multiple transit data sets from the same period by researchers. In many cases, only a single data source is available and the studies

[1]Department of Transport and Planning, Faculty of Civil Engineering and Geosciences, Delft University of Technology, GA Delft, The Netherlands
[2]ENTPE, University of Lyon, Lyon, France

Corresponding Author:
Address correspondence to Ding Luo: d.luo@tudelft.nl

primarily developed methods to infer missing information. For example, Chu and Chapleau early on presented how spatiotemporal bus load profiles could be estimated based on AFC data only (*6*). In the absence of real bus trajectory information, they managed to estimate the spatiotemporal paths of vehicles by combining the first and last transaction times at each stop and corresponding timetable. Their work is one of the pioneering studies that revealed the power of AFC data on load profile construction. Sun et al. subsequently investigated a similar problem, however, in the context of a metro system (*7*). With only AFC data available (both tap-in and tap-off information recorded), they developed a methodology for estimating trains' trajectories and linked individual passenger rides to these trajectories, which results in a spatiotemporal density of metro vehicles. More recently, Moreira-Matias and Cats proposed a novel method for estimating on-board loads of buses using AVL data only (*8*). Passenger loads are built by applying machine learning algorithms to smoothen the load profile based on actual dwell time records. In addition, a web-based application to visualize bus load profiles, called BusViz, has also been developed based on the AFC data in Singapore (*9*). Despite the progress on the visualization work, their approach to derive bus trajectories has several constrained assumptions. For instance, the arrival time of a bus at a stop is equated to the earliest entry time of the first passenger who boards or alights at that stop, while the departure time is set equal to the greatest of the card entry times of passengers who board or alight at the stop. More advanced vehicle trajectory inference techniques based on AFC data only, (see for example Min et al. and Zhou et al. [*10*, *11*]), can be adopted and extended to address these limitations.

Although multiple transit data sets that are comparable among each other have become increasingly available to researchers, most research effort has focused on a selected number of fields, such as transit origin–destination (OD) estimation (*12*, *13*), travel time reliability analysis (*14*, *15*), and passenger assignment modeling in urban rail systems (*16–19*). Few existing studies have comprehensively examined how spatiotemporal load profiles of transit vehicles can be constructed using multiple data sources. This study is therefore devoted to bridging this gap, which can benefit both researchers and practitioners. The study's contribution is twofold, including specific identification of the issues pertaining to a single or a combination of data sets (AFC, AVL, and GTFS), and the development of a methodology for addressing these issues and generating spatiotemporal load profiles of transit vehicles. The methodology consists of four steps through which raw data are processed and integrated to generate the passenger load profiles over space and time. These profiles allow service providers to analyze vehicle trajectories and demand patterns, and further investigate service utilization and the propagation of delays and crowding. The data collected from the urban transit network in The Hague, The Netherlands are utilized for demonstrating the methodology. A series of inference and matching steps are employed. This analysis results in profiles of vehicle trajectories and passenger loads which are further visualized through space-time occupancy graphs. Analogously to how space-time graphs of speed and flow enable traffic engineers to study spatiotemporal congestion patterns along routes in car traffic, these space-time occupancy graphs enable transit operators to study and inspect spatiotemporal on-board crowding patterns along transit service lines.

The remainder of this study is organized as follows. The next section describes the transit data sets available in the Dutch context, including AFC, AVL, and GTFS data, along with the related issues. Following this the methodology is described, with an overview and descriptions of all the steps as well as implementation. The results are then presented, followed by the conclusions and discussion of future research directions.

Data Description and Issue Identification

In this study three different transit data sets are used, all of which were collected from the urban transit system in The Hague, The Netherlands. These comprise AFC, AVL, and GTFS data. All three data sets cover the period of March 2015, which includes 22 normal working days and 9 weekend days. In this period, the system consisted of 12 tram lines and 8 bus lines, and the incumbent transit operator was HTM.

Clarification of Terminology

In order to avoid ambiguity, all terms used in this study are specified as follows based on what Robinson et al. proposed (*20*).

- (*Passenger*) *Ride*: This describes the movement of a passenger on a single vehicle, that is, a bus or tram. The ride begins at the stop where the passenger boarded the vehicle and ends at the stop where the passenger alighted the vehicle.
- (*Vehicle*) *Trip*: This describes the movement of a transit vehicle through a pre-defined sequence of stopping points. It is sometimes called "run" in the literature.

AFC Data

The Dutch smart card system (*OV-chipkaart* in Dutch) is a nationwide fare-collection system (*21*, *22*). An important feature of the Dutch smart card system is that passengers are required to check in and check out for every single ride of a journey for fare calculation, except when transferring within the national railway system and metro systems. Missing the check-out will result in a much higher fare, therefore making passengers have a strong incentive to finish a ride by correctly checking out. Consequently, information of travelers' origin and destination is fairly complete and accurate without the need to infer passenger alighting stops. However, individual rides still need to be combined to obtain an OD matrix, which is an important input for a variety of offline applications.

The smart card data set from The Hague for the whole month of March 2015 contains close to 8 million validated records. Each record characterizes a single transit passenger ride with anonymous card ID, line/route ID, date, stop ID, and time of check-in and check-out. The information of trip ID and vehicle ID, which would allow for the connection to the AVL data, is not available in this data set.

AVL Data

The AVL data of transit vehicles in The Netherlands are well stored and managed at a national level (*23*). The one-month data set used in this study contains over 22 million records in an event-based format. Individual trips are distinguished by a unique trip ID within an operating day. Besides information of line ID, stop ID, and vehicle ID, each row of data is specified as an event and its corresponding timestamp(s), such as the start of a trip, on-route, arrival, dwell, departure, and end of a trip. In addition, deviation from the scheduled time is indicated in the data under "punctuality." The entire fleet of HTM is equipped with the AVL system.

GTFS Data

GTFS, as one of the most prevailing formats for transit schedules and associated geographic information, has been widely adopted in the past decade by transit agencies and operators to share information with the public (*4*). In The Netherlands, GTFS data that cover most transit services nationwide can be accessed via a website called *OVapi* (*24*). These feeds are created from the open data files published by local transit operators under an open license. The website is well maintained with frequent updates of GTFS and GTFS-Realtime data.

The GTFS data used in this study contain all trips scheduled by HTM for March 2015 in The Hague. The GTFS data can be matched with the AVL data set using the trip ID.

Data Issues

Several issues pertaining to a single or a combination of data sets can be identified and need to be resolved for the current application, that is, constructing the spatiotemporal load profiles of transit vehicles. These issues are summarized and presented in Figure 1 in relation to their sources. Basically, issues specific to each individual data set are illustrated in the respective oval, whereas issues that arise when two or more data sources are combined are positioned at their intersections. The following issues have been identified:

- **(1) Indirect availability of line information about stops**

  Information in GTFS data is stored in a trip-based manner, meaning that what can be directly obtained from it are only individual vehicle trips that consist of all the stops per trip. Complete stop sets of service lines, which are crucial for subsequent analytics, are not directly available from the GTFS data.

- **(2) Erroneous AFC check-in/out records**

  AFC data contain different types of errors, including unrealistic alighting times or alighting locations, as well as missing and unrealistic trip numbers. The issues are the result of improper passenger behaviors and system malfunctions.

- **(3) Multiple timestamps for one arrival/departure event**

  Multiple timestamps can be occasionally found for one arrival or departure event in the raw AVL data set. It is unclear why this happens, but it jeopardizes the global consistency of vehicle trajectories.

- **(4) Timestamps missing for one arrival/departure event**

  Arrival and/or departure timestamps at a stop can be missing. The size of missing timestamps ranges from one event (arrival/departure) at a stop to an entire trip. Note that issue #3 and this one can happen to the same stop (e.g., two arrival events, missing departure event), which makes the issue even worse.

- **(5) Inconsistency in trip ID**

  The vehicle trip ID indices in AVL and GTFS data do not always match. This inconsistency causes problems in matching trip and trajectory when combining GTFS (scheduled trajectories) and AVL (recorded trajectories).

- **(6) Lack of information for matching rides to scheduled vehicle trips**

  Since the AFC data set used in this study does not contain vehicle trip ID, it is impossible to directly match

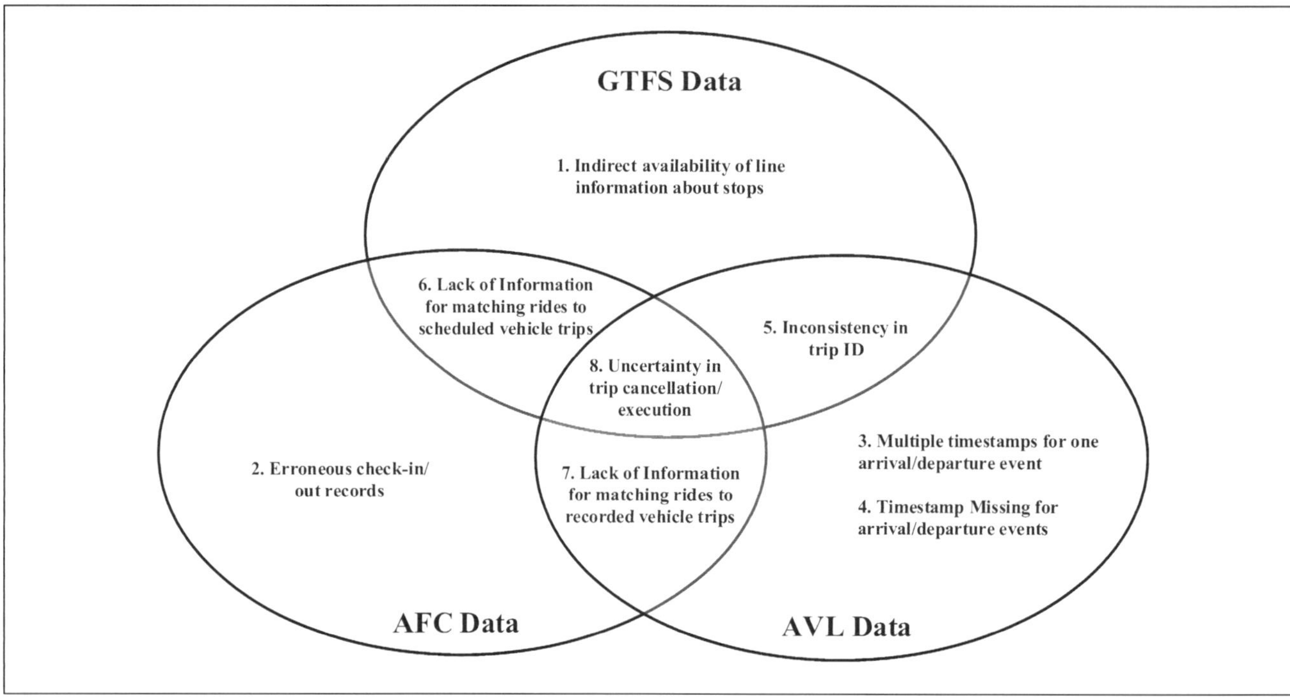

Figure 1. Identification of the issues pertaining to a single or a combination of data sets for constructing the spatiotemporal load profiles of transit vehicles. Issues specific to each individual data set are illustrated in the respective oval, whereas issues that arise when two or more data sources are combined are positioned at their intersections.

individual rides to scheduled vehicle trips that are extracted from the GTFS data.

- **(7) Lack of information for matching rides to recorded vehicle trips**

  The same issue as #6 holds for this situation too. There is no direct way to match rides to the recorded vehicle trips from AVL data.

- **(8) Uncertainty in trip cancelation/execution**

  The GTFS data contain all scheduled trips of a day. However, this does not provide conclusive evidence that all these trips are indeed executed. In many cases, the number of trips found in the AVL data set is smaller than the scheduled number of trips. It is uncertain whether this is a result of trip cancelation or AVL system malfunction without any additional information. AFC data may be used here to settle the discrepancy.

These issues are illustrated for a given day and line in Figure 2, which visualizes the recorded trajectories from AVL data; scheduled trajectories from GTFS data; and check-in/out records. Figure 2*a* first displays all the recorded trajectories from the AVL data. There are many gaps in this plot, which indicates that there is either a missing timestamp or multiple timestamps for the arrival or departure event at that stop. Figure 2*b* adds the layer of all scheduled trajectories (blue lines) underneath the recorded ones (red lines). It can be observed that overall vehicle trips adhere to the timetable very well. Next, check-out (blue circle points) activities are added in Figure 2*c*. An important finding from this plot is that when there is a trajectory gap, check-in/out activities also do not exist, or are very sparse, which implies that in the case where the arrival timestamp is missing but departure has at least one timestamp, the vehicle probably drives through the stop without serving passengers. Figure 2*d* displays a zoom-in plot to allow for a more detailed inspection. Check-in activities are clustered close to the vehicle arrival time, unlike check-out activities, because it is customary for passengers to check out in the segment directly upstream of the alighting stop.

Methodology

In this section, a methodology for constructing the spatiotemporal loads of transit vehicles based on aforementioned data sources is described. An overview is first provided, followed by subsections dedicated to each step. The final subsection describes how this was implemented.

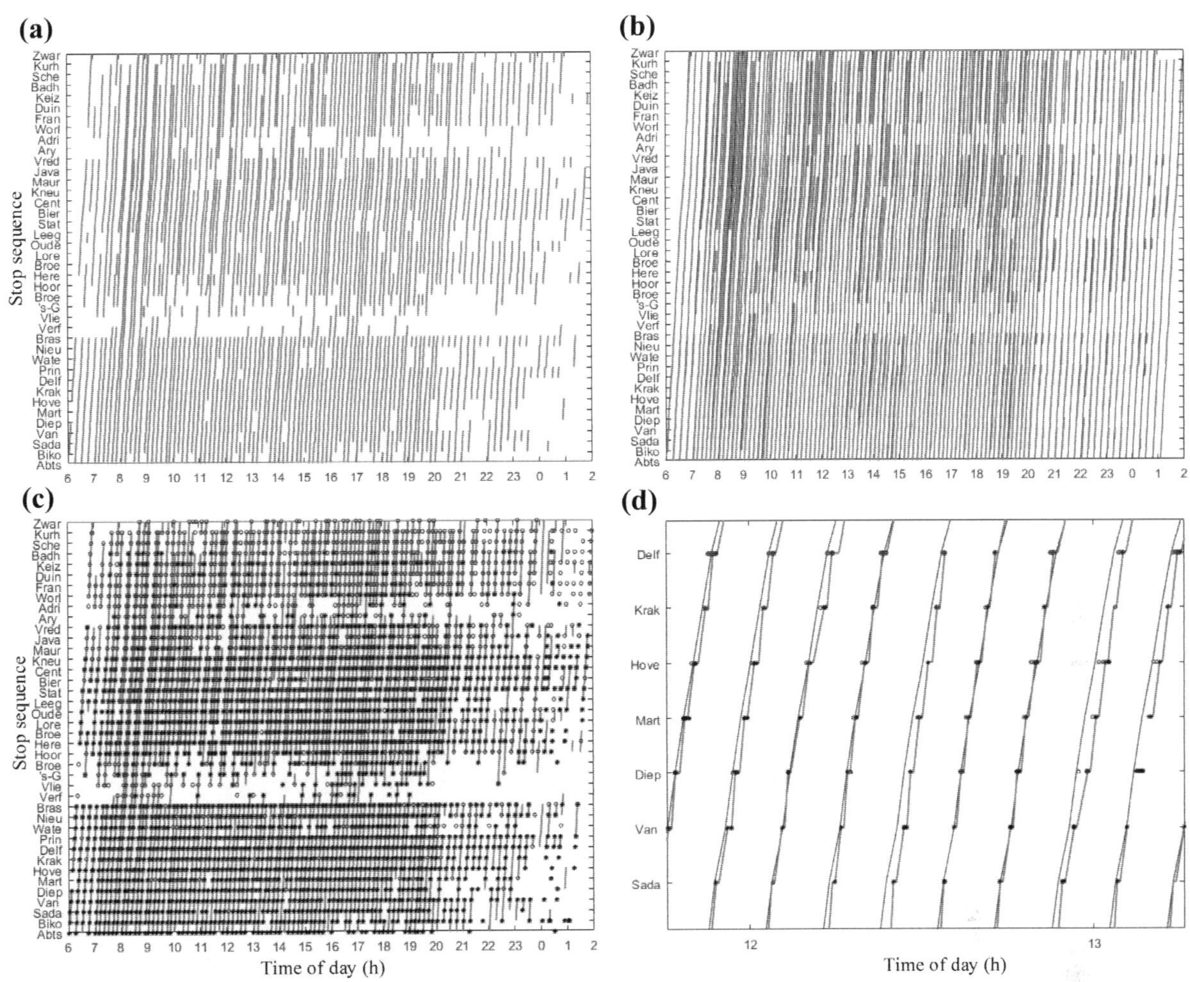

Figure 2. Visualization on how different data sources characterize 1-day services. The example is from line 1 from Delft Tanthof to Scheveningen Noorderstrand on March 5, 2015. (*a*) Recorded trajectories (red lines) obtained from the AVL data set; (*b*) Recorded trajectories (red lines) on top of all the scheduled trajectories obtained from the GTFS data set; (*c*) Recorded trajectories (red lines) on top of all the check-in (black star points) and check-out (blue circle points) activities; (*d*) Zooming-in for a selected hour (12–13) of the data presented in (*c*).

Overview

The methodology consists of four steps as shown in Figure 3. Raw data from three independent sources are used throughout the four steps. All three individual data sets are initially stored in separate databases. In step 1, the raw data are first obtained from the databases with all the information restructured at a daily level. By doing so, the subsequent workflow is made more computationally efficient. Issues #1 and #2 (described earlier) are addressed at the first step, resulting in data files respectively containing passenger rides; recorded vehicle trajectories, and scheduled vehicle trajectories. The integration of recorded and scheduled vehicle trajectories is then performed at the second step. Issue #5 is solved, resulting in data files that contain both scheduled and recorded vehicle trajectories. In step 3, passenger rides are matched with vehicle trips and trajectories (solving issues #6 and #7). In the last step, all scheduled trips are first labeled either "canceled" or "executed" based on the validation results. The data files from step 3 are used to perform this validation task, addressing issue #8. Finally, vehicle trajectories of validated trips are corrected by fusing multiple types of information.

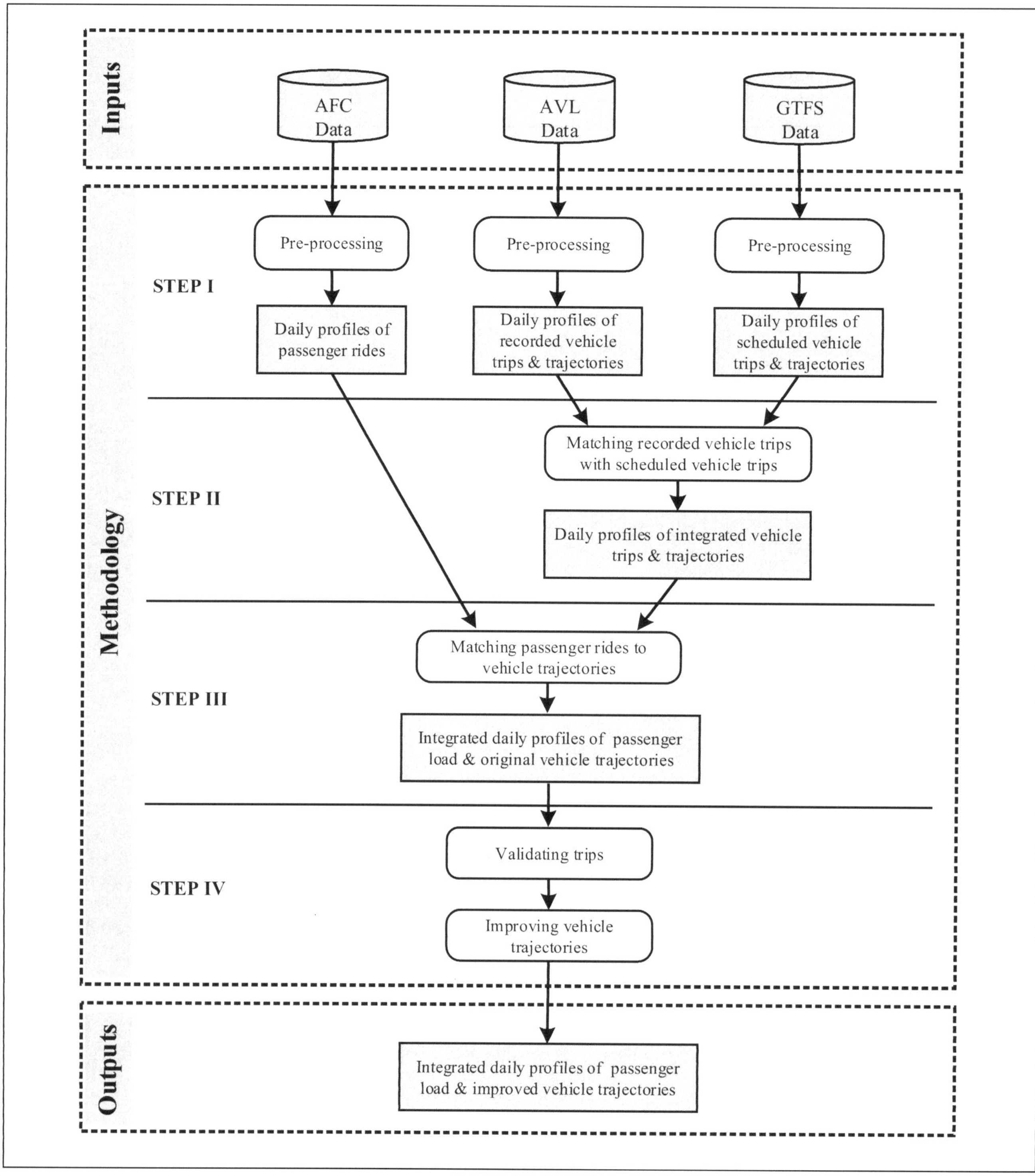

Figure 3. Overview of the four-step methodology. Inputs are raw information from individual data sets, and the final outputs are integrated profiles containing vehicle trajectories and passenger loads.

Consequently, all trajectories have complete trajectory and load information and are globally consistent. Issues #3 and #4 are thus solved in this final step. This sequential method decomposes the process into small sub-tasks with each step solving one or several of the identified issues.

Step 1: Data Pre-Processing

All three types of data are preprocessed in the first step at a daily level. For AFC data, single transaction records are first linked to generate individual passenger rides with both check-in and check out information (stops and times). Erroneous rides with unrealistic travel time and origin or alighting stops are identified and removed in this process. In this case, the travel time threshold was set to 90 min, which exceeds the maximal travel time between any pair of stops in the case study network. In addition, the indices of stops were also transformed to be consistent with those of the AVL and GTFS data.

Every single scheduled trajectory—characterized by the arrival and departure time at every stop along a line—is extracted from the GTFS data. This process is not straightforward because the GTFS standard does not contain direct information about regular stop sequences of individual transit routes. Designed originally for the purpose of route planning, the GTFS data make it quite handy to obtain arrival and departure times at each stop of individual trips by storing information based on trips. A trip is recorded in the trip.txt file and further detailed in the stop_times.txt file with its sequence of stops. All trips, including sub-lines and partial trips with some stops skipped, are thus easily stored in the data. However, this becomes an obstacle when information about the most regular and fullest stop sequence of transit lines is required. To overcome this problem, a brute-force approach is adopted. All trips of a transit route from a normal working day are scanned in order to acquire the complete set and right sequence of stops on this line and this direction. This is a straightforward yet effective solution to this problem. AVL data, however, do not need to be much processed since the information is already organized based on stop sequences of trips. If there is a missing record, a "Not A Number" label is added.

Step 2: Matching Trips in GTFS and AVL

This step is dedicated to matching all the trips recorded in the AVL data set to all the scheduled trips contained in the GTFS data. Ideally, the two data sources should share the same trip ID indexing scheme so that the matching is very straightforward. However, inconsistencies do exist, as illustrated in the previous section. To address this issue, the recorded arrival time as well as the delay at stops from the AVL data are used to compute the probable scheduled arrival times as follows:

$$\tilde{\pi}^s_{n,k} = \pi^r_{n,k} - d^r_{n,k} \tag{1}$$

where $\tilde{\pi}^s_{n,k}$, and $\pi^r_{n,k}$, respectively, denote the probable scheduled and recorded arrival times of trip n at stop k. $d^r_{n,k}$ denotes the recorded delay of trip n at stop k from the AVL data. The scheduled trip that has the closest arrival time at a stop to this "estimated" scheduled arrival time $\tilde{\pi}^s_{n,k}$ is then found, and its trip ID from the GTFS data is temporarily labeled to this stop visit. After applying this process to all the stops of this trip, the GTFS trip ID that has been most frequently labeled is adopted and assigned to the entire trip. The matching of the recorded trips (AVL) to scheduled trips (GTFS) is performed so that those trips that cannot be found in the AVL data will be later checked to assess whether they were really executed by taking the AFC data into consideration. In addition, headways based on AVL and GTFS are also computed and added to the trajectory profiles at the end of this step.

Step 3: Matching Passenger Rides to Vehicle Trajectories

The objective of this step is to match all individual passenger rides to the vehicle trips that these passengers traveled with. However, since the trip ID information is missing in the current AFC data set, a trip ID inference algorithm for all the rides is first developed, as shown in Figure 4.

Let $t^{\text{in}}_{i,k}$ denote the check-in time of passenger i at stop k. Let $\pi^r_{n,k}$ and $\pi^s_{n,k}$, respectively, denote the recorded and scheduled arrival times of trip n at stop k. Essentially, the algorithm attempts to find the trip ID for a single ride by the ith passenger so that his or her check-in time at the stop k, $t^{\text{in}}_{i,k}$, is closest to the vehicle arrival time $\pi^r_{n,k}$ and $\pi^s_{n,k}$ at the very same stop k along the trip n. Recorded arrival times π^r are used as the major benchmark because delays can introduce a significant bias when performing such inference. If the following condition is satisfied, then this ride was labeled with the trip ID of trip n:

$$\pi^r_{n,k} - \varepsilon^- < t^{\text{in}}_{i,k} \leq \pi^r_{n,k} + \varepsilon^+ \tag{2}$$

where ε^- and ε^+, respectively, represent the lower and upper bounds of the searching time window. In this case study, ε^- and ε^+ were empirically set to be 20 and 50 seconds, respectively, after scrutinizing the data. When the recorded arrival time of a trip at this stop π^r, nevertheless, is missing or has multiple values, the scheduled arrival time π^s is then employed with a larger time window as follows:

$$\pi^s_{n,k} - \frac{h^s_{n,k}}{2} < t^{\text{in}}_{i,k} \leq \pi^s_{n,k} + \frac{h^s_{n+1,k}}{2} \tag{3}$$

where $h^s_{n,k}$ denotes the scheduled headway between the current trip n and previous trip n–1 at stop k. Finally, if

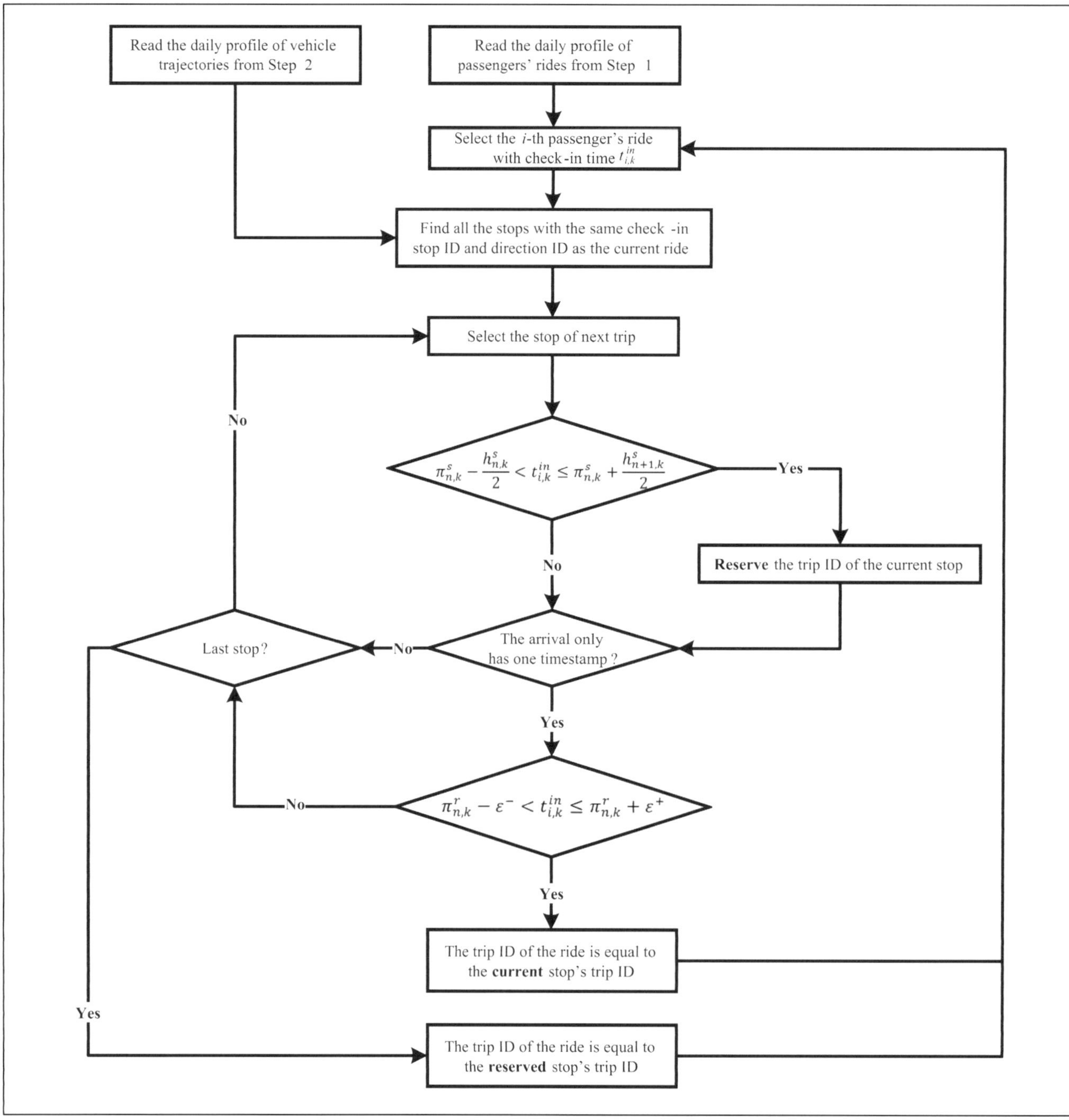

Figure 4. Algorithm for infering the trip ID of individual passenger rides.

the inference based on the recorded arrival time does not yield a result, the result based on the scheduled arrival time will be adopted.

Step 4: Improving Vehicle Trajectories

The last step is dedicated to the correction of vehicle trajectory profiles based on the results from step 3. It attempts to resolve the issues identified in the AVL data (#3, #4, and #8). The central idea is that the integrated information from three different data sources at the same detail level, including the recorded (AVL) and scheduled (GTFS) arrival/departure times, as well as check-in/out times at stops (AFC), will allow (1) whether a vehicle trip was indeed executed to be inferred, and (2) actual vehicle trajectories to be restored to the maximal extent. In the current study, practical solutions consisting of several rules were developed and applied.

To infer whether a trip is executed or canceled in reality, a series of rules are proposed based on the practical investigation. The inference takes into account (1) whether there is sufficient information about recorded timestamps for a trip; and (2) whether there are enough check-in activities that can be reasonably associated with this trip. Eventually, all scheduled trips are labeled as either "executed" or "canceled."

The restoration of vehicle trajectories is performed using a practical approach which attempts to fill in all the gaps in the recorded trajectories (exemplified in Figure 2*a*). Two rules and one assumption are made in this process:

- **Rule 1**: The arrival time cannot be later than the departure time at a given stop.
- **Rule 2**: The arrival time at any given stop cannot be earlier than the departure time from the last stop.
- **Assumption 1**: When only a timestamp for the departure is available, the arrival time should be equal to the departure time because the vehicle presumably did not have to serve this stop.

By applying these rules and this assumption, it is ensured that the vehicle trajectory of a trip is globally consistent in a sense that vehicles can never move backward. The assumption is made based on the practical investigation into the data. As Figure 2*c* illustrates, when there is a gap in the recorded trajectory, very few check-in/out activities can be spotted. Therefore, it can be safely assumed that vehicles skipped the stops when only departure times are logged in the database.

Implementation

All three raw data sets are stored in a PostgreSQL 9.3 database. A series of indices on date, line ID, stop ID, and so forth, were created to improve the SQL query performance. All of the abovementioned steps were coded in MATLAB.

Results

For simplicity and tractability, only the results of tram line 1 for the entire month of March 2015 are described in this section. Line 1 connects Delft, a mid-size old university city, to The Hague, the main city in its urban agglomeration, serving 41 stops per direction, including three major train stations. The service is frequent, with 208 trips on a normal weekday and up to 8 trams per hour in the peak on each direction. More than 670,000 passenger rides are recorded for line 1 in the AFC database over the case study period. After applying the pre-processing to

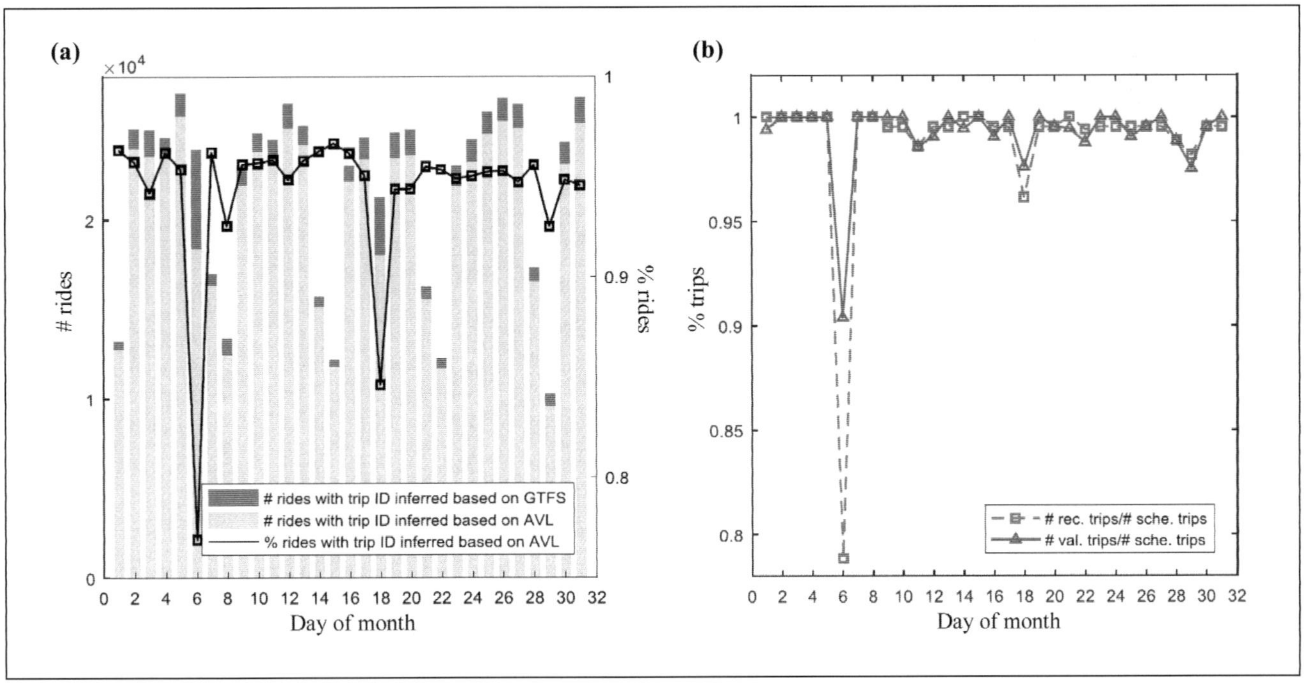

Figure 5. Results for trip ID inference of rides and trip validation. (*a*) Illustration of trip ID inference for rides that are based on recorded trajectories (AVL) and scheduled trajectories (GTFS), respectively. The line shows the percentage of rides of which trip IDs are inferred based on the recorded trajectories (AVL); (*b*) Comparison of numbers of scheduled trips, recorded trips and validated trips.

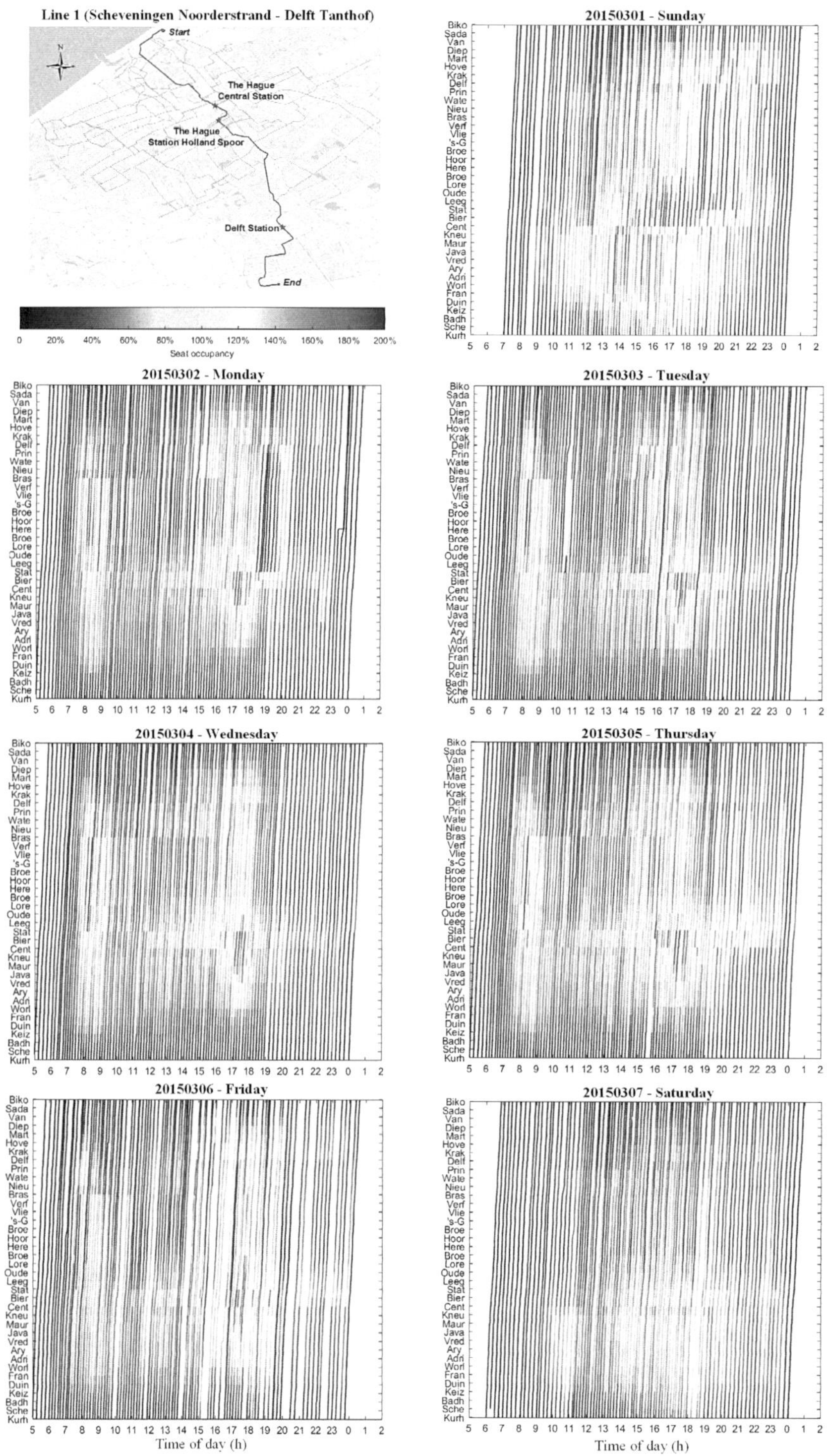

Figure 6. Illustrations of spatiotemporal seat occupancy of line 1 from Scheveningen Noorderstrand to Delft Tanthof over the first week of March 2015.

these rides, around 0.03% (around 200) rides were discarded because of the issues described earlier.

Figure 5*a* shows the result of vehicle trip ID inference for passenger rides. For most of the days, over 90% of the rides' trip ID can be successfully inferred based on the recorded trajectories owning to the relatively high quality of AVL data on these days. Two particular days, that is, the 6th and 18th of March, are noticeable because of their low AVL-based inference percentage. This is arguably caused as a result of the significant loss of AVL data on those two days, which implies that many inferences rely instead on just the scheduled trajectories.

The percentages of recorded trips in the AVL data and the executed trips that are eventually validated compared with the total number of scheduled trips are calculated and displayed in Figure 5*b*. Both directions are considered at the same time. It can be seen that on many days, not all scheduled trips were executed. A significant plunge appears on the 6th of March, where fewer than 80% of scheduled trips are recorded in the AVL data, whereas over 90% trips could be validated when also using information from check-in/out activities on that day. In addition, the 18th and 29th also yield results that are clearly worse than the average. As a result of inconsistent and incomplete trajectories from the AVL data, trip ID inference of rides on these days become more unreliable, resulting in a stronger reliance on GTFS-based inference as shown in Figure 5*a* for March 18th and 29th.

The final output profiles are visualized by plotting so-called space-time vehicle seat occupancy graphs (Figure 6). These seat occupancy graphs relate the vehicle occupancy to the seating capacity. In the color schemes in Figure 6, 100% occupancy means that the number of passengers on board is equal to the number of seats (76 for all the vehicles running on line 1). Significant crowdedness is thus easily identified when the seat occupancy is higher than 100% (warmer color). The upper bound is set to 200%, corresponding to the maximal vehicle capacity (around 150 people). It should be emphasized that this visualization technique has great potential for decision support in transit planning and operations, ranging from timetable optimization, network and fleet scheduling, and designing sub-services running over partial lines, to name just a few. The graphs provide a single, information-rich, and intuitive global view of service quality.

Although detailed analyses are beyond the scope of this exploratory study, example visualizations presented in Figure 6 (line 1 from Scheveningen Noorderstrand to Delft Tanthof over the first week of March 2015; March 1st is a Sunday) allow for identification of crowdedness over space and time. Overall, similar crowding patterns can be recognized on weekdays with clearly visible morning and afternoon peak-hour flows, while there are more variations during weekends. For instance, severe on-board crowding always occurs from Station Hollands Spoor to Badhuiskade in the morning, and from Vredespaleis to Station Hollands Spoor in the afternoon during this week.

Conclusions and Future Research

Obtaining on-board load profiles of transit vehicles has remained as a difficult task for operators in recent years as a result of technical and financial constraints. In this paper, a new methodology for constructing such profiles with multiple transit data sources is presented, including AFC, AVL, and GTFS. Difficulties of utilizing these data are discussed with the issues arising from a single or a combination of data sets specifically identified. The methodology consists of four steps through which the raw information from individual data sources is processed and corrected. The output profiles can convey integrated information regarding both vehicle trajectories and passenger demand on a large spatiotemporal scale. The methodology is demonstrated with data collected from the urban transit system in The Hague, The Netherlands. A key output is so-called space-time seat occupancy graphs, which provides operators with a compact and powerful reference to intuitively examine the on-board crowding patterns over time and space, thus helping to improve their services, such as timetable optimization, network and fleet scheduling, designing sub-services running over partial lines, and so forth.

The contribution of this study is twofold. First, the study aimed to integrate different transit data sources for obtaining state estimations for passenger loads. In this process, the issues related to each and the combination of different data sets, namely AFC, AVL, and GTFS, are specifically identified. Although based on data available in the Dutch context, most of their properties are universal, and this way of presenting all the issues can be beneficial for researchers and practitioners with different data formats but similar difficulties. Second, a methodology that solves these issues in a sequential manner is described and yields service profiles containing both vehicle trajectories and passenger loads. The complexity of approaches and algorithms in each step can vary depending on the availability of information.

On the basis of the current study, several research directions can be further explored. For example, more advanced techniques for correcting vehicle trajectories while assuring global coherence can be developed to replace the current rule-based approach. In addition, the resulting data set can significantly benefit network-wide transit flow analysis to deepen the understanding of transit system dynamics.

Acknowledgments

The authors thank HTM and Stichting OpenGeo for providing the AFC and AVL data sets, respectively. We also acknowledge the support of the SETA project funded by the European Union's Horizon 2020 research and innovation program.

Author Contributions

The authors confirm contribution to the paper as follows: study conception and design: Ding Luo, Oded Cats, Hans van Lint and Loïc Bonnetain; data collection: Ding Luo; analysis and interpretation of results: Ding Luo, Loïc Bonnetain and Oded Cats; draft manuscript preparation: Ding Luo, Oded Cats, and Hans van Lint. All authors reviewed the results and approved the final version of the manuscript.

References

1. Moreira-Matias, L., J. Mendes-Moreira, J. F. de Sousa, and J. Gama. Improving Mass Transit Operations by Using AVL-Based Systems: A Survey. *IEEE Transactions on Intelligent Transportation Systems*, Vol. 16, No. 4, 2015, pp. 1636–1653.
2. Pelletier, M.-P., M. Trépanier, and C. Morency. Smart Card Data Use in Public Transit: A Literature Review. *Transportation Research Part C: Emerging Technologies*, Vol. 19, No. 4, 2011, pp. 557–568.
3. Utsunomiya, M., J. Attanucci, and N. Wilson. Potential Uses of Transit Smart Card Registration and Transaction Data to Improve Transit Planning. *Transportation Research Record: Journal of the Transportation Research Board*, 2006. 1971: 119–126.
4. Google. *GTFS Static Overview*. https://developers.google.com/transit/gtfs/.
5. Wong, J. Leveraging the General Transit Feed Specification for Efficient Transit Analysis. *Transportation Research Record: Journal of the Transportation Research Board*, 2013. 2338: 11–19.
6. Alfred Chu, K., and R. Chapleau. Enriching Archived Smart Card Transaction Data for Transit Demand Modeling. *Transportation Research Record: Journal of the Transportation Research Board*, 2008. 2063: 63–72.
7. Sun, L., D.-H. Lee, A. Erath, and X. Huang. Using Smart Card Data to Extract Passenger's Spatio-temporal Density and Train's Trajectory of MRT System. *Proc., ACM SIGKDD International Workshop on Urban Computing*, ACM, New York, NY, 2012, pp. 142–148.
8. Moreira-Matias, L., and O. Cats. Toward a Demand Estimation Model Based on Automated Vehicle Location. *Transportation Research Record: Journal of the Transportation Research Board*, 2016. 2544: 141–149.
9. Anwar, A., A. Odoni, and N. Toh. BusViz: Big Data for Bus Fleets. *Transportation Research Record: Journal of the Transportation Research Board*, 2016. 2544: 102–109.
10. Min, Y.-H., S.-J. Ko, K. M. Kim, and S.-P. Hong. Mining Missing Train Logs from Smart Card Data. *Transportation Research Part C: Emerging Technologies*, Vol. 63, 2016, pp. 170–181.
11. Zhou, Y., L. Yao, Y. Chen, Y. Gong, and J. Lai. Bus Arrival Time Calculation Model Based on Smart Card Data. *Transportation Research Part C: Emerging Technologies*, Vol. 74, 2017, pp. 81–96.
12. Gordon, J., H. Koutsopoulos, N. Wilson, and J. Attanucci. Automated Inference of Linked Transit Journeys in London Using Fare-transaction and Vehicle Location Data. *Transportation Research Record: Journal of the Transportation Research Board*, 2013. 2343: 17–24.
13. Nassir, N., A. Khani, S. Lee, H. Noh, and M. Hickman. Transit Stop-level Origin-Destination Estimation Through Use of Transit Schedule and Automated Data Collection System. *Transportation Research Record: Journal of the Transportation Research Board*, 2011. 2263: 140–150.
14. Ma, Z.-L., L. Ferreira, M. Mesbah, and A. T. Hojati. Modeling Bus Travel Time Reliability with Supply and Demand Data from Automatic Vehicle Location and Smart Card Systems. *Transportation Research Record: Journal of the Transportation Research Board*, 2015. 2533: 17–27.
15. Ma, Z., S. Zhu, H. N. Koutsopoulos, and L. Ferreira. Quantile Regression Analysis of Transit Travel Time Reliability with Automatic Vehicle Location and Farecard Data. *Transportation Research Record: Journal of the Transportation Research Board*, 2017. 2652: 19–29.
16. Zhao, J., F. Zhang, L. Tu, C. Xu, D. Shen, C. Tian, X.-Y. Li, and Z. Li. Estimation of Passenger Route Choice Pattern Using Smart Card Data for Complex Metro Systems. *IEEE Transactions on Intelligent Transportation Systems*, Vol. 18, No. 4, 2017, pp. 790–801.
17. Hörcher, D., D. J. Graham, and R. J. Anderson. Crowding Cost Estimation with Large Scale Smart Card and Vehicle Location Data. *Transportation Research Part B: Methodological*, Vol. 95, 2017, pp. 105–125.
18. Kusakabe, T., T. Iryo, and Y. Asakura. Estimation Method for Railway Passengers' Train Choice Behavior with Smart Card Transaction Data. *Transportation*, Vol. 37, No. 5, 2010, pp. 731–749.
19. Zhu, Y., H. N. Koutsopoulos, and N. H. M. Wilson. A Probabilistic Passenger-to-Train Assignment Model Based on Automated Data. *Transportation Research Part B: Methodological*, Vol. 104, No. C, 2017, pp. 522–542.
20. Robinson, S., B. Narayanan, N. Toh, and F. Pereira. Methods for Pre-processing Smartcard Data to Improve Data Quality. *Transportation Research Part C: Emerging Technologies*, Vol. 49, 2014, pp. 43–58.
21. Luo, D., O. Cats, and H. van Lint. Constructing Transit Origin–Destination Matrices with Spatial Clustering. *Transportation Research Record: Journal of the Transportation Research Board*, 2017. 2652: 39–49.
22. Van Oort, N., T. Brands, and E. de Romph. Short-Term Prediction of Ridership on Public Transport with Smart Card Data. *Transportation Research Record: Journal of the Transportation Research Board*, 2015. 2535: 105–111.
23. Van Oort, N., D. Sparing, T. Brands, and R. M. Goverde. Data Driven Improvements in Public Transport: The Dutch Example. *Public Transport*, Vol. 7, No. 3, 2015, pp. 369–389.
24. OVapi. http://gtfs.ovapi.nl/. Accessed July 1, 2018.

The Public Transportation Group (AP000) peer-reviewed this paper (18-02399).

Article

TRR
JOURNAL OF THE TRANSPORTATION RESEARCH BOARD

Transportation Research Record
2018, Vol. 2672(8) 187–198
© National Academy of Sciences:
Transportation Research Board 2018
Article reuse guidelines:
sagepub.com/journals-permissions
DOI: 10.1177/0361198118793273
journals.sagepub.com/home/trr
SAGE

A Study of Bus High-Resolution GPS Speed Data Accuracy

Miguel A. Figliozzi[1] and Nicholas B. Stoll[1]

Abstract

The recent availability of high-frequency transit (HFT) GPS data for buses has allowed the estimation of detailed bus-travel speed profiles between bus stops. HFT data are defined as data comprising GPS vehicle trajectory recorded at or less than 5 s intervals. With HFT data it is now possible to measure changes in bus speed at specific locations of interest, such as intersections, ramps, crosswalks, and so on. Previous research efforts have not compared GPS-based bus speeds with general traffic speeds at specific locations between stops. This research fills this knowledge gap, utilizing accurate stationary radar speed data as a baseline or ground truth data to estimate GPS-based bus speed accuracy. A thorough data analysis of the bus and traffic speed data indicates that HFT speed estimations between stops are accurate and highly correlated with traffic speeds. Time–space speed profiles and regression analysis are utilized to quantify factors that affect HFT speed estimation accuracy. The relative advantages and limitations of the HFT data are presented and discussed. This study concludes that large HFT datasets can be utilized to accurately monitor speeds between transit stops. HFT datasets are suitable to cost-effectively monitor recurrent arterial speed performance for both passenger and transit vehicles.

There are several well-established technologies for collecting speed and travel time data, including vehicle identification sensors—like Bluetooth readers, radar devices, loop detectors, and probe car data. Loop detectors and vehicle identification sensors, once installed, continuously record most (or a share of in the case of Bluetooth readers) vehicles passing specific road sections. Bluetooth readers sample a fraction of the traffic and can estimate average speeds between two readers, but cannot produce speed profiles between the reader locations. The density of Bluetooth, radar, or loop detectors is typically very low on most arterials and most non-freeway network links do not have speed profile data.

This research utilizes high-frequency transit (HFT) bus (probe) vehicle data to estimate arterial street speed and produce speed profiles along a designated arterial. The GPS data are denoted as high frequency when GPS points are obtained at or less than 5 s intervals; the vast majority of transit agencies in the USA are currently storing 30 s interval GPS data or not storing any GPS data outside transit stops. Dedicated probe vehicles can be utilized to collect travel time and other data for designated routes in the network. However, because of cost considerations, government agencies cannot run probe vehicles continuously; moreover, privacy concerns make it problematic to gather data from private vehicles. Private data vendors can provide crowdsourced probe vehicle data, but these datasets can be expensive, and in many cases the data are already aggregated or preprocessed.

The recent availability of HFT data for buses has allowed the estimation of bus-travel speed profiles between bus stops. HFT data are defined as data comprising GPS vehicle trajectory records at or less than 5 s intervals. With HFT data it is now possible to measure changes in bus speed at specific locations such as intersections, ramps, crosswalks, and so on. No previous research efforts have compared GPS-based bus speeds with general traffic speeds at specific locations between stops. This research fills this knowledge gap. This research utilizes accurate stationary-radar speed data as a baseline or ground truth data to estimate GPS-based bus speed accuracy. More specifically, this research is original because it aims to answer two novel research questions: (1) how accurate are the speed measurements obtained utilizing HFT data when compared to state-of-the-art stationary speed sensors; and (2) what are the factors likely to affect the accuracy of HFT speed estimations at specific locations between bus stops?

[1]Transportation Technology and People (TTP) Lab, Department of Civil and Environmental Engineering, Portland State University, Portland, OR

Corresponding Author:
Address correspondence to Miguel A. Figliozzi: figliozzi@pdx.edu

To analyze the accuracy of the HFT speed data, exploratory data analysis, correlations, time–space speed profiles, and regression analysis are utilized. The final sections of this study analyze the impact of transit vehicle frequency and accuracy, as well as advantages and limitations to the HFT speed data. The next section provides a brief literature review.

Literature Review

Previous research efforts with GPS vehicle data have tried to utilize data publicly available from taxi fleets or ambulances; these efforts have focused on estimating speeds between segments or at the network level. Traffic speeds from taxis or ambulances are not calculated at a specific location because of the irregularity of most taxi or ambulance trips. Researchers have used these data from ambulances (*1*) or taxi fleets (*2, 3*) to estimate hourly average link or path travel times. Other researchers have utilized private data sources from crowdsourced mobile phone data (*4*) or Bluetooth sensors that read and track devices carried via vehicles or people, such as smartphones (*5*). Other researchers have focused on datasets that are sparse concerning low-frequency GPS sampling but use many probe vehicles (*3, 6*).

Unlike previous research efforts, the HFT data analyzed in this paper are sparse concerning the number of probe (bus) vehicles per hour, per link—the flow of transit vehicles per hour—instead of sparse concerning the GPS sampling frequency. Research indicates that probe vehicle flow rates significantly affect travel time estimation accuracy. Results show that a significant reduction in speed estimation variance is observed when the number of observations per hour increases from 12 to 36 probe vehicles per hour. Minor gains in estimation accuracy are obtained beyond 48–60 probe vehicles per hour (*7*).

Using public buses as probe vehicles *between bus stops* has been studied in the past (*8, 9*). These early research efforts revealed that when automobiles experience long delays, buses on the same facility are also likely to be delayed but the reverse relationship is not always true, such as when buses dwell at stops because they are ahead of schedule. Previous research efforts in the Portland region have utilized stop-to-stop bus-travel data to assess arterial performance and transit performance (*10*). However, all of these studies (*8–10*) were severely limited by the lack of GPS coordinates between bus stops; the datasets utilized in these studies (*8–10*) were sparse concerning both probe vehicle flow rates and GPS sampling frequency.

The recent availability of 5 s GPS data for buses has removed much of the guesswork involved in estimating bus-travel speed profiles between bus stops (*11*); it is now possible to measure *relative* changes in bus speed at intersections, ramps, crosswalks, and so on (*11*). "Relative" is emphasized because previous research efforts have not compared the actual speeds of transit vehicles and general traffic. For example, Figure 1 shows a time–space

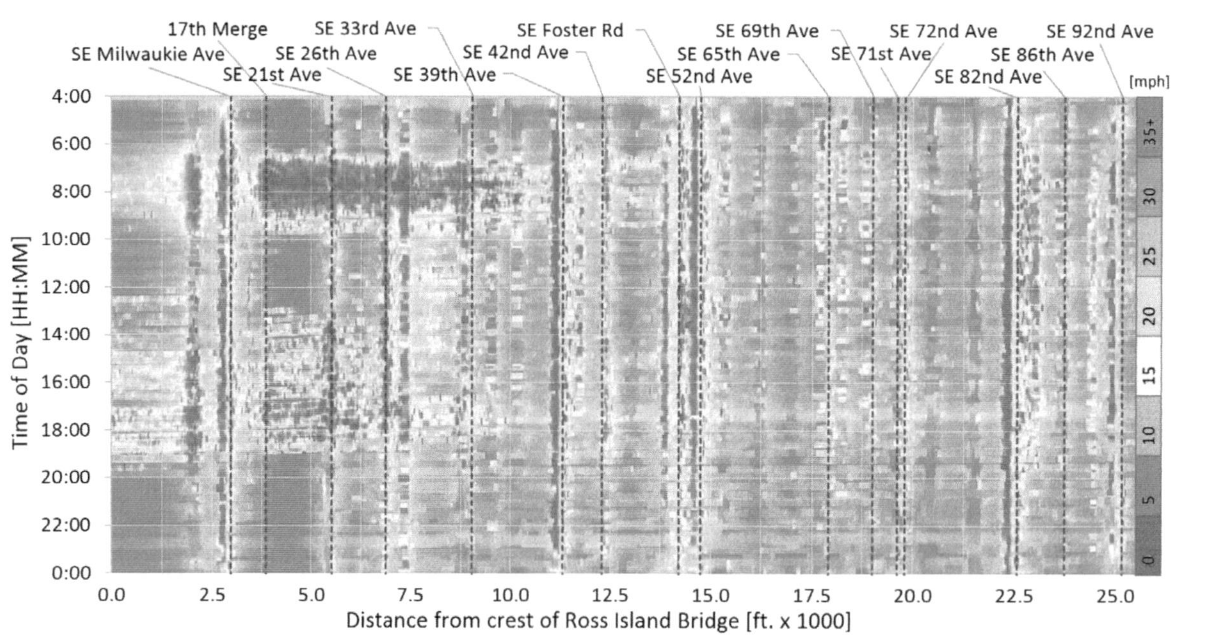

Figure 1. Space–time speed heatmap without dwell times, travel left-to-right, adapted from Glick and Figliozzi (*11*).

speed diagram constructed utilizing HFT data. In Figure 1, signalized intersections are labeled and it is possible to observe delays at some busy intersections like SE 39th or SE 82nd, which are major four-lane arterials. It is also possible to observe peak-hour morning congestion between SE 17th and SE 39th (caused by a bottleneck at a congested downstream bridge access). The most significant limitation of the utilized HFT speed data is that buses do not reflect traffic speeds while servicing customers at a bus stop. However, a recent publication (*11*) showed that *between stops* buses can be utilized to detect recurrent congestion and analyze traffic trends; dwell times or stop service times were removed to create Figure 1—see Glick and Figliozzi for more details (*11*).

Past research has compared speed estimates for link travel times (*9*) but not speeds at a specific location using stationary sensors as a baseline or as ground truth data. No previous research efforts have compared HFT speeds with stationary-speed-sensor data. This research fills this knowledge gap, utilizing Wavetronix Digital Wave Radar (DWR) sensor data as a baseline or ground truth data by measuring on location and with high accuracy. Mid-block Wavetronix sensors are very accurate when properly installed, measuring speeds and volumes with a 1–2% error range (*12–14*) or a spread of ± 1–2 mph when compared against speed guns (*14, 15*).

HFT bus speed data can even detect bus speed changes when approaching intersections, as well as bus speed changes that are caused by on-ramps and crosswalks (*16*). This research also takes advantage of recent algorithms and data processing strategies to estimate speed percentiles for spot speeds at specific locations (*11*) that may remove the bias that is generated by bus stops (*17*) or traffic signal stops because the buses cannot take advantage of signal progression (*18*). These improvements are key to eliminating the differences between transit and private vehicle speeds by utilizing only transit data from vehicles that are not stopped by an upstream traffic signal and intermediate transit stop. Better and more accurate transit data can then be utilized to visualize transit operations (*19*), transit delay (*20*), and forecast travel times (*21*).

Available HFT Data and Study Location

The HFT data utilized in this research are provided by TriMet, the public transportation agency in the Portland region. On most arterials outside the downtown transit mall, the TriMet HFT data are relatively *sparse* because the flow of vehicles (buses) is, at most, 6–10 vehicles per hour in the majority of arterials and collector streets outside downtown Portland. However, the data have high frequency because GPS data are recorded in 5 s intervals. Reliable data are available for over 220 weekdays per year if holidays and abnormal weeks are removed from the dataset. Therefore, millions of GPS coordinates can be easily obtained over a 6 month period. Other advantages of TriMet's dataset include: (1) the data are freely available; (2) buses travel across congested areas and cover all major arterials and collector streets in the region (even some freeways for express routes); (3) bus data are linked to a physical sensor in the vehicle (measuring wheel motion), making it possible to detect congestion with more accuracy than with GPS data alone; (4) bus stop activity is recorded and GPS data around bus stops can be removed if necessary; and (5) unlike taxis or ambulances, bus-travel paths tend to be extremely regular.

The study location was chosen based on three factors: (1) availability of high-quality Wavetronix data; (2) availability of HFT data; and (3) familiarity with the area. The chosen location, southeast Powell Boulevard, meets the three criteria. The accuracy of the two Wavetronix sensors was previously analyzed as part of a research effort along the same corridor (*22*). Nonetheless, as part of this research effort more recent Wavetronix speed–volume data were reanalyzed to evaluate accuracy of the sensors: no problems or errors were found. Along SE Powell runs the TriMet frequent-service route 9, with approximately 10 min headways during peak hours and 15 min headways during off-peak hours. The researchers are thoroughly familiar with the research area and traffic patterns. This roadway is part of an instrumented corridor where the research team has recently studied the performance of the adaptive traffic signal system SCATS (*22*).

Along SE Powell Boulevard, two Wavetronix sensors are located mid-block near cross streets southeast 24th Avenue and southeast 35th Avenue. The locations of these Wavetronix sensors were chosen to ensure high-quality data and also to best capture free-flow traffic during off-peak hours and severe queuing during peak hours (*22*). At SE Powell and 24th, the distance to the nearest signalized intersections is approximately 655 ft (200 m); at SE Powell and 35th, the distance to the nearest signalized intersections is approximately 1050 ft (320 m). At SE Powell Boulevard and 24th, a Wavetronix sensor measures speed and volumes across a section comprising two through westbound lanes, two through eastbound lanes, and a middle left-turn lane; a similar configuration is present at the intersection of SE Powell Boulevard and 35th. The distances between bus stops vary per location and direction of travel. Distances between the Wavetronix sensors and the bus stops (in feet) are included in Figure 2. Details for each location and travel direction are summarized below:

- Eastbound 24th: The Wavetronix sensor is approximately 50 ft (15.3 m) upstream of the

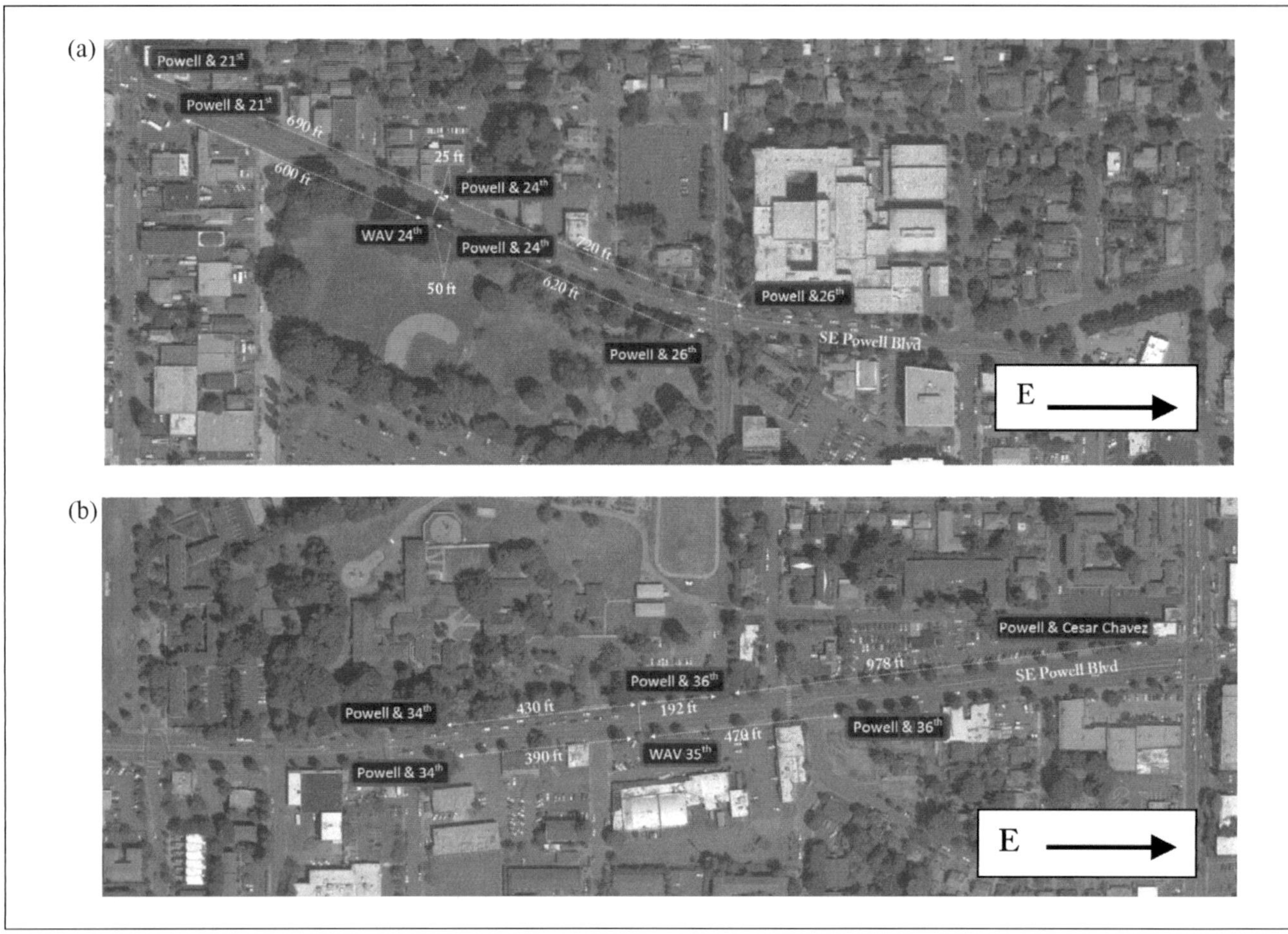

Figure 2. Sensor and bus stop locations: (*a*) SE Powell and 24th, distance between Wavetronix sensor (red) and bus stops (blue); (*b*) SE Powell and 35th, distance between Wavetronix sensor (red) and bus stops (blue).

nearest bus stop (stop ID 4625) and over 600 ft (183 m) downstream of the SE Powell Boulevard and 21st bus stop (stop ID 4622). It is also within close proximity of a crosswalk. At the SE Powell Boulevard and 24th bus stop (stop ID 4625), less than 20% of bus trips service this stop; conversely, more than 80% of bus trips service the stop at the SE Powell Boulevard and 21st (stop ID 4625).

- Westbound 24th: The nearest bus stop is roughly 25 ft (7.5 m) upstream of the sensor (stop ID 4626), but this bus stop is not heavily serviced; roughly 30% of bus trips service this stop.
- Eastbound 35th: There are no bus stops within close proximity to the Wavetronix sensor. The closest upstream bus stop is located 350 ft away at SE Powell Boulevard and 34th (stop ID 4647).
- Westbound 35th: The upstream bus stop is located 190 ft upstream of the sensor, at SE Powell and 36th (stop ID 4649); roughly 50% of the buses service this stop.

Data Processing

TriMet supplied four different datasets: cyclic data, stop event data, stop data, and block data. Cyclic data are recorded at 5 s intervals. This is the probe (bus) vehicle GPS data denominated as HFT data or 5 s resolution data (5-SR data); see Table 1 for a small data sample. The first column, trip number, is used to identify a bus trip. The second column, stop number, is updated whenever the bus services a stop. For example, in Table 1 the bus serviced four different stops. The third and fourth columns record odometer distance in meters and time after midnight in seconds, respectively. The fifth and sixth columns record the latitude and longitude of the GPS locations, respectively.

The stop event and stop datasets contain information on instances where the bus stops—for example, type of stop, dwell time, estimated bus load, boardings, alightings, and so on. Stop event and stop data can be

Table 1. Sample of HFT or 5-SR Dataset

Trip number	Stop number	Distance	Time	Longitude	Latitude
247917030	247917070	65169	30088	45.49713	−122.537
247917030	247917070	65210	30093	45.49712	−122.537
247917030	247917070	65252	30098	45.49711	−122.538
247917030	247917071	65313	30123	45.49707	−122.539
247917030	247917071	65375	30128	45.49704	−122.539
247917030	247917071	65453	30133	45.497	−122.54
247917030	247917072	65531	30138	45.49697	−122.541
247917030	247917072	65608	30143	45.49694	−122.542
247917030	247917072	65685	30148	45.49691	−122.543
247917030	247917073	65763	30153	45.49685	−122.544
247917030	247917073	65832	30158	45.49677	−122.545

combined with the 5-SR data for a given trip number and stop number, but require the bus block dataset to merge all the datasets. The Wavetronix sensors along SE Powell Boulevard record data in 10 s intervals. Wavetronix data provide traffic speed, volume, and occupancy data per lane, aggregated in consecutive 10 s intervals.

The data processing steps, at a conceptual level, can be succinctly described as four major steps:

1. Extract right lane Wavetronix data for the time intervals of interest (only right lane data because buses travel along the right lane on Powell Boulevard).
2. Tag all HFT GPS points *before* reaching and *after* passing a Wavetronix sensor for each direction of travel and location.
3. Match Wavetronix 10 s interval data to the closest *before* and *after* HFT GPS point timestamps.
4. HFT speeds were calculated employing a methodology described in a recently published article (*11*).

Exploratory Data Analysis

The literature review section indicated that stationary sensors, especially Wavetronix radar sensors, are considered to be highly accurate devices for measuring roadway speeds and traffic volumes. Here, Wavetronix data are considered ground truth data and are used as a baseline to measure the quality of HFT speed data.

Figure 3 compares Wavetronix and HFT speed profiles when data are aggregated in 15 min, 30 min, 45 min, and 60 min intervals. The *y*-axis indicates average travel speed at a given interval and the *x*-axis is time of day (TOD) between 5:00 a.m. and 22:00 p.m. It is possible to observe that Wavetronix and HFT speed profiles do not overlap at all times, but there seems to be a strong correlation. In almost all cases, HFT speeds are lower than general traffic speeds. A clear gap can be seen at the westbound 24th and westbound 35th locations.

Speed distribution kernel plots indicate that when HFT speeds are lower—for example, at the westbound 24th and westbound 35th locations—the speed distributions are clearly bimodal (see Figure 4). If Wavetronix and HFT speed data are plotted utilizing different colors for specific bus events, it is possible to find some noteworthy patterns. For example, Figure 4 shows the relationship between Wavetronix speeds and HFT 5-SR data, where colors are utilized to differentiate events when the bus door opens (i.e., services a bus stop) at the nearest bus stop downstream from the Wavetronix sensor. The small secondary peaks seen in blue at the westbound locations in Figure 4 are matched by a cloud of low-speed blue points (opened bus doors) that are shown in Figure 5.

Regression Analysis

To analytically identify the key variables affecting HFT speed quality, a regression analysis was undertaken. The following variables were included in the analysis:

- Dependent variable: Error or speed difference (i.e., Wavetronix speed minus HFT speed) for each bus trip. It is important to note that transit vehicle speed is estimated at the location of the Wavetronix sensor and that some locations are near heavily used bus stops, while other locations are farther away from bus stops.
- Independent variables: Variables associated with the vehicle and vehicle activity (e.g., bus load, whether an upstream or downstream stop was serviced, whether bus doors opened, stop service time, and so on), variables associated with traffic conditions (Wavetronix data for traffic speeds, volumes, sensor occupancy, vehicle classification, and so on), and variables related to the nearest GPS coordinates (distance to Wavetronix sensor, time of the day, and so on).

A regression analysis with these dependent and independent variables has not been performed before. To time-effectively find patterns, backward linear regression models were estimated utilizing all the regression variables that were not significantly correlated. A summary of the regression results per location and direction of travel are reported in Table 2. Only coefficients for significant variables are reported, and the coefficient's relative importance is also reported. Relative importance was measured by the contribution of each independent variable to explained variance. The R package "relaimpo" was utilized to estimate all statistics.

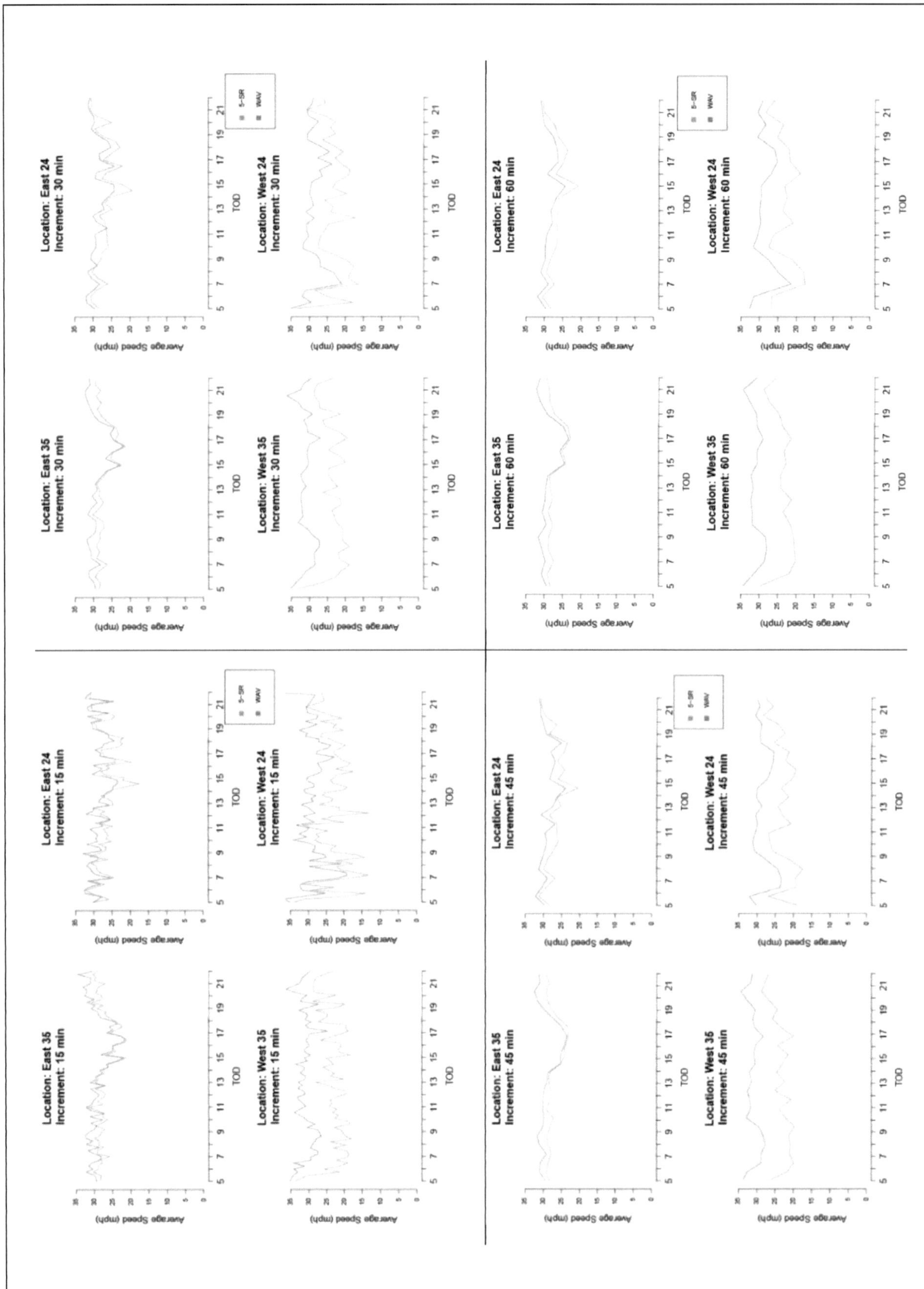

Figure 3. Comparison of average speed profiles by TOD.

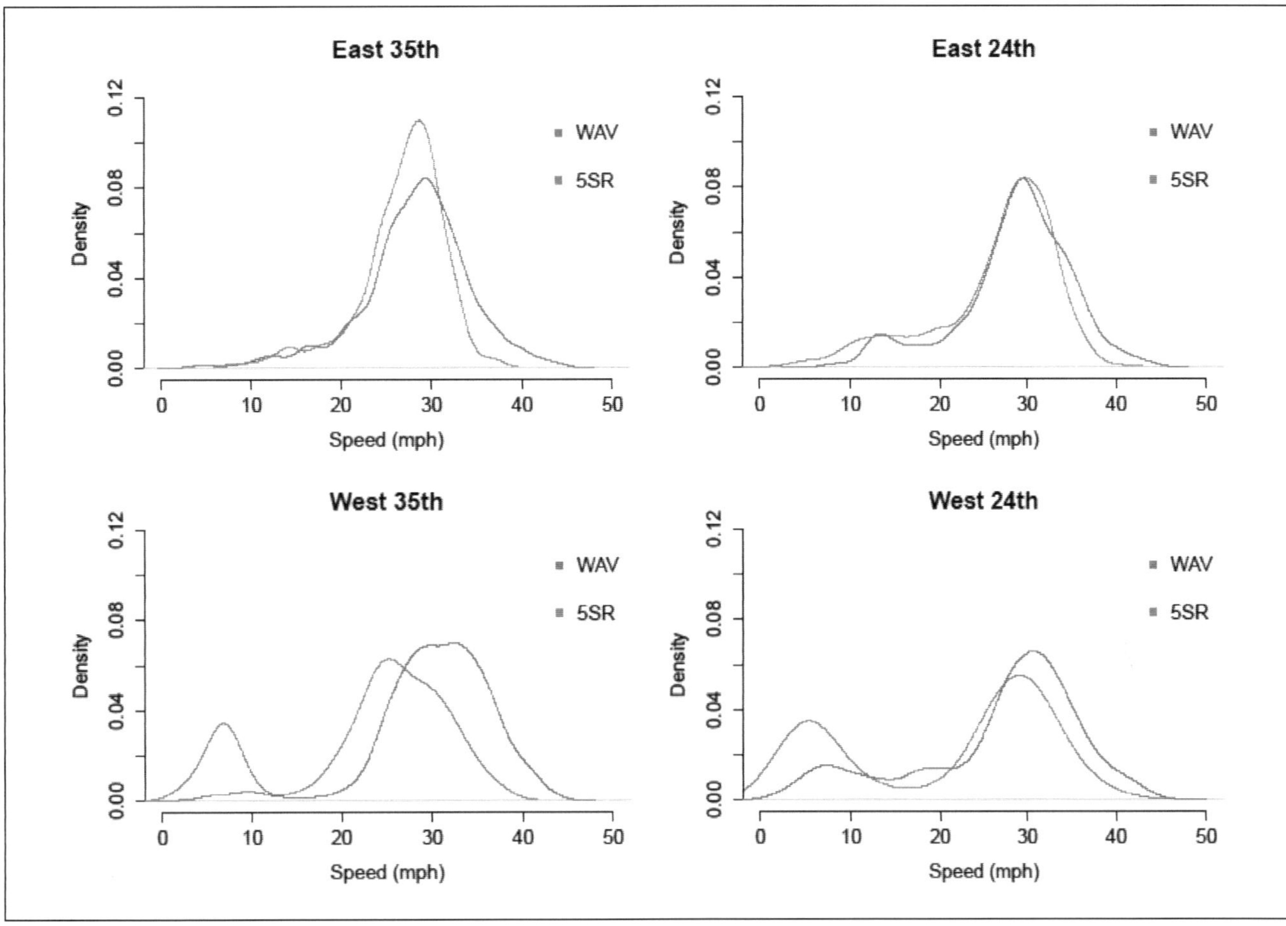

Figure 4. Kernel density plots.

The results of the regression analysis were highly encouraging: (1) All the models had relatively high adjusted R^2 values and (2) there was sign consistence among the key significant variables. The potential interpretation of the key variables is mostly intuitive. Across the board, Wavetronix speed was a highly significant variable, with high relative importance, and a positive sign. The Wavetronix speed variable indicates that, all things being equal, the higher the traffic speed, the larger the error or difference between Wavetronix and HFT speed. This is intuitive because buses are large and heavy vehicles with limited maneuverability and a slower acceleration profile than most passenger vehicles. The speed limit is 35 mph along this section of SE Powell.

The Wavetronix data also show that during free-flow conditions, a significant portion of regular vehicles travel at or faster than the 35 mph speed limit (see Figure 4); owing to their safety training and continuous video monitoring, bus drivers are more likely to obey speed limits than are regular drivers. When traffic conditions are congested, the difference between traffic and HFT speed decreases. This is favorable for the analyst who is interested in finding congestion in specific areas or TODs.

Important differences can also be observed in Table 2. If the bus door opened after passing the Wavetronix sensor, the error or speed difference increases significantly. This variable has a high relative importance only for westbound 24th and 35th; for eastbound 24th and 35th locations, the stop variables are less important. This finding is consistent with the plots shown in Figure 5 (see the difference between red and blue dots at some locations). The other variables—distance between GPS points and the Wavetronix sensors—have mostly negative values.

Error or speed difference tends to decrease as distance increases, holding all the other variables constant. This finding is intuitive because HFT speeds and distance between GPS points are positively correlated. The GPS data are recorded in 5 s intervals; the faster the bus travels, the greater the distance between GPS points. The locations near 24th Avenue have significantly lower R^2 values than the locations near 35th Avenue These bus stops are close to the Wavetronix sensor at 24th Avenue

Table 2. Regression Summary and Comparison

Variable	Westbound 24th		Eastbound 24th		Westbound 35th		Eastbound 35th	
	Coefficient	Relative importance	Coefficient	Relative importance	Coefficient	Relative importance	Coefficient	Relative importance
Intercept	-13.404^{***}	na	-8.189^{***}	na	-14.451^{***}	na	1.202^{***}	na
Sensor speed	0.774^{***}	26.6%	0.655^{***}	49.4%	0.814^{***}	27.1%	0.950^{***}	58.2%
Distance: sensor to point (before)	-0.664^{***}	12.1%	-0.745^{***}	31.7%	0.356^{***}	8.2%	-1.400^{***}	20.8%
Distance: sensor to point (after)	-0.056^{**}	3.9%	-0.141^{***}	6.2%	-0.617^{***}	20.2%	-1.399^{***}	20.9%
Bus door opened (before)	-2.199^{***}	0.9%	-3.158^{***}	2.1%	-7.659^{***}	2.2%	0.200^{**}	0.1%
Bus door opened (after)	15.638^{***}	56.6%	6.150^{***}	10.6%	13.105^{***}	42.3%	na	na
Adjusted R^2	0.687		0.519		0.734		0.914	

$^{*}p < 0.1$. $^{**}p < 0.05$. $^{***}p < 0.01$.

and this may explain the drop in adjusted R^2 values. It is likely that the available variables failed to capture the changes in errors or speed variability at 24th Avenue.

Traffic performance along an arterial is a complex spatial and temporal phenomenon. The regression results are useful but it is unlikely that a linear regression model can capture all the spatial and temporal subtleties of arterial traffic. Space–time speed profiles are typically more useful to describe traffic patterns, and these are analyzed in the next section.

Time–Speed Profiles and Speed Correlations

The time–speed profiles for each location and direction of travel are presented in Figures 6 and 7. The distances at the bottom of the graph are centered on the location of the Wavetronix sensor. At the bottom of each profile, a graph depicts estimated bus delay and the location of the Wavetronix sensor and nearby bus stops. It is possible to observe significant delays around heavily used bus stops. In addition, the speed profile for westbound 24th shows congestion in the morning hours (see Figure 6*b*); this is realistic as commuter traffic moves toward downtown Portland and generates morning queues at a downstream bridge that propagate all the way up to 24th Avenue and beyond. The same congestion appears in Figure 1 between 3 and 8 miles and between 7:00 and 9:00 a.m.

Table 3 shows correlations between Wavetronix speeds and HFT speeds. The top portion of the table shows the correlations when speeds are measured or estimated at the Wavetronix sensor. At eastbound 24th and eastbound 35th there is a high degree of correlation; at westbound 24th and westbound 35th the correlations are significantly lower, and this can be explained by the proximity of the Wavetronix sensor to heavily used nearby bus stops (see Figures 6 and 7 and regression analysis results). To test this hypothesis, HFT speeds were calculated 200 ft away from the sensor—see Figures 6*b* and 7*b* with a marker for "New Location." At westbound 24th (Figure 6*b*) and westbound 35th (Figure 6*a*) bus stop delays do not influence the HFT speed estimations. Table 3 shows that the correlations increased significantly at the new locations 200 ft away from the Wavetronix sensor at westbound 24th and westbound 35th.

However, moving the location of the HFT data speed estimation does not necessarily improve correlations in all cases. For example, at eastbound 24th (Figure 7*a*), if the HFT speed is estimated upstream, the correlation value decreases. In this example, the Wavetronix sensor is likely capturing delays from peak-hour afternoon queuing that originate at the downstream traffic signal (SE 26th) that are not captured when the HFT speed estimation location is moved 200 ft upstream. Figure 4 shows a slight bimodal distribution for the Wavetronix

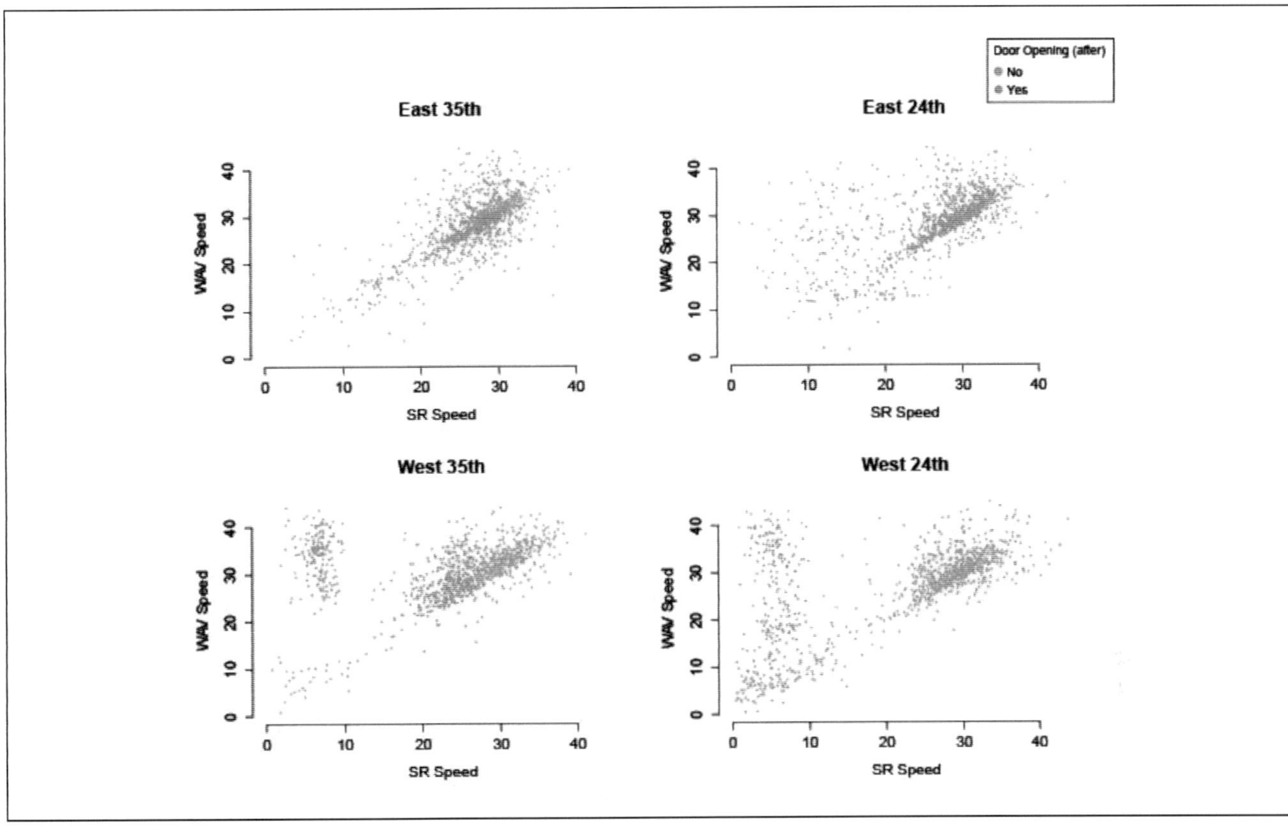

Figure 5. Wavetronix and 5-SR speed conditional on door opening (after point).

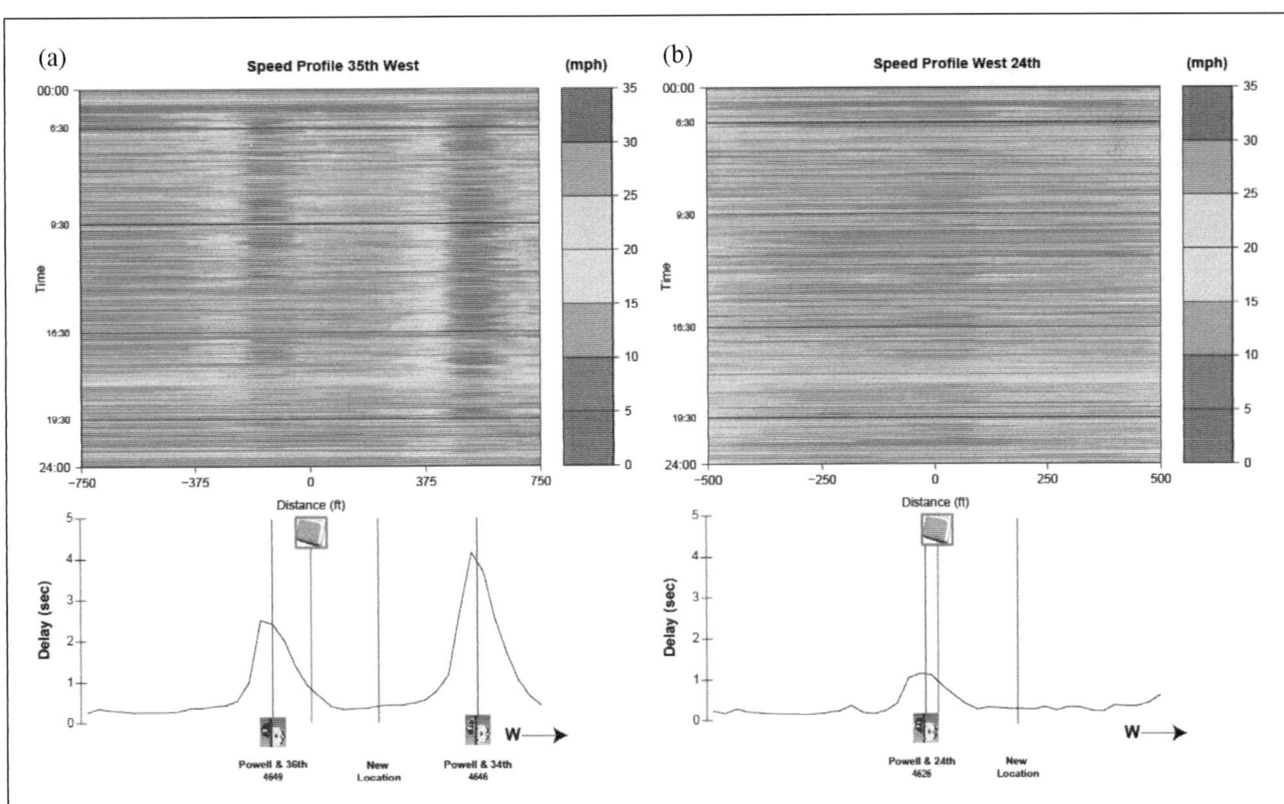

Figure 6. Speed heatmaps and sensor locations for westbound travel: (a) westbound 35th; (b) westbound 24th.

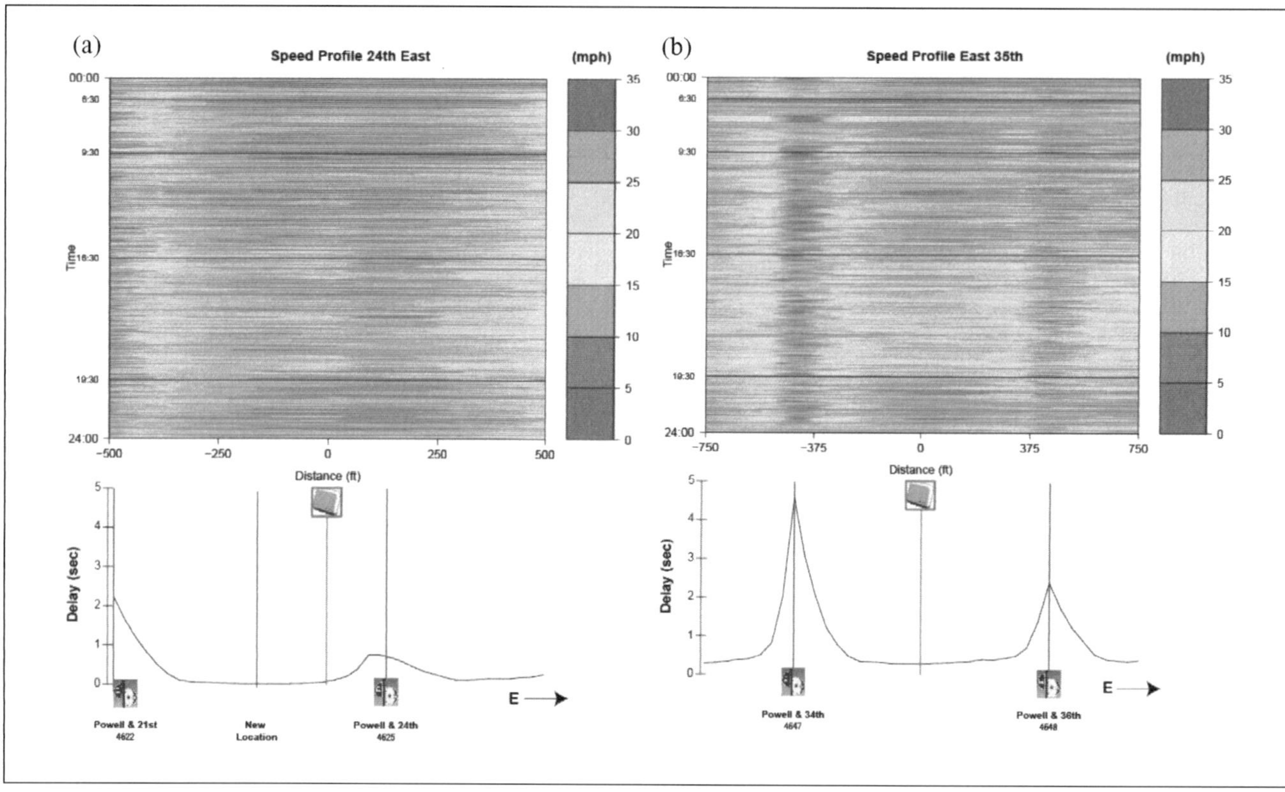

Figure 7. Speed heatmaps and sensor locations for eastbound travel: (a) eastbound 24th; (b) eastbound 35th.

Table 3. Correlations at Wavetronix Sensor and New Locations

		Time intervals—Bin sizes			
Proximity to Wavetronix sensor	Location	15 min	30 min	45 min	60 min
At the Wavetronix sensor	Eastbound 35th	0.92	0.94	0.96	0.96
	Eastbound 24th	0.80	0.90	0.90	0.94
	Westbound 35th	0.60	0.76	0.68	0.79
	Westbound 24th	0.70	0.74	0.70	0.84
Away from the Wavetronix sensor	Eastbound 24th	0.53	0.60	0.62	0.75
	Westbound 35th	0.84	0.92	0.91	0.94
	Westbound 24th	0.83	0.93	0.92	0.96

speeds at 24th. The low speeds of the second, smaller peak are likely to be associated with delays that originate at the downstream intersection when traffic is heavy. Yellow areas that propagate beyond the Wavetronix sensor can be observed in Figure 7a.

HTF Vehicle Flow Rates and Accuracy

The regression analysis and study of the time–speed profiles and correlations indicate that it is not accurate to include transit vehicles that service stops. It is possible to remove all the trips that service a stop location to increase the accuracy of the HFT speed estimation. It is recommended that HFT speed data are not utilized to evaluate arterial speeds along a segment that is ± 200 ft (± 60 m) around a heavily utilized bus stop without first removing GPS points that are near the bus stop. The bus stop impact area can be readily seen in a time–space speed profile (see Figures 6 and 7).

A reduction in the flow of HFT vehicles has to be offset by a longer data-collection period. The relationships between the frequency of probe data and variance of travel time estimates have been studied in the past (*7*); a significant reduction in variance is observed when the number of observations per hour increases from 12 to 36 probe vehicles per hour (*7*). Minor gains in estimation accuracy are obtained beyond 48–60 probe vehicles per hour. Therefore, it is desirable to obtain a sample size of 50 or more trips per hour. An example can be used to illustrate the tradeoffs. Assuming that nine out of 10 buses service a stop and bus headways are 12 min, to obtain 50 "undisturbed" weekday observations per hour it is necessary to analyze 20 weeks of data. In this case, "undisturbed" means vehicles that do not service a stop and therefore have a useful speed profile. Another example: If bus (vehicle) flow rate is low and equal to one vehicle per hour, in 10 weeks (or less than 3 months), it is possible to collect over 50 observations for a given hour of the day.

The minimum vehicle flow rate to obtain reasonable travel time estimation accuracy has other implications. It is unlikely to find arterials where the HFT flow rate is greater than 20 vehicles per hour per route. Assuming a minimum sample size of 50 more trips per hour, it is not possible to use HFT data to provide accurate real-time speed estimations unless five or six high-frequency bus routes are along an arterial. However, even if the hourly rate is below 50 trips per hour, the HFT data can provide accurate temporal depictions of average-speed travel times—that is, a good indicator for *recurrent* traffic conditions. If the analyst is interested in recurrent traffic conditions, the HFT data can provide accurate estimations of speed profiles along the arterial instead of speeds at specific locations where stationary sensors are located.

Conclusion

The exploratory data analysis, the time–space speed profiles, and the speed correlation graphs indicate that HFT bus speed data have a high degree of correlation with stationary Wavetronix sensors. HFT speed data can be used to cost-effectively analyze bus operations and monitor arterial recurrent speeds and performance with a high degree of accuracy. Better and more accurate transit data can then be utilized to visualize transit operations (*20*), estimate transit delay (*21*), and forecast travel times (*22*).

The high regularity of the HFT data means that in 3–6 months it is possible to accumulate large datasets with enough observations to analyze or detect locations and/or times with recurrent congestion along any section of an arterial. For bus HFT data, the results of this research show that a much shorter HFT data-collection period is necessary to obtain high correlations (greater than 0.90) with Wavetronix speed data when HFT speeds are not influenced by nearby heavily serviced bus stops. Around heavily utilized bus stops, the GPS data can be removed to avoid the bias introduced by slowed buses (*17*, *18*). Likewise, to eliminate differences between transit and private vehicle speeds, only transit data from vehicles that are not stopped by an upstream traffic signal and intermediate transit stop must be utilized (*19*).

HFT bus speed data are reasonably accurate and extremely economical when compared to the cost of purchasing, installing, and maintaining a network of stationary sensors. It is not argued here that large datasets of HFT speed data should replace stationary sensors; rather, HFT speed data can be used to complement stationary sensors and/or reduce the number of stationary sensors that are necessary to cost-efficiently monitor the performance of an arterial street or provide ground truth data and traffic volumes. Stationary sensors are useful to obtain data for a specific roadway section or point, whereas HFT speed data provide speed profiles that can be used to fill the gaps between stationary sensors. HFT data can also be used in real-time applications if the data are used to complement data from stationary sensors. Because of its potential cost-effectiveness, the joint utilization of HFT and stationary sensor data to monitor real-time or recurrent traffic conditions along arterial corridors is a subject that deserves additional research efforts.

Practical applications of this research include that knowing that HFT speed data are accurate allows the direct comparison of locations and the creation of a ranking of delays. Rankings are useful to prioritize locations that need improvements (e.g., transit signal priority or queue jump treatments at signalized intersections) and to produce speed heatmaps that show the impact of general traffic congestion on transit performance.

Acknowledgments

The authors would like to thank TriMet staff members Steve Callas and Miles J. Crumley for graciously providing the datasets used in this analysis and for their support in understanding the intricacies of how the data are structured. The authors would also like to thank Ellen Bradley for carefully editing and proofreading the paper. The authors would also like to acknowledge the support of NITC (National Institute for Transportation and Communities) Transportation Center for funding this research effort.

Author Contributions

The authors confirm contribution to the paper as follows: study conception and design: MAF, NBS; data collection: Tri County Metropolitan District of Oregon (TriMet); analysis and interpretation of results: MAF, NBS; draft manuscript preparation: MAF, NBS. All authors reviewed the results and approved the final version of the manuscript.

References

1. Westgate, B. S., D. B. Woodard, D. S. Matteson, and S.G. Henderson. Large-Network Travel Time Distribution

Estimation for Ambulances. *European Journal of Operational Research*, Vol. 252, No. 1, 2016, pp. 322–333.
2. Zhan, X., S. Hasan, S. V. Ukkusuri, and C. Kamga. Urban Link Travel Time Estimation Using Large-Scale Taxi Data with Partial Information. *Transportation Research Part C: Emerging Technologies*, Vol. 33, 2013, pp. 37–49.
3. Jenelius, E., and H. N. Koutsopoulos. Travel Time Estimation for Urban Road Networks Using Low Frequency Probe Vehicle Data. *Transportation Research Part B: Methodological*, Vol. 53, 2013, pp. 64–81.
4. Bar-Gera, H. Evaluation of a Cellular Phone-Based System for Measurements of Traffic Speeds and Travel Times: A Case Study from Israel. *Transportation Research Part C: Emerging Technologies*, Vol. 15, No. 6, 2007, pp. 380–391.
5. Wasson, J., J. Sturdevant, and D. Bullock. Real-Time Travel Time Estimates Using Media Access Control Address Matching. *ITE Journal*, Vol. 78, No. 6, 2008, pp. 20–23.
6. Herring, R., A. Hofleitner, P. Abbeel, and A. Bayen. Estimating Arterial Traffic Conditions Using Sparse Probe Data. *Proc., 13th International IEEE Conference on Intelligent Transportation Systems*, Funchal, Portugal, IEEE, New York, NY, pp. 929–936.
7. Sen, A., P. Thakuriah, X. Q. Zhu, and A. Karr. Frequency of Probe Reports and Variance of Travel Time Estimates. *Journal of Transportation Engineering*, Vol. 123, No. 4, 1997, pp. 290–297.
8. Hall, R., and N. Vyas. Buses as a Traffic Probe: Demonstration Project. *Transportation Research Record: Journal of the Transportation Research Board*, 2000. 1731: 96–103.
9. Cathey, F., and D. Dailey. Transit Vehicles as Traffic Probe Sensors. *Transportation Research Record: Journal of the Transportation Research Board*, 2002. 1804: 23–30.
10. Bertini, R., and S. Tantiyanugulchai. Transit Buses as Traffic Probes: Use of Geolocation Data for Empirical Evaluation. *Transportation Research Record: Journal of the Transportation Research Board*, 2004. 1870: 35–45.
11. Glick, T. B., and M. A. Figliozzi. Traffic and Transit Travel Time Reliability Indices and Confidence Intervals. *Transportation Research Record: Journal of the Transportation Research Board*, 2017. 2649: 28–41.
12. Gates, T., S. Schrock, and J. Bonneson. Comparison of Portable Speed Measurement Devices. *Transportation Research Record: Journal of the Transportation Research Board*, 2004. 1870: 139–146.
13. Mohammad, H., S. Schrock, and E. Fitzsimmons. Accuracy of Traffic Speed and Volume Data Detected Using Radar Technology. Presented at 95th Annual Meeting of the Transportation Research Board, Washington, D.C., 2016.
14. Minge, E., J. Kotzenmacher, and S. Peterson. *Evaluation of Non-Intrusive Technologies for Traffic Detection*. No. MN/RC 2010-36. Minnesota Department of Transportation, Research Services Section, Saint Paul, MN, 2010.
15. Sanchez, G. *Evaluation of the Accuracy of Approach Volume Counts and Speeds Collected by Microwave Sensors*. Thesis. Brigham Young University, Provo, UT, 2016.
16. Stoll, N., T. B. Glick, and M. A. Figliozzi. Utilizing High Resolution Bus GPS Data to Visualize and Identify Congestion Hot-Spots in Urban Arterials. *Transportation Research Record: Journal of the Transportation Research Board*, 2016. 2539: 20–29.
17. Glick, T., and M. A. Figliozzi. Measuring the Determinants of Bus Dwell Time. *Transportation Research Record: Journal of the Transportation Research Board*, 2017. 2647: 109–117.
18. Thornton, D., and B. Coifman. Signal Progression Impacts on Transit Buses as Travel Time Probes. *Journal of Transportation Engineering*, Vol. 141, No. 8, 2015, 04015009.
19. Kurkcu, A., F. Miranda, K. Ozbay, and C. T. Silva. Data Visualization Tool for Monitoring Transit Operation and Performance. *Proc., 2017 5th IEEE International Conference on Models and Technologies for Intelligent Transportation Systems (MT-ITS)*, Naples, Italy, IEEE, New York, NY, 2017, pp. 598–603.
20. Sidhu, B. S., R. L. Bertini, and A. Pande. Exploiting New High Resolution Transit Data Sources: Developing Improved Data Driven Models of Transit Travel Time and Delay. Presented at 96th Annual Meeting of the Transportation Research Board, Washington, D.C., 2017.
21. Hans, E., N. Chiabaut, L. Leclercq, and R. L. Bertini. Real-Time Bus Route State Forecasting Using Particle Filter and Mesoscopic Modeling. *Transportation Research Part C: Emerging Technologies*, Vol. 61, 2015, pp. 121–140.
22. Slavin, C., W. Feng, M. Figliozzi, and P. Koonce. Statistical Study of the Impacts of SCATS Adaptive Traffic Signal Control on Traffic and Transit Performance. *Transportation Research Record: Journal of the Transportation Research Board*, 2013. 2356: 117–126.

The Public Transportation Group (AP000) peer-reviewed this paper (18-05735).

Any errors or omissions are the sole responsibility of the authors.

Article

TRR
JOURNAL OF THE TRANSPORTATION RESEARCH BOARD

Transportation Research Record
2018, Vol. 2672(8) 199–209

DOI: 10.1177/0361198118793519
journals.sagepub.com/home/trr
SAGE

Evaluation of Route Changes Utilizing High-Resolution GPS Bus Transit Data

Travis B. Glick[1] and Miguel A. Figliozzi[1]

Abstract
This study applies high-resolution archived transit data to study the effect of roadway changes using data collected before and after the completion of a project affecting transit. Methodologies are presented to compare percentile and time-of-day performance measures before and after the project. In addition, differences in travel time and travel-time variability are examined over the altered route. The case study examines one heavily used route in Portland that was recently diverted onto a newly built transit-only bridge to examine the claims that travel times would decrease and reliability would increase. The results of this study indicate that travel times increased for the majority of trips but travel-time variability during the peak period was sharply reduced.

Methodologies to quantify transit performance influence transportation planning and subsequent decisions; those decisions may affect operating speeds, travel times, ridership, costs, and efficiency (*1*). Transit operations research is continually evolving with the introduction of new or improved data-collection systems. Onboard global position systems (GPS) are opening up new research opportunities to visualize and quantify transit behaviors hidden by legacy data-collection systems. High-resolution (HR) GPS data-collection is expanding analysis options when implemented. However, these data-collection systems are new, not widespread, and as a result, understudied.

This study applies HR data to quantify the impact of changes to roadways as a before-and-after study. This study expands on existing systems to show applicable methodologies that quantify transit performance changes following a roadway modification in locations where transit has been traditionally excluded.

The case study for this paper examines the effect of the Tilikum Crossing. This new bridge in Portland, OR is the largest vehicle-free bridge in the United States (*2*). Although it is designed for light rail, streetcar, bikes, buses, and pedestrians, personal vehicles are not permitted. TriMet, Portland's public transportation provider, claimed the new bridge would reduce travel times and improve efficiency on routes 9 and 17 (*3*). This paper examines those claims for Route 9. The bridge cost was estimated at US$134 million paid for by federal grants, OR state lottery, and TriMet revenue.

Background and Literature Review

Before-and-after studies are not new, and frequently seek to quantify effects of roadway modifications. These studies exist across transportation modes and roadway applications. Their widespread implementation is one of the main ways new systems are adopted.

These studies typically use Bayesian statistics or are descriptive in nature. Studies that use Bayesian methodologies for before-and-after analysis reach back more than 30 years. The literature on Bayesian methodologies up to 2006 is well summarized in a paper that claims that a lack of sufficient data, such as traffic or crash counts, leads to high levels of uncertainty (*4*). Due to the natural variation in count data, crash rates fluctuate wildly month-to-month and year-to-year. Improvements for count usage have been proposed by newer papers that have added observations to reduce uncertainty by including additional counts estimated from models, which are based on locations with physical similarities to a given study area. These counts are then modified using a known prior distribution (*5*, *6*).

[1]Transportation Technology and People Lab, Department of Civil and Environmental Engineering, Maseeh College of Engineering and Computer Science, Portland State University, Portland, OR

Corresponding Author:
Address correspondence to Miguel A. Figliozzi: figliozzi@pdx.edu

Traffic assessment often uses descriptive before-and-after approaches. For example, a study looking at the change in fatal crashes following the introduction of anti-lock brakes compared results using the difference in an estimated risk ratio and its associated 95% confidence interval (CI) calculated from the collected data (*7*). Researchers have also examined the effect of traffic-calming systems across the United States for pedestrian wait times, using *t*-tests to measure differences in mean values (*8*). Descriptive analyses are appropriate for large sample sizes and appear across many fields.

This study did not use counts, which therefore limited the application of Bayesian methodologies. Variance of speed data was calculated using asymptotic variance and a probability mass function (9) described in the methodology section of this paper. Additionally, this research employed recently published performance metrics methodologies that use percentile travel-time, travel speed, and CI estimates to identify and rank low-performance hotspots (*10*). Percentiles are more general and suited for both symmetric and skewed distributions; for example, off-peak travel follows lognormal distributions, which are asymmetric.

Travel times play a key role in the overall ridership of a city. Travel-time elasticity is not readily available for all cities, as noted by Ayvalik et al. for Chicago (*11*); however, Portland, OR is a city with a full travel-forecasting model (*12*). This model indicates that decreased transit travel-times will increase overall ridership. When estimating travel times and trajectories, researchers have used proxies. Early research using buses as probes showed that buses experience the same types of delays as automobiles but that the reverse relationship is not always true. For example, buses will dwell at designated locations (bus stops) that regular vehicles do not (*13*, *14*). TriMet buses have been used widely to quantify arterial performance for both automobiles and transit (*15*, *16*).

Stop event (SE) data collects information at each bus stop including, but not limited to, actual and scheduled times of arrival and departure, dwell times, door usage, and passenger movements. When combined with data from loop detectors, signal cycle lengths, and other proprietary data-collection systems, SE data has been used to estimate travel times of other factors that may influence service reliability at the point, stop-to-stop, and route levels (*13, 17–19*). Each new study adds useful information, but the representation remains an average between stops due to the nature of SE data. Stop disturbance (SD) data, which includes information about any location where the wheels of the bus have stopped, such as the type (i.e., scheduled or unscheduled), location, and time of stop, may be combined with SE data to provide insights into stopping behavior associated with intersections or locations with heavy congestion. The integration of HR data with SE data may be used to mitigate some of the shortcomings of SE or SD data. For example, one study using GPS data found that for narrow ranges of departure times, travel-time distributions are best characterized by normal distributions. The same is true for peak-hour travel windows, but off-peak travel-times follow lognormal distributions (*20*).

Data Sources

This study relied on data provided by the Tri-County Metropolitan Transportation District of Oregon (TriMet), which has archived automatic vehicle location (AVL) data for all trips since 1997. TriMet updated their bus dispatch system with HR data collection in 2013. SE, SD, and HR data are the foundation of TriMet's collection systems. TriMet also creates onboard video recordings of all trips; however, TriMet, as with many agencies, erases video on a weekly cycle unless an incident occurs or if requested for a specific date (*21*).

At the authors' request, TriMet provided three sets of AVL data: SE, SD, and HR. Each of the AVL data sets represent the same buses, routes, and times; this allows for comparisons and integration of the data sets. The visuals and comparisons obtained using a combined data set provide additional details, hidden or not included in each of the sets individually.

For the before case, there were 1,227 and 1,158 buses in the eastbound and westbound directions, respectively. The after case had 1,453 and 1,691 buses in the eastbound and westbound cases, respectively.

Study Areas

The changes to bus operations on Southeast Powell Boulevard (Powell) resulted from the completion of the Tilikum Crossing, a new bridge over the Willamette River. This bridge was built to carry light rail, streetcars, and buses. Route 9, which historically crossed the Ross Island Bridge, was diverted to the new crossing. The changes examined are for the section where the two different routes converge on the east side of the river. In addition to the travel speed percentile comparisons, Powell was also examined statistically for changes by time of day to isolate when the roadway saw increases in speed.

Typically, this area of Powell is highly congested and carries upwards of 40,000 vehicles daily. However, no traffic counts are available for this stretch of Powell from 2016. Data collected at the macroscopic level (i.e., Portland region), mostly on freeways, indicate the peak periods are lasting longer across the metro area and that daily vehicle miles traveled, daily congestion hours, and daily vehicle hours of delay increased by

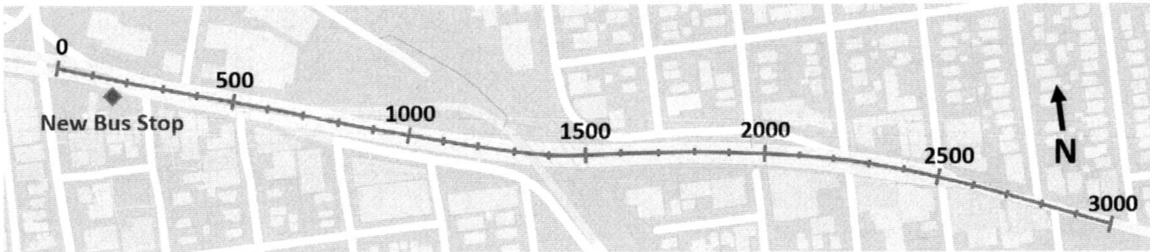

Figure 1. Map of study area for Southeast Powell Boulevard. Measurements correspond to x-axis of results.

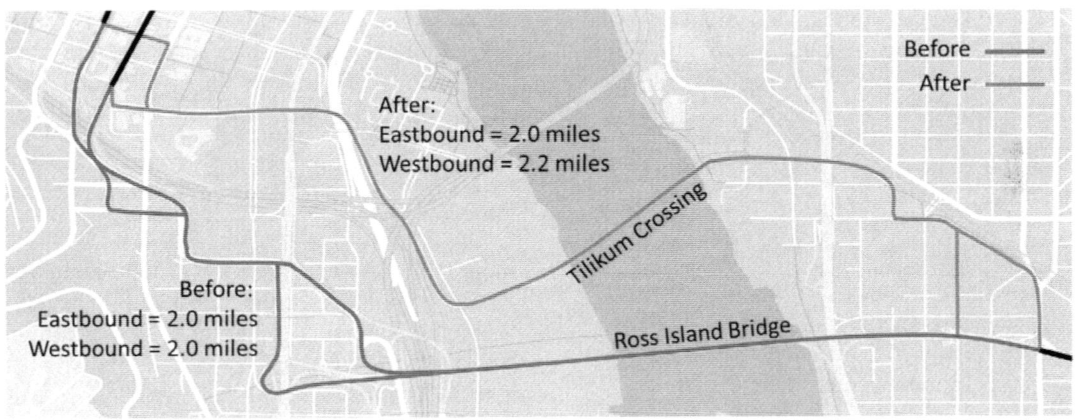

Figure 2. Map of route before and after change. Red shows the previous configuration. Green shows the new configuration.

approximately 1%, 11%, and 17%, respectively from 2014 to 2016 (*22*).

Figure 1 shows the study area leading up to the changed route on the east side of the river. This segment saw no physical changes but is examined for performance changes that may be the result of the new route.

Figure 2 shows the change in route. The distance traveled by eastbound buses was unchanged. Westbound buses now travel an additional 0.2 miles (~322 meters).

Methodology

The three data sets provided by TriMet cover the same time period, routes, and buses, with some exceptions caused by different data-collection parameters. These data sets must be filtered and cleaned to provide a uniform set of trips, which requires a unique identification number across the sets. All data sets include a bus number, trip number, and date; these serve as a starting point. Individual trips are separated based on route and direction included in the SE and SD data sets. The HR data, which does not include these identifiers, must be compared with the other sets to separate out individual trips based on time. This process culminates in a unique identification number for each bus and trip across all the data.

Using the unique identification number, all three data sets are integrated starting with the SE and SD data sets. When a bus dwells at a bus stop, these locations are considered SDs, so these events are duplicated, whereas pass-by stops would not be recorded. The integration of the data sets ensures that pass-by stops, dwell events, and other places where the wheels of the bus stop moving are all included with as much detail as can be provided by the combination of both sets.

Following this step, the HR data is interwoven by timestamps with the SE and SD data to provide a complete picture of the bus's trajectory, an example of this is shown in Figure 3. The merged data sets provide a means to quantify roadway behaviors that account for and can exclude bus-stopping behavior, which allows for estimates of general traffic behavior.

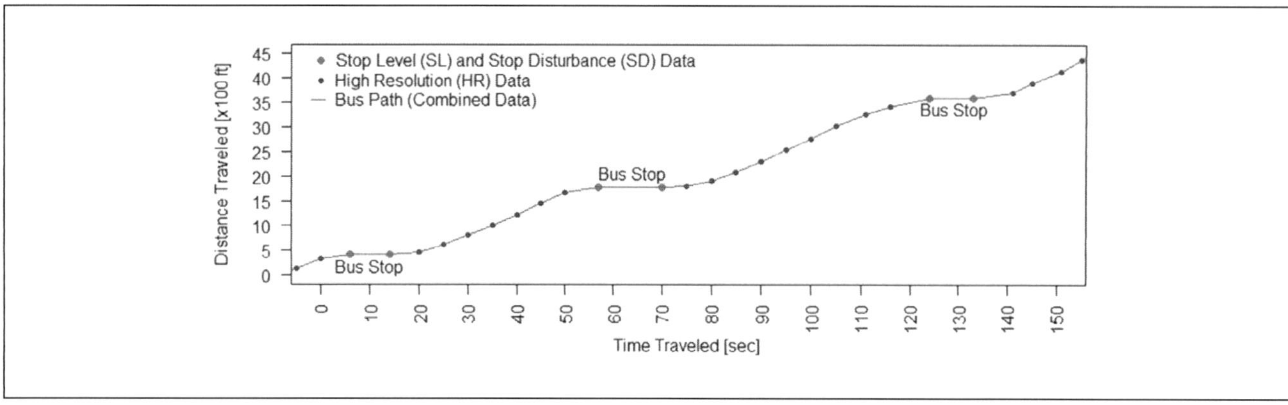

Figure 3. Bus trajectory using the combined data set.

Each study area is divided into a system of equal length and non-overlapping segments, s_i. Each unique segment shall be given an index of i. The total number of segments shall be denoted as n_I.

$$i \in I = \{1, 2, 3, \ldots, n_I\}$$

The combined data will have a set of j bus trip that pass through each segment, s_i. The consecutive GPS coordinates that surround each centerpoint of a given segment, s_i, are extracted to make up a pair of GPS coordinates (or vehicle trips) for each s_i. The total number of vehicle trips in each segment is denoted as n_{J_i}.

$$j \in J_i = \{1, 2, 3, \ldots, n_{J_i}\}$$

The data used in the analysis deviates from normal distributions. As such, this study relies heavily on percentiles and their associated percentile variances. The methodology for calculating speeds stems from a previously published paper using HR data (*10*), but with a procedure that allows the estimation of CIs when distributions are not necessarily normal. Any s_i has a set of percentile travel speeds found by ordering the data and extracting a number that correspond to a percentile, p. The value of the speed percentile in segment$i$ is denoted $v_{i,p}$. The standard deviation of the estimated $v_{i,p}$ is denoted $\sigma_{i,p}$.

The process to estimate each $\sigma_{i,p}$ is described in Figure 4. For example, Figure 4 upper left shows the speed histogram for any segment i. A cumulative distribution function (CDF) is estimated from the data (upper right) and later spline smoothing is applied to the CDF to create a continuous function that approximates the CDF (lower left). From this generated spline-smoothed function, the probability of a given point can be calculated by taking the derivative of the CDF to produce an estimate of the probability density function (PDF) – see lower right. This process can be applied to any probability distribution.

The estimate of the variance for any percentiles of univariate data is estimated through a CDF and its derivative, the PDF. To estimate $\hat{\sigma}_{v_{i,p}}{}^2$, the variance of a speed percentile, the following equation is utilized (*23*):

$$\hat{\sigma}^2_{v_{i,p}} = \frac{p(1-p)}{f(\hat{v}_{i,p})^2 \cdot n_{J_i}}$$

where $f(\hat{v}_{i,p})$ is the probability of the PDF given the input velocity, $\hat{v}_{i,p}$, and n_{J_i} is the number of observations in each segment i. Assuming the number of observations is large (>160) (*23*), this estimate of variance may be used to estimate the CIs for each $\hat{v}_{i,p}$. For a confidence level α and its associated z-score, $z(\alpha)$, the range of percentile values that may represent an estimated percentile is found.

$$CI_{\hat{v}_{i,p}} = \left[\hat{v}_{i,p} - \hat{\sigma}_{v_{i,p}} \cdot z(\alpha), \hat{v}_{i,p} + \hat{\sigma}_{v_{i,p}} \cdot z(\alpha)\right]$$

This interval provides the indices for the extremes of the CI around $v_{i,p}$.

A speed variability $\Delta\hat{v}_i$, is used to identify segments that are more heavily congested during the peak hour (*21*). It is calculated by subtracting the 15th percentile travel speed from the 85th percentile travel speed. When divided by the median travel-time, a speed variability index ($\hat{\mu}_i$) is obtained for each segment (*21*).

$$\Delta\hat{v}_i = \hat{v}_{i,85} - \hat{v}_{i,15}$$

$$\hat{\mu}_i = \frac{\Delta\hat{v}_i}{\hat{v}_{i,50}} = \frac{\hat{v}_{i,85} - \hat{v}_{i,15}}{\hat{v}_{i,50}}$$

The standard deviation of the estimated Δv_i can be estimated as follows:

$$\hat{\sigma}_{\Delta v_i} = \sqrt{\hat{\sigma}^2_{v_{i,85}} + \hat{\sigma}^2_{v_{i,15}} - 2 \cdot cov(\hat{v}_{i,85}, \hat{v}_{i,15})}$$

Note that this formula for $\hat{\sigma}_{\Delta v_i}$does not require that the distributions be normal for large enough sample sizes.

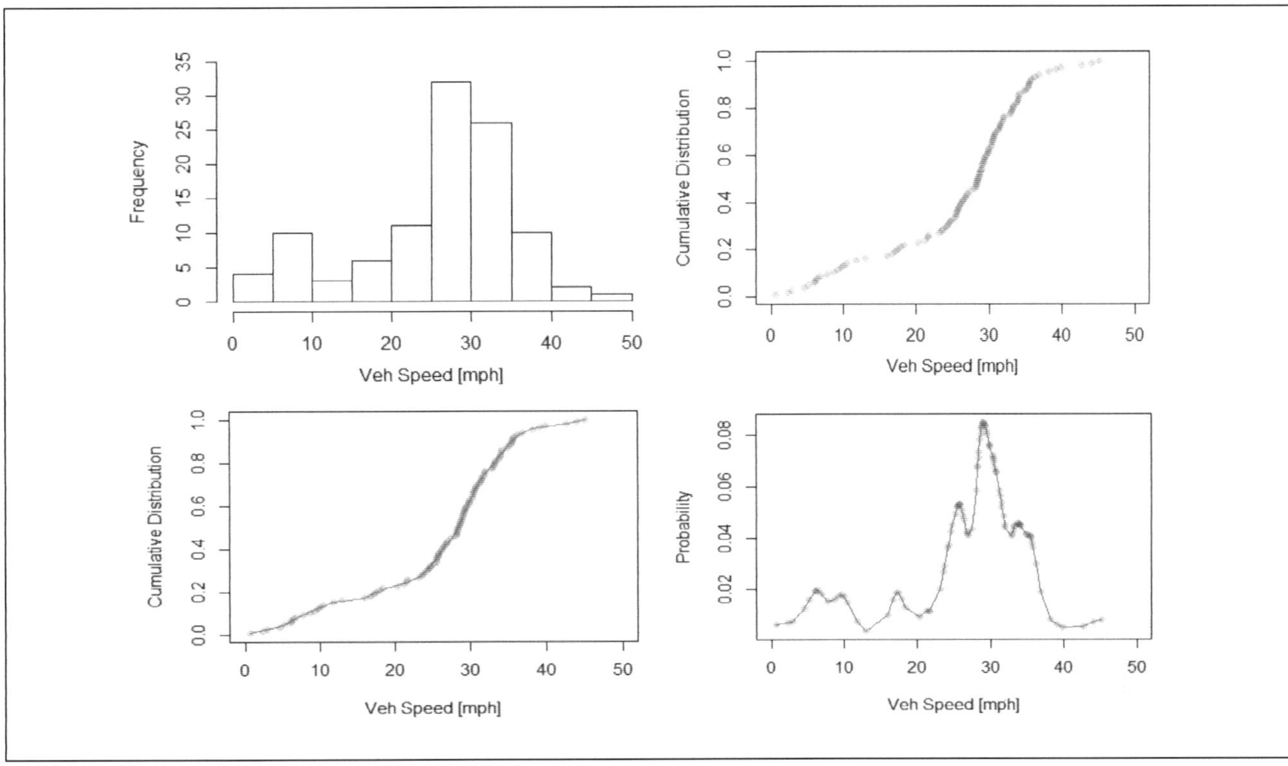

Figure 4. Upper left: histogram of randomly generated data in 5-mph (~8-kph) bins. Upper right: actual cumulative distribution of points. Lower left: cumulative distribution function creating through spline smoothing. Lower right: probability mass function for data.

As the speed percentiles are not independent, the covariance increases the estimation CI:

$$CI_{\Delta v_i} = [\Delta\hat{v}_i - \sigma_{\Delta\hat{v}_i} \cdot z(\alpha), \Delta\hat{v}_i + \sigma_{\Delta\hat{v}_i} \cdot z(\alpha)]$$

If 0 falls within the estimated interval, the null hypothesis $\Delta\hat{v}_i = 0$ cannot be rejected using a confidence level based on α.

The standard deviation of the estimated μ_i can be approximated as follows utilizing a well-known formula for propagation of uncertainties (*24*) assuming uncorrelated speed percentiles:

$$\frac{\hat{\sigma}_{\mu_i}}{|\hat{\mu}_i|} \sim \sqrt{\left(\frac{\hat{\sigma}_{\Delta v_i}}{\Delta\hat{v}_i}\right)^2 + \left(\frac{\hat{\sigma}_{v_{i,50}}}{\hat{v}_{i,50}}\right)^2} =$$

$$\sqrt{\frac{\hat{\sigma}_{v_{i,85}}{}^2 + \hat{\sigma}_{v_{i,15}}{}^2 - 2 \cdot cov(\hat{v}_{i,85}, \hat{v}_{i,15})}{\hat{v}_{i,85} - \hat{v}_{i,15}{}^2} + \left(\frac{\hat{\sigma}_{v_{i,50}}}{\hat{v}_{i,50}}\right)^2 + \left(\frac{\hat{\sigma}_{v_{i,50}}}{\hat{v}_{i,50}}\right)^2}$$

As the speed percentiles are not independent, the estimated σ_{μ_i} using the previous equation is a lower bound. The value $\hat{\sigma}_{\mu_i}$ can be used to estimate a lower bound CI for $\hat{\mu}_i$:

$$CI_{\mu_i} = \left[\hat{\mu}_i - \sigma_{\hat{\mu}_i} \cdot z(\alpha), \hat{\mu}_i + \sigma_{\hat{\mu}_i} \cdot z(\alpha)\right]$$

If 0 falls within the estimated interval, the null hypothesis $\hat{\mu}_i = 0$ cannot be rejected using a confidence level based on α.

The methodologies described provide the foundation to compare the initial and changed conditions of the study area. For all the data, before-and-after data will be designated by sub-indexes 0 and 1, respectively. A δ added before a variable will denote the difference between after and initial conditions. The differences in the means and percentile speeds and travel times and peak-hour performance metrics are denoted as follows:

$$\delta\hat{v}_{i,p} = \hat{v}_{i,p,1} - \hat{v}_{i,p,0}$$

$$\delta\Delta\hat{v}_i = \Delta\hat{v}_{i,1} - \Delta\hat{v}_{i,0}$$

$$\delta\hat{\mu}_i = \hat{\mu}_{i,1} - \hat{\mu}_{i,0}$$

The previously estimated standard deviations can be used to estimate standard deviations and CIs for any of the difference statistics denoted by δ. For example, the estimated standard deviation for $\delta\hat{v}_{i,85}$ is calculated as follows:

$$\delta\hat{\sigma}_{v_{i,85}} = \sqrt{\hat{\sigma}^2_{v_{i,85},1} + \hat{\sigma}^2_{v_{i,85},0} + Cov\left((\hat{v}_{i,85})_1, (\hat{v}_{i,85})_0\right)}$$

Travel times, denoted as t_j, between two points are extracted from the data where j is a single bus.

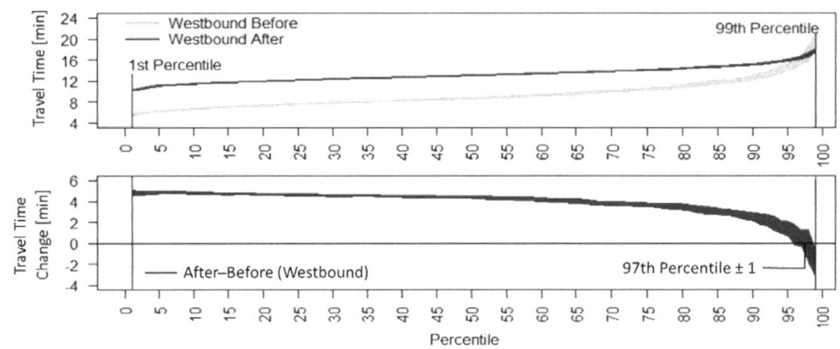

Figure 5. Top: Westbound travel-times. Bottom: travel-time difference (after minus before).

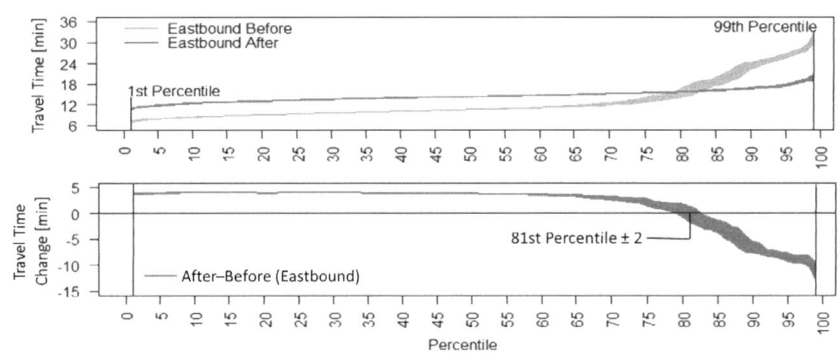

Figure 6. Top: Eastbound travel-times. Bottom: travel-time difference (after minus before).

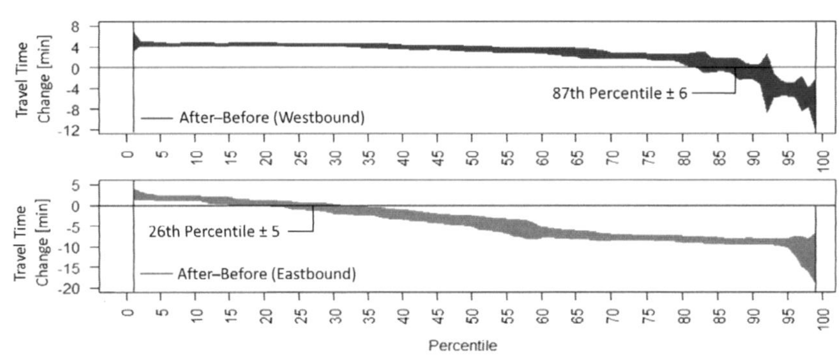

Figure 7. Travel-time differences (after minus before) during the PM peak. Top: westbound. Bottom: eastbound.

Percentiles and CIs are calculated using the same methodology as above.

$\hat{t}_p$ = percentile travel-time, and

$\hat{\sigma}_{tp}$ = standard deviation of percentile travel-time.

Average travel-time, $\bar{t}$, can be found by summing each percentile travel-time as if it were an individual bus then correcting for the number of buses n_{J_i}. On average each percentile travel-time should be seen an equal number of times.

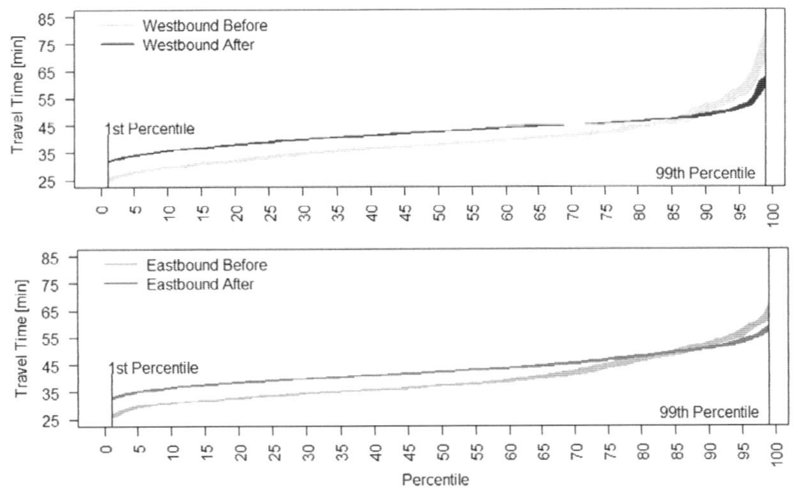

Figure 8. Travel times over 7.2 miles (11.6 km) of Powell for all trips. Top: westbound. Bottom: eastbound.

Table 1. Mean Hourly Travel-Time, Standard Deviation, Passenger Load, and Scheduled Trips per Day

	EB		WB		EB		WB		EB		WB			
	bef.	aft.	bef.	aft.	bef.	aft.	bef.	aft.	bef.	aft.	bef.	aft.		
Time	Mean travel-time (min)				Travel-time standard deviation (min)				Mean passenger load				EB trips per day	WB trips per day
6:00	31.4	36.2	34.2	38.9	2.5	1.4	3.3	2.9	4	2	10	9	3	6
7:00	34.1	40.1	47.8	48.5	2.8	2.7	10.2	5.7	4	4	11	12	3	8
8:00	35.1	40.2	50.5	51.2	2.4	2.4	10.7	7.7	4	4	12	14	5	6
9:00	34.9	40.2	40.9	44.0	3.0	2.5	6.7	3.5	5	4	11	11	4	4
10:00	33.0	40.9	36.4	42.8	19.9	1.8	2.8	3.3	5	5	7	11	4	4
11:00	35.6	42.3	38.2	44.4	2.5	2.3	2.9	4.0	5	6	7	11	4	4
12:00	37.8	44.6	38.6	44.1	2.3	3.6	3.2	2.7	6	8	9	8	4	4
13:00	39.3	44.3	39.0	43.8	2.5	2.3	3.7	3.0	9	7	7	11	4	4
14:00	39.1	45.8	41.0	43.9	2.3	2.8	5.5	3.7	8	11	7	9	5	4
15:00	46.4	49.2	41.8	45.1	7.0	4.4	4.6	4.2	12	12	8	10	6	5
16:00	50.4	51.4	42.0	44.8	5.7	4.5	6.6	2.2	13	12	8	10	8	5
17:00	52.9	50.4	47.4	44.8	6.0	4.2	11.8	3.4	10	14	7	8	9	4
18:00	45.4	45.5	38.3	42.2	7.7	3.9	9.0	3.1	12	12	9	7	5	5
19:00	34.3	41.9	33.4	40.9	14.5	2.9	2.4	2.6	10	10	5	6	4	3
20:00	33.3	40.8	30.8	38.2	2.2	3.8	2.3	2.4	8	8	7	7	4	3

Note: EB = eastbound; WB = westbound; aft. = after; bef. = before.
Travel times are for the same distance covered in Figure 8.

$$\bar{t} = \frac{n_{J_i}}{99}\sum_{p=1}^{99}\hat{t}_p$$

Results

Travel-Time Changes

Travel times in both directions of travel across the Ross Island Bridge—the former river crossing for this route—increased for the vast majority of trips. For westbound travel (Figure 5), it is likely that only a few percent of trips saw a decrease in travel times. For eastbound trips (Figure 6), approximately 20% of trips saw a reduction in travel times. However, although travel times increased for most trips, both directions experienced a large decrease in travel-time variability (i.e., the difference between the 99th and 1st percentile).

The top panels of figures 5 and 6 show the actual travel-times by percentiles with the 95th percentile CI shown. The bottom panel of each figure shows the

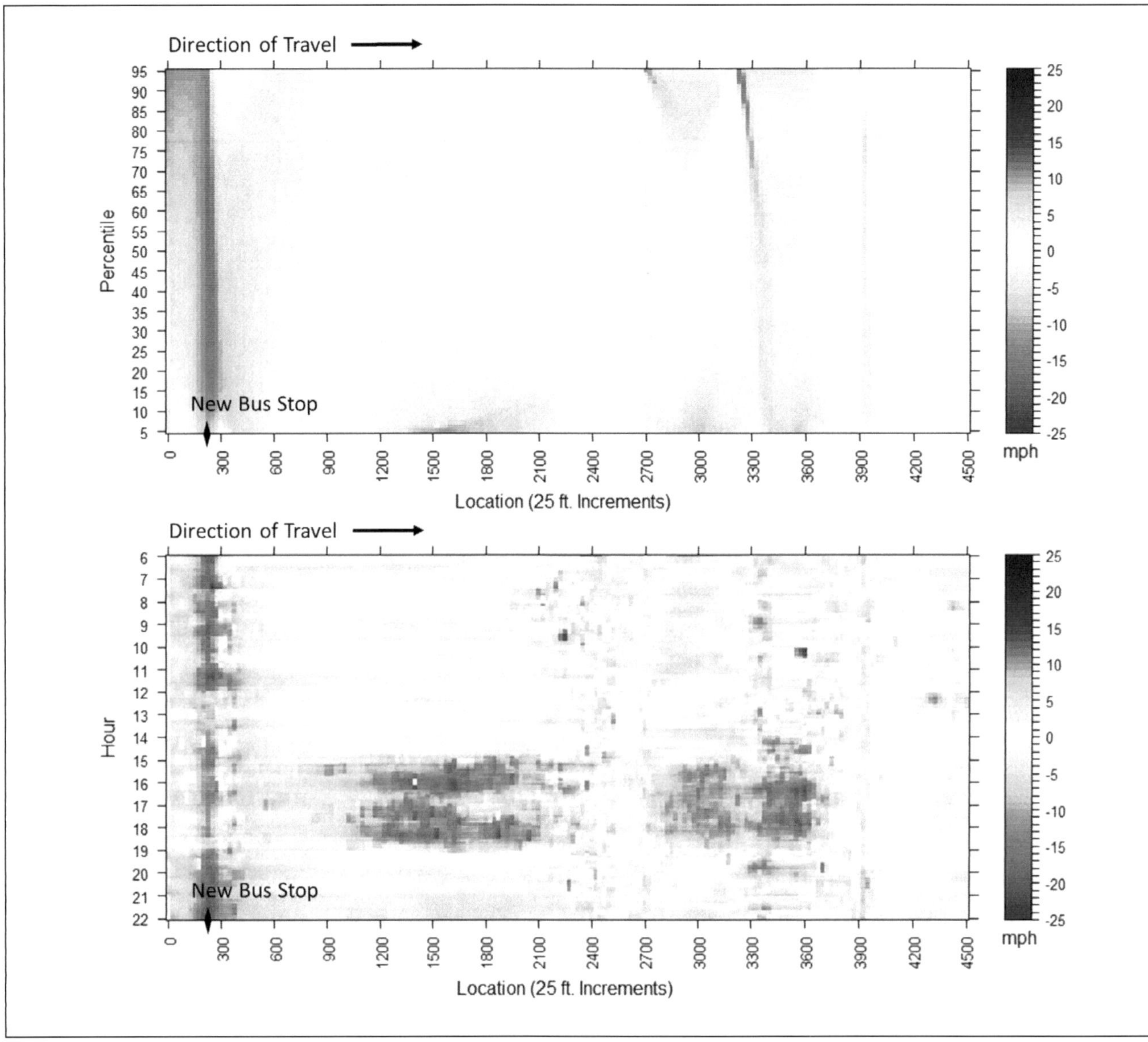

Figure 9. Top: difference in the 5th through 95th percentile travel speeds between before-and-after cases along Powell in the eastbound direction. Bottom: difference in the travel speeds between before-and-after cases along Powell in the eastbound direction by time of day. Each value is created using a 30-min moving average.

estimated difference with positive values indicating longer travel-times between the 1st and 99th percentile. In Figure 5, the travel-time change does not consider the 0.2-mile increase in travel distance for westbound trips. Assuming slow transit speeds of just 15 mph (~24 kph), that distance increase accounts for less than a minute of the total change.

The same type of analysis was conducted for the AM and PM peak periods in both directions. Although, the AM peak showed nearly identical results as the all-day case in both directions, the PM peak saw significant changes in travel times. Figure 7 shows the PM peak travel-time changes (after minus before) for westbound trips (top) and eastbound trips (bottom).

Estimates of travel times were expanded to include an additional 5 mi of Route 9; the resulting 7-mi segment is the most heavily congested part of Powell. As shown in Figure 8, travel times increased for approximately 85% of all trips; however, the range of possible travel times (99th minus 1st percentile) decreased by 23 and 17 min in the westbound and eastbound directions, respectively, despite increases in peak-hour congestion in the Portland metro.

Additionally, as indicated by Table 1, average travel-time became much more consistent by time of day after

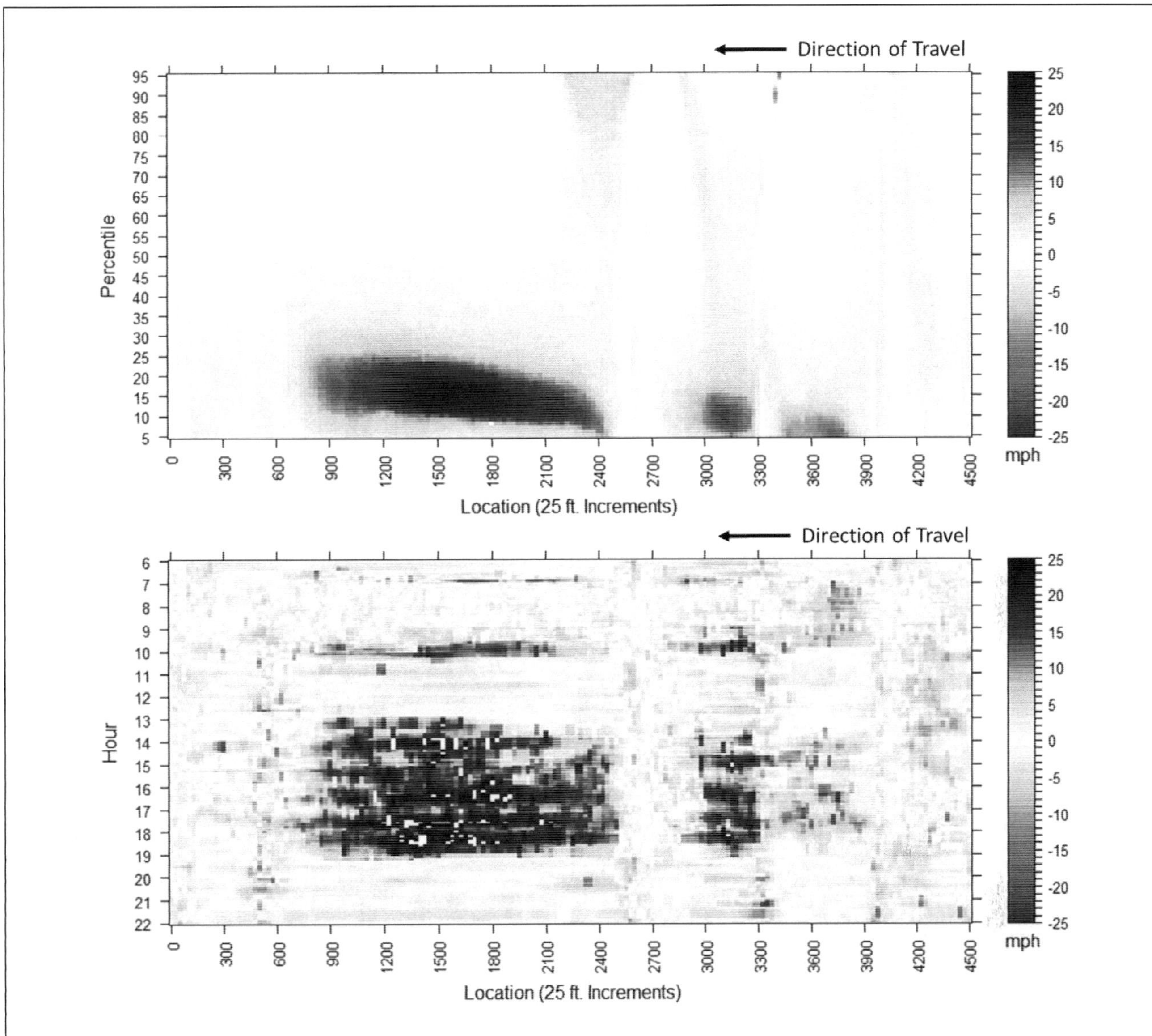

Figure 10. Top: Difference in the 5th through 95th percentile travel speeds between before-and-after cases along Powell in the westbound direction. Bottom: Difference in the travel speeds between before-and-after cases along Powell in the westbound direction by time of day. Each value is created using a 30-minute moving average.

the change. For example, the sharp increase in travel times in eastbound travel at 17:00 is notably reduced after the route change. Passenger loading remains similar.

Eastbound Travel Speeds

The travel speed differences following the changes to Powell indicate substantial changes to bus performance. However, the study area (excluding where route changed) was not significantly altered. The road was repainted without altering the location of lines. The decrease in speed in Figure 9 at $x = 250$ ft (~76.2 m) is accounted for by the addition of a new bus stop. Prior to the change in the route, that bus stop was a nearside stop approximately 50 ft (~15 m) before the start of the analysis area. The before and after routes do not overlap. As such, the previous bus stop and the new left turn on the route were excluded. By percentile, much of eastbound travel appears to have decreased in speed. By time of day, the majority of the decreased speed is seen to be in the peak period.

Westbound Travel Speeds

For westbound travel, travel speeds increased significantly for the 10th through 25th percentiles. This

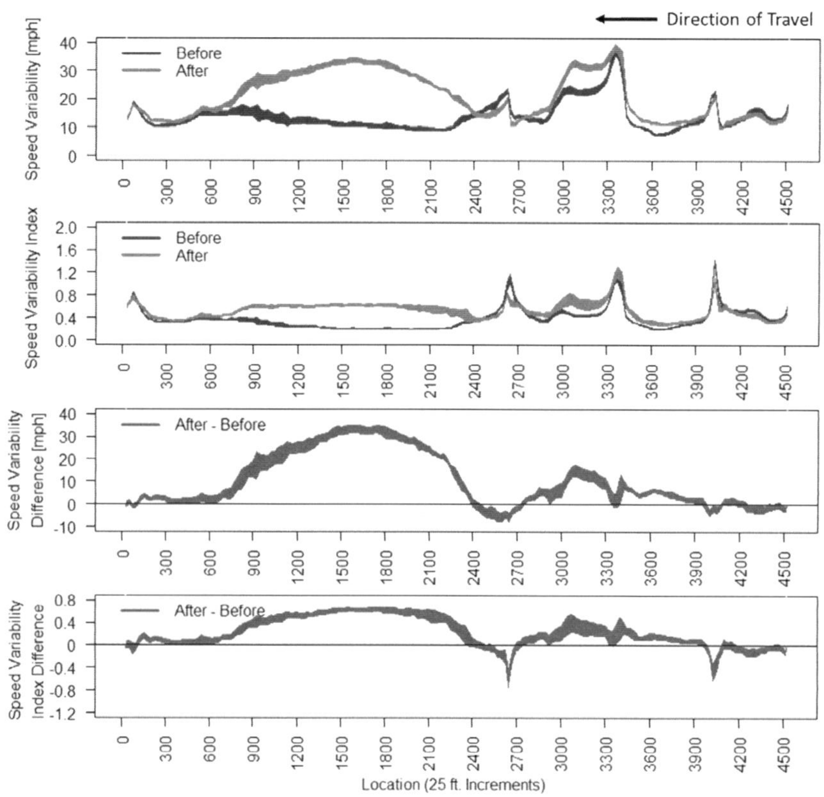

Figure 11. All graphs: westbound direction on Powell. Thickness of line is the 95th percentile confidence interval. Top: speed variability (85th minus 15th percentile travel speed) in miles per hour. Upper middle: speed variability index (ratio between the speed variability and 50th percentile travel speeds). Lower middle: difference between speed variability. Bottom: difference between speed variability index.

decreased travel-time is concentrated during the evening commute. Traffic patterns suggest that, typically, vehicles are attempting to leave the city center. The time-of-day plot in Figure 10 shows that there is less congestion heading into the city after the route change than there was before.

This major change in traffic patterns is supported by the plots of speed percentile difference and the speed variability index in Figure 11, which shows different traffic patterns before and after the change, as well as a large non-zero difference between them. In the eastbound direction, the differences in speed variability and the speed variability index show near zero and statistically insignificant changes.

Conclusion

The methodologies outlined in this paper provide a means to quantify differences in transit performance on roadway segments from before and after a change to that roadway. HR GPS data can provide detailed information between stops and highlight areas and times-of-day in which speeds or travel times differ and can be applied to different types of roadway reconfigurations. Furthermore, the use of percentile CIs can provide insights into how travel-time variability changes at different locations and travel speeds.

TriMet's claim that travel times would decrease is not clear from the results of this data. For the majority of trips, travel times actually increased for Route 9. One contributing factor to the increased travel-times may be the 25-mph (~40-kph) speed limit for buses and trains (*3*). This limit is a product of line-of-sight requirements that are limited by the grade of the bridge, which, to provide enough clearance for ships, is just under 5% for the majority of its length (*25*). Another factor may be the wait time at traffic signals around the new bridge. These signals give priority to light rail and force buses to wait to keep space between the two travel modes. In terms of efficiency, travel appears significantly more predictable due to the dramatic difference in travel-time variability during the peak period. This predictability may reduce the need for added and unscheduled buses during peak congestion, which may reduce long-term costs.

Acknowledgments

The authors thank Steve Callas and Miles J. Crumley of TriMet for providing the data sets used in this analysis and for their support in understanding the intricacies of how the data are structured. The authors also acknowledge the support of National Institute for Transportation and Communities transportation center for funding this research effort.

Author Contributions

The authors confirm contribution to the paper as follows: study conception and design: Glick, Figliozzi; data collection: Tri-County Metropolitan District of Oregon (TriMet); analysis and interpretation of results: Glick, Figliozzi; draft manuscript preparation: Glick, Figliozzi.

References

1. Levinson, H. S. Analyzing Transit Travel Time Performance. *Transportation Research Record: Journal of the Transportation Research Board*, 1983. 915: 1–6.
2. Portland Bureau of Transportation. *Tilikum Crossing: Bridge of the People*. The City of Portland [Online]. 2015. Accessed: May 2017. https://www.portlandoregon.gov/transportation/68548.
3. TriMet. *Tilikum Crossing Bridge of the People*. TriMet [Online]. 2015. Accessed: May 2017. https://trimet.org/tilikum/.
4. Persuad, B., and C. Lyon. Empirical Bayes Before–After Safety Studies: Lessons Learned from Two Decades of Experience and Future Directions. *Accident Analysis & Prevention*, Vol. 39, No. 3, 2007, pp. 546–555.
5. Hauer, E. *Observational Before-After Studies in Road Safety*. Elsevier Sciences Ltd., Oxford, 1997.
6. Hauer, E., and B. Persaud. Common Bias in Before-and-Bfter Accident Comparisons and Its Elimination. *Transportation Research Record: Journal of the Transportation Research Board*, 1983. 905: 164–174.
7. Farmer, C. M., A. K. Lund, R. E. Trempel, and E. R. Braver. Fatal Crashes of Passenger Vehicles Before and After Adding Antilock Braking Aystems. *Accident Analysis & Prevention*, Vol. 29, No. 6, 1997, pp. 745–757.
8. Huang, H. F., and M. J. Cynecki. Effects of Traffic Calming Measures on Pedestrian and Motorist Behavior. *Transportation Research Record: Journal of the Transportation Research Board*, 2000. 1705: 26–31.
9. Ringner, B. Law of the Unconscious Statistician. *Unpublished Note*, 2009.
10. Glick, T. B., and M. A. Figliozzi. Traffic and Transit Travel Time Reliability Indexes and Confidence Intervals. *Transportation Research Record: Journal of the Transportation Research Board*, 2017. 2649: 28–41.
11. Ayvalik, C. K., and Jotin Khisty, C. Heuristic Analysis of Impacts of Commuter Rail Station Consolidation on Pedestrian Access. *Tranportation Research Record: Journal of the Transportation Research Board*, 2002. 1793: 47–54
12. Kim, K. "Metro Travel Forcasting 2008 Trip-Based Demand Model Methology Report" Planning Department - Transportation Research and Modeling Services, 2008.
13. Hall, R., and N. Vyas. Buses as a Traffic Probe: Demonstration Project. *Transportation Research Record: Journal of the Transportation Research Board*, 2000. 1731: 96–103.
14. Cathey, F., and D. Dailey. Transit Vehicles as Traffic Probe Sensors. *Transportation Research Record: Journal of the Transportation Research Board*, 2002. 1804: 23–30.
15. Bertini, R., and S. Tantiyanugulchai. Transit Buses as Traffic Probes: Use of Geolocation Data for Empirical Evaluation. *Transportation Research Record: Journal of the Transportation Research Board*, 2004. 1870: 35–45.
16. Berkow, M., M. Wolfe, C. Monsere, and R. Bertini. Using Signal System Data and Buses as Probe Vehicles to Define the Congested Regime on Arterials. Presented at 87th Annual Meeting of the Transportation Research Board, Washington D.C., 2008.
17. Skabardonis, A., and N. Geroliminis. Real-time Estimation of Travel Times on Signalized Arterials. *Proc. 16th International Symposium on Transporation and Traffic Theory*, College Park, MD, 2005, pp. 387–406.
18. Bertini, R. L., and A. M. El-Geneidy. Generating Transit Performance Measures with Archived Data. *Transportation Research Record: Journal of the Transportation Research Board*, 2003. 1841: 109–119.
19. Chakroborty, P., and S. Kikuchi. Using Bus Travel Time Data to Estimate Travel Times on Urban Corridors. *Transportation Research Record: Journal of the Transportation Research Board*, 2004. 1870: 18–25.
20. Mazloumi, E., G. Currie, and G. Rose. Using GPS Data to Gain Insight into Public Transport Travel Time Variability. *Journal of Transportation Engineering*, Vol. 136, No. 7, 2010, pp. 623–631.
21. Crumley, M. Interviewee, Operational TriMet Questions [Interview]. 2017.
22. *Portland Region 2016 Traffic Performance Report*. Oregon Department of Transportation, 2016.
23. Brown, M. B., and R. A. Wolfe. Estimation of the Variance of Percentile Estimates. *Computational Statistics & Data Analysis*, Vol. 1, 1983, pp. 167–174.
24. Taylor, J. R. *An Introduction to Error Analysis The Study of Uncertainties in Physical Measurements*. University Science Books, Sausalito, Calif., 1997.
25. Anderston, M. *Tilikum Crossing: A Few Words of Advice for Folks Working to Conquer Hills*. Bike Portland [Online], 2015. Accessed. May 2017. https://bikeportland.org/tag/tilikum-crossing.

The Public Transportation Group (AP000) peer-reviewed this paper (18-06530).

Any errors or omissions are the sole responsibility of the authors.

Article

TRR
JOURNAL OF THE TRANSPORTATION RESEARCH BOARD

Transportation Research Record
2018, Vol. 2672(8) 210–219

DOI: 10.1177/0361198118758284
journals.sagepub.com/home/trr
SAGE

Private Shuttles and Public Transportation: Effects of Shared Transit Stops on Travel Time and Reliability in Seattle

Elyse O'C. Lewis[1], Don MacKenzie[1], and Regina Clewlow[2]

Abstract

An Employer Shared Transit Stop (ESTS) pilot was introduced in April, 2017 by the Seattle Department of Transportation (SDOT) and King County Metro (KCM). Eleven bus stops within the City of Seattle were identified for stop-sharing with private shuttles that serve employees of and are operated by Microsoft and Seattle Children's Hospital, respectively. This study utilizes real-time transit performance data that was generated by KCM, then collected and aggregated by the company Swiftly to assess the impacts of the ESTS program on bus transit reliability. Data analyses based on visualization tools on the Swiftly Insights platform of route-level schedule adherence as well as fixed panel regressions at the mean and at the 0.90 and 0.95 quantile of stop-level schedule adherence are considered. On average, bus transit reliability has not been affected by the ESTS pilot program. Statistically significant results indicating a potential delay in bus arrival times as a result of ESTS stop-sharing were found at only one of the nine stops considered in the study. An increase in participating companies and stops is recommended.

As a city with significant geographical constraints and a rapidly growing population (*1*), the City of Seattle looks for ways to manage roadway congestion and maintain equitable, quality access for citizens (*2*). To these ends, goals for reducing drive-alone rates during commute hours have been adopted at a state level through the Commute Trip Reduction (CTR) Law (*3*). CTR requires cities and counties in Washington State "to reduce the number and length of drive-alone commute trips" by working with major employers in their jurisdictions to develop and implement employee commute programs (*3*). Such measures include, but are not limited to, shuttle services for employees, particularly shuttle services for first mile/last mile links to transit and for routes underserved by transit. Additionally, public transit service providers such as King County Metro (KCM) that share the congested streets of Seattle with personal vehicles have established service guidelines to measure and pursue the goal of quality transportation alternatives for all Seattle residents (*4*).

In an attempt to improve private shuttle services provided by major employers for their employees in the City of Seattle, the Seattle Department of Transportation (SDOT) and KCM began the Employer Shared Transit Stop (ESTS) pilot to "test the feasibility of allowing employer-provided shuttles to use public transit stops while minimizing impacts to public transit operations" (*5*). On April 24th, 2017 the pilot began, allowing private shuttles operated by Microsoft and Seattle Children's Hospital (SCH) to share nine of eleven bus stops throughout the city of Seattle that were identified for the pilot program (*5*). Employers pay a monthly fee for each stop in exchange for special signage and permission to use of the public right of way (ROW) as a pick-up and drop-off for employees (*6*).

Microsoft and SCH both provide private shuttles to employees through their CTR programs. Shuttle fleets are comprised of community buses and motor coaches as defined by the Transportation Capacity and Quality of Service Manual (TCQSM) (*7*). Microsoft Connector Buses provide an alternative to personal vehicle commuting to their campus in Redmond, Washington located 13 miles from downtown Seattle (*8*). SCH's 22-shuttle system provides both first mile/last mile connections to the University of Washington (UW) LINK light rail transit hub as well as along underserved routes (*9*). SCH's CTR program has been particularly aggressive, earning multiple awards for excellence (*10*).

The ESTS pilot program was met with mixed public opinion. The potential positive benefits include: (1) cost-sharing in the maintenance and upgrades to transit stops, which could result in better quality stops overall; (2) employees using

[1]University of Washington, Seattle, WA
[2]Research Affiliate, University of California, Davis, CA

Corresponding Author:
Address correspondence to Elyse O'C. Lewis: elyseoc@uw.edu

shuttle buses can more easily transfer between private and public transit; (3) formal loading and unloading is safer for all users of the public ROW; (4) existing transit facilities would have higher overall utilization and improve pedestrian-oriented investments nearby; and (5) sharing facilities could put less pressure on limited curb space throughout the transportation network (*11, 12*). Negative concerns are mostly related to the existence of such shuttles in general, namely their potential to enable gentrification (*13, 14*). Beyond this, concerns about the shuttles' impact on transit have also been raised in response to the announcement of the ESTS pilot program specifically (*6*).

The ESTS pilot is a trial to determine what the impacts of allowing private shuttles to utilize public transit ROW will have on transit services. In a review of literature related to bus transit service reliability, reliability was consistently held as a critical passenger priority, arguably only second in importance to "arriving safely at destinations," and passenger patronage of bus transit is directly correlated to service reliability (*15*). Moreover, from an operator's perspective, low reliability not only affects patronage but can contribute "to increased operating costs" as it "impacts the schedule recovery component of cycle time" (*7*).

Reliability, however, is an ambiguous attribute that is consequently difficult to quantify and measure. In current practice, reliability is typically quantified as on-time performance (OTP). However, while OTP is an easy metric to calculate, it does not adequately represent reliability as perceived from a passenger's perspective. The TCQSM recognizes not only reliability explicitly in its five key concepts but that it also identifies it as a passenger-perspective quality of service performance measure (*7*). To address this, recent efforts by innovative transit agencies are currently seeking to identify more effective ways to measure reliability from a passenger's perspective (*16*).

Though identifying a single metric for reliability is difficult, it generally relates to time (*15*). Many agencies nationally, including KCM, maintain access to General Transit Feed Specification (GTFS) and APIs of real-time transit performance data (*17*). The company Swiftly has collected this data and aggregated it into a format that can be easily queried by tools through its advanced analytics platform as well as downloaded in CSV format for further analysis.

To determine the ESTS pilot impacts, this paper first reviews practical metrics and analysis methods for reliability, then applies these methods to the Swiftly data set of KCM bus schedules and adherence. The three analyses presented in this paper include a standard OTP runtime performance assessment per KCM guidelines, a fixed effects panel regression of schedule adherence at the stop level, and a quantile regression of schedule adherence at the stop level.

Measuring Reliability

While the definition of reliability varies, it consistently involves time, from how actual bus times relate to scheduled times, to consistency of travel time and minimized waiting times (*15*). The most basic and most commonly used metric for reliability is on-time performance (OTP) (*15*). While "most agencies define reliability in terms of OTP," the exact time ranges that constitute an on-time bus arrival/departure vary widely depending on the operator, though, on average, operators consider a bus on-time if it arrives within the window of 1 min early to 5 mins late (*18*). At KCM, schedule reliability is measured with an OTP window that only considers late bus arrivals; routes that experience arrivals >5 min late 35% of the time during PM peak periods or 20% of the time on average are considered non-compliant (*19*). Any early arrivals are considered on-time. In a 2016 OTP service review, KCM found 60 routes that needed "service-hour investments to improve reliability" (*19*).

Percent on-time OTP, though widely used by operators, is considered a sub-par metric throughout the research literature because it does not adequately capture time reliability from a passenger perspective. For passengers, waiting time at bus stops is valued "more than any other time component of their trip" (*15*). In fact, studies have found that passengers can value waiting time at a rate of 3–5 times their time in-vehicle and that the quality of waiting locations can also be valued in in-vehicle time equivalents up to 1.3 min for a stop shelter (*7*). Furthermore, passengers tend to overestimate waiting time, particularly if that waiting time is unpredictable (*15*). Therefore, beyond seeking to reach a set percentage of trips/arrivals that are on-time, looking at the arrivals that fall outside of the "on-time" range and the severity of deviation better captures the passenger's perception of reliability. Additionally, analyzing schedule adherence at the stop level rather than at the route runtime level further focuses on the most critical time component of the trip from a passenger's perspective of reliability.

Based on this, multiple analyses of reliability were considered to determine the impact of the ESTS pilot on KCM service reliability. These methods attempt to capture the impact of company shuttles on transit performance not only from a planning perspective, but also from a user perspective. In a percent on-time OTP analysis (based on KCM standards), the dependent variable is schedule adherence of runtime for each of the affected routes. For the panel regression comparisons, the dependent variable is schedule adherence at the stop level.

Data

Swiftly's Data and Analytics Platform

Swiftly, Inc. is a software company that specializes in developing accurate real-time passenger information and robust data analytics for transit operators. Swiftly Insights is a cloud-based platform that processes and archives GPS data from the computer-aided dispatch (CAD) and automated vehicle location (AVL) systems of transit agencies. The

Table 1. Treated Stops and Routes Identification and Labeling for the Study

Study label	KCM ID#	Stop identification KCM Street	KCM cross-street	Direction of travel	Company shuttle	Implem. date	Routes that utilize the stop				
A	2180	Queen Anne Ave N	W Harrison St	S	Microsoft	24/04/2017	1	2	8	13	32
B	10562	Sand Point Way NE	40th Ave NE	N	SCH	24/04/2017	75				
C	11420	15th Ave E	E Mercer St	S	Microsoft	24/04/2017	10				
D	12340	E Madison St	25th Ave E	SW	Microsoft	24/04/2017	8	11			
E	13250	19th Ave E	E Harrison St	S	Microsoft	24/04/2017	12				
~~F~~	~~25200~~	~~NE 45th St~~	~~Union Bay Pl NE~~	~~W~~	~~SCH~~	~~Later~~	~~31~~	~~32~~	~~67~~	~~75~~	~~78~~
G	25765	Montlake Blvd NE	NE Pacific Pl	N	SCH	24/04/2017	65	78			
H	29720	NW Market St	20th Ave NW	W	Microsoft	24/04/2017	44				
~~I~~	~~29920~~	~~NE 45th St~~	~~Mary Gates Memorial~~	E	~~SCH~~	~~Later~~	65	~~75~~			
J	31970	California Ave SW	SW Spokane St	S	Microsoft	24/04/2017	50	55	128		
K	37920	NE 65th St	39th Ave NE	W	Microsoft	24/04/2017	62	71	76		

platform allows transit agencies, planners, and researchers to quickly access and analyze GPS data points, both for real-time monitoring and for analysis of historical data.

Swiftly Insights generates aggregate variables such as schedule adherence in addition to presenting the basic data pulled from the agency AVLs. Schedule adherence is calculated as the difference between scheduled and actual bus departure times. These variables can then be analyzed either through the visualization tools provided within the Swiftly Insights dashboard or through analyses of CSV downloads of the data.

Data-Set Development

To determine appropriate date and time ranges for the data, KCM press releases and route revisions since the start of Swiftly data records (3/23/2016) were reviewed to determine potential scheduling/routing impacts. For analyses considering before and after effects of the ESTS program, data was considered from the Monday of 3/13/17 after the implementation of a 3/11/17 semi-annual KCM schedule change and until 6/04/17, providing six weeks of data on either side of the 4/24/17 pilot start date. Multiple study stops fall near the UW Seattle campus, so it should be noted that classes were in session throughout the study period with the exception of week 2 of the study when classes were out for spring break.

To determine the routes needed for analysis, the OneBusAway online browser was used to identify affected stops and related routes based on the SDOT press release announcing the ESTS pilot program (summarized in Table 1). According to this release, all but two Children's stops were introduced on the 4/24/2017 pilot start date (*5*). These two stops were removed from the study (as shown by the strike-throughs in Table 1), resulting in a total of nine shared stops considered as the treatment stops in the study. As shown in the Table 1 summary of treated stop and route information for the study, all of the study stops are low-volume, with only one initial stop (A) serving five routes. The majority (7/9) of study stops only serve one or two routes and all of the study stops are simple curb-side (rather than bus bay) stops.

To develop the control group, schedule adherence reports from the previous year were avoided given the rapid growth in Seattle. Instead, two types of stops were identified as potential control stops: upstream (US) stops and different route (DR) stops. US stops are stops located upstream (or before) a treated stop whereas DR stops are located in the same geographic area but not along the same route as a treated stop. US stops were selected because, among a wide variety of factors that have been found to affect bus reliability, many are directly tied to a given route; factors include driver experience, length and complexity of route, operating environment, and other route conditions such as on-street parking, signalized intersections, and direction of travel (*20*). For both control group stop types, only low-volume, curb-side bus stops were considered.

While stops along nearby streets with comparable surrounding land use and parking patterns as well as number and type of routes utilizing the stops were held as the ideal for DR control stop selection, such stops exist for only two of the nine treated stops. As a result, the majority of DR stops identified were nearby stops in the opposite direction of travel of treated routes/stops, that is, stops across the street from treated stops. To try to control for the impact of the differing directionality, DR stops in the opposite direction of travel from treated stops were only selected if they experienced similar weekday peak congestion per the Google Maps Typical Traffic™ function.

Based on these requirements, the CSV data download feature in the Swiftly Insights platform was used to extract the historical KCM data for all potentially affected routes. The data was disaggregated into AM and PM peak sets per KCM guidelines: AM peak from 6 AM to 9 AM and PM peak from 3:30 PM to 6:00 PM (*21*). All routes in the vicinity of the treated stops operate within mixed traffic environments as defined by the TCQSM (*7*). Cumulative Distribution Function

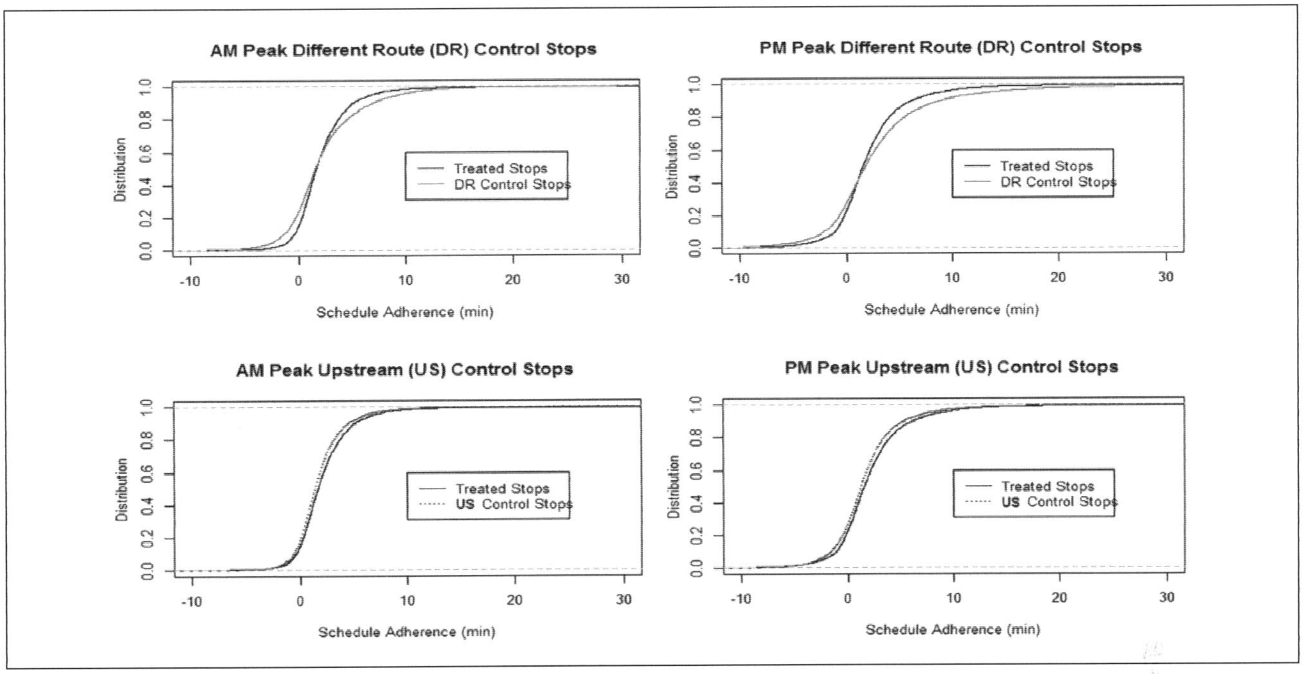

Figure 1. CDF plots comparing the schedule adherence distributions of potential collections of control stops upstream (US) or along different routes (DR) to treated stops before the ESTS program began.

(CDF) plots were used to compare the distributions before the implementation of the ESTS program of the potential control stops versus the treated stops to determine the best control group for the study. As shown in Figure 1, the DR stops have a different distribution of schedule adherence compared to the treated stops, especially in the PM peak. In contrast, the US stops have a similar distribution before the start of the ESTS stop sharing pilot in both the AM and PM peak as shown in Figure 1. Based on this, US stops at a ratio of 3:1 control:treated stops were selected as control stops for the analysis with one exception.

As shown in Figure 2, treated stop B is downstream of treated stop G. Additionally, stops along the west and southwest edges of the UW serve as major transfer hubs for buses within the region. As a result, control stops for either stop B or G could not be identified immediately upstream. Control stops for treated stop G were selected from US stops west (or further upstream of) the transfer hubs that border the University campus. Because stop B is located near the start of route 75, a route that runs through multiple bus transfer hubs on UW's campus, only one US stop could be identified. As a result, stop B has a ratio of 1:1 control:treated stops. However, because the US control stop serves a second route in addition to the 75, observations are roughly 2:1 control stop:stop B.

Analysis Methods

Four analysis methods were selected to determine ESTS impacts on KCM reliability from the Swiftly schedule adherence data set. The first analysis most closely adheres to the KCM guidelines for reliability measurement. It considers the route runtime and analyzes the data using the tools provided by the Swiftly platform. From this, the percentage of late run times along each route before and after the ESTS start date is identified.

The second analysis method utilizes panel regression which considers all treated stops, along with US control stops, along the same affected route. Panel regression is a widely used method in social sciences and econometrics to analyze a data set in terms of identity and time to consider cross-sectional and longitudinal relationships and impacts (*22*). Specifically, a fixed effects panel regression was utilized to determine the average difference between treated and control stop types, between weeks within the study, and between stops affected by the ESTS pilot shuttles and non-treated stops. A total of 12 study weeks were considered with the ESTS pilot beginning at the start of the seventh study week. Stops affected by the ESTS pilot shuttles are the treated stops during weeks 7–12, and non-treated stops are all stops in the control group and stops in the treated stop group during weeks 1–6. This model ultimately identifies the average waiting time at the stop caused by the introduction of ESTS pilot shuttles. Given that "unreliable transit service will increase average waiting time," the model provides a good metric to determine whether the ESTS pilot affected transit reliability (*7*).

The third analysis method considers quantile panel regressions of the 0.90 and 0.95 quantile of schedule adherence. Because quantile regression focuses on a user-specified

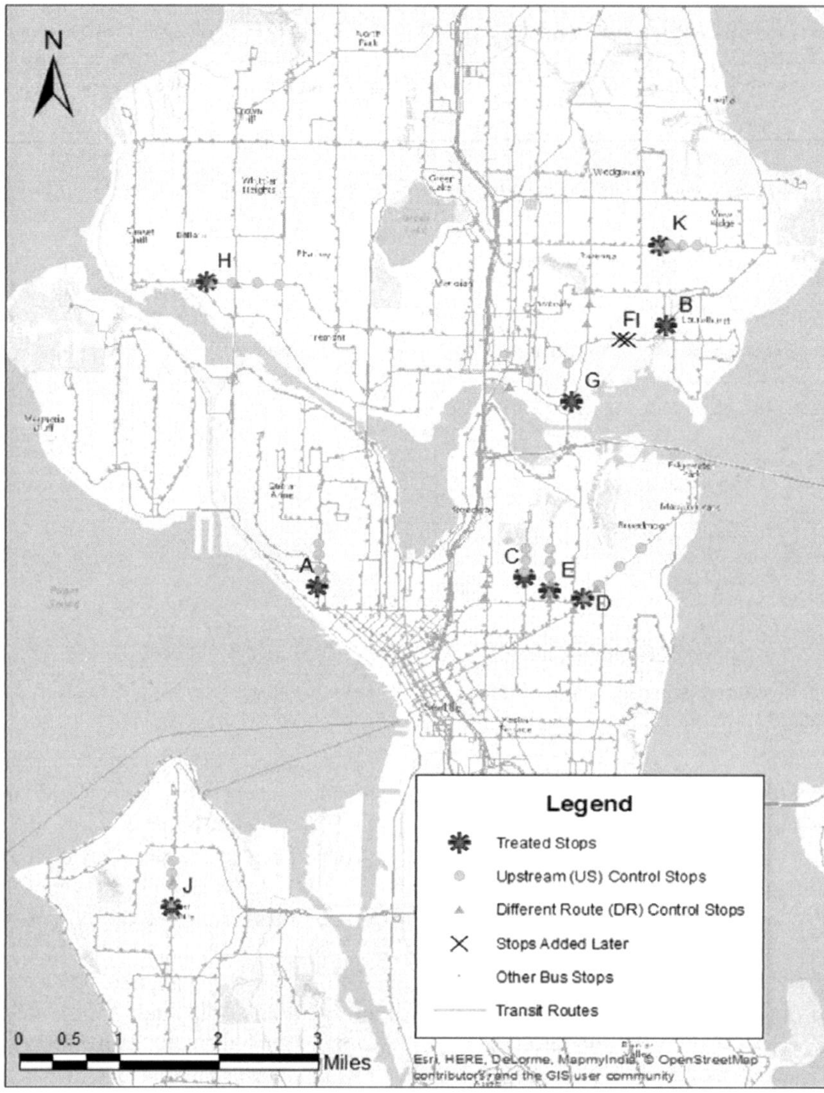

Figure 2. ESTS pilot study stops and control stops considered within the analysis.

portion of a data distribution rather than fitting an average to the distribution as a whole, it generates models that more robustly estimate the effect of outliers in a data set (*23*). It is therefore ideal for assessing the frequency of severe bus delays and thus more closely approximates the frequency of low reliability from a passenger perspective. The quantile regression R package is "quantreg" (*24*).

Finally, to separate and further investigate individual treated stops, CDF plots were utilized to identify different distributions of schedule adherence before and after the ESTS pilot start date. First, the quantiles of schedule adherence at each of the treated stops before and after the ESTS pilot start date were compared to identify a subset of stops with increased arrival delays after the start of the ESTS pilot. Quantile regression at the median (0.50 quantile) as well as in the 0.90 and 0.95 quantiles were fitted to the subset of stops to determine if the increase in bus delays was the result of the ESTS shuttles or some other factor.

Discussion of Results

The results of the four analysis methods are discussed below.

Analysis 1: OTP Measurement Following KCM Guidelines

KCM guidelines for reliability identify routes experiencing >35% late arrivals (i.e., arrivals more than 5 mins past the scheduled time) in PM peak as unreliable and in need of

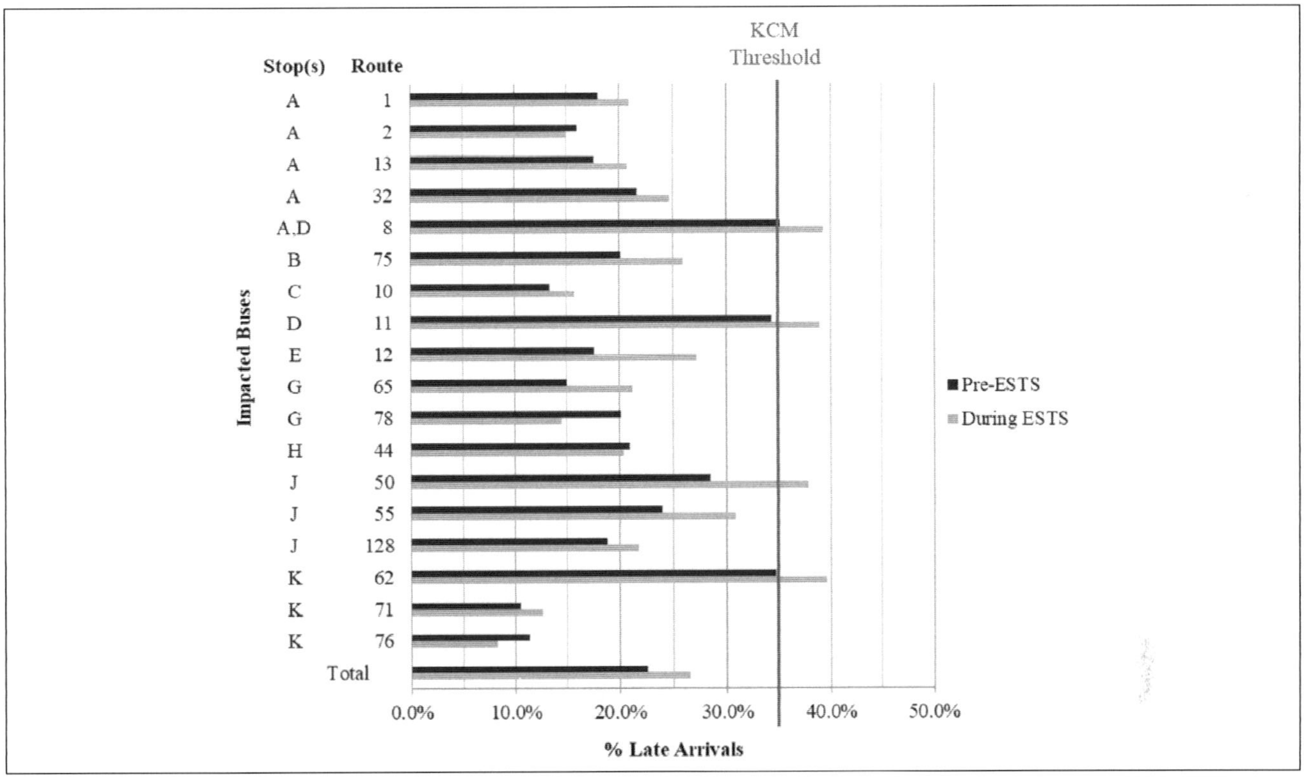

Figure 3. Schedule adherence by route for routes with shared stops before and after the ESTS pilot start date during the PM peak.

remedial attention (*19*). Per these guidelines, Figure 3 was generated using the Schedule Adherence by Route feature on the Swiftly Insights platform to look at the percent on-time OTP of routes affected by ESTS shuttle stops before and after the pilot start date (4/24/17). As Figure 3 shows, the percentage of late arrivals increased in most study routes with four (rts 8, 11, 50, 62 which stop at A and D, D, J, and K, respectively) of the 18 routes passing the KCM threshold of 35% late arrivals. Comparatively, only one (rt 8 which stops at A) of the 18 routes passed the 35% threshold before the ESTS start date.

While this suggests a potential impact on routes, it does not indicate whether these increased delays are the result of the ESTS shuttles or some additional factors. For example, the 8 is the only bus that runs along a highly congested east-west thoroughfare near the Amazon campus with a portion of the roadway under construction during the study dates. Additionally, the 62 has been receiving remedial attention since it was created as a combination of two routes discontinued after the LINK light rail extension in March of 2016. The 50, however, has remained constant in schedule with no published concerns since early 2016, as has the 11, beyond minor routing changes in September 2016. Therefore, while this percent on-time OTP route-level analysis suggests the possibility that ESTS shuttles are affecting OTP, additional analysis is necessary.

Analysis 2: Fixed Effects Panel Regression

To better identify whether the ESTS shuttles affect bus performance, this second analysis not only focuses more acutely on the point of impact (the stops), but also introduces a set of control observations. Additionally, both AM and PM peak periods are considered. Once US control stops were identified (as explained in the "DATA" section), control and treated stops were compared in a fixed effects panel regression as specified in Equation 1.

$$\begin{gathered} y_{iwt} = \alpha_i + \gamma_w + \beta D_{iw} + \epsilon_{iwt} \\ i = \text{stops} \\ w = \text{weeks} \\ t = \text{observations} \end{gathered} \tag{1}$$

The model specification includes fixed effects both for stops (α_i) and for weeks (γ_w). The stop-level fixed effects capture average schedule adherence at each stop, while the week-level fixed effects capture area-wide variations between weeks such as those caused by seasonality and UW sessions. A treated dummy variable (D_{iw}) is used to specify when treated stops are subjected to the ESTS pilot program treatment, and the coefficient β represents the average treatment effect. The dummy variable D_{iw} has a value of 1 from weeks 7 to 12 at the treated stops (since week 7 marks the start of the ESTS pilot program) and is 0 otherwise. Table 2 shows

Table 2. Fixed Effects Panel Regression of Schedule Adherence in the AM and PM Peak at All ESTS Study Stops

	Dependent variable: Schedule adherence (min)	
	AM peak	PM peak
Week 2	−0.256** (0.063)	−0.582** (0.094)
Week 3	−0.116 (0.066)	0.542** (0.091)
Week 4	−0.015 (0.066)	−0.107 (0.092)
Week 5	−0.653** (0.066)	−0.837** (0.092)
Week 6	−0.365** (0.067)	−0.016 (0.093)
Week 7	−0.270** (0.069)	−0.401** (0.095)
Week 8	−0.063 (0.070)	1.235** (0.098)
Week 9	−0.246** (0.069)	0.016 (0.096)
Week 10	−0.227** (0.068)	0.174 (0.096)
Week 11	−0.054 (0.069)	0.532** (0.097)
Week 12	−0.245** (0.070)	0.426** (0.100)
Treated dummy	0.001 (0.060)	0.102 (0.087)
Constant	2.534** (0.069)	1.880** (0.095)
Stop fixed effects	Yes	Yes
Observations	37,477	37,155
R^2	0.1	0.077
Adjusted R^2	0.099	0.076
Residual std. error	2.641 (df = 37431)	3.763 (df = 37109)
F statistic	92.904** (df = 45; 37431)	69.227** (df = 45; 37109)

Note: *$p < .05$; **$p < .01$.

Table 3. Fixed Effects Quantile Regression in the AM and PM Peak for All ESTS Study Stops

	Dependent variable: Schedule adherence (min)			
	AM peak		PM peak	
	0.9	0.95	0.9	0.95
Treated dummy	−0.147 (0.085)	−0.141 (0.135)	−0.376* (0.171)	−0.226 (0.229)
Constant	5.846** (0.142)	7.314** (0.164)	5.512** (0.186)	7.798** (0.240)
Weekly fixed effects	Yes	Yes	Yes	Yes
Stop fixed effects	Yes	Yes	Yes	Yes
Degrees of freedom	140116 total; 140041 residual		149170 total; 149095 residual	

Note: *$p < .05$; **$p < .01$.

the results of the regression analysis. The estimated treatment effects of the ESTS pilot program are small and not statistically significant. This suggests that, on average, sharing stops with private shuttles did not affect schedule adherence for KCM buses at the studied stops. Further, the "Constant" for the models represents the average schedule adherence in the data set; for both AM and PM peaks, the average arrival time for buses occurs on-time, only ~2 mins after the scheduled arrival time.

Analysis 3: Quantile Regressions of Stop Schedule Adherence

While the fixed effects panel regression considered the average impact of ESTS shuttles on the treated stops, or impact at the mean, quantile regressions provide a shifted focus. Quantile regressions of the fixed effects panel data were performed on all stops at the 0.90 and 0.95 quantiles (i.e., the 90th and 95th percentiles) to look more closely at the severity of later bus arrivals. Models were fitted per the same specifications given in Equation 1 and the resulting coefficients for the intercept and treatment dummy variable are presented in Table 3. As shown, results in the 0.95 quantile are not statistically significant, while results for the 0.90 quantile have a *p*-value of .028 in the PM peak. The estimated treatment effect implies that schedule adherence improved as a result of the ESTS pilot. This is an unreasonable causal impact and suggests that additional factors not present in this analysis affected the results. To seek further

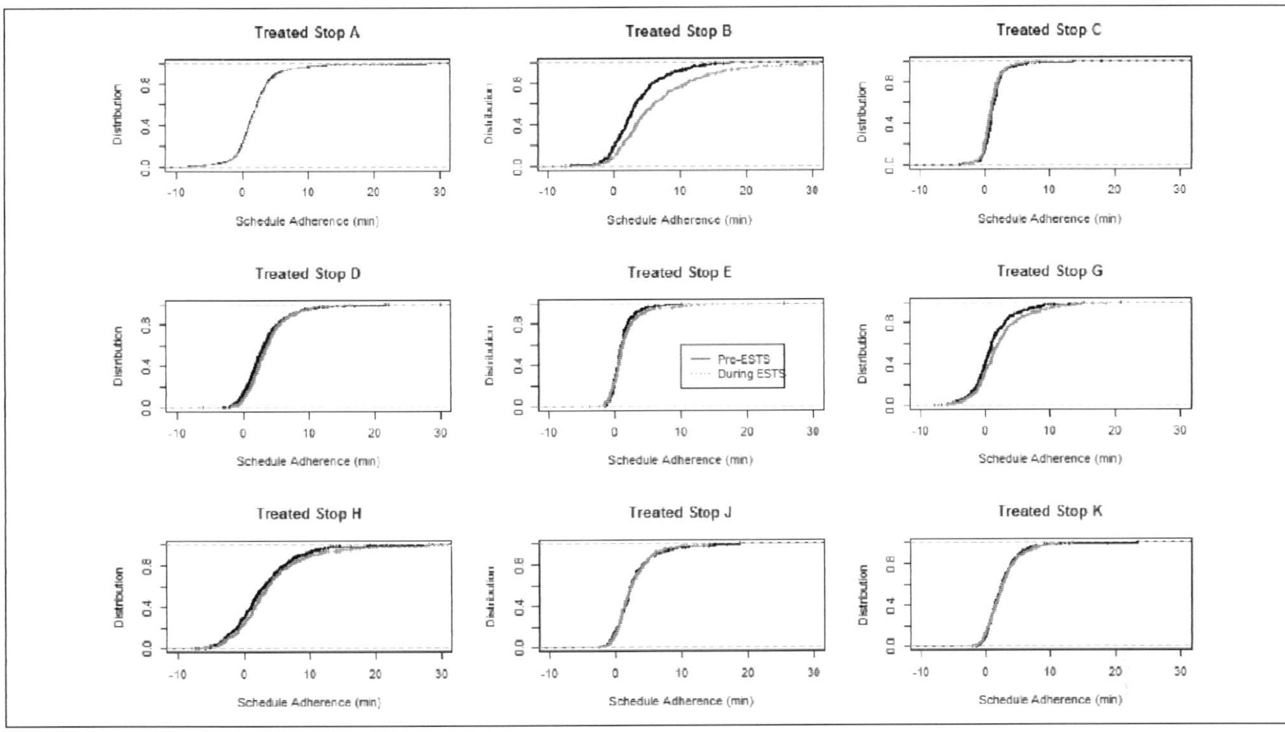

Figure 4. CDF plots of schedule adherence for treated stops before and after the start of ESTS.

clarification, a fourth analysis of each stop individually was considered.

Analysis 4: Quantile Comparisons and Regressions of Individual Stop Schedule Adherence

While the results at both the mid and upper quantiles did not indicate an effect of the ESTS program on transit performance, an additional analysis was conducted to determine whether or not individual stops have been affected. Rather than relying exclusively on the first analysis of an increase in late runtimes after the start of the ESTS program, schedule adherence before and after the start of the ESTS program at the stop level was assessed by comparing distributions. Figure 4 presents the CDF plots of schedule adherence times before and after the start of ESTS at the treated stops. Most of the stops show little to no difference, as the CDF curves overlap almost perfectly. Plots B and G, however, represent the stops that have experienced clear disparities between the distribution of schedule adherence at the stop before and after the start of the ESTS program. At both stops, buses are consistently later after the start of the ESTS program, with significantly later arrivals occurring at stop B after the start of the ESTS program. Based on these variations, treated stops B and G were identified as stops that required additional attention.

To determine whether these disparities are the result of the presence of ESTS shuttles or simply the result of other factors not considered in the analysis, fixed effects panel regressions at the median (0.50), 0.90, and 0.95 quantile comparing the treated stops to their corresponding control stops were considered. Table 4 presents the estimated treatment effects for stops B and G. Quantile regression models for stop G failed to reject the null hypothesis that the ESTS shuttles had no impact on bus schedule adherence during both peak periods and at all quantiles tested.

The models for stop B, however, reject the null hypothesis at the median (0.50) and 0.90 quantile during the PM peak. As Figure 5 shows, an increase in delays at the treated stop B compared to the US control stop occurs after the ESTS pilot begins. The differences between stops are very statistically significant, many with p-values $<.0001$. Based on this, the model suggests that, while arrival times at stop B are consistently later than arrival times at the control stop throughout the study period, buses began arriving even later at stop B after the start of the ESTS pilot. In the aggregate, these results suggest that the introduction of ESTS shuttles at stop B may have affected schedule performance.

Conclusions and Recommendations

Many factors affect bus schedule adherence. This study attempts to control for a multitude of factors through a robust control group development effort to study the impact of the introduction of private shuttles sharing bus transit stop ROW through the ESTS program with schedule adherence (time) information. While standard KCM reliability analysis methods looking at OTP at the route level were considered, they

Table 4. Treated Dummy Coefficients for Schedule Adherence at the 0.50 (Median), 0.90, and 0.95 Quantile of Stops B and G during the AM and PM Peak

	Treated dummy values for dependent variable: Schedule adherence (min)					
	AM peak			PM peak		
Models	0.5	0.9	0.95	0.5	0.9	0.95
Stop B	0.395 (0.311)	0.791 (1.130)	1.382 (0.957)	1.864** (0.457)	2.756* (1.386)	1.644
Stop G	−0.165 (0.125)	−0.629 (0.378)	−0.748 (0.585)	0.178 (0.226)	−0.889 (0.764)	−1.874 (1.284)
Weekly fixed effects	Yes	Yes	Yes	Yes	Yes	Yes
Stop fixed effects	Yes	Yes	Yes	Yes	Yes	Yes
Deg. of freedom stop B	5187 total; 5173 residual			7146 total; 7132 residual		
Deg. of freedom stop G	18039 total; 18023 residual			17946 total; 17930 residual		

Note: $*p < .05$; $**p < .01$.

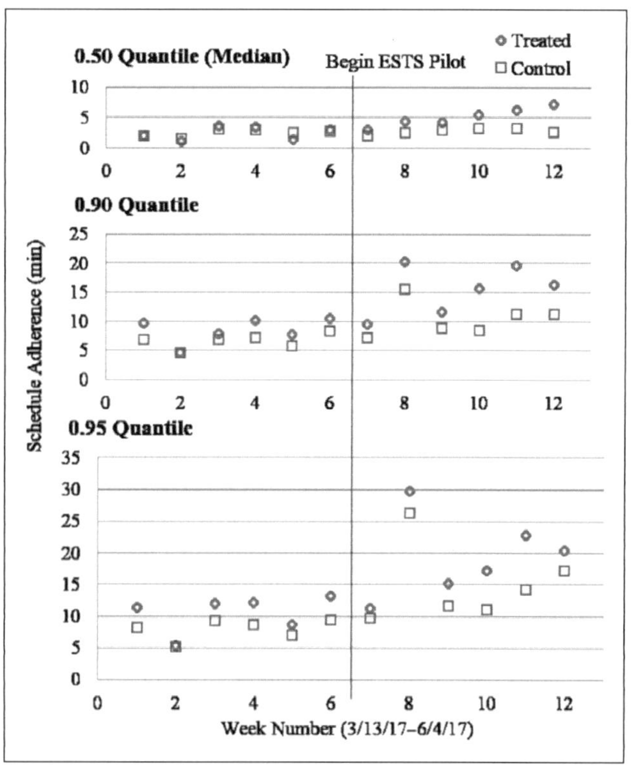

Figure 5. Stop B treated versus control stop schedule adherence in the PM peak by quantile.

were ultimately too zoomed out from the study locations. Moreover, discrepancies were found at the stop level that did not appear at the route level and vice versa. Furthermore, stop-level analysis is more likely to capture the perceived reality of impact caused by ESTS shuttles by better representing what passengers experience and would consider to be unreliable service.

An aggregate analysis of all nine stops selected for the ESTS pilot study paired against control stops and considered at the mean as well as at the 0.90 and 0.95 quantile in fixed effects panel regressions did not suggest a negative impact on public transit performance. At the stop level, however, one model suggested a potential relationship. Stop B that serves only one route (route 75) may have been affected. Stop B is a low capacity bus facility with only one loading area. Given that capacity and reliability are inherently linked, it is possible that the addition of shuttles stopping at this low capacity location is the sole cause of the increased bus arrival delays (cite TCQSM pp. 1–7). However, given that none of the other low-volume, single loading zone stops exhibit statistically significant impacts, the presence of multiple loading areas at a stop should not necessarily be a prerequisite for shared stop selection.

Based on this analysis, further investigation is needed to determine whether ESTS shuttle sharing is the sole cause of these delays, and, if so, why this stop is adversely affected while others are not. Further, because low-volume pilot stops were selected, results from these stops may not provide a clear indication of the potential impacts if a wider-spread program made use of busier stops. However, given the apparent minimal to no impact caused by the ESTS shuttle stops sharing ROW with transit buses at lower-volume stops, an expansion of the program at similar, lower-volume stops is recommended at this time. A more detailed analysis that considers more factors beyond time is necessary to fully isolate and determine impacts from ESTS. However, the potential for positive utilization of public curb space ROW and increased investment in transit services that revenue from a stop-sharing program such as ESTS provides outweigh the minimal negative impacts and warrant further investigation into expansion options.

Acknowledgments

Thank you to Swiftly for support and data for this project.

References

1. Puget Sound Regional Council. *Population Grows at Fast Pace in 2017 [Internet]*. Puget Sound Regional Council, 2017. https://www.psrc.org/whats-happening/blog/population-grows-fast-pace-2017. Accessed July 13, 2017.

2. Puget Sound Regional Council. *Transportation 2040 Executive Summary [Internet]*. Puget Sound Regional Council, 2010. https://www.psrc.org/sites/default/files/4web_finalt2040es.pdf. Accessed May 26, 2017.
3. King County Metro. *About the Commute Trip Reduction (CTR) Law [Internet]*, 2017. http://www.kingcounty.gov/transportation/CommuteSolutions/About.aspx. Accessed July 18, 2017.
4. King County Metro. *King County Metro Strategic Plan: Service Guidelines [Internet]*. King County Metro Transit, 2013. http://metro.kingcounty.gov/planning/pdf/Metro_ServiceGuidelines_2013.pdf. Accessed June 23, 2017.
5. Seattle Department of Transportation. *Seattle Department of Transportation: Employer Shared Transit Stop Pilot [Internet]*, 2017. http://www.seattle.gov/transportation/SharedTransitStopPilot.htm. Accessed June 23, 2017.
6. Bruce Nourish. *Pragmatism, Efficiency, and Who Shares [Internet]*. Seattle Transit Blog, 2017. https://seattletransitblog.com/2017/05/04/pragmatism-efficiency-and-who-shares/. Accessed July 18, 2017.
7. Kittelson & Associates, Inc., Parsons Brinckerhoff, K. F. H. Group, Texas A&M Transportation Institute, Arup. *Transit Capacity and Quality of Service Manual [Internet]*. Report No.: 165. Transportation Research Board of the National Academies, Transit Cooperative Research Program, 2013. https://www.nap.edu/download/24766#. Accessed July 31, 2017.
8. Brian Crockford. *Microsoft's Commute Program Keeps Employees Moving|NAIOP [Internet]*, 2016. http://www.naiop.org/en/Magazine/2016/Winter-2016-2017/Development-Ownership/Microsofts-Commute-Program-Keeps-Employees-Moving.aspx. Accessed July 13, 2017.
9. Association for Commuter Transportation. *Getting to Work: Spotlight on Employer-Sponsered Commuter Programs [Internet]*, 2017. http://actweb.org/wp-content/uploads/2017/01/Getting-to-Work-Jan-2017-Final.pdf. Accessed July 13, 2017.
10. Seattle Department of Transportation. *Seattle Commute Trip Reduction Employer Profile [Internet]*, 2015. http://www.seattle.gov/waytogo/docs/CTR_Profile_Boards_seattlechildrens.pdf. Accessed July 13, 2017.
11. Fesler, S. *Seattle Launches Shared Transit Stops with Microsoft And Children's Hospital » The Urbanist [Internet]*. The Urbanist, 2017. https://www.theurbanist.org/2017/04/24/seattle-launches-shared-transit-stops-microsoft-childrens-hospital/. Accessed July 13, 2017.
12. Lee, J. *A First-of-Its-Kind Program: Microsoft, Seattle Children's Shuttles Are Now Using Metro Stops*. The Seattle Times [Internet], 2017. http://www.seattletimes.com/seattle-news/transportation/a-first-of-its-kind-program-microsoft-seattle-childrens-shuttles-are-now-using-metro-stops/. Accessed July 18, 2017.
13. Crucchiola, J. *SF's Tech Bus Problem Isn't About Buses. It's About Housing [Internet]*, 2016. WIRED. https://www.wired.com/2016/02/sfs-tech-bus-problem-isnt-about-buses-its-about-housing/. Accessed July 13, 2017.
14. Grande, A. *Microsoft Connector Shuttles Targeted by Protestors [Internet]*. KIRO, 2014. http://www.kiro7.com/news/microsoft-connector-shuttles-targeted-protestors/81833193. Accessed July 13, 2017.
15. Diab, E. I., M. G. Badami, and A. M. El-Geneidy. Bus Transit Service Reliability and Improvement Strategies: Integrating the Perspectives of Passengers and Transit Agencies in North America. *Transport Reviews*, Vol. 35, No. 3, 2015, pp. 292–328.
16. TransitCenter. *To Improve Transit, Be Smart About Delays [Internet]*. TransitCenter, 2016. http://transitcenter.org/2016/05/17/improve-transit-smart-delays/. Accessed August 2, 2017.
17. King County Metro. *Developer Resources [Internet]*, 2010. http://www.kingcounty.gov/transportation/kcdot/MetroTransit/Developers.aspx. Accessed August 2, 2017.
18. Bates, J. W. Definition of Practices for Bus Transit On-Time Performance: Preliminary Study. *Transportation Research Circular [Internet]*, No. 300, 1986. https://trid.trb.org/view.aspx?id=276035. Accessed July 13, 2017.
19. King County Metro. *2016 System Evaluation: Annual Service Guidelines Report Attachement A [Internet]*. King County Metro Transit, 2016. http://metro.kingcounty.gov/planning/pdf/2011-21/2016/service-guidelines-full-report.pdf. Accessed July 18, 2017.
20. Strathman, J. G., and J. R. Hopper. Empirical Analysis of Bus Transit On-Time Performance. *Transportation Research Part A: Policy Practice*, Vol. 27, No. 2, 1993, pp. 93–100.
21. King County Metro. *Fares & ORCA Passes: What to Pay [Internet]*, 2015. http://kingcounty.gov/depts/transportation/metro/fares-orca/what-to-pay.aspx. Accessed June 23, 2017.
22. Maddala, G. S., and K. Lahiri. *Introduction to Econometrics*, 4th ed. Wiley, Chichester, UK, 2009, p. 654.
23. Ijsmi, E. Application of Quantile Regression in Clinical Research: An Overview with the Help of R and SAS Statistical Package. *International Journal of Statistical Medical Informatics*, Vol. 2, No. 1, 2017, pp. 1–6.
24. Koenker, R., S. Portnoy, P. T. Ng, A. Zeileis, P. Grosjean, and B. D. Ripley. *Package "quantreg." Quantile Regres In [Internet]*, 2017. ftp://ftp.ussg.iu.edu/pub/CRAN/web/packages/quantreg/quantreg.pdf. Accessed August 2, 2017.

The Standing Committee on Passenger Intermodal Facilities (AP045) peer-reviewed this paper (18-05768).

Article

TRR
JOURNAL OF THE TRANSPORTATION RESEARCH BOARD

Transportation Research Record
2018, Vol. 2672(8) 220–229

DOI: 10.1177/0361198118791377
journals.sagepub.com/home/trr
SAGE

Transforming the Transportation Customer Experience through Contemporary Artistic Practice: A Case Study in Strategic Implementation

Zipporah Lax Yamamoto[1]

Abstract

Institutionalized government support for artwork incorporated into transportation environments has a history spanning more than 80 years in the United States. Although the Federal Transit Administration has a history of policy guidelines, and the American Public Transportation Association, and Americans for the Arts have developed best practices for the successful integration of artworks during the planning and construction phases of transportation projects, little research exists concerning the public life of those artworks during the operational phase, beyond maintenance concerns. This paper examines an artist-led tour program initiated by the Los Angeles County Metropolitan Transportation Authority (Metro), as a case study, to demonstrate one way contemporary artistic practices are strategically implemented to further community engagement, by extending the reach of existing, integrated artworks during a capital project's operational phase. The study begins with a context for public art in transportation environments, including a brief overview of the tradition of government support for public art in the United States. Artworks integrated within the Metro system are examined with a focus on features that engage customers as landmarks, inspire responses from participants rooted in the memory of historical events or popular culture, and respond to pre-programmed cues. Lastly, this research discusses the transformational value offered by contemporary artistic practices rooted in relational aesthetics, approaches that prioritize transitory experience over material product, by providing continued engagement with communities beyond the planning and construction phases, as examined through an artist-led tour program developed by Metro.

There is a highly-regarded and widely documented tradition of integrating art into transportation environments in the United States. Since the 1970s, when the United States Department of Transportation (USDOT) began formally supporting art in federally funded transit projects, a significant body of publications has emerged regarding best practices for the successful integration of artworks during the planning and construction phases of transportation capital projects. However, little research exists concerning the public life of those existing artworks after capital projects have become operational. This paper examines an artist-led tour program initiated by the Los Angeles County Metropolitan Transportation Authority (Metro), as a case study, in order to demonstrate one approach to strategically employ contemporary artistic practices rooted in relational aesthetics, approaches that prioritize transitory experience over material product, in order to further community engagement in the transit environment, and to cultivate ridership as a lifestyle choice. The study begins with a context for public art in transportation environments, including a brief overview of the tradition of codified government support for public art in transportation environments in the United States. Artworks integrated within the Metro system are examined with a focus on features that engage customers as landmarks, inspire responses from participants rooted in the memory of historical events and popular culture, and respond to cues initiated by passers-by. Lastly, this research discusses the transformational value contemporary artists' practices offer for providing continued engagement with communities beyond the planning and construction phases, as examined through an artist-led tour program developed by Metro. The study

[1]Los Angeles County Metropolitan Transportation Authority (Metro), Los Angeles, CA

Corresponding Author:
Address correspondence to Zipporah Lax Yamamoto:
yamamotoz@metro.net

utilizes qualitative data gathered through customer surveys conducted by Metro and discusses findings from a customer experience perspective.

Art in Transportation and Public Facilities

Documented artworks incorporated into transportation environments in the United States date as far back as the late 1800s, with a tradition of national government programs specifically targeted at support for art in public facilities that began in the mid-1930s. At that time, President Franklin D. Roosevelt created the Works Progress Administration (WPA), which was primarily designed to create jobs during the Great Depression. The administration invested in infrastructure projects nationwide. A division of the program, the Federal Art Project (FAP), hired artists in financial distress to create artworks, including murals for public facilities. According to Norwalk Transit, which today houses several WPA-era murals, 2,566 murals in public buildings were created during this time (*1*).

As Tess Thackara writes in her article, *What We Can Learn from the Brief Period When the Government Employed Artists*, "Artists were employed with specific goals in mind: to help the government communicate with the rest of the country, to inspire pride in a nation that had been brought to its heels, and to document the country's recovery effort" (*2*).

Thackara continues to explain how unusually equitable the program appeared when compared with the traditional art market of the time. "The Federal Art Project's non-discrimination clause meant that it attracted, and hired, not just white men but also artists of color and women who received little attention in the mainstream art world of the day. The artists created posters, murals, paintings, and sculptures to adorn public buildings" (*2*).

Cities began to institutionalize support for art in public places nearly a decade and a half after the FAP ended, with Philadelphia adopting the first "percent for art" program in 1959. As Ellen Dissanayake writes in *New Land Marks: Public Art, Community, and the Meaning of Place*, the book published in 2001 as an accompaniment to an exhibition at the Pennsylvania Academy of the Fine Arts, and celebrating the Fairmount Park Art Association's longstanding partnerships focused on public art in Philadelphia,

> The arts in traditional societies from the Paleolithic era to the present were rarely made for private delectation, nor were they expressive of the private ruminations of a solitary artist. Prehistoric paintings on rock walls may have been seen only by certain people, such as hunters, or initiates, or males – no one knows. Their subject matter, however, reflected group belief, needs, and lore (*3*).

The human desire to record collective mythology in prominent, visible locations is a motivation with deep, documented, and well-preserved cultural roots.

The USDOT began funding artworks in transportation in 1977. The Federal Transit Administration (FTA) developed Circular 9400.1, Design and Art in Public Transportation Projects, in 1981, which was updated by Circular 9400.1A, Design and Art in Transit Projects, in 1995. In the latter document, "FTA articulates its commitment to fund quality design and art in mass transit projects" (*4*). The policy states, "The visual quality of the nation's mass transit systems has a profound impact on transit patrons and the community at large. Mass transit systems should be positive symbols for cities, attracting local riders, tourists, and the attention of decision makers for national and international events" (*4*). The document continues to stress the value that design and art contribute to the transit environment, specifically, "Good design and art can improve the appearance and safety of a facility, give vibrancy to its public spaces and make patrons feel welcome. Good design and art will also contribute to the goal that transit facilities help to create livable communities" (*4*).

In 2013, the American Public Transportation Association (APTA) released *Best Practices for Integrating Art into Capital Projects*, which addresses the importance of art program development, "at the earliest stages of a capital project's development to ensure integration of art into the project" (*5*). The document continues to outline, "Art is not a one-time undertaking; once art is installed, it must be maintained, and ongoing programs should introduce the public to the art on the system" (*5*). The publication does not elaborate about the potential for new and emerging artistic practices to contribute to ongoing programs of that nature.

Similarly, in 2016 the Americans for the Arts Public Art Network (PAN) developed updated *Proposed Best Practices for Public Art Projects*. Americans for the Arts PAN is the national professional organization for public art practitioners nationwide. Although the best practices developed by PAN are not specific to art in transportation, like the documents developed by FTA and APTA, the PAN white paper does not go into detail regarding practical approaches to extend the reach of integrated artworks once a capital project has moved into the operational phase. This document explores the added value contemporary artistic practices rooted in relational aesthetics offer in extending the reach of integrated artworks in the operational transit environment.

Los Angeles Metro

Metro is the transportation planner and coordinator, designer, builder and operator for Los Angeles (L.A.)

County. Nearly ten million people, almost one-third of California residents, live, work and play in the service area. L.A. may be known for its car culture, however with the desire for expanded mobility options expressed by over two-thirds of county voters through the passage of two local sales tax initiatives in the past decade – Measure R in 2008, and Measure M in 2016 – the agency is rapidly executing a plan to develop transportation alternatives countywide. Perceptions about transit in L.A. have radically shifted since the agency's inception.

Over the years, more than one thousand artists have been commissioned to develop site-specific integrated artworks, temporary exhibitions, onboard posters, art tours, and live performances for the agency, through Metro Art. As APTA articulated in *Best Practices for Integrating Art into Capital Projects*, "A successful art program is one more way for transit agencies to project an overall image of competence and desirability" (*5*).

Metro's integrated artwork assets and cultural programming work to advance the agency's core business goals by enhancing the customer experience, cultivating ridership, and connecting people and communities throughout L.A. As Metro's Deputy Executive Officer Maya Emsden, who oversees the agency's art program, was quoted as saying in *Artists take Metro Commuters on Another Kind of Journey*, "The customers' experience is essential. Art is a wonderful, engaging way to transform their journeys" (*6*). The broad range of artists who have created works for Metro over the years represents the diversity of the county. The agency is currently executing the most extensive public works program in the country and boasts one of largest transit art collections in the nation.

Integrated Public Artwork

Metro's collection of integrated artworks enhances the customer environment for daily riders, and acts as a stand-alone cultural attraction. Thousands of visitors have been lured to transit annually since 1999, through a series of free, docent-led art tours offered through Metro Art. All artworks in the system are designed specifically for their transit locations. In keeping with best practices identified by APTA, "Selection of artists [is] based on high quality art works and appropriateness of their work to the scope and scale of the project" (*5*). Every artwork is tailored to the needs of the specific project, and artworks draw on a variety of engagement strategies after they are installed.

Integrated artworks may engage customers as cultural landmarks, they may inspire responses from participants rooted in the memory of historical events or popular culture, and they may respond to a set of cues initiated by riders. *For Your Intellectual Entertainment*, by Daniel

Figure 1. *For Your Intellectual Entertainment* by artist Daniel Martinez.

Martinez, (Figure 1) is an example of one artwork in the Metro system that acts as a landmark and visually links the station with the surrounding community. This artwork takes advantage of the station's elevated location to display a highly visible, playful nod to the aviation industry in the local neighborhood. This monumental scale and placement is possible when an artist is included in the early design phases of a project, allowing architects and engineers to design for the integration of a unique station feature. As Craig Amundsen wrote in *Public Art and Public Transportation*, "The inclusion of artists in the early architectural programming for a transit-related space has become popular as a better way to achieve more humanized public spaces" (*7*). Marking a place, acting as a physical locator, is one tremendously valuable yet primarily fixed engagement role for artworks during the operational phase.

Metro Art has commissioned a wide range of artworks that draw on cultural memory to appeal to people from all walks of life. As Edmund N. Bacon, the long-time planner for the Philadelphia City Planning Commission wrote in *Design of Cities*, "The city is a people's art, a shared experience, the place where the artist meets the greatest number of potential appreciators" (*8*). *The*

Figure 2. The Movies: Fantasies and Spectacles by artist Joyce Kozloff (detail).

Movies: Fantasies and Spectacles, by Joyce Kozloff, includes hand-glazed ceramic tiles that reference hundreds of Hollywood films on underground station mezzanine walls. In Figure 2, children excitedly point to scenes they recognize. This type of casual interaction triggered by memory and experience is another way artworks engage customers during their travel through the system.

The Metro Art collection also features integrated artworks that engage the public by responding to cues. *A Train*, a light and sound piece by artist Bill Bell, (Figure 3) includes a plaque with a built-in microphone. When a person speaks a code word into the plaque, the artwork produces a series of pre-programmed audible and visual returns. This approach embedded into the artwork's design presents a third, responsive, customer-initiated opportunity for direct interaction.

APTA notes, "Public transit agencies continually work to improve service reliability, safety, competence, and desirability. A positive customer experience is critical for maintaining and increasing transit ridership. Art is a cost-effective way of heightening the customer experience and associating public transit with positive cultural images and aspirations" (*5*). Though anyone might enjoy the automatic light and sound components of *A Train* that run when the piece is in a latent state, how might a causal transit rider know that it is possible to activate the artwork without insider knowledge? Similarly, for a fixed artwork, are there visual or cultural codes that might enhance the customer experience of the system, and if so, what can agencies do to deepen the relationship between the rider and the transit landscape?

The transit customers of today bring a different level of expectation to their experiences than the customers of years past. As Clayton Lord notes in the executive

Figure 3. *A Train*, by artist Bill Bell is an integrated, responsive, light and sound artwork.

summary to Constance Y. White's essay, "Arts, Transportation and Infrastructure," in *Arts and America: Arts, Culture, and the Future of America's Communities*,

> As the country continues to transition to a customized economy, where consumers demand deep, personalized user experiences even in circumstances where utility was previously enough, one area that will be deeply impacted is mass transportation. Art interventions in stations and transportation modalities will help raise the quality of the end-user experience, while advances in design and other creative workforce applications will meld deeper engagement with efficient construction and design (*9*).

Customer expectation for personalized and positive interaction is growing. It is no longer enough for a transit system to deliver a person from one place to another. Customers place a high value on the overall quality of differentiated, yet deeply integrated experiences. This awareness has led to a national focus on creative place-making, which is defined by Transportation for America as, "An approach that deeply engages the arts, culture, and creativity, especially from underrepresented communities, in planning and designing projects so that the resulting communities better reflect and celebrate local culture, heritage and values" (*10*).

Artistic practices have also evolved in tandem with this social focus. As Nicolas Bourriaud writes in his groundbreaking book, *Relational Aesthetics*, "It is no longer possible to regard the contemporary work as a space to be walked through … . It is henceforth presented as a period of time to be lived through, like an opening to unlimited discussion" (*11*). Many contemporary artists engaged in this approach are not primarily focused on producing objects. They are primarily concerned with artwork as ephemeral engagement, in a specific place, at a specific time. Bourriaud continues, describing, "Those artists proposing as artworks: a/ moments of sociability [and] b/ objects producing sociability" (*11*). Artists whose practice centers on relational aesthetics work with a social medium. Their work focuses on the interactions between people, objects, and places. They may draw on a wide variety of tools to achieve situational, social relationships, including the visual arts, poetry and spoken word, music, dance, theater, and movement.

When utilized strategically, are contemporary artistic practices that privilege experience over object well-suited to provide meaningful overlays onto the customer experience? The case study that follows examines the implementation of a relational aesthetics approach explored primarily through a field of art commonly known as "social practice." This application provides unique opportunities for interactions with physically integrated works of art, imbues existing artworks with extended meaning created through positive, associative transit experiences, and provides heightened engagement that contributes to a comfortable level of familiarity with the overall Metro system.

Contemporary Artistic Practices as Transformational Overlays

In 2012, Metro Art invited contemporary artists to use a relational aesthetics approach to amplify the highly successful, docent-led public art tour program that had been in existence since 1999. The popular guided tours exemplified one of the best practices advocated by APTA as, "An ongoing program to introduce the public to art on the system" (*5*). At present, a team of over 30 volunteer docents continue to offer regular, public, monthly tours of integrated artworks countywide. Custom tours are also available for groups, by request.

Many of the tour leaders are professional docents, and have volunteered their services for years – even decades – at prominent, local museums. They are highly trained specialists who engage a diverse group of participants. Tour attendees are general members of the public from all demographics, ranging in age from school children to senior citizens. The tours attract both tourists and locals who discover new ways to engage with transit. As one L.A. resident commented, "Everyone raved about the [tour] experience as being educational, enlightening and entertaining. [We] enjoyed it immensely and want to bring other friends and family members on future tours. Most of the attendees had no idea such a rich art collection was waiting underground." English- and Spanish-language tours are offered regularly, with interpretive services in additional languages available by advance request. All tours are free.

When the docent tour program began in 1999, 80% of tour participants were first-time Metro riders. However, much has changed in L.A. over the last two decades. The arts and cultural scene has gone from incubatory to thriving; riding transit has become a lifestyle choice for many, and discretionary riders cite quality of experience as a key contributing factor in their mobility choices. As Constance Y. White writes, "In a shift from the more utilitarian nature of public planning prevalent in the second half of last century, public agencies have now started prioritizing user experience in consideration of planning, and are building more livable communities with a focus on improved quality of life" (*9*). Adding tours led by artists allowed the agency to further diversify the range of arts and cultural experiences that promote a lifestyle intertwined with transit.

Goals. APTA best practices affirm, "Art can entice the choice rider, a major target audience for increasing ridership, as well as attract new riders" (*5*). The goals of adding contemporary artists to use the artistry of engagement to invigorate experiences with the Metro system are fourfold:

- Cultivate ridership as a lifestyle choice;
- Familiarize the public with the expansive breadth of the countywide Metro system;
- Educate people about transit etiquette;
- Boost perceptions about public transportation.

In a county that covers a 1,433-m^2 service area, in which transit is somewhat regarded as a novelty, one challenge is to acclimate the general public to the overall culture of transit. This includes guiding people to weave transit and other mobility options into their lives in ways that are convenient, personally fulfilling, and qualitatively more attractive than driving alone.

Strategies. A series of strategies were designed to capture a demographic that seeks to consume experiences and information, by linking places where people live, work, and play with arts and cultural outings that offer special opportunities to engage with the Metro system. Those strategies include:

- Adding artists who specialize primarily in engagement practices to tours;
- Linking places people live, work, and play with cultural outings;
- Creating special opportunities to engage with the Metro system;
- Partnering with local businesses to offer cross-promotion and post-tour offers and discounts;
- Creating positive, associative transit experiences.

Implementation. Artists lead the public through existing transit spaces and use their crafts – visual arts, poetry and spoken word, music, dance, theater, movement, and act as facilitators of social experiences, to engage participants. Peirce Lewis, a geographer who has written extensively about the American landscape and cultural geography, describes acclimating his students to "read" a landscape. He writes, "So our first job, if we want to teach students how to read a landscape, is to help them to develop the habit of thoughtful looking and asking questions about what they see in the ordinary landscape they inhabit" (*12*). Artists in the Metro system use their honed engagement skills to guide tour goers in a similar fashion. Participants slow down and interact with integrated artworks in choreographed moments that are different from the ways they might during a hurried commute. As Nita Lelyveld noted in the *L.A. Times*, "They make you notice details you might otherwise walk past. They also provide, for those who need it, generous and judgment-free instruction in TAP cards [fare media], turnstiles and train-traveling etiquette" (*13*).

In Figure 4, a tour participant uses her body to respond to part of the *People Portraits: in Creativity, Performing, Sports and Fashion* series of mosaic artworks by Faith Ringgold. By physically investigating the landscape of a station through the art of performance, riders create personal, familiar, embodied relationships with the built environment.

Ed Fuentes, a writer for the broadcast network KCET, observed the responsive nature of the artist-led engagement. "This tour suggests public art isn't just about revitalizing an underserved downtown, or reinventing the city. It responds to what is already there and helps prove that decades of strategies aimed to make the city a creative center point have been constant" (*14*).

Relational aesthetics utilize engagement as a critical tool, which is an ideal medium for ongoing programs that connect the public with the art in the transit system. Metro's strategic implementation focuses on providing a transformational experience by giving customers the time, the opportunity, and a special set of arts-based tools to use to activate existing integrated artworks, and to interact with the Metro system. Artists are supported by docents and staff who answer questions about Metro and the deep history of artworks integrated throughout the system. Tour goers have the opportunity for a fresh perspective, transit through the eyes of artists, in addition to the institutional and practical knowledge that career staff and docents with decades of experience provide.

Many Metro Art contemporary artist-led tours use transit to link places where people work with seasonal celebrations and cultural outings. Participants learn about county neighborhoods, explore various cultural traditions, and have the opportunity to celebrate holidays with new friends. In honor of Día de los Muertos (Day of the Dead), Metro has organized tours that encouraged festive dress and led participants to the heart of the county's most well-known Día de los Muertos celebration, at Self-Help Graphics, a non-profit organization in East L.A. (Figure 5) As Sandra Bloodworth writes in *Along the Way: MTA Arts for Transit*, "The best public art eloquently and effectively links the place where it is located to the people who come there; both the place and the people are altered, however imperceptibly, by the dialogue between art and the public" (*15*).

Figure 4. A participant strikes a pose with an artwork by Faith Ringgold while on a tour that includes a movement artist as a guide.

Figure 5. Participants point to platform-integrated artwork during a Día de los Muertos (Day of the Dead) tour, led by Consuelo Flores.

Figure 6. Over a hundred participants gather at Union Station during the inaugural Fourth of July art tour.

A Fourth of July artist-led tour traveled through several downtown L.A. rail stations (Figure 6). Participants began the tour with a series of warm-up exercises designed to introduce tour goers to the often close confines in the transit system, and to build a level of comfort between group members. Next, they searched for embedded visual clues to unlock the storyline in a mosaic artwork. Later, they explored choreography as a tool for navigating a maze of embedded symbols in a paving installation. The tour ended at a new civic park, which featured special programming, including live music, food trucks, and fireworks – the first for the downtown area.

Metro has also worked with artists to design engagement-focused tours for Valentine's Day. One year, Mike the Poet created original poetry that responded to a series of artworks in the system and encouraged tour goers to "Fall in love with Metro." Feedback from participants indicated the value of the tour for first-time riders, with written notes such as, "An excellent first-time experience, showing lots of hidden gems, art and architecture." Comments specifically noted the transformation overlay provided through artist-led experience, such as, "Excellent interpretation of L.A. history on the Metro through the lens of a poet!" Attendees also left requests for additional artist-led programs, noting, "Mike is a gem. Please keep expanding your program to the other lines."

In response to the enthusiastic feedback, Metro continued to develop the offerings. The Independent Shakespeare Company staged a series of acts at locations near rail stations in subsequent years. Participants experienced the system and surrounding areas through a contemporary application of Shakespeare's writing, performed by local artists (Figure 7). An Earth Day tour led by the artist team Fallen Fruit began with a lemonade stand in front of Union Station (Figure 8). Participants created self-portraits on lemons, drank lemonade, and prepared to examine several artworks focusing on nature, the earth, and time. Through the series of artist-led tours, there is the potential for participants to engage with the same artwork multiple times, and for each experience to be uniquely perceived through a different contemporary artist's practice. These opportunities allow existing integrated artworks to be experienced in a multiplicity of transformational ways over time.

Partnerships with local businesses offer cross-promotion and post-tour offers and discounts by leveraging existing programs at the agency. People are able to show their TAP cards at participating locations in order to save money on admission, obtain discounts on meals, and receive free gifts.

While on the tours, people learn how to use a TAP card, often through an artist-led activity – a game. The fare system can be confusing to first-time riders, but people work together through an interactive experience to figure it out together. Participants associate Metro with having a good time, and they will not stumble the next time they ride Metro their own. "They [the tours] are free. They are fun," noted Nita Lelyveld in the *L.A. Times*. "Not everyone comes for the help. Art is the lure. But the

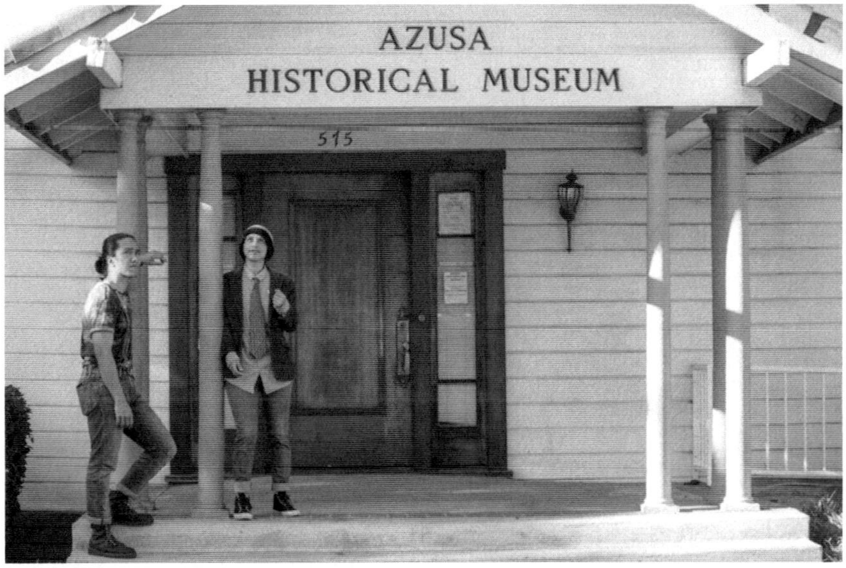

Figure 7. The Independent Shakespeare Company enlivens destinations near rail stations through a series of love-related scenes during a recent Valentine's Day tour.

Figure 8. Participants on an Earth Day tour led by artist collective Fallen Fruit hold self-portraits they have drawn on lemons, as they explore artworks in the Metro system that are connected with nature, the earth, and time, including *The Wheels of Change*, by Chusien Chang, which lies beneath their feet.

aim is to seduce and thus inspire further public transit use" (*13*).

Participant Responses. Opt-in surveys were utilized to learn more about contemporary artist-led tour program participants. Hard copy survey cards were distributed at the end of each tour. Participants were able to return the forms immediately, or choose to mail the forms to the agency. The information that follows was received from a self-selecting group of artist-led tour participants who chose to respond to hard copy surveys, over a 5-year period spanning 2012–2017.

When the volunteer docent-led tour program began in 1999, 80% of those tour participants were first-time Metro riders. The tours focused heavily on the practicalities of riding Metro. Survey responses gathered between 2012 and 2017 indicate a high percentage of repeat riders (85%), and a relatively lower percentage of riders being introduced to the system through artist-led tours (15%). This increase in repeat rider participation indicates a greater number of tour goers are choosing to ride transit. It also suggests confidence in the Metro system may be growing.

Tours are more commonly engaging repeat riders who are there to experience the system in new and interesting ways. As one respondent wrote, "Awesome experience; lived in L.A. for 24 years and I can't believe I hadn't experienced it [Metro] before. Definitely would recommend it to others!! Made me appreciate L.A. and public transport so much more!!"

Although the slim majority (53%) of tour goers who responded to the survey indicated they arrived by car, a close percentage (47%) indicated they chose transit. In the last 5 years Metro has opened extensions to three rail lines and one bus rapid transitway, which may be a contributing factor to the high number of transit patrons who responded to the surveys. The system now extends further throughout the county and provides additional linkages, which makes transit more practical and convenient.

The tours have received favorable ratings overall: 84% of respondents rated the tour they attended "excellent"; 13% of respondents rated the tour they attended "good"; 3% of respondents rated the tour they attended "fair." In addition, 100% of respondents over the past 5 years indicated they would take a tour again in the future, and 100% of respondents indicated they would recommend a tour to a friend. Many comments spoke specifically to the transformational nature of the artist-led engagement, with comments such as, "The tour was informative, made Pasadena areas come alive. Please do MORE!!"

Conclusion

The transportation customer of today is not fully satisfied by the stand-alone, utilitarian value of the primary service provided. They seek personalized, qualitative value from their experience. As observed through the L.A. Metro artist-led tour program case study, contemporary artistic practices utilizing engagement methodologies are well-suited to engage this customer after a capital project moves into the operational phase, after base transit services are provided.

The landscape of transportation, and the multitude of ways in which people communicate within and identify a lifestyle surrounding those spaces, are in a state of continual flux. Peirce Lewis writes about the concept of the landscape as a "selectively incomplete document" (*12*). It is a partial and transitory text. The transportation landscape includes fixed physical locators, such as a longstanding tradition of integrated artworks, through which customers identify place in relationship with the broader landscape. When implemented strategically, artist-led engagement activities, in this case study structured through art tours, provide fresh and varied opportunities for customers to revisit well-traveled public spaces, and guide extended interactions with existing artworks to create positive, associative transit experiences that promote ridership as an attractive and viable lifestyle choice.

Author Contributions

The author confirms contribution to the paper as follows: study conception and design: ZY; data collection: Metro Art; analysis and interpretation of results: ZY; draft manuscript preparation: ZY. All authors reviewed the results and approved the final version of the manuscript.

References

1. Norwalk Transit. *Public Transportation in Southwest CT.* https://www.norwalktransit.com/wpamurals/. Accessed July 25, 2017.
2. Thackara, T. What We Can Learn from the Brief Period When the Government Employed Artists. *Artsy*, February 1, 2017. https://www.artsy.net/article/artsy-editorial-government-paid-working-class-artists. Accessed July 25, 2017.
3. Dissanayake, E. Why Public Art is Necessary. In *New Land Marks* (Penny Balkin Bach, ed.), Editions Ariel, and imprint of Grayson Publishing, Washington, D.C., 2001, pp. 25–34.
4. FTA Circular 9400.1A. Design and Art in Transit Projects. *National Transportation Library*, June 1995. https://ntl.bts.gov/DOCS/9400-1.html. Accessed July 25, 2017.
5. American Public Transportation Association. *Best Practices for Integrating Art into Capital Projects.* APTA SUDS-UD-RP-007-13. American Public Transportation Association, Washington, D.C., 2013.
6. Bradner, L. Artists Take Metro Commuters on another Kind of Journey. *Los Angeles Times*, January 6, 2013. http://articles.latimes.com/2013/jan/06/entertainment/la-et-cm-los-angeles-metro-art-20130106. Accessed July 26, 2017.
7. Amundsen, C. *Public Art and Public Transportation.* Transportation Research Board of the National Academies, Washington, D.C., 1995. http://onlinepubs.trb.org/Onlinepubs/trbconfproc/vol.8/8-008.pdfß.
8. Bacon, Edmund N. *Design of Cities.* Penguin Books, New York, 1967.
9. Lord, C. (ed.). *Arts and America: Arts, Culture, and the Future of America's Communities.* Americans for the Arts, Washington, D.C., 2015.

10. Transportation for America. *What is Creative Placemaking?* http://creativeplacemaking.t4america.org/what-is-creative-placemaking/. Accessed July 26, 2017.
11. Bourriaud, N. *Relational Aesthetics.* Les Presses du Réel, France, 2002. (English translation).
12. Lewis, P. The Monument and the Bungalow. In *Everyday America: Cultural Landscape Studies After J.B. Jackson.* University of California Press, Berkeley, Los Angeles and London, 2003, pp. 85–108.
13. Lelyveld, N. Teaching New Metro Riders the Subway of Doing Things. *Los Angeles Times*, March 9, 2014. http://articles.latimes.com/2014/mar/09/local/la-me-adv-beat-metro-art-tour-20140310. Accessed July 25, 2017.
14. Fuentes, Ed. Metro's Summer Public Art Tour Connects Rail, Artists, and Communities. *KCET*, June 13, 2013. https://www.kcet.org/history-society/metros-summer-public-art-tour-connects-rail-artists-and-communities. Accessed July 25, 2017.
15. Bloodworth, S., and W. Ayes. *Along the Way: MTA Arts for Transit.* The Monacelli Press, New York, 2006.

The Standing Committee on Passenger Intermodal Facilities (AP045) peer-reviewed this paper (18-05497).

Article

TRR
JOURNAL OF THE TRANSPORTATION RESEARCH BOARD

Transportation Research Record
2018, Vol. 2672(8) 230–242

DOI: 10.1177/0361198118781395
journals.sagepub.com/home/trr
SAGE

Empirical Analysis of Traveling Backwards and Passenger Flows Reassignment on a Metro Network with Automatic Fare Collection (AFC) Data and Train Diagram

Ruihua Xu[1], Yanan Li[1], Wei Zhu[1], and Sijie Li[1]

Abstract

Metro (subway) systems are becoming overcrowded in some of China's mega-cities, such as Beijing, Shanghai, and Guangzhou. As many passengers are unable to board trains on an overcrowded metro network during peak periods, some of them are willing to spend more time and energy by traveling backwards in order to secure a seat or even room for standing. Traditional studies on travel behavior analysis and transit assignment models seldom deal with this situation. We propose a methodology including the affinity propagation cluster method with between-within-proportion (BWP) index and an adaptive "0-1" model named the traveling backwards model (TBM) to identify the phenomenon of "traveling backwards" and to reassign passenger flows on a metro network using automatic fare collection (AFC) data and actual train diagrams. As a numerical example, this integrated approach is applied to the Beijing metro system. Our research shows that the affinity propagation cluster method with BWP index and implicit enumeration algorithm for TBM work well. TBM is a good replacement for the existing assignment model and travel behavior analysis, especially for those mega-cities' networks in peak hours.

Metro (subway) is playing an increasingly important role in public transportation, especially in large cities, owing to its faster velocity, higher reliability, and larger capacity. However, these benefits attract large numbers of passengers to the metro system which creates challenges for management. The metro system is becoming overcrowded in some mega-cities in China. According to a passenger flow study in Beijing, the capacity loading rate of several sections is up to 150% in peak hours (*1*). During peak hours, many passengers have to stand in the trains or even fail to board trains (left behind) because of limited capacity. This results in a phenomenon called traveling backwards (TB) where passengers purposely board trains in an outbound direction for an inbound trip, changing direction near the end of the line for a seat or just a place to stand. (A simple representation of this travel behavior is shown in Figure 1.) The passenger flow study shows that the proportion of these TB passengers is higher than 10% in certain suburban metro stations of the metro that extend beyond the circle line into the outer fringes of the suburban area (*1*).

TB seems to go beyond traditional modes in route choice decisions which are modeled and evaluated according to cost and elements related to travel time (*2*). Recent research on public transport systems shows that service reliability, riding comfort, seat availability, and perceptions of crowding levels also have great impacts on route choice behaviors (*3–5*). Thanks to the development of automated fare collection (AFC) systems, researchers can understand passenger flow characteristics better with large quantities of real-world data from individual passengers' tap-in/tap-out records. Compared with conventional field surveys that yield small data (*6*), when the sample size increases greatly, the data is more objective and can cover different situations. In order to describe how passengers travel in spatiotemporal dimensions, researchers have tried to connect passenger travel time with train diagrams (*7–10*). Crowding at stations and on platforms can result in delays, potential safety issues, and passengers left behind, which can significantly degrade level of service in terms of wait times, comfort,

[1]The Key Laboratory of Road and Traffic Engineering, Ministry of Education, Tongji University, Shanghai, P.R. China

Corresponding Author:
Address correspondence to Wei Zhu: zhuweimail@tongji.edu.cn

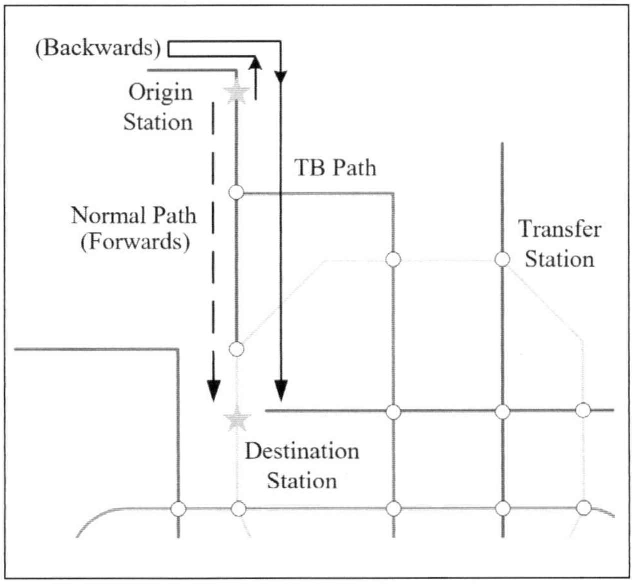

Figure 1. Illustration of behavior of passengers "traveling backwards" in an overcrowded metro network.

and reliability (*10*). Passengers being left behind leads to AFC records of long journey times (*10*), which may be regarded as abnormal data if it is very much longer than the general traveling time (for example, if a record corresponds with the passenger having boarded the fifth train after he came to the platform). Therefore, the corresponding AFC data of TB passengers would always be regarded as abnormal data and eliminated from the data set during data cleaning. Traditional passenger behavior analysis and passenger flow assignment models have ignored this kind of behavior, which may cause spatio-temporal bias and errors in passenger flow distribution.

In this paper we propose a methodology for identifying the phenomenon of TB and reassigning passenger flows in a metro network based on AFC data and actual train diagrams. The main contributions of this work are as follows:

(a) We first identify the phenomenon of TB. The difficulty is that "one-ticket transfer" AFC data only records origin–destination (OD) information without its chosen path. It is hard to identify whether the passenger is traveling backwards or not, and which station is the turning point. Therefore, a novel clustering algorithm is introduced to classify the passengers.
(b) We build a "traveling backwards model" (TBM) to track the actual routes of TB passengers and an implicit enumeration (IE) algorithm is proposed.
(c) We establish the TB path to reassign the TB passenger flows back to the network, which will reflect more actual passenger flow distribution.

The remainder of this paper is organized as follows. In the next section the relevant literature is reviewed. The third section describes the data for better understanding of the research work. The fourth section presents the methodology framework and the corresponding algorithms are developed in the fifth section. In the sixth section the proposed methodology is applied to the Beijing metro network as a case study. The final section summarizes the paper and indicates areas for further exploration.

Literature Review

A metro system operates under the constraints of the train diagram and the flexibility of passengers. Compared with public bus transit and private cars, route choice behavior in a metro system is much harder to describe. On a bus, smart cards can record not only boarding and alighting time, but also the corresponding ID of the vehicle (*11*). However, metro passengers tap their card at the entry to the station and the interval between trains is short, which makes it harder to identify which train the passenger actually boarded. In addition, a "one-ticket transfer" metro system cannot record the transfer information. Compared with private cars, the cost of tracking each individual passenger by sensing technologies such as inductance loop detector or video is too costly and too imprecise to measure crowds in peak hours (e.g., the passenger inflow can be up to 200 per minute at a station in rush hours).

In recent years, AFC data has been used in metro systems to analyze passenger behavior (*12*), for operational management (*13*), service adjustments (*14*), passenger flow assignment (*15, 16*), etc. Pelletier et al. (*17*) make a comprehensive review of using smart card data at different levels of management. Barry et al. (*18*) first use smart card data to estimate OD demand. Kusakabe et al. (*19*) develop a methodology for estimating the exact train which an individual passenger takes (*7, 8*), combining AFC data and train diagrams to identify the passenger's route choice and assign passenger flow more accurately. Zhu et al. (*20*) present a calibration approach for metro assignment models using a genetic algorithm-based framework and make a validation of the rail transit assignment models with AFC data. Sun et al. (*11*) propose an integrated Bayesian statistical inference framework to characterize the passenger flow assignment model. Hong et al. (*15*) develop a novel clustering algorithm to connect traveling time with route choices and an interesting approach for determining the maximum accuracy of the assignment model.

However, there has been little research published on metro passenger "traveling backwards" behavior and passenger flow assignment in peak hours. Li et al. (*21*)

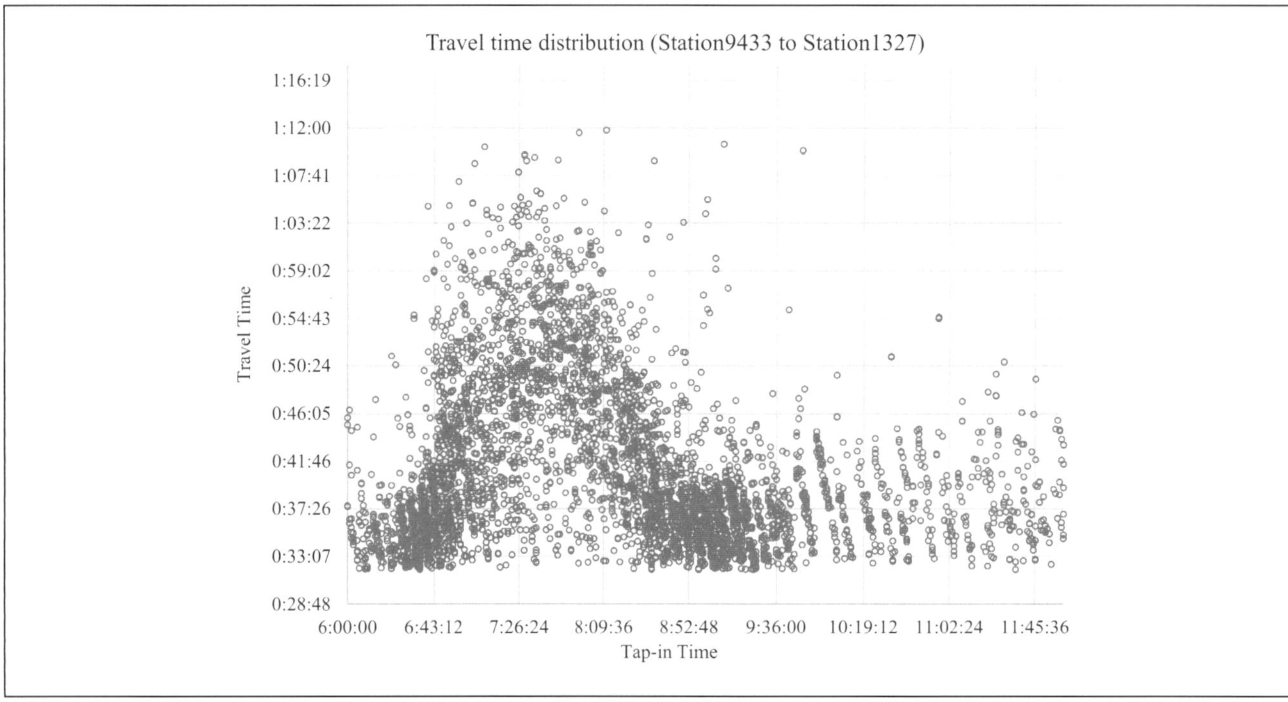

Figure 2. Travel time distribution of OD pair from six days' AFC data.

developed an affinity propagation method to cluster AFC travel time to identify passenger behavior in an overcrowded metro system. Tirachini et al. (*12*) estimated the differences in the valuation of sitting and standing during public transport trips using revealed preferences of TB users in Singapore with the simplified situation that origin and destination stations were on the same line, but they did not do further studies on route choice and flow distribution. Sun et al. (*8*) analyzed each component of a rail transit trip including passengers left behind (FtB – Fail to Board) and Sun (*9*) proposed a schedule-based passenger's path-choice estimation model and a weighted assignment technique based on FtB parameters.

To our knowledge, little attention has been paid to dealing with TB behavior in a metro system, which is becoming a frequent phenomenon in peak hours of mega-metro network. This paper focuses on how to identify TB behavior and reassign passenger flow in peak hours using AFC data and train diagram to improve the accuracy of the existing transit assignment models.

Description of Data Used

AFC Data

AFC systems provide large quantities of data, recording passengers' activities with origin station ID, destination station ID, tap-in time, and tap-out time. We abstract the traveling time from AFC data. Figure 2 shows travel time distribution of an OD pair (Station 9433 to 1327) from six days' AFC data. The horizontal axis represents tap-in time and the vertical axis represents traveling time.

In general, at each station, passengers will always ride the first train to arrive after they reach the platform in order to minimize their waiting time. There is a special situation where passengers give up the first train and choose the following one in order to be seated. However, few commuters would waste their time waiting for another if the current train still has available capacity, especially in peak periods. Nevertheless, we cannot define all travelers who fail to board the first train as TB passengers directly, as overcrowding is not the only reason for overtime traveling. Apart from the passengers left behind due to overcrowded trains, there are some other possibilities, for example, passengers waiting for their friends on the platform or going to the restroom. Therefore, AFC records need more exact trip-chaining analysis.

Travel time from AFC data is the total journey time, which should be divided into the following elements: Element (1) and element (4) have been investigated by the staff of metro company; element (2) and element (3) will be illustrated in the next section (Train Diagram).

(1) Enter/exit time between fare gate and platform at station A_i ($t_{A_i}^{in}/t_{A_i}^{out}$);
(2) Waiting time at platform of station A_i ($t_{A_i}^{w}$);

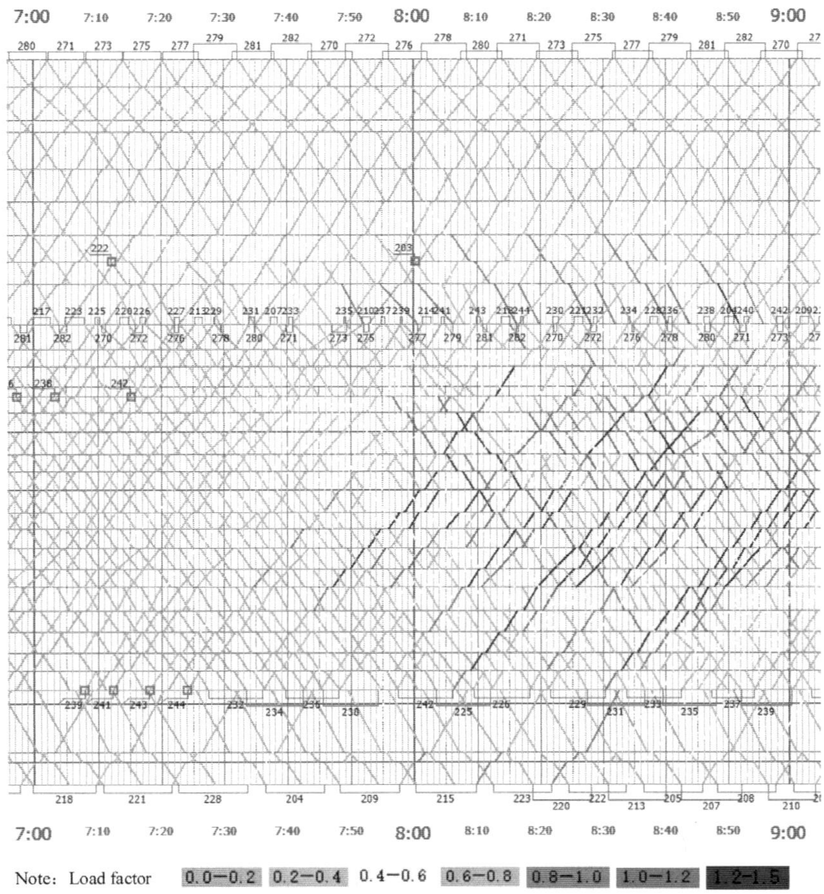

Figure 3. Example of actual train diagram with load factor.

Table 1. Example of Train Diagram Data

Date	Train ID	Direction	Destination station ID	Stop station ID	Stop order	Arrival time	Departure time
20161018	1017	1	9441	9433	8	07:13:05	07:14:18
20161018	1018	1	9441	9433	8	07:18:57	07:20:00
20161018	2012	2	9419	9433	5	07:14:23	07:15:30
20161018	2075	2	1321	1327	13	09:09:54	09:10:37

(3) Pure travel time from A_O to A_D (T_{A_O,A_D});
(4) Transfer time between platforms at A_i and A_j ($t_{A_i,A_j}{}^{transfer}$).

Train Diagram

The train diagram illustrates the relationship between space and time for train operation (Figure 3). Necessary elements for the model formula are summarized in Table 1.

The following are the necessary parameters:

$t_{A_O}^{(I)}/t_{A_D}^{(O)}$	Tap-in/Tap-out time at origin station A_O/destination station A_D.
$t_{A_i}^{(A)}/t_{A_i}^{(D)}$	Arrival/Departure time at A_i.
$t_{A_i,A_{i+1}}^{(R)}$	Running time from A_i to A_{i+1}.
$t_{A_i}^{(S)}$	Dwelling time at A_i.
$t_{A_i}^{(H)}$	Transfer time at A_i, from the opposite direction to the forward direction.

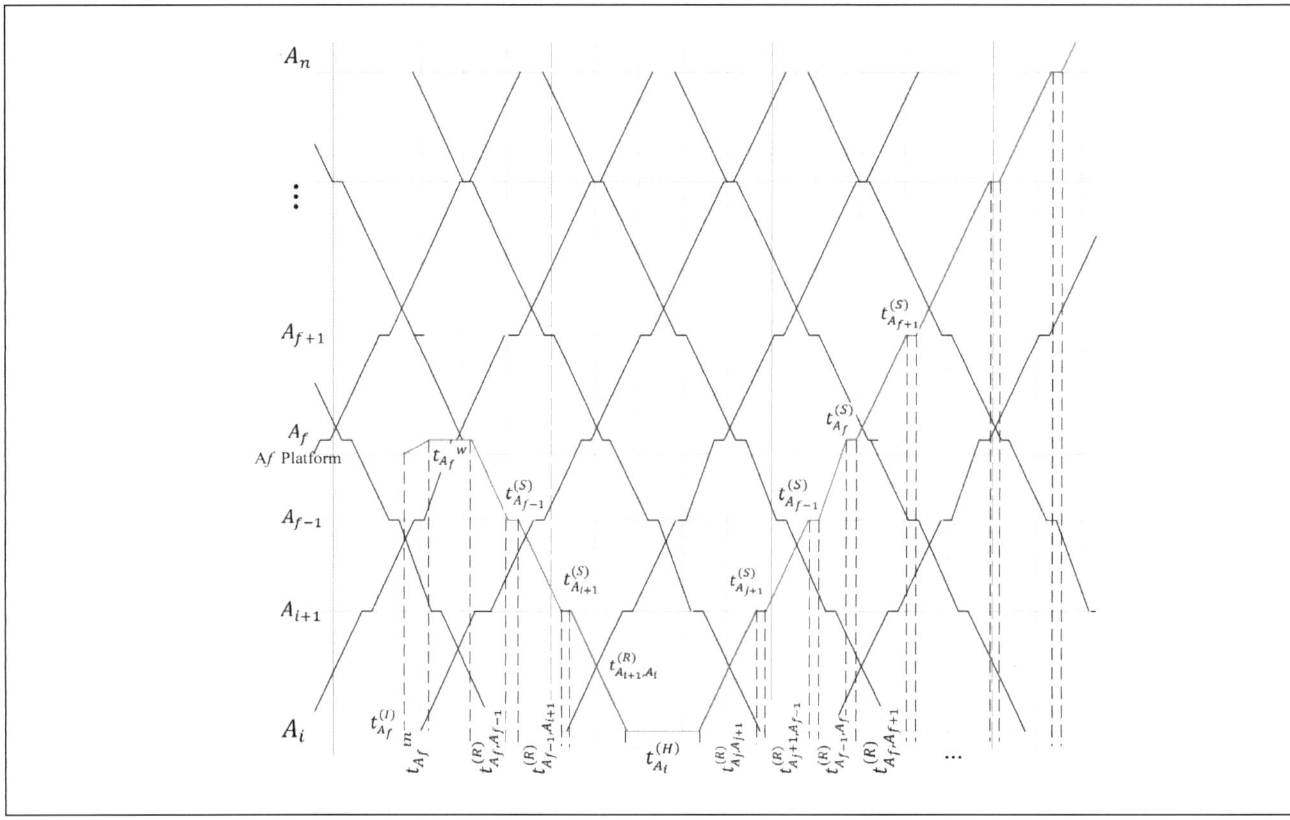

Figure 4. Illustration of the "traveling backwards" phenomenon by train diagram.

Therefore, the running time for each section (A_i – A_{i+1}) can be depicted as:

$$t^{(R)}_{A_i,A_{i+1}} = t^{(A)}_{A_{i+1}} - t^{(D)}_{A_i} \tag{1}$$

And the dwelling time for each station can be derived as:

$$t^{(S)}_{A_i} = t^{(D)}_{A_i} - t^{(A)}_{A_i} \tag{2}$$

Here, $t^{(H)}_{A_i}$ is equal to transfer walking time and waiting time. As the turning point for traveling backwards should be an island station, we assume $t^{(H)}_{A_i}$ is equal to the headway of the trains.

The waiting time at station $A_i(t_{A_i}{}^{w})$ is related to the passenger arrival time at platform ($t^{(PA)}_{A_i}$), $t_{A_i}{}^{w} = t^{(A)}_{A_i} - t^{(PA)}_{A_i}$, and $t^{(PA)}_{A_i}$ can be estimated by the method of Zhou and Xu (*7*) based on AFC data and train diagram.

The pure travel time for an OD pair is $T = t^{(A)}_{A_D} - t^{(D)}_{A_O} = (t^{(O)}_{A_D} - t^{(I)}_{A_O}) - (t_{A_i}{}^{in} + t_{A_i}{}^{out})$, and T can also be described as Equation 3:

$$T_{A_O,A_D} = t^{(A)}_{A_D} - t^{(D)}_{A_O} = \sum_{i=O}^{D-1} t^{(R)}_{A_i,A_{i+1}} + \sum_{i=O+1}^{D-1} t^{(S)}_{A_i} \tag{3}$$

However, T is always beyond the normal value during peak hours because of the limited train capacity (Figures 2 and 3). Traditional models may regard these overtime data as abnormal data and hence ignore TB behavior. The TB path can be described as in Figure 4. (The origin station is A_f, and the destination station is A_n. The planned path is A_f-A_{f+1}-...A_n, while the TB path is A_f-A_{f-1}-...A_{i-1}-A_i-A_i-A_{i+1}...A_f...A_n.) Here, A_i is the turning point in this TB path.

To identify the TB phenomenon and reassign this part of the overtime data back to the network, we propose a TBM as a supplement to the current models under the overcrowded situation.

Traveling Backwards Model

Overview

To identify TB passengers, we make the following assumptions:

(a) Passengers choose a "non-left behind" station as the turning point in their backward ride;

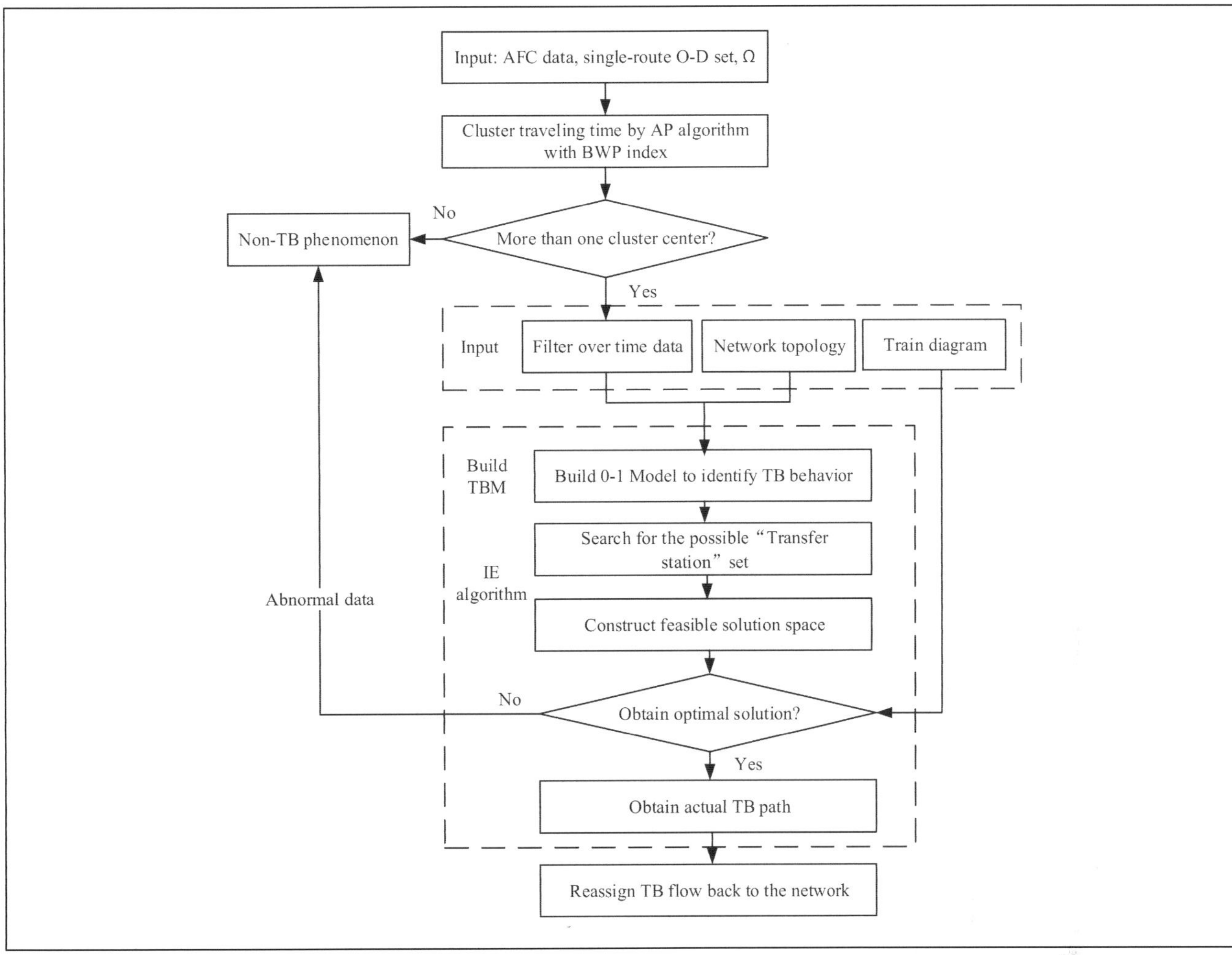

Figure 5. Methodology overview.

(b) Passengers try to choose the nearest turning point as the "transfer station" and turn back into the forward direction;

(c) The origin station at which the TB phenomenon takes place is a suburban station, which is ahead of the first transfer station in the direction toward the urban district, defined as $A_O \in \Omega$;

To manage TB identification and passenger flow reassignment, we propose a methodology framework (Figure 5):

(a) Cluster traveling time to filter overtime data which could be related to TB;

(b) Build the TBM to search for the actual paths of TB passengers based on train diagram and AFC data in order to confirm that overtime data is caused by TB, but not the abnormal data;

(c) Reassign TB passenger flow back to the network.

Filter Overtime Data

Clustering is an unsupervised learning method that tries to find some obvious distributions and patterns in unlabeled data sets by maximizing the similarity of the objects in a common cluster and minimizing the similarity of the objects in different clusters. It is crucial to determine optimal number of clusters for the quality of clustering in cluster analysis.

Affinity propagation (AP) is a clustering algorithm based on the concept of "message passing" between data points (*22*). Unlike clustering algorithms such as k-means or k-medoids, AP does not require the number of clusters to be determined or estimated before running the algorithm. In addition, it is crucial to determine the optimal number of clusters for the quality of clustering. Studies on clustering validity index include Davies-Bouldin, in-group proportion, and between-within-proportion (BWP). Theoretical research and experimental results

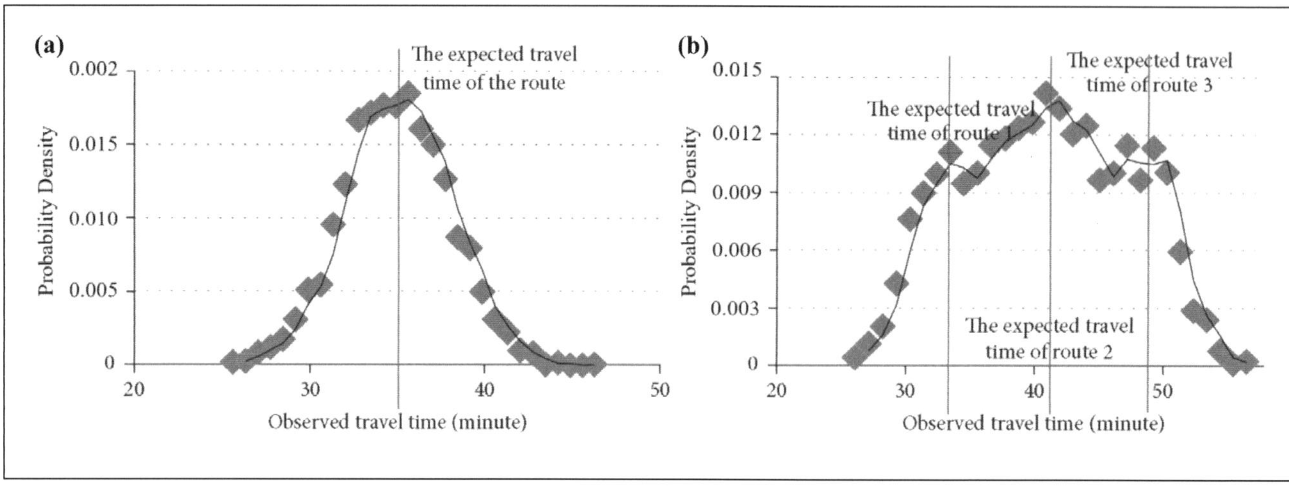

Figure 6. Relationship between the number of clusters and the number of routes (*16*).

show that BWP is suitable for determining the optimal number of clusters based on AP clustering algorithm (*23*). Therefore, we develop the AP method under BWP index to cluster AFC data with single-path OD.

In common sense, the number of cluster centers for single-path OD should be only one (Figure 6). If the results turn out to be more than one, it means that a considerable proportion of passengers are traveling for longer times and then we can choose from these overtime data where A_O stations belong to Ω.

Traveling Backwards Model

The necessary parameters of the TBM are as follows:

The decision variables of the proposed model are:
The objective of the TBM is to find a traveling plan whose traveling time is the most similar to the pure time extracted from AFC data. In other words, the aim is to minimize the difference between AFC pure travel time and the tracking path's travel time:

$$\min Z = T - \left(\left(\sum_{i=0}^{f-1} x_i \cdot t^{(R)}_{A_{i+1},A_i} + \sum_{i=1}^{f-1} x_i \cdot t^{(S)}_{A_i} \right) + \left(\sum_{j=0}^{n-1} y_j \cdot t^{(R)}_{A_j,A_{j+1}} + \sum_{j=1}^{n-1} y_j \cdot t^{(S)}_{A_j} \right) + \sum_{i}^{f-1} z_i \cdot t^{(H)}_{A_i} \right) \quad (4)$$

Constraints:

T	Pure travel time for each AFC data except from $t_{A_i}{}^{in}/\ t_{A_i}{}^{out}$(mentioned in "Description of Data" section, above).
$t^{(R)}_{A_{i+1},A_i}$	Running time from A_{i+1} to A_i in the backward direction.
$t^{(R)}_{A_j,A_{j+1}}$	Running time from A_j to A_{j+1} in the forward direction.
$t^{(S)}_{A_i}$	Dwelling time at A_i in the backward direction.
$t^{(S)}_{A_j}$	Dwelling time at A_j in the forward direction.
$t^{(H)}_{A_i}$	Transfer time at turning point A_i, from the opposite direction to the forward direction.
i, j, f	Index of metro station where i=0,...,f-1; j=0,...,n; f is the index of origin station.

x_i	$x_i \in \{0, 1\}$: 1 if a train stop/departure at A_i; and 0 otherwise.
y_j	$y_j \in \{0, 1\}$: 1 if a train stop/departure at A_j; and 0 otherwise.
z_i	$z_i \in \{0, 1\}$: 1 if a passenger turns back at A_i; and 0 otherwise.

(1) If $x_i = 1$, then $x_{i+1} = x_{i+2} = \ldots = x_{f-1} = 1$, and $x_{i-1} = x_{i-2} = \ldots = x_0 = 0$. Therefore, x_i can be described as:

$$x_{i+1} \geq x_i \tag{5}$$

(2) In the forward direction, the passing stations include stations in both directions (Figure 4). Thus, y_j can be determined by Equation 6:

$$y_j = \begin{cases} x_i, j = i = 0, \cdots, f-1 \\ 1, j = f, \cdots n \end{cases} \tag{6}$$

(3) TB passengers choose only one station in the backward direction as the turning point (called "transfer station"), and this chosen station will be the farthest station the passengers reach, which means that z_i represents a turning point. Therefore, we declare z_i as Equations 7 and 8.

$$z_i = \begin{cases} x_0, i = 0 \\ x_i - x_{i-1}, i = 1, \cdots, f-1 \end{cases} \tag{7}$$

$$\sum_{i=0}^{f-1} z_i = 1 \tag{8}$$

(4) As T is an overtime data for identifying TB passengers, T must be longer than the tracking path time, the object value should be constrained as: $Z \in R^+$

After obtaining the actual path of overtime data, this part of AFC data can be reassigned to the network as a replacement for the current passenger flow assignment model.

Solution Algorithm

Affinity Propagation with BWP Index

Affinity propagation takes as input a collection of real-valued similarities between data points, where the similarity $s(i, k)$ indicates how well the data point with index k is suited to be the exemplar for data point i. When the goal is to minimize squared error, each similarity is set to a negative squared error. There are two kinds of message exchanged between data points, and each takes into account a different kind of competition: "responsibility" $r(i, k)$ and "availability" $a(i, k)$.

Algorithm 1. AP algorithm with BWP index

Data: AFC data and Single-path OD dataset
for each pair of single-path OD **do**
  Select the search range for the cluster number [k_{min}, k_{max}]
  for $k = k_{min}$ to k_{max} **do**
    call AP algorithm (*21*);
    calculate BWP value for each AFC data (*23*);
    calculate average BWP
  end
end
return Cluster number and cluster center for each single-path OD

The BWP index is to used estimate the number of clusters in a dataset. The goal is to attain high intra-cluster similarity and low inter-cluster similarity.

Implicit Enumeration Algorithm

The IE algorithm is a special branch and bound technique for problems with binary variables (*24*). IE searches the solution space by creating a sequence development tree. The tree is developed by branching on the next station to be added to the sequence. There are f possible branches out of the first node, assuming f is the number of stations that are active to be the "transfer station". As a 0-1 model (TBM), the solution space will be 2^f-vectors by enumeration. However, we are able to use a combination of function and constraint information to implicitly enumerate large numbers of feasible f-vectors, thereby significantly speeding up the solution of the problem.

According to Equations 5–8, one feasible solution can be described as (taking $f = 4$ as an example, the origin station is the fifth station from the farthest "transfer station" (A_0)):

$$\begin{array}{c} \\ x_i \\ y_j \\ z_i \end{array} \begin{array}{c} \begin{array}{ccccccccc} A_0 & A_1 & A_2 & A_3 & A_4 & A_5 & A_j & \ldots & A_n \end{array} \\ \begin{pmatrix} 0 & 0 & 1 & 1 & - & - & - & \ldots & - \\ 0 & 0 & 1 & 1 & 1 & 1 & 1 & \ldots & 1 \\ 0 & 0 & 1 & 0 & - & - & - & \ldots & - \end{pmatrix} \end{array}$$

The above solution means that the actual path for this passenger is "A_4-A_3-A_2-A_3-A_4-A_5-…-A_n". Then connect the feasible solution with train diagram to calculate the tracking path's travel time:

$$\begin{pmatrix} 0 & 0 & t^{(R)}_{A_3,A_2} + t^{(S)}_{A_2} & t^{(R)}_{A_4,A_3} + t^{(S)}_{A_3} & - & - & - & \ldots & - \\ 0 & 0 & t^{(R)}_{A_2,A_3} + t^{(S)}_{A_2} & t^{(R)}_{A_3,A_4} + t^{(S)}_{A_3} & t^{(R)}_{A_4,A_5} + t^{(S)}_{A_4} & t^{(R)}_{A_5,A_6} + t^{(S)}_{A_5} & t^{(R)}_{A_j,A_{j+1}} + t^{(S)}_{A_j} & \ldots & t^{(R)}_{A_{n-1},A_n} + t^{(S)}_{A_{n-1}} \\ 0 & 0 & t^{(H)}_{A_2} & 0 & - & - & - & \ldots & - \end{pmatrix}$$

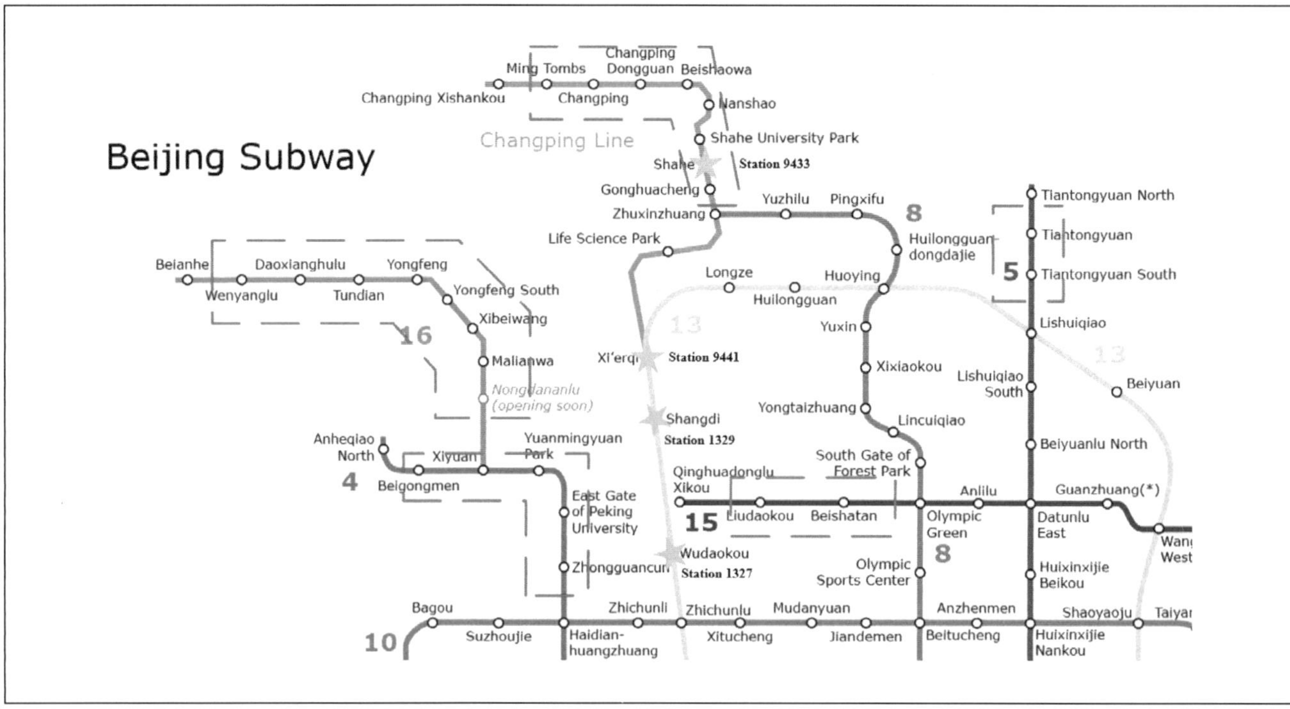

Figure 7. Map of Beijing metro (excerpt).

As the object of the model is to minimize the difference, so we can define $x_i = z_i = 0$ $(i = f, \cdots, n)$. What is more, train diagram data should obey the actual train diagram, which means that the values of Equation 9 are heterogeneous. All elements of the train diagram have contained the adjustment in the actual situation such as delay at the platform, speeding up in the section and so on.

$$\left(\left(\sum_{i=0}^{f-1} x_i \cdot t^{(R)}_{A_{i+1},A_i} + \sum_{i=1}^{f-1} x_i \cdot t^{(S)}_{A_i}\right) + \left(\sum_{j=0}^{n-1} y_j \cdot t^{(R)}_{A_j,A_{j+1}} + \sum_{j=1}^{n-1} y_j \cdot t^{(S)}_{A_j}\right) + \sum_{i} z_i \cdot t^{(H)}_{A_i}\right) \quad (9)$$

However, checking each path using a train diagram to determine the path validity still needs much storage space and calculation. Using the assumption mentioned above, we can continue to reduce the solution space:

(a) As passengers will choose a "non-left behind" station as the "transfer station" in the opposite direction, A_i should be a station with sufficient capacity (non-left behind station) when $z_i = 1$. The set of A_i can be obtained by calculating the proportion of passengers left behind *(7)*;

(b) Passengers try to choose the nearest "transfer station" to turn back, therefore, A_0 should not be very far away from A_f. Here, we can set a threshold like $f \leq 4$, which means that the origin station is the fifth station counting from the farthest "transfer station". The threshold can be determined by the common experience. Considering that travel time for 4*2 sections is nearly half an hour, we can say that the threshold is enough to cover all the solutions.

After determining the feasible solution space X, each solution should be tested to obtain the minimum Z in Equation 4. Based on the ideas of the IE algorithm, the testing sequence can also be set to improve the efficiency of computing. As the object is to minimize Z, and maximize Equation 9, in other words, the testing sequence can start from the feasible solution where $x_0 = 1$.

In a word, the IE algorithm is efficient for the TBM and reduces the computation greatly.

Case Study

In this section, the quality of methodology will be illustrated by using a real case of the Beijing metro system. All the results are calculated by C#. net and PL/SQL database programming.

Basic Information

Beijing metro is a system with 79,654 OD pairs and the highest passenger flow has been up to 12.69 million per

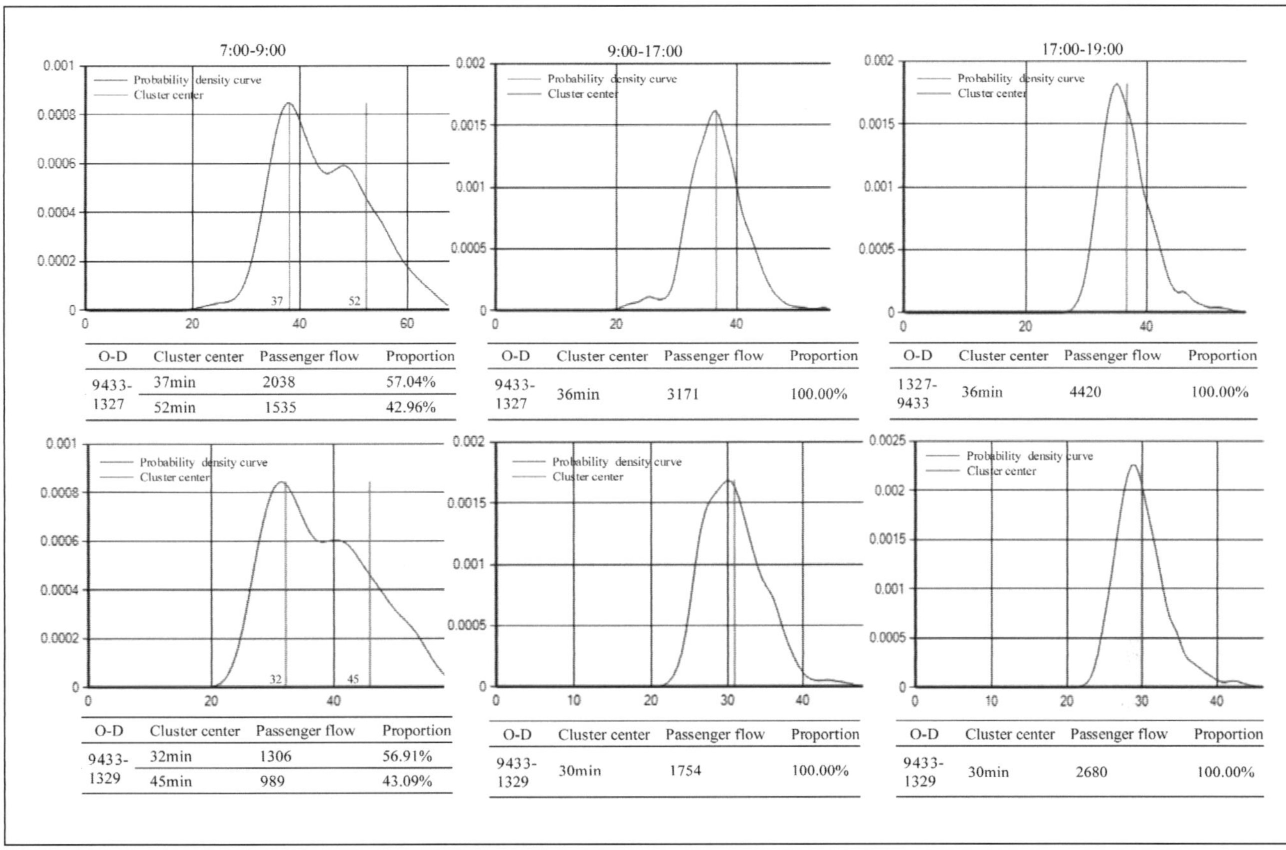

Figure 8. Cluster results for Station 9433 to Station 1327/1329.

day. In this paper, we use a total of 39,453,138 AFC records in 2016. To make a clear statement of TB behavior, we only search for the OD pairs with single path (26,509 pairs). These account for 33.28% of all the OD pairs and the single-path passenger flow accounts for 53.83% of the total passenger flow in the six days studied. The OD pairs of stations can belong to different paths, but there is only one path from the origin station to the destination station. In Figure 7, stations in the red dashed box belong to **Ω**.

Cluster Results

After the calculation with the AP algorithm under the BWP index, we can easily find the overtime single-path OD pairs having more than one cluster center. We cluster six days of AFC data into three periods: morning peak hours (tap-in time: 7:00~9:00), off-peak period (9:00~17:00) and evening peak hours (17:00~19:00).

Here, Station 9433–Station 1327/Station 1329 is taken as an example (Figure 7). There is more than one cluster center for single-path OD pairs (9433–1327 and 9433–1329) in the morning peak hours (Figure 8). Considering that morning and evening peak passenger flow are in the opposite directions, we choose 1327/1329–9433 for analysis in evening peak hours. The TB phenomenon is much more obvious in the morning peak hours, which is consistent with the fact that travel time in the morning is more concentrated compared with the other periods. The limited capacity of the metro cannot meet the demand of morning peak commuters. However, the clustering method can only make an initial filter, and not all the overtime journey data are caused by TB. Data classified to the second cluster needs further analysis by the TBM.

TB Path and Passenger Flow Reassignment

Taking one overtime AFC record as an example (Figure 7, $t_{Station9433}{}^{in}$ = 07:14:00, $t_{Station9441}{}^{out}$ = 07:57:30), station 9433 belongs to Ω, and it is the first suburban station with capacity dilemma in the direction toward the urban district. We define $f = 4$, so Station 9433 is A_4, and the AFC record can be described as $t_{A_4}{}^{in}$ = 07:14:00 and $t_{A_8}{}^{out}$ = 07:57:30). Above all, the feasible solution space $\boldsymbol{X}$ can be reduced into Equation 4 by the IE algorithm:

$$X_1 = \begin{pmatrix} 1 & 1 & 1 & 1 & 0 & 0 & 0 & 0 & 0 \\ 1 & 1 & 1 & 1 & 1 & 1 & 1 & 1 & 1 \\ 1 & 0 & 0 & 0 & 0 & 0 & 0 & 0 & 0 \end{pmatrix},$$

$$X_2 = \begin{pmatrix} 0 & 1 & 1 & 1 & 0 & 0 & 0 & 0 & 0 \\ 0 & 1 & 1 & 1 & 1 & 1 & 1 & 1 & 1 \\ 0 & 1 & 0 & 0 & 0 & 0 & 0 & 0 & 0 \end{pmatrix},$$

$$X_3 = \begin{pmatrix} 0 & 0 & 1 & 1 & 0 & 0 & 0 & 0 & 0 \\ 0 & 0 & 1 & 1 & 1 & 1 & 1 & 1 & 1 \\ 0 & 0 & 1 & 0 & 0 & 0 & 0 & 0 & 0 \end{pmatrix},$$

$$X_4 = \begin{pmatrix} 0 & 0 & 0 & 1 & 0 & 0 & 0 & 0 & 0 \\ 0 & 0 & 0 & 1 & 1 & 1 & 1 & 1 & 1 \\ 0 & 0 & 0 & 1 & 0 & 0 & 0 & 0 & 0 \end{pmatrix}.$$

Matching each feasible solution to the corresponding operational train diagram. According to the idea of the IE algorithm, we calculate the value objective function (Equation 4) by the sequence of X_1, X_2, X_3, and X_4. After testing, both $Z(X_1)$ and $Z(X_2)$ are negative, and disobey Equation 4. X_3 is well satisfied, and the tracking path is shown as in Figure 9.

As TB passengers are matched to the paths, the problem of TB passenger flow reassignment is solved at the same time. In this case study, the proportion of TB passenger flow is up to 12.92% during morning peak hours (the number of AFC records in this case is 2305 in total), and two-thirds of TB passengers choose Nanshao as their "transfer station." A TB passenger flow survey was also carried out along Line Changping (*1*) by counting the number of TB passengers at a certain station. The result shows that the TB phenomenon mostly happens at Shahe University Park and Nanshao, and the total proportion of TB passengers (where the origin station belongs to Line Changping) is about 10%.

According to this research, the TB phenomenon is common in the morning peak hours in suburban stations of the Beijing metro system. In addition, the traditional assignment model has exposed the spatiotemporal bias and errors in passenger flow distribution of the previous model (*1*). Our proposed methodology can effectively help to reassign the overtime data back to the network and rectify the deviation. Analysis of the TB phenomenon is a way to learn more about travel behavior, and it is valuable to guide the operation and planning of the

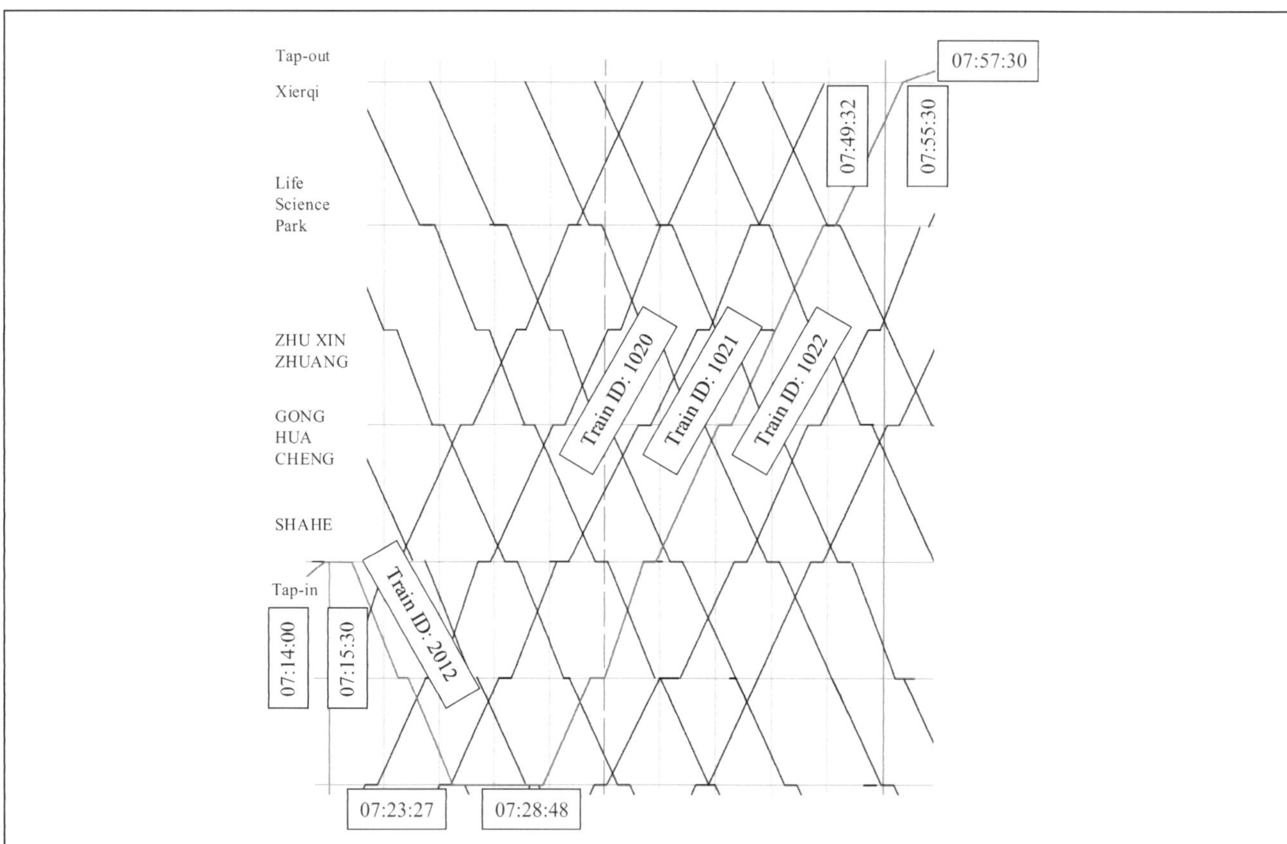

Figure 9. Traveling backwards path for Shahe (station 9433) to Xi'erqi (station 9441).

metro in the following ways: (a) Improve operational capacity to ease overcrowding such as shortening the travel interval; (b) Carry out passenger flow limitation measures to slow down the current flow; (c) Carry out passenger flow guidance measures such as staggered shifts or reward for choosing another, less crowded path; (d) Tighten security at stations with the worst overcrowding.

Conclusion and Future Research

This paper contributes a method for identifying the "traveling backwards" phenomenon, tracking the actual path and passenger flow reassignment in the overcrowded metro system:

(1) The TB phenomenon, which is a result of limited capacity in the overcrowded metro system, is identified by the AP cluster algorithm with the BWP index and a TBM. The TB phenomenon is more serious in the morning peak periods.
(2) The IE algorithm is efficient in reducing solution space for TBM. TBM is a 0-1 model developed for searching for the actual TB path and the TB passenger flow can be reassigned.
(3) The TB methodology framework is a supplement for the current researches on passenger route choice behavior and passenger flow assignment, especially under the crowded situation. Suggestions have been given for better operation and passenger flow organization.

A possible future research direction is to expand the methodology framework to multi-path OD pairs, and work with operations to minimize TB. In addition, more efforts are necessary to achieve a comprehensive understanding of overcrowding externalities on user behaviors, wellbeing, and choices.

Acknowledgments

The study was financially supported by the National Natural Science Foundation of China (71271153 and 71701152), Fundamental Research Funds for the Central Universities of China, and Shanghai Philosophy and Social Science Foundation (2015EGL006). The authors acknowledge Beijing Metro Co., Ltd for providing basic data during the research, and thank the anonymous referees for their valuable comments and suggestions.

Author Contributions

The authors confirm contribution to the paper as follows: study conception and design: Ruihua Xu, Yanan Li, Wei Zhu; data collection: Yanan Li, Sijie Li; analysis and interpretation of results: Yanan Li, Wei Zhu; draft manuscript preparation: Yanan Li. All authors reviewed the results and approved the final version of the manuscript.

References

1. Zhu, W., Y. N. Li, and Y. L. Zhou. *Calibration and Eva luation of Rail Transit Assignment Models with Passenger Flow Investigation and Big Data Analysis.* Tongji University and Beijing Metro Co., Ltd, China, 2017, pp. 28–30.
2. de Dios Ortúzar, J., and L. G. Willumsen. *Modelling Transport.* John Wiley & Sons, Inc., New York, 2011.
3. Kim, J. K., B. Lee, and S. Oh. Passenger Choice Models for Analysis of Impacts of Real-Time Bus Information on Crowdedness. *Transportation Research Record: Journal of the Transportation Research Board*, 2009. 2112: 119–126.
4. Tirachini, A., D. A. Hensher, and J. M. Rose. Crowding in Public Transport Systems: Effects on Users, Operation and Implications for the Estimation of Demand. *Transportation Research Part A Policy & Practice*, Vol. 53, 2013, pp. 36–52.
5. Batarce, M., J. C. Muñoz, J. de Dios Ortúzar, S. Raveau, C. Mojica, and R. A. Ríos. Use of Mixed Stated and Revealed Preference Data for Crowding Valuation on Public Transport in Santiago, Chile. *Transportation Research Record: Journal of the Transportation Research Board*, 2015. 2535: 73–78.
6. Chen, C., J. Ma, Y. Susilo, Y. Liu, and M. Wang. The Promises of Big Data and Small Data for Travel Behavior (aka Human Mobility) Analysis. *Transportation Research Part C: Emerging Technologies*, Vol. 68, 2016, pp. 285–299.
7. Zhou, F., and R. H. Xu. Model of Passenger Flow Assignment for Urban Rail Transit Based on Entry and Exit Time Constraints. *Transportation Research Record: Journal of the Transportation Research Board*, 2012. 2284: 57–61.
8. Sun, Y., and R. H. Xu. Rail Transit Travel Time Reliability and Estimation of Passenger Route Choice Behavior: Analysis Using Automatic Fare Collection Data. *Transportation Research Record: Journal of the Transportation Research Board*, 2012. 2275: 58–67.
9. Sun, Y., and P. M. Schonfeld. Schedule-Based Rail Transit Path-Choice Estimation using Automatic Fare Collection Data. *Journal of Transportation Engineering*, Vol. 142, No. 1, 2016, p. 04015037.
10. Zhu, Y., H. Koutsopoulos, and N. Wilson. A Probabilistic Passenger-to-Train Assignment Model Based on Automated Data. *Transportation Research Part B: Methodological*, Vol. 104, 2017, pp. 522–542.
11. Sun, L., Y. Lu, J. G. Jin, D. H. Lee, and K. W. Axhausen. An Integrated Bayesian Approach for Passenger Flow Assignment in Metro Networks. *Transportation Research Part C: Emerging Technologies*, Vol. 52, 2015, pp. 116–131.
12. Tirachini, A., L. Sun, A. Erath, and A. Chakirov. Valuation of Sitting and Standing in Metro Trains Using Revealed Preferences. *Transport Policy*, Vol. 47, 2016, pp. 94–104.
13. Sun, L., J. G. Jin, D. H. Lee, K. W. Axhausen, and A. Erath. Demand-Driven Timetable Design for Metro Services. *Transportation Research Part C: Emerging Technologies*, Vol. 46, 2014, pp. 284–299.

14. Li, W., R. Xu, Q. Luo, and S. Jones. Coordination of Last Train Transfers Using Automated Fare Collection (AFC) System Data. *Journal of Advanced Transportation*, Vol. 50, No. 8, 2016, pp. 2209–2225. https://onlinelibrary.wiley.com/doi/abs/10.1002/atr.1455.
15. Hong, L., W. Li, and W. Zhu. Assigning Passenger Flows on a Metro Network Based on Automatic Fare Collection Data and Timetable. *Discrete Dynamics in Nature and Society*, 2017:1–10.
16. Zhu, W., F. Zhou, J. Huang, and R. Xu. Validating Rail Transit Assignment Models with Cluster Analysis and Automatic Fare Collection Data. *Transportation Research Record Journal of the Transportation Research Board*, 2015. 2526: 10–18.
17. Pelletier, M. P., M. Trépanier, and C. Morency. Smart Card Data Use in Public Transit: A Literature Review. *Transportation Research Part C: Emerging Technologies*, Vol. 19, No. 4, 2011, pp. 557–568.
18. Barry, J., R. Newhouser, A. Rahbee, and S. Sayeda. Origin and Destination Estimation in New York City with Automated Fare System Data. *Transportation Research Record. Journal of the Transportation Research Board*, 2002. 1817: 183–187.
19. Kusakabe, T., T. Iryo, and Y. Asakura. Estimation Method for Railway Passengers' Train Choice Behavior with Smart Card Transaction Data. *Transportation*, Vol. 37, No. 5, 2010, pp. 731–749.
20. Zhu, W., H. Hu, and Z. Huang. Calibrating Rail Transit Assignment Models with Genetic Algorithm and Automated Fare Collection Data. *Computer-Aided Civil and Infrastructure Engineering*, Vol. 29, No. 7, 2014, pp. 518–530.
21. Li, Y. N., F. Zhou, L. Hong, and W. Zhu. Empirical Analysis of Failing to Board and Traveling Backwards in an Overcrowded Urban Rail Transit System. *CICTP 2017: Transportation Reform and Change - Equity, Inclusiveness, Sharing, and Innovation - Proceedings of the 17th COTA International Conference of Transportation Professionals*, Vol. 2018-January, 2018, pp. 1979–1989.
22. Frey, B. J., and D. Dueck. Clustering by Passing Messages between Data Points. *Science*, Vol. 315, No. 5814, 2007, pp. 972–976.
23. Zhou, S. B., Z. Y. Xu, and X. Q. Tang. Method for Determining Optimal Number of Clusters Based on Affinity Propagation Clustering. *Control and Decision*, Vol. 26, No. 8, 2011, pp. 1147–1152, 1157.
24. Brinton, C. R. An Implicit Enumeration Algorithm for Aircraft Arrival Scheduling. *Proceedings IEEE/AIAA 11th Digital Avionics Systems Conference*, Seattle, WA, 1992, pp. 268–274.

The Standing Committee on Rail Transit Systems (AP065) peer-reviewed this paper (18-03644).

Article

TRR
JOURNAL OF THE TRANSPORTATION RESEARCH BOARD

Transportation Research Record
2018, Vol. 2672(8) 243–254

DOI: 10.1177/0361198118772958
journals.sagepub.com/home/trr
SAGE

Timetable Synchronization and Optimization Considering Time-Dependent Passenger Demand in an Urban Subway Network

Pan Shang[1], Ruimin Li[1], Zhiyong Liu[1], Kai Xian[2], and Jifu Guo[2]

Abstract

In an urban subway network, a synchronized and optimized timetable can provide a good service to all passengers. Therefore, the timetable is typically the most important factor for an operator. This study builds a time-dependent passenger demand-driven timetable synchronization and optimization (TDTSO) model as a mixed-integer programming model to minimize passenger total travel time in an urban subway network by adjusting departure times, running times, stopping times, and headways of all trains on each line. To solve the model, a binary variable determination (BVD) method is developed to calculate the binary variables of the TDTSO model, and a genetic algorithm based on the BVD method is proposed. The proposed TDTSO model is applied to the Beijing subway network in a case study. Several performance indicators are presented to verify the efficiency of the proposed model. The results can provide the operator of an urban subway network with additional options to produce a synchronized and optimized timetable for real-world situations.

The problem of timetable synchronization and optimization in an urban subway network has recently attracted considerable interest. Non-transferring passengers go through only one line to arrive at their destination, whereas transferring passengers typically make at least one transfer during their trips. For non-transferring passengers, an optimized timetable for one line can improve the operation of the service. For transferring passengers, synchronizing a timetable between different lines to improve transfer efficiency is an important issue in urban subway network operation.

Smart card automated fare collection (AFC) systems, which produce large quantities of detailed data on onboard transactions (e.g., departure/arrival time and station of each passenger), are being used increasingly by public transit agencies, particularly in urban subway networks. By aggregating the smart card data obtained from the AFC system by origin–destination (OD) pair and by arrival time interval, the historical time-dependent passenger demand in typical days can be obtained, which contains the time-dependent passenger demand information and can be adopted to improve the performance of the timetable of an urban subway network (*1*).

In the following section, previous literature is reviewed and the contribution of this study is presented.

Literature Review

Literature on the timetable synchronization and optimization problem in an urban subway network is abundant. Cacchiani et al. and Ibarra-Rojas et al. surveyed existing studies that primarily dealt with the planning, operation, and control problems of transport systems (*2, 3*). In an urban subway network, passengers typically complain at transfer stations when there is a long wait for connecting trains (*4*). Thus, minimizing passenger transfer waiting time in an urban subway network has been studied for years. Domschke proposed a model to minimize passenger transfer waiting time by optimizing train dispatch time without considering passenger demand (*5*). Considering the timetable synchronization problem of the Hong Kong mass transit railway, Wong et al. proposed a mixed-integer programming optimization model that minimized the transfer waiting time of all passengers in the subway system (*6*). Wu et al. and

[1]Department of Civil Engineering, Hang Lung Center for Real Estate, Tsinghua University, Beijing, China
[2]Beijing Transport Institute, Beijing, China

Corresponding Author:
Address correspondence to Ruimin Li: lrmin@tsinghua.edu.cn

Shafahi and Khani considered the number of transfer passengers, which was defined by the number of passengers who transferred from one train on one line to another train on a connecting line; they then proposed a genetic algorithm (GA) to solve the model (*4, 7*).

Apart from passenger transfer waiting time, resolving passenger total waiting time is another objective that researchers have focused on. Vansteenwegen and Van Oudheusden proposed a standard linear programming model to improve passenger service with a waiting cost function that weighted different types of waiting times and late arrivals; they then applied the programming model to a small part of the Belgian railway network (*8*). Wu et al. considered a bus timetabling problem with stochastic travel times and added slack time to mitigate randomness in bus travel times (*9*). They also developed a stochastic integer programming model to minimize total waiting time cost for three passenger types (i.e., transferring, boarding, and through passengers). Sun et al. and Barrena et al. constructed models to minimize passenger total waiting time in a single-line transit railway based on time-dependent OD demand (*10–12*).

Researchers have studied other objectives in addition to passenger transfer waiting time and total waiting time. Ceder et al. proposed a mixed-integer linear programming model to maximize the number of simultaneous bus arrivals at transfer nodes by considering the satisfaction and convenience of users of the system (*13*). Castillo et al. proposed a model to minimize the sum of relative travel times by considering user preferences (*14*). Lin and Ku proposed a model to maximize the profit made by a rail company by optimizing stopping patterns using GA (*15*). Wang et al. proposed an event-driven model that involved three types of events, that is, departure, arrival, and passenger arrival; they studied how rates had changed events for the train scheduling problem in an urban rail transit network (*16*). The model included the routing of arriving passengers at transfer stations and passenger transfer behavior. Kang et al. proposed models for last trains to optimize transfer connection headways (*17*). Other researchers have studied the integrated problem of simultaneously preparing a timetable and passenger routes based on an event activity network to minimize passenger total travel time that considered constant OD demand (*18–20*).

Several researchers have studied the timetable optimization problem with multi-objectives. Zhou and Zhong proposed a multi-mode resource-constrained project scheduling formulation by considering variations in interdeparture times and train travel time in double-track train timetabling applications (*21*); they designed several dominance rules to find Pareto optimum schedules in an intercity passenger corridor. Hadas and Ceder introduced synchronized timetables to reduce the waiting time caused by transfers, and developed a dynamic programming model for minimizing total travel time (*22*). Ho et al. formulated a railway timetabling process as a multi-objective optimization problem that considered infrastructure capacity and customer satisfaction (*23*); they proposed intelligent solution-searching techniques to provide a possible solution. Liu et al. considered three objectives, namely, total waiting time, total in-vehicle travel time, and total operating cost, and then built an optimization model to minimize the weighted sum of the three objectives to establish a bus stop-skipping scheme with random travel time (*24*). Heydar et al. considered the railway scheduling problem with two types of trains, namely, express and local, to minimize dispatching cycle and total train delay (*25*). Yang et al. proposed a two-objective cooperative scheduling model to increase regenerative energy utilization and shorten passenger waiting time (*26*). Sun et al. presented a multi-objective optimization model that considered the average travel time of trains, energy consumption, and user satisfaction (*27*). An improved GA was designed to solve the problem.

Most of the previous papers addressed the timetable synchronization and optimization problem by focusing mainly on minimizing passenger transfer and total waiting times. Recently, passenger demand in urban subway networks has increased significantly, and oversaturation usually occurs both at the station platform and in trains in peak hours, meaning that passengers are increasingly concerned about the total travel time of their trips. Therefore, this study considers the total travel time of all passengers in an urban subway system as the objective, and formulates a time-dependent passenger demand-driven timetable synchronization and optimization (TDTSO) model using historical time-dependent passenger demand on typical days. The contributions made in this study are summarized as follows:

(a) This study builds a TDTSO model, which is a mixed-integer programming model in an urban subway network that considers historical time-dependent passenger demand on typical days. Passenger total travel time is set as the objective of the TDTSO model, whereas other objectives, such as transfer waiting time (*4, 6, 7*) and total waiting time (*28*), are formulated based on the TDTSO model, as a result of its flexibility.

(b) A binary variable determination (BVD) method is developed to calculate binary variables (e.g., passenger feasible loading variable and passenger actual loading variable), which can significantly reduce the constraints of the TDTSO model, and thus passenger waiting time and riding time can be easily and rapidly obtained.

(c) A BVD-based GA is proposed to solve the TDTSO model, and the efficiency of GA is compared with those of other algorithms [e.g., pattern

Table 1. Notation Used in this Study

Notation	Description
Indices and sets	
t	Index of time intervals, $t \in T$, where T is the set of time intervals
$\boldsymbol{l}, \boldsymbol{l}'$	Index of lines, $l, l' \in L$, where L is the set of lines
$\boldsymbol{s}, \boldsymbol{s}'$	Index of stations, $s, s' \in S_l$, where S_l is the set of stations on line l
$\boldsymbol{w}(\boldsymbol{t})$	Index of time-dependent OD pairs, $w(t) \in W(t)$, where $W(t)$ is the set of time-dependent OD pairs in time interval t
j	Index of sub-trips, $j \in \{1, 2, \ldots, g_{w(t)}\}$, where $g_{w(t)}$ is the number of sub-trips for passengers of time-dependent OD pair $w(t)$
$\boldsymbol{q}, \boldsymbol{q}'$	Index of trains, $q \in Q_l$, where Q_l is the set of trains on line l
Timetable notations	
$\boldsymbol{t}_{sl,q}^{A}$	Arrival time of train q at station s on line l
$\boldsymbol{t}_{sl,q}^{D}$	Departure time of train q at station s on line l
$\boldsymbol{t}_{sl,q}^{R}$	Running time of train q from station $s-1$ to station s on line l
$\boldsymbol{t}_{sl,q}^{S}$	Stopping time of train q at station s on line l
$\boldsymbol{t}_{sl,q}^{H_A}$	Arrival headway between train $q-1$ and train q at station s on line l
$\boldsymbol{t}_{sl,q}^{H_D}$	Departure headway between train $q-1$ and train q at station s on line l
$\boldsymbol{t}_{sl}^{R,\min}\left(\boldsymbol{t}_{sl}^{R,\max}\right)$	Minimum (maximum) running time of train q from station $s-1$ to station s on line l
$\boldsymbol{t}_{sl}^{S,\min}\left(\boldsymbol{t}_{sl}^{S,\max}\right)$	Minimum (maximum) stopping time of train q at station s on line l
$\boldsymbol{t}_{sl}^{H,\min}\left(\boldsymbol{t}_{sl}^{H,\max}\right)$	Minimum (maximum) headway between train $q-1$ and train q at station s on line l
Passenger notations	
$\boldsymbol{f}_{w(t)}$	Number of passengers of time-dependent OD pair $w(t)$
$\boldsymbol{P}_{w(t)}$	Set of sub-trips for passengers of time-dependent OD pair $w(t)$, $P_{w(t)} = \{s_j l_j \rightarrow s_j' l_j' \mid j = 1, 2, \ldots, g_{w(t)}\}$, where $s_j l_j \rightarrow s_j' l_j'$ records the first and last stations for passengers of time-dependent OD pair $w(t)$ in their j th sub-trip
$\boldsymbol{x}_{w(t),j,q}$	Passenger feasible loading variable is equal to 1 if train q can possibly load passengers of time-dependent OD pair $w(t)$ at station s_j on line l_j in their j th sub-trip and is equal to 0 otherwise
$\boldsymbol{y}_{w(t),j,q}$	Passenger actual loading variable is equal to 1 if train q can actually load passengers of time-dependent OD pair $w(t)$ at station s_j on line l_j in their j th sub-trip and is equal to 0 otherwise
$\boldsymbol{\tau}_{w(t),j}$	Arrival time of passengers of time-dependent OD pair $w(t)$ at the first station in their j th sub-trip
$\boldsymbol{T}_{s_j' l_j', s_{j+1} l_{j+1}}^{\text{transfer}}$	Transfer time of passengers from station s on line l to station s' on line l'
$\boldsymbol{T}_{w(t),j}^{\text{wait}}$	Waiting time of passengers of time-dependent OD pair $w(t)$ in their j th sub-trip
$\boldsymbol{T}_{w(t),j}^{\text{ride}}$	Riding time of passengers of time-dependent OD pair $w(t)$ in their j th sub-trip
Energy notation	
$\boldsymbol{e}\left(\boldsymbol{t}_{sl,q}^{R}\right)$	Energy cost for train q running from station $s-1$ to station s on line l using time $t_{sl,q}^{R}$

search (PS), simulated annealing (SA), and branch-and-bound (BnB)] via a numerical test.

(d) The TDTSO model and GA are applied to the Beijing subway network. Adjustments of the operational parameters are analyzed based on the case of the Beijing subway network.

Formulation

Definitions

Synchronization: Synchronization denotes minimum transfer waiting time and maximum transfer efficiency between different lines in an urban subway network.

Trip: A trip denotes completed travel for passengers from their origin stations to destination stations in an urban subway network.

Sub-trip: A sub-trip denotes part of travel for passengers traveling on a specific line, and several sub-trips comprise a complete trip if passengers need to transfer in an urban subway network.

Time-dependent OD pair: A time-dependent OD pair denotes a group of passengers with the same physical OD pair (i.e., origin station and destination station) and departure time at origin station.

All notations and key concepts used in the formulation of TDTSO model are stated in Table 1.

Passenger Waiting and Riding Times

In an urban subway network, passengers are concerned with their total travel time, which is composed of waiting, riding, and transferring, which collectively indicate

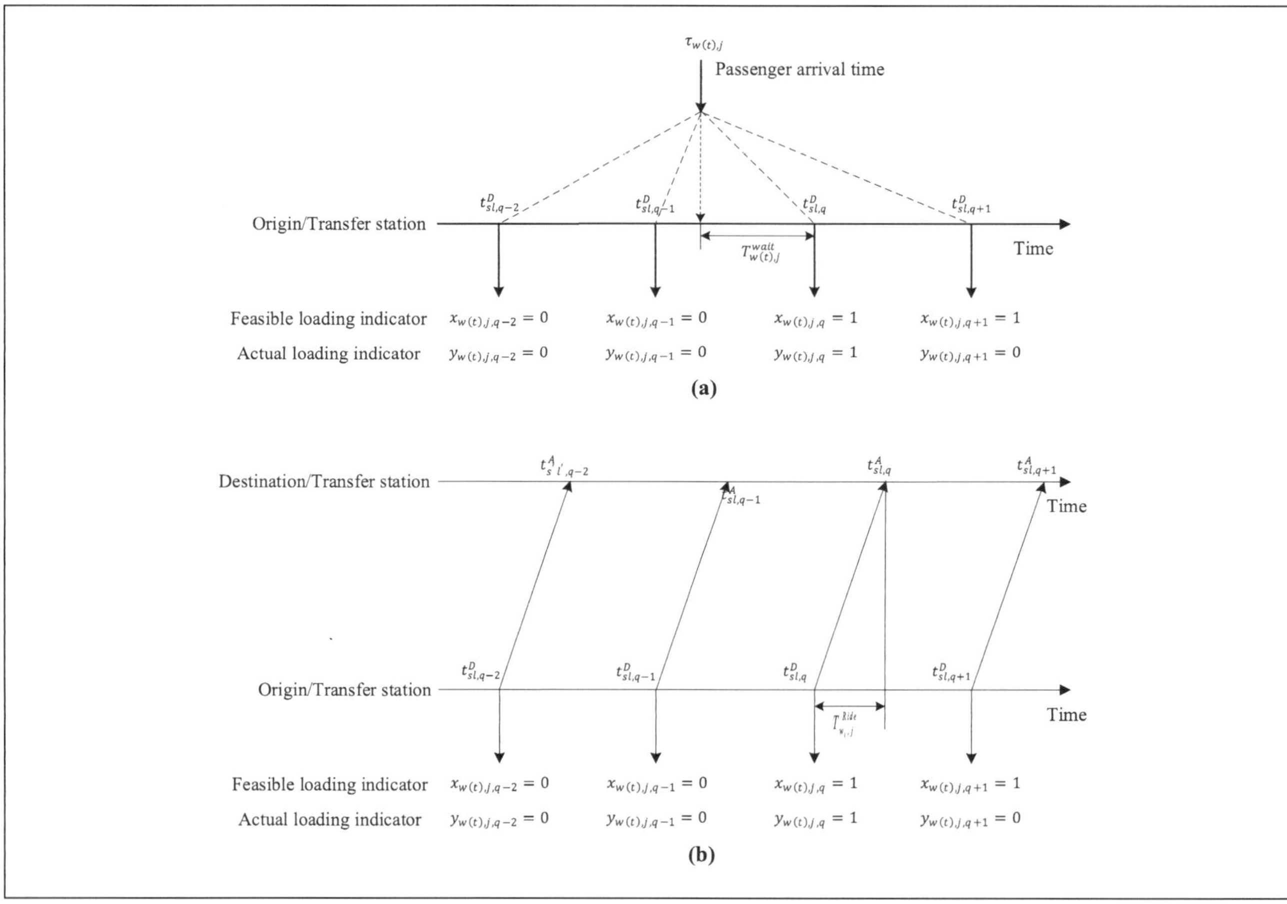

Figure 1. Passenger waiting time and riding time; (*a*) waiting time, (*b*) riding time.

the time cost during the trips of a passenger from origin to destination (*16*). This section presents the calculation methods for passenger waiting and riding times. Assuming that all passengers always take the first arriving train after they reach the platform at a station to reduce their waiting time, to illustrate, Figure 1*a* provides the description of waiting time $T^{\text{wait}}_{\text{w(t),j}}$ in the j th sub-trip for passengers of OD pair $w(t)$.

In Figure 1*a*, the arrow above the timeline is passenger arrival time $\tau_{\text{w(t),j}}$ at the origin station in the j th sub-trip. For passengers of OD pair $w(t)$, the arrival time $\tau_{\text{w(t),1}}$ at the origin station in the first sub-trip is time t, and the following arrival time $\tau_{\text{w(t),j}}$ at transfer station during the j th sub-trip can be calculated by Equation 1. The arrow below the timeline represents the departure times of $(q-2)$ th, $(q-1)$ th, q th, and $(q+1)$th trains. In addition, the dotted lines indicate that passengers will get on the trains. Evidently, getting on $(q-2)$th and $(q-1)$th trains is impossible for passengers; however, they can possibly get on q th and $(q+1)$th trains. Thus, the feasible loading indicators for these four trains in this case are $x_{\text{w(t),j,q-2}} = 0$, $x_{\text{w(t),j,q-1}} = 0$, $x_{\text{w(t),j,q}} = 1$, and $x_{\text{w(t),j,q-1}} = 1$. To obtain the feasible loading indicator $x_{\text{w(t),j,q}}$, Constraint 2 is proposed to determine $x_{\text{w(t),j,q}}$ in the TDTSO model.

$$\tau_{\text{w(t),j}} = \tau_{\text{w(t),j-1}} + T^{\text{wait}}_{\text{w(t),j-1}} + T^{\text{ride}}_{\text{w(t),j-1}} + T^{\text{transfer}}_{s'_{j-1}l'_{j-1},s_jl_j} \quad (1)$$

$$M\left(x_{\text{w(t),j,q}} - 1\right) \le t^{D}_{s_jl_j,q} - \tau_{\text{w(t),j}} < Mx_{\text{w(t),j,q}} \quad (2)$$

Subsequently, we assume that all passengers always take the first arriving train after they reach the platform at a station to reduce their waiting time. In this case, passengers will ultimately get on q th train. Thus, the actual loading indicators for the four trains in this case are $y_{\text{w(t),j,q-2}} = 0$, $y_{\text{w(t),j,q-1}} = 0$, $y_{\text{w(t),j,q}} = 1$, $y_{\text{w(t),j,q+1}} = 0$. The actual loading indicator $y_{\text{w(t),j,q}}$ can be calculated using Equation 3, which presents the exclusive train for passengers to get on.

$$y_{\text{w(t),j,q}} = x_{\text{w(t),j,q}} - x_{\text{w(t),j,q-1}} \quad (3)$$

Passenger waiting time (Figure 1*a*), which is the time span between passenger arrival time and actual loading train departure time, can be easily calculated using Equation 4.

$$T^{\text{wait}}_{\text{w(t)},\text{j}} = \sum\nolimits_{\text{q}\in\text{Q}_\text{l}} y_{\text{w(t)},\text{j},\text{q}} \times \left(t^{D}_{\text{s}_\text{j}\text{l}_\text{j},\text{q}} - \tau_{\text{w(t)},\text{j}}\right) \quad (4)$$

Passenger riding time (Figure 1*b*) refers to the time span between passenger departure time and the actual loading train arrival time at the destination/transfer station. After the actual loading indicator $y_{\text{w(t)},\text{j},\text{q}}$ is obtained when passenger waiting time is calculated, passenger riding time can be easily calculated using Equation 5.

$$T^{\text{ride}}_{\text{w(t)},\text{j}} = \sum\nolimits_{\text{q}\in\text{Q}_\text{l}} y_{\text{w(t)},\text{j},\text{q}} \times \left(t^{D}_{\text{s}'_\text{j}\text{l}'_\text{j},\text{q}} - t^{D}_{\text{s}_\text{j}\text{l}_\text{j},\text{q}}\right) \quad (5)$$

Timetable for an Urban Subway Network

The timetable for an urban subway network determines the arrival and departure times of trains, which can indicate the efficiency of the urban subway network. A timetable should satisfy the following constraints to ensure its feasibility.

Arrival and Departure Times of Trains. For any line $l \in L$, Equations 6 and 7 are given to calculate departure time $t^{D}_{\text{sl},\text{q}}$ and arrival time $t^{A}_{\text{sl},\text{q}}$, respectively, as follows:

$$t^{D}_{\text{sl},\text{q}} = t^{A}_{\text{sl},\text{q}} + t^{S}_{\text{sl},\text{q}} \quad (6)$$

$$t^{A}_{\text{sl},\text{q}} = t^{D}_{(\text{s}-1)\text{l},\text{q}} + t^{R}_{\text{sl},\text{q}} \quad (7)$$

Headway of Trains. Equations 8 and 9 define the arrival and departure headways between adjacent trains at each station on each line, which is the minimum time between two adjacent trains.

$$t^{\text{H}_\text{A}}_{\text{sl},\text{q}} = t^{A}_{\text{sl},\text{q}} - t^{A}_{\text{sl},\text{q-1}} \quad (8)$$

$$t^{\text{H}_\text{D}}_{\text{sl},\text{q}} = t^{D}_{\text{sl},\text{q}} - t^{D}_{\text{sl},\text{q-1}} \quad (9)$$

To ensure operational safety and fulfill service requirements, the minimum and maximum headways on line l are denoted as $t^{\text{H},\min}_{\text{sl}}$ and $t^{\text{H},\max}_{\text{sl}}$, respectively. Thus, Constraints 10 and 11 ensure that the headways are within the given range.

$$t^{\text{H},\min}_{\text{sl}} \le t^{\text{H}_\text{A}}_{\text{sl},\text{q}} \le t^{\text{H},\max}_{\text{sl}} \quad (10)$$

$$t^{\text{H},\min}_{\text{sl}} \le t^{\text{H}_\text{D}}_{\text{sl},\text{q}} \le t^{\text{H},\max}_{\text{sl}} \quad (11)$$

Running Time and Stopping Time. In real operation, each train has a limited speed range, which leads to the limitations on running times between adjacent stations. The minimum stopping times must guarantee that all passengers waiting at the platform can get on the train, whereas the maximum stopping times can guarantee the efficiency of urban subway service. Thus, setting limitations on the running times and stopping times of all trains is required. Constraint 12 ensures that running times are between the lower and upper bounds. Constraint 13 is the stopping time constraint, which uses the lower bound to ensure that all waiting passengers can board the train and stay safe, as well as the upper bound to ensure operational efficiency.

$$t^{\text{R},\min}_{\text{sl}} \le t^{R}_{\text{sl},\text{q}} \le t^{\text{R},\max}_{\text{sl}} \quad (12)$$

$$t^{\text{S},\min}_{\text{sl}} \le t^{S}_{\text{sl},\text{q}} \le t^{\text{S},\max}_{\text{sl}} \quad (13)$$

TDTSO Model Construction

On the basis of the formulation of passenger waiting and riding times, as well as the timetable for an urban subway network, the TDTSO model can be constructed as follows.

Objective function:

$$\begin{aligned}\min c(\textit{travel time}) = &\sum_{t\in T}\sum_{w(t)\in W(t)}\sum_{j\in\{1,\dots,g_{w(t)}\}} f_{w(t)} \times T^{\text{wait}}_{w(t),j} + \sum_{t\in T}\sum_{w(t)\in W(t)}\sum_{j\in\{1,\dots,g_{w(t)}\}} f_{w(t)} \times T^{\text{ride}}_{w(t),j} \\ &+ \sum_{t\in T}\sum_{w(t)\in W(t)}\sum_{j\in\{1,\dots,g_{w(t)-1}\}} f_{w(t)} \times T^{\text{transfer}}_{s'_j l'_j, s_{j+1} l_{j+1}}\end{aligned} \quad (14)$$

Subject to: Equations 1 to 13.

The objective function (*14*) aims to minimize the total travel time of all passengers in an urban subway network, which contains passenger waiting time, riding time, and transfer time, and the TDTSO model subject to Equations 1 to 13 stated above.

Modified Objective Function for Analyzing Train Energy Cost, Transfer Waiting Time, and Total Waiting Time

Previous studies considered the energy cost as part of the objective function (*26, 27*), to analyze the tradeoff between total passenger travel time and energy cost. Based on the energy cost curve, there is an empirical energy cost function for train q running from station $s-1$ to station s on line l using time $t^{R}_{\text{sl},\text{q}}$, which can be presented with Equation 15. Therefore, the total energy cost of all trains can be calculated by Equation 16.

$$e_{\text{sl},\text{q}} = E\left(t^{R}_{\text{sl},\text{q}}\right) \quad (15)$$

$$c(energy) = \sum_{l \in L} \sum_{s \in S_l} \sum_{q \in Q_l} e_{sl,q} = \sum_{l \in L} \sum_{s \in S_l} \sum_{q \in Q_l} E\left(t^R_{\mathrm{sl,q}}\right) \quad (16)$$

The objective function can be further modified as Equation 17 to analyze the tradeoff between total passenger travel time and energy cost for operators of an urban subway system, where α and β are weights for passenger total travel time and energy cost in the objective function.

$$\min \alpha \times c(travel\ time) + \beta \times c(energy) \quad (17)$$

Besides the total travel time presented by Equation 14, transfer waiting time (Equation 18) and total waiting time (Equation 19) can also be adopted to design the timetable in an urban subway network, thanks to the flexibility of the TDTSO model.

$$\min c(transfer\ waiting\ time) = \sum_{t \in T} \sum_{w(t) \in W(t)} \sum_{j \in \{2,\ldots,g_{w(t)}\}} f_{\mathrm{w(t)}} \times T^{\mathrm{wait}}_{\mathrm{w(t),j}} \quad (18)$$

$$\min c(total\ waiting\ time) = \sum_{t \in T} \sum_{w(t) \in W(t)} \sum_{j \in \{1,\ldots,g_{w(t)}\}} f_{\mathrm{w(t)}} \times T^{\mathrm{wait}}_{\mathrm{w(t),j}} \quad (19)$$

Solution Algorithm

Various algorithms have been applied to solve the timetable synchronization and optimization problem in an urban subway network, which belongs to the NP-hard class of problems. Recently, artificial intelligence techniques have been used to solve this problem. For example, GA, SA, Tabu-searching algorithm, and artificial neural network algorithm are widely adopted. Among these, GA is an extensively applied stochastic optimization procedure (*26*).

In this section, we propose a BVD method to calculate feasible passenger loading indicator $x_{\mathrm{w(t),j,q}}$ and passenger actual loading indicator $y_{\mathrm{w(t),j,q}}$, which are key variables in obtaining passenger waiting time and riding time, respectively. The BVD method can be utilized to calculate fitness in GA. GA is presented based on the BVD method to solve the TDTSO model.

BVD Method

The BVD method is presented to calculate passenger feasible loading variable $x_{\mathrm{w(t),j,q}}$ and passenger actual loading variable $y_{\mathrm{w(t),j,q}}$. For example, to capture waiting time $T^{\mathrm{wait}}_{\mathrm{w(t),j}}$, we should calculate $\{y_{\mathrm{w(t),j,1}}, y_{\mathrm{w(t),j,2}}, \ldots, y_{\mathrm{w(t),j,|Q_l|}}\}$. Variables $\{y_{\mathrm{w(t),j,1}}, y_{\mathrm{w(t),j,2}}, \ldots, y_{\mathrm{w(t),j,|Q_l|}}\}$ are restrained by Equations 2 and 3, which enables the TDTSO model to obtain a large number of constraints and variables. Therefore, in terms of operating efficiency, the BVD method is demonstrated to determine $x_{\mathrm{w(t),j,q}}$ and $y_{\mathrm{w(t),j,q}}$ quickly and easily. The BVD method for calculating $x_{\mathrm{w(t),j,q}}$ and $y_{\mathrm{w(t),j,q}}$ $(j \in \{1, 2, \ldots, g_{\mathrm{w(t)}}\}, q \in Q_l)$ for each OD pair $w(t) \in W(t)$ is described as follows.

Step 1. Initialize.
Input the initial parameters: set of sub-trip $P_{\mathrm{w(t)}} = \{s_j l_j \rightarrow s'_j l'_j | j = 1, 2, \ldots, g_{\mathrm{w(t)}}\}$; *current timetable* $t^A_{\mathrm{sl,q}}, t^D_{\mathrm{sl,q}}, t^R_{\mathrm{sl,q}}, t^S_{\mathrm{sl,q}}$ *for* $s \in S_l$, $l \in L$, *and* $q \in Q_l$. *Set* $j = 1$ *and* $\tau_{\mathrm{w(t),j}} = t$.
Step 2. Determine variables.
Step 2.1. Set $q = 1$.
Step 2.2. If $t^D_{\mathrm{s_j l_j, q}} \geq \tau_{\mathrm{w(t),j}}$, *then* $x_{\mathrm{w(t),j,q}} = 1$; *otherwise,* $x_{\mathrm{w(t),j,q}} = 0$.
Step 2.3. Set $y_{\mathrm{w(t),j,q}} = x_{\mathrm{w(t),j,q}} - x_{\mathrm{w(t),j,q\text{-}1}}$.
Step 2.4. Set $q = q + 1$ *and return to Step 2.2 until* $q = |Q_l|$.
Step 3. Calculate passenger waiting time and riding time.
Step 3.1. Calculate passenger waiting time using Equation 4.
Step 3.2. Calculate passenger riding time using Equation 5.
Step 3.3. Calculate passenger arrival time at the first station of the next sub-trip using Equation 1.
Step 4. Update.
Set $j = j + 1$ $j = j + 1$ *and return to Step 2 until* $j = g_{\mathrm{w(t)}}$.

By adopting this method, passenger feasible loading variable $x_{\mathrm{w(t),j,q}}$ and passenger actual loading indicator $y_{\mathrm{w(t),j,q}}$ can be obtained for passenger OD pair $w(t) \in W(t)$, whereas passenger waiting time and riding time can be determined using the BVD method.

GA

GA is a heuristic algorithm that can solve models for large-scale networks. GA is used here to solve the model after considering the characteristics of the TDTSO model and computing efficiency. In GA, the train arrival time at the first station of each line ($t^A_{\mathrm{1l,q}}$), running time ($t^R_{\mathrm{sl,q}}$), and stopping time ($t^S_{\mathrm{sl,q}}$) are selected to represent the genes of chromosomes. The chromosome has three parts: a series of arrival times at the first station, a series of running times, and a series of stopping times. In this study, the first population is created according to the initial timetable. Passenger total travel time, which is the objective function in the TDTSO model, is selected to be the fitness function and captured using the BVD method.

Three common operators used in GA are selection, crossover, and mutation. The selection operator chooses the parents for the next generation based on their scaled values from the fitness function. The stochastic uniform selection method chooses each parent that corresponds to the length proportional to the wheel value sequence, which can provide a low objective value with an increased possibility to be chosen and a high objective value with less possibility to be chosen, to conform to the rule of survival of the fittest. Scattered crossover is adopted in GA. This crossover creates a random binary vector, which has the same length as the chromosome. Then, the genes where the vector is 1 from the first parent and where the vector is 0 from the second parent are selected and combined to produce a child. The mutation operation produces slight random changes in the individuals of the population, thereby providing genetic diversity and enabling GA to escape local optima. A uniform mutation is adopted in GA. The uniform mutation first selects a fraction of the vector points of an individual for mutation. A random value α is then produced in $\alpha \in [0, 1]$, which replaces the chosen gene with $\alpha \times (Upper - Lower)$.

Given that crossover and mutation operations may result in infeasible solutions, the minimum and maximum boundaries of decision variables as well as the headway constraints are used as the input. Thus, the algorithm will delete infeasible solutions and produce new ones until a feasible solution is generated.

The last issue is the termination of GA. Three criteria for GA termination are typically used (*7*), namely, (a) stall test (the best solution does not change after a given number of iterations), (b) function tolerance (the difference between the best and worst solutions in a population is less than a given value), and (c) generation limit (a maximum number of iterations is reached). In the proposed GA, the generation limit is selected as the unique criterion to stop the algorithm.

In summary, the detailed algorithmic steps are described as follows.

Step 1. Initialize.
Step 1.1. *Initialize GA parameters: population size N, initial generation k = 1, and maximum number of generations K.*
Step 1.2. *Initialize a parent chromosome, including train arrival time at the first station of each line ($t^A_{1l,q}$), running time ($t^R_{sl,q}$), and stopping time ($t^S_{sl,q}$) according to the initial timetable.*
Step 1.3. *Take a mutation operation to the parent chromosome and generate another chromosome. Apply a crossover operation to these two chromosomes and generate a chromosome until the number of feasible chromosomes reaches N.*
Step 1.4. *Produce the initial population P(1) with these N feasible chromosomes.*
Step 2. Create.
Step 2.1. *Calculate the fitness function value using the BVD method and return the fitness function value of each chromosome in population P(1).*
Step 2.2. *Perform selection operation and produce a new population $P_n(k)$.*
Step 3.2. *Perform crossover operation and produce crossover children from $P_n(k)$.*
Step 3.3. *Perform mutation operation and produce mutation children from $P_n(k)$.*
Step 3.4. *Combine the elite children, crossover children, and mutation children to form a new population.*
Step 3. Stop.
Step 3.1. *Update k = k + 1.*
Step 3.2. *If k = K, then stop. Otherwise, return to Step 2.*
End.

Numerical Test

Test Network Description

In this section, the TDTSO model is implemented and the efficiency of the GA is compared with those of other common algorithms (i.e., PS, SA, and BnB) via the test network. As shown in Figure 2*a*, the test network has two lines (Lines 1 and 2), and three stations are found on each line. Transfers can occur between Station 2 on Line 1 and Station 2 on Line 2 at a transfer time of 1 min.

Three trains on each line are considered. The running and stopping times of the train are fixed (running time: 4 min, stopping time: 1 min), while the minimum and maximum headways are 3 min and 8 min, respectively. The time-dependent passenger demand is shown in Figure 2*b*.

Optimization Results and GA Performance

The numerical test was run on an Intel® Core™ i5-4460 Processor CPU 3.40 GHz 16.00 GB RAM in MATLAB 2014b. GA was convergent after 30 iterations within 0.52 s for this simple network test. The objective value on line 1, objective value on line 2, and total travel time can be reduced by 13.87%, 1.79%, and 9.11%, respectively.

The results of the TDTSO model from different algorithms are shown in Table 2. GA, PS, SA, and BnB obtain similar results, but CPU time is different. GA takes 0.52 s to obtain an optimal solution after 30 iterations or steps. However, PS, SA, and BnB take 0.95, 5.30, and 24.90 s, respectively, to solve the TDTSO model. Evidently, GA is more efficient than the other methods.

Two parameters of GA related to the size of the problem are tested in this study. Experience shows that better results tend to come from the proposed model with a larger generation (*K*) and population (*N*). As shown in Figure 3, the number of generations increases

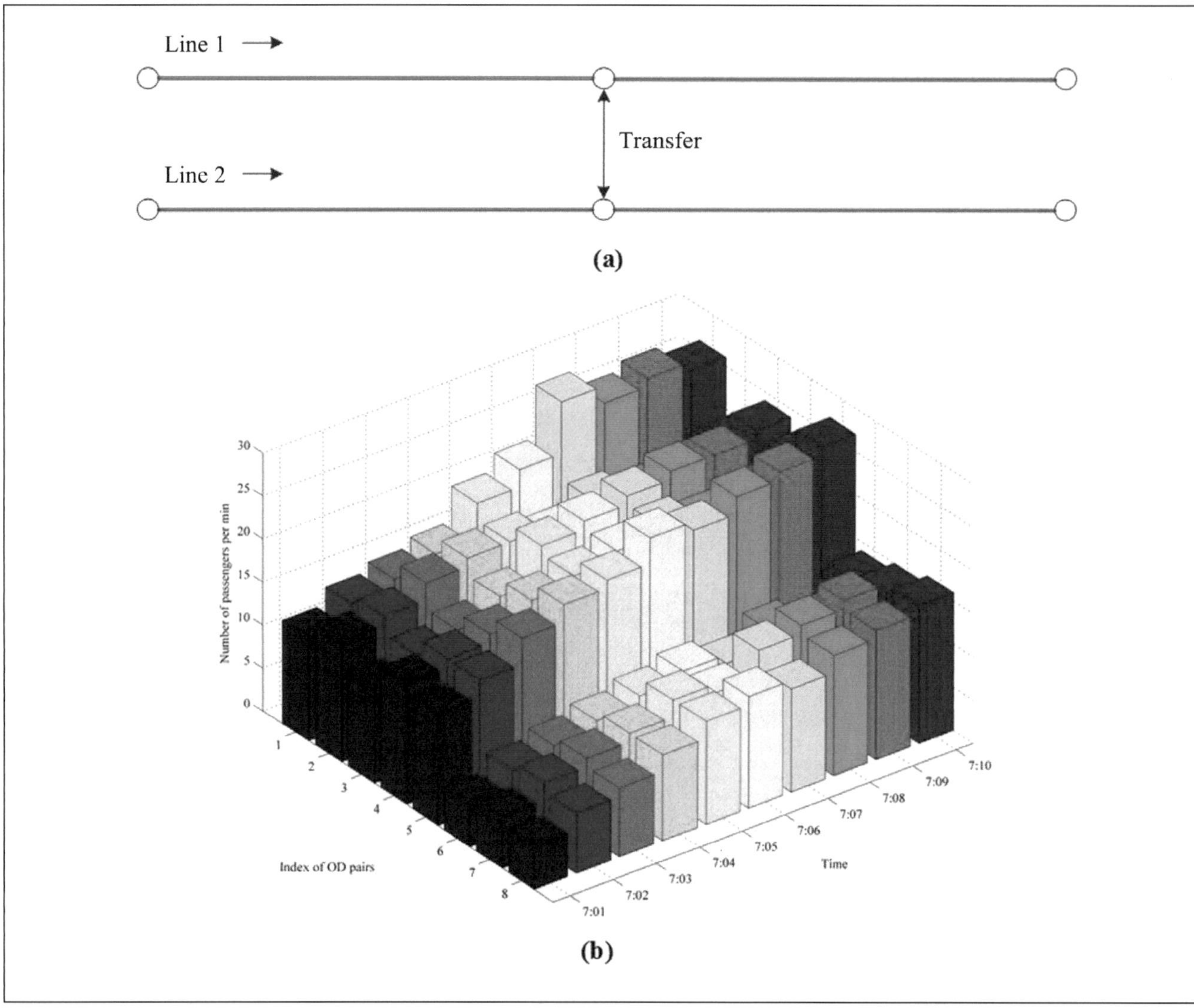

Figure 2. Test network and time-dependent passenger demand; (*a*) simple network, (*b*) time-dependent passenger demand.

Table 2. Results of the TDTSO Model from Different Algorithms

Algorithm	Initial value	Optimization value	Percent	Iteration/Step	CPU time
GA	12,357 min	11,231 min	9.11	30	0.52 s
PS	12,357 min	11,231 min	9.11	36	0.95 s
SA	12,357 min	11,231 min	9.11	822	5.30 s
BnB	12,357 min	11,231 min	9.11	2,084	24.90 s

from 0 to 50, and population size rises from 0 to 50. Each objective value can be obtained for each group of generation and population. The test of GA parameters demonstrates that larger population and generation lead to better results and longer computing time in the numerical test.

Case Study

Network Description and Basic Parameters

In this section, the performances of the TDTSO model and GA are illustrated through the real-world case of the Beijing subway network. Figure 4 shows a portion of the

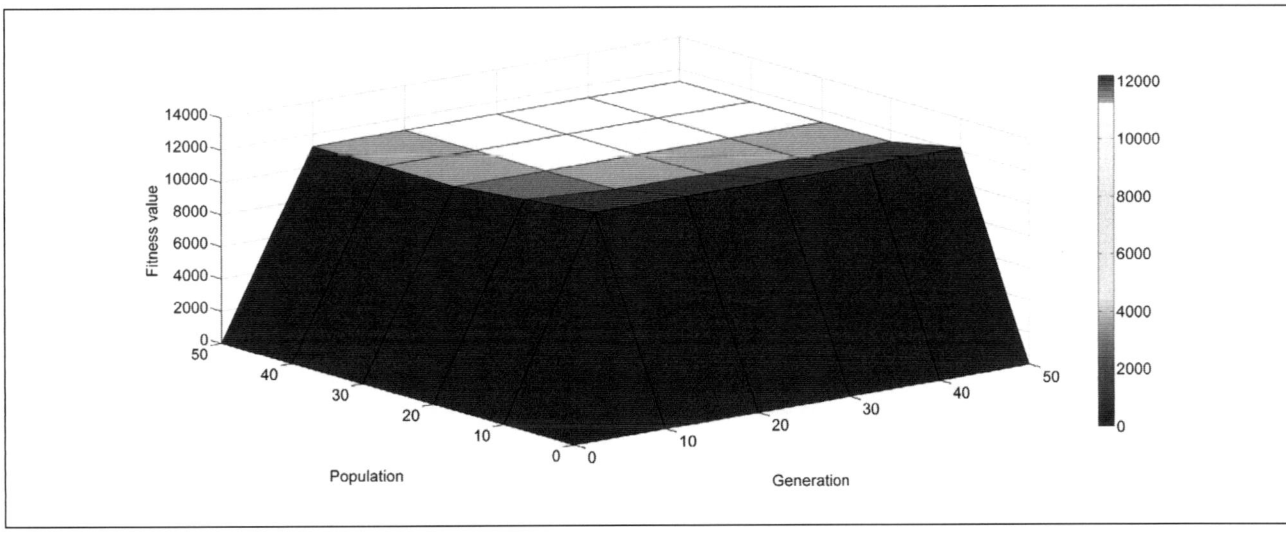

Figure 3. Analysis of GA parameters.

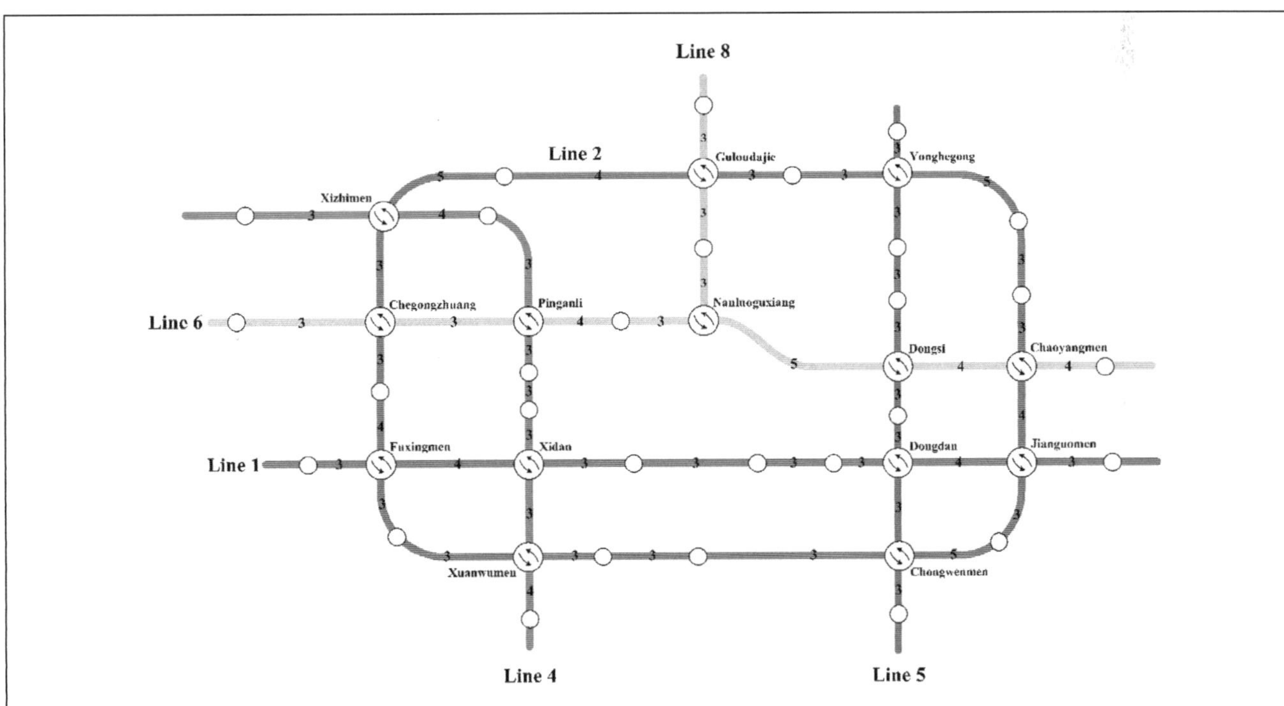

Figure 4. Beijing subway network.

Beijing subway network, focusing on the busiest area of Beijing network, containing 12 lines (six two-direction lines), 14 transfer stations, and 28 normal stations.

The research time period is from 09:30 to 10:30, which is a normal hour in the Beijing subway network; the minimum time interval in the case study is 30 s. During the research time period, 144 trains (12 trains in each up and down direction on one line) are scheduled along 120 time intervals. The stopping times are fixed at 30 s, and the running time between adjacent stations is shown in Figure 4. The allowable adjustments of running time and headway are ± 1 min and ± 1 min 30 s, respectively. The time-dependent passenger demand can be obtained from the AFC data of Beijing subway network.

Table 3. Optimization Results

Item	Initial value	Optimized value	Percent reduction
Total travel time	296,701 min	275,513 min	7.14
Waiting time	92,134 min	85,783 min	6.89
Transfer waiting time	26,630 min	23,551 min	11.56

Table 4. Three Situations with Different Adjustments of Operational Parameters and Optimization Results

	Parameter adjustments		Percent reduction		
Situation	Running time	Headways	Total travel time	Waiting time	Transfer waiting time
Strict situation	± 30 s	± 1 min	5.92	7.48	11.57
Basic situation	± 1 min	± 1 min 30 s	7.14	6.89	11.56
Relaxed situation	± 1 min 30s	± 2 min	7.46	10.58	12.91

The objective of this study is to design a synchronized and optimized timetable during the preplanning period. A population size of 2000 is selected, as the population size should be larger than the number of variables of TDTSO model to keep the individual diversity in GA. Crossover, mutation, and elite fractions are set to be 0.8, 0.15, and 0.05, respectively.

Optimization Results

The GA was run for 500 generations within less than 3600 min. The detailed optimization results are presented in Table 3.

Table 3 shows that passenger total travel time, which is the objective function of the TDTSO model, is minimized by 7.14%. When passenger total travel time is minimized in the case of the Beijing subway network during the research time period, passenger total waiting, riding, and transfer waiting times are minimized by 6.89%, 7.38%, and 11.56%, respectively. The optimized results show that the TDTSO model proposed in this study minimizes passenger total waiting, riding, and transfer waiting times when passenger total travel time is the objective function chosen to be minimized.

Analysis of Adjustments of Operational Parameters

In the actual operation of an urban subway network, the running time between adjacent stations and the headway is restrained by various factors (e.g., signal pattern, communication pattern, safety consideration, and others). The allowable adjustments of operational parameters are mostly determined by running time and headway, which significantly affect the performance of the synchronized and optimized timetable. To analyze the effect of adjusting operational parameters, three situations with different adjustments of operational parameters were designed (Table 4). The models of these situations are solved using GA with the same parameters.

Table 4 shows the optimized results of the three situations. It indicates that the strict situation presents the minimum percent reduction for total travel time, which is the objective function of the TDTSO model, whereas the relaxed situation exhibits the maximum percent reduction. The total waiting and transfer waiting times are reduced to approximately 6–11% and 11–13%, respectively. The comparison among the three situations proves that larger adjustments of operational parameters can better adapt to time-dependent OD demand. For a limited number and capacity of trains, the operator can offer a better service by improving the operation level, which can determine the adjustments of operational parameters.

Conclusions

This study proposes a TDTSO model that considers time-dependent OD demand in an urban subway network to minimize passenger total travel time. This model considers all types of passengers (transferring and non-transferring passengers) in an urban subway network that is adapted to time-dependent passenger demand when producing a synchronized and optimized timetable. A BVD method is developed to calculate the binary variables of the TDTSO model, and a GA based on this method is proposed. GA efficiency is compared with those of other algorithms (PS, SA, and BnB) via a numerical test. The TDTSO model and GA are applied to the Beijing subway network for a case study. The case study shows that the proposed model can be easily implemented and is capable of obtaining a synchronized and optimized timetable that minimizes passenger total travel time in an urban subway network. Three situations with different adjustments of operational parameters (running time and headway) are implemented. It was found that larger adjustments of operational parameters can better adapt to time-dependent passenger demand, thereby providing operators with a method to improve services under a limited number and capacity of trains. The optimization results of the proposed model with three objectives were analyzed. The three objectives can offer operators of urban subway networks additional options to produce a synchronized and optimized timetable that can be adopted in actual situations.

Acknowledgment

The study is support by National Natural Science Foundation of China (71361130015). The authors would like to thank the Key Lab of Beijing for Urban Transport Simulation and Decision-making.

Author Contributions

The authors confirm contribution to the paper as follows: study conception and design: Pan Shang, Ruimin Li; data collection: Kai Xian, Jifu Guo; analysis and interpretation of results: Pan Shang, Zhiyong Liu; draft manuscript preparation: Pan Shang, Ruimin Li, Zhiyong Liu. All authors reviewed the results and approved the final version of the manuscript.

References

1. Pelletier, M.-P., M. Trépanier, and C. Morency. Smart Card Data Use in Public Transit: A Literature Review. *Transportation Research Part C: Emerging Technologies*, Vol. 19, No. 4, 2011, pp. 557–568.
2. Cacchiani, V., D. Huisman, M. Kidd, L. Kroon, P. Toth, L. Veelenturf, and J. Wagenaar. An Overview of Recovery Models and Algorithms for Real-Time Railway Rescheduling. *Transportation Research Part B: Methodological*, Vol. 63, 2014, pp. 15–37.
3. Ibarra-Rojas, O., F. Delgado, R. Giesen, and J. Muñoz. Planning, Operation, and Control of Bus Transport Systems: A Literature Review. *Transportation Research Part B: Methodological*, Vol. 77, 2015, pp. 38–75.
4. Wu, J., M. Liu, H. Sun, T. Li, Z. Gao, and D. Z. Wang. Equity-Based Timetable Synchronization Optimization in Urban Subway Network. *Transportation Research Part C: Emerging Technologies*, Vol. 51, 2015, pp. 1–18.
5. Domschke, W. Schedule Synchronization for Public Transit Networks. *Operations-Research-Spektrum*, Vol. 11, No. 1, 1989, pp. 17–24.
6. Wong, R. C., T. W. Yuen, K. W. Fung, and J. M. Leung. Optimizing Timetable Synchronization for Rail Mass Transit. *Transportation Science*, Vol. 42, No. 1, 2008, pp. 57–69.
7. Shafahi, Y., and A. Khani. A Practical Model for Transfer Optimization in a Transit Network: Model Formulations and Solutions. *Transportation Research Part A: Policy and Practice*, Vol. 44, No. 6, 2010, pp. 377–389.
8. Vansteenwegen, P., and D. Van Oudheusden. Developing Railway Timetables Which Guarantee a Better Service. *European Journal of Operational Research*, Vol. 173, No. 1, 2006, pp. 337–350.
9. Wu, Y., J. Tang, Y. Yu, and Z. Pan. A Stochastic Optimization Model for Transit Network Timetable Design to Mitigate the Randomness of Traveling Time by Adding Slack Time. *Transportation Research Part C: Emerging Technologies*, Vol. 52, 2015, pp. 15–31.
10. Sun, L., J. G. Jin, D.-H. Lee, K. W. Axhausen, and A. Erath. Demand-Driven Timetable Design for Metro Services. *Transportation Research Part C: Emerging Technologies*, Vol. 46, 2014, pp. 284–299.
11. Barrena, E., D. Canca, L. C. Coelho, and G. Laporte. Exact Formulations and Algorithm for the Train Timetabling Problem with Dynamic Demand. *Computers & Operations Research*, Vol. 44, 2014, pp. 66–74.
12. Barrena, E., D. Canca, L. C. Coelho, and G. Laporte. Single-Line Rail Rapid Transit Timetabling Under Dynamic Passenger Demand. *Transportation Research Part B: Methodological*, Vol. 70, 2014, pp. 134–150.
13. Ceder, A., B. Golany, and O. Tal. Creating Bus Timetables with Maximal Synchronization. *Transportation Research Part A: Policy and Practice*, Vol. 35, No. 10, 2001, pp. 913–928.
14. Castillo, E., I. Gallego, J. M. Ureña, and J. M. Coronado. Timetabling Optimization of a Mixed Double-and Single-Tracked Railway Network. *Applied Mathematical Modelling*, Vol. 35, No. 2, 2011, pp. 859–878.
15. Lin, D. Y., and Y. H. Ku. Using Genetic Algorithms to Optimize Stopping Patterns for Passenger Rail Transportation. *Computer-Aided Civil and Infrastructure Engineering*, Vol. 29, No. 4, 2014, pp. 264–278.
16. Wang, Y., T. Tang, B. Ning, T. J. van den Boom, and B. De Schutter. Passenger-Demands-Oriented Train Scheduling for an Urban Rail Transit Network. *Transportation Research Part C: Emerging Technologies*, Vol. 60, 2015, pp. 1–23.
17. Kang, L., J. Wu, H. Sun, X. Zhu, and Z. Gao. A Case Study on the Coordination of Last Trains for the Beijing Subway Network. *Transportation Research Part B: Methodological*, Vol. 72, 2015, pp. 112–127.
18. Siebert, M., and M. Goerigk. An Experimental Comparison of Periodic Timetabling Models. *Computers & Operations Research*, Vol. 40, No. 10, 2013, pp. 2251–2259.
19. Borndörfer, R., H. Hoppmann, and M. Karbstein. *Timetabling and Passenger Routing in Public Transport*. Technical Report 15–31, ZIB Report, 2015. CASPT, 2015.
20. Schmidt, M., and A. Schöbel. Timetabling with Passenger Routing. *OR Spectrum*, Vol. 37, No. 1, 2015, pp. 75–97.
21. Zhou, X., and M. Zhong. Bicriteria Train Scheduling for High-Speed Passenger Railroad Planning Applications. *European Journal of Operational Research*, Vol. 167, No. 3, 2005, pp. 752–771.

22. Hadas, Y., and A. A. Ceder. Optimal Coordination of Public-Transit Vehicles using Operational Tactics Examined by Simulation. *Transportation Research Part C: Emerging Technologies*, Vol. 18, No. 6, 2010, pp. 879–895.
23. Ho, T., C. W. Tsang, K. H. Ip, and K. Kwan. Train Service Timetabling in Railway Open Markets by Particle Swarm Optimisation. *Expert Systems with Applications*, Vol. 39, No. 1, 2012, pp. 861–868.
24. Liu, Z., Y. Yan, X. Qu, and Y. Zhang. Bus Stop-Skipping Scheme with Random Travel Time. *Transportation Research Part C: Emerging Technologies*, Vol. 35, 2013, pp. 46–56.
25. Heydar, M., M. E. Petering, and D. R. Bergmann. Mixed Integer Programming for Minimizing the Period of a Cyclic Railway Timetable for a Single Track with Two Train Types. *Computers & Industrial Engineering*, Vol. 66, No. 1, 2013, pp. 171–185.
26. Yang, X., B. Ning, X. Li, and T. Tang. A Two-Objective Timetable Optimization Model in Subway Systems. *IEEE Transactions on Intelligent Transportation Systems*, Vol. 15, No. 5, 2014, pp. 1913–1921.
27. Sun, Y., C. Cao, and C. Wu. Multi-Objective Optimization of Train Routing Problem Combined with Train Scheduling on a High-Speed Railway Network. *Transportation Research Part C: Emerging Technologies*, Vol. 44, 2014, pp. 1–20.
28. Niu, H., X. Zhou, and R. Gao. Train Scheduling for Minimizing Passenger Waiting Time with Time-Dependent Demand and Skip-Stop Patterns: Nonlinear Integer Programming Models with Linear Constraints. *Transportation Research Part B: Methodological*, Vol. 76, 2015, pp. 117–135.

The Standing Committee on Rail Transit Systems (AP065) peer-reviewed this paper (18-03741).

TRR
JOURNAL OF THE TRANSPORTATION RESEARCH BOARD

Article

Transportation Research Record
2018, Vol. 2672(8) 255–265

DOI: 10.1177/0361198118777629
journals.sagepub.com/home/trr
SAGE

Multi-Objective Optimization of Train Timetable with Consideration of Customer Satisfaction

Aris Pavlides[1] and Andy H. F. Chow[2]

Abstract

This paper presents a multi-objective optimization framework for mainline train timetabling with consideration of customer satisfaction. The customers' perspectives considered include their perceptions of service punctuality, waiting times, and comfort of journeys. The optimization formulation is subject to constraints representing operational requirements and signaling systems. The optimization model is applied to the Brighton main line network in south-east England as a case study, and the results demonstrate how the proposed optimization framework can help government and train operators to derive more effective and equitable timetables with consideration of customer satisfaction.

Owing to tight fiscal, physical, and environmental constraints, continuous construction of new tracks will not always be a sustainable solution (*1*). It is, hence, important to explore effective train timetabling techniques for maximizing the utilization of existing infrastructure (*2*–*4*). There have been a number of studies looking at effective timetabling or scheduling techniques for mainline train services, of which a comprehensive review can be found in Fang et al. (*5*). In general, mainline train services refer to connections between cities as opposed to local metro services, whereas timetabling is regarded as the process of deriving a feasible schedule for a given set of train services over a specific route by specifying the associated arrival and departure times at each designated point. Mainline services are generally heterogeneous, that is, they consist of passenger and freight trains, slow and express services, domestic and international connections, etc. The feasibility of a timetable is also subject to a number of factors including availability of trains and crew, infrastructure capacity, and travel demand (*6*). As a consequence, deriving a timetable that satisfies different stakeholders, including passengers, train operators, and infrastructure managers, is always a challenge (*7*).

Many of the previous studies discussed above focus on the operator's perspective, which aims to maximize the service throughput and/or minimize the operating cost such that the revenue gained from running the train services can be maximized. Other customer-related service indicators, including journey times and the associated reliability, waiting times, and crowdedness, do not appear to have received the same level of attention in the literature compared with the operator's interest. As recognized in many national railway sectors, such as the Rail Safety Standard Board (RSSB) in the UK (*6*), it is vital to take customers' perspectives into account when deriving a train timetable for long-term benefits such as ensuring a sustainable and viable patronage.

This study contributes to the literature by looking at the issue of timetabling mainline train services for improving the overall efficiency and equity of the rail systems from the customer perspective. We demonstrate the formulations of cost functions reflecting the various factors of customer-concerned performance, which include journey times, customer waiting times, punctuality, and crowdedness, as our evaluation and optimization objectives (*6*). The customer waiting time, which is related to the service throughput (i.e., number of train runs given a period of time), as well as the service punctuality will also be of interest from the perspective of operators who aim to maximize their profits. The cost functions are then used to formulate a multi-objective optimization formulation (see, e.g., *8*–*10*) that is solved by an integrated genetic algorithm (GA)–Dijkstra

[1]Centre for Transport Studies, University College London, Gower Street, London, UK
[2]Systems Engineering and Engineering Management, City University of Hong Kong, Kowloon Tong, Hong Kong

Corresponding Author:
Address correspondence to Andy H. F. Chow: andychow@cityu.edu.hk

heuristics for a train timetable that can minimize the total cost associated with all objectives considered herein. The derivation of the timetable is subject to a set of predefined operational constraints including signal blocking, minimum train headways, minimum dwell times, and speed limits. The cost functions and optimization framework is applied to the Brighton main line (BML) in south-east England as a case study. The results reveal how the current timetable is assessed and able to improve with the proposed cost functions and optimization framework.

The paper is organized as follows: we start with presenting the model specification of a timetable and its associated constraints. Then, we formulate a set of cost functions for assessing different performances from both operators' and customers' perspectives. The cost functions are used to formulate a multi-objective optimization problem for train timetabling with the proposed solution heuristics. Next, the cost functions and optimization framework are applied to a case study of Brighton main line to demonstrate the proposed method and the results are discussed. Finally, the paper provides some final remarks and suggestions for future work.

Specification of Timetable and Constraints

In our study, a train timetable is represented through a defined series of times of train arrivals $\tau_{n,s}$ and departures $\sigma_{n,s}$ of each train n over a set of control points s (which can be a station, junction, etc.) along its service route (*11*, *12*). Given $\sigma_{n,s}$ and $\tau_{n,s}$, we can derive the running time $T_{n,s}$ of train n between station s and $s+1$ as

$$T_{n,s} = \tau_{n,s+1} - \sigma_{n,s} \tag{1}$$

and also the dwell time $D_{n,s}$ of train n at station s

$$D_{n,s} = \sigma_{n,s} - \tau_{n,s} \tag{2}$$

The setting of the variables $\sigma_{n,s}$ and $\tau_{n,s}$ will be subject to a set of operational constraints. We first have the minimum sectional running time constraints to reflect the speed limit imposed on each track section $(s, s+1)$:

$$\tau_{n,s+1} \geq \sigma_{n,s} + \frac{\Delta_{s,s+1}}{v^*_{n,s}} \tag{3}$$

where $\Delta_{s,s+1}$ is the distance between stations s and $s+1$, $v^*_{n,s}$ is the maximum admissible speed limit for train n traveling from station s toward $s+1$. Moreover, we also have the minimum dwell time constraints that define the minimum time that has to be spent by each train n at station s:

$$\sigma_{n,s} - \tau_{n,s} \geq d^*_{n,s} \tag{4}$$

The minimum dwell time $d^*_{n,s}$ imposed here is a predefined constant that will typically be determined by a number of factors on the demand side such as demand level of passengers or freight for that specific train at that specific station, or the consideration of connectivity where it is necessary to ensure a long enough dwell time for passengers or goods to transfer from one train to another at the station or interchange.

Finally, to ensure safety, each track section is further disaggregated into a series of blocks, as shown in Figure 1. Under the current fixed block signaling systems in practice, each block can only accommodate up to one train at a time to ensure safe operations. To model the fixed block signaling system, we denote the departure and arrival times of train n at block j between station pair $(s, s+1)$ as $\sigma_{n,s,j}$ and $\tau_{n,s,j}$, respectively, as shown in Figure 2. The shaded region in the figure represents the location and time period (during times t_{in} and t_{out}) that is occupied by the train of interest during which other trains are prohibited from entering. Following the specification in the current UIC (International Union of Railways) operational code (*13*), we have

$$t_{in} \geq \tau_{n,s,j} - \frac{\delta_{n,j}}{v_{n,s,j}} \tag{5}$$

where $\delta_{n,j}$ is the visual distance of train n to the entrance of block j; $v_{n,s,j}$ is the nominal speed of train n traveling through block j, for all blocks j in between stations s and $s+1$. The time t_{in} represents the time when the driver of train n observes the signal aspect at block j and starts to take action accordingly. Moreover,

$$t_{\text{out}} \geq \sigma_{n,s,j} + \frac{L_n}{v_{n,s,j}} \tag{6}$$

where L_n is the length of train n, which is around 20–25 m per train coach. The time t_{out} represents the time when the tail of the train n clears from the block section. Owing to the signaling system, it is expected that congestion will occur when the train volume on a track section is high (*14*). The signal blocking system requires that the entry time of train $n+1$ into block j at $\tau_{n+1,s,j}$ should be after (i.e., greater than or equal to) the existing of its predecessor train n at t_{out}. Following the constraint in Equation 6, this will give mathematically the corresponding signal blocking constraint for all station pairs $(s, s+1)$ and signal blocks j as

$$\tau_{n+1,s,j} \geq \sigma_{n,s,j} + \frac{L_n}{v_{n,s,j}} \tag{7}$$

in which train $n+1$ is the train following immediately after train n.

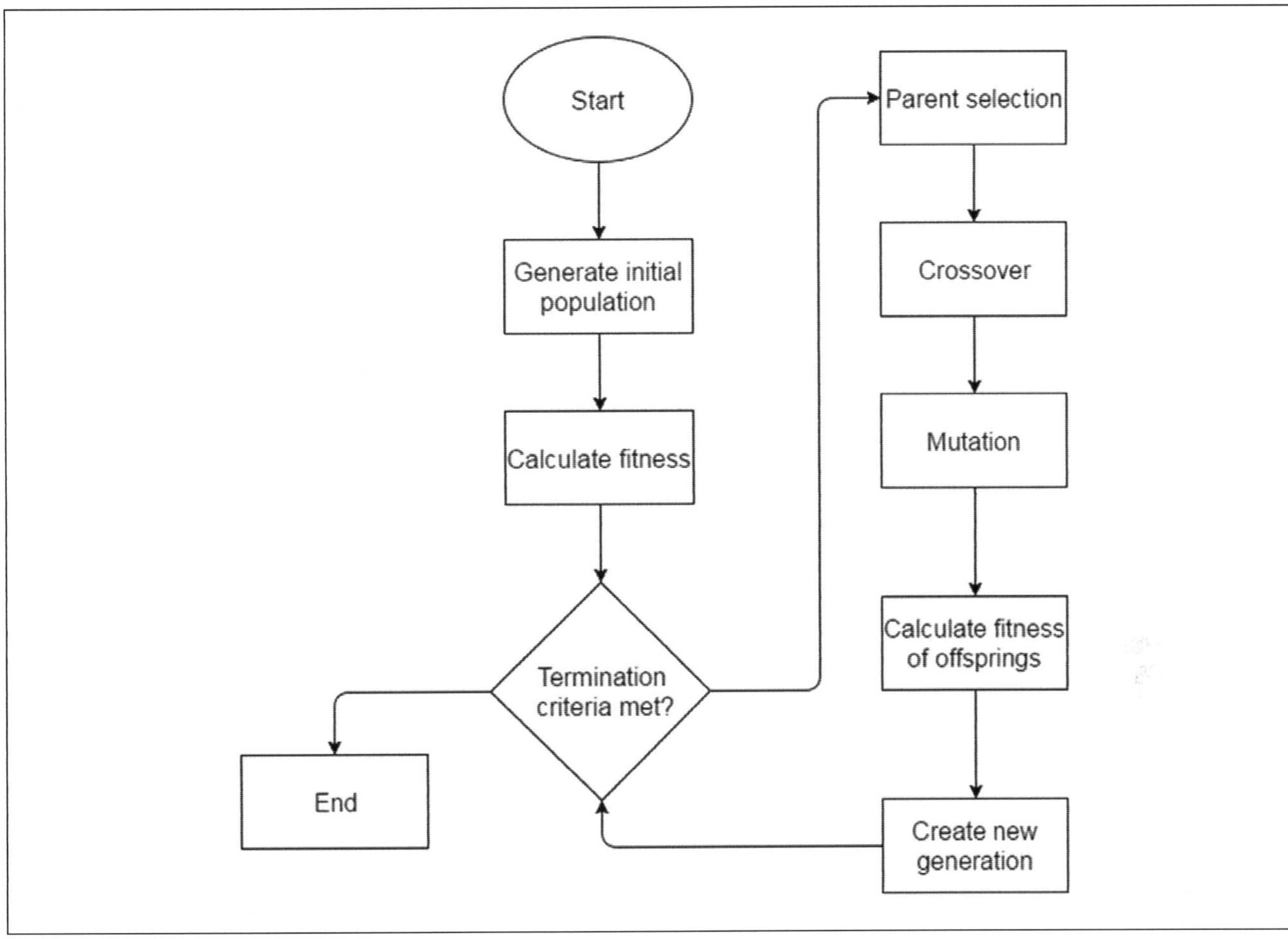

Figure 1. Flowchart of the GA.

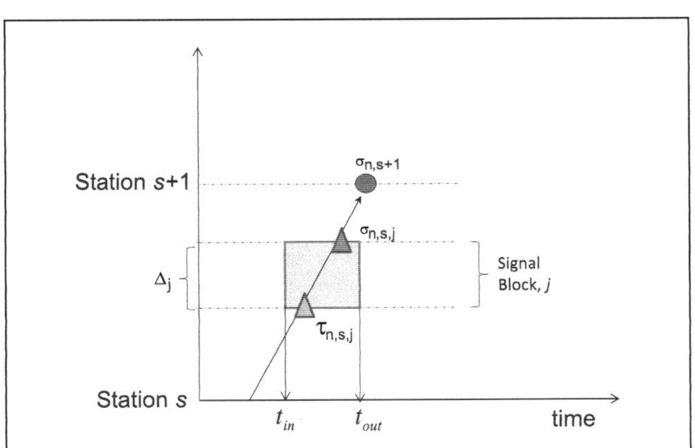

Figure 2. Mathematical representation of the fixed block signaling system.

Cost Functions

We now formulate the cost functions to evaluate different performances in railway timetabling and operations. Following the comprehensive review in (6), we focus on four indicators in the railway industry: train journey times, customer waiting times, service punctuality, and the crowdedness of trains that are of great interest and importance from the customers' and social perspective.

Journey Times

Journey times $T_{n,s}$ of customers on trains n over all section $(s, s+1)$ can be obtained from Equation 1 following the specification of timetable variables $\sigma_{n,s}$ and $\tau_{n,s}$. Given all journey times $T_{n,s}$, we define the following cost function

$$C_T = \hat{c}_T \sum_{s=1}^{S} \sum_{n=1}^{N} T_{n,s} p_{n,s} \tag{8}$$

where N and S represent the total number of trains and stations in the system, respectively. The variable $p_{n,s}$ is a quantity associated with the demand that represents the number of passengers (or amount of goods) on train n

running between stations s and $s+1$. Determining this $p_{n,s}$ will require origin–destination (OD) demand information. It should be emphasized that incorporating the passenger information $p_{n,s}$ is crucial to reflect the importance of customer satisfaction and consideration. Dropping the quantity $p_{n,s}$ (say, if OD information is not available) could result in the evaluation and timetabling process being determined by the traditional operators' approach, which treats each train n equally regardless of the actual passengers' demand and expectation, and hence defeating the purpose of improving customer satisfaction through the process. Applying the cost function (Equation 8) to the timetabling process will ensure the corresponding timetables give higher priority to trains carrying more passengers or goods through $p_{n,s}$.

To compare and integrate this cost function with other cost functions to be presented later in the paper, we need to convert all the costs into a common unit (e.g., monetary unit). For this purpose, the notation $\hat{c}_T$ in Equation 8 represents the monetary cost associated with the journey times. Some existing examples of this conversion factor, which is known as "value of time," can be found in (*15–17*). We have further discussion on the choice of this $\hat{c}_T$ and other monetary cost coefficients in the latter sections.

Waiting Times of Customers

To derive the cost function associated with waiting times of customers (i.e., passengers or freight operators), we first assume $\lambda_{n,s}(t)$ be the number of customers wishing to board train n at station s at time t. This $\lambda_{n,s}(t)$ can be derived readily from a standard OD matrix. Given a standard OD matrix (in which rows represent "origins" and columns represent "destinations"), we can obtain this $\lambda_{n,s}(t)$ by summing each row over its columns for all rows s at time interval t. Moreover, fundamental queueing analysis (e.g., *18*) gives the total waiting time W of all customers within the analysis period (in the unit of [persons-time] or [goods-time]) as

$$W = \sum_{s=1}^{S} \sum_{n=1}^{N_s - 1} \int \int_{\tau_{n,s}}^{\tau_{n+1,s}} \lambda_{n,s}(t) dt dt \qquad (9)$$

where N_s is the total number of trains serving station s over the study time period. The time interval between $\tau_{n,s}$ and $\tau_{n+1,s}$ specifies the headway of train service at station s.

Following the calculation of W, a monetary cost conversion factor is added:

$$C_W = \hat{c}_W W \qquad (10)$$

where $\hat{c}_W$ is the monetary cost associated with waiting times. The purpose of incorporating the waiting time into the optimization framework is to ensure that there are adequately frequent services for number of passengers or goods at the station without creating excessive waiting times. Empirical studies conducted by the UK Department for Transport (*17*) suggest that this $\hat{c}_W$ will be around two or three times larger than $\hat{c}_T$ as the waiting time is generally regarded as a dead loss.

Punctuality of Service

Punctuality is measured as the time discrepancy between the scheduled and the actual arrival times of the train services. To quantify the punctuality in monetary units (see *19*, *20*), we adopt a linear schedule cost function. Let τ^* denote the ideal arrival time of the train service and let Φ denote a time allowance for lateness (e.g., Φ is considered to be 3 min under the UK railway operational regulations). If the corresponding train is delayed by more than Φ from the ideal arrival time τ^*, a schedule delay cost will be imposed on the train operator by the infrastructure manager for lateness. It is considered here that this schedule delay cost increases linearly with a slope of $\hat{c}_P$ over arrival time τ, where $\tau \geq \tau^* + \Phi$. This penalty rate $\hat{c}_P$ represents the value of lost time of customers (passengers or freight companies) per unit lateness in time. Following this linear specification, the total schedule delay cost associated with punctuality can be determined, taking the arrival of passengers and/or goods into account, as

$$C_P = \hat{c}_P \sum_{s=1}^{S} \sum_{n=1}^{N_s - 1} \int_{\tau_{n,s}}^{\tau^*_{n+1,s}} \lambda_{n,s}(t) \left(\tau_{n+1,s} - \tau^*_{n+1,s} - \Phi \right)^+ dt \qquad (11)$$

where $\tau^*_{n+1,s}$ is the scheduled arrival time for train $n+1$ at station s, $\left(\tau_{n+1,s} - \tau^*_{n+1,s} - \Phi \right)^+ = \max\left(\left(\tau_{n+1,s} - \tau^*_{n+1,s} - \Phi \right), 0 \right)$.

It is noted that this punctuality cost analysis is generally applicable to other schedule cost functions apart from the linear assumption, by revising the cost function term '$\left(\tau_{n+1,s} - \tau^*_{n+1,s} - \Phi \right)^+$' in Equation 11 accordingly.

Crowdedness of Trains

The crowdedness cost function measures the difference between traveling in crowded and uncrowded trains and is an important aspect for passengers when assessing the attractiveness of public transport (*21*). The difference is derived from the fact that passengers value their time

Table 1. Values of $R_{n,s}$ in Crowdedness Cost Function (*21*)

Sitting penalties			
Load factor	Leisure	Business	Commute
60%	0	0	0
70%	0.02	0.04	0
80%	0.04	0.08	0
90%	0.07	0.11	0.11
100%	0.09	0.15	0.22
110%	0.15	0.21	0.32
120%	0.22	0.27	0.43
130%	0.28	0.34	0.54
140%	0.35	0.4	0.65
150%		Linear extrapolation of the previous values	
160%			
Standing penalties			
Load factor	Leisure	Business	Commute
60%	0	0	0
70%	0	0	0
80%	0	0	0
90%	0	0	0
100%	2.12	2.86	1.76
110%	2.33	2.93	1.89
120%	2.54	3.01	2.03
130%	2.75	3.08	2.16
140%	2.96	3.15	2.3
150%		Linear extrapolation of the previous values	2.43
160%			2.57

higher when they are traveling in crowded trains. We formulate the corresponding cost function as

$$C_D = \hat{c}_D \sum_{s=1}^{S} \sum_{n=1}^{N} (R_{n,s} - 1) T_{n,s} p_{n,s} \tag{12}$$

where $R_{n,s}$ is a parameter depending on the loading factor of the train, which is defined as the number of passengers on board divided by the seating available on the train.

Following Wardman and Whelan (*21*), Table 1 shows the relationship between $R_{n,s}$ and the corresponding loading factor for both sitting and standing passengers. When the loading factor is below 0.6, the level of crowdedness is regarded as low and $R_{n,s}$ is taken as zero, meaning that no penalty is imposed for crowding (i.e., $C_D = 0$). When the loading factor goes beyond 0.6, $R_{n,s}$ and, hence, the penalty increases with the train loading while it varies over different kinds of passengers (*21*). When $R_{n,s}$ goes beyond 0.9, the penalty associated with standing passengers starts to apply as some passengers will start to be unable to find a seat. It is noted that the parameter $R_{n,s}$ for standing passengers is much higher than that for seated passengers to reflect the extra dissatisfaction of passengers when they are unable to find a seat, as supported by empirical study (*21*).

Multi-Objective Train Timetabling

The cost functions are now applied to formulate a multi-objective train timetabling framework. The optimization process aims to determine the optimal train sequence and schedule such that the following total cost is minimized:

$$C = C_T + C_W + C_P + C_D \tag{13}$$

The total cost derived in Equation 13 is in monetary units in which the components are integrated through the monetary cost coefficients as discussed in previous sections. The cost minimization problem is subject to the operational constraints in Equations 3, 4, and 7.

The train timetable optimization problem is combinatoric in that it involves different feasible combinations of $\tau_{n,s}$ and $\sigma_{n,s}$ representing different sequencing and scheduling of trains (see, e.g., *22–24*), as well as the total number of trains in operation. Considering a scenario where there are N trains to schedule, the number of possible sequences for scheduling these trains will be $N!$. This has not included the infinite number of ways of setting the departure and arrival times of these trains along the service route given a sequence, as well as the potential adjustment in the total number of trains N running the service.

Moreover, the cost function shown in Equation 13 involves consideration of uncertainties in the cost components owing to the stochastic variations in train running times $T_{n,s}$ and passenger arrivals $\lambda_{n,s}(t)$. Consequently, the timetabling problem must be considered as a stochastic optimization problem. We adopt a Monte Carlo or scenario-based approach (*25*) for solving this. Following general practice (*19*, *20*), the running times of the trains are first assumed to follow an exponential distribution with respect to their corresponding nominal values. Regarding the passenger arrival pattern, given the hourly arrival volume, the hourly passenger volume will first be divided by the total number of train runs within an hour. This gives an expected number of passengers boarding when a train arrives. For example, if the hourly passenger volume is 100 passengers/hour and if the train service headway is 15 mins (i.e., 4 trains/hour), then one can expect that on average there will be 25 passengers boarding on each train arrival. Following determination of this average number of boarding passengers to each train, we assume that the passenger arrivals will follow a Gaussian distribution with a mean value equal to this average number, associated with a coefficient of variation of 10%. For each feasible operational plan (including sequence, schedule, and number of trains) to be considered in the timetabling process, the cost function components will be evaluated over 100 Monte Carlo simulation runs over the running time and passenger arrival distributions through the train network. The corresponding performance of the timetable plan in the optimization process will be represented by the mean value of the costs calculated over these 100 Monte Carlo runs.

To derive a solution within a reasonable time, the Monte Carlo-based timetable solution process is implemented in C through a hybrid GA–Dijkstra calculation procedure. The first part (GA) of the solution procedure is to derive the sequence of trains departing from their corresponding origins. Given the departing train sequence, the second part (Dijkstra) of the procedure is to derive the combination of service routes, and the corresponding arrival and departure times of the trains over each station and other specified control points that would give the lowest cost based upon the estimation of the train running times and dwell times.

GA (see Figure 1) is formulated to mimic the concept of natural selection and is widely used in combinatorial optimization problems. In our setting, GA starts with generating an initial (random) set of timetables, where in our case the population size of this set is 200. Each timetable contains a specific sequence of trains departing from their origins in which feasible train sequences are encoded into a set of "chromosomes." Different from the traditional binary (0–1) encoding scheme (*26*, *27*), here we adopt a permutation encoding scheme in which each chromosome represents a specific train sequence. For example, consider there are eight trains (A, B,..., H) with different service paths and characteristics to schedule, which gives a total of 8! = 40,320 possible sequences. Each of these 40,320 possible combinations can be represented by an 8-bit chromosome. Two of these possible chromosomes in this case are

ABCDEFGH
ACBDGHFE

Given each of the chromosomes which represents a specific sequence of trains, they will further be complemented with the corresponding departure and arrival times of each train. The departure and arrival times are computed by using Dijkstra's algorithm. Dijkstra's algorithm (see *28*) is a widely known and adopted algorithm for determining shortest (or cheapest) paths through a given network for each specific OD (source–sink) pair. Each train is associated with a predefined list of stops en route from its origin toward its destination that the train will have to serve. Based upon the train departure sequence from each origin (terminal) determined by the GA, Dijkstra's algorithm is used to determine the route, and the corresponding arrival and departure times that each train can reach while passing all the required intermediate stops with lowest cost while subject to the constraints in Equations 3, 4, and 7. Specifically, Dijkstra's algorithm is summarized as follows: the trains are loaded one by one from their origins into the network following the sequence as determined in GA. Dijkstra is used to determine the shortest (cheapest) path from the origin toward the train's required stop, and repeat the process until it reaches the destination after passing all required stops. The running times between each pair of stops will have to satisfy Equations 3, 4, and 7 (i.e., the fixed block signaling constraints), with consideration of the existence and influence of other existing trains in the system.

Following the determination of both train sequences and schedules, the total cost associated with each train sequence and the schedule is calculated using Equations 8–12 and combining them as shown in Equation 13. With both sequences and schedules specified, the optimization algorithm "reproduces" chromosomes proportional to their "fitness" values in the next iteration. The fitness value is calculated based upon the value of total cost (Equation 13) associated with the train sequence and departures specified in the chromosome. Essentially, a higher fitness value will be assigned to a chromosome if the chromosome achieves lower total cost, and the fitness function FIT_i for each chromosome i is defined as

$$FIT_i = 1 - \frac{C_i}{C_{\max}} \tag{14}$$

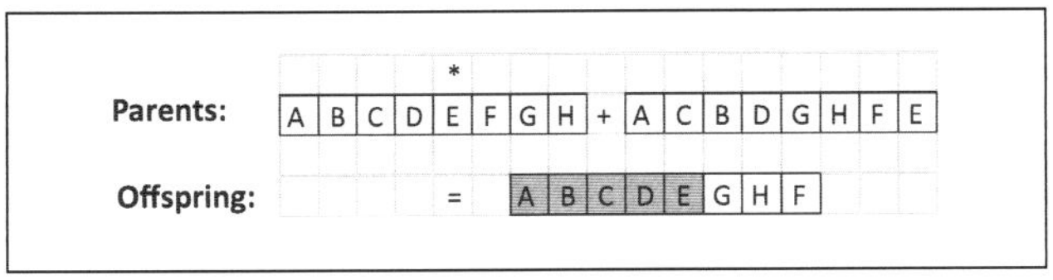

Figure 3. Example of a crossover operation in sequence optimization.

in which C_i is the mean value of total costs calculated from Equation 13 over 100 Monte Carlo simulations on the running time distributions based on the sequence and schedule of trains specified in chromosome i, $C_{\max}$ is the maximum cost value identified in the current iteration. The top 5% of the population's chromosomes will be maintained and reproduced in the next iteration. This is to ensure that the chromosomes in the next iteration will be at least as good as the previous one. The remaining 95% will be replaced by new ones created by the "cross-over" and "mutation" processes as described in the following.

The "crossover" operation will then randomly select and "mate" two chromosomes (regarded as "parents"). Figure 3 shows an example of the crossover operation. Given the pair of chromosomes, a crossover point is first randomly selected (e.g., the 5th bit in Figure 3). The chromosome of the "offspring" of this pair of parents is first copied from the first parent (ABCDEFGH) until the crossover point (i.e., ABCDE, the shaded part in the offspring). Then the other parent is scanned from the left; if an alphabet is found in the other parent that is not yet in the offspring, it will be added. The crossover process is for generating the next set of population with some entirely new characteristics with respect to the previous population and, hence, avoiding the optimization process from trapping into local optima.

Finally, the "mutation" process randomly selects a pair of bits in an arbitrary chromosome, and swaps the two bits. The probability at which a bit is selected is predefined, which is 0.02 in our case. This is again to prevent the optimization process from getting trapped into a local optimum. The GA optimization process above (reproduction–crossover–mutation) will continue until the predefined maximum number of iterations (which is set to be 100 in our application) is reached.

The GA–Dijkstra optimization process described above will continue until the predefined maximum number of iterations (which is set to be 100 in this study) is reached. Eventually, the set of train sequences and schedule of trains which has the lowest cost value of Equation 13 through the process will be identified as the final optimal solution.

Case Study

The cost functions and the multi-objective optimization framework are applied to the BML in southeast England (Figure 4). The BML is approximately 80 km long and is

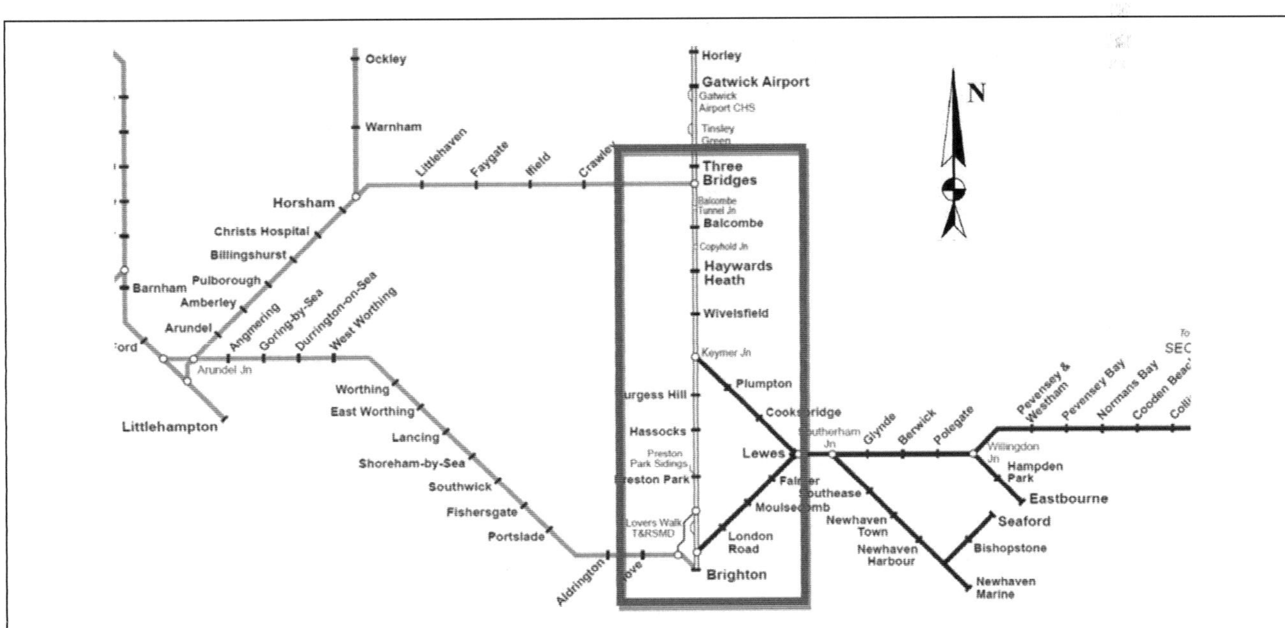

Figure 4. Test network, Brighton main line (UK).

Table 2. Coefficients in Cost Function

Components	Coefficient	Travel purpose (unit: £/hour)		
		Business	Commute	Leisure
Journey times	$\widehat{c_T}$	47.18	6.46	5.71
Waiting times	$\widehat{c_W}$	117.95	16.15	14.28
Punctuality	$\widehat{c_P}$	117.95	16.15	14.28
Crowdedness	$\widehat{c_D}$	47.18	6.46	5.71

an electrified connection linking London Victoria and London Bridge stations with Brighton via East Croydon and Gatwick Airport. The line itself has a complex structure with a variable number of tracks (four tracks from London down to Balcombe Tunnel Junction and two tracks thereafter), different speed limits along the line, multiple branch lines (e.g., at Junctions Horsham, Lewes), and sidings (e.g., along Ardingly, Lovers Depot). Passenger operators that operate on the BML include Southern and Govia. We selected the section between Gatwick Airport and Brighton as highlighted in Figure 5. This is one of the busiest sections along the BML. The study period is 08:00–10:00 a.m., which is regarded as the morning peak, on weekdays. During the study period there is currently a total of 22 trains running from Brighton toward Gatwick and, hence, central London (the "up" direction) and 18 trains running from Gatwick toward Brighton (the "down" direction). We derive this "base case" train timetable with information obtained from Network Rail. There are two different train classes running through the section during the study period: Classes 375 (total: 18 trains) and 442 (total: 22 trains) with Class 375 used for the express connection. Each train consists of eight coaches during the study period and each coach is 21 m long.

The coefficients in the cost function (*13*) are set according to the official guideline set by the Department for Transport in the UK, which are summarized in Table 2. The figures are set according to the webTAG Unit 3.5.6 guidance (*17*) which specifies the values of time of different kinds of travelers. The kinds of passengers considered here include passengers traveling for work purpose during office hours (Business), passengers traveling to and from work (Commute), and passengers traveling for leisure purpose (Leisure).

The cost functions also need specification of the OD matrix for deriving passengers on-board ($p_{n,s}$) and passengers waiting to board for each train n at each station ($\lambda_{n,s}(t)$). The OD matrix is constructed from field observations and annual surveys conducted by Network Rail. In addition, it is estimated from in-house surveys conducted by Network Rail that 20% of passengers are traveling for Business purpose, 60% for Commute, and

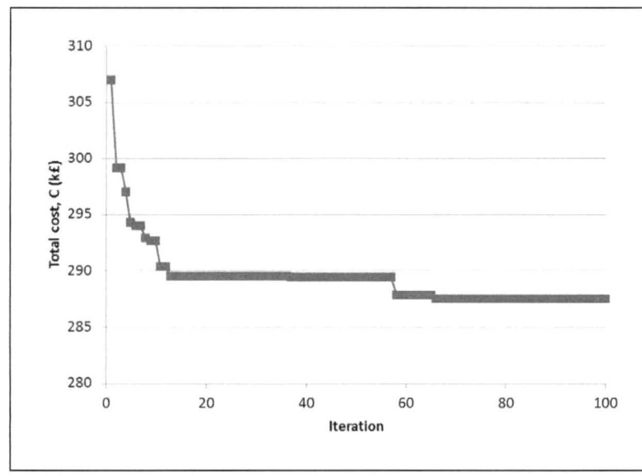

Figure 5. Train diagrams before and after optimization. Dotted line = fast trains; solid line = slow trains.

20% for Leisure. Capturing the passenger composition is needed for deriving an equitable timetable as different passengers value their time spent differently (see Table 2). We adopt the following way to allocate the seats among different passengers: when a train arrives, we allow business (Work) passengers to first take a seat. If any remaining seats are available, commuters (Commute) will then follow, and finally leisure passengers (Leisure). This is because it is found that business and commuting passengers are more likely to schedule their trip in advance and, hence, be able to find a seat, provided pre-booking is available.

Basic Results

Given the network configuration and cost functions, the optimization algorithm is implemented in C on a standard Windows 7 (64-bit) desktop computer. There is a total of (20 + 22 =) 42 trains in the system under the base case. Each of these 42 trains is characterized by its type (e.g., fast, slow, freight, etc.), list of stops, origin, and destination. We set the total number of optimization iterations to be 100 and the whole optimization process takes about an hour to complete on the computer. The

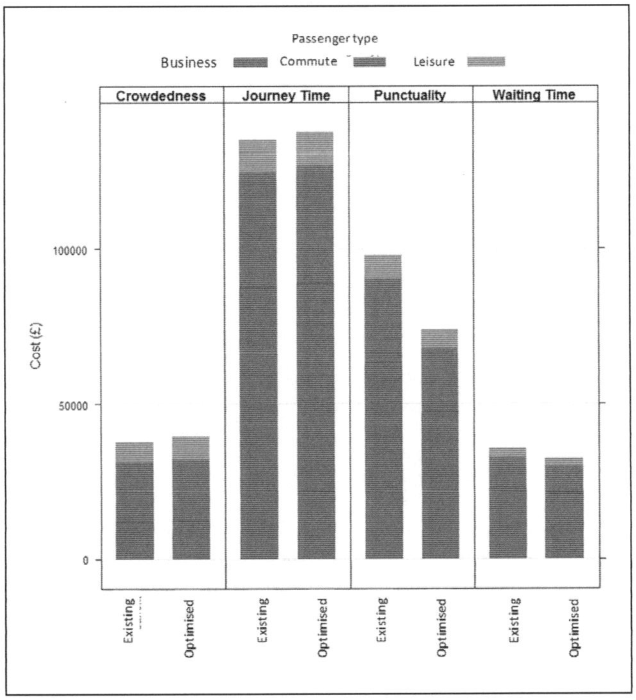

Figure 6. Optimization process.

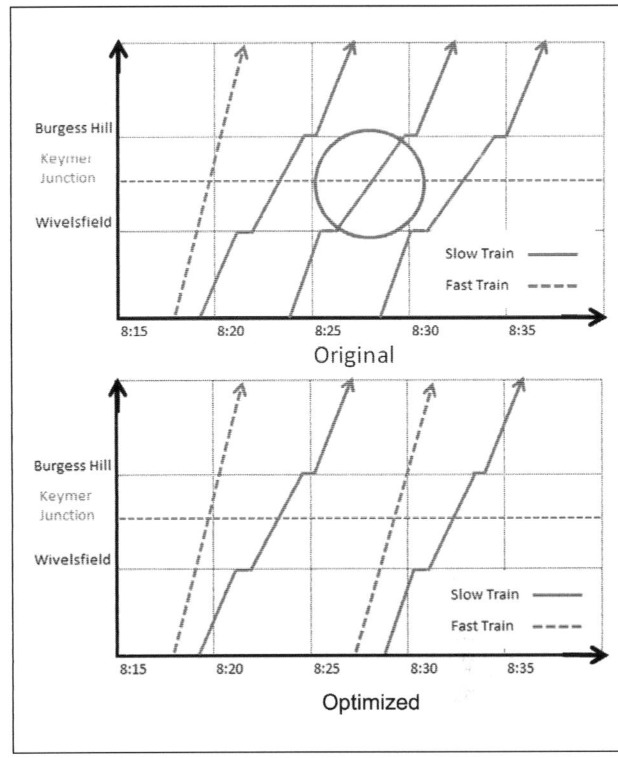

Figure 7. Cost components before and after optimization.

total number of train runs is fixed as its existing value. Consequently, the improvement shown here is from re-sequencing and re-scheduling the trains with respect to the existing plan. The idea is to derive an optimized timetable from the proposed optimization framework with the same number of trains within the same study period. We then compare the "optimized" timetable with this "base case" timetable to see how much improvement, in terms of reduction in costs, can be achieved in different aspects through re-sequencing and re-scheduling.

Figure 5 shows the optimization process. Similar to other similar applications (see *26*, *27*), the most significant improvements are observed in the first few generations while the optimization process gradually converges slowly to the ultimate final solution at later iterations. As shown in the figure, the total cost associated with the existing timetable is decreased from the original £307,000 to £288,000 by the optimizer over 100 iterations. Figure 7 further shows the breakdown of reductions in different cost components. The results reveal that the optimizer would choose to improve waiting times and punctuality at the expense of journey times and crowdedness. This is due to the differences in the associated cost coefficients adopted as shown in Table 2. Hence, we can expect that the optimization framework could be more beneficial when applied to high-demand scenarios (e.g., peaks) when train services are subject to high passenger volumes and congestion.

Further investigation into the train diagram (Figure 5; time period: 08:15–08:35) suggests that the reduction in costs associated with punctuality and waiting time, and increases in journey times and crowdedness, is due to the congestion caused by changes in train runs around the location of Keymer Junction. The punctuality, and hence the waiting time, of the train service is improved due to the more effective sequencing of the train services by the optimizer. Specifically, the improvement is achieved by scheduling trains (which carry relatively few passengers) departing from or heading to Three Bridges station along the Crawlay branch to later times to give way to the trains (which carry more passengers) traveling along the mainline between Gatwick and Brighton.

Conclusion

This paper started with formulating a set of cost functions for evaluating different performance associated with the train timetabling process. With special consideration given to the perspective of customers, the performances considered here include journey times of trains, waiting times of customers for service, punctuality, and crowdedness onboard. The cost functions are further applied to formulation of a multi-objective framework which derives timetables for mainline train services. Given the definition of the timetable-based decision variables (arrival and departure schedules of trains), we also formulate a set of operational constraints representing the physical and safety requirements on the train running times and dwell times, as well as the fixed block signaling system. This paper also presents a solution heuristics that consists of a hybrid GA–Dijkstra algorithm for determining the sequence and schedule of train runs. The stochasticity in train running times and passenger arrivals and hence their implications on the cost function components are also captured in the optimization process through an underlying Monte Carlo simulation.

The algorithms are applied to the BML in southeast England as a case study. The optimizer is shown to be able to reduce the cost of operations in the aspects of crowdedness, waiting times, and punctuality by improving the sequencing and scheduling of train runs. With the assumed seat allocation scheme following conversations with professionals, it is also found that commuters would have benefited most from the optimized timetable due to the significant reduction in Crowdedness cost when compared with the other kinds of passengers. This study highlights the importance of incorporating passenger demand information in the timetabling process, and reveals how cost functions and optimization techniques can be adopted to assist government and train operators to derive more effective and equitable timetables with consideration of customers. Finally, it is noted that the focus of the present paper lies on the formulation of cost functions and their application to timetabling instead of developing advanced optimization algorithms. We agree that it will be worth conducting further research on alternative optimization algorithms and behavioral studies for improving the quality and reliability of the optimal solutions. Another possible research topic is to find the dynamic changes pattern and corresponding train schedule considering the dynamic equilibrium. We will leave these for future studies.

Acknowledgments

The authors thank all the anonymous referees for the constructive comments. This study is funded by an Industrial CASE (iCASE) studentship awarded by UK Engineering and Physical Sciences Research Council (EPSRC, Ref: EP/K504610/1) and Rail Safety and Standard Board (RSSB). The authors thank Catherine Baker at RSSB and Meena Dasigi at Network Rail for their valuable comments.

References

1. Chow, A. H. F., A. Santacreu, I. Tsapakis, G. Tanaksaranond, and T. Cheng. Empirical Assessment of Urban Traffic Congestion. *Journal of Advanced Transportation*, Vol. 48, No. 8, 2014, pp. 1000–1016.
2. Schwanhausser, W. The Status of German Railway Operations Management in Research and Practice. *Transportation Research Part A: Policy and Practice*, Vol. 28, No. 6, 1994, pp. 495–500.
3. Zhou, X., and M. Zhong. Bicriteria Train Scheduling for High Speed Passenger Railroad Planning Applications. *European Journal of Operational Research*, Vol. 167, No. 3, 2005, pp. 752–771.
4. Lee, Y., and C. Chen. A Heuristics for Train Pathing and Timetabling Problem. *Transportation Research Part B: Methodological*, Vol. 43, No. 8–9, 2009, pp. 837–851.
5. Fang, W., S. Yang, and X. Yao. A Survey on Problem Models and Solution Approaches to Rescheduling in Railway Networks. *IEEE Transactions on Intelligent Transportation Systems*, Vol. 16, No. 6, 2015, pp. 2997–3016.
6. Chen, L., and C. Roberts. *Defining the Optimization Function for Timetabling as Part of Future Traffic Regulation (FuTRO)*. Report submitted to Railway Safety and Standards Board, United Kingdom, 2012.
7. Ibarra-Rojas, O., R. Giesen, and Y. Rios-Solis. An Integrating Approach for Timetabling and Vehicle Scheduling Problems to Analyze the Trade-Off Between Level of Service and Operating Costs of Transit Networks. *Transportation Research Part B: Methodological*, Vol. 70, 2014, pp. 35–46.
8. Albrecht, T. Automated Timetable Design for Demand-Oriented Service on Suburban Railways. *Public Transport*, Vol. 1, No. 1, 2009, pp. 5–20.
9. Sama, M., C. Meloni, A. D'Ariano, and F. Corman. A Multi-Criteria Decision Support Methodology for Real-Time Train Scheduling. *Journal of Rail Transport Planning and Management*, Vol. 5, No. 3, 2015, pp. 146–162.
10. Chow, A. H. F., S. Li, and R. Zhong. Multi-Objective Optimal Control Formulations for Bus Service Reliability with Traffic Signals. *Transportation Research Part B: Methodological*, Vol. 103, 2017, pp. 248–268.
11. Caprara, A., M. Fischetti, and P. Toth. Modeling and Solving the Train Timetabling Problem. *Operations Research*, Vol. 50, No. 5, 2002, pp. 851–861.
12. Dorfman, M. J., and J. Medanic. Scheduling Trains on a Railway Network Using a Discrete Event Model of Railway Traffic. *Transportation Research Part B: Methodological*, Vol. 38, No. 1, 2004, pp. 81–98.
13. International Union of Railways (UIC, 2004). *UIC Code 406: Capacity*, 2004.
14. Dingler, M. H., Y. Lai, and C. P. L. Barkan. Population Synthesis for Microsimulating Travel Behavior. *Transportation Research Record: Journal of the Transportation Research Board*, 2009. 2117: 41–49.
15. Wardman, M. The Value of Travel Time: A Review of British Evidence. *Journal of Transport Economics and Policy*, Vol. 32, No. 3, 1998, pp. 285–316.
16. Wardman, M. Public Transport Values of Time. *Transport Policy*, Vol. 11, No. 1, 2004, pp. 363–377.
17. DfT. *TAG UNIT 3.5.6: Values of Time and Vehicle Operating Costs*. Department for Transport, UK, 2012.
18. Daganzo, C. F. *Fundamentals of Transportation and Traffic Operations*. Pergamon, New York, 1997.
19. Carey, M. Reliability of Interconnected Scheduled Services. *European Journal of Operational Research*, Vol. 79, No. 1, 1994, pp. 51–72.

20. Carey, M. Ex Ante Heuristic Measures of Scheduled Reliability. *Transportation Research Part B: Methodological*, Vol. 33, No. 7, 1999, pp. 473–494.
21. Wardman, M., and G. Whelan. Twenty Years of Rail Crowding: Evidence and Lessons from British Experience. *Transport Reviews*, Vol. 31, No. 3, 2010, pp. 379–398.
22. Cacchiani, V., A. Caprara, and P. Toth. Scheduling Extra Freight Trains on Railway Networks. *Transportation Research Part B: Methodological*, Vol. 44, No. 2, 2010, pp. 215–231.
23. Cacchiani, V., and P. Toth. Nominal and Robust Timetabling Problems. *European Journal of Operations Research*, Vol. 219, No. 3, 2012, pp. 727–737.
24. Kuo, A., E. Miller-Hooks, and H. S. Mahmassani. Freight Train Scheduling with Elastic Demand. *Transportation Research Part E: Logistics and Transportation Review*, Vol. 46, No. 6, 2010, pp. 1057–1070.
25. Chow, A. H. F., and Y. Li. Robust Optimization of Dynamic Motorway Traffic via Ramp Metering. *IEEE Transactions on Intelligent Transportation Systems*, Vol. 15, No. 3, 2014, pp. 1374–1380.
26. Lo, H., and A. H. F. Chow. Control Strategies for Over-Saturated Traffic. *ASCE Journal of Transportation Engineering*, Vol. 130, No. 4, 2004, pp. 466–478.
27. Chow, A. H. F., C. Y. Cheung, and H. T. Yoon. Optimization of Police Facility Locationing. *Transportation Research Record: Journal of the Transportation Research Board*, 2015. 2528: 60–68.
28. Ahuja, R. K., T. L. Magnanti, and J. B. Orlin. *Network Flows: Theory, Algorithms, and Applications*. Prentice Hall, New Jersey, 1993.

The Standing Committee on Rail Transit Systems (AP065) peer-reviewed this paper (18-04675).

The contents of this paper reflect the views of the authors, who are responsible for the facts and the accuracy of the data presented herein. The contents do not reflect the official views or policies of RSSB, Network Rail, or any other organization. This paper also does not constitute a standard, specification, or regulation.

Article

TRR
JOURNAL OF THE TRANSPORTATION RESEARCH BOARD

Transportation Research Record
2018, Vol. 2672(8) 266–274

DOI: 10.1177/0361198118784386
journals.sagepub.com/home/trr
SAGE

Operational Impacts of Platform Doors in Metros

Alexander Barron[1], Shane Canavan[1], Richard Anderson[1], and Judith Cohen[2]

Abstract

Platform doors are increasingly installed by metros, primarily to improve safety. However, they have the potential for both positive and negative operational impacts, mostly by affecting dwell times at stations. Using data from the CoMET and Nova international metro benchmarking consortia of 33 metro systems, this paper seeks to understand and quantify these operational impacts. Overall, platform doors have a net negative impact on dwell times, leading to between 4 and 15 s of extra time per station stop. This is due to the additional time required for the larger doors to open and close, slower passenger movements due to the additional distance between platforms and trains, and, most importantly, extended departure delays after both sets of doors are closed caused by the need to ensure safety (that no one is trapped in the gap between the two sets of doors). This is a particular problem in mainland China, where metros conduct manual safety checks that require drivers to step out of trains onto platforms. However, despite longer dwell times, platform doors have a net positive impact on metro operations, largely due to the many safety benefits that also reduce delays and thereby improve service performance. There are also potential benefits regarding energy and ventilation. To mitigate the negative impacts, metros should seek to refine procedures and improve technology to reduce dwell time delays caused by platform doors. Reducing or eliminating these extra delays are essential to delivering efficient service and maximum capacity, provided that safety can be assured.

Platform doors, whether platform screen doors (typically full height), platform edge doors (typically half- or three-quarters height), or automatic passenger gates (typically shorter than half-height), have become commonplace elements of metro infrastructure in recent years. Although bringing marginal extra operating and maintenance costs (and a potential for failure), they are often recommended because of the wide range of benefits they bring – most notably safety (preventing suicides and trespassing on the track). Other potential benefits include improving station air-conditioning efficiency and/or supporting tunnel ventilation.

Platform doors have simultaneously both positive and negative operational impacts on dwell times at stations. They are wider than train doors so can take longer to open and close, and can lead to slower movement rates as passengers must cross two doorways to enter and leave trains. Their presence creates a risk that passengers could become trapped between the platform doors and train doors, in many cases leading to additional safety checks, which can delay trains. However, platform doors remove the risk of an incoming train striking a passenger, enabling faster platform entry speeds. In 2007, the net impact was estimated by Harris and Anderson (*1*) as increasing dwell times by 1 to 2 s. In the intervening years a larger and more diverse survey dataset has been accumulated; this study aims to investigate and quantify the impact of platform doors on station dwell times while also looking at their other overall operational impacts on metros.

The paper is structured as follows. The next section provides an overview of the relevant literature. This is followed by a description of the data and methodology. Discussion of the main results is then provided, followed by the main conclusions.

Literature Review

The most recognizable consequence of installing platform screen doors (PSDs) is the obvious improvements they bring to safety, reducing suicides and accidents by limiting access to the track. Evidence from Hong Kong (which examined the effectiveness of installing PSDs on suicide rates) found that over an 11-year period, PSDs were responsible for a 60

[1]Railway and Transport Strategy Centre, Department of Civil and Environmental Engineering, Imperial College London, London, UK
[2]Strategy and Network Development, London Underground, Transport for London, London, UK

Corresponding Author:
Address correspondence to Alexander Barron: alexander.barron@imperial.ac.uk

reduction in railway suicides (*2*). Remarkably, no significant displacement of potential attempted suicides to other platforms without platform doors was observed. However, there may be circumstances where PSDs could have a negative impact on passenger safety, for example emergency situations in which passengers are evacuated. This is particularly true during peak periods. Qu and Chow tested several scenarios with different passenger loading and opening conditions of PSDs (*3*). The study confirms that emergency evacuation time for trains is prolonged when PSDs are present. This is particularly sensitive to the stopping point of the train; when a train does not stop at the correct position, passenger movement can be disrupted. Reductions in exit width can lead to longer evacuation times, and in serious situations can lead to passenger panic. In addition, evidence from Chow et al., who examined incidents on fire and ventilation provision in metros in Hong Kong, further confirms the issue of aligning the train with PSDs (*4*). The study also refers to reports of glass panes of PSDs breaking suddenly, despite being made of toughened glass intended for extended exposure to fire, introducing an additional hazard.

PSDs do not only limit the movement of passengers; by also limiting air flow from tunnels into stations they also help control heating, ventilation, and air conditioning, reducing energy use. Evidence from Taipei Mass Rapid Transit (MRT) simulations indicates that PSDs can significantly decrease the average and peak cooling load, thus reducing the capacity/size of cooling equipment and allowing the chiller cooling load to be abridged (*5*). This in turn leads to significant savings in energy consumption. However, electricity consumption by ventilation equipment increases notably when PSDs are used.

By limiting air flow, PSDs can also shorten exposure to hazardous air pollutants. Passengers on metros can have significant exposure to particulate matter (PM10 and PM2.5) (*6*). Evidence provided by Lee et al. (*7*) suggests that after installing PSDs on the Seoul metro system, the average PM10 concentration on the platform without PSDs was 1.4 times higher than platforms with PSDs.

Other hazards examined include polycyclic aromatic hydrocarbons (PAH) (*8*), as well as potential exposure to a range of metals such as iron, manganese, chromium, aluminum and copper (*9–12*). Jeon et al. (*13*) examined the levels of radon, a naturally-occurring radioactive gas which can cause lung cancer. The study of the Seoul metro system surveyed changes in radon concentrations before and after PSD installation on 54 stations across six lines. After implementation of PSDs the mean radon concentrations at platforms decreased by approximately 56%, from 121.7 Bq/m^3 to 54.0 Bq/m^3, with the number of platforms where radon concentration was 74 Bq/m^3 decreasing from 38 to 12.

Additional benefits of PSDs include the effects on diffusing train noises (*14*). The study found that noise levels emitted by trains was reduced by PSDs, particularly blocking the lower frequency components of train noises. In doing so this greatly improves passenger comfort.

Dwell times are by their nature random processes that depend on the number and behavior of the passengers, as well as the action of closing door and departure/take-off (*15*). With regards to the impact of PSDs on boarding and alighting times, de Ana Rodríguez et al. suggest that they do not have a detrimental impact on times (*16*). The findings, from both laboratory experiments as well as analysis of video footage recorded on the London Underground, indicate that the presence of the doors affects passenger behavior at the platforms, inducing a more organized boarding and alighting process. Passengers were found to wait in a more organized fashion with less clustering. Waiting beside the doors rather than in front of them allows for more room to give way to alighters more often than without PSDs. Current literature, which mostly focuses on optimizing dwell time, has examined a range of additional factors. These include, but are not limited to, Zhang et al., who considered individual desire, pressure from passengers behind, personal activity and tendencies (dependent on gender, age, etc.) (*17*); Yalçınkaya and Mirac Bayhan, who considered average travel time, carriage fullness and headways (*18*); Assis and Milani, who considered the waiting time of passengers at stations, onboard passenger comfort, train trip duration and number of trains in service (*19*); and Martinez et al., who considered the variability of dwell times by considering peak and off-peak hours, incidents, and other operational variables (*15*).

Data and Methodology

We used data from the CoMET and Nova benchmarking consortia, comprising 33 metro operators in 31 cities worldwide. CoMET and Nova have been in existence since 1994, with the key objectives of comparing performance and sharing practices between metros in order to improve. The groups function within a framework of confidentiality, and a confidentiality agreement therefore governs the provided data that is referenced in this paper. This means that research results must be anonymized for publication. CoMET metros included in this study are: Beijing BMTROC, Berlin BVG, Delhi DMRC, Guangzhou GMC, Hong Kong MTR, London Underground, Metro de Madrid, Mexico City STC, Moscow Metro, New York NYCT, Paris RATP (Metro and RER, two systems), Metro de Santiago, Metro Sao Paulo, Shanghai SSMG, Singapore SMRT, and Taipei TRTC. Nova metros included in this study are: Barcelona TMB, Brussels STIB, Buenos Aires Metrovias, Bangkok BEM, Istanbul Ulasim, Kuala Lumpur RapidKL, Lisbon Metro, London DLR, Montreal STM, Nanjing Metro, Newcastle Nexus, Oslo Sporveien, Metro Rio, Shenzhen SZMC, Sydney Trains, and Toronto TTC.

This study refers to data from two principal sources: 1) a large database of survey results collected on a rolling basis over the past 15 years, and 2) questionnaire responses from a detailed case study conducted by the groups in 2013 about dwell time management. The Railway and Transport Strategy Centre (RTSC) has developed a significant database of dwell

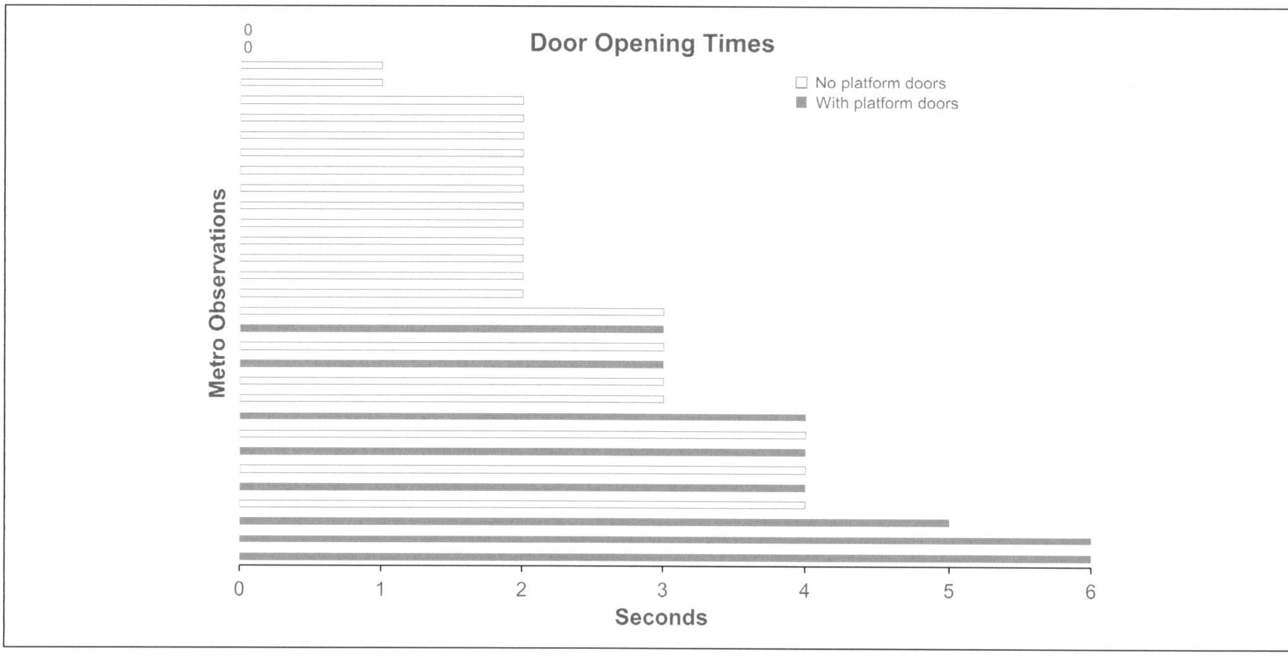

Figure 1. Observed door opening times (based on RTSC surveys at busy metro stations worldwide, 2000–2013).

time surveys carried out on a rolling basis since 2000. The full survey database includes surveys for every CoMET and Nova metro, as well as additional surveys from other metros and railways worldwide. A subset of the survey data, comprising at least one survey from each CoMET and Nova metro, was used for this analysis.

The surveys follow a standard methodology that enables direct comparisons between metros, with surveys typically carried out at the busiest station on the busiest line as the operational performance at this location determines how the overall system performs. Detailed timings are made of each element of the station stop, including wheel stop, door opening, "door closing" sound, actual train door closure, and wheel start. The survey also captures passenger movements at the busiest (or most critical) door, which is the door most likely to delay the train and therefore determines the overall dwell time and, ultimately, the capacity of the line. Passenger movements recorded include the number of passengers alighting, boarding, left behind/unable to board, and those traveling in the train vestibule who may be obstructing passenger flow. Due to the fact that these surveys have been carried out over many years, they accurately reflect a snapshot of on-the-ground passenger conditions and are useful for comparative purposes.

The survey data has been supplemented by questionnaires completed by many participating metros as part of a 2013 case study on dwell time management. The questionnaire responses include further details of train and station operations for these busy locations with long dwell times, and also include metros' own explanations of performance, operational experiences, and perceptions. Responses from 28 metros were provided confidentially to the consortia for the case study.

Results and Discussion

The key operational impacts of platform doors on metros with regard to dwell times are (station stop times), safety, and energy. We find that platform doors have a net negative impact on station dwell times, adding 4 to 15 s on average. However, despite the longer dwell times, we find that platform doors have a net positive overall impact on metros, primarily because of the dramatic reduction in track intrusions (to near zero) and the potential to save energy. These other benefits are especially relevant where track intrusion is a problem, such as high rates of suicides or trespassers, which are more common in Europe and North America, or where stations are air-conditioned, which is common for underground stations in most Asian metros. Each of the constituent impacts will be discussed in the subsections below.

Impacts on Dwell Times

The most important impact of platform doors on metro operations is dwell times (the time that trains are stopped in stations). Dwell time can be defined as the total time from the wheels stopping to the wheels starting to move again. Platform doors affect dwell times in several different ways, with the net impact not entirely clear.

Opening and Closing Times. Because platform doors are larger and slower than train doors, they typically take longer to open and close. A sample of door-width data for six metros with platform doors from the survey shows that platform doors ranged from 22% to 48% wider than the corresponding

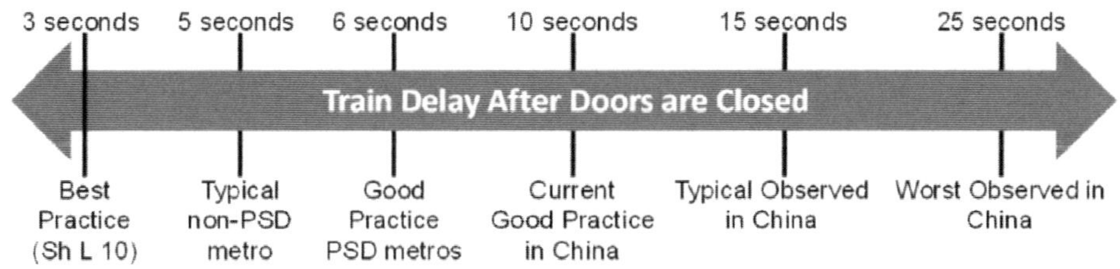

Figure 2. Observed departure delay worldwide at busy stations during peak hours (based on RTSC surveys).

train doors, and were above 2 m wide in all but one of the metros. Although this is necessary to allow for the stopping tolerance of the train, it does also mean that the doors are bigger and slower (and also easier to block).

Figure 1 below shows the observed door opening times at busy metros stations worldwide from the surveys. Although there is overlap of door opening times of between 3 and 4 s for both with and without PSDs, it is clear that the presence of PSD generally adds to door opening times. Opening times of 2 to 4 s are typical without platform doors, whereas times are typically 3 to 5 s with platform doors. In some cases, platform doors open and close after train doors, which causes the process to take even longer. For the metros with very short door opening times (e.g., zero or 1 s), the doors may begin opening before the train wheels come to a complete stop – this can sometimes be found on older trains in older metros. Based on these observations, we estimate that the longer opening and closing times add approximately 1 to 2 s to station dwell times with platform doors.

Passenger Flow Rates. Passenger flow rates (for both boarding and alighting from trains) are influenced by a wide variety of factors, including the precise geometry of each station and train surveyed, passenger volumes, and cultural factors. These boarding and alighting rates have been observed at metros worldwide through the survey process. Statistical tests using regression analysis were conducted, but the results relating to platform doors were inconclusive. This was partly due to the relatively small number of observations with platform doors in the entire sample, but also because of the diversity of the sample. For example, platform doors are more prevalent in Asian metros, where passenger volumes are higher and passenger flow rates may be proportionally higher than elsewhere (with or without platform doors).

However, anecdotal observations from surveys and individual survey results suggest that passenger movements are generally slightly slower with platform doors. This is primarily because of the slightly greater distance from the platform to the train, and the related need for passengers to pass through two sets of doors and thresholds instead of just one. On average, boarding and alighting rates are both approximately 1.1 passengers per second per door. Even if these passenger flow rates are only 10% slower (i.e., 1.0 passengers per second rather than 1.1), the aggregate effect on dwell times can be significant. With a hypothetical flow of 30 passengers (combining boarders and alighters) at a busy interchange station, a 10% reduction in the movement rate would lead to an additional 2 to 3 s. It is also important to note that this situation only has to occur at one door of the train (the busiest door) to cause the dwell time increase.

Departure Delay. The most significant impact of platform doors on dwell times is due to departure delay, which is the time between the final door closing and the train starting to move (after both the train and platform doors are closed). These delays have been observed to be very significant in some metros, as illustrated in Figure 2 below. All of the examples highlighted in the figure use automatic train operation (ATO), with the "best practice" example of Shanghai Line 10 now using unattended train operation (UTO).

It is normal for there to be some small amount of post-door closing delay for all metros. This time is spent confirming that all doors are closed and locked and that it is therefore safe to depart. First, there can be some technical delay during which the door status is communicated throughout the train. On lines with drivers, there can also be delay while the driver observes this doors closed/locked notification and/or makes their own observations, such as using cameras or platform mirrors. In some cases, the design of the cab may contribute to longer dwell time if the operation requires the driver to move away from their seat. In more traditional circumstances, with a driver and no platform doors, we have observed this delay to be, on average, approximately 5 s at busy stations during peak hours, although it can range from as little as 2 to 3 s up to 10 to 12 s.

With platform doors, this procedure is complicated by having a second set of doors that need to be confirmed as closed and locked, but also – most importantly – by having a gap between the platform doors and the train. With platform doors, we have observed the post-door closure delay to be on average 12 s worldwide, but with a very wide range of between 3 and 25 s. However, the variation observed has been clearly divided between observations in mainland China and elsewhere.

Outside of China, the observed delay has ranged from approximately 5 to 10 s, with 6 s suggested to be good

Figure 3. Example of Chinese metro driver checking procedure (taken in Guangzhou Metro by RTSC, 2014).

practice. In these cases, the presence of platform doors only adds approximately 1 to 2 s to station dwell times compared with stations without platform doors. In mainland China, however, the observed delay has ranged from 10 to 25 s, with 15 s being the most typically observed value. The reason for these significant delays is the extensive manual checking procedures in place for the gap between the platform doors and the train. Typically, the driver must physically step out of the train onto the platform and conduct a series of maneuvers to check no one is trapped in the gap between the platform doors and the train (before then re-entering the train and preparing to drive off), as shown in Figure 3 below. This procedure was introduced to ensure safety, and is practiced partly because of two known cases in China (2007 in Shanghai and 2014 in Beijing) where a passenger became trapped between a closed train and platform doors and was ultimately killed.

It is important to consider the implications of this practice. Although it may help to ensure safety, the resulting delays also have significant costs, both in terms of metro capacity and passenger time. If the departure delay in Chinese metros could be reduced from the current typical value of 15 s to the 6 s observed good practice value from elsewhere in the world – thus saving 9 s per station stop – an additional one or two trains per hour could be operated, providing valuable extra capacity for very crowded lines. Alternatively, if a typical line has 30 stations, then saving 9 s per station would save 4.5 min from one end of the line to another, which for many lines would be sufficient in peak hours (when frequencies may be 2 to 3 min) to enable the same service to operate with fewer resources (for example, one fewer driver or train, or both one fewer driver and train).

However, the greater impact may be in the value of lost time for passengers. If we consider a hypothetical example in Shanghai, we can attempt to calculate a value for the time lost. If the average passenger travels through approximately eight stations and experiences the same extra time loss of 9 s per station as explained above, then 72 s would be lost per journey. With an average of nearly 5 million passenger journeys per day, this suggests about 100,000 h of customer time may be lost per day. If the average value of time is assumed to be equal to the average net city wage in Shanghai (US$5.21 in 2015 US$ purchasing power parity terms [*20*]), then the value of time lost per day is US$521,000, or nearly US$200 million per year.

This hypothetical example is used to illustrate the cumulative effect that such seemingly small impacts have. It is also important to consider the broader impacts of these costs on cities. Any passengers discouraged from traveling by metro (either because of the lost time or, more likely, the lack of available capacity) who therefore travel by other modes are likely to reduce overall safety, because other transport modes are almost certainly less safe than metros. Those same potential passengers may also create additional costs by contributing to road congestion. It is hoped that this example illustrates how calculating the economic cost of delays can produce powerful data showing exactly how much the value of time lost could amount to. This data could be shown to senior management and governments, and could potentially help persuade them to change policies or regulations, or to invest in technology that would reduce or even eliminate the problem.

However, one observed exception to these significant departure delays has been observed in China – on Line 10 in Shanghai. Line 10 was originally designed for future fully automated (UTO/GoA4) operations, and in August 2014 Shanghai Metro began live testing of a more automatic mode. This new operating mode includes fully automated dwell times, where the station stop is managed completely by the computer, and the system ensures safe departure using a series of interlocked systems and sensors for train doors,

Figure 4. Interlocked gap detection system using lasers on Shanghai Metro Line 10 (photo taken by author, 2014).

platform doors, and gap detectors (see Figure 4). If ultimately successful, this solution has great potential to eliminate the long delays experienced in China, which would both improve capacity and save significant sums of money in passenger time. However, it is still under evaluation, and the costs to implement similar systems on other metro lines may be significant. Still, in the future as infrastructure and equipment ages and ultimately is refurbished or replaced, installing such systems may be advisable.

Other Factors. Although issues related to the opening and closing of doors, passenger flow rates, and departure delays are the most important impacts, platform doors can also affect dwell times by impacting passenger behavior and the consistency of dwell times. Some metros – particularly those with platform doors on only small proportions of their network, or only on new lines – have reported better passenger behavior with platform doors. Although we do not have any concrete data to support or refute this, it is believed that this may be a reflection of the different equipment. For example, passengers may perceive the need to pass through (or hold) two sets of doors to be more intimidating than just one, and therefore be more timid about taking disruptive actions. Furthermore, passengers may (even subconsciously) associate new platform doors with more technology and automation and consider that it is advisable to not challenge them. Conversely, if passengers know that a driver is manually closing the doors and looking at them from the front of the train, they may believe that the driver will wait or reopen the doors for them.

In addition, there could be positive impacts associated with passengers being sure of where the train doors will be and lining up accordingly, which may actually improve passenger flow when trains arrive. Without platform doors, passengers will not know exactly where the train doors will be located (and train stopping positions may be more variable, particularly with manual driving), which may lead to some delay as passengers re-align themselves with the door. However, this situation may actually be less relevant for very busy stations, where passengers who could not board a previous train will already be lined up by the door positions, and could also be mitigated by using platform markings. Anecdotal observations from surveys suggest that this characteristic generally has a positive, but very marginal, impact on dwell times.

Finally, there is evidence to suggest that platform doors may make station dwell times more consistent, which in turn improves the overall regularity of train service. In the survey dataset, the coefficient of variation of station stop times was about 37% better with platform doors than without. However, it is important to recognize that this association is likely also linked to age of equipment, level of automation, and the fact that more platform doors are present in Asian metros, where service is significantly more reliable overall. Nonetheless, greater regularity of dwell times would be consistent with better passenger behavior, and some metros that are planning to retrofit platform doors have reported that they are expecting to achieve more consistent (if slightly longer) dwell times.

Impacts on Safety

Improving safety is typically a key reason to implement platform doors in metros, and there is no doubt that platform doors improve safety significantly. Based on a previous safety study conducted by the CoMET and Nova groups in 2011, it is clear that with platform doors there is a dramatic reduction in track intrusions to nearly zero. This reflects a reduction not only in suicides and suicide attempts but also passengers hit by trains, passengers falling onto the track, and passengers trespassing on the track. The physical separation of the track right-of-way from the platform, strongest with full-height PSDs, is crucial for eliminating both intentional and unintentional intrusions. There can also be improvements in other related issues, such as track fires (as significantly less litter from passengers can actually reach the track).

The only related safety indicator that did not improve with platform doors was the number of passengers caught in doors. In fact, this indicator actually increased, which can be

Figure 5. Passengers queuing at platform doors in London, showing how queues lead to inefficient use of platform space (photo taken by author, 2013).

explained by the presence of two sets of doors for passengers to get caught in, and perhaps in some cases the big and slow nature of platform doors attracting passengers to try to hold them or pass through them as they are closing.

The other safety-related issue to mention is the potential for passengers to get trapped in the gap between the train and the platform doors. Although very dangerous, the number of instances appears to be very low. There seems to be only two reported occurrences worldwide – one in Shanghai in 2007 and one in Beijing in 2014 (both of which unfortunately resulted in the death of the passenger). This risk can be mitigated by design, ensuring that the gap is as small as possible and that platform doors have taper plates that prevent closure if someone is standing in the gap, and by technology, using systems such as laser detectors (like the ones on Shanghai Line 10 shown in Figure 4) to confirm that the gap is clear before the train departs.

Apart from the safety benefits of platform doors, it is important to recognize the benefits of reduced safety-related incidents on train service. The negative impacts of delays associated with these incidents are massive – for example, entire metro lines may be shut down for hours after a suicide, and trespassers may require power to be turned off and police to search tunnels. By virtually eliminating these incidents and the related delays, the metro and all passengers benefit significantly. In fact, the costs associated with retrofitting platform doors may be justifiable on the savings linked to fewer disruptions (when also considering the value of time lost by passengers).

Impacts on Energy

Although not the focus of the surveys referenced in this paper, it has been observed that platform doors can lead to energy savings for metros. This is especially true where underground stations are air-conditioned, as platform doors retain the cool air in the station and avoid leakage into the tunnels. To achieve these savings, the stations need to be completely sealed off from the tunnels with full-height PSDs. Evidence from Hu and Lee, who carried out a case study on the Taipei MRT system, estimates that in the peak periods cases without PSDs can experience as much as 50% more electricity consumption than cases with PSDs (5).

Even for metros without air-conditioned stations, platform doors may help achieve some savings – for example, by contributing to a metro's overall approach to ventilation. However, this is very dependent on the metro's existing ventilation setup, and therefore the impact of platform doors may vary significantly.

Other Impacts

Finally, this section summarizes other operational impacts of platform doors in metros:

- If metros have manual driving, then platform doors can enable faster train speeds when entering stations. Anecdotal evidence from surveys suggests that this could potentially save 2 s per station. However, it is important to note that this could also be done with automatic driving (i.e., ATO) without platform doors.
- Increased investment and maintenance costs are required to purchase, install, and maintain platform doors. In particular, retrofitting platform doors can be very expensive, as platforms may need structural works to support the additional weight and cantilevered loads, and there may be challenges associated with overhead obstructions or curved platforms.
- Platform doors introduce another piece of equipment that can fail, which can cause delays – although long-term data from CoMET and Nova suggest that the number of delays caused by platform doors is very small.
- Platform doors can help to prevent objects, including trash, from reaching the track. This in turn reduces the delays associated with retrieving or removing objects, as well as the prevalence of track fires.
- Space may be better utilized on platforms with platform doors, as passengers can safely use all of the space up to the edge. This may in turn increase platform capacity and reduce any need to restrict access to stations due to crowding. However, anecdotal evidence, including that from Figure 5, suggests that perhaps the queuing encouraged by platform doors also means that platform space between doors is actually less used.

Conclusion

We have seen that platform doors have significant operational impacts for metros, most importantly relating to dwell times

Table 1. Summary of Key Dwell Time Factors Influenced by Platform Doors

Factor	Average impact
Slower door opening and closing times	+ 1–2 seconds
Slower passenger flow rates (boarding and alighting)	+ 2–3 seconds
Longer departure delay after all doors closed	+1–10 seconds
Platform doors	+4–15 seconds

and safety, but also to other factors such as energy and possibly passenger behavior. Overall, we found that platform doors have a net negative impact on dwell times, adding between 4 and 15 s on average to the time trains are stopped at each station. This additional time adds up to significant lost time for passengers, and could also lead to reduced total capacity provided. The key factors leading to longer dwell times with platform doors and the average impact compared with stations without platform doors are summarized in Table 1.

However, despite these longer dwell times, we found that platform doors have a net overall positive impact on metros, primarily due to the significant safety benefits. These include both the reduction in safety incidents themselves, in which platform doors virtually eliminate track intrusions (and therefore nearly all suicides and cases of passengers being hit by trains), but also the significant reductions in delays associated with safety incidents. It is also important to acknowledge that platform doors may lead to more consistent, regular dwell times, and help to enable higher degrees of automation, both of which improve overall service performance.

To retain the positive benefits and mitigate the negative impacts, metros should seek to refine procedures and improve technology to reduce dwell time delays caused by platform doors. The main example of this is the significant departure delay experienced primarily in mainland China, where delays of typically 15 s, but as high as 25 s, are experienced after both train and platform doors are closed and before the train begins to move. Reducing or eliminating these extra delays is essential to delivering efficient service and maximum capacity, provided that safety can be assured. For metros considering platform doors – either for new lines or retrofitting for existing lines – it is important to consider these impacts at the planning stage, noting that the significant benefits of platform doors may justify the not insignificant costs of installation.

Acknowledgments

We thank the metros of the CoMET and Nova consortia for supporting this research, including by their responses to the case study questionnaire and participation in the on-site surveys.

Author Contributions

The authors confirm contribution to the paper as follows: study conception and design: AB, RA; data collection: AB, JC; analysis and interpretation of results: AB, SC, RA, JC; draft manuscript preparation: AB, SC. All authors reviewed the results and approved the final version of the manuscript.

References

1. Harris, N. G., and R. J. Anderson. An International Comparison of Urban Rail Boarding and Alighting Rates. *Proceedings of the Institution of Mechanical Engineers, Part F: Journal of Rail and Rapid Transit*, Vol. 221, No. 4, 2007, pp. 521–526.
2. Law, C. K., P. S. Yip, W. S. Chan, K. W. Fu, P. W. Wong, and Y. W. Law. Evaluating the Effectiveness of Barrier Installation for Preventing Railway Suicides in Hong Kong. *Journal of Affective Disorders*, Vol. 114, No. 1, 2009, pp. 254–262.
3. Qu, L., and W. Chow. Platform Screen Doors on Emergency Evacuation in Underground Railway Stations. *Tunnelling and Underground Space Technology*, Vol. 30, 2012, pp. 1–9.
4. Chow, W., L. Qu, and E. Pang. Incidents on Fire and Ventilation Provision in Subway Systems in Hong Kong. *International Journal On Engineering Performance-Based Fire Codes, Accepted To Publish*, Vol. 10, No. 3, 2011, pp. 41–47.
5. Hu, S. C., and J. H. Lee. Influence of Platform Screen Doors on Energy Consumption of the Environment Control System of a Mass Rapid Transit System: Case Study of the Taipei MRT System. *Energy Conversion and Management*, Vol. 45, No. 5, 2004, pp. 639–650.
6. Johansson, C., and P. Å. Johansson. Particulate Matter in the Underground of Stockholm. *Atmospheric Environment*, Vol. 37, No. 1, 2003, pp. 3–9.
7. Lee, J. G., S. H. Byeon, and J. H. Lee. The Effect of Platform Screen Door (PSD) for Fine Particles at Subway Train in Seoul, Korea. *Proc., ICCAS-SICE, 2009*, Fukuoka, Japan, IEEE, New York, 2009.
8. Fromme, H., A. Oddoy, M. Piloty, M. Krause, and T. Lahrz. Polycyclic Aromatic Hydrocarbons (PAH) and Diesel Engine Emission (Elemental Carbon) Inside a Car and a Subway Train. *Science of the Total Environment*, Vol. 217, No. 1, 1998, pp. 165–173.
9. Chillrud, S. N., D. Epstein, J. M. Ross, S. N. Sax, D. Pederson, J. D. Spengler, and P. L. Kinney. Elevated Airborne Exposures of Teenagers to Manganese, Chromium, and Iron from Steel Dust and New York City's Subway System. *Environmental Science & Technology*, Vol 38, No. 3, 2004, pp. 732–737.
10. Crump, K. S. Manganese Exposures in Toronto during Use of the Gasoline Additive, Methylcyclopentadienyl Manganese Tricarbonyl. *Journal of Exposure Analysis & Environmental Epidemiology*, Vol. 10, No. 3, 2000, p. 227.
11. Pfeifer, G., R. Harrison, and D. Lynam. Personal Exposures to Airborne Metals in London Taxi Drivers and Office Workers in 1995 and 1996. *Science of the Total Environment*, Vol. 235, No. 1, 1999, pp. 253–260.

12Sitzmann, B., M. Kendall, J. Watt, and I. Williams. Characterisation of Airborne Particles in London by Computer-Controlled Scanning Electron Microscopy. *Science of the Total Environment*, Vol. 241, No. 1, 1999, pp. 63–73.

13. Jeon, J.-S., J. C. Yoon, H. C. Lee, S. W. Eom, and and Y. Z. Chae. A Noticeable Change in Indoor Radon Levels after Platform Screen Doors Installation in Seoul Subway Station. *Journal of Korean Society for Atmospheric Environment*, Vol. 28, No. 1, 2012, pp. 59–67.
14. Socta, Y., and R. Shimokura. Change of Acoustic Characteristics Caused by Platform Screen Doors in Train Stations. *Applied Acoustics*, Vol. 73, No. 5, 2012, pp. 535–542.
15. Martinez, I., B. Vitoriano, A. Fernandez, and A. P. Cucala. Statistical Dwell Time Model for Metro Lines. *Proceedings of Urban Transport XIII*, edited by C.A. Brebbia, WIT Press, 2007, pp. 223–232.
16. de Ana Rodríguez, G., S. Seriani, and C. Holloway. Impact of Platform Edge Doors on Passengers' Boarding and Alighting Time and Platform Behavior. *Transportation Research Record: Journal of the Transportation Research Board*, 2016. 2540: 102–110.
17. Zhang, Q., B. Han, and D. Li. Modeling and Simulation of Passenger Alighting and Boarding Movement in Beijing Metro Stations. *Transportation Research Part C: Emerging Technologies*, Vol. 16, No. 5, 2008, pp. 635–649.
18. Yalçınkaya, Ö., and G. Mirac Bayhan, Modelling and Optimization of Average Travel Time for a Metro Line by Simulation and Response Surface Methodology. *European Journal of Operational Research*, Vol. 196, No. 1, 2009, pp. 225–233.
19. Assis, W. O., and B. l. E. A. Milani. *Generation of Optimal Schedules for Metro Lines Using Model Predictive Control. Automatica*, Vol. 40, No. 8, 2004, pp. 1397–1404.
20. UBS. *Prices and Earnings*. UBS AG, 2015. https://www.ubs.com/microsites/prices-earnings/prices-earnings.html, 2015.

The Standing Committee on Rail Transit Systems (AP065) peer-reviewed this paper (18-02825).

TRR
JOURNAL OF THE TRANSPORTATION RESEARCH BOARD

Article

Transportation Research Record
2018, Vol. 2672(8) 275–287

Article reuse guidelines:
sagepub.com/journalsPermissions.nav
DOI: 10.1177/0361198118791628
journals.sagepub.com/home/trr
SAGE

Integrating Dynamic Signaling Commands Under Fixed-Block Signaling Systems into Train Dispatching Optimization Problems

Xiaojie Luan[1], Bart De Schutter[2], Francesco Corman[3], and Gabriel Lodewijks[4]

Abstract

In railway operations, when a disruption occurs, train dispatchers aim to adjust the affected schedule and to minimize negative consequences during and after the disruption. As one of the most important components of the railway system, railway signals are used to guarantee the safety of train services. We study the train dispatching problem with consideration of railway signaling commands under the fixed-block signaling system. In such a system, signaling commands dynamically depend on the movement of the preceding trains in the network. We clarify the impact of the signaling commands on train schedules, which has so far been neglected in the literature on railway train dispatching, and we propose an innovative set of signaling constraints to describe the impact. The determination of the signal indicators is presented using "if-then" constraints, which are further transformed into linear inequalities by applying two transformation properties. Activation of the train speed limits that result from the signaling commands is the core purpose of the signaling constraints, and this is implemented by using the signal indicators. Moreover, we formulate the Greenwave (GW) policy, which requires that trains always proceed under green signals, and we further investigate the impact of the GW policy on delays. In numerical experiments, the proposed signaling constraints are employed within a time-instant optimization problem, which is a mixed-integer linear programming (MILP) problem. The experimental results demonstrate the effectiveness of the proposed signaling constraints and show the impact of the signaling commands and GW policy on the train dispatching solution.

Railways are crucial to address the ever-increasing mobility of people and goods, due to their positive characteristics of high capacity, high speeds, and eco-friendliness. A negative characteristic of railway services, however, is their limited reliability and punctuality, which hinders attainment of the modal share sought by policy makers and researchers. Train movements on a railway network are regulated by timetables. In daily operations, perturbations (caused by bad weather, infrastructure failures, extra passenger flows, etc.) unavoidably occur, which may affect the normal operations and cause a deviation from the planned timetable. In such cases, the task of a train dispatcher (controller) is to take proper measures to adjust the schedules affected, in order to reduce the negative consequences (delays). This is the train dispatching problem. Due to the high interdependency between trains utilizing the available capacity, train dispatching work is usually complex, especially when the railway network is operated close to saturation, in densely urbanized zones, or during peak hours. An ineffective train dispatching decision could result in a snowball effect with consecutive delays, downgrading the reliability and punctuality of train services. Fast and effective decisions for the train dispatching problem are always desired.

Railway signaling is one of the most important parts of the railway system. There is a wide variety of railway signals and many signaling systems with different

[1]Section Transport Engineering and Logistics, Delft University of Technology, Delft, The Netherlands
[2]Delft Center for Systems and Control, Delft University of Technology, Delft, The Netherlands
[3]Institute for Transport Planning and Systems (IVT), ETH Zürich, Zürich, Switzerland
[4]School of Aviation, Faculty of Science, University of New South Wales, Sydney, Australia

Corresponding Author:
Address correspondence to Xiaojie Luan: x.luan@tudelft.nl

principles all over the world, for example, the moving-block signaling systems commonly used in high-speed railway networks and the fixed-block signaling systems commonly used by conventional railways. The core function of the signaling systems is same, however—indicating the state of the block section(s) ahead for the train driver in order to guarantee the safety of train services. A signaling command can be indicated by a single aspect or by multiple aspects. In the United States and in many European countries, the signaling command provides an additional requirement for train operations, namely indicating the maximum allowed speed to the driver. The driver has to control the train to prevent it from exceeding this speed to ensure safety. Otherwise, a worse situation may happen, where the available distance is not enough to stop the train.

An extensive body of studies is available in the literature that addresses the train dispatching problem, having different focuses, such as: considering multiple classes of running traffic (*1*), passenger connections (*2*), speed management (*3*), and maintenance plans (*4*), and using different approaches (e.g., linear/nonlinear optimization and heuristics). However, a gap still exists with regard to the signaling commands. Train speed limits that result from signaling commands have been neglected in the literature (*5*), although they are indeed required in real train operations. To the best of our knowledge, no study is yet available for generating optimal train dispatching solutions that integrate precisely the actual signal aspect shown to train drivers, and that guarantee no violation of the signaling commands, based on the fixed-block signaling system. The reality of train operations and the gap in the scientific literature have motivated us to include the signaling commands while addressing the train dispatching problem.

We therefore study the train dispatching problem with consideration of railway signaling commands, focusing on railway networks with a fixed-block signaling system, as is common in the United States and in many European countries. As the signaling commands dynamically depend on the movement of the preceding trains in the network, we use binary variables (namely signal indicators) to indicate the signaling commands. These signal indicators are determined by a set of "if-then" constraints, which could be further transformed into linear inequalities by applying two transformation properties. These constraints could be generalized to other signaling systems. Train speed limits that result from the signaling commands are restricted in the signaling constraints by employing the signal indicators. In addition, we formulate the Greenwave (GW) policy and explore its impact on the train dispatching solution (i.e., the train delays). Basically, the aim of the GW policy is to require trains to follow their scheduled speed profile exactly, thus avoiding the need for speed profile adjustments. In the numerical experiments, a time-instant optimization approach (mixed-integer linear programming, MILP) proposed in our previous work (*6*) is used to apply the proposed signaling constraints, aiming at delivering a train dispatching solution with minimization of train delays. The experimental results demonstrate the effectiveness of the proposed model, including signaling constraints, and show the impact of the signaling commands and GW policy on the train dispatching solution.

Literature Review

An extensive body of literature is available for the railway train dispatching problem; interested readers might refer to the surveys (*5*, *7*, *8*). This section briefly reviews the state of the art for the train dispatching problem, especially focusing on the studies that use operations research-based techniques and on how signaling issues are treated.

Advances in scheduling theory have made it possible to solve real-life train dispatching instances in which not only departure/arrival times (*9*), but also train orders, routes, speeds, and further operational freedom were considered as variables (*10–14*).

Several operations research-based techniques are now available for addressing the train dispatching problem. A particularly popular model is the alternative graph-based model, which uses a combination of job shop and alternative graph techniques (*9*). This alternative graph-based formulation method considers microscopic details and is further employed in many studies, for example, dealing with the train rerouting problem by developing meta-heuristics, including a tabu search algorithm (*11*) and a variable neighborhood search algorithm (*15*), and investigating the impact of the details and the number of operational constraints on the applicability of models, in terms of solution quality and computational efficiency (*16*).

Another stream of studies addresses the train dispatching problem at a macroscopic level, which allows for faster resolution and a larger geographical scope. Schöbel (*17*) proposed an event-activity based integer programming model to solve the delay management problem. The model was further extended to address a discrete time/cost trade-off problem of maintaining service quality and reducing passengers' inconvenience (*18*), and by including headways and capacity constraints and testing multiple pre-processing heuristics in order to fix integer variables and to speed up the computation (*19*). In the proposed problems, connections are decided to be maintained or dropped by minimizing the number of missed connections, while minimizing the delays of all events.

Other approaches have also been proposed to solve this problem. Luan et al. (*4*) employed the flag variables-based formulation method to address the integration of train scheduling and preventive maintenance planning. Pellegrini et al. developed a heuristic algorithm, named RECIFE-MILP, based on an extended version of the MILP formulation they had proposed (*20*, *21*). Samà et al. (*22*) further investigated how to select the most promising train routes among all possible alternatives, through developing an ant colony optimization meta-heuristic. Some dispatching decision support systems have been developed by researchers, and a few of them have been used in practice. Mazzarello and Ottaviani (*23*) developed a traffic management support system that is able to optimize traffic flow in large railway networks equipped with either fixed- or moving-block signaling systems. D'Ariano et al. (*24*) developed an advanced decision support system, known as ROMA (railway traffic optimization by means of alternative graphs), for dispatching trains based on microscopic details. Decomposition technologies were used for handling large areas in this system. It is verified that the system is able to find feasible and efficient schedules quickly. An exact approach and a master-slave solution algorithm (based on decomposition) were presented by Lamorgese and Mannino (*25*), based on which a dispatching decision support system was developed. This system has been in operation in Norway since February 2014 and represents one of the first operative applications of mathematical optimization to train dispatching.

Most approaches in the literature neglect the impact of signaling issues while rescheduling the traffic. A recent study (*3*) takes the impact of signaling into account, focusing on high-speed traffic based on a quasi-moving-block system. The authors proposed an alternative graph-based optimization problem to reschedule the high-speed traffic, which integrates the modeling of traffic management measures and the supervision of speeds, braking, and headways. The current paper fills the knowledge gap regarding the signaling issues in the train dispatching problem, which are mostly neglected in the literature. The main contribution is thus the consideration of different signaling commands, not only "Clear" and "Stop" (commonly satisfied by a track capacity constraint), but also "Approach (Limited)" (mostly neglected in the literature), depending dynamically on the traffic state. The proposed approach is suitable for inclusion in optimization schemes, in time-instant (or time-continuous, in the terminology of [*5*]) formulations, and can be extended to virtually any other fixed-block signaling system.

Problem Description

This section introduces fixed-block signaling systems, followed by the relevant signaling commands that should be respected during train operations. We then describe the formulation method of the time-instant optimization approach used in our previous study (*6*) for addressing the train dispatching problem. This formulation method will be used to construct the signaling constraints later on.

Railway Signaling Systems

A railway signaling system is used to direct railway traffic and to keep trains clear of each other at all times (*26*, *27*). A railway signal shows whether the track is clear ahead and also indicates to train drivers how far ahead the track is clear. Figure 1 illustrates three fixed-block signaling systems that are basic and widespread all over the world. In such fixed-block signaling systems, each track is divided into a sequence of block sections, and each block section is protected by a fixed signal placed at its entrance and displayed to the driver of an approaching train.

Figure 1*a* illustrates a two-aspect signaling system, which has a basic signal with a red and a green aspect. A green aspect indicates that the block section is accessible for trains, and a red aspect rejects the access of trains. As stated in (*28*), this two-aspect signaling system works well for trains with speeds less than 50 km/h; however, for a train traveling at a higher speed (e.g., over 50 km/h), the train driver needs a warning of a red aspect ahead to allow time to stop. Therefore, multi-aspect signals have appeared, as shown in Figure 1*b* and *c*, for a three-aspect signal and a four-aspect signal respectively. The three-aspect signal has a red, a yellow, and a green aspect. If a block section is occupied by a train, the signal placed at its entrance will display a red aspect, and the two signals behind will show a yellow aspect and a green aspect respectively. The yellow aspect provides an advance warning of the red aspect ahead, and the driver then knows that there is only one clear block section ahead. The green aspect indicates that there are at least two clear block sections ahead, and the driver can maintain the design speed until facing a yellow aspect. The four-aspect signaling system in Figure 1*c* works similarly to the three-aspect signaling system, except that two advance warnings are provided before a red aspect: a single-yellow and a double-yellow aspect. One purpose of doing this is to provide an earlier warning for higher-speed trains, and another purpose is to allow better track occupancy by shortening the length of the block sections.

In the United States and in most European countries, the signal indicates the maximum allowed speed to the driver, and the driver has to control the train to avoid exceeding this speed to ensure safety (otherwise, the driver will be penalized). The blue curves in Figure 1 indicate the maximum speed profiles of train f_1 at the

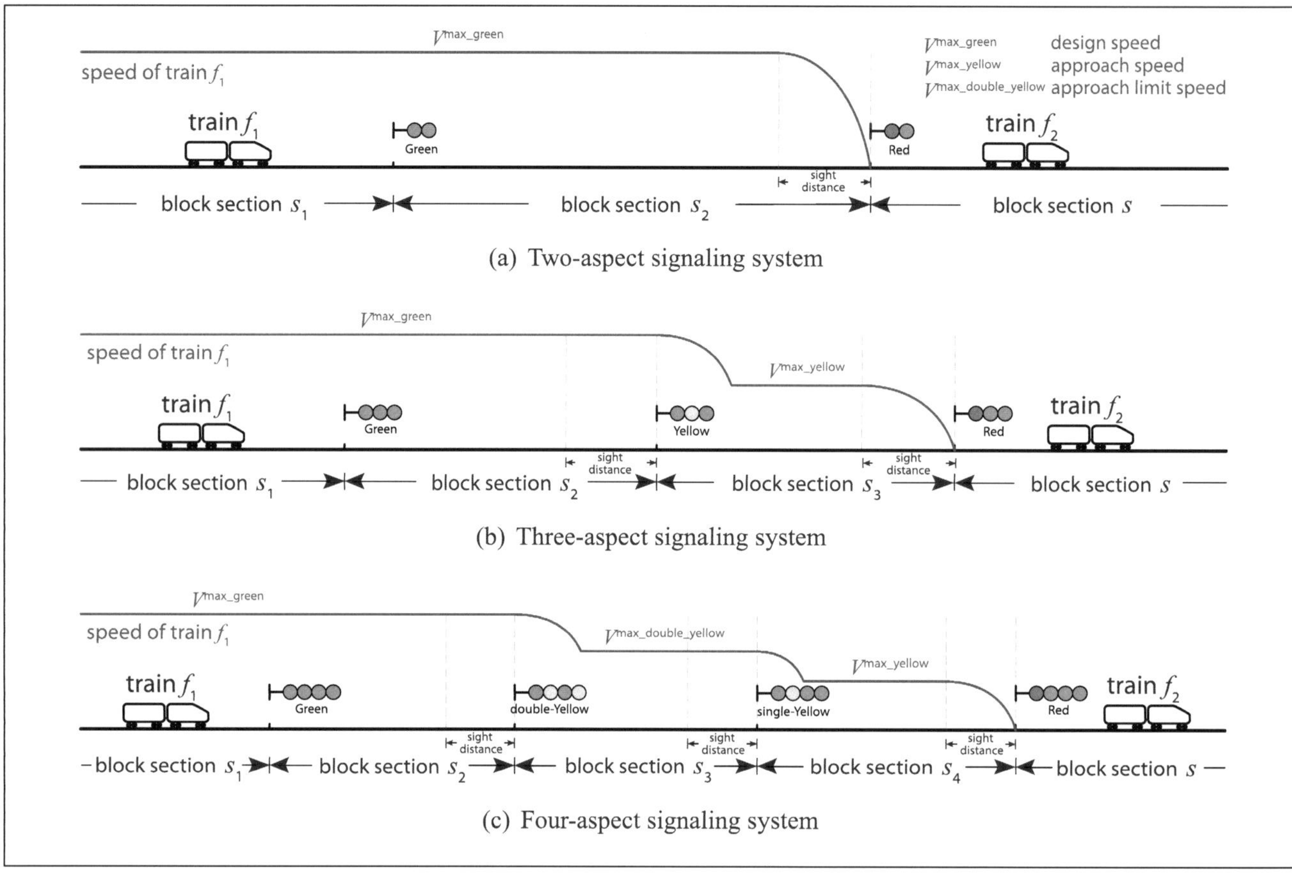

Figure 1. Three fixed-block signaling systems.

current moment, in the case of different signal aspects at sight distance.

Let us consider the three-aspect signaling system as an example to explain the impact of signal aspects on train speeds. Figure 2 presents the time-space-speed graphs for a train f_1 on five consecutive block sections. Different scenarios are illustrated, depending on the movement of the preceding train f_2. The orange horizontal line marked for each train on each block section directs the signal aspect faced by the train at the corresponding sight distance. In the case of always facing a green aspect, the driver can proceed with the design speed ($V^{max\text{-}green}$), as illustrated in the planned scenario of Figure 2*a*. The preplanned train paths are further indicated by the dashed black lines in the cases of disruptions, illustrated in Figure 2*b*–*f*, and the train paths after adjusting are indicated by the solid black lines. In the upper portion of each subfigure, we sketch the train paths and the signal aspects (in color) displayed as a function of the time; in the lower portion, we present the maximum allowed speed for train f_1. In the case of facing a yellow aspect at the sight distance of block section s_2, the driver of train f_1 has to decrease the train speed from the design speed ($V^{max\text{-}green}$) to the approach speed ($V^{max\text{-}yellow}$), which is maintained until the following signal becomes visible. When the train reaches the next sight point, there are five possible scenarios, as shown in Figure 2*b*–*f*, which are labeled as scenario C1,..., C5 respectively:

- In scenario C1, the signal placed at the sight distance of block section s_3 shows a green aspect, and the driver is allowed to accelerate the train from the approach speed to the design speed.
- In scenario C2, the signal stays yellow, and the train is controlled to enter the next block section at the approach speed.
- In scenario C3, the signal stays red until the train completely stops, and the signal becomes green only after a certain waiting time.
- In scenario C4, the signal is yellow and then switches to green in the sight time (i.e., the running time of a train over the sight distance on a block section). The driver possibly accelerates the train to the design speed after a small reaction time.
- In scenario C5, the signal is red and then switches to yellow (or green directly) during the braking phase. In this scenario, the driver stops braking

and possibly accelerates the train to the approach speed (or the design speed) after a small reaction time.

We can conclude that the maximum allowed speed of a train on a block section depends on the signaling command (i.e., the signal aspect displayed), and the signaling command dynamically depends on the movement of the preceding train in the railway network. Note that hereafter we make an assumption of considering the most conservative case in terms of safety—the movement of each train on each block section respects the signaling command shown at the moment that the train is reaching sight distance, that is, the possibility of scenarios C4 and C5 in Figure 2 is ignored.

The Time-Instant Formulation Method for the Train Dispatching Problem

In a time-instant (or time-continuous, in the terminology of [5]) optimization approach for addressing the train dispatching problem, we use arrival time variables a and departure time variables d to describe train movements on block sections. More specifically, $a_{f,s}$ indicates the arrival time of train f at block section s, and $d_{f,s}$ indicates the departure time of train f from block section s. The arrival and departure safety headway time intervals $g_{f,s}$ and $h_{f,s}$ can be either pre-determined as parameters (*6*) or considered as variables (*10*). For determining the section blocking time, the occupancy time of block section s for the arrival of train f is formulated as

$$\sigma_{f,s} = a_{f,s} - g_{f,s}, \quad \forall f \in F, s \in E_f, \tag{1}$$

and the release time of block section s for the departure of train f is formulated as

$$\delta_{f,s} = d_{f,s} + h_{f,s}, \quad \forall f \in F, s \in E_f, \tag{2}$$

where F is the set of trains, E_f is the set of block sections that train f may use, and $\sigma_{f,s}$ and $\delta_{f,s}$ indicate the starting and ending time of blocking section s for train f.

Figure 3 illustrates the movement of train f on block section s by using arrival and departure time variables. More specifically, train f arrives at time $a_{f,s} = 4$ and departs at time $d_{f,s} = 7$. As we have the safety headway times $g_{f,s} = 2$ and $h_{f,s} = 1$, block section s is blocked for train f from time $\sigma_{f,s} = 2$ to time $\delta_{f,s} = 8$.

To generate a conflict-free train dispatching solution, the block section capacity constraint is proposed by avoiding the overlap between any pair of trains on the same block section, formulated as follows:

$$\sigma_{f_2,s} + \left(1 - \theta_{f_1,f_2,s}\right) \cdot M \geq \delta_{f_1,s}, \quad \forall f_1 \in F, f_2 \in F, s \in E_{f_1} \cap E_{f_2} \tag{3}$$

$$\sigma_{f_2,\bar{s}} + \left(1 - \theta_{f_1,f_2,s}\right) \cdot M \geq \delta_{f_1,s}, \quad \forall f_1 \in F, f_2 \in F, s \in E_{f_1}, \bar{s} \in E_{f_2}. \tag{4}$$

where $\theta_{f_1,f_2,s}$ is a binary train order variable, with $\theta_{f_1,f_2,s} = 1$ if train f_2 arrives at block section s or block section $\bar{s}$ after train f_1, and otherwise $\theta_{f_1,f_2,s} = 0$, and M is a sufficiently large positive number. Note that we indicate bi-directional block section on a single-track segment as s and $\bar{s}$, which refer to one physical block section in the opposite direction. Thus, the model can be applied to single-track, double-track, or N-track networks. Interested readers may refer to references (*6*), (*10*), and (*14*) for more details.

Mathematical Formulation of the Signaling Commands

This section formulates the signaling constraints to implement the signaling commands, which are innovative in comparison with previous studies. We first clarify the impact of the signaling commands on the train schedules, as revealed by the train travel times. Then, we formulate signaling constraints by applying signal indicators to represent signaling commands, which dynamically depend on the condition of the block sections ahead, that is, the relative position of the preceding train in the network. The time-instant formulation method is considered for constructing the signaling constraints. Moreover, we consider the three-aspect signaling system in Figure 1*b* as an example to formulate the signaling constraints. For the four-aspect signaling system and other signaling systems that are not detailed in this paper, a similar approach can be followed.

Impact of Signaling Commands on Train Schedules: Additional Train Travel Time

A non-green signal requires a reduction in train speed. To implement the speed reduction, a direct way is to restrict train speed. However, few train dispatching models consider the train speed as variable; as a result, we cannot add restrictions to the train speed. As the reduction in train speed can be reflected by an increased travel time of a train on a block section, we can formulate the reduction in train speed by requiring additional train travel time. Therefore, this section interprets the impact of the signaling commands on train schedules, as revealed by train travel time.

Figure 4 is a three-layer figure that illustrates the possible scenarios of a case that train f traverses two adjacent block sections s_1 and s_2. The sight point and the end point of block sections s_1 and s_2 are labeled as P1,..., P4 respectively, as shown in Layer (a) of Figure 4. A speed-time graph is used to show the relation between the speed

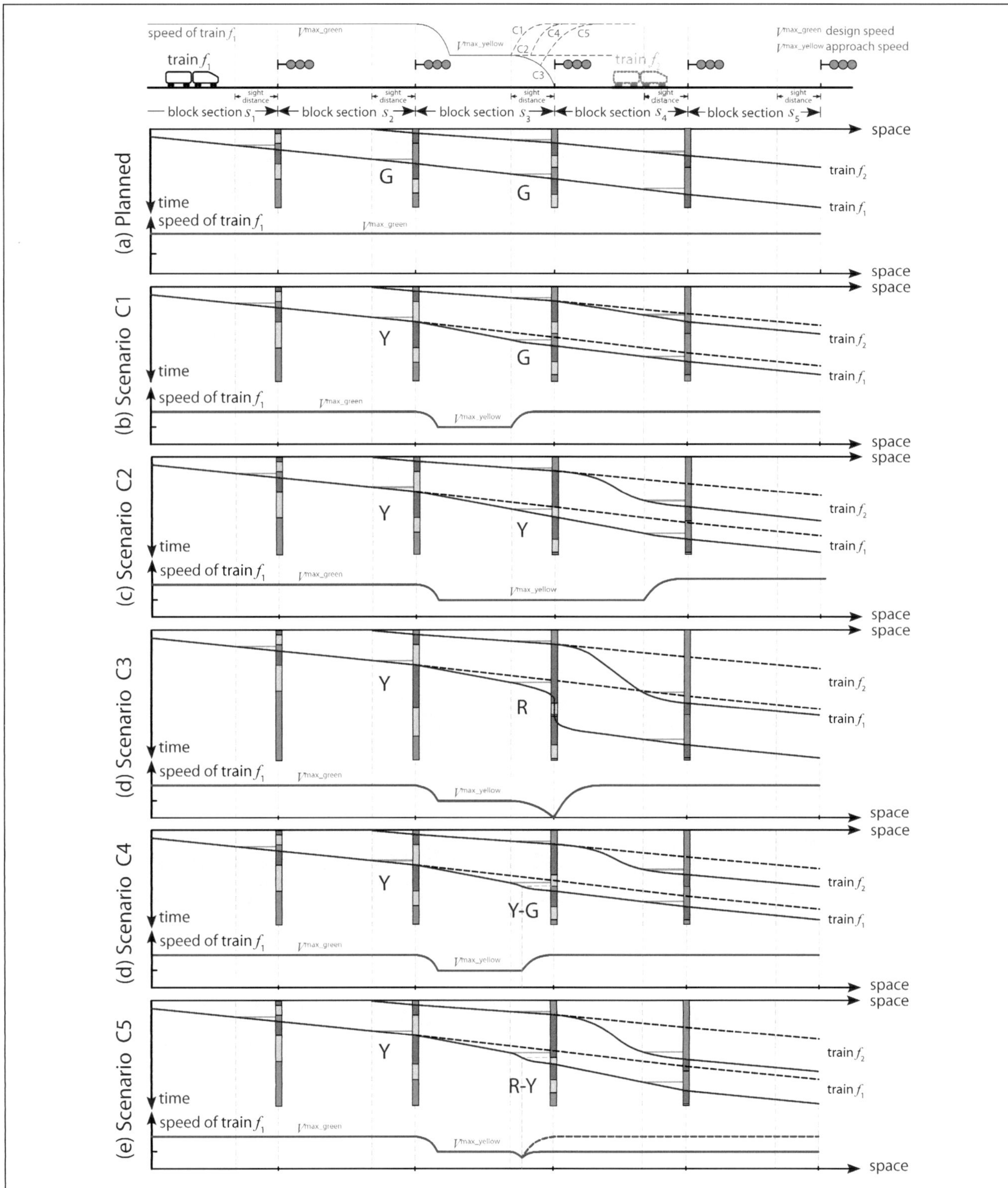

Figure 2. Time-space-speed graphs of the possible scenarios for train operations.

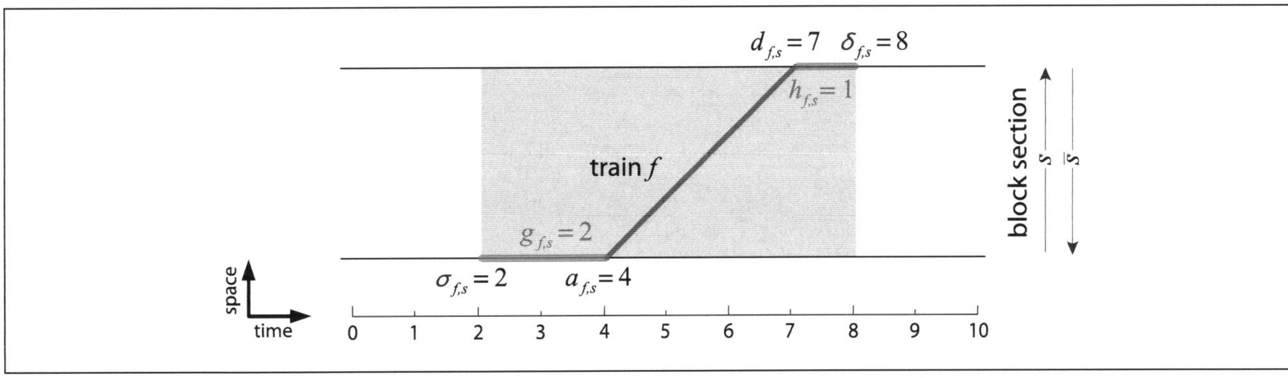

Figure 3. The time-instant optimization method.

reduction and the additional travel time. In Layer (b), seven speed-time graphs for train f from the position P1 to P4 are sketched. A single or double capital letter(s) is used to indicate the signal aspects shown at position P2 and P4 respectively, for example, "GG" means two green aspects. The X-axis indicates the time at which train f arrives at the positions P1,..., P4. An integrated graph of these separate speed-time graphs is provided in Layer (c), in which the effects of the signaling commands (speed reduction) on the train travel times are graphically interpreted. In this layer, the X-axis indicates the time at which train f arrives at a position under a given scenario, for example, the time point labeled "P4_YG" is the arrival time of train f at position P4, if the two signals at positions P2 and P4 show the yellow and green aspects, respectively.

In Layer (b) of Figure 4, the colored area of the graphs (1–6) indicates the distance from position P1 to P4, that is, the distance traveled by a train equals its speed multiplied by the elapsed time; thus these areas should be equal to each other. As sketched in graphs (3) and (7), train f faces the red aspect directly after the green aspect, which should be prevented in the three-aspect signaling system for safety purposes; therefore, these two scenarios will not occur. The other five graphs correspond to the scenarios C1,..., C3 presented in Figure 2. Recall that we make an assumption of considering the worst case; so the possibilities of scenarios C4 and C5 in Figure 2 are neglected. Moreover, we use two binary variables (speed indicators) $\lambda^{\text{red}}_{f,f_2,s}$ and $\lambda^{\text{yellow}}_{f,f_2,s}$ to indicate whether train f_2 results in a red aspect and a yellow aspect for train f on block section s. More specifically, $\lambda^{\text{red}}_{f,f_2,s} = 1$ if train f_2 is occupying the block section one ahead of block section s, which makes train f face the red aspect on block section s, otherwise, $\lambda^{\text{red}}_{f,f_2,s} = 0$. Thus, the scenarios presented in Layer (b) of Figure 4 can be represented by using these speed indicators, as marked.

Let us focus on Layer (c) of Figure 4. Note that $\tau_{f,s}$ indicates the free-flow travel time of train f on block section s under the normal condition (i.e., a green aspect), and $\Delta^{\text{Y}}_{f,s}$, $\Delta^{\text{YG}}_{f,s}$, and $\Delta^{\text{YR}}_{f,s}$ indicate the additional/decreased travel time caused by the train speed reduction (i.e., by the "Approach" and "Stop" signaling commands). If the signal placed at position P2 is green, as in graphs (1) and (2) in Layer (b), no train speed reduction is required. If the signal placed at position P2 is yellow, as in graphs (4) to (6) in Layer (b), then train f reduces its speed by following the yellow curve, and its travel time on block section s_2 increases. The amount of the increased travel time further depends on the display of the signal placed at position P4:

- If the signal at position P4 is green, i.e., graph (4) in Layer (b), then train f is allowed to accelerate by following the green curve, and the additional travel time (compared with the free-flow travel time) in this case is shorter than that in the case of facing a yellow aspect, which is calculated by $\Delta^{\text{Y}}_{f,s_2} - \Delta^{\text{YG}}_{f,s_2}$.
- If the signal at position P4 is yellow, i.e., graph (5) in Layer (b), then train f maintains its speed by following the yellow line, and the additional travel time on block section s_2 is denoted as $\Delta^{\text{Y}}_{f,s_2}$, compared with the free-flow travel time in the normal scenario of facing green aspects. We consider that a second yellow aspect allows the train to pass at the approaching speed.
- In the remaining case that train f faces a red aspect at position P4, corresponding to graph (6) in Layer (b), train f has to decelerate by following the red curve until it stops completely, and the additional travel time is measured by $\Delta^{\text{Y}}_{f,s_2} + \Delta^{\text{YR}}_{f,s_2}$.

Formulations of the Signaling Constraints

This section formulates the signaling constraints, dynamically determining the signal indicators and implementing the reduction in train speed by requiring additional

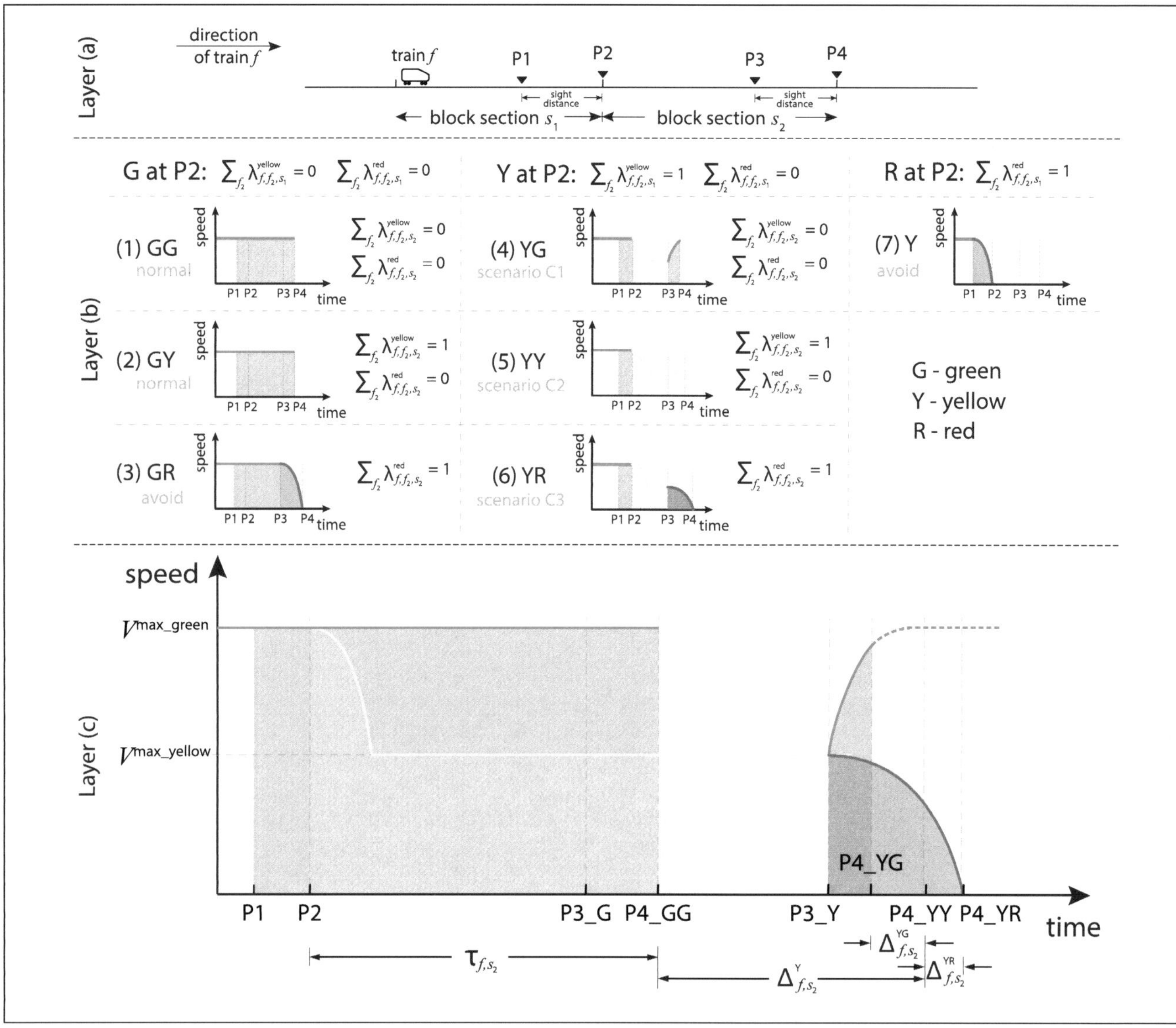

Figure 4. Speed-time graphs to illustrate the relation between the speed reduction and the additional travel time.

travel time. As discussed, for each train on each block section, the signaling command received by the driver depends on the movement of the preceding train in the network. Therefore, we need to identify the signaling commands by determining the condition (i.e., being occupied or released) of the block section(s) ahead.

Figure 5 illustrates how to identify the "Stop" signaling command; for the "Approach" command, a similar approach can be followed. In Figure 5*a*, train f_1 meets a red aspect in block section s_1 at the sight distance, because train f_2 is occupying block section s_2 at that moment. In such cases, the blocking time of train f_1 on block section s_1 and the blocking time of train f_2 on block section s_2 are "overlapping" in the sight time as shown in the timetable; we then have $\lambda^{\text{red}}_{f_1,f_2,s_1} = 1$. For other cases shown in Figure 5*b*, train f_1 faces a non-red aspect in block section s_1, and no "overlap" happens in the sight time between the blocking time of train f_1 and train f_2. As a result, we have $\lambda^{\text{red}}_{f_1,f_2,s_1} = 0$.

To determine the red signal indicator $\lambda^{\text{red}}_{f_1,f_2,s}$, let us consider the binary variables $\gamma^{\text{r1}}_{f_1,f_2,s}$ and $\gamma^{\text{r2}}_{f_1,f_2,s}$ to satisfy the conditions

$$\left[d_{f_1,s_1} \leq \sigma_{f_2,s_2}\right] \Leftrightarrow \left[\gamma^{\text{r1}}_{f_1,f_2,s_1} = 1\right],$$

$$\forall f_1 \in F, f_2 \in F, f_1 \neq f_2, s_1 \in E_{f_1}, s_2 \in E_{f_2} \cap E^{\text{adj2}}_{s_1}, \quad (5)$$

$$\left[\delta_{f_2,s_2} \leq d_{f_1,s_1} - \iota^{\text{sight}}_{f_1,s_1}\right] \Leftrightarrow \left[\gamma^{\text{r2}}_{f_1,f_2,s_1} = 1\right],$$

$$\forall f_1 \in F, f_2 \in F, f_1 \neq f_2, s_1 \in E_{f_1}, s_2 \in E_{f_2} \cap E^{\text{adj2}}_{s_1} \quad (6)$$

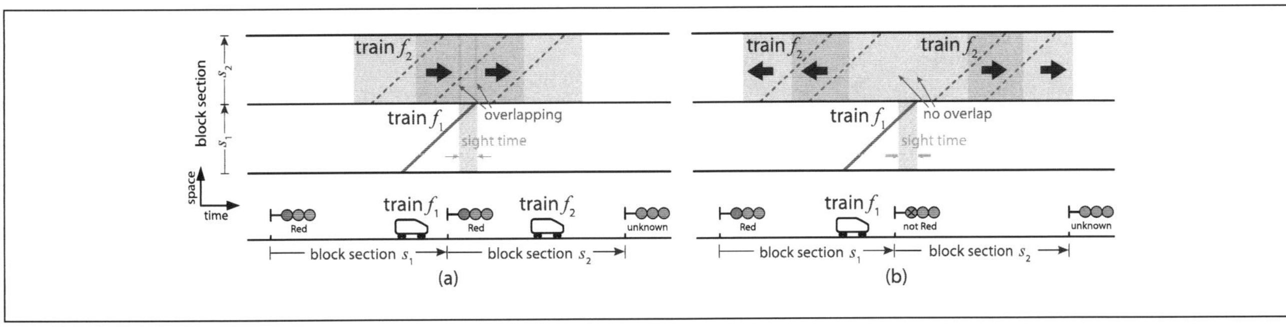

Figure 5. Identifying the "Stop" signaling command.

if train f_1 and train f_2 run in the same direction, or

$$[d_{f_1,s_1} \le \sigma_{f_2,\bar{s}_2}] \Leftrightarrow \left[\gamma^{\mathrm{r1}}_{f_1,f_2,s_1} = 1\right],$$
$$\forall f_1 \in F, f_2 \in F, f_1 \ne f_2, s_1 \in E_{f_1}, \bar{s}_2 \in E_{f_2} \cap E^{\mathrm{adj2}}_{s_1}, \quad (7)$$

$$\left[\delta_{f_2,\bar{s}_2} \le d_{f_1,s_1} - \iota^{\mathrm{sight}}_{f_1,s_1}\right] \Leftrightarrow \left[\gamma^{\mathrm{r2}}_{f_1,f_2,s_1} = 1\right],$$
$$\forall f_1 \in F, f_2 \in F, f_1 \ne f_2, s_1 \in E_{f_1}, \bar{s}_2 \in E_{f_2} \cap E^{\mathrm{adj2}}_{s_1} \quad (8)$$

if train f_1 and train f_2 run in opposite directions. Note that $\iota^{\mathrm{sight}}_{f_1,s_1}$ indicates the running time of train f_1 over the sight distance on block section s_1, and $E^{\mathrm{adj2}}_{s_1}$ indicates the set of the adjacent block sections of block section s_1. As a result, we have $\gamma^{\mathrm{r1}}_{f_1,f_2,s_1} = 0$ and $\gamma^{\mathrm{r2}}_{f_1,f_2,s_1} = 0$ for the case in Figure 5*a*, and we have either $\gamma^{\mathrm{r1}}_{f_1,f_2,s_1} = 1$ or $\gamma^{\mathrm{r2}}_{f_1,f_2,s_1} = 1$ for the case in Figure 5*b*. Then, the red signal indicator $\lambda^{\mathrm{red}}_{f_1,f_2,s}$ can be formulated as

$$\lambda^{\mathrm{red}}_{f_1,f_2,s} = (1 - \gamma^{\mathrm{r1}}_{f_1,f_2,s}) \cdot (1 - \gamma^{\mathrm{r2}}_{f_1,f_2,s}),$$
$$\forall f_1 \in F, f_2 \in F, f_1 \ne f_2, s \in E_{f_1}. \quad (9)$$

By applying the transformation properties proposed in (*29*), the if-then constraints Equations 5 to 8 and the nonlinear constraint (Equation 9) can be reformulated as linear inequalities. Moreover, formulations similar to Equations 5 to 9 can also be constructed for the yellow signal indicator $\lambda^{\mathrm{yellow}}_{f_1,f_2,s}$. For the sake of compactness, we do not provide those details here.

According to the above analysis of the additional train travel time, which reflects the train speed reduction required by the signaling commands, the travel time constraint is given as follows:

$$d_{f,s_2} - a_{f,s_2} \ge w^{\min}_{f,s_2} + \tau_{f,s_2} + \Delta^{\mathrm{Y}}_{f,s_2} \cdot \sum_{f_2 \in F} \lambda^{\mathrm{yellow}}_{f,f_2,s_1} + \Delta^{\mathrm{YR}}_{f,s_2} \cdot \sum_{f_2 \in F} \lambda^{\mathrm{red}}_{f,f_2,s_2} -$$
$$\Delta^{\mathrm{YG}}_{f,s_2} \cdot (1 - \sum_{f_2 \in F} \lambda^{\mathrm{red}}_{f,f_2,s_2}) \cdot (1 - \sum_{f_2 \in F} \lambda^{\mathrm{yellow}}_{f,f_2,s_2}),$$
$$\forall f \in F, s_1 \in E_f, s_2 \in E_f \cap E^{\mathrm{adj2}}_{s_1}, \quad (10)$$

where $w^{\min}_{f,s_2}$ indicates the minimum dwell time of train f on block section s_2. Equation 10 ensures that the actual travel time of train f on block section s is not less than the sum of the minimum dwell time, the free-flow travel time, and the additional travel time caused by the non-"Clear" signaling commands.

Moreover, the GW result would not violate the speed limits that result from the signaling commands, and the GW policy is proven to be effective in saving energy consumption and dealing with delays (*30*). We can employ the GW policy by setting all signal indicators to zero, as follows:

$$\sum_{f \in F} \sum_{s \in E_f} \sum_{f_2 \in F: f_2 \ne f} \left(\lambda^{\mathrm{yellow}}_{f,f_2,s} + \lambda^{\mathrm{red}}_{f,f_2,s}\right) = 0. \quad (11)$$

As a result, all trains will always only encounter green aspects when traversing the network.

As the signaling constraints proposed here can be transformed into linear inequalities, we can still use the solution approach of the original models to solve the dispatching problem that integrates the signaling commands. If we consider the train dispatching model proposed in (*6*), which is an MILP model, a standard MILP solver can be used. For more details, interested readers may refer to the description in reference (*6*).

Numerical Experiments

We conducted the experiments on a rail network with two main tracks, composed of 48 nodes and 53 block sections, as depicted in Figure 6. The two tracks in different directions are independent, so only one direction is considered. We consider 1.5 hours of traffic with 20 trains in three train categories: eight intercity, eight sprinter, and four freight trains. Sprinter trains stop at all stations; intercity and freight trains stop only at the origin and destination stations. We consider 20 delay cases of the primary delays following a three-parameter Weibull distribution, as investigated in (*31*). We adopt the CPLEX optimization studio 12.6.3 with default settings to solve MILP problems. The following experiments are

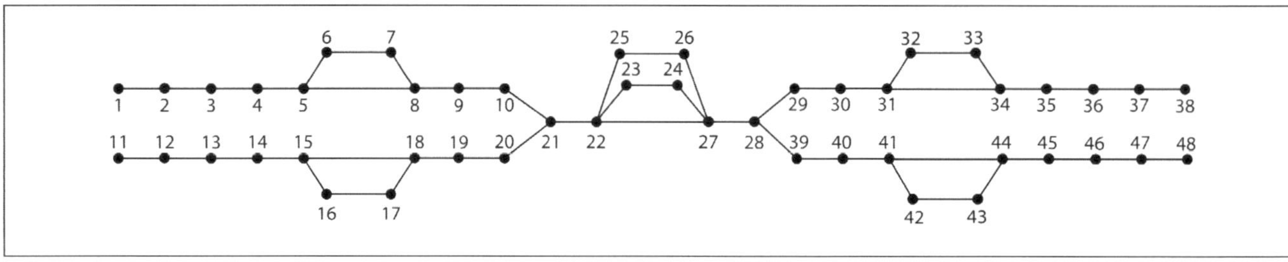

Figure 6. A rail network.

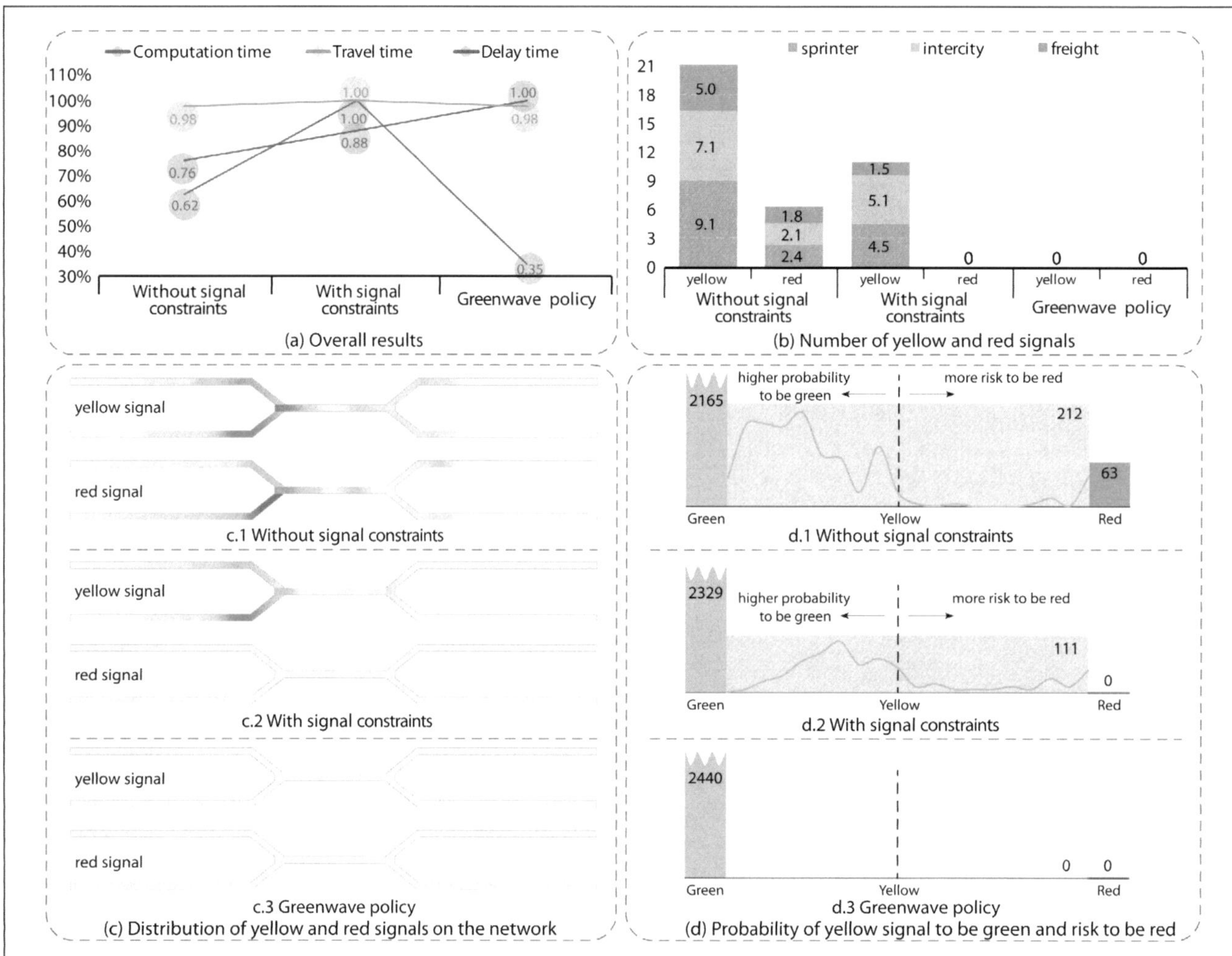

Figure 7. Illustration of the experimental results.

performed on a computer with an Intel®Core™ i7 @ 2.00 GHz processor and 16 GB RAM.

In Figure 7*a* and *b*, the average results of the 10 delay cases are given, in terms of the train delay times, the train travel times, and the number of the red and yellow signals faced by the trains. The results in Figure 7*a* are normalized in the range 0 to 1. As shown, when integrating the signaling constraints, the total delay time and the total travel time increase by 16% and 2% respectively, and the number of yellow aspects that the trains meet decreases by 47%. As the speed limits required by the signaling commands are activated, the model optimizes the train departure and arrival times by avoiding trains facing non-green aspects (as much as possible). However, some trains still face yellow aspects on some block sections, that is, 11 times on average. In such cases, the speed reduction (i.e., the additional travel time) is enforced in our model; therefore, the train travel time becomes larger, which further results in the increased total delay time and travel time. *Considering the signaling constraints enables us to avoid the full stop caused by the red signal, as the times that the trains confront the red signal are reduced to zero.* By applying the GW policy, where the trains can always proceed under green signals, the total delay time further increases by 13%, compared with the results with the signaling constraints; however, the total travel time decreases to become similar to the results without the signaling constraints. The computation time for obtaining the results with consideration of the signaling constraints is longer, but still within 300 seconds.

We further explore more details regarding the signals. In Figure 7*c* and *d*, we present the results of all the 10 delay cases by summing up the number of the red/yellow signals that are faced by the trains. Figure 7*c* illustrates the distribution of the red and yellow signals on the network, of which the layout originates from Figure 6. A darker color implies that a larger number of red/yellow signals occurred at the position. As shown, the red and yellow signals are mostly faced by trains before entering the merging area of the two lines and also in the merging area. This results from the increased train interactions in these areas. Figure 7*d* presents the probability of the yellow signals becoming green and the risk of being red, which is obtained by analyzing the overlapping time of the two blocking times, as interpreted in Figure 5. The bars indicate the numbers of the green, yellow, and red signals, and the exact values are labeled. The Y-axis of the yellow line (within the yellow bar) indicates the number of the yellow signals, and the X-axis is the probability of becoming green or red. The black dashed line is a benchmark line, indicating that the probability to become green equals the risk to be red. When neglecting the signaling constraints, lots yellow signals occurred have big chance to become green. By considering the signaling constraints, the total number of the yellow signals decreases from 212 to 111, and the number of the yellow signals that have larger probability to become green is reduced. *Considering the signaling constraints enables us to reduce the number of the yellow signals faced by the trains, i.e., taking the chance of letting the yellow signals become green as much as possible.*

We can conclude that the consideration of the signaling commands leads to larger delay times and larger travel times, but increases the realism of the dispatching solution obtained and the safety of the train services. The GW policy hardly affects the train travel time, but results in a significant increase of train delays.

Conclusions

This paper integrates dynamic signaling commands under the fixed-block signaling system into train dispatching optimization problems. We have investigated the impact of the signaling commands on train schedules, and we have implemented this impact in our model by presenting the signaling constraints. In these constraints, the signaling commands (represented by the signal indicators) are dynamically determined by the train movements, and the train speed limits are activated according to the signal indicators. In our experimental results, consideration of the signaling commands results in larger delay times (16%) and longer travel times, but reduces the number of the non-green signals that are met and increases the realism of the train dispatching solution and the safety of the train services, as the real operational requirements (speed limits) caused by the signaling commands are included. In fact, the 16% gap can be thought of as an approximation error of currently available fixed speed dispatching models neglecting signaling aspects. Moreover, we consider the Greenwave (GW) policy to dispatch trains, which results in longer train delay times; however, the total train travel time does not change much, and it is similar to the result that neglects the signaling impact.

Our future research will focus on the following main extensions. First, as the proposed approach for constructing the signaling constraints can be extended to other signaling systems, we will study different signaling systems and compare their performance, in terms of delay, energy, capacity, robustness, etc., from the planning and operational perspectives. Second, when applying the GW policy, the energy consumption is expected to be reduced. We will evaluate the energy consumption by applying the model we proposed in reference (*10*), instead of that in reference (*6*). Finally, in order to extend the applicability of the model to large-scale networks, we will focus on developing heuristic algorithms and distributed optimization methods for improving the computational efficiency of the model.

Acknowledgments

This work is supported by China Scholarship Council under Grant 201507090058.

Author Contributions

The authors confirm contribution to the paper as follows: study conception and design: Xiaojie Luan, Bart De Schutter, Francesco Corman, and Gabriel Lodewijks; data collection: Xiaojie Luan, Francesco Corman; analysis and interpretation of results: Xiaojie Luan, Bart De Schutter, and Francesco Corman; draft manuscript preparation: Xiaojie Luan, Bart De Schutter, Francesco Corman, and Gabriel Lodewijks. All authors reviewed the results and approved the final version of the manuscript.

References

1. Corman, F., A. D'Ariano, I. A. Hansen, and D. Pacciarelli. Optimal Multi-Class Rescheduling of Railway Traffic. *Journal of Rail Transport Planning & Management*, Vol. 1, No. 1, 2011, pp. 14–24.
2. Dollevoet, T., F. Corman, A. D'Ariano, and D. Huisman. An Iterative Optimization Framework for Delay Management and Train Scheduling. *Flexible Services and Manufacturing Journal*, Vol. 26, No. 4, 2014, pp. 490–515.
3. Xu, P., F. Corman, Q. Peng, and X. Luan. A Train Rescheduling Model Integrating Speed Management During Disruptions of High-Speed Traffic Under a Quasi-Moving Block System. *Transportation Research Part B: Methodological*, Vol. 104, 2017, pp. 638–666.
4. Luan, X., J. Miao, L. Meng, F. Corman, and G. Lodewijks. Integrated Optimization on Train Scheduling and Preventive Maintenance Time Slots Planning. *Transportation Research Part C: Emerging Technologies*, Vol. 80, 2017, pp. 329–359.
5. Cacchiani, V., D. Huisman, M. Kidd, L. Kroon, P. Toth, L. Veelenturf, and J. Wagenaar. An Overview of Recovery Models and Algorithms for Real-Time Railway Rescheduling. *Transportation Research Part B: Methodological*, Vol. 63, 2014, pp. 15–37.
6. Luan, X., F. Corman, and L. Meng. Non-Discriminatory Train Dispatching in a Rail Transport Market with Multiple Competing and Collaborative Train Operating Companies. *Transportation Research Part C: Emerging Technologies*, Vol. 80, 2017, pp. 148–174.
7. Corman, F., and L. Meng. A Review of Online Dynamic Models and Algorithms for Railway Traffic Management. *IEEE Transactions on Intelligent Transportation Systems*, Vol. 16, No. 3, 2015, pp. 1274–1284.
8. Fang, W., S. Yang, and X. Yao. A Survey on Problem Models and Solution Approaches to Rescheduling in Railway Networks. *IEEE Transactions on Intelligent Transportation Systems*, Vol. 16, No. 6, 2015, pp. 2997–3016.
9. D'Ariano, A., D. Pacciarelli, and M. Pranzo. A Branch and Bound Algorithm for Scheduling Trains in a Railway Network. *European Journal of Operational Research*, Vol. 183, No. 2, 2007, pp. 643–657.
10. Luan, X., Y. Wang, B. De Schutter, L. Meng, G. Lodewijks, and F. Corman. Integration of Real-Time Traffic Management and Train Control for Rail Networks-Part 1: Optimization Problems and Solution Approaches. *Transportation Research Part B: Methodological*, Vol. 115, 2018, 41–71.
11. Corman, F., A. D'Ariano, D. Pacciarelli, and M. Pranzo. A Tabu Search Algorithm for Rerouting Trains During Rail Operations. *Transportation Research Part B: Methodological*, Vol. 44, No. 1, 2010, pp. 175–192.
12. Corman, F., A. D'Ariano, D. Pacciarelli, and M. Pranzo. Bi-Objective Conflict Detection and Resolution in Railway Traffic Management. *Transportation Research Part C: Emerging Technologies*, Vol. 20, No. 1, 2012, pp. 79–94.
13. Törnquist, K. J. Design of an Effective Algorithm for Fast Response to the Re-Scheduling of Railway Traffic During Disturbances. *Transportation Research Part C: Emerging Technologies*, Vol. 20, No. 1, 2012, pp. 62–78.
14. Meng, L., and X. Zhou. Simultaneous Train Rerouting and Rescheduling on an N-Track Network: A Model Reformulation with Network-Based Cumulative Flow Variables. *Transportation Research Part B: Methodological*, Vol. 67, 2014, pp. 208–234.
15. Samà, M., A. D'Ariano, F. Corman, and D. Pacciarelli. A Variable Neighbourhood Search for Fast Train Scheduling and Routing During Disturbed Railway Traffic Situations. *Computers & Operations Research*, Vol. 78, 2017, pp. 480–499.
16. Kecman, P., F. Corman, A. D'Ariano, and R. M. Goverde. Rescheduling Models for Railway Traffic Management in Large-Scale Networks. *Public Transport*, Vol. 5, No. 1–2, 2013, pp. 95–123.
17. Schöbel, A. *Integer Programming Approaches for Solving the Delay Management Problem*. Heidelberg, HR: Springer, 2007.
18. Ginkel, A., and A. Schöbel. To Wait or Not to Wait? The Bicriteria Delay Management Problem in Public Transportation. *Transportation Science*, Vol. 41, No. 4, 2007, pp. 527–538.
19. Schachtebeck, M., and A. Schöbel. To Wait or Not to Wait—and Who Goes First? Delay Management with Priority Decisions. *Transportation Science*, Vol. 44, No. 3, 2010, pp. 307–321.
20. Pellegrini, P., G. Marlière, R. Pesenti, and J. Rodriguez. RECIFE-MILP: An Effective MILP-Based Heuristic for the Real-Time Railway Traffic Management Problem. *IEEE Transactions on Intelligent Transportation Systems*, Vol. 16, No. 5, 2015, pp. 2609–2619.
21. Pellegrini, P., G. Marlière, and J. Rodriguez. Optimal Train Routing and Scheduling for Managing Traffic Perturbations in Complex Junctions. *Transportation Research Part B: Methodological*, Vol. 59, 2014, pp. 58–80.
22. Samà, M., P. Pellegrini, A. D'Ariano, J. Rodriguez, and D. Pacciarelli. Ant Colony Optimization for the Real-Time Train Routing Selection Problem. *Transportation Research Part B: Methodological*, Vol. 85, 2016, pp. 89–108.
23. Mazzarello, M., and E. Ottaviani. A Traffic Management System for Real-Time Traffic Optimisation in Railways. *Transportation Research Part B: Methodological*, Vol. 41, No. 2, 2007, pp. 246–274.
24. D'Ariano, A., F. Corman, D. Pacciarelli, and M. Pranzo. Reordering and Local Rerouting Strategies to Manage Train Traffic in Real Time. *Transportation Science*, Vol. 42, No. 4, 2008, pp. 405–419.
25. Lamorgese, L., and C. Mannino. An Exact Decomposition Approach for the Real-Time Train Dispatching Problem. *Operations Research*, Vol. 63, No. 1, 2015, pp. 48–64.
26. Theeg, G. *Railway Signalling & Interlocking: International Compendium*. Eurailpress, Hamburg, Germany, 2009.
27. Hansen, I. A., and J. Pachl. *Railway Timetabling & Operations: Analysis, Modelling, Optimisation, Simulation, Performance Evaluation*. Eurailpress, Hamburg, Germany, 2014.
28. The Railway Technical Website. http://www.railway-technical.com/, 2017.
29. Williams, H. P. *Model Building in Mathematical Programming*. Chichester, England: John Wiley & Sons, 2013.
30. Corman, F., A. D'Ariano, D. Pacciarelli, and M. Pranzo. Evaluation of Green Wave Policy in Real-Time Railway

Traffic Management. *Transportation Research Part C: Emerging Technologies*, Vol. 17, No. 6, 2009, pp. 607–616.

31. Corman, F., A. D'Ariano, M. Pranzo, and I. A. Hansen. Effectiveness of Dynamic Reordering and Rerouting of Trains in a Complicated and Densely Occupied Station Area. *Transportation Planning and Technology*, Vol. 34, No. 4, 2011, pp. 341–362.

The Standing Committee on Rail Transit Systems (AP065) peer-reviewed this paper (18-01197).

Article

TRR
JOURNAL OF THE TRANSPORTATION RESEARCH BOARD

Transportation Research Record
2018, Vol. 2672(8) 288–296

DOI: 10.1177/0361198118794547
journals.sagepub.com/home/trr
SAGE

Urban Metro Rail Demand: Evidence from Dynamic Generalized Method of Moments Estimates using Panel Data

Shane Canavan[1], Daniel J. Graham[1], Richard J. Anderson[1], and Alexander Barron[1]

Abstract

This paper estimates elasticities of demand for metro services with respect to fares, income, quality of service, population and network length. Data for 32 world metro systems covering the period from 1996 to 2015 are analyzed within a dynamic panel data specification. Three key contributions are made. First, we collate a database for estimation that is more extensive than that used in previous studies. Second, the quality of the data we have available allows us to more accurately represent quality of service than has been possible previously. Lastly, we estimate and compare two different measures of demand. Our analysis finds a statistically significant negative fare elasticity of −0.25 in the long run for a passenger km specified model and −0.4 in the long run for a passenger journey specified model, and a positive long run income elasticity of 0.17 and 0.18 for the passenger km and passenger journey models respectively. Regarding quality of service, we find positive long run elasticities of 0.56 and 0.47 for the passenger km and passenger journey models respectively. Income levels, population, and the size of the network are also found to be statistically significant and positive in nature. The results suggest passenger km and passenger journeys will increase more in response to changes in service (here represented by increased capacity) than to changes in fares, with the difference in elasticities of service and fares being more pronounced for passenger km.

Introduction

Urban metro rail systems, hereafter referred to as metros, provide one of the most successful solutions to transportation challenges in large metropolitan areas. To plan metro services effectively and achieve service performance targets, it is necessary to understand demand at a fundamental level and how it responds to changes in operational and economic characteristics. In this paper we estimate the effects of fares, income, quality of service, population, and the scale of the system on metro demand.

The literature argues that the factors affecting the demand for public transport are well established, and include fares, income levels, a measure of the service provided, car ownership levels as well as the price of petrol. However, variation in elasticities is considerable, and much of the existing evidence is based on older data sources and econometric methods. In this paper we make three key contributions to the existing literature. First, we collate a database for estimation that is more extensive in terms of the number of observations than that used in previous studies, and which contains a larger number of systems over a longer period of time. Second, the quality of the data we have available allows us to more accurately represent various dimensions of quality of service than has been possible previously. Lastly, we estimate and compare two different measures of demand. The study makes use of conventional demand economics, and it is hoped that the additional evidence provided may enhance results provided by traditional transport models such as the four-step model.

The structure of the paper, consisting of six sections, is as follows. Following this introductory section, we provide a review of the literature. We then describe the advantages of using panel data along with the technical challenges that need to be addressed. From this, the dynamic panel Generalized Method of Moments (GMM) approach is presented as an appropriate approach to overcome these challenges and its application is described. We then explain the data that have

[1]Railway and Transport Strategy Centre, Centre for Transport Studies, Department of Civil and Environmental Engineering, Imperial College London, London, UK

Corresponding Author:
Address correspondence to Shane Canavan: shane.canavan@imperial.ac.uk

been collated for estimation along with hypotheses based on findings from the literature. The penultimate section presents our main results. The final section draws together the main findings of the research and suggests some policy implications and suggestions for future work.

Literature Review

The study benefits from several comprehensive and high-quality reviews and guides in the literature that have already been conducted. Perhaps the most notable of these are the two companion papers of Goodwin (*1*) and Oum et al. (*2*). The former focused on the empirical estimation of demand elasticities while the latter also deals with theoretical and empirical issues in estimating transport demand elasticities.

Starting with Goodwin (*1*), which provides a review of demand elasticities with special reference to short and long run effects of price changes, the survey finds that for metros the fare elasticities vary significantly with a clear pattern for long-term elasticities to be between 50% higher and three times higher than the short term for all modes. However, as Greening et al. (*3*) point out, caution should be exercised as the studies reviewed were predominantly carried out using old aggregated data and that inferring meaning from the results may be misleading. Furthermore, the amount of metro evidence found by the study was sparse in comparison to bus and rail evidence provided. The issue of variance of estimates was further emphasized by Oum et al. (*2*), which examined over 60 studies that reported estimates of price elasticities of transport demand. The survey found demand elasticities for urban transit with regard to price vary significantly, between −0.01 and −0.78 with most values ranging between −0.1 and −0.6. Further evidence of variability amongst demand elasticities is found in Luk and Hepburn (*4*) who reviewed Australian travel demand elasticities.

In addition to the reviews that have been carried out, the Transport Research Laboratory (TRL) has produced a guidance manual, TRL (*5*), along with a corresponding academic paper, Paulley et al. (*6*). The manual is targeted at public transport operators, planning authorities, as well as academics and other researchers. Included in the manual are factors that have been identified as influencing the demand for public transport, as well as quantitative indicators for how these factors influence demand. The high degree of variation in elasticities highlighted by the reviews are also emphasized in this guidance manual. More recently, the meta-analysis provided by Holmgren (*7*) investigates this issue further and provides an in-depth discussion of some of the underlying reasons. In summary, variation in estimates is viewed as being dependent on the quality and characteristics of the data used, along with the modeling techniques applied. Namely, the structures of the models, whether they are static or dynamic, the estimators used, as well as the specification of the variables. The study also provides recommendations of the variables that demand models should include, which are broadly similar to the TRL guidance.

A summary table of the key literature is presented below in Table 1.

Methodology

As explained above, our study analyzes a panel dataset consisting of demand variables for metros, as well as a set of explanatory factors that help to explain changes in the levels of demand. The advantages of using a panel dataset, which comprises both cross-sectional (between) and time-series (within) variation, include increased

Table 1. Summary Table of Key Literature

Author(s)	Title	Comment
Goodwin (*1*)	A review of new demand elasticities with special reference to short and long run effects of price changes	Review of empirical estimation of demand elasticities
Oum et al. (2)	Concepts of price elasticities of transport demand and recent empirical estimates: an interpretative survey	Companion paper to Goodwin (*1*) dealing with theoretical and empirical issues in estimating transport demand
Greening et al. (*3*)	Energy efficiency and consumption — the rebound effect — a survey	Survey of empirical evidence from the US
Luk and Hepburn (*4*)	New review of Australian travel demand elasticities	Report reviewing and summarizing information empirically determined in Australia
TRL (*5*)	The demand for public transport: a practical guide	Transport Research Laboratory (TRL) guidance manual
Paulley et al. (*6*)	The demand for public transport: The effects of fares, quality of service, income and car ownership	Corresponding academic paper to TRL's guidance manual
Holmgren (*7*)	Meta-analysis of public transport demand	Meta-analysis

precision in estimation owing to more data and variability (8, 9). Importantly, panel data allow for estimation of models that account for unobserved time-invariant heterogeneity which may be correlated with the error term, and thus for differences between metros that we know to exist but cannot be measured. In addition, panel data allow us to investigate dynamics and adjustments in behavior.

Our underlying dynamic panel model is

$$D_{it} = \gamma X'_{it} + \beta D_{i,t-1} + u_{it} \text{ where } u_{it} = \mu_i + v_{it} \quad (1)$$

where demand, represented by D for the ith metro ($i = 1,\ldots,N$) in year t of a panel with T years ($t = 1,\ldots,T$), is explained by a set of explanatory variables (X'_{it}) as well as the lagged value of D itself (the dynamic component). In this case, γ is the set of coefficients for each corresponding explanatory variable included in the econometric model, while β represents the coefficient for the lagged variable D. All variables are in natural logarithms, therefore, the γ coefficients represent the corresponding demand elasticity for a particular explanatory variable.

The error component, u_{it}, is composed of an unobserved time-invariant individual-specific effect μ_i and a random disturbance v_{it}. The former is interpreted as the metro-specific effect not included in the regression which accounts for differences known to exist between the metros that have not been measured.

A rudimentary first approach includes that of the Pooled Ordinary Least Squares (OLS). This approach specifies constant coefficients and applies the usual OLS Gauss–Markov assumptions used for cross-sectional analysis. This is an appropriate approach when μ_i is thought to be constant across systems (i.e., $\mu_i = \mu$). Pooled OLS is carried out by stacking the data over i and t and estimating using OLS. Effects unique to a metro are then subsumed in the error term.

Two main approaches to the treatment of time-invariant individual-specific effects (μ_i) include Fixed Effects (FE) and Random Effects (RE) models. When the metro-specific effects are correlated with the regressors, a FE model should be used, while when they are not correlated, a RE model can be used with the μ_i assumed independent and identically distributed $\mu_i \sim i.i.d.\left(0, \sigma_\mu^2\right)$, and uncorrelated with the predictor variables and the random disturbance $\varepsilon_i \sim i.i.d.\left(o, \sigma_v^2\right)$.

The assumption that regressors are uncorrelated with the errors in the model is hard to define in the case of a dynamic model because typically $Cov(D, u) \neq 0$.

In this context the Generalized Method of Moments (GMM) estimator, stemming from Hansen (10), offers a possible solution. GMM allows for estimation of model parameters in the presence of endogenous variables by deriving instruments from the nature of the panel data itself. This is done by estimating parameters based on the information in population moment conditions, which refers to the expected values of powers of a variable with population average under consideration. Two GMM approaches are available, these are commonly known as the difference and the system GMM (11).

The difference GMM estimator, as proposed by Arellano and Bond (11), is used to estimate Equation 1 based on the following orthogonality conditions:

$$E(D_{it-s}\Delta u_{it}) = 0 \text{ for } t = 3, \ldots, T \text{ and } 2 \leq s \leq T - 1 \quad (2)$$

where D_{it-s} are suitable lags of the dependent variable. The approach uses the time series nature of the data itself as an instrument for the residual of Equation 1 in differences. System GMM is an augmented version and is outlined by Arellano and Bover (12) and fully developed by Blundell and Bond (13), which uses lagged first differences as instruments for equations in levels as well as the lag variable instruments for first-difference equations.

In the presence of data with little variation over time, the difference GMM has been shown to suffer from weak instrument bias (13). In this context, the system GMM (12, 13) is preferable to the difference GMM and proceeds by estimating regressions in differences and in levels simultaneously.

The consistency of the dynamic GMM estimator depends on two crucial assumptions: no first-order serial-autocorrelation in the error term of the levels equation, and instrument exogeneity. To evaluate these assumptions, we consider the Arellano and Bond autocorrelation tests and the Sargen/Hansen tests. The Arellano and Bond (14) test tests for serial autocorrelation and evaluates the hypothesis that there is no second-order serial correlation in the first differenced residuals, which in turn implies that the errors from the levels equations are serially uncorrelated (15). If there is serial correlation, then the instruments used are not valid and there is need for another set of instruments starting with deeper time lags. The Sargan/Hansen test of over-identifying restrictions is the standard test for the validity of the instrument (15).

The individual parameters of these models are difficult to interpret when considered alone; as a result, we compute long run elasticity estimates related to respective results. Considering Equation 1 above, the long run coefficient is calculated as follows:

$$\gamma_{LR} = \frac{(\gamma_1 + \gamma_2 + \ldots + \gamma_{m)}}{(1 - (\beta_1 + \beta_2 + \ldots + \beta_p))} \quad (3)$$

where $\gamma_{\rm p}$ and $\beta_{\rm p}$ are the coefficients of variables described in Equation 1 lagged mth periods.

Data

In this section we describe the variables that were tested along with their respective sources and hypotheses. In general, the majority of the data originate from two consortia of metro system operators, namely the Community of Metros (CoMET) and the Nova Group of Metros (Nova) (https://cometandnova.org/), managed by the Railway and Transport Strategy Centre (RTSC) based at Imperial College London. The consortia focus on benchmarking, and the work developed by the RTSC over the last two decades is supported by a comprehensive database of key performance indicators related to different areas of metro operation. Accordingly, the data allow us to compile an unbalanced panel dataset consisting of 32 metro systems over 20 years between 1996 and 2015.

The CoMET metros included in this study were: Beijing BMTROC, Berlin BVG, Delhi DMRC, Guangzhou GMC, Hong Kong MTR, London Underground, Metro de Madrid, Metro de Santiago, Metro Sao Paulo Mexico City STC, Moscow Metro, New York NYCT, Paris RATP (Metro and RER, two systems), Shanghai SSMG, Singapore SMRT and Taipei TRTC. Nova metros included in this study: Bangkok BEM, Barcelona TMB, Brussels STIB, Buenos Aires Metrovias, Istanbul Ulasim, Kuala Lumpur RapidKL, London DLR, Metro Rio, Montreal STM, Nanjing Metro, Newcastle Nexus, Oslo Sporveien, Shenzhen SZMC, Sydney Trains, and Toronto TTC.

Our analysis is supported by a high-standard level of data quality, as during the years of benchmarking work carried out within the groups, a series of systematic data cleaning processes have been developed and maintained which carry out verification and validation tests. This is complemented by ongoing contact with CoMET and Nova members through phone calls, face-to-face meetings, and site visits. It is also worth noting here that owing to the sensitive commercial nature of the RTSC data, an existing confidentiality agreement requires results to be presented in an anonymized form.

For the purpose of this specific study, we use data for demand, as well as a set of explanatory variables which have been identified in the literature as having an influence on the levels of demand for metros.

For demand, two metro patronage measures are considered. These are passenger journeys and passenger km and are sourced from the CoMET and Nova databases. Passenger journeys refers to the annual total of all passengers recorded or estimated as entering the system at the origin station to exiting at the destination, irrespective of the number of different lines used or the number of interchanges between lines. The measure includes all journeys irrespective of the fare paid. This includes fare evaders and passengers travelling on reduced fares such as students, the elderly, and children of all ages, but does not include passengers travelling on other modes on behalf of the metro such as replacement bus services. For passenger km, this measure refers to the distance travelled by each passenger on metro trains, summed for all passengers. The inclusions and exclusions are consistent with the passenger journeys measure.

A proxy for metro fares is estimated as annual revenue from fares divided by the annual number of passengers. Again, sourced from the CoMET and Nova databases. It is worth noting that this variable represents an average fare per journey, and excludes seasonal variation, changes in conditions relating to ticket types, and special travel offers such as railcards. The value does not account for fare evaders. Data for fare evaders is available, but with too few observations to use. For the metros with data, the average percentage of passenger journeys attributed to evaders is 1.5% (SD = 1.8). As the rates appear to be relatively low and consistent amongst the groups it is unlikely that accounting for fare evasion will adjust the elasticity estimates dramatically. As the data improve it is hoped that this can be tested in the future. The values that were tested are 2015 equivalent US dollars based on Purchasing Power Parity (PPP). This accounts for differences in exchange rates as well as inflation. From the literature, metro fare elasticities have been found to be on average approximately −0.3 in the short run and −0.6 in the long run (*5*), and these elasticities have been found to be steadily increasing over time.

Income figures are proxied using Gross Domestic Product (GDP) per capita, again based on 2015 equivalent US dollars based on PPP rates. The values have been sourced from the World Bank's International Comparison Program Database, specifically from the World Development Indicators. GDP at purchaser's prices is the sum of gross value added by all resident producers in the economy plus any product taxes and minus any subsidies not included in the value of the products. Evidence of the effect of income on demand for metros has been found to be fairly substantial, with the most recent study finding a short run elasticity of 0.03 and long run of 0.18 (*16*). As this study also used GDP per capita as a proxy it is expected that these results will be broadly similar. It should be noted that using GDP per capita does have its shortcomings, as it does not capture differences between urban and rural areas, or indeed differences between cities in the same state. An alternative option considered was to employ the UBS price and earnings data at the city level. However, this brings its own short comings. For a start, this is based on survey data on a limited sample of professions. The data are derived based on numerous assumptions, obtained from third parties, which are believed to be accurate but this is based on good faith, and no representation or warranty, express or implied, is made or provided as to the accuracy or completeness of the data. As such, GDP per capita was deemed the most appropriate at this stage.

To account for the quality of service provided by the metro, the operated capacity provided by all actual car km operated in revenue service is considered and is normalized by the network length. By using capacity km, which refers to spaces for people to travel, and by normalizing by the size of the network the measure reflects both overall capacity and frequency provided and accounts for factors such as the size of the cars and their design, as well as the train length. This can be considered as providing an indication of the intensity of the service. The measure accounts for both seating and standing capacity, and is a standardized measure of capacity whereby standing density has been normalized to four people per square meter. The measure includes all lines in the network as well as turnaround and reversal at terminals. Excluded from the measure are empty stock movements, movement of trains to and from depots, movements of trains in depots, engineering trains, driver training runs (without passengers), train scheduled to run but which did not actually operate (e.g., cancellations), and rail replacement bus services.

Owing to the number of different definitions of quality of service found in the literature and the fact that fewer studies have been carried out specifically on metros than on railway or bus services, evidence is less clear. The most recent metro demand study found that estimated elasticities, Graham et al. (*16*), constructed a quality of service variable consisting of car km per route km, finding a short run elasticity of 0.07 and long run of 0.51. However, by using car km instead of capacity km as proposed by this study, the measure does not account for car size or design, with results that may be skewed by varying train lengths between and amongst metros (for example, consider a line running four-car trains compared with a line running eight-car trains. Both lines may offer the same total capacity per train but the latter would exhibit 50% more car km). As such, it is evident that the measure constructed for this paper is more appropriate and reflects the level of service provided by metros much more accurately. The improvement of this variable forms a major contribution to this work.

Population figures have primarily been sourced from the United Nations internet-based data service (http://data.un.org). These include a combination of official census figures along with estimates during years where a census was not carried out. Where reasonable, interpolation was carried out along with supporting figures from OECD statistics. Additional sources include the National Bureau of Statistics of China, Guangzhou Statistical Yearbook, Nanjing Statistical Yearbook, Department of Civil Affairs for Taipei City Government, and the Census and Statistics Department for the Government of Hong Kong. It is expected that as population increases so too do the number of trips being made. It is expected that the elasticities for the passenger journeys model to be highly elastic.

Network length, again from CoMET and Nova, refers to the length of the network used by trains operating in passenger service. The total length of the network is used, which is the sum of all lines of length of line (end to end, including any/all branches). This excludes tracks in depots, yards, and sidings, as well as the tracks used only for turning around trains. The measure accounts for lines with multiple tracks and for parallel lines, as the length is only counted once for double track lines (one track in each direction). Four-track lines/sections of line, often used to allow express trains to pass stopping services, are counted as twice the length of double track lines, for the length. Single track sections, including loops, count as 0.5 of double track length, while section of route where two or more lines share the same tracks are counted once. No evidence has been found for the influence of network length on demand, but it is reasonable to assume that as the network increases, so too do the number of journeys and average lengths of the journeys. As such, it is expected that the elasticities for both models to be highly elastic and positive in nature.

In terms of car ownership and the price of petrol, the remaining major factors thought to influence the demand for public transport, an inadequate level of data were compiled. Leaving out factors which we believe to have an effect on demand leads to estimation of parameters that are likely to be biased; formally this is referred to as omitted variable bias. However, despite this shortfall, as described in the methodology, this is addressed to a first approximation by an adequately specified panel data regression which can account for unobserved time-invariant heterogeneity.

Table 2 provides an overview of the variables that have been used and gives their descriptive statistics. Table 3 provides the correlation matrix of the set of explanatory variables. The values indicate that there are no issues with multicollinearity with the exception of the lagged dependent variable which, as discussed in the methodology section, can be adequately dealt with using the GMM approach described.

Results

Table 4 reports the results of estimation of Equation 1 using Pooled OLS, random effects, fixed effects, difference GMM, and system GMM models respectively. The econometric models present demand elasticities for the main demand drivers of urban metro rail systems. The coefficients represent the elasticities, while the values in brackets represent the corresponding standard errors.

Table 2. Overview of Variables for 32 Metro Rail Systems 1996–2015

Variables	Obs	Mean	SD	Min	Max
Metro patronage					
Passenger km (millions)	454	6,964.04	8,440.75	229.16	41,608
Passenger journeys (millions)	461	703.19	679.70	32.51	3,241.50
Fare					
Annual fare revenue from fares divided by annual passenger journeys (2015 US$ PPP)	423	1.17	0.51	0.22	2.63
Income					
GDP per capita (2015 US$ PPP)	726	22,280.34	15,230.29	1,378.50	86,128.17
Intensity of service					
Capacity km per network length	457	165.33	90.52	27.94	440.23
Population					
Population	517	9,359,276	5,880,417	1,012,051	25,700,000
Network length					
Network length (km)	482	156.32	134.56	19.74	588

Table 3. Correlation Matrix of Explanatory Variables

	Fare	Income	Intensity of service	Population	Net. length
Fare	1.0000	–	–	–	–
Income	0.0864	1.0000	–	–	–
Intensity of service	0.0543	0.0034	1.0000	–	–
Population	0.0239	–0.4615	0.2826	1.0000	–
Network length	0.1864	0.1475	0.0432	0.3871	1.0000

The levels of significance are also reported for 10%, 5%, and 1% levels.

As noted, the Pooled OLS assumes homogeneity across the metros. While this is not a reasonable assumption, the results help to identify an appropriate model. Likewise, the random and fixed effects panel data estimators are inappropriate as they are subject to bias via endogeneity owing to the lagged demand variable. For the GMM models, to deal with the endogeneity several specifications were tested. The choice of model was based on the significance of the estimates, the Arellano–Bond tests for AR(1) and AR(2) in first differences, as well as the Sargan and Hansen tests of over-identifying restrictions. In each case, the tests confirm that GMM is appropriate and that the instruments used are valid.

On examining the results, it is evident that the system GMM estimates appear to be biased. The elasticities appear very similar to the Pooled OLS estimates with similar magnitudes throughout and evidence of counter-intuitive signs for the income estimates for both the passenger km and passenger journey models. We know the Pooled OLS estimates are biased and inconsistent as they assume homogeneity across the metros, and so we assume the System GMM is also inappropriate.

As the data do not appear to be burdened by persistence, the difference GMM does not appear to suffer from weak instrument bias. As such, the difference GMM models are considered to be the most appropriate, with a specification that treats the intensity of service provided as endogenous, prices as predetermined but not strictly exogenous, and income as strictly exogenous. In both cases, the explanatory variables are statistically significant at a 99% level, with signs and magnitudes that are intuitive. From these, long run elasticity estimates are calculated and are presented in Table 5 along with a summary of the short run elasticity estimates.

The fare results imply that a 10% increase in fares corresponds to 1.1% decrease in passenger km and 2% decrease in passenger journeys in the short run, and 2.5% decrease in passenger km and 4% decrease in passenger journeys in the long run.

As expected, the results regarding income, which are proxied by GDP per capita, are consistent with Graham et al. (*16*), with values of 0.17 and 0.18 for the long run estimates for the passenger km and passenger journey models respectively. This confirms previous findings that indicate that metros are perceived as normal goods.

Again, consistent with Graham et al. (*16*), the values for the service levels are larger than the absolute values of the fare elasticities, reaffirming the implication that a more effective approach to increasing metro patronage stems from service improvements rather than from

Table 4. Results of the Dynamic Panel Model

	Passenger km					Passenger journeys				
Variable	Pooled OLS	RE	FE	Diff GMM	Sys GMM	Pooled OLS	RE	FE	Diff GMM	Sys GMM
ln (Fare)	−0.00	−0.07***	−0.09***	−0.11***	−0.00	−0.03**	−0.13***	−0.20***	−0.20***	−0.03***
	(0.01)	(0.02)	(0.03)	(0.02)	(0.01)	(0.01)	(0.02)	(0.02)	(0.02)	(0.01)
ln (Income)	−0.06***	−0.02	0.09**	0.08***	−0.06***	−0.04	0.02	0.09***	0.09***	−0.04***
	(0.01)	(0.02)	(0.03)	(0.03)	(0.01)	(0.01)	(0.01)	(0.03)	(0.02)	(0.01)
ln (Intensity of service)	0.10***	0.22***	0.21***	0.25***	0.10***	0.07***	0.16***	0.24***	0.24***	0.07***
	(0.03)	(0.04)	(0.05)	(0.04)	(0.02)	(0.02)	(0.03)	(0.03)	(0.03)	(0.01)
ln (Population)	0.03***	0.13***	0.21*	0.18**	0.03***	0.03***	0.12***	0.41***	0.41***	0.03***
	(0.01)	(0.02)	(0.11)	(0.09)	(0.01)	(0.01)	(0.02)	(0.09)	(0.07)	(0.01)
ln (Network length)	0.10***	0.27***	0.33***	0.37***	0.10***	0.08***	0.18***	0.25***	0.25***	0.08***
	(0.02)	(0.03)	(0.04)	(0.03)	(0.01)	(0.01)	(0.02)	(0.03)	(0.02)	(0.01)
ln ($Demand_{t-1}$)	0.90***	0.72***	0.58***	0.55***	0.90***	0.91***	0.72***	0.50***	0.50***	0.91***
	(0.02)	(0.03)	(0.03)	(0.02)	(0.02)	(0.01)	(0.02)	(0.03)	(0.02)	(0.01)
AB AR (1)	–	–	–	0.000	0.000	–	–	–	0.000	0.000
AB AR (2)	–	–	–	0.755	0.281	–	–	–	0.841	0.429
Sargan	–	–	–	0.000	0.000	–	–	–	0.000	0.000
Instruments	–	–	–	310	344	–	–	–	312	346
Adj. R^2	0.99	0.99	0.98	–	–	0.99	0.99	0.94	–	–
Observations	344	344	344	310	344	346	346	346	312	346

Note: Significance at 10%, 5%, and 1% levels is indicated by *; **; and ***; respectively.

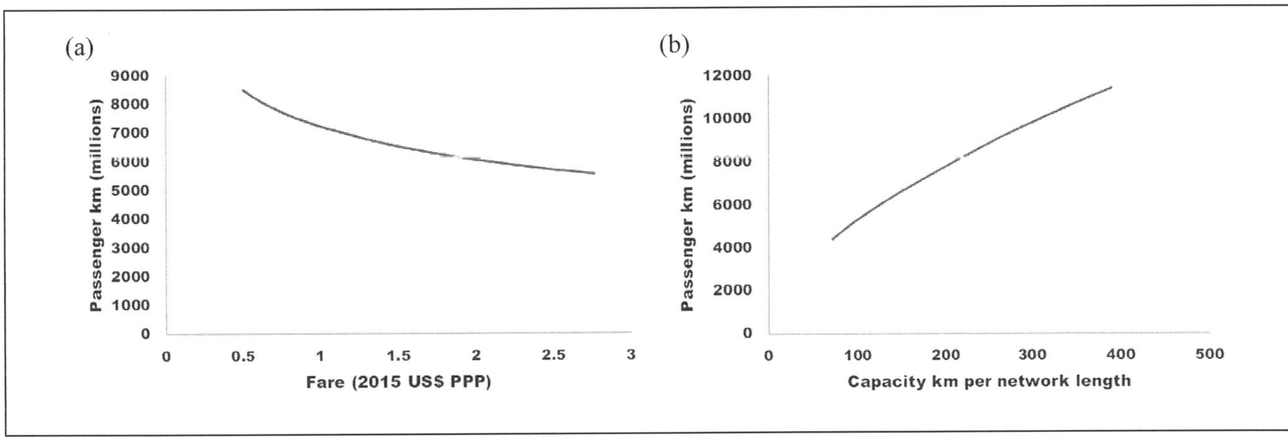

Figure 1. Visualizing an example of changes in passenger km given changes in fare levels and capacity km per network length based on the estimated elasticities.

Table 5. Summary of Elasticity Estimates from Difference GMM Model

	Passenger km		Passenger journeys	
Variable	Short run	Long run	Short run	Long run
Fare	–0.11	–0.25	–0.20	–0.40
Income	0.08	0.17	0.09	0.18
Intensity of service	0.25	0.56	0.24	0.47
Population	0.18	0.40	0.41	0.83
Network length	0.37	0.83	0.25	0.50

changes in fares. Owing to the improved variable, an updated interpretation may be described as a 10% increase in capacity km per network length may lead to a 5.6% increase in passenger km and a 4.7% increase in passenger journeys in the long run.

As expected, results for population and network length are significant and positive in nature. Interestingly, the magnitudes between the passenger km and passenger journey models differ. However, these differences appear to be intuitive. As the network increases in size, it is reasonable to assume that many journeys will increase in length, with greater distances leading to longer trips. Thus, it is unsurprising that, in the long run, estimates regarding network length should be larger when explaining passenger km than for explaining passenger journeys. This is indeed the case and the results can be interpreted as a 10% increase in network length corresponding with an 8.3% increase in passenger km and a 5% increase in passenger journeys. Alternatively, the estimates for population are found to be the reverse with a 10% increase in population corresponding with a 4% increase in passenger km and an 8.3% increase in passenger journeys. This can perhaps be explained by higher populated cities being denser in nature, and suggests that as the population increases the number of journeys increases, however with journey lengths that are shorter on average.

Overall, on average the long run estimates appear to be slightly above double that of their short run counterparts. This is unsurprising and appears to be consistent with studies examined in the review provided by Goodwin (*1*) which finds long run estimates ranging between 50% and three times higher. The long run estimates being more elastic is intuitive as in the long run consumers are better placed to adjust their behavior according to the changes in the factors which influence their demand, such as by changing their home or work locations.

Of the variables tested, metro operators have direct control over the fare and the level of service provided. Figure 1 below illustrates how the demand may vary given different levels of fare and service according to the elasticities that have been estimated. The graphs demonstrate that demand is more sensitive to the service provided (with a steeper curve) than changes in fare levels. Naturally, decision makers should take into consideration that adjusting fares will affect fare revenue, while adjusting the level of service will affect the operational cost.

Conclusion

Urban rail operators seek to understand how demand may rise and fall. This paper has estimated metro demand functions to identify factors that affect urban metro rail demand. It is clear from the literature that due to the variation in estimates found in previous studies, which have been based on weak and often deficient data, it is timely to derive a set of elasticities that use high quality up-to-date data and appropriate robust methods.

With more observations and a generally higher-quality dataset available, along with variables that more accurately reflect the services provided by metros, our results hopefully provide more reliable elasticities for metro service planning and forecasting in the absence of city information

The results confirm that demand is indeed more sensitive to the service provided by metros rather than changes in fare levels. This suggests that changes in the capacity provided by metros will have a greater effect on the patronage of a metro than changes in fares. By estimating the likely change in demand (and, therefore, revenue) from additional capacity, these elasticities can also be used by metros to evaluate potential changes to service levels and test their financial viability. In addition, the discrepancy between the two models perhaps emphasizes a need for distance based fare structures to be in place to encourage continued sustainability.

The income elasticities confirm that metros are perceived as normal goods, suggesting that as the standard of living increases so too does the level of trip making. The relationship between commuting by metro with population and network length is also of key importance, and should play an important role in making decisions related to investments into metro capacity.

Given that the long run estimates are generally double those of the short run estimates, it is important to bear in mind that metro customers take time to adjust their behavior in response to the explanatory variables examined here. Considerations need to be made for the customers' knowledge of adjustments and how entrenched their travel habits may be.

Naturally, a key limitation of this study is that the elasticities provided are averages across all metros and all circumstances. In the future, data permitting, it may be possible to carry out further evaluations by reconciling the results with disaggregated data and examine how demand varies over different parts of a metro network at different times of day. In doing so, the results would be beneficial to metro operators' strategies in terms of capacity utilization and provision given the improved understanding of how exactly the demand of service varies spatially and temporally.

Author Contributions

The authors confirm contribution to the paper as follows: study conception and design: Shane Canavan, Daniel J. Graham, Richard J. Anderson, Alexander Barron; data collection: Shane Canavan; analysis and interpretation of results: Shane Canavan, Daniel J. Graham, Richard J. Anderson, Alexander Barron; draft manuscript preparation: Shane Canavan.
All authors reviewed the results and approved the final version of the manuscript.

References

1. Goodwin, P. B. A Review of New Demand Elasticities with Special Reference to Short and Long Run Effects of Price Changes. *Journal of Transport Economics and Policy*, Vol. 26, 1992, pp. 155–169.
2. Oum, T. H., W. G. Waters, and J.-S. Yong. Concepts of Price Elasticities of Transport Demand and Recent Empirical Estimates: An Interpretative Survey. *Journal of Transport Economics and Policy*, Vol. 26, No. 2, 1992, pp. 139–154.
3. Greening, L. A., D. L. Greene, and C. Difiglio. Energy Efficiency and Consumption — The Rebound Effect — A Survey. *Energy Policy*, Vol. 28, No. 6, 2000, pp. 389–401.
4. Luk, J., and S. Hepburn. *New Review of Australian Travel Demand Elasticities*. ARRB Research Report 249. Australian Road Research Board, Victoria, 1993.
5. Transport Research Laboratory. *The Demand for Public Transport: A Practical Guide*. TRL Report, TRL593. Transport Research Laboratory, 2004.
6. Paulley, N., R. Balcombe, R. Mackett, H. Titheridge, J. Preston, M. Wardman, J. Shires, and P. White. The Demand for Public Transport: The Effects of Fares, Quality of Service, Income and Car Ownership. *Transport Policy*, Vol. 13, No. 4, 2006, pp. 295–306.
7. Holmgren, J. Meta-analysis of Public Transport Demand. *Transportation Research Part A: Policy and Practice*, Vol. 41, No. 10, 2007, pp. 1021–1035.
8. Wooldridge, J. M. *Econometric Analysis of Cross Section and Panel Data*. MIT Press, Cambridge, Mass., 2010.
9. Baltagi, B. *Econometric Analysis of Panel Data*. John Wiley & Sons, Hoboken, NJ, 2008.
10. Hansen, L. P. Large Sample Properties of Generalized Method of Moments Estimators. *Econometrica: Journal of the Econometric Society*, Vol. 50, 1982, pp. 1029–1054.
11. Arellano, M., and S. Bond. Some Tests of Specification for Panel Data: Monte Carlo Evidence and an Application to Employment Equations. *The Review of Economic Studies*, Vol. 58, No. 2, 1991, pp. 277–297.
12. Arellano, M., and O. Bover. Another Look at the Instrumental Variable Estimation of Error-Components Models. *Journal of Econometrics*, Vol. 68, No. 1, 1995, pp. 29–51.
13. Blundell, R., and S. Bond. Initial Conditions and Moment Restrictions in Dynamic Panel Data Models. *Journal of Econometrics*, Vol. 87, No. 1, 1998, pp. 115–143.
14. Arellano, M., and S. Bond. Some Tests of Specification for Panel Data: Monte Carlo Evidence and an Application to Employment Equations. *Review of Economic Studies*, Vol. 58, No. 2, 1991, pp. 277–297.
15. Graham, D. J., P. S. Melo, P. Jiwattanakulpaisarn, and R. B. Noland. Testing for Causality between Productivity and Agglomeration Economies. *Journal of Regional Science*, Vol. 50, No. 5, 2010, pp. 935–951.
16. Graham, D. J., A. Crotte, and R. J. Anderson. A Dynamic Panel Analysis of Urban Metro Demand. *Transportation Research Part E: Logistics and Transportation Review*, Vol. 45, No. 5, 2009, pp. 787–794.

The Standing Committee on Rail Transit Systems (AP065) peer-reviewed this paper (18-04076).

TRR
JOURNAL OF THE TRANSPORTATION RESEARCH BOARD

Article

Transportation Research Record
2018, Vol. 2672(8) 297–306

DOI: 10.1177/0361198118781646
journals.sagepub.com/home/trr
SAGE

Urban Services on Regional Rail: Local and Network-Based Methodologies for Evaluating New Stations in Toronto

Becca Nagorsky[1] **and Richard Borbridge**[1]

Abstract

The Greater Toronto and Hamilton Area's regional GO Rail network is undergoing a substantive upgrade to provide a two-way, all-day, electrified, 15-min service. These investments have led to discussions about the nature of the service and calls to add new stations, particularly within the City of Toronto. This paper explores the methodology used to evaluate specific new station locations in the City of Toronto and beyond, along with a concurrent analysis of the impacts of a package of new stations on the network performance and hierarchy. Both assessments used a multi-faceted business case framework. Travel-time impacts, land use, costs, feasibility, and other factors influenced the recommendation of sites intended to optimize the substantial investment in network infrastructure, and address city-building objectives like encouraging development and providing access to underserved communities. Network analysis points to new stations within the city shifting the traditional three-tier network hierarchy comprised of regional rail, urban rapid transit, and local transit to a hybrid network in which regional rail serves some intermediate functions.

Like many growing urban regions around the world, the Greater Toronto and Hamilton Area's (GTHA) transit network is challenged by high ridership, pressures for expansion, and financial constraints. The GO Rail regional network runs through the City of Toronto and into downtown, but, as a largely suburban commuter service, has limited stations. A 10-year, 13.5 billion Canadian dollars ($CA) program to significantly upgrade the GO Rail network served as a catalyst for discussion about adding new stations in Toronto and beyond, requiring concurrent analysis of both individual station locations as well as the broader implications of the entire package of new stations on the network hierarchy.

In 2017 GO Rail celebrated its 50th year in operation as the commuter rail service for the GTHA. The system consists of seven operating corridors, 280 mi (450 km) of track, 65 existing stations across 11 municipalities, and over 225,000 daily riders. Operations on five corridors consist of hourly-or-better trips during peak and shoulder periods on weekdays, and two corridors operating a half-hourly service all-day, every day. The broader public transit network includes three subway lines operated in the City of Toronto by the Toronto Transit Commission (TTC), and an extensive network of streetcars and buses. Surrounding municipalities have been planning for and building out bus rapid transit and light rail transit routes, particularly over the last decade.

In 2014 the provincial government of Ontario announced a plan to electrify the GO Rail network and provide 15-min frequencies and bi-directional service throughout the day. The transition to new fleet, new service patterns, and major infrastructure investment evoked consideration for optimizing GO Rail to play a larger role in local service, partly through the provision of new stations.

The GO Rail network employs a hub and spoke typology with downtown Toronto's Union Station the primary destination for passengers. In recent years, increasing employment and residential densities in outer Toronto (within city limits but outside the downtown) and in surrounding municipalities has diversified the centers around the region, resulting in new travel patterns such as inter-municipal travel between growing suburbs. The GO Rail expansion program strengthens links between different nodes of the regional transportation system.

This paper reviews the parallel and coordinated consideration of the individual new station locations and the overall resulting change to the regional transit network. It introduces the methodology developed to evaluate the impact of new stations added within the extents of the existing heavy-rail commuter network and considers the resulting shift in functionality from a three-tier network hierarchy within the City of Toronto to a hybrid network. It outlines the concurrent evaluation processes for both new station locations and the

[1]Metrolinx, Toronto, ON, Canada

Corresponding Author:
Address correspondence to Becca Nagorsky:
becca.nagorsky@metrolinx.com

consequent change to the network and highlights results of the analysis. Given the funding constraints and crowding challenges faced by transit networks in many large cities, the emerging Toronto model of a move toward using the regional transit network to fill urban transit needs is likely to be relevant to other jurisdictions. The analytic methodology outlined in this paper may serve as a basis for consideration of both individual new line stations on regional rail services, coupled with analysis of the broader networks.

We will begin with a review of relevant literature followed by an overview of the parallel site-specific and network-wide evaluation of the impact of new stations. We conclude with the results of this process and recommendations for further study.

Literature Review

A great deal of research has been conducted on optimizing network structure and hierarchy and, in particular, on the relationship between regional commuter transit service and local commuter service. This research is instructive in highlighting advantages and disadvantages of strictly hierarchical networks as compared with hybrid networks. Ultimately, the research suggests that, under certain conditions, a hybrid network can be efficient to operate and meet travel needs. This research helps to inform thinking on the changes to the GTHA's network and places those changes into the North American and global context.

Hermann Orth et al. set out a standard three-level transit network to serve three different trip lengths – short, medium, and long – but note that many cities deviate from this structure (*1*). Typically, systems that deviate from the structure operate two-level hybrid networks. In hybrid networks, local service is characterized by frequent stops and serves short trips, as it does in a three-level network, but both medium- and long-distance trips are served by regional service, with medium- to long-distance stop spacing. Two-level networks typically lack an intermediate rapid transit layer. Station spacing in the regional network might be longer in the outer reaches of the network and grow more frequent closer to the central city. The potential for express service on the regional transit layer can speed travel times for longer-distance passengers on a hybrid network. This network typology may serve medium-sized cities well, but has an inherently lower capacity than a three-level network. Orth et al. cite the example of Zurich, Switzerland where a two-level network functions well but is reaching capacity (*1*).

Karlsruhe, Germany offers another example of a hybrid, two-level network with its well-studied tram-train service. Karlsruhe's tram-trains run as streetcars within the inner city, adhering to streetcar/light rail regulations, and as trains outside the city limits, adhering to heavy-rail operations regulations (*2*). This network design is the result of a number of constraints unique to the Karlsruhe context, and vehicles were developed specifically for Karlsruhe to meet these operational needs. In particular, Karlsruhe's tram-trains are appropriate for an urban structure where the central train station is located a short distance from the city's downtown, such that conventional regional rail servicing a single central train station would be impractical for most passengers (*3*). Tram-train service allows for longer-distance travelers to benefit from higher speeds and longer station spacing outside the city along with finer-grained access inside the city. The structure has been successful in Karlsruhe and other small European cities but would be challenging to translate to a larger city where at-grade light rail could not provide sufficient capacity as the top tier of a transit network.

John Nash and Herbert Levinson explore the concept of regional rapid transit with a North American focus, providing analysis of the role that this type of transit can fill (*4*). Nash and Levinson trace the origins of regional rapid transit to the systems that were built in the late 1960s and 70s, relying on advanced technologies like automatic train operation that allowed for a single operator on each train and thus operational efficiencies. According to their description, regional rapid transit functions between commuter rail and urban rapid transit, characterized by reliance on park and ride lots and long stop spacing like traditional commuter rail, but providing more extensive urban service than a typical commuter rail service. Nash and Levinson's regional rapid transit systems are comparable to the two-level, hybrid networks that Orth et al. observe in the European context and rely on a single rail service to provide both suburban and urban transit (*1*,*4*).

The above examples of hybrid networks each focus on transit systems intentionally developed with the goal of providing both suburban and urban service. The following example, most comparable to the Toronto context, addresses the possibility of adapting commuter rail service to better provide urban service. In the late 1990s, Boston's Massachusetts Bay Transportation Authority (MBTA) studied converting the underutilized Fairmount Line into rapid transit service as well as other options for altering the service. The Fairmount Line is Boston's shortest commuter rail line at 9.2 mi (14.8 km) and does not extend into Boston's suburbs. Interestingly, the study found that the population in the Fairmount Line's catchment area was highly transit dependent and that local bus services from the area into Boston's downtown were slow and unreliable, yet ridership on the Fairmount Line itself remained low (*5*). The study recommended adding four new stations along the route to increase access to the rail service, along with frequency and service span upgrades, fare structure changes, and state of good repair investments. Stations were recommended at locations that would draw both additional boardings from local residents as well as function as destinations. Notably, the proposed improvements were not expected to reduce auto travel, but were instead forecast to reduce demand for local bus service.

Ultimately, three of the four recommended stations were built, with a fourth under construction. Service has been

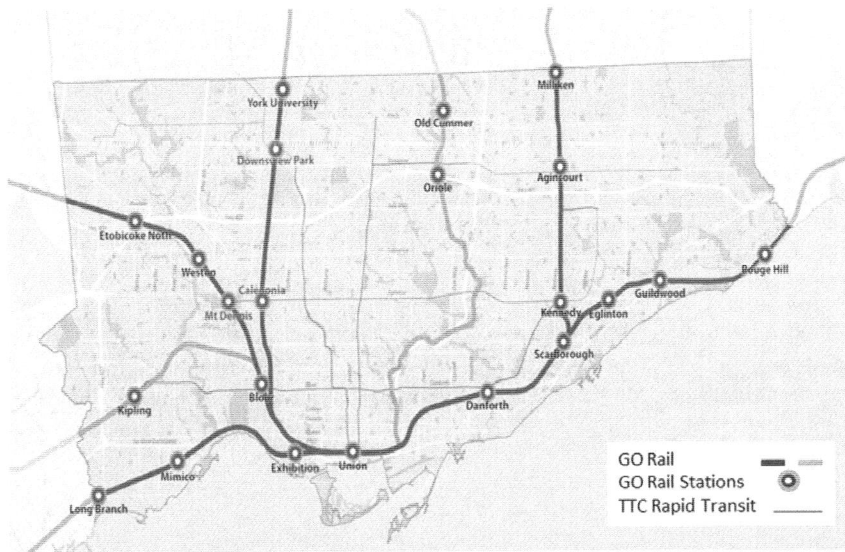

Figure 1. Existing GO Rail stations within the City of Toronto.

increased to accommodate growing demand and stations have been upgraded. In terms of transit service typology, the line is now a hybrid regional and commuter service; station infrastructure mirrors that of other commuter rail stations but station spacing is about 1 mi (1.6 km) compared with 2 to 5 mi (3.2 to 8 km) on other MBTA commuter rail lines. The Fairmount Line attracts considerably higher off-peak ridership than other commuter rail lines, consistent with its identity as a hybrid service (*6*).

It should be noted, however, that a 2016 ridership count on the line suggests ridership is lower than the initial 1990's forecasts. Barriers to higher ridership include the lack of an integrated fare structure with Boston's urban transit service and stations, which are less integrated into the urban fabric. Still, the changes to the Fairmount Line are regarded as a success in terms of improving travel choice and reducing crowding on the local bus lines.

Globally, a number of examples exist of hybrid transit networks in which transit lines or services provide both urban and suburban service. These services are successful in achieving operational efficiencies and in making best use of transit facilities and funding. They are, however, limited in the capacity they can provide. Most hybrid services were designed as hybrid, rather than retrofitted, as is the plan in Toronto. Additionally, little research has looked at the impacts of a hybrid service on long-distance commuters.

At the station level, Hamacher et al. examine two modes for adding new stations to an existing network: optimizing location to maximize accessibility of a station for new users, and the change in travel time for existing passengers imposed by adding the station (*7*). This work employs an extensive modeling exercise to identify optimal patterns on this basis but recognizes the value of local and network-wide considerations when locating sites. They conclude that building and operating costs, and the change in passenger volumes are among important additional considerations for selecting new station sites (*7*).

Methodology for Evaluating Individual Stations

Along with the substantial upgrades to GO Rail corridor infrastructure, planners and policy makers began discussing the changes that may become possible with a frequent, electrified service. In particular discussions focused on how GO Rail could better serve City of Toronto travelers in addition to the traditional suburban customer base. Toronto's subway is focused on the city's downtown and equity analyses often note the dearth of rapid transit service in the outer areas of the city, home to higher concentrations of vulnerable populations. The GO Rail network runs through many of these neighborhoods not served by other rapid transit but faces obstacles to ridership including wide station spacing, locations outside population or employment centers, and higher fares compared with the TTC (see Figure 1). Nonetheless, a move to a hybrid network structure offers opportunities to better serve some of these areas.

New stations have been previously added by extending further into suburban areas to serve new commuters. The increased frequency and reduced dwell times that would be provided by electrified vehicles allowed for consideration of new stations within the existing network, between stations, to serve new markets or align better with areas of density or intensification that were not previously possible. New subway and light rail transit services also intersected existing rail corridors, yielding candidate sites for modal interchange.

Two types of new stations were assessed – locations in largely residential areas that would serve new AM peak

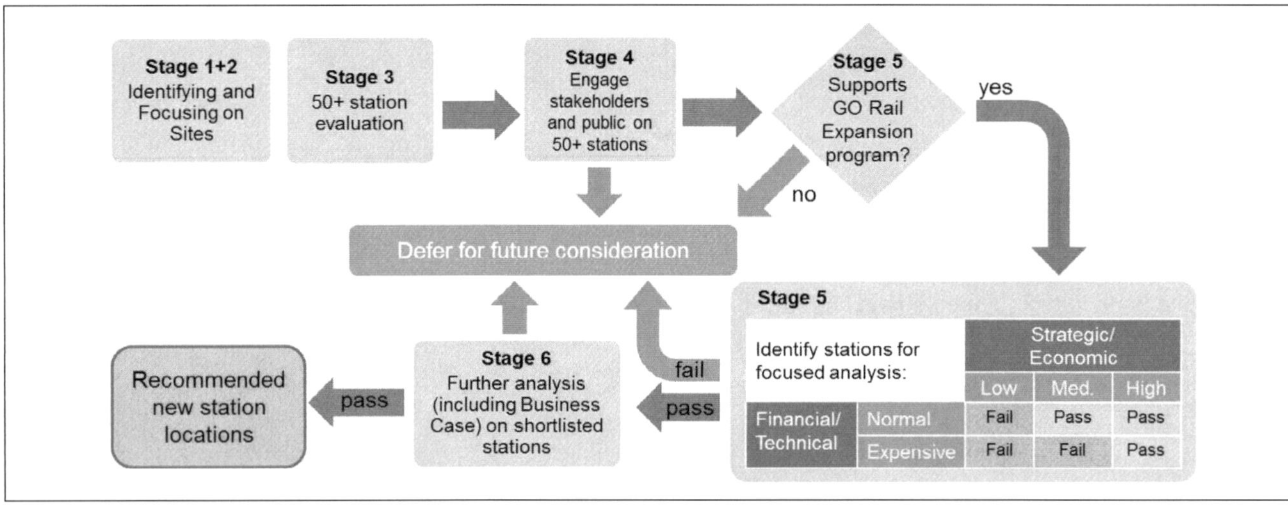

Figure 2. Decision-making process.

boardings, and locations in secondary employment areas at the periphery of Toronto's downtown that could serve AM peak alightings. New boarding stations would expand the reach of GO service to new areas but reinforce the current network pattern of boardings along the lines and alighting at Union Station in the downtown, whereas new alighting stations would shift the GO system from a single downtown hub to a system serving more dispersed destinations.

Following a comprehensive evaluation of individual station locations, and parallel evaluation of the network effects of new stations, 12 sites were recommended and approved, including four in suburban municipalities and eight within the City of Toronto (*8*).

Five objectives were established that focused the approach to evaluating new stations, with the intent to maximize planned investment in new corridor infrastructure. These objectives reflect both local- and regional-serving characteristics of a two-level network structure, and attempt to incorporate the parallel considerations of access to transit and travel-time optimization, along with the strategic objectives of city building. New stations should:

- Improve service and add riders
- Minimize impact on trip time for existing customers
- Maintain appropriate station spacing for the vehicle technology
- Support existing regional and municipal plans
- Consider the different roles and needs of each location and adapt to the urban or suburban context. At different stations this may have included
    - Minimize impacts on stable residential neighborhoods
    - Deeper integration with emerging urban fabric and so forth

Methodology

The Metrolinx business case framework is a mandated exercise for major projects. It uses a four-chapter approach for project evaluation consisting of the deliverability and operations case, the strategic case, the economic case, and the financial case. Each case looks at alternatives from a different perspective and each must be supportive in order for an alternative to be recommended. While Metrolinx has codified its own business case framework, the methodology captures the range of factors common to most transit evaluations across jurisdictions and should have broad cross-applicability.

Individual Station Evaluation

Process. The study of potential new station locations consisted of six stages, each broadening and deepening the criteria for analysis. Figure 2 illustrates the decision-making process used to focus the analysis. Initially, a list of over 120 potential sites was compiled for consideration. This list was sifted based on connectivity, land use, and feasibility criteria to arrive at 24 of the best performing sites on an individual basis, structured to provide a consistent comparison of disparate sites.

Strategic Case Summary. The following key criteria guided strategic case evaluation:

- Policy alignment
- Development potential and intensification
- Real estate market demand
- Natural environment
- Operational system
- Connectivity and ridership drivers
- Station access
- Social inclusivity and accessibility

The strategic chapter of the business case qualitatively assessed and documented the planning and policy considerations at a provincial and municipal level. They were ranked high, moderate, or low, relative to their alignment with a selection of policy statements and plans encompassing land use, transportation, environmental, and social policy. Station locations were preferred where they aligned with existing land use plans, and avoided where municipal official plans and infrastructure would not support it or limit the potential to concentrate growth.

Current and potential ridership is a central consideration for any new station location. Although estimates are typically derived from a comprehensive transportation model, in this case the level of effort was prohibitive given the wide geographic area and the number of unknowns related to infrastructure buildout that were under development simultaneously. Rather, this work estimated relative ridership using boarding and alighting data at comparable existing stations, populations, and jobs within catchment areas, and growth potential within the vicinity through a review of "soft sites" and surrounding developable land.

Ridership informed a travel-time savings calculation, which estimated the upstream impacts of additional journey time on riders passing through against the potential time saved by new users to the GO network and the station. Connectivity and ridership drivers included the estimated number of trips starting and ending at the station, connections to existing and planned high-order transit routes, number of nearby destinations, and places of interest.

Broader policy objectives were captured by proxy where detailed analysis was not available. Social inclusivity was gauged by proximity to lower-income communities or City of Toronto-designated Neighborhood Improvement Areas, and high-level environmental impacts were evaluated by considering proximity and potential impact to designated protection areas, parks, and woodlots.

Financial Case Summary. The following key criteria guided financial case evaluation:

- Incremental GO ridership (millions of trips)
- Fare revenue (A)
- Total costs (B) (capital + operating costs)
- Operating costs (C)
- Net present value (A-B)
- Revenue to cost ratio (A/B)
- Operating cost recovery ratio (A/C)

The financial case was based on a high-level estimate of construction costs and extrapolated approximate operating costs. As an early examination of sites, real costs were estimated roughly and for comparative purposes. A standard costing table was produced from precedent station construction and improvement projects, with consultant input. The variety of station configurations yielded capital costs ranging from CA$25 million to CA$250 million.

Operating costs for new stations were derived from those of existing facilities through a pre-existing calculation model. Comparative station estimates ranged from CA$10 million to CA$45 million annually.

Calculation of potential fare revenue was derived from ridership estimated in the strategic and economic analyses. Consideration of whether users were new passengers for the GO Rail system, or simply redistributed from an alternate existing station also informed the relative value of a site and potential cost recovery.

Economic Case Summary. The following key criteria guided economic case evaluation:

- Travel-time savings (millions of person-hours)
- Auto distances saved (millions of vehicle-kilometers traveled; VKTs)
- Benefits, costs, net present value, benefit–cost ratio
- Transportation user impacts:
  - Travel-time savings
  - Vehicle operating cost savings
  - Decongestion on road network
  - Safety impacts
  - Environmental impacts

The economic analysis of sites was aligned with internal Metrolinx guidance documents developed based on independent research and best practices. These manuals established assumptions for economic costs and values, such as VKTs (CA$0.74/km), the economic benefit of safety improvement, accident reduction benefit (CA$0.08/VKT), value of time (CA$16.71/h), and greenhouse gas emission reduction, with the full set of assumptions published as an appendix to the station evaluations. Costs and benefits associated with transportation user benefits, change in automobile use, and travel-time impact on customers was calculated as economic net present value (NPV) over a 60-year period.

Deliverability and Operations Case Summary. The following key criteria guided deliverability and operations case evaluation:

- Constructability
- Stakeholder impacts
- Room for growth
- Approvals/permits required
- Operating impacts
- Other key risks and impact

At the initial business case stage, deliverability was gauged at a high level. Many factors required supplementary engineering and design to provide meaningful consideration, and operational issues arising from network and track design were outstanding at the time of analysis. Issues such as the full extent of property requirements and ease of acquisition, soil quality, and community response to the project could not be more than assumed. Feasibility was qualitatively categorized from "normal," such as greenfield sites with few impediments; to "difficult": tightly constrained urban conditions with major operational impacts, local tracks and extensive subsurface remediation required, for instance.

Sensitivities. The business case reports also incorporated a set of sensitivity scenarios to evaluate key emerging issues for which there was insufficient data for more thorough evaluation or direct consideration in the business case. Evaluation of the sites was generally conducted on the basis of current or known conditions. Identifying sensitivities helped to respond to anticipated questions about the magnitude of impact that different assumptions or significant network changes might have on the performance of a station. The sensitivities considered, for example, whether heightened development potential would significantly affect the surrounding population and employment and resulting ridership, or if changes to the regional fare structure would facilitate inter-modal transfer and change the value proposition of transit for the users. These were judged as having "high" or "low" potential to alter the relevant key criteria, and thereby the relative performance of the station against the base condition. This helped to focus further work and served as a starting point for discussions with project stakeholders.

Individual Station Ranking and Selection

The results of individual business cases allowed for a better understanding of specific sites and a comparison of potential station impacts. They did not, however, provide robust guidance on the relative merits of one site against another given their widely varying contexts. The initial framework was not equipped to address the additive effects of multiple new stations on a corridor, or adequately highlight the influence of criteria only measured qualitatively. For this, a ranking approach was undertaken as a follow-up process to prioritize recommendations for each corridor, and bridge the network impacts discussed in the next section. Four additional screenings were undertaken that examined these issues.

Since the economic NPV served to combine a host of financial and economic factors including travel-time impact, it provided a first screen to assess stations against several of the original objectives. The screening process began with a ranking of NPV from highest to lowest, and into three categories: positive, marginal, or poor.

Rankings were then adjusted based on the capability of the station to impose positive or negative strategic or deliverability factors that were not captured in the NPV. This screen emerged through consultation with stakeholders, and acknowledgment that several of the qualitative considerations were of high value to municipalities. Station selection needed to reflect city building and social objectives beyond delivery of transit. Stations located in areas of intense development focus, or which had access to lower-income neighborhoods were elevated in the ranking. Locations that posed concerns related to local environmental factors such as disrupting protected areas reduced a station's ranking.

A "network fit" screen considered a station's proximity to existing or committed rapid transit lines, recognizing the importance of building an interconnected network in serving the largest area and audience. This screen further considered the proximity of high performing stations to each other, and impact on network expansion where a new station today could preclude a potentially better performing station in the future.

The final screen recognized and attempted to preserve the operational and journey time benefits of new GO Rail infrastructure. With an established objective of not significantly increasing end-to-end journey time for each line, the decrease in journey time resulting from electrification was compared against the increase in journey time imposed by stopping at a new station. This screen posited outer stations as having minimal impacts, as compared to stations closer to downtown Toronto. The relative impact of adding a station to the ends of the line, where relatively few passengers are affected, was compared against the significant effect of adding travel time for every passenger on full trains near the core of the network. The introduction of up to three stations on a corridor was deemed to provide a balance.

The results of this work identified five locations that generally performed well, five stations that were identified as contingent on further refining details of the GO Rail investment and upgrade program, and seven that were comparatively low performing. As capital costs are better understood through design, and planning is advanced for the areas around stations, further work is planned to update the business cases over time.

Methodology for Evaluating Network Effects of New Stations

Network Considerations

In addition to evaluating the costs and benefits of each individual station, an analysis was conducted of the network impacts of adding packages of new stations (*9*). At present, the GTHA has a traditional three-level transit network comprised of local bus and streetcar networks providing frequent stop service, a subway providing grade-separated intermediate service, largely within the City of Toronto, and commuter rail providing service from Toronto's suburbs into the downtown.

As noted above, several structural barriers exist to better use of GO Rail among Toronto residents, including higher fares and absence of free transfers between GO and the local

Toronto transit service. In January 2018, a discounted transfer fare will be introduced, better integrating the Toronto and regional transit networks and a follow-up study will be required to understand the ridership impacts. In assessing the network impacts of new stations, although they may have substantial influence, fare and transfer policies were assumed to be out of the scope of the analysis.

Methodology

The network analysis was carried out concurrently with the individual site analysis. As with individual stations and consistent with Metrolinx's Business Case policy, alternatives were assessed using the four-chapter business case framework.

Network Alternatives

Four alternatives were analyzed with each capturing a different means of intensifying service along the GO Rail system, within the City of Toronto, through a mix of service upgrades (beyond those planned through the regional express rail investment) and new stations. Alternatives were developed using the planned and funded upgrades to the GO Rail network as the base case and looked at how service could be further improved within the City of Toronto.

- Alternative A proposed adding five new stations, focusing on emerging employment areas at the periphery of the downtown and doubling planned train frequencies to 5-min peak service and 7.5-min off-peak service.
- Alternative B proposed adding eight new stations in a mix of employment and residential locations and changing the service pattern to local and express service, thus allowing the new stations to be served without lengthening journey times for longer-distance passengers.
- Alternative C proposed adding seven to eight new stations in a mix of employment and residential locations throughout Toronto without any changes to the planned regional express rail service.
- Alternative D proposed adding four to five new stations with a focus on serving employment areas at the periphery of Toronto's downtown without any changes to the planned service.

Alternatives Evaluation through the Business Case

Deliverability and Operations Case Screen Summary. The deliverability and operations case assessed whether the alternatives were technically feasible, assuming a baseline of infrastructure planned for GO regional express rail service that includes track and signaling upgrades as well as electrification and new fleet. The individual station business cases discussed in Methodology for Evaluating Individual Stations evaluated the constructability of each individual station and thus only stations deemed to be feasible were included in the network phase of the evaluation. Given the complexity of changes to a rail network, the deliverability and operations case was used as a screening tool to eliminate infeasible alternatives from further consideration.

The analysis was carried out by translating each alternative into a sample timetable and then mapping that timetable onto the track layout in a string-line diagram. These string-line diagrams created a visual representation of train speeds and track geography and enabled an understanding of where trains would meet and thus require multiple tracks to safely accommodate the service.

This analysis revealed that alternatives A and B would require additional tracks beyond what is already planned, entailing substantial property acquisition of between 800 and 1,000 individual commercial and residential properties abutting the rail corridor. For Alternative A, the substantially higher frequencies were a challenge for corridors where a mix of diesel and electric services are planned due to the differential speeds achieved. In these corridors, even small increases in frequency beyond what is planned resulted in train-meets. For Alternative B, the express and local stopping pattern resulted in numerous locations where an express train needed to pass a local train, resulting in the need for passing track or an additional track altogether. Beyond track infrastructure, the analysis suggested that the additional service considered in alternatives A and B would also require added platform capacity at Union Station, Toronto's central rail terminus. Union Station is a constrained site and additional platforms would likely entail a below- or above-grade solution which would add considerable complexity and cost.

For these reasons, alternatives A and B were screened from further consideration through the business case framework. Alternatives C and D were found to be operationally feasible through the string-line diagram analysis and were carried forward for full evaluation.

Strategic Case Summary. The strategic case was used to explore how the different alternatives implied changes to the objectives and character of transit and affected the ways the passengers travel. It also looked at the extent to which the alternatives meet policy objectives, including growing transit ridership, decreasing transit travel times, and achieving social equity and city-building goals.

The two remaining alternatives have strategic advantages and disadvantages. Alternative C blurs the network hierarchy to a greater extent by adding more stations within Toronto, it also goes further to increase access to rapid transit service within Toronto. This effect is concentrated in the city's periphery, which is less well-served by rapid transit. The addition of more new stations would also lengthen travel times for medium- and long-distance passengers. In contrast, Alternative D has a smaller impact on network hierarchy by adding only four to five new stations. The addition of a

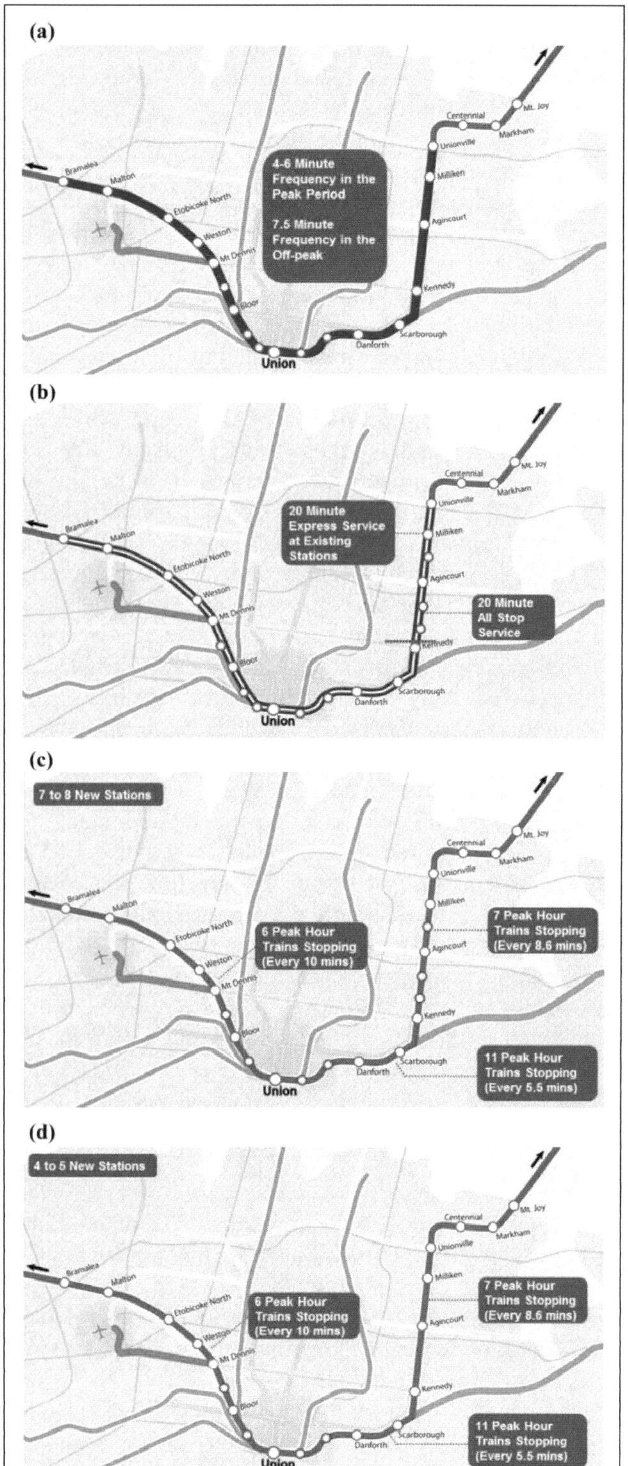

Figure 3. Network alternatives: (*a*): alternative A; (*b*) alternative B; (*c*) alternative C; (*d*) alternative D.

smaller package of new stations would not expand travel choice as significantly for Toronto passengers, but also would have less impact on journey times for medium- and long-distance passengers.

In terms of policy objectives, analysis looked at ridership, city-building goals, and social equity. Both alternatives were forecast to increase overall ridership on the GO Rail system by approximately 10%. Furthermore, forecasts for both alternatives suggested a net increase in overall travel time, and erosion of some of the travel-time savings resulting from electrification. Alternative C would increase transit access for low-income residents and would also further advance city-building goals because the higher number of new stations provides more opportunities for redevelopment and neighborhood uplift.

The most interesting aspect of the strategic case analysis comes through the discussion of travel behavior in light of the changes the two alternatives would bring about to the network hierarchy. The ridership forecasting suggested that the new stations proposed in emerging employment centers at the periphery of the downtown are more promising than those in residential areas in attracting new boardings. This finding suggests that the traditional network hierarchy in which commuter rail passengers must transfer to local transit for their last mile could be optimized through the addition of new stations at the periphery of downtown enabling direct access to a broader range of final destinations beyond the downtown.

The last strategic case finding emerged through a close reading of the ridership forecasting, While alternatives C and D are forecast to attract similar numbers of new riders, they are not the same new riders. Because Alternative C would lengthen travel times more significantly, it is projected to deter some medium- and long-distance passengers from using GO Rail. Although Alternative C attracts more new passengers through a higher number of new stations, it also deters passengers through longer journey times. Alternative D is forecast to attract a smaller number of new passengers, but avoids discouraging existing passengers because the impact on overall travel times is relatively insignificant. This finding suggests that there may be a tipping point in blurring network hierarchy in which the traditional commuter rail market is deterred from using transit because the commuter rail layer is not sufficiently differentiated from the intermediate layer.

Economic Case Summary. As in the individual new stations' business cases, the economic case quantifies and monetizes the benefits of the alternatives with a focus on travel-time savings and VKTs. Travel-time savings were assessed by forecasting the passengers who would use the new station and save time either at their origin or destination point compared with the number of passengers who would lose a small amount of time due to the added travel time associated with an additional stop.

The base case for this analysis was the GO regional express rail program, currently under construction, which is expected to bring about significant benefits in travel time and expanded ridership. Because both alternatives C and D increase overall travel time compared with the base case, both have negative net benefits and both would decrease the benefit–cost ratio of the regional express rail program. Alternative C is forecast to have net benefits of -CA$2.2 billion and Alternative D is forecast to have net benefits of -CA$1.1 billion.

In terms of network hierarchy, the most interesting economic finding was in looking at the change in VKT and the corresponding savings in auto operating costs. Because Alternative C goes further to hybridize the regional system than Alternative D, it was forecast to shed some long-distance commuters, while attracting new shorter-distance commuters. Pushing long-distance commuters to auto travel substantially erodes benefits related to auto operating cost savings, compared with Alternative D (CA$20 million compared with CA$110 million for Alternative D).

Financial Case Summary. The financial case looks at the forecast capital and operating costs, to the Province of Ontario, of the two alternatives. The projected capital cost for Alternative C is CA$1.1 billion to CA$1.7 billion and the projected operating cost is CA$8.7 million. The projected capital cost for Alternative D is CA$0.7 billion to CA$1.0 billion and the projected annual operating cost is CA$5.3 million.

Additionally, because GO Rail operates on a fare by distance formula and Alternative C would deter some long-distance passengers and attract slightly fewer new passengers, Alternative C is projected to increase annual revenue to GO by CA$11.8 million. Alternative D is projected to increase annual GO revenue by CA$15.1 million. Note that these revenue projections do not capture revenue lost to local transit caused by a switch in passengers from one system to the other.

Business Case Conclusions. Based on the business case analysis, the addition of a small number of new stations focused on those in secondary employment areas emerged as the optimal direction. In terms of network hierarchy implications, this conclusion suggests that there is room to soften the network hierarchy by allowing the regional system to serve some local functions, particularly in providing a more fine-grained downtown service. However, going too far in hybridizing the regional network to serve both local and regional functions risks jeopardizing the medium- and long-distance level of service that the regional level of the network is best suited to serve.

Converging, Conclusions, and Future Study

The parallel station and network analysis provided insights into the opportunities and potential challenges of adding new stations to the GO Rail network. The business case framework provided a tool for multi-faceted evaluation of each, with the local and network lenses providing greater depth to a question that is more often examined through one lens or the other: where should we build a new station?

Both studies served as complementary exercises, bringing to light a broader array of considerations. With tight budgets and ever-increasing demands to optimize transit investment and maximize value-for-money, an upfront exercise that examines multiple scales also builds a stronger constituency of support. The transition of GO Rail from a regional commuter service to a hybrid model is a lengthy process with a planned 10-year rollout and budget over CA$15 billion. Extensive planning, design, and construction will follow this work, creating opportunities to bring new data and perspectives to the multi-faceted process of network expansion.

In the case of GO Rail, the transition from long-distance to more locally oriented services provided the opportunity to re-examine the role of rail service on city-building goals while simultaneously raising and considering questions about how best to optimize operations. The expected shifts in passenger demographics will not just be realized by bringing service to new neighborhoods. Rather, with a 15-min service, fundamental shifts in regional travel behavior and mode choice may arise. Trips previously only possible via automobile can be taken up by new combinations of regional, local, and active transportation if new stations maximize connectivity and minimize journey time.

This exercise is also an important step in expressly connecting the decision-making processes for land use and transportation with implications for the role of a hybrid model on land use patterns and adjacent development. It inspires many additional avenues of inquiry: How is the role of specific new sites measured, how do they change the travel patterns of users in the vicinity and beyond? By extension, do cities grow more densely or do the added stations contribute to an overall disaggregation of development patterns?

Does the network effect of new destinations ultimately outweigh the travel-time impacts of adding the stations? Is there a clear benefit to regional sustainability in moving toward a hybrid model that would not have been realized by holding fast to GO Rail as a third tier of transportation for regional passengers?

With increasing opportunities and desire to maximize infrastructure investment, the push to add stations to operating rail corridors is only likely to increase. This paper suggests that it is important to define at the outset the desired role of the "new" system within the context of the existing transportation network. The examples of multi-tiered systems illustrate the need to develop the right tool for the job. This paper has helped to identify the challenges and benefits of inserting stations into a network while the function of the network remains under development.

Acknowledgments

The authors acknowledge a number of Metrolinx staff who were critical to this work. Joshua Engel-Yan's guidance of the work and his thoughtful review of this paper were invaluable. Elana Horowitz was instrumental in guiding the new stations work and her attention to detail and commitment to urbanism were a credit to the work.

Anna Pace's long lens of experience and dedication to improving the regional transit network were critical in guiding the network-level analysis. Lastly, David Prichard developed the ridership forecasting technique used for both sets of analyses and his ability to understand and address novel challenges with technical rigor were essential to advancing this work.

Author Contributions

The authors confirm contribution to the paper as follows: study conception and design: BN, RB; data collection: BN, RB; analysis and interpretation of results: BN, RB; draft manuscript preparation: BN, RB. All authors reviewed the results and approved the final version of the manuscript.

References

1. Orth, H, A. Nash, and U. Weidmann. A Level-Based Approach to Public Transport Network Planning. *Transportation Research Record: Journal of the Transportation Research Board,* 2015. 2537: 1–12.
2. Brandl, P. G., and K. W. Axhausen. Karlsruhe 1975-1995: A Case Study of Light Rail Transit Development. *Transportation Research Record: Journal of the Transportation Research Board*, 1998. 1623: 155–164.
3. Walker, J. Karlsruhe: A Ride on the "Tram-Train". http://humantransit.org/2009/10/karlsruhe-the-tramtrains.html. Accessed July 14, 2017.
4. Nash, J., and H. Levinson. Regional Rapid Transit: Past, Present, and Future. *Transportation Research Record: Journal of the Transportation Research Board*, 2011. 2219: 69–77.
5. Nelson, D., Y. Duse-Anthony, and S. Friemann. Improving Inner-City Mobility with Commuter Rail Service Case Study of Fairmount Line in Boston, Massachusetts. *Transportation Research Record: Journal of the Transportation Research Board*, 2005. 1930: 44–53.
6. Increasing Ridership on the Fairmount Line Final Report, April 2017. https://www.tbf.org/~/media/TBFOrg/Files/Reports/Increasing%20Ridership%20on%20the%20Fairmount%20Line.pdf, The Boston Foundation. Accessed July 16, 2017.
7. Hamacher, H., A. Liebers, A. Schobel, D. Wagner, and F.F. Wagner, Locating New Stops in a Railway Network. *Electronic Notes in Theoretical Computer Science*, Vol. 50, No. 1, 2001, pp. 1–11. http://www.sciencedirect.com/science/article/pii/S1571066104001628
8. New Stations. Metrolinx, Toronto, http://www.metrolinx.com/en/regionalplanning/newstations/default.aspx. Accessed July 29, 2017.
9. GO RER-SmartTrack Integration Options Initial Business Case, June 2016. http://www.metrolinx.com/en/docs/pdf/board_agenda/20160628/20160628_BoardMtg_Appendix_3_SmartTrack_EN.pdf. Metrolinx, Accessed July 25, 2017.

The Standing Committee on Rail Transit Systems (AP065) peer-reviewed this paper (18-00076).

Article

TRR
JOURNAL OF THE TRANSPORTATION RESEARCH BOARD

Transportation Research Record
2018, Vol. 2672(8) 307–315
© National Academy of Sciences:
Transportation Research Board 2018
Article reuse guidelines:
sagepub.com/journals-permissions
DOI: 10.1177/0361198118782027
journals.sagepub.com/home/trr
SAGE

Experimental Study for Estimating the Passenger Space at Metro Stations with Platform Edge Doors

Sebastián Seriani[1,2] and Taku Fujiyama[1]

Abstract

Platform edge doors (PEDs) are used in various metro stations to improve safety, comfort, and ventilation conditions; however, limited research has been done to estimate the passenger space (PS) in the boarding and alighting process when PEDs are installed. The objective of this paper is to estimate the PS needed for alighting at metro stations. For this purpose, laboratory experiments have been performed at University College London's Pedestrian Accessibility Movement Environmental Laboratory. The experiments consisted of a mock-up of a carriage and the relevant portion of the platform, in which different load scenarios of boarding and alighting were conducted. The scenarios were based on a preliminary analysis observed at Westminster Station (with PEDs) and Green Park Station (without PEDs). To obtain the position of each passenger on the platform a tracking tool was used. The results show that the PS for alighting passengers can be represented as an asymmetrical ellipse, in which the longitudinal and lateral radii change according to the negotiations with other passengers alighting or waiting on the platform to board the train. Therefore, there is a relationship between the PS and the level of interaction, which suggests that passengers adjust their PS to avoid collision. This research can be used in pedestrian models by traffic engineers to estimate the PS of passengers boarding and alighting when PEDs are used. This in turn can help in designing the platform–train interface and platforms at transport infrastructures.

The platform–train interface (PTI) is considered the space where the majority of interactions of passengers occur. This space is defined as the area between the train doors and the corresponding adjacent platform (*1*). For example, in the UK railway network more than 3 billion interactions take place each year, and 21% of the total safety risks (injuries and fatalities) and 48% of the total fatality risks to passengers are produced in this space (*2*).

To reduce these risks at the PTI, platform edge doors (PEDs) have been implemented in metro stations worldwide (*3–5*). PEDs can also influence ventilation conditions and act as smoke detectors in case of a fire (*6*, *7*). In the case of London Underground (LU), PEDs work as sliding barriers simultaneously with the train doors (see Figure 1, left).

There is a common assumption that the use of PEDs increases the dwell time, which is defined as the time the train remains on the platform to transfer passengers. However, recent studies have shown that PEDs has no significant impact on the boarding and alighting time, which corresponds to the dynamic component of the dwell time (*8*, *9*). However, these elements change the behavior of boarding passengers insofar as they tend to wait beside the doors rather than in front of them. For those studies, the authors used a mock-up train and PTI in a controlled environment at University College London's Pedestrian Accessibility Movement Environmental Laboratory (PAMELA), based on observation from video footage at two LU platforms (see Figure 1, right). However, because of the absence of an automatic tracking tool, the authors did not obtain the exact position of each passenger, therefore the results obtained left the question of the effect of PEDs on the interaction of passengers boarding and alighting open for further research.

This unresolved question was partially solved by the study of the level of interaction (LOI) (*10*). The authors used the same experiments and observations in De Ana Rodríguez et al. (*8*) to track the exact position of passengers boarding and alighting. However, as a result of some technical faults in the tracking process and lack of resources, the authors in Seriani et al. (*10*) could only consider the space between two passengers alighting (e.g., pair A_1–A_2) or two passengers boarding (e.g., pair B_1–B_2), and not the space between each passenger alighting and boarding (e.g., pair A_1–B_1), espe-

[1]University College London, London, UK
[2]Faculty of Engineering and Applied Science, Universidad de los Andes (Chile), Santiago, Las Condes, Región Metropolitana

Corresponding Author:
Address correspondence to Sebastián Seriani:
sebastian.seriani.14@ucl.ac.uk; sseriani@miuandes.cl

Figure 1. PEDs used in Westminster Station (left) and Pedestrian Accessibility Movement Environmental Laboratory (PAMELA) (right).

cially when boarding passengers are considered an "obstacle" for alighting.

Therefore, the aim of this study is to fill that knowledge gap and to expand the analysis and results in earlier studies (*8–10*), estimating the passenger space (PS) for alighting at metro stations with PEDs. The specific objectives are: (a) to mock-up a carriage and the relevant portion of the platform at PAMELA; (b) to conduct different load (flow) scenarios of boarding and alighting at PAMELA; (c) to measure the PS for each scenario.

This paper is composed of six sections, including this one. In section two a literature review is presented. Next, the method followed by this work is explained. Section four shows the results of PS measurements. In section five a discussion is provided. Finally, in the last section the conclusions are delivered.

Literature Review

According to Fruin, in any standing area (e.g., metro stations and surroundings) a pedestrian can be represented as an ellipse of area 0.30 m^2 comprising a body depth of 50 cm and a shoulder breadth of 60 cm (*11*). However, when the pedestrian starts to walk, this area increases to 0.75 m^2 because there is extra space used for leg and arm movements, and therefore a "face to face" distance between pedestrians of less than 0.5 m will be considered as intimate (*12*). Therefore, if the distance between two pedestrians' heads is less than 1 m (taking into account 0.5 m plus two times half the body depth of [*11*]), then pedestrians will feel that their space is being invaded.

In psychology the term "personal space" is used to represent the "virtual" space needed for each passenger to feel comfortable in different situations. The effect of intimacy was first studied by Hall (*13*). The author classified intimate distance into four groups: (a) intimate (when the distance is less than 0.5 m and pedestrians have a special relationship); (b) personal zone (between 0.5 m and 1.2 m, and pedestrians know each other); (c) social consultative zone (between 1.2 m and 4 m, and pedestrians do not know each other but they have permitted communication); and (d) public distance (between 4 m and 10 m, and pedestrians do not know each other). In the case of metro stations, Sommer (*14*) studied the social behavior of passengers and used three groups to classify personal space: (a) intimate (<0.5 m); (b) personal (0.5–1.2 m); and (c) social (>3.0 m). However, this feeling of invasion is based on perception rather than physical space, which is difficult to calibrate at the PTI zone. In addition, acceptable densities vary by situation and duration; for example, people stand closer together on escalators or stairs than when walking, and are more tolerant of passing closely when walking in opposite directions than in the same direction.

From another perspective, Schmidt and Keating introduced the concept of personal control which is less subjective and is categorized into three forms: behavioral, cognitive, and decisional (*15*). The behavioral form is related to crowding situations which are produced when the density interferes with the behavioral sequence or blocks the goal of pedestrians, or when pedestrians feel that they lose control or freedom of their space, in which collision avoidance is a way to return to a non-crowded situation. With respect to cognitive aspects of traveling, the personal control depends on the way each pedestrian anticipates and interprets the event or impending condition. Finally, the decisional control is related to the desired situation of each pedestrian when selecting outcomes.

In the case of railway and metro systems, a crowded situation at the PTI can occur when passengers are walking where there is a density of more than two passengers per square meter, or there are more than 5 pass/m^2 in a waiting area (*16*). To evaluate the degree of congestion in an area, the Level of Service (LOS) reported in Fruin can be used based on general parameters such as speed, density or flow (*11*). The LOS goes from Level A (free flow with no conflicts) to Level F (critical density, sporadic flow, frequent stops, and physical contact), where Level E is equal to capacity. However, there is a difference between density (physical characteristics of the environment) and crowding (psychological phenomenon), because a high-density situation is not always perceived as crowded with a high level of stress (*17*). The authors proposed a model with a high level of density and perception of crowding and stress level, and also identified the relationship between crowding and risk safety. Similarly, Evans and Wener studied high density and stress while commuting in trains where passengers have to sit next to others (*18*). The authors found that when the density increased, passengers perceived a high stress level.

Therefore, the PS is related to situations in terms of physical measurements, that is, as a function of density and capacity on the platform and train, but also as a psychological dimension which is more about the perception of crowding (*19*). To capture crowding in railway and metro systems Lam et al. proposed a binary logit model to represent discomfort of passengers (*20*). Similarly, to study the effect on the level of stress and feeling of exhaustion, Mahudin et al. proposed a model to measure crowds based on psychological aspects of crowds, evaluation of the environment where the crowd is situated, and how crowds react in specific situations (*21*). Recently, Kim et al. identified that people avoid delays caused by crowding and the stress caused by crowding such as lack of availability of seats, and avoid other passengers, or worry about sexual harassment (*22*).

With respect to microscopic pedestrian models, the space used by pedestrians differs in shape and size. For example, in cellular automata models each pedestrian is represented as a square cell with a side length of 0.3 m or 0.4 m (*23*, *24*), while in force-based models pedestrians are represented as particles (or circles) (*25*). These representations simplify each pedestrian as a rigid body without leg or arm movement, therefore they present a PS which is not entirely realistic in overlapping or overtaking interactions between pedestrians. This has been improved by looking for other ways (e.g., shapes) of representing pedestrian interaction (*26*, *27*). However, not all of these models are calibrated and validated to represent boarding and alighting; therefore, there is a lack of microsimulation models to represent a more realistic PS at metro stations.

In term of experiments, according to Gérin-Lajoie et al. each pedestrian needs a space represented as an elliptical area 0.96 m wide by 2.11 m deep, which is smaller when overtaking a static rather than a moving obstacle (*28*). In addition, Gérin-Lajoie et al. demonstrated that this space can be asymmetrical in shape and side (left and right) during the circumvention of a cylinder (or column) as an obstacle, in which the longitudinal axis of the ellipse is related to speed, whereas the lateral axis is related to the avoidance of contact with other pedestrians or obstacles (*29*). Recent laboratory experiments have also used static obstacles to define the space used by pedestrians as a function of the vision field, in which an angle of 45 degrees was reached and a distance of less than 1.5 m was perceived as difficult to avoid and react to (*30*). However, these studies are related to the concept of sensory zone, in which a person can perceive, evaluate, and react to approaching hazards. For example, at a normal walking speed the sensor zone can be estimated as an elliptical area of 1.06 m wide by 1.52 m deep (*31*). Similarly, Fruin calculated that the sensory zone reached a distance of 1.48 m for a normal walking speed of 1.37 m/s (*32*). Another experiment found that the number of lanes in a bottleneck depends on the door width but also on the distance between pedestrians and their behavior in which a "zipper effect" (overlapped) is reached (*33*). In relation to metro stations, different experiments have been performed to identify the effect of train design features on the behavior and flow of passengers at the doors (*34–36*).

Despite the research related to personal space, personal control, and crowding, little research has been done to estimate the PS for alighting at metro stations when PEDs are used. Therefore this paper, which extends earlier research (*8–10*), analyzes the PS based on the behavior at LU stations, which can then be expanded to other metro systems.

Methods

The exact train loadings were defined as well as the different situations to be tested, which were based on the observation of two weeks of CCTV footage at Green Park (without PEDs) and Westminster (with PEDs). Both stations presented the same platform layout and similar demand profiles.

The operational variables at Green Park and Westminster were recorded during the most congested hours of the day (8:15 to 9:15 am and 5:15 to 6:15 pm), reaching a flow of 30 train/h. To do this, 15 days (November 2014) of data were collected using the software Observer XT 11 and the videos were converted into AVI format.

For this study, the loads used were as described in Table 1, in which high-density situations are reached. Three scenarios of ratio (R) between boarding and alighting were defined (R = 4, R = 1, R = 0.25). Each of these scenarios was tested with PEDs and without PEDs. The LC_0 and LC_1 loads were only tested to prepare passengers for each day and to check initial values or boundaries of the experiment when there were no passengers inside the train or on the platform. In the case of LC_5 this scenario was used to calculate the total load of the train.

These scenarios were simulated at PAMELA using a mock-up of an underground tube carriage and a portion of the platform with similar characteristics to Green Park (without PEDs) and Westminster (with PEDs). The mock-up was 10.00 m long and 2.65 m wide, with 20 seats (12 fixed seats and 8 tip-up seats), and two double doors 1.6 m wide. This produced a total floor area of 17.46 m^2, which allowed a capacity of 90 passengers (for a density of 4 pass/m^2) or 142 passengers (for a density of 7 pass/m^2) inside the train. The horizontal gap between the train and the platform was equal to 90 mm, while the vertical gap was 170 mm (without

Table 1. Loads used in the Experiment at PAMELA

Load condition code	Boarding passengers per door	Alighting passengers per door	On-board passengers per door	Ratio (boarding/alighting)	Number of runs per scenario
LC_0	55	0	0	–	2
LC_1	0	55	0	–	2
LC_2	40	10	5	4	10
LC_3	10	40	5	0.25	10
LC_4	20	20	15	1	10
LC_5	55 + crush	0	0	–	2

PEDs) and 0 mm (with PEDs). The platform was 10.00 m long and 3.30 m wide.

The PTI was defined as the space between the train doors and PEDs (similar to Westminster), and in the case without PEDs (similar to Green Park) it was the space between the train doors and the yellow safety line on the platform.

The analysis focused on the period between the train doors opening and closing (i.e., after the train arrived). For this simulation, 110 participants were recruited to form 11 groups of 10 passengers each. In addition, boarding passengers used red hats and alighting passengers used white hats, and each set of 10 passengers wore different colored bibs on which each passenger had a unique number. Therefore, each passenger was identified by their bib color, hat color, and number. This produced an input density on the platform of 3.3 pass/m^2 (when all passengers were standing on the platform) and 5.15 pass/m^2 inside the car (when all passengers were inside the train).

Passengers at the experiment were instructed to walk "naturally" as if they were boarding and alighting from a train in the LU. To make sure that this behavior was represented over time, random groups were chosen to board, alight, or remain inside the carriage. In addition, a complete sound system was provided in order to make the experiment feel real for the participants. The sound included the train arriving, braking, door opening alarm, door closing alarm, and departure.

The cameras were located in the ceiling (4 m height), which enabled the recording of a space on the platform of only 2.4 m wide by 5.0 m long in front of each train door (which produced an observed area on the platform A_p = 12 m^2).

Using the tracking software Petrack (*37*), the position (x, y) of each alighting passenger A_i was recorded each time he/she exited the PTI zone. Therefore, the time step ($\Delta t = i - (i-1)$) was defined as the difference in seconds between two consecutive alighters (A_i and A_{i-1}) who exited the PTI zone.

In this work the interaction between the first passenger alighting and the first passenger boarding was not considered, therefore i = 2,..,N_a (N_a = total number of passengers who alighted per door).

In addition, Petrack was used to track the number of passengers around A_i. Each alighter A_i had at least 4 passengers around him/her (front, back, left and right) who could be

Figure 2. Example of Petrack used to track the position of A_i (i = 3) when R = 1 at PAMELA.

alighting or boarding. The following criteria were used to select those passengers X_i who were around A_i:

- Passenger A_i should have a clear view of passengers X_i, that is, if the angle between A_i and X_i is smaller than five degrees then X_i is not tracked; and
- Passenger A_i should be closer to passengers X_i, that is, if the distance between A_i and X_{i+1} is double the distance between A_i and another passenger X_i then X_{i+1} is not tracked.

For example, Figure 2 shows the position of A_i (1) and seven other passengers around him/her. Position 5 and 8 were alighting passengers located in front and at the back of A_i, respectively, while positions 2, 3, 4, 6, and 7 represented boarding passengers around A_i.

Following the example in Figure 2, the position of passengers around each A_i was plotted to represent the space used of each alighter A_i, which represented an asymmetrical ellipse. The area of each asymmetrical ellipse was calculated using an approximation of triangles between the position of A_i and the surrounding passengers X_i who were boarding (B_i or B_{i+1}) or alighting (A_{i+1} or A_{i-1}). According to Heron's Formula, the area of each triangle i can be obtained using Equation 1. The sum of all triangles will be the area of the asymmetrical space (AS) for A_i (see Equation 2). The distance between A_i and A_{i+1} is defined as longitudinal front radius. The longitudinal back radius is the distance between

Table 2. Average Dimensions of Asymmetrical Ellipse for Each Alighter (A_i) Surrounded by Passengers X_i when PEDs are Used at PAMELA

R	Number A_i tracked	Number X_i tracked	Longitudinal front radius (cm)	Longitudinal back radius (cm)	Lateral right radius (cm)	Lateral left radius (cm)
4	68	227	63.23	61.29	89.02	78.86
1	150	523	76.74	59.80	85.35	77.46
0.25	232	714	79.45	61.65	95.05	82.41

A_i and A_{i-1}. The distance between A_i and B_i (or B_{i+1}) is defined as the lateral right or left radii.

$$A_{\text{triangle}} = \sqrt{(t \cdot (t-a) \cdot (t-b) \cdot (t-c)}, \quad \text{where } t = (a+b+c)/2 \tag{1}$$

$$AS = \sum_{i=2}^{N_a} \left(A_{\text{triangle}} \right)_i \tag{2}$$

where a, b, and c are the length of the sides of each triangle i, obtained using the Euclidian method between A_i and the surrounding passengers tracked with Petrack. The number of triangles is equal to the number of passengers around each Ai.

The AS was compared with the overall space (OS), which was obtained as the total platform area divided by the total number of passengers on the platform. In other words the AS only considered those passengers surrounding alighters, while the OS was obtained by counting all passengers on the platform. This comparison was done for each value of R.

Finally, an ANOVA test single factor (significance level of 5%) was done to compare if there were significant differences in the PS over the different values of R, comparing the case with PEDs and without PEDs. The null hypothesis was defined as the samples having the same mean.

Results

Passenger Demographics

The subjects used in PAMELA were volunteers: 46% men and 54% women; 78% of them were regular users of the LU and most were under 45 years old (15% were under 24 years, 26% 25–34, 19% 35–44, 27% 45–59, 7% 60–64, and 7% more than 65 years old). The total passenger load tested in the scenario LC_0 and LC_1 was 8221 kg (including seated passengers). The average height of passengers was 170 cm with a deviation standard of 8 cm.

Dimensions of Asymmetrical Ellipse

Table 2 shows the average dimensions of the asymmetrical ellipse for each passenger alighting (A_i) in the different scenarios of ratio between boarding and alighting (R) at PAMELA when PEDs are used. All cases (total tracked of 450 alighters) of R presented smaller longitudinal back radius than the longitudinal front radius, reaching up to a 22.4% difference when R = 0.25. The standard deviation of the longitudinal front radius was about 26 cm for all cases of R, and the longitudinal back radius reached a standard deviation in the range of 14 cm to 19 cm.

With respect to lateral radii, Table 2 shows that passengers alighting maintained more distance from the left side than from the right side, reaching up to 15% in difference when R = 0.25. This was produced in all scenarios of R (a total of 1464 passengers X_i tracked around the total of 450 passengers A_i). The standard deviation of the lateral left radius was around 25 cm; the lateral right radius in R = 1 reached almost 10 cm lower standard deviation compared with R = 0.25 and R = 4.

An ANOVA test single factor was used with a significance level of 5% (α = 0.05) to compare the longitudinal front radius and longitudinal back radius for each value of R. The null hypothesis (H_0) was defined as the samples having the same mean. The results of the ANOVA showed that the p-value was lower than 0.05. This means that the null hypothesis is rejected, that is, there is significant difference between the longitudinal front radius and the longitudinal back radius in each case of R. The same statistical test was performed to compare the lateral right radius and lateral left radius for each value of R. The results showed significant differences for these two variables in each case of R.

Space Used by Alighting Passengers

Figure 3 shows the longitudinal and lateral radii plotted for each scenario of R when PEDs are used at PAMELA. The coordinate (0,0) represents the alighting passenger A_i, who is surrounded by passengers X_i (who were boarding or alighting). The shape of the PS changed with respect to each value of R.

Figure 4 shows the average AS (using Equation 2) for each passenger alighting (A_i) when PEDs are used. The x-axis shows the number of passengers alighting when they came out from the doors (i = 2,..,N_a). The variable AS followed a "U" curve approximated by polynomial equations of order 3, reaching a coefficient of determination of 0.87 (R = 4), 0.85 (R = 1) and 0.79 (R = 0.25).

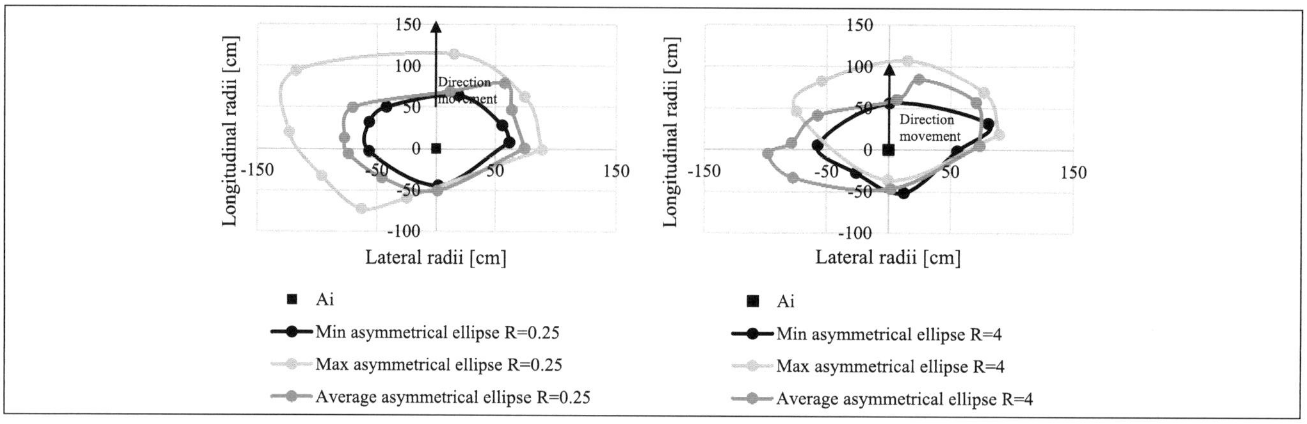

Figure 3. Shape of the asymmetrical ellipse for R = 0.25 (left) and R = 4 (right) when PEDs are used at PAMELA.

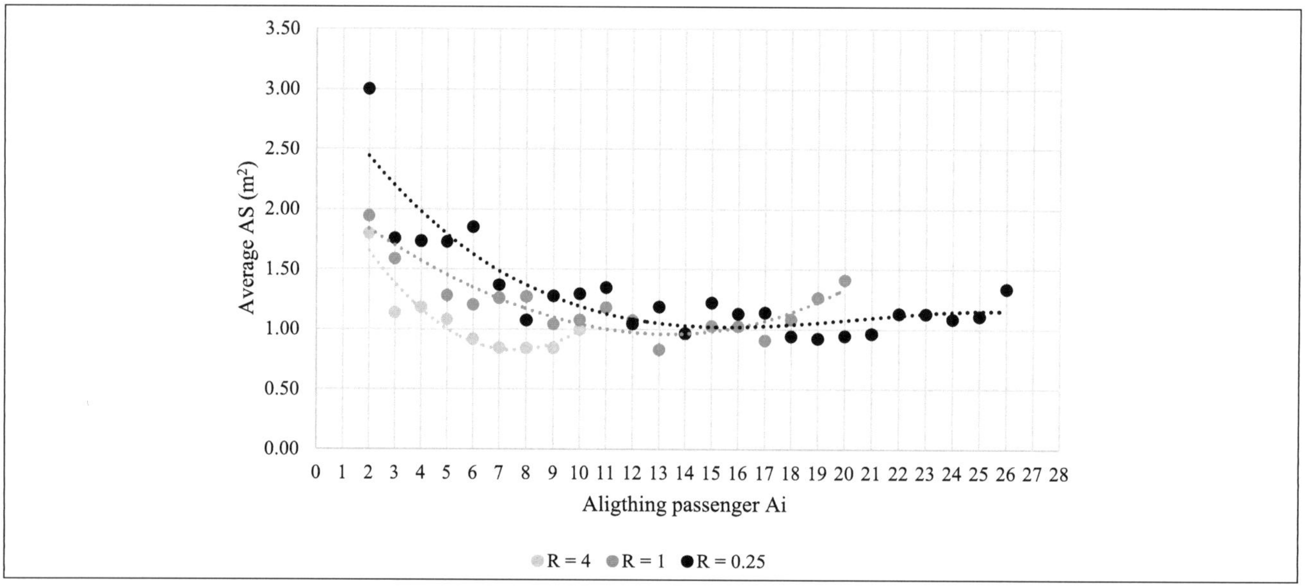

Figure 4. Average asymmetrical space (AS) of each passenger alighter (A_i) according to each R when PEDs are used at PAMELA.

With respect to minimum values of AS in Figure 4, when R = 1 there are 0.83 m^2/pass or LOS E (passenger A_{13} from a total of 20 alighters). When R = 4 and R = 0.25 the minimum values were slightly higher, reaching 0.84 m^2/pass (passenger A_7 from a total of 10 alighters) and 0.92 m^2/pass (passenger A_{19} from a total of 26 alighters), respectively. In all the cases of R, the minimum values of AS presented a LOS = E.

Regarding maximum values of AS, Figure 4 shows that passengers alighting reached a LOS C in the case of R = 1 (1.94 m^2/pass) and R = 4 (1.80 m^2/pass). However, in the case of R = 4 a LOS B was obtained with 3.0 m^2/pass on average, which is 70.45% higher with respect to the following passenger alighting. These values are presented in the early stages of the alighting process (second passenger alighting).

The AS can be compared with the overall passenger space (OS). On average, the AS for alighters presented a LOS D for all cases of R; however, the OS reached up to LOS F for R = 4. In other words, the OS reached up to 0.60 m^2/pass difference compared with the AS when R = 4. In the case of R = 1 this difference is slightly lower, reaching 0.56 m^2/pass, whereas in R = 0.25 the difference is reduced to 0.30 m^2/pass.

Similar "U" curves were found in the case without PEDs. An ANOVA test (one factor and significance level of 5%) was performed to compare the AS with PEDs and without PEDs for each value of R. The results show that the only case that presented significant differences was the comparison between R = 0.25 and R = 1, in which the *p*-value was lower than $\alpha = 0.05$. In the case of R = 4 there were no significant differences.

Discussion

In this paper a new method is presented to estimate the PS needed for alighting when PEDs are used. Significant differences in the dimensions of the asymmetrical ellipse were reached for each scenario at PAMELA. The average values for all the three cases of R (4, 1, and 0.25) showed that the lateral left radius was smaller than the lateral right radius.

The difference between the lateral radii could be caused because passengers preferred to maintain a certain distance to avoid collision. This distance can be considered as intimate when a value lower than 1.0 m is reached between the heads of two passengers (2 times the body depth defined in Fruin [*11*], plus 0.5 m defined elsewhere [*12–14*]).

Therefore, the results of this work showed that, on average, the lateral distance between passengers alighting and boarding (pair A_i–B_i) was around 80 cm, which means that passengers felt a high LOI, extending the study in Seriani et al. (*10*). However, this distance could be influenced by the behavior of passengers boarding and the location of the exit gate on the platform, because alighting formed lanes of flow; this effect could be considered for further research.

Similarly, it seems that passengers alighting preferred to maintain a greater distance in front of them than behind them as a result of collision avoidance techniques. On average, the longitudinal front and back radius reached a value lower than 100 cm, showing a high LOI, extending the research in Seriani et al. (*10*). The results also showed that the value of R had an impact on the longitudinal front radius. In contrast, in all the cases of R passengers maintained a similar distance from behind.

With respect to the area used by passengers alighting when PEDs are used, the AS is more representative of the space used by passengers than the OS, reaching a bigger difference when R = 4. The differences could be because the OS considered all passengers on the platform, whereas the AS considered only those who were surrounding each alighter. Therefore, AS can be used to design the PTI area and OS to design the whole platform.

In terms of shape, the AS followed "U" curves over time. This may be because the first passengers who alighted had more space on the platform than the rest of the passengers. In addition, toward the end of alighting, alighters could have more space between them as the supply of alighters from the seating sections of the carriage decreases. This could be related to the field of vision of each passenger, which was not covered in this work. However, further experiments can be carried out at PAMELA and the results can be compared with existing laboratory studies (*30*), in which participants used an eye camera to identify their space.

Compared with existing studies, on average the AS reached a lower value than obtained by Gérin-Lajoie et al. (*28*) and Templer (*31*) in walkways. This could be because the number of passenger alighting increased over time, producing congestion in the PTI.

Overall, and bearing in mind the methodological limitations, there seems to be no impact of PEDs on the AS when R = 4. However, in the other two cases (R = 1 and R = 0.25) the presence of PEDs produced a significant effect on the AS. However, this impact could be related to the level of demand at the PTI rather than the use of PEDs. When R = 4 the PTI is totally congested, allowing little space for passengers to get off the carriage, and forming only one narrow single lane of flow, while in the cases R = 1 and R = 0.25, because of the lower demand level (compared with R = 4) at the PTI, the number of alighting lanes of flow varied between one and two lanes, affecting the AS. Therefore, further research is needed to study the relationship between the formation of lanes and the presence of PEDs.

Conclusions

This paper has studied the estimation of the PS for alighting at metro stations with PEDs by means of laboratory experiments.

The results show that the PS can be considered as an asymmetrical ellipse for passengers alighting, in which the longitudinal and lateral radii are affected by the collision avoidance of a person in front of him/her and by the interaction with other passengers alighting or waiting to board the train.

In contrast to microscopic pedestrian models, the results show that the shape and value of PS varies over time, forming "U" curves in which the first passengers reach a higher PS than the last passengers.

There is no clear advantage or disadvantage of PEDs versus no PEDs in terms of PS, where demand plays a more important role. In any case, there seems to be no clear negative impact of PEDs on PS.

The results in this paper are based on two LU stations. However, they could be considered as a starting point to study the use of PEDs in other transport systems.

Further research should deepen the understanding of the relationship between PEDs and formation of lanes of passengers alighting or boarding. It would also be interesting to collect more data to identify the impact of passengers with encumbrances (luggage, shopping, buggy, backpacks) or mobility aids (wheelchair, pram) on the PS.

Acknowledgments

The authors would like to thank all the participants and the PAMELA team for helping with the experiments. Also the authors are grateful to London Underground Limited for providing the data used in this research.

Author Contributions

Study conception and design: S. Seriani, T. Fujiyama; data collection: S. Seriani; analysis and interpretation of results: S. Seriani, T. Fujiyama; draft manuscript preparation: S. Seriani. All authors reviewed the results and approved the final version of the manuscript.

References

1. Seriani, S., and R. Fernandez. Planning Guidelines for Metro-bus Interchanges by Means of a Pedestrian Microsimulation

Model in Chile. *Transportation Planning and Technology*, Vol. 38, No. 5, 2015, pp. 569–583.
2. *Platform Train Interface Strategy*. Rail Safety and Standards Board (RSSB). London, 2015.
3. Law, C. K., and P. S. Yip. An Economic Evaluation of Setting Up Physical Barriers in Railway Stations for Preventing Railway Injury: Evidence from Hong Kong. *Journal of Epidemiology and Community Health*, Vol. 65, No. 10, 2011, pp. 915–920.
4. Kyriakidis, M., R. Hirsch, and A. Majumdar. Metro Railway Safety: An Analysis of Accident Precursors. *Safety Science*, Vol. 50, No. 7, 2012, pp. 1535–1548.
5. Kroes, E., M. Kouwenhoven, L. Debrincat, and N. Pauget. Value of Crowding on Public Transport in Île-de-France, France. *Transportation Research Record: Journal of the Transportation Research Board*, 2014. 2417: 37–45.
6. Le Clech, A. Heat Extraction in Underground Railway Tunnels. *Eurotransport Transport Research Laboratory*, Vol. 3, No. 1, 2005, pp. 35–40.
7. Qu, L., and W. K. Chow. Platform Screen Doors on Emergency Evacuation in Underground Railway Stations. *Tunnelling and Underground Space Technology*, Vol. 30, 2012, pp. 1–9.
8. De Ana Rodríguez, G., S. Seriani, and C. Holloway. Impact of Platform Edge Doors on Passengers' Boarding and Alighting Time and Platform Behavior. *Transportation Research Record: Journal of the Transportation Research Board*, 2016. 2540: 102–110.
9. Seriani, S., G. De Ana Rodríguez, and C. Holloway. The Combined Effect of Platform Edge Doors and Level Access on the Boarding and Alighting Process in the London Underground. *Transportation Research Record: Journal of the Transportation Research Board*, 2017. 2648: 60–67.
10. Seriani, S., T. Fujiyama, and C. Holloway. Exploring the Pedestrian Level of Interaction on Platform Conflict Areas at Metro Stations by Real-scale Laboratory Experiments. *Transportation Planning and Technology*, Vol. 40, No. 1, pp. 100–118.
11. Fruin, J. J. Designing for Pedestrians: A Level-of-service Concept. *Highway Research Record*, Vol. 377, 1971, pp. 1–15.
12. Pushkarev, B., and J. Zupan. *Urban Space for Pedestrians*. The MIT Press, Cambridge, Massachusetts, 1975.
13. Hall, E. *The Hidden Dimension*. Doubleday, Garden City, N.Y., 1966.
14. Sommer, R. *Personal Space: The Behavioral Bases of Design*. Prentice Hall, N.J., 1969.
15. Schmidt, D. E., and J. P. Keating. Human Crowding and Personal Control: An Integration of the Research. *Psychological Bulletin*, Vol. 86, No. 4, 1979, pp. 680–701.
16. *Station Planning Standards and Guidelines*. London Underground Limited, London, 2012.
17. Cox, T., J. Houdmont, and A. Griffiths. Rail Passenger Crowding, Stress, Health and Safety in Britain. *Transportation Research Part A*, Vol. 40, No. 3, 2006, pp. 244–258.
18. Evans, G. W., and R. E. Wener. Crowding and Personal Space Invasion on the Train: Please Don't Make Me Sit in the Middle. *Journal of Environmental Psychology*, Vol. 27, No. 1, 2007, pp. 90–94.
19. *Health and Safety Effects of Rail Crowding: Hazard Identification (T307)*. Rail Safety and Standards Board (RSSB), London, 2005.
20. Lam, W. H., C. Y. Cheung, and C. F. Lam. A Study of Crowding Effects at the Hong Kong Light Rail Transit Stations. *Transportation Research Part A*, Vol. 33, No. 5, 1999, pp. 401–415.
21. Mahudin, N. D. M., T. Cox, and A. Griffiths. Measuring Rail Passenger Crowding: Scale Development and Psychometric Properties. *Transportation Research Part F*, Vol. 15, No. 1, 2012, pp. 38–51.
22. Kim, K. M., S. P. Hong, S. J. Ko, and D. Kim. Does Crowding Affect the Path Choice of Metro Passengers? *Transportation Research Part A*, Vol. 77, 2015, pp. 292–304.
23. Zhang, Q., B. Han, and D. Li. Modeling and Simulation of Passenger Alighting and Boarding Movement in Beijing Metro Stations. *Transportation Research Part C*, Vol. 16, 2008, pp. 635–649.
24. Davidich, M., F. Geiss, H. G. Mayer, A. Pfaffinger, and C. Royer. Waiting Zones for Realistic Modelling of Pedestrian Dynamics: A Case Study Using Two Major German Railway Stations as Examples. *Transportation Research Part C*, Vol. 37, 2013, pp. 210–222.
25. Helbing, D., L. Buzna, and A. Johansson. Self-organized Pedestrian Crowd Dynamics: Experiments, Simulation, and Design Solutions. *Transportation Science*, Vol. 39, No. 1, 2005, pp. 1–24.
26. Langston, P. A., R. Masling, and B. N. Asmar. Crowd Dynamics Discrete Element Multi-circle Model. *Safety Science*, Vol. 44, No. 5, 2006, pp. 395–417.
27 Bandini, S., M. Mondini, and G. Vizzari. Modelling Negative Interaction Among Pedestrians in High Density Situations. *Transportation Research Part C*, Vol. 40, 2014, pp. 251–270.
28. Gérin-Lajoie, M., C. L. Richards, and B. J. McFadyen. The Negotiation of Stationary and Moving Obstructions During Walking: Anticipatory Locomotor Adaptations and Preservation of Personal Space. *Motor Control*, Vol. 9, No. 3, 2005, pp. 242–269.
29 Gérin-Lajoie, M., C. L. Richards, J. Fung, and B. J. McFadyen. Characteristics of Personal Space During Obstacle Circumvention in Physical and Virtual Environments. *Gait & Posture*, Vol. 27, No. 2, 2008, pp. 239–247.
30. Kitazawa, K., and T. Fujiyama. Pedestrian Vision and Collision Avoidance Behavior: Investigation of the Information Process Space of Pedestrians Using an Eye Tracker. In *Pedestrian and Evacuation Dynamics 2008*, Springer Berlin Heidelberg, 2010, pp. 95–108.
31. Templer, J. A. Human Territoriality and Space Needs on Stairs. In *The Staircase: Studies of Hazards, Falls, and Safer Design*. MIT Press, Cambridge, MA, 1992, pp. 61–70.
32. Fruin, J. J. *Designing for Pedestrians: A Level-of-Service Concept*. PhD dissertation, Polytechnic Institute of Brooklyn, 1970.
33. Hoogendoorn, S. P., and W. Daamen. Pedestrian Behaviour at Bottlenecks. *Transportation Science*, Vol. 39, No. 2, 2005, pp. 147–159.
34. Daamen, W., Y. Lee, and P. Wiggenraad. Boarding and Alighting Experiments: An Overview of the Set Up and

Performance and Some Preliminary Results on the Gap Effects. In *Transportation Research Record: Journal of the Transportation Research Board*, 2008. 2042:71–81.

35. Fernandez, R., P. Zegers, G. Weber, and N. Tyler. Influence of Platform Height, Door Width, and Fare Collection on Bus Dwell Time. Laboratory Evidence for Santiago de Chile. *Transportation Research Record: Journal of the Transportation Research Board*, 2010. 2143: 59–66.

36. Fujiyama, T., R. Thoreau, and N. Tyler. The Effects of the Design Factors of the Train-platform Interface on Pedestrian Flow Rates. *Pedestrian and Evacuation Dynamics 2012*, 2014, pp. 1163–1173.

37. Boltes, M., and A. Seyfried. Collecting Pedestrian Trajectories. *Neurocomputing*, Vol. 100, 2013, pp. 127–133.

The Standing Committee on Rail Transit Systems (AP065) peer-reviewed this paper (18-02096).

Article

TRR
JOURNAL OF THE TRANSPORTATION RESEARCH BOARD

Transportation Research Record
2018, Vol. 2672(8) 316–326

DOI: 10.1177/0361198118773185
journals.sagepub.com/home/trr
SAGE

Demand-Driven Transfer Coordination of Last Trains in Metro Network

Feng Zhou[1], Yanan Li[1], and Ruihua Xu[1]

Abstract
Under the condition of metro network operation with one-ticket transfer, coordination of last trains is especially important because it is the last chance for many travelers to arrive at their destination. Considering passenger flow demand and operating time threshold, this paper proposes a methodology for the effective coordination of last trains, which can not only maximize the total number of passengers transferring successfully in the last-train period, but also improve network accessibility and shorten the time span of network accessibility. The hierarchical progressive algorithm and double-level evaluation indexes are put forward and applied to a case study of the Guangzhou Metro network. The results show that the number of passengers transferring successfully has increased by 10.4% and the method improves accessibility to the network. The research achievements can be applied to Apps and websites to provide passengers with travel information guidance, and can also be used to control the automatic vending machines to sell tickets.

The rapid development of urbanization and motorization has aggravated the crowding of road traffic in urban areas. Metro, as an efficient transport system, is now the mainstay of urban passenger transport in many megacities (*1*). Metro is characterized by a high level of safety, wide accessibility, and high energy performance, and offers reliable service with adequate punctuality (*2*). However, the last-train coordination problem is a significant challenge for metro system planners and operators to improve the level of service. As most of the metro systems do not provide passengers with 24-hour service (typically for track or fleet maintenance), the last trains are the last chances for many passengers to reach their final destinations. The metro system is under network operation, and the accessibility of the metro network greatly decreases as metro lines cease to operate one by one in the evening. Therefore, the coordination of last trains is especially important to meet the passengers' travel demand.

The last-train transfer coordination problem is especially complicated as it involves transfers among numerous trains, stations, and different coordination directions. There have been many works on the study of the non-last-trains scheduling problem, and various studies have attempted to address the issue of planning coordinated train schedules. Cevallos and Zhao, Wong et al., and Shafahi and Khani proposed optimized models to minimize transfer times (*3–5*), and Xu et al. built a multi-objective subway timetable optimization approach with minimum passenger time (*6*). However, the current methods and models cannot solve the last-trains coordination problem well. First, the effect of transfer failure in the last-trains period is more serious than normal/peak hours, as it is the final chance for passengers to take trains or transfer. The second difference is that the passenger flow is relatively small and the capacity of last trains is generally sufficient in the last-train period. Moreover, the trade-off involves level of service and operating costs. It is unnecessary to extend operating time and compress the interval of trains in the last-train period beyond the passengers' demand.

Some research has addressed last-train coordination but primarily attempted to reduce passenger waiting time (*7*) or transfer connection time (*8–10*) based on physical and operational characteristics. Little research has paid attention to the passenger flow demand during the last-train period (*9*), or the demand of passengers was always assumed or forecasted (*11–13*). Thus, much existing research may not be practical for application in the field.

This paper proposes a methodology for effective coordination of last trains considering passenger flow demand and operating time threshold. The corresponding

[1]Key Laboratory of Road and Traffic Engineering of the Ministry of Education, College of Transportation Engineering, Tongji University, Shanghai, P.R. China

Corresponding Author:
Address correspondence to Ruihua Xu: rhxu@hotmail.com

Table 1. Summary of Previous Last Train Coordination Studies

Researcher	Methodology	Passenger demand data	Operational constraints	Algorithm
Xu et al. (2008)	Calculated departure time domains for multiple-direction transfers	Not considered	Objective function	Hierarchical coordination
Xu and Li (2012)	Optimized the connection scheme of the last trains	Forecasted data	Time span mentioned	Reference to Kruskal algorithm
Li et al. (2013)	Compiled network-level timetables based on operator-specific policies	Assumed by operators	Mentioned	Pick-and-filter algorithm
Zhou et al. (2013)	Reduced transfer waiting time and improved passengers' accessible reliability	Randomly generated	Mentioned	Genetic algorithm
Kang et al. (2014)	Maximized average transfer redundant time	Not considered	Minimum-maximum operational time constraints	GSA algorithm
Kang et al. (2015)	Minimized passenger transfer connection headways	Not considered		Genetic algorithm
Li et al. (2017)	Maximized number of passengers able to board with genetic algorithm	Actual AFC data	Mentioned	Genetic algorithm

algorithm is developed to guarantee more traveling demand in the last-trains period. Specifically, actual passenger flow demand data are acquired by automatic fare collection (AFC) system, and the method has been applied to the Guangzhou metro system. Moreover, evaluation indexes are put forward including: transfer-level index (Volume of transfer passengers and Number of coordinated directions in last-trains period) and network-level index (Time span for last trains, Network accessibility distribution, and Time span of network accessibility).

The remainder of the paper is organized as follows. In the next section, relevant literature on transfer coordination and present a summary of last-train coordination studies is reviewed. The methodology is then introduced in detail, including the model for transfer-coordinated optimization of last trains and the evaluation approaches. The corresponding algorithm is developed, and the proposed methodology is then applied on the Guangzhou metro network as a case study; and in the final section, conclusions and research extensions are discussed.

Literature Review

On the research of coordination for operation plan, scholars put emphasis on optimization of the headway (*14*), the departure time (*9*), or dwell time (*10*), making connections of bus routes and metro lines (*15, 16*). Most of the objective function is to minimize the total passenger waiting time. Some other research has aimed at minimizing the transfer time at transfer stations (*17, 18*). The preceding problems are usually solved by proposing mixed integer programming models (*4, 5*). Artificial intelligence algorithms are also developed to search for optimal solutions, such as genetic algorithm (*10, 14*) and gravitational search algorithm (GSA) (*9, 18*).

The last-trains coordination problem has attracted increasing attention from researchers. Most of the studies have attempted to reduce passenger waiting times during transfers, following the previous methodology. However, transfer coordination in last-train periods is different from other periods, as mentioned earlier. Table 1 summarizes previous last-train coordination studies, reflecting the lack of consideration in terms of demand and an algorithm that is less practical for field application. Li et al. (*19, 20*) took passenger demand into consideration and proposed a methodology to optimize the number of passengers who can reach their final destination, using AFC data, but the algorithm checks each AFC data point with a corresponding path and train diagram, leading a higher space complexity and time complexity.

This paper focuses on how to maximize the passengers' demand during the last-train period also considering the demand of metro operators. In addition, a practical and effective algorithm is introduced.

Methodology

A demand-driven model for coordinated schedule of last-train transfers is proposed. The double-level evaluation indexes are also illustrated in this section.

Transfer Coordination Model for Last Trains

Basic Assumption. The metro system is complex and dynamic. To facilitate model formulation without losing generality, several assumptions are made throughout the paper.

(1) The capacity of each of the last trains is sufficient to carry all passengers as there are fewer passengers on the last trains;
(2) All trains run accurately according to the schedules in the last-trains period. As the timetables for last trains are published to the public in advance, so all trains need to run on time in the last-trains period. In addition, there are few disturbances for last-train operation based on assumption (1). Thus, assumption (2) is considered to be reasonable;
(3) The last trains stop at each station, and short routing is not considered.

Definition of Parameters.

(1) Train diagram

The train diagram illustrates the relationship between space and time for train operation. The main information conveyed by the train diagram is section running times, station dwelling times, and station arrival times and departure times. Necessary elements for the model formulation are summarized.

The following are the necessary parameters:

$t^{i,d}_{A,s}/t^{i,d}_{D,s}$	Arrival time/Departure time at the sth station in the *d* direction of Line *i* (*d*= 0, if the direction is upward; *d*= 1, if the direction is downward).
$t^{i,d}_{R,s,s+1}$	Running time from the sth station to the (s + 1)th station in the *d* direction of Line *i*.
$t^{i,d}_{S,s}$	Dwelling time at the sth station in the *d* direction of Line *i*.

Therefore, the arrival time of each station can be depicted as Equation 1 and the departure time of each station can be described as Equation 2, where $t^{i,d}_{D,0}$ is the departure time at the origin station. We define $t^{i,d}_{S,0}=0$ and s starts from 1.

$$t^{i,d}_{A,s} = t^{i,d}_{D,0} + \sum_{m=0}^{s-1} t^{i,d}_{S,m} + \sum_{m=0}^{s-1} t^{i,d}_{R,m,m+1} \quad (1)$$

$$t^{i,d}_{D,s} = t^{i,d}_{D,0} + \sum_{m=1}^{s} t^{i,d}_{S,m} + \sum_{m=0}^{s-1} t^{i,d}_{R,m,m+1} \quad (2)$$

(2) Passenger transfer waiting time

The last train is usually the last chance for many passengers to transfer, therefore, successful transfer coordination is a hard constraint. For example, if a passenger needs to transfer at transfer station *H* from Line *i* downward to Line *j* upward, Equation 3 must be satisfied:

$$t^{j,0}_{D,H} \geq t^{i,1}_{A,H} + t^{k}_{iHj} \quad (3)$$

where k is an array of coordination direction depicted as (the direction of feeder train in Line *i*, the direction of the preceding connecting train Line *j*). Therefore, k= (1,0) in Formula 3; t^{k}_{iHj} is defined as transfer walking time at station *H* from Line *i* to Line *j* in coordination direction k.

Thus, the passenger transfer waiting time defined as $t^{W,k}_{iHj}$ can be determined by Equation 4:

$$t^{W,k}_{iHj} = t^{j,d}_{D,H} - t^{k}_{iHj} - t^{i,d}_{A,H} \quad (4)$$

In addition, the value of $t^{W,k}_{iHj}$ for each *H* should be positive if passengers travel through several transfer stations.

Demand-Driven Transfer Coordination Modeling. The basic aim of transfer coordination is to meet the traveling demand of passengers in the last-train period. In other words, the objective function optimizes the number of passengers who can transfer successfully during the last-train period. Considering the demand of operators, operation time domains must be controlled in order to reserve sufficient time for maintenance and crew rostering.

The necessary parameters of the model are as follows.

The decision variables of the proposed model are:

T	The set of transfer stations in a network, $H \in \mathbf{T}$.
L	The set of lines in a network, $i \in \mathbf{L}, i=1, 2, \ldots$.
d	*d* = 0, if the direction is upward; *d* = 1, if the direction is downward.
n	The number of coordination directions at transfer station *H*, $n = (N_e)^2 + 4(N_t)^2 + 4(N_t)(N_e - 1) - N_e$ where N_t is the number of lines passing through *H* and N_e is the number of lines ending up at *H*. For example, if two lines come across at *H* ($N_e = 0, N_t = 2$), then $n = 8$.
Q^k_H	The transfer passenger flow at transfer station *H* in the *k* th coordination direction, *k* = 1,..., *n*.
$t^{W,k}_H$	Transfer waiting time at transfer station *H* in the *k* th coordination direction.
$t^{i,d}_{D,O}$	Departure time at the origin station O in the *d* direction of Line *i*.
$t^{i,d}_{A,D}$	Arrival time at the destination station D in the *d* direction of Line *i*.
$t^{i,d}_D/t^{i,d}_A$	Departure/Arrival time in the *d* direction of Line *i* according to operational policy.

Thus, the demand-driven transfer coordination model in the last-train period is built as follows:

x_H^k	$x_H^k \in \{0, 1\}$: 1 if the k th coordination direction at station H is successful; and 0 otherwise.
X	The set of decision variables, $\{X \mid x_H^k, H \in T, k = 1, \ldots, n\}$

$$\max F(X) = \sum_{H \in T} \sum_{k=1}^{n} x_H^k Q_H^k \tag{5}$$

Subject to:

(1) Transfer coordination constraints:

$$x_H^k = \begin{cases} 1, t_H^{W,k} \geq 0 \\ 0, t_H^{W,k} < 0 \end{cases} \tag{6}$$

Here, we define M is a sufficient large positive number. Thus, Equation 6 equals Equation 7:

$$M(x_H^k - 1) \leq t_H^{W,k} < M t_H^{W,k} \tag{7}$$

(2) Operator demand constraints:

$$\min\left(t_D^{i,d}\right) \leq t_{D,O}^{i,d} \leq \max\left(t_D^{i,d}\right) \tag{8}$$

$$\min\left(t_A^{i,d}\right) \leq t_{A,D}^{i,d} \leq \max\left(t_A^{i,d}\right) \tag{9}$$

Double-Level Evaluation Index

Evaluation Indexes can be classified into two levels: transfer-level indexes focus on coordination, and network-level indexes pay more attention to network accessibility. Each index is defined as follows.

Transfer-Level Indexes

(1) Volume of transfer passengers: The total number of passengers that successfully transfers for each transfer station in all the coordination direction in last-train period.
(2) Number of coordinated directions: The maximum number of coordinated transfer directions of each transfer station.

Network-Level Indexes

(1) Time span for last trains: The time span between the latest arrival time and the earliest departure time of the last trains of each line in the network.
(2) Network accessibility distribution: As network accessibility gradually decreases in the last-train period, so the distribution of network accessibility with time can describe its change rule.
(3) Time span of network accessibility (τ): The time span between τ'' (the last origin-destination [OD] pair from reachable to unreachable) and τ' (the first OD pair from reachable to unreachable). In other words, $\tau = \tau'' - \tau'$.

Algorithm

Transfer coordination in the metro network has been proved to be an NP-hard problem (*9*). The problem is usually solved by heuristic algorithms or artificial intelligence algorithms. However, existing studies always ignore actual passenger demand, or are computationally costly, which is not suitable for field-practical applications. Therefore, a "Hierarchical progressive algorithm" was developed here to improve the applicability and effectiveness of the algorithm.

Identify Principal Coordination Relation

Considering the contradiction among transfer stations in a network and the contradiction among different coordination directions in a transfer station, it is hard to achieve bidirectional coordination successfully because of the transfer walking time. For example, Line1 (upward)→Line2 (upward) and Line2(upward)→Line1(upward) can hardly be coordinated at the same time, especially for the channel transfer (*10*). Therefore, it is important to identify principal coordination relation first. That is to say, the priority of coordination lines and directions should be determined:

Step 1: Determine the sequence of transfer station.

The transfer station is graded by the transfer passenger flow, geographical position, and the number of coordinated lines. The weight of each scoring factor is ω_1, ω_2, and ω_3, where $\omega_1 + \omega_2 + \omega_3 = 1$.

(1) Score of transfer passenger flow (a_H)

Sort the transfer passenger flow of each transfer station H descending, set the station with the maximum transfer passenger flow to 100 points, and the other transfer stations' score is determined by the percentage of the largest passenger flow:

$$a_H = \frac{Q_H}{\max(Q_H)} \times 100 \; H \in T \tag{10}$$

Here, Q_H is the transfer passenger flow at transfer station H in the last-train period.

(2) Score of geographical position (b_H)

b_H is scored by the H's location in the city area and its location in the line: an urban transfer station gets a higher score than a suburban transfer station, and the midpoint of the line gets higher score than the line endpoint. According to the above principles, first-class transfer station is 100 points, two-class transfer station is 75 points, descending successively.

(3) Score of coordination (c_H)

Set the station with the maximum coordinated lines to 100 points, and the other transfer stations' score is determined by the percentage of the largest number of coordination lines:

$$c_H = \frac{n_H}{\max(n_H)} \times 100 \quad H \in T \tag{11}$$

Above all, the final score of each transfer station H is $g_H = \omega_1 \cdot a_H + \omega_2 \cdot b_H + \omega_3 \cdot c_H$. Thus, the sequence of transfer station can be determined by the g_H in descending order.

Step 2: Determine the sequence of coordinated lines.

Among the set of lines belonging to the prioritized station (determined in Step 1), the prioritized line is determined according to the total number of transfer passengers, connection lines, and transfer stations belonging to the line.

Step 3: Determine the prioritized coordination direction referring to the transfer passenger flow of each coordination direction at the transfer station H.

Taken together, the transfer station priority is first determined, then the coordination line and the coordination direction can be determined progressively.

Hierarchical Progressive Algorithm

Step 1: Calculate arrival time/departure time of each station belonging to the prioritized line (baseline) chosen by the principle described above. Put the baseline into the determined line set **V**.

Step 2: Calculate arrival time/departure time of each station. If Line i and Line j transfer at station H, there are two situations according to the coordination direction: Line i (feeder train of baseline)→Line j (coordinated train of coordinated line) and Line j (feeder train of baseline)→Line i (coordinated train of coordinated line).

(1) Line i→Line j:

The departure time at the H th transfer station in Line j can be determined by Equation 12:

$$t_{\mathrm{D,H}}^{\mathrm{j,d}} \geq t_{\mathrm{A,H}}^{\mathrm{i,d}} + t_{\mathrm{iHj}}^{k} = t_{\mathrm{D,0}}^{\mathrm{i,d}} + \sum_{m=0}^{H-1} t_{\mathrm{S,m}}^{\mathrm{i,d}} + \sum_{m=0}^{H-1} t_{\mathrm{R,m,m+1}}^{\mathrm{i,d}} + t_{\mathrm{iHj}}^{k} \tag{12}$$

Then the departure time at origin station of Line j can be calculated by Equation 13:

$$t_{\mathrm{D,0}}^{\mathrm{j,d}} \geq t_{\mathrm{D,0}}^{\mathrm{i,d}} + \sum_{m=0}^{H-1} t_{\mathrm{S,m}}^{\mathrm{i,d}} + \sum_{m=0}^{H-1} t_{\mathrm{R,m,m+1}}^{\mathrm{i,d}} + t_{\mathrm{iHj}}^{k} - \sum_{m=0}^{H} t_{\mathrm{S,m}}^{\mathrm{j,d}} - \sum_{m=0}^{H-1} t_{\mathrm{R,m,m+1}}^{\mathrm{j,d}} \tag{13}$$

(2) Line j→Line i:

The arrival time at the H th transfer station in Line j can be determined by Equation 14:

$$t_{\mathrm{i,H}}^{\mathrm{j,d}} \leq t_{\mathrm{D,H}}^{\mathrm{i,d}} - t_{\mathrm{jHi}}^{k} = t_{\mathrm{D,0}}^{\mathrm{i,d}} + \sum_{m=0}^{H} t_{\mathrm{S,m}}^{\mathrm{i,d}} + \sum_{m=0}^{H-1} t_{\mathrm{R,m,m+1}}^{\mathrm{i,d}} - t_{\mathrm{jHi}}^{k} \tag{14}$$

Then the departure time at origin station of Line j can be calculated by Equation 15:

$$t_{\mathrm{D,0}}^{\mathrm{j,d}} \leq t_{\mathrm{D,0}}^{\mathrm{i,d}} + \sum_{m=0}^{H} t_{\mathrm{S,m}}^{\mathrm{i,d}} + \sum_{m=0}^{H-1} t_{\mathrm{R,m,m+1}}^{\mathrm{i,d}} - t_{\mathrm{jHi}}^{k} - \sum_{m=0}^{H-1} t_{\mathrm{S,m}}^{\mathrm{j,d}} - \sum_{m=0}^{H-1} t_{\mathrm{R,m,m+1}}^{\mathrm{j,d}} \tag{15}$$

To optimize the transfer waiting time, the above inequalities are given " = " to simplify the calculation.

Step 3: Determine the last-train operation time span of the coordinated line. If not exceeding the time threshold (Equation 8), the line can be stored in **V** and go to Step 4; If so, let the departure time equal to the threshold value of the threshold: if $t_{\mathrm{D,O}}^{\mathrm{j,d}} > \max\left(t_D^{\mathrm{j,d}}\right)$, then $t_{\mathrm{D,O}}^{\mathrm{j,d}} = \max\left(t_D^{\mathrm{j,d}}\right)$; else if $t_{\mathrm{D,O}}^{\mathrm{j,d}} < \min\left(t_D^{\mathrm{j,d}}\right)$, then $t_{\mathrm{D,O}}^{\mathrm{j,d}} = \min\left(t_D^{\mathrm{j,d}}\right)$. Then recalculate the departure/arrival time of each station and store the line in **V**.

Step 4: Determine the next transfer station, coordinated lines, and coordinated direction as benchmarks according to the steps in section "Identify Principal Coordination Relation". Return to Step 2, until all the lines in L have been put in **V**.

The "Hierarchical progressive algorithm" is described in Figure 1.

Case Study

In this section, the quality of the transfer coordination model and the algorithm will be illustrated by using the real case of the Guangzhou metro system. The Guangzhou metro is a system with nine lines and 156 stations (including 20 transfer stations) by 2016 (see Figure 2). Actual passenger flow demand data is acquired by the AFC system from Auto Fare Clearing Center. The results are calculated by C#. net and PL/SQL database programming on a Intel(R) Core(TM) i5-4590 CPU 3.3 GHz and 12.0 GB RAM PC. The time to obtain the result is 9.30 s. The results will be analysis according to the double-level evaluation indexes.

Calculation

First, the sequence of transfer stations is calculated (Step 1 in section "Identify Principal Coordination Relation"), and the results are presented in Figure 3.

Tiyu Xilu Station is the first base transfer station. This station connects with three lines, including Line 1, Line 3, and Line 3 North. The scores of the connected lines are calculated (Step 2 in section "Identify Principal Coordination Relation") in Table 2. Thus, Line 1 becomes the prioritized coordinated line (baseline). Calculate the time schedule including arrival time/departure time of each station belonged to the Line 1, and put Line 1 into the determined line set **V** (Step 1 in section "Hierarchical progressive algorithm").

According to the transfer passenger flow of each coordination direction (Step 3 in section "Identify Principal Coordination Relation"), the prioritized coordination

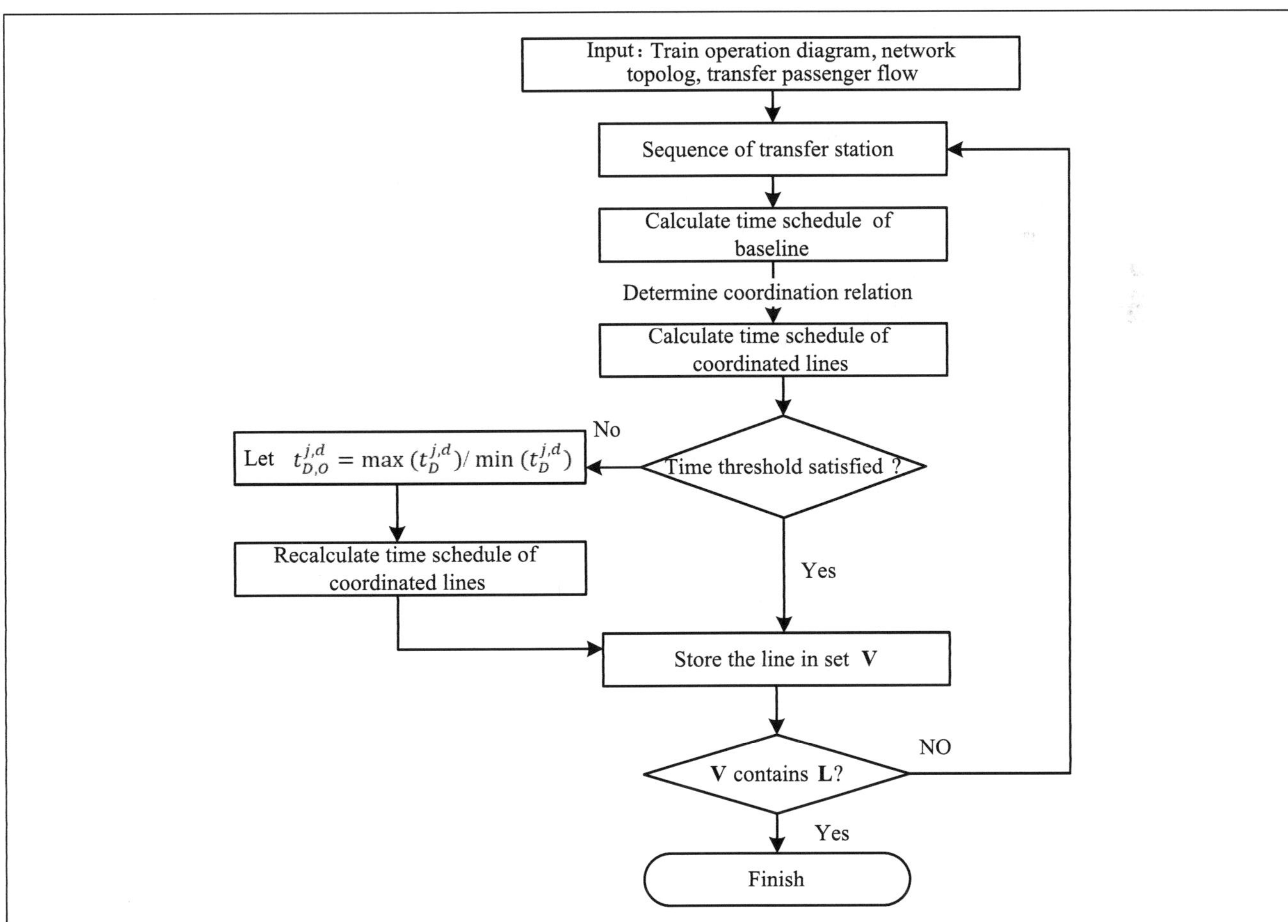

Figure 1. Flow chart.

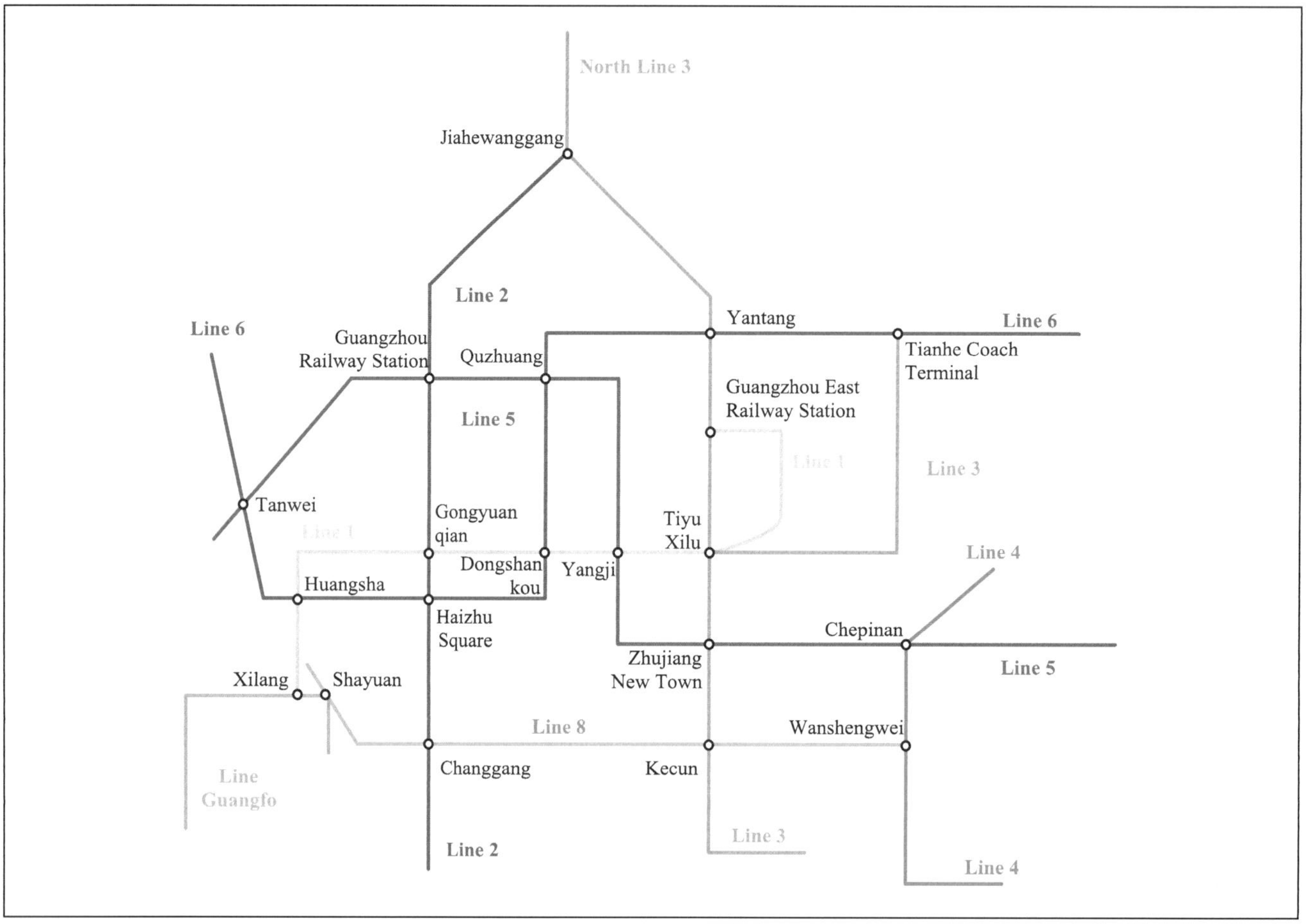

Figure 2. A sketch network based on the Guangzhou metro network.

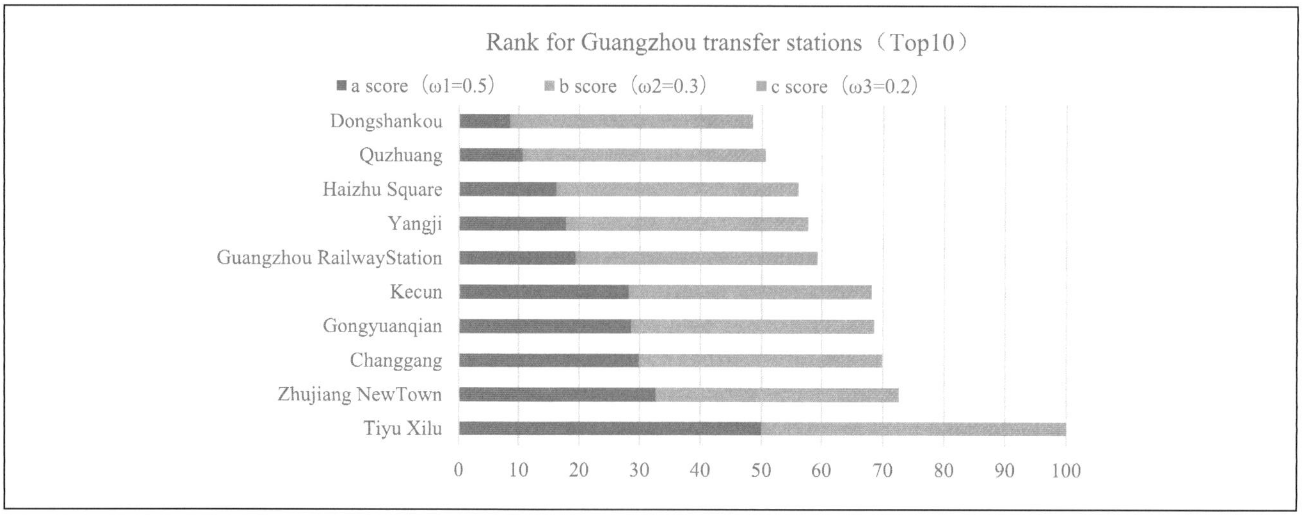

Figure 3. Rank for Guangzhou transfer stations.

Table 2. Scores for Connected Lines of TiyuXilu Station

Line	Transfer passenger flow	Score (0.5)	Connected lines	Score (0.3)	Transfer stations	Score (0.2)	Total score
1	13054	90.85	6	100.00	7	100.00	95.43
3	14368	100.00	5	83.33	5	71.43	89.29
North 3	8639	60.13	4	66.67	5	71.43	64.35

Table 3. Results of Hierarchical Progressive Adjustment (Excerpts)

Transfer station	Line (direction)	Baseline (direction)	Baseline time	Transfer walking time	Presumption time	Previous time	Adjusted time
Tiyu	3 (down)	1 (up)	23:21:40	+ 0:03:22	23:25:02	23:24:19	23:25:02
Xilu	3 (up)	1 (up)	23:21:40	+ 0:03:22	23:25:02	23:42:10	23:32:10
	North 3 (down)	3 (down)	23:25:02	+ 0:03:02	23:28:04	23:30:00	23:28:04
	North 3 (up)	3 (up)	23:32:10	+ 0:03:02	23:35:12	23:36:46	23:35:12
Zhujiang	5 (down)	3 (down)	23:25:02	+ 0:02:14	23:27:16	23:21:07	23:27:16
New Town	5 (up)	3 (up)	23:39:10	+ 0:02:14	23:41:24	23:57:26	23:47:26

direction is Line 1 (upward) to Line 3 (downward) and Line 3 (upward). Then, calculate time schedule for coordinated line (Step 2 in section "Hierarchical progressive algorithm") and determine the last-train operation time span (Step 3 in section "Hierarchical progressive algorithm").

In Guangzhou, the time threshold is set in the following principles: $\min\left(t_D^{i,d}\right) = t_{D,O}^{i,d}$-10, and $\max\left(t_D^{i,d}\right) = t_{D,O}^{i,d} + 15$, which mean that the results cannot be 10 minutes earlier or 15 minutes later than the previous time. In addition, the transfer walking time at each transfer station of different coordinated directions is provided by Guangzhou metro operators. As the data were investigated by the staff at the station, the investigation data by were multiplied by 120% to simulate real transfer walking time.

When calculating the last-train operation time span for Line 3(up), the presumption time at the transfer station (23:25:02) is much earlier than the previous time. Therefore, the presumption time should be adjusted to 23:32:10 (23:42:10–00:10:00). Recalculate the time schedule of Line 3 (up), and store Line 3 into **V**. Then, determine the time schedule of Line 3 North, the other coordinated line based on Tiyu Xilu. After that, Zhujiang New Town was set as the base transfer station according to the sequence (Figure 3). Search for the prioritized coordinated lines and directions. If the coordinated lines have been put into **V**, then skip this transfer station. The calculation will be finished when all nine lines in the Guangzhou metro system have been put into **V** (see Table 3).

Results Evaluation

(1) Transfer-level indexes

Compared with the calculation results and the previous results, the volume of transfer passengers in the last-train period and the number of coordinated directions are listed in Table 4.

The results show that the number of coordinated directions for all the transfer stations has been significantly improved, as has the transfer passenger flow in the last-train period. The number of successful transfer passengers has increased by 2046, with the growth rate of 10.4%.

(2) Network-level indexes

The time span of the adjusted time schedule calculated by the "Hierarchical progressive algorithm" is the same as the previous coordination plans (Figure 4). The time span of last trains is 00:21:55–22:15:00 = 2:06:55. The departure time span is from 22:15:00 to 23:40:00, and the arrival time span is from 23:08:54 to 00:21:55.

The network accessibility can be analyzed in Figure 5. The first unreachable OD appeared at 21:01:24 in the previous coordination plan, whereas the optimized plan delays it to 21:16:54. The time period of the fastest declining rate was 22:30:00–23:00:00 in the previous plan, whereas the calculated one delays it for half an hour. That is to say, more OD pairs are still accessible when these OD pairs cannot be reached in the original plan.

Table 4. Comparison of Successful Coordinated Passengers and Directions

Station	Connected directions (line)	Coordinated direction successful		Transfer passenger flow	
		Previous	Optimized	Previous	Optimized
Changgang	8 (2, 8)	4	4	1,637	1,637
Chepinan	8 (4, 5)	4	4	510	510
Gongyuanqian	8 (1, 2)	4	4	1,832	2,324
Tiyu Xilu	16 (3, North 3, 1)	6	7	4,695	4,991
Guangzhou East Station	4 (1, North 3)	2	2	512	512
Guangzhou Railway Station	8 (2, 5)	3	2	1,411	916
Haizhu Square	8 (2, 6)	3	4	539	1,083
Yangji	8 (1, 5)	4	4	1,735	1,735
Dongshankou	8 (1, 6)	3	2	171	59
Huangsha	8 (1, 6)	3	4	483	573
Jiahewanggang	8 (2, North 3)	1	1	549	549
Kecun	8 (3, 8)	4	4	1,092	1,092
Quzhuang	8 (5, 6)	3	4	462	637
Tanwei	8 (5, 6)	4	3	508	111
Tianhe Coach Terminal	4 (3, 6)	1	2	155	191
Wanshangwei	4 (4, 8)	0	0	0	0
Xilang	4 (1, Guangfo)	0	0	0	0
Yantang	8 (North 3, 6)	3	4	781	1,133
Zhujiang New Town	8 (3, 5)	3	4	2,159	2,630
Shayuan	8 (8, Guangfo)	4	4	450	1,044
Total	148	59	63	19,681	21,727

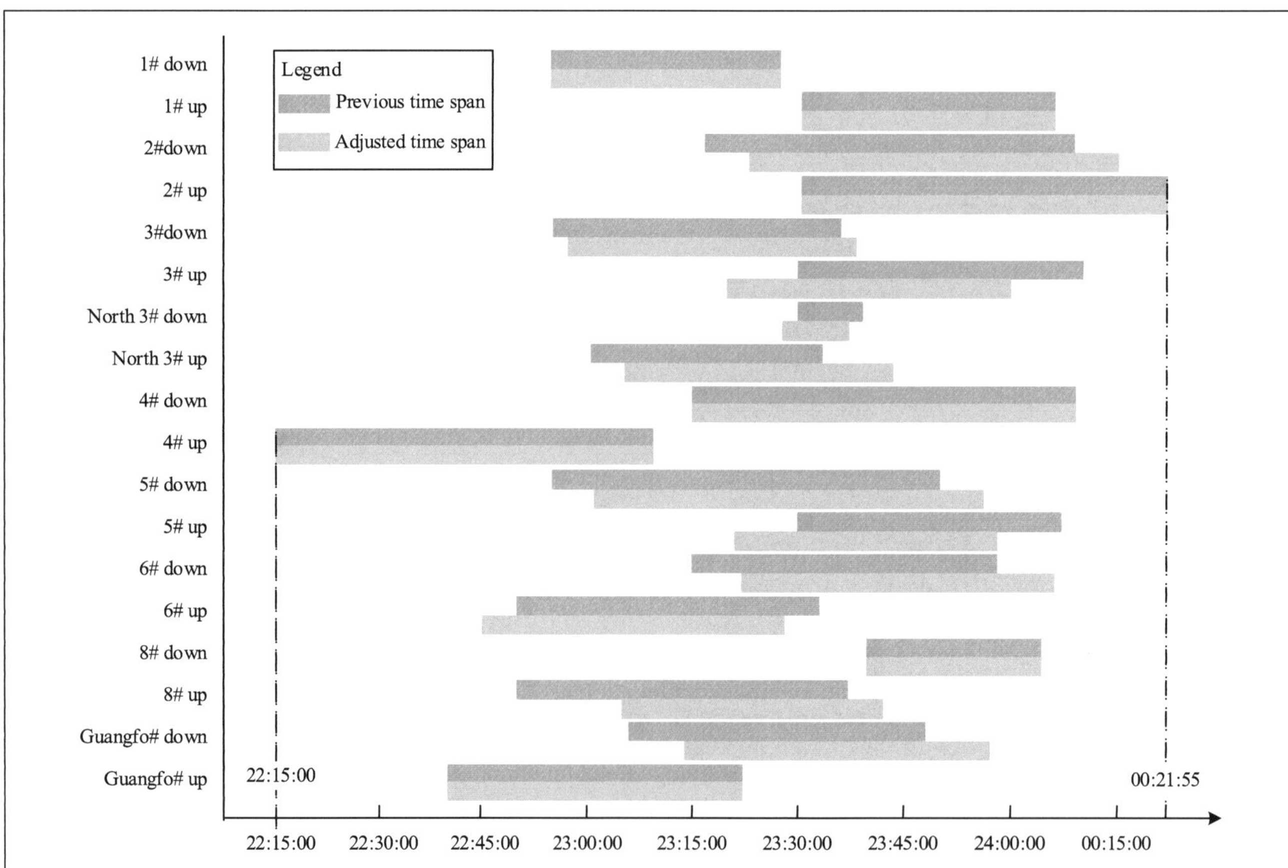

Figure 4. Time span of last trains.

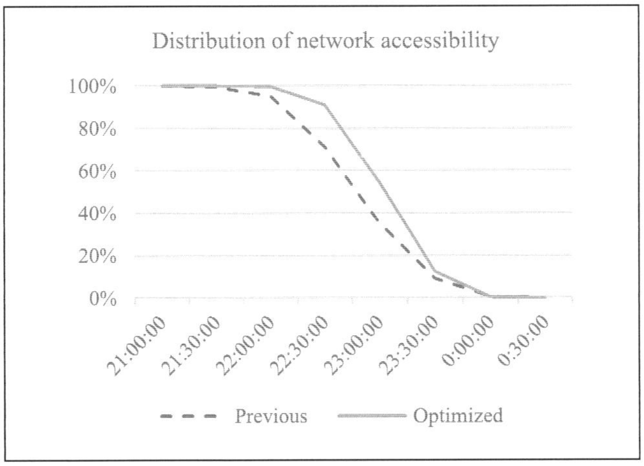

Figure 5. Distribution of network accessibility.

The time span of network accessibility of the optimized results is $\tau =$ 24:14:15–21:16:54 = 2:57:21, and the previous span is $\tau_0 =$ 24:19:50–21:01:24 = 3:18:26. The span of network accessibility is shortened by nearly 20 minutes, which means that operation work is more compact at night and the capacity utilization is more efficient.

In conclusion, the optimized results prove that the demand-driven coordination model not only enables more people to transfer to their destination in the last-train period, but also improves network accessibility and shortens the time span of network accessibility. The methodology and the corresponding algorithm have been put into practical applications in Guangzhou Metro Co., Ltd.

Conclusions

Most previous studies on the coordination of last-train schedules paid little attention to passenger flow, although meeting the demand of passengers is core to upgrading the level of service.

This paper contributes a methodology for transfer coordination of last trains in a metro network, aiming to maximize the number of passengers who can complete their trips in the last-train period. With the constraints of the operating time threshold, the model can also consider the demand of operators, and the optimized plan is favorable to organization and management. The "Hierarchical progressive algorithm" based on transfer stations is very applicable and effective for field application. In addition, double-level evaluation indexes are proposed to evaluate the results systematically.

The research results have been applied to Guangzhou Metro Co., Ltd. The results can be used for Apps and websites to provide passengers with travel information guidance, and can also be used to prevent automatic vending machines selling tickets when passengers would miss the last trains at transfer stations.

Acknowledgments

The study was financially supported by the Fundamental Research Funds for the Central Universities (22120170239 and 2016kj032), and Shanghai Science and Technology Commission Foundation (16DZ1203802). The authors wish to acknowledge Guangzhou Metro Co., Ltd, for collaboration during the research, and thank the anonymous referees for their valuable comments and suggestions.

Author Contributions

The authors confirm contribution to the paper as follows: study conception and design: Feng Zhou, Ruihua Xu, Yanan Li; data collection: Feng Zhou, Yanan Li; analysis and interpretation of results: Feng Zhou, Yanan Li; draft manuscript preparation: Yanan Li, Feng Zhou. All authors reviewed the results and approved the final version of the manuscript.

References

1. Jin, J. G., K. M. Teo, and L. Sun. *Disruption Response Planning for an Urban Mass Rapid Transit Network*. Presented at 92nd Annual Meeting of the Transportation Research Board, Washington, D.C., 2013.
2. Krasemann, J. T. Design of an Effective Algorithm for Fast Response to the Re-scheduling of Railway Traffic during Disturbances. *Transportation Research Part C*, Vol. 20, No. 1, 2012, pp. 62–78.
3. Cevallos, F., and F. Zhao. Minimizing Transfer Times in Public Transit Network with Genetic Algorithm. *Transportation Research Record: Journal of the Transportation Research Board*, 2006. 1973: 74–79.
4. Wong, R. C., T. W. Yuen, K. W. Fung, and J. M. Y. Leung. Optimizing Timetable Synchronization for Rail Mass Transit. *Transportation Science*, Vol. 42, No. 1, 2008, pp. 57–69.
5. Shafahi, Y., and A. Khani. A Practical Model for Transfer Optimization in a Transit Network: Model Formulations and Solutions. *Transportation Research Part A: Policy and Practice*, Vol. 44, No. 6, 2010, pp. 377–389.
6. Xu, X., K. Li, and X. Li. A Multi-objective Subway Timetable Optimization Approach with Minimum Passenger Time and Energy Consumption. *Journal of Advanced Transportation*, Vol. 50, No. 1, 2016, pp. 69–95.
7. Vansteenwegen, P., and D. Van Oudheusden. Decreasing the Passenger Waiting Time for an Intercity Rail Network. *Transportation Research Part B*, Vol. 41, 2007, pp. 478–492.
8. Feng, Z., S. Jungang, and P. Hanchuan. Optimization Method for Last Train Coordination Plan of Urban Rail Transit Based on Network Operation. *Procedia - Social and Behavioral Sciences*, Vol. 96, 2013, pp. 2706–2712.
9. Kang, L., X. Zhu, J. Wu, H. Sun, S. Siriya, and T. Kanokvate. Departure Time Optimization of Last Trains in

Subway Networks: Mean-Variance Model and GSA Algorithm. *Journal of Computing in Civil Engineering*, Vol. 29, No. 6, 2015, p. 04014081.

10. Kang, L., J. Wu, H. Sun, X. Zhu, and Z. Gao. A Case Study on the Coordination of Last Trains for the Beijing Subway Network. *Transportation Research Part B: Methodological*, No. 72, 2015, pp. 112–127.
11. Xu, R., and X. Li. Comprehensive Optimization for Connection Scheme of Last Trains in Urban Mass Transit Network. *Journal of Tongji University (Natural Science)*, Vol. 10, 2012, pp. 1510–1516.
12. Li, X., and Y. Shi. Optimization for the Last Train Timetable of Urban Rail Transit Networks Based on Transfer Coordination. *Fourth International Conference on Transportation Engineering*, 2013, pp. 190–195.
13. Zhou, W., L. Deng, M. Xie, and X. Yang. Coordination Optimization of the First and Last Trains' Departure Time on Urban Rail Transit Network. *Advances in Mechanical Engineering*, Vol. 5, 2013, pp. 1–12.
14. Kang, L., J. Wu, H. Sun, X. Zhu, and B. Wang. A Practical Model for Last Train Rescheduling with Train Delay in Urban Railway Transit Networks. *Omega*, Vol. 50, 2015, pp. 29–42.
15. Lee, K. K., and P. Schonfeld. Optimal Slack Time for Timed Transfers at a Transit Terminal. *Journal of Advanced Transportation*, Vol. 25, No. 3, 2001, pp. 281–308.
16. Lee, K. T. *Optimization of Timed Transfers in Transit Terminals*. University of Maryland, College Park, Md., 2003.
17. Domschke, W. Schedule Synchronization for Public Transit Networks. *OR Spektrum*, Vol. 11, No. 1, 1989, pp. 17–24.
18. Zhao, F., and Zeng, X. Optimization of Transit Network Layout and Headway with a Combined Genetic Algorithm and Simulated Annealing Method. *Engineering Optimization*, Vol. 38, No. 6, 2006, pp. 701–722.
19. Li, W., S. Jones, and R. Xu. Coordination of Last Trains for Metro Networks with AFC System Data: A Case Study of Shanghai Metro Network. Presented at 95th Annual Meeting of the Transportation Research Board, Washington, D.C., 2016.
20. Li. W., R. Xu, Q. Luo, and S. Jones. Coordination of Last train Transfers Using Automated Fare Collection (AFC) System Data. *Journal of Advanced Transportation*, Vol. 50, No. 8, 2017, pp. 2209–2225.

The Standing Committee on Rail Transit Systems (AP065) peer-reviewed this paper (18-04620).

Article

TRR
JOURNAL OF THE TRANSPORTATION RESEARCH BOARD

Transportation Research Record
2018, Vol. 2672(8) 327–338

DOI: 10.1177/0361198118792116
journals.sagepub.com/home/trr
SAGE

A Real-Time Service Management Decision Support System for Train Dispatching at New York City Transit

Daniel Wood[1], Timon Stasko[1], Lauren Tarte[1], Darian Jefferson[1], and Alla Reddy[1]

Abstract

With ridership near modern highs, New York City Transit's (NYCT) subway network frequently operates at or near capacity. This makes maintaining a high-quality service both challenging, due to the lack of "slack," and exceptionally important, due to the large number of riders affected by disruptions. To this end, train dispatchers constantly monitor the network and adjust service to respond to delays. This paper presents a decision support system developed by NYCT which uses real-time train movements and historical ridership information to provide dispatchers with recommendations for holds and station skips in real time. The system uses heuristic headway criteria to determine hold or skip candidate trains, and then estimates the net passenger time savings of each potential hold or skip using estimated origin–destination flows and basic assumptions about passenger behavior. Potential actions that meet a passenger benefit threshold are recommended, and communicated to dispatchers with a simple dashboard. A pilot implementation of the system has been in use at NYCT's Rail Control Center (RCC) for several months, though many details of the system are still in development. Initial observations indicate the system is helping dispatchers manage train service more effectively, producing large passenger time savings.

In most high-frequency, high-ridership transit systems, maintaining reliable service is a constant challenge, and New York City's subway system is no exception. Small delays, even from something as seemingly inconsequential as a passenger holding doors, can easily grow as the delayed train experiences increased boardings and alightings at following stations, often causing longer dwells that push the train further behind schedule. Furthermore, short headways (New York City Transit's [NYCT] minimum scheduled headway is 2 min) make a delay on one train likely to affect following trains; if a disruption causes service to become uneven, the impact can cascade through the system. As a crowded NYCT subway train can be carrying well over 1,000 people, the impacts on passengers' time can be significant. For this reason, it is important to actively manage service, closing problematic gaps in service to reduce passenger delays and prevent gaps from growing larger.

Like many subway systems, NYCT's subway service management is done by train dispatchers, who continually monitor train locations and decide whether, and what, interventions are needed to improve service. Typically, a dispatcher will select between holding (extending a train's station dwell time) or skipping (having a train skip station stops) if they decide a service intervention is needed, and communicate the action over radio to the train's operator. During severe disruptions dispatchers may undertake more extreme service actions, such as rerouting trains over different lines or short-turning ahead of a blockage. These service management decisions are generally based on dispatchers' intuition and experience.

In "regular" service, the task of selecting which actions to take might seem simple. If a train has a small gap in front of it and a large gap behind, it should be held. If a train has a large gap in front of it and a small gap behind, some stations should be skipped. However, these principles are far from guaranteed. Some large gaps exist intentionally for upcoming merges or put-ins from a train yard, whereas others are not worth closing because too many passengers would be delayed in the process. Furthermore, even where these principles are

[1]Data Research & Development, Operations Planning, New York City Transit Authority, New York, NY

Corresponding Author:
Address correspondence to Daniel Wood: Daniel.Wood@nyct.com

correct, they do not yield complete actions: Where should the train be held, and for how long? Which station(s) should be skipped? Answering these questions requires a much more nuanced analysis, as well as knowledge of line ridership patterns. Despite these complexities, dispatchers need to decide on actions quickly, or risk missing the opportunity.

In this challenging environment, opportunity exists for software tools to enhance dispatchers' ability to manage service. Such tools, with a vast and precise memory for ridership patterns, the capacity to perform complex calculations nearly instantly, and the resources to remain entirely focused on service management (whereas dispatchers are sometimes required to shift their attention to other tasks), could evaluate the passenger impacts of hundreds of possible service actions in real time, something beyond the capability of any human dispatcher, while incorporating the best practices of experienced dispatchers. The results of such analysis, made available by the tool in the form of actionable recommendations, could empower dispatchers to provide more responsive and passenger-centered service management.

Toward this goal, NYCT is developing a software tool, named the Service Intervention Recommendation Engine (SIRE), which draws on real-time train movements and historical ridership to provide its subway train dispatchers with actionable, near-real-time service intervention recommendations. This tool is envisioned to be especially useful for NYCT's new train dispatchers, who have less experience to rely on. In particular, SIRE is envisioned to play several roles:

- Assist dispatchers in making nuanced decisions where historical ridership is key, such as selecting which stops to skip,
- Highlight developing issues and the actions needed to fix them before they grow large enough to be easily detected,
- Provide a prioritized list of recommended actions for dispatchers to refer to,
- Help "standardize" dispatching practice, effectively sharing best practices of experienced dispatchers among all dispatching staff.

The next section outlines the previous work on transit service management strategies. Section 3 lays out SIRE's design—its components, data sources, and logic used to generate service intervention recommendations. Section 4 describes how SIRE has been developed in collaboration between technical staff and train dispatchers, and the initial findings on SIRE's impact on service management. The paper concludes with directions for future work.

Previous Work

Active transit service management has been a research topic for several decades. Early work relied on manual headway observation at control points, and simple control schemes, such as Turnquist's "prefol" strategy of holding buses at a few control points until preceding and following headways are similar (*1*). More recent models have used real-time vehicle location data to analyze the entire route in real time, and produce frequently updated service control actions for all vehicles en route. Many of these models use heuristic rules designed to produce even headways; such models have been developed for both bus (*2*) and rail (*3*) operations, incorporating holding, expressing, and short-turning strategies. Other models (*4, 5*) take an optimization approach, generating a set of control actions that minimize some reliability cost function, often total passenger journey time, over a forecast rolling horizon. With recent computing power gains, these optimization models have become feasible for use in real time. Both heuristic and optimization service control models have featured models of dwell times as a function of headways and passenger demand, modeling the dynamics that lead to "bunching" if unchecked.

Service control strategies in the literature have typically been evaluated by comparing the reliability of simulated service(s) with or without the proposed control strategy. However, a few studies have partnered with bus operators to test service control strategies on actual bus routes, and evaluate their impact on real-world reliability. An early study was done by TriMet in 2000, using a field supervisor to communicate terminal holds and short-turns for several routes sharing a common terminal (*6*). More recently, Lizana et al. in 2014 tested an optimization-based system on several high-frequency, high-ridership bus routes in Santiago (*7*), whereas Argote-Cabanero et al. in 2015 tested a heuristic-based system on several interlined routes in San Sebastian, Spain (*2*). These two studies used real-time bus locations to generate control actions (hold, slow down, or speed up), which were transmitted to bus operators via Android tablets mounted in each bus. Significant reliability gains were observed in all three studies when service control systems were used.

No similar field tests of service control systems have been done for rail transit systems, particularly those relying on manual dispatching like NYCT. "Passive" tools for train dispatchers—that is, tools not producing recommended actions—have been developed and field tested, however. Tribone developed a daily line reliability "scorecard" for the Massachusetts Bay Transportation Authority (MBTA) subway operations staff that prompted changes to dispatching practices (*8*). NYCT has previously developed and field tested a real-time train time-space diagram tool (*9*), and later a real-time "gap table" tool showing the largest gaps between trains in a dispatcher's territory, highlighting trains that might need service interventions. More recently, NYCT developed a dashboard providing real-time predictions of platform crowding levels at certain stations, informing dispatcher decisions about where to skip stations (*10*).

This study's work is thus a significant departure from the prior work on transit service management. To the authors' knowledge, NYCT's SIRE is the first real-time tool providing service management actions developed and implemented on a high-frequency, high-ridership rail transit system. This work also differs from prior active service control systems in being explicitly designed for train dispatchers and their environment, rather than communicating directly with vehicle operators, or neglecting the communication issue in general (effectively the case with simulated service control strategies).

Methods

Design Overview

NYCT's service management decision support system for train dispatchers was developed with the following main design objectives:

- Generate service control actions that will produce substantial passenger journey time savings;
- Be easy for train dispatchers to use—recommendations should be clear and actionable, and not so numerous as to overwhelm dispatchers;
- Indicate higher-benefit actions, so dispatchers can prioritize which to implement when time is limited;
- Update with near-real-time frequency (every 30 s or less);
- Ensure basic logic is understandable and justifiable for non-technical staff, to promote users' confidence in the recommendations.

The resulting SIRE has the following general structure. First, train schedules and real-time train locations are processed to generate arrival time predictions for every train's downstream stations, which are then used to estimate the headways between each train. SIRE then loops through all trains currently en route, checking if each train's leading and following headways meet the criteria for potentially needing a hold or skip. If a train's headways meet the criteria, SIRE estimates the net passenger time savings of a hold or skip, using historical ridership data to estimate the number of passengers who gain and lose from the action. If the net passenger benefit meets a minimum threshold, the action is recommended, and relayed to train dispatchers at the Rail Control Center (RCC) using a web dashboard accessible from every dispatcher's computer.

A heuristic, rather than optimization, approach was chosen mainly to ensure fast computation time and to speed up the development process, but also to avoid creating a "black box" whose recommendations would be challenging to verify and troubleshoot, and whose logic would be difficult for non-technical staff to understand, which could hinder buy-in from SIRE's intended users. SIRE is designed only for centrally-dispatched NYCT subway services for which full real-time train location data is available—currently the 1, 2, 3, 4, 5, and 6 lines—but could be expanded as more services get real-time data (see Future Work). The rest of this section describes each part of SIRE in more detail.

Train Trajectory and Headway Generation

SIRE pulls in real-time train location and schedule data from two sources. The primary source is a direct feed from the automatic train supervision system for the 1 to 6 services, which provides real-time train locations at the track circuit level (roughly every 400 to 800 ft), and train schedules roughly 20 min before each train's departure (updated en route for skips and reroutes). From a train's current location and scheduled stopping pattern, forecast arrival times for the trip's remaining stops are then calculated using historical running times. If a train is between stations, the time to the train's next stop arrival is estimated as the historic median running time between the train's current circuit location and the next stop.

Given a train's predicted arrival time at its next stop, the expected running time to each following stop is the sum of the median running times (including dwells) between all pairs of adjacent stations between the next stop and the target following stop. The historic median stop-to-stop running times are pre-calculated and tabulated by hour to capture time of day trends in speed and dwell time.

Once arrival time forecasts are generated for all trains, it becomes straightforward to estimate each train's leading and following headway at all downstream stations. It should be noted that this method does not model the dynamic interaction between headways and dwell times—that is, the dynamics that cause bunching—as has been done in many prior systems. This was done partly for tractability reasons, but mainly because NYCT has not yet been able to create a workable subway dwell time model, with enough predictive power to be useful for operations applications. Tackling this problem is a potential area for future SIRE development (see Future Work).

Generating List of Potential Hold and Skip Trains

Once headways have been estimated for all trains, SIRE then checks each train's current leading and following headways against a set of headway thresholds to be considered for a hold or skip. These thresholds' main purpose is to filter out trains with fairly even headways that do not need corrective action. Their specific values, essentially defining when headways are uneven enough to potentially merit intervention, were derived from dispatchers' past practices and academic literature. These rules quickly remove the large majority of trains from analysis, speeding computation time significantly.

Both quantitative and qualitative methods were used to translate dispatchers' experience into concrete headway criteria. Initial skip headway thresholds were derived from analysis of historic train movement data, which indicated there were rough cutoffs for absolute difference between headways before and after, the ratio of headway before to headway after, and absolute headway before size, below which trains were rarely skipped. The skip results informed the hold criteria, as historical data for holds was not easily obtainable (distinguishing holds from delays in NYCT's data is quite challenging). This statistical analysis was combined with interviews with dispatchers about how they decided when to take actions, as well as many hours spent with dispatchers at work, observing when they decided to hold or skip trains; these interactions yielded many refinements to the rules. Thresholds drawn from historical dispatcher actions were compared with theoretical estimations from the literature where possible. In particular, the gap ratio threshold findings were similar to findings by Eberlein (*11*).

Through the above process, sets of hold and skip headway criteria were developed. For holds, the criteria are

$$h_a - h_b \geq v_H$$
$$h_a / h_b \geq \alpha_H$$
$$h_a \geq g_H$$

where

h_b is the headway before;
h_a is the headway after;
v_H is the minimum gap difference parameter for holds;
α_H is the minimum gap ratio parameter for holds; and
g_H is the minimum gap after parameter for holds.

For skips, the criteria are

$$h_b - h_a \geq v_S$$
$$h_b / h_a \geq \alpha_S$$
$$h_b \geq g_S$$
$$h_a \leq h_{\max,S}$$

where

v_S is the minimum gap difference parameter for skips;
α_S is the minimum gap ratio parameter for skips;
g_S is the minimum gap before parameter for skips; and
$h_{\max,S}$ is the maximum headway after for skips.

The first three skip criteria are essentially the same as the three hold criteria, but with the positions of h_a and h_b reversed, reflecting that the "big gap" to be closed is h_a for holds and h_b for skips. The last skip rule, which sets a maximum headway after for skips ($h_{\max,S}$), is intended to prevent skips that would significantly delay passengers boarding or alighting at skipped stations. Because such delays are likely to be perceived as especially frustrating (since passengers can see the skipped train passing them, or have to alight and take another train), and are clearly the result of dispatcher actions, preventing long delays caused by skips was thought worthwhile from a customer-perception perspective, even if unnecessary from a utilitarian passenger time savings standpoint.

Initially, single values of each parameter were used. However, early SIRE testing revealed issues with this. SIRE's headway estimates are constantly changing, due to the various dynamics of the subway system, with even fairly "stable" headways fluctuating by around 30 s over the course of a few minutes. This often caused the headways of a train being evaluated for a hold or skip to fluctuate above and below one of the hold/skip thresholds, in turn resulting in "unstable" hold/skip recommendations that would appear, disappear, and reappear. This behavior frustrated dispatchers, as it was unclear whether such unstable recommendations should be acted on. To address the issue, most parameters were changed to have two values: one to keep an action recommended, and another, 10% to 20% higher, to initially recommend an action. Each parameter's two values were set so that if a train's headways are uneven enough to clear the "add" threshold, normal headway variation should not cause the train's headways to fall below the "keep" threshold. This change significantly improved recommendation stability.

If a train meets the headway unevenness thresholds, SIRE then checks it for possible merge issues. If two trains on different tracks are forecast to arrive at a merge point at roughly the same time, their order through the merge becomes somewhat unpredictable, and merge conflicts can occur, in which one of the trains gets delayed while waiting for the other to pass through the merge. If not accounted for, merge problems can produce unstable recommendations, and recommendations that are ineffective (or even delay-causing) when acted on. To prevent this, SIRE requires trains near merges to satisfy the requirement

$$\min\{h_a, h_b\} \geq m_j$$

where m_j is the estimated minimum train headway at junction stop j through which order is reliably maintained. Parameter m_j varies among different subway junctions, depending on their observed "predictability." Furthermore, for all trains SIRE looks for merges upcoming in the next 25 min, and blocks skipping or holding a train if the action would cause a severe merge conflict.

For potential skip trains near the end of their trip (generally 10 or fewer stops remaining), SIRE also checks if a skip could cause a terminal holdout. Terminal holdouts are delays when a train cannot enter its terminal station because all the terminal's tracks are occupied by trains laying over; they usually occur when several trains arrive at the terminal in close succession, which can be triggered by a skipped train "catching up" to its leader. To prevent this, SIRE monitors the speed of trains entering terminals. If a train takes significantly longer to enter a terminal than normal, an indicator of a terminal holdout taking place, the maximum time savings for skips near the terminal is temporarily restricted to z_{ho}, calculated as

$$z_{ho} = h_b - \delta_T - g_{ho}$$

where δ_T is the recent holdout delay time observed at terminal T, and g_{ho} is a "safety factor" parameter (currently 150 s) to ensure a sufficient headway between a skip train and its leader at the terminal. If z_{ho} is under 45 s, the train is removed from skip consideration. This method is based on the observation that terminal holdouts usually occur in groups, so if a terminal experienced a holdout recently, the leader of a potential skip train nearing that terminal is more likely to also experience a terminal holdout delay, reducing the skip train's effective headway before.

Generating Actions and Estimating Passenger Benefit

If a train meets the headway criteria for a hold or skip consideration, and has no apparent merge or holdout issues, a proposed hold or skip is generated and the net passenger time saving is estimated. If the net passenger benefit exceeds a threshold, it will be recommended. For both holds and skips, there are both "winners" who save time and "losers" who get delayed, and tradeoffs between maximizing passenger time saved and reducing the perceived inconvenience of service interventions. The process for generating actions and estimating benefit is fairly different for holds and skips, so they will be discussed separately. However, both depend on the same input historical ridership data, which merits discussion before the specifics of passenger benefit estimation.

Input Historical Ridership Data. SIRE's historical ridership input data is derived from the output of NYCT's subway ridership model (*12*). The subway ridership model takes as input a day's historical automated fare collection entry swipe data and train movement data, and outputs an estimate of every unlinked trip's origin, destination (inferred from a passenger's next entry swipe station), train used, begin time, and end time.

To generate the input data for SIRE, the subway ridership model is run for a set of 15 to 20 recent weekdays, and the results aggregated and averaged together to get train-trip-level boardings, alightings, and leave load information for a "typical" weekday. This process is repeated every 3 to 4 months, to keep the ridership input up-to-date with respect to seasonality and ridership trends. Efforts to create typical Saturday or typical Sunday datasets have so far been unsuccessful, due to the frequent maintenance-related weekend service changes on the New York City (NYC) subway, which vary greatly from weekend to weekend; for this reason, SIRE to date has only been used for daytime weekday service.

The typical weekday train-trip-level data is then aggregated by route, direction, and hour to generate the following ridership input data, for any stop s:

- Average train leave load $L_{avg}(s)$;
- Average passenger alightings per train $A_{avg}(s)$;
- Average route-independent passenger boarding rate, $\lambda_I(s)$. The subset of boarding passengers whose destinations can be served by any train serving the same station/track/direction, regardless of route;
- Average train route-dependent passenger boarding rate, $\lambda_D(s)$. The subset of boarding passengers whose destinations can only be served by the route in question.

L_{avg} and A_{avg} are converted into train-level leave load and alightings estimates L and A using a headway before correction factor β, defined as

$$\beta = \left(\frac{h_{b,s}}{h_s^*}\right)^{0.5}$$

where h_s^* is the scheduled headway at s, and $h_{b,s}$ is the train's forecast headway before at s. L and A are then estimated as

$$L(s) = \beta L_{avg}(s)$$
$$A(s) = \beta A_{avg}(s)$$

The factor β models the property that, assuming random passenger arrivals, a train's expected passenger load is roughly proportional to its headway before. The 0.5 exponent attempts to account for the uncertainty in the train's prior headways—just because a train has a large/small prior headway now, does not mean it had one 15 stops previously.

Hold Generation and Passenger Benefit Calculation. If a train meets the criteria for hold consideration, SIRE will then evaluate the net passenger benefit of holding for duration y at the next stop, for all y in the set (in minutes) $\{0.5, 1.0, 1.5, \ldots, y_{max}\}$. The overall maximum hold time $y_{\max}$, currently 3 min, was added to prevent very long holds that would frustrate passengers and train crews, even if justified from a pure passenger time savings perspective.

Excluding dynamics such as denied boardings and headway-dwell interaction, SIRE considers three groups of passengers affected by a hold:

- Passengers on the train while it is held, each delayed by the hold time y, with total delay D_{train};
- Downstream boarding passengers who would have made the train without the hold, each having to wait y longer for the train, with total delay D_{wait};
- Downstream boarding passengers who would have missed the train without the hold, each saving wait time equal to the headway after at the downstream station, with total time savings B_{wait}.

This breakdown of hold delay and benefit is illustrated in Figure 1. Given a train's route, direction, time period, and

- set of train's scheduled stops $S = \{s_1, s_2, \ldots, s_n\}$, where s_1 is the origin terminal and s_n is the destination terminal,
- S_k, the hold stop, and
- h_b^i and h_a^i, forecast headways before and after at stop s_i (not accounting for hold).

The total passenger time delay/savings for the three groups, $D_{\text{train}}, D_{\text{wait}}, B_{\text{wait}}$, can be described for a non-interlined route by

$$D_{\text{train}} = y(L(s_{k-1}) - A(s_k))$$
$$D_{\text{wait}} = \sum_{i=k}^{n} h_b^i \lambda(s_i) y$$
$$B_{\text{wait}} = \sum_{i=k}^{n} y\lambda(s_i)\left(h_a^i - y\right)$$

The total net passenger benefit for the hold is then

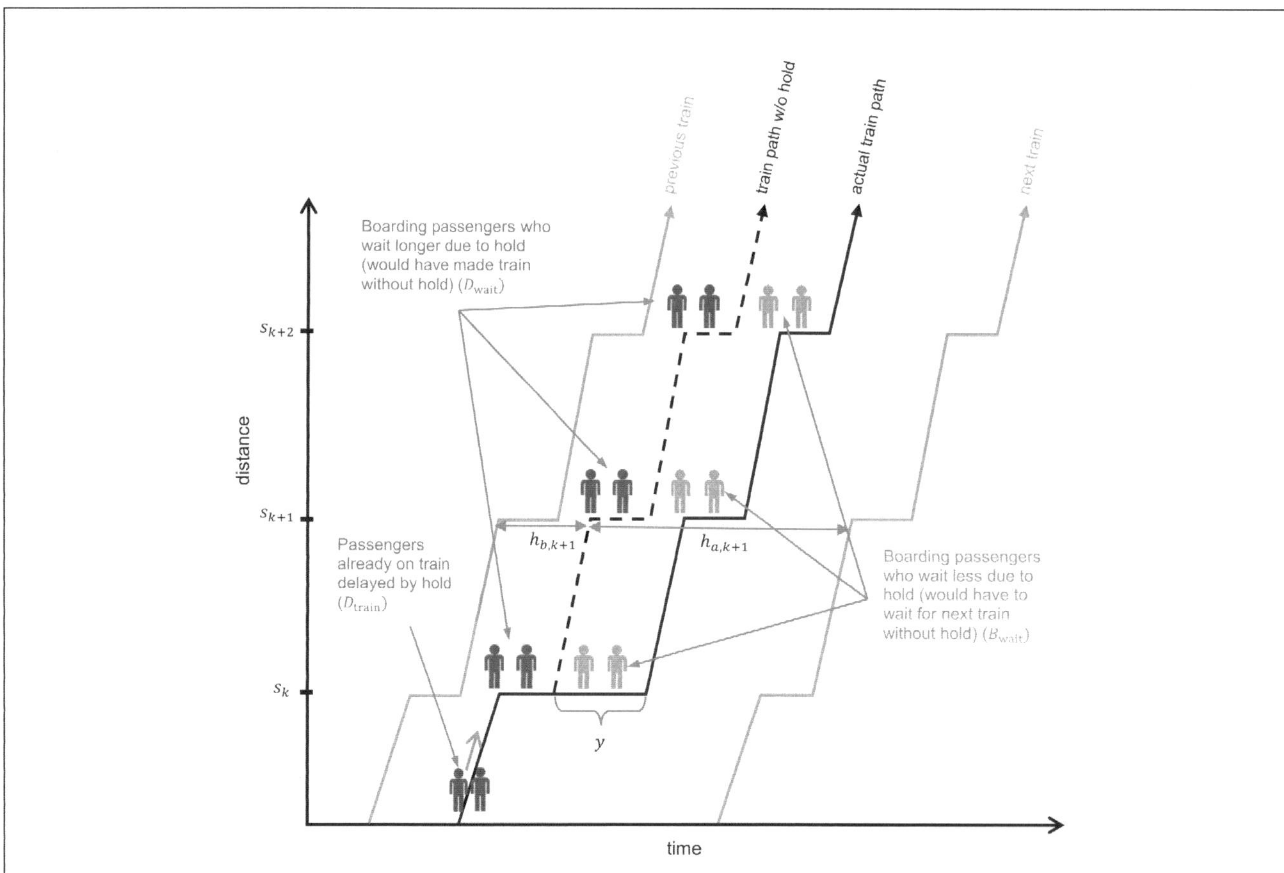

Figure 1. Breakdown of passenger benefit and delay for a hold of duration y at station s_k (no interlining), into waiting benefit $\boldsymbol{B_{\text{wait}}}$, on-train delay $\boldsymbol{D_{\text{train}}}$, and waiting delay $\boldsymbol{D_{\text{wait}}}$.

$$NPB_H = B_{\text{wait}} - D_{\text{wait}} - D_{\text{train}}$$
$$= \sum\nolimits_{i=k}^{n} y\lambda(s_i)\left(h_a^i - y\right)$$
$$- \sum\nolimits_{i=k}^{n} h_b^i \lambda(s_i) y - y(L(s_{k-1}) - A(s_k))$$

which simplifies to

$$NPB_H = -\,y(L(s_{k-1}) - A(s_k)) + y\sum\nolimits_{i=k}^{n} \lambda(s_i)\left(h_a^i - h_b^i - y\right)$$

For a hold at an interlined station, downstream passengers need to be split into route-dependent and route-independent boardings. This is expressed as

$$NPB_H = -\,y(L(s_{k-1}) - A(s_k)) + y\sum\nolimits_{i=k}^{n} \lambda_I(s_i)$$
$$\left(h_a^i - h_b^i - y\right) + \lambda_D(s_i)\left(h_{a,r}^i - h_{b,r}^i - y\right)$$

where $h_{b,r}^i$ and $h_{a,r}^i$ are the route headways at stop s_i. This equation describes the calculation done by SIRE to estimate a hold's net passenger benefit. If the route-level headways happen to be the same as the track-level headways, or $\lambda_D = 0$, the route-dependent equation reduces to the route-independent equation.

After evaluating NPB_H for all potential hold times y, SIRE selects the "best" hold time y^* that has the highest NPB_H. If this maximum net passenger benefit exceeds a minimum threshold, the hold of time y^* at the next stop will be recommended.

Skip Generation and Passenger Benefit Calculation. SIRE's skip recommendation generation starts with creating a list of the potential skip train's upcoming stops that could be skipped. An upcoming stop is considered a skip candidate if it:

- is forecast to be reached within the next 12 to 15 min,
- is in the next 10 stops,
- has average boardings plus alightings less than 4% of the average route/direction total (4% roughly corresponds to the threshold for "major" stations),
- was not skipped by the previous train,
- is not forecast by NYCT's real-time platform crowding model *(10)* to experience high levels of platform crowding in the next 15 min.

The first two rules are meant to prevent skip recommendations very far in the future. The others are intended to prevent skipping stops with especially high numbers of boardings and alightings, or which would likely especially frustrate passengers (e.g., skipping the same stop twice in a row).

After the list of potential skip stops is generated, SIRE then generates all possible combinations of skip stops, and calculates their total average boardings plus alightings, and estimated skip time savings z. The time saved z is estimated using the model

$$z = a\sum\nolimits_{sX}\left(d_{\text{avg},s} - d_{\text{skip},s}\right) - bN_X$$

where

X is the set of skipped stops;

$d_{\text{avg},s}$ is the average dwell time at stop s;

$d_{\text{skip},s}$ is the average dwell time at stop s when the station is skipped;

N_X is the number of groups of consecutive stations in skipped station set X; and

a and b are the model parameters.

The model was estimated using linear regression on about 5 months of historical train skip time savings data, and was found to estimate skip time savings reasonably well, with $R^2 \approx 0.45$ and SE $\approx$ 30 s.

These potential skip stop combinations are then filtered on two requirements: (1) the estimated time saved z is less than the holdout maximum z_{ho}, and overall skip maximum z_{max} (currently 5 min); and (2) the skipped stations' total average boardings plus alightings are less than P_{max} (currently 9% of the route/direction total). These restrictions prevent skip recommendations that would cause delays or severely affect service, which although possibly justified from a pure passenger time savings view would likely elicit customer complaints and are contrary to dispatchers' historical practice.

Next, the total net passenger time savings is estimated for every potential skip stop combination that meets the above requirements. SIRE in effect weighs the tradeoff between the benefit to passengers on the skipped train and those waiting downstream who save time because of the skip, and the delay to passengers whose origins or destinations are skipped, or who miss the train because of the skip. SIRE breaks down these passengers into five categories:

- Passengers on the train before the skip whose destination is after the skip, each saving in-vehicle time z, with total time savings B_{train};
- Passengers on the train whose destination is one of the skipped stations, who must alight at the stop before the skip and wait for the next train to their destination, producing total delay D_{alight};
- Passengers who were going to board the train at one of the skipped stations, who must wait for the next train to their destination, producing total delay D_{pass};
- Passengers after the skip who wait z less time because the train comes sooner, producing total time savings B_{after};

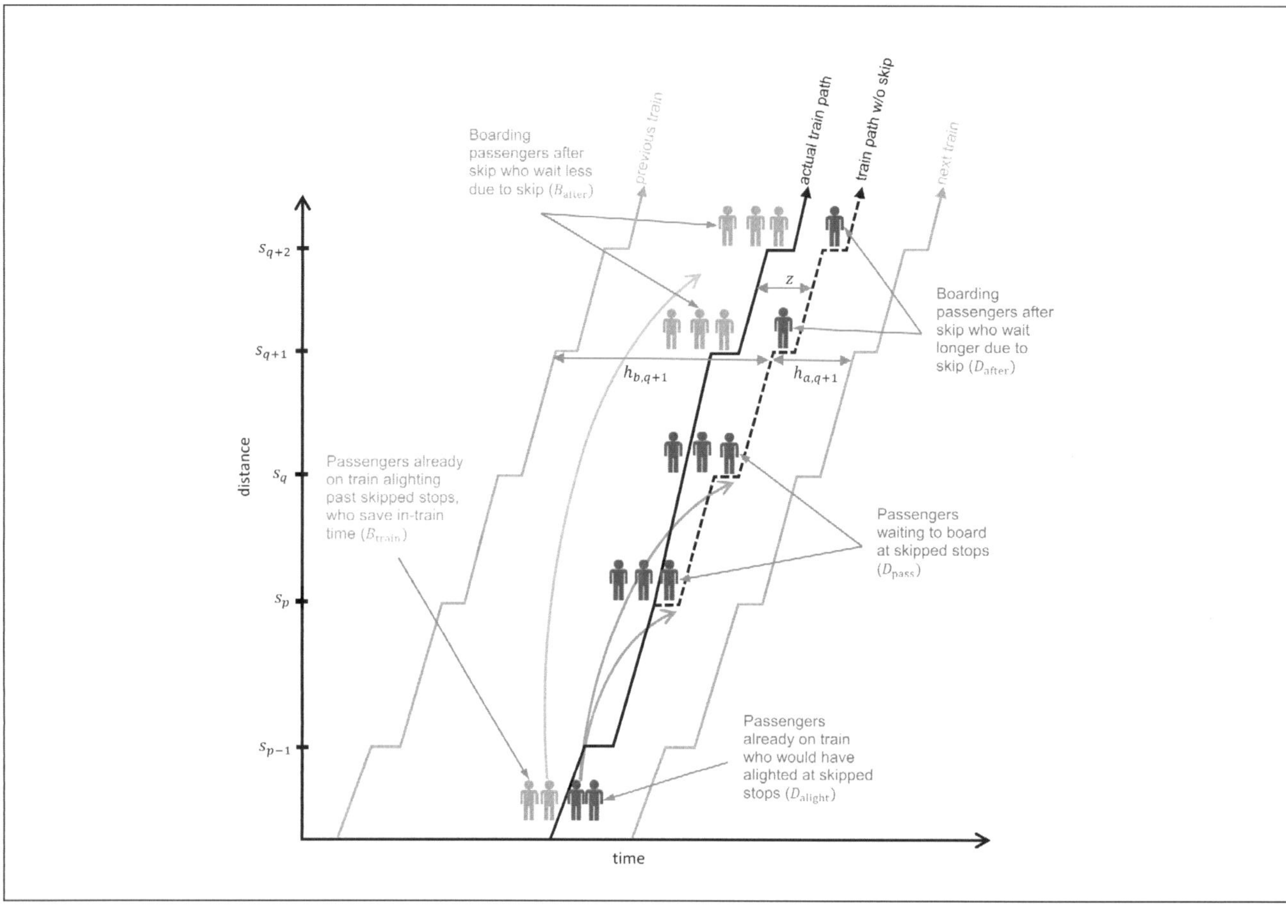

Figure 2. Breakdown of passenger benefit and delay for skipping stations s_p and s_q with time savings z (no interlining or next trip benefit), into on-train benefit B_{train}, delay to passengers alighting at skipped stations D_{alight}, delay to passengers boarding at skipped stations D_{pass}, and benefit and delay to passengers boarding after the skip, B_{after} and D_{after}.

- Passengers after the skip who miss the train because the train comes sooner, who must wait for the next train to their destination, producing total delay D_{after}.

This breakdown of passenger delay and benefit for a skip is illustrated in Figure 2. For the case of a single contiguous set of skip stations, that is, $N_x = 1$, affecting only the current trip, these passenger time savings and delays are estimated for a non-interlined route by:

$$B_{train} = z\beta\left(L(s_{p-1}) - \sum_{i=p}^{q} A(s_i)\right)$$
$$D_{alight} = \beta \sum_{i=p}^{q} A(s_i)h_a^i$$
$$D_{pass} = \sum_{i=p}^{q} \lambda(s_i)h_b^i h_a^i$$
$$B_{after} = z\sum_{i=q+1}^{n} \lambda(s_i)(h_b^i - z)$$
$$D_{after} = z\sum_{i=q+1}^{n} \lambda(s_i)h_a^i$$

where p and q are the stop indices of the first and last skipped stations; all other notation follows from the hold benefit equations. From this, the total skip net passenger benefit can be defined

$$\begin{aligned} NPB_S &= B_{train} - D_{alight} - D_{pass} + B_{after} - D_{after} \\ &= z\beta\left(L(s_{p-1}) - \sum_{i=p}^{q} A(s_i)\right) - \beta\sum_{i=p}^{q} A(s_i)h_a^i \\ &\quad - \sum_{i=p}^{q} \lambda(s_i)h_b^i h_a^i \\ &\quad + z\sum_{i=q+1}^{n} \lambda(s_i)(h_b^i - z) - z\sum_{i=q+1}^{n} \lambda(s_i)h_a^i \end{aligned}$$

which simplifies to

$$\begin{aligned} NPB_S &= z\beta L(s_{p-1}) - \sum_{i=p}^{q} \beta A(s_i)(z + h_a^i) \\ &\quad + \lambda(s_i)h_b^i h_a^i + z\sum_{i=q+1}^{n} \lambda(s_i)(h_b^i - h_a^i - z). \end{aligned}$$

The formulation above is then extended to the interlined route scenario as follows

$$NPB_S = z\beta L(s_{p-1}) - \sum_{i=p}^{q} \beta A(s_i)(z + h_a^i) + \lambda_I(s_i)h_b^i h_a^i + \lambda_D(s_i)h_{b,r}^i h_{a,r}^i + z\sum_{i=q+1}^{n} \lambda_I(s_i)(h_b^i - h_a^i - z) + \lambda_D(s_i)(h_{b,r}^i - h_{a,r}^i - z)$$

When a train is near the end of its trip (specified the same as for holdout detection), SIRE extends B_{after} and D_{after} to include early stops on the train's next trip, up to 10 stops or the first merge point. The additional next trip benefit and delay, B_{nt} and D_{nt}, are estimated as

$$B_{nt} = z\sum_{i=1}^{w_{nt}} \lambda_I(s_i^{nt})(h_b^{nt} - z) + \lambda_D(s_i^{nt})\left(h_{b,r}^{nt} - z\right)$$
$$D_{nt} = z\sum_{i=1}^{w_{nt}} \lambda_I(s_i^{nt})h_a^{nt} + \lambda_D(s_i^{nt})h_{a,r}^{nt}$$

where

s_i^{nt} is the ith station of the train's next trip;

h_b^{nt} and h_a^{nt} are the next trip's forecast track headways at the origin;

$h_{b,r}^{nt}$ and $h_{a,r}^{nt}$ are the next trip's forecast route headways at the origin; and

w_{nt} is the index of the next trip's merge point (or the tenth stop, if sooner).

The next trip delay and benefit are only added toward the end of the trip because before then the next trip headway forecasts are too uncertain to be relied on. Likewise, when the next trip benefit is added, the calculation only extends to the first merge point or tenth station because headway forecasts past there are unreliable or unavailable.

The skip net passenger benefit equations above are extended in SIRE's logic to apply to skips with multiple, non-consecutive groups of stations (i.e., $N_X>1$). For the sake of brevity, the extended formulations are not shown, but essentially they are derived by treating non-consecutive skips as a series of cascading single-group skips.

By cycling through all possible skip stop combinations, SIRE is able to find the skip stop combination with the highest estimated net passenger benefit. As with holds, if this "best" skip stop combination's benefit exceeds a minimum threshold, it will be recommended.

User Interface and Computer Setup

The hold and skip actions recommended by SIRE are communicated to train dispatchers via a dashboard (Figure 3) that dispatchers can view on their desktop computer. Dispatchers can filter recommendations on line and dispatcher territory, to focus only on recommendations relevant to them. The recommendations are color-coded black, orange, or red based on estimated passenger benefit (red highest, black lowest), clearly and quickly indicating to dispatchers which recommendations are highest priority. Based on dispatcher suggestion, the recommendations are split by railroad direction (northbound/southbound).

SIRE's main program, handling train movement data processing, train forecast generation, and recommendation generation, was written entirely in Python. The program takes only 2 to 3 s to run for the 1 to 6 lines combined on an average desktop computer, allowing for frequent recommendation updates (currently every 15 s). The user interface was built as a web application (eliminating the need to install software on users' computers) using Bootstrap and Node.js. All software components used were open-source, and all software development was done in-house, reducing development costs and increasing development speed (since no procurement was required).

Development and Implementation

Development and Implementation Process

An important aspect of the SIRE project is the development and implementation process used. Rather than using a traditional "waterfall" design process, with formal requirements writing, design period, testing, and acceptance, SIRE was developed using an agile, iterative process. Starting from a simple, bare-bones version of SIRE, technical staff would test each iteration of SIRE with train dispatchers at the RCC, spending hours observing SIRE's functioning, and getting feedback from dispatchers on its recommendations (Do the recommendations make sense? Are there too many/not enough? Are they stable?) and general user-friendliness. The technical staff would then incorporate these findings into a new version, and repeat the testing process. Through dozens of rounds of testing, feedback, and revision over the course of more than a year, SIRE was incrementally built up into its current form.

This approach produced a tool that, though built by technical staff, was significantly shaped by dispatchers' direct input. Many key features of SIRE, including merge conflict prevention, holdout prevention, measures to improve stability, and non-revenue train handling, were the direct result of dispatcher feedback bringing to attention issues not noticed by the technical staff. Furthermore, many needs probably would not have been anticipated by a formal requirements process, given the complexity of train operations and lack of precedent for SIRE-like tools. But besides producing a better tool, the collaborative development process also promoted dispatchers' ownership in SIRE—it was something they helped to create, rather than something arbitrary imposed on them. This ownership helped build dispatchers' trust in SIRE's recommendations. The SIRE technical staff continues to solicit feedback from dispatchers

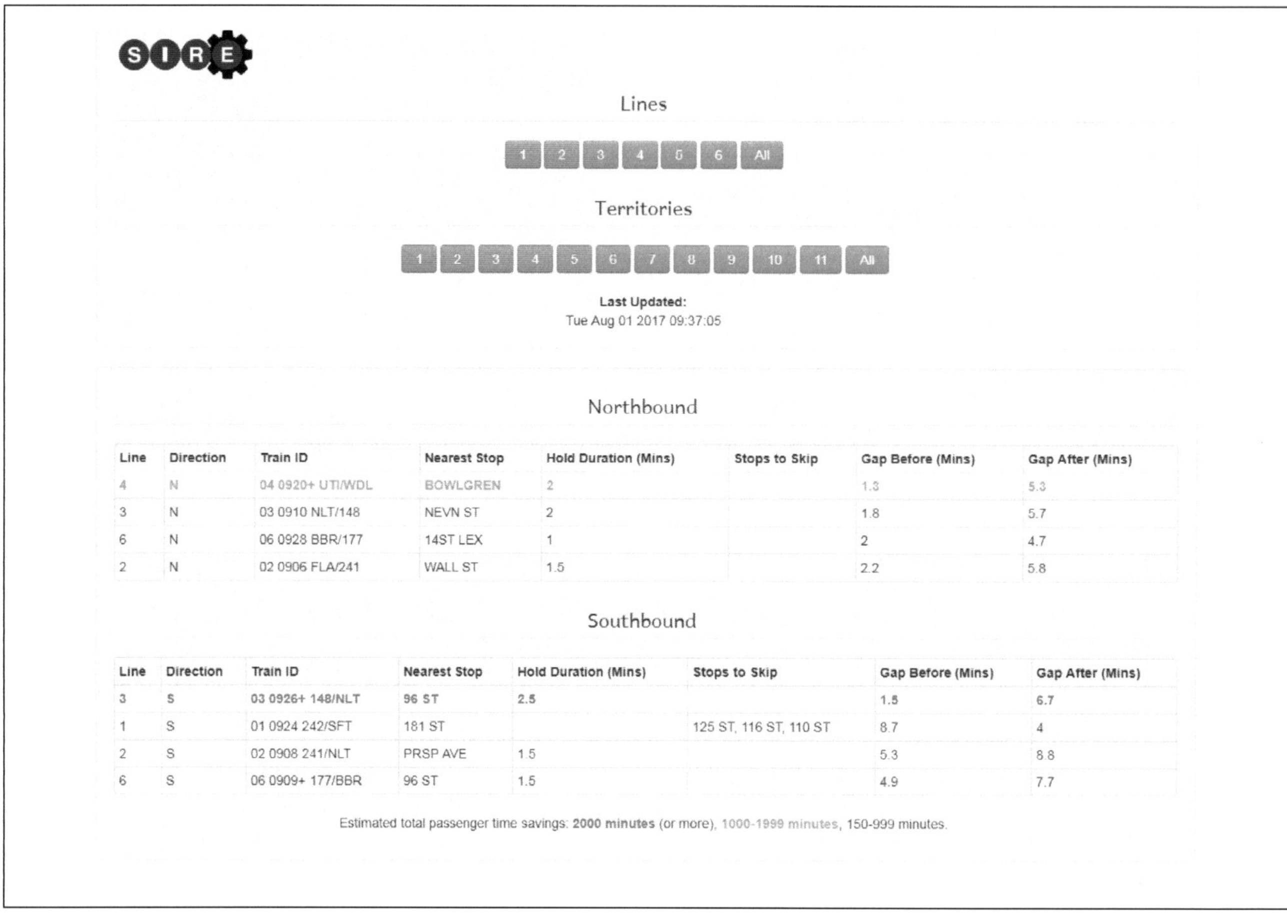

Figure 3. SIRE user interface.

and make improvements in response—SIRE remains a work in progress, not a finished product.

SIRE's implementation proceeded in tandem with the field testing process. Implementation began with a pilot phase on the 6 line during the p.m. shift with only hold recommendations. Gradually, this was expanded to holds and skips, then both a.m. and p.m. shifts, then other lines. During this rollout, technical staff conducted one-on-one training sessions with dispatchers new to the tool, as well as several group training sessions. This interactive training not only facilitated dispatchers' adoption of SIRE, but also provided an opportunity to get feedback on SIRE from new users, which informed further SIRE iterations.

SIRE is currently used by most 1, 2, 3, 4, 5, and 6 line dispatchers during daytime weekday service. The goal is for dispatchers to implement as many SIRE recommendations as possible, though dispatchers are allowed to modify or disregard SIRE recommendations, particularly if they foresee the action causing operational problems. Giving dispatchers this flexibility was crucial for dispatchers' acceptance of using SIRE, especially at the beginning of implementation when SIRE's logic was significantly less sophisticated, and input data quality problems were more common. Even now, dispatchers have qualitative information relevant to service management (e.g., equipment problems, police investigations) that is unavailable to SIRE. Dispatchers often need to ignore SIRE when they are responding to a major disruption, which usually requires all of their attention and involves more drastic actions to restore service (reroutes, short-turns, etc.).

Results and Impact

In-person observations and discussions with RCC staff have (qualitatively) indicated that when dispatchers are actively using SIRE, they are able to act on more gaps in service, and act on them sooner. SIRE recommendations have also revealed new hold and skip strategies—new skip stop patterns, for example—that appear to be effective in improving service quality. Furthermore, SIRE has revealed types of service actions that, although they even trains' spacing, are likely to produce minimal or negative passenger time savings. For example, at first glance holding a train toward the end of its trip seems like a reasonable action to even trains' spacing. However, with such end-of-line holds there are usually many more on-train passengers delayed during the hold than remaining boarding passengers who will benefit; thus, end-of-line holds usually have negative passenger benefit, and therefore are not recommended by SIRE.

When dispatchers act on SIRE recommendations, they produce significant passenger journey time savings. Using SIRE records and historical train movement data, it was estimated that in May 2017, SIRE-recommended holds and skips by dispatchers on the 1 to 6 lines saved passengers 90,000 h of travel time, with around 180,000 passengers per day benefiting from SIRE actions. Individual SIRE-recommended actions were often estimated to save more than 20 h of passenger journey time. However, these numbers are only rough estimates of SIRE's impact; the total time savings do not distinguish between SIRE-recommended actions that were prompted by SIRE, and those that dispatchers would have done without SIRE recommendations. The passenger journey time savings estimates also do not include the benefit of preventing denied boardings and bunching, and the benefit from SIRE preventing "bad" actions that would have produced net passenger delay (e.g., end-of-line holds). Overall, the development, implementation, and results of the SIRE project indicate that a computer-driven active service management control system can be successful for high-frequency rail transit with manual dispatching.

Future Work

There are many potential directions for enhancing the sophistication of SIRE's logic. SIRE's passenger benefit estimation could potentially be significantly enhanced by incorporating:

- Estimates of passenger time saved by preventing denied boardings due to overcrowding;
- Time saved by preventing gaps in service growing larger due to longer dwell times (i.e., bunching dynamics);
- Better estimation of the time savings of a particular skip, perhaps in the form of a "self-teaching" algorithm that would monitor skips' effectiveness and adjust accordingly;
- Value of time weights and/or penalties, to account for how passengers perceive the inconvenience of different kinds of delay differently (e.g., on-train delay vs. having to transfer).

Also beneficial would be more sophisticated methods of predicting and preventing merge conflicts and terminal holdouts, which would permit more aggressive service management near junctions and terminals.

There is also opportunity to significantly expand the scope of SIRE. At NYCT, SIRE could be expanded to serve the rest of the NYC subway network as the availability of real-time train location data grows. Another version of SIRE could be built for NYCT's terminal dispatchers who are responsible for deciding whether to hold trains at terminals, or send trains out early (something outside the purview of dispatchers at RCC). Looking further, SIRE could be expanded to NYCT's bus network; in many respects SIRE for buses should be simpler than SIRE for subways, due to their more flexible operation (no merge conflicts, no terminal holdouts, ability to pass, etc.). Now restricted to holds and skips, SIRE could also theoretically be extended to recommend actions like reroutes, short-turns, and deadheading.

Looking outside NYCT, SIRE could be adapted for systems without central dispatching, using tools developed in prior studies *(2, 7)* to communicate actions directly to bus or train operators. Other systems may also find certain service management issues and objectives are more or less important (fewer merge issues, for example, or greater importance of timed transfers with less frequent service).

Acknowledgments

The authors thank the following at NYCT: Shri Iyer and Anne Halvorsen for their help developing and validating SIRE, the RCC dispatchers and management for their invaluable input and feedback, and Tuan Huynh, Jameel Alverez, and Joseph Fong for their help in building the SIRE dashboard.

Author Contributions

The authors confirm contribution to the paper as follows: study conception and design: Daniel Wood, Timon Stasko, Lauren Tarte, Darian Jefferson, Alla Reddy; data collection: Daniel Wood, Timon Stasko, Lauren Tarte, Darian Jefferson; analysis and interpretation of results: Daniel Wood, Timon Stasko, Darian Jefferson; draft manuscript preparation: Daniel Wood, Timon Stasko. All authors reviewed the results and approved the final version of the manuscript.

References

1. Turnquist, M. A. Strategies for Improving Reliability of Bus Transit Service. *Transportation Research Record: Journal of the Transportation Research Board*, 1981. 818: 7–13.
2. Argote-Cabanero, J., C. F. Daganzo, and J. W. Lynn. Dynamic Control of Complex Transit Systems. *Transportation Research Part B*, Vol. 81, 2015, pp. 146–160. http://dx.doi.org/10.1016/j.trb.2015.09.003
3. Song, W. *Real Time Dispatching Control in Transit Systems*. Master's thesis. Massachusetts Institute of Technology, 1998. http://hdl.handle.net/1721.1/50512

4. Delgado, F., J. C. Munoz, and R. Giesen. How much can Holding and/or Limiting Boarding Improve Transit Performance? *Transportation Research Part B*, Vol. 46, 2012, pp. 1201–1217.
5. Sanchez-Martinez, G. E. *Real-time Operations Planning and Control of High-Frequency Transit*. Doctoral thesis. Massachusetts Institute of Technology, 2015. http://hdl.handle.net/1721.1/99550
6. Strathman, J. G., et al. Bus Transit Operations Control: Review and an Experiment Involving Tri-Met's Automated Bus Dispatching System. *Journal of Public Transportation*, Vol. 4, 2001, pp. 1–26. https://doi.org/10.5038/2375-0901.4.1.1
7. Lizana, P., J. C. Munoz, R. Giesen, and F. Delgado. Bus Control Strategy Application: Case Study of Santiago Transit System. *Procedia Computer Science*, Vol. 32, 2014, pp. 397–404. https://doi.org/10.1016/j.procs.2014.05.440
8. Tribone, D. *Making Data Matter: The Role of Information Design and Process in Applying Automated Data to Improve Transit Service*. Master's thesis. Massachusetts Institute of Technology, 2013. http://hdl.handle.net/1721.1/81640
9. Suchkov, B., M. Boguslavsky, and A. Reddy. Development of a Real-Time Stringlines Tool to Visualize Subway Operations and Manage Service at New York City Transit. *Transportation Research Record: Journal of the Transportation Research Board*, 2015. 2538: 19–29.
10. Caspari, A., B. Levine, J. Hanft, and A. Reddy. Real-Time Estimation of Platform Crowding for New York City Subway: Case Study at Wall St 2/3 Station in Financial District. *Transportation Research Record: Journal of the Transportation Research Board*, 2017. 2648: 68–78.
11. Eberlein, X. J. *Real-Time Control Strategies in Transit Operations: Models and Analysis*. Doctoral thesis. Massachusetts Institute of Technology, 1995. http://hdl.handle.net/1721.1/11387
12. Stasko, T., B. Levine, and A. Reddy. Time-Expanded Network Model of Train-Level Subway Ridership Flows Using Actual Train Movement Data. *Transportation Research Record: Journal of the Transportation Research Board*, 2016. 2540: 92–101.

The Standing Committee on Rail Transit Systems (AP065) peer-reviewed this paper (18-03239).

Responsibility for errors or omissions remains entirely with the authors. Opinions expressed or implied are the authors' and do not necessarily reflect official polices of MTA New York City Transit or any other organization.

Article

TRR
JOURNAL OF THE TRANSPORTATION RESEARCH BOARD

Transportation Research Record
2018, Vol. 2672(8) 339–350

DOI: 10.1177/0361198118790096
journals.sagepub.com/home/trr
SAGE

Streetcars and Economic Development: Do Streetcars Stimulate Employment Growth?

Sarah Jack Hinners[1], Arthur C. Nelson[2], and Martin Buchert[3]

Abstract

This study reports economic development outcomes—defined as change in employment—for areas within 0.25 mi of three streetcar stations along four lines in each of four cities: Portland OR; Salt Lake City UT; Seattle WA; and New Orleans LA. Employment 3 years prior to the start of the streetcar construction was compared with employment within the three streetcar station areas in 2013. Before–after comparisons were also applied to 10 bus stops within each urbanized area selected as "controls" because they were quantitatively comparable to each streetcar station area studied. Shift-share analysis is used to compare each streetcar and control station area's employment trends with respect to central county economic performance. Portland's system showed strong economic development consistency followed by New Orleans. The Salt Lake City and Seattle lines, however, did not perform as consistently. These results indicate that although streetcar investment may support economic development, it is not alone a driver of employment growth.

Prior to the opening of Portland, Oregon's North–South streetcar line in 2001, streetcars in North America were novelties – a charming relic of a bygone era found only in a handful of cities. Streetcars are now popping up all over North America, with about a dozen in operation and many more being planned. Until recently, it seemed that a true streetcar renaissance was taking place that would reshape public transportation in American cities. Streetcars have come under scrutiny lately, which led in part to the cancellation of two streetcar projects in Arlington, Virginia in late 2014. The debate rages between those who see streetcars as transit projects and those who see streetcars as economic development or urban revitalization investments, or both. This dichotomy of perspective is more than nuance, for it is essentially what shut down the Arlington streetcar project and "destroyed 15 years of a joint effort … It set back transit options in this part of the region for at least a generation or more" (*1*).

Those in the anti-streetcar camp argue that as a way of moving people around, streetcars are excessively expensive and inefficient. They point out that streetcars are subject to getting stuck in traffic, do not move much faster than a person on foot or bicycle, yet require multi-million dollar infrastructure investments. The transportation needs of communities, they argue, would be better met by increasing bus service or by improving roads (*2*, *3*). In a recent comparative study, Brown, Nixon and Ramos note that with respect to operations costs, streetcars are a less efficient transit mode than buses (*4*).

The pro-streetcar camp argues that streetcars can be more than just transportation projects as they are highly visible investments that serve as a catalyst for economic development. For instance, Brown, Nixon and Ramos note that Portland's streetcar system stood alone as a successful economic development and urban revitalization investment perhaps because it was planned to facilitate revitalization in addition to transportation (*4*). An 2014 article in *The Guardian* speaks of the "emotional and aesthetic allure" of streetcars (*5*), and Brown, Nixon and Ramos offer that streetcars "have now taken on an iconic role completely separate from any transportation function they possess" being, notably, a catalyst for economic development (*4*).

[1]Ecological Planning Center, Department of City and Metropolitan Planning, University of Utah, Salt Lake City, UT
[2]School of Landscape Architecture and Planning, University of Arizona, Tucson, AZ
[3]Global Change and Sustainability Center, University of Utah, Salt Lake City, UT

Corresponding Author:
Address correspondence to Arthur C. Nelson:
acnelson@ArthurCNelson.com

This article is the first to systematically assess the economic development outcomes to streetcar investments. It does so through analysis of comparisons between streetcar lines built in Portland, Salt Lake City, Seattle, and New Orleans. It concludes with an interpretation of the applications and implications of the results for understanding the relationship between streetcars and employment.

Streetcars and Economic Development

Because they are so new, there are very few comparative, objective studies on the linkage between streetcars and economic development. Existing studies are either single-city case studies (*6*, *7*), or non-peer-reviewed reports commissioned and published by interested parties such as transit agencies, or both (*8*).

As the relationship between transportation investments and economic development is one of the most researched topics in urban planning, we focus on a few recent perspectives relevant to this article. Nelson et al. lay out a series of economic arguments in favor of transit investments such as facilitating agglomeration economies, increasing real estate values, and enhancing connectivity between people and jobs (*9*). Higgins and Kanaroglou note that although transit is assumed to generate positive economic outcomes such as increasing real estate values near transit stations, 40 years of research on the topic shows considerable variation across cities, transit modes, and even stations (*10*). Kolko surmises that, "The relationship between transit and surrounding land values and densities depends both on how businesses and residents value proximity to transit and on public-sector decisions about zoning, land use, and other incentives for transit-oriented development" (*11*). As a case in point, a study of projected outcomes to streetcar investments in Cincinnati found that impacts were spatially limited without accompanying policies designed to stimulate economic development (*12*).

To assess directly the extent to which jobs are attracted to transit investments, Nelson et al. studied the Eugene-Springfield, Oregon Bus Rapid Transit (BRT) system, finding enhanced growth in employment relative to the rest of the metropolitan area within 0.25 mi of BRT stations (*13*). This growth was not consistent across job sectors however, a finding that is supported by Belzer et al.'s analysis of preferences of different employment sectors to locate near transit infrastructure (*14*). In contrast, Kolko examined 204 rail transit stations in California and found that growth in employment near stations was variable, with more station areas losing jobs than gaining (*11*).

These and other studies show transit investments do not automatically lead to economic development. And outcomes may vary by transit mode. Notably, compared with light rail transit (LRT) and BRT systems, streetcars are much slower as they tend to move in traffic at the speed of traffic—often little faster than walking. Streetcars also stop more frequently than BRT and LRT. As Hovee quips, "If light rail systems function as highways and arterials, streetcar systems function as the *local streets*" (*8*). They also tend to be found in downtowns, the only notable exception being Portland's extension of its downtown streetcar looping across the Willamette River.

It is against this backdrop that our study addresses the question of whether streetcars promote economic development, measured by employment change. It is the first comparative study of employment patterns associated with modern streetcars in the United States, examining four modern streetcar systems that had completed construction and initiated service at the time of this study.

Methodology

Economic development may be assessed in various ways. In the literature on transportation and land use there are three common metrics. One approach measures economic development in terms of financial investment in real estate development (*6*), another as changes in real estate value (*10*), and a third in terms of employment (*9*, *11*). It is difficult in any case to make a direct causal link between a transportation project and the real estate development metric; the relationship is generally inferred by examining spatial patterns and temporal trends in the data. A further challenge is presented in studies that compare more than one city, because local data associated with real estate development, building permits, and so forth, are collected and maintained locally and in different ways from one jurisdiction to the next. Thus no objective, comparative, multi-city studies of economic development associated with streetcar projects have been completed to date.

In order to avoid the challenges of obtaining comparable real estate investment data for all four cities in our study, employment was chosen to be the metric for economic development. This metric has the advantage of being available nationwide in a common format (US census). However, it has the arguable disadvantage of being perhaps less responsive to streetcar development in the sense that real estate investment generally must occur first, in order for the employment to follow. For this reason, and because the streetcar lines in question are so new, this analysis can only lay the groundwork for detecting employment change for the new lines studied in this paper. The processes of change taking place may well be far from having run their course and only show small indications of things to come.

From a city planning perspective, forward-looking investments in public transit such as light rail are expected to increase density (residential, employment, or both residential and employment) near transit stations, as rail infrastructure is too expensive to cost-efficiently serve low-density land uses. There are complications in studying the impacts of rail transit investment, however, because changes in density associated with rail transit projects may either reflect new, net economic growth occurring preferentially in those locations, or it may reflect spatial redistribution of pre-existing jobs. In addition, there is a chicken-or-egg dilemma in assessing economic impacts, as many transit investments are intentionally located to serve already dense or growing population or job centers [see (*11*) for trends in California]. That is, transit projects may be built specifically in locations that are already on a growth trajectory.

In this project, four streetcar systems were studied that were new streetcars – not "heritage" lines – carrying modern trains and developed primarily to serve transit functions for the local community rather than tourists. The four study systems were Portland, Oregon's Central Loop; Salt Lake City, Utah's S Line; Seattle, Washington's South Lake Union line; and New Orleans, Louisiana's Rampart-St. Claude Line. Changes in employment around focal streetcar stops were tracked over time against a set of comparable non-streetcar control sites, with the central county as a whole, and within economic groups. We use central counties for our region because streetcar systems are designed to serve only the most centralized and built out area of the central county. For each study system, a "before" year, specified as 3 years prior to the beginning of construction, and an "after" year, specified as 2013, being the most recent year for which data were available, were determined (see Appendix for dates). The before date was considered to reflect the date of the project funding announcement, as such an announcement might trigger real estate speculation, however as only employment is being studied, not real estate investment or land values, the date of 3 years prior to the construction start was considered sufficiently early as to represent a "before" condition. The after dates are the primary limitation of this study: service on two of these streetcar systems began so recently [Salt Lake City: 08 December 2013 (*15*); New Orleans: 28 January 2013 (*16*)] that the most recent available data at the time of this study are less than 1 year after start of service.

Study Locations and Controls

For each streetcar line studied, three focal stops were selected to serve as experimental "treatments." The primary selection criterion for streetcar stops was spacing, that they be far enough apart that quarter-mile buffers around each stop would not overlap significantly with one another and that they represent as much of the full length of the line as possible. Each treatment site consists of a quarter-mile radius walkshed centered on the selected stop. Guerra, Cervero and Tischer find support for a quarter-mile catchment as appropriate for employment–transit studies (*17*). Each walkshed had a different socioeconomic profile and to detect the effect of streetcar development, a set of control sites with similar profiles but without streetcars, were selected for comparison. A multidimensional similarity measure was used to select 10 control sites for comparison with each study stop; these control sites were located within the same metropolitan area, were served by existing bus routes, and shared similar socioeconomic characteristics in terms of population, employment, housing units, households, and median household income as reported at the census block group level prior to the streetcar project development's "before" date. A similar conceptual approach, with slightly different methodology, may be found in Kolko's analysis (*11*). The process of selecting control station areas consisted of two steps: first, creating a large set of candidate points, and second, assessing each candidate point for similarity with the actual streetcar locations. For the first step, it was assumed that bus service follows corridors that are similar to those considered by planners when designing streetcar alignments, and a large number of points ($n = 1000$) randomly distributed along the network of arterial streets was created.

The second step used census measures of population, employment, housing units, households, and median household income at the census block group level from the identified "before" year. These variables were chosen because of their availability at annual intervals going back to the earliest year considered in our study (2003 for the Seattle line) and their universal availability from the US census; the block group is dictated as the spatial unit of reporting as this is the finest unit for which the Census Bureau reports the median household income variable.

For each streetcar station (treatment) point and candidate control station point, values were estimated for each of the five variables for the area within 0.25 mi of the point, allocating block group level measures proportionately to the fraction of the block group falling within this area. The distributions of the five measures were checked for normality and, if necessary, transformed to normalize them; individual measures were then standardized to z-scores. Taking each streetcar station (treatment) point individually, the distance in the standardized five-dimensional space was computed between the treatment point and each candidate control point. This distance is proportionate to dissimilarity, and the final 10 control sites were selected starting with the smallest dissimilarity distances from each streetcar station (treatment) site and rejecting candidates that overlapped in geographic space with already-selected control sites.

Descriptive Analysis

Each system was evaluated for the change in overall employment and employment in economic groups over the study periods. The streetcar station areas and the central county were compared, as well as control station areas with the central county – in both cases using z-scores to test whether differences over time are statistically different from null. Although we did not analyze the impact of planning policies associated with each streetcar project, these policies are listed in the Appendix.

Employment Analysis

With this study design, employment within each treatment walkshed is compared with itself over time (before–after), and with its respective set of control walksheds, also over time. In addition, as in the BRT study by Nelson et al., shift-share analysis was used to examine employment trends at each of the study sites and controls (*9*). This approach looks at employment within each walkshed at the before and after dates, compared with broader trends in the central county as a whole, and within industry sectors in the central county, and allows the determination of changes in employment at each studied treatment or control site as attributable to broader trends within the central county (as a whole, within industry sectors or both) or as due to local factors at the study site. Nelson et al. developed the shift-share approach for transit-related economic development that was used in this study (*9*).

Longitudinal employer household dynamics (LEHD) data from the US Census Bureau was used for the before and after years at the census block and central county levels. LEHD economic sectors were combined into economic groups reflecting roughly comparable employment, land use, and building characteristics [see (*18*) for details]. Economic sectors were not included for which workers do not occupy building spaces on a regular basis, such as farming, forestry, fishing, mining, or construction.

Application and Results

The results are presented for each streetcar line in the following format: 1) overview of the system history and focal station areas; 2) employment change for the streetcar station areas and for the central county as a whole with individual sector assessments; and 3) apportionment of shift-share results with respect to control station areas and station areas. At the end of the section, the results are broadly summarized for comparative purposes. In addition, all four cities enacted zoning adjustments to facilitate development along the new streetcar corridors. Although we did not analyze the impacts of these policies, they are listed for reference in the Appendix.

Portland, OR, Central Loop Line

The Portland Phase II streetcar system, known as the Loop, opened in 2012, although elements of the system, such as the Tilikum Crossing Bridge, were not operational until September 2015. As discussed previously, the first phase of Portland's streetcar system, the North–South Line, opened in 2001 and represented the first modern streetcar project in the United States. The success of Phase I in stimulating economic development has already been explored, so the Loop portion was the focus of the current research. The Central Loop Line was chosen because 1) it serves an area outside of, and across the Willamette River from, downtown and 2) the original North–South Line has already been studied. Land use policies were put in place around the line, as described in the Portland Comprehensive Plan, to support development around, and the use of, the streetcar. The policies describe the potential use of transit investments, support of economic development goals with transportation planning, and rezoning permitted uses and density maximums to further foster economic growth. The Portland stations were all located along the "B Loop" and include these stations:

- Northeast Grand at Holladay
- Grand and Stark
- Southeast Grand and Hawthorne

For brevity, results are reported for the combination of treatment and control sites. Table 1 reports descriptive results as well as shift-share results, focusing only on the control station area and streetcar station area effects. Discussion follows.

Overall, the control areas increased jobs by 6% during the study period whereas the central county (Multnomah) grew by 8%, which is not a significant difference in change based on z-scores. In contrast, the streetcar station areas added 22% to their cumulative job base between 2006 and 2013, which is significant. Shift-share analysis shows that whereas control areas lost share of change in jobs during the study period, streetcar station areas gained share. Cumulatively, both streetcar station areas and controls had statistically significant gains in jobs in several economic groups, though the control areas lost jobs in the light industrial and

Table 1. Pooled Descriptive and Shift-Share Results for Portland Streetcar Economic Development Analysis

Economic group	Control station area 2006	Control station area 2013	Central county 2006	Central county 2013	Control station area change	Central county change	z	Control station area share
Manufacturing	29	54	40,166	36,988	86%	–8%	*	27
Light industrial	310	166	28,403	26,919	–46%	–5%	*	(128)
Retail-lodging-food	3,490	3,402	90,257	97,573	–3%	8%	*	(371)
Knowledge	687	750	58,714	70,044	9%	19%		(70)
Office	2,365	2,818	164,032	172,446	19%	5%	*	332
Education	126	200	19,002	23,652	59%	24%	*	43
Health	5,210	5,605	60,270	68,912	8%	14%	*	(352)
Arts-entertainment-recreation	242	271	15,432	18,482	12%	20%		(19)
Total	12,459	13,266	476,276	515,016	6%	8%		(537)
Economic group	**Streetcar station area 2006**	**Streetcar station area 2013**	**Central county 2006**	**Central county 2013**	**Streetcar station area change**	**Central county change**	**z**	**Streetcar station area share**
Manufacturing	863	904	40,166	36,988	5%	–8%	*	109
Light industrial	1,767	2,017	28,403	26,919	14%	–5%	*	342
Retail-lodging-food	2,038	2,522	90,257	97,573	24%	8%	*	319
Knowledge	958	1,736	58,714	70,044	81%	19%	*	593
Office	5,978	7,052	164,032	172,446	18%	5%	*	767
Education	76	106	19,002	23,652	39%	24%		11
Health	878	1,070	60,270	68,912	22%	14%		66
Arts-entertainment-recreation	173	188	15,432	18,482	9%	20%		(19)
Total	12,731	15,595	476,276	515,016	22%	8%	*	2,189

*z-scores are $p < .05$.

retail-lodging-food groups. Education and health care economic groups added significant numbers of jobs in the control areas, whereas the streetcar station areas did not. On the whole, streetcar station areas performed better during the study period in attracting jobs in more economic groups than the controls.

It is clear that the employment mix differs among streetcar station areas, and among control areas as well as between the streetcar station areas and their matched control station controls. The three consistent employment growth areas for streetcar stations in Portland were knowledge, retail-lodging-food, and health.

Salt Lake City, UT, S Line

The S Line is the first modern streetcar line in Utah. It joins the main TRAX light rail system at its western end, and runs east for 2 mi through a dedicated right-of-way that previously carried freight rail, ending in Salt Lake City's "second downtown," the Sugar House neighborhood. Future planned phases will extend the line deeper into Sugar House and eventually create a loop, connecting back to the TRAX Red Line, however no funding has yet been secured for future phases. The line was built rapidly, with construction initiated in April 2012 and the opening of the line in December 2013. Redevelopment activity along the line has been quite variable, with high levels of investment in the Sugar House area and less at other locations. A form-based code was put in place around the streetcar line to support redevelopment activity. The form-based code allows a variety of land uses and building types that increase economic potential along the line. Streetcar stations used in the analysis include:

- Fairmont Station
- 500 East Station
- South Salt Lake Station

The S Line is unique in several ways. In addition to running through an abandoned rail corridor instead of in a street, the project also includes a linear park and bike path that represents a critical linkage in the regional bicycle system. Moreover, the South Salt Lake Station is substantially surrounded by a park-and-ride lot serving the TRAX light rail system. With the exception of the Sugar House end of the line, the areas served by the S Line were not on a growth or redevelopment trajectory prior to the construction of the streetcar. Table 2 reports pooled descriptive and shift-share results for streetcar station areas and control areas, and for Salt Lake County as a whole.

During the study period—extending from the depths of the Great Recession well into recovery—jobs increased by 25% in the control station areas but decreased by 2% in

Table 2. Pooled Descriptive and Shift-Share Results for Salt Lake City Streetcar Economic Development Analysis

Economic group	Control station area 2009	Control station area 2013	Central county 2009	Central county 2013	Control station area change	Central county change	z	Control station area share
Manufacturing	79	108	52,503	54,956	37%	5%	*	25
Light industrial	32	20	33,170	34,713	–38%	5%	*	(13)
Retail-lodging-food	753	1,404	117,648	125,702	86%	7%	*	599
Knowledge	267	230	71,808	81,006	–14%	13%	*	(71)
Office	445	507	242,008	254,659	14%	5%		39
Education	195	200	17,421	21,496	3%	23%	*	(41)
Health	935	950	60,907	65,460	2%	7%		(55)
Arts-entertainment-recreation	92	70	13,757	14,420	–24%	5%	*	(26)
Total	2,798	3,489	609,222	652,412	25%	7%	*	457
Economic group	**Streetcar station area 2009**	**Streetcar station area 2013**	**Central county 2009**	**Central county 2013**	**Streetcar station area change**	**Central county change**	**z**	**Streetcar station area share**
Manufacturing	402	141	52,503	54,956	–65%	5%	*	(280)
Light industrial	285	207	33,170	34,713	–27%	5%	*	(91)
Retail-lodging-food	1,441	1,492	117,648	125,702	4%	7%		(48)
Knowledge	338	361	71,808	81,006	7%	13%		(20)
Office	1,376	1,563	242,008	254,659	14%	5%	*	115
Education	91	64	17,421	21,496	–30%	23%	*	(48)
Health	75	101	60,907	65,460	35%	7%	*	20
Arts-entertainment-recreation	81	92	13,757	14,420	14%	5%		7
Total	4,089	4,021	609,222	652,412	–2%	7%	*	(345)

*z-scores are $p < .05$.

the streetcar station areas, both significant relative to the Salt Lake central county increase of 7%. Shift-share analysis further shows that by controlling for regional and industrial mix influences, the control station areas gained share of jobs while streetcar station areas lost share. The control areas gained jobs in most economic groups, notably retail-lodging-food. In contrast, the streetcar station areas lost jobs in many more groups than they gained, though they did gain substantially more and statistically significant jobs in the office economic group.

When considered separately, there is no consistency among streetcar stations in Salt Lake with regard to employment sector dynamics. What growth there is at streetcar stations is driven primarily by growth at the Sugar House end of the line, although there is some office growth at South Salt Lake. The 500 East Station area experienced a hemorrhaging of jobs over the study period.

Short-term economic development outcomes of the S Line should not be surprising. Not only does the line not travel along a regular street, but a key station (South Salt Lake) is mostly a park-and-ride lot. Much of the length of the line is also currently dominated by relatively low-density and low-to-moderate-income single-family housing that is unlikely to redevelop in the near future. This project was built opportunistically, with a view to the long-term infrastructure of the city, not as a local economic development stimulus. The economic growth that has occurred around the Sugar House terminus is in a regional center that was poised to take off regardless of the addition of a new rail transit line.

Seattle, WA, South Lake Union Streetcar

The South Lake Union Streetcar in Seattle started operation in 2007, connecting the South Lake Union neighborhood to downtown. The neighborhood was approved to be developed as a biotechnology and biomedical research center and those sectors contributed substantial funding for the streetcar. In 2010, Amazon built a new campus building in South Lake Union, which increased ridership considerably. The area is substantially developed into offices and other nonresidential uses though new high-density residential development has dominated new construction on parking lots and the redevelopment of older, low-rise structures. All of the development occurred after rezoning the area from industrial to land uses that would create a productive urban center. The effect of the change in policies was major redevelopment, economic growth, and a willingness on the part of developers to invest in projects around the streetcar line. The streetcar stations evaluated include the following:

- Westlake Hub
- Westlake Ave and Thomas Street

Table 3. Descriptive and Shift-Share Results for Seattle Streetcar Economic Development Analysis

Economic group	Control station area 2003	Control station area 2013	Central county 2003	Central county 2013	Control station area change	Central county change	z	Control station area share
Manufacturing	508	624	112,737	112,807	23%	0%	*	116
Light industrial	8,850	9,041	68,725	68,425	2%	–0%	*	230
Retail-lodging-food	7,139	8,155	223,327	246,110	14%	10%	*	288
Knowledge	18,553	24,263	194,368	254,175	31%	31%		1
Office	38,163	51,432	386,941	419,973	35%	9%	*	10,011
Education	7,666	8,291	27,731	36,057	8%	30%	*	(1,677)
Health	1,283	2,458	114,410	159,481	92%	39%	*	670
Arts-entertainment-recreation	1,363	1,517	36,577	44,137	11%	21%	*	(128)
Total	83,525	105,781	1,164,816	1,341,165	27%	15%	*	9,511
Economic group	**Streetcar station area 2003**	**Streetcar station area 2013**	**Central county 2003**	**Central county 2013**	**Streetcar station area change**	**Central county change**	**z**	**Streetcar station area share**
Manufacturing	751	396	112,737	112,807	–47%	0%		(355)
Light industrial	7,874	7,824	68,725	68,425	–1%	–0%		(16)
Retail-lodging-food	10,227	10,491	223,327	246,110	3%	10%	*	(779)
Knowledge	15,521	18,356	194,368	254,175	18%	31%	*	(1,941)
Office	22,032	15,358	386,941	419,973	–30%	9%	*	(8,555)
Education	609	818	27,731	36,057	34%	30%		26
Health	2,281	4,633	114,410	159,481	103%	39%	*	1,453
Arts-entertainment-recreation	858	1,805	36,577	44,137	110%	21%	*	770
Total	60,153	59,681	1,164,816	1,341,165	–1%	15%	*	(9,397)

*z-scores are $p < .05$.

- Fairview Ave and Campus Drive

What follows is a discussion of the key interpretations of Table 3, which reports descriptive and shift-share results.

Spanning 10 years, this line's before–after study period was the longest of the cases evaluated. Unlike Portland, the Seattle streetcar system has not been expanded beyond its small downtown footprint. As seen in Table 3, it has also not performed well relative to the central county. During the study period, the control station areas increased jobs by 27% whereas King County's increased by 15%, and streetcar station areas actually lost 1% of jobs, mostly in the office sector group. Otherwise, statistically significant changes among economic groups around streetcar stations were positive. During much of the study period and especially since the Great Recession, residential development in downtown Seattle has outstripped office and other nonresidential development. Results shown in Table 3 confirm this. The effect of residential development near streetcar stations should be the subject of future research.

Shift-share analysis shows that whereas control station areas gained share of jobs overall, streetcar station areas lost share. Indeed, streetcar station areas lost share of jobs in six of eight economic groups compared with just one for the control station areas. Health and arts-entertainment-recreation appear to be the most consistent positive growth sectors for streetcar station areas but again, there is little consistency between locations.

New Orleans, LA, Rampart-St. Claude Line

New Orleans' Rampart-St. Claude Streetcar line opened in 2013, one of the newest systems in this study. It begins at a transit hub that connects Amtrak, Greyhound, and Regional Transit Authority (RTA) bus lines, and runs 0.8 mi into the New Orleans Central Business District, running through the historic arts and theater district. The streetcar station areas are:

- Union Passenger Terminal
- Rampart-St. Claude Avenue and Poydras Street
- Canal Street and Elk

As this is the newest system studied, there may not have been enough time for the market to respond to the streetcar system. Land use policies were established to help with economic development, such as rezoning with overlays that support mixed use and multimodal options and aim for 24-h programming. On the other hand, among all the systems, this may be the one with the most advanced planning. Construction and initiation of service on this line follows a sustained wave of recovery-related growth for Orleans Parish (the central county)

Table 4. Descriptive and Shift-Share Results for New Orleans Streetcar Economic Development Analysis

Economic group	Control station area 2008	Control station area 2013	Central county 2008	Central county 2013	Control station area change	Central county change	z	Control station area share
Manufacturing	313	116	6,970	4,766	−63%	−32%	*	(98)
Light industrial	1,212	882	5,707	4,904	−27%	−14%	*	(160)
Retail-lodging-food	5,256	5,149	41,972	53,293	−2%	27%	*	(1,524)
Knowledge	3,559	3,424	25,205	28,858	−4%	14%	*	(652)
Office	4,036	4,571	77,590	77,023	13%	−1%	*	565
Education	53	84	15,249	19,540	58%	28%		16
Health	253	176	18,591	21,868	−30%	18%	*	(122)
Arts-entertainment-recreation	205	84	9,534	9,627	−59%	1%	*	(123)
Total	14,887	14,485	200,818	219,879	−3%	9%	*	(2,098)
Economic group	**Streetcar station area 2008**	**Streetcar station area 2013**	**Central county 2008**	**Central county 2013**	**Streetcar station area change**	**Central county change**	**z**	**Streetcar station area share**
Manufacturing	121	97	6,970	4,766	−20%	−32%		14
Light Industrial	603	432	5,707	4,904	−28%	−14%	*	(86)
Retail-lodging-Food	3,744	7,309	41,972	53,293	95%	27%	*	2,555
Knowledge	3,880	5,086	25,205	28,858	31%	14%	*	644
Office	8,281	8,877	77,590	77,023	7%	−1%	*	657
Education	33	131	15,249	19,540	297%	28%	*	89
Health	3,757	3,172	18,591	21,868	−16%	18%	*	(1,247)
Arts-entertainment-recreation	2,844	2,147	9,534	9,627	−25%	1%	*	(725)
Total	23,263	27,251	200,818	219,879	17%	9%	*	1,900

*z-scores are $p < 0.05$.

initially from Hurricane Katrina and then the Great Recession. Finally, it is in an area dominated by surface parking lots and older properties, prime for redevelopment. Table 4 reports economic development outcomes for the streetcar and control station areas. The results are broadly summarized in Table 5 for comparative purposes.

Table 4 shows that overall, the streetcar station areas outperformed Orleans Parish and the control areas, gaining jobs at 17% compared with the Parish at 7%, whereas the control areas lost 3%. Moreover, where the control areas lost jobs in six of the eight economic groups, the streetcar station areas lost jobs in just three. The shift-share results show further that the streetcar station areas gained share of central county jobs overall whereas the control areas lost job share.

Overall, relative to their matched controls, the streetcar station areas are all clearly more dynamic with regard to employment than control areas in New Orleans, however the dynamism does not follow consistent trends across station areas. The only consistent growth areas for streetcar station areas were office and retail-lodging-food.

Interpretations and Implications

The primary outcome desired from these analyses is consistency in the response of employment to streetcars. Does each city streetcar line behave in an internally consistent manner, or is the difference between station areas greater than the difference between cities? If local employment effects are consistent among the sites within a single streetcar line, and in their differences with control sites, it is likely that there is a streetcar effect for that city. If streetcars may be said to reliably promote economic development (as measured here by employment), then there should be: 1) across all sites and cities, a greater positive change in employment at treatment sites than at controls, and 2) positive local effects in the shift-share results. It should be noted that all four cities enacted zoning changes and design standards that facilitated development along the streetcar corridor, including mixed-use, transit-oriented development, and increased permitted densities (Appendix).

Among the study streetcar lines, Portland's Central Loop Line exhibits by far the greatest internal consistency in employment response. As expected, there is positive employment growth among the streetcar station areas that exceeds the growth at control sites. There is an economic development pattern that extends the pattern established with the original Portland North–South streetcar line. This confirms that Portland has an established streetcar economic-development strategy that works for Portland. The city and its business community have "bought in" to the streetcar paradigm and effective planning and design have created a streetcar system that

Table 5. Summary and Comparison of Four Streetcar Systems

Name	Location	"Before" date; date opened	Stations	Control area employment growth	Streetcar area employment growth	Employment growth sectors	Surrounding influential land use
Central Loop (Phase II)	Portland, OR	2006; 2012 (bridge opened 2015)	(1) NE Grand + Holladay (2) Grand + Stark (3) SE Grand Hawthorne	8%	22%	Knowledge, retail-lodging-food, health	Crossing bridge, convention center, Lloyd District, aging industrial, county administration
South Line	Salt Lake City, UT	2009; 2013	(1) Fairmont (2) 500 East (3) South Salt Lake	25%	−2%	Office	Linear park, bike path, park-and-ride lots, major commercial ctr, aging industrial
South Lake Union	Seattle, WA	2003; 2007	(1) Westlake Hub (2) Westlake Ave + Thomas St (3) Fairview Ave + Campus Dr	27%	−1%	Health, arts-entertainment-recreation	Amazon campus, other corporate headquarters, biotechnology/ biomedical research, museums
Rampart-St. Claude	New Orleans, LA	2008; 2013	(1) Union Passenger Terminal (2) Rampart-St. Claude Ave + Poydras St (3) Canal St + Elk	−3%	17%	Office, retail-lodging-food	Bus terminal, central business district, surface parking lots, entertainment district

functions well in both its transportation and economic functions. Although the raw scale and industry mix of employment varies from one station to the next, the results of the analysis reflect heavy economic development and job growth locating preferentially at streetcar stations, well ahead of control areas and the metropolitan area as a whole, even though the Loop line only became fully connected in 2015.

Of the other streetcars included in this study, it might be predicted that Seattle would be the most likely to follow the Portland model. Both are Pacific Northwest cities with similar culture and feel, although Seattle is much larger. Seattle's streetcar has also been in operation longer than any of the other lines considered in this study, second only to the original Portland line. However, this prediction was not borne out in analysis. Certainly, considerable investment in real estate redevelopment by major employers such as Amazon has occurred within and near the streetcar station areas, but job performance has been surprisingly weak, actually declining. The Seattle example is lacking the internal consistency that was expected, showing patchy development along the streetcar line. However, it is also known from current ongoing work that whereas job production may be weak, residential development has dominated the streetcar station areas since the Great Recession.

The New Orleans streetcar was built within the unique context of recovery from Hurricane Katrina. As noted by Guthrie and Fan, there has been significant commercial investment along the streetcar line during the post-Katrina years, but this has occurred within the context of strong investment and recovery activities citywide (*6*). New Orleans also has a long history of streetcars, making it more likely that investors will feel comfortable with streetcars as a valuable amenity. In fact, all three streetcar stations in this study showed positive employment growth and outperformed the control sites in both direct comparison and shift-share analysis. Furthermore, it is evident that all of the streetcar sites in New Orleans are more dynamic than their matched controls. Could streetcar sites in New Orleans become the centers of focused development that we have seen in Portland? This is left to future research, but there is certainly strong potential.

The streetcar with perhaps the least internal consistency is the S Line serving the Sugar House area of Salt Lake City. The streetcar concept is new to Salt Lake City (although it had a streetcar network in the past), and the corridor through which the S Line runs is in the early stages of what will probably be decades of change. Wildly differing and rapidly changing employment numbers along the line indicate possibly several processes at work: relocation of jobs and activities along the corridor, vacating of older buildings and businesses in preparation for demolition, redevelopment, renovation, and simply very different uses of land and land densities at the different stops. Furthermore, the S Line does not run within a street, but through its own right of way that also supports a linear park, bike trail, and easy pedestrian movement. It is a novel and unique amenity, and it still remains to be seen how employment activities might orient themselves in relation to it. Among all streetcar lines studied and perhaps even all such lines nationally, the S Line's design may be the most unique, though also, potentially, least conducive to stimulating economic development.

Overall, Portland has continued to follow in its own footsteps as this analysis confirms previous observations of Portland's streetcar success story. All of the other cities lacked the consistency of Portland. As the literature review above showed, Portland's streetcar itself is only part of a mosaic of planning and development policies designed to reinforce streetcar use. Our policy inventory (Appendix) indicated that all four cities did implement zoning changes to facilitate development along the streetcar line. It seems likely that in the case of the newer systems, employment development will follow, eventually, as there is evidence of employment change occurring. Portland may have been in a better position to develop quickly, based on its recent experience with the earlier streetcar line. Our study also did not have a way to gauge the degree of buy-in on the part of the local business community and the dynamics of public–private sector relationships in each city. A final point is that despite the short distance between streetcar stations, from the perspective of supporting business development and therefore employment growth, a streetcar line does not behave as a corridor but as a series of discrete points associated with individual station areas.

Appendix. Streetcar Planning Policies

City	Streetcar line	Policies	Intended effect of policies	Additional incentives	Source
Salt Lake City	S Line	Zoning: high density mixed use and residential	High density near transit stops to increase accessibility and transit ridership		East Streetcar Neighborhood Form-Based Code
		Design standard: implementation of greenway along streetcar line	A multi-use corridor to increase access and attract new businesses, residents, and visitors		Downtown South Salt Lake Zoning Ordinance and Design Standards
		East Streetcar Neighborhood Form-Based Code	Transit-oriented redevelopment near streetcar corridor		East Streetcar Neighborhood Form-Based Code
				Suggested incentive programs: first-time home buyer assistance; financing for public improvements, such as lighting or streets; financial partnership to build housing that the city desires such as senior or workforce housing; financial partnership on shared parking; rental unit buy back	East Streetcar Master Plan
		Streetcar Community Development Area (CDA)	Negotiate participation terms with taxable entities in CDA		South Salt Lake Economic Development Report
Portland	Central Loop (Phase II)	Central Employment Zone	Mixed use of commercial and industrial buildings in a central location with transit access that matches existing development		Portland Zoning Code

(continued)

Appendix. (continued)

City	Streetcar line	Policies	Intended effect of policies	Additional incentives	Source
		Central Commercial Zone	Intense development with high building coverage, large buildings, close spacing, and pedestrian oriented with attractive streetscapes		Portland Zoning Code
				Streetcar installation expected as development catalyst	Portland Streetcar Loop Development Strategy
Seattle	South Lake Union	Incentive zone overlay	Provide affordable housing or other public amenities	Developers gain extra floor area within their project by providing external amenities	Seattle Municipal Code; Incentive Zoning Overview, City of Seattle Office of Housing
		Seattle Mixed Zone	Higher density and transit-supportive residential, commercial, and office uses will redevelop industrial areas		Seattle Municipal Code
		Downtown Mixed Zone	Urban center with mixed uses supporting transit and pedestrian orientation		Seattle Municipal Code
		Downtown Office Core Zone	High-density office and commercial land uses due to higher building heights take the place of older office and light manufacturing areas		Seattle Municipal Code
New Orleans	Rampart-St. Claude	Current zoning: commercial streetcar corridor now exists for the buildings fronting the streetcar	High-density commercial buildings along streetcar line will boost economic development in the area		City of New Orleans Property Viewer and Comprehensive Zoning Ordinance
		Design standard for central commercial and historic commercial	Promotes historic preservation and tourist traffic in the area		City of New Orleans Vieux Carre Design Guidelines
				Streetcar implementation expected to be a catalyst for commercial/private development and to bring economic profit to the area by increasing property taxes and promoting tourist traffic	Mayor Landrieu and RTA's press release 10/03/16: https://www.nola.gov/mayor/press-releases/2016/20161003-pr-rampart-st-claude-line-opening/

Acknowledgments

The authors acknowledge support for research reported in this article from the U.S. Department of Housing and Urban Development's Office of Policy Development and Research with assistance from the University of Utah, the National Institute for Transportation and Communities, and the University of Arizona. We recognize especially contributions from Richard Decker, Dejan Eskic, Emily Guffin, Michael Larice, Kate Morrell, and Cassie Younger, all at the University of Utah.

Author Contributions

The authors confirm contribution to the paper as follows: study conception and design: SJH, ACN, MB; data collection: MB; analysis and interpretation of results: SJH, ACN; draft manuscript preparation: SJH, ACN. All authors reviewed the results and approved the final version of the manuscript.

References

1. Sullivan, P., and A. Olivo. Arlington Officials Halt Efforts on Long-Planned Streetcar Projects for Columbia Pike, Crystal City. *Washington Post*, November 18, 2014. https://www.washingtonpost.com/local/virginia-politics/arlington-officials-major-announcement-on-columbia-pike-streetcar-project-at-noon/2014/11/18/ce2a8170-6f38-11e4-8808-afaa1e3a33ef_story.html? noredirect=on&utm_term=.c59f11411be8.
2. O'Toole, R. *The Great Streetcar Conspiracy*. Cato Institute Policy Analysis no. 699, 2012.
3. The Economist. *Streetcars and Urban Renewal: Rolling Blunder*, August 9, 2014. http://www.economist.com/news/united-states/21611123-federal-subsidies-have-inspired-some-silly-transit-projects-rolling-blunder.
4. Brown, J., H. Nixon, and E. Ramos. *The Purpose, Function, and Performance of Streetcar Transit in the Modern U.S. City: A Multiple-Case-Study Investigation*. Mineta Transportation Institute Report 12-39, 2015.
5. Marshall, S. Streetcars of Desire: Why are American Cities Obsessed with Building Streetcars? *The Guardian*. February 20, 2015. http://www.theguardian.com/cities/2015/feb/20/streetcars-of-desire-why-are-americans-obsessed-with-building-trams. Accessed December 10, 2015.
6. Guthrie, A., and Y. Fan. Streetcars and Recovery: An Analysis of Post-Katrina Building Permits around New Orleans Streetcar Lines. *Journal of Planning Education and Research*, Vol. 33, No. 4, 2013, pp. 381–394.
7. Hovee, E. D., and R. Gustafson. *Streetcar-Development Linkage: The Portland Streetcar Loop*. Transportation Research Board Conference Publication, 2012. http://onlinepubs.trb.org/onlinepubs/conferences/2012/LRT/RGustafson.pdf.
8. Hovee, E. D. *Streetcar-Development Linkage: The Portland Streetcar Loop*. E. D. Hovee & Company, LLC, City of Portland Office of Transportation, Portland, OR, 2008.
9. Nelson, A. C., G. Anderson, K. Bartholomew, P. Perlich, T. W. Sanchez, and R. Ewing. *The Best Stimulus for the Money: Briefing Papers on the Economics of Transportation Spending*. Smart Growth America, 2009. http://www.smartgrowthamerica.org/documents/thebeststimulus.pdf.
10. Higgins, C. D., and P. S. Kanaroglou. 40 Years of Modelling Rail Transit's Land Value Uplift in North America: Diverse Methods, Differentiated Outcomes, Debatable Assumptions, and Future Directions. Presented at 94th Annual Meeting of the Transportation Research Board, Washington, D.C., 2015.
11. Kolko, J. *Making the Most of Transit: Density, Employment Growth, and Ridership around New Stations*. Public Policy Institute of California, 2011. http://www.ppic.org/main/publication.asp?i=947.
12. Mokadi, E., D. Mitsova, and X. Wang. Projecting the Impacts of a Proposed Streetcar System on the Urban Core Land Redevelopment: The Case of Cincinnati, Ohio. *Cities*, Vol. 35, 2013, pp. 136–146.
13. Nelson, A. C., R. Ewing, M. Miller, S. Kannan, and B. Appleyard. Bus Rapid Transit and Economic Development. *Journal of Public Transportation*. Vol. 16, No. 3, 2013, pp. 41–57.
14. Belzer, D., S. Srivastava, and M. Austin. *Transit and Regional Economic Development*. Oakland, CA: Center for Transit-Oriented Development, Oakland, CA, 2011. http://www.reconnectingamerica.org/resource-center/browse-research/2011/transit-and-regional-economic-development/.
15. Wikipedia. Utah Transit Authority, https://en.wikipedia.org/wiki/S_Line_(Utah_Transit_Authority). Accessed March 3, 2016.
16. Wikipedia. Rampart-St. Claude Streetcar Line, https://en.wikipedia.org/wiki/Rampart-St._Claude_Streetcar_Line. Accessed March 3, 2016.
17. Guerra, E., R. Cervero, and D. Tischler. 2011. *The Half-Mile Circle: Does It Best Represent Transit Station Catchments?* UCTC-FR-2011-09. University of California Transportation Center, Berkeley, CA, 2011.
18. Hinners, S. J., and M. A. Larice. *Assessing the Impacts of Streetcars on Economics, Equity, and Quality of Life*. Metropolitan Research Center, University of Utah, Salt Lake City, Utah, 2016. http://mrc.cap.utah.edu/wp-content/uploads/sites/8/2015/12/Assessing-the-Impacts-of-Streetcars-on-Economics-Equity-and-Quality-of-Life-HUD-Sustainable-Communities-Grant-Deliverable-August-2016.pdf.

The Standing Committee on Light Rail Transit (AP075) peer-reviewed this paper (18-00337).

The views expressed here do not necessarily reflect those of the research sponsor or the authors' universities.

Article

TRR
JOURNAL OF THE TRANSPORTATION RESEARCH BOARD

Transportation Research Record
2018, Vol. 2672(8) 351–362

DOI: 10.1177/0361198118797219
journals.sagepub.com/home/trr
SAGE

The Impact of Various Streetcar Types on Passenger Activity and Running Times

Wen Xun Hu[1], Ehab Diab[2], Aya Aboudina[1,3], and Amer Shalaby[1]

Abstract

Transit agencies that operate streetcars use vehicles of varied sizes and characteristics to accommodate demand and service requirements. While some studies have been conducted on the influence of different bus types on service operations, there is limited research on streetcar types. Therefore, this paper examines the impacts of three different types of streetcar operating in the City of Toronto on passenger service time and running time using statistical models. Field data was collected in July 2017 for three types of streetcars serving the busiest corridor in Toronto. Descriptive statistics and two linear regression models were employed to compare their performance. The analysis results support the use of low-floor vehicles with wider doors as well as larger number of doors. Results also indicate a negative impact of the current stop design on the usage of longer streetcars. This study offers transit agencies new insights into the effects of distinct types of streetcars on door usage, passenger service time, and travel time, which are important components of transit service efficiency and attractiveness.

Transit agencies operate vehicles of various sizes and designs to satisfy fluctuating demand and accommodate operational needs. The vehicle's design affects the boarding experience of passengers as well as dwell time. For example, low-floor vehicles are effective in reducing dwell times and serving passengers with reduced mobility (*1*), and more and wider vehicle doors are also beneficial for reducing dwell times. Therefore, it is valuable to understand the advantages and disadvantages of using distinct types of transit vehicles and their impact on passenger activities and travel time.

The Toronto Transit Commission (TTC) operates North America's largest and busiest streetcar network of 11 routes, mostly in downtown Toronto. With the introduction of the new streetcar (Flexity Outlook) in 2014, three types of streetcars are now serving the transit network, providing a unique opportunity to compare their service performance and usage by the public. It should be noted that a proof-of-payment (POP) scheme is employed across the network, which means passengers can board via any door. Onboard payment—in all three streetcar types—includes tokens, PRESTO cards (smart cards), credit cards, and cash. TTC passes, validated e-tickets, and paper transfers can be used as POP. This wide range of payment methods provides several convenient options for users, improving their experience with transit. The objective of this paper is to study the impact of different vehicle designs, associated with the three streetcar types, on service performance.

Literature Review

Previous studies on streetcars have investigated the effects of various factors and control strategies on streetcar performance. For example, a study in Melbourne and Toronto indicates that transit signal priority (TSP) can reduce streetcar travel time by 6% to 10%, although improvements in reliability of streetcar service due to TSP are inconsistent (*2*). The design of tram stops also affects the boarding experience, travel time performance, safety, and accessibility (*3*, *4*). Low-floor vehicles are recognized for reducing dwell times and improving accessibility for passengers with limited mobility (*1*). However, few studies have focused on how different streetcar types with distinct combinations of design features perform with respect to dwell times and travel times.

The number of boarding and alighting passengers, crowding, fare type, bus design, and relative height of vehicle floor and platform levels have been cited in the literature as major factors influencing bus dwell times (*5*, *6*). Similar factors, in addition to streetcar stop design,

[1]Department of Civil and Mineral Engineering, University of Toronto, Ontario, Canada
[2]Department of Geography and Planning, University of Saskatchewan, Saskatoon, SK, Canada
[3]Engineering Math and Physics Department, Cairo University, Giza, Egypt

Corresponding Author:
Address correspondence to Wen Xun Hu: wenxun.hu@mail.utoronto.ca

also affect the dwell time of streetcars (*7*, *8*). Bus dwell time studies are more commonly found in the literature, in which automatic vehicle location (AVL) and automatic passenger counting (APC) data were the main data sources. However, AVL and APC data tend to overestimate passenger service (or activity) time compared with manual data (*9*). The first empirical study that investigated the impact of crowding and stop design on tram dwell time was carried out in Melbourne (*3*). Dwell time was defined as the total of passenger service time and time waiting for departure (*10*). This study instead uses the passenger service time, which is measured from the time the door opens until the last boarding or alighting passenger passes the door.

Running time is the time taken by a transit vehicle to travel between two stops along a route. Running time (or travel time) models are common in the literature, and they have been widely estimated using AVL and APC data for investigating the effects of various strategies and factors (*11*). Running/travel time is strongly affected by the segment length, density of signalized intersections and utilization of control methods such as TSP, type of right of way, type of vehicle, and time of day (*2*, *12*, *13*).

A limited number of studies have investigated the impacts of different transit vehicle types on service performance. Evaluation of the running time of articulated buses has demonstrated that they experience longer delays than regular buses (*14*). Multi-unit streetcars, which are streetcars coupled to provide higher capacity, reduce bunching and headway variability by providing lower service frequency (*15*, *16*). Since various streetcar systems use different types of streetcars, it is valuable to understand how these differences might influence the service quality. This paper presents the first empirical study that explores the influence of streetcar types with distinct design factors on passenger service and travel time, providing transit agencies with a reference for streetcar design updates and rolling stock upgrades.

Methodology

To understand the performance of the three streetcar types, ride and spot checks were conducted during the morning and afternoon peak hours on three consecutive weekdays when the weather was mild and there was minimal road construction activity. GPS tracking data was also recorded using smart phones at one-second intervals. Ride check data were collected by research assistants riding the streetcars, where a research assistant was positioned at each door to collect accurate information about door operation times and passenger activity. The collected data was processed to investigate passengers' choices of door usage, the acceleration rate of streetcars, and differences in the passenger service time and travel time between the Flexity Outlook (FLX), Canadian Light Rail Vehicle (CLRV) and Articulated Light Rail Vehicles (ALRV). All the CLRVs and ALRVs currently serving the streetcar network are scheduled to be replaced by FLXs by 2019. The specifications of streetcars are outlined in Table 1.

Two streetcar routes were selected, namely the 504 King and 514 Cherry. They operate on the same King Street corridor and share 25 eastbound and westbound stops (Figure 1). This minimized the influence of non-captured factors (like traffic volume) on the analysis results. King Street is the busiest surface transit corridor in the entire city, carrying over 65,000 riders daily. Routes 504 and 514 operate in a mixed right of way; passengers need to walk across a vehicle lane to board. Vehicles and bicycles on this lane are required to stop to let people get on and off streetcars safely. Route 504 runs CLRVs throughout the day and uses ALRVs to accommodate extra demand in peak hours. On the other hand, all vehicles that serve route 514 are FLXs. People normally board the first approaching vehicle to reach one of the two subway stations at two mid-points along the corridor. The study section, including 21 overlapping curbside stops, begins and ends at time points. A preliminary survey confirmed that these two routes have comparable passenger demand.

Descriptive statistics as well as two statistical models at the stop and segment levels were involved in the data analysis. Table 2 includes a list of variables incorporated in the analysis with detailed descriptions. At the stop level, the descriptive data was used to summarize and compare the general patterns of door usage on the streetcars. Linear regression analysis was performed to investigate the passenger service time at the door most heavily used at each stop and its relationship with the number of passengers, stop location, and streetcar type. In this paper, passenger service time is defined as the time from door opening until the last boarding/alighting passenger passes the door. Dummy variables were included to segregate vehicle types and reveal the impacts of the three types on passenger service time and travel time. Interaction variables were utilized to capture the influence of FLXs on passenger activities.

At the segment level, the model investigated the segment travel time and its relationship with the number of passengers, direction, stop location, distance, time period, and streetcar type. This was done to provide transit agencies with aggregated information about the impact of different streetcar types on the service performance. A time-point segment (or a segment) was defined as the route section between two consecutive time points. The time points are important locations for transit agencies' schedule building and operational control. Squared terms of some independent variables were used to

Table 1. Specifications of Streetcars

	CLRV	ALRV	FLX
			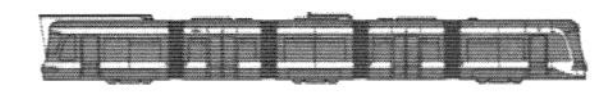
Crush load (passengers)	132	205	251
Seating	46	61	70
Car length (m)	15.44	23.16	30.20
Width (m)	2.54	2.54	2.54
Height (m)	3.36	3.36	3.84
Weight (kg)	22,685	36,745	48,200
Floor	High	High	Low
Stairs	4 steps	4 steps	1 step
Doors	2	3	4
First (front) door	Dual bi-fold, two channels, about 1 m to the vehicle front line	Dual bi-fold, two channels, about 1 m to the vehicle front line	One-leaf slide door, one channel, about 1 m to the vehicle front line
Second door	Double leaf, two channels, about 8 m to the vehicle front line	Double leaf, two channels, about 8 m to the vehicle front line	Double-leaf slide door, two channels, about 8 m to the vehicle front line
3rd door	NA	Double leaf, two channels, about 16 m to the vehicle front line	Double-leaf slide door, two channels, about 19 m to the vehicle front line
4th door	NA	NA	One-leaf slide door, one channel, about 27 m to the vehicle front line
Passenger boarding/alighting channels	4	6	6
Fare box	At the front door, accepts tokens and exact cash	At the front door, accepts tokens and exact cash	NA
PRESTO card reader	Two in total, one on the left-hand side at the front door, one in the middle of the two channels at the second door	Three in total, one on the left-hand side at the front door, one in the middle of the two channels at the second and third doors	Six in total, one at each of the front and fourth doors, two on both sides at the each of the second and third doors
Fare and transfer machine	NA	NA	Two in total, located across at each of the central double doors, accept cash, tokens, and credit/debit card
Ticket validator	NA	NA	Two in total, located beside the fare and transfer machines
Fleet size	112	15	31

Note: The head of each of the streetcars in the illustrations is on the right. NA = not applicable.

account for possible non-linear relationships. An interaction variable of ALRV/FLX and the number of boarding and alighting passengers were considered to capture the effects of the total number of people boarding and alighting ALRV/FLX vehicles on the travel time. In addition, a non-linear relationship was observed between the FLX passenger activity and the dependent variables; therefore, its squared term was used in the models. The effect of bunching, reflected by the variable named Number of blocked stops (Table 2), was also explored since it is a severe issue in Toronto causing substantial delays to streetcars. Blocked stops are identified when a dwelling streetcar is observed at a stop, while an incoming streetcar attempts to serve the same stop. GPS tracking data was used to analyze the acceleration rate of each streetcar after serving users at a given stop. A sensitivity analysis was also conducted, using the travel time model, to examine how travel times of different streetcar

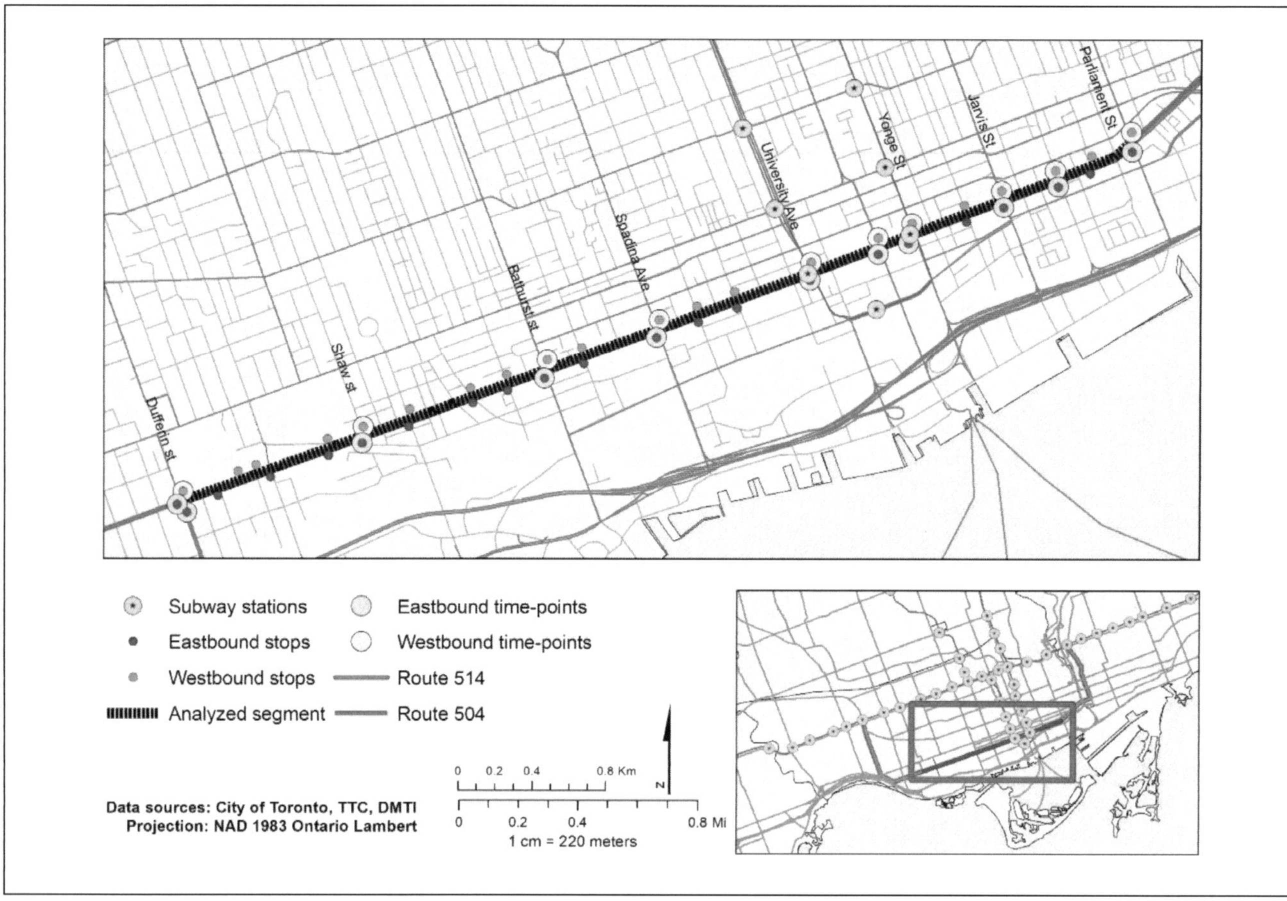

Figure 1. Map of the selected streetcar routes and segments.

types would change with the increase of boarding and alighting volumes.

Results and Analysis

Door Usage

The collected data consisted of 32 trips in total, among which 11 used FLXs, 14 used CLRVs, and 7 used ALRVs. In total, 570 stop-level observations were collected in the study period. The breakdown of door usage by streetcar type for all passengers, boarding and alighting, are illustrated in Figure 2. In general, the usage of doors can be attributed to factors including the door location, door width, and payment methods available at the door. Alighting passengers used the front door less often, possibly to avoid conflicts with the long boarding queues and because passengers have traditionally been directed to exit via the rear doors.

All the analyzed stops are near-side, which means that the streetcar drivers usually open their vehicle doors either when the streetcar's front end is at the stop post/shelter/stop line of the intersection or upstream of that point if passenger cars are stopped in front of the streetcar. Generally, the boarding passengers preferred using the front doors, which are closer than the other doors to the streetcar stop areas where passengers wait for the streetcars. These doors include the first and second doors of the FLXs, and the front doors of the CLRVs and ALRVs. The farther doors were used less frequently for boarding, likely because of the longer walking distance from the stops.

Alighting passengers, who are spread out inside the vehicles, used the doors closest to their standing or seating spots. Because the distances to the central doors (second and third doors of the FLXs, second door of the CLRVs, second and third doors of the ALRVs) are expected to be the shortest for most riders, they were used for alighting more than the other doors at the ends (e.g., first and fourth doors of the FLXs, first door of the CLRVs, first door of the ALRVs).

In addition to door locations, door width also has an impact on door usage. This is obvious on the FLXs where the second door with two channels was used more

Table 2. Description of Variables in the Regression Models

No.	Variable name	Description
Passenger service time model		
1	Passenger service time	The longest passenger service time among all doors at a stop (dependent variable)
2	$Boarding_{max}$	The number of boarding passengers at the door with the longest (maximum) passenger service time
3	${Boarding_{max}}^2$	The square term of variable # 2 in this table
4	$Alighting_{max}$	The number of alighting passengers at the door with the longest (maximum) passenger service time
5	${Alighting_{max}}^2$	The square term of variable # 4 in this table
6	Passenger load	The number of onboard passengers before serving the boarding and alighting passengers at a stop
7	At subway station	A dummy variable that is equal to 1 if the stop is located at a subway station
8	ALRV	A dummy variable that is equal to 1 if the trip used ALRV
9	FLX	A dummy variable that is equal to 1 if the trip used FLX
Interaction terms		
10	$FLX*(boarding\ and\ alighting)_{max}$	An interaction variable that is the product of variable # 9 and the sum of variables # 2 and 4 in this table
11	$(FLX*[boarding\ and\ alighting]_{max})^2$	The square term of variable # 10 in this table
Segment travel time model		
12	Segment Travel Time	The difference between departure time from the last stop of a time-point segment and the departure time of the first stop of a time-point segment for a designated (dependent variable)
13	$(Boarding\ and\ alighting)_{seg}$	The total number of boarding and alighting passengers at all doors and all stops (except the first stop) in the time-point segment
14	${(Boarding\ and\ alighting)_{seg}}^2$	The square term of variable # 13 in this table
15	At subway station	A dummy variable that is equal to 1 if the trip was made along a time-point segment that includes a subway station
16	Eastbound	A dummy variable that is equal to 1 if it is an eastbound trip
17	Number of blocked stops	The segment's total number of blocked stops, where a blocked stop incident is identified if the incoming streetcar is not able to serve a stop because of another streetcar dwelling at the same stop
18	Distance (m)	The length of the time-point segment in metres
19	Skipped stops	The total number of skipped stops in the time-point segment
20	PM peak	A dummy variable that is equal to 1 if the trip was made in the afternoon peak
21	FLX	A dummy variable that is equal to 1 if the trip used FLX
Interaction terms		
22	$FLX*(boarding\ and\ alighting)_{seg}$	An interaction variable that is the product of the dummy variable # 21 and variable # 13 in this table
23	${(FLX*boarding\ and\ alighting)_{seg}}^2$	The square term of variable # 22 in this table
24	ALRV	A dummy variable that is equal to 1 if the trip used ALRV
25	$ALRV*(boarding\ and\ alighting)_{seg}$	An interaction variable that is the product of the dummy variable # 24 and variable # 13 in this table

often by both boarding and alighting passengers than the front door with one channel only.

For boarding passengers, another crucial factor that influenced their usage of doors is the payment types available at the door. In addition to the door width, the second door on the FLX is more attractive than the front door because it is equipped with a fare payment machine. This is also a reason for the unbalanced door usage by boarding and alighting passengers observed on the old streetcars. Passengers using tokens and cash must pay at the fare box at the front door and receive a transfer from the driver. Approximately 19% of riders on old streetcars used the fare box to pay. As the first doors of CLRVs and ALRVs were used more by boarding passengers, it was more common to alight from the rear doors.

Table 3. Average Passenger Service Time per Door per Person

	FLX	CLRV	ALRV
Door 1	3.49	3.55	4.18
Door 2	1.87	2.46	2.23
Door 3	2.17	NA	2.74
Door 4	4.06	NA	NA
Average	2.58	3.01	3.28

Note: Values expressed in seconds. NA = not applicable.

Table 3 reports the average time taken by one passenger to get on or off a streetcar door. As can be noticed from the FLX column in Table 3, the times of the first and last doors are almost double those of the second and

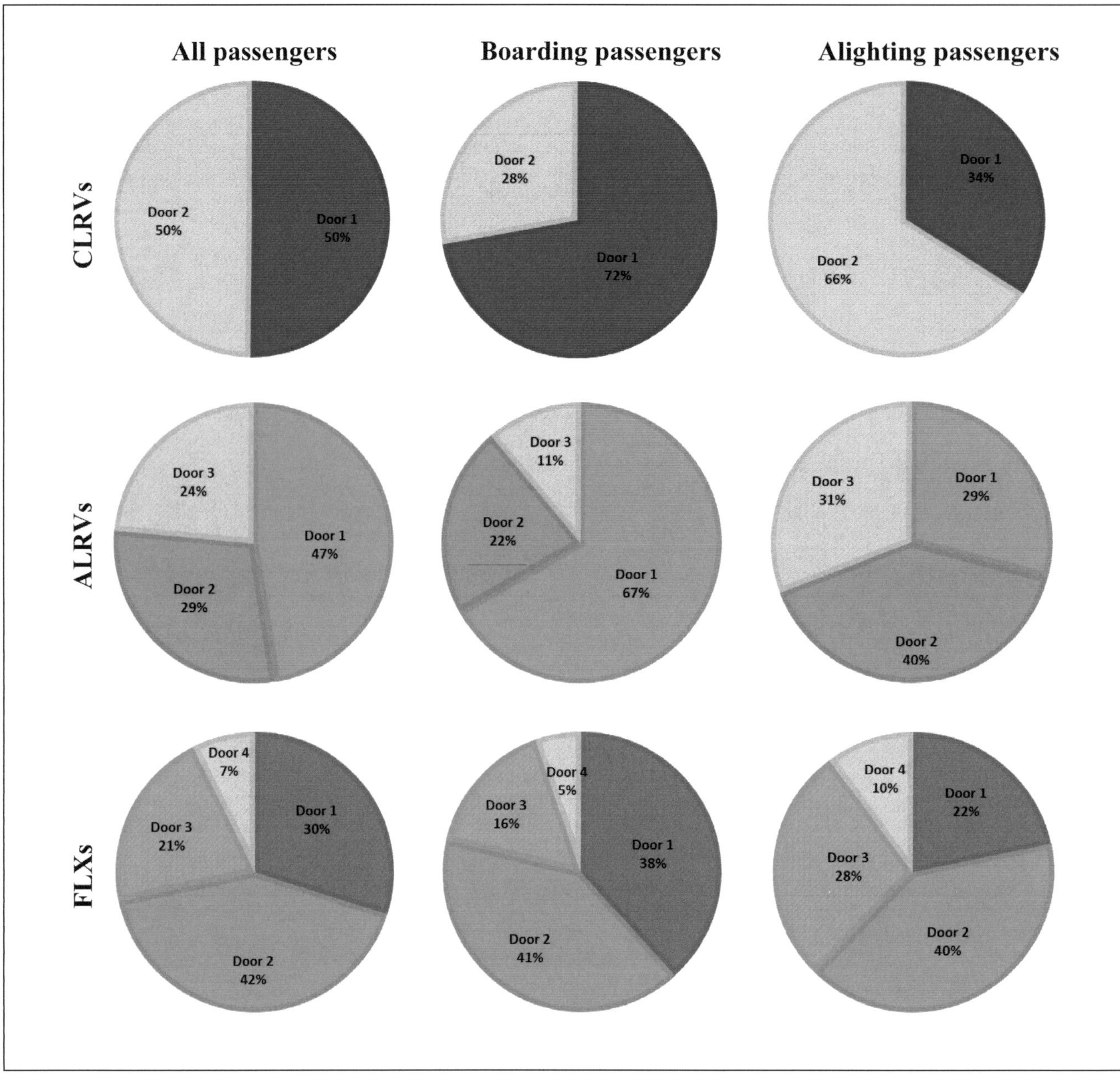

Figure 2. Breakdown of door usage by streetcar type for all passengers, boarding passengers, and alighting passengers.

third doors. This is because the first doors have one channel, hence can only serve one passenger at a time, unlike the latter doors having two channels. It can also be observed that the time at the fourth door is approximately 0.6 second longer than that at the first door, due to the time spent reaching it. Although the first doors of CLRVs and ALRVs have the same size as the other doors, they had longer passenger service time, which can be attributed to more boarding passengers (as observed on Figure 2). Boarding passengers normally take more time than alighting passengers for several reasons, including the time taken to approach the stopped vehicle and payment of fare (if needed). The average service time per passenger, T_a, should be calculated for each streetcar type as:

$$T_a = \sum_{n=1}^{N} (P_n)(t_{an}) \quad (1)$$

where N is the total number of streetcar doors, n is the door number, P_n is the percentage door usage at the nth door (which is illustrated in Figure 2 for different streetcar types), and t_{an} is the average service time per passenger at the nth door (listed in Table 3). As is clear in the table, the average time taken to board or alight CLRVs and ALRVs are longer due to stairs. This is consistent with the findings of (6). As low-floor vehicles are more accessible, seniors and passengers with strollers were more often observed on the FLXs. Based on our observation, they would choose to wait for a FLX rather than taking a CLRV or ALRV even if it arrived first. To confirm that passenger service times are shorter on low-floor vehicles, a passenger service model is estimated and presented in the following section.

Since the doors are not evenly used and have varied sizes, the time spent on boarding and alighting varied depending on which doors were used. Theoretically, the optimal (i.e., minimum) service time ($T_{optimal}$) can be achieved if passengers are redistributed to doors until equal total passenger service times are reached at all doors. Assuming the total number of boarding and alighting passengers at the stop to be equal to M, the optimal situation described can hence be expressed as follows:

$$\begin{aligned} T_{optimal} &= t_{a1} \cdot p_1 \cdot M = t_{a2} \cdot p_2 \cdot M \\ &= \ldots = t_{an} \cdot p_n \cdot M = \ldots = t_{aN} \cdot p_N \cdot M \end{aligned} \tag{2}$$

where p_n is the percentage door usage at the nth door in the optimal situation. Using Equation 2 and the values of t_{an} reported in Table 3 for different streetcar types, the optimal percentage usage per door, hence the minimal service time $T_{optimal}$, can be determined for each streetcar type. The average percentage saving in passenger service time, achieved in the optimal situation, can be calculated as follows for each streetcar type:

$$\begin{aligned} &\% \text{ Saving of Passenger Service Time} \\ &= \left(\frac{1}{m_2 - m_1 + 1} \sum_{M = m_1}^{M = m_2} \frac{T_{observe} - T_{optimal}}{T_{observe}} \right) * 100 \end{aligned} \tag{3}$$

where $T_{observe}$ is the average observed passenger service time when the number of boarding and alighting passengers reported per stop equals M; m_1 and m_2 are the lower and upper observed values of M; and $T_{optimal}$ is the optimal passenger service time calculated through Equation 2, at a given value of M. Based on the data collected during the ride checks, m_1 and m_2 were found to be 2 and 40, respectively. This range is also used in the "Sensitivity Analysis" section presented below. According to Equation 3, the percentage saving in passenger service time (in the optimal situation) was found to be 45.37% for FLXs, 41.74% for CLRVs and 37.99% for ALRVs (i.e., 40% on average for all three streetcar types). The aggregated impact of the number of doors on passenger service time and travel time is analyzed using the segment travel time model.

Descriptive Statistics of Regression Models

Table 4 presents the summary statistics of the variables used in the passenger service and segment travel time models. As shown in the table, statistics are differentiated according to the streetcar type. For the passenger service time model, a total of 458 records (i.e., stops) were included in the analysis. Stops with no passenger activity, skipped stops, and cases of door reopening were removed from the analysis. For the segment travel time model, a total of 230 trip segments were included. These segments had at least one passenger activity at a stop. As observed in Table 4, the passenger service times and travel times of CLRVs were the highest, while ALRVs and FLXs were much faster. To understand these results better—while controlling a set of influential variables—two statistical models, namely "passenger service time model" and "segment travel time model," are presented in the following two sections.

Passenger Service Time Model

A linear regression model was developed using passenger service time in seconds as the dependent variable. The output of this model is reported in Table 5. The model contains 458 stop observations and explains 52% (R^2 = 0.52) of the variation in passenger service time. This R^2 value is comparable to previous research that used similar models (*17*).

As shown in Table 5, each passenger boarding the streetcar system in Toronto adds about 2.96 seconds to the total passenger service time. The squared term of boarding passengers indicates that each additional passenger added less time to overall passenger service time than the passenger before her/him. More specifically, each additional boarding is expected to take 0.06 seconds less time. Therefore, the more boarding passengers, the less time it takes per passenger to board, which is consistent with the findings of previous research (*14*, *17*). Each passenger alighting the streetcar system in Toronto adds 1.10 seconds to the total passenger service time, while the alighting square term indicates that each additional alighting passenger is expected to take 0.02 seconds less time. This suggests a faster service time per person for alighting than boarding, which is expected due to the time spent in crossing a traffic lane and approaching the streetcar in the boarding process. This is also in agreement with the Transit Capacity and Quality of Service Manual (TSCQM) (*10*) which introduced the boarding lost time factor to account for the time required for boarding passengers to approach transit vehicles. In addition, climbing the four steps of CLRVs and ALRVs takes more effort than descending these steps for alighting passengers.

Every additional onboard passenger, before arriving at the stop, adds about 0.08 seconds to the total passenger

Table 4. Summary Statistics of Variables Used in the Models

	FLXs		CLRV		ALRV	
	Mean	SD	Mean	SD	Mean	SD
Variables of the passenger service model						
Passenger service time	13.89	11.29	17.14	16.20	12.51	11.47
$Boarding_{max}$	3.18	3.75	2.90	4.00	1.80	3.03
$Boarding_{max}^2$	24.11	57.67	24.33	67.16	12.31	51.83
$Alighting_{max}$	2.57	4.99	2.94	4.60	2.16	2.51
$Alighting_{max}^2$	31.34	162.0	29.71	130.2	10.92	23.74
At subway station	0.10	0.30	0.11	0.32	0.11	0.32
Passenger load	58.48	29.74	45.38	21.89	48.38	25.78
CLRV	0.00	0.00	1.00	0.00	0.00	0.00
ALRV	0.00	0.00	0.00	0.00	1.00	0.00
FLX	1.00	0.00	0.00	0.00	0.00	0.00
FLX*$(\text{boarding and alighting})_{max}$	5.75	6.44	0.00	0.00	0.00	0.00
$(\text{FLX*[boarding and alighting]}_{max})^2$	74.30	243.5	0.00	0.00	0.00	0.00
Observations (N)	168		192		98	
Variables of the segment travel time model						
Segment travel time	214.40	104.82	226.69	115.64	181.31	82.91
Eastbound	0.58	0.50	0.29	0.45	0.22	0.42
Number of blocked stops	0.26	0.62	0.46	0.76	0.11	0.37
Distance (m)	611.6	301.5	614.0	309.5	606.8	306.7
Skipped stops	0.06	0.28	0.09	0.28	0.04	0.19
PM peak	0.60	0.49	0.71	0.45	0.22	0.42
$(\text{Boarding and alighting})_{seg}$	29.86	24.43	18.77	15.17	19.58	16.59
$(\text{Boarding and alighting})_{seg}^2$	1481	2986	580.0	1026	653	1176
At subway station	0.24	0.43	0.24	0.43	0.20	0.40
FLX	1.00	0.00	0.00	0.00	0.00	0.00
FLX*$(\text{boarding and alighting})_{seg}$	29.86	24.43	0.00	0.00	0.00	0.00
$(\text{FLX*[boarding and alighting]}_{seg})^2$	1481.13	2985.94	0.00	0.00	0.00	0.00
ALRV	0.00	0.00	0.00	0.00	1.00	0.00
ALRV*$(\text{boarding and alighting})_{seg}$	0.00	0.00	0.00	0.00	19.58	16.59
Observations (N)	85		91		55	

Table 5. Passenger Service Time Model

				95.0% confidence interval	
Independent variables	Coeff.	t-stat	Sig.	Lower bound	Upper bound
Constant	**3.14**	**2.61**	**0.01**	**0.78**	**5.50**
$Boarding_{max}$	**2.96**	**10.33**	**0.00**	**2.40**	**3.53**
$Boarding_{max}^2$	**−0.06**	**−3.81**	**0.00**	**−0.09**	**−0.03**
$Alighting_{max}$	**1.10**	**4.25**	**0.00**	**0.57**	**1.54**
$Alighting_{max}^2$	**−0.02**	**−2.07**	**0.04**	**−0.04**	**0.00**
Passenger load	**0.08**	**4.13**	**0.00**	**0.04**	**0.12**
At subway station	**7.18**	**4.06**	**0.00**	**3.70**	**10.66**
ALRV	−1.85	−1.53	0.13	−4.23	0.53
FLX	−1.52	−0.90	0.37	−4.84	1.79
Interactions					
FLX*$(\text{boarding and alighting})_{seg}$	**−0.70**	**−2.17**	**0.03**	**−1.34**	**−0.07**
$(\text{FLX*[boarding and alighting]}_{seg})^2$	0.01	1.39	0.16	−0.01	0.03
Adjusted R^2	0.52				
N	458				

Note: Bold values indicate statistical significance.

service time. Therefore, heavily loaded streetcars are expected to have longer passenger service times, which was also concluded in the literature (*3*). The passenger service times at transfer stops to subway stations are about 7.20 seconds longer, which can be attributed to continuous flow of passengers approaching from distant subway exits.

Table 6. Segment Travel Time Model

Independent variables	Coefficients	t-stat	Sig.	95.0% confidence interval	
				Lower bound	Upper bound
Constant	−23.78	−1.48	0.14	−55.43	7.86
(Boarding and alighting)$_{seg}$	**3.52**	**3.71**	**0.00**	**1.65**	**5.39**
(Boarding and alighting)$_{seg}^2$	**−0.04**	**−2.84**	**0.00**	**−0.06**	**−0.01**
At subway station	**43.29**	**4.02**	**0.00**	**22.08**	**64.51**
Eastbound	**17.02**	**1.96**	**0.05**	**−0.13**	**34.16**
Number of blocked stops	**49.50**	**7.83**	**0.00**	**37.04**	**61.97**
Distance (m)	**0.25**	**16.63**	**0.00**	**0.22**	**0.27**
Skipped stops	**−28.85**	**−1.82**	**0.07**	**−60.04**	**2.35**
PM peak	**26.81**	**3.20**	**0.00**	**10.31**	**43.31**
FLX	18.80	1.00	0.32	−18.29	55.90
ALRV	12.40	0.77	0.44	−19.27	44.07
Interactions					
FLX*(boarding and alighting)$_{seg}$	**−2.84**	**−2.52**	**0.01**	**−5.05**	**−0.62**
(FLX*[boarding and alighting)$_{seg}$]^2	**0.04**	**2.71**	**0.01**	**0.01**	**0.07**
ALRV*(boarding and alighting)$_{seg}$	**−1.22**	**−1.97**	**0.05**	**−2.45**	**0.00**
Adjusted R^2	0.71				
N	230				

Note: Bold values indicate statistical significance.

The key policy variables, ALRV and FLX, have negative signs, but were insignificant. In other words, the passenger service times of ALRVs and FLXs are not significantly different than CLRV. However, the model includes an interaction variable "FLX*(Boarding and Alighting)$_{seg}$" to capture the impact of FLXs on passenger activity, which shows a negative and statistically significant effect on passenger service time. This indicates that every passenger boarding/alighting a FLX reduced the passenger service time by 0.70 seconds compared with CLRVs and ALRVs. This finding confirms the benefit of using low-floor FLXs. The FLX passenger activity squared term shows a positive sign indicating that the observed savings by passengers decrease with the increase of number of users; however, the significance level of the variable is less than 90%. It should be noted that the type of fare payment did not have a significant impact in our models. This is may be due to the fact that all front doors of ALRVs and CLRVs are wide doors (two-channel doors) which allow users to bypass any user depositing the exact change or tokens. For FLXs, which have one-channel doors, the payment machines are located inside the vehicle (not near the door) and allow passengers to pay using exact change, token, or credit card. To better understand the previous findings at the aggregated level, the following section presents the segment travel time model.

Segment Travel Time Model

A linear regression model was developed using time-point segment travel time in seconds as the dependent variable. The output of this model is reported in Table 6, which contains 231 segment level observations and explains 71% of the variation in travel time. This proportion of explained variance is comparable to previous results (*18*).

As seen in Table 6, each passenger boarding/alighting a streetcar adds 3.52 seconds to the total segment travel time, while keeping all other variables at their mean values. Each additional boarding/alighting passenger is expected to take 0.04 seconds less time than the passenger who boarded before her/him. The differences between the estimates of this model and the previous model were expected, since the previous one did not include the extra door open time without passenger activity. Travel times at segments with a subway station were longer by 43.3 seconds, which is also expected due to higher passenger volumes at these locations and continuous arrivals and more frequent door reopening. Eastbound trips were faster by 17.0 seconds on average, which could be explained by traffic conditions and the more diverse and vibrant land uses on the east side of the corridor. The number of blocked stops has a positive significant coefficient, indicating that bunching at stops increased the streetcar travel time by 49.5 seconds (about 70% of a signal cycle length). This is expected since the following streetcar has to wait for the leading one's dwell time as well as their stopping at traffic lights.

As expected, distance has a positive and significant impact on travel time. For every metre in length, total travel time increases by 0.25 seconds, which equals an average speed of 14.4 km/h. Every skipped stop saves about 28.9 seconds of the segment travel time. Trips during the afternoon peak were slower by 26.8 seconds than morning peak, as observed in many other cities (*11*, *19*).

While FLXs and ALRVs have a positive sign in terms of their impact on segment travel time compared with CLRVs, their coefficients are not significant. This suggests that FLXs and ALRVs were not consistently slower than CLRVs. Nevertheless, the interaction between the FLXs and their passenger activity shows a significant impact on the segment travel time. Boarding/alighting FLXs reduced travel time by 2.84 seconds per passenger. This demonstrates the aggregated benefit on travel time of using FLXs. Moreover, a non-linear relationship was observed using the square term of the FLX passenger activity. It is shown that every additional boarding/alighting passenger on FLX adds 0.04 seconds to the segment travel time. This can be attributed to several factors. First, boarding passengers had to walk further to approach the back doors. Therefore, while the passengers using the first and second doors of FLXs benefited from the low-floor, it took more time per person for those who walked to the third and fourth doors. Second, FLXs include a mix of two-channel and one-channel doors which were not equally used (as indicated in the "Door Usage" section). With more passenger activities, users at the first door of an FLX (single-channel door) would have to wait longer to board, or otherwise shift to other doors, which increases the total service time. For boarding and alighting passengers on ALRVs, represented by the "ALRV*(Boarding and Alighting)$_{seg}$" variable, travel time was reduced by 1.22 seconds, with no observed non-linear relationship. This demonstrates less saving per boarding/alighting passenger for ALRVs than FLXs. However, these savings were more consistent across different volumes of passengers (a linear relationship). Therefore, changing the streetcar stop location along the corridor, while adding some marks on the ground showing the location of each door of the FLXs, is highly recommended to maximize their benefits.

The acceleration rate of the streetcar after dwelling might also affect the travel time. To get a statistically significant estimate of the acceleration rate of each streetcar type, speed values were calculated at more than 30 different situations in which streetcars (of the same type) departed from stops along their routes (*20*). The acceleration rate was calculated in our analysis by dividing the distance travelled from the time of departure (t_d) until a certain time interval (T) elapsed by the duration of that time interval (i.e., by T). A suitable time interval T was found to be 10 seconds, after which the streetcar might decelerate due to traffic congestion ahead of it. The distance travelled during T was calculated based on the streetcar's location (identified from the GPS data) at time instances t_d and $t_d + T$. To overcome possible inaccuracies of GPS data, the streetcar location at certain time was determined by averaging the locations captured by GPS at the same time instant, one second before, and one second after. The acceleration rate of each streetcar type was hence estimated as the statistical mean of the speeds calculated for that type, over more than 30 stops. The average acceleration rate intervals estimated for the three streetcar types, under 90% confidence level, are as follows:

- CLRV: 15.29 ± 1.77 km/hr
- ALRV: 12.83 ± 1.50 km/hr
- FLX: 11.83 ± 1.87 km/hr

The numbers obtained indicate that the acceleration rate of CLRVs was higher than those of FLXs and ALRVs. Since CLRVs and ALRVs have the same engine, the heavier ALRVs have lower acceleration rate. Although FLXs have the largest weight, they also use more advanced engines, so the difference between FLXs and ALRVs is not significant. It should be emphasized, however, that having higher acceleration rate does not necessarily imply faster streetcar travel time, since other factors affect the latter quantity, as clarified above in the "Segment Travel Time Model" discussion.

Sensitivity Analysis

A sensitivity analysis was conducted using the coefficients of the segment travel time model (presented in Table 6). This was done to demonstrate the changes in travel times of different streetcar types with respect to different passenger volumes, keeping all other variables at their mean values. The estimates were constructed for the eastbound trips in the afternoon peak without blocked stops or skipped stops. The travel distance was set at a constant value of 609.5 metres, which is the average value of the analyzed segments.

The streetcars travel time is shown in seconds in Table 7. The travel time increases with the number of passengers, because more passengers entail longer passenger service time at stops. The travel time of CLRVs is the longest in all scenarios regardless of the number of boarding and alighting passengers, but the difference between CLRVs and ALRVs is small when passenger numbers are low. In other words, at segments with passenger activity less than 20 passengers, CLRVs are on average 5% and 9% slower than ALRVs and FLXs, respectively. This is because ALRVs and FLXs have more doors than CLRVs. In addition, FLXs are faster since they are low-floor vehicles which saves more time on boarding and alighting.

The percentage of savings generally increases with the number of boarding and alighting passengers between CLRVs and the two other types. In scenarios of passenger-volume between 20 and 40, CLRVs are on average 15% and 20% slower than ALRVs and FLXs,

Table 7. Sensitivity Analysis of Travel Time

Total boarding and alighting passengers	2	4	6	8	10	15	20	25	30	35	40
TT of CLRV	177	183	190	196	201	214	226	235	242	248	252
TT of ALRV	174	178	182	186	189	196	201	204	206	205	203
TT of FLX	171	173	174	175	177	180	184	188	191	195	199
ALRV[a]	1.4%	2.7%	3.9%	5.0%	6.1%	8.6%	10.9%	13.0%	15.2%	17.3%	19.4%
FLX[a]	3.1%	5.9%	8.3%	10.4%	12.2%	15.9%	18.5%	20.2%	21.1%	21.4%	21.1%

[a]Percentage of savings by taking ALRVs and FLXs compared with CLRVs.

respectively. This indicates the significant benefits of the latter two types. However, the percentage savings of FLXs decline with the increase of passenger activities compared with ALRVs, which has been explained by several factors in the "Travel Time Model" section. At passenger-volume level of around 40, both ALRVs and FLXs have similar impacts on travel time, compared with CLRVs. Therefore, it is highly recommended to review streetcar stop location and implementing policies to increase users' awareness about the door locations on FLX vehicles.

Conclusion

The main objective of this paper was to analyze the impact of streetcar types on passenger service time and segment travel time. Ride and spot checks were performed to gather data for analysis. Descriptive statistics and regression models were used to investigate the factors that influence the door usage, passenger service time, and travel time.

The descriptive analysis of door usage depicts the underlying factors affecting passengers' door choice. Boarding passengers were seemingly affected by the distance of doors from the stops, the payment methods available, and the door width. Alighting passengers tended to avoid using the front door and alight via the closest door. The passenger service time and travel time models confirmed the benefit of using vehicles with low-floor and larger number of doors. More specifically, the passenger service time model suggested that, with the same number of boarding and alighting passengers, low-floor FLXs had a shorter service time per passenger than CLRVs and ALRVs.

At the segment level, CLRVs had the slowest travel times compared with ALRVs and FLXs. This difference in travel times was affected considerably by the number of boarding and alighting passengers. In other words, with a total number of boarding and alighting passengers less than 20 passengers per segment, CLRVs were about 5% and 9% slower than ALRVs and FLXs, respectively. The difference in travel times increased dramatically with the increase of passenger activity. For segments with a total passenger activity between 20 and 40 passengers, CLRVs were slower by 15% and 20% than ALRVs and FLXs, respectively. Although CLRVs have better acceleration performance, this difference demonstrates a challenge in operating CLRVs along busy corridors with considerable passenger activity. Along these corridors streetcar schedules should be adjusted to accommodate the additional travel time needed by CLRVs to serve passengers in comparison with the two other types.

The analysis also shows that the benefits exhibited by ALRVs and FLXs were not consistent, with FLXs taking more time as the passenger activity increases. This was attributed to the unbalanced use of doors on FLXs in comparison with ALRVs. Therefore, to reduce the travel times for longer streetcars (such as FLXs in Toronto) and maximize their benefits, it is important to take measures that encourage even usage of all doors. The near-side stops can be relocated a few metres upstream of the intersection instead of at the stop line, so that when the streetcar dwells at the stop line, the stop aligns with the centerline of the streetcar. Also, dedicated streetcar platforms can be used more commonly such that waiting passengers can spread themselves more evenly along the platform. This would encourage boarding passengers to use the central and back doors more. Signs indicating the location of doors would also help passengers line up in front of different doors as they do normally at the Metrorail platform.

This study investigates the impact of streetcar types on service operations on one of the busiest streetcar corridors in North America. Three types of streetcars were considered, covering distinct vehicle designs with respect to the number of doors (ranging from two to four), level of boarding, payment method (fare box, POP, and payment machine), door width, streetcar length (ranging from 15 to 30 metres), and door location. Even though cities around the world may use other streetcar types, they would share some features with CLRVs, ALRVs, or FLXs, as the former two are built based on standard design and more than 3,000 Flexity vehicles are serving light rail systems globally. Therefore, this paper offers transit planners and decision makers valuable insights into the streetcar design and infrastructure improvements of new or existing systems. Although this research explores solely the impact of streetcar design on transit service operations, another important aspect is to understand the users' perceptions of the benefits of distinct types of vehicles, which is left for future research.

Acknowledgments

The authors gratefully acknowledge Junbo Liang and Ryan Williams from the University of Toronto for their help in data collection and for their comments which helped to improve the study. This research was funded partially by the Natural Sciences and Engineering Research Council of Canada (NSERC).

Author Contributions

The authors confirm contribution to the paper as follows: study conception and design: WXH, ED, AS; data collection: WXH, ED; analysis and interpretation of results: WXH, ED, AA, AS; draft manuscript preparation: WXH, ED, AA, AS. All authors reviewed the results and approved the final version of the manuscript.

References

1. Levine, J. C., and G. Torng. Dwell-Time Effects of Low-Floor Bus Design. *Journal of Transportation Engineering*, Vol. 120, No. 6, 1994, pp. 914–929. http://ascelibrary.org/doi/abs/10.1061/(ASCE)0733-947X(1994)120:6(914).
2. Currie, G., and A. Shalaby. Active Transit Signal Priority for Streetcars. Experience in Melbourne, Australia, and Toronto, Canada. *Transportation Research Record: Journal of the Transportation Research Board*, 2008. 2042: 41–49.
3. Currie, G., A. Delbosc, S. Harrison, and M. Sarvi. Impact of Crowding on Streetcar Dwell Time. *Transportation Research Record: Journal of the Transportation Research Board*, 2013. 2353: 100–106.
4. Currie, G., and P. Smith. Light Rail and Major Activity Center Circulation Systems: Innovative Design for Safe and Accessible Light Rail or Tram Stops Suitable for Streetcar-Style Conditions. *Transportation Research Record: Journal of the Transportation Research Board*, 2006. 1955: 36–46.
5. Fletcher, G., and A. El-Geneidy. Effects of Fare Payment Types and Crowding on Dwell Time: Fine-Grained Analysis. *Transportation Research Record: Journal of the Transportation Research Board*, 2013. 2351: 124–132.
6. Milkovits, M. Modeling the Factors Affecting Bus Stop Dwell Time: Use of Automatic Passenger Counting, Automatic Fare Counting, and Automatic Vehicle Location Data. *Transportation Research Record: Journal of the Transportation Research Board*, 2008. 2072: 125–130.
7. Lin, T. M., and N. H. M. Wilson. Dwell Time Relationships for Light Rail Systems. *Transportation Research Record: Journal of the Transportation Research Board*, 1992. 1361: 287–295.
8. Currie, G., A. Delbosc, and J. Reynolds. Modeling Dwell Time for Streetcars in Melbourne, Australia, and Toronto, Canada. *Transportation Research Record: Journal of the Transportation Research Board*, 2012. 2275: 22–29.
9. Grisé, E., and A. El-Geneidy. Identifying the Bias: Evaluating the Effectiveness of Automatic Data Collection. *Transportation Research Record: Journal of the Transportation Research Board*, 2017. 2647: 33–40.
10. Kittelson, P., K. Quade, and K. Hunter-Zaworski. *TCRP Report 165: Transit Capacity and Quality of Service Manual*. Transportation Research Board of the National Academies, Washington, D.C., 2003.
11. Diab, E. I., and A. M. El-Geneidy. Understanding the Impacts of a Combination of Service Improvement Strategies on Bus Running Time and Passenger's Perception. *Transportation Research Part A: Policy and Practice*, Vol. 46, No. 3, 2012, pp. 614–625. https://doi.org/10.1016/j.tra.2011.11.013.
12. Hu, W. X., and A. Shalaby. Use of Automated Vehicle Location Data for Route- and Segment-Level Analyses of Bus Route Reliability and Speed. *Transportation Research Record: Journal of the Transportation Research Board*, 2017. 2649: 9–19.
13. Abkowitz, M. D., and I. Engelstein. Factors Affecting Running Time on Transit Routes. *Transportation Research Part A: General*, Vol. 17, No. 2, 1983, pp. 107–113. https://doi.org/10.1016/0191-2607(83)90064-X.
14. El-Geneidy, A. M., and N. Vijayakumar. The Effects of Articulated Buses on Dwell and Running Times. *Journal of Public Transportation*, Vol. 14, No. 3, 2011, pp. 63–86.
15. Currie, G., and A. Shalaby. Success and Challenges in Modernizing Streetcar Systems: Experiences in Melbourne, Australia, and Toronto, Canada. *Transportation Research Record: Journal of the Transportation Research Board*, 2007. 2006: 31–39.
16. Shalaby, A. S., K. Ling, and J. Sinikas. Multiple-Unit Streetcar Operation: Evaluation using Microsimulation Modeling. Presented at 85th Annual Meeting of the Transportation Research Board, Washington, D.C., 2006.
17. Diab, E. I., and A. M. El-Geneidy. The Farside Story: Measuring the Benefits of Bus Stop Location on Transit Performance. *Transportation Research Record: Journal of the Transportation Research Board*, 2015. 2538: 1–10.
18. Diab, E. I., and A. M. El-Geneidy. Understanding the Impacts of a Combination of Service Improvement Strategies on Bus Running Time and Passenger's Perception. *Transportation Research Part A: General*, Vol. 46, No. 3, 2012, pp. 614–625. https://doi.org/10.1016/j.tra.2011.11.013.
19. Strathman, J., K. Dueker, T. Kimpel, R. Gerhart, K. Turner, P. Taylor, S. Callas, D. Griffin, and J. Hopper. Automated Bus Dispatching, Operations Control, and Service Reliability: Baseline Analysis. *Transportation Research Record: Journal of the Transportation Research Board*, 1999. 1666: 28–36.
20. Montgomery, D. C., and G. C. Runger. *Applied Statistics and Probability for Engineers*. John Wiley & Sons, Inc., Hoboken, NJ, 2010.

The Standing Committee on Light Rail Transit (AP075) peer-reviewed this paper (18-06740).

TRR
JOURNAL OF THE TRANSPORTATION RESEARCH BOARD

Article

Transportation Research Record
2018, Vol. 2672(8) 363–373

DOI: 10.1177/0361198118794534
journals.sagepub.com/home/trr
SAGE

Improving High-Frequency Transit Performance through Headway-Based Dispatching: Development and Implementation of a Real-Time Decision-Support System on a Multi-Branch Light Rail Line

Joshua J. Fabian[1,2], Gabriel E. Sánchez-Martínez[1], and John P. Attanucci[1]

Abstract

Service reliability is a major concern for public transportation agencies. Transit services experience natural variability in operations performance, due to factors such as congestion, changes in demand, and operator behavior. This variability leads to irregular headways, resulting in longer passenger waits and decreased effective capacity as gaps in service form. Real-time control strategies allow controllers to regulate service and improve performance. This research tested the effectiveness of a headway-based dispatching strategy at a terminal on the Massachusetts Bay Transportation Authority (MBTA) Green Line in Boston, a complex, four-branch light rail line. Terminal personnel were provided with tablet computers showing departure times optimized by an even-headway policy. When optimized departure times were adhered to, peak period headway variability was reduced by 40%. The average wait was shortened by 15% (30 sec), and the 90th percentile wait was shortened by 21% (90 sec). The results show that adopting headway-based dispatching at terminals of high-frequency lines promises significant benefits to service and passengers if operational changes are accompanied by improved supervision.

Service quality is critical to a transit system's ability to serve its passengers traveling to and from their homes, places of work, schools, and other destinations and origins. In the view of agencies and passengers, reliability—the degree to which actual service reflects scheduled service—is a major part of overall service quality (*1*). Reliability can be affected by factors such as weather, operator behavior variability, passenger demand spikes, converging branches, and right of way interference. For high-frequency services, passengers typically do not consult schedules for timing their arrival at origins, but rather gauge experiences on the set headway (*2*). Real-time control actions, including holding, expressing, and short-turning, can be implemented in attempts of correcting headway variability as it occurs. Considering holding-type actions, previous simulation-based research shows that dispatching on headways at terminals yields the shortest and most reliable passenger wait times of any single control strategy (*3*).

This research piloted headway-based dispatching on the Massachusetts Bay Transportation Authority (MBTA) Green Line in Boston, a four-branch light rail service. Trains are currently dispatched by schedule from the outer terminals. An algorithm combining two commonly-studied headway-based holding strategies was formulated, utilizing real-time vehicle location data. During the weeklong experiment, supervisory staff, or inspectors, at one terminal, Riverside, were provided with tablet computers running a decision-support system that displays recommended train departure times from the algorithm, along with other vehicle and operator information.

The goal of this research is to implement headway-based dispatching in the field and determine its effectiveness on operations and passenger experience. This study is novel in that it extends a headway-based dispatching

[1]Massachusetts Institute of Technology, Cambridge, MA
[2]Massachusetts Bay Transportation Authority, Boston, MA

Corresponding Author:
Address correspondence to Joshua J. Fabian: jfabi@alum.mit.edu

strategy to a complex, multi-branch light rail line and has inspectors implement it using a decision-support tool. Various service-based and passenger-based measures of reliability are compared prior to and during the experiment to evaluate the benefits of the headway-based dispatching strategy.

Following this introduction is a literature review and brief description of existing operations. The decision-support system is then described, along with the field experiment. The pilot results are presented and discussed, and recommendations are made.

Literature Review

Past research has explored headway-based holding and dispatching strategies as well as decision-support systems which implement these control strategies. The MBTA Green Line has also been the case study for previous research.

Barnett (*4*) and Turnquist (*2*) investigated holding methods to minimize waiting and traveling times, with the latter proposing the *Prefol*, or even-headway, holding strategy. Cats et al. (*5*) simulated different types of holding strategies on a trunk bus rapid transit (BRT) line and found the even-headway strategy to minimize passenger cost. Bartholdi and Eisenstein (*6*) trialed a strategy using limited terminal holding as the only control method on a bus line. Rolling horizon optimization models have been used by Delgado et al. (*7*) and Sánchez-Martínez et al. (*8*) to simulate holding and dispatching times considering capacity constraints. Fabian and Sánchez-Martínez, using a simulation of the Green Line, found that headway-based dispatching at terminals resulted in shorter and more reliable passenger wait times than en route holding alone and holding by schedule (*3*).

Public transit applications of decision-support systems have been previously proposed and tested. Fellows (*9*) conceptualized a user interface (UI) for a system to control operations on the MBTA Green Line. The UI takes in real-time automated vehicle location (AVL) data and shows estimated costs and benefits for possible control actions. Pangilinan et al. (*10*) provided supervisors on a bus line with a handheld tablet displaying a map with real-time vehicle locations; its use reduced headway variability by 21% during the study period. In-vehicle displays on Stockholm buses, displaying headway adherence information and instructions based on an even-headway policy, reduced passenger travel times by 10% (*11*). In Santiago, Lizana et al. (*12*) applied the optimization model by Delgado et al. (*7*) to a BRT route. The mean cycle time was reduced by 8% and the standard deviation of cycle time was halved. Maltzan (*13*) provided bus inspectors with a tablet that recommended departure times based on a target-headway strategy; the experiment reduced waiting times on a high-frequency route during the rush hour. Both Lizana et al. (*12*) and Maltzan (*13*) experienced issues relating to operator and inspector compliance with recommendations, suggesting that the benefits of real-time control strategies cannot be realized until compliance can be reasonably assured.

This research adds to existing literature by conducting a field test of a headway-based dispatching strategy enforced by terminal inspectors carrying portable decision-support tools, which displays real-time information, and it is carried out at one branch terminal of a multi-branch light rail line.

Existing Operations

Dispatchers based in the Operations Control Center (OCC) and inspectors based in the field make real-time service decisions. Control actions on the Green Line are currently based on either scheduled departure times or the scheduled headway, depending on the location of the action and personnel ordering it. Peak-hour trains are scheduled every 5–7 min per branch, leading to a combined headway of 90 sec in portions of the trunk.

Terminal inspectors are primarily responsible for dispatching trains and ensuring that operators report for scheduled runs. They are located at the dispatched end of each of the four branches: Boston College, Cleveland Circle, Riverside, and Lechmere, all locations with vehicle storage space (*14*). Figure 1 shows a diagram of the Green Line (the E-branch is controlled from Lechmere instead of Heath). When reporting for duty, operators sign in with the inspector at their assigned terminal. If an operator is absent, the inspector must either find a substitute, typically drawing from a list of extraboard operators available at the terminal (who are scheduled to stand by to substitute regular operators who are absent), or operate the trip with one fewer car (each Green Line car must be staffed to control the doors), or cancel the trip.

Trains are dispatched from terminals primarily using the schedule. Inspectors refer to paper sheets that list the day's scheduled departures and arrivals, along with the scheduled operators corresponding to each trip, and fill out actual times on the sheets as they occur, noting operator changes as needed. Inspectors aim for on-schedule departures, and can take actions to improve punctuality, such as pulling out trains from the yard, reordering operators, or drawing from the extraboard list. In certain situations, departure times are adjusted to even out headways, especially if another trip is canceled. Such decisions are made considering fleet and crew availability. Terminal inspectors also decide when trains should be pulled into the yard (increasing headways, as scheduled following a peak period).

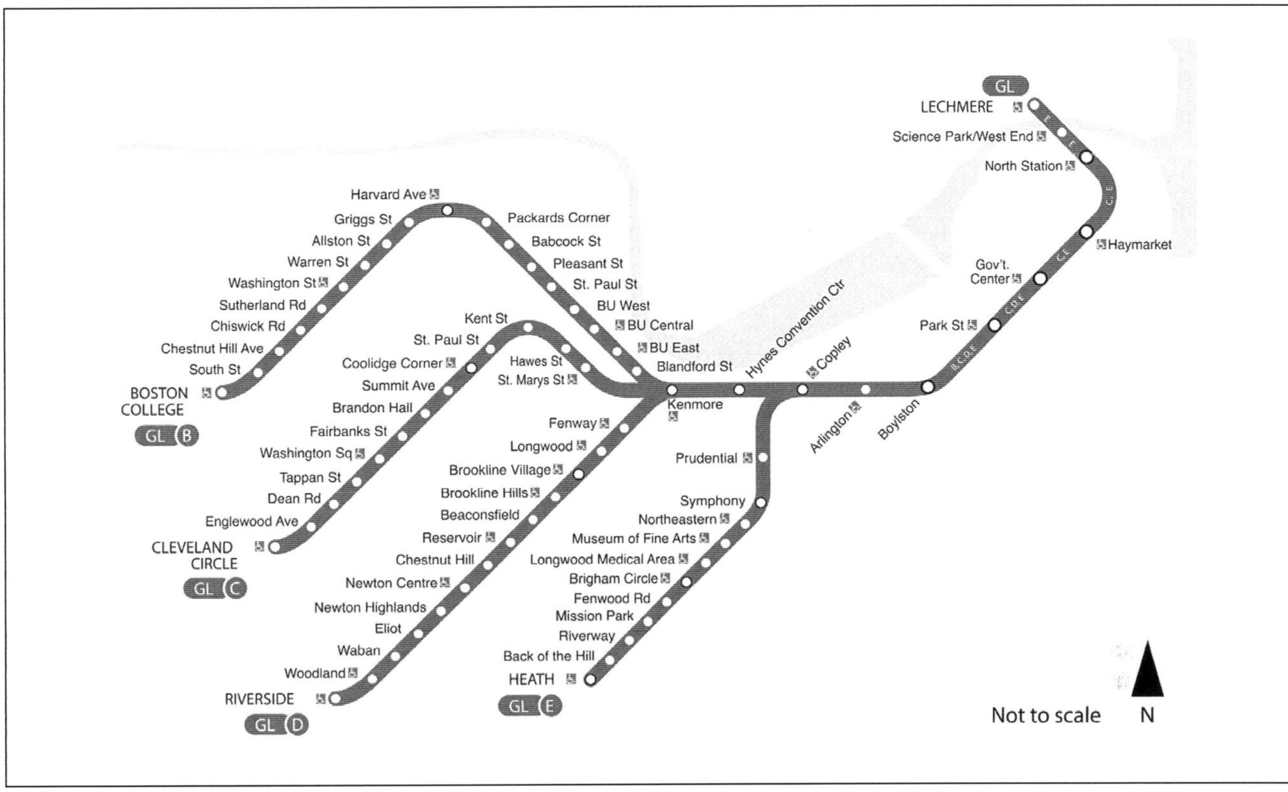

Figure 1. Schematic of MBTA Green Line.

At terminals, updated knowledge on train arrivals is important for coordinating trip allocations and yard movements. Until recent years, inspectors have estimated arrival times by communicating with train operators via radio. With GPS tracking now available for most trains, a public-facing application programming interface (API) provided by the MBTA outputs train locations and arrival predictions in real time. But even with modern train-tracking technology, service decisions are often made in an ad hoc manner, with inconsistencies among inspectors and dispatchers. Many dispatching decisions are not made in coordination with, or with notification of, dispatchers or other inspectors downstream, who may apply control based on different criteria.

Experiment Control Strategy

A policy combining the target-headway strategy (*2*, *15*) and the even-headway strategy (*2*) was formulated for use in this research:

$$d_i = \max\left(\frac{d_{i-1} + a_{i+1} + l}{2}, a_i + l, d_{i-1} + h_t\right) \quad (1)$$

where

d_i = departure time of trip i;
a_i = arrival time of trip i;
l = minimum required layover time; and
h_t = target headway at time of day t.

This policy builds upon the even-headway strategy, adding two constraints: a train cannot depart until the minimum layover, l, and target headway, h_t, have both elapsed. Passenger-related dwell at terminals is not incorporated because it is assumed that passengers can alight and board during the minimum layover time. The target headway is usually equal to the scheduled headway, h_t^*. Following large service gaps, defined by the MBTA as 1.5 times the scheduled headway (*16*), the target headway (h_t) is reduced to 0.65 h_t^* for the following trip. Following gaps longer than twice the scheduled headway, the adjustment is made for two subsequent departures. These adjustments are meant to alleviate reduced capacity if the first train following a gap leaves behind passengers on the platform. The adjustments were rarely applied during the experiment, however, because long gaps were often filled by adding extra trains and using extraboard operators.

Decision-Support System

A decision-support system was developed to implement the aforementioned strategy in a front-end application for a tablet computer. It combines scheduled train information from the General Transit Feed Specification (GTFS), scheduled crew information from internal sources, and real-time train locations and arrival predictions from the MBTA's real-time API. A decision engine assigns trains traveling toward the terminal to future scheduled trips and recommends terminal departure times based on initial arrival time and dispatching strategy.

Terminal inspectors make modifications on the tablets to, for instance, cancel a trip or substitute an operator. They also indicate if a trip will be served by a new train from the yard or if an arriving train will be pulled into the yard. Departures are automatically optimized after modifications. Figure 2*a* shows a screenshot of the front-end application in its primary view, with train arrivals on the left half and the departures of the same trains on the right. (Operator identifiers are redacted to safeguard privacy.) Each row represents one car of a two-car train. Figure 2*b* shows an example of a dialog box from which various attributes of a trip can be edited.

The front-end application is supported by two back-end applications and a database. The first program acts as a real-time data interpreter, processing train location updates and user inputs, and as a decision engine, applying the control policy and recommending terminal departure times. Train data is cleaned in real time, before interpretation, to maintain the integrity of trip and operator information. A database stores records of each trip and operator, scheduled and operated, and is accessed by the back-end applications. These records provide a log of all train movements, control actions, and inspector decisions. The second application is an API which acts as an interface between the database and the front-end client.

Experiment Description

An experiment controlling departing headways from Riverside was conducted over the entirety of five weekdays from opening until close of service: Wednesday, March 1 through Tuesday, March 7, 2017. Preceding the experiment were five full weekdays of training personnel on site.

Terminal inspectors scheduled to work at Riverside were the primary users of the decision-support tool. During the experiment, they had the added responsibility of using the tablet computer and were instructed to dispatch trips at the times given by the headway-based algorithm rather than the schedule. A ring-off bell was installed at Riverside, operable from the tablet. The bell, similar to ones used at the MBTA's heavy rail terminals, rings for 15 sec through the platform's public address system, signaling operators to depart. Inspectors were told to activate the bell when indicated in the UI (30 sec before recommended departures). In general, recommended departure times were communicated to operators verbally, mirroring existing practice for schedule adjustments, even when the bell was in use. Nine different officials had shifts at Riverside during the experiment.

Assistants from the MBTA Office of Performance Management and Innovation (OPMI) and Massachusetts Institute of Technology (MIT) provided on-site support continuously during the experiment. Assistants shadowed inspectors and ensured necessary adjustments were made through the UI. When an inspector encountered difficulties using the application, the assistant on duty provided technical support. Throughout the experiment, assistants took notes on recommendation adherence, inspector and operator actions, feedback from all parties involved, and service-related events.

Results

When evaluating the benefits of the dispatching experiment, several service-oriented performance metrics, including headway, cycle time, and recommendation compliance, and passenger-oriented metrics, namely passenger wait time and wait reliability, are considered. The following analysis focuses on the morning peak period, from 06:00 to 09:00, in the peak direction (eastbound). The morning peak is analyzed because the greatest number of passengers travel from Riverside at this time, commuting toward downtown Boston.

The effectiveness of the decision-support system in improving performance was measured relative to observed performance on the Riverside branch during a baseline period consisting of the weekdays of three weeks preceding the experiment, Monday, January 30 through Friday, February 17, 2017. Four weekdays within this period were excluded from the baseline due to significant winter storms or large-scale events disrupting typical service and ridership patterns. Several shorter periods with severe disruptions were also excluded.

Compliance with Recommendations

A total of 782 departures were observed at Riverside during the experiment. Table 1 shows that 36.1% of all departures left within 30 sec of the recommendation and that just fewer than half of all trips left within 45 sec. Comparing adherence to the departure recommendations against schedule punctuality, trains were more likely to leave at the scheduled time.

Based on the results in Table 1, a more detailed analysis of when trips departed early or late is conducted. Figure 3 summarizes the share of early, on-time, and late departures by the difference between scheduled and

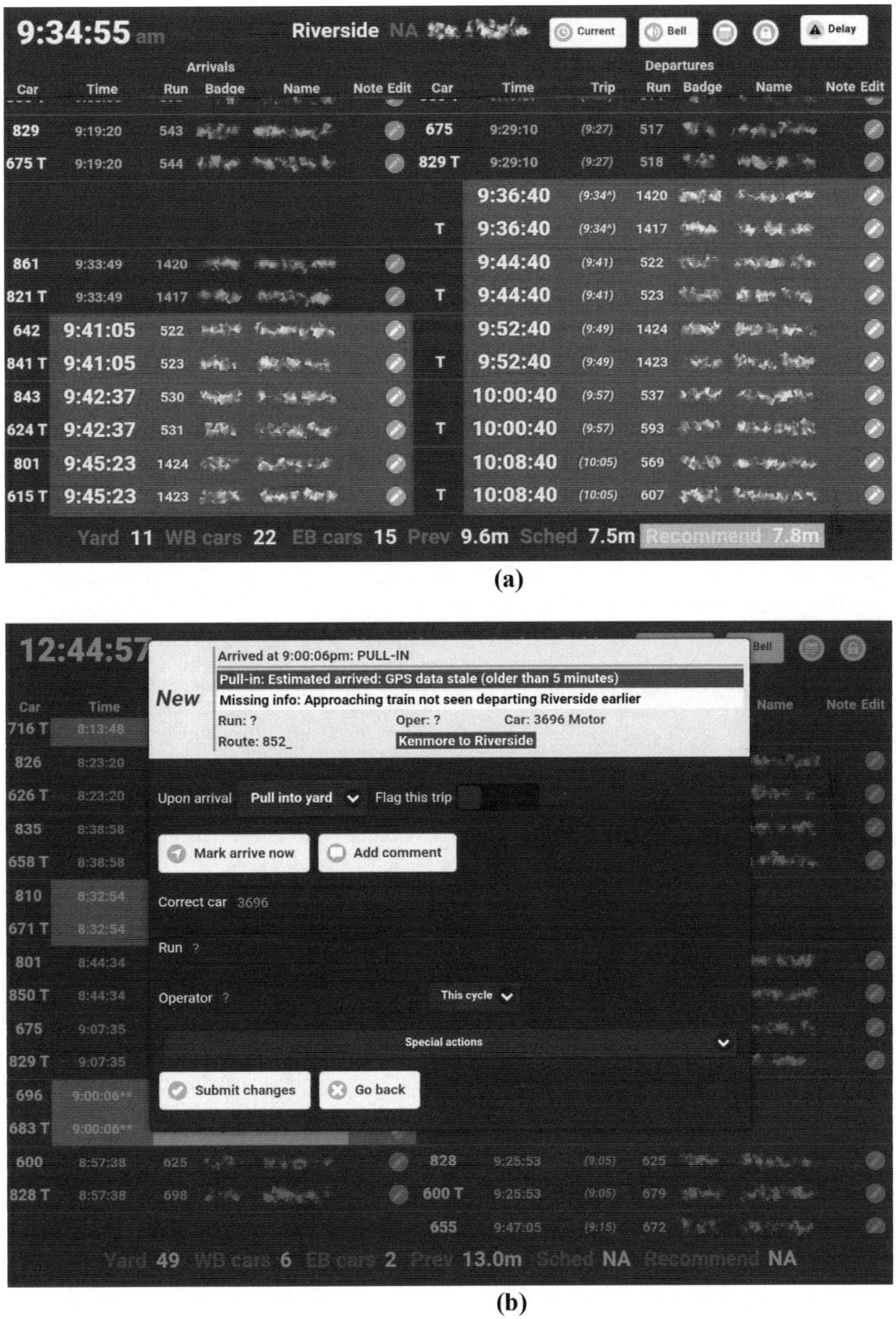

Figure 2. Screenshots from front-end application.

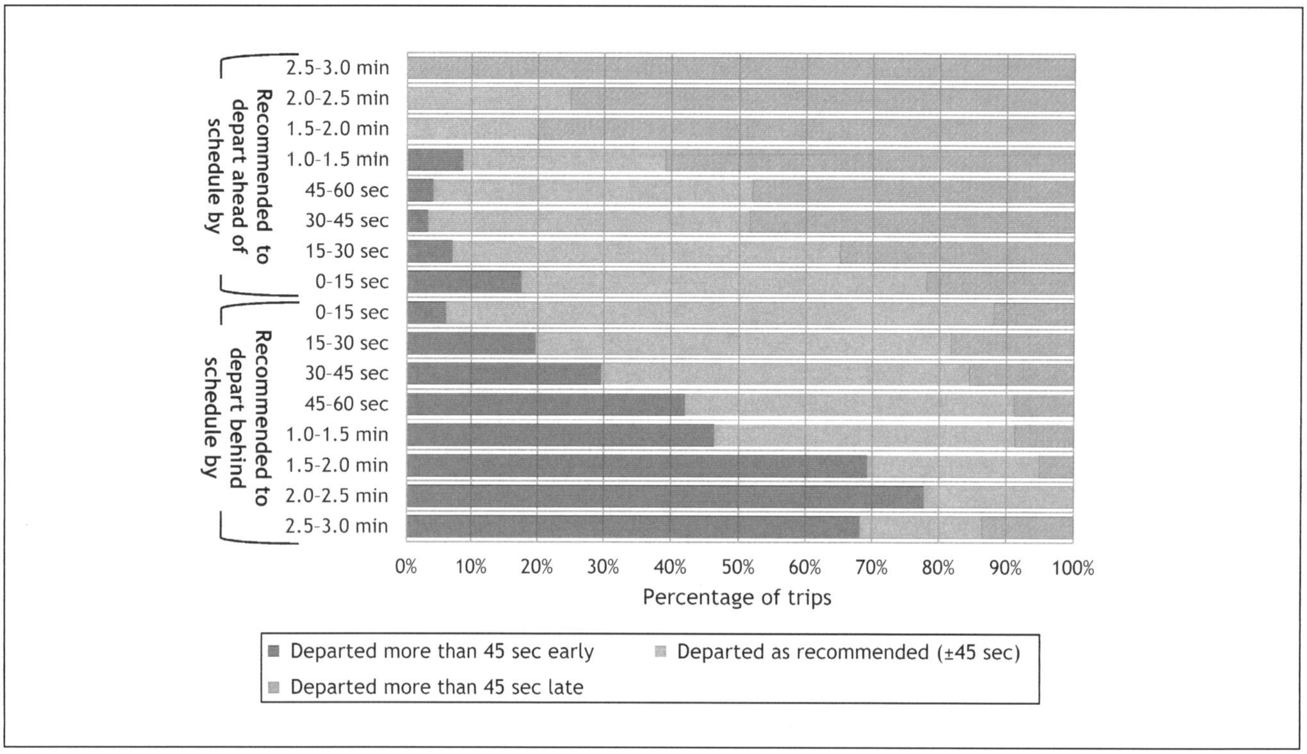

Figure 3. Shares of trips departing early, on time, and late (to recommended).

Table 1. Adherence to Recommendation and Schedule by Compliance Threshold

Threshold	Trips compliant with recommendation	Trips compliant with schedule
Within 30 sec	36.1%	48.4%
Within 45 sec	49.3%	62.7%
Within 60 sec	60.6%	72.4%
Within 90 sec	77.6%	83.4%

recommended departure times. When recommendations are made for 90 sec or more after the scheduled time, for example, trains leave early more often than on time. For recommendations 30 sec or more ahead of schedule, trains leave late at least as often as on time. These results suggest that compliance was a major issue during the experiment, with the problem being slightly worse for recommendations made ahead of the schedule. Even when recommended times were within 30 sec of the schedule, however, only 65.5% of trips left on time. This relatively low compliance rate suggests the opportunity for improved operations control procedures.

Terminal departures are classified as either compliant or non-compliant to the recommendation, based on a threshold of $\pm$45 sec. This is done to separate the effects of using the holding strategy from those of using existing dispatching procedures, as the recommendations were not followed consistently. Since a train's running time and headway are often impacted by the vehicles ahead (leading to uneven boardings) and behind (leading to en route control actions), a trip is only considered compliant if the adjacent trains had also departed within the 45 sec threshold. During the experiment, 15.6% of Riverside-based trips met the criteria. When broken by time of day, the share of fully compliant trips varied between 12.8% (midday) and 20.3% (morning peak). For passenger-based metrics, the definition of compliance is relaxed in that the following train does not need to have met the threshold, because it is assumed that passengers board the first available train. For example, if two trains pass a station in close succession, a waiting passenger is assumed to board the first, unless it is full; the second train has no effect on the passenger's wait time.

Headway Variability

Figure 4 shows the coefficient of variation of headway, a measure of vehicle spacing, along the Riverside branch, for compliant and non-compliant trips during the experimental and baseline periods. The coefficient of variation of headway is commonly used to measure headway irregularity because it is mathematically related to expected passenger waiting time. For each station, the coefficient

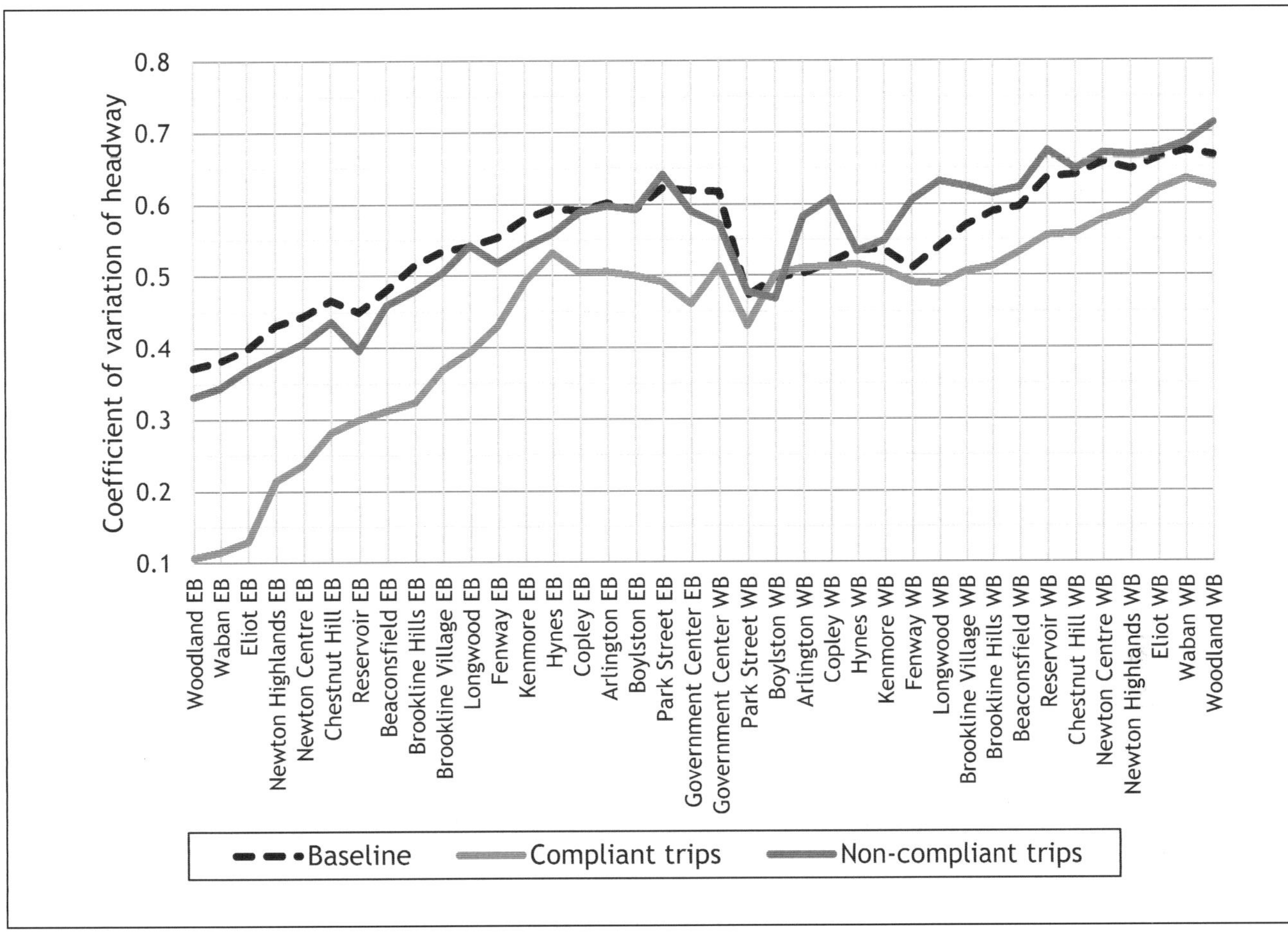

Figure 4. Coefficients of variation of headways.

of variation of headway is the standard deviation of the headway at that location divided by the mean headway. The left half shows headway variability in the eastbound (EB) direction and the right half shows the westbound (WB) direction.

Headway variability generally increases as trains progress from the outer terminal, the experiment site (the first station after departing Riverside is Woodland EB), to the inner terminal (Government Center) and back. The headway variability of compliant eastbound trips begins much lower than the baseline, and the difference decreases as trains approach and merge with other branches at Kenmore EB, though the improvements are still statistically significant. Notable decreases in the coefficient of variation between stops, such as at Reservoir EB or Park Street WB, happen where other control is applied. The magnitude of the decreases in the coefficient around the eastern terminal, at Government Center and Park Street, suggests that any existing control there (outside the scope of this experiment) is limited in effectiveness. Westbound service is more variable than eastbound service largely due to layover and track constraints within the trunk. Across both directions, the headway variability of compliant trips decreased by 41% compared to the baseline when the coefficients of variation are weighed by boardings at each station. This indication of overall improvement reflects the different waits experienced by passengers.

Cycle Time

Table 2 lists summary statistics for observed cycle times (excluding the layover time at Riverside). The mean cycle time decreased by 3.4 min for compliant trips, a statistically significant improvement from the baseline. In practice, however, vehicle schedules are typically set between the 85th and 95th percentiles of observed cycle times over a survey period, (*17*). Consistently dispatching trains based on headway resulted in a 5.9% (or 6.4 min) decrease in the 90th percentile cycle time. By adjusting the scheduled cycle time likewise, the current number of trains could be scheduled to serve additional trips, or

Table 2. Vehicle Cycle Times

Service scenario	Median (min)	Mean (min)	Standard deviation (min)	90th percentile (min)	Scheduled (min)
Baseline	98.1	97.8	9.7	107.4	100.0
Non-compliant trips	96.9	97.9	7.1	107.0	
Compliant trips	94.2	94.4	5.4	101.0	

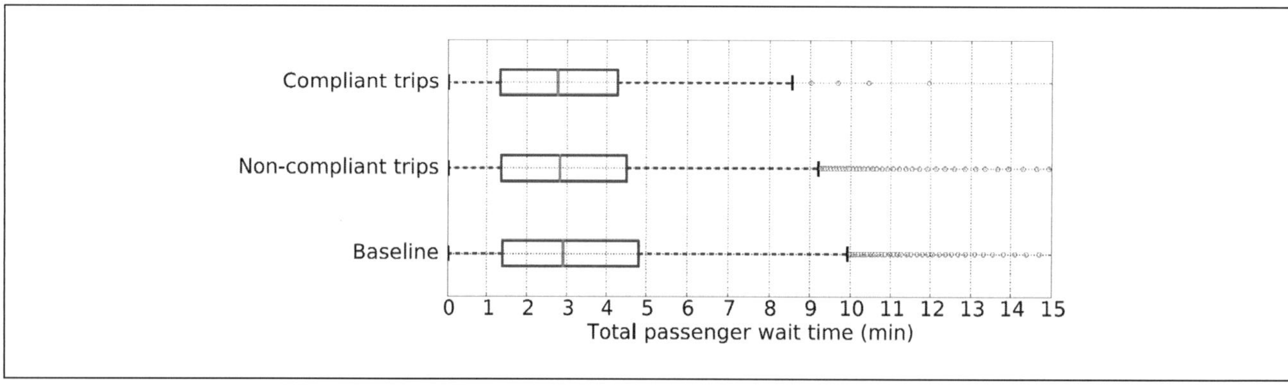

Figure 5. Range of eastbound passenger wait times.

fewer trains could serve the current number of scheduled trips. A 6.4 min cut from the scheduled cycle time could require one fewer train, although the scheduled cycle time is shorter than what was observed during the experiment at the 90th percentile, meaning that trains were still returning to Riverside late.

Passenger Wait Time

A passenger's wait time is the time between the arrival at the origin station and the vehicle boarding time. Wait times are estimated by simulating passenger activity using typical demand derived from a model that infers passenger activity from automated fare collection and AVL data (*18*, *19*). The simulated passengers are matched to actual service operated each day using AVL data. Passengers in the analysis include all those who traveled between the surface-level branch portion and the trunk. Assuming that passengers arrive at their origins randomly and are able to board the next train to arrive, the mean wait time for the ideal transit route equals one-half the scheduled headway. In reality, service does not operate perfectly on schedule, and expected wait times increase with the variability of headway (*20*).

Peak direction passengers on compliant trips experienced a 15% decrease in mean wait time as well as a 32% reduction in the standard deviation of wait time. The average passenger waits 30 sec less for a train. More importantly, the change in standard deviation means that passengers experience a smaller range of waiting. Even in the westbound, reverse commute direction, the wait time variability decreased by 10% for those riding trains which originated on compliant trips from Riverside.

The box plots in Figure 5 show the wait time ranges of eastbound passengers. The ends of each box represent the first and third quartiles, and the range between the end of the box and the closest whisker is 1.5 times the interquartile range. The third quartile (75th percentile) is reduced by over 30 sec for those on compliant trips. The 90th percentile wait time decreases by 90 sec (21%). This percentile is important to note because it may represent an unusually long trip for the average commuter, experienced perhaps twice a month, but one that they must anticipate and plan around daily.

Passenger Wait Reliability

Wait time reliability, as defined by Tribone et al. (*21*), is a measure of customer experience. It provides the proportion of passengers with wait times shorter than the scheduled headway. The MBTA, as a performance measure on its rail services, calls for at least 90% of passengers to experience reliable waits. Table 3 shows the wait time reliabilities for passengers riding between the trunk and branch portions of the Riverside branch.

Wait reliability in the eastbound direction increases from 81.2% to 89.2% for passengers boarding compliant

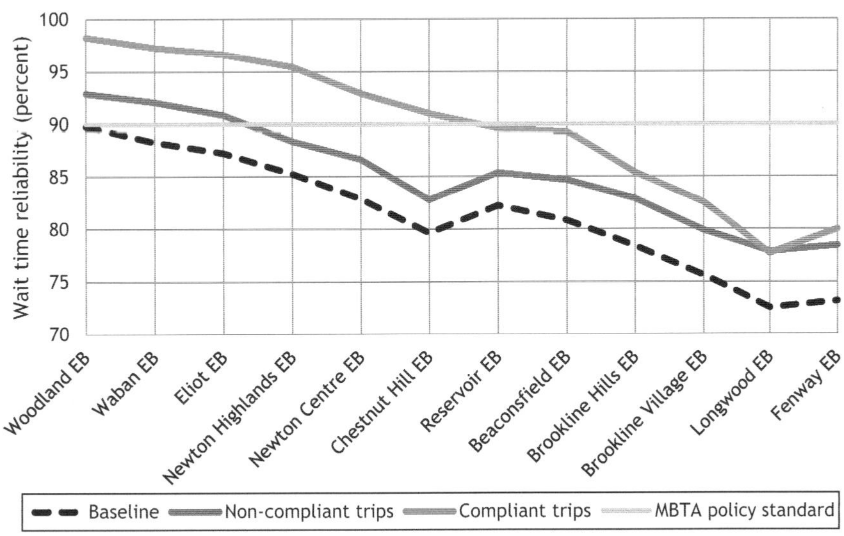

Figure 6. Stop-level wait time reliability along the surface portion of the Riverside branch.

Table 3. Wait Time Reliability

Direction	Service scenario	Morning peak (%)
Eastbound	Baseline	81.2
	Non-compliant trips	84.9
	Compliant trips	89.2
Westbound	Baseline	77.6
	Non-compliant trips	78.6
	Compliant trips	80.4

trains, meaning that an additional 8.0% of all passengers would experience acceptable wait times with terminal headway-based control. In the westbound, or non-peak direction, wait reliability increases by a more modest 2.8%. Those passengers who boarded non-compliant trips during the experiment also experienced some increase in wait reliability; this can be attributed to the minimum headway requirement reducing vehicle bunching throughout the service day.

Figure 6 compares wait time reliability by station. Under headway-based control from Riverside, the 90% standard is met at six out of twelve surface-level stations (at two more stations, 89% of passengers have reliable waits). Under existing operations, only the first station departing Riverside, Woodland, achieves a reliability of 89%. Some benefits are observed for non-compliant trips during the experiment period, partially because of less en route holding following decreased bunching. Another possible reason for the improvement in non-compliant trips is the increased supervision operators and inspectors experienced during the experiment—some non-compliant trips operated closer to headway regularity than they may have during the baseline period.

Discussion

The results show that both average passenger wait time and wait time variability decrease in both directions when trains are dispatched by headway as recommended by the decision-support system. In the morning peak direction, the average wait is reduced from the baseline by 30 sec, or 15%. At the 90th percentile, wait times decrease by 90 sec. Similarly, wait time reliability increases. These improvements mean that commuters and other passengers may, over time, feel comfortable budgeting less buffer time to their trips, increasing productivity. Positive effects in the return direction suggest that operational changes at a small number of targeted locations—outer terminals, in this experiment—can result in significant improvements. This being said, a low compliance rate was experienced throughout the experiment period. Non-compliance with the recommended terminal departure time is correlated with the deviation of the recommendation from the time in the normal schedule, but further work is needed to determine the different reasons, such as human factors, for this result.

Decreases in observed cycle times mean that trains make their journeys faster, as do passengers when combined with wait time decreases. Faster trips may permit scheduled cycle times to be reduced or, maintaining the current schedule, decrease the number of trips returning late and thus the amount of overtime incurred by operators.

Despite the observed improvements, certain issues remained after implementing headway-based dispatching at Riverside. Wait times still became more unreliable, and headways more variable, further along the branch due to factors such as the variability of operator speeds and of dwell times—from unequal demand within a short time, an unequal distribution of passengers boarding the two cars of a train, or even passengers interacting with the onboard farebox with cash. As shown in Figure 6, the 90% wait reliability target was still not met at the branch stations furthest from Riverside in the controlled direction. This means that those passengers traveling from near the control point to the trunk were more likely to experience benefits during the experiment than those originating further downstream. In order to meet the reliability standard at every station, supplementing headway-based terminal dispatching with coordinated headway control at some stations en route could be considered.

Future Work and Recommendations

The results of this research support the use of headway-based dispatching at terminals. Improvements in real-world train regularity and passenger experience are realized even though the light rail service used for the experiment shares a trunk section with three other branches. The implementation of a headway-based dispatching policy benefits from the use of a decision-support system to calculate and communicate optimal departure times as train positions are updated in real time. Transit agencies wishing to improve reliability on high-frequency services should first evaluate terminal policies and procedures and determine if operations are run efficiently.

Future research may focus on testing more sophisticated dispatching policies that respond to historical demand, platform crowding (from fare gate or video-based counts), and predicted or actual vehicle loads, in addition to headways, to determine if additional service and passenger benefits can be obtained. An algorithm incorporating load or demand may, for example, recommend that a train be dispatched with a short headway if it is predicted that the previous train left behind passengers on the platform (*8*). Other control action types, such as en route holding, short-turning, or expressing could also be recommended with a decision-support system. In the case of the Green Line, adding headway-based control at more than one terminal may be studied, since service on one branch is affected by activity on other branches within the trunk.

Successful adoption of decision-support tools for managing service controls is dependent on two conditions: high-quality real-time data and increased rates of departure compliance at terminals. A low-latency, highly-granular automated data collection system is required in order to know the locations of vehicles—and their upcoming arrival times—with high confidence. This requires a high density of collection points, if wayside equipment is used, or high vehicle polling frequencies, if a wireless system, potentially involving GPS, is used. The MBTA Green Line uses a combination of these two systems. Such a system could also transmit real-time passenger load information, from weight sensors or automated passenger counting systems, which may be incorporated into dispatching algorithms as described above. Data collected in real time must also be adequately cleaned before being used by a decision-support system.

Although trips complying with the optimized departure times from the decision-support system improved service quality, only half of the trips were dispatched close to the recommended times. Improving location predictions could help predict service deterioration and thus improve the effectiveness of a control policy, especially in very high-frequency services (*22*, *23*). The human factors contributing to non-compliance, including those related to technology and culture, should also be studied in depth. Implementation of dispatching decision-support tools must be paired with initiatives to improve cooperation among operators and inspectors, since non-compliance significantly reduces the benefits that are realized. Initiatives may include training sessions focused on reliability, sign-in policy changes, performance-based bonuses or perks, audible cues, in-station displays showing operators updated vehicle assignments and departure times, and in-vehicle consoles alerting operators of en route control actions.

Acknowledgments

This research was conducted through the MIT Transit Lab with financial support from the Massachusetts Bay Transportation Authority (MBTA). The authors thank the many individuals at the MBTA who provided their time and assistance for this experiment.

Author Contributions

The authors confirm contribution to the paper is as follows: study conception and design: JJF, GESM and JPA; data collection: JJF; analysis and interpretation of results: JJF, GES-M, and JPA; draft manscrupt preparation: JJF. All authors reviewed the results and approved the final version of the manuscript.

References

1. Iseki, H., and B. D. Taylor. Style versus Service? An Analysis of User Perceptions of Transit Stops and Stations. *Journal of Public Transportation*, Vol. 13, No. 3, 2010, pp. 39–63. https://doi.org/10.5038/2375-0901.13.3.2.

2. Turnquist, M. A. Strategies for Improving Reliability of Bus Transit Service. *Transportation Research Record: Journal of the Transportation Research Board*, 1981. 818: 7–13.
3. Fabian, J. F., and G. E. Sánchez-Martínez. Simulation-Based Comparison of Holding Strategies for a Multi-Branch Light-Rail Service. *Transportation Research Record: Journal of the Transportation Research Board*, 2017. 2648: 23–32.
4. Barnett, A. On Controlling Randomness in Transit Operations. *Transportation Science*, Vol. 8, No. 2, 1974, pp. 102–116. https://doi.org/10.1287/trsc.8.2.102.
5. Cats, O., A. N. Larijani, H. N. Koutsopoulos, and W. Burghout. Impacts of Holding Control Strategies on Transit Performance: A Bus Simulation Model Analysis. *Transportation Research Record: Journal of the Transportation Research Board*, 2011. 2216: 51–58.
6. Bartholdi, J. J. III, and D. D. Eisenstein. A Self-Coordinating Bus Route to Resist Bus Bunching. *Transportation Research Part B: Methodological*, Vol. 46, No. 4, 2012, pp. 481–491. https://doi.org/10.1016/j.trb.2011.11.001.
7. Delgado, F., J. C. Muñoz, and R. Giesen. How Much Can Holding and/or Limiting Boarding Improve Transit Performance? *Transportation Research Part B: Methodological*, Vol. 46, 2012, pp. 1202–1217. https://doi.org/10.1016/j.trb.2012.04.005.
8. Sánchez-Martínez, G. E., H. N. Koutsopoulos, and N. H. M. Wilson. Real-Time Holding Control for High-Frequency Transit with Dynamics. *Transportation Research Part B: Methodological*, Vol. 83, 2016, pp. 1–19. https://doi.org/10.1016/j.trb.2015.11.013.
9. Fellows, R. *Using and Enhancing Automatic Vehicle Identification to Improve Service Control on the MBTA Green Line*. Master's thesis. Massachusetts Institute of Technology, Cambridge, MA, 1990.
10. Pangilinan, C., N. H. M. Wilson, and A. Moore. Bus Supervision Deployment Strategies and Use of Real-Time Automatic Vehicle Location for Improved Bus Service Reliability. *Transportation Research Record: Journal of the Transportation Research Board*, 2008. 2063: 28–33.
11. Cats, O. *A Brief Summary of Real-Time Bus Regularity Control: From a Simulation Study to Full-Scale Implementation*. https://odedcats.weblog.tudelft.nl/files/2016/08/Real-time-control-weblog-summary.pdf. Accessed April 15, 2017.
12. Lizana, P., J. C. Muñoz, R. Giesen, and F. Delgado. Bus Control Strategy Application: Case Study of Santiago Transit System. *Procedia Computer Science*, Vol. 32, 2014, pp. 397–404. https://doi.org/10.1016/j.procs.2014.05.440.
13. Maltzan, D. *Using Real-time Data to Improve Reliability on High-Frequency Transit Services*. Master's thesis. Massachusetts Institute of Technology, Cambridge, MA, 2015.
14. Massachusetts Bay Transportation Authority. *Ridership and Service Statistics*, 14th ed. July 2014. http://www.mbta.com/uploadedfiles/documents/2014%20BLUEBOOK%2014th%20Edition.pdf. Accessed May 24, 2017.
15. Daganzo, C. F., and J. Pilachowski. Reducing Bunching with Bus-to-Bus Cooperation. *Transportation Research Part B: Methodological*, Vol. 45, No. 1, 2011, pp. 267–277. https://doi.org/10.1016/j.trb.2010.06.005.
16. Massachusetts Bay Transportation Authority and IBI Group. *MBTA-Realtime API for Performance Data Documentation (v0.9.1)*. http://realtime.mbta.com/Portal/Content/Documents/MBTA-realtime_Performance_APIDocumentation_v0_9_1_2016-04-14_public.pdf. Accessed April 15, 2017.
17. Muller, T. H. J., and P. G. Furth. Integrating Bus Service Planning with Analysis, Operational Control, and Performance Monitoring. In *Proc., ITS America 10th Annual Meeting and Exposition: Revolutionary Thinking, Real Results*, Washington, D.C., 2000.
18. Gordon, J. B., H. N. Koutsopoulos, and N. H. M. Wilson. Automated Inference of Linked Transit Journeys in London using Fare-Transaction and Vehicle Location Data. *Transportation Research Record: Journal of the Transportation Research Board*, 2013. 2343: 17–24.
19. Sánchez-Martínez, G. E. Inference of Public Transportation Trip Destinations by Using Fare Transaction and Vehicle Location Data: Dynamic Programming Approach. *Transportation Research Record: Journal of the Transportation Research Board*, 2017. 2652: 1–7.
20. Wilson, N. H. M., D. Nelson, A. Palmere, T. H. Grayson, and C. Cederquist. Service-Quality Monitoring for High-Frequency Transit Lines. *Transportation Research Record: Journal of the Transportation Research Board*, 1992. 2652: 3–11.
21. Tribone, D., L. Riegel, R. Warade, and D. Barker. Measuring Transit Agency Performance using Open Realtime Data. Presented at 95th Annual Meeting of the Transportation Research Board, Washington, D.C., 2016.
22. Alvarez, A., F. Merchan, F. Poyo, and R. George. A Fuzzy Logic-Based Approach for Estimation of Dwelling Times of Panama Metro Stations. *Entropy*, Vol. 17, No. 5, 2015, pp. 2688–2705. https://doi.org/10.3390/e17052688.
23. Hans, E., N. Chiabaut, L. Leclercq, and R. L. Bertini. Real-time Bus Route State Forecasting using Particle Filter: An Empirical Data Application. *Transportation Research Procedia*, Vol. 6, 2015, pp. 434–447. https://doi.org/10.1016/j.trpro.2015.03.033.

The Standing Committee on Light Rail Transit (AP075) peer-reviewed this paper (18-01706).

TRR
JOURNAL OF THE TRANSPORTATION RESEARCH BOARD

Transportation Research Record
2018, Vol. 2672(8) 374–383
© National Academy of Sciences: Transportation Research Board 2018
Article reuse guidelines: sagepub.com/journals-permissions
DOI: 10.1177/0361198118777080
journals.sagepub.com/home/trr
SAGE

Biking to Ride: Investigating the Challenges and Barriers of Integrating Cycling with Regional Rail Transit

Léa Ravensbergen[1], Ron Buliung[1], Meaghan Mendonca[2], and Naren Garg[2]

Abstract

Integrating bicycling with public transport can potentially benefit cyclists and transit operators. Successfully coordinating these transport modes, however, can be a difficult task when so little is known about the social and environmental barriers to this type of multimodal travel in the North American context. Using data derived from a survey of regional train service patrons in the Greater Toronto and Hamilton regions of Ontario, Canada, this study examines the challenges faced by those who cycle to/from the train, the barriers that keep passengers from commuting to/from the train by bicycle, and the sociodemographic characteristics of those cycling—and not cycling—to/from the train. Safety concerns, worries about bicycle security, and rules restricting when bicycles are permitted on trains were among the top challenges identified by individuals currently cycling to and/or from train stations. Among those who do not cycle to or from the train, appearance and comfort were the two primary concerns. Results also indicate that certain groups were more likely to cycle to/from the train than others. Notably, a large gender gap exists, approximately two-thirds (67%) of those cycling to their local train station were male. Results from this study may inform policy makers on how to successfully, and equitably, integrate cycling with regional rail transit.

In recent years, the bicycle has received increasing attention from planners, decision makers, and researchers (*1*). Owing to the bicycle's associated efficiency, environmental, and health benefits, it is promoted as a potential key transport mode in producing sustainable urban mobility (*2*). Transportation research tends to focus on single transport modes independent of others. For example, most research on city cycling focuses solely on trips that begin and end by bike (*2*). However, cyclists often exhibit multimodal travel behavior combining the bicycle with other modes such as buses or trains (*2*). In a study of Danish travel behavior, it was found that few cyclists used a bicycle as their sole travel mode; the majority combined bicycle trips with other modes (*3*). Furthermore, scholars have argued that bicycles and public transit can be allies with their joint use potentially competing with the private car (*4, 5*). Planning for sustainable transport could mean taking a closer look at multimodal passenger travel, including cycle-transit integration.

Integrating cycling with public transit can be a mutually beneficial combination. Bicycles can extend the catchment area of transit stations far beyond the pedestrian range (*6–8*) and at a lower cost than cars, which require expensive parking facilities (*6*). Access to public transit not only allows cyclists to make longer trips that are difficult to make by bike alone, but can also provide alternatives when cyclists encounter bad weather, mechanical problems, or gaps in the bicycle network (*7*). While cycling rates remain low in many U.S. and Canadian cities, cycling levels, public transit use, and the integration of bicycles with public transit have grown over the past decade (*7*).

This paper focuses on the challenges and barriers to bicycle–rail integration of the GO Transit system in the Greater Toronto and Hamilton areas in Ontario, Canada. GO Transit is a division of Metrolinx, a transportation planning agency of the Government of Ontario, and consists of an extensive network of commuter train lines and bus routes serving 65 million passengers a year (*9*). The system's rail network, the focus of this study, encourages cycling by providing bicycle parking facilities at each station and by allowing passengers to bring their bicycles on trains during off-peak hours. However, the number of individuals reaching the service by bicycle remains low, averaging approximately 1% across the network of stations (*10*). This study answers the following research questions through an analysis of a survey distributed to passengers of GO's train service: What challenges do integrated riders report? What barriers keep rail passengers

[1]University of Toronto at Mississauga, Mississauga, ON, Canada
[2]Metrolinx, Toronto, ON, Canada

Corresponding Author:
Address correspondence to Léa Ravensbergen: lea.ravensbergen@mail.utoronto.ca

from biking to rail? And what is the demographic profile of those who do—and those who do not—cycle to a regional train station? The paper begins with a brief review of articles examining bicycle–train integration. Sections detailing methods, results, and a concluding section with policy recommendations follow the review.

Literature Review

Previous research has found that few studies explore bicycle–public transport integration (*2*, *8*, *11*). One article identified five main approaches to aid in promoting bicycle–public transit integration in U.S. and Canadian cities including: provision of bicycle parking facilities at bus stops and rail stations; multi-function bike stations providing a range of services such as bicycle rental, repairs, lockers; bicycle racks on buses; allowing bicycles on board public transit vehicles; and coordinating the bicycle network with public transit (*7*). In a study of bike-and-ride trip characteristics and user profiles in the UK, Netherlands, and Germany, the authors found that more people cycled to faster, higher-order types of public transit, for example, trains or intercity buses, than slower types of public transport such as local buses or streetcars (*4*). Research out of the Netherlands suggests that two improvements could potentially increase ridership: the quality of bicycle routes and bicycle parking (*6*). Additionally, a countrywide program that upgraded bicycle parking at train stations in the Netherlands reported increased use and greater user-satisfaction (*12*). Finally, through the examination of the effect of passenger attitudes on the decision to commute by public transport and bicycle, it has been found that those who commute solely by public transport have more positive attitudes toward commuting by car and less positive attitudes toward commuting by bike, than those who combine bicycle and public transport for commuting (*11*). Bicycle–rail integration has not been studied extensively, especially in the North American context. Given that European cities tend to differ greatly from North American ones regarding their planning and political histories, built form, and transit, car, and bicycle networks and ridership, research on bicycle–train integration in North America is needed. Furthermore, we could not locate work examining, as this study does, the challenges faced by those currently combining bicycle and train travel, nor the barriers preventing rail passengers from doing so.

In addition, mode choice, trip purpose, and social characteristics have been found to influence travelers' needs, desires, and priorities when it comes to their transport journey (*13*). For example, punctuality and reliability are important to rush hour commuters, whereas safety, reliability, and service frequency have been found to be more important to female commuters (*13*). A substantial literature on the social characteristics of different kinds of travelers exists. For example, we know that approximately two-thirds of commuter cyclists are men in Canada, the United States, the UK, and Australia (*14*). Although less studied than the gender gap in cycling, discrepancies between other demographic groups have been observed as well. For instance, in a study set in London, England, most cyclists identified as White, male, and came from more affluent social groups (*15*). In the city of Toronto, utilitarian cyclists are predominantly male (66%), are between the ages of 25 and 54 (59%), and maintain a downtown address (*16*).

Few studies have assessed the social characteristics of those who combine cycling and train travel. Research set in Nanjing, China, found that women, older commuters, and low-income commuters were more likely to use a private bicycle, and less likely to use public bicycle/bike share, to access rail transit (*17*). Given that there is little accessible data on how bicycles combine with transit in most countries (*2*, *4*), especially countries with low cycling rates such as Canada (*4*), the lack of knowledge on the social characteristics of multimodal travelers is not surprising. This paper explores the challenges faced by those who combine bicycle and rail, the barriers preventing people from doing so, and the social characteristics of those who do, and those who do not, combine cycling with rail in the Greater Toronto and Hamilton areas, Canada. In doing so, this paper addresses two research gaps identified in this literature review: a lack of studies on the challenges and barriers to bicycle–rail integration in the North American context, and limited data on the social characteristics of multimodal travelers.

Data and Methods

Sample Characteristics

A bike policy survey was administered to GO Train passengers between November 17th and 30th 2014. It was sent to 6,483 panelists of Metrolinx's online customer advisory community called the "Let GO Know" panel. Panelists are recruited annually through advertisements on many channels (e.g., newspaper ads, website, train posters, train schedule departure screens, etc.). Let GO Know panelists are compared to a larger system wide GO Transit rail survey, with approximately 17,000 respondents, to validate its member make-up against the population of passengers. Non-panelists were also requested to participate in the survey through postcard invitations at specific stations (i.e., small printed invitations with links to the survey were distributed by hand to passengers during rush hour) and through a link posted to a Toronto-based cycling advocacy group's newsletter. Survey respondents were not directly compensated, but were automatically entered in a draw to win credit on the GO Transit system.

The survey was completed by 1,323 Let GO Know panelists (19.87% panelist response rate). Of those, 1,247 had ridden a GO Train in the past year. Furthermore, 33 non-panelists successfully completed the survey, resulting in a sample of 1,280 individuals. The primary goal of the study was to understand passengers' knowledge and use of the train system's bike policy, that is, the restriction of bicycles on GO Trains traveling through the most central and connecting station in the network during rush hour. However, the survey

Table 1. Summary Statistics of the Survey Sample

Sample characteristics	%	*n*
Own a bicycle (*n* = 1025)	62.54	641
Have biked to the train in the past year (*n* = 1280)	20.00	256
Very interested in cycling to the train (*n* = 641)	5.62	36
Somewhat interested in cycling to the train (*n* = 641)	21.84	140
Neither interested or uninterested in biking to the train (*n* = 641)	11.54	74
Somewhat uninterested in cycling to the train (*n* = 641)	16.69	107
Very uninterested in cycling to the train (*n* = 641)	44.31	284
Gender (*n* = 1,280)		
Male	46.80	599
Female	53.20	681
Age group (*n* = 1,280)		
Under 18	0.47	6
18–24	7.34	94
25–34	17.34	222
35–44	23.28	298
45–49	14.69	188
50–54	17.58	225
55–64	17.27	221
65+	2.03	26
Annual income in Canadian dollars (*n* = 1,247)		
Under $25,000	3.29	41
$25,000–$49,999	6.26	78
$50,000–$74,999	10.67	133
$75,000–$99,999	12.99	162
$100,000–$124,999	12.11	151
$125,000–$149,999	7.70	96
$150,000 or more	14.11	176
Prefer not to answer	32.88	410

also asked the participants about their cycling behaviors and attitudes as well as their age, gender, and income.

Most participants owned a bicycle (62.54%) (Table 1). One in five survey respondents stated that they had biked to or from a train station that year. Approximately one-third (27%) of the sample indicated they were somewhat to very interested in riding a bicycle to their home train station. Few respondents were under 18, older than 65, or earned less than CAD$25,000 yearly.

Variables and Units of Analysis

The survey included multiple-choice and open-ended questions. Responses were assessed through quantitative analysis and qualitative reporting. Specifically, to answer the first research question, what challenges do those who combine cycling with rail face?, responses to the open question "What are some of the challenges you experience combining cycling and GO Transit travel?" (*n* = 237) were assessed. Recurrent themes were then identified. To answer the second research question, what barriers keep rail passengers from reaching train stations by bicycle?, responses to the multiple-choice question "What are the main reasons you don't ever ride your bicycle to the GO Station" were examined. Participants could select more than one option resulting in 1,416 responses from the 641 participants. Finally, to answer the last research question, what is the demographic profile of those who do—and those who do not—cycle to reach the regional train station?, the gender, age, and income profiles of those who combine biking and rail were established. Then, responses to the questions: "What are the main reasons you don't ever ride your bicycle to the GO Station?" and "How interested are you in riding a bicycle to [your home station]?" were compared across income, age, and gender categories. Chi-squared tests were performed to check for statistically significant differences between male and female responses to these two questions. The names of train stations were removed to preserve respondent anonymity.

Results

What Challenges Do Those Who Combine Cycling with Rail Face?

Most passengers who had biked to GO at least once in the past year, specifically 237 of the 256, responded to the open question "What are some of the challenges you experience combining cycling and GO Transit travel?" Four factors were consistently cited by the respondents: unsecure and inconveniently designed bicycle parking at stations, travel safety concerns for the cycle part of the trip, unclear and restrictive rules regarding when bicycles are allowed on trains, and impracticality of navigating stations with bicycles. Other reported challenges included weather (n = 16), and lack of changing rooms at stations (n = 3).

Bicycle Parking at Stations. The most frequently mentioned challenge—identified by 33.76% of cyclists (80 out of 237)—was station bicycle parking. Two types of bicycle parking issues emerged: the security of locked bicycles and poor/inconvenient bicycle parking design. Regarding the inconvenience of the bicycle parking at stations, reported difficulties included distance between racks and station platforms, inadequate shelter for locked bicycles, and proximity between smoking and bicycle parking areas.

As one participant explained:

> [some] bike racks … are only accessible if you get off your bike IN THE ROAD and lift it onto a curb. You cannot bike from the sidewalk because a bench is in the way. These bike racks were not placed with a biker in mind [emphasis original].

Multiple participants described their own or acquaintances' experiences of bike theft:

> I've had my bike vandalized (and/or parts stolen) at the train station on a couple of occasions. During the day, thieves have a good, long window between trains when there is nobody around.

Safety Concerns Cycling to/from Stations. Following bicycle parking concerns, the most frequently cited issue was traffic safety while riding to/from train stations (22.36%). Many cyclists blamed this lack of perceived safety on driver behaviors; others expressed their desire for dedicated bicycle lanes to attenuate conflict.

Many cyclists vividly described these safety issues:

> Where to start? Have you actually tried to bike from the road to where the bike parking is at a GO Station? There are no bike lanes. I'm fighting for space with drivers trying to park. We need dedicated lanes! You have the space.

Another expressed the need for infrastructure:

> Very limited bicycle lanes to and from the GO station at [X station] make for a VERY dangerous ride [emphasis original].

GO Train Bicycle Policy. Bicycles are not allowed on regional trains during rush hour (due to potential overcrowding) if the train travels through the downtown station, called Union Station. Many individuals, however, still carry their bicycles on board during this period. Almost a quarter (specifically 22.36% or 53 out of 237) of cycling respondents identified the rush hour policy as a key challenge to integration. Folding bicycles are permitted on regional trains at all times.

The rules/policies were perceived as restrictive and lacking clarity:

> The hours and rules are too restrictive. Sometimes going against rush hour traffic (e.g., on [X] line), there is no need to restrict bikes during rush hour. Also, until recently there were no trains outside of rush hour on the [Y] line, so it wasn't even possible to use a bike on some lines.

With regard to clarity:

> The policy about bicycles ON the train is not clear at all. SOME GO employees say it's only on peak hours; SOME others say I can take my bike with me at any time. Then to make things more confusing, almost ALL GO train coaches have a sign by the door clearly showing that IT IS ALLOWED to take up to two bikes on the train. I looked for clear rules on the GO website or on the stations and there's nothing in writing that can tell the user what the real policy is [emphasis original].

Train Station and Rolling Stock Design Issues. Approximately 15% of people who cycled (35 out of 237) reported encountering difficulties navigating the stations with their bicycles. Identified issues included: lack of space for bicycles on train cars, awkwardness when using elevators, and difficulty getting onto train cars. Many participants listed issues they have encountered:

> Handlebars getting caught on the vertical post in the doorway, carrying bicycle up and down stairs at the stations, no dedicated bicycle car.

Another stated:

> No obvious elevators in stations to facilitate bringing the bike to and from the platform. No ramp into the trains to facilitate bringing in a bike (accessibility ramp could be used for this purpose). … The doors have almost closed on my bike several times because it took me some time to lift it in.

The responses made clear that the train stations and coach designs were not designed with cyclists in mind. These qualitative observations suggest that bicycle parking at train stations, perceived traffic risks, restrictive and unclear bicycle policy, and the station and train's bicycle-unfriendly designs are key service integration challenges.

What Are the Barriers Keeping Rail Passengers from Reaching Train Stations by Bicycle?

Respondents who did not cycle to the train were asked, "What are the main reasons you don't ever ride your bicycle to the train station?" Appearance and comfort were the majority concerns with 42% of patrons who do not cycle to the train, indicating concern about maintaining a professional appearance, and 38% reporting they feared they would be sweaty during their train ride (Figure 1).

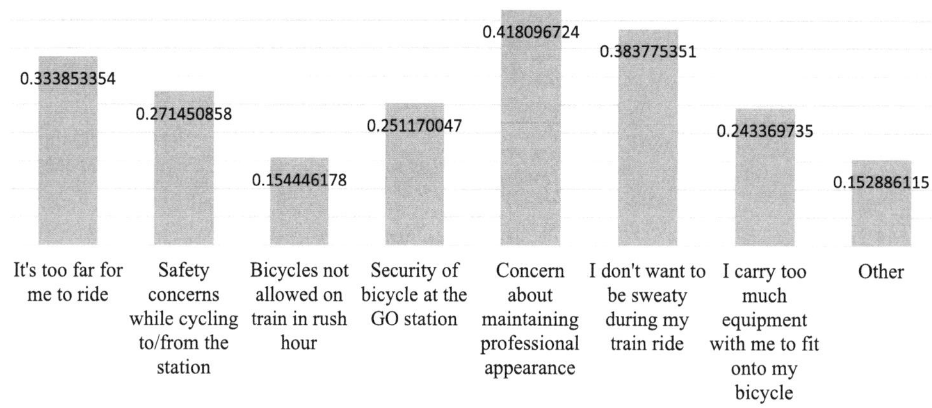

Figure 1. Primary reasons passengers do not ride their bicycle to the train station.
Note: Respondents could select multiple options.

Thirty-three percent of those who do not cycle indicated that train stations were too far from their home to bike. After distance, key reported barriers included: personal safety concerns while cycling to/from stations (27%), concerns about bicycle security at stations (25%), finding cycling impractical due to the amount of equipment they carry to/from the train (24%), and bike policy (15%) (Figure 1). While only 25% of passengers who do not bike to/from the train station identified bicycle parking as a barrier, 44% of those who do not cycle to/from the train, but who indicated being somewhat to very interested in biking to/from the train, cited concerns about the security of their parked bicycles as the main deterrent. Finally, 15% of respondents selected "other" responses including living close enough to walk ($n = 40$), coordinating commutes with household responsibilities ($n = 8$), weather concerns ($n = 4$), medical reasons ($n = 3$), and a lack of lockers/storage at stations ($n = 2$).

What Are the Social Characteristics of Those Who Combine Cycling with Rail and Those Who Do Not?

Gender. Consistent with North American bicycle commuting trends, the data suggest that two-thirds of regional rail cyclists are men (67% male vs. 33% female). A "female" risk aversion hypothesis has been suggested as an explanation for this type of gender gap in cities with low cycling rates (*18–20*). This study, however, does not support the gendered safety risk hypothesis (p-value > 0.1) (Table 2). It has also been hypothesized that women cycle less than men due to the greater social pressures women face to look feminine and attractive (*14*). However, while women indicated concern about professional appearance more than men, this observation was not statistically significant (p-value > 0.01). Men more frequently cited not wanting to be sweaty during the train ride than women (p-value < 0.1).

Others have suggested that women may not like cycling as much as men (*14*). This study suggests that women are equally interested in cycling to/from the train (p-value > 0.1). Previous work has found that women are more likely to trip-chain than men due to greater household responsibilities (*14*, *21*). This disproportionate unpaid household-serving burden limits their time available for other activities and makes it more difficult to commute by bicycle. Of those who selected "other" as a response, six women, compared with two men, specified that household duties, specifically picking up their children ($n = 5$) or running errands ($n = 1$) were among the main reasons they do not bike to/from the train. One woman stated:

> [I] need to pick up my children at the end of the day and can't do that with a bike.

Furthermore, women in this survey indicated carrying too much equipment as a barrier to cycling to/from the train more frequently than men (p-value < 0.1) (Table 2). Although the survey did not specify, this extra equipment may also relate to trip chaining (i.e., extra equipment for groceries, childcare, etc.). Survey questions did not directly investigate trip chaining.

Income. Most current cyclists to the train earn more than \$150,000 yearly (Figure 2*a*). Looking within each income category, a greater proportion of lower-income respondents cycle to the train. In fact, intra-income category rates decline as income increases and then rise within the highest income category (Figure 2*b*). Those earning \$25,000–\$49,999 yearly are most interested in cycling to the train (39% were somewhat to very interested).

Many barriers to cycling vary by income. Respondents within the lowest income group reported greater concern about their safety while cycling to the station as well as

Table 2. Survey Responses by Gender

	Men (n = 302)	Women (n = 339)
What are the main reasons you don't ever ride your bicycle to the train station? (multiple choice)		
It's too far for me to ride	108	106
Safety concerns while cycling to/from the station	77	97
Bicycles not allowed on the train in rush hour	59***	40
Security of bicycle at the GO Station	84	77
Concern about maintaining a professional appearance	117	151
I don't want to be sweaty during my train ride	127*	119
I carry too much equipment with me to carry onto my bicycle	61	95*
How interested are you in riding a bicycle to [your home station]? (multiple choice)		
Very interested in riding a bike to the train station	18	18
Somewhat interested in riding a bike to the train station	63	77
Neither interested nor uninterested in riding a bike to the train station	39	35
Somewhat uninterested in riding a bike to the train station	58	49
Very uninterested in riding a bike to the train station	124	160

***p-value < 0.01. *p-value < 0.1.

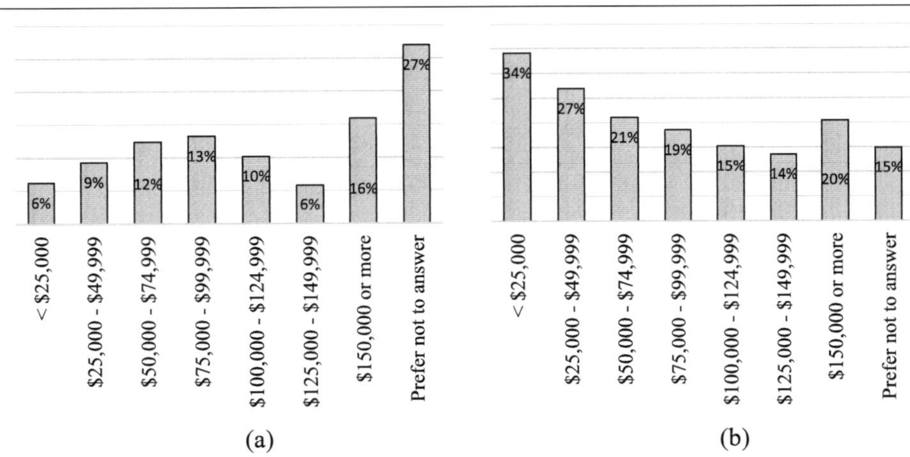

Figure 2. Income profile of those who cycle to the train: (*a*) income distribution of sample who bike to/from train; (*b*) proportion of income group who bike to/from train.

bicycle parking security more than all other income groups. Furthermore, those in the middle-income bracket (i.e., earning $50,000–$74,000 yearly) cite the system's bicycle policy as a deterrent more frequently than other income groups (Figure 3).

Finally, a positive correlation between income and both concerns over maintaining a professional appearance and wanting to avoid being sweaty was found (Figure 4).

Age. Most respondents currently cycling to/from the train station are between the ages of 25 and 44 (Figure 5*a*). However, a greater proportion of young passengers cycle to/from the train. In fact, these rates decline with age (Figure 5*b*).

Before the age of 35, similar proportions of commuters are somewhat to very uninterested and somewhat or very interested in cycling to/from the train station (approximately 40%). This interest drops dramatically from the age of 35 until 64 (Figure 6).

Discussion and Conclusion

This study examined the challenges and barriers reported by those who currently do and do not combine cycling with rail travel. Limitations include potential self-selection bias of survey respondents. This study did not consider how these challenges and barriers vary by season. Future work is needed to consider potential seasonal variation, particularly in the winter months, in rider participation in bicycle–rail integration programs. Furthermore, this case study focused on bike–rail integration in the context of the Greater Toronto and Hamilton areas. Therefore, while the study contributes to the broader discussion on bicycle–rail integration, the policy recommendations

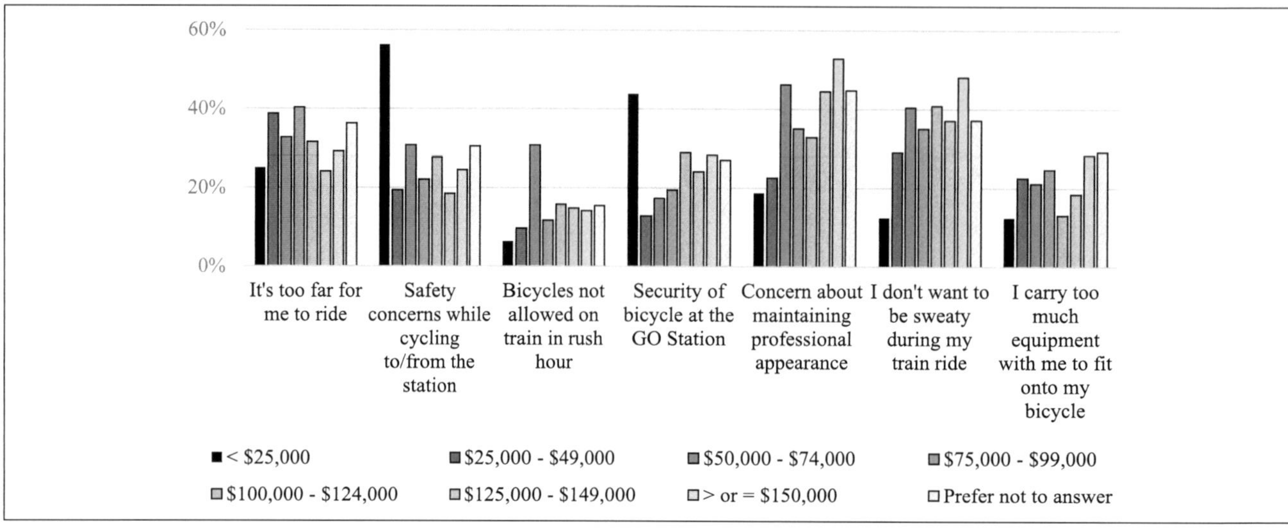

Figure 3. Reasons not to bike to/from the train station organized by income.

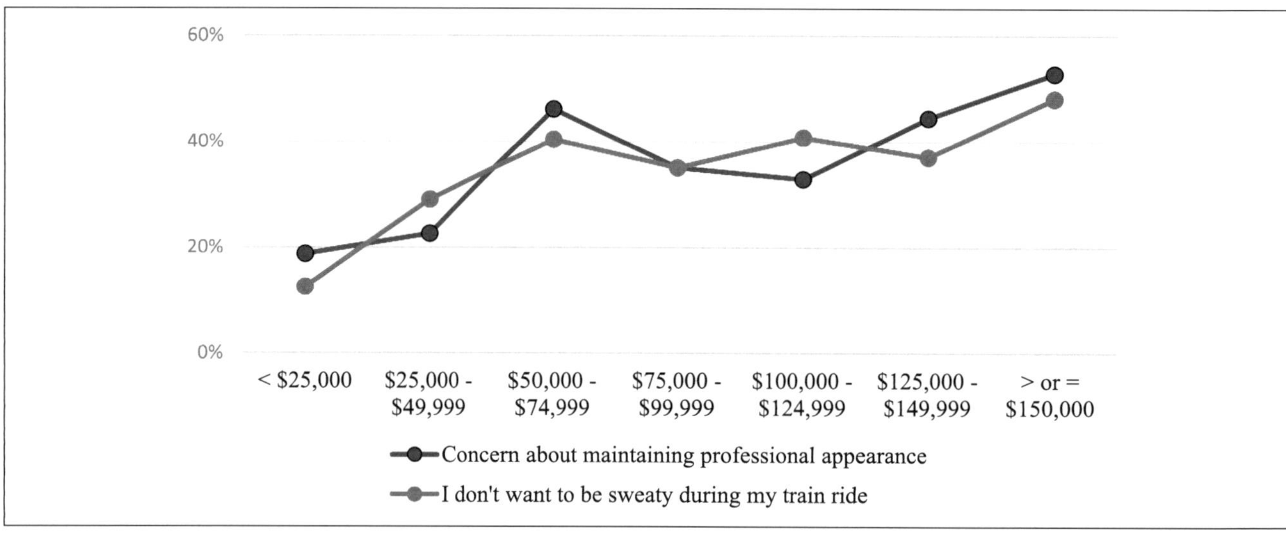

Figure 4. Relationship between concerns over maintaining a professional appearance, sweating, and annual income.

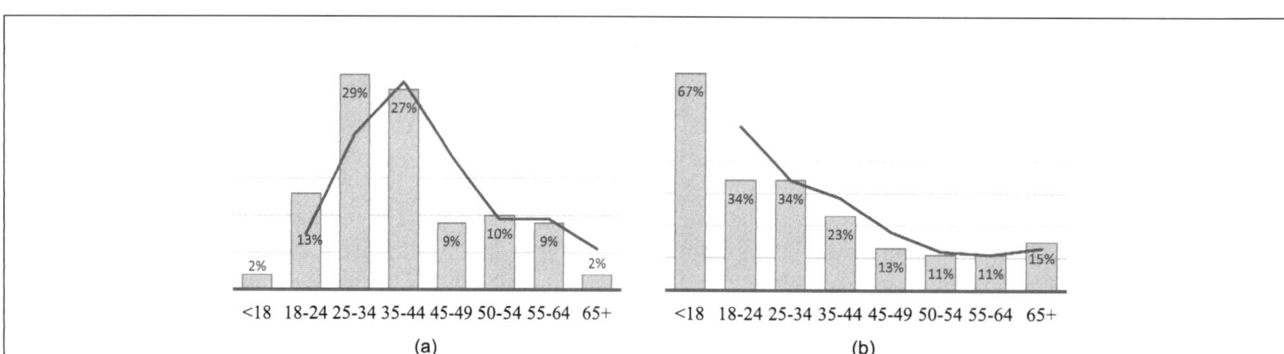

Figure 5. Age profile of those combining cycling with rail: (a) age distribution of sample who bike to/from train; (b) proportion of age group who bike to/from train.

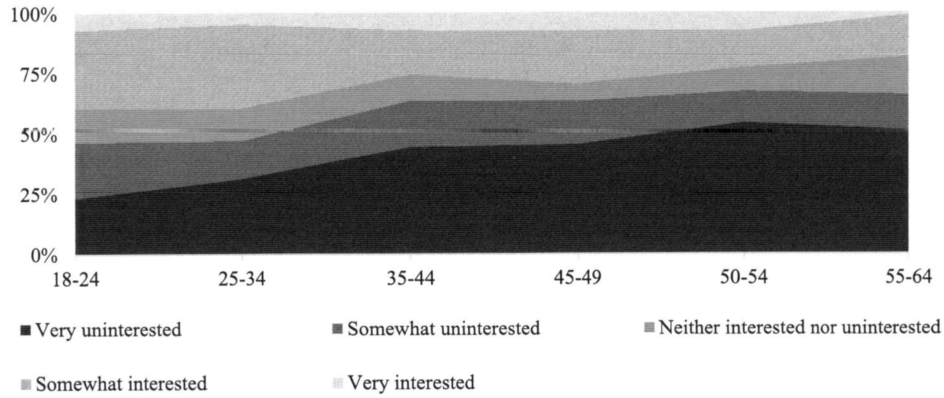

Figure 6. Interest in cycling to the train station across the life course.

emerging from this research may not be transferrable to other rail networks. This concluding section includes a discussion of key findings and policy recommendations.

Both cyclists (34%) and those not cycling to the train (25%) identified station bicycle parking as a barrier. Transit agencies wanting to encourage cycling, particularly the one studied here, should provide secure and convenient bicycle parking at every station. The majority of cyclists (69%) stated that they were interested in using a shared bicycle parking area with key access only, if such facilities were available at train stations. This type of infrastructure, which can be found at select stations on the Caltrain service in California (*22*) and in Melbourne, Australia (*23*), takes up the space of three parking spots, but can fit 26 bicycles though costing less than half the price of a single parking space (*23*). One in four cyclists also stated that video surveillance would encourage them to park their bicycle at the train station. Given that 44% of passengers who do not bike to the train, but are interested in doing so, cited concerns about bicycle security, solving the bicycle parking security challenge could weaken the impact of this key barrier.

Of course, improved bicycle parking would encourage cycling for only one side of the multimodal trip: their bikes cannot be used at their destination stations. Previous work has found that public transit users are more likely to use their own bicycles to get from their (often suburban) home to their origin station than to their final destination (often downtown) from their destination station (*24*). Given that destination stations may be at or near the city center, ensuring adequate bike share programs at stations in the city center may facilitate cycle–train integration. Downtown Toronto is served by 270 public bicycle stations (*25*). The fares for this Bike Share service and the regional rail service are not integrated, however, GO Train passengers who own a PRESTO Card, the electronic fare payment system available across many transit agencies in the Greater Toronto and Hamilton areas, receive a 40% discount on the annual Bike Share membership (*25*). While some, of course, prefer the convenience of bringing their own bicycle on the train to use it for both the first and last mile of their trip, coordinating Bike Share and rail systems could still encourage cycling to/from train stations. Indeed, others have similarly recommended that rail agencies provide multimodal offers such as access to shared bicycle systems, or integrated multimodal fare cards, to encourage cycle–train integration (*26*). Future research could specifically examine how best to coordinate public bicycle sharing systems with rail.

Approximately one-quarter of those who combined bikes and rail, and 27% of those who do not bike to/from the train station cited safety concerns as a challenge or barrier. This result highlights the need for cycling infrastructure to provide cyclists with safe access to train stations across the transit network. Dedicated bicycle lanes within the train station property, with clear orientation to bike parking, as well as lanes connecting train stations to the municipalities they serve are recommended. Improving cyclist safety while accessing stations is also in line with the 2016 GO Rail Station Access Plan that includes the guideline to "create safe and direct pedestrian and cycling routes to transit stations that are complemented with clear way-finding and potential bike share programs within the station area" (*10*). Given that the lowest income group of this study was most concerned with their personal safety while cycling to/from train stations as well as station bicycle security, policies aimed at improving rider safety and bicycle security could be particularly beneficial to lower-income train commuters. Safety concerns are not necessarily limited to those associated with traffic; future work could explore the different sources of cycling-related safety concerns such as fear of being mugged, assaulted, or harassed in general.

Navigation within the station envelope was also reported as a barrier. New stations should be designed and built with all passengers, including cyclists, in mind. For stations that have already been built, modifications can be made to accommodate cyclists. For example, stairways in stations can be modified to include a ramp to roll bicycles up and/or down, bicycle stickers can be placed on elevators to indicate that cyclists and their bicycles are invited to use these facilities, and a dedicated train car for bicycles would also make it easier for cyclists to navigate train stations. Within the train coaches, a more flexible design of the bike area (e.g., folding seats, more spacious standing zone) may also support bikes on trains, and could also be used for strollers, luggage, etc.

The GO system's bike policy was also identified as a challenge or barrier for both cyclists (22%) and those who do not cycle to/from the train (15%). The two main critiques from cyclists were that the policy lacked clarity and that it was restrictive. Bicycle-related policies should be simple and easy to follow. Many respondents (including those who do or do not cycle) support the bike policy and expressed some acrimony toward the idea of letting bicycles onto rush hour trains. Adding a separate car dedicated to bicycles during rush hour periods, as suggested above, could allow cyclists to incorporate cycling into their daily commute without inconveniencing other passengers. For example, Caltrain includes bike cars that can accommodate 24–72 bicycles (*23*). Rail operators must also weigh the benefits of increased bicycle capacity against the costs of increased dwell times, initial capital investment, maintenance, and other ongoing costs when deciding to implement this policy (*8*).

The two main barriers preventing passengers from combining bike and train were concerns about maintaining a professional appearance and comfort. This result may have to do with perceptions around cycling; some see biking as a high intensity sport that should be "performed" while wearing specialized clothing. Utilitarian cycling, however, does not need to involve heavy exertion resulting in sweating or an unprofessional appearance. Even the frame geometry of many utilitarian bikes enables a more relaxed riding position and pace. In many places, people of all ages and fitness levels cycle for transport. Other than providing showers or lockers at train stations, or in the workplace, little can be done in the short term to encourage people with these concerns to cycle, but over the long term, a change in the culture of mobility could attenuate such concerns.

Other than identifying key policies that could encourage bicycle–train integration, this study also contributes knowledge on the social characteristics of those combining cycling with train travel. Specifically, approximately two-thirds (67%) of those cycling to the train were male. Furthermore, this study does not offer evidence supporting gendered hypotheses regarding risk, appearance, or a general dislike of cycling, as an explanation for the reported gender gap in cycle rates. Some evidence points to women cycling less due to the need to trip-chain in support of unpaid household labor. However, further research is needed to investigate this hypothesis. It is important to note that although this study identified behavioral and perception-based patterns across social groups, why these patterns and rates exist was not studied in detail. Further research should investigate how the transportation behaviors, needs, and experiences of people of different ages, genders, and incomes are shaped by, and shape, the context within which they live.

Acknowledgment

Léa Ravensbergen received the Joseph-Armand Bombardier Graduate Scholarship from the Social Sciences and Humanities Research Council of Canada (SSHRC).

Author Contributions

The authors confirm contribution to the paper as follows: study conception and design: L.R., M.M, N.G.; data collection: M.M, N.G., R.B.; analysis and interpretation of results: L.R.; draft manuscript preparation: L.R., R.B., M.M. All authors reviewed the results and approved the final version of the manuscript.

References

1. Börjesson, M., and E. Jonas. The Benefits of Cycling: Viewing Cyclists as Travelers Rather than Non-Motorists. In *Cycling and Sustainability* (J. Parkin, S. Ison, and J. Shaw, eds.), Turpin Distribution Services Limited, Biggleswade, 2012, pp. 83–108.
2. Kuhnimhof, T., B. Chlond, and P. C. Huang. Multimodal Travel Choices of Bicyclists. *Transportation Research Record: Journal of the Transportation Research Board*, 2010. 2190: 19–27.
3. Olafsson, A. S., T. S. Nielsen, and T. A. Carstensen. Cycling in Multimodal Transport Behaviours: Exploring Modality Styles in the Danish Population. *Journal of Transport Geography*, Vol. 52, 2016, pp. 123–130.
4. Martens, K. The Bicycle as a Feedering Mode: Experiences from Three European Countries. *Transportation Research Part D: Transport and Environment*, Vol. 9, No. 4, 2004, pp. 281–294.
5. Keijer, M. J. N., and P. Rietveld. How do People Get to the Railway Station? The Dutch Experience. *Transportation Planning and Technology*, Vol. 23, No. 3, 2000, pp. 215–235.
6. Geurs, K. T., L. La Paix, and S. Van Weperen. A Multi-Modal Network Approach to Model Public Transport Accessibility Impacts of Bicycle-Train Integration Policies. *European Transportation Research Review*, Vol. 8, No. 4, 2016, p. 25.
7. Pucher, J., and R. Buehler. Integrating Bicycling and Public Transport in North America, *Journal of Public Transportation*, Vol. 12, No. 3, 2009, pp. 79–104.
8. Krizek, K. J., and E. W. Stonebraker. Bicycling and Transit: A Marriage Unrealized. *Transportation Research Record: Journal of the Transportation Research Board*, 2010. 2144: 161–167.
9. Metrolinx. *Go Transit*, 2015. http://www.gotransit.com/publicroot/en/default.aspx. Accessed October 9, 2017.
10. Metrolinx. *GO Rail Station Access Plan*, 2016, p. 5. http://www.metrolinx.com/en/regionalplanning/projectevaluation/studies/GO_Rail_Station_Access_Plan_EN.pdf. Accessed July 20, 2017.
11. Heinen, E., and W. Bohte. Multimodal Commuting to Work by Public transport and Bicycle: Attitudes toward Mode Choice. *Transportation Research Record: Journal of the Transportation Research Board*, 2014. 2468: 111–122.
12. Martens, K. Promoting Bike-and-Ride: The Dutch Experience. *Transportation Research Part A: Policy and Practice*, Vol. 41, No. 4, 2007, pp. 326–338.

13. Susilio, Y. O., and O. Cats. Exploring Key Determinants of Travel Satisfaction for Multi-Modal Trips by Different Traveler Groups. *Transport Research Part A: Policy and Practice*, Vol. 67, 2014, pp. 366–380.
14. Garrard, J., S. Handy, and J. Dill. Women and Cycling. In *City Cycling* (J. Pucher and R. Buehler, eds.), MIT Press, Cambridge, Mass., 2012, pp. 211–234.
15. Steinbach, R., J. Green, J. Datta, and P. Edwards. Cycling and the City: A Case Study of How Gendered, Ethnic and Class Identities Can Shape Healthy Transport Choices. *Social Science and Medicine*, Vol. 72, No. 7, 2011, pp. 1123–1130.
16. Toronto Cycling Think & Do Tank. *Mapping Cycling Behaviour in Toronto,* 2013. http://www.torontocycling.org/uploads/1/3/1/3/13138411/mapping_cycling_behaviour_in_toronto_final_23_may_printer_tl.pdf. Accessed May 10, 2018.
17. Yi, Y., Y. Fan, A. Ermagun, X. Cao, W. Wang, and K. Das. Public Bicycle as a Feeder Mode to Rail Transit in China: The Role of Gender, Age, Income, Trip Purpose, and Bicycle Theft Experience. *International Journal of Sustainable Transportation*, Vol. 11, No. 4, 2017, pp. 308–317.
18. Agarwal, A., and A. North. Encouraging Bicycling Among University Students: Lessons from Queen's University, Kingston, Ontario. *Canadian Journal of Urban Research*, Vol. 21, No. 1, 2012, pp. 151–168.
19. Garrard, J., G. Rose, and S. Lo. Promoting Transportation Cycling for Women: The Role of Bicycle Infrastructure. *Preventive Medicine*, Vol. 46, 2008, pp. 55–59.
20. Heesch, K. C., S. Sahlqvist, and J. Garrard. Gender Differences in Recreational and Transport Cycling: A Cross-Sectional Mixed-Methods Comparison of Cycling Patterns, Motivators, and Constraints. *International Journal of Behavioral Nutrition and Physical Activity*, Vol. 9, No. 1, 2012, p. 106.
21. Heinen, E., B. van Weeb, and K. Maat. Commuting by Bicycle: An Overview of the Literature. *Transport Reviews: A Transnational Transdisciplinary Journal*, Vol. 30, No. 1, 2010, pp. 59–96.
22. Caltrain. *Bicycle General Info.* http://www.caltrain.com/rider-info/Bicycles/Bicycle_General_Info.html. Accessed October 5, 2017.
23. Bicycle Network. *Parkiteer*, 2016. https://www.bicyclenetwork.com.au/general/programs/370/. Accessed January 17, 2017.
24. Pucher, J., and R. Buehler. Cycling for Everyone: Lessons from Europe. *Transportation Research Record: Journal of the Transportation Research Board*, 2008. 2074: 58–65.
25. Bike Share Toronto. *Bike Share Toronto*, 2016. https://www.bikesharetoronto.com/. Accessed October 5, 2017.
26. Bachand-Marleau, J., J. Larsen, and A. El-Geneidy. Much-Anticipated Marriage of cycling and Transit. *Transportation Research Record: Journal of the Transportation Research Board*, 2011. 2247: 109–117.

The Standing Committee on Commuter Rail Transportation (AP070) peer-reviewed this paper (18-02701).

Article

TRR
JOURNAL OF THE TRANSPORTATION RESEARCH BOARD

Transportation Research Record
2018, Vol. 2672(8) 384–403

DOI: 10.1177/0361198118783888
journals.sagepub.com/home/trr
SAGE

Exploring the Complex Relationship between Railroad Infrastructure, Operating Constraints, Maximum Speeds, and Public Schedules

Alex Lu[1], Janek Kozlowski[2], Sean Fahey[1,3], and Shaun Whitman[4]

Abstract

Outside the railroad industry, passenger trip times are often treated as a performance measure or yardstick of the industry's health. However, many factors not known or well understood by casual observers affect trip times—indeed, some are recognized techniques utilized by infrastructure owners to deliver journey time reductions. Factors triggering running time modifications include: track (design, maintenance, layout); infrastructure (structures, signal, power, grade crossings); cant deficiency, tractive effort, braking rates; rules, operating practices, timetable speeds; and schedulers' decisions on train performance modeling, en-route adjustments, and dwell times. Historical operating documents issued by railroads can be used to reconstruct events associated with trip timing changes, but their ephemeral nature makes it difficult. Because some information is never written down, the rationale for certain modifications may be unknowable. Public timetables offer a general sense of trends in service offerings but don't show what was really happening. Opposing industry trends can actually give rise to similar schedule impacts, e.g. deferred maintenance and track rehabilitation both lead to longer journey times. Applying forensic analysis methodology to Penn Central's Mohawk/Buffalo Divisions revealed that some time degradations are correlated to downgraded infrastructure and vehicle issues, but long-term changes were balanced. Metro-North's New Haven Line case study reveals how new stations, necessary safety modifications, increasing congestion, and temporary construction delays all contributed to recent timing changes. To maximize system performance, operators must balance time, capacity, and reliability subject to an overarching constraint of safe operations.

This paper serves three basic purposes:

1. Explain for a non-rail-industry audience many factors behind the determination of published public timetable trip times between station pairs on conventional railroad networks.
2. Show that changes in public timetable running times alone is insufficient evidence or information to infer or explain happenings in rail industry investment, maintenance management, infrastructure stewardship, etc.
3. Demonstrate how operating documents (where available) can be utilized to reconstruct reasons for changes to advertized journey time and to feel the industry's pulse, while also demonstrating that they might not tell the entire story.

The paper is divided into four major sections:

1. Technical and operational factors that affect running times
2. Operating documents that describe these factors
3. Methods for constructing a schedule that properly accounts for them
4. Case studies in forensic schedule analysis:
   (a) Penn Central's Mohawk and Buffalo Divisions (1956–2010): utilizing only operating documents and modeling methodologies
   (b) Metro-North's New Haven Line (2009–2016): utilizing institutional knowledge of section schedulers and published sources

[1]ADD-Strategic Operating Initiatives and Field Administration, Metro-North Railroad, Ossining, NY
[2]DD-Operating Capital Projects—New York State, Metro-North Railroad, New York, NY
[3]ADD-Operations Analysis & Logistics, Metro-North Railroad, New York, NY
[4]Rules Examiner, Metro-North Railroad, New York, NY

Corresponding Author:
Address correspondence to Alex Lu: me@lexciestuff.net

Motivation

Standard methodologies for rail transit planners and schedulers to create service specifications, to establish public timetables, and to create vehicle assignments, traincrew diagrams (pairings), routing plans, and terminal docking plans are well understood and documented (e.g. [*1–3*]). However, running times between stations are sometimes seen as empirical observations obtained from operational experience, running time checks, or track occupancy records. Although specialized software is available (e.g. [*4*, *5*]) to estimate the impacts of numerous factors on running times, delays, and network congestion (e.g. [*6*]), considerations affecting published running times are less familiar to outside stakeholders. Indeed, recent work (*7*) attempted to infer what was generally happening in the rail industry through retrospective inspection of public timetables. Some outside pressure groups (*8*), trade press articles (*9*), and even industry strategy documents (*10*) cite trip times as a performance measure or goal.

Advertized times, even in aggregate, should not be seen as a performance yardstick to measure the general state of passenger rail. The relationship between system performance capability and published journey times is many-to-one where different confluences of multiple events (some good, some bad) could all result in similar running time improvements or degradations. Just as it is widely acknowledged that many factors independently cause reduction in on-time performance or line capacity (trains per hour, tph), multiple reasons cause scheduled time modifications; indeed, some factors trade-off with one another. Maximizing on-time performance (*11*, *12*) or line capacity is not a reasonable long-term strategy—and good express train running times is only one of multiple factors indicating healthy and well-managed railway systems. The "railway races" on Anglo-Scottish grouse trains is an early example (*13*), which ultimately ended tragically at Salisbury in 1906 when London & South Western Railway wrecked its premier overnight express due to relentless emphasis on journey time.

Planners performing commuter rail feasibility studies (e.g. [*14*, *15*]) are familiar with assumptions and existing conditions that must be defined prior to generating journey time estimates. Beyond infrastructure capabilities and specifications, medium-term operational considerations like track maintenance, train stopping patterns, and rostered equipment all have an effect on advertized trip times. Short-term temporary speed restrictions (TSR), capacity-related delays, or dispatching considerations also add extra minutes, communicated to customers via supplemental schedules or special timetables.

Railroad Track Factors

Track geometry and technologies are fundamental to passenger train operations, ensuring safety at speed. Right-of-way curvature constraints and track design and maintenance parameters are translated to maximum authorized speeds (MAS)—the highest speed at which one given train may operate on the given track segment at that specific time—which in turn directly affect published journey times.

Track Geometry

Figure 1*a* shows an excerpt from a typical Track Chart & Maintenance Program (*16*) from a major Northeastern railroad, showing constraints affecting passenger train running times. Diamonds above the track are mileposts (MPs). Design track geometry is shown at the bottom. Curve at MP 225 is ~0.3 miles long (horizontal scale measurement) and has a degree of curvature 1°50". Design superelevation on Tracks 2&1 is 4". Assuming track geometry is properly maintained within design limits, allowable speeds are given by formulae in 49 CFR 213.329 Appendix A.

When contemplating new track construction, track geometry is rarely known before right-of-way design. However, constraining curve radius can be approximated by inspecting aerial photographs or maps. The best-case curve could be sketched out (Figure 1*b*), curve radius measured, then converted to best-possible geometry assumption. (curve radius ≈ 5,730 feet ÷ degrees.) A common misconception is that if track geometry is legal at certain speeds, then services could operate at that speed. Track speed is the first step in figuring out MAS; many other factors can result in MAS being revised downwards (rarely, upwards). Track speeds provide a lower bound for running time.

Track Maintenance

Economic factors like intended traffic and labor/material costs drive railroad operators' choices of track maintenance levels. FRA's "class of track" designation describes the maximum permissible deviation from design geometry in parameters like gauge, alinement, and crosslevel. Whenever track inspectors discover deviations approaching FRA limits, maintenance gangs must be called to restore track profile.

Track class is an additional constraint over-and-above design geometry. Although a curve geometry might be designed for 75 mph, if track is maintained only to FRA Class 3 standards, passenger trains would be limited to 60 mph. Maintaining to Class 3 standards means, amongst other items, track gauge must be kept between 4'8" and 4'9¾" (*18*). To understand if track maintenance constraints contribute to longer journey times, we can check if FRA track class was downgraded. Some railroads' track charts report this information (Figure 1*a*).

Special Trackwork

Taking diverging routes through switches and crossings usually requires lower speeds. In North America this is typically shown as a signal speed. Once designed, switch sizes

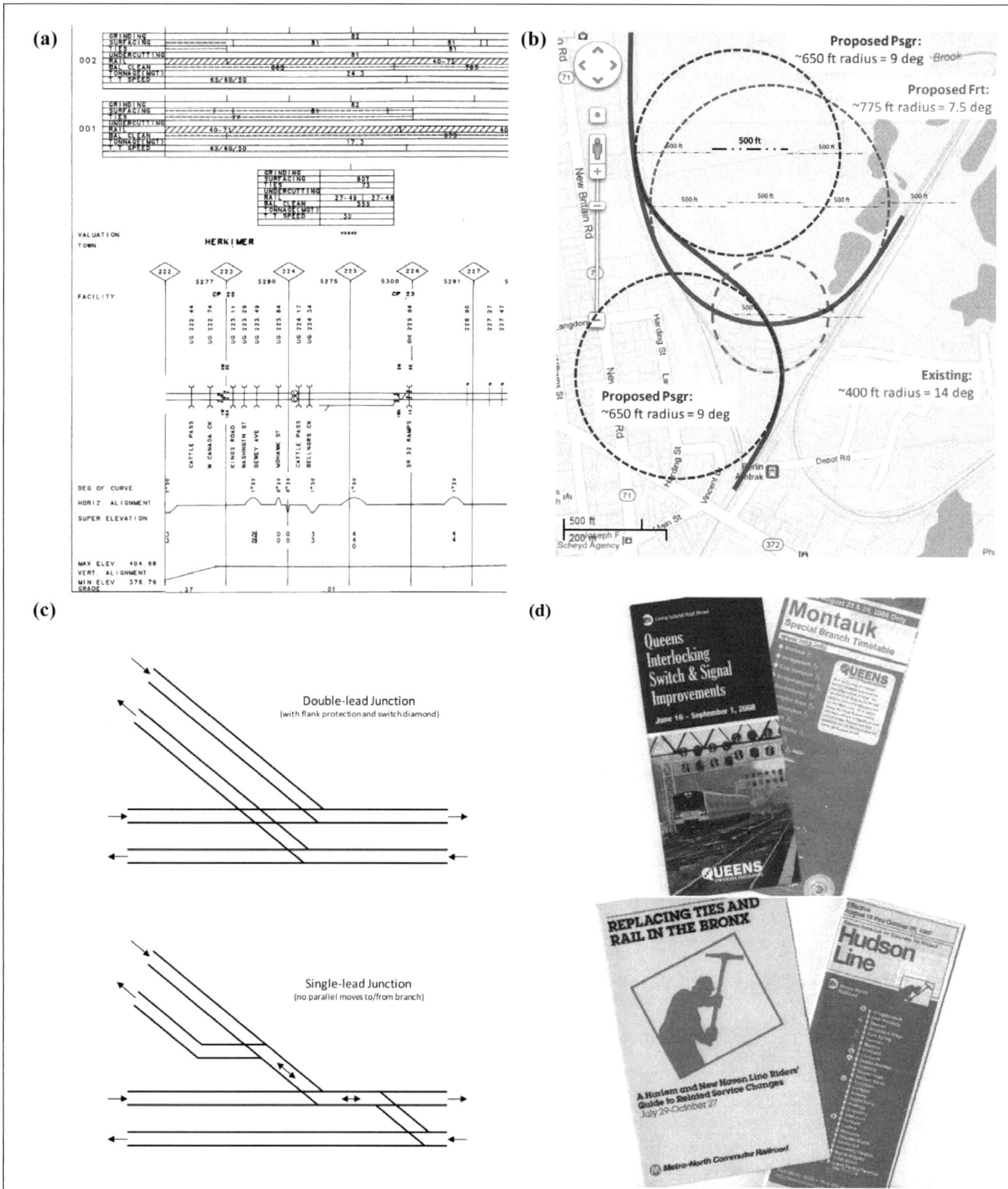

Figure 1. Tools for speed research: (*a*) Conrail track chart from the vicinity of Herkimer, NY; (*b*) sketch curve radius measurements in the vicinity of Berlin, CT; (*c*) speed impacts of junction layouts (*17*); (*d*) special timetables issued by railroads showing specifically the impacts on journey time of rehabilitation projects.

are not normally changed, for example, No.20 turnouts have a 45 mph MAS.

Three situations arise where turnout speeds do change: (a) in track rationalization or normal switch replacement, if space constraints or parts availability dictates, turnouts might be replaced with smaller ones, for example, No. 20 replaced with No. 15, with MAS reducing to 30 mph; (b) for speed improvements, straight points may be replaced with curved points, for example, "curved" No. 20 are good for 60 mph while occupying the same footprint; (c) turnouts improperly adjusted, worn, or having incorrect geometry results in TSRs. Advertized running times are affected in each case.

Special trackwork layouts affect timing in less obvious ways. Removal of critical switches or wyes may result in trains making reverse moves to access platform areas or major terminals, adding several minutes. Some railroads have eliminated maintenance-heavy switch diamond by replacing standard junctions with single-leads (Figure 1*c*), preserving mainline MAS but requiring diverging moves at lower speeds. Single-lead junctions effectively introduce short single-track segments, additional delays occur when two opposing branch trains need to access the mainline simultaneously.

Infrastructure Considerations Other than Track

Non-track infrastructure components have an impact on MAS (and trip time capabilities). Normally, designs maximize latent achievable right-of-way speeds (limited by curvature and terrain). As railroads move through rebuild cycles, components become MAS constraints in some situations.

Structures

Dynamic forces exerted by moving trains on structures are related to equipment axle weights and train speeds. When installed, designs generally support intended maximum load (based on the Cooper rating—representation of two typical locomotives and its train) at prevailing line speeds. Situations resulting in structural MAS constraints include: (a) over time, degraded structures are "de-rated," having their dynamic loads reduced to maintain safety due to compromised members; (b) vehicle weights may increase, resulting in higher loads, thus permissible speeds are reduced to compensate; (c) area linespeed increased due to civil improvements, but a bridge remaining qualified only for original speed. Industry references (*19*, *20*) on relationships between bridge ratings and speeds are available.

Signal Braking Distances

Standard signaling textbooks (*21*) or railroad documentation (*22*) show braking curves for computing minimum signal spacing from design variables like entry speed, vehicle characteristics, number of signal system aspects, and required capacity (minimum permissible headway in tph). Lines are initially equipped with track circuits and signal blocks at correct distances such that all trains (with train-type-specific speed restrictions) can come to a safe stop prior to encountering a Stop signal. The system provides sufficient sighting distances and appropriate aspects for anticipated traffic patterns. For given aspects, an inverse relationship exists between capacity and signal design speed. Once installed, track circuit, interlocking, and signal locations are normally fixed. Design speeds are normally higher than civil speeds and therefore do not require restrictions.

Situations where signal systems require permanent speed restrictions (PSR) below civil speeds include: (a) traffic pattern changes after signal installation, resulting in inappropriate signal spacing; (b) vehicle characteristics assumptions (e.g. braking rates) in initial design rendered obsolete by vehicle changes (e.g. additional weight), resulting in speed restrictions to match new braking curves to existing signal spacing; (c) track reconfigurations (e.g. siding extension, interlocking expansion) require signal relocations, but desired locations do not work well with existing blocks (expensive to relocate), resulting in lower MAS to provide sufficient braking distance.

Signal modifications can also require operating speed changes. Removal/addition of automatic intermediate signals results in block length changes, thus changing distance traveled at lower speeds when signal indication is not the most favorable aspect possible. Changes in capability to display aspects results in signals displaying the next-most-restrictive aspect, changing the speed at which moves could legally be made.

Systems maintained per original design generally have no impact on operating speeds. However, when signals are modified or simplified to reduce maintenance costs or support different traffic patterns, they could reduce capacity or inadvertently lengthen trip times. References on the speed implications of signal design are available (*21*, *23*).

Cab Signal Systems (CSS)

Cab signals are normally designed as overlays to conventional multi-aspect signals. A cab signal system (CSS) may merely repeat and enforce wayside aspects, provide additional aspects allowing higher speeds, or replace automatic wayside signals with code change points, providing cost savings. Suitably equipped trains can generally access higher MAS through additional aspects or improved safety from aspect enforcement. CSS does not usually affect trip times once installed, however, where CSS is removed or had capabilities modified (e.g. aspects removed/added), changes to operating speeds can occur after initial introduction.

U.S. passenger railroads relied on CSS modifications to enforce curve speeds prior to full implementation of positive train control (PTC) at high-risk locations, like Frankford

Junction and Spuyten Duyvil. CSS aspects are enforced at speeds (e.g. 15/30/45 mph) that may not exactly match civil restriction (specified to nearest 5 mph). Signal engineers use next-most-restrictive aspect at nearest-code-change point, resulting in, for example, 30 mph enforced for 2,200 ft for a 40 mph restriction requiring 1,400 ft braking distance. This has a measurable impact on trip times compared with locomotive engineers manually managing deceleration.

Level Crossings

Where roads and railways cross at grade, speed restrictions may be required in high-speed territory (>79 mph in U.S.A.). FRA guidelines (*24*) recommend different protection levels based on train speeds, although it is not clear if this applies to existing grade crossings or generates new PSRs if crossing protection has not been upgraded. In the U.K., enhanced level-crossing protection is required where trains operate at >100 mph.

Electrical Supply Infrastructure

On electrified railroads, short-term substation power ratings can limit locomotive power demand (e.g. maximum throttle or "dial down" requirements), tonnage, train length, headway, and acceleration, although they do not normally generate speed restrictions. However, resultant achievable speeds could affect running times.

Catenary system designs can have subtle and complex impacts on maximum speeds: e.g., catenary pole spacing in central N.J. limits speeds to 140 mph even with new constant tension designs (*25*). Railroads impose TSRs relating to pantograph-catenary contact instability in windy, hot, or cold conditions (typically in variable-tension territory). Third rail systems are normally limited to 100 mph maximum speed; difficulties exist in designing nosepieces to accept shoes at higher speeds. Industry resources (*43*, *44*) further explain these constraints.

Vehicle Factors

Many speed constraints are vehicle dependent and relate specifically to vehicle performance.

Cant Deficiency (Underbalance)

Limiting speeds (above which derailment occurs) on continuously welded rail (CWR) with modern fasteners can be much higher than the speed at which passengers feel nauseous or unsteady from curving forces (*26*). Since passenger comfort constrains MAS, vehicle-based tilting systems were developed to improve comfort, thereby increasing speeds.

Conventional U.S. passenger equipment normally operatea at 3" (76 mm) of underbalance; active-tilt equipment operates at 7" (178 mm). Each imbalance level gives rise to different MAS on the same physical curve. Higher underbalance results in faster rail wear. If railroads qualify equipment at a different underbalance level, MAS is changed on most curves, and affects advertized trip times. Figure 2*d* shows differential MAS (*27*) for United Aircraft's TurboTrain on the Shore Line.

In the U.K., vehicle size, weight, and suspension design allows conventional equipment to operate at 150 mm (5.9") of cant deficiency; at one time Railtrack was exploring operating non-tilt passenger equipment at 165 mm (6.5") of "exceptional" cant deficiency. In Europe, tilting equipment can operate at between 225 mm and 300 mm of underbalance (*28*).

Installed Power

On diesel-powered trains, engine horsepower/power-to-weight ratio directly impacts acceleration rates and MAS. Acceleration rates and loading times make tremendous differences to journey times, particularly on curvaceous routes peppered with PSRs or local services with many stops. If railroads improve locomotive horsepower or assign fewer cars per train, published running times are affected. Indeed, British Rail had timing profiles for each consist type down to whether high-speed trains (HSTs) had seven, eight, or nine coaches, each class of electric locomotives, and trailing tons hauled (Figure 2*j*). When Britain's InterCity 125 HST trainsets were introduced on regional lines with 75–90 mph maximum speeds, trip time reductions of 10–15% were nevertheless achieved (compared with previous coaches and locomotives) due to improved acceleration and braking capabilities.

Braking Rates

Advances in vehicle technology give rise to differential braking rates. Most modern U.S. passenger coaching stock utilizes two disc and two tread brakes per axle; Amtrak's Acela utilizes three disc brakes per axle to achieve 26% better braking rates (*29*).

On HSTs, disc-and-tread combinations delivered 9%*g* (0.88 m/s^2, 2.0 mphps) of service brake, enabling operations at 125 mph on legacy lines equipped with signal systems designed for 100 mph (*31*). This allowed for tremendous time reductions, not only from the 125 mph capability but also reduced deceleration time upon approach to every PSR and station stop. Hydrostatic brakes designed to deliver 12%*g* (12% of acceleration due to gravity, 1.17 m/s^2 or 2.6 mphps) braking rates on British Rail's Advanced Passenger Train were not commercially successful, but electro-pneumatic brakes delivered more predictable and responsive braking performance in suburban multiple-units and heavy freight trains.

Operational Practices

Operating practices also impact published journey times.

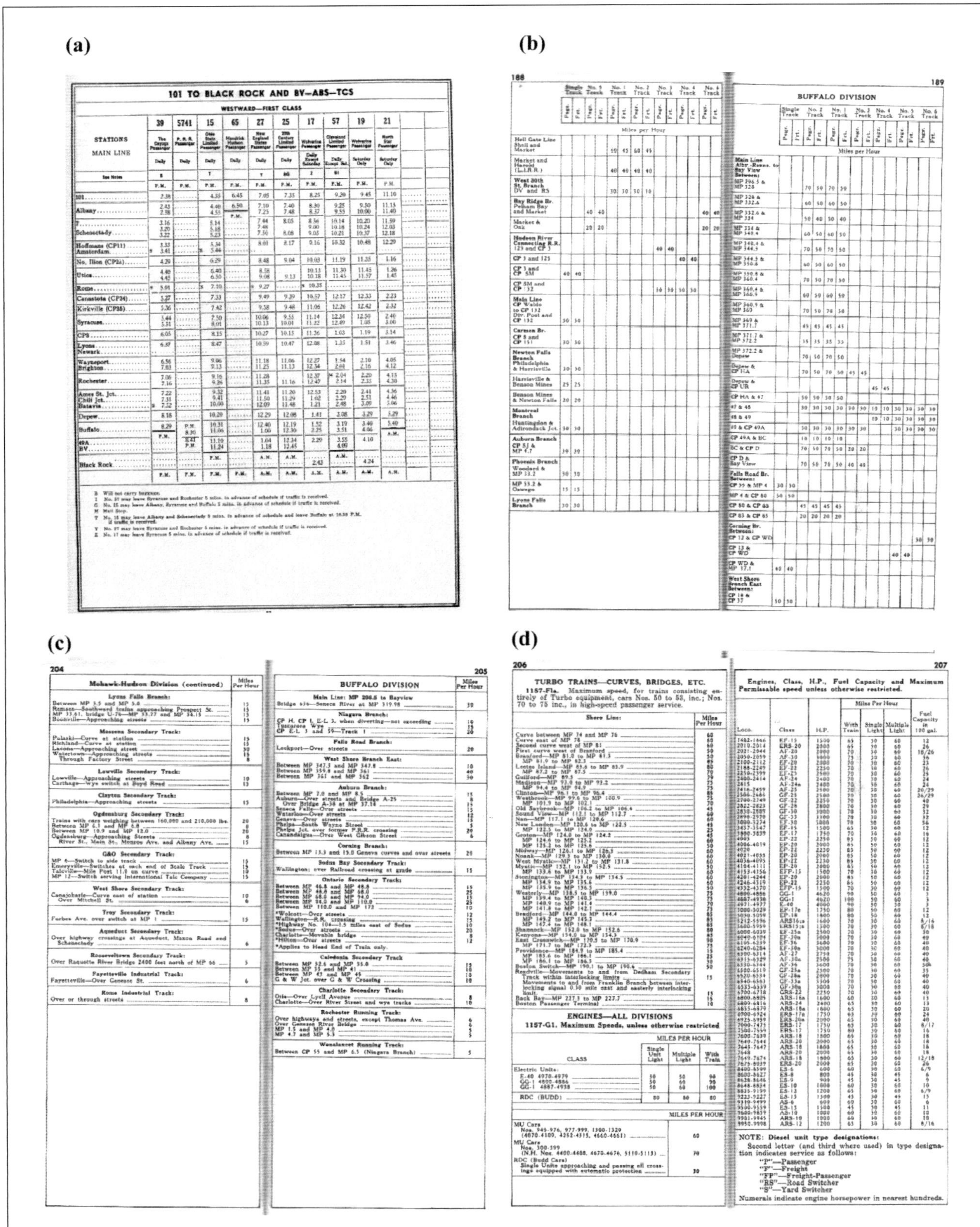

Figure 2. (Continued)

(e)

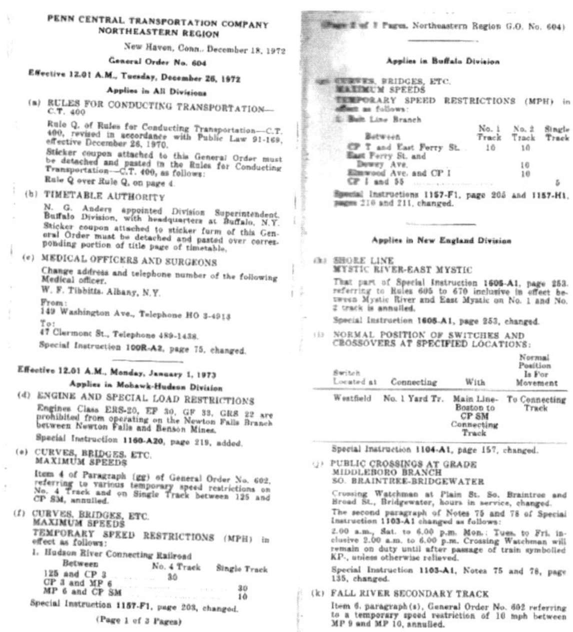

PENN CENTRAL TRANSPORTATION COMPANY
NORTHEASTERN REGION

New Haven, Conn., December 18, 1972

General Order No. 604

Effective 12.01 A.M., Tuesday, December 26, 1972

Applies in All Divisions

(a) RULES FOR CONDUCTING TRANSPORTATION—C.T. 400

Rule Q, of Rules for Conducting Transportation—C.T. 400, revised in accordance with Public Law 91-169, effective December 26, 1970.

Sticker coupon attached to this General Order must be detached and pasted in the Rules for Conducting Transportation—C.T. 400, as follows:

Rule Q over Rule Q, on page 4.

(b) TIMETABLE AUTHORITY

N. G. Anders appointed Division Superintendent, Buffalo Division, with headquarters at Buffalo, N.Y. Sticker coupon attached to sticker form of this General Order must be detached and pasted over corresponding portion of title page of timetable.

(c) MEDICAL OFFICERS AND SURGEONS

Change address and telephone number of the following Medical officer.

W. F. Tibbitts, Albany, N.Y.

From:
149 Washington Ave., Telephone HO 3-4913

To:
47 Clermont St., Telephone 489-1438.

Special Instruction 100R-A2, page 75, changed.

Effective 12.01 A.M., Monday, January 1, 1973

Applies in Mohawk-Hudson Division

(d) ENGINE AND SPECIAL LOAD RESTRICTIONS

Engines Class ERS-20, EF 30, GF 33, GRS 22 are prohibited from operating on the Newton Falls Branch between Newton Falls and Benson Mines.

Special Instruction 1160-A20, page 219, added.

(e) CURVES, BRIDGES, ETC.
MAXIMUM SPEEDS

Item 4 of Paragraph (gg) of General Order No. 602, referring to various temporary speed restrictions on No. 4 Track and on Single Track between 125 and CP SM, annulled.

(f) CURVES, BRIDGES, ETC.
MAXIMUM SPEEDS
TEMPORARY SPEED RESTRICTIONS (MPH) in effect as follows:

1. Hudson River Connecting Railroad

Between	No. 4 Track	Single Track
125 and CP 3	30	
CP 3 and MP 6		30
MP 6 and CP SM		10

Special Instruction 1157-F1, page 203, changed.

(Page 1 of 3 Pages)

(Page 2 of 3 Pages, Northeastern Region G.O. No. 604)

Applies in Buffalo Division

(g) CURVES, BRIDGES, ETC.
MAXIMUM SPEEDS
TEMPORARY SPEED RESTRICTIONS (MPH) in effect as follows:

1. Belt Line Branch

Between	No. 1 Track	No. 2 Track	Single Track
CP T and East Ferry St.	10	10	
East Ferry St. and Dewey Ave.		10	
Elmwood Ave. and CP I		10	
CP I and 55			5

Special Instructions 1157-F1, page 205 and 1157-H1, pages 210 and 211, changed.

Applies in New England Division

(h) SHORE LINE
MYSTIC RIVER-EAST MYSTIC

That part of Special Instruction 1605-A1, page 253, referring to Rules 605 to 670 inclusive in effect between Mystic River and East Mystic on No. 1 and No. 2 track is annulled.

Special Instruction 1605-A1, page 253, changed.

(i) NORMAL POSITION OF SWITCHES AND CROSSOVERS AT SPECIFIED LOCATIONS:

Switch Located at	Connecting	With	Normal Position Is For Movement
Westfield	No. 1 Yard Tr.	Main Line-Boston to CP SM Connecting Track	To Connecting Track

Special Instruction 1104-A1, page 157, changed.

(j) PUBLIC CROSSINGS AT GRADE
MIDDLEBORO BRANCH
SO. BRAINTREE-BRIDGEWATER

Crossing Watchman at Plain St. So. Braintree and Broad St., Bridgewater, hours in service, changed.

The second paragraph of Notes 75 and 78 of Special Instruction 1103-A1 changed as follows:

2.00 a.m., Sat. to 6.00 p.m. Mon.; Tues. to Fri. inclusive 2.00 a.m. to 6.00 p.m. Crossing Watchman will remain on duty until after passage of train symbolled KP-, unless otherwise relieved.

Special Instruction 1103-A1, Notes 75 and 78, page 135, changed.

(k) FALL RIVER SECONDARY TRACK

Item 6, paragraph (s), General Order No. 602 referring to a temporary speed restriction of 10 mph between MP 9 and MP 10, annulled.

(f)

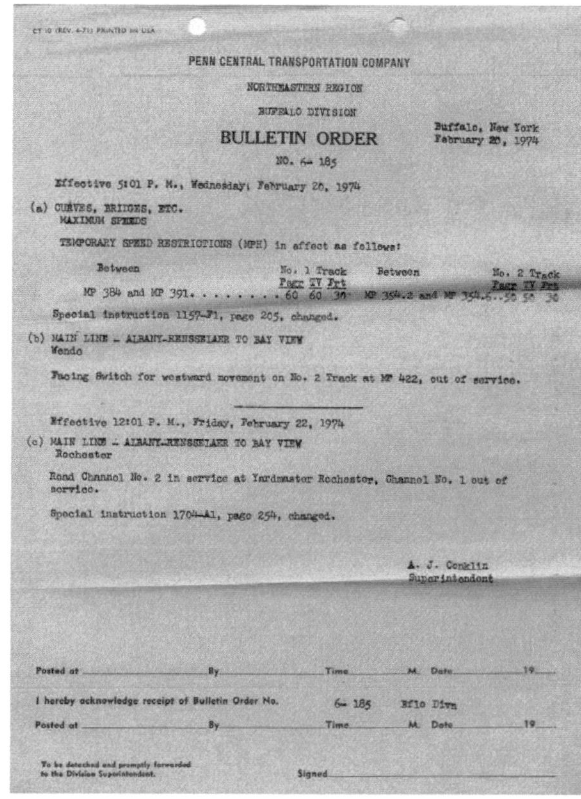

CT 10 (REV. 4-71) PRINTED IN USA

PENN CENTRAL TRANSPORTATION COMPANY
NORTHEASTERN REGION
BUFFALO DIVISION

BULLETIN ORDER

NO. 6-185

Buffalo, New York
February 20, 1974

Effective 5:01 P. M., Wednesday, February 20, 1974

(a) CURVES, BRIDGES, ETC.
MAXIMUM SPEEDS

TEMPORARY SPEED RESTRICTIONS (MPH) in affect as follows:

Between	No. 1 Track Pagr	No. 1 Track TV	No. 1 Track Frt	Between	No. 2 Track Pagr	No. 2 Track TV	No. 2 Track Frt
MP 384 and MP 391	60	60	30	MP 354.2 and MP 354.6	50	50	30

Special instruction 1157-F1, page 205, changed.

(b) MAIN LINE - ALBANY-RENSSELAER TO BAY VIEW
Wendo

Facing Switch for westward movement on No. 2 Track at MP 422, out of service.

Effective 12:01 P. M., Friday, February 22, 1974

(c) MAIN LINE - ALBANY-RENSSELAER TO BAY VIEW
Rochester

Road Channel No. 2 in service at Yardmaster Rochester, Channel No. 1 out of service.

Special instruction 1704-A1, page 254, changed.

A. J. Conklin
Superintendent

Posted at ________ By ________ Time ____ M. Date ________ 19__

I hereby acknowledge receipt of Bulletin Order No. 6-185 Bflo Divn

Posted at ________ By ________ Time ____ M. Date ________ 19__

To be detached and promptly forwarded to the Division Superintendent. Signed ________

(g)

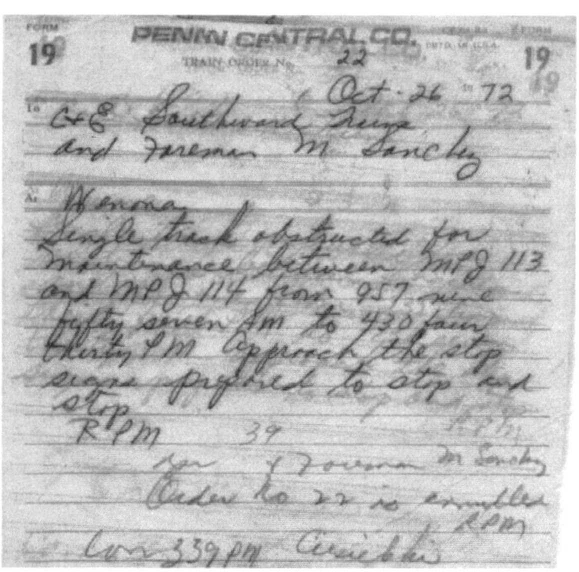

FORM 19
PENN CENTRAL CO.
TRAIN ORDER No. 22
19

Oct. 26 '72

To C+E Southbound Trains and Foreman M. Sanchez

At Wenona

Single track obstructed for maintenance between MPJ 113 and MPJ 114 from 957 nine fifty seven Am to 430 four thirty PM Approach the stop signs prepared to stop and stop.

RPM 39

Foreman M. Sanchez

Order No 22 is complied RPM

Com 339 PM

(h)

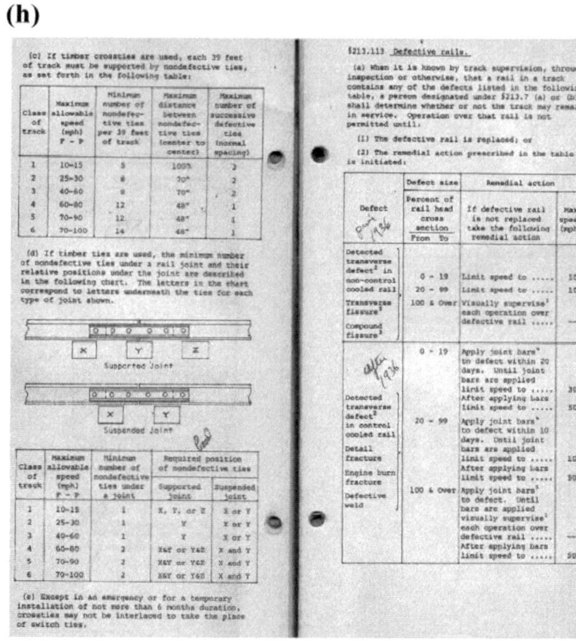

(c) If timber crossties are used, each 39 feet of track must be supported by nondefective ties, as set forth in the following table:

Class of track	Maximum allowable speed (mph) F - P	Minimum number of nondefective ties per 39 feet of track	Maximum distance between nondefective ties (center to center)	Maximum number of successive defective ties (normal spacing)
1	10-15	5	100"	3
2	25-30	8	70"	2
3	40-60	8	70"	2
4	60-80	12	48"	1
5	70-90	12	48"	1
6	70-100	14	48"	1

(d) If timber ties are used, the minimum number of nondefective ties under a rail joint and their relative positions under the joint are described in the following chart. The letters in the chart correspond to letters underneath the ties for each type of joint shown.

X Y Z
Supported Joint

X Y
Suspended Joint

Class of track	Maximum allowable speed (mph) F - P	Minimum number of nondefective ties under a joint	Required position of nondefective ties: Supported joint	Required position of nondefective ties: Suspended joint
1	10-15	1	X, Y, or Z	X or Y
2	25-30	1	Y	X or Y
3	40-60	1	Y	X or Y
4	60-80	2	X&Y or Y&Z	X and Y
5	70-90	2	X&Y or Y&Z	X and Y
6	70-100	2	X&Y or Y&Z	X and Y

(e) Except in an emergency or for a temporary installation of not more than 6 months duration, crossties may not be interlaced to take the place of switch ties.

§213.113 Defective rails.

(a) When it is known by track supervision, through inspection or otherwise, that a rail in a track contains any of the defects listed in the following table, a person designated under §213.7 (a) or (b) shall determine whether or not the track may remain in service. Operation over that rail is not permitted until:

(1) The defective rail is replaced; or

(2) The remedial action prescribed in the table is initiated:

Defect	Defect size: Percent of rail head cross section (From – To)	Remedial action: If defective rail is not replaced take the following remedial action	Max. speed (mph)
Detected transverse defect[2] in non-control cooled rail; Transverse fissure[3]; Compound fissure[3]	0 - 19	Limit speed to	15
	20 - 99	Limit speed to	10
	100 & Over	Visually supervise[1] each operation over defective rail	—
Detected transverse defect[2] in control cooled rail; Detail fracture; Engine burn fracture; Defective weld	0 - 19	Apply joint bars[4] to defect within 20 days. Until joint bars are applied limit speed to	30
		After applying bars limit speed to	50
	20 - 99	Apply joint bars[4] to defect within 10 days. Until joint bars are applied limit speed to	10
		After applying bars limit speed to	50
	100 & Over	Apply joint bars[5] to defect. Until bars are applied visually supervise[1] each operation over defective rail	—
		After applying bars limit speed to	50

Figure 2. (Continued)

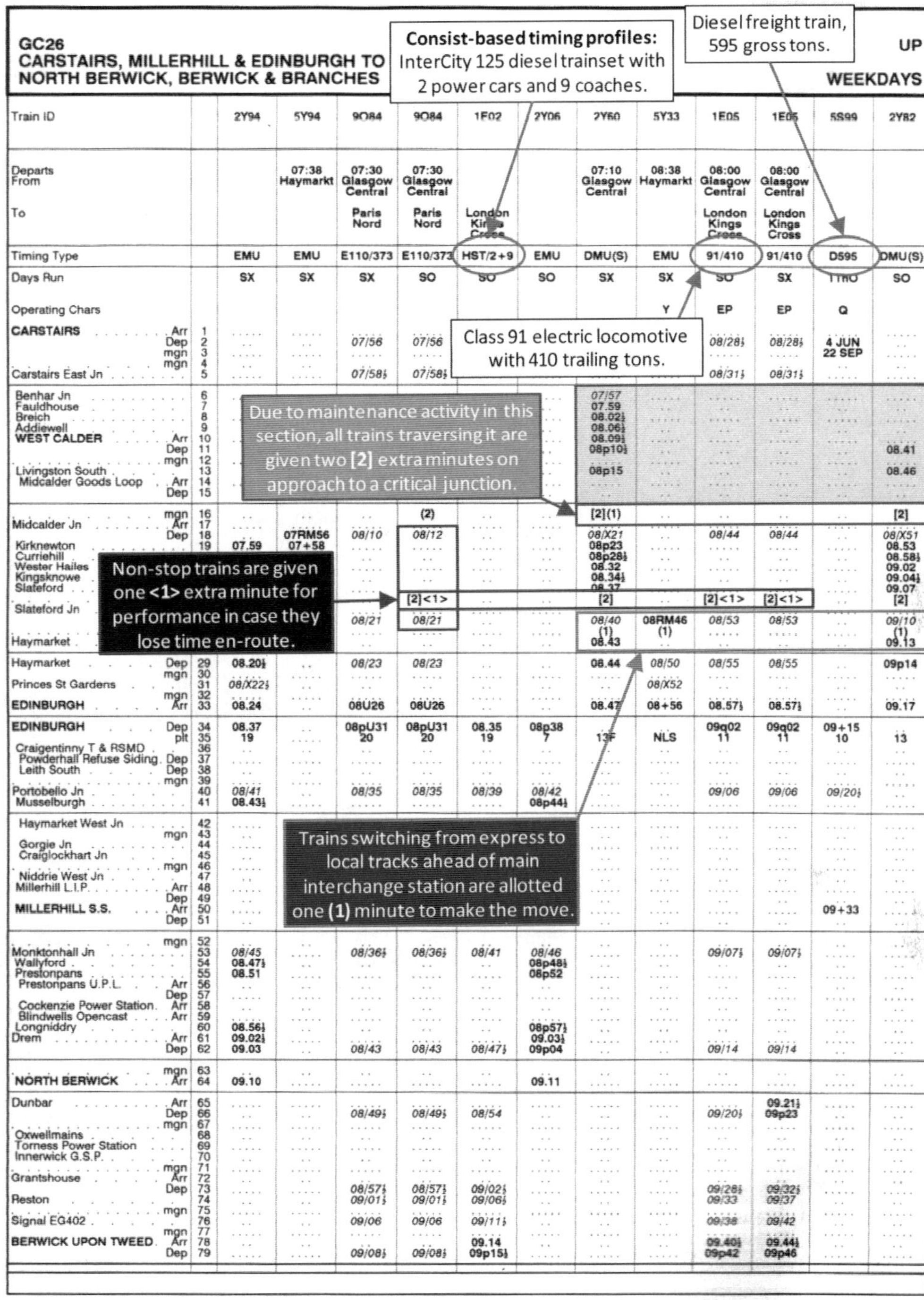

(j)

Figure 2. Operating documents relating to railway speed restrictions. Permanent speed restrictions: (*a*) schedule pages from New York Central Railroad Timetable Mohawk-Syracuse-Buffalo Divisions No. 19, p. 53, effective 1966-04-24; (*b*) Penn Central Transportation Company (PCRR) Northeastern Region Timetable No. 6, pp. 188-189, effective 1972-10-29—Line Speeds; (*c*) Civil Speed Restrictions; (*d*) Equipment Restrictions. Temporary speed restrictions: (*e*) Penn Central Transportation Company (PCRR) Northeastern Region General Order No. 604, effective 1972-12-26; (*f*) PCRR Northeastern Region, Buffalo Division Bulletin Order No. 6-185, effective 1974-02-20; (*g*) PCRR Train Order ("Form 19") advising southbound trains of a track obstruction at Milepost 113-114; (*h*) PCRR Manual for Construction and Maintenance of Track (MW-4) detailing speed restrictions for each type of track defect. Schedule detail resulting from permanent and temporary speed restrictions: (*j*) Railtrack Working Timetable showing use of Square, Circle, and Triangle times and Consist-based Timing Loads on the East Coast Mainline in Scotland.

Rules or Staffing-Based Delays

Changed operating rules or staffing requiring existing infrastructure to be operated differently (typically for increased safety margin) can lengthen trip times. Examples are: (a) some railroads require trains to hold outside station if another train is already platformed (especially at low platforms), resulting in running time degradation for trains in the opposite direction; (b) reduction in yard personnel leads to more through trains stopping to operate hand-throw switches; (c) reduction in trainman or station positions leads to longer intermediate station dwell times; (d) reduction in towerman positions or hours leads to reduced en-route flexibility, making it difficult to pass slower trains or requiring routing via slower tracks.

Work Zone Delays

Track conditions affect speeds, but so do track maintenance activities. As train volumes increase due to growth or traffic consolidation from duplicate lines, maintenance windows became narrower, making it difficult to perform work effectively and clear up prior to each train's arrival. It is now common for express trains to stop on the mainline while track gangs clear up, and then pass the work site at reduced speeds. Delays from "slow zones" can be so severe that railroads issue special construction timetables (Figure 1*d*) reflecting extended timings.

Single Tracking Delays

On double-track lines, major projects require trains to be single-tracked past the work site when one track is out of service. Schedules are written such that opposing trains do not meet near single-tracked sections, so trains do not have to wait to clear. However, moves through interlockings (at either end) are made at reduced speeds; in densely trafficked areas several minutes' wait for opposing trains to clear is inevitable. If track rehabilitation projects are planned in advance, these delays are written into published schedules.

Approach Control Delays

When remotely-controlled switches are thrown from high-speed routes to slower routes, most signal systems display restrictive aspects to approaching traffic, to enforce speeds through junctions, even though blocks ahead maybe unoccupied (*31*, *32*). In congested locales, this can add minutes (colloquially, "burn time") to running times. When routing plans are modified, scheduled times can change substantially.

Train Slot Delays

At junctions or single-track territory, trains are planned to operate within designated timetable slots to avoid occupying key interlockings simultaneously. Slots do not always line up with unconstrained running time between them, thus extra time waiting for slots is written into timetables (see "Circle Time" discussion, below).

Maximum Authorized Speeds

Operationally speaking, railroad MAS is generally given by the Employee Timetable. The Timetable is a living document issued to operating employees, periodically updated by proper authority. It is not the only document that governs MAS for a given location at a given time. It is said that, "MAS is a long math problem where the answer is a single number." Although speeds are shown on other sources like track charts, diagrams, etc., speeds shown therein are necessarily simplified, and not authoritative. For simplicity, typical practice in Northeastern U.S.A. is discussed here.

Hierarchy of Documents

Employees in charge of trains (or trackcars) must be familiar with all operating documents. In descending order of generality, these are:

1. Operating Rule Book
2. Employee Timetable: Schedules & Special Instructions
3. General Orders
4. Bulletin Orders
5. Train Orders (issued by Dispatchers)

Some railroads also utilize General Notices and Operating Notices, of informational nature.

Operating Rule Book

Rule Books provide instructions governing all aspects of operations and define how other documentation should be interpreted. All rules therein apply unless superseded explicitly by more specific instructions. MAS are not discussed here, because speeds are necessarily specific to one location. Geographically specific information is given in the Timetable.

Employee Timetable—Schedules

The Employee Timetable contains the authority for train movements under "timetable and train order operation." Station pages define station locations. Schedule pages (Figure 2*a*) show authority to occupy track by each train at each location at specific times, subject to superiority of trains. Although Employee Timetables have more timing points (e.g. interlockings, employee stops) than shown in public schedules, they do not typically provide rationales for timing, thus they are only a little better than public schedules for understanding why running times change. City-to-city running times divided by rail mileage are not good indicators of MAS, or even average speeds while underway.

Employee Timetable—Special Instructions

Special Instructions are where MAS are defined. MAS are defined under three headings (*27*):

- Maximum Speeds Unless Otherwise Specified (i.e. "Line Speeds", Figure 2*b*),
- Permanent Speed Restrictions—Curves & Bridges (also "Civil Restrictions", Figure 2*c*), and
- Special Maximum Speeds (more specifically, "Equipment Restrictions," relating to train equipment operated, Figure 2*d*).

To find MAS for a specific location, first look up Line Speeds, then check Civil Restrictions and lower MAS if necessary, and finally determine Equipment Restrictions—including blanket and location-specific restrictions. The MAS thus determined is the base under normal operations, to which time-specific conditions may be applied.

General Orders

General Orders (GO) are periodically issued documents that make permanent changes to the Operating Rule Book, Employee Timetable, or Timetable Special Instructions under the authority of the Superintendent of Operations. GO (Figure 2*e*) may be issued in sticker form; employees are expected to moisten and paste over relevant timetable pages (like stamps), such that new information covers over superseded information. GOs may also be issued in loose-leaf format, replacing superseded pages.

When the Engineering Department makes permanent changes to infrastructure, resulting in MAS changes or new infrastructure being put "in service," these are described in GOs (*34*). If TSRs are in effect for prolonged periods, the Superintendent may elect to print TSRs in GOs.

GOs could help track the general state of railroad maintenance, because they contain a partial history of TSRs. When analyzing GOs, each and every single one must be obtained. They are issued with sequence number and effective date, so completeness of the records is easily determined. Each GO is issued with a summary page describing changes made; thus a reasonably complete history (and corresponding impacts on journey times) is usually readily reconstructable.

Bulletin Orders

Bulletin Orders (BOs, e.g. Figure 2*f*) are frequently issued documents making temporary changes to operating documents. Most railroads have a daily BO issued under different names, e.g. Daily Train Operations BO, TSR Bulletin, etc. This document is where most TSR information could be found.

However, a TSR by definition occurs due to temporary conditions, and thus should not appear as permanent changes to operating documents. Certain TSRs, for example, allowing tracks to settle for 72 hours after tamping, are lifted in three days. Others, like "mud spot" conditions, might persist for several weeks until track gangs can effect repairs. Other TSRs could be associated with poor tie conditions being unable to hold gauge, leading to downgraded track class, which might not be repairable until rotted ties are replaced by traveling production gangs and hence could persist for months. These might stay in the BO and never make it into GOs, as long as authorities intend to capitally rebuild the track when scheduled.

BOs carry information on working limits, requiring trains to contact track foremen to obtain permission through work sites. This generates delays and prescribes an MAS past the work site, although it is technically a signal speed, and not a TSR (Working Limit Stop Sign is considered a "signal.")

BO data is perhaps the best resource to understand the state of maintenance relating to MAS and consequently published speeds. The challenge is that BOs are temporary changes and therefore are never pasted into timetable books. Although operating employees are required to have all BOs in effect on their person while on duty, when they are superseded by Summary BOs, GOs, or Timetable Reprints, all superseded BOs are typically discarded to avoid confusion. Therefore, the railroad state of maintenance is ephemeral information: unless you were there at the time, it is highly unlikely that a reliable, accurate, and complete account of factors affecting travel speeds can be reconstructed.

Despite computerization, records that change daily are often not retained electronically for long periods, because business justification for their retention usually is not there. It is said, "In Operations, you're only as good as your last rush hour; nobody remembers great work you did last Thanksgiving."

BOs also promulgate train schedule revisions, typically for special movements—when necessary to convey to all operating employees one-day-only schedule, e.g., for circus trains. Normally, schedules (and consequently advertized times) are not modified in response to TSRs, even for those lasting several months. It is therefore difficult to infer state of maintenance using public running times.

BOs identify general speed restrictions in effect for one-day only, for example, Heat Restriction, or Reduced Rail Adhesion. Most railroads utilizing CWR issue speed restrictions on exceptionally hot days to guard against risks from track buckling (e.g. 20 mph MAS reduction when highest forecast temperature is >85°F or a 25°F change in temperature (*34*), although cut-off and restriction severity varies by railroad and region).

Train Orders

Train Orders (TOs), for example, “Form D,” “Form M,” or “Form 19” (Figure 2*g*), are issued to specific trains and may contain instructions superseding anything discussed above. Typical uses include:

- **As Track Warrants to convey movement authority**, e.g. when a superior train is running so late that the Dispatcher allows an opposing inferior train to make progress down the line and meet it at a different siding. Dispatcher issues TOs to both trains superseding timetable authority, instructing superior train to hold at a non-scheduled location, and inferior train to meet it there.
- **To Remove Track or Signal System from Service**, e.g. when Maintenance Foremen require exclusive use of blocks to effect repairs.
- **To Issue Emergency Speed Restrictions (ESRs)**, e.g. when routine track patrols find defects like broken joint bars, heat kinks, or pull-aparts such that, to reduce risk of derailment, trains must travel at lower than MAS. Engineering documents like MW-4 (Figure 2*h*) normally prescribe train speeds over each defect type.

Unfortunately, TOs containing ESR records are also perishable, most railroads requiring retention for only one to seven days. TO is also not the only way to communicate ESRs to train crews. Some Northeastern railroads allows daily BOs to be amended via radio with additional speed restrictions. Even if historical TOs were available, at best they represent an incomplete record of speed restrictions.

Computerized records seldom deliver sufficient detail for useful forensics. Centralized Traffic Control (CTC) playback “tapes” or recorded radio voice records are typically expunged after 2–4 weeks, and systems are normally designed for specific known-time event investigation access, rather than extensive data trending analysis.

Track defect prevalence can correlate with state of maintenance (but it is not an absolute indicator, because maintenance strategy can also drive this number). Broken joint bars rarely occur if track structure is properly supported. Heat kinks are unusual with correctly adjusted CWR, properly ballasted shoulders, and functioning fastening systems. Pull-aparts can occur during extremely cold weather, particularly at spots with internal rail defects, head cracks, engine burns, or defective welds; they are minimized by frequent rail grinding and prompt rail repair. Although track inspection records must be kept pursuant to 49 CFR 213.241 and are subject to inspection by the FRA, they are not generally available and are expunged after several years.

Defining Public Timetable Running Times

To understand published journey times, it is worth pondering how train schedulers arrive at trip times from the constraints discussed above. For clarity, British practice is discussed here. North American passenger schedulers utilize similar concepts (e.g. [*6*]), but no industry standard exists.

Train Performance Calculator (TPC)

A train performance calculator (TPC) models locomotive performance over specified terrain with defined speed restrictions. Basic inputs are: physical characteristics (speed restrictions, gradients, curvature, station locations), traction characteristics (wheel adhesion, horsepower, braking rate), and load characteristics (tonnage, load, rolling resistance, wind resistance, etc.). The output is unconstrained running time—before considering station dwell times, congestion, signal checks, TSRs, etc. (i.e. theoretical minimum under ideal conditions). Normally, four types of sectional running times (SRT) are given for each station-to-station segment: stop-to-stop, stop-to-pass, pass-to-stop, and pass-to-pass. “Stop-to-pass” means train stops at first station but by-passes second station; acceleration from standing stop at first station is included, required braking to stop when arriving at second station is excluded (but required braking to comply with any PSRs in second station’s limits is included).

TPC models are provided by industry vendors (*35*) and typically validated by running time checks requiring operations of special test trains. To build speed profiles, schedulers traditionally utilize onboard speedometers and record mph readings every few seconds. To understand traction output, another scheduler records throttle and brake positions and times actuated. (Today, this work is performed with computerized dataloggers.) Schedulers also note any signal checks or unusual conditions. Speed profiles and control manipulations are compared with TPC outputs and internal parameters are tweaked until the model reproduces recorded traction performance under various conditions.

Square, Circle, and Triangle Times

Building a public schedule from raw running times necessitates addition of extra allowances to ensure reliability and achievability of train operations under field conditions (i.e. to make schedules “robust”). The three types of en-route time adjustments are:

- **Engineering Adjustment [Square Time]** or “Recovery Time”: Extra minutes or fractional minutes inserted wherever major track engineering work or TSRs are in effect. They are location-specific and typically derived by comparing TPC runs under unconstrained condition with TPC runs with TSRs in force.
- **Routing Time (Circle Time)** or “Pathing Time”: Extra time inserted upon approach to major junctions, to absorb delays resulting from signal checks, waiting for conflicting movements to clear, or as insurance against missing critical slots at busy junctions. They

are junction-and-time specific and derived from inspection of stringline charts and junction utilization/clearing times. They are also inserted for trains running on close headways where a following train may operate for long stretches at less than MAS due to adverse signal indications (called "running on double yellows"). Accuracy and adequacy is critical, as trains missing assigned slots on busy lines can have knock-on effects far beyond the immediate locale.

- **Performance Allowance <Triangle Time>** or "Pad": Time inserted upon approach to major stations whose purpose is to absorb unanticipated en-route delays not explicitly allowed for in Square and Circle times, like individual variations in driver performance (i.e. operating techniques and skills of different locomotive engineers). For example, experienced operators typically shut off power and coast at an earlier point if approaching major terminals with allowance to spare.

Published departure-to-arrival times are the sum of all appropriate station-to-station times plus any Square, Circle, and Triangle times inserted en route (Figure 2*j*). Taken together, en-route adjustments should account for all operational and infrastructure factors discussed in this paper's first half. It is impossible to back-out en-route adjustments from public schedules without access to rail industry documentation. Fortunately, in the U.K., Network Rail's Working Timetables (WTT) are published for all to see (*37*). However, the U.S. passenger rail industry does not generally publish en-route adjustments in schedule documents, not even in Employee Timetables. Adjustments are typically domains of railroad scheduling departments and are known under names like "schedule skeleton." In some cases, these bases for running times are carried in individual section schedulers' heads and never written down.

Understanding scheduling rationale and the reasoning behind each adjustment requires either interviewing the section scheduler when the timetable was written, or determining all contemporaneous constraints schedulers should have taken into consideration when determining the published running time.

Station Dwell Times

En-route adjustments do not cover the time required for station work, typically defined as the period between the time when a train is fully berthed in the station (i.e. wheels stop moving), to the time when the doors are closed and "okay to go" is given.

Dwell times are open to debate even amongst train schedulers because they can include time required for passengers to make transfer connections, to service trains at major terminals (typically, food and baggage, more rarely, fuel, water, change of gauge, change of host railroad, etc.), for hand-off between incoming and relief crews, and for combining or separating a train's different sections. On commuter lines it also accounts for platform congestion and passengers holding doors. On mixed freight-and-passenger trains it can include required time for switching freight cars. Frailey (*37*) discusses North American passenger train switching operations extensively, including W.E. Deming's landmark study for the Burlington Northern.

Fortunately, station times are easier to understand; they are often published in public timetables for major terminals. However, suburban carriers do not typically publish dwell times at intermediate stops and may even have flag stops or situations (e.g. evening outbound drop-offs) where trains are permitted to operate ahead of published schedule. They can also be hidden, as mixed freight-and-passenger trains can make freight stops with unpublicized dwell times, and trains often have unadvertized operational stops requiring dwell times—even on commuter lines, employee stops or time required to receive TOs are omitted from public schedules.

Forensics of Running Time Degradations

Research has thus far established that modifications in public timetable running times result from changes taking place in these general categories:

- Track design geometry, maintenance, layout, and special work
- Infrastructure factors like structures, signals, power, and grade crossings
- Wheel–rail interaction issues like cant deficiency
- Vehicle factors like tractive effort and braking rates
- Operating rules, practices, and ongoing projects to rehabilitate infrastructure
- MAS per Timetable (from above constraints)
- Scheduler's decisions:
  1. Train performance modeling
  2. En-route adjustments
  3. Station dwell times

It follows, therefore, that to fully understand *What Happened to Speed* (*7*), specifics of what changed in each category during the study period for the services in question should be examined. Although public timetables offer a general sense of trends in service offerings, they cannot tell us whether these changes resulted from schedulers' decisions (for either marketing or operational reasons), deterioration of physical infrastructure, ongoing projects to restore the plant, or permanent system upgrades. Indeed some factors counteract one another, or indicate completely opposite industry trends. For instance, deferred track maintenance and increases in track rehabilitation both lead to longer journey times, and passengers would be none the wiser.

It is possible for published journey times to remain static after substantial upgrades that increased MAS, if schedulers elect to utilize the upgraded capabilities to improve reliability or provide additional dwell time for connections at intermediate stations. Trip time capabilities may be improved even while physical plant is deteriorating by changing operating practices, or reducing congestion by reprioritizing and canceling trains.

Although advertized trip times and the industry's health may not be all that well correlated, longitudinal analysis of journey times can be a high-level screening tool to identify corridors of interest where in-depth forensic analysis could reveal reasons behind running time changes. Secondary sources written by contemporary observers (e.g. [*38*]) can sometimes provide helpful insight into management actions and decision rationales.

This Happened to Speed: Case Study of Penn Central's Mohawk and Buffalo Divisions

Exploring this hypothesis a little further, we examine a case study utilizing some sources discussed above. It provides a more specific (but still incomplete) explanation as to *why* public timetables were revised during the study timeframe. Limited to the former Penn Central Northeastern Region, this methodology can be applied to any corridor of interest.

When correlating passenger corridor performance with infrastructure investment, three dimensions should be considered: service speed, capacity (trains per day operated at design speed), and reliability (probability that planned timing is actually achieved). In the service design of mixed traffic corridors, published service speeds often result from trade-offs along these dimensions. MAS and resulting SRTs are practically the only elements where infrastructure owners exercise complete control.

In 1966 the *20th Century Limited* (16:00) ran 14% faster than the *Chicagoan* (18:30), chiefly because of shorter scheduled dwell times (0:35 versus 1:48) at intermediate stations. Today's westbound *Lake Shore* (19:05) has less dwell time (1:22) built into its schedule, even when marshalling time at Albany is included, but nonetheless takes 3% longer than the *Chicagoan* in 1966.

Scheduled Dwell Time

Figure 3*a* shows dwell times in the Penn Central Northeastern Region of westbound New York–Chicago trains leaving Grand Central Terminal in late evening (*Chicagoan*, PC *#63*, *Lake Shore*), collated from contemporary public and employee timetables (*27*, *39–42*). A few unpublished times were estimated. Beginning in 1961, dwell times mushroomed, especially during the Penn Central era. Amtrak stopped this in 1971, "making trains worth riding again." Dwell time growth began again at Albany in 1979; however, it was minimized at other locations.

Figure 3*b* shows that NY Central assigned dwell times quite deliberately and methodically, with overnight trains having longer dwell times. In 1963, all trains had extra time at Buffalo, even important trains like *20th Century Limited* were booked for 11 minutes. Significant dwell time was also included at intermediate points like Utica and Rochester.

Dwell times have critical impacts on train operations. When delays occur outside the operator's control (e.g. weather, passenger action, etc.), extra dwell time can help absorb impacts. Typically, schedulers use average running time, but recovery time at major terminals can be derived from 95th-percentile time. Departing westbound, by maximizing the probability of leaving Buffalo on-time (95th-percentile ensures trains leaving Buffalo have only 5% chance of lateness), it minimizes downstream impacts by not having trains operate out of slot.

Amtrak added extra recovery time at stations where host railroad change is necessary (e.g. Cleveland), and at crew change points (e.g. Toledo), to allow time required for coordination between different dispatching offices and relief paperwork.

Sectional Running Times

Figure 3*c* presents a rather complex picture of those same trains' SRTs. Relative periods of stability existed 1983–1997, but gentle upwards trends nonetheless existed, likely contributing to the perception that trains were getting slower. Marked deterioration occurred 1968–1971, adding 35 minutes (12%) of running time. Another turbulent period was 1999–2003, likely associated with the Conrail split when major operational changes occurred on the Chicago Line. As passenger train performance became political at the Federal level circa 2008, public timetables became more of a contractual commitment, rather than a reflection of infrastructure capabilities or a quantitative description of operational intent.

During 1976–1980 SRTs were abnormally long and likely not entirely due to deterioration of infrastructure. The relevant special instruction (*40*) states:

> AMTRAK Engines, Class SDP-40F, in number series 540 to 649, are restricted as follows—trains with one SDP-40F Unit alone [...] must not exceed 40 MPH on curves of 1 degree 30 minutes or greater. (PCRR Rule 1157-G1b)

It then lists 45 and 22 such curves in the Mohawk and Buffalo Divisions, respectively. These 67 severe speed restrictions applying to passenger trains partly explain increased running times.

Interestingly, the introduction of RoadRailers in 1993, and the cancelation of Amtrak Mail in 2004 and ExpressTrak in 2006, did not have noticable impacts on scheduled times—at

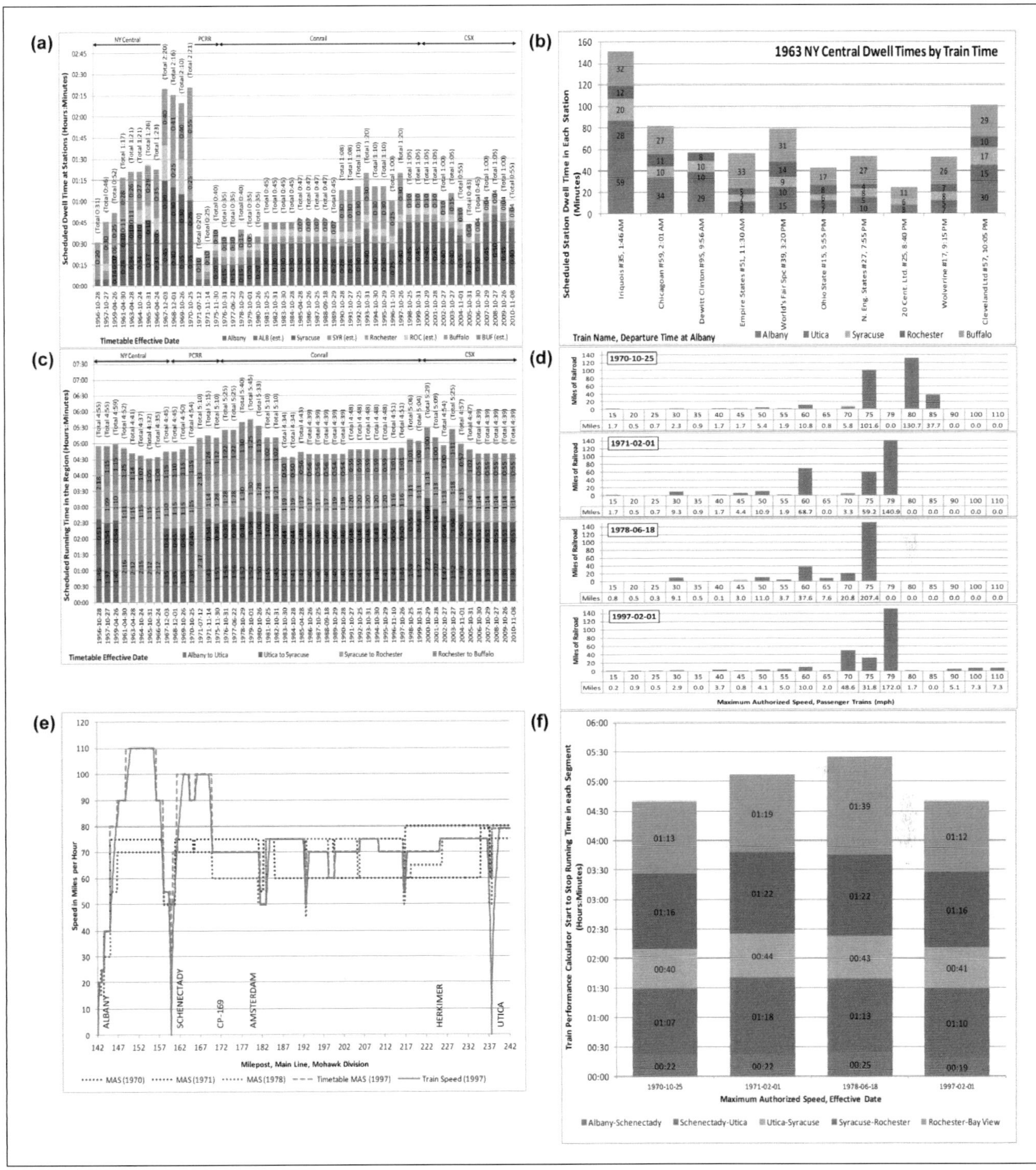

Figure 3. Longitudinal analysis of schedules and speeds on the Penn Central Northeastern Region, Mohawk and Buffalo Divisions, 1956–2010: (*a*) *Lake Shore Limited* dwell time analysis; (*b*) 1963 NY Central Railroad dwell time by train; (*c*) sectional running time analysis; (*d*) maximum authorized speeds (MAS); (*e*) train performance curve (TPC) from Albany to Utica; (*f*) minimum achievable running time per simplified train performance calculator.

least not in this Region. Since the mid-2000s, Amtrak has utilized TPC to derive pure running times and has produced standardized schedules. Different categories of scheduled time allowances are explicitly documented internally.

Changes in Infrastructure

Figure 3*d* shows histograms of MAS for passenger trains. In October 1970, despite rail industry issues in the Northeast, significant portions of mainline were available for 75–80 mph operations, with almost 40 miles qualified for 85 mph, primarily west of Seneca River (*27*). However, the infrastructure was degrading fast during the first years of Penn Central's bankruptcy:

> Applies in Buffalo Division: Passenger Trains—79 mph over the entire Division. (PC GO 407 (aa), 1/1/71)

> Intermittent inductive Automatic Train Stop [ATS] System on the entire region, out of service. (PC GO 409 (a), 2/1/71)

ATS was deactivated as a maintenance cost-saving measure. By February 1971, numerous TSRs had noticeably increased Slow orders in 30, 50, and 60 mph categories (*33*). ESRs may have been even more numerous. By 1978, the de-facto speed limit over the entire Region was 75 mph. As infrastructure slowly returned to a state of good repair, speed profiles mostly returned to normal by February 1997, along with substantial new segments of 90, 100, and 110 mph running. Higher-speed running, which requires cab signals and higher superelevation on curves, occurred east of CP-169 where passenger traffic dominates. West of CP-169, where heavy freight trains from Alfred E. Perlman Yard join the mainline, the highest MAS was 79 mph.

Figure 3*e* shows simplified TPC runs using timetable MAS, indicating infrastructure speed capabilities did indeed degrade, beginning in 1971, and reached a low point in 1978, but recovered by 1997. This correlates well with the SRT findings. Running at 110 mph contributed a 3-minute saving between Albany and Schenectady, but Figure 3*f* shows time was lost en route to Utica due to new or more severe restrictions.

Nothing Happened to Speed (At Least Not Around Here)

This is not a definitive history of train speeds in the Mohawk and Buffalo Divisions, but it demonstrates some factors already discussed. Passenger train scheduling is a complex discipline and a multitude of issues are involved, all of which affect public timetable end-to-end trip times. Financial ramifications and difficult public sentiments are consequences of inappropriate advertized running times.

In a study period of over 50 years, running times from Albany to Buffalo have basically remained at about 4 hours 45 minutes, punctuated by periods when specific technical issues (only some of which are known from this analysis) have elongated travel times followed by recovery once issues were addressed. One can either rejoice in the successful achievement of state-of-good-repair, or regret that no true speed improvements were evident from this data.

It's Not About Speed: Case Study of Metro-North's New Haven Line

Scheduling is a delicate balancing act. Here, we get behind the scenes a little to understand decisions and analytics supporting these actions performed every day in railroad scheduling offices nationwide.

The Long View: 1940–2009

Figure 4*a* shows running times for one New Haven Line early morning express train (Grand Central Terminal arrival at ~08:30) over the last 75 years. Journey times did not change from 1940 to 1970, excepting one minor revision in stopping pattern between 1940 and 1955. Ownership and sponsorship of commuter services turned over to the public sector on October 27, 1970, under a purchase-of-service contract, leading to dramatic changes. Train paths formerly serving intercity clientele originating from far as Springfield, Mass. were truncated at New Haven, Conn. and extra stops were added for the benefit of commuters. In 1973, new "Metropolitan" M-2 electric multiple unit equipment with better acceleration and top speed was introduced, together with high-level platforms, which enabled additional stops to be added while trip time was further reduced. Timings remained relatively stable until 2009, although increasing customer demand and expectations for service reliability caused slight upward trends. Nonetheless, express trains were still scheduled 14% faster in 2009 than in 1940.

Service Reliability: 1983–2012

Customer demand and operational achievements for more reliable service is borne out in Figure 4*b*. From 1983–2012, trade-offs moved towards scheduling longer running times to assure better reliability and to account for expected delays due to track congestion. This was a period of tremendous growth in train volumes and ridership, particularly during peak hours at Mott Haven Junction in the Bronx (MO) which came to be operated near design capacity. Despite MO having been rebuilt in a new layout with higher-speed switches and CTC cutover in June 1993, it remained a flat junction. Figure 4*c* shows all trains approaching MO from all three lines during typical morning peak. Higher throughputs required more precision in operations (because each train slot was now open for a shorter time window), which translated into incremental needs for additional recovery time on the approach to MO, adding 2 minutes. Schedulers carefully balanced trip time, capacity, and reliability but it resulted in slightly longer running time.

Impacts of Extraordinary Incidents: 2013

Figure 4*d* shows the impacts of three major incidents in 2013. In May, a joint-bar failure resulted in changes to track inspection procedures, which required more on-track time

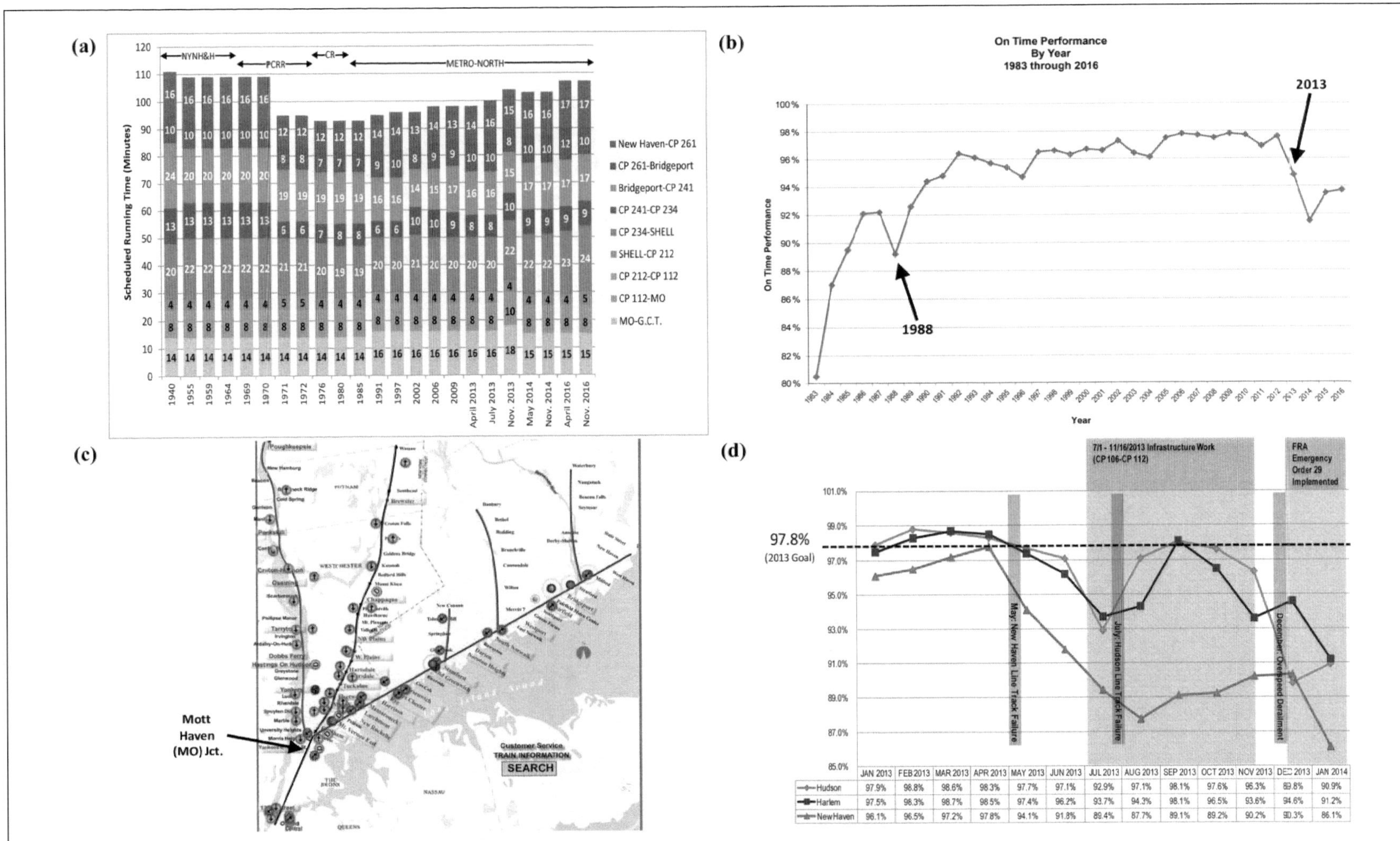

Figure 4. Case study of schedules and on-time performance (OTP) on the Metro-North New Haven Line: (*a*) express train running times by line segment 1940–2016; (*b*) OTP history 1983–2016; (*c*) train tracker showing peak utilization at around 7:45 a.m. weekdays; (*d*) monthly OTP in 2013; (e) "heat maps" showing OTP by train by location in 2013; (*f*) running time revisions 2002–2016 with reasons for changes; (*g*) maximizing system performance requires train operators to balance trip time, capacity, and reliability subject to overarching constraint of safe operations.

(e)

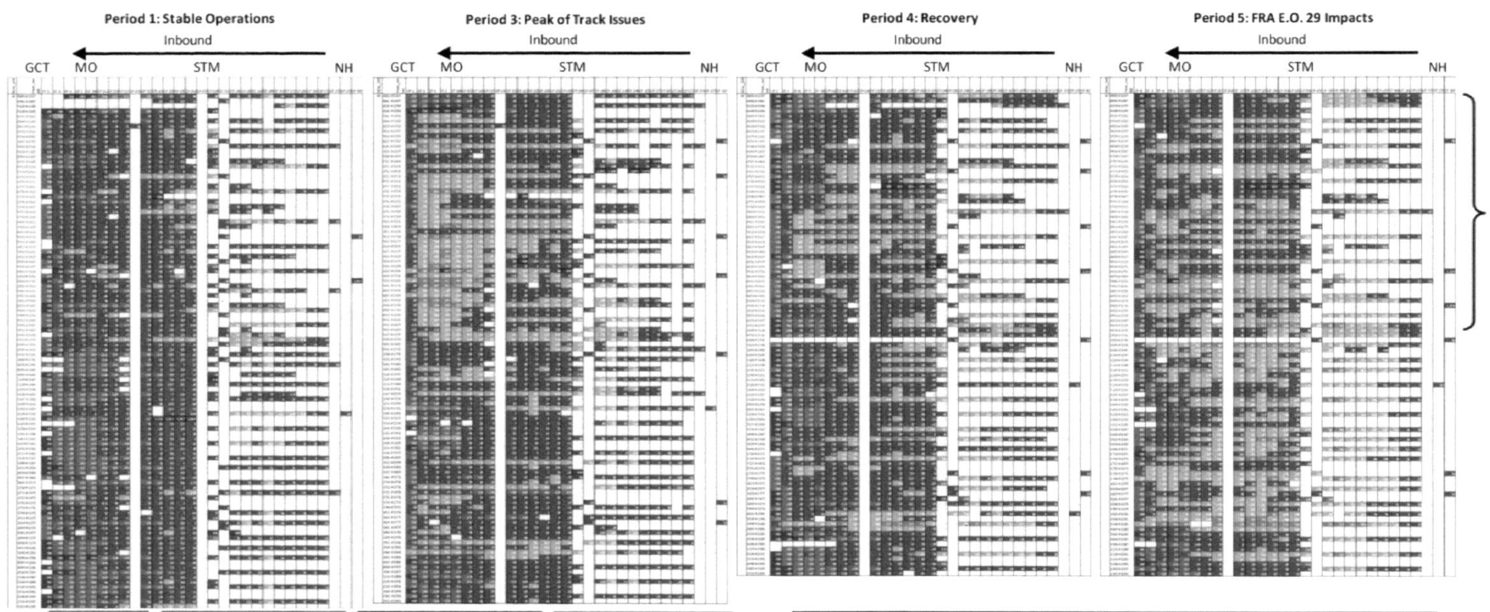

(f)

Fairfield Metro passenger station opened December 5, 2011.

West Haven passenger station opened August 18, 2013. Continuing adjustments reflect fine-tuning based on practical experience.

Cab signal dropdowns provided for MAS reductions of more than 20 mph. Running time added to reflect new infrastructural constraint.

Additional two minutes needed to accommodate Devon Transfer project. Upon project completion, extra time was removed.

Effective Date	10/27/02	10/01/06	10/18/09	04/07/13	07/01/13	11/17/13	05/11/14	11/09/14	04/03/16	10/02/16
Train #	*1527*	*1527*	*1527*	*1527*	*1527*	*1527*	*1523*	*1523*	*1523*	*1523*
N.H.-CP 271	3	3	3	3	3	3	3	3	3	3
CP 271-CP 266	5	5	5	5	(+2) 7	(-1) 6	(+1) 7	7	7	7
CP 266-CP 261	5	(+1) 6	(-1) 5	(+1) 6	6	6	6	6	(-1) 7	7
CP 261-CP 257	5	(+1) 6	6	(+1) 7	7	(-2) 5	(+2) 7	7	(+2) 9	(-2) 7
CP 257-CP 255	4	[illegible]	4	(+1) 5	5	(-1) 4	4	4	4	4
CP 255-CP 248	6	6	6	(+1) 7	7	7	7	7	7	7
CP 248-CP 244	4	4	4	4	4	4	(+1) 5	5	5	5
CP 244-CP 241	3	(+1) 4	(+2) 6	(-3) 3	3	3	(+1) 4	4	4	4
CP 241-CP 234	10	10	(-1) 9	(-1) 8	8	(+2) 10	(-1) 9	9	9	9
CP 234-CP 229	8	(-1) 7	7	7	7	(+1) 8	8	8	(+1) 9	9
CP 229-CP 223	6	6	6	6	6	6	(+1) 7	7	7	7
CP 223-CP 217	7	(-1) 6	6	6	6	(+1) 7	(-1) 6	6	6	(+1) 7
CP 217-CP 215	1	(+1) 2	2	2	2	2	2	2	2	2
CP 215-CP 212	3	3	3	3	3	3	3	3	3	(+1) 4
CP 212-CP 112	2	2	2	2	2	2	2	2	2	2
CP 112-MO	8	8	8	8	8	(+2) 10	(-2) 8	8	8	8
MO-G.C.T.	16	16	16	16	16	(+2) 18	(-3) 15	15	15	15
Total Trip Time	**96**	**98**	**98**	**98**	**100**	**104**	**103**	**103**	**107**	**107**

Running time studies showed that trains are short on time in this area due to increasing congestion from additional movements in Stamford.

Additional time provided to accommodate track remediation work in the Bronx. Removed when project complete.

Time needed for crossing moves due to bridgeplates being needed at local stations to support track and catenary renewal projects requiring continuous track outages.

stable operating environment
(programmed engineering adjustments move with work location)

fast-changing operating environment
(additional repair work occurring in many locations simultaneously)

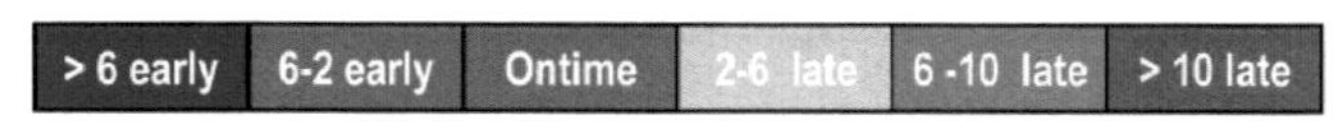

(g)

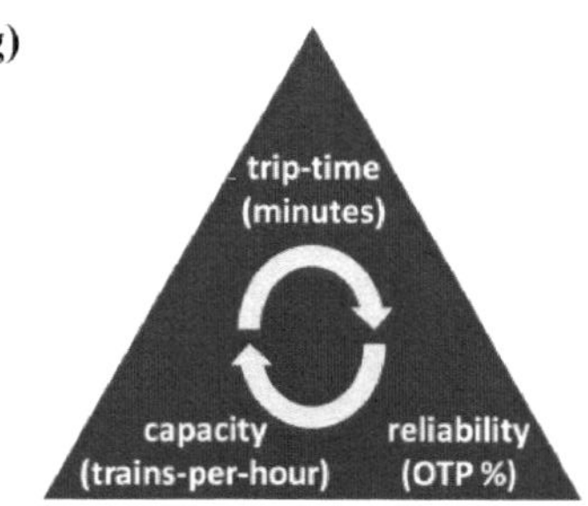

Figure 4. (Continued)

(*11*); in July, a freight-train derailment resulted in an emergency four-month program to eliminate "mud spots" within a busy track section, which introduced TSRs and impacted reliability. In December, an overspeed accident resulted in FRA Emergency Order (EO) 29 requiring a second headend crewmember at braking distance from where MAS decreases by more than 20 mph unless civil speed protection via cab signal modifications was in place.

The impacts of these changes are visualized in Figure 4*e*, a color chart of minimum observed (i.e. best-case achieved) inbound running times. In Period 1 (before May), peak trains had correct SRTs to maintain reliable service. During Period 3, when TSRs and track outages were in effect for infrastructure remediation, normal recovery times were simply insufficient to absorb delays incurred particularly near MO during morning peak hours. There was a brief reprise during Period 4 after work completion, but during Period 5, due to operational changes required by EO 29, actual running times lengthened again. These impacts were not reflected in public timetables due to their varying nature.

Addressing the Maintenance/Operation Balance: 2009–2016

Figure 4*f* shows differential SRT analysis from one control point (CP) to another drawn from recent operating schedules. We discussed the rationale behind each change with section schedulers responsible for the territory. Two changes (green) relate to permanent infrastructure change (two new stations opened); the blue change accounts for EO 29 cab signal modifications to provide civil speed protection; and two changes (orange) relate to color chart work (Figure 4*e*) that identified increasing congestion around Stamford for which insufficient SRT were previously allocated. The majority of trip time increases (red) were due to planned temporary construction conditions, which were restored when maintenance work was complete.

The main reasons behind apparent cumulative increases in trip times were in fact specific changes happening at accelerated pace due to simultaneous infrastructure improvements taking place on the New Haven Line. Many projects—like catenary replacement, track maintenance, bridge reconstruction, drainage improvement, and PTC installation—are absolutely vital to state-of-good-repair and usually invisible to commuter ridership. The combination of construction volume and train frequency simply got to the point where moving the few available minutes of engineering allowance around was insufficient to cover all the work required, resulting in net journey time increases.

The Nuances of Speed on the Railroad

Through technical material presented and case studies, we have seen that right-of-way, track, infrastructure, vehicle characteristics, and operating practices can all affect scheduled times. To properly understand all speed constraints, it is necessary to study timetable special instructions, GOs, BOs, and potentially even specific TOs. When constructing timetables, schedulers add engineering, routing, and performance allowances, and station dwell times to pure run time to arrive at a realistic schedule. Therefore, it is impossible to infer what was happening on the railroad by reading a public timetable.

Speed, Capacity, and Reliability Trade-Off

Figure 4*g* represents one way to consider these issues. To maximize system performance, train operators must balance trip time, capacity, and reliability subject to an overarching constraint of safe operations. Minimizing advertized journey time is no more of an appropriate goal for the passenger rail industry than maximizing tph or OTP. Infrastructure investments, when complete, can typically improve all three variables in absolute terms, but when projects seek to "maximize" one variable, they typically do so at the expense of the other two equally important service attributes.

The authors hope that the complex relationship between published journey time and state of the industry is better understood in the public sphere from considerations outlined in this paper.

Acknowledgments

The authors thank J.G. Allen for providing the impetus for this paper. The authors also acknowledge P. Held, J-P. Fiorentino, M. Corbett, G. Raguseo, J. Durousseau, J. Grumblatt, F. Gimenez, J. Wang, J. Garofalo (Sono Switch Tower Museum), A.H. Dreyer (Railway & Locomotive Historical Society), and F.N. Nowell, III (B&M Historical Society) for their assistance in this research. The authors thank M. Shiffer for permission to use Figure 4*d*; and are indebted to G.E. Hayden and J.E. Kesich for their helpful comments. A. Lu expresses his appreciation to Railtrack Scotland's E. Boyle, B. Gardiner, K. Brown, K. White, and N. Wunsch, and ScotRail's A. Rattary and F. Grey for their teaching and mentoring on the art of journey-time reduction and scheduling.

Author Contributions

The authors confirm contribution to the paper as follows: study conception and design: Lu; data collection: Lu; analysis and interpretation of results: Lu, Fahey, Kozlowski, Whitman; draft manuscript preparation: Lu. All authors reviewed the results and approved the final version of the manuscript.

References

1. Boyle, D. K. Chapter 7: Rail Scheduling. *Controlling System Costs: Basic and Advanced Scheduling Manuals and Contemporary Issues in Transit Scheduling*, TCRP Report No. 135, Washington, D.C., 2009.
2. Japanese National Railways. Chuo West Line/Shinonoi Line Train Diagram (中央本線 篠ノ井線列車ダイヤグラム), Effective 1975-03-10. In *鉄道ファン*, Vol. 15/173, Nagoya, Japan, 1975.
3. Nock, O. S. Chapter 12: The Working Time-Table. In *Single Line Railway*, David & Charles, Newton Abbot, UK, 1966.

4. Berkeley Simulation Software, L.L.C. *Rail Traffic Controller*. http://www.berkeleysimulation.com/rtc.php. Accessed June 13, 2017.
5. Vantuono, W.C. SYSTRA: Investments, improvements for RAILSIM®. In *Railway Age*, Simmonds-Boardman Publishers, Omaha, Neb., 2013.
6. Oricchio, V., J. Fox, and R. Trepal. Commuter Rail Implementation Study for New Haven and Hartford, Connecticut, and Springfield, Massachusetts. *Transportation Research Record: Journal of the Transportation Research Board*, 2010. 2146: 127–132.
7. Allen,J.G.,andLevinson,H.S.WhatHappenedtoSpeed?Scheduled Speeds and Travel Times of North American Passenger Trains, 1965 to 2015. *Transportation Research Record: Journal of the Transportation Research Board*, 2016. 2546: 60–68.
8. Business Council of Fairfield County. *Metro-North Timetable Study*, Stamford, CT, 2014. http://www.businessfairfield.com/wp-content/uploads/2014/12/TIMETABLE-STUDY-FINAL.pdf. Accessed June 8, 2017.
9. Steffee, D. M. The Year Amtrak Abandoned Racetracks, the Conventional Failed to Top 80 mph. *TRAINS Magazine*, June 1972.
10.Railtrack PLC. Chapter 6: Route Strategies. In *Network Management Statement for Great Britain*, Euston, England, 1997.
11. National Transportation Safety Board. *Organizational Factors in Metro-North Railroad Accidents*, Special Investigation Report NTSB/SIR-14/04, Washington, D.C., 2014. https://www.ntsb.gov/safety/safety-studies/Documents/SIR1404.pdf. Accessed February 2, 2018.
12. Federal Railroad Administration. *Operation Deep Dive: Report to Congress*. https://www.fra.dot.gov/Elib/Document/3586. Accessed July 26, 2017.
13. Nock, O. S. Chapter 10: The Aftermath. In *Railway Races to the North*, Ian Allan, London, 1959.
14.Maine Department of Transportation. *Portland North Alternative Modes Transportation Project—Alternative Analysis*, Auburn, ME, 2011. http://www.maine.gov/mdot/planning/docs/portlandnorth/AlternativeAnalysis082011.pdf. Accessed June 8, 2017.
15.Florida Department of Transportation. *Final Conceptual Alternatives Analysis/Environmental Screening Report, South Florida East Coast Corridor Transit Analysis: Miami-Dade, Broward, and Palm Beach Counties*, Fort Lauderdale, FL, 2006. http://tri-railcoastallinkstudy.com/docs/FINAL_%20AA-ESR_SFECCTA_Phase1%20-%2001-07-09.docx_test4.pdf. Accessed June 8, 2017.
16.Consolidated Rail Corporation. Conrail Main Line Albany-Buffalo, MP 220-MP230 NY. In *Maintenance Program and Track Chart, Northeastern Region, Mohawk Division*, Philadelphia, PA, 1983.
17. Hall, S. Chapter 10: Purley and Glasgow Bellgrove—Accident or Design? In *Danger on the Line*, Ian Allen, Shepperton, England, 1989.
18. Keefe, K. P. *Track Classifications: Classifying and Inspecting Railroad Track*. http://trn.trains.com/railroads/abcs-of-railroading/2006/05/track-classifications. Accessed June 10, 2017.
19.Sorgenfrei, D. F., and W. N. Marianos, Jr., Chapter 23: Railroad Bridges. In *Bridge Engineering Handbook*, CRC Press, Boca Raton, FL, 2000.
20.American Railway Engineering and Maintenance-of-Way Association (AREMA). Chapter 15. In *Manual for Railway Engineering*, AREMA, Landover, Md., 1997.
21. Nock, O. S. Chapter 1: Philosophy of Railway Signalling. In *Railway Signalling*, Institution of Railway Signal Engineers, London, U.K., 1980.
22. Consolidated Rail Corporation. *Train Stopping Distance Curves* (September 19, 1981—Revised May 7, 1986), Philadelphia, PA, 1986.
23. Collins, G. E. *Elements of Railway Signaling*, General Railway Signal, Rochester, N.Y., 1975.
24. Federal Railway Administration. Appendix: Potential Tier Structure for Passenger Systems. In *Highway-Rail Grade Crossing Guidelines for High-Speed Passenger Rail*. https://www.fra.dot.gov/eLib/Details/L03536. Accessed June 11, 2017.
25. Fazio, A. E. HSR in the Northeast: What Next? In *Railway Age*, November 2015. http://www.railwayage.com/index.php/passenger/high-performance/hsr-in-the-northeast-what-next.html. Accessed October 30, 2017.
26. Marquis, B. *Cant Deficiency, Curving Speeds and Tilt, U.S. DOT Volpe Center*. http://ntl.bts.gov/lib/52000/52700/52789/Cant_Deficiency_Curving_Speeds_and_Tilt.pdf. Accessed June 11, 2017.
27. Penn Central Transportation Company. *Northeastern Region, Timetable No.4*, in Effect 4.01 A.M., Eastern Standard Time, Sunday, October 25, 1970.
28. Klauser, P. Operating at High Cant Deficiency. In *Interface: The Journal of Wheel-Rail Interaction*, October 2005. Retrieved from http://interfacejournal.com/archives/581. Accessed June 11, 2017.
29. Maloney, B., and D. Phillips All Aboard Amtrak's Acela. In *The Washington Post Online*. http://www.washingtonpost.com/wp-srv/travel/features/acelaflash.htm. Accessed October 30, 2017.
30. Nock, O. S. *Two Miles a Minute: Story Behind the Conception and Operation of Britain's High Speed and Advanced Passenger Trains*, Patrick Stephens Ltd., Yeovil, U.K., 1983.
31. Olver, P. M. *Report on the Collision that Occurred on 19 September 1986 at Colwich Junction in the London Midland Region of British Railways*, Ministry of Transport, Her Majesty's Stationery Office, London, UK, 1988.
32. Hall, S. Colwich Junction Accident. In *Danger Signals: An Investigation into Modern Railway Accidents*. Ian Allen, Shepperton, England, 1989.
33. Penn Central Transportation Company. *Northeastern Region, General Order No. 409*, Effective 12:01 A.M., Monday, February 1, 1971.
34.Maryland Transit Administration. *What Are Heat Orders?* https://mta.maryland.gov/advisories/what-are-heat-orders. Accessed July 24, 2017.
35.Howard, S. M., L. C. Gill, and P. J. Wong. Review and Assessment of Train Performance Simulation Models. *Transportation Research Record: Journal of the Transportation Research Board*, 1983. 917: 1–6.
36. Network Rail. *The Working Timetable (WTT) is the Rail Industry's Version of the Public National Timetable*. https://www.networkrail.co.uk/running-the-railway/timetabling/working-timetable/. Accessed June 13, 2017.
37. Frailey, F. W. Zephyrs Against a Head Wind. In *Twilight of the Great Trains*, Indiana University Press, Bloomington, Ind., 2010.

38. Pinkepank, J. A. Why New Haven Re-Electrified—So Long as the Wires are Up, it's Cheaper to Use 'em. *TRAINS Magazine*, August 1964.

39. New York Central Railroad Company. *Boston & Albany, Syracuse, Buffalo Divisions, Time Table No. 14 for Employees Only*. Effective 2:00 AM E.S.T., Sunday, Oct. 27, 1963.

40. Consolidated Rail Corporation. *Timetable No.1*, in Effect 12.01 A.M., Eastern Standard Time, Sunday, June 18, 1978.

41. Conrail. *Albany Division, Station Pages and Division Special Instructions for System Timetable No.4*, Effective 12:01 A.M., Eastern Standard Time, Saturday, February 1, 1997.

42. Museum of Railway Timetables. *The Amtrak System Timetables, 1971-2010*. http://www.timetables.org/. Accessed February 23, 2016.

43. Oura, Y., Y. Mochinaga, H. Nagasawa. Railway Electric Power Feeding Systems. In Japanese Railway Technology Today, East Japan Railway Culture Foundation, Shibuya, Tokyo, Japan, 2001.

44. Keenor, G. Overhead Line Electrification for Railways, 4th Edition. http://www.railwaysarchive.co.uk/OLE/. Accessed June 13, 2018.

The Standing Committee on Commuter Rail Transportation (AP070) peer-reviewed this paper (18-01516).

Article

TRR
JOURNAL OF THE TRANSPORTATION RESEARCH BOARD

Transportation Research Record
2018, Vol. 2672(8) 404–410

DOI: 10.1177/0361198118777063
journals.sagepub.com/home/trr
SAGE

Strategic Planning and Prospects of Rail-Bound Demand Responsive Transit

Oded Cats[1] and Jesper Haverkamp[1,2]

Abstract

Fully automated services allow for greater flexibility in operations and lower marginal operational costs. In this study we examined the strategic planning implications of a novel service concept – an automated rail demand responsive transit (DRT) system that offers a direct non-stop service. The objective of this study was to determine the capacity requirements of the envisaged service and discuss its prospects and feasibility. A cost minimization approach for determining the optimal track and station platform capacities for a rail-DRT system so that passenger, infrastructure, and operational costs are minimized is described. The macroscopic model allows for studying the underlying relations between technological, operational and demand parameters, optimal capacity settings, and the obtained cost components. The model was applied to a series of numerical experiments to test its implications for different network structures and demand distributions. The results of the numerical experiments indicate that – unlike conventional rail systems in which stations often are capacity bottlenecks – link capacity properties are more critical for the performance of automated rail-DRT systems than station capacity. A series of sensitivity analyses was performed to test the consequences of various cost and capacity specifications, as well as the characteristics of future automated rail-DRT systems.

The rapid advancements in the developments of fully-automated vehicles have led to an increasing interest in the concept of demand responsive transit (DRT) systems. Automated services will allow for greater flexibility in operations and lower marginal operational costs. The area of application has almost exclusively been limited to road-bound systems. Although automation on rail networks is not new, to the best of the authors' knowledge, applying DRT as a substitute for current heavy rail services has not been considered in the literature or practice to date. This paper presents a first step into the relatively unknown area of rail DRT.

In this work, a DRT system that has been designed as a full replacement of scheduled heavy rail for a given (sub-) network is envisaged. The automated rail-bound vehicles offer a direct, non-stop service and are in the rail network, in response to passenger requests, with no predefined routes and schedules. Vehicles transport passengers that share the same origin and destination stations. Vehicles can be sized according to the operator's preference, but they are considerably smaller than existing trains.

The objective of this study was to determine the capacity requirements of rail DRT. Unlike road-bound DRT services, which operate in rural areas or cater for special user groups, rail DRT is designed to serve a large geographical area with relatively high demand that could result in an operation constrained by congestion and capacity limitations. The strategic planning objective of this study constitutes a major difference from most of the earlier models such as Anderson (*1*) and Winter et al. (*2*), which considered microscopic operational models with a stochastic passenger arrival.

The development of technological and service concepts that will facilitate rail-DRT systems are still in their early stages. It is therefore not surprising that literature on rail DRT is limited. An early research identified the challenges of short vehicle headways and limited station capacity in the context of dense urban operations (*3*). The vehicle engineering RailCab project developed technical and mechanical solutions for small driverless rail-bound traffic (*4*). Vehicle design solutions for operating at short headways in an automated guideway transit system were studied by Choromanski and Kowara (*5*), whereas capacity in relation to station layout has been analyzed in more detail by Greenwood et al. (*6*). Although these studies provide preliminary insights into anticipated advanced in-vehicle technology, there is a lack of knowledge on the strategic planning aspects of such operations, such as the capacity requirements they inflict on railway network infrastructure and related system performance and level-of-service.

[1]Department of Transport and Planning, Delft University of Technology, Delft, Netherlands
[2]Nederlandse Spoorwegen, Department of Operations Control, Section Business Intelligence, Utrecht, Netherlands

Corresponding Author:
Address correspondence to Oded Cats: o.cats@tudelft.nl

This paper presents an optimization approach for determining the optimal track and station platform capacities for a rail-DRT system so that passenger, infrastructure, and operational costs are minimized. The model is formulated as a cost minimization problem in which vehicle flow distribution attains system-optimum conditions. The model allows for studying the underlying relations between technological, operational and demand parameters, optimal capacity settings, and the obtained cost components. The model is applied to a series of numerical experiments to illustrate and explore the strategic planning aspects of possible rail-DRT network-wide implementations. The performance under various scenarios is assessed, the implications and feasibility of which are discussed.

Methodology

Modeling Approach

Current railway models with timetables as their cornerstone are unsuitable for rail-DRT applications. The main challenge in modeling network-wide long-term planning for rail DRT is that the model needs to capture flow distribution and capacity constraints without representing system dynamics in microscopic detail. Other rail-DRT challenges include the need to handle large numbers of hourly passenger requests, strict routing constraints due to the nature of rail infrastructure, and highly heterogeneous service characteristics compared with traditional rail systems. The approach taken in this study was to develop a novel macroscopic model by considering rail DRT as a special case of the network flow problem.

The deterministic and static optimization problem is solved for a given network topology and passenger demand distribution (DD), both are thus exogenous to the model. Other inputs to the model include vehicle size (homogenous fleet), cost units, and track and node flow-speed and -delay functions. Model outputs are the optimal infrastructure capacity per network element (i.e., rail segments and stations), vehicle flow distribution, and the value of passenger, infrastructure, and operational costs.

Network Transmission Cost Functions

This section describes how the link speed-density function and node delay-density function are defined. The speed-density function is assumed to follow a logistic function (*7*) albeit with an abrupt transition from free-flow to jam, which is more suitable for automated systems. Link travel times are then determined based on link length, free-flow speed, and a logistic term with the volume over capacity ratio.

Vehicle arrivals are assumed to follow the Poisson distribution, implying that service requests of all vehicles can be represented as a joint Poisson process. Vehicles either drive through or call at origin and destination stations. Considering each platform as a server, and assuming that all vehicles have the same mean service time with an exponential service time distribution and that all vehicles can use all platforms, the DRT station is characterized as a non-pre-emptive multi-server queuing model where servers are governed by a Poisson process and job service times are distributed exponentially denoted as a M/M/c system (similarly to metro stations in for example Xu et al. [*8*]). If the station has more than two platforms, it is assumed that through-going vehicles can overtake dwelling vehicles, otherwise the station is governed by non-prioritized M/M/c queues. The expected waiting time in the non-pre-emptive M/M/c queue depends on delay probability (*9*) with the corresponding expected delays in prioritized and non-prioritized queuing systems being determined according to the formulations provided in Wagner (*10*). This function assigns different values for through-going vehicles and dwelling vehicles at stations in which overtaking is possible, otherwise no distinction is made.

Cost Minimization Problem

Considering rail DRT as a special case of the network flow problem, the decision variables are link capacity, node capacity, and the share of vehicle flow routed via each route alternative per origin–destination (OD) pair. The objective is to minimize the cumulative value of infrastructure capacity costs, passenger travel time costs, and operational costs. The rail-DRT planning objective is to balance between the costs of adding infrastructure capacity and the costs of delay or detours caused by insufficient infrastructure capacity.

The cost minimization problem is then formulated as follows: (i) passenger travel times calculated over all links and stations; (ii) track capacity investment costs; (iii) station capacity investment costs; and (iv) variable operational costs. Each of the objective function components is associated with a weight that is the corresponding monetary cost.

Link travel times and the expected delay at stations are calculated as described in the previous section. Additional constraints include demand satisfaction, flow conservation, and non-negative decision variables. Decision variables include the capacity of each link and each station and the number of vehicles traversing each link that travel between a certain OD pair. Hence, the cost minimization problem entails the simultaneous solution of setting the capacity per rail-segment and station and obtaining the corresponding system-optimum solution of network flow distribution while minimizing user, investment, and operational costs.

Service frequency is specified in this study as approximately proportional to passenger demand for a given OD pair unless passenger demand is not sufficient to justify a predefined minimum service frequency. Hence, waiting times are determined in the initialization phase, independent of the decision variables, and can be left out of the analysis of alternative solutions.

Model Implementation and Specification

Model specification involves setting values for a series of technological and service parameters. In the following, base

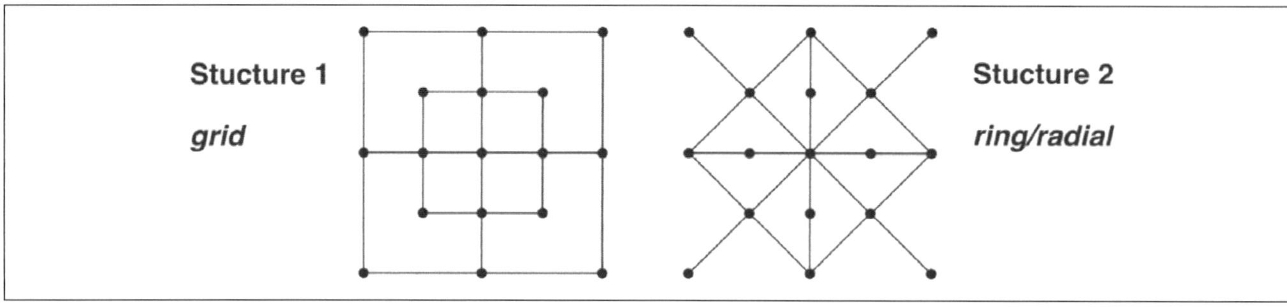

Figure 1. The two network structures considered in the numerical experiments: grid (left) and ring/radial (right). All lines represent bidirectional arcs.

case values designed to reflect the prevalent conditions in the Netherlands – for which the authors have access to relevant information – are specified.

Objective function weights were set as follows: the Dutch value-of-time was €10/h. Based on estimates from the Dutch rail industry, investment costs were estimated to amount to €50 million per km of track and €123.5 million per station platform assuming a 38-m long platform required for the envisaged system operations. Both were assumed to have depreciated over a 30-year period. Operational cost per seat-km was estimated at €0.02.

The speed-density function involves specifying the free-flow speed and track vehicle capacity (VC). The former was set to 100 km/h based on current conditions in the dense Dutch network. Track capacity (TC) was set to 180 vph in line with operational people-mover systems, which operate at 20-s headways.

Station platform operation is governed by the queuing servers. Based on observations in an existing automated system (i.e., Rivium Parkshuttle), dwell time was set to 20 s. Mean service rate was set to 5 s for non-stop vehicles and 80 s for dwelling vehicles. The latter is based on an estimation of the time required for setting switches, pulling into the station, dwelling, and clearing the platform.

Finally, the base case VC was set to 24 passengers. Vehicles are designed so that all passengers are seated. Service frequency per OD pair was determined to satisfy the demand with a load factor of at least 0.7. OD-pairs for which demand does not justify at least one departure per hour (i.e., 17 passengers) remain unserved by the DRT service. The optimization problem was solved in MATLAB.

Numerical Experiments

The cost minimization model described in the previous section was applied to a series of numerical experiments to analyse the generic properties of rail-DRT systems, underlying relations between model variables, and the sensitivity of model outputs to variations in input parameters.

Experimental Set-up

The numerical experiments were performed using a graph composed of 17 nodes and 48 unidirectional links. Two distinct network structures were considered: a grid and a ring/radial structure, as illustrated in Figure 1. The base case passenger demand corresponds to 34,000 hourly requests distributed over the network based on a gravity model using the Euclidean distance between all origin and destination nodes (an average of 125 trips per OD pair). Each optimization problem was solved within 15 s on a standard PC.

A series of sensitivity analyses pertaining to service design, cost parameters, technological capabilities, and demand scenarios were performed. In the following, the results for variations in vehicle passenger capacity, TC, and DD pattern are reported and discussed. In the case of VC, it is also varied in conjunction with corresponding changes in operational costs and minimum service frequency threshold.

Results and Analysis

Table 1 summarizes the results of the base case scenario along with the key sensitivity analysis scenarios. The base case is set with the parameter values specified in the Model Implementation and Specification section – including a VC of 24 passengers and TC of 180 vph – and with the Euclidian gravity DD. The table reports results for other VC and TC scenarios noted with the corresponding value, as well as DD scenarios as detailed below. Table 1 includes the four cost components for the optimal solution expressed in hourly terms (columns 2–5), followed by two indicators of system resources – fleet size and total seat-km offered – and the resulting service speed.

Overall, the grid network was found to outperform the ring/radial network, yielding lower values for all cost components in the optimal solution with the existing DD. Careful investigation revealed that this difference stemmed primarily from the fact that in the grid network flows can often be rerouted at constant mileage, whereas in the ring/radial network rerouting always comes at the price of increased travel distance. Consequently, flow rerouting was common in the grid network, whereas all vehicles took the shortest route in the ring/radial scenario.

Figure 2 presents the unrounded (left) and rounded (right) allocated infrastructure capacity and corresponding utilization in the grid network. All but four arcs that require a double track have a single track. Hourly arc capacity costs increase to €73,000 compared with €32,000 in the unrounded

Table 1. Numerical Experiments Scenario Results for Grid; Ring/Radial Networks

Scenario	Passenger costs (€1000)	Link costs (€1000)	Node costs (€1000)	Operation-al costs (€1000)	Fleet size (vehicle)	Offered seat-km (1000 km)	Service speed (km/hr)
Base	47.1; 49.1	31.8; 33.5	46.1; 46.3	9.9; 10.4	370; 386	495.4;521.6	74; 74
VC_12	58.2; 60.3	69.8; 72.9	83.3; 83.2	11.7; 12.2	915; 948	583.5;609.2	70; 71
VC_48	32.4; 36.4	14.0; 13.5	27.0; 26.8	7.0; 7.8	127; 143	348.7;390.1	75; 75
VC_96	26.5; 27.5	8.3; 7.6	18.1; 18.0	5.5; 5.9	52; 54	276.6;294.6	73; 75
TC_30	60.6; 63.3	160.5;169.0	46.4; 46.3	9.9; 10.4	476; 497	495.4;521.6	57; 58
TC_45	55.6; 58.0	111.0;116.9	46.4; 46.3	9.9; 10.4	437; 456	495.4;521.6	62; 63
TC_120	48.7; 50.7	45.8; 48.2	46.4; 46.3	9.9; 10.4	382; 399	495.4;521.6	71; 72
DD_CC	80.1; 75.2	57.9; 53.9	52.7; 52.4	18.0; 16.8	630; 591	902.1;839.4	79; 78
DD_U	78.6; 74.1	56.9; 53.4	49.5; 49.5	17.7; 16.7	618; 582	887.1;832.5	79; 79

Note: VC = vehicle capacity; TC = track capacity; DD = demand distribution; CC = closeness centrality; U = uniform.

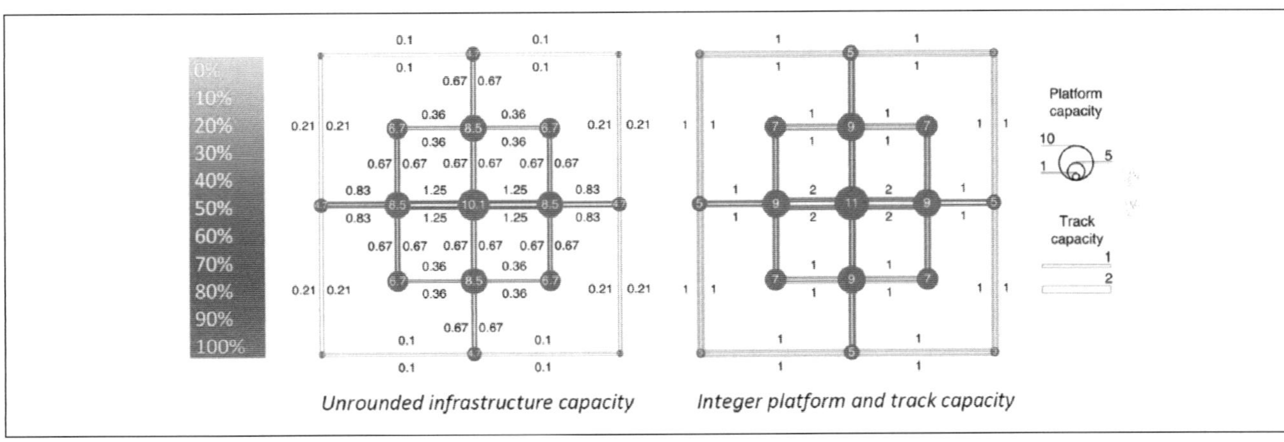

Figure 2. Allocated capacity (numbers) and corresponding utilization (indicated in color) in the grid network.

case. Hourly node capacity costs increase to €50,000 up from €46,000. Notably, arc costs are considerably more sensitive to the integer capacity criterion. The utilization level is close to 70% on most arcs and in excess of 80% for nodes. These values hold for the major part of the network. The cost minimization solution results with a low variability in capacity utilization across the network (67% to 72%).

Solution sensitivity to VC was tested for the following values: 12, 24, 48, and 96. These capacities correspond to the number of seats in van, base case, bus, and intercity train car unit, respectively. Station capacity costs were adjusted to correspond to changing platform length requirements. A pronounced trend can be observed for both network structures in the results reported in Table 1 – the smaller the VC, the higher all cost components become. This is also clearly visible in Figure 3, which displays cost component per VC scenario for the grid network. Reducing VC from the base case value of 24 to 12 results in an increase of approximately 120% and 80% in link and node capacity costs, respectively. This increased capacity is required to handle a fleet size that is 2.5 times larger, but nevertheless results in a 25% rise in passenger costs. Part of this difference was attributed to the increasing number of passengers that were not offered a direct DRT service as vehicle size increased. This share increased dramatically from negligible levels to a sizable minority of passenger demand in the scenario with a capacity of 96. With the exception of the lowest capacity scenario, service speed remained at a relatively constant level.

Smaller rail vehicles are associated with higher operational costs, because the seat-to-engine ratio is lower. It might therefore be argued that to ensure a fair comparison, a change in vehicle size needs to be accompanied with an opposite change in operational costs. The following scenarios are studied for the grid network: 12 seats at €0.08 per seat km, 24 seats at €0.02 per seat km (base case), 48 seats at €0.015 per seat km, and 96 seats at €0.0125 per seat km. As shown in Figure 4, the simultaneous change in seat km price and unit operational costs has most influence on the component of operational costs. Nevertheless, the results in Table 1 show that a change in unit operational costs has no major influence on the decision variables. Differences between the various vehicle size scenarios are in line with the results reported in Table 1.

To overcome the deficiency of large shares of unserved demand when vehicle size increases, the second set

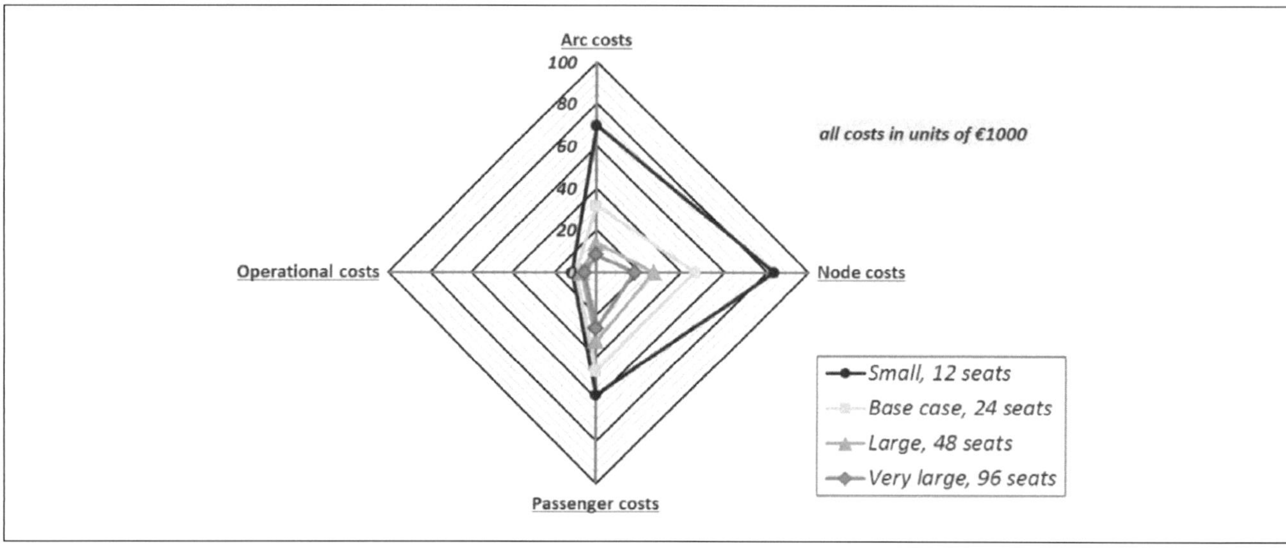

Figure 3. Cost component sensitivity to vehicle capacity in the grid network.

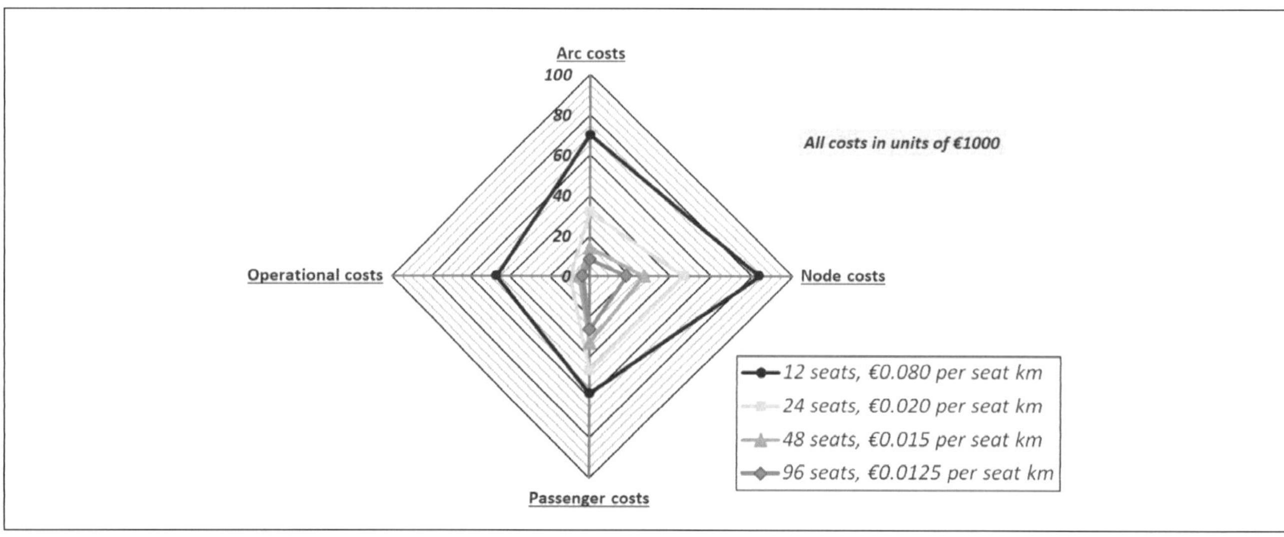

Figure 4. Cost component sensitivity to combined changes in vehicle capacity and unit operational costs in the grid network.

of combinatorial scenarios studied a decrease of minimum service frequency at an increase in VC for the grid network. The largest vehicle was associated with a frequency threshold of one departure per hour. The threshold values for smaller vehicles were set such that the minimum capacity of 96 seats per hour was obtained in all scenarios: 12 seats and at least eight departures per hour; 24 seats and 4 hourly departures; 48 seats and 2 hourly departures; and 96 seats with at least one departure per hour. In this way, the share of unserved demand was fixed in all cases (12%). Though served demand was equal in all scenarios, passenger costs still decreased with increasing VC, albeit more modestly (i.e., 46,940 and 40,560 for 12 seats and 96 seats, respectively) than reported in Table 1. Link and node capacity costs were significantly higher in low VC scenarios with costs more than five and three times as high for link and node costs, respectively, when comparing 12-seat and 96-seat scenarios, yet resulting in a lower operational speed. Notwithstanding, smaller vehicles may still be preferred because they offer higher service frequency.

The automated rail TC assumed in the base case scenario to be 180 vph, is very high compared with existing heavy rail systems. To test the consequences of technological developments that result in lower values, the model was run with TC values of 30, 45, and 120 vph. The most conservative value corresponds to the maximum frequency in classical heavy rail with European Railway Traffic Management System (ERTMS) signaling. A reference value of 45 is taken from automated metro systems. An intermediate value of 120 can

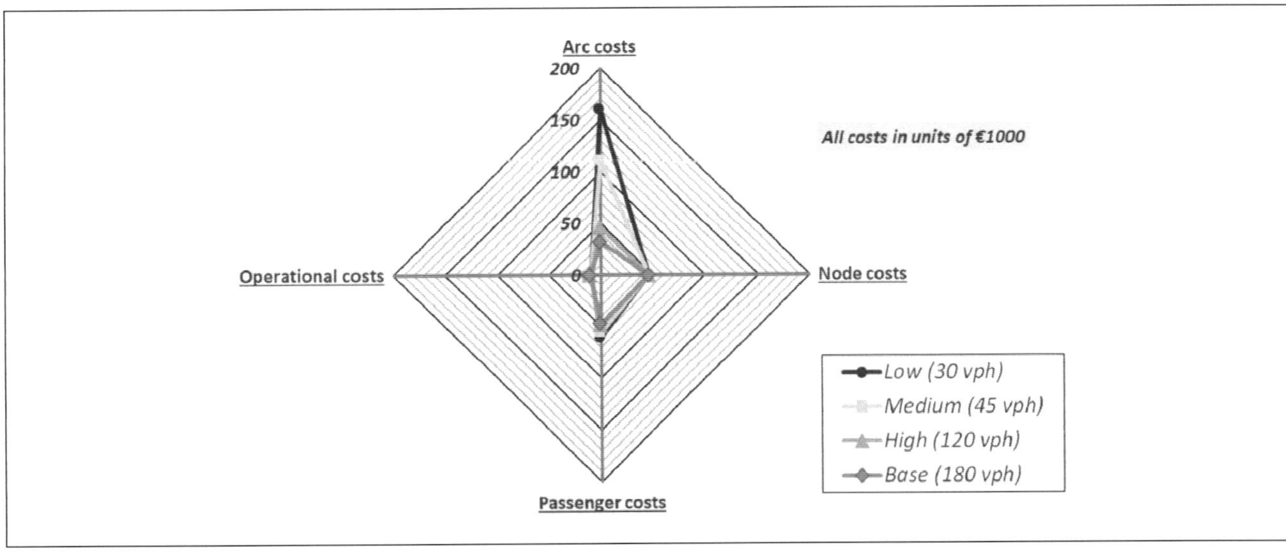

Figure 5. Cost component sensitivity to changes in track capacity in the grid network.

be indicative of more limited technological advancements. The results of the sensitivity analysis are visualized in Figure 5. As might be expected, link costs are most affected by changes in TC as they exercise a linear relation. In the scenario of 30 vph/track, link costs increased by over 400% compared with the base case. In contrast, the impact of TC on passenger costs exhibited a logistic relation through the operational speed resulting in an increase of 29% when comparing the low capacity scenario of 30 vph with the base case scenario. Node costs and operational costs remained unaffected. Link costs dominated the cost function for low TC values and decreased to levels similar to station capacity and passenger costs for TC values of 120 vph and then were exceeded by the latter two for a capacity of 180 vph (i.e., base case).

Finally, DD was expected to have profound effects on system performance. Two DD scenarios were considered in addition to the base case, which had demand oriented toward the network's center of gravity: the opposite case of uniform demand distribution (DD_U) and an intermediate case where demand is proportional to the node closeness centrality (DD_CC) metric (i.e., average distance to all other nodes). Note that unlike the uniform and gravity distributions, DD in the latter scenario is not independent from network connectivity and hence results in a different OD-matrix for the two network structures. In contrast to the base case results, link costs in the optimal solutions for the two alternative DD scenarios surpassed station costs. A more uniformly distributed demand required a larger fleet and led to higher mileage and thus TC on more network links, albeit with lower congestion levels as reflected in the increase in average speed for both network structures. Operational costs were lower for the ring/radial network than for the grid network in the average closeness and uniform scenarios due to its better connectivity, whereas the grid network performed better when demand was concentrated at the center where it offered shorter routes.

Discussion and Conclusion

This study presents a network cost minimization model for determining the optimal infrastructure capacity and flow distribution of rail-DRT systems. The results of the numerical experiments indicate that – unlike conventional rail systems in which stations often are capacity bottlenecks – link capacity properties are more critical for the performance of automated rail-DRT systems than station capacity. Despite the highly heterogeneous service characteristics of rail DRT, it is concluded that only a limited number of stations need to offer overtaking capacity. Hence, DRT stations in general do not have to be positioned off the main line.

As this study is pioneering in the public transport system it envisaged, model specification required making a series of assumptions about plausible characteristics of future automated rail-DRT systems. A series of sensitivity analyses was performed to test the impacts of even extreme deviations from the assumed values on model outputs. Future developments are expected to allow finer specifications of technological and service features such as the link speed-density function, the station platforms queuing regime, and related assumptions.

The automated rail TC assumed in the base case scenario to be 180 vph, is very high compared with existing heavy rail systems. From the sensitivity analysis, it is concluded that link costs are most affected by changes in TC. Link costs dominated the cost function for the lowest TC values examined and decreased to levels similar to station capacity and passenger costs for TC values of 120 vph. From a conventional heavy rail point of view, rail-DRT TC characteristics are challenging and require vehicle switching on fixed rails, communications-based train control, and automated dispatching. In addition to the technological exploration of this study, there are station management aspects to offering rail-DRT services that need to be addressed in future research. These design and technical aspects include the coordination of passenger travel requests, and informing passengers and vehicles about the dwelling and loading positions within the station.

Under the base case parameter settings, the optimal link and station capacity allocation is comparable to those currently available in some heavy rail networks. Link infrastructure utilization is however higher, in the order of 70% to 85%, compared with approximately 65% with today's system. The need to invest in link capacity strongly depends on vehicle characteristics of minimum headway and the speed-density relation. Trading-off link costs against passenger costs is possible in the form of either inducing longer travel time or reducing service frequency by increasing VC.

The sensitivity of the overall results to changes in unit operational costs is negligible within any reasonable range of unit operational cost values. Choices on infrastructure capacity and vehicle flow routing can therefore be made regardless of unit operational costs. This is an important conclusion with respect to the development of rail-DRT vehicles. Even in the unfortunate case when rail-DRT vehicles are more expensive to operate than anticipated, the strategic choices on available infrastructure are still valid.

The optimal solution corresponded to the system-optimum flow distribution. Even though model formulation allows for rerouting vehicles to attain global system-optimum conditions, only in extreme cases is a minority of the vehicle flow not assigned to the shortest route. The implications of competing DRT service providers can be studied by adjusting the objective function used in this study to guarantee user-equilibrium conditions.

Although certain network structures appeared to be better equipped in reducing the objective function value for certain DDs, the overall pattern is that the denser a network becomes, the lower the costs per passenger kilometer. This relation is particularly pronounced when the higher connectivity is offered in the area of highest demand. In practical terms, this implies that also in the context of rail DRT, stations should be higher in number and closer together in areas of high demand, such as urban regions. In low demand zones, such as rural areas, the number of stations should be more conservative.

This study examined the long-term planning consequences of offering a new transportation technology and service concept. To test its potential to substitute the existing rail service, current demand levels were tested in the application. Future research is needed to assess travelers' perceptions and preferences toward such services and potential changes in demand patterns that may be caused by the introduction of rail-DRT services. For example, the service might have consequences for station attractiveness due to the increased correlation between service frequency and demand for specific connections. This may also have ramifications for network structure design and service availability, including equity considerations.

Acknowledgments

The authors thank Dennis Huisman for his suggestions and critical input. Dutch Railways (NS) kindly provided input on operational settings.

Author Contributions

The authors confirm contribution to the paper as follows: study conception and design: O.C.; data collection: J.H.; analysis and interpretation of results: J.H. and O.C.; draft manuscript preparation: O.C. and J.H. All authors reviewed the results and approved the final version of the manuscript.

References

1. Anderson, J. E. Simulation of the Operation of Personal Rapid Transit Systems. *Proc., 6th International Conference on Railway Engineering Design*, Lisboa, Portugal, WIT Press, 1998, pp. 523–532.
2. Winter, K., O. Cats, G. H. A. Correia, and B. Van Arem. Designing an Automated Demand-Responsive Transport System: Fleet Size and Performance Analysis for the Case Study of a Campus Train Station Service. *Transportation Research Record: Journal of the Transportation Research Board*, 2016. 2542: 75–83.
3. Bendix, T., and R. Hesse. Cabinentaxi: A Personal Public Transport System. *Transport*, Vol. 1, No. 3, 1972, pp. 322–329.
4. Henke, C., M. Tichy, T. Schneider, J. Böcker, and W. Schäfer. Organization and Control of Autonomous Railway Convoys. *Proc., 9th International Symposium on Advanced Vehicle Control (AVEC 08)*, Kobe, Japan, 2008, pp. 1–6 .
5. Choromanski, W., and J. Kowara. PRT – Modeling and dynamic simulation of track and vehicle. *13th International Conference on Automated People Movers and Transit Systems*, Paris, France, ASCE, 2011, pp. 294–306.
6. Greenwood, J., L. Mantecchini, and J. Schweizer. Analytical Capacity Limits of Personal Rapid Transit Stations. *Proc., 13th International Conference on Automated People Movers and Transit Systems*, Paris, France, ASCE, 2011, pp. 326–338.
7. Rotoli, F., E. N. Cawood, and A. Soria. *Capacity Assessment of Railway Infrastructure: Tools, Methodologies and Policy Relevance in the EU Context*. JRC Technical Report. European Commission, 2016. ISBN 1831-9424.
8. Xu, X.-Y., J. Liu, H.-Y. Li, and J.-Q. Hu. Analysis of Subway Station Capacity with the Use of Queueing Theory. *Transportation Research Part C: Emerging Technologies*, Vol. 38, 2014, pp. 28–43.
9. Kella, O., and U. Yechiali. Waiting Times in the Non-Preemptive Priority M/M/c Queue. *Stochastic Models*, Vol. 1, No. 2, 1985, pp. 257–262.
10. Wagner, D. Analysis of Mean Values of a Multi-Server Model with Non-Pre-Emptive Priorities and Non-Renewal Input. *Stochastic Models*, Vol. 13, No. 1, 1997, pp. 67–84.

The Standing Committee on Automated Transit Systems (AP040) peer-reviewed this paper (18-02339).

Article

TRR
JOURNAL OF THE TRANSPORTATION RESEARCH BOARD

Transportation Research Record
2018, Vol. 2672(8) 411–420

DOI: 10.1177/0361198118794286
journals.sagepub.com/home/trr
SAGE

Predicting Bus Operator Retention Based on Employee Characteristics and Work History

Eric Lind[1] and Joel Huting[1]

Abstract

Turnover of employed bus operators increases cost of service for transit agencies, meaning interventions to reduce turnover can be a common goal. To be effective, interventions should be based on an understanding of the types, timelines, and underlying reasoning for operators separating from employment. However, to date there has been little quantitative research into bus operator separation dynamics. To better understand these dynamics, a hierarchical survival model was used to quantify monthly probability of separation for individual bus operators, based on a sample of two-year work histories for over 1300 bus operators employed by Metro Transit (Minneapolis–St. Paul, MN). Characteristics of operators at hire, such as age, gender, and driving experience, as well as work history events such as absences, incidents, and accidents, were used to understand what influences separation. Separation rate differed predictably with time, with the highest risk of both voluntary and involuntary separation in the first ten months of operator employment. Of bus operator characteristics at hire, prior commercial driving experience lowered the probability of separation. The number of absences from work, and accidents classified as the responsibility of the operator, increased the separation risk probability per month. The highest probability of separation was in operators hired without a commercial driver's license, whose work history included multiple accidents. Alterations to hiring and training practices may ameliorate risk of bus operator separation. Additionally, the hierarchical modeling approach utilized here has promise for predicting relative impacts of characteristics of employees and their work history on multiple aspects of employment.

Effective public transportation relies in great part on the everyday labor of bus operators, who can fulfill the scheduled service safely and on time. The job demands are unique: one must be able to operate a large vehicle in mixed traffic while adhering to a schedule, monitoring electronic communications, and providing retail-level interaction with customers. Because of these demands, not every employee who begins as a part-time bus operator will spend a career as a transit employee; some will leave voluntarily, and a smaller fraction will be involuntarily separated from employment. These separations are costly to transit agencies, as they include not only the sunk costs of recruitment and training of the departing employee, but also lost productivity during an interim period before the employee can be replaced (*1*).

Because of the high cost of bus operator turnover, transit agencies have an incentive to understand the factors that might predict operator success and retention. Previous research has focused mainly on differentiating potential operators at the time of hire. For instance, Jacobs et al. (*2*) surveyed bus drivers across nine organizations using a battery of over 350 questions, which were then combined into scores and compared to supervisory ratings, absences, and accident data. The scores were able to differentiate the top performers, but not distinguish among the bottom 2/3 of operators, and thus may not be useful in predicting turnover. In contrast, Kompier et al. (*3*) modeled separation as a cumulative process that arose from medical hardship on the job leading eventually to disability. In this way, the absence rate predicted the eventual separation by the operator, either by choice or medical necessity (*3*). Case studies (*4*) have also been valuable in understanding bus operator hiring and turnover dynamics in particular geographic and economic situations. However, a quantitative statistical exploration of factors commonly measured at transit agencies and the probability of a bus operator to stay on

[1]Metro Transit, Minneapolis, MN

Corresponding Author:
Address correspondence to Eric Lind: eric.lind@metrotransit.org

the job has not been attempted. Such a model would be especially valuable if it could predict the time course and risk factors involved in separation, based on commonly collected data.

There are two categories of characteristics of employees that might predict separation (and by its inverse, retention). Characteristics of employees at hire are of use for two reasons. First, personal characteristics such as gender and age can be indicative of diversity in expectations for, and approaches to, the job. Significant differences in retention predicted by these characteristics may indicate that an organization needs to focus training or other programs differently. Other characteristics at hire might be indicative of job success. Previous research identified 11 separate dimensions of performance required of bus operators, including stress management, multitasking, and customer service, in addition to safe and timely bus operation (*5*). Because experience with commercial driving can build these skills, a primary desire of transit agencies is to hire qualified individuals who hold a commercial driver's license (*4*). Additional education beyond the minimum job requirement (e.g., a high school diploma) may correlate with some of the non-driving skills identified by Hausdorf and Risavy (*5*). If found to be influential in an analysis of operator retention, recruitment can be tuned to favor these characteristics (where appropriate and legal to do so).

Characteristics of a bus operator's work history can indicate the stresses or experiences on the job that might increase the risk of separation regardless of an operator's individual characteristics, or that might indicate a predisposition to separation. Non-crash incidents and on-road accidents are a regular feature of bus operation but can differ by the conditions and experience of the operator (*6*). Incidents which are not the responsibility or fault of the driver may accumulate as a simple function of time behind the wheel, while accidents classified as the responsibility of the operator may indicate a skill deficiency that could lead to separation. Other work conditions can influence the probability that a driver will separate, such as the work site location (home garage). The aspect of work history over which the operator has most control is the rate of absences from the job. Absenteeism is a vexing problem in transit operations due to the need to replace the work, often on short notice, in order to maintain service (*2*, *7*). However, beyond the immediate impact of absenteeism, the accumulation of missed work may in some cases, though not all, be an indicator of an impending separation from employment. Resolving the predictability of separation from absenteeism rate is thus an important goal for bus operations administrators.

To obtain a full understanding of the factors that might predict bus operator retention and separation, it is necessary to combine data across levels or scales into one coherent framework. A person's age at hire does not change during the course of their employment, while their garage assignment changes multiple times; additionally, the influence of a garage on operator retention, through the supervisors and culture, may differ across operators of different ages. Thus, the individual characteristics and work history predictors may interact in complex ways. To resolve these interactions in a useful way, it is necessary to construct a model of operator separation at the individual level, rather than using aggregate regression approaches in which the variation within individual work history experiences is not resolvable. The goal of these analyses is to understand and predict separation at the level of the individual, which is the optimal resolution of the data model.

Data and Methods

Metro Transit, the public transportation organization serving the cities of Minneapolis–St Paul, Minnesota, operates over 900 buses using a bus operator workforce that numbers around 1500 people. Operator characteristics and work history are stored in a variety of bus-operations and human-resources datasets, with reliably complete data from 2008. Data were extracted from these datasets and joined based on employee identification numbers, which were then randomly masked before analysis and reporting to prevent identification of individuals. The dataset for which all the characteristics of the operators described below were available, which included 1318 employees, comprised 90% of the total operator workforce employed from 2008 to 2016.

When operators leave employment, they are classified as having left voluntarily or involuntarily. In the case of involuntary separation, a cause is assigned, and is typically the result of a violation of negotiated work rules, whether during or after probation. With voluntary separation, while exit interviews are offered, they are not commonly submitted, and in general there is little to no information stated by the exiting employee about the reasons behind the separation. Nonetheless, voluntary separations are the majority of operator separations (more than half of our dataset), thus understanding the factors that contribute to, and might even predict, the decision to separate could lead to specific actions to reduce operator turnover.

For each bus operator in the available dataset, characteristics of the operators and their work experiences were assembled as indicated in Table 1. Variables were of one of two types: individual or work history. Individual characteristics were descriptors of the operator at time of hire, including age, self-reported gender, self-reported education history, the year the employee was hired, and whether the employee had a commercial driver's license

Table 1. Variables Characterizing 1318 Metro Transit Bus Operators Employed during 2008–2016

Variable	Type	Data mean (SD), mode, or frequency
Gender	Individual	25.2% female
Age at hire	Individual	41.8 (11.04) years old at hire
Year hired	Individual	2013
CDL at hire?	Individual	1041 had CDL, 277 no CDL
Highest education	Individual	Some College
Training	Work history	
Probation	Work history	
Cumulative incidents	Work history	1.38 (1.6) events by month 12
Cumulative responsible accidents	Work history	0.35 (0.6) events by month 12
Cumulative absence occurrences	Work history	6.63 (5.4) absences by month 12
Assigned garage	Work history	
Separated	Work history	318 total separations
Separation type	Work history	138 involuntary, 180 voluntary

(CDL) at hire. The education and CDL variables were used to understand the influence of prior experience on the potential to separate once hired as an operator. Any level of CDL was grouped together in the variable (Minnesota has four levels of driver's license, three of which are commercial. The majority of operators in the dataset held Class B at time of hire, which indicates certification to operate single-unit vehicles over 26,000 pounds gross vehicle weight rating, with a passenger endorsement).

Work history variables captured different aspects of the course of employment hypothesized to be important for understanding propensity to separate from the job. Binary indicators were used for training and probation to highlight times when operators were undergoing a high degree of supervision or were operating under more stringent work rules regarding absences and accidents. After probation, some work rules lead to forced separation (e.g., at Metro Transit 13 unexcused absences in a rolling 12-month window) but are rarely triggered. The model included all absence occurrences, whether excused or not, under the hypothesis that increased levels of missed work could be a precursor to voluntary separation.

In-service incidents and accidents are classified by supervisors as "responsible" or "not responsible" depending on the operator's role in the event. Work rules address the number of responsible accidents that lead to involuntary separation. We included both responsible and not-responsible incidents and accidents in the analysis, under the hypothesis that increased incident activity could be a precursor to voluntary separation as well. All reported events (ranging from side mirror scrapes to serious collisions) were included.

For each bus operator, a time series of employment by month was constructed, aligning each time series so all 1318 operators began as employees simultaneously in the model. Individual characteristics of the operator, which do not change over the course of employment, were not included in the time series. There was interest within Metro Transit in examining the influence of being assigned to a particular work site (garage), so the work history variables included this assignment, which can change by month during an operator's employment.

Because initial data visualization indicated more rapid turnover in the first two years of employment than thereafter, and because there has been a higher rate of hiring in Metro Transit since 2015, the first 24 months of employment were chosen to be included in the model (Figure 1). For each employee-month, the following was scored: whether the employee was in training, or on probation; the rolling sum of non-responsible incidents and accidents; the rolling sum of responsible accidents; the rolling sum of occurrences of absences from work (for any reason, and any duration); and the garage to which the operator reported for work. More than one garage could be recorded for one operator in a given month, so binary indicator variables were used for each location.

Statistical Approach

The prediction of interest in the analysis is a separation event; at some month of employment, an employee leaves, either voluntarily or involuntarily, reducing the workforce. The dataset was combined of employees who had separated involuntarily; had separated voluntarily; and had not separated. We classified terminated operators as involuntary separators, and all other operators as voluntary separators. Some unknown number of the employees that have not at the time of analysis separated, will do so before 24 months of employment. Thus the "losses" depicted in the timeseries in Figure 1 includes employees who have separated, but also employees who are still employed but have worked less than 24 months. The analysis must account for the characteristics of these operators as well.

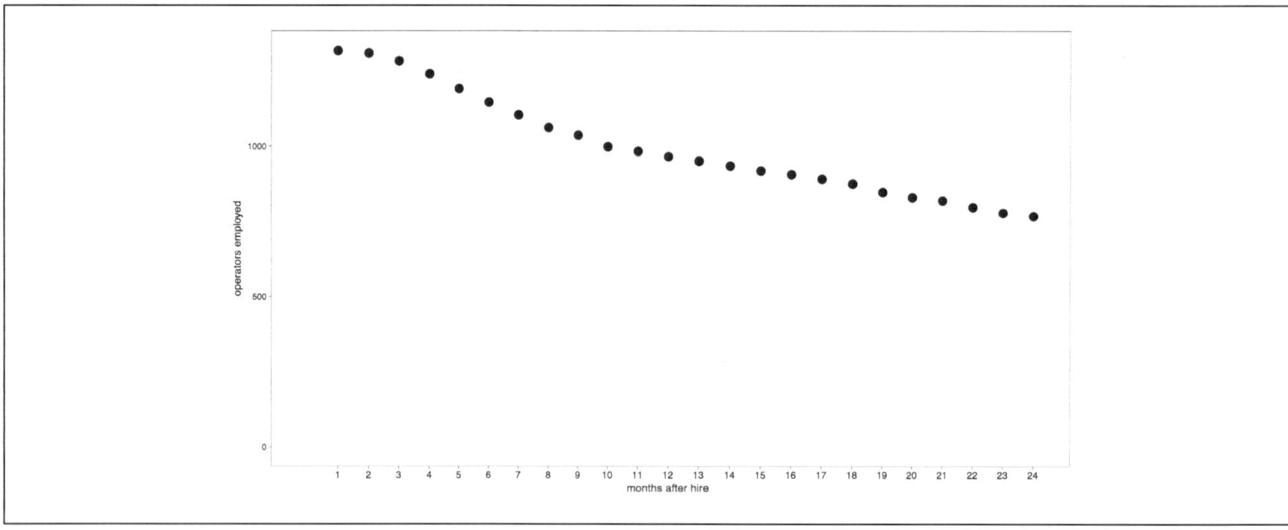

Figure 1. Time series of employment for 1318 operators at Metro Transit 2008–2016. Operator employment was aligned so that all began at month = 0. The time series shows the pattern of operator separation over the first 24 months of employment, regardless of year initiated.

To use both individual and work history variables in the same model to understand and predict separation, a probabilistic hierarchical survivorship model was constructed. In this case the "survival" to time t (in months) represents continued employment to that month, which integrates survivorship at that month as well as all previous months. The risk of separation is modeled as a Poisson random variable which varies by month,

$$\text{separation}_t \sim \text{Poisson}(\lambda_t) \tag{1}$$

The expected separation risk is a combination of two possible rates, based on the distinction between operators who separate voluntarily and those who separate involuntarily. Operators are categorized as voluntary separators according to probability θ, and as involuntary separators according to $(1-\theta)$, based on the individual characteristics of the operator. The expected separation risk λ for operator i at time t is modeled as

$$\lambda_{i_t} = \begin{cases} \lambda_{V_t} & \text{if } \theta_i = 1 \\ \lambda_{I_t} & \text{if } \theta_i = 0 \end{cases} \tag{2}$$

where λ_{V_t} is the rate of separation among voluntary operators in month t, and λ_{I_t} is the involuntary separation rate in month t. These rates are drawn from a normal distribution centered on a common hierarchical baseline rate which varies with time, with a pooled variance as

$$\left.\begin{matrix} \lambda_{V_t} \\ \lambda_{I_t} \end{matrix}\right\} \sim N(\lambda_\mu, \sigma_p) \tag{3}$$

That is, the different separation rates for voluntary or involuntary classification at each time point are connected through a common baseline λ_μ at that timepoint, and the pooled variance across timepoints σ_p. Thus, the risk of separating voluntarily λ_V and involuntarily λ_I both change through time, but in potentially different ways. The probability of an operator being in the voluntary category is drawn from a Bernoulli distribution (a single trial of a binomial distribution with probability p of being classified as voluntary).

$$\theta_i \sim \text{Bernoulli}(p) \tag{4}$$

The probability p of voluntary classification is linked by logit transformation to the effects of the individual characteristics,

$$\text{logit}(p) = \beta_{\text{ind}} x_i \tag{5}$$

where x_i are the individual characteristics of the operators (age, gender, CDL, education, and year hired) and β_{ind} is a vector of slopes representing the influence of those characteristics on the probability of an operator being classified as a voluntary separator.

To incorporate the work history time series, the baseline risk probability of each type in each month was scaled by the work history factors hypothesized to alter the risk of separation. In generic form, the scaling is an exponentiated function of the work history observations,

$$\lambda_{\text{scaled}}(t) = \lambda_t e^{\beta_w x_t} \tag{6}$$

where x_t are the work history variables in month t, and β_w is a vector of slopes representing the influence of that work history on scaling of separation risk at time t.

The work history scaling component of the risk model was allowed to vary with voluntary classification as well, so that the full expected risk model for operator i at time t becomes

$$\lambda_{i_t} = \begin{cases} \lambda_{V_t} e^{\beta_{w_V} x_t} & \text{if } \theta_i = 1 \\ \lambda_{I_t} e^{\beta_{w_I} x_t} & \text{if } \theta_i = 0 \end{cases} \tag{7}$$

where β_{w_V} and β_{w_I} represent the coefficients for the effects of the work history variables in the voluntary and involuntary group, respectively.

To summarize, operator retention was predicted by first probabilistically classifying operators as belonging to the voluntary or involuntary separator group, based on the individual characteristics of the operator. This represents a way to assign different risk pools based on these characteristics. Second, the risk in each month is estimated separately for voluntary and involuntary categories, but from a common pooled baseline risk that varies across types through time. Finally, the work history of an operator scales these baseline expectations, so that events or work locations that increase or decrease risk can be included in each time period for each operator. The probablility of a separation event is given by equation, and the probability that an operator i is still employed (has "survived") at month t is

$$S(i,t) = \exp\left(-\int_0^t \lambda_i(t)dt\right) \tag{8}$$

This formulation of the model allows diverse sources of information to be incorporated in one coherent framework, as well as allowing the estimation of the probabilities of a full set of parameters of interest. In addition to a predictive framework for operator separation based on characteristics and work history, the model estimates the following probabilities: the typical rate of operator separation overall across the dataset; how these rates differ over time, and between operators separating voluntarily and involuntarily; the relative influence of individual characteristics on probability of separating voluntarily or involuntarily; how work history variables contribute to overall separation probability; and how the influence of work history differs by separation type. Each of these parameters of interest are probabilistic, meaning the estimates are a distribution of values. Following McElreath (*8*) distributions were characterized as means and 89% credible intervals (denoted "89% CI"), which capture the density of the effect size distributions, but avoids prompting unconscious, binary "significance" assessment (which is not applicable to these models).

Implementation

The following numeric variables were centered (values subtracted from the mean) before analysis to allow for comparison across values on different scales: age, year hired, and self-reported highest education completed (coded as an ordered factor as follows: 1 = Some High School, 2 = High School, 3 = Some College, 4 = Technical College, 5 = Associate's Degree, 6 = Bachelor's Degree, 7 = Master's Degree or Doctorate).

The statistical model was implemented in Stan, a programming language for Bayesian probabilistic models (*9*). The RStan package in R version 3.3.3 (*10*) was used to interface with Stan (*11*). Noninformative, regulating priors were used for the λ and β variables, normally distributed with mean = 0, SD = 2. These priors allow for conservative estimation of effects based on the data while keeping effects within reasonable bounds (*12*). Half-Cauchy priors with mean = 0, SD = 2 were used for σ_p; the Cauchy is a student-t distribution with degrees of freedom = 1, restricted to have positive values (the "half" above zero) to ensure a positive estimate for σ_p (*12*). Four chains sampling 2000 times each after warmup of 1000 iterations were run, with the "adapt.delta" control option set to 0.9 to account for divergent transitions. Resulting model fits were examined for proper chain mixing, evidence of divergent transitions, and covariance among posterior estimates using the shinystan package in R. The full Stan model code can be found at https://metrotransitmn.shinyapps.io/operator-retention/.

Results

The Stan model converged successfully, and there were no divergent transitions. Mixing of MCMC chains was verified through visual inspection, and all reported estimates had good convergence based on Gelman-Rubin $\hat{R}$ statistics < 1.1.

Temporal Patterns of Separation Risk

Operators in the dataset separated from employment at an overall mean rate of 3.7% per month (89% CI: 2.7%, 4.9%). However, this overall rate hid variation both in types of operators and when they separate. The retention of operators had a clear time signal across months of employment (Figure 2). There was little to no probability of separation in months one and two following hire, corresponding to the typical training period at Metro Transit. Starting in month three, probabilities of separation increased markedly, with the highest probability of separation in months 4–8. Despite sharing a common baseline risk hazard, there is clear evidence for differing rates of retention among employees classified as voluntary and involuntary separators (Figure 2).

Operators voluntarily separated from employment at higher rates in month three to month eight of employment, than the remaining 16 months of employment, which indicated steady probability of turnover of about 3% per month. In contrast, involuntary separation

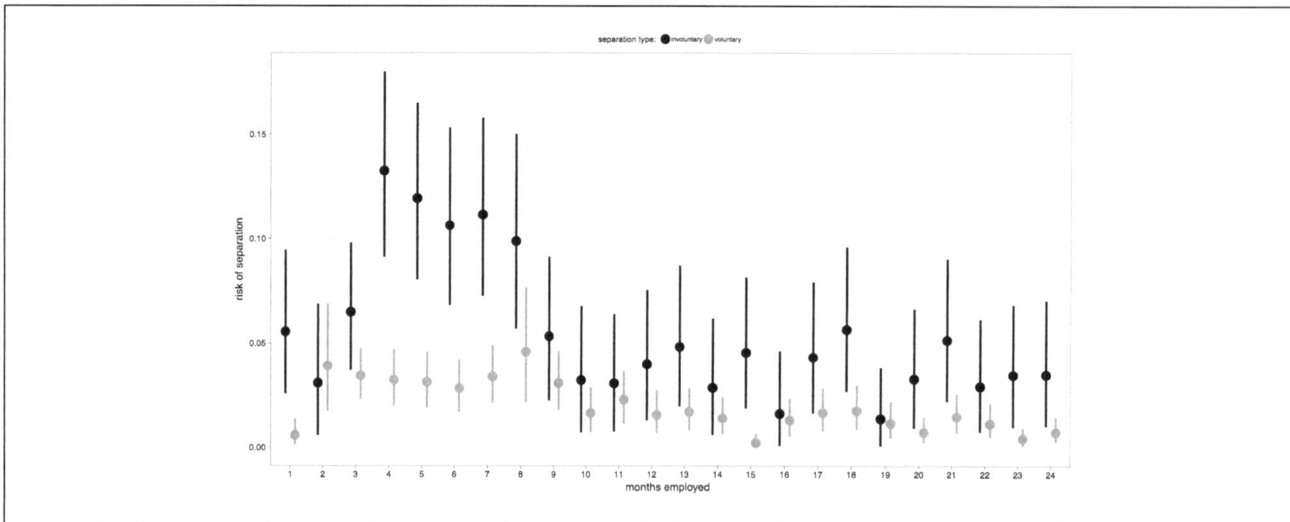

Figure 2. Risk rates of separating through time. Overall baseline probability of separation for bus operators at each month in the involuntary group (black) and the voluntary group (gray), depicted as the mean and 89% CI of the posterior probability distribution. Rates can be thought of as percent turnover during the month indicated.

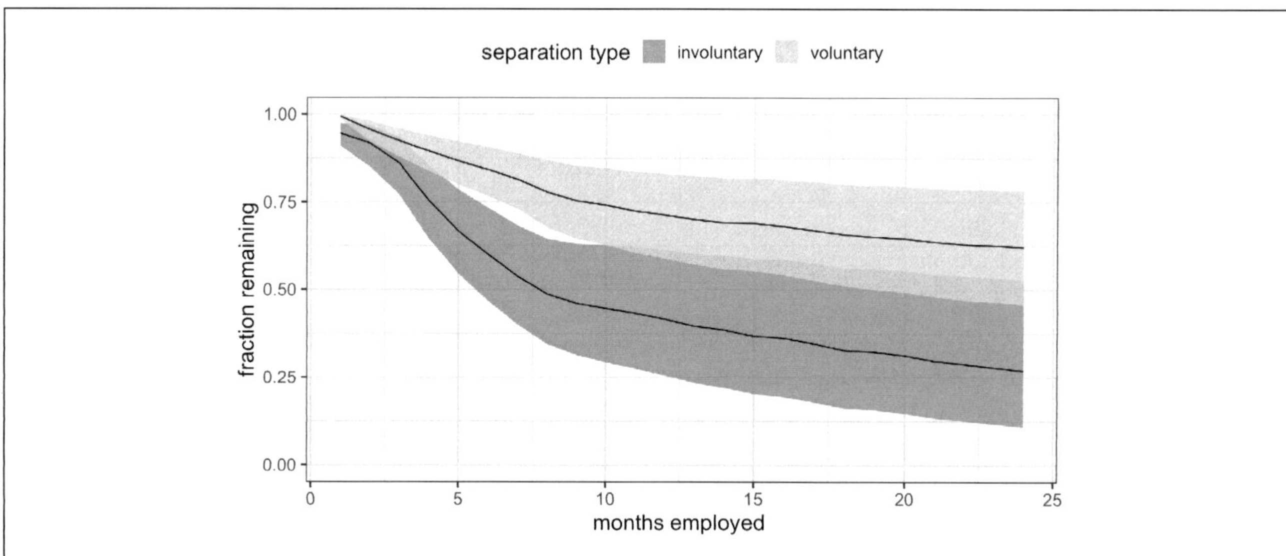

Figure 3. Fraction operators remaining employed through time. Line depicts mean and shading indicates 89% CI of the posterior probability of the fraction remaining after given number of months.

resulted in a much steeper decline in retained employees, starting from month two until month nine of employment. These higher separation rates early on result in large differences in retention over two years of employment (Figure 3). Of the employees predicted to be involuntarily separated, fewer than 50% were likely to be still employed after 10 months.

Individual Characteristics Predicting Separation

Because of the strong divergence in baseline rates, the predictors of classification of bus operators as voluntary or involuntary separators (θ_i) thus become extremely influential on the overall separation rate. The indicator of whether operators already held a CDL at hire was found to be the single most influential predictor of being classified as voluntary, and thus having a lower separation rate (Figure 4). Operators not holding a CDL at hire were predicted to be involuntary separators at 5.2 times the rate of CDL holders (derived from exponentiating the log-odds ratio; 89% CI: [4.4, 6.04]).

Age and highest education completed at hire had no measurable effect on classification as voluntary and thus

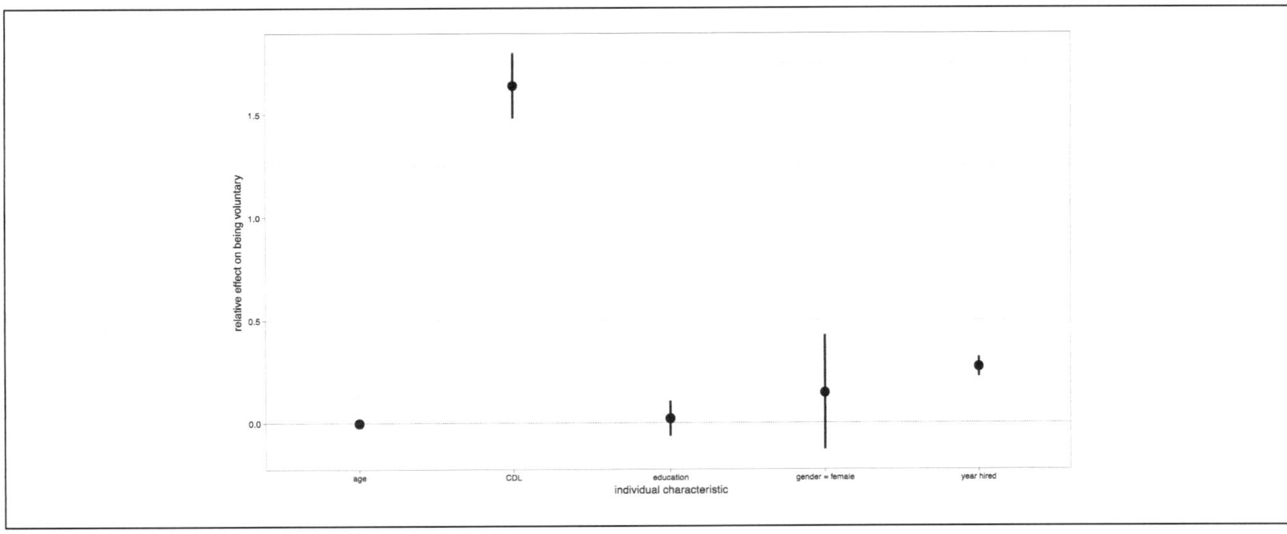

Figure 4. Standardized effects of characteristics on being voluntary. Points are mean, and bars depict 89% CI of parameter estimate, which is a linear predictor on the log-odds scale.

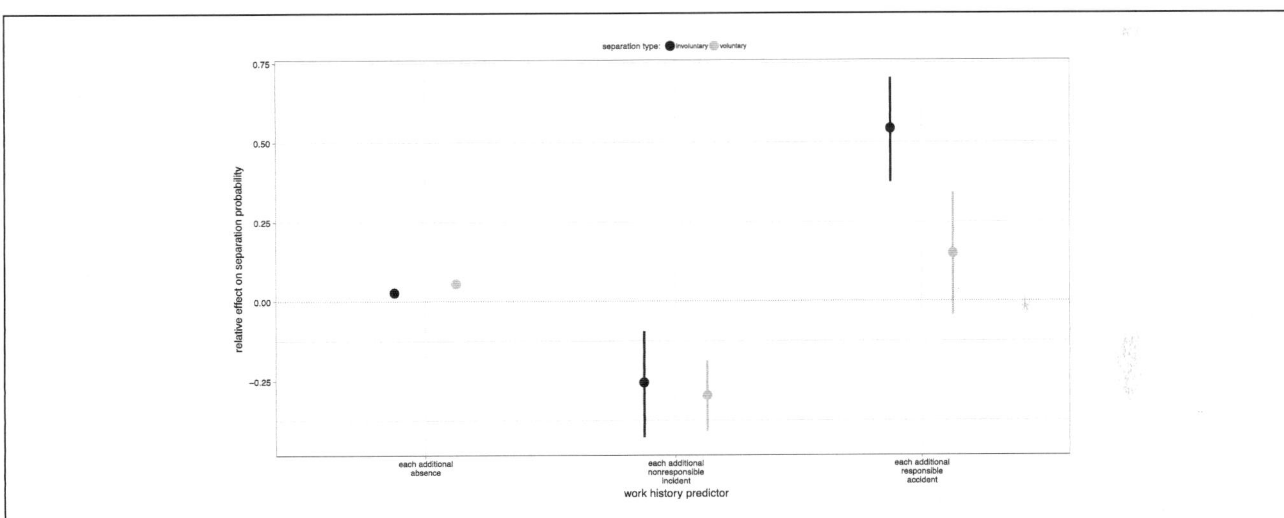

Figure 5. Standardized effects of work history variables on separation probability. Points are mean and bars depict 89% CI of parameter estimate, which is a linear predictor on the log scale.

separation rate. There was a slightly higher probability of females to be classified as voluntary, and a hire year where operators hired in more recent years had a higher probability of being classified as voluntary, and thus higher rates of retention (Figure 4).

Work History Variables Predicting Separation

In general, work history variables had similar influences on separation probablility across bus operators who voluntarily and involuntarily separated (Figure 5). Operators who were on probation in a given month had a much lower probability of separation, and operators in training had a slightly lower probability of separation, than the long-term trend. These results confirm the intuition that higher supervision and contact rates associated with the early phase of an operator's career do reduce the probability of separation. However, operators who were in training did not have a significantly lower risk of leaving voluntarily, indicating that at the time of training the dynamics of decision making leading to a voluntary separation were similar to the remainder of an operator's career.

Absences had small but predictable impacts which differed by separation type. For involuntary separators, each additional occurrence of an absence (a missed work day for any reason and for any duration) increased the monthly probability of separation by 0.027 (89% CI: [0.014, 0.04]). The effect was significantly higher when leading to voluntary separation, such that each additional absence increased the monthly probability of separation by 0.055 (89% CI: [0.042, 0.067]). Therefore, each additional absence by an operator indicated a substantially higher risk of voluntary separation through time.

Accident and incident work history variables differed in sign and strength of the impact on overall separation probability, as well as by separation type. The cumulative number of incidents classified upon review as not the responsibility of the operator predicted a lower separation probability, of either type. The per-incident decrease of the monthly probability of separation was by a mean of −0.255 (89% CI: [−0.428, −0.093]) for involuntary separators and −0.296 (89% CI: [−0.409, −0.188]) for voluntary separators.

In contrast, each additional accident for which the operator was responsible led to an increasing probability of separation, though this increase differed by separation type. Operators in the involuntary category increased their risk of separation by 0.543 (89% CI: [0.374, 0.703]) with each additional accident. The impact on operators separating voluntarily was significantly lower, and some of the parameter density was not greater than zero (mean 0.151, 89% CI: [−0.044, 0.34]).

The impacts of work location (garage assignment) were similar across separation types, with some garages leading to an increased risk of separation more than other locations. While informative for internal use these effects are not generalizable and are not reported here.

Discussion of Results

The separation of bus operators from their employment with Metro Transit followed a predictable temporal pattern, differing by voluntary and involuntary pathways, and was predictable from characteristics of individuals and their work history. The highest rates of separation were concentrated in months 3–10 after hire, which is after operators have finished training, and spanned the typical post-training probationary period (8 months at Metro Transit). This pattern was true of both involuntary and voluntary separation (Figure 2), though rates of involuntary separation were significantly higher during this period. By the end of the two-year time series used in this study, voluntary separation risk was very low and was equivalent to longer-term rates of turnover. Thus, one implication of our results is that interventions designed to reduce separation risk should be implemented from the beginning of the bus operator's work history, especially during the first six to eight months of on-street operation. By the end of the first year, operators had a much lower risk of separation, so while interventions throughout the course of employment are valuable, they have less room for improvement after this point in time.

The possession of a CDL at hire was the single most important factor in differentiating involuntary and voluntary separators. The significantly higher monthly risk observed in the baseline time series means that interventions to lower overall bus operator turnover should focus on reducing the pool of operators who are classified as candidates for involuntary separation. This does not mean that all operators beginning work without a CDL will separate involuntarily, but that they are more likely to be subject to the higher risk probability of involuntary separation, especially in the first 8 months of employment.

The occurrence of at-fault accidents magnifies this effect, as each additional accident for which the operator is deemed responsible has a large increase in separation risk per month (Figure 5). Two factors are at play here. First, a separate analysis of this dataset showed operators without a CDL were about 40% more likely to experience an accident or incident. Second, the influence of accidents on separation risk is disproportionately high in the involuntary separator group. This might indicate an ability for experienced operators to "bounce back" from accidents in a way that non-CDL-holding operators cannot. The pool of bus operators most at risk of separation are those hired without a CDL, who have an accident within the first six months of on-street operation. Interventions and training designed to reduce operator turnover overall, should prioritize training and other opportunities to reduce at-fault accident occurrence, especially in operators without commercial driving experience, and should act quickly to evaluate and improve driving skills for these operators in the early part of their work history.

There was an interesting positive recency effect of year of hire on operator retention, in which operators hired in the last few years had lower risk of separation (Figure 4). Partially this effect could be due to improved training and mentorship activities implemented within Metro Transit, though it could also be that separation probabilities for more recently hired operators are underestimated as many have not yet worked two full years and may separate before doing so. Nonetheless, the year effect suggests the cohort of bus operators hired at a given time may not act like those hired another time, which aligns with understanding of external forces such as local employment rates and other economic conditions.

In contrast to the direct experience of CDL holders, there was no effect of education level prior to hire on the probability of separation. However, education level in and of itself has a diffuse connection to the actual skills and abilities needed for daily operation of an in-service bus. Indication of previous experience in retail or other customer-facing jobs may provide predictive factors of the success of operators in the customer interaction portion of the job, analogous to the CDL predicting safe operation. Metro Transit has recently begun collecting such indicators from prospective applicants, which may provide interesting data for future modeling efforts.

The work history variables that increased separation risk per month included cumulative number of absences, and cumulative number of responsible accidents. These variables have long been used to evaluate bus operators (*2*), though use of these same metrics to predict ultimate separation is more novel. Surprisingly, the effect of cumulative absences was higher for operators separating voluntarily than involuntarily (Figure 5). While the per-absence effect is small, the cumulative nature of the probability of separation means small number of increases of absences can indicate significant increases in risk. For instance, an operator in the voluntary category with 16 absences after 16 months has a 5% higher cumulative probability of separation in that month than the same operator with only 9 absences to that point (the median of the dataset). Absences are known to cause immediate challenges to effective service operation, but these results demonstrate how as absences accumulate they can be a significant predictor of voluntary separation from employment. One previous study came to similar conclusions based on the accumulation of work-related injury leading to eventual disability (*3*), but all types of voluntary separation are included here. It is not possible to resolve with these data whether the mechanism of connection between absenteeism and eventual separation is physical, psychological (*7*), due to personal conflicts with schedules, or some other reason. Nonetheless, transit agencies would do well to use accumulations of missed work to serve as a warning sign of eventual separation and intervene accordingly.

Because the work history variables were modeled as having a constant effect throughout the operator's work history, in some cases the effects were smaller or of opposite sign than might be expected. For instance, while increasing absences from work did predict higher separation rates, the increase was modest when applied across 24 months. A more refined version of the model might examine how the influence of absences changes through time, so that multiple absences in month 4 of employment might increase separation probability more than multiple absences in month 22 of employment.

Similarly, the surprising result that operators decreased the risk of separation with each additional non-responsible incident might be explained by considering that as operators increase their service time, they will accumulate non-responsible incidents proportionally (*6*). Thus, the lowering of separation risk with increased non-responsible incidents may reflect the longevity of the operator. Again, a model with time-varying coefficients for these incident effects on separation would allow a better understanding of these dynamics.

The statistical approach used here is critical to aiding the understanding of bus operator retention. By using a hierarchical model which allows for both individual characteristics and aspects of work history to inform predictions of separation from employment, a single model captures distinct processes that contribute to employee turnover. By utilizing a probabilistic framework, the generative process of separation that includes unpredictable (or "random") factors not easily quantified in studies of human decision-making behavior is more accurately portrayed. In all, the flexibility of the modeling framework indicates useful extensions may be possible into similar operational questions, such as predicting absenteeism or the importance of work assignments to eventual separation.

Conclusion

Bus operator separation was successfully modeled as a combination of two distinct groups: high-risk bus operators who were terminated involuntarily, and voluntary separators who had lower turnover throughout the course of employment. Thus, increasing the pool of bus operators with characteristics matching the voluntary group should be a priority for transit agencies with similar dynamics. The most important characteristic of bus operators in this respect was the possession of a CDL, indicating that this should be a priority for hiring, and that agencies should evaluate training and mentorship for those bus operators hired without a CDL. Absences from the job, in addition to causing immediate operational challenges, can be an indicator of potential separation by bus operators in the long term. Accidents fed back to increase risk of separation more strongly in the already at-risk group, indicating that commercial driving experience may increase resilience to accidents during bus operation. A hierarchical model as used here is able to capture both static and dynamic characteristics of an employee and their work history and has promise for understanding other aspects of transit operations.

Acknowledgments

We thank Metro Transit staff, especially Donathan Brown and Scott Cady, for assisting with data compilation.

Author Contributions

The authors confirm contribution to the paper as follows: study conception and design: EL, JH; data collection: EL; analysis and interpretation of results: EL; draft manuscript preparation: EL, JH. All authors reviewed the results and approved the final version of the manuscript.

References

1. Cronin, C. B., A. Alexander, B. Cronin, C. Riches, J. Stern, R. Lazaro, and V. Lazaro. *TCRP Report 162: Building a Sustainable Workforce in the Public Transportation Industry—A Systems Approach*. Transportation Research Board of the National Academies, Washington, D.C., 2013.
2. Jacobs, R. R., J. M. Conte, D. V. Day, J. M. Silva, and R. Harris. Selecting Bus Drivers: Multiple Predictors, Multiple Perspectives on Validity, and Multiple Estimates of Utility. *Human Performance*, Vol. 9, No. 3, 1996, pp. 199–217. https://doi.org/10.1207/s15327043hup0903_2.
3. Kompier, M., H. Mulders, T. Meijman, M. Boersma, G. Groen, and R. Bullinga. Absence Behaviour, Turnover and Disability: A Study among City Bus Drivers in the Netherlands. *Work & Stress*, Vol. 4, No. 1, 1990, pp. 83–89. https://doi.org/10.1080/02678379008256968.
4. Cook, T., and J. Lawrie. Recruitment, Selection, and Retention of Personnel at North Carolina Community Transportation Systems: Findings and Recommendations from a Study. *Transportation Research Record: Journal of the Transportation Research Board*, 2004. 1884: 75–82.
5. Hausdorf, P. A., and S. D. Risavy. Predicting Training and Job Performance for Transit Operators. *International Journal of Selection and Assessment*, Vol. 23, No. 2, 2015, pp. 191–195. https://doi.org/10.1111/ijsa.12107.
6. Huting, J., J. Reid, U. Nwoke, E. Bacarella, and K. E. Ky. Identifying Factors That Increase Bus Accident Risk by Using Random Forests and Trip-Level Data. *Transportation Research Record: Journal of the Transportation Research Board*, 2016. 2539: 149–158.
7. Strathman, J., J. Broach, and S. Callas. Analysis of Short-Duration Unscheduled Absences of Transit Operators. *Transportation Research Record: Journal of the Transportation Research Board*, 2009. 2111: 185–194.
8. McElreath, R. *Statistical Rethinking: A Bayesian Course with Examples in R and Stan*. CRC Press, Taylor & Francis Group, Boca Raton; London; New York, 2016.
9. Carpenter, B., A. Gelman, M. D. Hoffman, D. Lee, B. Goodrich, M. Betancourt, M. Brubaker, J. Guo, P. Li, and A. Riddell. Stan: A Probabilistic Programming Language. *Journal of Statistical Software*, Vol. 76, No. 1, 2017. https://doi.org/10.18637/jss.v 076.i01.
10. Stan Development Team. *RStan: The R Interface to Stan*. 2016.
11. R Core Team. *R: A Language and Environment for Statistical Computing*. R Foundation for Statistical Computing, Vienna, Austria, 2017.
12. Gelman, A., J. B. Carlin, H. S. Stern, D. B. Dunson, A. Vehtari, and D. B. Rubin. (eds.). *Bayesian Data Analysis*. CRC Press/Chapman & Hall, Boca Raton, Fla., 2013.

The Standing Committee on Transit Management and Performance (AP010) peer-reviewed this paper (18-00853).

Article

TRR
JOURNAL OF THE TRANSPORTATION RESEARCH BOARD

Transportation Research Record
2018, Vol. 2672(8) 421–430

Article reuse guidelines:
sagepub.com/journals-permissions
DOI: 10.1177/0361198118786671
journals.sagepub.com/home/trr
SAGE

Assessing and Improving Operational Strategies for the Benefit of Passengers in Rail-Bound Urban Transport Systems

Anne Durand[1], Niels Van Oort[1], and Serge Hoogendoorn[1]

Abstract

Unplanned disruptions in transit can have consequent impacts on passengers. The more inconvenienced passengers are, the more likely operators will be negatively impacted. Yet so far, operators and researchers have addressed the rescheduling problem during disruptions mainly with a supply-side focus – timetable, crews, and vehicles – and not with a passenger perspective. Urban rail transit particularly lacks insights in terms of passenger-focused rescheduling. Being able to assess the inconvenience experienced by passengers during disruptions compared with what they normally experience, and being able to compare how different rescheduling strategies affect them are therefore two major challenges. The framework developed in this study precisely aims at tackling these challenges. A case study of the Rotterdam Metro is used to test the framework developed in this paper. Alternative strategies are developed focusing on the incident phase (from the beginning of the incident until its cause is resolved). The application of the framework reveals that a regularity-focused rescheduling strategy would be beneficial for high-frequency service users. Realistically, yearly savings could amount to around €900,000 in terms of societal passenger costs for the operator in the Rotterdam area alone. However, the omnipresence of the punctuality paradigm, through which most operators plan and analyze operations, makes the implementation of passenger-focused strategies a challenging task for traffic controllers. The results of the study are valuable for transit operators worldwide, and the framework could provide decision makers with insights on the performance of different strategies, bringing to light trade-offs between the supply and passenger sides during disruptions.

Passengers can be seriously inconvenienced by unplanned disruptions occurring in transit, that is, major events with "a beginning and an end in time and a location where its effects can be felt" (*1*). At the beginning of a disruption, traffic controllers decide on and implement a service control strategy, one of the final goals being to restore operations as originally planned. These strategies consist of service control measures that directly affect vehicles and thus passengers. When the implemented strategy results in unfavorable conditions for passengers, rippling adverse publicity and revenue loss may follow for the operator. It is therefore crucial for operators to take a closer look at control strategies and their implementation in non-recurrent conditions.

Rescheduling Research and Practices in Transit

Disruption management is usually divided into three categories in the rail transport domain: timetable adjustment, crew rescheduling, and rolling stock rescheduling (*2*). Only in the past few years has the focus started to shift toward passengers (*3*). A disruption consists of two distinct phases (*4*):

- The incident phase, from the start of the incident until the cause of the disruption is resolved. Capacity is scarce but demand peaks.
- The service recovery phase, when traffic controllers work to bring the system back to a target state.

Passengers and Transit Rescheduling Research. The lack of rescheduling research focusing directly on passengers can be explained by vehicle data having been available longer than passenger data. Technology has evolved in the past few years though, and smartcard data for instance have

[1]Department of Transport and Planning, Faculty of Civil Engineering and Geosciences, Delft University of Technology, The Netherlands

Corresponding Author:
Address correspondence to Anne Durand:
AnneLouise.Durand@gmail.com

tangible applications at the operational level (*5*), opening up new opportunities to focus more on passengers.

The first steps in the shift from a supply to a passenger focus in research pertain mostly to heavy rail (*3*), with the development of optimization algorithms including variables such as passenger travel and waiting time in objective functions (*6*). Although relevant, such studies do not necessarily fully acknowledge the complexity of the traffic controllers' environment and may therefore lack applicability (*1*). Unlike heavy rail, fewer studies taking the passenger perspective into account were produced for non-recurrent conditions in urban rail transit. This may be due to the apparent lower complexity of operations compared with heavy rail. Consequently, there is a gap in knowledge regarding passenger-centered rescheduling strategies for non-recurrent conditions in urban rail transit.

Rescheduling in Urban Rail Transit. To support traffic controllers, control strategies can be planned. In rail urban transit systems, partial pre-plans are acknowledged as reasonable (*1, 4*), providing elements like control measures and potential capacity issues at different times of the day, while leaving traffic controllers a degree of freedom (*4*). Many operators have already adopted pre-planning. The transit operator in Rotterdam in the Netherlands, RET (Rotterdam Electric Tram), designed pre-plans for each potential partial or full blockage in the metro network, consisting of lists of control measures. Traffic controllers at RET acknowledge however that the passenger perspective has never been formally considered in the design of these pre-plans.

Measures typically used in urban rail transit can be grouped in three main categories (see [*7, 8*] for more details):

- *Speed control measures*, including holding – delaying the departure of a vehicle to reduce headway variance – speeding up, and slowing down. The two latter have little application in high-frequency urban transit systems.
- *Station-skipping control measures*. These include expressing, deadheading, short-turning, diverting, and cancelling. Expressing is when a vehicle with passengers skips one or multiple stations that it was supposed to serve. Deadheading is similar, but with an empty vehicle. Short-turning means allowing a vehicle to turn and run in the opposite direction before it has reached its terminal. Diverting can occur when branches exist in the network. The ultimate station-skipping measure is the cancellation of a service. In all cases, the aim is to restore regularity, punctuality, or both. Because of the mismatch between supply and demand during the incident phase, an aggressive use of station-skipping measures is usually not recommended and short-turning is typically used, along with diversion when possible (*7*).
- *Other measures*. Service addition measures like adding a gap vehicle and implementing a shuttle service may be used. Furthermore, the single-track operations measure may be used when one of the two tracks of a rail system is unavailable and if crossover tracks are available on both sides of the disrupted segment. This measure is typically used during the incident phase.

Most research on these measures in rail-bound urban transit systems has been produced either with a passenger focus but with recurrent conditions in mind only (*9*) or for non-recurrent conditions but without acknowledging passenger impacts, typically for single-track operations (*10*). A recent study provides an exception though (*11*). Because passengers are considerably affected during the incident phase (*4*), and because of the lack of research on single-track operations, this study focuses on the incident phase.

Main Challenges When Investigating Passenger-Centered Rescheduling Strategies During Disruptions

There are two main challenges in this study. First, it is necessary to be able to quantify the inconvenience experienced by passengers in non-recurrent conditions with a clear framework, similar to what Van Oort et al. and Fadaei and Cats did for recurrent conditions (*9,12*). Second, the impacts of different strategies on passengers need to be compared for the same disruption, a challenge already highlighted in literature (*1*). To overcome this challenge, simulation is used.

Motivation of the Study and Outline

The aim of this study was to determine *how service control strategies used in non-recurrent conditions could be improved and developed when the passenger perspective is considered.* To this end, a framework was developed based on literature and analytical thinking. The framework was then tested using a case study of the Rotterdam Metro network. This framework allows the assessment, development, and comparison of strategies. The study demonstrates that there is room for improvement in terms of passenger-oriented rescheduling in the Rotterdam Metro and sheds light on the types of actions that benefit passengers during disruptions in high-frequency transit systems in general.

This paper fulfills the aforementioned objective by first presenting the development of the framework. The subsequent section presents the case study and the results. The final section then provides general conclusions, recommendations, and future research directions.

Development of the Framework

In this section, the development of the framework is described. Following a clearly defined objective, data needs are established. The structure of the framework is presented.

Framework Objective

The aim of the framework is to allow for multiple service control strategies to be assessed, developed, and then compared from a passenger perspective for one given disruption. To allow for a suitable quantification of the costs of a disruption, strategies should be assessed with recurrent conditions as a reference, and not ideal conditions as defined by the operations plan. Indeed, because of the variability of travel times inherent in transit, recurrent conditions are often different from what was planned. As inspired by Wilson et al., the assessment should allow distinguishing between different (groups of) OD (origin–destination) pairs to provide meaningful insights into trade-offs between groups of passengers (*13*).

Data Needs

In previous studies where passenger inconvenience is assessed in recurrent conditions, a combination of passenger and vehicle data is used (*9*, *11*). This is also the chosen approach here. As the passenger trip chain and vehicle processes are intertwined (*9*), supply-side (automatic vehicle location, AVL) data combined with demand-side (automatic passenger count, APC) data can facilitate derivation of passenger impacts (*14*). It is best to make use of APC data that easily allows reconstitution of the full passenger journey, as the way passengers are affected depends on their full journey (*13*). Data from smartcards that need to be tapped in at the beginning of the journey and tapped out at the end are thus needed.

Organization of the Framework

Using a combination of AVL and APC data led Van Oort et al. to define a three-step approach, providing a good basis for the assessment developed here (*9*). *Supply-side indicators* are computed based on AVL data and then translated into *passenger impacts* based on APC data. It is assumed that passengers do not cancel their trip or reroute; this assumption will be discussed in the Reflection section. Then, passenger impacts are turned into *monetary impacts* based on values of travel time components, thereby allowing for comparisons.

To complete the framework, passenger impacts need to be defined and a method must be chosen to generate alternative strategies. Ideally, to compare the impacts on passengers of various strategies, one would have the AVL data of multiple situations with the same disruption and identical circumstances but various strategies. As this is unlikely to ever happen, AVL data corresponding to the use of alternative strategies must be simulated.

Passenger Impacts. Quality indicators are required to properly quantify the quality of service (*15*), in this instance, the performance of a transit system in non-recurrent conditions from the passenger perspective. The chosen impacts are loosely based on research by Van Oort et al. (*9*). To derive each passenger impact, AVL data (from historical data or simulation) were used and supply-side indicators were computed (Step 1 of the three-step approach), such as headways and percentage regularity deviation mean (PRDM) (*16*). Let n be 0 or a positive integer, designating respectively the case when the strategy used by traffic controllers or an alternative strategy is used. For a single passenger traveling on the OD pair (y, z) in case n, the following impacts and their translation into costs (Steps 2 and 3 of the three-step approach) were chosen:

- Average additional effective in-vehicle time $\Delta t_{in,eff}^{n,(y,z)}$. The associated cost is found by multiplying this impact by a value of time (VoT, expressed in €/hour).
- Average additional waiting time at the first boarding $\Delta t_{wait}^{n,y}$. The associated cost is found by multiplying this impact by a value of waiting time ($VoWT$, expressed in €/hour).
- Average additional perceived in-vehicle time $\Delta t_{in,per}^{n,(y,z)}$. This is computed by multiplying in-vehicle time by factors that depend on the level of crowding, based on a meta-study (*17*). The associated cost is found by multiplying this impact by the VoT.
- Average additional denied boarding occurrences, with their associated waiting times. The average denied boarding cost for a passenger, $db^{n,y}$, already contains the $VoWT$ and assumes that denied boarding is equally distributed over passengers at station y. A crush capacity (maximum capacity, seating plus standing) assuming two people standing per square meter was chosen as it is the current approach at RET. It is noted cc_2.
- Average amount of unplanned transfers, with their associated waiting times. With $\lambda^{n,s}$ the probability of having to make an unplanned transfer at

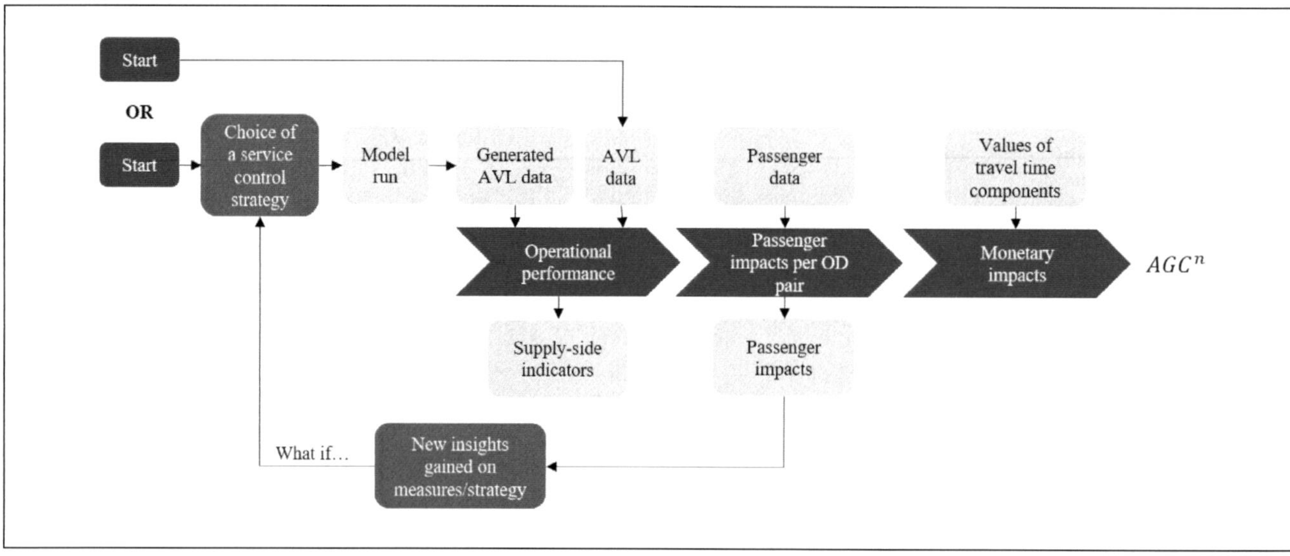

Figure 1. Overview of the developed framework to assess service control strategies and design alternative strategies in non-recurrent conditions from a passenger perspective.

a stop s on the route r from y to z (noted as $r_{(y,z)}$) in case n, and P the penalty of having to make such a transfer, the cost of this impact is the sum over all unplanned transfers of P and the additional waiting time multiplied by the $VoWT$.

The VoT for the Dutch situation was based on Kouwenhoven et al. (*18*). The VoWT was determined based on a trade-off between values from studies by Wardman, and by Van der Waard (*19*, *20*). The transfer penalty was based on Bovy and Hoogendoorn-Lanser (*21*). Using the variables described above, the additional generalized costs (AGC) in Euros in case n and for all passengers p are described by Equation 1.

$$AGC^{n} = \sum_{(y,z)\in S} AGC^{n,(y,z)} = \sum_{(y,z)} p^{(y,z)} \times \left(\left(\Delta t^{n,(y,z)}_{in,eff} + \Delta t^{n,(y,z)}_{in,per} \right) \times VoT + \Delta t^{n,y}_{wait} \times VoWT + db^{n,y} + \sum_{\forall s \in r_{(y,z)}} \lambda^{n,s} \times \left(\Delta t^{n,s}_{wait} \times VoWT + P \right) \right) \quad (1)$$

where S is the chosen spatial scope. Note that only average values are used for passenger impacts. This is a choice linked to the type of simulation used; more explanations can be found in the Reflection section.

Generation of Alternative Strategies. As previously mentioned, simulation is used to generate alternative strategies. The inputs of the simulation are the initial situation (i.e., location of the vehicles in the network) and a service control strategy. The latter consists of dispatching times from terminals, holding times at stops, number of dispatched vehicles, and routing of vehicles. The outputs of the simulation are AVL data corresponding to the use of alternative strategies: times are recorded at the arrival and the departure of each vehicle at each stop so that a file comparable to an AVL file can be generated (Figure 1).

Discrete-event simulation (using the software package Arena) was chosen because it allows for incremental changes to be easily implemented and queues to be easily modeled. In the simulation, each stop is represented as a resource. Each entity (train) successively seizes and releases each resource on its path. An entity can only seize a resource if it is available, and tracks and switches are available: this is the safety system. Otherwise, it queues. Queues follow a FIFO (first in, first out) discipline. Due to time constraints, modeling choices were made: passenger interactions were not included but APC data were used to estimate boarding and alighting rates at each stop. The time during which each resource is seized (dwell time) can be based on AVL data from previous disruptions, including expected crowding, and on when the next resource is available to seize (bunching time). Deterministic running times can be chosen if available AVL data show that this is a reasonable assumption.

Once the list of measures to be included in the assessment is established (see Rescheduling in Urban Rail Transit section) and the strategy used by traffic controllers is assessed, alternative strategies can be generated based on a "what-if" approach. The objective is to reduce AGC while making sure that improvements (on OD pairs) are not obtained at the expense of other passengers. Indeed, extreme inconvenience is likely to drive passengers away from urban transit (*22*). For each modification of a strategy, the variable inputs of the simulation model are incrementally modified by the analyst based on the results from the previously assessed strategies (passenger impacts for each OD pair). For instance, once the strategy used by traffic controllers is assessed, the analyst may decide to add, modify, or remove a measure (e.g., hold vehicles at the station s_1 to prevent congestion at station s_2) to mitigate the effects of the disruption on the most heavily impacted OD pairs. A new control strategy is then crafted. Testing traffic controllers' suggestions can also offer interesting insights. In the end, it is expected that sufficient knowledge on the situation will be gained to derive an ideal strategy. The generation of alternatives ceases when marginal improvements are no longer significant. This semi-automated process is a pragmatic approach using scientific techniques while still taking into account the constraints of traffic controllers (i.e., four trains cannot receive instructions at the same time with three traffic controllers). The ideal strategy is not guaranteed to be mathematically optimal, but this is not necessarily required here since the study predominantly aims to understand how each strategy that could *realistically be implemented by traffic controllers in their current work environment* affects passengers.

The discrete-event simulation model used within the framework was verified, calibrated, and validated before use. The selected disruption (see Presentation of the Case Study section) was reproduced with the simulation, yielding generated AVL data. These were translated into a time–distance diagram that was qualitatively validated by experienced traffic controllers. Generated AVL data were then compared with AVL data from the disruption by calculating impacts and AGC. Generated AVL data were found to slightly overestimate AGC, by 3%. This was deemed sufficient. The reader can refer to Durand for further details (*23*).

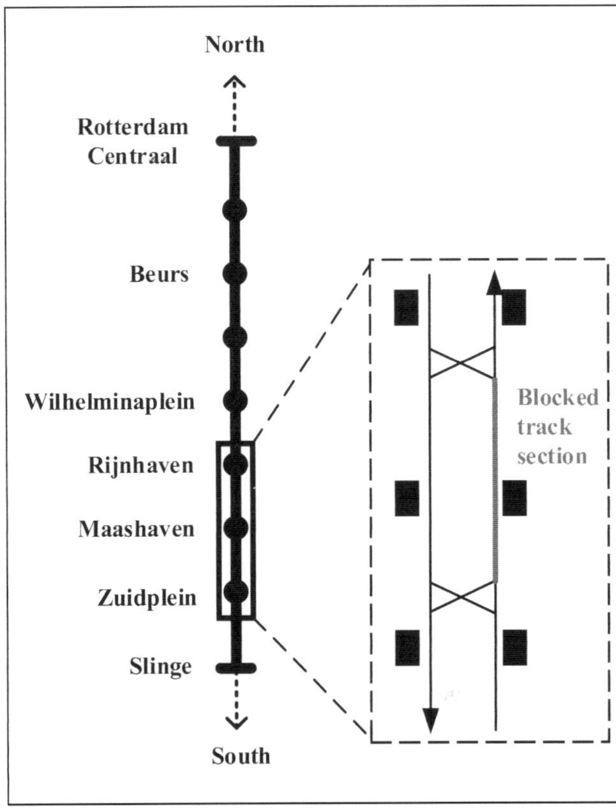

Figure 2. Slinge–Rotterdam Centraal track segment and zoom on the disrupted area.

Case Study

In this section, the case study is presented and the framework applied. Results are then described and implications for transit operators and conclusions pertaining to the framework itself are formulated.

Presentation of the Case Study

The framework was tested with a comprehensive case study, dealing with a disruption that occurred in Rotterdam Metro. Rotterdam is the second-largest city of the Netherlands, with approximately 630,000 inhabitants (*24*). RET operates the metro, all trams, and most buses. In 2017, the metro network comprised five lines and 53 stations. APC data as required in the Data Needs section are available due to the use of contactless smartcards by all passengers; see Van Oort et al. for more details (*25*). In 2016, 54% of all check-ins at RET were carried out in the metro and the total number of trips by metro amounted to 88 million (*26*). AVL data are available via the Dutch nationwide AVL system (*27*).

The most concentrated period in terms of disruptions is 6:00–8:00 a.m., whereas 8:00–9:00 a.m. is the busiest hour in terms of check-ins. Eighty-two percent of the morning trips in Rotterdam Metro are undertaken by commuters (*28*). It is likely that these frequent travelers will be affected in some way by a disruption at least once a month. A generic disruption occurring during the morning peak of a working day was therefore chosen as the case study: a vehicle was blocked from roughly 8:00 a.m. to 9:00 a.m. on the northbound track of Maashaven Station. Two lines share the tracks between Rotterdam Centraal and Slinge; one extends further north whereas the other extends further south (Figure 2). During peak hours (7:00–9:00 a.m. and 3:00–6:00 p.m.), the planned frequency is 18 vehicles per h.

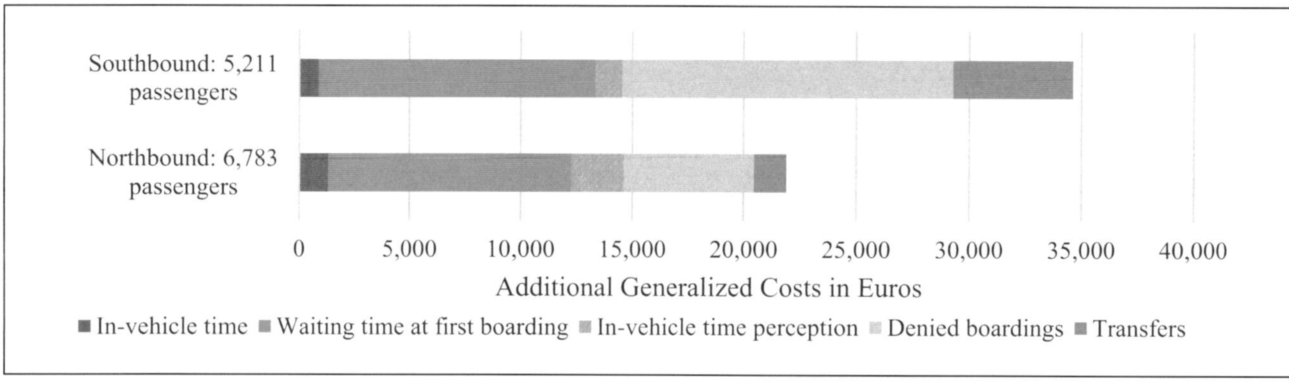

Figure 3. Assessment of the inconvenience experienced by passengers due to the disruption, with the strategy implemented by traffic controllers, strategy (A).

The corresponding pre-plan for such a disruption states that:

- All southbound vehicles from the line that extends north of Rotterdam Centraal should short-turn in Rotterdam Centraal, thereby leaving no train from this line between Rotterdam Centraal and Slinge. This reduces the frequency to 12 trains per hour between Rotterdam Centraal and Slinge.
- Vehicles from the other line should drive as planned, preferably alternating two by two on the single-track segment, the bottleneck section.

On the day of the disruption, traffic controllers started by applying this plan. However, since 12 vehicles in each direction on the single-track segment create congestion on both sides of the bottleneck, they also decided after several minutes to short-turn some northbound vehicles in Slinge, to relieve the bottleneck.

Results of the Framework Application

In the assessment, only passengers starting at, stopping at, or crossing at least one station between Rotterdam Centraal and Slinge were considered. This is a good starting point given that the pre-plan aims at "protecting" frequencies beyond this segment. The assessment is carried out for 1 h, the duration of the partial blockage, that is, the incident phase, the scope of this study. Extending beyond this phase involves further quantitative considerations on vehicles and crews, which are not yet part of this research.

Performance of the Strategy used by Traffic Controllers. Smartcard AVL data were used to assess the performance of the strategy used by traffic controllers. Although there were 30% more passengers traveling northbound than southbound, the latter suffered greater inconvenience on average, especially in terms of denied boarding (Figure 3). The observed imbalance is probably due to a significant headway gap in Rotterdam Centraal, around 18 min between two southbound trains. The corresponding AGC, using the base strategy (A), amounted to €57,000 of passenger societal costs.

Generation of Alternative Strategies. In total, 10 alternative strategies were generated with the procedure presented in the Generation of Alternative Strategies section. The three most important alternative strategies are presented here. These strategies are named (B), (C), and (D), respectively, the latter being the best-performing strategy.

- In strategy (B), the sequence of trains in the bottleneck for single-track operations was modified compared with strategy (A). During the transition phase, that is, from steady operations in recurrent conditions to steady operations in non-recurrent conditions, trains need to be sent into the bottleneck in a way that anticipates gaps in headway created by the unplanned event. Implementing this change alone reduced AGC by around 12%.
- Strategy (C) consists of strategy (B) to which holding at stations for regularity purposes was added. A combination of holding upstream and downstream of the bottleneck was found to be most efficient. Around half of the vehicles were held between 30 s up to 5 min. In strategy (C), an 18% reduction in AGC was achieved compared with strategy (A).
- Strategy (D) uses strategy (C) as a basis, and the implementation of short-turning in Rotterdam

Table 1. Bandwidth for Additional Generalized Costs

Strategy	(A)	(B)	(C)	(D)
Short description of the strategy	Base strategy: Traffic controllers' strategy: short-turning + single-track operations.	(A) + different sequence in the bottleneck.	(B) + holding for regularity purposes.	Best-performing strategy: (C) + A redirected train.
Bandwidth for AGC in €	57,000–43,000	50,000–37,000	47,000–35,000	35,000–28,000

Note: AGC = additional generalized costs.

Centraal as described in the pre-plan is delayed. This way, a vehicle can fill the gaps in headway created by the unplanned event. A 35% reduction in AGC was achieved compared with strategy (A) and no further impacts on the branch north of Rotterdam Centraal were expected. Denied boarding and waiting time at the first stop were the two components most significantly reduced with strategy (D), with a 61% and 29% decrease in costs, respectively, compared with strategy (A).

Furthermore, structural changes such as implementing short-turnings within the Rotterdam Centraal–Slinge segment led to substantially higher AGC than with strategy (A), indicating that pre-plans designed by RET most likely form a good basis. Nevertheless, the assessment shows that there is room for improvement regarding passenger-centered strategies. With strategy (A), a passenger traveling from Beurs to Wilhelminaplein is affected more than three times as much as a passenger traveling from Zuidplein to Beurs (€10.6 versus €3.4). This is paradoxical because the first passenger is traveling upstream of the blockage and never crosses it, whereas the second passenger does cross the bottleneck. This is probably due to traffic controllers focusing their efforts on the single-track segment. With strategy (D), the inconvenience for both OD pairs is reduced (€4.2 versus €2.9 respectively). In general, more focus on regularity led to a substantial decrease in AGC for OD pairs that did not cross the bottleneck, more than for OD pairs that did. Because of the ripple effects of a disruption, some OD pairs that do not cross the bottleneck remain more affected than OD pairs using the single-track segment. Nonetheless, a positive correlation was found between a stricter control of regularity – making sure that headway gaps were not becoming too large – and a decrease in total AGC.

Sensitivity Analysis. The starting point of this study was the current approach at RET, which is why the crush capacity cc_2 was used. In general, crush capacity is difficult to determine. It depends on the vehicle, stops on the line, interpersonal distance in culture, time of day, etc. However, research conducted at another Dutch urban transit operator (with relatively similar vehicles) shows that it is closer to 3 or 3.5 (*25*). A sensitivity analysis using these two numbers, noted respectively as cc_3 and $cc_{3.5}$, was therefore undertaken to evaluate the influence of this parameter.

Even with increased crush capacities, trends remained similar. There was still a decrease by around 35% in AGC with strategy (D) compared with strategy (A) and denied boarding still occurs with all strategies, although to a lesser extent with strategy (D). This analysis allowed derivation of bandwidths for the results (Table 1). Such bandwidths were even more relevant, as simulation outputs were shown to slightly overestimate AGC (see Generation of Alternative Strategies section).

An estimate for yearly AGC savings can be computed based on these bandwidths. In 2016, approximately 60 disruptions like the case study regarding time, duration, and cause occurred in the RET metro network. Assuming around €15,000 of savings in AGC per incident, savings for passengers could amount to approximately €900,000 if every disruption like the case study was addressed with a strategy similar to (D) (a saving of €1 of AGC means reducing the waiting time of one passenger by 5 min).

The final step of the framework test is the validation of results. No quantitative validation was done because of the constraints of the project, however multiple interviews were conducted. Technical experts at RET all found both the qualitative and the quantitative results to be reasonable and realistic; moderate changes in strategy led to moderate – yet non-negligible – improvements. The authors therefore consider the framework to be currently valid and are working on an improved validation process.

Implications in Practice for Operators and Authorities. Although a significant part of this research was a case study, multiple insights are transferable to other operators and authorities worldwide; a global lack of passenger focus has been observed by Van Oort (*29*). In non-recurrent conditions, traffic controllers need to focus more on regularity at a network scale. Focusing on regularity has already been found to be beneficial for passengers (*30*). However, traffic controllers are used to working with a punctuality paradigm as their daily goal is to ensure that deviations from the operations plan (timetable, crew, and vehicle schedules), built on punctuality, are as small as possible. This is the main dilemma. Nevertheless, authorities and operators should strive to make it easier for traffic controllers to focus on passengers when needed. Such organizations should therefore ask themselves the following questions:

- *Are traffic controllers sufficiently aware of passenger issues related to a lack of regularity?* Busy with daily work, some may not be completely aware of the punctuality paradigm. Furthermore, they may have been metro train drivers in the past, resulting in a particularly acute awareness of crew and vehicle issues.
- *Is the work environment of traffic controllers conducive to a focus on regularity when needed?* For instance, at RET, controllers can see train delays on control screens, but no regularity-related indicator is displayed, such as the time elapsed since the last train at each station. This makes it almost impossible for them to spot or anticipate gaps in headway. Pre-plans and their application should also be questioned.
- *Are there elements outside of the traffic control center (tactical, strategic levels) that may indirectly contribute to traffic controllers' limited focus on passengers?* For instance, operations at RET are being analyzed with indicators that are typically not passenger-oriented, a conclusion based on a comparison between the metrics used at RET and those suggested by Barron et al. (*31*). Moreover, the incentives defined by the authority that grants RET the right to operate Rotterdam Metro as a concession are mostly punctuality-based. These incentives do afford traffic controllers some freedom to focus on regularity without the risk of being fined, but this involves strict registration and justification of every action, which is currently materially impossible.

Modifications of the work environment and procedures are likely to increase traffic controllers' workload, at least on a short-term basis. Changes should be gradual and consider the existing work culture. For instance, RET could start by implementing regularity-related indicators in their monitoring system and by modifying pre-plans. They could include peak-hour variants and estimated capacity for a given bottleneck for each potential partial blockage. Such changes would already allow for traffic controllers to be more proactive.

Reflection

In the assessment, it is assumed that passengers do not reroute or cancel their trips. The no-rerouting assumption may not always hold. In 2017, 79.2% of the Dutch population had a smartphone and used it to access the Internet at least once every three months (*32*). Passengers are likely to use a journey planner to quickly reroute. Some may simply know the network and find alternatives routes themselves. Parts of the network in the case study offer limited alternatives, but the assessment may still have overestimated denied boarding and waiting time costs. The no-cancellation assumption may be more correct as, during the morning peak, 82% of the passengers on Rotterdam Metro are commuting for work and study (*28*). Thus, it would be difficult for most people to postpone their trips. Further research on these assumptions would be needed.

If these passengers were to experience frequent and serious inconveniences, they may be driven out of the system. Therefore, an additional impact in the assessment could be the "long-term effect" of a disruption. Reliability buffer time (RBT) could be used for this purpose. Uniman et al. define this as the difference between the 95th percentile travel time and the 50th percentile travel time (*33*). An RBT of 5 min therefore means that if a commuter plans 5 min of buffer time for their journey, they will reach their destination on time 95% of the time. A high RBT in non-recurrent conditions might lead passengers to readjust their departure time, extending the impacts of the disruption to multiple days or weeks. A long-term effect can thus be computed. In general, using extreme-value-based impacts instead of average impacts is recommended, because they better reflect passengers' perceptions of traveling and thus inconvenience (*22*). However, a simulation with a greater level of detail would have been required to make the use of extreme-value-based impacts relevant. Average values were deemed more likely to be accurate for this study.

Conclusions and Recommendations

This study aimed to determine how service control strategies used in urban rail systems could be improved and developed when the passenger perspective is considered. The framework allowed assessment of the inconvenience experienced by passengers during the incident phase of a disruption. It also allowed for the development and comparison of the performance of various service control strategies in response to one specific disruption, each strategy yielding a different quality of service. In the

Rotterdam case study, societal passenger costs were reduced by up to €7 per passenger (e.g., one unplanned transfer and 30 min of waiting time) for certain OD pairs with the best-performing strategy. Yearly savings in societal passenger costs could amount to approximately €900,000 if every disruption like the case study is addressed with an approach similar to the best-performing strategy.

Based on the study results, insights were gained into passenger-centered disruption management in high-frequency, rail-bound transit systems in general that could apply to transit operators and authorities worldwide. In such systems, pre-plans should exist for potential full or partial blockages and provide variants for peak hours, highlighting potential capacity issues. In addition, results suggest that real-time decisions favoring regularity considerations could benefit passengers considerably. However, instead of attributing by default the lack of passenger focus to traffic controllers, operators should first question whether traffic controllers are able to focus on regularity. Moreover, all levels of a company (operational, tactical, and strategic) could potentially influence the amount of attention passengers are given during disruptions. Naturally, all changes come at a cost. Therefore, it is suggested that decision makers use the outputs of the framework developed in this study as inputs for a cost-benefit analysis, to determine the extent to which being passenger-oriented is worthwhile.

Furthermore, insights were gained into the single-track operations measure. In rail-bound transit systems, this measure allows avoidance of cutting a full segment because of a partial blockage. There were two main findings: first, planners and traffic controllers should have an idea of the capacity the single-track segment can reasonably sustain and, second, single-track operations need to be associated with proactive holding upstream of the bottleneck to prevent major irregularities. Nonetheless, conducting more research to gain a better understanding of the different variables that influence the performance of this measure is suggested. Three other areas of research are also suggested for the medium term. First, more research regarding the behavior of urban transit passengers during unplanned non-recurrent conditions could be useful, to better understand how, why, and which passengers choose to reroute or give up on their trips. It would also be useful to know what might drive passengers away from urban transit. Once there is more knowledge in these areas, passengers could be modeled as entities in future simulations and passenger interactions could be properly taken into account. Second, undertaking more research on in-vehicle crowding is recommended. Indeed, capacity was found to be a major variable of the assessment and thus deserves attention. Further studies could use AVL and APC data to determine crush capacities more precisely. In general, it would be interesting to know which factors influence crush capacity and how. Third, extending the assessment with some non-passenger-related impacts, such as the impact on crews is recommended. Ultimately, this would allow the framework to be applied to the recovery phase. The framework would then provide even more meaningful insights into the possible trade-offs that could be made by traffic controllers during a disruption.

Acknowledgments

This research was made possible owing to working closely with RET, which provided the necessary data, resources, and support to complete this study. The authors thank the anonymous reviewers for their valuable comments.

Author Contributions

The authors confirm contribution to the paper as follows: study conception and design: AD, NVO, SH; data collection: AD; analysis, interpretation of results, and draft manuscript preparation: AD, NVO, SH. All authors reviewed the results and approved the final version of the manuscript.

References

1. Carrel, A., R. Mishalani, N. Wilson, J. Attanucci, and A. Rahbee. Decision Factors in Service Control on High-Frequency Metro Line. *Transportation Research Record: Journal of the Transportation Research Board*, 2010. 2146: 52–59.
2. Jespersen-Groth, J., D. Potthoff, J. Clausen, D. Huisman, L. Kroon, G. Maróti, and M. N. Nielsen. Disruption Management in Passenger Railway Transportation. *Robust and Online Large-Scale Optimization*, Vol. 5868, 2009, pp. 399–421. http://dx.doi.org/10.1007/978-3-642-05465-5_18.
3. Van der Hurk, E., H. N. Koutsopoulos, N. H. M. Wilson, L. G. Kroon, and G. Maroti. Shuttle Planning for Link Closures in Urban Public Transport Networks. *Transportation Science*, Vol. 50, 2016, pp. 947–965. http://doi.org/10.1287/trsc.2015.0647.
4. Moore, A. *A Process for Improving Transit Service Management During Disruptions*. Master thesis. Massachusetts Institute of Technology, 2003.
5. Pelletier, M.-P., M. Trépanier, and C. Morency. Smart Card Data Use in Public Transit: A Literature Review. *Transportation Research Part C: Emerging Technologies*, Vol. 19, 2011, pp. 557–568. http://dx.doi.org/10.1016/j.trc.2010.12.003.
6. Veelenturf, L, L. Kroon, and G. Maróti. Passenger-oriented Railway Disruption Management by Adapting Timetables and Rolling Stock Schedules. *Transportation Research Part C: Emerging Technologies*, Vol. 80, 2017, pp. 133–147. http://doi.org/10.1016/j.trc.2017.04.012.
7. Carrel, A. *Diagnosis and Assessment of Operations Control Interventions*. Master thesis. Massachusetts Institute of Technology, 2009.

8. Van Oort, N. *Service Reliability and Urban Public Transport Design Service*. Doctoral thesis. Delft University of Technology, 2011.
9. Van Oort, N., T. Brands, E. De Romph, and A. Flores. Unreliability Effects in Public Transport Modelling. *International Journal of Transportation*, Vol. 3, 2015, pp. 113–130. http://dx.doi.org/10.14257/ijt.2015.3.1.08.
10. Louwerse, I., and D. Huisman. Adjusting a Railway Timetable in case of Partial or Complete Blockades. *European Journal of Operations Research*, Vol. 235, 2014, pp. 583–593. http://dx.doi.org/10.1016/j.ejor.2013.12.020.
11. Cats, O., M. Yap, and N. Van Oort. Exposing the Role of Exposure: Public Transport Network Risk Analysis. *Transportation Research Part A: Policy and Practice*, Vol. 88, 2016, pp. 1–14. http://dx.doi.org/10.1016/j.tra.2016.03.015.
12. Fadaei, M., and O. Cats. Evaluating the Impacts and Benefits of Public Transport Design and Operational Measures. *Transport Policy*, Vol. 48, 2016, pp. 105–116. http://dx.doi.org/10.1016/j.tranpol.2016.02.015.
13. Wilson, N. H., R. Macchi, R. Fellows, and A. Deckoff. Improving Service on the MBTA Green Line through Better Operations Control. *Transportation Research Record: Journal of the Transportation Research Board*, 1992. 1361: 296–304.
14. Furth, P. G., B. Hemily, T. H. J. Muller, and J. G. Strathman. *Using Archived AVL-APC Data to Improve Transit Performance and Management*. Transit Cooperative Research Program (TCRP) Report 113. Transportation Research Board, Washington D.C., 2006. http://dx.doi.org/10.17226/13907.
15. Kittelson & Associates Inc., Parsons Brinckerhoff, KFH Group Inc., and Texas A&M Transportation Institute, Arup. *TCRP Report 165: Transit Capacity and Quality of Service Manual*. Transportation Research Board of the National Academies, Washington D.C., 2013.
16. Van Oort, N. Incorporating Enhanced Service Reliability of Public Transport in Cost-Benefit Analyses. *Public Transport*, Vol. 8, 2016, pp. 143–160. http://dx.doi.org/10.1007/s12469-016-0121-3.
17. Wardman, M., and G. Whelan. Twenty Years of Rail Crowding Valuation Studies. *Transport Reviews*, Vol. 31, 2011, pp. 379–398. http://dx.doi.org/10.1080/01441647.2010.519127.
18. Kouwenhoven, M., G. De Jong, P. Koster, V. A. C. Van den Berg, E. T. Verhoef, J. Bates, and P. M. J. Warffemius. New Values of Time and Reliability in Passenger Transport in The Netherlands. *Research in Transportation Economics*, Vol. 47, 2014, pp. 37–49. http://dx.doi.org/10.1016/j.retrec.2014.09.017.
19. Wardman, M. Public Transport Values of Time. *Transport Policy*, Vol. 11, 2004, pp. 363–377. http://dx.doi.org/10.1016/j.tranpol.2004.05.001.
20. Van der Waard, J. The Relative Importance of Public Transport Trip-time Attributes in Route Choice. *Proc., PTRC*, London, 1988, pp. 67–78.
21. Bovy, P. H. L., and S. Hoogendoorn-Lanser. Modelling Route Choice Behavior in Multi-modal Transport Networks. *Transportation*, Vol. 32, 2005 pp. 341–368. http://dx.doi.org/10.1007/s11116-004-7963-2.
22. Furth, P. G., and T. H. J. Muller. Service Reliability and Hidden Waiting Time. *Transportation Research Record: Journal of the Transportation Research Board*, 2006. 1955: 79–87.
23. Durand, A. *Managing Disruptions in Public Transport from the Passenger Perspective*. Master thesis. Delft University of Technology, 2017.
24. City of Rotterdam. *Rotterdam in Numbers* (in Dutch), 2017. http://rotterdam.buurtmonitor.nl. Accessed July 2017.
25. Van Oort, N., T. Brands, E. de Romph, and M. Yap. Ridership Evaluation and Prediction in Public Transport by Processing Smart Card Data: A Dutch Approach and Example. In *Chapter 11, Public Transport Planning with Smart Card Data* (F. Kurauchi, and J. D. Schmöcker, eds.), CRC Press, 2016. http://doi.org/10.1201/9781315370408-15.
26. RET. *Annual Report 2016* (in Dutch), 2017. http://retjaarverslag.nl. Accessed July 2017.
27. Van Oort, N., D. Sparing, T. Brands, and R. M. Goverde. Data-driven Improvements in Public Transport: The Dutch Example. *Public Transport*, Vol.7, 2015, pp. 369–389. http://dx.doi.org/10.1007/s12469-015-0114-7.
28. RET. *Customer Satisfaction Survey, Metro Passengers*. Survey conducted between November 2015 and April 2016.
29. Van Oort, N. Incorporating Service Reliability in Public Transport Design and Performance Requirements: International Survey Results and Recommendations. *Research in Transportation Economics*, Vol. 48, 2014, pp. 92–100. http://dx.doi.org/10.1016/j.retrec.2014.09.036.
30. Schmöcker, J., S. Cooper, and W. Adeney. Metro Service Delay Recovery: Comparison of Strategies and Constraints Across Systems. *Transportation Research Record: Journal of the Transportation Research Board*, 2005. 1930: 30–37.
31. Barron, A., P. Melo, J. Cohen, and R. Anderson. A Passenger-Focused Management Approach to the Measurement of Train Delay Impacts. *Transportation Research Record: Journal of the Transportation Research Board*, 2013. 2534: 75–83.
32. CBS (Netherlands Statistics). *Internet: access, use and devices* (in Dutch), 2016. http://statline.cbs.nl/Statweb/publication/?DM = SLNL&PA = 83429NED&D1 = 0,2-5,20&D2 = 0,3-6&D3 = 0&D4 = l&HDR = T&STB = G1,G2,G3&VW = T. Accessed June 2018.
33. Uniman, D. L., J. P. Attanucci, R. G. Mishalani, and N. H. Wilson. Service Reliability Measurement Using Automated Fare Card Data. *Transportation Research Record: Journal of the Transportation Research Board*, 2010. 2143: 92–99.

The Standing Committee on Transit Management and Performance (AP010) peer-reviewed this paper (18-00524).

TRR
JOURNAL OF THE TRANSPORTATION RESEARCH BOARD

Article

Transportation Research Record
2018, Vol. 2672(8) 431–442

DOI: 10.1177/0361198118772702
journals.sagepub.com/home/trr
SAGE

Application of a Cost-Allocation Model to Swiss Bus and Train Lines

Marc Sinner[1], Ulrich Weidmann[2], and Andrew Nash[3]

Abstract

Public transport costs are an important decision factor when making system design choices and network or timetable planning. Although many cost models of varying degrees of sophistication have been proposed in literature, practitioners often use relatively simple models such as purely vehicle-kilometer-based cost estimations. This paper applies a cost-allocation model to Swiss bus and train lines. The proposed model is based on operating statistics including 'productive hours', vehicle-kilometers, vehicle number and size. The cost model is then applied to a case study for a generic public transport line with a typical demand distribution. The case study calculates the demand level at which train service becomes more effective than bus service for various combinations of minimal frequency, average speed and maximum number of coupled Electric Multiple Units.

This section presents a brief literature review of research in the domain of cost models for public transport.

In the early 1980s, tightening public funds increased interest in cost models for both researchers and practitioners. Cherwony et al. (*1*) present an overview of the most commonly used models, their assumptions as well as the parameters needed for use. The most used of these models, the cost-allocation model, assigns cost categories to specific operating statistics like vehicle-hours, vehicle-km or number of vehicles. The unit cost per operating statistic depends on the type of vehicle or the time period (peak or off-peak).

Subsequently these models were refined by investigating specific aspects in greater detail and applied to different regions. For instance, Peskin (*2*) presents a detailed cost-allocation model for bus- and rail-based public transport in Houston, Texas, whereas Stopher et al. (*3*) use data from Southern California. The latter also formulate general requirements for cost models. Cherwony and Porter (*4*) analyze how marginal schedule changes affect driver costs. Similarly, Carter et al. (*5*) compare how different models predict the delta-costs resulting from service expansions or reductions. Parody et al. (*6*) address how to allocate capital costs for infrastructure, vehicles, and fare revenues to peak and off-peak services. The unit cost differences between peak and off-peak periods were also investigated by Cherwony and Mundle (*7*) as well as Cervero (*8*). Nuworsoo (*9*) presents a method to include pull-out and pull-in trips to the cost calculations, which helps making decisions on the optimal location of the bus storage facility. Most of these studies focus on bus operation. Only a few also consider rail services. For instance, Bruun (*10*) presents a parametric cost model for urban transport by comparing bus rapid transit (BRT) and light rail transit (LRT) operations.

At the same time, European researchers focused especially on comparing costs between bus and rail, both in urban and rural context. Whereas in the U.S. commuter rail networks are mostly concentrated around the larger cities, European countries also have extensive railway networks with regional passenger traffic outside the major agglomerations. Allport (*11*) compares the costs of buses, trams and metros in Rotterdam. The study found that labor was the largest cost segment for bus services, whereas infrastructure was largest for metros. Aberle et al. (*12*) estimate the costs of different bus types and small diesel trains for rural lines in Germany. Leuthard completed similar research (*13*), and later published several studies on bus and tram line cost structure (*14, 15*).

More recent publications by the German Forschungsgesellschaft für Strassen- und Verkehrswesen (FGSV) (*16*) or by Hass-Klau et al. (*17*) focus on city

[1]ETH Zurich, Institute for Transport Planning and Systems, Zurich, Switzerland
[2]ETH Zurich, Vice President Human Resources and Infrastructure, Zurich, Switzerland
[3]Emch+Berger AG Bern, Bern, Switzerland

Corresponding Author:
Address correspondence to Marc Sinner: marc.sinner@ivt.baug.ethz.ch

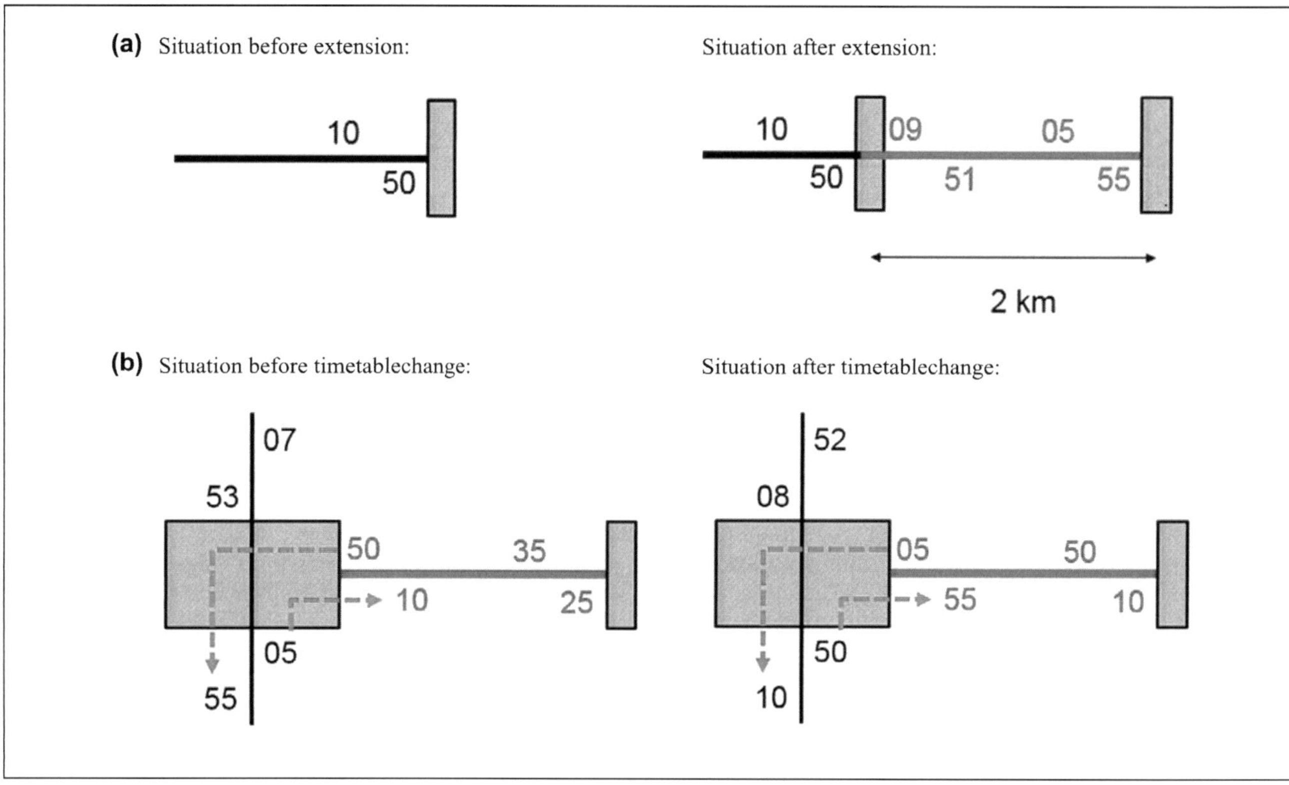

Figure 1. Examples of service changes.

traffic and the system choice between buses and trams. Frank et al. (*18*) developed a method to estimate the annual cost of buses based on several input parameters. It can also calculate the costs of single lines or marginal line changes by comparing 'before' and 'after' scenarios. However, the method requires a large amount of input data.

Despite the general research conclusion that simple cost estimation procedures linking costs to a single operating statistic (e.g. vehicle-km) are not sufficiently accurate (*4, 19*), many practitioners are reluctant to use more sophisticated models, such as proposed by Harmatuck (*20*). Instead many still use kilometer-based values (*21, 22*) to estimate costs. Taylor et al. (*23*) point out a gap between research and practice by operators and public authorities. In most European countries, public tenders use cost per vehicle-Km as a measure for efficiency (see Avenali et al. (*24*) for an Italian example). Often, the lack of the required data for more elaborate models is a problem, as Bösch et al. (*25*) point out for Switzerland.

Figure 1 shows two extreme cases in which cost estimation based solely on distance falls short. The respective timetable settings are displayed as netgraphs (commonly used to describe periodic timetables). They concisely illustrate the timetable of periodic lines by only showing the arrival and departure minutes (which recur every hour). The numbers closer to the gray boxes (representing stations) correspond to the arrival minute and those further away correspond to the departure minute. Here, the arrival and departure minutes are always on the right-hand side of the line in the direction of travel they refer to.

Example 1

There is an hourly regional bus line with a zero-symmetric periodic timetable having a layover time of 20 min (see Figure 1*a* left). The bus arrives at the layover at HH:50 and leaves at HH:10. As shown on the right side of Figure 1*a*, the route could be extended by 2 km, thus reducing the layover to roughly 10 min (arrives at new layover HH:55, leaves HH:05). A kilometer-based calculation using estimates published by the Swiss Federal Office of Transport (*26*) would yield the following additional costs for this extension:

$$2 \cdot 2\,\text{km/trip} \cdot 6.86\,\text{CHF/km} = 27.44\,\text{CHF/trip}$$

For each trip, the additional cost would be 27.44 CHF, although vehicle and driver are available (as they are currently just laying over). The additional cost is thus overestimated.

Example 2

The kilometer-based calculation can also underestimate the operating costs. Figure 1*b* (left) shows a situation in which an hourly bus with 15 min travel time (marked in red) connects to a train (marked in black) with hourly service, too. The timetable is assumed to be periodic with network-wide symmetry at the full hour. The connection times for the main demand relation of the bus line (shown with green arrows) are short (bus arrives at HH:50, train leaves at HH:55; opposite direction analogously). The bus line can be served with one vehicle (layovers of 20 min and 10 min respectively).

If the train departure and arrival times are shifted by 15 min (in both directions of travel) while upholding the full hour symmetry, the bus schedule also needs to be shifted to maintain the service quality in the main demand relation (green arrows) (Figure 1*b* right). If one wants to keep the short connections in both directions (train -> bus, bus -> train) of the main demand relation, the bus service cannot be covered with one vehicle. The layovers increase to 50 min and 40 min respectively. A second vehicle and driver are needed, although no additional trips are added and no additional kilometers are driven. A purely kilometer-based calculation would thus yield no additional cost, although they rise significantly when an additional driver and vehicle are needed.

Research Goal

This paper thus proposes a simple cost estimation model for Swiss public transport designed to address the problems outlined above. It is based on the cost-allocation method proposed by Cherwony et al. (*1*), Taylor et al. (*23*) or Small and Verhoef (*27*). Based on the findings of Stopher et al. (*3*), the proposed model shall fulfill the following requirements:

- The model shall be able to provide costs on a line-detail basis.
- Marginal cost effects as illustrated by the above examples shall be captured accurately.
- Operating statistics shall rely as far as possible on easily available data, in this case vehicle-hours, vehicle-kilometers, the number of vehicles and vehicle size.

In the second part of this paper, the model is used in a case study to determine the break-even points between bus and train operation and to evaluate the influence of different parameters on the location of these break-even points.

Methodology

A sample with real cost data of 294 mostly outer-urban bus lines and 96 train lines in two Swiss cantons (cantons are the federated states of Switzerland) was used to estimate the operating costs for buses and trains. The sample includes data from 15 bus companies and 13 train companies. The reporting standard across the sample was not uniform: on the one hand companies assigned their respective cost items to a varying number of categories; on the other hand, the same cost items were sometimes assigned by different companies to different categories. The costs therefore needed to be aggregated into major categories: the chosen categories and their dependent operating statistics are summarized in Table 1.

The term ‘productive hours’ is the sum of the trip hours and the layovers (*28*) as, according to Swiss law, the layovers count as paid worktime. Different legal provisions in other countries could require the use of a different operating statistic like trip-hours. Vehicle costs include depreciation and interest for vehicle ownership, fuel, maintenance, and so forth. The category ‘Operational side costs’ groups various smaller costs including dispatching, maintenance of bus stops, and so forth. The category ‘Distribution’ includes ticketing costs, such as staffed ticket counters and ticket vending machines at stations or on-board vehicles.

Most regional trains in Switzerland run without a train attendant. The only staff member on-board is the train operator. Passengers must buy tickets before boarding. The category ‘train attendant’ thus covers selective ticket controls and accompanied trains (often used for evening service). The model estimates both the hourly rate and the average percentage of service hours the rate

Table 1. Cost Categories

Cost category	Function of ...
Bus lines	
Driver cost	Productive hours
Vehicle cost	Type of bus, #vehicles, Km
Operational side costs	Productive hours, type of bus, #vehicles, Km
Overhead	supplement in % of the 3 above
Distribution	supplement in % of the 4 above
Train lines	
Operator cost	Productive hours
Train attendant	Productive hours, % of accompanied hours
Vehicle cost	#vehicles, Km, vehicle size
Operational side costs	Productive hours, #vehicles, Km, vehicle size
Infrastructure	Km, vehicle size
Overhead	supplement in % of the 5 above
Distribution	supplement in % of the 6 above

is applied to. Some lines in the sample are known to have fully accompanied service. In this case, the corresponding percentage was set to 100%.

The infrastructure costs are the access fees paid by the train operating company (TOC) to the network owner. In Switzerland, TOCs are charged on a non-discriminatory basis according to the provisions fixed by the Federal Office for Transportation (*29*). In the proposed model, the calculation method was simplified by having it only depend on train-km and seat-km.

For buses, discrete vehicle sizes with specified passenger capacities exist: minibus, standard bus, articulated bus. For train lines, the rolling stock is more diverse; vehicle sizes vary significantly from one-car trains to twelve-car double deck trains. A proxy variable thus needed to be used to capture this variation. The model approximates vehicle size by the number of seats. In addition to the cost data, the sample included the following operating statistics: number of productive hours, vehicle-km and seat-km. For trains, the average vehicle size was obtained by dividing the seat-km by the vehicle-km. For bus lines, the type of vehicle was indicated in the sample.

The number of vehicles required to cover all rotations was approximated by the formula

$$\#\text{vehicles} = \frac{\text{productive hours}}{\frac{365}{7} \cdot \sum_{\text{days}} \left(h_{\text{service hours}} \cdot \frac{f_{\text{off-peak}}}{f_{\max}} + h_{\text{peak hours}} \cdot \frac{f_{\text{peak}} - f_{\text{off-peak}}}{f_{\max}} \right)}$$

f_{peak} and $f_{\text{off-peak}}$ designate the peak and off-peak frequencies expressed in trains or buses per hour. $h_{\text{peak hours}}$ is the number of hours during which the peak frequency is operated. $h_{\text{service hours}}$ is the total number of service hours. This data was collected from the publicly available timetable for each line and day type (Monday–Friday, Saturday and Sunday).

Results

Model Validation

The sample was randomly split into two parts of equal size. Parameters were estimated separately for each part using the method of least squares and were subsequently applied to the other part for validation (cross-validation). Table 2 summarizes the quality of calibration and validation by indicating the R^2-value of the respective match between real and modeled costs.

For the bus lines we note that the R^2-values are consistently better for sample part 2 than for sample part 1. The latter thus contained more outliers than the former. For the train lines, we see that the quality of the model is particularly poor when it comes to the operational side costs (in terms of volume, a secondary cost category), whereas the model fits well for the more important categories of vehicle cost and infrastructure charges. It is also interesting to note that in the validations the R^2-values of the total costs are consistently higher than all R^2-values of the individual cost categories, with the only exception of train infrastructure cost. Inaccuracies of the different

Table 2. Quality of Calibration and Validation

	Relative cost volume	Calibration on part 1 Validation on part 2		Calibration on part 2 Validation on part 1	
	(average over entire sample)	R^2 of the calibration	R^2 of the validation	R^2 of the calibration	R^2 of the validation
Bus lines					
Driver cost	49%	0.903	0.940	0.941	0.903
Vehicle cost	30%	0.924	0.935	0.944	0.893
Operational side costs	8%	0.453	0.536	0.591	0.417
Overhead	10%	0.609	0.691	0.697	0.593
Distribution	3%	0.547	0.547	0.624	0.344
Total cost	100%	----------	0.952	----------	0.919
Train lines					
Operator cost	22%	0.840	0.823	0.830	0.830
Train attendant	5%	0.751	0.710	0.761	0.663
Vehicle cost	36%	0.935	0.890	0.916	0.902
Operational side costs	3%	0.249	0.066	0.265	−1.412
Infrastructure	20%	0.976	0.957	0.959	0.972
Overhead	8%	0.757	0.784	0.799	0.753
Distribution	7%	0.490	0.557	0.563	0.496
Total cost	100%	----------	0.921	----------	0.949

cost categories thus seem to balance each other out when estimating the total cost. The reason lies most probably in the inhomogeneous reporting standards of the different operating companies.

Given the good fits of the overall total costs, as well as of the most important cost categories (by cost volume), we consider the model to be sufficiently good for onward use. Now that the model itself and the calibration method have been validated, the unit cost parameters used for all onward calculation are estimated on the entire sample.

Unit Cost Parameters

Bus Costs. The unit costs of the bus cost model are summarized in the top part of Table 3. For the operational side costs, the model shows distance-dependent costs to be equal for minibuses and standard buses, whereas they are higher for articulated buses. A possible explanation could be that articulated buses generally operate in regions with higher demand and thus also more car traffic, which in turn increases the need for dispatching activities. Lines with higher demand also tend to have more advanced infrastructure at bus stops.

Train Costs. The unit cost parameters of the train cost model are summarized in the lower part of Table 3. Unlike the bus cost model, the operational side costs do not have a time-dependent cost component. Overhead is smaller than for bus lines, which can be explained by the fact that it is applied to a higher base cost level. Distribution is more expensive, which can be explained by the higher number of ticket machines, whereas on outer-urban bus lines tickets are often sold by the driver.

Comparison with Data from Literature and Practice

Bus Costs. The obtained unit cost parameters are tested by applying them to a regional and an urban bus line respectively. For the regional bus line we assume a standard bus driving approximately 70,000 km/year (from Leuthard (*13, 21*) and Weidmann (*30*)) at an average speed of 25 km/h (*26*). Turnaround times are assumed to account for 15% of the productive hours. On first glance,

Table 3. Unit Cost Parameters

Cost Category	Unit costs		
Bus Unit Costs	Minibus	Standard Bus	Articulated Bus
Driver cost	66.61 CHF/H		
Vehicle cost			
fixed part	7,446 CHF/a	27,767 CHF/a	50,247 CHF/a
variable part	0.97 CHF/Km	1.61 CHF/Km	2.03 CHF/Km
Operational side costs			
time-dependent	2.35 CHF/H		
distance-dependent	0.33 CHF/Km		0.52 CHF/Km
Overhead	supplement of 11.08% of the 3 categories above		
Distribution	supplement of 5.13% of the 4 categories above		
Train Unit Costs			
Operator cost	115.01 CHF/H		
Train attendant			
hourly cost	114.86 CHF/H		
applied to	24.4% of productive hours (unless fully accompanied)		
Vehicle cost			
fixed part	153,727 CHF/a		
Km-variable part	2.25 CHF/Km		
seat-Km-variable part	0.01411 CHF/seat-Km		
Operational side costs			
Km-variable part	0.33 CHF/Km		
seat-Km-variable part	0.00048 CHF/seat-Km		
Infrastructure			
Km-variable part	1.63 CHF/Km		
seat-Km-variable part	0.00873 CHF/seat-Km		
Overhead	supplement of 9.53% of the 5 categories above		
Distribution	supplement of 8.22% of the 6 categories above		

Table 4. Examples of Cost Repartition

Cost Category	Absolute cost	Relative part
Regional bus line		
Driver cost	3.13 CHF/Km	48%
Vehicle cost		
fixed part	0.40 CHF/Km	31%
variable part	1.61 CHF/Km	
Operational side costs		
time-dependent	0.11 CHF/Km	7%
distance-dependent	0.33 CHF/Km	
Overhead	0.62 CHF/Km	9%
Distribution	0.32 CHF/Km	5%
Total	6.52 CHF/Km	100%
Urban bus line		
Driver cost	4.11 CHF/Km	53%
Vehicle cost		
fixed part	0.46 CHF/Km	27%
variable part	1.61 CHF/Km	
Operational side costs		
time-dependent	0.15 CHF/Km	6%
distance-dependent	0.33 CHF/Km	
Overhead	0.74 CHF/Km	9%
Distribution	0.38 CHF/Km	5%
Total	7.78 CHF/Km	100%
Regional train line		
Operator cost	2.13 CHF/Km	14%
Train attendant	0.52 CHF/Km	3%
Vehicle cost		
fixed part	1.02 CHF/Km	41%
Km-variable part	2.25 CHF/Km	
seat-Km-variable part	2.82 CHF/Km	
Operational side costs		
Km-variable part	0.33 CHF/Km	3%
seat-Km-variable part	0.10 CHF/Km	
Infrastructure		
Km-variable part	1.63 CHF/km	23%
seat-Km-variable part	1.75 CHF/Km	
Overhead	1.20 CHF/Km	8%
Distribution	1.13 CHF/Km	8%
Total	14.87 CHF/Km	100%

this may seem high, but in regional traffic, with strong timetable constraints dictated by connecting railway lines, turnaround times of 15% of total service time are actually rather optimistic. With these assumptions, the model yields the km-based costs shown in Table 4 (top). The final cost of 6.52 CHF/km matches well with values from literature (*26, 30*).

The urban case considers the same bus with average operating speed of approximately 18 km/h (from Leuthard (*21, 22*)). Turnaround time is reduced to 10% of the total service time, as frequency is higher and there are fewer timetable constraints. In the urban context, the turnaround time merely serves as buffer against imported delays. On the other hand, the annual mileage drops to around 60,000 km/year. The results are shown in Table 4 (middle). The operating speed is a very sensitive parameter, which has a significant influence on the total costs. This finding is supported by the work of Avenali et al. (*24*), whose results also confirm the inverse proportional relationship between speed and cost.

It is also interesting to note that the fixed vehicle costs per year of standard and articulated buses given in Table 3 are in the same order of magnitude as the capital cost of buses found by Taylor et al. (*23*). They found annual capital costs of $34,371.

Train Costs. The train cost model was tested assuming a regional train (electric multiple unit) with a capacity of 200 persons, travelling 150,000 km/year, running at an average speed of 60 km/h and having a turnaround time of 10% of the total service time. Results are presented in Table 4 (bottom). The obtained total cost matches well with data from practice and literature (*26, 30*). The share of operator cost is only around 15%, and infrastructure charges are roughly 25%. Again, one should note that these charges do not cover the full infrastructure cost. In Switzerland, they only cover around 40% (*31*). The remaining costs are mostly covered by federal subsidies. Given the comparably low share of the operator cost, train automation does not yield direct savings as large as for buses. The main interest in railway automation is more efficient use of existing infrastructure made possible by more precise driving trajectories.

The relative cost volumes of the original data sample in Table 2 show a higher share of the train operator costs (22%) than the calculations here (14%, Table 4). The 10% share of turnaround time we assumed here is rather optimistic. Among the sample, most lines have a higher share of turnaround time (due to less efficient rotations), which subsequently increases the cost share of the train operator. Furthermore, the average speed has an influence, too.

Limitations

One should note that the above results are valid for Switzerland. Their transposition to other countries is possible, but the specific unit cost parameters need recalibration to fit the local conditions, for instance, wages.

Case Study: Annual Operating Costs as a Function of Demand

Assumptions and Parameters

The proposed model was applied in a case study for a generic public transportation line. In this study the yearly operating costs for both a bus and a train service were calculated for different scenarios and boundary conditions. The line is assumed to be 20 km long and to have a standard temporal demand distribution with strong

Table 5. Parameters of the Case Study Scenarios

	Scenario 1	Scenario 2	Scenario 3	Scenario 4
Bus speed	25 km/h	**30 km/h**	25 km/h	25 km/h
Train speed	40 km/h	**60 km/h**	40 km/h	40 km/h
Min. bus turnaround time	5 min			
Min. train turnaround time	10 min			
Bus vehicle size	articulated bus, 50 seats			
EMU vehicle size	200 seats	200 seats	**100 seats**	200 seats
Lower min. frequencies in the early morning and late evening	No	No	No	**Yes**
Min. bus rotation time	*1h 46 min*	***1h 30 min***	*1h 46 min*	*1h 46 min*
Min. train rotation time	*1h 20 min*	***1 h***	*1h 20 min*	*1h 20 min*

peak-period demand oriented towards a city center (*32*). Table 5 summarizes the basic case study parameters. In all scenarios the train service is operated with Electric Multiple Units (EMU); for scenarios 1, 2 and 4 the vehicle is a standard Flirt (built by Stadler Rail) shown in Figure 2*a*; scenario 3 considers a smaller vehicle.

Scenario 1 serves as the reference case. In scenarios 2 to 4, we change one type of parameter:

- Scenario 2 (higher operating speed): importantly operating speed does not only concern the travel time itself, but also the required number of vehicles. In scenario 1, the minimal rotation time of the train is 1 hour 20 min. In scenario 2, it is 1 hour, which saves 1 or 2 vehicles depending on the frequency.
- Scenario 3 (smaller vehicles): evaluates the effect of smaller EMU vehicles (Figure 2*b* GTW 2/6 built by Stadler Rail). In off-peak hours, smaller vehicles help reduce costs, whereas in the peak-hours their reduced capacity requires additional trains.
- Scenario 4 (reduced frequencies): investigates to what extent lower minimal frequencies in the early morning (between 5:00 and 6:00) and late evening (after 21:00) helps reduce costs.

The scenarios are calculated for each of the following combinations:

- Minimal frequencies of 60 min, 30 min and 15 min respectively. For scenario 4, the standard frequency of 30 min is reduced to 60 min in the early morning and late evening while the standard frequency of 15 min is lowered to 30 min. The standard frequency of 60 min is assumed to be the minimal acceptable level of service which cannot be lowered any further.
- For trains, we investigate how the maximum number of EMUs that can be coupled together

(a) Image source: Marc Sinner

(b) Image source: Marc Sinner

Figure 2. Photos of train types.

influences total cost. The calculation was done once for up to 2 EMUs and once for up to 3 EMUs.

When demand during a particular hour exceeds the available capacity, additional capacity is added according to the following rules:

- For buses, additional buses are put in service. The evaluation takes into account that some of these buses are required for more than one hour or in both directions, thus requiring them to return to the origin. Others are only needed for one trip and do not return to the origin. That way, a simplified rostering process is simulated.
- For trains, the procedure is similar, with the notable exception that existing trains are lengthened by attaching one or two additional EMUs (depending on the maximum number allowed) before operating additional trains. The demand level when the first additional peak-hour trains are required depends on the available capacity of one EMU and the number of EMUs being coupled together.

Results

Figure 3*a–d* shows the annual operating costs as a function of the daily demand for each scenario. Inside each scenario, the combinations of mode, frequency and (as applicable) train length described above are represented by separate curves. The blue colored curves represent 15 min frequency, orange 30 min and green 60 min frequency. Buses are represented by dotted lines and trains by solid (maximum 3 EMUs) or dashed lines (maximum 2 EMUs).

Table 6 summarizes the break-even points of bus and train cost functions (i.e. in which the cost of operating bus service is equal to train service). For trains, we assume the more favorable case with up to three EMUs coupled. The order of magnitude of these break-even points matches with the required levels of demand identified by the Swiss Federal Office for Transportation for the implementation of 30 min and 15 min frequencies (2,000 and 8,000 people/day/direction respectively) (*33*). The results in Table 6 are valid for Swiss conditions in which comparably high wages generally play in favor of trains which have a significantly lower share of staff costs than buses. In countries with lower wages, break-even generally occurs at higher demands.

Discussion of Results

From these results, we can draw the following conclusions:

1) In all scenarios, the higher the desired minimal frequency, the higher the demand needed for a train service to be more economical than a bus service. Trains require bundling of demand due to their higher capacity, which conflicts with the attractiveness of higher frequencies.
2) In all scenarios, the gap between the cost functions of 60 min minimal frequency and 30 min minimal frequency closes progressively, both for buses and trains. This is because additional trains or buses need to be added to the 60 min frequency schedule also in off-peak hours as demand increases. Thus, the schedule becomes closer and closer to an all-day 30 min frequency schedule. Eventually both curves meet. The same holds for the 30 min and 15 min curves with the difference that their meeting point is at a higher demand level.
3) The higher the speed difference between buses and trains (cf. comparison between scenarios 1 and 2), the lower the break-even point. In addition to the direct effect of speed, the possibility for efficient vehicle rotations is crucial. Although scenario 1 allows efficient rotations for 30 min and 15 min frequencies, it does not for the 60 min frequency. The vehicle has long layovers. On the other hand, scenario 2 does allow very efficient rotations for all frequencies.
4) Smaller rail vehicles can help reduce the break-even point between buses and trains as capacity can be tailored to demand. On the other hand, in high demand ranges, the use of small vehicles increases costs (compare scenarios 1 and 3 in Figure 3*a* and *c*).
5) As expected, the possibility of coupling more EMUs helps reduce costs. If the vehicle size is small, the effect is greater.
6) The acceptance of lower frequencies in early mornings and late evenings reduces costs significantly for both buses and trains. However, given their lower capacity, the demand range in which these savings can be seen for buses is more reduced and the resulting savings are lower. Table 6 shows that the break-even point can be shifted in favor of trains.

Conclusions

This paper proposes a cost-allocation model for Swiss bus and train lines using real cost data from 294 bus and 96 train lines. It is based on easily available operating statistics such as productive hours, vehicle-km and number and size of the employed vehicles. The model's resulting unit cost parameters match closely existing data from

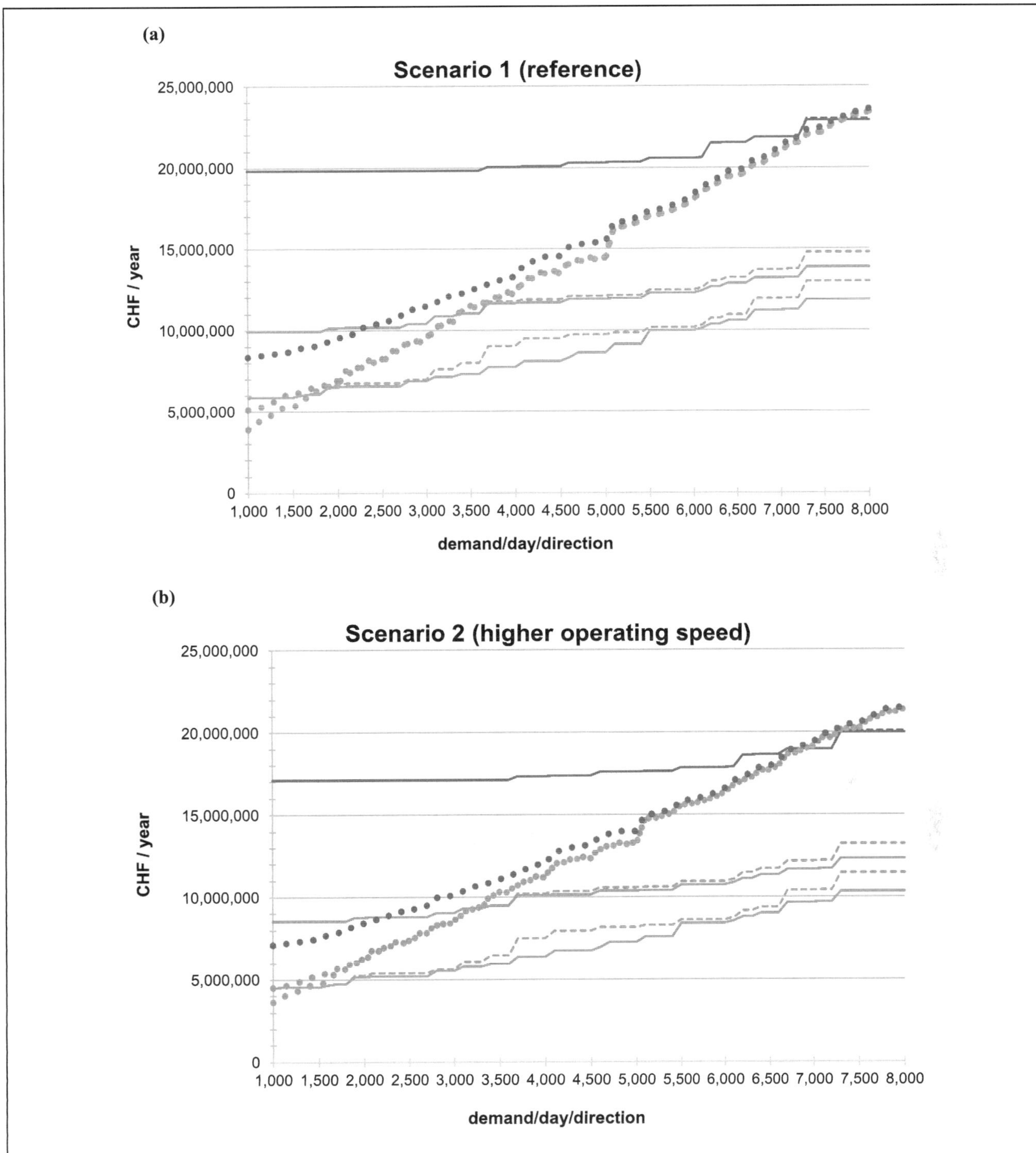

Figure 3. (Continued)

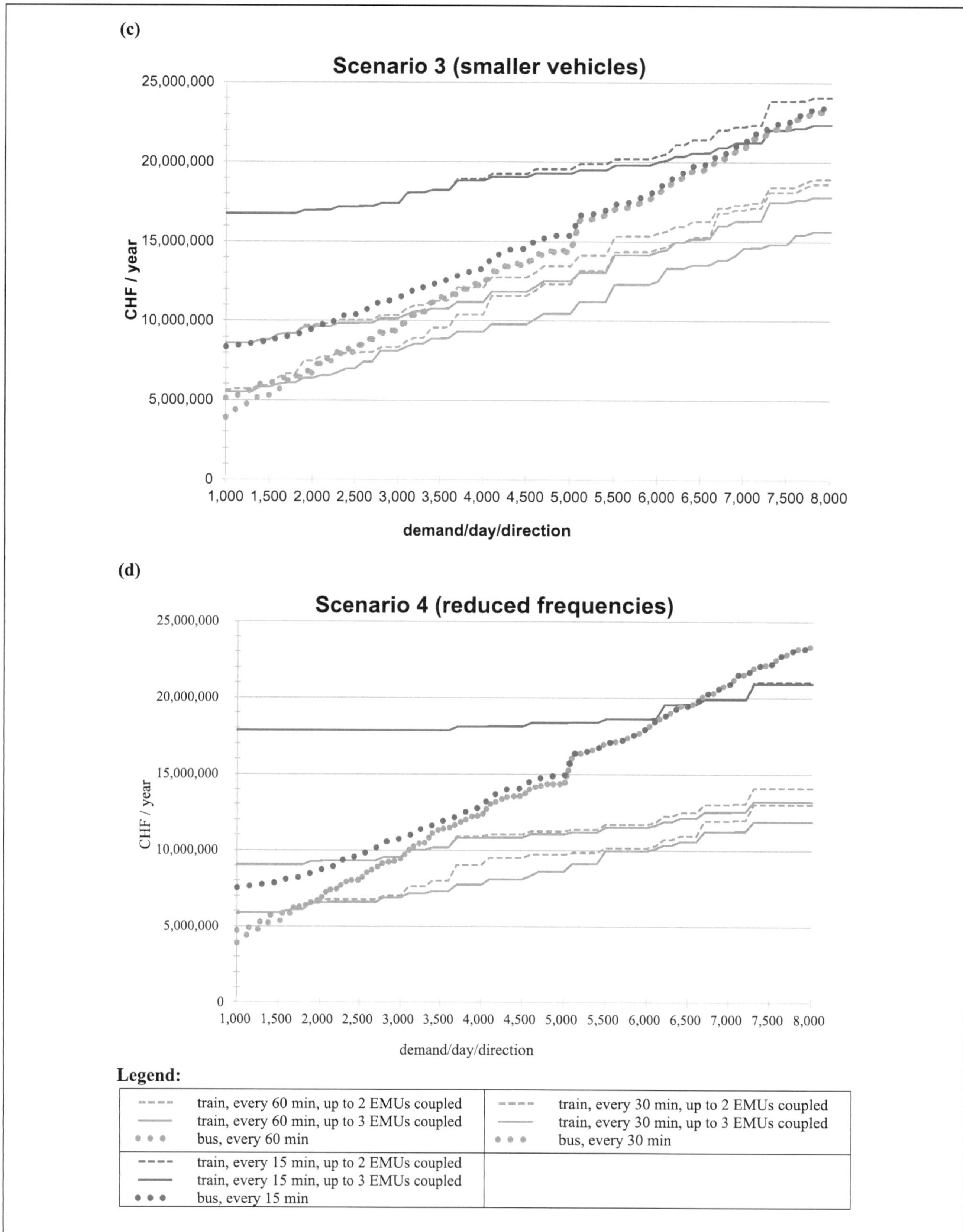

Figure 3. Cost vs demand plots.

Table 6. Break-Even Points

	Break-even between bus and train at … [people / day / direction]		
	… 60 min frequency	… 30 min frequency	… 15 min frequency
Scenario 1 (reference)	1700	3400	7600
Scenario 2 (high operating speed)	1400	3200	6900
Scenario 3 (smaller vehicles)	1700	3400	7100
Scenario 4 (reduced frequencies)	1700	3200	6600

literature and practice. This model provides public transport planners with a tool they can use to more accurately estimate the real cost of network changes such as line extensions (cf. Figure 1*a*) or timetable shifts (cf. Figure 1*b*).

The model was applied in a case study for a generic public transport line to estimate costs for various scenarios regarding minimal frequency, average speed or the maximum number of EMUs that could be coupled. The results showed the importance of speed and more specifically the ability for efficient rotations on operating costs and the break-even point between buses and trains. With minimal frequencies of 60 min or 30 min, a train service can be more cost-efficient than a bus from comparably low to medium demand ranges of 1,400–1,700 or 3,200–3,400 people/day/direction respectively upwards. These numbers are valid for Swiss cost level and would be higher in countries with lower wages for bus drivers and train operators.

So far, the question of the availability of slots on the railway network was not taken into account. In further research, this issue shall be looked at in combination with a broader variety of train types. In fact, the limited availability of slots might require different types of trains such as double deck trains in order to cover the capacity in the peak hours, although their capacity is not needed during off-peak hours.

Acknowledgments

The authors express their gratitude to the two Swiss cantons and the companies who provided the real data for this research. Furthermore, they would like to thank the anonymous reviewers for their valuable comments.

Author Contributions

The authors confirm contribution to the paper as follows: study conception and design: Marc Sinner, Ulrich Weidmann; data collection: Marc Sinner, Ulrich Weidmann; analysis and interpretation of results: Marc Sinner, Ulrich Weidmann; draft manuscript preparation: Marc Sinner, Andrew Nash. All authors reviewed the results and approved the final version of the manuscript.

References

1. Cherwony, W., S. Mundle, B. Porter, and G. Gleichman. Review of Bus Costing Procedures. *Transportation Research Record: Journal of the Transportation Research Record*, 1982. 854: 54–60.
2. Peskin, R. Development and Testing of a Cost-Allocation-Based Cost-Estimating Method. *Transportation Research Record: Journal of the Transportation Research Record*, 1982. 862: 40–47.
3. Stopher, P., L. Brandrup, B. Lee, and S. Parry. Development of a Bus Operating Cost Allocation Model Compatible with UMTA Urban Transportation Planning System Models. *Transportation Research Record: Journal of the Transportation Research Record*, 1987. 1108: 31–41.
4. Cherwony, W, and B. Porter. Scheduling-Based Marginal Cost-Estimating Procedure. *Transportation Research Record: Journal of the Transportation Research Record*, 1982. 862: 35–39.
5. Carter, D., S. Mundle, and B. McCollom. Comparative Evaluation of Bus Route Costing Procedures. *Transportation Research Record: Journal of the Transportation Research Record*, 1984. 994: 1–8.
6. Parody, T., M. Lovely, and P. S. Hsu. Net Costs of Peak and Offpeak Transit Trips Taken Nationwide by Mode. *Transportation Research Record: Journal of the Transportation Research Record*, 1990. 1266: 139–145.
7. Cherwony, W, and S. Mundle. Peak-Base Cost Allocation Models. *Transportation Research Record: Journal of the Transportation Research Record*, 1978. 663: 52–56.
8. Cervero, R. Multistage Approach for Estimating Transit Costs. *Transportation Research Record: Journal of the Transportation Research Record*, 1982. 877: 67–75.
9. Nuworsoo, C. Model To Minimize Nonrevenue Costs in Bus Transit Operations. *Transportation Research Record: Journal of the Transportation Research Record*, 1999. 1666: 14–22.
10. Bruun, E. Bus Rapid Transit and Light Rail: Comparing Operating Costs with a Parametric Cost Model. *Transportation Research Record*, Vol. 1927, 2005, pp. 11–21.
11. Allport, R. The costing of bus, light rail transit and metro public transport systems. *Traffic Engineering & Control*, Vol. 22, No. 12, 1981, pp. 633–639.
12. Aberle, G., K. Kindsvater, C. Aufderbeck, and D. Trost. *Schienen- und Busverkehr in der Region: einzelwirtschaftliche und gesamtwirtschaftliche Kostenvergleichsrechnung*. Deutscher Verkehrs-Verlag, Hamburg, 1998.

13. Leuthardt, H. Bus oder Bahn in der Region? *Der Nahverkehr*, Vol. 14, No. 11, 1996, pp. 12–16.
14. Leuthardt, H. Kostenstrukturen von Stadt-, Überland- und Reisebussen. *Der Nahverkehr*, Vol. 16, No. 6, 1998, pp. 19–23.
15. Leuthardt, H. Betriebskosten von Stadtbahnen. *Der Nahverkehr*, Vol. 18, No. 10, 2000, pp. 14–17.
16. Forschungsgesellschaft für Strassen- und Verkehrswesen (FGSV). *Hinweise zu Systemkosten von Busbahn und Strassenbahn bei Neueinführung*. Forschungsgesellschaft für Strassen- und Verkehrswesen (FGSV), Köln, 2008.
17. Hass-Klau, C., G. Crampton, C. Biereth, and V. Deutsch. *Bus or Light Rail: Making the Right Choice*. Environmental & Transport Planning, Brighton, 2003.
18. Frank, P., M. Friedrich, and J. Schlaich. Betriebskosten von Busverkehren schnell und genau ermitteln. *Der Nahverkehr*, Vol. 26, No. 11, 2008, pp. 15–22.
19. Cohen, H., J. Stowers, W. Cherwony, and L. Polin. *NCTRP Report 16: Estimating Incremental Costs of Bus Route Service Changes*. TRB, National Research Council, Washington, D.C., 1988.
20. Harmatuck, D. Light Rail Cost Functions and Technical Inefficiency. *Transportation Research Record: The Journal of the Transportation Research Record*, 2008. 2042: 58–70.
21. Leuthardt, H. Betriebskosten von Linienbussen im systematischen Vergleich. *Der Nahverkehr*, Vol. 23, No. 11, 2005, pp. 20–25.
22. Leuthardt, H. Die Wirtschaftlichkeit von Gelenkbussen und Buszügen. *Der Nahverkehr*, Vol. 28, No. 5, 2010, pp. 30–34.
23. Taylor, B., M. Garrett, and H. Iseki. Measuring Cost Variability in Provision of Transit Service. *Transportation Research Record: Journal of the Transportation Research Record*, 2000. 1735: 101–112.
24. Avenali, A., A. Boitani, G. Catalano, T. D'Alfonso, and G. Matteucci. Assessing Standard Costs in Local Public Bus Transport: Evidence from Italy. *Transport Policy*, Vol. 52, 2016, pp. 164–174.
25. Bösch, P., F. Becker, H. Becker, and K. Axhausen. Cost-Based Analysis of Autonomous Mobility Services. *Transport Policy*, Vol. 64, 2018, pp. 76–91.
26. Bundesamt für Verkehr (BAV), Kennzahlen Regionaler Personenverkehr, 2016. https://www.bav.admin.ch/bav/de/home/das-bav/aufgaben-des-amtes/finanzierung/finanzierung-verkehr/personenverkehr/rpv-mit-erschliessungsfunktion/rpv-kennzahlen.html. Accessed 12 June, 2017.
27. Small, K, and E. Verhoef. *The Economics of Urban Transportation*. Routledge, Abingdon, 2007.
28. Bundesamt für Verkehr (BAV). Leitfaden Kennzahlen RPV, 2017. https://www.bav.admin.ch/bav/de/home/themen/alphabetische-themenliste/vollzugshilfen/leitfaeden/leitfaden_kennzahlen_rpv.html. Accessed 12 June, 2017.
29. SBB Infrastruktur, Leistungskatalog Infrastruktur 2018, 2017. https://company.sbb.ch/content/dam/sbb/de/pdf/sbb-konzern/sbb-als-geschaeftspartner/angebote-fuer-evus/onestopshop/Leistungskatalog_2018_DE.pdf. Accessed 12 June, 2017.
30. Weidmann, U. System- und Netzplanung, Institut für Verkehrsplanung und Transportsysteme (IVT), ETH Zürich, 2013.
31. SBB Infrastruktur, Das Trassenpreissystem, 2015. https://company.sbb.ch/content/dam/sbb/de/pdf/sbb-konzern/sbb-als-geschaeftspartner/angebote-fuer-evus/onestopshop/Flyer_A4_Trassenpreissystem_de.pdf. Accessed 12 June, 2017.
32. Weidmann, U. Systemdimensionierung und Kapazität, Institut für Verkehrsplanung und Transportsysteme (IVT), ETH Zürich, 2011.
33. Bundesamt für Verkehr (BAV). Dokumentation Planungsgrundlagen Ausbauschritt 2030 (Anlageband), 2014. https://www.bav.admin.ch/dam/bav/de/dokumente/themen/fabi-step/dokumentation_planungsgrundlagenstepas2030anlageband.pdf.download.pdf/dokumentation_planungsgrundlagenstepas2030anlageband.pdf. Accessed 9 June, 2017.

The Standing Committee on Transit Management and Performance (AP010) peer-reviewed this paper (18-00903).

Article

TRR
JOURNAL OF THE TRANSPORTATION RESEARCH BOARD

Transportation Research Record
2018, Vol. 2672(8) 443–454

DOI: 10.1177/0361198118791665
journals.sagepub.com/home/trr
SAGE

Subway Service Down Again? Assessing the Effects of Subway Service Interruptions on Local Surface Transit Performance

Ehab Diab[1] and Amer Shalaby[2]

Abstract

Cities around the world are keen to offer modern urban transit systems that connect various locations in an efficient and reliable manner with the highest degrees of riders' experience. These systems may consist of various modes such as buses, streetcars (or trams), and rapid transit systems (e.g., subways). In both research and practice, the quality of transit service has traditionally been measured and investigated on a mode-by-mode basis. Therefore, it is rare to find studies that investigate the impacts of poor performance or breakdown of one transit mode on other functioning modes in multimodal integrated transit systems. This research aims at understanding the impact of subway service interruptions on the speed performance of surface transit in Toronto, Ontario. To do that, a detailed data set of subway service interruptions collected in 2013 by the Toronto Transit Commission (TTC), the public transit provider in the City of Toronto, was used. In addition, another data set was obtained from the TTC's automatic vehicle location system for 51 bus and streetcar routes that are within a short walking distance of some subway stations. Using two statistical models, the paper results indicate that subway service interruptions have a statistically significant negative impact on bus and streetcar service operations in terms of slower speeds, with more immediate and intense impacts on streetcar service.

Cities around the world are keen to offer modern urban systems that connect various locations in an efficient and reliable manner with the highest degrees of riders' experience. These systems may consist of various modes such as buses, streetcars (or trams), and rapid transit systems (e.g., subways). A sizable body of literature has focused on understanding the impacts of general factors or various transit improvement strategies on transit service performance and reliability (*1–3*). These studies generally focused on one transit mode at a time, such as the bus system, the streetcar system, or the subway system (*1–8*), while ignoring the impacts of one system's disruption on the other systems in multimodal networks. For example, Diab and El-Geneidy (*1*, *9*), using bus data collected from the automatic vehicle location (AVL) and automatic passenger count (APC) systems in Montreal, tested the impact of a set of improvement strategies (e.g., articulated buses, bus lanes) on the bus service running time and variation. In another study, they investigated the impact of bus stop location on service quality (*10*). Mesbah et al explored the impacts of weather conditions on the tram system's travel times (*4*). Currie and Reynolds (*6*) investigated the impacts of two different fare collection methods on streetcar service performance, whereas Currie et al (*5*) evaluated the impact of passenger crowding on tram dwell times. Regarding the subway related studies, Weng et al (*8*) examined the impacts of causal variables on subway delay duration, such as power/vehicle failures and switch malfunction, whereas Louie et al (*7*) investigated the impacts of both causal and non-causal factors, such as passenger related incidents.

Other researchers focused on understanding, modeling and proposing conceptual frameworks for rail (or subway) disruption management (i.e., emergency response plans), while testing different approaches or categorizing rail incidents according to some criteria (*11–14*). For

[1]Department of Geography and Planning, University of Saskatchewan, Saskatoon, SK, Canada
[2]Department of Civil Engineering, University of Toronto, Toronto, ON, Canada

Corresponding Author:
Address correspondence to Ehab Diab: ehd361@mail.usask.ca

example, Pender at al. proposed a method to evaluate the economic viability of providing a dedicated bus service reserved for bus bridging purposes for rail system disruption (*15*). Darmanin at al (*16*) and van der Hurk et al (*17*) developed mathematical approaches with the aim of minimizing commuter discomfort following a service disruption and estimating the required shuttle line frequencies, respectively.

Despite the previous efforts, it is rare to find studies that focused on understanding the impacts of service failures of one transit mode on the performance of other functioning transit modes in multimodal integrated networks. An improved understanding of the various causes behind the changes in transit systems' performance that users experience on a daily basis would help transit agencies develop more appropriate policies to maintain and improve transit systems' resilience and performance in the future. This research aims at understanding the impact of the subway system's interruptions (caused by incidents) on the speed performance of surface transit routes in Toronto. Using Toronto as a case study provides a unique opportunity to understand the effects of subway service interruptions on the service performance of both bus and streetcar lines in a highly integrated multimodal transit system.

Study Context and Methodology

Toronto, Ontario, is the most populous city in Canada with more than 2.8 million inhabitants in 2015 (*18*). The Toronto Transit Commission (TTC) operates the transit service within the City of Toronto, serving about 2.7 million daily passengers (*19*). It operates a multimodal transit system, consisting of four subway lines, 11 streetcars lines, and 141 bus routes. The TTC subway system extends to a total length of 68 km serving 69 stations (*20*).

Two data sets were used in this study. The first data set includes detailed subway system interruption data collected in 2013 by the TTC. In this data set, each record includes the incident's date, time, subway station, direction of travel, amount of delay (in minutes), vehicle number/type, as well as a brief description of the incident and a code representing the incident type. Around 12,000 subway incidents were reported at the stop level in 2013 along the two main subway lines (i.e., Yonge–University–Spadina [YUS] and Bloor–Danforth lines) operated by the TTC in Toronto (Figure 1*a*). The incident causes, for example, included: debris/intrusions at track level; track circuit/track down; signal/switch/power problems, and disabled train/train body problems as well as incidents related to passenger activity issues and security aspects. A complete list of incident types at the stop level can be found at Louie et al (*7*). The second data set was acquired from the TTC's AVL system for several bus and streetcar routes that are within a short walking distance of some subway stations. This data set includes information about bus and streetcar locations (x- and y-coordinates) recorded every 20 s as well as other information related to the time of the record, and route and run number.

The study time frame of interest is all the weekdays of May 2013. This month was selected for analysis because it experienced the greatest number of incidents with the largest amount of delays, and the lowest standard deviations for both at the system level. This captures the impacts of a good number of subway incidents with varying degrees of delays. Other criteria were chosen to identify the subway stations to be included in the analysis. These criteria were based on generating a composite indicator (or index) for each subway station using the 2013 daily ridership data and the TTC subway incident data to identify the most vulnerable stations in the subway system. This indicator consists of three equally weighted variables: daily ridership at the subway station; and total number of incidents and total amount of delay per weekdays of the month. Each variable was normalized using a *Z*-score, then a total was generated by summing all three *Z*-scores (Figure 1*b*). This index was used to rank all the subway stations in May 2013. The results of this method helped identify the subway stations that suffered most from delays in May 2013. Most of these stations were located along the YUS Line. Therefore, this study placed its focus on this line.

Furthermore, a subway station should satisfy the following conditions. It should be neither the first or last station of the subway line nor a transfer station, where two subway lines are intersecting. This is because delays in one subway line can encourage passengers to use the other line. Finally, the top ranked 24 subway stations, in terms of Z-score value, along the YUS Line (out of 32 stations) were included in the analysis. This allowed for analyzing the impacts of around 388 incidents with a total delay of 1,702 min, ranging from 2- to 73-min incidents.

Over 100 million individual observations were acquired from the TTC's AVL system for 41 bus routes and 10 streetcar routes for the weekdays between May 1, 2013 and May 31, 2013. These routes have stops within 200 m of the selected subway stations, representing a viable option for subway users when a subway service interruption occurs. The selected month had mild and clear weather with no recorded snow on the ground that could impact the bus and streetcar route operations.

To isolate the effects of the subway service interruptions on surface transit performance, two statistical models were developed. They capture the impacts of subway service interruptions on the streetcar and bus service speeds, respectively, at the segment level of analysis. The segment was defined as the route section between two consecutive

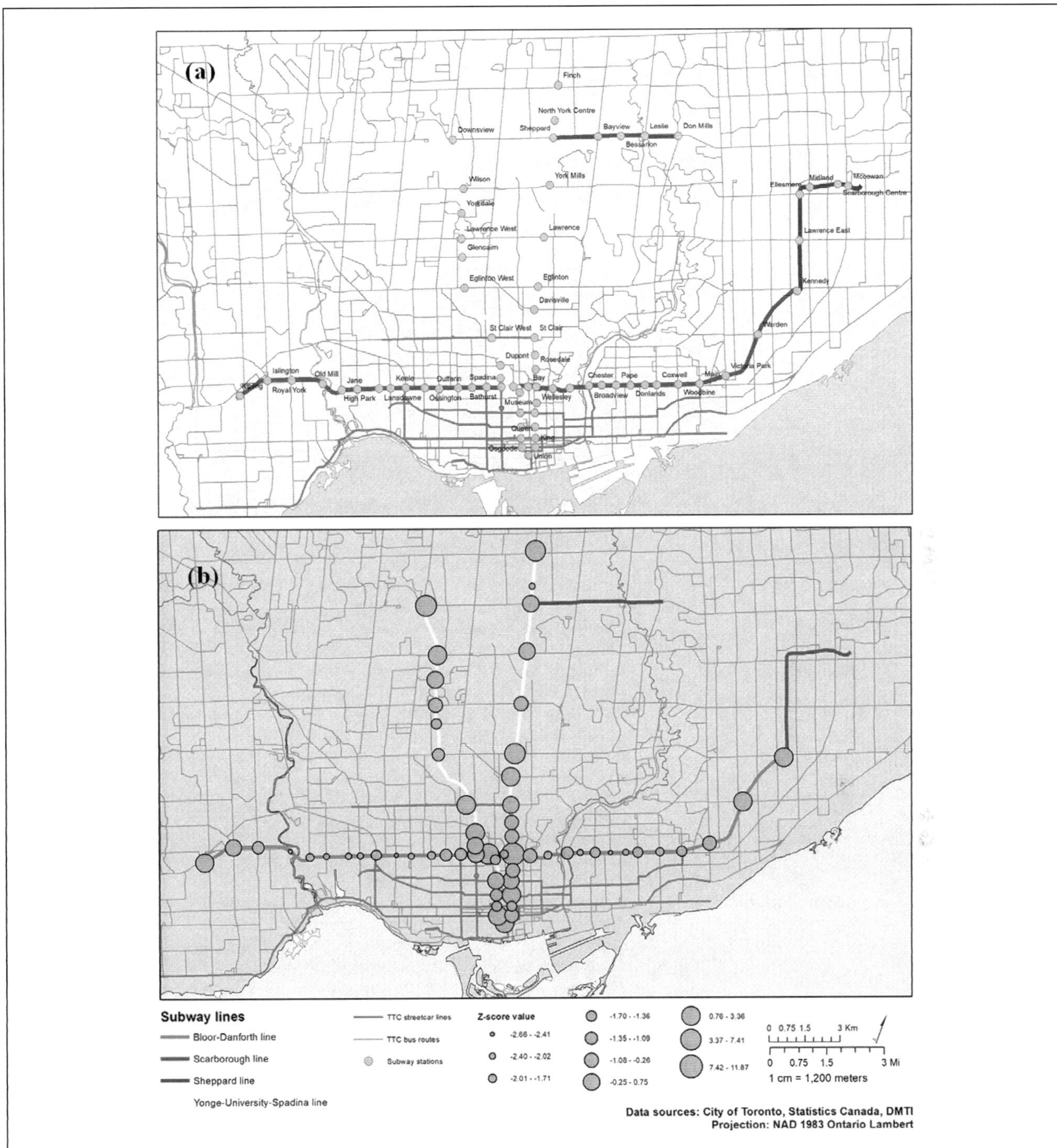

Figure 1. (*a*) Toronto transit system and (*b*) subway system total *Z*-score by subway station.

time points along a route. The time points are important locations for transit agencies' schedule building and operational control (*21*, *22*). Subsequently, all the variables in this study were defined according to that. For example, the average speed per trip-segment was computed as the average speed of all GPS points of a given trip within a given segment. In the analysis, we kept only trip-segments that start within a 3-km straight-line distance buffer of the YUS Line to reduce the number of analyzed trip-segments. Data preparation involved processing the raw AVL data

Table 1. Description of Variables Used in the Model

Variable name	Description
Trip speed (km/h)	Trip speed over a segment. The segment is defined as the route section between two consecutive time points along a route (dependent variable)
Segment sequence	A number that represents the sequence of the time-point segment along a trip calculated from the terminal
Number of scheduled stops	Number of scheduled stops within a time-point segment
Adjacent segment	A dummy variable equals 1 if the segment is within 200 m of subway station
Segment with a layover	A dummy variable equals 1 if the segment includes a layover stop
Segment distance (km)	Segment length between the two consecutive time points
Distance to Union Station (km)	Distance measured from Union Station to the nearest point of the time-point segment
Distance to Union Station $(km)^2$	Distance to Union Station squared. The squared term is used to account for the observed non-linear relationship between the variable and the dependent variable
Morning peak	A dummy variable equals 1 if the trip over a time-point segment started between 6:00 and 9:00 a.m. and zero otherwise
Afternoon peak	A dummy variable equals 1 if the trip over a time-point segment started between 3:00 and 7:00 p.m. and zero otherwise
Early evening	A dummy variable equals 1 if the trip over a time-point segment started between 7:00 and 10:00 p.m. and zero otherwise
Late evening	A dummy variable equals 1 if the trip over a time-point segment started between 10:00 pm and 1:00 am and zero otherwise
Subway station ridership (000s)	Daily ridership of the subway station
Subway station $ridership^2$	Subway station daily ridership squared
Trips starting within 5 min of an incident	A dummy variable equals 1 if the trip started between 0 and 5 min after a nearby subway service interruption took place
Trips starting within 5 to 10 min of an incident	A dummy variable equals 1 if the trip started between 5 and 10 min after a nearby subway service interruption took place
Trips starting within 10 to 20 min of an incident	A dummy variable equals 1 if the trip started between 10 and 20 min after a nearby subway service interruption took place
Trips starting within 20 to 30 min of an incident	A dummy variable equals 1 if the trip started between 20 and 30 min after a nearby subway service interruption took place
Trips starting within 30 to 60 min of an incident	A dummy variable equals 1 if the trip-segment started between 30 and 60 min after a nearby subway service interruption took place
Trips starting within 60 + min of an incident	A dummy variable equals 1 if the trip-segment started 60 min or more after a nearby subway service interruption took place
Trips starting after a cleared incident time (min)	Time between the trip starting time (after an incident was cleared) and the incident clearance time. This time should range between 1 and 30 min (i.e., within the 30-min recovery buffer)
Trip-segment downstream a segment adjacent to interruption	A dummy variable equals 1 if the trip-segment was downstream of a trip-segment adjacent to the location of subway service interruption
Bus line *i*	A dummy variable equals 1 when the trip-segment is along bus line *i*, where *i* represents a bus line; it equals zero otherwise
Streetcar–bus	A dummy variable equals 1 if the trip-segment used buses along the streetcar routes
Streetcar–CLRV	A dummy variable equals 1 if the trip-segment used a Canadian light rail vehicle (CLRV)
Streetcar–ALRV	A dummy variable equals 1 if the trip-segment used an articulated light rail vehicle (ALRV)
Streetcar–Flexity	A dummy variable equals 1 if the trip-segment used a Flexity Outlook
Streetcar line *i*	A dummy variable equals 1 when the trip-segment is along streetcar line *i*, where *i* represents a streetcar line; it equals zero otherwise

in ArcGIS using an automated script written in Python, and removing system recording errors, duplicate records, segment trips with less than two GPS records, and holiday and weekend trips. Subsequently, the trip-segment's average speed, direction, starting time, and route and trip numbers were obtained.

Table 1 includes a list of variables incorporated in the statistical analysis with a detailed description. The squared terms of some independent variables are used to account for a possible non-linear relationship between them and the dependent variable. Other variables were tested in the initial model specification but were eliminated later from the study because of their insignificance (p-value ≥ 0.1) and/or correlation with other used variables (with a Pearson coefficient ≥ 0.80). The removed variables include *bus/streetcar headway, express routes*, and *total number of nearby bus/streetcar lines*. According to previous studies, the general factors affecting bus and streetcar travel times include stop sequence along a trip, time of day, segment distance, and number of stops (*22–24*).

Regarding the control variables, dummy variables for the direction of travel and time of the day were included in the models to isolate the impacts of traffic congestion on the service performance. A dummy variable, *Adjacent segment*, was included in the analysis to control for the trip-segments that start within 200 m of a subway station. These segments are expected to have lower speeds because of higher passenger activity at these locations. Another variable that was used in the models is the distance between Union Station and the segment in question. Union Station, the country's largest intermodal transit hub, is located in the heart of downtown Toronto.

Several variables were included in the models to examine the potential effect of subway service interruptions on surface transit speed. The first is a set of dummy variables that captures whether a trip has taken place after the occurrence of a subway service incident along a bus/streetcar time-point segment located within 200 m of a subway station (from the created *Adjacent segment* variable). These dummy variables were utilized to capture both the spatial and temporal connections between the subway incidents and the neighboring bus and streetcar segments. These dummy variables were categorized into six groups, namely trips starting within 0 to 5, 5 to 10, 10 to 20, 20 to 30, and 30 to 60 min of an incident, as well as trips starting 60 min after an incident. Theses dummy variables represent trips that have started after a subway incident starting time, while the incident is still taking place, or within a 30-min buffer of the incident clearance time. These 30 min were considered as the maximum recovery time for the system to return to its normal operation. Accordingly, another variable, *Trips starting after a cleared incident time (min)*, representing the difference between the trip's starting time and the incident clearance time, was used to control for the changes in speed of trips that start after the incidents' clearance time and within the 30-min buffer. Another variable was used to understand in more detail the spatial and temporal impacts of subway system interruptions. This variable, *Trip-segment downstream a segment adjacent to interruption*, distinguishes the trip-segments downstream of a trip-segment adjacent to the location of subway service interruption. In other words, this variable captures the change in speed along the bus/streetcar segments that are not immediately adjacent to the subway but lie within the 3-km buffer of the YUS Line. It is important to note that the TTC's subway service interruption data are recorded at the station level, indicating the type and amount of delay, with no clear information about the extent of its spatial effect on other subway stations that had also experienced no service. Therefore, another cleaning step was done . This step was based on removing all the bus/streetcar segments that are within the interruption impact window (incident duration + 30 min of recovery time) and within 200 m of all subway stations that are not currently analyzed. This was implemented to make sure that we are capturing the subway service interruptions' impact at only one location (i.e., the first location).

In this paper, we used two models to reveal the potential effects of the subway service interruptions. The first is a streetcar speed model, which uses the streetcar speed (in km/h) over a time-point segment as the dependent variable. A negative value for the coefficient in the model means that a streetcar had a slower speed. In this model, additional variables are used to control for the type of vehicles. TTC uses more than one type of vehicle along each single streetcar route, including the short Canadian light rail vehicle (CLRV) and the long articulated light rail vehicle (ALRV). Furthermore, in 2014, the TTC began introducing new low-floor and articulated streetcar vehicles, named Flexity Outlook, to offer more capacity (*25*). Also, sometimes, and because of the shortage of streetcar service capacity, the TTC uses regular buses on streetcar routes to provide a supplementary service alongside the streetcar service. The second model is the bus speed model, which uses the bus speed over a segment as the dependent variable. In this model, 41 dummy variables of the bus route numbers were generated to isolate the unique impacts of each individual bus route on service speed.

Analysis

Descriptive Statistics

Table 2 presents the summary statistics of the variables used in the streetcar and bus models. The results are differentiated according to the subway service operations: trips during subway service normal operations; and trips during subway service interruptions. For the streetcar system, a total of 753,820 records were included in the analysis, with about 2,800 records occurring after a subway incident and within 200 m of the subway station. In contrast, a total of 1,074,497 records were included in the bus model, with about 10,300 records occurring after a subway incident took place and within 200 m of the subject subway station.

As observed in Table 2, the overall speed of all streetcar trips during normal operations was about 12.8 km/h with a standard deviation of 5.5 km/h, whereas the

Table 2. Streetcar and Bus Route Descriptive Statistics

	Streetcar trip-segments				Bus trip-segments			
	Subway normal operations		After subway incident		Subway normal operations		After subway incidents	
	Mean	SD	Mean	SD	Mean	SD	Mean	SD
Segment average speed (km/h)	12.81	5.55	9.64	3.92	18.56	9.26	14.23	9.56
Outbound direction	0.50	0.50	0.54	0.50	0.50	0.50	0.51	0.50
Time-point sequence	5.73	3.35	6.61	1.54	6.24	4.23	6.60	5.26
Number of scheduled stops	4.78	9.76	3.51	2.87	3.63	2.48	3.07	2.51
Adjacent segment	0.18	0.38	1.00	0.00	0.23	0.42	1.00	0.00
Segment with a layover	0.19	0.49	0.13	0.39	0.16	0.37	0.29	0.45
Segment distance (km)	1.00	0.64	0.88	0.41	1.20	0.79	1.27	0.92
Streetcar–bus	0.10	0.30	0.06	0.23	0.00	0.00	0.00	0.00
Streetcar–CLRV	0.79	0.41	0.86	0.35	0.00	0.00	0.00	0.00
Streetcar–ALRV	0.11	0.31	0.09	0.28	0.00	0.00	0.00	0.00
Streetcar–Flexity	0.00	0.00	0.00	0.00	0.00	0.00	0.00	0.00
Average headway	3.84	1.61	3.89	2.02	7.00	5.77	7.19	6.12
Distance to Union Station (km)	2.22	1.48	1.95	1.85	7.91	3.35	8.22	2.82
Distance to Union Station $(km)^2$	7.11	9.82	7.22	10.2	73.7	54.1	75.4	40.6
Morning peak	0.18	0.39	0.23	0.42	0.20	0.40	0.29	0.45
Afternoon peak	0.24	0.42	0.25	0.43	0.23	0.42	0.37	0.48
Early evening	0.13	0.34	0.15	0.35	0.13	0.34	0.08	0.26
Late evening	0.13	0.33	0.06	0.23	0.13	0.33	0.06	0.24
Trips starting within 5 min of an incident	0.00	0.00	0.14	0.35	0.00	0.00	0.15	0.36
Trips starting within 5 to 10 min of an incident	0.00	0.00	0.15	0.36	0.00	0.00	0.16	0.37
Trips starting within 10 to 20 min of an incident	0.00	0.00	0.30	0.46	0.00	0.00	0.30	0.46
Trips starting within 20 to 30 min of an incident	0.00	0.00	0.27	0.44	0.00	0.00	0.29	0.45
Trips starting within 30 to 60 min of an incident	0.00	0.00	0.12	0.33	0.00	0.00	0.10	0.30
Trips starting within 60 + min of an incident	0.00	0.00	0.01	0.10	0.00	0.00	0.00	0.03
Trips starting after a cleared incident time (min)	0.00	0.00	12.63	9.56	0.00	0.00	13.34	9.42
Trip-segment downstream a segment adjacent to interruption	0.03	0.17	0.00	0.00	0.03	0.18	0.00	0.00
Number of records	751,016		2,804		1,064,197		10,300	

Note: SD = standard deviation.

average speed of streetcar trips that occurred after a subway service interruption was 9.6 km/h with a standard deviation of 3.9 km/h. For the bus routes, a similar pattern can be observed with regard to the difference between trips occurring after a subway incident and those taking place during normal subway service operations. Nevertheless, to better understand how subway service interruptions affect surface transit operations while controlling a set of relevant variables, two statistical models are presented in the following section.

Streetcar Speed Model

A linear regression model was developed using streetcar speed per trip-segment in km/h as the dependent variable. The output of this model is reported in Table 3. Only significant variables were kept in the model. The model contains 753,820 observations and explains 32% of the variation in speed. As seen in Table 3, generally, the key policy variables, *Trips starting within (…) min of an incident*, accounting for the trip's starting time after the occurrence of a subway service interruption and before resumption of subway service, have negative significant coefficients. Although this indicates a general negative impact of subway incidents on the streetcar speed, the impacts vary according to when the trip starts relative to the incident's start time. More specifically, trips starting within 0 to 5 min of a subway incident that has not been cleared yet do not suffer any significant change in their speed. This indicates that there is no

Table 3. Streetcar Segment Speed Model

	Coeff.	Z	95% conf. interval	
			Lower bound	Upper bound
(Constant)	**9.73**	**288.4*****	**9.67**	**9.80**
Outbound direction	**−0.40**	**−37.4*****	**−0.42**	**−0.38**
Time-point sequence	**0.01**	**2.68*****	**0.00**	**0.01**
Number of scheduled stops	**−0.03**	**−44.35*****	**−0.03**	**−0.03**
Adjacent segment	**−2.85**	**−152.8*****	**−2.89**	**−2.81**
Segment with a layover	**−1.90**	**−146.5*****	**−1.93**	**−1.88**
Segment distance (km)	**2.24**	**213.8*****	**2.22**	**2.26**
Streetcar–bus	**−0.06**	**−1.73*****	**−0.13**	**0.01**
Streetcar–ALRV	**−0.19**	**−4.58*****	**−0.28**	**−0.11**
Streetcar–Flexity	0.42	0.20	−3.60	4.43
Distance to Union Station (km)	**1.30**	**76.5*****	**1.27**	**1.33**
Distance to Union Station $(km)^2$	**−0.18**	**−57.1*****	**−0.18**	**−0.17**
Morning peak	**1.51**	**96.8*****	**1.48**	**1.54**
Afternoon peak	**−0.72**	**−50.6*****	**−0.75**	**−0.70**
Early evening	**1.16**	**67.6*****	**1.13**	**1.20**
Late evening	**4.00**	**226.9*****	**3.96**	**4.03**
Subway station ridership	**−0.05**	**−17.54*****	**−0.05**	**−0.04**
Subway station $ridership^2$	**0.00**	**18.55*****	**0.00**	**0.00**
Trips starting within 5 min of an incident	0.07	0.28	−0.40	0.53
Trips starting within 5 to 10 min of an incident	**−0.68**	**−2.83*****	**−1.16**	**−0.21**
Trips starting within 10 to 20 min of an incident	**−0.57**	**−2.36****	**−1.05**	**−0.10**
Trips starting within 20 to 30 min of an incident	**−0.81**	**−2.23****	**−1.53**	**−0.10**
Trips starting within 30 to 60 min of an incident	**−1.01**	**−2.37****	**−1.84**	**−0.18**
Trips starting within 60 + min of an incident	**−1.47**	**−1.74***	**−3.12**	**0.19**
Trips starting after a cleared incident time (min)	**0.035**	**2.23****	**0.00**	**0.06**
Route 501	**−0.69**	**−15.10*****	**−0.78**	**−0.60**
Route 502	**−0.66**	**−13.41*****	**−0.76**	**−0.56**
Route 503	−0.10	−1.33	−0.24	0.05
Route 504	**0.07**	**3.72*****	**0.03**	**0.11**
Route 506	**0.15**	**6.94*****	**0.11**	**0.20**
Route 509	**1.23**	**25.45*****	**1.14**	**1.33**
Route 510	**−2.75**	**−107.6*****	**−2.80**	**−2.70**
Route 511	**−1.51**	**−54.77*****	**−1.56**	**−1.45**
Route 512	**1.07**	**20.84*****	**0.97**	**1.17**
N			753,820	
Adjusted R^2			0.32	
F statistics			(34, 753819) 10430	
F significance (Prob > F)			0	

Note: Coeff. = coefficient; conf. = confidence. Bold font indicates statistical significance.
***Significant at 99%.
**Significant at 95%.
*Significant at 90%.

immediate impact of subway incidents on streetcar performance within the first 5 min. However, streetcar trips starting within 5 to 10 min of a subway incident that is still persisting start to see some delays, experiencing a reduction in speed by 0.68 km/h. This may reflect the combined effect of people coming to use the subway only to learn the service was interrupted and deciding to use the streetcar lines instead, and subway users off-loaded from the interrupted subway trains and deciding immediately to use the nearby streetcar lines instead of waiting for the resumption of subway service.

The speed reduction persists at about 0.57 km/h when the streetcar trip starts in 10 to 20 min after a subway incident start time and while it is still in effect. The trips starting within 20 to 30 min of a subway incident that is still taking place suffer a reduction of 0.81 km/h in speed. The decrease in speed grows to reach 1.01 km/h and 1.47 km/h for trips starting within 30 to 60 min of an incident and more than 60 min after an incident, respectively. This means that more delays and lower speeds are expected for streetcar trips starting while longer subway incidents are persistent, while keeping all other variables at their mean values. This can be because of the current policies followed by TTC of providing fewer shuttle buses at disrupted subway locations served by nearby streetcar services in downtown Toronto (*26*). Therefore, transit agencies should provide more resources and different disruption management strategies in responding to subway incidents in downtown areas to reduce the adverse impacts on nearby surface transit that are typically characterized with overcrowding issues. After the incident clearance time, every minute difference between the streetcar trip's starting time and the incident's clearance time is expected to improve the streetcar service speed by 0.035 km/h. Using this coefficient value, this indicates that trips starting after 10 min of the subway incident clearance time could expect about 0.35 km/h recovery in their average speed, whereas trips starting after 20 and 30 min could expect a 0.70 and 1.05 km/h recovery in their average speed, respectively. This indicates (with comparison to the model's coefficients in Table 3) that most of the trips will recover within the 30-min window to their original speed, except for the trips that start at 60 min or later after an incident that is still taking place. However, this window still represents a good recovery approximation as the longest subway incident duration in our sample is 73 min.

In accordance with previous work, the remaining control variables in the model perform as expected with regard to the expected sign and significance (*22*, *23*, *27*, *28*). Trips during the afternoon peak were slower by 0.7 km/h than midday trips, whereas trips made during other time periods were faster than midday trips. This observation is not specific to the context of Toronto as it has also been reported in two other North American cities (*9*, *23*, *27*). Segment distance increases the service speed. For every additional kilometer, streetcar speed is expected to increase by 2.85 km/h. This suggests that longer distances between time-point stops may allow streetcar drivers to maintain higher speeds, which is consistent with previous research (*23*). For every additional kilometer in distance to downtown, the streetcar speed is expected to increase by 0.13 km/h. This is expected because of less traffic congestion and demand at locations that are farther from downtown. Nonetheless, the distance to Union Station squared term had a statistically significant positive association with the speed, indicating that after a certain distance no negative association between the distance to Union Station and streetcar speed can be expected.

Route speed also varies by direction, likely resulting from traffic conditions and other unseen factors not explained by the model's variables. The increase in the segment sequence within a trip is expected to increase the streetcar operational speed slightly by 0.006 km/h for every additional segment. The number of scheduled stops within a segment is expected to decrease the speed by 0.03 for every additional stop, which is expected because of passenger activity and streetcar acceleration and deceleration at stops. Any segment within 200 m of a subway station and any segment that includes a layover stop are slower by 2.85 and 1.9 km/h compared with other segments, respectively. This is expected because of the high volume of passengers at these locations. For every additional 1,000 passengers per subway station, the streetcar service speed starting at segments that are within 200 m is expected to be slower by 0.05 km/h. Nevertheless, the subway ridership squared term shows that after a certain ridership level, no decrease in speed can be expected. Finally, the type of streetcar vehicles and route number also have an impact on streetcar average speed, which can be related to the vehicle offered capacity and specifications, traffic conditions, route lengths, and TTC scheduling and route management practices.

Bus Speed Model

A linear regression model was also developed using bus speed per trip-segment in km/h as the dependent variable. The output of this model is reported in Table 4. Only the significant variables were kept in the model. The model contains 1,074,496 observations and explains 37% of the variation in bus service's speed. As seen in the table, a different pattern can be observed with respect to the variable signs and magnitude of coefficients in comparison to the previous model. The key policy variables, *Trips starting within (…) min of an incident*, have a negative significant coefficient for only two cases, namely for trips starting within 20 to 30 min and 30 to 60 min of a subway incident start time. More specifically, trips starting within 20 to 30 min and 30 to 60 min of a subway incident while it is still in effect suffer a reduction of 0.25 km/h and 0.53 km/h in their speed, respectively.

Nevertheless, the observed reductions in speed are relatively minor compared with the previous model, particularly considering that bus speeds are normally higher than streetcar speeds. Therefore, the following section presents a sensitivity analysis to better understand the relative impacts on both systems. Bus trips starting more than 60 min after a subway incident do not suffer a statistically significant reduction in their service speed. This indicates an improved situation, which can be related to the TTC's practice of employing shuttle buses at these locations, relieving the impacts of subway incidents. However, bus speeds over the same trip's segments following an impacted segment by the subway service interruption is expected to be slower by 0.47 km/h, which could be explained by induced passenger activity along these segments and/or by traffic conditions along the bus corridor. Traffic conditions could be worsened by

Table 4. Bus Segment Speed Model

	Coeff.	Z	95% conf. interval	
			Lower bound	Upper bound
(Constant)	**13.11**	**150.6***	**12.9**	**13.3**
Outbound direction	**0.28**	**19.24***	**0.25**	**0.30**
Time-point sequence	**0.02**	**6.92***	**0.02**	**0.03**
Number of scheduled stops	**−0.21**	**−47.0***	**−0.22**	**−0.20**
Adjacent segment	**−3.98**	**−208.5***	**−4.02**	**−3.94**
Segment with a layover	**−7.13**	**−295.2***	**−7.18**	**−7.09**
Segment distance (km)	**2.85**	**195.8***	**2.82**	**2.88**
Distance to Union Station (km)	**0.42**	**29.63***	**0.39**	**0.45**
Distance to Union Station (km)^2	**−0.01**	**−18.36***	**−0.02**	**−0.01**
Morning peak	**0.55**	**26.49***	**0.51**	**0.59**
Afternoon peak	**−2.11**	**−108.0***	**−2.15**	**−2.07**
Early evening	**2.00**	**85.68***	**1.96**	**2.05**
Late evening	**6.42**	**266.4***	**6.37**	**6.46**
Subway station ridership	**−0.04**	**−12.39***	**−0.05**	**−0.03**
Subway station ridership^2	**0.001**	**13.63***	**0.00**	**0.00**
Trips starting within 5 min of an incident	0.13	0.66	−0.25	0.51
Trips starting within 5 to 10 min of an incident	0.19	0.98	−0.19	0.56
Trips starting within 10 to 20 min of an incident	0.07	0.52	−0.21	0.35
Trips starting within 20 to 30 min of an incident	**−0.25**	**−1.70***	**−0.53**	**0.04**
Trips starting within 30 to 60 min of an incident	**−0.53**	**−2.28****	**−0.98**	**−0.07**
Trips starting within 60 + min of an incident	−2.98	−1.14	−8.07	2.12
Trip-segment downstream a segment adjacent to interruption	**−0.47**	**−11.7***	**−0.55**	**−0.39**
40 dummy variables to control for the route number				
N			1,074,496	
Adjusted R^2			0.37	
F statistics			(62,1,074,495)10220	
F significance (Prob > F)			0	

Note: Coeff. = coefficient; conf. = confidence. Bold font indicates statistical significance.
***Significant at 99%.
**Significant at 95%.
*Significant at 90%.

additional taxi and Uber-like services which may be requested by some subway riders affected by subway disruption events. The stopping maneuvers of taxis and Uber vehicles in the vicinity of bus and streetcar paths may create conflicts with transit flow and travel times.

Regarding the other remaining control variables in the model, it is clear that the model coefficients follow the same signs and magnitude as the previous model. Afternoon peak trips are slower than all other trips during the day. The distance to Union Station increases the bus service speed up to a certain point, beyond which no positive impact is observed with longer distance. The segment distance and segment sequence have a positive impact on speed, whereas the number of stops within a segment, trips that are within 200 m of a subway station, and trips with layovers have negative impacts on bus speed. Finally, the route direction and route number (40 dummy variables but not reported in Table 4) also showed a significant impact on bus speeds, which can be attributed to each route's specifications, traffic conditions, and scheduling.

Sensitivity Analysis

Using the previous models' coefficients, it was possible to conduct a sensitivity analysis to demonstrate the relative impacts of subway incidents on streetcar and bus systems' speed while keeping all variables constant at their mean values. In fact, this sensitivity analysis enables a better understanding of the changes and a meaningful comparison of the previous models' results. Table 5 shows the expected average speed for streetcars and buses during the different time periods of the day. For the streetcar system, the speed estimates were made for the eastbound trips of Route 501 and using CLRVs. For the bus system, the speed estimates were made for the westbound trips of Route 96. Both routes included in the sensitivity analysis operate at the median speed of their

Table 5. Streetcar and Bus Service Speed (km/h) Sensitivity Analysis

	Morning peak		Midday		Afternoon peak		Early evening		Late evening		Average
	Streetcar service speed (km/h)										
During normal operations	**11.0**		**9.6**		**8.8**		**10.7**		**13.6**		**10.7**
During subway disruption	Speed	%	Speed	%	Speed	%	Speed	%	Speed	%	%
Trips starting within 5 min of an incident	11.0	**0.0**	9.6	**0.0**	8.8	**0.0**	10.7	**0.0**	13.6	**0.0**	**0.0**
Trips starting within 5 to 10 min of an incident	10.3	**−6.2**	8.9	**−7.2**	8.1	**−7.8**	10.0	**−6.4**	12.8	**−5.0**	**−6.5**
Trips starting within 10 to 20 min of an incident	10.5	**−5.2**	8.9	**−6.0**	8.2	**−6.5**	10.1	**−5.4**	12.9	**−4.2**	**−5.5**
Trips starting within 20 to 30 min of an incident	10.2	**−7.4**	8.7	**−8.5**	8.0	**−9.2**	9.9	**−7.6**	12.7	**−6.0**	**−7.7**
Trips starting within 30 to 60 min of an incident	10.0	**−9.1**	8.5	**−10.6**	7.8	**−11.4**	9.7	**−9.4**	12.5	**−7.5**	**−9.6**
Trips starting within 60 + min of an incident	9.60	**−13.3**	8.1	**−15.4**	7.3	**−16.6**	9.2	**−13.7**	12.1	**−10.8**	**−13.9**
	Bus service speed (km/h)										
During normal operations	**16.5**		**17.0**		**15.9**		**21.0**		**26.4**		**19.4**
During subway disruption	Speed	%	Speed	%	Speed	%	Speed	%	Speed	%	%
Trips starting within 5 min of an incident	16.5	**0.0**	17.0	**0.0**	15.9	**0.0**	21.0	**0.0**	26.4	**0.0**	**0.0**
Trips starting within 5 to 10 min of an incident	16.5	**0.0**	17.0	**0.0**	15.9	**0.0**	21.0	**0.0**	26.4	**0.0**	**0.0**
Trips starting within 10 to 20 min of an incident	16.5	**0.0**	17.0	**0.0**	15.9	**0.0**	21.0	**0.0**	26.4	**0.0**	**0.0**
Trips starting within 20 to 30 min of an incident	16.3	**−1.5**	16.7	**−1.4**	15.6	**−1.6**	20.7	**−1.2**	26.1	**−0.9**	**−1.3**
Trips starting within 30 to 60 min of an incident	16.0	**−3.2**	16.4	**−3.1**	15.3	**−3.3**	20.4	**−2.5**	25.8	**−2.0**	**−2.8**
Trips starting within 60 + min of an incident	16.5	**0.0**	17.0	**0.0**	15.9	**0.0**	21.0	**0.0**	26.4	**0.0**	**0.0**

Note: % refers to the percentage of change in speed = (trip speed during an incident category – trip speed during normal operations)/trip speed during normal operations. Bold font shows this percentage.

respective systems. In Table 5, the percentage of change in average speed because of subway service interruptions compared with normal operations is reported.

As shown in Table 5, the average speed for streetcars over an adjacent segment within 200 m of a subway station and during normal operations is 10.7 km/h. This average decreases by 6.5% for trips starting 5 to 10 min after a subway incident start time and while it is still in effect. The reduction in speed remains at about 5.5% for trips that start within 10 to 20 min of a subway incident's start time. Afterwards, the speeds further decline gradually by 7.7%, 9.6 %, and 13.9% for trips starting within 20 to 30 min and 30 to 60 min of an accident, and more than 60 min after an incident, respectively. This reflects a higher magnitude of impact of longer subway incidents on the streetcar service speed. The maximum delay for streetcars occurs during the afternoon peak period, with a 16.6% reduction in speed for trips starting more than 60 min after an incident. This highlights a challenge during this time period that requires more attention from the transit agency.

The average speed for buses over an adjacent segment within 200 m of a subway station and during normal operations is 19.4 km/h. This average decreases slightly by 1.3% and 2.8% for trips starting within 20 to 30 min and 30 to 60 min of a subway incident's start time and while it is still in effect, respectively. Similar to the streetcar system, the peak delay occurs during the afternoon peak period, with a 3.3% reduction in speed for trips starting within 30 to 60 min of a subway incident's start time. This highlights a maximum of 13.3% of additional delay for streetcars because of subway service interruptions in comparison with buses. This delay means an increased waiting time, travel time, and overcrowding experienced by the streetcar's users, which would have a direct negative effect on their perceptions of the service quality. Therefore, developing more appropriate policies to maintain and improve the streetcar system's performance during these situations is highly recommended.

Conclusion

The objective of this study is to evaluate the impact of the subway system's interruptions on the surface transit system's performance in Toronto. To do that, the study used TTC's subway system interruption data and AVL system data for 51 surface transit routes within a short walking distance of a subway station. The selection of subway stations to be included in the analysis was based on a composite indicator generated to identify the most vulnerable stations in the subway system and to capture the exceptional events of subway interruptions. Two models were utilized using bus and streetcar speeds at the disaggregate level of time-point segment and trip as the dependent variable.

The models indicate that subway service interruptions have a statistically significant impact on streetcar and bus operations, contributing to longer passenger waiting times at downstream locations and in-vehicle passengers' total travel time. Nevertheless, the intensity of delay varies according to the mode and the starting time of the trip category in comparison with the starting time of the incident. More specifically, subway incidents have more immediate and long lasting negative impacts on the streetcar service speed. This reduction in speed ranged between 5.5% and 13.9% of the average speed during normal operations. In contrast to the streetcar service, the bus service experienced less significant delays in its operational speed because of subway service interruptions. This reduction in speed ranged only between 1.3% and 2.8% of the average speed during incident-free operations.

This difference in impacts between the two systems may reflect the TTC's use of disruption management protocols for dealing with subway service disruptions. These protocols mainly deploy fewer shuttle buses along the southern section of the subway system (U-shaped section) where the nearby streetcar service is available, which could be the reason for the sharper decline in streetcar service speed. Thus, a detailed investigation with regard to the shuttle service level of employment that is required to minimize the subway incidents' impact on local streetcar service is recommended. We were not able to investigate this here because of missing accurate information about shuttle service arrival and departure times during our study period.

Given the dearth of studies that explore the impact of the service disruption of one transit mode on other transit modes in a multimodal environment, one of the key contributions of this research is to develop a methodology that can be used by other agencies and researchers to understand these impacts at different settings and locations. It should be noted that APC system data were not available for this study because the TTC currently does not equip its streetcars and most of its buses with such a system. Not equpping all the fleet with an APC system is commonly the case of large transit agencies in North America (*29*). It is recommended that another case study of a similar multimodal transit system is undertaken in the future while using other sources of passenger demand data such as APC to further investigate the impact of disruptions across transit modes in a different context. Finally, although this research explores the impact of subway system interruptions on the speed performance of neighboring (within 200 m) bus and streetcar services, it is recommended that this study is expanded to investigate the impacts on other performance indicators such as variability of service, headway delays, and user delays. It is also recommended to expand this study spatially to investigate subway disruption impacts on surface transit routes not in the immediate vicinity of affected subway stations, according to the used case study context.

Acknowledgments

The authors gratefully acknowledge Kenny Ling and Francis Li from the TTC for providing the data used in the paper. This research was funded by the Natural Sciences and Engineering Research Council of Canada (NSERC) and the Ontario Research Fund (ORF).

Author Contributions

The authors confirm contribution to the paper as follows: study conception and design: Ehab Diab, Amer Shalaby; data collection: Ehab Diab; analysis and interpretation of results: Ehab Diab, Amer Shalaby; draft manuscript preparation: Ehab Diab, Amer Shalaby. All authors reviewed the results and approved the final version of the manuscript.

References

1. Diab, E., and A. El-Geneidy. Variation in Bus Transit Service: Understanding the Impacts of Various Improvement Strategies on Transit Service Reliability. *Public Transport*, Vol. 4, No. 3, 2013, pp. 209–231.
2. Yetiskul, E., and M. Senbil. Public Bus Transit Travel-Time Variability in Ankara (Turkey). *Transport Policy*, Vol. 23, 2012, pp. 50–59.
3. Kimpel, T., J. Strathman, R. Bertini, P. Bender, and S. Callas. Analysis of Transit Signal Priority Using Archived TriMet Bus Dispatch System Data. *Transportation Research Record: Journal of the Transportation Research Board*, 2005. 1925: 156–166.
4. Mesbah, M., J. Lin, and G. Currie. "Weather" Transit is Reliable? Using AVL Data to Explore Tram Performance in Melbourne, Australia. *Journal of Traffic and Transportation Engineering (English Edition)*, Vol. 2, No. 3, 2015, pp. 125–135.
5. Currie, G., A. Delbosc, S. Harrison, and M. Sarvi. Impact of Crowding on Streetcar Dwell Time. *Transportation Research Record: Journal of the Transportation Research Board*, 2013. 2353: 100–106.
6. Currie, G., and J. Reynolds. Evaluating Pay-on-Entry versus Proof-of-Payment Ticketing in Light Rail Transit. *Transportation Research Record: Journal of the Transportation Research Board*, 2016. 2540: 39–45.
7. Louie, J., A. Shalaby, and K. Habib. Modelling Disruption Duration for Toronto's Subway System: An Empirical Investigation Using Lognormal Regression and Hazard Models. Presented at 95th Annual Meeting of the Transportation Research Board, Washington, D.C., 2016.
8. Weng, J., Y. Zheng, X. Yan, and Q. Meng. Development of a Subway Operation Incident Delay Model Using Accelerated Failure Time Approaches. *Accident Analysis & Prevention*, Vol. 73, 2014, pp. 12–19.

9. Diab, E. I., and A. M. El-Geneidy. Understanding the Impacts of a Combination of Service Improvement Strategies on Bus Running Time and Passenger's Perception. *Transportation Research Part A: Policy and Practice*, Vol. 46, No. 3, 2012, pp. 614–625.
10. Diab, E. I., and A. M. El-Geneidy. The Farside Story: Measuring the Benefits of Bus Stop Location on Transit Performance. *Transportation Research Record: Journal of the Transportation Research Board*, 2015. 2538: 1–10.
11. Schmöcker, J., S. Cooper, and W. Adeney. Metro Service Delay Recovery: Comparison of Strategies and Constraints across Systems. *Transportation Research Record: Journal of the Transportation Research Board*, 2005. 1930: 30–37.
12. Kepaptsoglou, K., and M. Karlaftis. The Bus Bridging Problem in Metro Operations: Conceptual Framework, Models and Algorithms. *Public Transport*, Vol. 1, No. 4, 2009, pp. 275–297.
13. Wang, Y., J. Guo, G. Currie, W. Dong, and B. Pender. Bus Bridging Disruption in Rail Services with Frustrated and Impatient Passengers. *IEEE Transactions on Intelligent Transportation Systems*, Vol. 15, No. 5, 2014, pp. 2014–2023.
14. Codina, E., and A. Marin. A Design Model for the Bus Bridging Problem. *Proc., World Conference on Transport Research Society*, 12th World Conference on Transport Research, Lisbon, Portugal, 2010.
15. Pender, B., G. Currie, N. Shiwakoti, and A. Delbosc. Economic Viability of Bus Bridging Reserves for Fast Response to Unplanned Passenger Rail Disruption. *Transportation Research Record: Journal of the Transportation Research Board*, 2015. 2537: 13–22.
16. Darmanin, T., C. Lim, and H. Gan. Public Railway Disruption Recovery Planning: A New Recovery Strategy for Metro Train Melbourne. *Proc., The 11th Asia Pacific Industrial Engineering and Management Systems Conference; The 14th Asia Pacific Regional Meeting of International Foundation for Production Research*, Asia Pacific Industrial Engineering and Management Systems Conference, Melaka, 2010.
17. van der Hurk, E., H. Koutsopoulos, N. Wilson, L. Kroon, and G. Maróti. Shuttle Planning for Link Closures in Urban Public Transport Networks. *Transportation Science*, Vol. 50, No. 3, 2016, pp. 947–965.
18. Ontario Ministry of Finance. *Ontario Population Projections Update, 2015–2041*. Ontario Ministry of Finance, Toronto, Canada, 2016. http://www.fin.gov.on.ca/en/economy/demographics/projections/. Accessed July 24, 2016.
19. American Public Transportation Association. *APTA Transit Ridership Report, Second Quarter, 2013*. American Public Transportation Association, Washington, D.C., 2013.
20. TTC. *TTC Operating Statistics 2013*. Toronto, Canada. https://www.ttc.ca/About_the_TTC/Operating_Statistics/2013.jsp. Accessed May 2, 2016.
21. Vuchic, V. *Urban Transit: Operations, Planning, and Economics*. John Wiley & Sons, Hoboken, NJ, 2005.
22. Kimpel, T.. Time Point-Level Analysis of Transit Service Reliability and Passenger Demand. In *Urban Studies and Planning, No. Doctor of Philosophy*, Portland State University, 2001, p. 154.
23. El-Geneidy, A., J. Horning, and K. Krizek. Analyzing Transit Service Reliability Using Detailed Data from Automatic Vehicular Locator Systems. *Journal of Advanced Transportation*, Vol. 45, No. 1, 2011, pp. 66–79.
24. Figliozzi, M., W. Feng, and G. Lafferriere. A Study of Headway Maintenance for Bus Routes: Causes and Effects of "Bus Bunching" in Extensive and Congested Service Areas. In *Civil and Environmental Engineering Faculty Publications and Presentations, No. Paper 96*, Transportation Research and Education Center, Portland, Ore., 2012.
25. TTC. *New streetcars*. TTC, Toronto, Canada. http://www.ttc.ca/About_the_TTC/Projects/New_Vehicles/New_Streetcars/index.jsp. Accessed July 16, 2017.
26. Diab, E., W. Feng, and A. Shalaby. Breaking into Emergency Shuttle Service: Aspects and Impacts of Retracting Buses from Existing Scheduled Bus Services. Canadian Journal of Civil Engineering, 2018, pp. 647–658.
27. Strathman, J., K. Dueker, T. Kimpel, R. Gerhart, K. Turner, P. Taylor, S. Callas, D. Griffin, and J. Hopper. Automated Bus Dispatching, Operations Control, and Service Reliability: Baseline Analysis. *Transportation Research Record: Journal of the Transportation Research Board*, 1999. 1666: 28–36.
28. Strathman, J., T. Kimpel, and S. Callas. *Headway Deviation Effects on Bus Passenger Loads: Analysis of Tri-Met's Archived AVL-APC Data*. Center for Urban Studies, College of Urban and Public Affairs, Portland State University, Seattle, Wash., 2003.
29. Boyle, D. K. *TCRP synthesis 77: Passenger Counting Systems*. Transportation Research Board, Washington, D.C., 2008.

The Standing Committee on Transit Management and Performance (AP010) peer-reviewed this paper (18-05031).

The ideas and findings presented in this paper represent the authors' views in an academic exercise.

TRR
JOURNAL OF THE TRANSPORTATION RESEARCH BOARD

Article

Transportation Research Record
2018, Vol. 2672(8) 455–463

DOI: 10.1177/0361198118787363
journals.sagepub.com/home/trr
SAGE

Defining TransMilenio Users' Value and Satisfaction through the Lean Thinking Approach

Carlos Garcia-Suarez[1], Andres L. Rivera-Perez[1], and Alvaro Rodriguez-Valencia[2]

Abstract

Bogota is witnessing a steady decline in the user satisfaction for its Bus Rapid Transit (BRT) system. On a one to five point scale, general satisfaction with the system, known as TransMilenio, has fallen from 4.60 in 2001 to 2.62 in 2016. This paper analyzes what is happening to TransMilenio user satisfaction by defining the "Value Concept" within the principles of the "Lean thinking" approach. Qualitative research techniques were used to identify factors that are important to users with regard to what they expect to receive from the system. Quantitative data were collected through an in-person survey of a representative random sample of TransMilenio riders. These data were used in a linear regression model to gain meaningful insights into factors that determine levels of satisfaction for TransMilenio users. The results show that six factors explain user satisfaction. From these results, this investigation proposes that the main efforts of TransMilenio should be in line with riders' expectations. Our findings can help TransMilenio understand its shortcomings and shed light on how to prevent service declines in other BRT systems modeled on that of Bogota.

Bogota is witnessing a rapid and steady decline in its Bus Rapid Transit rider satisfaction. According to the latest satisfaction survey, TransMilenio users have an average general satisfaction level of 2.62, on a one to five point scale on which 5 represents the maximum level of satisfaction (*1*). Figure 1 shows annual ratings from four different sources, which have very good matches in overlapping years. The results of the survey conducted in 2016 by *Bogota Cómo Vamos*, a citizen effort to track and monitor changes in the quality of life in Bogota (*2*), show that only 18% of TransMilenio users were satisfied with its service (*3*).

Several authors have offered possible explanations for decreasing user satisfaction with TransMilenio. The most commonly mentioned factors are rush hour overcrowding, resulting in discomfort and disorder within the System (*4–7*). Other authors attribute perceptions of poor service to deterioration of infrastructure and insecurity (*4*, *7*). Factors such as failure to expand TransMilenio infrastructure (*5*, *6*), delays in feeder bus services (*5*), and a need to increase bus speed to reduce travel times (*8*) affect rider satisfaction. In addition, TransMilenio's decision to reduce operational costs by only operating a portion of its entire fleet at any given moment, has had a negative impact on bus frequencies (*7*, *9*).

In light of the previously published literature, this study uses the lean thinking philosophy to uncover dimensions that may contribute to declining satisfaction with TransMilenio. Elsewhere, this approach has produced outstanding results in industries such as construction, transportation, and logistics. Qualitative research techniques were used to identify dimensions that users value to evaluate the service they receive. Quantitative data were collected through an in-person survey of a representative random sample of TransMilenio riders to identify dimensions that explain users' levels of satisfaction. Linear regression techniques were applied to that end. This paper is devoted to a study of the first of the five lean thinking principles and is preliminary to examine the application of the remaining four principles to TransMilenio.

Literature Review

Value and Satisfaction

Understanding preferences and perceptions of public transportation passengers is relevant for evaluating transit service performance (*10*) and travelers' choices (*11*). According to Lai and Chen, satisfaction with a service and its perceived

[1]Empresa de Transporte Tercer Milenio, TransMilenio S.A., Bogota, Colombia
[2]Department of Civil and Environmental Engineering, Universidad de los Andes, Bogota, Colombia

Corresponding Author:
Address correspondence to Alvaro Rodriguez-Valencia: alvrodri@uniandes.edu.co

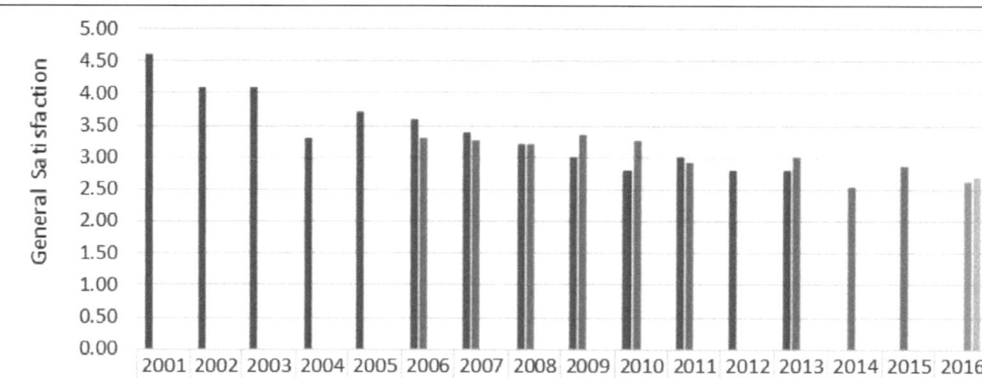

	2001	2002	2003	2004	2005	2006	2007	2008	2009	2010	2011	2012	2013	2014	2015	2016
Bogotá Cómo Vamos	4.60	4.10	4.10	3.30	3.70	3.60	3.40	3.20	3.00	2.80	3.00	2.80	2.80			
Observatorio de Movilidad						3.31	3.26	3.22	3.37	3.27	2.91		3.01	2.53	2.86	
Centro Nacional de Consultoría																2.62
García, Rivera																2.69

Figure 1. Results of the satisfaction surveys since 2001.

value affect behavioral intentions of passengers which may be a prime determinant of long-term financial performance of transit operations (*12*). In addition, positive customer satisfaction translates into greater numbers of passengers through retention of current riders and attraction of new ones. It also results in a more positive public image (*13*).

Tam (*14*) provides a broad review of the literature to define *perceived service quality, perceived value* and *customer satisfaction* while simultaneously providing insightful empirical evidence about how to connect them. In general, all three concepts relate an 'input' and an 'output'. *Perceived service quality* is a perception of the discrepancy between customers' expectations and perceptions of service performance. *Perceived value* refers to the ratio between service obtained and costs incurred by users. To this point, both monetary and non-monetary costs are involved in the assessment. The trade-off in the case of perceived value is input defined as the service received and output defined as what riders give up to obtain service. Finally, *customer satisfaction* is an emotional response that results from a cognitive process of evaluating perceived value. According to Tam "as customers' perceptions of the quality of the service increase, they perceive higher value in the service and feel more satisfied" (*14*). Tam shows that all three concepts are directly and causally correlated.

By understanding users' perspectives, service providers can identify areas that require immediate intervention so that the difference between consumers' expectations and the service they actually receive is increasingly smaller. Through this process, customers' perception of value and levels of satisfaction can be expected to increase. Imam states that service providers should give priority to those dimensions users evaluate as poor because improvements in these areas are likely to increase levels of satisfaction (*15*).

Quite a lot of factors that influence users' satisfaction can be found in the literature. Eboli and Mazzulla conducted a stated preference experiment to measure service quality in public transportation and found that waiting time and fares are the two strongest factors explaining quality (*16*). They found that other important attributes include reliability, cleanliness and bus stop facilities (*16*). Factors that negatively influence urban service quality perceptions are travel time, walking time, time between services (headway) and delays, and those that positively influence perceptions are comfort, the attitude of the driver, information, and cleanliness.

The study from Tyrinopoulos and Antoniou of the transit systems in two Greek cities found that the relevant dimensions depend on the type of transit system (*17*). They found that the most important satisfaction dimensions for metro systems are vehicle cleanliness, behavior of staff other than drivers, and ticketing systems, but for bus and trolley systems service frequency, vehicle cleanliness, network coverage, waiting conditions, and punctuality are the key factors (*17*). As metro systems are very reliable, and bus system reliability depends on traffic congestion, relevant quality dimensions change according to the mode.

Lean Thinking: A Management Business Philosophy Based on Value Concept

Lean thinking is a management philosophy that seeks to maximize efficiency of companies' functioning through continuous improvement in the organization of activities and individuals that intervene in the fabrication of a product or in the provision of a service (*18*). This philosophy was born in the 1950's at Toyota. Although it was originally related solely to the manufacturing industry, "over the years Lean Thinking has gained force as a strategic model of management" (*19*) with successful application in diverse sectors including construction, healthcare, and logistics.

Application of lean thinking requires collaboration from everyone involved in the provision of service. It is based on five principles: value, value stream, flow, pull, and perfection. The methodologically correct way to apply those principles is through a process that can be summarized as a "method to specify value, line up value-creating actions in the best sequence, conduct these activities without interruption whenever someone requests them, and perform them more and more effectively" (*20*).

For lean thinking, the reason companies exist is to create value. The starting point for correct implementation of lean thinking is the effort to define what value is from the final consumer's perspective rather than from the point of view of the producer (*20*). This is true as value consists of the attributes of a product or service that lead a client to pay for it, or what generates a customer's preference for one product or service over other equivalent alternatives (*19*). The value concept makes sense when it is expressed as a result that satisfies a client's requirement at the right moment at a concrete price (*20*).

Certainly, the identification of value in a service industry such as passenger transportation is a challenging task. While the value of goods produced through manufacturing is specified with regard to tangible characteristics including size, weight, and materials, the value of a service can include an endless number of variables. The case of FedEx has been recognized for documenting the application of lean thinking in the transport and logistics sector by defining what factors were important for satisfying their clients' desires and necessities (*21*). These requirements are equivalent to what Tam refers to as "expectations" and that under the lens of lean thinking is nothing other than the concept of "value" (*14*). The company consulted its clients to identify the value that they are willing to pay for. As a result, FedEx has established a list of "Eight Rights" which are the dimensions that must be considered for correct provision of the service: materials, quantity, time, place, resources, price, quality, and service. Through offering the correct rights in the correct quantities at the right time, FedEx guarantees customer satisfaction without incurring costs for which customers are not willing to pay (*21*).

Context

TransMilenio revolutionized the world's mass transportation systems with regard to passengers moved per hour and the public and legal framework that was created for the system. TransMilenio accounts for 15.96% of the total trips in Bogota, Colombia, more than 2.3 million passengers per day in 2015 (*22*). The institutional and business models were part of the success of this BRT. It was designed in such a way that the private sector could be involved in fleet investment, bus operation, technology, and fare collection. For years, TransMilenio's sole source of income was fare collection. Fares must suffice to pay for operational costs of the private bus companies, the fare collector, and the TransMilenio bureaucracy. Infrastructure construction and maintenance is the responsibility of the city of Bogota.

TransMilenio's operational principles focus on achieving standards of quality, dignity and comfort while reaching goals of financial and environmental sustainability to improve riders' quality of life (*23*). To follow those principles and reach its goals, the company has implemented measures of inner functioning such as its strategic plan which is based on the statutes of the TransMilenio society (*24*), an action plan composed of commitments, indexes, percentage advantage, and actors involved (*25*), and parameters including indicators of progress in each part of the company (*26*).

Methodology

The methodology adopted for this study has two parts: one qualitative and another quantitative. The qualitative part of the study aims to complement what is reported in the FedEx case by establishing dimensions that define "value" through interviewing TransMilenio riders. The quantitative part of the study determines which dimensions are explanatory variables for riders' levels of satisfaction based on the results of 384 surveys.

Qualitative Stage: Uncovering Value in TransMilenio Service

A qualitative approach is useful as a complement to existing secondary sources to determine dimensions relevant to the service provided by TransMilenio. FedEx's eight rights as explained by Martichenko are the starting points for uncovering dimensions that determine value (*21*). The in-depth interviews of local users aim to identify important dimensions that the literature does not consider.

The qualitative part of this study used semi-structured interviews. Between October 3 and October 7, 2016, one TransMilenio rider was randomly selected at each of six TransMilenio stations: Portal 80, Portal Americas, Portal Usme, Portal Norte, Portal Dorado, and Universidades Station. We asked them to freely answer four questions:

- How is a typical trip in TransMilenio?
- What aspects of TransMilenio do you consider to be positive?
- What aspects of TransMilenio do you consider to be negative?
- What is missing in TransMilenio that you would add?

To integrate answers from interviewees, researchers completed a trip stage evaluation matrix, based on the life history method (Figure 2). Life history consists of chronological narrations of the experiences of participants about one or more relevant issue (*27*). To report the relevant factors (second and

	Planning the trip	Access to the station	Ticket purchase	Circulation to the platform	Boarding	In vehicle- travel	Egress
+							
-							

Figure 2. Trip stage evaluation matrix.

Table 1. Variables Evaluated in the Survey

Variable	Description	Range
Section 1: General satisfaction (dependent variable)		
Satisfaction	Satisfaction level: 1 = Very dissatisfied, 5 = Very satisfied	1 To 5
Improvement	What attributes would you change in TransMilenio (TM) to improve the service?	Open ended
Section 2: Service dimensions		
Price	TM ticket: 1 = Expensive, –1 = Not expensive	–1 To 1
Comfortable	TM System makes me feel: 1 = Comfortable, –1 = Uncomfortable	–1 To 1
Safety	TM System makes me feel: 1 = Safe, –1 = Unsafe	–1 To 1
Frequency	TM buses: 1 = Pass often, –1 = Do not pass often	–1 To 1
Cleanness	The TM System: 1 = Is clean, –1 = Is not clean.	–1 To 1
Speed	TM routes: 1 = Are fast, –1 = Are not fast.	–1 To 1
Staff	TM staff: 1 = Do their job well, –1 = Does not do their job well.	–1 To 1
Coverage	TM coverage: 1 = Works, –1 = Does not work.	–1 To 1
Infrastructure	TM infrastructure: 1 = Is good, –1 = Is not good.	–1 To 1
Information	Information tools: 1 = Are useful, –1 = Are not useful.	–1 To 1

third questions), interviewers wrote down the positive and negative factors in each of the stages of the trip. The stages can be seen in Figure 2. The final output of this qualitative stage is a comprehensive list of dimensions that define what gives value to the service provided by TransMilenio.

Quantitative Stage: Explanatory Variables for Service Satisfaction

This stage deals with identifying which items on the list of dimensions defined from the qualitative stage explain users' satisfaction in TransMilenio. A survey of TransMilenio riders was conducted in November 2016. The survey inquired about service quality, perceptions, and behaviors of riders. Our aims were to obtain a 95% confidence level and a 5% margin of error from the random sample (N = 384). Survey administration included random user selection within stations. The stations selected for the survey are on different TransMilenio corridors. The administration time varied according to peak and off-peak hours, day and night hours, and the day of the week.

The questionnaire was divided into five sections: (a) general satisfaction and improvement actions; (b) rating service dimensions; (c) rating other aspects of service; (d) behavioral questions; and (e) socio-demographic questions. Questions about general satisfaction with the service were answered on a 5-point Likert scale. An open ended question asking about system attributes that could be modified to improve service allowed us to see what was at the top of users' minds. In section two, the core of the survey, we asked respondents to rate agreement with statements related to each of the dimensions established in the qualitative stage. Table 1 summarizes sections 1 and 2. The third section of the survey was aimed at learning riders' perceptions of very specific aspects of each of the variables of section 2.

Survey results were processed with R Software to determine significant variables using the ordinary least squares (OLS) method. OLS allows estimation of parameters of a linear regression model when the relation between the dependent variable and the explanatory variable is a hypothesis that needs to be proven (*28*). The linear regression will then show which dimensions (among those that define value) are explanatory variables of the dependent variable (user's satisfaction level). This relation can be presented as:

$$S = \beta_0 + \beta_1 x_1 + \beta_2 x_2 \ldots + \beta_n x_n + \varepsilon \qquad (1)$$

S refers to the satisfaction level and β_0, β_1, β_2, $\ldots\beta n$ are unknown coefficients. When this relation is plotted, β_0 represents an intercept term, $\beta 1, \beta 2 \ldots \beta n$ are coefficients of slope, and the remaining terms are random errors in the model (*28*).

Results

The Determinants of Value in TransMilenio

Together with lean thinking methodology and the literature, the interview process allowed us to determine dimensions that explain the value in TransMilenio's service. Question 1 (in the in-depth interview) aims to find the dimensions of TransMilenio that cause users to prefer this mode of transport over other available options. The response given by the user at the "Universidades" station sums up what was said by the other people interviewed: "*It is safe; its buses are clean; it facilitates access to vulnerable populations of disabled people, the elderly, pregnant women; it has good coverage and nearby stations, and its price is affordable.*" Interview answers were mostly related to coverage and speed. Responses to this question also highlighted security, cleanliness, price, infrastructure (for service to vulnerable people) and clear information about stops and routes.

Answers to the question about the positive aspects of TransMilenio (What aspects of TransMilenio do you consider to be positive?) provide a glimpse into dimensions that currently contribute to levels of rider satisfaction with the system. Coverage again appeared to be relevant, but the greatest advantage of TransMilenio today appears to be the speed of its buses. Four out of the six respondents highlighted this together with coverage. According to their answers, TransMilenio's ability to move buses at much higher speeds than ordinary buses is because of its use of dedicated lanes (infrastructure) which are exclusively used by BRT buses.

Answers about negative aspects of TransMilenio's service indicate dimensions that probably affect current perceptions of service. The main problem for those interviewed is inadequate bus frequency and reliability. These, in turn, are causes of other negative aspects such as discomfort resulting from failure of passengers to queue up for boarding and overcrowding of buses.

Next, participants filled out the "life history" form for each stage of a typical TransMilenio trip as shown in Figure 2. We found that these riders perceive that route information tools are not clear for the planning stage and that the staff in charge of giving information are not friendly and cannot be easily identified. These riders think that there is no order in lines at ticket machines and at turnstiles for entering and leaving a station. They perceive that overcrowding in the stations, especially at peak hours, makes it difficult to circulate within stations. In addition, they feel that areas around stations are not safe and that maintenance of infrastructure is insufficient. During on-platform waiting time, interviewees recalled disorderly and malfunctioning boarding/exiting processes. They also commented on lack of staff to organize waiting riders and to provide information to promote the proper movement of passengers within the station. Finally, the people interviewed detected problems within the buses which were mostly related to comfort. Buses are too full, seats are uncomfortable, venders work inside the buses without authorization or regulation, and drivers are sometimes careless. Nevertheless, these people once again emphasized that the speed of the BRT buses is one of the best attributes of the system.

Based on the interview answers, the literature review, and FedEx's eight rights, we elaborated a list of 10 dimensions that define value for TransMilenio service:

- **Price** evaluates the user's perception of cost in relation to the level of service received.
- **Comfort** combines aspects related to density of passengers, organization of riders and good citizenship.
- **Security** measures how safe users feel about crime in and around the system.
- **Frequency** relates to waiting times before boarding buses.
- **Cleanliness** evaluates neatness, orderliness and cleanliness of system components including stations and buses.
- **Speed** integrates factors that affect or contribute to travel time.
- **Staff** measures the service provided by employees who have direct contact with riders including drivers, ticket clerks and information staff.
- **Coverage** evaluates sufficiency of existing trunk lines and ticketing locations to meet the demands of existing system users.
- **Infrastructure** inquiries into the current state of the infrastructure.
- **Information** assesses clarity and relevance of existing resources for planning trips, supporting the choice of route, and providing in-station directions.

Only five out of the eight rights of FedEx have equivalents in the service provided by TransMilenio: price, time (frequency), place (coverage), resources (infrastructure), and service (staff). Nevertheless, the interview results provided five new dimensions that complement what was considered in the FedEx case, which results in a total of 10 dimensions for the case of TransMilenio.

Explanatory Variables for Service Satisfaction

The general satisfaction level of TransMilenio's riders was 2.69/5.0 with a standard deviation of 0.95 (recall that the April 2016 official TransMilenio's survey satisfaction level was 2.62/5.0). Considering that the dimensions were rated on a range between –1 and 1, results of the survey show that speed of travel (0.61), coverage (0.59), information (0.56) and staff (0.39) have positive averages; cleanliness (0.09) and infrastructure (0.09) have neutral averages; and comfort (–0.58), frequency (–0.57), safety (–0.54) and price (0.46) show negative averages when it is taken into account that for this last dimension a positive answer indicates that the price is considered to be high.

According to the averages above, people consider ticket prices to be expensive. The survey shows that 70% of the passengers surveyed considered that TransMilenio ticket prices are high in relation to the service that they receive.

Table 2. Results of the Linear Regression

	Estimate	p-value	Sig.
Price	–0.1193	0.0143	*
Comfort	0.2630	0.0000	***
Security	0.0720	0.1870	
Frequency	0.1795	0.0014	**
Cleanliness	0.0707	0.1358	
Speed	0.1467	0.0086	**
Staff	0.1936	0.0001	***
Coverage	0.0857	0.1097	
Infrastructure	0.0336	0.4565	
Information	0.1664	0.0019	**
Intercept	2.7179	0.0000	***
R2 = 0.328	R2 adj. = 0.309		

Note: Significance codes: 0 = ***; 0.001 = **; 0.01 = *.

Also, they consider that bus routes are fast, but they must wait a long time to get on board buses. Of those surveyed, 76% consider that they spend too much time waiting to board buses. Users rate system coverage, information, and the work of TransMilenio's staff as good. Finally, on average their opinions of the infrastructure and cleanliness of the system are neither favorable nor unfavorable.

Users feel neither comfortable nor safe while using this system. In percentage terms, 77% of the users surveyed said that they are dissatisfied with current comfort conditions while only 19% consider that TransMilenio is comfortable. According to survey results, the principal negative aspect of this variable is overcrowding: a near unanimous 94% of riders surveyed feel that the buses are overcrowded. The presence of musicians and street vendors affects riders comfort and is disliked by 73% of the surveyed passengers. Space is important for comfort, and 58% of the sample consider that the stations are spacious. Nevertheless, comfort within these spaces is affected by the lack of order in waiting lines, and 90% of those surveyed argued that waiting lines are disorganized and, in any case, not respected.

The results of the multivariate linear regression model show that customer satisfaction, the dependent variable, is related to some of the dimensions that define value. Specifically, the results of the model (Table 2) prove that comfort, staff, frequency, information, speed, and price are dimensions that affect users' satisfaction.

If vehicles could pass continuously with short intervals between them, and if passengers could travel fast, these changes might lead to increased levels of user satisfaction. As in the previous dimensions, the pertinence of information about trip planning and the service provided by TransMilenio station workers is positively related to general satisfaction. Nevertheless, improvements made in favor of the comfort of the passengers will have greater impact on satisfaction than any improvements in other dimensions. On the other hand, the higher the fare, the lower the level of user satisfaction, as established by several authors.

Analysis of Results

The literature provides evidence that relevant dimensions in transit user satisfaction vary from case to case. TransMilenio's explanatory dimensions, namely comfort, staff, frequency, information, speed, and price appear in other studies. Relevance of time-related variables (i.e. speed and frequency) has appeared recurrently in the literature and is also the case in our research. Cleanliness, which was reported to be important by Eboli et al. and Tyrinopoulos et al. (*16*, *17*), appears within the lowest level of priority for action in our study (highly rated and not-significant). Similar to results found by Román et al., comfort and information positively affect customer satisfaction (*29*). It should be noted that other studies of transit satisfaction in Bogota considered insecurity to be a key factor (*4*, *7*). Here however, in spite of being very poorly rated, it is not significant as an explanatory variable for satisfaction in TransMilenio.

Comfort is especially important in this study because the magnitude of its coefficient is the highest of all those estimated in the regression analysis taking into account that they were all measured on the same scale. The value given for this dimension means that it is the most relevant with regard to survey results. Comfort contains within it several aspects related to users' satisfaction. For instance, culture orderliness of riders waiting in line, space available in stations and articulated buses, and the presence of informal vendors.

Price is the only variable that has an inverse relation to satisfaction levels. Fare increases result in reductions in the level of satisfaction. TransMilenio has two criteria for definition of fares: financial sustainability and user affordability (*30*). As transit subsidies are not easily justifiable within the Colombian regulatory framework, incentives favor reduction of operation expenditures to keep fares affordable for users. As private operators gain revenue according to kilometers traveled, operating efficiently with small fleets is the shadow objective of TransMilenio. In fact, survey results show that 70% of the passengers surveyed consider TransMilenio ticket prices to be high in relation to the service that they receive.

After price and comfort, frequency is the third significant dimension which is poorly rated and which deserves careful analysis. Frequency of service is indirectly associated with other dimensions such as comfort, price, information, and infrastructure. Infrequent buses may cause overcrowded buses and longer waiting times which affect comfort. Crowds of people waiting within the stations may decrease perceptions of infrastructure quality. It should be noted that many authors who used other methods, found that peak hour overcrowding is the cause for negative perceptions of TransMilenio's service (*4–7*). In addition, the decision by TransMilenio to operate only part of its fleet at any given moment to reduce operational costs negatively affects bus frequencies (*7*, *9*).

Figure 3 shows aspects of TransMilenio service that users identified in response to the request in section 1 of the survey (open-ended question) to state three attributes that would improve service. Frequency is the most commonly recalled

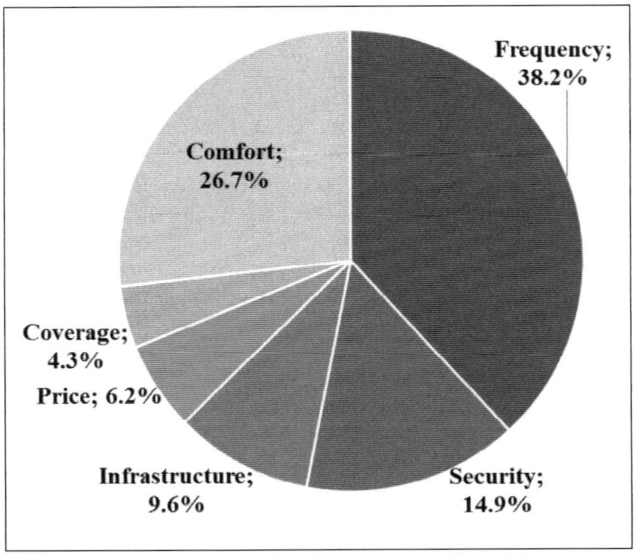

Figure 3. Factors that TransMilenio's riders would improve.

problem. This coincides with the model and the direct ratings of users. Security and users' comfort were each mentioned by respondents as relevant areas within which actions for improvement should be taken. However, security does not appear to be significant. Improvements in security, as suggested by riders would have uncertain effects on overall satisfaction.

Our results could send a constructive message to transit authorities in Bogota if the objective is to set priorities for service improvements. As established by Imam, the average service dimensions ratings (section 2 of the survey) indicate action on poorly rated dimensions (*15*). Nevertheless, significant variables explaining satisfaction (linear regression model) point to specific dimensions between those which are poorly evaluated for service improvements.

Table 3 provides a cross tabulation between significant and not significant variables and poorly rated and highly rated dimensions. Cross tabulation between significance and ratings is a starting point for continuous improvement within the orientation of lean thinking principles. This categorization can aid company prioritization of variables to directly impact riders' levels of satisfaction beginning with those that are classified as significant and poorly rated.

High bus frequency is one of the most attractive features of BRT systems because it reduces travelers' waiting times (*31*). If running more buses were the solution, assuming financial constraint, it would not necessarily solve the problem in Bogota's BRT. TransMilenio has peak hour platform capacity deficiencies at busy terminals and trunk stations. There are stations at which a single bus position at a platform serves up to seven different high frequency bus routes. In peak periods, boarding and exiting require more time resulting in longer bus queues. This means that running more buses to improve frequency would be futile and could even be counterproductive because it could cause more bus congestion at stops. However, under the lean thinking philosophy, the problem of bus frequency may also be related to management of the current fleet, and further research would be required (conducted on the remaining four lean principles).

Table 3. Classification of Variables to Prioritize Them

	Poorly rated	Highly rated
Significant	• Price • Comfort • Frequency	• Speed • Staff • Information
Not significant	• Security	• Coverage, • Infrastructure and • Cleanliness

To put the lean thinking principles of value, value stream, flow, pull, and perfection into practice in a process to improve TransMilenio, this study has begun with an analysis of what constitutes value within this system. Before proceeding to putting the second principle, value stream, into practice, we propose to prioritize significant dimensions that are poorly rated by users. In this regard, bus frequency is a significant poorly rated dimension that users identified as in urgent need of improvement. It should be noted that the determination of frequencies is a harrowing process within this company, we propose to use value mapping which will start the application of the second lean principle.

Conclusion

This research uses the lean thinking philosophy to uncover dimensions that may explain declining satisfaction with TransMilenio. Through the use of mixed qualitative and quantitative methods, this research provides empirical evidence of the determinants of value for TransMilenio's users and explanatory dimensions for their levels of satisfaction. Previous lean thinking examples in transportation, other literature sources, and semi-structured interviews were used to identify a broad list of determinants of value. Linear regression analysis based on a representative random sample (N = 384) permitted selection of those determinants which most strongly contribute to an explanation of levels of costumers' satisfaction with this service.

Although users value ten dimensions of the service, the regression model contributed to identification of comfort, frequency, price, service from staff, information, and swiftness as the dimensions that are relevant for satisfaction. This analysis based on the lean thinking philosophy focuses the discussion on what customers who use the service expect and how these expectations match up with what providers do in their daily activities. This is a clear example of how transportation services are seen through different lenses. While TransMilenio users place high values on comfort, frequency, price, service from staff, information, and swiftness, the company has spent efforts on many aspects, sometimes in many other dimensions, not necessarily the crucial ones.

Nevertheless, the attributes in need of improvement and the ratings are other valuable sources that transit agencies can use to develop improvement strategies. The cross tabulation of significance and ratings presented in the analysis section of this paper can contribute to identification of priorities for improvement. As the users' suggestions for improvements of TransMilenio in response to the open ended question in Section 1 of the survey and the dimensions established in the qualitative stage coincide, service providers can create more value and satisfaction for their customers by matching their daily operations with users' expectations. Even though other significant dimensions were also poorly rated and identified by users as in need of urgent improvement, frequency has been targeted as a good starting point for service improvement.

This case is an example of lean thinking application in transportation, specifically for urban passenger public transportation. Understanding mismatches between users' value and what suppliers provide is the next phase of the application of lean thinking to urban transit systems.

Acknowledgments

We want to thank the Urban and Regional Sustainability Research Group (SUR) and the Engineering and Construction Management Research Group (IN2GECO) of Universidad de los Andes for their valuable support during our research. In addition, we thank the Department of Civil and Environmental engineering for the resources provided to make the surveys possible. Finally, we are grateful to Daniel R. Rosas, who helped us with the data base, graphs, and management.

Author Contributions

The authors confirm contribution to the paper as follows: study conception and design: Carlos Garcia-Suarez, Leonardo Rivera-Perez, and Alvaro Rodriguez-Valencia; data collection: Carlos Garcia-Suarez, Leonardo Rivera-Perez, and Alvaro Rodriguez-Valencia; analysis and interpretation of results: Carlos Garcia-Suarez, Leonardo Rivera-Perez, and Alvaro Rodriguez-Valencia; draft manuscript preparation: Carlos Garcia-Suarez, Leonardo Rivera-Perez, and Alvaro Rodriguez-Valencia. All authors reviewed the results and approved the final version of the manuscript.

References

1. Transmilenio, S.A. *Encuesta de Satisfacción TransMilenio S.A.* Centro Nacional de Consultoría S.A. http://www.transmilenio.gov.co/Publicaciones/Ãndice_de_satisfaccion_del_usuario_con_el_sistema_transmilenio_abril_2016. Accessed November 1, 2016.
2. Bogota Como Vamos. *Acerca de Bogota Como Vamos*. http://www.bogotacomovamos.org/acerca/. Accessed July 17, 2016.
3. Bogota Como Vamos. *Encuesta de Percepción Ciudadana 2016*. http://www.bogotacomovamos.org/documentos/encuesta-de-percepcion-ciudadana-2016/. Accessed November 6, 2016.
4. Gilbert, A. Bus Rapid Transit: Is Transmilenio a Miracle Cure? *Transport Reviews*, Vol. 28, No. 4, 2008, pp. 439–467.
5. Hidalgo, D., and P. Graftieaux. Bus Rapid Transit Systems in Latin America and Asia: Results and Difficulties in 11 Cities. *Transportation Research Record: Journal of the Transportation Research Board*, 2008. 2072: 77–88.
6. Muñoz, J. C. *¿El Modelo de Transmilenio Sigue Siendo Válido Para Transantiago?* Departamento de Ingeniería de Transporte y Logística, Escuela de Ingeniería, Pontificia 5 Universidad Católica de Chile, Santiago de Chile. https://www.ing.uc.cl/transporte-y-logistica/el-modelo-transmilenio-sigue-siendo-valido-para-transantiago/. Accessed November 29, 2016.
7. Batarce, M., J. C. Munoz, J. Ortuzar, S. Raveau, C. Mojica, and R. A. Rios. *Evaluation of Passenger Comfort in Bus Rapid Transit Systems*. Inter-American Development Bank, Washington D.C. 2015, pp. 1–44.
8. Kash, G., and D. Hidalgo. The Promise and Challenges of Integrating Public Transportation in Bogota, Colombia. *Public Transport*, Vol. 6, No. 1–2, 2014, pp. 107–135.
9. Velásquez, J. M. Análisis de Factores Que Inciden En La Demanda Del Sistema TransMilenio En Bogota, Colombia. *Revista de Ingenieria*, Vol. 30, 2009, pp. 56–64.
10. Eboli, L., and G. Mazzulla. How to Capture the Passengers' Point of View on a Transit Service through Rating and Choice Options. *Transport Reviews*, Vol. 30, No. 4, 2010, pp. 435–450.
11. Cirillo, C., L. Eboli, and G. Mazzulla. On the Asymmetric User Perception of Transit Service Quality. *International Journal of Sustainable Transportation*, Vol. 5, No. 4, 2011, pp. 216–232.
12. Lai, W. T., and C. F. Chen. Behavioral Intentions of Public Transit Passengers-The Roles of Service Quality, Perceived Value, Satisfaction and Involvement. *Transport Policy*, Vol. 18, No. 2, 2011, pp. 318–325.
13. van Lierop, D., and A. El-geneidy. Public Image: Determining the Effects of Users' Image of Public Transit on Loyalty. Presented at 96th Annual Meeting of the Transportation Research Board, Washington, D.C., 2017.
14. Tam, J. L. M. Customer Satisfaction, Service Quality and Perceived Value: An Integrative Model. *Journal of Marketing Management*, Vol. 20, No. 7–8, 2004, pp. 897–917.
15. Imam, R. Measuring Public Transport Satisfaction from User Surveys. *International Journal of Business and Management*, Vol. 9, No. 6, 2014, pp. 106–114.
16. Eboli, L., and G. Mazzulla. A Stated Preference Experiment for Measuring Service Quality in Public Transport. *Transportation Planning and Technology*, Vol. 31, No. 5, 2008, pp. 509–523.
17. Tyrinopoulos, Y., and C. Antoniou. Public Transit User Satisfaction: Variability and Policy Implications. *Transport Policy*, Vol. 15, No. 4, 2008, pp. 260–272.
18. Womack, J. P. *The Machine That Changed the World*. McMillan, New York, 1990.
19. Riadi, P. *Lean Healthcare Y Su Aplicación En El Laboratorio Clínico*. Santiago de Chile, 2013.
20. Womack, J. P., and D. T. Jones. From Lean Production to the Lean Enterprise. *Harvard Business Review*, Vol. 72, No. 2, 1994, pp. 93–103.
21. Taylor L., and R. Martichenko. *Lean Transportation - Fact or Fiction?* 2006, pp. 1–9.
22. Cámara de Comercio & U. de los Andes. *Observatorio Anual de Movilidad de Bogota*, 2017.

23. Transmilenio, S. A. *Estructura Orgánica Y Talento Humano*. Transparencia y acceso a la información pública TransMilenio: Misión y Visión. http://www.transmilenio.gov.co/Publicaciones/la_entidad/transparencia_y_acceso_a_la_informacion_publica_transmilenio/3_estructura_organica_y_talento_humano/mision_y_vision. Accessed October 4, 2016.
24. Transmilenio, S.A. *Mejorar El Servicio Por El Cual Se Paga El Precio Mas Alto de Latinoamerica, Aumentando El Numero de Buses*. Bogota D.C., 2017.
25. Transmilenio, S.A. *Plan de Acción de Transmilenio S.A.* http://www.transmilenio.gov.co/Publicaciones/la_entidad/transparencia_y_acceso_a_la_informacion_publica_transmilenio/6_planeacion/plan_de_accion. Accessed July 11, 2016.
26. Transmilenio, S.A. *Indicadores de Gestión Transmilenio S.A.* http://www.transmilenio.gov.co/Publicaciones/la_entidad/transparencia_y_acceso_a_la_informacion_publica_transmilenio/6_planeacion/indicadores_de_gestion. Accessed Jul. 10, 2016.
27. R. Hernandez Sampieri, R., C. Fernandez Collado, and P. Baptista Lucio. *Metodología de la Investigación*. 6th ed. McGraw Hill, México D.F., 2010.
28. Chumney, E., and K. Simpson. *Methods and Designs for Outcomes Research*. American Society of Health-System Pharmacists, Charleston, South California, 2006.
29. Román, C., J. C. Martín, and R. Espino. Using Stated Preferences to Analyze the Service Quality of Public Transport. *International Journal of Sustainable Transportation*, Vol. 8, No. 1, 2014, pp. 28–46.
30. Transmilenio, S.A. *Informe de Gestión 2017, Transmilenio S.A., Bogota;* 2017, p. 34.
31. Hook, W., and L. Wright. Bus Rapid Transit Planning Guide. *Institute for Transportation & Development Policy*, New York, 2010, p. 37.

The Standing Committee on Transit Capacity and Quality of Service (AP015) peer-reviewed this paper (18-05019).

TRR
JOURNAL OF THE TRANSPORTATION RESEARCH BOARD

Transportation Research Record
2018, Vol. 2672(8) 464–474

DOI: 10.1177/0361198118796940
journals.sagepub.com/home/trr
SAGE

Bus Load Inference and Crowding Performance Evaluation through Disaggregate Analysis of Fare Transaction, Vehicle Location, and Passenger Count Data

Gabriel E. Sánchez-Martínez[1], Laurel Paget-Seekins[2], Christopher W. Southwick[1], and John P. Attanucci[1]

Abstract

Comfort is an important aspect of the transit passenger experience. Crowding can significantly decrease passenger comfort and disrupt service delivery, causing passenger travel times to increase and even resulting in passengers being unable to board an arriving vehicle. This research explores the use of automatically collected vehicle location data, fare transaction data, and passenger origin–destination inference to measure crowding on buses. Three model components are involved: scaling vehicle trip-level origin–destination transfer data, measuring crowding as perceived by passengers through performance measures defined for this purpose, and determining the sources of crowding. The latter is important to identify the most effective means of addressing crowding in each case. The models are tested on data from the Massachusetts Bay Transportation Authority, and examples of graphical applications already being used by planners are presented.

Together with trip time, cost, and reliability, ride comfort is among the factors that influences people's choice of mode for commuting and other urban transportation needs (*1–6*). Comfort is especially relevant in the context of urban public transportation systems, which may struggle to provide the required capacity at peak times. Riders of public transportation services operating near capacity experience highly crowded vehicles, and are less likely to find a seat or ride with sufficient space to read or use a mobile device, making their ride not only uncomfortable but also less productive. It has been shown that increasing frequency to decrease crowding can increase social welfare, in spite of the additional operating costs and traffic that additional vehicles can cause (*7*).

When vehicles reach capacity, people waiting at a stop may not be able to board the next vehicle to arrive, resulting in additional waiting time and inconvenience that is not captured by traditional performance measures. High loads also slow the boarding and alighting process, resulting in longer stop dwell times, slower operating speeds, and longer cycle times, which increase operating costs, as well as increased headway irregularity, which further increases waiting times and decreases the reliability as experienced by riders (*8–13*).

Given the importance of crowding to passengers' experience, mode choice, capacity, reliability, and operating cost, it is important for transit agencies to continuously monitor vehicle loads, identify routes that need service improvements, determine what changes are needed to mitigate high crowding, and implement those changes. Measuring loads is a long-established practice in transit agencies (*14–16*), but the aggregation level at which this is typically done, spatially at the route level and temporally for the whole day or by time periods, which may suffice for reporting and meeting the requirements of a service delivery policy, may hide how passengers experience their trips and thus the real impact of high loads on demand and performance (*17*). For example, the 2010 MBTA Service Delivery Policy (*18*) specifies that the average maximum load for a bus route over a peak time period should not exceed 1.4 times the number of seats on a vehicle. It is probable that among the

[1]Massachusetts Institute of Technology, Cambridge, MA
[2]Massachusetts Bay Transportation Authority, Boston, MA

Corresponding Author:
Address correspondence to Gabriel E. Sánchez-Martínez: gsanmar@mit.edu

routes satisfying this requirement are ones with full vehicles at the peak of the peak. The nature of the problem is such that most passengers take the most crowded vehicles, so the experience of the average passenger may be one of much higher crowding (along with longer waiting time, longer trip time, and lower reliability) than what is reflected by load in the average vehicle.

Even when loads are measured at higher spatiotemporal resolutions, the causes of high loads are not apparent from load measurement alone. Unacceptably high loads may be caused by insufficient scheduled frequency, running fewer trips than scheduled, demand variability at various timescales, headway irregularity in high-frequency routes, or a combination thereof. The manner in which each of these causes can be addressed is different. For example, high loads caused by high demand must be addressed through service planning. High loads caused by dropped trips must be addressed by improving vehicle maintenance, implementing policies and initiatives to reduce driver absence, and improving the agency's ability to operate trips with standby drivers. High loads caused by headway irregularity must be addressed through real-time operations control. Each of these causes has different implications on an agency's budget, required attention from the management, and opportunities to improve performance through technology. It is therefore necessary to identify the causes of crowding to make the load performance measures actionable and have a tangible impact on service quality.

The issues outlined above present a challenge for transit agencies to measure crowding as experienced by passengers and at high spatiotemporal resolution. The current state of practice leaves many agencies relying more on passenger complaints than on their own data to address crowding. The increasing availability of automatically collected data, especially from automated fare collection (AFC), automated vehicle location (AVL), and automatic passenger counting (APC) systems, has enabled the development of data-driven approaches to service planning (*19*), demand analysis (*20–24*), and performance measurement (*25*, *26*) in public transportation agencies. While data alone seldom provide the information required to solve key transportation problems, the combination of various data sources with mathematical models can (*27*). There has also been a movement toward measuring performance as perceived by passengers, especially in the context of trip times and reliability (*25*), and to making performance measures actionable (*26*), but not in the context of bus crowding.

This research advances the state of the art by (a) formulating a model to scale up origin–destination matrices inferred from disaggregate fare transaction data, (b) introducing measures of crowding that reflect passenger experience, and (c) formulating an approximate crowding source allocation model to make measures of crowding actionable. The implementation of these models and measures in the Massachusetts Bay Transportation Authority (MBTA) is discussed in the application section.

Framework

Figure 1 illustrates the framework of this research. The first step is to measure or estimate loads at a disaggregate level, that is, at each stop of each trip. Many transit agencies have only a portion of their fleet equipped with APC, so loads must be inferred. Disaggregate AFC and AVL data can be combined through an origin–destination inference algorithm (ODX) such as the ones developed by Trepanier et al. (*20*), Gordon et al. (*21*), Munizaga and Palma (*22*), and Sánchez-Martínez (*23*). These algorithms cannot infer destinations for all fare transactions, such as transactions of cards appearing once in a day, so the output must be scaled up to the full demand. Scaling can also account for the flow of passengers that does not interact with the fare system, be it because of fare evasion or because pass holders are allowed to board without interacting. Non-interaction factors can be derived from manual surveys or from a combination of ODX output and APC data. Loads can be determined from the resulting scaled trip-level origin–destination matrices. Those agencies having all or a great majority of their bus fleet equipped with APC can bypass this modeling step (as what is needed are loads rather than origin–destination matrices) and measure loads directly with APC. Manual ride-check surveys could be used to measure loads in the absence of APC, but it could be prohibitively expensive at the scale required by this application.

Once disaggregate load data are available, a model calculates crowding performance measures that reflect passenger experience, which can be based on loads in excess of a threshold that depends on vehicle type, as well as the duration of the excess load. These performance measures can feed reports and power online dashboards that enable planners to learn when and where vehicles are the most crowded. To make the information actionable, however, the quantified crowding is classified according to the likely underlying cause of crowding. This empowers the management of the transit agency to decide where to focus their attention and what directions to give. For example, if the crowding is mostly caused by headway irregularity, the agency can save the operating resources that would have gone toward adding vehicles to the route to increase frequency, and instead focus on improving dispatch discipline at terminals and implementing real-time control decision support systems. Once the appropriate action comes into effect, loads

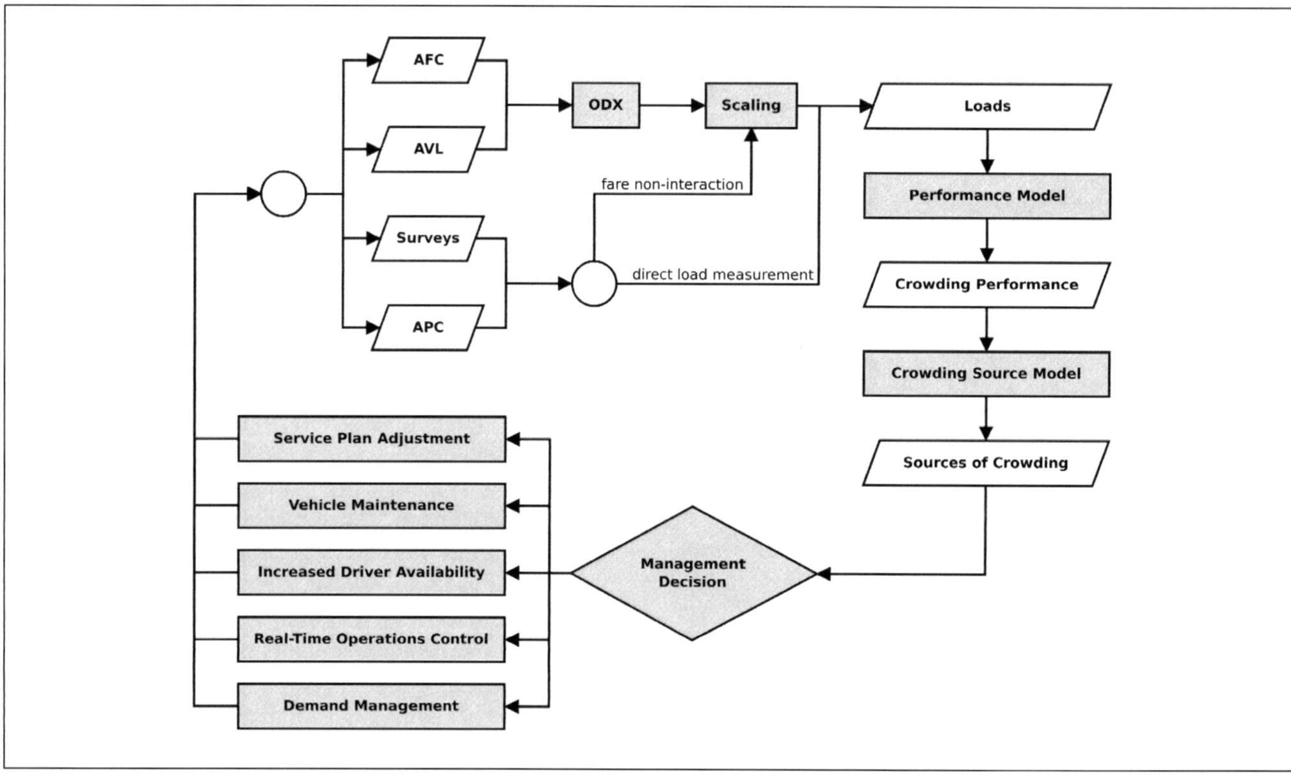

Figure 1. Research framework.

should decrease. The monitoring process continues to confirm the expected result and identify new sources of high crowding as they develop.

Methodology

This section presents the model to scale up incomplete origin–destination matrices to obtain loads, introduces performance measures to capture crowding as experienced by passengers, and describes the method through which the sources of crowding are identified.

Estimating Loads from Unscaled Origin–Destination Matrices

The ODX model combines AFC and AVL data to infer the origins and destinations of each trip and transfers between legs (or stages) of a journey. Origins are inferred by matching transaction timestamps and farebox identifiers in AFC data to AVL records, to find the location of a vehicle at the time of the transaction. Destinations are inferred by trip chaining, assuming that the destination of each trip is near the location of the next origin of that card, and finding a stop downstream of the origin in the same bus trip from which the next origin can be reached with minimum disutility (in-vehicle time and walking time). Transfers are inferred when there is a short time and distance between successive trips of different lines (*23*).

In the MBTA system, origins are inferred for 94% of bus fare transactions. The remaining transactions can sometimes be matched to a vehicle, but not to a particular stop of a particular trip, sometimes because the time difference between AFC and the closest AVL record is too large, and more often because of AVL equipment malfunction or data errors. Destinations are inferred for 64% of bus fare transactions, and cannot be inferred in the remaining cases for several reasons: (a) there was only a single trip made with the fare card that day or the origin of the next trip was not inferred (17% of transactions), (b) the most likely destination is too far from the next origin (11% of transactions), (c) the origin of the current transaction was not inferred (5% of transactions), or (d) one of various logical verification conditions was not met (3% of transactions).

Thus from the output of ODX it is possible to obtain unscaled bus trip-level origin–destination matrices, accounting for about two-thirds of the total demand. The matrices have one row and column per stop, and have all of the flow in the upper triangle, that is, the diagonal and the lower triangle are populated with structural zeros. The scaling process increases the values of the cells in this matrix to account for the full demand, expressed mathematically as follows. Let T be the true origin–destination matrix, I be the unscaled matrix derived from fare transactions with inferred origins and destinations, U be unknown matrix of demand with inferred origin but uninferred destination, and N be the unknown matrix of demand with uninferred origin or of non-interaction trips without a fare transaction. Then:

$$T = I + U + N \tag{1}$$

and because I is known, we can estimate T by adding our estimates of U and N to I as follows:

$$T \approx I + \tilde{U} + \tilde{N} \tag{2}$$

addressing first the estimation of U, the objective is to assign destinations probabilistically to fare transactions with inferred origin but uninferred destination, represented by a vector u containing the number of boardings at each stop (of trips with uninferred alighting stop). It is assumed that the large sample of passenger trips with inferred origin and destination (over many bus trips and dates and under homogenous conditions, such as morning peak) is generally representative of the destinations of all trips. However, by construction, passenger trips followed by a transfer (thus inferred, to a second trip with inferred origin) must have an inferred destination, and are therefore always included in I. As these trips involve a fare transaction close in space and time to the destination, it is highly likely that the destination is inferred. This implies that I contains most or all trips followed by transfers, and U contains few or no trips followed by transfers. Therefore, scaling u to I would yield an estimate of T that overestimates the number of alightings at common transfer locations. To avoid this bias, we instead scale u to $L \subseteq I$ of passenger trips with inferred origin and destination that are not followed by a transfer. More precisely, we scale to an aggregate version $\hat{L}$ of L over many vehicle trips in homogenous conditions (e.g., morning peak) to obtain a large sample and a more accurate distribution of destinations by origin, and normalize $\hat{L}$ so that each row sums to 1, such that each cell value is the probability of an alighting stop not followed by a transfer, given a boarding stop:

$$\tilde{U} = u\hat{L} \tag{3}$$

Then $I + \tilde{U}$ is an estimate of the origin–destination matrix of trips that interact with the farebox and have an inferred origin. To estimate N, we assume that the distribution of destinations by origin in $I + \tilde{U}$ is representative of all trips, and in particular of trips with uninferred origin and non-interaction trips. By comparing the boardings registered in AFC data against APC data (for the sample of trips run on vehicles equipped with APC) or manual ride-check surveys, non-interaction factors can be estimated. These factors can be at the trip, segment, or stop level, and could be different by time period. For instance, research into non-interaction in the MBTA's Green Line found that non-interaction is significantly higher at the peak load stops in the peak period, when vehicles loads are high, as passengers may have an excuse to board through the back door (*24, 28*). Non-interaction factors by stop, contained in vector n, can then be applied to estimate N, as follows:

$$\tilde{N} = n\left(I + \tilde{U}\right) \tag{4}$$

Note that if the factors are determined at the trip level, n has the same value for each row. The extent to which it is important to accurately estimate N depends on the extent of fare non-interaction in the transit service being analyzed and on the potential consequences of not accounting for the additional demand when changing service based on estimated crowding.

Combining Equations 2, 3, and 4 we obtain:

$$T \approx (1 + n)\left(I + u\hat{L}\right) \tag{5}$$

Loads can be calculated from the scaled matrix by taking the difference of cumulative boardings and alightings. The load of a vehicle as it departs the stop s is:

$$l_s = \sum\nolimits_{b=1}^{s} \sum\nolimits_{a=b}^{z} T_{\text{ba}} - \sum\nolimits_{a=1}^{s} \sum\nolimits_{b=1}^{a} T_{\text{ba}} \tag{6}$$

where b and a are boarding and alighting stop indices, respectively, z is the number of stops on the trip, and T_{ba} is the number of people boarding at b and alighting at a. Note that, by construction, $T_{\text{ba}} = 0$ when $b \geq a$. The first term in the equation is the number of passengers boarding at s or upstream and alighting downstream of s. The second term is the number of passengers alighting at s or upstream, regardless of boarding location.

In order to validate the model, the number of boardings and alightings inferred at each stop were compared with those automatically counted in the sample of vehicles equipped with APC in the MBTA. Figure 2 shows the distribution of the difference between inferred and counted boardings (blue) and alightings (red) over several months. The distributions peak at zero (no error) and drop quickly as the error increases, suggesting that both boardings and alightings can be estimated accurately.

Crowding Performance Measures

Passengers do not feel that a vehicle is crowded until a certain threshold is reached. After this threshold, crowding could be experienced differently by different passengers, for example depending on whether they are sitting or standing. The experience of crowding intensifies not

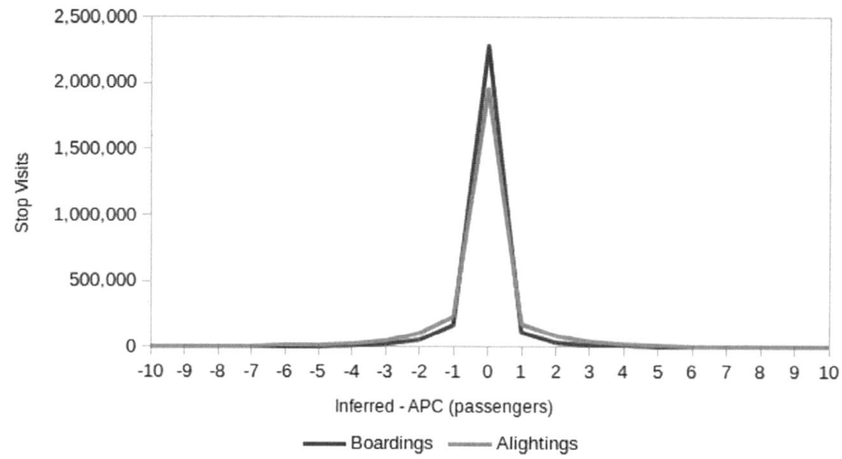

Figure 2. Distribution of difference between inferred and automatically counted boardings and alightings.

only with the number of passengers in the vehicle, but also with the duration of crowding, as passengers might more easily tolerate a short segment in a crowded vehicle than most of their trip in the same conditions. An occasional crowded trip is more easily tolerated by the agency than consistent crowding, the latter more strongly nudging riders to consider alternative routes or modes. Likewise, crowded trips in the peak might be more tolerable than crowded trips off-peak, as passengers have fewer non-crowded transit alternatives and the road network might be congested. The crowding performance measure defined below considers these factors. For a given bus trip t and stop s, we define *uncomfortable passenger time* (UPT) as follows:

$$c_{t,s} = f_t(l_{t,s}) \cdot r_{t,s} \tag{7}$$

where $l_{t,s}$ is the load as the vehicle departs stop s, $r_{t,s}$ is the running time to the next stop, and f_t is a function that gives the number of uncomfortable passengers for a particular load in a particular trip. The latter function, which may be nonlinear and discontinuous, reflects what the agency regards as a crowded vehicle, which can depend on vehicle type, time of day, day of the week, and so forth. For instance, the Southeastern Pennsylvania Transportation Authority aims to have load factors of 1.0 (all passengers can sit) off-peak, but tolerates load factors between 1.31 and 1.74 in peak period service, recognizing that some passengers may have to stand in order to meet the high peak period travel demand with available resources; the variation between 1.31 and 1.74 depends on how vehicles are configured and the relationship between load factor and density (*29*). The Los Angeles County Metropolitan Transportation Authority (LA Metro) tolerates load factors of 1.4 for routes with headways under 10 min, but aims for load factors no greater than 1.0 for low-frequency routes, recognizing that the additional waiting time of passengers who cannot board the first bus they see arrive because it is full is much more onerous with long headways; the aim is to guarantee being able to board vehicles of low-frequency service (*30*).

Once the number of uncomfortable passengers is multiplied by the segment running time in Equation 7, the number of passenger minutes in crowded conditions is obtained. This can be aggregated to the trip level by adding each stop's contribution over each trip, and to time periods or the whole day by adding each trip's contribution over all trips in a date range. Spatially, UPT can be aggregated to the segment level instead of trip level, enabling the exploration of where and when the bus network is most crowded. The level of aggregation should be chosen according to the analysis being carried out.

In addition to UPT, we define *comfort fraction* (CF) as the percentage of passenger time spent riding in comfortable conditions. This is calculated by dividing UPT by the total number of passenger minutes and subtracting the quotient from 1. As with UPT, it can be aggregated spatially to the stop, segment, or trip level, and temporally to the trip, time period, or day level. Routes with higher frequency and ridership may have higher UPT than other routes even when the percentage of passengers experiencing crowding is not as high. CF is useful because it normalizes UPT by a route's ridership, making it easier to compare crowding across routes serving different levels of demand.

Identifying Sources of Crowding

Measuring UPT across all routes in a bus network enables the management of a transit agency to focus their attention on routes with most crowding, but before taking action it is important to identify the causes of crowding in each route. Omitting this step exposes the agency to the risk of suboptimal spending of resources, by adding vehicles to a route that has sufficient frequency in the schedule, but is poorly managed at the terminal, for example. The causes of crowding can be identified by comparing UPT in actual service with the same performance measure in several hypothetical cases:

1. Ridership spread across days: The total ridership in a certain time period is evenly distributed across all the trips over all days (of the same type, e.g., weekdays) in the analysis period (e.g., afternoon peak). The resulting UPT can be attributed to insufficient frequency of service, even for average demand. To decrease this UPT, the agency should increase the scheduled frequency.
2. Ridership spread across scheduled trips each day: Each day's ridership is evenly distributed across the scheduled trips of each time period. The sum of UPT across days is because of insufficient scheduled frequency. UPT in this case when there is no UPT in the first case indicates that the scheduled frequency suffices for average demand but not for busier-than-average days. To decrease this UPT, the agency should increase the scheduled frequency, considering that demand varies across days and looking at higher percentiles of daily ridership.
3. Ridership spread across operated trips each day: Each day's ridership is evenly distributed across the operated trips of each time period. The sum of UPT across days is attributable to insufficient delivered frequency. UPT in this case in excess of UPT in the second case indicates that crowding is being caused by the agency operating fewer trips than scheduled. To decrease this UPT, the agency should address the underlying problem, such as inability to pull out vehicles under repair, or driver absences.
4. Ridership assigned to operated trips on schedule: Ridership is reassigned to vehicles of operated trips as if each trip departed each stop on schedule. For high-frequency routes, which tend to have passengers arriving at stops randomly, more passengers arrive during longer headways than shorter headways. By simulating punctual departures, we estimate UPT with regular headways. The difference in UPT between actual operations and this case quantifies the amount of crowding caused by headway irregularity. For example, if actual UPT is significant while this case has no UPT, the crowding is not caused by insufficient scheduled frequency or dropped trips, but by inadequate real-time operations control. To decrease this UPT, the agency should improve the dispatching discipline at terminals and employ holding control en route to maintain regular headways.

Application

The framework and models of this research were implemented with performance data from the MBTA, focusing on non-holiday weekdays of Fall 2015.

Setting the Standard

The MBTA sets its crowding (or comfort) standard for bus service as part of its Service Delivery Policy (*18*). In January 2017, the MBTA governing board adopted an updated Service Delivery Policy that took advantage of new data sources to better measure the passenger experience. Over the course of 2015 and 2016, MBTA staff worked with internal and external stakeholders to determine the policy objectives and technical definitions for specific performance measures.

The MBTA sought both qualitative and quantitative input on passenger crowding. An online survey showed passengers pictures of transit vehicles at six different levels of occupancy and asked if the crowding level was acceptable, wasn't acceptable but they would still board, or was unacceptable and they wouldn't board. Survey respondents were also asked about the importance of getting a seat and whether they ever changed the time of their trip or station they used in order to get a seat. In addition, the MBTA held workshops with a diverse set of passenger constituency groups to gather feedback. Workshop leaders marked off a square meter on the floor and asked participants to enter the square until an acceptable level of crowding was achieved. After the exercise, participants discussed the factors that contribute to discomfort onboard, including the ability to easily disembark and length of time experiencing crowding.

The MBTA used this feedback, along with Massachusetts Department of Public Utilities regulations on bus vehicle loading, to set its crowding/comfort standard and define f in Equation 7. Different functions were defined for peak periods and off-peak periods, and the CF for each bus route was calculated:

- High-volume time periods: The maximum comfortable passenger-to-seat ratio for high-volume travel periods is 140%. At loads of 140% or less of seated capacity, all passengers are considered comfortable. No passengers are considered comfortable when the vehicle load exceeds 140% of seated capacity (*31*).

- Low-volume time periods: The maximum comfortable passenger-to-seat ratio for lower-volume travel periods is 125%. At loads up to 125% of seated capacity, all passengers are considered comfortable; above 125% and up to 140% of seated capacity, only seated passengers are considered comfortable; and no passengers are considered comfortable when the vehicle load exceeds 140% of seated capacity (*31*).

In order to use this service measure, the MBTA defined a threshold of CF below which a route is considered to not meet the standard. Ideally, all bus routes would have a CF of 100%, but in practice there are resource constraints and a mix of considerations in the service planning process that make it necessary to tolerate uncomfortable service a small fraction of the time. In order to set a realistic goal, the MBTA examined the distribution of CF over the agency's bus routes, shown in Figure 3. The majority of bus routes had over 95% of passenger minutes in comfortable conditions. There were 12 routes having CF under 92%, and most key bus routes had CF between 92% and 96%. Based on this, and in order to allow the consideration of other factors in the service planning process, the MBTA set the minimum threshold at 92%, but also established a target CF of 96%.

Analyzing the Data

Having defined performance measures and established a threshold, a detailed analysis of crowding on routes with high crowding was carried out. UPT and CF were calculated for 30 min windows for each route, separate by direction and for each segment, and the sources of crowding were identified. The computation was performed in a relational database, and an API and a web application dashboard were built to visualize the results over time and space. Figure 4 shows a screenshot of the dashboard, displaying the crowding performance of Route 9. A map of the city and its bus routes is shown on the left pane. The routes displayed on the map can be filtered by CF to focus on crowded routes, using the slider at the top left. Routes can be selected by clicking on this map or by selecting from a drop-down menu at the top. The middle pane shows graphs of CF and UPT by time of day or for the whole day, for the selected direction. CF for Route 9 in the inbound direction is 100% at most times of day, but it decreases to about 50% in the morning peak. UPT rises to its peak at the same time, as expected. The right pane shows a schematic of the route with segments color-coded according to crowding intensity at the selected half-hour time period. The most crowded segments are shown in deep orange. Selecting a segment displays UPT and CF for that segment at the selected time. The dashboard shows that only a third of passengers ride comfortably in the peak segment at the

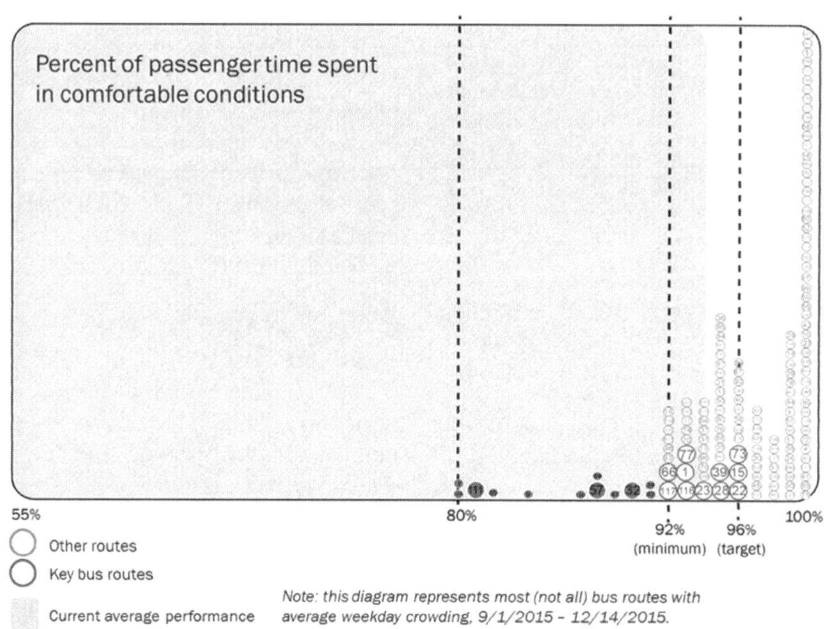

Figure 3. Distribution of comfort fraction across bus routes, Fall 2015.

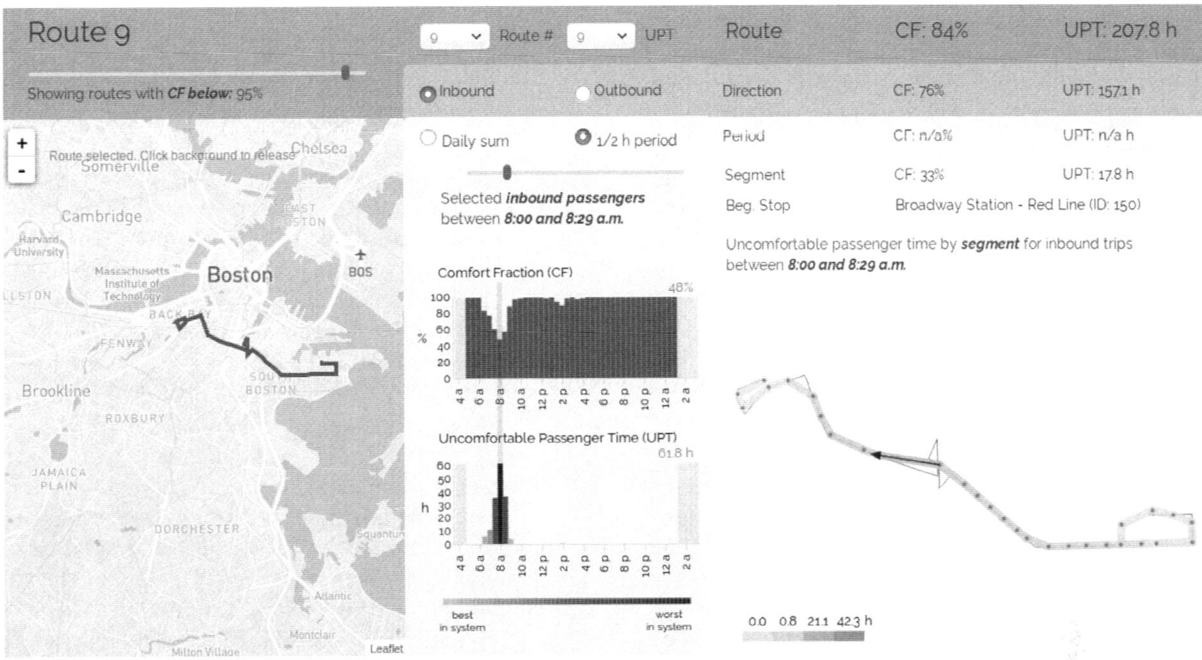

Figure 4. Crowding dashboard displaying information for Route 9, Fall 2015.

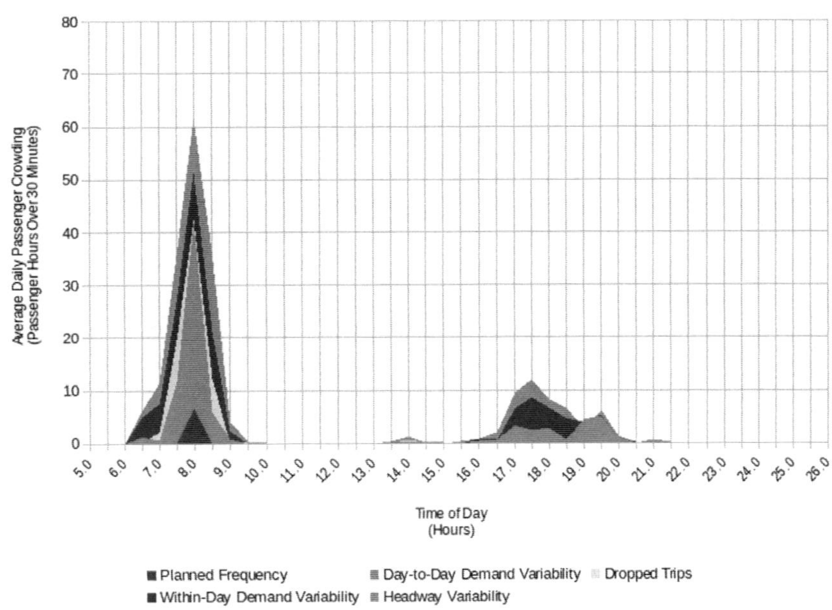

Figure 5. Sources of crowding in Route 9, Fall 2015.

peak time. This type of dashboard allows planners to quickly identify the times and places with most crowding on the system.

The final analysis examined the sources of crowding by time period. Figure 5 shows crowding on both directions of Route 9, by time of day and by source.

Crowding is only present in the peak periods, and is much greater in the morning peak than in the afternoon peak. The service plan provides sufficient frequency of service to meet the average daily demand at all times except the busiest half hour in the morning. However, there are significant fluctuations in ridership across days, and the service plan does not provide adequate frequency for days with higher-than-average demand. Dropped trips (scheduled but not operated) and demand variability across trips of the same time period in the same day also cause crowding in the morning, to a lesser extent. Headway irregularity in the peaks also causes a significant amount of crowding. While it appears that most of the crowding must be addressed by increasing the scheduled frequency and adding vehicles to the route, a meaningful portion could be decreased by dispatching trips from the terminal on regular headways and by implementing holding strategies where possible to keep headways even throughout the route.

Informing Service Changes

The MBTA has defined crowding using a passenger experience measure, developed the modeling tools to quantify this measure, and carried out the analysis to determine where crowding is taking place in time and space, and its causes. Now the transit agency is starting to use the analysis to propose service changes or operational improvements to better meet the standard. This will take place in the context of a new service planning process, which started in Fall 2017.

Conclusions

Comfort is an important aspect of the transit passenger experience. Crowding can significantly decrease passenger comfort and disrupt service delivery, causing passenger travel times to increase and even resulting in passengers being unable to board an arriving vehicle. Reducing crowding can encourage ridership growth in the most attractive parts of the transit network. This research explores the use of automatically collected vehicle location data, fare transaction data, and passenger origin–destination inference to measure crowding on buses, and makes three key contributions:

1. A scaling model for disaggregate origin–destination inference data of bus service is developed, which avoids biasing the scaled demand structure toward alightings at transfer locations, and which captures fare non-interaction factors at the resolution they are known.
2. Two bus crowding performance measures are defined, which can be computed from scaled disaggregate demand data, and which reflect passenger experience. These are based on a general function that transforms loads to crowding, which the agency can define reflecting tolerances and expectations that vary by time period, frequency of service, and vehicle configuration. *Uncomfortable passenger time* (UPT) measures the total time passengers spend riding buses in crowded conditions. *Comfort fraction* (CF) measures the percentage of time passengers spend in comfortable conditions. Both of these can be analyzed at different resolutions in space and time, to identify where and when crowding occurs.
3. A method for allocating crowding to various sources is developed. Crowding is measured for several hypothetical demand reassignment cases, and the comparison between these cases and real operations attributes crowding to insufficient frequency of service, fluctuations in demand between days or between trips within a day (which must also be addressed by increasing frequency), dropped trips, or headway irregularity. This points managers to the most effective measures that their agency can adopt to decrease crowding.

The framework has been tested at the MBTA. Several graphic applications, including an interactive web dashboard, were developed to assist planners in measuring crowding as experienced by passengers in a systematic manner. The MBTA has begun implementing service changes based on these tools.

The Transit Capacity and Quality of Service Manual (*6*) states that comfort is an important determinant of service quality and that demand responds to crowding. The methodology presented in this research could inform future revisions of the manual to include guidance on how agencies can leverage automatically collected data to calculate crowding performance measures and allocate them to specific sources in the interest of making them actionable.

Future research will track the progress made on reducing crowding based on the types of operational changes made. Opportunities to build upon the methodological components of this research include modeling crowding in corridors shared by multiple routes, measuring when passengers are left behind using vehicle-mounted cameras and inference algorithms based on the stops skipped on heavily loaded vehicles, and investigating how crowding relates to route and mode choice. The load estimation process could be refined with more sophisticated modeling of travel patterns by passengers without an inferred destination and those not interacting with the fare system. The distribution of destinations could be segmented by fare product (e.g., stored value, monthly pass, senior student) or fare medium (e.g., paper ticket, smartcard). The methods presented here could be applied in other transit agencies.

Acknowledgments

This research was funded by the Massachusetts Bay Transportation Authority (MBTA). The Massachusetts Green High Performance Computing Center (MGHPCC) supported this research by hosting the computer server on which data analysis was carried out. The authors thank Erica Gunn and Arthur Prokosch for their work designing and building the dashboard illustrated in Figure 4.

Author Contributions

The authors confirm contribution to the paper as follows: study conception and design: Sánchez-Martínez, Paget-Seekins, Southwick, Attanucci; data collection: Sánchez-Martínez, Paget-Seekins, Southwick, Attanucci; analysis and interpretation of results: Sánchez-Martínez, Paget-Seekins, Southwick, Attanucci; draft manuscript preparation: Sánchez-Martínez, Paget-Seekins. All authors reviewed the results and approved the final version of the manuscript.

References

1. Zorn, L., E. Sall, and D. Wu. Incorporating Crowding into the San Francisco Activity-based Travel Model. *Transportation*, Vol. 39, No. 4, 2012, pp. 755–771.
2. Batarce, M., J. C. Muñoz, J. D. Ortúzar, S. Raveau, C. Mojica, and R. A. Ríos. Use of Mixed Stated and Revealed Preference Data for Crowding Valuation on Public Transport in Santiago, Chile. *Transportation Research Record: Journal of the Transportation Research Board*, 2015. 2535: 73–78.
3. Cantwell, M., B. Caulfield, and M. O'Mahony. Examining the Factors that Impact Public Transport Commuting Satisfaction. *Journal of Public Transportation*, Vol. 12, No. 2, 2009, pp. 1–21.
4. Tirachini, A., D. A. Hensher, and J. M. Rose. Crowding in Public Transport Systems: Effects on User's Operation and Implications for the Estimation of Demand. *Transportation Research Part A: Policy and Practice*, Vol. 53, 2013, pp. 36–52.
5. Thompson, K., L. Hirsch, S. Muller, and S. Rainbird. *A Socioeconomic Study of Carriage and Platform Crowding in the Australian Railway Industry*. Technical Report. CRC for Rail Innovation, Brisbane, Australia, 2012.
6. Kittelson & Associates, Inc., Parsons Brinckerhoff, Inc., KFH Group, Inc., Texas A&M Transportation Institute, and Arup. *TCHRP Report 165: Transit Capacity and Quality of Service Manual*, 3rd ed. Transportation Research Board of the National Academies, Washington, D.C., 2013.
7. Tirachini, A., D. A. Hensher, and J. M. Rose. Multimodal Pricing and Optimal Design of Urban Public Transport: The Interplay Between Traffic Congestion and Bus Crowding. *Transportation Research Part B: Methodological*, Vol. 61, 2014, pp. 33–54.
8. Adebisi, O. A Mathematical Model for Headway Variance of Fixed-route Buses. *Transportation Research Part B: Methodological*, Vol. 20, No. 1, 1986, pp. 59–70.
9. Lin, T., and N. Wilson. Dwell Time Relationships for Light Rail Systems. *Transportation Research Record: Journal of the Transportation Research Board*, 1992. 1361: 287–295.
10. Tirachini, A. Bus Dwell Time: The Effect of Different Fare Collection Systems, Bus Floor Level and Age of Passengers. *Transportmetrica A: Transport Science*, Vol. 9, No. 1, 2013, pp. 28–49.
11. Milkovits, M. Modeling the Factors Affecting Bus Stop Dwell Time: Use of Automatic Passenger Counting, Automatic Fare Counting, and Automatic Vehicle Location Data. *Transportation Research Record: Journal of the Transportation Research Board*, 2008. 2072: 125–130.
12. Strathman, J., T. Kimpel, and S. Callas. *Headway Deviation Effects on Bus Passenger Loads: Analysis of Tri-Met's Archived AVL-APC Data*. Technical Report. United States Department of Transportation, Washington, D.C., 2003.
13. Bunker, J.M. Assessment of Transit Quality of Service with Occupancy Load Factor and Passenger Travel Time Measures. *Transportation Research Record: Journal of the Transportation Research Board*, 2015. 2535: 45–54.
14. Ryus, P., M. Connor, S. Corbett, A. Rodenstein, L. Wargeling, L. Ferreira, Y. Nakanishi, and K. Blume. *TCRP Report 88: A Guidebook for Developing a Transit Performance-Measurement System*. Transportation Research Board of the National Academies, Washington, D.C., 2003.
15. Furth, P. G., B. Hemily, T. H. Muller, and J. G. Strathman. *TCRP Report 113: Using Archived AVL-APC Data to Improve Transit Performance and Management*. Transportation Research Board of the National Academies, Washington, D.C., 2006.
16. Boyle, D. *TCRP Synthesis 77: Passenger Counting Systems*. TRB, National Research Council, Washington, D.C., 2008.
17. Li, Z., and D. Hensher. Crowding in Public Transit: A Review of Objective and Subjective Measures. *Journal of Public Transportation*, Vol. 16, No. 2, 2013, pp. 107–134.
18. Massachusetts Bay Transportation Authority. *Service Delivery Policy*. Massachusetts Bay Transportation Authority, Boston, MA, 2010.
19. Vanderwaart, C., J. P. Attanucci, and F. P. Salvucci. Applications of Inferred Origins, Destinations, and Interchanges in Bus Service Planning. *Transportation Research Record: Journal of the Transportation Research Board*, 2017. 2652: 70–77.
20. Trepanier, M., N. Tranchant, and R. Chapleau. Individual Trip Destination Estimation in a Transit Smart Card Automated Fare Collection System. *Journal of Intelligent Transportation Systems*, Vol. 11, No. 1, 2007, pp. 1–14.
21. Gordon, J., H. Koutsopoulos, N. Wilson, and J. Attanucci. Automated Inference of Linked Transit Journeys in London Using Fare-Transaction and Vehicle Location Data. *Transportation Research Record: Journal of the Transportation Research Board*, 2013. 2343: 17–24.
22. Munizaga, M. A., and C. Palma. Estimation of a Disaggregate Multimodal Public Transport Origin–destination

Matrix from Passive Smartcard Data from Santiago, Chile. *Transportation Research Part C: Emerging Technologies*, Vol. 24, 2012, pp. 9–18.

23. Sánchez-Martínez, G. E. Inference of Public Transportation Trip Destinations by Using Fare Transaction and Vehicle Location Data: Dynamic Programming Approach. *Transportation Research Record: Journal of the Transportation Research Board*, 2017. 2652: 1–7.
24. Sanchez-Martinez, G. E. Estimating Fare Non-interaction and Evasion Using Disaggregate Fare Transaction Data. *Transportation Research Record: Journal of the Transportation Research Board*, 2017. 2652: 98–105.
25. Uniman, D. *Service Reliability Measurement Framework Using Smart Card Data: Application to the London Underground.* Master's thesis. Massachusetts Institute of Technology, Cambridge, MA, 2009.
26. Tribone, D., D. Block-Schachter, F. Salvucci, J. Attanucci, and N. Wilson. Automated, Data-driven Performance Regime for Operations Management, Planning, and Control. *Transportation Research Record: Journal of the Transportation Research Board*, 2014. 2415: 72–79.
27. Sánchez-Martínez, G. E., and M. Munizaga. Workshop 5 Report: Harnessing Big Data. *Research in Transportation Economics*, Vol. 59, 2016, pp. 236–241.
28. Prokosch, A., and A. Gartsman. All-Door Boarding Without Proof-of-Payment: Revenue Impacts and Operational Implications. Presented at 96th Annual Meeting of the Transportation Research Board, Washington D.C., 2017.
29. Southeastern Pennsylvania Transportation Authority. *Service Standards and Process*. Technical Report. Southeastern Pennsylvania Transportation Authority, Philadelphia, PA, 2014.
30. Los Angeles County Metropolitan Transportation Authority. *Transit Service Policy*. Technical Report. Los Angeles County Metropolitan Transportation Authority, Los Angeles, CA, 2015.
31. Massachusetts Bay Transportation Authority. *Service Delivery Policy*. Massachusetts Bay Transportation Authority, Boston, MA, 2017.

The Standing Committee on Transit Capacity and Quality of Service (AP015) peer-reviewed this paper (18-04503).

TRR
JOURNAL OF THE TRANSPORTATION RESEARCH BOARD

Article

Transportation Research Record
2018, Vol. 2672(8) 475–484

DOI: 10.1177/0361198118781398
journals.sagepub.com/home/trr
SAGE

Understanding Human Perception of Bus Fullness: An Empirical Study of Crowdsourced Fullness Ratings and Automatic Passenger Count Data

Xidong Pi[1], Zhen (Sean) Qian[1,2], Aaron Steinfeld[3], and Yun Huang[4]

Abstract
Bus fullness, namely bus crowding level, is one of the most critical metrics used to quantify public transportation service quality and consumer satisfaction. Often it is simply represented by the head count within a bus vehicle that is measured using the prevailing automatic passenger counter (APC)., Little is known, however, about how the precise passenger count reflects the human perception of bus fullness. This study examines the linkage between APC data and crowdsourced fullness ratings data from riders in Pittsburgh, U.S.A., to understand how riders' perception of bus fullness is related to spatial, temporal, and demographic factors in addition to the passenger counts. We found that human perception of bus fullness, matched with passenger counts, can vary substantially by time of day, bus vehicle seat capacity, and passengers' income. Last but not least, we proposed a statistical model that estimates riders' perception of bus fullness based on APC data and bus route characteristics. This model can be used by transit agencies who possess APC data to assess service quality better, and ultimately to enhance the design and operation of transit systems.

Bus fullness, namely bus crowding level, is one of the most critical metrics used to quantify the quality of and customer satisfaction with public transportation services (*1*, *2*). Bus crowding results in not only uncomfortable travel experience but also low efficiency of bus service. It increases dwell time at bus stops (*3*), and overcrowding and discomfort on buses can double the average cost of trips (*4*). Many public transit agencies use the loading level of a bus vehicle to represent the bus fullness, defined by the ratio of current loads (passenger counts) over the vehicle seat capacity. It is common practice to use automatic passenger counter (APC) systems to compute this ratio as the metric of bus fullness. The prevailing APC systems (*5*, *6*), usually accompanied by automatic vehicle location (AVL) systems, help automate the measurements of passenger counts in every trip and at every stop, which was traditionally collected by hired observers at a much smaller coverage of spatial and temporal scale. By deploying APC systems, public transit operators and planners are now receiving and archiving considerably more passenger count measurements than in the past.

However, this definition of bus fullness does not take into account riders' perceptions, which may depend on various factors. First, rider perception of bus fullness varies by individual. Some users may regard the same level of fullness as more crowded than others. Thus, they may feel more discomfort than others on the same vehicle. The perception of bus fullness may also vary by time of day and location. Users may psychologically feel more uncomfortable at certain times and locations, and thus perceive the effect of fullness very differently. For instance, peak hours may often cause a higher crowding level than off-peak hours, but passengers are also more used to overcrowding in peak hours. Given the same ridership over seat capacity ratio, a peak-hour bus may be perceived as less crowded than an off-peak bus.

To understand bus fullness better in order to quantify service quality and customer satisfaction, we examine human perception of bus fullness instead of simply using the ratio of passenger count over seat capacity. In particular, we answer the following questions: How much is there a gap between machine measurements (APC data) and human perception of bus fullness? Is there any discrepancy in the perception among different bus riders? What factors are associated with human perception of fullness? Can we infer human perception from APC measurements combined with other factors?

[1]Department of Civil and Environmental Engineering, Carnegie Mellon University, Pittsburgh, PA
[2]Heinz College, Carnegie Mellon University, Pittsburgh, PA
[3]Robotics Institute, Carnegie Mellon University, Pittsburgh, PA
[4]School of Information Studies, Syracuse University, Syracuse, NY

Corresponding Author:
Address correspondence to Zhen (Sean) Qian: seanqian@cmu.edu

Thanks to the Tiramisu project (*7–10*), a crowdsourced transit platform providing real-time bus location and fullness in the Pittsburgh region, we have access to bus fullness ratings from app users that reflect their perceptions. Users of the Tiramisu application submit their perceived ratings of bus fullness at their boarding bus stops. The ratings are categorized by "many seats", "few seats", "stand," and "full". This paper examines both the crowdsourced fullness ratings data from Tiramisu and APC data to understand how riders' perceptions of bus fullness are related to spatial, temporal, and demographic factors in addition to the measured passenger counts. Few transit agencies outside of Pittsburgh are able to measure fullness perception directly from riders. Therefore, they can use APC data to infer the perceived fullness to assess the service quality better, and ultimately to enhance system design and operation.

This research contributes to the literature in the following aspects:

- Provides an empirical study of the relation and comparison between APC data and crowdsourced ratings of bus fullness (a proxy of human perception of bus fullness).
- Shows the effect of several factors on human fullness perception. Those factors include time of day (e.g., peak hours versus off-peak hours), bus seat capacity (e.g., 40-seats versus 60-seats), household income level.
- Develops an estimation model to infer passengers' perception of fullness from APC measurements combined with other spatial, temporal, and demographic factors.

The paper is organized as follows. In the next section, we first provide a literature review on studies of bus load and bus fullness, followed by an introduction to the characteristics of our two data sets: APC data and crowdsourced fullness ratings, as well as the data preprocessing methods. Then, we present the statistical experiment results and analyses. We also provide a brief discussion of our findings. The final section presents the conclusions from this research.

Literature Review

Research on bus fullness is fairly intensive. Concerning passengers' perceptions, teams have studied how bus crowding is related to passenger satisfaction using case studies and survey data (*2*, *11*). From a quantitative perspective, other research has investigated how bus crowding impacts the transit service quality in terms of the dwelling time and service reliability (*12*, *13*), and predicted the bus fullness and passenger demand (*14*, *15*). The research gap is that the measured bus fullness (or ridership to capacity ratio) can differ from passengers' perceptions of fullness, and the relationship is still unclear.

Li and Hensher (*16*) define two types of measures of crowding: measured passenger ridership as an "objective" measure and passenger's perceived crowding as a "subjective" measure. These two measures are not identical, and the perceived crowding is a compound of actual passenger density and many other factors, such as personal preferences and environmental characteristics. The authors also provide a comprehensive discussion about the discrepancy between measured and perceived crowding. Others have shown that this discrepancy is significant through two recent surveys from a rail operator in Australia (*17*). Rail passengers' perceptions of crowding are also related to psychological factors. The authors proposed a self-rating questionnaire to study the users' perceptions of crowding on rail and found a complex connection among passenger stress, negative feelings, and crowding. Nevertheless, most transit agencies and public transit literature still use the "objective" bus fullness measures, directly from APC data (*18*, *19*). Therefore, it is critical to study the relations between "objective" and "subjective" bus fullness quantitatively.

In this paper, we clearly identify two measures of bus fullness: APC-based passenger ridership to capacity and crowdsourced user ratings on fullness. The crowdsourced user fullness ratings from mobile applications enable us to study the subjective and objective bus fullness measures statistically. We also build an estimation model (a coarse mapping between the two dimensions) based on a large volume of data.

Data Description

Crowdsourced Fullness Data

Tiramisu (*7–9*) is a mobile phone application which was designed to crowdsource and share bus information, such as real-time bus location, arrival time, bus fullness, issues, and riding stories in the community. Users of the Tiramisu application share their GPS location and submit the fullness ratings in four categories: "many (seats)," "few (seats)," "stand," or "full", when they are boarding. This application is able to provide the estimated vehicle arrival time and user-perceived bus fullness information in real-time to other users. It helps to improve the accessibility of the transit service and the user's experience. Tiramisu was the only way for passengers to get real-time transit information in Pittsburgh before the AVL/APC systems were deployed at the end of year 2012.

In this research, we use the crowdsourced fullness ratings data of all weekdays from September 2012 to October 2015 from Tiramisu app (except for August 2013, October 2013–January 2014, and November 2014, because either the APC system or Tiramisu app was off-line during those time periods). Each rating represents the user's perception of bus fullness at the boarding stop. Each rating also comes with a trace in the data set that contains information regarding trip date, trip ID, route ID, boarding location, bus stop, and time-stamp, as well as the user's fullness rating. These fullness ratings covered all major bus routes operated by the Port Authority of Allegheny County (PAAC), the sole transit provider of Allegheny County.

APC Data

The APC data used in this research were collected and provided by PAAC. PAAC upgraded and installed the APC system on all buses in 2012. It measures the number of passengers boarding and disembarking by means of laser sensors at each bus door, and then archives the data in on-board computers. The system is activated immediately when a bus departs its depot and starts a trip. The APC data contains trip information (date, route ID, direction [inbound or outbound], trip ID, the number of seats), stop information (stop ID, arrival and departure time), and passenger counts (boarding, alighting, and in-vehicle counts). We compute the bus fullness for any bus trip at any location and time following:

$$\text{Fullness} = \frac{\text{Number of Passengers}}{\text{Number of seats}}$$

We refer to this metric as "APC-based fullness" or "passenger-over-capacity ratio" throughout this paper. In Pittsburgh, it is a real number spanning (0.0, 2.0) in most cases, with the very rare occurrence of being greater than 2.0. This implies that almost all bus loads in Pittsburgh were less than 200% of seat capacity.

Data Preprocessing

The crowdsourced data were first pre-processed to remove the records with empty fullness ratings. We then filtered out users with less than or equal to three contributions to eliminate fraudulent and inaccurate submissions from inactive users. By matching the date, route ID, direction, trip ID, and stop ID, we joined both APC and fullness ratings data, which essentially created one-to-one mapping from fullness ratings to APC-based fullness for bus trips and stops. After this mapping, we consolidated the two crowdsourced fullness rating classes "stand" and "full" to one class "full", due to the possible linguistic ambiguity between "stand" (a bus has no available seats) and "a bus is full (full capacity)". In addition, the number of records for these two classes was much less than for the other two classes "many seats" and "few seats." Consolidating the two classes may enhance statistical significance. Finally, we filtered out "irresponsible" users, which were defined as users with more than 15% of "outlier" ratings. An outlier rating was a rating with its associated APC-based fullness value smaller than 5 quantiles or greater than 95 quantiles of all samples at this rating. For example, if the mean APC-based fullness of all samples rated as "many seats" is 0.30, and the 5 and 95 quantiles were 0.03 and 0.80 respectively, then samples that were rated as "many seats" records and their associated APC-based fullness value less than 0.03 or greater than 0.80 were identified as outlier ratings.

After data preprocessing, there were in all 42,238 valid rating traces from 1,529 users, plotted in Figure 1. It showed that the rating data covered all main corridors and streets along bus routes. For each of the three fullness rating classes:

- Many seats: 23,092 valid records
- Few seats: 13,569 valid records
- Full: 5,577 valid records

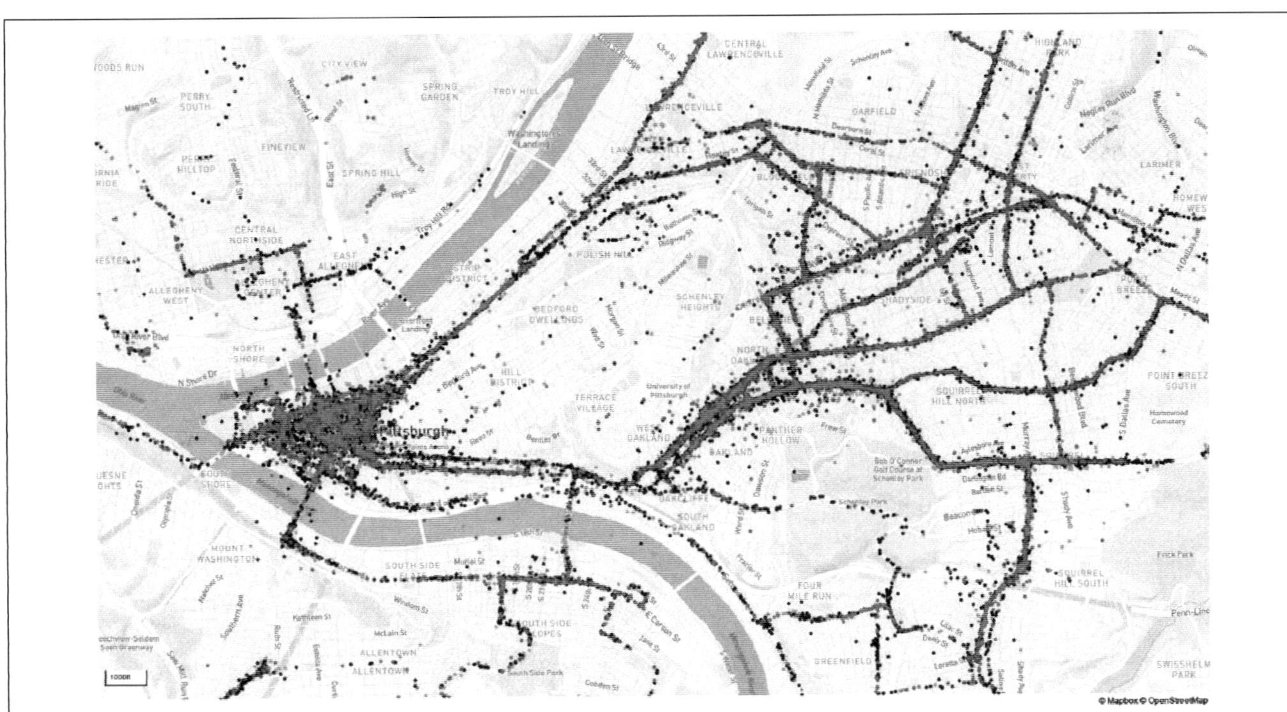

Figure 1. Crowdsourced bus fullness ratings data: spatial distribution in central districts of Pittsburgh. Year 2012= orange; 2013= black; 2014= pink; 2015= cyan. Created with ©Mapbox (*20*).

Experiments and Results

Statistical Difference between Crowdsourced Fullness Ratings and APC-Based Bus Fullness

The first question we wanted to answer was whether user fullness ratings were statistically consistent with the actual bus load. We created a hexbin plot of time of day (x-axis) and APC-based measurements (y-axis) in Figure 2(a–c) for each of the three rating classes: "many", "few," and "full", along with marginal distributions. The boxplot of APC measurements for the three rating classes is in Figure 2(d). On average, users perceived around 31% of bus loads over capacity to be "many seats," around 52% to be "few seats," and around 76% to be "full." However, the variance increases when the fullness is rated as high. This indicates human perception is generally more consistent when a bus is empty, whereas the perception is more diversified when the bus is almost full. Also notice that the marginal distributions of APC-based fullness by time of day are more concentrated towards morning and

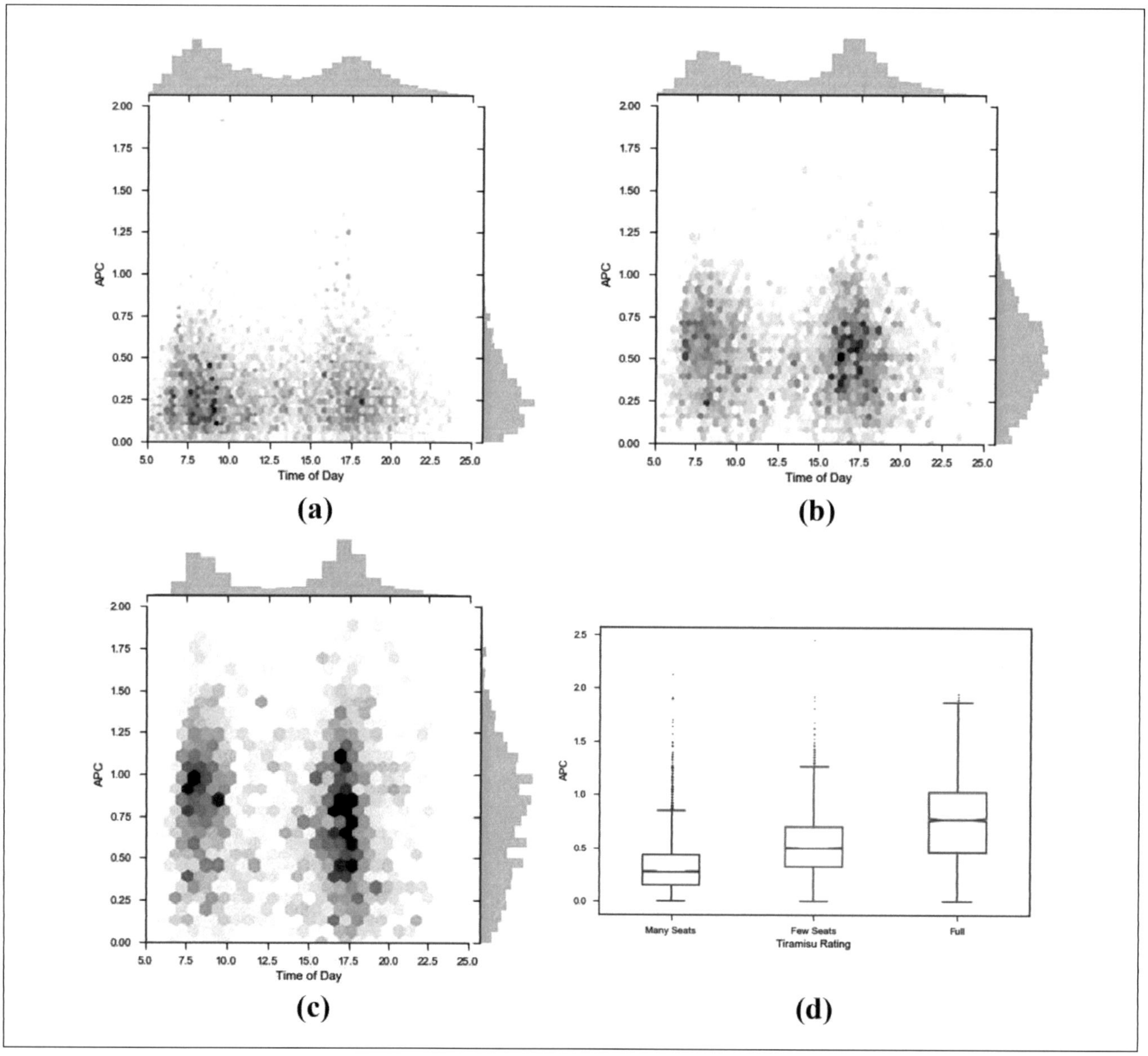

Figure 2. The hexbin plot and box plot of time of day and APC-based fullness. (a) "many seats", mean = 0.315, SD = 0.218; (b) "few seats", mean = 0.517, SD = 0.270; (c) "full", mean = 0.758, SD = 0.382; (d) box plot of APC fullness at each rating level.

Table 1. Statistical Tests for the Three Rating Groups

	One-way ANOVA test		Kruskal–Wallis *H*-test	
Groups	Statistics	*p*-value	Statistics	*p*-value
All 3 groups	5014.4	<0.005*	7114.0	<0.005*
"Few" versus "many"	4213.2	<0.005*	3946.9	<0.005*
"Many" versus "full"	9328.1	<0.005*	5172.4	<0.005*
"Few" versus "full"	2437.7	<0.005*	1719.3	<0.005*

afternoon peak hours at the higher rated fullness, implying more consistent human perception/ratings for a full bus than an empty bus during peak hours. However, the box plot also implies that aggregating over all time of day, human perception is generally more consistent among all users when the bus is empty than it is full.

We performed a one-way ANOVA on the three groups of data: "many seats", "few seats," and "full", to test if the means of the APC-based fullness among the three groups are different statistically. The one-way ANOVA hypothesis test requires three main assumptions including: (a) samples are independent; (b) each group of samples follows a normal distribution; and (c) homoskedasticity: the standard deviation of samples within each group is equal. We also performed a nonparametric version of one-way ANOVA test: Kruskal–Wallis *H*-test, which requires no explicit assumption on the distribution of samples. Both hypothesis tests gave extremely large statistics and extremely small *p*-values, see Table 1. Next, we investigate other factors that are also associated with human perception of fullness.

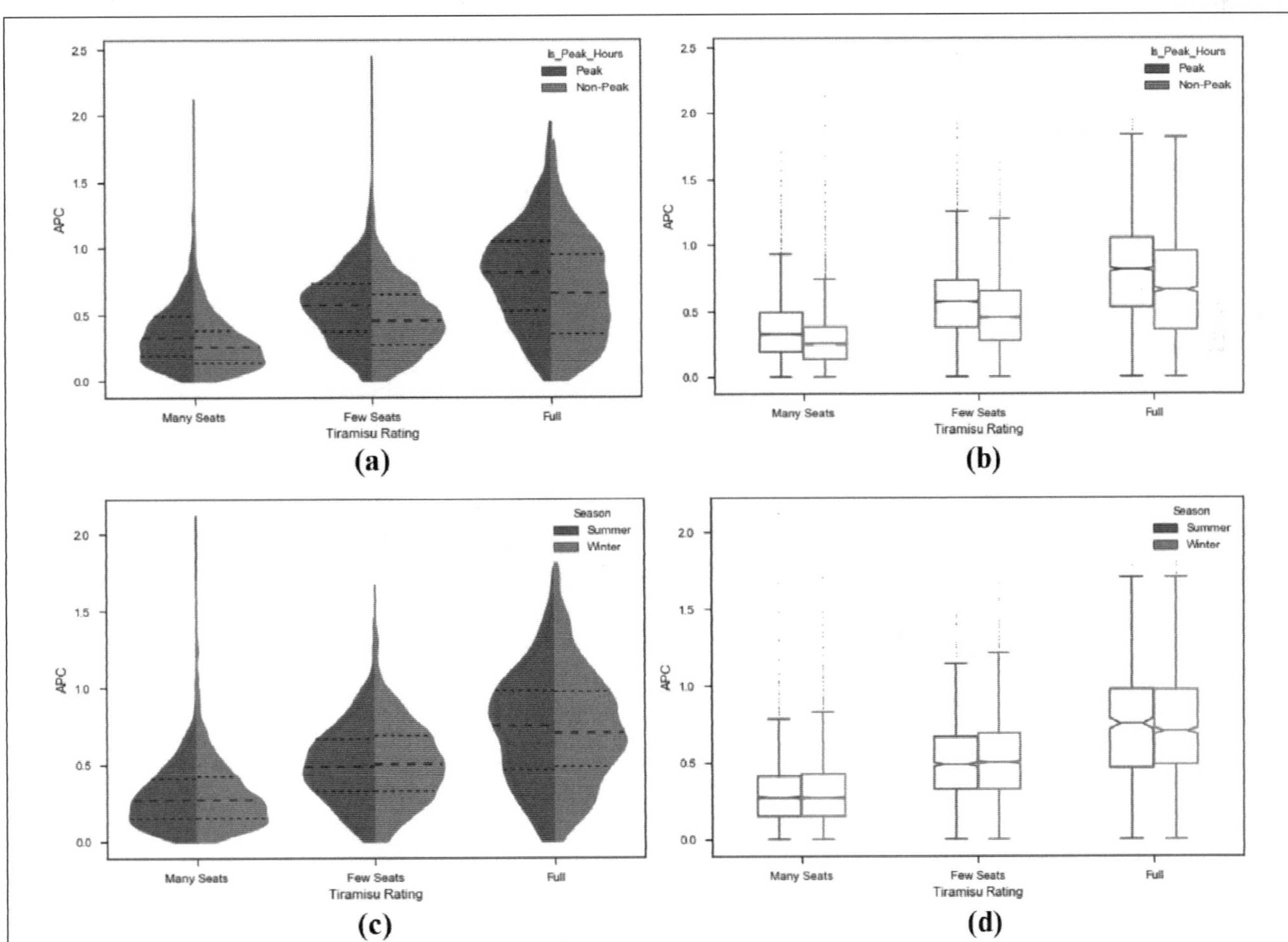

Figure 3. Violin plots and box plots of: peak hours versus off-peak hours (*a*) and (*b*); winter versus summer (*c*) and (*d*).

Table 2. Statistical Tests for the Influence of Time of Day and Seasons

Groups	Mann–Whitney rank test	
	Statistics	*p*-value
Time of day:		
"Peak-many" versus "off-peak-many"	5.02e+07	7.08e-161*
"Peak-few" versus "off-peak-few"	1.83e+07	3.60e-96*
"Peak-full" versus "off-peak-full"	2.84e+06	4.72e-30*
Seasons:		
"Winter-many" versus "summer-many"	1.05e+07	0.312
"Winter-few" versus "summer-few"	3.27e+06	0.090
"Winter-full" versus "summer-full"	3.78e+05	0.399

Factors Associated with Human Perception of Fullness

For this part, we hoped to find the external factors that significantly affect passengers' perceptions of bus fullness. Since personal information was not collected in our data set, we focus on the most prominent temporal and spatial impact factors, as well as the bus size.

Temporal analysis. First, we examined the effect of time of day (i.e. "peak hours" versus "off-peak hours") on the human perception of bus fullness. The "peak hours" are defined as inbound trips (directions from suburbs to downtown) in the morning [7 AM, 10 AM] and outbound trips (directions from downtown to suburbs) in the afternoon/evening [4 PM, 7 PM]. All other trips are treated as "off-peak."

Figure 3(*a* and *b*) shows the distribution of APC for trips of peak hours and off-peak hours, as well as the box plot implying the mean, standard deviation, and outliers of data. We also performed a nonparametric hypothesis test, the Mann–Whitney rank test (Table 2). Passengers' perceptions of fullness were statistically different in peak hours and off-peak hours, provided with the passenger counts. In peak hours, passengers were more likely to tolerate overcrowding in buses. For all three fullness ratings: "many," "few," and "full", the average APC measurements are higher in peak hours than in off-peak hours.

We also studied seasonal effects, in particular winter and summer, on users' perceptions of bus fullness. In Pittsburgh, due to the cold climate, people wear bulkier clothes during the winter. We considered December to February as winter, and June to August as summer. Similar to the study of the time-of-day effect, the violin plot and box plots for groups of winter and summer are shown in Figure 3(*c–d*). Surprisingly, the distributions of APC are quite similar between summer and winter. The Mann–Whitney rank test results in Table 2 also show their group mean is statistically insignificant. Clearly, the data shows that passengers can tolerate the loss of space from thick clothing well.

Spatial Analysis. Next, we examined the spatial variation of fullness perception. It is speculated that geographic regions with different demographics may have effects on users' perceptions. We divided the city of Pittsburgh into census tracts, and acquired the median annual household income data for each census tract from 2011 to 2015 using American Community Survey (ACS) 5-Year Estimates. Three income classes are applied.

- Low income: lower than $40,000
- Middle income: from $40,000 to $100,000
- High income: higher than $100,000

The violin plot and box plot are shown in Figure 4 (*a–b*). The mean APC measurements for all three rating groups increase along with the income level. These results are interesting, implying that trips in high-income tracts can probably accommodate more passengers for the same fullness rating than trips in low-income tracts. This may be a region-specific result since peak-hour routes serving workers in the finance and legal professions are very busy in Pittsburgh due to constrained parking in the civic center. Hypothesis test results listed in Table 3 support this conclusion. The mean of APC measurements varies substantially among tracts with different income levels, provided with the same fullness ratings.

Bus Seat Capacity. Bus seat capacity may also affect users' perceptions of fullness. There are two types of buses running in Pittsburgh: small-size buses have 37, 40, or 42 seats, while large-size buses have 56, 57, 60, or 63 seats. The violin plot and box plot are shown in Figure 4(*c–d*), paired with hypothesis test results in Table 3. The results show that large-size buses have a smaller mean of passenger-over-capacity ratios in each of the rating groups. This effect is statistically significant, and most prominent for those trips rated at "full." Users with frequent experience of riding small-size buses are more likely to tolerate overcrowding.

Estimating Human Perception of Bus Fullness

We found that human perception of bus fullness is statistically related to time of day (peak hours or off-peak hours), spatial variation (represented by income level) and bus seat

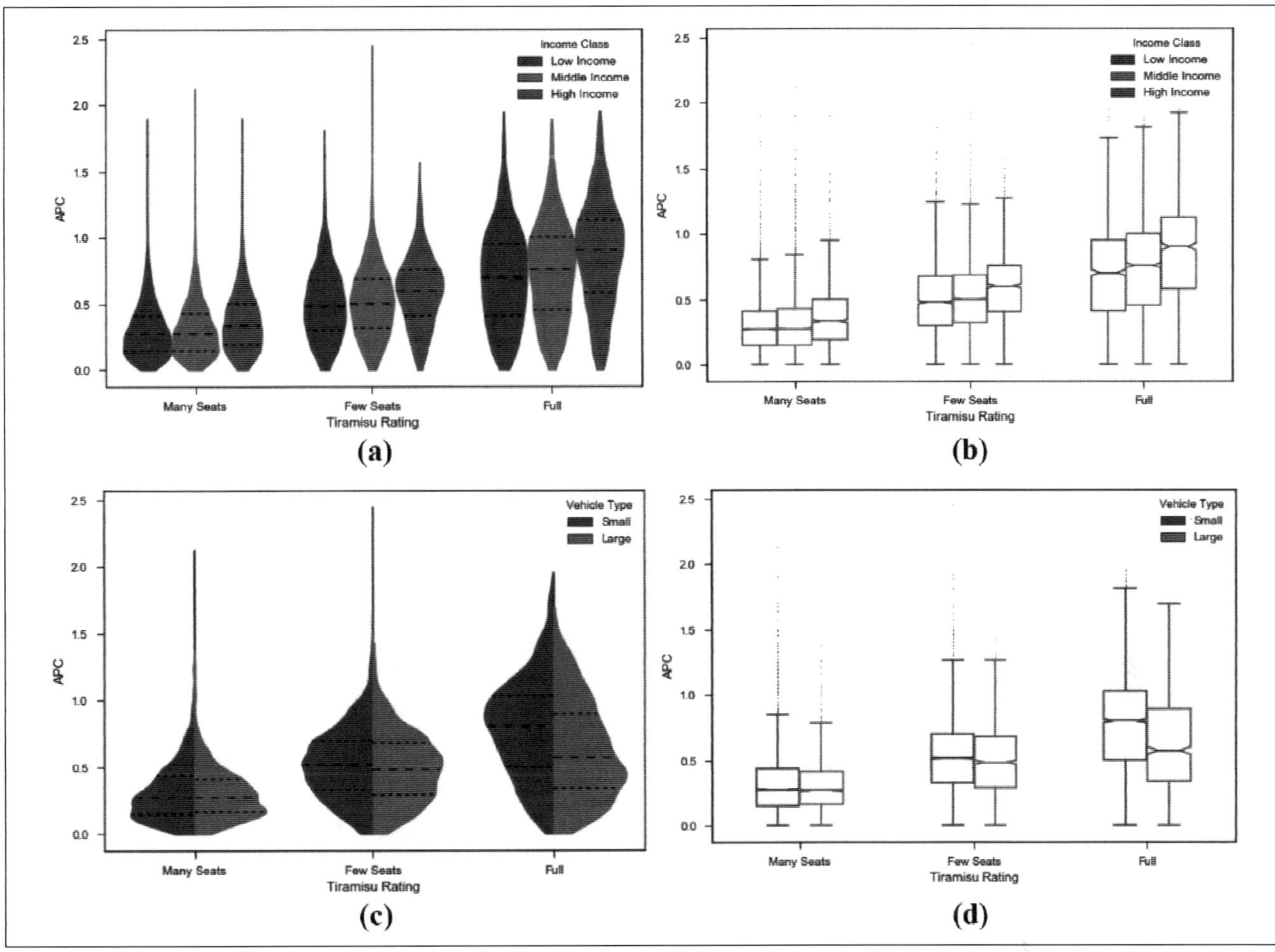

Figure 4. Violin plots and box plots of: different income levels (*a*) and (*b*); different vehicle sizes, (*c*) and (*d*).

capacity, in addition to the actual passenger-over-capacity ratios measured by APC. This section develops models to estimate human-rated bus fullness using APC data, so that they can be used for transit agencies to assess service quality better without necessarily having crowdsourced ratings data. We defined two classes of models:

- Baseline model: APC-based fullness being the sole predictor
- Improved model: The predictors include APC-based fullness, bus route ID, hour of day, trip direction, median income of the tract, bus seat capacity

Table 3. Statistical Tests for the Influence of Different Income Levels and Vehicle Sizes

Groups	Mann–Whitney rank test *p*-value
income levels:	
"Low income-many" vs. "middle income-many"	5.14e-03*
"Middle income-many" vs. "high income-many"	2.20e-27*
"Low income-few" vs. "middle income-few"	3.81e-04*
"Middle income-few" vs. "high income-few"	1.07e-31*
"Low income-full" vs. "middle income-full"	2.13e-05*
"Middle income-full" vs. "high income-full"	4.41e-19*
Vehicle sizes:	
"Small-many" vs. "large-many"	0.081
"Small-few" vs. "large-few"	2.96e-04*
"Small-full" vs. "large-full"	4.80e-33*

Table 4. Estimation Accuracy with Different Hyperparameters in a Random Forest Classifier

Min-split	2			3			5			10		
n-trees	B	I	Imp	B	I	Imp	B	I	Imp	B	I	Imp
20	64.0	75.6	18.1%	64.2	75.5	17.7%	64.1	75.2	17.4%	64.0	74.0	15.5%
50	64.0	76.2	19.0%	64.0	76.0	18.7%	64.1	75.4	17.7%	64.0	74.0	15.5%
100	64.0	76.5	19.4%	64.0	76.5	19.4%	64.1	76.1	18.7%	64.1	74.4	16.1%
200	64.1	76.4	19.0%	64.0	76.6	19.6%	64.1	76.1	18.8%	64.0	74.3	16.0%
300	64.0	76.4	19.4%	64.0	76.5	19.4%	64.0	76.1	18.8%	64.1	74.3	15.9%
500	64.0	76.4	19.3%	64.1	76.6	19.5%	64.0	76.3	19.1%	64.1	74.4	16.1%

Note: B = baseline model; I = improved model; Imp = improvement in estimation accuracy.

We randomly selected 50% of the data as the training set for a random forest classifier. The remaining 50% of the data was used as the test data set. We did a grid search of hyperparameters for a random forest classifier: (a) *n-trees*: the number of trees in the forest; and (b) *min-split*: the minimum number of data samples required to split an internal node. The estimation accuracies are shown in Table 4. Figure 5 shows the confusion matrix of baseline and improved model for a random forest classifier with the best hyperparameters found in grid search (*n-trees* = 200, *min-split* = 3).

Generally, the non-linear classification model can accurately estimate the users' fullness ratings 77% of the time. This substantially improves on the baseline model that simply regards the fullness ratings to be equivalent to categorized passenger-over-capacity ratios, where the number is 64%. This also implies that humans' perception of fullness can be influenced by many factors other than the passenger counts.

Bus passengers are heterogeneous, and their perceptions can also be influenced by their personal characteristics. This kind of information is not available and thus cannot be included in our model. The model has an error rate of about 23%. Its accuracy may be further improved by incorporating personal demographic data.

Discussion

Analysis of APC data and crowdsourced fullness rating data revealed that they are highly correlated. On average, users perceive the same fullness that is largely determined by passenger-over-capacity ratios. However, there exist large variance of perception of fullness among users, especially when the bus is crowded. Such variance of perception can be partially attributed to peak hours, bus size, and riders' income level. However, there is no clear seasonal effect on the perception. We summarize additional insights as follows:

- Fullness ratings by users, namely their perception of fullness, can vary quite substantially among users. Therefore, averaged ratings may not reveal user diversity. Generally, this variation of perception is relatively small when rating a crowded bus during peak hours or an empty bus.
- Riders are more likely to "tolerate" overcrowding during peak hours than off-peak hours. This is probably because riders are used to and prepared for crowded buses during peak hours. For transit agencies, the implication is that when planning and operating bus trips during peak hours, more passengers may be tolerated to ensure the same service quality to users than during off-peak hours.
- Bus trips in high-income areas can probably accommodate more passengers for the same fullness rating than trips in low-income areas. This is probably due to two reasons: self-selection of high-income riders and more space requirement for low-income riders. People with high income are more likely to travel using private cars, so whoever chooses to ride a bus is more likely to tolerate in-vehicle crowding. Low-income riders are likely to be captive to buses. Thus, they may travel with more bags and carry-ons.
- On average, passengers perceive around 90% of seat capacity to be "full" for small-size buses, and 60–70% of capacity to be "full" for large-size buses. Large-size buses have a smaller mean of passenger-over-capacity ratios for each of the three rating groups. This effect is the most prominent for buses perceived as "full." Therefore, there is declining benefit in perceived fullness when extending the number of seats in larger, articulated buses. One possible explanation is that passengers can only get on board at the front door of the bus and they tend not to move to the far back of large-sized buses. Thus, the perceived bus fullness from passengers who have just boarded could be "full" even when the APC measure is only 60–70%.

We built an effective model for estimating passengers' perceptions of fullness using spatial and temporal factors in addition to APC data. Estimating human perceptions of bus fullness solely based on passenger counts is generally inaccurate (64% in our study). If we add trip information (time of day, direction, number of bus seats, route ID) and demographic

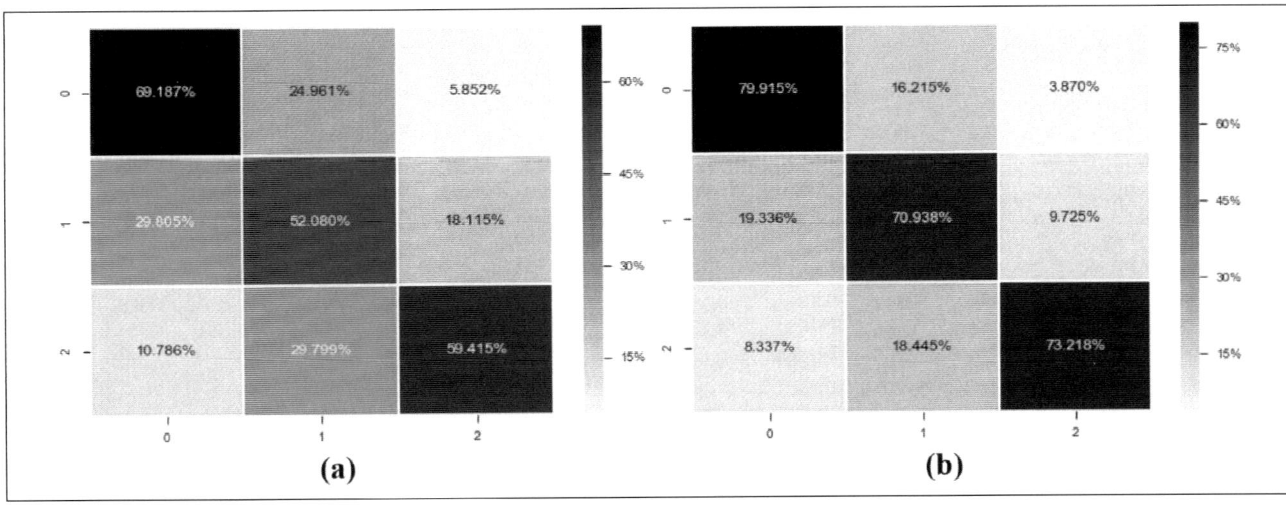

Figure 5. Confusion matrices with heat-maps of baseline and improved models for random forest: (*a*) baseline model, (*b*) improved model. Row = prediction label; column = true label; 0 = "many seats"; 1 = "few seats"; 2 = "full".

information (district average annual income) to the predictor, the estimation accuracy can be substantially improved (77% in our study). However, there still exists quite a bit of variability of perception that cannot be explained by the estimation model. This is likely due to individual variability in perception by humans and possibly reckless (false) ratings by some users we were unable to filter out. This noise and error would be extremely difficult to eliminate.

Conclusion and Future Work

This paper examines the linkage between APC data and crowdsourced fullness ratings data from users (acquired from the Tiramisu application) in the Pittsburgh region, to understand better how riders' perceptions of bus fullness are related to spatial, temporal, and demographic factors in addition to passenger counts. We found that human perception of bus fullness, provided with the same passenger counts, can vary substantially by time of day (e.g., peak hours versus off-peak hours), bus vehicle seat capacity (e.g., 40-seats versus 60-seats), and area (differentiated by income level). A model was therefore developed to estimate passengers' fullness ratings using those factors. This model estimates the fullness perception, as a metric for service quality, better than using passenger count as the sole estimator. This model could be used by transit agencies which have APC data to assess service quality better, without necessarily having perception data, and ultimately to enhance system design and operation. The output could also be used to populate fullness rating in real-time feeds for rider information systems. For example, riders who use wheelchairs may opt to alter their trip preemptively if their desired bus is full (*7*, *10*).

Our work has two major limitations. First, we only used transit data from Pittsburgh to build and test the model. Future work should focus on testing this model in other cities and validation using state-preference surveys to assess causality. Second, our findings suggest that a variety of factors are associated with human perception of bus fullness, but we cannot claim the causality. Future work should apply qualitative methods to investigate the causal relations further. Third, this work could be extended from bus to light rail trains or subways. This study is focused on bus transit due to the limited crowdsourced passenger rating data for light rail transit collected by Tiramisu in Pittsburgh, but this method is applicable if future data for light rail trains becomes available.

Acknowledgments

This research is funded in part by Traffic 21 Institute and Carnegie Mellon University's Mobility 21, a National USDOT University Transportation Center for Mobility which is sponsored by the US Department of Transportation. Parts of this work are also funded under a grant from the National Institute on Disability, Independent Living, and Rehabilitation Research (NIDILRR grant number 90RE5011-01-00). NIDILRR is a Center within the Administration for Community Living, Department of Health and Human Services. The authors also thank Amy Silbermann from the Port Authority of Allegheny County (PAAC) for providing all the APC data for the experiments and analysis.

Author Contributions

The authors confirm contribution to the paper as follows: study conception and design: Xidong Pi, Zhen (Sean) Qian, Yun Huang, Aaron Steinfeld; data collection: Yun Huang, Aaron Steinfeld; data analytics: Xidong Pi, Zhen (Sean) Qian; draft manuscript preparation: Xidong Pi, Zhen (Sean) Qian, Yun Huang, Aaron Steinfeld. All authors reviewed the results and approved the final version of the manuscript.

References

1. dell'Olio, L., A. Ibeas, and P. Cecin. The Quality of Service Desired by Public Transport Users. *Transport Policy*, Vol. 18, No. 1, 2011, pp. 217–227.

2. Eboli, L., and G. Mazzulla. Service Quality Attributes Affecting Customer Satisfaction for Bus Transit. *Journal of Public Transportation*, Vol. 10, No. 3, 2007, p. 2.
3. Lin, T. M., and N. H. Wilson. Dwell Time Relationships for Light Rail Systems. *Transportation Research Record: Journal of the Transportation Research Board*, 1992. 1361: 287–295.
4. Litman, T. Valuing Transit Service Quality Improvements. *Journal of Public Transportation*, Vol. 11, No. 2, 2008, p. 3.
5. Hodges, C. C. *Automatic Passenger Counter Systems: The State of the Practice*. Technical Report No. DOT-1-87-36. Urban Mass Transportation Administration, US Department of Transportation, 1985.
6. Attanucci, J., and D. Vozzolo. Assessment of Operational Effectiveness, Accuracy, and Costs of Automatic Passenger Counters. *Transportation Research Record: Journal of the Transportation Research Board*, 1983. 947: 15–23.
7. Zimmerman, J., A. Tomasic, C. Garrod, D. Yoo, C. Hiruncharoenvate, R. Aziz, N. R. Thiruvengadam, Y. Huang, and A. Steinfeld. Field Trial of Tiramisu: Crowd-Sourcing Bus Arrival Times to Spur Co-Design. *Proc., SIGCHI Conference on Human Factors in Computing Systems*. ACM, 2011, pp. 1677–1686.
8. *Tiramisu: The Real-Time Bus Tracker*. http://www.tiramisutransit.com/. Accessed April 20, 2017.
9. Tomasic, A., J. Zimmerman, C. Garrod, Y. Huang, T. Nip, and A. Steinfeld. The Performance of a Crowdsourced Transportation Information System. Transportation Research Board 94th Annual Meeting Compendium of Papers, 2015.
10. Steinfeld, A., J. Zimmerman, A. Tomasic, D. Yoo, and R. Aziz. Mobile Transit Information from Universal Design and Crowdsourcing. *Transportation Research Record: Journal of the Transportation Research Board*, 2011. 2217: 95–102.
11. Cantwell, M., B. Caulfield, and M. O'Mahony. Examining the Factors that Impact Public Transport Commuting Satisfaction. *Journal of Public Transportation*, Vol. 12, No. 2, 2009, p. 1.
12. Milkovits, M. Modeling the Factors Affecting Bus Stop Dwell Time: Use of Automatic Passenger Counting, Automatic Fare Counting, and Automatic Vehicle Location Data. *Transportation Research Record: Journal of the Transportation Research Board*, 2008. 2072: 125–130.
13. Fritz, M. S. Effect of Crowding on Light Rail Passenger Boarding Times. *Transportation Research Record: Journal of the Transportation Research Board*, 1983. 908.
14. Ma, Z., J. Xing, M. Mesbah, and L. Ferreira. Predicting Short-term Bus Passenger Demand Using a Pattern Hybrid Approach. *Transportation Research Part C: Emerging Technologies*, Vol. 39, 2014, pp. 148–163.
15. Tirachini, A., D. A. Hensher, and J. M. Rose. Crowding in Public Transport Systems: Effects on Users, Operation and Implications for the Estimation of Demand. *Transportation Research Part A: Policy and Practice*, Vol. 53, 2013, pp. 36–52.
16. Li, Z., and D. A. Hensher. Crowding in Public Transport: A Review of Objective and Subjective Measures. *Journal of Public Transportation*, Vol. 16, No. 2, 2013, p. 6.
17. Mahudin, N. D. M., T. Cox, and A. Griffiths. Measuring Rail Passenger Crowding: Scale Development and Psychometric Properties. *Transportation Research Part F: Traffic Psychology and Behaviour*, Vol. 15, No. 1, 2012, pp. 38–51.
18. Furth, P. G., B. Hemily, T. H. Muller, and J. G. Strathman. *Using Archived AVL-APC Data to Improve Transit Performance and Management*. Transit Cooperative Research Program (TCRP) Report 113, published by Transportation Research Board of the National Academies, Washington, D.C.,. 2006.
19. Kittelson & Associates, United States. *Federal Transit Administration, Transit Cooperative Research Program, & Transit Development Corporation*. Transit Capacity and Quality of Service Manual (No. 100). Transportation Research Board of the National Academies, Washington, D.C., 2003.
20. *Mapbox: Build Experiences for Exploring the World*. https://www.mapbox.com/. Accessed April 20, 2017.

The Standing Committee on Transit Capacity and Quality of Service (AP015) peer-reviewed this paper (18-05900).

TRR
JOURNAL OF THE TRANSPORTATION RESEARCH BOARD

2018 FRED BURGGRAF AWARD: Public Transportation

Transportation Research Record
2018, Vol. 2672(8) 485–496

DOI: 10.1177/0361198118780834
journals.sagepub.com/home/trr
SAGE

Measuring Spontaneous Accessibility for Iterative Transit Planning

Matthew Laquidara[1]

Abstract
Public transit planners rely on measurements of network performance to anticipate the impact of changes to a transportation system. Accessibility-based measurements emphasize how well a transportation system allows individuals to reach desired opportunities, rather than maximizing network properties such as capacity. This paper presents a measurement of accessibility for transit customers making unanticipated, spontaneous trips. Measuring this spontaneous accessibility was facilitated by developing an open-source software tool that can analyze a transit network throughout an entire day, over a complete municipal boundary or transit agency service area, at fine spatial granularity, and without some of the simplifying assumptions made by previous studies. The tool is used to study spontaneous accessibility within the city of Seattle over a one-year period featuring the opening of a light rail extension and restructure of bus services. Studies of this nature require only limited data sources but produce precise results, and thus can be utilized to measure iterative refinement of the transit network. Furthermore, techniques from the discipline of information theory provide insight into ways to reduce the computational demands, giving planners the ability to consider more alternatives.

Individuals with use of a car enjoy an ease of access to distant destinations that those without vehicles do not. This difference in accessibility is especially evident when making unanticipated trips, such as going to a grocery store to pick up a single item needed for a recipe, visiting an urgent care clinic, or responding to a family emergency. Those with vehicles can start their journeys immediately in response to the spontaneous need or desire, and proceed to their destination in the most direct way. Those limited to walking and taking transit can make relatively short trips spontaneously by walking, but for longer trips they must contend with the fact that public transit generally runs on a schedule, transit lines may be indirect, and transfers may be required to reach destinations. All of these factors constrain the ability to complete unanticipated trips in a timely fashion, making private vehicle ownership an attractive proposition.

In the city of Seattle, the number of households without access to a car has been growing over the last five years (*1*). For commuting trips originating within the Seattle city limits and terminating in the downtown Central Business District (CBD), transit mode share dominates the use of single occupancy motorized vehicles (*2*). In spite of this, Seattle's rate of households with at least one vehicle exceeds that of cities including Milwaukee, Detroit, and San Francisco (*3*). The contrast between these observations could indicate that the desire for private vehicle ownership in Seattle is not tied solely to its use in getting to and from work. Thus, transit planners at agencies tasked with supporting a trend of decreased car ownership may want to prioritize the network's capability to support trips beyond those representative of standard commuting patterns.

This paper examines the ability of a public transit system, specifically that which exists in the city of Seattle, to support unanticipated trips. It defines the capability of a transit network to allow such trips as its spontaneous accessibility and describes comparable measurements for it that differ from accessibility measurements presented previously in the literature. Creating this measurement was facilitated by developing an open-source software tool that measures the transit network in ways that mirror a customer making an unexpected transit trip. The tool must evaluate the network over a full day period, as the need to make a transit trip may occur at any time. It must consider journeys from many origins to many destinations. A large set of origins ensures that the analysis is relevant to people throughout the studied area. Since the destinations of unexpected trips are by definition unpredictable, the analysis considers all destinations on the map. Furthermore, the tool must not measure the network in an overly abstract way; it should not make approximations that result in allowing transit trips that are infeasible in actuality. These requirements, however, work in opposition to the desire for transit planners to compute this measurement whenever considering service changes. Slow computation

[1]Public Transit Analytics, Seattle, WA

Corresponding Author:
Address correspondence to Matthew Laquidara:
matt@publictransitanalytics.com

would preclude spontaneous accessibility measurements from being a part of periodic service evaluation, limiting them to being long-range planning tools. In support of finding balance between these priorities, this paper makes use of techniques from information theory to quantify the amount of error introduced by simplifying the analysis. As a demonstration of these techniques, spontaneous accessibility measurements are used to show the extent to which a rider's ability to take unanticipated trips within Seattle changed over a one-year period that included a light rail line extension and restructure of bus services.

Literature Review

Path Finding

Almost any measurement of a transportation network requires an understanding of the time expended to reach destinations. In the context of public transit, this requires knowledge of the transit routes one will take to their destination. Dial (*4*) describes the pathfinder program, which converts stops and the scheduled duration between them to a graph and uses a tree solving algorithm to find minimal paths from starting points. The algorithm assumes that starting locations are transit hubs and that connections between multiple lines only occur where transit lines share a stop. No walking is accounted for in this model and the transfer wait time is always half the frequency of the transit lines servicing a stop. For modeling transit networks abstractly, these limitations may be acceptable.

For more precise modeling of the transit network that is available to riders, path finding algorithms must have a notion of absolute time. Tong and Richardson (*5*) implement a method of finding minimal paths by converting a set of transit stops and schedules into a format where the minimal paths can be found by Dijkstra's graph algorithm. A graph models the transit network, as in earlier approaches, but the algorithm keeps track of the current time, and changes the durations between stops based on that time. It also models walking between certain stops. Crisalli and Rosati (*6*) present DY-RT (DYnamic Regional Transit), which uses the schedules of a larger regional bus and rail network, but makes performance measurements at the level of municipalities. The heuristic approach taken by Ayed et al. (*7*) allows larger and more detailed representations of transit networks, at the cost of absolute accuracy. Though the approaches vary in focus on urban or interurban travel, they provide a more accurate model of the transit network.

Though many path finding solutions use graph-theoretic approaches, other strategies exist. RAPTOR, described by Delling et al. (*8*), is not based on Dijkstra's algorithm; instead it uses dynamic programming to find best paths. The authors use it on a complete transit map of London with over 20,000 stops. As a result of its non-graph-theoretic construction, it more easily supports parallel processing and calculating best paths for times in a range.

Transit Network Measurement

Many large-scale transportation planning decisions make use of the urban transportation modeling system, which uses the four-step model consisting of trip generation, trip distribution, mode choice, and traffic assignment (*9*). This approach assumes a certain geographic area will generate a number of trips based on its underlying characteristics. Destination points have some measurement of gravity by which trips are attracted to them. The nature of the transit network influences the path that will be taken, and thus the expected ridership of a route within the network can be approximated. In the context of transit planning, route changes are tested by exploring how they affect the riderships projected by the four-step model.

The four-step model involves considerable data aggregation; trips are formed out of the aggregate properties of origin and destination areas. Hägerstand (*10*) asserts that such an approach does not take into account the fact that any trip occurs because of an individual's desires and needs, not out of the natures of regions. Such reasoning informs approaches to transit network evaluations that focus on what Hanson (*11*) describes as "personal accessibility," the ability of a person to reach sites where activities occur, or opportunities, from their home. Counting these opportunities and weighting them by their distance creates an "accessibility index" whereby the personal accessibility provided by the transportation system can be contrasted by origin point. Handy and Niemeier (*12*) argue that there are many dimensions to accessibility measurement and, while there is no best approach, a study's goal informs the proper choices to make. Geurs and van Wee (*13*) assert that accessibility measures can be useful for making transportation and land use decisions, provided that a set of criteria are met. They further contend that person-based accessibility measurements, those which focus on what an individual can access given their time and space constraints, are effective for evaluating transportation network changes, but complexity of calculation and large data requirements make them difficult to perform. Work by Bertolini et al. (*14*) uses the concept of an isochrone to determine which neighborhoods can reach job centers within a time frame. The study only considers trips using a single transit route. In spite of its limitations, it demonstrates numerically-measured accessibility as a viable transit planning tool.

Geographic Information Systems

Understanding the ability of an individual to access destinations via transit is hindered when the walking paths that a person can take are not accurately known. Nyerges (*15*) describes the use of geographic information systems (GIS) in deducing an accurate representation of locations in Seattle that can reach public transportation lines within a fixed walking distance. Tribby and Zandbergen (*16*) evaluate a new bus service by producing an index that measures change in time to reach a single destination given transit need. GIS enables accurate modeling of the walking distance from individual homes to buses, as well as transfers between them. Averages are used for transit ride and wait times. Work by Mavoa et al. (*17*) has similar characteristics, but measures access to a variety of destination types, such as schools, from individual property parcels. Silva and Pinho (*18*) propose the structured activity layer, which expresses the diversity of opportunities accessible to an individual as an index measurement. O'Sullivan et al. (*19*) use GIS as a way to produce more accurate isochrones, rather than an index. Though a similar set of assumptions is made about transit travel time, GIS enables measurement at a scale befitting an individual's interaction with a transit network.

Schedule Data

Accurate measurements of accessibility must incorporate schedules, since real transit vehicles neither move at a constant rate, nor arrive at stops after a rider waits exactly half the headway. Though schedule-based path finding algorithms have long existed, the lack of easily accessible schedule data limited applications. The transit accessibility planning analyst described by Lei and Church (*20*) creates isochrones that use a single starting time, schedule data from contemporary timetables, walking time calculations from GIS, and a Dijkstra-like path finding algorithm. The authors propose evaluating transit networks using these isochrones by starting at a set of origins and summing the destinations reachable over a set of starting times, but do not suggest a mechanism of selecting either. The availability of transit schedules in general transit feed specification (GTFS) format (*21*) enabled considerable expansion in accessibility studies that use full schedule data. Work by Anderson et al. (*22*) suggests that, for some transit stops, it is critical to consider all starting times in a day, as accessibility varies both within and between hours. This is employed in a study of access to jobs in the Twin Cities area (Minnesota) by Owen and Levinson (*23*), wherein a RAPTOR-like algorithm is used to compute the number of jobs that can be reached from every bus station in the network, over a two-hour peak-time commute period. The study performs aggregation at the census block level; initial walks to transit stops use an estimated walking time based on the straight-line distance. The accessibility value of a resource is weighted by a decay function to make distant opportunities less appealing than close ones. Owen and Levinson (*24*) use an unlimited transfer model and accurate walking routes, with census block groups as the unit of spatial granularity for measuring access to jobs. The accessibility values that they find correlate with observed transit mode share, a testament to the feasibility of using advanced accessibility-based measurements as a transit planning technique. The urban access tool, proposed by Blanchard and Waddell (*25*), uses similar data sources, but takes the approach of averaging headway times for a time period rather than computing reachability at every minute. Gillespie and Fahrenwald (*26*) add the dimension of automobile travel to transit to produce a time-weighted access to jobs measurement. It is notable for its use of a regular grid. The methodology described by Conway et al. (*27*) is most similar to the approach of this paper. The authors model the spatial environment with a high-resolution grid, use the ranged-RAPTOR algorithm for path finding, incorporate full schedules, and use accurate walking paths. However, because of their focus on access to jobs, the measurement of value for reaching a given area emphasizes the expected behaviors of commuters. The transit accessibility studies reviewed in this section are set out in Table 1 for comparison.

Methodology

The purpose of this study is to construct a measurement of a transit network's capability to provide spontaneous accessibility—the ability of individuals to make unanticipated trips to unexpected destinations. The measurement is designed as comparable and dimensionless, allowing the evaluation of alternatives as well as the contrasting of the accessibility within areas. Accurately modeling the transit network as seen by riders is a critical component of any accessibility study. However, the literature survey reveals that the most advanced modeling has only been used on studies that focus on home-to-work trips. Studies that consider accessibility to more destinations have lacked the same precision. Therefore, this section discusses the requirements for a spontaneous accessibility measurement and how they are fulfilled.

A spontaneous accessibility measurement is an isochrone: it measures a proportion of opportunities that can be reached within a fixed duration. In this case, it is difficult to define what opportunities are. Given that riders are taking trips in response to immediate needs, the value of any destination is potentially unknowable in advance. Rather than attempting to evaluate what areas of the map are of high value, all destinations are defined to provide equal opportunity. The measurement applies this same reasoning to selecting starting points for the isochrone. While it is possible to weight more populous areas as having greater value, such a measurement devalues the burden placed on an individual who finds that they must start their trip from an uncommon origin, and thus are not the default. The measurement also considers every minute of the entire day equally. Increasing the value given to times when the most people travel assumes that unanticipated events occur more often during these periods. The resulting measurement, though expressible as a single dimensionless quantity, is somewhat complex: it convolves every origin point, destination point, and time of day. The use of a strict isochrone, rather than a decay-based measurement, helps limit complexity. Eschewing this dimension preserves "interpretability and communicability," which Geurs and van Wee (*13*) assert is essential for a practical accessibility measurement.

Table 1. Comparison of Transit Accessibility Studies

Work	Transit model	Walking model	Origins & destinations	Time frame	Measurement
Bertolini et al. (*14*)	Unimodal, avg. speed	N/A	Neighborhoods, job centers	N/A	Population, jobs
Tribbly and Zandbergen (*16*)	Avg. ride time, wait time	Paths	Residences, single point	Range	Index: Time change given transit need
Mavoa et al. (*17*)	Avg. ride time, fixed wait time	Paths	Land parcels, various destination points	Range	Index: Time thresholds to destinations
Silva and Pinho (*18*)	Avg. speed	Unspecified	Census tract	N/A	Index: Variety of activities
Lei and Church (*20*)	Full schedule	Paths	Single point, unspecified destinations	Single time	Reachable area
Owen and Levinson 2012 (*23*)	Full schedule	Straight line	Census block	Range	Time-weighted access to jobs
Owen and Levinson 2015 (*24*)	Full schedule	Paths	Census block group	Range	Access to jobs
Blanchard and Waddell (*25*)	Avg ride time, wait time	Paths	Census block	Range	Access to jobs
Gillespie and Fahrenwald (*26*)	Unspecified	Paths	Grid (0.5 × 0.5 mi.)	Range	Time-weighted access to jobs
Conway et al. (*27*)	Full schedule	Paths	Grid (78 × 78 m.)	Range	Access to jobs
Laquidara	Full schedule	Paths, elevation	Grid (176 × 281 m.)	Full day	Reachable area

Spatial Environment

An important consideration in designing a spontaneous accessibility measurement is appropriately modeling the space that is being evaluated. Unanticipated trips, by their definition, may have origin or destination points anywhere. Practically, however, a spontaneous accessibility calculation is made over a bounded area. Furthermore, it would be impossible to determine paths between every point in the area. Therefore, the bounds are divided into uniformly-sized, non-overlapping sectors. While prior studies have used census or property divisions, their lack of uniformity is problematic. Using non-uniform divisions overestimates the accessibility of areas of the map within large sectors, causing the measurement to systematically undervalue transit that serves these areas.

Constructing the spatial environment can be done with a very small collection of resources. GTFS files of the transit agencies operating in the region are required to model bus service. For the available walking routes, OpenStreetMap (*28*) data is used. However, the route-finding component is entirely separable, allowing substitution of commercial data sources if available and desired. Water body data, in GeoJSON format (*29*), is used to eliminate entirely-water sectors. The ease of collecting these resources minimizes the preparation needed to make a spontaneous accessibility measurement, enabling it to be performed frequently.

Path Finding

Path finding is used to determine which sectors fall within the isochrone of a given duration, time of day, and center location. The software preprocesses transit data to support the data access patterns used by the path finder component. A single run of a transit vehicle is called a trip and is a list of pairs of stops and times. Only trips that are inside the time span and spatial environment are considered. Additionally, a table of entry points has rows of stops and sorted columns of times, allowing reference of the trip arriving at that stop, at that time.

Walks can be initiated from the starting point or a transit stop and can end at another transit stop or the border of any sector. All points outside of the bounds are eliminated. To further reduce the number of walking trips that must later be tested for feasibility, the straight-line distance of each potential walking trip is calculated and converted into an estimated time. All estimates less than the maximum duration are retained and sorted by the estimated time. When more precise distances are later needed, the destinations to test are limited to those where the estimated time is less than the allowed time. Final measurement of these distances is done by a separate subsystem, currently an instance of the GraphHopper software (*30*), though commercial products can be substituted.

The path finder itself is intended to provide a highly accurate model of trips that the transit network permits. The algorithm that it uses is a dynamic programming algorithm similar to RAPTOR (*8*) that takes advantage of the preprocessed data. The least-time path is always found; no preference is given to minimizing initial waiting time, waiting time during transfers, or walking distance, though the number of mode transfers is minimized as a consequence of the algorithm. While these properties may conflict with rider preferences, they are intended to describe the network's capability.

Measurement

Every spontaneous accessibility measurement is determined by finding paths for a set of tasks. A task contains the parameters for performing path finding: a starting time, starting location, and isochrone threshold. Executing a task yields the sectors that were reached given the task's parameters and the best path to each sector. Consequently, each sector maintains a task count: the number of tasks wherein that sector was reached. A single executed task allows the computation of the simplest measurement known as the time-qualified point accessibility ratio (TQPAR). This is the ratio of sectors reached to the total number of sectors. The results of Anderson et al. (*22*) indicate that accessibility can vary considerably, even within a restricted range of times. Thus, tasks are executed for every minute of an entire day and the ratios derived from each task are averaged to create the point accessibility ratio (PAR). The path finder uses techniques similar to ranged-RAPTOR (*8*) to compute these tasks more efficiently than running them individually. Executing tasks for the product of every minute of the day and each sector center, accounts for transit riders taking trips that originate from locations that they cannot anticipate. The ratios from each task are averaged to give the network accessibility ratio (NAR). Formal descriptions of each calculation are given in Equation 1 where T is the set of all times, S is the set of valid sectors, s_0 is a chosen starting point, t_0 is a chosen starting time, and "reached" is a function that computes the number of sectors reached in a fixed duration for a center point and starting time.

$$
\begin{aligned}
TQPAR_{duration}^{time} &= \frac{reached(t_0, s_0)}{|S|} \\
PAR_{duration} &= \frac{\sum_{t \in T} reached(t, s_0)}{|T| \cdot |S|} \\
NAR_{duration} &= \frac{\sum_{t \in T} \sum_{s \in S} reached(t, s)}{|T| \cdot |S|^2}
\end{aligned}
\tag{1}
$$

Measuring Sampling Error with Kullback-Leibler Divergence

Reducing the number of tasks comprising a NAR provides an opportunity to measure spontaneous accessibility at a considerably lower cost. The theoretical validity of this approach is intuitive. Measuring the spontaneous accessibility from the center of one sector also reveals information about nearby sectors, as the close proximity makes it likely that some of the same transit stations can be reached at similar times of day. Thus, it may be possible to use a sample of sector centers as starting points rather than every one. Before accepting this approach as valid, it is useful to quantify the impact of this sampling.

In the discipline of information theory, Kullback and Leibler (*31*) describe a technique for measuring the divergence between two random variables where one is an approximation and one is known to be true. The formulation of this, for a true distribution p and an assumed one q, is presented in Equation 2. Using such a technique for spontaneous accessibility measurements would require viewing them as random variables.

$$
D(p \| q) = \sum_{x \in X} p(x) \log \frac{p(x)}{q(x)} \tag{2}
$$

While each of the spontaneous accessibility ratio measurements yields a single value, they are formulated from what can be thought of as a collection of observations. Thus, given a set of sectors and their task counts, it is possible to construct a probability distribution out of the likelihoods of observing task counts. To allow comparison of non-sampled and sampled results, where task counts will be lower, the ratio of a sector's task count to the total number of tasks is used. These ratios are divided into bins, and the bins must be defined such that every bin with values in the true distribution also has values in the sampled distribution. Otherwise, the Kullback-Leibler Divergence will be infinite, preventing reasonable attempts to compare samplings. Using this strategy, a network accessibility calculation and an empirically chosen number of bins is used to generate the distribution for p while the sampled network accessibility and the same number of bins yield q.

Since measuring the divergence from a distribution requires knowing that true distribution, sampling is not used for measuring the present spontaneous accessibility of a transit network. However, once a sample with an acceptably low Kullback-Leibler divergence has been found, planners can test their desired changes using only the sampled points as starting points rather than the centers of every sector. When calculated with a sample of origins rather than every sector center, the spontaneous accessibility measurement is known as a sampled network accessibility ratio (SNAR), presented in Equation 3, where S' is the set of sampled sectors. By performing the SNAR calculation rather than a full NAR, the computation time is reduced, allowing more experimentation.

$$
SNAR_{duration} = \frac{\sum_{s \in S'} \sum_{t \in T} reached(t, s)}{|T| \cdot |S| \cdot |S'|} \tag{3}
$$

Application and Results

To demonstrate the properties and uses of spontaneous accessibility measurements, this paper presents two analyses. The first highlights a practical process that planners can use to make decisions using spontaneous accessibility. In it, NARs quantify the impact of transit network changes. Specifically, it considers one year's worth of transit changes, a light rail extension and bus network restructures, and evaluates whether these changes have been successful in improving the ability to make unexpected trips within Seattle. Planners can use this process to examine historic changes in their transit network. With small modifications, they can use it to evaluate the impact of proposed changes or speculated improvements. The second analysis pertains to the nature of SNARs. Though planners are unlikely to replicate this analysis, its results reveal advisable practices when sampling. As such, it is of use to planners who want to test several modifications to the transit network, but are operating under time constraints.

Environment

Seattle is served by three transit agencies: King County Metro operates an all-day bus service within the city bounds as well as some commute-focused suburban buses; Sound Transit is a multi-county agency that provides the high-frequency Link light rail line as well as some commuter buses; and the Seattle Department of Transportation manages two streetcar lines. These agencies make service changes at six-month intervals. In March of 2016, Sound Transit opened an extension to the Link line, with stations in the Capitol Hill neighborhood and near the University of Washington. As a result, King County Metro restructured its bus services to eliminate redundancies and provide expanded access to the new light rail stations as part of its six-month periodic restructure process. Many bus lines in neighborhoods surrounding the light rail stations received higher frequencies as a result of the restructure, at the cost of eliminating services considered to be redundant.

Schedule data is available through King County Metro's GTFS files. Though three different agencies control transit in Seattle, King County Metro operates all of the Seattle Department of Transportation streetcars, the Link light rail, and most of the Sound Transit buses that provide transit service between points within Seattle. As such, they publish GTFS files including this infrastructure, allowing the use of King County Metro's files alone to provide the data needed for this study. Because the analyses span multiple service changes, two sets of GTFS files are needed as these files typically represent service over a fixed date range. Planners doing historic analyses can use this approach; those conducting speculative analysis can construct modified GTFS files reflecting the proposed network changes.

In these analyses, Seattle is physically represented by a bounding rectangle of the city's borders. As a consequence, some areas outside of the actual city boundary are included; for the purposes of this study, they are considered to be a part of Seattle. In their analyses, planners may choose bounds that reflect the totality of their transit network or choose smaller bounds, such as an individual city or neighborhood, if they wish to measure spontaneous accessibility for individuals traveling between points in that area. The bounding box is divided into a grid of one hundred by one hundred sectors. This number is chosen arbitrarily; higher counts more accurately model the experience of individual riders as the smaller sectors convolve the experiences of fewer individuals, but increase the computational power required to conduct an analysis. In this case each sector has dimensions of 176 meters by 281 meters (577 feet by 921 feet). These are slightly larger than the typical census blocks in the area, but smaller than all but the smallest census block groups. Sectors entirely of water are eliminated, leaving 6,063 sectors reachable.

Geographic data on Seattle is freely available. An OpenStreetMap extract of Washington State provides routing data for the entire bounding box. To detect water, a hydrographic map of the region is freely available from the county and was reprojected and converted to GeoJSON with open-source software.

Analyses

To perform a comparative spontaneous accessibility analysis of the impact of the year's network changes, NARs are selected as the measurement type. This choice reflects that the goal of the analysis is to judge the impact of the transit changes on Seattle as a whole. In almost all transit planning cases, this is an appropriate measurement, though planners evaluating the impact of transit network modifications on a particular location would compare PARs centered at the location. In this retrospective analysis, comparable dates must be chosen. Two non-holiday Mondays at approximately the same time of year, January 25, 2016 and January 30, 2017, represent service before and after the transit network changes took place, while controlling for seasonal variability. A time span of an entire day is required to account for the unpredictability of needs arising. Planners may choose an isochrone duration arbitrarily, but it should reflect data or perception of the amount of time individuals will travel to fulfill an unexpected need. These analyses use a 30-minute duration.

To evaluate the efficacy of using sampled network accessibility in place of network accessibility, several sampling variants are contrasted using otherwise the same parameters as the January 25, 2016 network accessibility measurement. This study chose sample sizes of 1,000, 2,000, 3,000, and 4,000 to reflect considerable differences in the amount of sampling used. Five different random samples at each size allow a basic analysis of the distribution of sample quality. For computing the Kullback-Leibler divergence, planners must empirically choose a range and bin number such that none of the bins from the sampled distribution have zero items. After computing the NAR and the 20 SNARs, it was found that a range from zero to 0.2 with twelve bins fulfilled this requirement.

Discussion

Table 2 summarizes the results of both analyses. The NAR measurement from January 25, 2016 of 0.06017 is fairly

Table 2. Summary of Spontaneous Accessibility Measurements

Date	Measurement	Samples	Value	K-L divergence (bits)
2016-01-25	NAR_{30}	6063 (all)	0.06017	0
2016-01-25	$SNAR_{30}$	1000	0.06009	0.00442
2016-01-25	$SNAR_{30}$	1000	0.06080	0.01407
2016-01-25	$SNAR_{30}$	1000	0.06053	0.00155
2016-01-25	$SNAR_{30}$	1000	0.05945	0.00692
2016-01-25	$SNAR_{30}$	1000	0.06109	0.00863
2016-01-25	$SNAR_{30}$	2000	0.05922	0.00420
2016-01-25	$SNAR_{30}$	2000	0.06057	0.00527
2016-01-25	$SNAR_{30}$	2000	0.05987	0.00192
2016-01-25	$SNAR_{30}$	2000	0.05992	0.00219
2016-01-25	$SNAR_{30}$	2000	0.05884	0.00641
2016-01-25	$SNAR_{30}$	3000	0.06050	0.00081
2016-01-25	$SNAR_{30}$	3000	0.06006	0.00197
2016-01-25	$SNAR_{30}$	3000	0.06017	0.00120
2016-01-25	$SNAR_{30}$	3000	0.05980	0.00089
2016-01-25	$SNAR_{30}$	3000	0.05977	0.00196
2016-01-25	$SNAR_{30}$	4000	0.06039	0.00195
2016-01-25	$SNAR_{30}$	4000	0.06013	0.00063
2016-01-25	$SNAR_{30}$	4000	0.05986	0.00171
2016-01-25	$SNAR_{30}$	4000	0.06006	0.00136
2016-01-25	$SNAR_{30}$	4000	0.06018	0.00022
2017-01-30	NAR_{30}	6063 (all)	0.06140	0

abstract on its own. It indicates that if an independent selection of a random origin point, destination sector, and starting time is made, there is a 6.017% chance that that selection will correspond to a trip that can be made in 30 minutes. The value of this number is considerably clarified when viewed comparatively. As a result of the year's transportation changes, the NAR on January 30, 2017 was 0.06140. This represents a 2.0% increase in the number of trips with random origin, destination, and starting time combinations that can be made in 30 minutes. The absolute value of this increase is small, but it is taken over the entire bounds of the city at all times of day and corresponds to approximately 65,109,250 additional feasible trips. The relatively small change in value also reflects that the major network changes were localized to the light rail stations and bus service restructures in limited areas. Nevertheless, a planner asked to ascertain whether historical changes have improved the network can use the comparative result to determine that they have increased spontaneous accessibility overall. The same process can be employed for comparing several hypothetical alternatives and, before putting one into effect, choosing one with the greatest increase in NAR.

In addition to making single-value measurements, planners can spatially decompose network accessibility measurements for insight into the distribution of spontaneous accessibility. Figure 1 shows the map for the January 25, 2016 NAR calculation. Each sector is colored according to the proportion of tasks in which that sector can be reached within 30 minutes. The average of these ratios equals the NAR. The map indicates that several patterns of service can create areas with elevated spontaneous accessibility. Many bus routes converge on the CBD (A), which is also served by light rail. Light rail stations (B, C, D, E, F, G) show pockets of increased spontaneous accessibility, though the extent varies, as a result of the amount of connecting bus services. Areas where frequent buses converge from perpendicular directions (H, J), or where several frequent bus lines converge on a common street (K, M) have accessibilities that match or exceed those of light rail stations outside of the CBD. Such a map can also indicate to planners areas for future transit expansion by highlighting areas with low present spontaneous accessibility.

With the network accessibility map serving as a baseline, Figure 2 shows the impact of the changes made between January 25, 2016 and January 30, 2017. Sectors are colored based on the ratio of improvement or degradation between the earlier and later measurement. The greatest improvement is seen at the terminal of a new frequent bus route where only limited service was available in the past (N). The University of Washington light rail station (O) shows a considerable increase in accessibility in its immediate vicinity. While restructure of the bus service connecting to the station has made unanticipated trips to northeast Seattle (P) more viable, restructures in the University District (Q) and Montake (R), to the west and south of the station, respectively, have not been successful in this regard, perhaps the result of more journeys taking an indirect path through the light rail station. Capitol

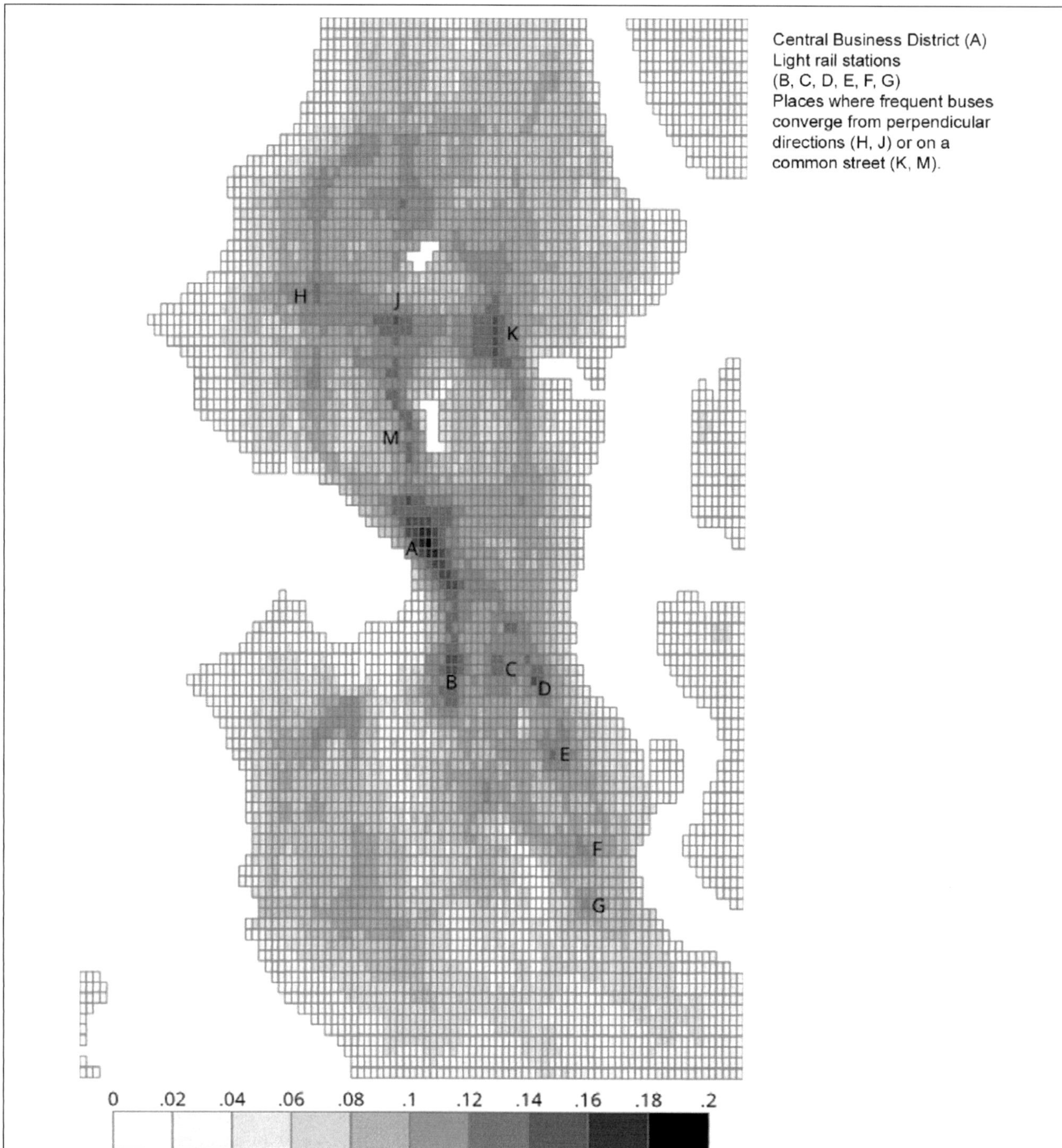

Figure 1. Network accessibility in graphical form. Each sector is colored based on the proportion of tasks that allow it to be reached within a 30-minute period.

Hill (S) shows noted improvement, but not to the extent of the University of Washington station and northeast Seattle. Bus service restructures in this area were limited. In southeast Seattle (T) a large swath of modest improvement is the result of the extension of a bus line through the area. This came at a cost to Georgetown (U), where service through the neighborhood remains equally frequent, but originates from fewer other locations. Viewing the map provides a more nuanced view of the 2.0% gain in spontaneous accessibility. A planner can use such a map to qualify the success of a network change,

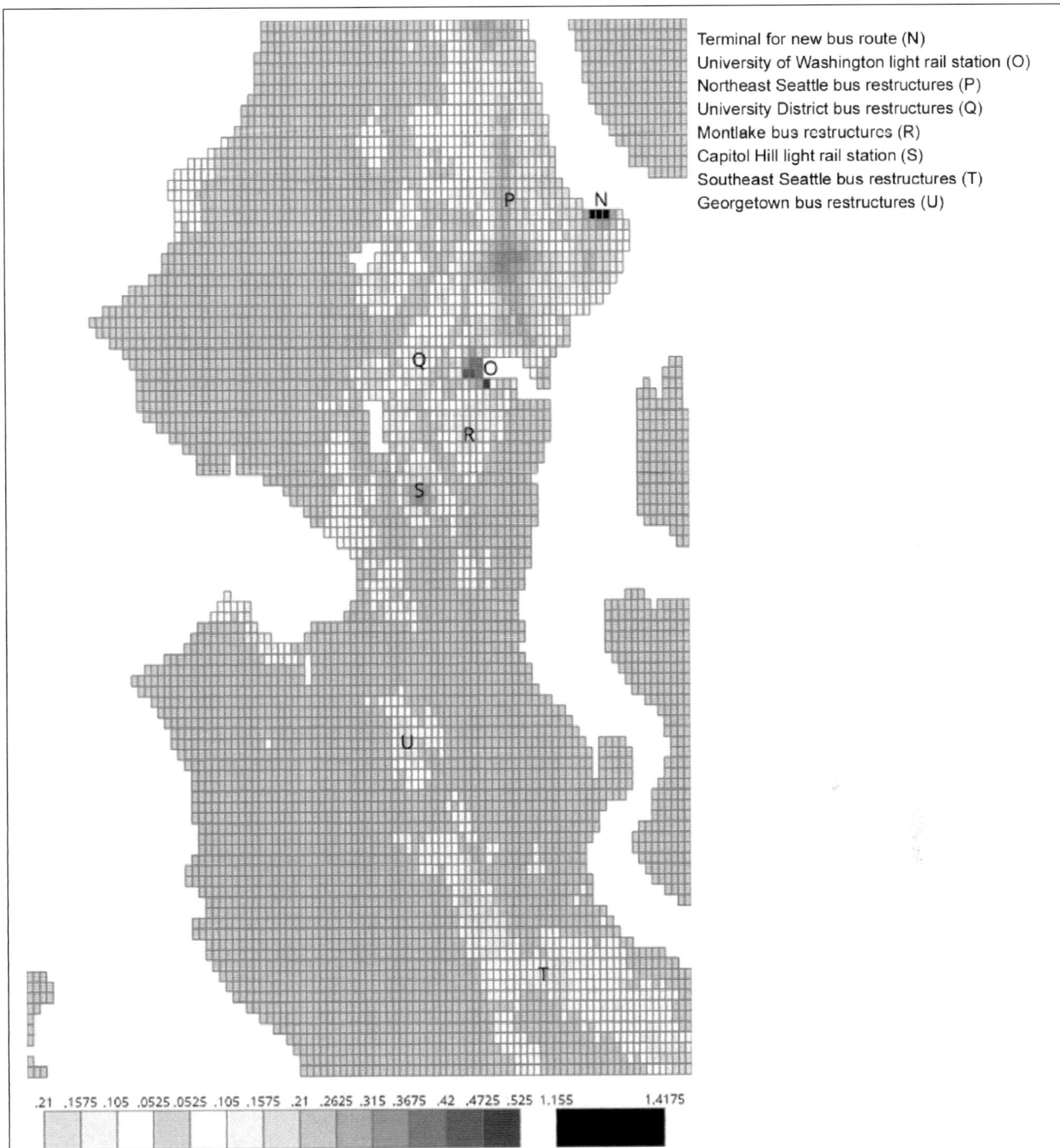

Figure 2. Comparative network accessibility in graphical form. Each sector is colored based on the ratio of change between January 25, 2016 and January 30, 2017 in the number of tasks that allow the sector to be reached within 30 minutes. Sectors more strongly orange are reachable under more circumstances on the latter date, those more strongly blue being reachable under fewer.

in this case being able to report that benefit has been realized unevenly despite the overall improvement. Furthermore, bus service restructures, intended only to reduce redundancy, harm spontaneous accessibility in some areas.

The spatial decomposition of the comparative network accessibility also allows planners to construct customized measurements. Though spontaneous sccessibility is intended to model trips with arbitrary origin and destination points,

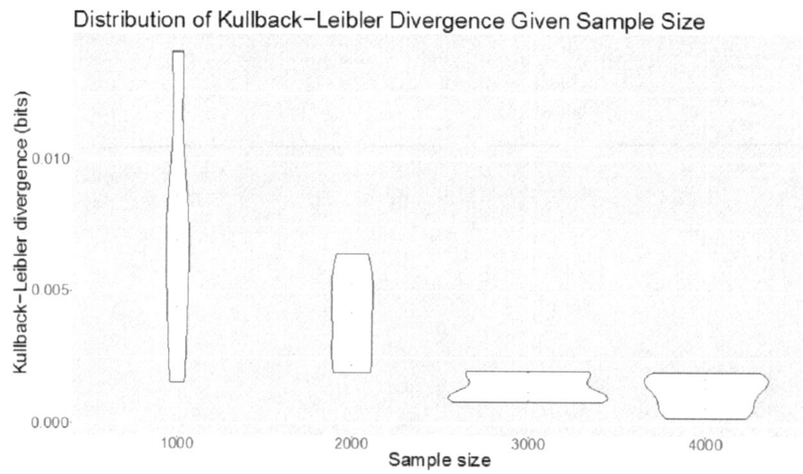

Figure 3. Violin plot showing the distribution of Kullback-Leibler divergence given the number of random sectors with which to calculate the sampled network accessibility.

planners managing limited resources may wish to prioritize service in areas that have historically generated many trips. In this case, planners can weigh the amount of change at individual sectors by a factor, such as relative measures of ridership, population density, or jobs, before averaging the amount of change. More simply, proposed changes that impact spontaneous accessibility negatively for a given set of vital sectors can be rejected on those grounds. Using decomposed measurements in conjunction with the overall change in network accessibility can balance planners' goals of improving the network for unexpected trips while maintaining its suitability for existing, predictable ones.

The results of the second analysis give planners guidance on using sampled network accessibility measurements. Figure 3 graphs the distribution of the Kullback-Leibler divergence of each of five samples at each sampling level. Kullback-Leibler divergence in all cases ranges from zero to one bit of information, but lacks an intuitive sense of scale for making value judgments of the quality of a distribution. For it to be of value, it is used comparatively. This analysis does not find a number of samples beyond which no benefit is gained from adding more. However, the primary benefit of increasing the sample count is lower variability, not necessarily a better sample. The lowest of the 1000-sector sample rivals the means of the 3,000- and 4,000-sector samples and has a value lower than any of the 2,000-sector samples. This result demonstrates that the efficacy of a sampled network accessibility for approximating the network accessibility has a strong dependency on sector selection. Producing a random sample of sectors and measuring the Kullback-Leibler divergence is a computationally inexpensive operation compared to a sampled network accessibility calculation. The result of this analysis suggests that a time-effective way to compare many network changes is to take several samples at a relatively low sample count and use the sample with the lowest Kullback-Leibler divergence regardless of its absolute value. Confirming this suggestion by examining the distributions of divergence using a different location or alternative sector sizes is a topic of further research.

The results also suggest that there is value in finding sampling techniques that select a small number of sectors in a way that minimizes the Kullback-Leibler divergence. This would allow much faster analysis of transportation network changes and thus more experimentation. Intuitively, such an algorithm would avoid the selection of sectors whose the reachability can be inferred from adjacent sectors. A sector selector that prioritizes sectors distant from already-selected ones accomplishes this to some extent, but does not account for sectors that differ from their neighbors considerably for geographic reasons. From a theoretical perspective, variational autoencoders (*32*) may provide a way to find such samples. This is a potential topic of future research.

Conclusion

By extending the concepts presented in a variety of previous accessibility-based studies of transit networks, this paper builds precise measurements for spontaneous accessibility—the ability to make unanticipated trips using public transit. Network accessibility captures the all-day, full-network accessibility of a public transit system. Its ability to make precise statements about incremental changes such as the Link extension and bus restructures in Seattle demonstrates that it possesses the granularity to be part of periodic evaluations of a transit network, rather than being saved for long-term planning. Though this study analyzes a transit network change that occurred in the past, planners can use a similar process to evaluate transit network modifications that they may be considering. When planners judge the merits of several alternatives, sampled network accessibility can reduce the time needed by allowing less computationally demanding but still sufficiently representative evaluations. Spontaneous accessibility measurements are not a wholesale replacement for existing transit planning technologies—they do not address vehicle capacity or incorporate observations of real riders. However, a measurement that encodes the ability of transit customers to make unanticipated trips has value as a supplement, as contemporary research has not emphasized it. By incorporating spontaneous accessibility into their planning process, planners can design transit networks that help those who do not have personal vehicles to enjoy the advantages of those who do.

Author Contributions

The author confirms contribution to the paper as follows: study conception and design; data collection; analysis and interpretation of results; and draft manuscript preparation. The author reviewed the results and approved the final version of the manuscript.

References

1. United States Census Bureau. *American FactFinder, B08201: Household Size by Vehicles*, https://factfinder.census.gov/bkmk/table/1.0/en/ACS/15_1YR/B08201/1600000US5363000, 2015.
2. Commute Seattle. *2016 Center City Commuter Mode Split Survey*. Commute Seattle, 2017.
3. Sivak, M. *Has Motorization in the U.S. Peaked? Part 4: Households without a Light-Duty Vehicle*. The University of Michigan Transportation Research Institute, Ann Arbor, MI, 2014.
4. Dial, R. B. Transit Pathfinder Algorithm. *Highway Research Record*, Vol. 205, 1967, pp. 67–85.
5. Tong, C. O., and A. J. Richardson. A Computer Model for Finding the Time-Dependent Minimum Path in a Transit System with Fixed Schedules. *Journal of Advanced Transportation*, Vol. 18, No. 2, 1984, pp. 145–161.
6. Crisalli, U., and L. Rosati. DY-RT: A Tool for Schedule-Based Planning of Regional Transit Networks. In *Schedule-Based Dynamic Transit Modeling Theory and Applications* (N. H. M. Wilson and A. Nuzzolo, eds.), Springer Science+Business, New York, NY, 2004, pp. 135–158.
7. Ayed, H., C. Galvez-Fernandez, Z. Habbas, and D. Khadraoui. Solving Time-Dependent Multimodal Transport Problems Using a Transfer Graph Model. *Computers & Industrial Engineering*, Vol. 61, 2011, pp. 391–401.
8. Delling, D., T. Pajor, and R. F. Werneck. Round-Based Public Transit Routing. *Transportation Science*, Vol. 49, No. 3, 2014, pp. 591–604.
9. Pas, E. I. The Urban Transportation Planning Process. In *Geography of Urban Transportation* (S. Hanson, ed.), 2nd ed. Guilford Press, New York, NY, 1995, pp. 53–77.
10. Hägerstand, T. What about People in Regional Science? *Papers in Regional Science*, Vol. 24, No. 1, 1970, pp. 7–24.
11. Hanson, S. The Context of Urban Travel Concepts and Recent Trends. In *The Geography of Urban Transportation* (S. Hanson and G. Giuliano, eds.), 3rd ed. Guilford Press, New York, NY, 2004, pp. 3–29.
12. Handy, S. L, and D. A. Niemeier. Measuring Accessibility: An Exploration of Issues and Alternatives. *Environment and Planning A*, Vol. 29, No. 7, 1997, pp. 1175–1194.
13. Geurs, K. T, and B. van Wee, Accessibility Evaluation of Land-Use and Transport Strategies: Review and Research Direction. *Journal of Transport Geography*, Vol. 12, 2004, pp. 127–140.
14. Bertolini, L., F. le Clercq, and L. Kopoen, Sustainable Accessibility: A Conceptual Frame-work to Integrate Transport and Land Use Plan-Making. Two Test-Applications in the Netherlands and a Reflection on the Way Forward. *Transport Policy*, Vol. 12, 2005, pp. 207–220.
15. Nyerges, T. L. GIS in Urban-Regional Planning. In *The Geography of Urban Transportation* (S. Hanson, ed.), 2nd ed. Guilford Press, New York, NY, 1995.
16. Tribby, C. A, and P. A. Zandbergen. High-Resolution Spatio-Temporal Modeling of Public Transit Accessibility. *Applied Geography*, Vol. 34, 2012, pp. 345–355.
17. Mavoa, S., K. Witten, T. McCreanor, and D. O'Sullivan. GIS Based Destination Accessibility via Public Transit and Walking in Auckland, New Zealand. *Journal of Transport Geography*, Vol. 20, 2012, pp. 15–22.
18. Silva, C, and P. Pinho. The Structural Access Layer (SAL): Revealing how Urban Structure Constrains Travel Choice. *Environment and Planning*, Vol. 42, No. 11, 2010, pp. 2735–2752.
19. O'Sullivan, D., A. Morrison, and J. Shearer. Using Desktop GIS for the investigation of Accessibility by Public Transport: An Isochrone Approach. *International Journal of Geographic Information Science*, Vol. 14, No. 1, 2000, pp. 85–104.
20. Lei, T. L, and R. L. Church. Mapping Transit-Based Access: Integrating GIS, Routes and Schedules. *International Journal of Geographical Information Science*, Vol. 24, No. 2, 2010, pp. 283–304.
21. Google Developers. *GTFS Static Overview*. https://developers.google.com/transit/gtfs/, 2016.
22. Anderson, P. A., A. Owen, and D. M. Levinson. The Time Between: Continuously-Defined Accessibility Functions for Schedule-Based Transportation Systems. *Presented at 92nd Annual Meeting of the Transportation Research Board*, Washington, D.C., 2013.
23. Owen, A, and D. Levinson. *Access to Destinations: Annual Accessibility Measure for the Twin Cities Metropolitan Region*. Final Report 2012-34. Minnesota Department of Transportation, St. Paul, MN, 2012.
24. Owen, A, and D. M. Levinson. Modeling the Commute Mode Share of Transit Using Continuous Accessibility to Jobs. *Transportation Research Part A*, Vol. 74, 2015, pp. 110–112.
25. Blanchard, S. D, and P. Waddell. Urban Access: Generalized Methodology for Measuring Regional Accessibility with an Integrated Pedestrian and Transit Network. *Transportation Research Record: Journal of the Transportation Research Board*, 2017. 2653: 35–44.
26. Gillespie, W, and P. Fahrenwald. Transit Access Measure: Incorporating Walk and Drive Access. *Transportation Research Record: Journal of the Transportation Research Board*, 2017. 2653: 82–92.
27. Conway, M. W., A. Byrd, and M. van der Linden. Evidence-Based Transit and Land Use Sketch Planning Using Interactive Accessibility Methods on Combined Schedule and Headway-Based Networks. *Transportation Research*

Record: Journal of the Transportation Research Board, 2017. 2653: 45–53.

28. Haklay, M., and P. Weber. OpenStreetMap: User-Generated Street Maps. *IEEE Pervasive Computing*, Vol. 7, No. 4, 2008, pp. 12–18.
29. Butler, H., M. Daly, A. Doyle, S. Gillies, and T. Schaub. *The GeoJSON Format Specification*, 2008.
30. Karich, P., S. Schröder, and M. Zilske. *GraphHopper*. https://github.com/graphhopper/graphhopper/tree/0.9, 2017.
31. Kullback, S., and R. A. Leibler. On Information and Sufficiency. *The Annals of Mathematical Statistics*, Vol. 22, No. 1, 1951, pp. 79–86.
32. Doersch, C. Tutorial on Variational Autoencoders. *arXiv:1606.05908 [stat.ML]*, 2016.

The Standing Committee on Transit Capacity and Quality of Service (AP015) peer-reviewed this paper (18-01522).

TRR
JOURNAL OF THE TRANSPORTATION RESEARCH BOARD

Article

Transportation Research Record
2018, Vol. 2672(8) 497–504

DOI: 10.1177/0361198118794291
journals.sagepub.com/home/trr
SAGE

Estimation of Passengers Left Behind by Trains in High-Frequency Transit Service Operating Near Capacity

Eli Miller[1], Gabriel E. Sánchez-Martínez[1], and Neema Nassir[1]

Abstract

Measuring rail system crowding is important to transit agencies. Crowding data has implications for safety, operations control, service planning, performance measurement, and customer information. This paper proposes a bi-level regression model that transit agencies can use to estimate the number of passengers left behind on a platform by high-frequency trains operating at capacity. Inputs to the model include the number of passenger arrivals between trains and train departure times, which are derived from automatic fare collection and vehicle location data. The data are used to calculate the proposed measure of cumulative capacity shortage, which is shown to have high correlation with the number of passengers left behind. A bi-level regression approach is introduced and applied to calibrate the model parameters based on manual counts of passengers left behind. A case study using data from the Chicago Transit Authority's Blue Line demonstrates promising results, with an adjusted coefficient of determination of 0.81. The model could be used for post-hoc analysis of crowding performance or, in the context of real-time operations monitoring, for near-term predictions of passengers left behind.

Public transit agencies are interested in measuring crowding on their rail systems for safety, operations control, service planning, performance measurement, and customer information. While many agencies currently have automated fare collection (AFC) systems installed in their rail stations, few have automated passenger counting systems on their rail cars. This configuration allows agencies to calculate the rate of passenger arrivals at platforms, but not crowding levels without conducting expensive manual surveys. Information on crowding levels is particularly important when trains running at capacity can leave passengers behind. The model proposed in this paper overcomes this gap, and could be employed by transit agencies to compute crowding levels using automated vehicle location (AVL) data and AFC data.

In this paper, we present a platform-specific left-behind (PSLB) linear regression model that can be used to estimate left-behinds using two simple inputs: automatically collected train departure and passenger entry data. The model can be applied to any rail system where this type of automatically collected data are available, including those with tap-in only fare collection systems. Additionally, the model can be applied to systems with station platforms that serve multiple lines and directions.

Crowding data has important implications for customer safety. This data is particularly critical for systems where excessive platform crowding necessitates metering passenger entries. Between March 2014 and February 2016, Transport for London reported temporarily preventing passengers from entering stations 1,078 times due to overcrowding. During this time period, passenger entry to Oxford Circus station was closed 219 times (nearly once every three days) (*1*). In 2016, New York City Transit (NYCT) developed a model to forecast platform loads over concerns that some of their platforms were becoming dangerously overcrowded (*2*). NYCT's platform loading model assumes sufficient train car capacities, and as a result, does not estimate passengers left behind on the platform due to high train car loads.

Most transit agencies' performance metrics focus on delays experienced by trains rather than customers. However, train and platform crowding data has implications for evaluating performance as perceived by passengers. Passengers may experience longer waiting times as the result of being left behind from a crowded train, or they may experience inconvenience as the result of on-board or platform congestion. Crowding data can help

[1]Massachusetts Institute of Technology, Cambridge, MA

Corresponding Author:
Address correspondence to Eli Miller: esm2af@mit.edu

agencies develop metrics that better capture how customers experience their system, by evaluating how often customers are left behind on the platform due to overcrowding and how long customers need to wait at a station before they can board a train with available space. Several agencies with tap-in/tap-out fare collection systems, such as Bay Area Rapid Transit (BART) and Washington Metropolitan Area Transit Authority (WMATA), have developed loading models that estimate train loads (*3*, *4*). However, none of these models estimate the number of passengers left behind on the platform due to excessive train car loads. Additionally, neither of these models is compatible with tap-in only rail systems.

Crowding data is also critical for service planning and scheduling. Information on passenger loads and crowding can help agencies make more informed changes to supplied capacity, be it through frequency or train length, to better meet demand and relieve crowding. Also, as a means for travel demand management, the train and platform crowding data can be released to passengers to allow them to make more informed choices when they use the system. Users with more flexible schedules could use this data to make their trips when or where the system is less congested, helping to decrease peak crowding. Transport for London recently developed a signage system that informs customers when their rail system experiences peak crowding levels (*5*).

Literature Review

Previous work on train crowding focuses on establishing excess crowding standards, guidelines on how to measure crowding, and the effects of overcrowding (*6–10*). Travel demand models and particularly transit network assignment models are conventionally used to estimate train loads and platform crowding conditions for planning purposes. These models are based on the economics principle of user equilibrium, which assumes that passengers optimize their path choices in a day-to-day learning process, based on their previous day's travel, including travel time, delay, convenience, etc. of the experienced path alternatives. Capacity-constrained transit assignment models can effectively capture the overcrowding and left-behind phenomena along with the associated delay or inconvenience, to incorporate in the prediction of passengers' path choices (*11–14*). However, conventional transit assignment models, despite their effectiveness for long-term and strategic planning applications, lack the desired precision that is necessary to generate short-term predictions or passenger advisory information. In addition, these models generally produce predictions for typical days (e.g., weekday peak hours) and are not designed for special event situations, disruption scenarios, or any kind of abnormality in the system which could result in excessive crowding and passengers left behind. AFC and AVL data are valuable sources of data that can be utilized to fill this gap. AFC data not only reflect the travel demand at a very granular level of space and time, but also present an unprecedented opportunity to analyze special events and incidents, anytime and anywhere in the network.

NYCT has built upon the existing transit assignment methodologies in the literature and utilized historic AFC and AVL data to estimate train loads in the New York City subway system. NYCT developed a network flow transit assignment model that estimates historic train and platform loads (*15*). However, this model is not designed to estimate crowding conditions in real time, due to the computational complexity of the model. In 2016, NYCT developed a simplified version of that model to forecast platform loads. However, the loading model assumes sufficient train car capacities, and as a result, does not estimate passengers left behind on the platform due to high train car loads (*2*). Zhu et al. developed a method for estimating the probability of being left behind at a specific station and time (*16*). This method is only compatible with a tap-in/tap-out fare system.

The PSLB model proposed in this paper offers a simpler and more general method for estimating passengers left behind that can easily be adapted to any rail system. Unlike the high computational demands required to run some network flow models, the computational simplicity of the PSLB model makes it amenable for producing near-term crowding predictions. The key contribution of this research is developing a general framework with a regression model to overcome the identified gap in the literature. The proposed model estimates the number of passengers left behind on the platform using automatically collected data of train departures and passenger entries.

The next section outlines the framework of the PSLB model. The details of the PSLB model are described in the model section. An example application of the model is then presented and its results are discussed. The last section concludes with a summary of contributions and a discussion of future work.

Framework

This research utilizes AFC and rail AVL data to estimate the number of passengers left behind on a specific platform by full trains. Both AVL and AFC systems store information in databases that can later be accessed for analysis. Rail AVL entries include a timestamp and signal block or location identifier. This data is used to determine when trains depart each station. AFC transaction records include a timestamp and faregate identifier. This

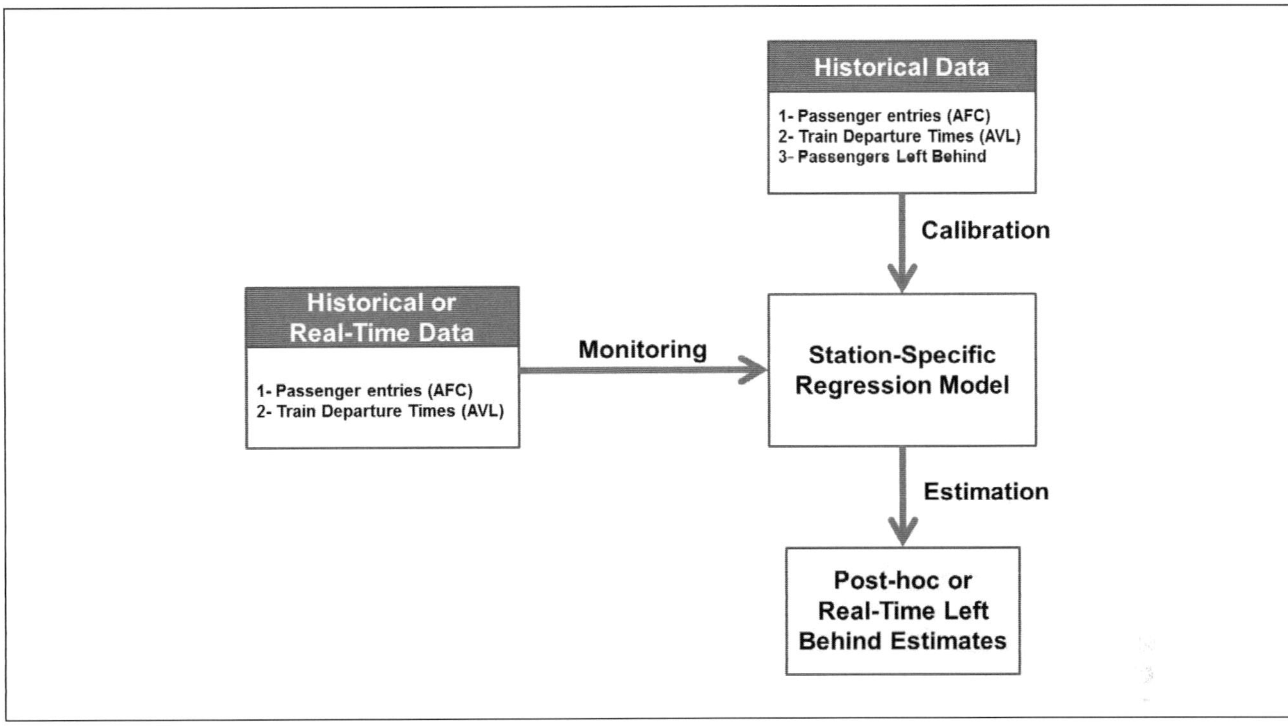

Figure 1. Platform-specific left-behind regression model framework.

data is used to determine the number of passengers that enter a particular station between train departures. The framework proposed in this paper feeds AVL and AFC data into a model that outputs the number of passengers left behind on the platform.

Figure 1 describes the framework for the PSLB regression model. There are two inputs for the model:

1. **Train departure times:** AVL data are used to determine when trains depart the station of interest and optionally upstream stations.
2. **Passenger entries:** AFC data are used to determine how many passengers enter a station between train departures. Survey data or an origin–destination–transfer inference model such as that by Sánchez-Martínez (*17*) can be used to split entries by direction. As with AVL, passenger entries at the station of interest and optionally stations upstream should be obtained.

The PSLB model is independently calibrated for each platform using manually collected left-behind counts, passenger entry data from AFC, and train departure times from AVL. The calibration involves estimating platform-specific model parameters, as described in the next section. The dependent variable in the model estimates the number of passengers left behind on the platform due to insufficient remaining capacity in the trains serving the platform. Once the model is calibrated, historic entries and departure times can be fed into the model to estimate historic numbers of passengers left behind on the platform. Alternatively, near-term forecasts of train departures and historic or real-time entries can be fed into the model to predict near-term numbers of passengers left behind on the platform.

Model

The PSLB regression model is a bi-level regression model, where the dependent variable is the number of passengers left behind by departing trains and the explanatory variable is the cumulative capacity shortage (CCS) for every train, as defined later in this section. This variable is calculated based on passenger arrival data, train departure data, and a proxy capacity parameter that must also be estimated. A bi-level optimization program minimizes the model's estimation error by finding the value for the platform-specific proxy capacity, in the upper level, and finding the linear regression parameter values in the lower level. This is illustrated in Figure 2.

In its simplest form, the PSLB model is a single-variable model, but it can also be expanded and improved by including one CCS variable for each upstream station. Both forms are formulated in this section and tested in the application section. The remainder

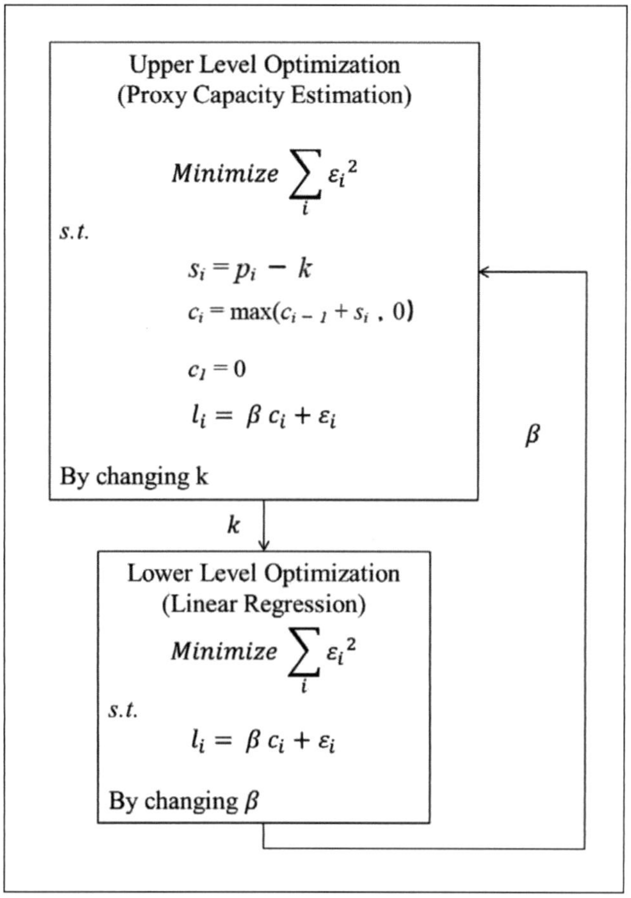

Figure 2. Bi-level regression model flowchart.

of this section introduces the concept of the CCS variable and describes how to calibrate the PSLB model using AVL data, AFC data, and manually collected observations of passengers left behind.

Cumulative Capacity Shortage

CCS captures the cumulative effect of trains arriving with insufficient remaining capacity to serve all passengers waiting to board on the platform. When the number of passengers intending to board a train exceeds the available capacity of the train, passengers are left behind on the platform, and the CCS increases. When the number of passengers intending to board a train is less than the available capacity, no passengers are left behind on the platform, and the CCS decreases. Capacity shortages accumulate because passengers left behind by a train increase the boarding demand for following trains. However, capacity surpluses do not accumulate, since any space for additional passengers on departing trains cannot be utilized later to reduce platform crowding. Therefore, CCS is always non-negative.

The available capacity of a train upon arriving at a platform and after alightings have taken place is an unknown quantity. Knowing the available capacity would allow the calculation of loads and left-behinds directly from platform arrivals. Instead, the model incorporates a proxy of available capacity, which is calibrated in the upper level of the estimation process.

These concepts can be defined mathematically as follows. Order train departures chronologically and refer to them by an index $i = 1, 2, \ldots, n$. Let p_i be the number of passengers arriving between the departures of trains $i-1$ and i, and intending to board train i. Let k be the available capacity proxy of trains, defined for each platform. The capacity shortage s_i of train i is

$$s_i = p_i - k \quad (1)$$

where
s = capacity shortage
i = train departure identifier
p = number of passengers arriving the platform between train departures
k = available capacity proxy of trains

The CCS for train i, denoted by c_i, can be calculated recursively as follows:

$$c_i = \max(c_{i-1} + s_i, 0) \quad \forall i \subset [2, n] \quad (2)$$

where
c = cumulative capacity shortage
n = number of train departures

$$c_1 = 0 \quad (3)$$

Because CCS is constructed to reflect the physics of train loading and the interactions between supply and demand under constrained capacity, it provides a nonlinear signal that is highly correlated with the actual number of passengers left behind, l_i. This can be exploited in a linear model to estimate platform crowding.

$$l_i = \beta c_i + \varepsilon_i \quad (4)$$

where
l = passengers left behind
β = regression coefficient
ε = error term

The regression parameter scales the signal to maximize estimation accuracy. The intercept is constrained to zero so that no left-behinds are predicted when CCS is zero.

Figure 3 illustrates how the CCS and its components change in response to the availability or shortage of capacity of each train. In this example, the available capacity proxy k, shown in black, is set to 313 passengers. The number of station entries, shown in yellow,

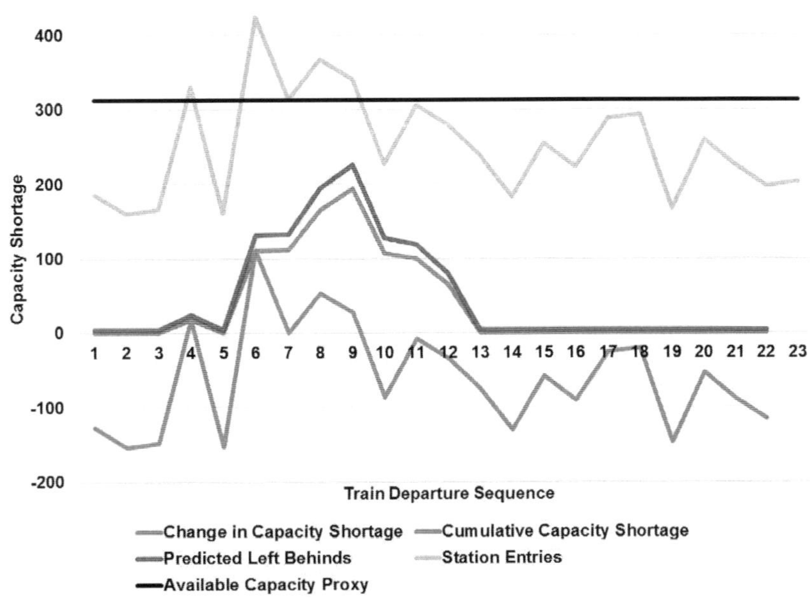

Figure 3. Station entries, change in capacity shortage, cumulative capacity shortage, and predicted left-behinds.

mostly falls below this capacity, but exceeds capacity on the 4th, 6th, 8th, and 9th train departures. The difference between the number of arriving passengers and the proxy capacity, shown in blue, drives the changes in CCS, shown in green. Negative changes in capacity shortage (or capacity surpluses) decrease CCS when it is positive, reflecting trains picking up passengers left behind by previous trains, but CCS never becomes negative. Positive changes in capacity shortage always increase CCS. The CCS value and the predicted number of passengers left behind, shown in red, become positive when the number of station entries between successive departures exceeds 313 passengers, but they do not necessarily return to zero once the number of station entries between successive departures drops below 313 passengers. This feature of the CCS parameter captures what is commonly observed after a platform accumulates a significant number of passengers left behind: it often takes several train departures to clear the platform. The CCS parameter captures passengers who may be denied boarding and incorporates them as additional boardings for subsequent trains. The regression coefficients are responsible for the difference in scale between the red and green lines in Figure 3.

Parameter Calibration

The PSLB model contains two parameters that require calibration based on observations of passengers left behind by full trains: the available capacity proxy k and the linear scale parameter of β. Calibration requires a manual data collection effort to count the number of passengers left behind by each train. These observations can later be paired with the number of passenger entries in AFC. Although the model can be estimated using the total number of station entries, its accuracy can be increased by splitting station entries by line and direction, either through manual observation or with an origin–destination–transfer model, and only considering passengers arriving at the platform of interest and destined for stations served by each train downstream. The parameters are calibrated to left-behind observations with the proposed bi-level regression. When recording left-behind observations, it is critical to distinguish (even for a platform serving a single branch and direction) between passengers who were left behind and those who are waiting on purpose. When a platform serves multiple branches, lines, or directions, this distinction becomes more important because far more passengers will choose not to board because they are waiting for a train of a different branch, line, or direction.

Once calibrated, the model estimates the number of passengers left behind at the platform by each train, given CCS data, which can be calculated recursively from AFC and AVL data. This can be done offline for post-hoc analysis or in real time for operations monitoring. If AFC data is not available in real time, it can be substituted by the typical number of station entries by time of day, which when combined with real-time AVL data allows estimation of the number of passengers arriving during each headway. The model could be used predictively if fed near-term forecasts of train movements.

Incorporating Upstream Information

The PSLB model can be extended to include information from upstream stations, which can increase accuracy by capturing how congested upstream stations may result in full trains arriving at the platform of interest. This requires collecting the number of passenger arrivals at each upstream platform for each train (from AFC data), and calculating CCS for each platform. A separate capacity proxy must also be estimated for each station in regression. The number of passengers left behind by each train is only required at the station of interest. For example, the PSLB model for station D including information about upstream stations A, B, and C would take the following form:

$$l_{Di} = \beta_A c_{Ai} + \beta_B c_{Bi} + \beta_C c_{Ci} + \beta_D c_{Di} + \varepsilon_{Di} \quad (5)$$

where
D = station D
A = upstream station A
B = upstream station B
C = upstream station C

Application

This section demonstrates the utility of the PSLB model with an application on the Chicago Transit Authority's (CTA) Blue Line. This line was selected because of its high ridership and reported high level of crowding. In 2016, the Blue Line provided 46.2 million rides or 150,000 rides daily (*18*). Clark/Lake was chosen as the station of interest due to its high entries, high dwell time, and high levels of crowding. Left-behind data was collected using platform video feeds for 15 weekday evening rush hour periods (16:30–18:30) in January 2017. Overall, data was collected for 465 O'Hare-bound train departures from Clark/Lake. Of these, 268,train departures left passengers behind on the platform due to lack of available capacity on the train.

Passenger entry data was collected using the agency's AFC system, Ventra. The entries were split by intended line and direction using an origin–destination–transfer model (*17*). Splitting the entries by line and direction is particularly critical at Clark/Lake station, as the station serves six different heavy rail lines during peak hours. Tables 1 and 2 summarize the regression results for the bivariate and multivariate PSLB models, respectively. The multivariate PSLB model includes information from Clark/Lake along with the two stations upstream from Clark/Lake for northbound Blue Line trains: Washington and Monroe. The model parameters β and k were estimated individually for Clark/Lake, Washington, and Monroe stations by regressing the CCS for each run (as it departed Clark/Lake, Washington, or Monroe) against the observed passengers left behind by the run at Clark/Lake. The adjusted R^2 increases from 0.75 to 0.81 when information from upstream stations (Washington and Monroe) is included.

Table 1. Regression Parameters and Statistics for Bivariate PSLB Model

Parameter	Value	Standard error	*T*-statistic
$\beta_{Clark/Lake}$	0.90	0.02	37.22
$k_{Clark/Lake}$	165		
Adjusted R^2	0.75		
N	464		

Table 2. Regression Parameters and Statistics for Multivariate PSLB Model

Parameter	Value	Standard error	*T*-statistic
$\beta_{Clark/Lake}$	0.43	0.05	9.20
$\beta_{Washington}$	0.64	0.06	10.69
β_{Monroe}	0.08	0.06	1.25
$k_{Clark/Lake}$	165		
$k_{Washington}$	139		
k_{Monroe}	134		
Adjusted R^2	0.81		
N	464		

Figure 4 compares the observed number of passengers left behind with the estimates generated by the multivariate PSLB model. The model is effective in estimating the number of passengers left behind. Differences between the model's estimated values and reality can be attributed to changes in effective train capacity due to the number of passengers with luggage, passengers using wheelchairs, and varying willingness to board a crowded train, which may depend on the displayed headway to the next train, time of day, and person-specific attitudes toward crowding.

Conclusion

The number of passengers left behind at platforms by full trains is the result of two factors: the number of passengers on the platform intending to board and the amount of space available in the train after alightings. The key contribution of this research is a platform-specific left-behind (PSLB) model that estimates the number of passengers left behind based only on the first of these two

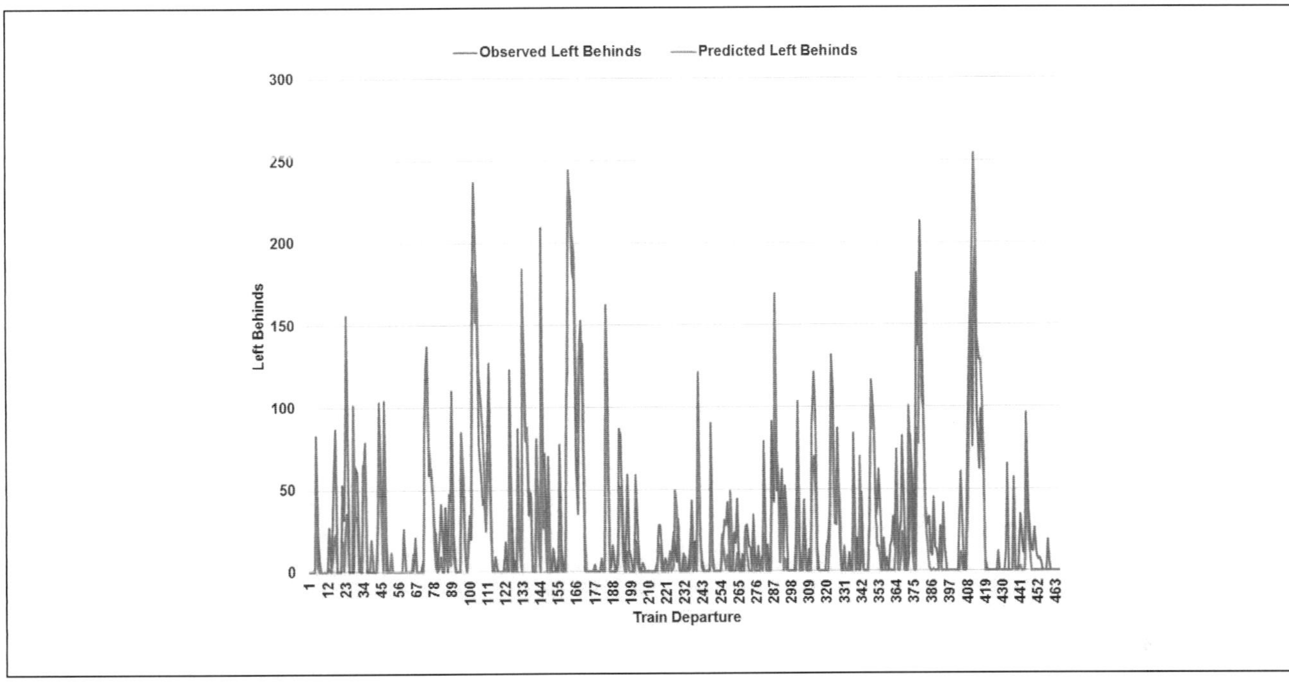

Figure 4. Left-behind passengers observed and estimated by the PSLB model.

factors. This is significant because, while many transit agencies have an AFC system that records station entries, very few agencies have automated passenger counting systems on their rail cars. The model can be applied to rail lines in which the fare system requires tapping in but not out.

This is accomplished by introducing the concept of cumulative capacity shortage (CCS), which reflects the capacity reduction effects that long headways and headway variability have on a transit line. The model is calibrated to the platform of interest through a bi-level regression of several parameter values to observations of passengers left behind (manually collected) paired with the number of passengers arriving at the platform or entering the station (from AFC data). Once calibrated, the model only requires automatically collected data to estimate the number of passengers left behind on the platform due to high train loads.

The PSLB model was tested with data from the CTA's Blue Line, producing relatively accurate estimates, considering the simple model structure. A modest improvement in estimation accuracy can be gained from including information from upstream stations. The model can be used for post-hoc analysis or for short-horizon predictions. The latter could be accomplished by substituting historic vehicle departure data with vehicle movement predictions, and historic passenger arrival data with typical passenger arrival rates. During special events, the model can be used with real-time AFC data, if available, or possibly with historic demand that is adjusted to the level of expected demand for the special event.

While the PSLB model is effective at predicting the number of passengers left behind on the platform at high-demand stations, the model is not sensitive enough to produce reliable left-behind estimates at low-demand stations where passengers are nonetheless sometimes left behind. These types of stations are often just downstream of high-demand stations that experience a greater number of passengers left behind. Since low-demand stations have few passengers left behind, the probability of being left behind depends more strongly on train loads than the number of passengers arriving at the platform during each headway. Research to develop a state-tracking model that explicitly estimates train loads at each station in parallel with estimating the number of passengers left behind is underway.

Acknowledgments

This research was funded by the Chicago Transit Authority (CTA). The authors thank the CTA for their generous support of the project.

Author Contributions

The authors confirm contribution to the paper as follows: study conception and design: Miller, Sánchez-Martínez, Nassir; data collection: Miller; analysis and interpretation of results: Miller,

Sánchez-Martínez, Nassir; draft manuscript preparation: Miller, Sánchez-Martínez, Nassir. All authors reviewed the results and approved the final version of the manuscript.

References

1. Proto, Laura. Revealed: London's Most Overcrowded Tube Stations. *London Evening Standard*, June 22, 2016. http://www.standard.co.uk/news/transport/revealed-londons-most-overcrowded-tube-stations-a3277706.html. Accessed July 17, 2017.
2. Caspari, A., B. Levine, J. Hanft, and A. Reddy. Real-Time Estimation of Platform Crowding for New York City Subway: Case Study at Wall St 2/3 Station in Financial District. *Transportation Research Record: Journal of the Transportation Research Board*, 2017. 2648: 68–78.
3. *BART.gov Launches New Estimated Train Crowding Feature*. Bay Area Rapid Transit, April 23, 2013. https://www.bart.gov/news/articles/2013/news20130423. Accessed July 17, 2017.
4. Jia, W., and M. Chow. How Crowded is Crowded? A Practitioner's Tool to Assessing Rail Congestion. Presented at 95th Annual Meeting of the Transportation Research Board, Washington, D.C., 2016.
5. *Improving Station Information*. Transport for London. https://tfl.gov.uk/corporate/about-tfl/how-we-work/planning-for-the-future/improving-station-information. Accessed July 17, 2017.
6. Transportation Research Board. *TCRP Report 165: Transit Capacity and Quality of Service Manual*, 3rd ed. Transportation Research Board of the National Academies, Washington, D.C., 2013.
7. Li, Z., and D. Hensher. Crowding in Public Transport: A Review of Objective and Subjective Measures. *Journal of Public Transportation*, Vol. 16, No. 2, 2013, pp. 107–134. http://dx.doi.org/10.5038/2375-0901.16.2.6
8. Veitch, T., J. Partridge, and L. Walker. Estimating the Costs of Over-crowding on Melbourne's Rail System. *Proc., Australian Transport Research Forum*, Brisbane, Queensland, Australia, 2013.
9. Berkovich, A., A. Lu, B. Levine, and A. Reddy. Observed Customer Seating and Standing Behavior and Seat Preferences on Board Subway Cars in New York City. *Transportation Research Record: Journal of the Transportation Research Board*, 2013. 2353: 33–46.
10. O'Malley, T., and M. Vaishnav. Is This Seat Taken? Multifaceted Research Study To Inform the Chicago, Illinois, Transit Authority's Future Rail Car Seating Design. *Transportation Research Record: Journal of the Transportation Research Board*, 2014. 2419: 33–39.
11. Poon, M., S. Wong, and C. Tong. A Dynamic Schedule-Based Model for Congested Transit Networks. *Transportation Research Part B: Methodological*, Vol. 38, No. 4, 2004, pp. 343–368. https://doi.org/10.1016/S0191-2615(03)00026-2
12. Hamdouch, Y., H. Ho, A. Sumalee, and G. Wang. Schedule-Based Transit Assignment Model with Vehicle Capacity and Seat Availability. *Transportation Research Part B: Methodological*, Vol. 45, No. 10, 2011, pp. 1805–1830. https://doi.org/10.1016/j.trb.2011.07.010
13. Leurent, F., E. Chandakas, and A. Poulhes. A Passenger Traffic Assignment Model with Capacity Constraints for Transit Networks. *Procedia-Social and Behavioral Sciences*, Vol. 54, 2012, pp. 772–784. https://doi.org/10.1016/j.sbspro.2012.09.794
14. Nuzzolo, A., U. Crisalli, and L. Rosati. A Schedule-Based Assignment Model with Explicit Capacity Constraints for Congested Transit Networks. *Transportation Research Part C: Emerging Technologies*, Vol. 20, No. 1, 2012, pp. 16–33. https://doi.org/10.1016/j.trc.2011.02.007
15. Stasko, T., B. Levine, and A. Reddy. Time-Expanded Network Model of Train-Level Subway Ridership Flows Using Actual Train Movement Data. *Transportation Research Record: Journal of the Transportation Research Board*, 2016. 2540: 92–101.
16. Zhu, Y., H. Koutsopoulos, and N. Wilson. Inferring Left Behind Passengers in Congested Metro Systems from Automated Data. *Transportation Research Procedia*, Vol. 23, 2017, pp. 362–379. http://dx.doi.org/10.1016/j.trpro.2017.05.021
17. Sánchez-Martínez, G. Inference of Public Transportation Trip Destinations by Using Fare Transaction and Vehicle Location Data: Dynamic Programming Approach. *Transportation Research Record: Journal of the Transportation Research Board*, 2017. 2652: 1–7. http://dx.doi.org/10.3141/2652-01
18. *Annual Ridership Report, Calendar Year 2016*. Ridership Analysis and Reporting, Chicago Transit Authority, February 1, 2017. http://www.transitchicago.com/assets/1/ridership_reports/2016_Annual_-_Final.pdf. Accessed July 17, 2017.

The Standing Committee on Transit Capacity and Quality of Service (AP015) peer-reviewed this paper (18-01557).

Article

TRR
JOURNAL OF THE TRANSPORTATION RESEARCH BOARD

Transportation Research Record
2018, Vol. 2672(8) 505–510

DOI: 10.1177/0361198118777078
journals.sagepub.com/home/trr
SAGE

Modeling the Influence of Weather on Transit Ridership: A Case Study from Brisbane, Australia

Ming Wei[1], Jonathan Corcoran[1], Thomas Sigler[1], and Yan Liu[1]

Abstract

This paper examines the influence that weather conditions exert on transit ridership in sub-tropical Brisbane, Australia. Drawing on transit smart card and weather station records covering a period of six months, the weather-transit ridership relationship across three different travel modes is modeled. Our results reveal that weather does impose an effect on transit ridership and that its influence varies by both transit mode and across weekdays and weekends. On weekdays, the heat index, an indicator of human-perceived temperature, was found to exert a negative effect on bus and train ridership, while ferry ridership was shown to be most sensitive to rainfall. On weekends, the ridership across all three transit modes decreased on rainy days. Results from quantile regression suggest that on weekdays during periods of higher ridership the influence of weather on transit ridership becomes stronger, while on weekends the influence of weather is lessened. We argue that our findings are important for transit agencies in their capacity to unveil how an important predictor of ridership variation, weather, exerts an influence across a transit system. These preliminary findings mark the first step in assembling the necessary evidence to build a more weather-resilient transit system.

It is well established in the literature that the weather can significantly affect our daily travel behavior (*1*). Extreme weather, such as heavy precipitation or extreme temperatures, has been shown to have diverse effects such as reducing travel demand, increasing traffic congestion and elevating the risk of traffic accidents (*2*, *3*). Set against the background of climate change, which points to the intensification of extreme weather with increasing frequency collectively, understanding the relationship between weather events of various kinds and travel behavior has become increasingly salient.

In recent decades, increasing attention has been paid to the influence of weather on travel behavior. In one study by Keay and Simmonds (*4*), a negative relationship between rain and traffic volume was highlighted, and that more intense rainfall led to a larger reduction in traffic volume. In another study, Thomas et al. showed that most daily fluctuations in cycling demand could be explained by weather conditions and that recreational demand was much more sensitive to weather than utilitarian demand (*5*).

Despite this growing body of literature on the influence of weather on travel behavior, there appears to be several inconsistencies in their findings. For example, some studies support the notion that rain exerts a negative influence on the traffic volume (*6*–*8*), whereas others find that rain is positively associated with traffic volume (*9*, *10*). One plausible reason for the inconsistent findings could be that individuals in different climatic contexts exhibit differing behavioral patterns. For example, precipitation may induce commuters to drive instead of taking mass transit where moisture poses an inconvenience, but the opposite may be true where precipitation poses a hazard on the roadways, such as when flooding or ice are present.

A relatively large body of work has focused on the influence of weather on private and non-motorized transport modes (*11*), with comparatively less focus being placed on the way that weather exerts an effect on public transport usage. Given the imperative progress toward more sustainable urban forms of transport, we would argue that there is a need to better understand the way in which weather influences our use of public transport, and this is the focus of this paper.

This study draws on Brisbane, Australia as the case study context, and aims to redress several research gaps, and to extend our understanding on the influence of weather on transit ridership. We draw on transit smart card data and local weather station records from November 2012 to April 2013 and apply a suite of negative binomial regression models coupled with negative binomial quantile regression models to capture the weather-transit ridership relationship.

The remainder of the paper is organized as follows: in the next section, a brief review of the extant weather-transit ridership relationship research is discussed. Then the case

[1]School of Earth and Environmental Sciences, The University of Queensland, QLD, Australia

Corresponding Author:
Address correspondence to Ming Wei: m.wei@uq.edu.au

study context, data and empirical approach are described. Subsequently, the modeling results are presented followed by a discussion of their importance before offering some tentative conclusions and charting some avenues for future research.

Background Literature

Evidence has highlighted that certain weather conditions (e.g., heatwaves and storms) exert an important effect on public transport usage (*1*, *12*). In previously published research, rain, temperature and wind have been the most examined weather variables in relation to their effects on transit ridership (*1*, *2*). The following sections discuss the body of evidence assembled to date as it relates to the influence of these three weather parameters on transit ridership.

Rain

In most previous studies, rain has been found to exert a negative influence on transit ridership. Changnon examined the effect of summer rainfall on daily bus ridership in Chicago and identified a ridership reduction on rainy days (*13*). Stover and McCormack investigated the impact of rain on bus ridership in Pierce County, Washington using ordinary least squares (OLS) regression (*14*). It was found that rain negatively influenced bus ridership across all four seasons. One inch of rain can lead to a decrease in ridership of 3,650 people (9.73%) for spring, the highest of the seasons. Guo et al. examined the effect of five weather elements (temperature, rain, snow, wind, and fog) on daily bus and train ridership using linear regression models (*15*). Rain has shown a negative effect on transit ridership and the influence is stronger on weekends than on weekdays, and stronger for bus than for train. However, some studies argue that the effect of rain on transit ridership is negligible. For example, Kashfi et al. explored the relationship between bus ridership and weather conditions in Brisbane for a one-year period and found that rain has insignificant influence on the ridership (*16*). Similar empirical results can also be found in a recent study based in Shenzhen, China (*11*).

Temperature

Previous research has mainly reported two different roles of temperature in influencing transit ridership, namely positive effects and insignificant effects. In a case study from Chicago, Guo et al. (*15*) reported that temperature exerted a positive influence on bus and train ridership. A 1°C increase of temperature resulted in a weekday ridership increase of between 400 and 700 people for bus, and between 300 and 500 people for rail. For Gipuzkoa, Spain, Arana et al. suggested that a one-degree temperature rise could increase bus ridership by 474 people on a Saturday and 491 on a Sunday (*17*). In a case study from Washington, Stover and McCormack suggested that temperature only had positive effect in winter and no significant effect had been identified in other seasons (*14*). In New York City, Singhal et al. (*18*) found that daily subway ridership did not have a significant response to the temperature variation. Two recent studies developed in a subtropical climate context both reported that temperature had negligible influence on transit ridership (*11*, *19*).

Wind

Although wind was found to exert negative effects on active transportation, its influence on transit ridership are rather more complex (*2*). Guo et al. reported that wind speed is negatively associated with bus ridership and a 1 km/h increase in wind speed could result in 1290–2010 decrease in bus ridership per day, while the influence on train ridership is negligible (*15*). Arana et al. found that in Gipuzkoa, Spain, 1 km/h increase in wind speed could cause decreases in bus ridership of 115 people on a Saturday, but the influence on Sunday turned out be quite small (*17*). Singhal et al. claimed that at a hourly level the wind had negative effects on bus ridership while at a daily level the impacts are insignificant (*18*). In addition, some researchers claimed that there was no association between wind speed and transit ridership (*16*, *20*). From the empirical study in Washington state, results showed that wind speed was significant in decreasing bus ridership in winter, spring and autumn, but not in summer (*14*) . By applying the transit smart card data from Shenzhen, China, Zhou et al. reported that wind had different influences between weekdays and weekends (*11*). On weekdays, wind was shown to negatively influence bus and subway ridership; on weekends, the wind effect was found to be positive.

Overall, the weather-transit ridership literature has yielded several inconsistent findings. To this end, there is need for further studies to add to this growing literature in an attempt to solidify our understanding of this complex relationship. Developing this more robust understanding will be the first step to assembling the necessary evidence on which transit operators and policy makers will be able to (re)design more weather-resilient transit systems.

Study Context and Data

Brisbane, located in the southeast corner of Queensland, Australia, has a resident population in excess of 2 million inhabitants, and is the state capital and the most populous city in the State, and the third largest city in Australia. Brisbane is situated in the maritime subtropical climate region and has a mild winter and a hot humid summer. January is the hottest month with an average temperature of 26 °C (79°F) and the coldest is July at 15 °C (59°F). From November 2012 to April 2013, the average temperature was

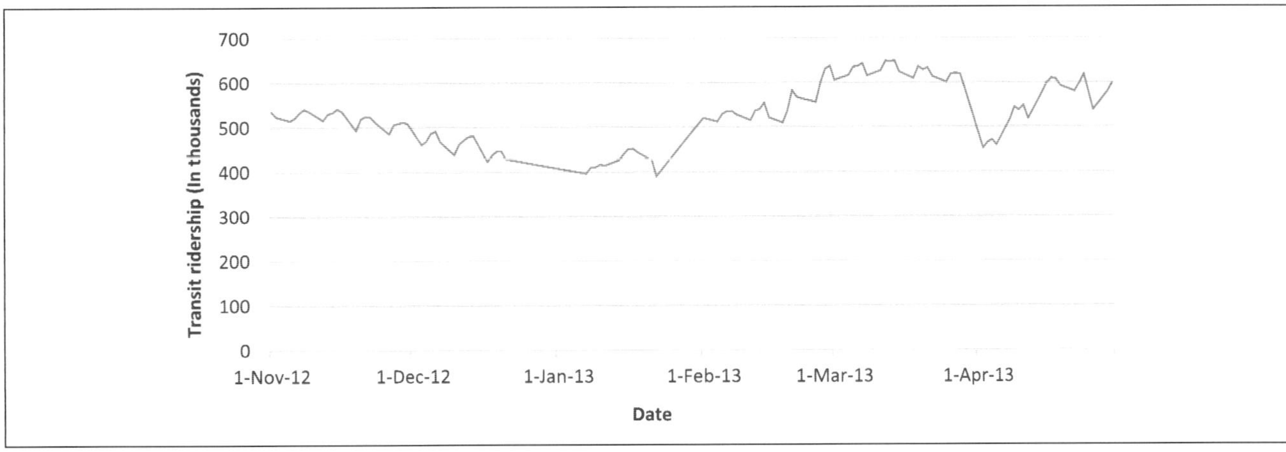

Figure 1. Weekday transit ridership from November 2012 to April 2013.

24.3 °C (75.7°F), the average relative humidity is 68.6% and the cumulative rainfall is around 560 mm. A similar climate can be observed in south-eastern parts of Brazil, China, and the United States.

To investigate the weather-transit ridership relationship, two main datasets are employed in this study, including transit smart card data and weather station records. Both datasets were collected from November 2012 to April 2013. To compare the influence of weather on transit ridership between weekdays and weekends, 109 weekdays and 50 weekend days in the six months were identified.

Transit ridership across three modes at the daily level were obtained from transit smart card data. In Brisbane, the transit smart card was implemented in 2008 across all public transit modes including bus, train and ferry. As an automatic fare collection method, a transit smart card requires passengers to touch their card at a card reader each time they start and finish a trip (*21*). The convenience and price advantage over paper tickets means that transit smart cards record over 80% trips using public transit services in Brisbane (*22*). From November 2012 to April 2013, on every weekday nearly 600,000 trips were generated and more than 250,000 passengers used their transit smart card. To ensure passenger privacy, the card type is not available in this study. Figure 1 presents the weekday transit ridership recorded by transit smart card across the six months.

The weather information required from weather station records contains rain (mm), temperature (°C), wind speed (km/h) and relative humidity (%). In this study, four weather parameters were selected to examine their influence on transit ridership, including rain, temperature, wind speed and heat index. It is worth noting that heat index, an indicator of apparent temperature capturing the temperature equivalent perceived by human, is calculated based on the heat index equation considering the combined effects of temperature and relative humidity. The heat index equation is as follows:

$$\begin{aligned} HI = & -42.379 + 2.04901523*T + 10.14333127*RH \\ & - 0.22475541*T*RH - 0.00683783*T*T \\ & - 0.05481717*RH*RH + 0.00122874*T*T*RH \\ & + 0.00085282*T*RH*RH - 0.00000199*T*T*RH*RH \\ & - Adjustment\ 1 - Adjustment\ 2 \end{aligned}$$

$$Adjustment\ 1 = \left[(13 - RH)/4\right] * \sqrt{(17 - |T - 95|)/17}$$

$$Adjustment\ 2 = \left[(RH - 85)/10\right] * \left[(87 - T)/5\right]$$

In which T is the temperature in degrees F and RH is relative humidity in percent. HI is the heat index expressed as an apparent temperature in °F. Adjustment 1 exists on the condition that the RH is less than 13% and the temperature is between 80 and 112 °F. Adjustment 2 is valid on the condition that the RH is greater than 85% and the temperature is between 80 and 87 °F.

Methodology

Negative Binomial Regression

For the purposes of investigating the weather-transit ridership relationship, we applied regression analysis, taking the transit ridership as the dependent variable and weather parameters as independent variables. Given transit ridership is a kind of count data, the negative binomial model and Poisson model are the ideal model candidates for modeling those count variables (*23*). By testing and comparing the two models, it was found that the negative binomial regression model showed a better fit in all three transit modes across weekdays and weekends. The negative binomial regression model is formulated as follow ($Y_{1,2,3}$ denotes the ridership for three different transit modes):

$$\begin{aligned} Y_{1,2,3} = & \exp(\alpha + \beta_1 Rain + \beta_2 Temperature \\ & + \beta_3 Wind + \beta_4 HeatIndex) \end{aligned}$$

Table 1. Modeling Results on the Influence of Weather on Weekday Transit Ridership

Transit mode	Weather parameter	Negative binomial (mean)	Negative binomial (quantile) 0.25	0.5	0.75
Bus	Rain	−0.0013	0.0010	−0.0005	−0.0020
	Temperature	0.0112	−0.0242	0.0119	0.0376
	Wind	−0.0021	−0.0078	−0.0094	0.0005
	Heat index	−0.0270***	−0.0249	−0.0249***	−0.0394***
Train	Rain	−0.0012	−0.0017	−0.0006	−0.0013
	Temperature	0.0141	0.0137	0.0133	0.0276*
	Wind	−0.0002	−0.0023	−0.0014	−0.0018
	Heat index	−0.0217***	−0.0174***	−0.0202**	−0.0269***
Ferry	Rain	−0.0023*	−0.0019*	−0.0026*	−0.0044*
	Temperature	−0.0116	−0.0100	−0.0014	−0.0160
	Wind	−0.0038	−0.0055	0.00154	−0.0036
	Heat index	−0.0060	−0.0049	−0.0098	0.0001

Note: ***, **, *denote the significance at the 1%, 5% and 10% level, respectively.

Table 2. Modeling Results on the Influence of Weather on Weekend Transit Ridership

Transit mode	Weather parameter	Negative binomial (mean)	Negative binomial (quantile) 0.25	0.5	0.75
Bus	Rain	−0.0015**	−0.0026**	−0.0009*	−0.0007*
	Temperature	−0.0580	−0.0548	0.0119	−0.0614
	Wind	0.0283*	0.0323*	0.0533	0.0246
	Heat index	0.0228	0.0273	0.0359	0.0310
Train	Rain	−0.0005*	−0.0008**	−0.0006*	−0.0004
	Temperature	−0.0427	−0.0569	−0.1348	−0.0451
	Wind	0.0422*	0.0446*	0.0437*	0.0460
	Heat index	0.0230	0.0251	0.0670	0.0222
Ferry	Rain	−0.0072**	−0.0089**	−0.0069**	−0.0056**
	Temperature	−0.0586	−0.0428	−0.0672	−0.0309
	Wind	0.0277	−0.0105	0.00254	0.0282
	Heat index	0.0630	0.0849	0.0198	0.0102

Note: ***,**, *denote the significance at the 1%, 5% and 10% level, respectively.

Negative Binomial Quantile Regression

Quantile regression is a well-established complement to regression at the mean in its capacity to explore how independent variables explain variations in the dependent variable at different points in the dependents' distribution (*24*, *25*). When considering the dependents' distribution, quantile regression offers a more comprehensive characterization of the relationships between the dependent and independent variables and captures the different effects of the independent variables. In this study, the use of quantile regression is applied to examine how the influence of the various weather variables explains variations in transit ridership at different levels of the ridership. This is especially important given that transit ridership is known to experience large fluctuations across a day and week (Figure 1).

Results

From the negative binomial regression results, it is apparent that weather exerts different influences on transit ridership by mode and the influence also varies between weekdays (Table 1) and weekends (Table 2). On weekdays, the heat index is found to negatively influence bus and train ridership, rain has been shown to only exert a negative effect on ferry ridership. While on weekends, it shows that the ridership of bus, train and ferry all decrease in the presence of rainfall, wind is positively associated with bus and train ridership, and heat index becomes insignificant in influencing the transit usage. As rain has shown significant influence on ferry ridership on both weekdays and weekends, by comparing the coefficients, it is apparent that the influence of rain is stronger on weekends than that on weekdays, and on weekends ferry ridership is most sensitive to rainy conditions, and train ridership is the least affected. In addition, in this study case, air temperature does not show a significant impact on transit ridership on both weekdays and weekends.

In this study, quantile regression is based on the negative binomial regression and presents the results for the 0.25, 0.5 and 0.75 quantiles. From the results, firstly, it is apparent that the quantile regression coefficients are quite different from the negative binomial coefficients at the upper and lower quantiles, indicating that the influence of weather on transit ridership is not consistent in response to the ridership fluctuations. Secondly, on weekdays, the coefficients of the heat index become larger (in a negative direction) from the lower quantile to the upper quantile, indicating that the influence of heat index tends to become stronger with higher bus and train usage. In addition, an insignificant result of heat index on bus ridership at the 0.25 quantile indicates that on a low bus usage day the heat index is negligible in influencing the ridership. As for the coefficients of rain on ferry ridership, it follows the same trend as with the heat index on the bus and train ridership. This demonstrates that on weekdays the influence of weather becomes stronger when the ridership becomes higher. Thirdly, on weekends, the trend of the coefficients shows a different direction. The coefficients of rain are larger at the 0.25 quantile than the upper quantiles, indicating that the negative rain influence is stronger on the days with low transit usage. The coefficients of wind on bus ridership and train ridership show that positive wind influence is only significant at lower quantiles. The results reveal another important difference in the weather-transit ridership relationship between weekdays and weekends.

Conclusion

This paper investigates the influence of weather on transit ridership in Brisbane, Australia, using weather station records and transit smart card data. Employing negative binomial regression at the mean and three quantiles (25th, 50th and 75th), the role of weather on transit ridership and its variation by mode and weekday/weekend were modeled.

The results from the modeling exercise reveal that the influences of weather on transit ridership differ between weekdays and weekends and vary by transit mode. On weekdays, bus and train passengers are shown to be sensitive to variations of perceived temperature, hot and wet conditions, resulting in marked decreases in bus and train ridership. On weekends, bus and train ridership are shown to increase on windy days. One plausible reason for the increase could be the consequence of modal substitution in which individuals switch from active to motorized modes, as we know from previous research that active transport is particularly vulnerable to weather conditions (*26–28*). In this study, rain was found to be the most influential weather parameter in shaping variations in transit ridership with the most marked effects being on weekdays for ferry-based travel and for all modes on weekends.

The quantile regression results show that on weekdays, the influence of weather on transit ridership becomes stronger at higher levels of ridership. However, the results on weekends are in the inverse, in which weather was found to exert a stronger influence at lower levels of ridership. Given that on weekdays, transit ridership is highly related to the academic year, high levels of ridership normally appear during the academic term. The result, stronger weather influence at higher levels of ridership on weekdays, indicates that the response of passengers to weather change may vary by passenger type and student passengers could be more sensitive to weather conditions. As for the results on weekends, these could be explained by major events such as fireworks, festivals, sporting events, as high transit ridership is less sensitive to weather change.

There are two lines of enquiry that appear fruitful to progress in future research. The first is related to the temporal scale. Common to our study, most extant research examines the weather-transit ridership relationship using aggregate daily transit records. Investigating the weather-transit ridership relationship at a finer temporal scale could be an avenue for future studies to unpack the way in which weather might impose an impact on transit ridership at say an hourly level, for example, how would a dry morning, but a wet afternoon impact transit ridership. The second relates to the theoretical framework used to explain the influence of weather on transit ridership. So far, many studies have presented the results of empirical analyses without a comprehensive consideration for the mechanisms controlling travel behavior and the way in which weather influences these mechanisms. Some preliminary work has begun in this are (*29*). However, there remains a need to develop a theoretical framework that has the agility to capture the complex weather-transit ridership relationship from which future weather-transit studies can be grounded.

To conclude, the findings of this study enrich our understanding of the weather-transit ridership relationship and contribute to the assembly of a new evidence base with the capacity to inform the (re)design of a more weather-resilient transit system.

References

1. Böcker, L., M. Dijst, and J. Prillwitz. Impact of Everyday Weather on Individual Daily Travel Behaviours in Perspective: A Literature Review. *Transport Reviews*, Vol. 33, No. 1, 2013, pp. 71–91.
2. Liu, C., Y. O. Susilo, and A. Karlström. Weather Variability and Travel Behaviour – What We Know and What We Do Not Know. *Transport Reviews*, Vol. 37, No. 6, 2017, pp. 715–741.
3. Lam, W. H., H. Shao, and A. Sumalee. Modeling Impacts of Adverse Weather Conditions on a Road Network with Uncertainties in Demand and Supply. *Transportation Research Part B: Methodological*, Vol. 42, No. 10, 2008, pp. 890–910.

4. Keay, K., and I. Simmonds. The Association of Rainfall and Other Weather Variables with Road Traffic Volume in Melbourne, Australia. *Accident Analysis and Prevention*, Vol. 37, No. 1, 2005, pp. 109–124.
5. Thomas, T., R. Jaarsma, and B. Tutert. Exploring Temporal Fluctuations of Daily Cycling Demand on Dutch Cycle Paths: The Influence of Weather on Cycling. *Transportation*, Vol. 41, No. 1, 2013, pp. 1–22.
6. Edwards, J. B. Speed Adjustment of Motorway Commuter Traffic to Inclement Weather. *Transportation Research Part F: Psychology and Behaviour*, Vol. 2, No. 1, 1999, pp. 1–14.
7. Maze, T. H., M. Agarwal, and G. Burchett. Whether Weather Matters to Traffic Demand, Traffic Safety, and Traffic Operations and Flow. *Transportation Research Record: Journal of the Transportation Research Board*, 2006. 1948: 170–176.
8. Kalkstein, A. J., M. Kuby, D. Gerrity, and J. J. Clancy. An Analysis of Air Mass Effects on Rail Ridership in Three US Cities. *Journal of Transport Geography*, Vol. 17, No. 3, 2009, pp. 198–207.
9. Sabir, M., M. J. Koetse, and P. Rietveld. *The Impact of Weather Conditions on Mode Choice: Empirical Evidence for the Netherlands*. Department of Spatial Economics, VU University, Amsterdam, 2008.
10. Akin, D., V. P. Sisiopiku, and A. Skabardonis. Impacts of Weather on Traffic Flow Characteristics of Urban Freeways in Istanbul. *Procedia - Social and Behavioral Sciences*, Vol. 16, 2011, pp. 89–99.
11. Zhou, M., D. Wang, Q. Li, Y. Yue, W. Tu, and R. Cao. Impacts of Weather on Public Transport Ridership: Results from Mining Data from Different Sources. *Transportation Research Part C: Emerging Technologies*, Vol. 75, 2017, pp. 17–29.
12. Chapman, L. Transport and Climate Change: A Review. *Journal of Transport Geography*, Vol. 15, No. 5, 2007, pp. 354–367.
13. Changnon, S. A. Effects of Summer Precipitation on Urban Transportation. *Climatic Change*, Vol. 32, No. 4, 1996, pp. 481–494.
14. Stover, V. W., and E. D. McCormack. The Impact of Weather on Bus Ridership in Pierce County, Washington. *Journal of Public Transportation*, Vol. 15, No. 1, 2012, p. 6.
15. Guo, Z., N. H. M. Wilson, and A. Rahbee. Impact of Weather on Transit Ridership in Chicago, Illinois. *Transportation Research Record: Journal of the Transportation Research Board*, 2007. 2034: 3–10.
16. Kashfi, S. A., J. M. Bunker, and T. Yigitcanlar. Understanding the Effects of Complex Seasonality on Suburban Daily Transit Ridership. *Journal of Transport Geography*, Vol. 46, 2015, pp. 67–80.
17. Arana, P., S. Cabezudo, and M. Peñalba. Influence of Weather Conditions on Transit Ridership: A Statistical Study Using Data from Smartcards. *Transportation Research Part A: Policy and Practice*, Vol. 59, 2014, pp. 1–12.
18. Singhal, A., C. Kamga, and A. Yazici. Impact of Weather on Urban Transit Ridership. *Transportation Research Part A: Policy and Practice*, Vol. 69, 2014, pp. 379–391.
19. Kashfi, S. A., J. M. Bunker, and T. Yigitcanlar. Modelling and Analysing Effects of Complex Seasonality and Weather on an Area's Daily Transit Ridership Rate. *Journal of Transport Geography*, Vol. 54, 2016, pp. 310–324.
20. Böcker, L., M. Dijst, J. Faber, and M. Helbich. En-route Weather and Place Valuations for Different Transport Mode Users. *Journal of Transport Geography*, Vol. 47, 2015, pp. 128–138.
21. Wei, M., and J. Zhou. Central Places of Transit Riders: A Visual Story of Brisbane, Australia. *Regional Studies, Regional Science*, Vol. 3, No. 1, 2016, pp. 531–533.
22. Wei, M., Y. Liu, and T. J. Sigler. An Exploratory Analysis of Brisbane's Commuter Travel Patterns Using Smart Card Data. *Proc., State of Australian Cities National Conference, 2015*, Gold Coast, Queensland, 2015.
23. Liu, C., Y. Susilo, and A. Karlstrom. Measuring the Impacts of Weather Variability on Home-based Trip Chaining Behaviour: A Focus on Spatial Heterogeneity. *Transportation*, Vol. 43, No. 5, 2016, pp. 843–867.
24. Faggian, A., J. Corcoran, and F. Rowe. Evaluating the Effects of Australian Policy Changes on Human Capital: The Role of a Graduate Visa Scheme. *Environment and Planning C: Government and Policy*, Vol. 34, No. 1, 2016, pp. 151–170.
25. Kc, K., and J. Corcoran. Modelling Residential Fire Incident Response Times: A Spatial Analytic Approach. *Applied Geography*, Vol. 84, 2017, pp. 64–74.
26. Corcoran, J., T. Li, D. Rohde, E. Charles-Edwards, and D. Mateo-Babiano. Spatio-temporal Patterns of a Public Bicycle Sharing Program: The Effect of Weather and Calendar Events. *Journal of Transport Geography*, Vol. 41, 2014, pp. 292–305.
27. Nankervis, M. The Effect of Weather and Climate on Bicycle Commuting. *Transportation Research Part A: Policy and Practice*, Vol. 33, No. 6, 1999, pp. 417–431.
28Helbich, M., L. Böcker, and M. J. Dijst. Geographic Heterogeneity in Cycling Under Various Weather Conditions: Evidence from Greater Rotterdam. *Journal of Transport Geography*, Vol. 38, 2014, pp. 38–47.
29. Ahmed, F., G. Rose, and C. Jakob. Commuters' Cycling Behavior in Response to Weather Variation: Insight from an Extended Theory of Planned Behavior. Presented at 95th Annual Meeting of the Transportation Research Board, Washington, D.C., 2016.

The Standing Committee on Transit Capacity and Quality of Service (AP015) peer-reviewed this paper (18-02618).

Article

TRR
JOURNAL OF THE TRANSPORTATION RESEARCH BOARD

Transportation Research Record
2018, Vol. 2672(8) 511–522
© National Academy of Sciences:
Transportation Research Board 2018
Article reuse guidelines:
sagepub.com/journals-permissions
DOI: 10.1177/0361198118772725
journals.sagepub.com/home/trr
SAGE

Transit Vehicle Performance Analysis for Service Continuity/Termination: A Data Envelopment Analysis Approach

S. Kiavash Fayyaz S.[1], Xiaoyue Cathy Liu[1], and Ran Wei[2]

Abstract

Public transit agencies aim to improve services while reducing operating costs. Transit performance analysis, as the main approach used to assess operating cost and revenue, has received much attention in recent decades. Most of such studies focus on macro-level performance analysis by comparing across transit agencies or within a transit agency across different parts of its operation. This macro-level analysis assumes that bus drivers and vehicles have identical performance in terms of production and resource consumption, yet they can vary significantly and the variations directly influence service reliability and operational efficiency. As a result, micro-level vehicle performance analysis is needed for operation optimization. This paper introduces an innovative and effective use of the data envelopment analysis (DEA) approach to estimate, project, and compare the operational efficiency of each transit vehicle. Using the paratransit fleet of Utah Transit Authority (UTA) as a case study, the study demonstrates the varying cost structures and operational efficiencies over time associated with different vehicle types. It shows that such variations and time series analysis can be used to guide prioritization of vehicle procurement and service continuity/termination, which further leads to significant cost savings and improvement in reliability of service. The proposed approach is replicable for any transit fleet with available maintenance and operation data. The proposed method provides transit agencies with data-driven analytics to facilitate the decision-making process.

Public transit represents an integral and important part of the transportation ecosystem, particularly in metropolitan areas. With the decline in government funding allocation, public transit agencies have been aiming to improve services with minimal operating costs (*1*). This goal is usually addressed from multiple angles and perspectives. A wealth of literature has focused on macro-level efficiency improvement, where the performance of transit agencies is evaluated based on system revenue and operating costs (*2–8*). System revenue is usually measured by ridership or vehicle-miles, and operating cost is measured by a combination of labor, capital, energy consumption, and operating time. As a result, the macro-level performance evaluation can help identify inefficiencies in transit services by comparing across transit agencies or within a transit agency across different parts of its services (e.g., bus routes).

For public transit systems, several inputs and outputs are jointly influential on system performance, which makes it challenging to evaluate the underlying relationship between them (*9*). Data envelopment analysis (DEA), as a non-parametric approach, addresses this issue and has been widely used in recent years to compare the operational efficiencies of multiple decision-making units (DMUs) such as transit agencies (*10*) and bus routes (*11*). For example, Karlaftis (*12*) selected labor (as total number of employees), capital (as total number of vehicles), and energy consumption (as total fuel consumption) as inputs, and vehicle-miles and passenger-miles as outputs, to measure and compare the efficiency and effectiveness of all U.S. transit systems. Lao and Liu (*2*) used operation time, round-trip distance, and number of bus stops as input variables and total number of passengers as output to measure and compare the operational efficiencies of bus routes operated by Monterey-Salinas Transit. Wei et al.

[1]Department of Civil and Environmental Engineering, University of Utah, Salt Lake City, UT
[2]School of Public Policy, University of California at Riverside, Riverside, CA

Corresponding Author:
Address correspondence to Xiaoyue Cathy Liu: cathy.liu@utah.edu

(*13*) assessed the operational efficiencies of bus routes within the Utah Transit Authority (UTA) network. Total operation time per day, number of operating buses per day, and total operating mileage per day were the inputs and average number of passengers per day was the output in their DEA model. They proposed a comprehensive assessment method that evaluates both operational efficiency and access equity.

Macro-level transit performance analysis generally assumes consistent performance of entities (e.g., drivers and vehicles) within a transit system or DMU (*14*). For example, analysis of the operational efficiency of bus routes assumes that all bus drivers and vehicles have identical performance in terms of production (e.g., driving speed and mileage) and resource consumption (e.g., wages and fuel consumption) (*2*). Yet in reality bus drivers' characteristics such as age, training, and driving style could vary. This results in different driving speeds and stop management, for example, and consequently varying operational efficiencies of the service (*15, 16*). Vehicle characteristics, such as fuel economy, fuel type, and required maintenance, can also influence the mileage output and operational efficiency of the service (*17, 18*). Micro-level analysis of transit performance can address this issue, but it requires detailed data collection for entities. In recent years there has been growing adoption of sensing technologies that facilitate and automate collection of high-resolution data, such as automatic vehicle location (AVL), automatic passenger count (APC), and fuel consumption information. Researchers have started to use such data sources for various micro-level performance assessments. For example, Fayyaz et al. (*19*) analyzed the bus dwell time and fare payment structure using AVL and APC data. They have shown that characteristics of bus stops such as stop placement and design, and the built environment, influence bus dwell time and consequently the operational efficiency of the bus route.

Each transit vehicle is an important micro-part in a public transit system, with vehicles accounting for a significant portion of capital, maintenance, and operating costs (*20*). Vehicle managers face several decision-making challenges every day with regard to transit vehicles, including selection of vehicles for service continuity/termination, selection of type of vehicle for procurement, and estimating the future maintenance and operating costs associated with each vehicle. Vehicle selection for service continuity is the process of selecting one or multiple vehicles from the fleet to run extra miles due to insufficient supply (i.e., of vehicles). Vehicle selection for service termination concerns selecting one or more vehicles from the fleet for retirement. The retired vehicles may be replaced with newly procured ones. Currently, most of the decisions regarding vehicle selection for service continuity/termination are made based on planning/engineering judgements rather than data-driven analytics.

Mounting evidence suggests that vehicles with different characteristics (e.g., build, model, fuel economy, and age) tend to demonstrate varying efficiency patterns over time (*21–23*). These variations in efficiency over life cycle are an important factor in the selection of vehicle type for procurement. For example, assume type *A* vehicles have higher fuel efficiency compared with type *B* vehicles. The maintenance costs of type *B* vehicles are gradually rising over time, while type *A* vehicles have almost constant maintenance costs over their entire life cycle. Type *A* vehicles are beneficial for transit agencies with a mature network targeting operation of their vehicles for relatively longer mileage and expecting roughly constant maintenance expenses over time. Type *B* vehicles are more suited for transit agencies aiming to generate high revenue in a short time period and expecting to replace their fleet more frequently.

The aforementioned issues can be summarized into the following questions: (a) Which vehicles should be given priority for service continuity/termination? (b) What are the expected maintenance and operating costs of each vehicle in the near future? and (c) What type of vehicles to purchase? In order to answer these questions, a data-driven method must be developed to measure and monitor vehicles' performance over time. The results will help guide transit agencies in the decision-making process of vehicle procurement and in effective cost control. To the best of the authors' knowledge, no study to date has focused on comparing vehicles' performances for such purposes.

This paper introduces an innovative and effective application of the DEA method to measure and compare operational efficiencies across transit vehicles. The contribution of this paper is twofold. First, the DEA method is applied to compare the operational efficiency of transit vehicles based on their life-time operational and maintenance records. We demonstrate that the aggregated historical vehicle performance data can be used to extract surrogate measures for vehicle efficiency and reliability, and guide future vehicle fleet procurement. Second, time series DEA are attempted to compute vehicles' operational efficiency over time. Such underlying operational efficiency trends can be used for predicting the future maintenance and operational costs associated with each vehicle. The vehicles are then ranked based on the predicted operational efficiency for service continuity/termination.

The rest of the paper is structured as follows. We provide a detailed description of the proposed DEA models both for vehicles' entire operating time span and consecutive time sequences, and explain how these DEA models help to reveal the operational efficiency of each vehicle.

The method is then applied to the paratransit fleet operated by UTA in the State of Utah. Their high operating costs and wide range of vehicle mileage make paratransit vehicles a suitable choice for our study. The paper concludes with a discussion of the results and implications of the study findings.

Methodology

This study aims to address challenges in transit vehicle management and decision making. To this end, we propose a DEA analysis to measure the operational efficiency of each vehicle in the fleet. The operational efficiency of each vehicle is compared with that of other vehicles over their operating period. The vehicle's operated mileage is used as the output, and maintenance labor costs, maintenance part costs, maintenance overhead costs, fuel consumption, and number of preventive maintenance measures (PMs) performed are used as inputs in the DEA model (*2, 12, 24*). This computational procedure is described at length in the "DEA Model" section.

Note that the operational efficiency of transit vehicles varies over time. Efficiency might decay at a faster rate for some vehicles compared with the rest of fleet. To capture this temporal fluctuation, each vehicle's operational efficiency can be traced for consecutive time sequences (i.e., each month) over the entire operating period. Such temporal trends differentiate vehicles and can guide decisions related to vehicle procurement, maintenance, and service continuity/termination. The computational procedure for estimating time series of operational efficiency for each vehicle is described in the "Time Series DEA" section.

DEA Model

DEA is a non-parametric method to evaluate the relative performance (efficiency) of DMUs when the presence of multiple inputs (resources) and outputs (work, goods, or service) makes comparison difficult (*25, 26*). A DEA model estimates the production frontier by establishing an empirical relation between inputs and outputs. The operational efficiency of a DMU is then measured by comparing its own production frontier with the estimated optimal production frontier. In other words, an output (input) DEA model measures operational efficiency by comparing the real outputs (inputs) with the optimal outputs (inputs) given a certain combination of inputs (outputs). As a result, efficient DMUs determined from the DEA model may not necessarily form the best production frontier possible, but rather represent the best-practice frontier (*27*). The major advantages of DEA are: (a) not requiring a specified functional form to relate inputs and outputs, and (b) independence of unit and scale of input and output variables.

For the specific problem of transit vehicle performance, vehicles serve as DMUs. The miles traveled (in miles) as the amount of work produced by a DMU (vehicle) functions as the output. Maintenance labor, parts, and overhead costs (in dollars), fuel consumption (in gallons), oil changes (in gallons), and number of PMs associated with each DMU (vehicle) serve as inputs in the input-oriented DEA model. The input-oriented DEA model measures the operational efficiency of DMUs by evaluating the difference between the DMU's inputs and minimum inputs while keeping the outputs at their current level (*28*). As a result, the input-oriented model reflects transit agencies' goal to minimize vehicle maintenance and operating costs. Let k represent the index of a set of vehicles (n), r and i represent the index of outputs (s) and inputs (m), respectively. The operational efficiency of DMU_k is determined by solving the following model (*29*):

$$Maximize\ \theta_k = \frac{\sum_{r=1}^{s} u_r y_{rk}}{\sum_{i=1}^{m} v_i x_{ik}} \tag{1}$$

subject to

$$\frac{\sum_{r=1}^{s} u_r y_{rj}}{\sum_{i=1}^{m} v_i x_{ij}} \leq 1\ \forall j = 1, \ldots, n \tag{2}$$

$$u_r, v_i \geq 0 \quad \forall r = 1, \ldots, s; i = 1, \ldots, m \tag{3}$$

where θ_k is the relative efficiency of DMU_k, y_{rk} is the rth (i.e., $r = 1$: miles traveled) output of DMU_k, x_{ik} is the ith input of DMU_k (i.e., $i = 1$: labor cost, $i = 2$: parts cost, $i = 3$: overhead cost, $i = 4$: preventive maintenance, $i = 5$: oil changes, and $i = 6$: fuel consumption), and u_r, v_i are decision variables reflecting the weights associated with outputs and inputs, respectively. The objective (Equation 1) is to maximize the relative efficiency θ_k of the kth vehicle (DMU_k). Constraints (Equation 2) ensure that each of the vehicles, using the same weights, has a relative efficiency that is less than or equal to one. Constraint (Equation 3) imposes non-negativity restrictions for weights. This optimization model can be converted into a linear programming model as shown in (*25*). In order to derive relative efficiency for each of the n vehicles, this optimization model must be solved n times, once for each vehicle (k = 1, 2, ..., n). The "rDEA" package in R was used to solve the DEA model.

The operational efficiency of each vehicle, θ_k, ranges from zero to one. An efficiency score of one indicates that the vehicle is performing best compared with other vehicles based on inputs and outputs. In other words, these vehicles are minimizing resource consumption (i.e., inputs) given their miles traveled (i.e., outputs). It is important to mention that both low- and high-cost vehicles can be operationally efficient because the DEA model evaluates how well these resources are used to generate the outputs. On the other hand, the operational efficiency of a vehicle with a score close to zero performs the worst compared with its peers. These vehicles are usually producing almost the same output (i.e., mileage and age) by consuming more resources (inputs).

The aggregated inputs and outputs for a specific time period can be used in the DEA model to measure vehicles' relative operational efficiencies and compare their performances during the selected period. Yet the operational efficiency of any vehicle evolves over time. This is caused by aging and a myriad of random events (e.g., vehicle failure) that might impose additional maintenance and operating costs. Trend analysis is thus necessary to estimate the expected operational efficiency of each vehicle in the future.

Time Series DEA

To understand better the productivity and costs associated with different vehicles over time, the DEA analysis must be applied to time series of inputs and output measures. For this purpose, the cumulative inputs and outputs of each vehicle were calculated for consecutive time sequences (i.e., every month) over their entire operating time span. Each time the sequence data of each vehicle is treated as a DMU. As a result, the DMU set n in the DEA model presented above is replaced with the set of vehicles over each time sequence (n_t). Everything else remains the same in Equations 1, 2, and 3. Note that u_r and v_i are estimated for each vehicle at each time sequence separately. The temporal fluctuation of operational efficiency associated with each vehicle can then be traced to assess the underlying trends.

A decreasing operational efficiency trend indicates that vehicle performance is decaying over time. In other words, the cost associated with each additional mile operated exceeds the operational cost for previous miles. A constant time series operational efficiency indicates that the cost associated with operating each additional mile is almost the same as the operational cost of previous miles. The vehicle performance is consistent as it ages. An increasing operational efficiency shows that the vehicle performance is improving over time. In this case, age has a positive effect on vehicle performance. This usually happens when the cost associated with each additional mile operated is less than the costs for previous miles. For example, if a vehicle underwent a major maintenance event (e.g., engine failure) in time sequence t_1, it would experience a sudden jump in costs (inputs) while the miles traveled (output) would increase at the normal rate. At the next time sequence t_2, this vehicle would experience a normal growth rate for both costs and miles traveled. This results in higher operational efficiency at time sequence t_2 than time sequence t_1. Consequently, the operational efficiency of this vehicle shows an increasing trend during that period.

Operational efficiency trends can guide transit agencies in their future decision making regarding vehicle procurement, maintenance, and service continuity/termination. The cost associated with each vehicle type during its life cycle is an important guiding factor for vehicle procurement. Public transit agencies can select the vehicle type that best suits their particular needs. The operational efficiency trend of each vehicle can also be used for predicting its future maintenance and operating costs. This helps the budget allocation process. Finally, the expected operational efficiencies of vehicles are measured relative to each other. Thus, they can be used for prioritizing vehicles for service continuity/termination with the objective of minimizing costs.

Data

UTA is the primary transit provider in the Wasatch Front Region of the State of Utah. UTA's services cover six counties: Salt Lake, Utah, Davis, Weber, Box Elder, and Tooele. The transit network consists of 125 fixed transit routes encompassing bus, light rail, and commuter rail. In this study, we chose a paratransit fleet operated by UTA to demonstrate the applicability of our method. Paratransit is in general a demand-responsive service (*30*). Unlike fixed-route transit, paratransit operates according to demand and has no fixed routes or schedules (*31*). Paratransit vehicles are usually moving in mixed traffic. Each paratransit trip includes dispatching a vehicle to the passengers' points of origin, picking up passengers, transporting passengers to their destinations, and returning to the base. These trips can thus vary significantly in distance. As a result, paratransit vehicles have a wider range of mileage compared with fixed-route transit vehicles. Another important difference between paratransit and fixed-route transit is its higher operating cost per passenger trip. This is mainly caused by two factors: (a) a paratransit service is on-demand; and (b) a paratransit service is designed to address passengers' specific needs (including those with disability or carrying large baggage), and thus requires special equipment (e.g., lift) (*30, 32*). These distinctive characteristics of paratransit (high operating cost per passenger trip and wide range of vehicle mileages) make it a suitable choice for this study. UTA's paratransit service is a curb-to-curb (origin-to-destination) service provided for people with physical, cognitive, or visual disabilities who are functionally unable to use the fixed-routes service independently either all of the time, temporarily, or only under

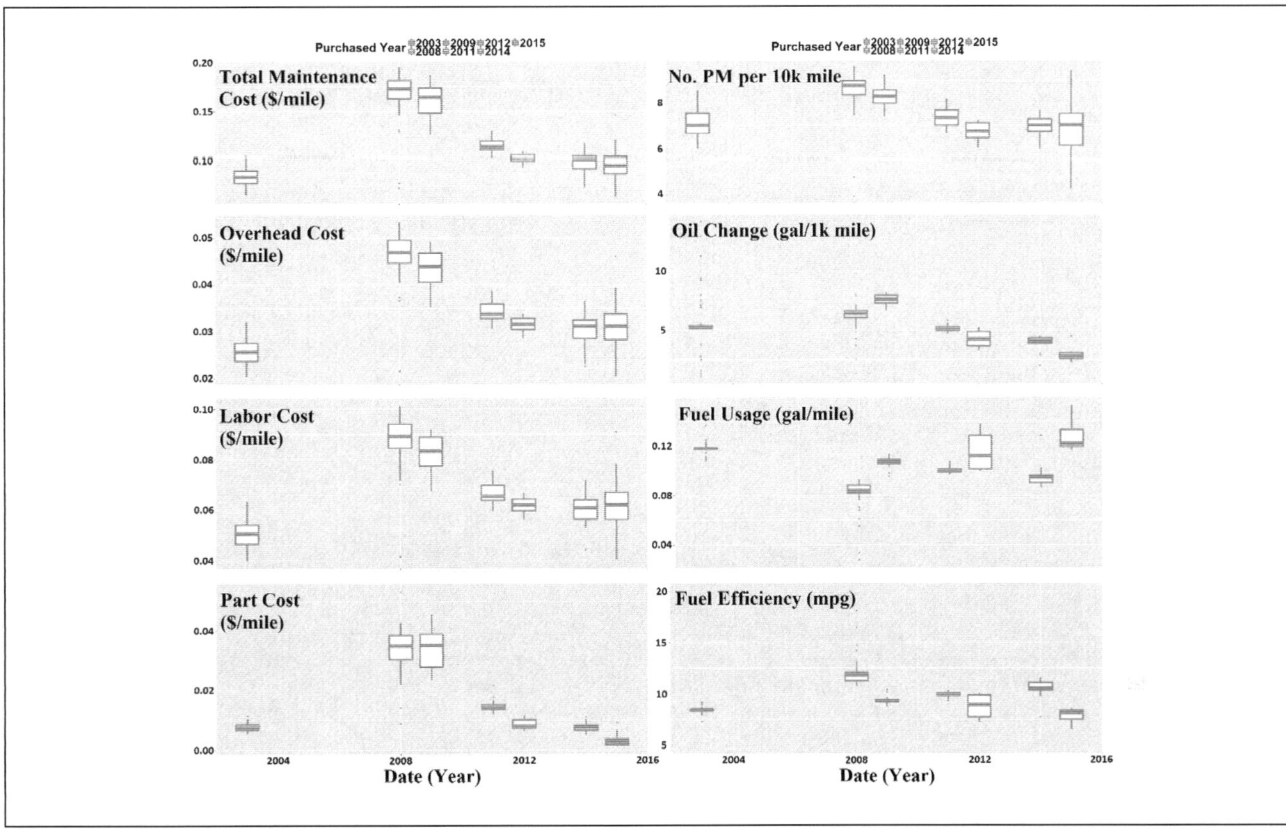

Figure 1. Paratransit vehicles: maintenance and operation data break-down according to year purchased.

certain circumstances. UTA's paratransit fleet includes accessible buses and vans. Because paratransit is provided through reservations and requires specific equipment (e.g., lift), the service operates at a much higher cost per passenger trip than regular UTA services.

In this study the maintenance and operation data including oil changes, fuel usage, miles driven, preventive maintenance, and maintenance costs including labor, parts, and overhead costs were collected daily for 223 paratransit vehicles from May 2003 to May 2017. All the 223 paratransit vehicles are Glaval vans with identical passenger capacity and equipment. A total of 53, 47, 28, 32, 20, 26, and 18 vehicles were purchased in years 2003, 2008, 2009, 2011, 2012, 2014, and 2015, respectively. The daily records were further aggregated into monthly and through the entire operation period for each paratransit vehicle. Figure 1 presents a summary of the collected data over mileage by the procurement years. Note that vehicles purchased in the same year are of identical type.

The visualization in Figure 1 represents the distribution of aggregated maintenance and operation data collected for each vehicle type. It shows that for different vehicle types the maintenance and operating costs vary, as do their distribution. For example, the group of vehicles purchased in 2008 has higher total maintenance costs than the group of vehicles purchased in 2003. On the other hand, the group of vehicles purchased in 2008 has better fuel economy and lower fuel usage than the group of vehicles purchased in 2003. These facts seem to counterbalance each other, which makes performance comparison challenging. For instance, on average, the group of vehicles purchased in 2008 incurs $13,000 higher total maintenance costs at 200,000 mileage than the group of vehicles purchased in 2003. However, the 2008 vehicles have better fuel economy which leads to nearly 8,500 gallons (~$13,000) less fuel usage than the 2003 group at 200,000 mileage.

The bottom and top of the box in each boxplot presented in Figure 1 represents the 25th and 75th percentiles, and the band in between shows the median value of the respective maintenance and operating cost variables for each vehicle type. The graph shows the level of consistency of various maintenance and operating costs for specific types of vehicles. For example, the parts cost boxplot for the group of vehicles purchased in 2008 is larger than the parts cost boxplot for the group of vehicles purchased in 2003. This indicates that the parts costs of vehicles purchased in 2003 were relatively consistent

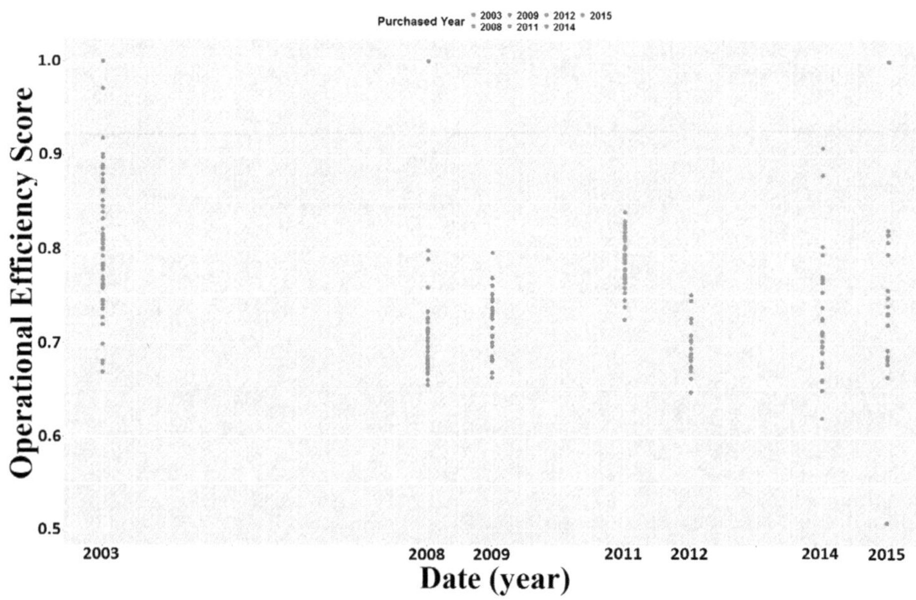

Figure 2. Paratransit vehicles: operational efficiency score distribution by year purchased.

(and thus were more predictable), whereas the parts costs of vehicles purchased in 2008 varied significantly (and thus were less predictable).

Results and Discussion

As described in the "Methodology" section, the operational efficiencies of paratransit vehicles are computed using the DEA model. For this purpose, the inputs and outputs were aggregated over the entire data collection period for each paratransit vehicle. Note that the collected data cover the entire life cycle for some vehicles (i.e., vehicles purchased in 2003, 2008, and 2009) and only part of the life cycle for others (i.e., vehicles purchased in 2011, 2012, 2014, and 2015). Figure 2 displays the operational efficiency distribution of the paratransit fleet by year purchased. There are several interesting facets to these results. First, the groups of vehicles purchased in 2003 and 2011 have the highest average operational efficiency score (˜0.8). This indicates that those vehicles are performing the best compared with other groups in terms of minimizing costs (e.g., maintenance cost and fuel consumption) given their productivity (i.e., miles traveled). Second, the variance of operational efficiency scores for vehicles purchased in 2003, 2014, and 2015 are the highest compared with other groups. This shows that these vehicles demonstrate less consistent performance. In other words, the expected costs of these vehicles vary more compared with other groups, which makes estimations of future costs and efficiency more volatile. It is important to note that the high variations in operational efficiency scores of vehicles purchased in 2014 and 2015 may be due to their relatively early stage of life cycle, as explained in Figure 2. Third, the variation in efficiency scores of vehicles within the same group illustrates the difference in observed performance across vehicles of the same type. This information helps guide transit agencies in service continuity/termination planning, particularly if such planning is conducted in phases. For example, it is logical to give higher priority for service termination to vehicles with low operational efficiency scores (closer to 0) among the vehicles purchased in 2003. On the other hand, vehicles with high operational efficiency score (˜1) can be the candidates to remain in service for a longer period, if needed.

Figure 3 shows the operational efficiency score of paratransit vehicles by their mileage. The operational efficiency of vehicles purchased in 2003 decayed after 200,000 miles. In this group, the vehicles that remained in service for a longer period (e.g., mileage greater than 200,000) have an average efficiency score of 0.75. However, UTA could save about 10% in costs by selecting the vehicles with highest operational efficiency score for continuing in service. For instance, selecting vehicles with the highest operational efficiency scores for longer service (i.e., mileage less than 200,000 and operational efficiency score greater than 0.85) could result in at least 10% (i.e., 0.85–0.75) cost reduction for the same service. Clearly, this issue has been considered for groups of vehicles purchased in 2008 and 2009 as vehicles with the

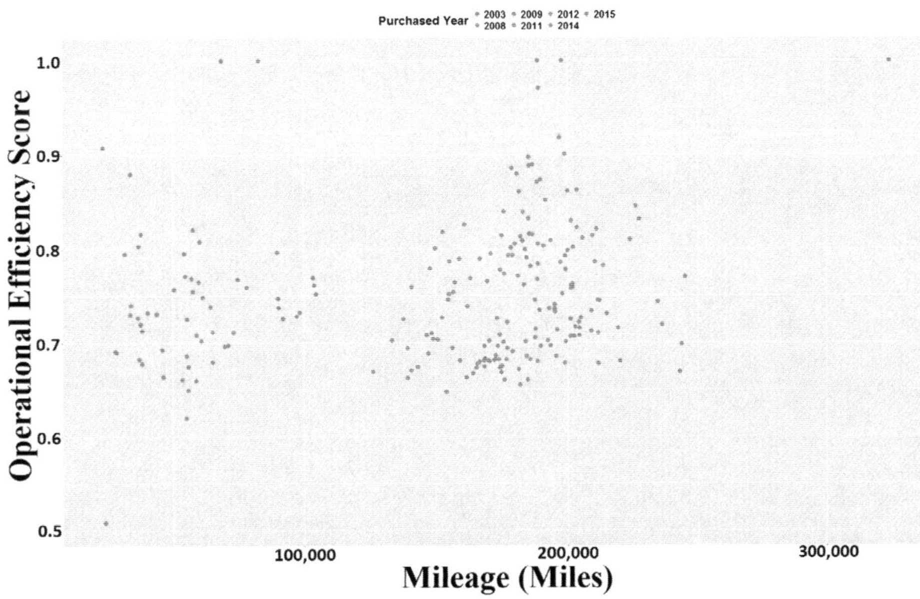

Figure 3. Paratransit vehicles: operational efficiency score by mileage.

highest efficiency scores remained in service for longer periods.

In order to predict the expected future costs associated with each paratransit vehicle, time series DEA analysis was applied to groups of vehicles purchased in 2003, 2008, and 2009. The aggregated inputs and outputs for each month of the service years were computed for each vehicle. Figure 4 shows the operational efficiency score trajectories of these vehicles over time and mileage. Note that the scores fluctuate significantly in the first 25,000 miles of service. This is due to the fact that even small changes in cost structure would influence the input values in the DEA model at the early phase of service. As mentioned earlier, the time series DEA uses cumulative inputs, which are monotonically increasing over time. At the early stage in service, those input values are much smaller and thus more sensitive to minor changes in cost structure. After 25,000 miles, the operational efficiency trend becomes more stable. Note that a sudden drop in the operational efficiency score after 25,000 miles implies significant cost increase during that month. This is usually caused by a major maintenance event such as engine failure.

The operational efficiency of each vehicle differs from its peers. For example, the range of scores at 50,000 miles is 0.15, 0.25, and 0.15 for groups of vehicles purchased in 2003, 2008, and 2009, respectively (as shown in Figure 4*b*). The range further reduces and stabilizes after 150,000 miles at about 0.10 for all groups. Correspondingly, the expected future costs among the same vehicle type can also vary significantly.

In order to visualize and compare the operational efficiency scores of the three groups of vehicles over time better, Figure 4 is smoothed using locally weighted polynomial regression (LOESS) method (*33*). LOESS is a non-parametric regression method that measures each local smoothed value by a weighted quadratic least square function over the span of values (i.e., operational efficiency score). The main advantage of LOESS is that it does not require function specification to fit the data. The result of smoothed graph is shown in Figure 5. It shows that vehicles purchased in 2003 have low operating costs in the first 100,000 miles of service. Afterwards, operating costs start increasing and operational efficiency starts to drop by 1.7% for each 25,000 additional miles traveled. As a result, this type of vehicle is a more desirable procurement choice for transit agencies that aim to replace their fleet more frequently (e.g., every five years or 150,000 miles). It is also a good choice for transit agencies that project low revenue in the first five years followed by a high revenue period. The high revenue period can help cover the high operating costs of this type in the final years of service. On the other hand, the groups of vehicles purchased in 2008 and 2009 demonstrate steady operating costs and operational efficiencies. Transit agencies with relatively stable budgets that plan to replace their fleet less frequently (e.g., every 8–10 years and 250,000 miles) will be interested in these types. These fleets are also more reliable given that the

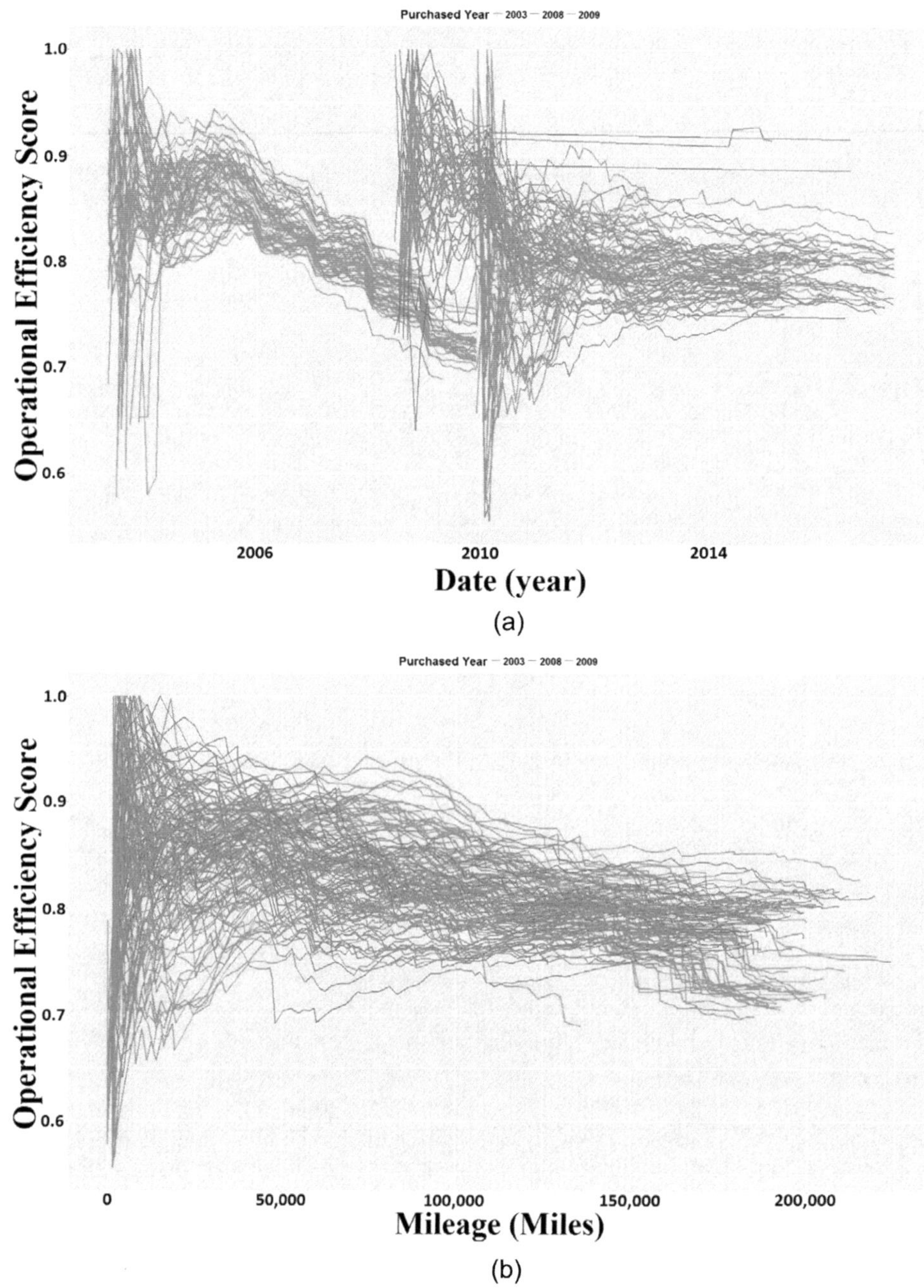

Figure 4. Time series DEA analysis: operational efficiency score by (*a*) operation time and (*b*) mileage.

probability of sudden major maintenance events (and additional operating costs) is smaller compared with the 2003 fleet (as shown in Figure 2). Thus, transit agencies running tight fleet schedules which require vehicles with high reliability will be more interested in groups of vehicles purchased in 2008 and 2009.

Time series of operational efficiency scores can be further analyzed to project the expected future operating costs. For better visualization, the operational efficiency of three specific vehicles (one from each group of vehicles purchased in 2003, 2008, and 2009) and their LOESS smoothed time series are shown in Figure 6*a*. As expected, Vehicle No. 3216, purchased in 2003, experiences low operating costs for the first 100,000 miles followed by an increasing pattern in operating costs, to the point where it becomes too costly after 175,000 miles.

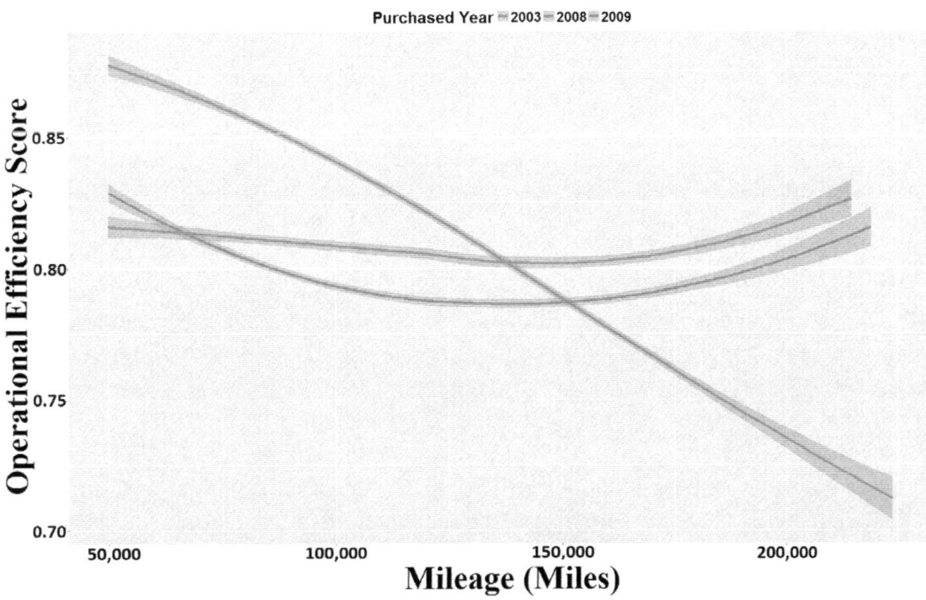

Figure 5. DEA trend analysis: LOESS for groups of vehicles purchased in 2003, 2008, and 2009 over mileage.

On the other hand, Vehicles No. 8222 and No. 9202, purchased in 2008 and 2009 respectively, show relatively consistent performance over time (mileage). These vehicles can provide service at reasonable costs for longer mileage (e.g., 250,000) compared with Vehicle No. 3216.

The same comparison can be made between vehicles of the same type (e.g., purchased in 2008) for service continuity/termination prioritization. Figure 6*b* shows operational efficiency scores time series of three sample vehicles purchased in 2008. Assume that the decision for service continuity/termination was made at 150,000 miles. Since almost all three vehicles show a steady operational efficiency pattern between 100,000 to 150,000 miles, the one vehicle with the highest/lowest operational efficiency score has the priority for continuity/termination. Thus, Vehicle No. 8244 is given the top priority for service continuity followed by Vehicles No. 8222 and 8202, respectively. Similarly, Vehicle No. 8202 is given the top priority for service termination followed by Vehicles No. 8222 and 8244, respectively. These results show that UTA has made a sound decision as Vehicles No. 8244 and 8222 remained in service for longer with a higher mileage than Vehicle No. 8202.

Conclusion

Vehicle performance has a significant impact on public transit maintenance and operating costs, as its failure has a direct influence on transit reliability and operational efficiency. Even though the performance of transit agencies and/or service operation has been studied before, analysis of transit vehicle performance has been overlooked in past literature. A practical data-driven approach is needed to determine quantitatively and to project vehicle performance over time (or mileage) to assist with decision making. The growing availability of maintenance and operation data in the majority of transit systems supports such a data-driven method.

This paper proposes a DEA approach to assess the performance (operational efficiency) of each vehicle that can support transit agencies' decision-making processes. Maintenance and operation data were collected for various types of paratransit vehicles purchased in different years. The data were aggregated for each vehicle's entire operating time span and consecutive time sequences (i.e., each month). Our input-oriented DEA model uses maintenance labor, parts, and overhead costs (in dollars), fuel consumption (in gallons), oil changes (in gallons), and number of PMs associated with each vehicle as inputs, and miles traveled (in miles) as output. The results show that maintenance and operating costs of different vehicle types can vary significantly (e.g., up to 35%). The operational efficiency scores vary notably among vehicles of the same type indicating that: (a) the relatively higher variance of operational efficiency scores within the same vehicle type means more unexpected patterns in future maintenance and operating costs. As a result, such a vehicle type has lower reliability and is less appealing to mature transit agencies that aim for reliability and seek high predictability; and (b) vehicles of the same type may

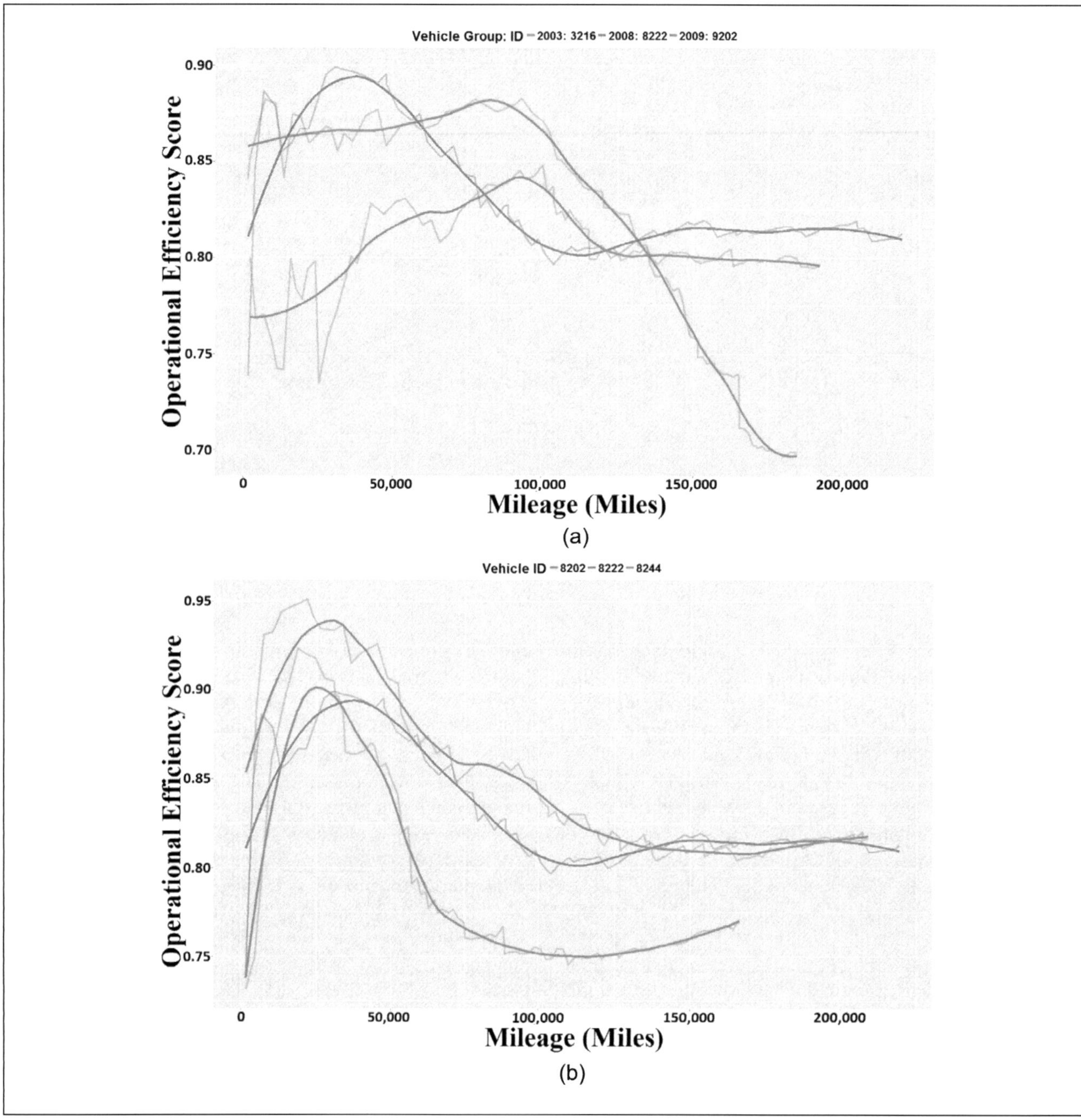

Figure 6. DEA trend analysis: (*a*) LOESS for three random vehicles over mileage and (*b*) LOESS for three random vehicles purchased in 2008 over mileage.

have significantly different maintenance and operating costs (e.g., the highest and lowest operational efficiency of vehicles purchased in 2003 differed by about 35%). This indicates that prioritizing vehicles for service continuity/termination can significantly reduce costs.

In order to analyze the operational efficiency trend of each vehicle, DEA was applied to time series data (aggregated monthly). The results indicate that some vehicle types have lower maintenance and operating costs (higher operational efficiency) in the initial service miles (e.g., before 100,000 miles). Afterwards, their maintenance and operating costs start to increase (operational efficiency decreases) and they become less reliable. These vehicle types are desirable for: (a) transit agencies planning to replace their fleet more frequently since their fleet will be replaced before reaching the tipping point; and (b) smaller transit agencies that project low revenue in the near future followed by a high revenue period. Some types of vehicles show fairly consistent maintenance and operating costs and high reliability for longer mileage (e.g., 200,000 miles). Mature transit agencies with steady revenue will be more interested in these vehicles since their high predictability (reliability) is required for operating to tight schedules.

Finally, the operational efficiency trends of vehicles of the same type were compared to provide a basis for prioritizing vehicles for service continuity/termination. Vehicles were ranked for service continuity/termination based on their current and expected future operational efficiency scores. The results show that expected operational efficiency for vehicles of the same type can vary significantly (e.g., 10% for group of vehicles purchased in 2008). Considering the fact that a mid-size transit agency like UTA is making hundreds of service continuity/termination decisions each year, the cost savings by the proposed method will be significant. Assuming such a decision was made on 100 vehicles to remain in service for an additional 50,000 miles where total maintenance and operating costs of vehicles are $ 0.27 per mile and the cost saving by using our method is 10%, the agency would save approximately $150,000 in costs.

The proposed method should not be the final word on the matter, rather it inspires the need for the next logical step. This next step is to tailor the method further for the specific needs of agencies, such as selection of input variables and output variables, linear/non-linear estimation of expected future operational efficiency, estimation of the expected value of each input variable, and development of platforms for service continuity/termination prioritization, all of which can improve transit efficiency and reduce maintenance and operating costs. The results are potentially useful for optimizing transit services, especially for paratransit operation.

Acknowledgments

The authors would like to thank the staff of UTA for providing the data, and for their support, and feedback on this study.

Author Contributions

The authors confirm contribution to the paper as follows: study conception and design: S. Kiavash Fayyaz S., Xiaoyue Cathy Liu; data collection: Kiavash Fayyaz, Xiaoyue Cathy Liu; analysis and interpretation of results: Kiavash Fayyaz, Xiaoyue Cathy Liu, Ran Wei; draft manuscript preparation: Kiavash Fayyaz, Xiaoyue Cathy Liu, Ran Wei. All authors reviewed the results and approved the final version of the manuscript.

References

1. Tsamboulas, D. A. Assessing Performance under Regulatory Evolution: A European Transit System Perspective. *Journal of Urban Planning and Development*, Vol. 132, No. 4, 2006, pp. 226–234.
2. Lao, Y., and L. Liu. Performance Evaluation of Bus Lines with Data Envelopment Analysis and Geographic Information Systems. *Computers, Environment and Urban Systems*, Vol. 33, No. 4, 2009, pp. 247–255.
3. Fielding, G. J., T. T. Babitsky, and M. E. Brenner. Performance Evaluation for Bus Transit. *Transportation Research Part A: General*, Vol. 19, No. 1, 1985, pp. 73–82.
4. Fielding, G. J. *Managing Public Transit Strategically: A Comprehensive Approach to Strengthening Service and Monitoring Performance*. Jossy-Bass, San Francisco, Calif., 1987.
5. Benn, H. *TCRP Synthesis 10: Bus Route Evaluation Standards*. Transportation Research Board of the National Academies, Washington, D.C., 1995.
6. De Borger, B., K. Kerstens, and A. Costa. Public Transit Performance: What Does One Learn from Frontier Studies? *Transport Reviews*, Vol. 22, No. 1, 2002, pp. 1–38.
7. Taylor, B., M. Garrett, and H. Iseki. Measuring Cost Variability in Provision of Transit Service. *Transportation Research Record: Journal of the Transportation Research Board*, 2000.1735: 101–112.
8. Triantis, K. P. Engineering Applications of Data Envelopment Analysis. In *Handbook on Data Envelopment Analysis*, Springer, Boston, MA, 2004, pp. 401–441.
9. Barnum, D. T., S. McNeil, and J. Hart. Comparing the Efficiency of Public Transportation Subunits Using Data Envelopment Analysis. *Journal of Public Transportation*, Vol. 10, No. 2, 2007, p. 1.
10. Chu, X., G. J. Fielding, and B. W. Lamar. Measuring Transit Performance Using Data Envelopment Analysis. *Transportation Research Part A: Policy and Practice*, Vol. 26, No. 3, 1992, pp. 223–230.
11. Hawas, Y. E., M. Khan, and N. Basu. Evaluating and Enhancing the Operational Performance of Public Bus Systems Using GIS-Based Data Envelopment Analysis. *Journal of Public Transportation*, Vol. 15, No. 2, 2012, p. 2.
12. Karlaftis, M. G. A DEA Approach for Evaluating the Efficiency and Effectiveness of Urban Transit Systems. *European Journal of Operational Research*, Vol. 152, No. 2, 2004, pp. 354–364.
13. Wei, R., X. Liu, L. Wang, A. Golub, and S. Farber. Evaluating Public Transit Services for Operational Efficiency and Access Equity. *Journal of Transportation Geography*, Vol. 65, pp. 70–79.
14. Novaes, A. G. Rapid-Transit Efficiency Analysis with the Assurance-Region DEA Method. *Pesquisa Operacional*, Vol. 21, No. 2, 2001, pp. 179–197.
15. Vangi, D., and A. Virga. Evaluation of Energy-Saving Driving Styles for Bus Drivers. *Proceedings of the Institution of Mechanical Engineers, Part D: Journal of Automobile Engineering*, Vol. 217, No. 4, 2003, pp. 299–305.
16. Zarkadoula, M., G. Zoidis, and E. Tritopoulou. Training Urban Bus Drivers to Promote Smart Driving: A Note on

a Greek Eco-Driving Pilot Program. *Transportation Research Part D: Transport and Environment*, Vol. 12, No. 6, 2007, pp. 449–451.
17. Gintner, V., N. Kliewer, and L. Suhl. Solving Large Multiple-Depot Multiple-Vehicle-Type Bus Scheduling Problems in Practice. *OR Spectrum*, Vol. 27, No. 4, 2005, pp. 507–523.
18. Tzeng, G. H., C. W. Lin, and S. Opricovic. Multi-Criteria Analysis of Alternative-Fuel Buses for Public Transportation. *Energy Policy*, Vol. 33, No. 11, 2005, pp. 1373–1383.
19. Fayyaz, S. K., X. Liu, and R. J. Porter. Genetic Algorithm and Regression-Based Model for Analyzing Fare Payment Structure and Transit Dwell Time. *Transportation Research Record: Journal of the Transportation Research Board*, 2016. 2595: 1–10.
20. Karlaftis, M. G., and P. McCarthy. Cost Structures of Public Transit Systems: A Panel Data Analysis. *Transportation Research Part E: Logistics and Transportation Review*, Vol. 38, No. 1, 2002, pp. 1–18.
21. Ally, J., and T. Pryor. Life-Cycle Assessment of Diesel, Natural Gas and Hydrogen Fuel Cell Bus Transportation Systems. *Journal of Power Sources*, Vol. 170, No. 2, 2007, pp. 401–411.
22. Clark, N. N., F. Zhen, W. S. Wayne, and D. W. Lyons. *Transit Bus Life Cycle Cost and Year 2007 Emissions Estimation*. Final Report, FTA-WV-26-7004.2007.1. Federal Transit Administration, US Department of Transportation, Washington, D.C., 2007.
23. Cooney, G., T. R. Hawkins, and J. Marriott. Life Cycle Assessment of Diesel and Electric Public Transportation Buses. *Journal of Industrial Ecology*, Vol. 17, No. 5, 2013, pp. 689–699.
24. Fielding, G. J. Managing Transit Strategically. *ITS Review: Institute of Transportation Studies*, Vol. 11, No. 2, 1988, pp. 4–6.
25. Charnes, A., W. W. Cooper, and E. Rhodes. Measuring the Efficiency of Decision Making Units. *European Journal of Operational Research*, Vol. 2, No. 6, 1978, pp. 429–444.
26. Banker, R. D., A. Charnes, and W. W. Cooper. Some Models for Estimating Technical and Scale Inefficiencies in Data Envelopment Analysis. *Management Science*, Vol. 30, No. 9, 1984, pp. 1078–1092.
27. Cook, W. D., K. Tone, and J. Zhu. Data Envelopment Analysis: Prior to Choosing a Model. *Omega*, Vol. 44, 2014, pp. 1–4.
28. Zhu, J. *Quantitative Models for Performance Evaluation and Benchmarking: Data Envelopment Analysis with Spreadsheets*. Springer, Berlin, Vol. 213, 2014.
29. Cooper, W. W., L. M. Seiford, and J. Zhu. Data Envelopment Analysis. In *Handbook on Data Envelopment Analysis*, Springer, Boston, MA, 2004, pp. 1–39.
30. Lave, R., and R. Mathias. State of the Art of Paratransit. *Transportation in the New Millennium*, Vol. 478, 2000, pp. 1–7.
31. Joewono, T. B., and H. Kubota. User Satisfaction with Paratransit in Competition with Motorization in Indonesia: Anticipation of Future Implications. *Transportation*, Vol. 34, No. 3, 2007, pp. 337–354.
32. Fu, L., and G. Ishkhanov. Fleet Size and Mix Optimization for Paratransit Services. *Transportation Research Record: Journal of the Transportation Research Board*, 2004. 1884: 39–46.
33. Cleveland, W. S., and S. J. Devlin. Locally Weighted Regression: An Approach to Regression Analysis by Local Fitting. *Journal of the American Statistical Association*, Vol. 83, No. 403, 1988, pp. 596–610.

The Standing Committee on Transit Capacity and Quality of Service (AP015) peer-reviewed this paper (18-01838).

The work presented in this paper remains the sole responsibility of the authors.

Article

TRR
JOURNAL OF THE TRANSPORTATION RESEARCH BOARD

Transportation Research Record
2018, Vol. 2672(8) 523–535

DOI: 10.1177/0361198118795002
journals.sagepub.com/home/trr
SAGE

Moving Minds: The Next Generation of Real-Time Transit Information in San Francisco

Jason Lee[1], William Tyner[1], and Nealay Vasavda[1]

Abstract

In 2001, the San Francisco Municipal Transportation Agency (SFMTA) revolutionized the transit riding experience by implementing its first-generation customer information system to predict when its Muni transit vehicles would arrive in real time. Since then, mobile technologies have ushered in on-demand ride-hailing services and raised expectations for transit service quality. Seeking to stabilize and increase ridership, transit operators like SFMTA are working to adapt and innovate in this rapidly changing environment. SFMTA is now embarking on the next generation of real-time customer information. To guide this multi-million dollar investment, SFMTA conducted extensive quantitative and qualitative research to design new system features and uncover how they could shape travel behavior. An in-depth survey of 5,856 Muni customers revealed that income is a primary mode-choice driver, coupled with the relative availability of ride-hailing services. As income rises, the propensity to wait for Muni decreases, particularly when nearby ride-hailing options are plentiful. Applicable to other cities, this research also affirmed that better real-time information such as alternative routes and transfer connections could increase ridership overall, potentially reducing demographic influences on mode choice and fostering a more equitable and sustainable transportation system. Informed by this research, SFMTA's new system will focus on improving prediction accuracy, keeping customers informed throughout their journey, and developing a transit-focused mobile app.

The San Francisco Municipal Transportation Agency (SFMTA) is developing a new real-time vehicle arrival and service update system for its Muni public transportation network. Known formally as the Next Generation Customer Information System, the project will empower Muni customers to take transit to their destinations quickly and reliably. Muni is the backbone of the city's multimodal transportation network, providing over 700,000 average weekday trips with a fleet of electric trolley coaches, motor coaches, light rail vehicles, historic streetcars, and cable cars.

Following a successful pilot program on one route in 1999, the SFMTA launched the nation's first large-scale real-time transit information system in 2001. Faced with poor schedule adherence, Muni used to publish only general schedule information to avoid disappointing customers if their vehicle did not arrive on time. Timetable booklets did not list specific times for most trips and instead provided scheduled service frequency. With neither specific timetables nor a reliable service, customers would venture out to their stop and hope that Muni would come shortly. In reality, they could end up waiting a highly variable amount of time.

The first-generation system revolutionized how people used transit by giving them an expectation of when Muni would arrive. This basic model, still largely employed throughout North American transit systems, delivers vehicle arrival predictions through countdown signs located at waiting shelters and stations, and now through mobile apps and online.

Times have changed, however. People have grown accustomed to a state-of-the-art technological landscape that continues to redefine the transportation industry through on-demand services and real-time information both before and during one's trip. The explosion of app-based car sharing, bike sharing, private commuter vans, scooters, and ride-hailing services has raised traveler expectations, making it more challenging for transit to meet customer needs with its current information model.

Perhaps the greatest impact to San Francisco's transportation environment has come from transportation

[1]The San Francisco Municipal Transportation Agency, San Francisco, CA

Corresponding Author:
Address correspondence to Jason Lee: jason.lee@sfmta.com

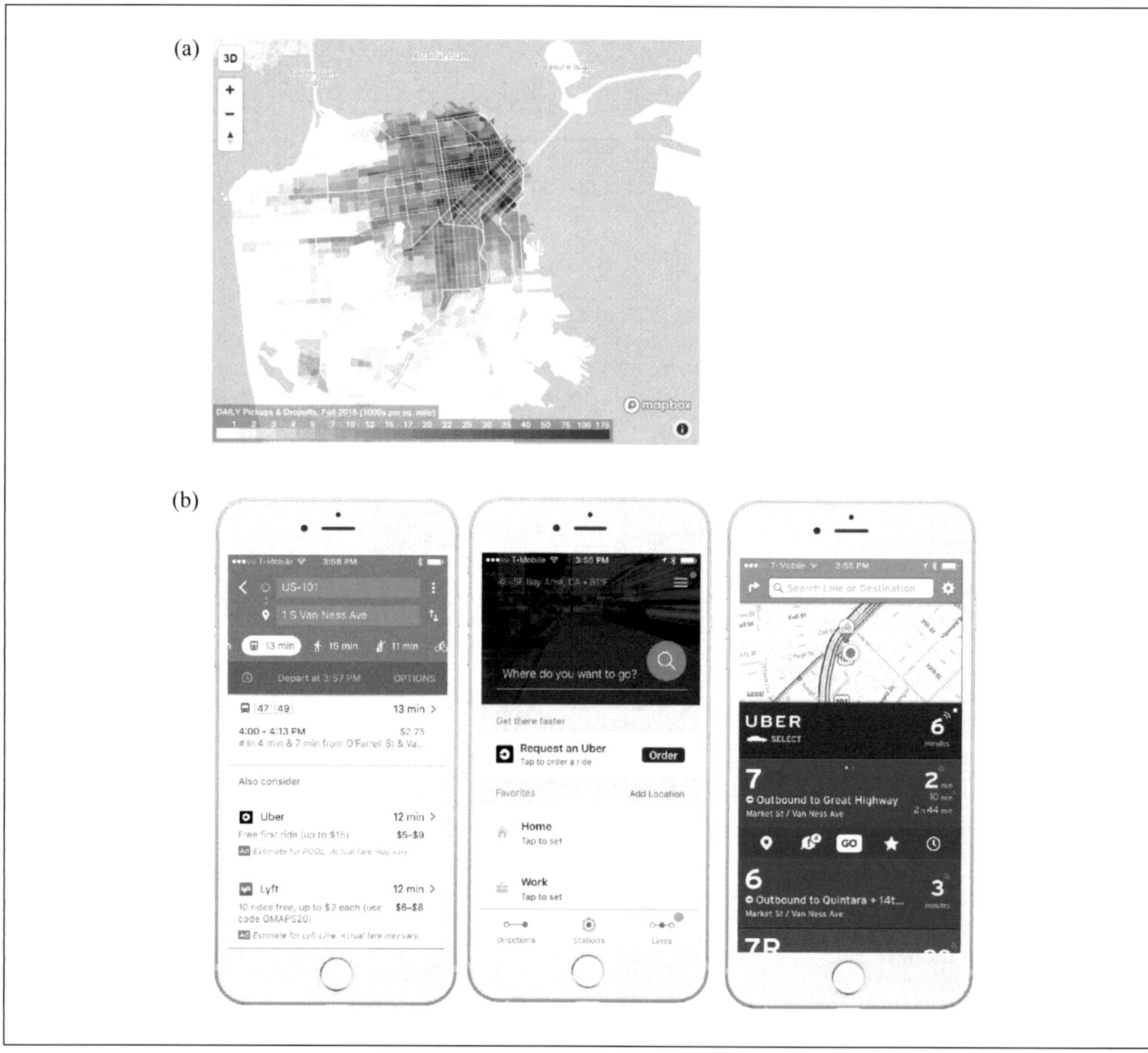

Figure 1. Transportation Network Company (TNC) impacts: (*a*) TNC trips are concentrated in parts of San Francisco where Muni service is the most robust but also vulnerable to congestion (Source: San Francisco County Transportation Authority [1]); (*b*) third-party apps are featuring TNC ads alongside real-time transit information provided through SFMTA open data.

network company (TNC) ride-hailing services such as Uber and Lyft. TNCs allow people to travel to their destination with minimal trip planning—offering personal mobility, but with large-scale consequences from the transportation system perspective. Typically carrying just one or two passengers at a time, TNCs now generate 170,000 vehicle trips, or 570,000 VMT, every weekday in San Francisco. At peak times, they comprise 20% to 26% of vehicle trips in Downtown and South of Market, areas where Muni service is the most robust but also vulnerable to congestion (Figure 1*a*, [*1*]). Furthermore, the SFMTA has observed that third-party mobile apps are increasingly advertising TNCs alongside real-time transit information provided through SFMTA open data (Figure 1*b*).

In addition to traditional factors such as gas prices and service levels, these technological changes may be affecting transit. In 2016, national ridership fell by 2.3% from the previous year according to the American Public Transportation Association. Bus ridership plunged 4.1% when the economy and transportation demand grew. Shifting people from transit to lower-capacity TNCs could impact the livability and sustainability of heavily congested, dense cities like San Francisco.

SFMTA is investing in the next generation of real-time customer information to make Muni the preferred travel choice. The new system will focus on improving prediction accuracy, keeping customers continually informed, retaining those who might otherwise use less sustainable transportation modes, and using data analytics to improve service and operational planning. Examples of potential innovations include:

- Developing a more sophisticated and accurate vehicle arrival prediction algorithm;
- Implementing alternatively powered signage to expand customer access to information at unpowered shelters and stops;
- Reducing travel times by showing nearby alternative routes with shorter waits on digital signage at stops;
- Balancing ridership loads by providing crowding alerts and suggesting parallel services with space available;
- Strengthening network connectivity by showing transfer connection times on board vehicles;
- Communicating service delays and disruptions on board vehicles;
- Providing real-time stop accessibility information and elevator/escalator outage alerts to facilitate travel for seniors and persons with disabilities;
- Using data from mobile technologies to understand customer preferences and improve service and operational planning.

The transit industry's current models and understanding of mode choice must evolve as customer behavior adapts to new transportation options. To assess the potential for the next generation system to build transit ridership, SFMTA fielded an extensive survey and conducted qualitative research to identify how customers would react to the new features being contemplated. Through this effort, SFMTA believes that providing more and better information at different points along a customer's journey could help transit recapture ridership.

To support this conclusion, this paper will first explore prior real-time transit information research. It will then detail the project team's methodology and study findings, discuss implications, and finally describe opportunities for further research.

Literature Review

Extensive research confirms the importance of real-time information, validating its potential to grow ridership in cities like New York and Chicago (*2*, *3*). For example, Brakewood et al. employed a fixed effects model of average weekday unlinked bus trips in New York to reveal a median ridership increase of 1.7% system-wide and 2.3% on the largest quartile of routes (*2*). Other research confirms that real-time information also increases customer satisfaction and feelings of safety and security (*4*–*6*). In another paper, Brakewood, Barbeau, and Watkins revealed that real-time information decreased anxiety and frustration when waiting for the bus (*7*). Much of this work, however, acknowledges that ridership increases attributed to real-time information have been nonetheless modest.

Research also concludes that real-time information reduces actual and perceived waiting times (*8*, *9*). For example, Chow, Block-Schachter, and Hickey conducted in-station surveys before and after the Massachusetts Bay Transportation Authority deployed real-time information to measure changes in satisfaction and wait time expectations. Their results revealed that after the introduction of countdown signs, people reduced their overestimation of wait time by 50% (*10*). Similarly, research commissioned by OneBusAway in Seattle found that real-time information decreases the perceived average wait time by 0.7 min or about 13% (*6*). Brakewood et al. also detected a larger decrease of nearly 2 min for riders using real-time information (*7*).

Others have explored the relationship between real-time crowding information and ridership. Research conducted by Kim, Lee, and Oh revealed that the impact of crowding information varied by group such as students, homemakers, older people, and by trip purpose such as commute and non-commute (*11*). Recently, Zhang, Jenelius, and Kottenhof found crowding information helped redistribute passengers from more to less crowded rail cars by approximately 4 percentage points (*12*). There is an opportunity to supplement these studies by exploring the qualitative attitudes supporting the observed behavior.

Other research explored real-time information's psychological effects on ridership. Tang and Thakuriah concluded that real-time information would be most successful in increasing ridership when paired with programs that increase the public's exposure to transit. They also found that real-time transit information might intervene to break non-transit users' travel habits and increase their transit use (*13*).

Carrel, Halvorsen, and Walker investigated the relationship between riders' past experiences with transit unreliability and their corresponding adaptation strategies. They concluded that the most significant negative experiences that reduced transit use were agency-attributable delays, long transfer waits, and boarding prevention due to crowding (*14*).

Although many have validated the benefits of real-time information, some argue that these benefits come with risks. Gooze, Watkins, and Borning highlighted the importance of error-reporting systems when implementing real-time information. They learned that riders assume a margin of error that negatively impacts ridership. Nevertheless, transit agencies can mitigate this loss by enabling customers to easily report quality control issues (*15*).

In 2017, Song, Peeta, and Hsu modeled how cognitive burden, cognitive decisiveness, and emotional relief influenced customers' route choice decision-making process. They found that stress from information overload or information-induced confusion can weaken real-time information's effectiveness (*16*).

This study builds upon existing research in key ways. Its mixed quantitative and qualitative methodology, large sample size, and focus on underrepresented populations provides a comprehensive and rich data set to analyze how the next generation of real-time information could influence customer perceptions and their resulting mode choices. With survey responses representing a diverse ridership base, the research also provides further empirical evidence into how demographic factors and other variables shape travel decisions among disaggregated subpopulations. Most importantly, this research demonstrates how new types of real-time information—such as alternative routes and transfer connections—could help transit providers retain and increase ridership. Finally, the findings will shape and justify an imminent multi-million-dollar investment in San Francisco's next generation of customer information, while also assisting transit professionals and researchers elsewhere facing similar issues.

Research Goals and Methodology

With the Muni customer experience in mind, SFMTA embarked on an extensive public engagement initiative to determine what information is critical to customers making informed travel decisions. This outreach has allowed the agency to incorporate key findings into the next generation system requirements while revealing the new system's potential to increase mode share. Some key questions SFMTA's research helps answer include:

- How long are customers willing to wait for transit?
- How do customers perceive that transit is "on time" versus "late"?
- What are the salient factors influencing travel choice?
- Can providing real-time information help retain transit customers who might otherwise use another mode?
- How might third-party apps influence travel choice?
- How can real-time information improve willingness to transfer between routes?
- How can real-time information reduce behavioral differences across demographic groups?

In-Depth Survey and Statistical Analysis

From May to October 2017, SFMTA conducted a comprehensive online survey to capture customer travel patterns, inner reasoning, reactions, and preferences (Table 1). The key objective was to assess how customer information could influence travel behavior toward transit.

To publicize the survey, SFMTA leveraged press releases, social media, agency listservs and notices on real-time arrival signs. To increase participation among underrepresented populations, SFMTA also provided Chinese and Spanish translations, and distributed paper versions of the survey. With 5,856 people completing the 10-to-15-min survey, this sample size reflects a margin of error of $\pm$ 1.3% at a 95% confidence level.

To capture diverse feedback and to isolate key demographic factors influencing mode choice, the survey asked voluntary questions on age, ethnic background, gender identity, household income, and disability status. These demographic questions—each of which had at least a 90% completion rate—came at the end to avoid response bias. Based on these responses, the survey has a geographically representative sample from San Francisco and nearby counties that reflects Muni ridership. Results included a robust quantity of samples with regards to ethnicity, age, income, gender and residential ZIP code. SFMTA then applied a series of binary logistic regressions to the data to determine which demographic factors most affected transit mode share with respect to real-time information.

Qualitative Research

Supplementing its quantitative research, the SFMTA conducted focus groups to ensure diverse representation and input. Stakeholder groups included Senior and Disability Action, the Independent Living Resource Center, LightHouse for the Blind and Visually Impaired, and the Youth Commission, as well as low-income communities of color in the Bayview, Chinatown, the Mission, Potrero Hill, and Visitacion Valley. In addition, the project team held "ride-along" immersive field research sessions with customers to observe and better understand their travel behavior.

Findings

Service Reliability

Continuous, accurate real-time information is paramount to meeting customer expectations for service reliability. Transit systems that operate in mixed traffic—the vast

Table 1. Survey Topics

Survey topic	Description
Transportation system usage	
Transportation mode usage	Frequency of use of Muni, other transit systems, walking, bicycling, transportation network companies, taxis, employer shuttles, driving alone, and carpooling
Willingness to wait	Maximum desired wait time without any real-time information during the day, during the evening, and when transferring
Direct vs. transfer trips	Frequency of riding a direct route vs. transfers
Fare payment	Fare media usage (for example, unlimited-ride passes vs. pay-as-you-go)
Muni route(s)	Regularly-ridden Muni routes
Customer information	
Information tools usage	Usage of different information tools (for example, shelter signs, mobile apps, website trip planners, information hotline)
Mobile apps	Usage of different mobile apps
Opinions of existing system	Five-point scale rating of different elements of the existing system
Desired features	Free-form response
Situational questions (stated preference)	
Mode share direct trip	Mode preference in four scenarios when going from work or school: • Customer arrives at stop. Countdown sign predicts a 20-min wait • Countdown sign displays an earlier-arriving alternative route three blocks away • Customer checks smartphone before walking to stop, which predicts a 20-min wait • Customer's smartphone app also advertises for-profit TNC services
Mode share transfers	Mode preference in two scenarios when going from work or school: • Smartphone app shows two routes to reach the destination but not the connection time • Smartphone app also shows estimated connection time
Crowding	Mode preference when countdown sign shows a currently full vehicle followed by a vehicle currently with space available
Demographic questions	
Demographics	Residential ZIP code, age, ethnic background, gender identity, annual household income, and disability
Motor vehicles	Number of motor vehicles in household

majority of service in the United States—face challenges maintaining scheduled service frequency and consistent headways. Actual service differs from schedules because of traffic congestion, variable passenger loads, the deployment of wheelchair ramps, and other factors. All Muni routes, including the Muni Metro light rail system, operate all or portions of their services on surface streets, making them likely to encounter headway variability.

As per City Charter Section 8A.103, "a (Muni) vehicle is considered on time if it is no more than one minute early or four minutes late as measured against a published schedule that includes time points." Focus groups revealed that SFMTA customers rarely consult the official schedule because service is generally frequent and timetables are only available online. When asked about vehicle "lateness," focus groups reported that they define a late vehicle as one that does not arrive according to real-time predictions.

On a scale of 1 to 5 (1 = poor, 3 = good, 5 = excellent), survey respondents rated current prediction accuracy a 2.5. This suggests that improving prediction accuracy could enhance perceptions that Muni is "reliable" and "on time," even without any changes to officially measured on-time performance.

When service disruptions occur, the existing customer information system is unable to provide real-time information about the incident. Not surprisingly, respondents ranked the current communication of service delays and reroutes just 2.0. By providing real-time updates on unplanned service changes and offering alternatives, SFMTA anticipates that the next generation system will reduce the impacts of delays and disruptions for customers, which in turn will improve perceptions of system reliability.

Service Frequency

Although Muni offers some of the highest levels of transit service in the country, there are also weaker parts of the network. In some outer neighborhoods, or in general during evenings and weekends, many routes operate every 20 to 30 min (Figure 2).

Even with this high level of service, customer expectations of maximum waiting times are also high. The survey asked respondents to indicate how long they would be willing to wait if they had just arrived at a stop without real-time information. As indicated in Table 2, the median respondent reported a willingness to wait between 10 and 15 min. Few people are willing to wait

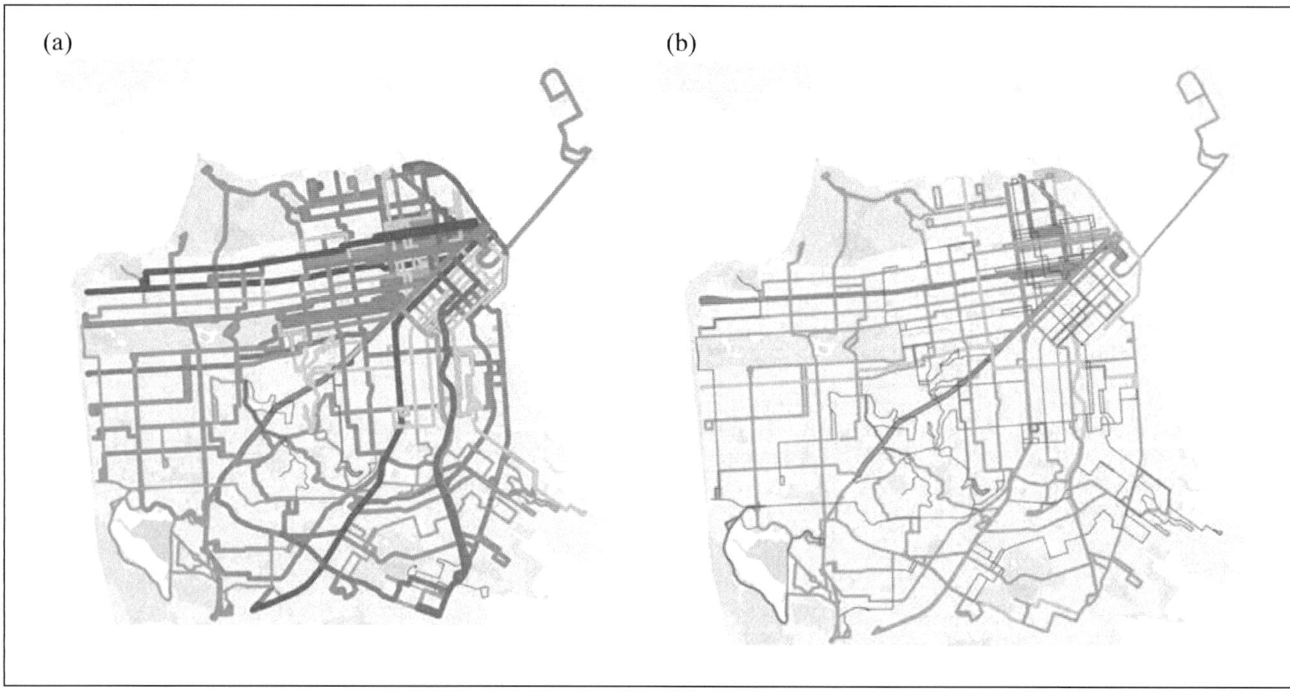

Figure 2. Muni service frequency: (*a*) during rush hours (generally every 15 min or better); (*b*) during late evening (generally every 20 to 30 min).

Table 2. Percentage of Customers Willing to Wait for Transit Without Real-Time Arrival Information

Waiting time until next Muni vehicle	During the day (*n* = 5,856)	During the evening or at night (*n* = 5,856)	When transferring (*n* = 5,856)
5 min	97%	94%	93%
10 min	73%	67%	59%
15 min	35%	34%	22%
20 min	14%	15%	8%
30 min	5%	5%	3%

20 min or more, which is how often many routes operate at night or at other off-peak periods. The willingness to wait also declines noticeably for transfers.

While Muni's frequent service during the daytime aligns well with customer expectations, the discrepancy between willingness to wait and service frequency underscores the importance of having real-time information, particularly during off-peak periods and for transfers.

Transit Customer Information and Mode Choice

To measure how real-time information delivered at different times and places could affect mode choice, SFMTA designed situational questions to test the following:

- Can providing nearby alternative routes with shorter waits help retain transit customers who might otherwise use another transportation mode?
- How do customers respond to transit information disseminated through third-party mobile apps that often advertise private ride-hailing services?
- Does real-time transfer information increase the willingness to transfer between routes?

The survey asked respondents to imagine scenarios where they were going home from work or school and had to wait 20 min, which is not uncommon if there is a service gap or during evenings and weekends. Suspecting that few respondents would be willing to wait that long, which survey findings confirmed (Table 2), the project team designed questions to determine whether different types of real-time information could effectively extend one's willingness to wait.

In Scenario 1 (the base case), respondents arrived at their stop and saw a digital sign predicting a 20-min wait. The survey asked customers what they did the last time they encountered a similar situation. In Scenario 2, the sign suggested an alternate route three blocks away arriving sooner. In Scenario 3, customers consulted their smartphone before walking to their stop and saw their wait would be 20 min. Finally, Scenario 4 was identical to Scenario 3 except that a TNC advertisement accompanied the wait prediction.

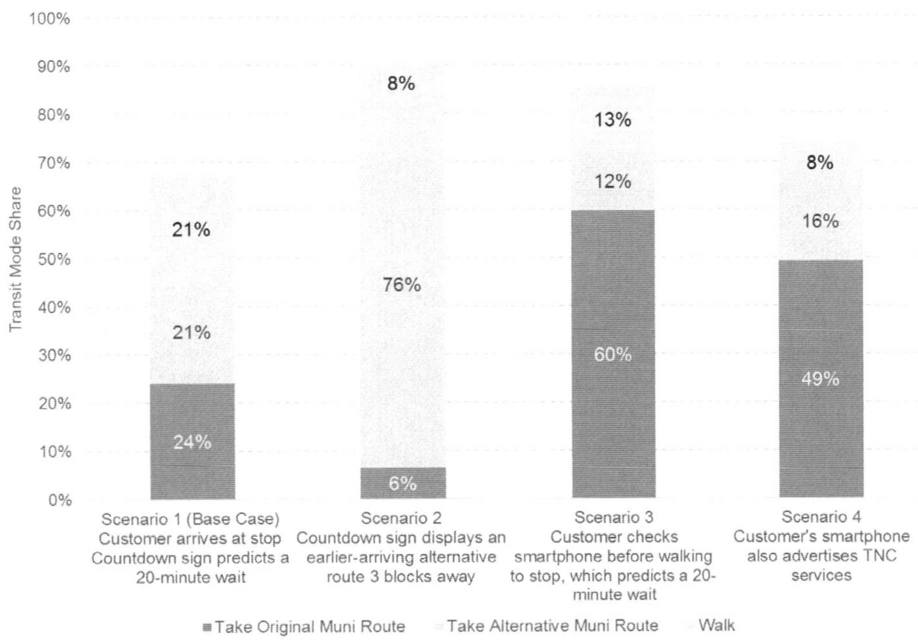

Figure 3. Stated preference mode choice for scenarios involving a predicted 20-min wait ($n = 5{,}856$). Varying types of real-time information presented at different times and places significantly impact the mode share of sustainable transportation options including Muni and walking.

As shown in Figure 3, Muni mode share varied greatly based on the scenario. In Scenario 1, only 45% of respondents took Muni, either waiting the entire 20 min or finding an alternative transit route on their own. Proactively suggesting an alternative route (Scenario 2) boosted Muni's mode share to 82%. When respondents checked a mobile app before walking to their stop (Scenario 3), 72% chose Muni. Finally, when the mobile app also advertised TNC services (Scenario 4), Muni's mode share fell to 65%.

Demographic Variables Influencing Mode Choice

To determine how customer demographics influence mode share, SFMTA conducted binary logistic regressions for each scenario. Table 3 details model results, whereas Table 4 shows whether such variables (all else held equal) influence mode share across scenarios (p-values were statistically significant at or below the 0.05 level). Model highlights include:

- Income: The propensity to ride Muni declines as income grows, but more and better real-time information at strategic times and places could mitigate this.
- TNC density by ZIP: People in ZIP codes with high TNC density are less likely to ride Muni.
- Fare payment method: Customers with prepaid passes are more likely to ride Muni than those who pay per ride.

Income. In recent years, rapid employment and population growth have contributed to widening income inequality and gentrification in the San Francisco Bay Area. Coinciding with these broader economic and social trends, the proliferation of private employer shuttles, commuter vans, and TNCs has also raised the prospect of a two-tiered, income-based transportation system. The analysis confirms that income strongly influences transportation choices—in the base case (Scenario 1), up to a 29-percentage-point mode share difference exists between the lowest- and highest-income brackets (29% vs. 58%, Figure 4*a*).

Figure 4, *a* and *b* show how the next generation of real-time information might alter customer behavior to sharply reduce or even virtually eliminate the influence of income disparities on transit mode share. Faced with waiting 20 min at a stop (Scenario 1), people are far more likely to shift away from Muni as their income increases. In contrast, when the stop's countdown sign offers a nearby alternative (Scenario 2) or customers can consult their smartphones before walking to their stop (Scenario 3), income disappears as a statistically significant variable. The income gap, however, reappears when

Table 3. Binary Logistic Regression Analysis

Variable	Coefficient	Standard error	*p*-value	Odds ratio
Scenario 1 (base case) model results: customer arrives at stop and countdown sign predicts a 20-min wait				
Fare payment method	0.493	0.065	*< 0.001*	1.637
TNC density by ZIP	−0.470	0.067	*< 0.001*	0.625
Age	0.164	0.021	*< 0.001*	1.178
Vehicles in household	0.046	0.034	0.182	1.047
Ethnicity	0.336	0.136	*0.014*	1.399
Gender (male)	0.034	0.065	0.595	1.035
Gender (non-binary)	−0.275	0.450	0.541	0.759
Income	−0.125	0.019	*< 0.001*	0.882
Disability status	0.098	0.120	0.413	1.103
Ethnicity by income	< 0.001	0.028	0.995	1.000
Constant	−0.392	0.142	0.006	0.676
Scenario 2 model results: countdown sign displays an earlier-arriving alternative route three blocks away				
Fare payment method	0.435	0.085	*< 0.001*	1.545
TNC density by ZIP	−0.303	0.090	*0.001*	0.738
Age	−0.012	0.028	0.668	0.988
Vehicles in household	−0.023	0.044	0.601	0.977
Ethnicity	−0.405	0.177	*0.022*	0.667
Gender (male)	−0.025	0.084	0.767	0.975
Gender (non-binary)	1.450	1.028	0.158	4.264
Income	−0.022	0.024	0.375	0.979
Disability status	−0.617	0.141	*< 0.001*	0.540
Ethnicity by income	0.047	0.035	0.182	1.048
Constant	1.928	0.190	< 0.001	6.875
Scenario 3 model results: customer checks smartphone before walking to stop, which predicts a 20-min wait				
Fare payment method	0.414	0.072	*< 0.001*	1.513
TNC density by ZIP	−0.288	0.075	*< 0.001*	0.750
Age	−0.074	0.023	*0.001*	0.929
Vehicles in household	−0.047	0.037	0.202	0.954
Ethnicity	−0.394	0.149	*0.008*	0.674
Gender (male)	−0.069	0.071	0.331	0.934
Gender (non-binary)	−0.188	0.487	0.700	0.829
Income	0.004	0.020	0.860	1.004
Disability status	−0.040	0.130	0.762	0.961
Ethnicity by income	0.054	0.030	0.074	1.055
Constant	1.409	0.159	< 0.001	4.093
Scenario 4 model results: customer's smartphone also advertises TNC services				
Fare payment method	0.341	0.066	*< 0.001*	1.407
TNC density by ZIP	−0.369	0.070	*< 0.001*	0.692
Age	0.065	0.022	*0.003*	1.067
Vehicles in household	0.043	0.036	0.229	1.044
Ethnicity	0.191	0.143	0.182	1.211
Gender (male)	0.052	0.066	0.428	1.054
Gender (non-binary)	0.229	0.488	0.639	1.258
Income	−0.054	0.019	*0.004*	0.947
Disability status	0.146	0.131	0.263	1.157
Ethnicity by income	−0.023	0.028	0.423	0.978
Constant	0.631	0.147	< 0.001	1.879

Note: Italics indicate statistically-significant *p*-values (at or below the 0.05 level). Variable is likely to influence transit mode share, all else held equal.

customers see predictions on a mobile app that also advertises TNCs (Scenario 4).

TNC Vehicle Density. As shown in Figure 1*a*, TNC vehicle trips are most highly concentrated in San Francisco's northeastern quadrant, followed by outlying commercial districts. To gauge whether TNC availability influences transit ridership, the analysis assigned San Francisco ZIP codes into two categories based on TNC density.

Combining both income and TNC density, Figure 4, *c* and *d* show that people living in TNC-dense neighborhoods are statistically less likely to ride Muni in all scenarios, and that this gap generally holds across all income brackets. Most significantly, when faced with a

Table 4. Statistical Significance (*p*-Value) of Demographic Variables

Statistical significance (*p*-value) across cases				
Variable	Scenario 1 (base case)	Scenario 2	Scenario 3	Scenario 4
Fare payment method	Yes	Yes	Yes	Yes
TNC density by ZIP	Yes	Yes	Yes	Yes
Age	Yes	No	Yes	Yes
Vehicles in household	No	No	No	No
Ethnicity	Yes	Yes	Yes	No
Gender (male)	No	No	No	No
Gender (non-binary)	No	No	No	No
Income	Yes	No	No	Yes
Disability status	No	Yes	No	No
Ethnicity by income	No	No	No	No

20-min wait (Scenario 1), highest-income earners living in high-density TNC ZIP codes were 16 percentage points less likely to take Muni than the highest-income earners living in low-density TNC ZIP codes (22% vs. 38%, Figure 4*c*).

Although these results suggest that SFMTA currently might be losing significant ridership in high-density TNC areas, the next generation system could alter this dynamic. Most strikingly, suggesting transit alternatives to avoid a 20-min wait (Scenario 2) lifted transit mode share by 25 to 58 percentage points (55% vs. 80% at the lowest-income level and 22% vs. 80% at the highest-income level, Figure 4*c*). Because high-density TNC areas also overlap with the densest and most frequent parts of Muni's network, there could easily be occasions in which someone could take a nearby alternative Muni route rather than a TNC if informed of that option.

Other Notable Findings

Fare Payment Method. SFMTA enables customers to pay for individual transit rides (with cash at the farebox or with electronic cash value on a smartcard or app) or in bulk (with an unlimited-ride pass). Likely due to SFMTA fare discounts for low- and moderate-income San Franciscans, including the Lifeline Pass and the Free Muni for Youth, Seniors, and People with Disabilities programs, only 2% of survey respondents cited expensive fares as a deterrent to riding transit.

Nevertheless, as indicated in Figure 5, the regression analysis uncovered a correlation between fare payment method and mode choice. Of respondents stating they rode Muni at least 4 days per week, 34% reported paying-as-you-go, despite likely riding enough to make purchasing an unlimited-ride pass financially worthwhile. Though the causality is uncertain, SFMTA hypothesizes that by not making an upfront financial commitment to transit, pay-as-you-go customers might be more likely than pass holders to consider and choose other transportation modes for each trip. Consequently, the number of pay-as-you-go respondents choosing transit was 5 to 14 percentage points lower than for pass holders, depending on the scenario.

Fare policy is an area for further research. In addition to encouraging monthly passes, agencies may wish to consider other fare options that not only accommodate flexibility but also encourage more frequent transit use.

Transfers. Two questions asked customers how they would return home from work or school if their trip required transferring between two Muni vehicles, with and without a real-time prediction of the connection time. Respondents could choose whether they would take Muni all the way, take Muni until the transfer point and then find another transit mode to their destination, call a TNC, or use a taxi. Providing transfer time predictions (assuming a hypothetical 6-min wait) boosted the percentage of respondents who would take Muni for at least the first portion of their trip from 75% to 90% and for the entire length of their trip from 48% to 83%.

Crowding. State-of-the-art automatic passenger counters installed on all buses and light rail vehicles purchased after 2014 give SFMTA the technical ability to report ridership loads and crowding alerts in real time through its new radio communications system. To determine whether such information would be useful, the survey asked what customers would do if the countdown sign showed that a currently full vehicle would be arriving in 2 min followed by a vehicle with space available in 6 min. In addition, the survey asked participants why they made their choice in an open-ended question.

While 68% chose to wait for the following vehicle, 23% opted to try to board the first vehicle despite it being full. In open-ended responses, customers explained that they chose the first vehicle because they did not have

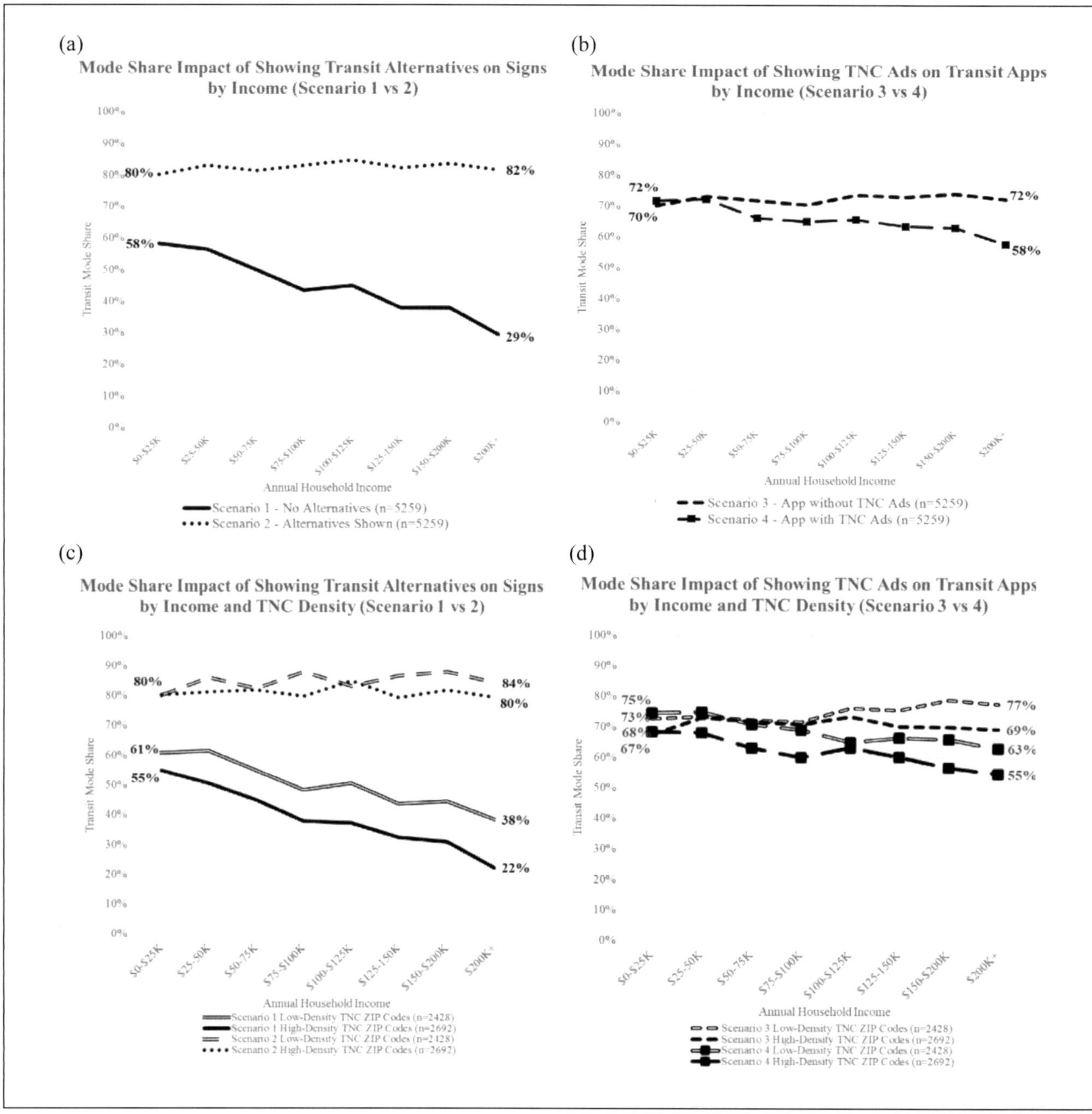

Figure 4. Stated preference transit mode share by income bracket and ZIP code-based TNC vehicle density: (*a*) mode share impact of showing transit alternatives by income; (*b*) mode share impact of showing TNC ads on transit apps by income; (*c*) mode share impact of showing transit alternatives by income bracket and ZIP code-based TNC vehicle density; (*d*) mode share impact of showing TNC ads on transit apps by income bracket and ZIP code-based TNC vehicle density.
Note: The number of respondents in each income bracket ranges from 424 to 845, representing a margin of error of ± 3– ± 5% at a 95% confidence level. Data exclude the 10 percent of respondents who did not indicate their income level.

time to wait for the second vehicle or lacked confidence in the predictions. Many respondents felt the second vehicle would not arrive in the predicted 6 min, would have filled up before reaching their stop, or would have to turn back before the end of the line. Though they might have felt less comfortable on the first vehicle, at least their departure was guaranteed. The extra 4-min wait represented an additional time investment, which would leave customers feeling "burned" if the second vehicle failed expectations. Considering this feedback,

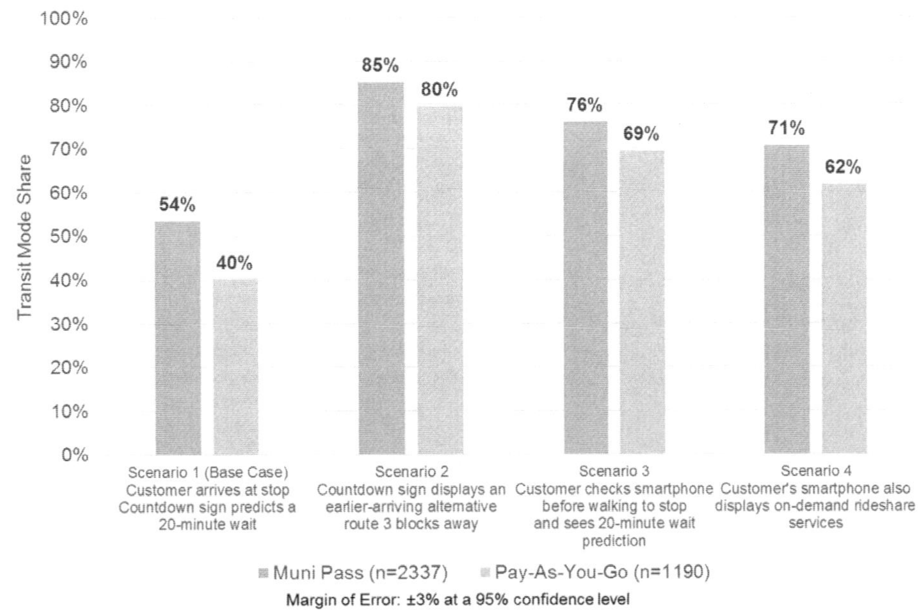

Figure 5. Fare payment, customer information and mode choice. Among survey respondents riding Muni at least 4 days per week, pass users are more likely than pay-as-you-go customers to choose Muni across all tested scenarios for a 20-minute predicted wait.

SFMTA will conduct further research to explore how best to present vehicle occupancy information.

Deterrents to Transit Ridership and How the Project Would Address Them

Finally, SFMTA's survey asked customers to recall the last time they chose another form of transportation over Muni, and to select up to two factors that influenced their decision. As indicated in Table 5, the vast majority were service-related. The next generation system could mitigate some of these factors and turn negative experiences into neutral or even positive ones. Over the long term, improved customer satisfaction would translate into ridership retention and growth.

Further Research

To test how certain enhancements might impact customer mode-choice behavior, this research necessarily relies on stated preferences to hypothetical scenarios. Though this methodology does not always reflect revealed preferences, the results in the aggregate and the high number of respondents imply the next generation system has significant potential to increase Muni ridership.

Qualitative feedback indicates that customers weigh multiple factors when making travel choices. Through the new system's mobile app and website, SFMTA would be able to collect behavioral user data with customer permission in order to gauge real-time information's impact and improve service and operations planning. Once implemented, actual usage data would permit future research into topics such as wait tolerance, latent demand, crowding and comfort, origin–destination patterns, and ridership elasticity.

Conclusion

In the new millennium, San Francisco has emerged as a real-life laboratory to explore how technology can radically alter the transportation landscape. More than 15 years ago, this high-tech hub helped revolutionize the transit riding experience by informing customers when their vehicle was coming in real time. Since then, this first-generation real-time transit information system has remained relatively static while the city has seen an explosion of car sharing, bike sharing, employer shuttles, mobile app-based private commuter vans, and TNCs. San Francisco residents and visitors seemingly have more transportation options than ever, which raises consumer expectations.

SFMTA's Next Generation Customer Information System must adapt to this reality. Traditional demand and mode-choice models must also evolve in response to new transportation options. San Francisco is fortunate

Table 5. Deterrents to Transit Ridership and How the Project Would Address Them

Factor	Percent (n = 5,856)	Potential next generation customer information system benefits
Scheduled service was too infrequent.	32%	Although the new system would not directly increase service, it could reduce the negative impacts of infrequent service by allowing customers to time their arrival at a stop or take a nearby alternative route coming sooner.
Ride on board Muni would take too long.	32%	Though the new system would not directly address vehicle speeds, it could potentially recommend faster route options if available, spread demand, and reduce total travel time by shortening customer waits.
Muni did not arrive when predicted.	31%	By providing more accurate predictions, the new system would improve customer perceptions that Muni is "on time" and lower the chances that customers would give up on Muni while waiting.
There was a service delay.	23%	Though the new system would not directly eliminate service delays, it would notify customers of service disruptions or reroutes during their journey so they can seek alternatives and/or inform those waiting for them at their destination that they will be late.
Muni was too crowded.	21%	Although the new system does not directly increase resources to reduce overcrowding, it could provide crowding alerts to allow customers to wait for a later trip or find a nearby less crowded route.
Transfers were required.	8%	By providing real-time transfer connection times online and on board vehicles, the new system would make transferring more predictable and give customers confidence they can take multiple Muni routes to more destinations.
I did not feel safe or secure.	8%	By providing arrival predictions that allow customers to reduce waiting times, the new system may improve perceptions of safety and security, at least while waiting for Muni.
Muni did not stop for me.	7%	Vehicles often do not stop for customers because they are full. By providing crowding alerts to allow bus customers to wait for a later trip or find a nearby less crowded route.

that Muni still attracts a ridership base representative of its ethnic and socioeconomic diversity. The project team's research findings suggest that the status quo could intensify inequities by dividing transportation services into two income-based systems. Losing transit customers to private or ride-hailed automobiles could lead to a downward spiral: Muni could become slower and less attractive as TNC-induced traffic congestion grows, while less fare revenue due to lower ridership could lead to service cuts and fare increases.

With these challenges also come opportunities. Using public input, the project will focus on improving prediction accuracy, keeping customers informed throughout their journey particularly with respect to service disruptions and transfers, leveraging mobile technology and offering alternatives, and other supplementary information. This research has affirmed that providing more and better information at different points along a customer's journey could potentially increase transit ridership across all demographics, leading to a more equitable and sustainable transportation system. The lessons learned from implementing the Next Generation Customer Information System will in turn help other transit systems prepare for the future.

Author Contributions

The authors confirm contribution to the paper as follows: study conception and design: JL, WT; data collection: JL, WT; analysis and interpretation of results: JL, WT, NV; draft manuscript preparation: JL, WT, NV. All authors reviewed the results and approved the final version of the manuscript.

References

1. *TNCs Today A Profile of San Francisco Transportation Network Company Activity*. San Francisco County Transportation Authority, San Francisco, CA, 2016.
2. Brakewood, C., G. Macfarlane, P. Brinckerhoff, and K. Watkins. The Impact of Real-Time Information on Bus Ridership in New York City. *Transportation Research Part C: Emerging Technologies*, Vol. 53, 2015, pp. 59–75.
3. Tang, L., and P. Thakuriah. Ridership Effects of Real-Time Bus Information System: A Case Study in the City of Chicago. *Transportation Research Part C*, Vol. 22, 2012, pp. 146–161.

4. Zhang, F., Q. Shen, and K. J. Clifton. Examination of Traveler Responses to Real-time Bus Arrival Information Using Panel Data. *Transportation Research Record: Journal of the Transportation Research Board*, 2008. 2082: 107–115.
5. Watkins, K., B. Ferris, and A. Borning. OneBusAway: Behavioral and Satisfaction Changes Resulting from Providing Real-Time Arrival Information for Public Transit. Presented at 90th Annual Meeting of the Transportation Research Board, Washington, D.C., 2011.
6. Watkins, K., B. Ferris, A. Borning, G. Scott Rutherford, and D. Layton. Where is My Bus? Impact of Mobile Real-Time Information on the Perceived and Actual Wait Time of Transit Riders. *Transportation Research Part A: Policy and Practice*, Vol. 45, No. 8, 2011, pp. 839–848.
7. Brakewood, C., S. Barbeau, and K. Watkins. An Experiment Evaluating the Impacts of Real-Time Transit Information on Bus Riders in Tampa, Florida. *Transportation Research Part A: Policy and Practice*, Vol. 69, pp. 409–422.
8. Cats, O., W. Burghout, T. Toledo, and H. Koutsopoulos. Effect of Real-Time Transit Information on Dynamic Path Choice of Passengers. *Transportation Research Record: Journal of the Transportation Research Board*, 2011. 2217: 46–54.
9. Cats, O., W. Burghout, T. Toledo, and H. Koutsopoulos. Modeling Real-Time Transit Information and its Impacts on Traveler Decisions. Presented at 91st Annual Meeting of the Transportation Research Board, Washington, D.C., 2012.
10. Chow, W., D. Block-Schachter, and S. Hickey. Impacts of Real-Time Passenger Information Signs in Rail Stations at the Massachusetts Bay Transportation Authority. *Transportation Research Record: Journal of the Transportation Research Board*, 2014. 2419: 1–10.
11. Kim, J., B. Lee, and S. Oh. Passenger Choice Models for Analysis of Impacts of Real-Time Bus Information on Crowdedness. *Transportation Research Record: Journal of the Transportation Research Board*, 2009. 2112: 119–126.
12. Zhang, Y., E. Jenelius, and K. Kottenhoff. Impact of Real-Time Crowding Information: A Stockholm Metro Pilot Study. *Public Transport*, Vol. 9, No. 3, 2016, pp. 483–499.
13. Tang, L., and P. Thakuriah. Will Psychological Effects of Real-time Transit Information Systems Lead to Ridership Gain? *Transportation Research Record: Journal of the Transportation Research Board*, 2011. 2216: 67–74.
14. Carrel, A., A. Halvorsen, and J. Walker. Passengers' Perception of and Behavioral Adaptation to Unreliability in Public Transportation. *Transportation Research Record: Journal of the Transportation Research Board*, 2013. 2351: 153–162.
15. Gooze, A., K. Watkins, and A. Borning. Benefits of Real-Time Transit Information and the Impacts of Data Accuracy on the Rider Experience. *Transportation Research Record: Journal of the Transportation Research Board*, 2013. 2351: 95–103.
16. Song, D., S. Peeta, and Y. Hsu. Psychological Effects of Real-Time Travel Information on Traveler Route Choice Decision-Making Process. Presented at 96th Annual Meeting of the Transportation Research Board, Washington, D.C., 2017.

The Standing Committee on Public Transportation Marketing and Fare Policy (AP030) peer-reviewed this paper (18-03781).

Article

TRR
JOURNAL OF THE TRANSPORTATION RESEARCH BOARD

Transportation Research Record
2018, Vol. 2672(8) 536–545

DOI: 10.1177/0361198118798242
journals.sagepub.com/home/trr
SAGE

Data-Driven Opportunities from an Account-Based Fare Payment System

Raymond Chan[1], Maulik Vaishnav[1], Scott Wainwright[1], Paul Murray[1], and Alex Cui[1]

Abstract

The development of an account-based transit card has provided an opportunity for the Chicago Transit Authority (CTA) to examine transit usage differently. This new automated fare card/smart card system integrates many parts of a transit fare system, such as physical card, fare product, purchase information, and transit use location. The data are also integrated on a much longer timeframe than the previous generation transit card system. Increased data longevity allows the CTA to embark on a longitudinal study of transit card usage by separating cards in a priori groups. In a period between May 2016 and May 2017, a slight decrease in cards used in the system led to a small ridership drop, but a larger drop in ridership was found in passengers who used the system from one year to the next. This longitudinal application of passenger utilization of the transit network provides the agency with additional information that can be used to adjust fare policies and start marketing campaigns based on customer usages. This allows the agency to improve decision-making ability with greater insight on customer behavior.

The advancement of automated fare cards (AFC) has rapidly progressed in last two decades. At the Chicago Transit Authority (CTA), the introduction of the Chicago Card in 2003, a stored-value smart card fare system, and increased data from the magnetic stripe fare cards led to a development of an integrated data analysis suite. These tools allow planning and scheduling staff to use cross-sectional analyses of bus and train ridership information to design and implement service to the half-hour level (*1*). In 2013, CTA implemented a new contactless, account-based smart card fare system, Ventra, which provides numerous advantages to the organization. Increased reporting speeds, data fusion with automated vehicle location (AVL) technology, and a centrally managed electronic purse (compared with a stored-value card) increases the ability for the agency to better understand customer patterns.

An account-based system provides a steady stream of transit user data over time, and improved understanding of changing fare product and transit use behavior. The ability to link transit usage and fare product sales over long periods of time provides detailed passenger behaviors, helping the agency understand customer characteristics such as turnover. The results from these analyses have tremendous modeling and marketing potential for transit agencies.

This paper starts with a research objectives statement outlining the aims of this work followed by a literature review of academic advances in AFC and smart card use as well as some practitioner ready advances in AFC use in planning and market research. Then, background information on the Ventra account-based system is presented along with the longevity of the dataset. The data are then applied to examine how transit usage has changed over a year, which is used in two different policies and marketing ideas. This paper concludes with agency impacts and future opportunities.

Research Objectives

The goal of this research is to provide useful and practitioner ready data for decision makers to better understand customers at the CTA with the use of currently available data and new techniques. This research utilizes the account-based, transactional fare card data at the CTA to develop a longitudinal ridership model. This piece explores the possibilities with these data by

[1]Chicago Transit Authority, Chicago, IL

Corresponding Author:
Address correspondence to Raymond Chan: rchan@transitchicago.com

providing a case study with enhanced ridership trends. It also provides a next step, illustrating further use of this method to other customer dimensions such as fare value loading, and demonstrates a potential action for agencies with targeting marketing communication.

Background

The academic literature on AFC analyses for planning purposes can be categorized into three different decision-making horizons: strategic (long term), operational (medium term), and tactical (short term). This differs slightly from Pelletier et al., which focuses on planning and technical aspects of smart card and AFC usages in transit agencies (*2*).

Strategic uses of AFC and smart card data focus on identification of different classes of behaviors to separate out different types of users of the system. This activity modeling is based on the idea that transit agencies have different types of passengers, such as traditional transit commuters, visitors, and infrequent transit users. Such applications have uses, from targeted marketing campaigns to improved service design based on passenger types.

Different clustering algorithms, such as *K*-means and density-based spatial clustering of application with noise (DBSCAN) are used to group and separate users. Both temporal (*3*) and spatial-temporal pattern analysis (*4, 5*) are used to classify users. A priori classification of different users is applied to the groups to cluster passengers (*6*). Inferred activities and passenger classification are validated with traditional survey to collect missing information. Demographic information (*7*) and trip purpose (*8*) are inferred through behavioral patterns derived from AFC data and validated with a survey sent out to a small sample.

Operational and tactical decision making are augmented with smart card and AFC data. Many systems that have AFC validation at network entry ("Tap on systems") and barrierless transfers between routes and/or modes infer alightings through additional data about the next transaction (*9*). To capture passengers that are not using smart cards, inferences are made based on smart card users (*10*). Passenger alighting and journey information aids in the development of origin–destination (OD) models, which are applied in transit microsimulation models (*11, 12*) to analyze passenger movements throughout a public transit network.

By relating AVL data to fare transaction data and automated passenger counters, information on performance metrics, such as running time and on-time performance, are assessable by using the fare's transaction time as a probe for vehicular time (*13*). Agencies can use this methodology to understand how different types of users, based on fare classes, are impacted by service changes to a route (*14*).

Transit Agency Implementation

The literature shows that although AFC technology was not developed explicitly for planning purposes, agencies are increasingly using these data for increased accuracy. TCRP Report 126 details how transit agencies can leverage AFC and other intelligent transportation system (ITS) technologies, such as automated passenger counters (APC) and AVL, for market research and other planning activities (*1*). MTA's New York City Transit utilizes AFCs, combining this with AVL information to create an OD model for route planning purposes (*15*).

Although much of the literature has focused on operational and tactical-level decision support tools, the literature lacks models and analyses of long-term transit trends over months and years utilizing primarily smart card data. This is potentially because of the AFC data limitations, the smart cards penetration on a transit system, or AFC fare media longevity. Furthermore, many analyses focus on statistical models on short periods of time, such as transfer analysis or journey analysis on a daily basis or the stability of travel patterns over a few days or weeks.

Most agencies perform cross-sectional analyses, where all behaviors are measured repeatedly. For example, FTA's National Transit Database data is a cross-sectional dataset of passenger behaviors at a transit agency. While this dataset can detect if ridership has dropped on a transit system, it cannot discern between different events that could lead to this outcome, such as "many passengers took fewer rides" versus "new passengers took fewer rides than passengers who left the system." Understanding how transit passengers change their behaviors is important in developing better transit systems. A long-range longitudinal analysis, examining changes to a cohort over a long period of time, focuses on the behavior of individuals over a period of time, which can help understand how people change their habits.

Background on CTA's Open Fare System: Ventra

The Ventra system is an account-based, open fare payment system. Instead of a stored-value system, transit value is a centrally stored electronic purse, with the physical card providing account details, validated when a user requests a payment (entry) into the transit network. This works similar to a debit card, where the value is stored centrally in a bank, rather than encoded on the physical media. Therefore, when a user loses their card, another card is issued to that user, linked directly to their old account. The physical card is an abstraction of the fare mechanism (i.e., unlimited 7-day pass, college, student, etc).

CTA implemented Ventra in fall of 2013. Every fare payment ("transaction") is centrally recorded on a database after each fare payment at a farebox, rather than stored in a farebox and transmitted at a centralized vehicular facility ("probed"). At each transaction, AVL information is also provided, allowing researchers to examine the data for both spatial and temporal patterns. Lastly, the Ventra system also can provide researchers information on the type of mechanism and where the value was loaded (i.e., cash at a ticket vending machine or a credit card on the Ventra app). As of 2017, over 95% of all fares collected at CTA are on a smart card. Although rail stations only accept Ventra cards, CTA busses still allow cash payments.

The open fare payment system allows any credit/debit card or mobile payment system that is compatible with contactless payment technologies (*16*) to pay at the faregate or on a bus without having to purchase a Ventra card. This includes most mobile payment mechanisms such as Apple Pay, Android Pay, and Samsung Pay that include a compatible near field communication (NFC) technology.

Cards versus Customers

The transition to Ventra provided more robust transactional data. In the past, the percentage of registered stored-value smart cards was low compared with the magnetic stripe cards. Given their non-account-based nature, the potential to segment and identify those behaviors was relegated to a smaller sample of the transactions occurring in the Chicago Card smart card. In the new environment, almost all cards are account based. In addition, the Chicago Card smart card only had two main fare products: pay per use and the 30-day unlimited ride pass. Some 3% of the fare media used on the system are limited-use disposable tickets and are not included in this analysis.

Systems that employ disposable magnetic stripe cards can have a difficult time using a card's identifier as a proxy for a customer. While Ventra's account-based system does not perfectly model the customer, it reduces the ratio of cards per customer. For example, the unlimited passes distributed by the CTA prior to Ventra's introduction could only be used for the life of the pass. Counting each physical fare card individually would oversample the shorter duration unlimited passes compared with longer duration passes as the analysis period expanded. For example, a customer could use three 1-day unlimited passes and one 7-day unlimited pass in a month. On the magnetic stripe system, this results in four different physical cards use. A single Ventra card is used for different fare products.

Yet, more cards have been produced and accounts have been created compared with the unique customers using the CTA. Customers still lose their cards. Old cards are found at the bottom of the drawer. There are often gaps of at least one month in card usage use even among frequent transit card users. Furthermore, a customer could use different physical cards. For example, a user who forgets their Ventra card could use a non-Ventra credit card to enter the system. Each new card is assigned its own unique account number. Mobile devices create a virtual credit card for each physical card they replicate, potentially leading to one credit card account assigned two or more Ventra accounts. The increased tokenization from mobile devices and card number changes are expected to grow.

One method to link disparate cards together could be to use the registration statuses of cards. Ventra allows a customer to link multiple Ventra cards to the same registration. Customers are incentivized to register because they are charged a $5 fee which is refunded on registration. Just over half of all cards used in March–May 2017 were registered, including 64% of standard issue Ventra card, the *full fare card*. However, two-thirds of new cards (activated in 2017) are not registered, suggesting that many consider the registration process too burdensome. This presents a challenge given the increasingly competitive nature of urban transportation, with new entrants into the market. New and non-regular customers regularly face the decision to register their new card, pay the $5 fee for their new card, or take another mode altogether which likely offers seamless payment. Even so, the registration rates among many long-term customers strengthen this analysis.

Customers that register their cards are assigned a patron identifier in the dataset which can be used to connect multiple cards. However, many of these registrations are also created at a household level, with dependents or elderly parents are registered together with the primary transit user. This ambiguity around registrations at a household or individual level makes inferring transit behaviors difficult. Approximately 6.8 million Ventra transit accounts were created since its inception, which is approximately 1.4 Ventra cards per registered account.

Life of a Card

To utilize the account-based smart card data in a longitudinal study, CTA needed to understand the longevity of the Ventra accounts and cards that are used. In this natural experiment setup, the longevity of the cards is reliant on many factors, such as the natural turnover rate and the cohort availability in an analysis period. *Full fare*

Table 1. Longevity of Cards by Rider Class

Rider class	2013.5	2014.0	2014.5	2015.0	2015.5	2016.0	2016.5	2017.0	Total
Full fare	18%	13%	8%	7%	9%	9%	13%	22%	100%
O'Hare employee	31%	21%	9%	7%	8%	7%	9%	9%	100%
College	25%	3%	16%	4%	19%	4%	24%	5%	100%
Student	14%	5%	22%	4%	19%	4%	22%	9%	100%
Student child	4%	6%	10%	7%	17%	14%	25%	18%	100%
Disabled	47%	20%	9%	5%	6%	4%	5%	5%	100%
Paratransit	27%	17%	11%	8%	8%	8%	8%	13%	100%
Senior	34%	28%	13%	5%	5%	4%	5%	5%	100%
Disabled free ride	51%	25%	10%	5%	3%	2%	2%	1%	100%
Senior free ride	41%	28%	16%	5%	3%	2%	2%	2%	100%
Active military	16%	24%	18%	12%	10%	7%	7%	6%	100%
Disabled military	32%	30%	15%	7%	6%	4%	4%	3%	100%
Total	23%	14%	11%	6%	9%	8%	13%	17%	100%

cards make up approximately 66% of all cards distributed. Of the 5.5m *full fare cards* used since inception, 70% of the then were active for 12 months or less and 31% were active for 1 month or less

The rate of turnover on *full fare cards* is non-trivial; 22% of these cards used from March–May 2017 were new in 2017 and an additional 22% were new in 2016, as seen in Table 1. Even so, nearly 40% of all *full fare cards* have been used for 3 years or more.

Disabled, Paratransit, Senior, Disabled Ride Free and *Senior Ride Free* cards have a much longer life. Users must meet certain requirements set by Chicago's Regional Transit Authority to apply for the cards. This higher barrier to obtain these cards is potentially responsible for the longer card life. Over 85% of the cards active in the March–May 2017 timeframe were first used in 2013 or 2014.

Overall, 224,000 cards that have been used every month over the last 2 years. These accounts represent 17–20% of the cards used in any one month and account for 28% of total monthly ridership. The longevity of these uses provides an opportunity for longitudinal study.

Methodology

Customer usage of transit was analyzed over a 5-week period starting April 30, 2017 through June 3, 2017 and compared with May 1, 2016 through June 4, 2016. This methodology approximates a month and accounts for variations in a calendar month and seasonality over a year. The CTA uses the same month–last year method, year-over-year analysis (YoY), to compare many metrics in a cross-sectional analysis. A unique customer identifier is used to provide longitudinal analysis.

Seasonality is naturally accounted for in a YoY analysis, allowing analysts to focus on impacts from other factors. Although this works well for many cross-sectional analyses reported on a monthly basis, utilizing this methodology for longitudinal customer studies has numerous limitations. One of the limitations of the data is the longevity of physical cards and the presence of a card on the transit network from one year to the next. In any given month, about half of all cards used were not used in the same month the previous year. Figure 1 illustrates the amount of Ventra data that can be used in a YoY methodology based on data from May 2017.

Out of the cards that were not available to be longitudinally analyzed in the YoY methodology, three classes of users were identified: continuous cards, non-continuous cards, and new cards in May 2017. A continuous card is defined as a card that saw no calendar gap in ridership—these cards took at least one ride every month from the card's first use through May 2017. Non-continuous cards had ridership patterns that were difficult to define on a monthly basis. For example, some cards were not active in the month of December 2016, which would suggest that these were college students who did not depend on CTA outside of the school year. A new card in May 2017 was a card that was never used prior to May 2017.

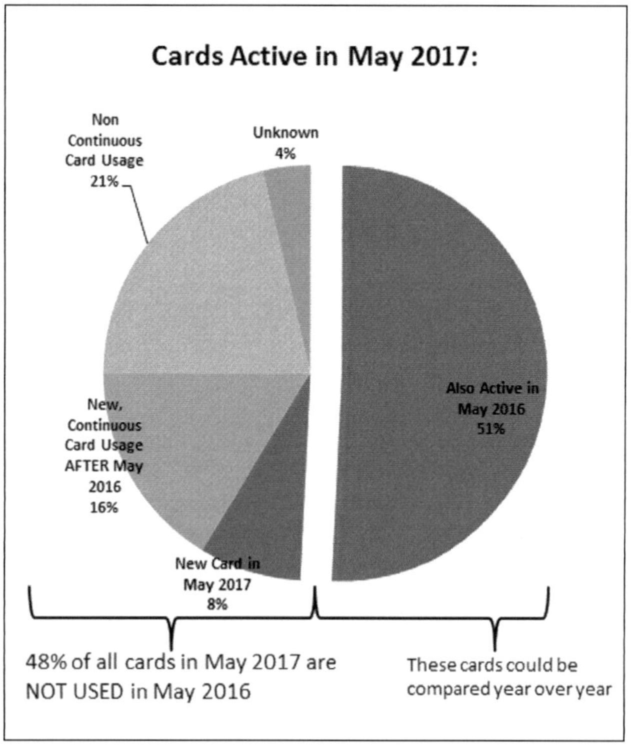

Figure 1. Analytical penetration of longitudinal year-over-year analysis.

Analysis: Frequency of Transit Use

Understanding the how customers change their transit use patterns is important for the agency to answer complex questions. CTA uses the YoY methodology to specifically compare ridership in one month with the same period 1 year prior to account for seasonality in transit ridership in a cross-sectional study. However, the use of a longitudinal analysis to support the cross-sectional reporting helps answer questions about the *source* of the ridership loss.

CTA used the frequency dimension to create groups based on different types of transit usage. These groupings are used to illustrate the types of passengers on the system. Figure 2 shows the amount of cards that took a specific number of transit trips over a 5-week period in 2017 in a column chart.

The columns are a histogram of the number of cards that have taken a specific number of rides over the 5-week period between April 30 and June 3, 2017. The coloring of the columns shows the frequency groups that were established a priori based on typical travel patterns. The line plots the total number of rides taken by all binned users defined by the histogram (cards * transit rides). As a passenger's usage increases, the total number of rides increases as well until around 50 rides, approximately 10 rides per week. This demonstrates that the

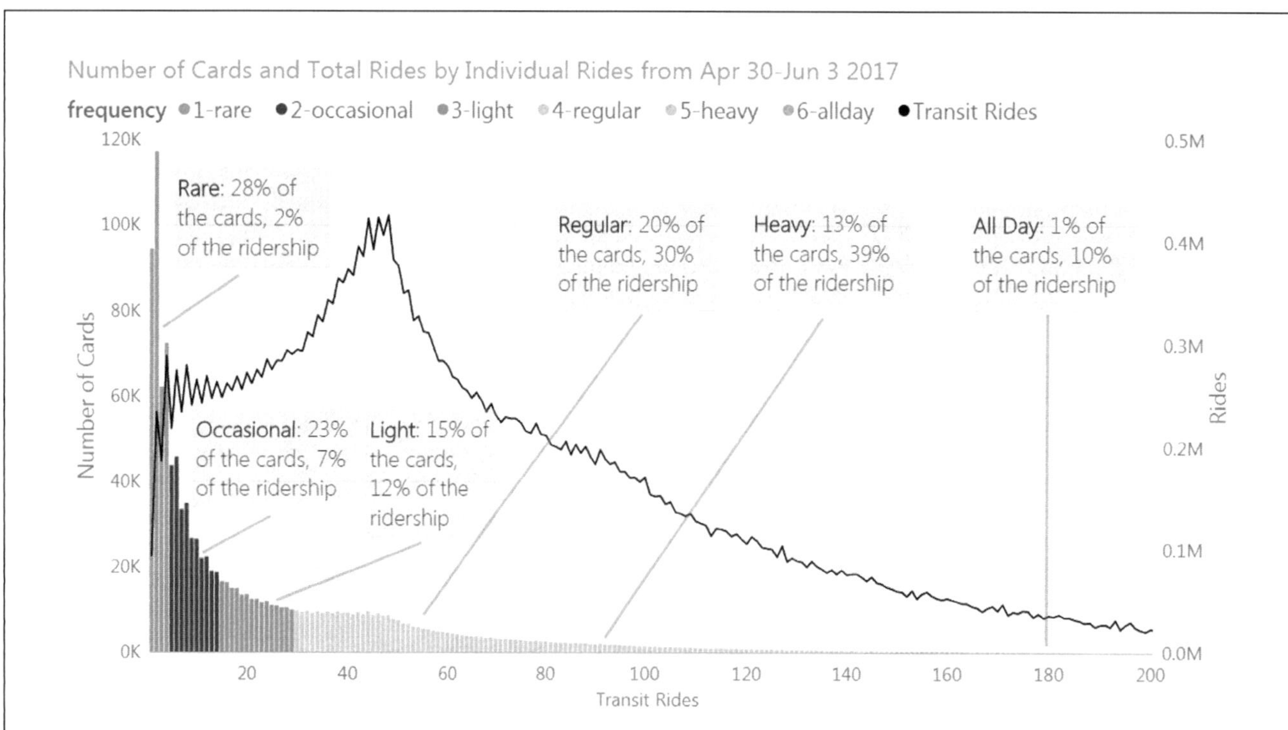

Figure 2. Distribution of cards and aggregate rides by number of transit trips.

Table 2. Cards Grouped by Transit Usage (May 2016 vs. 2017, in Thousands)

		Current period (4/30/2017–6/3/2017)						
		Rare (1–4 rides)	Occasional (5–14)	Light (16–29)	Regular (30–59)	Heavy (60–149)	All day (>150)	
Prior period (5/1/2016–6/4/2016)	Rare	80	37	10	8	1	0.0	
	Occasional	45	49	19	11	1	0.0	
	Light	14	22	27	18	2	0.0	
	Regular	10	13	22	71	6	0.0	
	Heavy	1	2	2	8	8	0.1	
	All day	0.0	0.0	0.0	0.0	0.1	0.1	
		Rare	Occasional	Light	Regular	Heavy	All day	Total
Prior period	YoY cards	136	126	83	122	20	0.2	487
	Non YoY cards	280	154	66	61	11	0.1	572
	Total cards	416	280	149	183	31	0.3	1,059
		Rare	Occasional	Light	Regular	Heavy	All day	Total
Current period	YoY cards	150	123	80	116	18	0.2	487
	Non YoY cards	262	145	65	64	11	0.1	547
	Total cards	412	268	145	180	29	0.3	1,034

regular (30–59 rides) and *heavy* (60–149 rides) riders are responsible for a larger amount of transit trips. Furthermore, while there are many individuals who have used transit during this period, their transit usage was lower. These cards are categorized in the *rare* (1–4 rides) or *occasional* (5–14 rides) frequency groups.

Then, the 2017 data are compared against a similar dataset from May 1 through June 4, 2016 to understand the shift in transit usage by cards over the course of a year. Even though approximately half of the cards appear in both time periods, this can provide insight on changing travel patterns. Table 2 illustrates the shift in usage pattern among cards that were present in both May 2016 and 2017. In addition, the table provides information on the cards that were only used in one of the two periods.

The top part of the table shows the shift in transit habits from one year to the next. For instance, 45,000 riders in the prior period which were *occasional* riders in 2016 are now *rare* riders in 2017. The counterfactual example shows that 37,000 riders increased their ridership, elevating them from the *rare* group in 2016 to the *occasional* group in 2017. Overall, 235,000 (48%) of cards stayed in the same group (along the diagonal in green), while 113,000 (23%) of cards moved to a more frequent group (in the top triangle in blue) and 139,000 (29%) of the cards moved to a less frequent group (in the bottom triangle in red).

1.03 million cards were used in May 2017 which was 25,000 or 2.3% less than 2016. Only 47% (487,000) of the cards used in May 2017 was also used in May 2016. The non YoY cards also declined in all the defined groups, except for the *regular* frequency group. Whereas approximately 64,000 more new cards used transit compared with the 61,000 that left, nearly 6,000 of the returning cards moved out of the *regular* group in 2017.

Similarly, Table 3 looks at the number of transit trips taken by all cards based on the usage grouping established earlier. Table 3 is organized similarly to Table 2, with differences in rides from the cards per group that appear in both years on top, followed by a summary of all riders' usage below.

According to Table 3, 17.7 million rides were taken on the CTA in May 2017, which was 0.6 million or 3.6% less than May 2016. Whereas the cards that returned in 2017 collectively saw fewer rides taken on transit, the cards that only appeared in 2017 took more rides than the cards that did not appear in 2017. Furthermore, cards which used transit less collectively took 2.4 million fewer rides (red shaded region), compared with cards that used transit more, which collectively took 2.0 million more rides (blue shaded region).

The *regular* frequency group from 2016 collectively decreased ridership by 1.1 million rides. However, the *regular* frequency group in 2017 only saw an increase of 0.8 million rides, leading to a net drop in 0.3 million rides from customers who rode the CTA in both years. Similar conclusions can be derived from the other groups. These data support the theory that the decline in ridership from the prior period to the current period is made up mainly of transit passengers who used the system consistently from year to year, but chose to take fewer rides.

The power of a more robust smart card transactional data environment is evident when looking at frequency by card or usage. The ability to extract even more insight into the behavior of transit passenger is enriched with other characteristics such as fare product, mode, location, time of day, or even differentiating between weekdays and weekends. This can assist a transit agency in understanding its transit passengers to better market and educate the effectiveness and value of public transit.

In addition, in a traditional cross-sectional analysis, analysts can characterize the overall travel patterns in each period, but cannot readily explain them. When ridership drops, decision makers and policymakers quickly point to other related events. For example, many will point to the continued population losses in Chicago's south and west sides as the culprit for CTA's ridership loss. This longitudinal analysis shows that although fewer customers use the CTA, most ridership loss can be attributed to returning customers using public transit less often. The ability to surgically specify transit usage change by individual behavior has important policy implications for decision makers, given the increasingly competitive urban transportation environment.

Next Steps

The Ventra dataset has opened up many opportunities at CTA for different applications based on more detailed and longitudinal data. While analyses such as the one above provide answers to questions and insights onto existing models, the strength of this integrated dataset provides new opportunities that were not possible with a traditional stored-value magnetic strip or smart card-based system. For example, CTA can utilize these data to better understand how and where customers purchase different fare products and in a targeting marketing campaign or in traditional route and service planning.

Purchase Habits: Sales Channel, Fare Product, and Transit Use

With the integrated Ventra dataset, CTA is able to connect customer transit usage directly to the fare products purchased. Before, CTA's ability to understand customer's fare product purchase patterns was limited to which distributors (and which locations) products were purchased. The introduction of Ventra provides the CTA the ability to observe the entire activity pattern, from physical card purchase location, value load mechanism (cash, credit), the fare product purchased (none, unlimited pass), to usage location.

From the data, customers who use retail partners to load value onto Ventra are more likely to buy a 7-day unlimited ride pass and use the bus network more often. They also use the transit network throughout the day and weekends. On the contrary, passengers who have setup autoload via a credit card or load value through an employer pre-paid benefits program rely more on 30-day passes, ride rail more often, and use transit during traditional commute hours. Although information on the older magnetic stripe fare system provided similar information based on retail sales locations, the ability to examine these data on an individual basis provides more insight on the types of fare products customers may shift between.

This methodology can also be used with the introduction of new fare purchase mechanisms. With the introduction of new account management mechanisms, such as managing from a webpage from Ventra's inception to the introduction of the Ventra app in 2015, consumers have increasing options to load value and purchase fare products. This change has dramatically increased the share of sales at internet and ticket vending machine (TVM) sales channels, reducing customer reliance on the retail network, as seen in Figure 3.

This cross-sectional analysis shows that the retail network has lost some market share whereas online channels and TVM have gained. However, sales at the retail network have also declined because people are loading less value per transaction and buying fewer passes. Longitudinal analysis allows CTA to compare the flows between fare purchase mechanisms. The net migration change is also negative, suggesting fewer transit riders who depend on the retail network. Continuing to understand the changes in customer behavior can help guide future investment in providing transit fare passes to customers.

The results from the longitudinal sales analysis have many implications for fare modeling. These longitudinal data can provide insight into how transit usage is linked

Table 3. Usage by Transit Usage Groups (May 2016 vs. 2017, in Thousands)

		Current period (4/30/2017–6/3/2017)						
		Rare (1–4 rides)	Occasional (5–14)	Light (16–29)	Regular (30–59)	Heavy (60–149)	All day (>150)	
Prior period (5/1/2016–6/4/2016)	Rare	-7	199	185	310	60	2	
	Occasional	-253	-7	211	367	80	2	
	Light	-253	-249	-6	312	82	1	
	Regular	-398	-429	-379	-48	145	3	
	Heavy	-74	-102	-102	-189	-12	5	
	All day	-0.3	-0.8	-1	-3	-4	-162	
		Rare	Occasional	Light	Regular	Heavy	All day	Total
Prior period	YoY rides	322	1,079	1,769	5,208	1,506	1,903	11,787
	Non YoY rides	614	1,279	1,387	2,523	788	21	6,612
	Total rides	936	2,358	3,156	7,731	2,294	1,924	18,399
		Rare	Occasional	Light	Regular	Heavy	All day	Total
Current period	YoY rides	346	1,050	1,714	4,915	1,303	1,746	11,074
	Non YoY rides	573	1,201	1,367	2,676	828	23	6,668
	Total rides	919	2,251	3,081	7,591	2,131	1,769	17,742

to different fare product purchases. For example, without these data, many fare models might have modeled switching between unlimited 7-day passes and 30-day passes. However, account-level data have suggested that there may not be much overlap between the two fare products, as the populations using them are significantly different. Pricing pass discounts differently for the two products may not lead to as much switching between them as previously thought. These behavioral characteristics can be used to infer other information that would otherwise be difficult to capture without survey information. This can help inform and develop new fare product policies, pricing, and strategies.

Marketing Application: Targeted E-mail Communication

On-going account-level analyses found that many new customers were riding CTA regularly but were not buying unlimited ride pass products, even though it would benefit them to do so. Moreover, some other customers had stopped buying passes and opted for pay-per-use but continued to ride regularly. To encourage unlimited pass sales, CTA reached out to some of these customers via a targeted email campaign.

Customers who register their Ventra cards can provide their email address and opt in to receive promotional emails from Ventra. CTA has used this communication tool to send out promotional messages as well as market research surveys. With longitudinal customer patterns, CTA could embark on a marketing-specific communication to encourage more efficient pass purchasing behavior.

The CTA identified more than 11,000 accounts which had taken transit rides valued at more than $20/week for at least 15 out of 20 weeks in 2017. A subset of about 3,600 accounts received an email on CTA unlimited passes, pre-tax transit benefits, and the transit value autoload feature. The goal was to see if simple and low-cost targeted marketing efforts can change purchase

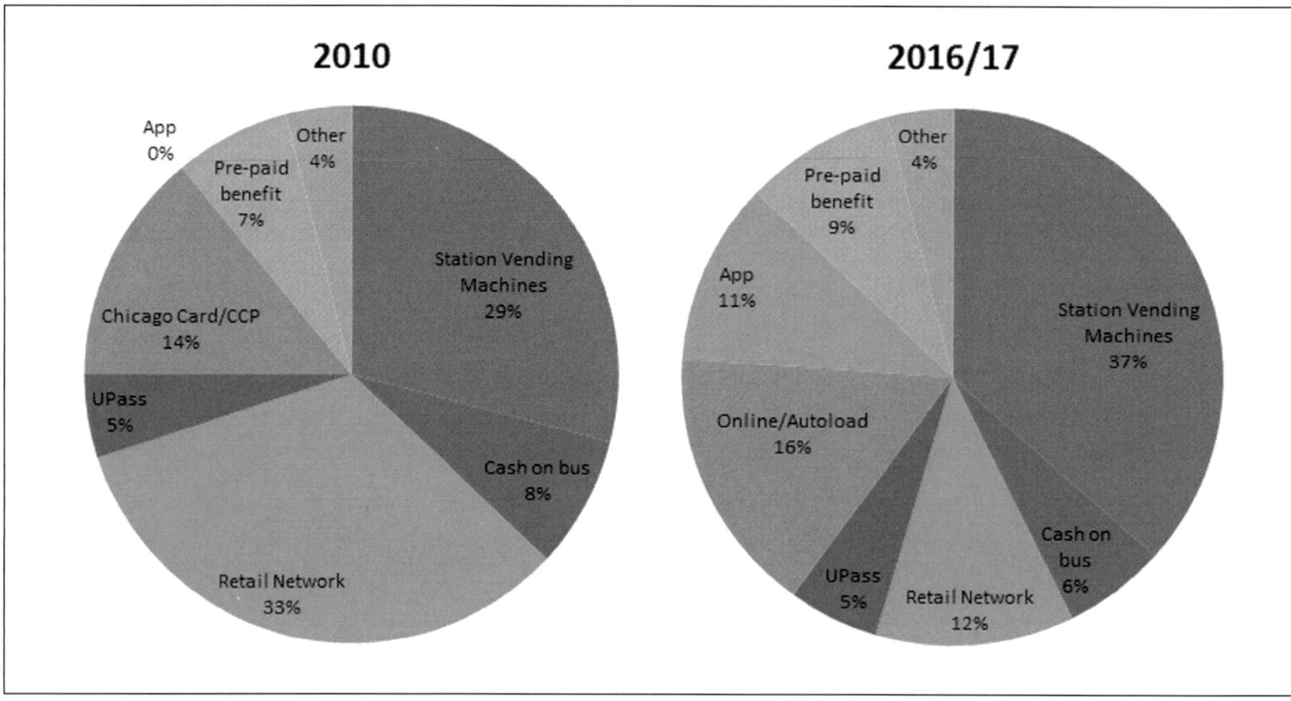

Figure 3. Shifts in value loading mechanism between 2010 and 2017.

behavior, and if that change in behavior induces ridership.

Initial results show that the treatment group bought more passes, especially 7-Day Passes, within a few weeks compared with a control group, which received no communication. The number of accounts that loaded passes grew 60 percent in the treatment group compared with 24 percent in the control group. Two months later, almost 3 percent of the nudge group was buying pass products compared with 1 percent of the no nudge group. However, changes in ridership per account had remained fairly similar between the two groups.

Conclusion

CTA has emphasized knowledge of customer behavior as one tool to make informed policy decisions. This longitudinal analysis shows that CTA can study the patterns of many of its customers at a detailed level. Using this type of analysis can help understand behavior changes over time.

Although this case study focuses on a YoY method to minimize any seasonality with ridership trends, this method is applicable to trend analyses over multiple consecutive months (month-to-month) as well. For instance, the same question and analysis on ridership gains for an unusual event such as sporting championship series could be used by examining multiple time periods around events. Is the increase in ridership a result of increased activity from current riders, or is it an influx of new riders?

There are still challenges CTA faces in making conclusions about the individual behavior of riders. Not all cards offer longitudinal study, account turnover is present, and there appear to be multiple accounts per user. Registration and long-term transit users mitigate some of these characteristics. Understanding these differences is important in highlighting the uncertainties in this analysis. Yet, without perfect customer information, CTA's ability to better understand consumer behavior has allowed the agency to embark on interesting and complex analyses of systems without fragile data fusion tactics needed to integrate AVL data to AFC data. The agency has gained better insights on specific customer behaviors over time.

Reframing problems such as traditional ridership analyses as cross-sectional and longitudinal studies helps structure and identify interesting research problems in public transportation. However, traditional longitudinal studies are significantly more costly and require years before the experimental data have statistical power and a large cohort. Therefore, using techniques similar to natural experiments can help mimic experimentally controlled longitudinal studies.

CTA's decision to implement and save transactional, account-based, fare card data in prior years may not have had a traditional 'return on investment' decision process. Such investments in data represent a risk based on promises that the data could potentially provide use in the future with better techniques that are more accessible to public agencies, like big data analysis and cloud computing. Future flexibility and decision-making ability are only possible with organizational investment in data storage, organization, and analysis as a strategy.

The underlying data powering longitudinal study methods can inform fare policies decisions and create more effective marketing campaigns. The Ventra transit card and its related account information make these things possible through an integrated dataset with long-lasting cards. From this new information, CTA can increase the effectiveness of traditional planning and market research applications, leading to better models and policies for future decisions.

Author Contributions

The authors confirm contribution to the paper as follows: study conception and design: R. Chan, A. Cui, M. Vaishnav, S. Wainwright, P. Murray; data collection: R. Chan, A. Cui, M. Vaishnav, S. Wainwright, P. Murray; analysis and interpretation of results: R. Chan, A. Cui, M. Vaishnav, S. Wainwright, P. Murray; draft manuscript preparation: R. Chan, A. Cui, M. Vaishnav, S. Wainwright, P. Murray. All authors reviewed the results and approved the final version of the manuscript

References

1. National Academies of Sciences, Engineering, and Medicine. *Leveraging ITS Data for Transit Market Research: A Practitioner's Guidebook*. The National Academies Press, Washington, D.C., 2008.
2. Pelletier, M.-P., M. Trépanier, and C. Morency. Smart Card Data Use in Public Transit: A Literature Review. *Transportation Research Part C: Emerging Technologies*, Vol. 19, No. 4, 2011, pp. 557–68.
3. El Mahrsi, M. K., E. Côme, J. Baro, and L. Oukhellou. Understanding Passenger Patterns in Public Transit Through Smart Card and Socioeconomic Data. *Proc., 3rd International Workshop on Urban Computing*. Chicago, IL, 2014.
4. Ma, X., C. Liu, H. Wen, Y. Wang, and Y.-J. Wu. Understanding Commuting Patterns Using Transit Smart Card Data. *Journal of Transport Geography*, Vol. 58, 2017, pp. 135–45.
5. Ghaemi, M. S., B. Agard, M. Trépanier, and V. P. Nia. A Visual Segmentation Method for Temporal Smart Card Data. *Transportmetrica A: Transport Science*, Vol. 13, No. 5, 2017, pp. 381–404.
6. Kieu, L. M., A. Bhaskar, and E. Chung. Passenger Segmentation Using Smart Card Data. *IEEE Transactions on Intelligent Transportation Systems*, Vol. 16, No. 3, 2015, pp. 1537–548.
7. Goulet-Langlois, G., H. N. Koutsopoulos, and J. Zhao. Inferring Patterns in the Multi-week Activity Sequences of Public Transport Users. *Transportation Research Part C: Emerging Technologies*, Vol. 64, 2016, pp. 1–16.
8. Kusakabe, T., and Y. Asakura. Behavioural Data Mining of Transit Smart Card Data: A Data Fusion Approach. *Transportation Research Part C: Emerging Technologies*, Vol. 46, 2014, pp. 179–191.
9. Nassir, N., M. Hickman, and Z.-L. Ma. Activity Detection and Transfer Identification for Public Transit Fare Card Data. *Transportation*, Vol. 42, No. 4, 2015, pp. 683–705.
10. Hong, S.-P., Y.-H. Min, M.-J. Park, K. M. Kim, and S. M. Oh. Precise Estimation of Connections of Metro Passengers from Smart Card Data. *Transportation*, Vol. 43, No. 5, 2015, pp. 749–769.
11. Ali, A., J. Kim, and S. Lee. Travel Behavior Analysis Using Smart Card Data. *KSCE Journal of Civil Engineering*, Vol. 20, No. 4, 2015, pp. 1532–539.
12. Gaudette, P., R. Chapleau, and T. Spurr. Bus Network Microsimulation with General Transit Feed Specification and Tap-in-Only Smart Card Data. *Transportation Research Record: Journal of the Transportation Research Board*, 2016. 2544: 71–80.
13. Trépanier, M., C. Morency, and B. Agard. Calculation of Transit Performance Measures Using Smartcard Data. *Journal of Public Transportation*, Vol. 12, No. 1, 2009, pp. 79–96.
14. Alfred, K. K., and A. Lomone. Reproducing Longitudinal In-Vehicle Traveler Experience and the Impact of a Service Reduction with Public Transit Smart Card Data. *Transportation Research Record: Journal of the Transportation Research Board*, 2016. 2541: 81–89.
15. Hanft, J., S. Iyer, B. Levine, and A. Reddy. Transforming Bus Service Planning Using Integrated Electronic Data Sources at NYC Transit. *Journal of Public Transportation*, Vol. 19, No. 2, 2016, pp. 89–108.
16. Ventra. *Ventra How-To: Paying with Contactless Bankcards*. https://www.ventrachicago.com/how-to/contactless-bankcards/. Accessed May 15, 2017.

The Standing Committee on Public Transportation Marketing and Fare Policy (AP030) peer-reviewed this paper (18-05611).

Article

TRR
JOURNAL OF THE TRANSPORTATION RESEARCH BOARD

Transportation Research Record
2018, Vol. 2672(8) 546–556

DOI: 10.1177/0361198118790839
journals.sagepub.com/home/trr
SAGE

Examining the Provision of Railway Transit Information to Foreign Visitors in the Tokyo Metropolitan Area and Strategies for Improvement

Xiao Yun Chang[1,2], Yoshiki Ikeda[1], Sohei Tsujimura[3], and Kiyoshi Sakamoto[1]

Abstract
Well known for its extensive network and state-of-the-art technology, the Japanese railway system has become a must-see attraction for foreign tourists and yet has not ceased to perplex them. Despite the growing use of multilingual signs, real-time train information, and information centers equipped with English-speaking staff, visitors from abroad could still find themselves lost when navigating the railway system. This research aims to examine the provision of railway transit information to foreign visitors in the Tokyo Metropolitan Area, and focuses on the railway system of Japan Railway (JR) East—the largest railway company in Japan and a global railway leader—as a case study. We identified four areas of research that are critical to the ease of use of railway systems, namely wayfinding, route planning, ticketing, and staff assistance. We examined the factors contributing to successful information provision and quantified the influences of sign design on the cognitive cost of wayfinding through interviews and surveys. We found that sign legibility, inadequate explanation for those with little background knowledge, and station staff's language skills were pressing concerns among foreign passengers, and transit information needs to be addressed systematically. This research will contribute to the understanding of tourists' use of public transit and provide a practical guide to improving the usability of urban railway systems not only in Japan but also worldwide.

Overview of the Railway System in the Tokyo Metropolitan Area

A country well known for its state-of-the-art railway system, Japan has been facing a spike of inbound tourism as a result of the Tokyo Olympics in 2020 and aggressive inbound travel stimulation policies (*1*). In response to the increase in the mobility demands of foreign tourists, the topic of foreigner-friendly design in transportation systems has gained much more attention over the past few years. However, the conceptual difficulties in realizing this goal are multifold given the complexity of the Japanese railway system.

A brief introduction of the Japanese railway system shall set the background for readers' understanding. The development of railway infrastructure in Japan could be backdated as far as the 19th century: a mix of private and public railway projects in the past century have contributed to railways becoming widespread in Japan, resulting in a total of 27,268 km of rail nationwide. The six Japan Railway (JR) companies, products of the privatization of the public entity named Japanese National Railways, operate 19,470 km of rail as of 2014; the rest is operated by other public and private organizations. The Tokyo Metropolitan Area is home to most of the lines operated by JR East, the largest JR company among the six, more than ten private commuter lines, the Tokyo subway network, and other suburban railway networks. Now, the railway system in the Tokyo Metropolitan Area stands among the most complicated systems in the world. It is different from most railway systems in the world in the following ways:

[1]Frontier Service Laboratory, R&D Center for JR East Group, Saitama Prefecture, Japan
[2]University of California, Berkeley, CA
[3]Department of Urban and Civil Engineering, Ibaraki University, Ibaraki Prefecture, Japan

Corresponding Author:
Address correspondence to Xiao Yun Chang: xy.chang@berkeley.edu

- Multiple agencies: Most cities in the world have two or three railway operators at most.
- Complex network: JR East and Tokyo Metro both have fairly complicated networks composed of more than 10 lines and hundreds of stations operated on complex schedules.
- Complex fare policies: Railway fares in the Tokyo Metropolitan Area are mostly distance based and sometimes speed based. Even for the same pair of origin and destination stations, a speed-based premium charge could apply.
- Mega-sized, multipurpose station buildings: Due to the network complexity, several major train stations in the Tokyo Metropolitan Area can run more than 10 platforms and comprise lines from multiple agencies. These huge train stations are often crowded and hard to navigate without proper wayfinding systems. Moreover, transit-oriented development has brought large-scale shopping zones and ubiquitous advertisements to these stations.

These characteristics, although distinguishing the railway system in the Tokyo Metropolitan Area from most railway systems in the world, also have perplexed foreign passengers who are not yet used to such complexity. There have been attempts to study the usability of transit; however, very few literature accounts of passenger navigation with a focus on foreign passengers in large-scale train stations exist (*2–4*). Kinsella et al studied the usability of Dublin's public transit for new users (*5*). Le-Klähn et al studied the usability of public transit in Munich for new users, frequent users, and non-users (*6–8*). Studies by Dziekan on the railway system in Munich, the scale of which is comparable to the subject of this report, show that tourists who arrive at a new city experience a time-consuming learning process as they try to explore the railway system (*9–13*). According to Dziekan's works, the three stages are: starting with the knowledge of the existence of public transportation options; advancing into the identification of appropriate modes of transport; and, finally, learning how to navigate successfully within the system. Short-term tourists usually do not have the luxury of time to learn completely, and we cannot estimate the hardship they experience purely from the eyes of frequent, local users who have learned the system by heart. Furthermore, we need to anticipate their possible confusion and mistakes by understanding their expectations, background knowledge about railways, and the process through which they learn to use the railway system.

Information plays a key role in the interaction between users and the system. The complexity of information provided to the railway users is determined by the structure of the railway system, and users' understanding of the system is dependent on the quality of information. Owing to the aggressive promotion by JNTO and the upcoming Olympic Games in Tokyo, the inbound tourism has guaranteed the need for understandable information on the railway system.

The term "understandable" is challenged by another trend in inbound tourism, that is, diversification of visitors by geographical regions. Traditionally, railway agencies provided information in Japanese and English to accommodate their users until the late 2000s. Then, information in Chinese and Korean started to become available as inbound tourism diversified and railway agencies realized that the bilingual standard was insufficient. Further, the diversifying inbound tourism over recent years has brought in more non-speakers of any of these four languages. Some railway agencies have been producing information in Thai in an attempt to further accommodate passengers from Thailand. At the same time, we see a potentially perilous slippery slope: when do we stop adding languages to the sign boards?

This uncontrolled addition of information would also result in increasingly clumsier, messier information, which could diminish the legibility of information and make the learning process harder for foreign tourists. This problem needs to be handled delicately, potentially with the development of translation technology on mobile devices.

Comprehensive Transit Information

A successful ride on railway requires one to obtain travel information, understand the information, and make the right decisions at critical steps in the trip (*14*). As a starting point, we can consider the following process. We need information about the routes we take on the network (route planning), information about how to move within the system (wayfinding), and information about tickets, assuming that the users obtain the information on their own. When users fail to obtain or understand the information on their own, staff assistance comes into the picture to inform passengers (Figure 1).

Figure 1's traditional flow diagram can be modified to include passengers purchasing railway visitor passes, who tend to only use routes covered by their passes. These passes are often valid on certain routes operated by selected railway agencies rather than on all routes in a region, and such exclusivity compels the passengers to plan their routes more carefully.

With the framework illustrated in Figure 2, we now consider issues with information provision. First, the lack of integration among these components could lead to obstacles for using the system smoothly. Second, staff assistance acts only as a support of other means of obtaining information, and we cannot rely solely on

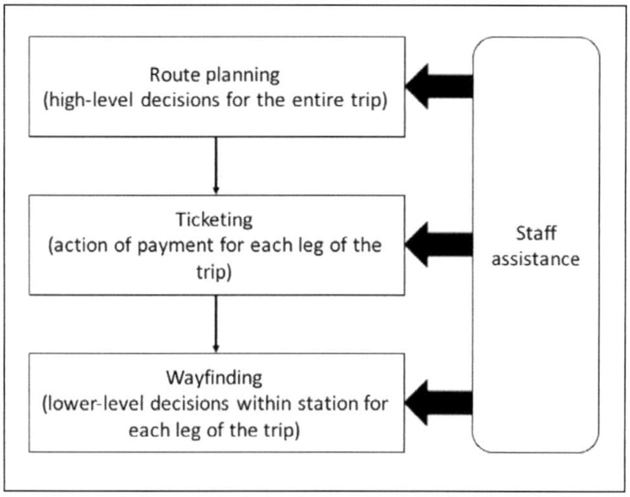

Figure 1. Staff assistance supports decision making at every stage.

having more staff. Third, we must appreciate the difference in information needs and thus the variability in informational retrieval methods. Users who have greater familiarity with the system are less likely to look up fundamental information (e.g., location of the station closest to home), whereas newcomers still need knowledge that local users might deem as common sense. It was also found that foreign passengers in Japan are less concerned with real-time delay information than static information such as network, routes, and signs in stations. Foreign passengers also rely on tools with which they are familiar, which results in the wide use of Internet-based tools, and the reluctance to switch to tools which they do not know well. We found that Google Maps, despite occasional errors and incomplete information with regard to its data sources in Japan, has been the most popular tool among the railway users we surveyed.

We face a dilemma of presenting easily understood information that lowers the cognitive cost for foreign users while confronted by the complexity of the system and the need for multilingual information, which could worsen the information overload. We must also acknowledge the inadequacy of existing solutions, for they do not entirely answer the fundamental questions foreign passengers may have. Strategic information design that organizes information in an organic, efficient way, ensuring correct understanding at every step of a trip on the railway system, is the answer to the dilemma; it requires a systematic analysis of the status quo, a thorough understanding of customers' needs, and a set of implementable strategies delivering the design philosophy in practice.

Objectives

The goals of this project are as follows:

- Study the information needs of foreign passengers using railway in the Tokyo Metropolitan Area.
- Quantitatively and qualitatively understand the gap between information needs and information provision for foreign passengers.
- Formulate strategies to bridge the gap through improved information on all steps of transit usage.

Summary of the Project

This project was carried out in June to December 2016 as a preliminary step toward a satisfactory solution to the problems we have found. From quantitative and qualitative investigations regarding four aspects of

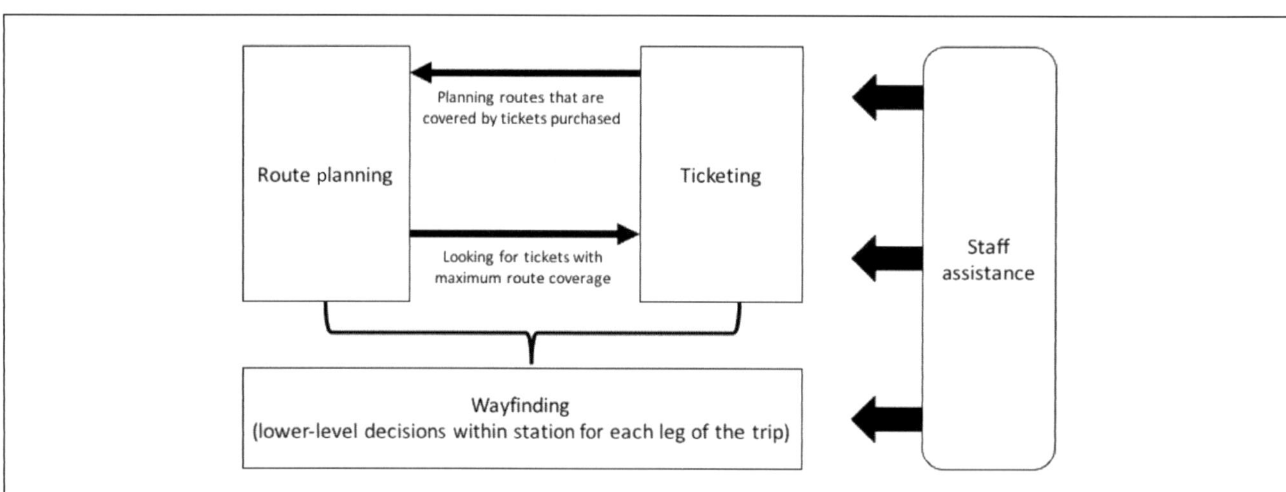

Figure 2. Decision flow capturing interaction between route planning and ticketing.

transit information provision—wayfinding design, route planning information, ticketing information, and staff assistance—we have come to understand the status quo and persistent problems, and formulated strategies toward improvement.

The next section is a holistic discussion of the methods of investigation. We then present an overview of foreign passengers' information needs that could provide significant insights in the project. We explore the four aspects of information provision and conclude the report with a summary of strategies and areas of further research.

Methodology

Field Investigations

Study of Signs at Major JR East Stations. In July 2016, we visited five JR East train stations in the Tokyo Metropolitan Area: Ueno, Shinjuku, Tokyo, Suidobashi, and Narita Airport Terminal 1. During these visits, we took photos of signs in stations and simulated the experience of passengers looking for certain exits of and destinations outside the stations.

Visits at the Narita Airport JR Travel Service Centers. In November 2016, we visited the two JR Travel Service Centers, located in Narita Airport Terminal 1 and Terminal 2 respectively, for a total of 9 h. During these visits, we interacted with and assisted customers at the centers and observed other staff at the centers serving customers.

Evaluation Grid Method

This interview method was used to understand the characteristics of desirable signs in the context of transportation and wayfinding, especially outside one's home country. The participants were five U.S. college student volunteers. Through the categorization of many photos of signs in train stations worldwide that are sufficiently different in their order of desirability, we collected participants' stated preferences. Then we extracted the students' evaluation frameworks from their descriptions of the reasons for which they make such classifications, how these reasons make signs more desirable, and features of signs leading to those reasons. We then generalized the results into a single evaluation framework.

Caption Evaluation Method

In December 2016, we carried out this experiment in Tokyo Station (one of the most complicated railway stations in Tokyo) with five native speakers of English and five native speakers of Chinese, mostly in their 20s, who had stayed in Japan for less than 6 months. Participants were prompted to walk toward a destination point in the station from a given starting point; while trying to reach the destination within a time limit, participants were told to take photos of objects and scenes that caught their attention and to take note of their impressions of these objects and scenes. A total of 101 photos along with impressions was aggregated to show passengers' attitudes toward various elements of train stations and implicitly their preferences with regard to functionality of train stations.

Survey at the Tokyo Station

In September 2016, we conducted a survey at the Sobu Underground Platform 4 in Tokyo Station, which operates the airport-bound Narita Express, a JR express train known to be mostly used by foreign passengers. The questionnaire consisted of three sections: 1) Internet access in Japan, 2) trip planning tools used in Japan, and 3) satisfaction with railway in the Tokyo Metropolitan Area. We collected 254 samples featuring customers from Asia, North America, Europe, South America, and Oceania.

Online Survey

In December 2016, we conducted another survey targeting foreign users of the Japanese railway system on the Internet. The samples consisted of 30 native speakers of Chinese residing in the Greater China Area, 30 native speakers of Korean residing in South Korea, 30 native speakers of English, and 30 European and Southeast Asian residents who are not native speakers of Chinese, Korean, or English, who had visited the Tokyo Metropolitan Area in the past year and stayed less than 30 days. The survey was comprised of questions about satisfaction with railway in the Tokyo Metropolitan Area and a series of A/B testing of sign design.

Overall Information Needs

From survey responses and interviews with foreign passengers, we have found that foreign visitors' experiences of getting lost, confused, or disoriented are often related to unaddressed information needs that are different from those of local passengers. By examining the results, we have found five representative information needs currently not adequately fulfilled and which deserve further attention.

Emphasis of Fundamental Information

As shown in Figure 3, foreign passengers are mostly concerned with fundamental information required for a transit trip, that is, route, ticket, and wayfinding

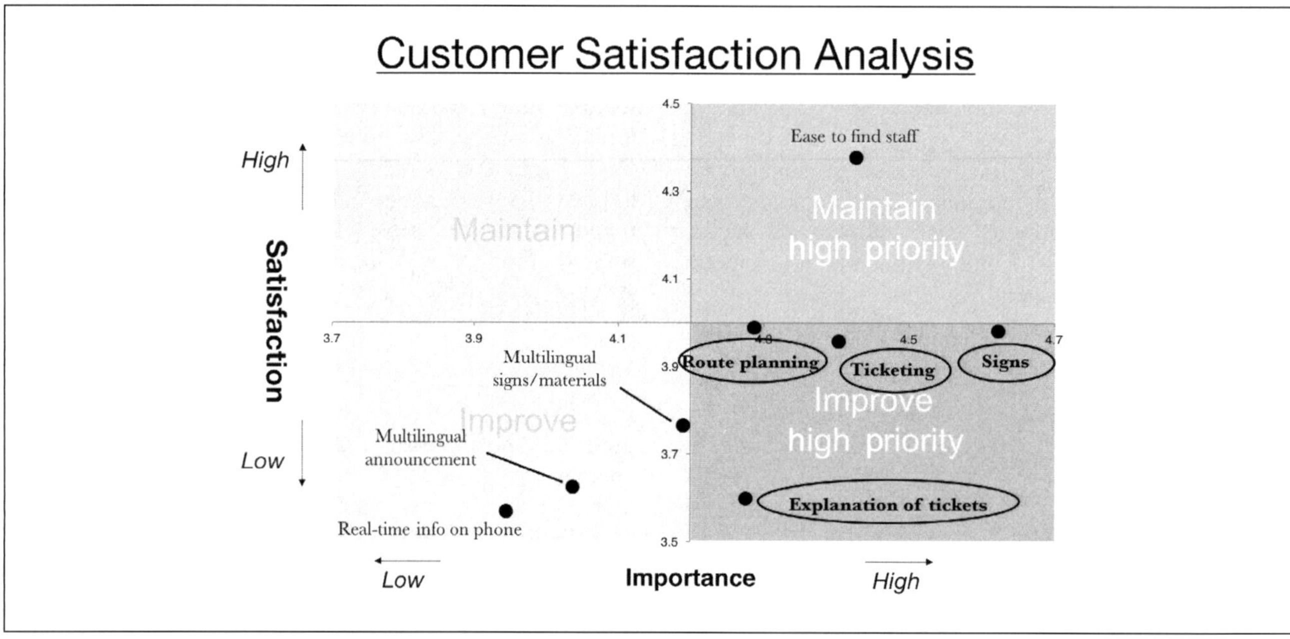

Figure 3. Customer satisfaction portfolio analysis for foreign passengers.
(Source: Survey at Tokyo Station.)

information; real-time delay information is less important in contrast. However, based on railway agencies' attempts to make real-time delay information widely available, real-time delay information is a higher priority for local passengers. This distinction indicates that short-term passengers have not yet mastered basic knowledge about the system, unlike local passengers who are more familiar with the static information and demand more accurate real-time information to further perfect their trip planning process.

Simplified Ticketing Process

A comprehensive railway system gives rise to a more complicated fare system and more complicated ticketing procedures; according to Figure 4, foreign passengers raise concerns about the ticketing process—where to buy, what to buy, and how to use tickets once purchased. Without clearing up these types of confusion altogether, foreign passengers will continue to find it difficult to purchase tickets.

Consistent Information for Easy Identification

According to survey responses and interviews, foreign passengers often use their own apps and tools for trip planning and wayfinding. However, the difficulty of distinguishing lines, platforms, and areas within maze-like stations renders these tools useless. For example, passengers who are informed from maps to take northbound trains at a station would prefer seeing signs specifying the directions of the lines rather than a cluster of the lines' final destinations.

Clarification of Commonly Misunderstood Terms

We have found that foreign passengers can have systematic misunderstanding on some concepts, and these could be prevented if the railway agencies were more proactive about educating their customers rather than expecting customers to properly self-educate before using the system.

Multilingual Information

Figures 3 and 4 reveal that multilingual information is a significant area for improvement. This could range from conversations with station staff, static and dynamic signs, announcements, and so on. Currently in JR East's railway system, general station staff are not specifically trained in foreign languages; not only does this prevent station staff from effectively helping non-Japanese-speaking passengers, this creates deeper communication issues when makeshift signs are made because of indoor construction and temporary announcements are needed because trains are delayed. The lack of a robust multilingual information system has profound implications.

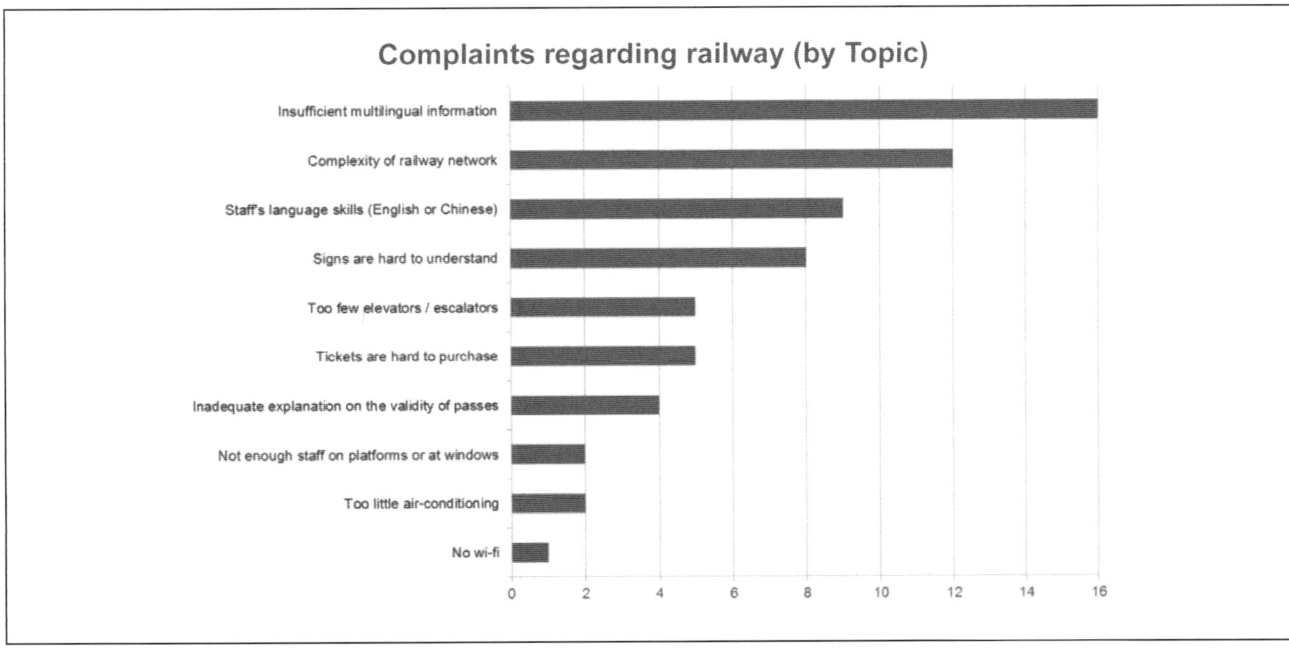

Figure 4. Customer complaints with regard to the railway system in the Tokyo Metropolitan Area. (Source: Survey at Tokyo Station.)

The Wayfinding System

Definition

Conventionally, wayfinding systems refer to the guiding signs in a physical environment, and most wayfinding design literature is concerned with these visual guides. In the context of railway transport, we would like to enlarge the scope of the definition to include a broader range of elements that help passengers navigate in the railway system.

Referring to Figure 1, we first identify the possible occasions for which wayfinding elements are needed. Movement occurs not only in stations but also during boarding and alighting. On a moving train, passengers could rely on cues including in-train announcements, network maps in train cars, and running-in boards located along platforms to identify their current location in the network and whether they should get off at the next stop. These cues are part of the wayfinding system because they enable passengers to make correct movements.

Another dimension of enlargement has to do with the senses used to obtain wayfinding information. This includes visual signs, audio announcements, and tactile cues related to wayfinding. However, given the limited time span of the project, we will focus mainly on static signs in stations, the most commonly seen type of wayfinding tools. These signs are heavily relied on and closely related to users' levels of satisfaction and trustworthiness with regard to railway.

Static Signs in Stations

Signs are significant in wayfinding, especially in large, complicated train stations with a limited view to the outside where it is almost impossible to identity directions without proper guidance. The following types of signs are related to wayfinding (*15*):

- Identification signs: signs that act as identifiers allowing people to confirm their current locations.
- Directional signs: signs pointing people toward specific directions.
- Operational signs: signs providing information about the use and operation of a physical environment (e.g., map signs and train timetables).

We found three categories of issues (visibility, content, installation) related to these signs in the JR East stations as elaborated below. These issues were synthesized from various sign design manuals, interviews, surveys, and authors' observations (*15–17*). A critical review of the JR East Sign Design Manual that follows revealed the reasons behind the prevalence of these issues.

Visibility. The visibility of signs in stations is critical given the visual nature of signs, and it can be affected by the following:

- Density of installation
- Lighting

- Distinguishability from the surroundings
- Density of passengers in the station
- Physical elements blocking the view
- Size of sign boards

Content. Signs containing ambiguous information or information that takes longer than usual to understand could undermine users' confidence in navigating around the system and waste users' time unnecessarily. The factors influencing the legibility of a single sign are as follows:

- Density of information on signs
- Font size from a certain distance
- Layout and hierarchy of information
- Availability of information in languages other than Japanese
- Universally recognizable symbols
- Color coding

Installation. The configuration of signs in a station affects passengers' movement macroscopically. Common sign design standards stress the importance of installing signs that match passengers' lines of movement, that is, having the right signs at the right places. The need to proactively prevent possible confusion because of the complexity of the stations and the railway systems requires us to go beyond the general lines of movement and be more specific at each critical decision point. The issues with the installation at a macroscopic level are described below:

- Assumption of background knowledge
- Overuse of makeshift signs
- Untimely removal of obsolete signs
- Lack of consistency
- Missing signs at critical decision points

The Sign Design Manual. The JR East Sign Design Manual provides standards of making and installing information signs in JR East train stations. It lays out guidelines of sign configuration in train stations and guidelines for both contents and formats of various types of signs. For sign contents, the manual also provides standardized translations for most commonly seen information, such as station names, "Central Exit," and "Restroom for Women," in English for all phrases and in Chinese and Korean for selected phrases. The manual is updated every few years to incorporate changes with respect to all aspects of train operation, for example, stations on a particular line, translation of certain phrases, and the level of detail to which certain terms are explained.

However, the scope of the manual is limited and does not cover the entire process of sign making. Each station is in charge of managing its signs from the design stage to the final implementation, that is, installation. As a result, the sign manual acts as a reference to the station for how signs should look and where they should be installed when new signs belonging to the categories described in the manual are introduced. Nonetheless, the manual has no bearing on the removal of old signs and the installation of signs outside the scope of the manual, for example. We discuss the issues with the current implementation of the manual below:

- It does not fully ensure the visibility of information on the signs: Specifying signboard sizes, font sizes, and height of placement alone cannot adequately ensure the visibility of the information on the signs or even the signs themselves. The guidelines provided in the manual cannot cover all the complexities caused by station environments; nor do they provide approaches to systematically evaluate the visibility of signs.
- Its treatment of translations does not meet the needs of foreign passengers: A huge step toward universal design, most information on signs is available in English and key terms such as "exit" and "fare adjustment machine" are available also in simplified Chinese and Korean under the stipulation of the manual. However, the number of signs that are solely in Japanese is still striking, and the translations are merely clones of existing information in Japanese. The quantity and the quality of information that non-Japanese speakers receive from signs is lower than that received by the Japanese-speaking passengers.
- It lacks provision to cases of makeshift signs: The manual covers the most regular types of signs but not makeshift signs, which are often set up to explain temporary situations, leaving the design to the hands of staff in charge at each station. As a result, these signs lack consistency and sometimes fail to communicate effectively with the intended viewers.
- It does not require active solicitation of customers' opinions, especially those of foreign passengers: Currently, the manual is managed by the marketing department, with little input from other departments of the company, and changes, unless executive, are mostly motivated by complaints filed by passengers on JR East's website. This approach is fragile in the context today. Those who are less likely to successfully navigate in stations—foreign passengers, first-timers, and the visually impaired—happen to be those who are less likely to make their voices heard. This

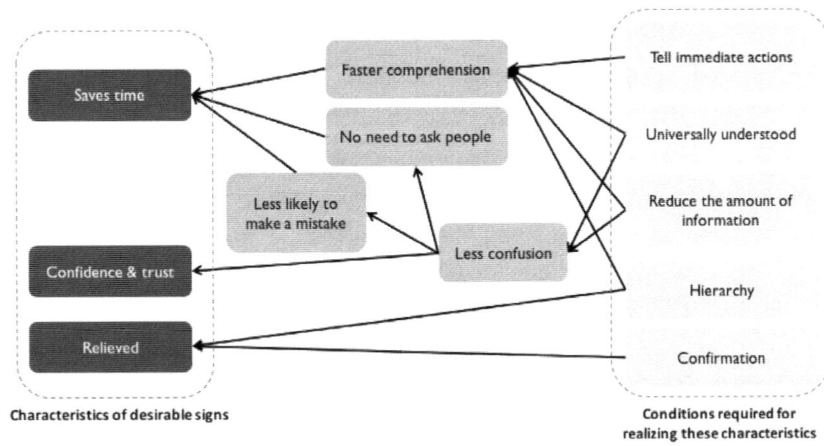

Figure 5. Aggregated result of evaluation grid method interviews.

prevents the authors of the manual from adjusting the manual accordingly.

- It does not provide standards for systematic renovation; nor does it maintain consistency across stations: Since the design and installation of signs is implemented by each station, the manual cannot dictate the outcomes of the implementation and has no further control over actions that stations take on existing signs and makeshift signs. Consequently, we still see unintegrated, inconsistent, and confusing signs in stations today.

Desirable Signs

We have found through our evaluation grid method interviews, summarized in Figure 5, that signs in train stations should meet certain conditions (Figure 5, right-hand column) that eventually save users' time, allow users to build confidence and trust, and help users feel relieved when navigating in the system (Figure 5, left-hand column). These conditions interact through the mechanism laid out between the two columns in Figure 5, and can inform sign designers about design-related decisions. Most importantly, many of the conditions point at "faster comprehension," which is equivalent to reduction in cognitive costs; unfortunately, we have observed many overloaded signs that do not meet this criterion for foreign passengers.

Route Planning and Ticketing

Route Planning

Route planning on a complicated railway network has irreplaceable significance; it allows users to navigate the system with some degree of certainty and learn the system in the longer run. According to our surveys, foreign passengers rely almost equally on traditional route planning tools such as paper maps and Internet-based tools such as Google Maps. Figure 6 shows the result of the survey conducted in the Tokyo Station. We found that foreign passengers are much more likely to use route planning apps that they already know, for example, Google Maps, rather than apps that are more popular in Japan and are known to be more accurate. This highlights the importance of having a global view when railway agencies design information. Besides navigation apps, foreign passengers rely mostly on materials published by the railway agencies and station staff to obtain route planning information.

Ticketing

Ticketing does not necessarily happen after route planning. We will discuss both the information about tickets and the actual process of obtaining tickets in this section.

- Validity of tickets and passes unclear: We are aware that foreign passengers frequently have questions or doubts about the validity of their passes and tickets, either before or after the purchase. Although railway agencies often try to explain the validity of these tickets and passes, possibly to increase sales, there still seems to be a lack of effective communication. As an example, passengers who hold a Japan Rail Pass are not always aware that they need to reserve seats for the Narita Express before getting on board. There could be a few reasons leading to the bafflement, for instance, not knowing about seat reservation

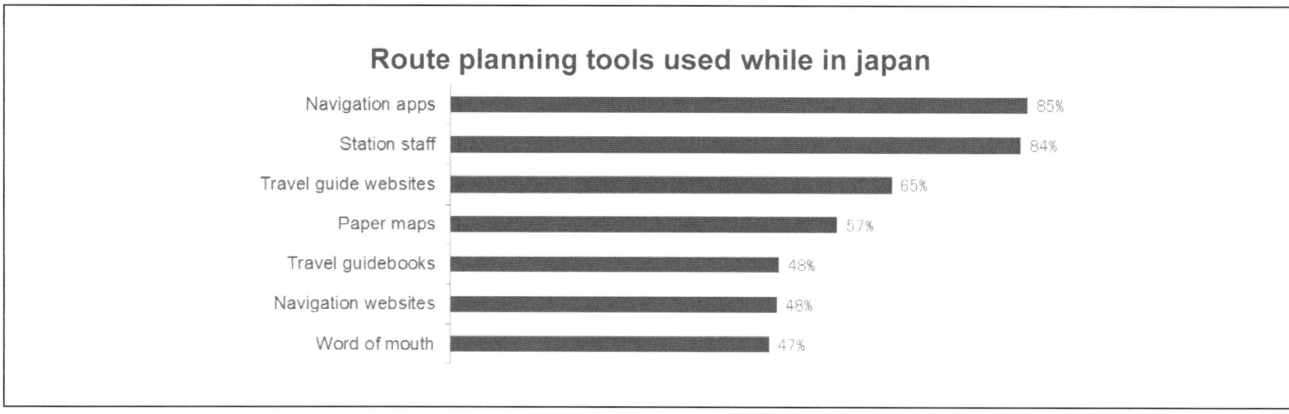

Figure 6. Route planning tools used by foreign passengers in the Tokyo Metropolitan Area.
(Source: Survey at Tokyo Station.)

rules, not being aware that Narita Express only has reserved seats, or equating Narita Express with Shinkansen where unreserved seats are available. Another common misunderstanding among foreign passengers is the distinction between basic fare ticket and express fare ticket, which is the result of a long-standing historical fare system.

- Hard to purchase online: Online ticketing systems can be a very helpful tool for foreign passengers to decide the types of tickets to obtain, and a highly effective channel for the railway agencies to inform passengers about particular rules and validity of tickets. Currently, there does not seem to be a railway agency in the Tokyo Metropolitan Area that has a well-established, foreigner-friendly online ticketing system.
- Multiple types of ticket machines: For historical reasons, JR East has separate ticket machines for different types of tickets. The basic ticket machines do not sell Shinkansen tickets, whereas the express ticket machines do not sell local train tickets. These machines have distinct features, and even for a single type of machine, several generations with slightly different looks can coexist.

Staff Assistance

Status Quo and Issues

The current state of staff assistance as sampled across JR East's stations and information centers shows several patterns that call for strategic changes. The level of satisfaction with regard to service etiquette and manner has been high whereas with regard to language skills it remains low.

JR East has been setting up Travel Service Centers, staffed with speakers of English and other languages, in key stations since 2007. However, staff outside these information centers tend to be less equipped with the language skills necessary for interaction with foreign passengers, as our surveys suggest.

In response, translation tools including tablets and apps are being used in some information centers, the results of which are mixed. Despite JR East's resolution in developing such tools, we have witnessed cases in which communication failed in the presence of translation software. Besides the limits of translation software, there are at least two other reasons for not solely relying on these tools. First, the outputs of these tools are prone to errors, especially in the case of homonyms; second, passengers might not know the names of the concepts they had in mind when talking to staff, for example, visitor pass products, railway lines to take. In these cases, these tools alone cannot solve the communication problems because of the lack of language skills.

At the organizational level, there has been little effort of joining the dots of customer-facing service at individual stations and information booths. This can be easily understood by looking at the variety of forms in which identical or similar information is displayed at different stations. In comparison, questions that were repeatedly asked at a single location could suggest unaddressed needs or unresolved problems that require changes at a higher level.

Conclusions and Strategies

We have examined the extent to which foreign railway users' information needs are met in the Tokyo Metropolitan Area. The lack of proper understanding of what foreign passengers do and do not know, the passiveness in soliciting feedback and post-design evaluation, and unsystematic information provision are the three common themes that have diminished the usability of the transit system. We propose the following strategies to improve transit information.

Strategies on Wayfinding

We have explored that signs have not adequately addressed foreign passengers' needs, and our strategies aim to ensure that these needs are taken into consideration. The first step is to utilize some of the findings from this project and in the longer run more macroscopic efforts will be required. We strongly recommend that our findings be circulated with sign designers of railway agencies and put to use as soon as possible.

Visibility issues still present challenges for foreign passengers; we recommend that the agencies develop proper measures for sign visibility and reconsider the relationship between signs and profit-generating elements including, but not limited to, store banners and advertisement.

In the long run, the agencies should make sure that foreign passengers' needs are met and this requires the establishment of validation mechanisms; the agencies should actively reach out to the foreign users of their systems and solicit feedback on a regular basis.

Strategies on Route Planning and Ticketing

As foreign passengers tend to use a different set of route planning tools from local passengers, there is value in learning more about their experience with these tools. For example, if a navigation app uses exits as locational identifiers whereas a certain station frequently uses landmarks as locational identifiers, it would be beneficial for this station to install more signs about exits. Railway agencies can cooperate with route planning tool companies (e.g., Google) to ensure the quality of information provided to customers.

Realizing the close tie between route planning and ticketing, we recommend that the two be integrated through convenient, easily accessible means, possibly through mobile apps. Deutsche Bahn's DB Navigator app is a successful case that provides navigating and ticket booking services simultaneously. Railway agencies could design a section in the app specifically for visitor passes that explains the validity of visitor passes and allows users to plan trips by only using routes covered by their visitor passes.

To simplify ticketing, we could either improve the ticketing facilities or ultimately simplify the fare system. Considering the challenges of revolutionizing the entire fare system in Japan, we recommend that railway agencies upgrade ticket machines, develop user-friendly online ticketing systems, and make ticketing information more accessible.

Strategies on Staff Assistance

The agencies should centralize the disaggregate experience of customers and address foreign passengers' underlying needs. At the most basic level, station staff's lack of foreign language skills should be improved systematically; there have been attempts to develop better translation tools as a response to this issue, and agencies could hire more staff who speak languages other than Japanese.

The agencies should collect repeated passenger questions and come up with unified responses to save human capital at the station level and to provide consistent answers. However, repeated questions are indicators of underlying issues that await bigger changes. We suggest that the agencies incorporate these issues into resolving the fundamental issues that cause distress and confusion.

Author Contributions

The authors confirm contribution to the paper as follows: study conception and design: Xiao Yun Chang, Kiyoshi Sakamoto; data collection: Xiao Yun Chang, Yoshiki Ikeda; analysis and interpretation of results: Xiao Yun Chang, Yoshiki Ikeda, Sohei Tsujimura; draft manuscript preparation: Xiao Yun Chang. All authors reviewed the results and approved the final version of the manuscript.

References

1. インバウンドの現状と、受入環境の整備について. 日本政府観光局 (Japan National Tourism Organization), 2016.
2. Thompson, K. Tourists' Use of Public Transport Information: What They Need and What They Want. Presented at the Association for European Transport, Strasbourg, France, 2004.
3. Tsao, Y. C., S. Y. Huang, K. K. Chang, and C. H. Lai. Universal Design with Cognitive Psychology Applied on Sign and Guidance Information Systems of Transportation Public Station. Presented at the Asian Conference on Psychology & the Behavioral Sciences, Osaka, Japan, 2012.
4. Tyrinopoulos, Y., and C. Antoniou. Public Transit User Satisfaction: Variability and Policy Implications. *Transport Policy*, Vol. 15, No. 4, 2008, pp. 260–272. https://doi.org/10.1016/j.tranpol.2008.06.002.
5. Kinsella, J., and B. Caulfield. An Examination of the Quality and Ease of Use of Public Transport in Dublin from a Newcomer's Perspective. *Journal of Public Transport*, Vol. 14, No. 1, 2011, pp. 1–20.
6. Le-Klähn, D. T., R. Gerike, and C. Hall. Tourist Use of Public Transport at Destinations - A Review. *Current Issues in Tourism*, Vol. 18, No. 8, 2015, pp. 785–803. https://doi.org/10.1080/13683500.2014.948812.
7. Le-Klähn, D. T., R. Gerike, and C. Hall. Visitor Users vs. Non-Users of Public Transport: The Case of Munich, Germany. *Journal of Destination Marketing & Management*, Vol. 3, No. 3, 2014, pp. 152–161. https://doi.org/10.1016/j.jdmm.2013.12.005.
8. Le-Klähn, D. T., C. Hall, and R. Gerike. Analysis of Visitor Satisfaction with Public Transport in Munich. *Journal*

of Public Transportation, Vol. 17, No. 3, 2014, pp. 68–85. http://dx.doi.org/10.5038/2375-0901.17.3.5.
9. Dicke-Ogenia, M., and K. Dziekan. Reducing Uncertainty and Supporting Cognitive Maps in Travel Information for Public Transport. *World Review of Intermodal Transportation Research*, Vol. 3, No. 1/2, 2010, pp. 73–90. https://doi.org/10.1504/WRITR.2010.03158.
10. Dziekan, K. *Ease-of-Use in Public Transportation – A User Perspective on Information and Orientation Aspects*. PhD thesis. Royal Institute of Technology, Stockholm, Sweden, 2008.
11. Dziekan, K. The Transit Experience of Newcomers to a City - Learning Phases, System Difficulties and Information Search Strategies. Presented at 87th Annual Meeting of the Transportation Research Board, Washington, D.C., 2008.
12. Dziekan, K. What Do People Know About Their Public Transport Options? *Transportation*, Vol. 35, No. 4, 2008, pp. 519–538. https://doi.org/10.1007/s11116-008-9164-x.
13. Dziekan, K., and A. Scholz. How to Measure “Ease-of-Use” in Public Transportation?: Scale Construction and Testing. Proceedings of the Biennial Conference on Environmental Psychology, Bayreuth, Germany, 2007.
14. Smallman, C., and K. Moore. Process Studies of Tourist Decision Making. *Annals of Tourism Research*, Vol. 37, No. 2, 2010, pp. 397–422. https://doi.org/10.1016/j.annals.2009.10.014.
15. Calori, C., and D. Vanden-Eynden. *Signage and Wayfinding Design*. John Wiley & Sons, Inc., Hoboken, N.J., 2015.
16. *Station Design Principles for Network Rail*. Network Rail, 2015.
17. *Wayfinding at Stations A Good Practice Guide*. Rail Safety & Standards Board, 2006.

The Standing Committee on Public Transportation Marketing and Fare Policy (AP030) peer-reviewed this paper (18-04476).

Article

TRR
JOURNAL OF THE TRANSPORTATION RESEARCH BOARD

Transportation Research Record
2018, Vol. 2672(8) 557–565

DOI: 10.1177/0361198118792765
journals.sagepub.com/home/trr
SAGE

BART Perks: Using Incentives to Manage Transit Demand

Ryan Greene-Roesel[1], Joe Castiglione[2], Camille Guiriba[2], and Mark Bradley[3]

Abstract

BART Perks, offered by Bay Area Rapid Transit (BART) and the San Francisco County Transportation Authority (SFCTA), was a six-month test program that provided incentives to riders for traveling during the shoulder hours of the morning peak period. The main goals of the Perks program were to: (1) reduce peak period and peak direction crowding, and improve person throughput; (2) improve BART customer satisfaction; (3) increase employer support for flexible work schedules; and (4) identify implementation challenges and solutions to provide lessons learned. About 17,800 participants enrolled, and about 2,600 of them were frequent weekday peak hour travelers. During the trial, an average of 250 participants, or about 10% of those who previously traveled during the peak hour, shifted their ride to one of the shoulder hours. Recommendations for future programs include recruiting a larger number of riders in the targeted travel market, structuring it to better reward behavior change rather than preexisting behavior, and considering more efficient methods of participant recruitment and retention.

Ridership on the San Francisco Bay Area Rapid Transit (BART) system has increased rapidly in recent years following population and employment growth. Between 2004 and 2016, ridership grew by about 40% overall and 75% in the Transbay corridor connecting San Francisco's financial district and East Bay cities via the underwater Transbay Tube (*1*). According to Federal Transit Administration thresholds for available space per passenger (*2*), the corridor is over capacity.

Increased demand can be accommodated through expanded physical capacity, or through more efficient use of existing capacity. Although BART is working to expand physical capacity in the Transbay Tube, it will be several years before significant expansion can be realized. More efficient use of existing capacity can be an effective short-term solution for alleviating capacity constraints while infrastructure improvements are brought online. Use of price signals, such as charging higher tolls or fares during the most congested times, is sometimes used for efficiently matching demand with capacity. Peak pricing is not currently under consideration by BART's Board of Directors, so staff sought to test another approach of managing peak congestion using incentives.

BART Perks was a six-month test program that provided incentives for traveling during the shoulder hours of the morning peak period. The first program of its kind in North America, BART Perks was modeled after a similar program deployed in Singapore. Perks was a joint project of BART and the San Francisco County Transportation Authority (SFCTA), the congestion management agency for San Francisco, and was funded by the Federal Highway Administration, BART, and the SFCTA.

The paper is organized as follows. First, a brief review of relevant literature is provided. The goals and structure of BART Perks are then described. Enrollment, participation, and demand-management outcomes are summarized. Finally, the paper provides guidance on lessons learned for public agencies interested in pursuing similar programs.

Literature Summary

Transit agencies around the world have tried several demand-management techniques to reduce peak crowding, including peak pricing or off-peak discounts, better information, or employer partnerships to encourage

[1]San Francisco Bay Area Rapid Transit District, Oakland, CA
[2]San Francisco County Transportation Authority, San Francisco, CA
[3]Resource Systems Group, San Diego, CA

Corresponding Author:
Address correspondence to Ryan Greene-Roesel: rgreene@bart.gov

more schedule flexibility among workers (*3*). Pricing is considered a particularly effective method of reducing peak period congestion (*4*), but may be politically difficult to implement.

Another approach, that of using incentives and gamification to reduce peak travel, is a potentially more palatable alternative (or complement) to peak pricing because, rather than penalizing people for traveling during peak times, it rewards them for traveling during less congested times. One prominent example is the Singapore Land Transportation Authority's (LTA) INSINC program (now TSR—Travel Smart Rewards). INSINC was implemented as part of a suite of peak period demand-management measures that include discounted off-peak rides and employer financial incentives to promote employer schedule flexibility (*5*).

Participating commuters earned points for all travel on the LTA system, and additional points for travel during fixed shoulder periods. Points could be redeemed for a fixed dollar value or by playing an online game. Users also received personalized offers ("Magic Boxes") for additional points. An early evaluation of INSINC found a 7.5% reduction in the average share of trips made during peak periods during the program compared to beforehand among participants; most of this shift seems to have been to the few minutes just before the peak hour. The evaluation also concluded that shift effects were greater for those with friends in the program, and that the special offers had a positive effect in encouraging additional shift. The evaluation noted that commuters most likely to sign up were those already traveling in the shoulder hours, but stated that these committed off-peak commuters played a crucial role in the program by inviting more friends to join, and influencing the behavior of their peers (*6*).

The Tokyo Metro has also implemented an incentives program to encourage off-peak commuting. During certain times of year, travelers earn points based on how many hours they take the train before or after peak hour. Once the traveler has earned sufficient points, they can enter a lottery to win cash rewards (*7*). Although the effects of incentives had been tested previously in Singapore and Tokyo, the applicability of the incentives model to the North American rail context is not considered a given in the development of BART Perks. Henn et al. (*8*) found that urban rail demand-management techniques applied successfully in one context can fail in another.

Incentives have also been used to encourage off-peak commuting among drivers. The Stanford CAPRI (Congestion and Parking Relief Incentives) program aimed to reduce peak hour traffic in Stanford University by rewarding commuters who drive during off-peak hours. CAPRI users earned points for off-peak trips and could redeem rewards in an online game. Approximately 60% of users self-reported that they had shifted their travel times outside of peak periods during the program, and results suggest that users shifted by the minimum amount necessary to earn a reward (*9*).

Program Goals

BART Perks was modeled after the Singapore TSR program and used the same underlying strategies and software implementation to shift travel behavior. The main goals of the Perks program were to:

- Reduce peak period and peak direction crowding, and improve person throughput: The program sought to test the effects of offering riders incentives for shifting travel outside the morning rush, especially among peak period Transbay travelers.
- Improve BART customer satisfaction: Crowding has led to historic declines in BART customer satisfaction. It was hoped that a travel incentives program could help improve satisfaction among participants.
- Increase employer support for flexible work schedules: BART Perks aimed to provide employers with technical resources to encourage adoption of flexible work hour policies. Workplace constraints have been identified as a top barrier to shifting travel out of peak periods (*10*).
- Identify implementation challenges and solutions: Provide lessons learned for subsequent phases of the program, and as guidance to other regions considering similar programs.

Program Overview

This section describes the structure of the Perks program in terms of recruitment and employer outreach, earning points, and redeeming rewards.

Recruitment and Employer Outreach

Participants enrolled in the program through a mobile-friendly website (www.bartperks.com, now inactive). At sign-up they accepted a user agreement and were prompted to enter their transit smart card ID number. Linking each user with their smart card was necessary to award them points based on the frequency, timing, and length of their trips, and to evaluate program effectiveness. Participants were recruited in four ways:

- Direct outreach: For three days at the launch of the program, the marketing team distributed fliers to riders at the Embarcadero and Montgomery stations in San Francisco's financial district during the morning peak. There were also banners advertising the program hung at the two stations for most of the duration of the pilot period.

- Earned media coverage/word-of-mouth: Approximately 20 news outlets covered the launch.
- Employer partnership program: Employers who signed up as a Perks partner and promoted the program among their employees were offered a $20 gift certificate for the first 10 sign-ups from their company or organization, and employees received a sign-up bonus of 200 points. Employer partners were also offered resources and technical assistance from a consultant specializing in flexible work hours.
- Friend referral: Participants were offered 250 points when a friend registered for the program. Additionally, in October 2016, all participants received a special offer ("bonus box") to earn 500 points by recruiting a friend.

Earning Points

In the initial four months of the program, participants earned one point per mile for all travel on BART and between three and six points per mile for travel initiated during morning peak shoulder periods, also known as Bonus Hours (Table 1), depending on their status in the program. Program status was determined by the number of peak shoulder trips made in a given time period (Table 2). For the last two months of the program, the design was changed so that participants earned 17 points per trip (equivalent to the average points earned per trip during the first part of the program), rather than one point per mile traveled. The intent was to compare the effectiveness of the two approaches, especially for certain travel markets, such as longer distance origin–destination pairs where participants would receive relatively more incentives under the initial scheme, and relatively fewer under the later scheme.

In addition to earning points per mile or per trip, randomly selected participants also received occasional special offers ("bonus boxes") of additional points during the program. The purpose of these bonus boxes was to test the sensitivity of participants to different types and levels of targeted incentives, and to encourage further shift among the target market. Three offers ranging from 20 to 500 points were made during the program duration to encourage additional shift among regular Transbay Tube participants.

Redeeming Rewards

Participants had three options for exchanging points for cash rewards:

- Autoplay: For those enrolled in Autoplay, all points earned in the previous week were automatically entered into a game (effectively, a random rewards generator) once per week. Depending on the outcome, the participant could receive nothing, or a reward ranging between $1 and $100.
- Spin-to-Win game: Participants could turn off Autoplay and use points to play an online chutes-and-ladders game to win more points or cash rewards. Like Autoplay, this game was essentially a random rewards generator, but included an interactive element.

Table 1. Perks Program Points Rules

	Bonus hour 6:30–7:30 a.m.	Peak hour 7:30–8:30 a.m.	Bonus hour 8:30–9:30 a.m.	All other times
Points earned	3, 4, 5, or 6 points/mile[a] (depending on status)	1 point per mile[a]	3, 4, 5, or 6 points/mile[a] (depending on status)	1 point per mile[a]

[a]In January 2017, the scheme was changed to award 17 points per trip instead of 1 point per mile.

Table 2. BART Perks Program Status Rules

Status	Bronze	Silver	Gold	Platinum
Bonus hour point multiplier	3 points/mile[a]	4 points/mile[a]	5 points/mile[a]	6 points/mile[a]
Maximum reward (in game play)	$10	$20	$50	$100
Number of bonus hour trips required for this status	0	2/week, for at least 2 weeks	3/week for at least 2 weeks	4/week for at least 2 weeks

[a]In January 2017, the scheme was changed to award 17 points per trip instead of 1 point per mile.

- Cash buyout: Participants could exchange points at a rate of $1 per 1,000 points; 1,000 points was equivalent to 1,000 miles traveled on BART—a typical commuter might travel this distance in about a month and a half.

Initially, the project team hoped to pay rewards as value added back onto participant transit smart cards, but this proved infeasible. Instead, BART paid out cash rewards once per month via PayPal. However, this arrangement was suboptimal, because participants needed to have an active PayPal account that used the same email address that was used to register for BART Perks.

Limitations of the Approach

This program design replicated the approach implemented in Singapore, because it had been shown to be successful in shifting travel behavior, and because it was felt that the design was simple and understandable to the public. However, the design had several limitations, most notably with respect to the use of a static, universal definition for the shoulder peak period (6:30–7:29 a.m. and 8:30–9:29 a.m.) and peak (7:30–8:29 a.m.) travel. In fact, the peak travel period on BART varies significantly by line, day of week, season, and in response to erratic delays—therefore a single overall peak period may or may not apply on any given day.

Additionally, a key goal of the program was to reduce crowding in the Transbay Tube in the peak hour in the peak direction. However, what constitutes a "peak departure" varies based on the individual's origin station and distance from the Transbay Tube. For example, someone departing from the end of a line (for example, the Pittsburg/Bay Point station) during the incentivized period of 6:30–7:29 a.m. will travel about 60 min before reaching the Transbay Tube, and could reach the tube during the exact period (7:30–8:29 a.m.) that they were incentivized to avoid. The project team conducted analysis to confirm that this situation would occur for only a minority of participants, and decided to prioritize the simplicity/understandability of the test program over technical precision.

Results

This section describes the program results in terms of enrollment, engagement/rewards redemption, travel behavior change, participant satisfaction, and the employer partnership program.

Enrollment

Active enrollment peaked at about 17,800 participants (an additional 2,400 enrolled but never activated their account). Most enrollment was obtained through direct outreach and earned media coverage in the first few days of the program. A minority of participants (6%) signed up for the program through a friend referral or through their employer. Enrollment trends were also affected by the short duration of the program. Beginning in January 2017, program staff began messaging that the program would end within two months, and enrollment was formally closed in mid-January.

The project team hoped that targeting marketing efforts at the Embarcadero and Montgomery stations during the morning peak period would result in high program enrollment among peak period Transbay travelers. Sweepstakes law, which applied to BART Perks because of the Spin-to-Win component, precluded more direct methods of recruitment such as qualifying participants based on their travel history. Sweepstakes law dictates that any California resident over the age of 18 must be allowed to participate in a sweepstakes program.

Ultimately, peak period Transbay travelers were a small share of program participants. Table 3 summarizes participants by their "usual" commute time (this was determined by the timing of their commute travel before program initiation). It shows that approximately 50% of program participants were regular morning commuters. However, morning peak hour Transbay commuters comprised at most 13% of all program participants. This indicates that significant numbers of program participants were not regular users of the congested Transbay Tube, which was the focus of the decongesting efforts.

Travel Behavior

Perks' effect on participant travel behavior was evaluated by comparing behavior before and during the program, estimating the number of trips shifted, and examining whether shifts persisted after the program ended.

Share of Trips Made during Peak Hour, before and during the Program. The share of participant trips that began during

Table 3. Participants by Regular Morning Commute Travel Pattern

Commute	Number of participants	Percent of participants
Regular morning commute		
Usual peak hour	1,737	10
Sometimes peak hour	1,917	11
Not peak hour	5,272	30
Total	8,926	51
Regular morning Transbay commute		
Usual peak hour	1,155	6
Sometimes peak hour	1,230	7
Not peak hour	3,403	19
Total	5,788	32

Note: Total participants = 17,800.

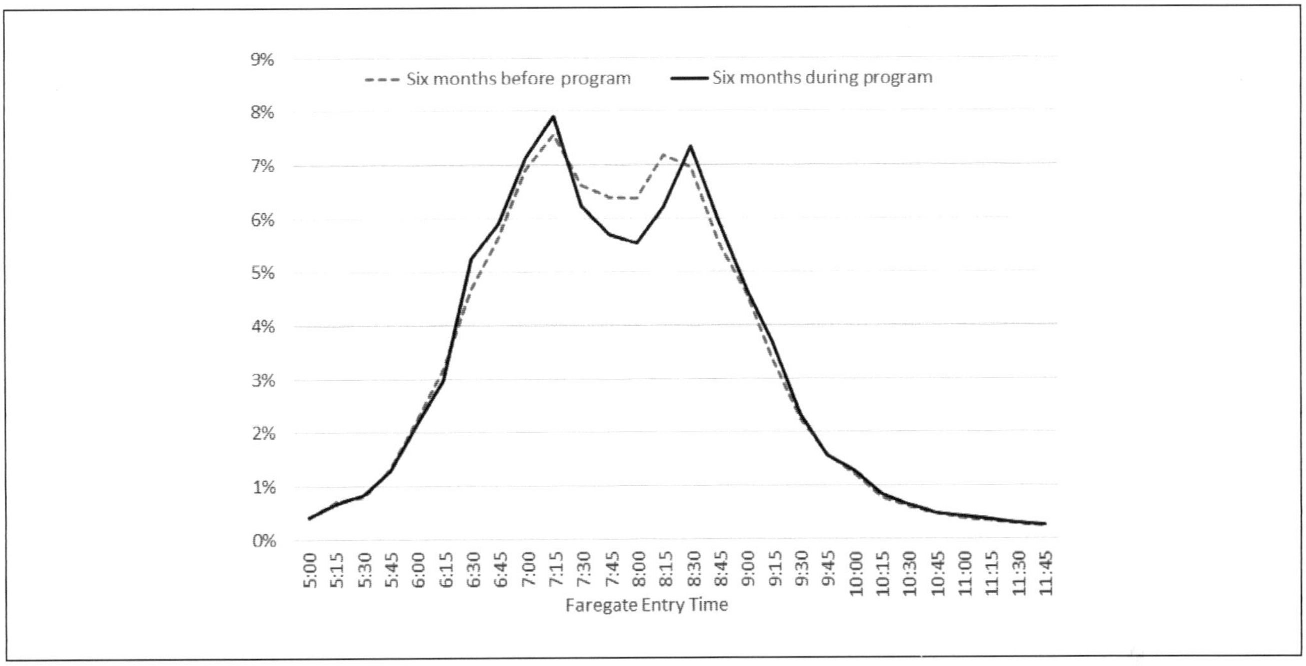

Figure 1. Inbound morning Transbay participant trips before and during BART Perks program.

Table 4. Peak Hour Travel Shares by Market

	System		Transbay inbound	
	All travelers	BART Perks participants	All travelers	BART Perks participants
Daily trips by market	420,300	18,200	113,900	6,300
Peak hour share—before (%)	11.4%	14.3%	20.5%	26.6%
Peak hour share—during (%)	11.4%	12.9%	20.3%	23.7%
Peak hour share—difference	0.0%	−1.4%	−0.2%	−2.9%
Percentage reduction	0.3%	−9.8%	−0.9%	−10.9%
Total reduction	121	−250	−214	−180

the peak hour (7:30–8:29 a.m.) before the program began was compared with the share of participant trips that began during the peak hour during the program. Figure 1 illustrates the result by comparing the faregate entry times of Transbay inbound morning trips made by participants on average during weekdays (Tuesday–Thursday) in the six months before the program to the six months during the program. It illustrates that the percentage of entries made during the peak hour (7:30–8:29 a.m.) fell during the program compared with the period before the program. It also illustrates that many participants already traveled in the peak shoulder period before the program, and that shifters primarily moved to just before or just after the peak hour, similar to results from the Singapore program (*6*). The degree of participant shift did not vary significantly between the mileage-based and trip-based schemes.

Number of Trips Shifted. To estimate the number of trips that were shifted by the program, the shares of daily participant trips that entered the system in the peak hour before Perks and during Perks were calculated using participants' smart card data. These shares were then applied to the average daily trips made by participants. The difference is the estimate of shifted trips. This can be calculated systemwide for the Transbay inbound market, or for any other market segment.

As shown in Table 4, it is estimated that systemwide during the six-month trial period about 10% of all peak riders, or an average of 250 participants (roughly two full train cars of passengers), shifted their ride to either before or after the peak morning hour each weekday. Inbound Transbay peak hour trips were reduced by 10.9% or 180 trips. The table also shows that the share of travel occurring in the peak hour remained approximately the same

Table 5. Share of Regular Commuter Trips by Morning Peak Hour

	5:00–6:30 a.m.	Bonus hour 6:30–7:29 a.m.	Peak hour 7:30–8:29 a.m.	Bonus hour 8:30–9:29 a.m.	9:30–11:59 a.m.
Pre-program	9.1%	27.4%	29.9%	24.0%	9.5%
During program	8.7%	29.1%	26.2%	26.2%	9.7%
Post-program	9.4%	27.1%	28.6%	24.5%	10.3%
Change from pre-program to during program	–0.4%	1.7%	–3.7%	2.2%	0.2%
Change from during program to post-program	0.7%	–2.0%	2.4%	–1.7%	0.6%
Change from pre-program to post-program	0.3%	–0.3%	–1.3%	–0.5%	0.4%

during this period for all BART riders and Transbay inbound travelers. This suggests that the reductions in peak travel among Perks participants are attributable to the Perks program and not background trends affecting the BART system.

Persistence of Shift after the Program. The BART Perks program was in effect for six months, from late August 2016 through late February 2017. On termination of the program, it was possible to observe whether the behavioral shifts from the peak hour persisted. Table 5 shows a summary of changes in trip-making by peak hour for "regular commuters," which was defined as those travelers who made at least 15 morning peak trips before the program was initiated, at least 30 morning peak trips during the program, and at least 15 trips after the program ended. Note that the average number of trips made by regular commuters during the program phases was significantly higher—approximately 60 trips per commuter during the three-month pre- and post-program phases, and nearly 90 trips per commuter during the six-month program. This table shows that regular commuters reduced their share of peak hour trip-making from 29.9% before the program to 26.2% during the program (slightly more change than the reduction from 26.6% to 23.7% of all Transbay inbound trips). However, after the program ended, this share then rose back to 28.6%, suggesting that, for the target market of morning peak hour inbound commuters, the program effects partially, but not completely, persisted after the program ended.

Participant Game Engagement and Reward Redemption

Points accrued automatically, and most users retained the default Autoplay option, meaning that their points were entered into a random rewards generator every week. About 86% of program cash rewards were paid out this way, compared to 13% paid out through manual online play of the Spin-to-Win game, and less than 1% through the cash buyout. Manual game play was associated with higher levels of shift: 27.8% of participants who only actively played the game shifted, compared to 8.7% who used Autoplay.

The average participant made about one trip per weekday, and earned a total of \$13, or a little over \$2 per month, for the duration of the program. However, the distribution of rewards varied considerably by participant (Figure 2).

Participant Satisfaction with BART

Perks participants were surveyed in December 2016 and again in February 2017. Nearly 10,000 participants responded. Table 6 summarizes participant responses to questions about satisfaction with Perks, BART overall, and with the degree of crowding on BART, and compares them to responses to the same questions included in BART's biannual Customer Satisfaction Survey, which was conducted around the same time. Responses from the latter survey reflect only those BART riders who use the smart card, since this group is more comparable to the Perks participants, who were also smart card users.

Table 6 shows that participants generally supported the Perks program (nearly 70% said they were very or somewhat satisfied). However, their satisfaction with BART remained lower than BART riders overall who use smart cards. This is likely because they appear to experience worse crowding than typical BART riders, as shown by their lower ratings for the availability of seating and standing room (Table 6). Results from BART's general Customer Satisfaction Survey indicate that crowding is one of the top factors that influences customers' overall satisfaction with BART (*11*).

Participants also provided about 6,100 open-ended comments about the program. Top comments were

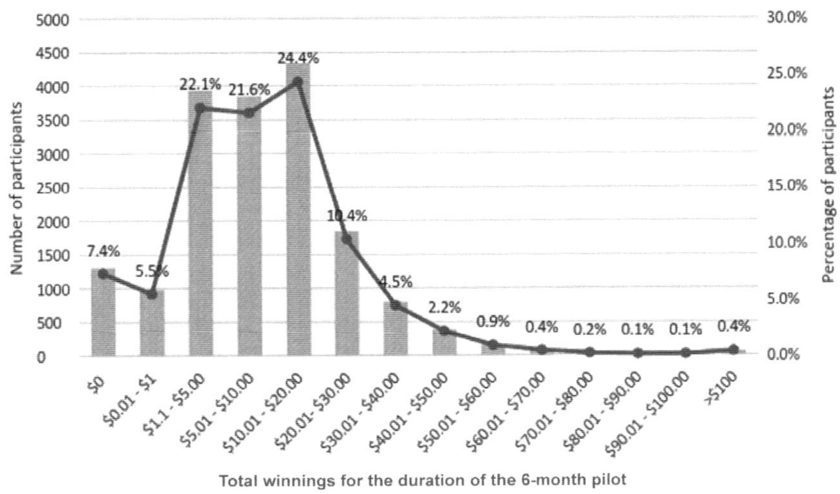

Figure 2. Frequency distribution of total winnings by participant for the six-month pilot.

Table 6. Perks Participant Satisfaction Compared to All BART Riders

	Perks participants	All BART riders who use smart cards
Satisfaction with Perks program (percentage very or somewhat satisfied)	69%	N/A
Satisfaction with BART overall (percentage very or somewhat satisfied)	59%	67%
Rating for availability of standing room on BART (1–7 scale, 7 is best)	3.93	4.27
Rating for availability of seats on BART (1–7 scale, 7 is best)	3.18	3.66
Source	Perks Participant Survey December 2016 and February 2017 (n = 9,865)	BART Customer Satisfaction Study, September and October 2016 (n = 3,793)

requests for additional/higher levels of rewards/points, requests for expanded or different payout options (besides PayPal), and general positive feedback about the program. Participants rated the level of rewards offered lowest of all the program attributes, with only 44% rating the level of rewards Good or Excellent.

Employer Partnership Program

Fifteen employers signed up as BART Perks Partners and were committed to promoting the program among their employees, including UCSF Benioff Children's Hospital, Alliant International University, CRI, and Integral Group. Two employers requested to receive technical assistance from the flexible work schedules consultant.

Evaluation against Program Goals

Reduce Peak Period and Peak Direction Crowding, and Improve Person Throughput

The Perks program demonstrated that incentives can be successfully used to shift the departure times of peak period travelers. Among all program participants there was a reduction in peak hour inbound Transbay travel of 10.9%, and a reduction in peak hour overall system travel of 9.8%, which is slightly greater than the 7.5% reduction in peak hour trips made by participants in the TSR program in Singapore (*6*).

While promising, this shift was on a small base of peak period trips and therefore had little impact on crowding. The 17,800 active participants only generated about 1,700 Transbay morning peak hour, peak direction trips on a typical weekday, and about 10.9% of these trips shifted during the program, representing a shift of approximately 170 trips. For context, BART moves about 26,000 people through the inbound Transbay Tube in the peak hour. About 1,300 people, or one 10-car train worth, would need to shift for there to be a 5% reduction in overall inbound Transbay morning peak hour volume.

Improve BART Customer Satisfaction

Perks program participants were generally positive about their experience with the program, and about 70% indicated that they were somewhat or very satisfied. However, participation in the Perks program does not appear to have impacted participant satisfaction with BART overall. In fact, because Perks participants experienced more crowded conditions than BART riders on average, they were less satisfied with BART. Future programs could have a greater effect on satisfaction by significantly reducing peak crowding.

Increase Employer Support for Flexible Work Schedules

Perks included an employer partnership program that was aimed at increasing Perks program enrollment while providing resources to increase employer support for greater adoption of flexible work schedules. While many employers signed up to be partners in promoting the program, few were interested in receiving technical assistance. This may be because many of the employers who signed up as partners already offered flexible work schedules, and that offers of technical assistance may not have provided a compelling reason for employers to initiate a change in employee policies. This is consistent with past pilot programs we have undertaken, in which we found it difficult to interest employers in making employee policy changes through offers of technical assistance (*12*). Future programs may wish to explore different employer engagement strategies.

Key Recommendations

The authors offer the following recommendations for future programs:

- Obtain sufficient peak travelers. We found that despite efforts to target recruitment at peak riders, the majority of participants did not regularly travel in the peak hour before the start of the program. To significantly increase peak shift, future programs will need to recruit larger number of peak riders, such as by scaling up enrollment to ensure that sufficient numbers of peak riders are participating, providing higher levels of rewards to entice more peak riders to sign up, or qualifying participants based on their travel history.
- Focus rewards on behavior change. We found strong self-selection into the program among participants who were already traveling during the desired times before the program. Future programs should ideally be structured to reward behavior change rather than preexisting behavior, such as by establishing a behavior baseline (e.g., frequent peak travel) and rewarding change from that baseline.
- Consider more efficient methods of participant recruitment and retention. BART Perks rewarded all BART riders to some degree for traveling on BART, and this likely was an important factor in explaining the appeal of the program and rapid enrollment of nearly 18,000 participants within a few weeks of launch. However, the broad appeal of the program resulted in a high level of participation among individuals who were not frequent BART commuters. Future programs will need to explore more efficient means of recruiting and engaging participants.
- Consider the trade-off between a sweepstakes component and participant engagement. While the team experienced legal and technical complications in including the sweepstakes feature, we found that participants who were more engaged in the Spin-to-Win game also had higher levels of shift. Future programs should consider the trade-off of sweepstakes complications with the potential for greater engagement and shift.

Next Steps

BART has received a grant from the Federal Transit Administration to conduct further research on the potential to use incentives to reduce crowding in the Transbay corridor. BART and SFCTA staff are working to scope a more targeted, smaller-scale trial of an incentive program that builds on the lessons learned from BART Perks.

Acknowledgments

The authors would like to thank the former staff of Urban Engines (Dr. Balaji Prabhakar, Dr. Christopher Pluntke, Dave Parker, Kim de Rose, and Tom Yue) for creating and deploying the BART Perks program, and Allen Greenberg of FHWA for providing essential funding and expertise to inform the program design. Staff would like to also thank the consultant team that supported the program deployment and analysis, including

Civic Edge Consulting, Resource Systems Group, Seasons Productions, Barretto-Co, and AECOM, and the staff of SFCTA and BART who played key roles in the program development. BART Perks was funded by a generous grant from the FHWA Value Pricing Program, BART, and San Francisco's half-cent sales tax for transportation. The full BART Perks Evaluation Report for FHWA is available on the SFCTA website: www.sfcta.org.

Author Contributions

The authors confirm contribution to the paper as follows: study conception and design: RGR, JC, CG; data collection: RDR, JC, CG; analysis and interpretation of results: RGR, JC, CG, MB; draft manuscript preparation: RGR, JC, CG. All authors reviewed the results and approved the final version of the manuscript.

References

1. Bay Area Rapid Transit District. Total Annual Exits FY1973–FY2017. https://www.bart.gov/about/reports/ridership. Accessed July 28, 2017.
2. Federal Transit Administration. Final Interim Policy Guidance: Federal Transit Administration Capital Investment Grant Program, June 2016. https://www.transit.dot.gov/sites/fta.dot.gov/files/docs/FAST_Updated_Interim_Policy_Guidance_June%20_2016.pdf. Accessed July 29, 2017.
3. Halvorsen, A. Improving Transit Demand Management with Smart Card Data: General Framework and Applications. Submitted to the Department of Civil and Environmental Engineering at the Massachusetts Institute of Technology in partial fulfillment of the requirements for the degree of Master of Science in Transportation, 2015. http://hdl.handle.net/1721.1/99543.
4. Cervero, R. Transit Pricing Research. *Transportation*, Vol. 17, No. 2, 1990, pp. 117–139.
5. Singapore Land Transportation Authority. Travel Smart. https://www.lta.gov.sg/content/ltaweb/en/public-transport/mrt-and-lrt-trains/travel-smart.html. Accessed July 18, 2017.
6. Pluntke, C., and B. Prabhakar. INSINC: A Platform for Managing Peak Demand in Public Transit. *JOURNEYS, Land Transport Authority Academy of Singapore*, 2013, pp. 31–39. https://web.stanford.edu/~balaji/papers/13INSINC.pdf
7. Van der List, B. How the Tokyo Metro Handles Rush Hour to Operate On Time. *Metro Magazine*, April 13, 2017.
8. Henn, L., G. Karpouzis, and K. Sloan. A Review of Policy and Economic Instruments for Peak Demand Management in Commuter Rail. Presented at the 33rd Australasian Transport Research Forum Conference, Canberra, Australia, 2010.
9. Zhu, C., J. S. Yue, C. V. Mandayam, D. Merugu, H. K. Abadi, and B. Prabhakar. *Reducing Road Congestion through Incentives: A Case Study*. Transportation Research Board, Washington, D.C., 2015.
10. Litman, T. Transit Price Elasticities and Cross-Elasticities. *Journal of Public Transportation*, Vol. 7, No. 2, 2004, pp. 37–58.
11. Bay Area Rapid Transit Marketing and Research Department, Corey, Canapary & Galanis Research. 2016 Customer Satisfaction Survey. https://www.bart.gov/sites/default/files/docs/CustSat2016Report_fnl.pdf. Accessed March 25, 2017.
12. Greene-Roesel, R., K. Singa, and J. Karlin-Resnick. The Challenge of Soliciting Voluntary Participation in Parking Cashout: Lessons Learned from San Francisco. Presented at 94th Annual Meeting of the Transportation Research Board, Washington, D.C., 2015.

The Standing Committee on Public Transportation Marketing and Fare Policy (AP030) peer-reviewed this paper (18-04803).

Article

TRR
JOURNAL OF THE TRANSPORTATION RESEARCH BOARD

Transportation Research Record
2018, Vol. 2672(8) 566–577

DOI: 10.1177/0361198118797485
journals.sagepub.com/home/trr
SAGE

Differences in Control and Regulatory Structures of Public Transport within the United Kingdom and Ireland: Implications for Quality and Effectiveness of Service Delivery

Austin Smyth[1] and Luke Kelleher[2]

Abstract

Regulation and control of public transport is the subject of continuing debate in the United Kingdom and Ireland. In Great Britain, e.g., this stems from dissatisfaction with market liberalization. Elsewhere the concerns stem from pressures to liberalize the market and funding constraints. Such concerns and calls for further reform pose questions about the relative effectiveness of greater or reduced state involvement in public transport. This paper reviews evidence on the performance of public transport operators under a variety of control and regulations informed by benchmarking alternatives as well as funding under different regulatory systems. Various secondary data sources informed development of performance indicators for both bus and rail including level of service (vehicle kilometers) by per capita population, by region, population size, and population density; quality of service by fares and level of comfort; passenger satisfaction by operator across a suite of level of service parameters; and passengers per capita population, region, population size, and population density. This paper provides a commentary on the relative performances of the different types of public transport operator and informs an understanding of possible reasons for variations in performance under differing regulatory and control "models."

Ever since deregulation of buses in Great Britain (GB) (outside London) was proposed in 1984 there has been controversy concerning the merits of regulated and deregulated systems (*1–3*). In recent years there have been increasing calls for bus re-regulation and rail re-nationalization. This paper examines the effectiveness and efficiency of different forms of control and regulation covering public transport provision in the United Kingdom and Ireland (UK) by benchmarking alternative regimes with particular reference to areas outside the Greater London area and away from the major intercity corridors. It draws on evidence collated and analyzed by the Northern Ireland Audit Office (NIAO). In 2015 NIAO published the results of an extensive investigation into the effectiveness of public transport in Northern Ireland (*4*).

Effectiveness in the organization and delivery of public transport is defined as the extent to which the policies achieve their intended objectives in addition to any unanticipated side benefits. Key indicators encompass consumer satisfaction ratings and market performance among providers. Efficiency refers to the costs of delivery, in particular financial costs. Overall cost-effectiveness refers to the benefits generated as a function of the costs to the public purse and the internal costs of service providers.

The focus here is on intermediate output indicators of the performance of providers supplemented by quality of service as assessed by consumers together with the costs of service delivery. In conjunction with these quality-of-service ratings and market performance, detailed consideration of efficiency informs an assessment of the overall effectiveness of various models of control and regulation outside Greater London and beyond the major intercity rail corridors.

[1]Centre for Sustainable Communities, School of Life and Medical Sciences, University of Hertfordshire, Hatfield, UK
[2]University College Dublin, Planning and Environmental Policy, Dublin, Ireland

Corresponding Author:
Address correspondence to Austin Smyth: a.smyth2@herts.ac.uk

Regulation, implying interference in the workings of the market, appears to conflict with the norms of economic theory (*5*). According to this line of thought, the state can best achieve allocative efficiency by minimizing its role and allowing markets to become perfectly contestable. Inspired by this principle, governments have sought to privatize a range of activities. The UK was among the first countries to pursue this policy. From the 1980s until the mid-1990s, privatization of public transport was an important component of economic policy in the UK, except for Northern Ireland. This contrast in policy and practice facilitates benchmarking public and private ownership of public transport within the same state. Given the scale of such changes in public policy it is appropriate to place benchmarking in the context of a full understanding of developments in UK policy on control and regulation of public transport.

Key Developments in the Control and Regulation of Public Transport in the UK

The form and scale of intervention by UK governments in transport in the UK has changed with time. According to Smyth, historically, many of the regulatory controls in the UK were aimed specifically at containing the potential power of railway monopolies, whereas the inter-war period witnessed controls over highly competitive road-based modes such as buses and road haulage (*5*). Post-Second World War policy saw nationalization of the main transport supply industries to improve internal efficiency by facilitating greater co-ordination within and between modes.

Less than four decades later however, public ownership was no longer regarded as the optimum solution and emphasis was placed on market forces. In this new situation, economic regulation was intended to replicate efficient market outcomes rather than being seen as a straightforward replacement for market processes. More recently, UK governments have tended to withdraw from direct control of transport with the introduction of wide-ranging privatization. Private investment and management in public transport was espoused by the Conservative government that assumed power in 1979. Deregulation and competition were seen as central to arresting the decline in bus services, avoiding rising costs and minimizing subsidies (*6*). Deregulation of local bus services was enacted under the Transport Act 1985.

From October 1986, road service licensing in GB, with the exception of London, was abolished and replaced by a system of registration. The duties of local authorities to coordinate public transport ended.

The government's plan for the reorganization of the British Railways Board was reflected in the Railways Act 1993. It provided for a complex system of privatization of the GB network (*7*). The rail system was split into three structures: a stock market listed public limited company (Railtrack) responsible for infrastructure, train operating companies (TOCs) bidding for passenger franchises, with rolling stock (engines and carriages) leased from rolling stock companies (*8*). The election of the "New Labour" government in 1997 did not see a significant reversal of transport policies. However, under the Transport Act 2000 both the Director of Passenger Rail Franchising and the British Railways Board were abolished and their functions transferred to the Strategic Rail Authority.

Under these arrangements rail services, including fares, are partly specified by government or its regulatory bodies. In contrast, outside London bus services are largely left to the market to determine, with the notable exceptions of concessionary fare provisions and a limited number of defined and publicly subsidized routes. In contrast to other parts of GB, bus services in London, though operated by private companies, are specified and tightly regulated by Transport for London and answerable to the Mayor of Greater London.

In recent years, there have been increasing calls for bus re-regulation and rail re-nationalization across GB (*9*). In 2015 the Labour party election manifesto announced plans to regulate the bus and rail industry (*10*). At the 2017 election Labour called for a rolling program of re-nationalization as existing franchises came to an end.

Continuity in Control and Regulation of Public Transport in the UK: Northern Ireland—A Place Apart

In contrast to the widespread privatization and deregulation of public transport operations in the rest of the UK from the 1980s, public transport in Northern Ireland remains state-owned and wholly regulated and gives the operators a near monopoly on scheduled services. Privatization of public transport did not occur in Northern Ireland, largely attributable to the consequences of 30 years of political instability and violence.

Public transport in Northern Ireland is controlled and regulated by the Department for Infrastructure (DFI). It sponsors the Northern Ireland Transport Holding Company (NITHC) and provides financial and administrative support to help ensure the operational delivery of transport policy. This extends to the planning, delivery, and governance of public transport. NITHC is responsible for the operation of Translink, a brand name introduced to cover the integrated services of Ulsterbus, Citybus (now rebranded Metro), and Northern Ireland Railways (NIR), all of them subsidiary companies of

NITHC. NIR is the only state-owned vertically integrated rail company providing passenger train services in the UK. These operating companies determine routes, timetables and fares with DFI exercising a light touch oversight of such matters, while at the same time providing strong control over major capital investments. A feature of the latter is the significant underfunding of capital investment in public transport in Northern Ireland compared with the UK as a whole.

Benchmarking Process

The benchmarking process encompassed defining a series of effectiveness and efficiency indicators for bus and rail services within the UK separately, and collating performance for selected operators, modes, or both operators and modes serving specified cities, areas, and jurisdictions. For both modes of transport, the findings are grouped into discrete indicators classified under the effectiveness of the system and its efficiency.

Effectiveness of the bus sector is measured as follows:

- Frequency of service (vehicle kilometers, VKM) per capita, by areal unit and where available in relation to population size of urban area and population density for each operator
- Bus fares indicator: Average cost per passenger journey
- Bus comfort indicators: Vehicle design and seating availability; age of the fleet and load factor

Effectiveness is also measured in terms of customer satisfaction ratings across a battery of quality-of-service indicators as follows:

- Punctuality
- Cleanliness
- Comfort
- Seating availability
- Service information provided
- Driver helpfulness
- Value for money

Further, it can be argued effectiveness is also reflected in patronage and market-share indicators.

Efficiency for the bus sector is defined by a series of supply-side indicators, including operating costs and staff productivity. The indicators include:

- Staff per bus
- Operating staff to management staff ratios
- Annual wage cost per employee
- Operating cost per bus
- Turnover per employee

Investigation of comparable data available for the rail sector yielded the following indicators of effectiveness:

- Service frequency for various distance bands
- Journey times for various distance bands
- Fares for various distance bands

As with the bus sector, effectiveness of the different rail systems can be measured in relation to customer satisfaction across various quality-of-service indicators as follows:

- Punctuality
- Cleanliness
- Comfort
- Seating availability
- Service information provided
- Driver helpfulness
- Value for money

As with buses, effectiveness is also reflected in patronage trends and market-share indicators.

For the rail sector, efficiency is defined in terms of a series of supply-side indicators including:

- Operating cost per passenger km
- Operating cost per timetabled train

The data on which the assessment was based was drawn from a variety of published sources including the annual TAS Partnership Business Monitor and Passenger Focus Survey data. Passenger Focus is the independent passenger watchdog, set up by the UK government. The research was also supplemented by information obtained from a variety of consultancy studies undertaken on behalf of DFI, NITHC, or both. In the absence of direct indicators, author calculations were derived from census data, transport statistics, and TAS Partnership Business Monitor data (see Figure 1, *a* and *b*). In addition, various travel planning tools including Traveline, Traveline Scotland and Ireland's National Transport Authority tools together those of National Rail, Irish Rail, Bus Eireann, Dublin Bus, Translink, and various GB operators were utilized (See Figure 3). Journey times by car were drawn from various journey planning tools.

Operator Performance by Benchmark Indicator

Effectiveness of Bus Services by Control and Regulatory Model

VKM provided by Northern Ireland's state-owned bus services were benchmarked against operators in England, Scotland, and Wales. The available data permitted

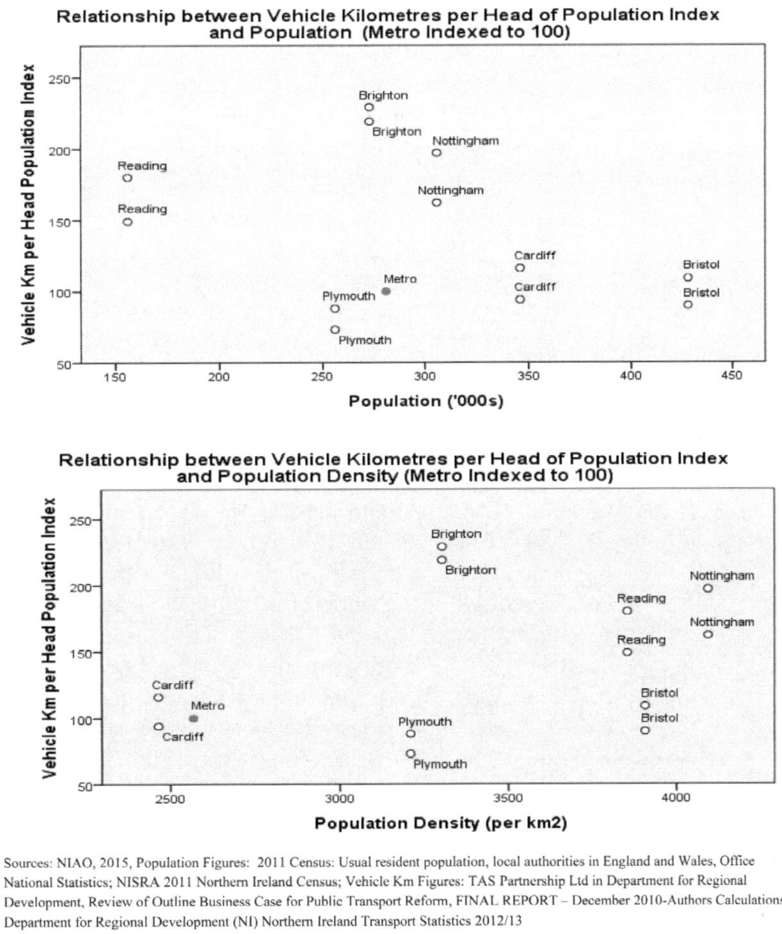

Figure 1. (*a*) Vehicle kilometers per head of population and population for Translink Metro Service (Belfast) and six other UK urban areas, and (*b*) vehicle kilometers per km^2 and population density for Translink Metro Service (Belfast) and six other UK urban areas.
Sources: NIAO, 2015, Population Figures: 2011 Census: Usual resident population, local authorities in England and Wales, Office National Statistics; NISRA 2011 Northern Ireland Census; Vehicle Km Figures: TAS Partnership Ltd in Department for Regional Development, Review of Outline Business Case for Public Transport Reform, FINAL REPORT December 2010-Authors Calculations; Department for Regional Development (NI) Northern Ireland Transport Statistics 2012/13

comparable service-level indicators to be investigated at four spatial scales: jurisdiction (country level), Passenger Transport Executive (PTE) areas (major conurbations outside London) and non-PTE areas aggregated, and for selected non-metropolitan urban areas.

Jurisdiction Level: "Country". Table 1*a* provides index values for VKM per head of population and VKM per hectare for Northern Ireland, Scotland, Wales, and England. Northern Ireland's VKM per head of population and per hectare is given a base value of 100.

Overall bus service provision per head of population in Northern Ireland is lower than all other parts of the UK but markedly lower than in Scotland. Service levels are likely to be directly related to population density. Taking this into account however, overall Northern Ireland's bus service levels per hectare are still lower than Scotland, markedly lower than England but on a par with Wales. To investigate the extent to which these differences reflect background demographic profiles or geographical conditions the next stage in the benchmarking exercise involved splitting the analysis between major PTE and non-PTE areas as reflected in the national statistics. Non-PTE areas can range from smaller urban areas to very remote, low-density locations with small total populations. In the absence of more disaggregate figures this was adopted as a pragmatic approach to compare service levels by area type before pursuing a

Table 1.
(a) Bus Vehicle Kilometers per Head of Population and per Hectare for Northern Ireland, England, Scotland and Wales

	BVK per head of population	BVK per hectare
Northern Ireland	100	100
England	109	323
Scotland	194	118
Wales	110	95

(b) Bus Vehicle Kilometers per Head of Population for Translink Ulsterbus and Non-PTE Area

	BVK per head of population
Translink ulsterbus	100
Non-PTE areas	86

(c) Bus Vehicle Kilometers per Head of Population and per Hectare for Translink Metro (Belfast), City Regions (PTE Areas) and London

	BVK per head of population	BVK per hectare
Translink metro	100	100
City regions (PTE Areas)	140	130
London	167	471

Note: Table 1 uses an index scale with Northern Ireland's bus vehicle kilometers per head of population and per hectare given a base value of 100. BVK = bus vehicle kilometers; PTE = Passenger Transport Executive. Sources: Northern Ireland Audit Office 2015, TAS Partnership Ltd in Department for Regional Development, *Review of Outline Business Case for Public Transport Reform: Final Report*: December 2010 (Chapter 7, p. 85) (*11*).

more in-depth investigation for selected local areas for which data could be obtained.

Ulsterbus and Non-PTE Areas. Ulsterbus operates almost all bus services outside the Greater Belfast area, serving many smaller towns as well rural Northern Ireland. Table 1*b* benchmarks Ulsterbus's level of service against GB non-PTE areas. The index scale allocates a base value of 100 for Ulsterbus in the case of VKM per head of population and per hectare. The evidence indicates that Northern Ireland (outside Belfast) enjoys a significantly higher level of service than comparable areas within GB. Other aspects of effectiveness and the issue of efficiency and overall cost-effectiveness are considered elsewhere.

Translink Metro Bus, PTE Areas and London. The Belfast Metropolitan Area represents some one-third of Northern Ireland's population, although the city and its hinterland are significantly smaller than any of the PTE areas. Table 1*c* depicts VKM per head of population and VKM per hectare for Translink Metro (Belfast) and City Regions (PTE Areas). Figures for London are provided for completeness.

When we consider Belfast, separately from other areas in Northern Ireland, bus service levels per head of population are significantly lower than overall average service levels in GB city regions and London. Moreover, Belfast's bus service levels per hectare are also somewhat lower than overall figures for GB city regions and barely 20% of that provided in London. However, Belfast is markedly smaller than the city regions for which aggregate data on service levels is readily available. Therefore the next stage was to benchmark Translink Metro against companies operating in other similar-sized UK cities.

Benchmarking Translink Metro with Selected Bus Company Operators in Medium-Sized Cities. The bus companies chosen for more detailed analysis are based in: Bristol, Brighton, Cardiff, Nottingham, Plymouth, and Reading. Metro is broadly comparable with these in size and operating territory. The aggregate phase of the analysis investigated the relationship between VKM per head of population and total population for Belfast (Translink Metro Service) and the six other UK urban areas. Figure 1*a* uses an index scale with Translink Metro's VKM per head of population given a base value of 100.

In the absence of direct indicators for such operators two VKM per head of population values were estimated and plotted for each of the six UK urban areas; one was based on PTE areas and one based on non-PTE areas. This enabled upper and lower boundaries to be placed on the "true" VKM per head of population for each of the six urban areas. The same protocol was adopted to investigate the VKM per Km^2. Figure 1*a* depicts the pattern of VKM per head of population in relation to population size for the urban areas analyzed. Figure 1*b* plots the VKM per Km^2 and population density for Translink Metro Service (Belfast) and the six other UK urban areas using the same values. Figure 1*a* indicates bus service frequencies per head are lower in Belfast than for most cities with similar populations. This appears consistent with the aggregate findings reported above for Belfast compared with the major city regions. Figure 1*b* shows bus frequencies per km^2 are typically lower in Belfast than many cities with similar populations, but typical of urban areas with relatively low population densities. In relation to service frequency Belfast performs poorly in absolute terms but in line with lower-density urban areas.

Buses Fares Indicator: Average Cost per Passenger Journey. Quality of bus service under state ownership and for deregulated systems also encompassed benchmarking bus fares and levels of on-board comfort for Belfast and the six other operations assessed for frequency variations. The average cost per passenger journey in Belfast at £1.58 is the fourth highest of the seven areas with Brighton and Hove £2.40, Reading £2.20, Bristol £1.70, Nottingham £1.45, Cardiff £1.20, and Plymouth £1.00 (£1 = US$1.52 April 25th, 2015; £1 = US$1.30 September 11th, 2018). Belfast is higher than the less prosperous benchmarked urban areas.

Bus Comfort, Vehicle Design, and Seating Availability Indicators: Age of the Fleet and Load Factor. Two industry indicators, age of the fleet and load factor (seating availability), for which data could be derived for major PTE areas and Metro (Belfast) as well as non-PTE areas from published and other secondary sources were chosen as indicative of comfort. In 2008–2009 Ulsterbus and Metro had the youngest average fleet age, with the exception of London. By 2012 to 2013, this was still the case despite a rise in Metro's average fleet age. Translink Metro services operate with marginally lower load factors than other city regions in the UK, but markedly lower than in London. Together these indicators suggest both Translink Metro and Ulsterbus passengers enjoy relatively good levels of comfort. The analysis of effectiveness reported above, relates to output or intermediate indicators.

Bus Customer Satisfaction. The definition of effectiveness however goes further to encompass the extent to which the policies achieve their intended benefits. It was noted in the introduction that key indicators of the realization of benefits include consumer satisfaction ratings and market performance among the relevant providers. Therefore an important element of benchmarking public transport is how the consumer views the service on offer. In Northern Ireland, Translink undertakes surveys of customer satisfaction in conjunction with the area's official consumer representative body. Similarly, passenger focus coordinates/undertakes user satisfaction surveys periodically among bus users in GB. Drawing on both sources, comparative satisfaction scores were compiled for the seven benchmarked urban areas. In addition to Belfast (Metro bus) the areas included: Brighton and Hove, Bristol, Cardiff, Nottingham, Plymouth, and Reading. The seven areas were measured across seven quality-of-service parameters.

Overall, Metro customers were less satisfied than customers in other parts of GB. Metro compared poorly in the areas of punctuality, cleanliness, seating availability, information provision, and driver helpfulness. They were average in terms of value for money, but had the highest satisfaction levels in terms of comfort. These findings seem to be broadly consistent with the output indicator performance reported above. In contrast, Translink's Ulsterbus operations achieve higher levels of consumer satisfaction than those enjoyed by Metro. This might be expected given the assessment of objective output indicators referred to above.

Efficiency of Bus Operations

Overall cost-effectiveness also reflects the relationship between the benefits generated and the costs to the public purse and internal costs of providers. The next stage benchmarked the efficiency of Northern Ireland's regulated bus service with deregulated counterparts in England, Wales, and Scotland. The same comparator operators that were selected for the effectiveness benchmarking exercise were chosen (with the exception of Nottingham for which no data was available). Various performance indicators were generated to measure operator efficiency across the different areas. The same performance indicators were used to measure the cost-effectiveness of the Ulsterbus service. This was benchmarked against five bus operators in non-metropolitan areas of England, Wales, and Scotland. These were Arriva Cmryu, First Devon and Cornwall, Stagecoach in Lincolnshire, Stagecoach Devon, and Stagecoach Scotland.

Examination of each of the comparators found that Translink Metro and Ulsterbus performed relatively well in terms of operating cost per bus and staff per bus, but they lay at the lower end of the range on turnover per employee. The single largest components of the cost base for bus operators are wages and pensions costs. Between 2006 and 2011, annual wage cost per employee was highest in both the Metro service and Ulsterbus service compared with the other benchmarked UK areas. Higher overall costs for road staff than elsewhere in the UK tend to reflect the burden of a final salary pension scheme to which Translink staff belong. In GB most operators have switched to the cheaper defined benefits scheme.

Figure 2 illustrates the operating staff to management staff ratio between 2006 and 2011 for Northern Ireland's Metro service and bus operators in the comparable UK cities. It is apparent that the lowest ratio was found in the Metro service. Ulsterbus also had the lowest ratio of operating staff to management staff in comparable regions. The benchmarking data also indicated that the Metro and Ulsterbus services are relatively "over-managed" compared with other comparator public transport companies.

Toward an Assessment of the Relative Overall Effectiveness of Bus Operations by Control/Regulatory Model

With regard to the effectiveness of the different control/regulatory models, the evidence is conflicting. For areas

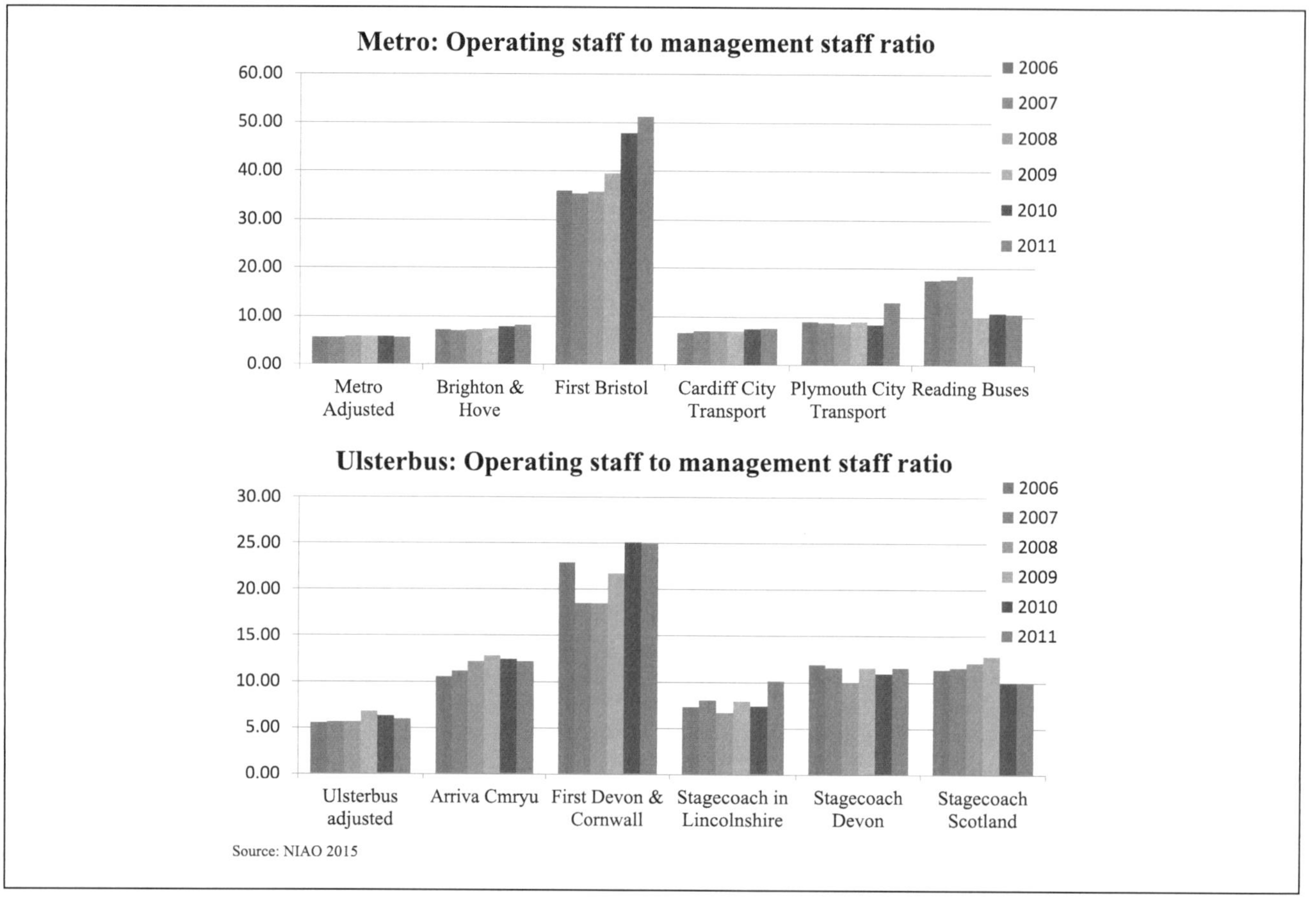

Figure 2. Metro and Ulsterbus operating staff to management staff wages benchmarked to GB comparators.

outside the major cities and conurbations, service levels are higher under the Northern Ireland state-owned model. In contrast, frequency of service in urban areas for the deregulated systems is superior to the state-owned fully regulated system. However, bus comfort levels are better under the equivalent state-owned regulated model. For fares in general, the cost of travel is broadly similar under both models. Consumer satisfaction scores also exhibit this variable pattern of performance across a range of quality-of-service indicators.

Although public transport use has continued to rise since 2001, bus use in Northern Ireland remains lower than in England, Scotland, or Wales. These overall figures, however, mask significant differences between the state-owned operators. Up to 2011-2012 overall bus patronage fell slightly, with a decrease at Ulsterbus balanced by an increase in Metro ridership in the Greater Belfast area. However, Belfast's performance remained toward the lower end of the range exhibited by these six cities. Turning to the efficiency of the different systems, performance is variable. Whereas the regulated system in Northern Ireland performed better with regard to operating cost per bus and staff per bus, the deregulated systems operated with lower wages per employee and a superior operating staff to management ratio.

Effectiveness of Rail Services by Control and Regulatory Model

Journey Time, Journey Fare, and Service Frequency for Various Distance Bands. Assessment of rail focused on service frequency, journey time, and fares for various distance bands. As with the bus sector, effectiveness of the different rail systems is also measured in terms of customer satisfaction across a battery of quality-of-service indicators. The assessment examined levels of service for NIR in comparison with journeys on selected TOC routes in GB. To facilitate comparison, a sample of journeys was chosen that is representative of different journey types, that is, 0 to 50 mi, 51 to 100 mi, and above 100 mi.

The findings suggest that the frequency and journey services provided by NIR are on a par with similar routes in GB in the case of shorter journeys, but typically slower and less frequent for longer journeys. Turning to fares, "walk on" rail fares – fares payable on the day, rather than purchasing a ticket in advance over the web or

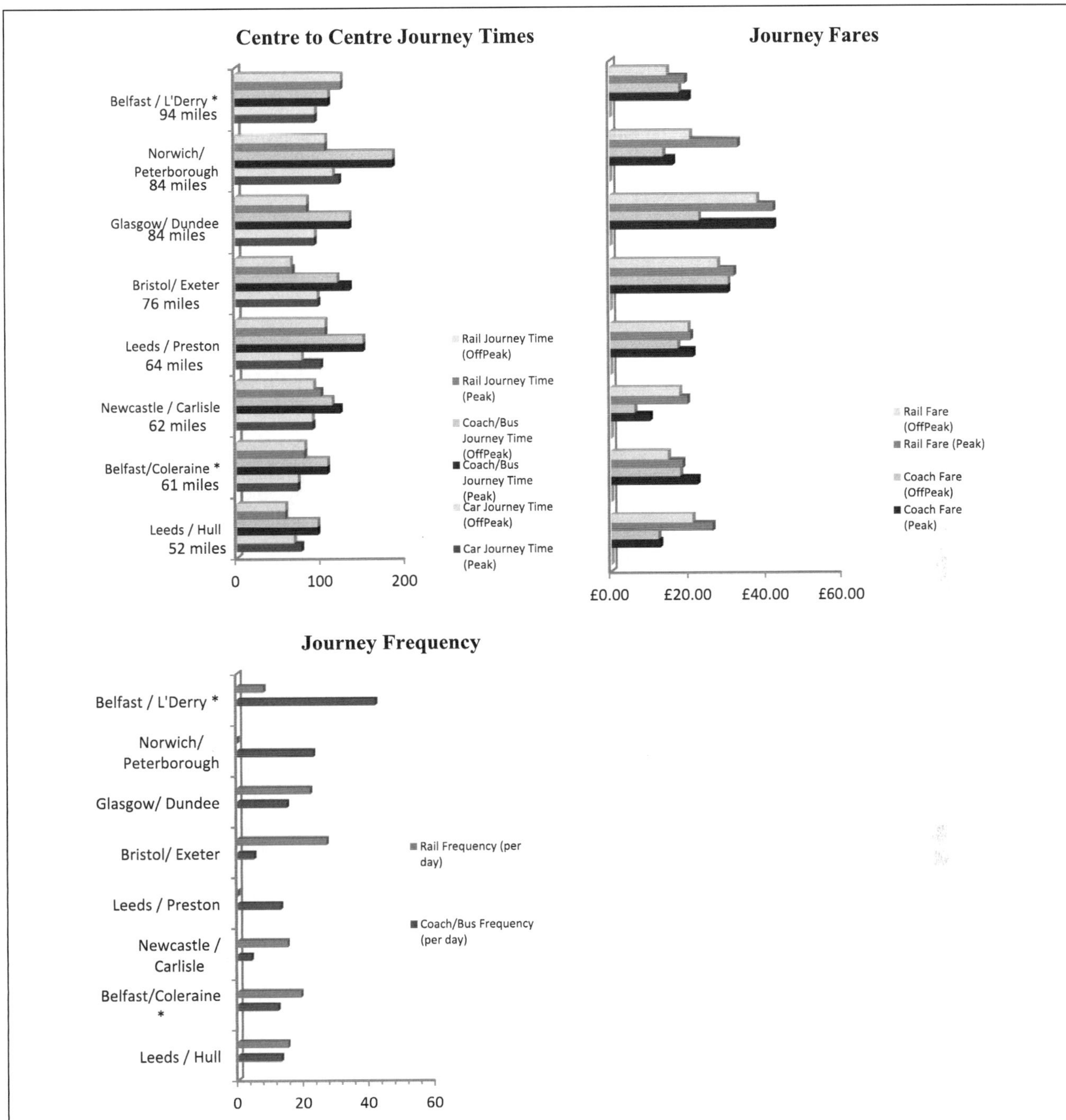

Figure 3. Journey times, journey fares, and journey frequency (51 to 100 mi).

through a discounted scheme – were collated for the same routes. Additional comparative information was also collated for journeys within the Republic of Ireland. The results tend to show that for shorter distance journeys, rail fares in Northern Ireland are in line or slightly higher than local fares in Scotland and the Republic of Ireland, but lower than elsewhere in GB. For all journeys above 50 mi, NIR fares are typically cheaper and in the case of longer distances, markedly lower than fares in the same distance band in GB (Figure 3).

Rail: Passenger Satisfaction Over a Suite of Level of Service Parameters. Rail passenger satisfaction was assessed in a similar way to bus passenger satisfaction using data from NIR and eight other UK TOCs. These were chosen for the similarity of their operation types and markets with those of NIR. The eight rail services selected for comparison were: Arriva Wales, Chiltern, East Midlands, First Hull, First TransPennine, Grand Central, Northern Rail, and Scotrail. The eight areas were measured across seven different quality-of-service parameters. Overall, NIR customers' satisfaction ratings compared very favorably with other customers' satisfaction ratings of their local rail companies. NIR was best for information provided, comfort, and staff helpfulness. It was also ranked second best for value for money, cleanliness, and punctuality. It was ranked as the third best company for availability of seating.

As for bus transport, ultimately effectiveness of policy for rail is determined by the extent to which the intended beneficiaries take advantage of the benefits available to them by the relevant policy. This reflects not only the absolute level and quality of the service provided but also how competitive it is compared with alternatives, and in the final analysis, patronage and market share secured. In contrast to bus transport, NIR passenger numbers more than doubled in the period 2002–2003 to 2013–2014 (110%). In relation to modal share, rail has achieved a substantial increase in market share on shorter and mid-distance commuter routes (see Figure 4). These patterns reflect market conditions, changes in the competitiveness of rail compared with alternatives, or both market conditions and changes incompetitiveness. The competitiveness of alternatives is in turn reflected in the fares that rail can levy. Figure 4*a* shows that the Belfast/Bangor revenue yield is the highest compared with all other journeys benchmarked. For longer journeys of over 100 mi, market share has fallen and in the case of the cross-border route, dropped significantly in absolute terms since 2000 (Figure 4*b*). This is vividly illustrated in the case of the Belfast to Dublin cross-border route. Its declining market performance and the constraints on the potential to raise revenue reflects, in part, a massive upgrade in the parallel road network, with journey time by car and bus reduced by a third, rail track capacity constraints and congestion, and a failure to maintain rail infrastructure and journey times.

Efficiency of Rail Services by Control and Regulatory Model

The next stage of the benchmarking analysis focused on the efficiency of the regulated NIR system compared with the TOCs elsewhere in GB. This analysis was limited by availability of comparator information and the differing operating models of British train companies. However, relevant data was obtained for six comparator companies included in the analysis of passenger satisfaction.

Operating Cost Per Passenger km. The most up-to-date figures available were from 2011-2012. The operating cost of the selected TOCs was measured and compared with both passenger kilometers traveled and timetabled train kilometers. Under both comparators, NIR is relatively efficient. It has the third lowest cost per passenger kilometer traveled and the second lowest cost per timetabled train kilometer.

Toward an Assessment of the Relative Overall Effectiveness of Rail Operations by Control/Regulatory Model

The evidence suggests Northern Ireland's state-owned rail operator offers service levels at least on a par with the best performing TOCs, and delivered as efficiently as the best performing TOCs operating similar service types. A notable exception in relation to journey times relates to journeys over 50 mi with respect to which NIR services tend to be slower than TOCs offering similar types of service. In contrast, however, fares are almost always lower for NIR customers compared with those faced by rail users in GB on routes offering a similar service type. These are, in turn, typically cheaper on a per mile basis than intercity routes. This objective evidence for output indicators is supported by the high satisfaction ratings achieved by NIR for a range of quality-of-service indicators. NIR achieves ratings in the top three, generally first or second, of the 10 operators surveyed.

In contrast to the modest growth in bus transport, overall NIR passenger numbers more than doubled in the period 2002-2003, to 2013-2014 (by 110%). Turning to modal share, NIR has achieved a substantial increase in market share on shorter and mid-distance commuter routes, whereas for longer journeys of over 100 mi, market share has fallen and, in the case of the cross-border route, dropped significantly in absolute terms since 2000. In the case of the Belfast to Dublin cross-border route, declining market performance and the constraints on its potential to raise fares and revenue reflects how uncompetitive rail has become in recent years. This loss of competitiveness demands greater revenue support to sustain the service.

Toward an Assessment of the Relative Overall Cost-Effectiveness of Public Transport Operations in the UK by Control/Regulatory Model

Space constrains the depth to which the review of the overall cost-effectiveness of Northern Ireland's public

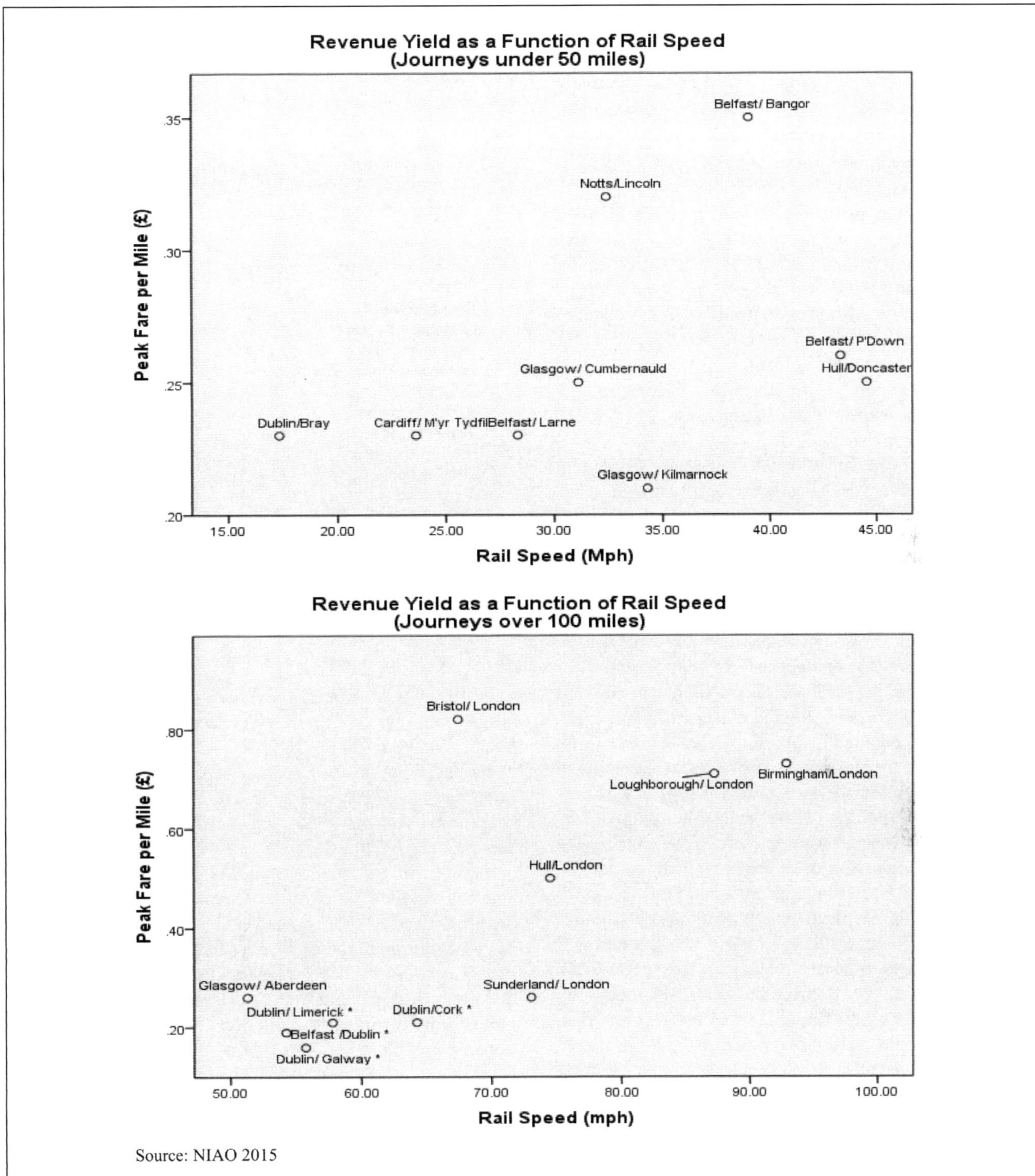

Figure 4. (*a*) Revenue yield as a function of rail speed (journeys under 50 mi) and (*b*) revenue yield as a function of rail speed (journeys over 100 mi).

transport system, on which this paper is based, can be presented here. The evidence does suggest that public policy for rail services in Northern Ireland is cost-effective overall, compared with the control and regulatory arrangements that apply elsewhere in the UK in relation to non-intercity services and commuter services beyond London and the South East England. In the case of bus services the position is less clear cut. Compared with areas outside the major urban areas in GB, overall Northern Ireland bus users enjoy more comprehensive coverage at comparable fares. However, for Belfast, the main urban area, the frequency of service is certainly poorer than larger urban centers in GB or more densely populated medium-sized urban areas. It is similar but at the lower end of the provision enjoyed by lower-density, medium-sized centers in England and Wales. For other indicators the pattern is mixed. The operator recognises local political considerations influence bus service provision geographically, with cross subsidy of rural dwellers by bus users in Belfast. In relation to efficiency, the state-owned operator achieved similar performance, providing the costs related to final salary pension schemes that apply to the state owned operators were excluded from such comparisons.

Ultimately, effectiveness is determined by the extent to which the intended beneficiaries take advantage of the benefits afforded to them by public policy interventions in the public transport sector. This will reflect not only the absolute level and quality of the "service" provided but also how competitive it is compared with alternative choices potentially available to the user and as reflected in patronage and market share. Overall bus use in Northern Ireland is lower than in England, Scotland, or Wales. Variations may reflect demographic conditions and geography. However, Northern Ireland also experienced an unprecedented drop in public transport use during the 1970s, attributable to the onset of the political violence that engulfed the area after 1969. This was particularly true of Belfast where ridership dropped by around three-quarters in a period of 5 years during the 1970s. With the cessation of violence and improvements to services, the late 1990s witnessed something of a modest rebound. Public transport use (including rail: 16% of total journeys) has continued to rise since 2001. Total passenger journeys by all modes/operators rose 5.8% from 2002–2003 to 2011–2012. Patronage increased by 11% between 2002–2003 and 2013–2014. However, passenger journeys per head of population have remained virtually unchanged over 13 years. These modest changes are also reflected in the broad stability in market share enjoyed by public transport for work travel, with 8% in 2001 and 7% in 2011. This trend is similar to figures for Scotland: 16% in 2001 and 2011, and 16% (2001) and 18% (2011) in the cases of England and Wales, respectively.

These figures mask significant differences between the state owned operators. In the case of bus transport up to 2011 to 2012 overall bus patronage fell slightly, with a decrease at Ulsterbus (rural/small town and inter-urban services) balanced by an increase in Metro ridership in the Greater Belfast area. By way of comparison, bus passenger journeys in GB outside London fell by between 5.83% between 2000–2001 and 2011–2012. Ulsterbus experienced an overall reduction of 12% in passenger numbers over the period 2002–2003 to 2013–2014. It can be noted Ulsterbus passenger journeys were already in decline before 2002. In contrast, Metro achieved a 33% increase in ridership between 2002–2003 and 2013–2014. The peak modal share for car travel to Belfast city center dropped by 17% between 2011 and 2013. Notwithstanding these achievements, Belfast demonstrates an absolute level of ridership at the lower end of the range for benchmarked medium-sized urban areas. Belfast's performance is consistent with the strong relationship exhibited between ridership and population density. The massive reduction in the city's population from the 1970s accounts for some of the modest benchmarked performance of the city.

This paper has assessed the effectiveness and efficiency of different forms of control and regulation for public transport within the UK by benchmarking performance across various supply- and demand-side measures as summarized in Table 2. The evidence summarized in this paper, together with the wider findings of the NIAO 2015 report (*4*) offer a sound basis for concluding a robust case can be made for public transport services to be regulated and delivered under public ownership.

Table 2. Relative Effectiveness and Efficiency of State Owned Compared to Privately Owned Public Transport in the UK: Summary of Indicators

Indicators	Bus	Rail
Effectiveness		
Operator indicators		
Frequency of service	✓	✓
Fares indicator	✓	✓
Comfort indicators	✓	✓
Journey time		✓
Consumer satisfaction indicators		
Punctuality	✓	✓
Cleanliness	✓	✓
Comfort	✓	✓
Seating availability	✓	✓
Service information provided	✓	✓
Driver helpfulness	✓	✓
Staff helpfulness		
Value for money	✓	✓
Efficiency		
Operating costs	✓	✓
Staff productivity		✓
Overall cost-effectiveness		
Market performance		

Note: ✓ = state owned public transport operator achieves performance equal to or better than peer group privately owned public transport operator for specified indicator.

Author Contributions

The authors confirm contribution to the paper as follows: study conception and design: AS; data collection: LK, AS; analysis and interpretation of results: AS, LK; draft manuscript preparation: AS, LK. All authors reviewed the results and approved the final version of the manuscript.

References

1. Banister, D. Deregulating the Bus Industry in Britain—(A) the Proposals. *Transport Reviews: A Transnational Transdisciplinary Journal*, Vol. 5, No. 2, 1985, pp. 99–103.
2. Gwilliam, K. M., C. A. Nash, and P. J. Mackie. Deregulating the Bus Industry in Britain—(B) the Case Against. *Transport Reviews*, Vol. 5, No. 2, 1985a, pp. 105–132.
3. Beesley, M. E., and S. Glaister. Deregulating the Bus Industry in Britain: A Reply. *Transport Reviews*, Vol. 5, No. 3, 1985, pp. 223–224.
4. NIAO. DRD: The Effectiveness of Public Transport in Northern Ireland: Report by the Comptroller and Auditor General, Belfast 21 April 2015. ISBN 978-1-910219-85-0. Available at https://www.niauditoffice.gov.uk/publication/drd-effectiveness-public-transport-northern-ireland (Last accessed September 11 2018)
5. Smyth, A. Competition and Regulation: The Role of Ownership, Published in Issue 16 of International Symposium on Theory and Practice in Transport Economics, 2003. ISSN 2311-827X. Available at https://www.oecd-ilibrary.org/transport/16th-international-symposium-on-theory-and-practice-in-transport-economics_9789282103494-en (Last accessed September 11 2018)
6. Butcher, L. 'Buses: Deregulation in the 1980s, House of Commons Library Standard Note SN/BT/ 1534, 2010.
7. White, P. Competition in Public Transport in Great Britain. *European Transport*, Vol. 33, 2006, pp. 69–82.
8. Bowman, A. An Illusion of Success: The Consequences of British Rail Privatisation. *Accounting Forum*, Vol. 39, 2015, pp. 51–63.
9. Rowney, M., and W. Straw. *Greasing the Wheels: Supporting and Improving Britain's Rail and Bus Services*. Report for Institute for Public Policy Research, 2014. Available at https://www.ippr.org/files/publications/pdf/greasing-the-wheels_Aug2014.pdf
10. Labour Party. *Britain can be Better: The Labour Party Manifesto*. The Labour Party, London, 2015.
11. FGS McClure Watters and the TAS Partnership 'Review of Outline Business Case for Public Transport Reform commissioned by the Department for Regional Development, 2010.

The Standing Committee on Strategic Management (ABC10) peer-reviewed this paper (18-05149).

Article

TRR
JOURNAL OF THE TRANSPORTATION RESEARCH BOARD

Transportation Research Record
2018, Vol. 2672(8) 578–587

DOI: 10.1177/0361198118769902
journals.sagepub.com/home/trr
SAGE

Improving Pathways to Fixed-Route Transit: Transit Agency Practices to Expand Access for All Users

Stephanie DiPetrillo[1], Andrea Lubin[1], Anastasia Loukaitou-Sideris[2], Carla Salehian[3], Stephen C. Gibson[4], Kristine M. Williams[5], and Theodore Trent Green[6]

Abstract

This paper discusses challenges to and promising practices for making infrastructure improvements that enable persons with disabilities and all system users to safely reach public transportation stations and stops to access fixed-route transit options. Although tremendous strides have been made in national transit system accessibility since the 1990 passage of the Americans with Disabilities Act, far less progress has been made in ensuring that pathways to accessible transit stations and stops are in place. This reality poses a critical mobility concern as the nation's aging population continues to rise, as does the percentage of the population with disability who may need to utilize public transit to access life enhancing and sustaining services. The study uses a national survey and targeted case study analysis to identify effective best practices for facilitating improvement to pathways to transit. These include approaching cost and funding creatively; pursuing partnerships; initiating and maintaining open communication and collaboration; exploring the use of data and new technologies to reduce cost and improve outcomes; and thinking holistically to identify and address pathway to transit improvement needs. Useful to both practitioners and policymakers, the findings presented offer insights on how agencies can successfully achieve infrastructure improvements that extend beyond their own facilities and that allow people with disabilities and others to more easily access public transit.

Access to fixed-route public transit options contributes to increased mobility. For transit services to be used effectively and fully by persons with disabilities, infrastructure and built environment components, that is, pathways to transit, must permit individuals to reach stops and stations without difficulty. Transit agencies have an incentive to promote the use of fixed-route services by persons with disabilities, as these services are generally less costly to operate than Americans with Disabilities Act (ADA) paratransit services, and increased ridership usually contributes to more cost-effective operations. This research uses case studies to explore efforts and promising practices undertaken by U.S. transit agencies to improve the pathways to transit.

Transit agencies and their partners have made significant progress since the 1990 passage of the ADA in increasing system accessibility, with the U.S. public transit vehicle fleet now reaching close to total accessibility for those using wheelchairs and with other disabilities (*1*). However, less progress has been made on pathway to transit accessibility—even though both agencies and riders benefit when exemplary pathways exist permitting riders to reach stations and stops more easily. Transit agencies find themselves in a difficult position though, as they often cannot make these improvements themselves, but must collaborate with others (governments, agencies, private property owners, etc.) to improve the built environment beyond their properties. To this end, many transit agencies nationwide have worked with local governments and private partners, as well as made changes to their policies and procedures in an effort to improve the physical connections that allow persons with disabilities and others to access regularly scheduled transit options.

This paper shares successful strategies utilized by select U.S. transit agencies in their pursuit of initiatives to improve

[1]Alan M. Voorhees Transportation Center, Edward J. Bloustein School of Planning and Public Policy, Rutgers, The State University of New Jersey, New Brunswick, NJ
[2]Luskin School of Public Affairs, University of California Los Angeles, Los Angeles, CA
[3]University of California Los Angeles, Los Angeles, CA
[4]Faculty of Arts, Business, Law and Education, UWA School of Design, University of Western Australia (M433), Crawley (Perth), Australia
[5]Center for Urban Transportation Research, University of South Florida, Tampa, FL
[6]School of Architecture & Community Design, University of South Florida, Tampa, FL

Corresponding Author:
Address correspondence to Stephanie DiPetrillo: sdipetrillo@ejb.rutgers.edu

pathways that enable persons with disabilities to access transit services. The strategies and practices were identified and explored by an interdisciplinary research team assembled for a study funded by the Mineta Transportation Institute that was completed in 2015 (*2*). The team conducted a brief national survey to identify prospective case study locations and then carried out interviews and site visits with the selected transit agencies and their partners. The core objective of the case study investigation was to learn promising practices from agencies actively improving the pathways that connect transit consumers—particularly persons with disabilities—to transit stations and stops.

Background

Pathways to Transit: Definition and Importance

All public transit riders must travel along some path to reach and access transit. Pathways to transit include all of the elements encountered during a journey from origin until aboard a transit vehicle and from vehicle to destination. These elements include sidewalks, street crossings, curb cuts, bus shelters, and access to the vehicle.

Researchers recognize that conditions existing between origin/destination and station/stop affect one's willingness to travel by transit, and how one chooses to travel to the station/stop. Distance, safety, and density effects on traveling to/from a station/stop have been well explored by researchers concerned with land use–transportation interactions (*3–6*). Other researchers have explored ways to traverse the "last mile" to the station/stop and how this influences the willingness of individuals to use transit (*7*). Park recognized that the quality of path walkability and walking distance affect mode choice in traveling to stations (*8*). More recently, Tilahun and Li explored how sidewalk accessibility affects the propensity to travel by transit (*9*).

None of these studies, however, explicitly recognize that for a portion of the population considered transportation disadvantaged, the option to travel by some other means (e.g., private auto) is more limited and makes the question of mode choice largely irrelevant. For this population, use of transit, enabled through sufficient infrastructure, may provide the only ready transportation option and may be the primary means to transportation independence (*10*), despite the prevalence of physical limitations that may affect their ability to reach transit easily (*11*).

The population living with disability in the U.S. is large and growing. According to the 2015 American Community Survey, one in five Americans reports having a disability (*12, 13*). The aging of the U.S. population contributes to the growth in persons with disabilities, as individuals frequently acquire disability through the aging process, and as average lifespans increase allowing individuals to live with disability for longer periods (*14*). As of 2014, there were an estimated 44.6 million people aged 65 or older in the U.S., or about 14 percent of the population. The Census projects the number of older Americans will nearly double to 88 million by 2050, when they will comprise 22 percent of the population (*12*).

Negative Outcomes without Transit Access

Inaccessible transit stops prevent these persons with disability and older adults from using fixed-route bus transit services, thereby limiting their mobility and access to employment, goods and services, and other activities. Lynott and Figueiredo reported in their highlights of the 2009 National Household Travel Survey that more than half of persons 65 and older who reported not taking a trip outside their home in the past week indicated they wanted to get out more often (*15*). Both Loprest and Maag (*16*) and Rosenbloom (*17*) recognized the vital role transportation plays in facilitating the successful community integration of persons with disabilities by enabling them to reach employment, medical, and other destinations. Lubin and Deka found people with disability actively seeking employment and cited deficits in pedestrian infrastructure between home and nearest station/stop as a concern (*18*). Theis and Furner (*19*) reported declining community participation resulting from perceived environmental barriers, and Maynard (*20*) recognized that fear of failure while making a trip can result in individuals choosing not to make the trip at all. In total, because of the significant percentage of the U.S. population currently living with disability, the anticipated increase of those coping with disability in the near future, and evidence from the literature demonstrating the importance of access to transit for these persons, it is strikingly obvious that increased accessibility of pathways to public transit must become a nationwide priority.

Gap in Research

Most accessibility studies omit examination of the needs of people with disabilities traveling to transit stations/stops, instead limiting their focus to stop or vehicle characteristics. Koppa et al. found that most studies on physical barriers for people with disabilities using transit focused on adaptive equipment and techniques such as on-board audible information and wheelchair securement (*21*). In a rare exception, Lavery et al. explored three components that contribute to an ability to travel—person, vehicle, and built environment—and suggested a design approach that addressed deficiencies in sidewalk surface pavement as a means to mitigate infrastructure barriers (*22*).

Rosenbloom suggested that people with disabilities would be well served by increased examination of best practices that address sidewalk maintenance, curb ramp installation, and related enforcement issues. She cited an environment of weak regulation and enforcement that perpetuated significant barriers in the pedestrian network, including a paucity of "accessible route[s] to ... accessible bus stop[s]." To create safe, secure, and accessible transit environments for all users, she recommended the development and maintenance of accessible and fully lit pedestrian paths and improved parking, safety, and security enforcement strategies (*17*).

Of particular concern for people with disabilities and older adults traveling by transit is whether they can count on the trip being "seamless." Breaks in the trip chain, no matter how seemingly small or insignificant to other riders, can have devastating effects on this population and can perpetuate their disuse of transit long after the incident occurs. Maynard offers that the "devil is in the detail" and that access requires not only a commitment to creating an unbroken "journey chain" but an ongoing commitment to management (*20*).

Challenges for Transit Agencies

Notwithstanding a desire to address pathway barriers to their services, public transit agencies face critical funding concerns that limit their ability to make these enabling improvements. Notably, ADA mandates do not provide additional funding, constraining support for additional or non-mandated services. Transit agencies cite budget limitations as a primary constraint to increasing the number of transit stops for ADA improvements, though researchers offered that the largest improvement can be achieved by a methodical selection of transit stops designed to optimize the overall benefits to individuals with disabilities. Wu et al. suggested that rather than focusing on improvements by selecting transit stops based on staff experiences or requests from elected officials, selection based on optimization has the ability to serve as a beneficial platform for identifying ADA requirements and other related improvements (*23*).

Infrastructure supporting pedestrian accessibility has not witnessed improvements parallel to the rest of the transportation network—partly because of a lack of enforceable regulations mandating the pedestrian (or public right-of-way) system be made accessible to travelers with either physical or visual impairments. Limitations to making these improvements often result from land boundaries that are in effect design boundaries, because entities (transit agencies and governmental bodies) rarely effect change beyond their prescribed territory or in coordination with adjacent territories. This condition governs how many transit agencies enact accommodating improvements and how most city and county governments have interpreted the ADA mandate (*20*).

Two Trends Contributing to Improvements

Despite the above-noted limitations, two trends offer promise and support for improved pathways to transit for people with disabilities: Universal Design (UD) and Complete Streets. Audirac described UD as a school of thought governed by a comprehensive set of design principles that focus on creating a barrier-free, accessible, inclusive, and transgenerational environment that incorporates the use of assistive technology with the goal of serving all users regardless of age and ability (*24*). Illustrating the difference between ADA-inspired improvements in transit and those evoked by UD, Audirac contrasted ADA-accessible buses equipped with lifts for people with disabilities and stairs for those without disability with low-floor, ramp-based, and stairless buses. Both provide accessibility. The former improvement, however, highlighted difference whereas the latter epitomized inclusion and optimized access for all.

Conceptually, Complete Streets also supports better connections for all transit users by way of more comprehensive pedestrian networks. A national movement, Complete Streets promotes "streets for everyone" through the adoption of local policies and support for improvements that provide access to all users of all modes, ages, and abilities (*25*). Complete Streets design recommendations accommodate older pedestrians through pavement maintenance and materials (flat surfaces), curb ramps (in the direction of crosswalks), median refuges (for wider crossings), pedestrian crossing signals (with countdowns and audible signals) and other pedestrian amenities (pedestrian lighting, street trees, resting spots, and wide sidewalks) (*26*). The adoption of Complete Streets policies is typically an indication that governmental partners—local, county, and state governments and departments within those governments—are amenable to working with local transit providers to improve access to and conditions at stations/stops.

Research Design and Methodology

The question the research team sought to answer was: What agency practices and policies contribute to improving the accessibility of pathways to transit? To begin this inquiry, the researchers conducted a brief (10-question) survey to identify transit agencies actively pursuing policies and projects to improve pathways to transit stations and stops, receiving responses from a total of 152 transit agencies in 37 states and the District of Columbia.

From the data collected, the researchers identified agencies, partners, locations, and innovative projects, programs, and practices for further investigation. Following inquiries with potential case study sites, the research team selected five agencies for in-person interviews and site visits: Broward County Transit (BCT), Link Transit Washington (Link), Memphis Area Transit Authority (MATA), NJ TRANSIT (NJT), and Tri-County Metropolitan Transportation District of Oregon (TriMet). Although case studies cannot be representative, the selection of these five agencies did offer the research team the opportunity to explore pertinent issues and innovative solutions implemented in diverse parts of the country, by agencies working in rural, suburban, and urban locations and responsible for transit in both small and large service areas—ranging from a single county, to a multi-county region, to a statewide agency. See Table 1.

Table 1. Case Study Agency Characteristics

Agency	Service area (jurisdiction)	Service area (mi^2)	Population	No. of vehicles (all modes)	No. of annual unlinked trips	Improvements undertaken
Broward County Transit, Florida (BCT)	County	410	1,869,235	274	37,809,246	Self-evaluation process Bus stop upgrades Shelters and amenities program Transit stops retrofit Community Design & Complete Streets Travel training
Memphis Area Transit Authority, Tennessee (MATA)	Metro area	338	755,141	150	8,097,475	Bus shelter program Trolley station ADA improvement program Complete Streets Travel training
NJ TRANSIT, New Jersey (NJT)	Statewide	5,325	10,594,013	3,151	276,498,375	Station area improvements Complete Streets Arts in transit Travel training
Tri-County Metropolitan Transportation District of Oregon, Oregon (TriMet)	Multicounty (urban)	534	1,560,803	638	101,380,984	Pedestrian network analysis Bus stop improvements Pedestrian improvement project Complete Streets Travel training
Link Transit, Washington	Multicounty (rural)	197	108,660	44	1,063,333	Rural bus stop improvement program Travel training

Note: No. = Number.
Sources: National Transit Database, 2015; interviews.

The five agencies undertook actions and programs that improved pathways to transit. Practices that contributed to improvements included ongoing systematic self-evaluation of existing infrastructure conditions (BCT, TriMet), bus and transit stop improvement programs (BCT, Link, MATA, TriMet), major station improvements (NJT), integration of community design and Complete Streets principles (BCT, MATA, NJT, TriMet), usage of innovative materials, design, or public art (Link, NJT, TriMet), and implementation of supportive policies such as travel training (BCT, Link, MATA, NJT, TriMet).

Findings

Limitations Faced by Agencies

Through both the survey and the subsequent interviews, the team sought out information about the barriers agencies faced that prevented or impaired their ability to improve pathways to stations/stops. Survey analysis demonstrated that although several agencies indicated they faced no challenges or that they only or primarily served their riders via demand-response services, most agencies affirmed that improving pathways to transit facilities was within their agency's purview—even if not directly under their sole control. A majority of respondents (65%) indicated that financial limitations constrained their ability to improve accessibility to transit facilities. See Table 2.

A small share of respondents (18%) stated that jurisdictional or bureaucratic hurdles posed a significant barrier to making these kinds of accessibility improvements. These challenges elucidate the complicated process that was often required to address deficiencies in locations that were beyond agency control, including:

- Acquisition of state and local permits (right-of-way or encroachment);
- Requirement to purchase land for the placement of shelters/stops and ramps;
- Policies that prevent a transit agency from building improvements on unincorporated lands;
- Lack of coordination between other city divisions, or between city division and utility providers;
- Lack of jurisdiction/inconsistent rules, regulations, and processes across multiple jurisdictions; and
- Lack of policy requiring sidewalks for existing development.

Survey respondents, who cited jurisdictional barriers, explained that difficulties arose when working with other

Table 2. Survey Results Highlights

Question	Count	Percent
Does your agency have any practices or programs that have been developed (or adopted) to address physical barriers faced by riders with disabilities?		
Yes	107	70%
No	45	30%
Is your transit agency aware of infrastructural barriers to accessible transit stops and/or stations that are prevalent in your service area? (Check all that may apply)		
Lack of sidewalks	118	78%
Lack of curb cuts or wheelchair accessible ramps	93	61%
Lack of bus shelters/shade	93	61%
Broken or cracked sidewalks	90	59%
Non-signalized street crossing/intersections	86	57%
Inadequate lighting	72	47%
Utility poles or other barriers on sidewalks	69	45%
Lack of signage/wayfinding information	69	45%
Insufficient maintenance	22	14%
Lack of elevators/escalators	12	8%
Other	13	9%
Has your agency taken any action or measures to address these barriers independently or in cooperation with local government and/or other partners? (Select one)		
Yes	112	74%
No	40	26%
What obstacles (bureaucratic, financial, etc.) prevented or hindered your agency from addressing physical barriers faced by transit riders with disabilities? (Fill in)		
Financial	99	65%
Jurisdictional/Bureaucratic	27	18%
Lack of community support	10	7%
Lack of information or expertise within agency	3	2%
Other	17	11%

agencies, governmental units, and private landowners, often resulting in increased cost and less frequent project success. Depending on the jurisdiction, agencies may be required to secure funds (through federal, state or other funding), acquire an easement (determine legal property boundaries, identify and negotiate with property owner(s), address liability concerns, and overcome citizen concerns), prepare a design and specifications (address issues of meeting ADA and other standards), acquire appropriate permits (work with multiple government bodies and departments), and execute projects (working with in-house or consultant construction teams). One survey respondent succinctly described the obstacles faced while coordinating with various state and local governing bodies and controlling departments:

> For one shelter to be built we must deal with the property owner, possibly the DOT rights-of-way section, local zoning approvals, local Police Department review, state and local traffic control construction or modification, State construction standards review, and the local building department. For a small agency, this can be daunting and extremely time consuming. The time lag from conception, grant writing and approval to the actual completion can take up to and sometimes more than three years.

Thus, the issue of financial limitations is closely tied to that of jurisdictional and bureaucratic barriers. It may be a truism but time costs and complexity costs more.

Survey respondents also offered "other" conditions that limited their ability to address physical barriers to transit, which included issues of agency philosophy, lack of high-level agency management support, general disinterest in improving transit services, and a lack of clarity on appropriate length of pathway to best enhance accessibility to transit.

Exploring Best Practices

Case study analysis yielded various best practices that enabled agencies to improve pathways to transit facilities. Five thematic approaches arose and are discussed below: a) approaching cost and funding creatively, b) pursuing partnerships, c) initiating and maintaining communication and collaboration, d) exploring the use of data and new technologies to reduce cost and improve outcomes, and e) thinking holistically to identify and address needs. None of these practices exist in isolation—for example, successful partnerships rely on open and consistent communication—but each is sufficiently distinct and contributed enough to overall project success to warrant discussion.

Approach Cost and Funding Issues Creatively. As previously noted, 65 percent of participants who responded to this study's national survey reported that financial factors posed a significant impediment to addressing infrastructure barriers to transit. Case study interviewees confirmed this finding. Recognizing that state and local government entities do typically operate in a restricted funding environment does not, however, preclude ability to successfully pursue such improvements. Case study analysis demonstrated that pursuit and usage of multiple funding sources and openness to seeking non-traditional funding, as well as project partnership opportunities, can yield success.

Federal grants served as one significant source of funding, and most case study agencies utilized federal grant programs, such as the Federal Transit Administration New Freedom/Section 5310 initiative, to at least partially support their improvements. For example, MATA used New Freedom funds to implement sidewalk and curb ramp improvements and to install bus shelters along various routes. However, most successful pathway projects utilized funding from a variety of sources that included a combination of federal and state programs and grants, local funding, passenger revenue (e.g., fares), and local taxes.

On the subject of local taxes as a funding source, 70 percent of Washington's Link Transit special project funding is generated through a local sales tax. This relatively sustainable tax source was endorsed by the citizens of the region and is levied through Washington State's Public Transportation Benefit Area, which is a special taxing district established by the state for the purpose of providing public transportation. Oregon's TriMet utilizes another local tax—payroll—to pursue its pathways to transit improvements and other transportation initiatives.

Pursue Partnerships. Despite agencies reporting the difficulties in working with other agencies, partnerships with governmental entities and private partners did provide a means to creatively reduce costs and achieve goals that might not otherwise have been feasible. Case study agencies frequently partnered in operations. For example, BCT collaborates with 18 Broward County municipalities in its Community Bus service, a local circulator system. Municipalities lease the buses from BCT and then either contract with a private vendor to provide services or provide the service internally. BCT and its partners fund the service through a combination of a local option gas tax, county, and federal sources. The municipalities supplement revenue through advertising and by charging a fare, which five cities opt to do. This case of intergovernmental partnering has been critical to the service's success. BCT took a similar approach to address its pathway to transit infrastructure needs. In 2006, the agency accelerated its bus stop improvement efforts when the Broward County Commission committed $10 million over five years to retrofit stops for ADA compliance. BCT collaborated with local governments, the Broward County Metropolitan Planning Organization, and the Florida Department of Transportation. Municipalities with existing sidewalk improvement programs split costs with BCT. Cities unable to match BCT funding contributed by streamlining the permitting process, allowing BCT to more easily execute improvements in the municipal right-of-way.

TriMet's work on the comprehensive Line 57-Tualatin Valley (TV) Highway/Forest Grove pedestrian and corridor improvement project in Portland, Oregon offered another example of creative partnering that surmounted financial constraints. As TriMet explained, cost savings were achieved when the agency pursued relationships that harnessed the capabilities, resources, and expertise of each partner. In the Line 57 TV Highway project, the City of Portland and TriMet developed a plan and an intergovernmental cost sharing agreement that resulted in halving projected project costs. The City, with its expertise in construction, managed the building process and covered construction labor costs but the agency paid for project material costs. As the case study interviewee explained:

> The reason why our agency is different is the time we save on our short construction. Funding is the least of my worries, especially when we can get surrounding jurisdictions involved. They already have the tools and the skills. When we are able to have them act as our contractors, then I can just put the plans together, transfer it over, and get twenty of these improvements built in the time it would normally take to build one (*2*).

With its work in the cities of Newark and New Brunswick, NJT demonstrated how some projects could only be achieved through partnership. In the Newark project, key stakeholders included the City, other transportation providers, local and agency police, consultants, and adjacent landowners. This partnership resulted in improvements to the street crossings and surrounding properties and enhanced vehicular access to the site. In New Brunswick, NJT worked closely with the local development corporation, DEVCO, and other public and private partners to make a wide range of improvements and enhancements that include a pedestrian bridge to an adjacent street, access to 24-hour elevators, and sidewalk and ramp upgrades on both agency and private property.

Communicate with Users and Partners. The agencies often emphasized the role communication played in advancing projects that improved pathways to transit. Both communication between the agency and users and communication between the agency and its partners affected project design and execution.

Successful interaction between the agency and end users and other community members contributed to meeting project goals in three ways: 1) understanding the needs of riders, 2) shaping and refining projects, and 3) soliciting project and agency support by demonstrating benefits. To determine and prioritize improvements, Link Transit interviewed its riders to supplement its evaluation of existing service routes.

Similarly, BCT incorporated consumer feedback when choosing among potential locations identified through the agency's bus stop self-evaluation process. The other agencies also employed consumer input to fine-tune project design and implementation. TriMet and NJT worked with their respective consumer stakeholder groups (Citizens for Accessible Transportation and the ADA Task Force) to identify consumer needs and to discuss projects under consideration. In the work that NJT completed to improve access to Newark and New Brunswick Station areas, the ADA Task Force reviewed designs and offered input. MATA worked with a community advisory committee composed of members of several local advocacy groups to inform its improvements plans. A community advisory committee and other stakeholders helped the agency determine the best locations for bus shelters.

The agencies found that open and frequent communication with internal and external partners expedited funding and facilitated project design, implementation, and completion. BCT noted that frequent communication between the agency and partner municipalities during land use planning activities and development review allowed the agency to align improvements with need and promoted better use of limited funds. Similarly, the aforementioned improvements to New Brunswick Station area, as well as the improvements to the Newark Station area, arose out of frequent and ongoing discussions between NJT and adjacent property owners. These projects benefited when well-established channels of communication inspired trust between partners and thus offered a means to overcome incipient difficulties. NJT assumed a quality assurance role on work completed on properties contiguous to the station, thus guaranteeing that improvements would meet the needs of people with disabilities. Implementation of these activities by the agency required field observation and communication with multiple partners including the municipalities, private property owners, and construction personnel. NJT observed that although working closely with external partners requires a high level of commitment, collaborations can permit an agency to undertake much more extensive improvements and stretch its limited resources over a much greater area.

Employ Data, Technologies, and Physical Design Innovations to Achieve Goals. Consideration and usage of data, various technologies (including green materials), and physical design innovations served as a vital means to increase project efficiency and generate long-term cost savings. Both BCT and TriMet undertook programs of self-evaluation to determine and monitor locations in need of improvement. Spurred by a desire to better monitor agency performance, BCT established a database and collected information that identified non–ADA compliant bus stop locations, documented needed improvements, and monitored progress. Information collected included the presence of a bus shelter, bench, landing pad, and sidewalk connection to the stop. These data were associated with boardings and ridership figures, locations (municipality and geocoordinates), and before, during, and after improvement photographs.

Similarly, TriMet's Pedestrian Network Analysis project facilitated pathway improvement by developing an objective and quantifiable model to prioritize locations with greatest need and potential effect, based on usage. Data collected included density, mix of uses, proximity of stop to high-need destinations, and place characteristics that adversely affect walking conditions such as high-speed vehicular traffic and missing sidewalks (*27*).

With respect to the use of technologies to achieve pathway improvements, Washington's Link Transit solved one of its pathway deficiencies by using an innovative low-cost material to improve 70 of its rural bus stops, a water-soluble soil stabilization material known as "Rhino Snot." With the material, the agency created flat, stable, and durable landing pads instead of more costly concrete landing pads. Link Transit found that the durability of the pads exceeded the manufacturer specified 3-year life expectancy with the same pads safely used over a 7-year period.

TriMet in Portland also discussed the efficiency and other benefits of using technologies and physical design innovations. In relation to the former, the agency developed fare and real-time arrival information applications for smart phones. They also installed a blinking signalization device at certain bus stops to inform the bus driver that a rider is waiting at the stop. In relation to design innovations, the agency opted to use the innovative design offered through installation of Simme-seats, a space-saving seating option that requires little to no maintenance, at selected bus stops. TriMet installed these seats in locations with limited space, enabling the agency to more easily maintain ADA sidewalk width requirements. They also opted to use plastic guards at the curb or concrete edge of bus stops to protect bus tires and sidewalks from damage and redesigned their bus stop sign shape to increase visibility from any angle and improve wayfinding. Finally, TriMet opted to sandblast vandalized glass at bus shelters into artistic design patterns instead of replacing the damaged glass—thus, creating aesthetically pleasing public art while incurring repair savings estimated to be $100k per year.

One potential challenge in pursuing new technologies was that a given agency not familiar with utilizing said technology may need to partner with one or more other entities to achieve success. For example, when NJT sought to incorporate adaptive signal technology into its Newark project, the agency benefited from partnership assistance from the local power company and the city itself to facilitate the integration of the technology into the existing infrastructure.

Think Holistically to Identify and Address Needs. Finally, discussions with the case study agencies repeatedly returned to the question of how to achieve their primary goal of providing transportation to persons with disability and others with

Figure 1. Examples of pathway to transit improvements: (a) Link Transit = Rural bus stop with "Rhino Snot" pad; (b) MATA = Trolley station lift; (c) NJT = Pedestrian ramp connecting street and local businesses to platform; (d) BCT = Bus stop reconstruction; (e–f) TriMet = Simme seating and bus shelter art.

mobility impairments. The agencies stated that improving access to transit by improving infrastructure best achieved this goal when it was undertaken as part of a comprehensive strategy. For example, TriMet argued that limiting accessibility initiatives to vehicle accessibility was ultimately shortsighted and thus the agency considered ways to enhance walkability and access to transit facilities in its efforts to serve customers with disabilities. Agencies (and their partners) found success when they applied multiple approaches and actions concurrently to address three distinct areas: rider knowledge (the know-how to use the fixed-route system), system attributes (the accessibility of vehicles and facilities such as pathways to stations/stops), and public awareness (an understanding of how to access and use the transit system).

Link Transit advocated for careful review of the conditions and the impediments that riders may experience along every segment of a trip as an initial step toward overcoming these barriers. As a result, Link Transit employed a multipronged plan that addressed each of the identified areas of concern. The agency offered free individual and group travel training for all its passengers, targeting the program toward local older adults. They also implemented several fixed-route system improvements. In addition to making physical improvements to stop locations, and where appropriate to the pathway to the stop, the agency purchased low-floor buses and created service routes that better aligned with the needs of transportation-disadvantaged riders. They also sought public awareness and support through an aggressive marketing campaign designed to reach both persons with disabilities and the general public, informing them of the system improvements. The campaign communicated the benefits of using the accessible fixed-route system and the social equity importance of making transport services and access to those services available to all users.

Conclusions and Takeaways

Although significant improvements to vehicles, stations, and stops have allowed people with disabilities and others with limited mobility to more easily use public transit to meet their needs, the relative neglect of pathways to stations and stops continues to pose accessibility barriers. To better serve these riders, the infrastructure that surrounds transit facilities, and the pathways that lead to transit facilities, must also be accessible. As difficult as achieving vehicle and station/stop improvements may be, improving the pathways to those stations and stops can be more problematic because the pathways to transit generally lie outside direct agency control. Thus, agencies wanting to address this gap in service often must work outside their organizations and engage with riders and the public as well as with public and private partners to achieve these goals.

Many agencies surveyed for this research reported that insufficient funding posed a significant barrier to pursuing these kinds of improvements. The case study agencies used a variety of funding sources to achieve project goals including federal, state, and local funds, passenger revenues, and local, special project funding (local sales tax). From their riders, agencies gained valuable insights into the location and nature of infrastructure gaps that limit access. Engagement with riders with disabilities helped agencies define needs, plan improvements, refine designs, and evaluate progress. Open and consistent communication with all stakeholders as well as collaboration with public and private partners allowed agencies to expand their reach and effect improvements that extended well beyond agency properties. Agency efforts to improve pathways benefited from dedicated programs to collect and evaluate information about the conditions surrounding their facilities. BCT's self-evaluation process and TriMet's pedestrian network analysis afforded each agency the means to assess existing conditions and to optimize the location of improvements. Use of newer technologies, materials, and design innovations contributed to project success and often enabled agencies to achieve cost savings.

Finally, these agencies all have embraced the idea that people with disabilities are best served when consideration is paid to the entire trip—from origin to destination—and have understood that seemingly small gaps in the path of travel can impose significantly negative effects. Collectively, the measures employed by these agencies (illustrated in Figure 1) permitted them to substantially improve access to their services and better serve their riders with and without disabilities. Agencies found that improving pathways to transit improved access for all riders and yielded the added benefit of promoting inclusion. As one agency interviewee noted, "Operating a truly inclusive transit system enabled riders to learn about one another and from one another, contributing to increased awareness for folks with disabilities and their needs."

Acknowledgments

The authors thank all of the agencies and project partners who have worked so diligently to improve pathways to transit for persons with disabilities and for the benefit of all transit users. This research was supported by the Mineta Transportation Institute.

References

1. Neff, J., and M. Dickens. *2015 Public Transportation Fact Book*, 66th ed. American Public Transportation Association, Washington, D.C., 2015.
2. DiPetrillo, S., A. Lubin, A. Loukaitou-Sideris, C. Salehian, S. Gibson, K. Williams, and T. T. Green. *Improving Pathways to Transit for Persons with Disabilities*. MTI Report 12-59, San José State University, San José, Calif., 2016. http://transweb.sjsu.edu/PDFs/research/1233-transit-access-for-persons-with-disabilities.pdf.
3. Canepa, B. Bursting the Bubble: Determining the Transit-Oriented Development's Walkable Limits. *Transportation Research Record: Journal of the Transportation Research Board*, 2007. 1992: 28–34.
4. Loukaitou-Sideris, A., R. Liggett, and H. Iseki. The Geography of Transit Crime: Documentation and Evaluation of Crime Incidence on and around the Green Line Stations in Los Angeles. *Journal of Planning Education and Research*, Vol. 22, No. 2, 2002, pp. 135–151. http://dx.doi.org/10.1177%2F0739456X02238443.

5. Cervero, R. Walk-and-Ride: Factors Influencing Pedestrian Access to Transit. *Journal of Public Transportation*, Vol. 7, No. 3, 2001, pp. 1–23.
6. Cervero, R., and R. Ewing. Travel and the Built Environment. *Journal of the American Planning Association*, Vol. 76, No. 3, 2010, pp. 265–294. https://doi.org/10.3141/1780-10.
7. Brons, M., M. Givoni, and P. Rietveld. Access to Railway Stations and Its Potential in Increasing Rail Use. *Transportation Research Part A: Policy and Practice*, Vol. 43, No. 2, 2009, pp. 136–149. https://doi.org/10.1016/j.tra.2008.08.002.
8. Park, S. *Defining, Measuring, and Evaluating Path Walkability, and Testing Its Impacts on Transit Users' Mode Choice and Walking Distance to the Station*. ProQuest Dissertations Publishing Database. (Accession No. 3331760). University of California, Berkeley, Calif., 2008.
9. Tilahun, N., and M. Li. Walking Access to Transit Stations: Evaluating Barriers with Stated Preference. *Transportation Research Record: Journal of the Transportation Research Board*, 2015. 2534: 16–23.
10. Burkhardt, J., A. T. McGavock, C. A. Nelson, and C. G. Mitchell. *Improving Public Transit Options for Older Persons: TCRP Report 82*. Transit Cooperative Research Program. Transportation Research Board of the National Academies, Washington, D.C., 2002. http://onlinepubs.trb.org/onlinepubs/tcrp/tcrp_rpt_82v2a.pdf.
11. Casas, I. Social Exclusion and the Disabled: An Accessibility Approach*. *The Professional Geographer*, Vol. 59, No. 4, 2007, pp. 463–477. http://dx.doi.org/10.1111/j.1467-9272.2007.00635.x.
12. U.S. Census Bureau. American Community Survey, 2015 American Community Survey 5-Year Estimates, Table S1810: Disability Characteristics. 2015. https://factfinder.census.gov/.
13. Brault, M. W. *Americans With Disabilities: 2010. Current Populations Reports*, Vol. P70-131, U.S. Census Bureau, Washington, D.C., 2012, pp. 1–23. https://www2.census.gov/library/publications/2012/demo/p70-131.pdf
14. He, W., and L. J. Larsen. *Older Americans with a Disability: 2008–2012*. American Community Survey Reports, Vol. ACS-29, U.S. Census Bureau, Washington, D.C., 2014, pp. 1–30. https://www.census.gov/content/dam/Census/library/publications/2014/acs/acs-29.pdf.
15. Lynott, J., and C. Figueiredo. *How the Travel Patterns of Older Adults Are Changing: Highlights from the 2009 National Household Travel Survey*. Fact Sheet 218, AARP Public Policy Institute, Washington, D.C., 2011.
16. Loprest, P., and E. Maag. *Barriers to and Supports for Work among Adults with Disabilities: Results from the NHIS-D*. Urban Institute, Washington, D.C., 2001.
17. Rosenbloom, S. The Transportation Patterns and Needs of People with Disabilities. In *The Future of Disability in America* (M. J. Field and A. Jette, eds.), National Academies Press, Washington, D.C., 2007, pp. 519–560.
18. Lubin, A., and D. Deka. Role of Public Transportation as Job Access Mode: Lessons from Survey of People with Disabilities in New Jersey. *Transportation Research Record: Journal of the Transportation Research Board*, 2012. 2277: 90–97.
19. Theis, K. A., and S. E. Furner. Shut-in? Impact of Chronic Conditions on Community Participation Restriction among Older Adults. *Journal of Aging Research*, Vol. 2011, 2011, Article ID 759158. http://dx.doi.org/10.4061/2011/759158.
20. Maynard, A. Can Measuring the Benefits of Accessible Transport Enable a Seamless Journey? *Journal of Transport and Land Use*, Vol. 2, No. 2, 2009, pp. 21–30.
21. Koppa, R. J., B. Davies, and K. Rodriguez. *Barriers to Use of Transportation Alternatives by People with Disabilities*. Publication SWUTC/98/467402-1, Texas Transportation Institute, Southwest Region University Transportation Center, Texas A & M University System, 1998.
22. Lavery, I., S. Davey, A. Woodside, and K. Ewart. The Vital Role of Street Design and Management in Reducing Barriers to Older Peoples' Mobility. *Landscape and Urban Planning*, Vol. 35, 1996, pp. 181–192. https://doi.org/10.1016/0169-2046(96)00309-X.
23. Wu, W., A. Gan, F. Cevallos, and L. D. Shen. Selecting Bus Stops for Accessibility Improvements for Riders with Physical Disabilities. *Journal of Public Transportation*, Vol. 14, No. 2, 2011, pp. 133-149. http://scholarcommons.usf.edu/jpt/vol14/iss2/7/.
24. Audirac, I. Accessing Transit as Universal Design. *Journal of Planning Literature*, Vol. 23, No. 1, 2008, pp. 4–16. https://doi.org/10.1177/0885412208318558.
25Smart Growth America. What Are Complete Streets? Washington, D.C., 2017. https://smartgrowthamerica.org/program/national-complete-streets-coalition/what-are-complete-streets/. Accessed June 8, 2017.
26. Lynott, J., J. Haase, K. Nelson, A. Taylor, H. Twaddell, J. Ulmer, B. McCann, and E. R. Stollof. *Planning Complete Streets for an Aging America*. No. 2009-02. AARP Public Policy Institute, Washington, D.C., 2009.
27. TriMet. *Technical Memo #1: Targets, Methodology, and Data Inventory*. Pedestrian Network Analysis, Portland, 2010. http://trimet.org/pdfs/pednetwork/trimet-ped-network-technical-memo-1.pdf.

The Public Transportation Group (AP000) peer-reviewed this paper (18-03153).

The contents of this paper reflect the views of the authors who are responsible for the facts and accuracy of the information presented.

Article

TRR
JOURNAL OF THE TRANSPORTATION RESEARCH BOARD

Transportation Research Record
2018, Vol. 2672(8) 588–598

DOI: 10.1177/0361198118758048
journals.sagepub.com/home/trr
SAGE

Applying a Model for Trip Assignment and Dynamic Routing of Automated Taxis with Congestion: System Performance in the City of Delft, The Netherlands

Xiao Liang[1], Gonçalo Homem de Almeida Correia[1], and Bart van Arem[1]

Abstract

This paper proposes a method of assigning trips to automated taxis (ATs) and designing the routes of those vehicles in an urban road network, and also considering the traffic congestion caused by this dynamic responsive service. The system is envisioned to provide a seamless door-to-door service within a city area for all passenger origins and destinations. An integer programming model is proposed to define the routing of the vehicles according to a profit maximization function, depending on the dynamic travel times, which varies with the ATs' flow. This will be especially important when the number of automated vehicles (AVs) circulating on the roads is high enough that their routing will cause delays. This system should be able to serve not only the reserved travel requests, but also some real-time requests. A rolling horizon scheme is used to divide one day into several periods in which both the real-time and the booked demand will be considered together. The model was applied to the real size case study city of Delft, the Netherlands. The results allow assessing of the impact of the ATs movements on traffic congestion and the profitability of the system. From this case-study, it is possible to conclude that taking into account the effect of the vehicle flows on travel time leads to changes in the system profit, the satisfied percentage and the driving distance of the vehicles, which highlights the importance of this type of model in the assessment of the operational effects of ATs in the future.

In recent years, technology development has accelerated the future roll-out of vehicle automation. This technology is expected to have a beneficial impact on travel efficiency, especially on interurban roads. Studies have mostly used micro-simulation tools or mathematical models to estimate the changes in road capacity and congestion under different levels of vehicle automation and cooperation (*1–3*).

Regarding the use of automated vehicles (AVs) as transit systems, there are currently several projects which are testing the implementation automated pods or buses in pilot experiments. One of the most relevant examples being the CityMobil2 project (*4*) in which different field experiments are being run in several European cities. In 2015 the province of Gelderland in the Netherlands started developing a project named WEpods with two self-driving vehicles, which are being tested within the Wageningen University campus, and between the campus and Ede-Wageningen train station (*5*).

Several approaches have been proposed to test the effect of transit automation on mobility, especially looking at the combination between traditional taxis and shared use taxi fleets. The automated taxis (ATs) may provide a new type of door-to-door service competing with traditional taxis or even public transport because these new systems would be able to avoid extra manpower costs and make the ATs service cheaper. Two methods are widely used to test the impact of AVs as public transport, agent-based simulation and mathematical optimization. Martinez et al. used agent-based simulation to build a model to test the introduction of 100% autonomous fleets of taxis to satisfy transport demand in a city (*6*). Spieser et al. used an analytical mathematical formulation to estimate the number of shared AVs to replace all modes of personal transportation in the case-study city of Singapore (*7*). Liang et al. proposed two integer programming models to optimize the service area of ATs for the last mile of train trips. Applying the methods to the real case-study city of Delft, the Netherlands, they concluded that fleet size influences the profitability of the taxi system (*8*). Despite

[1]Department of Transport & Planning, Faculty of Civil Engineering and Geosciences, Delft University of Technology, The Netherlands

Corresponding Author:
Address correspondence to Xiao Liang: x.liang@tudelft.nl

these models and their application, they all seem to be limited in having taken into account the effect that the taxis themselves can have on the traffic flow, and hence the delays in the network.

A recent study addressed the traffic assignment of privately owned AVs from the user optimum perspective with the objective of minimizing each family's own transport costs (*9*). An equilibrium convergence was proposed in which households with similar trips should have similar transport costs. Since traffic assignment is a nonlinear problem, this was tackled by an iterative process in which travel times do not change during each assignment of vehicles to trips and to the network. Another study also considered the traffic congestion when studying the routing problem of a large number of shared AVs, in which the traffic flow is modeled through link-transmission model (*10*). Then the model is applied to a small network as the considerable computation time means this formulation is not ready to solve the AVs' dynamic routing problems on a real size network.

In this paper, a model is proposed to address the problem of assigning the vehicles to clients and to paths on the urban road network. The model should consider dynamic travel times, which vary with the flow of the ATs. Moreover, it does not only serve reserved requests, but also some real-time requests. A rolling horizon scheme is proposed in which an optimization problem is solved over several horizons with updated demand. This approach advantage is two-fold: it helps reduce the problem size and it allows for adaption of the supply to the real-time demand.

Rolling horizon is a standard solution methodology for the dynamic vehicle routing problem (*11*). It is achieved by repeatedly solving a static problem over the events that have occurred from the current time t, to $t+L$, where L is the length of a horizon. Double-horizon based heuristics were proposed by Mitrović-Minić et al., incorporating both the long-term and short-term horizons (*12*). Yang et al. presented a rolling horizon scheme for truck fleet assignment and scheduling, when dealing with the real-time information (*13*). Luo and Schonfeld compared the rolling horizon strategy with the immediate strategy when inserting requests implemented in dial-a-ride problem heuristics. They concluded from their computation results that when satisfying all of the demand, the rolling horizon strategy reduced the vehicles by up to 10%, compared with the immediate strategy. This is because the rolling horizon strategy benefits from the advance information available (*14*).

The paper is organized as follows. Firstly, it introduces the mathematical model for assigning the passengers' requests to ATs and searching the optimal vehicle routing solutions. Secondly, the rolling horizon scheme is described to deal with the reserved and real-time requests. Then the application to the case-study city of Delft is presented, which is followed by the results of comparing the running of the model with and without congestion. The paper ends with the main conclusions that can be taken from the model application.

Integer Programming Model

This section describes a linear programming model for which the objective is to determine the optimal vehicle routing of the ATs system. A private taxi service can serve any pair of nodes within the city road network in which it is assumed that there is no background traffic flow, meaning the flow is generated only by the ATs themselves. Clients use an online app to book ATs providing travel information that includes the origin, the destination and the desired departure time. The model works on the assumption that the taxi company can achieve total control of the system by being free to accept or reject requests according to a profit maximization objective. It is assumed that the system accepts only individual trips for each vehicle. The model sets, data, parameters and mathematical formulation are shown below.

Sets

$\boldsymbol{N} = \{1,\ldots,i,\ldots N\}$: set of the nodes in the network, in which N is the total number of nodes.
$\boldsymbol{T} = \{0,\ldots,t,\ldots,T\}$: set of time instants in the service period, in which T is the total number of time steps in the operation time.
$\boldsymbol{T'} = \{-T^b,\ldots 0,\ldots,t,\ldots,T,\ldots,T+T^b\}$: set of time instants in the operation time, including the service period $\{0\ldots T\}$ and the buffer period $\{-T^b\ldots 0\}$ and $\{T\ldots T+T^b\}$, in which T^b is the number of time steps in the buffer period.
$\boldsymbol{E} = \{1,\ldots,e,\ldots,E\}$: set of trips, in which E is the total number of all the trips in the operation time.
$\boldsymbol{V} = \{1,\ldots,v,\ldots,V\}$: set of vehicles, in which V is the total number of taxis in the system.
$\boldsymbol{K} = \{0,\ldots,k,\ldots,K\}$: set of break-points when calculating the travel time based on the traffic flow, in which K is the total number of the break-points.
$\boldsymbol{A} = \{\ldots,(i,j),\ldots\}$: set of connected links of the road network, in which vehicles move, $i,j \in \boldsymbol{N}$, $i \neq j$
$\boldsymbol{B} = \{\ldots,(i_{t_1},j_{t_2}),\ldots\}$: set of links in the time-space network, $\forall (i,j) \in \boldsymbol{A}, \forall t_1,t_2 \in \boldsymbol{T'}, t_1 < t_2, \delta_{ij}^{\min} \leq t_2 - t_1 \leq \delta_{ij}^{\max}$

Data

a^e : desired departure time for the e th trip, $e \in \boldsymbol{E}$
b^e : latest arrival time for the e th trip, $e \in \boldsymbol{E}$
c^e : earliest departure time for the e th trip, $e \in \boldsymbol{E}$
wait^e : maximum waiting time for departing on the e th trip, $e \in \boldsymbol{E}$
ori^e : the origin node for the e th trip, $e \in \boldsymbol{E}$
des^e : the destination node for the e th trip, $e \in \boldsymbol{E}$
$\delta_{ij}^{\max}$: maximum travel time from node i to node j in time steps, $\forall (i,j) \in \boldsymbol{A}$
$\delta_{ij}^{\min}$: minimum travel time from node i to node j in time steps, $\forall (i,j) \in \boldsymbol{A}$

δ_{ij}^{k}: travel time from node i to node j in time steps at break-point k, $\forall(i,j)\in \boldsymbol{A}, k\in \boldsymbol{K}$
d_{ij}: length of the link (i,j) in kilometers, $\forall(i,j)\in \boldsymbol{A}$
Opt_{ij}: shortest travel time from node i to node j in free-flow conditions, $\forall i,j\in \boldsymbol{N}$
Cap_{ij}: capacity of each link (i,j) which is the number of vehicles that can go through the link at the highest travel time, $\forall(i,j)\in \boldsymbol{A}$

Parameters

Pr: price rate of a taxi, € / time step
C_{fuel}: fuel cost, €/ km
C_{p}: parking cost, € / time step
C_{d}: vehicle depreciation cost, € / day
ρ: penalty cost if a trip is rejected by the system, € / trip
C_{w}: penalty for passengers' waiting, € / time step
C_{del}: penalty for delivery delay, where the delay is defined as the real travel time minus the shortest travel time for a request, € / time step
μ: expansion coefficient, represents the number of taxis with the same characteristics in the population

Decision Variables

$x_{i_{t_1},j_{t_2}}^{v}$: binary variable equal to 1 if vehicle v drives from i to j from time instant t_1 to t_2, $\forall(i_{t_1},j_{t_2})\in \boldsymbol{B}, \forall v\in \boldsymbol{V}$
$y_{i_t}^{v}$: binary variable equal to 1 if vehicle v parks at i from time instant t to $t+1$, $\forall i\in \boldsymbol{N}$, $\forall v\in \boldsymbol{V}$, $\forall t\in \boldsymbol{T}', t<T$
S_{ij}^{ev}: binary variable equal to 1 if trip e from node $i=\text{ori}^{e}$ to node $j=\text{des}^{e}$ is done using vehicle v, $\forall e\in \boldsymbol{E}$, $\forall v\in \boldsymbol{V}$
P_{ij}^{evt}: binary variable equal to 1 if trip e from node $i=\text{ori}^{e}$ to node $j=\text{des}^{e}$ done by vehicle v starts at time instant t, $\forall e\in \boldsymbol{E}$, $\forall v\in \boldsymbol{V}$, $\forall t\in \boldsymbol{T}, c^{e}\le t\le a^{e}+\text{wait}^{e}$
A_{ij}^{evt}: binary variable equal to 1 if trip e from node $i=\text{ori}^{e}$ to node $j=\text{des}^{e}$ done by vehicle v ends at time instant t, $\forall e\in \boldsymbol{E}$, $\forall v\in \boldsymbol{V}$, $\forall t\in \boldsymbol{T}', c^{e}+\text{Opt}_{ij}\le t\le b^{e}$
ϕ_{ij}^{ev}: integer variable of the difference between the desired pick-up time and real pick-up time of trip e from node $i=\text{ori}^{e}$ to node $j=\text{des}^{e}$ done by vehicle v, $\forall e\in \boldsymbol{E}$, $\forall v\in \boldsymbol{V}$
Load_{t}^{v}: binary variable equal to 1 if vehicle v is transporting an individual passenger from time instant t to $t+1$, $\forall v\in \boldsymbol{V}$, $\forall t\in \boldsymbol{T}', t<T$
F_{ij}^{t}: integer variable of the vehicle flow on link (ij) starting from time instant t, $\forall(i,j)\in \boldsymbol{A}$, $\forall t\in \boldsymbol{T}', t<T'$
$\lambda_{ij}^{t,k}$: binary variable equal to 1 if travel time on link (ij) starting from t is at break-point k, $\forall(i,j)\in \boldsymbol{A}$, $\forall t\in \boldsymbol{T}', t<T'$, $\forall k\in \boldsymbol{K}$
δ_{ij}^{t}: integer variable of travel time when travelling on link (ij) starting from time instant t in time steps, $\forall(i,j)\in \boldsymbol{A}$, $\forall t\in \boldsymbol{T}', t<T'$

Objective Function

$$\max(\text{Profit})=\text{Pr}\cdot\sum_{\substack{e\in E,v\in V\\ i=\text{ori}^{e}\\ j=\text{des}^{e}}} S_{ij}^{ev}\cdot\text{Opt}_{ij}-C_{\text{fuel}}\cdot\sum_{\substack{(i_{t_1},j_{t_2})\in A\\ v\in V}} x_{i_{t_1},j_{t_2}}^{v}\cdot d_{ij}-C_{d}\cdot V-C_{p}\cdot\sum_{\substack{i\in N,v\in V\\ t\in T,t<T}} y_{i_t}^{v}-\rho\cdot\sum_{e\in E}\left(1-\sum_{\substack{v\in V\\ i=\text{ori}^{e}\\ j=\text{des}^{e}}} S_{ij}^{ev}\right)-C_{w}\cdot\sum_{\substack{e\in E,v\in V\\ i=\text{ori}^{e}\\ j=\text{des}^{e}}}\phi_{ij}^{ev}-C_{\text{del}}\cdot\sum_{\substack{e\in E,v\in V\\ i=\text{ori}^{e}\\ j=\text{des}^{e}}}\left(\sum_{t\in T'}(A_{ij}^{evt}*t)-\sum_{t\in T}(P_{ij}^{evt}*t)-\text{Opt}_{ij}\cdot S_{ij}^{ev}\right) \tag{1}$$

The objective function is considered from the perspective of both the ATs company and the passenger. For the ATs company, it aims to maximize the total profit which includes the total revenue from the customers, the vehicle fuel costs, the vehicle depreciation costs and the parking costs. The revenue is only charged for the shortest travel time of each request to avoid extra detours to charge more to the client. Regarding the passengers, when a request is rejected, passengers are assumed to use public transport as an alternative to finish their trip and the ATs system should pay a fixed charge for the users as a penalty. Late arrival and delivery delay are also penalized to consider the level of service offered to the clients.

Subject to the Following Constraints

The constraints given in Equation 2 describe that trip e from node i to node j can only be satisfied by vehicle v if that vehicle has passed through node i between the earliest departure time c^{e} and the latest arrival time b^{e}:

$$S_{ij}^{ev}\le\sum_{\substack{i_{t_1},j_{t_2}\in \boldsymbol{B}\\ t_1\ge c^{e},t_2\le b^{e}}} x_{i_{t_1},j_{t_2}}^{v}\ \forall e\in \boldsymbol{E}, v\in \boldsymbol{V}, i=\text{ori}^{e}, j=\text{des}^{e} \tag{2}$$

The constraints given in Equation 3 describe that trip e from node i to node j can only be satisfied by vehicle v if that vehicle has passed through node j at time instant $a^{e}+\delta_{ij}$:

$$S_{ij}^{ev}\le\sum_{\substack{i_{t_1},j_{t_2}\in \boldsymbol{B}\\ t_1\ge c^{e},t_2\le b^{e}}} x_{i_{t_1},j_{t_2}}^{v}\ \forall v\in \boldsymbol{V}, e\in \boldsymbol{E}, i=\text{ori}^{e}, j=\text{des}^{e} \tag{3}$$

The constraints given in Equation 4 assure that a satisfied trip must have only one departure time or this trip is not satisfied at all:

$$\sum_{\substack{t\in T' \\ c^e \le t \le a^e + \text{wait}^e}} P_{ij}^{evt} = S_{ij}^{ev}\ \forall e \in \boldsymbol{E}, v \in \boldsymbol{V}, i = \text{ori}^e, j = \text{des}^e \tag{4}$$

The constraints given in Equation 5 guarantee that one trip can only have one departure time when it is served or this trip is not satisfied by any vehicle:

$$\sum_{\substack{t\in T' \\ c^e \le t \le a^e + \text{wait}^e}} P_{ij}^{evt} \le 1\ \forall e \in \boldsymbol{E}, v \in \boldsymbol{V}, i = \text{ori}^e, j = \text{des}^e \tag{5}$$

The constraints given in Equation 6 assure that a satisfied trip must have only one arrival time or this trip is not satisfied at all:

$$\sum_{\substack{t\in T' \\ c^e + \text{Opt}_{ij} \le t \le b^e}} A_{ij}^{evt} = S_{ij}^{ev}\ \forall e \in \boldsymbol{E}, v \in \boldsymbol{V}, i = \text{ori}^e, j = \text{des}^e \tag{6}$$

The constraints given in Equation 7 guarantee that one trip can only have one arrival time when it is served or this trip is not satisfied by any vehicle:

$$\sum_{\substack{t\in T' \\ c^e + \text{Opt}_{ij} \le t \le b^e}} A_{ij}^{evt} \le 1\ \forall e \in \boldsymbol{E}, v \in \boldsymbol{V}, i = \text{ori}^e, j = \text{des}^e \tag{7}$$

The constraints given in Equation 8 compute the difference between the desired departure time and the real departure time, which is the passengers' waiting time:

$$\phi_{ij}^{ev} = \sum_{t\in T} P_{ij}^{evt} \cdot t - a^e \cdot S_{ij}^{ev}\ \forall e \in \boldsymbol{E}, v \in \boldsymbol{V}, i = \text{ori}^e, j = \text{des}^e \tag{8}$$

The constraints given in Equation 9 impose that the departure time instant of a satisfied trip must happen before the arrival time instant:

$$\sum_{\substack{t\in T \\ c^e \le t \le a^e + \text{wait}^e}} \left(P_{ij}^{evt} \cdot t\right) \le \sum_{\substack{t\in T' \\ c^e + \text{Opt}_{ij} \le t \le b^e}} \left(A_{ij}^{evt} \cdot t\right) \forall e \in \boldsymbol{E}, v \in \boldsymbol{V}, \quad i = \text{ori}^e, j = \text{des}^e \tag{9}$$

The constraints given in Equation 10 ensure that a trip can only be served by one vehicle:

$$\sum_{v\in V} S_{ij}^{ev} \le 1\ \forall e \in \boldsymbol{E}, i = \text{ori}^e, j = \text{des}^e \tag{10}$$

The constraints given in Equation 11 impose that for a specific time step t to $t+1$ each one of the ATs should only park at one node or not park at all:

$$\sum_{i\in N} y_{i_t}^v \le 1\ \forall v \in \boldsymbol{V}, t \in \boldsymbol{T}', t < T' \tag{11}$$

The constraints given in Equation 12 guarantee that one trip with a specific origin and departure time can only have one destination and one arrival time:

$$\sum_{i_{t_1}, j_{t_2} \in \boldsymbol{B}} x_{i_{t_1}, j_{t_2}}^v \le 1\ \forall i \in \boldsymbol{N}, v \in \boldsymbol{V}, t_1 \in \boldsymbol{T}' \tag{12}$$

The constraints given in Equation 13 ensure that each vehicle can only have one status at time instant t: either going to the next destination or parking at that place:

$$\sum_{\substack{i_{t_1}, j_{t_2} \in \boldsymbol{B} \\ t_1 \le t, t_2 > t}} x_{i_{t_1}, j_{t_2}}^v + \sum_{i\in N} y_{i_t}^v = 1\ \forall v \in \boldsymbol{V}, t \in \boldsymbol{T}', t < T' \tag{13}$$

The constraints given in Equation 14 are flow conservation constraint which make sure that the number of taxis leaving from node i and parking there from time instant t is equal to the number of vehicles arriving at node i and parking at the same place until t:

$$\sum_{i_t, j_{t_1} \in \boldsymbol{B}} x_{i_t, j_{t_1}}^v + y_{i_t}^v = \sum_{i_{t_2}, j_t \in \boldsymbol{B}} x_{i_{t_2}, j_t}^v + y_{i_{t-1}}^v\ \forall i \in \boldsymbol{N}, \quad t \in \boldsymbol{T}', t < T', v \in \boldsymbol{V} \tag{14}$$

The constraints given in Equation 15 describe the initial status of the ATs fleet. At the beginning of the operation period, all vehicles are stocked at the central parking station:

$$\sum_{\substack{i_0, j_{t_1} \in B \\ v\in V}} x_{i_0, j_{t_1}}^v + \sum_{v\in V} y_{i_0}^v = V \qquad i = \text{parking station} \tag{15}$$

The constraints given in Equation 16 compute the flow of vehicles in each road link (i, j) from time instant t_1:

$$F_{ij}^{t_1} = \left(\sum_{v\in V} x_{i_{t_1}, j_{t_2}}^v\right) \cdot \mu\ \forall \left(i_{t_1}, j_{t_2}\right) \in \boldsymbol{B} \tag{16}$$

The constraints given in Equation 17 limit the flow to the capacity of each link:

$$F_{ij}^t \le Q_{ij}\ \forall (i, j) \in \boldsymbol{A}, \forall t \in \boldsymbol{T}', t < T' \tag{17}$$

The constraints given in Equations 18 and 19 impose a break-point k to describe the congestion level of each link:

$$F_{ij}^t = \sum_{k\in K} \lambda_{ij}^{t,k} \cdot k \cdot \mu\ \forall (i, j) \in \boldsymbol{A}, \forall t \in \boldsymbol{T}', t < T' \tag{18}$$

$$\sum_{k\in K} \lambda_{ij}^{t,k} = 1\ \forall (i, j) \in \boldsymbol{A}, \forall t \in \boldsymbol{T}', t < T' \tag{19}$$

The constraints given in Equation 20 define the travel time on road link (i, j) from time instant t based on the break-point k:

$$\delta_{ij}^t = \sum_{k\in K} \delta_{ij}^k \cdot \lambda_{ij}^{t,k}\ \forall (i, j) \in \boldsymbol{A}, \forall t \in \boldsymbol{T}', t < T' \tag{20}$$

The constraints given in Equations 21 and 22 impose that the travel time is not greater than the maximum travel time or smaller than the minimum travel time. Meanwhile, they also guarantee that when vehicle v is travelling on link (i_{t_1}, j_{t_2}), the travel time must be $\delta_{ij}^{t_1}$:

$$\delta_{ij}^{t_1} \leq (t_2 - t_1) \cdot x_{i_{t_1}, j_{t_2}}^{v} + t_{ij}^{\max} \cdot \left(1 - x_{i_{t_1}, j_{t_2}}^{v}\right) \forall \left(i_{t_1}, j_{t_2}\right) \in \boldsymbol{B}, v \in \boldsymbol{V} \quad (21)$$

$$\delta_{ij}^{t_1} \geq (t_2 - t_1) \cdot x_{i_{t_1}, j_{t_2}}^{v} + t_{ij}^{\min} \cdot \left(1 - x_{i_{t_1}, j_{t_2}}^{v}\right) \forall \left(i_{t_1}, j_{t_2}\right) \in \boldsymbol{B}, v \in \boldsymbol{V} \quad (22)$$

The constraints given in Equation 23 guarantees link consistency. It is assumed that vehicles entering on a link (i, j) later from node i should not leave earlier from node j, which means the ATs do not pass one another. These constraints were established by Kaufman et al. (*15*):

$$t_1 + \delta_{ij}^{t_1} \leq t_2 + \delta_{ij}^{t_2} \; \forall (i, j) \in \boldsymbol{A}, t_1, t_2 \in \boldsymbol{T}', t_1, t_2 < T', t_1 < t_2 \quad (23)$$

The constraints given in Equation 24 compute the number of passengers loaded on each vehicle during time step t to $t+1$:

$$\text{Load}_t^v = \sum_{\substack{i=\text{ori}^e, j=\text{des}^e \\ e \in \boldsymbol{N}, \forall t_1 \in \boldsymbol{T}, t_1 \leq t}} P_{ij}^{evt_1} - \sum_{\substack{i=\text{ori}^e, j=\text{des}^e \\ e \in \boldsymbol{N}, \forall t_2 \in \boldsymbol{T}, t_2 \leq t}} A_{ij}^{evt_2} \; \forall t \in \boldsymbol{T}, t < T', v \in \boldsymbol{V} \quad (24)$$

The constraints given in Equation 25 enforce that only one passenger can be in each vehicle. In addition, when the vehicle is transporting passengers, it should not park at any node of the road network:

$$\text{Load}_t^v + \sum_{i \in \boldsymbol{N}} y_{i_t}^v \leq 1 \forall t \in \boldsymbol{T}, t < T', v \in \boldsymbol{V} \quad (25)$$

Rolling Horizon

The mathematical model described in the previous section entails knowing all the demand before the optimization. This means all taxi system users should reserve this transport service and provide all the travel information including the origin, the destination, and the departure time beforehand. Therefore, it would be impossible to change the movement of the ATs and serve real-time demand. Using a rolling horizon scheme, it is possible to consider those requests and at the same time reduce the complexity of the model that was proposed above.

Rolling Horizon Scheme

A rolling horizon scheme is used to divide one day into several horizons in which both the real-time and the pre-booked demand are considered. Model constraints given in Equations 1–25 are run for every horizon with updated demand from the previous. The horizon rolls forward with a specific rolling length and arrives at the next horizon with updated ATs requests. In this section, $\boldsymbol{H} = \{1, \ldots, h, \ldots H\}$ is the set of horizons in a day, in which H is the total number of horizons. t_r is the time length for rolling, and t_h is the time length of a horizon, $t_r < t_h$.

The rolling horizon scheme is shown in Figure 1. For horizon 1, the system collects the reserved requests which have the desired departure time within the time horizon 0 to t_h as input I and optimizes the ATs movements using these requests. After obtaining the optimal results as output I, the ATs system only executes the routing results for each vehicle from 0 to t_r (output II) and rolls forward to the next horizon. For the second horizon, two types of demand are both entered into the system. In addition to the reserved requests from t_r to $t_r + t_h$ (input I), the ATs company also receives real-time demand to be served as soon as possible from 0 to T_r as input II (these cannot be served at the rolling interval in which they were generated due to the vehicles' routes already being optimized). The optimization process for horizon 2 will be performed just before the time instant T_r, and all the ATs only perform the optimal results for T_r to $t_r + t_r$ (output II). Similarly, when it rolls to horizon h, the system input will be the reserved requests for $(h-1) \cdot t_r$ to $(h-1) \cdot t_r + t_h$ and the real-time requests for $(h-2) \cdot t_r$ to $(h-1) \cdot t_r$. Then the system will only implement the routing results for $(h-1) \cdot t_r$ to $h \cdot t_r$ as output II out of all the potential results for $(h-1) \cdot t_r$ to $(h-1) \cdot t_r + t_h + t_b$ which is the output I of horizon h.

The rolling length and the horizon length can be explained from a practical perspective. This rolling horizon scheme is able to optimize the ATs' routing time after time with real-time information. The horizon length determines how far the system will include the pre-known information: the reserved taxi service requests. The rolling length indicates how often the system will input the new requests, update the optimal results and execute the routing plan. The rolling length is set shorter than the horizon length because the system can plan for the future requests and update once new requests enter. If the horizon is as short as a rolling length, the system cannot relocate for the next horizon trip, which may decrease the system efficiency. If the rolling is as long as the horizon, the real-time request clients may wait for a whole horizon to be calculated for optimization.

Demand Initialization

Different time components for reserved and real-time requests are set to initialize the system demand. The reserved requests arrive in advance before the horizon start time. No waiting time is allowed to provide a better service for these clients. This means the system should serve them as they require. As a result, the desired departure time is also the earliest departure time for this kind of demand. For the real-time requests, the time they make requests is the desired departure time. Therefore, this model allows some waiting time for each request to be served by ATs as these requests are only analyzed in the next horizon. The earliest departure time is the start time of the current horizon. In this model, the maximum waiting time is set equal to the time length for rolling.

For horizon h, the desired departure time $a^{e,h}$ for each request should be transformed from an absolute time to a

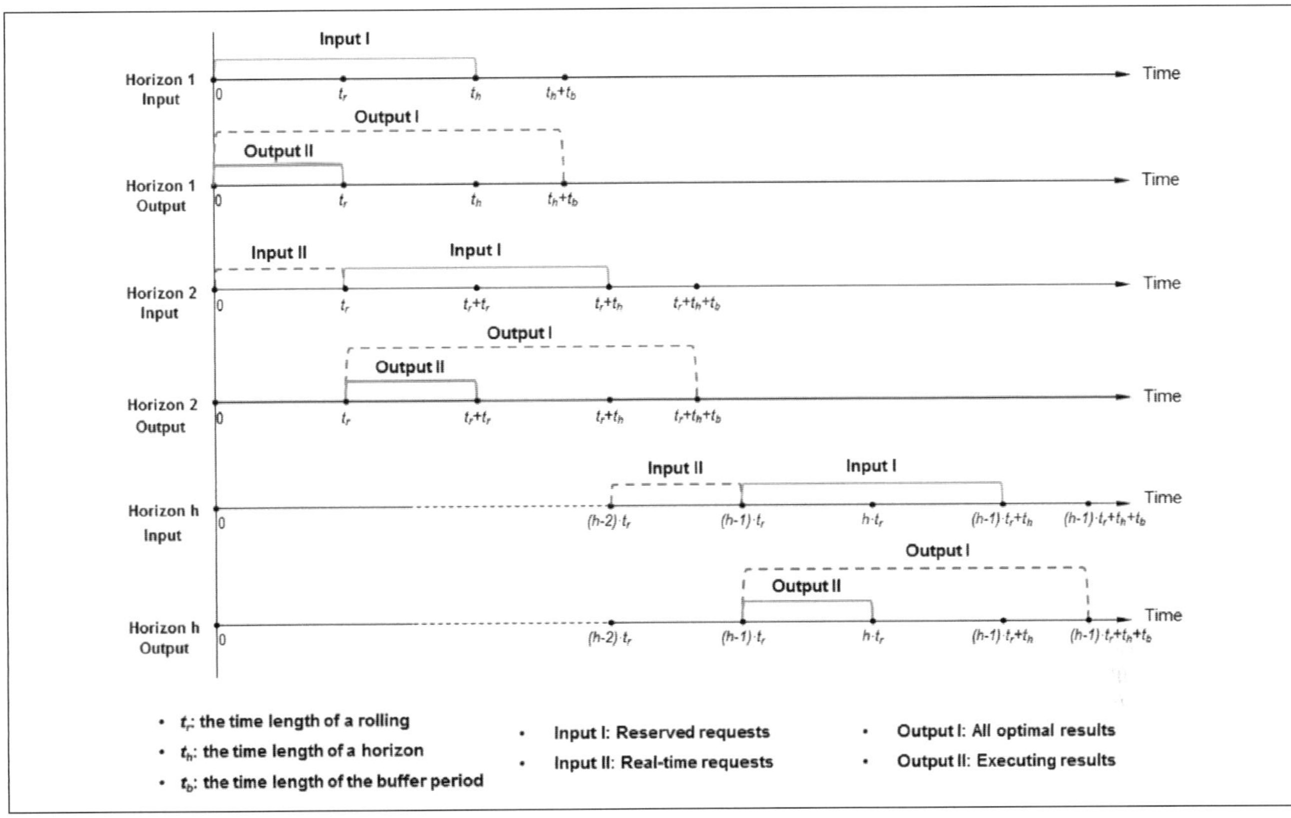

Figure 1. Rolling horizon scheme.

relative time: $a^{eh} = a^e - (h-1)*t_r$. The earliest departure time for reserved requests in horizon h is: $c^{eh} = a^{e,h}$. The earliest departure time for real-time requests in horizon h is: $c^{eh} = 0$. The latest arrival time for reserved requests in horizon h is: $b^{eh} = a^{eh} + \text{Opt}_{ij} \cdot \delta$. The latest arrival time for real-time requests in horizon h is: $b^{eh} = wait^e + \text{Opt}_{ij} \cdot \delta$, where δ is the scale factor between the maximum travel time and the shortest travel time.

Continuity

There are three types of satisfied trips in each optimization horizon which can be handled in different ways in the rolling horizon scheme (Figure 2). If the optimal departure time and the arrival time for each request are both within the executing period (from $(h-1)*t_r$ to $h*t_r$), the system will fully execute the routing results for that request (trip type 1). If the departure and arrival time are both after the end of the executing period t_r, the system will leave it in the demand pool and perform the optimization later in the next horizon (trip type 2). If the departure time is before t_r and the arrival time is afterward (trip type 3), then the system will execute the routing results from the departure time until t_r. Based on the optimal solution, the system will know which vehicle satisfies this request and the location of it at time instant t_r. This location will be the new origin of that request and this time is the new departure time in the next horizon. In some cases, at time instant t_r the vehicle v is in the middle of a road link. The cut-off time for this request will be the one closest to t_r after t_r, and the new origin of it will be the location of vehicle v at that time instant.

When obtaining the optimal results from the previous horizon, some new notations are used to save the system status information, and set them as a starting point for the next horizon:

loc^{vh} : the location of vehicle v at or most close to the beginning of the current horizon, $\forall v \in \boldsymbol{V}, h \in \boldsymbol{H}$

et^{vh} : the first available time when vehicle v finishes its trip starting from the last horizon, $\forall v \in \boldsymbol{V}, h \in \boldsymbol{H}$

vs^{eh} : which vehicle satisfies trip e in the previous horizon, $\forall e \in \boldsymbol{E}, h \in \boldsymbol{H}$

$vo_{i_t}^{vh}$: equal to 1 if vehicle v is occupied by an unfinished trip from time instant t to $t+1$ from the previous horizon, $\forall v \in \boldsymbol{V}, t \in \boldsymbol{T'}, t < T', i = loc^v, h \in \boldsymbol{H}$

$vo\_last_{i_t}^{vh}$: equal to 1 if vehicle v is occupied by an unfinished trip from time instant t to $t+1$ from the previous horizon, and will end this trip at time instant $t+1$, $\forall v \in \boldsymbol{V}, t \in \boldsymbol{T'}, t < T', i = loc^{vh}, h \in \boldsymbol{H}$

The equations below are used to save the vehicle status information from the previous horizon:

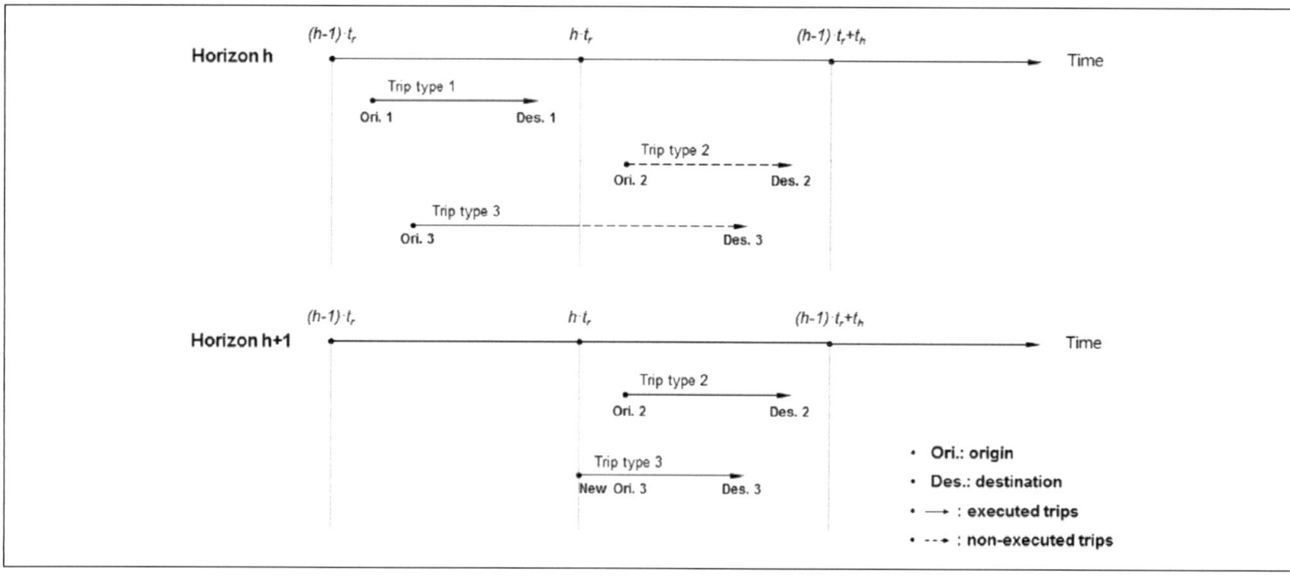

Figure 2. Trips between horizons.

$$loc^{v(h+1)} = \sum_{\substack{i_{t_1}, j_{t_2} \in \boldsymbol{B} \\ t_1 < T', t_1 \geq T'}} x\_result^{vh}_{i_{t_1}, j_{t_2}} \cdot j + \sum_{\substack{\forall i \in \boldsymbol{N}, \forall t \in \boldsymbol{E}' \\ t = T_r - 1}} y\_result^{vh}_{i_t} \cdot i \, \forall v \in \boldsymbol{V} \tag{26}$$

Equation **(26)** computes the final location of vehicle v at the last time instant of the execute period T_r, where $x\_result^{vh}_{i_{t_1}, j_{t_2}}$ is the value of decision variable $x^v_{i_{t_1}, j_{t_2}}$ from horizon $h-1$, and $y\_result^{vh}_{i_t}$ is the value of decision variable $y^v_{i_t}$ from horizon $h-1$. If this vehicle is on an unfinished trip, then the location will be the destination of that trip.

$$et^{v(h+1)} = \sum_{\substack{i_{t_1}, j_{t_2} \in \boldsymbol{B} \\ t_1 < T', t_1 \geq T'}} x\_result^{vh}_{i_{t_1}, j_{t_2}} \cdot (t_2 - T_r) + \sum_{\substack{\forall i \in \boldsymbol{N}, t \in \boldsymbol{T}' \\ t = T_r - 1}} y\_result^{vh}_{i_t} \cdot (t + 1 - T_r) \forall v \in \boldsymbol{V} \tag{27}$$

Equation **(27)** is equal to 0 if vehicle v is free at the last time instant of the execute period T_r. If it is not, it is the first available time when vehicle v finishes its trip starting from the last horizon.

$$vs^{e(h+1)} = \sum_{\substack{v \in \boldsymbol{V} \\ i = ori^e, j = des^e}} S\_result^{evh}_{ij} \cdot v \forall e \in \boldsymbol{E} \tag{28}$$

Equation **(28)** indicate which vehicle serves the eth request. If it is 0, this request is not satisfied by the system, where $S\_result^{evh}_{ij}$ is the value of the decision variable S^{ev}_{ij} from horizon $h-1$.

The value of $vo^{v(h+1)}_{i_t}$ is not from an equation. If $et^{v(h+1)}$ is bigger than 0 and the location i is equal to $loc^{v(h+1)}$, then $vo^{v(h+1)}_{i_t} = 1, \forall i \in \boldsymbol{N}, \forall t \in \boldsymbol{T}', t \leq et^v$.

New desired information should be loaded into the half executed requests. Firstly, the system should assign a new origin to request e in horizon h: $ori^{eh} = loc^{vh}$, when $v = vs^{eh}$, which is a half-executed request. Then this updated request needs a new desired departure time, which is the time instant when the vehicle finished the previous trip.

Based on the model constraints in Equations 1–25 and the system status of the previous horizon, this section renovates the model to guarantee the continuity between sequential horizons.

The constraints given in Equation 29 are modified from those given in Equation 13, which indicate that ATs can only have one out of three statuses: going to the next destination, parking at the current node or driving on a trip from the previous horizon,

$$\sum_{\substack{i_{t_1}, j_{t_2} \in \boldsymbol{B} \\ t_1 \leq t, t_2 > t}} x^v_{i_{t_1}, j_{t_2}} + \sum_{i \in \boldsymbol{N}} y^v_{i_t} + \sum_{i \in \boldsymbol{N}} vo^{vh}_{i_t} = 1 \forall v \in \boldsymbol{V}, t \in \boldsymbol{T}, h \in \boldsymbol{H} \langle T', h \rangle 1 \tag{29}$$

The constraints given in Equation 30 are an updated version of those given in Equation 14. Vehicles arriving at i_t are not only from the trip in the current horizon, but also from the previous horizon:

$$\sum_{i_t, j_{t_1} \in \boldsymbol{B}} x^v_{i_t, j_{t_1}} + y^v_{i_t} = \sum_{j_{t_2}, i_t \in \boldsymbol{B}} x^v_{j_{t_2}, i_t} + y^v_{i_{t-1}} + vo\_last^{vh}_{i_{t-1}1} \forall i \in \boldsymbol{N}, t \in \boldsymbol{T}', v \in \boldsymbol{V}, h \in \boldsymbol{H}, t \langle T', h \rangle 1 \tag{30}$$

The constraints given in Equation 31 impose that all the vehicles must start from the same location as they end in the previous horizon:

$$\sum_{j\in N, t_2 \in T'} x^{v}_{i_{t_1} j_{t_2}} + y^{v}_{i_{t_1}} = 1 \forall v \in \boldsymbol{V}, h \in \boldsymbol{H}, \quad i = loc^{vh}, t = et^{vh}, h > 1 \tag{31}$$

The constraints given in Equation 32 guarantee that the half executed requests in the last horizon must be served by the same vehicle in the current horizon:

$$S^{ev}_{ij} = 1 \quad \forall e \in \boldsymbol{E}1, h \in \boldsymbol{H}, i = \mathrm{ori}^{eh}, \quad j = \mathrm{des}^{e}, v = vs^{eh}, h > 1 \tag{32}$$

Case Study and Results

The model is applied to the case-study city of Delft, which is located in the Netherlands, having a total area of 24 km^2, and a population of about 100,000.

Demand

To apply the model for studying the ATs system in the city of Delft, the following data are needed:

1. The information of the requests;
2. The price and cost of running the system;
3. The road network information of Delft.

The mobility data were obtained from the Dutch mobility dataset (MON 2007/2008) and transformed in the study of Correia and van Arem (*9*). These data include the origin and destination, travel mode, departure and arrival time, which is the required information to apply the model in this paper and can be freely obtained online (*16*).

The survey sample includes 68,640 requests which were made by residents who travelled only within the city of Delft during the surveyed day. For this application, only the trips done by cars and taxis are considered. A total of 22,240 trips resulted which are represented by 1112 requests in the operation day (expansion coefficient $\mu = 20$). At the same time, each one of the ATs in our system represents 20 vehicles therefore a taxi satisfying a trip represents 20 taxis satisfying 20 requests. In these experiments, 50% of all the requests are considered to be booked in advance (reserved requests), while 50% are real-time requests. Monte Carlo simulation is used to decide which requests are pre-booked and which are real-time. The original origin and destination, as well as the desired departure time recorded in the data base are used as preferred in the experiments.

According to the time distribution of all the requests, the earliest trip started at 06:48 and the latest trip at 23:35. 06:30–24:00 is chosen as the operation period of the ATs service. The time step of the optimization is 2.5 min. Each horizon contains 12 time steps ($t_h = 12$) and the rolling length is 6 time steps ($t_r = 6$). In practice this means that at each time the system looks forward to the requests in the next 30 min and updates the ATs' status every 15 min. This also means the real-time requests will wait 15 min as the maximum waiting time.

The cost values considered are as follows:
$\mathrm{Pr} = 10$ € / time step.
$C_{\mathrm{fuel}} = 0.1$ € / km.
$C_d = 17.5$ € / day.
$\rho_1 = 5$ € / trip
$\rho_2 = 2$ € / trip
$C_w = 0.1$ € / time step

The penalty for unsatisfied reserved and real-time requests are given different values, which is mainly because the system should give priority to reservations.

The parking outside of Delft city center is free of charge. ATs cannot park in the city center (nodes 28, 4, 3, 44 and 9), but they are able to deliver passengers there and leave immediately. This is based on the parking regulations in the city of Delft.

Road Network

The network has 66 road links and 46 nodes. Some of the links have one lane per way and some have two lanes. The capacities were considered as 1600 and 3200 vehicles per hour per direction based on the number of lanes on that link. The maximum speeds were assumed to be 50 and 70 km/h, respectively, for the lower and higher capacity links.

The minimum travel time $\delta^{\min}_{ij}$ on link (i, j) is obtained from the free flow speed. The maximum travel time $\delta^{\max}_{ij}$ in time step is computed based on the speed of 5 km/h. Travel time at break-point k in time step δ^{k}_{ij} is considered as a function of the traffic flow, which is reflected by the value of break-point k. There are three break-points for the road links with lower capacity and six break-points with higher capacity. These travel times are obtained based on the Bureau of Public Roads function (*17*) in equation $t = t_0\left(1 + a \times \left(\frac{V}{Q}\right)^{b}\right)$, where t_0 is the minimum travel time, $\frac{V}{Q}$ is the value of break-point k, a is $\left(\delta^{\max}_{ij} / \delta^{\min}_{ij}\right) - 1$ and b is 4.

The shortest travel time between the nodes (only for which there is demand) is computed by the shortest path method with free flow travel time (Opt_{ij}). The travel distance on each link is the link length in reality.

Results

The optimization model was run in an i5 processor @3.10 GHz, 12.00 GB RAM computer under a Windows 7 64-bit operating system by Xpress, an optimization tool that uses advanced branch-and-cut algorithms for solving integer programming problems.

Two kinds of scenarios are considered. The first one, which is called dynamic travel time, considers the travel time as a function of the taxis traffic flow on each link, as formulated in Equations 18–20. The second scenario assumes that the travel times do not change and are thus always equal to the minimum travel time on that link, even if the traffic flow varies (static travel times). Therefore, the travel time constraints are turned off. Another key variable for creating these scenarios is the fleet size which represents the number of ATs available in the system.

A comparison of the routing optimization of ATs between different scenarios was performed and is presented in Table 1. The fleets with 100, 200, 300, 400, and 500 ATs are represented by 5, 10, 15, 20, and 25 ATs in the optimizing process, with the expansion coefficient $\mu = 20$. Congestion delay is calculated as the time difference between the real travel time and the shortest travel time of a request. The waiting time is only generated by the real-time requests, which is the difference between the desired departure time and the departure time that the traveler experienced.

In the dynamic travel time scenarios, the fleet size is a critical parameter with respect to the system performance. The daily profit is sensitive to the number of requests satisfied by ATs. When there are only 100 vehicles in the taxi system 1840 requests are satisfied in total. In this case, the satisfied rate is the lowest among all the scenarios. When the fleet size increases to 200, the satisfied demand almost doubled, which also leads to a big increase in the system benefits. This trend continues in scenarios 3, 4 and 5 with 300, 400, and 500 ATs. However, the growth rate is falling in these three scenarios, especially when the fleet size reaches 500. Even though, with more vehicle depreciation cost and less working efficiency, scenario 5 with 500 ATs is the most profitable one among these scenarios with dynamic travel time.

As the penalty of reserved requests is higher than the real-time ones, the system decides to serve reserved requests as a priority. For example, when there are 300 ATs available, 32% of reserved requests are served, while 10% real-time requests are satisfied. As the value of penalty costs has a big influence on deciding which requests are to be served. It also reveals that the system can use this method to adjust the service strategy for different kinds of demand.

The average number of satisfied trips indicates the working efficiency of each AT. In the first five scenarios with dynamic travel time, the most efficient is the one with 100 ATs, with 18.4 requests per vehicle. This also can be seen from the column "average travel time," "average idle time," "idle rate," and "average travel distance." Scenario 5 has the highest idle time and idle rate. This means more than half of the total operation time ATs are parking somewhere without any movement. The average travel distance is 149.5 km per AT, which is also lower than in any other scenario with the dynamic travel time condition. Actually, it does not conflict with the unsatisfied request rate. The demand is not uniformly distributed during the day which brings out the peak horizons and the off-peak horizons. In the off-peak horizons, 500 ATs are redundant and generate idle time, while in peak horizons there are not enough to serve all the requests.

In the static travel time scenarios, the same trend can be found from scenarios 6 to 10 with regards to the total profitability. Scenario 10 with 500 vehicles has the highest profit, which serves the most requests with a 37% satisfied rate. When the fleet size is 100, each of the ATs are more efficient, serving 19.0 requests in average. The travel distance is also the highest, 236.9 km per vehicle. When the number of ATs decreases, the transport efficiency decreases.

Comparing the static travel time scenarios with dynamic ones, it is easy to find that the ATs system is always more profitable when considering the minimum travel times. When the fleet size is 100, the system obtains 52,200 € under the dynamic travel time condition, which is 6.1% lower than the static travel time with 55,600 €. In addition, the number of satisfied trips in the dynamic travel time scenarios is also less than the one from the static travel time scenarios. The differences in service performance are more obvious in other fleet size scenarios. With 300 ATs, scenario 3 has 19.9% less profit, while with 500 ATs the dynamic travel time scenario has 36.0% less daily profit. This is because when the travel time is always the minimum value, the system would take less time than the dynamic travel time scenario, which makes it possible for the ATs to serve more trips and receive more profit. However, it cannot reflect the real traffic status and the impact of ATs on travel time. This can be seen from the column "Total congestion delay" and the "Avg. congestion delay" in Table 1, which represents the difference between the real travel time and the shortest travel time. These time steps are caused by the traffic congestion and decrease the available service time of ATs.

Conclusion

This paper proposed a mathematical model to study a dynamic travel time based ATs system to provide transport service within the city area. This model involves both the reserved requests and real-time requests by the rolling horizon scheme and optimizes the vehicle routing horizon by horizon. The contribution of this paper is to consider the travel time on each road link according to the number of vehicles travelling on it and perform system optimization with the objective to maximize the total profit of such a system by deciding on each ATs routing selection. It also establishes a rolling horizon scheme to deal with the dynamic vehicle routing problem including real-time information. The model was then applied to a case study in the city of Delft, the Netherlands with a 17.5 hours service period and 23,340 travel requests generating from the road network with 46 nodes and 66 links. From that application it was possible to draw the following conclusions.

Table 1. Optimization Model Results

	Scenario	Fleet size	Daily profit ($\times 10^3$ €)	Satisfied requests	Satisfied rate	Satisfied rate of reserved requests	Satisfied rate of real-time requests	Avg. satisfied requests per vehicle	Avg. idle time min/veh.	Avg. travel time min/veh.	Idle rate	Avg. travel distance km/veh.	Avg. congestion delay min/satisfied requests	Avg. waiting time min/satisfied real-time requests
Dynamic travel time	1	100	52.2	1840	8%	13%	3%	18.4	459.0	591.0	44%	257.0	3.5	13.8
	2	200	172.4	3640	16%	28%	5%	18.2	506.3	543.8	48%	229.3	2.2	13.5
	3	300	218.6	4660	21%	32%	10%	15.5	535.0	515.0	51%	210.3	3.7	12.7
	4	400	256.9	5440	24%	35%	14%	13.6	624.1	425.9	59%	172.3	2.9	12.2
	5	500	260.2	5760	26%	35%	16%	11.5	672.4	377.6	64%	149.5	3.2	12.9
Static travel time	6	100	55.6	1900	9%	14%	3%	19.0	510.5	539.5	49%	236.9	-	13.7
	7	200	170.8	3700	17%	28%	5%	18.5	521.3	528.8	50%	225.2	-	12.9
	8	300	273.0	5360	24%	41%	7%	17.9	519.2	530.8	49%	222.1	-	13.2
	9	400	333.6	6680	30%	47%	13%	16.7	560.8	489.3	53%	204.9	-	13.3
	10	500	406.5	8220	37%	55%	19%	16.4	578.6	471.4	55%	196.9	-	12.7

Note: Avg = average; veh. = vehicle; - = not applicable.

1) The fleet size is an important factor on system profitability, regardless of the dynamic or static travel time. The reason is that more ATs are able to serve more requests which bring more income for the ATs system, even though the depreciation costs are also increasing and the service efficiency is decreasing.
2) Knowing that real-life travel conditions have dynamic travel time, application of a static minimum travel time vehicle routing model will overestimate the system profitability. Traffic congestion happens when vehicles are travelling on the network, and it leads to less profit, arrival delay and fewer satisfied requests than in the case of the static travel time. However, it reflects the real traffic situation and makes the routing results more realistic.
3) The influence of considering travel time as dynamic in ATs vehicle routing problems can be seen from two aspects: the time delay on the congested road links and the extra travel distance caused by detouring.

Acknowledgments

The authors would like to thank FICO (Xpress supplier) for establishing a research agreement with the Department of Transport & Planning at TU Delft. This work was supported by ProRail, the train infrastructure manager of the Netherlands. Thanks go also to the Chinese Scholarship Council for sponsoring the first author.

References

1. Bose, A., and P. A. Ioannou. Analysis of Traffic Flow with Mixed Manual and Semiautomated Vehicles. *IEEE Transactions on Intelligent Transportation Systems*, Vol. 4, No. 4, 2003, pp. 173–188.
2. Calvert, S. C., T. H. A. Van Den Broek, and M. Van Noort. Modelling Cooperative Driving in Congestion Shockwaves on a Freeway Network. *Proc., 14th IEEE Intelligent Transportation Systems Conference, ITSC*, Washington, D.C., 2011, pp. 614–619.
3. van Arem, B., C. J. G. Van Driel, and R. Visser. The Impact of Cooperative Adaptive Cruise Control on Traffic-Flow Characteristics. *IEEE Transactions on Intelligent Transportation Systems*, Vol. 7, No. 4, 2006, pp. 429–436.
4. CityMobil2. *CityMobil2 - Cities Demonstrating Automated Road Passenger Transport*. http://www.citymobil2.eu/en/.
5. WEpods. *WEpods project DAVI – Dutch Automated Vehicle Initiative*. http://davi.connekt.nl/wepods-project/.
6. Martinez, L. M., G. H. A. Correia, and J. M. Viegas. An Agent-Based Simulation Model to Assess The Impacts of Introducing a Shared-Taxi System: An Application to Lisbon (Portugal). *Journal of Advanced Transportation*, Vol. 49, No. 1, 2015, pp. 475–495.
7. Spieser, K., K. Treleaven, R. Zhang, E. Frazzoli, D. Morton, and M. Pavone. Toward a Systematic Approach to the Design and Evaluation of Automated Mobility-on-Demand Systems: A Case Study in Singapore. In *Road Vehicle Automation* (G. Meyer and S. Beiker, eds.), Springer, New York, 2014, pp. 229–245.
8. Liang, X., G. H. de Almeida Correia, and B. van Arem. Optimizing the Service Area and Trip Selection of an Electric Automated Taxi System Used for the Last Mile of Train Trips. *Transportation Research Part E: Logistics and Transportation Review*, Vol. 93, 2016, pp. 115–129.
9. de Almeida Correia, G. H., and B. van Arem. Solving the User Optimum Privately Owned Automated Vehicles Assignment Problem (UO-POAVAP): A Model to Explore the Impacts of Self-Driving Vehicles on Urban Mobility. *Transportation Research Part B: Methodological*, Vol. 87, 2016, pp. 64–88.
10. Levin, M. W. Congestion-Aware System Optimal Route Choice for Shared Autonomous Vehicles. *Transportation Research Part C: Emerging Technologies*, Vol. 82, 2017, pp. 229–247.
11. Psaraftis, H. N. Dynamic Vehicle Routing Problems. In *Vehicle Routing: Methods and Studies* (B. L. Golden and A. A. Assad, eds.), North Holland, Amsterdam, The Netherlands, 1988, pp. 223–248.
12. Mitrović-Minić, S., R. Krishnamurti, and G. Laporte. Double-Horizon Based Heuristics for the Dynamic Pickup and Delivery Problem with Time Windows. *Transportation Research Part B: Methodological*, Vol. 38, No. 8, 2004, pp. 669–685.
13. Yang, J., P. Jaillet, and H. Mahmassani. On-Line Algorithms for Truck Fleet Assignment and Scheduling Under Real-Time Information. *Transportation Research Record: Journal of the Transportation Research Board,* 1999. 1667: 107–113.
14Luo, Y., and P. Schonfeld. Online Rejected-Reinsertion Heuristics for Dynamic Multivehicle Dial-a-Ride Problem. *Transportation Research Record: Journal of the Transportation Research Board*, 2011. 2218: 59–67.
15. Kaufman, D. E., J. Nonis, and R. L. Smith. A Mixed Integer Linear Programming Formulation of the Dynamic Traffic Assignment Problem. *Proc., IEEE International Conference on Systems, Man and Cybernetics*, Chicago, Illinois, Vol. 1, 1992, pp. 232–235.
16. Correia, G. H. de A., and B. van Arem. Trips and network of the case-study city of Delft. https://www.researchgate.net/publication/309134999_Trips_and_network_of_the_case-study_city_of_Delft
17. Dafermos, S., and F. Sparrow. Traffic Assignment Problem for a General Network. *Journal of Research of the National Bureau of Standards B*, Vol. 73, No. 2, 1969, pp. 91–118. http://www.scopus.com/inward/record.url?eid=2-s2.0-0014491807&partnerID=tZOtx3y1.

The Standing Committee on Emerging and Innovative Public Transport and Technologies (AP020) peer-reviewed this paper (18-01038).

Article

TRR
JOURNAL OF THE TRANSPORTATION RESEARCH BOARD

Transportation Research Record
2018, Vol. 2672(8) 599–607
© National Academy of Sciences:
Transportation Research Board 2018
Article reuse guidelines:
sagepub.com/journals-permissions
DOI: 10.1177/0361198118758674
journals.sagepub.com/home/trr
SAGE

Personalized Menu Optimization with Preference Updater: A Boston Case Study

Xiang Song[1], Mazen Danaf[1], Bilge Atasoy[1], and Moshe Ben-Akiva[1]

Abstract

This paper presents a personalized menu optimization model with preference updater in the context of an innovative Smart Mobility system that offers a personalized menu of travel options with incentives for each incoming traveler in real time. This Smart Mobility system can serve as a major travel demand management system that encourages energy-efficient travel options. The personalized menu optimization is built on a logit mixture model that captures each individual traveler's choice behavior. The personalized menu optimization model is enhanced with a preference updater that can update the estimates of individual traveler's preference parameters when new choice data is received. To illustrate the advantages of the proposed methodology, a case study is presented based on real travelers and trips in the greater Boston area from the Massachusetts Travel Survey data. The case study consists of two parts. In the first part, the personalized menu optimization with preference updater is tested in a setting where the travelers are new to the system and their preferences are updated through preference updater. A comparative analysis of the performance of the proposed method with preference updater is presented against the method without preference updater. In the second part, the benefit of using individual level preference parameters instead of population level preference parameters in the personalized menu optimization model is analyzed. The case study shows that the proposed method can outperform the hit rates of its two counterparts.

Smart Mobility systems are emerging to provide innovative app-based mobility solutions that can deal with real-time requests from heterogeneous travelers. Companies including Uber, Lyft, and Zipcar have been successful in addressing travelers' needs and have therefore been attracting many travelers every day (*1*). The recommendations from Smart Mobility systems need to be personalized and optimized to attract mobility consumers and to operate efficiently, which can be enabled by the personalized menu optimization. The personalized menu optimization integrates travel behavioral modeling and optimization techniques, which can capture heterogeneous travelers' behaviors while accomplishing an operator's objectives such as maximizing traveler benefits or a system's total revenue (*2–4*). In a Smart Mobility system called Flexible Mobility on Demand (FMOD), the personalized menu optimization selects a subset of travel options, including taxi, shared-taxi, and mini-bus, to be presented on the menu that can maximize the expected revenue, or consumer surplus, or both (*2, 3*).

In this paper, a personalized menu optimization is presented with preference updater in the context of a research project, called Tripod: Sustainable Travel Incentives with Prediction, Optimization and Personalization. The idea is to incentivize travelers based on energy savings to increase the utilization of more energy efficient options. This framework brings together various methodologies including system and user level optimizations, travel behavior models, traffic simulation with prediction capabilities, energy simulation, and agent-based simulation for evaluating the performance of the incentive control mechanisms. The travel incentives are in the form of tokens for energy reduction and are decided by system optimization. By accepting and executing a specific travel option, a traveler earns tokens that depend on the system-wide energy saving he or she creates, encouraging them to consider not only their own energy cost, but also the impact of their choice on the system. Tokens can then be redeemed for services and goods from participating vendors and transportation agencies. For example, a cup of coffee or a movie ticket can be exchanged with different amounts of tokens from the Tripod system.

[1]MIT, Civil and Environmental Engineering, Cambridge, MA

Corresponding Author:
Address correspondence to Xiang Song: bensong@mit.edu

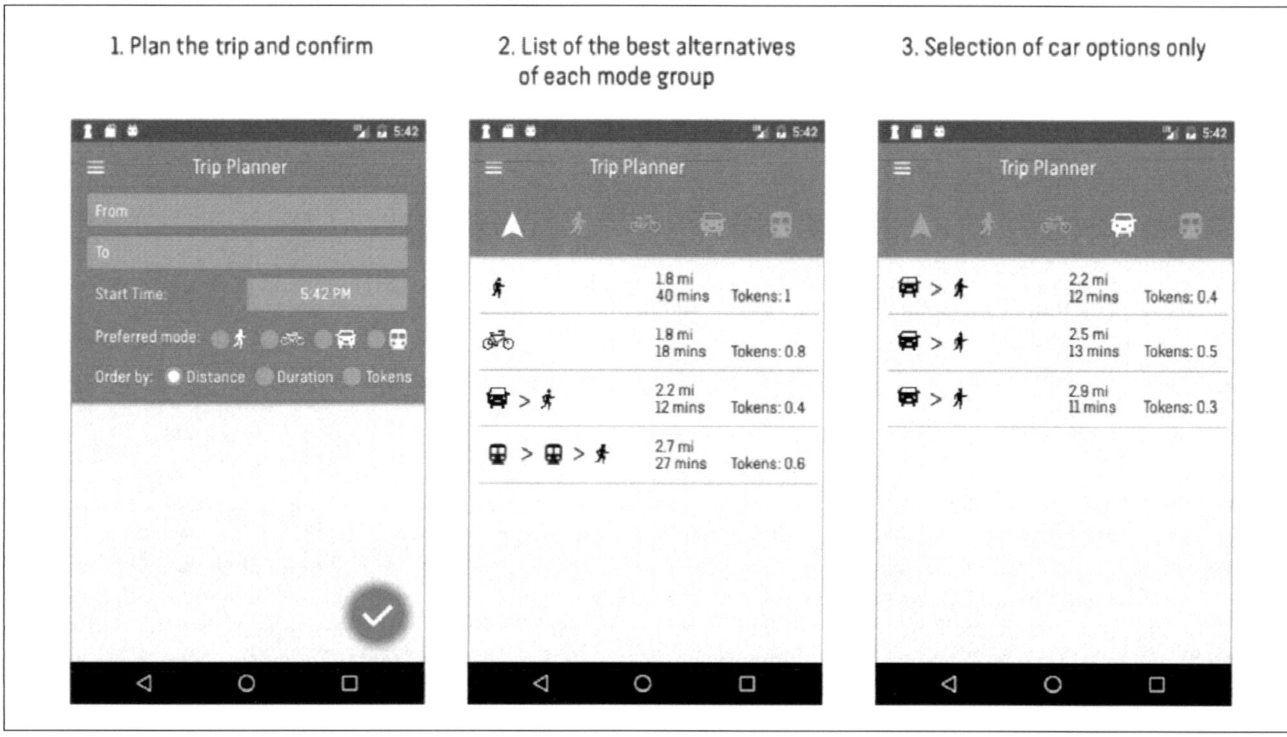

Figure 1. Tripod menu.

In Tripod, personalized menu optimization generates a personalized and optimal menu that maximizes consumer benefits in the form of a log-sum formula (*4*). Travelers make trip requests on the Tripod app, and the user-level optimization (User Optimization) generates personalized menus (see Figure 1) as a list of travel options including mode, departure time, route, energy usage, and travel incentives in the form of tokens to incentivize the user toward green travel options.

User Optimization has been presented and tested before with Monte Carlo experiments and it was shown that a logit mixture model which captures well the heterogeneity among the synthetic population will significantly improve the performance of a personalized menu. In this paper, there are several extensions of methodology and evaluation experiments. First, all the evaluations of methodology are based on real travelers and their real trips in the greater Boston area from Massachusetts Travel Survey (MTS). Second, the personalized menu optimization is enhanced with a preference updater which can incorporate the new choice data and update efficiently the estimates of preference parameters of these associated individuals. The enhancement with preference updater is tested against an option without such enhancements. Third, the proposed menu optimization with individual-level preference parameters is tested against the one with population-level preference parameters, and the added value of personalization is discussed.

The remainder of the paper is organized as follows. In the second section, relevant technical literature is reviewed including recommender systems based on discrete choice models, personalized menu optimization, and recommender systems with preference updates. In the third section, the methodology is presented including a personalized menu optimization model and a preference updater. In the fourth section, a case study based on Boston travelers and their travel history is presented to illustrate the benefits of the proposed method against its counterparts. Finally, the paper is concluded and future research directions are provided in the last section.

Literature Review

In this section, existing studies relevant to this paper are introduced. These include discrete choice models in recommender systems, assortment optimization in operations management and personalized menu optimization applied in Smart Mobility systems, and recommender systems with preference updates.

Discrete Choice Models and Recommender Systems

Recommender systems vary in different underlying techniques including widely used collaborative filtering, content-based filtering, data mining techniques, and knowledge-based and context-aware methods (*4*). With rich individual data becoming available in the era of Big Data, the ability of discrete choice models to predict individual choices has attracted interest in recommender systems (*5–7*).

Jiang et al. (*7*) used a multi-level nested multinomial logit model to measure users' preferences toward an entire recommendation list (*7*). The recommendation problem was formulated as a nonlinear binary integer programming problem. The authors pointed out that discrete choice models could incorporate product diversity in the proposed recommendation, which differed from traditional recommender systems (*7*). Choice models that can account well for consumer heterogeneity are crucial for recommender systems. There exist various recommendation techniques that account explicitly for heterogeneity in user preferences (*4, 8–10*).

Assortment Optimization and Personalized Menu Optimization

Personalized menu optimization is adapted from assortment optimization which is an emerging topic in operations management and becomes popular in many practical settings such as retailing and online advertising (*11*). The goal of assortment optimization is to select a subset of items to offer from a universe of substitutable items to maximize the expected revenue when consumers exhibit a random choice behavior. Different discrete choice models are often used to model the random choice behaviors of consumers including multinomial logit, nested logit, and logit mixture (*12–14*). The reader is referred to Kök et al. (*15*) for more details of assortment optimization literature and industry practice (*15*).

In both FMOD and User Optimization in Tripod, assortment optimization models were used to offer a real-time personalized menu (or assortment) to each individual, which was slightly different from most assortment optimization algorithms where they offered a fixed menu for all individuals to achieve maximum expected revenue. Therefore, the proposed method is called personalized menu optimization, which is the underlying recommendation method for the Tripod system in this paper.

Recommender Systems with Preference Updates

Recommender systems built on consumer preferences can update their assessments of consumer preferences as more choice data becomes available. The menu optimization or recommendation can be endogenous or exogenous with preference updates. Endogenous recommendation means that the generation of a recommended list of products is related to learning the uncertain preferences. For example, if we know little about users' preferences for a particular product, we might recommend that product in order to gain a better understanding of the associated users' preferences.

A multi-armed bandit approach deals with the trade-off between exploration (learning uncertain consumer preferences of some products) and exploitation (maximizing the reward such as hit probability in the current period) where a recommendation is endogenous with preference updates. There are different types of multi-armed bandit problems including stochastic, adversarial, and Markovian depending on the assumed nature of the reward process (*16*). Distinct strategies have been shown to be effective for each specific bandit model: the upper confidence bound (UCB) algorithm in the stochastic case, the exponential-weight algorithm for exploration and exploitation (Exp3) algorithm in the adversarial case, and the so-called Gittins indices in Markovian or Bayesian cases (*16*). For stochastic and adversarial bandit models, the reader is referred to Bubeck and Cesa-Bianchi (*16*). For Markovian bandit or Bayesian bandit models, the book by Gittins et al. (2011) is the main reference (*17*). Different bandit algorithms have been applied in many real world applications such as personalized news, display advertising, and personalized digital coupon display (*18–20*).

In this paper, we present personalized menu optimization as the recommendation engine for the Tripod smartphone app. Personalized menu optimization is built on a discrete choice model, namely logit mixture. The estimation and update of preferences are implemented through an innovative preference updater which is exogenous to the recommendation. Preference updater uses the Hierarchical Bayes estimation procedure for logit mixture as developed by Danaf et al. (*21*). The update procedure includes an offline update (estimates individual-level as well as population-level parameters) and an online update (estimates individual-level parameters using previous offline estimates of population-level parameters and leaves them unchanged) (*21*). The online update is quite efficient as it does not re-estimate population level parameters and therefore it is suitable for real-time recommendation models such as User Optimization in Tripod. More details of the preference updater applied in Tripod will be described in the methodology section.

The personalized menu optimization proposed in this paper contributes to the literature in various ways. First, the proposed method offers a real-time personalized solution while existing assortment optimization studies offer assortments for a group of individuals. Additionally, the proposed method is enhanced with a preference updater which keeps learning preferences, while existing studies often assume known preferences. Third, a logit mixture model is used in the proposed method while existing assortment optimization methods typically do not capture inter-consumer heterogeneity across the population. Fourth, this paper presents an innovative application of personalized menu optimization in the context of a Smart Mobility system with numerical experiments based on real trip data of travelers from the greater Boston area.

Methodology

In this section, the personalized menu optimization model used in Tripod is described along with its associated logit mixture model. The methodology underlying the preference updater is then introduced.

Personalized Menu Optimization

As outlined above, the personalized menu optimization is built on choice models of consumer behavior. When Tripod is in implementation, the user choice and trip behavior data can be collected through capabilities of Future Mobility Sensing (FMS) which will support the implementation of the proposed method. The data used in this case study is from MTS not FMS. The reader is referred to Zhao et al. (2015) for more details of FMS (*22*).

With the choice data from Tripod users, one can estimate a logit mixture model. Based on that, the estimates of individual level preference parameters can be obtained. Each travel option on the menu can be thought of as a combination of attributes. Let z_i denote the vector of attributes associated with travel option i. Let ζ_n denote the vector of parameters of individual n associated with corresponding attributes. The utility (U_{ni}) of option *i* for individual *n* is given by

$$U_{ni} = \zeta_n z_i \tag{1}$$

Let $v_{ni} = \exp(U_{ni})$. The probability that individual *n* will choose option *i* given the choice set C (all the options individual *n* faces) is given as follows:

$$P_i(\zeta_n) = \frac{v_{ni}}{\sum_{i' \in C} v_{ni'} + v_{opt}} \tag{2}$$

where v_{opt} is the utility of the opt-out option (opt-out means that the individual rejects all the options on the menu). With probability $\frac{v_{opt}}{\sum_{i' \in C} v_{ni'} + v_{opt}}$, individual *n* does not choose any option on the menu. The utility of the opt-out option can be individual-specific.

For User Optimization in Tripod, the objective is to maximize the consumer surplus associated with the presented menu which can be represented by the expected maximum utility of travel options on the menu. Ben-Akiva (1973) has shown that it can be represented as the log-sum formula (*23*). The app-based menu is also restricted in menu size M. The personalized menu optimization problem is

$$\max_{x_{ni} \in \{0,1\}, i \in C, \sum_{i \in C} x_{ni} \le M} \log\left(\sum_{i \in C} x_{ni} * v_{ni}\right) \tag{3}$$

where x_{ni} is the binary decision variable denoting whether the option *i* is chosen by individual *n* and v_{ni} is the exponential utility of option *i* for individual *n*.

Note that the scale parameter (marginal utility of income) is omitted in the log-sum formula as it will not affect the result of menu optimization. This model can be efficiently solved through a sorting algorithm (*4*). If the estimates of individual level preference parameters are not available, one would instead use $U_i = \mu z_i$ as the utility of each individual for travel option *i* where μ is the vector of population-level parameters. As a result, the menu optimization model in Equation 3 will generate a non-personalized menu.

Preference Updater

The preference updater is built on the Hierarchical Bayes (HB) procedure for estimating the logit mixture model (*24*). Using this procedure, the estimates of population-level as well as individual-level parameters can be obtained (*24*). The HB procedure is a "data borrowing" technique that stabilizes individual-level preferences for each individual using information from not only his/her past choices, but also those of other individuals within the same data set (*25*).

There are three sets of preference parameters in a typical logit mixture model including the individual-level mean ζ_n, population-level mean μ, and population-level covariance matrix Ω. The posterior distribution of these parameters for *n* individuals is

$$K(\mu, \Omega, \zeta_n \, \forall n | d_n \, \forall n) \propto \left\{ \prod_{n=1}^{N} \left[\left[\prod_{i=0}^{I_n} P_i(\zeta_n)^{d_{in}} \right] \mathcal{N}(\zeta_n | \mu, \Omega) \right] \right\} k(\mu) k(\Omega) \tag{4}$$

where $k(\mu), k(\Omega)$ denote the prior distribution of population-level parameters. d_{in} is a binary variable denoting that individual *n* chooses option i and I_n denotes the total number of options for individual *n*. Posterior mean is the estimate of each parameter. Since this posterior distribution does not have a closed form, Markov chain Monte Carlo is used to sample from the posterior distribution in the HB procedure.

The HB procedure includes three Gibbs sampling/ Metropolis-Hasting steps as follows.

(1) Drawing $\mu | \Omega, \zeta_n$ using a Normal Bayesian update with unknown mean and known variance.
(2) Drawing $\Omega | \mu, \zeta_n$ using a Normal Bayesian update with known mean and unknown variance.
(3) Drawing $\zeta_n | \mu, \Omega$ using a Metropolis-Hastings (MH) procedure.

After a sufficient number of Gibbs iterations, individual-specific coefficients can be estimated as the posterior means of ζ_n. The reader is referred to Ben-Akiva et al. (2016) for more details of the so-called Allenby-Train HB procedure (*26*).

Recommender systems require updating individual preferences continuously and in real time. However, re-estimating all individual and population level coefficients frequently using the Allenby-Train HB procedure might not be feasible because of computational limitations. To update individual preferences in real time, two interacting and repeated steps are applied: the offline and the online procedures (*21*).

The offline estimation procedure updates individual as well as population-level parameters for all individuals. Periodically (e.g., every week), data are pooled and all coefficients (μ, Ω, and ζ_n) are updated to reflect the effects of all choices made by all individuals since the last update. This is done by applying the full HB procedure and obtaining draws from all posterior distributions.

The online estimation procedure updates individuals' preferences in real time as they make repeated choices. The individual specific parameters (ζ_n) are updated after every choice using the same MH procedure (i.e., step 3 of Allenby-Train procedure), assuming that the population parameters μ and Ω are fixed. This update is computationally efficient, and it can be done for the specific individual who has made a new choice. Danaf et al. (2017) showed that this method yields results that are close to those obtained from a full HB procedure (*21*). Therefore, the online update is important and efficient for improving the individual-level parameters that will be used for the next choice of the user.

Boston Case Study

In this case study, the added value of using the proposed methodology is illustrated against its counterparts. The case study is built on real travelers and their travel histories provided by MTS. MTS includes activity diaries for 33,000 individuals belonging to 15,000 households. The data was collected between June 2010 and November 2011. Individuals were asked to fill out all activities they performed on a designated weekday (24 H, Monday to Friday) in chronological order, and to provide the activity location, the transport mode used to arrive at this location, the arrival and departure times, and accompanying household or non-household members.

The user population in the case study is constructed from MTS data with individuals who have made at least five tours in the specified day. There are 246 such individuals in total. Their travel choice sets for five choice situations are also constructed by using Google Map API with origin, destination, and departure times from MTS data. There are five travel modes considered in each individual's choice set including walk, bike, car, car-pool, and transit. The sizes of choice set can vary from 4 to 16 representing different routes and modes (some modes are not always available). With this data-generation procedure, 246 Tripod users are obtained and each of them is using Tripod for five tours. In this dataset, most users prefer to travel by car (>80% of total tours). Few tours are made using green transportation modes, 6% were walking. In the population, most trips (around 72%) are by car or car-pool. Few trips are made using green modes, 10% were walking. In general, the modal share is similar in the selected sample and the population.

With the choice sets and revealed choices from MTS data, one can estimate a logit mixture model capturing individuals' preferences of travel choices. The utility function is in linear specification with preference parameters including alternative specific constants (ASC) for bike, car, car-pooling, and transit, and the coefficient for travel time:

$$U_{ni} = ASC_{n,bike} * 1_{i,bike} + ASC_{n,car} * 1_{i,car} + ASC_{n,carpool} * 1_{i,carpoool} + ASC_{n,transit} * 1_{i,transit} - \exp\left(\zeta_{n,TravelTime}\right) * TravelTime_{ii} \quad (5)$$

where $1_{i,bike}$ denotes if this option i is a bike option. All ASCs are normally distributed while the travel-time coefficient is transformed to be log-normally distributed to ensure that travel time enters the utility with a negative sign as expected.

For the case study, it is assumed that Tripod user travel choice behavior follows a logit mixture model and it is also assumed that the offline update of logit mixture estimates based on choice situations from 1 to 4 are the "true" preference parameters, which are used to simulate their hit behavior. The performance of different menu optimization algorithms with different inputs of preference parameters are evaluated through a hit experiment. The hit experiment simulates 10,000 repeated choice situations with a specific generated menu for a specific individual. The individual can choose one option on the menu or opt-out, that is, reject all options on the menu. The (average) hit rate is defined as the proportion of the number of hits over all the repeated choice situations. The opt-out utility is defined as the expected maximum utility of all the options that are not on the menu, which is given by the log-sum as follows:

$$P_{opt} = \frac{\sum_{i' \in C \setminus C_M} v_{ni'}}{\sum_{i \in C} v_{ni}} \quad (6)$$

where C_M denotes the set of indices that are presented on the menu and C denotes the full choice set. Note that with FMS there will be information on which travel option is chosen by the Tripod user even if he or she opts out once the system is fully implemented.

The case study consists of two parts which represent two different stages of the Smart Mobility system. In the first part, it is considered at the beginning stage of Tripod where the travel preferences have not been fully understood. An online update is applied for estimates of individual-level preference parameters and the estimates are used for the menu optimization. Specifically, an offline update with choice data from the first two choice situations (i.e., a typical

Table 1. Population Level Parameters

	Offline12		Offline123		Offline1234	
Parameters	Mean	StDev	Mean	StDev	Mean	StDev
ASC_{bike}	−1.25	8.51	−0.82	3.03	−0.67	3.67
ASC_{car}	3.50	5.28	3.51	3.76	3.60	3.74
$ASC_{carpool}$	−1.36	3.49	−1.58	3.33	−1.54	3.90
$ASC_{transit}$	−0.41	5.92	−1.19	3.73	−0.51	4.13
$\mu_{Travel\ time}$	1.72	1.84	1.40	4.44	1.82	4.72

Note: StDev = standard deviation.

HB procedure of estimation logit mixture model) is conducted and its estimates are denoted as Offline12. Then an online update of individual level-preference parameters based on Offline12 with choice data from the third-choice situations is conducted and estimates are denoted as Online3. An offline update with choice data from the first three choice situations is also conducted and denoted as Offline123. Similarly, the "true" individual-level preference parameters are denoted as Offline1234 which is from the offline update with the first four choice situations. To evaluate the performance of personalized menu optimization with preference updater, the hit rates of generated menus based on all these different preference parameters are compared. It is believed that that with more choice data the performance will be better. In other words, Offline1234 will perform better than Offline123 and Offline12. For Online3, it is expected that the performance would be close to Offline123 and in general better than Offline12. It is not expected that for every individual Online3 will always perform better than Offline12 since Online3 does not update the population-level parameters. Table 1 shows the values of population level mean and standard deviations of all the offline updates. Note that the coefficient for travel time is log-normally distributed. It is observed that the car is much preferred by the selected population as the ASC mean for car is positive, and the heterogeneity across individuals is large as the standard deviations are high in general. The population means are similar across three offline updates especially for ASC_{car}, $ASC_{carpool}$ and $\mu_{Travel\ Time}$. The standard deviations for ASCs of Offline12 are higher than those of Offline123 and Offline1234. But the standard deviation of travel time is lowest at Offline12. The standard errors of these population-level parameters are small, ranging from 0.11 to 0.38.

In the second part, the mature stage of Tripod where the travel preferences are well understood is considered. As in part 1, the individual level parameters from Offline1234 are assumed to be the "true" preference parameters and menu optimization with individual-level parameters are considered as the personalized menu optimization. The performance of personalized menu optimization is compared with the menu optimization with population-level preference parameters from Offline1234 which is considered as non-personalized menu optimization on the fifth-choice situation. This comparison illustrates the added value of personalization.

Case Study Results

Under the described case study set-up, the comparative results are presented in this section. Figure 2 shows the comparison between Offline12, Online3, Offline123 and Offline1234. For each type of parameter estimate and each individual trip request, a menu optimization model with varying menu size constraints (from 1 to 10) is run to generate a menu. Hit experiments are then conducted on these menus. The hit rates of menu sizes from 1 to 10 are shown in Figure 2.

It is observed that menu optimization performs better with a better hit rate when there is more choice data. Offline1234 performs better than Offline123, and Offline123 performs better than Offline12. It is also observed that Online3 performs better than Offline12 in general. Online3 performs similarly to Offline123 (for menu size 1, Offline123 is slightly better and for other sizes like menu size 2, Online3 is slightly better). Note that the idea behind Online3 is to achieve similar estimates of individual-level parameters to those from Offline123 without the need for estimating the whole set of population-level parameters. Note that for menu sizes of 1, 8, 9, and 10, it is observed that Offline12 outperforms Offline123. Since we are testing menus on tour data 4. This could happen since individuals behave similarly in tours 1, 2, and 4 but a bit differently in tour 3. It is expected that choice data menu optimization with more trips would outperform that with just the first few trips.

As menu size increases, the gaps between different methods decrease since the menu starts to include more options that are also preferable and the hit rates increase. This shows that when a small subset of options is selected from the full choice set to be presented on the menu, the parameter estimates are critical. In the case of transportation, the number of options grows quickly with route, mode, departure time, and driving style dimensions. Therefore, it is crucial to have a good set of parameters for selecting a subset of options to be presented on a Smart Mobility app. As Online3 is real-time efficient and achieves similar parameters as Offline123, the proposed method is promising for getting good estimates in real-time.

Figure 3 presents the comparison between personalization and non-personalization. It can be observed from the figure that the personalized menu optimization always performs better than non-personalized menu optimization and the gaps are in general larger when the menu sizes are relatively small. The hit rates increase with the menu sizes increase. Note that even with a menu size of 10, non-personalized menu optimization has around a 3–4% possibility of missing the preferred choice of the user. By conducting two sample t tests, the hit rates between personalized and non-personalized menu optimization are not statistically significant across menu sizes 1–10. But it is statistically significant at 0.1 level across menu sizes 2–6. On the other hand, personalized menu optimization achieves a hit rate close to 1 after a menu size of 6

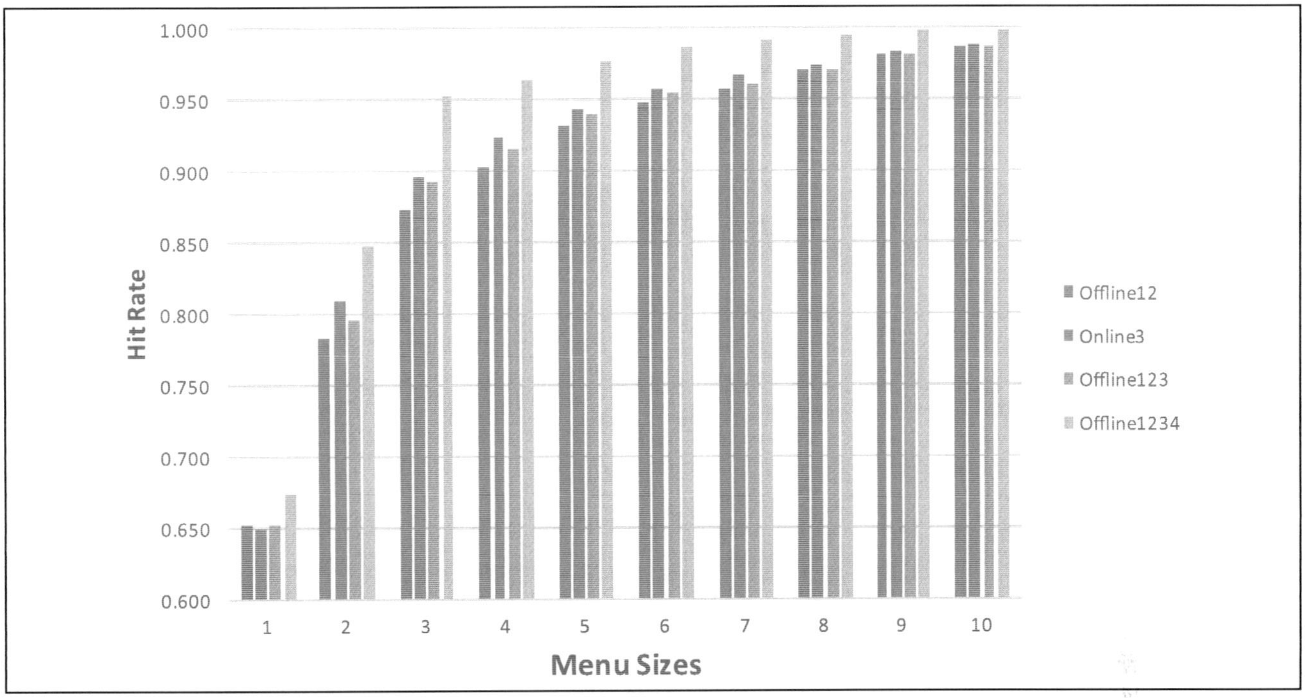

Figure 2. Added value of preference updater.

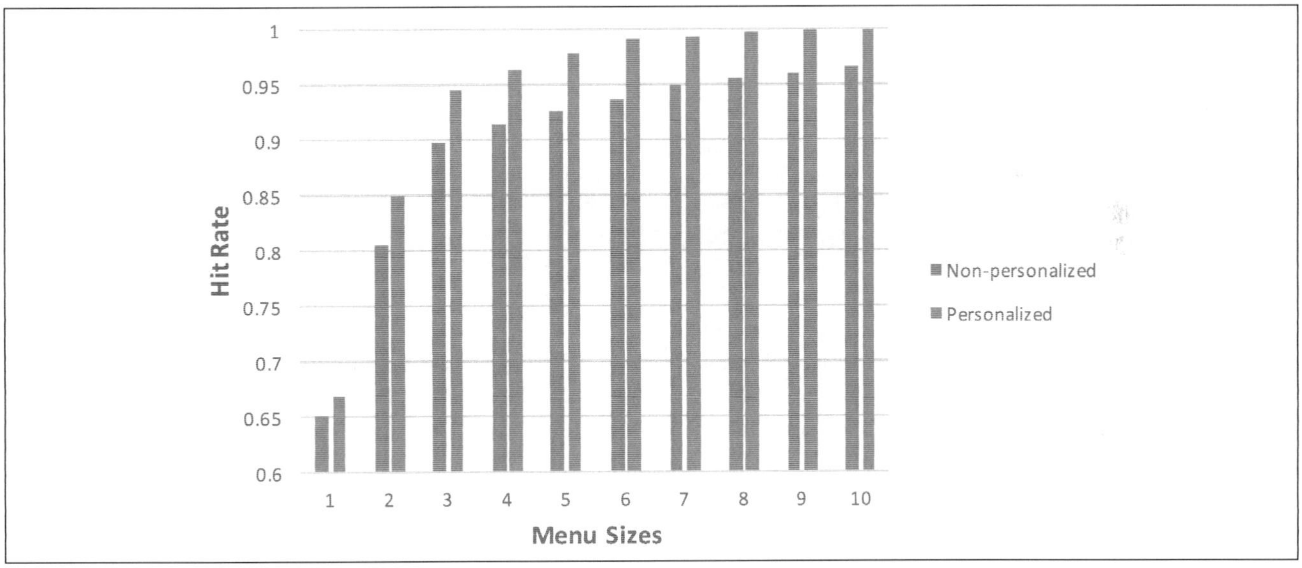

Figure 3. Added value of personalization.

in this case study. The discussion is similar to the previous case: as the selection of the menu is more critical (small menu size) the added value of personalization is higher.

To further illustrate the influence of personalized menu optimization, we analyze the hit result with menu size 1 and true choice to see how they match. For trip 4, there are 32 out of 246 individuals whose hit options are different with or without personalized menu optimization. The matching result indicates that the personalized menu optimization can capture the preferred option even with menu size 1. The result is expected since maximizing consumer surplus is similar to maximizing the hit rate. We, therefore, expect that the hit result with personalized menu optimization would be close to true choice. If the objective is to maximize something else such as energy saving, the matching result could be quite different.

Conclusion and Future Work

This paper presents a methodology of personalized menu optimization with continuously updated preference parameters through a preference updater. The methodology is flexible and can be applied in various Smart Mobility systems to offer personalized travel solutions that evolve with the traveler. The proposed methodology is evaluated with a case study based on real data from the greater Boston area and the benefits of preference updater and personalization have been shown.

This paper has contributed to the literature in several ways. First, the added value of personalization versus non-personalization has been shown which supports its applications in personalized recommendations for Smart Mobility systems. Second, the integration of preference updater with personalized menu optimization makes personalized menu optimization more sustainable. Namely, in the initial stages of Smart Mobility systems in real life, preferences are not estimated with confidence as travelers' preferences toward travel options are uncertain to system operators. Similarly, the preferences of travelers evolve as conditions change in the transportation network or based on changes in their own needs. Therefore, sustainable systems need to represent such changes. Third, the evaluation is achieved through a case study based on real trip data.

In the future, the performance of personalized menu optimization needs to be evaluated in a more comprehensive way and it should be compared with other benchmark personalized recommender systems. First, the evaluation of the proposed methodology will be conducted with pooled stated preference and revealed preference data specific to the Tripod system. Second, further investigations of the performance of the proposed methodology will be done with larger sample sizes and with different menu sizes for individuals. Third, it would be worthwhile comparing the proposed recommendation algorithm with preference updates to the multi-armed bandit approach where the recommended menu is endogenous with the preference learning. Fourth, more performance evaluations can be done with the proposed methodology supported with more complicated discrete choice models such as logit mixture model with inter- and intra-consumer heterogeneity (*27*). Fifth, how Smart Mobility Systems given personalized menu optimization and token incentives would influence travelers' choice preferences should be investigated, especially when the objective is not to maximize consumer surplus or hit rate.

Acknowledgments

This work is in the context of a research project, called Sustainable Travel Incentives with Prediction, Optimization and Personalization (Tripod), sponsored by the U.S. Energy Department's Advanced Research Projects Agency-Energy (ARPA-E). It was awarded by ARPA-E's one of the latest programs, the Traveler Response Architecture using Novel Signaling for Network Efficiency in Transportation (TRANSNET).

References

1. Rayle, L., S. Shaheen, N. Chan, D. Dai, and R. Cervero. *App-Based, On-Demand Ride Services: Comparing Taxi and Ridesourcing Trips and User Characteristics in San Francisco. Presented at* 94th Annual Meeting of the Transportation Research Board, Washington, D.C., USA, 2015.
2. Atasoy, B., T. Ikeda, X. Song, and M. Ben-Akiva. The Concept and Impact Analysis of a Flexible Mobility on Demand System. *Transportation Research Part C: Emerging Technologies*, Vol. 56, 2015, pp. 373–392.
3. Atasoy, B., T. Ikeda, and M. Ben-Akiva. Optimizing a Flexible Mobility on Demand System. *Transportation Research Record (TRR)*, Vol. 2536, 2015, pp. 76–85.
4. Song, X., B. Atasoy, and M. Ben-Akiva. Smart Mobility Through Personalized Menu Optimization. *Presented at 96th Annual Meeting of the Transportation Research Board,* Washington, D.C., USA, 2017.
5. Chaptini, B. *Use of Discrete Choice Models with Recommender Systems*. PhD dissertation. Department of Civil and Environmental Engineering, MIT, 2005.
6. Polydoropoulou, A., and M. Lambrou. Development of an e-Learning Recommender System Using Discrete Choice Models and Bayesian Theory: A Pilot Case in the Shipping Industry. In *Security Enhanced Applications for Information Systems*, 2012, ISBN: 978-953-51-0643.
7. Jiang, H., X. Qi, and H. Sun. Choice-Based Recommender Systems: A Unified Approach to Achieving Relevancy and Diversity. *Operations Research*, Vol. 62, No. 5, 2014, pp. 973–993.
8. Allenby, G., and J. Ginter. Using Extremes to Design Products and Segment Markets. *Journal of Marketing Research*, Vol. 32, 1995, pp. 392–403.
9. Rossi, P., R. McCulloch, and G. Allenby. On the Value of Household Purchase History Information in Target Marketing. *Marketing Science*, Vol. 15, No. 4, 1996, pp. 321–340.
10. Ansari, A., S. Essegaier, and R. Kohli. Internet Recommendation Systems. *Journal of Marketing Research*, Vol. XXXVII, 2000, pp. 363–375.
11. Désir, A., V. Goyal, D. Segev, and C. Ye. Capacity Constrained Assortment Optimization Under the Markov Chain Based Choice Model. Technical report, Columbia University, 2015. http://www.columbia.edu/~vg2277/constrained_MC.pdf.
12. Davis, J., G. Gallego, and H. Topaloglu. *Assortment Planning Under the Multinomial Logit Model with Totally Unimodular Constraint Structures*. Department of IEOR, Columbia University, 2013. http://www.columbia.edu/gmg2/logit_const.pdf.
13.Davis, J. M., G. Gallego, and H. Topaloglu. Assortment Optimization Under Variants of the Nested Logit Model. *Operations Research*, Vol. 62, No. 2, 2014, pp. 250–273.
14. Feldman, J., and H. Topaloglu. Bounding Optimal Expected Revenues for Assortment Optimization Under Mixtures of Multinomial Logits. *Production and Operations Management*, Vol. 24, No. 10, 2015, pp. 1598–1620.
15. Kök, A. G., M. L. Fisher, and R. Vaidyanathan. Assortment Planning: Review of Literature and Industry Practice. In *Retail Supply Chain Management*, Springer US, 2008, pp. 99–153.
16Bubeck, S., and N. Cesa-Bianchi. Regret Analysis of Stochastic and Nonstochastic Multi-Armed Bandit Problems.

Foundations and Trends in Machine Learning, Vol. 5, No. 1, 2012, pp. 1–122.

17.Gittins, J., K. Glazebrook, and R. Weber. *Multi-Armed Bandit Allocation Indices*. John Wiley & Sons, New York, 2011.

18Li, L., W. Chu, J. Langford, and R. E. Schapire. A Contextual-Bandit Approach to Personalized News Article Recommendation. *Proc., 19th International Conference on World Wide Web*, ACM, 2010, pp. 661–670.

19. Schwartz, E. M., E. T. Bradlow, and P. S. Fader. Customer Acquisition Via Display Advertising Using Multi-Armed Bandit Experiments. *Marketing Science*, Vol. 36, No. 4, 2017, pp. 200–522.

20. Song, X. *A Bayesian Bandit Approach to Personalized Online Coupon Recommendations*. Master thesis, Massachusetts Institute of Technology, 2016.

21. Danaf, M., F. Becker, X. Song, B. Atasoy, and M. Ben-Akiva. *Personalized Recommendations using Discrete Choice Models with Inter-and Intra-Consumer Heterogeneity*. Working Paper, MIT, 2017.

22. Zhao, F., F. C. Pereira, R. Ball, Y. Kim, Y. Han, C. Zegras, and M. Ben-Akiva. Exploratory Analysis of a Smartphone-Based Travel Survey in Singapore. *Transportation Research Record: Journal of the Transportation Research Board*, 2015. 2494: 45–56.

23.Ben-Akiva, M. E. *Structure of Passenger Travel Demand Models*. Doctoral dissertation, Massachusetts Institute of Technology, 1973.

24.Allenby, G. M., and P. E. Rossi. Marketing Models of Consumer Heterogeneity. *Journal of Econometrics*, Vol. 89, No. 1, 1998, pp. 57–78.

25.Orme, B., and G. Baker. *Comparing Hierarchical Bayes Draws and Randomized First Choice for Conjoint Simulations*. Sawtooth Software Research Paper Series, 2000.

26. Ben-Akiva, M., D. McFadden, and K. Train. *Foundations of Stated Preference Elicitation, Consumer Choice Behavior and Choice-Based Conjoint Analysis*. Working Paper, University of California, Berkeley, 2015.

27. Becker, F., M. Danaf, X. Song, B. Atasoy, and M. Ben-Akiva. *Hierarchical Bayes Estimator of a Logit Mixture with Inter- and Intra-Consumer Heterogeneity*. Working paper, MIT, 2016.

The Standing Committee on Emerging and Innovative Public Transport and Technologies (AP020) peer-reviewed this paper (18-05753).

Article

TRR
JOURNAL OF THE TRANSPORTATION RESEARCH BOARD

Transportation Research Record
2018, Vol. 2672(8) 608–618

DOI: 10.1177/0361198118758630
journals.sagepub.com/home/trr
SAGE

Automated Mobility-on-Demand vs. Mass Transit: A Multi-Modal Activity-Driven Agent-Based Simulation Approach

Rounaq Basu[1], Andrea Araldo[1], Arun Prakash Akkinepally[1], Bat Hen Nahmias Biran[2], Kalaki Basak[2], Ravi Seshadri[2], Neeraj Deshmukh[2], Nishant Kumar[2], Carlos Lima Azevedo[1], and Moshe Ben-Akiva[1,2]

Abstract

Among the new transportation services made possible by the introduction of automated vehicles, automated mobility-on-demand (AMoD) has attracted a lot of attention from both industry and researchers. AMoD provides a service similar to taxi or ride-sharing services, while being driverless. It is expected to attract a huge fraction of travelers currently using mass transit or private vehicles and will have a disruptive effect on urban transportation. While most studies have focused on the operational efficiency of the technology itself, our work aims to investigate its *impact* on urban mobility. Our contribution is two-fold. First, we present a flexible AMoD modeling and simulation framework developed within a multi-modal agent-based urban simulation platform (SimMobility). The framework allows the detailed simulation and assessment of different AMoD operations together with an activity-based framework that accounts for changes in demand, such as activity participation, trip making, mode, destination, or route choice decisions. Second, we focus our attention on the role of mass transit in a futuristic urban system where AMoD is widely available. Mass transit is already challenged by current ride-sharing services, for example, Uber and Lyft, which provide comparatively better and cheaper services. This trend will plausibly be exacerbated with the introduction of AMoD, which may indirectly act as a replacement to mass transit. Our simulation results show that mass transit is irreplaceable, despite the high efficiency of AMoD, in order to avoid congestion and maintain a sustainable urban transportation system with acceptable levels of service.

New technologies and the ubiquitous use of smartphones have opened the possibilities for more convenient, affordable, fast, and safe options in urban transportation. This has led to the emergence of mobility-on-demand (MoD) systems, such as Uber and Lyft, which aim to provide fast and reliable mobility that is catered to individualistic needs. At the same time, automated vehicle (AV) technology has advanced at an impressive pace. Corporations, such as Google and Tesla (*1*), have been in a race to develop a fully automated vehicle. The combination of these two promising technologies, known as automated mobility-on-demand (AMoD), has recently attracted interest among both researchers and industry (for example, Uber (*2*) has started testing AV programs in several states in the US).

The term AMoD (*3*) designates a service similar to MoD or taxi, with the difference that vehicle operations are driverless. AMoD combines the benefits of MoD and AVs in several aspects. First, operational cost is drastically reduced, given the complete removal of driver labor costs and superior energy efficiency of AVs. Furthermore, negative externalities, such as emissions, travel time uncertainty, and accidents, may also reduce, as already observed for MoD (*4*) and for AVs (*5*). The latter also observes that AMoD will increase road network utilization, making it possible to transport more passengers with less congestion, with respect to privately owned cars. Fagnant and Kockelman (*6*) found that AV benefits would amount to between $2,690 and $3,900 annually per vehicle, incorporating decreases in insurance, parking costs, and traffic congestion.

It is clear that AMoD is a disruptive technology that will deeply impact the transportation system. Most of the literature (*7–11*) has focused on the efficiency of AMoD and AVs, in terms of road movement and fleet management. However, only a handful of recent studies have shown the importance

[1]Intelligent Transportation Systems Lab, Massachusetts Institute of Technology, Cambridge, MA
[2]Singapore-MIT Alliance for Research and Technology, Singapore

Corresponding Author:
Address correspondence to Rounaq Basu: rounaq@mit.edu

of considering changes in demand when designing such disruptive technologies. Azevedo et al. (*12*) assessed the sensitivity of AMoD fleet sizes and parking station configurations on individual mobility patterns, specifically with regard to modal shares, routes, and destinations. Martinez and Viegas (*13*) and ITF (*14*) focused on the mobility, mode performances, and environmental changes that might result from large-scale adoption of shared and self-driving vehicle fleets in Lisbon and Helsinki, respectively.

Cities are increasingly focused on developing sustainable urban transportation systems, with attempts to increase quality and access to public transport and non-motorized modes. In light of these initiatives, the question of the hour is: *How will AMoD impact an urban transportation system?* Merely studying efficiency of AMoD operations is not sufficient to answer this question. The demand for AMoD depends on both its attractiveness compared with other existing modes (such as private automobiles and mass transit) and on the creation of new trips due to enhanced accessibility. *Would AVs lead to extreme on-road congestion if mass transit were to be discarded?* The performance of the overall transportation supply system is of primary interest to urban policymakers. *Will the ridership (and therefore revenue) of mass transit services decrease and, as a consequence, potentially lead to a downward spiral of reduced investment and level of service?*

Questioning the role of mass transit in the future thus becomes increasingly relevant. Debates are ongoing on whether current MoD will replace mass transit (*15,16*). Researchers and transit authorities indicate ride-sharing as one of the possible causes for decline in transit ridership (*17–19*). Polzin (*20*) warns that the loss in revenue resulting from this decline would create a vicious cycle of decreased investment in public transport, which would in turn slow down infrastructure improvement and hamper the quality of service. Consequently, mass transit would experience an even larger decline in ridership. If these concerns are relevant now (with MoD), they will be even more so with the introduction of AMoD, which is expected to lead to a rapid improvement of service performance and user convenience. These concerns have been confirmed in the literature as well (*21*).

To answer questions like these, we follow an *integrated* approach, which embraces the operational aspects of the new technology while considering the response of individuals to their availability. To that effect, this work provides a twofold contribution. First, we introduce a flexible framework, developed in the open-source platform SimMobility (*22*), for the detailed simulation of the operation of AMoD services. Designed to be modular and easily extensible, we expect the research community to greatly benefit through rapid development, testing, and comparison of results. Current research works focusing on AMoD simulations lack in this aspect, which makes repeatability of experiments and verification of results almost impossible.

Second, we apply this framework to a case study, where we investigate the role of mass transit in future urban transportation systems. *Will mass transit have the same vital importance as it currently does in the future when AMoD will be easily deployable?* Using our simulation framework in a case study, we show that, in dense urban environments with infrastructural constraints, a mass transit system is still necessary to cope with high mobility demand. However, AMoD can also contribute greatly in improving urban mobility if integrated with mass transit.

Methodology and Framework

Along with a brief introduction to SimMobility, we provide an outline for the design and implementation of AMoD services in our simulation framework.

Overview of SimMobility

SimMobility is a multi-scale agent-based simulation platform that incorporates time-scale dependent behavior modeling through activity-based frameworks. Considering land-use, transportation and communication interactions, SimMobility can be used for a variety of applications ranging from implementation of intelligent transportation systems and estimating vehicular emissions to evaluation of alternative future scenarios and generation of innovative policy and investment strategies. SimMobility encompasses three major components.

1. *Long-Term (LT):* A macroscopic land-use transport simulator that involves creation of a synthetic population along with housing location choice, job location choice, and car ownership choice at the temporal scale of days to months to years (*23,24*);
2. *Medium-Term (MT):* A mesoscopic supply simulator coupled with a microscopic demand (daily activity) simulator that involves mode choice, route choice, activity pattern, and incident-sensitive (re)scheduling at the temporal scale of seconds to minutes (*25*);
3. *Short-Term (ST):* A microscopic traffic simulator that involves lane changing, gap acceptance, route choice, and acceleration–braking behavior at the temporal scale of a fraction of a second (*26*).

This paper will focus on the MT simulator only, as the AMoD framework has been designed for MT-level decisions. While it would be interesting to observe the effect of inclusion of such services on LT decisions, it serves as a direction for future research. The MT modeling framework is an integration of activity-based demand modeling systems with dynamic traffic assignment for modeling supply decisions. There are three components in the MT framework that interact with each other leading to feedback mechanisms for agent decisions.

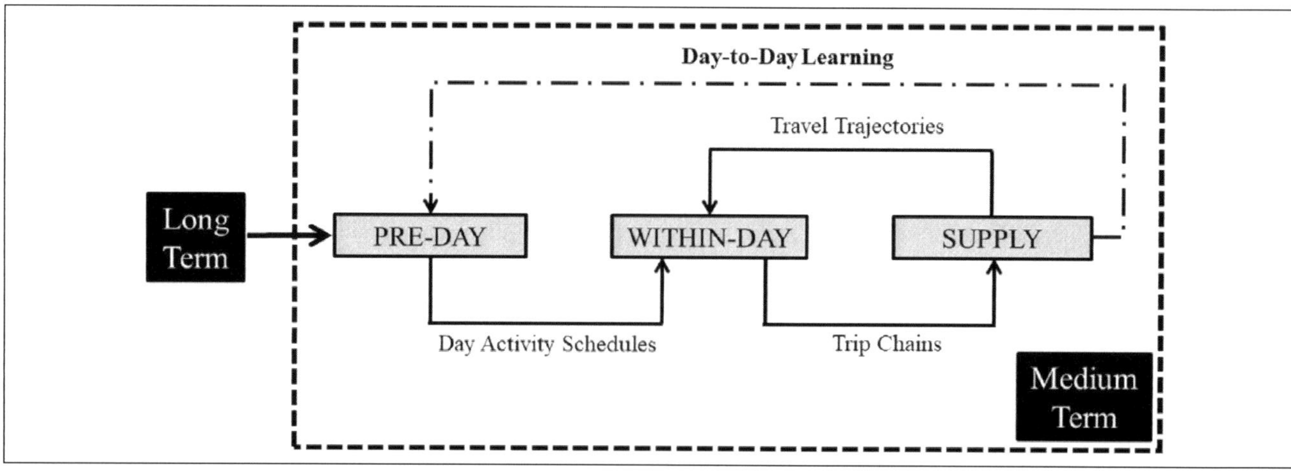

Figure 1. Feedback loop mechanism in SimMobility Medium-Term (adapted from Lu et al. [25]).

1. *Pre-day* models daily activity travel patterns at the individual level for a synthetic population;
2. *Within-day* simulates departure time choice and route choice decisions incorporating en-route decisions such as re-scheduling;
3. *Supply* provides network attributes and supply-based models for both private and public transportation modes.

AMoD Framework

After providing a brief overview of how AMoD fits into SimMobility, we proceed to explain the design and implementation in further detail.

Overview. The AMoD service is first made available in the pre-day, that is, individuals from the synthetic population have access to it in their choice set while constructing their activity schedules. At the pre-day level, each individual makes a mode choice which is modeled using the random utility framework (*27*). The systematic utility of each mode is computed as a weighted sum of the following parameters: *total travel time*, which consist of in-vehicle travel time, waiting time, and walking time; *travel cost*, which consists of road tolls, ticket prices, service prices, and so forth; *number of transfers in public transport, a dummy variable for the CBD, vehicle ownership, age* and *gender*. A probability is computed for each mode depending on their utility, assuming a known distribution for the error term. Then, each individual makes mode choices based on these probabilities and constructs an activity schedule for the day, including the movement between locations as well as their preferred mode. We make assumptions on the generalized travel cost of AMoD (see Experimental Design section) based on literature since relevant data are not available for this mode. The reader should refer to Li and Biran (*28*) for further details about the underlying behavioral models.

Travel times and waiting times are historical values, which are updated with the use of the day-to-day learning component shown in Figure 1, when "permanent or recurrent" changes on the system status quo are observed. This module feeds back supply outputs in the form of zone-to-zone travel times and costs to the pre-day level to update travelers' knowledge of the system. After a simulation is complete, the simulated costs, waiting, and travel times are recorded, aggregated at the zone-to-zone level and fed back to the pre-day module for its next run (day), that is, weighted averages of values from the previous and current iterations are used to update the historical values in the database. The pre-day simulation is then re-launched, in which the individual re-plans the activities and trips for the next simulation iteration (day). Assuming that there is indeed a change on travelers' knowledge of the system, multiple iterations of the day-to-day learning are expected until convergence of these transferred values is achieved.

Before expanding on the implementation, it is worth emphasizing the benefits of this approach. By using such a framework, we relax the common assumption that demand is fixed and pre-defined. In fact, in a multi-modal urban transportation system, such static conditions are hardly observed and it is crucial to consider partial behavioral shifts (based on differential utility), as our approach does, to provide a more realistic approach for studying the adoption and impact of AMoD at the system level. Furthermore, our framework is flexible enough to allow investigations under different hypotheses. For instance, we can determine a feasible price for AMoD operators by testing scenarios with different levels of price. The major contribution of our framework is to allow for the evaluation of impact of performance metrics on user adoption and vice-versa. For example, a certain AMoD operational strategy may be widely accepted leading to large demand but is only able to maintain a required level of service till a certain level of demand. Therefore, the target price can be adjusted accordingly through experiments. We cannot

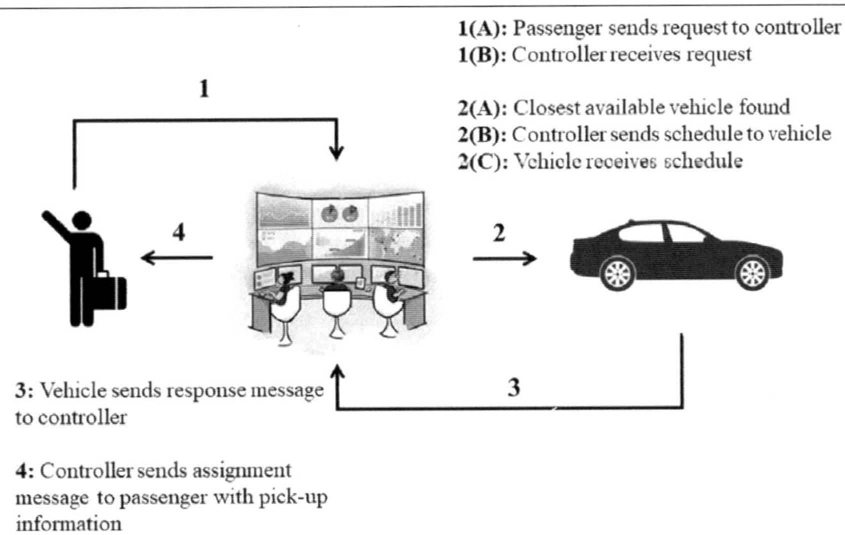

Figure 2. Control flow diagram for the AMoD framework.

present such analyses in this paper for brevity, but we emphasize that our framework enables this kind of research.

Model and Design of AMoD in the Supply Simulator. In the supply simulator, the AMoD framework is composed of three components: the *passenger*, the *vehicle*, and the *controller*. The different interactions between them are shown in Figure 2 through a control flow diagram. Our framework allows for simulating different AMoD services, each controlled by a separate controller, at the same time. The passenger sends a request to an AMoD service, specifying the intended pick-up and drop-off locations. The requests directed to an AMoD service are periodically processed by the relative controller. The processing consists of the creation of a *schedule* for each vehicle subscribed to that service. A schedule is a sequence of commands to instruct the vehicle. They can be of type *pick-up a passenger*, *drop-off a passenger*, *cruise in a certain zone*, and *park*. Each schedule is then sent to the intended vehicle, which will execute the sequence of commands.

It is noteworthy that this mechanism is naturally suited to implement ride-sharing, that is, serving two or more users with the same vehicle at the same time. For example, if we want passenger 1 to share a ride with passenger 2, it suffices to send to a vehicle a schedule containing the following commands: *pick-up passenger 1*, *pick-up passenger 2*, *drop-off passenger 1*, *drop-off passenger 2*. Since the passengers, vehicles, and the controller are interacting asynchronously through messages, the simulation is made fully parallelizable, leading to significant savings in computation time.

While unable to include a detailed description of the ride-matching algorithm here for brevity, we refer the reader to Araldo et al. (*29*). The algorithm of the AMoD controller tries to assign each incoming request to an available vehicle periodically. The match is feasible if adding the pick-up and the drop-off of the new passenger to the schedule of the vehicle satisfies the following constraints.

1. The *waiting time* of all the passengers already included in the schedule, as well as of the new passenger, must be below a certain threshold.
2. For each passenger, we first estimate the *minimum time* that the passenger would spend for their trip if a vehicle were only serving them, going from their origin to their destination directly. Next, we compute the estimate of the time between their pick-up and drop-off, as specified in the schedule, which we call *serving time*. The difference between serving time and minimum time is the *additional delay*, which accounts for detours made due to the fact that the trip is not exclusively serving a single user, but is shared with others. We also impose that the additional delay be below a certain threshold.

We consider 10 minutes to be the threshold for both parameters and the sharing capacity to be 2 passengers in this study.

Case Study: AMoD and Mass Transit

We now apply the framework described above to consider the role of mass transit in future urban mobility. *Will mass transit remain as relevant when AMoD is widely available?* The answer to this question will drive future investment in transportation infrastructure, which is the major motivation behind this case study.

Experimental Design

We designed three different scenarios which are separately simulated and subsequently compared. We consider a *Base Case* scenario where smart mobility services have not been introduced yet. The available modes are single occupancy car (*Car*), pooling with one extra passenger (*Carpooling 2*), sharing with two extra passengers (*Carpooling 3*), public bus (*Bus*), Mass Rail Transit (*MRT*), traditional taxis (*Taxi*), motorcycle (*Motorcycle*), and walking (*Walk*). The modal availabilities are in accordance with our study area, which we describe in the following section.

In the *Without Mass Transit* scenario we introduce AMoD in lieu of mass transit modes such as Bus and MRT. We assume that the behavioral preferences toward AMoD are the same as that of taxi and that the AMoD price is reduced by 40% (w.r.t. taxi), based on the literature *(30)*. Network performance in the form of travel times were passed to the day-to-day learning module, which feeds back to the pre-day model to update individual choices.

A final scenario, termed *With AMoD*, is constructed with the inclusion of AMoD along with availability of all modes from the base case scenario. Apart from the traditional on-demand door-to-door service, AMoD can also augment the rail service by providing first–last mile connectivity. This would cater only to trip legs that have trip-ends at MRT stations, in which ride-sharing is also heavily encouraged. The initial parameter values for AMoD remain the same; however, the waiting time is increased by 5 minutes for each trip leg compared with walking to reflect the delay passengers would experience while waiting for the vehicle to arrive. Overall travel time for the first and last mile of the trip will be reduced by 87.5% as compared with walk, which is as a result of our modification of the access/egress time value in the utility equation for MRT from a generic walking speed (4 km/hr) to an average AMoD speed (40 km/hr). Note that we use the average AMoD speed for the access/egress time in the utility equation only for the first run of the process (as an a-priori value). Subsequent runs have updated AMoD access/egress times obtained from the supply simulation. The feedback methodology described earlier is utilized in this scenario as well.

We conducted 24-hour simulations for each scenario in SimMobility using the study area mentioned in the following sub-section. The initial simulation was used to generate real-time parameter estimates such as travel times for different modes, which were then used as feedback for the choice models in the pre-day component (recall the *feedback loop* described earlier). The results of subsequent simulations, which represent the actual experiments, are summarized below.

Study Area

This case study was tested using a prototypical city, *Virtual City*, which consists of a moderately sized network and population, generated so as to resemble land-use patterns, travel behavior, and activity patterns observed in Singapore. The total population in our Virtual City is 351,000 (~10% of Singapore) with an average tour rate of 1.14 per individual. The road network consists of 95 nodes (intersections), 286 segments (road sections with homogeneous geometry), and 254 links (groups of one or more segments with similar properties). There are 12 bus lines, each having a constant service headway that ranges from 3 minutes to 9 minutes, spanning the region with 86 bus stops. Virtual City also has 4 MRT lines with a total of 20 subway stations, and 24 Traffic Analysis Zones (TAZs) overall. Relevant maps are shown in Figure 3.

Results for Virtual City

As mentioned in previous sections, a strong suit of SimMobility is the flexibility to adapt demand models as a part of the pre-day simulation component (that generates the day activity schedule). Therefore, we are able to observe the changes in mode choice while comparing the pre-day simulation results for the three scenarios. The temporal distribution of total trips, obtained through the time-of-day (TOD) choice model is shown in Figure 4. Since the TOD model is largely dependent on individual socio-economic characteristics as independent variables, we do not witness significant changes in the pattern across the three scenarios, i.e., 4(*a*), 4(*b*), and 4(*c*).

From Figure 4, we can see that the demands for Taxi and AMoD during both morning and evening peak periods are significantly higher compared with the remainder of the day. Therefore, we adopt a dynamic fleet sizing strategy, whereby the fleet size is selected to match or be slightly higher than the demand for the remainder of the day but will only match about 10–15% of the demand during peak periods. Despite this seemingly low fleet size, the network experiences considerable levels of congestion, as we discuss in the following sub-sections, making addition of further vehicles to the fleet counter-productive.

We observe from Figure 5 that the sharpest increase in mode share for the *Without Mass Transit* scenario occurs in the case of AMoD. AMoD is also found to be more preferable to Taxi because of lower tariffs for both the futuristic scenarios. The consistently high Walk share is in agreement with a study by the Land Transport Authority (LTA) in Singapore that reported several South-East Asian hubs like Bangalore, Delhi, Mumbai, Beijing, Osaka, Shanghai, Tokyo, and Singapore itself witnessing high mode shares for walking, ranging from 20% to 30% (*31*). Considering the *With AMoD* scenario, it is clear that the drop in Bus share is compensated by AMoD. An interesting point of note is a rise in the choice of MRT with the introduction of AMoD, thereby indicating that availability of first–last mile connectivity does play a role in enhancing the choice of mass transit.

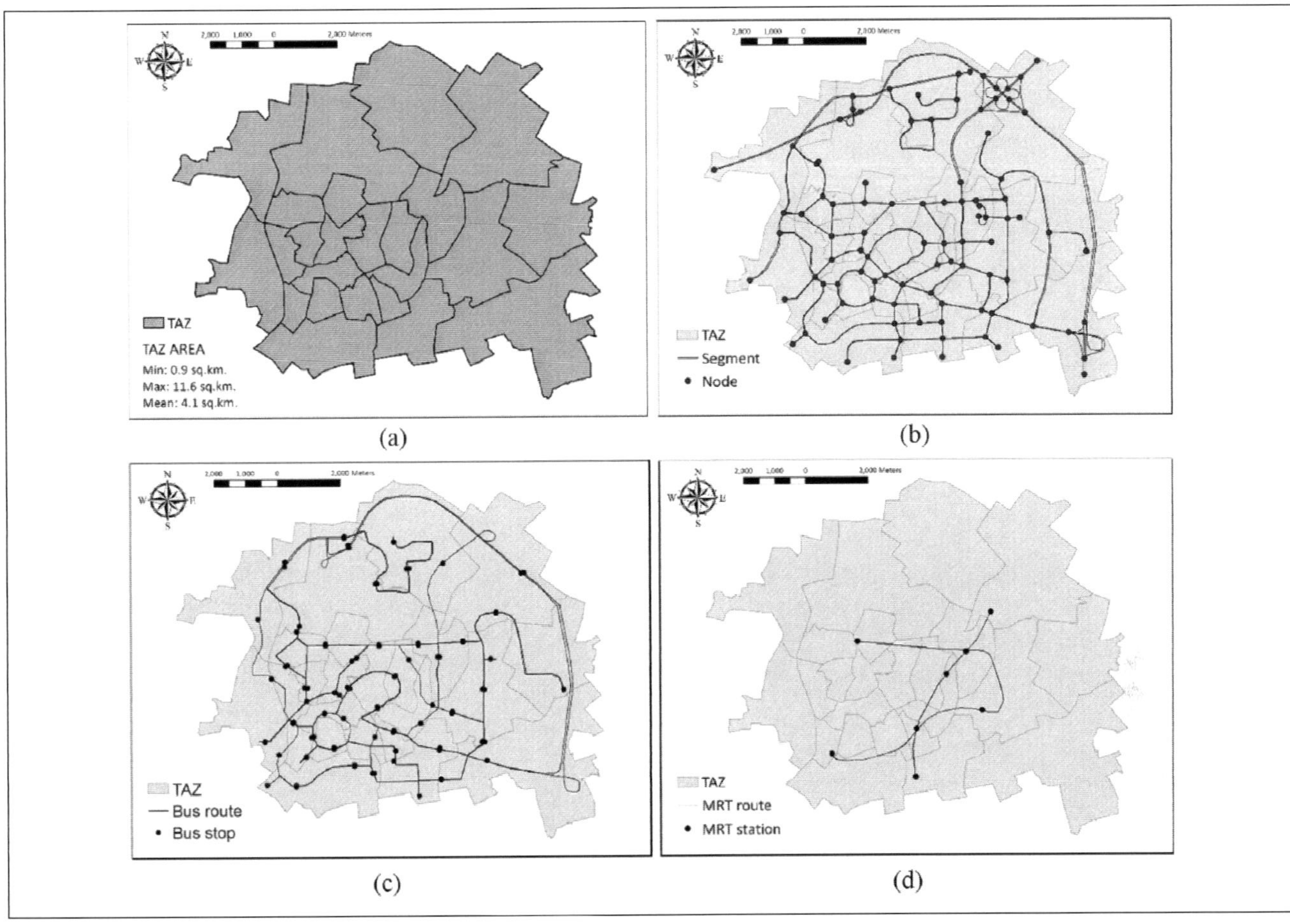

Figure 3. Maps of the Virtual City showing: (a) TAZ boundaries, (b) road network, (c) bus network, and (d) MRT network.

To evaluate the performance of the whole transportation system on a typical day, three categories of metrics are defined.

1. *System performance metrics* reflect the system's overall behavior for a particular scenario. We are specifically interested in identifying the levels of congestion. To this aim, we compute segment density and queuing. Segment density is defined as the number of aggregated vehicles on a road segment over its length, which can be expressed in veh/lane/km. As indicators of queuing in the network, we investigate proportion of segments experiencing queuing along with their average queue lengths.
2. *User experience metrics* reflect an individual's quality of travel in the network. We compute parameters of interest such mode-specific in-vehicle travel times (IVTTs) and waiting times (WTs) at the trip level. Longer IVTTs might point toward high levels of congestion in the network, and longer WTs would indicate that the current supply parameters such as fleet size are not adequate for the existing demand. The total travel time (TT) is also computed as the sum of IVTT and WT, except for Car which does not have WT, meaning that TT is equal to IVTT.
3. *AMoD performance metrics*: Introduction of AMoD warrants a separate examination of its performance. A preliminary evaluation of the ride-sharing algorithm can be made through observing the average vehicle occupancy and the proportion of shared trips. We would also like to explore the vehicle kilometers of travel (VKT) (which acts as a proxy indicator of emissions) by AVs while in different states.

System Performance Metrics. As the morning peak has a higher trip volume than the evening peak (Figure 4) and we wish to examine the overall system in its most critical state, we drop the evening peak period from further analysis. Figure 6 shows segment density maps for Virtual City during the morning peak period and the off-peak period across the three scenarios. It is easy to observe that congestion for *With AMoD* scenario is comparatively higher than the base case, while the *Without Mass Transit* scenario experiences extremely high levels of congestion. It is interesting to note in the *Base Case* and *With AMoD* scenarios that congestion

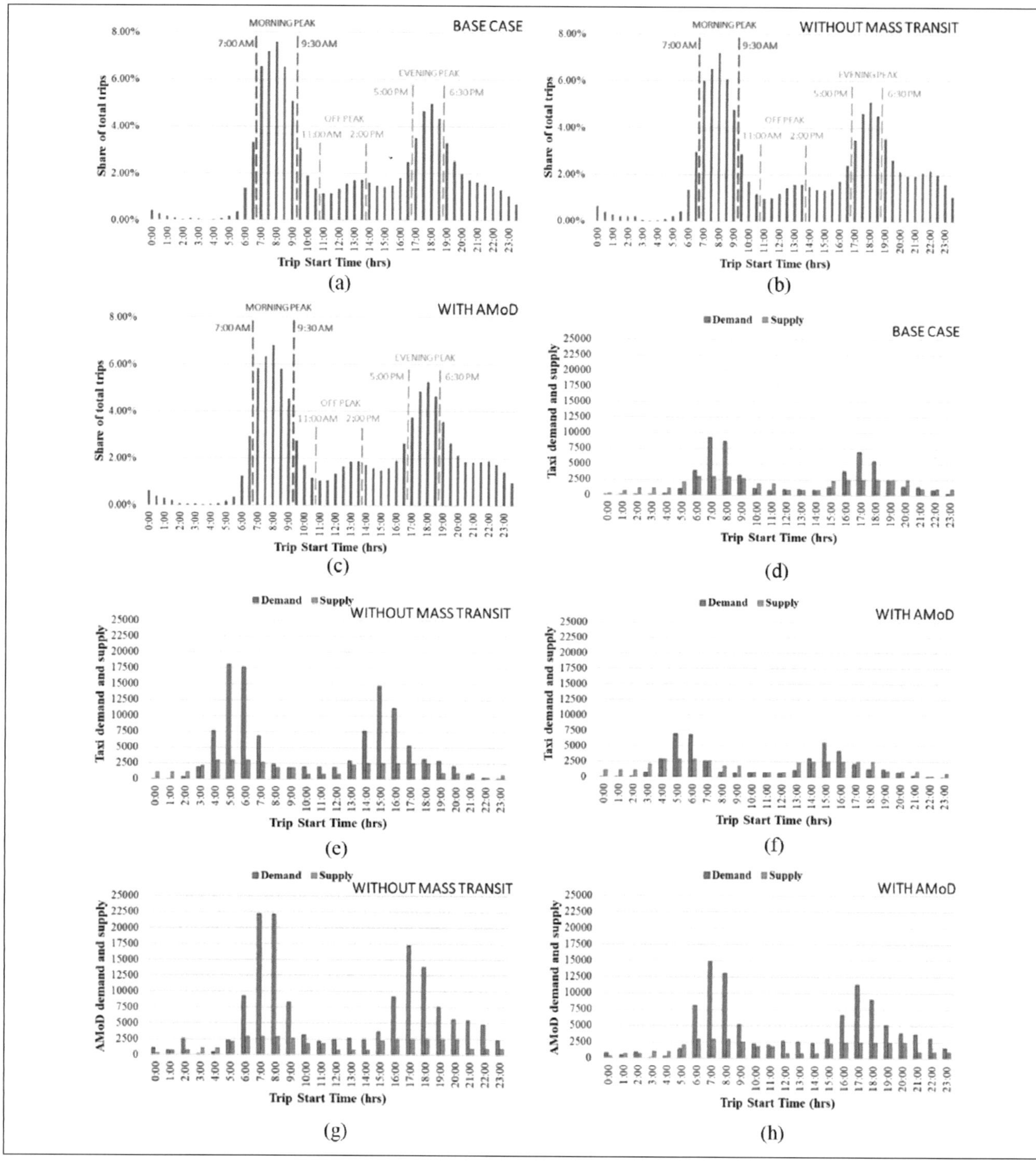

Figure 4. Temporal distribution of total trips (a, b, c), taxi demand and supply (d, e, f) and AMoD demand and supply (g, h) for all possible scenarios.

drains away during the off-peak period, as one would expect. However, in the *Without Mass Transit* scenario, congestion deteriorates considerably.

To examine the queuing patterns, the number of queued vehicles and total queue length were averaged over the morning peak and off-peak periods, following which we focused only on the segments which had a queue. For the morning peak period, 64% of segments experienced queuing with an average queue length of 6.6 m in *Base Case*. This increased to 69% with a massive jump in queue length to 229 m for *Without Mass Transit*, while *With AMoD* had a queuing rate of 67% with an average queue length of 49 m.

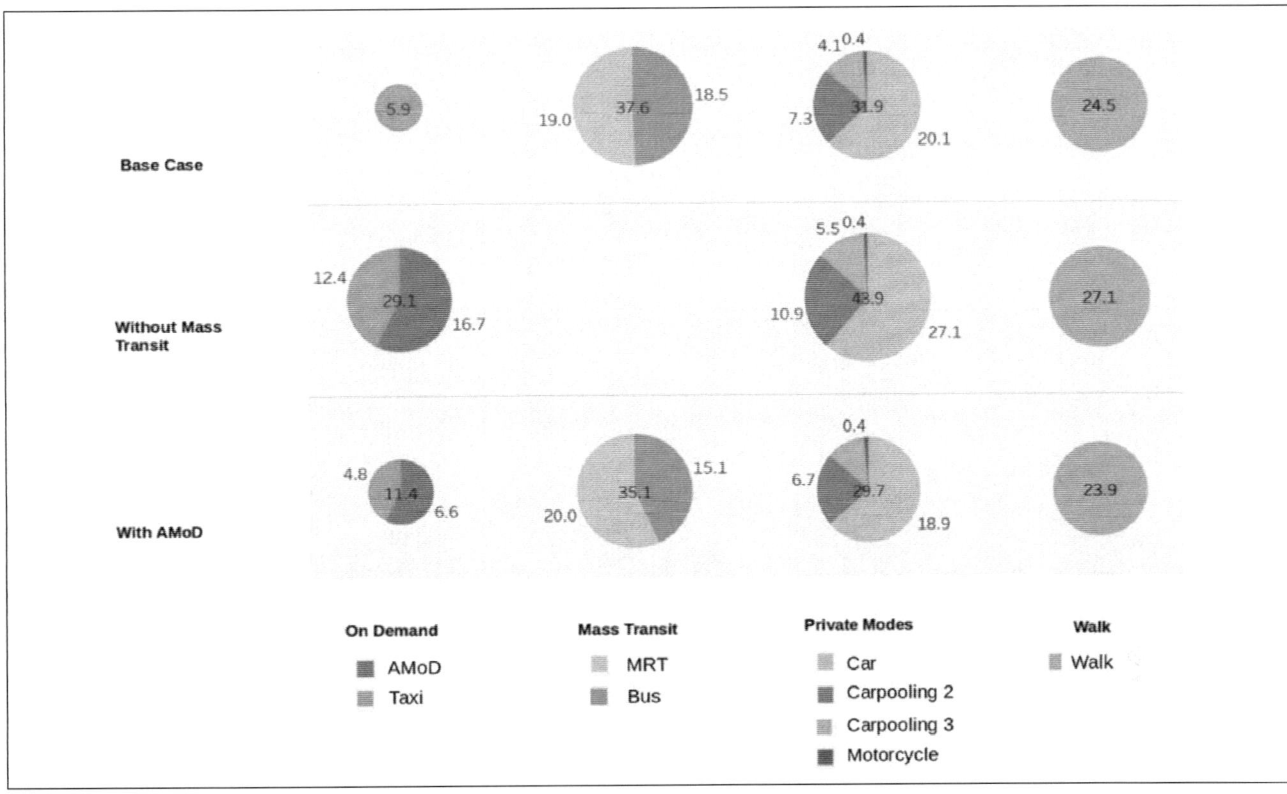

Figure 5. Changes in modal shares across the three scenarios.

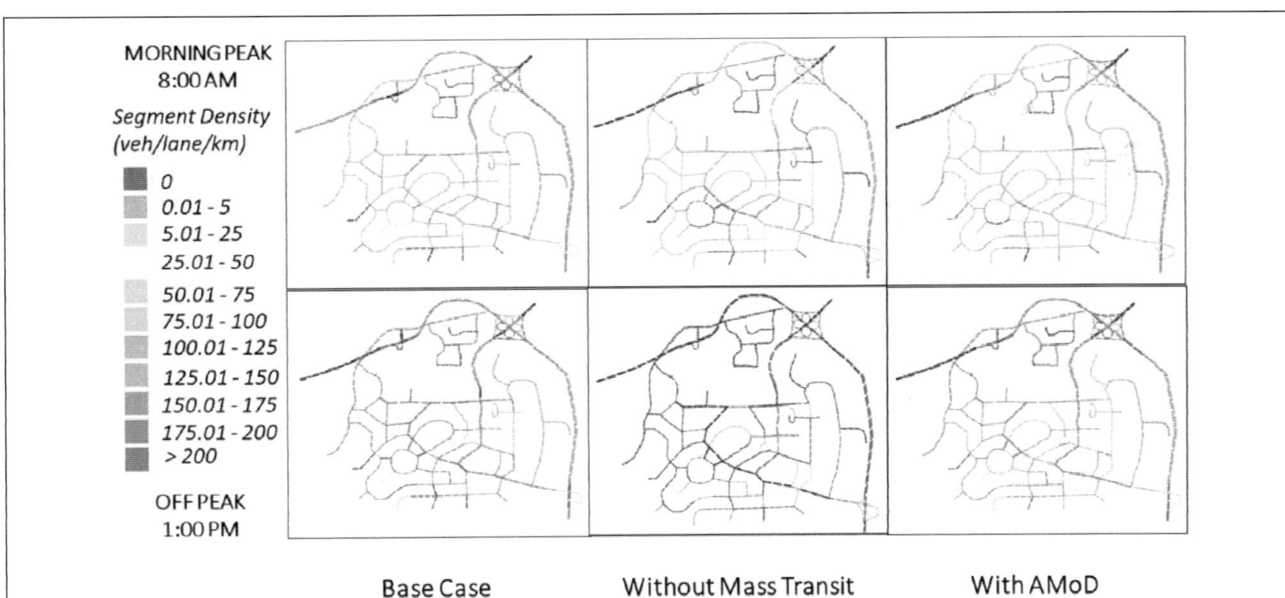

Figure 6. Congestion maps for peak and off-peak periods.

Similar trends, although of lower magnitudes, were observed for the off-peak period as well. *Base Case* had 28% segments with 0.1 m as the average queue length, indicating minimal levels of congestion. As observed from the congestion maps, *With AMoD* also experienced minimal congestion with 49% of segments experiencing queuing but with an average queue length of only 0.2 m. The extent of unexpectedly high congestion in the off-peak period for *Without Mass Transit* can be witnessed through a queuing rate of 41% with a staggeringly large average queue length of 1100 m.

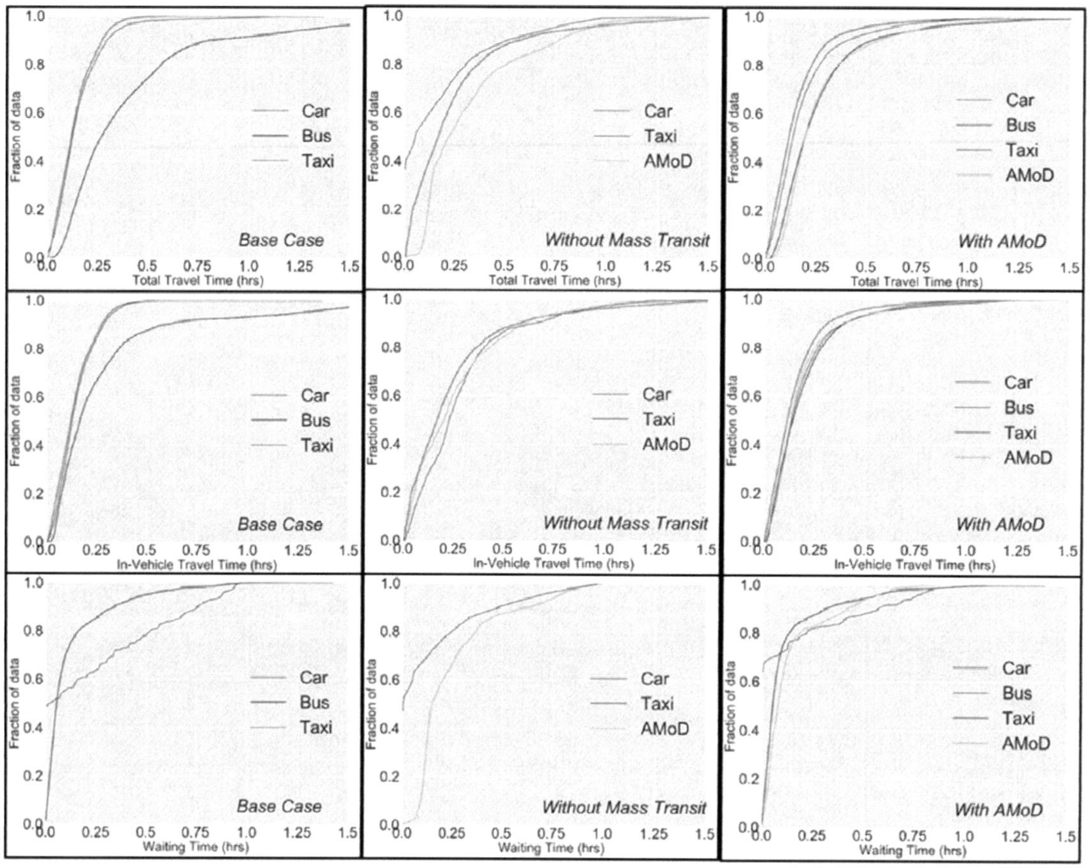

Figure 7. Overall user experience metrics for different modes across scenarios.

User Experience Metrics. An important parameter that define a user's travel experience is travel time, which has two components: IVTT and WT. We present cumulative distribution functions for mode-specific travel times in Figure 7. We keep MRT out of this analysis, since it operates on a fixed schedule-based service and its user experience is not affected by on-road congestion. Although increased demand would require an increase in train capacity or a more frequent service, this investigation is an avenue for future research.

Bus users have almost similar waiting experiences in both the *Base Case* and *With AMoD* scenarios because of the constant headway-based bus service. However, users in the *With AMoD* scenario have a higher IVTT due to more on-street congestion. While 65% of Taxi users have to wait for less than 15 minutes in *Base Case*, the same is experienced by 80% in the other two scenarios indicating a good level of service. However, because of congestion, the IVTT increases considerably with the 95th percentile being around 15, 45 and 30 minutes, respectively, for the three scenarios. Car users experience a similar trend in IVTT as well. The dynamic fleet sizing strategy provides a lower WT for AMoD users in the *With AMoD* case. This scenario also witnesses lower IVTT and, therefore, lower total travel time. The higher congestion in the *Without Mass Transit* scenario has resulted in the comparatively largest travel times for AMoD users, leading to the most unfavorable travel experience.

AMoD Performance Metrics. We now examine the performance of AMoD specifically to gain deeper insights into their impact on the system as a whole. Considering the considerably higher demand for AMoD in the *Without Mass Transit* scenario, we observe the average vehicle occupancy to decrease from 1.80 to 1.47 in the *With AMoD* scenario. The proportion of shared trips also decreases by around 30%, thereby indicating that AMoD adjusts to incorporate more ride-sharing as demand increases.

For the *With AMoD* scenario, around 60% of total VKT is spent while traveling with a passenger, 35% while going for pick-up or parking, and 5% for empty vehicle cruising. The same observations for the *Without Mass Transit* scenario are 55%, 40%, and 5%, respectively. The considerably high non-trip contribution to VKT has been pointed out earlier in our discussions, which can be decreased through future explorations of better ride-matching and routing algorithms. However, an interesting observation is that higher demand may lead to an increased contribution to VKT while going for a pick-up.

Discussion

Advancements in AV technology and MoD services herald the arrival of smart urban mobility in the next decade. In the light of concerns about what this might bode for the future of mass transit, it is important to investigate the impact of AMoD on an urban multi-modal transportation system. We undertake a simulation-based approach to examine hypothetical scenarios for different policy implementations with regard to the role of AMoD.

While *With AMoD* encourages the introduction of AMoD and suggests that it might be useful for boosting mass transit ridership by providing first–last mile connectivity, *Without Mass Transit* represents a scenario where mass transit has been completely discarded. Although we obtain mode-specific insights about individual travel experience, all three categories of scenario evaluation metrics point out that *completely replacing mass transit with AMoD might not be possible without adversely affecting user experience and level of service*. Making mass transit unavailable would indeed force a modal shift to on-road modes such as cars, taxis, and AMoD. However, the resulting congestion would make for a terrible travel experience and lead to a complete breakdown of the system because of extremely heavy congestion. Mass transit would be irreplaceable in areas with dense demand because of its high passenger/space ratio: one bus equals about 30 cars in terms of capacity, whether the car is automated or not.

A key contribution of this research has been to clearly identify how policy-makers should encourage the growth of AMoD and attempt to incorporate it in the existing transportation system. Results from *With AMoD* look promising enough to perhaps consider providing subsidies to services that encourage use of mass transit, such as provision of access/egress connectivity. Therefore, future research should focus on the scenarios in which future urban mobility shapes AMoD not as a competitive but complementary service to mass transit.

We acknowledge that this is a preliminary investigation and the trends we observe in our case study cannot be generalized. Therefore, several research avenues remain open for further exploration. Supply decisions such as optimal fleet sizing and more refined algorithms for re-balancing and ride-sharing would be important contributions. Further scenarios can be designed to test different policy implementations of AMoD, such as a modified AMoD price scenario for subsidized AMoD services providing connectivity to mass transit. The effects of preferential traffic treatments such as transit signal priority or high-occupancy vehicle lanes should be evaluated. The interplay between mass transit and AMoD deserves further investigation. Redesigning mass transit services (headways, routes, capacities, and so forth) based on changes in vehicle occupancies on different transit lines due to availability of AMoD is a particularly promising avenue.

Finally, the next step would be to test these scenarios on different urban scenarios, including mixed urban and suburban areas, contrasting a transit-friendly city like Singapore with automobile-dependent regions such as Boston, and different levels of car ownership and income. All these aspects are expected to impact the mode share and the performance of the transportation system. The authors do not expect AMoD to be a panacea for all the urban scenarios; we rather reinforce that urban policies must be driven by inherent transportation cultures.

Conclusion

AMoD is unanimously considered as a game-changer in urban transportation. In order to fully exploit the benefits and limit the potential negative effect, systematic investigation is needed. Focusing on the performance of the operations of AMoD only is not sufficient. To correctly capture its *impact* on future urban transportation, an integrated approach is needed which also encompasses how travelers will respond to the availability of this technology. To this aim, we presented an AMoD framework in SimMobility that allows us to evaluate the adoption of the new service by travelers, as well as its operations. Furthermore, we apply the framework to a case study which shows that mass transit may play an irreplaceable role in urban transportation, especially under scenarios of wide adoption of AMoD. Having proposed a method to quantify the implications of discarding mass transit and demonstrated it in a specific AMoD-inclusive urban scenario, the next question we ask is: *When are we better off without mass transit, and when does mass transit complement AMoD?*

Acknowledgments

This research was supported in part by the National Research Foundation (NRF) Singapore through the Singapore MIT Alliance for Research and Technology's Future Urban Mobility IRG research program, the Ministry of National Development (MND) Singapore through the Land and Livability National Innovation Challenge (L2NIC) grant, and by the U.S. Department of Energy's Advanced Research Projects Agency-Energy (ARPA-E) under its TRANSNET Program.

References

1. Tesla. *Launch of Tesla Model 3*. 2017. https://www.tesla.com/model3
2. Uber. *Uber: Advanced Technologies Group*. 2017. https://www.uber.com/info/atg/
3. Kornhauser, A. *Uncongested Mobility for All: New Jersey's Area-wide aTaxi System*. Princeton University, Princeton, N.J., 2013.
4. Martin, E. W., and S. A. Shaheen. Greenhouse Gas Emission Impacts of Carsharing in North America. *IEEE Transactions on Intelligent Transportation Systems*, Vol. 12, No. 4, 2011, pp. 1074–1086.
5. Milakis, D., B. Van Arem, and B. Van Wee. Policy and Society Related Implications of Automated Driving: A Review of Literature and Directions for Future Research. *Journal of Intelligent Transportation Systems*, Vol. 21, No. 4, 2017, pp. 324–348.
6. Fagnant, D., and K. Kockelman. Preparing a Nation for Autonomous Vehicles: Opportunities, Barriers and Policy

Recommendations. *Transportation Research Part A: Policy and Practice*, Vol. 77, 2015, pp. 167–181.
7. Fernandes, P., and U. Nunes. Platooning of Autonomous Vehicles with Intervehicle Communications in SUMO Traffic Simulator. *Proc., 13th International IEEE Conference on Intelligent Transportation Systems (ITSC)*, Madeira Island, Portugal, IEEE, New York, 2010, pp. 1313–1318.
8. Santi, P., G. Resta, M. Szell, S. Sobolevsky, S. H. Strogatz, and C. Ratti. Quantifying the Benefits of Vehicle Pooling with Shareability Networks. *Proceedings of the National Academy of Sciences*, Vol. 111, No. 37, 2014, pp. 13290–13294.
9. Fagnant, D., K. Kockelman, and P. Bansal. Operations of Shared Autonomous Vehicle Fleet for Austin, Texas market. *Transportation Research Record: Journal of the Transportation Research Board*, 2015. 2536: 98–106.
10. Boesch, P. M., F. Ciari, and K. W. Axhausen. Autonomous Vehicle Fleet Sizes Required to Serve Different Levels of Demand. *Transportation Research Record: Journal of the Transportation Research Board*, 2016. 2542: 111–119.
11. Alonso-Mora, J., S. Samaranayake, A. Wallar, E. Frazzoli, and D. Rus. On-Demand High-Capacity Ride-Sharing Via Dynamic Trip-Vehicle Assignment. *Proceedings of the National Academy of Sciences*, Vol. 114, No. 3, 2017, pp. 462–467.
12. Azevedo, C. L., K. Marczuk, S. Raveau, H. Soh, M. Adnan, K. Basak, H. Loganathan, N. Deshmunkh, D. H. Lee, E. Frazzoli, and M. Ben-Akiva. Microsimulation of Demand and Supply of Autonomous Mobility on Demand. *Transportation Research Record: Journal of the Transportation Research Board*, 2016. 2564: 21–30.
13. Martinez, L. M., and J. M. Viegas. Assessing the Impacts of Deploying a Shared Self-Driving Urban Mobility System: An Agent-Based Model Applied to the City of Lisbon, Portugal. *International Journal of Transportation Science and Technology*, Vol. 6, No. 1, 2017, pp. 13–27.
14. International Transport Forum. *Shared Mobility Simulations for Helsinki*. The International Transport Forum, OECD, Paris, 2017.
15. Rayle, L., D. Dai, N. Chan, R. Cervero, and S. Shaheen. Just a Better Taxi? A Survey-Based Comparison of Taxis, Transit, and Ridesourcing Services in San Francisco. *Transport Policy*, Vol. 45, 2016, pp. 168–178.
16. Hall, J. D., C. Palsson, and J. Price. *Is Uber a Substitute or Complement for Public Transit?* University of Toronto, 2017. https://ideas.repec.org/p/tor/tecipa/tecipa-585.html
17. Fitzsimmons, E. G. Subway Ridership Declines in New York: Is Uber to Blame? *The New York Times*, 2017.
18. Lazo, L. Ripple Effect of Metro's Troubles: Plummeting Bus Ridership Across the Region. *The Washington Post*, 2016.
19. Sadowsky, N., and E. Nelson. *The Impact of Ride-Hailing Services on Public Transportation Use: A Discontinuity Regression Analysis*. Economics Department, Bowdoin College, 2017. https://digitalcommons.bowdoin.edu/econpapers/13/
20Polzin, S. E. *Implications to Public Transportation of Automated or Connected Vehicles*. National Center for Transit Research, University of Florida, 2016. https://www.transitavi.org/speculative-research/implications-to-public-transportation-of-automated-or-connected-vehicles/
21. Litman, T. *Autonomous Vehicle Implementation Predictions: Implications for Transport Planning*. Victoria Transport Policy Institute, Victoria, Canada, 2014.
22. Adnan, M., F. C. Pereira, C. M. L. Azevedo, K. Basak, M. Lovric, S. Raveau, Y. Zhu, J. Ferreira, Z. Christopher, and M. Ben-Akiva. *SimMobility: A Multi-scale Integrated Agent-based Simulation Platform*. Presented at 95th Annual Meeting of the Transportation Research Board, Washington, D.C., 2016.
23. Zhu, Y., and J. Ferreira. Synthetic Population Generation at Disaggregated Spatial Scales for Land Use and Transportation Microsimulation. *Transportation Research Record: Journal of the Transportation Research Board*, 2014. 2429: 168–177.
24.Le, Y., G. Cernicchiaro, C. Zegras, and J. Ferreira, Jr. Constructing a Synthetic Population of Establishments for the Simmobility Microsimulation Platform. *Transportation Research Procedia*, Vol. 19, 2016, pp. 81–93.
25. Lu, Y., M. Adnan, K. Basak, F. Pereira, C. Carrion, V. H. Saber, H. Loganathan, and M. Ben-Akiva. *Simmobility Mid-Term Simulator: A State of the Art Integrated Agent Based Demand and Supply Model*. Presented at 94th Annual Meeting of the Transportation Research Board, Washington, D.C., 2015.
26. Azevedo, C. L., H. Soh, S. Oh, K. Basak, N. M. Deshmukh, T. Toledo, B. Marimuthu, L-S. Peh, K. Marczuk, and M. E. Ben-Avika. *SimMobility Short-Term: An Integrated Microscopic Mobility Simulator*. Presented at 96th Annual Meeting of the Transportation Research Board, Washington, D.C., 2017.
27.Ben-Akiva, M., and S. Lerman. *Discrete Choice Analysis: Theory and Application to Travel Demand*. MIT Press, Cambridge, Mass., 1985.
28. Li, S., and B. N. Biran. *Technical Report for Mode Choice for Work Tour to Usual Location SimMobility MT*. 2017. http://fm.smart.mit.edu/technical-report-for-mode-choice-for-work-tour-to-usual-location-simmobility-mt/. Accessed on September 30, 2017. SMART, Singapore.
29. Araldo, A., N. Deshmukh, K. Basak, R. Seshadri, C. L. Azevedo, and M. Ben-Akiva. *An On-Line Scheduling Algorithm for Ride-Sharing in On-Demand Mobility Services*. 2017. https://www.researchgate.net/publication/321038260_An_on-line_scheduling_algorithm_for_ride-sharing_in_on-demand_mobility_services. Accessed on September 30, 2017.
30. Spieser, K., et al. Toward a Systematic Approach to the Design and Evaluation of Automated Mobility-on-Demand Systems: A Case Study in Singapore. In *Road Vehicle Automation* (G. Meyer and S. Beiker, eds.), Springer Springer International Publishing, 2014, pp. 229–245.
31Land Transport Authority (LTA). *Passenger Transport Mode Shares in World Cities*. LTA, 2011. http://studylib.net/doc/7972629/passenger-transport-mode-shares-in-world-cities

The Standing Committee on Emerging and Innovative Public Transport and Technologies (AP020) peer-reviewed this paper (18-06186).

TRR
JOURNAL OF THE TRANSPORTATION RESEARCH BOARD

Article

Transportation Research Record
2018, Vol. 2672(8) 619–628

DOI: 10.1177/0361198118783597
journals.sagepub.com/home/trr
SAGE

Incomplete Hub Location Model for Designing Van-Taxi Networks

Seyed Sina Mohri[1] and Meisam Akbarzadeh[1]

Abstract

Compared with ordinary taxis, van taxis offer higher vehicle capacity. Therefore, where possible, accommodating taxi trips by van would enhance the productivity of urban transportation systems. The difficulty of boarding and alighting vans could be mitigated by designing limited stop services. Studying the spatial pattern of taxi trips is a prerequisite for designing effective van routes. In this study, an incomplete p-hub median location model in combination with the concordance analysis method is presented to select the most suitable locations of van-taxi hubs of a real urban network. Several scenarios are defined according to the desired number of hubs. The model satisfies the system and passengers' expectations simultaneously. The objective function to be minimized is the vehicle-kilometers traveled by taxi and van. For satisfying the passengers' utility, the origin–destination (OD) travel time for all OD pairs is bounded with respect to existing OD travel time. The number of hub stations, the number of required vans, and the reduction in vehicle-kilometers traveled in the van-taxi network are model outputs. These are used as inputs for evaluating the scenarios using the concordance analysis approach. The proposed model is implemented on the Isfahan (Iran) network. Data were extracted from the smart card system for Isfahan taxi fare collection. The results show that construction of two or three van-taxi stations is the best scenario and decreases the vehicle-kilometers traveled in the taxi network up to 25% and 19% for morning and evening peak period, respectively.

Increasing the vehicle occupancies enhances the productivity of urban transportation systems and helps in mitigating the adverse consequences of traffic congestion. Offering incentives to high-occupancy vehicles and promoting transit and paratransit modes are examples of this approach. Relative difficulty of boarding and alighting would be an impediment in using van-taxi services. Therefore, van taxis are suitable as shuttles; that is, having one origin and one destination.

This paper proposes a method for designing the most suitable van-taxi network based on taxi trip patterns extracted from a real city. The major part of the proposed method is a new approach for designing an incomplete hub network for urban taxi trips. In the hub network for urban taxi trips, van service is considered for moving people between hub nodes and the shared taxi service is used between hub and non-hub nodes. The shared taxi service is common in many cities throughout the world. The incomplete shape of the hub network depends on the network shape. For instance, limited-access and arterial facilities are preferred for van services. Therefore, the shape of the urban highway network imposes the shape of the hub network. In addition to travel time and travel distance matrices for taxi trips based on common urban taxi routes, two other travel time and travel distance matrices are considered for van trips based on the existence of limited-access and arterial facilities in the proposed model. The proposed mixed-integer programming model for locating hub stations in a van-taxi network aims to increase the benefit of the system as well as utility for the passengers. The objective of the system is reducing the total vehicle-kilometers. Compared with taxi cars, vans have higher capacity, and this could help in achieving the goal of reduction in vehicle-kilometers in the hub network. In order to satisfy passengers' expectations, the model does not allow the total origin–destination (OD) travel time on hub networks to become longer than the existing travel time on taxi network. The most important advantage of the new hub network for users is to reduce travel time and waiting time. The travel time of passengers would reduce as a result of the removal of stops between two hubs, and also by passing through limited-access and arterial facilities with higher speed limits. The removal of direct taxi

[1]Department of Transportation Engineering, Isfahan University of Technology, Isfahan, Iran

Corresponding Author:
Address correspondence to Seyed Sina Mohri: s.mohri@te.iut.ac.ir

trips in the city makes the taxi transportation system work in two parts—collecting passengers at the first hub, and distributing them from the second hub—that will lead to an increase of network coverage and decrease of passengers' waiting time in origin and second hub nodes. The presented model is implemented on the network of taxi trips in Isfahan (Iran) and its results are evaluated.

Literature Review

In network design, connecting all origins and destinations via direct links is prohibitively costly and causes complexity and congestion throughout the network. One of the strategies to resolve this problem is to set some network nodes as hubs. "Hubs" refers to the centers of collection and distribution that are used for a direct relationship between the two points. In a simple oriented network with n nodes, a direct relationship among all nodes with each other $n(n\text{-}1)$ to oriented arc (one pair of source-destination) is needed, while in case of the presence of a hub in the network, the number of all arcs will be equal to $2(n\text{-}1)$ (*1*).

The selection of the optimal situation of hub nodes in the network is determined by the hub location problem (HLP). Based on objective function performance, hub location models can be classified into five categories: hub medium location problem, hub center location problem, hub covering location problem, multi-objective HLP and continuous HLP (*2*). The most common method in hub location is using hub median location models. O'Kelly proposed a single allocation hub location model with a quadratic objective function (*3*). Then, Campbell proposed the first linear integer programming for HLP and named it the "p-hub median location problem" (*4, 5*). The p-hub median location problem is divided into two problems of single allocation p-HLP and multiple allocation p-HLP (*4, 5*). In single allocation p-HLP each non-hub node is allowed to connect only to one hub, but in multiple allocation p-HLP, non-hub nodes can be connected to several hubs. After Campbell, Skorin-Kapov et al., Sohn and Pak, Ernst and Krishnamoorthy, Ebery, Hamacher et al., and Marin et al. modified the primary model and presented new models to reduce the variables of decision and increase the problem-solving speed for larger networks (*6–11*). For example, Ernst and Krishnamoorthy took the idea of the multi-commodity flow problem and presented a new simple model for the hub median location problem. In this model, each commodity shows the flow of the produced commodity in a specific node. Ernst and Krishnamoorthy's model of has $n^3 + n^2$ decision variables including n^2 binary variables (*8*).

Hub location models are employed in various transportation applications such as air transportation (*12–14*), freight transportation (*15–17*), maritime transportation (*18–20*), and public transportation (*21–24*). The classic HLPs considered four assumptions:

- Assumption 1: the hub network is a complete graph.
- Assumption 2: a discount factor (α) is considered for hub edges.
- Assumption 3: no direct connection between non-hub nodes is allowed.
- Assumption 4: the movement cost follows the triangle inequality rule.

The HLPs presented in the context of urban transportation have usually relaxed the first assumption. For designing an incomplete network graph usually a tree (*25–29*), ring (*30–32*), or general (*21, 33, 34*) network structure is considered in the literature. In tree-shape networks, there is only one path for traveling between the hubs. In ring-shape networks, each hub is connected to exactly two other hubs. In the general shape, the structure of hub network forms after solving the model and no ex-ante structure is assumed.

In urban transportation networks, faster services are used for connecting hubs and slower services are used between non-hub and hub nodes. Because of the high implementation costs of rapid transit services, previous studies have designed an incomplete hub network in the urban area. To the authors' best knowledge, Nickel et al. presented the first model for HLP in public transportation network (*21*). In this study, fixed costs were considered for hub nodes and all arcs, and a general shape for hub network was considered. In this study, PT (public transportation) and GPT (generalized public transportation) were presented, in which the first model the assumptions (1) and (4) and in the second model the assumptions (1), (3) and (4) were relaxed. In the second model, because of the possibility of connecting two non-hub nodes, n^2 variable were added to the problem. The proposed models were tested on the CAB dataset (*21*). Gelareh presented a base model for hub location in public transportation network called HLPPT (HLP in public transportation) (*35*). Furthermore, he proposed several extensions and additions to HLPPT under non-restrictive network policy, considering delay (DHLPPT), tree shape (TSHLPPT), and polygonal HLPPT (PHLLPT), and multi-period HLP. In this study several heuristic algorithms were proposed for solving the problems. Also, in this study, the real tendencies of passengers in using public transportation systems were considered. For example, passengers who enter the hub network do not tend to change their travel type, and continue their trip in this network to the last hub node near to their final destination (*35*). Gelareh and Nickel presented a bender decomposition algorithm for solving HLPPT (*36*). Gelareh and

Nickel dealt with HLPs in a multi-period context. When a project includes considerable construction and investment costs, a planning horizon with several decision-making periods are considered for its implementation. Thus, they considered a limited budget for constructing the hub network (*33, 36*).

Methodology

The presented model is a multiple allocation p-HLP model which aims to locate the stations of a van-taxi network by considering system and passenger benefits. The model increases the system's benefits (stated as the total traveled vehicle-kilometers) and prevents the reduction of the benefits to passengers (stated as the OD travel times). The reduction in users' benefits was evaluated by comparing OD travel times on the primary and the proposed hub networks.

The stages of combing hub location model and concordance analysis method for selecting the optimal scenario are as follows:

- For all OD pairs, calculate the travel distance and travel time matrices of van and taxi trips.
- Implement the hub location model for all scenarios.
- For each scenario, calculate:
    - the number of hubs
    - the fleet size
    - the reduced vehicle-kilometers
- Run the multi-criteria decision-making (MCDM) model and select the optimal scenario.

Before presenting the model, the procedure used for calculating travel time and travel distance matrices for van and taxi modes is explained.

Figure 1 shows an example network. The links of this network are either limited access, for example highway, or urban, for example local and collector/distributor. Limited-access links are illustrated by red solid lines and urban links are illustrated by blue dotted lines. This network includes 26 nodes and 11 demand nodes, that is, centroids. Trips are produced in centroids. It is assumed that in order to prevent excess traffic interruption and delay in urban links, the van services are only allowed to pass through limited-access links.

In this network, any node may potentially be selected as a hub. Therefore, two states can accrue for computing the van travel time and travel distance between two hub nodes:

- State (a): The first and second hub nodes are placed on limited-access links.
- State (b): At least one of the two hub nodes are not located on limited-access links.

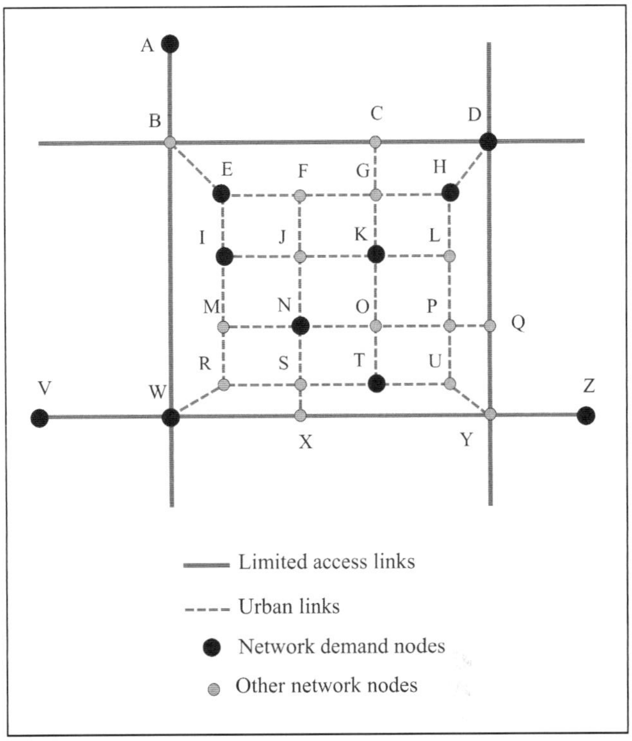

Figure 1. An urban network example.

In state (a), the van travel distance and travel time between two hubs are easy to compute and are estimated based on the shortest path time on limited-access links. For example, to travel from hub *A* to *V* the path *ABWV* exists. In state (b), if the two considered hub nodes can connect to limited-access links with at most one urban link, the van travel time and travel distance can be computed, otherwise, a large positive value is assigned to these two parameters. For example, traveling from *H* to *R* can be done using van service because two links *HD* and *RW* connect these two points to limited-access links. As another example, consider two points *V* and *K*. Although *V* is on limited-access links, *K* cannot connect to limited-access links with maximum one urban link, and therefore *K* cannot be a hub node. Also, if more than one path exists between two hub nodes, the path with shortest travel time is considered.

Now the model is presented based on the abovementioned assumptions. Maximizing the benefit of the system is addressed in the objective function. Improving the travel time of the passengers and feasibility requirements are addressed in the constraints:

$$Min \sum_{i} \sum_{j} \sum_{k} \sum_{m} \left(\frac{d_{\mathrm{ik}}}{C_t} + \frac{d'_{\mathrm{km}}}{C_v} + \frac{d_{\mathrm{mj}}}{C_t} + \epsilon \right) h_{\mathrm{ij}} X_{\mathrm{ij}}^{\mathrm{km}} + \sum_{i} \sum_{j} \frac{h_{\mathrm{ij}} d_{\mathrm{ij}}}{C_t} \left(1 - \sum_{k} \sum_{m} X_{\mathrm{ij}}^{\mathrm{km}} \right) \quad (1)$$

Subject to:

$$t_{ij}^{km} \le (t^t{}_{ij} + t^d{}_{ij})X_{ij}^{km} + M\left(1 - X_{ij}^{km}\right) \quad \forall i,j,k,m \in D_{all} \tag{2}$$

$$\sum_k Z_k = P \tag{3}$$

$$\sum_k \sum_m X_{ij}^{km} \le 1 \quad \forall i,\ j \in D_{all} \tag{4}$$

$$t_{ij}^{km} = T^{move} + T^{wait} \quad \forall i,j,k,m \in D_{all} \tag{5}$$

$$T^{move} = Z_i t_{ik}^{t(hub)} + (1 - Z_i)t^t{}_{ik} + t_{km}^{t(hub)} + Z_j t_{mj}^{t(hub)} + \left(1 - Z_j\right)t^t{}_{mj} \quad \forall i,j,k,m \in D_{all} \tag{6}$$

$$T^{wait} = Z_i \beta t^d{}_{ik} + (1 - Z_i)\alpha t^d{}_{ik} + \beta t^d{}_{km} + Z_j \beta t^d{}_{mj} + \left(1 - Z_j\right)\gamma t^d{}_{mj} \quad \forall i,j,k,m \in D_{all} \tag{7}$$

$$X_{ij}^{km} \le Z_k \quad \forall i,j,k,m \in D_{all} \tag{8}$$

$$X_{ij}^{km} \le Z_m \quad \forall i,j,k,m \in D_{all} \tag{9}$$

$$X_{ij}^{km} \in \{0,1\} \quad \forall i,j,k,m \in D_{all} \tag{10}$$

$$Z_k \in \{0,1\} \quad \forall k \in D_{all} \tag{11}$$

Where:

- Sets /input parameters

d_{ij}: The travel distance between nodes i and j in the urban network (km)

d'_{km}: The travel distance between nodes i and j (km) in the limited-access and arterial facilities

h_{ij}: Demand flow from node i to node j (passenger)

C_t: Capacity of taxi vehicle (passenger/ vehicle)

C_v: Capacity of van vehicle (passenger/ vehicle)

D_{all}: A set of network nodes

M: A large positive constant

α: A reducing coefficient of the existing passengers waiting time for taxi in origin nodes for moving toward van stations

β: A coefficient for converting the existing passengers waiting time to the waiting time for van service at the first hub station

γ: A reducing coefficient of the existing passengers waiting time for taxi at van stations for traveling to the destination

ϵ: A very small number to make the node k and m non-uniform ($k \neq m$)

t_{ij}^{km}: The total travel time of passengers in the van-taxi network

T^{move}: The travel time of passengers in van-taxi network

T^{wait}: The waiting time of passengers for taxi and van services in van-taxi network

t_{ik}^t: Taxi travel time in the urban network from node i to node k

$t_{km}^{t(hub)}$: Van travel time in the limited-access and arterial network from node k to node m

$t^d{}_{ik}$: The existing passengers waiting time for taxi in node i for traveling from node i to node k. *if* $i = k$ *then* $t^d{}_{ik} = 0$

- Decision variables

X_{ij}^{km}: A binary variable, equal to one if the demand flow from node i to node j passes through hub k and m, and zero otherwise

Z_k: A binary variable, equal to one if node k is determined as hub, and zero otherwise

Equation 1 is the objective function and calculates the total vehicle-kilometers traveled in the network. The second term of this expression states the possibility of direct trips (without using van). If the value of the binary decision variable X_{ij}^{km} is 1, the first term of the objective function will be used and the amount of vehicle-kilometers will be calculated in the van-taxi network. If the total vehicle-kilometers of the network demands in the system with a van is less than the direct taxi trip, the system will select this service and locate its stations. If nodes k and m are the same, parameter d'_{km} in the first term of this equation will become equal to zero and the vehicle-kilometers traveled in the first term would be equal to the second term of Equation 1. In this situation, the objective function of the problem will not make any certain decisions for passing demand from i to j through the van-taxi service or direct service. Therefore, the value ϵ in such conditions makes the amount of vehicle-kilometers traveled in the van-taxi network be more than the direct taxi trip so that the demand can flow directly without using the hub network.

In addition to the systemic decision, the travel time of passengers in the network with van should not be more than the direct taxi trip. This limitation is designed in Equation 2 and provides the users' expectations in addition to providing the system's objectives. Equation 3 limits the number of van stations (hubs) in the network, which is considered as a policy value and therefore is an input of the problem. Equation 4 ensures that if van-taxi is selected, the flow between OD points (i.e., i to j) is not more than total demand from node i to node j. Equation 5 calculates the travel time of passengers in the van-taxi network and is composed of two parts, travel time and waiting time. Equation 6 calculates the travel time of passengers in the van-taxi network. In the case of using van service, the travel time is considered. In this equation, if the origin or destination nodes are considered as van stations, the travel time from the origin to the second van station or the travel time from the van station to the destination should be equal to the travel time on limited-access links, respectively. Equation 7 calculates the total passengers' waiting time in the van-taxi network. As a result of the removal of taxi service in the path between two hubs and shifting the excess taxis to the collection and distributing legs of travels, the coefficients of α and γ will reduce the passengers' waiting time for taxi in origin and second hub stations. Also, converting the waiting time for taxi to the waiting time for van is carried out using β reducing coefficient. Equations 8 and 9 state that if any flow directly passes from hub nodes k to m, both nodes must be selected as hub stations. Equations 10 and 11 show the type of the decision variables (binary variable). Since the model does not limit the capacity of hub nodes, the type of variable X_{ij}^{km} is a binary variable. If the model changes to a capacitated incomplete HLP, X_{ij}^{km} will be considered a continuous variable that is less than 1, and one other equation for limiting the capacity of hub nodes is added to the presented model.

The optimization may yield several optimal solutions. Selecting among these solutions requires consideration of some other criteria. In order to solve this problem, a MCDM method was used, which is explained in the following section.

Concordance Analysis Method

MCDM problems are those in which a decision maker wants to select among a set of n alternatives (A) or rank them with considering p criteria. The score of each alternative with regard to each criteria is organized as a matrix $Z_{n\times p}$. Criteria may be either the benefits achieved from alternatives or the costs incurred by the decision maker for selecting that alternative. The first criteria are called positive and the second criteria are called negative criteria. In the general case, the importance of criteria is not necessarily equal. Therefore, a weighting vector (W) is also considered of which element w_j states the importance of criteria j for the decision maker.

Concordance analysis is a method used for MCDM. The general technique of concordance analysis method is called the ELECTRE method and has been used in various studies (*22*). When the decision-making criteria are contradictory, or when an MCDM problem cannot be simply converted into an equivalent criterion, concordance analysis would be an appropriate tool. One of the main advantages of this technique is the possibility of considering different weight vectors for the criteria, so that each weight vector may emphasize a different subset of the decision criteria. In the end, the alternative selected under majority of the weight vectors will be selected as the superior alternative. Concordance analysis method consists of five steps, as follows (*37–39*).

As the values and variation range of the columns of $Z_{n\times p}$ are not the same, normalization of the matrix elements is necessary (step 1). The linear normalization method is used in concordance analysis: for positive criteria column k, each element z_{ij} is transformed to r_{ij} which is equal to $z_{ij}/\max_k z_{kj}$. For negative criteria k, each element z_{ij} is transformed to r_{ij} which is equal to $1 - z_{ij}/\max_k z_{kj}$. With this normalization, the matrix values are all transformed to benefits.

In the next step (step 2), the concordance and discordance sets and indices are calculated. The concordance set of alternative i and alternative i' ($C_{ii'}$) consists of the alternatives in which i has a better normalized score than i'. The discordance set of alternative i and alternative i' ($D_{ii'}$) consists of the alternatives in which i' has a better normalized score than i. In other words:

$$C_{ii'} = \{j|r_{ij} \geq r_{i'j}\} \tag{12}$$

$$D_{ii'} = \{j|r_{ij} \leq r_{i'j}\} \tag{13}$$

Based on the concordance and discordance sets, the indices of concordance and discordance are calculated (step 3). Concordance index of alternatives i and i' ($c_{ii'}$) is the sum of the weights of the criteria members of their concordance set. Discordance index of alternatives i and i' ($d_{ii'}$) is the average of weighted and normalized values of differences scores of i and i' in criteria which are members of discordance set. In other words:

$$c_{ii'} = \sum_{j\in C_{ii'}} w_j \quad \forall i \neq i' \tag{14}$$

$$d_{ii'} = \frac{\sum_{j\in D_{ii'}} \left\{\frac{w_j|r_{ij}-r_{i'j}|}{d^{\max}}\right\}}{m} \forall i \neq i' \tag{15}$$

$$d^{\max} = \max_{i,i',j} w_j|r_{ij} - r_{i'j}| \tag{16}$$

$$m = \max_{i,i'}\{number\ of\ elements\ in\ D_{ii'}\} \tag{17}$$

The net values of concordance and discordance indices are calculated as the difference of the sum of concordance (discordance) indices of i and i'(step 4):

$$c_i = \sum_{i\neq i'} c_{ii'} - \sum_{i\neq i'} c_{i'i} \tag{18}$$

$$d_i = \sum_{i\neq i'} d_{ii'} - \sum_{i\neq i'} d_{i'i} \tag{19}$$

For each weight vector, the alternatives with $c_i>0$ and $d_i<0$ are placed in the selection set of that weight vector. Finally, the option that is more frequently repeated in the selection sets of the weight vectors are selected as the best alternative (step 5).

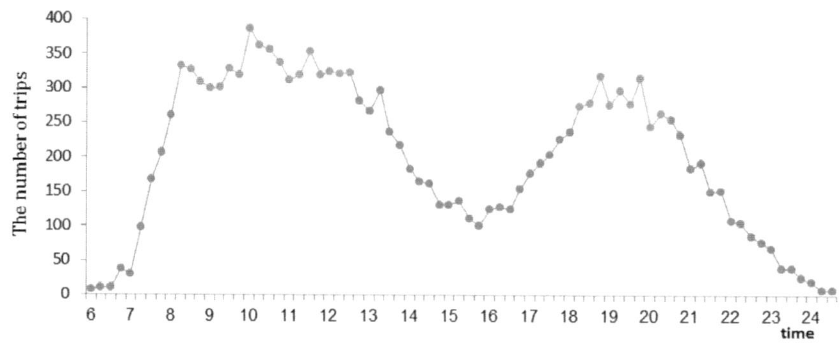

Figure 2. Number of taxi trips during the day.

Data Preparation and Analysis

Data of taxi trips from Isfahan (Iran) during two weeks of May 2015 were used for the analysis. Isfahan is the third largest city of Iran, located in the central part of the country and with 1.7 million inhabitants. The database included the coordinates of origin and destination of each trip, trip travel time, and real travel distance. It included more than 118,000 taxi trips by 2,700 unique taxis. The following steps were taken for data preparation:

Step 1: Finding the morning and evening design time interval in a day.

Taxi trips were grouped into 15-minute time intervals for all mentioned days. Next, the day that had the most taxi trips during its morning and evening rush hours, selected as day design and the average paths travel time during the rush hours, was computed. Finally, each rush hour extended to a design time interval so that all paths' travel times during the interval were mostly equivalent. As shown in Figure 2, time intervals 8:15 to 12:15 and 18:15 to 20:15 of May 5, 2015 were selected as the design time intervals. The number of trips in each time interval of May 5 is shown in Figure 2.

Step 2: Extracting a taxi network and its OD matrix demand from the taxi smart card system information.

Because of the advantage of van-taxi service for long trips, the origin and destination of trips longer than 7 km carried out within the design time intervals were extracted and analyzed. Areas in the city with high taxi trip production and attraction were detected from the data. These areas were mainly located at major intersections and shopping or medical centers. In order to build up the OD matrix, these points were selected as OD matrix points, and the trips originated or destined in their vicinity (less than 200 meters distance) were allocated to them. Consequently, a 30 $\times$ 30 OD matrix was established which included more than 80% of the total taxi trips in Isfahan. The considered network nodes are illustrated in Figure 3.

Step 3: Computing the travel distance and travel time matrices of urban and limited-access facilities.

In order to compute the travel distance and travel time matrices of urban and limited-access facilities, first the nodes that could potentially be selected as hub based on states (a) and (b) were selected. Then, the approximate travel distance on limited-access and arterial facilities of the network were extracted using Google Map. By comparing the estimated travel approximate distance and calculated trip length by the smart system of taxies in each OD pair, the limited-access and arterial trips among these nodes were separated from the set of trip data and their average travel times were calculated. By removing the data of limited-access and arterial trips from the data set, the average travel time of urban trips among the network's nodes was calculated.

Step 4: Computing the existing passengers waiting time for taxi ($t^d{}_{ik}$) for network nodes.

In network nodes with high production and attraction of taxi trips, it is expected that the quality of taxi service will be suitable and the waiting time for a taxi will be less than other nodes. Therefore, instead of a unique waiting time for all network nodes, a proper waiting time in relation to the point's production and attraction of trips is considered for each network node. By assuming the maximum and minimum waiting time for a taxi as 10 and 2 min, the existing passengers waiting time for taxi ($t^d{}_{ik}$) in each problem's node was computed as a linear function of the total point's production and attraction of trips.

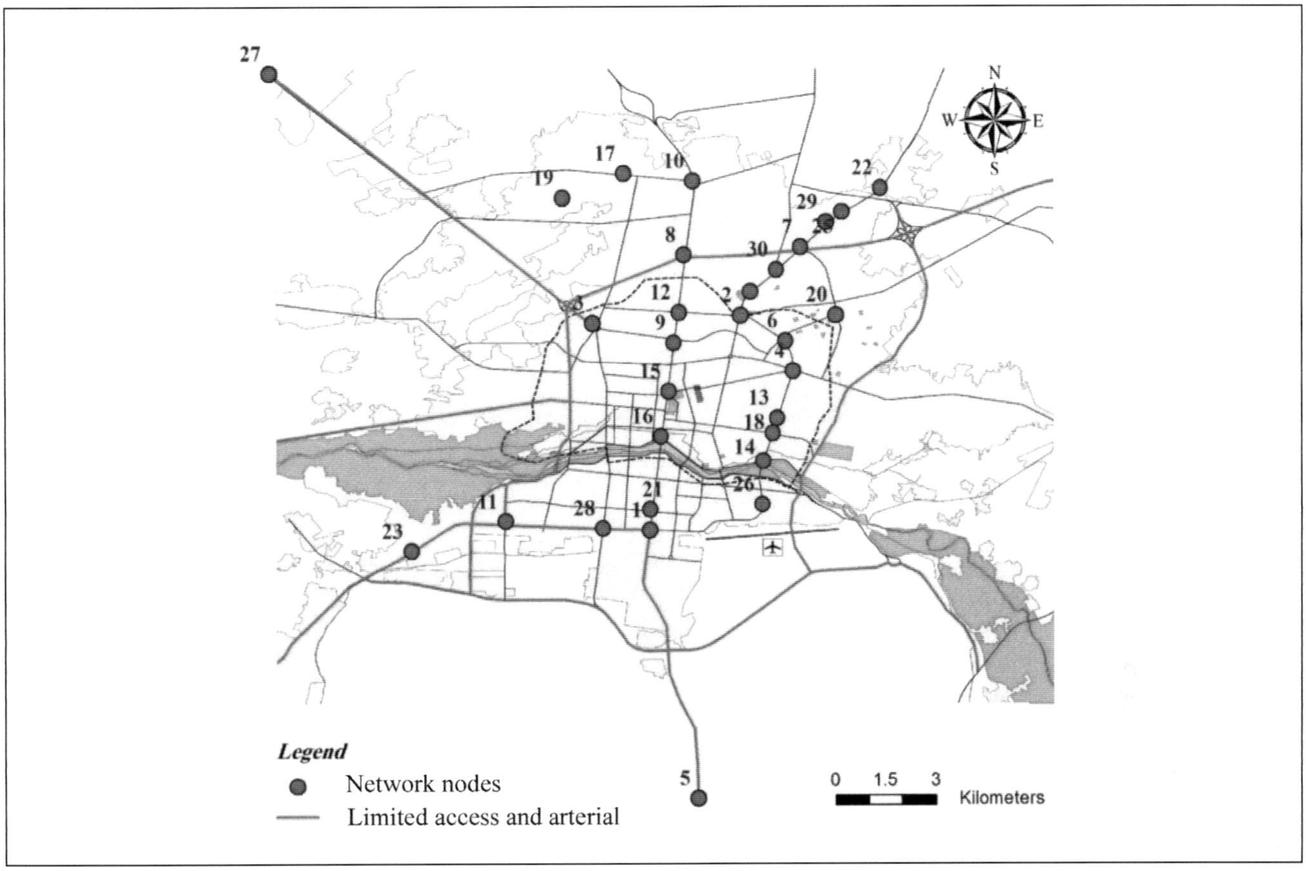

Figure 3. The network nodes of taxi trips in Isfahan city.

Results

The model was solved for defined time intervals and five scenarios with two to six hubs for each time interval. Commercial software IBM ILOG CPLEX 12.6.1 was used for solving the model in a device with Intel(R) Core(TM) i7 CPU @2.13 GHz and 6 Gbytes of RAM under the 64-bit Windows 7 operating system. α and γ coefficients for Isfahan network were considered as 0.7, and β coefficient is equal to 0.5. Also, the taxi capacity is assumed as four passengers and van capacity as 12 passengers. In Table 1, the results of the scenarios in addition to the do-nothing scenario (the absence of hub in the network) are presented. In other words, the do-nothing scenario represents the existing situation, that is, the regular taxi service.

A comparison between the results of five considered scenarios and do-nothing scenario (traditional taxi service) for morning or evening period shows the establishing a van-taxi service without increasing the users' travel time decreases the value of vehicle-kilometers in the taxi network. Although all scenarios reduce the total vehicle-kilometers in the taxi network, selecting the most economic scenario among them is a challenging task.

According to the similar stations determined at both scenarios of morning and evening, and also the more crowded interval of the morning, the selection of the best and the most economic scenario was carried out according to the feature of morning interval. To select the optimal scenario based on the concordance analysis method, three criteria of decision-making were considered as follows:

- Criterion 1: The number of hubs

The increase in the number of hubs in the network increases the construction costs of van service stations, and thus is considered as a negative criterion in evaluating the scenarios.

- Criterion 2: The number of vehicles operating on line

Another decision-making criterion for selecting the optimal scenario is the number of vehicles operating on line for each scenario. Calculation of an accurate number of vehicles operating on line for each scenario is not

Table 1. Comparing the Scenarios of Van-Taxi Network in Morning and Evening Periods

Time interval	Scenario ID	Number of hubs	Objective function value (vehicle-kilometers)	Solution time (sec)	Hub nodes	OD[a]	Attracted demand (%)[b]	Length (km)[c]
Morning	0	0	2622	-	-	0	0	0
	1	2	2094	3.3	1-5	4	17	8.6
	2	3	1979	3.3	1-5-14	20	29	12.6
	3	4	1934	4.5	1-4-5-14	31	32	14.7
	4	5	1829	3.8	1-3-5-11-14	35	40	22.4
	5	6	1710	4.6	1-3-5-11-14-27	37	45	32.8
Evening	0	0	887	-	-	0	0	0
	1	2	765	3.5	1-5	3	9	8.6
	2	3	718	3.1	1-5-14	12	16	12.6
	3	4	691	4.2	1-4-5-14	23	21	14.7
	5	5	665	5.4	1-3-5-11-14	27	24	22.4
	6	6	627	6.2	1-3-5-11-14-27	30	26	32.8

[a]The number of OD attracted to hub network.
[b]The percentage of demand attracted to hub network.
[c]The length of hub network (km).

possible, but it is estimated through the Equation 20 according to the method proposed in Vuchic (*40*):

$$N_{\mathrm{TU}} = \frac{2L \times P_d(1+\gamma)}{\alpha \times n \times C_v \times V_c} = L \times P_d \times \left(\frac{2(1+\gamma)}{\alpha \times n \times C_v \times V_c}\right) = L \times P_d \times K_1 \tag{20}$$

In the above equation, L is the service path length, P_d is the current demand, and K_1 is the product of the system features that are considered the same in all comparative scenarios. Thus, the number of vehicles operating on line at each scenario was dependent on the product of the two features of path length and path demand that were used as the representative of the number of van vehicles in comparing the scenarios. This criterion is considered as a negative criterion.

- Criterion 3: The reduced vehicle-kilometers

The transfer of taxi trips to van caused the reduction of vehicle-kilometers traveled in the network, which will lead to the reduction of fuel consumption, reduction of air pollution, and reduction of travel time of taxi trips in the network. Thus, this criterion is a positive one when comparing the scenarios.

By considering these three criteria in the selection of the best scenario, four different weight systems were considered. An important step in the concordance analysis method is determining the weight of criteria. In this study, the weight of all criteria was considered the same according to the theory of concordance method in weight system number 1. In three other weight systems, one of the criteria was emphasized, so that its weight was considered twice the other criteria (*39*). The weights of criteria in four considered weight systems are as follows:

- Weighting system 1 equal emphasis on all criteria:
  - Criterion 1, 2, and 3 = 0.33
- Weighting system 2 emphasis on criterion 1:
  - Criterion 1 = 0.50
  - Criterion 2 and 3 = 0.25
- Weighting system 3 emphasis on criterion 2:
  - Criterion 1 and 3 = 0.25
  - Criterion 2 = 0.50
- Weighting system 4 emphasis on criterion 3:
  - Criterion 1 and 2 = 0.25
  - Criterion 3 = 0.50

By determining the weight of each criterion in weight systems, the concordance analysis method was established for each weight system. Table 2 shows the results of the method (values of concordance (c_i) and discordance (d_i) indices for each scenario in quadratic weight systems.

According to the values of concordance and discordance indices in all weight vectors, scenarios 1 and 2 were placed in the selection set. Scenarios 1 and 2 had the bigger positive concordance index and the smaller negative discordance index in all weight systems. Therefore, scenarios 1 and 2 will be the optimal scenarios of van-taxi network implementation in this case, and stations 1, 5, and 14 will be the three optimal hub stations of the van-taxi network. According to Table 1, considering stations 1, 5, and 14 as hub nodes (scenario 3) decreases by 25 and 19% total vehicle-kilometers in morning and evening periods, respectively.

Table 2. The Values of Concordance and Discordance Indices in Each Scenario

	Concordance index (c_i)				Discordance index (d_i)			
Scenario	Weighting system 1	Weighting system 2	Weighting system 3	Weighting system 4	Weighting system 1	Weighting system 2	Weighting system 3	Weighting system 4
1	1.3	2	2	0.1	−1.2	−1.5	−1.56	−0.52
2	0.67	2	2	0.05	−0.73	−0.86	−0.94	−0.41
3	0	0	0	0	−0.27	−0.2	−0.45	−0.15
4	−0.67	−1	−1	0	0.57	0.74	−0.71	0.26
5	−1.3	−2	−2	0	1.63	1.84	2.2	0.81

Conclusion

In this paper, an incomplete van-taxi hub location model was proposed by van system to cover long and high-demand taxi OD pairs. The presented model is capable of increasing the system (social) and users' benefits. The system attempts to reduce the vehicle-kilometers traveled in the existing taxi network by offering a van service. Maximizing the benefit of the system is addressed in the model's objective function. For satisfying the passenger utility, the OD travel time for all OD pairs is bounded with respect to existing taxi travel time for each OD pair. Improving the travel time of the passengers and feasibility requirements are addressed in the model's constraints. Solving the presented optimization model, several optimal scenarios result based on variation of number of hubs as a model parameter. In order to find the optimal scenario, an MCDM method, concordance analysis approach, was established on the results of scenarios. Three important criteria—the number of hub stations, the number of vans required, and the reduction in vehicle-kilometers traveled in the van-taxi network—were extracted from scenario's results. These criteria were considered as inputs for evaluating the scenarios based on the concordance analysis method.

The model was implemented on the taxi network in Isfahan city. The final results showed that the scenarios of implementing two and three van stations in the network were the optimal scenarios for implementing a van-taxi network in Isfahan city. Implementing these two and three van stations in Isfahan city can reduced the amount of vehicle-kilometers traveled in the network of taxi trips to 20 and 25% in the morning and 14 and 19% in the evening. The proposed method could be applied to other cities and provide mitigation of the adverse consequences of overcrowding, in addition to the promotion of transit and paratransit services.

Author Contributions

The authors confirm contribution to the paper as follows: Study conception and design: SSM and MA. Data collection: MA. Analysis and interpretation of results: SSM and MA. Draft manuscript preparation: SSM and MA. All authors reviewed the results and approved the final version of the manuscript.

References

1. Hekmatfar, M., and M. Pishvaee. *Hub Location Problem in Facility Location*. Physica-Verlag, Berlin, 2009, pp. 243–270.
2. Farahani, R. Z., M. Hekmatfar, A. B. Arabani, and E. Nikbakhsh. Hub Location Problems: A Review of Models, Classification, Solution Techniques, and Applications. *Computers & Industrial Engineering*, Vol. 64, No. 4, 2013, pp. 1096–1109.
3. O'Kelly, M. E. A Quadratic Integer Program for the Location of Interacting Hub Facilities. *European Journal of Operational Research*, Vol. 32, No. 3, 1987, pp. 393–404.
4. Campbell, J. F. Location and Allocation for Distribution Systems with Transshipments and Transportion Economies of Scale. *Annals of Operations Research*, Vol. 40, No. 1, 1992, pp. 77–99.
5. Campbell, J. F. Integer Programming Formulations of Discrete Hub Location Problems. *European Journal of Operational Research*, Vol. 72, No. 2, 1994, pp. 387–405.
6. Skorin-Kapov, D., J. Skorin-Kapov, and M. O'Kelly. Tight Linear Programming Relaxations of Uncapacitated P-Hub Median Problems. *European Journal of Operational Research*, Vol. 94, No. 3, 1996, pp. 582–593.
7. Sohn, J., and S. Park. Efficient Solution Procedure and Reduced Size Formulations for P-Hub Location Problems. *European Journal of Operational Research*, Vol. 108, No. 1, 1998, pp. 118–126.
8. Ernst, A. T., and M. Krishnamoorthy. Efficient Algorithms for the Uncapacitated Single Allocation P-Hub Median Problem. *Location Science*, Vol. 4, No. 3, 1996, pp. 139–154.
9. Ebery, J. Solving Large Single Allocation P-Hub Problems with Two or Three Hubs. *European Journal of Operational Research*, Vol. 128, No. 2, 2001, pp. 447–458.
10. Hamacher, H. W., M. Labbé, S. Nickel, and T. Sonneborn. Adapting Polyhedral Properties from Facility to Hub Location Problems. *Discrete Applied Mathematics*, Vol. 145, No. 1, 2004, pp. 104–116.
11. Marín, A., L. Cánovas, and M. Landete. New formulations for the Uncapacitated Multiple Allocation Hub Location

Problem. *European Journal of Operational Research*, Vol. 172, No. 1, 2006, pp. 274–292.

12. Jaillet, P., G. Song, and G. Yu. Airline Network Design and Hub Location Problems. *Location Science*, Vol. 4, No. 3, 1996, pp. 195–212.
13. Kuby, M. J., and R. G. Gray. The Hub Network Design Problem with Stopovers and Feeders: The Case of Federal Express. *Transportation Research Part A: Policy and Practice*, Vol. 27, No. 1, 1993, pp. 1–12.
14. Eiselt, H. A., and V. Marianov. A Conditional P-Hub Location Problem with Attraction Functions. *Computers & Operations Research*, Vol. 36, No. 12, 2009, pp. 3128–3135.
15. Alumur, S. A., B. Y. Kara, and O. E. Karasan. The Design of Single Allocation Incomplete Hub Networks. *Transportation Research Part B: Methodological*, Vol. 43, No. 10, 2009, pp. 936–951.
16. Campbell, J. F. Hub Location for Time Definite Transportation. *Computers & Operations Research*, Vol. 36, No. 12, 2009, pp. 3107–3116.
17. Racunica, I., and L. Wynter. Optimal Location of Intermodal Freight Hubs. *Transportation Research Part B: Methodological*, Vol. 39, No. 5, 2005, pp. 453–477.
18. Aversa, R., R. C. Botter, H. E. Haralambides, and H. T. Y. Yoshizaki. A Mixed Integer Programming Model on the Location of a Hub Port in the East Coast of South America. *Maritime Economics & Logistics*, Vol. 7, No. 1, 2005, pp. 1–18.
19. Hsu, C.-I., and Y.-P. Hsieh. Routing, Ship Size, and Sailing Frequency Decision-Making for a Maritime Hub-And-Spoke Container Network. *Mathematical and Computer Modelling*, Vol. 45, No. 7, 2007, pp. 899–916.
20. Choong, S. T., M. H. Cole, and E. Kutanoglu. Empty Container Management for Intermodal Transportation Networks. *Transportation Research Part E: Logistics and Transportation Review*, Vol. 38, No. 6, 2002, pp. 423–438.
21. Nickel, S., A. Schöbel, and T. Sonneborn. Hub Location Problems in Urban Traffic Networks. *Mathematical Methods on Optimization in Transportation Systems*, Vol. 48, 2001, p. 95.
22. Nijkamp, P., and A. van Delft. *Multi-Criteria Analysis and Regional Decision-Making*. Springer Science & Business Media, Amsterdam, the Netherlands, 1977.
23. Gelareh, S., and S. Nickel. A Benders Decomposition Algorithm for Single Allocation Hub Location Problem. *Annual International Conference of the German Operations Research Society*, Saarbrücken, Germany 2007.
24. Gelareh, S., and S. Nickel. New Approaches to Hub Location Problems in Public Transport Planning. Technical report, Department of Optimization, Fraunhofer Institute for Industrial Mathematics (ITWM), D 67663 Kaiserslautern, Germany, 2007.
25. Kim, J.-G., and D.-W. Tcha. Optimal Design of a Two-Level Hierarchical Network with Tree-Star Configuration. *Computers & Industrial Engineering*, Vol. 22, No. 3, 1992, pp. 273–281.
26. Lee, Y., B. H. Lim, and J. S. Park. A Hub Location Problem in Designing Digital Data Service Networks: Lagrangian Relaxation Approach. *Location Science*, Vol. 4, No. 3, 1996, pp. 185–194.
27. Zhang, L. The Three-Hub Center Allocation Problem. *Proc., 47th Annual Southeast Regional Conference*, Clemson, SC, 2009.
28. Contreras, I., E. Fernández, and A. Marín. Tight Bounds from a Path Based Formulation for the Tree of Hub Location Problem. *Computers & Operations Research*, Vol. 36, No. 12, 2009, pp. 3117–3127.
29. Contreras, I., E. Fernández, and A. Marín. The Tree of Hubs Location Problem. *European Journal of Operational Research*, Vol. 202, No. 2, 2010, pp. 390–400.
30. Klincewicz, J. G. Hub Location in Backbone/Tributary Network Design: A Review. *Location Science*, Vol. 6, No. 1, 1998, pp. 307–335.
31. Lee, C.-H., H.-B. Ro, and D.-W. Tcha. Topological Design of a Two-Level Network with Ring-Star Configuration. *Computers & Operations Research*, Vol. 20, No. 6, 1993, pp. 625–637.
32. Wang, Z., C. Lin, and C.-K. Chan. Demonstration of a Single-Fiber Self-Healing CWDM Metro Access Ring Network with Unidirectional OADM. *IEEE Photonics Technology Letters*, Vol. 18, No. 1, 2006, pp. 163–165.
33. Gelareh, S., and S. Nickel. Hub Location Problems in Transportation Networks. *Transportation Research Part E: Logistics and Transportation Review*, Vol. 47, No. 6, 2011, pp. 1092–1111.
34. Yoon, M.-g., Y.-H. Baek, and D.-w. Tcha. Optimal Design Model for a Distributed Hierarchical Network with Fixed-Charged Facilities. *Management Science and Financial Engineering*, Vol. 6, No. 2, 2000, pp. 29–45.
35. Gelareh, S. Hub Location Models in Public Transport Planning. PhD thesis, Vom Fachbereich Mathematik der Technischen Universität Kaiserslautern, Germany, 2008.
36. Gelareh, S., and S. Nickel. A Benders Decomposition for Hub Location Problems Arising in Public Transport. *Annual International Conference of the German Operations Research Society*, Saarbrücken, Germany Springer, Berlin, Heidelberg, Germany, 2007, pp. 129–134.
37. Rietveld, P. *Multiple Objective Decision Methods and Regional Planning*. North-Holland, Amsterdam, the Netherlands, 1980.
38. Roy, B., and P. Vincke. Multicriteria Analysis: Survey and New Directions. *European Journal of Operational Research*, Vol. 8, No. 3, 1981, pp. 207–218.
39. Giuliano, G. A. Multicriteria Method for Transportation Investment Planning. *Transportation Research Part A: General*, Vol. 19, No. 1, 1985, pp. 29–41.
40. Vuchic, V. R. *Urban Transit Systems and Technology*. Hoboken, NJ, John Wiley & Sons, 2007.

The Standing Committee on Paratransit (AP060) peer-reviewed this paper (18-03001).

TRR
JOURNAL OF THE TRANSPORTATION RESEARCH BOARD

Article

Transportation Research Record
2018, Vol. 2672(8) 629–638

DOI: 10.1177/0361198118794537
journals.sagepub.com/home/trr
SAGE

Analysis of Paratransit Feeder-Service Pilot: Projected versus Actual Ridership and Cost-Benefit Results

Todd Hansen[1]

Abstract

The research presented examines projected and actual operational cost savings from the implementation of a paratransit feeder-service pilot in Houston, Texas. Transit agencies can offer feeder service by utilizing demand–response paratransit vehicles to connect to fixed-route rail and buses for the remaining portion of the trip. Feeder service has the potential to create savings in operating costs for transit agencies along with additional travel options for ambulatory paratransit customers. At the direction of Houston METRO, researchers projected the number of trips that would transfer to feeder service from standard paratransit service to support the agency in designing a pilot. Subsequently, the researchers analyzed the actual feeder-service trip data at the conclusion of the pilot period to assist the agency in studying the final pilot results. The presented research compares projected trip-demand and operational costs conducted before the pilot period to actual trip and cost savings during the 6-month pilot to determine the rate of trip transfer between regular paratransit service and combination feeder-service connecting paratransit to fixed-route service. The research analyzes the cost per revenue hour for feeder-service trips compared with the pre-pilot projections and the actual cost of the same trip using entirely paratransit service to show the net benefit for the transit agency. The results show lower operational costs for paratransit feeder-service compared with regular paratransit service during the pilot period. Transit agencies interested in implementing feeder service can use the research methodology to estimate the impacts of potential paratransit feeder trips.

METROLift is the paratransit division of the Metropolitan Transit Authority of Harris County (METRO), the designated public transit authority in Houston, Texas. Since 1979, METROLift has provided scheduled demand–response service for persons with disabilities to and from requested origins and destinations in Harris County. METROLift initially provided service within the central area of Houston and gradually expanded its service area to keep up with increasing trip demand and legislation from the Americans with Disabilities Act (ADA) of 1990 (*1*). METRO created the *METROLift Moving Forward* initiative in 2013 with a series of community outreach workshops to gain public feedback on policy topics related to quality of service and sustainability amid rising operational costs (*2*). One of the policy areas discussed during the workshops was travel training/feeder service. METROLift had not operated a formal feeder-service program to that point and was interested in customer perceptions of a potential program. The workshop series had a total of 334 participants, 52 of which were METROLift users. Seventy-two percent of workshop respondents either Strongly Agreed or Agreed that METROLift should develop feeder service to fixed-route stations and transit centers (*1*, *3*). METRO requested that the researchers study the potential impacts of feeder service on passenger trips and operational costs to assist the agency in designing a feeder-service pilot.

Within this research, the term "feeder service" means demand–response service designed to connect customers to fixed-route service, rather than linking fixed routes to other fixed routes. The concept of the METROLift feeder-service pilot was to provide an option for riders to use fixed-route service for a portion of their trip rather than using paratransit demand–response service for the entire trip. METROLift designed the pilot to have feeder-service trips exclusively connect to fixed routes at METRO facilities rather than sidewalk bus stops to ensure better accessibility features at the service transfer

[1]Texas A&M Transportation Institute, Houston, TX

Corresponding Author:
Address correspondence to Todd Hansen: t-hansen@tti.tamu.edu

point. Feeder service has the benefit of greater travel options for paratransit riders by utilizing frequent, scheduled fixed-route service. Paratransit feeder-service operates similarly to regular ADA paratransit service in terms of reserving trips with a scheduler at the transit agency. Depending on the service design, feeder service could also benefit paratransit riders through faster travel times, closer reservation windows, and fare discounts. Feeder service provides the transit agency the opportunity for operating cost savings by transporting paratransit riders part of the way on a fixed-route at no additional cost for a portion of the overall trip.

This research assisted METRO in understanding the cost and ridership impacts of trips shifting from the METROLift paratransit service for the entire trip over to regular fixed-route service for a portion of the trip. The success of the pilot's impacts were to be determined through the ability to coordinate paratransit and fixed-route schedules, training, and outreach to paratransit customers, and the amount of trip destinations near METRO fixed-route facilities. The research accounted for conservative trip transfer rates assumed for the feeder-service pilot due to it being an optional service, meaning customers could choose to schedule feeder trips to take advantage of incentives built into the pilot program. The research used METROLift trip manifest data before and during the pilot period to calculate the number of trips using feeder service and quantify operational cost savings of trips connecting to fixed-route service. In conjunction with the projected impacts analysis before the pilot, researchers conducted a peer review of feeder-service practices at other transit agencies in the United States and assisted METRO in determining the structure of service for the pilot (*4*). The feeder-service pilot began on January 1, 2017 and concluded on June 30 of the same year.

Literature Review

The researchers reviewed previous literature and studies about feeder-service concepts as well as peer transit agencies' practices on feeder service, including the topics of pricing, policies, and performance measurement. Paratransit feeder-service as a combination of demand–response service and general public fixed-route transit has been summarized within Transit Cooperative Research Program (TCRP) *Synthesis 76: Integration of Paratransit and Fixed-Route Transit Services*. Paratransit feeder-service is typically more prosperous in situations with successful marketing to customers and use of available technology to schedule trips (*5*). The TCRP synthesis found that geographically elongated service areas can be well-suited to feeder service due to longer trip lengths, and that feeder service is better received in areas with a frequent fixed-route service. The synthesis also found that conditional eligibility practices for feeder service can result in significant cost savings, whereas voluntary policies for feeder-service trips as requested by customers typically generate little interest in the service (*5*).

Benefits of paratransit feeder-service include operational cost savings, shorter passenger trip lengths, increased fixed-route ridership, and increased independence for persons with disabilities (*2*). TransLink in Vancouver, Canada, reported that the feeder-service trips it operates cost the agency less than half as much as a similar trip taken by demand–response paratransit (*6*). In general, feeder-service legs of the trip on paratransit average about 41 min (*5*), whereas the average travel time for demand–response paratransit is around 90 min (*7*). ACCESS in Pittsburgh reported increases in fixed-route ridership after implementing feeder service (*6*). For paratransit customers, feeder service provides an additional travel option to take advantage of fixed-route service through a safe connection and transfer while reducing personal dependence on ADA paratransit (*6*, *8*, *9*). Amenities at the transfer point such as benches and shelters can be especially important for ADA paratransit customers using feeder service, and long travel times and transfer windows should be avoided by transit agencies scheduling feeder-service trips (*9*).

Previous research conducted with Houston METRO found that the transit agency could reduce operational costs and allocate resources more effectively if some paratransit riders completed part of their trip using feeder service, and fixed-route stops could be easily accessed on the ground (*2*). The research used an October 2012 sample of trips to identify trip points within a ¼-mi of METRORail stations, longer than 30 min in travel time, and occurring during peak morning and afternoon hours. Researchers found that 4.2% of trips in the sample qualified for potential feeder-service and applied a cost savings average per trip based on 5%, 10%, and 15% trip transfer rates from paratransit service to feeder-service (*2*). Researchers used this method as a starting point to begin the analysis process for the feeder-service pilot with more recent data and to expand potential feeder trips to connect more types of METRO facilities (*2*).

Data and Methods

The researchers calculated the average costs of METROLift paratransit service to project the passenger trip and cost impacts of feeder service given the data available. Researchers focused on METROLift manifest data to look at where METROLift trip points occurred around fixed-route facilities to project trip transfer rates to feeder service for the pilot period. This section discusses the research setting, pilot structure, and data used in the analysis.

Research Setting

The research setting was within the service area of Houston METRO located in Harris County, Texas. The research involved both the fixed-route and paratransit services of Houston METRO to facilitate the scheduled feeder-service connections for trips during the pilot period.

METROLift Paratransit and METRO Fixed-Route Services. METROLift provides curb-to-curb and door-to-door services within the required ¾-mi area around fixed routes (as determined by the ADA) and beyond the required area to provide persons with disabilities the opportunity to make the trips they desire. As of July 3, 2013, METROLift had 17,340 active customers using their demand–response service (*2*). METRO provides fixed-route service chiefly through local bus routes, commuter bus routes, and light rail lines. In August 2015, METRO began operations of its new bus network schedule, which reconfigured local bus routes to have more direct travel paths, increased service frequencies, and created more transfer points between routes.

Pilot Design. The researchers assisted METRO with looking at peer practices of other transit agencies operating feeder services in the United States to help determine how agencies decided to schedule feeder service, including trip length, types of transfer points, eligibility for feeder service, fare policies, and scheduling systems. METRO used the peer information in determining policies and scheduling for its feeder-service pilot along with consideration of the best interests of its paratransit customers. METRO decided that the best practice for the pilot was to allow customers to choose to schedule feeder-service trips rather than creating a decision process to determine which trips would be mandatory for feeder service. In other words, scheduling feeder-service trips would be completely voluntary for customers. METRO would only determine eligibility for feeder service by confirming customer eligibility for ADA paratransit and the connection of a requested feeder trip to a METRO fixed-route facility. The METROLift feeder-service pilot setup and subsequent projection analysis before the pilot period used the following assumptions on feeder-service design:

- METROLift vehicles will transport customers for part of the trip length; METRO fixed-route buses and rail will be used for the rest of the trip.
- METROLift vehicles will connect to fixed-route service at METRO transit centers, park and rides, and light rail stations only, as opposed to any bus stop in the system.
- Feeder-service trips are scheduled so that customers will not need to transfer between different fixed-route services.

Through background research of other feeder-service programs around the country, METROLift decided on two incentives during the pilot period for customers to schedule feeder trips. The major incentive was that any METROLift-eligible customer who took a feeder-service trip would be able to take the paratransit portion of the trip free of charge. Eligible METROLift customers can already ride fixed-route vehicles for free with the Freedom Q Card program, which means that both legs of the feeder trip would be entirely fare free. The other incentive during the pilot was an allowance for repeat feeder-service users to schedule feeder-service trips on the same day, up to a few hours before the trip, depending on vehicle capacity. The same day scheduling incentive allowed feeder-service customers to schedule trips within the normal 24-h reservation deadline and provided more flexibility for customers to manage their personal schedules. Normally METROLift customers must schedule a trip at least 1 day in advance, with a time window up to 5 p.m. the day before.

The feeder-service program aimed to take advantage of METRO's accessible fixed-route facilities and fleet while making the connection to the fixed-route system from areas in METRO's service area that do not have a strong fixed-route service or sidewalk accessibility (*1*). The feeder-service pilot included the following policies:

- Feeder trips would be determined by customer self-selection, never mandatory for any ADA paratransit customers.
- Only ADA paratransit customers eligible to use METROLift would be eligible to schedule feeder-service trips.
- Any trips taken by feeder service would be entirely fare-free for METROLift customers on all METROLift and METRO fixed-route vehicle legs of the trip.
- Customers scheduling feeder-service trips can call into METROLift on the same day of the desired travel to schedule the trip—METROLift would meet the scheduling request as capacity is available.
- Feeder service would be available 7 days a week between the hours of 6 a.m. and 9 p.m.—METROLift created a new telephone number specifically for feeder-service trip reservation.

- METROLift would extend its guaranteed-ride program to cover feeder-service customers in case any connections to fixed-route service are missed, ensuring customers are never stranded.
- Customers would be dropped by the METROLift paratransit vehicle directly at the bus bay or near the rail platform entrance needed to transfer to fixed-route service. METROLift would then pick the customer up at the same location for the return trip.

METRO trained METROLift staff members and drivers on scheduling feeder-service trips with the fixed-route schedule, dispatching vehicles with both feeder service and ADA paratransit riders, and picking up/dropping off customers at fixed-route facilities. METRO provided outreach sessions to customers prior to the pilot to talk about how feeder service works and how to utilize travel training and schedule information to take fixed-route service for a portion of the trip.

Fixed-Route Facilities. METRO determined that fixed-route stations/facilities, totaling 79 possible transfer locations, would be the best fit for connecting customers to fixed-route service using feeder service. At the time of the pilot, METRO operated services at 44 light rail stations, 21 transit centers, and 27 park-and-ride stations; some facilities share multiple services such as a rail station and bus transfer center. Figure 1 depicts the locations of the fixed-route facilities and light rail lines in METRO's service area using Houston METRO's online geographic information system (GIS) maps. The transit centers are depicted by the light purple bus icons, while the park-and-ride stations are depicted by blue parking icons.

Study Design

Researchers used the METROLift manifest data from the January to June 2017 period to determine the travel times and trip distances between METROLift feeder-service trips and all other METROLift trips during the 6-month pilot period. Researchers utilized tools in Microsoft Excel and ESRI ArcGIS® software to organize the data and calculate trip data and operational cost per trip. These data were then compared with the initial analysis projecting trip transfer rates to feeder service and associated operational impacts that assisted METROLift in putting together the pilot.

Projection Analysis. The initial projection analysis conducted aimed to assess the potential rate of trips that would transfer from regular METROLift service to feeder service based on assumptions of trip locations, trip lengths, and customer preferences. The analysis used a week's sample of METROLift manifest data from April 10 to 16, 2016, during which 27,751 total completed METROLift trips occurred. METRO asked researchers to look at METROLift trips during the sample period that occurred in proximity to METRO rail stations, transit centers, and park-and-ride stations. The objective of the feeder-service pilot was to schedule connections between METROLift vehicles at METRO fixed-routes at transfer facilities rather than standard sidewalk bus stops, allowing feeder service to utilize the better accessibility features present at the facilities and the frequent service connections from the fixed routes (*4*).

The researchers geolocated trip addresses from the April 2016 week sample and identified trip pickups and dropoffs that occurred within a ¼-mi of METRO fixed-route facilities. Most research on transit user behavior uses a ¼-mi as the maximum acceptable distance a person is comfortable walking to a transit stop from their trip origin point. In North American cities from the late 1960s through the 1980s, between 75% and 80% of transit passengers walked no farther than a ¼-mi to get to bus stops (*10*). Some research uses a ½-mi buffer area as a comfortable walking distance for transit, particularly in the case of walking to rail stations. Researchers kept the ¼-mi area for all fixed-route facilities regardless of whether METRO light rail service was present at the facility because of feeder service dealing with ADA paratransit customers. The equidistant ¼-mi buffer area used to identify the geolocated trip address is imperfect in reality due to potentially unknown physical barriers or disconnected street networks that may exist on the way to or from a fixed-route facility, preventing some customers from being able to access their ultimate trip destination from a METRO facility were they to use feeder service (*4*).

Researchers used GIS software tools to clip the pickup and dropoff points from the METROLift April 2016 week trip sample with ¼-mi buffer zones around METRO facilities, factoring out any trip points that were outside the buffer zones. A trip point within a buffer zone was assigned to the METRO facility as part of the stop's potential total of feeder-service trips that could be connected to fixed-route transit. If a trip point fell within two or more buffer zones, which typically only occurred at light rail station platforms in the Houston central business district, researchers duplicated the point for each additional buffer zone (*4*).

Researchers then used the XY to Line toolbox feature in ArcGIS® to determine the trip lengths between the pickup and dropoff points geolocated within METRO's facilities. The concept of the upcoming feeder-service pilot was for METRO to serve paratransit customers choosing the feeder option by connecting the customer

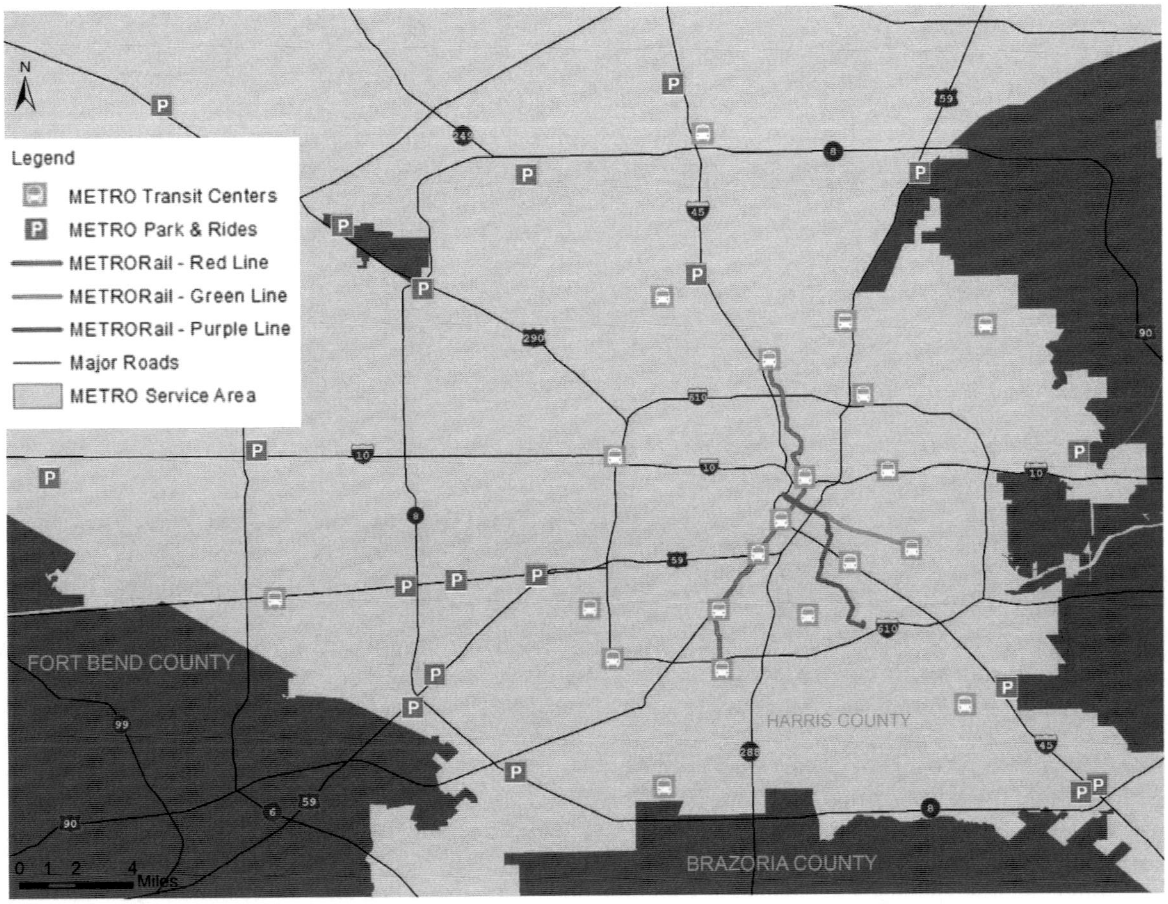

Figure 1. Locations of METRO fixed-route facilities.

to the nearest METRO facility that would then get them to their destination by transferring a single fixed-route. The transfer between a paratransit vehicle and a fixed-route vehicle is most useful when the overall trip distance is longer and the fixed-route can transport the customer for a significant length of the overall trip. If the overall trip distance is relatively short, it is more useful for METROLift to transport the customer by the paratransit vehicle for the entire trip length rather than the customer having to spend time transferring to a different vehicle.

Researchers used these assumptions of a ¼-mi proximity of either a pickup or dropoff point to a METRO facility and an overall paratransit trip length of at least 5 mi between the pickup and dropoff to identify the number of trips in the sample period that would be potential candidates for using feeder service. Researchers determined 5 mi as a suitable distance for a feeder paratransit trip to be operationally feasible, as feeder trip lengths with shorter distances would make more sense on paratransit alone without a transfer to fixed-route service. Researchers identified a subtotal of 2,683 total pickups and dropoffs meeting these criteria in the April 2016 week sample, approximately 9.7% of all METROLift trips. Table 1 shows the geolocation and trip point results for the week sample along with daily and 6-month calculations (*4*).

For the projection analysis, researchers used ArcGIS to determine the nearest METRO facility for the trip point not within a ¼-mi of any facility. For example, if the dropoff point was located within a ¼-mi of the METRO facility, the distance between the corresponding pickup point and the METRO facility was calculated. This distance represented the trip length for feeder service to connect to the fixed-route facility so that the customer can reach their final destination. Researchers created two groups of trips out of the total of 2,683 trip points found: one group with all of the trips and another subgroup of long trips with a greater than 5-mi difference between the feeder trip portion and the true

Table 1. Projected Trips Transferring to Feeder Service

Trip points and METROLift facilities	Week sample of trips	Calculated daily trips	Calculated 6-month trips
Rail station pickups	795	114	20,729
Rail station dropoffs	877	125	22,867
Transit center pickups	393	56	10,247
Transit center dropoffs	415	59	10,821
Park-and-ride pickups	105	15	2,738
Park-and-ride dropoffs	98	14	2,555
Subtotal of trips meeting criteria	2,683	383	69,957
Total METROLift trips	27,751	3,964	723,588
% of total	9.7%	9.7%	9.7%
10% transfer of subtotal trips to feeder service	268	38	6,996

distance between the pickup and dropoff points (what would be the standard paratransit trip). The distinction between all trips and long trips in the sample was made to highlight how feeder service would be particularly useful for the agency to operate in cases when the feeder-service trip is significantly shorter than a paratransit trip for the whole trip length.

The researchers used a cost per revenue mile average of US$2.55 that was provided through the METROLift annual budget in October 2015. The cost figure represents variable operating costs and does not include fixed costs such as administrative or facility rental expenses. Researchers applied cost per revenue mile to the trip lengths for the paratransit trips in the sample and the calculated trip distance of using feeder service for the same trip. The researchers used cost per revenue mile during the initial projection analysis because data was measured in trip distance as opposed to travel time. The researchers used a 10% trip transfer rate of the 2,683 trips in the week sample to represent a conservative number of customers choosing to take feeder service for their trip (*4*). Given a 10% transfer rate of trips meeting these criteria, the projected amount of feeder-service trips over a 6-month period using an average of 30.42 days per month would be 6,996. Figure 2 illustrates the process used in the projected cost–benefit analysis, the results of which are discussed later in this research paper.

METROLift Manifest Data. For the analysis of the actual pilot data, researchers requested manifest data from METRO for all METROLift trips that occurred between January 1 and June 30, 2017. Before the pilot, METRO created a code in the scheduling system to indicate which trips within ADA paratransit were designated and scheduled as feeder-service trips. The scheduling code allowed researchers to easily determine which trips in the manifest should be analyzed as feeder service and which ones were regular ADA paratransit trips. The manifest data included addresses for the pickup and dropoff points, codes indicating whether the scheduled trip was

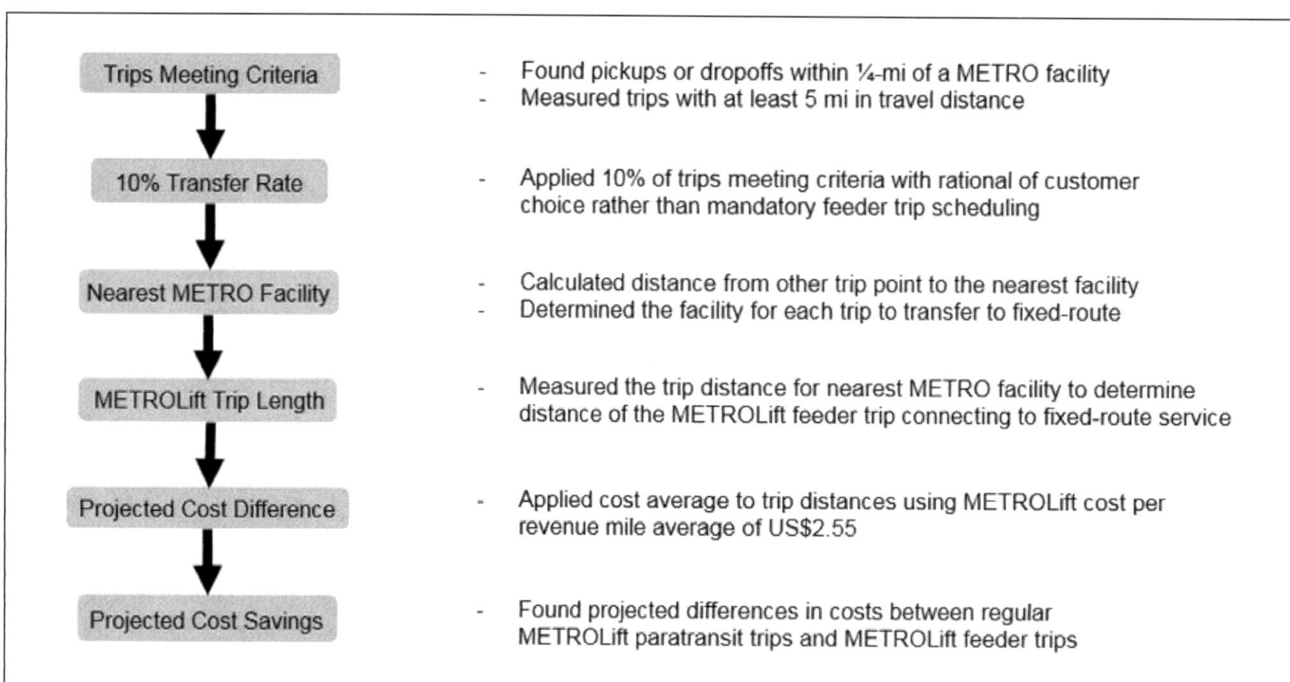

Figure 2. Process of projected feeder-service cost savings.

completed without a no show or cancellation, scheduled and actual arrival times, and actual mileage driven by the vehicle during the trip. METROLift completed 756,810 paratransit trips and 1,844 feeder-service trips during the 6-month feeder pilot period.

Paratransit Feeder Trip Segments versus Fixed-Route Trip Segments. One data restriction that limited the pilot period analysis was the unavailability of the corresponding boardings and alightings data for the fixed-route portion of the trip that feeder service connected to. Fixed-route data were not available for customer privacy reasons, protecting information from electronic fare media, and due to the differences in software systems between the fixed-route and METROLift services. With such data, the analysis could have potentially mapped out the arrival time of the customer to their final destination using the address/stop point at the boarding of the return trip starting with fixed-route service. In this scenario, the analysis would have compared the time traveled by the customer from their origin to their destination using a combination of feeder service and fixed-route in comparison with what the travel time would have been using paratransit service alone.

Researchers could identify the addresses of METRO facilities in the METROLift manifest data, but without the corresponding address of the other facility at the end of the trip, the analysis was incapable of determining the final destinations of feeder trips.

Cost Savings of Feeder Service

An appealing aspect about feeder service is that it utilizes the existing fixed-route system to complete the remaining length of the trip with any additional cost imposed on the fixed-route schedule. Researchers looked at feeder trip manifest data to calculate the cost of providing a feeder-service trip, utilizing the measures of average cost per revenue hour and revenue mile. The cost averages were determined from the METROLift October 2015 budget, which showed average cost rates for METROLift van and minivan services at US\$39.57 per revenue hour or US\$2.55 per revenue mile. The revenue cost figures are weighted averages of METROLift variable costs for service performed by taxis and vans. The cost per revenue mile figure was used for the initial projection analysis because trip distance was the determining factor in whether the trips might be scheduled as feeder service.

For the analysis of the actual feeder-service pilot data, researchers used the actual travel time calculated from the METROLift manifest data along with the average cost figure of US\$39.57 per revenue hour. Researchers used the cost per revenue hour figure for the pilot analysis because the travel time data were a better measure of the effort of operational cost for METROLift service. Researchers added up the total amount of actual travel time for the completed feeder trips and other METROLift trips from January through June, then multiplied the totals by the average cost figure. The method yielded a cost per trip of US\$13.30 for feeder-service trips, based on an average trip length of 20.2 min, and US\$25.67 for other METROLift trips during the pilot period. Table 2 shows the results of the analysis.

Discussion of Results

Researchers considered the initial analysis of projected feeder-service utilization and operational costs before the feeder-service pilot and actual feeder-service utilization during the pilot. Both analyses found a benefit for

Table 2. Cost Per Trip Comparison During Pilot Period

Month	Number of trips	Total travel time (minutes)	Total cost (US\$)	Cost per trip
METROLift feeder service (January through June 2017)				
January	100	2,061	\$1,359	\$13.59
February	161	3,086	\$2,035	\$12.64
March	346	6,998	\$4,615	\$13.34
April	372	8,760	\$5,777	\$15.53
May	438	8,771	\$5,784	\$13.21
June	427	7,516	\$4,957	\$11.61
Grand total	1,844	37,192	\$24,528	\$13.30
METROLift paratransit service (January through June 2017)				
January	119,067	4,721,100	\$3,113,565	\$26.15
February	116,170	4,632,156	\$3,054,907	\$26.30
March	130,719	5,229,991	\$3,449,179	\$26.39
April	115,543	4,710,142	\$3,106,339	\$26.88
May	139,961	5,113,594	\$3,372,415	\$24.10
June	135,350	5,049,552	\$3,330,180	\$24.60
Grand total	756,810	29,456,535	\$19,426,585	\$25.67

Table 3. Feeder-Service Trips by METROLift Customers

Number of feeder trips per customer (January–June 2017)	Number of feeder trip customers	Average trip distance (miles)	Average trip time (minutes)
Fewer than 5 trips	83	3.4	17.4
Between 5 and 10 trips	24	3.4	20.0
Between 11 and 50 trips	16	4.0	20.5
More than 50 trips	4	4.6	20.7

Table 4. Difference in Feeder Service and Paratransit Service Costs

Feeder service and paratransit service from January–June 2017

Trip type	Total travel time (minutes)	Total travel time (hours)	Multiplied by METROLift cost per revenue hour (US$)	Divided by number of completed trips	Equals cost per trip (US$)
Feeder service	37,192	619.9	$39.57	1,844	$3.30
Regular paratransit	29,456,535	490,942.3		756,810	$24.42

Scenario: Feeder service versus no feeder service

Trip type	Number of completed trips	Multiplied by average travel time (minutes)	Total travel time (hours)	Multiplied by METROLift cost per revenue hour (US$)	Equals total operating cost (US$)
Feeder service	1,844	20.2	619.9	$39.57	$24,528
No feeder service		38.9	1,196.2		$47,334

METRO in operational cost from using fixed-route service for a portion of the trip length, however utilization rates were not as anticipated. This section of the paper discusses the findings.

Feeder-Service Utilization

In the initial projected feeder-service analysis, researchers used a sample of manifest data from a week with fair, non-precipitous weather to control for any adverse environmental effects on travel. The data sample was then extrapolated to daily and 6-month calculations of projected feeder trips based on the distance and a 10% transfer rate criteria. The 6-month subtotal of projected trips shown in Table 1 would translate to 6,996 potential trips over the period (*4*).

However, the projected numbers were higher than the actual number of feeder-service trips during the 6-month pilot. The total number of feeder-service trips from January to June 2017 was 1,844. Feeder-service utilization during the pilot was lower than anticipated; it may have been lower due to reasons such as the introduction of an unfamiliar service type, gradual dissemination of information about the pilot program, the given trip destinations not being located near fixed-route facilities, customer preferences about using the fixed-route system, and some types of disabilities being incompatible with using fixed-route service comfortably, safely, or both. The number of feeder-service trips by month did gradually increase as the pilot progressed, reaching over 400 trips per month by the end of May 2017, as shown in Table 2.

Researchers calculated the amount of repeat feeder-service customers from manifest data to get a sense of how frequently customers utilized the pilot program. Most customers, 83 persons, used feeder service to fewer than five trips during the pilot period. Four customers used feeder service over 50 times during the pilot, representing the most frequent feeder-service customers. Researchers did not find dramatic differences in trip distances and travel times between the utilization frequencies. Table 3 shows the ranges of feeder-service trips taken by number of customers.

Differences in Operational Costs

Both the initial projected analysis and the final pilot period analysis used cost averages data from the October 2015 METROLift budget data to calculate the operational costs of feeder-service trips compared with if the trips in question were completed entirely with paratransit vehicles.

Table 5. Comparison of Projected Analysis and Actual Pilot Cost Savings (US$)

Projected analysis (calculated 6-month trips)			Actual pilot (January–June 2017)		
Projected paratransit service cost	$183,311	Percent savings from total	Actual paratransit service cost	$47,334	Percent savings from total
Minus projected feeder-service cost	$90,634		Minus actual feeder-service cost	$24,528	
Equals operating cost savings	$85,132	46%	Equals operating cost savings	$22,806	48%

Actual Pilot Period. Table 4 shows the analysis method and calculations of the actual feeder-service pilot data. Researchers used the total travel time in minutes from the METROLift manifest data, translated it into travel time hours by dividing by 60, then multiplied it by the average cost per revenue hour figure US$39.57. Dividing the total cost per hour by the number of completed trips, researchers found that the cost per trip for feeder service, US$13.30, was nearly half of the cost per trip for all other METROLift service.

Researchers used the average travel time differences between the service types to determine the operating cost for the feeder trips had they been completed as standard paratransit trips. In this scenario, the average feeder trip travel time is adjusted from 20.2 min to 38.9 min, then multiplied by the same cost per revenue hour figure to find the total cost differences. The 1,844 feeder trips cost METRO US$24,528 as feeder service but would have cost US$47,334 if delivered as ADA paratransit service for the entire trip length, resulting in an operational savings of US$22,806 for the agency.

Comparison with Projected Cost Savings. The previous analysis of the projected impacts of feeder service was conducted with specific distance parameters and a conservative 10% transfer rate of trips from standard paratransit service to feeder service. The analysis projected that the availability of feeder service for 6,996 trips over a 6-month period would provide a semi-annual net cost saving of US$85,132 at existing trip-demand levels, as shown in Table 5 (*4*). The projected cost savings from the initial analysis resulted in a 46% estimated operating cost saving, whereas the analysis of actual pilot data resulted in a 48% operating cost saving from operating feeder service for a portion of trips instead of regular paratransit service.

The calculated cost savings assumes that all trips taken using feeder service during the pilot period would have been taken with standard METROLift paratransit service instead had the pilot not been in place, as only ADA paratransit-eligible customers were entitled to schedule feeder service. However, some of the feeder trips taken may have been in addition to the usual demand for other METROLift trips rather than displacing those trips; additional data would be required to determine this.

Suggestions for Future Research

The research for this analysis focuses on the utilization of the METROLift manifest data from the feeder-service pilot period. The data are helpful for measuring the amount of service delivered and determining costs, but they do not answer more specific questions about customer motivations and other potential factors influencing utilization of feeder service. Some of these factors may include the effects of weather, type of disability, use of mobility aides, age, impact on travel time, trip purpose, and socioeconomic variables. Further analysis efforts could be conducted with additional data resources and methodology steps, such as comparing feeder-trip reservations with frequent paratransit trips as a measure of trip transfer between services, filtering home origin data through census geographies to determine area demographics, or matching customer database information with requested trips to discover specific customer factors. Future research of feeder-service utilization with respect to these factors would be useful for a more thorough assessment of the potential of paratransit feeder-service.

Conclusions

The analysis of manifest data from Houston METRO's feeder-service pilot shows lower operational costs per trip on average for feeder-service trips than other METROLift ADA paratransit trips. The lower costs are the result of reduced trip travel times and distances for feeder-service customers on paratransit vehicles and no additional operational cost for the transit agency to transport the customer on already scheduled fixed-route service. The pilot did not generate as many trips on feeder service as projected, but the transit agency still saved on operational costs for each feeder-service trip on average during the 6-month period. The cost savings were achieved by utilizing existing staff, technology systems, and vehicles by implementing policies that would allow staff to assist customers in scheduling trips on feeder service and planning trips to connect with the fixed-route service schedule.

More notably, the transit agency did not implement a mandatory program but instead offered customers the choice to schedule trips using feeder service, consistent with agency goals for feeder service to be an additional option for customers rather than required feeder trip scheduling based on conditional eligibility. Any future efforts with feeder service for the agency could use similar feeder-service scheduling and incentive policies in conjunction with increased awareness of the program to give more customers the knowledge to consider feeder service as a travel option. The resulting cost savings of operating feeder service for a portion of its overall service will allow the agency to apply those savings toward providing more trips for ADA paratransit customers.

Author Contributions

The authors confirm contribution to the paper as follows: study conception and design, data collection, analysis and interpretation of results, and draft manuscript preparation: TH. All authors reviewed the results and approved the final version of the manuscript.

References

1. Hansen, T. *History of Houston METROLift White Paper*, Metropolitan Transit Authority of Harris County, Houston, 2017.
2. Tan, S., and T. Hansen. Impacts of Coordination between Fixed-Route and Paratransit on Service and Sustainability. Presented at 94th Annual Meeting of the Transportation Research Board, Washington, D.C., 2015.
3. Edrington, S., L. Cherrington, and T. Hansen. *METROLift Moving Forward: Workshop and Community Feedback Outcome Report*. Texas A&M Transportation Institute, College Station, TX, June 2013.
4. Hansen, T. Cost-Benefit Analysis of Estimated Paratransit Feeder Service Demand. Presented at 96th Annual Meeting of the Transportation Research Board, Washington, D.C., 2017.
5. Weiner, R. *TCRP Synthesis 76: Integration of Paratransit and Fixed-Route Transit Services*. Transportation Research Board of the National Academies, Washington, D.C., 2008.
6. *Issue Paper: Transit Sustainability and Integration for Alameda County Transportation Commission (ACTC)*. Alameda County Transportation Commission. http://www.alamedactc.org/files/managed/Document/2418/Transit_Sustainability_and_Integration.pdf. Accessed July 8, 2014
7. *A Guide to Paratransit Services*. Dallas Area Rapid Transit. https://www.dart.org/riding/paratransitguide.asp. Accessed July 8, 2014.
8. *Atlanta Regional Coordinated Human Services Plan*. Atlanta Regional Commission, Atlanta, GA, February 2007. http://documents.atlantaregional.com/tcc/HST/2012-2013_HST_Plan_Limited_Update_No_Appendices.pdf. Accessed July 8, 2014.
9. Thatcher, R., and C. Ferris. *TCRP Report 163: Strategy Guide to Enable and Promote the Use of Fixed-Route Transit by People with Disabilities*. Transportation Research Board of the National Academies, Washington, D.C., 2013.
10. Kittelson & Associates, Inc., Parsons Brinckerhoff, KFH Group, Inc., Texas A&M Transportation Institute, Arup. *TCRP Report 165: Transit Capacity and Quality of Service Manual*, 3rd ed. Transportation Research Board of the National Academies, Washington, D.C., 2013.

The Standing Committee on Paratransit (AP060) peer-reviewed this paper (18-01752).

TRR
JOURNAL OF THE TRANSPORTATION RESEARCH BOARD

Article

Transportation Research Record
2018, Vol. 2672(8) 639–648

DOI: 10.1177/0361198118797829
journals.sagepub.com/home/trr
SAGE

Probabilistic Fleet Sizing and Routing Problem to Minimize Mobility Disparities

Marouane Zellou[1] and Mahdieh Allahviranloo[2]

Abstract

The objective of this paper is to develop an analytical model to analyze the mobility behavior of a target population and to minimize the disparities. The analysis is conducted using household travel survey data collected from 7993 residents of five boroughs of New York City, where citizens with mobility impairments are considered as our target population. The study starts by quantifying the existing gap in the patterns of the mobility-impaired individuals and the base population followed by the generation of new activities in the agenda of the target population using the patterns of their nearest neighbors in the base population. Activities are generated based on copula sampling method. Attributes of activities forming the new agenda of the target population are the inputs that will be used for the design of an optimum ridesharing system. By integrating probabilistic models with the existing routing models, we develop a methodology to identify the optimum fleet size, optimum route, and optimum schedule such that all requests of the physically impaired travelers are fulfilled within a reasonable wait time for ride requesters.

Time-use behavior of individuals and whether individuals of different groups spend time in the same types of places are important for quality of life, access to opportunities, and segregation studies. Mobility is a critical factor generating many inequalities in our society, which reinforces the importance of the development of methods to quantify the geographic scale and proximity, as well as temporal variations in daily behavior of mobile population. Variation in local contexts, both in terms of the availability of resources and social connections—e.g., where do people travel during the day? With whom do they interact? What kind of facilities do they have access to?—can produce inequalities that have long-term impacts on the wellbeing of urban dwellers. An accessible transportation system directly improves the quality of life by bringing opportunities for all segments of society to participate in activities (*1*), and lack of such system increases the risk of social exclusion (*2*). Church et al., identifies seven categories of social exclusion in relation to transportation and city formation (*3*). Furthermore, transportation and accessibility are considered as influential factors in determining social interaction potential (SIP) (*4*) and betterment of urban design relies on the methodological improvements in the spatial–temporal analysis of mobility behavior (*5*, *6*).

With the goal of minimizing mobility disparities in different sectors of the population and providing an equal and accessible mobility system, we propose a methodology to analyze mobility patterns and maybe provide solutions by bridging machine learning, probabilistic analysis, and optimization models together. In the literature, analysis of mobility in the dimension of space has been studied from different perspectives. Use of ellipses to portray mobility behavior and dynamics of urban travel behavior is one of the most common methods (*7*). Earlier models of accessibility or space–time prism analysis (*8*) were based on computing the total volume of the prisms or accessible locations within the prism. Kim and Kwan introduced an accessibility measure that designates the accessible opportunities reachable by individuals while taking into account the distribution of the opportunities as well as the time budget constraints imposed by activity–space–time prisms (*9–12*).

In this paper, first, built on the methods developed by Le Priol and Allahviranloo (*13*), disparity in the mobility patterns—time use, location, and activity purpose—of a target population and the rest of the population, named as "base population", is measured. Once the disparities

[1]Ecole Nationale des Travaux Publics de l'Etat, Vaulx-en-Velin, France
[2]The City College of New York, New York, NY

Corresponding Author:
Address correspondence to Mahdieh Allahviranloo:
mallahviranloo@ccny.cuny.edu

are measured new activities are added to the agenda of the target population using multidimensional copula sampling method (*14*, *15*). Copula sampling is widely used in different domains of engineering, economics, mathematics, etc. (*16*). In this paper, we use a multivariate Gaussian copula to sample multiple attributes of activities and insert new sets of activities in the agenda of the target population (*17*, *18*).

After generation of new patterns, a fleet-sizing, routing, and scheduling model is solved to estimate the optimum number of vehicles on the span of the day that can address the mobility demand of the target population. As one of the oldest models in transportation, fleet sizing has been investigated in the literature for the past 40 years. The first fleet-sizing problem was introduced by Ball et al. (*19*) where they use route-first, cluster-second strategy, and a greedy insertion heuristic. Golden et al. (*20*) made the fleet size term more explicit and expanded the indices to consider multiple classes of vehicles, and modified the Clarke-Wright savings method to solve the problem. Beaujon and Turnquist (*21*) developed stochastic dynamic version of fleet-sizing model, and Sayarshad and Ghoseiri (*22*) solved a fleet-sizing model over multiple periods. Ropke and Cordeau (*23*), Sexton and Bodin (*24*, *25*), and Sexton and Choi (*26*) use decomposition methods to solve fleet-sizing problem for large-scale studies.

Providing an exact solution for large-scale fleet-sizing model is a challenging mathematical problem. A variety of decomposition methods or heuristics have been proposed as solution techniques to handle the complexity of these problems. Unlike network-based formulations for fleet-sizing models, Diana et al. (*27*) optimized the fleet-sizing problem for a many-to-many demand-responsive system using a probabilistic approach. The proposed approach is time efficient for short periods of planning. Therefore, in this paper, we integrate the approach proposed by Diana et al. to estimate the optimum number of vehicles that need to serve the demand in short periods of the time. After finding the optimum fleet size in each time slice, we proceed with routing using pickup and delivery problem with time windows (PDPTW) (*28*) to provide ride for the requesters. We define our target population to be the people with mobility disabilities. Indeed, the study is very critical for cities like New York where only 112 of the 472 subway stations (24%) are equipped to accommodate mobility demand of disabled or elderly population (*29*). The infrastructure maintenance or redesign is costly and would disturb the daily operations of the subway system; therefore, designing a more flexible mobility solution for this sector of society using the existing infrastructure would be a viable solution for both short-term and long-term planning purposes. It should be noted that the authors are well aware of the existence of rich literature on the subjects of mobility pattern recognition, fleet-sizing, and vehicle routing problems, however we limit our review to the works that are directly relevant to the topic for the sake of simplicity of the paper. In the following sections, we first present some definitions used in the paper, then describe the methodology, followed by presenting the results of the empirical analysis and conclusions.

Definitions

Prior to proceeding to the next section, here, we present some of the terms as following:

a) Base population (BP): refers to the entire population in the surveyed data.
b) Target population (TP): is a subset of base population where we want to analyze and compare their patterns to the base population.
c) Nearest neighbor population (NNP): subset of base population who are not in the target population but are demographically similar to the individuals in the target population.
d) Activity patterns (Act): sequence of trips and activities made by individuals during the day. It is defined by five attributes of activity location (x), activity duration (d), activity purpose (p), activity start time (T), and trip duration (tr); $\text{Act} = \langle x, d, p, T, \text{tr} \rangle$.
e) Fixed activities: defined as a set of activities that are robust to any change in their attributes. In this study pickup/drop off, work, and school activities are categorized as fixed set of activities.
f) Flexible activities: defined as a set of activities that can be rescheduled/replaced/relocated.
g) Time slice (TS): shorter periods used to estimate the required fleet size in the probabilistic method. The duration of the time slices depends on the total number of the demand and their spatial–temporal distribution.

Methodology

The work presented here is composed of four main steps: (I) activity generation and insertion in the agenda of the individuals in the target population, (II) development of a probabilistic fleet-sizing model within each time slice, (III) routing and scheduling the fleet of vehicles within each time slice using PDPTW, and (IV) chaining the generated routes in each time slice to the routes in other time slices using vehicle routing (VRP) models. Each step will be described thoroughly in the following sections.

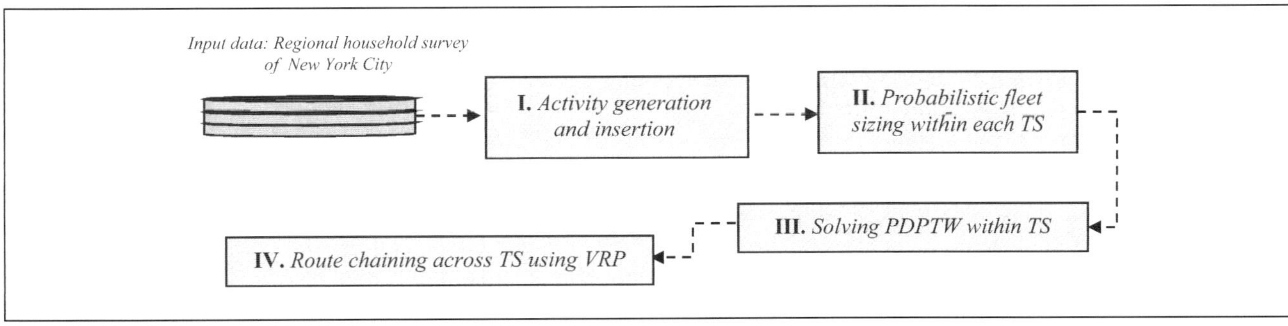

Figure 1. Flowchart of the proposed methodology.

Activity Generation and Insertion

With the goal of minimizing the disparity among mobility patterns of the people in the target population with their nearest neighbors, this step generates and inserts some feasible activities to the agenda of the target population, considering the limitations caused by their original agenda. The task is conducted by detecting the NNP for each person in the target population using their demographic attributes (e.g., age, gender, ethnicity, etc.). Once the NNP sets are generated for each individual in the target set, new activities for the target population will be generated based on the patterns of NNP. As described earlier, each activity is defined by five correlated attributes of $\langle x, d, p, T, \mathrm{tr} \rangle$. We use multidimensional copula sampling method to capture the inherent correlation among these attributes. Copula contains the information on the dependence structure between the components of a random vector, which is used to generate random samples with the same dependency structure among its components. As stated by Zou and Zhang (*18*), a nice feature of copula sampling is that the model used to describe the random variables is independent of the choice of the marginal distributions. For n dimensional variables, the algorithm for execution of the copula sampling is described as follows:

Algorithm for the generation of random samples using copulas

I. for $i \leftarrow 1$ to N_{TP} do:
II. for $j \leftarrow 1$ to n do:
III. $w^T \leftarrow$Compute inverse CDF $\left(F_1^{-1}(u_1), \ldots, F_n^{-1}(u_n)\right)$.
IV. $\mathbf{P} \leftarrow$ Compute matrix of linear correlation

$$\begin{pmatrix} 1 & \cdots & \rho_{1n} \\ \vdots & \ddots & \vdots \\ \rho_{n1} & \cdots & 1 \end{pmatrix}_{n \times n}.$$

V. $C(u_1, \ldots, u_n) \leftarrow$ Compute

$$\int_{-\infty}^{F_1^{-1}(u_1)} \cdots \int_{-\infty}^{F_n^{-1}(u_n)} (2\pi)^{-0.5n} |\mathbf{P}|^{-0.5} \exp\left(-\tfrac{1}{2} w^T \mathbf{P}^{-1} w\right) dw.$$

N_{TP} refers to the size of the target population. As described in the algorithm, the parameters are sampled from the inverse of the correlated cumulative distribution functions, as represented by w^T, and matrix $\mathbf{P}$ represents the correlations among different attributes. For our analysis, the copula has five dimensions of $\langle x, d, p, T, \mathrm{tr} \rangle$. For every individual in the target population, sampling is conducted using the probability distribution of the patterns for the corresponding NNP set selected from the base population. Let i be the index of the person in the target population and $\mathrm{act}^{(i)}$ be the sampled number of activities for person i. For each activity $a \in \mathrm{act}^{(i)}$ five attributes of the activity $\langle x_a, d_a, p_a, T_a, \mathrm{tr}_a \rangle$ are sampled using copula method. The sampled activity is added to the agenda of the target population depending on the availability of the free time. Also, it will depend on the start and end time of the fixed activities of person's original agenda and the attributes of the sampled activity. Insertion of the activity will be in a sequential order, where after each insertion, the free time windows in the agenda of the target population is adjusted accordingly, as described by the following insertion rule:

$\mathrm{Agenda}^{(i)} = \left\{ \mathrm{Agenda}^{(i)} \cup \mathrm{act}^{(i)} \right\}$ if $\mathrm{act}^{(i)}$ is not conflicting with $\mathrm{Agenda}^{(i)}$; if not, discard $\mathrm{act}^{(i)}$.

Design of the Shared Mobility System

Step II, III, and IV presented in Figure 1 will produce the optimum size of the required fleet of the vehicles to accommodate the mobility requests of individuals in the target population. Solution to the optimum fleet sizing was inspired by the works of Diana et al., where a probabilistic approach is used. Due to the combinatorial nature of the approach, the proposed method works best for short periods. Therefore, we slice the span of the day to shorter periods represented by TS. After finding the optimum fleet size in each time slice, the optimum routing and scheduling of vehicles inside each time slice will be found using PDPTW. Finally, generation of the optimum routes and schedules inside each time slice will be followed by chaining routes across different time slices based on the space–time feasibility constraints among the routes through a simple vehicle routing optimization model.

Probabilistic Fleet Sizing within Each Time Slice

This is the step II of the proposed methodology presented in Figure 1. For the purpose of clarity, we follow the same notations used in the original paper of Diana et al. (*27*). As described earlier, the activity agenda of the target population will identify the spatial–temporal distribution of the demand for service in our study area. Target population can request one or multiple rides containing origin and destination of the trips alongside the earliest and latest pickup time (EPT, LPT) and earliest and latest desired drop-off time (EDT, LDT). For each request, we assume that passengers can wait up to a maximum wait time (WS) to be served—a fixed maximum wait state is considered for all the customers in the system. From the updated activity patterns for the target population, we can compute the arrival and departure time for each activity. In order to be able to use the proposed methodology described in Diana et al. (*27*) the following assumptions are made in our analysis:

Assumption 1: Requesters do not have any preferences over the vehicles, the only service constraints are space and time reachability. Clearly, this assumption is very strong and it does not represent the reality. In fact, depending on the type of impairment, different restrictions on the type of vehicles should be considered. For instance, some travelers might need visual aids, some might need wheelchair to move around, etc. Despite this assumption, the proposed methodology is still valid, and it can be easily integrated with mode choice models. After identifying the trips that need to be accomplished using each mode, we can apply our methodology to dispatch fleet of vehicles. However, the focus of the current paper is the development of analytics to find the optimum fleet size and routing which can minimize mobility disparities; mode choice is beyond our scope. Furthermore, we did not have enough observations for different modes of transportation among physically impaired citizens in our dataset and the information regarding the type of disability or preferences on using alternative transportation modes is not properly recorded. Lack of sufficient information would cause identification problems, and for these reasons we excluded mode choice from the study.
Assumption 2: Vehicles travel at uniform speed. We execute five experiments with different average speed for vehicles varying in the range of 15–55 mph.

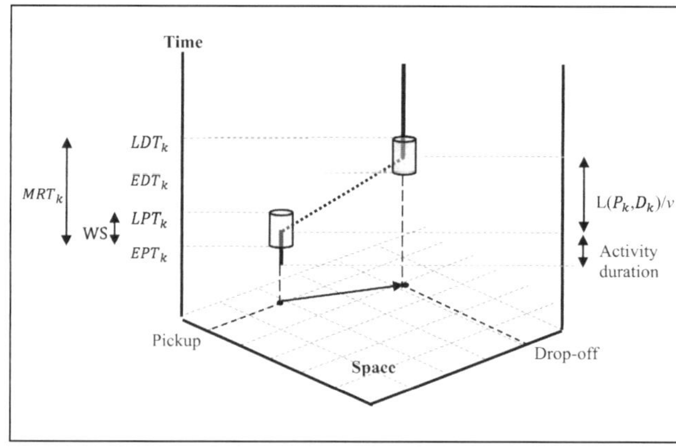

Figure 2. Time windows definition.

Assumption 3: Vehicles should pick up or drop off passengers within the requested time windows.
Assumption 4: Service time at each stop is set to 5 minutes.
Assumption 5: The maximum wait state for every person before each pick up is set to 20 minutes.
Assumption 6: Vehicles are allowed to idle between services.
Assumption 7: Vehicles' capacity is limited to three persons.
Assumption 8: We set the length of each time slice to 5 minutes. This duration was identified after a preliminary analysis on the data, and 5 minutes interval results in the least number of fleet size.

Let $L(A,B)$ be the Euclidean distance between point A and B. Assuming an average speed of $\mathbf{v}$, the ride time between location of pickup point P and drop off point of D for the request number k is equal to $L(P_k, D_k)/v$. Maximum ride time (MRT) for request k can be stated by a linear function of the ride time described as $\text{MRT}_k = a + b.(L(P_k, D_k)/v)$, where $(a \geqslant 0,\ b \geqslant 1)$. Equations 1 to 3 can be stated between earliest and latest pickup time (EPT, LPT) and earliest and latest desired drop-off time (EDT, LDT) for request k:

$$\text{LPT}_k = \text{EPT}_k + \text{WS} \tag{1}$$

$$\text{EDT}_k = \text{EPT}_k + \frac{L(P_k, D_k)}{v} \tag{2}$$

$$\text{LDT}_k = \text{EPT} + \text{MRT}_k \tag{3}$$

Figure 2 demonstrates the relationship between activity time windows and the WS (maximum wait time) graphically.

We define r_i as the probability of serving a set of i requests out of the n total requests by one vehicle. According to Diana et al. (*27*) it can be proven that the expected number of vehicles, denoted by $E(z)$, is computed using

$$E(z) = n\sum_{i=1}^{n}\frac{r_i}{i}\prod_{j=i+1}^{n}(1-r_j) \quad (4)$$

However, the value of r_i decreases as the value of i increases—probability of serving passengers with one car decreases as the number of passengers increases. We therefore only need to compute the first m values of r, where m is the index from which computing the next value of r will not change the expected number of vehicles.

Let dp be the probability of success of visiting a drop-off point (d) and a subsequent pickup point (p)—similarly we can define pp (pickup/pickup), pd (pickup/drop off), dd (drop off/drop off). The values of different combinations of probabilities are computed through Equations 5 to 8, where $f(.)$ denotes the corresponding probability density function:

$$dp = \int_{0}^{v.WS} f\left(L(D_i,P_j) + L(P_i,\mathrm{D}_i)\right)d\left(L(D_i,P_j) + L(P_i,\mathrm{D}_i)\right) \quad (5)$$

$$pp = \int_{0}^{v.WS} f\left(L(D_i,P_j)\right)d\left(L(D_i,P_j)\right) \quad (6)$$

$$pd = \int_{-v.WS}^{v.WS} f\left(L(D_i,P_j) - L(P_i,\mathrm{D}_i)\right)d\left(L(D_i,P_j) - L(P_i,\mathrm{D}_i)\right) \quad (7)$$

$$dd = \int_{-v.WS}^{v.WS} f\left(L(D_i,P_j) + L(P_i,D_i) - L(P_j,D_j)\right)d\left(L(D_i,P_j) + L(P_i,D_i) - L(P_j,D_j)\right) \quad (8)$$

The probability of serving m passengers with one vehicle, r_m, can be stated as

$$r_m = pd.\max_{i=1,\ldots,m}\left(pp^{i-1}, pd^{m-i}, dp^{m-i}, dd^{i-1}\right) \quad (9)$$

This equation allows us to compute the expected number of vehicles, $V(\mathrm{TS})$, for each time slice using

$$V(\mathrm{TS}) = n\sum_{j=1}^{n}\frac{pd.\max_{i=1,\ldots,m}\left(pp^{i-1}, pd^{m-i}, dp^{m-i}, dd^{i-1}\right)_i}{j}\times \prod_{k=j+1}^{n}\left(1 - pd.\max_{i=1,\ldots,m}\left(pp^{i-1}, pd^{m-i}, dp^{m-i}, dd^{i-1}\right)_k\right) \quad (10)$$

Optimum Routing and Scheduling in Each Time Slice

Once the optimum fleet size is estimated for each time slice, represented by $V(\mathrm{TS})$, the optimum routing and scheduling of requests inside each time slice is found through the classic capacitated PDPTW. We set the objective function of each time slice ($\mathrm{Obj}^{\mathrm{TS}}$) to be minimize the deviation of the actual service time from passengers requested time, and minimizing the total travel time for fleet of vehicles. The objective function for this problem is

$$\mathrm{Obj}^{\mathrm{TS}} = \min_{X,T}\left\{\sum_{i\in n(\mathrm{TS})}|T_i - \mathrm{TD}_i| + \sum_{\substack{i,j\in n(\mathrm{TS}),\\ v\in V(\mathrm{TS})}} t_{ij}X_{ij}^{v}\right\} \quad (11)$$

In this equation, $n(\mathrm{TS})$ refers to the number of nodes requested to be served within time slice TS, T_i refers to the time that the node i is served by a vehicle, TD refers to the desired service time requested by passenger at location i, t_{ij} is the travel time between pairs of nodes (i,j), and X_{ij}^{v} is the binary variable which takes the value of 1 if node j is being served right after node i by vehicle v.

This step generates the optimum route, $R_v^{(\mathrm{TS})} = \left\{X_{ij}^{v}\right\}$, for vehicle $v \in V$ at time slice TS, which is constructed based on the nodes visited by vehicle v. Also it generates the corresponding optimum scheduling for the route, $\mathrm{Sch}_v^{(\mathrm{TS})} = \left\{T_i^{v}\right\}$, constructed based on the time that vehicle v serves the (pickup or drop-off) request at each node i. For time slices with large number of requests we route vehicles using decomposition methods. The problem is decomposed to two problems of routing and scheduling (*23–26*).

Route Chaining

The last step of our proposed methodology is to find the optimum number of the vehicles and their routing/scheduling specification over the entire day. After the completion of stage III in Figure 1, which finds the optimum routing of each vehicle necessary to serve all the requests within a time slice which generated $\left\{R_v^{(\mathrm{TS})}, \mathrm{Sch}_v^{(\mathrm{TS})}\right\}$, we chain the routes of vehicles from one time slice to another conditional on the feasibility of the chain and the reachability of one route to another within the constraints generated from the spatial and temporal attributes of the routes. To check the feasibility of the route chaining, we must check if the travel time from one sequence to another is sufficient for the on-time service in both matching routes. The relevant objective function for

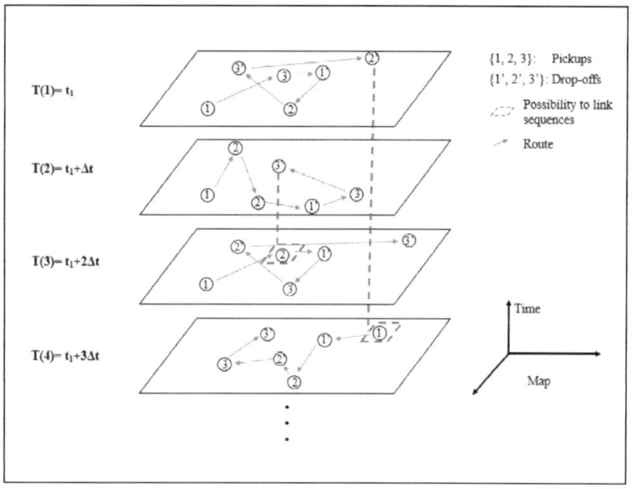

Figure 3. Route chaining between time slices.

route chaining is minimizing the idle time of the vehicles and the travel time from the last visited point of one route to the initial point of the other route. In fact, the problem is again a classic vehicle routing problem. We can cluster all nodes inside each route and represent them by one dummy node, where service time for each dummy node is the duration of the route, travel time from other nodes in the network to each dummy node is adjusted based on travel time to the first node of the route, and travel time from each dummy node to the other nodes is adjusted based on travel time from the last node of the route to others. For instance, consider patterns presented in Figure 3. Let us assume that the optimum duration for the sequence presented in the first time slice T(1) for vehicle V1 is computed to be 15 minutes. The entire route in this time slice can be represented by a new dummy node, DN, with the service duration of 15 minutes. The travel time from other nodes to DN is equal to travel time to node 1 in T(1), and travel time from DN to other nodes is equal to travel time from node 2′ in T(1) to other nodes. The location of the last service in T(1) is closest to the start location of the sequence of the fourth time slice T(4); these two routes can be chained and served by one vehicle if the travel time from the sequence one to four does not violate the time windows constraints for the first node of the route T(4).

Results of Empirical Analysis

The 2011 Regional Household Travel Survey of New York (*30*) is used for this study. The data is comprised of person files, trip files, household files, and vehicle file describing the person and household characteristics and the detailed information of the trips made during the day. A previous work (*13*) quantified disparities between mobility patterns of disabled citizens as target population using pattern recognition and clustering techniques. According to this study, 604 disabled persons have participated in a total number of 3337 trips during the day of the survey and 7993 non-disabled persons have executed total number of 51,421 trips on the day of the survey. Using copula sampling a total of 648 new activities (1296 trips) were generated and successfully inserted to the agenda of mobility-impaired travelers, resulting in a total number of 4633 trips. In order to compare the updated trips of individuals in the target population to their original trips, and to the trips made by the base population, we created a surface based on the mobility behavior of 604 disabled population in the data. Setting home location as the reference, intensity, spread, and length of trips completed by the target population is measured. 3D surfaces presented in Figure 4, show surface boundaries generated prior and after the insertion of new activities to the agenda of the target population. Figure 4*a* compares the patterns of the target population before (presented by red surface), and after the activity insertion (presented by white surface). Figure 4*b* compares updated patterns of the target population to the base population, illustrated by white and green surfaces respectively. As it is shown in the figure, after insertion of new activities to the original agenda of mobility-impaired individuals, the absolute difference in the boundary surface between patterns of the base population and updated patterns of the target population reduces from 29.9% to 15.1%. The comparison assists us to have a better visualization of the disparities in the mobility behavior of the target versus the base population in an aggregate level.

Figure 5 demonstrates the spatial distribution of the original and newly added activities to the agenda of the target population in three periods of the time. Figure 5*a*, *b*, and *c* illustrate the spatial distribution of these activities in the morning, midday, and evening time periods, respectively. Figure 5*d*, *e*, and *f* represent the spatial distribution of the generated activities (resulting in 1296 trips) in three time periods as illustrated in the figure. As we progress to later times of the day, the probability of insertion of new activities to the agenda reduces due to availability of free time.

We executed our model under six scenarios created based on the average speed of vehicles. The estimated fleet size for a total number of 4633 trips made by the target population, are 859, 648, 583, 482, and 439 vehicles for the average speed of 15, 25, 35, 45, and 55 mph, respectively. Depending on the fleet size, average number of trips per vehicle varies in the range of 5.4 trips, when the average speed is 15 mph, to 10.55 trips, when the average speed is 55 mph. We might conclude that the system is not operating very efficiently, for instance vehicles are occupied only about 19% of the time for the case of

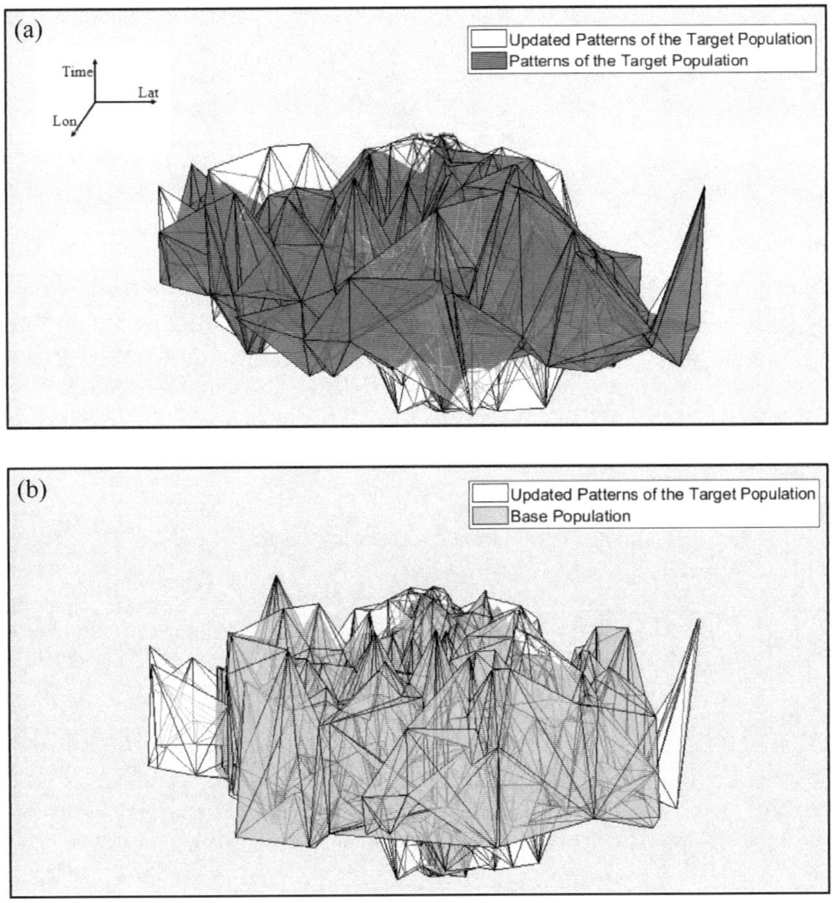

Figure 4. (*a*) Comparing the boundaries of the mobility patterns of the target population prior and after insertion of the activities; and (*b*) comparing the boundaries of the mobility patterns the target population after insertion of the activities to the patterns of the base population.

25 mph; The main contributing reason for large fleet sizes is the occurrence of simultaneous requests spread over large spatial area, which forces the model to allocate one vehicle to serve just a few rides.

The following figures shows the distribution of the number of vehicles required to serve all the requests in each time slice of the day (Figure 6*a*), and the total travel time spent by each vehicle to serve the request (Figure 6*b*). These graphs present the results for the case when the average speed is 25 mph.

Conclusion and Future Work

Livable cities and communities are defined as places where every individual has almost equal access to facilities and resources. However, this goal often is not fulfilled due to physical and financial limitations that agencies or cities are faced with. Emergence of new modes of transportation and adaptation to shared mobility systems could be a solution to eliminate mobility-related barriers in communities. In the work presented here, we developed a methodology to learn the mobility disparities and provided analytical solution for the design of a shared mobility system. Inspired by pattern recognition methods generated to measure mobility disparities among a target population and the rest of the society (*13*), we first quantified the existing mobility gap and then provided solutions to reduce it. In this study, we defined the target population to be mobility-impaired citizens, which include 604 individuals out of 7993 in our data. Then using the patterns of the individuals with similar demographic attributes, nearest neighbors, we sampled and inserted set of feasible activities to the agenda of the target population, where a total of 648 activities were added to their agenda. Copula sampling is used to sample different attributes of activities—activity

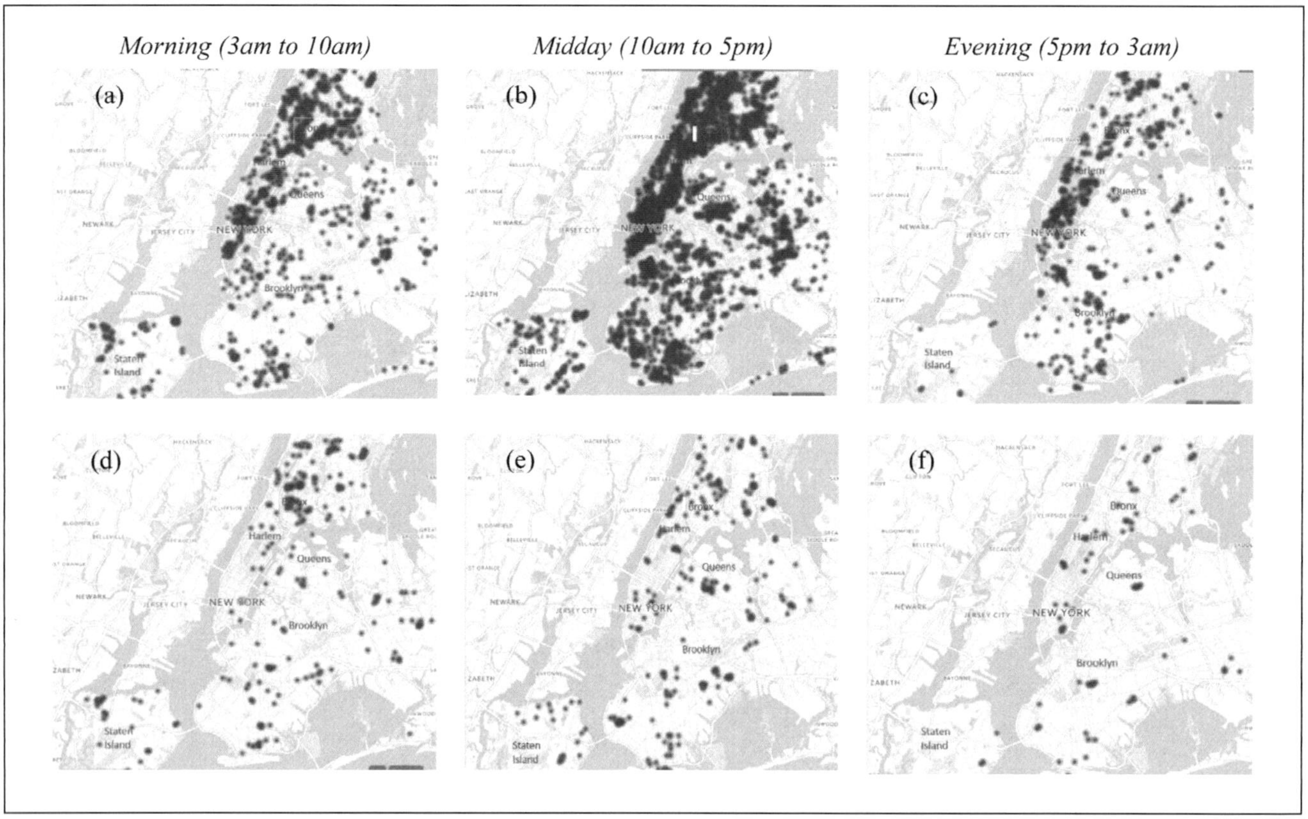

Figure 5. (*a, b, c*) Spatial distribution of activities in the agenda of target population in three time periods of morning, midday, and evening and (*d, e, f*) illustrating the spatial distribution of inserted activities in the agenda of target population in three time periods of morning, midday, and evening.

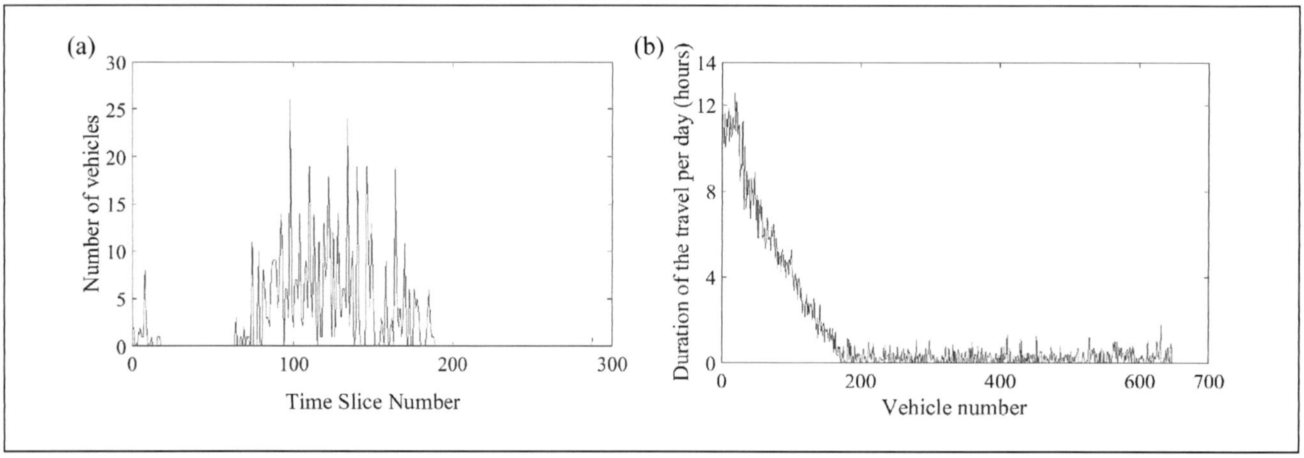

Figure 6. (*a*) Expected number of fleet size per time slice; and (*b*) total travel time spent by each vehicle to serve the requests—the average speed assumed to be 25 mph.

type, duration, start time, location, and trip duration—simultaneously. The optimum design of the ridesharing system starts with segmenting the span of the day to different time slices, where within each time slice we first found the optimum fleet size that can serve the requested demands. The optimum duration of time slice is found to be 5 minutes. Optimum routing of the vehicles in each time slice is based on the solution to PDPTW. Finally, the routes within each time slice were merged using vehicle routing problem, generating the fleet sizes and routing specifications under different speed ranges. The large dimension of study area which covers five boroughs of Manhattan, Bronx, Brooklyn, Staten Island, and Queens along with disproportionate distribution of the requests during the day impacts the analysis results, which at the end estimated a relatively large optimum fleet size compared with the total number of requests. The proposed methodology is time efficient and can be easily adapted to any target population, and even increase in the number of population and requests will result in an efficient use of the fleet in the system since it will increase the probability of inserting more requests in routes.

The study can be extended to consider variations in different types of vehicles, however due to lack of such information in our dataset, pursuit of this goal was not possible for this paper. The work would benefit if a direct survey was conducted from the target population to capture their actual willingness to participate in activities, the attributes of the preferred activities, and their preferences over the vehicle types and equipment. Additionally, we did not consider any dynamic modeling approach in this paper and we assumed that the activities would be planned ahead of time; however, inclusion of real-time ride request and considering uncertainty in the requested patterns is worthy for further explorations.

Author Contributions

The authors confirm contribution to the paper as follows: study conception and design: M. Allahviranloo; data collection: M. Zellou; analysis and interpretation of results: M Zellou, M.Allahviranloo; draft manuscript preparation: M Zellou, M.Allahviranloo. All authors reviewed the results and approved the final version of the manuscript.

References

1. Koppa, R., B. Davies, and K. Rodriguez. *Barriers to Use of Transportation Alternatives by People with Disabilities*. Texas Transportation Institute Texas A&M University System, Southwest Region, College Station, TX, 1998.
2. Pyrialakou, V. D., K. Gkritza, and J. D. Fricker. Accessibility, Mobility, and Realized Travel Behavior: Assessing Transport Disadvantage from a Policy Perspective. *Journal of Transport Geography*, Vol. 51, 2016, pp. 252–269. https://doi.org/10.1016/j.jtrangeo.2016.02.001.
3. Church, A., M. Frost, and K. Sullivan. Transport and Social Exclusion in London. *Transport Policy*, Vol. 7, No. 3, 2000, pp. 195–205. https://doi.org/10.1016/S0967-070X(00)00024-X.
4. Farber, S., M. O'Kelly, H. J. Miller, and T. Neutens. Measuring Segregation Using Patterns of Daily Travel Behavior: A Social Interaction Based Model of Exposure. *Journal of Transport Geography*. Vol. 49, 2015, pp. 26–38. https://doi.org/10.1016/j.jtrangeo.2015.10.009.
5. Næss, P., and N. Vogel. Sustainable Urban Development and the Multi-Level Transition Perspective. *Environmental Innovation and Societal Transitions*, Vol. 4, 2012, pp. 36–50. https://doi.org/10.1016/j.eist.2012.07.001.
6. Hong, J., Q. Shen, and L. Zhang. How Do Built-Environment Factors Affect Travel Behavior? A Spatial Analysis at Different Geographic Scales. *Transportation*, Vol. 41, No. 3, 2014, pp. 419–440. https://doi.org/10.1007/s11116-013-9462-9.
7. Newsome, T., H. W. A. Walcott, and P. D. Smith. Urban Activity Spaces: Illustrations and Application of a Conceptual Model for Integrating the Time and Space Dimensions. *Transportation*, Vol. 25, No. 4, 1998, pp. 357–377. https://doi.org/1005082827030.
8. Lenntorp, B. *Paths in Space–Time Environment, a Time Geographic Study of Possibilities of Individuals*. Lund Studies in Geography. Department of Geography, The Royal University of Lund, Lund, Sweden, 1976.
9. Kim, H., and M. P. Kwan. Space-Time Accessibility Measures: A Geo-Computational Algorithm with a Focus on the Feasible Opportunity Set and Possible Activity Duration. *Journal of Geographical Systems*, Vol. 5, No. 3, 2003, pp. 71–91.
10. Kwan, M. P. GIS Methods in Time-Geographic Research: Geocomputation and Geovisualization of Human Activity Patterns. *Geografiska Annaler: Series B Human Geography*, Vol. 86, No. 4, 2004, pp. 267–280. https://doi.org/10.1111/j.0435-3684.2004.00167.x.
11. Xiao, N., M. P. Kwan, M. F. Goodchild, and S. Shekhar. Geographic Information Science. *Proc., 7th International Conference, GIScience*, Columbus, OH, 2012. https://doi.org/10.1007/978-3-642-33024-7.
12. Li, R., and D. Tong. Constructing Human Activity Spaces: A New Approach Incorporating Complex Urban Activity-Travel. *Journal of Transport Geography*, Vol. 56, 2016, pp. 23–35. https://doi.org/10.1016/j.jtrangeo.2016.08.013.
13. Le Priol, E., and M. Allahviranloo. A Methodology on Quantifying Spatial-Temporal Disparities in Travel Behavior of Mobility-Impaired Travelers with Non-Disabled Travelers: Case Studies of London and New York. *Journal of Transport & Health*, Vol. 5, 2017, p. 3. https://doi.org/10.1016/j.jth.2017.05.315.
14. Navarro, J., and F. Spizzichino. On the Relationships between Copulas of Order Statistics and Marginal Distributions. *Statistics and Probability Letters*, Vol. 80, No. 5–6, 2010, pp. 473–479. https://doi.org/10.1016/j.spl.2009.11.025.
15. Gräler, B., and E. Pebesma. The Pair-Copula Construction for Spatial Data: A New Approach to Model Spatial Dependency. *Procedia Environmental Sciences*, Vol. 7, 2011, pp. 206–211. https://doi.org/10.1016/j.proenv.2011.07.036.
16. Dowd, K. Copulas in Macroeconomics. *Journal of International and Global Economic Studies*, Vol. 1, No. 1, 2008, pp. 1–26.
17. Zilko, A. A., D. Kurowicka, and R. M. Goverde, Modeling Railway Disruption Lengths with Copula Bayesian Networks. *Transportation Research Part C: Emerging*

Technologies, Vol. 68, 2016, pp. 350–368. http://dx.doi.org/10.1016/j.trc.2016.04.018.

18. Zou, Y., and Y. Zhang. A Copula Based Approach to Accommodate the Dependence among Microscopic Traffic Variables. *Transportation Research Part C: Emerging Technologies*, Vol. 70, 2016, pp. 53–68. http://dx.doi.org/10.1016/j.trc.2015.11.003.
19. Ball, M. O., B. L. Golden, A. A. Assad, and L. D. Bodin. Planning for Truck Fleet Size in the Presence of a Common-Carrier Option. *Decision Science*, Vol. 14, No. 1, 1983, pp. 103–120. http://dx.doi.org/10.1111/j.1540-5915.1983.tb00172.x.
20. Golden, B. L., A. Arjang, L. Levy, and F. Gheysens. The Fleet Size and Mix Vehicle Routing Problem. *Computers & Operations Research*, Vol. 11, No. 1, 1984, pp. 49–66. https://doi.org/10.1016/0305-0548(84)90007-8.
21. Beaujon, G. J., and M. A. Turnquist. A Model for Fleet-Sizing and Vehicle Allocation. *Transportation Science*, Vol. 25, No. 1, 1991, pp. 19–45. https://doi.org/10.1287/trsc.25.1.19.
22. Sayarshad, H. R., and K. Ghoseiri. A Simulated Annealing Approach for the Multi-Periodic Rail-Car Fleet-Sizing Problem. *Computers & Operations Research*, Vol. 36, No. 6, 2009, pp. 1789–1799. https://doi.org/10.1016/j.cor.2008.05.004.
23. Ropke, S., and J.-F. Cordeau. Branch-and-Cut-and-Price for the Pickup and Delivery Problem with Time Windows. *Transportation Science*, Vol. 43, No. 3, 2016, pp. 267–286. https://doi.org/10.1287/trsc.1090.0272.
24. Sexton, T. R., and L. D. Bodin. Optimizing Single Vehicle Many-To-Many Operations With Desired Delivery Times: I. Scheduling. *Transportation Science*, Vol. 19, No. 4, 1985, pp. 378–410. https://doi.org/10.1287/trsc.19.4.378.
25. Sexton, T. R., and L. D. Bodin. Optimizing Single Vehicle Many-to-Many Operations with Desired Delivery Times: II. Routing. *Transportation Science*, Vol. 19, No. 4, 1985, pp. 411–435. https://doi.org/10.1287/trsc.19.4.411.
26. Sexton, T. R., and Y. M. Choi. Pickup and Delivery of Partial Loads with "Soft" Time Windows. *American Journal of Mathematical and Management Sciences*, Vol. 6, No. 3–4, 1986, pp. 369–398. http://dx.doi.org/10.1080/01966324.1986.10737200.
27. Diana, M., M. M. Dessouky, and N. Xia. A Model for the Fleet-Sizing of Demand Responsive Transportation Services with Time Windows. *Transportation Research Part B: Methodological*, Vol. 40, No. 8, 2006, pp. 651–666. https://doi.org/10.1016/j.trb.2005.09.005.
28. Solomon, M. M. Algorithms for the Vehicle Routing and Scheduling Problem with Time Window Constraints. *Operations Research*, Vol. 35, No. 2, 1987, pp. 254–265. https://doi.org/10.1287/opre.35.2.254.
29. *MTA Guide to Accessible Transit*. http://web.mta.info/accessibility/stations.htm. Accessed June. 13, 2017.
30. *2010–2011 Regional Household Travel Survey*. New York Metropolitan Transportation Council, North Jersey Transportation Planning Authority, Newark, NJ, 2013.

The Standing Committee on Accessible Transportation and Mobility (ABE60) peer-reviewed this paper (18-05493).

Article

TRR
JOURNAL OF THE TRANSPORTATION RESEARCH BOARD

Transportation Research Record
2018, Vol. 2672(8) 649–661

DOI: 10.1177/0361198118794290
journals.sagepub.com/home/trr
SAGE

Analytical Method to Determine Transportation Service Gaps for Transportation Disadvantaged Populations

Ilir Bejleri[1], Soowoong Noh[1], Zongni Gu[1], Ruth L. Steiner[1], and Sandra M. Winter[2]

Abstract

Transportation disadvantaged (TD) populations, which include elderly people, people with disabilities, and people who do not own a vehicle, face mobility challenges because alternative transportation services are limited. To identify specific transportation service gaps for TD populations, an assessment of both transportation demand and transportation accessibility is necessary. Previous studies have focused on a single aspect of transportation accessibility or on studying geographic transportation gaps at the state or county level. This study develops a methodology aimed at identifying TD populations with major gaps in alternative transportation services. Using spatial overlays and network analysis, we develop a model for calculating transportation supply through a comprehensive transportation accessibility measure that considers relevant trip destinations and three alternative transportation services: public transportation, on-demand services, and taxi services. Gaps in service are identified by overlaying the TD population demand volume with transportation accessibility. The resulting supply–demand matrix highlights transportation deficient areas and populations, based on overlays of high demand and low supply. We apply this methodology to Alachua County, Florida. Results expose areas of deficiency, some in unexpected places. The findings highlight the need to assess alternative transportation for TD populations, to determine gaps to be addressed to improve TD populations' mobility and reduce personal auto trips and related traffic injuries. This method can guide policy makers in fostering equitable transportation provision across users.

Mobility benefits include access to people and places, psychological health, and increased community participation. In contrast, limited transportation accessibility has detriments, especially for transportation disadvantaged (TD) persons—the elderly, people with disabilities, and people who do not own vehicles. TD individuals need more convenient mobility options than individual automobile users, due to their physical limitations and limited financial means. Despite the issue's importance, research determining deficiencies in transportation accessibility and transportation needs of TD populations is limited (*1–4*).

This study contributes a methodology using a case study to identify such deficiencies, by developing a comprehensive measure of transportation accessibility and transportation demand for TD populations. This methodology uses geographic information systems (GIS) to pinpoint spatial gaps with high transportation demand and low transportation accessibility at a fine geographic scale. The map-based results provide an effective way to aid planners and policy makers in reducing transportation deficiencies and expanding alternative transportation for TD populations via future plans and policy implementations. Our case study applies the methodology to Alachua County, Florida.

Literature Review

In this section, we review existing research on transportation accessibility measuring TD-related factors. The definition of transportation accessibility varies by research area and has been applied to a wide range of topics. In transportation planning, transportation accessibility has

[1]Department of Urban and Regional Planning, University of Florida, Gainesville, FL
[2]Department of Occupational Therapy, University of Florida, Gainesville, FL

Corresponding Author:
Address correspondence to Soowoong Noh: nswscott@gmail.com

been used in vehicle routing and scheduling, travel demand forecasting, and transportation system planning (*5, 6*). In land-use planning, transportation accessibility has been applied to topics such as population allocation and facility location (*7, 8*).

In this study, we define transportation accessibility as ease of access to available relevant transportation services (*9*). Based on the literature, we consider three categories of services: public transportation with fixed routes and schedules; on-demand services with flexible routes and schedules and requiring advance request; and taxi services with flexible routes and schedule available by immediate request.

Due to convenience and flexibility, TD populations, especially the elderly, largely prefer automobile travel (*10*). However, numerous studies have demonstrated the significant role public transportation plays in helping senior adults meet daily needs including case studies of Sydney, Australia, and London, UK (*11–17*). Other research efforts calculate public transportation *served areas* as a ratio of bus stops walking catchment area divided by the area of the analysis zone (*1–3, 18, 19*). Additionally, several scholars illustrate that identical spatial coverage does not mean identical service availability for all users. They argue that measurements should consider not only spatial availability but also service frequency and span (*20, 21*). Moreover, researchers have studied public transportation supply and demand gaps for specific user groups, such as the elderly, people with disabilities, and low-income populations (*22–24*).

Although public transportation has many benefits, persons with physical disabilities typically limit their use of public transit (*13–15, 25*). To address this limited use, *on-demand services* have emerged as an alternative. On-demand services provide flexible routes including "route-deviation services, demand responsive bus transport, dial-a-ride buses and flex-route buses" (*13*). Cervero found these services have several benefits: increased mobility options, improved environmental conditions, and stimulation of advanced transit technologies (*26*). However, on-demand services require advance request and need a minimum number of users to operate, therefore previous studies have focused mainly on optimal fares and numbers of users (*27, 28*). Many studies have measured on-demand service accessibility as the ratio of population to the number of available facilities (*29–31*).

Similar to on-demand services, *taxi services* provide flexible routes, without an advanced request. Because these services are based on immediate requests, response time is a key variable in evaluating accessibility (*32*). However, response time can vary daily or weekly based on traffic. The service cost is another important variable to consider, due to lack of affordability for segments of population.

Prior transportation accessibility studies present some limitations. Little research has been done to develop comprehensive measures of transportation accessibility for TD populations. Prior research is primarily limited to public transit and emergency vehicles. In addition, most prior studies use census boundaries (e.g., census blockgroup or census tract in the U.S.) as the spatial unit of analysis to measure transportation accessibility. However, although census boundaries are useful because they are accompanied by population demographic profiles, the actual population distribution is typically not uniform within each census area (*3*) and assumption of even population distribution is a known limitation. Moreover, most studies have used walking (or bicycling) distance from transit stops to measure accessibility, but this may overlook challenges faced by persons with physical disabilities and the elderly who may not be able to walk to stops (*15*). Last, more research is needed for on-demand services to develop quantitative measures, as well as qualitative studies, such as evaluation of services via user satisfaction surveys (*26*).

Methodology

This study presents a geospatial methodology for identifying geographic areas and respective TD populations inadequately served with alternative transportation options. The spatial analysis unit is at the census blockgroup level given the available demographic profiles of the TD populations. The method is applied to Alachua County, Florida, identified as a high-priority county for transportation disadvantage by the Florida Department of Transportation (*33*).

The methodology takes a supply–demand approach. A conceptual model of the methodology is presented in Figure 1. *Transportation supply* is measured by quantifying transportation accessibility based on three transportation service provider categories: *public transportation* accessibility including network walking distance, and number of residential units; *on-demand service* accessibility is measured by applying origin–destination (OD) network analysis within boundaries of the areas served by each service provider; *taxi service* accessibility is measured by applying OD network analysis to destinations. *Transportation demand* is computed by calculating the volume of TD users: elderly people, people with disabilities, and people who do not own an automobile. Deficient areas or gaps are found by identifying spatial differences between transportation supply and demand.

Data and Variables

The necessary data to implement this methodology involve locations and attributes of service providers,

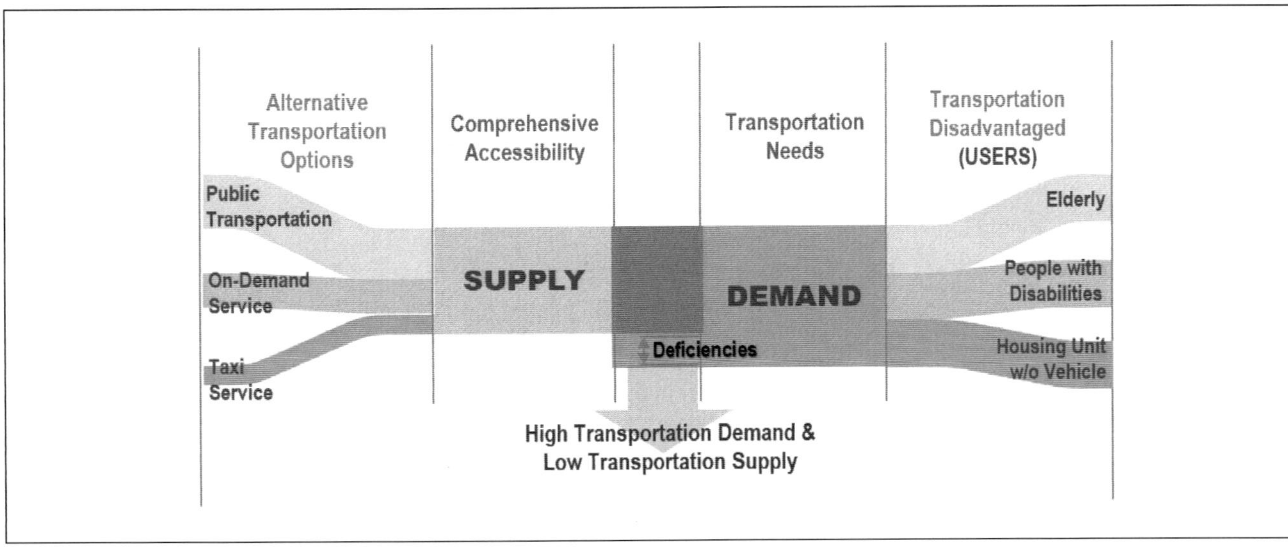

Figure 1. Conceptual model.

travel origins, and destinations, street network, and demographic profiles of TD populations. Figure 2 shows data maps for our case study.

Travel Origins, Destinations, and Street Network. Travel origins are established at the centroid point of each census blockgroup boundary. Alachua County has 155 blockgroups. Blockgroup information is obtained from the Florida Geographic Data Library (FGDL). The Florida unified GIS street network is used to determine transportation service areas and calculate the OD matrix based on origins and destinations defined below.

Possible destination categories and trip volumes to each destination are determined from the 2009 National Household Travel Survey (NHTS) including the Florida add-on. NHTS provides the most comprehensive survey examining travel patterns and behavior (*34*). We determine the destinations from the NHTS trip purpose categories, for this study, selecting destinations relevant to the TD population based on Florida Safe Mobility for Life Coalition priorities. Destinations include shopping, recreational, meals, educational, and medical. We create maps for each destination category using FGDL locations. Destinations include 866 locations for shopping, 138 for recreation, 94 for meals, 609 for education, and 223 for medical purposes.

Service Providers. Transit stops and route information for the public transportation provider—Regional Transit System (RTS)—which has fixed routes and schedules, were obtained from the General Transit Feed Specification (GTFS), freely available from Google. On-demand services include service providers that provide paratransit services, senior transportation, disability transportation, and medical transportation. They typically offer flexible routes, require advance requests, and provide door-to-door service. Provider information is obtained from the Florida Senior Safety Resource Center (FSSRC) database. Their locations were geocoded using GIS address matching technique and the Florida unified GIS streets basemap. Variables for each service included eligible trip purpose, user eligibility, and service area geographic boundary. For this case study, five service providers, predominantly within the boundaries of the city of Gainesville, supported all eligible trip purposes (shopping, recreation, meals, education and medical), and are exclusively available to people with disabilities, elderly, indigent populations, and veterans. Two other service providers support trips within city boundaries but only for medical purposes for people with disabilities, elderly people, and veterans. Two taxi services are available with flexible routes and immediate requests. While taxi service, in general, is an available option, it may not be suitable for all TD populations due to limited wheelchair accessibility and high costs.

User Demographics Profile. We obtained demographic information for TD populations—elderly people, people with disabilities, and persons in housing unit without a vehicle, from FGDL based on the 2010 U.S. Census and the 2010–2014 American Community Survey (2014 ACS) blockgroup-level data. In the study there are 29,074 elderly people (age 65 +), 13,573 people with disabilities, and 8,200 housing units without a vehicle. The number of people with disabilities, provided by 2014 ACS, applies to people between the ages of 20 and 64. We extracted

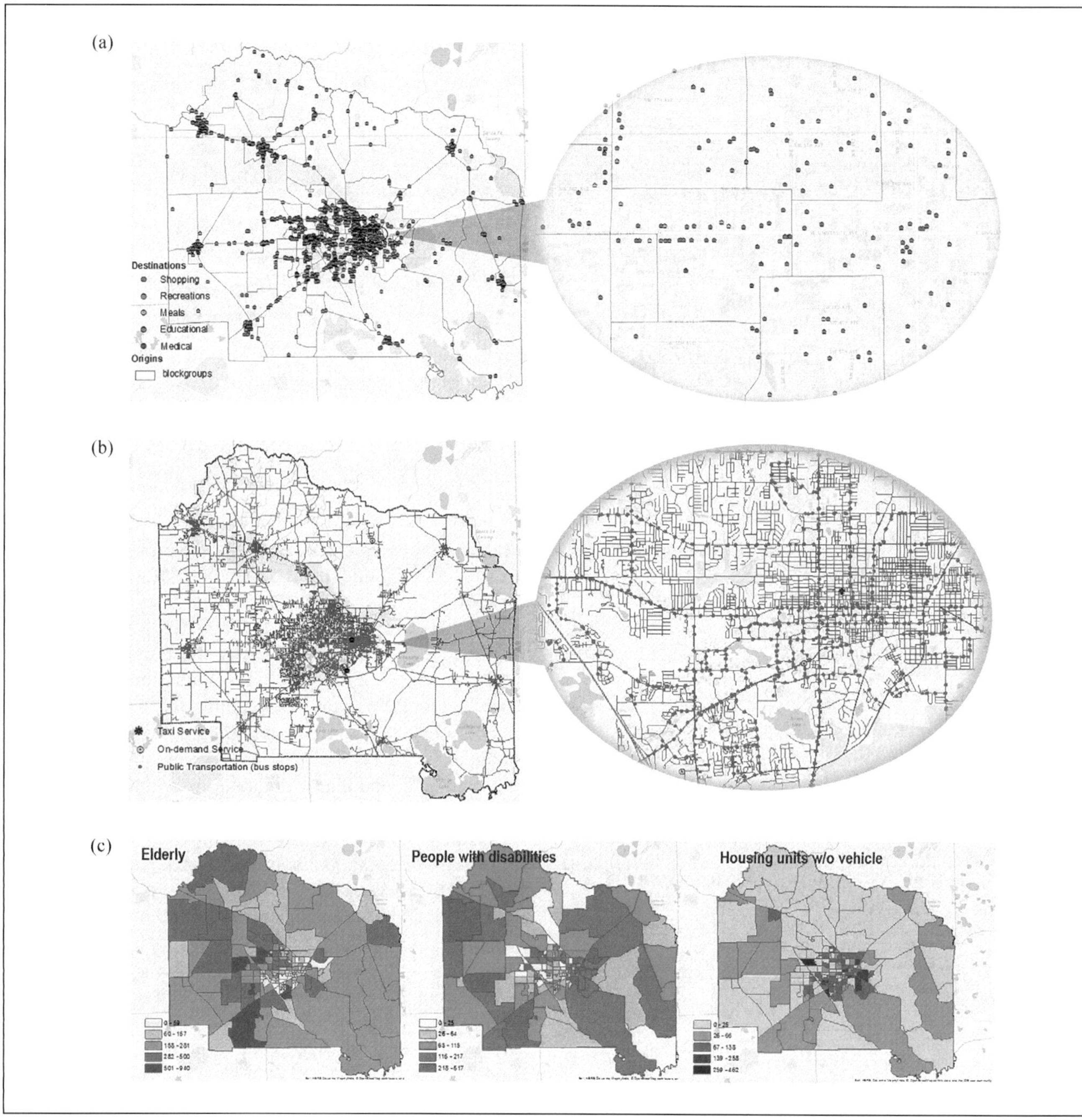

Figure 2. Map data: (*a*) travel origins and destinations, (*b*) service providers and street network, (c) TD populations.

the location and the number of residential units for each blockgroup from local government property parcel data.

Calculation of Transportation Supply

To determine transportation service supply for TD, we assessed each service provider category and quantified users' potential to access these services. Importantly, while these services may be *available*, actual user access may or may not be possible (e.g., public transportation is neither accessible, nor useful, if a user cannot walk to the nearest transit stop). Therefore, in this section, we assess transportation supply as a measure of service accessibility.

Measuring Access to Public Transportation Services. To measure public transportation access, we operate under the assumption that for public transportation to be useful, transit stops must be within a reasonable walking distance. Although there is variance, typically, a five-minute walk is considered a feasible distance to and from transit stops (*1–3*). Specifically, for elderly people, a walking speed between 2.59 and 3.58 mph, sustained for five to six minutes of walking, is considered acceptable (*25*). Therefore, for this study, we use a five-minute distance assuming an average walking speed of 3.0 mph.

To create walking catchment areas, we employ GIS network analysis around each bus stop using street network data and considering only streets designated as walkable. Next, we identify *served areas*, the areas of residential parcels served by public transportation. We calculate served areas by overlaying the walking catchment areas with the residential property parcels. Last, by spatially joining served residential parcels with blockgroups, we compute blockgroup public transportation accessibility:

$$TA_{pt} = \frac{RU_S}{RU_T} \tag{1}$$

where TA_{pt} is the public transportation accessibility for the blockgroup; RU_S is the number of served residential units, and RU_T is the total number of residential units. This formula measures the level of accessibility to public transportation for each blockgroup. The values range from 0 to 1, with 0 representing no access at all and 1 representing full blockgroup access.

This method improves on previous studies that measure the served areas as a ratio of the *service area* to the *total area* of the blockgroup. Given the uneven distribution of residential units in a blockgroup, the use of the residential parcel locations increases the accuracy by measuring this variable at a finer geography.

Measuring Access to On-Demand Services. Since on-demand services vary by cost, we group them into two categories: flat and variable fee services.

Flat fee: These services are accessible to all residential areas falling within their service area. Therefore, served residential parcels are determined by overlapping the residential parcels and service areas. Next, by spatial joining served residential parcels and blockgroups, we use Equation 1 to compute the ratio of residential units served to the total number of residential units in each blockgroup.

Variable Fee: For these services, providers charge more for longer travel distances. Therefore, distance to destinations must be considered. To accomplish this, we use cumulative opportunity scores, which measure the potential accessibility of each blockgroup to all eligible destinations, diminished by trip distance (*35–37*). Using GIS OD matrix analysis, we first calculate the network distance from each blockgroup centroid (origin) to each destination point. In the equation, network distance acts as impedance since longer distances indicate less opportunity for access to destinations. Next, the cumulative opportunity score to all destinations is calculated as follows:

$$S_{co} = \sum \frac{W_i}{d_{ij}} \tag{2}$$

where S_{co} is the cumulative opportunity score, d_{ij} is the network distance from each blockgroup ($_j$) to each destination ($_i$) and W_i is the weight for each destination category. The weights represent the relative trip frequency to each destination category. Based on NHTS, they are 0.44 for shopping, 0.27 for recreation, 0.15 for meals, 0.08 for education, and 0.06 for medical destinations. We applied these weights accordingly to calculate accessibility provided by each on-demand service.

Measuring Access to Taxi Services. Taxi service is similar to variable fee, on-demand service, with the benefit of responding to immediate requests. Therefore, the lack of nearby destinations translates into higher costs to access distant destinations. This could decrease the likelihood of using these services, resulting in lower user accessibility, despite abundant service availability. To calculate taxi service accessibility, we use Equation 2 above to compute cumulative opportunity scores of variable fee on-demand services.

Calculation of the Combined Transportation Accessibility (Supply). To measure comprehensive accessibility to transportation services that considers all three service categories—public transportation, on-demand services, and taxi services—measures for each category need to be standardized before they can be combined, since each measure is in different units. The values are standardized in a scale of 0 to 1 based on the relationship of the score to the highest value in its series using the formula below:

$$SS = \frac{D_i - \min(D_i)}{\max(D_i) - \min(D_i)} \tag{3}$$

where SS = standardized score; Di = original value of dataset i; min (Di) = minimum value of dataset i; and max (Di) = maximum value of dataset i.

Comprehensive transportation accessibility is finally calculated by applying a weighted sum of the three standardized scores. We adopted the method from Currie (*2, 3*) since this method is useful to generate a single supply score by combining accessibility of different transportation modes. The comprehensive transportation accessibility for TD population for each blockgroup is calculated as:

$$CTA = \left(SS_p \times W_p\right) + \left(SS_o \times W_o\right) + \left(SS_t \times W_t\right) \quad (4)$$

where SS_p is the standardized score for public transportation accessibility; SS_o is the standardized score for on-demand service accessibility; SS_t is the standardized score for taxi service accessibility; W_p is the weight for public transportation; W_o is the weight for on-demand service; and W_t is the weight for taxi service. For this case study, we select weights that reflect the relative use of each service category by TD populations using NHTS information for Florida. They are 0.55 for public transportation, 0.35 for on-demand services and 0.1 for taxi service. The values of the comprehensive transportation accessibility range from 0 to 1, with zero indicating no accessibility and 1 representing maximum accessibility. These values are used as measures of transportation supply.

Calculation of Transportation Demand

To calculate transportation demand, we adopt the method developed by Currie (*2, 3*). This method enables demand calculation based on multiple demographic datasets. We apply this method to the TD populations as follows:

$$DS = (SS_e \times W_e) + (SS_d \times W_d) + (SS_v \times W_v) \quad (5)$$

where DS = demand score; SS_e = standardized score for the number of elderly; SS_d = standardized score for the number of people with disabilities; SS_v = standardized score for the number of housing units without vehicle; W_e = weight for elderly population; W_d = weight for people with disabilities; and W_v = weight for housing units without vehicle.

Currie (*2, 3*) used weights based on government statistics, however, here we use equal weights because government statistics are unavailable. Transportation demand values range from 0 to 1, with zero indicating no transportation needs and 1 representing maximum transportation needs. Note the distance to destinations is not used in the calculation of demand, because it is considered in the supply/accessibility equations.

Gap Analysis

Gap analysis aims to determine spatial deficiencies between transportation supply and demand. To enable this analysis, using natural breaks classification, we reclassify demand and supply scores that range from 0 to 1, into seven categories: very low, low, medium low, medium, medium high, high, and very high. Natural breaks classification clusters the data in categories with similar values by exploiting natural gaps in the data, and thus reducing the variance within classes and maximizing the variance between classes (*38, 39*). By spatially comparing newly categorized demand and supply layers, we identify transportation service deficiencies as areas with high demand and low supply.

Results

Transportation Supply

Figure 3 shows the spatial distribution of transportation supply and populations affected. In the study area, 45.92% (13,351 people) of elderly people have poor accessibility (categories "very low" and "low"). Poor accessibility also applies to 37.72% (5,119 people) of people with disabilities and 12.39% (1,016 units) of housing units without access to a vehicle.

On the contrary, the city of Gainesville and surrounding areas have better accessibility. Areas of "medium" to "very high" accessibility are concentrated in central Gainesville. These areas cover 40.74% (11,845 people) of elderly people and 54.33% (7,374 people) of people with disabilities. Fortunately, these well-served areas include 82.39% (6,756 units) of housing units without a vehicle. The majority of the housing units without access to vehicle are located in Gainesville and probably these are the residences of college students.

Transportation Demand

Figure 4 shows the spatial distribution of transportation demand and populations affected. In the study area, 44.18% (12,844 people) of elderly people have high transportation need (categories "very high" and "high"). High need also applies to 42.03% (5,705 people) of people with disabilities and to 43.88% (3,598 units) of housing units without access to a vehicle. Geographically, high transportation demand is located in small communities outside Gainesville.

Transportation Gaps

Figure 5*a* shows the distribution of demand and supply scores in each category, as well as the thresholds used to define the categories using natural breaks classification. Overall scores are scattered, showing no obvious relationship between supply and demand. This pattern indicates that gaps are dispersed over the study area. These results are consistent with another study (*3*). Figure 5*b*, which we refer to as the demand–supply matrix, shows TD population affected for each combination of supply and demand categories. The most deficient areas, identified from the scatter plot and the demand–supply matrix

(a)

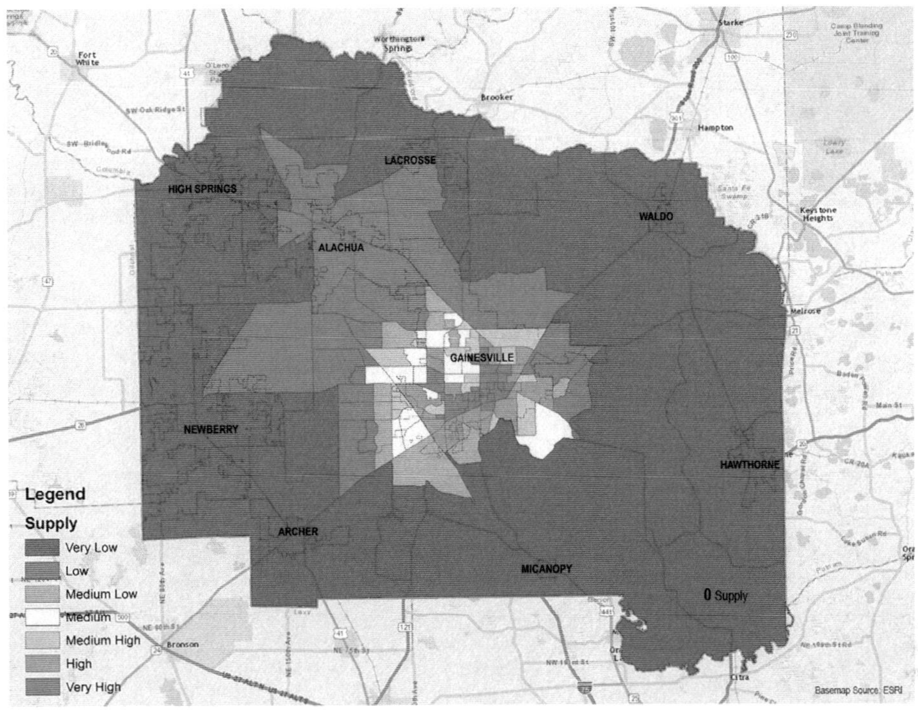

(b)

Supply	Number of blockgroups	Number of elderly	Number of people with disabilities	Number of housing units without vehicle
Very Low	32	7,478	3,294	585
		25.72%	24.27%	7.13%
Low	21	5,873	1,825	431
		20.20%	13.45%	5.26%
Medium Low	12	3,878	1,080	428
		13.34%	7.96%	5.22%
Medium	17	4,337	1,367	603
		14.92%	10.07%	7.35%
Medium High	18	3,142	1,680	1,808
		10.81%	12.38%	22.05%
High	28	2,566	2,256	2,414
		8.83%	16.62%	29.44%
Very High	27	1,800	2,071	1,931
		6.19%	15.26%	23.55%
Grand Total	155	29,074	13,573	8,200
		100.00%	100.00%	100.00%

Figure 3. Spatial distribution of transportation supply (*a*) and affected TD populations (*b*).

are those with "very high" or "high" demand and "very low" or "low" supply scores.

Figure 6 shows the spatial distribution of these areas. In the category of highest deficiency, we see one blockgroup with "very low" supply and "very high" demand where the population includes 2.17% (632 people) of elderly people, 1.73% (235 people) of people with disabilities, and 0.45% (225 units) of housing units without vehicle. In the category of second highest deficiency, four blockgroups rank as "very low" in supply and "high" in demand. They have 4.29% (1,246 people) of elderly people, 5.50% (746 people) of people with disabilities, and 1.66% (136 units) of housing units without vehicle. The two areas of lower deficiency contain six blockgroups with "low" supply but "very high" and "high" demand. They comprise 10.94% (3,181 people) of elderly people, 7.24% (983 people) of people with disabilities, and 3.74% (307 units) of housing units without a vehicle.

(a)

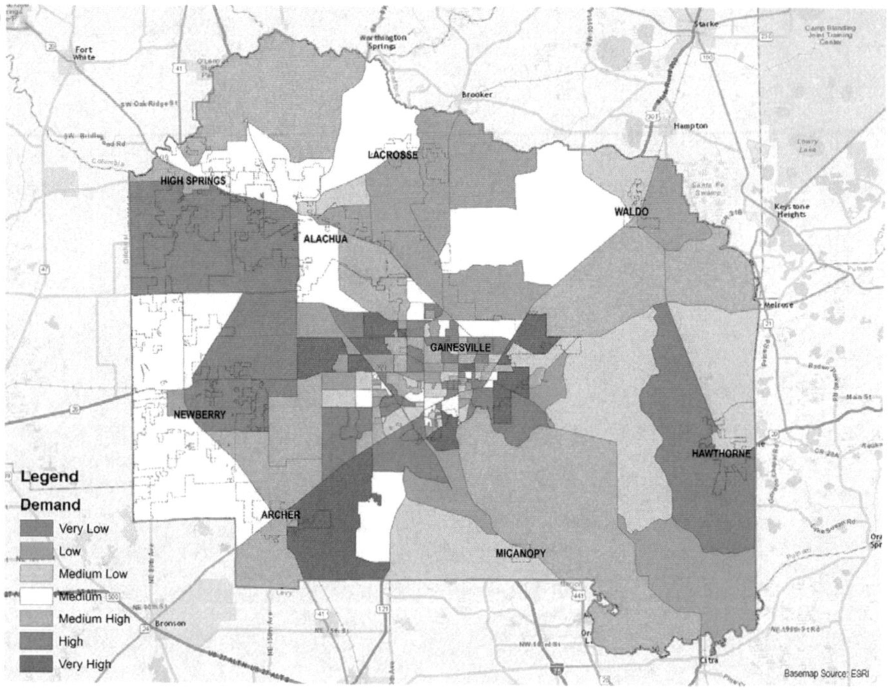

(b)

Demand	Number of blockgroups	Number of elderly	Number of people with disabilities	Number of housing units without vehicle
Very Low	14	247	135	257
		0.85%	0.99%	3.13%
Low	30	3,487	1,174	498
		11.99%	8.65%	6.07%
Medium Low	25	2,348	1,827	1,114
		8.08%	13.46%	13.59%
Medium	25	3,959	2,069	1,337
		13.62%	15.24%	16.30%
Medium High	26	6,189	2,663	1,396
		21.29%	19.62%	17.02%
High	25	7,480	3,282	2,291
		25.73%	24.18%	27.94%
Very High	10	5,364	2,423	1,307
		18.45%	17.85%	15.94%
Grand Total	155	29,074	13,573	8,200
		100.00%	100.00%	100.00%

Figure 4. Spatial distribution of transportation demand (*a*) and demand by TD populations (*b*).

Discussions and Conclusion

This paper presented a methodology for identifying gaps in alternative transportation services for TD populations. We applied this methodology to Alachua County, Florida, and found 11 deficient areas, representing 17.40% (5,059 people) of elderly people, 14.47% (1,964 people) of people with disabilities, and 5.85% (480 units) of housing units without a vehicle. Our *a priori* assumption was that deficiencies would be mostly in the unincorporated parts and small cities of the county. The city of Gainesville, which has the largest population concentration in the county, has a wide variety of transportation options. As expected, gaps were found in the unincorporated areas of the county or inside smaller cities. However, one gap is located within the Gainesville city limits, in the eastern part of the city, due to lack of transportation supply and concentration of TD population. Several rural areas have transportation gaps with low population density and large land areas. Residential land use can be overlaid on such gaps to gain a more accurate understanding of where people live in order to

(a)

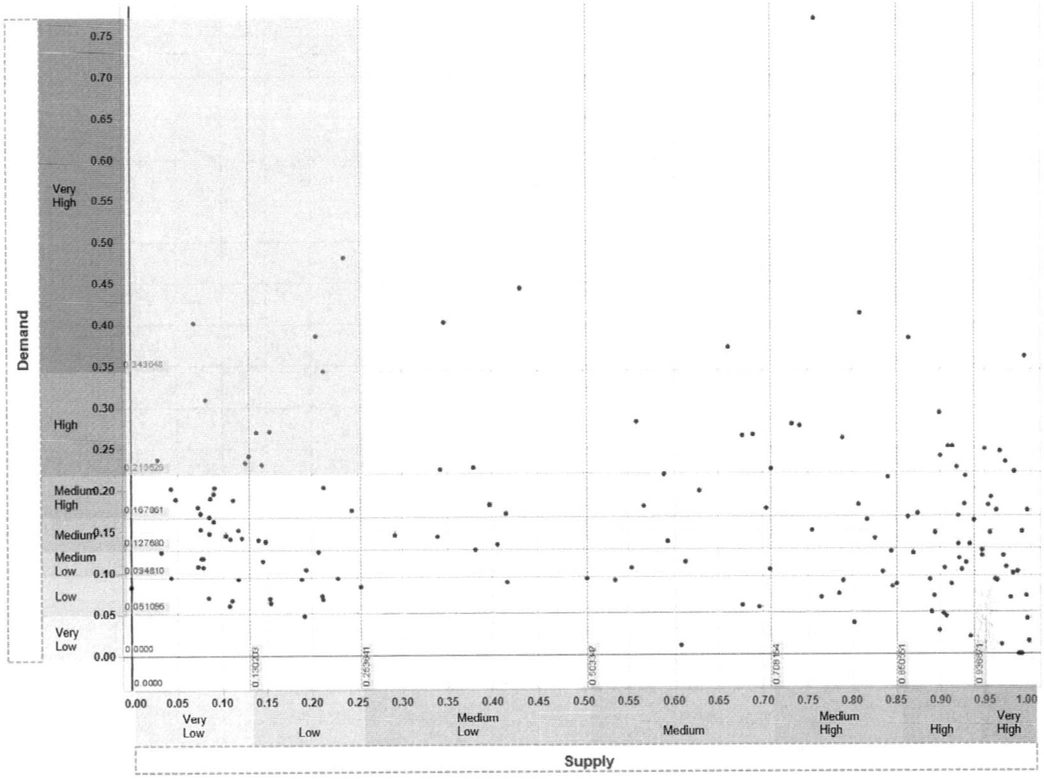

(b)

Demand	Supply	Very Low	Low	Medium Low	Medium	Medium High	High	Very High	Grand Total
Very High	Number of blockgroups	1	2	2	1	2	1	1	10
	Elderly	632	1,160	1,418	413	1,355	214	172	5,364
	People with disabilities	235	609	403	348	374	185	269	2,423
	Housing units w/o vehicle	37	88	116	0	634	258	174	1,307
High	Number of blockgroups	4	4	2	4	3	5	3	25
	Elderly	1,246	2,021	748	1,473	562	694	736	7,480
	People with disabilities	746	374	147	373	631	642	369	3,282
	Housing units w/o vehicle	136	219	126	379	291	828	312	2,291
Medium High	Number of blockgroups	8	2	2	4	2	3	5	26
	Elderly	2,255	817	309	1,408	263	722	415	6,189
	People with disabilities	935	132	216	299	236	288	557	2,663
	Housing units w/o vehicle	162	5	145	111	206	172	595	1,396
Medium	Number of blockgroups	8	2	3	1	3	6	2	25
	Elderly	1,590	333	939	133	300	582	82	3,959
	People with disabilities	791	233	115	86	162	513	170	2,069
	Housing units w/o vehicle	191	14	21	48	335	512	216	1,337
Medium Low	Number of blockgroups	5	3	1	3	2	5	6	25
	Elderly	884	387	263	528	59	131	96	2,348
	People with disabilities	344	273	31	183	105	403	488	1,827
	Housing units w/o vehicle	59	41	20	22	188	344	440	1,114
Low	Number of blockgroups	6	7	2	3	5	3	4	30
	Elderly	871	1,033	201	382	583	153	264	3,487
	People with disabilities	243	198	168	62	166	123	214	1,174
	Housing units w/o vehicle	0	64	0	43	117	155	119	498
Very Low	Number of blockgroups		1		1	1	5	6	14
	Elderly		122		0	20	70	35	247
	People with disabilities		6		17	6	102	4	135
	Housing units w/o vehicle		0		0	37	145	75	257
Grand Total	Number of blockgroups	32	21	12	17	18	28	27	155
	Elderly	7,478	5,873	3,878	4,337	3,142	2,566	1,800	29,074
	People with disabilities	3,294	1,825	1,080	1,367	1,680	2,256	2,071	13,573
	Housing units w/o vehicle	585	431	428	603	1,808	2,414	1,931	8,200

Figure 5. Supply–demand results: (*a*) scatter plot: supply–demand scores; (*b*) supply–demand matrix.

delineate gaps more closely, but data on the actual distribution of TD populations based on residential location is not available.

Practitioners can use this methodology to identify objectively areas to target for resource allocation, as part of policy implementations to improve TD population transportation accessibility. For Alachua County in particular, problematic areas identified in this study can be prioritized by county planners and policy makers, informing steps toward narrowing service gaps.

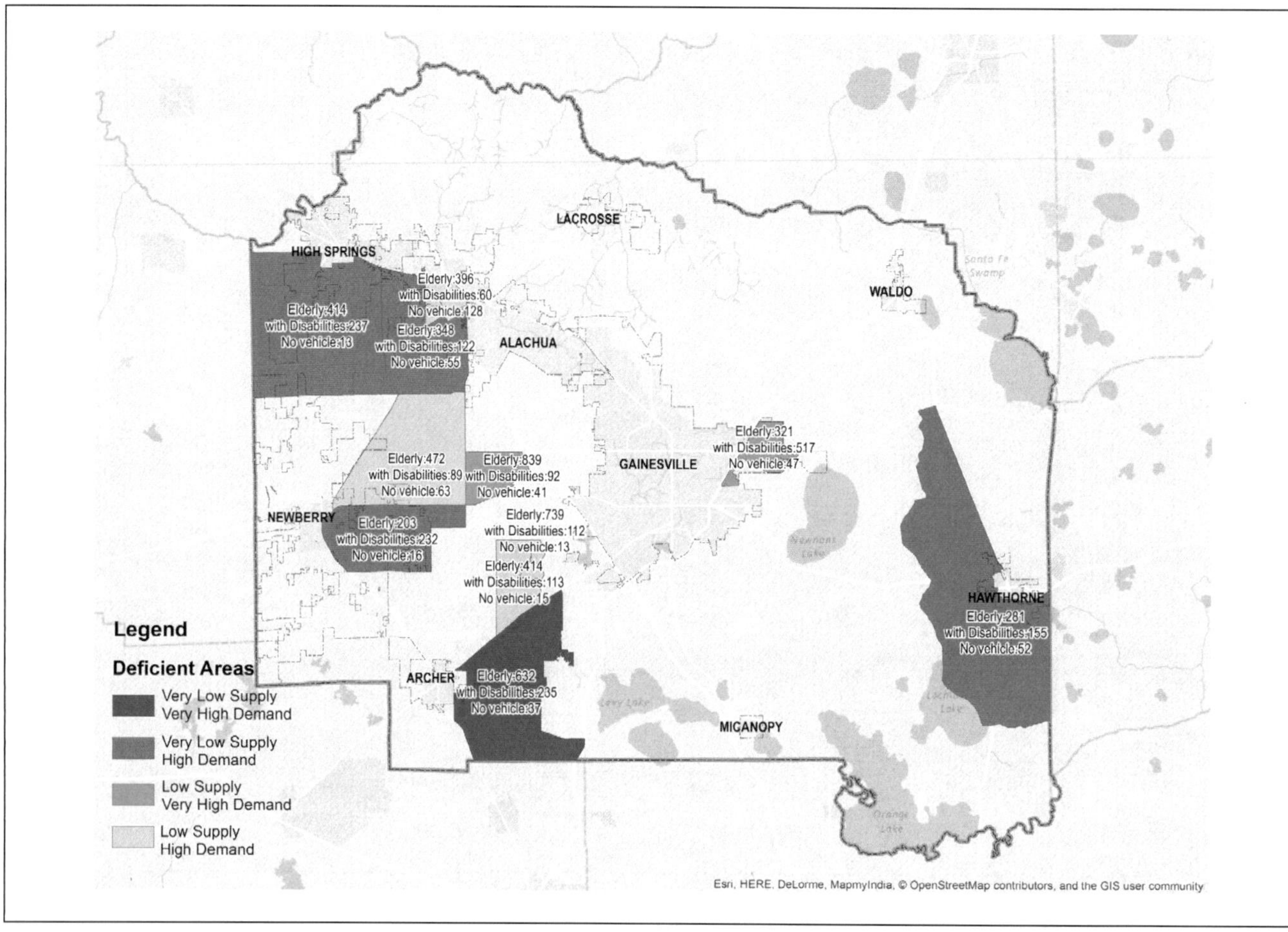

Figure 6. Spatial distribution of transportation service gaps.

As a methodological research tool, the model developed in this study provides improvements compared with previous studies. First, the methodology develops a comprehensive assessment of three transportation options: transit service, on-demand service, and taxi service. Second, in the calculation of accessibility, this method considers only relevant destinations for TD, and it weights the accessibility based on the relative travel frequency to these destinations and based on the transportation service category used. All the weights are derived from NHTS travel survey. Third, to determine the areas served, accessibility ratios are calculated by using the residential parcels rather than the area of the blockgroup, thus providing higher accuracy, especially for those blockgroups in which the distribution of residential areas is highly uneven. A recent case study in North Dakota evaluates gaps between supply and demand to prioritize public transit investment (*40*). However, that study uses the county as the analysis unit and measures the level of service; in this study, we use blockgroups and measure accessibility to destinations.

For broader professional practice, our methodology provides several benefits. First, it highlights the power of map-based overlay analysis. The deficient areas delineated on the map can be easily understood by planners, policy makers, and the public alike, and can be used as a guide for improvements and resource allocation. Second, these maps can be overlaid on other long-range transportation plans that address population growth, leading to synergies that efficiently integrate alternative transportation options into larger plans and policies. Third, the method developed here can be easily replicated. The model is constructed using a technique called Model Builder that can be packaged to be shared with other researchers and practitioners. Most of the required data—census, street networks, and destinations—are broadly available from the U.S. Census, departments of transportation, metropolitan planning organizations, planning departments, and property appraiser offices. Some effort may be needed to collect the information about private service providers for on-demand services. It should be noted that, since we used Florida results from NHTS data, these weights might not apply to other states and could be changed accordingly. Fourth, the method can be applied to larger geographic areas such as a multi-county region, i.e., MPO, DOT district, or even an entire state. Additionally, the method can be customized to determine gaps for specific users and service providers selectively. For example, to find the gaps just for people with disabilities, the method can only use the relevant service providers to calculate the supply and percentage of people with disabilities to calculate the demand. Fifth, the supply–demand matrix developed here can be used as a prioritization tool to sequence interventions, by incrementally considering additional areas defined by the cells of the matrix at the intersection of "demand" and "supply" categories. For example, practitioners can study additional areas with high demand defined as "medium high" and low supply defined as "medium low" and so on.

We acknowledge some data-related limitations in this study. First, since the focus of this study is to determine spatial gaps of transportation services, we do not consider the temporal dimension of public transportation such as the service hours and service frequency. These items will be the subject of future research. Second, application of the method in our case study used only services located in a single county. We may not have captured services offered in surrounding counties, a potential resource for residents, especially those living at the edge of the county. This can be addressed by a regional, or ideally, a statewide, application of the methodology. We are currently working on applying the same model to the entire state of Florida. Third, we acknowledge the potential for "double-counts" when calculating transportation demand between potentially overlapping demand segments of elderly, persons with disabilities, and household units without vehicles. Although 2014 ACS data specifies the populations of the elderly (age 65 +) and people with disabilities (between the ages of 20 and 64), due to the sample size, it is difficult to assess the potential of inflating demand values for those housing units without a vehicle that may contain elderly people or people with disabilities. Fourth, although using residential parcels could be a more accurate method to assess supply, the distribution of TD populations at the property parcel level is unavailable. However, point sources such as large care facilities can be used to improve the accuracy of demand calculation. Fifth, to calculate combined transportation accessibility, we use weights relevant to Florida, but the weights can change by population type and transportation mode if data is available. In our case, the weights for transportation mode based on NHTS were only available for elderly people. In the absence of this information for other TD populations, we apply the same weight to each transportation mode. Another option to calculate combined accessibility is to compute the spatial accessibility by considering the benefits provided to an individual based on spatial choices (*9*). This log-sum measurement could summarize with one index the desirability of all possible choices of the set (*41*). However, since this method assumes all transportation modes have the same degrees of substitutability (i.e., cross-elasticities are constant), it is not realistic in transportation studies (*9*). Therefore, since we use three different transportation options, we compute the accessibility of each option and combine them into a single index based on Currie's method (*2, 3*).

Another issue regarding access to services is affordability of paid services, which may prevent TD individuals from using some services, regardless of their geographic access. Although cost estimations are not in the scope of this study, we address the issue of the limited use of taxi service due to unaffordability by assigning taxi service a very low weight. The cost of service will be incorporated in future research.

Recently, transportation network companies (TNC), which connect passengers with drivers via mobile apps, have become popular alternatives to taxi services. Given the proliferation of TNCs like Uber or Lyft, local government could consider future experimentation with subsidizing such services as a potentially affordable transportation option to reduce service gaps. In fact, a study of a nine-month experimental program implemented by the City of Gainesville found that Uber improved the accessibility of older adults (*42*). If such a service were to be applied consistently, it could eliminate the East Gainesville gap area identified in our study. More broadly, the gap identification method presented here can include TNC services as additional service providers in the taxi category and test the effects of these services to reduce gaps and improve transportation accessibility for TD populations.

Finally, more studies like this are necessary to address deficiencies and improve mobility for elderly and other TD populations. Such improvements also have the potential toreduce personal auto trips and related traffic injuries. The phenomena of longer life expectancy, increasing numbers of elderly persons, and the serious issue of traffic safety for elderly persons, indicate that the demand for solutions to support alternative transportation options, especially on-demand services for TD populations, will only increase.

Author Contributions

The authors confirm contribution to the paper as follows: study conception and design: Ilir Bejleri and Soowoong Noh; literature review: Soowoong Noh and Sandra M. Winter; data collection: Soowoong Noh and Zongni Gu; methodology development: Ilir Bejleri, Soowoong Noh, and Zongni Gu;

analysis and interpretation of results: Ilir Bejleri, Soowoong Noh, and Ruth L. Steiner; draft manuscript preparation: Ilir Bejleri, Soowoong Noh, Ruth L. Steiner, and Sandra M. Winter. All authors reviewed the results and approved the final version of the manuscript.

References

1. Al Mamun, S., and N. Lownes. Measuring Service Gaps. *Transportation Research Record: Journal of the Transportation Research Board*, 2011. 2217: 153–161.
2. Currie, G. Gap Analysis of Public Transport Needs—Measuring Spatial Distribution of Public Transport Needs and Identifying Gaps in the Quality of Public Transport Provision. *Transportation Research Record: Journal of the Transportation Research Board*, 2004. 1895: 137–146.
3. Currie, G. Quantifying Spatial Gaps in Public Transport Supply Based on Social Needs. *Journal of Transport Geography*, Vol. 18, No. 1, 2010, pp. 31–41.
4. Currie, G., and A. Delbosc. Exploring Public Transport Usage Trends in an Ageing Population. *Transportation*, Vol. 37, No. 1, 2010, pp. 151–164.
5. Morris, J. M., P. L. Dumble, and M. R. Wigan. Accessibility Indicators for Transport Planning. *Transportation Research Part A: General*, Vol. 13, No. 2, 1979, pp. 91–109.
6. Wang, G., Y. Zhong, C.-P. Teo, and Q. Liu. Flow-Based Accessibility Measurement: The Place Rank Approach. *Transportation Research Part C: Emerging Technologies*, Vol. 56, 2015, pp. 335–345.
7. Hansen, W. G. How Accessibility Shapes Land Use. *Journal of the American Institute of Planners*, Vol. 25, No. 2, 1959, pp. 73–76.
8. Yoshida, N., and U. Deichmann. Measurement of Accessibility and Its Applications. *Journal of Infrastructure Development*, Vol. 1, No. 1, 2009, pp. 1–16.
9. Miller, H. J. Place-Based Versus People-Based Accessibility. In Levinson, D. M. and Krizek, K. J.(Eds.), *Access to Destinations*, Elsevier, Oxford, 2005, pp. 63–89.
10. Metz, D. H. Mobility of Older People and their Quality of Life. *Transport Policy*, Vol. 7, No. 2, 2000, pp. 149–152.
11. Golob, T. F., and D. A. Hensher. The Trip Chaining Activity of Sydney Residents: A Cross-Section Assessment by Age Group with a Focus on Seniors. *Journal of Transport Geography*, Vol. 15, No. 4, 2007, pp. 298–312.
12. Schmöcker, J.-D., M. A. Quddus, R. B. Noland, and M. G. H. Bell. Mode Choice of Older and Disabled People: A Case Study of Shopping Trips in London. *Journal of Transport Geography*, Vol. 16, No. 4, 2008, pp. 257–267.
13. Broome, K., L. Worrall, J. Fleming, and D. Boldy. Evaluation of Flexible Route Bus Transport for Older People. *Transport Policy*, Vol. 21, 2012, pp. 85–91.
14. Mercado, R., A. Páez, and K. B. Newbold. Transport Policy and the Provision of Mobility Options in an Aging Society: A Case Study of Ontario, Canada. *Journal of Transport Geography*, Vol. 18, No. 5, 2010, pp. 649–661.
15. de Boer, E. Introducing and Sustaining Accessible Transport: Social and Physical Challenges. *Transportation Research Record: Journal of the Transportation Research Board*, 2004. 1885: 15–20.
16. Rye, T., and D. Scotney. The Factors Influencing Future Concessionary Bus Patronage in Scotland and their Implications for Elsewhere. *Transport Policy*, Vol. 11, No. 2, 2004, pp. 133–140.
17. Banister, D., and A. Bowling. Quality of Life for the Elderly: The Transport Dimension. *Transport Policy*, Vol. 11, No. 2, 2004, pp. 105–115.
18. Cheung, C. L., and A. W. Agrawal. TTSAT: A New Approach to Mapping Transit Accessibility. *Journal of Public Transportation*, Vol. 13, No. 1, 2010, p. 55.
19. Polzin, S. E., R. M. Pendyala, and S. Navari. Development of Time-of-Day-Based Transit Accessibility Analysis Tool. *Transportation Research Record: Journal of the Transportation Research Board*, 2002. 1799: 35–41.
20. Neutens, T., M. Delafontaine, D. M. Scott, and P. De Maeyer. A GIS-Based Method to Identify Spatiotemporal Gaps in Public Service Delivery. *Applied Geography*, Vol. 32, No. 2, 2012, pp. 253–264.
21. Schoon, J. G. Accessibility Indices: Pilot Study and Potential Use in Strategic Planning. *Transportation Research Record: Journal of the Transportation Research Board*, 1999. 1685: 29–38.
22. Delbosc, A., and G. Currie. Using Lorenz Curves to Assess Public Transport Equity. *Journal of Transport Geography*, Vol. 19, No. 6, 2011, pp. 1252–1259.
23. Hess, D. B. Walking to the Bus: Perceived Versus Actual Walking Distance to Bus Stops for Older Adults. *Transportation*, Vol. 39, No. 2, 2012, pp. 247–266.
24. Lubin, A., and D. Deka. Role of Public Transportation as Job Access Mode. *Transportation Research Record: Journal of the Transportation Research Board*, 2012. 2277: 90–97.
25. Steffen, T. M., T. A. Hacker, and L. Mollinger. Age- and Gender-Related Test Performance in Community-Dwelling Elderly People: Six-Minute Walk Test, Berg Balance Scale, Timed Up & Go Test, and Gait Speeds. *Physical Therapy*, Vol. 82, No. 2, 2002, pp. 128–137.
26. Cervero, R. *Paratransit in America: Redefining Mass Transportation*. Praeger, Westport, Conn., 1997.
27. Bearse, P., S. Gurmu, C. Rapaport, and S. Stern. Paratransit Demand of Disabled People. *Transportation Research Part B: Methodological*, Vol. 38, No. 9, 2004, pp. 809–831.
28. Franklin, J. P., and D. A. Niemeier. Discrete Choice Elasticities for Elderly and Disabled Travelers between Fixed-Route Transit and Paratransit. *Transportation Research Record: Journal of the Transportation Research Board*, 1998. 1623: 31–36.
29. Case, C., and T. L. Hawthorne. Served or Unserved? A Site Suitability Analysis of Social Services in Atlanta, Georgia using Geographic Information Systems. *Applied Geography*, Vol. 38, 2013, pp. 96–106.
30. Sathisan, S., and N. Srinivasan. Evaluation of Accessibility of Urban Transportation Networks. *Transportation Research Record: Journal of the Transportation Research Board*, 1998. 1617: 78–83.
31. Tong, D., W.-H. Lin, J. Mack, and D. Mueller. Accessibility-Based Multicriteria Analysis for Facility Siting.

Transportation Research Record: Journal of the Transportation Research Board, 2010. 2174: 128–137.
32. Carr, B. G. Access to Emergency Care in the United States. *Annals of Emergency Medicine*, Vol. 54, No. 2, 2009, pp. 261–269.
33. Safe Mobility for Life Coalition. *Aging Road User Strategic Safety Plan*. 2017 http://www.safeandmobileseniors.org/pdfs/flagingroaduserstrategicsafetyplan.pdf. Accessed Mar. 15, 2018.
34. Florida Department of Transportation. *2009 National Household Travel Survey Florida Data Analysis*. Trends and Conditions Report. Florida Department of Transportation, 2013.
35. Yang, H., and S. C. Wong. A Network Model of Urban Taxi Services. *Transportation Research Part B: Methodological*, Vol. 32, No. 4, 1998, pp. 235–246.
36. Van Meter, E., A. B. Lawson, N. Colabianchi, M. Nichols, J. Hibbert, D. Porter, and A. D. Liese. Spatial Accessibility and Availability Measures and Statistical Properties in the Food Environment. *Spatial and Spatio-Temporal Epidemiology*, Vol. 2, No. 1, 2011, pp. 35–47.
37. Barnes, T. L., N. Colabianchi, J. D. Hibbert, D. E. Porter, A. B. Lawson, and A. D. Liese. Scale Effects in Food Environment Research: Implications from Assessing Socioeconomic Dimensions of Supermarket Accessibility in an Eight-County Region of South Carolina. *Applied Geography*, Vol. 68, 2016, pp. 20–27.
38. Jaramillo, C., C. Lizárraga, and A. L. Grindlay. Spatial Disparity in Transport Social Needs and Public Transport Provision in Santiago De Cali (Colombia). *Journal of Transport Geography*, Vol. 24, 2012, pp. 340–357.
39. Baz, I., A. Geymen, and S. N. Er. Development and Application of GIS-Based analysis/synthesis Modeling Techniques for Urban Planning of Istanbul Metropolitan Area. *Advances in Engineering Software*, Vol. 40, No. 2, 2009, pp. 128–140.
40. Mattson, J., R. Godavarthy, and J. Hough. Method for Estimating Statewide Transit Needs and Investment Priorities for Rural and Small Urban Areas. *Transportation Research Record: Journal of the Transportation Research Board*, 2016. 2543: 25–33.
41. Handy, S. L., and D. A. Niemeier. Measuring Accessibility: An Exploration of Issues and Alternatives. *Environment and Planning A*, Vol. 29, No. 7, 1997, pp. 1175–1194.
42. Leistner, D. L., and R. L. Steiner. Uber for Seniors? Exploring Transportation Options for the Future. *Transportation Research Record: Journal of the Transportation Research Board*, 2017. 2660: 22–29.

The Standing Committee on Accessible Transportation and Mobility (ABE60) peer-reviewed this paper (18-01491).

Article

TRR
JOURNAL OF THE TRANSPORTATION RESEARCH BOARD

Transportation Research Record
2018, Vol. 2672(8) 662–674

DOI: 10.1177/0361198118772952
journals.sagepub.com/home/trr
SAGE

Data Quality of Travel Behavior Studies: Factors Influencing the Reporting Rate of Self-Reported and GPS-Recorded Trips in Persons with Disabilities

An Neven[1], Ine De Schutter[2], Geert Wets[1], Peter Feys[3], and Davy Janssens[1]

Abstract

In persons with disabilities, the accuracy of self-report data collection methods in travel behavior studies may be influenced by disease-related dysfunctions. The present study determines in detail whether disease-related physical, cognitive or psychosocial impairments contribute to the reporting rate of subjective self-report travel diaries and objective GPS tracking devices, besides socio-demographic and trip-related characteristics. The reporting rate of both data collection methods was analyzed in 108 persons with Multiple Sclerosis with various disability severities during a seven-day data collection period. The results demonstrated that there was only limited influence of disease-related dysfunctions on the reporting rate of both data collection methods, as well as only limited significant differences between subgroups with various disability severity. Overall, the data quality of the diary was higher than the quality of the GPS data: 66% of the trips were reported in both data collection methods, while overall more than one fifth of all trips were forgotten and not registered by GPS, and 11% were forgotten and not reported in the diary. Self-report travel diaries seemed to be more suitable for persons with a higher disability severity, as these persons more often forgot to take their GPS device with them when making a trip because of several organizational issues related to their mobility limitations.

Physical activity (PA) levels are often measured to identify current levels and changes in PA of specific population groups, and to evaluate the effectiveness of interventions designed to increase activity levels (*1*) which is an important part of health promotion research (*2*). The measurement method may have a significant impact on the observed PA levels, as self-report PA measures generally differ from direct, objective measures of PA (*1*). While self-report methods like questionnaires, interviews, or diaries are often preferred as they are cheap to administer and offer additional information on the context of activities, objective measurement devices like pedometers or accelerometers provide more accurate measurement of PA and are not prone to certain biases in self-report methods such as recall bias (*3*). Especially in research studies of persons with disabilities (PWD), the accuracy of self-report surveys for measuring PA can be influenced by cognitive dysfunctions like memory or emotional problems (*4*). Commonly used activity questionnaires and diaries rely on recall and honest reporting, and require individuals to have no or limited cognitive deficits to reduce potential bias in reporting results (*5*). According to research conducted by (*6*), (*8*) and (*7*), self-report subjective PA questionnaires show only limited reliability and validity on the frequency of different activity types (e.g. respondents tend to indicate their highest recent frequency of participation) and the duration and amount of PA. On the other hand, data from objective monitors may also sometimes be incomplete if, for example, participants forget to wear their monitors, or if the monitors suffer external limitations such as low battery life, signal interference, or malfunction (*9*).

Recently, travel behavior studies have been increasingly used in public health studies to assess PA levels, as interventions to discourage car use and substitute active

[1]Hasselt University, Transportation Research Institute (IMOB), Diepenbeek, Belgium
[2]Hasselt University, Diepenbeek, Belgium
[3]Hasselt University, REVAL Rehabilitation Research Institute, BIOMED Biomedical Research Institute, Faculty of Medicine and Life Sciences, Diepenbeek, Belgium

Corresponding Author:
Address correspondence to An Neven: An.Neven@uhasselt.be

travel alternatives are motivated by concerns to improve health through increased PA (*10*). In this respect, traditional travel behavior studies make use of self-report activity-related travel diaries, typically requiring participants to provide information about activity purpose, travel mode, start and end times, location of trip origin and destination, and company of the trip. Over the past decade, recent advances in GPS technology and mobile positioning data have resulted in conventional paper self-report activity-related travel diaries being supplemented (and sometimes replaced) with GPS tracking devices as a collection method to record travel behavior (*11–14*). In the context of health promotion, accurate measurement of travel behavior by means of this combination of self-report and objective estimates allows, for example, a detailed assessment of the time spent using active modes of transport (*15*), or the classification of accelerometer-derived PA bouts into walking behavior (*16*).

Similar to problems with self-report PA questionnaires, it has been documented previously that traditional travel behavior survey data suffer from incompleteness and inaccuracies of reported trips (*17*), because of the detailed required trip-level information. Under-reporting of trips in travel studies has being documented by comparing self-reported trips with travel data collected by GPS tracking devices, to identify those trips recorded in the GPS data that were not reported by respondents in the travel diaries. Several key *demographic factors* were found (*18–21*) to significantly contribute to the misreporting of trips: younger respondents, men, individuals with a lower educational degree, unemployed individuals, low-income households, households with ≥3 vehicles, or households with ≥3 workers were more likely to under-report their trips. In addition, *trip-related characteristics* were found to significantly influence trip under-reporting (*22*). Short duration activities or trips, discretionary trips, trips made at the end of the travel day, among others, were more often forgotten or omitted in paper travel diaries (*17, 20, 23*). Individuals who made many trips in one day, were also more likely to underreport their travel (*24*).

On the other hand, there are also some (but fewer) studies that focused on situations, and/or associated factors, where fewer trips were recorded by GPS tracking devices than were reported in the travel diary. Data missing from GPS devices may result from device limitations such as signal drop inside buildings or low battery life, as well as user errors like forgetting to carry the device (*9*). Previous studies showed that trips occurring before respondents went to work or after they left work, as well as trips on Saturdays or Thursdays, were more often forgotten and not recorded by GPS (*25*).

While there are studies available comparing diaries and GPS tracking for the estimation of the travel behavior of healthy persons, there is relatively little research available about the *reliability of travel behavior studies in PWD* (*26–28*). It may be expected that the accuracy of self-report data collection methods in travel behavior studies of PWD may be influenced by disease-related cognitive dysfunctions, such as problems in remembering (details of) all performed trips.

Therefore, the present study contributes to the field by determining in detail whether *disease-related physical, cognitive, or psychosocial impairments* also contributed significantly to changes in reporting in activity-related travel diaries and recording by GPS tracking devices by PWD.

Participants in this study were diagnosed with Multiple Sclerosis (MS), a progressive inflammatory and neurodegenerative chronic disease of the central nervous system, and had various degrees of disability. As MS is characterized by a combination of physical, cognitive, and psychosocial symptoms, the cognitive status of persons with MS (PwMS) may interfere with their ability to accurately recall and report the locations, distances, and frequencies of their community mobility (*4*). In MS, cognitive impairment occurs in 40% to 65% of its patients (*29*). In this study, the differences between both travel behavior data collection methods (travel diaries and GPS tracking devices) were analysed during a seven-day data collection period. Many physical, cognitive and psychosocial function tests were included to investigate their specific and relative impact to the misreporting. The study is part of a large-scale research project into influencing factors in travel behavior in MS.

Methodology

Participants

There were 108 persons with clinical definite MS who gave written informed consent and participated. The study was approved by ethical committees of Hasselt University, Antwerp University Hospital, and National MS Centre Melsbroek. PwMS were recruited based on databases of the REVAL Rehabilitation Research Center, by neurologists of the rehabilitation centers, and after information sessions in an MS-specialized fitness center and support groups of MS Society Flanders. PwMS were included if they made only one trip a week, and were excluded if they were bedridden, or had a relapse or related corticosteroid treatment in the month before the study.

Participating PwMS were further divided in three subgroups, according to their Disease Steps (DS) (*30*) describing ambulatory dysfunction. The DS is a simple brief clinical rating scale, based on a general physical examination and the assistive devices needed to walk 25 feet (7.62 meters). Persons in the 'mild' subgroup (DS $\leq$ 2, n = 51) experienced no to mild limitations or might

Table 1. Socio-demographic Characteristics of PwMS by Subgroup

Variable	Mild MS (n = 51)	Moderate MS (n = 27)	Severe MS (n = 30)	Total MS (n = 108)
Age				
22–34	7	1	0	8
35–44	15	4	5	24
45–54	20	8	6	34
55–64	8	12	7	27
>64	0	2	12	14
Gender				
Male	16	13	12	41
Female	34	14	18	66
Education				
Primary	3	1	4	8
Secondary	25	17	19	61
Higher	22	9	7	38
Work				
Not working	31	22	28	81
Half-time	10	5	1	16
Full-time	9	0	1	10
Driving ability				
No	7	5	18	30
Uncertain	4	4	2	10
Yes	39	18	10	67
Household size				
1 person	5	3	10	18
2 persons	17	13	13	43
>2 persons	28	11	7	46
Household income				
<€1000	0	0	1	1
€1000–2500	20	10	12	42
€2500–5000	21	10	5	36
>€5000	1	0	0	1
Unknown	8	7	12	27
Cars in household				
0	4	2	7	13
1	24	22	16	62
2 or more	22	3	7	32

have a visibly abnormal gait, but did not require ambulation aids. Persons in the 'moderate' subgroup (DS 3–4, n = 27) required intermittent or continuous unilateral support to walk more than 25 feet; while persons in the 'severe' subgroup (DS 5–6, n = 30) required bilateral support or were confined to a wheelchair.

The division into subgroups was based on ambulatory (physical) dysfunction, because it was shown previously that changes in activity and travel behavior in MS were (mainly) related to increasing ambulatory dysfunction (*26*). This made it possible to examine the influence of trip-related factors (e.g., number of trips) on the reporting rate, as well as to determine to what extent the agreement between objective GPS devices and self-reported diaries would differ between disability status. Socio-demographic characteristics of PwMS by subgroup are summarized in Table 1 (missing socio-demographic data of 1 PwMS in mild subgroup).

Experimental Design and Outcome Measures

A detailed overview of the experimental design can be found in Neven et al. (*26*). During the first individual contact moment with the PwMS, measures of physical, cognitive and psychosocial functioning were taken. Socio-demographic data was collected by means of a questionnaire, and activity-related travel behavior measures were thoroughly explained. Then, during seven consecutive days, activity-related travel behavior was measured by completing a travel diary and wearing a GPS logger. These devices were additionally explained in a self-written manual and a permanent helpline was available. In the second meeting, the self-report indices and the GPS logger were returned.

Travel Behavior Outcome Measures. Both subjective self-reported activity-related travel diaries and objective GPS tracking devices were used to obtain a comprehensive

Table 2. Clinical Characteristics of PwMS by Subgroup

Disability variable	Mild MS (*n* = 51) Subgroup 1	Moderate MS (*n* = 27) Subgroup 2	Severe MS (*n* = 30) Subgroup 3	Kruskal–Wallis anova	Mann–Whitney post-hoc comparison between MS Subgroups		
					1–2	1–3	2–3
Disease steps	1.4 ± 0.6 (0–2)	3.4 ± 0.5 (3–4)	5.4 ± 0.5 (5–6)	<0.01	<0.01	<0.01	<0.01
Disease duration	10.6 ± 7.6 (1–30)	15.3 ± 7.7 (5–32)	21.4 ± 11.1 (2–48)	<0.01	<0.01	<0.01	<0.05
MSFC ↑	0.5 ± 0.4 (−0.4–1.3)	−0.1 ± 0.6 (−2.3–0.9)	−2.9 ± 2.5 (−6.6–0.3)	<0.01	<0.01	<0.01	<0.01
Physical functioning							
T25FW (m/s) ↑	1.4 ± 0.3 (0.9–2.0)	0.8 ± 0.3 (0.3–1.3)	0.2 ± 0.3 (0.0–0.9)	<0.01	<0.01	<0.01	<0.01
MSWS-12 ↓	27.4 ± 11.2 (12–58)	47.1 ± 9.7 (21–60)	46.6 ± 21.1 (0–60)	<0.01	<0.01	<0.01	ns
9HPT (s) ↓	23.4 ± 5.2 (15.0–39.9)	29.5 ± 10.4 (17.5–60.2)	41.1 ± 15.8 (17.2–76.0)	<0.01	<0.01	<0.01	<0.01
SF36 physical ↑	39.6 ± 10.2 (15.9–59.1)	28.7 ± 8.1 (12.40–43)	22.3 ± 9.0 (9.1–51.5)	<0.01	<0.01	<0.01	<0.01
Cognitive functioning							
PVSAT ↑	51.8 ± 9.7 (22–60)	47.8 ± 13.2 (0–60)	38.5 ± 16.0 (0–60)	<0.01	ns	<0.01	<0.05
TMT (s) ↓	47.5 ± 20.3 (18.9–106.6)	58.2 ± 21.2 (32.8–137.0)	106.6 ± 66.0 (32.6–257)	<0.01	<0.05	<0.01	<0.01
Psychosocial functioning							
MFIS ↓	31.7 ± 17.9 (2–74)	46.7 ± 15.0 (13–72)	36.3 ± 19.9 (0–70)	<0.01	<0.01	ns	<0.05
HADS ↓	10.2 ± 7.9 (1–33)	14.0 ± 6.9 (1–27)	11.7 ± 7.6 (0–25)	0.04	<0.05	ns	ns
FAI ↑	29.0 ± 7.2 (12–42)	24.2 ± 7.2 (7–35)	17.9 ± 7.8 (5–31)	<0.01	<0.01	<0.01	<0.01
SF36 mental ↑	50.4 ± 10.6 (20.8–66.1)	47.1 ± 12.9 (20.2–63.7)	53.5 ± 10.2 (25–73.1)	ns	/	/	/

understanding of the activity-related travel behavior of PwMS. In the diaries, participants had to indicate all information about their outdoor activities (e.g., activity type, start time and location) and resulting trips (e.g., travel mode and company). Participants were asked to carry the GPS logger for each outdoor trip in a handbag or trouser pocket, by which the current location could be determined and saved in the memory. The GPS logger, 747Pro 66-Channel GPS Trip Recorder with motion sensor (TranSystem Inc.) allowed the recording of more accurate information about routes travelled, as well as detecting trips that were not filled out in the travel diaries (see further for the description of detecting trips by GPS). A trip means an outdoor displacement which is identified by a clear activity motive (e.g., working), and can consist of one or several travel modes.

Activities were divided into one of the following categories: working/education, shopping/ services, social (e.g., visiting friends), leisure (e.g., hobbies), bring or get activities (bringing or getting relatives or things to somewhere), personal care (e.g., rehabilitation and doctor), other (not belonging to one of the previous), and home activities. Walking (either independently or using any assistive device), driving, or cycling around without destination was also seen as a trip, and was classified as 'travelling'. The outcome measures that were analyzed in this study were: research day, week day, departure time, trip duration, activity type, transport mode, planning of trips, chain of trips.

Clinical Outcome Measures. Clinical characteristics of PwMS by subgroup are shown in Table 2. Physical, cognitive, and psychosocial functions were assessed to investigate their specific and relative impact on the reporting rate of the data collection methods in patient profiles with various disability severity. The multidimensional *Multiple Sclerosis Functional Composite* (MSFC) measured the ambulation/leg function by the Timed 25-Foot Walk test (T25FW), the arm/hand function by the 9-Hole Peg test (9HPT) and cognition by the Paced Visual Serial Addition Test (PVSAT) (*31*).

Physical functioning:

- *Timed 25-Foot Walk test* (T25FW): measured the time PwMS needed to walk 25 feet as quickly as possible, using their usual assistive devices (*31*).
- *Multiple Sclerosis Walking Scale* (MSWS-12): measured the impact of MS on walking ability by means of a questionnaire consisting of 12 items (*32*).
- *9 Hole Peg test* (9HPT): measured the time PwMS needed to put nine pegs in holes in a plastic board, and remove them again (*31*).
- *36-item Short-Form Health Survey* (SF-36): measured the self-reported health related quality of life (QoL), resulting in a physical health summary score (*33*).

Cognitive functioning:

- *Paced Visual Serial Addition Test* (PVSAT): measured working memory, arithmetic capabilities and attention, by which PwMS were shown a number every three seconds and asked to say aloud the sum of the second last (*31, 34*).
- *Trail Making Test* (TMT): measured visual attention and task switching by recording the time PwMS needed to connect 25 consecutive dots on a sheet of paper (numbers in part A, numbers and letters in part B) (*35*).

Psychosocial functioning:

- *Modified Fatigue Impact Scale* (MFIS): measured the impact of fatigue on daily functioning (*36*).
- *Hospital Anxiety and Depression Scale* (HADS): measured the level of depression and anxiety (*37*).
- *Frenchay Activities Index* (FAI): measured instrumental activities of daily living which required some initiative from the patient in the last three and six months (*38*).
- *36-item Short-Form Health Survey* (SF-36): measured the self-reported health related QoL, resulting in a mental health summary score (*33*).

In Table 2, values are presented as mean ± SD (range). Ns means not significant. Upward arrows indicate better performance with higher scores (e.g., a higher score on T25FW (m/s) means that the participant has a better ambulation/leg function); while downward arrows indicate worse performance with higher scores (e.g., a higher score on MSWS-12 means that the participant's MS has a greater (negative) impact on his walking ability).

Data Processing and Statistical Analysis

As a first step, the data of the travel diaries were implemented in a database in Access. Next, all individual trips were manually compared by the researcher, by viewing the GPS traces of participants in the geographic information system software package ArcGIS and visually comparing them with the information from the travel diary (i.e., the data which were inserted in the database). If the trip was forgotten in the diary (and thus not in the database), the database was complemented with the trips detected by the GPS data. As such, the final database consisted of all trips made during the data collection period. Every trip was labelled if it was either (i) registered in the diary alone, (ii) by means of the GPS tracking device alone, of (iii) by both data collection instruments. If a trip was forgotten by both instruments (e.g., a homeward trip) but could be logically deduced (e.g., if the next trip again started from the home location), it was complemented in the database as 'forgotten by GPS and diary.' Other examples of these discovered trips are: a trip starting at location C, while the previous trip was from location A to location B (in this example the trip B–C was forgotten); or if the first trip of the day started from the work location while the person arrived home the day before (in this example the trip home–work was forgotten).

Numerical data were analyzed using SPSS Statistics ($p < 0.05$). Descriptive analyses were used for the questionnaires and travel diaries. The Shapiro–Wilk test indicated non-normal distributions of most variables and, therefore, non-parametric Kruskal–Wallis analysis of variance (anova), and post-hoc Mann–Whitney tests for independent samples were used to examine differences between MS subgroups (e.g., 'Is the reporting rate by GPS better in the mild subgroup compared with the moderate subgroup?'). The Wilcoxon test was applied to determine differences in outcome measures for each data collection method (e.g., 'Was research day 1 better filled in in the diary than research day 2?'). Bivariate Spearman correlation coefficients were calculated to assess the level of association between variables, by which a correlation was considered as poor (<0.30), low (0.30–0.50), moderate (0.50–0.70), high (0.70–0.89), and very high (>0.90). Multiple regression analyses with a stepwise selection procedure were performed to determine the most predictive outcome measures for the reporting rate of the travel diary and the GPS. Multicollinearity was checked for all models.

Results

Differences between MS Subgroups: Clinical Characteristics

Table 2 shows that the overall significant disparity among MS subgroups justified the selected DS cut-off

Table 3. Reporting Rate by Data Collection Method of PwMS by Subgroup

Reporting method (%)	Total MS (*n* = 108)	Mild MS (*n* = 51) Subgroup 1	Moderate MS (*n* = 27) Subgroup 2	Severe MS (*n* = 30) Subgroup 3	Kruskal–Wallis anova	Mann–Whitney Post-hoc Comparison between MS Subgroups		
						1–2	1–3	2–3
Diary	87.65 ± 18.09 (0–100)	84.94 ± 21.39 (0–100)	87.78 ± 15.93 (50–100)	92.29 ± 12.28 (50–100)	ns	/	/	/
GPS	76.95 ± 23.00 (0–100)	78.59 ± 20.58 (0–100)	80.56 ± 16.35 (41–100)	70.70 ± 30.66 (0–100)	ns	/	/	/
Diary & GPS	66.31 ± 25.46 (0–100)	65.35 ± 26.24 (0–97)	69.50 ± 20.36 (25–100)	65.04 ± 28.76 (0–100)	ns	/	/	/
Only diary	21.34 ± 21.77 (0–100)	19.59 ± 20.12 (0–97)	18.28 ± 15.79 (0–59)	27.26 ± 28.07 (0–100)	ns	/	/	/
Only GPS	10.64 ± 16.60 (0–100)	13.24 ± 20.36 (0–100)	11.06 ± 14.19 (0–47)	5.66 ± 8.84 (0–35)	ns	/	/	/
Forgotten by GPS & diary	1.71 ± 5.49 (0–50)	1.82 ± 3.16 (0–14)	1.16 ± 3.14 (0–15)	2.05 ± 9.32 (0–50)	<0.05	ns	<0.05	ns

scores 2 and 5 for differentiating between patients with various ambulatory dysfunctions. Significant differences were found in disease duration, MSFC, all physical and cognitive functioning measures outcomes, and almost all outcomes on psychosocial functioning measures, meaning that the moderate and, especially, the severe MS subgroups suffered more from physical, cognitive, and psychosocial symptoms than the mild MS subgroup.

Baseline Results: Reporting Rate of Data Collection Methods

The 108 individuals reported a total of 2,562 trips in their diaries, while the GPS tracking devices carried by these individuals recorded a total of 2,311 trips. In general, 71 individuals (66%) reported more trips in the diary than were detected by GPS; eight individuals (8%) reported the same number of trips in both data collection methods; and 28 individuals (26%) reported fewer trips in the diary. More information about the travel behavior (number of trips, modal split, activities, etc.) can be found in (*26*).

Table 3 shows that the majority of trips (66%) were reported in both data collection methods. In general, more than one fifth of all trips (21%) were only reported by diary and thus forgotten and not registered by GPS (e.g., forgotten to carry or to switch on the GPS); compared with 11% of trips that were only registered by GPS and thus forgotten and not reported in the diary. Overall, as well as across MS subgroups, the data quality of the diary was higher than the quality of the GPS data. In the severe MS subgroup, more trips were forgotten and not registered by GPS, while no significant difference was found between subgroups. Values are presented as mean % ± SD (range), and ns means not significant.

Reporting Rate of Data Collection Methods across Research Days and Travel Outcome Measures

Wilcoxon signed-rank tests were applied to determine if the travel outcome measures influenced the reporting rate of the different data collection methods. Figure 1 shows that trips were less registered by GPS on the first research day compared with the subsequent days, which appeared to be significant (Wilcoxon signed-rank test, $p < 0.05$). There were no significant differences between reporting on weekdays or weekend days, neither in the diary nor by GPS. Registration by GPS was more often forgotten in trips in the afternoon (16–20 H) compared with other time periods, in trips of short duration (<10 min) compared with longer trip durations, and in trips that did not occur as part of a trip chain. Reporting in the diary was less complete in homeward trips compared with other activity types, and in trips by bicycle or public transport compared with trips by car. It is possible that homeward trips were more often forgotten because participants forgot that return home trips were also part of their outdoor activities. Trips by bicycle or public transport may be forgotten because they were more often part of a travel chain (for which participants may have forgotten to indicate all travel modes used). The planning of the trips did not influence the reporting rate of the data collection methods.

Reporting Rate of Data Collection Methods by Travel Outcome Measure and MS Subgroup

Table 4 shows the reporting rate per outcome measure, indicating only limited significant differences between MS subgroups. Persons with mild MS more often forgot

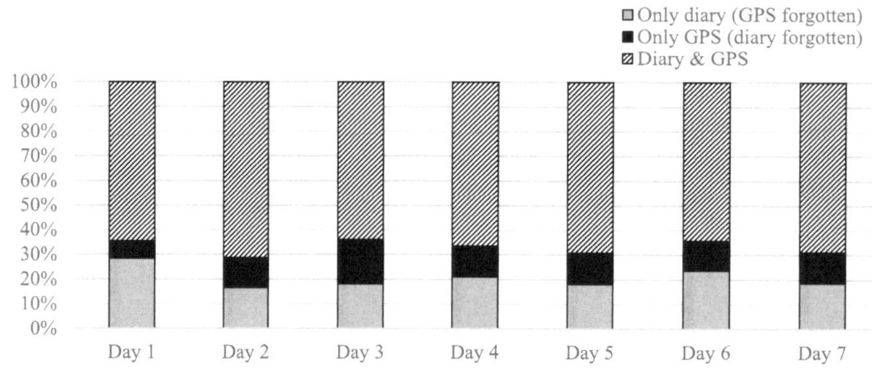

Figure 1. Reporting rate of data collection method per day.

to carry their GPS device for short trips or for single trips, compared with other subgroups. In addition, they more often forgot to report trips in the morning (6–12 H), trips as car passenger, and spontaneous trips in their diary. On the other hand, persons with mild MS reported their leisure trips better in both data collection methods compared with other subgroups. Persons with severe MS reported trips better in the morning and afternoon in their diary, and reported their trips as car passenger better compared with the other subgroups. There was no difference between the reporting rate of MS subgroups depending on the research day or the type of day (weekday versus weekend day). It was notable that none of the work or educational trips was forgotten and not filled in the travel diary in any of the subgroups. In the table, values are presented as mean ± SD (range), and ns means not significant.

Influencing (Disease-Related) Clinical, Socio-demographic and Trip-Related Factors

Table 5 shows the correlation coefficients between the reporting rates of both data collection methods, and clinical, socio-demographic, and trip-related outcome measures by MS subgroup. Within the total sample, associations with clinical measures were absent for both data collection methods. In the mild MS subgroup, persons with a higher MSFC (lower overall disability severity) score better at filling out their travel diary. In the moderate MS subgroup, there was a low positive correlation between the DS and the reporting rate of the GPS, while the T25FW correlated moderately negative. Persons in the moderate MS subgroup who had more ambulatory (walking) problems, seemed thus to register their trips better by GPS. There was a low correlation between the physical component of the SF36 in the severe MS subgroup and the reporting rate of the diary, indicating that persons with severe MS who experienced a higher (physical) QoL, performed worse at reporting their trips in the diary.

In the case of socio-demographic factors, the educational level correlated poorly with the reporting rate of the diary within the total sample, and moderately in the moderate and severe MS subgroups. Persons with a higher educational level seem to report their trips better in the diary. The work situation was positively associated with the reporting rate in the diary (PwMS who worked filled out their diary better), while the household size seemed to correlate negatively (PwMS with a larger household size filled out their diary worse). In the moderate MS subgroup, the registration of trips by GPS was positively influenced by a higher age. Overall, PwMS who performed more trips reported their trips worse in their travel diary. In the mild MS subgroup, planned trips were reported better in the diary but worse when registered by GPS. The correlation coefficients between trips by public transport and the reporting rate of the GPS were dissimilar between the subgroups. However, it should be noted that a bad registration by GPS when travelling by public transport, can also result from external factors like signal interference (e.g., in trains). In the table, significant correlation coefficients are indicated with * if $p < 0.05$ and with † if $p < 0.01$. Ns means not significant.

Results of the multiple regression analyses, performed with the total sample and the subgroups separately, are presented in Table 6. Overall, the models better explained variability in the mild subgroup regarding the reporting rate of the diary, and in the moderate subgroup regarding the reporting rate by GPS (respectively 57.7% and 38.6%), the models which also had clinical variables as significant predictors. In the mild MS subgroup, the overall disability severity (MSFC) and trip-related factors like the share of planned trips and trips performed as a car passenger, determined the reporting rate in the diary. In the moderate MS subgroup, the reporting rate

Table 4. Reporting Rate of Data Collection Instrument by Travel Outcome Measure and MS Subgroup

Reporting rate (%)	Only Diary (GPS Forgotten)				Only GPS (Diary Forgotten)				Diary & GPS				Mann–Whitney Comparison between Subgroups								
													Only Diary			Only GPS			Diary & GPS		
MS Subgroup	Tot	Mild	Mod	Sev	Tot	Mild	Mod	Sev	Tot	Mild	Mod	Sev	1–2	1–3	2–3	1–2	1–3	2–3	1–2	1–3	2–3
Research day																					
Day 1	31.7	24.2	35.3	42.7	6.6	7.0	5.1	7.3	61.6	68.7	59.6	50.0	ns	ns	ns	ns	ns	ns	ns	ns	ns
Day 2	18.2	16.5	15.6	25.0	9.2	11.2	9.9	3.7	72.7	72.3	74.5	71.2	ns	ns	ns	ns	ns	ns	ns	ns	ns
Day 3	21.5	16.0	22.6	33.3	13.3	17.1	12.4	5.3	65.2	67.0	65.0	61.3	ns	ns	ns	ns	ns	ns	ns	ns	ns
Day 4	20.7	22.0	11.8	27.5	11.3	11.2	14.4	8.2	68.0	66.8	73.8	64.3	ns	ns	ns	ns	ns	ns	ns	ns	ns
Day 5	19.7	22.5	7.4	23.1	11.9	13.3	15.1	6.6	68.4	64.2	77.5	70.3	ns	ns	ns	ns	ns	ns	ns	ns	ns
Day 6	23.1	25.6	21.6	17.7	11.1	10.9	14.4	7.0	65.8	63.5	64.0	75.3	ns	ns	ns	ns	ns	ns	ns	ns	ns
Day 7	21.6	23.3	22.1	17.6	11.1	14.0	12.9	3.8	67.3	62.7	65.0	78.6	ns	ns	ns	ns	ns	ns	ns	ns	ns
Week day																					
Weekday	22.3	21.7	19.8	25.5	10.2	12.9	9.6	5.9	67.7	65.4	70.6	69.3	ns	ns	ns	ns	ns	ns	ns	ns	ns
Weekend day	22.0	20.0	18.7	28.6	12.6	12.4	19.8	6.6	65.4	67.6	61.5	64.8	ns	ns	ns	ns	ns	ns	ns	ns	ns
Departure time																					
6–12 H	24.2	22.4	21.1	30.5	7.0	10.0	4.6	3.7	68.8	67.6	74.4	65.8	ns	ns	ns	ns	< 0.05	ns	ns	ns	ns
12–16 H	20.6	17.0	15.4	31.9	9.9	13.1	11.3	3.2	69.5	70.0	73.3	65.0	ns	ns	ns	ns	ns	< 0.05	ns	ns	ns
16–20 H	26.3	22.4	24.8	35.1	12.8	15.8	13.6	6.3	61.0	61.9	61.6	58.6	ns	ns	ns	ns	< 0.05	ns	ns	ns	ns
20–6 H	19.7	14.9	17.5	33.1	18.2	23.1	18.9	6.4	62.1	62.0	63.6	60.5	ns	ns	ns	ns	ns	ns	ns	ns	ns
Trip duration																					
<10 min	21.0	23.0	15.7	22.6	13.2	14.9	14.5	8.2	65.9	62.2	69.8	69.3	< 0.05	ns	ns	ns	ns	ns	ns	ns	ns
11–20 min	18.4	14.7	19.9	25.0	12.8	13.1	16.6	7.7	68.8	72.2	63.6	67.3	ns	ns	ns	ns	ns	< 0.05	ns	ns	ns
>20 min	15.2	14.0	11.7	21.0	7.7	10.4	6.8	3.3	77.1	75.6	81.6	75.7	ns	ns	ns	ns	ns	ns	ns	ns	ns
Activity type																					
Work/Education	26.3	31.7	0.0	33.3	0.0	0.0	0.0	0.0	73.7	68.3	100.0	66.7	< 0.05	ns	< 0.01	ns	ns	ns	< 0.05	ns	< 0.01
Shops/Services	19.6	16.5	12.5	34.2	1.5	2.6	0.0	1.1	78.9	80.9	87.5	64.7	ns	ns	ns	ns	ns	ns	ns	ns	< 0.05
Leisure	21.6	16.2	25.9	28.8	2.8	1.4	6.1	2.1	75.7	82.4	68.0	69.2	ns	ns	ns	ns	ns	ns	< 0.05	ns	ns
Social	27.2	30.0	22.0	26.7	2.9	1.8	4.6	3.5	69.9	68.2	73.4	69.8	ns	ns	ns	ns	ns	ns	ns	ns	ns
Personal care	29.5	24.0	31.2	37.4	0.9	0.9	1.7	0.0	69.6	75.0	67.2	62.6	ns	ns	ns	ns	ns	ns	ns	ns	ns
Walking	23.0	31.2	12.5	17.2	4.0	3.0	10.0	0.0	73.0	65.8	77.5	82.8	ns	ns	ns	ns	ns	ns	ns	ns	ns
Bring/Get	27.3	27.1	24.6	37.5	3.1	4.6	0.0	0.0	69.6	68.3	75.4	62.5	ns	ns	ns	ns	ns	ns	ns	ns	ns
Other	35.6	50.0	0.0	60.0	0.0	0.0	0.0	0.0	64.6	50.0	100.0	40.0	ns	ns	ns	ns	ns	ns	ns	ns	ns
Home	22.8	19.3	20.5	31.1	10.6	13.1	11.4	5.2	66.7	67.6	68.1	63.7	ns	ns	ns	ns	ns	ns	ns	ns	ns
Transport Mode																					
Car driver	19.4	22.1	16.9	14.2	2.5	1.6	54.	1.1	77.9	76.0	77.3	84.7	ns	ns	ns	ns	ns	ns	ns	ns	ns
Car passenger	24.3	21.7	20.8	32.5	2.4	0.0	9.2	0.8	72.7	78.3	70.0	64.2	ns	ns	ns	< 0.01	ns	< 0.05	ns	ns	ns
Cyclist	26.3	26.5	31.7	0.0	16.4	13.6	8.3	100.0	53.2	54.6	60.0	0.0	ns	ns	ns	ns	ns	ns	ns	ns	ns
Pedestrian	28.3	31.2	25.3	21.4	5.4	4.3	2.2	20.0	65.9	64.2	71.5	58.6	ns	ns	ns	ns	ns	ns	ns	ns	ns
Public transport	35.9	41.7	4.8	66.7	13.6	18.2	11.7	0.0	49.6	40.2	80.2	33.3	ns	ns	ns	ns	ns	ns	ns	ns	ns
Wheelchair	22.9	0.0	0.0	33.5	7.5	0.0	16.7	3.3	69.5	0.0	83.3	63.2	ns	ns	< 0.01	ns	ns	ns	ns	ns	< 0.05
Adapted	43.0	0.0	50.0	47.5	0.0	0.0	0.0	0.0	57.0	100	50.0	52.5	ns	ns	ns	ns	ns	ns	ns	ns	ns
Unknown	18.1	7.7	22.2	40.6	31.0	28.4	48.1	18.8	11.5	14.9	0.0	15.6	ns	ns	ns	ns	ns	ns	ns	ns	ns

(continued)

Table 4. Continued

Reporting rate (%)	Only Diary (GPS Forgotten)				Only GPS (Diary Forgotten)				Diary & GPS				Mann–Whitney Comparison between Subgroups								
													Only Diary			Only GPS			Diary & GPS		
MS Subgroup	Tot	Mild	Mod	Sev	Tot	Mild	Mod	Sev	Tot	Mild	Mod	Sev	1–2	1–3	2–3	1–2	1–3	2–3	1–2	1–3	2–3
Planning trips																					
Planned	24.6	20.6	29.9	26.2	4.9	7.1	2.6	3.3	70.5	72.2	67.5	70.5	ns	ns	ns	ns	ns	ns	ns	ns	ns
Spontaneous	21.2	17.2	26.8	22.5	4.3	7.6	0.4	2.5	74.6	75.2	72.9	75.0	ns	ns	ns	< 0.05	ns	ns	ns	ns	ns
Unknown	8.9	7.6	9.8	10.3	62.6	62.7	50.7	73.6	28.5	29.7	39.4	16.1	ns	ns	ns	ns	ns	ns	ns	ns	ns
Trip chaining																					
Single	26.0	24.7	33.9	23.9	8.7	12.6	5.3	5.1	64.5	62.7	60.8	71.0	< 0.05	ns	ns	ns	ns	ns	ns	ns	ns
Chain	14.8	13.3	13.7	18.5	16.9	21.7	9.7	15.2	68.3	65.1	76.6	66.3	ns	ns	ns	ns	ns	ns	ns	ns	ns

by GPS was influenced by the perceived QoL (SF-36) and the age of the participants. Within the total sample, only trip-related factors were of importance, while clinical or socio-demographic factors did not contribute in explaining the variability regarding the reporting rate in the diary. In the table, R^2 is predictive value, β is estimate, SE is standard error and t is t-value. Significant regression coefficients are indicated with * if $p < 0.05$ and † if $p < 0.01$.

Discussion

The present study investigated the influence of disease-related physical, cognitive, and psychosocial characteristics, besides socio-demographic and trip-related variables, on the data quality of different data collection methods in a travel survey in PWD (more specific PwMS). This study demonstrated that clinical variables had only limited influence on the reporting rate (and associating data quality) of subjective travel diaries and objective GPS tracking devices. In addition, there were only limited significant differences between subgroups with various disability severities. 66% of the trips were reported in both data collection methods, while overall more than one fifth of all trips were forgotten and not registered by GPS, and 11% were forgotten and not reported in the diary.

Within the total sample, it was interesting (but unexpected) that no clinical variable correlated with the reporting rate of either the travel diary or the GPS device. Since there was overall significant disparity among subgroups in the cognitive tests (Table 2), we expected more potential bias in reporting results by PwMS with severe dysfunction. Persons in this subgroup more often forgot to carry their GPS, but still reported their trips better in their travel diary (although not significantly different between subgroups). This could be explained by family members assisting in recroding trips in the diaries. However, it seemed that the household size negatively influenced the reporting rate in the diary, possibly because of other household activities with higher priority. In the moderate MS subgroup, persons with a higher perceived health-related QoL (mental component of SF-36) registered trips better by GPS. Persons with a lower perceived QoL might have more difficulties in making an (independent) trip, and therefore might well be less inclined to carry the GPS device with them.

Confirming previous literature (*22*), the educational level and work situations correlated positively. However, there were dissimilarities among subgroups with the reporting rate of the travel diary. Education seemed to correlate moderately in the more advanced disease stages, while the work situation was only of importance in the mild MS subgroup and the total sample. It was

Table 5. Bivariate Spearman Correlation Analysis between Reporting Rate and Clinical, Socio-demographic and Trip-related Variables by MS Subgroup

Reporting rate (%)	Diary				GPS			
MS Subgroup	Total	Mild	Mod	Sev	Total	Mild	Mod	Sev
Clinical								
Disease steps	ns	ns	ns	ns	ns	ns	0.44*	ns
Disease duration (yrs)	ns	ns	ns	ns	ns	ns	ns	ns
MSFC ↑	ns	0.33*	ns	ns	ns	ns	ns	ns
Physical functioning								
T25FW (m/s) ↑	ns	ns	ns	ns	ns	ns	−0.51†	ns
MSWS-12 ↓	ns	ns	ns	ns	ns	ns	ns	ns
9HPT (s) ↓	ns	ns	ns	ns	ns	ns	ns	ns
SF36 physical ↑	ns	ns	ns	−0.39*	ns	ns	ns	ns
Cognitive functioning								
TMT (s) ↓	ns	ns	ns	ns	ns	ns	ns	ns
PVSAT ↑	ns	ns	ns	ns	ns	ns	ns	ns
Psychosocial functioning								
MFIS ↓	ns	ns	ns	ns	ns	ns	ns	ns
HADS ↓	ns	ns	ns	ns	ns	ns	ns	ns
FAI ↑	ns	ns	ns	ns	ns	ns	ns	ns
SF36 mental ↑	ns	ns	ns	ns	ns	ns	0.47*	ns
Sociodemographic								
Age	ns	ns	ns	ns	ns	ns	0.56†	ns
Gender	ns	ns	ns	ns	ns	ns	ns	ns
Education	0.28†	ns	0.50†	0.61†	ns	ns	ns	ns
Work situation	0.23*	0.36*	ns	ns	ns	ns	ns	ns
Driving ability	ns	ns	ns	ns	ns	ns	ns	ns
Household size	−0.22†	ns	ns	ns	ns	ns	−0.41*	ns
Household income	ns	ns	ns	ns	ns	ns	ns	ns
Cars in household	ns	ns	ns	ns	ns	ns	ns	ns
Trip-related								
Number of trips	−0.25†	−0.31*	ns	ns	ns	ns	ns	ns
Car passenger	ns	0.29*	ns	ns	ns	ns	ns	ns
Public transport	ns	ns	ns	ns	−0.25†	−0.28*	0.44*	ns
Planned trips	0.25*	0.37†	ns	ns	ns	−0.30*	ns	ns
Trips to work	ns	0.37†	ns	ns	ns	ns	ns	ns
Trips to shops/services	0.29†	ns	0.54†	ns	ns	ns	ns	ns
Trips for personal care	0.20*	ns	ns	ns	ns	ns	ns	ns
Trips 12–16 H	ns	ns	ns	ns	0.26†	ns	ns	ns

somewhat surprising that, in the moderate subgroup, older persons registered their trips better by GPS, in contrast with literature (*22*) stating that GPS should be used for younger generations. The number of trips negatively influenced the reporting rate of the travel diary in the total sample and mild subgroup, likely because it is too difficult to remember a high number of trips, or perhaps because of the amount of time involved in reporting each trip. As well in the mild subgroup, and within the total sample, persons who planned more trips in advance reported their trips better, probably because it is easier to recall trips that were planned before.

Overall, using only GPS tracking devices (the current standard method) for collecting data in PwMS and more generally PWD, should be undertaken with caution, as more than one fifth of all trips were not registered in this study, and as much as 29% in the severe subgroup. Participants were allowed to keep the device switched on (only at night were they asked to switch it off), as the logger automatically switched to the stand-by function when no movement was detected for more than five minutes (by a motion sensor) and, similarly, started recording when the person moved again. However, participants had to remember to take the device with them when they started travelling, which often appeared to be problematic. Besides possible cognitive problems, PWD are also likely to have to think about several organizational issues when making a trip (e.g., carrying assistive mobility devices) because of their mobility limitations, which may explain why they forget to take their GPS with them more often than participants without limitations. A (straightforward) solution may be to attach the GPS device to another attribute they usually carry while moving outdoors. Current innovations in GPS technology

Table 6. Multiple Linear Regression: Clinical, Socio-demographic and Trip-related Factors Related to Reporting Rate

Reporting rate (%)	Diary									GPS		
	Total			Mild			Severe			Moderate		
MS Subgroup	β	SE	t	β	SE	t	β	SE	t	β	SE	t
Clinical factors												
MSFC				0.16	0.05	3.37†						
SF-36 mental										0.01	0.00	2.25*
Socio-demographic factors												
Age										0.01	0.00	2.71*
Education							0.04	0.02	2.40*			
Trip-related factors												
Number of trips	−0.02	0.01	−2.35*									
Car passenger				0.29	0.10	2.94†						
Planned trips	0.22	0.06	4.01†	0.51	0.08	6.13†						
Trips to shops/services	0.20	0.08	2.61*									
Overall model												
R	0.224			0.577			0.176			0.386		
Adjusted *R*	0.201			0.550			0.146			0.335		
β constant	0.757			0.381			0.775			0.154		
Standard error	0.056			0.065			0.065			0.173		
p	<0.001			<0.001			<0.05			<0.01		

and mobile positioning data, by which trips and stops can be automatically extracted from GPS traces and presented on a map to the traveler (prompted recall) (*12*), may lower the respondent burden. If the GPS acquisition could become (almost) completely automated, it is expected that the majority of trips would be registered by GPS, leading to a higher data quality. Although we assumed in our study that we captured the complete travel behavior of participants by combining the travel diary and the GPS data, it is possible that some trips were not logged because of GPS device problems (and thus not user error), as we did not control the GPS data quality at the individual point level or for each trip.

The results suggest that traditional (subjective) paper and pencil diaries may be more suitable for PWD with more advanced stages of the disease, the difference being that participants would not have to think about filling in the diary immediately, but would be able to do this at a later time when they had less organizational things on their minds. On the other hand, this longer time period may cause delayed memory problems with more potential bias in reporting results. Informing family members or caregivers about the research could help to achieve more complete results.

In summary, we recommend the complemented use of both subjective self-report and objective GPS tracking devices to obtain detailed information about the actual activity-related travel behavior of PWD. Despite recent advances in GPS technology, self-report diaries remain an important source of information for understanding why and how people make their trips (e.g., for which purpose, with whom) at least for persons for whom it is possible to use this method. While in this study the missing diary data were complemented by visually checking the logged GPS data, this is not feasible in a larger population, suggesting that an automatic checking approach is necessary. Participants suffering from problems with physical problems like fine hand motor skills, as well as those with intellectual impairments, may not be able to fill in diaries or questionnaires. For these people, GPS devices may provide a convenient solution to capture their travel behavior, but it must be ensured that this device will not be forgotten when leaving on trips. This could be achieved by, for example, sending a daily message to participants to remind them to carry their GPS. In addition, new technologies that enable researchers to collect objective data about underlying activity types and accompanying persons, like visual life-logging technologies capturing every day activities, may potentially make self-report diaries redundant (*39*). To make a substantiated choice of data collection instrument in travel behavior studies in PWD, the (disease-related) capabilities of the participants should be taken into account and the selected method should be targeted toward the disability level of the target group. The present study may give initial indications about how this can be done.

Author Contributions

The authors confirm contributions to the paper as follows: study conception and design: Neven, Wets, Feys, Janssens; data collection: Neven, De Schutter; analysis and interpretation of results: Neven, De Schutter, Wets, Feys, Janssens; draft manuscript preparation: Neven. All authors reviewed the results and approved the final version of the manuscript.

References

1. Prince, S. A., K. B. Adamo, M. E. Hamel, J. Hardt, S. C. Gorber, and M. Tremblay. A Comparison of Direct versus Self-Report Measures for Assessing Physical Activity in Adults: A Systematic Review. *International Journal of Behavioral Nutrition and Physical Activity*, Vol. 5., No. 56, 2008.
2. Bauman, A., P. Phongsavan, S. Schoeppe, and N. Owen. Physical Activity Measurement - A Primer for Health Promotion. *Promotion & Education*, Vol. 13, No. 2, 2006, pp. 92–103.
3. Haskell, W. L. Physical Activity by Self-Report: A Brief History and Future Issues. *Journal of Physical Activity & Health*, Vol. 9, 2012, S5–S10.
4. Gosney, J. L., J. A. Scott, E. M. Snook, and R. W. Motl. Physical Activity and Multiple Sclerosis. Validity of Self-report and Objective Measures. *Family and Community Health*, Vol. 30, No. 2, 2007, pp. 144–150.
5. Hale, L., K. Williams, C. Ashton, and T. Connole. Reliability of RT3 Accelerometer for Measuring Mobility in People with Multiple Sclerosis: Pilot Study. *Journal of Rehabilitation Research & Development*, Vol. 44, No. 4, 2007, pp. 619–628.
6. Shephard, R. J. Review. Limits to the Measurement of Habitual Physical Activity by Questionnaires. *British Journal of Sports Medicine*, Vol. 37, 2003, pp. 197–206.
7. Dewulf, B., T. Neutens, D. Van Dyck, I. De Bourdeaudhuij, and N. Van de Weghe. Correspondence Between Objective and Perceived Walking Times to Urban Destinations: Influence of Physical Activity, Neighbourhood Walkability, and Socio-demographics. *International Journal of Health Geographics*, Vol. 11, No. 43, 2012, pp. 1–10.
8. Helmerhorst, H. J. F., S. Brage, J. Warren, H. Besson, and U. Ekelund. Review. A Systematic Review of Reliability and Objective Criterion-related Validity of Physical Activity Questionnaires. *International Journal of Behavioral Nutrition and Physical Activity*, Vol. 9, No. 103, 2012, pp. 1–55.
9. Dunton, G. F., E. Dzubur, K. Kawabata, B. Yanez, B. Bo, and S. Intille. Development of a Smartphone Application to Measure Physical Activity using Sensor-assisted Self-report. *Frontiers in Public Health*, Vol. 2, No. 12, 2014, pp. 1–13.
10. de Nazelle, A., M. J. Nieuwenhuijsen, J. M. Antó, M. Brauer, D. Briggs, C. Braun-Fahrlander, N. Cavill, A. R. Cooper, H. Desqueyroux, S. Fruin, G. Hoek, L. I. Panis, N. Janssen, M. Jerrett, M. Joffe, Z. J. Andersen, E. van Kempen, S. Kingham, N. Kubesch, K. M. Leyden, J. D. Marshall, J. Matamala, G. Mellios, M. Mendez, H. Nassif, D. Ogilvie, R. Peiró, K. Pérez, A. Rabl, M. Ragettli, D. Rodríguez, D. Rojas, P. Ruiz, J. F. Sallis, J. Terwoert, J. F. Toussaint, J. Tuomisto, M. Zuurbier, and E. Lebret. Improving Health Through Policies that Promote Active Travel: A Review of Evidence to Support Integrated Health Impact Assessment. *Environment International*, Vol. 37, No. 4, 2011, pp. 766–777.
11. Bellemans, T., D. Janssens, G. Wets, H. J. P. Timmermans, and B. Kochan. In the Field Evaluation of the Impact of a GPS-enabled Personal Digital Assistant on Activity-travel Diary Data Quality. *Transportation Research Record: Journal of the Transportation Research Board*, 2007. 2049: 136–143.
12. Cich, G., L. Knapen, T. Bellemans, D. Janssens, and G. Wets. TRIP/STOP Detection in GPS Traces to Feed Prompted Recall Survey. *Proc., 6th International Conference on Ambient Systems, Networks and Technologies*, London, 2015, Elsevier, pp. 262–269.
13. Raza, A., L. Knapen, K. Declercq, T. Bellemans, D. Janssens, and G. Wets. Diary Survey Quality Assessment Using GPS Traces. *Proc., 6th International Conference on Ambient Systems, Networks and Technologies*, London, 2015, Elsevier, pp. 600–605.
14. Ahas, R., H. J. Miller, and F. Witlox. From the Guest Editors: Mobility, Communication, and Urban Space. *Journal of Urban Technology*, Vol. 21, No. 2, 2014, pp. 1–7.
15. Panter, J., S. Costa, A. Dalton, A. Jones, and D. Ogilvie. Development of Methods to Objectively Identify Time Spent Using Active and Motorised Modes of Travel to Work: How do Self-reported Measures Compare? *International Journal of Behavioral Nutrition and Physical Activity*, Vol. 11, No. 116, 2014.
16. Kang, B., A. V. Moudon, P. M. Hurvitz, L. Reichley, and B. E. Saelens. Walking Objectively Measured: Classifying Accelerometer Data with GPS and Travel Diaries. *Medicine & Science in Sports & Exercise*, Vol. 45, No. 7, 2013, pp. 1419–1428.
17. Bricka, S., and C. R. Bhat. A Comparative Analysis of GPS-Based and Travel Survey-Based Data. *Transportation Research Record: Journal of the Transportation Research Board*, 2006. 1972: 9–20.
18. Zmud, J., and J. Wolf. Identifying the Correlates of Trip Misreporting - Results from the California Statewide Household Travel Survey GPS Study. *Proc., 10th International Conference on Travel Behaviour Research*, Lucerne, 2003, IATBR, pp. 1–16.
19. Pierce, B., J. Casas, and G. Giaimo. Estimating Trip Rate Under-Reporting: Preliminary Results from The Ohio Household Travel Survey. Presented at 82nd Annual Meeting of the Transportation Research Board, Washington, D.C., 2003.
20. Forrest, T. L., and D. F. Pearson. Comparison of Trip Determination Methods in Household Travel Surveys Enhanced by a Global Positioning System. *Transportation Research Record: Journal of the Transportation Research Board*, 2005. 1917: 63–71.
21. Houston, D., T. T. Luong, and M. Boarnet. Tracking Daily Travel: Assessing Discrepancies Between GPS-derived and Self-reported Travel Patterns. *Transportation Research Part C*, Vol. 48, 2014, pp. 97–108.
22. Bricka, S. G., et al. An Analysis of the Factors Influencing Differences in Survey-reported and GPS-recorded Trips. *Transportation Research Part C*, Vol. 21, 2012, pp. 67–88.
23. Wolf, J. *Using GPS Data Loggers to Replace Travel Diaries in the Collection of Travel Data*. Georgia Institute of Technology, Atlanta, 2000.

24. Stopher, P. R., C. FitzGerald, and M. Xu. Assessing the accuracy of the Sydney Household Travel Survey with GPS. *Transportation*, Vol. 34, No. 6, 2007, pp. 723–741.
25. Stopher, P., and L. Shen. An In-depth Comparison of GPS and Diary Records. *Transportation Research Record: Journal of the Transportation Research Board*, 2011. 2846: 32–37.
26. Neven, A., D. Janssens, G. Alders, G. Wets, B. Van Wijmeersch, and P. Feys. Documenting Outdoor Activity and Travel Behaviour in Persons with Neurological Conditions Using Travel Diaries and GPS Tracking Technology: A Pilot Study in Multiple Sclerosis. *Disability and Rehabilitation*, Vol. 35, No. 20, 2013, pp. 1718–1725.
27. Risser, R., S. Iwarsson, and A. Stahl. How do People with Cognitive Functional Limitations Post-stroke Manage the Use of Buses in Local Public Transport? *Transportation Research Part F*, Vol. 15, 2012, pp. 111–118.
28. Shoval, N., G. Auslander, K. Cohen-Shalom, M. Isaacson, R. Landau, and J. Heinik. What can We Learn About the Mobility of the Elderly in the GPS Era? *Journal of Transport Geography*, Vol. 18, 2010, pp. 603–612.
29. Jongen, P. J., A. T. Ter Horst, and A. M. Brands. Cognitive Impairment in Multiple Sclerosis. *Minerva Medica*, Vol. 103, No. 2, 2012, pp. 73–96.
30. Hohol, M. J., E. J. Orav, and H. L. Weiner. Disease Steps in Multiple Sclerosis: A Simple Approach to Evaluate Disease Progression. *Neurology*, Vol. 45, 1995, pp. 251–255.
31. National Multiple Sclerosis Society. *Multiple Sclerosis Functional Composite - Administration and Scoring Manual*. 2001, UNITECH Communications.
32. Hobart, J. C., A. Riazi, D. L. Lamping, R. Fitzpatrick, and A. J. Thompson . Measuring the Impact of MS on Walking Ability. The 12-Item MS Walking Scale (MSWS-12). *Neurology*, Vol. 60, No. 31, 2003, pp. 31–36.
33. Ware, J. E., and C. D. Sherbourne. The MOS 36-item Short-form Health Survey (SF-36). *Medical Care*, Vol. 30, No. 6, 1992, pp. 473–483.
34. Nagels, G., L. Geentjens, D. Kos, L. Vleugels, M. B. D'hooghe, P. Van Asch, K. Vuylsteke, and P. P. De Deyn. Paced Visual Serial Addition Test in Multiple Sclerosis. *Clinical Neurology and Neurosurgery*, Vol. 107, 2005, pp. 218–222.
35. Tombaugh, T. N. Trail Making Test A and B: Normative Data Stratified by Age and Education. *Archives of Clinical Neuropsychology*, Vol. 19, No. 2, 2004, pp. 203–2014.
36. Kos, D., G. Nagels, M. B. D'hooghe, M. Duportail, and E. Kerckhofs . A Rapid Screening Tool for Fatigue Impact in Multiple Sclerosis. *BMC Neurology*, Vol. 6, No. 27, 2006, pp. 1–8.
37. Bambauer, K. Z., S. E. Locke, O. Aupont, M. G. Mullan, and T. J. McLaughlin. Using the Hospital Anxiety and Depression Scale to Screen for Depression in Cardiac Patients. *General Hospital Psychiatry*, Vol. 27, 2005, pp. 275–284.
38. Schuling, J., R. de Haan, M. Limburg, and K. H. Groenier. The Frenchay Activities Index. Assessment of Functional Status in Stroke Patients. *Stroke*, Vol. 24, No. 8, 1993, pp. 1173–1177.
39. Mavoa, S., M. Oliver, K. Witten, and H. M. Badland. Linking GPS and Travel Diary Data using Sequence Alignment in a Study of Children's Independent Mobility. *International Journal of Health Geographics*, Vol. 10, No. 64, 2011, pp. 1–10.

The Standing Committee on Accessible Transportation and Mobility (ABE60) peer-reviewed this paper (18-02900).

Article

TRR
JOURNAL OF THE TRANSPORTATION RESEARCH BOARD

Transportation Research Record
2018, Vol. 2672(8) 675–685

DOI: 10.1177/0361198118787082
journals.sagepub.com/home/trr
SAGE

Revisiting Clear Floor Area Requirements for Wheeled Mobility Device Users in Public Transportation

Aravind Bharathy[1] and Clive D'Souza[1]

Abstract

Current accessibility standards in the U.S. prescribe minimum dimensions for "clear floor area" to accommodate wheeled mobility device (WhMD) users on transportation vehicles. Prior research on the anthropometry of WhMD users (n = 500) indicates that these dimensions are too small to accommodate the size of many occupied WhMDs, especially power chairs and scooters. This paper describes a development project designed to update the evidence for these technical criteria and communicate them to the vehicle designers and accessibility standards developers in a manner that would facilitate making good decisions. An interactive web-based design tool was developed for determining the dimensions of clear floor area to achieve a user-specified level of physical accommodation based on occupied device length and width measurements taken on 500 WhMD users. The web-based design tool is now available to practitioners who seek to accommodate a wider range of WhMD users than the minimum standards required by regulations. The design tool is also intended as a visual evidence base for regulatory activity and universal design practice with higher ambitions. The advent of driverless automated vehicles will increase the importance of accessibility and usability to accommodate the diversity of riders with disabilities. Clear floor space to enable independent ingress, interior circulation, and egress among WhMD users will be a foremost concern. The transportation industry, standards developers, disability advocates, mobility device manufacturers and prescribers, need to understand the limitations of current accessibility standards and work to address these limitations through updated vehicle design standards and policies.

Increases in life span, improved medical care, and increasing availability of assistive technologies that promote independence have resulted in a growing need for accessible public transportation that provides efficient and safe services for those with disabilities, including users of wheeled mobility devices (WhMDs). The U.S. population of non-institutionalized (i.e., not in nursing homes or hospitals) WhMD users is currently at 3.4 million (*1*) and continues to grow at an annual rate of 5% (*2*). However, over 40% of WhMD users living in areas served by public transit experience significant environmental and design barriers when using public transit services (*3*). WhMD users present a unique set of challenges for physical accessibility in public transportation because of concerns over large spatial requirements for floor area and maneuvering (*4*, *5*).

Currently, accessibility standards and codes are used throughout the U.S. and in other countries to implement laws that mandate accessibility to transportation (e.g., buses, vans, trains, etc.) and buildings (e.g., public restrooms, bus stops, transit facilities, etc.). The Americans with Disabilities Act Accessibility Guidelines (ADAAG) for Transportation Vehicles (*6*) is the key document used for the design of accessible facilities and transportation systems. A primary feature of the ADAAG is the specification of a "clear floor area" for wheelchairs, a space 760 mm (30 in.) wide by 1,220 mm (48 in.) in length. The clear floor area is used to determine the size of the area planned for wheelchair securement, the width of doorways and minimum interior clearances for maneuvering a WhMD, and the space needed to accommodate people in WhMDs in bus shelters, bus stop pads, and in terminals – therefore, a critical component for ensuring physical accessibility.

The technical criteria in these standards are based on body sizes and functional abilities (i.e., anthropometry) of adults and sometimes children with disabilities. To understand the spatial implications of contemporary wheeled mobility technology and user populations, the second author (CD) collaborated with researchers at the IDeA Center at the University at Buffalo to develop an extensive anthropometry database of WhMDs in the U.S. with the intention of improving accessibility to buildings, facilities, and public

[1]Inclusive Mobility Research Laboratory, Center for Ergonomics, Department of Industrial and Operations Engineering, University of Michigan, Ann Arbor, MI

Corresponding Author:
Address correspondence to Clive D'Souza: crdsouza@umich.edu

transportation vehicles (*7*, *8*). A key finding from an early analysis on the anthropometry of 369 WhMD users indicated that the minimum clear floor area dimensions for transit buses are too small to accommodate the size of many occupied WhMDs, especially power chairs and scooters (*9*). Key reasons for this are worth revisiting. Foremost was the realization that research conducted in the 1970s was the basis for the current criteria for accommodating WhMD users (*10*). Since that time, the sizes and characteristics of WhMDs and their users have changed considerably. By the late 1990s, anecdotal evidence was emerging that the standards were no longer adequate.

Today, the range of WhMDs available is much more diverse. The size, weight, and maneuvering characteristics of WhMDs vary considerably between device type (e.g., manual wheelchairs, powered wheelchairs, and electric scooters), make, and model (*7–9, 11*). Both manual and powered wheelchairs offer some customizability to fit user needs in postural support, comfort, and ease of use (e.g., leg rests and footrests, tilt and recline), which also affects space requirements. The growing availability and use of scooters and power chairs are expected to further change the mobility device user demographics in the coming years (*2*). Power chairs have allowed people with more severe disabilities to become mobile in the community. The choice of drive configuration (i.e., front-, mid-, and rear-wheel-drive) and powered seating options influence maneuverability of powered wheelchairs (*12*). People are larger due to the obesity epidemic, resulting in a growth in bariatric wheelchairs.

The increasing use of scooters among the growing older adult population has put greater emphasis on scooter access in public transit systems. Scooters typically have a larger turning radius and are less maneuverable than wheelchairs because their configuration and chassis length need to accommodate the drive controls (e.g., tiller) and foot placement (*12–14*). These devices tend to be designed and prescribed for outdoor use. They are intended to assist individuals with limited ambulation and good trunk control and sufficient upper extremity function to control the steering tiller, but who also lack the upper extremity stamina or range of motion necessary to use a manual wheelchair. These problems are not unique to the U.S.; research studies from the UK suggest similar concerns regarding increased WhMD sizes and challenges to accommodation on public transit vehicles (*7*, *15*, *16*).

Over time, the emphasis of our research effort evolved to creating data-driven graphical design tools to provide an evidence base to evaluate existing accessibility standards and to support designers and standards developers who seek to accommodate a wider range of WhMD users than the minimum standards required by regulations (*17*, *18*). Facilitating this outcome was the production of statistical analyses and visualizations with results in a graphical format (e.g., the percentage of WhMD users that could fit, turn, reach, and see, in specific conditions). In particular, multivariate accommodation models were very effective for communication and standards development activities by providing a sophisticated way to visualize the impact of revisions to accessibility standards as "*what if*" analyses (*17*, *18*). This shift was informed by lessons learned from our engagement with the design and accessibility standards community (*18*) that:

- Research results are best presented to decision makers in a form that is familiar to them, in this case, two-dimensional illustrations that resemble or can be overlaid with illustrations used in the existing standard, and,
- Graphic presentations that illustrate the impact of alternative decisions, such as interactive accommodation models, can greatly facilitate the adoption of findings.

Study Objectives

The objectives of this study were to:

1. Develop a data-driven algorithm for determining the dimensions of clear floor area to achieve a user-specified level of physical accommodation based on occupied device length and width measurements taken on 500 WhMD users,
2. Implement the algorithm as a web-based accommodation model with a graphical and interactive user interface to support standards development, regulatory action, and design,
3. Evaluate existing clear floor area requirements prescribed in the ADAAG using the developed accommodation model and where necessary generate recommendations for improving the minimum clear floor area requirements.

Implications of these findings are discussed.

Methods

The analysis described in this study used anthropometry measurements previously taken on 500 adults in the U.S. who relied solely on WhMDs for mobility (*8*). WhMD users were recruited through many sources, including a local independent living center, a United Cerebral Palsy Association, a geriatric day care center, and local hospitals, including Veterans Affairs Medical Centers in Buffalo and Pittsburgh. In addition, advertisements were posted in local newspapers and flyers placed in local organizations and stores. The university's institutional review board approved the study and all participants provided written informed consent.

The research team made a deliberate attempt to recruit a diverse group of users, rather than just individuals who possessed a specific set of physical capabilities so that the results obtained could be extended to the broader WhMD

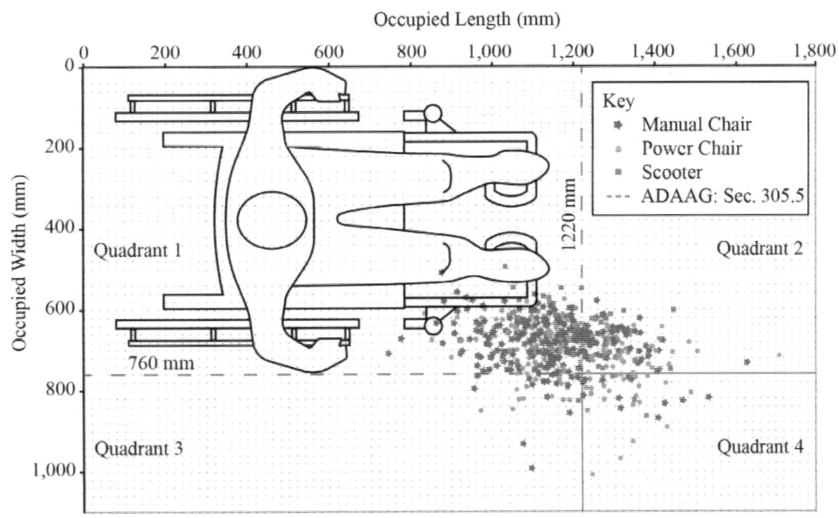

Figure 1. Scatterplot of occupied length vs. occupied width, overlaid with the ADAAG Sec. 305.5 requirement for minimum clear floor area of size 760 mm × 1,220 mm (30 in. × 48 in.).

user population in the U.S. The measurement protocol included the collection of demographic information, mobility device specifications, structural anthropometric measurements, and functional anthropometric information from each participant (*8*).

A three-dimensional digitizing technique was developed to collect detailed data on static anthropometric variables (*19*, *20*). A set of over 130 standardized three-dimensional data points for each person and device were collected that could describe the three-dimensional volume of the occupant and device, while occupants held a comfortable or typical seated posture. Three-dimensional coordinates of the landmarks were then used to derive estimates of widths, heights, and depths of mobility device characteristics and body dimensions relevant to built environment design. The current study used three computed dimensions:

1. Occupied length: computed as the horizontal distance from the extreme rearmost and forward-most point of the combined occupant and mobility device.
2. Occupied width: computed as the horizontal distance between the extreme lateral right-most and left-most points on the body or mobility device with the participant seated in a comfortable, relaxed posture.
3. Occupied floor area: defined as the estimated rectangular floor space required for each occupant-mobility device in the study sample and computed as a product of the occupied length and occupied width.

Web Database and User Interface

This development project advances our anthropometry research by implementing a new web-based framework to support multivariate analysis of specific aspects of WhMD user interaction with the built environment (e.g., clearance, approach, reach capability, line of sight, and turning space requirements). This framework consists of an interconnected, hierarchical set of modules (using open source scripting languages PHP and JavaScript) to query the underlying anthropometry of wheeled mobility database implemented in MySQL and to perform multivariate anthropometry analyses. A visual interface was developed using D3.js, which is a JavaScript library for producing dynamic, interactive data visualizations in web browsers by combining SVG for graphics, HTML5 for content, CSS for aesthetics, PHP for data retrieval, and JavaScript for data manipulation.

Displaying Clear Floor Area

The visual interface allows users to retrieve and display information on occupied width and occupied length as individual datapoints on a two-dimensional scatterplot (Figure 1). A scatterplot is very useful for depicting the association between two variables, that is, occupied length and width in our case. Data can be displayed for all 500 WhMD users in the study sample or stratified by a defined sub-group (e.g., type of WhMD, sex, age range, etc.).

To provide context to designers and standards developers, a plan view of a WhMD user and ADAAG minimum requirement for clear floor area (*6*) is superimposed on the display (Figure 1). Datapoints enclosed by the ADAAG required minimum clear floor area (Quadrant 1) represent individuals accommodated, while datapoints outside represent individuals excluded either due to inadequate clear floor length (Quadrant 2), width (Quadrant 3), or both length and width (Quadrant 4).

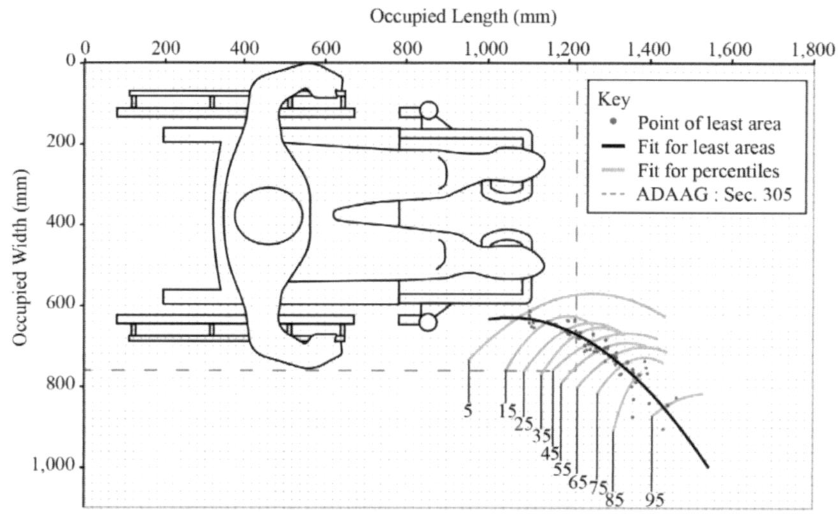

Figure 2. Quadratic fit for least areas by percentile for manual wheelchair users, overlaid with the ADAAG Sec. 305.5 requirement for minimum clear floor area of size 760 mm × 1,220 mm.

Accommodation Model for Clear Floor Area

An accommodation model was developed for determining the minimum dimensions of clear floor area to achieve a user-specified level of physical accommodation based on occupied device length and width measurements taken on 500 WhMD users. In the interactive web-based display two modes for conducting "*what if*" analyses are implemented. First, end-users can *drag* the boundary of the clear floor area (dotted lines in Figure 1) to increase/decrease the clear floor area with the interface display updating in near real-time to communicate the proportion of the sample accommodated (i.e., number of datapoints in Quadrant 1 divided by the total number of datapoints for that group of users). Alternately, the end-user can specify on a slider-bar the percentile of users to be accommodated and the boundary of the clear floor area on the 2-D display updates in near real-time to depict the corresponding minimum required clear floor area length and width.

The latter percentile analysis of occupied floor area provides a more useful starting point for determining suitable dimensions for clear floor area. In contrast to our previous report (*9*), this study presents a more refined method for computing the minimum required clear floor area length and width to accommodate the desired percentile of WhMD users. The need for this stems from the fact that the Pearson's correlation coefficients between occupied width and occupied length were weak, that is, 0.21 for manual chairs, 0.30 for power chairs, and 0.50 for scooters, suggesting that neither occupied width nor occupied length can necessarily be used to reliably predict the other. It also implies that the specific combination of clear floor area length and width to accommodate the desired percentile is not trivial.

Procedure for Calculating Minimum Clear Floor Space Requirements by Percentile

Summarized below is the procedure developed for calculating by WhMD type the minimum clear floor area length and width to accommodate a user-specified percentile accommodated (Figure 2):

1. For a given percentile of accommodation, say 5%, the WhMD users in the study sample whose corresponding occupied length and occupied width accommodate 5% ± 2% of users in the sample are identified. We opted for a tolerance range of ± 2% due to the relatively small sample sizes, particularly for scooter users. This step produces multiple possible lengths and widths for a given value of percentile accommodation.
2. Next, we determine the specific combination of length and width that would produce the smallest clear floor area from these identified points. This is done by fitting or approximating a quadratic curve to the identified data points (i.e., a yellow curve at 5% in Figure 2). Any point on this curve would accommodate 5% of users – some combinations with large occupied widths, some with large occupied lengths, with others in between. A quadratic curve was chosen due to low mean squared error values compared with a linear fit and other higher-order polynomial fit. The equation for a quadratic curve takes the following general form,

$$f(x) = y = ax^2 + bx + c \quad (1)$$

3. The clear floor area, A, obtained by any point on the curve can be represented as,

$$Area(A) = xy \tag{2}$$

where
x = occupied length
y = occupied width.
Substituting for 'y' from Equation (1),

$$A = x\left(ax^2 + bx + c\right) \tag{3}$$

The point (x_{min}, y_{min}) on this quadratic curve that would yield the smallest clear floor area can be obtained by computing the solution to the first derivative, namely,

$$\frac{d}{dx}(A) = \frac{d}{dx}\left(x\left(ax^2 + bx + c\right)\right) = 0 \tag{4}$$

$$3ax^2 + 2bx + c = 0 \tag{5}$$

4. The solution to this equation is:

$$x_{min} = \frac{2b \pm \sqrt{4b^2 - 12ac}}{6a} \tag{6}$$

Substituting for x in Equation (1), we obtain:

$$y_{min} = a\left(x_{min}\right)^2 + b\left(x_{min}\right) + c \tag{7}$$

5. Steps 1 to 4 are repeated for all percentiles from 5 to 95% in increments of 1%. Figure 2 shows the corresponding quadratic curves in increments of 5%.
6. Next, a quadratic curve is fit through all the pairs of minimum occupied length and occupied width obtained from each percentile value. Again, a quadratic curve is chosen due to low mean squared error. Each point on this curve provides a clear floor area with a corresponding length and width, and this area will be close to the minimum for a given accommodation value. This is depicted in Figure 2 by a solid-black curve passing through the different percentile-based combinations of points (x_{min}, y_{min}). The curve shows the trajectory by which the occupied length and width should change to increase/decrease accommodation. Clearly, this trajectory is non-linear, that is, occupied length and width do not change proportionately. Separate trajectories were produced for each type of WhMD.

Data Analysis

Three types of analyses were performed stratified by the type of WhMD:

1. Descriptive analyses to evaluate the distributional characteristics of occupied length, width, and area. The mean, standard deviation, range, 5th and 95th percentile values for occupied length, width, and area were calculated using SPSS Statistics v23.0 (*21*).
2. Occupied lengths and widths in the sample were compared with the ADAAG minimum requirement for clear floor area to determine the proportion of the sample accommodated and excluded.
3. Using the graphical accommodation model described in the previous section, the minimum required clear floor area lengths and widths for commonly used percentiles accommodated were computed and summarized in a graphical and tabular format for use by accessibility designers and standards developers.

Results

Sample Demographics

The study sample consisted of 264 (52.8%) men and 236 (47.2%) women. Among men, 147 (55.7%) used a manual wheelchair, 100 (37.9%) used power chairs, and 17 (6.4%) used scooters. Among women, 130 (55.1%) used a manual chair, 89 (33.7%) used a power chair, and 17 (7.2%) used a scooter. The mean (standard deviation) age of the sample was 56.4 (18.8) years with a range of 18 to 100 years.

Regarding self-reported medical conditions that led to dependence on mobility devices, central nervous system disorders (e.g., multiple sclerosis, cerebral palsy, etc.) were the most frequently occurring (34%), followed by spinal cord injuries (26%), orthopedic injuries/deformities (13%), and cerebral vascular diseases such as stroke (10%). Instances of medical conditions such as amputations (3%), traumatic brain injuries (2%), or respiratory and cardiovascular diseases (2%), were less frequent, with the remaining cases listed under "other" (10%).

Univariate Analysis of Occupied Width, Occupied Length, and Occupied Floor Area

Descriptive statistics for occupied width, occupied length, and occupied floor area stratified by mobility device type are tabulated (Table 1).

Comparisons with the ADAAG Minimum Clear Floor Space Requirement

Table 2 provides the frequency count and percentages in parentheses (percentages are computed based on the

Table 1. Summary Statistics for Occupied Width (mm), Occupied Length (mm), and Occupied Area (m^2), Stratified by Type of WhMD

Manual chairs (n = 277)

					Percentile		
Dimension	Mean	SD	Min	Max	5th	50th	95th
Occupied width (mm)	685.0	61.4	508.0	992.0	595.7	677.3	786.4
Occupied length (mm)	1,150.5	129.8	743.0	1,625.0	934.7	1,154.8	1,362.0
Occupied area (m^2)	0.79	0.125	0.44	1.25	0.60	0.78	1.00

Power chairs (n = 189)

					Percentile		
Dimension	Mean	SD	Min	Max	5th	50th	95th
Occupied width (mm)	706.7	72.4	574.0	1,008.0	606.9	695.3	827.5
Occupied length (mm)	1,196.3	133.7	831.0	1,709.0	977.0	1,183.4	1,414.5
Occupied area (m^2)	0.85	0.145	0.50	1.27	0.63	0.84	1.11

Scooters (n = 34)

					Percentile		
Dimension	Mean	SD	Min	Max	5th	50th	95th
Occupied width (mm)	650.0	89.6	488.0	857.0	526.7	617.3	829.3
Occupied length (mm)	1,208.5	110.0	1,025.0	1,439.0	1,039.3	1,202.8	1,433.1
Occupied area (m^2)	0.79	0.158	0.50	1.17	0.55	0.77	1.12

Table 2. Percentage of Wheeled Mobility Device Users in the Study Sample (n = 500) with Lengths and Widths That Are Accommodated (Quadrant 1) or Excluded (Quadrants 2, 3, 4) Compared with the ADAAG Required Minimum Clear Floor Area of Size 760 mm × 1,220 mm

	Quadrant				
Mobility device type	Q1 (Included) L < 1,220 & W < 760	Q2 (Exceed length) L >= 1,220 & W < 760	Q3 (Exceed width) L < 1,220 & W >= 760	Q4 (Exceed length & width) L >= 1,220 & W >= 760	Total
Manual	185 (66.8%)	65 (23.5%)	15 (5.4%)	12 (4.3%)	277 (100%)
Power	94 (49.7%)	55 (29.1%)	16 (8.5%)	24 (12.7%)	189 (100%)
Scooter	18 (52.9%)	12 (35.3%)	1 (2.9%)	3 (8.8%)	34 (100%)
Combined	297 (59.4%)	132 (26.4%)	32 (6.4%)	39 (7.8%)	500 (100%)

corresponding mobility device category) for cases in each of the four quadrants depicted in Figure 1. Only 59.4% of combined mobility device users in the sample were accommodated in the ADAAG minimum clear floor space requirement (Quadrant 1). Comparisons of occupied length and occupied width of WhMD users in the study sample with the ADAAG prescribed minimum clear floor length (1,220 mm) and minimum clear floor width (760 mm) showed that a larger percentage of users exceeded on occupied length (Quadrant 2) as compared with occupied width (Quadrant 3) or both, occupied length and width (Quadrant 4).

Percentile-Based Accommodation Model for Clear Floor Area Length and Width

Figure 3 shows the quadratic fit for the data points at which least area for a given percentile was observed. The coefficient of determination (r^2) or the fit of the quadratic curve was determined to be 0.86 for all mobility devices combined, 0.89 for manual chairs, and 0.80 for power chairs. Scooters had a very low r^2 value due to insufficient data points that result from the small number of scooter users in the sample. Table 3 gives the same values by device type.

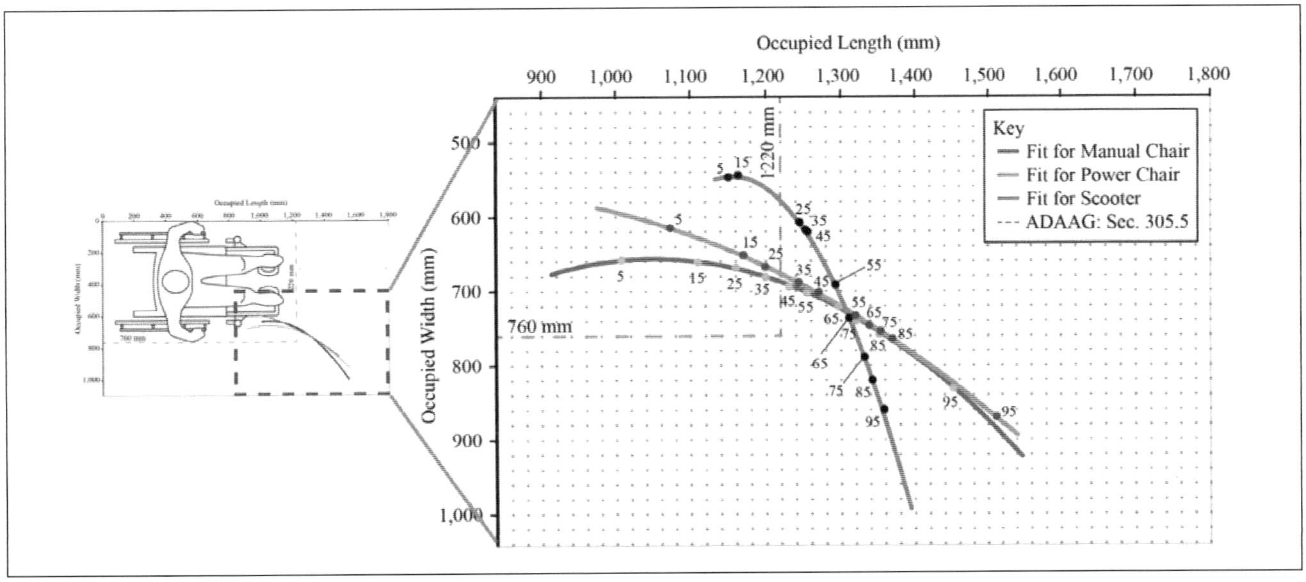

Figure 3. Accommodation model for clear floor area by type of WhMD. Close-up view on the right.

Discussion

Contemporary wheeled mobility technology can support greater independence, better utilization of public transportation, and increased social participation of individuals with disabilities into mainstream society (e.g., employment, shopping/doctor visits, recreational travel). However, the full potential of this technology cannot be realized unless device size and shape, and the size of occupants are accommodated by transportation systems to insure access to safe, comfortable, and timely travel.

Implications for Accessibility Design Standards and Policies

Standards serve as the foundation for the evidence-based design of the built environment because architects, engineers, product designers, and other design professionals rely on experts in accessibility, safety, and public health to translate knowledge into guidelines for practice. Like product development, standards development is a long-term, expensive, and time-consuming activity. It requires the creation of supporting evidence and translation of that evidence into realistic standards.

The findings of this research provide a guide to policy makers and accessibility standards developers for evaluating and revising current transportation accessibility standards. The results suggest that the current standards are not adequate to accommodate a large proportion of contemporary WhMD users. Three design strategies could address this finding in practice. The standards for minimum clear floor area could be increased. This would have limited, if any, implications for new construction in the built environment but could have major implications for the design of some vehicle types, especially low floor buses where the space between wheel wells is very limited. The use of a four-passenger longitudinal flip-up seat in place of a three-passenger longitudinal flip-up seat in one wheelchair securement area would provide enough space for larger scooters and wheelchairs, reducing overall seating capacity by only one person. Finally, the location of boarding ramps at entries in the middle of low floor vehicles instead of front door access could increase the maneuvering space available by avoiding the need for passage between wheel wells. This last strategy would require using a curbside fare collection or an alternate method that does not require face-to-face driver interaction or supervision. It also may have, at present, a limited application to boarding from raised platforms because in most low floor buses, the kneeling feature, which enables a lower ramp slope, is only installed on the front of the bus. These strategies above are not mutually exclusive. Together, they provide policy makers with some options that could lead to a reasonable solution with minimum cost and operational impact. The alternative is to enact operating policies that restrict access on mass transport vehicles to devices that have footprints that fall within specified limits, but such policies based on size could be considered discriminatory and require drivers to make difficult judgments in the field.

Implications for Accessible Design of Transportation Vehicles

Small increases to the minimum securement area to better accommodate contemporary WhMDs on occupied length and width also lead to reduced entry and positioning times and difficulty, depending on the specific bus design, without

Table 3. Percentile Accommodated and the Corresponding Estimated Minimum Clear Floor Area Length and Width Stratified by Mobility Device Type

	Percentile Accommodated for Users of Manual Wheelchairs (n = 277)									
Dimension	5%	15%	25%	35%	45%	55%	65%	75%	85%	95%
Length (mm)	1,008	1,111	1,163	1,195	1,230	1,255	1,295	1,319	1,352	1,449
Width (mm)	658	660	670	679	692	702	721	734	754	828
Area (m^2)	0.66	0.73	0.78	0.81	0.85	0.88	0.93	0.97	1.02	1.20
	Percentile Accommodated for Users of Power Chairs (n = 189)									
Dimension	5%	15%	25%	35%	45%	55%	65%	75%	85%	95%
Length (mm)	1,076	1,168	1,204	1,244	1,267	1,320	1,338	1,355*	1,372	1,508*
Width (mm)	616	652	669	689	702	732	744	755*	766	870*
Area (m^2)	0.66	0.76	0.80	0.86	0.89	0.97	0.99	1.02	1.05	1.31
	Percentile Accommodated for Users of Scooters (n = 34)									
Dimension	5%	15%	25%	35%	45%	55%	65%	75%	85%	95%
Length (mm)	1,148	1,161	1,246	1,252	1,253	1,291	1,310	1,327*	1,343	1,360*
Width (mm)	549	548	609	617	619	687	729	785*	817	860*
Area (m^2)	0.63	0.64	0.76	0.77	0.77	0.89	0.95	1.04	1.10	1.17

*Indicates values that were interpolated (if in between) or extrapolated (if at the end) from the best-fit quadratic curve due to few observations at those percentiles.

affecting seating capacity (*22, 23*). However, a space increase in existing public transit environments may be prevented by technological constraints.

The results presented here also help to identify design goals for future transportation and wheeled mobility technology. For example, the data provide a goal for the design of new suspension systems for low floor buses that could increase the available space between wheel wells, for new securement systems that might allow additional available room in the securement area, or alternative vehicle interior layouts that offer better access and circulation spaces. Also important are innovations in mobility device design that improve device maneuverability making them more adaptable to constrained environments, such as on buses, or technologies that increase the comfort and postural support for the occupant while possibly reducing the occupied floor area.

Designing the interior of buses and other transit vehicles for accessibility is a multi-dimensional design problem. Multiple adjacent design features in addition to the size of the securement area (such as its location in relation to the front or rear doorway, forward- vs. rear-facing orientation) and seat layout and orientation (transverse vs. longitudinal) influence an individual WhMD user's ability to enter, situate in, and exit the securement area in an efficient manner. In some instances, it can require simultaneous consideration for additional variables such as shoulder widths and heights, knee and toe clearance widths and heights, and functional reach capability, to name a few. Future research activities will explore the development of multivariate analyses tools that can factor in additional variables besides width and length of the occupant and device. Additionally, in laboratory experiments using a full-scale mock-up of a bus with reconfigurable interiors, we demonstrated quantifiable relationships between occupied device dimensions and actual performance in dynamic activities such as boarding, interior circulation, and disembarking (*22–24*).

The aging of the population, rising social expectations of accessible mobility options, and the increased cost of vehicle ownership will lead to increased utilization and demand for inclusive, reliable transportation systems. The advent of automated vehicles will present opportunities for reducing mobility barriers for individuals with disabilities. But little attention is being given to the design of these vehicles to serve the disability community. The lack of drivers will also introduce new problems. On the one hand, the design of automated vehicles will necessitate a higher standard of usability to ensure independent use without driver assistance. On the other hand, such vehicles will have more space due to the removal of obsolete components such as driver stations and fare machines. The new mobility industry and technology developers have yet to embrace the need and value of designing these vehicles to be accessible and inclusive. Standards developers, disability advocates, mobility device manufacturers and prescribers need to understand the limitations of current accessibility standards and work to address these limitations through updated vehicle design standards and policies. As demonstrated in the case example of clear floor area, human factors and ergonomics research can improve our knowledge of how best to accommodate the diverse population of passengers.

Methodological Limitations

The proportion of wheelchair users mentioned above might not represent the same proportion expected in the adult U.S. population of wheelchair users. For example, the adult U.S. population of WhMD users comprises an estimated 84% manual chair users, 8% power chair users, and about 8% scooter users (*2*, *3*). The current study sample had a much larger proportion of power wheelchair users and more severe physical limitations compared with the U.S. population. However, power chair users were intentionally oversampled to obtain a better understanding of the functional abilities of this user group, which typically has more severe physical limitations and is therefore more sensitive to design restrictions.

Measurements in this study were taken with participants seated in a comfortable posture that they could maintain for the duration of the measurement process, typically lasting 15 to 20 minutes. Participants often chose to rest their arms on the armrest and/or extend the footrests and leg rests to support the lower extremities. These data do not take into account the ability of some individuals to move their limbs inboard of their devices, swing leg rests and footrests out of the way, or adjust backpacks and bags, which may extend beyond the boundaries of their devices. It is also worth noting, however, that traveling on public transportation is an activity that requires remaining relatively stationary for extended periods of time. While these measurement conditions and postures may tend to overestimate the occupied width and occupied length dimensions for very short time durations, they do reflect postures that WhMD users consider comfortable for times that are commensurate with the vast majority of transit trips. Further, the samples were drawn largely from cold weather cities which may introduce some bias toward larger and more durable equipment, but data were collected all year round in an attempt to minimize any bias in recruitment associated with the seasons.

The analyses presented in this paper are based on the assumptions of a rectangular area required for static positioning of a WhMD. The findings suggest that the current minimum requirements of a 760 mm (30 in.) wide by 1,220 mm (48 in.) length of clear floor area are clearly inadequate for a sizeable proportion of WhMD users measured in this study, especially those with the most serious disabilities, namely, power chair users. Wheeled mobility anthropometry dimensions from this study have also been compared with existing accessibility standards and anthropometry research findings from other countries such as Australia, Canada, and the UK (*7*). These international comparisons also suggest a trend toward an increase in occupied floor area among mobility devices, and the inability of accessibility standards to reflect the space requirements of WhMD users in the built environment. One approach to increasing the level of accommodation would be to provide a larger clear floor area than the existing standards. The findings described here are intended to provide guidance for modifying standards, particularly in applications where the amount of space available in some vehicles is very limited. Many individuals in the study were excluded by the minimum clear floor area requirements due to larger occupied lengths (Quadrant 2 in Figure 1), suggesting that the greatest improvements in accommodation could be first achieved by increasing the length dimension of the clear floor area. It is worth noting that, based on our research findings and other considerations from stakeholders, the ICC/ANSI A117.1 Committee, which publishes a voluntary building standard for accessibility, has revised its minimum requirements for clear floor area length in all new construction, increasing it from 1,220 mm (48 in.) to 1,320 mm (52 in.) (*18*, *25*). Also, space requirements for maneuvering into and out of these spaces inside vehicles were not considered in the current analysis and might suggest the need for additional clearances (*22*).

Conclusion

This report describes a data-driven graphical and interactive accommodation model that can be used by architects and designers to visualize and understand the impact of physical design features on accommodation levels. The accommodation model presented here shows the importance of linking occupied width and length of WhMDs when establishing suitable dimensions for minimum clear floor area in accessibility standards, including those used in the transportation sector. The findings from the analysis of clear floor area need to be the subject of considerable dialogue in the transportation community to determine what level of accommodation is realistic and practical with regard to contemporary vehicle technology. Involving WhMD manufacturers, mobility device vendors, rehabilitation engineers and therapists, and representatives from transit agencies and organizations serving individuals with disabilities in this dialogue is also critical to arriving at acceptable and easily implementable solutions. For now, given prevailing standards provisions, rehabilitation engineers and therapists prescribing wheelchairs need to consider the transportation needs of clients and the occupied device size relative to the space available for wheelchair circulation and securement areas on contemporary public transportation vehicles.

The report also demonstrates the utility of the assembled anthropometry database in helping standards developers and designers improve accessibility, independent mobility, and safety for travelers using WhMDs on buses, other public transportation vehicles, and potentially future autonomous driverless vehicles. The web-based implementation provides the accessible design and standards community worldwide with the capability to query our anthropometry database and perform multivariate analyses without requiring users to have programming or statistical expertise. Usability testing of the website and implementation of other such multivariate accommodation models is ongoing. The accommodation model can be accessed via the website for the University of Michigan's Inclusive Mobility Research Laboratory at http://dsouzalab.engin.umich.edu/research/anthro/index.php.

Acknowledgments

The contents of this manuscript were developed under a grant from the National Institute on Disability, Independent Living, and Rehabilitation Research (NIDILRR) through the Rehabilitation Engineering Research Center on Universal Design at Buffalo (RERC-UD; Grant # 90RE5022-01-00). Research underlying the anthropometry database on WhMD users presented in this paper was supported under a grant from NIDILRR through the RERC-UD (Grant # H133E990005) and from the U.S. Access Board (contract # TDP-02-C-0033). NIDILRR is a center within the Administration for Community Living (ACL), Department of Health and Human Services (HHS).
The authors thank the principal investigators of the RERC-UD and the Anthropometry of Wheeled Mobility project, Drs. Edward Steinfeld, Victor Paquet, and James Lenker at the IDeA Center, University at Buffalo for anticipating the need and initiating this important research effort more than a decade ago.

Author Contributions

The authors confirm contributing to the paper as follows: study conception and design: CD; analysis and interpretation of results: CD, AB; draft manuscript preparation: AB, CD. All authors reviewed the results and approved the final version of the manuscript.

References

1. Brault, M. W. Americans with Disabilities: 2010. In *Current Population Reports*. U.S. Census Bureau: U.S. Census Bureau, Washington, D.C., 2012, pp. 70–131.
2. Flagg, J. Wheeled Mobility Demographics. In *Industry Profile on Wheeled Mobility* (S. Bauer and M. E. Bunning, eds.), RERC on Technology Transfer, Buffalo, NY, 2009.
3. LaPlante, M. P. H. S. Kaye. *Demographics and Trends in Wheeled Mobility Equipment Use and Accessibility in the Community. Assistive Technology: The Official Journal of RESNA*, Vol. 22, No. 1, 2010, pp. 3–17.
4. Goldman, J. M., and G. Murray. *TCRP Synthesis 88: Strollers, Carts, and Other Large Items on Buses and Trains*. Transportation Research Board of the National Academies, Washington, D.C., 2011.
5. Nelson/Nygaard Consulting Associates. *Status Report on the Use of Wheelchairs and Other Mobility Devices on Public and Private Transportation*. Nelson/Nygaard Consulting Associates, Editor. Easter Seals Project ACTION, Washington, D.C., 2008.
6. U.S. Access Board. *Federal Register 36 CFR Part 1192*. ADA Accessibility Guidelines for Transportation Vehicles, 1998. http://www.access-board.gov/transit/html/vguide.htm. Accessed February 17, 2010.
7. Steinfeld, E., J. Maisel, D. Feathers, and C. D'Souza. Anthropometry and Standards for Wheeled Mobility: An International Comparison. *Assistive Technology*, Vol. 22, No. 1, 2010, pp. 51–67.
8. Steinfeld, E., V. Paquet, C. D'Souza, C. Joseph, and J. Maisel. *Final Report: Anthropometry of Wheeled Mobility Project. Prepared for the U.S. Access Board*. IDeA Center, Buffalo, NY, 2010.
9. D'Souza, C., E. Steinfeld, V. Paquet, and D. Feathers. Space Requirements for Wheeled Mobility Devices in Public Transportation: Analysis of Clear Floor Space Requirements. *Transportation Research Record: Journal of the Transportation Research Board*, 2010. 2145: 66–71.
10. Steinfeld, E., S. Schroeder, and M. Bishop. *Accessible Buildings for People with Walking and Reaching Limitations*. U.S. Department of Housing and Urban Development, Washington, D.C., 1979.
11. Fitzgerald, S. G., A. Kelleher, E. Teodorski, D. M. Collins, M. Boninger, and R. A. Cooper. The Development of a Nationwide Registry of Wheelchair Users. *Disability and Rehabilitation: Assistive Technology*, Vol. 26, No. 2, 2007, pp. 358–365.
12. Koontz, A., E. Brindle, P. Kankipati, D. Feathers, and R. Cooper. Design Features that Impact the Maneuverability of Wheelchairs and Scooters. *Archives of Physical Medicine and Rehabilitation*, Vol. 91, 2010, pp. 759–764.
13. Dutta, T., E. C. King, P. J. Holliday, S. M. Gorski, and G. R. Fernie. Design of Built Environments to Accommodate Mobility Scooter Users: Part I. *Disability and Rehabilitation: Assistive Technology*, Vol. 6, No. 1, 2011, pp. 67–76.
14. King, E. C., T. Dutta, S. M. Gorski, P. J. Holliday, and G. R. Fernie. Design of Built Environments to Accommodate Mobility Scooter Users: Part II. *Disability and Rehabilitation: Assistive Technology*, Vol. 6, No. 5, 2011, pp. 432–439.
15. Mitchell, C. The Size of the Reference Wheelchair for Accessible Public Transport. In *11th International Conference on Mobility and Transport for Elderly and Disabled Persons (TRANSED)*, Montreal, Canada, 2007.
16. Steinfeld, E., C. D'Souza, and J. Maisel. Clear Floor Space For Contemporary Wheeled Mobility Users. In *12th International Conference on Mobility and Transport for Elderly and Disabled Persons (TRANSED 2010)*, Hong Kong, 2010.
17. D'Souza, C., V. Paquet, E. Steinfeld, and D. Feathers. Anthropometric Data Visualization Tools to Improve Accessibility of Built Environments. In *3rd International Conference on Applied Human Factors and Ergonomics, 2010 AHFE International*. Taylor and Francis, Ltd., Miami, Florida, 2010.
18. Steinfeld, E., C. D'Souza, and J. White. *Developing Evidence-Based Standards: A Case Study in Knowledge Translation*. In *International Conference on Universal Design, June 2014* (P. H. H. Caltenco, A. Larsson, K. Rassmus-Gröhn, and B. Rydeman, eds.), IOS Press, Lund, Sweden. 2014, pp. 89–98. doi:10.3233/978-1-61499-403-9-89.
19. Feathers, D., V. Paquet, and C. Drury. Measurement Consistency and Three-dimensional Electromechanical Anthropometry. *International Journal of Industrial Ergonomics*, Vol. 33, 2004, pp. 181–190.
20. Paquet, V., and D. Feathers. *An Anthropometric Study of Manual and Powered Wheelchair Users. International Journal of Industrial Ergonomics*, Vol. 33, No. 3, 2004, pp. 191–204.
21. SPSS Inc. *SPSS for Windows*, 2007, SPSS Inc., Chicago.
22. D'Souza, C. V. Paquet, J.A. Lenker, and E. Steinfeld. Effects of Transit Bus Interior Configuration on Performance of Wheeled

Mobility Users During Simulated Boarding and Disembarking. *Applied Ergonomics*, Vol. 62, 2017, pp. 94–106. doi: 10.1016/j.apergo.2017.02.008.

23. D'Souza, C., V. Paquet, J. A. Lenker, and E. Steinfeld. Self-Reported Difficulty and Preferences of Wheeled Mobility Device Users for Simulated Low Floor Bus Boarding, Interior Circulation, and Disembarking. *Disability and Rehabilitation: Assistive Technology*, Vol. 1, 2017, pp. 1–14. DOI: 10.1080/17483107.2017.1401128.
24. Steinfeld, E., V. Paquet, J. Lenker, C. D'Souza, and J. Maisel. Universal Design Research on Boarding and Using Buses. In *12th International Conference on Mobility and Transport for Elderly and Disabled Persons (TRANSED 2010)*, Hong Kong, 2010.
25. ICC/ANSI. *ICC A117.1-2017: Accessible and Usable Buildings and Facilities*. American National Standards Institute/International Code Council, New York, 2017.

The Standing Committee on Accessible Transportation and Mobility (ABE60) peer-reviewed this paper (18-02377).

The contents of this manuscript do not necessarily represent the policy of NIDILRR, ACL, HHS, U.S. Access Board, and you should not assume endorsement by the Federal Government.

Article

TRR
JOURNAL OF THE TRANSPORTATION RESEARCH BOARD

Transportation Research Record
2018, Vol. 2672(8) 686–697

DOI: 10.1177/0361198118796732
journals.sagepub.com/home/trr
SAGE

Evaluating Potential Demand and Operational Effects of Coordinated Americans with Disabilities Act Paratransit and Taxi Service

Vincent Turmo[1], Mahour Rahimi[2], Eric J. Gonzales[3], and Price Armstrong[4]

Abstract

This paper presents an evaluation of the potential implications of using taxis along with conventional paratransit to serve customers eligible for demand responsive service through the Americans with Disabilities Act (ADA). Subsidizing a taxi service to carry some paratransit demand could potentially reduce the total cost of the service if diverted trips are inexpensive to serve by taxi and the remaining paratransit operations remain efficient. This study makes use of ADA paratransit operations and demand data from the Pioneer Valley Transit Authority (PVTA) in the greater Springfield, Massachusetts, region to develop an algorithm for optimally allocating paratransit demand to conventional paratransit services and taxis. First, a model is presented to optimize the spatial extent of the ADA paratransit service area, assuming that all trips outside the service region can be incentivized to use taxis. A more detailed investigation makes use of survey data from PVTA to estimate which trips could potentially be diverted to taxis or transportation network companies (TNCs) based on the type of disability that a customer has. The result is a model that shows which trips could be incentivized to switch to taxi and the magnitude of total operating cost savings that may be achieved from such a program.

Demand responsive transit (DRT) services are common in cities across the United States, especially because the Americans with Disabilities Act (ADA) requires transit agencies to provide paratransit services to people with disabilities as a requirement for federal funding eligibility. This service is also important for agencies to meet requirements of Title VI of the Civil Rights Act of 1964 prohibiting discrimination in federally-assisted programs. The challenge of managing a system of DRT services for people with disabilities is that the operation of the system depends on the demand-side considerations related to traveler behavior and supply-side considerations related to the system structure and costs. ADA paratransit services must provide curb-to-curb service for people with disabilities whose trip origin and destination lie within 0.75 mi of fixed route bus service and who want to travel during the hours of fixed route operation. In some cases, paratransit services are offered to a broader geographic region or to a more general population of eligible users. A consistent feature of these systems is that the operating cost per trip is very high, requiring large subsidies to sustain.

Because of the high cost of providing DRT services, it is important to understand the factors that affect these costs and to be able to predict the effect of changes in demand and service alignment on total agency costs. The benefits to the agency will be to lower the cost of providing services, which will allow services to continue even as demand in some regions is increasing. These systems often divide their service area into smaller regions to simplify management and operations. The objective for the transit agency is to identify efficient ways to provide this service, especially the required service for ADA paratransit customers, who have disabilities.

In an effort to reduce costs, the Massachusetts Bay Transportation Authority (MBTA) in Boston recently implemented a new program allowing customers to use the Ride (existing paratransit service), a taxicab, Uber, or Lyft. This program is an effort to make use of alternative and emerging mobility service providers to reduce the costs of provided transportation to people with

[1]Ecole Nationale des Travaux Publics de l'Etat, Vaulx-en-Velin, France
[2]Massachusetts Bay Transportation Authority, Boston, MA
[3]Department of Civil and Environmental Engineering, University of Massachusetts, Amherst, MA
[4]Pioneer Valley Transportation Authority, Springfield, MA

Corresponding Author:
Address correspondence to Eric J. Gonzales: gonzales@umass.edu

disabilities. MBTA provides a subsidy for taxi, Uber, and Lyft trips covering the fare after the first $2 up to a total subsidy of $13 per trip. This provides an incentive for users to choose an alternative mode for short trips that can be served less expensively than by the existing paratransit service. This program raises questions about the implication of new service options on demand and supply and the potential for reducing the total cost of service.

The aim of this paper is to develop and identify optimal service structures to coordinate conventional paratransit operations and taxis for serving paratransit demand. Using operations and demand data from the Pioneer Valley Transit Authority (PVTA) in Western Massachusetts, the proposed algorithm demonstrates how the combined operations of paratransit and taxis can be optimized by strategically incentivizing some trips to divert from conventional paratransit to a subsidized taxi trip. The organization of the paper is as follows. First, the existing literature is reviewed on modeling paratransit services and alternative methods for serving customers with disability. Next, the research methodology is presented for the models of paratransit and taxi costs as well as the proposed algorithm to define an optimal zoning structure for the service. A section then presents the available data for application to PVTA service areas.

Literature Review

There are two important bodies of literature that are relevant to this work. First are the studies related to modeling DRT operations and costs. Many models have been developed to optimize the routing and zoning structures for paratransit services. The approaches to model and study demand responsive services can be broadly classified as detailed simulation models and approximate mathematical models. Simulation has been employed to assess the effects of zoning strategies as well as time window variations on productivity measures of DRT services (*1*). Applications include evaluations of trip miles, deadhead miles, and fleet size for Los Angeles County, California (*2*). Studies of paratransit in Harris County, Texas, include comparison of centralized and decentralized operating strategies (*3*) and the effect of zoning with transfers (*4*, *5*). Variations in travel time and scheduling reliability can also be modeled with detailed methods (*6*, *7*). However, developing an accurate simulation is expensive because it requires detailed data inputs and a significant amount of time for model construction and calibration.

In contrast, approximate mathematical models are simpler to construct and require readily available data inputs. These types of models show the connection between data inputs and cost estimates using analytical models. Several aggregate analytical models based on continuous approximations geometric probability have been developed to represent various operational strategies for DRT, and to predict the total distance, operating time, and fleet size required to serve a given demand (*8*). This analytical approach has been used in a mathematical problem to optimize the total time and fleet size in the system subject to quality of service constraints (*9*–*12*). Stochastic models have also been developed that more realistically account for demand uncertainty to relate the fleet size to demand distribution, service area, and time window (*13*). Recent studies have shown that aggregate methods provide good estimates of annual operating costs using data from New Jersey Transit's ADA paratransit service (*14*, *15*).

A second relevant body of research evaluates policies to reduce the cost of ADA paratransit operations by coordinating with other service providers. Transit agencies have made agreements to partner with taxi companies to provide to provide transit services for decades. Many studies have considered the role that taxis can play to complement or supplement other public transportation services for passengers with disabilities (*16*–*19*). The potential to reduce the costs of operating paratransit services by partnering with taxis has led to successful taxi voucher programs in many cities, including San Francisco Municipal Transportation Authority (SFMTA) in San Francisco, California; Pace Transit in suburban Chicago, Illinois; Metropolitan Transit Authority of Harris County (METRO) in Houston, Texas; and Washington Metropolitan Area Transportation Authority (WMATA) in greater Washington, D.C. (*20*). TCRP Report 121 presents a spreadsheet toolkit for comparing the effectiveness of non-dedicated fleets (primarily taxis) for augmenting peak demand, handling long trips, or providing service in low-demand periods (*21*). The toolkit provides estimates of cost savings resulting from integrating service with non-dedicated vehicles, but the tool does not explicitly consider the geographic distribution of origins and destinations within the service region.

In recent years, transportation network companies (TNCs) have emerged as an alternative provider of transportation services in many metropolitan regions. TNCs are rapidly changing urban mobility by using digital platforms to allow customers to request on-demand service from suppliers. Several different business models have been developed. Some services essentially mimic exclusive-ride point-to-point taxicab service, distinguished primarily by the fact the customers use a smartphone to hail or request service. Other services seek to group riders into demand responsive shared rides using sedans or vans that can resemble jitneys or even dial-a-ride shuttles, aiming to achieve economical operations by simultaneously serving many passengers. With the rapid expansion of TNCs and their adoption by the general public, there have been several studies evaluating their relationship to public transit and extent to which TNCs operate in competition with transit or as a supplement to existing services (*22–24*). Although the potential of TNCs to substitute conventional transit trips and reduce transit efficiency is a cause for concern (*25–28*), ADA paratransit services are so costly to provide that substitution may have the desirable effect of lowering total operating costs.

With increasing interest among transit agencies throughout the United States in considering cooperative arrangements between ADA paratransit services and taxis or TNCs, there is a pressing need for quantitative methods to assess the potential value of these programs. To date, the quantitative modeling of demand responsive systems has focused exclusively on paratransit operations as a stand-alone system. The research on collaborative arrangements with taxi companies and other service providers has focused more on the policy issues and quality of service considerations for users. This paper connects these two bodies of research to provide a quantitative method to model and optimize a program to shift some paratransit trips from conventional ADA paratransit operations to taxis or TNCs.

Model of Paratransit and Taxis

This study seeks to quantify the impact of changes in paratransit service on the total cost of operations. Because the detailed origin–destination pattern of trips varies from day to day, it works with aggregate models that provide estimates of required resources and costs associated with monthly or annual demand patterns. The benefit of working with models and data over this long timeframe is that many observations can be used to characterize the distribution of travel demand across the region. This study makes use of an aggregate model of paratransit that was originally developed to quantify costs in NJ Transit's Access Link system (*14*) and later extended to consider zoning strategies (*15*) and dynamic pricing strategies (*29*). The model provides estimates of the required vehicle fleet size, vehicle miles of operations, and vehicle hours of operations based on the total demand, λ trips/time, and the total service area,A. For this study, taxi costs are estimated based on the local published taxi fare rates and the distance and time associated with average trips for each origin–destination pair in the region.

Modeling ADA Paratransit Operations and Costs

To calculate annual costs of operating paratransit for λ trips in a service area of size A, the following formulas from (*14*) are used for vehicle miles traveled (VMT), vehicle hours traveled (VHT), and required fleet size (M):

$$\text{VMT} = \frac{1}{2}\left(\frac{1}{\sqrt{\lambda T}} + \frac{1}{\sqrt{n}}\right) r_2 \lambda \sqrt{A} \tag{1}$$

$$\text{VHT} = Mt_p = \lambda t_p \left(b_1 + b_2 + \frac{1}{2v}\left(\frac{1}{\sqrt{\lambda T}} + \frac{1}{\sqrt{n}}\right) r_1 \sqrt{A}\right) \tag{2}$$

where

t_p = the daily time period of ADA operations during which trips are served;
λ = the annual demand;
b_1 and b_2 = the boarding time and the drop-off time per trip;
T = the time window for picking up a passenger;
A = the service area;
v = the average travel speed of vehicles in the network;
n = the average occupancy of paratransit vehicles; and
r_1 and r_2 = specific travel factors which change with the transportation system.

The annual operating cost of the paratransit system is based on the estimates of annual VHT and VMT. These operations parameters are associated with dollar costs, which can be estimated from agency operations and cost data. The total cost model takes the following form:

$$\text{TC}_p = \alpha_0 + \alpha_1 \text{VMT} + \alpha_2 \text{VHT} + \alpha_3 M \tag{3}$$

where α_0 are the fixed costs associated with setting up a paratransit operation in a region, and α_1, α_2, and α_3 are the incremental cost of each vehicle hour, vehicle mile, and vehicle in the fleet.

Taxi Costs

This study considers taxis as an alternative to the conventional ADA paratransit service provided by a transit agency. Taxis are typically operated as a regulated industry in which many private operators provide services, charging fares according to a regulated formula based on distance and time. Therefore, the cost to serve a trip by taxi can be estimated based on the distance traveled and the time required to complete the trip, both of which are data that are available from records of ADA paratransit trips. The average cost of a single taxi trip, (AC_t) is given by:

$$\text{AC}_t = \beta_0 + \beta_1 l + \beta_2 d \tag{4}$$

where

β_1 = the fixed cost to enter a taxicab;
β_2 = the cost per distance traveled;
l = the length of the trip;
β_3 = the cost per delay; and
d = the delayed time experienced.

An estimate of annual costs can be made by multiplying the cost of a trip of average length and delay by the total number of trips served. Delay is defined with the average speed and a free flow speed. When the average speed is lower than the free flow speed, it means that there is congestion. In this study, the cost of taxi trips associated with individual origin–destination pairs is considered in order to identify specific trips that may be most cost-effectively diverted from conventional paratransit to taxis. In this paper, it is supposed that TNCs would follow a similar cost structure to taxis, and no distinction is made between the services although actual TNC cost structures may differ depending on the company.

Combined System of Paratransit and Taxis

Now consider a region of size A in which a transit agency provides a demand responsive service to an annual demand of λ ADA-eligible trips. If all ADA trips are served by the paratransit system, the total cost of serving a single region would be given by Equation 3. If the expressions for VMT, VHT, and M from Equations 1 and 2 are substituted into 3, the total annual cost can be expressed as a function of the total ADA demand λ:

$$\begin{aligned} \mathrm{TC}_p(\lambda) &= \alpha_0 + \alpha_1 \frac{r_2\sqrt{A}}{2}\left(\sqrt{\frac{\lambda}{T}} + \frac{\lambda}{\sqrt{n}}\right) \\ &\quad + \left(\alpha_2 t_p + \alpha_3\right)\left[\lambda(b_1 + b_2) + \frac{r_1\sqrt{A}}{2v}\left(\sqrt{\frac{\lambda}{T}} + \frac{\lambda}{\sqrt{n}}\right)\right] \end{aligned} \tag{5}$$

Note that this result is a combination of constants and terms that increase with $\sqrt{\lambda}$ and λ. Therefore, the total cost of ADA paratransit is strictly increasing with demand.

The objective of this paper is to identify the trips that can be diverted to an alternative service, such as taxis, in order to reduce the total cost of the combined system—that is, the total cost of paratransit and combined cost of all trips diverted to taxis. If the paratransit service must continue to operate over the entire service area to serve those users who are unable or unwilling to use taxis, then the value of A can be considered constant. The marginal cost (MC) of the paratransit service is the first derivative of Equation 5 with respect to λ:

$$\begin{aligned} \mathrm{MC}_p(\lambda) &= \alpha_0 + \alpha_1 \frac{r_2\sqrt{A}}{2}\left(\frac{1}{2\sqrt{T\lambda}} + \frac{1}{\sqrt{n}}\right) + \left(\alpha_2 t_p + \alpha_3\right) \\ &\quad \left[b_1 + b_2 + \frac{r_1\sqrt{A}}{2v}\left(\frac{1}{2\sqrt{T\lambda}} + \frac{1}{\sqrt{n}}\right)\right] \end{aligned} \tag{6}$$

Note that the marginal cost is decreasing with demand, so that the lowest marginal cost is associated with serving all trips by paratransit.

If a trip can be served at lower cost by taxi, the total cost can be reduced by diverting it from the paratransit service. To minimize the total cost of the combined paratransit and taxi system, this process should be repeated until there are no taxi trips that can be served for less than the marginal cost of the taxi service. A graphical representation of this solution method is presented with the application to PVTA data in which taxi trips are ranked in order from least to greatest cost and plotted against the corresponding marginal cost of the paratransit service. The result is either an optimized system with both paratransit and taxi used in coordination or one mode used in isolation:

1. If the taxi curve includes trips that cost less than the marginal cost of paratransit serving the total demand and increases to equal the marginal cost at some point, then the cost is locally minimized with a mix of taxi and paratransit service. This solution is a global optimum if at least some ADA demand cannot be served by a taxi. Otherwise, it is necessary to check the cost associated with the boundary point that all trips are served by taxi.
2. If no such intersection point (local minimum) exists, then the optimal solution is either to serve all feasible trips by taxi or by paratransit, but not with a combined system.

The following sections present an application of this modeling and analysis method to the PVTA to quantify the potential for reducing operating costs by coordinating ADA paratransit operations with taxis.

Study Data

The paratransit operations and demand data used for this study are from a 2-year set of trip records from the PVTA. From June 2015 to June 2017, PVTA's ADA paratransit system served 432,830 trips. The raw data included no personally identifying information.

For each trip, 22 columns contain the following information:

1. Date and time of each of the following: vehicle arrival at a trip origin for pick-up, vehicle departure from trip origin, vehicle arrival at trip destination for drop-off, and vehicle departure from trip destination. These data can then be used to calculate the duration of each pick-up and drop-

off stop as well as the duration of the trip from origin to destination.

2. Location of each trip origin and destination. The locations are available as detailed latitude/longitude coordinates, which are also classified in geographic aggregations by zip code and community (i.e., town or city).
3. Distance traveled on the network by paratransit vehicles between stops for pick-up or drop-off. From this distance, along with the trip durations, it is possible to calculate the average network speed associated with each trip.
4. Vehicle ID and corresponding garage ID and garage location for the vehicle serving each trip. From this it is possible to string together the sequence of vehicle routes that operate each day, so operations with and without passengers onboard are accounted for in the total vehicle miles and vehicle hours of operations.
5. Type of disability for each customer. The dataset includes some details about the types of devices that customers used, but the relevant information for this research is identification of which customers are ambulatory, implying that they would be physically able to use a taxi without equipment for loading a wheelchair.
6. Anonymous customer ID. This number is used to link customer trip records with a survey that was conducted to collect additional data.

The raw data included no personally identifying information, such as names or origin/destination addresses; rather, they were aggregated to the zip code level.

The raw data represent a total ADA demand of approximately 215,000 trips per year originating in a 627 mi^2 area that includes 24 communities served by PVTA, including Springfield, Massachusetts. This region is covered by 44 zip codes, which is the unit of spatial aggregation used for the analysis in this paper. Figure 1 shows the annual number of trips originating in each zip code, which shows that the demand is largely concentrated in the urban core of the region around Springfield, Chicopee, and Holyoke but with several trips also being made in outlying areas.

In addition to the data that is logged for each trip, PVTA has recently conducted a survey of paratransit users that received 381 responses from a random sample of 500 customers who were contacted for participation. The high response rate was the result of a program to follow up with individuals to encourage participation, and the respondents were representative of the distribution of ages and home locations known for the full population of eligible ADA customers. The survey gathered additional demographic data that is not normally known about paratransit customers: race/ethnicity, household income, access to other means of transportation, and access to a smart phone. The survey responses are linked to the anonymous customer IDs in the trip log so that these additional demographic characteristics can be

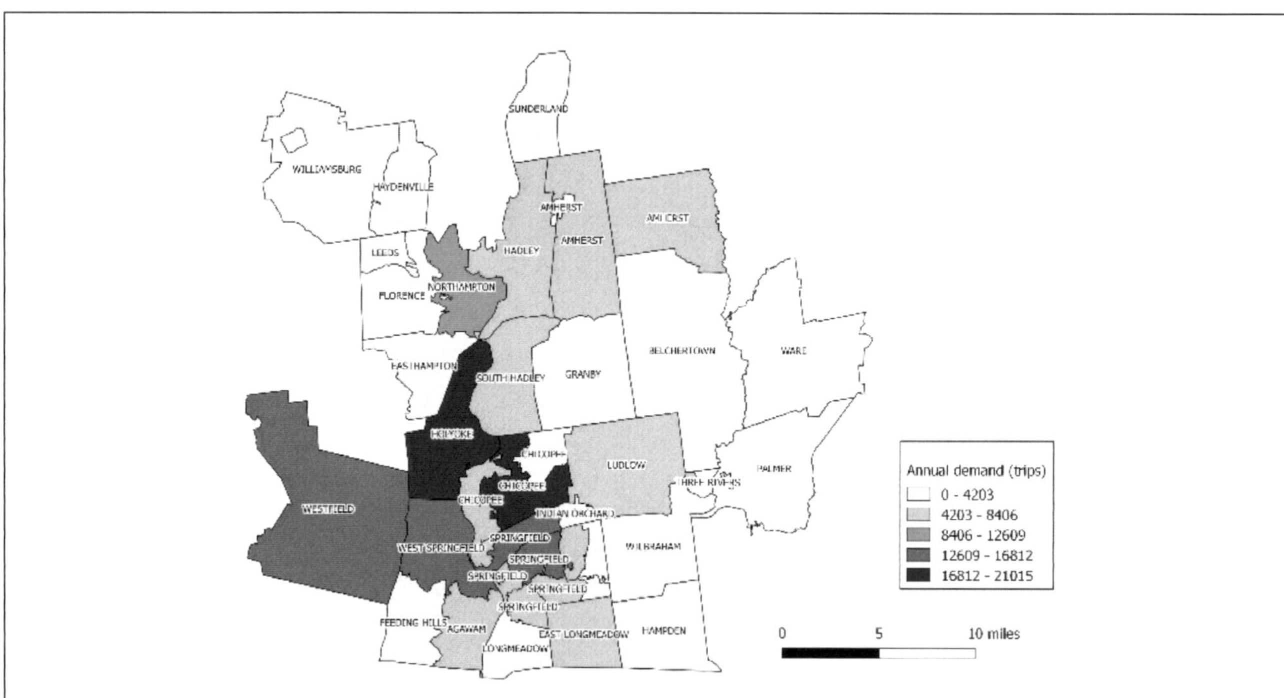

Figure 1. Annual ADA trip origins by zip code.

Table 1. Parameters of Paratransit and Taxi Models

Parameter	Value	Units	Fit statistic (R^2)
Paratransit operations parameters			
Area, A	627	mi^2	na
Time window, T	40	min	na
Vehicle occupancy, n	1.2	pax/veh	na
Passenger loading/unloading time, $b_1 + b_2$	10	min	na
Average vehicle speed, v	20.5	mph	na
Calibration parameter for VHT and M, r_1	0.90	na	0.96–0.99
Calibration parameter for VMT, r_2	0.82	na	0.97–0.99
Paratransit cost parameters			
Fixed annual cost, α_0	480,123	\$	na
Cost per vehicle mile, α_1	0.518	\$	0.80
Cost per vehicle hour, α_2	19.89	\$	0.86
Annual cost per vehicle, α_3	55,044	\$	0.93
Taxi cost parameters			
Fixed cost per trip, β_0	2.40	\$	na
Cost per distance, β_1	2.50	\$/mi	na
Cost per delay time, β_2	21.00	\$/h	na

Note: na = not applicable; pax/veh = passengers per vehicle.

related to the actual trips served by the system. This data source is useful for assessing the potential equity implications of a program to shift some ADA paratransit demand to taxis or alternative providers.

Application of Model

Operation and Cost Model Parameters

Using data from PVTA's ADA paratransit service, a cost model for the paratransit was developed that reflects the relationship between demand and service area with required vehicle miles, vehicle hours, and fleet size for operations. This model is then linked with cost parameters to create a total cost function. PVTA currently organizes paratransit service in six service zones. The observable parameters of the operations models were estimated using aggregate data for the six regions combined, and the values are reported in Table 1. Calibrations of the operations models were done using the linear regression method described in (*14*) to identify the values of r_1 and r_2 in each region, which fit the observed data with R^2 values between 0.96 and 0.99, indicating a good fit. For this analysis, the paratransit service in the Pioneer Valley is treated as a single region, so an average of the parameter values is used. Aggregated system-wide cost data from PVTA was used to estimate the cost coefficients also using linear regression; cost values and fit statistics are also included in Table 1.

In addition to the model data for paratransit, the cost parameters for taxi trips are based on published fare rates for yellow cabs in Springfield, Massachusetts. Although there are number of taxi companies operating in communities throughout the Pioneer Valley, and Uber has recently launched a service in the region as well, this published cab rate is used for this analysis under the assumption that prices are similar for competing services.

Demand Data and Taxi Cost Estimation

To simplify the analysis, the original data set containing 432,830 trip records over the course of 2 years was aggregated into an origin–destination (OD) matrix using the zip code as the unit of spatial aggregation. Trips were observed for 1,306 of the 1,936 possible OD pairs. The average observed distance and travel time were calculated for each OD pair. To estimate the average cost of a taxi trip for each OD, this data was used for the inputs into Equation 4. The trip length, l, is simply the average observed distance. The delay, d, is the calculated positive difference between the observed travel time and the free flow travel time that would have been experienced if the vehicle traveled at an average free flow speed of 25 mph. If the average trip was completed at a speed faster than 25 mph, the delay is considered to be zero. A sensitivity analysis of the effect of free flow speed on the cost calculation showed that, within a range of reasonable values, this had far less effect than the distance traveled. Figure 2 shows a spatial comparison of the average trip length and corresponding estimated taxi trip cost from each origin zip code. The figure confirms that shorter trips in the urban core around Springfield are typically cheaper to serve than longer trips from communities with less dense demand and population at the edge of the region.

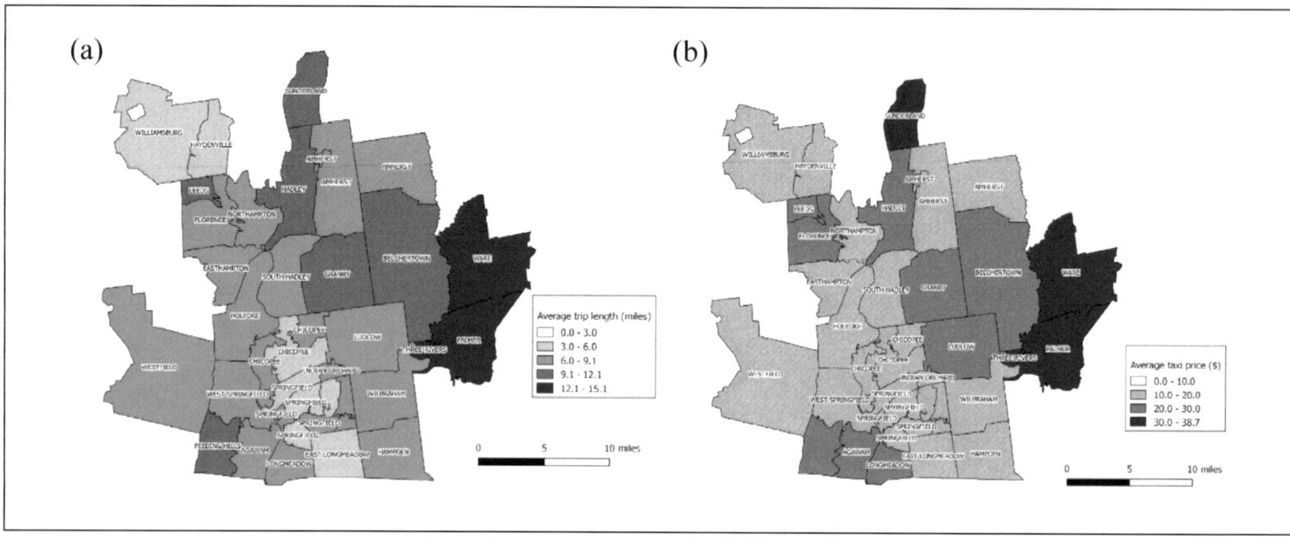

Figure 2. Spatial distribution of (*a*) average trip length and (*b*) estimated taxi fare by origin zip code.

In considering more carefully which trips could potentially be shifted to taxis, the total OD demand matrix was split into parts: an OD matrix for ambulatory trips that could potentially use taxis, and an OD matrix for non-ambulatory trips (i.e., customers using a mobility device such as a wheelchair or scooter) that would require service by a specially equipped vehicle in the existing paratransit fleet.

Single Provider

Implementing the model for the existing conditions, in which all trips are served by PVTA's paratransit service, the estimated annual operating cost is \$5,858,000. This is equivalent to an average cost of \$27.07 per trip which is slightly less than the actual reported cost, but the model approximates demand as a constant rate over the whole day. The existing marginal cost of a PVTA paratransit trip is \$26.58, as estimated by the model. By comparison, serving all trips by taxi would equate to taxi fares adding up to \$4,401,000. The following cases consider the opportunity to further reduce costs by optimizing the allocation of demand to paratransit and taxis.

Case 1: Potential for All Users to Use Taxis

The first case for consideration is to consider all paratransit customers as potential taxi passengers. In this case, each OD pair is ranked in order from lowest estimated trip cost to highest estimated trip cost. The marginal cost of using taxis to serve demand can then be represented by plotting this cost per trip against the cumulative taxi demand for trips at equal or lower price. Plotting this function appears as the increasing blue curve in Figure 3*a*, with the annual number of trips diverted to taxi, λ_t, represented on the horizontal axis. This is the cost that should be compared with the marginal cost of the ADA paratransit service for the remaining demand, $\lambda_{\text{total}} - \lambda_t$. In Figure 3*a*, the marginal cost appears as an increasing function shown by the orange curve, because the demand for paratransit is decreasing as the number of trips diverted to taxis increases. To interpret the figure, the left side ($\lambda_t = 0$) represents the existing policy in which all ADA-eligible trips are served by the paratransit service. Moving to the right, trips in OD pairs associated with the lowest estimated taxi cost per trip are shifted away from paratransit. As more trips are shifted to taxis, the marginal cost of paratransit increases because the system serves fewer customers, and the marginal cost of taxi trips increases because the cheaper trips have already been diverted. Where the two curves intersect, the total cost is a minimum as shown in Figure 3*b*.

For the cost parameters and demand pattern in the Pioneer Valley, the policy that would minimize the combined cost of paratransit and taxi trips would be to divert all trips to taxi that could be served with a fare of less than \$26.61, which is here called the optimal cost threshold. This corresponds to 157,400 trips annually from 528 OD pairs, representing 73% of the total demand. The annual combined cost of this optimized system would be an estimated \$3,945,000, which is 33% less than serving all trips by paratransit and 10% less than serving all trips by taxi. The optimal cost threshold is only a little higher than the marginal cost of \$26.58 associated $\lambda_t = 0$, because the costs per paratransit trip are changing very slowly over a wide range of demand levels.

The proposed optimal policy applies if the transit agency bears the direct cost of operating the paratransit

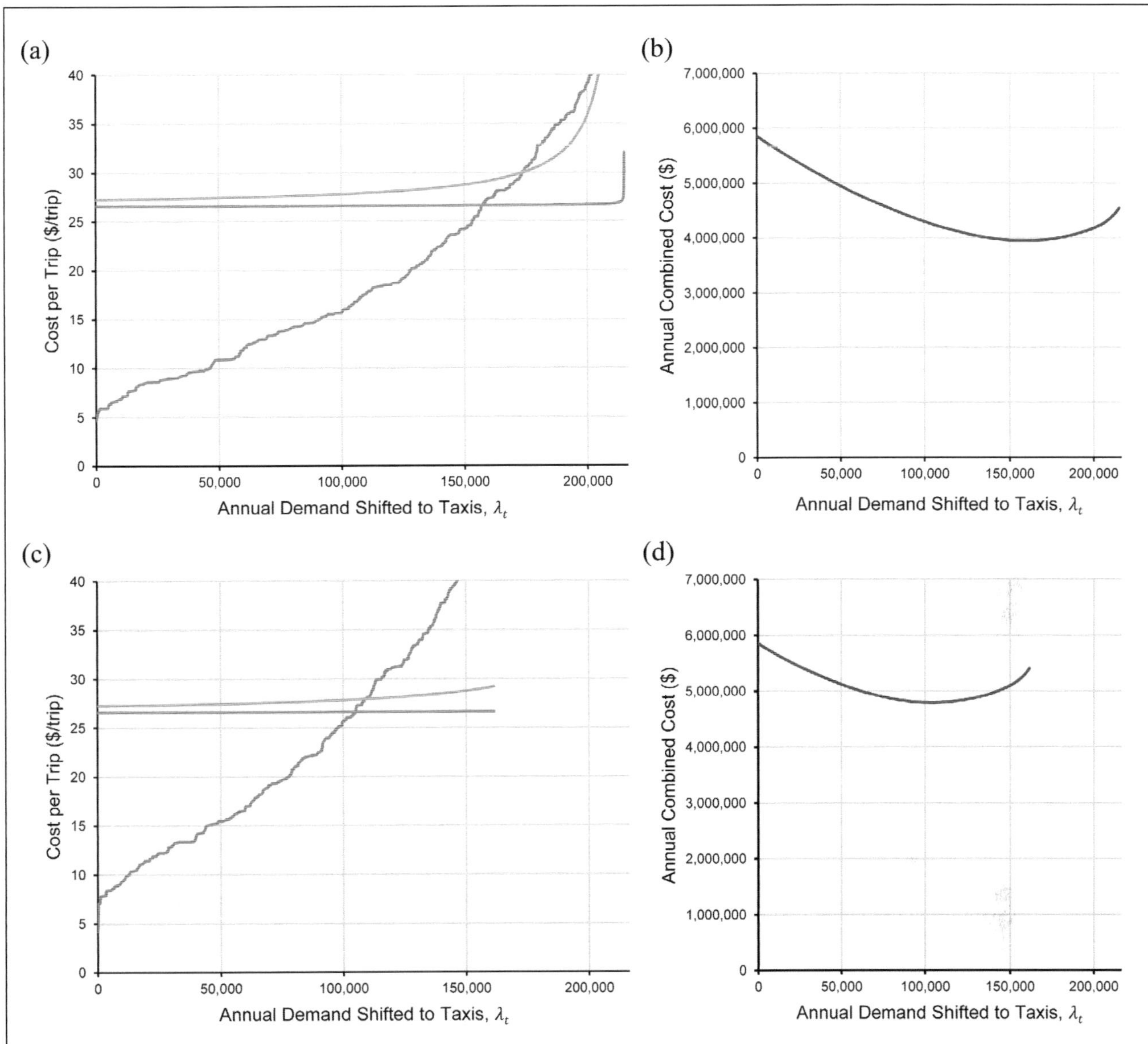

Figure 3. Effect of shifting demand to taxis on cost per trip and total combined system cost for Case 1 and Case 2 (*a*) cost per trip for Case 1 (all potential users); (*b*) total cost for Case 1; (*c*) cost per trip for Case 2 (ambulatory users); (*d*) total cost for Case 2.

service in addition to the taxi fares for diverted trips. Many transit agencies, including PVTA, contract out operations of their ADA paratransit service to a private operator. The terms of the contract specify how the operations of paratransit translate into fiscal obligations from the transit agency. One common way of structuring such a contract is for the transit agency to pay a fixed average cost per trip served based on an expected level of paratransit demand. Under this contracting model, the transit agency minimizes costs by comparing the taxi fare to the average cost rather than the marginal cost. Note, as in Figure 3*a*, that the taxi cost curve always intersects the average cost curve to the right of the marginal cost curve, because paratransit exhibits economies of scale, and the average cost always exceeds the marginal cost. The average cost threshold in this case is $29.96 resulting in a combined system cost of $3,971,000.

Case 2: Considering That Only Ambulatory Customers Potentially Use Taxis

Customers can become eligible for ADA paratransit service with a variety of disabilities, many of which do not impede a customer from being able to ride in a regular car. An important distinction among customer disability data is the classification of ambulatory customers (those who are able to walk) and non-ambulatory customers who rely on a mobility device that often requires a vehicle with special equipment, such as a lift. A realistic constraint on a program to divert trips from paratransit to taxis is to consider that such a mode shift may only be feasible for 165,550 annual trips by ambulatory passengers (75% of the total ADA paratransit demand).

To evaluate this case, the cost function for the paratransit service is the same as before, because the paratransit system will continue to serve non-ambulatory customers and could potentially serve all ambulatory customers as well. The difference is that the taxi cost function is scaled against the cumulative demand of ambulatory customers only, because those are the customers who could potentially switch modes. Figure 3*c* shows the increasing taxi fare against the number of trips diverted to taxis, λ_t, along with the marginal cost and average cost of the corresponding paratransit service. Note that the taxi cost appears to rise much faster than in Figure 3*a*, so the taxi fare equals the marginal cost of paratransit at a much lower demand corresponding to the minimum combined cost shown in Figure 3*d*.

If only ambulatory customers can potentially use taxis, then the combined cost of paratransit operations and taxi fares is minimized when trips are diverted to taxi up to an optimal cost threshold of $26.59. This corresponds to 105,000 trips from 424 OD pairs representing 63% of ambulatory demand and 49% of total ADA paratransit demand. The estimated annual combined cost is $4,789,000, which is 18% less than the cost of serving all demand by paratransit. This anticipated benefit is smaller than the result from Case 1, because there are fewer potential trips to divert to taxi and therefore fewer cost savings.

A transit agency that pays a contractor for paratransit operations on a fixed average cost basis would now minimize costs by keeping trips with taxi fare greater than $27.91 on the paratransit system with the non-ambulatory customers. The result would be 108,900 annual taxi trips and a combined annual cost of $4,792,000.

Case 3: Differentiating Demand by Time of Day

Most of the ADA trips served by PVTA occur between the hours of 6 a.m. and 6 p.m. By breaking the day into time periods, it is possible to identify the effect of temporal variations in demand on the effectiveness of a taxi policy. This analysis takes the demand profile of Case 1 and breaks down the time period into four 3-h blocks: 6 to 9 a.m., 9 a.m. to 12 noon, 12 noon to 3 p.m., and 3 to 6 p.m. These time periods account for 96% of the total demand. Three analyses are compared in Table 2.

First, the cost of serving all trips by paratransit is calculated by accounting for each of the time periods separately. Note that the required fleet size varies from time period to time period, but the annual cost requires that the maximum fleet size be paid for. The annual cost of $6,341,000 is greater when temporal variations are considered compared with the aggregated results presented above, because the variations mean that resources cannot be used with uniform intensity (and efficiency) throughout the day. Second, the temporal variation is considered for the optimal cost threshold of $26.59, identified for

Table 2. Effect of Temporal Variation in Demand on Cost

Period	Threshold cost	Taxi demand	Annual taxi cost	Paratransit demand	VMT	VHT	*M*
All trips served by paratransit							
6–9 a.m.	na	na	na	45,165	425,958	30,333	28
9 a.m.–12 noon	na	na	na	43,899	414,051	29,485	27
12 noon–3 p.m.	na	na	na	44,474	419,462	29,870	27
3–6 p.m.	na	na	na	72,560	683,411	48,683	44
Total annual combined cost							$6,341,000
Single threshold cost as identified in Case 2							
6–9 a.m.	$26.59	19,968	$317,037	25,197	238,139	16,949	15
9 a.m.–12 noon	$26.59	24,356	$372,220	19,543	184,913	13,157	12
12 noon–3 p.m.	$26.59	23,891	$378,280	20,583	194,705	13,855	13
3–6 p.m.	$26.59	37,128	$604,325	35,432	334,431	23,811	22
Total annual combined cost							$4,845,000
Optimized threshold for each time period							
6–9 a.m.	$27.69	22,421	$381,923	22,744	215,055	15,305	14
9 a.m.–12 noon	$21.80	21,158	$294,901	22,741	215,017	15,302	14
12 noon–3 p.m.	$23.46	21,786	$326,000	22,688	214,533	15,267	14
3–6 p.m.	$35.68	49,795	$1,001,929	22,765	215,248	15,318	14
Total annual combined cost							$4,572,000

Note: na = not applicable.

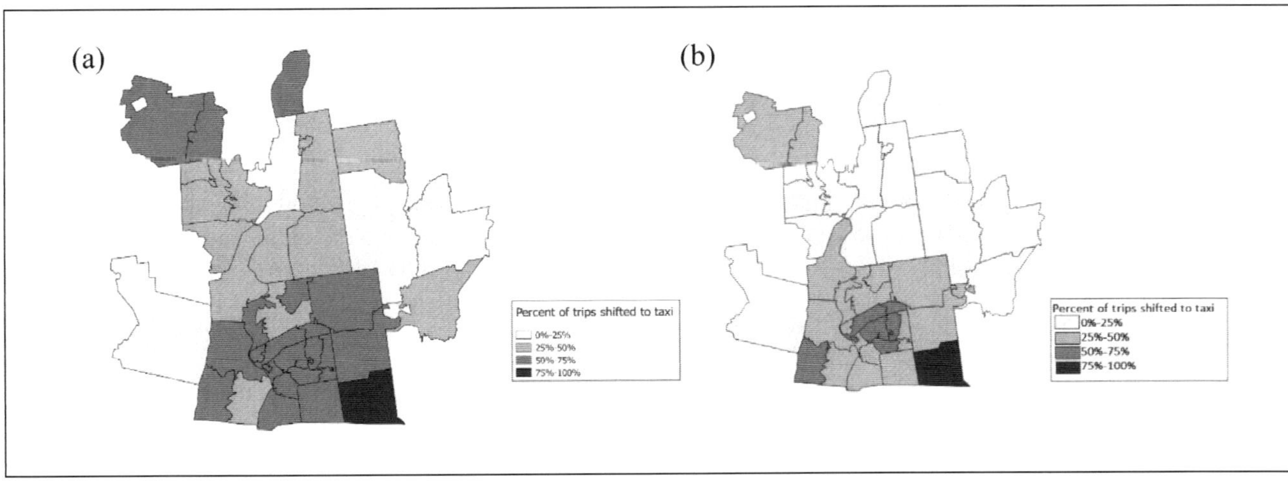

Figure 4. Share of total paratransit trips diverted to taxis for (*a*) Case 1 and (*b*) Case 2.

Case 2. The resulting cost of \$4,853,000 achieves a 24% reduction in system cost. Third, the cost threshold is optimized for each time period by accounting for the cost of VMT and VHT within each period and attributing the cost of M to the maximum fleet required across all periods. For the PVTA, the cost thresholds vary by time to level the paratransit demand throughout the day so that the required fleet is 14 vehicles at all times. The marginal cost when only VMT and VHT are considered is \$18.20, but the demand is high enough even in off-peak hours that this threshold would require a larger fleet. The resulting cost \$4,572,000 adds an additional 4% cost saving over the use of a single threshold cost.

Implications for Demand and Policies

Impacts on Different Users

The method presented in this paper for reducing the system cost of serving ADA-eligible paratransit customers by diverting some trips to taxis has shown that potentially large cost savings can be accomplished, at least with the example of PVTA. A constraint on diverting paratransit trips to taxis or TNCs is that the federally-required ADA service must provide an equivalent service to all eligible customers, so a customer cannot be offered only an inaccessible taxi. A more flexible way to coordinate paratransit services with other providers is to continue to offer conventional next-day paratransit service to all customers, and then give all customers to option to use taxis or TNCs if they choose to contact the taxi company or TNC on their own (e.g., as the MBTA program is implemented in Boston). Such an option could be perceived as offering a high quality of service, because reservations do not need to be made at least a day in advance, and the pick-up time for taxis and TNCs is typically much shorter than the 20-min time window guaranteed by PVTA paratransit services. If the transit agency subsidizes fares only up to the threshold, then customers will have less incentive to use taxis or longer and costlier trips.

The method for identifying trips to give incentives to shift to taxis is based on calculating the threshold taxi fare below which trips should be served by taxi instead of paratransit. As a result, the characteristics of trips and customers that are switched to taxis can be compared with those who remain on paratransit to gain insights about how different customers are served by the combined system. This is important for ensuring compliance with Title VI of the Civil Rights Act, which prohibits discrimination by race, color, or national origin for federally-assisted programs.

One relevant comparison is to consider how the policy would affect different geographic locations in the region. Figure 4 shows maps for Case 1 and Case 2 of the percentage of trips originating in each zip code that would be diverted to taxis in an optimally managed combined system. Not surprisingly, the greatest percentages of diverted trips are in the locations that have the lowest estimated taxi trip costs as shown in Figure 2. The percentage of diverted trips is also noticeably lower in Case 2, because fewer customers are considered to be able to use taxis. There remains a concentration of eligible trips centered on Springfield.

Implications for Policies

One concern associated with policies that incentivize the use of taxis or TNCs for paratransit services is that improvements in the perceived quality of service from alternative modes may increase demand. For example, if it becomes more convenient to make trips once taxis are an option for service, it is likely that some increase in discretionary travel will result. Regulations prohibit transit agencies from discriminating ADA paratransit service on the basis of trip purpose, and this has the potential to undermine projected cost savings. The analysis presented in this paper is based on the assumption that travel demand is inelastic. If trips are induced, the effect on cost would be equivalent to increasing the taxi cost curve proportionally. For example, if eligibility for taxi service doubles the number of trips that are made, the effect would be the same as if taxi fares were doubled. Further research is necessary to consider how more sophisticated demand models can be implemented to account for changes in demand across demographic groups, geographic locations, and times of day.

The analysis presented in this paper is based on the cost structure for paratransit and taxicab operations in the Pioneer Valley, around Springfield, Massachusetts. The potential cost savings from a taxi or TNC policy could be much greater in regions where paratransit is costly to operate but the development is dense enough that customers are making short trips. Conversely, such a policy may not make sense in more rural areas where taxi service is costly if it is available at all. In this paper, taxis were considered to be available in all communities throughout the Pioneer Valley, but some outlying communities, such as Ware, Massachusetts, do not have taxi services available, and this may leave transit agencies with an obligation to operate conventional paratransit services out to the periphery of the service region. In the face of uncertainty about demand response, it may make sense to start with a lower threshold to incentivize a smaller number of trips, and then increase the threshold as more information about demand patterns for a new policy becomes available.

Conclusion

This paper has presented a method to quantify the potential benefits of implementing a program to allow ADA-eligible customers to choose taxis or TNCs instead of the conventional paratransit service to complete their trips. The method uses aggregate models of paratransit operations to estimate the marginal cost of service with conventional vans compared with the cost per trip of using a taxi.

Actual demand patterns and operating costs for the PVTA's ADA paratransit service were used as a case study. Accounting for temporal variations in demand and the potential for ambulatory passengers to use taxis, the results show that by subsidizing taxi fares, the annual cost for PVTA could potentially be reduced from $6,341,000 to $4,845,000 (23.5% reduction) if a fixed cost threshold is defined or $4,572,000 (27.9% reduction) if a time-varying cost threshold is used. Costs remain higher if some customers are unable or unwilling to use taxis, because the potential diversion of trips is constrained, and the remaining paratransit users have a higher likelihood of using mobility devices and requiring more time for boarding and alighting. The benefit is based on optimizing the threshold for subsidizing taxi fares to account for time-varying demand and assuming that customers always switch to taxi in response to incentives. Cost reductions would be lower if some customers choose to remain on the paratransit service or if new trips are induced. Although there are risks to coordinating service between public and private providers, there are potentially large benefits for maintaining or improving quality of service for ADA-eligible users while reducing costs for transit agencies.

Author Contributions

The authors confirm contribution to the paper as follows: study conception and design: EJG, MR, and PA; data analysis and interpretation of results: VT, MR, and EJG; draft manuscript preparation: VT and EJG. All authors reviewed the results and approved the final version of this manuscript.

References

1. Fu, L. A Simulation Model for Evaluating Advanced Dial a Ride Paratransit Systems. *Transportation Research Part A*, Vol. 36, No. 4, 2002, pp. 291–307.
2. Dessouky, M., L. Quadrifoglio, and F. Ordez. A Simulation Study of Demand Responsive Transit System Design. *Transportation Research Part A*, Vol. 42, No. 4, 2008, pp. 718–737.
3. Quadrifoglio, L., and C. Shen. Evaluating Centralized versus Decentralized Zoning Strategies for Metropolitan ADA Paratransit Services. *Journal of Transportation Engineering*, Vol. 139, No. 5, 2013, pp. 524–532.
4. Shen, C., and L. Quadrifoglio. The Coordinated Decentralized Paratransit System: Formulation and Comparison with Alternative Strategies. Presented at 90th Annual Meeting of the Transportation Research Board, Washington, D.C., 2011.
5. Shen, C., and L. Quadrifoglio. Evaluation of Zoning Design with Transfers for Paratransit Services. *Transportation Research Record: Journal of the Transportation Research Board*, 2012. 2277: 82–89.
6. Fu, L. Improving Paratransit Scheduling by Accounting for Dynamic and Stochastic Variations in Travel Time. *Transportation Research Record: Journal of the Transportation Research Board*, 1999. 1666: 74–81.
7. Fu, L., and S. Teply. On-Line and Off-Line Routing and Scheduling of Dial-a-Ride Paratransit Vehicles. *Computer-Aided Civil and Infrastructure Engineering*, Vol. 14, No. 5, 1999, pp. 309–319.
8. Daganzo, C. F. An Approximate Analytic Model of Many-to-Many Demand Responsive Transportation

Systems. *Transportation Research Record: Journal of the Transportation Research Board*, 1978. 12: 325–333.
9. Daganzo, C. F. The Distance Traveled to Visit *n* Points with a Maximum of *c* Stops per Vehicle: An Analytic Model and an Application. *Transportation Science*, Vol. 18, No. 4, 1984, pp. 331–350.
10. Fu, L. Analytical Model for Paratransit Capacity and Quality-of-Service Analysis. *Transportation Research Record: Journal of the Transportation Research Board*, 2003. 1841: 81–89.
11. Figliozzi, M. A. Planning Approximations to the Average Length of Vehicle Routing Problems with Varying Customer Demands and Routing Constraints. *Transportation Research Record: Journal of the Transportation Research Board*, 2008. 2089: 1–8.
12. Figliozzi, M. A. Planning Approximations to the Average Length of Vehicle Routing Problems with Time Window Constraints. *Transportation Research Part B*, Vol. 43, No. 4, 2009, pp. 438–447.
13. Diana, M., M. M. Dessouky, and N. Xia. A Model for the Fleet Sizing of Demand Responsive Transportation Services with Time Windows. *Transportation Research Part B*, Vol. 40, No. 8, 2006, pp. 651–666.
14. Rahimi, M., M. Amirgholy, and E. J. Gonzales. System Modeling of Demand Responsive Services: Evaluating Cost Efficiency of Service and Coordinated Taxi Usage. *Transportation Research Part E*, Vol. 112, 2018, pp. 66–83.
15. Rahimi, M., and E. J. Gonzales. Systematic Evaluation of Zoning Strategies for Demand Responsive Transit. Presented at 94th Annual Meeting of the Transportation Research Board, Washington, D.C., 2015.
16. Chia, D. *TCRP Synthesis 74: Policies and Practices for Effectively and Efficiently Meeting ADA Paratransit Demand: A Synthesis of Transit Practice*. Transportation Research Board of the National Academies, Washington, D.C., 2006.
17. Burkhardt, J. E. *Potential Cost Savings from Taxi Paratransit Programs*. Social Research in Transport, Institute of Transport Studies, Monash University, Melbourne, Australia, 2010.
18. Tuttle, D., and K. Eaton. *Optimizing Accessible Taxi Service to Augment Traditional Public Transit Services in Delaware*. DCT 222. Institute of Public Administration, University of Delaware, Newark, DE, 2012.
19. Ellis, E. *Use of Taxis in Public Transportation for People with Disabilities and Older Adults*. TCRP Synthesis 119. Transportation Research Board, Washington, D.C., 2016.
20. Burkhardt, J. E., J. Doherty, J. M. Rubino, and J. Kim. *Accessible Community Transportation in Our Nation: A Survey on the Use of Taxis in Paratransit Programs*. National Aging and Disability Transportation Center, Washington, D.C., 2008.
21. Nelson/Nygaard Consulting Associates. *TCRP Report 121: Toolkit for Integrating Non-Dedicated Vehicles in Paratransit Service*. Transportation Research Board of the National Academies, Washington, D.C., 2007.
22. Taylor, B., R. Chin, M. Crotty, J. Dill, L. Hoel, M. Manville, S. Polzin, B. Schaller, S. Shaheen, D. Sperling, M. Zafar, and S. Zielinski. *Special Report 319: Between Public and Private Mobility: Examining the Rise of Technology-Enabled Transportation*. Transportation Research Board of the National Academies, Washington, D.C., 2015.
23. Feigon, S., and C. Murphy. *TCRP Research Report 188: Shared Mobility and the Transformation of Public Transit*. Transportation Research Board of the National Academies, Washington, D.C., 2016.
24. Shin-pei, T., Z. Accardi, B. Schaller, and K. Hovenkotter. *Private Mobility, Public Interest: How Public Agencies Can Work with Emerging Mobility Providers*. TransitCenter, New York, NY, 2016.
25. Martin, E., S. Shaheen, and J. Lidicker. Impact of Carsharing on Household Vehicle Holdings: Results from North American Shared-Use Vehicle Survey. *Transportation Research Record: Journal of the Transportation Research Board*, 2010. 2143: 150–158.
26. Henao, A. *Impacts of Ridesourcing – Lyft and Uber – on Transportation Including VMT, Mode Replacement, Parking, and Travel Behavior*. PhD dissertation. Civil Engineering Program, University of Colorado at Denver, Denver, CO, 2017.
27. Schaller, B. *Unfinished Business: A Blueprint for Uber, Lyft and Taxi Regulation*. Schaller Consulting, New York, NY, 2016.
28. Schaller, B. *Unsustainable? The Growth of App-Based Ride Services and Traffic, Travel and the Future of New York City*. Schaller Consulting, New York, NY, 2017.
29. Amirgholy, M., and E. J. Gonzales. Demand Responsive Transit Systems with Time Dependent Demand: User Equilibrium, System Optimum, and Management Strategy. *Transportation Research Part B*, Vol. 92, 2016, pp. 234–252.

The Public Transportation Group (AP000) peer-reviewed this paper (18-04159).

Article

TRR
JOURNAL OF THE TRANSPORTATION RESEARCH BOARD

Transportation Research Record
2018, Vol. 2672(8) 698–707

DOI: 10.1177/0361198118791391
journals.sagepub.com/home/trr
SAGE

Transport Policy Optimization with Autonomous Vehicles

Patrick M. Bösch[1], Francesco Ciari[2], and Kay W. Axhausen[1]

Abstract

Autonomous vehicles (AVs), here self-driving and driverless vehicles, SAE levels 4 and 5 are becoming more clearly a reality. Potential services based on AVs and their consequences for the transport system are of increasing importance. This paper investigates policy combinations for a world with such services. The policy measures investigated are pricing of public transport (through subsidies), pricing of private motorized transport (through taxation or mobility pricing), and the organization of AV services (monopoly vs. oligopoly, with or without ride-sharing). Further, the perception of travel times for autonomous private cars is considered. All combinations of policies (respectively two to four levels each) were implemented in a simulation to determine their synergies. The applied model was the agent-based transportation simulation MATSim. The scenario employed for the tests was the agglomeration of Zug, Switzerland. The results suggest that, given the current spatial distribution of the demand and the current transport system, AV systems are only able to reduce travel times at the cost of substantial mode shifts and additional vehicle kilometers driven. Of the tested policy measures, although all showed the expected causality, only the organizational form of the AV service had a statistically significant effect. Therefore, this paper suggests that policy makers should be cautious when confronted with the promises of future transport services. To invest the benefits of automation into an improvement of the existing transport system (e.g., automation of public mass transit or complementing public mass transit with ride-sharing AVs in low-demand areas) might be a good alternative.

Autonomous vehicles (AVs), in this paper driverless and self-driving vehicles (SAE levels 4 and 5 [*1*]), promise to revolutionize the transport system. The possibility of driverless relocations in shared vehicle systems, a substantial cost reduction in public transport operations (*2*), and driving transformed into productive time are just a few of the revolutionary features expected from AVs (for a comprehensive overview see [*3*]).

Such fundamental changes of the transport system were topics of early papers on AVs (*3–5*). Recently however, the focus has shifted to more detailed questions. Although insight on the required fleet sizes to serve a city (*6–10*), or the organization of new services (*11–14*) is background knowledge on the possibilities offered by AVs, they are years, maybe decades away from implementation.

Thanks to this background knowledge however, it is now possible to return to the fundamental question of transport system organization: given all these new opportunities, the current system as a starting point, and given the financial and political constraints, how should the future transport system best be organized?

This paper aims to assess different possibilities AVs allow for future transport services given their benefits and costs for society. It evaluates policy measures available to policy makers to influence and shape the transport system to make the most of the benefits AVs could possibly bring.

Finding a preferred combination of policy measures is a classic optimization problem (*15–17*). A target function is evaluated in a multi-dimensional space of possible measures and their respective implementation ranges. It is evaluated in an appropriate model of the transport system, meaning that it has to be able to represent the system dynamics and responses to the proposed policy measures. At the same time, it should be fast enough to test many policy combinations.

Although an appropriate model is certainly most important, another fundamental part is the definition of the possible policies (*15*). Policy makers have the following possibilities to influence the transport system:

[1]IVT, ETH Zurich, Zurich, Switzerland
[2]LIFE, Joanneum Research, Graz, Austria

Corresponding Author:
Address correspondence to Patrick M. Bösch:
patrick.boesch@ivt.baug.ethz.ch

through direct management they can optimize the usage of existing infrastructure (e.g., traffic management), or, assuming sufficient financial means, they can provide new or extend existing infrastructures or public transport services, or both. Using taxes (including mobility pricing) and subsidies, they can change the costs of certain modes versus others and of the transport system overall, and using legislation, they can regulate the organization and usage of the transport system (e.g., speed limits, priority lanes, etc.). Finally, yet importantly, using advertising campaigns, they can (try to) influence general attitudes toward different modes.

The third element is the definition of the target function. The goal of any transport system should be to move people and goods rapidly, cheaply, easily, safely, and sustainably. An optimal transport system maximizes all these targets at the same time. However, as long as safe and cheap beaming is not possible, trade-offs need to be assessed and priorities need to be set. Though endless variations in target weighting exist and any choice can be debated, the list of goals per se is manageable and indicators can be found (*18*).

In summary, as literature shows (*15–17*), combining the available possibilities to influence the system with a comprehensive target function and an appropriate transport model effectively allows identification of high-performing transport policy strategies.

Although the above is an established methodology, attempts to apply it to investigate how a socially beneficial AV-based transport system could and should look have so far been very limited. To date, literature has been mostly restricted to either describing the system qualitatively or focusing on detailed, mostly operational questions.

This study is an attempt to fill this gap. Starting from the current system and considering financial and legislative constraints, this paper focuses on policies to influence the price of existing transport services and the organization of future AV-based transport services. Given the importance of the value of travel time in AVs, but also its uncertainty, these policies are tested against three assumed perceived travel times in AVs.

This set is evaluated against two performance indicators (instead of a single target function). These are the total travel time (TT) of all trips in the area, representing the "output" of the transport system, and total vehicle kilometers traveled (VKT) representing the costs required and externalities produced.

The methodology is applied to the region of Zug, Switzerland. The region is modeled with an agent-based transport model (MATSim, [*19*]), which, given its ability to represent single individuals (agents), is particularly suitable to investigate the impact of policy measures. In fact, the impact at the systemic level is the consequence of individual reactions to the policies.

In this paper, the next section describes the methodology, the chosen policy measures, and the objectives in more detail. The Transport Model section introduces MATSim and describes the Zug model. Next, specific sections present and discuss the results. The Conclusion section presents policy recommendations based on the results and concludes the paper.

Methodology

The methodology followed was proposed by May et al. for the development of optimal integrated transport policy strategies (*15*). They applied it to different European cities to evaluate combinations of transport policies. Here, given that it is unknown when AVs will be available and to account for the many other uncertainties on the future transport system, a simplified version of the original methodology was used. The temporal aspects of the policy measure staging were excluded and the policies were less detailed.

The methodology consisted of three parts: first, development of possible policy measures including a respective range for each policy; second, definition of an appropriate objective function that evaluated how well different combinations work; and third, the application of these measures and the objective function in a model of the transport system.

With an analysis of the full policy ranges, not only could the preferred strategy be identified, but also and evaluation of the transport system's sensitivity to the different individual policy measures. The model used was a MATSim scenario of the Swiss area of Zug. An introduction to MATSim and a description of the scenario follows in the next section.

Policy Measures

The selection of policy measures depends not only on the system characteristics, but also on external restrictions. On top of the obvious ones, such as physical feasibility and financial restrictions, the required political support is also a major condition, if not the most important one. The policy measures proposed here were designed and selected with this in mind.

As mentioned earlier, the number of possible ways for policy makers to influence the transport system are limited. Investments in services or infrastructure, influencing price through taxes (including mobility pricing) or subsidies, legislative measures, direct traffic management, and advertisement campaigns are the main ones. The policies investigated were selected from this set. Most policy measures allow for a continuous or near-continuous range in their application. For simplicity however, only discrete levels were investigated here.

Policy Measures for Existing Modes. Existing modes include mass transit public transport (PT), the slow modes (SM) walk and bike, and motorized individual transport (MIT). For PT and MIT, the respective autonomous version was assumed (aPT and aMIT).

The two policy measures aPT pricing and aMIT pricing were selected. Other possible measures were not further investigated either for their political or financial feasibility (e.g., infrastructure projects), or for their impact being difficult to quantify (e.g., advertisement campaigns). A closer investigation of other possible measures will constitute part of future work.

These two policy measures were complemented by different assumptions on the possible comfort changes through automation.

- *Pricing of aPT* represents any policy measures increasing or decreasing the user price of aPT. The main policy lever is the level of subsidies. The automation of aPT (buses) was estimated to halve its production cost (*2*). As, today, subsidies cover 50% of the cost of Swiss PT (*20*), the following three levels of aPT subsidies were investigated: no subsidies, which results in the same price for aPT as for PT today (0.27 Swiss Francs [CHF]/km [*2*]); the same relative level of subsidies (50%) as today, which results in half the price for aPT as for PT today (CHF 0.13/km); and the same absolute level of subsidies as today, which results in a free at the point of use aPT.
- *Pricing of aMIT* aims to increase or decrease the average cost per distance for aMIT. The main policy instruments to achieve this are taxes (e.g., on fuel or vehicles) or mobility pricing (for areas or road categories). Bösch et al. (*2*) found the cost of aMIT to be similar to today's MIT costs. Therefore, two possibilities were assumed here: first, a similar level of taxes or mobility pricing, or both as today, which results in the same marginal cost of aMIT as MIT today (CHF 0.18/km [*2*]); and second, new taxes or mobility pricing for aMIT in the range of 25% of today's cost of MIT, resulting in CHF 0.22/km.
- *Comfort changes of aMIT* is not actually a policy measure, but represents the expected benefit of autonomous driving technology to transform driving into productive time. It thus reduces the negative value of travel time (VOT) in aMIT. Three levels were investigated here: the same VOT as today, that is as if driving (CHF 23.29/h [*21*]); the same VOT for aMIT as for PT (CHF 14.43/h [*21*]); and, given that other passengers represent for most people a negative factor of traveling with PT (*22*), a 25% lower negative VOT for the individual aMIT as for PT, resulting in CHF 10.82/h.

Organizational Form of Future Modes. Future modes represented here are all based on autonomous taxis, which can be operated as a traditional taxi service (aTaxi) with exclusive single user service or as a ride-sharing service (aRS) which can carry multiple passengers at the same time. They can be operated by a public agency or by a private company, which can provide different comfort and price levels. Other models and forms of future modes, such as for example autonomous mini-buses, point-to-point shuttles, and so forth, were omitted here.

The future form of organization of such services is an important question policy makers should start to think about. If they wait too long before taking action, the market will organize itself. This might result in a suboptimal system from a societal point of view.

To represent these different forms of organization, the following services were proposed as "policy measures": a monopoly aTaxi service, a monopoly aRS service, and an oligopoly in which six different services compete (3 aTaxi and 3 aRS).

Following the above assumptions, negative VOT was assumed to be the same for aRS as for PT, whereas for aTaxis it was assumed to be 25% less negative (more comfortable). The monetary prices per passenger kilometer (PPKM) for the services follow Bösch et al. (*2*). The fleet sizes for the services were estimated based on Bösch et al. (*10*). They found that for a good level of service, one aTaxi could replace four private cars. Here, the monopolist's fleet was expected to serve 25% of the population with such a level of service. Therefore, 25% of 25% of the current car fleet of Zug (96,000 [*23*]) results in 6,000 aTaxis. For aRS, a 33.3% smaller fleet was assumed (4,000 vehicles).

In the competitive situation, each of the services was assumed to have a fleet one-third of the respective monopolist (rounded up to the next 500 vehicles). This resulted in 6,000 aTaxis and 4,500 aRS AVs in the area, which increased the total fleet by 75%, respectively 162.5% compared with the monopolistic cases. This is realistic, as each service requires a substantial fleet to offer a good service in the area.

This resulted in the following four cases:

1. A monopoly service offering 6,000 aTaxis for individual transport (VOT: CHF 10.82/h, PPKM: CHF 0.46/km).
2. A monopoly service offering 4,000 AVs for ride-sharing (VOT: CHF 14.43/h, PPKM: CHF 0.30/km).
3. An oligopoly of services, represented here by six services, three different experiences (VOT as

Table 1. Overview of the Policy Measures Investigated in this Paper

Policy measure	Level 0	Level 1	Level 2	Level 3
Pricing of aPT	No change	–50%	–100%	na
Pricing of aMIT	No change	+ 25%	na	na
Comfort changes of aMIT	No change	as PT	as PT - 25%	na
Future modes	None	Monopoly aTaxi	Monopoly aRS	Oligopoly (6 services: 3 comfort–price levels, aTaxi/aRS)

above, –25%, and + 25%) and matching prices (price as above, + 25%, and −25%) and each with aRS (1,500 vehicles per service) or as aTaxis (2,000 vehicles per service).

4. No AV-based service (base case).

This resulted in 3 × 2 × 3 × 4 = 72 different transport scenarios. The policy measures and their ranges, as well as the assumed levels of comfort of aMIT are summarized in Table 1.

Performance Indicators

Instead of a single objective function, two performance indicators were used. The first, TT, represents the performance of the system in providing access to activities. The second, total VKT, represents monetary costs and externalities of the transport system (*18*). Although a more detailed analysis is required for the assessment of individual solutions, the reduction to these two indicators allows comparison of the solutions without politically influenced weights (as would be the case for single target function values).

Total VKT is often used in transport studies as the direct and single indicator to calculate various costs of the system. Examples range from pure monetary costs (fuel and vehicle costs per VKT), health costs (accidents per VKT), negative externalities (noise per VKT), to ecologic costs (e.g., CO_2 per VKT) (*18*). Here, for each case, the total VKT were provided directly as a proxy for the overall cost of the transport system. They were calculated as the sum of VKT of all modes.

The total TT of all trips in the analysis area represents the transport system's performance in providing access to activities. It serves as an illustrative and direct indicator of the performance as the number of trips and their origins and destinations remain constant across all scenarios.

For the monopolist scenarios, profitability represents a third factor. This was calculated by multiplying the total passenger kilometers of the AV service with the passenger price and comparing this to the cost per VKT multiplied by the total VKT of the service. Profitability indicates a requirement for subsidies under the proposed pricing scheme.

Transport Model

MATSim

The transport model used here was a MATSim model. MATSim, an agent-based transport model (*19*), was chosen for its suitability for the evaluation of transport policies targeting individuals and their traveling decisions, and because of its computational performance.

MATSim uses a co-evolutionary, iterative optimization process to identify the user equilibrium of a transport system. A population of agents with daily plans, listing activities to be executed, and routes and modes to get from one activity location to the next, represents transport demand.

Each iteration, a random sample of agents can modify their plans (change modes, routes, or departure times); the transport simulation then simulates a full day with all agents executing their daily plan. A queue model is used to simulate traffic (*24*). After the simulation, all agents score their plan with a scoring function. It rewards activity time and punishes TT and cost. During the iterative process, plans with good scores are kept, whereas plans with bad scores are discarded.

The MATSim functionality particularly important for this study was the simulation of AVs (*14*, *25*). In the basic configuration used here, it simulated AV-based taxi services organized by a central dispatcher. Agents, which would like to use a taxi, place a request at the central dispatcher, which looks for the closest free taxi and assigns it to the agent. The taxi serves the agent and waits at the agent's destination for the next assignment. Relocation was not included. The initial placing of the taxis was based on population density.

MATSim also allows for the simulation of ride-sharing (*25*). With ride-sharing, as long as the first passenger's additional TT remains less than a given threshold (here 800 s), the dispatcher can pool another passenger.

Region of Zug

The city of Zug is a mid-size town located about halfway between Zurich and Lucerne. It has 29,000 inhabitants and is the capital of the canton of Zug with 120,000

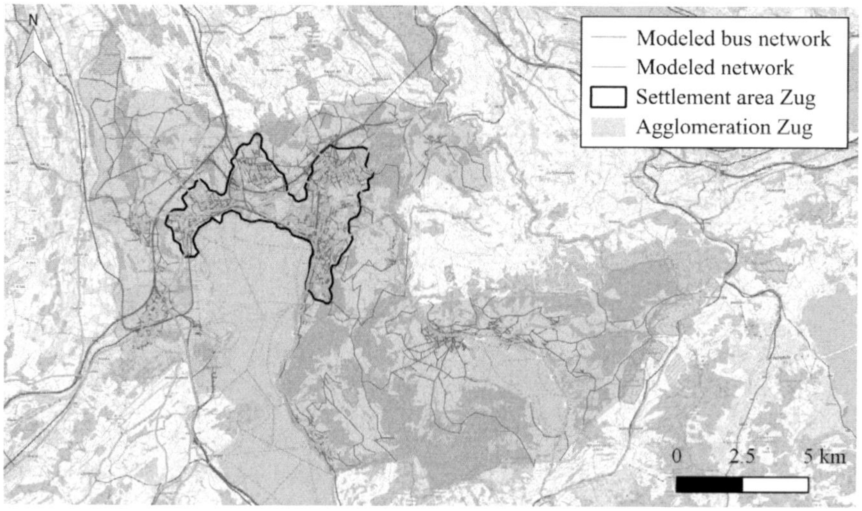

Figure 1. Zug area.

inhabitants (*23*). The canton approximately represents the agglomeration of Zug (*26*).

From a simulation point of view, Zug is very suitable for this study as it is a large enough town to have its own agglomeration and its own PT system (*27*) densely covering the main settlement area (Figure 1), but is also small enough to allow for quick computation times even if the full population is represented by agents. This representation was required to get realistic results on the usage of PT and taxi services (*28*).

Furthermore, Zug has been attracting increasing attention for transport experiments in Switzerland. It was selected for an AV shuttle experiment by the Federal Swiss Railways (*29*). Starting summer 2018, an AV shuttle will connect Zug's main train station with a nearby research campus. Recently, the canton of Zug was also selected by the Swiss Federal Government for a study on the potential and the possible effects of mobility pricing (*30*). A study on future transport systems and suitable transport policies for Zug is therefore a good fit with these events.

MATSim Model of Zug

The MATSim model of Zug used for this study was cut from a recently developed 2015 MATSim model for Switzerland (see Bösch et al. [*31*] for a detailed description). The model covers all agents having their home within the area, that is the agglomeration of Zug (Figure 1), and all agents of the full Switzerland scenario, which have an activity in the area or pass through the area. Within the area, the modeled infrastructure (street network, PT, facilities) was fully detailed as described in Bösch et al. (*31*). Outside of the area, the street network was modeled on the level of arterial roads (capacity min. 1,000 vph) and PT was reduced to rail lines only. AV services were restricted to trips within the area and empty rides with private AVs were excluded.

The scenario represents the full population, which means that every agent in the simulation represents one real person (assuming no error in the available statistics). In 2015, the population of canton Zug consisted of 117,695 persons (*23*). To avoid border effects however, reporting of transportation system performance was focused only on the densely populated main settlement area of Zug and its inhabitants, outlined in Figure 1 (*32*).

In the baseline scenario, the trip-based modalsplit of these agents was 14% PT, 42% MIT, and 44% SM. Compared with the official statistics (14% PT, 37% MIT, 48% SM, [*33*]), there was a slightly higher use of MIV at the expense of SM. This was balanced however, by the average distance traveled per agent per day with 8 km/d PT, 21 km/d MIT, and 3.1 km/d SM compared with the official 7 km/d PT, 26.6 km/d MIT and 3.5 km/d SM (*34*). This leads to the conclusion that the model fits well and that the deviations are likely due to the different sampling processes. This is further supported by the reasonably fitting average speeds (model: 17.3 km/h PT, 48.7 km/h MIV, and 2.6 km/h SM, compared with the official 20.2 km/h PT, 46.9 km/h MIV, 5.9 km/h SM, [*34*]).

Results

In total 72 scenarios were simulated. Each simulation was run for 250 iterations. This is a low number of iterations for MATSim runs but was considered sufficient for an indication of the system development. Nevertheless,

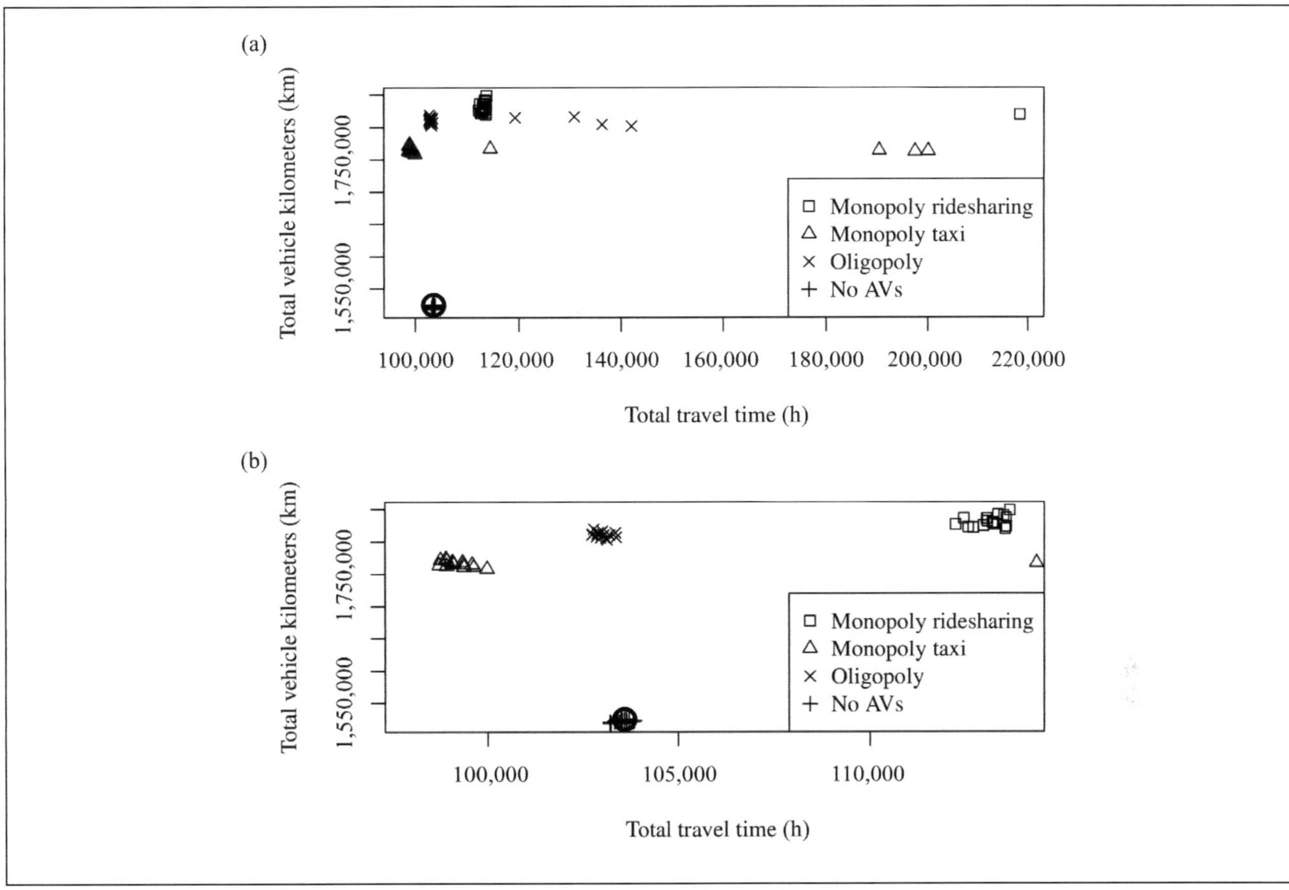

Figure 2. TT–VKT trade-off by market configuration (reference scenario, i.e., no AV service and no policy implemented, highlighted with a circle): (*a*) complete set of configurations, and (*b*) focus on main configuration clusters.

results indicate that AV-based services might gain more mode share until user equilibrium. The following results should therefore be seen as conservative indications on the user potential of AV-based services.

Organizational Forms of Future Modes

This subsection compares all scenarios. It focuses on the performance of the different organizational forms of the AV-based services. For each scenario, Figure 2 presents the total TT of all agents versus the total VKT in the area, differentiating the scenarios based on the organizational form of the AV-based service.

Figure 2, especially Figure 2*b*, shows that monopoly aTaxi services have the potential to reduce TT. However, this improvement comes at a great cost: a 4.1% reduction in TT for a 16.0% increase in VKT (averaged across all combinations without outliers). The other policy configurations showed a similar increase in VKT, but they were not able to reduce the overall TT. With aTaxis, the expected pick-up and drive time were 2.6 min and 2.4 min, whereas for aRS these were 12.4 min and 10.9 min, despite similar served average Manhattan trip distances (aTaxi: 1.28 km, aRS: 1.32 km). The reason is that, in its current implementation, aRS always gives priority to bundling of trips vs. single trips, even if this means substantial detours. Oligopolies, consisting of both aTaxi and aRS services, lie in between the other two clusters.

The tight clustering of configurations shows that the chosen organizational model is the most important determinant for future performance of the system. Other policy measures (price of aMIT and aPT, VOT of aMIT) appear to have only a minor effect within the cluster.

Several AV scenarios produced substantially more TT than the rest of their service's cluster. These were considered outliers. Analysis showed that major traffic congestion was the reason for these outliers. No clear common reason could be identified however, which indicates that, independent of the chosen policy measures, the additional VKT are bringing the system generally closer to breakdown (major congestions) with occasional tipping.

Effect of Policy Measures on Total Travel Time

To analyze the effect of the different policy measures, Table 2 presents a multiple linear regression analysis of

Table 2. Multiple Linear Regression Analysis of Total Travel Time with Policy Measures as Variables (Outliers Excluded from Analysis)

Coefficients	Estimate	SE	*t* value
(Intercept)	103665.31	373.51	277.54***
Price of aPT	85.11	97.24	0.87
Price of aMIT	−247.46	322.63	−0.76
VOT of MIT like PT	109.63	102.65	1.06
VOT of MIT better than PT	64.68	98.64	0.65
AV organization – RS monopoly	9672.90	107.73	89.79***
AV organization – taxi monopoly	−4276.57	114.02	−37.50***
AV organziation – oligopoly	−492.51	113.87	−4.32***

Note: (a) PT = (automated) public transport; (a) MIT = (automated) motorized individual transport; VOT = value of time; AV = autonomous vehicle; RS = ride-sharing service.
Signif. codes: ****p* = .001.

all scenarios except the outliers. System performance, that is, total TT, was the response variable of the model. The model shows that, as expected from Figure 2, the influence of the organizational models was highly significant, whereas the other measures had no statistically significant effect. This is surprising as the interventions were substantial. It seems that, at least within the degrees of freedom of the model (e.g., no activity location adaptation), neither a substantial price reduction of aPT (−50% and −100%) nor a substantial increase in the price of aMIT (+ 25%) could significantly reduce the total TT.

Concerning the observed relations, the model represents expected associations. A reduction of the price of aPT seems to lead to lower travel times (positive estimate), whereas a reduction in the price of aMIT increases total TT (negative estimate). Furthermore, an increase in attractiveness of aMIT (i.e., decrease in VOT) leads to the expected decrease in total TT (positive estimate).

Monopolist aTaxi Strategies

Under the chosen pricing scheme and the provided fleet sizes, none of the aTaxi monopolists could operate profitably. Excluding the outliers from the analysis, the average passenger trip with an aTaxi was 1.84-km long. To serve this trip, the average pick-up distance (driving distance from the closest free aTaxi to the passenger) was 2.13 km, leading to a productive to total VKT ratio (= 1.84 km /(1.84 km + 2.13 km)) of 46%.

Here, for a more detailed picture, the neutral aTaxi monopolist scenario is compared with the reference scenario. The neutral scenario was chosen to avoid random influences of other policy measures (see above). By multiplying the TT of each mode with its VOT, the TT improvement of the aTaxi scenario vs. the reference scenario can be monetarized. This approach results in perceived travel cost savings of CHF 185,000 per day. These savings could be obtained by accepting an increase of 15.7%, that is, 240,000 km, in VKT per day, which cause additional costs (before externalities) of CHF 103,000 per day (costs per VKT based on Bösch et al. [*2*]). The modalsplit of the aTaxi scenario was 32.9% aTaxi, 35.5% car, 11.7% PT, and 19.9% SM. Compared with the reference scenario (47.9% car, 16.0% PT, and 36.0% SM), this means aTaxi gains substantially from all three traditional modes, but most from SM (16.1 percentage points).

Discussion

The results presented in this paper are surprising in the sense that the literature (e.g., [*9*, *11*]) usually assumes – more or less explicitly – that shared AV fleets improve the transportation system. Here, comparing different possible systems with shared AVs and, especially, comparing them with the case without such services, reveals that such an improvement depends heavily on the chosen implementation. Furthermore, although positive effects of shared AV fleets can be expected, for example, 12.4 percentage points of mode share switching from private cars to aTaxis, which would result in a substantial fleet size reduction (*10*), they might also come at great cost: in additional VKT, and an increased risk of congestion, as suggested by the presence of outliers. The additional VKT originate from empty rides, but also from substantial mode share changes from the VKT-neutral modes aPT and SM to aTaxis (16.1 percentage points from SM and 4.3 from aPT).

The example of Uber in Manhattan indicates that this is indeed a valid observation (*35*). There, additional empty miles by Uber taxis combined with the attraction of former pedestrians or PT users, led to an increase of VKT in a system already operating at its limits and thus a worsening of the overall situation.

Another surprise from the results is that policy measures (the pricing of aMIT and aPT and the reduction of VOT of aMIT), although showing the expected relations, seem to have had no statistically significant effect on the total TT – or at least a negligible one compared with the chosen implementation of AVs. This was certainly unexpected and is the subject of forthcoming research (e.g., [*36*]).

The same applies to the observation that aRS services reached substantial mode shares (on average 13%) despite being very inefficient (8.4 km of average driving distance for trips with an average Manhattan origin–destination distance of only 1.3 km). Although this indicates the remarkable attractiveness of these services, it also shows the importance of careful system and pricing scheme designs for the operators of such systems. The latter further emphasized by the non-profitability of the aTaxi services.

This leads to a discussion of the transport model. Even though the transport model scenario reproduced the existing transport situation well, parameter choices and the design of the offered AV-based services indicate future work. The fleet sizes were estimated based on an educated guess. As required fleet sizes heavily depend on local context (e.g., car ownership rates, spatial structure, PT infrastructure), a more in depth analysis of the required fleet size will be part of future work. The same applies for the chosen level of prices. Although based on a detailed estimation (*2*), the assumptions leading to the estimated prices did not fully apply to the scenario at hand and were, together with the suboptimal fleet sizes, a suspected reason for the non-profitability of the aTaxi services. The simulation of the AV fleets themselves has also been – up until now –rather simple. Relocation to minimize pick-up distances and more complex assignment algorithms are in development and might be applied to future studies.

These limitations should be considered when interpreting the results. Nevertheless, the results represent valuable indications that the introduction of services based on shared AVs might come with considerable additional VKT and not always with an improvement in total TT. Having systematically reached this conclusion certainly adds to the present discussion on possible AV-based services.

Conclusion

In this paper, different policy measures for future transport systems were investigated. They included different levels of subsidies for aPT, of pricing of aMIT, and different organizational frameworks (monopoly vs. oligopoly) for AV-based services (Table 1). These were complemented with different assumptions on the future VOT of aMIT, which means how comfortable private AVs will be.

Following May et al. (*15*), different possible combinations of these policies and assumptions were simulated in a scenario of the agglomeration of Zug, Switzerland, using MATSim (*19*). Zug is small enough to allow simulation of a substantial number of scenarios, but large enough to produce relevant outcomes.

The results of these simulations showed that scenarios including AV services, independent of their organizational form, increased total VKT substantially (Figure 2). More VKT means more externalities, more intense infrastructure usage, and a less SM-friendly environment. Reasons for this increase, apart from empty pick-up rides, are substantial mode shifts from aPT and SM to AVs. If these additional VKT (costing CHF 103,000 per day before externalities) and mode shifts are accepted however, AVs have the potential to reduce total TT. Even without sophisticated pricing, vehicle relocation, and so forth, here, CHF 185,000 per day in reduced TT costs could be achieved with aTaxis.

Focusing on the accompanying measures, the results did reveal the expected effects: a price reduction in aPT and a price increase in aMIT, each leading to a decrease in total TT, whereas a reduced travel time burden in aMIT led to an increase in TT. These effects however, were only observed in a statistically non-significant range (Table 2).

In terms of policy recommendations, these results suggest a need to be careful with new AV-based services. Policy makers would be well advised to be critical about the promises of such new services and to evaluate in detail how they might fit into their specific transport system. Although having the potential to reduce total TT and costs (as also shown, for example, by Merlin [*37*] for Ann Arbor), they would likely increase traffic by adding, for example, pick-up rides. This increases the risk of traffic jams and additional waiting times for customers and other traffic participants. A real-life example for this is what happened with Uber in Manhattan (*35*). In this sense, the results also suggest that policy makers and society need to prepare for an "it will get worse before it gets better" scenario.

Small-scale experiments with AVs and the development of new services are to be encouraged as long as they do not cause too much additional traffic and do not disturb the existing system. When the day for large-scale introduction comes, policy makers should be aware, however, that such a system is likely to represent a very attractive competitor for existing PT and SM.

Until this day comes, however, it is suggested that policy makers use automation to improve the existing system (such as the automation of public mass transit, complementing it in low-demand areas with AV-based ride-sharing services, or investing road capacity gained by more efficient use of road space in more generous infrastructure for SM) – a suggestion also supported by other recent studies (*38*, *39*).

To conclude, it might not be as clear as is often suggested that services based on shared AVs will actually improve the overall performance of the transport system. The existing system has grown and evolved during the past century. It is about 100 years since affordable private cars came on the market and, thus, the last major "game changer" in transport was introduced. Since then the system has had a central role in the economic growth of several countries, especially in the midst of the last century, and the emergence of obvious limitations and issues has been addressed with efforts trying to find a balance among externalities, affordability, and accessibility. This should make us aware that replacing such a transportation system from one day to the next might not be as straightforward as it seems neither an obvious gain.

Acknowledgments

The authors thank the four anonymous reviewers for their valuable inputs, which helped to improve the paper. This research was funded by the Swiss National Science Foundation (project no. 200021_159234, Autonomous Cars).

Author Contributions

The authors confirm contribution to the paper as follows: study conception and design: PB, KA; data collection: PB; analysis and interpretation of results: PB, FC; draft manuscript preparation: PB, FC, KA. All authors reviewed the results and approved the final version of the manuscript.

References

1. SAE International. *Taxonomy and Definitions for Terms Related to On-Road Motor Ve3 hicle Automated Driving Systems*. Warrendale, PA, 2014. http://standards.sae.org/j3016_201401/.
2. Bösch, P., F. Becker, H. Becker, and K. Axhausen. *Cost-based Analysis of Autonomous Mobility Services*. Working Paper, 1225. Institute for Transport Planning and Systems (IVT), ETH Zurich, Zurich, 2017.
3. Fagnant, D.J., and K.M. Kockelman. Preparing a Nation for Autonomous Vehicles: Opportunities, Barriers and Policy Recommendations. Transportation Research Part A: Policy and Practice, Vol. 77, 2015, pp. 167–181.
4. Lutin, J. M., A. L. Kornhauser, and E. Lerner-Lam. The Revolutionary Development of Self-Driving Vehicles and Implications for the Transportation Engineering Profession. *ITE Journal*, Vol. 83, No. 7, 2013, pp. 28–32.
5. Burns, L. D. Sustainable Mobility: A Vision of Our Transport Future. *Nature*, Vol. 497, 2013, pp. 181–182.
6. Burns, L. D., W. Jordan, and B. Scarborough. *Transforming Personal Mobility*. Technical Report, The Earth Institute, Columbia University, 2013.
7. Fagnant, D. J., and K. M. Kockelman. The Travel and Environmental Implications of Shared Autonomous Vehicles, Using Agent-based Model Scenarios. *Transportation Research Part C: Emerging Technologies*, Vol. 40, 2014, pp. 1–13.
8. Spieser, K., K. Treleaven, R. Zhang, E. Frazzoli, D. Morton, and M. Pavone. Toward a Systematic Approach to the Design and Evaluation of Automated Mobility-on-Demand Systems: A Case Study in Singapore. In *Road Vehicle Automation* (Meyer, G., and S. Beiker, eds.), Springer, 2014, pp. 229–245.
9. International Transport Forum. *Urban Mobility System Upgrade: How Shared Self-driving Cars Could Change City Traffic*. Technical Report. OECD, Paris, 2015.
10. Bösch, P. M., F. Ciari, and K. W. Axhausen. (2016) Required Autonomous Vehicle Fleet Sizes to Serve Different Levels of Demand. *Transportation Research Record: Journal of the Transportation Research Board*, 2016. 2542: 111–119.
11. Zachariah, J., J. Gao, A. Kornhauser, and T. Mufti. Uncongested Mobility for All: A Proposal for an Area Wide Autonomous Taxi System in New Jersey. Presented at 93rd Annual Meeting of the Transportation Research Board, Washington, D.C., 2014.
12. Chen, D. T., K. M. Kockelman, and J. P. Hanna (2015) Implications of a Shared, Autonomous, Electric Vehicle (SAEV) Fleet. Presented at 14th International Conference on Travel Behavior Research, Windsor, 2015.
13. Zhang, R., K. Spieser, E. Frazzoli, and M. Pavone. Models, Algorithms, and Evaluation for Autonomous Mobility-On-Demand Systems. Presented at the American Control Conference (ACC), Chicago, 2015.
14. Bischoff, J., and M. Maciejewski. Autonomous Taxicabs in Berlin – A Spatiotemporal Analysis of Service Performance. Presented at International Scientific Conference on Mobility and Transport Transforming Urban Mobility, mobil.TUM, Munich, 2016.
15. May, A. D., S. P. Shepherd, and P. M. Timms. Optimal Transport Strategies for European Cities. *Transportation*, Vol. 27, No. 3, 2000, pp. 285–315.
16. May, A. D., S. P. Shepherd, G. Emberger, A. Ash, X. Zhang, and N. Paulley. Optimal Land Use - Transport Strategies - Methodology and Application to European Cities, *Transportation Research Record: Journal of the Transportation Research Board*, 2005. 1924: 129–138.
17. Zhang, X., N. Paulley, M. Hudson, and G. Rhys-Tyler. A Method for the Design of Optimal Transport Strategies. *Transport Policy*, Vol. 13, No. 4, 2006, pp. 329–338.
18. Kockelman, K., S. Boyles, P. Avery, C. Claudel, L. Loftus-Otway, D. Fagnant, P. Bansal, M. W. Levin, Y. Zhao, J. Liu, L. Clements, W. Wagner, D. Stewart, G. Sharon, M. Albert, P. Stone, J. Hanna, R. Patel, F. Hagen, T. Choudhary, T. Li, A. Nichols, K. Sharma, and M. Simoni. *Bringing Smart Transport to Texans: Ensuring the Benefits of a Connected and Autonomous Transport System in Texas*. Final Report - Report No. 0-6838-2. Center for Transportation Research, The University of Texas at Austin, Austin, Texas, 2016.
19. Horni, A., K. Nagel, and K. W. Axhausen (eds.) *The Multi-Agent Transport Simulation MATSim*. Ubiquity, London, 2016.
20. Laesser, C., and S. Reinhold. *Finanzierung des OeV in der Schweiz: Was zahlt der Nutzer, was die Allgemeinheit?* Report. Schriftenreihe SBB Lab, St. Gallen, 2013.

21. VSS. *Kosten-Nutzen-Analysen (KNA) bei Massnahmen im Strassenverkehr: Zeitkosten im Personenverkehr*. Norm, SN 641 822a. Swiss Association of Road and Transport Professionals (VSS), Zurich, 2009.
22. Schmid, B., S. Schmutz, and K. W. Axhausen. Explaining Mode Choice, Taste Heterogeneity and Cost Sensitivity in a Post-car World. Presented at 95th Annual Meeting of the Transportation Research Board, Washington, D.C., 2016.
23. Zuger Kantonalbank. (2016) *Der Kanton Zug in Zahlen - Ausgabe 2016*. Report. Zuger Kantonalbank and Fachstelle für Statistik des Kantons Zug, Zug, 2016.
24. Rieser, M., K. Nagel, and A. Horni. Qsim. In *The Multi-Agent Transport Simulation MATSim* (Horni, A., K. Nagel, and K. W. Axhausen, eds.), Ubiquity Press, London, 2016, pp. 77–80.
25. Hörl, S. Agent-based Simulation of Autonomous Taxi Services with Dynamic Demand Responses. Presented at 6th International Workshop on Agent-based Mobility, Traffic and Transportation Models, Methodologies and Applications (ABMTrans 2017), Madeira, 2017.
26. Schuler, M., P. Dessemontet, and D. Joye. *Raumgliederung der Schweiz*. Technical Report. Swiss Federal Statistical Office (BFS), Neuenburg, 2005.
27. Zugerland Verkehrsbetriebe. *Startseite*. Webpage, http://www.zvb.ch/en/home/. Accessed August 1, 2017.
28. Bösch, P. M., and F. Ciari. MacroSim - A Macroscopic Mobsim for MATSim. *Procedia Computer Science*, Vol. 109, 2017, pp. 861–868.
29. Tagesanzeiger. SBB testen selbstfahrende Busse. *Online Article*, March 2017, http://www.tagesanzeiger.ch/wirtschaft/unternehmen-und-konjunktur/SBB-testen-selbstfahrende-Busse/story/29231859. Accessed August 1, 2017.
30. Bundesrat. Bundesrat vertieft Mobility Pricing mit kantonaler Wirkungsanalyse. *Online Press Release*, July 2017, https://www.admin.ch/gov/de/start/dokumentation/medienmitteilungen.msg-id-67431.html. Accessed July 14, 2017.
31. Bösch, P., K. Müller, and F. Ciari. The IVT 2015 Baseline Scenario. Presented at 16th Swiss Transport Research Conference, Ascona, 2016.
32. Menzi, R. *Die Landschaften der Schweiz - Landschaftstypologie Schweiz, Manual*. Bundesamt für Raumentwicklung (ARE), Bundesamt für Umwelt (BAFU), Bundesamt für Statsitik (BFS), Bern, 2000.
33. Perret, C., J.-L. Muralti, C. Siegenthaler, C. Seewer, M. Kowald, A. Corpataux, and A. Danalet. *Verkehrsverhalten der Bevölkerung - Ergebnisse des Mikrozensus Mobilität und Verkehr 2015*. Technical Report. Bundesamt für Statistik (BFS) and Bundesamt für Raumentwicklung (ARE), Neuchatel and Bern, 2017.
34. Bundesamt für Statistik (BFS), Bundesamt für Raumentwicklung (ARE). Verkehrsverhalten der Bevölkerung, Kenngrössen 2015 - Kanton Zug. *Online Data*, 2017, https://www.zg.ch/behoerden/baudirektion/statistikfachstelle/themen/11-mobilitaet-und-verkehr/verkehr-1/downloads/Kenngroessen-MZ-2015-A2_ZG.xls/download. Accessed August 1, 2017.
35. Miller, S. Uber's Own Numbers Show it's Making Traffic Worse. *Online Article*, 2015, http://www.streetsblog.org/2015/07/22/ubers-own-data-reveals-it-slows-manhattan-traffic-9-percent/. Accessed April 20, 2016.
36. Bösch, P. M. *Autonomous Cars – The Next Revolution in Mobility (Working Title)*. PhD thesis. ETH Zurich, Zurich, (forthcoming).
37. Merlin, L. Comparing Automated Shared Taxis and Conventional Bus Transit for a Small City. *Journal of Public Transportation*, Vol. 20, No. 2, 2017, pp. 19–39.
38. Meyer, J., H. Becker, P. M. Bösch, and K. W. Axhausen. Autonomous Vehicles: The Next Jump in Accessibilities? *Research in Transportation Economics*, Vol. 62, 2017, pp. 80–91.
39. Winkler, T. *Selbstfahrende Fahrzeuge auf der ersten und letzten Meile des ÖVs*. Master thesis. ETH Zurich, Zurich, 2017.

The Standing Committee on Automated Transit Systems (AP040) peer-reviewed this paper (18-06253).

Article

TRR
JOURNAL OF THE TRANSPORTATION RESEARCH BOARD

Transportation Research Record
2018, Vol. 2672(8) 708–718

DOI: 10.1177/0361198118756883
journals.sagepub.com/home/trr
SAGE

Who is Interested in Carpooling and Why: The Importance of Individual Characteristics, Role Preferences and Carpool Markets

Yujin Park[1], **Na Chen**[2], **and Gulsah Akar**[1]

Carpooling is an informal form of commuter ride-sharing that can mitigate traffic congestion and parking demand. While ride-sharing and carpool formation require both a driver and at least one passenger, there has been a lack of studies that explore potential users of carpooling and their role preferences (driver versus passenger); perceived incentives, attitudes and current mode choices may affect their willingness to carpool and their role preferences. This study examined the factors affecting individuals' carpooling decisions and carpool role preferences using data from Ohio State University's 2012 Campus Travel Pattern Survey. Based on individuals' stated interests and choices, three market segments were identified: current carpool users, potential carpool users, and not-interested commuters. The analyses revealed that incentives and deterrents to carpooling significantly differ by role preference. People favoring the passenger role emphasized safety, flexibility and parking cost-savings, whereas those favoring the driver role tended to find value in the convenience and opportunities for socializing through a carpool trip. People receptive to both roles emphasized flexibility and all types of cost reductions. Although most people did not elect technological aids as a meaningful intervention, potential carpool users, particularly those favoring the passenger role, were more likely to seek these aids. Using probit and bivariate probit regressions, carpool interest and choice were found to be associated with commute distance, university affiliation, marital status, current travel mode, and personal attitude. Concern for the environment, and the propensity to make stops while commuting were positively associated with both carpooling interest and decision to carpool.

Since the 1970s, planners and policy makers have adopted transportation demand management (TDM) strategies to deal with transport-related urban problems (e.g., traffic congestion, energy consumption, environmental impacts, social segregation, etc.) (*1–4*). Among the TDM strategies, promoting carpooling is an approach emphasizing travel behavior modification using existing transportation facilities and resources. Since its peak usage in the 1970s due to the energy crisis in terms of fuel shortages and rising fuel costs, carpooling experienced a continuous decline in mode share during the following three decades (*5,6*). Compared with other TDM strategies such as promoting public transit, walking, and cycling, carpooling provides an alternative traveling option without asking people to give up their automobiles and without requiring as significant an infrastructure investment (*5*). Empty seats in every commuter's automobile constitute a great resource for policy makers to explore. This therefore calls for a better understanding of carpooling behavior and the markets to help policy makers and practitioners promote this underutilized mode of transport effectively.

Many existing carpooling studies focus on the decision to carpool given a set of transportation alternatives available to a trip maker (*3,5–8*). While some studies have examined the behavior of potential carpool users (*9*), there has been a lack of consideration of the preference for a carpool role: driver versus passenger. Limited knowledge is available regarding what types of commuters consider carpooling to be a viable option and which role they might prefer. Neoh et al. suggest that enlarging the pool of potential carpool users is essential for the success of matching like-minded carpool users (*9*). In their study, however, potential carpoolers generally referred to workers in organizations that were spatially close to one another, regardless of their interest in carpooling (*9*). A few

[1]City and Regional Planning, Ohio State University, Columbus, OH
[2]School of Planning, College of Design, Architecture, Art, and Planning, University of Cincinnati, Cincinnati, OH

Corresponding Author:
Address correspondence to Yujin Park: park.2329@osu.edu

studies have provided insights into carpooling interest and role preferences. Buliung et al. (*10*) and Shaheen et al. (*11*) revealed that those who prefer a single role (driver or passenger) are less likely to engage in carpooling in the long run compared with individuals who share the driving responsibility. Role preferences of current and potential carpool users are expected to influence carpool choices in general. Role preferences could also be indicative of individuals' various motivations for and deterrents to carpooling. Therefore, examining carpooling interest and role preferences could assist policy makers and planners in understanding different carpooling behaviors, preferences and motivations, thus facilitating the development of targeted programs and policies.

This study aims to enhance the understanding of varying carpooling interest and choices, taking the driver versus passenger choice into account. Our study analyzes: (i) carpooling interest as a driver, a passenger, or both, and (ii) its relationship with actual carpool choices of diverse socio-demographic groups, using data from Ohio State University's (OSU) 2012 Campus Travel Pattern Survey.

University campuses can be ideal study areas to examine potential carpool use for several reasons (*1,8,12*). First, a university is often the largest employer in an area and therefore has a diverse pool for ride-sharing with the same destination (campus) (*1*). Second, a university can serve as a test site for trialing transportation policies that aim to increase the use of alternative travel modes (*8,12*). Research on modeling carpooling interest and choice in university campuses is limited. This study evaluated the main motivations for and deterrents to carpooling for people with different carpool interests and role preferences. Commuter characteristics, commute factors and personal attitudes associated with role preferences were assessed, as was whether carpool interest relates to actual increases in the decision to carpool among campus commuters.

Literature Review

In this paper the term carpooling is broadly defined as the casual grouping of travelers into common commute trips by a single vehicle (*2*). Determinants that affect carpooling choices examined by existing studies can be categorized in several ways (*5–9,13*). Neoh et al. undertook a meta-analysis to evaluate internal and external factors that encourage people to carpool (*9*). Internal factors include demographics (e.g., gender, age, etc.) and judgmental factors (e.g., perceptions and incentives). External factors refer to third-party interventions (e.g., extra parking charges for single occupancy vehicles (SOVs), and partner matching services) and situational factors (e.g., travel time, and distance). Vanoutrive et al. analyzed the determinants of carpooling to workplaces in Belgium, and found that having more employees with regular work schedules, lower accessibility by automobile and train, and a guaranteed ride home positively contributed to the success of carpooling (*13*). Wang and Chen examined three groups of variables (socio-demographics, attitudes, and the built environment) that affected modal shifts between SOVs and carpooling in the Puget Sound Region (*7*). The authors found that only commute time and personal attitudes led to modal shifts. Population and employment densities were not significant.

Some studies have evaluated pilot programs, such as carpool clubs, or have performed experimental modeling to investigate how policy interventions might improve these programs (*5,10,11,14–16*). Owing to the rapid development of a series of carpooling programs, researchers are increasingly interested in identifying who uses these programs and why. Correia and Viegas found that introducing carpool clubs did not guarantee greater system flexibility (*14*). They demonstrated that people only trust and interact with their colleagues for carpooling, thus carpool use does not necessarily increase as the system scale increases with carpool club programs. Habib et al. employed a hybrid discrete-choice model that conditions mode choice set formation based on the probability of a commuter considering carpooling as a viable option (*5*). They found that once carpooling is considered a viable mode by an individual, having other TDM-related options, such as flexible office hours, increases the propensity of carpooling. Otherwise, the same TDM strategy would have little influence on an individual's carpool choice (*5*). The results suggest that people who think about carpooling as an alternative and people who are not interested in carpooling behave and respond to the same TDM policies quite differently.

In the context of large-campus commuters, Deakin et al. analyzed the size and characteristics of the market for dynamic ride-sharing by focusing on commuters to the University of California, Berkeley (central campus) and downtown Berkeley (*17*). Using focus group interviews and surveys, the authors found that Berkeley commuters' motivations and interest in dynamic ride-sharing were affected by the perceived degree of savings in time and money, indicating that increasing parking charges or providing discount parking at convenient locations would be effective means for the promotion of dynamic ride-sharing. Zhou examined car-sharers' characteristics at the University of California, Los Angeles with a sample of employees who commute using an alternative mode to driving alone (*8*). The study revealed that university employees' participation rates in the car-sharing program was generally high for bus commuters, females and low-income employees. Targeting the same campus as this study (OSU), Shay used a stated preference survey of the OSU community to explore whether people were willing to participate in a real-time employer-based ride-sharing program, and their individual characteristics and travel behavior (*18*). The study revealed that those who usually traveled by car, whether shared or alone, were more likely to participate in a real-time ride-sharing program than those who did not (*18*).

Most of the studies discussed above focus on the determinants of choosing to carpool or participating in carpooling programs. Only a few have found that willingness to carpool has an association with role preference (*10,11*). Shaheen et al. (*11*) considered a binary variable indicating a casual carpooling role ("only a rider" or "sometimes a driver") in a multinomial logit model, and this variable turned out to be the largest contributing factor that affected the resulting mode choice. The limitation is that this study's sample mostly consisted of passengers waiting for a ride at casual pick-up locations, thus it does not reflect driver perspectives sufficiently.

Policies to spur carpooling as a TDM strategy should be preceded by a better understanding of role preferences to facilitate targeting the right market. The present study adds to the existing literature through emphasizing stated interests in carpooling in relation to different role preferences. Using OSU's 2012 Campus Travel Pattern Survey data, this study provides in-depth analyses on the variations in carpooling interest, role preference and carpool choice across a variety of socio-demographic groups. The estimated models quantify the effects of hypothesized factors on carpooling interest and the link to actual carpooling choices.

Data and Methodology

OSU 2012 Campus Travel Pattern Survey

The data set used for this study is OSU's 2012 Campus Travel Pattern Survey. The survey was conducted online between May 3, 2012 and May 19, 2012 by the OSU's Transportation and Parking Services in conjunction with a research team from the City and Regional Planning Program. Invitation emails with a link to the survey questionnaire were sent to a sample of 21,500 randomly selected members of the university community (about 25% of the campus population). Approximately 3% of the campus population (2,527 respondents) completed the survey. After checking for inconsistencies in the data and removing cases with incomplete information, the sample comprised 1,856 off-campus residents. About 98.7% of the respondents lived within 40 miles of the campus and 93.3% lived within 20 miles of the campus (Figure 1). Each respondent's network-based commute time from their residence to the center of the campus was calculated using Google Drive API. There are a range of public transit agencies in central Ohio, but only a few of them operate fixed routes, and a limited number of these routes serve the OSU campus. There are very few exclusive high occupancy vehicle (HOV) lanes (3.25 miles) or equivalents (0.38 miles) in operation in Ohio, and few of these are accessible to central Ohio commuters (*19*). A website for a ride-share matching service (www.rideshareohio.com) exists, sponsored by regional planning agencies. It allows commuters in central, northwest and southwest Ohio to find carpool partners and share schedules. However, public awareness about this website is very low.

The survey data provide various pieces of information on OSU commuters' travel behavior: the number of times per week that a person uses each travel mode (e.g. carpool, bicycle, transit, walking, and driving alone), significant factors affecting mode choice, whether a person is interested in participating in carpooling as a driver, passenger, or both, and influential factors that encourage or deter carpooling. In this study, respondents were classified into three carpool market segments:

- Current carpool users (N=252): commuters who carpool at least once a week
- Potential carpool users (N=723): commuters who do not currently carpool but have indicated an interest
- Not-interested commuters (N=881): commuters who neither use nor consider carpooling

Figure 1 presents the spatial distribution of the survey respondents by role preference and carpool market segment. No particular spatial pattern is found. A summary of the sample characteristics based on carpooling role preferences and choices is presented in Table 1.

Modeling Approach

Two binary probit models and a seemingly unrelated bivariate probit regression model were developed to examine factors that affect carpool interest and carpool choice. The odds of having an interest in carpooling and being a current carpooler may vary considerably among commuters in response to a range of factors.

To analyze an individual's interest in a given carpool role, two binary probit models were estimated. The first model estimated the probability of indicating an interest in a driver role and the second one estimated the probability of indicating an interest in a passenger role.

It is reasonable to hypothesize that people who are interested in carpooling may be more active in finding company and adjusting their schedules, thus having a higher probability of successfully forming a carpool group. As such, being interested in carpooling and choosing to commute by carpool are closely and positively correlated. One potential approach to account for this relationship in modeling is to include carpool interest as an explanatory variable in univariate probit modeling for carpool choices. However, univariate probit results could be potentially biased if there is an overlap in unobserved characteristics that determine both carpool interest and carpool choice (*20*). This unobserved heterogeneity could result in the error term in the univariate model being correlated with the explanatory variables that predict carpool choices, thus violating the exogeneity assumption (*21*). Another approach, which was used in this study, was to acknowledge the link between willingness to carpool and actual choice by using bivariate probit regression. This

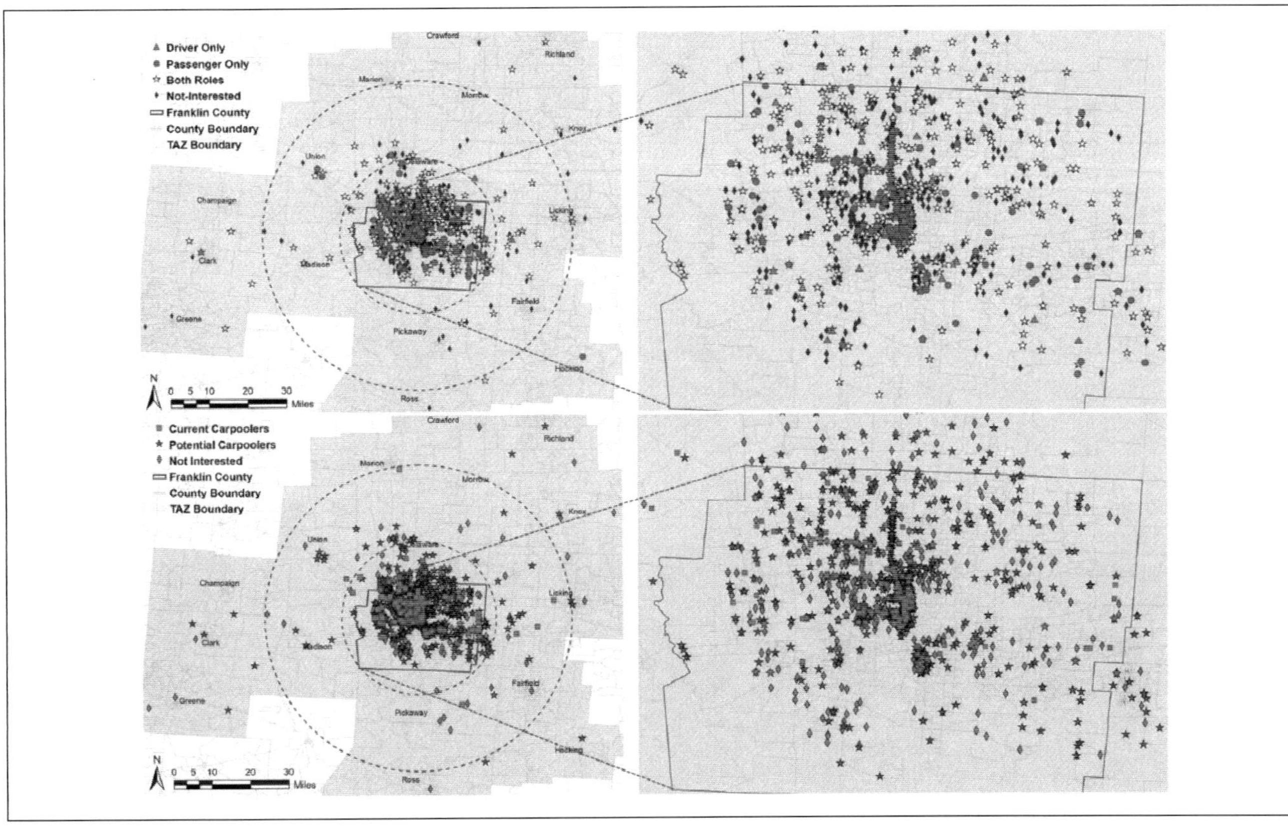

Figure 1. Spatial distribution of the survey respondents' residences (N = 1,856).
Note: The circles represent 20-mile and 40-mile buffer areas of the campus.

approach fixes the bias in the previous approach through measuring the probability that carpool interest and decision to carpool are jointly determined, allowing the error terms to be jointly distributed as bivariate normal with means 0, variances 1 and correlation ρ :

$$y_{1j}^* = x_j\beta + \epsilon_{1j} + \mathbf{offset}_j^{\beta}$$
$$y_{2j}^* = z_j\gamma + \epsilon_{2j} + \mathbf{offset}_j^{\gamma}$$
$$E(\epsilon_1) = E(\epsilon_2) = \mathbf{0}$$
$$\mathbf{Var}(\epsilon_1) = \mathbf{Var}(\epsilon_2) = \mathbf{1}$$
$$\mathbf{Cov}(\epsilon_{1j}, \epsilon_{2j}) = \rho$$

where y_{1j}^* and y_{2j}^* are the unobserved latent variables for carpool interest and carpool-choice models, respectively; the outcome y_{ij}, which is equal to 1, is only observed if the latent variable y_{ij}^* is positive, or $y_{ij} = 0$ if otherwise (for i = 1, 2). Error terms, ϵ_{1j} and ϵ_{2j}, are assumed to follow bivariate normal distribution ($\rho : \epsilon_{1j}, \epsilon_{2j} \sim N(\mathbf{0}, \mathbf{0}, \mathbf{1}, \mathbf{1}, \rho)$) (*20,21*).

One of the key advantages of employing bivariate probit regression compared with standard probit regression to analyze the association between carpooling interest and decision is that an estimate of the interrelatedness (error covariance) of the two decisions is obtained (*20,21*). A significant error covariance estimate suggests that the decisions are significantly interrelated. Therefore, a bivariate probit model was an effective approach to estimate the two equations (one for carpool interest and one for carpool choice) at the same time while controlling for suspected biases.

Discussion of Results

Carpool Interest and Role Preferences

The results in Table 1 show that 32.6% of the respondents are interested in being a carpool driver and 44.1% are interested in being a passenger (not mutually exclusive). In total, 47.8% of the respondents were interested in either role, and 28.9% both roles. About 13.6% of all respondents were current carpoolers. This rate is slightly higher than the U.S. carpool commute mode share of 10.7% and 9.4% based on the 2006 and 2013 American Community Survey (ACS) data respectively (*22*). OSU is one of the largest employers in the region, thus it allows for a relatively higher probability of finding carpool partners for commuters. Younger commuters had higher rates of carpooling interest and were more likely to be actual carpoolers. Single individuals were more likely to be interested in carpooling than married people. Respondents taking more than 30 min to commute to campus were most likely to be interested in carpooling, although their rate of

Table 1. Characteristics of the Survey Sample based on Role Preferences in Carpooling (before Imputation for Missing Data)

	Total		Role Preference for Carpooling[1]			
			Driver	Passenger	Either Role	Current Carpooler
Variables	N	%	%	%	%	%
Gender						
Male	744	40.4	33.3	40.5	44.8	14.1
Female	1,099	59.6	32.2	46.7	50.0	13.4
Age						
<25	685	37.0	37.5	49.5	55.8	18.5
25–35	418	22.6	31.6	44.5	50.7	15.3
35–45	268	14.5	26.9	37.7	42.2	9.3
>45	479	25.9	23.4	33.0	40.5	7.5
Marital status						
Single	948	53.0	35.4	47.9	52.1	14.6
Married, with children	421	23.6	32.1	40.4	43.5	10.7
Married, no children	269	15.1	31.6	45.0	48.3	18.2
Unmarried partner	149	8.3	32.2	46.3	51.0	12.1
University affiliation						
Faculty	179	9.7	27.4	31.3	34.6	14.0
Staff	784	42.4	30.0	41.1	44.5	8.7
Graduate/professional students	289	15.6	38.1	50.2	55.4	20.8
Undergraduate students	598	32.3	35.1	49.2	52.5	16.6
Commute time[2]						
<10 min	618	33.3	27.5	40.0	43.4	18.0
10–20 min	745	40.1	34.1	44.4	49.0	13.4
20–30 min	383	20.6	35.8	48.0	50.7	8.4
>30 min	110	5.9	40.9	50.9	54.5	8.2
Car ownership						
Yes	1757	95.0	33.1	43.4	47.2	13.2
No	67	3.6	0.0	50.7	50.7	19.4
No, but I am thinking of purchasing one	25	1.4	44.0	76.0	80.0	24.0
Frequent commute mode[1] *(≥ 3 times/week)*[1]						
Carpool	109	5.9	51.4	69.7	74.3	100.0
Transit	337	18.2	31.8	53.7	57.3	17.2
Automobile (driver alone)	1,126	60.7	35.2	43.3	47.6	6.7
Bicycle	142	7.7	24.6	35.9	36.6	13.4
Walking	334	18.0	26.3	41.9	44.0	17.1
Important factors in mode choice[3]						
Shorter commute time	1,448	78.0	34.5	46.2	50.0	14.4
More Flexibility in when I depart from campus	1,416	76.3	33.1	44.0	48.2	13.4
The ability to make stops on the way to/from campus	956	51.5	34.0	42.4	46.3	14.0
Cost	1,221	65.8	35.1	49.5	52.3	14.7
Concern for the environment	777	41.9	35.9	50.5	53.4	15.6
I have no option but to drive to campus	836	45.0	35.3	42.6	46.8	10.0
Total	1,856	100.0	32.6	44.1	47.8	13.6

[1]The survey question allows the selection of multiple outcomes. The percentages do not add up to 100%. For instance, taking females as an example, 32.2% are interested in being a driver, 46.7% in being a passenger and 50.0% are interested in either role.
[2]Commute time is calculated by Google Map API from each respondent's residence to the center of the campus.
[3]Measured at a binary scale (1 if important, 0 if otherwise).

actual carpool use was smaller than other groups with shorter commute times. People who did not own a car but had considered purchasing one were far more likely to be interested in carpooling, particularly as a passenger (76%), than those who owned a car (43.4%) or those who neither owned nor had thought of buying one (50.7%). Bicycle commuters showed the lowest interest in carpooling (36.6%), whereas transit users reported the highest interest (57.3%). Around 47.6% of

Table 2. Encouraging and Deterring Factors for Carpooling by Role Preference

	Driver Role Only	Passenger Role Only	Both Roles	None	Total	*p*-Value[1]
Encouraging factors						
Website to find a carpool group	8.8	21.2	17.1	3.3	10.0	0.08
More convenient parking	**33.3**	22.6	27.5	10.9	18.1	0.19
Cost of regular car parking	29.8	**36.4**	36.2	12.0	22.9	0.62
Cost of gasoline	29.8	35.3	**45.2**	**17.3**	**28.4**	0.09
Finding good company to ride with	**34.8**	28.7	36.4	**19.5**	27.0	0.18
Less expensive than driving alone	29.8	35.0	**46.6**	16.6	**28.1**	0.00
A guaranteed ride home if you were unable to return by carpool	**38.6**	**56.7**	**56.8**	**23.0**	**37.9**	0.03
A variety of pick-up and drop-off times	22.8	**47.9**	38.9	16.1	27.3	0.00
Deterring factors						
I need my auto for off-campus trips	50.9	39.6	46.1	45.6	45.1	0.17
I have a constrained or irregular schedule	**61.4**	**59.0**	**67.6**	**62.0**	**63.1**	0.08
I need to pick-up/drop-off my children	7.0	7.4	13.0	14.5	12.8	0.06
I do not have a way to find a carpool group	10.5	28.6	27.8	5.8	15.4	0.02
I do not have time to wait on others	50.9	50.7	**53.9**	**59.1**	**56.2**	0.75
I do not like to depend on others I do not know well	**71.9**	**64.5**	**60.9**	**58.1**	**60.3**	0.23
I am concerned about my safety	8.8	24.4	13.3	9.5	12.7	0.00
I prefer to ride alone	21.1	14.3	11.1	26.9	20.5	0.08
I am concerned about becoming stranded on campus	**52.6**	53.9	53.1	36.4	44.2	0.96
Number of observations	57	217	414	807	1495	-

Note: Respondents were asked to select the three most encouraging factors and the three most deterring factors.
Note: The numbers in bold indicate the three most frequently selected factors in each group.
[1]: This indicates the *p*-value of the joint significance chi-squared test among three groups (driver, passenger, and both roles). The joint significance tests among four groups (driver, passenger, both roles, and none) were all significant ($p < 0.05$).

people who usually drove alone indicated an interest in carpooling either as a driver or passenger, although they were least likely to actually choose this (6.7%).

Carpooling Motivations and Deterrents

In this subsection, the results for carpooling motivations and deterrents are discussed, focusing on the differences between groups preferring different roles. The respondents were asked to select the three most influential factors. The significance of the differences was tested using the chi-squared test of independence. The level of significance is indicated by *p*-values (Tables 2 and 3).

Table 2 shows that people who reveal an interest in taking both roles (driver and passenger) tend to consider the cost-reduction aspects of carpooling more strongly than those interested in a single role. Specifically, a higher portion of those interested in both roles were likely to select "cost of gasoline" (45.2%) and "less expensive than driving alone" (46.6%) as one of the top three motivations for carpooling. People who preferred to be a driver only focused on the convenience and entertaining aspects of carpooling: "more convenient parking" (33.3%) and "finding good company to ride with" (34.8%). Meanwhile, people who favored only being a passenger considered flexibility of carpool formation, "a variety of pick-up and drop-off times" (47.9%), and parking cost-savings (36.4%) as the most important factors. Overall, "a guaranteed ride home" is one of the factors chosen most frequently by all three groups. The availability of alternative options for getting back home may positively impact an individual's carpooling propensity. "Finding good company to ride with" is considered important by all groups, although relatively more important to those favoring a driver role only, and those who did not indicate carpool interest. However, technical assistance to facilitate carpool formation, for instance, "a website to find a carpool group," was only chosen by a few respondents. This result suggests that people prefer more casual carpool formation compared with a matching program. It is notable that individuals only interested in being a passenger consider this factor as a relatively influential one (21.2%) compared with other groups.

Unlike the varied findings on the encouraging factors for carpooling, a rather uniform pattern is observed in terms of deterring factors across people with different role preferences. Arrangement barriers, such as "I do not want to depend on others I do not know well" and "I have a constrained or irregular schedule" were selected by more than half of the people within each group. Although safety concerns did not receive many votes, people preferring only a passenger role showed more concern about their safety (24.4%) than those preferring a driver role (8.8%), or both roles (13.3%). Individuals preferring a passenger role were less likely to select the need for trip chaining (i.e., off-campus trips) as a deterrent. People willing to take both roles showed the least tolerance in terms of waiting time. They

Table 3. Encouraging and Deterring Factors for Carpooling by Carpool Market

	Current Carpooler	Potential Carpooler	Not Interested	Total	*p*-Value
Encouraging factors					
Website to find a carpool group	6.6	19.6	3.4	10.0	0.00
More convenient parking	35.3	23.7	10.0	18.1	0.00
Cost of regular car parking	**41.3**	33.5	10.9	22.9	0.00
Cost of gasoline	40.1	**40.5**	**16.8**	**28.4**	0.00
Finding good company to ride with	34.1	34.4	**18.9**	26.5	0.00
Less expensive than driving alone	**46.1**	39.5	15.6	**28.1**	0.00
A guaranteed ride home if you were unable to return by carpool	**43.1**	**57.5**	**21.9**	**37.9**	0.00
A variety of pick-up and drop-off times	25.7	**43.0**	15.8	27.3	0.00
Deterring factors					
I need my auto for off-campus trips	40.1	45.4	45.9	45.1	0.39
I have a constrained or irregular schedule	**65.3**	**64.6**	**61.5**	**63.1**	0.43
I need to pick-up/drop-off my children	12.6	10.2	14.8	12.8	0.45
I do not have a way to find a carpool group	15.6	28.2	5.7	15.4	0.00
I do not have time to wait on others	**53.5**	52.3	**59.8**	**56.2**	0.02
I do not like to depend on others I do not know well	**62.3**	**63.0**	**57.9**	**60.3**	0.15
I am concerned about my safety	12.6	17.4	9.2	12.7	0.00
I prefer to ride alone	10.2	14.4	27.3	20.5	0.00
I am concerned about becoming stranded on campus	43.7	**54.7**	36.4	44.2	0.00
Number of observations	167	570	758	1495	-

Note: Respondents were asked to select the three most encouraging factors and the three most deterring factors. The numbers in bold indicate the three most frequently selected factors in each group.

were more likely to select irregular schedules (67.6%) and lack of time to wait on others (53.9%) as deterrents than others. People who state an interest in both roles may be those who are seriously considering carpooling despite perceived time constraints, and this may partly explain why they are more open to switching roles depending on circumstances.

Table 3 summarizes the encouraging and deterring factors with regard to the three carpool market segments: current carpoolers, potential carpoolers and not-interested commuters. Current and potential carpoolers put a lot of emphasis on cost-saving advantages and chose such factors as "less expensive than driving alone" and "cost of gasoline," whereas these benefits hardly motivated commuters without any interest in carpooling. Arrangement barriers, such as irregular schedules and reluctance to wait on others, were found to be equally influential for current and potential carpool users. Notably, concerns about being stranded on campus and personal safety were the distinctive factors significantly associated with potential carpoolers (54.7% and 17.4%, respectively) rather than with current carpoolers (43.7% and 12.6%). Potential carpool users emphasized system flexibility ("a variety of pick-up and drop-off times," 43.0%) more than current carpool users (25.7%). Potential carpoolers turned out to be the group that most needed technical assistance or relevant policies that could help them find company for carpooling. More potential carpoolers selected "a website to find a carpool group" (19.6%) and "I don't have a way to find a carpool group" (28.2%) than current carpoolers (6.6% and 15.6%) and not-interested individuals (3.4% and 5.7%). As such, efforts to increase carpool usage among commuters could be more effective if they were accompanied with strategies to increase the availability of alternative transportation options, such as public transit, and its connectivity with other modes. Furthermore, strategic technological assistance that facilitates ride-matching and carpool formation has the potential to help those people who are actually considering carpooling.

Factors Affecting Carpool Interest and Carpool Choice

Two binary probit models and a seemingly unrelated bivariate probit regression model were employed to examine the factors affecting carpool interest and carpool choice. The modeling results are reported in Tables 4 and 5.

The two binary probit models considered binary alternatives in the choice set. The first model examined the choice of a driver role (1, if interested in being a driver; 0 if not). The second model examined the choice of a passenger role (1, if interested in being a passenger; 0 if not).

The seemingly unrelated bivariate probit regression model accounts for two decisions:

(i) interested in carpooling as either a driver or passenger (= 1) or neither (= 0);

Table 4. Model Estimation Results: Binary Probit Regression

Variables	Driver Interest Model[1]		Passenger Interest Model[1]	
	Estimated Coefficients			
Socio-demographics	Coefficient	(z-stat)	Coefficient	(z-stat)
(Base: female, aged 25–45, single, graduate student)				
Male	**0.147**	**2.02**	−0.084	−1.18
Aged under 25	**0.470**	**3.87**	**0.341**	**2.89**
Aged over 45	**−0.249**	**−2.47**	*−0.176*	*−1.82*
Married, with children	0.048	0.43	0.045	0.42
Married, no children	−0.021	−0.18	0.103	0.93
Unmarried partner	−0.010	−0.08	0.005	0.04
Faculty	−0.260	−1.57	**−0.432**	**−2.66**
Staff	**−0.246**	**−2.08**	*−0.193*	*−1.68*
Undergraduate student	−0.109	−0.87	−0.001	0.00
Commute time by car (Base: less than 10 min)				
10–20 min	**0.454**	**4.04**	**0.413**	**3.82**
20–30 min	**0.476**	**3.44**	**0.565**	**4.21**
More than 30 min	**0.576**	**3.03**	**0.680**	**3.67**
Modal choice				
Frequent use of bike (more than 3 times/week)	**−0.588**	**−3.47**	**−0.660**	**−4.22**
Frequent use of automobile (more than 3 times/week)	0.056	0.56	−0.126	−1.28
Frequent use of transit (more than 3 times/week)	*−0.235*	*−1.81*	−0.116	−0.94
Frequent use of walking (more than 3 times/week)	**−0.335**	**−2.54**	**−0.354**	**−2.85**
Factors affecting mode choice				
Shorter commute time	**0.226**	**2.31**	**0.287**	**3.07**
More flexibility in when I depart from campus	−0.096	−1.01	−0.076	−0.82
The ability to make stops on the way to/from campus	*0.135*	*1.74*	−0.083	−1.10
Cost	0.098	1.18	**0.262**	**3.29**
Concern for the environment	**0.250**	**3.24**	**0.285**	**3.83**
I have no option but to drive to campus	*−0.166*	*−1.77*	**−0.279**	**−3.02**
Constant	**−1.058**	**−5.71**	**−0.611**	**−3.47**
Log likelihood	−849.86		−923.70	
Likelihood-ratio chi-squared	109.32		143.70	
Number of observations	1,495		1,495	

Note: Bold coefficients: $p<0.05$, coefficients in italics: $p<0.1$.
[1]Each model is estimated by binary probit regression.

(ii) currently use carpooling for commute more than once a week (=1) or not (=0).

The explanatory variables in our models included gender, age, marital status, university affiliation (faculty, staff, graduate and undergraduate student), commute time, frequency of used travel modes, and factors affecting mode choice. The effects of several land-use variables at respondents' places of residence were tested (such as population, employment, and housing densities, and median household income). However, these factors were not statistically significant. This is consistent with Wang and Chen's findings that population densities at residential traffic analysis zones (TAZs), and employment densities at destination TAZs are not significant in determining the shift from SOV to carpool (*7*). These variables were excluded from the final models.

One of the important results is the estimate of the covariance (ρ) of the error terms in the bivariate model (Table 5). If the estimated ρ is not significantly different from 0, then the two choices considered are independent of each other, and model coefficients would not be biased if two variables representing these two choices were included in the same equation. In our model estimation, the estimate of ρ was 0.406 with a *z*-value of 5.83, indicating that these two choices were correlated and the seemingly unrelated bivariate probit model therefore needed to be used.

Males were more likely to be interested in being a carpool driver, consistent with men's active propensity for driving. Respondents who were younger than 25 were more likely to be interested in being a driver or a passenger than others. Being older than 45 decreased the probability of having an interest in either role. This finding supports the advantage of exploring carpooling behavior in university campuses where

Table 5. Model Estimation Results: Bivariate Probit Regression

Variables	Estimated Coefficients			
	Carpool Interest Model[1]		Carpool-Choice Model[1]	
Socio-demographic	Coefficient	(z-stat)	Coefficient	(z-stat)
(Base: female, aged 25–44, single, graduate student)				
Male	−0.056	−0.80	−0.019	−0.21
Aged under 25	**0.527**	**4.45**	**0.433**	**2.78**
Aged over 45	**−0.246**	**−2.63**	**−0.290**	**−2.06**
Married, with children	0.064	0.62	**0.524**	**3.27**
Married, no children	0.086	0.79	**0.768**	**5.20**
Unmarried partner	0.032	0.25	0.124	0.71
Faculty	**−0.405**	**−2.58**	−0.098	−0.49
Staff	−0.184	−1.62	**−0.366**	**−2.51**
Undergraduate student	−0.132	−1.07	*−0.245*	*−1.65*
Commute time by car (Base: less than 10 min)				
10–20 min	**0.429**	**4.03**	−0.009	−0.07
20–30 min	**0.478**	**3.63**	**−0.424**	**−2.39**
More than 30 min	**0.666**	**3.65**	−0.300	−1.13
Modal choice				
Frequent use of bike (more than 3 times/week)	**−0.506**	**−3.22**	-	-
Frequent use of autom obile (more than 3 times/week)	*0.199*	*1.87*	-	-
Frequent use of transit (more than 3 times/week)	−0.008	−0.06	-	-
Frequent use of walking (more than 3 times/week)	**−0.245**	**−1.97**	-	-
Factors affecting mode choice				
Shorter commute time	**0.222**	**2.45**	0.169	1.33
More flexibility in when I depart from campus	−0.033	−0.36	−0.141	−1.18
The ability to make stops on the way to/from campus	−0.051	−0.69	**0.228**	**2.29**
Cost	**0.236**	**3.02**	**0.317**	**2.85**
Concern for the environment	**0.295**	**4.02**	*0.186*	*1.92*
I have no option but to drive to campus	**−0.299**	**−3.30**	**−0.248**	**−2.11**
Constant	**−0.704**	**−4.03**	**−1.652**	**−7.50**
ρ	-	-	**0.406**	**5.83**
Log-likelihood	−1,412.68			
Likelihood-ratio chi-squared	218.37			
Number of observations	1,495			

Note: Bold coefficients: $p < 0.05$; coefficients in italics: $p < 0.1$.
[1]The two equations are jointly estimated by seemingly unrelated bivariate probit regression.

lots of commuters are young students. Graduate students were more likely to be interested in carpooling in either role, compared with faculty and staff. Undergraduate and graduate students were not significantly different from one another in terms of their carpool interest and role preferences. However, the carpool-choice model revealed that undergraduate students were less likely to carpool than graduate students. This may be because undergraduate students usually live closer to campus and are more likely to take advantage of other travel modes, such as bicycles, transit or walking. Many of them may not actually choose carpools despite their interest. Marital status had a significant influence on carpool choice, but not on carpool interest. Married people, with or without children, were more likely to carpool compared with unmarried individuals with or without a partner. This may in part relate to the joint travel behavior of a couple.

The significant effects and the signs of the commute time variables suggest that people living farther from campus are more likely to be interested in carpooling. This is reasonable since commuters with longer travel times would benefit most from the advantages of carpooling, such as cost savings and convenience. However, in the bivariate model for carpool choice, the coefficients for the commute time variables were not significant, except for the "20–30 min," which were negative. This suggests that people who live less than 10 min away from campus are more likely to carpool than those who live within 20–30 min from campus. This mixed finding implies that the influence of commuting time on carpool behavior is not strictly linear. As commute time and distance increase, the utility of using carpooling to save costs and have a pleasant journey with company also arises, motivating people to explore carpool opportunities. However, if someone lives far from campus, the availability of alternative transport modes, such as public transit and cycling as back-up options, and the size of the carpool, is likely to diminish, making this option less feasible. It would be useful for policy makers to consider the clashing impact of distance on carpool interest and choice to strengthen the advantages of carpooling for long commute trips and expand the availability of alternative modes for remote commuters.

Results for respondents' frequent commute modes show that if someone regularly commutes by cycling or walking (at least three times/week), his/her probability of being interested in a driver or a passenger role significantly decreases. This implies that promoting carpool behavior will not counteract the choice of non-motorized modes, which are also promising components of TDM strategies. The coefficient of frequent transit-users was significant and negative in the driver model but not in the other models. Although transit users were found to be the largest portion of those interested in carpooling (Table 1), being a frequent transit-user does not have a significant impact on carpool interest after other attributes are accounted for. Therefore being a transit user does not necessarily translate into higher carpooling interest. Socio-demographic characteristics, commute distance and personal attitudes all come into play. Driving alone at least 3 days/week to campus showed a significantly positive effect only in the aggregated interest model.

Respondents who considered "shorter commute time" as an important factor in their mode choice decisions were more likely to have an interest in either role. Thus, it can be deduced that people who are interested in carpooling expect to reduce their commuting time after they start carpooling. The factor "the ability to make stops on the way to/from campus" was found to be positively associated with having an interest in being a driver, but not in a passenger role; this factor was also associated with being a current carpooler. This result points to the possibility that having a range of destinations on the way to campus and home increases one's chances or intention to find company to share a ride. In addition, people who want to make stops during their commute significantly prefer a driver role because it gives more autonomy in steering a trip chain while carpooling. Cost-related factors have been found to be important reasons for carpooling in the literature (*6,9,11*). Our study provides further insight that people who value travel costs have a higher likelihood of being interested in a passenger role. This cost factor had a significantly positive effect in the carpool-choice model, showing that cost is quite important for both interest and actual choice. Environmental concern was found to be positively associated with carpool interest and use. The perception of limited transport options ("I have no option but to drive to campus") negatively impacted on carpool choice as well as carpool interest.

Overall, the model results reveal that the influence of socio-demographics, commute time, and individual perceptions vary across the stated carpool interests for being a driver, or a passenger, or both and revealed choices. The results demonstrate that policies for fostering a carpooling culture should not be a single good-enough policy, but rather a bundle of targeted policies that can meet the diverse needs of people favoring different carpool roles. The joint-model results support our hypothesis that, all things being equal, having an interest in carpooling is associated with a higher probability of actual carpool use.

Conclusions

Using data on university commuters' interest in carpooling, this study explored the characteristics of people in different market segments for carpooling and the significance of role preferences in motivating people to carpool. The proportion of people willing to take both roles (driver and passenger) in carpooling was larger than that of those interested in a passenger role only (28.9% versus 15.1%, respectively). The portion of people only interested in being a driver was fairly small (3.7%). Most people who revealed an interest in carpooling were willing to change roles flexibly to sustain carpooling. People interested in both roles tended to underline the financial benefits of carpooling, while individuals interested in only being a passenger showed more concerns about safety and flexibility in trip planning. A guaranteed ride home was a universally crucial factor across all groups.

Age, marital status, university affiliation, commute distance, modal split, and personal attitudes were key factors associated with carpool interest and choice. Younger people and graduate students were more willing to participate in carpooling and actually used it more. Married people, with or without children, tended to carpool at least once a week compared with unmarried people, while there was little difference in the likelihood of having an interest in carpooling between them. A longer commute distance was positively associated with carpool interest. However, this trend did not exist in revealed carpool choices, suggesting a non-linear relationship between commute distance and carpool choice. After other attributes were controlled for, frequent transit-users tended to avoid a driver role, and frequent cyclists and walkers were likely to reject an interest in both roles. Frequent automobile-users were more likely to have an interest in carpooling generally. The need for trip chaining or off-campus trips increased the likelihood of having an interest in a carpool driver role, but not in a passenger role. Environmental awareness was associated with both carpool interest and choice.

The overall results underscored the importance of financially tangible incentives and the availability of other transport options complementary to a carpool commute for current and potential carpool users, and for people interested in both carpool roles. The results also suggested that a website or a matching system would have a minimal effect in attracting uninterested people into carpooling and encouraging current users to keep carpooling. Nevertheless, potential carpool users, particularly those preferring a passenger role, were found to be more in need of technical assistance to find a carpool member.

The findings from our comparative analysis and statistical modeling indicate that there are opportunities to increase carpool use through focusing on individuals showing an interest in carpooling who prefer a particular role with specific incentives in mind.

As a limitation, our study did not differentiate the potentially varying intensities of interest in carpooling. Future research could benefit from the findings in this study by focusing on individuals who do not carpool despite their significant interest in doing so. Finally, although this study focuses on a specific population from a large university setting, the results still inform general policy making and the methodology developed could be applied elsewhere.

References

1. Akar, G., C. Flynn, and M. Namgung. Travel Choices and Links to Transportation Demand Management: Case Study at Ohio State University. *Transportation Research Record: Journal of the Transportation Research Board*, 2012. 2319: 77–85.
2. Chan, N. D., and S. A. Shaheen. Ridesharing in North America: Past, Present, and Future. *Transport Reviews*, Vol. 32, No. 1, 2012, pp. 93–112.
3. Hasnine, M. S., A. Weiss, and K. N. Habib. Development of an Employer-Based Transportation Demand Management Strategy Evaluation Tool with an Advanced Discrete Choice Model in its Core. *Transportation Research Record: Journal of the Transportation Research Board*, 2016. 2542: 65–74.
4. Litman, T. The Online TDM Encyclopedia: Mobility Management Information Gateway. *Transport Policy*, Vol. 10, No. 3, 2003, pp. 245–249.
5. Habib, K. N., Y. Tian, and H. Zaman. Modelling Commuting Mode Choice with Explicit Consideration of Carpool in the Choice Set Formation. *Transportation*, Vol. 38, No. 4, 2011, pp. 587–604.
6. Li, J., P. Embry, S. Mattingly, K. Sadabadi, I. Rasmidatta, and M. Burris. Who Chooses to Carpool and Why?: Examination of Texas Carpoolers. *Transportation Research Record: Journal of the Transportation Research Board*, 2007. 2021: 110–117.
7. Wang, T., and C. Chen. Attitudes, Mode Switching Behavior, and the Built Environment: A Longitudinal Study in the Puget Sound Region. *Transportation Research Part A: Policy and Practice*, Vol. 46, No. 10, 2012, pp. 1594–1607.
8. Zhou, J. An Analysis of University Employee Car-Sharers in Los Angeles. *Transportation Research Part D: Transport and Environment*, Vol. 17, No. 8, 2012, pp. 588–591.
9. Neoh, J. G., M. Chipulu, and A. Marshall. What Encourages People to Carpool? An Evaluation of Factors with Meta-Analysis. *Transportation*, Vol. 44, No. 2, 2017, pp. 423–447.
10. Buliung, R. N., K. Soltys, R. Bui, C. Habel, and R. Lanyon. Catching a Ride on the Information Super-Highway: Toward an Understanding of Internet-Based Carpool Formation and Use. *Transportation*, Vol. 37, No. 6, 2010, pp. 849–873.
11. Shaheen, S. A., N. D. Chan, and T. Gaynor. Casual carpooling in the San Francisco Bay Area: Understanding User Characteristics, Behaviors, and Motivations. *Transport Policy*, Vol. 51, 2016, pp. 165–173.
12.Lue, A., and A. Colorni. A Software Tool for Commute Carpooling: A Case Study on University Students in Milan. *International Journal of Services Sciences*, Vol. 2, No. 3–4, 2009, pp. 222–241.
13.Vanoutrive, T., E. Van De Vijver, L. Van Malderen, B. Jourquin, I. Thomas, A. Verhetsel, and F. Witlox. What Determines Carpooling to Workplaces in Belgium: Location, Organisation, or Promotion? *Journal of Transport Geography*, Vol. 22, 2012, pp. 77–86.
14.Correia, G., and J. Viegas. Carpooling and Carpool Clubs: Clarifying Concepts and Assessing Value Enhancement Possibilities through a Stated Preference Web Survey in Lisbon, Portugal. *Transportation Research Part A: Policy and Practice*, Vol. 45, No. 2, 2011, pp. 81–90.
15. O'Brien, J., and A. Dunning. Community Rideshare for a Small City: Preliminary Results for the Lawrence, Kansas, OnBoard Prototype Program. *Transportation Research Record: Journal of the Transportation Research Board*, 2014. 2469: 32–40.
16Ko, J., and D. Kim. Employer-Based Travel Demand Management Program: Employer's Choice and Effectiveness. *Transport Policy*, Vol. 59, pp. 1–9.
17. Deakin, E., K. Frick, and K. Shively. Markets for Dynamic Ridesharing? Case of Berkeley, California. *Transportation Research Record: Journal of the Transportation Research Board*, 2010. 2187: 131–137.
18. Shay, N. *Investigating Real-Time Employer-Based Ridesharing Preferences Based on Stated Preference Survey Data*. MS thesis. Ohio State University, O.H., 2016.
19.Office of Highway Policy Information. *High Occupancy Vehicle Lanes by State*, 2008. https://www.fhwa.dot.gov/policyinformation/tables/03.pdf.
20. Fleming, C., and P. Kler. I'm too Clever for this Job: A Bivariate Probit Analysis on Overeducation and Job Satisfaction in Australia. *Applied Economics*, Vol. 40, No. 9, 2008, pp. 1123–1138.
21Rayton, B. A. Examining the Interconnection of Job Satisfaction and Organizational Commitment: An Application of the Bivariate Probit Model. *The International Journal of Human Resource Management*, Vol. 17, No. 1, 2006, pp. 139–154.
22McKenzie, B. Who Drives to Work? Commuting by Automobile in the United States: 2013. *American Community Survey Reports*, 2015.

The Standing Committee on Transportation Demand Management (ABE50) peer-reviewed this paper (18-05121).

TRR
JOURNAL OF THE TRANSPORTATION RESEARCH BOARD

Article

Transportation Research Record
2018, Vol. 2672(8) 719–730

DOI: 10.1177/0361198118801346
journals.sagepub.com/home/trr
SAGE

The Relation between Train Access Mode Attributes and Travelers' Transport Mode-Choice Decisions in the Context of Medium- and Long-Distance Trips in the Netherlands

Peter van der Waerden[1] and Jaap van der Waerden[2]

Abstract
This paper presents the findings of a study that investigated travelers' transport mode-choice behavior in the context of medium- and long-distance trips, with special attention given to attributes of train access modes. The goal of the paper is to provide greater insight into the contribution of both access and main travel-mode attributes to the travelers' decision to use the car or the train for medium- and long-distance trips. Based on data collected through a stated choice experiment, a mixed logit model is estimated to identify the contributions of all included attributes. In total, 32 attributes were included in the experiment describing the main transport modes, train and car, and the access modes: bicycle, bus, and drop-off/car. Based on a fractional factorial design, the attributes and corresponding levels were combined into 81 different mode-choice situations. The stated choice experiment was included in an online questionnaire that was distributed among members of a marketing panel. Panel members were invited to choose the preferred travel-mode for a medium- and long-distance trip given a detailed description of both main and access modes. Each member evaluated nine choice situations. In total, 415 panel members completed the questionnaire. Combined, panel members evaluated 3,735 choice situations. The mixed logit model analysis shows that time and cost-related attributes significantly contribute to the attractiveness of transport modes. However, these effects differ considerably between the investigated modes. Conversely, safety-related attributes, chance of delay, and transfer time from access mode to train platform play a minor role.

The introduction of the car in the late 1800s and the increase in car ownership changed the spatial structure of urban settlements considerably (*1*). People were no longer forced to live close to their work or to any other activity (e.g., social, leisure, and shopping). One of the consequences of living some distance away was that people started to make longer-distance trips. The most frequently used transport mode to make medium- and long-distance trips in the Netherlands is the car, followed by the train (*2*). In many cases both transport modes are considered by travelers when making a choice. Van Exel and Rietveld found that 45% of the train travelers on a major corridor in the Netherlands have a car in their choice set, whereas 47% of car travelers have public transport in theirs (*3*).

Looking at the use of cars, the growth of the population in and around cities is currently causing major problems such as congestion, pollution, and parking problems (*4*). The accessibility of urban locations is worsening, and the livability of residents in cities is under pressure. To solve the problems caused by increased car use, transportation planners are diligently looking for opportunities to stimulate travelers to change from the car to more sustainable transport modes. For medium- and long-distance trips in the Netherlands, the train is considered a viable alternative to the car (*2, 3*). To encourage car drivers to change to the train, the service provided by the train has to approach the level of service provided by cars (*5, 6*). In this context, the mode of access to and egress from train stations is acknowledged as an essential part

[1]Urban Planning Group, Department of the Built Environment, Eindhoven University of Technology, Eindhoven, the Netherlands
[2]Social Psychology Group, Department of Psychology, Radboud University, Nijmegen, the Netherlands

Corresponding Author:
Address correspondence to Peter van der Waerden:
p.j.h.j.v.d.waerden@tue.nl

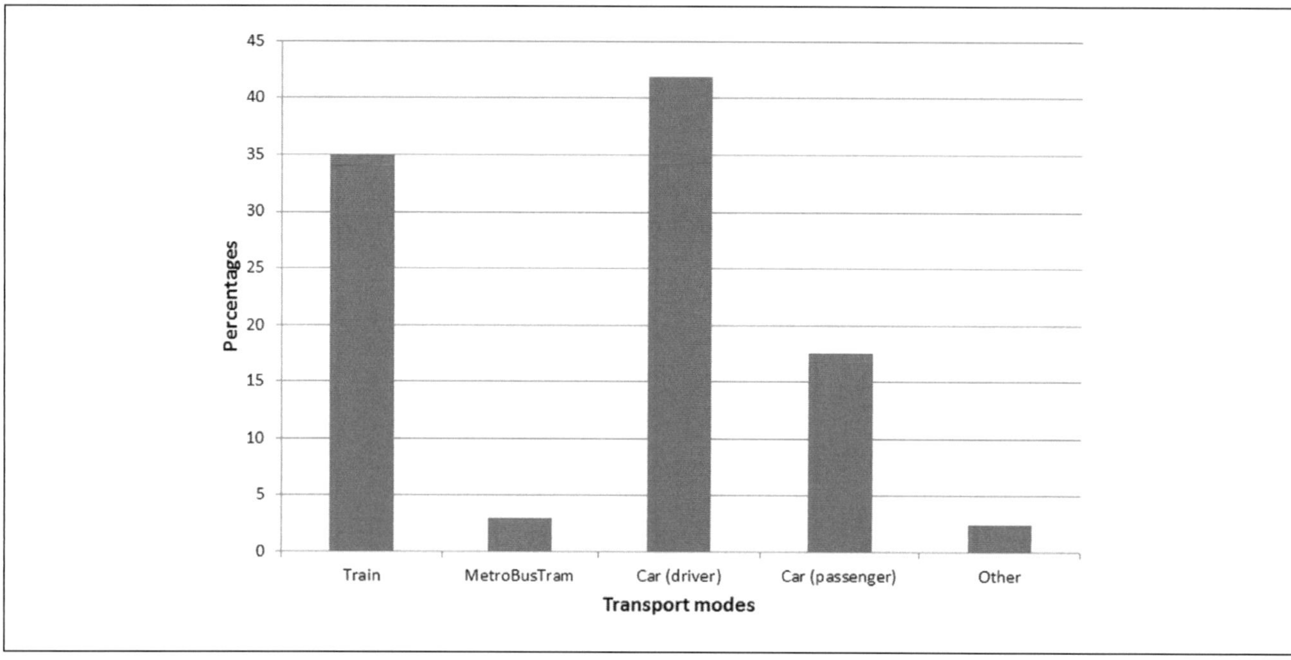

Figure 1. Transport mode share: medium- and long-distance trips in the Netherlands (2).

of the train service resulting in higher use of public transport (7). Detailed insights regarding the role of access and egress modes in travelers' mode-choice decisions are still limited. This is especially true when looking at the competition between car and train and considering attributes besides travel time and cost (7, 8).

The aim of this study is to provide more insight into the factors that contribute to travelers' mode-choice decisions for medium- and long-distance trips with special attention paid to the competition between the transport modes of car and train. In the context of the Netherlands, medium- and long-distance trips cover a distance of approximately 50 km or more, mostly between major cities (2). In the current study, special attention is paid to the role played by the various attributes of train access modes in travelers' mode-choice decisions. In the Netherlands, the train is a feasible alternative to the car when considering medium- and long-distance trips (Figure 1). A train journey can be subdivided into different parts that all need to be considered when exploring travelers' transport mode-choice decisions, as it is not a simple door to door form of transport. Rietveld stated that "the market potential of railways services depends on the quality of the total chain from residence to place of activity" (9). In this study, the focus is on the access mode because it is an essential element in the competitive position of the train (9). Changing the mode of access to stations is also seen as an effective way to improve public transport access to low-wage jobs (7).

In the Netherlands, the bike is the most frequently used train access mode followed by walking and public transport (Figure 2). Figure 2 shows that, with 11%, the car (driver and passenger) also plays a considerable role as an access mode.

The remainder of the paper is organized as follows. First, a brief overview of previous studies is given to clarify the background of the current study. In the section that follows that, the adopted research approach is outlined. Next, the data collection methods are presented including a brief description of the research sample. The subsequent section illustrates the experiences of the respondents regarding traveling over medium- and long-distances and their use of transport modes. After this, the results of the model analyses are presented. The paper closes with conclusions, and makes some recommendations for transportation planners, and suggestions for future research.

Literature

In the past, several studies regarding the characteristics of access journeys to public transport in general and trains in particular have been carried out. Based on these studies, an overview of insights regarding the adopted research focuses and included attributes is presented. Regarding the research focus a distinction can be made between studies dealing with travelers' mode-choice behavior in general, studies exploring customer satisfaction with public transport services, and those dealing specifically with train journeys.

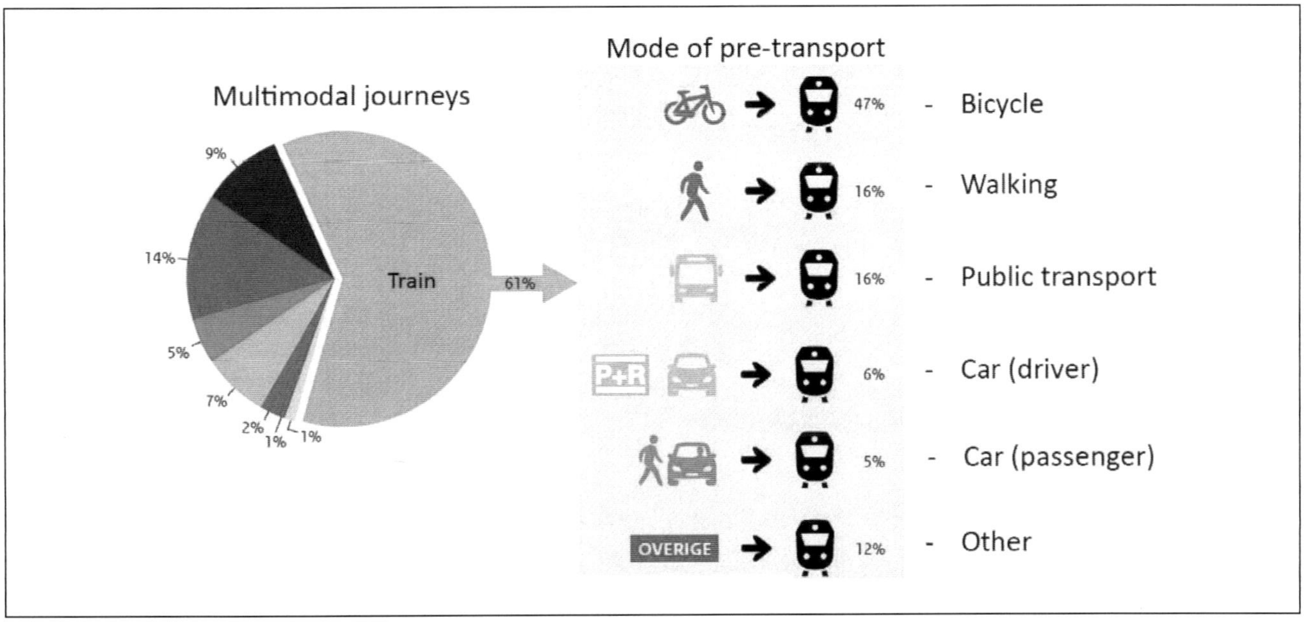

Figure 2. Access mode use in the Netherlands (*10*).

In the first types of study, researchers are looking for attributes that influence travelers' choices mostly between car, public transport, bike, and walking. Some studies extend the set of choice alternatives with carpool and telecommuting (*11*). Others specifically focus on the competition between car and train (*3*, *5*). Goel and Tiwari investigated factors that contribute to the choice of access mode (walk, cycle, cycle rickshaw, auto rickshaw, motorized two-wheeler, bus, and car) of metro users in Delhi (*12*). They found that trip length, vehicle ownership, location of metro station, and population density are significantly associated with the choice of access mode. Such studies investigate a variety of attributes covering physical environment, transport modes, trip-maker and trip characteristics, and psychological factors. Most studies are based on revealed choice behavior. Almost all studies use a general specification of the public transport access mode, for example, by including walk and bike as access modes (*3*) or distance to public transport service (*5*). Brands et al. argued that mode-choice models do not sufficiently consider the car and bike as access modes to public transport (*8*). They investigated travelers' public transport route choices with all kinds of single-mode and multimodal travel options without paying attention to the attributes of the various access modes.

The second group of studies mainly focuses on attributes that make public transport more attractive. For example, Redman et al. investigated various public transport quality attributes that might inspire car drivers to use public transport more often (*6*). Based on an extensive literature review, they concluded that attributes such as reduced fare promotions will attract car drivers but that other attributes related to accessibility, reliability, and mobility provision will keep the car drivers using public transport. An alternative approach is presented by Dell'Olio et al. who investigated the effect of variables on public transport users' desired quality of service (*13*). Based on a stated choice experiment they found that waiting time at the bus stop, cleanliness of the vehicle, and comfort of the buses are the most valued variables. Variables such as driver kindness, bus occupancy, and journey time were found to be less valued. More recently, Mouwen investigated customers' satisfaction with public transport services (*14*). Based on an extensive data set containing the satisfaction scores of urban and regional Dutch public transport users, he concluded that the service attributes of on-time performance, travel speed, and service frequency are the most significant in determining customer satisfaction levels. A study by Salas evaluated non-motorized access ratings in relation to increasing non-motorized access to commuter rail stations (*15*). He found that only providing bike-lanes and sidewalks is not sufficient to improve a station's access rating. Attention also has to be paid to the connectivity of bike and pedestrian routes, the quality of facilities maintenance, and policies affecting non-motorized transportation projects. This topic was also investigated by Chowdhury et al.: they looked at how the ease of access to terminals influences public transport users' attitudes (*16*). Based on the reactions of public transport users, they concluded that ease of access to terminals and accessibility to various destinations has a significant effect on users' ridership satisfaction.

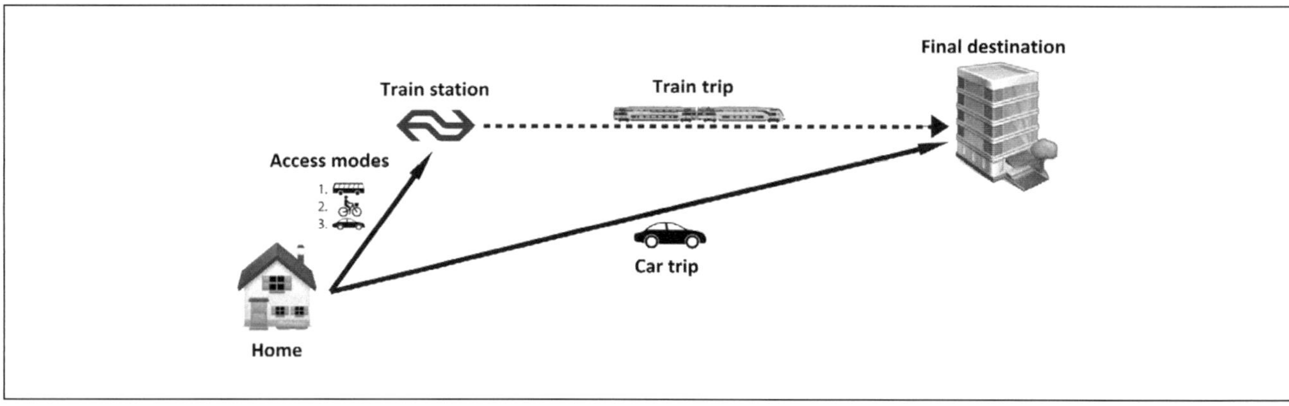

Figure 3. Four different combinations of transport modes (*20*).

The final group of studies focuses on travelers' train use and the access of train stations. Some studies investigate customer satisfaction with a train journey in combination with train access mode. Givoni and Rietveld investigated the contribution of several attributes (price/quality ratio, overall station evaluation, connection with public transport, and car and bicycle parking) on a traveler's perception of train station access (*17*). They found that the quality of the station and the access/egress facilities have a significant effect on the general perception of traveling by train. In addition, Brons et al. (*18*) related the number of daily trips made by rail with more or less the same characteristics as used by Givoni and Rietveld (*17*). They found that the quality of the access to the station is an important element in explaining rail use. Another very interesting finding of the study was that improving the journey to the station seems to be more important than facilitating a smooth connection between access and main mode through better parking facilities.

Looking at approaches and findings in previous studies, it appears that only limited attention has been paid to attributes of different access (bicycle, bus, and car/drop-off) and main transport modes (train and car). Several researchers acknowledged that detailed insights into the contribution of access modes to the attractiveness of public transport are essential (*5*, *6*, *8*, *15*, *17*). Therefore, a more detailed investigation into individual attributes was adopted as the main topic of the current study.

Research Approach

To get insight into travelers' preferences regarding different attributes of both access and main transport modes for medium- and long-distance trips, a stated choice experiment was developed. A stated choice experiment provides a much better opportunity to investigate the competition between car and train than a revealed choice experiment (*19*). Potential travelers can be confronted with many more details of both alternatives even when they rarely use these. In the experiment four different transport mode combinations were included, namely bus–train, bicycle–train, drop-off/car–train, and solo car (Figure 3). The figure was also included in the questionnaire to clarify the choice options.

For each separate mode various attributes were selected to represent the appearance of the separate transport (access) modes. The selection was based on the findings in the literature review and, for each transport mode, covers the travel time and costs, connected safety measures and levels, and chance of delay. In addition, some mode specific attributes were selected such as walking time to bus stop (bus) and cost of parking (bicycle and car). Finally, one attribute was specified to represent the connectivity between the various access modes and the train. In total, 32 attributes were included in the stated choice experiment. Table 1 gives an overview of the selected attributes and their corresponding attribute levels. The table contains a short description of each attribute that was also included in the questionnaire. The levels of the travel time attributes set were based on a valid situation for an average trip between a respondent's home location surrounding Eindhoven and one of the cities in the Netherlands for which car and train are competing options.

The attribute levels were combined into choice situations using a fractional factorial design of 3^{32} (*19*, p. 115 onwards). In total, 81 different choice situations were generated. One example of a choice situation is presented in Figure 4. Respondents were invited to indicate what (combination of) transport mode they preferred when traveling from their home location in the vicinity of Eindhoven to one of the major cities in the Netherlands (see next section for included cities). There were no specific travel characteristics specified such as travel purpose, travel day, or travel party. Each respondent was invited to evaluate nine different choice situations.

Table 1. Overview of Attribute and Attribute Levels (*20*)

Attribute	Attribute levels	Explanation to respondent
Attributes: Train		
In-vehicle time (IVT)	45 min, 50 min, 55 min	Train in-vehicle time is the time you spend in the train traveling from the beginning station to the end station.
Waiting time (WT)	3 min, 9 min, 15 min	The waiting time is the time you have to wait on the railway platform until you can get on the train.
Cost of travel	€10.00, €15.00, €20.00	The cost of train travel.
Chance of delay	5%, 10%, 15%	The chance that the train will be delayed (travel time longer than scheduled) during the trip.
Safety	No additional measures; additional security cameras; additional security cameras and staff	The safety measures present at the railway station.
Possibility to work	No additional facilities; personal work tables; special work cabins	The train can have facilities that improve the possibility of working in the train.
Attributes: Car		
In-vehicle time (IVT)	50 min, 60 min, 70 min	Car in-vehicle time is the time you spend in the car traveling from home to the end destination.
Parking search time (PST)	3 min, 9 min, 15 min	The time you need to find a parking spot at the destination.
Cost of travel	€10.00, €14.00, €18.00	The cost of traveling by car.
Parking costs	€2.00 per hour, €4.00 per hour, €6.00 per hour	The costs per hour of the parking facility at the end destination.
Chance of congestion	10%, 20%, 30%	The chance that you will experience congestion during car travel.
Safety	No additional measures; additional security cameras; additional security cameras and staff	The safety measures present at the parking facility where your car is parked.
Attributes: Bus		
Time to get to bus stop	1 min, 4 min, 7 min	The time you need to get from home to the bus stop.
In-vehicle time (IVT)	15 min, 20 min, 25 min	Bus in-vehicle time is the time you spend in the bus traveling from the beginning stop to the end stop.
Time to get to railway platform	1 min, 3 min, 5 min	The time you need to get from the bus to the railway platform.
Cost of travel	€1.00, €2.00, €3.00	The cost of traveling by bus.
Chance of delay	0%, 15%, 30%	The chance that the bus will be delayed (travel time longer than scheduled) during the trip.
Type of bus	Regular bus, express bus, shuttle bus	The regular bus stops frequently. The express bus has fewer stops, and the shuttle bus travels directly to the railway station.
Safety measures	No additional measures; additional security cameras; additional security cameras and staff	The safety measures present at the bus stops.
Safety level of the environment	Low, average, high	The safety level of the environment refers to the presence of sufficient light, escape routes, and the physical openness at the bus stops.
Attributes: Bicycle		
On-vehicle time	20 min, 25 min, 30 min	The on-vehicle time of the bicycle is the time you need to cycle from home to the railway station.
Time to get to railway platform	2 min, 4 min, 6 min	The time you need to get from the bicycle parking facility to the railway platform.
Cost of parking bicycle	€0.50 per time, €1.00 per time, €1.50 per time	The cost of the parking facility at the railway station.
Chance of delay	0%, 5%, 10%	The chance that you suffer from any form of delay (travel time longer than usual) during the travel with the bicycle.
Safety measures	Poor presence of cycling facilities; moderate presence of cycling facilities; good presence of cycling facilities	The safety measures for the bicycle are about the cycling facilities that are present on the cycling route. Good cycling facilities include separate cycle paths and priority intersections.

(continued)

Table 1. (continued)

Attribute	Attribute levels	Explanation to respondent
Safety level of the environment	Low, moderate, high	The safety level of the environment means the presence of enough light and physical openness on the bicycle paths.
Attributes: Car/drop-off		
In-vehicle time (IVT)	10 min, 15 min, 20 min	Car in-vehicle time is the time you spend in the car traveling from home to the railway station.
Time to get to railway platform	2 min, 5 min, 8 min	The time you need to get from the parking facility or drop-off place to the railway platform.
Cost of parking car	€3.00 per day, €3.50 per day, €4.00 per day	The cost of the parking facility at the railway station per day.
Chance of delay	0%, 10%, 20%	The chance that you will experience any form of delay (travel time longer than usual) while traveling by car.
Safety measures	No additional measures; additional security cameras; additional security cameras and staff	The safety measures present at the parking facility or the place you are dropped off by car.
Safety level of the environment	Low, moderate, high	The safety level of the environment means the presence of enough light, flight routes, and physical openness at the parking/drop-off facilities.

Example choice task

Characteristics	Train			Car
In-vehicle travel time	50 minutes			60 minutes
Parking wait & search time	3 minutes			9 minutes
Costs train / car trip	14 euros			15 euros
Costs car parking				4 euros per hour
Chance on delay	10 percent			20 percent
Safety	No specific measures			Extra camera's and guards
Possibilities to work	Special work spaces			
	Pre-transport			
	Bus	Bicycle	Car drop-off/parking	
Time to bus stop	4 minutes			
In-vehicle travel time	20 minutes	25 minutes	15 minutes	
Time to platform	3 minutes	4 minutes	5 minutes	
Travel and storage costs	1 euro	1.50 euro per use	3.00 euro per day*	
Chance on delay	15 percent	5 percent	10 percent	
Type of bus	Express bus			
Safety measures	No specific measures	Spacious bicycle facilities	Extra security camera's	
Safety level environment	High	Average	Low	
Your preference	○	○	○	○

* This parking tariff holds only if you park the car in the vicinity of the train station.

Figure 4. Example transport mode-choice task.

Data Collection

The data collection took place in the spring of 2015. The respondents that participated in this study were recruited from the panel of PanelClix Online Market Research (www.panelclix.co.uk). Because of the setting of the stated choice experiment, only panel members living in Eindhoven or in the vicinity of Eindhoven were invited to participate (for spatial distribution, see Figure 5). In total, 415 respondents completed the questionnaire. Certain personal details of the respondents are presented in Table 2. The table shows that the sample is very diverse with an acceptable respondent distribution across all distinct levels.

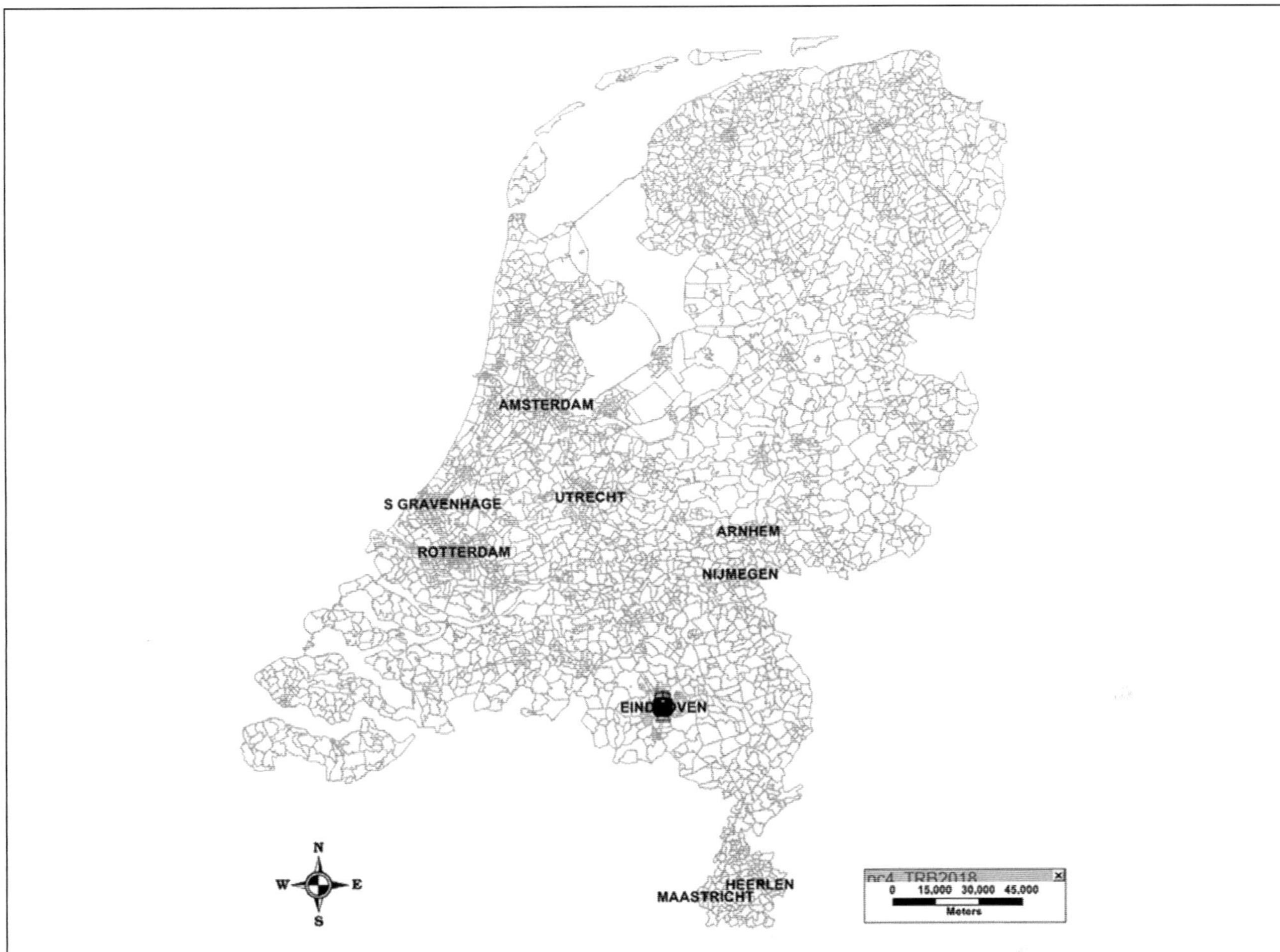

Figure 5. The spatial distribution of respondents and major cities in the Netherlands.

Descriptive Analyses

The first part of the questionnaire was set up to gain greater insight into the respondents' experiences of traveling from Eindhoven to major cities in the Netherlands (Figure 5). In total, eight cities were included in the questionnaire: Utrecht (by car 90 km from Eindhoven), Amsterdam (130 km), Nijmegen (70 km), Arnhem (80 km), Heerlen (90 km), Maastricht (90 km), Rotterdam (110 km), and 's-Gravenhage (140 km).

These questions were also included to prepare the respondent for the stated choice part in which they had to assume that they were going to make a trip from their home location in Eindhoven or surroundings to a major city in the Netherlands. In this paper two major insights are highlighted: the respondents' visit frequency to major cities and the mode used when a respondent visits a major city. Regarding visit frequency, the respondents were asked to indicate for the aforementioned eight cities how often they have visited each (1 = Never, 2 = Sometimes, 3 = Often, 4 = Very often). For presentation purposes, the numbers connected to the answer categories are summed, creating a form of "visiting index." The values of the index range between 8 (none of the eight cities have been visited) and 32 (all eight cities have been visited "Very often"). Subsequently, the indices are classified into categories for which a higher category means more experience visiting a major city. Figure 6 shows the percentages of respondents assigned to each category. There was no respondent that had "Never" visited all of the included cities (value 8). There was only one respondent who frequently visited all eight cities (value 32). From Figure 6 it can be concluded that all respondents have some experience of visiting at least one of the major cities in the Netherlands. This makes it easier for the respondent to understand the provided choice situation in the stated choice experiment.

Regarding the transport mode used when visiting a city, a distinction is made between the use of car, train, and other (including taxi, electric bike, and express bus). Figure

Table 2. Personal Characteristics of the Respondents

Characteristics	Levels	Frequency	Percentage
Gender	Male	189	45.5
	Female	226	54.5
Age	49 years and younger	221	53.3
	50 years and older	194	46.7
Educational level	Medium (professional)	192	46.3
	High (college and university)	223	53.7
Family composition	Single	114	27.5
	Family without children	181	43.6
	Family with children	120	28.9
Total number of respondents		415	100.0

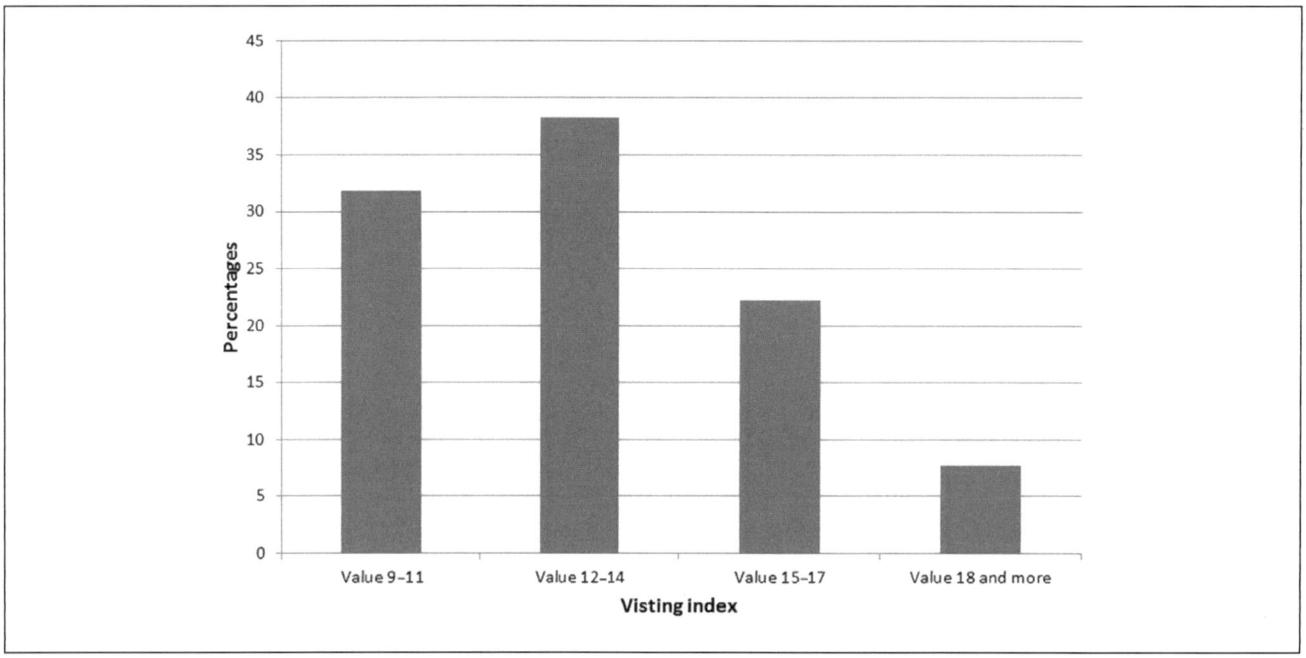

Figure 6. Sample visiting index values (n = 415).

7 shows the percentages for respondents who have visited a city at least "Sometimes" (indicated by N). It is possible that the sum of the percentages exceeds 100% because some respondents indicated that they use various transport modes. From the figure, it can be concluded that for certain cities (e.g., Utrecht and Amsterdam) the majority of the respondents use the train when visiting the city. In the case of the other cities it is clear that the car is preferred but that the train attracts a considerable share.

Model Analysis

The individual choices were investigated using a mixed logit (ML) model with panel effects. This model takes into account random taste variation in the population of decision makers and the fact that a decision maker can make more than one decision. In the current experiment each respondent evaluated nine choice situations. The model was flexible enough to completely relax the independence and identically distribute the error structure of the standard multinomial logit (MNL) model (*21*). Several examples show that ML models perform better than the standard MNL models (*22, 23*).

The 415 respondents evaluated a combined total of 3,735 choice situations. The software package NLOGIT 5.0 (*24*) was used to estimate the model parameters. To represent the effects of the attribute levels on the probability of a certain transport mode (combination), effect coding was used. In the current paper, the model with linear effects only is presented because it appeared to be

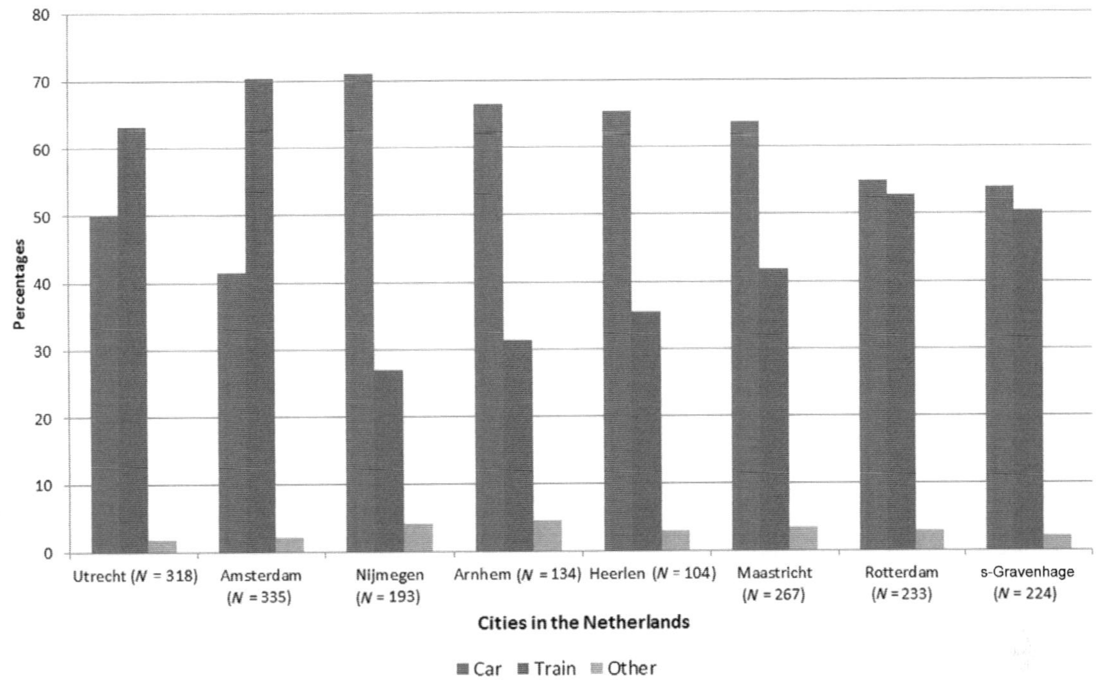

Figure 7. Transport mode use to major cities in the Netherlands.

the best performing model. A model with both linear and non-linear effects showed no increase in performance. The same was true for a model with context effects (effects based on age, gender, education, etc.) This means that the estimated parameter represents the part-worth utility of the first attribute level (see Table 1). The part-worth utility of the second level is always equal to 0, and the part-worth utility of the third level is equal to $-1 \times$ the estimated parameter. In addition, only parameters for the standard deviations of the access mode specific constants were included in the model to look for taste variation.

The estimated model was tested against the model with all coefficients equal to 0 (null model) using the log-likelihood ratio statistic (LRS). The LRS-value indicates that the estimated model performs significantly better than the null model (for details see Table 3). The model is also perfectly able to manage back forecasting the observed choices (see McFadden's Rho squared values).

Looking in more detail at the estimated parameters, it appears that several attributes significantly contribute to the total utility of the included transport mode alternatives. A positive sign means that the total utility increases, which means that the probability of the alternative increases. A negative value means that the total utility decreases. The parameters of the alternative specific constants show that, if no other attributes are considered, respondents prefer the car over the train with any of the access modes. However, the significant parameters for the standard deviations show that there are taste variations between respondents (heterogeneity).

The attractiveness of the train increases when the travel time decreases (e.g., from 55 min to 45 min). The same is true for a decrease in travel costs (e.g., from €20.00 to €10.00). The attractiveness of the train is intensified when the parameters of the train access modes (bus, bicycle, and drop-off/car) are positive. In the case of the bus, the attractiveness of the train increases when walking time to the bus stop is short (approximately 1 min), in-vehicle time is brief (approximately 15 min), cost of travel is low (approximately €1.00), the chance of a delay is small (approximately 0%), safety measures are present, and the environment safety-level is high. The following attributes of the bicycle contribute to the attractiveness of the bike itself and, in addition, to the attractivity of the train significantly: on-vehicle time, chance on delay, and safety measures. The following drop-off/car attributes contribute to the attractiveness of the drop-off car and train alternative: in-vehicle time, time to railway platform, and cost of travel. The attractiveness of the car is significantly related to in-vehicle time, cost of travel, and parking costs.

The comparison between the contributions of all attributes can best be presented by the range in part-worth utility that an attribute covers (see Figure 8). It appears that the attribute in-vehicle time of the access mode

Table 3. Parameter Estimates of Mixed Logit Model, Linear Effects Only

Attributes	Attribute levels	Mean	Standard deviation
Constants	Bus	–0.2495	4.5470***
	Bicycle	–0.6264**	4.6069***
	Drop-off/car	–0.8796***	3.6758***
Train			
In-vehicle time	45 min	0.1462*	
Waiting time	3 min	0.0484	
Cost of travel	€10.00	0.1868**	
Chance of delay	5%	0.0510	
Safety measures	No additional measures	–0.0650	
Possibility to work	No additional facilities	0.0132	
Bus			
Time to bus stop	1 min	0.2693***	
In-vehicle time	15 min	0.3084***	
Time to railway platform	1 min	0.0438	
Cost of travel	€1.00	0.3067***	
Chance of delay	0%	0.3523***	
Type of bus	Regular bus	0.0572	
Safety measures	No additional measures	–0.1411*	
Safety levels: environment	Low	–0.1526**	
Bicycle			
On-vehicle time	20 min	0.4488***	
Time to railway platform	2 min	0.0182	
Cost of bicycle parking	€0.50 per time	0.1172	
Chance of delay	0%	0.1573**	
Safety measures	Poor presence of cycling facilities	–0.2363***	
Safety levels: environment	Low	–0.0755	
Drop-off/car			
In-vehicle time	10 min	0.4650***	
Time to railway platform	2 min	0.2439***	
Cost of travel	€3.00 per day	0.1699**	
Chance of delay	0%	0.1024	
Safety measures	No additional measures	–0.0281	
Safety levels: environment	Low	–0.0791	
Car			
In-vehicle time	50 min	0.3797***	
Parking search time	3 min	0.0785	
Cost of travel	€10.00	0.2959***	
Parking costs	€2.00 per hour	0.3466***	
Chance of congestion	10%	0.0021	
Safety	No additional measures	–0.0480	
Goodness of fit			
Loglikelihood (β)		–5177.8094	
Loglikelihood (*0*)		–3026.8471	
LRS (test value for 38 df: 53.384)		4301.9247	
Rho^2		0.415	
Adjusted Rho^2		0.413	

Note: ***Significant at 99%. **Significant at 95%. *Significant at 90% confidence level. df = degrees of freedom.

drop-off/car has the highest contribution to this transport mode context, followed by on-/in-vehicle times of bikes and cars. The limited effect of in-vehicle time for the train, and the safety measures and safety levels of the bus is remarkable.

Conclusions

This study was initiated because of the limited availability of insights into the influence of train access modes on the competitive position of trains versus the car. A stated choice experiment was set up to investigate various

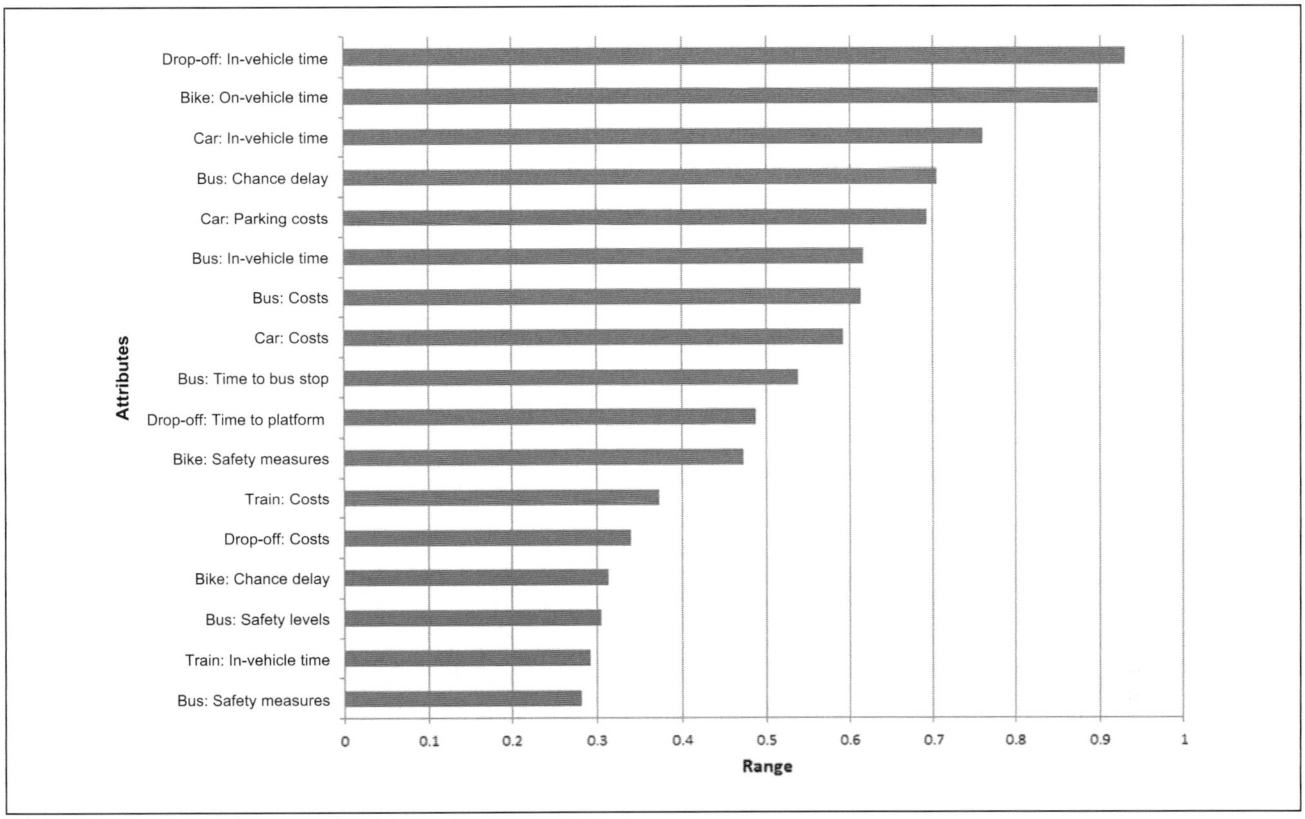

Figure 8. Range in part-worth utilities (only the significant parameters are illustrated).

attributes of train access modes in more detail. The results of the model estimation showed that diverse attributes of train access modes play a role when travelers compare the train and car. The model estimation results show that there are considerable differences in the effects of these attributes. The most influential attributes are related to in-vehicle travel time and travel costs, including parking costs. In the context of the access modes, bus and bicycle safety measures play an influential role. The insights of this study can be used to set up a strategy to motivate car drivers to switch from car use to the train for medium- and long-distance trips. The focus of improvement plans should be on cost reduction, travel time and delay minimization, and safety measures. The insights could also be used to improve the access modes for train users by offering fast and safe routes for bicyclists. This is especially relevant in the context of transit-oriented development in which the focus is on stimulating sustainable transport use by developing optimal urban development and public transport conditions.

The study included some limitations that would be worth exploring in the near future. As mentioned before, the stated choice experiment did not include travel purpose, travel day, or travel party. These kinds of trip characteristics could also contribute to travelers' transport mode decisions because of differences in time pressure, levels of crowdedness in trains, congestion on roads, and traveling with, for example, children. For each trip characteristic, the challenge is to find sufficient respondents who are familiar with the tradeoff between train and car. Even though a large set of attributes was used, other attributes might be considered in the future covering, for example, dynamic information provision and smart navigation possibilities. Finally, it is important to validate the findings from this study with real-world data to identify whether the stated preferences are in line with revealed preferences.

Acknowledgments

The authors thank Wim Schaefer and Koen Sanders of the Construction Management & Engineering group of the Eindhoven University of Technology for their contribution to the research.

Author Contributions

The authors confirm contribution to the paper as follows: study conception and design: PvdW, JvdW; data collection: PvdW, JvdW; analysis and interpretation of results: PvdW, JvdW;

draft manuscript preparation: PvdW. All authors reviewed the results and approved the final version of the manuscript.

References

1. Rodrigue, J. P., C. Comtois, and B. Slack. *The Geography of Transport Systems*. Routledge Taylor & Francis Group, London, UK, 2006.
2. Central Bureau of Statistics. *Dutch National Travel Survey 2015*. Central Bureau of Statistics, Heerlen, the Netherlands, 2015.
3. Van Exel, N. J. A., and P. Rietveld. Could You Also Have Made This Trip by Another Mode? An Investigation of Perceived Travel Possibilities of Car and Train Travelers on the Main Travel Corridors to the City of Amsterdam, the Netherlands. *Transportation Research Part A*, Vol. 43, 2009, pp. 374–385.
4. Van Wee, B. The Unsustainability of Car Use. In *Handbook of Sustainable Travel*, edited by T. Gärling, D. Ettema, and M. Friman., Springer, Dordrecht, the Netherlands, 2014, pp. 69–83.
5. Beirão, G., and J. A. S. Cabral. Understanding Attitudes towards Public Transport and Private Car: A Qualitative Study. *Transport Policy*, Vol. 14, No. 6, 2007, pp. 478–489.
6. Redman, L., M. Friman, T. Gärling, and T. Hartig. Quality Attributes of Public Transport that Attract Car Users: A Research Review. *Transport Policy*, Vol. 25, 2013, pp. 119–127.
7. Boarnet, M. G., G. Guiliano, Y. Hou, and E. J. Shin. First/Last Mile Transit Access as an Equity Planning Issue. *Transportation Research Part A*, Vol. 103, 2017, pp. 296–310.
8. Brands, T., E. de Romph, T. Veitch, and J. Cook. Modeling Public Transport Routes Choice with Multiple Access and Egress Modes. *Transportation Research Procedia*, Vol. 1, 2014, pp. 12–23.
9. Rietveld, P. The Accessibility of Railway Stations: The Role of the Bicycle in the Netherlands. *Transportation Research Part D*, Vol. 5, 2000, pp. 71–75.
10. Kennisinstituut voor Mobiliteitsbeleid/KIM. *Ministry of Infrastructure and Ecology, Mobility Vision 2014* (in Dutch). Kennisinstituut voor Mobiliteitsbeleid/KIM, The Hague, the Netherlands, 2014.
11. Zhou, J. Sustainable Commute in a Car-Dominant City: Factors Affecting Alternative Mode Choice among University Students. *Transportation Research Part A*, Vol. 46, 2012, pp. 1013–1029.
12. Goel, R., and G. Tiwaro. Access-Egress and Other Travel Characteristics of Metro Users in Dehli and Its Satellite Cities. *IATSS Research*, Vol. 39, 2016, pp. 164–172.
13. Dell'Olio, L., A. Ibeas, and P. Cecin, The Quality of Service Desired by Public Transport Users. *Transport Policy*, Vol. 18, 2011, pp. 217–227.
14. Mouwen, A. Drivers of Customers Satisfaction with Public Transport Services. *Transportation Research Part A*, Vol. 78, 2015, pp. 1–20.
15. Salas, R. D. Comparison of Station Access Quality Ratings and Travel Behavior: An Examination of Commuter Rail Access Station Mode Access. Presented at 94th Annual Meeting of the Transportation Research Board, Washington, D.C., 2015.
16. Chowdhury, S., K. Zhai, and A. Khan. The Effect of Access and Accessibility on Public Transport Users' Attitudes. *Journal of Public Transportation*, Vol. 19, 2016, pp. 97–113.
17. Givoni, M., and P. Rietveld. The Access Journey to the Railway Station and Its Role in Passengers' Satisfaction with Rail Travel. *Transport Policy*, Vol. 14, 2007, pp. 357–365.
18. Brons, M., M. Givoni, and P. Rietveld. Access to Railway Stations and Its Potential in Increasing Rail Use. *Transportation Research Part A*, Vol. 43, 2009, pp. 136–149.
19. Hensher, D. A., J. M. Rose, and W. H. Greene. *Applied Choice Analysis: A Primer*. Cambridge University Press, Cambridge, UK, 2005.
20. Sanders, K. *The Role of Pre-Transport in Travelers' Transportation Mode Choice: A Study Conducted in the Region of the Railway Station of Eindhoven*. Master thesis in Construction Management & Engineering. Eindhoven University of Technology, Eindhoven, the Netherlands, 2015.
21. Bhat, C. R., N. Eluru, and R. B. Copperman. Flexible Model Structure for Discrete Choice Analysis. In *Handbook of Transport Modeling*, 2nd ed., edited by D. Hensher, and K. J. Button Elsevier Ltd, Oxford, UK, 2008, pp. 75–104.
22. Hess, S., and J. W. Polak, Mixed Logit Estimation of Parking Type Choice. In *Transport Statistics*, edited by B. W. Sloboda. J. Ross Publishing Inc., Fort Lauderdale, FL, 2009, pp. 77–102.
23. Borgers, A., A. Kemperman, L. Toll, and H. Timmermans. Measuring Preferences for Parking Facilities in Old Residential Areas. *Proc., 7th International Symposium on City Planning and Environmental Management in Asian Countries*, Fukuoka, Japan, 2010, pp. 1–14.
24. Economic Software Inc. *NLOGIT Version 5*, Economic Software Inc., Plainview, TX, 2012.

The Standing Committee on Transportation Demand Management (ABE50) peer-reviewed this paper (18-01017).

Article

TRR
JOURNAL OF THE TRANSPORTATION RESEARCH BOARD

Transportation Research Record
2018, Vol. 2672(8) 731–741

DOI: 10.1177/0361198118773195
journals.sagepub.com/home/trr
SAGE

Modal Share Change following Implementation of Travel Demand Management Strategies

Francesco Piras[1], Eleonora Sottile[1], and Italo Meloni[1]

Abstract

The demand for urban mobility has seen a marked increase over the last few decades. The expansion of road networks, the increase in distance traveled and in the number of trips made has created car-dependent cities. One solution for achieving the switch from private car to public transport is the introduction of a new public transport line that improves the accessibility and frequency of service. However, an important behavioral process underpins travel choices and often a change in choice context is not sufficient to evoke behavior change. Voluntary travel behavior change (VTBC) programs were developed for heightening individuals' awareness of the consequences of their travel choices and of the benefits to be reaped from using sustainable alternatives, through information provision and communication strategies. The objective of the present paper is to analyze the effect on travel mode choice of introducing a new light railway line into the choice set (hard measure) when implementing a VTBC program on a large scale, but with a high degree of personalization (soft measure). Although numerous studies have demonstrated the need to implement a (hard) measure that acts on the choice context, in combination with a (soft) measure that acts in a personalized way on demand, as far as the authors are aware, few have put this into practice. The first results have shown that a combination of hard and soft measures is more effective in evoking travel behavior change and personalized travel plans produce better results than generalized soft measures.

The demand for urban mobility has grown substantially over the last few decades, to the extent that projections based on data for 2000 predict a threefold to fourfold increase in passenger kilometers traveled by the year 2050 (*1*). The reason for this is the rapid population growth, especially in the urban population as a consequence of urbanization. This, in turn, has led to the haphazard and uncontrolled development of cities and towns with major negative impacts on the environment. The proliferation of road infrastructure, the increase in the distance traveled and number of trips has created a strong dependence on the car. In Italy, 65.3% of journeys are made by car (mostly as driver), an increase of nearly eight points over the last 15 years (*2*).

To reduce car dependence in favor of more sustainable alternative modes of travel requires promoting appropriate policies that also ensure an efficient organization of the transportation system. Public transport has a major role to play in reducing private car use (*3*). The introduction of new routes for improving accessibility, density, and frequency of public transport services, so as to be able to compete effectively with the private car, may be one solution for reducing vehicle kilometers traveled. Service quality is perceived as an important determinant of users' travel demand (*4*).

Therefore, analyzing travel behavior is not a simple task for car use (*5*), as it embraces individuals' attitudes, perceptions, habits, and knowledge. The so-called voluntary travel behavior change (VTBC) programs were conceived to take these aspects into account (*6*). These programs use information and communication tools to encourage people to change their travel behavior, and their implementation has proven to be effective in reducing private car use (*7*). Analysis of several projects conducted over the past 20 years has pinpointed the strengths and weaknesses of these measures (*8–10*). In particular, information provision is just one of the elements of the process, and alone, without any intervention in the choice context, it may well be ineffective.

For these reasons, the implementation of an integrated system of structural and information measures

[1]University of Cagliari, Cagliari, Italy

Corresponding Author:
Address correspondence to Francesco Piras: francesco.piras@unica.it

could yield better results and enhance the effect of each measure.

The aim of this paper is to analyze the effect on travel mode choice of introducing a new sustainable form of transport into the choice set (hard measure) when implementing a VTBC program (soft measure). The transport context chosen for this experiment is a corridor linking the city center of Cagliari (Italy) to a university/hospital complex, where a new light rail route went into service in February 2015. A survey called "Cittadella Mobility Styles" was designed that included the new light rail line in the choice set of a large-scale VTBC program conducted in the metropolitan area of Cagliari. The survey included the following steps: (1) interception of individuals traveling to the university complex and collection of information on their travel habits before the new light rail service became operational, on the psychological aspects underpinning their travel choices and life style, and on socio-economic characteristics; (2) design and delivery of a personalized travel plan (PTP), providing a detailed description of the tailored sustainable trip for reaching the university campus with the new light rail service, as an alternative to the private car; (3) monitoring of the post-PTP phase for detecting any travel behavior changes and thus the effect of combining information (soft) and hard measures. In particular, an attempt was made to overcome the critical issue concerning evaluation of the measure by creating a control group, so as to disentangle the effect of the structural measure from that of personalized information provision.

The paper is organized as follows. We first provide a literature review of the different travel demand management measures and applications implemented worldwide. Then we describe the application. Next, we present the results as behavioral change achieved. Conclusions and future research directions are presented in the final section.

Literature Review

Travel demand management (TDM) is defined as any initiative with the object of reducing the negative impact of car use (*11*).

Structural measures, also referred to as hard measures, aim to change the external context in which choices are made (*12*). Hard measures include, for instance, improvements to infrastructure and management of public transport services, increasing costs for car use, and bans which restrict car use (*13*).

However, hard measures do not always produce the desired results as they do not succeed in making car use less appealing or triggering behavior change. Traveling by car offers many personal benefits, both instrumental (flexibility, shorter travel times, comfort) and psychological (feeling of freedom, etc.) (*14*), that make it difficult to change a choice that has become habitual, even though other alternatives, objectively more advantageous in time, cost, and so on, do exist (*15*). Furthermore, people who behave habitually are less likely to be interested in information about available alternatives and tend to make decisions that involve few cognitive resources (*16*). For all these reasons a structural measure may not be sufficient to influence car drivers to change their travel behavior: they need to be informed and educated to enhance their knowledge (for example of transport alternatives), heightening their awareness (for example of environmental impacts) and modifying attitudes, thereby strengthening the propensity to adopt non-motorized forms of travel (*17*). On the basis of the studies of these process, soft measures, known as VTBC programs (*6*), were developed. They aim to change individuals' perceptions, motivations, knowledge and norms (*12*), so as to steer them toward a more pro-environmental behavior. Informational strategies aim to enhance car users' knowledge of behavioral alternatives and their pro and cons (*12*). They may include social modeling (*e.g.* preeminent public figures using alternative travel modes) to strengthen social norms as well as individualized marketing (providing people with customized information about their travel options) (*18*).

Cairns et al. (*19*) identified and defined 10 different soft measures, including:

- Workplace/school travel plans, in which a package of measures is introduced at an individual school/workplace to encourage students/employees to travel more sustainably;
- Personalized travel planning (PTP), in which individuals are provided with personalized information, to make them aware of available sustainable means of transport and encourage their use;
- Public transport information and marketing (PTIM), which includes advertising campaigns and the provision of information in more accessible formats;
- Travel awareness campaigns, in which several media are involved and attempt to enhance general public awareness of the negative consequences of car use.

Numerous PTP programs have been implemented worldwide over the last 20 years adopting different methodological approaches. Some examples are IndiMark and TravelSmart (*7*), Travel Blending (*20*), Travel Feedback Programs (*21*), PTP (*19*), PTP in Sweden (*22*) and Casteddu Mobility Styles (*5*). Analysis of all these experiences demonstrated that PTP can reduce car usage by around 10% and lead to significant increases in walking, cycling, and public transport use.

Other experiences have shown that "soft" measures, such as PTP, generally work better when used in conjunction with relevant infrastructure improvements (new cycle paths, new bus routes, new pedestrian crossings, etc.) or with the introduction of "sticks" such as parking charges or reduced capacity for road traffic or parked vehicles (*10*). Cairns et al. report that plans which included parking management measures achieved an average reduction in car driver trips of >24%, compared with 10% for those that did not (*23*). A study by Transport for London shows that the implementation of smarter choice measures combined with a road pricing measure achieved a reduction in total traffic levels of 8–17%, compared with 2–4% if they had been applied alone (*19*).

TDM strategies have also been implemented during special events. An application of Travel Blending was implemented in Sydney as a part of the efforts to reduce car travel in the lead up to the Olympics in 2000 (*24*). During the Olympic Games in London in 2012 a series of TDM strategies, which included information measures, were implemented (*25*).

Analysis of the numerous projects conducted over the past 20 years identified the key factors distinguishing a successful VTBC program (*9*):

- Target mobility context: if the context in which the measure is to be implemented is inappropriate, it may be ineffective.
- Target behavior and population: selecting clear and measurable behavioral goals facilitates behavior change (*26*). It is also essential to target a suitable segment of the population for the alternative to be promoted. This implies an in-depth knowledge of participants' socio-economic and psychological characteristics.
- Removing barriers to behavior change. Barriers can be identified as:
  - External barriers, associated with the transport and social-economic context;
  - Internal barriers, associated with the psycho-motivational nature of the travel choice and reflecting the individuals' point of view on the travel choice made (negative attitude toward a certain means of transport, poor awareness of automobile externalities, etc.);
  - Barriers caused by habitual behavior.
- Personalization: this is the most effective tool for breaking down internal barriers to a sustainable means of transport (*21*). The greater the level of information customization of a VTBC program, the greater its effectiveness will be (*9*). The proposed alternative must take into account individuals' limits and needs (*27*). Another important element is the contact between the VTBC program team and participants, to exchange opinions and offer advice (*21*).
- Information: this is the core of a VTBC program. The information provided must be able to influence those psychological aspects crucial to mode choice. It should be useful, usable, used and readily acquired (*28*). Feedback serves to compare the disadvantages of actual behavior with the advantages of sustainable behavior, as individuals are unable to quantify them.
- Persuasion: information, by itself, may be ineffective in evoking behavior change (*29*). Information needs to be integrated with communication and persuasion techniques, based on the knowledge of individuals' psychology (*9*).
- Use of one of three behavioral approaches or a combination thereof: 1) behavioral theories and models such as the Theory of Planned Behavior (*30*), the Norm Activation Model (*31*), the Value-Belief-Norm theory (*32*), or the Theory of Interpersonal Behavior (*33*); 2) behavior change theories such as Lewin's Theory of Change, Transtheoretical Model (*34*), Self-Regulation Theories (*35*), or the Model of Implementation Intention (*36*); 3) DEFRA social marketing (*37*) and persuasion techniques (*38*).
- Evaluating the effectiveness of interventions and monitoring: evaluation of a soft measure is important for researchers and policy-makers (*12*). In general, these are assessed in terms of car use reduction (time traveled, distance, number of trips) and benefits for the single individual (kilocalories [kcal] burned]) or community (e.g., carbon dioxide [CO_2] emissions reduction). Monitoring is another key factor for observing any changes in travel behavior and whether changes are long lasting.

Notwithstanding the above results and the well-known procedures for conducting a successful VTBC program some unsolved issues remain. Tertoolen found that although information can affect attitudes, it does not affect travel behavior (*39*). Eriksson et al. found that information campaigns are considered by car drivers to be a relatively ineffective measure compared with improving public transport and increasing fuel taxes (*40*). One problem associated with information is the level of personalization, as it affects the effectiveness of the measure. Eriksson et al. found that less than 2% of respondents believed that improving timetable information would persuade them to use public transport (*41*).

However, a high degree of personalization can be costly and difficult to achieve in large scale programs (*9*).

Another problem is how to evaluate the effectiveness of a measure, particularly in the absence of a control group. VTBC programs are often conducted within broader campaigns launched to reduce car travel, with a focus on the overall effect (*42*). So, without a control group, it is difficult to assess the effectiveness of the implementation of a single measure and disentangle it from a combination of effects. A study by Fuji et al. shows that outcomes of 15 Travel Feedback Programs conducted in Japan differ depending on whether suitable control groups have been included in the study or not (*43*). Another problem that affects evaluation of VTBC programs is the variability in travel from day to day, from season to season, and in response to external stimuli (*44*). Multi-day self-report surveys, with the compilation of a 2-day activity diary among those who agreed to participate, may be the best option to overcome this problem (*44*). Another criticism is that the evaluation is highly dependent on self-reporting (*10*). A solution could be the use of passive GPS devices, which require almost no action on the part of respondents (*44*).

Application

The VTBC program described here was implemented by CRiMM research center (Centro di Ricerca Modelli di Mobilità) at the University di Cagliari between 2013 and 2015 in the metropolitan area of Cagliari, Italy. The program, called "Cittadella Mobility Styles" focused on the different travel patterns of individuals going to the university's hospital/scientific complex (Cittadella Universitaria) situated in the suburbs.

The number of people potentially attracted daily to the Cittadella amounts to a little more than 10,200: 1,784 university and hospital employees (17.5%), 7,872 students (77.2%) and 580 (5.7%) for hospital admissions, medical examinations, visiting patients, and so on.

Parking in the Cittadella is free and unrestricted and students are entitled to reduced fares on public transport.

In 2008 the first light rail line began operating between the center of Cagliari and the municipality of Monserrato, along a route with seven stops between the two terminals. During implementation of the VTBC program, the line was extended by 2 km up to the hospital/university complex providing a direct connection from the city center, so users no longer had to travel the last segment by bus. The service became operational in February 2015. Journey time along the entire route is around 22 minutes and trains run every 10 minutes at off-peak times, more frequently during rush hours.

In light of the above, a decision was made to promote the new light rail line as a sustainable travel alternative with a view to determining the effect of combining hard and soft measures. The survey targeted car drivers as public transport users travel sustainably.

The Cittadella Mobility Styles program consisted of three macro-phases:

- First wave survey: a preliminary survey to capture the travel patterns of individuals reaching the Cittadella;
- PTP: creation and delivery of a PTP containing suggestions for a sustainable travel alternative to the Cittadella (designed on the basis of the information collected with the first wave survey);
- Second wave survey: assessment of the effectiveness of the measure and monitoring by administering a second questionnaire after deliver of the PTP.

First Wave Survey

The first survey was implemented in April 2013. The aim was to intercept as many university students, university and hospital staff, and visitors to the hospital for medical examinations, admissions, and so on, as possible. Potential candidates were contacted both via mailing lists provided by the university and hospital (8,847 invitation mails were sent) and through a promotional campaign conducted via traditional communication channels (postcards, press conferences, TV and daily newspapers) and social media, inviting them to complete an online questionnaire hosted on the project's web site (the web site recorded 6,402 visits during the promotional campaign period). The possibility of winning an iPad or €100 gift voucher was offered as an incentive to filling in the questionnaire.

The objective of the first wave survey was to gather detailed information about commuters and individuals traveling occasionally to the Cittadella, so as to be able to define, with a high degree of personalization, the information to be presented in the PTP delivered to the selected target. In particular, the questions concerned:

- Trip description:
  - Trip origin and frequency,
  - Mode used,
  - Trip characteristics (walking time, waiting time, time aboard, etc.),
  - Fare;
- Personal perceptions, propensities, attitudes, beliefs:
  - Attachment to car,
  - Dislike of public transport,
  - Propensity to use the light rail,
  - Proenvironmental behavior,
  - Environmental awareness,

- Level of confidence with information systems;
- Personal information:
  - Age,
  - Gender,
  - Education,
  - Occupation,
  - Income,
  - Household structure,
  - Car ownership.

The questionnaire was filled in by 2,886 individuals; 2,163 questionnaires were completed (74.9%) and 723 partially completed (25.1%).

Sample characteristics are shown in Table 1. The most widely used mode of travel reported by respondents was public transport with a percentage of around 52.7% (which reflects the large number of students in the sample), followed by car as driver, 37.7%, car as passenger, 7.7%, and motorbike/moped, 1.9%. Public transport is the preferred option of students (88.2%), as the majority of them (83.9%) do not own a car (the reason being low income) and they are entitled to reduced fares. Individuals traveling for purposes other than study prefer alternative means of transport, in spite of public transport being, on average, cheaper. In particular, car drivers who had the option to travel by public transport paid, on average, € 2.18 for a single car trip, versus an average of € 0.89 for a single trip on public transport.

The PTP

The PTP is one of the key components of a VTBC program. The choice to opt for a personalized approach was based on the findings of several works that demonstrated personalized information to be more effective in evoking travel behavior change than mass communication, as car drivers cannot simply ignore it (*15*).

Analysis of the questionnaire responses, particularly place of residence and means of transportation used to reach the Cittadella, made it possible to identify the target population. All car drivers living within the metropolitan area of Cagliari, along the light rail corridor, were considered as potential users. These numbered 544 out of a total of 2,163 car drivers. The PTPs were not delivered to all 544 car drivers identified, as a sub-sample of 20% (109 users) was randomly chosen as a control group for evaluating the effectiveness of the VTBC program. In this way it was possible to disentangle the effect of the information (soft) measure from the structural (hard) measure and vice versa and at the same time assess the combined effect of the hard and soft measures.

The travel alternative was created on the basis of the travel patterns reported by respondents to the preliminary questionnaire. All the travel information and quantitative feedback on weekly use of the car, distance traveled, and journey time were analyzed. To create the PTP using the light railway as an alternative means of travel, the analysis took into account the departure station, walking time and distance to the nearest station, waiting time at the station, journey time and distance, and fare paid. The alternatives were processed using Citilabs CUBE software.

Special attention was devoted to the graphics of the PTP and to the information contained therein as it has been demonstrated that accuracy is important and that it matters that the information is conveyed in a manner that is easily understood (*45*).

The PTP contained the following information:

- A practical map showing the route to reach the light railway station;
- Personalized slogans and other useful information on sustainable travel in general and specifically on use of the light railway;
- Links that provide useful information on mobility;
- A detailed description of the actual individual and environmental effects of the travel behavior adopted.

In fact, the PTPs presented a daily, weekly and annual evaluation of the key factors that come into play in travel choice: monetary cost, journey time, calories burned, and CO_2 emissions. The disadvantages associated with car use with respect to the light rail were also indicated, so as to offer new viewpoints and food for thought about mobility and its effects, with a view to evoking travel behavior change;

The PTP was e-mailed to the participants at the address indicated in the preliminary survey. The information contained therein could be read from any mobile device (pc, tablet, smartphone), so that participants always had the information on hand.

The cost of the marketing campaign, staff, and creation of the PTP was around € 30,000.00.

Second Wave Survey

Once the light rail service had become operational (February 2015), PTPs were delivered to the selected target. A few months later a second wave survey was implemented to give participants time to process the information and try the proposed alternative. The aim of the second survey was to:

- Monitor users' behavior after PTP delivery and the introduction of the light rail service;
- Obtain feedback on the design and layout of the PTP and on the information contained;

Table 1. Data Collected in the First Wave Survey and Socio-Economic Characteristics

Variables	Total	By car as driver	By car as passenger	By motorcycle	By bus
Questionnaires completed	2,163	816 (37.7%)	166 (7.7%)	42 (1.9%)	1,139 (52.7%)
Socio-economic characteristics					
Gender (percent male)	40.6%	49.4%	30.7%	73.8%	34.5%
Age					
18–30	71.4%	47.1%	71.1%	47.6%	89.7%
31–40	10.3%	17.5%	7.8%	14.3%	5.3%
41–80	18.3%	35.4%	21.1%	38.1%	5.0%
Level of education					
Low (high school and lower)	71.7%	56.6%	74.7%	64.3%	82.2%
Medium (undergraduate and master degree)	20.2%	26.5%	18.7%	19.0%	16.0%
High (higher than master degree)	8.1%	16.9%	6.6%	16.7%	1.8%
Employment status					
Student	69.4%	45.0%	68.1%	40.5%	88.2%
Employed	28.1%	52.8%	22.9%	57.1%	10.0%
Unemployed or retired	2.5%	2.2%	9.0%	2.4%	1.8%
Has children	13.9%	27.6%	15.7%	26.2%	3.4%
Monthly personal income					
Low (<1,000 euro)	61.9%	43.8%	72.3%	42.8%	74.1%
Medium (1,000–2,000 euro)	25.1%	35.7%	19.9%	28.6%	18.1%
High (>2,000 euro)	13.0%	20.6%	7.8%	28.6%	7.8%
Trip characteristics					
Purpose					
Study	69.5%	48.4%	59.0%	47.6%	87.0%
Work	18.0%	36.3%	9.7%	40.5%	5.3%
Medical treatment	6.3%	7.5%	16.3%	4.8%	4.1%
Visiting patients	3.6%	5.4%	7.8%	0.0%	1.8%
Other	2.6%	2.4%	7.2%	7.1%	1.8%
Time and distance					
Travel distance (average) [km]	–	10.72	14.04	7.62	9.2
Travel time aboard (average) [min]	–	17.68	19.04	11.93	27.54
Walking time from origin to bus-stop (average-bus only) [min]	–	–	–	–	5.19
Number of changes (bus only)					
0	–	–	–	–	80.1%
1	–	–	–	–	17.2%
2	–	–	–	–	2.7%

- Collect the necessary information for modeling the phenomenon.

The questions in the second wave survey concerned:

- Trip description:
  - Trip origin and frequency,
  - Mode used,
  - Trip characteristics (walking time, waiting time, time aboard, etc.),
  - Fare;
- The light rail service:
  - Level of knowledge about the light rail service,
  - Motivation for using or not using the light rail service;
- Information about PTPs:
  - Design and layout of the PTP,
  - Quality of information provided,
  - Quantity of information provided.

The second questionnaire was sent to all individuals who accurately completed the first one, not simply to the potential light rail users in the target. The reason for this was not to preclude the possibility of analyzing the behavior of those users who were not included in the target. So, the results of the second wave survey were analyzed distinguishing three different categories:

- PTP: participants who received the PTP;
- CG: participants who formed part of the control group;
- No PTP: participants who did not receive the PTP and were not included in the CG (these were users of bus services or car drivers who did not benefit from using the light rail service as they did not live along the corridor served by light railway).

Fully completed questionnaires numbered 740, 34.2% of the total sent. Unfortunately, with surveys comprising different phases spread over time, many of the initial participants may fall by the wayside (*45*). On the other hand, in this way it is possible to collect detailed information from those who made the effort to fill in both questionnaires, and to examine the phenomenon in depth.

Analysis of the socio-economic characteristics of respondents to the second wave survey, limited to those who traveled to the Cittadella at least once (516 individuals), revealed that the sample reproduces both the first wave survey sample and the population gravitating toward the Cittadella. Table 2 summarizes individuals' socio-economic characteristics, split into the three categories mentioned above.

Results

Analysis of the results was limited to those users who participated in the first wave survey and reported traveling to the Cittadella at least once (516 individuals), the reason being that only these individuals could be compared to understand whether they had changed their travel habits after implementation of the hard and soft measures. Note that the hard measure involved all

Table 2. Data Collected in the Second Wave Survey and Socio-Economic Characteristics

Variables	PTP	CG	No PTP	Total
Questionnaires sent	435	109	1,619	2,163
Questionnaires completed	169 (38.9%)	45 (41.3%)	526 (32.5%)	740 (34.2%)
Participants traveling to the Cittadella	113 (66.9%)	29 (64.4%)	374 (71.1%)	516 (69.7%)
Socio-economic characteristics				
Gender (percent male)	42.5%	41.4%	36.4%	38.0%
Age				
18–30	39.8%	48.3%	78.6%	68.4%
31–40	17.7%	17.2%	9.4%	11.6%
41–80	42.5%	34.5%	12.0%	20.0%
Level of education				
Low (high school and lower)	44.3%	51.8%	78.3%	69.4%
Medium (undergraduate and master degree)	32.7%	24.1%	16.1%	20.2%
High (higher than master degree)	23.0%	24.1%	5.6%	10.4%
Employment status				
Student	38.1%	37.9%	78.3%	67.2%
Employed	59.3%	62.1%	19.5%	30.6%
Unemployed or retired	2.7%	0.0%	2.1%	2.1%
Has children	27.4%	31.0%	8.8%	14.1%
Monthly personal income				
Low (<1,000 euro)	42.5%	48.3%	65.2%	59.3%
Medium (1,000–2,000 euro)	34.5%	34.5%	24.1%	26.9%
High (>2,000 euro)	23.0%	17.2%	10.7%	13.8%

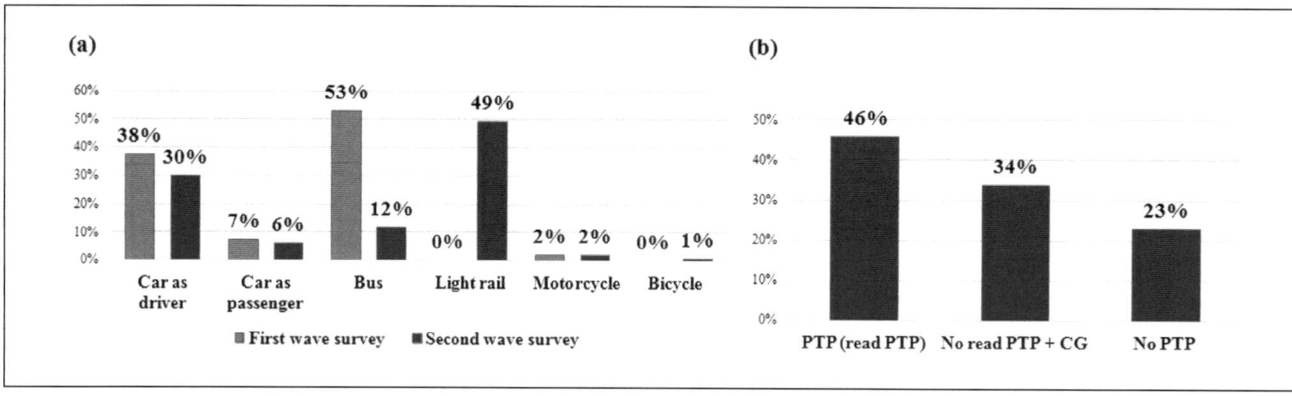

Figure 1. (*a*) Comparison of modal share first and second survey waves, and (*b*) behavior change (switch from car as driver to light rail).

individuals (PTP, CG and No PTP), whereas the effect of the soft measure is divided into two categories: 1) one related to mass communication and advertising campaign of the light rail service, which involved all individuals 2) the second related to the personalized measure represented by the PTP, which involved all those who had received and read the PTP. Therefore, it is worth noting that all the individuals, in addition to the hard measure, were intercepted by a generalized soft measure which could have had an effect, though not quantifiable, on travel behavior.

PTP Results

Note that 54 out of the 113 participants (47.8%) in PTP group who had traveled to the Cittadella at least once after the introduction of the light rail service stated they had not received/read the PTP. Therefore, from now on, they will be included in the control group as they have the same characteristics (they belonged to the potential light rail users' target but did not receive/read any personalized information).

The overall effect of evaluating the measure was not distorted by this assumption as not having received/read the PTP neutralizes the effect of the personalized measure. Thus, these individuals are only liable to be influenced by mass communication and the hard measure. Treating these individuals as PTP recipients would, conversely, have resulted in erroneously evaluating the effect of the personalized measure as it was not actually delivered.

It was decided to send the PTP via e-mail as most people are now familiar with modern technology and it enabled the measure to be implemented on a large scale. Moreover, as 65% of the sample of the first wave survey possessed a smartphone, the PTP was always on hand making it possible to test the proposed alternative without any prearrangement. However, this was not sufficient to lead people to read the PTP: direct contact, though costlier and affecting sample size, seems to still be the best way to attract people's interest.

Special attention was devoted to assessing the effectiveness of the PTP, both its design and layout and the quality of the information contained. In particular, analysis of the responses of participants who read the PTP revealed that the information was well-organized (97%), not excessive (85%) and the colors used facilitated reading (95%). The suggested alternative was explained clearly and links to the maps were considered useful for understanding it. Therefore, it can reasonably be claimed that the objective of providing participants with a graphically pleasing personalized plan and a clear description of the suggested alternative has been achieved.

As far as the PTP is concerned, the respondents stated that it also provided information on aspects they were unaware of or unable to quantify such as CO_2 emissions and kcal burned (75%), so they were able to appreciate the benefits of using the light rail (71%). This shows that the PTP was successful in delivering precise information about the alternative available to car drivers: in fact, 65% stated that the PTP provided them with information that enabled them to make a decision.

Analysis of the second wave questionnaire indicated that 47.5% had tried the travel alternative suggested in the PTP, 28.8% stated they would try it in the future, but 23.7% stated they had not tried it and did not intend to do so in the future.

Behavior Change Analysis

To evaluate any changes in travel behavior the modal share observed in the second wave survey was analyzed (Figure 1*a*). First of all, 49% of respondents had chosen to use the light railway, of whom 41% already traveled by public transport, 7% were car drivers and 1% used other means of transport.

The diagram (*b*) of Figure 1 quantifies the effect of each measure implemented on the switch from car driver to light rail user. Roughly half (46%) the participants who read the PTP (27 individuals) changed their travel behavior, against just 34% of the control group, which also included those who had not received/read the PTP, demonstrating the greater effectiveness of a personalized measure compared with mass communication. Another significant aspect is the duration of the use of the light rail: 27.9% who read the PTP and 21.7% of the control group became a frequent user (the ratio of light rail use frequency to total frequency was greater than 0.5). Considering this result, it can reasonably be claimed that, all other conditions being equal (hard measure and mass communication), the personalized measure can contribute by 12% to behavior change for those participants in the potential light rail user target. This is in line with the findings of the few studies reported in the literature that quantify the contribution of VTBC programs to behavior change as between 5 and 15% (*19*). However, it should be stressed that in these cases the effects of behavior change may well have been overestimated as monitoring was limited to a very short period of time.

Nevertheless, it is important to note that mass communication and introduction of a new line of light service also played an important role in travel behavior change.

A moderate percentage (23%) of participants who were not included in the potential light rail user target (No PTP) also changed their behavior, but the percentage is lower than the other two cases, as was to be expected.

Lastly, evaluation took place of which information contained in the PTP had the greatest effect on travel behavior change. Respondents' answers were measured on a 5-point Likert scale. Participants considered all the information provided important for deciding whether to change their travel behavior: CO_2 emissions (3.67 AVG), journey time (3.56 AVG), cost (3.44 AVG), calories burned (3.26 AVG) in that order. Though the results for journey time and costs were predictable, surprisingly CO_2 emissions were considered the most important. As mentioned previously, people are usually unaware of this aspect or are unable to quantify emissions. This information was considered significant by 51.7% of those who changed their travel behavior.

Discussion and Conclusions

This work focuses on the attempt to assess in quantitative terms the role that a combination of hard (i.e., new mode of sustainable transport—light railway line in the choice set) and soft measures (differentiated with respect to the degree of personalization—mass communication versus personalized travel planning) can have in travel mode choice (switch from car driver to light rail). A comparison was carried out of the data collected in two survey waves before and after a new light railway line went into service. The paper analyzes the additional effect of the personalized soft measure versus the combination of hard and soft (mass communication) measures usually implemented when a new service becomes operational.

Additionally, it was possible to evaluate the overall effect of all three measures. The paper describes in detail the methodology and procedure adopted for implementing the sensitization campaign (soft measure). Validation of the results depends both on the methodological framework and on how the different phases of the campaign are conducted, so as to be able to distinguish the contribution of hard from soft measures and a combination thereof. For the design of the personalized soft measures (VTBC program) the key factors that distinguish successful VTBC programs were used, in an attempt to overcome those problems that generally affect this kind of measure (e.g., lack of a control group). For assessing changes in travel behavior a comparison was made of the modal shares observed in the first and second wave surveys. Considering the results, it can reasonably be claimed that the awareness campaign (soft measure + VTBC program) in the specific context contributes by 12% to behavior change. This is in line with the findings of the few studies reported in the literature that quantify the contribution of VTBC programs to behavior change as 5 to 15 % (*19*).

Further, the combination of hard and soft measures achieved a change in travel behavior of 34%, when the measure is not personalized, and 46% with the VTBC program. This confirms the importance of adopting TDM measures on a large scale, with a combination of measures that changes the choice context and information measure, both generalised and personalized.

The work highlights, as reported in the literature, the difficulty in gathering sufficient data for evaluating the effectiveness of a single measure and in combination with others. In fact, even though the original sample size was relatively large, the size of the sample for the second wave survey was rather small to produce conclusive evidence. This is a shortcoming but represents a good starting point and an additional contribution to the research in this area. One possible solution to this problem could be keeping in regular contact with the participants, to increase their involvement and encourage participation.

A key to the success of VTBC programs is the high degree of personalization and how the information is conveyed (it needs to be useful and readily understandable). Implementing these kinds of measures on a large scale can be expensive. However, technology, especially smartphones, can benefit the implementation of VTBC programs, making it possible to send personalized information to users in real time about available sustainable travel options in a given moment.

Future research will focus on conducting an in-depth analysis of travel behavior change, examining in detail individuals' propensities and psycho-attitudinal characteristics and constructing demand forecasting models using panel data (first and second wave surveys).

Acknowledgments

The authors wish to express their thanks to the University of Cagliari, the Azienda Ospedaliera Universitaria and the Azienda Regionale Sarda Trasporti SpA for their collaboration.

Author Contributions

The authors confirm contribution to the paper as follows: study conception and design: Francesco Piras, Eleonora Sottile; data collection: Italo Meloni; analysis and interpretation of results: Francesco Piras, Eleonora Sottile, Italo Meloni; draft manuscript preparation: Francesco Piras, Eleonora Sottile. All authors reviewed the results and approved the final version of the manuscript.

References

1. Un-Habitat Core Team. *Planning and Design for Sustainable Urban Mobility: Global Report on Human Settlements 2013*. UN-HABITAT, Routledge, 2013.
2. ISFORT. *14° Rapporto sulla mobilità in Italia*. ISFORT, www.isfort.it, 2016.
3. EU. *WHITE PAPER Roadmap to a Single European Transport Area: Towards a Competitive and Resource Efficient Transport System*, COM(2011), 144 final. European Commission, 2011.
4. Prioni, P, and D.A. Hensher. Measuring Service Quality in Scheduled Bus Services. *Journal of Public Transportation*, Vol. 3, No. 2, 2000, pp. 51–74.
5. Sanjust, B., I. Meloni, and E. Spissu. An Impact Assessment of a Travel Behavior Change Program: A Case Study of a Light Rail Service in Cagliari, Italy. *Case Studies on Transport Policy*, Vol. 3, No. 1, 2015, pp. 12–22.
6. Ampt, E. *Voluntary Household Travel Behaviour Change: Theory and Practice*. Paper presented at the 10th International Association of Travel Behaviour Research Conference, Lucerne, Switzerland, August 15–16 2003.
7. Brög, W., E. Erl, I. Ker, J. Ryle, and R. Wall. Evaluation of Voluntary Travel Behaviour Change: Experiences from Three Continents. *Transport Policy*, Vol. 16, No. 6, 2009, pp. 281–292.
8. Richter, J., M. Friman, and T. Gärling. Soft Transport Policy Measures: Gaps in Knowledge. *International Journal of Sustainable Transportation*, Vol. 5, No. 4, 2011, pp. 199–215.
9. Meloni, I., and B. S. di Teulada. I-Pet Individual Persuasive Eco-Travel Technology: A Tool for VTBC Program Implementation. *Transportation Research Procedia*, Vol. 11, 2015, pp. 422–433.
10. Bonsall, P. Do We Know Whether Personal Travel Planning Really Works? *Transport Policy*, Vol. 16, No. 6, 2009, pp. 306–314.
11. Taylor, M. A., and E. S. Ampt. Travelling Smarter Down Under: Policies for Voluntary Travel Behaviour Change in Australia. *Transport Policy*, Vol. 10, No. 3, 2003, pp. 165–177.
12. Steg, L., and C. Vlek. Encouraging Pro-Environmental Behaviour: An Integrative Review and Research Agenda. *Journal of Environmental Psychology*, Vol. 29, No. 3, 2009, pp. 309–317.
13. Bamberg, S., S. Fujii, M. Friman, and T. Gärling. Behaviour Theory and Soft Transport Policy Measures. *Transport Policy*, Vol. 18, No. 1, 2011, pp. 228–235.
14. Steg, L. Car Use: Lust and Must. Instrumental, Symbolic and Affective Motives for Car Use. *Transportation Research Part A: Policy and Practice*, Vol. 39, No. 2, 2005, pp. 147–162.
15. Gärling, T., and S. Fujii. Travel Behavior Modification: Theories, Methods, and Programs. *The Expanding Sphere of Travel Behaviour Research*, 2009, pp. 97–128.
16. Verplanken, B., H. Aarts, A. Knippenberg, and A. Moonen. Habit versus Planned Behaviour: A Field Experiment. *British Journal of Social Psychology*, Vol. 37, No. 1, 1998, pp. 111–128.
17. Steg, L., and G. Tertoolen. Sustainable Transport Policy: The Contribution from Behavioural Scientists. *Public Money and Management*, Vol. 19, No. 1, 1999, pp. 63–69.
18. Gärling, T., and G. Schuitema. Travel Demand Management Targeting Reduced Private Car Use: Effectiveness, Public Acceptability and Political Feasibility. *Journal of Social Issues*, Vol. 63, No. 1, 2007, pp. 139–153.
19. Cairns, S., L. Sloman, C. Newson, J. Anable, A. Kirkbride, and P. Goodwin. Smarter Choices: Assessing the Potential to Achieve Traffic Reduction Using "Soft Measures". *Transport Reviews*, Vol. 28, No. 5, 2008, pp. 593–618.
20. James, B., and W. Brög. TravelSmart: Large-Scale Cost-Effective Mobility Management. Experiences from Perth, Western Australia. *Proceedings of the Institution of Civil Engineers: Municipal Engineer*, 2002, Vol. 151, No. 1, pp. 39–48.
21. Fujii, S., and A. Taniguchi. Determinants of the Effectiveness of Travel Feedback Programs: a Review of Communicative Mobility Management Measures for Changing Travel Behaviour in Japan. *Transport Policy*, Vol. 13, No. 5, 2006, pp. 339–348.
22. Friman, M., L. Larhult, and T. Gärling. An Analysis of Soft Transport Policy Measures Implemented in Sweden to Reduce Private Car Use. *Transportation*, Vol. 40, No. 1, 2013, pp. 109–129.
23. Cairns, S., A. Davis, C. Newson, and C. Swiderska. *Making Travel Plans Work: Research Report*. Department for Transport, London, 2002.
24. Rose, G., and E. Ampt. Travel Blending: An Australian Travel Awareness Initiative. *Transportation Research Part D: Transport and Environment*, Vol. 6, No. 2, 2001, pp. 95–110.
25. Currie, G., A. Jones, and J. Woolley. The Travel Demand Management Program for the London 2012 Olympic Games: Impacts and Lessons. *Proc., World Conference on Transportation Research (WCTR)*, Rio de Janeiro, Brazil, 2013.
26. Davies, N. What are the Ingredients of Successful Travel Behavioural Change Campaigns? *Transport Policy*, Vol. 24, 2012, pp. 19–29.

27. Stopher, P. R. Voluntary Travel Behavior Change. In *Handbook of Transport Strategy, Policy and Institutions* (K. J. Button, and D.A. Hensher ed.), Emerald Group Publishing Limited, Bingley, 2005, pp. 561–579.
28. Fogg, B. J. Persuasive Technology: Using Computers to Change What We Think and Do. *Ubiquity*, Vol. 5, 2002, pp. 89–120.
29. Abrahamse, W., L. Steg, C. Vlek, and T. Rothengatter. A Review of Intervention Studies Aimed at Household Energy Conservation. *Journal of Environmental Psychology*, Vol. 25, No. 3, 2005, pp. 273–291.
30. Ajzen, I. The Theory of Planned Behavior. *Organizational Behavior and Human Decision Processes*, Vol. 50, No. 2, 1991, pp. 179–211.
31. Schwartz, S. H. Normative Influences on Altruism. *Advances in Experimental Social Psychology*, Vol. 10, 1977, pp. 221–279.
32. Stern, P. C., T. Dietz, T. Abel, G. A. Guagnano, and L. Kalof. A Value-Belief-Norm Theory of Support for Social Movements: The Case of Environmentalism. *Human Ecology Review*, Vol. 6, 1999, pp. 81–97.
33. Triandis, H. C. Values, Attitudes, and Interpersonal Behavior. In *Nebraska Symposium On Motivation* (Howe, H.E. Jr., ed.), University of Nebraska Press, Lincoln, NE, 1979, pp. 195–259.
34. Prochaska, J. O., and C. C. DiClemente. Transtheoretical Therapy: Toward a More Integrative Model of Change. *Psychotherapy: Theory, Research & Practice*, Vol. 19, No. 3, 1982, pp. 276.
35. Carver, C. S., M. F. Scheier, and J. K. Weintraub. Assessing Coping Strategies: A Theoretically Based Approach. *Journal of Personality and Social Psychology*, Vol. 56, No. 2, 1989, pp. 267.
36. Gollwitzer, P. M. Goal Achievement: The Role of Intentions. *European Review of Social Psychology*, Vol. 4, No. 1, 1993, pp. 141–185.
37. Defra, A. *Framework for Pro-Environmental Behaviours*. Department for Environment, Food and Rural Affairs, London, 2008.
38. Cialdini, R. B. Harnessing the Science of Persuasion. *Harvard Business Review*, Vol. 79, No. 9, 2001, pp. 72–81.
39. Tertoolen, G., D. Van Kreveld, and B. Verstraten. Psychological Resistance against Attempts to Reduce Private Car Use. *Transportation Research Part A: Policy and Practice*, Vol. 32, No. 3, 1998, pp. 171–181.
40. Eriksson, L., J. Garvill, and A. M. Nordlund. Acceptability of Travel Demand Management Measures: The Importance of Problem Awareness, Personal Norm, Freedom, and Fairness. *Journal of Environmental Psychology*, Vol. 26, No. 1, 2006, pp. 15–26.
41. Eriksson, L., J. Garvill, and A. M. Nordlund. Acceptability of Single and Combined Transport Policy Measures: The Importance of Environmental and Policy Specific Beliefs. *Transportation Research Part A: Policy and Practice*, Vol. 42, No. 8, 2008, pp. 1117–1128.
42. Tørnblad, S. H., S. Kallbekken, K. Korneliussen, and T. K. Mideksa. Using Mobility Management to Reduce Private Car Use: Results from a Natural Field Experiment in Norway. *Transport Policy*, Vol. 32, 2014, pp. 9–15.
43. Fujii, S., S. Bamberg, M. Friman, and T. Gärling. Are Effects of Travel Feedback Programs Correctly Assessed? *Transportmetrica*, Vol. 5, No. 1, 2009, pp. 43–57.
44. Stopher, P., E. Clifford, N. Swann, and Y. Zhang. Evaluating Voluntary Travel Behaviour Change: Suggested Guidelines and Case Studies. *Transport Policy*, Vol. 16, No. 6, 2009, pp. 315–324.
45. Gaker, D., and J. L. Walker. Insights on Car-Use Behaviors from Behavioral Economics. In *Auto Motives: Understanding Car Use Behaviours* (K. Lucas, E. Blumenberg, and R. Weinberger eds.), Emerald Group Publishing Limited, Bingley, 2011, pp. 107–120.

The Standing Committee on Transportation Demand Management (ABE50) peer-reviewed this paper (18-05513).

Article

TRR
JOURNAL OF THE TRANSPORTATION RESEARCH BOARD

Transportation Research Record
2018, Vol. 2672(8) 742–753

DOI: 10.1177/0361198118781659
journals.sagepub.com/home/trr
SAGE

New and Unique Aspects of University Campus Transportation Data to Improve Planning Methods

Christopher R. Cherry[1], William Riggs[2], Bruce Appleyard[3], Nirbesh Dhakal[1], Alexander Frost[3], and Saman Tabrizizadeh Jeffers[1]

Abstract

University campuses have unique characteristics that influence travel demand. They benefit from controlled planning frameworks, can influence almost all aspects of pricing and on-campus transportation infrastructure, benefit from simplified many-to-one travel patterns, and maintain extensive data on nearly the entire campus community. Campuses represent millions of trips daily, yet little research has been conducted that focuses on developing tailored frameworks to assess transportation demand, target travel demand management strategies, or to assess environmental or other impacts. This paper describes how two sets of commonly collected data can be leveraged to provide new insights into travel behavior, incentives, and environmental policy. Specifically, we illustrate the fusion of standard travel demand survey data with disaggregate and precise household address data to provide new inferences into appropriate strategies to improve transportation sustainability. We applied three use-cases—carpool potential, walk and bike incentives, and mode shift to reduce greenhouse gas emissions—to the data from three universities, University of Tennessee, Cal Poly San Luis Obispo, and San Diego State University, respectively. We found that the richer dataset provided substantially better data resolution that allowed for transportation strategies to be more precisely targeted and expansion of the potential impact of transportation demand management strategies.

University campuses provide unique mixed land use, multimodal, and walkable environments to students, faculty, and staff. Often, these campuses are islands of sustainable transport within mostly auto-oriented cities. Campuses are often internally balanced, with most amenities needed by the campus community located within walking distance. Campus populations vary widely but can range from several thousand to tens of thousands, the size of a small town. Universities can also be the largest single employer in many urban areas, providing significant leverage when developing transportation planning strategies. Despite millions of university students and employees, little research has been conducted on best practices and unique methods of transportation and environmental analysis for these unique cases. Universities are increasingly developing comprehensive transportation and environmental plans with little guidance on the best approaches that cater to the unique features of university-related travel.

Universities have relatively unique many-to-one travel patterns and tend to have central control over transportation and land-use policies (e.g., parking) for the campus community. Moreover, universities tend to keep some household data on all or most campus users (e.g., home address). These features make some traditional transportation and environmental planning and travel demand management (TDM) strategies ill-fitting for campuses.

This paper presents a unique methodology with specific cases outlined for exploiting some of the data and travel behaviors that are available to universities. We focus on pairing specific origin (home location) data provided by universities with traditional travel and activity surveys to identify travel demands and opportunities for improved TDM. This is a novel contribution relative to mostly descriptive university-related travel research—allowing more precise policy interventions. We do this in the context of traditional TDM plans and emerging campus greenhouse gas (GHG) inventories and climate action plans.

The paper is organized as follows. First, we describe the existing work on campus transportation planning and identify gaps in the literature. Next, we propose a method

[1]The University of Tennessee, Knoxville, TN
[2]University of San Francisco, San Francisco, CA
[3]San Diego State University, San Diego, CA

Corresponding Author:
Address correspondence to William Riggs: wriggs@usfca.edu

that fuses precise spatial location information (e.g., home coordinates) with survey data and existing transportation infrastructure variables to identify campus transportation demand and GHG impacts. Last, we illustrate three use-cases of such data to show the unique aspects of this work. The three case studies focus on carpool and ridesourcing potential, walk and bicycle potential, and parking and transit policy, applied at three universities: University of Tennessee, Knoxville (UTK), San Diego State University (SDSU), and California Polytechnic State University, San Luis Obispo (Cal Poly), respectively.

Background and Literature

Universities are major activity generators requiring advanced TDM strategies (*1*, *2*). However, few studies have focused on methods to improve campus-planning outcomes. The mix of housing concentrated within a short distance from campuses, along with compactness and density make them ideal for walking and bicycling, but the commute to university campuses often relies on car-commuting (*3*). An online survey has been used to study mode switching of urban campus populations in places like Virginia, Berkeley, and Perth (*4–9*). Most campus travel demand surveys mimic conventional travel surveys to aid planning, but tend not to leverage unique features of campuses to inform tactical transportation decisions. New spatial tools are being used to evaluate travel behavior including precise geographic information system (GIS) methods to inform travel demand and environmental policies (*10*).

In recent years, environmental and climate plans have pushed campuses to more carefully consider transportation's role in GHG emissions and air pollution (*11*). In order to do so, a systematic environmental management approach is needed to reduce negative impacts on the environment, with a particular focus on transportation (*12*, *13*).

Parking pricing is one of the powerful policies being used frequently by campuses (*14*). Universities rarely provide free parking. Moreover, some universities develop parking cash-out programs. For example, Stanford through its "Clean Air Cash" program pays 2,500 employees who do not purchase a parking permit during the year (*11*). Likewise, University of Colorado, Boulder uses a fleet of small buses with an "Ecopass," for which employers buy passes for their employees and pass holders ride the buses free of charge with valid ID. After this policy was enacted with the city of Boulder and the University of Colorado, total transit use in the last 5 years has increased by 400% (*15*). University of California, Berkeley's Class Pass has had similar success, providing substantial discounts to students in exchange for unlimited marginally free transit rides (*16*).

Travel choice is tied both to market norms as well as social norms, and this is true in the campus environment. On campuses, travel behavior is related not only to personal health but to the environmental sustainability and social capital of communities. Sustainable transport can also result in substantially more economically resilient communities (*17*, *18*), but the practical reality is that this type of environment is not the case in many university campuses and communities across the United States (*19*).

Built environment factors play a role in active travel behavior, and policy and incentives play a key role in transportation decisions. Transportation choice is tied to financial and social factors as well as to public policies (*20–24*). Yet the interrelationship between these forces, particularly in the university context, remains under-explored, especially in the face of new technology and mobile proliferation. Policies and incentives can encourage or deter driving behaviors and influence auto ownership (*16*, *21–27*). Some communities use a "carrot" approach, offering incentives such as free transit passes, cash back ("cash-out") parking programs, or social media nudge tools (*28*) to reward alternatives to driving (*29*). Other communities prefer to use the "stick" approach, with high prices for parking, tolls, or other usage fees. Universities have a unique opportunity to balance these approaches.

Unique Methods and Data

Universities tend to rely on campus travel surveys, in the tradition of conventional travel surveys that often inform travel demand modeling efforts. One of the main advantages of university travel survey deployment is that university employee and student databases tend to be comprehensive, allowing for 100% sampling rates. This could potentially remove some sampling bias (*30*). However, the survey could still suffer from response bias.

Here, we rely on three different travel surveys, developed for different purposes (i.e., transportation demand modeling and GHG inventories), but the three surveys include similar travel-related data. Specifically, they include home location and trip-level mode choice questions (in standard travel diary format) relevant to this study, along with some attitudinal questions regarding travel demand strategies. All surveys were conducted online and the campus population was sampled using the university-supplied email recruitment. The online surveys allowed branching and tailored surveys for different members of the campus communities (e.g., on-campus student residents, faculty, staff etc.).

In this aspect, these surveys are not just conventional travel surveys in a university context such as those which have been featured in publications by Riggs et al., Wang et al., Khattak et al., and others (*7–9*). Although in the three cases evaluated, these surveys are relatively standard, they offer new ways of modeling university communities as small microcosms with the capacity for more robust analysis than larger regional travel studies—perhaps the next frontier of travel studies, unique to universities. One key point of difference in the three cases presented is that the compact campuses are small relative to the distance between home and campus, and in almost all cases, travel demand is generated at home (off campus) and destined for campus, creating a many-to-one travel pattern. This eliminates one of the key challenges in travel demand modeling, especially for college or corporate campuses with large employment or student bodies—developing defensible trip distribution patterns (i.e., origin and destination pairs). Here, we know that (almost) all respondents of our survey, for commute trips, travel from a home origin to a single campus destination. This leads to the main unique innovation for campus travel demand studies.

Beyond surveys of the campus community, which have a high sample rate but generally a low response rate, universities tend to have near-perfect and precise disaggregate data on home location for all members of the campus community. This campus population can number in the tens of thousands, and the implications of this resolution of campus data have been scarcely studied. In the three cases described here, we have addresses for all students, faculty, and staff (Figure 1). This allows stratification of survey data to more closely represent true travel patterns of the campus community.

Moreover, with precise data on home locations, and a sample of mode choice and trip making behavior, we can, with some precision, estimate total commute travel, mode split, GHG and other travel metrics for the entire campus community. These data can be used to develop targeted TDM interventions for specific subgroups (either spatially, or by employee/student classification) based on location and travel patterns nearby. These new datasets can inform many policies, as outlined below.

Potential Data Applications

Pricing and TDM in the Campus Environment

Pricing is one of the key factors that conceals the social and environmental costs of driving decisions (*27*, *31*), and keeping rates low potentially encourages the purchase of the annual permit, increasing the drive-alone rate and GHG emissions (*32*). Based on other decisions one can assume that the cost of parking in general ascribes an arbitrary coherence to the cost of other types of travel, for example via walking and biking, as the full costs are paid upfront usually on a monthly basis and it therefore appears "free" on a daily basis when compared with the cheaper (and healthier) options of walking, biking, or taking transit—with people making what are effectively irrational decisions (*22*, *27*, *31–34*).

GHG Estimation Methods

Data about mode shift, vehicle class, and vehicle miles traveled (VMT) can provide an estimation of GHG reduction by considering all the commuters and people willing to change their mode, housing location, or vehicle (*35*). Furthermore, survey data in our surveys includes make, model, and year of vehicle used to commute. Coupled with sampled household locations and paired with full population household locations and estimated fuel economy of vehicles, we can estimate, with some certainty, the amount of fuel and thus GHG emissions from single-occupant vehicles (SOVs).

Carpooling or App-Based Ridesourcing

There is abundant capacity in existing personal vehicles to accommodate a large portion of the campus community. University carpool programs can be very effective at reducing SOVs. At University of California, Santa Barbara, within four years of implementing a carpool program, carpooling rates increased by 30% for staff and increased by 25% among students (*36*). The primary barrier to carpooling is ride matching (spatially and temporally). Universities can overcome these challenges with common destinations and fewer work-schedule challenges. Moreover, the administrative ability to match coworkers is greater at universities than distributed employers, with many successful examples. Transportation network companies aim to provide real-time ride matching with many-to-many trip distribution patterns. The university case is dramatically simplified by many-to-one trip distribution patterns, a feature that enables earlier nontechnology-enabled ridesharing like casual carpool or "slugging" (*37*).

Walk and Bike Access to Campus with Supportive Infrastructure

With survey data alone, bike and walk trips could be underreported, and the total population that lives within bicycle or walking distance is unknown. Understanding the spatial distribution of actual bicycle and walk trips (from surveys) and campus populations (from home address data) allows targeted interventions, either through marketing or infrastructure, and allows policies that could expand those areas (e.g., promoting electric bikes).

Transit Access

Transit is among the most efficient users of road space, has low marginal emissions, and requires little parking or terminal space. Many universities offer transit benefits to employees (in addition to parking benefits). Again, strategic investment in transit infrastructure, incentives, and other policies is important for universities. Here, it is important to determine the proportion of the campus population that is inclined to use, or is even within reach of transit service. In the case of University of Tennessee, only two-thirds of students and a quarter of staff and faculty are within walking distance of the urban transit system, discouraging broad-based programs like deep-discount fares (*16*).

Case Studies

Here we illustrate possible uses of these data with three case studies. These are small data use-cases that are not exhaustive, but show some of the possibilities of how such data could be

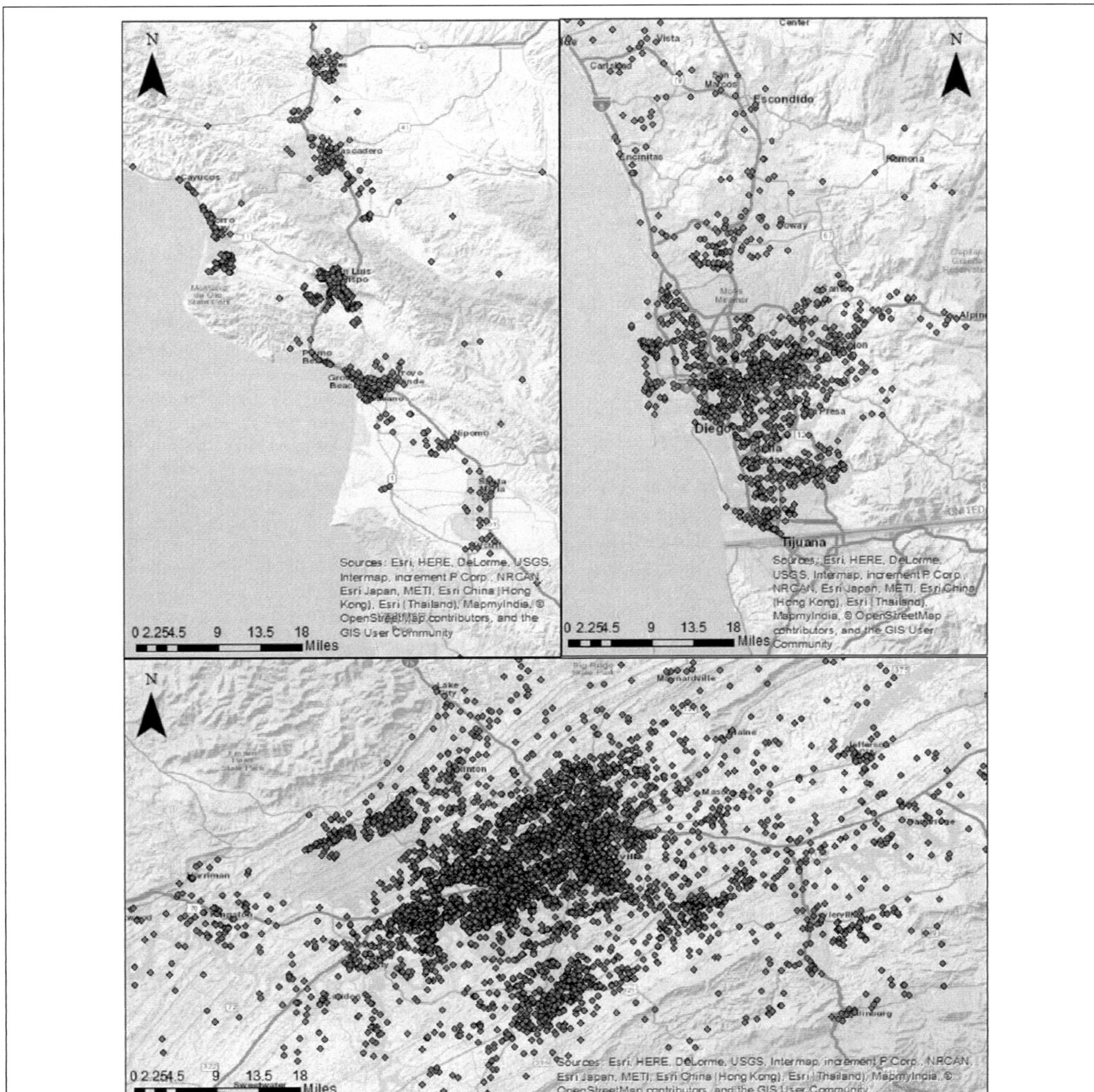

Figure 1. Population distribution at different campuses. All scales are the same. Top left is poly-centric survey respondents from San Luis Obispo (Cal Poly); top right is all dense urban staff home locations in San Diego State University (SDSU); bottom is all faculty, staff, and students in sprawling Knoxville (UTK).

used by campus transportation or environmental planners. The three case studies reflect slightly different types of data, collected for different purposes, with different spatial resolution. We start with UTK, located in a medium-sized city. In Knoxville, we focus on ridesharing potential with a combination of household location data and survey data. Next, we discuss walk and bike potential at Cal Poly, San Luis Obispo, whose campus is in a small city on California's central coast. Last, we investigate policy at SDSU, located in the third-largest metropolitan area in California.

University of Tennessee: Carshare and Ridesource Potential

UTK is an urban campus built on a peninsula bounded by the Tennessee River and adjacent to downtown Knoxville. The neighborhood of Fort Sanders borders UTK to the north, which is the most densely populated area in the city, with a population of over 15,000 living in approximately 3,100 residential units, in a 50-block 474-acre area (*32*). UTK is at the crossroads of West Knoxville's transit system (Knoxville Area Transit, KAT) and the university operates a circulator bus system that serves campus and surrounding neighborhoods. An important note is that the only fixed route transit system (KAT) that operates in the region only serves the limits of the city, and not the surrounding urbanized area, where a large portion of the campus community lives.

UT also has relatively good expressway access from all directions. The campus is well-served by off-road greenways and other bicycle infrastructure, and the city of Knoxville has a League of American Bicyclist "Bronze" rating. The UT student body is about 28,000 students (~three-quarters undergraduate). UT also employs 9,800 faculty and staff and is the largest employer in Knoxville.

UTK has developed a Climate Action Plan to achieve carbon neutrality by 2060 (*38*). Transportation is part of that goal and the transportation sector emits 14% of total GHGs on UTK campus (including commute and on-campus transportation). Of that 14%, 13% GHG is contributed by SOVs. Therefore, the Climate Action Plan at UTK has set an intermediate goal of reducing SOV commuting miles per year to 25% below 2007 levels by 2020 (*39*).

Moreover, part of the UTK masterplanning process is to move most parking to the periphery of campus and use other modes of transportation such as transit, walking, or biking—making campus a more pedestrian- and bicycle-friendly area. This strategy could make driving less convenient, or at least diminish SOV's door-to-door advantage compared with other modes. These initiatives require more diverse travel patterns and this study investigated many TDM strategies to assess and reduce energy and GHG-intense modes. Here we focus on rideshare and carpool approaches.

Methodology: Surveys, Home Location, and Cluster Analysis for Targeted Intervention. UTK, like most other campuses, keeps household data on most of their campus users including home addresses. This dataset is composed of identifiable and geocoded home addresses of students and faculty, and classification (e.g., graduate or undergraduate student), without any other characteristics or mode choice information. This information is initially collected for non-transportation purposes. To complement this information, a detailed preference travel demand survey was conducted capturing the attributes that may be associated with transportation mode choice decisions of commuters and non-commuters.

The survey garnered 721 respondents, 103 non-commuters (living on campus) and 618 commuters. Commuters included faculty, staff, and students living outside campus. Of off-campus commuters 85% of the respondents traveled by car and the rest walked, biked, and used transit. Most non-commuters (55) walked to campus although 70% have a vehicle and 10% have a bike on campus. The unique aspect of this application is that student and staff address data (collected for other purposes) were merged with conventional travel data to identify opportunities for spatially nuanced TDM strategies. Administrative units on campus do not suffer from the privacy concerns of the conventional private sector or public sector datasets, as the data are generated and analyzed within the institution.

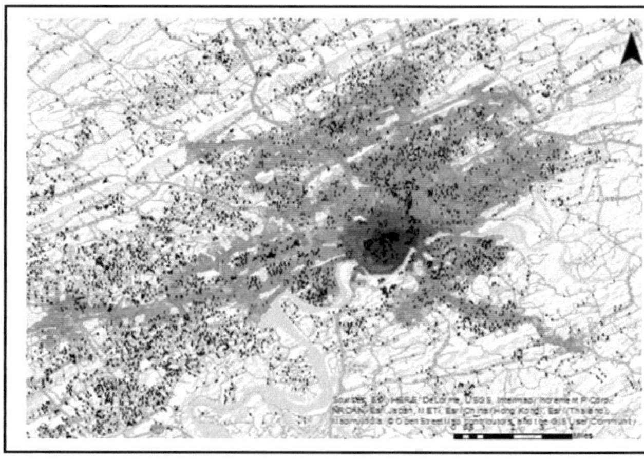

Figure 2. Campus population overlaid on transportation zones. Darkest zone is walk distance; gray zone is bicycle distance; and green zone is transit distance. Black dots represent home locations of students, faculty, and staff. This map is scaled about five times larger than the map in Figure 1.

In this research, the addresses of the campus users were mapped using ArcGIS Pro's online address geocoding engine. A network for Knoxville was developed using the Network Analyst extension comprising public data for roads, counties, and zip codes (from data.gov). Network analyst was used to determine trips from each location to campus. The average home-to-campus trip distance for students' trips is 6.1 mi, for faculty is 9.5 mi, and for staff is 10.6 mi.

Using Network Analyst in ArcGIS, we mapped the zones that were within walking and biking distance to campus and within walking distance to transit stops. A 1-mi distance was considered walkable to campus, 2 mi considered bikable, and a quarter mile for walk to transit. We added one more zone a half-mile distance from bus stops as bike to transit zones. Collectively all these zones were identified as transport alternative zones as shown in Figure 2. About 62% of students, 32% of faculty, and 29% of staff lived in that zone. The black dots outside of the green buffer represent those who are effectively out of reach of transit or other alternative modes. Our survey responses indicate that almost 100% drive alone or carpool in those areas. Absent major expansions in the transit system, or significant land-use changes and movement to the urban core, inducing a shift toward alternative modes is unfeasible for nearly half the campus population. As such, we focus (in this illustration) on carpool and rideshare options for car-dependent suburbs. Knowing specific home locations enables precise analysis of transit or

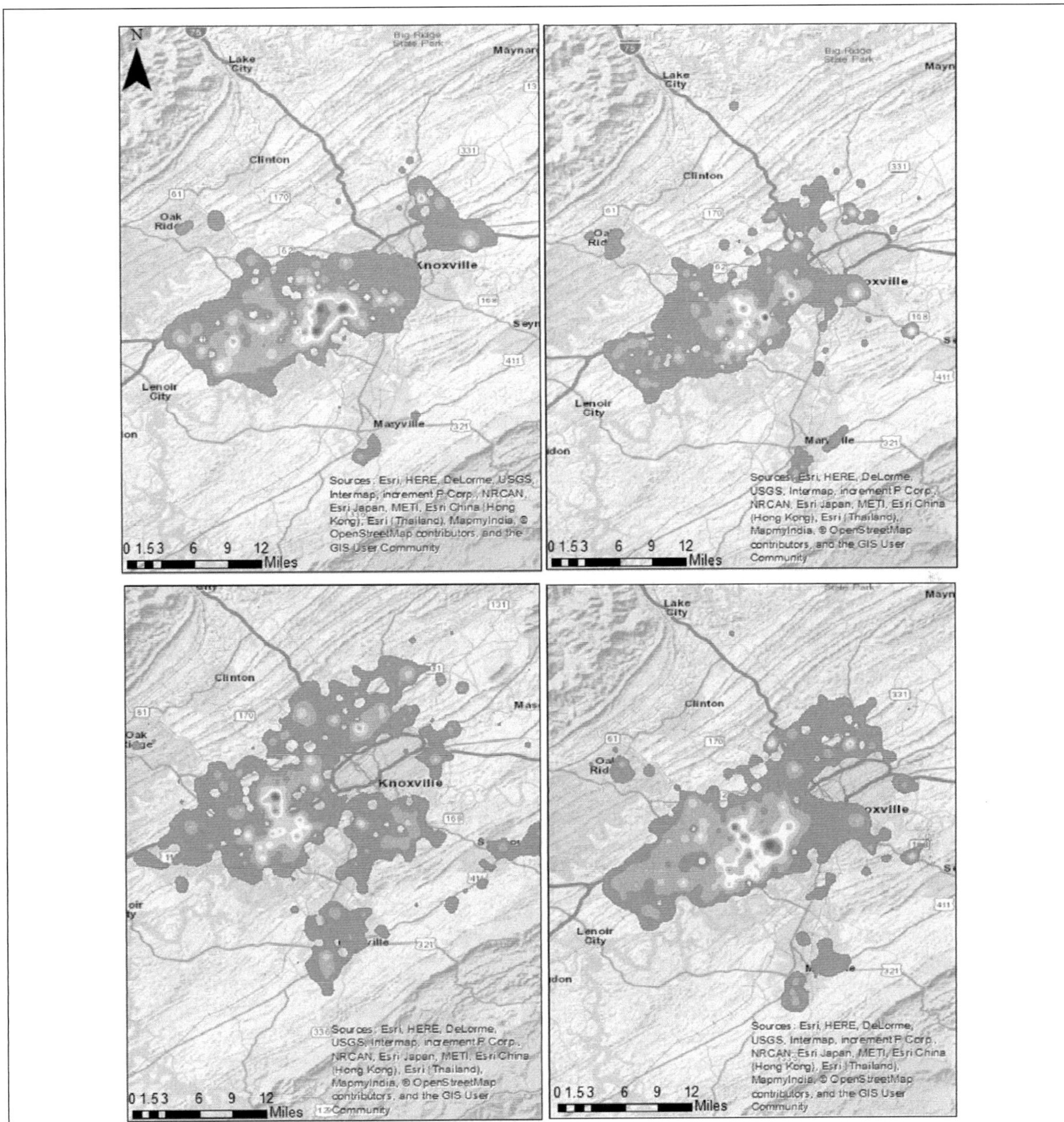

Figure 3. Cluster zones for faculty (top left), students (top right), staff (bottom left), and overall (bottom right).

other non-motorized modes, for which conventional traffic analysis zones do not provide the precision needed to assess short trips.

Rideshare or Carpool as a Travel Demand Strategy. Segregated zoning and dispersion of population away from commercial corridors results in most of the population living outside zones with alternative modes of transportation. Here, we focus on those areas outside of transit's reach. We visualized the population dispersion to search for clusters, mapping the density of home locations. The population clusters were completed separately for students, faculty, and staff as employees have a relatively fixed schedule. Figure 3 shows the population clusters for each population.

Campus-user populations outside of alternative-mode zones were given priority in consideration of carpooling, though some of the densest clusters of the campus population have transit access, which is promising (and could be a target for transit-oriented incentives). The travel demand survey asked about mode choice of commute trips for the past week. During weekdays, around 7% of the respondents car-pooled while 70% drove alone. Of those carpooling, 36% were from zip codes 37920 and 37909. Although relatively close to campus, those areas have low transit access. This strengthened our belief that the population outside of the transit accessibility zone would benefit from carpool.

Carpooling can be proposed in two different ways, either as a point-to-point park and ride model or by picking up persons along a route. The latter would be more feasible in terms of capital cost of facilities and ease for commuters, and is enabled by some of the corridor-oriented residential distributions in Knoxville. There are already several websites and forums (Tennessee Carpool Center, RTA Carpool ride match, Knoxville Smart Trips, Carpool World, etc.) connecting many carpoolers to plan and share their rides. An app could be developed to further connect the riders, or other apps could be used, mimicking ridesourcing apps that provide carpool service (e.g., UberPool and LyftLine).

To promote this practice, UTK could further incentivize the riders or partner service providers. Like other universities, the incentive could be free and convenient parking for carpoolers, or designated lots for carpool cars. The university would benefit from this program, as this would reduce demand for parking on campus, reducing capital expenses needed to support their current structured parking initiatives. The western part of Knoxville has a larger population of students, staff, and faculty, and based on that, we selected a test corridor in this general vicinity to demonstrate the applicability of a carpool strategy. In all, 23 faculty, 75 staff, and 146 students live within 2.5 mi from this location. We found each rider choosing to carpool would save about 15.6 mi each way. Furthermore, we looked at a route running through a higher density suburban area. On this corridor, 55 faculty, 60 staff, and 151 students are within a ¼ mi buffer from this street. A driver agreeing to carpool would not have to make a significant detour to pick up other riders. As stated before, an app or website could enable trip matching and incentives could be tested to generate sufficient demand.

This case study aims to showcase a specific application of using an internal database (enrollment and human resource records) to inform specific TDM strategies and tactics. The home-location data was paired with a conventional survey-based travel demand analysis to target specific strategies to reduce SOV travel to campus. Prior to this, like many campuses, TDM strategies were not targeted to populations (e.g., transit benefits for all). This study shows how, using a spatially rich dataset, transportation administrators could target and directly market different TDM incentives to individuals based on their home location. This could, if well executed, be more cost effective than blanket incentives, marketing, and education campaigns.

Cal Poly San Luis Obispo: Encouraging Biking and Walking

Cal Poly is the largest employer in its respective county and occupies a large share of the urban center. As such, it could be likened to either a small town or a large corporate campus, and transportation plays a large role in the campus's functionality—both in terms of trips generated and facilitating accessibility and the interactions that make it a vibrant place of intellectual exchange and innovation. At the same time, it poses land-use, sustainability, and costs challenges, especially as the campus is an urban environment with constraints on space and budgets.

The Cal Poly campus has ample surface parking that in the past limited the need for more aggressive alternative transportation programs. The monthly permit system requires payment upfront for monthly parking priced below market rates. Based on this, demand is high, and the daily decision on commute alternatives is embedded into the parking permit decision. The low cost of parking has affected demand and increased circling for oversubscribed spaces and spillover into residential neighborhoods and on-street spaces. Due to these challenges related to housing, traffic, and environment issues, the campus has chosen to experiment with pricing and behavioral strategies.

In terms of travel choices, many students walk to campus. Walking accounts for 41% of the students' commutes (*40, 41*). Eight percent of faculty and 19% of staff walk to campus. Ten percent of students, and around 5% of staff and faculty, commute by public transit. Despite this, the average round trip commute length to campus is 17.4 VMT per day per commuter, because many students live far away from the campus—on average, further from campus than faculty do. Although some may find this trend counterintuitive, as many students do live close to campus, in actuality the second largest share of both students (and of staff) live more than than 10 mi from the campus.

Transportation currently accounts for 50% of the GHG emissions at Cal Poly. This sector consists of automobiles, public transportation, bicycles, and walking. The major contributor to GHG emissions within this sector is single automobile commute. Only 24% of students commute by this method, as well as 68% of faculty and staff, however many of the students that commute by car have longer commutes than faculty and staff. This makes for a total of 38% of commuters driving alone to campus.

Methodology: Surveying with a Focus on Incentives. In this context, the campus engaged in a Climate Action Plan and Travel Survey in 2015 with recognition that travel behavior is complex. Researchers recognized that knowledge about and attitudes toward transit and driving, self-image, and travel choices all play a role in transportation behavior. Prior experience and habit have also been found to enter into transportation mode choice decisions and to shape what alternatives are considered, as well as how alternatives are assessed.

The planning process began with a survey issued to all campus constituents. The survey received a total of 3,961 responses, 17% of the entire campus population of roughly 23,000. Unsurprisingly, the majority of respondents were students, totaling 68.6%, whereas the rest were made up of faculty, staff, and visitors. Respondents answered questions about the closest intersection to their home and travel patterns. Like UTK's efforts, these address locations were mapped using the ArcGIS online geocoding engine, and Network Analyst was used to determine trip location and distance.

Focus on Biking, Walking, and Housing. In addition to establishing the modal split breakdown presented earlier, the Cal Poly survey focused on testing incentives to bike or walk, asking a segment of respondents to answer a series of structured questions about how they might behave when presented with an incentive to walk or bike. This included randomly assigned incentives: a) a US$5 monetary incentive; b) a free cup of coffee or juice; c) a free cup of coffee or juice, with a specified value of US$2; or d) a social request to give up their parking pass for altruistic reasons—in this case of benefit to the environment.

The rationale for this was to encourage a culture of biking and walking and to test a new TDM approach focused on behavioral economics. This focus was rational given the temperate climate of the city and the active bicycling community that already exists. The City received a Gold Award from the League of American Bicyclists. The City of San Luis Obispo, other local agencies, and cycling advocates regularly host workshops, lessons, school assemblies, and community events to continue to promote cycling.

As a result, a pilot of both financial and social incentives was conducted finding that social incentives that tapped into altruistic tendencies had a strong pull on campus travel habits. The program found that such incentives were much more effective than financial incentives at influencing individuals living beyond the 10-mi threshold to explore transit, carpooling, and cycling (*22*). This provides impetus for the campus to explore transport-related social programs including features like a social application that could facilitate group connections; incentives such as a commute club in which campus travelers are entitled to a free cup of coffee or juice when they travel via walking or biking; or a free monthly gym membership to enable access to showers before work and support a holistic healthy lifestyle/workplace.

The survey also revealed a glaring housing issue for students and that by increasing the number of housing units on campus, the university could begin to decrease the percentage of students who live in off-campus housing and subsequently commute using SOVs. As a result, the Climate Action Plan suggested increasing the number of housing units for students by an average of 500 units every 5 years for students, faculty, and staff.

San Diego State University (SDSU): Commute Mode Sheds

SDSU Commute/Policy Catchment Zones and the Standard Deviation Ellipse. In 2014 and 2015, as SDSU began to explore emissions reduction strategies, commuting was expected to be responsible for a large portion of emissions. A comprehensive online travel demand survey was undertaken, gathering information regarding various aspects of university transportation behaviors. The survey was deployed via email to faculty, staff, and auxiliaries in the fall of 2015, and students in the spring of 2016 (*42*). The results of the survey help understand travel behavior and the incentives necessary for creating a significant mode shift. A total of 2,801 survey responses were gathered, representing a 7.55% sampling rate, with 2,122 student responses, 202 faculty responses, and 477 staff responses. The survey revealed that 80% of staff and 63% of students drive to campus. These responses were then geocoded in ArcGIS, yielding a total of 2,297 valid responses (as shown in Table 1). The data were made non-identifiable for privacy reasons and ArcGIS was used to determine trip distance. The resulting map provided the spatial distribution of the campus population and created spatial context for proposed policy strategies. This was created using the standard deviation ellipse function as shown in Figure 4.

The survey respondents were divided on the basis of mode chosen—similar to both UTK and Cal Poly cross-tabulation methods. The points for each mode were used in the standard deviation ellipse function in ArcGIS to find the commute zone for each mode as shown in Figure 4. The inclination of the zones explains the distribution of each mode. Most people walking and biking live close to campus, especially toward the east and west of the campus. From registrar data, about 41% of the student population was found to be living in the biking zone. The first transit zone contained 32% of the population, and the walking zone represented 6% of students. These data can be used to identify which modes are viable for each population group, based on the mode choice of their neighbors.

Emissions Calculations. The trip distance calculated by Network Analyst was used to calculate the student emissions in each zone. Overall, 29,845.6 million tonnes of carbon dioxide equivalent ($MtCO_2e$) (94.76%) of total student emissions were distributed throughout five catchment zones. The other 5% of emissions were created by students who reside outside the catchment zones. Transit Zone 2 was found to constitute 53.03% (15,828 $MtCO_2e$) of student emissions; 7,340.9 $MtCO_2e$, or 24.6%, of emissions came from Transit Zone 1. Another 18.93% (5,649 $MtCO_2e$) were produced within the car zone. Only 325.4 $MtCO_2e$ and 702.3 $MtCO_2e$ were produced in the walk and bike zones, respectively. The biking zone contained 40.92% of the student data but contributed only 702.3 $MtCO_2e$ in emissions. Given this, a strategic

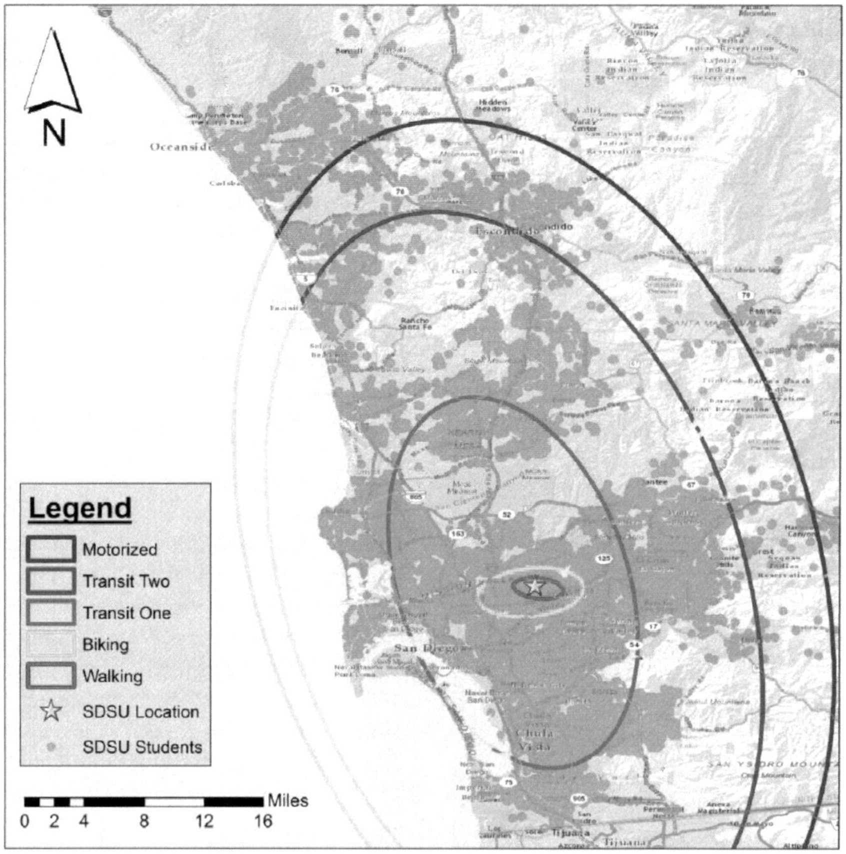

Figure 4. SDSU standard deviation ellipse commute zones for walk, bike, bus, and car trips.

policy was proposed that focused on providing incentives to 1) shift students to the biking zones closer to campus through housing, or 2) expand the biking zone through better infrastructure or improved technology (e.g., e-bikes).

Policy Strategy 1: Changes in Student Housing. This policy scenario is based on the possibility of SDSU realizing a total of 10,000 beds on campus by 2030, as indicated in the SDSU master planning processes. Currently over 4,385 students live in on-campus housing, leaving over 30,000 commuting students. Expanding the on-campus population by 5,615 would reduce commuting-related emissions by approximately 6,392.8 $MtCO_2e$ a year (16.4%) by 2030 from the projected business-as-usual scenario, highlighting the importance of university housing policy and land-use planning in the reduction of the campus carbon footprint and expanding multimodal transport.

Policy Strategy 2: Encourage more Bicycling and Walking. Another proposed strategic policy is to promote better walking and biking infrastructure, instead of parking facilities. This focuses on car-free commuting and has the potential to reduce emissions by 1027.7 $MtCO_2e$, or 2.71% from the baseline scenario. Given the terrain surrounding SDSU, these findings suggest potential benefits from not only improving bike lanes, but also making student e-bike subsidies available.

Conclusion

This study aimed to highlight more advanced approaches to using data that is often generated by universities, sometimes for alternate reasons, that can inform travel demand studies. In short, universities can develop like small cities—with large captive populations, top-down planning processes and, in many cases, significant power to develop campuses in very sustainable ways. As summarized in Table 2, our cases provide lessons for many growing corporate and technology campuses that are producing increasingly regional trips and impacts. Both university and corporate campus typologies have developed as progressive mixed-use developments with relatively low auto-reliance and highly walkable and livable spaces, often in the middle of car-centric cities. This presents significant opportunities for campus transportation- and environmental planners as they seek to address commute-related issues.

Table 1. Distance by Cohort

		Less than 1.5 miles	1.5–5 miles	5–10 miles	More than 10 miles	Total
Student	Count	357	805	17	453	1,632
	% within user type	22%	49%	1%	28%	100%
	% within distance cohorts	70%	72%	68%	71%	71%
	% of total	16%	35%	1%	20%	71%
Faculty	Count	63	105	2	59	229
	% within user type	28%	46%	1%	26%	100%
	% within distance cohorts	12%	9%	8%	9%	10%
	% of total	3%	5%	0%	3%	10%
Staff/other	Count	94	209	6	127	436
	% within user type	22%	48%	1%	29%	100%
	% within distance cohorts	18%	19%	24%	20%	19%
	% of total	4%	9%	0%	6%	19%
Total	Count	514	1119	25	639	2297
	% within user type	22%	49%	1%	28%	100%
	% within distance cohorts	100%	100%	100%	100%	100%
	% of total	22%	49%	1%	28%	100%

Table 2. Summary of Cases

Location	Method	Policy recommendations	Opportunities and outcomes
University of Tennessee, Knoxville	• Surveys • GIS/location • Cluster analysis	Rideshare/carpool as a travel demand management strategy	Pricing for parking Opportunity to target (primarily carpool) interventions using specialized location-based matching
Cal Poly, San Luis Obispo	• Surveys • GIS/location • Regression and segment analysis	Focus on social norms to encourage biking and walking	Decision to address parking pricing Opportunity to target campus patrons by individual social characteristics Parallel decision to institute more on-campus housing
San Diego State University	• Surveys • GIS/location • Zonal analysis	Strategies to incentivize housing on campus and biking/walking	Pricing for parking Decision/commitment to construct 10,000 beds on the campus Zone/location-based incentives to encourage biking and walking

As is summarized in the table, a key lesson is that each campus used a sophisticated data approach, starting with a survey and then moving to GIS and more robust statistical modeling. This is important because universities have the ability to harness specific high-resolution data for their entire campus population, to assess current conditions and target populations for interventions. This is unprecedented in conventional public sector transportation planning methods—unique over other more traditional jurisdictional boundaries. To leverage this advantage, we argue that university transportation planners and administrators should be well versed in more advanced data sciences and economics to induce more sustainable behavior. As can be shown in our examples, every campus used an advanced modeling method. Some of this expertise could be developed through interactions with faculty and students in those areas.

Moreover, as integrated data provides more opportunities for improving campus operations, improvements in communication between campus units could allow for better analysis. In these cases, campus transportation, sustainability, enrollment, and human resources departments could share data to improve specific transportation applications. Improved data transparency between campus units could enable these opportunities. This provides a key recommendation for further study—that common GHG and sustainability metrics begin to be applied across campuses, allowing for better benchmarking and comparisons.

Further, while the opportunities and outcomes at each campus varied slightly, each had the similarity in that policy was directed at nudging individual behavior away from driving through hyper-targeted (and data enabled) marketing and incentives, and parking pricing. This policy highlights an inherent power campuses have that is important to harness—that they control almost all parking and even some transit to their entire community, which often is all destined to one location (many origins to one destination). This allows significant control over economic or other incentives to push behavior.

While these provide important lessons for other campuses they also highlight ongoing research and study should continue on this topic. In that light, the following items also warrant more analysis in the future.

- Research should evaluate how and if the lessons from campus planning apply to many growing corporate and technology campuses.
- Evaluation should be done to judge the efficacy of policy and transportation demand management measures developed to address GHG emissions and if they are having longitudinal impacts on reductions.
- As all campuses used climate solutions that involved student housing, the land use–transportation connection should be investigated with greater purpose to demonstrate the reduction in trips from changes in land use at these locations.

To conclude, this paper highlights a few examples of how data can be used to target policies or develop scenarios that can shift behavior to reduce parking demand, reduce the adverse effects of drive-alone traffic, or reduce the environmental footprint of a university. The applications of such data are broad and this paper only highlights a few of the applications we applied, specifically focused on carpool or rideshare strategies, targeted bike and walk incentives, and mode shift to reduce GHGs. While not a comprehensive view of how these datasets can be applied, or even a comprehensive view of all data sources that can be mined and fused from universities for transportation purposes, this does illustrate how data can be gathered and applied in novel ways to address the complicated issues of campus commute and climate-related emissions.

Acknowledgments

The authors thank the many people who assisted in data collection and processing at each of the three universities in the case studies, including many students and staff who assisted in developing survey instruments and compiling appropriate data. The authors also acknowledge the institutional funding sources at UTK, Cal Poly, and SDSU for funding this data collection effort.

Author Contributions

The authors confirm contribution to the paper as follows: study conception and design: CC, WR, BA; data collection: CC, WR, BA, ND, AF, STJ; analysis and interpretation of results: CC, WR, BA, ND, AF, STJ; draft manuscript preparation: CC, WR, BA. All authors reviewed the results and approved the final version of the manuscript.

References

1. Bustillos, B. I., J. Shelton, and Y.-C. Chiu. Urban University Campus Transportation and Parking Planning Through a Dynamic Traffic Simulation and Assignment Approach. *Transportation Planning and Technology*, Vol. 34, No. 2, 2011, pp. 177–197.
2. Riggs, W., and J. Kuo. The Impact of Targeted Outreach for Parking Mitigation on the UC Berkeley Campus. *Case Studies on Transport Policy*, Vol. 3, No. 2, 2015, pp. 151–158.
3. Delmelle, E. M., and E. C. Delmelle. Exploring Spatio-Temporal Commuting Patterns in a University Environment. *Transport Policy*, Vol. 21, 2012, pp. 1–9.
4. Shannon, T., B. Giles-Corti, T. Pikora, M. Bulsara, T. Shilton, and F. Bull. Active Commuting in a University Setting: Assessing Commuting Habits and Potential for Modal Change. *Transport Policy*, Vol. 13, No. 3, 2006, pp. 240–253.
5. Zhou, J. Proactive Sustainable University Transportation: Marginal Effects, Intrinsic Values, and University Students' Mode Choice. *International Journal of Sustainable Transportation*, Vol. 10, No. 9, 2016, pp. 815–824. https://doi.org/10.1080/15568318.2016.1159357
6. Rybarczyk, G., and L. Gallagher. Measuring the Potential for Bicycling and Walking at a Metropolitan Commuter University. *Journal of Transport Geography*, Vol. 39, No. Suppl. C, 2014, pp. 1–10. https://doi.org/10.1016/j.jtrangeo.2014.06.009
7. Riggs, W. Dealing with Parking Issues on an Urban Campus: The Case of UC Berkeley. *Case Studies on Transport Policy*, Vol. 2, No. 3, 2014, pp. 168–176. https://doi.org/10.1016/j.cstp.2014.07.009
8. Wang, X., A. Khattak, and S. Son. What Can Be Learned from Analyzing University Student Travel Demand? *Transportation Research Record: Journal of the Transportation Research Board*, 2012. 2322: 129–137.
9. Khattak, A., X. Wang, S. Son, and P. Agnello. Travel by University Students in Virginia: Is This Travel Different from Travel By the General Population. *Transportation Research Record: Journal of the Transportation Research Board*, 2011. 2255: 137–145.
10. Kamruzzaman, M., J. Hine, B. Gunay, and N. Blair. Using GIS to Visualise and Evaluate Student Travel Behaviour. *Journal of Transport Geography*, Vol. 19, No. 1, 2011, pp. 13–32.
11Balsas, C. J. Sustainable Transportation Planning on College Campuses. *Transport Policy*, Vol. 10, No. 1, 2003, pp. 35–49.
12Alshuwaikhat, H. M., and I. Abubakar. An Integrated Approach to Achieving Campus Sustainability: Assessment of the Current Campus Environmental Management Practices. *Journal of Cleaner Production*, Vol. 16, No. 16, 2008, pp. 1777–1785.

13. Emanuel, R., and J. Adams. College Students' Perceptions of Campus Sustainability. *International Journal of Sustainability in Higher Education*, Vol. 12, No. 1, 2011, pp. 79–92.
14. Bond, A., and R. Steiner. Sustainable Campus Transportation Through Transit Partnership and Transportation Demand Management: A Case Study from the University of Florida. *Berkeley Planning Journal*, Vol. 19, No. 1, 2006, pp. 125–142.
15. Toor, W., and S. Havlick. *Transportation and Sustainable Campus Communities: Issues, Examples, Solutions*. Island Press, Washington, D.C., 2004.
16. Nuworsoo, C. Deep Discount Group Pass Programs: Innovative Transit Finance. *Berkeley Planning Journal*, Vol. 18, No. 1, 2005, pp. 151–165.
17. Dugundji, E., and J. Walker. Discrete Choice with Social and Spatial Network Interdependencies: An Empirical Example Using Mixed Generalized Extreme Value Models with Field and Panel Effects. *Transportation Research Record: Journal of the Transportation Research Board*, 2005. 1921: 70–78.
18. Blume, L. E., and S. N. Durlauf. *The Economy as an Evolving Complex System, III: Current Perspectives and Future Directions*. Oxford University Press, New York, 2005.
19. Marchal, F., and K. Nagel. Modeling Location Choice of Secondary Activities with a Social Network of Cooperative Agents. *Transportation Research Record: Journal of the Transportation Research Board*, 2005. 1935: 141–146.
20. Heyman, J., and D. Ariely. Effort for Payment a Tale of Two Markets. *Psychological Science*, Vol. 15, No. 11, 2004, pp. 787–793.
21. Schwartz, H. Predictably Irrational: The Hidden Forces That Shape Our Decisions. *Business Economics*, Vol. 43, No. 4, 2008, p. 69.
22. Riggs, W. Painting the Fence: Social Norms as Economic Incentives to Non-Automotive Travel Behavior. *Travel Behaviour and Society*, Vol. 7, 2017, pp. 26–33. https://doi.org/10.1016/j.tbs.2016.11.004
23. Riggs, W. Testing Personalized Outreach as an Effective TDM Measure. *Transportation Research Part A: Policy and Practice*, Vol. 78, 2015, pp. 178–186. https://doi.org/10.1016/j.tra.2015.05.012
24. Amir, O., D. Ariely, A. Cooke, D. Dunning, N. Epley, U. Gneezy, B. Koszegi, D. Lichtenstein, N. Mazar, and S. Mullainathan. Psychology, Behavioral Economics, and Public Policy. *Marketing Letters*, Vol. 16, No. 3–4, 2005, pp. 443–454.
25. Deakin, E. Sustainable Development and Sustainable Transportation: Strategies for Economic Prosperity, Environmental Quality, and Equity. *Institute of Urban & Regional Development Research Report*, UC Berkeley, 2001, pp. 1–42.
26. Deakin, E., A. Bechtel, A. Crabbe, M. Archer, S. Cairns, A. Kluter, K. Leung, and J. Ni. Parking Management and Downtown Land Development in Berkeley, California. *Transportation Research Record: Journal of the Transportation Research Board*, 2004. 1898: 124–129.
27. Tudela-Rivadeneyra, M. S., E. D. Aldo, M. Shirgoakar, E. A. Deakin, and W. W. Riggs. The Cost Versus Price for Parking Spaces at Major Employment Centers: Findings from UC Berkeley. Presented at 94th Annual Meeting of the Transportation Research Board, Washington, D.C., 2015.
28. Shoup, D. C. *The High Cost of Free Parking*. Planners Press, Chicago, 2005.
29. Cervero, R., and M. Duncan. Walking, Bicycling, and Urban Landscapes: Evidence from the San Francisco Bay Area. *American Journal of Public Health*, Vol. 93, No. 9, 2003, pp. 1478–1483.
30. Flyvbjerg, B. Measuring Inaccuracy in Travel Demand Forecasting: Methodological Considerations Regarding Ramp Up and Sampling. *Transportation Research Part A: Policy and Practice*, Vol. 39, No. 6, 2005, pp. 522–530.
31. Gärling, T., and G. Schuitema. Travel Demand Management Targeting Reduced Private Car Use: Effectiveness, Public Acceptability and Political Feasibility. *Journal of Social Issues*, Vol. 63, No. 1, 2007, pp. 139–153.
32. Saelens, B. E., J. F. Sallis, and L. D. Frank. Environmental Correlates of Walking and Cycling: Findings from the Transportation, Urban Design, and Planning Literatures. *Annals of Behavioral Medicine*, Vol. 25, No. 2, 2003, pp. 80–91.
33. Ariely, D., and G. Loewenstein. Coherent Arbitrariness: Stable Demand Curves Without Stable Preferences. *Technology*, Vol. 73, 2003, pp. 73–105.
34. Helbing, D., and P. Molnar. Social Force Model for Pedestrian Dynamics. *Physical Review E*, Vol. 51, No. 5, 1995, p. 4282.
35. Cervero, R., and K. Kockelman. Travel Demand and the 3Ds: Density, Diversity, and Design. *Transportation Research Part D: Transport and Environment*, Vol. 2, No. 3, 1997, pp. 199–219.
36. Schmitt, A. *Five Ways Colleges Are Coaxing Students Out of Their Cars*. http://usa.streetsblog.org/2014/02/05/five-ways-colleges-are-coaxing-students-out-of-their-cars/. Accessed January 8, 2016.
37. Kelly, K. L. Casual Carpooling-Enhanced. *Journal of Public Transportation*, Vol. 10, No. 4, 2007, pp. 119–130.
38. Harris, W. *The Cumberland Avenue Corridor and the University of Tennessee: The Relationship Between a University and Its Neighborhood*. University of Tennessee, 2012.
39. Tennessee, U. O. *Climate Action Plan for University of Tennessee, Knoxville*. http://rs.acupcc.org/cap/483/. Accessed January 8, 2016.
40. Lord-Farmer, K., W. Riggs, and A. Greve. Transportation Policy for Campus Climate Action Planning: Process and Policy Implications. Presented at 96th Annual Meeting of the Transportation Research Board, Washington, D.C., 2017.
41. Greve, A., C. Clark, W. Riggs, J. Carpentier, and C. Lord-Farmer. *Cal Poly Climate Action Plan*. City & Regional Planning Studios and Projects, 2016, pp. 1–124.
42. Appleyard, B., J. Mckinstry, and A. Frost. Calculating the Campus Carbon Footprint: Measuring University Associated Greenhouse Gas Emissions from Transport. Presented at 96th Annual Meeting of the Transportation Research Board, Washington, D.C., 2017.

The Standing Committee on Transportation Demand Management (ABE50) peer-reviewed this paper (18-00651).

The authors take full responsibility for the findings and they do not reflect views of other people or institutions associated with this project.

Article

TRR
JOURNAL OF THE TRANSPORTATION RESEARCH BOARD

Transportation Research Record
2018, Vol. 2672(8) 754–763

DOI: 10.1177/0361198118799031
journals.sagepub.com/home/trr
SAGE

Designing a Transit-Feeder System using Multiple Sustainable Modes: Peer-to-Peer (P2P) Ridesharing, Bike Sharing, and Walking

Daisik Nam[1], Dingtong Yang[1], Sunghi An[1], Jiangbo Gabriel Yu[1], R. Jayakrishnan[1], and Neda Masoud[2]

Abstract

Peer-to-peer (P2P) ridesharing is a relatively new concept that aims to provide a sustainable method for transportation in urban areas. Previous studies have demonstrated that a system that incorporates both P2P ridesharing and transit would enhance mobility. We develop schemes to provide travel alternatives, routes and information across multiple modes, which includes P2P ridesharing, transit, city bike-sharing and walking, within the network. This study includes a case study of the operation of the multimodal system that includes P2P ridesharing participants (both drivers and riders), the Los Angeles Metro Red line subway rail, and the Los Angeles downtown bike-share system. The study conducts a simulation, enhanced by an optimization layer, of providing travel alternatives to passengers during morning peak hours. The results indicate that a multi-modal network expands the coverage of public transit, and that ride- and bike-sharing could be effective transit feeders when properly designed and integrated into the transit system.

Metropoles such as Los Angeles are encountering serious congestion issues owing to high demand for transportation and the limited capacity of the street networks. In such cities, public transportation plays a significant role in alleviating congestion on the street network. However, the problem of transporting people to and from public transport stations, also known as the first- mile/last-mile problem, remains an issue. Commuters who would have otherwise used public transportation choose to drive their vehicles owing to the difficulty of access to public transportation stations.

Introducing sustainable transportation alternatives to provide access to public transportation allows for the reduction of congestion and its side-effects. These alternatives include Peer-to-peer (P2P) rideshare, bike share, walk, and transit. First, these alternatives encourage people to reduce the usage of personal low-occupancy vehicles, which would reduce greenhouse gas emissions. Second, mode combination allows the sustainable modes to complement each other, overcoming the weaknesses of each one, were they to be used as the main mode of transportation. Last but not least, the combination of modes may even reduce travel time and improve reliability. This research proposes a transit-feeder system that combines several modes of transportation to provide door-to-door transportation.

In P2P ridesharing, drivers who are traveling to perform activities use empty seats in their vehicles to transport passengers who have spatiotemporal proximity with them. Masoud et al. proposed a transit feeder system to promote transit ridership by connecting riders to transit using P2P ridesharing (*1*, *2*). Their matching algorithm has a multi-hop property in which a passenger can transfer between multiple vehicles/modes of transport. They also allow for each vehicle to carry multiple passengers at the same time. The system will take over the routing of drivers to place them in spatiotemporal proximity with passengers.

This research extends works of Masoud et al. by integrating multiple shared-mobility alternatives (*1*). In this study bike sharing will also be integrated into the transit

[1]Department of Civil and Environmental Engineering, Anteater Instruction and Research Building (AIRB), Institute of Transportation Studies, University of California, Irvine, CA
[2]Department of Civil and Environmental Engineering, University of Michigan, Ann Arbor, MI

Corresponding Author:
Address correspondence to Daisik Nam: nds1027@gmail.com

feeder system, along with P2P ridesharing, in an attempt to increase accessibility to transit stations and improve transit ridership. Bikes offer several advantages compared to normal vehicle usage: (i) they are not affected significantly by the street traffic conditions, and (ii) while drivers' pre-specified schedules and the transit system's fixed routes and schedules constrain the potential for matches, the route and schedule of bikes are flexible as long as bikes are available at stations. By guiding riders to walk some distance to the nearby bike stations and P2P ridesharing go-points and therefore aggregating the demand (*3*), the ride matching rate could potentially increase as well.

Integrating multiple modes into the transit-feeder system is accompanied by certain challenges in the design and operational management, which this paper attempts to address. First of all, we introduce a comprehensive multi-modal platform in which each transportation alternative is allocated a separate layer in a multi-layer network. This multi-modal platform can be regarded as a super network. A layer dedicated to a single mode of transportation can contain the mode's specific characteristics, reduce the computing time to find the shortest path, and provide the basis for efficient management of the network database (*4*).

This research evaluates the proposed transit feeder system by applying it to P2P ridesharing participants (both drivers and riders), the Los Angeles Metro Red line and the bike-sharing program in downtown LA. The reason why we selected this area is because the ridership of the Metro Red line has declined in recent years (*1*). In addition, a recently launched bike-sharing program in downtown LA has experienced low usage. The goal of the transit system feeder is to increase the ridership of both.

Related Literature

A lot of the past research results show that there is a significant correlation between the accessibility and transit ridership (*5–9*). The goal of transit feeder services is to improve transit ridership by improving accessibility to the transit system. Transit feeder systems can use multiple modes of transportation to connect travelers to transit, and have been receiving more attention during recent years.

Several studies have focused on integrating bikes and transit to promote transit ridership (*10–13*). These studies have concluded that combining the two modes of transportation can be beneficial to travelers by increasing their accessibility, and to the transit and bike-share systems by increasing their utilization rates as a result of establishing a mutually complementary relationship between them. Riders in this multi-modal network can take advantage of the benefits of both modes: transit is reliable and affordable, and can be faster than the average private vehicle in some regions; bikes can enlarge the catchment area of transit with low costs, reducing congestion, and provide environmental and health benefits (*14*). Integrating bike-sharing into the transit-feeder system reduces the number of passengers who drive to, or are dropped-off at metro stations by having them switch to bikes.

Personal vehicles or taxi and ride-share vehicles can also be utilized as a transit feeder. According to Mo, 10% of Metro rail passengers in LA are dropped off by another personal vehicle (*7*). Nam et al. and Wang and Ross analyzed taxi trip characteristics using GPS data and found that a meaningful portion of transit passengers utilize taxis for their last mile (*15*, *16*). Regue et al. proposes Car2work for integrating ridesharing into existing public transit systems (*17*). Masoud et al. focuses on designing a system to feed the LA Metro Red line using peer to peer ridesharing and found that ridesharing can improve transit ridership (*1*).

A multi-modal transit feeder system can benefit from the different characteristics that each mode has to offer. For example, a multi-modal transit-feeder system can promote health through bike-sharing, and make public transportation more attractive and accessible by covering the proverbial first/last mile. This paper introduces a system to better integrate the components of multi-modal transportation systems by proposing desirable itineraries to travelers. The system provides easier access to public transit, partially by using non-motorized transport options, and partially by having individuals share motorized transportation options, therefore decreasing the number of vehicles on the street network. Lastly, the proposed transit-feeder system provides better accessibility to public transportation, specifically by addressing the last-mile challenge faced by public transit.

This system requires an elaborate ride-matching algorithm that is capable of processing multimodality. Finding the best route for a passenger in a multimodal network can be formulated as an optimization problem. Several ride-matching techniques have been formulated recently, based on time-expanded network models (*1*, *2*, *18–20*). We use the special case for the formulation proposed in Masoud and Jayakrishnan that uses a dynamic programming algorithm to find the optimal route for a passenger in a matter of a few seconds (*2*). This algorithm has been applied in Masoud et al., showing reasonable matching results between riders, drivers and transit (*1*). We further enhance the dynamic programming algorithm proposed in Masoud and Jayakrishnan to include bike-sharing and walking (*2*).

Multi-Hop and Multi-Modal Ride-Matching System

In this paper we devise a multi-hop and multi-modal ride-matching algorithm. The proposed algorithm provides a traveler with an itinerary with multiple potential connections, such as walk-ridesharing-transit-bike. This

Table 1. Types of Locations in the Multi-Modal Shared-Ride Network

Station type	Symbol	Description
Go-points-vehicles	S_V	Points where a rider starts/ends their journey by taking a ridesharing vehicle $\forall S_V \in S_G$
Go-points-bike	S_B	Points where a rider starts/ends their journey by bike-sharing $\forall S_B \in S_G$
Go-points-connection points	S_C	Points where an individual can connect to other drivers or modes (bike, transit) $S_C \in S_V, S_C \in S_B, \forall S_C \in S_G$
Go-points-transit stations	S_T	Points where an individual can transfer to/from a transit station $\forall S_T \in S_C, \forall S_T \in S_G$
Riders' origin/destination	S_O, S_D	Points where a rider starts/end their journey - is connected to go-points by walking

information provides travelers with door-to-door guidelines on how to combine several modes of transportation for their trips. The goal of the ride-matching algorithm is to find passengers' itineraries that can provide them with the highest utility, in which utility is defined as a weighted combination of mode travel time, travel cost, waiting time and transfer penalty. Each passenger will be asked to provide the trip origin (S_O) and the trip destination (S_D), along with the earliest starting time (ES) and the latest arrival time (LA) of the trip. Passengers are also encouraged to state their preferences about the maximum number of connections (between different modes of transportation, or different vehicles of the same mode), modes of transportation, and characteristics of the vehicles on which they travel. It is possible to even elicit preferences on the type of individuals with whom they may share rides. Based on user input and available modes of transportation, the system will devise itineraries within the travel time windows specified by passengers, and propose these to them.

The dynamic programming algorithm proposed by Masoud and Jayakrishnan has a flexible, multi-hop routing scheme (*21*), and Masoud et al. apply the algorithm to design P2P ridesharing to be a transit feeder (1). This algorithm, however, only considers the combination of P2P ridesharing and transit. We reformulated the algorithm to include bike-sharing and walking to access the transit-feeder system. With algorithm enhancement and network expansion, the proposed method allows a rider's itinerary to include as many modes of transportation as desired. We redefined a network structure to efficiently manage multiple modes of transportation and introduced methods to improve the matching rate and provide utility-maximizing itineraries to travelers.

To model the transit feeder system, we discretize the study time horizon into short time periods (5-min periods in this study). Furthermore, we define locations in the network where travelers can start and end their trips, and/or transfer between transportation alternatives. Note that in this study we have several types of locations with different functionalities, elaborated in Table 1. The proposed algorithm has a node-link network structure. Let us define a node n_i to be a tuple of the time period (t_i) and the station (s_i), that is, $n_i = (t_i, s_i)$. A link is denoted as (t_i, s_i, t_j, s_j), such that it can be interpreted as a trip that starts from the station s_i at time t_i, and ends at the station s_j at time t_j. We define a set of "go-points", denoted by S_G, as pre-specified locations where riders can start or end a (leg of a) trip in a driver's vehicle, start or end a shared-bike ride, or transfer between modes. Our research does not assume, however, that every rider starts their journey from a go-point, as was done in past research. We allow riders to walk a certain distance between a go-point and their actual start/end points (S_O and S_D). This additional level of flexibility introduces advantages over the earlier schemes in which trips were assumed to start and end only at go-points: (1) it reflects the actual behavior of riders which can be extended to real mobile services: (2) it increases riders' route flexibility because they are not always restricted to one selected go-point; and (3) the ridesharing system can have a higher matching rate owing to this flexibility.

To allow multiple modes, we introduce a super network concept that utilizes an independent layer for each mode and integrates all modes using connections at mode-transfer stations. The locations (physical nodes in the network) are categorized into five types as shown in Table 1. To promote transit ridership, we restrict Go-points for bikes to be connected only to transit stations in this study.

In a multimodal system, we should consider the various characteristics of each mode. The network contains four different modes (P2P ridesharing, Bike-sharing, Transit, and Walking), which would result in four separate network sub-layers. In Figure 1, black, red and green lines represent the three different layers, and the blue lines represent the rider network (which contains the walking mode for the rider as well as the entire rider path). In this example, as the rider travels from his origin (O) to destination (D), the rider would walk to a bike station, ride a bike, then transfer to transit, and a ride-share vehicle will be used for his last mile. This ride matching can be accomplished through optimization on such a multimodal network, along the lines of those used in Masoud and Jayakrishnan (*2*).

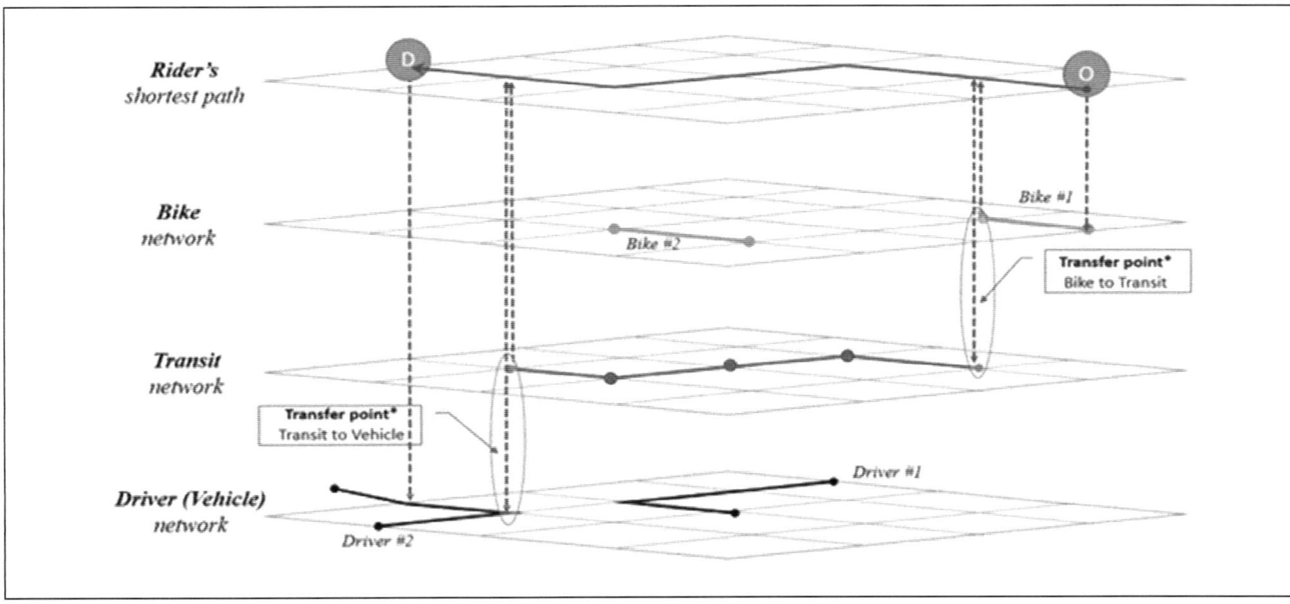

Figure 1. Layers of transportation alternatives in the super network of the transit-feeder system.

The P2P ridesharing network only contains vehicle related link-node information such as the travel routes of drivers, turn restrictions, tolls and travel times of vehicles, as it is designed only for vehicles. The bike network would only include availability of bikes, cost, and routes to nearby subway train stations. The transit network would only have information about frequencies, routes, and fares. The layer for each travel mode is independent except for the connection points of time and space. The riders' travel can be accomplished through a certain combination of several modes. Furthermore, riders should walk to a nearby go-point from an actual origin point and from a go-point to an actual destination, where the origins and destinations are typically homes or work/shopping locations. For simplicity, we use a straight line for any walk link, though it can also be represented as a separately found walking route.

Each connection go-point (S_C) is represented by separate tuples (t_i, s_i) corresponding to each mode because each mode has a different available time window. Our decomposed optimization algorithm for ride-matching in a ridesharing system can optimize a multi-modal system in a similar fashion, essentially by considering a bike or a transit vehicle as similar to a virtual "driver" in a ridesharing system.

The integrated multi-modal network improves efficiency in pre-processing, ride-matching and managing of the database. The optimization algorithm includes multiple shortest path calculations in pre-processing and ride-matching. The multimodal network structure reduces computing time for shortest path calculations by restricting the number of node explorations from any node to only those of the associated mode, except at the connection nodes between the mode layers (mode transfer nodes). For database management during path optimization, this network structure reduces the search time to find feasible drivers, as the user limits his/her preferences to only certain modes. The query process then searches for feasible drivers in only the preferred-mode set, which is a relatively small database when compared to one that includes all modes.

The demand generation process is designed to reflect a realistic spatial distribution among the riders. As explained above, previous research has assumed that riders' origin/destination are predefined locations, as shown with blue dots in Figure 2 (*1–3*). In reality, however, riders should access one of the go-points by walking unless their starting/ending points are exactly at the go-points. We design our ride matching network to reflect the riders' actual starting/ending points. Instead of assuming that demand exists at a representative point, we randomly disperse the travel demand for a transportation analysis zone (TAZ) within the associated TAZ, as red dots in Figure 2. In other words, we consider accessibility to/from a go-point.

To connect the randomly generated riders' origin and destination points to the network of go-points, we introduce dynamic walk links as connectors to the nearby go-points. They are indicated by the dashed line in Figure 3. We design the dynamic walk links to be connected to the nearest *n* go-points. These dynamic links are temporarily generated in our network when a rider requests a ride. After finishing the ride-matching process, the walk links are eliminated from the network for the next process.

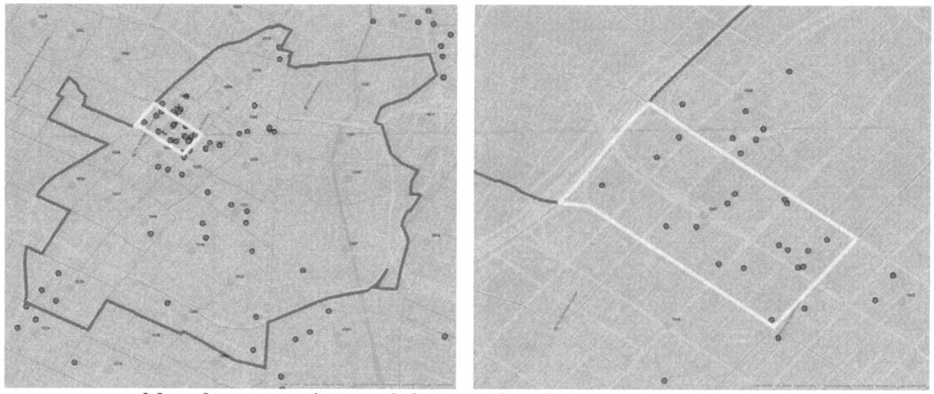

Figure 2. Example of demand generation.

This dynamic process allows a simpler network structure for further processing.

Multiple dynamic connectors from an origin to go-points, and to a destination from go-points have the potential to increase matching rates, as shown in Figure 3. Connecting a walk link to only one go-point has the limitation that the available drivers and possible routes are spatially restricted. Figure 3*a* and *b* show an example of a ride-matching failure with a single connection. In the example, Rider 1's origin is R1 in a circle and his/her destination is R1 in a rectangle. Figure 3*a* indicates that the rider's go-points are G1 for origin and G3 for destination.

There are two drivers in the sample network. Driver 1 traverses D1-G1-G2-D1 and Driver 2 travels to D2 through G4. If there is no driver going to the riders' solely connected go-point near his destination, the rider's trip cannot be matched. This problem could be solved if we find the second nearest go-point and search a route again. However, it is computationally expensive as the matching process should be repeated until the ride is matched. The concept of multiple connectors can solve this problem, as shown in Figure 3*c* and *d*, as multiple connected links can include drivers/modes near multiple go-points. Although introducing the connectors marginally increases the computational time owing to the increased network size, the better matching rate more than compensates for this.

Another advantage of dynamic links is to find better paths for riders. The two sample networks in Figure 3*e* and *f* showcase this property in the following example. Here we assume that there are three drivers: 1) D1-G1-G2-D1, 2) D2-G2-G3-D2, and 3) D3-G4-D3. Figure 3*e* shows the case when we connect only one walk link to the nearest station. Rider 1 will be guided to walk to G1 from the origin because the nearest go-point is go-point 1. She will then transfer to Driver 2 at go-point 2 (G2) and be dropped off at go-point 3. The total travel time for this itinerary is 20 minutes (5 min walk). In comparison, our proposed network structure, as shown in (f), can reduce the travel time to 15 minutes and with no transfer, although a rider must walk little longer (8 min). Here, three driver candidates are considered for Rider 1.

Case Study

Research Scope and Data

For a parametric study of the application of our ride matching system, we selected the city of Los Angeles, as in our earlier study, which developed a network based on the Southern California Association of Governments (SCAG) RTP 2040 Travel Demand Model 2016 Scenario 3 (*1*). We enhanced that network to include multi-modal layers and the current bike-sharing stations in LA downtown as shown in Figure 4. Actual coordinates of the bike stations were collected and used to build the network. We used various data and actual travel time information, such as the LA Metro time table, and the Google directions API (*22*) for automobiles, bikes, and walk modes. Spatial connections between bike stations and transit stations are included in the network design step to efficiently improve accessibility to transit stations. Bike stations near transfer points (including metro stations) were selected as connection points.

In the preprocessing step, participants (riders and drivers) are randomly selected from the vehicle origin–destination trip table of the SCAG demand model, as the interest is in finding how many vehicles travelers will change their mode to our system. Then, as discussed in the demand generation step, we randomly placed the selected-riders in the given TAZ, as shown in Figure 2.

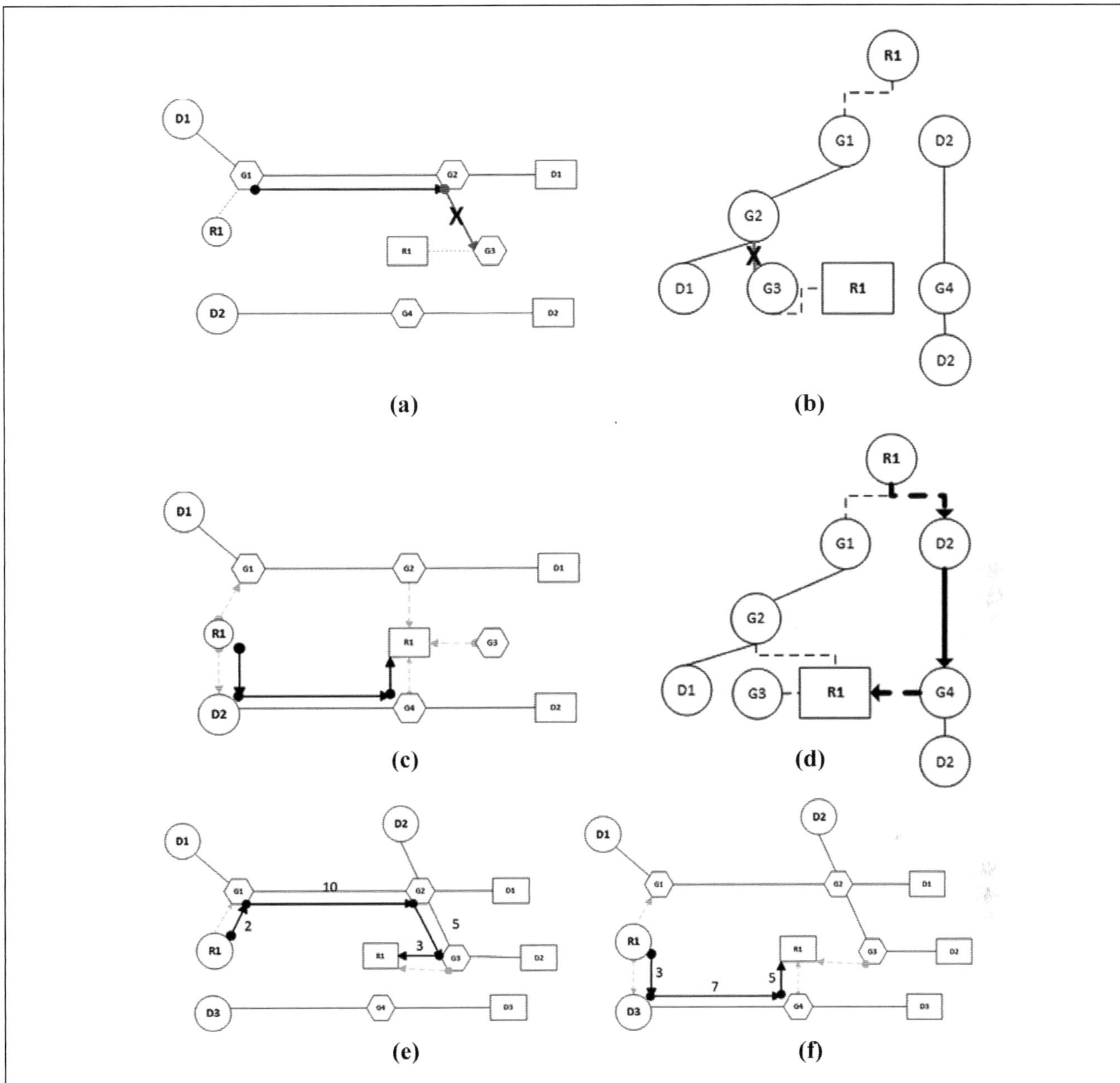

Figure 3. Advantage of multiple connectors: (*a*) ride-match failure case with single connector; (*b*) graph representation of ride-match failure case (single connectors); (*c*) ride-match success case with multiple connectors; (*d*) graph representation of ride-match success case (multiple connectors); (e) matched-rides result with a single connector (walk-D1-D2-walk); (*f*) matched-rides results with multiple connectors (walk-D3-walk).

The demand in the OD table is at an aggregated level containing the origin ID, destination ID, and the travel volume. SCAG also provides the geo-locations of each TAZ (both points and polygons). Points are generally located at the center of the associated polygons. A point and a related polygon are mapped by a key index. Using python programming and geographic information system (GIS) libraries, all riders' origin–destinations are randomly generated and located.

Within the constraints of the travel time budget, which is the difference between the earliest start (ES) and the latest arrival (LA) times that are specified by any rider, the dynamic matching algorithm finds the multi-hop paths maximizing each rider's utility. The utility function in our algorithm is expressed as a linear combination of the rider's cost components: travel time, mile-based travel cost, connection penalty, and bike-sharing cost. In this research, we use default values of \$20/hour for the value

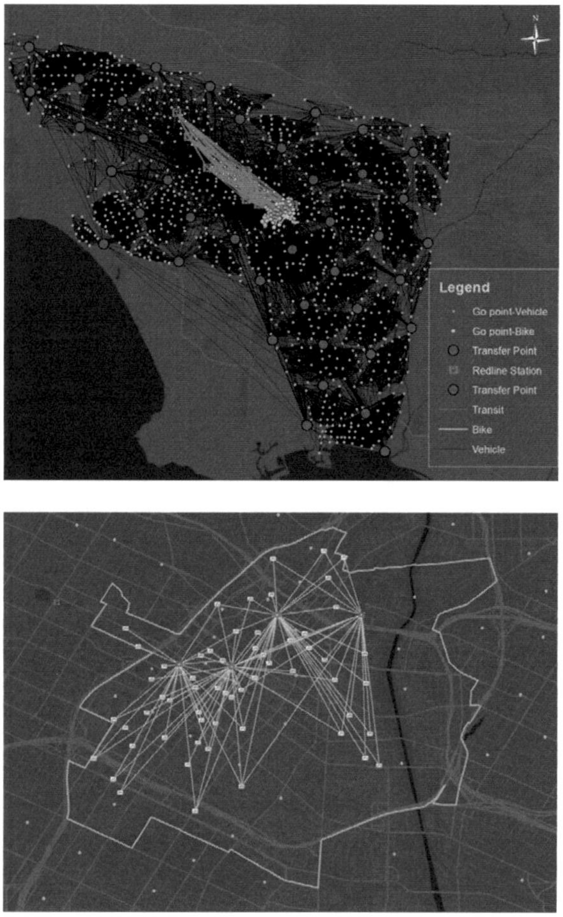

Figure 4. Node-link set and bike network expansion (Los Angeles region and the study area).

of time, and \$ 0.25 for each mile of ridesharing. Finally, we postulate a \$ 0.1 additional cost for each connection, and for each time period spent in waiting for a connection. In downtown LA, usage of a bike is priced at \$1.5 by a half hour period. When a bike is used in our matching system, it must be returned to another bike station.

Metro Rail Stations Accessibility

Reducing the first/last mile for the transit ridership implies improvements of accessibility to transit stations. Improved accessibility to the Metro increases transit ridership. Walking is a main mode to access to rail stations. Mo shows that about 52% of Metro users in LA County are willing to walk to Metro stations (*7*). Furthermore, they tend to walk when their walking time is less than 10 min. Therefore, we set the accessibility criteria as 10 min in our model. Our main expectation is that our proposed method attracts more riders by providing improved accessibility to Metro stations and reduced accessing time. In addition, it is evident that current riders, who walk to stations, can also benefit from the proposed method.

We examined how the proposed system improves accessibility to the Metro Red line subway stations in the morning peak. Assuming that riders only access the Red line stations if the access time is less than 10 minutes, we found the possible catchment region from where riders are willing to use our system. To identify catchment areas, this study applied a network analysis with our multi-modal network. To measure improvements to the catchment regions, we set access time to the Red line stations as an index. Access time to a metro station s is decomposed into access time by mode m from a go-point i to a metro station s, that is, ($tt^{m}_{i,s}$), and walking time to a go-point I, that is, (tt^{walk}_{i}), in Equation 1. Here $tt^{m}_{i,s}$ has three components: driving time from a go-point i to metro station s, denoted as ($tt^{m}_{inmode,i,s}$), waiting time for mode m at a go-point I, denoted as ($tt^{m}_{wait,i}$), and processing time for a mode at go-point I, denoted as ($tt^{m}_{process,i}$):

$$\text{Access Time}_s = tt^{m}_{i,s} + tt^{walk}_{i} \qquad (1)$$

where $tt^{m}_{i,s} = tt^{m}_{inmode,i,s} + tt^{m}_{wait,i} + tt^{m}_{process,i}$

From our multi-modal network, in which the actual travel times during the morning peak hour are found from the Google Directions API (*22*), a Dijkstra shortest path search identifies all possible go-points where mode travel time ($tt^{m}_{inmode,i,s}$) is less than a certain limit (in minutes). Each mode layer has mode-specific characteristics such as average wait time ($tt^{m}_{wait,i}$) for the mode m at a go-point i, and processing time ($tt^{m}_{process,i}$)of a mode m at a go-point i as presented in Equation 1. Average wait time for transit is calculated as half of the average transit headway (i.e., 1/frequency), which technically assumes, implicitly, that the scheduled headways are generally uniform and that there is no substantial schedule variance. The waiting time for bikes is set to zero and the processing time for bike rental is assumed to be 2 minutes. Our network does not include actual walk links, thus we again utilized a private API which provides walking level travel time and geographic boundaries from a point (*23*).

Figure 5 shows the accessible area to Red line stations, which indicates that our ridesharing system improves accessibility of Red Line stations in the morning peak. Red areas show the case when no feeder mode exists (except walking). Blue areas imply that more travelers can reach their nearest station by P2P ridesharing. We found the system to improve the area of accessibility from 8.64 square miles to 14.10 square miles. The proposed method with bike-sharing also has the potential to improve accessibility to 15.64 square miles. An interesting fact is that the bike-sharing system in the downtown area has more potential to increase the Red line's

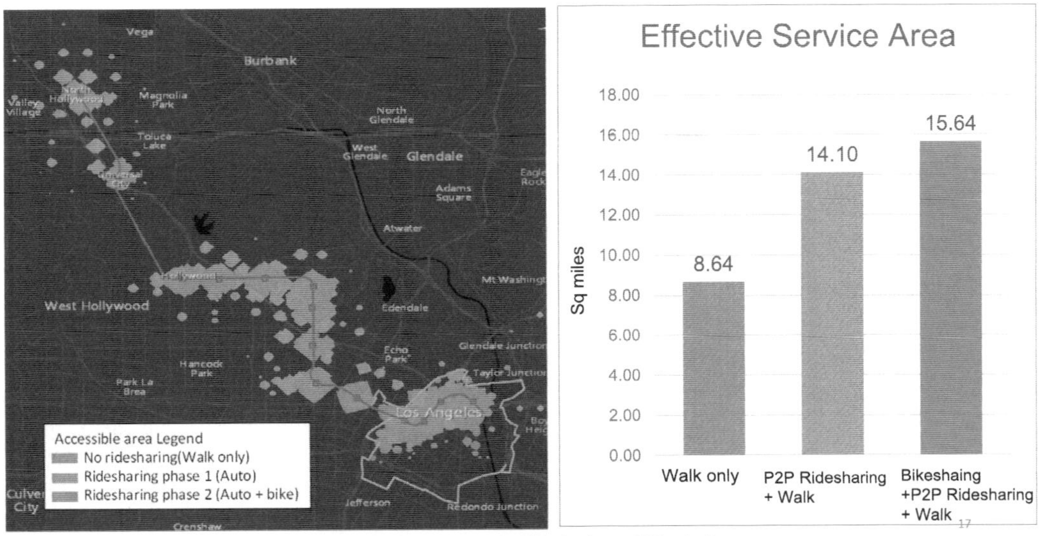

Accessibility improvement by Ridesharing and Bikesharing

Figure 5. Accessibility improvements by P2P ridesharing and bike-sharing: (*a*) accessible area to Red line stations; (*b*) effective service area.

catchment area than an automobiles-based ridesharing system as bikes are generally faster than automobiles that may become stuck in downtown congestion.

Parametric Study

Under the assumption that travelers will change their mode from personal automobiles, our simulations involve increasing the number of riders stepwise from 1,000 to 4,000, and increasing the number of drivers respectively from 1,000 to 10,000. The matching rate increases sharply at the beginning when we increase the number of drivers, as can be seen in Figure 6*a*. Then the marginal matching rate decreases. When we increase the number of drivers from 1,000 to 5,000, the matching rate triples. We may conclude that, the number of riders, drivers and the ratio between riders and drivers affect the matching rate. When the number of riders is small and the number of drivers increases to twice that of the riders, the matching rate would increase sharply. When the number of drivers is a relatively large number, doubling the number would not increase the matching rate to the same level. If the ratio continues to increase beyond two, the improvement in the matching rate is not as significant as before. The general pattern is similar to that reported in Masoud et al. (*1*), but we can see that the matching rate in our study is significantly higher. The network expansion, addition of bike-sharing, multiple-layer network, and dynamic walk connectors that we propose will enable riders to have more alternatives.

Increasing metro rail and bike usage among those who use automobiles is our main research interest. A series of simulations was designed, so as to understand how the usage numbers change when we increase the number of riders and drivers. We expect the number of Metro rail and bike users to also linearly increase when the number of total participants increases linearly. The usage results are shown in Figure 6*b* and *c*.

Firstly, for metro users, we can observe an approximately linear trend when we increase the number of participants. Out of the entire simulation sample (224,196 individuals), not all have direct access to the metro system (i.e., origins and destinations are far away from metro stations). Out of all of the individuals in the much larger area, around 8,400 individuals were potential metro users (as only Metro Red line is included in this study, and it covers only a small portion of the area). For a sampled 4,000-rider case, there are about 150 potential metro users, and 5 of whom are matched with a metro usage (3.3%). When the rider-to-driver ratio is high (which means the rideshare service may be in shortage, or not broad enough to cover all areas), the algorithm tends to match more riders with metro services. Therefore, designing a proper proportion of rider/driver ratio would help the usage of metro transit.

For bike users, the situation is similar. The trend is approximately linear when the total number of participants increases. As the study designs bikesharing only for a transit feeder, it does not include bike-only usage for temporary travel within the downtown region. In the 4,000-rider case, 75 riders have either the origin or the destination located in the downtown area. We identify this group of people as potential bike users. Among 75 riders, a single rider was matched with a bike (1.4%). If

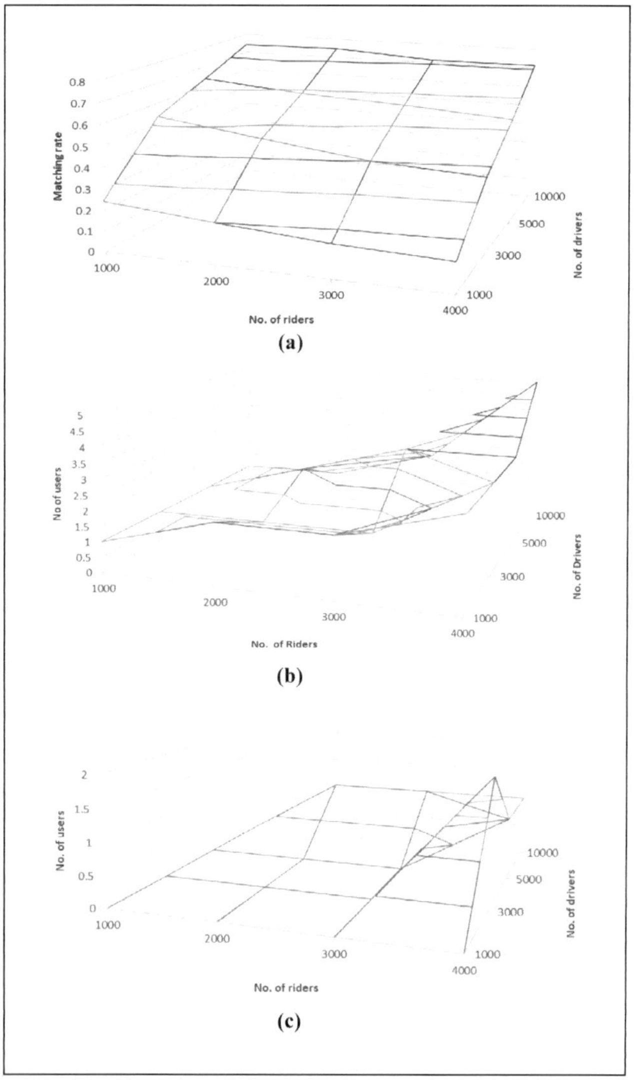

Figure 6. The results of the parametric study: (*a*) matching rate; (*b*) number of transit users according to the participants; (*c*) number of bike-sharing system users according to the participants.

we include bike-only travel, we will however see a higher bike usage rate in the downtown area.

Conclusion

This paper introduced schemes to study sustainable transportation alternatives that provide access to public transportation. Extending our earlier study (*1*), the paper contributes to the literature of ridesharing and transit feeder systems in the following ways.

First, we designed a transit feeder system that finds itineraries for riders using multiple sustainable travel modes: P2P ridesharing, bike-sharing, walk, and transit. For an efficient multi-modal model and easy data management, we introduced a super network concept that allocates each travel mode's network to a different layer that ensures each mode's operation is independent but at the same time accounts for transfers between modes. Riders are matched with different modes by moving across different mode layers. Green transportation modes in the system can be mutually beneficial for improving ridership, saving costs and increasing mobility.

Second, this paper proposes a scheme to achieve increased matching rates. Realizing that connecting a rider's origin to only one go-point (ridesharing/transit/bike station) can restrict the riders' mobility, a multiple-connector concept was proposed. These multiple connectors link the starting point to multiple go-points for each passenger, which contributes not only to increasing ride-matching rates for the system, but also to finding shorter travel-time itineraries for the riders.

The proposed system was tested in LA County. Target modes were the Metro Red line subway, Metro bike-sharing program in Downtown Los Angeles, walking, and P2P ridesharing. Geographical analysis for accessibility indicates that both P2P ridesharing and bike-sharing can enlarge the catchment area of the Red line stations. In the morning peak, bikes are more effective in Downtown LA towing to the existence of exclusive bike lanes, and because bikes are generally not affected by downtown street congestion. The parametric study indicates that our system generally improves matching rates when we compare it to our previous study on the rideshare system only being a feeder to transit (*1*).

The insights gained from our parametric study include the following. First, the matching rate is determined by the number of riders and drivers and the rider-to-driver ratio. When both the number of riders and drivers are small, the rate increases sharply at first and then remains relatively stable when the number of drivers is more than two times that of riders. When the number of drivers is a relatively large number (which means their routes are broad enough to cover whole areas), a further increase in the number of drivers would not improve the matching rate significantly. Besides, both P2P ridesharing and shared bikes are used as transit feeders by some riders. The usage would increase linearly when the availability of drivers and bikes increases.

One limitation, which is also a future research topic, is that this study only included travel demand from personal vehicle demands. In other words, this study focused on the rideshare matching potential from personal vehicle travel, and not the total travel demand (vehicle and transit). In future research, transit demand data could be included to study the potential improvements and mode shift. To reflect the actual travelers' behavior in the system, more elaborate behavior models, such as random utility models, should be implemented in our future research. Some sensitivity analysis could also be included in future research, so as to measure the effects of pricing, transfer and waiting time on riders' choice behavior.

Acknowledgments

The case study in this paper originated from a project funded by University of California Center on Economic Competitiveness in Transportation (UCConnect, the USDOT Region-9 UTC) and the California Department of Transportation for studying the impact of using peer-to-peer ridesharing and bike-share services as transit system feeders. We thank the sponsors for the support.

Author Contributions

The authors confirm contribution to the paper as follows: study conception and design: DN, DY, RJ, NM; data collection: DN, JY and SA; analysis and interpretation of results: DN, DY; draft manuscript preparation: DN, DY, RJ. All authors reviewed the results and approved the final version of the manuscript.

References

1. Masoud, N., D. Nam, J. Yu, and R. Jayakrishnan. Promoting Peer-to-Peer Ridesharing Services as Transit System Feeders. *Transportation Research Record: Journal of the Transportation Research Board*, 2017. 2650: 74–83.
2. Masoud, N., and R. Jayakrishnan. A Decomposition Algorithm to Solve the Multi-Hop Peer-to-Peer Ride-Matching Problem. *Transportation Research Part B: Methodological*, Vol. 99, 2017, pp. 1–29.
3. Stiglic, M., N. Agatz, M. Savelsbergh, and M. Gradisar. The Benefits of Meeting Points in Ride-Sharing Systems. *Transportation Research Part B: Methodological*, Vol. 82, 2015, pp. 36–53.
4. Liao, F., T. Arentze, and H. Timmermans. Supernetwork Approach for Multimodal and Multiactivity Travel Planning. *Transportation Research Record: Journal of the Transportation Research Board*, 2010. 2175: 38–46.
5. Cheng, Y. H., and K. C. Liu. Evaluating Bicycle-Transit Users' Perceptions of Intermodal Inconvenience. *Transportation Research Part A: Policy and Practice*, Vol. 46, No. 10, 2012, pp. 1690–1706.
6. Brons, M., M. Givoni, and P. Rietveld. Access to Railway Stations and Its Potential in Increasing Rail Use. *Transportation Research Part A: Policy and Practice*, Vol. 43, No. 2, 2009, pp. 136–149.
7. Mo, B. *GIS Network Analysis for Finding the Potential Metro Rail Ridership by Access Introduction.* 2010 AASHTO GIS for Transportation Symposium, 2009.
8. Moniruzzaman, M., and A. Páez. Accessibility to Transit, by Transit, and Mode Share: Application of a Logistic Model with Spatial Filters. *Journal of Transport Geography*, Vol. 24, 2012, pp. 198–205.
9. Chang, J. S., and J. Lee. Accessibility Analysis of Korean High-Speed Rail: A Case Study of the Seoul Metropolitan Area. *Transport Reviews*, Vol. 28, No. 1, 2008, pp. 87–103.
10. Ortúzar, J. D. D., A. Iacobelli, and C. Valeze. Estimating Demand for a Cycle-Way Network. *Transportation Research Part A: Policy and Practice*, Vol. 34, No. 5, 2000, pp. 353–373.
11. Rietveld, P. The Accessibility of Railway Stations: The Role of the Bicycle in the Netherlands. *Transportation Research Part D: Transport and Environment*, Vol. 5, No. 1, 2000, pp. 71–75.
12. Martens, K. The Bicycle as a Feedering Mode: Experiences from Three European Countries. *Transportation Research Part D: Transport and Environment*, Vol. 9, No. 4, 2004, pp. 281–294.
13. Lloret-Batlle, R., N. Masoud, and D. Nam. Peer-to-Peer Ridesharing with Ride-Back on High-Occupancy-Vehicle Lanes. *Transportation Research Record: Journal of the Transportation Research Board*, 2017. 2668: 21–28.
14. Krizek, K., and E. Stonebraker. Bicycling and Transit - A Marriage Unrealized. *Transportation Research Record: Journal of the Transportation Research Board*, 2010. 2144: 161–167.
15. Nam, D., K. Hyun, H. Kim, K. Ahn, and R. Jayakrishnan. Analysis of Grid Cell–Based Taxi Ridership with Large-Scale GPS Data. *Transportation Research Record: Journal of the Transportation Research Board*, 2016. 2544: 131–140.
16. Wang, F., and C. L. Ross. New Potential for Multimodal Connection: Exploring the Relationship between Taxi and Transit in New York City (NYC). *Transportation* (advance online publication), 2017, pp. 1–22. https://doi.org/10.1007/s11116-017-9787-x.
17. Regue, R., N. Masoud, and W. Recker. Car2work. *Transportation Research Record: Journal of the Transportation Research Board*, 2016. 2542: 102–110.
18. Agatz, N., A. Erera, M. Savelsbergh, and X. Wang. Optimization for Dynamic Ride-Sharing: A Review. *European Journal of Operational Research*, Vol. 223, No. 2, 2012, pp. 295–303.
19. Stiglic, M., N. Agatz, M. Savelsbergh, and M. Gradisar. Making Dynamic Ride-Sharing Work: The Impact of Driver and Rider Flexibility. *Transportation Research Part E: Logistics and Transportation Review*, Vol. 91, 2016, pp. 190–207.
20. Schreieck, M., H. Safetli, S. A. Siddiqui, C. Pflügler, M. Wiesche, and H. Krcmar. A Matching Algorithm for Dynamic Ridesharing. *Transportation Research Procedia*, Vol. 19, 2016, pp. 272–285.
21. Masoud, N., and R. Jayakrishnan. A Real-Time Algorithm to Solve the Peer-to-Peer Ride-Matching Problem in a Flexible Ridesharing System. *Transportation Research Part B: Methodological*, Vol. 106, 2017, pp. 218–236.
22. Google Directions API. 2017. https://developers.google.com/maps/documentation/directions/.
23. Walk Score API. https://www.walkscore.com/professional/api.php. Accessed on March 15, 2017.

The Standing Committee on Emerging and Innovative Public Transport and Technologies (AP020) peer-reviewed this paper (18-06518).

Article

TRR
JOURNAL OF THE TRANSPORTATION RESEARCH BOARD

Transportation Research Record
2018, Vol. 2672(8) 764–774

DOI: 10.1177/0361198118786644
journals.sagepub.com/home/trr
SAGE

Multi-Agent Simulation of a Demand-Responsive Transit System Operated by Autonomous Vehicles

Benedikt Jäger[1], Carsten Brickwedde[1], and Markus Lienkamp[2]

Abstract

Shared mobility solutions improve traffic efficiency in cities. This paper addresses the design of an urban demand-responsive transit system operated by a fleet of autonomous modules as a public transportation service. We introduce a discrete event-based multi-agent simulation to reflect system dynamics at an operational level. Vehicles, customers, and intelligent stops are modeled as software agents. Each entity has its own decision-making and planning component. Vehicles compete for a limited number of order requests, and decentralized stops manage the matching between a customer and a prospective vehicle. Scheduling and routing events take place highly dynamically. A simulation study evaluates a demand-responsive operation as a replacement for a fixed-route bus service. Results show that a total demand of 2.3 million rides can be fulfilled with a fleet of around 43,000 six-seater modules. In comparison with the current bus system, the total energy consumption and the system's utilization are improved, while transport times and distances result in values below the bus system's performance and therefore must be optimized in further research.

With the ongoing urbanization and continuous growth of the world's megacities, the demand for mobility increases. Urban transportation is becoming a significant challenge for the future. In order to reduce road congestion and commuting times as well as improve customer service, the focus is shifting toward new technologies and mobility concepts (*1*). Autonomous electric vehicles will transport customers efficiently and effectively in a connected environment (*2*). To further reduce the number of vehicles, car- and ridesharing approaches are being pursued, and mobility is becoming more and more of a service than an individual good.

While traditional public transit typically has fixed routes and schedules, demand-responsive transit (DRT) operates more flexibly (*3*). Schedules in a dial-a-ride application result from immediate or advance bookings (*4*). Route planning is done dynamically based on given requests. Until now, DRT has been focused on areas with low demand where fixed-route and fixed-schedule bus services face profitability issues. Common applications offer specialized transportation services for customers, feeder services to fixed-route systems, and flexible-route segment or route-deviation services (*5*). In the United States, DRT systems are mostly used as a paratransit service for people with limited mobility (*6*). Origin–destination patterns vary between many-to-many, many-to-few, many-to-one, few-to-many, and few-to-few relationships (*5*). Transport tasks can be served directly without any transfer or detour, or without transfers but with a possible detour due to additional stops, or by allowing transfers. The size of a DRT fleet depends on the (peak-period) demand, service and provider policies, passenger characteristics, trip pattern type, and service area characteristics (*5*).

The decision-making between fixed- and demand-oriented systems depends on the specific use case. Edwards and Watkins (*7*) introduce a methodology to compare fixed and DRT operation modes for a feeder system. Shen et al. (*8*) investigate an autonomous vehicle DRT system as a last mile solution. Schofer et al. (*6*) present a planning tool to derive fleet requirements for DRT applications. Noda et al. (*9*) evaluate the efficiency of a dial-a-ride system with that of a traditional bus system, looking at the structure of a town as well as the demand and fleet size. Mo et al. (*10*) compare various pricing policies and vehicle scheduling mechanisms for a shared dial-a-ride service. Jung and Jayakrishnan (*11*) explore a high-coverage point-to-point transit system operated by electric vehicles.

The aim of this paper is to design and analyze the performance of a public many-to-many DRT system with real-time scheduling operated on a city scale by fully-autonomous modules.

[1]TUM CREATE Ltd., Singapore
[2]Chair of Automotive Technology, Technical University of Munich, Garching, Germany

Corresponding Author:
Address correspondence to Benedikt Jäger: benedikt.jaeger@tum.de

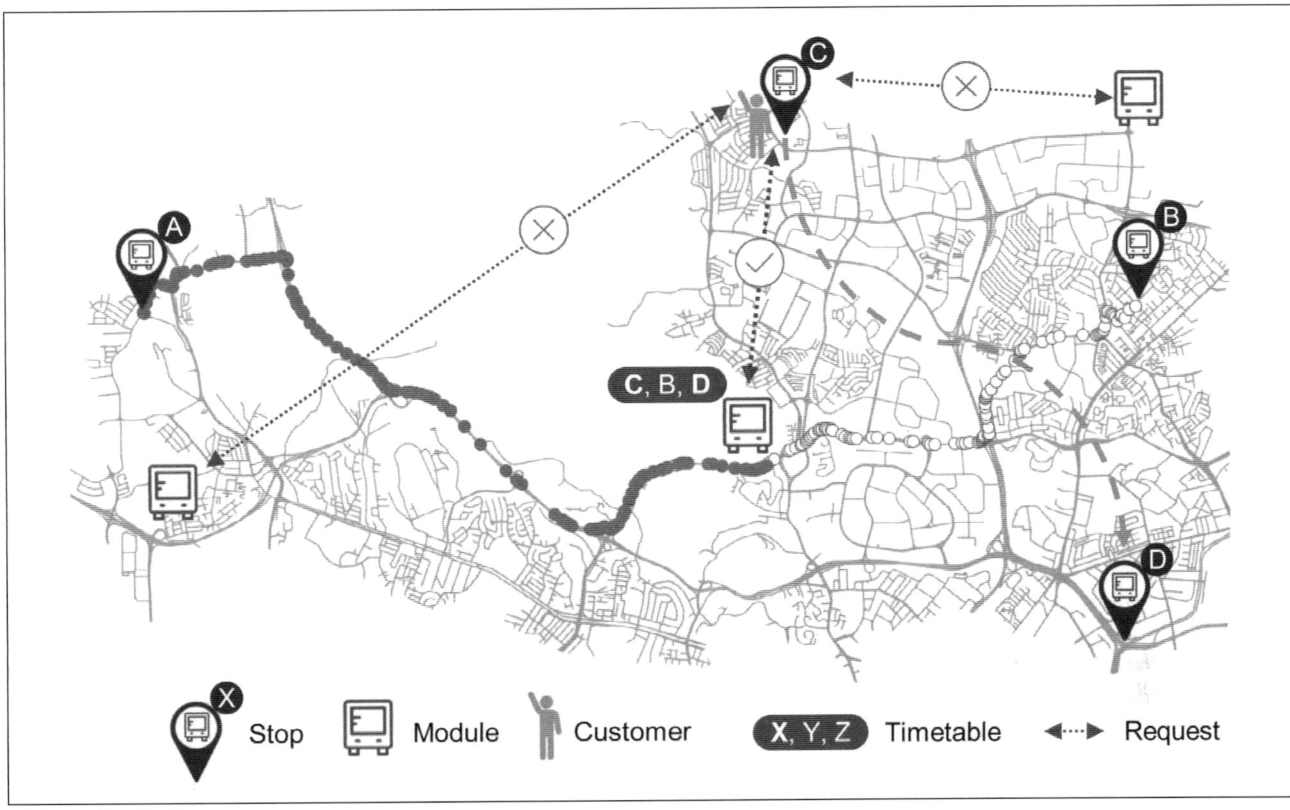

Figure 1. Customer–module matchmaking.

DRT in Terms of the Vehicle Routing Problem

Vehicles in a DRT system do not follow pre-defined routes. In Operations Research literature, the determination of the optimal set of routes for a fleet of vehicles that serves a set of distributed customers is described as the vehicle routing problem (VRP). Eksioglu et al. (*12*) provide a taxonomic overview including a categorization for the existing research on VRPs.

A DRT system can be categorized as a dial-a-ride problem with a time window, which is a specific form of the VRP. The objective is to minimize the total cost for a fixed number of vehicles with given capacities that must serve a set of dispersed customers within a specific time window (*13*).

The VRP can be addressed using a static or dynamic approach. The static case describes an optimization problem in which all requests for the entire time are known before the beginning of the service. Due to its non-deterministic polynomial-time harness in computational complexity theory, it is not possible to solve a larger problem within a given time frame with a reasonable computational effort (*14*).

The dynamic option describes situations with vehicles already executing transport tasks. Vehicle routing and scheduling must be conducted in real time. Existing plans must be taken into account, and routes may need to be adjusted accordingly. The system tries to find the optimal solution for each new request, which is usually dispatched by a control center. However, limited fleet capacity and temporal constraints can cause delays or refusals of bookings. Well-known solutions are based on classical heuristics, metaheuristics, and matheuristics (*15*).

Concept

The following sections cover the use case definition for the proposed DRT system, its conceptional design, and further simulation-related aspects.

Use Case Definition

To operate a DRT system, a vehicle fleet is required. Its main task is to transport customers between various pickup and drop-off points. We assume a service in which vehicles have the ability to move autonomously within a given road network. Boarding and alighting take place at defined stops. Customers, intelligent stops, and modules are connected and communicate to find the best transport option. Figure 1 illustrates a typical situation.

The initial action is a service request submitted by a customer at a stop without an advance-notice requirement. The stop organizes the ride for the customer by contacting several modules. It serves as a decentralized dispatcher and assigns the transportation task to exactly one module. This module

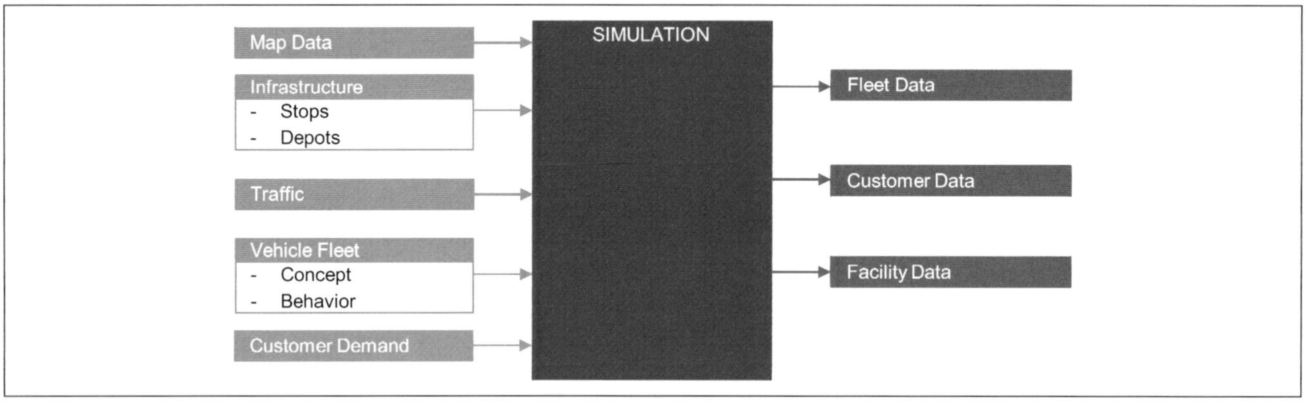

Figure 2. Input and output interfaces.

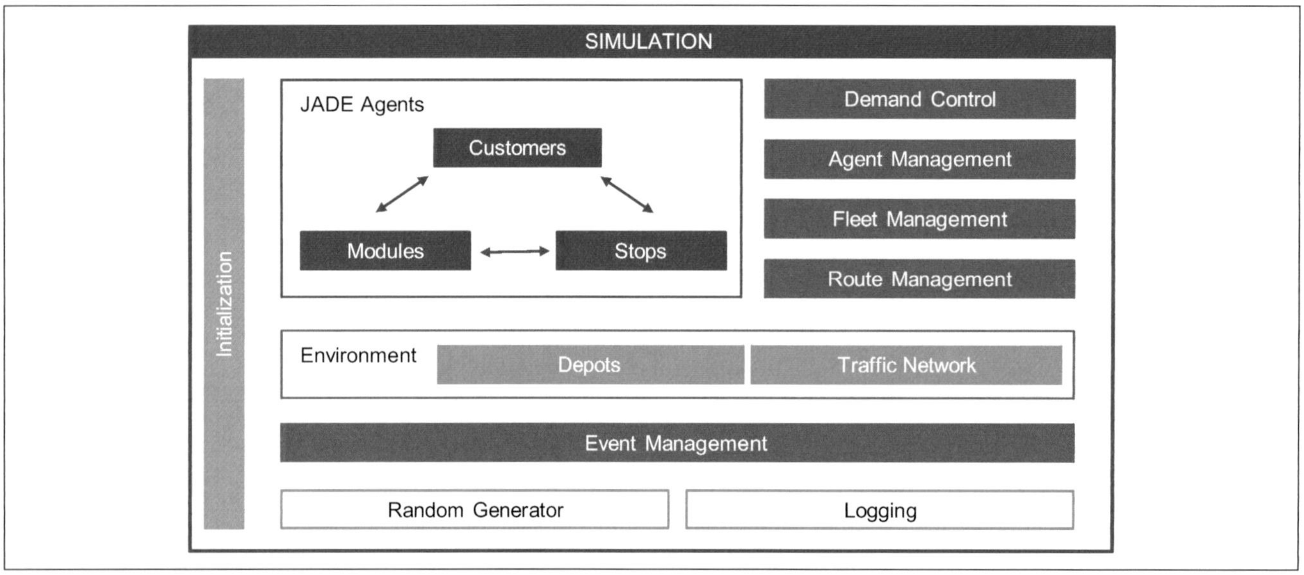

Figure 3. Design of the internal structure.

incorporates the job into its route planning. When a module has finished its schedule, it returns to a depot.

The decision to realize a stop-to-stop operation is made for several reasons. At first sight, it is more convenient for customers to order a door-to-door transportation service. However, including stops in a dial-a-ride system also has advantages. Stiglicet et al. (*16*) state that meeting points can increase the number of matched participants in a ridesharing application and reduce the total driven distance of the fleet. Additionally, stops allow several boarding and alighting actions to be aggregated to a single event. A proper stop design can decrease dwelling times and minimize the impact on other road users (*4*).

Simulation Approach

We introduce a multi-agent simulation to evaluate the performance of the proposed DRT system. Instead of trying to optimize the entire system at once, this approach follows the "divide and conquer" principle. Each transportation task defines a sub problem, which is solved step-by-step on an individual basis. This design has significant benefits when it comes to scalability, adaptability, and robustness.

The simulation itself is realized as a discrete event-based approach. It is written in Java, and its architecture supports multi-threading. JADE (*17*) is used as a middleware to incorporate multi-agent features. Figure 2 shows input and output interfaces.

The basic internal structure is illustrated in Figure 3. An agent management module handles the creation and destruction of agents. Since in JADE every agent is represented as an own thread, an event management module has to take over the synchronization task. A central fleet management module keeps a list of the latest agent states updated and provides this information to other entities. Navigation is organized via routing management. Additional modules are a

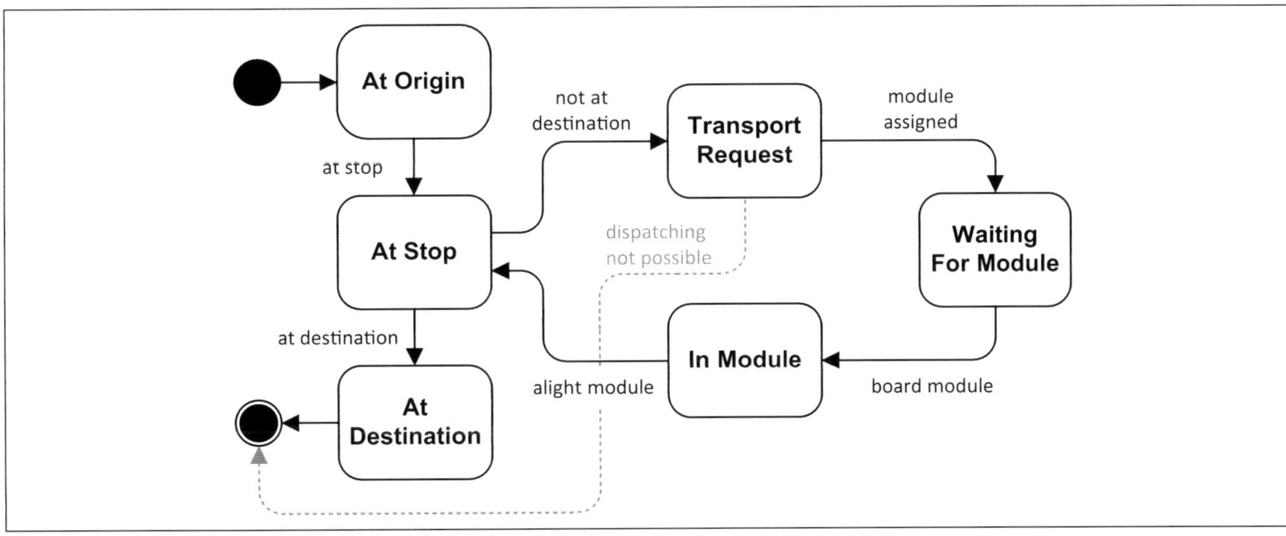

Figure 4. Customer finite state machine.

random number generator for stochastic calculations and a class for data logging. Further details about the simulation core have been published in a paper (*18*).

Customer, module, and stop agents represent active, intelligent entities. Their behavior is independent from each other and triggered by event or message calls. Dependent on the type, agents collaborate or compete for an open task. Modules are the most active components, whereas stops take over a mediator role to match customers with modules. All agents live in a shared environment consisting of depots and a traffic network. Depots provide space for the modules.

Customer: A customer agent is only temporarily in the system. It is created as soon as it submits a transport request and is removed after reaching the final destination. Its goal is to reach the target location as quickly as possible.

This requires one to request the transport at the stop and then follow the instructions of the system. A customer must wait for a module and board and alight at the required time. The customer's behavior is managed by a state machine, as shown in Figure 4.

Stop: A stop is primarily responsible for finding the best matching modules for a given transportation request. Its tasks include managing modules currently at a stop. Stops are facilities that provide transfer space but no parking space. The transfer space limits the number of vehicles that can simultaneously stop. A stop implements a timetable of planned incoming vehicles, including boarding and alighting customers. This information is used to address arriving vehicles during dispatching.

It further manages currently transferring vehicles as well as vehicles queuing at the stop. An arriving module contacts a stop which checks its available transfer space. It advises the module to either continue with the transfer or to wait in the queue. Once a previously transferring vehicle departs and the queue is not empty, the stop notifies the first waiting vehicle about the free space.

Modules: The module's main responsibility is to transport customers. This includes active bidding for offered transportation requests. Modules try to finish the remaining transportation tasks as quickly as possible by dropping all customers at their destinations and returning back to the closest depot.

Within the initialization, all modules are distributed to the depots. Whenever a module receives a request, it calculates the corresponding transportation effort and sends out a cost proposal. If a module is assigned a transport task, it incorporates it into its future planning. With the ability to transport multiple customers, the module can constantly add new customers as long as it has not reached its capacity limit.

To keep track of all customers and execute the tasks as desired, each module follows an internal plan realized as a timetable. This list includes all planned stops including arrival time and both boarding and alighting customers. Based on the timetable, the module knows exactly where and when to go next, as well as with whom to communicate. Knowing the number of boarding and alighting customers enables the module to estimate its dwelling time.

Whenever a new customer request is added to the timetable, the optimal positions for the new pickup and drop-off stop must be determined. The insertion of a stop element (Figure 5) is based on an optimization of the caused delay. It is required that the new pickup stop is added before the new drop-off stop and that the order of existing stops remains the same.

Similar to the customer agent, a module shows different behaviors and interactions depending on its state. To realize this, a finite state machine is implemented as shown in Figure 6.

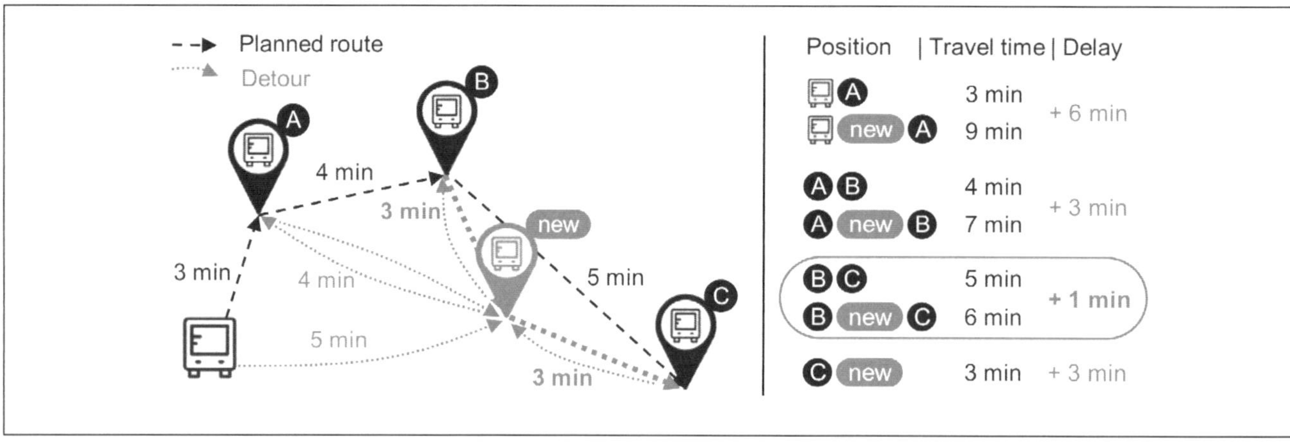

Figure 5. Decision-making to insert a new stop in given timetable.

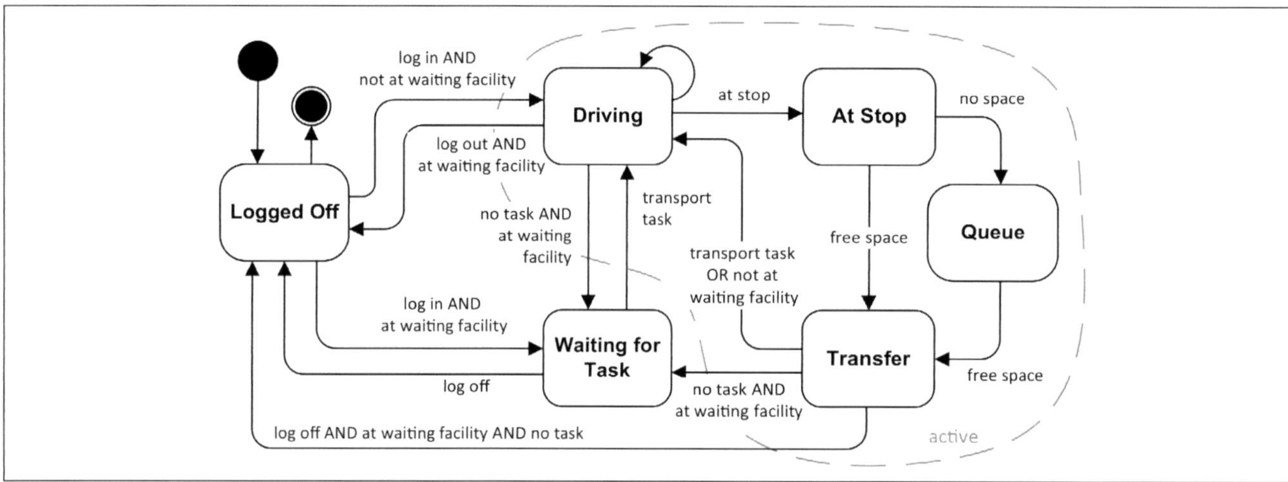

Figure 6. Module finite state machine.

The *Logged Off* state is the initial state. While *Waiting for task*, the module receives dispatching messages and participates in the competition for orders. When a job is assigned, the module becomes active. All movements happen within the state *Driving*. This includes customer transport, driving to a customer, and driving back to the depot. When arriving *At stop*, the module tries to register at the stop. Depending on the availability of free transfer space, it either directly continues with the transfer or enters the queue to wait until it receives a free space notification. The *Transfer* state defines the actual time with customers boarding and alighting.

The duration depends on vehicle, customers, and stop characteristics. At the beginning of a transfer, the module notifies all alighting passengers to exit. After completion, the vehicle deregisters from the stop. If further tasks need to be completed, it continues driving to the next planned location. Otherwise, the module checks if it can wait at the stop or if it needs to find a waiting facility.

Dispatching Communication. Dispatching causes an extensive message exchange. The communication act itself consists of two nested protocols as shown in Figure 7. An outer FIPA-Request (*19*) handles the messaging between customer and stop, and an inner FIPA Iterated Contract Net Interaction Protocol (*19*) specifies the communication between the stop and a set of modules.

First, a customer arrives at his departure stop and requests a transport to his desired destination. The stop agrees to conduct the dispatching and creates a transport job including stop-specific information. It determines modules to address and sends out a call for proposals. In order to guarantee the optimal dispatch result, all modules would need to be assessed. However, an increasing number of receivers significantly slows down the simulation performance. To minimize computational efforts and still achieve a sufficient quality, receivers are chosen heuristically in three categories, namely, surrounding vehicles, arriving vehicles, and empty vehicles.

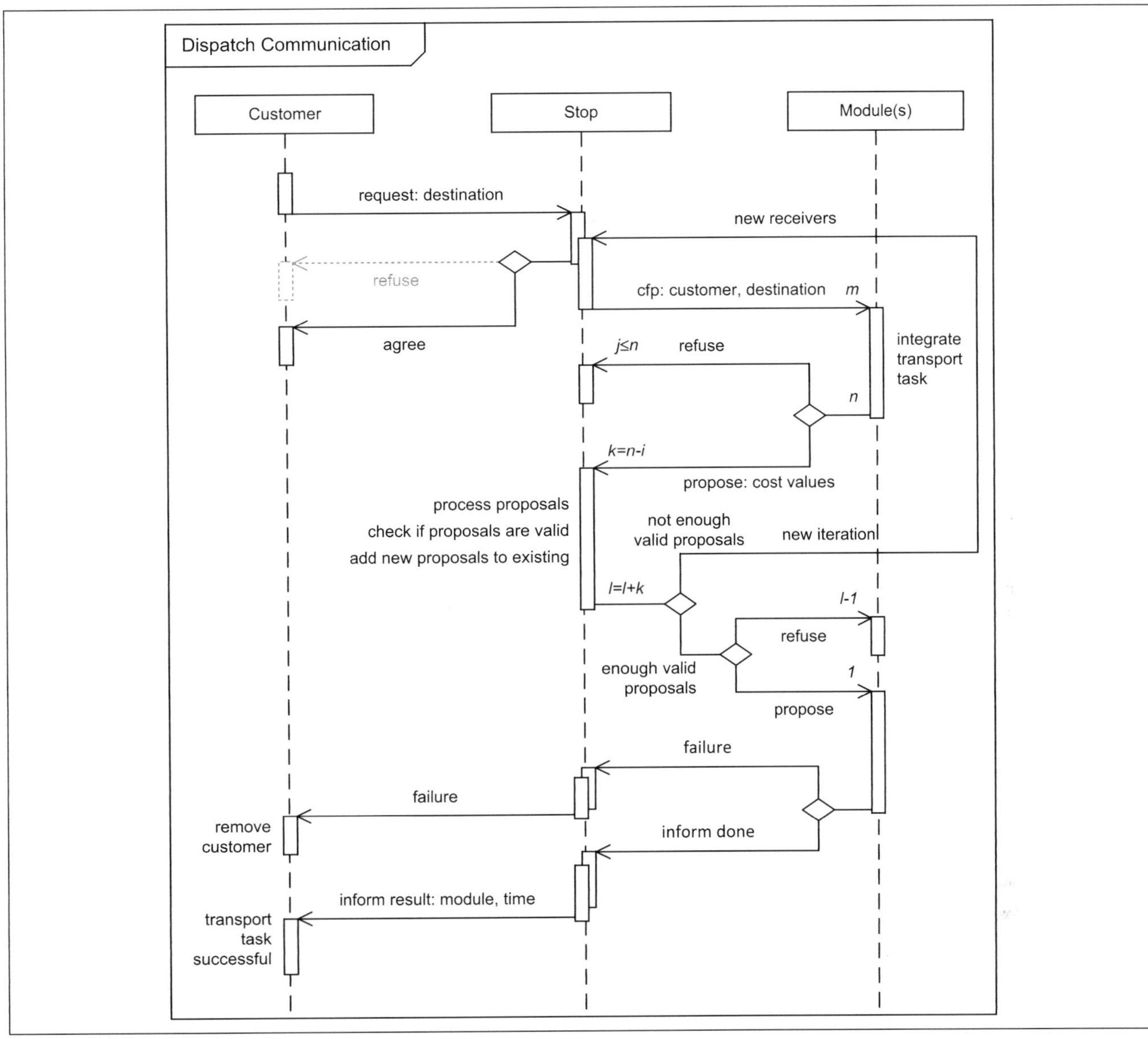

Figure 7. Dispatch communication.

This is done for several reasons. It is assumed that vehicles close to the departure stop reduce customer waiting times and empty trip distances. Modules with a stop already in their timetable might be suitable candidates as no additional detour is needed. Empty vehicles are added to the active fleet if needed. This ensures a minimum quality of service for a customer even if all surrounding modules are occupied.

Addressed modules have the option to accept or reject a given proposal. A direct rejection occurs when a module cannot provide any free space for the customer. Otherwise, they try to integrate the task into their plans and respond with a proposal that contains information about trip duration, expected waiting time, and additional energy consumption. The stop waits until it has received all responses and processes them. If the number of valid proposals is below a defined limit, the stop starts a new iteration, adding more receivers.

Having received enough proposals or when no additional vehicles remain, the stop performs a cost assessment. It calculates a final cost value for each vehicle and determines the best module. It replies to all proposals, accepting the best one and sending refusals to the others. The winning module integrates the transport job into its plan and informs the stop about its completion. If everything is successful, the stop informs the customer about the chosen module, estimated pickup time, and transport details. In case the stop is not able to find a suitable module, or a communication error occurs, the customer is informed about the failure instead.

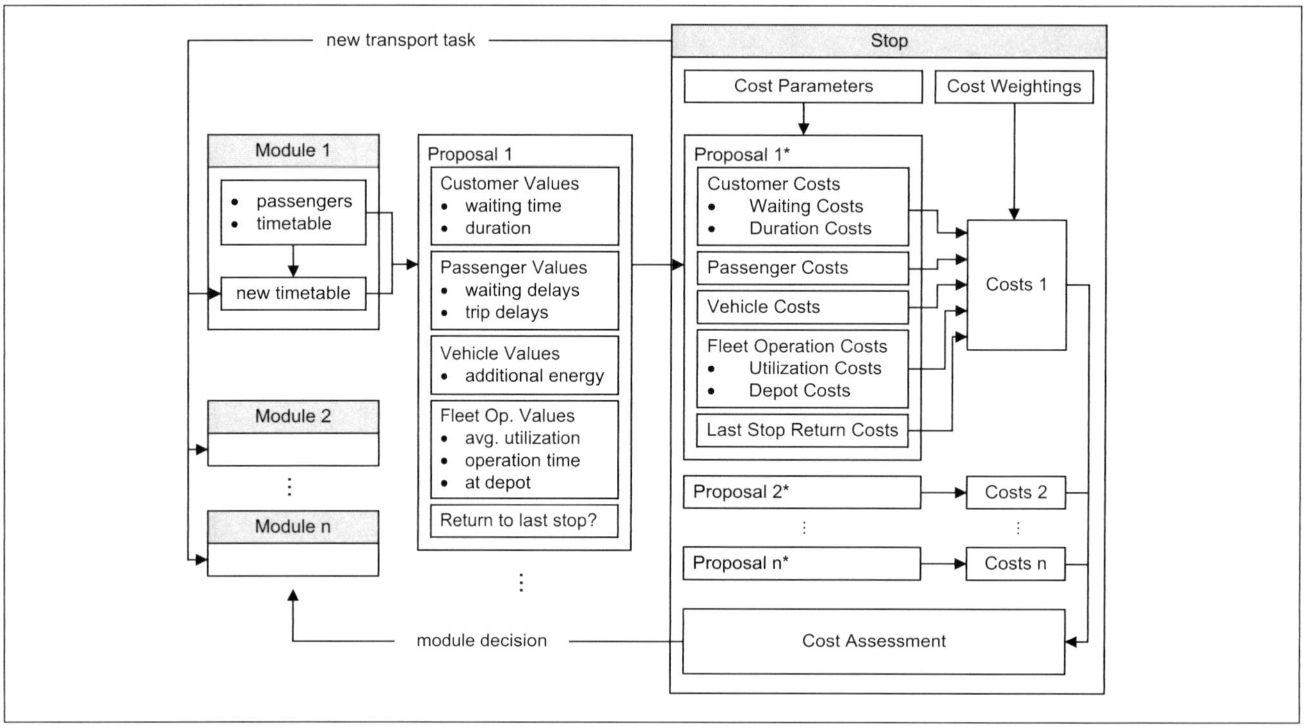

Figure 8. Cost calculation structure.

Cost Assessment

Since the cost calculation for dispatching is crucial for the DRT performance, its structure is explained in further detail. The proposed on-demand operation is only successful if it fulfills at least the minimum requirements regarding customer service and operation costs. Stakeholders have specific interests and needs. They typically ask different questions:

Customer: How long do I have to wait? How long is the travel time compared with other modes?

Passenger: How does a potential new customer affect my travel?

Vehicle: How much energy is needed to transport an additional customer?

Fleet Operation: Does the open task require the addition of a new vehicle to the active fleet?

In order to take these partially conflicting requirements into account, a trade-off has to be evaluated. The basic idea is to perform a weighted cost evaluation, as shown in Figure 8. *Customer costs* consider requirements regarding waiting time and the trip duration of a new customer. *Passenger costs* look at the same aspect, but for all passengers that are in the module or waiting at a stop. *Vehicle costs* reflect the additional energy needed for a new trip. *Fleet operation costs* aim to reduce the total number of active modules. This can be done by pulling as few new vehicles as possible from a depot and by getting active ones to return as quickly as possible. The *Last stop return costs* avoid situations in which vehicles return directly to their previous stops.

Simulation Study

To evaluate the feasibility of the proposed DRT system, a simulation study was conducted. We selected a use case in which the current bus system in Singapore is replaced by a DRT system and compared its performance. This subchapter presents the chosen input dataset and key findings.

Facilities and Road Network

The chosen infrastructure is based on current facilities in Singapore. It contains 14 bus depots and a total of 4,923 stops, including 35 hubs, 35 interchanges, and 23 terminals. The underlying road network was taken from Open Street Map and processed into a routable graph.

Demand

One of the key inputs is customer demand. To create a list of origin–destination relationships, Singapore's Land Transport Authority (LTA) provided smart card data from the year 2013. This dataset includes tap in and tap out information of

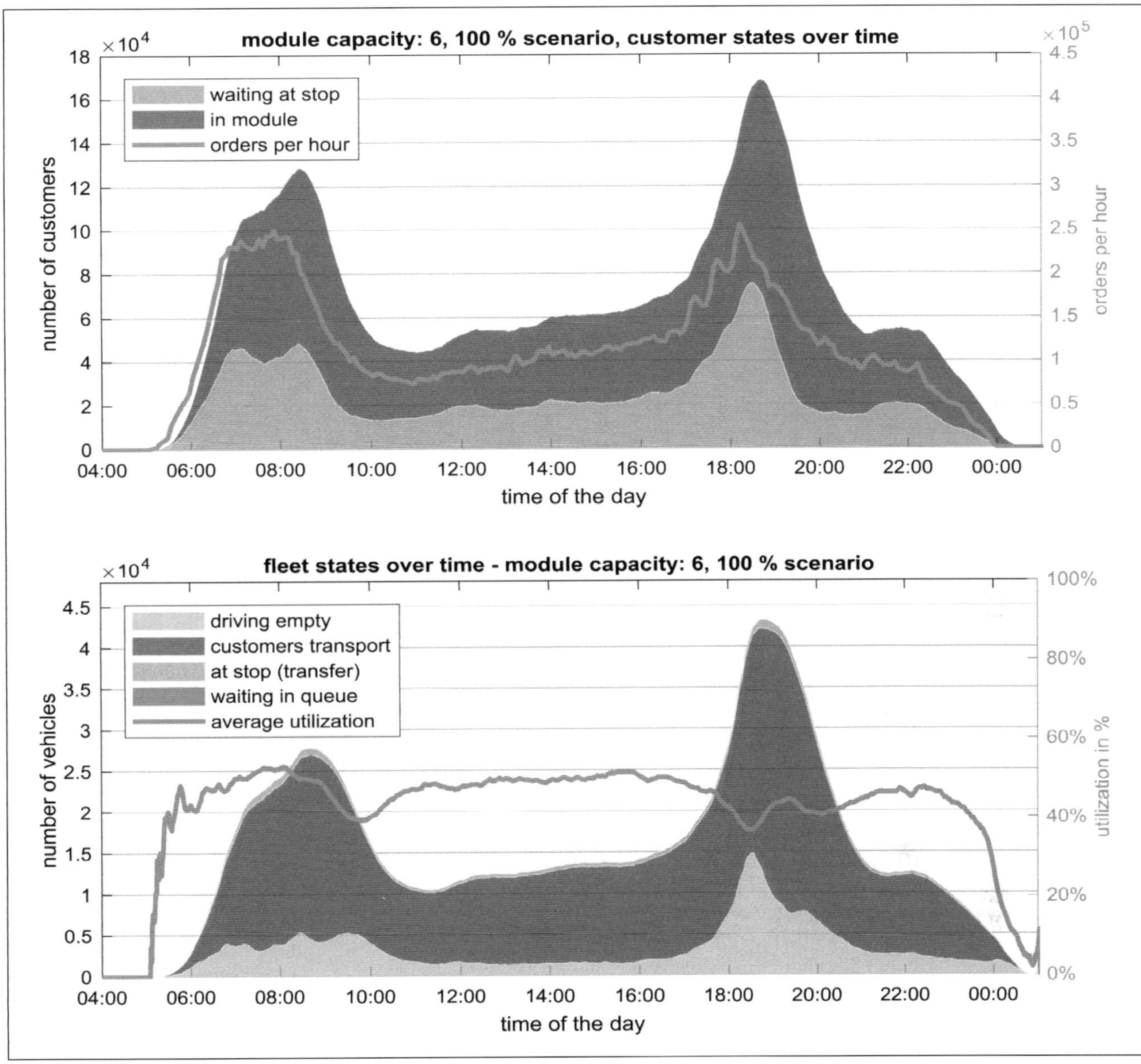

Figure 9. System analysis: vehicle states and customers over time.

all public transport journeys over several days. We filtered the data for only valid bus trips and extracted a weekday sample. Consecutive bus trips were combined into a single journey. Operation hours are from 5:00 to 12:00 a.m. This preprocessing step led to a list of 2.3 million demand events.

Traffic

The utilized routing engine provides travel times based on free-floating traffic. To consider real road conditions more accurately, a time-dependent traffic factor was introduced. This factor adopts the timing of routing results and is based on spatial-temporal fleet activity. Temporally, a speed factor distribution regarding the time of day was created using real traffic data. With a large fleet of simulated vehicles and the transport tasks representing real traffic demand, it is assumed that the known vehicles can be used as a good representation of the overall traffic. Assessing the vehicle's position and driving direction using a map grid, a speed factor for each position update event is determined by the time-based distribution. The traffic model was validated using the original trips' data with the average simulated speed being 4.6% slower than the original, and a standard deviation of 38.5%. Thus, the average travel times can be compared; however, due to the high standard deviation, single values can differ significantly.

Table 1. Performance Comparison

	Bus system		
	Diesel	Electric	DRT system
Fleet size	2,413		max. 43,165; avg. 16,073
Total seat capacity[a] in 1,000 persons	241		max. 259; avg. 96
Fleet distance per day in 1,000 km	547		5,393
Average utilization	18 %[a,b]		44%[c]
Fleet energy consumption per day in MWh	2,260[d]	985[e]	911[f]
Average trip time diff to bus	n/a		3.32 min (19.3%)
Average trip distance diff to bus	n/a		1.17 km (16.9%)

[a]estimated average capacity of 100 persons per bus (mix of single/double decker).
[b]average over distance.
[c]average over time.
[d]$E_{Consumption}$: 4.1 kWh/km (*21*).
[e]$E_{Consumption}$:1.8 kWh/km (*21*).
[f]$E_{Consumption}$: 0.169 kWh/km (22).

Vehicle Fleet

The chosen vehicle concept is a fully automated six-seater module with an electric powertrain. A variation of different seat capacities in a calibration phase showed that this module size is a good trade-off between operator- and customer-related needs. It benefits from a high utilization rate while providing a desirable quality of service.

System Behavior and Performance Analysis

Based on the described scenario, we evaluated the DRT system from a customer and operator point of view (Figure 9). The shift between the waiting customers and order-demand curve indicates that most customers are being picked up within a limited time frame. The mean customer waiting time is 12.1 min with a standard deviation of 10.9. In total, 0.07% of all trips are denied. In a comparison of the morning and evening peaks, a larger number of customers must wait during the evening peak. The diagram shows that at peak times, 170,000 customers are simultaneously active in the system.

By evaluating the fleet behavior, it is evident that the quantity of vehicles follows the demand. At the same time, the system is able to achieve a nearly constant utilization over time, indicating a limitation due to the dispatching. Many vehicles enter and leave the system at the beginning and at the end of peak times, indicated by the empty driving modules, either driving to the first customer or returning to the depot. This affects utilization, which decreases with a higher number of empty vehicles and increases again once they leave the system. With 43,165 modules active, the evening peak defines the required fleet size.

Discussion

With the demand being based on real bus trips, it is possible to compare the system performance between the existing bus and proposed DRT system. In 2013, LTA claimed the average daily ridership for buses to be 3.6 million passenger trips (*20*). Since this study simulated a demand of 2.3 million trips, the authors prorated the published data by a factor of 64%.

Table 1 summarizes the main results. In comparison with the bus system, the DRT system requires many more vehicles. The total seat capacity is just slightly higher (259,000 vs. 241,000), which is only required for a short period of time (peak demand), while, on average, a seat capacity of 96,000 is sufficient. With around 18 times more vehicles, the total driven distance increases as well and adds up to 5.4 million km per day, which is around 10 times more than the bus fleet. Analyzing the energy consumption, the DRT system beats the current system, even when replacing the entire bus fleet with electric buses.

Regarding utilization, smaller vehicles share the demand better than buses, and the average achieved utilization, at 44%, is more than twice as high. The system maintains a constant utilization by adding and removing vehicles to and from the fleet. Comparing the trip distance and duration, one can see that the DRT system is not able to beat the current bus system, and each trip is, on average, 19% longer than the original bus trip, while requiring 17% more time as well. This can be explained by detours caused by other passengers. Due to the large network of bus lines in Singapore, a customer is most likely to choose bus routes with a fast and direct connection. 76% of all bus journeys have no transfer to other buses; 21% have one transfer; and 3% have two or more transfers. Since a bus journey consists on average of 1.25 trips, a DRT system without transfers reduces the number of trips by 25%.

Conclusion

This paper presents a simulation model for a DRT system operated by automated modules on a city scale. The primary objective is to support decision-making in the early planning stage. Its operation is designed as a stop-to-stop service on flexible routes. It starts with a customer requesting a transportation task at a stop. The chosen stop is responsible for assigning the given order to a matching vehicle. The dispatch principle forms an auction mechanism realized by a contract net protocol. First, the solution space of transportation options is reduced by heuristically preselecting a subset of modules. Criteria are related to distances, planned activities, and vehicle states. The stop sends a call for a proposal message to the candidates and requests a cost estimation to fulfill the offered job. Each module calculates the effort. It is then the stop's responsibility to select the most suitable vehicle for the given transportation request.

The simulation approach itself is based on a discrete event-based multi-agent simulation. Agents are intelligent entities living independently in a shared environment. They solve the given dial-a-ride problem in competition.

A simulation study evaluated the potential to use a DRT system as a replacement for an existing bus system. The authors investigated a citywide operation with a total of 2.3 million requests per day. The results show a need for 43,165 six-seater modules. Due to its demand-driven design, the modules leave the system once no longer required and achieve constant utilization over time. Inactive vehicles are available for secondary use. On average, 16,073 modules are active.

The developed system is able to reduce total energy consumption, but service comfort cannot be improved at this stage of development.

The static bus route design outperforms the dynamic one, since passengers in a module may have different destinations, or a module may have to pick up new customers at additional stops.

Further research is required in order to create a superior system that is attractive to customers as well as operators. One idea is to improve the ridesharing potential by aggregating trip requests within a given time frame or by combining passengers with the same destination. Customer waiting time could be reduced by allowing parking along the street or at stop locations. To investigate the effects of additional congestion introduced by the DRT fleet, the model can be coupled with a traffic simulation. To examine different passenger responses, an extended order behavior model could be included.

Acknowledgments

To carry out the simulation study, the Land Transport Authority of Singapore provided helpful data. The authors would like to thank them for their support.

This project was funded by the Singapore National Research Foundation under its Campus for Research Excellence and Technological Enterprise (CREATE) programme and the Technical University of Munich.

Author Contributions

The authors confirm contributing to the paper as follows: study conception and design: Benedikt Jäger, Carsten Brickwedde; data collection: Benedikt Jäger, Carsten Brickwedde; analysis and interpretation of results: Benedikt Jäger, Carsten Brickwedde; draft manuscript preparation: Benedikt Jäger. All authors reviewed the results and approved the final version of the manuscript.

References

1. Dawes, M., and J. Zhao. User Identification of and Attitude Toward Dynamic Ridesourcing Services. Presented at 96th Annual Meeting of the Transportation Research Board, Washington, D.C., 2017.
2. Fagnant, D. J., and K. M. Kockelman. Preparing a Nation for Autonomous Vehicles: Opportunities, Barriers and Policy Recommendations. *Transportation Research Part A: Policy and Practice*, Vol. 77, 2015, pp. 167–181.
3. Koffman, D. *TCRP Synthesis of Transit Practice 53: Operational Experiences with Flexible Transit Services*. Transportation Research Board of the National Academies, Washington, D.C., 2004.
4. Vuchic, V. R. *Urban Transit Systems and Technology*. Wiley & Sons, Chichester, Hoboken, N.J., 2007.
5. *TCRP Report 165: Transit Capacity and Quality of Service Manual*, 3rd ed. Transportation Research Board of the National Academies, Washington D.C., 2013.
6. Schofer, J. L., B. L. Nelsol, R. Eash, M. Daskin, Y. Y. H. Wan, J. Yan, and L. Medgyesy. TCRP Report 98: *Resource Requirements for Demand-Responsive Transportation Services*. Transportation Research Board of the National Academies, Washington, D.C., 2003, pp. 366–371.
7. Edwards, D., and K. Watkins. Comparing Fixed-Route and Demand-Responsive Feeder Transit Systems in Real-World Settings. *Transportation Research Record: Journal of the Transportation Research Board*, 2013. 2352: 128–135.
8. Shen, Y., H. Zhang, and J. Zhao. Embedding Autonomous Vehicle Sharing in Public Transit System. An Example of Last Mile Problem. Presented at 96th Annual Meeting of the Transportation Research Board, Washington, D.C., 2017.
9. Shinoda, K., I. Noda, M. Ohta, Y. Kumada, and H. Nakashima. Is Dial-a-Ride Bus Reasonable in Large Scale Towns? Evaluation of Usability of Dial-a-Ride Systems by Simulation. *Proc., Multi-Agent for Mass User Support: International Workshop, MAMUS 2003*, Acapulco, Mexico, Springer, Heidelberg, 2004, pp. 105–119.
10. Mo, D. Y., Y. C. E. Lee, Y. Wang, and T. K. Y. Cheung. Shared Dial-a-Ride Service for More People with Lower Price. Presented at 96th Annual Meeting of the Transportation Research Board, Washington, D.C., 2017.
11. Jung, J., and R. Jayakrishnan. High-Coverage Point-to-Point Transit: Electric Vehicle Operations. *Transportation Research Record: Journal of the Transportation Research Board*, 2012. 2287: 44–53.
12. Eksioglu, B., A. V. Vural, and A. Reisman. The Vehicle Routing Problem: A Taxonomic Review. *Computers & Industrial Engineering*, Vol. 57, No. 4, 2009, pp. 1472–1483.
13. Tsubouchi, K., K. Hiekata, and H. Yamato. Scheduling Algorithm for On-Demand Bus System. *Proc., Sixth International Conference on Information Technology: New*

Generations, Las Vegas, Nev., IEEE, New York, 2009, pp. 189–194.
14. Larsen, A. *The Dynamic Vehicle Routing Problem*. PhD thesis. Department of Mathematical Modelling, Technical University of Denmark, Lyngby, 2000.
15. Pillac, V. *Dynamic Vehicle Routing: Solution Methods and Computational Tools*. PhD thesis. Department of Automation and Production, Ecole des Mines de Nantes, Nantes, 2012.
16. Stiglic, M., N. Agatz, M. Savelsbergh, and M. Gradisar. The Benefits of Meeting Points in Ride-sharing Systems. *Transportation Research Part B: Methodological*, Vol. 82, 2015, pp. 36–53.
17. Bellifemine, F., G. Caire, and D. Greenwood. *Developing Multi-Agent Systems with JADE*. John Wiley, Chichester, England, Hoboken, N.J., 2007.
18. Jäger, B., M. Wittmann, and M. Lienkamp. Agent-based Modeling and Simulation of Electric Taxi Fleets. *Proc., 6th Conference on Future Automotive Technology*, Fürstenfeldbruck, 2017.
19. Foundation for Intelligent Physical Agents. Fipa Abstract Architecture Specification. Geneva, 2002. www.fipg.org.
20. Land Transport Authority. *Singapore Land Transport: Statistics in Brief 2015*. Singapore, 2016. https://www.lta.gov.sg/content/ltaweb/en/publications-and-research.html.
21. Graurs, I., A. Laizans, P. Rajeckis, and A. Rubenis. Public Bus Energy Consumption Investigation for Transition to Electric Power and Semi-Dynamic Charging. *Proc., International Scientific Conference 2015*, Jelgava, Latvia University of Agriculture, 2015, pp. 366–371.
22. Schickram, S. *Derivation and Evaluation of Electric Vehicle Concepts for World Cities*. Dissertation, Institute of Automotive Technology, Technical University of Munich, 2016.

The Standing Committee on Emerging and Innovative Public Transport and Technologies (AP020) peer-reviewed this paper (18-00474).

Article

TRR
JOURNAL OF THE TRANSPORTATION RESEARCH BOARD

Transportation Research Record
2018, Vol. 2672(8) 775–784

DOI: 10.1177/0361198118801352
journals.sagepub.com/home/trr
SAGE

Potential of Taxi-Pooling to Reduce Vehicle Miles Traveled in Washington, D.C.

Yanshuo Sun[1] and Lei Zhang[2]

Abstract

This paper presents a pilot study of taxi-pooling in Washington, D.C. to demonstrate its potential for city-wide vehicle mileage reduction with real-world taxi trip data. In the taxi-pooling problem, riders specify their origins, destinations, earliest departure times, and latest arrival times. Taking all requests into consideration, the system dispatcher determines which requests can be grouped and thus served by one taxi, without violating any user-specified time windows and taxi capacity. This paper first presents a mathematical programming model whose objective is to minimize the system-wide vehicle miles traveled. Then an efficient solution approach to this problem is proposed. Results from analyzing the case of Washington, D.C. show that the total vehicle miles traveled can be reduced by 18% if the average buffer time in their travel schedule is 25% of their direct travel time.

A taxicab is usually dedicated to conveying a single rider or a small group of riders from the origin to the requested destination. Once a taxicab is occupied, it is not allowed, by regulation, to serve additional riders. Therefore, fighting over a cab is a long tradition in some city centers, such as New York City (NYC). Recently, this issue has been somewhat resolved as two ride-hailing apps in NYC, Via and Curb, began offering the option of sharing yellow taxis in NYC (*1*). The apps allow riders to opt for a shared taxi and each party only pays the discounted fare for their shared trip. From a systems perspective, sharing taxis can improve the utilization of the taxi fleet, reduce the number of taxi trips made, and alleviate traffic congestion. Riders can also benefit from it by reducing their wait time and paying a lower fare. In some states of Australia, such as New South Wales and Victoria, multiple hirings (two or more hirers use the same taxi simultaneously) are permitted with the consent of all parties involved. However, it is required that multiple hirings must start at the same time and at the same origin. All hires' destinations must be in the same general direction (*2*, *3*). As of 2017, taxi-sharing is not allowed in Washington, D.C. The purpose of this paper is to present a pilot study of taxi-pooling in Washington, D.C. to demonstrate its potential in city-wide vehicle mileage reduction with real-world taxi trip data.

The taxi-pooling problem of concern in this study is defined as follows: taxi riders who would like to participate in taxi-pooling use an app to specify their origins, destinations, earliest departure times, and latest arrival times. Taking all requests into consideration, the system dispatcher determines which requests can be grouped and thus served by one taxi, without violating any user-specified time windows and taxi capacity. If several riders are assigned to one taxi, the vehicle route should also be optimized to pick up and drop off riders in sequence. The objective is to minimize the system-wide vehicle mileage.

Similar problems have been studied in the literature, e.g., by Yan et al. (*4*) and by Santos and Xavier (*5*). Most studies formulate such problems as variants of the pickup and delivery problem with time windows, which is among the most challenging combinatorial optimization problems. Therefore, unless fast heuristics are designed, such models are too time consuming to solve and they cannot be used to analyze the taxi-pooling operations at a city level, as proved by detailed reviews of recent studies. The contribution of this study is to present an efficient computation tool to evaluate the impacts of large-scale taxi-pooling with real-world taxi trip data.

[1]Department of Industrial and Manufacturing Engineering, FAMU-FSU College of Engineering, Florida State University, Tallahassee, FL
[2]Department of Civil and Environmental Engineering, University of Maryland, College Park, MD

Corresponding Author:
Address correspondence to Lei Zhang: lei@umd.edu

The remainder of this paper is structured as follows. After reviewing relevant studies, this paper first presents a mathematical programming model for the taxi-pooling problem. Then, an efficient solution approach is designed to solve the optimization model. Case studies are conducted to demonstrate the efficiency of the proposed solution method as well as demonstrate the significant potential of taxi-pooling in reducing vehicle miles in Washington, D.C.

Literature Review

Some relevant studies on taxi-sharing are reviewed in this section. This review is by no means exhaustive. Interested readers are directed to survey articles (*6*, *7*) for additional reviews.

Yan et al. (*4*) employ a time–space network technique to study the taxi-pooling problem with case studies in Taiwan. Their model consists of several modules, such as the fleet routing and scheduling module, and passenger-taxi matching module. Nonetheless, it seems that their test cases are quite small. The Taipei metro area is divided into 40 zones and the taxi-pooling station is assumed to be located in the zone centroid. The size of the taxi fleet is only 25 and the number of passengers is 126 in the base analysis. Their largest test case only has 157 passengers. Their model can become problematic when a realistic problem instance consisting of thousands of passengers is considered.

D'Orey and Ferreira (*8*) conduct a simulation study of the taxi-sharing operations in the city of Porto, Portugal. Their taxi-sharing is distributed as rider requests are received by all vehicles; the routing algorithm is run separately at each individual vehicle. The rider then receives results from each vehicle and selects the best taxi to accept. Their simulation studies involving real-world taxi data show that taxi-sharing can benefit both the passengers and the taxi operators. In Porto's case, the total travel distance (distance travelled by all taxis) can be reduced by up to 9%.

Santi et al. (*9*) introduce the notion of a shareability network to study the taxi-sharing problem in NYC. With the shareability network, maximum matching algorithms are used to determine the best matches. Their case studies show that there is vast potential in NYC for such taxi trip-sharing services. They report that without increasing passenger discomfort significantly, the cumulative trip length can be reduced by at least 40%.

In the so-called Dial-a-Ride Problem with Money as an Incentive (DARP-M) proposed by Santos and Xavier (*5*), taxi-sharing is involved. The objective of this optimization problem is to maximize the total number of requests served and also to minimize the total fare paid by riders. To solve this complex combinatorial optimization problem, a greedy randomized adaptive search procedure is designed, and numerical experiments are conducted to evaluate the performance of the solution algorithm. They conclude that passengers pay 30% less than they would pay for exclusive rides.

Qian et al. (*10*) study a taxi group ride problem, where a group of riders with similar itineraries is assigned to professional drivers in a profit-driven taxi market. Grouped riders are picked up at the same meeting location and dropped off at the same delivery location, assuming that riders are willing to walk to and from these common points. This problem does not involve vehicle routing. The largest problem instance, solvable by an exact method within 15 minutes, consists of 88 riders. The problem becomes quickly intractable by exact methods when the problem size (represented by the number of riders) is doubled and heuristics are thus designed to find near-optimal solutions. Their numerical tests demonstrate significant vehicle mileage reductions, which exceed 47%.

Ma et al. (*11*) develop a dynamic taxi-sharing system and their extensive tests show that when some conditions are met, the taxi-sharing system can serve three times as many riders, as that in which no sharing is allowed. Meanwhile, the total travel distance can be reduced by 11% and the taxi fare per rider can be reduced by 7%.

Hosni et al. (*12*) present a Lagrangian decomposition approach in conjunction with two heuristics to solve the shared-taxi problem. Their computational results show that the proposed method can provide tighter bounds than CPLEX in shorter time. Nonetheless, their test cases are quite small. In simulation test one, 18 taxis are used and in the second test there are 31 taxis. Clearly, this approach is unlikely to be useable in evaluating the impact of taxi-sharing in a major city.

The taxi-sharing problem has also been studied by Jung et al. (*13*) with a case study in Seoul, Korea. Due to the complexity of the problem, hybrid-simulated annealing is proposed as the solution approach. In their simulations, 600 service vehicles are involved. They show that with the increase in riders' travel time, the proposed shared-taxi system can improve the system performance as compared with conventional taxi.

The reviews above show that many models of varying complexities for taxi-sharing have been developed and are tested extensively in simulation environments; some of the studies also involve real-world taxi trip data. Most studies report significant benefits of such services, with the reduction of vehicle miles ranging from 9% to 47%.

However, due to computational challenges, very few models are efficient enough to analyze city-scale problems, which makes it quite difficult to quantify the impacts of taxi-pooling operations in a city. Therefore, this study contributes to this line of research by

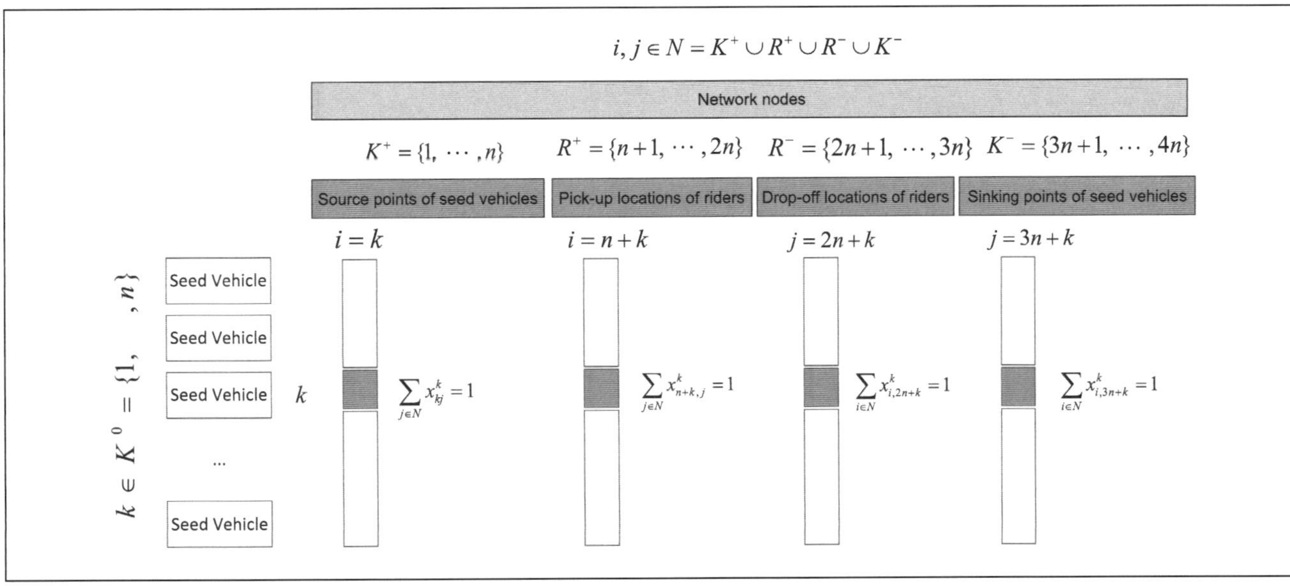

Figure 1. Illustration of some symbol definitions.

presenting an efficient computation tool to evaluate the potential of large-scale taxi-pooling with real-world taxi trip data in Washington, D.C.

Model of Taxi-Pooling Problem

Sets and Indices

For each rider request, a virtual taxicab—also called "seed vehicle" in this study—is created to serve this rider request exclusively. This vehicle has the same travel plan and schedule (i.e., origin, destination, earliest departure time at the origin, and latest arrival time at the destination) as the riders in the request. Suppose that the number of rider requests is denoted as n. The taxi-pooling problem is defined on a directed graph $G = (N, A)$, where $N = K^+ \cup R^+ \cup R^- \cup K^-$, $K^+ = \{1, \cdots, n\}$, $R^+ = \{n + 1, \cdots, 2n\}$, $R^- = \{2n + 1, \cdots, 3n\}$, $K^- = \{3n + 1, \cdots, 4n\}$. Subsets K^+, R^+, R^-, K^- contain source points of seed vehicles, pick-up locations of riders, drop-off locations of riders, and sinking points of seed vehicles, respectively, as shown in Figure 1. The starting node of seed vehicle $k \in K^0 = \{1, \cdots, n\}$ is k and ending node is $3n + k$. Each rider request $r \in R^0 = \{1, \cdots, n\}$ is associated with an origin node $n + r$ and a destination node $2n + r$.

The concept of a time window is illustrated in Figure 2. For example, for vehicle $k \in K^0$, its earliest departure time at the origin is e_k and its latest arrival time at the destination is l_{3n+k}. The direct travel time from the origin to the destination is t_k. The right time window at the origin and the left time window at the destination can be inferred with

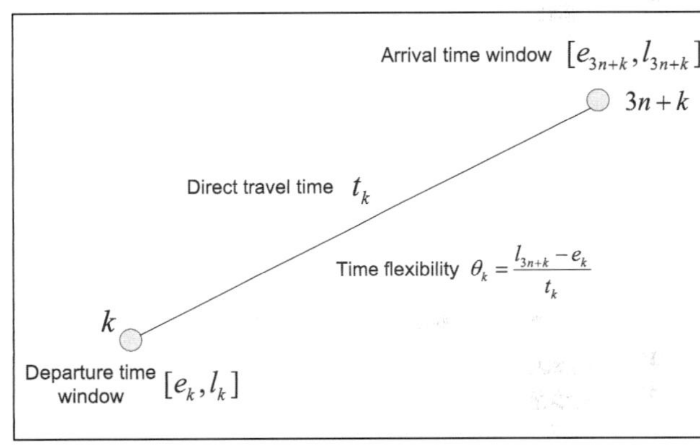

Figure 2. Inference of time windows and definition of time flexibility.

$$l_k = l_{3n+k} - t_k \quad \forall k \in K^0 \tag{1}$$

$$e_k + t_k = e_{3n+k} \quad \forall k \in K^0 \tag{2}$$

A ratio, which is calculated as the difference between the latest arrival time at the destination and the earliest departure time at the origin divided by the direct travel time, namely $\theta_k = \frac{l_{3n+k} - e_k}{t_k}$, is used to characterize the travel flexibility. This ratio is used in the literature (*14*). We note that the minimum value of θ is 1. When $\theta = 1$, at the origin $e_k = l_k$, meaning the width of the time window (calculated as the difference between the right time window and the left time window, $l_k - e_k$) is zero (namely no buffer or no flexibility at all). A larger value of θ corresponds to a large time flexibility.

A network node can be indexed by either i or $j \in N$. The set of arcs is $A = \{(i,j) : i,j \in N, i \notin K^-, j \notin K^+, i \neq j\}$, which means that for an arc (i,j) in the network G, the starting node could not be any destination of seed vehicles, the ending node could not be the origin of seed vehicles, and the starting and ending nodes should be different.

Parameters

$c_{ij}^k =$ cost of seed vehicle $k \in K^0$ traversing link $(i,j) \in A$;
$f_k =$ fixed cost (non-revenue mileage per taxi trip) of activating virtual vehicle $k \in K^0$;
$t_{ij} =$ travel time to traverse link $(i,j) \in A$;
$Q =$ the maximum number of rider requests to be served by each vehicle;
$q_i =$ load change at a location $i \in N$;
$d_i =$ service duration at a location $i \in N$;
$[e_i, l_i] =$ time window of node $i \in N$, where e_i is the earliest time to serve and l_i is the latest time;

At pick-up location $i \in R^+$, $q_i > 0$; at the paired drop-off location $(i+n) \in R^-$, $q_{i+n} = -q_i$, $i \in R^+$. At source and sink nodes of vehicles $i \in K^+ \cup K^-$, $q_i = 0$. It is also assumed that $d_i = 0$, $i \in K^+ \cup K^-$, due to no pickup or drop off of riders.

Decision Variables

x_{ij}^k = whether seed vehicle $k \in K^0$ is routed through arc $(i,j) \in A$;
z_k = whether seed vehicle $k \in K^0$ is activated as a sharing vehicle;
B_i^k = time at which vehicle $k \in K^0$ begins service at node $i \in N$

Formulation

The objective of the taxi-pooling problem is to minimize the system-wide vehicle miles traveled. This problem is formulated as follows:

$$\min_{\{x_{ij}^k, z_k, B_i^k\}} \sum_{k \in K^0} f_k z_k + \sum_{k \in K^0} \sum_{(i,j) \in A} c_{ij}^k x_{ij}^k \quad (3)$$

s.t.

$$\sum_{i \in N} \sum_{j \in N} x_{ij}^k \leq M z_k \quad \forall k \in K^0 \quad (4)$$

$$\sum_{j \in N} x_{ij}^k = 1 \quad \forall k \in K^0,\ i = n + k \quad (5)$$

$$\sum_{i \in N} x_{ij}^k = 1 \quad \forall k \in K^0,\ j = 2n + k \quad (6)$$

$$\sum_{k \in K^0} \sum_{j \in N} x_{ij}^k = 1 \quad \forall i \in R^+ \quad (7)$$

$$\sum_{i \in R^+} q_i \sum_{j \in N} x_{ij}^k \leq Q \ \forall k \in K^0 \quad (8)$$

$$\sum_{j \in N} x_{ij}^k - \sum_{j \in N} x_{n+i,j}^k = 0 \quad \forall i \in R^+,\ k \in K^0 \quad (9)$$

$$\sum_{j \in N} x_{kj}^k = 1 \quad \forall k \in K^+ \quad (10)$$

$$\sum_{j \in N} x_{ij}^k - \sum_{j \in N} x_{ji}^k = 0 \quad \forall i \in N/\{k, 3n+k\}, k \in K^0 \quad (11)$$

$$\sum_{i \in N} x_{i,3n+k}^k = 1 \quad \forall k \in K^+ \quad (12)$$

$$x_{ij}^k (B_j^k - B_i^k - t_{ij}) \geq 0 \quad \forall (i,j) \in A, k \in K^0 \quad (13)$$

$$e_i \leq B_i^k \leq l_i \quad \forall i \in N, k \in K^0 \quad (14)$$

$$x_{ij}^k \in \{0,1\} \quad \forall (i,j) \in A, k \in K^0 \quad (15)$$

$$z_k \in \{0,1\} \quad \forall k \in K^0 \quad (16)$$

$$B_i^k \geq 0 \quad \forall i \in N, k \in K^0 \quad (17)$$

Although we treat c_{ij}^k as a cost parameter, it does not necessarily represent monetary cost only; it can also represent vehicle mileage, fuel consumption, or emission. In this study, the units of c_{ij}^k are vehicle miles. Therefore, in the remainder of this paper, we do not explicitly distinguish cost and vehicle mileage. The objective has two components: fixed and variable. Here $\sum_{k \in K^0} f_k z_k$ is the total fixed cost of activating seed vehicles. The activation cost of a seed vehicle can be attributed to the distance from the current taxi location, when the dispatching decision is made, to the passenger pickup location. Alternatively, the activation cost of a seed vehicle can be regarded as the non-revenue mileage or deadheading mileage associated with each revenue trip. Here $\sum_{k \in K^0} \sum_{(i,j) \in A} c_{ij}^k x_{ij}^k$ is the total vehicle routing cost, which depends on the actual revenue mileage. The activation of a seed vehicle means the corresponding rider (based on whom the seed vehicle route is created) is the first to be picked up and the last to be dropped off. Once a seed vehicle is activated, other riders can be potentially served by this same vehicle. Constraint (4) indicates that unless a seed vehicle is activated, the vehicle cannot be routed. M is a sufficiently large constant. By definition, each seed vehicle serves a rider request exclusively. Constraints (5) and (6) ensure that the origin and also the destination are visited by the seed vehicle. For instance, seed vehicle $k \in K^0$ must visit origin $n + k$ and destination $2n + k$. Constraint (7) guarantees that each rider request is served and only once, i.e., the origin node is visited exactly once. Constraint (8) defines that the total number of rider requests served by vehicle $k \in K^0$ is within its capacity limit Q. Constraint (9) ensures the same vehicle visits the pickup and delivery locations of a rider request. Constraints (10–12) indicate that a vehicle must leave its source node; after visiting other nodes (neither source nor sink), it must leave for another destination; a vehicle must end at its sink node. The inequality (13) establishes the relation between service starting times at two adjacent nodes if the link (i,j) is used by a vehicle k. Constraint (13) imposes a unique travel direction, thus eliminating any sub-tours. This nonlinear constraint can be linearized into:

$$B_j^k \geqslant B_i^k + d_i + t_{ij} - M_B(1 - x_{ij}^k) \;\; \forall (i,j) \in A, k \in K^0 \quad (18)$$

where M_B is an arbitrary large number.

Constraint (14) affirms that the time window requirements are met. Equations 15–17 are feasibility constraints. In this formulation, if vehicle $k \in K^0$ is not routed through arc $(i,j) \in A$, definitions of B_i^k are void and their values are irrelevant.

Solution Approach

The formulation presented in the previous section is a vehicle-arc formulation, which is widely used in the vehicle routing literature. The classic vehicle routing problem (VRP) is to obtain the optimal set of routes for a fleet of vehicles to serve a given set of customers, which is a challenging combinatorial optimization problem. If cargos or passengers need to be picked up at certain locations and dropped off at other locations, the problem is called a vehicle routing problem with pickup and delivery (VRPPD). If a node can only be visited in a predetermined time window, the problem is called vehicle routing problem with time windows (VRPTW). VRP instances are computationally expensive and exact solution approaches can only solve small to medium-sized problems, e.g., involving dozens of customer requests. In the real world, thousands of rider requests can be received in a short time period and the matching must be optimized and announced shortly. Therefore, tradeoffs between solution quality and computation time must be resolved. Since the main objective of this paper is to explore the potential of taxi-pooling in the city of Washington, D.C., a fast heuristic is designed to solve the optimization problem efficiently, while there is no guarantee that the optimal solution is found.

Overall Procedure

Step 0: The set of vehicles that can serve additional riders is denoted as $\bar{K}$ and the set of unassigned rider requests is denoted as $\bar{R}$.

Step 1: If no vehicles can serve additional riders, i.e., $\bar{K} = \varnothing$, or all rider requests have been assigned, i.e., $\bar{R} = \varnothing$, go to Step 3; if there are vehicles that can serve additional riders and there are also unassigned rider requests, go to Step 2:

Step 2: For each vehicle $k \in \bar{K}$ and each unassigned rider request $i \in \bar{R}$, create a new vehicle route and check its time window feasibility (the checking procedure is provided later). If feasible, calculate how much cost (i.e., mileage) can be saved by accommodating request $i \in \bar{R}$ by vehicle $k \in \bar{K}$. The cost saving is denoted as π_{ki}.

Solve the binary program

$$\max_{\{y_{ki} \in \{0,1\}\}} \sum_{k \in K} \sum_{i \in R} \pi_{ki} y_{ki} \quad (19)$$

s.t.

$$\sum_{i \in R} y_{ki} \leqslant 1 \;\; \forall k \in \bar{K} \quad (20)$$

$$\sum_{k \in K} y_{ki} \leqslant 1 \;\; \forall i \in \bar{R} \quad (21)$$

$$\sum_{k \in K} y_{kj} + \sum_{i \in R} y_{ji} \leqslant 1 \;\; \forall j \in \bar{R} \cap \bar{K} \quad (22)$$

Objective (19) is to maximize the total cost saving by selecting some pairs of vehicle and rider requests. If a pair of vehicle k and request i is selected, $y_{ki} = 1, \forall i \in \bar{R}, k \in \bar{K}$. Constraint (20) ensures that at most one pair is selected for each seed vehicle and constraint (21) enforces a rider request appears at most in selected pairs. Constraint (22) determines a unique role for a rider request. Once a request is assigned to a certain vehicle, the seed vehicle associated with this rider request cannot be activated. If a seed vehicle route is activated, the rider request associated with the seed vehicle cannot be assigned to any other vehicles. To be more specific, if rider request $i \in \bar{R}$ is assigned to a vehicle, namely, $\sum_{k \in K} y_{ki} = 1$, by constraint (22), $\sum_{i \in R} y_{ki} = 0$, meaning that the corresponding seed vehicle route with $k = i$ cannot be selected. Similarly, for a vehicle k, if $\sum_{i \in R} y_{ki} = 1$, we obtain $\sum_{k \in K} y_{ki} = 0$, meaning the corresponding rider with $i = k$ cannot be assigned to any vehicle, because the rider is already covered by the route of seed vehicle k by definition. Because the problem is to maximize the total savings, only those pairs with positive cost savings ($\pi_{ki} > 0$) may be selected.

After solving the problem, vehicle set $\bar{K}$ and rider request set $\bar{R}$ are updated as follows: if $y_{ki} = 1, \forall i \in \bar{R}, k \in \bar{K}$, meaning the rider request i is assigned to vehicle k, i is no longer under consideration.

Then, i is also removed from $\bar{R}$ and the vehicle route of the relevant vehicle is updated. If the total number of assigned riders to a vehicle k is equal to the limit Q, this vehicle cannot serve additional riders, and k is removed from $\bar{K}$.

Go to **Step 1**.

Step 3: The algorithm terminates.

Insertion Heuristic

Figure 3 illustrates how a rider request can be inserted into a vehicle route and how the feasibility of the new route is checked. The origin and the destination of the rider are inserted separately and the destination is inserted after the origin. The insertion position is selected to minimize the incremental cost. For example, if origin E is inserted between A and C, the incremental cost is the cost of traversing arcs AE and EC minus the cost of traversing arc AC. The incremental cost can be calculated for each insertion position and the position with the smallest cost increase is the best position to insert.

If two positions have been selected for E and F, a new vehicle route, for example, is shown in Figure 3*b*. For the rider, its time window at origin E is $[e_E, l_E]$. For the vehicle to arrive within the time window $[e_E, l_E]$, the vehicle must depart between $\bar{e}_A$ and $\bar{l}_A$, where $\bar{e}_A = e_E - t_{AE}$ and

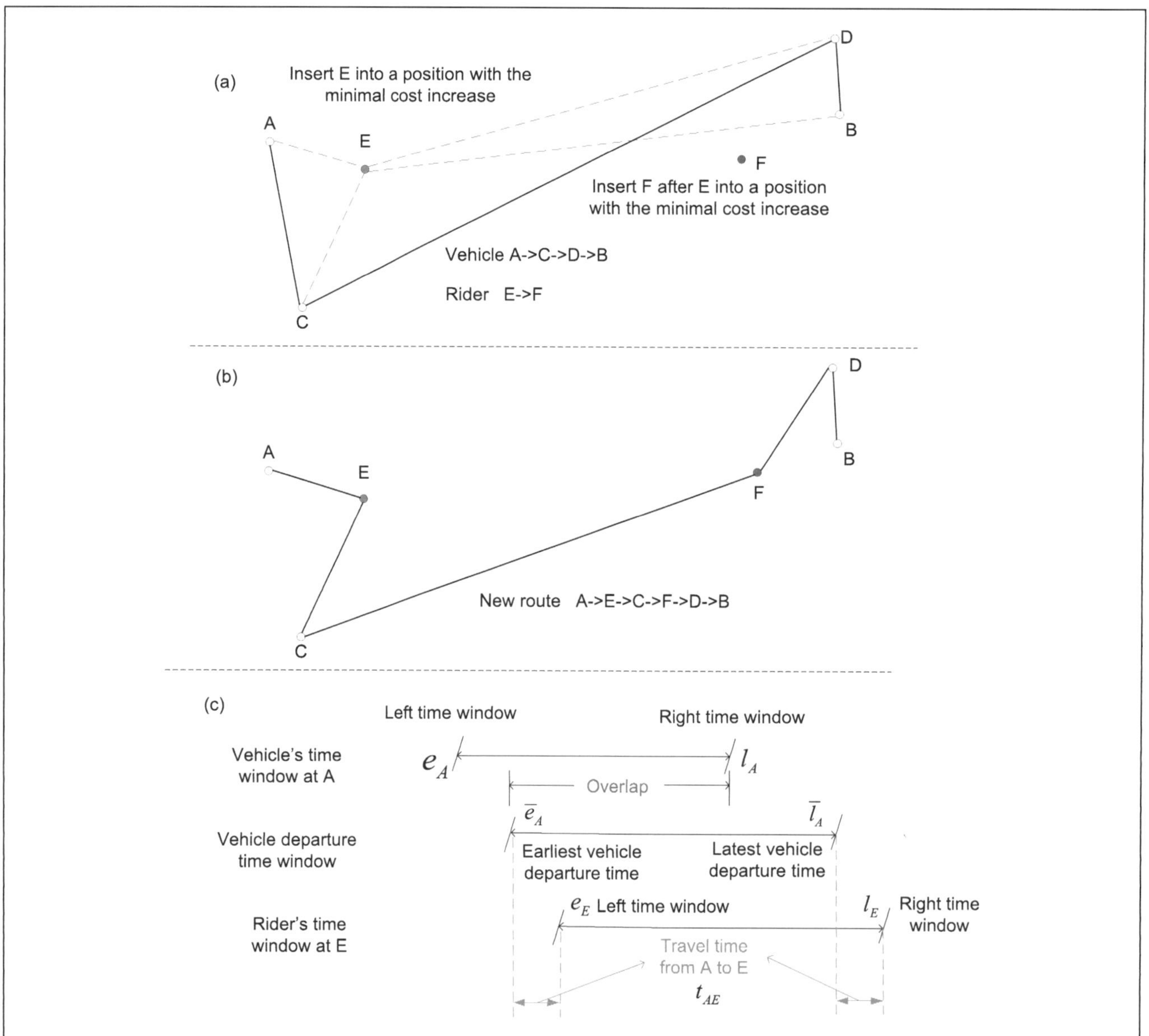

Figure 3. An example of rider insertion.

$\bar{l}_A = l_E - t_{AE}$. If $\max\{e_A, \bar{e}_A\} < \min\{l_A, \bar{l}_A\}$, the route is feasible; otherwise it is infeasible. If the new route is feasible, its cost saving can be calculated. π_{ki} is the cost of the original route plus the cost of serving request i exclusively minus the cost of the new route.

It should be noted that the heuristic described above can only produce a locally optimal solution, for several reasons. First, pickup and drop off locations of the new rider are separately inserted and the resulting route is not necessarily the best route leading to the maximum cost saving. Second, the binary problem (19-22) is to maximize the cost savings in each iteration of Step 2; however, this does not necessarily yield the most cost savings over several iterations of Step 2. For instance, it is possible that the cost savings is negative if either one of two requests is assigned to a taxi alone. However, if two rider requests are assigned to the taxi at the same time, the cost saving is positive.

A clear advantage of the proposed heuristic is its computational efficiency. It would take hours to solve the original vehicle-arc formulation with exact methods even for a medium size problem, e.g., with 100 rider requests.

Case Studies

Model Inputs

Taxi trip data in Washington, D.C. are made available by the D.C. Department of For-Hire Vehicles (DFHV) at http://opendata.dc.gov/datasets?q=taxi. Pickup and drop off locations are assigned to their nearest block locations based on the original latitude/longitude coordinates. Pickup and drop off times are also rounded to the nearest hour. To enrich the data, a uniform random number between 0 and 59 is chosen as the minute. In total, 3,670 taxi trips with pickup times between 8:00 am and 8:30 am on January 12, 2017, are selected for further analyses. Taxi trip data on other dates can be used as well.

Figures 4 and 5 show the pickup and drop-off locations of those taxi trips, respectively. As the route between two coordinates is not available, the distance of a trip is approximated by the straight-line distance from the pickup block to the drop-off block location multiplied by a ratio 1.3. We note that after routes between two coordinates become available with the aid of geographical information systems and the route length is measureable, such an approximation is no longer necessary. Figure 6 is a histogram of the taxi trip distance. It is found that most taxi trips are shorter than six miles according to the distance measure defined in this analysis.

Each taxi trip can be regarded as a rider request. Since the number of passengers in each rider request is unknown, we assume the number is 1 for convenience. Therefore, one passenger request is equivalent to one passenger. The value of time flexibility θ is assumed to

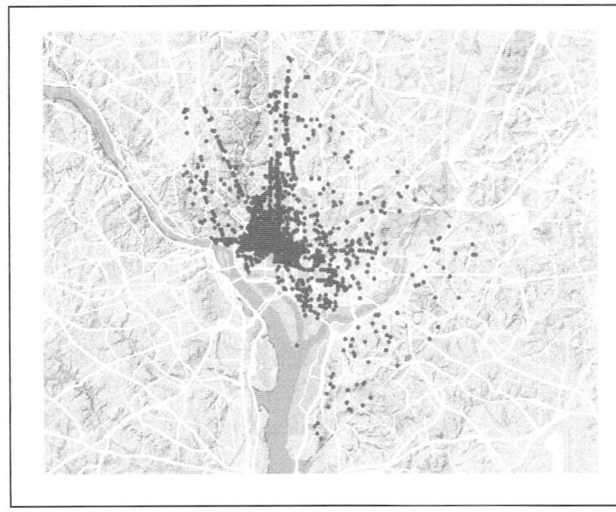

Figure 4. Distribution of pickup locations.

Figure 5. Distribution of drop-off locations.

be uniformly distributed between 1.0 and 1.5, unless specified explicitly otherwise. The mean value of θ is 1.25, meaning that on average, riders would like to spend 25% of their direct travel time in addition to their direct travel time in the taxi-pooling in exchange for a lower fare. Alternatively, the buffer time is 25% of the direct travel time.

The model is implemented in Python 2.7 and the mixed integer program solver is CPLEX v12.7.1.0. The program runs on a personal computer (Intel Core i-7 CPU 2.60 GHz, 16 GB RAM). It is expected that the computation time reported in this study can be reduced further if more powerful computers are used. Since the proposed solution algorithm can only generate locally optimal solutions, all cost savings reported are lower bounds on the

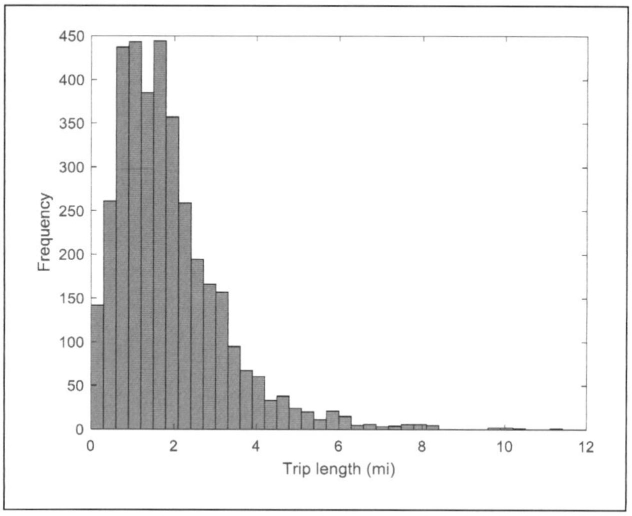

Figure 6. Histogram of the taxi trip length.

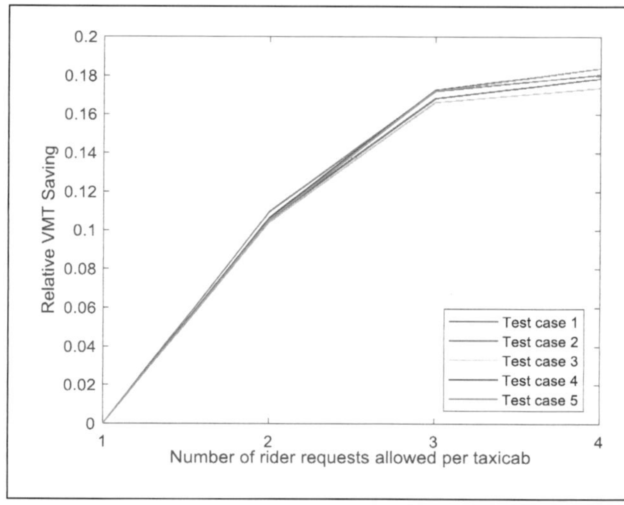

Figure 7. VMT saving made by taxi-pooling.

global optimal saving. In other words, the true savings are no smaller than those found in this study.

Benefits of Taxi-Pooling

The mileage of a taxi trip includes a fixed part and a variable part, as defined in objective (3). The fixed part is the non-revenue mileage from the current taxi location (when the dispatching decision is made) to the passenger pickup location. The variable part is the revenue mileage with at least one passenger on board. We introduce an example to illustrate the concept of relative VMT saving or cost saving. Assume that the non-revenue mileage associated with each revenue trip is 2. If taxi-pooling is not allowed, two riders are served by two different taxis. Actual VMTs of two taxis are assumed to be 5 and 7, respectively. With taxi-pooling, two riders are served by one vehicle; the total actual VMT by this shared taxi is 10. The total mileage without taxi-pooling is $(2 + 5) + (2 + 7) = 16$; the total mileage with taxi-pooling is $2 + 10 = 12$. The relative saving is $(16\text{-}12)/16 = 0.25$. Note that the relative saving depends on the value of non-revenue mileage per taxi trip.

Because some values, e.g., pickup time, are randomly generated, five runs of the model are conducted and the cost saving in each run is shown in Figure 7. The x-axis represents the number of rider requests allowed in each taxi trip. If the number is 1, a taxi serves each request exclusively, namely the status quo. If two requests are allowed to be served by one taxi, there would be some cost savings. As the number of allowed rider requests increases, the relative saving increases. The maximum number of requests allowed in this study is 4, considering that most taxicabs have 4 available passenger seats.

In Figure 7, when sharing is not allowed, the saving is 0. When two requests can share a vehicle, the relative saving increases to around 0.1. When four requests can share a vehicle, the relative saving grows to around 0.18. Although results from different runs vary, the trend is about the same.

Effect of Request Density

As expected, the relative cost savings significantly depend on the trip density, which is defined as the total number of requests per unit of time divided by the area of the study region. Since the study region is fixed in this paper, the trip density is higher when more trip requests are involved. In the base analysis, 3,670 requests with pickup times in a half-hour time period (8:00–8:30 am) are considered. To analyze the impact of trip density, 3,670 requests are shuffled. In the first run, the first 250 trips are considered. In the second, the first 500 trips are considered. In the last run, the first 3,000 trips are considered. The relative VMT saving in each run is shown in Figure 8. Clearly, the relative saving grows significantly with the number of requests in consideration. As the problem scale grows, the computation time also increases. Nonetheless, even if the number of requests is 3,000, the computation time is still manageable.

Effect of Schedule Flexibility

The reason why taxi-pooling can reduce VMT is that travelers are willing to spend more time in exchange for discounted fares. If each traveler requires his/her travel time to be exactly the same as the direct travel time (implying $\theta = 1$), there is no chance for taxi-pooling, unless

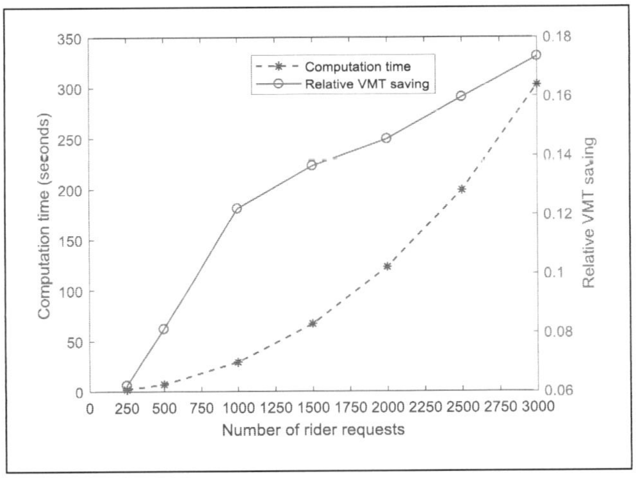

Figure 8. Impact of trip density.

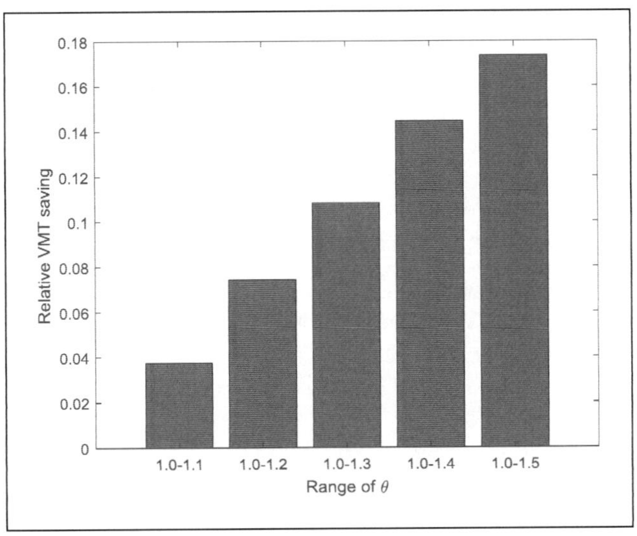

Figure 9. Impact of time flexibility.

travelers share exactly the same origin, destination, and trip start time.

Therefore, the cost saving is also highly dependent on time flexibility, as shown in Figure 9. As the time flexibility θ increases, the relative cost saving increases. Note that the x-axis is the range of θ. When the range is 1.0 to 1.5, implying that the average time flexibility θ is 1.25, the relative taxi mileage saving is around 0.17. In other words, if the average buffer time is 25% of the direct travel time, the system-wide cost saving is around 17%. Note that only the first 3,000 requests of the shuffled 3,670 requests are considered in generating the result in Figure 9. If all 3,670 requests are considered, the relative cost saving is expected to increase to around 18%, as already shown in Figure 7.

Conclusion

This paper evaluates how taxi-pooling can reduce the city-wide vehicle miles traveled. First, a vehicle-arc formulation for the taxi-pooling problem is provided and then a fast heuristic is proposed to solve the problem. To demonstrate the proposed model, case studies involving real-world taxi trip data are conducted. Results show that if on average the buffer time is 25% of the direct travel time, the system-wide vehicle miles traveled can be cut by 18%. It is also shown that the saving is highly dependent on the trip density. Computation tests show that the proposed solution method can handle relatively large problems (e.g., 3,000 requests) within a reasonable time (e.g., 5 minutes). In practice, rider requests are received over time and taxi-pooling dispatch decisions are made every few minutes. Assume that the decision interval is 5 minutes and there are 180 decision intervals (each hour has 12 intervals and there are 15 operating hours for example) in one day. A rough estimate is that a total of 540,000 taxi trips (3,000*180) can be handled in one day. Note that even NYC has no more than 500,000 daily taxi trips (*15*). Therefore, in terms of computational efficiency, the model should be able to realistically analyze a city-scale taxi-pooling problem. Although this paper presents a case study for Washington, D.C., the proposed methodology can be applied to other cities to evaluate the potential of taxi-pooling.

The present work can be improved in the following ways:

(1) Only the impact of taxi-pooling on vehicle miles traveled is examined. Therefore, the potential of taxi-pooling has not been fully explored. For example, taxi-pooling can reduce the service waiting time, which is defined as the time between the service request time and the pickup time.
(2) Local search heuristics can be introduced to further improve the solution quality and exact solution approaches should be designed to find the globally optimal solution.
(3) While some baseline values are used in this study, other model users are encouraged to calibrate the parameters (e.g., the non-revenue mileage per taxi trip) of the proposed model with their own data.

Acknowledgments

The financial support from the U.S. Department of Energy through its ARPA-E's TRANSNET Program is acknowledged.

Author Contributions

The authors confirm contribution to the paper as follows: study conception and design: Yanshuo Sun (YS), Lei Zhang (LZ);

data collection: YS; analysis and interpretation of results: YS, LZ; draft manuscript preparation: YS. All authors reviewed the results and approved the final version of the manuscript.

References

1. New Yorkers Try a Startling Idea: Sharing Yellow Cabs. *New York Times*. https://www.nytimes.com/2017/06/06/nyregion/new-york-yellow-taxis-ride-sharing.html. Accessed July 15, 2017.
2. *Multiple Hiring*. NSW Taxi Council, Australia. https://www.nswtaxi.org.au/passengers/multiple-hiring. Accessed July 15, 2017.
3. *Taxi Fares*. Taxi Services Commission, State Government of Victoria, Australia. http://taxi.vic.gov.au/passengers/taxi-passengers/taxi-fares. Accessed July 15, 2017.
4. Yan, S., C. Y. Chen, and C. C. Wu. Solution Methods for the Taxi Pooling Problem. *Transportation*, Vol. 39, No. 3, 2012, pp. 723–748.
5. Santos, D. O., and E. C. Xavier. Taxi and Ride Sharing: A Dynamic Dial-a-Ride Problem with Money as an Incentive. *Expert Systems with Applications*, Vol. 42, No. 19, 2015, pp. 6728–6737.
6. Agatz, N., A. Erera, M. Savelsbergh, and X. Wang. Optimization for Dynamic Ride-sharing: A Review. *European Journal of Operational Research*, Vol. 223, No. 2, 2012, pp. 295–303.
7. Furuhata, M., M. Dessouky, F. Ordóñez, M. E. Brunet, X. Wang, and S. Koenig. Ridesharing: The State-of-the-art and Future Directions. *Transportation Research Part B: Methodological*, Vol. 57, 2013, pp. 28–46.
8. d'Orey, P. M., and M. Ferreira. Can Ride-Sharing Become Attractive? A Case Study of Taxi-Sharing Employing a Simulation Modelling Approach. *IET Intelligent Transport Systems*, Vol. 9, No. 2, 2014, pp. 210–220.
9. Santi, P., G. Resta, M. Szell, S. Sobolevsky, S. H. Strogatz, and C. Ratti. Quantifying the Benefits of Vehicle Pooling with Shareability Networks. *Proceedings of the National Academy of Sciences*, Vol. 111, No. 37, 2014, pp. 13290–13294.
10. Qian, X., W. Zhang, S. V. Ukkusuri, and C. Yang. Optimal Assignment and Incentive Design in the Taxi Group Ride Problem. *Transportation Research Part B: Methodological*, Vol. 103, 2017, pp. 208–226.
11. Ma, S., Y. Zheng, and O. Wolfson. Real-time City-scale Taxi Ridesharing. *IEEE Transactions on Knowledge and Data Engineering*, Vol. 27, No. 7, 2015, pp. 1782–1795.
12. Hosni, H., J. Naoum-Sawaya, and H. Artail. The Shared-taxi Problem: Formulation and Solution Methods. *Transportation Research Part B: Methodological*, Vol. 70, 2014, pp. 303–318.
13. Jung, J., R. Jayakrishnan, and J. Y. Park. Dynamic Shared-Taxi Dispatch Algorithm with Hybrid-Simulated Annealing. *Computer-Aided Civil and Infrastructure Engineering*, Vol. 31, No. 4, 2016, pp. 275–291.
14. Masoud, N., and R. Jayakrishnan. A Decomposition Algorithm to Solve the Multi-Hop Peer-to-Peer Ride-matching Problem. *Transportation Research Part B: Methodological*, Vol. 99, 2017, pp. 1–29.
15. New York City Taxi & Limousine Commission. *2016 TLC Fact Book*. www.nyc.gov/html/tlc/downloads/pdf/2016_tlc_factbook.pdf. Accessed October 15, 2017.

The Standing Committee on Emerging and Innovative Public Transport and Technologies (AP020) peer-reviewed this paper (18-02130).

Article

TRR
JOURNAL OF THE TRANSPORTATION RESEARCH BOARD

Transportation Research Record
2018, Vol. 2672(8) 785–795

DOI: 10.1177/0361198118797462
journals.sagepub.com/home/trr
SAGE

Tangible Tools for Public Transportation Planning: Public Involvement and Learning for Bus Rapid Transit Corridor Design

Anson F. Stewart[1,2], P. Christopher Zegras[3], Phil Tinn[4], and Jeffrey L. Rosenblum[3]

Abstract

Open governance and open data have given rise to new collaborative tools for public involvement in transit planning. The research presented in this paper extends such tools, adding tangible and interactive features in an attempt to foster interaction, dialog, and social learning. Three tools, representing the impacts of bus rapid transit (BRT) projects at the street, neighborhood, and regional scales, were deployed at a series of public workshops in Boston. A pre-/post- survey design reveals substantive learning about BRT, supported by participants' general agreement with statements about social learning. The quality of dialog in the workshops may point to the potential for more in-depth, double-loop learning. Of the tools used in the workshop, participants judged the one representing the street scale to be the easiest to use, whereas they judged the touchscreen regional map to be the most relevant and credible. Further research could test such tools with more representative participants and in other settings.

The development of transportation improvements in any city poses a range of challenges, including meaningfully involving relevant stakeholders. With many members of the public skeptical of government's ability to generate solutions that will work for them, transport agencies and community organizations are increasingly seeking better ways to engage each other and the general public in developing project ideas. Among the various challenges is the fact that existing representational tools are not well-suited to allowing diverse stakeholders to understand, evaluate, and provide feedback on the benefits and tradeoffs of decisions. These interconnected decisions range widely, from local pedestrian flows around public transport stations, to parking provision on streets, to corridor alignment and priority schemes.

In recent years, new cooperative planning tools have emerged, made possible by the rapidly growing availability of open-source data platforms and interactive computing technology. These technologies promise to facilitate the inclusion of local knowledge in a way that could transform public participation processes. Emerging web-based tools, for example, can speed up data processing, spark new ways of representing and discussing potential projects, and improve interpretation among a broad spectrum of community members.

This paper describes efforts to explore these possibilities through the development, customization, and testing of several interactive tools to see how well they enable dialog. The tools are designed to allow workshop participants to explore impacts and alternatives at the street, neighborhood, and regional scales. In partnership with a local community development organization, they were deployed at a series of public workshops in Boston's Roxbury neighborhood. Though these workshops focused on the potential for implementing bus rapid transit (BRT) as a case study, their broader purpose was evaluating the extent to which new collaborative tools can encourage social learning among stakeholders. Such learning is a foundation for innovative co-creative transportation planning, whereby a range of stakeholders work together in problem framing and solution co-designing.

[1]MIT Interdepartmental Transportation Program, Cambridge, MA
[2]Conveyal, Boston, MA
[3]MIT Department of Urban Studies and Planning, Cambridge, MA
[4]MIT Media Lab, Cambridge, MA

Corresponding Author:
Address correspondence to Anson F. Stewart: astewart@conveyal.com

Background

Urban planning generally, and transportation planning specifically, has long grappled with meaningfully and effectively engaging the "public." In the United States, the top-down, engineering-driven urban highway program of the 1950s and 1960s came to a head in the late 1960s, culminating in the "freeway revolts" and national legislation requiring more public involvement (*1*). Around the same time and also in response to top-down initiatives such as urban renewal, participation "came into vogue" (*2*) among planning theorists and practitioners. Over the subsequent half century, as conceptions of planning have evolved, so have conceptions about the role of public participation. Lane (*3*) suggests that contemporary planning's inherently political dimensions and typically pluralistic and atomistic contexts make participation a "fundamental element" (p. 296). Innes and Booher call for "collaborative participation," through "multi-way interaction" among a range of stakeholders through a variety of mechanisms, formal and informal (*4*). Such arguments reflect the recognition that governance today implies a wider distribution of power and agency across society (*4*).

Paralleling the movement toward more open forms of governance, the ongoing revolution in information and communication technologies (ICTs) has ushered in new possibilities for planning and managing infrastructure and services and empowering citizens (*5*, *6*). "Open data"– borne of the basic idea that open government, transparency, and accountability can enhance public policy and service delivery (*7*) – epitomizes the potential. ICTs create new possibilities for distributed data generation, computation, and sharing, and also create new expectations among citizens about what information should be made available, by whom, to whom, under what conditions, and for what purposes. In the United States, these possibilities and expectations are increasingly manifest across the public sector. Public transportation provides a great example, through what Rojas terms "transit transparency" (*8*). Starting in the mid-2000s, transit agencies across the country began publicly releasing digital files of static transit information (e.g., stops and schedules) in a now de-facto standard (the general transit feed specification, GTFS) and creating ways (e.g., application programing interfaces, APIs) for third parties to access that information (*8*). An increasingly global phenomenon, transit transparency, exemplified by GTFS, is shifting how agencies interact with users and the general public, opening new methods of communication and service use, monitoring, assessment, and modeling (*9*).

Transportation modeling may represent the "last frontier" in unleashing the open data and open governance potentials for more collaborative participation. Born in, and subsequently codified during, the early years of urban highway planning in the 1950s (*10*), urban transportation models are necessarily complex – they aim to represent multi-modal supply–demand interactions for the purposes of predicting system responses to new interventions. Such tools have long been maligned as "black boxes," with only experts capable of understanding how model inputs actually produce travel forecasts (*11*). Yet, just as planning has shifted away from rigidity toward a more collective, inclusive, and communicative approach, advances in ICTs may change current urban modeling practices (*12*). GIS used in interactive participatory planning activities (in a "bottom-up" approach) can strengthen participation, helping citizens better articulate and understand perceptions relevant to local planning, especially by revealing spatial relationships, interconnections, and implications of values and strategies (*13*). More recent ICT and collaborative GIS advances may further strengthen these potentials, if they can synthesize disparate sources of information to improve users' spatial understanding of possible interventions (*14*).

A related advance is the shift from the traditional graphical user interfaces toward tangible user interfaces (TUIs). An early attempt to integrate TUIs into urban design and planning was the Luminous Planning Table: a projector casting light and digital shadows on a table top mounted with sensors monitoring physical blocks placed on it (*15*). Adjusting the location and orientation of the blocks changes shadows, lighting, and wind. Assessments of student groups using the tool suggested that those using the TUI, as opposed to traditional computer-design tools, were more likely to design through consensus (*15*). Other evidence has shown that designers using TUIs can perform multiple cognitive actions more quickly, iterate more frequently on design ideas, and better relate different spatial scales (*16*).

Opening the "black box" and making "planning support systems" more interactive and understandable may lead to greater social learning and better planning outcomes (*17*). Interactive, digitally enabled tools, operationalized in a context of collaborative governance, stand to transform stakeholder engagement in various realms of urban planning, including transportation. Sketch-planning tools can narrow the gap between conceptual design and traditionally expert-driven refinement and evaluation. Narrowing this gap, reducing the computational and cognitive burdens of analysis and evaluation, facilitates incorporating meaningful feedback throughout design, rather than seeking buy-in for already designed projects. Goodspeed (*17*) frames the potentials in terms of two forms of organizational learning theorized in Argyris (*18*): single-loop learning, the typical problem-solving learning style; and double-loop learning, which extends beyond immediate problem-solving, leading to a more in-depth reevaluation of goals and values. Organizations engaging in double-loop learning work not only to solve a fixed problem, but also to reconsider the problem itself; instead of merely completing a task, participants also refine new understandings and ways of working together. Double-loop learning may help build sustained community support for projects over the long-term, which would benefit transportation agencies with portfolios of multi-year projects. Though consensus, let alone ultimate outcomes, may be difficult to link to specific interactions between stakeholders that may have taken place years in the past, this paper takes as a premise that robust social learning in stakeholder engagement ultimately improves transportation projects, and that authentic dialog is a necessary, if not sufficient, condition for such learning. Building on this premise, this research explores to what extent digital tools may help planners realize Innes and Booher's call for more authentic dialog and multi-way interactions among stakeholders (*4*).

Research Objectives and Approach

To what extent can collaborative use of interactive planning tools encourage social learning in public engagement around transit planning? Might tangible, interactive planning tools, deployed in public engagement workshops, enhance single- and double-loop learning? Following Maher and Kim, tangibility can enhance learning by helping users think more rapidly, revisit ideas more frequently, better grasp spatial relations, and make new discoveries (*16*). Following Goodspeed, interactivity can improve individuals' imagination, alignment with others' perspectives, and engagement with others (*17*). Participants might learn by interacting with the tools directly, which would allow them to relate to different transit options based on personal experiences and specific relationships to place. They might also learn through group-based interaction, deliberation, and social learning, all enabled by rapid computation and quick and easy comparison of alternatives. Such learning might ultimately shift their attitudes and intentions.

To answer these questions and test these hypotheses, a series of pilot workshops was organized, involving a range of stakeholders and utilizing tangible and interactive tools designed to examine different public transport options. Workshop impacts were assessed through pre- and post-workshop surveys, observations, and transcriptions of participant interactions.

Context

Greater Boston has a long history of public involvement in transportation planning. It was a leader in the 1960s' "freeway revolt," becoming the first metropolitan area in the United States to have a nearly complete moratorium declared on highway construction in 1972. In the immediate aftermath of the moratorium came the Boston Transportation Planning Review (BTPR), a unique planning process that aimed to balance traditional technocratic transportation planning with broader participation to more meaningfully involve a range of interests, increase transparency, and more evenly consider a wider range of impacts (including social and environmental), at multiple scales (*19*). The result of this "open" planning process led to major transformations to the region's transit and highway plans and systems.

Today, Boston faces the challenge of expanding transit to serve current and future demand while struggling to operate and maintain its existing system. BRT provides one possibility for bus improvements, with separated rights of way (i.e., dedicated bus lanes) and payment of fares prior to boarding at high-quality stops (*20*). Despite its promise, BRT remains controversial. Exclusive bus lanes often require the removal of an existing car travel lane or parking. Boston's Washington Street Silver Line corridor, although promoted as BRT, has been characterized as "not BRT" according to an international standard (*21*). Subsequent efforts to improve bus service in Boston have faced strong community opposition. Factors contributing to the lack of public support include residents' unfamiliarity with the idea, planners' unfamiliarity with the communities, and memories of past and highly unwelcome "top-down" planning efforts (*22*). In addition, the association of BRT with inferior service provided by typical buses (slow, unreliable, with long, uncomfortable waits and rides) pervades the mindset of the public.

In 2013, a Boston-based civic philanthropy convened the Greater Boston BRT Study Group, a broad range of individuals from transportation, development, community, and environmental fields, to analyze the region's BRT potential. The study group recommended five corridors with particular promise for BRT implementation. Three of these corridors converge at Dudley Square in Roxbury, a key factor in the decision to hold workshops there. Transit service in nearby neighborhoods, the historic center of Boston's African American community and relatively distant from rail transit corridors, suffers especially low reliability and slow speeds.

The Tools

Three tools, drawing on open-source software and open data, were customized to provide users with a sense of the possible impacts of BRT corridor implementation at three scales: the region, the neighborhood, and the street. Each scale was represented in a different tangible tool at an interactive station in the workshop. The neighborhood and street scales were represented with CityScope, a platform that utilizes physical models (built from LEGO bricks) and projections of static and dynamic information. CityScope has been shown to encourage collaboration and communication among stakeholders (*23*). CityScope's BRT workshop versions allowed users to examine how different corridor types might perform in a neighborhood, or how different street elements (such as bus stop types or bike lanes in lieu of parking) might affect the streetscape and corridor performance. The region was represented through CoAXs, a web-based interactive platform, presented on a large horizontal touchscreen that allowed users to explore transit system performance region-wide and see how BRT projects might affect this performance (*24*, *25*).

CityScope: Neighborhood and Street Scales. Users interact with CityScope by moving tokens, tagged assemblies of LEGO bricks, to modify scenarios; effects of these modifications, such as changes in vehicle or pedestrian movements, are simulated. Simulation results can then be visualized, projected down onto the model, and summarized in other displays. For this project, CityScope was adapted to display predicted traffic performance (of buses and cars), based on the outputs of a Simulation of Urban Mobility (SUMO) model customized for the local area (*26*). One CityScope station represented a street segment and intersection, and the other represented approximately 1 km^2 of the neighborhood around Dudley Square.

For the street-scale CityScope station (Figure 1), users could place tokens representing bus lanes (curb, median), different bus stop types, bike lanes, and on-street parking onto a stylized street. Placing the tokens triggered summary statistics, corresponding to pre-run SUMO simulations, to be displayed on a large screen; animated vehicle movements were also projected on the physical model.

The neighborhood station (Figure 2) allowed users to place tokens representing different types of bus stops (e.g., regular or pre-pay boarding) and activate different corridor alignments. Based on these selections, pre-run traffic simulations (from SUMO) for the specific selections were used to generate summary statistics, showing, for example, bus speed changes or changes in estimated emissions levels. An adjacent screen displayed these statistics.

CoAXs: Regional Scale. CoAXs is an open-source, web-based visualization tool with two map-based modules: an accessibility view for displaying isochrones from user-selected origins and a corridor editor for modifying public transport routes (Figure 3). CoAXs uses Conveyal Analysis, an open-source cloud-hosted tool developed for professional planners to analyze land-use and transportation scenarios (*27*). It uses various open data sources, including GTFS feeds. Road and pedestrian network data are loaded from OpenStreetMap. Land-use data came from the EPA Smart Location Database (aggregated to the census block-group level). Stewart describes CoAXs and analyzes its use in these workshops in greater detail (*25*).

CoAXs builds on the basic idea that accessibility is an important measure of transport system performance (*28–30*). In this context, accessibility refers to the potential for individuals (in a given location) to reach destinations (such as certain types of jobs) by transit. For example, from a given location, a user could estimate how many healthcare jobs can be reached within 25 min, across the metropolitan area. After selecting a scenario of improvements from the corridor editor, the user could then see resulting changes in reachable healthcare jobs within the

Figure 1. Street-scale CityScope tool. (Source: Ariel Noyman of MIT Media Lab and authors.)

Figure 2. Neighborhood-scale CityScope tool. (Source: Elaine Hoffman of Goody Clancy and Ariel Noyman of MIT Media Lab.)

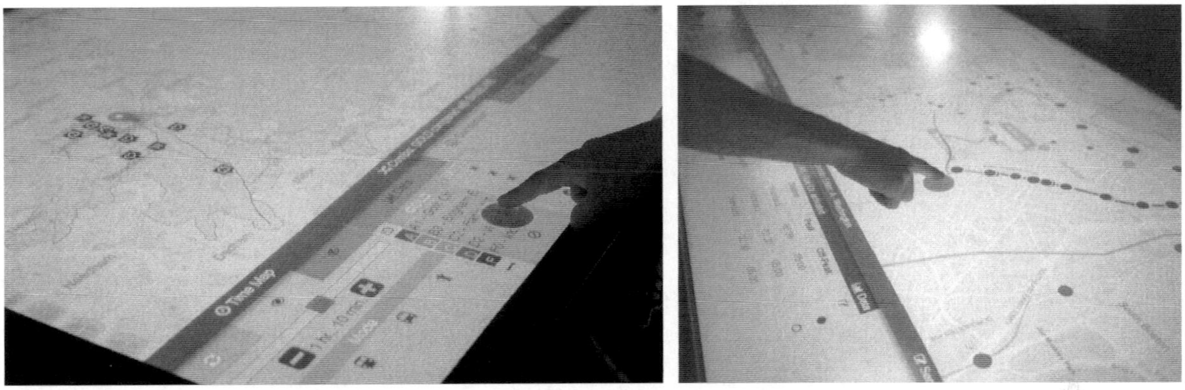

Figure 3. CoAXs interface, with accessibility view (left) and corridor editor view (right). (Source: authors.)

same 25-min cutoff. Essentially, the tool seeks to enable individuals to link accessibility maps and measures with travel patterns they deem important, thereby supporting dialog through a common, shared map.

Workshop Design

Six facilitated workshops took place over 4 days in October, 2015, each lasting approximately 2 h and following the same basic agenda and approach. Members from a local community development organization contributed to the workshop design and served as facilitators. Training prior to the workshops gave facilitators familiarity with the tools and provided valuable feedback on usability. Having community members as lead facilitators was an attempt to make the entire workshop experience peer-based and inclusive. Community facilitators led many of the tool-focused discussions, aiming to reduce participants' potential perception that "technical experts" controlled the technology, and therefore, the analyses. Facilitators had template scripts for guidance, but responded flexibly to different audiences as they saw fit.

Agenda. Each workshop followed the same overall agenda:

- Registration. Participants completed a pre-workshop survey then used a tablet to map points of interest to them (e.g., home, work, school, recreation, and healthcare locations), which were saved for later use in CoAXs.
- Orientation (30 min). Participants introduced themselves. One of the lead researchers then presented an overview of Boston's transportation

challenges, explained the workshop's purpose, and introduced key elements of BRT.

- Exploration using tangible tools (60 min total). Participants were divided into three groups. Each group started at one station and then rotated through the other stations in 20-min shifts. A member of the research team and a community facilitator were at the stations.
- Debrief discussion (30 min). All participants reconvened for a facilitated discussion of the experience. Participants sat in a circle, in an informal discussion facilitated by one of the researchers using open-ended questions to the participants.
- Post-workshop survey. Before leaving, participants were asked to fill out a post-workshop survey, which included space for open-ended feedback.

Data Collection. Data were collected through the pre- and post-workshop surveys, verbal feedback from the debrief session, and observations and transcriptions of audiovisual recordings. The pre-workshop survey collected basic demographics and measures of previous familiarity with digital media, public meetings, and BRT. The post-workshop survey included questions about learning and specific evaluations of the tools.

Participant Characteristics. A total of 51 individuals participated, from a range of advocacy, policy, neighborhood, and professional organizations. Six participants did not complete enough of the surveys for them to be useful for analysis.

Participants ranged in age from under 21 to over 75, although most were between 21 and 54 years old. Participants reported having a diverse range of professions, with 10% identifying as students and about 20% identifying as an engineer, architect, planner, transportation professional, or consultant. Based on these responses, 13 were classified as transportation "experts." Fifteen participants reported attending no public planning meetings within the preceding year. Almost three quarters answered yes to the question, "Are you familiar with the concept of bus rapid transit (BRT)?"

To categorize a participant as a "skeptic," the number of planning meetings they had attended in the preceding year was compared with the number of these meetings at which they declared learning something about the project in question. The difference between the former and the latter represented a measure of skepticism about learning in planning meetings. The nine participants who reported attending more total meetings than meetings at which they learned something were classified as "skeptics."

Findings

Mixed methods were used to analyze the workshops and the tools in an attempt to infer impacts, across participant types, on social learning, perspectives regarding transportation options, and communication. This section reports on participants' overall opinions of workshops, participants' learning in various dimensions, indicators of potential change in user behavior and attitude, and the comparative performance of the three tools used in the workshops.

Overall Learning

Single-loop learning is measured with a five-point Likert scale of (dis-)agreement with the statement: "I learned a great deal in the workshop." Almost three quarters of participants agreed or strongly agreed with this statement (Figure 4). There was also general agreement about social learning. More than three quarters of participants agreed or strongly agreed that they learned through observing others using the tools, learned through listening and conversing, and had helpful conversations. On the other hand, less than half agreed that they helped others learn.

Following Goodspeed, four survey questions were used to represent constructs (in parentheses below) important to enabling double-loop learning (*31*):

- I was able to get answers to the questions I asked. (Evidence seeking.)
- Workshop participants discussed issues in an open way. (Valid information.)
- Participants were open to differences in opinion. (Free and informed choice.)
- I would support recommendations created by the participants of the workshop. (Internal commitment to choice.)

The general agreement with these four statements (Figure 4) suggests that workshop conditions were conducive to double-loop learning.

Learning by Participant Type

Differences in reported learning were examined along two dimensions of individuals' past experiences: (1) whether one was a transportation planning "expert" and (2) whether one was a "skeptic" about public planning meetings, as described above. In general, non-experts reported more single-loop learning than experts. Non-skeptics also reported more single-loop learning than skeptics, and several skeptics reported that they did not learn anything in the workshops.

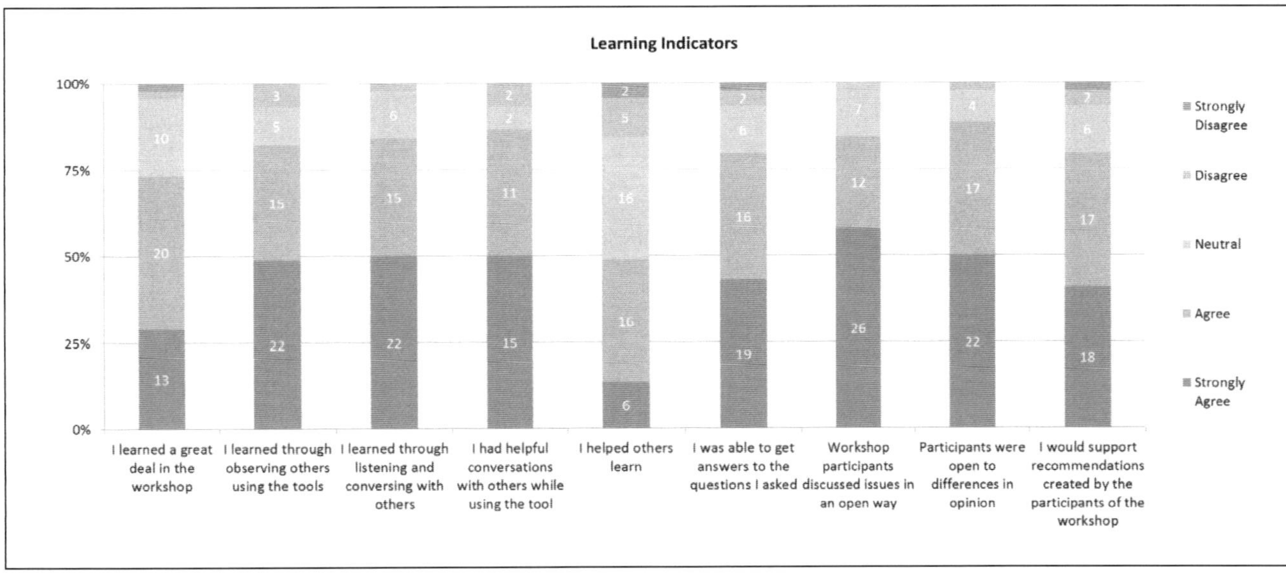

Figure 4. Learning indicators.

Substantive Learning

Because the workshops were organized around the topic of BRT, learning outcomes specific to BRT were also assessed, using responses to a question asking participants to list up to four elements of BRT in both the pre- and post-test surveys. Platform-level boarding, off-board fare collection, and dedicated bus lanes were mentioned more frequently in the post-test survey. This provides some evidence of learning about basic physical elements associated with successful BRT.

Metrics to quantify this substantive learning were developed using the Institute for Transportation and Development Policy's BRT Standard Scorecard, which includes about thirty corridor design elements that improve operations and service quality (*32*). The Scorecard's maximum possible score for each element was used to weight the elements identified by each participant in the pre- and post-test surveys. For example, if a participant listed "Off-board fare collection, designated bus lane, median busway, signal priority" in the post-test survey, his post-test BRT knowledge would be calculated as 8 (pre-board payment) + 8 (dedicated lane) + 8 (median alignment) + 7 (intersection treatment) = 31. Any difference when comparing total scores between the pre- and post-test surveys indicates learning about the elements of BRT. By this method, participants' scores increased by an average of nearly seven points per person, with over 60% of participants' responses indicating some amount of BRT knowledge gain. To illustrate the main BRT design elements "learned," Figure 5 shows the changes, between the pre- and post-workshop surveys, in the five design elements most commonly mentioned by participants in the post-workshop survey.

Impact of Learning on Attitudes and Intentions

To measure whether the workshops affected participants' attitudes and intentions, responses to three set questions asked in both surveys were compared. One question was the participant's degree of agreement with a statement related to agency: "I can play an active role in the planning of the community where I live." A second statement concerned the participant's belief in the impact of public involvement: "Public participation in planning advances the interests of my community." Finally, to gauge possible impacts learning might have on eventual behavior change – increased use of public transit – each participant's pre-test response to "In the last week, how many times did you travel by: car; subway/train; bus?" was compared with their post-test response to "If corridors like we discussed today are implemented, do you imagine yourself changing the way you travel? If so, how many times per week would you travel by: car; subway/train; bus?"

In each of the above three cases, the number of negative, zero, and positive changes was tallied. For the beliefs in agency and impact, nearly half of responses indicated no change, with the other half split almost equally across positive and negative changes. In terms of possible changes in future behavior, over half indicated they would increase their frequency of transit use if a BRT-type system were implemented in Boston. Of course, such stated intentions do not guarantee behavior would actually change.

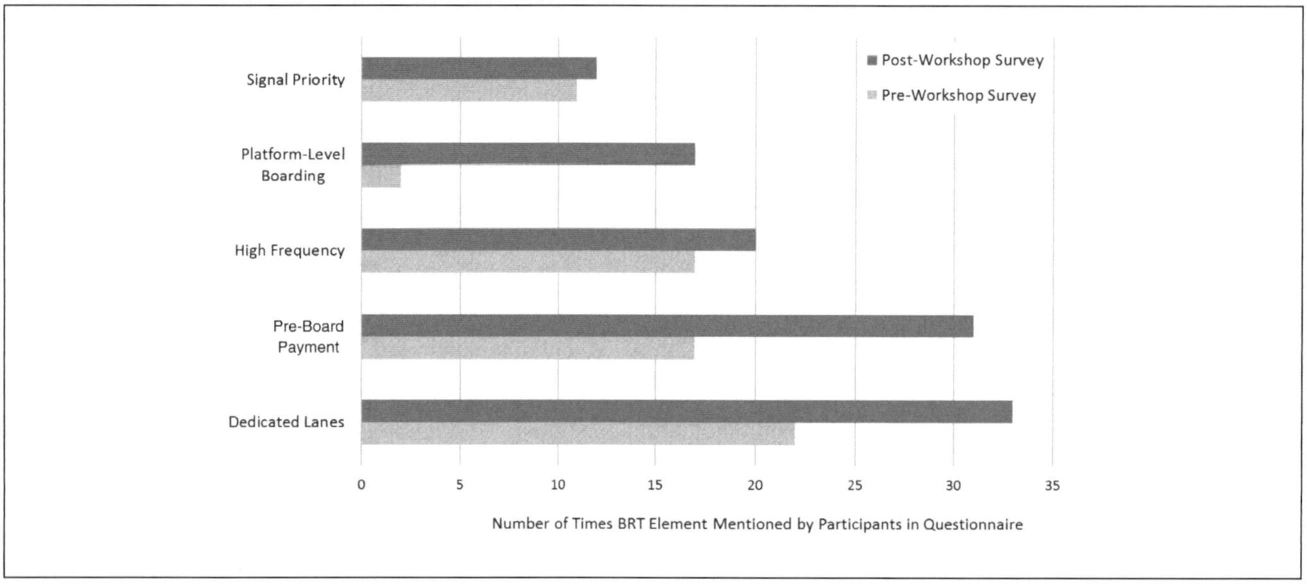

Figure 5. Changes in BRT design elements most commonly mentioned by participants.

Comparative Evaluation of Tools

To compare the tools, participants completed a series of questions for each station – street, neighborhood, and regional. Results for usability, usefulness, credibility, and relevance were generally positive, as detailed below (Figure 6).

Participants reported the street scale to be most usable. Over 50% strongly agreed that it was "easy to understand." The street-scale model was intentionally designed to represent BRT at its most concrete, human scale. Key features of BRT such as prepayment gates, elevated platforms, shelter, and dedicated lanes were physically represented and relatable at this scale, as compared to the neighborhood-scale and regional-scale models where BRT features become increasingly abstract. One participant commented "you can see the changes as you go," referring to the changes in the visualization immediately apparent upon changing the tangible tokens.

The tools were generally seen as useful for promoting discussion and teamwork. Very few participants disagreed that the tools "raised important issues for my team to discuss" and "helped my team work together." A majority indicated strong agreement with the former statement for the regional station.

Almost 90% of participants judged the regional scale as credible (scores of 4 or 5), whereas the other two tools obtained lower scores on this measure. Some participants commented on the relatability of the information presented relative to their personal experience. For example, a few participants mentioned how one could use personal commuting experience to familiar destinations as a reference for evaluating the accuracy of CoAXs' travel time map. For the neighborhood- and street-scale models, on the other hand, travel time appeared "hypothetical." Some focus group participants also pointed to the difference in breadth and depth of information across the three stations. One person, for example, attributed the credibility of the regional tool to its "limitless variations" of the travel time map in which people could input various destinations and "see all the places they want to go to," vis-à-vis the limited variations found in the street and neighborhood stations. Another pointed out that although CoAXs allowed each user to visualize their individual travel time map based on different transit service scenarios, in both neighborhood and street tools, there were "too many numbers to figure out which ones to trust" and the "tradeoffs were hard to see."

In relation to relevance, participants gave the highest scores to CoAXs (with almost 50% of responses giving it the highest score of 5). This finding may be attributable to participants' ability to place icons on the map to indicate their homes and frequent destinations as bookmarks, edit the service of a chosen corridor, and visualize accessibility from an origin of their choosing. The street and neighborhood tools did not have flexible geographic extents (zooming and panning) that enabled personalization through localization. Referring to the street scale, one participant commented, "It won't work on my street," whereas another reflected that "Every street block has different problems." Commenting on CoAXs, a participant pointed out the possibility to "see transit deserts," those parts of the city not easily accessed by transit service.

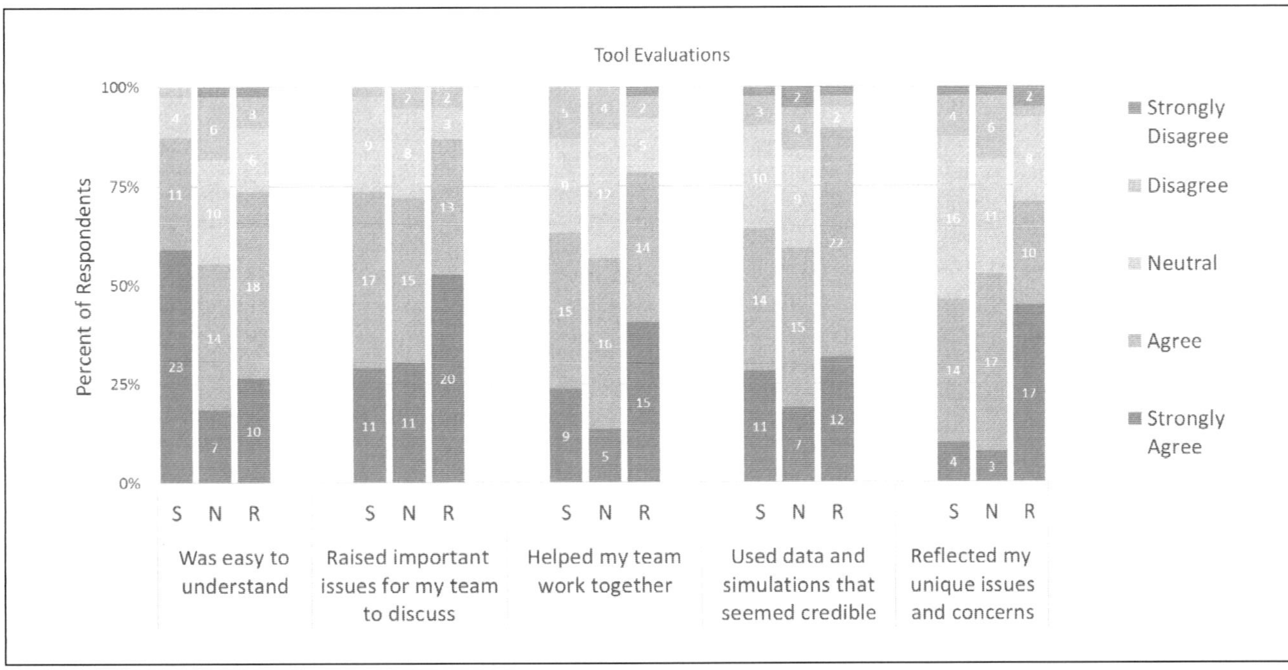

Figure 6. Comparison of tool usability, promotion of discussion and teamwork, credibility, and relevance for the street (S); neighborhood (N); and regional (R) scale tools.

Conclusions

The six workshops, held over 4 days with over 50 individuals from a range of backgrounds, reveal some evidence of learning among participants. Not only did participants report learning "a great deal," but they also showed the potential for more in-depth, longer-term double-loop learning. Participation seemed to generate substantive knowledge about BRT among the participants, although this effect dominated among "non-expert" participants, who generally had a lower baseline level of familiarity with BRT features. The evidence collected during the workshops hints at some mechanisms through which the tools contribute to learning, such as enabling users to interact with the tools and converse with others, question the tools and their assumptions, and relate easily to other points of view (*25*). Among the three different tools used in the workshop, participants judged the street-scale one to be the easiest to use, whereas they judged the regional-scale touchscreen map to be the most relevant and credible.

In the end, findings from this analysis should be viewed as suggestive. Assessing the impacts of interactive group activities, in which openness and local relevance are fundamental, is complex and difficult. Participants, recruited through various channels by community organizations, were not necessarily representative of the population at large. The workshops themselves evolved over time and were not consistently implemented; for example, flow, duration, and facilitation varied between the different sessions. Some of the interactions may be a result of the small group format (6 to 10 individuals) of each workshop. The lack of a control group hinders any conclusions about how such learning might have occurred in other ways and through other formats.

The three tools and their implementation had strengths and weaknesses. In one of the sessions, for example, CoAXs' basemaps failed, hindering user interaction. The LEGO-based models suffered from somewhat limited actual possibilities for "interacting" and "playing"– only a few tokens could be changed, constraining creativity. Overall, the 2-h duration of the workshops inevitably limited the depth of possible exploration and collaboration.

These various limitations and shortcomings point toward areas for future work. If these tools are to be used in the future, together or separately, questions immediately arise about scalability and portability. Building LEGO-based models to represent a large number of neighborhoods and bringing those models to such neighborhoods is not a minor undertaking. CoAXs can be relatively easily adapted to other places (with the right data available) and viewed by anyone with a web browser and an Internet connection; however, presumably much of the tool's value lies in inducing group interactions and common understanding while gathered around the touchscreen table. Additional research would be needed to better understand how these tools work together, or with other tools, to enhance stakeholder engagement for planning purposes. These innovative tools could be compared with more traditional media in terms of cost and effectiveness in achieving consensus. An ambitious longer-term research effort could also compare how tools affect the evolution and actual implementation of projects, though the difficulty of controlling for wider context would complicate such comparison. The tools may also have applications beyond community-based planning workshops, in, for example, the service planning work of public transit operators or to help train people about travel alternatives (e.g., taking the bus instead of driving). Ultimately, the workshops represent initial small steps toward leveraging new interactive tools for more inclusive and meaningful engagement in planning.

Acknowledgments

Facilitators included several members from *Nuestra Communidad.* Funding for the project was provided by the Barr Foundation, with additional support from the Volvo Research and Education Foundation through the BRT Centre of Excellence; the workshop space was provided in partnership with the Roxbury Innovation Center. Our participants made this research possible, and their insights and patience are appreciated. Open-source software is at the core of this research, including SUMO (DLR Institute of Transportation Systems); in their roles at Conveyal, Matt Conway, Trevor Gerhardt, Kevin Webb provided support for the open-source Analysis platform. Thanks also to Mary Skelton Roberts (Barr Foundation); Kent Larson, Ryan Chin, Ariel Noyman, Ira Winder (MIT Media Lab); Kuan Butts (Calthorpe Analytics); Allentza Michel (Powerful Pathways); and Deborah Perrotta and Marcos Paulo Schlickmann (MIT Portugal Program).

Author Contributions

The authors confirm contribution to the paper as follows: study conception and design: AFS, PCZ, PT, JR; data collection: PT, AFS, PCZ, JR; analysis and interpretation of results: PT, AFS, PCZ; draft manuscript preparation: PCZ, AFS, JR. All authors reviewed the results and approved the final version of the manuscript.

References

1. Napolitan, F., and P. C. Zegras. Shifting Urban Priorities? The Removal of Inner City Freeways in the United States. *Transportation Research Record: Journal of the Transportation Research Board*, 2008. 2046: 68–75.
2. Arnstein, S. A Ladder of Citizen Participation. *JAIP*, Vol. 35, No. 4, 1969, pp. 216–224.
3. Lane, M. B. Public Participation in Planning: An Intellectual History. *Australian Geographer*, Vol. 36, No. 3, 2005, pp. 283–299.
4. Innes, J. E., and D. E. Booher. Reframing Public Participation: Strategies for the 21st Century. *Planning Theory and Practice*, Vol. 5, No. 4, 2004, pp. 419–436.
5. Hollands, R. G. Will the Real Smart City Please Stand Up? *City*, Vol. 12, No. 3, 2008, pp. 303–320.
6. Allwinkle, S., and P. Cruickshank. Creating Smart-Er Cities: An Overview. *Journal of Urban Technology*, Vol. 18, No. 2, 2011, pp. 1–16.
7. Janssen, M., Y. Charalabidis, and A. Zuiderwijk. Benefits, Adoption Barriers and Myths of Open Data and Open Government. *Information Systems Management*, Vol. 29, No. 4, 2012, pp. 258–268.
8. Rojas, F. M. *Transit Transparency: Effective Disclosure through Open Data.* Taubman Center for State and Local Government, Harvard Kennedy School, Cambridge, MA, 2012.
9. Eros, E., S. Mehndiratta, C. Zegras, K. Webb, and M. C. Ochoa. Applying the General Transit Feed Specification to the Global South. *Transportation Research Record: Journal of the Transportation Research Board*, 2014. 2442: 44–52.
10. Weiner, E. *Urban Transportation Planning in the United States: An Historical Overview. Revised Edition.* Washington D.C., 1992.
11. Beimborn, E., R. Kennedy, and W. Schaefer. *Inside the Blackbox: Making Transportation Models Work For Livable Communities.* Citizens for a Better Environment and The Environmental Defense Fund, New York City, NY, 1996.
12. Batty, M. Planning Support Systems: Progress, Predictions, and Speculations on the Shape of Things to Come. In *Planning Support Systems for Cities and Regions* (Brail, R. K., ed.), Lincoln Institute of Land Policy, Cambridge, MA, pp. 3–30.
13. Talen, E. Bottom-Up GIS. *Journal of the American Planning Association*, Vol. 66, No. 3, 2000, pp. 279–294.
14. Chen, T., and A. Kratky. Touching Buildings – A Tangible Interface for Architecture Visualization BT - Universal Access in Human-Computer Interaction. In *Proc., Part I. Design Methods, Tools, and Interaction Techniques for eInclusion: 7th International Conference, UAHCI 2013, Held as Part of HCI International 2013* (Stephanidis, C., and M. Antona, eds.), Las Vegas, NV, USA, 21–26 July 2013, Springer, Berlin, Heidelberg, Germany, pp. 313–322.
15. Ben-Joseph, E., H. Ishii, J. Underkoffler, B. Piper, and L. Yeung. Urban Simulation and the Luminous Planning Table: Bridging the Gap between the Digital and the Tangible. *Journal of Planning Education and Research*, Vol. 21, 2001, pp. 196–203.
16. Maher, M. Lou, and M. J. Kim. Do Tangible User Interfaces Impact Spatial Cognition in Collaborative Design? In *Proc., Second International Conference, BT - Cooperative Design, Visualization, and Engineering: CDVE 2005* (Luo, Y., ed.), Palma de Mallorca, Spain, September 18–21, 2005. Springer, Berlin, Heidelberg, Germany, pp. 30–41.
17. Goodspeed, R. *Planning Support Systems for Spatial Planning Through Social Learning.* PhD dissertation. Department of Urban Studies and Planning, Massachusetts Institute of Technology, Cambridge, MA, 2013.
18. Argyris, C. Single-Loop and Double-Loop Models in Research on Decision Making. *Administrative Science Quarterly*, Vol. 21, No. 3, 1976, pp. 363–375.

19. Gakenheimer, R. A. *Transportation Planning as Response to Controversy: The Boston Case*. MIT Press, Cambridge, MA, 1976.
20. Stewart, A. F., J. P. Attanucci, and N. H. M. Wilson. Ridership Response to Incremental Bus Rapid Transit Upgrades in North America. *Transportation Research Record: Journal of the Transportation Research Board*, 2015. 2538: 37–43.
21. Weinstock, A., W. Hook, M. Replogle, and R. Cruz. *Recapturing Global Leadership in Bus Rapid Transit*. Institute for Transportation and Development Policy, New York City, NY, 2011.
22. MassDOT. *Roxbury-Dorchester-Mattapan Transit Needs Study*. Boston, MA, 2012.
23. Alrashed, T., A. Almalki, S. Aldawood, T. Alhindi, I. Winder, A. Noyman, A. Alfaris, and A. Alwabil. An Observational Study of Usability in Collaborative Tangible Interfaces for Complex Planning Systems. *Procedia Manufacturing*, Vol. 3, No. Ahfe, 2015, pp. 1974–1980.
24. Stewart, A. F., and P. C. Zegras. CoAXs: A Collaborative Accessibility-Based Stakeholder Engagement System for Communicating Transport Impacts. *Research in Transportation Economics*, Vol. 59, 2016, pp. 423–433.
25. Stewart, A. F. Mapping Transit Accessibility: Possibilities for Public Participation. *Transportation Research Part A*, Vol. 104, 2017, pp. 150–166.
26. SUMO. *Simulation of Urban MObility*. dlr.de/ts/sumo.
27. Conway, M. W., A. Byrd, and M. van der Linden. Evidence-Based Transit and Land Use Sketch Planning Using Interactive Accessibility Methods on Combined Schedule and Headway-Based Networks. *Transportation Research Record: Journal of the Transportation Research Board*, 2017. 2653: 45–53.
28. Handy, S., and D. Niemeier. Measuring Accessibility: An Exploration of Issues and Alternatives. *Environment and Planning A*, Vol. 29, No. 7, 1997, pp. 1175–1194.
29. Metz, D. The Myth of Travel Time Saving. *Transport Reviews*, Vol. 28, No. 3, 2008, pp. 321–336.
30. Zegras, C. Mainstreaming Sustainable Urban Transport: Putting the Pieces Together. In *Urban Transport in the Developing World: A Handbook of Policy and Practice* (Dimitriou, H., and R. Gakenheimer, eds.), Edward Elgar, Cheltenham, UK; Northhampton, MA, 2011, pp. 548–588.
31. Goodspeed, R. Sketching and Learning: A Planning Support System Field Study. *Environment and Planning B: Planning and Design*, Vol. 43, No. 3, 2015, pp. 444–463.
32. Institute for Transportation and Development Policy. *The BRT Standard*, 2014 ed. Institute for Transportation and Development Policy, New York City, NY, 2014.

The Standing Committee on Public Transportation Planning and Development (AP025) peer-reviewed this paper (18-05072).

Article

TRR
JOURNAL OF THE TRANSPORTATION RESEARCH BOARD

Transportation Research Record
2018, Vol. 2672(8) 796–806

DOI: 10.1177/0361198118796067
journals.sagepub.com/home/trr
SAGE

Segmenting the Market for New Modes using Stated and Revealed Preferences

Christoffel J. Venter[1]

Abstract

The paper describes a new method for segmenting the transport market on the basis of choice set heterogeneity, using a combination of revealed preference (RP) and stated preference (SP) data obtained from user surveys. The method refines previous definitions of mode captivity by differentiating between current automobile captives whose resistance to transit is enduring (due to lifestyle preferences and constraints), and those whose captivity is transient (i.e., they are willing to consider switching to transit if a suitably attractive new alternative is offered in the SP game). We test the methodology using data from Johannesburg, South Africa. By segmenting the market on this basis before estimating a mode choice model, the model fit is improved as compared with the conventional segmentation technique that defines captivity solely on automobile ownership and access to transit variables. It also delivers useful insights into preference heterogeneity between the captivity groups, which is especially helpful when planning for a new mode or service, as present patterns of subjective captivity may change when new options become available.

Market heterogeneity is an important issue in travel demand prediction. Understanding variation in the population in terms of tastes, attitudes and behavior helps not only to model choice behavior more accurately, but also to design services and modes to match the expectations of different users better (*1*, *2*). It is also important for project and policy evaluation, as ignoring heterogeneity could bias willingness to pay estimates and underestimate the welfare impacts derived from travel time reductions (*1*, *3*).

A large body of research has developed around analytical methods of dealing with heterogeneity. They can broadly be grouped into two groups depending on whether they deal with heterogeneity in preferences or in choice sets. Preference heterogeneity refers to variations in the preference structure across individuals or groups, and is typically expressed in terms of variations in the coefficients and/or variables present in the utility functions. Random parameter models allow coefficients to vary across individuals according to some distribution, and are commonly estimated using the mixed logit model (e.g., *2*, *4*). Alternatively, the population may be segmented into discrete sub-groups and different coefficients estimated for each. Segmentation may be done either a priori using socio-economic criteria (e.g., by income group), or during model estimation using endogenously defined and estimated segments (so-called latent class models [see *5*]). Attitudinal data have proven useful in helping identify latent classes through the use of structural equations (*6*, *7*) or hybrid choice models (e.g., *8*).

Choice set heterogeneity refers to situations where the set of alternatives considered by individuals varies across the population. Correct identification of choice sets is important during demand forecasting, since it is known that misspecification of the choice set can lead to biased and inconsistent estimates of demand (*9*, *10*). Market segmentation by choice set is commonly applied in transit services planning (*11*), typically by drawing a distinction between *captive* and *choice* passengers. Transit captives are limited to the use of one or more transit modes, typically because they do not have a driver's license or do not own a car (*9*); they are also sometimes called "transit dependent" (*12*). Automobile captives, on the other hand, are those who feel they have no other option but to use their car, due for instance to a lack of suitable transit services. Apart from modal availability, numerous personal and household constraints may contribute to transit or auto captivity, including disability, fear of crime, intra-household roles, and affordability (*13*–*15*). Travelers who

[1]Centre for Transport Development, Department of Civil Engineering, University of Pretoria, Pretoria, South Africa

Corresponding Author:
Address correspondence to Christoffel J. Venter: christo.venter@up.ac.za

are neither transit nor auto captive are typically defined as choice or discretionary passengers: they have various alternatives but choose their mode because they view it as superior to other options (*9*).

Captives comprise the majority of transit users worldwide; in the US this figure is about 70% on average (*12*). Captivity groups tend to differ demographically. Transit captivity, for instance, is more likely among people with low incomes, non-drivers, people with disabilities, high school and college students, and elderly people (*16*). Auto captives tend to be more heterogeneous than transit captives (*17*), comprising people of all income bands.

Segmentation by choice set often provides a useful a priori classification scheme for examining preference heterogeneity. Transit captive passengers have been found to be less sensitive to either price or service quality than choice riders (*16*, *18*), suggesting that transit improvements might be more effective at attracting new ridership among choice than captive populations. Thus transit agencies tend to direct efforts to attract new users at choice passenger segments, assuming auto captives are unreachable and that transit captives will continue to use their services in any case. Market segmentation has thus become a useful strategy for increasing transit ridership in the presence of market heterogeneity (*11*).

The interest of this paper lies with choice set based segmentation, particularly in the presence of new modes. We develop and test a refined method for undertaking a priori market segmentation, using a combination of revealed preference (RP) and stated preference (SP) data reflecting commuters' willingness to use transit under realistic travel conditions. The method responds to the shortcomings of existing market segmentation techniques which use either arbitrary deterministic rules (e.g., that all households without cars are transit captive), or purely subjective attitudinal responses. We refine the definition of automobile captives to distinguish between those who are persistent car users due to lifestyle constraints or immutable preferences (i.e., 'true' captives), and those whose captivity is transient due to the present unavailability of suitable transit, but who might become future choice users should sufficiently attractive transit options become available.

The method is tested in the context of Bus Rapid Transit (BRT) development in the city of Johannesburg, South Africa. We demonstrate that it provides a meaningful treatment of heterogeneity in the context of assessing the potential future market for the new service. Johannesburg's BRT is seen as underperforming in terms of passenger attraction and cost recovery (*19*). We argue that the new insights that are generated into the differential needs of various market segments could help to better target each segment and improve the overall patronage and performance of the system.

At a policy level, issues of captivity are particularly salient in the Global South, where transit captives comprise large proportions of ridership due to low automobile ownership levels. Captivity to transit and to non-motorized modes is of specific policy concern because it is often associated with high levels of transport deprivation and inequality (*20*), with poorer communities enjoying unequal access to affordable transport due to a combination of insufficient transit and spatial marginalization on the peripheries of cities. Reducing the extent or nature of captivity is thus sometimes seen as a goal of transport policy. The objective of this research is to develop a practical yet relevant way of measuring and, ultimately, predicting captivity in response to policy or system interventions.

This paper first presents a brief overview of current approaches to define market segments based on choice set heterogeneity, and then describes a method for using SP and RP responses to achieve this. Then the data and segmentation results as applied in the case study area are presented, followed by the results of a mixed logit estimation to test the impact of the segmentation on resulting mode choice models. Lastly conclusions are drawn on the applicability of the method to other cases.

Background: Choice Set Heterogeneity and Transit Captivity

Manski (*21*) formulated the choice problem as a two-stage process: in the first stage, alternatives are screened by some non-compensatory process (such as elimination-by-aspects) to construct a choice set; in the second, a choice is made using a compensatory process:

$$P_i = \sum_{C \in \Gamma} P(i|C)Q(C) \qquad (1)$$

The probability of choosing alternative i (P_i) is the product of $Q(C)$, the probability that the choice set is C (a subset of the "universal choice set" Γ), and $P(i|C)$, the conditional probability of choosing alternative i, if the choice set is C, which is typically a function of a set of utility equations that depend on a vector of estimable parameters.

The manner in which the function $Q(C)$ is defined determines the treatment of choice set generation and captivity in the model. There is ample evidence that assuming that $C = \Gamma$ (i.e., the universal choice set applies to all individuals), when in fact some individuals face restricted choice sets, leads to biased models and erroneous forecasting, particularly where changes in transport infrastructure or socio-demographics over time are expected to change the choice sets that decision-makers face (*22*, *23*).

The most common approach to modelling choice set generation is to reduce $Q(C)$ to a set of binary (0/1) indicators, related solely to socio-economic and spatial variables designed to capture the availability of specific modes to a decision-maker. It is common, for instance, to define individuals without access to cars as transit captive, and individuals with a car available but located further than a quarter mile (assumed to be a reasonable walking distance) from a transit stop as car captive (e.g., *9, 24*). We term this class of captivity models the rule-based approach.

The problem with the rule-based method is that deterministic rules are unlikely to capture real variations in captivity over time and from person to person (*25, 26*); nor do they include subjective factors such as lack of knowledge about alternatives, usability and security (*9*). In fact, Ben-Akiva and Boccara (*5*) argue that choice sets are actually unobserved and can never be computed with certainty on the basis of observable data.

A more suitable approach might thus be to treat choice set formation probabilistically, by estimating a parametrized function for $Q(C)$ that returns the probability that a certain choice set C is the actual choice set of an individual. Various formulations have been proposed and tested, including the Dogit (*27*) and logit-based models with latent choice sets (*26, 28*). The main difficulty with this approach is its combinatorial nature: the number of choice sets to be evaluated grows exponentially as the choice set space grows (*29*), making probabilistic choice set models difficult and time-consuming to estimate. As an alternative, single-stage models combine choice set generation and mode choice into a single model (e.g., *5, 29*). A variant of this approach integrates mode choice with vehicle ownership models (e.g., *30, 31*), such that the decision of whether to own a vehicle or not determines whether a vehicle is available during the mode choice.

Despite the theoretical appeal of probabilistic approaches to choice set prediction, their computational and data demands are such that they are not commonly used in practice. Predicting choice sets on the basis of a priori segmentation of the market remains a popular approach, and some research has gone into devising more context-sensitive segmentation strategies using a wider set of attitudinal and travel behavior variables.

Srinivasan et al. (*32*) add a category called semi-captive auto users, defined as individuals living in households with vehicles, but without drivers' licenses. Wilson et al. (*33*) expand both transit and private vehicle users into four categories, namely functional captive mode users, marginal captive mode users, marginal choice mode users and free choice users. Krizek and El-Geneidy (*17*) differentiate between car users who are auto captive and potential riders who might switch to transit in future. They also differentiate between regular and irregular users in each segment, arguing that the regularity of mode use affects the likelihood of switching to other modes. Jacques et al. (*34*) use information on user characteristics, satisfaction with current mode, and service variables in a cluster analysis technique to cluster travelers of all modes into four distinct groups based on the practicality and satisfaction derived from the mode, namely captivity, utilitarianism, dedication, and convenience. Of these, only the first are truly captive to their current modes. Recently van Lierop and El-Geneidy (*35*) identified a group called "captive by choice" to describe transit captives who choose not to have access to a car, rather than not being able to afford it. The work draws attention to the fact that what we think of as "captivity" may be due to a variety of reasons, including strong affective preference for a particular mode for environmental reasons.

What these approaches have in common is that they acknowledge that captivity is neither monolithic nor static. Market segments change; even captive users potentially have a choice in the long run if their situation changes (*32*). Some current car captives might become choice passengers if transit services expand; others will remain 'true' captives for whom transit is never subjectively an option (*9, 17*). Captivity is context dependent: the same person might be a choice passenger for the trip to work in town, but captive to the car for the evening trip to the theatre. While Krizek and El-Geneidy's (*17*) distinction between regular and irregular users implies as much, most captivity studies have focused on commute trips only.

The addition of subjective indicators of a respondent's willingness to consider currently unused modes is a clear advance on deterministic rule-based approaches using socio-economic and transit availability variables only. However, the way in which willingness to consider is measured is still problematic, as it elicits a general attitude rather than a context-specific response. For instance, Krizek and El-Geneidy (*17*) use responses to the question "How appealing, overall, is the idea of using the bus?" to help distinguish between auto captive and potential future transit users. Given the importance of mode-specific quality and trip-specific service variables to the question of whether an alternative is subjectively considered as a viable alternative, we argue that it is necessary to be more specific in the description of the alternative to elicit a realistic response. This is particularly important in the case of new modes, where users might have no experience on which to base such an answer.

This paper proposes an alternative to relying on deterministic rules or latent attitudinal variables for identifying choice-set based market segments. Instead, we make use of the actual and hypothetical mode choice made by the traveler in an actual choice situation. Stated and revealed preference surveys present a practical, intuitive method for collecting such choice data that can be tailored to local conditions.

Market Segmentation using Revealed and Stated Preferences

The proposed method depends on the collection of revealed preference (RP) and stated preference (SP) data, which often forms part of the planning for new modes or services. Although such data is suitable for estimating a wide range of models to deal with preference and choice set heterogeneity as described above, our interest lies in using it as an improved a priori market segmentation technique.

For this study four market segments were defined:

- Car captives: Trip makers with only the car mode available for the current trip. Following Krizek and El-Geneidy (*17*), car captives were further subdivided into two groups:
  - Persistent car captives: Car users whose captivity is due to personal, life cycle or activity-related factors—for instance, a worker who needs their car at work every day, or a parent whose trip patterns are too complex to undertake with public transport (called 'auto captives' by Krizek and El-Geneidy [*17*]); and
  - Transient car captives: Car users whose captivity is due to the current unavailability of alternatives, but who might be willing to switch to public transport in future, should an acceptable option become available (called 'potential riders' by Krizek and El-Geneidy [*17*]);
- Public transport captives: Trip makers with one or more public transport option but no car available for the current trip;
- Choosers: People with both a car and at least one public transport option available for their trip.

The distinction between persistent and transient car captives is important as it captures the level of persistence of car captivity: while the former are presumed insensitive to transit interventions, people in the latter category might become choosers if their set of options expands.

Revealed and stated preferences were used to identify these market segments, as follows. Firstly, the respondent completed an RP questionnaire in which details regarding a specific current trip (called the reference trip) were recorded, together with information regarding whether a feasible alternative mode existed for this trip, and, if so, its service data, such as expected costs and times. Then, a set of SP questions was constructed in which a specific hypothetical public transport option was offered as an alternative to the current mode for the reference trip, and the respondent's willingness to switch to the new alternative was recorded. At least one replication in the SP game contained a superior alternative in which all the hypothetical attributes were set at the best levels feasible. Market segments are then identified as follows:

- Car captives: Trip makers who chose the car for their current trip, and who reported no non-car alternatives in their RP survey.
  - Persistent car captives: Car captives who were non-traders in their SP survey (i.e., they declined to choose the transit option in any of the SP games offered, even the superior option).
  - Transient car captives: Car captives who were willing to switch to the transit option in at least one hypothetical SP game.
- Public transport captives: Current transit users who reported no car alternatives in their RP survey.
- Choosers: Current transit users with a car alternative, or current car users with a transit alternative reported in their RP survey.

Strictly speaking, market segmentation on this basis requires the SP games to be applied to car users only, as the rest of the segments can be identified from RP data alone. However execution of the SP survey among a sample of all users provides rich data for mode choice modelling as shown later in the paper.

By explicitly linking market segmentation to choice survey responses, we attempt to preserve external validity in at least three ways. Firstly, captivity—the absence of alternative modes—is implicitly self-identified rather than imputed by the analyst, making it more sensitive to the individual's (latent) perceptions and abilities. Secondly, captivity is defined for a *particular trip* made by the respondent, instead of being an unchangeable characteristic of the individual. The use of the reference RP trip, and pivoting the SP alternatives around this trip's characteristics, is a useful way of guaranteeing relevance of the SP responses and reducing hypothetical bias (*36*, *37*). For instance, if a car user carried luggage during their reference trip, this might influence their choice set by excluding hypothetical transit alternatives offered in the SP games. The impacts of both objective environmental factors and latent subjective attitudes are thus endogenized in the SP response. Thirdly, captivity is defined relative to a *particular transit option* with explicitly specified characteristics offered in the SP games. This is in line with an understanding of captivity as contingent on the specific alternatives on offer, rather than as a general, contextless characteristic of the decision maker.

Table 1. Sample Composition, City of Johannesburg Surveys, 2014

	Number (%) of interviews
Monthly household income	
0 – R2500 (low)	100 (8.3%)
R2501 – R8000 (medium)	469 (38.8%)
R8000 and more (high)	447 (37.0%)
Refused/unknown	192 (15.9%)
Current mode used (reference trip)	
Car (driver or passenger)	352 (29.1%)
Minibus-taxi	300 (24.8%)
Bus Rapid Transit	254 (21.0%)
Other bus	52 (4.3%)
Metrorail (commuter rail)	200 (16.5%)
Gautrain (premium rail)	50 (4.1%)
Trip purpose (reference trip)	
Work	762 (63.1%)
Non-work	446 (36.9%)
All trips	1,208 (100%)

Table 2. Attributes and Levels Tested in SP Experiment

Attribute	Levels
Mode constant	Car, Gautrain, Taxi, Bus, BRT, Train
Number of transfers (PT only)	No transfers; 1 transfer
Travel cost	current −30%; current; current +20%
In-vehicle travel time	current −25%; current; current +25%
Walk time to PT	5 mins; 10 mins; 30 mins
Wait time for PT	5 mins; 10 mins; 20 mins
Seat availability	Seat not available on BRT bus Seat is available on BRT bus

The following section demonstrates the operationalization of the segmentation approach using a real-world dataset.

Data

The data draws from a RP and SP survey conducted in 2014 in the City of Johannesburg. Johannesburg is South Africa's largest city, with a population of 4.4 million people of which about 60% use public transport. The City is currently rolling out a Bus Rapid Transit (BRT) system called *Rea Vaya*, of which two trunk corridors totaling 25 km and 16 km in length were operational in 2014. BRT ridership is relatively meagre, at about 40,000 passenger trips per day. Other public transport (PT) modes in the city include informal minibus-taxi paratransit (carrying about 70% of public transport trips), regular bus (9%), commuter rail (14%), and a new premium rail service called Gautrain (1%). About 30% of travelers walk to work and school (*38*).

The surveys were fashioned around understanding the choice behavior of potential users of BRT as a new mode. This was driven not so much by BRT's current patronage as by the key role this mode is expected to play in the future transformation of public transport in Johannesburg (*38*).

The surveys targeted 1,208 travelers selected using a stratified random sampling approach to ensure adequate coverage of the six main modes in Johannesburg (Table 1). Respondents were randomly selected at transit terminals (for the PT modes) and petrol stations and shopping malls (for car users), and surveyed using face-to-face computer-aided (CAPI) interviews. Tablet computers were used to capture responses and to dynamically generate SP questions. The RP section of the survey instructed respondents to provide information about a recent reference trip – either work or non-work – as well as one alternative (but unused) mode that was available for the trip (if any). In each of nine SP games, the respondent was offered a BRT alternative, carefully described to be similar to the current Rea Vaya offerings, but with varying service characteristics calculated as realistic variations on the current reference trip data. Table 2 shows the attributes and levels tested across the SP games. To reduce respondent burden, only three or four attributes were presented at a time using a block design.

Results

Market Segmentation

Application of the market segmentation methodology described above produced the results shown in Table 3. Car captives comprise about 14% of the sample, split evenly between persistent and transient captives. This was about half of all car users in the sample (the remainder being choice users), suggesting that one in two car users currently do not perceive any alternative to driving. However only one in four car users is persistently captive due to lifestyle preferences and constraints.

More than half of daily trips in the sample are captive to public transport. This high number is typical of lower-income countries with much lower car ownership than in developed countries.

Looking at the characteristics of each market segment, it is clear that considerable heterogeneity exists between groups. As expected, car captives consist overwhelmingly of high-income drivers who own cars, while public transport captives are predominantly from medium and low income households. Choice passengers tend to be from medium or high income households, with public transport available within walking distance from the home. Persistent and transient car captives seem to differ in two key respects. Firstly, while about a third of persistent captives could not answer the question of how far they

Table 3. Characteristics of Market Segments (Choice-Based Segmentation)

		Percentage of trips within segment complying with description			
Type of characteristic	Description	Persistent car captive	Transient car captive	Public transport captive	Choice
Mode availability	Public transport available within 10 minutes' walk	54%	45%	83%	84%
	Don't know how far to PT	32%	47%	30%	16%
	Households owns a car	100%	100%	26%	95%
Household demographics	Low-income household	4%	3%	10%	7%
	Medium-income household	1%	11%	51%	22%
	High-income household	95%	85%	39%	71%
	Household with children	65%	73%	65%	64%
Trip characteristics	Trip to/from work	63%	62%	63%	65%
	Trip origin or destination in CBD	6%	15%	28%	21%
All trips	Percentage in sample	9%	9%	57%	25%

live from the nearest public transport route, this percentage climbed to almost half for transient captives. This suggests that many car captives are in fact simply uninformed or unaware of transit, but that they will become conscious choice passengers once a transit option starts to appear on their mental horizon. Secondly, connectivity is important. Only about 6% of trips made by persistent car captives start or end in the Johannesburg Central Business District (CBD), as compared with 15% and 21% of trips by transient captives and choosers respectively. Transit services in Johannesburg are overwhelmingly radially oriented toward the CBD, making it very likely that travelers with suburb-to-suburb travel patterns do not have feasible transit alternatives at present. More important, though, is that this experience conditions their future likelihood of car captivity – even if direct suburb-to-suburb transit routes are offered, many suburban car users do not see themselves considering taking transit.

But there is also considerable heterogeneity *within* groups. For instance, a full 26% of people who considered themselves public transport captives live in households that own one or more cars. These might be tripmakers who do not have access to the car due to its use by another household member at the time of travel. Similarly, about half of car captives live within 10 minutes' walk of some public transport service, but evidently do not perceive it as a viable modal alternative for their specific trip.

To compare the choice-based segmentation with conventional deterministic methods, a rule-based segmentation was applied using the following car ownership and access to transit criteria:

- Respondents living in car-owning households, and within 5 minutes' walk of the nearest bus, taxi or train station, were classified as *choice users*.
- Respondents living in car-owning households, but further than 5 minutes' walk from the nearest bus, taxi or train station, were classified as *car captive*.
- Respondents living in households that do not own vehicles were classified as *PT captive*.

Table 4 shows the results. The choice-based market segmentation produces markedly different results than conventional deterministic segmentation. Overall, about 70% of the sample is classified the same using both approaches (the sum of the bold numbers in Table 4). The rule-based method allocates a much larger share of the sample to car captives (317 vs. 170), and lower shares to the other categories.

This leads to counterintuitive results when considering the willingness among car captives (identified by a rule-based method) to switch to future transit services. When offered a new bus service available within a 5-minute walk, about 70% of car captives chose the BRT option at least once in the SP game (last column of Table 4). This is a similar percentage to other rule-based segments. This is intuitively unappealing, as one would expect a larger proportion of current car captives to resist transit options due to lifestyle factors. The choice-based segmentation, not surprisingly, picks up the preference heterogeneity across segments much better, with only 52% of current car captives willing to switch to BRT as compared with 80% of choice users. This finding supports previous research indicating that simply allocating market segments on the basis of deterministic rules about household car ownership or transit access might lead to erroneous forecasts regarding the willingness to use alternative modes (*32*, *34*).

Segmented Mode Choice Models

The following section investigates the ability of choice-based segmentation to meaningfully distinguish between market segments in terms of their mode choice behavior, in the context of predicting the demand for a new transit

Table 4. Comparison of Market Segments: Choice-Based and Rule-Based Methods

		Choice-based segmentation						
		Persistent car captive	Transient car captive	Car captive (all)	Public transport captive	Choice	Total	Number (%) willing to choose BRT in SP game
Rule-based segmentation	Car captive	68	77	145	47	125	317	221 (70%)
	PT captive	0	0	0	434	13	447	315 (70%)
	Choice	13	12	25	62	103	190	141 (74%)
	Total	81	89	170	543	241	954	677
	Willing to choose BRT in SP game	0 –	89 (100%)	89 (52%)	396 (73%)	192 (80%)	677	

service. For the purposes of comparison, we estimated three models, namely:

- Model 1: Unsegmented model, without any a priori segmentation of respondents;
- Model 2: Rule-based segmentation model, using the conventional deterministic rules regarding car ownership and access to transit described in the previous section to segment the sample into three categories (car captive, PT captive, and choice); and
- Model 3: Choice-based segmentation model, using segmentation conditioned on RP/SP responses to segment the sample into four categories (persistent car captive, transient car captive, PT captive, and choice).

In each case, the combined SP and RP dataset was used in the estimation. A sequential estimation procedure was followed where the SP responses were used to estimate the coefficients of service variables like travel time and cost, while RP data were used to estimate the Alternative Specific Constants (ASC) in the utility functions (see *39*). Because of scale differences in the SP and RP data, an additional scale parameter was estimated that is used to rescale the SP coefficients relative to the ASCs.

We furthermore made provision for preference heterogeneity within each segment (or the entire dataset in the case of Model 1) by specifying a mixed logit (ML) model with the In-vehicle travel time and walk time variables as normally distributed random parameters. The ML model further allows for correlations in the errors between repeated observations for the same person, that is, panel effects stemming from the SP game (*4*).

In each model, separate coefficients were estimated for each market segment. Initial model runs for Models 2 and 3 indicated that coefficient estimates for the in-vehicle time and walk time variables did not vary significantly across segments, so they were constrained to be the same across all segments. A single ASC per mode was also estimated for each model as segment-specific constant produced poor results, probably due to the small sample sizes for some of the modes. In the case of Model 3, only two user segments were considered: public transport captives, and the combination of choice passengers and transient car captives. Since persistent car captives consistently chose the car mode in all RP and SP scenarios, their data contains no information on choice and no model could be estimated for them.

The results are given in Table 5. All models are highly significant and all coefficients are of the expected sign, while almost all are significant at the 99% level. The significant standard deviation coefficients indicate that the selected in-vehicle time and walk time parameters are indeed random, and that preference heterogeneity does exist. More importantly, the other coefficients for Models 2 and 3 vary significantly across segments, indicating that the segmentation employed generally distinguishes between groups with different preference structures.

Model 1 has the lowest log-likelihood value, suggesting that the two segmented models both outperform the unsegmented model. Market segmentation improves the model's ability to explain consumer choices, which is as expected as it makes more segment-specific variables available for doing so. More importantly, the log-likelihood value improves most from Model 2 to Model 3, suggesting that the proposed segmentation using observed RP/SP choices is an improvement on the deterministic rule-based approach. This is despite the fact that Model 3 is estimated on less data, as persistent car captives are excluded from the estimation. But the remaining data are more homogeneous, leading to better model performance.

It is easier to observe variations in preference structures by calculating willingness-to-pay (WTP) values for each model and segment. Table 6 shows the values, indexed against the value of travel time savings (VTTS) for in-vehicle time from Model 1. WTP values vary significantly across market segments. In most cases, PT

Table 5. Estimation Results: Segment-Specific Coefficients for Segmented Mode Choice Models and Comparison Model

Type of variable	Variable	Model 1 Unsegmented	Model 2 Rule-based segmentation: Car captive	Model 2 Rule-based segmentation: PT captive	Model 2 Rule-based segmentation: Choice	Model 3 Choice-based segmentation: PT captive	Model 3 Choice-based segmentation: Choice + transient car captives
Alternative specific constants	Bus	**−0.92 (−3.7)**	**−1.09 (−4.1)**	**−1.09 (−4.1)**	**−1.09 (−4.1)**	**−0.93 (−3.6)**	**−0.93 (−3.6)**
	BRT (reference category)	0.00	0.000	0.000	0.000	0.0000	0.0000
	Gautrain	+19.7 (0.0)	+19.8 (0.0)	+19.8 (0.0)	+19.8 (0.0)	+19.7 (0.0)	+19.7 (0.0)
	Minibus-taxi	**−0.96 (−6.7)**	**−0.86 (−5.9)**	**−0.86 (−5.9)**	**−0.86 (−5.9)**	**−0.92 (−6.4)**	**−0.92 (−6.4)**
	Train	+0.27 (1.2)	+0.21 (0.9)	+0.21 (0.9)	+0.21 (0.9)	+0.11 (0.5)	+0.11 (0.5)
	Car	**+1.77 (6.4)**	**+1.99 (6.9)**	**+1.99 (6.9)**	**+1.99 (6.9)**	**+1.76 (6.5)**	**+1.76 (6.5)**
Service variables[a]	Travel cost (Rands)	**−0.06 (−28.4)**	**−0.05 (−18.6)**	**−0.09 (−18.9)**	**−0.06 (−16.5)**	**−0.07 (−20.3)**	**−0.05 (−25.5)**
	In-vehicle travel time[b] (minutes)	**−0.005 (−5.7)**	**−0.006 (−6.0)**	**−0.006 (−6.0)**	**−0.006 (−6.0)**	**−0.005 (−4.9)**	**−0.005 (−4.9)**
	Walk time at start of trip[b] (minutes)	**−0.024 (−11.3)**	**−0.022 (−10.2)**	**−0.022 (−10.2)**	**−0.022 (−10.2)**	**−0.01 (−8.8)**	**−0.01 (−8.8)**
	Waiting time (minutes)	**−0.02 (−10.5)**	**−0.01 (−2.9)**	**−0.03 (−10.8)**	**−0.01 (−1.7)**	**−0.02 (−10.8)**	**−0.01 (−3.8)**
	Seat available on BRT (Yes = 1, only in BRT utility)	0.01 (0.3)	0.03 (0.7)	0.03 (0.7)	0.03 (0.7)	0.03 (0.7)	0.03 (0.7)
	Number of transfers	**−0.12 (−6.1)**	**−0.17 (−5.1)**	**−0.08 (−2.9)**	**−0.15 (−3.9)**	**−0.08 (−4.3)**	**−0.11 (−4.1)**
Standard deviation of random parameters[a]	SD of in-veh. travel time	**0.012 (5.9)**		**0.015 (7.5)**		**0.014 (8.3)**	
	SD of walk time	**0.053 (12.4)**		**0.054 (12.3)**		**0.034 (10.9)**	
Scale parameter	Scale parameter	**0.37 (8.5)**		**0.39 (9.1)**		**0.30 (8.6)**	
	No of observations	948		948 individuals		866 individuals	
	Log-likelihood	−4460.5		−4405		−4077	
	McFadden R^b	0.66		0.66		0.66	
	LR test: Chi^b (p-value)	**17052 (0.00)**		**17163 (0.00)**		**15666 (0.00)**	

Note: Bold indicates coefficients are significant at 1% or greater.
[a]Coefficient value with scale parameter already multiplied in.
[b]Coefficients estimated as random parameters; value shown is mean value.

Table 6. Mean Willingness-to-Pay Estimates (Indexed)

Variable	Model 1: Unsegmented	Model 2: Rule-based segmentation			Model 3: Choice-based segmentation	
		Car captive	PT captive	Choice	PT captive	Choice + transient car captives
In-vehicle travel time[a]	1.00	1.33	0.69	0.98	0.76	1.05
Walk time[a]	4.66	5.07	2.62	3.74	2.16	3.00
Waiting time	2.96	1.60	3.29	0.92	2.93	1.50
Value of each transfer	0.37	0.64	0.15	0.42	0.20	0.38

Note: All values are indexed against estimated VTTS for in-vehicle time from Model 1, R5.72 per hour. This equates to approximately a third of the proposed minimum wage in South Africa in 2017.
[a]Value shown is mean of normally distributed random coefficient.

captives have the lowest WTP values, consistent with constrained monetary budgets typical for this population. Choice passengers have values 40% to 200% higher, and car captives up to 300% higher (Model 2). An exception is the WTP for waiting time savings, where PT captives have the highest values in both segmented models. This is possibly a result of the fact that the actual experience of waiting for public transport among current PT users leads them to value it more negatively than does the hypothetical experience of infrequent or non-users.

WTP values in the choice-based model are generally lower than those of both the unsegmented model, and of comparable segments in the rule-based model. Removing 'true' captives from choice models tends to reduce the models' sensitivity to service variables, as less behavioral variation remains in the data. This matches previous findings in the literature that the inclusion of captives in choice models biases WTP estimates upwards and could lead to incorrect welfare evaluations (*3*, *23*).

The choice-based segmentation has the potential for more accurately predicting ridership change after the introduction of a new mode like BRT, because it explicitly treats each captivity segment differently. By lumping both types of car captives together into a single category, both undifferentiated and rule-based segmentation models might over- or underestimate the extent of mode-switching, particularly if the ratio of persistent to transient captives changes over time.

Lastly, the estimated VTTS values have implications for the design of new transit services like BRT. WTP values for in-vehicle travel in Model 3 are relatively low, at about a third of the minimum wage. This suggests firstly that potential BRT passengers in Johannesburg have a very limited willingness to pay for saving travel time due to affordability constraints, which can be expected to exert downward pressure on fares. Secondly, passengers value walk and wait time two to three times higher than in-vehicle time – a common finding in the literature (see *16*). Reasonable frequencies, connectivity, and network coverage to reduce walk distances are more important for new transit services than short travel times. This throws into question the prevailing design paradigm for BRT which tends to focus on providing dedicated bus lanes to raise travel speeds, rather than on promoting network connectivity and affordability. Choice passengers attach a higher value to speed and short walking times than do captives, a finding mirrored by studies in the US (*9*) and Australia (*40*). This suggests that denser feeder networks should be targeted at areas with higher proportions of present and potential choice passengers, where they will be more effective at attracting new passengers.

Conclusions

The paper describes a new method for segmenting the transport market using a combination of revealed and stated preference data obtained from user surveys. The method differentiates between current automobile captives whose resistance to transit is enduring (due to lifestyle preferences and constraints), named *persistent car captives*, and those who are willing to consider switching to transit if a suitably attractive service is offered – named *transient car captives*. The distinction is drawn on the basis of respondents' stated willingness to switch a specific recent trip to a specific transit alternative offered in an SP game, thus embedding the captivity model within the actual mode choice decision made by a traveler. We demonstrate that the model outperforms the more conventional deterministic segmentation technique based solely on automobile ownership and access to transit variables, in terms of explaining a greater portion of the observed behavior. It also endogenises the respondent's attitudes toward the alternative mode within the segmentation process, without the need for explicitly defining or measuring them. We see this as an advantage over recent segmentation-by-attitude approaches to market segmentation.

The method is likely to be especially helpful when planning for a new mode or service, as attitudes toward it may not yet be known or formed. The prerequisite is that it must be possible to adequately describe its service characteristics so that realistic responses may be elicited during the SP experiment. Although the case study presented here deals with Bus Rapid Transit as the new alternative in the presence of automobile and transit captivity, the method is completely generalizable and could be applied to other modes including non-motorized and shared modes. The fact that mode captivity is defined relative to the set of current and future modes on offer rather than as an immutable property of an individual, is seen as a key advantage as it reflects an understanding of captivity as a dynamic subjective concept that might change as the set of modes and their characteristics change.

Further refinement of the choice-based segmentation approach is needed to test its ability to improve mode choice predictions using independent samples. A potential shortcoming of the method is that conducting RP and SP surveys can be resource intensive, and although such surveys often form part of planning for new services, they may not be feasible for small jurisdictions or transit agencies. Further work is required to measure this extra cost against the benefits of enhanced predictive accuracy and perhaps even higher ridership due to better service design targeted at the needs of potential users. Wider application of the method may produce further insights into the characteristics and preferences of captivity groups in specific contexts, such that a better understanding of their distribution across space and over time can be gained. This might prove useful for policy analysis, especially in developing countries where much larger portions of the population face welfare-reducing modal captivity.

Acknowledgments

This work was partly funded by the City of Johannesburg, although the author takes sole responsibility for the analysis and interpretation. Marina Lombard contributed to the survey design and execution. Support from the BRT+ Centre of Excellence is also gratefully acknowledged.

Author Contributions

The author confirms sole responsibility for the following: study conception and design, data collection, analysis and interpretation of results, and manuscript preparation.

References

1. Bhat, C. R. Accommodating Variations in Responsiveness to Level-of-Service Measures in Travel Mode Choice Modeling. *Transportation Research Part A: Policy and Practice*, Vol. 32, No. 7, 1998, pp. 495–507.
2. Revelt, D., and K. Train. *Customer-Specific Taste Parameters and Mixed Logit*. University of California, Berkeley, CA, 1999.
3. Amador, F. J., R. M. González, and J. d. D. Ortúzar. Preference Heterogeneity and Willingness to Pay for Travel Time Savings. *Transportation*, Vol. 32, No. 6, 2005, pp. 627–647.
4. Hensher, D. A., and W. H. Greene. The Mixed Logit Model: The State of Practice. *Transportation*, Vol. 30, No. 2, 2003, pp. 133–176.
5. Ben-Akiva, M., and B. Boccara. Discrete Choice Models with Latent Choice Sets. *International Journal of Research in Marketing*, Vol. 12, No. 1, 1995, pp. 9–24.
6. Cheng, L., X. Chen, W. H. K. Lam, S. Yang, and D. Lei. Public Transit Market Research of Low-Income Commuters Using Attitude-Based Market Segmentation Approach: Case Study of Fushun, China. *Transportation Research Record: Journal of the Transportation Research Board*, 2017. 2671: 10–19.
7. Outwater, M., S. Castleberry, Y. Shiftan, M. Ben-Akiva, Y. S. Zhou, and A. Kuppam. Attitudinal Market Segmentation Approach to Mode Choice and Ridership Forecasting: Structural Equation Modeling. *Transportation Research Record: Journal of the Transportation Research Board*, 2003. 1854: 32–42.
8. Ben-Akiva, M., M. Ben-Akiva, K. Train, and D. McFadden. Hybrid Choice Models: Progress and Challenges. *Marketing Letters*, Vol. 13, No. 3, 2002, pp. 163–175.
9. Beimborn, E., M. Greenwald, and X. Jin. Accessibility, connectivity, and captivity: impacts on transit choice. *Transportation Research Record: Journal of the Transportation Research Board*, 2003. 1835: 1–9.
10. Williams, H., and J. d. D. Ortúzar. Behavioural Theories of Dispersion and the Mis-specification of Travel Demand Models. *Transportation Research Part B: Methodological*, Vol. 16, No. 3, 1982, pp. 167–219.
11. Elmore-Yalch, R. *A Handbook: Using Market Segmentation to Increase Transit Ridership*. TCRP Report 36. Transportation Research Board, National Research Council, Washington, D.C., 1998, p. 74.
12. Polzin, S., X. Chu, and J. Rey. Density and Captivity in Public Transit Success: Observations from the 1995 Nationwide Personal Transportation Study. *Transportation Research Record: Journal of the Transportation Research Board*, 2000. 1735: 10–18.
13. Chia, J., and B. Lee. *Variation in the Walking Time to Bus Stop by the Degree of Transit Captivity*. Australasian Transport Research Forum, 2015.
14. Habib, K. N. Household-level Commuting Mode Choices, Car Allocation and Car Ownership Level Choices of Two-Worker Households: The Case of the City of Toronto. *Transportation*, Vol. 41, No. 3, 2014, pp. 651–672.
15. Litman, T. Evaluating Transportation Choice. *Transportation Research Record: Journal of the Transportation Research Board*, 2001. 1756: 32–41.
16. Litman, T. Transit Price Elasticities and Cross-Elasticities. *Journal of Public Transportation*, Vol. 7, No. 2, 2004.

17. Krizek, K. J., and A. El-Geneidy. Segmenting Preferences and Habits of Transit Users and Non-users. *Journal of Public Transportation*, Vol. 10, No. 3, 2007, pp. 5.
18. Redman, L., M. Friman, T. Gärling, and T. Hartig. Quality Attributes of Public Transport that Attract Car Users: A Research Review. *Transport Policy*, Vol. 25, 2013, pp. 119–127.
19. Muñoz-Raskin, R., and H. Scorcia. Time for a Tailored Approach to South African BRTs. In *Connections - Transport and ICT. The World Bank*. Washington, D.C., 2017.
20. Lucas, K. Making the Connections Between Transport Disadvantage and the Social Exclusion of Low Income Populations in the Tshwane Region of South Africa. *Journal of Transport Geography*, Vol. 19, No. 6, 2011, pp. 1320–1334.
21. Manski, C. F. The Structure of Random Utility Models. *Theory and Decision*, Vol. 8, No. 3, 1977, pp. 229–254.
22. Stopher, P. R. Captivity and Choice in Travel-Behavior Models. *Transportation Engineering Journal of the American Society of Civil Engineers*, Vol. 106, No. 4, 1980, pp. 427–435.
23. Swait, J., and M. Ben-Akiva. Analysis of the Effects of Captivity on Travel Time and Cost Elasticities. *Behavioural Research for Transport Policy*, 1986, pp. 119–134.
24. Ben-Akiva, M., and T. Morikawa. Comparing Ridership Attraction of Rail and Bus. *Transport Policy*, Vol. 9, No. 2, 2002, pp. 107–116.
25. Cascetta, E., and A. Papola. Random Utility Models with Implicit Availability/Perception of Choice Alternatives for the Simulation of Travel Demand. *Transportation Research Part C: Emerging Technologies*, Vol. 9, No. 4, 2001, pp. 249–263.
26. Swait, J., and M. Ben-Akiva. Incorporating Random Constraints in Discrete Models of Choice Set Generation. *Transportation Research Part B: Methodological*, Vol. 21, No. 2, 1987, pp. 91–102.
27. Gaundry, M. J., and M. G. Dagenais. The Dogit Model. *Transportation Research Part B: Methodological*, Vol. 13, No. 2, 1979, pp. 105–111.
28. Swait, J., and M. Ben-Akiva. Empirical Test of a Constrained Choice Discrete Model: Mode Choice in Sao Paulo, Brazil. *Transportation Research Part B: Methodological*, Vol. 21, No. 2, 1987, pp. 103–115.
29. Swait, J. A Non-compensatory Choice Model Incorporating Attribute Cutoffs. *Transportation Research Part B: Methodological*, Vol. 35, No. 10, 2001, pp. 903–928.
30. Dissanayake, D., and T. Morikawa. Investigating Household Vehicle Ownership, Mode Choice and Trip Sharing Decisions Using a Combined Revealed Preference/Stated Preference Nested Logit Model: Case Study in Bangkok Metropolitan Region. *Journal of Transport Geography*, Vol. 18, No. 3, 2010, pp. 402–410.
31. Train, K. A Structured Logit Model of Auto Ownership and Mode Choice. *The Review of Economic Studies*, Vol. 47, No. 2, 1980, pp. 357–370.
32. Srinivasan, K., G. Pradhan, and G. Naidu. Commute Mode Choice in a Developing Country: Role of Subjective Factors and Variations in Responsiveness Across Captive, Semicaptive, and Choice Segments. *Transportation Research Record: Journal of the Transportation Research Board*, 2007. 2038: 53–61.
33. Wilson, F., A. Stevens, and J. Robinson. Identifying Mode Choice Constrained Urban Travel Market Segments. *Canadian Journal of Civil Engineering*, Vol. 11, No. 4, 1984, pp. 924–932.
34. Jacques, C., K. Manaugh, and A.M. El-Geneidy. Rescuing the Captive [mode] User: An Alternative Approach to Transport Market Segmentation. *Transportation*, Vol. 40, No. 3, 2013, pp. 625–645.
35. van Lierop, D., and A. El-Geneidy. A New Segmentation Approach: Evidence from Two Canadian Cities. *Journal of Public Transportation*, Vol. 20, No. 1, 2016.
36. Hensher, D. A. Hypothetical Bias, Choice Experiments and Willingness to Pay. *Transportation Research Part B: Methodological*, Vol. 44, No. 6, 2010, pp. 735–752.
37. Rose, J. M., M. C. J. Bliemer, D. A. Hensher, and A. T. Collins. Designing Efficient Stated Choice Experiments in the Presence of Reference Alternatives. *Transportation Research Part B: Methodological*, Vol. 42, No. 4, 2008, pp. 395–406.
38. City of Johannesburg. *Strategic Integrated Transport Plan Framework*. City of Johannesburg, Johannesburg, South Africa, 2013.
39. Swait, J., J. J. Louviere, and M. Williams. A Sequential Approach to Exploiting the Combined Strengths of SP and RP Data: Application to Freight Shipper Choice. *Transportation*, Vol. 21, No. 2, 1994, pp. 135–152.
40. Chia, J., J. Lee, and M. Kamruzzaman. Walking to Public Transit–Exploring Variations by Socio-economic Status. *International Journal of Sustainable Transportation*, Vol. 10, No. 9, 2016, pp. 805–814.

The Standing Committee on Public Transportation Planning and Development (AP025) peer-reviewed this paper (18-01407).

Article

TRR
JOURNAL OF THE TRANSPORTATION RESEARCH BOARD

Transportation Research Record
2018, Vol. 2672(8) 807–816

DOI: 10.1177/0361198118773897
journals.sagepub.com/home/trr
SAGE

Comparative Structural Evaluation of Transit Travel Demand using Travel Survey and Smart Card Data for Metropolitan Transit Financing

Tim Spurr[1], Antoine Leroux[2], and Robert Chapleau[2]

Abstract

Lrge metropolitan areas often comprise multiple municipalities and multiple transit-operating agencies that share infrastructure and passengers. In such cases, financing mechanisms are devised to share costs and revenues among the various jurisdictions. Using Montréal, Canada, as a case study, this paper investigates whether a large sample household travel survey (HTS) can provide sufficiently accurate and detailed information to form the basis for a metropolitan transit financing framework. The evaluation is made possible by the existence of a smart card (SC) fare collection system, deployed across the region, which provides, with some processing, an independent source of transit trip information. The structure of transit travel demand, as measured by SC and the HTS, were compared. The structural elements examined included the types of fare product used, the temporal distribution of trips during a typical weekday, and the spatial and temporal distribution of trips over the multiple networks serving the metropolitan area. The results of the comparison showed that the HTS constitutes a simplified portrayal of transit demand that over-represents symmetrical travel patterns prevalent during peak periods and under-represents other travel patterns. An important consequence of this bias is the over-representation of travel between the suburbs and downtown. Two theoretical allocation scenarios were designed to evaluate the potential effects of these differences on metropolitan-level cost- and revenue-sharing. A simple experiment showed the effects to be significant.

The global trend of increasing urbanization has caused many urban areas to agglomerate around multiple cities and towns spread over large geographic areas. These "metropolitan" urban regions that encompass multiple municipal jurisdictions present important challenges for the coordination, planning and financing of public transit services, especially when operating agencies share infrastructure and passengers. In such contexts, equitable and reliably objective mechanisms for sharing costs and revenues are pre-requisites for the harmonious functioning and development of public transit.

In parallel, the evolution of information technologies has made available unprecedented quantities of data describing the performance of complex systems, including urban transportation. In the domain of public transit, the smart card (SC) is progressively being adopted as the principal mode of payment. While SC technology has existed for decades, the data storage and processing capabilities required to extract meaningful insights from this technology are comparatively recent. Since the number of data points generated by an SC payment system can be of the order of millions per day, even in medium-sized transit systems, the analysis of SC data fits easily within the "Big Data" paradigm and has the potential to revolutionize urban transit planning, which has traditionally been based on the travel survey. While recent research has shown that telephone-based travel surveys suffer from significant response bias, travel surveys still provide dimensions of information that could not be obtained easily, if at all, from SC data.

The goal of this paper is to compare the structure of travel demand as reported by SC data and a telephone-based household travel survey (HTS) to assess the effects of using either data source as the basis for metropolitan financing mechanisms. The Montréal, Canada, region is used as a case study based on data from 2013. It is hoped that this improved understanding can inform the coordination and financing of public transit services in metropolitan urban agglomerations.

The structure of the paper is as follows. The next section provides the context and motivation for the study. The third section summarizes the data and methodology and provides some relevant information on the organization of public transit in the Montréal metropolitan area. The fourth section

[1]Autorité Régionale de Transport Métropolitain, Montréal, QC, Canada
[2]Department of Civil, Geological and Mining Engineering, École Polytechnique de Montréal, Montréal, QC, Canada

Corresponding Author:
Address correspondence to Tim Spurr: tspurr@artm.quebec

documents the comparison of transit trip structure as portrayed by the HTS and SC. The fifth section applies these differences to two theoretical allocation scenarios and the last section offers some concluding remarks.

Context of the Study

In most places, public transit systems require government subsidies to supplement the revenue obtained through fares that are insufficient to cover capital and operating costs. Frequently, especially in populous urban areas, a single transit system will serve multiple political territories, each of which might contribute to subsidizing the service. Where multiple agencies exist, integrated fare products (tickets or passes) are sold that allow access to combined services. Naturally, each participating agency will expect a share of the revenue from the sale of these products. In some cases, the equalization formulae governing payments between different municipal governments and transit agencies can lead to significant economic distortions. While the design of new sources of revenue (taxes, fees, tolls etc.) to pay for transit subsidies is a widely researched topic, the issue of how these subsidies should be shared among overlapping jurisdictions is only rarely addressed in the literature (*1*). Meanwhile, fare integration continues to be a subject of interest to metropolitan transit planning authorities (*2*).

One option for mitigating the distortions caused by a given fiscal scenario is to apply the principle that each participant pay costs in proportion to the benefit obtained (*3*). From a metropolitan planning perspective, this option is appealing as it forces local governments to take responsibility for their transit policies through their respective budgets. For many years in the Greater Montréal area, HTS have been used in conjunction with disaggregate network models to measure the number of passenger-kilometers traveled by transit users and associate this consumption with geopolitical and transit-operating entities. On the basis of these calculations, municipalities subsidize the metropolitan transit service proportional to the consumption of their residents, and transit agencies receive integrated fare revenue in proportion to the service provided to passengers who pay the integrated fare. Although this practice is perhaps unique to Montréal, at least one researcher (*4*) has questioned whether the "quality" of travel surveys generally is sufficiently high for revenue distribution purposes.

The Montréal HTS, undertaken roughly every 5 years since 1970, samples almost 5% of households in Greater Montréal area using telephone interviews and was most recently conducted in the fall of 2013, when 79,000 households were interviewed. The survey consists of a random sample, uniformly stratified over the metropolitan area and weighted based on age, sex, and household structure information contained in the national census. While the technical aspects of data collection for the survey have changed over time, one of the primary objectives has always been the observation of "revealed" public transit route choices. The itinerary information, recorded for each individual traveler, was necessary for the operational planning of bus and metro systems. Subsequently, as trips using multiple operators became more common, the information on itineraries was adopted as the basis for cost- and revenue-sharing between transit operators and municipalities. In order to maximize the accuracy of the route information provided by respondents, a process based on network simulation was developed to ensure that the declared transit itineraries were plausible. Now, with the implementation of an SC fare collection system in the region, a higher-level validation of the HTS measurement of transit trip structure becomes possible.

The problems inherent in telephone travel surveys have been extensively documented (*5–7*) and include non-representative samples, the time lag between the sample selection and the survey execution, under-reporting of trips by respondents (especially when responding on another household member's behalf) and inaccurate information on departure time and trip end locations. SC data, while addressing some of these issues (*8*), have other limitations, notably a lack of information on the socioeconomic characteristics of travelers and, importantly, their households (*9*).

A tap-in only SC fare collection system was implemented on all of the Montréal region's 14 transit agencies over the course of 2008. With some very minor exceptions, all transit users in the region must use an SC to board a transit vehicle. Consequently, the SC data represents a 95% sample of the transit-using population. The progressive implementation prevents a direct comparison with the data of the travel survey conducted that same year. Nevertheless, a comparison of the 2008 HTS with 2010 SC data on the Montréal metro (subway) system revealed the presence of significant sampling and reporting biases in the survey (*10*). A subsequent study attempted to match individual records from the 2013 HTS with the SC transactions recorded during the same period (*11*). That paper confirmed the biases suspected in the 2008 survey. The authors also found evidence that survey respondents may substitute their "typical" travel patterns for their actual trips on the day of the survey and that occasional transit users tend to under-report transit trips. Other studies comparing travel survey data with SC data were performed by Munizaga et al. (*12*), who found a good correspondence at the level of transit itineraries, and Trépanier (*13*) who found that the sampling rate of the HTS (5%, as in Montréal) was not large enough to accurately measure the ridership of specific bus lines and was unable to capture important temporal variations over the course of the day.

Taken together, these findings are at least sufficient to justify a validation of the structural elements of transit trips that are reported in an HTS.

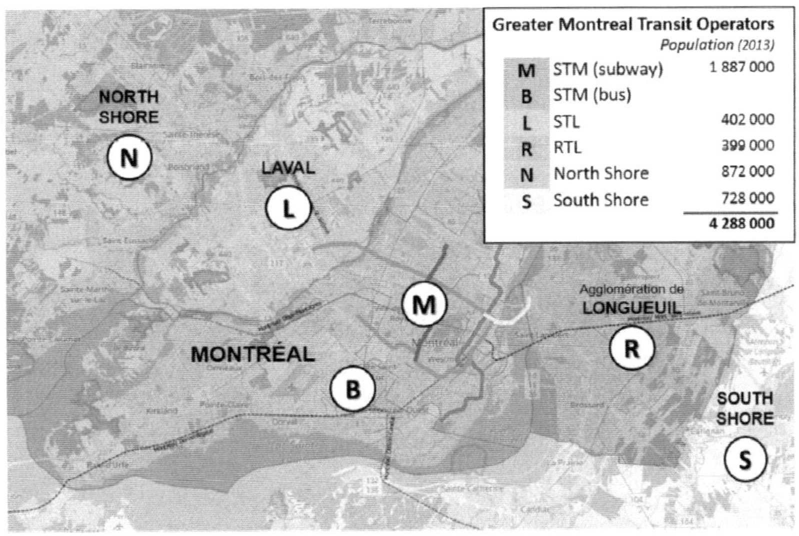

Figure 1. Map of the Greater Montréal area showing metro and commuter train lines.

Data and Methodology

The structure of transit demand is influenced by the organization of transit networks, the political geography of the territory they serve, and the fare structure. The Greater Montréal area is composed of 83 municipalities which, for analysis purposes, are grouped into five sub-regions, as shown in Figure 1. Until June 2017, the metropolitan region was served by 16 distinct transit agencies. To simplify the discussion in this paper, the 16 operators are grouped into 7 artificial operating entities, each associated with a region: STM metro (M) and STM bus (B) for Montréal central city, STL Laval (L), RTL Longueuil (R), North Shore bus (N) and South Shore bus (S). The STM operates the metro and bus services for the central city of Montréal. All other operators use only buses. The region is also served by a commuter train network but it does not require its users to validate their SCs when boarding and is therefore excluded from the analysis. Currently, passengers access transit services using local or integrated fare products. The integrated fares provide access to the services of multiple operators and the price of each fare varies according to the fare zone in which the boarding occurs. There are eight fare zones.

In the experiment described below, two independent data sets describing public transit travel demand in the Greater Montréal area during the fall of 2013 are compared. The first dataset is an extraction of the regional HTS, undertaken by telephone between September and December. The second data source is extracted from the archived SC transaction database for Thursday, October 24th 2013. While the survey data represent an "average weekday" created from an amalgam of days observed over a 4-month period, the SC data were collected on one specific day. In order to make meaningful comparisons with the survey data, the SC data for October 24th 2013 was analyzed in detail to ensure that the day was representative of fall 2013 as a whole (for partial confirmation, see Chapleau et al. (*14*), who use Montréal SC data from November 21st 2013). As shown in Figure 2, from Chapleau (*15*), the two datasets describe different objects and so direct comparison requires considerable data processing and some simplifying but plausible assumptions, which are described below.

The initial data extraction from the HTS included all trips made by public transit – 59,088 survey records representing 1.475 million daily trips for the entire population. These trips were made by 29,122 surveyed individuals representing 724,000 inhabitants of the Greater Montréal area.

Demand for the commuter train services is not measured accurately using the SC data, therefore trip segments made by train were removed from the survey dataset. Moreover, park-and-ride and kiss-and-ride account for 11% of daily public transit trips and, naturally, the segments of these trips made by car are not captured by the SC system. These trip segments were also removed from the survey dataset although the non-train transit segments of these trips were retained. Finally, connections between metro lines in survey trips were not considered transfers as they do not require SC validation. After these adjustments, the final survey dataset represented 1.4 million trips made by 690,000 travelers.

In the SC data, it is assumed that each card is used by a single person and that each trip involves at least one card validation. Care must be taken to ensure that transfers are not counted as separate trips. To make this distinction, three conditions were applied to consecutive validations of each SC:

1. A transfer validation cannot be more than 120 min after the previous validation.

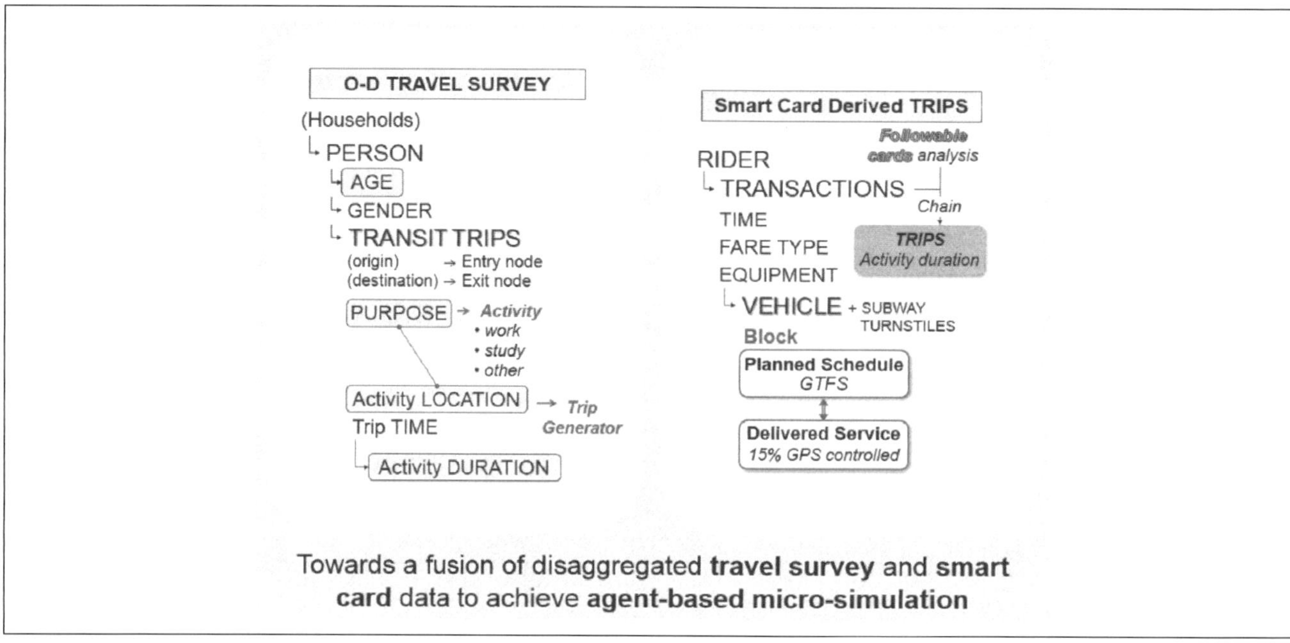

Figure 2. Schematic diagram comparing SC and HTS objects. GTFS = general transit feed specification.

2. A single trip cannot include consecutive metro entries.
3. A single trip cannot include consecutive validations using the same transit line.

This method allowed for the identification of 1.4 million trips made by 703,000 distinct SCs. Remarkably, these two quantities are very similar to the equivalent values derived from the HTS.

Transit Trip Structure: HTS versus SC

In this section, the structure of public transit demand is analyzed using both the SC data and the HTS, first from the perspective of the users of the transit system and then from the perspective of transit-operating agencies. In a metropolitan financing context, the relevant characteristics of transit system users are related to their use of different fare products. For the operating agencies, the analysis focuses on the proportion of passengers that are shared between multiple operating entities.

Fare Product Usage

In Figure 3, the total number of travelers detected by SC and by the travel survey are compared according to the fare product they used. The SC data contained 531 distinct fare products that could be grouped into four categories used by the HTS: regular monthly pass, reduced monthly pass, regular single fare, and reduced single fare. Over a 24-h period, the SC system and the travel survey record similar numbers of trips and travelers, however there are significant variations across fare categories. The travel survey overestimates the number of monthly pass-holders, underestimates the number regular single-fare users and overestimates the number of reduced single-fare users. In addition, the survey underestimates the average number of trips per day made by pass-holders (2.08 in the HTS vs. 2.30 or 2.28 with the SC). This finding indicates that the HTS simplifies travel behavior by over-reporting symmetrical travel (pairs of outbound and return home trips) and under-reporting intermediate trips. In fact, as the SC data indicate, pass-holders are more likely to make intermediate trips during the day, since doing so incurs no marginal cost. Ticket users, meanwhile, have an incentive to limit their transit usage since each additional trip consumes another ticket.

Meanwhile, part of the discrepancy observed in the average number of daily trips per traveler using individual tickets (1.86 and 1.76 with HTS vs. 1.32 and 1.62 with SC) can be attributed to the SC system's inability to detect multiple trips by the same traveler when the traveler uses tickets on a disposable SC.

A different perspective is provided by Figure 4, which shows the number of traveler fare products broken down by the number of daily trips per traveler. Even accounting for the SC system's inability to detect the same person using multiple single tickets, the HTS under-represents single-trip travelers and over-represents two-trip travelers, indicating a bias toward symmetrical travel behavior. This result is to be expected since each survey interviewee responds on behalf of all other household members and may not be aware of their intermediate trips. In addition, replying "yes" to the question "Did you use the same route to return home?" is

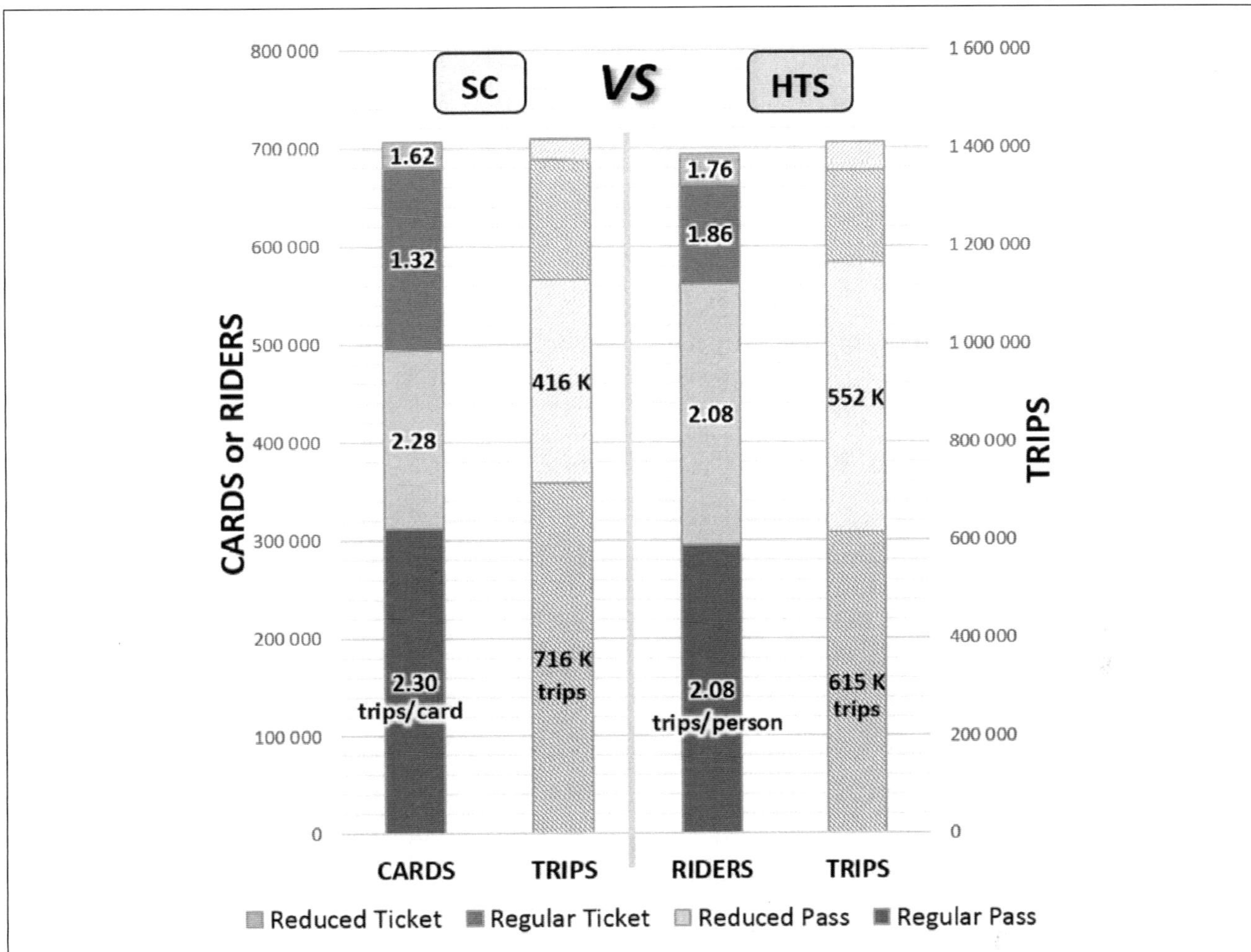

Figure 3. Comparison of traveler fare products using SC and HTS.

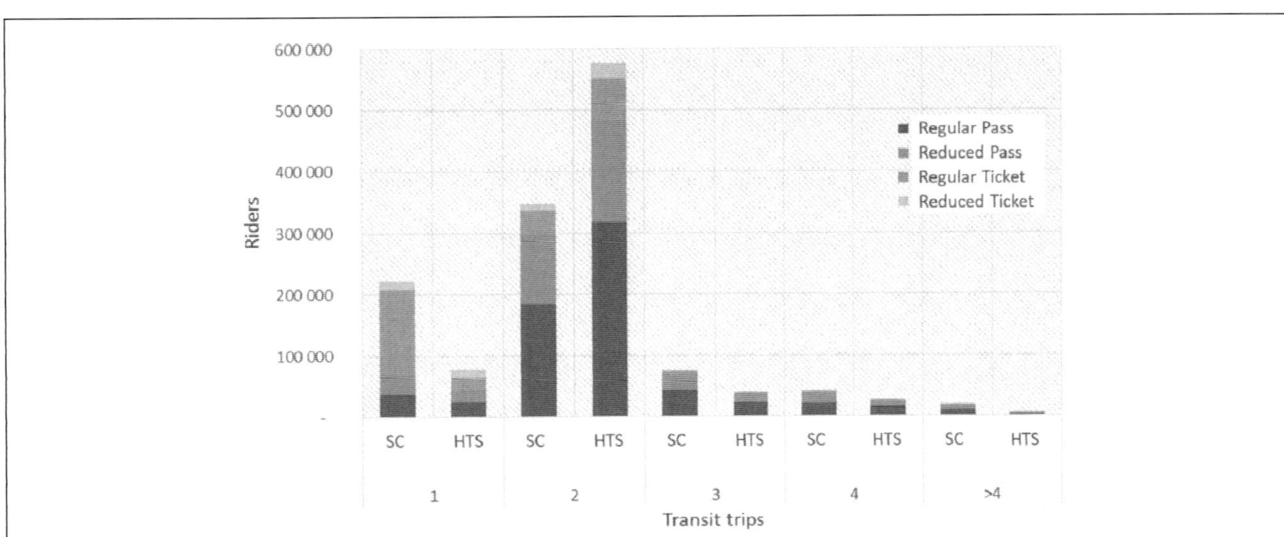

Figure 4. Comparison of traveler fare product use by number of trips per day.

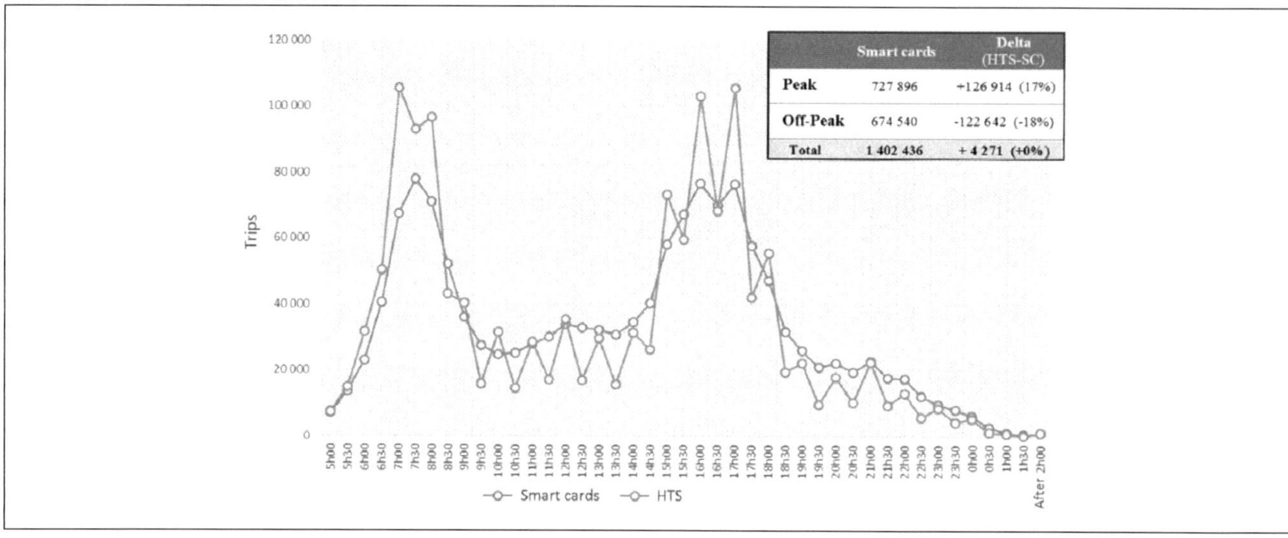

Figure 5. Comparison of temporal distributions of transit trips.

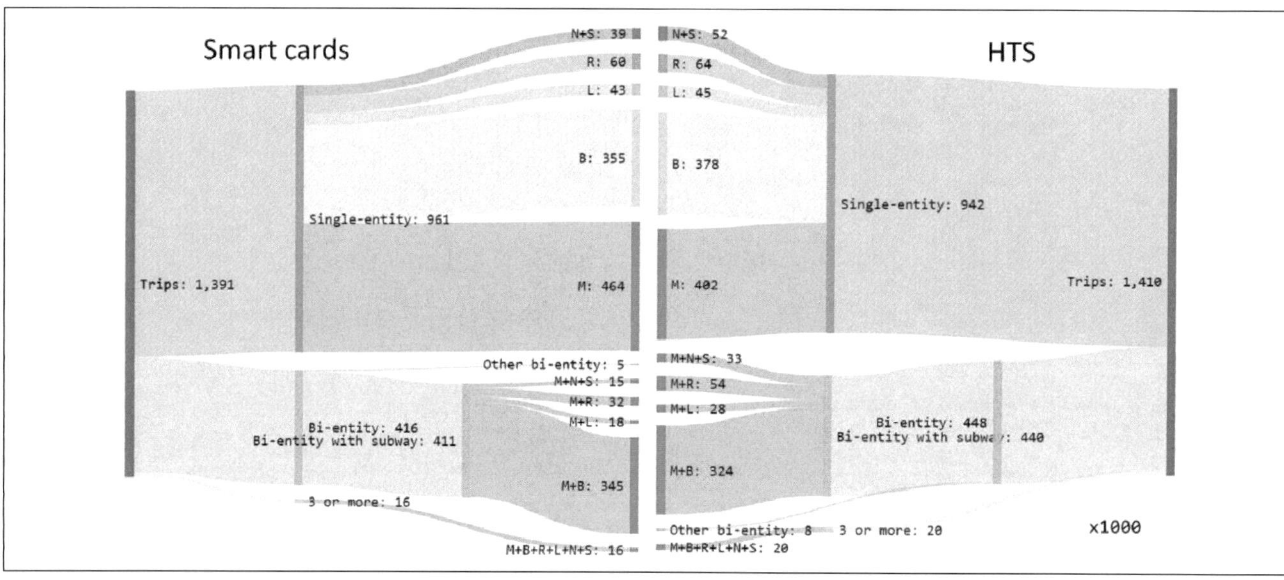

Figure 6. Comparison of operating entity combinations.

easier than remembering and describing an intermediate trip. Moreover, the large number of single-fare, single-trip travelers that are missing from the travel survey suggests that occasional transit users are under-represented.

Fare product usage patterns are related to the distribution of trips over the course of the day. Figure 5 shows the temporal distribution of trips by 30-min periods and illustrates that, globally, the survey overestimates peak period demand, which consists mostly of regular transit users having symmetric behavior, and underestimates off-peak demand generated from intermediate trips, complex trip chains, and occasional travelers. The graphic also illustrates the tendency of survey respondents to declare their departure times on the hour. It is worth emphasizing that over a 24-h period, the total number of survey trips matches almost exactly the number of SC trips. The pattern is very similar to the one described in Spurr et al. (*10*).

Single- and Multiple-Entity Transit Usage

From a metropolitan financing perspective, it is relevant to compare how SC and the HTS report multi-agency travel. Figure 6 compares how travelers use multiple entities (bus operators and the metro) over the course of a single trip, as measured by each data source. Based on the SC data, of the 1.4 million total daily transit trips, 69% use a single entity and

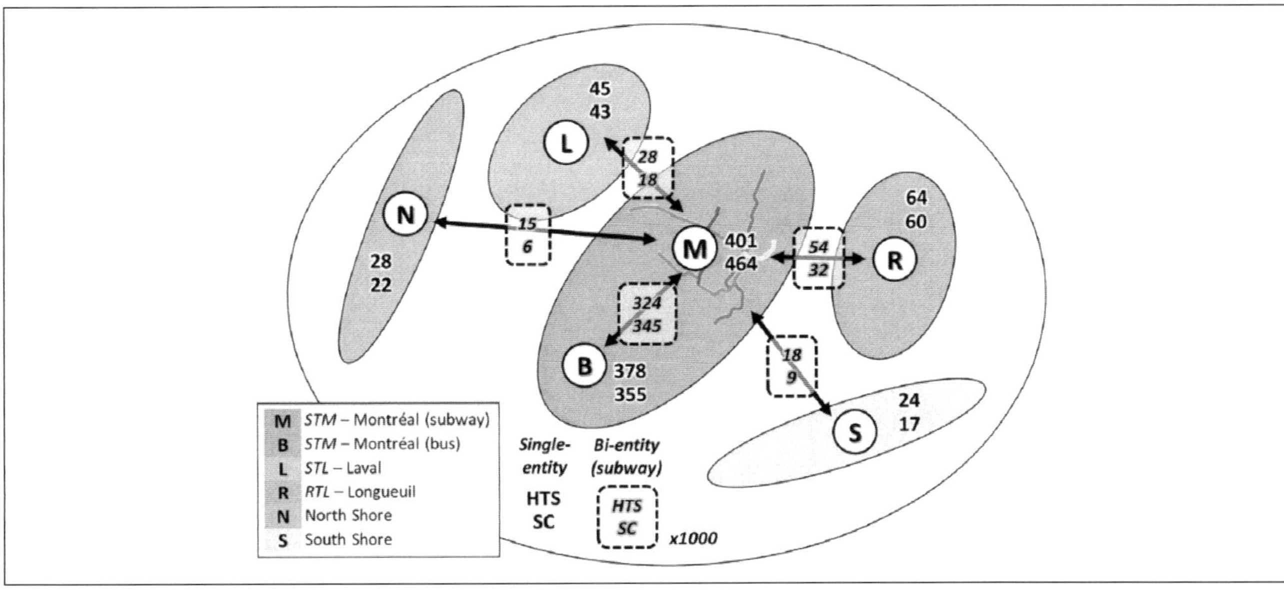

Figure 7. Comparison of geographic distribution of trips between and within regions.

30% use two entities. Only 1.2% of trips use three or more entities. Among trips using two entities, the metro is used 99% of the time (the commuter train is excluded from the analysis). A comparative assessment with the HTS of this market segment, subdivided according to which non-metro service was used, reveals a significant overestimation of all categories except those trips using entity (M + B), which it underestimates by 6% (324,000 trips vs. 345,000 trips), and metro only (M), which it underestimates by 14% (401,000 trips vs. 464,000 trips). Consistent with the findings above, trips involving B and M are more likely to be made off-peak for non-regular activities than trips using suburban bus services to access the metro. The largest overestimations (in percentage terms) of single-operator trips occur on the suburban bus services in the northern and southern suburbs (N and S).

The geographic distribution of the differences between the trip volumes estimated by the HTS and those estimated by SC is illustrated in Figure 7. In this figure, the numbers within each territory represent volumes of single-entity trips and the numbers on the arrows represent volumes of multiple-entity trips. Three points are worth emphasizing. First, the single-entity usage of all bus systems is overestimated in the HTS by 9.1%, although it varies by operator between 5% (L) and 35% (S). Second, bi-entity trips (almost all of which involve the metro) are overestimated by 7.1%, and those trips involving suburban bus services are overestimated by 77%. Third, metro trips that do not include suburban bus segments but may include central city bus (M and [M + B]) are underestimated by 10%. In summary, the HTS over-represents suburban trips, which, when made by public transit, are strongly associated with work and school activities. Central city travel by transit is motivated by a broader range of activities, many of which are irregular and occur off-peak, and is under-represented in the HTS. For revenue- and cost-sharing mechanisms, this discrepancy between the two data sources has some potential consequences, which are discussed in the next section.

Impact of Trip Structure Analysis on Transit Financing Calculation

To examine the possible effects of using HTS or SC as the basis of a metropolitan financing mechanism, two purely theoretical allocation scenarios were tested (Figure 8) for illustrative purposes. In the first scenario, transit consumption is measured at the operator level. In the second scenario, transit consumption is measured by entity, the effect of which is to separate the metro from the central city bus. In both scenarios, each transit trip generates a "score" which represents occasions where the traveler must pay a fare. For example, in the operator-based scenario, a trip using the central city bus (B) generates one unit of revenue for operator (M + B), whereas a trip using the central city bus, metro, and South Shore bus generates one unit of revenue for operator (M + B) and another unit of revenue for operator (R). In the entity-based scenario, the same trip generates one unit for the central city bus (B), one unit for the metro (M), and one unit for the South Shore bus (R). Here, consumption is measured solely on the basis of boarding volumes, although the distance traveled could be integrated into the calculation.

The experiment reveals significant differences in the allocation proportions depending on whether they are based on HTS or SC. In the operator-based scenario, the central city operator (M + B) is allotted 80.2% using HTS and 84.5% using SC. In the entity-based scenario, the corresponding figures for the central city (M) + (B) are 83.8% and 87.5%.

	HTS	SC 2013-10-24	SC week average	SC week s.d.
Scenario: by Transit Operators				
M+B	80.2%	84.5%	*84.4%*	*2.3%*
R	8.4%	6.8%	*6.7%*	*0.4%*
L	5.4%	4.7%	*4.8%*	*0.1%*
N+S	6.1%	4.0%	*4.1%*	*0.2%*
Scenario: by Operational Entity				
M	45.5%	48.4%	*48.0%*	*1.6%*
B	38.3%	39.1%	*39.5%*	*0.7%*
R	6.9%	5.4%	*5.4%*	*0.3%*
L	4.4%	3.8%	*3.8%*	*0.1%*
N+S	5.0%	3.2%	*3.3%*	*0.2%*

Scoring scheme (by Transit Operators):
B : M+B=1
BM : M+B=1
BMB : M+B=1
BMR : M+B=1, R=1

Scoring scheme (by Operational Entity):
B : B=1
BM : B=1, M=1
BMB : B=1, M=1
BMR : B=1, M=1, R=1

M *STM* - Montréal (subway) **B** *STM* - Montréal (bus) **R** *RTL* - Longueuil **L** *STL* - Laval **N** North Shore **S** South Shore

Figure 8. Measure of transit usage on several transit operators according to HTS and SC.

Scenario: by Transit Operators

HTS

rank 1 \ SEGMENT	B+M	L	R	N+S	Total
B+M	71.6%	0.4%	0.5%	0.2%	72.6%
L	2.2%	4.9%	0.0%	0.0%	7.2%
R	3.9%	0.0%	7.8%	0.1%	11.7%
N+S	2.6%	0.1%	0.1%	5.8%	8.5%
Total	80.2%	5.4%	8.4%	6.1%	100.0%

SC

rank 1 \ SEGMENT	B+M	L	R	N+S	Total
B+M	81.5%	0.9%	1.3%	0.6%	84.3%
L	0.9%	3.8%	0.0%	0.0%	4.8%
R	1.4%	0.0%	5.4%	0.0%	6.8%
N+S	0.7%	0.0%	0.0%	3.3%	4.1%
Total	84.5%	4.7%	6.8%	4.0%	100.0%

Scenario: by Operational Entity

HTS

rank 1 \ SEGMENT	B	M	L	R	N+S	Total
B	33.1%	15.2%	0.1%	0.1%	0.0%	48.6%
M	4.1%	23.5%	0.2%	0.2%	0.1%	28.1%
L	0.4%	1.7%	4.1%	0.0%	0.0%	6.2%
R	0.4%	3.1%	0.0%	6.4%	0.0%	9.9%
N+S	0.3%	2.0%	0.0%	0.1%	4.7%	7.2%
Total	38.3%	45.5%	4.4%	6.9%	5.0%	100.0%

SC

rank 1 \ SEGMENT	B	M	L	R	N+S	Total
B	30.5%	11.1%	0.2%	0.2%	0.1%	42.1%
M	8.1%	35.0%	0.5%	0.9%	0.4%	44.8%
L	0.2%	0.7%	3.1%	0.0%	0.0%	4.0%
R	0.2%	1.1%	0.0%	4.4%	0.0%	5.7%
N+S	0.1%	0.5%	0.0%	0.0%	2.7%	3.3%
Total	39.1%	48.4%	3.8%	5.4%	3.2%	100.0%

M *STM* - Montréal (subway) **B** *STM* - Montréal (bus) **R** *RTL* - Longueuil **L** *STL* - Laval **N** North Shore **S** South Shore

Figure 9. Distribution of transit consumption when the residential location of the rider is assigned to the first transit operator (HTS and SC).

In both scenarios, relative to the SC-based allocation, the HTS allocates more consumption to the suburban regions (R, L, N, S). Note that Figure 8 includes the same results calculated using the average SC transactions for the week of October 24th 2013 in addition to the results for that specific day. The differences between the two sets of results are negligible.

Often, the gap between the fare box revenues generated by a transit service and the costs of its operation is filled by municipal property taxes. In a metropolitan context, where the residents of one municipality use multiple operators for their transit trips, the task of calculating how much each municipality owes to each transit operator becomes complex.

Figure 9 presents an example calculation, based on the transit usage scoring mechanism described above. The figure portrays two equalization methods: the top one based on transit consumption by operator and the second based on transit consumption by entity. Two matrices are included for each scenario: one based on the HTS, the other on the SC data. The matrix format allows a simultaneous representation of transit consumption by territory distributed over multiple operating entities. The rows of each matrix represent the consumption of transit by municipal region. For every SC (or traveler in the HTS), the municipality of the first boarding of the day ("rank 1") is identified. This municipality is assumed to contain the residence of the traveler. The columns of each matrix represent the transit supplied, in terms of boarding volumes, per operator (or entity).

Starting with the operator-based scenario, the top left matrix, based on the HTS, indicates that residents of the central city (B + M) consume 72.6% of all operator trips in the region. Almost all of these trips (71.6% of the regional total) are completed using the central city transit operator (B + M). Meanwhile, the (B + M) operator carries 80.2% of all operator trips. The difference between transit consumed (72.6%) and transit supplied (80.2%) in the central city is due to the use of the (B + M) operator by residents of the suburban regions (L, R, and N + S). The residents of these regions consume 27.4% of operator trips. An examination of the top-right matrix, derived from SC data, gives a very different result. In the operator-based scenario, suburban residents consume 15.7% of operator trips. Similar results are obtained in the entity-based scenario (23.3% vs. 13.0%).

This discrepancy is the result of the difference between how HTS and SC data portray the structure of transit travel demand. It has been shown above that the HTS over-reports peak period travel, especially trips originating in the suburban regions. In addition, the HTS under-reports off-peak travel in the central city. The SC data, meanwhile, capture all trips made by pass-holders, including intermediate trips made during off-peak periods. The SC system is incapable, however, of accurately measuring consumption by single-ticket users, as it often cannot detect when the same traveler has used a different ticket. Some travelers are counted twice or more as a result. Since most intermediate and ticket-using trips occur in the central city, the SC trip data also produce a misleading representation of transit consumption.

Discussion and Conclusion

Trip data derived from archived SC validations facilitate, for the first time in the Montréal region, a detailed validation of the structure of transit trips reported in a large sample HTS. The goal of this paper was to compare the structure of transit travel demand, as revealed by SC and an HTS, to assess the extent to which each dataset yields different results when used to calculate cost- and revenue-sharing between transit operators and municipalities in a metropolitan area. A comparative analysis found that, despite the HTS's accurate measure of total trip volumes over an average weekday, the simplified structure of travel demand it portrays causes it to overestimate suburban transit consumption. The survey also over-reports multi-operator trips as well as the number of monthly pass-holders. The average use of monthly passes, as well as the number of single-ticket users, is under-reported.

The SC data have other shortcomings. Their accurate recording of discretionary travel by pass-holders means they measure transit trips that have no marginal cost for travelers and therefore generate no marginal revenue for operators. Thus it could be argued that these trips should not be included in the metropolitan financing framework. In addition, it is difficult to correctly measure travelers' use of single tickets using the SC system. An experiment using two purely theoretical allocation scenarios in the Montréal region reveals that the distorting effects of either data source are significant.

Consequently, at this stage in the research, it is apparent that neither the HTS nor the SC data are sufficiently representative of actual transit use to serve as the basis for a metropolitan transit financing mechanism without the application of corrective measures. The first such measure would be an adjustment of the weighting factors in the HTS to correct for the over-reporting of peak period suburb-based travel. A second measure would be some method for addressing the use of single tickets in the SC data. These topics will be the focus of future research.

Two limitations of this study are discussed here but are not believed to have influenced the general conclusions. First, the HTS represents a typical average weekday (an amalgam of many days of telephone interviews) whereas the SC data represents a single specific day. Although the total transaction volumes for October 24th, 2013 were determined to be "typical" for the season, local phenomena (at individual metro stations, for example) specific to that date could be significant. Second, the examination of alternative financing scenarios requires the imputation of a municipality of residence to each SC user. However, the "rank 1" hypothesis used in this paper does not necessarily hold for kiss-and-ride and park-and-ride trips.

Finally, the comparative analysis of transit trip structure using both HTS and SC suggests the possibility of developing a financing mechanism that considers transit to be a service rendered to people. In other words, with the continued development of automatic fare collection systems, the financing of public transit might be based on the number of people served, as opposed to the number of completed trips.

Acknowledgments

The authors gratefully acknowledge the resources provided by the Autorité Régionale de Transport Métropolitain as well as the collaboration of the Société de Transport de Montréal (STM). Special thanks to Phillippe Gaudette of Groupe MADITUC for his technical assistance.

Author Contributions

The authors confirm contribution to the paper as follows: study conception and design: A.L., R.C.; data collection: R.C.; analysis and interpretation of results: A.L., R.C., T.S.; draft manuscript preparation: T.S. All authors reviewed the results and approved the final version of the manuscript.

References

1. Rinks, D. B. Revenue Allocation Methods for Integrated Transit Systems. *Transportation Research Part A: General*, Vol. 20, No. 1, 1986, pp. 39–50.
2. Woo, L. *GTHA Fare Integration*. Metrolinx, Toronto, Canada, 2017, p. 23.
3. Chapleau, R. Measuring Impacts of Transit Financing Policy in Geopolitical Context: Montreal Case. *Transportation Research*

Record: Journal of the Transportation Research Board, 1995. 1496: 52–58.
4. Sammer, G. Identifying and Reconciling the Data Needs of Public Transit Planning, Marketing and Performance Measurement. In *Transport Survey Methods: Keeping up with a Changing World* (P. Bonnel, M. Lee-Gosselin, J. Zmud, and J. L. Madre, eds.), 2009, pp. 321–348.
5. Richardson, A. J., E. S. Ampt, and A. H. Meyburg. *Survey Methods for Transport Planning*. Eucalyptus Press, Parkville, VIC, Australia, 1995.
6. Ashley, D., T. Richardson, and D. Young. *Recent Information on the Under-Reporting of Trips in Household Travel Surveys*. Presented at 32nd Australian Transport Research Forum, Auckland, NZ, 2009.
7. Data Management Group. *Transportation Tomorrow 2011 TTS Version 1.0 Data Expansion & Validation*. Department of Civil Engineering, University of Toronto, Toronto, 2013.
8. Chapleau, R., M. Trépanier, and K. K. A. Chu. *The Ultimate Survey for Transit Planning: Complete Information with Smart Card Data and GIS*. Presented at 8th International Conference on Survey Methods in Transport: Harmonisation and Data Comparability, Annecy, France, 2008.
9. Pelletier, M. P., M. Trépanier, and C. Morency. Smartcard Data Use in Public Transit: A Literature Review. *Transportation Research Part C: Emerging Technologies*, Vol. 19, 2011.
10. Spurr, T., R. Chapleau, and D. Piché. Use of Subway Smart Card Transactions for the Discovery and Partial Correction of Travel Survey Bias. *Transportation Research Record: Journal of the Transportation Research Board*, 2014. 2405: 57–67.
11. Spurr, T., A. Chu, R. Chapleau, and D. Piché. A Smart Card Transaction "Travel Diary" to Assess the Accuracy of the Montréal Household Travel Survey. *Transportation Research Procedia*, Vol. 11, 2015, pp. 350–364.
12. Munizaga, M., F. Devillaine, C. Navarrete, and D. Sivla. Validating Travel Behavior Estimated from Smart Card Data. *Transportation Research Part C*, Vol. 44, 2014, pp. 70–79.
13. Trépanier, M., C. Morency, and C. Blanchette. Enhancing Household Travel Surveys Using Smart Card Data? Presented at 88th Annual Meeting of the Transportation Research Board, Washington, D.C., 2009.
14. Chapleau, R., P. Gaudette, and T. Spurr. Strict and Deep Comparison of Revealed Transit Trip Structure between CATI Household Travel Survey and Smart Cards. Presented at 97th Annual Meeting of the Transportation Research Board, Washington, D.C., 2018.
15. Chapleau, R. *Challenges in Integrating Smart Cards and GTFS Data to Perform Dynamic Assignment of Transit Demand*. Presented at UITP - Global Public Transport Summit, Montréal, 2017.

The Standing Committee on Public Transportation Planning and Development (AP025) peer-reviewed this paper (18-04398).

The numeric quantities presented in this paper are the result of analyses performed by the authors alone who assume full responsibility for any errors. The results presented are theoretical and do not impose any kind of obligation, financial or otherwise, on the transit agencies concerned.

TRR
JOURNAL OF THE TRANSPORTATION RESEARCH BOARD

Article

Transportation Research Record
2018, Vol. 2672(8) 817–826

DOI: 10.1177/0361198118798714
journals.sagepub.com/home/trr
SAGE

Factors Affecting Travel Demand by Bus: An Empirical Analysis at U.S. Metropolitan Statistical Area Level

Bhuiyan Monwar Alam[1], Hilary Nixon[2], and Qiong Zhang[3]

Abstract
Understanding the nature of transit travel demand is at the heart of transportation policy making and the success of transit systems. Unfortunately, most existing studies focus narrowly on a single or a few transit systems or metropolitan areas to analyze the determinants of transit travel demand. It is, therefore, difficult to generalize their findings over a broader geographic scale. This paper examines and explores the determinants of transit travel demand by bus generalizable to the entire U.S. by investigating internal and external factors at the U.S. metropolitan statistical area (MSA) level for 2010. The authors collected and used data for all 358 MSAs nationwide. Internal factors are those that the transit managers and operators have some control over (e.g., transit supply, transit coverage), whereas external factors are those that they do not have control over (e.g., population density, vehicles per household). Several variables came directly from the U.S. Census Bureau and National Transit Database, whereas the authors constructed others using data from these two sources. The paper employs a log-log regression model to address the issues of non-normality, non-linearity, and heteroscedasticity. Findings suggest that internal factors tend to be significant predictors of travel demand by bus mode compared with external ones. As such, transit authorities have the ability to attract ridership by adjusting the influential internal factors under their control.

Transit has been the least demanded transportation mode with reference to number of trips in the U.S. for last few decades, partly because it is not a convenient mode to many (*1*). Additionally, transit service is not always available in many regions of the country so riders who would like to choose transit as their mode may not be able to access it. Its demand, as a percentage of total trips made by all travel modes, has been essentially stable since 1970 despite various government initiatives to increase patronage. Such initiatives include increased subsidies for transit systems and maintaining relatively low fares (*2*). National Household Travel Survey (NHTS) data indicate that the share of person trips by each mode of transportation in 2001 and 2009 were 86.3% and 83.4% for private vehicles, 8.6% and 10.4% for walk, 3.4% and 4.2% for others, and 1.6% and 1.9% for transit, respectively (*3*). However, current trends in demographic and economic changes in the U.S. are positively affecting demand for both public transit and transit-oriented development (TOD) (*4*). As presented in Figure 1, transit ridership by bus mode was at a low level during the Great Depression in the early 1930s. Its demand experienced a rapid increase from 1934 and reached its peak in the late 1940s. Since then, the ridership by bus in both actual numbers and trips per capita steadily declined until 1972. Ridership numbers have been relatively stable since 1972 with unlinked passenger trips varying between 5 and 6 billion a year. However, per capita bus ridership shows a slight decline during the same period.

Although transit's market share is very low, typical transit users in small- and mid-sized metropolitan statistical areas (MSAs) are the most disadvantaged groups of U.S. society: people with disabilities, older adults, poor people, and women (*1*). It is important for transit planners and policy makers to know the appropriate demand functions of transit travel for better planning and policy making. Regrettably, most existing studies focus on the transit travel demand function on a single or a few transit systems (*5*). With such a narrow focus of the existing

[1]Department of Geography and Planning, The University of Toledo, Toledo, OH
[2]Department of Urban and Regional Planning, San Jose State University, San Jose, CA
[3]Department of Geography, Michigan State University, East Lansing, MI

Corresponding Author:
Address correspondence to Bhuiyan Monwar Alam:
bhuiyan.alam@utoledo.edu

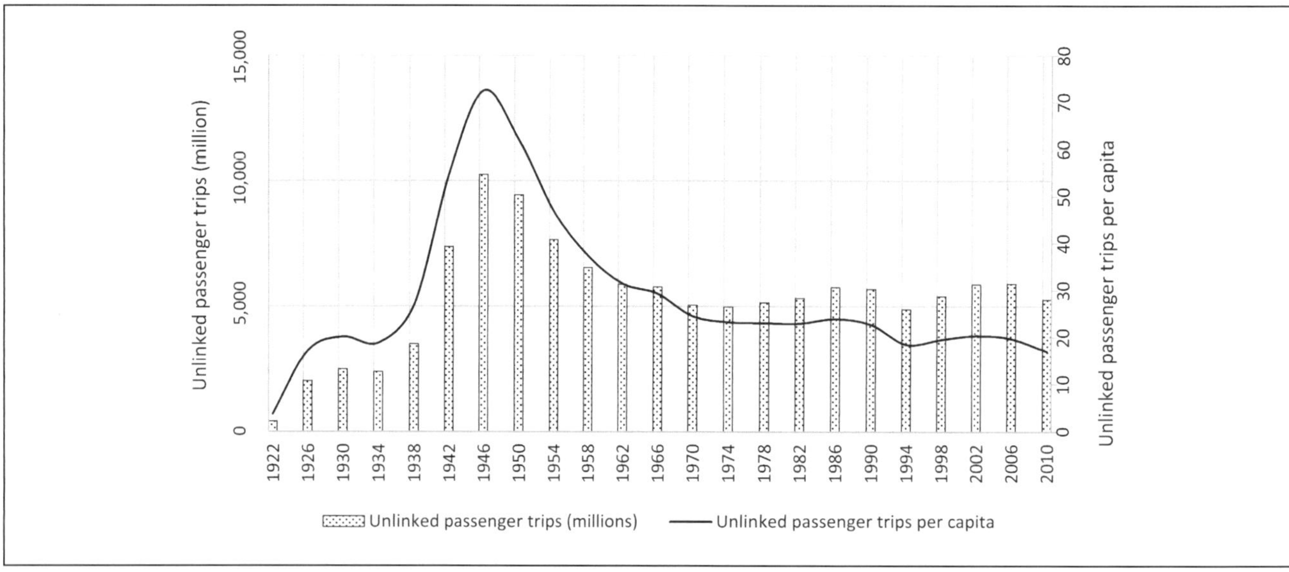

Figure 1. Annual transit ridership trend by bus mode in the U.S., 1922–2010.
(*Data Source*: 2012 Public transportation fact book, appendix A: Historical tables.)

studies, it is difficult to generalize their findings over the entire U.S. Although system-specific studies and their findings are important, having the knowledge of overall transit usage in the whole country is similarly important for planners and policy makers. This paper attempts to examine and explore the determinants of travel demand by bus that are generalizable to all of the U.S. It does so by investigating the factors at the U.S. MSA level in 2010. The study uses data on motor bus for both directly operated and purchased transportation types of services.

Existing Studies

Studies on public transit ridership analyze a range of economic, social, transit system, and other factors to identify significant determinants of ridership and transit mode share. The following two sections examine previous studies that analyze the relative influence of internal and external factors, while touching on the issue of transit demand-related public policy.

Internal Factors

Internal factors are system-specific that transit systems can adjust and control such as fare, operating hours, headways, and so forth. Studies of internal determinants assess the effects of service decisions made by transit agencies on ridership. Proponents of system determinants argue for transit systems' capacity to increase ridership through system design, fare adjustment, and service quality. Armbruster found reduction in miles covered by transit had a greater negative effect on ridership than a significant fare increase, which indicates that quality of service may be more important than fare (*6*). Other research found that positive incentives such as system expansion, service improvement, and stable fares best explained the success of certain transit systems (*4*, *7*, *8*).

Studies concerned with the effect of transit service on ridership may emphasize supply, extent, orientation pattern, and quality; the latter includes travel time, punctuality, comfort, safety, and so on. Studies tend to agree that service coverage/extent, frequency, and reliability are among the most significant determinants of transit mode choice. Sweet and Chen found evidence for travelers choosing transit over automobile commutes where transit was a faster and punctual alternative to traffic congestion (*9*). An international comparative study of transit use found that high levels of service (frequency, reliability, and destination proximity) were major drivers of light rail use in Europe, North America, and Australia (*10*). Thompson and Brown found that service coverage and frequency best explained ridership growth in western metropolitan service areas with populations from 1 to 5 million (*5*). Currie and Loader found that extending service in the evening hours had an unexpected and immediate positive effect on afternoon ridership, possibly indicating the viability of transit commutes for workers with evening shifts (*11*).

A few studies conducted over the last decade or so argue that radial transit systems, which involve transit lines radiating out from a central business district (CBD), fail to accommodate the needs of sprawling, low-density cities, many of them in the western U.S. (*12*). Studies argue that low-density metropolitan areas may achieve best results by using flexible bus routes to connect transit-dependent individuals with the dispersed employment sites characteristic of post-war auto-centric urban growth (*13, 14*).

External Factors

Unlike internal factors, external factors are those that transit managers and operators cannot exercise control over, such as population density, median household income, percentage of carless households, and so forth. Most research identifies external determinants as the most significant indicators of travel mode choice in the U.S., particularly within the wider context of long-term policy support for automobile use. Several studies found strong correlations between ridership, income, employment, real gross domestic product, and other economic factors (*2, 15*). Taylor and Fink found that economic factors, such as employment, income, and auto ownership, furnished the best immediate explanations for changes in transit mode shares within the wider context of policy (federal to municipal) that has consistently supported automobile as the dominant mode of travel nationwide (*16*).

In a comparative study of automobile dependency in 46 cities throughout the world, Kenworthy found that auto dependency was highest in the U.S., where it was associated with extremes in low-density, auto-centric land use patterns (*17*). Others found positive association between ridership, urban density, and transit accessibility (*18*). A subfield of studies examines the theory that travelers' mode choice may be influenced, and transit's share of travel increased, through investment in TOD, which situates planned, or pre-existing, sites of residential and employment density adjacent to transit, usually rail service (*19*). A dissenting body of work argues that many urban areas are best served by cheap, flexible bus transportation systems that connect transit-dependent riders with dispersed employment sites (*1*).

Guerra and Cervero found that population density increased the capital costs of transit, but also made systems cost effective because of increased ridership (*20*). Cervero and Kockelman demonstrate an association between low-density land use and travel by automobile on the one hand, and urban density, TOD, and viable transit on the other (*21*). Thompson and Brown explore that, contrary to popular belief, transit ridership grew most rapidly in the West, where bus systems had adapted to patterns of dispersed residence and employment, where the strength of the CBD is weaker (*5*). Controlling for factors thought to influence ridership, a 2008 study found no relationship between the strength of CBDs and ridership increase (*22*).

U.S. fuel prices have become increasingly unstable over the last decade, and so studies have considered gasoline price as a predictor for transit travel demand. Recent studies generally show a small but significant relation between rising fuel cost and increased ridership (*23, 24*).

Studies of socioeconomic determinants of transit ridership generally focus on the travel needs and behavior of demographic groups with very limited access to automobiles. Research shows that low-income groups lacking access to automobiles are most likely to rely on transit for access to employment, as well as fulfillment of household and other necessities (*1, 25*). Many recent immigrants have become public transit users since the mid-1990s (*26*). Studies of immigrant communities in "gateway" cities found low-income Latino immigrants to be most likely to use transit, with many also commuting by carpool (*27*). Analysis of nationwide data has found that foreign-born individuals are more likely to use transit, but has also indicated, again, that use tends to decrease over time, indicating the necessity for planning and policy initiatives, such as service extension and outreach to communities, to sustain or increase ridership levels (*28, 29*). Finally, a few studies examine the effects of offering discounted or free bus passes to university and college students who commute to campus, a policy that has produced notable ridership growth in particular localities (*8, 30*).

Data Sources and Collection Methods

The authors collected data on a wide array of variables from two major sources: the U.S. Census Bureau and the Integrated National Transit Database Analysis System (INTDAS) (*31*). INTDAS is a web-based database system designed for the retrieval and analysis of the National Transit Database (NTD). While most of the external factors for this study come from the Census Bureau, several internal factors come directly from NTD. Table 1 shows the factors that went into the final model, their sources, and construction methods.

The authors collected and used data on all 358 MSAs nationwide in 2010. Using data from the two sources, they constructed several other variables as well. For instance, they measured bus route coverage as the ratio of bus route miles to MSA population, and transit fare as the ratio of fare revenue to passenger trips by bus. Unlike previous studies that used a gas price index (*24*), the authors collected the absolute values of gas prices from gas charts provided by GasBuddy (*34*). The study calculated metropolitan sprawl index (MSI) for all 358 MSAs in the country in 2010 based on the procedures shown by Ewing et al (*32*) and Ewing and Hamidi (*33*).

Table 1. Variables Included in the Regression Model, Data Sources, and Construction Methods

Variable	Data source	Construction method
Travel demand by bus[a]	NTD	Passenger miles/MSA population
External factors		
Gas price	Gasbuddy.com	Estimated from gas price charts
Metropolitan sprawl index (MSI)	Reid Ewing	Estimated following the methods suggested by Ewing, et al (32) and Ewing and Hamidi (33)
Median household income	Census Bureau	Directly obtained from the Census Bureau
Percentage of carless households	Census Bureau	(Number of carless households in MSA/ total households in MSA) × 100
Vehicles per household	Census Bureau	Number of vehicles in MSA/total households in MSA
Percentage of college population	Census Bureau	Directly obtained from the Census Bureau
Percentage of immigrant population	Census Bureau	Directly obtained from the Census Bureau
MSAs in the South	NTD	MSAs not in the South = 0, MSAs in the South = 1
Population density	Census Bureau	MSA population/MSA size (in square miles)
Internal factors		
Transit orientation pattern	Interviews; visual inspection of bus transit network	CBD-oriented radial pattern = 0, multidestination oriented grid pattern = 1
Rail transit	NTD	No rail transit = 0, rail transit = 1
Transit fare (bus fare)	NTD	Fare revenue/passenger trips
Transit supply (by bus)	NTD	Vehicles miles/MSA population
Revenue hours (by bus)	NTD	Directly obtained from NTD
Average headway (of bus in minutes)	NTD	Directly obtained from NTD
Transit safety (no. of reported bus accidents)	NTD	Directly obtained from NTD
Bus route coverage	NTD	Bus route miles/MSA population

[a]Dependent variable.

MSI combines four factors: residential density; neighborhood mix of homes, jobs, and services; strength of activity centers and downtowns; and accessibility of the street network. Each of the factors, in turn, depends on other built environment and socioeconomic factors. Finally, this study estimated transit orientation patterns by telephone interviews of the transit managers and visual inspections of the network of each bus transit system to determine whether the system is of CBD-oriented radial pattern or multidestination-oriented grid pattern.

Explanation of the Variables

Informed by the existing literature and theories, the authors initially considered a wide array of variables to determine the factors that significantly influence transit travel demand by bus, measured as passenger miles per capita. NTD defines passenger miles as number of annual passenger trips multiplied by the system's average trip length in miles, which essentially is transportation service consumed with reference to total number of passenger miles. Initially, the authors constructed passenger trips per capita as well, which was obtained by dividing total passenger trips in an MSA by the population of the respective MSA. Originally, the authors thought of using passenger trips per capita as a measure of travel demand by bus. Their preliminary results, as reflected by the *F*-statistic and coefficients of multiple determination, indicated that passenger miles per capita was a more appropriate measure of transit demand than passenger trips per capita. Therefore, the authors used passenger miles per capita as the dependent variable in their final model. The use of passenger miles per capita as a measure of transit demand is also supported by literature (*5*).

Explanation of External Variables

Population Density. Both population and physical size of the MSA are important factors that influence transit ridership level. The study employs the relative influence of population and size, that is, population density, which is a ratio of MSA population to MSA size, and measures per square mile. The study expects that higher population density will attract more riders to bus transit.

Median Household Income. Median household income represents the economic condition of a geographic unit. Typically, this variable shows significant influence on transit ridership—the lower the income, the more the demand for transit ridership, and vice versa (*1*, *5*). Thus, the study expects a negative association between median household income and travel demand by bus.

Percentage of Carless Households. To some extent, the percentage of carless households reflects the socioeconomic condition of a geographic unit so it is important to include this variable in transit demand studies. It is expressed as the ratio of number of carless households to the total number of households in an MSA. The authors expect a higher percentage of carless households to cause an increase in transit travel demand. The initial regression model showed the presence of multicollinearity between percentage of carless households and percentage of minority population. Therefore, the authors dropped the percentage of minority population from the final model.

Percentage of College Population. There are two reasons to include the percentage of college population (defined as the total number of college-going population divided by the total MSA population) in the study. First, most college students are required or advised to live on or around campus, where their activities are usually concentrated. Consequently, it is not necessary for many of them to possess a car. Second, for most college students, the cost of having a car is expensive. While many colleges and universities have transit passes built into their student fees, many do not (*30*). Thus, it is important to include this variable in the study of transit demand analysis. The authors expect a positive relationship between percentage of college population and transit demand by bus.

Percentage of Immigrant Population. Despite a small proportion of highly skilled and educated people immigrating into the U.S. each year, a substantial proportion of immigrants are labor-force members who do not have advanced degrees or professional training. They usually have a relatively low income and often depend on transit for their daily travel. Thus, it is necessary to consider the immigrant population in the analysis of transit demand. One would expect that an increase in percentage of immigrant population would lead to more demand for transit, leading to a positive relationship between the two.

Gas Price. It is generally believed that many drivers switched to transit because of increased gas prices during the economic downturn in the late 2000s, indicating that a positive relationship is expected between gas price and transit travel demand by bus. The authors collected the absolute values of gas prices from a public website, GasBuddy.com, which contains free price charts that record the daily gas prices for major cities in each state. Although the Bureau of Labor Statistics has a consumer price index, most of its data are presented at the state level. Each MSA contains one socioeconomic core city or county. Therefore, it is reasonable to believe that the core city's gas price represents the corresponding MSA's gas price level.

Metropolitan Sprawl Index. Although MSI has not been employed in existing studies to investigate its impacts on demand for public transit, the authors included it because MSI measures the degree of urban sprawl not only from geographical perspectives but also from socioeconomic perspectives. The higher the MSI value, the less sprawled an MSA is. One would expect that the less sprawled an MSA is, the more demand for transit. Therefore, this study expects MSI to cause positive impacts on transit travel demand.

MSAs in the South. Studies indicate that apart from cities like New York, where transit supply and demand are still high, transit use is declining in the Midwest and Northeast relative to their population (*5*), while increasing in the South and West in recent years (*2, 35*). Such an increase in the South and West may be attributed to the sociodemographic change caused by a large influx of immigrants. This study includes a dummy variable named "South" to test if MSAs in the South make any difference in transit demand. Initially, it also included a dummy variable for MSAs in the West but kept only MSAs in the South in the final model.

Vehicles per Household. Vehicles per household is considered an important determinant of transit travel. It is expected that the demand for transit will go down with the increase in vehicles per household and vice versa. Thus, the authors expect a negative association between this variable and travel demand by bus.

Explanation of Internal Variables

Transit Supply. NTD estimates transit supply by dividing the number of annual vehicle miles, which includes revenue as well as deadhead miles, by the service area's population, which can be characterized as the number of miles of service provided to each person in the service area. Transit supply is a measure of the extensiveness of service provided in the service area. This study defines transit supply in the same way as NTD defines and estimates it by dividing vehicle miles by total MSA population. It assumes that the demand for travel by bus will increase if the transit systems increase transit supply; therefore, it expects a positive relationship between transit supply and transit demand by bus.

Revenue Hours. Revenue hours refers to total hours of operation along the routes where a transit system provides passenger service. An increase in revenue hours is expected to produce an increase in transit travel demand (*2*). As such, this study expects a positive association between revenue hours and transit travel demand by bus.

Average Headway. Transportation professionals measure average headway as the time the trailing bus takes to cover the distance between the front of the trailing bus and the front of the leading bus. NTD provides headway values for an entire system, in minutes. It first estimates the system size in terms of total non-directional route miles by doubling the directional route miles, which is then divided by the system's calculated average speed, that is, revenue miles per revenue hours to estimate the time it would take, in hours, to traverse all the system's total route miles. NTD divides this time figure by the system's number of peak vehicles to determine the number of hours it takes a vehicle to complete its portion of the total route miles once. Finally, this hour value is multiplied by 60 to obtain headway in minutes. A longer headway indicates a less frequent transit service. Traffic theory suggests that demand for transit travel will decrease with an increase in average headway (*36*); thus, the study expects a negative association between average headway and travel demand by bus. Average headway is a measure of transit frequency, which can be considered a measure of transit service efficiency as well (*2*).

Transit Safety. NTD defines transit safety by a surrogate variable, which is number of incidents or accidents reported. An incident or accident is counted if it causes casualty, collision, or property damage more than $1,000. More incident or accident reporting indicates that the transit system is less safe for the riders, which decreases their confidence in the transit system. One would expect that this decreased confidence in the safety feature of the transit system would eventually decrease the demand for transit. Consequently, the authors believe that a negative association exists between transit's safety and demand.

Transit Fare. General economic theory suggests that the demand for a commodity decreases if its price increases. Existing studies, like Taylor et al, have tested the influence of transit fare on patronage and found that when the transit fare goes up, ridership goes down (*2*). Transit fare is calculated as the ratio of passenger fare revenue to the total number of passenger trips.

Rail Transit. The presence of rail transit in an MSA might influence ridership by bus mode. Many riders may drive to the rail station, park at the station, and take the train to reach their destination. Similarly, bikers may use the train, and bike to and from the rail stations to reach their destinations. Others, however, may take the bus to and from the rail stations to meet their derived travel demand. Yet others may decide to only use the train and not ride the bus at all. This study includes a dummy variable that indicates the presence or absence of rail transit in an MSA: 0 indicates absence whereas 1 indicates presence of rail transit. The authors believe that the presence of rail transit in an MSA increases travel demand by bus, indicating that the presence of rail transit and travel demand by bus are positively associated.

Transit Orientation Pattern. There are two polarized route structures for transit orientation patterns—traditional CBD-oriented radial pattern and multidestination-oriented grid pattern (*12*). Transit managers who operate the former kind of system focus on attracting customers to the CBD, which they believe is the largest market for transit service (*14*, *37*). Transit managers who follow the latter type of system are usually dedicated to helping customers reach multiple destinations without transfer. Therefore, a distinguishing difference between the radial and grid pattern is the use of transfer (*14*). The authors coded 0 for CBD-oriented radial pattern and 1 for multidestination-oriented grid pattern. As multidestination-oriented transit networks provide direct services from origin to destination, they are generally expected to produce more transit demand. Thus, this study expects a positive association between transit orientation pattern and travel demand by bus.

Bus Route Coverage. Taylor et al (*2*) define route coverage as the ratio of route miles to land area, whereas Thompson and Brown (*5*) define service coverage as the ratio of population to route miles. Taylor et al (*2*) expect a positive relationship between transit demand and route coverage, whereas Thompson and Brown (*5*) expect a negative relationship. This study defines bus route coverage as the ratio of bus route miles to MSA population. One would expect that the greater the ratio of bus route miles to population, the larger the transit coverage by bus and, thereby, an increasing demand for transit by bus. Therefore, a positive association is expected between bus route coverage and travel demand by bus. This expectation is similar to that of Taylor et al (*2*).

Transit deserts (*38*, *39*), defined as areas lacking adequate public transit services but having a substantial proportion of population in need of such services, is another important variable that the authors thought of including in the model. Unfortunately, such data at MSA level are not available for the entire U.S. so the study could not include this variable.

Model Specification

Some believe that transit supply and transit demand are jointly produced and, therefore, they cause simultaneous effects on each other. This simultaneity between transit supply and demand can lead to inaccurate and biased coefficients of independent variables. Taylor et al argue that putting these two variables in one equation at the same time could produce a slope that is larger than it actually is (*2*). Thus, to address the simultaneous relationship between transit supply and demand, they constructed two-stage least squares (2SLS) simultaneous equation models in their cross-sectional analysis of transit use in 265 U.S. urbanized areas (UZA). McMullen and Eckstein apply this technique in their panel study of 87 U.S. urban areas to determine the vehicle miles traveled between 1982 and 2009 (*40*).

Contrarily, others believe that transit supply relatively quickly increases demand for transit, as measured by ridership, but that increased demand does not necessarily enhance supply in the context of a cross-sectional time frame. Transit planners would create more supply over a long period where they perceive greater demand, but that process generally takes years. Thompson and Brown argue that empirical results do not support that transit supply and demand are jointly produced and, therefore, might not have reciprocal impacts on each other (*5*). As a result, they use ordinary least squares (OLS) regression to estimate the transit travel demand change between 1990 and 2000. Controlling for possible simultaneity between transit supply and demand, Liu applied and compared 2SLS and OLS regression models to determine transit demand but found that applying the two techniques yielded no significant differences in the results (*41*).

The authors of this study believe, like Thompson and Brown, that transit supply and demand may have reciprocal relationships over a longer period but not for cross-sectional data over a short period of time (*5*). Therefore, they employed an OLS regression model to determine the factors that significantly influence travel demand by bus. Initial data screening indicated the presence of non-normality, non-linearity, and heteroscedasticity among the variables. To address these issues, the study transformed the continuous variables to natural logarithms while keeping three dummy variables untransformed, and employed a log-log regression model. It used variance inflation factor, tolerance, and correlation coefficients to determine multicollinearity among explanatory variables. Values for all these indicators were within acceptable limits (*42*).

Results and Analysis

Table 2 suggests that vehicles per household, percentage of college population, percentage of immigrant population, transit fare, transit supply, revenue hours, average headway, transit safety, and bus route coverage show statistically significant impacts on transit demand by bus. These variables affect travel demand by bus in their expected directions. Since the regression model is log-log in nature, the coefficients represent elasticities of change in travel demand by bus caused by a unit change in an explanatory variable. Transit supply, for instance, causes the highest impact on travel demand in the expected direction—a 10% increase in transit supply causes a 5.75% increase in travel demand by bus.

A higher number of revenue hours indicates better and more frequent service, which results in greater demand for transit. This study supports this hypothesis. Existing studies and general economic theory suggest that the demand for transit goes down as its fare goes up. For U.S. MSAs, at least, the study provides further proof: an inverse relationship exists between the transit's fare and its demand. A similar inverse relationship exists between average headway and travel demand by bus. Average headway is the time needed by a vehicle to cover the distance between the tip of the preceding vehicle to the tip of the following vehicle. The shorter the average headway, the greater the frequency of vehicles crossing a given point on a roadway segment per hour or day, which, all else being equal, leads to an eventual increase in transit travel demand. The current study supports this theory and indicates that travel demand by bus increases with the decrease in average headway, which, in turn, is a measure of transit service frequency (*2*).

Bus route coverage shows significant impacts on travel demand by bus in the expected direction. As explained earlier, the study defines bus route coverage as the ratio of route miles to MSA population. Thus, a larger ratio of route miles to MSA population indicates denser route miles per person that would lead to a more frequent and better quality bus service. This more frequent service, in turn, will cause increased demand for transit by bus. One would, therefore, expect a positive relationship between bus route coverage and its demand. The study supports this hypothesis and finds that bus route coverage is a major contributor to travel demand by bus: a 10% increase in bus route coverage will cause a 3.19% increase in travel demand by bus.

Transit safety, which is measured as the number of incidents or accidents reported per year, also shows significant impacts on travel demand by bus in the expected direction—more incident or accident reporting shows a decreasing demand for transit travel by bus. As discussed in the previous sections, existing studies indicate that the percentage of college population and percentage of immigrant population increase demand for transit. This study supports such findings of existing studies and indicates that the demand for travel by bus is significantly and

Table 2. Log-Log Regression Model Showing Impacts of Explanatory Variables on Travel Demand by Bus[a] and Their Expected and Demonstrated Behavior

Factors	Coefficient	SD	Expected behavior	Demonstrated behavior
Constant	–.739	2.403		
External factors				
Ln gas price	.292	.625	Positive	Positive
Ln metropolitan sprawl index	–.002	.116	Positive	Negative
Ln median HH income	.070	.209	Negative	Positive
Ln percentage of carless household	.073	.093	Positive	Positive
Ln vehicles per household	–.522	.245*	Negative	Negative
Ln percentage of college population	.170	.060**	Positive	Positive
Ln percentage of immigrant population	.094	.029**	Positive	Positive
MSAs in the South (0 = no; 1 = yes)	.002	.058	Positive	Positive
Ln population density	–.043	.036	Positive	Negative
Internal factors				
Transit orientation pattern (0 = radial; 1 = grid)	.062	.052	Positive	Positive
Rail transit (0 = no rail; 1 = rail)	–.050	.097	Positive	Negative
Ln transit fare (bus fare)	–.222	.037***	Negative	Negative
Ln transit supply (by bus)	.575	.082***	Positive	Positive
Ln revenue hours (by bus)	.385	.041***	Positive	Positive
Ln average headway (of bus in minutes)	–.227	.048***	Negative	Negative
Ln transit safety (no. of reported bus accidents)	–.080	.024**	Negative	Negative
Ln bus route coverage	.319	.065***	Positive	Positive

Note: SD = standard deviation; Ln = natural logarithm; HH = household; N = 358; R^2 = 0.809; Adjusted R^2 = 0.800; F-statistic = 84.978***.
[a]Dependent variable.
*$p < 0.05$. **$p < 0.01$. ***$p < 0.001$.

directly influenced by increases in the percentages of these two factors.

This study, however, found that certain variables that many transit planners and scholars view as important determinants of transit demand did not demonstrate significant impacts on travel demand by bus. Variables such as gas price, MSI, median household income, percentage of carless households, MSAs in the South, population density, transit orientation pattern, and presence of rail transit are expected to cause significant impacts on travel demand by bus, but were not proved so by this study. Among these, gas price, percentage of carless population, MSAs in the South, and transit orientation pattern behaved in their expected direction of relationships with travel demand by bus, whereas others showed opposite directions to what was expected.

Conclusions

The study regressed a total of 17 variables on travel demand by bus. The results show that the internal factors are the predominant significant predictors of travel demand by bus mode. In fact, six out of eight internal factors proved to be significant determinants. Only transit orientation pattern and presence of rail transit were insignificant internal variables. Additionally, vehicles per household, percentage of college population, and percentage of immigrant population are external factors that proved to be significant predictors for travel demand by bus. Among the remaining six insignificant external variables that the study tested, three demonstrated behavior in the expected direction whereas the other three demonstrated the opposite of the expected behavior.

Although this study does not support the conventional wisdom that external variables, such as population density and median household income, cause significant impacts on transit demand, it supports that special population groups, such as percentage of college population and percentage of immigrant population, demonstrate significant impacts on travel demand by bus. It indicates that although population density does not individually affect travel demand by bus, these population groups, who are often transit dependent, demonstrate a significant impact on travel demand by bus.

The findings of this study are important for transit policy makers, planners, operators, managers, and stakeholders. As the study finds that the factors internal to the transit systems predominantly determine the demand for travel by bus in the U.S. at MSA level, it largely falls onto transit systems to determine how to provide residents with efficient bus transit systems without depending on outside factors. In other words, they can attract and increase bus transit ridership by adjusting a few significant factors specific to their transit systems without being dependent on the outside world. Keeping this in mind, transit policy makers and planners may make appropriate plans and policies that will help transit authorities provide efficient transit systems to their users to increase ridership.

Overall, this study reveals that most of the socioeconomic factors that are beyond the control of transit systems do not necessarily contribute to transit's demand, and this simplifies the problem in a sense. It reveals that the job of building ridership belongs to, and is within reach of, transit policy makers, and that to achieve this goal, most efforts should focus on providing better transit systems that work more efficiently and effectively. However, one needs to be cautious in translating the findings of this research in a literal manner for two reasons. First, most decisions taken by transit operators and managers are dependent on funding availability. Often, such funding is insufficient for most transit systems. Adding adequate federal and state funding could be the first step to make transit, especially bus transit, a viable mode of transport. Of course, transit planners and policy makers need to continue to work on solving other policy and institutional barriers of service improvement. For instance, how regional transit operators and managers can provide efficient service amid widespread sprawling is a challenging task to accomplish. Second, NTD does not provide disaggregate data at an individual or household level but provides aggregate data at an MSA level. Therefore, one should be restrained in blindly applying the findings of this study to any specific MSA because the situation in an individual MSA could be different from the findings of this research, which uses aggregate-level data. The results, however, powerfully suggest that if transit providers and policy makers can muster the political will to improve service in certain critical respects, this will indeed grow the market for transit service.

Lastly, external factors like employment density, presence of millennial population (who are less reliant on cars), and presence of alternative transportation services (such as bike share and other transportation network companies like Uber and Lyft), which could compete with the bus service, would be valuable entries to the model. Regrettably, data on these variables are not yet available from the NTD data set at MSA level. Future research can model ridership demand by bus including these variables.

Author Contributions

The authors confirm contribution to the paper as follows: study conception and design: BMA; data collection: BMA and QZ; analysis and interpretation of results: BMA; draft manuscript preparation: BMA and HN. All authors reviewed the results and approved the final version of the manuscript.

References

1. Alam, B. M. Transit Accessibility to Jobs and Employment Prospects of Welfare Recipients without Cars: A Study of Broward County, Florida, using GIS and Econometric Model. *Transportation Research Record: Journal of the Transportation Research Board*, 2009. 2110: 78–86.
2. Taylor, B. D., D. Miller, H. Iseki, and C. Fink. Nature and/or Nurture? Analyzing the Determinants of Transit Ridership across U.S. Urbanized Areas. *Transportation Research Part A: Policy and Practice*, Vol. 43, No. 1, 2009, pp. 60–77.
3. National Household Travel Survey. *Summary of Travel Trends*. National Household Travel Survey, 2009. http://nhts.ornl.gov/2009/pub/stt.pdf.
4. Litman, T. *Evaluating Public Transit Benefits and Costs: Best Practices Guidebook*. Victoria Transport Policy Institute, Victoria, BC, Canada, 2017. http://www.vtpi.org/tranben.pdf.
5. Thompson, G., and J. Brown. Explaining Variation in Transit Ridership in U.S. Metropolitan Areas between 1990 and 2000: Multivariate Analysis. *Transportation Research Record: Journal of the Transportation Research Board*, 2006. 1986: 172–181.
6. Armbruster, B. P. *Factors Affecting Transit Ridership at the Metropolitan Level 2002–2007*. Master's thesis. Georgetown University, Washington, D.C., 2012. http://repository.library.georgetown.edu/handle/10822/553638.
7. Kain, J. F., and Z. Liu. Secrets of Success: Assessing the Large Increases in Transit Ridership Achieved by Houston and San Diego Transit Providers. *Transportation Research Part A: Policy and Practice*, Vol. 33, No. 7–8, 1999, pp. 601–624.
8. Litman, T. Transit Price Elasticities and Cross-Elasticities. *Journal of Public Transportation*, Vol. 7, No. 2, 2004, pp. 37–58.
9. Sweet, M. N., and M. Chen. Does Regional Travel Time Unreliability Influence Mode Choice? *Transportation*, Vol. 38, No. 4, 2011, pp. 625–642.
10. Currie, G., A. Ahern, and A. Delbosc. Exploring the Drivers of Light Rail Ridership: An Empirical Route Level Analysis of Selected Australian, North American and European Systems. *Transportation*, Vol. 38, No. 3, 2011, pp. 545–560.
11. Currie, G., and C. Loader. High Ridership Growth from Extended Transit Service Hours. *Transportation Research Record: Journal of the Transportation Research Board*, 2009. 2110: 120–127.
12. Thompson, G., and T. Matoff. Keeping Up with the Joneses: Radial vs. Multidestinational Transit in Decentralizing Regions. *Journal of the American Planning Association*, Vol. 69, No. 3, 2003, pp. 296–312.
13. Brown, J. R., and G. L. Thompson. Should Transit Serve the CBD or a Diverse Array of Destinations? A Case Study Comparison of Two Transit Systems. *Journal of Public Transportation*, Vol. 15, No. 1, 2012, pp. 1–18.
14. Brown, J. R., and G. L. Thompson. Examining the Influence of Multidestination Service Orientation on Transit Service Productivity: A Multivariate Analysis. *Transportation*, Vol. 35, No. 2, 2007, pp. 237–252.

15. Yoh, A., P. Haas, and B. Taylor. Understanding Transit Ridership Growth: Case Studies of Successful Transit Systems in the 1990s. *Transportation Research Record: Journal of the Transportation Research Board*, 2003. 1835: 111–120.
16. Taylor, B. D., and C. Fink. *The Factors Influencing Transit Ridership: A Review and Analysis of the Ridership Literature*. University of California Transportation Center, Berkeley, CA, 2003. http://www.reconnectingamerica.org/assets/Uploads/ridersipfactors.pdf.
17. Kenworthy, J. R. Patterns of Automobile Dependence in Cities: An International Overview of Key Physical and Economic Dimensions with Some Implications for Urban Policy. *Transportation Research Part A: Policy and Practice*, Vol. 33, No. 7–8, 1999, pp. 691–723.
18. Chakraborty, A. Land Use and Transit Ridership Connections: Implications for State-Level Planning Agencies. *Land Use Policy*, Vol. 30, No. 1, 2013, pp. 458–469.
19. Thompson, G., and J. Brown. Making a Successful LRT-Based Regional Transit System: Lessons from Five New Start Cities. *Journal of Public Transportation*, Vol. 15, No. 2, 2012, pp. 157–180.
20. Guerra, E., and R. Cervero. Cost of a Ride: The Effects of Densities on Fixed-Guideway Transit Ridership and Costs. *Journal of the American Planning Association*, Vol. 77, No. 3, 2011, pp. 267–290.
21. Cervero, R., and K. Kockelman. Travel Demand and the 3Ds: Density, Diversity, and Design. *Transportation Research Part D: Transportation and Environment*, Vol. 2, No. 3, 1997, pp. 199–219.
22. Brown, J. R., and D. Neog. *Urban Structure and Transit Ridership: A Reexamination of the Relationship in the United States*. Working Paper Series, WPS #07-01. Department of Urban and Regional Planning, Florida State University, Tallahassee, FL, 2008.
23. Currie, G., and J. Phung. Understanding Links between Transit Ridership and Gasoline Prices: Evidence from the United States and Australia. *Transportation Research Record: Journal of the Transportation Research Board*, 2008. 2063: 133–142.
24. Lane, B. W. The Relationship between Recent Gasoline Price Fluctuations and Transit Ridership in Major U.S. Cities. *Journal of Transportation Geography*, Vol. 18, No. 2, 2010, pp. 214–225.
25. Pucher, J., and J. L. Renne. Socioeconomics of Urban Travel: Evidence from the 2001 NHTS. *Transportation Quarterly*, Vol. 57, No. 3, 2003, pp. 49–77.
26. Blumenberg, E., and A. E. Evans. Planning for Demographic Diversity: The Case of Immigrants and Public Transit. *Journal of Public Transportation*, Vol. 13, No. 2, 2010, pp. 23–45.
27. Liu, C., and G. Painter. Travel Behavior among Latino Immigrants: The Role of Ethnic Concentration and Ethnic Employment. *Journal of Planning Education and Research*, Vol. 32, No. 1, 2011, pp. 62–82.
28. Chatman, D. G., and N. Klein. Immigrants and Travel Demand in the United States: Implications for Transportation Policy and Future Research. *Public Works Management and Policy*, Vol. 13, No. 4, 2009, pp. 312–327.
29. Blumenberg, E. Moving In and Moving Around: Immigrants, Travel Behavior, and Implications for Transport Policy. *Transportation Letters: The International Journal of Transportation Research*, Vol. 1, No. 2, 2009, pp. 169–180.
30. Brown, J. R., D. Hess, and D. Shoup. Fare-Free Public Transit at Universities: An Evaluation. *Journal of Planning Education and Research*, Vol. 23, 2003, pp. 69–82.
31. Integrated National Transit Database Analysis System. *Integrated National Transit Database Analysis System*. Florida International University, Miami, FL, 2012. http://www.ftis.org/intdas.html.
32. Ewing, R., R. Pendall, and D. Chen. *Measuring Sprawl and Its Impact*. *Smart Growth America*, Washington, D.C., 2003. http://www.smartgrowthamerica.org/documents/MeasuringSprawlTechnical.pdf.
33. Ewing, R., and S. Hamidi. *Measuring Urban Sprawls and Validating Sprawl Measures*. Research Geographic Information Systems and Science for Cancer Control, National Cancer Institute, National Institute of Health, Rockville, MD, 2014. http://gis.cancer.gov/tools/urban-sprawl/.
34. Gassbuddy. *Historical Price Charts*, 2017. http://gasbuddy.com/Charts.
35. Thompson, G., J. Brown, R. Sharma, and S. Scheib. Where Transit Use Is Growing? Surprising Results. *Journal of Public Transportation*, Vol. 9, No. 2, 2006, pp. 25–43.
36. TCRP Report 165: *Transit Capacity and Quality of Service Manual*, 3rd ed. Transportation Research Board of the National Academies, Washington, D.C., 1999.
37. Alam, B. M., G. L. Thompson, and J. R. Brown. Estimating Transit Accessibility with an Alternative Method: Evidence from Broward County, Florida. *Transportation Research Record: Journal of the Transportation Research Board*, 2010. 2144: 62–71.
38. Jiao, J., and M. Dilivan. Transit Deserts: The Gap between Demand and Supply. *Journal of Public Transportation*, Vol. 16, No. 3, 2013, pp. 23–39.
39. Jiao, J. Identifying Transit Deserts in Major Texas Cities Where the Supplies Missed the Demands. *The Journal of Transport and Land Use*, Vol 10, No. 1, 2017, pp. 529–540.
40. McMullen, B., and N. Eckstein. Determinants of VMT in Urban Areas: A Panel Study of 87 U.S. Urban Areas, 1982-2009. *Journal of Transportation Research Forum*, Vol. 52, No. 3, 2013, pp. 5–24.
41. Liu, Z. Determinants of Public Transit Ridership: Analysis of Post-World War II Trends and Evaluation of Alternative Networks. PhD thesis. John F. Kennedy School of Government, Harvard University, Cambridge, MA, 1993.
42. Gujarati, D. Basic Econometrics by Example. Palgrave McMillan, New York, NY, 2011.

The Standing Committee on Public Transportation Planning and Development (AP025) peer-reviewed this paper (18-06031).

Article

TRR
JOURNAL OF THE TRANSPORTATION RESEARCH BOARD

Transportation Research Record
2018, Vol. 2672(8) 827–839

DOI: 10.1177/0361198118786674
journals.sagepub.com/home/trr
SAGE

Transit Orientation: More Than Just Coverage—A New Method for the Assessment of Transit and Development Co-Location

Christian M. Marti[1], Luca Bertolini[2], and Ulrich Weidmann[1]

Abstract

This paper presents a new method for assessing transit and development co-location and applies it to a case study. Co-location is a core element of transit oriented development. It is currently assessed by designating activities within a given distance from transit as "close to transit" and calculating the activity density of these catchment areas or the share of activities within them. However, transit demand decreases with distance, so distribution of activities within transit catchment areas matters in addition to average density. The main contribution of the new method is explicitly assessing density distribution within transit catchment areas. It is based on the notion that density should not increase with distance from transit. Case study results demonstrate the method's ability to compare station areas based on aggregate indicator values, while also providing maps of disaggregate and spatially explicit co-location performance. This fine-grained analysis allows planners to identify potential future development areas. Results are compared to commonly used indicators for station area intensity and proximity of activities to transit. An important conclusion is that the new method should be used in combination with an intensity indicator.

Transit oriented development (TOD) is an important planning paradigm (*1–3*). A key goal of TOD – engrained in its very name – is orienting development toward transit. This co-location of transit and development can be achieved by locating activities where they are close to transit, or by creating transit where activities are located.

Co-location is important because the distance transit users are willing to travel (often by walking) to/from a transit stop is limited, and the share of people willing to walk to/from a stop decreases with distance (*4–10*). The exact form of this distance-decay of transit demand depends on factors including trip purpose, service characteristics, sociodemographic characteristics, and context. However, it is clear that the distance to/from a transit stop is one of the main factors determining the probability that potential transit users will actually use transit.

Co-location is commonly considered in TOD planning, land use transport integration analysis, or direct demand models for transit because of its importance. The most common approach is to define a catchment area – the area within a certain threshold distance (Euclidean or network) from a transit stop – and to designate all activities within this area as "close to transit", and all other activities as "not close to transit". This approach is inadequate because of the influence of distance on potential transit users' travel behavior.

This paper presents a new method for assessing transit and development co-location that accounts for the importance of proximity. It is organized into four parts. Part one describes current practices for co-location assessment and derives a research gap. Part two introduces a new indicator for co-location assessment. Part three applies the indicator in a case study. Part four presents the conclusions.

Co-Location Assessment

Application Scenarios

There are three applications for co-location assessment:

[1]ETH Zurich, Institute for Transport Planning and Systems (IVT), Zurich, Switzerland
[2]University of Amsterdam, Faculty of Social and Behavioural Sciences, Department of Human Geography, Planning and International Development Studies, Urban Planning Programme Group, Amsterdam, The Netherlands

Corresponding Author:
Address correspondence to Christian M. Marti:
christian.m.marti@gmail.com

1. planning support, by using the results as pointers for possible interventions;
2. normative or incentive policies, prescribing or encouraging a certain standard of co-location;
3. evaluation and learning.

As a planning tool, co-location indicators can help generate ideas and support decision-making when making changes to transit systems, the built environment, or both. Examples include analyzing existing or proposed co-location strengths and weaknesses or rating alternative proposals. As a policy tool, threshold values for co-location indicators could be developed and used as a condition for granting a building permit, for incentives such as a floor area bonus, or for transit subsidies. As an evaluation and learning tool, co-location assessment can shed light on how well co-location goals of executed projects have been achieved or highlight particularly successful cases that can serve as models for new developments.

Existing Approaches

Transit and development co-location assessment appears in the literature in different contexts. Most prominently, it is used as an independent variable in direct transit demand models (*7*, *11*), as an indicator in studies on the share of people with access to transit (*12*), and in TOD evaluation (*13*).

The most common approach for assessing co-location is to evaluate transit proximity by defining catchment areas within a fixed distance from stations. This is essentially an "all-or-nothing approach" (*7*, p. 1084) as any activity within the catchment area is considered to have transit access (and to potentially generate demand), while all activities beyond the threshold distance are considered to not have transit access (and not to generate any transit demand). The rationale for using a threshold is the notion of "maximum acceptable walking distance". The most common way of evaluating co-location is thus measuring intensity of catchment areas (e.g. density or number of activities), particularly for direct demand models. Other approaches capture proximity of activities to transit by calculating ratios, for example between the number of activities within two catchment areas defined using different distances such as 300 and 1200 m (*13*), or by analyzing the share of all activities located within a given threshold distance from transit (*12*). In any case, the exact distribution of activities or density is not considered in these methods for assessing co-location.

The problem with these "all-or-nothing" approaches is that transit demand decreases with increasing distance from transit (*4–10*). This means there is no fixed threshold distance people are willing to walk, but rather a continuous decay of transit demand with increasing distance. Thus, the influence of density on transit demand is determined by the exact distance of activities from transit rather than by the number of activities located within the threshold distance. In other words, transit demand depends not only on the average density of activities within transit catchment areas, but also on its distribution, or "articulated density" (*3*, p. 155).

In practical terms this means "concentrating housing and employment within several hundred feet of a rail station will produce far more riders than placing the same amount of development a half-mile away" (*11*, p. 286). In addition, it implies that existing approaches "cannot reflect the greater tendency to use public transport when the distance within the station catchment area is shorter. … They are not able to reflect the impact on travel of concentrating housing and employment at a longer/shorter distance from the station in cases where these developments are located within the station catchment area" (*7*, p. 1084).

Empirical evidence that density distribution within station catchment areas affects ridership is provided by Gutiérrez, Cardozo and García-Palomares (*7*). They developed a direct transit demand model that weights activities in transit station catchment areas by their distance from the station. Their model systematically produces better results than the all-or-nothing approach used by previous models. However, the distance-decay weighted regression applied by Gutiérrez et al. does not allow for the explicit assessment of co-location performance, such as comparing two different station areas or two alternative development scenarios for one station area based on their density distribution (*7*). Furthermore, it does not provide spatially explicit information about the causes of high or low co-location performance.

In summary, there is currently no approach that directly evaluates activity or density distribution around transit stations with regard to transit and development co-location.

Density Distribution Assessment

To address this gap, this research developed a method for assessing density distribution within transit catchment areas. One possible approach for assessing density distribution would be to recognize that transit demand decreases with increasing distance from stops, and therefore an ideal density distribution would locate all activities "as close as possible" to stops. An example is the indicator used by Provincie Noord-Holland and Vereniging Deltametropool (*13*). It calculates the share of activities located within 300 m of a station compared to all activities located within 1200 m from the station. This indicator could be further developed using several distances within the catchment area and averaging the share of activities within each distance. Any such "the closer the better" assessment would render the best result if all activities were located very close to transit – with either empty land between stops or very small stop spacing. However, using the notion of "the closer the better" as a general guideline for achieving co-location of transit and activities is problematic for three main reasons.

First, transit demand depends on many factors other than density and its distribution, among them the quality of the built environment. Concentrating all activities within a very small area around transit stops and leaving "gaps" in-between would lead to strong overcrowding in developed areas, and there is no evidence that this would create a high quality urban environment.

Second, proximity to transit can increase nuisances such as noise. Therefore, some potential transit customers might prefer to live a few hundred meters from a station instead of as close as possible.

Third, even if extreme concentration was generally desirable, it could still have negative impacts on overall transit demand, depending on future development demand and maximum densities (e.g. because of future building codes or developer behavior). There are two cases to be considered:

1. If all future development demand could be absorbed by the area directly around a transit stop (here called "core area", e.g. all land located within 300 m of a stop), it would be ideal from a transit perspective to locate all activities inside this core area. In this case, locating activities beyond the core area would redistribute density away from transit and thus decrease transit demand because of higher distances to transit.
2. However, if not all future development demand could be absorbed by the core area around a stop, limiting development to this core area would not be beneficial from a transit perspective. In such a case, development further away from transit is fully added to development within the core area, and thus still increases overall patronage, even if it creates less transit rides per activity.

This consideration needs to be undertaken at a corridor or even metropolitan level. However, as both future development demand and future maximum densities are not normally predictable, and the situation at the corridor or metropolitan level might be unknown when analyzing one specific station area, it seems unwise to generally limit development to the core areas very close to transit stops. However, this means that it is necessary to consider the importance of proximity in some other way.

In short, a better quality criterion than "the closer the better" is needed, one that considers the uncertainty of possible densities and demand for activities, as well as environmental problems stemming from extreme concentration of development around transit stops.

This paper proposes using the criterion that activity density should not increase with distance from a transit stop as a basis for creating a new method for assessing co-location. An assessment based on this notion would render the best result for any situation that has an even density distribution or increasing density toward transit stops. Thus, it is less prescriptive about concentration around transit stops than the "the closer the better" approach, but still penalizes concentration away from transit stops.

The proposed indicator assesses density distribution within a station area but does not consider the magnitude of density. This means that a station area could achieve a good indicator result even if it contained very few activities, as long as these activities were concentrated around the station. As density distribution and high average densities are both important for transit, the paper's proposed density distribution indicator should be used in combination with an intensity indicator such as average density or absolute number of activities in the station area. As these intensity indicators are commonly used, the remainder of this article focuses on density distribution indicators. The combination of density distribution with intensity indicators should be examined in further research.

Indicator for Density Distribution

Concept

This research translated the basic criterion that density at any point should not be larger than it is on average at points located closer to a transit stop into an indicator to assess density distribution for co-location of transit and activities. This was operationalized by comparing density at every point within the catchment area of a transit stop with the average density at all points located closer to that stop ("inwards density"), and with the average density at all points within the same catchment area that are located further away from that stop ("outwards density").

Density Representation

To operationalize the indicator, the catchment area of a transit stop is represented with raster cells for which both the network distance to the respective transit stop and the activity density are calculated. As data on activities is normally available for points (e.g. number of jobs or inhabitants per address), these data must be transformed into a continuous representation of density (essentially a density value per raster cell). Such a density surface is generated using kernel density estimation, which represents variation in population or services (*14–17*).

The basic idea is to estimate density at point s based on data points s_i (each point representing, e.g., an activity or a person) by weighting them with distance d_i between s and s_i so that data points further away from s contribute less to the estimated density at s. This analysis is conducted using all data points s_i with d_i smaller than a so-called bandwidth h. The distance weight is applied with a kernel function $k()$, which must integrate to one so that the overall count of activities estimated with the kernel within an analysis area remains the same as when simply summing up the number of points within that area.

A typical kernel for geographic density estimation (*14, 17, 18*) is the quartic function based on Silverman (*19*); it is the only available kernel in ArcGIS Spatial Analyst (*20, 21*) and one of the available options in QGIS (*22*), GRASS GIS (*23*), and SAGA (*24*). The kernel density estimation using a quartic kernel can be formulated as follows (*14*):

$$\hat{p}(s) = \sum_{d_i \leq h}^{n} \frac{3}{\pi h^2}\left(1 - \left(\frac{d_i}{h}\right)^2\right)^2 \quad (1)$$

in which $\hat{p}(s)$ is the estimated density at point s (in absolute terms, i.e., the sum of all $\hat{p}$ equals the total number of points) and n is the number of data points s_i that fulfil $d_i \leq h$. Thus, the summation is over all pairs s and s_i in which d_i does not exceed h. In this approach, data with counts (e.g. of activities or people per point) can simply be treated as multiple data points at a single location (*18*).

Indicator Formulation

The performance of density distribution with respect to co-location is measured for catchment area T of transit stop t. First, the entire developable surface of T is covered with raster grid cells m_k, and performance is evaluated for each cell. Subsequently, performance for T is obtained as the average performance of all cells $m_k \in T$.

For each cell m_k within T, activity density $\hat{p}(m_k)$ is estimated using kernel density estimation as described in Equation 1 and the pedestrian network distance between transit stop t and cell m_k, $d(t, m_k)$, is measured. Using this information, density $\hat{p}(m_k)$ is compared to the average "inwards" and "outwards" densities within T from the position of m_k, $\hat{p}_{in}(m_k)$ and $\hat{p}_{out}(m_k)$. The results of this evaluation, denoted $c_{in}(m_k)$ and $c_{out}(m_k)$, respectively, are within the range [0, 1], with the value of 1 representing the best outcome and 0 the worst. Results are computed using the ratio between the "inwards" density and $\hat{p}(m_k)$ and between $\hat{p}(m_k)$ and the "outwards" density:

$$c_{in}(m_k) = \begin{cases} 1, \hat{p}(m_k) \leq \hat{p}_{in}(m_k) \\ \dfrac{\hat{p}_{in}(m_k)}{\hat{p}(m_k)}, \text{otherwise} \end{cases} \quad (2)$$

and

$$c_{out}(m_k) = \begin{cases} 1, \hat{p}(m_k) \geq \hat{p}_{out}(m_k) \\ \dfrac{\hat{p}(m_k)}{\hat{p}_{out}(m_k)}, \text{otherwise} \end{cases} \quad (3)$$

with the "inwards" average density defined as

$$\hat{p}_{in}(m_k) = \frac{\sum_{l|d(t,m_l)<d(t,m_k)} \hat{p}(m_l)}{\sum_{l|d(t,m_l)<d(t,m_k)} 1} \quad (4)$$

and the "outwards" average density as

$$\hat{p}_{out}(m_k) = \frac{\sum_{l|d(t,m_l)>d(t,m_k)} \hat{p}(m_l)}{\sum_{l|d(t,m_l)>d(t,m_k)} 1} \quad (5)$$

The results of the two evaluations are averaged as

$$c_{total}(m_k) = \frac{1}{2}(c_{in}(m_k) + c_{out}(m_k)) \quad (6)$$

For the innermost cell, $\hat{p}_{in}$ has no value. Therefore, only c_{out} is computed, and $c_{total} = c_{out}$. For the outermost cell, this is reversed, that is, only c_{in} is computed, and $c_{total} = c_{in}$.

The density distribution within T is evaluated using the average performance of all cells m_k within T. The performance for T is thus:

$$C(T) = \frac{1}{K}\sum_{k=1}^{K} c_{total}(m_k) \quad (7)$$

in which K is the number of cells within T. $C(T)$ is in the range [0, 1], with 1 representing the best and 0 the worst performance of T with respect to co-location. The indicator value does not represent an absolute quantity; rather, it is a comparative measure.

Comparing density at raster cell m_k to both "inwards" and "outwards" average densities (Equations 4 to 6) is necessary to consistently capture the effects of "gaps" within the density distribution. For example, if we assume a situation in which density is very low and uniform within the catchment area of a stop, but there is one high-density activity concentration located far away from the stop. Comparing density of raster cells only to the "inwards" average density, all points located closer to the stop than the concentration, and all that are further away, would achieve very good results (their density is not higher than the average "inwards" density). Thus, the average result over all points would be good, which does not reflect the situation. By also comparing raster

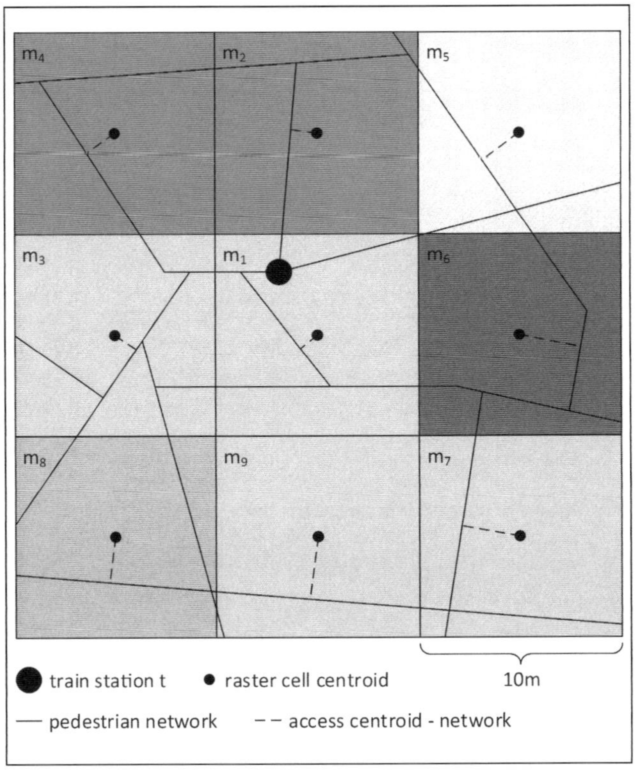

Figure 1. Generic example with nine raster cells of 10 x 10 m around a train station; darker shade of raster cell depicts higher density (see Table 1 for exact values).

cell density to average "outwards" density, results for all points located closer to the stop than the concentration are bad (because their density is lower than the average "outwards" density), and the effect of the isolated activity concentration is accounted for in the overall result.

Generic Example

Figure 1 depicts a generic example with one railway station and nine raster cells. Distance from a cell to the railway station is measured as the combination of the shortest path from the cell centroid to the pedestrian network and the network path from that access point to the railway station. Cells are denoted m_1 to m_9, with increasing distance to the station.

Density $\hat{p}(m_k)$ was assigned randomly to demonstrate the mechanism of the indicator. Table 1 contains the input values as well as the intermediate results for the indicator computation, as well as the indicator results. Note that the average of c_{total} in Table 1 is $C(T)$ for the generic example.

Indicator Interpretation

The indicator result shows how well transit and development are co-located within a transit stop area based on the distribution of activity density. At the stop area level, a maximum result value of 1.0 indicates that density is at no point within that area higher than the average density of all points located closer to the stop, and also at no point lower than the average density of all points within the stop area located further from the stop. In other words, a value of 1.0 is achieved if either there is uniform density within the stop area or there is a continuous decrease of density from the stop outwards. Any result value below 1.0 on the stop area level indicates some deviation from this "best case"– the closer to 0.0, the stronger the deviation. A "worst case" result value of 0.0 at the stop area level indicates a continuous increase in density with distance from the stop.

If the stop-level result is below 1.0, indicator values for individual raster grid cells are analyzed with result maps to examine how and where density distribution deviates from the "best case". Low indicator values in a specific area point at one of three possible states:

1. density is higher than on average closer to the transit stop;
2. density is lower than on average further away from the transit stop; or
3. both.

For an unambiguous interpretation, indicator result maps need to be complemented with density distribution maps. In general, the most common reasons for low indicator values in a specific area are either comparatively low density close to a stop or comparatively high density far from a stop. While some of these cases could be identified solely based on a density distribution map, the indicator result map allows for the systematic assessment of distribution – that is, it considers density at every point relative to density at all other points, thus highlighting some areas deviating from a "best case" density distribution that are not easily identifiable on the density map.

Case Study

Overview

The municipality of Zaanstad in the Netherlands was selected as a case study for applying the co-location indicator. Zaanstad is located north of Amsterdam and is part of its metropolitan region. It is expected to attract a significant share of metropolitan growth, because of its proximity to Amsterdam and high level of railway accessibility. Zaanstad has six railway stations with travel times of between 12 minutes (Zaandam) and 25 minutes (Krommenie-Assendelft) to Amsterdam central station and frequencies of between four and ten trains per direction and hour. Transit oriented development is a general goal of Dutch planning and is particularly pressing in

Table 1. Input Values and Indicator Results for Generic Example in Figure 1

k	$d(t, m_k)$[a]	$\hat{p}(m_k)$[b]	$\hat{p}_{in}$[b]	$\hat{p}_{out}$[b]	c_{in}	c_{out}	c_{total}
Cell no.	Distance from cell centroid to stop	Density of cell	Average inwards density	Average outwards density	Inwards evaluation result	Outwards evaluation result	Total evaluation result
1	8.00	1.50	–	2.40	–	0.63	0.63
2	8.50	3.80	1.50	2.20	0.39	1.00	0.70
3	11.00	1.30	2.65	2.35	1.00	0.55	0.78
4	14.50	3.60	2.20	2.10	0.61	1.00	0.81
5	17.50	0.50	2.55	2.50	1.00	0.20	0.60
6	23.50	4.60	2.14	1.80	0.47	1.00	0.73
7	27.00	1.20	2.55	2.10	1.00	0.57	0.79
8	29.00	2.60	2.36	1.60	0.91	1.00	0.95
9	29.50	1.60	2.39	–	1.00	–	1.00
Avg.	18.72	2.30	2.29	2.13	0.80	0.74	0.78

Note: 1 m = 3.281 ft; 1 m^2 = 10.764 ft^2; avg. = average; – = values that are not available because of the inner- and outermost cell.
[a]In m.
[b]In activities / 100 m^2; $\hat{p}(m_k)$ is chosen randomly.

the metropolitan areas such as Amsterdam, where roads are often at capacity.

To plan future development strategies, it is essential to know where co-location is already good and where it could be improved. Therefore, the case study compares co-location performance for station areas using the proposed indicator. Station areas are defined as the area around a station reachable within 1200 m using the pedestrian network. Results are compared based on $C(T)$ per station area as well as maps of the disaggregated raster cell results $c_{total}(m_k)$. The case study area with its six stations is depicted in Figure 2.

As the case study should illustrate the potential of using the proposed indicator for co-location assessment, indicator results are also compared to established indicators for co-location.

Data and Implementation

Railways. Railway stations were imported from the "Informatiesysteem Knooppunten" ("information system transport nodes") (*25*). Railway lines for visualization are based on cadaster data (*26*).

Pedestrian Network. The pedestrian network was extracted from OpenStreetMap data (*27*). Missing pedestrian crossings across roads represented with more than one line in the data or between offset sidewalks were added manually based on satellite images (*28*).

Activities. The number of residents was obtained from Statistics Netherlands (CBS) as hectare grid data for the year 2014 (*29*). The number of jobs was obtained from the job register LISA per four-digit-postcode for the year 2015 (*30*). Postcode areas for the year 2015 were obtained from ESRI Nederland (*31*).

Activities were distributed to addresses within the respective hectare grid cell (residents) or postcode area (jobs). Address data were obtained from the cadaster (*32*) for the year 2016, including usage category and floor area per address. Inhabitants were distributed to addresses with residential use, while jobs were distributed to all other addresses. Distribution weight was determined using the floor area per address.

Built-Up Area. In the context of the case study, large portions of land within a 1200 m network distance from a railway station are not actually developable – namely green spaces and water. To account for this, the analysis has been conducted for "built-up" areas only, that is, all land excluding green spaces and water. For this, public space distribution (*33*) and building footprints (*32*) have been obtained from cadaster data. The built-up area within the 1200 m station areas is depicted in Figure 2.

Computation. The proposed indicator was calculated using QGIS version 2.18.7 (*34*) for data management and visualization, GRASS GIS version 7.2.0 (*35*) for network analysis, and R version 3.2.2 (*36*) for data analysis. Additional indicators shown in Table 2 were computed with R using the same dataset. Density was estimated with SAGA tool "Kernel Density Estimation" (*24*) within the QGIS Processing Toolbox using a quartic kernel (as shown in Equation 1), a bandwidth of 100 m, and a raster resolution of 10 x 10 m.

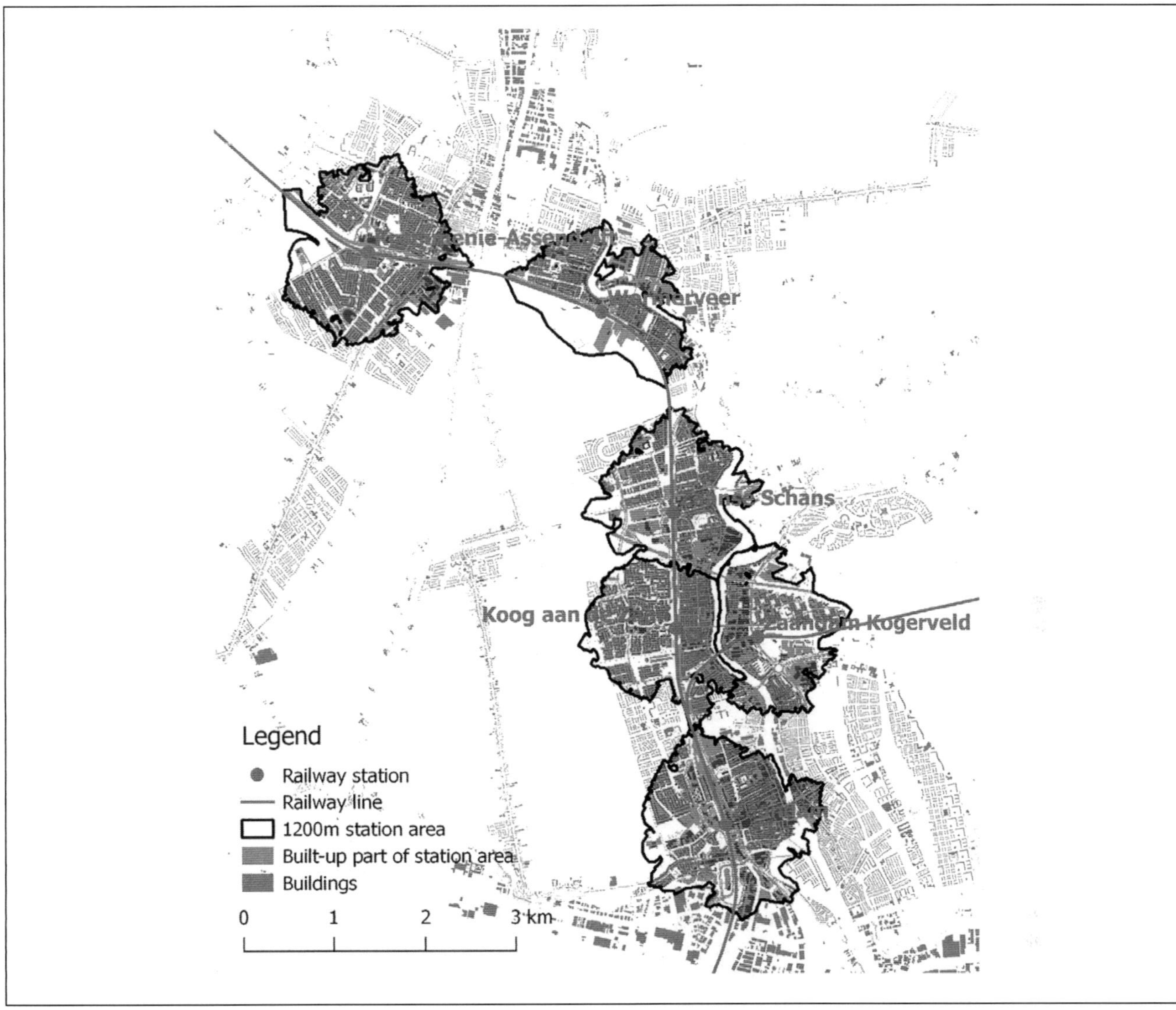

Figure 2 Overview of case study area.

Results

Table 2 contains co-location indicator results as well as additional indicators for the six case study station areas. The density distribution and respective indicator results on grid-cell level per station area are depicted in Figure 3.

The additional indicators in Table 2 reflect common approaches to co-location assessment:

- number of activities within the 1200 m station area—a measure of intensity;
- average density of the built-up area within the 1200 m station area—also a measure of intensity, considering differences in station area size;
- share of activities located within 300 m from the respective station in all activities located within the 1200 m station area—a measure of station proximity of activities;
- average distance from activities within the 1200 m station area to the station—also a measure of station proximity of activities.

Furthermore, Table 2 includes a simple ranking of the station areas according to the different indicators used.

Discussion

The proposed indicator clearly shows Koog aan de Zaan as the station area with the best co-location (indicator

Table 2. Overview of Results for Station Areas

	Krommenie-Assendelft	Wormer-veer	Zaanse Schans	Koog aan de Zaan	Zaandam-Kogerveld	Zaandam
New co-location indicator						
Co-location indicator result C(T)	**0.81**	**0.80**	**0.72**	**0.88**	**0.75**	**0.83**
Rank	*3*	*4*	*6*	*1*	*5*	*2*
Number of activities within station area						
Number of inhabitants in 1200 m[a]	14565	7388	7621	13222	6203	12653
Number of jobs in 1200 m[a]	3387	2567	4335	1162	4656	8564
Number of activities in 1200 m[a,b]	**17951**	**9955**	**11956**	**14384**	**10858**	**21217**
Rank	*2*	*6*	*4*	*3*	*5*	*1*
Average density of station area						
Built-up area within 1200m[a] [100 m²]	19390	11610	14153	14545	11613	19970
Average density of 1200 m station area (built-up[c]) [act. / 100m²]	**0.93**	**0.86**	**0.84**	**0.99**	**0.94**	**1.06**
Rank	*4*	*5*	*6*	*2*	*3*	*1*
Share of activities within 300m in station area						
Number of activities within 300 m[a]	553	346	415	1230	322	2641
Share of activities within 300 m in 1200 m[d]	**0.03**	**0.03**	**0.03**	**0.09**	**0.03**	**0.12**
Rank	*3*	*3*	*3*	*2*	*3*	*1*
Average distance to station						
Average distance to station from activities within 1200 m station area [m]	**840.67**	**836.27**	**801.39**	**720.73**	**785.02**	**726.03**
Rank	*6*	*5*	*4*	*1*	*3*	*2*

Note: 1 m = 3.281 ft; 1 m^2 = 10.764 ft^2.
[a]Pedestrian network distance from station.
[b]Activities = inhabitants + jobs.
[c]Only built-up area is counted for area computation.
[d]Share of activities within 1200m station area that are located within 300m station area.

value 0.88). The station areas of Zaandam (0.83), Krommenie-Assendelft (0.81), and Wormerveer (0.80) represent the middle field, while Zaandam-Kogerveld (0.75) and Zaanse Schans (0.72) have the lowest co-location performance.

The density map of the best-performing station area, Koog an de Zaan (Figure 3*g*), shows a very uniform density distribution, and a slight decrease in density with distance from the station. To understand why the station area has not achieved an indicator value of 1 despite this, it is necessary to consult the map of indicator results per grid cell (Figure 3*h*). This map reveals the locations of areas with low performance. Together with the density map, this allows for a more detailed interpretation: very low indicator values occur only in areas with very low density that are located southwest of the station, at the northwestern fringe of the station area, and along the railway infrastructure. These locations are also apparent on the density map. However, there are further areas with an indicator value below 1. For example, activity concentrations in the southwest and southeast of the station area exceed the average density of all points closer to the station – a conclusion not directly apparent from the density map.

The density distribution of the worst-performing station areas Zaandam-Kogerveld (Figure 3*i*) and Zaanse Schans (Figure 3*e*), shows both strong activity concentrations and low-density areas. However, given the highly irregular pattern, the density maps alone do not provide clear information about the main contributors to the overall low indicator values: which parts of the station areas perform particularly badly, and which do not? Indicator result maps (Figure 3, *f* and *j*) show that in Zaanse Schans, the lowest indicator values occur in areas with very low densities (at various distances from the station), whereas in Zaandam-Kogerveld, the lowest indicator values occur both in areas with high (far from the station) and low densities (also at various distances from the station). However, not all activity concentrations located toward the fringe of the two station areas (as apparent on the density distribution maps) entail particularly low indicator values. These are some of the additional insights provided by the indicator maps which would not be apparent simply from the density distribution maps.

The indicator result maps are particularly useful for identifying where interventions could have the highest potential effect on overall station area co-location performance. The specific type of intervention depends on boundary conditions (e.g., can transit or development be altered, or both?). However, there are some common

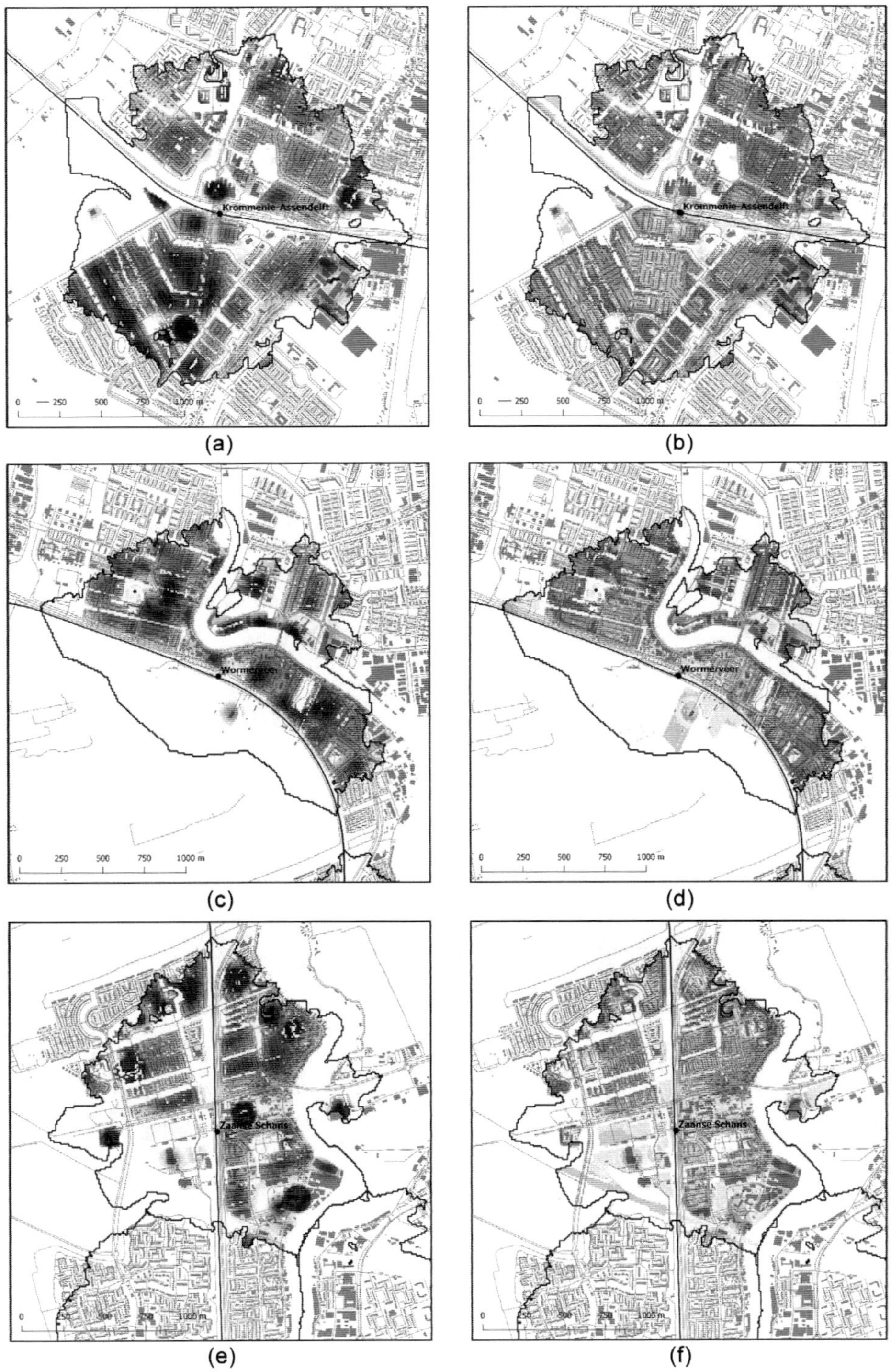

Figure 3. (continued)

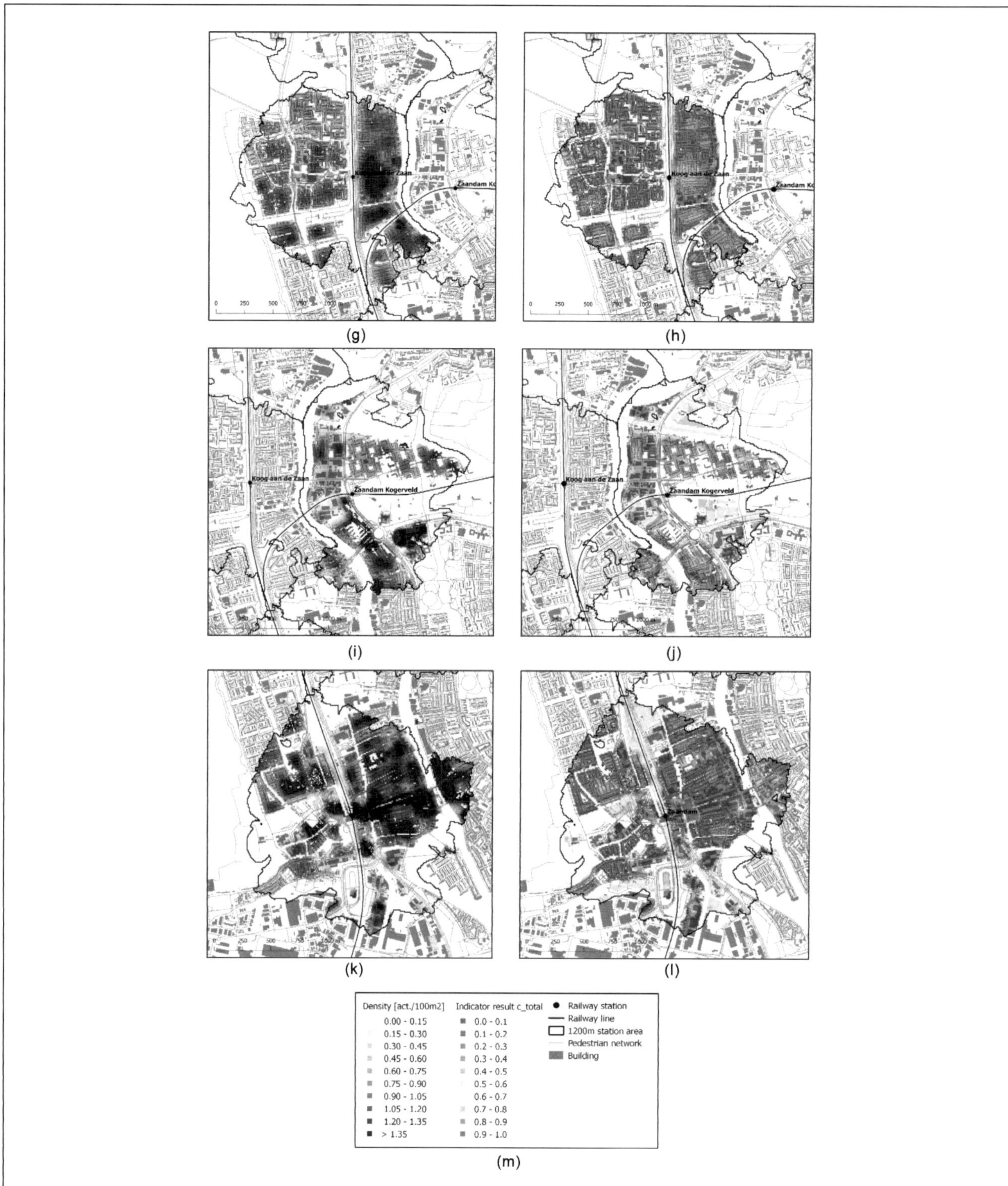

Figure 3. Density distribution $\hat{p}(m_k)$ and co-location indicator result $c_{total}(m_k)$ per grid cell within station areas: (*a*) Krommenie-Assendelft: density distribution, (*b*) Krommenie-Assendelft: co-location indicator, (*c*) Wormerveer: density distribution, (*d*) Wormerveer: co-location indicator, (*e*) Zaanse Schans: density distribution, (*f*) Zaanse Schans: co-location indicator, (*g*) Koog aan de Zaan: density distribution, (*h*) Koog aan de Zaan: co-location indicator, (*i*) Zaandam Kogerveld: density distribution, (*j*) Zaandam Kogerveld: co-location indicator, (*k*) Zaandam: density distribution, (*l*) Zaandam: co-location indicator, and (*m*) legend.

rules. Areas close to a station with a low co-location performance value should be considered for densification. For low-performing high-density areas far from stations, access and egress paths to/from transit should be made as direct and as comfortable as possible. If changes to transit are considered, these areas are possible focus points for future stops.

The other co-location indicators shown in Table 2 lead to a different station area ranking compared to the indicator developed in this research. For example, the indicator number of activities within the 1200 m station area reflects two aspects: the size of the station area (depending on pedestrian network and proximity of other stations) and the intensity of its land use. It shows large differences between station areas. The ranking order differs almost completely from the proposed indicator; this is expected since intensity does not consider activity distribution within station areas at all. The other intensity indicator, average station area activity density, shows much smaller differences between station areas because it accounts for their size. It leads to a third ranking order. Again, this is expected, since the three indicators reflect very different aspects of co-location.

The share of activities within 300 m of the station compared to all activities within the 1200 m station area reflects station proximity based on an arbitrary threshold distance. It leads to the same ranking for the first two places as average density but does not lead to differences between the remaining four stations. Last but not least, the average distance to the station within the 1200 m station area – also a station proximity indicator – is the only alternative indicator that has the same ranking order as the new indicator for the top two station areas, but it shows strong differences at the bottom end of the ranking. The two proximity indicators lead to a different ranking order than the proposed indicator and this reflects that, by accounting for density distribution, the proposed indicator considers more than simply the proximity to the station.

Conclusion

The proposed indicator developed in this research enables planners to evaluate the quality of station area transit and development co-location. The indicator can be complemented with maps of the results at a grid cell level to identify specific locations where co-location is strong or weak. This second analysis can be used to draw conclusions about locations for interventions such as densification or new stations. These conclusions can inform decision-makers about where to focus more detailed research.

Existing indicators for co-location assessment do not provide a similar level of detail – neither in their input for calculating results, nor in their possibilities for the presentation and interpretation of results. Intensity indicators (number of activities, average density) do not reflect density distribution within a station area at all. These indicators tend to simply show that the more centrally a station is located and the more urbanized the surrounding urban fabric, the better the value. Accordingly, they rank Zaandam station area first, which is located closest to central Amsterdam and is considered as the most urban station area. Proximity indicators consider density distribution to some extent. However, the 300 m share only assesses density based on one cut-off distance; therefore, inside the areas within 300 m and between 300 and 1200 m from the station, it does not assess distribution. Yet, there is an important difference whether development is concentrated at 400 or 1100 m from a station. Average distance to the station is ultimately the result of density distribution and thus considers distribution somewhat. However, it also depends on the extent of the station area (which can be reduced because of non-developable land or overlapping station areas), and it is hard to disentangle these influences. None of the other indicators allow for the spatially explicit interpretation of results, which means they do not help users understand the reasons why a certain station area performs well or badly.

On the other hand, all of the indicators discussed in this paper contain important information. In particular, the intensity indicators complement the proposed density indicator developed in this research. Importantly, the proposed indicator only reflects density distribution, regardless of the magnitude of density. Therefore, it is best used in combination with an intensity indicator.

The proposed indicator could be refined in three ways. First, the definition of built-up area used in the case study requires further development. In particular, large areas dedicated to transport infrastructure repeatedly show as performing badly in Figure 3. On the other hand, some transport infrastructure areas could possibly be developed in the future, and therefore clear criteria for their exclusion from analysis would be helpful. Furthermore, even the seemingly non-developable areas (such as open water or green space) excluded from the analysis might be considered for development at some point to exploit their proximity to transit. Therefore, the effects of undevelopable areas should be examined, for example by comparing the share and distribution of built-up area within station areas. Non-developable areas next to stations should also be considered in transit network design as potential barriers for transit access and as increasing distance between activities and transit. Second, more precise activity data would increase accuracy. Ideally, activities per building should be used as input data to avoid the potentially distorting activity disaggregation to addresses. Furthermore, visitor data for locations such as sports venues or shopping malls would be beneficial – currently, some of these show as areas of low density in Figure 3, which is not adequate. Third, experimentation with different bandwidths and cell sizes for kernel density estimation could shed light on the effect of extreme density concentration (single addresses with a very high number of activities) on indicator results.

In future research it would be interesting to use the proposed indicator combined with an intensity indicator in a direct transit demand model and compare its predictive power for transit demand with other indicators for co-location.

Acknowledgments

Five anonymous reviewers made helpful comments that improved the paper. This article is based on research supported by a Doc.Mobility grant from the Swiss National Science Foundation (SNSF). The authors are grateful to Andrew Nash for help in language editing and proofreading of the manuscript.

Author Contributions

The authors confirm contribution to the paper as follows: study conception and design: C.M. Marti, L. Bertolini, U. Weidmann; data collection: C.M. Marti; analysis and interpretation of results: C.M. Marti; draft manuscript preparation: C.M. Marti, L. Bertolini, U. Weidmann. All authors reviewed the results and approved the final version of the manuscript.

References

1. Calthorpe, P. *The Next American Metropolis, Ecology, Community, and the American Dream*. Princeton Architectural Press, New York, 1993.
2. Cervero, R. Transit-Oriented Development and the Urban Fabric. In *Sustainable Railway Futures: Issues and Challenges* (B. P. Y. Loo, and C. Comtois, eds.), Ashgate, Farnham, Surrey, UK and Burlington, Vt., 2015, pp. 75–93.
3. Suzuki, H., R. Cervero, and K. Iuchi. *Transforming Cities with Transit, Transit and Land-Use Integration for Sustainable Urban Development*. World Bank, Washington, D.C., 2013.
4. Walther, K. *Nachfrageorientierte Bewertung der Streckenführung im öffentlichen Personennahverkehr*. Dissertation, TH Aachen, 1973.
5. Zhao, F., L. F. Chow, M. T. Li, I. Ubaka, and A. Gan. Forecasting Transit Walk Accessibility - Regression Model Alternative to Buffer Method. *Transportation Research Record*: *Journal of the Transportatio Research Board*, 2003. 1835: 34–41.
6. Crowley, D. F., A. S. Shalaby, and H. Zarei. Access Walking Distance, Transit Use, and Transit-Oriented Development in North York City Center, Toronto, Canada. *Transportation Research Record*: *Journal of the Transportation Research Board*, 2009. 2110: 96–105.
7. Gutiérrez, J., O. D. Cardozo, and J. C. García-Palomares. Transit Ridership Forecasting at Station Level: An Approach Based on Distance-Decay Weighted Regression. *Journal of Transport Geography*, Vol. 19, No. 6, 2011, pp. 1081–1092.
8. Kittelson & Associates, Parsons Brinckerhoff, KFH Group, Texas A&M Transportation Institute, and Arup. *TCRP Report 165: Transit Capacity and Quality of Service Manual*, 3rd ed. Transportation Research Board of the National Academies, Washington, D.C., 2013.
9. El-Geneidy, A., M. Grimsrud, R. Wasfi, P. Tetreault, and J. Surprenant-Legault. New Evidence on Walking Distances to Transit Stops: Identifying Redundancies and Gaps Using Variable Service Areas. *Transportation*, Vol. 41, No. 1, 2014, pp. 193–210.
10. Chia, J., J. Lee, and M. Kamruzzaman. Walking to Public Transit: Exploring Variations by Socioeconomic Status. *International Journal of Sustainable Transportation*, Vol. 10, No. 9, 2016, pp. 805–814.
11. Cervero, R. Alternative Approaches to Modeling the Travel-Demand Impacts of Smart Growth. *Journal of the American Planning Association*, Vol. 72, No. 3, 2006, pp. 285–295.
12. Marks, M. *People near Transit: Improving Accessibility and Rapid Transit Coverage in Large Cities*. Institute for Transportation & Development Policy, New York, 2016.
13. Provincie Noord-Holland and Vereniging Deltametropool. *Maak Plaats! Werken aan Knooppuntontwikkeling in Noord-Holland*. Provincie Noord-Holland, Haarlem, 2013.
14. Gatrell, A. C., T. C. Bailey, P. J. Diggle, and B. S. Rowlingson. Spatial Point Pattern Analysis and Its Application in Geographical Epidemiology. *Transactions of the Institute of British Geographers*, Vol. 21, No. 1, 1996, pp. 256–274.
15. Porta, S., E. Strano, V. Iacoviello, R. Messora, V. Latora, A. Cardillo, F. Wang, and S. Scellato. Street Centrality and Densities of Retail and Services in Bologna, Italy. *Environment and Planning B: Planning and Design*, Vol. 36, No. 3, 2009, pp. 450–465.
16. Lewis, D. Kernel Density Estimation and Percent Volume Contours. In *Geocomputation: A Practical Primer* (C. Brunsdon, and A. Singleton, eds.), SAGE, Los Angeles, 2015, pp. 169–184.
17. Neutens, T. Accessibility, Equity and Health Care: Review and Research Directions for Transport Geographers. *Journal of Transport Geography*, Vol. 43, 2015, pp. 14–27.
18. O'Sullivan, D., and D. W. S. Wong. A Surface-Based Approach to Measuring Spatial Segregation. *Geographical Analysis*, Vol. 39, No. 2, 2007, pp. 147–168.
19. Silverman, B. W. *Density Estimation for Statistics and Data Analysis*. Chapman and Hall, New York, 1986.
20. *Quantitative Decisions*. Density Calculations. Quantitative Decisions, Rosemont, Pa. http://www.quantdec.com/SYSEN597/GTKAV/section9/density.htm. Accessed May 2, 2016.
21. *ESRI*. How Kernel Density works. Environmental Systems Research Institute, Redlands, Calif. http://pro.arcgis.com/

en/pro-app/tool-reference/spatial-analyst/how-kernel-density-works.htm. Accessed May 2, 2016.
22. *QGIS Development Team.* Heatmap Plugin. Open Source Geospatial Foundation Project. http://docs.qgis.org/2.0/en/docs/user_manual/plugins/plugins_heatmap.html. Accessed May 2, 2016.
23. Menegon, S., and R. Blazek. v.kernel. *GRASS Development Team.* https://grass.osgeo.org/grass70/manuals/v.kernel.html. Accessed May 2, 2016.
24. Conrad, O. *Module Kernel Density Estimation. System for Automated Geoscientific Analyses (SAGA).* http://www.saga-gis.org/saga_module_doc/2.1.3/grid_gridding_6.html. Accessed May 3, 2016.
25. *Vereniging Deltametropool.* Informatiesysteem Knooppunten. http://deltametropool.nl/nl/informatiesysteem_knooppunten. Accessed April 6, 2017.
26. Kadaster, N. L. *TOP10NL, State 2016-11-08.* Het Kadaster, Apeldoorn. https://www.kadaster.nl/-/top10nl, 2016. Accessed June 6, 2017.
27. OpenStreetMap Contributors. *OpenStreetMap.* http://www.openstreetmap.org, 2017. Accessed June 6, 2017.
28. Google Maps. *Google Maps.* http://maps.google.com, 2017. Accessed June 6, 2017.
29. CBS. *Statistische gegevens per vierkant.* Centraal Bureau voor de Statistiek, Den Haag.https://www.cbs.nl/nl-nl/dossier/nederland-regionaal/geografische%20data/kaart-met-statistieken-per-vierkant-van-100-bij-100-meter, 2014. Accessed June 6, 2017.
30. LISA. *Werkgelegenheidsregister van Nederland.* LISA, Enschede. www.lisadata.nl, 2015. Accessed June 6, 2017.
31. ESRI Nederland. *PC4 polygons (ESRI ArcGIS content), May 2015.* ESRI Nederland, Rotterdam. http://geoplaza.vu.nl/data/dataset/postcode/resource/3874f60f-e9a1-49b0-b0a8-a3e7d282c969, 2015. Accessed June 6, 2017.
32. Kadaster, N. L. *Basisregistraties Adressen en Gebouwen (BAG).* Het Kadaster, Apeldoorn. https://www.kadaster.nl/bag, 2016. Accessed June 6, 2017.
33. Kadaster, N. L. *Basisregistratie Grootschalige Topografie (BGT), State 16.05.2017.* Het Kadaster, Apeldoorn. https://www.kadaster.nl/bgt, 2017. Accessed June 6, 2017.
34. QGIS Development Team. *QGIS Geographic Information System.* Open Source Geospatial Foundation. http://qgis.osgeo.org, 2017. Accesssed February 14, 2017.
35. GRASS Development Team. *Geographic Resources Analysis Support System (GRASS) Software.* Open Source Geospatial Foundation Project. http://grass.osgeo.org, 2016. Accesssed February 14.
36. R Core Team. *R: A Language and Environment for Statistical Computing.* Foundation for Statistical Computing, Vienna, https://www.R-project.org/, 2015.

The Standing Committee on Public Transportation Planning and Development (AP025) peer-reviewed this paper (18-00611).

Article

TRR
JOURNAL OF THE TRANSPORTATION RESEARCH BOARD

Transportation Research Record
2018, Vol. 2672(8) 840–848

DOI: 10.1177/0361198118781150
journals.sagepub.com/home/trr
SAGE

Small Steps, Big Differences: Assessing the Validity of using Home and Work Locations to Estimate Walking Distances to Transit

Marie-Pier Veillette[1], Robbin Deboosere[1], Rania Wasfi[2,3], Nancy Ross[3], and Ahmed El-Geneidy[1]

Abstract

Walking to and from public transport can form a seamless way to integrate physical activity into our daily lives, thereby helping us achieve the recommended minutes of physical activity. To measure the link between physical activity and public transport use, it is critical to determine how far individuals are walking, and are willing to walk, to different modes of transit. Few planners, however, have access to detailed information on the exact public transport lines used by individuals, and therefore need to estimate distances by making use of only home and work locations. This study therefore compared two methods of calculating walking distances: one method using widely available home and work locations and a fastest route algorithm leveraging general transit feed specification data, and a second employing a detailed travel survey containing information on the real routes used by each respondent from Montreal, Canada, to generate more robust estimates of the distances individuals are walking to public transport stops. Results show that walking distances calculated from commonly available origin and destination information tend to underestimate real walking distances by 10%. Multilevel mixed-effect regression models indicate these differing results are mainly attributable to differences in travel behaviour and mode choice. Findings from this study provide a better understanding of how modelled and real walking routes to public transport stops differ, which could be of interest to professionals and urban decision makers wishing to correctly model walking to transit in their region when only limited information is available.

Obesity rates have skyrocketed in the last few decades, due to, among other reasons, an increasingly sedentary lifestyle, exacerbated by the widespread use of the private automobile as our main mode of transportation (*1–3*). In light of this epidemic, the World Health Organization stresses the importance of active transportation in determining physical health (*4*). Scholarly attention has been particularly paid to walking to public transport, and its potential to help individuals achieve the recommended minutes of weekly physical activity, and thereby contribute to a healthy lifestyle (*5–8*).

To measure the link between public transport and physical activity, it is critical to determine how far individuals are walking, and are willing to walk, to different modes of transit. While a large body of literature has examined this relationship by making use of detailed travel surveys, municipal planners often lack such information and only have access to home and work locations for individuals using public transport. Such a lack of detailed information might lead to errors in the calculation of walking distances, resulting in skewed and possibly incorrect findings, which then feed into public policy and municipal decision-making processes.

The goal of this study was therefore to assess the levels of walking to public transport for different modes, and to compare how these levels differ between two methods of calculation. The first method uses information from a detailed travel survey, which contains the exact route the commuter used, whereas the second method only uses origin and destination information, representing a common scenario in organisations where information is scarce. For this purpose, we employed the detailed 2013 origin–destination (OD) survey carried out by the Agence Métropolitaine de Transport in Montreal, Canada (*9*). This was obtained under a special agreement with the agency and contains detailed information on each person's route, such as the different bus, train, or subway lines that were used in a trip, in addition to the number of transfers, and the metro or train stations where

[1]School of Urban Planning, McGill University, Montreal, Canada
[2]Centre de Recherche du CHUM, Université de Montréal, Montreal, Canada
[3]Department of Geography, McGill University, Montreal, Canada

Corresponding Author:
Address correspondence to Ahmed El-Geneidy: ahmed.elgeneidy@mcgill.ca

each individual first boarded, transferred, and left the system. Walking distances are subsequently derived from this detailed OD Survey and compared with walking distances computed by using only origin and destination information and a shortest route algorithm based on data from the general transit feed specification (GTFS). By making use of GTFS and a fastest route algorithm in ArcGIS for these calculations, we employed information that is abundantly available to planners and urban decision makers in a variety of settings. By measuring how the two estimates for walking distances differ, we can construct adjustment factors that practitioners could use to correct walking distances calculated from only origins and destinations. The finding from this study are expected to be of relevance to professionals and decision makers wishing to measure the health benefits resulting from public transport trips in their region when only limited information is available.

The rest of the paper is organised as follows. Section 2 presents literature discussing walking to transit, how walking distances are measured, and their relation to public health. In section 3, the data and two methods used to derive walking distances are discussed; section 4 presents a multilevel mixed-effect regression model to disentangle the determinants of walking to transit. Section 5 then concludes the paper and provides recommendations to policy makers and professionals.

Walking to Transit

Only 15% of Canadian adults achieve the recommended levels of physical activity according to the 2007 to 2009 Canadian Health Measures Survey (*10*). While in the past the threshold for recommended physical activity was set to a minimum of 30 min/day, the rule of thumb was subsequently refined to 150 min/week of "moderate-to-vigorous physical activity" (*4*, *11*). In this context, research suggests that walking to public transport, as an activity carried out daily by individuals, can form a seamless way to integrate physical activity into their daily routine. It is therefore not surprising that previous research has extensively documented positive health benefits associated with the use of public transport; transit users who walk to or from stations experience increased levels of physical activity and can thereby achieve the recommended levels of daily and weekly physical activity when using certain modes (*1*, *5–8*, *11–17*).

Walking distance to public transport varies worldwide across cities, according to trip purpose, the built-environment, household and individual trip characteristics, and factors related to the specific mode used. The following section briefly reviews recent findings on these factors, and presents different methods used to measure walking distances.

Early research on the catchment area of public transport stops pinpoints the distance individuals are willing to walk to transit at 400 m for bus stops, and 800 m for heavy and light rail systems, indicating that the specific mode taken by individuals is a key determinant of walking distance (*18–20*). These catchment size areas have subsequently been refined to 524 m for bus stops and 1,259 m for commuter rail in a study of home-based trips in Montreal (*19*). Another Montreal study again demonstrates that commuter rail users walk longer distances to transit than subway and city bus users (*8*). In a non-western context, Wang, Chen and Xu (*21*) derive that 75% of survey respondents walk less than 623 m to access a subway station in Beijing. Moreover, besides mode factors, increases in transit service frequency and the number of transfers made during each trip have been demonstrated to reduce the total distance people are walking to transit (*8*, *19*, *22*).

In addition to the factors related to the mode and route used by individuals, socioeconomic characteristics have also been found to influence walking distance to transit (*19*, *21*, *23*). Men walk further to reach transit stops than women (*8*, *23*, *24*), whereas the probability of reaching a transit stop using an active mode of transportation (such as walking) decreases considerably as people age (*8*, *18*, *23*). Furthermore, individuals living in low-income households are more likely to spend 30 min or more walking to transit on an average day than those living in medium- and higher-income households (*11*); and university graduates accumulate less total walking time than individuals with lower levels of education (*11*).

Neighborhood characteristics, such as population density and the number of stops in close proximity to the home also affect walking distance to transit stops and stations (*21*, *25*). In California, the probability of reaching a transit stop with an active mode of transport decreases by 12% for every additional mile between the origin and the closest transit stop (*18*), whereas individuals living in more densely populated areas have an average daily walking time that is longer than those living in areas with a lower population density (*11*). In a similar vein, a study in China found that walking distances to bus rapid transit (BRT) increase by 75 m for every additional kilometre away from the city centre (*22*).

The results uncovered by the research presented above arguably depend on the type of data used to carry out the study. Most studies make use of travel surveys that report origins and destinations, and employ spatial analysis techniques in a geographic information system to calculate walking distances (*8*, *11*, *18*, *19*, *21–23*). Municipal planners and researchers, however, often lack detailed information on the actual transit lines used by respondents, and are therefore required to resort to different methods, such as shortest path algorithms between known origins and destinations, to estimate real walking distances. This study therefore compares walking distances generated by a detailed travel survey with estimates of walking distances derived from only origin and destination information.

Methodology

Our method consists of a two-step spatial and statistical analysis procedure. First, we calculated the levels of walking to five public transport modes (commuter train, subway, city bus, suburban bus, and peripheral bus) using ArcGIS for two different datasets: 1) the 2013 detailed OD Survey with full information on every respondent's route, and 2) only origin and destination information supplemented with GTFS data. The detailed OD Survey is conducted every 5 years and samples 5% of all households living in the Montreal Metropolitan Area, Canada (*9*). In addition to the origin and destination locations, it contains exclusive information on the respondents' socioeconomic and transit trip characteristics, such as the number of different transit modes used, the specific bus, commuter train, or subway lines taken, and the number of transfers made during one trip. To compare walking distances generated from this detailed dataset with walking distances estimated from commonly available information, we artificially modelled a scenario where planners only had access to origin and destination information: we extracted only the origins and destinations from the detailed OD Survey, and supplemented these with GTFS data to represent the public transport network and schedule in Montreal. We will refer to the first scenario (with detailed information) as "detailed OD," while the second scenario will be referred to as the "GTFS scenario," "GTFS calculations," or simply "GTFS." In a second step, the potential differences in the levels of walking to transit between the two datasets and methods were compared by conducting three multilevel mixed-effect regressions.

Sample

Our sample consisted of 9,588 home-based trips realised by adults (over 18 years old) who commute to work and live in the Montreal Metropolitan Area. Starting from the full OD Survey ($N = 410{,}741$ trips), we excluded all trips made with modes other than subway, commuter rail, and bus (city, peripheral, and suburban buses). We also omitted trips associated with unreported and unknown route numbers, and for purposes other than work. In addition, in order to retain only one trip per individual, we removed multiple trips made by the same respondents. Our sample represents 13.8% of all reported trips made for work purposes in the 2013 OD Survey.

Spatial Analysis

The two methodologies – detailed OD and GTFS – allow us to measure the daily walking distance to transit, the number of transfers made by each individual, the total distance traveled in-vehicle and the frequency of public transport service at the first stop or station. As the information provided in both scenarios was different, two methods of calculation were employed in ArcGIS.

Detailed OD Survey. To calculate the total daily walking distance with the detailed OD Survey, we used the Network Analyst "closest facility" tool. We first generated a ready-to-use pedestrian street network dataset and a file containing all stops served by the subway, train, city bus, peripheral bus, and suburban bus lines. When several bus stops were used by more than one transit line, we duplicated the number of stops according to the number of routes serving it. As a result, each bus stop was assigned to one transit line. Pedestrian routes were then generated from each home and work location to their respective nearest 200 transit stops. Among the routes from home to stop generated with Network Analyst, we then identified the shortest pedestrian route linking the home location to the closest stop if that stop matched the route number reported in the detailed OD Survey. A similar operation was undertaken to identify the shortest pedestrian route between the work locations and the closest stop serving the last line a respondent used. Note that, while for subway and train the exact first and last stops were reported, the method as explained above was also used for these two modes. The closest stops were compared with the real stops and no errors were noted. Both beginning and end-of-trip walking distances were subsequently summed to generate the total walking distance per trip, which was then multiplied by two to account for daily walking to and from public transport, assuming that the person commuting by public transport would be doing the same trip but in reverse order when going home.

To measure the distance traveled in-vehicle, we calculated the distance between the first and last transit stop, previously identified, with the Network Analyst "New Route" tool and a transit network dataset. The number of transfers made during one trip was provided in the detailed OD Survey and did not require any geographic information system computations.

Common Scenario using GTFS. To calculate walking distances in the common scenario in which detailed information is scarce, we first derived the fastest route between an individual's origin and destination using transit through the ArcGIS tool "Add GTFS to a network dataset." The fastest route algorithm using GTFS incorporates walking, waiting, boarding, and alighting, and in-vehicle time (based on the transit schedule) to compute total travel time. This resulted in a single line for every person, representing their fastest route at 8 a.m. on a regular Tuesday in 2013. To compute each person's walking distance, his/her route was intersected with the street network, after which the walking parts were extracted. The walking segments closest to the origin or destination were then said to represent the distance walked from home to the first stop taken and from the last stop taken to the destination, respectively. These two distances were summed to represent the total walking during a single trip and then multiplied by two to represent the total walking to and from public transport in a day.

The number of transfers in this second scenario was derived by adding the number of walking segments (minus the home-stop and stop-destination segments), combined with the number of different subway or rail lines the person took (computed by intersecting the rail and subway network with the route taken, coded by line). As the first line a person took was not computed in the calculation with GTFS, the frequency of the first route taken was derived as follows: for every stop, the number of different arrivals between 7 a.m. and 9 a.m. (whether a subway, a train, or a bus) was divided by the time interval of 2 h, resulting in the average time between two subsequent transit vehicles arriving at a stop during the morning peak. To ensure consistency, this method was also applied to calculate the frequency of the first stop an individual actually used, as derived from the detailed OD Survey.

Neighborhood Variables. Several neighborhood, variables were added to the dataset to supplement the trip and individual characteristics. Firstly, the percentage of people with a university degree in the origin census tract (CT), and population density in this same area, was gathered from Statistics Canada. Secondly, the number of transit stops within 1,000 m of the origin was calculated using ArcGIS tools, this to reflect the density of the public transport network in close proximity to each respondent's home. In addition, street connectivity was measured through the number of four-way street intersections within a 1,000-m buffer from the home, although this variable was not used as it was highly correlated with the number of bus stops within 1,000 m.

The data generated from the detailed OD Survey and the GTFS calculations were then appended to generate a bigger dataset ($N = 9{,}588*2$), and a dummy variable was included that equalled one if the GTFS method was used and zero if the detailed OD method was used in the calculations. When appending the two datasets, the individual identification number was retained to be used in the multilevel analysis, presented in the next section. The coefficient for the GTFS dummy variable will therefore show to what extent the total daily walking distance (as the dependent variable in the model presented below) is under- or overestimated by GTFS, whereas the coefficients of the other variables used in the model will offer explanations on the potential sources of the differences between the two methods.

Multilevel Mixed-Effect Regression

To disentangle the determinants of walking to transit, and to discern how the two methodologies of calculating walking distances differ, we conducted three multilevel mixed-effect regressions at the individual and CT levels. CTs are geographical areas that share similar characteristics, house between 2,500 and 8,000 people, and are used as a proxy for neighborhoods (*26, 27*). A multilevel mixed-effect regression has the ability to capture both neighborhood and individual effects on the variation of the total walking distance (*28*). As we can expect individuals living in the same neighborhoods to share some unobserved similarities determining their walking distance, a multilevel model is thus more suitable than a regular ordinary least squares regression. Our two-level data structure allows nesting individuals within their neighborhood, while also accounting for the two methods of calculation (detailed OD and GTFS) at the trip level (*29*).

Analysis

Table 1 shows descriptive statistics of the variables used in the three regression models. The average total daily walking distance, accumulated solely by accessing and egressing transit, is 2,021 m based on detailed OD Survey data, whereas the shortest path algorithm using GTFS data estimates this number as 1,816 m. The calculations through GTFS data thus underestimate the total daily walking distance by 205 m (or 10%) compared with the real distances walked by the OD Survey respondents. A *t*-test revealed that this difference in average daily walking distances is statistically significant. Other considerable differences between OD and GTFS calculations are also present: substantial dissimilarities in mode choice and in the number of transfers exist. More than twice as many people reported having used a commuter train in the detailed OD Survey compared with the modal split calculated through a shortest path algorithm using GFTS data. In addition, the GTFS calculations underestimated the modal split of the subway by 5%. In a similar vein, the number of transfers varied considerably between both methods of calculation; approximately three times more individuals realised three transfers or more according to the GTFS calculations. Note that the GTFS calculations assumed everyone left their home at 8 a.m., which artificially increases travel time for individuals taking e.g., the train at 8.30 a.m. This might cause a different route, for example using a bus, to be faster, partly explaining the large difference in train modal share. Although individuals would most likely stay at home for some time instead of waiting at the station, the GTFS application in ArcGIS did not incorporate these behavioural calculations. Similarly, an individual that would normally catch a train at 7.59 a.m. would be rerouted to a different stop, thereby resulting in a different walking distance. Ideally, the GTFS routes should be calculated every minute between 7 a.m. and 9 a.m., and the shortest route should be used, minimising any potential errors related to transit service variability. Nevertheless, the differences between the two methodologies are considerable.

Table 2 breaks down daily walking distances by their mean, median, 75th percentile, and 85th percentile for both methods of calculation, and presents the percentage of respondents achieving the recommended minutes of physical activity. For this purpose, a walking speed of 5.47 km/h was used, cf. Wasfi, Ross and El-Geneidy (*8*). According to the detailed OD Survey, around 71% of train users, 14% of subway riders, and 10% of city bus users meet the recommended

Table 1. Summary Statistics of Total Daily Walking Distance to and from Transit Stops or Stations in the Montreal Metropolitan Area, Canada using Two Methods of Calculation

	Detailed OD		GTFS	
Continuous variables	Mean	Standard deviation	Mean	Standard deviation
Total walking distance per day (m)	2,020.84	2,013.43	1,816.49	1,229.22
Trip characteristics				
Distance traveled in vehicle (km)	10.05	70.09	11.13	7.59
Frequency of service at first stop (min)	16.61	33.97	14.53	34.47
Neighborhood characteristics				
Percentage of people with university degree within origin census tracts	27.90	13.66	27.90	13.66
Population density within origin census tract (population/km^2)	7,473.37	5,091.67	7,473.18	5,091.44
Number of bus stops within a 1,000-m buffer of origin	79.66	32.41	79.66	32.40
Socioeconomic characteristics				
Age	43.15	12.24	43.15	12.25
Dummy variables	Proportion		Proportion	
Socioeconomic characteristics				
Gender – Male	0.45		0.45	
Low income (<CAN$30k)	0.11		0.11	
Medium income (CAN$30k–89k)	0.46		0.46	
Trip characteristics				
First mode taken				
Subway	0.30		0.25	
Commuter train	0.07		0.03	
Peripheral bus	0.05		0.08	
Suburban bus	0.14		0.15	
City bus	0.44		0.49	
Number of transfers				
No transfer	0.40		0.21	
One transfer	0.37		0.34	
Two transfers	0.19		0.30	
Three transfers or more	0.04		0.15	
Methodology				
GTFS	0 .00		1.00	

levels of physical activity solely by walking to transit. The calculations made with GTFS data severely underestimated the percentage of people achieving the recommended minutes of physical activity, by nearly 40% for train users and by almost 13% for subway users. Interestingly, both methods of calculation arrived at similar results for the percentage of city bus users meeting the recommended levels of physical activity. Note that the distances presented in the table represent total walking distance each day. The 85th percentile of walking distances from home to stop was 513 m for city bus stops, and 1,020 m for subway stations, which is consistent with findings from previous studies (*19*).

To determine the factors that drive the differences noted above, the results of the three regression models are presented (Table 3). The first model does not account for the mode taken and the number of transfers made by transit users. This model confirmed the summary statistics shown above: the total daily walking distance is underestimated by 199 m by the GTFS calculations, holding all other variables in the model constant.

When the first mode taken by a respondent is operationalised in the model, the coefficient for the dummy variable, representing a calculation based on GTFS data, decreases by 122 m to 77 m, see Model 2. Model 2 therefore suggests that the difference in the two walking distance calculations can be largely attributed to differing mode choice between the two modelling approaches. As OD Survey respondents are more likely to use the train than would be expected from the shortest route calculations with GTFS, they tend to walk more than predicted. Note that, as mentioned above, the difference in modal choice could be partly explained by the fixed start time (8 a.m.) for the GTFS calculations. Ideally,

Table 2. Total Daily Walking Distance by Mode and Method of Calculation

Total daily walking distance (m)	Train	Subway	City bus	Suburban bus	Peripheral bus
Detailed OD					
Mean	4,675	2,253	1,523	1,692	2,522
Median	3,822	1,986	1,197	1,346	2,025
75th percentile	5,477	2,706	1,832	2,068	3,172
85th percentile	6,610	3,163	2,316	2,606	4,195
Standard deviation	4,215	1,486	1,551	1,622	1,901
Percentage achieving recommended minutes of physical activity	70.58	24.08	9.88	13.39	32.57
GTFS					
Mean	2,369	1,761	1,588	1,941	2,922
Median	2,088	1,664	1,456	1,730	2,376
75th percentile	2,939	2,205	2,027	2,451	3,817
85th percentile	3,310	2,587	2,397	2,970	4,547
Standard deviation	2,143	858	900	1,235	2,324
Percentage achieving recommended minutes of physical activity	30.71	11.48	9.14	18.32	41.22

the GTFS routes should be calculated every minute during the morning peak and the start time resulting in the shortest route being used. Nevertheless, the coefficients in Model 2 demonstrate that differing mode choice between GTFS and the detailed OD Survey explain a large part of the variation between the two walking distance estimates. Planners could thus benefit from correctly modelling modal choice. This could be done, for example, by placing a priority on certain modes such as commuter trains, to take into account comfort and other factors not incorporated in our model. In practical terms, the GTFS calculations could artificially reduce the travel time for priority modes. More research is however needed to determine these priority modes and their weighting.

In Model 2, the coefficient for individuals living in low-income households is no longer statistically significant, which suggests that the shorter walking distances of low income groups in Model 1 could be explained by different mode choices: low income groups are more likely to use a bus than a subway or commuter rail and therefore walk less, which is consistent with previous research (*8*).

The third model, in addition to controlling for the first mode used by respondents, also controls for the number of transfers made during each trip. This model reveals that there are no longer statistically significant differences between the methods of calculation (the coefficient for the GTFS dummy is no longer significant). By comparing models 2 and 3, we can infer that the differing number of transfers accounts for around 77 m of the difference in total walking distance between the two approaches. A shortest route algorithm based on GTFS data, in order to minimise total travel time, routes an individual to a stop close to home, even if that route does not directly lead to the destination, thus resulting in more transfers and shorter walking distances. However, as most people are averse to transferring between routes (*30*), they are happy to walk further to access a direct route, thereby partly explaining the differences between the rational GTFS approach and real-life (irrational) travel behaviour. This shows that, in order to estimate robust walking distances to transit, planners could benefit from accurately modelling travel behaviour, in particular by limiting the number of transfers individuals are willing to make, and by correctly modelling modal choice. Making these two adjustments will allow for the calculation of accurate walking distances, even if only origin and destination information is available.

The other coefficients in Model 3 present interesting findings regarding how far transit users are willing to walk according to the first transit mode they used. All things being equal, trips made by a train are related to an increase in total daily walking distance of 1,750 m compared with a city bus, whereas OD Survey respondents walking to subway stops walk, on average, 892 m more per day than individuals using a city bus. Furthermore, a trip integrating two transfers is related to a reduction in the total walking distance of 485 m, all things being equal, and a trip with three transfers or more is associated with a decrease in the total walking distance of 656 m compared with a trip with no transfer, *ceteris paribus*.

Finally, the total daily walking distance is also influenced by socioeconomic, trip, and neighborhood characteristics. As expected, the coefficients for age, the number of university graduates in the origin CT, population density, and the number of bus stops within 1,000 m are statistically significant. Every year increase in age is related to a decrease in walking distance of 2.38 m. For every percent increase in the number of university graduates, total walking distance is predicted to decrease by 11.03 m. Furthermore, the total walking distance reduces by 0.05 m for each percent increase in population density, while one additional bus stop within 1,000 m of the origin is associated with a decrease in total walking distance of 8.08 m. The intra-class coefficients for Model 3 show that

Table 3. Multilevel Regression Models Predicting Total Daily Walking Distance

	Model 1				Model 2				Model 3			
Variables	Coef.	Sig.	Confidence int.†		Coef.	Sig.	Confidence int.†		Coef.	Sig.	Confidence int.†	
Socioeconomic characteristics												
Male	33.00		−11.73	77.72	20.77		−22.73	64.28	15.34		−28.36	59.03
Age	−2.65	***	−4.49	−0.81	−2.50	***	−4.29	−0.71	−2.48	**	−4.28	−0.68
Low income	−150.28	***	−229.39	−71.18	−75.41		−152.44	1.62	−52.77		−130.22	24.67
Medium income	−11.88		−60.70	36.94	12.26		−35.24	59.75	21.87		−25.84	69.58
Trip characteristics												
Distance traveled in vehicle	−0.001		−0.006	0.003	−0.01	***	−0.01	0.00	0.01	***	0.01	0.02
Frequency of service at first stop	4.08	***	3.16	5.00	0.72		−0.34	1.77	0.55		−0.50	1.60
First mode taken												
Subway					983.09	***	913.07	1053.12	891.58	***	820.90	962.27
Commuter train					1935.14	***	1782.80	2087.47	1749.78	***	1596.60	1902.96
Peripheral bus					−51.79		−245.46	141.88	−67.62		−258.93	123.70
Suburban bus					−35.36		−192.97	122.25	−15.70		−170.67	139.27
Number of transfers												
One transfer									−319.55	***	−374.30	−264.81
Two transfers									−485.20	***	−551.79	−418.62
Three transfers or more									−656.21	***	−751.18	−561.25
Neighborhood characteristics												
Percentage of people with university degree	−3.12		−9.81	3.57	−12.14	***	−18.88	−5.40	−11.03	***	−17.49	−4.57
Population density	−0.04	***	−0.06	−0.02	−0.06	***	−0.08	−0.04	−0.05	***	−0.07	−0.03
Number of bus stops	−5.68	***	−7.85	−3.50	−8.60	***	−10.75	−6.44	−8.08	***	−10.22	−5.95
Methodology												
GTFS	−198.68	***	−240.08	−157.28	−76.70	***	−116.49	−36.91	9.11		−31.66	49.88
Constant	3027.39	***	2780.20	3274.58	3376.51	***	3116.03	3636.99	3275.21	***	3019.73	3530.69
Random-effects parameters	Est.	SE	Confidence int.†		Est.	SE	Confidence int.†		Est.	SE	Confidence int.†	
Census tract												
var(Constant)	1530977	121135	1311050	1787797	1539675	11985	1321805	1793457	1393737	112403	1189961	1632409
Trip ID												
var(Constant)	80707	2342	45631	142747	128403	21729	92157.97	178902.3	164204	21930	126388	213334
var(Residual)	2077314	30389	2018598	2137737	1859207	27225	1806606	1913339	1808063	26579	1756713	1860914

Note: Dependent variable: total daily walking distance (m).
*95% significance level; ** 99% significance level; *** 99.9% significance level.
†95% confidence interval.

the origin CT and the individual explain 41.4% and 46.2% of the variance in total walking distances, respectively.

Conclusion

This paper contributes to the literature examining walking to transit by employing a highly detailed OD Survey containing the stops where each person first entered, transferred, and subsequently left the system. We compared robust estimates generated from this detailed survey to walking distances computed by making use of abundantly available origin and destination information and a fastest route algorithm based on GTFS data. We found that methods only making use of origin and destination information tend to underestimate the distance (and time) people are walking, or willing to walk, to transit. Our results show that walking distances calculated through GTFS data underestimate walking distances by 10% compared with real-life behaviour (1,816 compared with 2,021 m of daily average walking). Multilevel mixed-effect regression models indicate that these differing results can be mainly attributed to differences in travel behaviour and mode choice. Applying a shortest path algorithm to a transit network, though resulting in the fastest route, does therefore not accurately model travel behaviour: the routes calculated through GTFS data show over 15% of individuals taking three or more transfers, in reality only 3% of respondents transfer three times or more. We also found that more than twice as many respondents commute by train than is expected from calculations based on a fastest route with GTFS data, resulting in considerable differences in estimated walking distances. In order to generate robust estimates of walking distances to transit, practitioners could therefore benefit from correctly modelling modal choice and limiting the number of transfers an individual makes.

The difference in walking distance estimates could impose a bias on the number of individuals assumed not to meet the recommended minutes of daily physical activity through public transport. Therefore, if practitioners solely rely on calculations from home and work locations, the wrong interventions and policies might be proposed: practitioners might assume from their calculations that a large percentage of residents in a certain area are not meeting the recommended minutes of daily physical activity, though this might not be the case. This study therefore provides practitioners and researchers with a clear correction method when using GTFS data to calculate walking distances to public transport. Further research could improve on this study by using smartphone apps or wearable devices that could provide even more detailed information about walking paths and walking speeds. However, we do not expect large differences from our study in terms of total walking distances, except when an individual uses a shortcut through a private area or follows an irrational path to reach a public transport stop.

References

1. Jakicic, J., and K. Gallagher. Exercise Considerations for the Sedentary, Overweight Adult. *Exercise and Sport Sciences Reviews*, Vol. 31, No. 2, 2003, pp. 91–95.
2. Frank, L., M. Andresen, and T. Schmid. Obesity Relationships with Community Design, Physical Activity, and Time Spent in Cars. *American Journal of Preventive Medicine*, Vol. 27, No. 2, 2004, pp. 87–96.
3. Katzmarzyk, P. Physical Activity Levels of Canadian Children and Youth: Current Issues and Recommendations. *Canadian Journal of Diabetes*, Vol. 28, No. 1, 2004, pp. 67–78.
4. World Health Organization. *Global Recommendations on Physical Activity for Health*. World Health Organization, Geneva, 2011. http://www.who.int/dietphysicalactivity/physical-activity-recommendations-18-64years.pdf?ua=1. Accessed October, 2018.
5. Sallis, J., L. Frank, B. Saelens, M. Kraft, and E. Engelke. Active Transportation and Physical Activity: Opportunities for Collaboration on Transportation and Public Health Research. *Transportation Research Part A: Policy and Practice*, Vol. 38, No. 4, 2004, pp. 249–268.
6. Brown, B., and C. Werner. A New Rail Stop: Tracking Moderate Physical Activity Bouts and Ridership. *American Journal of Preventive Medicine*, Vol. 33, No. 4, 2007, pp. 306–309.
7. MacDonald, J., R. Stokes, D. Cohen, A. Kofner, and G. Ridgeway. The Effect of Light Rail Transit on Body Mass Index and Physical Activity. *American Journal of Preventive Medicine*, Vol. 39, No. 2, 2010, pp. 105–112.
8. Wasfi, R., N. Ross, and A. El-Geneidy. Achieving Recommended Daily Physical Activity Levels through Commuting by Public Transportation: Unpacking Individual and Contextual Influences. *Health and Place*, Vol. 23, 2013, pp. 18–25.
9. Agence Métropolitaine de Transport. Enquête Origine-Destination 2013. In Montréal, QC, 2013.
10. Statistics Canada. Physical Activity Levels of Canadian adults, 2007 to 2009, 2015. https://www.statcan.gc.ca/pub/82-625-x/2011001/article/11552-eng.htm. Accessed October, 2018.
11. Besser, L., and A. Dannenberg. Walking to Public Transit: Steps to Help Meet Physical Activity Recommendations. *American Journal of Preventive Medecine*, Vol. 29, No. 4, 2005, pp. 273–280.
12. Sallis, J., C. Spoon, N. Cavill, J. Engelberg, K. Gebel, M. Parker, C. Thornton, D. Lou, A. Wilson, C. Cutter, and D. Ding. Co-Benefits of Designing Communities for Active Living: an Exploration of Literature. *International Journal of Behaviorial Nutritional and Physical Activity*, 2015, pp. 12–30. https://ijbnpa.biomedcentral.com/track/pdf/10.1186/s12966-015-0188-2 the DOI is DOI 10.1186/s12966-015-0188-2
13. Brown, B., and C. Werner. Before and After a New Light Rail Stop: Resident Attitudes, Travel Behavior, and Obesity. *Journal of the American Planning Association*, Vol. 75, No. 1, 2008, pp. 5–12.
14. Warburton, D., C. Nicol, and S. Bredin. Health Benefits of Physical Activity: The Evidence. *Canadian Medical Association Journal*, Vol. 174, No. 6, 2006, pp. 801–809.
15. Sener, I. N., J. R. Lee, and Z. Elgart. Potential Health Implications and Health Cost Reductions of Transit-Induced Physical Activity. *Journal of Transport and Health*, Vol. 3, No. 2, 2016, pp. 133–140.

16. Bopp, M., V. V. Gayah, and M. E. Campbell. Examining the Link Between Public Transit Use and Active Commuting. *International Journal of Environnemental Research and Public Health*, Vol. 12, No. 4, 2015, pp. 4256–4274.
17. Lachapelle, U., and D. Gianini Pinto. Longer or More Frequent Walks: Examining the Relationship between Transit Use and Active Transportation in Canada. *Journal of Transport and Health*, Vol. 3, No. 2, 2016, pp. 173–180.
18. Durand, P. C., X. Tang, P. K. Gabriel, I. N. Sener, A. O. Oluyomi, G. Knell, K. A. Porter, M. D. Hoelscher, and W. H. Kohl, III. The Association of Trip Distance with Walking to Reach Public Transit: Daya from the California Household Travel Survey. *Journal of Transport & Health*, Vol. 3, No. 2, 2016, pp. 154–160.
19. El-Geneidy, A., M. Grimsrud, R. Wasfi, P. Tétreault, and J. Surprenant-Legault. New Evidence on Walking Distances to Transit Stops: Identifying Redundancies and Gaps Using Variable Service Areas. *Transportation*, Vol. 41, No. 1, 2014, pp. 193–210.
20. Zhao, F., L.-F. Chow, M.-T. Li, I. Ubaka, and A. Gan. Forecasting Transit Walk Accessibility: Regression Model Alternative to Buffer Method. *Transportation Research Record: Journal of the Transportation Research Board*, 2003. 1835: 34–41.
21. Wang, Z., F. Chen, and T. Xu. Interchange between Metro and Other Modes: Access Distance and Catchment Area. *Journal of Urban Developement*, Vol. 142, No. 4, 2016, p. 04016012.
22. Jiang, Y. P., C. Zegras, and S. Mehndiratta. Walk the Line: Station Context, Corridor Type and Bus Rapid Transit Walk Access in Jinan, China. *Journal of Transport Geography*, Vol. 20, No. 1, 2012, pp. 1–14.
23. Morency, C., M. Trépaniera, and M. Demers. Walking to Transit: An Unexpected Source of Physical Activity. *Transport Policy*, Vol. 18, No. 6, 2011, pp. 800–806.
24. García-Palomares, J. C., J. Gutiérrez, and O. D. Cardozo. Walking Accessibility to Public Transport: An Analysis Based on Microdata and GIS. *Environment and Planning B: Planning and Design*, Vol. 40, No. 6, 2013, pp. 1087–1102.
25. Ewing, R., and R. Cervero. Travel and the Built Environment. *Journal of the American Planning Association*, Vol. 76, No. 3, 2010, pp. 265–294.
26. Statistics Canada. Census tract: Detailed definition, 2015. https://www.statcan.gc.ca/pub/92-195-x/2011001/geo/ct-sr/def-eng.htm. Accessed October 2018
27. Ross, N. A., S. Tremblay, and K. Graham. Neighbourhood Influences on Health in Montréal, Canada. *Social Science & Medicine*, Vol. 59, No. 7, 2004, pp. 1485–1494.
28. Fuzhong, L., J. K. Fisher, R. C. Brownson, and M. Bosworth. Multilevel Modelling of Built Environment Characteristics Related to Neighbourhood Walking Activity in Older Adults. *Journal of Epidemiol Community Health*, Vol. 59, No. 7, 2005, pp. 558–564.
29. Sallis, J., B. Saelens, L. Frank, T. Conway, D. Slymen, K. Cain, J. Chapman, and J. Kerr. Neighborhood Built Environment and Income: Examining Multiple Health Outcomes. *Social Science & Medicine*, Vol. 68, No. 7. 2009, pp. 1285–1293.
30. Guo, Z., and N. Wilson. Assessing the Cost of Transfer Inconvenience in Public Transport Systems: A Case Study of the London Underground. *Transportation Research Part A: Policy and Practice*, Vol. 45, No. 2, 2011, pp. 91–104.

The Standing Committee on Public Transportation Planning and Development (AP025) peer-reviewed this paper (18-01807).

Article

TRR
JOURNAL OF THE TRANSPORTATION RESEARCH BOARD

Transportation Research Record
2018, Vol. 2672(8) 849–858

DOI: 10.1177/0361198118781182
journals.sagepub.com/home/trr
SAGE

Determining Effective Sample Size to Calibrate a Transit Assignment Model: A Bayesian Perspective

Mohadeseh Rahbar[1], Mark Hickman[1], Mahmoud Mesbah[1,2], and Ahmad Tavassoli[1,3]

Abstract

A transit assignment model is used to predict passenger loads in order to evaluate existing and future transit network scenarios. One fundamental issue affecting the calibration of a transit assignment model is the amount of data required. The present study is designed to determine the effect of different sample sizes on the accuracy level of the estimated passenger flow. A Bayesian model is adapted for transit assignment, and the sample size for three types of priors, namely: uninformative, informative, and overly informative, are examined. In order to assess the value of prior information on passenger flow, the root-mean-square error (RMSE) between each posterior estimate and the actual observation is computed. The posterior estimate that minimizes the %RMSE defines the effective sample size (ESS). This paper uses one day's automatic fare collection data from the South East Queensland (Australia) transit network to evaluate the effect of sample size and prior information on the posterior passenger flow estimates. The results show that it is not possible to determine the ESS for the Bayesian model with an uninformative prior. With an informative prior, the ESS is 50% of the population and, for the model with an overly informative prior, the ESS is 10% of the population. This means that the lack of prior information cannot simply be compensated by increasing the sample size in this Bayesian model. However, good prior information reduces the necessary sample size substantially.

The outcome of a transit assignment model is mainly affected by the assumption that a passenger wants to select a time- or cost-efficient path for his/her journey. This cost may be related to a single path, as in earlier models (*1–3*), or a hyperpath (*4*, *5*). Recently, some studies have employed Bayes' theory for detecting the preferences for used paths in a transit network. An important part of the Bayesian model is determining how to set the initial probability of parameters (known as "priors") using the data at hand (*6*).

One of the first applications of the Bayesian model in a transport context was by Maher (*7*) in estimating an origin–destination (OD) matrix. After that, the estimation of an OD matrix using the Bayesian approach has been more widely accepted by some other researchers such as Li (*8*), Perrakis et al. (*9*) and Alsger et al. (*10*). Given the paths used for each OD pair, Fu et al. (*11*) presented a Bayesian model to find the route shares and the mean and variance of travel times for each route. Lee and Sohn (*12*) proposed an approach to include the number of routes used as an unknown parameter into a Bayesian framework based on a reversible-jump Markov chain Monte Carlo (MCMC) algorithm. The proposed method showed flexibility in recognizing route-use patterns through the marginal posterior distribution of other unknown parameters. More pertinent to this study, a Bayesian statistical inference framework to model passenger assignment in a metro network was proposed by Sun et al. (*13*). By observing passenger travel times provided by smart card data and by including prior knowledge about the Singapore Metro network, they built the posterior density of path choice parameters. The results showed that the disutility of transfer time is about twice that of in-vehicle travel time in the Singapore Metro system. Rahbar et al. (*14*) developed a Bayesian statistical framework to model passengers' path choice behavior and to estimate travel time attributes in a multi-modal network. The travel time on each link, the transfer time, the impact of in-vehicle time, and the impact of transfer time on passenger path choice behavior were the important variables and parameters of their model. Results

[1]School of Civil Engineering, The University of Queensland, Queensland, Australia
[2]Department of Civil and Environmental Engineering, Amirkabir University of Technology, Iran
[3]Transurban, Melbourne, Australia

Corresponding Author:
Address correspondence to Mohadeseh Rahbar: m.rahbar@uq.edu.au

showed that passengers value transfer time more than in-vehicle time.

The amount of data to be used and the distribution of prior(s) are two fundamental questions in any Bayesian analysis (*15*). For many parametric Bayesian models, the answers are not straightforward. For example, with a small to moderate sample size, the prior's role is substantial. If the prior is a technically acceptable choice, then determining the sample size would become more important (*16*). On the other hand, by computing the sample size, the use of an overly informative prior that dominates the data may be avoided, in the sense that Bayesian inference is mostly affected by the prior rather than by the data.

In this paper, we propose an approach for computing the effective sample size (ESS) in a Bayesian model of transit assignment by adjusting a variety of settings. There are two main reasons for finding the ESS in this study. Firstly, this case study provides an opportunity to calibrate a Bayesian model on a transit network in which limited smart card data is available and additional data may needed to be collected. It would be more convenient to have an estimate of the required sample size to achieve an acceptable passenger flow forecast. Secondly, applying an unreasonably large sample size for the utilized model would increase the running time. Using this approach with enough data and an informative prior, acceptable results will be obtained in a reasonable time.

In this study, we select the transit assignment model suggested by Rahbar et al. (*14*) using a Bayesian inference approach. Second, we select different sample sizes from one weekday of smart card data from the South East Queensland (SEQ) transit network, including bus, train, and ferry modes. At each sample size, a number of repetitions are generated to overcome the bias of the randomly selected sample size. Third, priors are generated using three definitions: (*a*) uninformative, (*b*) informative, and (*c*) overly informative. The uninformative priors do not supply any particular information, while the informative priors might come from a literature review, an expert view, or explicitly from an earlier data analysis. We call the last type of priors overly informative because the source of information is strong and might dominate the likelihood. The combination of different sample sizes and priors gives different scenarios. The results of different scenarios are compared with the observations, and the sample size that minimizes the %RMSE is the ESS.

The remaining sections of this paper are organized as follows. The next section presents the methodology, including the application of Bayesian inference in transit assignment, the data description, sample size selection, and a definition of priors. The model applications are described in the results section, including a discussion comparing the model and actual observations. Conclusions are presented in the last section.

Methodology

Bayesian Model

Rahbar et al. (*14*) developed a Bayesian statistical framework to model passengers' path choice behavior and to estimate travel time attributes of a transit network. The travel time on each link c_a, the transfer time c_f, the impact of in-vehicle time Θ_1, and the impact of transfer time Θ_2 on passenger path choice behavior are the variables and parameters of their model, respectively. That model, with respect to its successful application in the transit network of Brisbane, Australia, is chosen for analysis in this paper. Mathematically, the model is written as follows:

$$\pi(c,\theta \mid T) \propto \prod_{od\in OD}\prod_{t\in T_{od}}\left[\sum_{r\in PT_{od}} \frac{\exp\left(\frac{-(t-\sum_{i\in r}c_i)^2}{2(\sum_{i\in r}c_i^2+\sigma^2)}\right)}{\sqrt{2\pi(\sum_{i\in r}c_i^2+\sigma^2)}} \times \frac{\exp\left(\theta_1\sum_{a\in r}c_a+\theta_2\sum_{f\in r}c_f\right)}{\sum_{r\in PT_{od}}\exp\left(\theta_1\sum_{a\in r}c_a+\theta_2\sum_{f\in r}c_f\right)}\right] \times \pi(c)\times\pi(\theta) \quad (1)$$

where $\prod$ represents the product of terms, and a and f are the indices of an in-vehicle link and a transfer link, respectively. Based on this model, the probability of parameters given the observed travel times, $\pi(c,\Theta|T)$, is equal to the probability of observing travel time t on path r from path set PT_{od}, multiplied by the probability of selecting path r, multiplied by the prior probability of parameters $\pi(c)\times\pi(\Theta)$. T_{od} represents the total travel time of a particular od pair. As can be seen in Equation 1, the likelihood function contains two parts, a multinomial logit model (Equation 2) which is a function of the route attribute (link travel time) and the probability of observing t given path r and other parameters (Equation 3).

$$p_r = \frac{exp\left(\theta_1\sum_{a\in r}c_a+\theta_2\sum_{f\in r}c_f\right)}{\sum_{\acute{r}\in PT_{od}} exp\left(\theta_1\sum_{a\in\acute{r}}c_a+\theta_2\sum_{f\in\acute{r}}c_f\right)} \quad (2)$$

$$p(t \mid r,c,\theta) = \frac{\exp\left(\frac{-(t-\sum_{i\in r}c_i)^2}{2(\sum_{i\in r}c_i^2+\sigma^2)}\right)}{\sqrt{2\pi(\sum_{i\in r}c_i^2+\sigma^2)}} \quad (3)$$

In this model, it is assumed that in-vehicle link travel times are independent of each other and follow a normal distribution with mean μ and standard deviation σ.

$$c \sim N\left(\mu,\sigma^2\right) \quad (4)$$

Data Preparation and Sample Size Selection

In this study, the main focus is on those journeys which start within a weekday morning peak (7 a.m.–9 a.m.) and finish within or after this period. Using the SEQ network's smart card fare payment system, the "*go*" card, all essential data such as date, time, boarding and alighting location, route number, and direction of each transaction (both "tap on" and

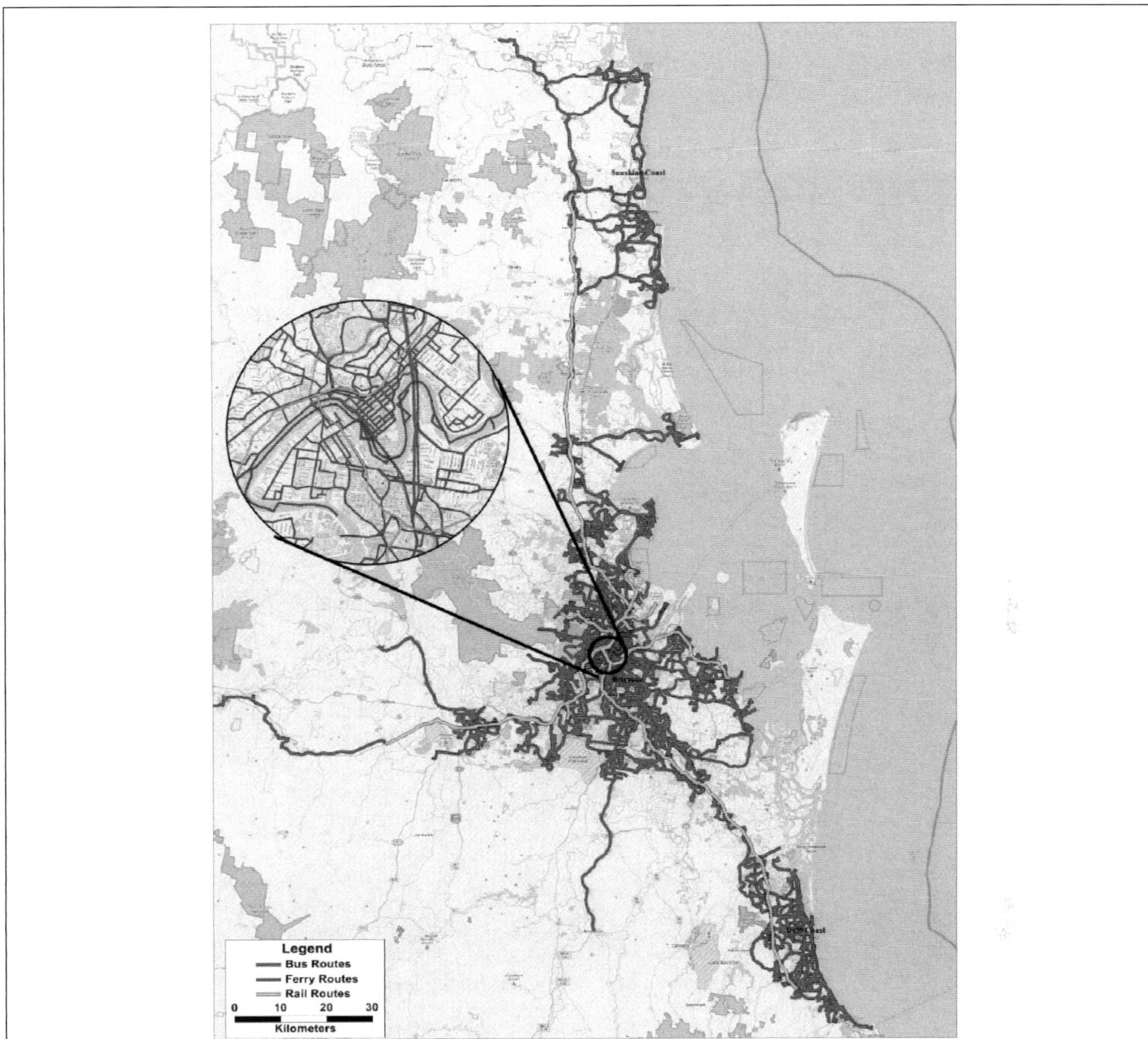

Figure 1. SEQ transit system adapted from Google Maps.

"tap off") are provided. Figure 1 shows the SEQ transit network.

Table 1 shows an example of four transactions of the selected data on March 20, 2013. For example, the passenger with card ID 2 has a transfer at Cultural Centre station.

Figure 2 illustrates the procedure of extracting the required information from the smart card data. In the first stage (data pre-processing), all records in which the boarding or alighting times are missing, the boarding or alighting stops are missing, the boarding stop is the same as the alighting stop, and/or the boarding time is later than the alighting time, are removed. At the data pre-processing step, about 30% of the data (including all records of a removed journey) was removed (*18*). Also, Australian Bureau of Meteorology data were consulted to ensure that the selected day of data is a typical day in terms of weather conditions and rainfall. In the second stage, based on the boarding time and alighting time of a unique card ID on the selected day, the sequence of all trip legs for that ID are extracted. Also, the network coordinates (nodes and links) from the strategic transport model are used to describe those trip legs. In the third stage, the in-vehicle time and the number of transfers for each unique card ID and trip leg are calculated. For ferry and bus, the difference between the alighting time and boarding time is defined as the in-vehicle time. For train, this difference in "boarding" and "alighting" times also includes waiting time and parts of access or egress time, as the smart card transactions occur on platforms, rather than on-board the vehicles.

Table 1. Sample Transactions from the Selected Data

Card ID	Boarding stop	Alighting stop	Boarding time	Alighting time	Route
1	Logan Rd at Swain St	Elizabeth St Stop 81 near George St	8:22	8:39	P189
2	RBWH station	Cultural Centre station	7:38	7:50	66
2	Cultural Centre station	UQ Lakes station	7:53	8:03	109
3	Maundrell Tce at John Goss Reserve	King George Square station	7:59	8:35	P343

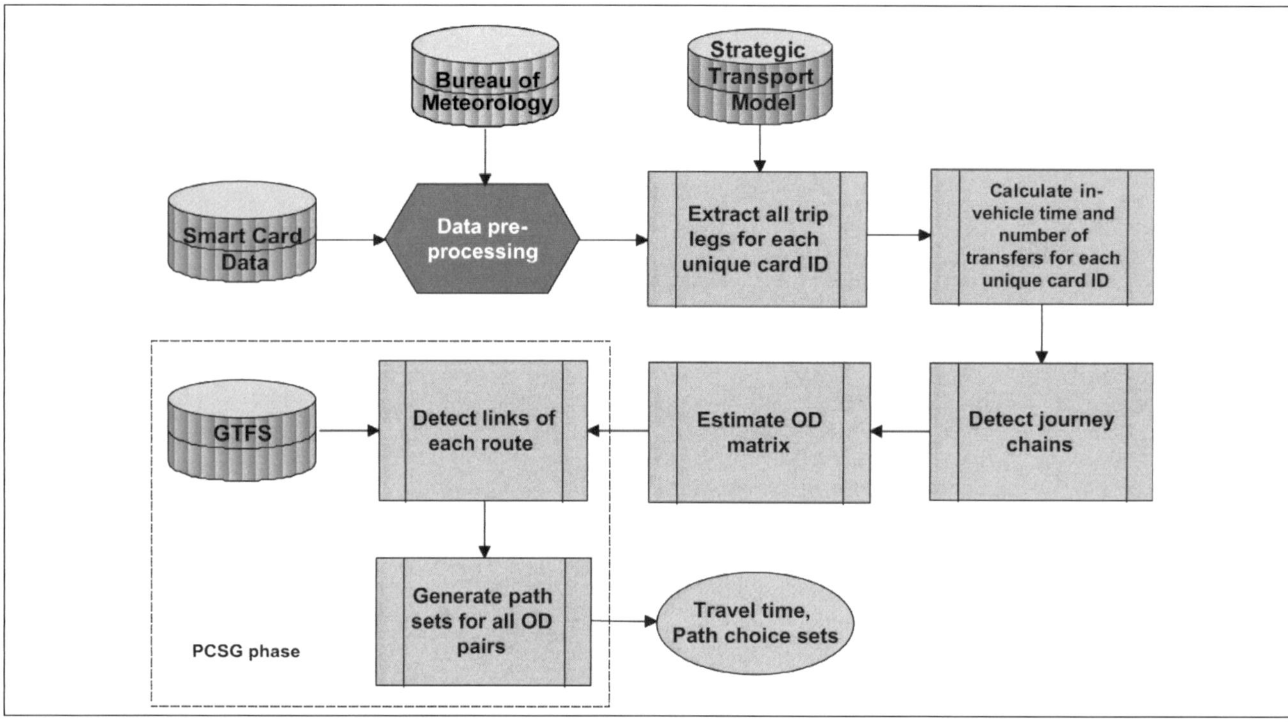

Figure 2. The data preparation algorithm.

For each passenger, the travel time for the same route will be slightly different due to the alighting and boarding times. In the fourth step, the trip leg sequence of each unique card ID helps us to create the journeys. In Brisbane, a 60-min transfer time is used to define a transfer for fare collection purposes (*14*). This means that, if the passenger starts the second trip within 60 minutes from the end of the first trip, he/she is still deemed to be on the previous journey and is not charged for the second trip. Considering this rule, the authors assume that if the calculated transfer time between two sequential trip legs is less than 60 minutes, these two trip legs are associated with one single journey (*19*). Then, using the journey's specifications and the number of passengers on each route, a stop-level (stop-to-stop) OD matrix is estimated. Using the route number, the links for passenger journeys are extracted from SEQ's GTFS (General Transit Feed Specification) (*20*). After detecting all the paths used by passengers, the path sets for all OD pairs are generated in the path choice set generation (PCSG) phase. Each path can include one or more than one route. Finally, the travel time for each passenger on each path, and the path choice sets, are the outputs of this algorithm.

For the selected typical day (March 20, 2013), there are 24,878 OD pairs and 80,767 journeys during the morning peak period. Table 2 shows the characteristics of the selected data set. For example, row 4 indicates that each path set has at least one path and at most 19 paths.

Table 3 shows the number of OD pairs, journeys, and used links among different sample sizes. For example, the 1% sample size contains 250 OD pairs and 840 journeys on 2,344 links. For each sample size, the OD pairs are selected randomly for five repetitions (i.e., five independent random draws).

The procedure of sample size selection, prior selection, and %RMSE calculation is illustrated in Figure 3. This procedure starts with the smart card data set. In this study, all journeys of the morning peak period for a weekday (March 20, 2013) are selected as the data set. The data preparation algorithm (Figure 2) is then called. The main functions of this algorithm are OD matrix estimation, the travel time calculation, and PCSG. Sampling is then performed

Table 2. Data Characteristics

Row	Characteristics	Value
1	Number of OD pairs	24,878
2	Number of trip legs	95,658
3	Number of journeys	80,767
4	Number of paths in each path set	[1,19]
5	Number of passengers on each path	[1,272]
6	Number of passengers on each OD pair	[1,560]
7	Number of used links	12,255

Table 3. Data Associated to the Selected Sample Sizes

Sample size (%)	No. of OD pairs	No. of journeys	No. of used links
1	250	840	2,344
5	1,250	3,620	5,907
10	2,490	7,270	8,037
30	7,464	25,578	10,454
40	9,952	31,413	11,152
50	12,450	39,270	11,390
70	17,415	57,283	11,933
100	24,878	80,767	12,255

randomly, based on the OD pair; for example, the samples include 1% of OD pairs, 5% of OD pairs, and so on. Since different OD pairs are selected for each sample size, specific path choice sets need to be generated. For this purpose, the PCSG method within the data preparation algorithm is called, for each sample. In the PCSG, the actual used paths by passengers are extracted from the dataset to generate the path choice sets (*14*). In the next stage, one type of prior information is selected. Using these priors, the Bayesian model (Equation 1) can be applied to the prepared data. For each sample size, a number of repetitions (independent random samples) are generated to overcome the bias of a single selected sample. The first outputs of this procedure are the posterior distributions of passenger flow on the network. The estimated passenger flows are compared with these distributions, and the %RMSE for these flows are computed. The value of the sample size that minimizes the %RMSE is used as the ESS.

Prior Information

To select the parameters of the prior (μ and σ for the normal distribution), three approaches are examined in this study: (*a*) uninformative, (*b*) informative, and (*c*) overly informative values.

In the first approach, it is assumed that there is no information on link travel times. We call this prior an "uninformative" prior and assume that $\mu = 5$ min and $\sigma^2 = 5$ min^2 for all of the links of the network.

In the second approach, some random observations from random OD pairs are selected, and the speeds of public transport vehicles are calculated using the observed travel time and distance. The average speed is 20 km/h. After calculating the distance between two successive stops, a unique travel time μ is provided for each link. For example, for one link with 1000 m length, the estimated μ would be 3 min. The second approach's prior is considered an "informative" prior.

Finally, in the third approach, GTFS data are used to provide link travel times as the prior information (so-called "overly informative" prior). Based on the GTFS data, the link travel times across the network vary from 1 to 37 min. Using the model-based clustering approach, links are categorized into seven groups, with μ = 1, 5, 9, 15, 23, 29, and 37 min. The model-based clustering approach is based on probability models wherein objects are assumed to follow a finite mixture of Gaussian probability distributions, such that each component distribution represents a cluster (*21*). The suggestion of the number of clusters and an appropriate model are the key advantages of the model-based approach, compared with standard clustering methods (k-means, hierarchical clustering, etc.). Fraley and Raftery (*22*) showed that the model-based approach, using the Bayesian information criterion, can provide much better performance than standard clustering methods. As an example, Figure 4 illustrates the clustering plots of the link travel times using the MCLUST package (*23*) in the R programming language (*24*). The main reason for clustering link travel times is the variety of travel times on rail, bus, and ferry links. For example, the seventh cluster with the mean of 37 min. contains railway links. Multimodality of the transit network thus indirectly affects the priors.

The variance values (σ^2) for all three travel time priors (uninformative, informative, and overly informative) are considered to be 5 min^2 to ensure that the selected distribution covers a greater interval around the mean value.

Results and Discussion

Using the R programming language, the Bayesian model (Equation 1) is implemented with the three priors on different sample sizes using the smart card data. The model was run for 108 scenarios (3 priors × 7 sample sizes × 5 repetitions, plus 3 priors × 100% data) on high performance computers at the University of Queensland. The estimated passenger flow on the whole network is compared with the actual observations. To show the associated errors, the %RMSE is used (*25*). The lower the value of %RMSE, the lower the differences between the predicted values and the actual observations. The %RMSE is:

$$RMSE = \sqrt{\frac{\sum_{i=1}^{N}\left(Model_i - Actual_i\right)^2}{N}} \quad (5)$$

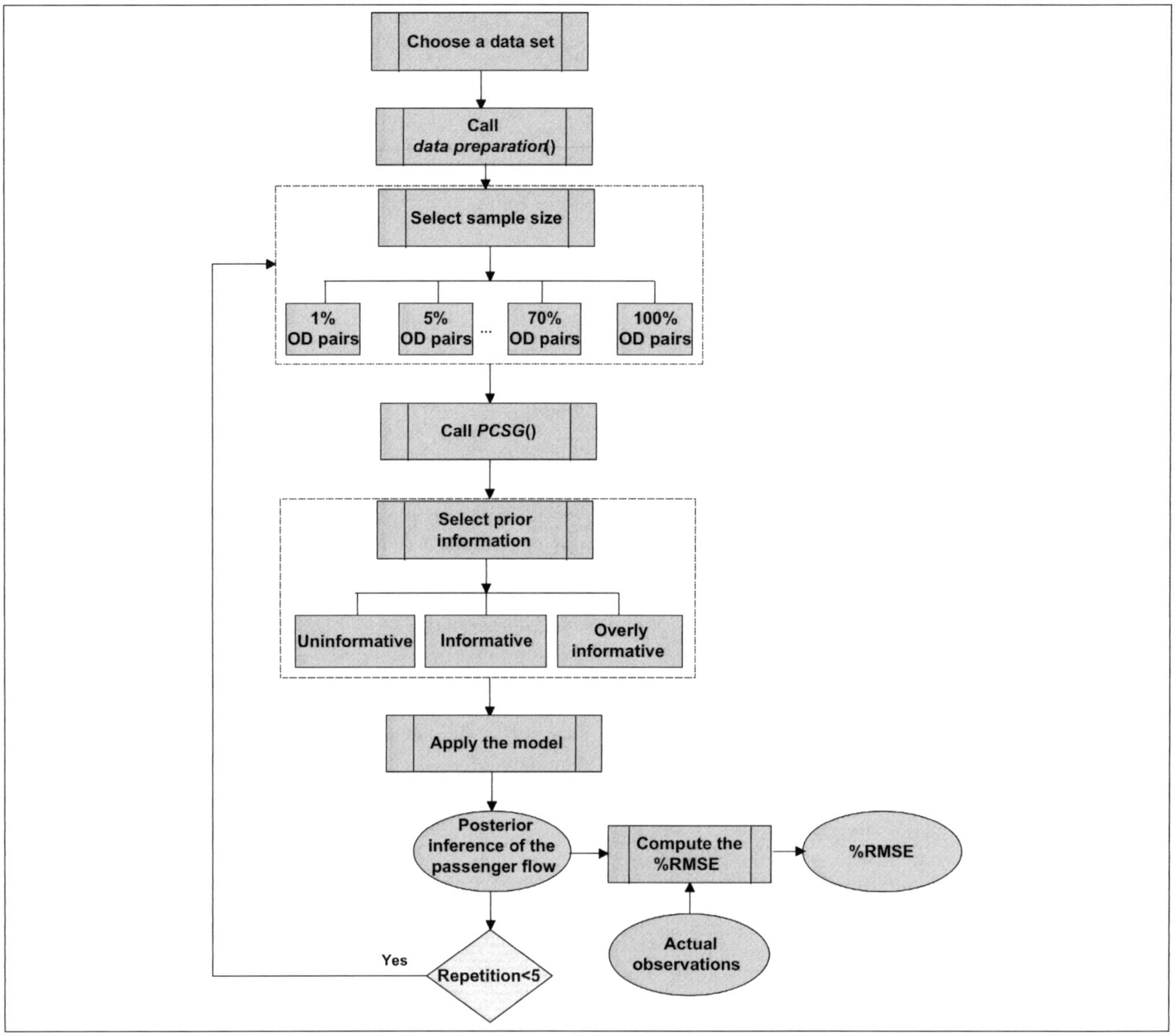

Figure 3. Sample size selection and Bayesian model estimation.

$$\%RMSE = \frac{RMSE}{\left(\frac{\sum_{i=1}^{N} Actual_i}{N}\right)} \times 100 \qquad (6)$$

where $Actual_i$ is the observed passenger flow from the "*go* card" data, $Model_i$ is the predicted value, and N is the number of predictions.

Table 4 presents the %RMSE between the estimated passenger flow from the Bayesian model and the actual observations for different sample sizes, as well as for different priors. For example, repetitions numbered one to five belong to the 1% sample size. As can be seen from the table, by increasing the sample size, the %RMSE decreases.

Graphical results are shown in Figure 5. All repetitions of each sample size are separated from other sample sizes by a solid gray vertical line. The best results belong to the model with an overly informative prior. In this case, the sample size does not have a strong influence on the posterior flow estimates. On the other hand, the worst results belong to the model with an uninformative prior. By increasing the sample size, a very gradual decrease in the %RMSE is observed. Even at the 100% level, the model with the uninformative prior cannot reach the level of an informative model with a 1% sample size. For the model with an informative prior, both the prior and the sample size have a significant effect on the posterior. This means that both the data itself and the prior information affect the estimated parameters. With an increase in the sample size, the %RMSE of an informative model approaches that of the overly informative priors.

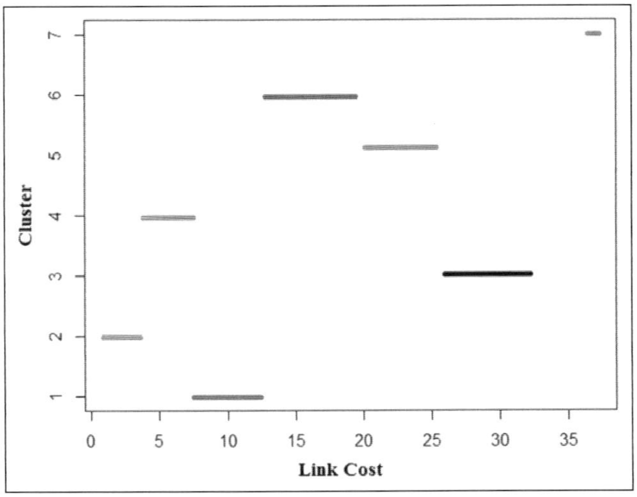

Figure 4. Sample GTFS clustering output.

As can be seen in Figure 5, even using 100% of the data, the estimated passenger flow using the uninformative prior is far away from the actual observations. It may be concluded that, in the presence of an uninformative prior, even a large set of data does not guarantee a close match on the posterior path flow estimates. Therefore, ESS identification using the uninformative prior is not meaningful; no sample size will result in useful posterior values.

For the model with an informative prior (red line in Figure 5), the errors are relatively high at low sample sizes and are reduced substantially as the sample size increases, particularly above the 50% sample size. The most surprising aspect of the results is in the range of the %RMSE. For example, 189% and 157% are the maximum and minimum range of the %RMSE for the model with an uninformative prior, while these values for the model with an informative prior are 116% and 46%, respectively. These represent significant improvements in the %RMSE.

Finally, for the model with an overly informative prior (green line in Figure 5), the posterior is strongly influenced by the prior. In this case, even with the smallest sample size, the %RMSE of the estimated values are low (52% in repetition 1). A possible explanation for this is that the GTFS, as the utilized prior, is the transit schedule based on which the transit services are operated. Hence, the actual travel times are very close to this overly informed prior. In summary, comparing these trends demonstrates that the prior's role is substantial in the accuracy of the final estimates. In the case of an overly informed prior, the posterior is mainly influenced by the prior and not much by the data. Stability is more visible by using 10% and larger sample sizes. Therefore, the ESS in this case would be a 10% sample size.

OD pair "1155" is used as an example to illustrate the output of the assignment model. Figure 6 shows three paths used for OD pair "1155." Based on the data, the demand for this OD pair is 26 passengers. In this example, the tuned parameters and random variables of the calibration phase are used.

Table 4. %RMSE Between Estimated Passenger Flow and Actual Observations

Sample size (%)	Repetitions	Uninformative	Informative	Overly informative
	1	169	113	52
	2	180	110	42
1	3	161	116	46
	4	158	110	43
	5	160	108	41
	6	169	96	45
	7	161	91	47
5	8	157	94	46
	9	162	95	44
	10	157	95	44
	11	158	92	41
	12	163	92	41
10	13	162	87	40
	14	158	87	41
	15	161	86	41
	16	158	50	40
	17	160	74	40
30	18	158	77	41
	19	159	63	40
	20	158	51	40
	21	157	62	46
	22	159	52	45
40	23	159	61	43
	24	159	52	44
	25	159	52	41
	26	158	51	44
	27	159	50	44
50	28	158	50	44
	29	159	51	44
	30	159	51	44
	31	159	48	43
	32	159	49	43
70	33	158	48	44
	34	158	49	43
	35	159	49	44
100	36	157	46	39

Using a 50% sample size and three different priors, the posterior estimates of passenger flows are shown in Table 5, with the last column containing the actual observations. The results of the model with an informative and with an overly informative prior are close to the observations.

Across all priors, however, there remains a substantial gap between the results and the actual observations. As one explanation, from the assignment model's perspective, passengers select their path only based on the estimated in-vehicle time because there are no transfer paths in this example. Thus, it is expected that most of the passengers choose either path 1 or

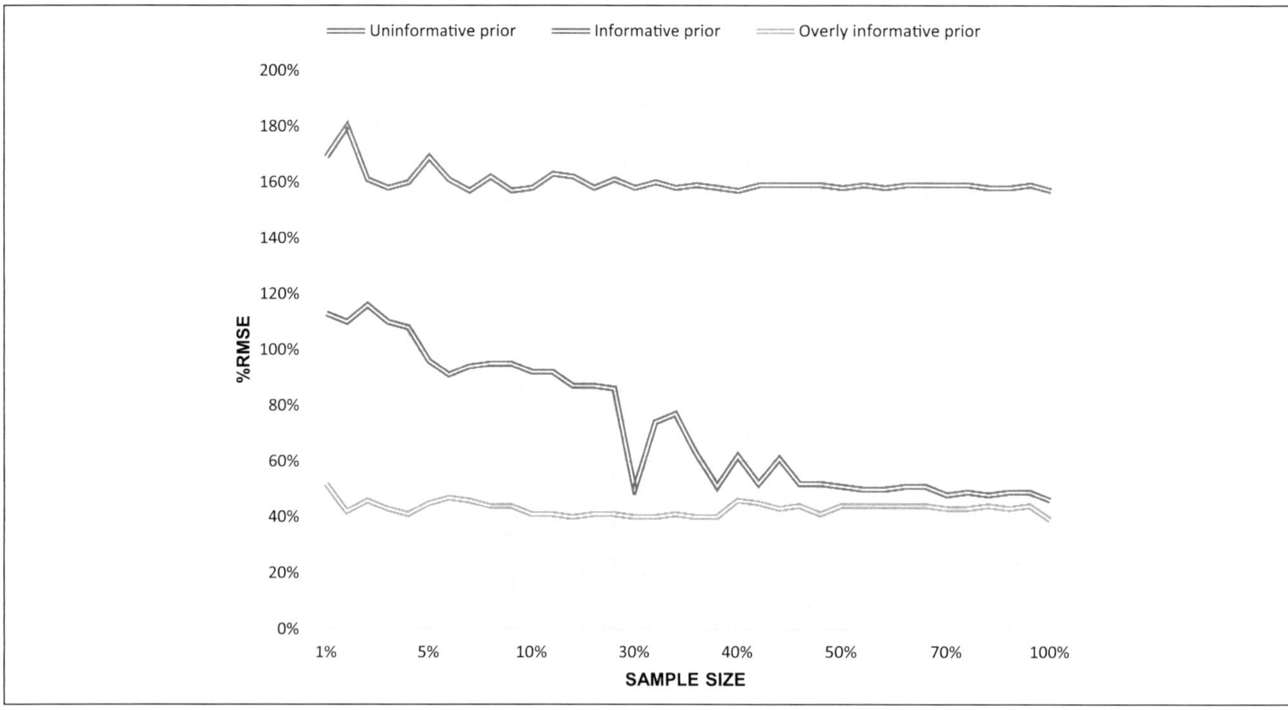

Figure 5. %RMSE between the model and actual observations.

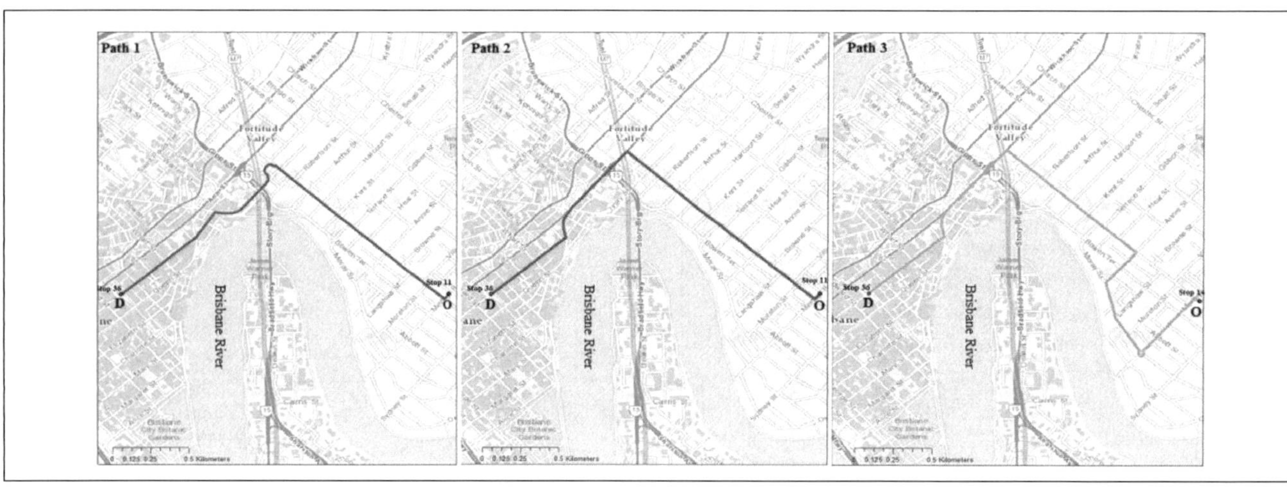

Figure 6. Three paths of OD pair "1155".

path 2, which have the lowest in-vehicle time. The actual observations show that path 1 is twice as desirable as path 2. This means there are more parameters in reality that affect passengers' decisions, such as waiting time and in-vehicle crowding.

Conclusions

Data collection is one essential prerequisite in calibration and validation of transit assignment models. The transit assignment model proposed by Rahbar et al. (*14*) is used for this study, which takes advantage of high-quality travel data provided by smart cards, and presents a transit assignment framework using a Bayesian inference approach. The main question in this analysis is, how much data is required to obtain a certain level of accuracy in the transit path flow estimates? In this paper, we conducted an experiment to compute the ESS in a variety of conditions. In addition, the accuracy level of model predictions is examined with different priors. Different sample sizes were chosen from one weekday data of the SEQ transit network including bus, train, and ferry modes. At each sample size, five repetitions (independent random samples) were generated to overcome the bias of a single randomly selected sample size.

Table 5. Results of the Model and Observations for OD Pair "1155"

Path	In-vehicle time (min)	Number of passengers			Observations
		Uninformative	Informative	Overly informative	
1	14	9	12	12	16
2	15	9	10	11	7
3	19	8	4	3	3

Then, travel time priors were generated using three definitions: (*a*) uninformative, (*b*) informative, and (*c*) overly informative priors. The combination of sample sizes with priors created 108 different scenarios. Finally, the estimated passenger flows of different scenarios were compared with the actual observations. Detecting the efficient sample size (ESS) for the model associated with an uninformative prior was not meaningful. For the model with an informative prior, the ESS is a 50% sample size (12,450 OD pairs and 39,270 journeys). For the model with an overly informative prior, the ESS is a 10% sample size (2,490 OD pairs and 7,270 journeys). The results highlight the effect of sample size and prior information on the accuracy level of the estimated passenger flow on the network.

This study has also shown that the model is sensitive to the prior information of the fundamental random variables such as link travel times. Further research should be undertaken to investigate the impact of different distributions for other parameters (Θ).

Since a normal distribution might produce negative values for link travel times, further research could explore other distributions for priors.

Acknowledgments

The authors thank TransLink (the public transport authority of Queensland, Australia) for providing the data. The authors also acknowledge the Australian Government Research Training Program for offering the lead author her scholarship for postgraduate study.

Author Contributions

The authors confirm contribution to the paper as follows: study conception and design, data collection, analysis and interpretation of results, and draft manuscript preparation: Mohadeseh Rahbar, Mark Hickman, Mahmoud Mesbah, and Ahmad Tavassoli. All authors reviewed the results and approved the final version of the manuscript

References

1. Dial, R. B. Transit Pathfinder Algorithm. *Highway Research Record*, Vol. 205, 1967, pp. 67–85.
2. Fearnside, K., and D. P. Draper. Public Transport Assignment. A New Approach. *Traffic Engineering & Control*, Vol. 13, 1971. pp. 298–299.
3. Le Clerq, F. *A Public Transport Assignment Method*, Vol. 23. Verkeerstechniek, Netherlands, 1972.
4. Nguyen, S., and S. Pallottino. Equilibrium Traffic Assignment for Large Scale Transit Networks. *European Journal of Operational Research*, Vol. 37, No. 2, 1988, pp. 176–186.
5. Spiess, H., and M. Florian. Optimal Strategies: A New Assignment Model for Transit Networks. *Transportation Research Part B*, Vol. 23, No. 2, 1989, pp. 83–102.
6. Congdon, P. *Bayesian Statistical Modelling*. Wiley, Chichester, New York, 2001.
7. Maher, M. J. Inferences on Trip Matrices from Observations on Link Volumes: A Bayesian Statistical Approach. *Transportation Research Part B: Methodological*, Vol. 17, No. 6, 1983, pp. 435–447.
8. Li, B. Bayesian Inference for Origin-Destination Matrices of Transport Networks Using the EM Algorithm. *Technometrics*, Vol. 47, No. 4, 2005, pp. 399–408.
9. Perrakis, K., D. Karlis, M. Cools, D. Janssens, K. Vanhoof, and G. Wets. A Bayesian Approach for Modeling Origin–Destination Matrices. *Transportation Research Part A: Policy and Practice*, Vol. 46, No. 1, 2012, pp. 200–212.
10. Alsger, A., B. Assemi, M. Mesbah, and L. Ferreira. Validating and Improving Public Transport Origin–Destination Estimation Algorithm Using Smart Card Fare Data. *Transportation Research Part C: Emerging Technologies*, Vol. 68, 2016, pp. 490–506.
11. Fu, Q., R. Liu, and S. Hess. A Bayesian Modelling Framework for Individual Passenger's Probabilistic Route Choices: A Case Study on the London Underground. Presented at 93rd Annual Meeting of the Transportation Research Board, Washington, D.C., 2014.
12. Lee, M., and K. Sohn. Inferring the Route-use Patterns of Metro Passengers Based Only on Travel-time Data Within a Bayesian Framework Using a Reversible-jump Markov Chain Monte Carlo (MCMC) Simulation. *Transportation Research Part B: Methodological*, Vol. 81, Part 1, 2015, pp. 1–17.
13. Sun, L. J., Y. Lu, J. G. Jin, D. H. Lee, and K. W. Axhausen. An Integrated Bayesian Approach for Passenger Flow Assignment in Metro Networks. *Transportation Research Part C: Emerging Technologies*, Vol. 52, 2015, pp. 116–131.
14. Rahbar, M., M. Mesbah, M. Hickman, and A. Tavassoli. Determining Route-choice Behaviour of Public Transport Passengers Using Bayesian Statistical Inference. *Road & Transport Research: A Journal of Australian and New Zealand Research and Practice*, Vol. 26, No. 1, 2017, pp. 64–72.
15. Morita, S., P. F. Thall, and P. Müller. Determining the Effective Sample Size of a Parametric Prior. *Biometrics*, Vol. 64, No. 2, 2008, pp. 595–602.
16. Berger, J. O., E. Moreno, L. R. Pericchi, M. J. Bayarri, J. M. Bernardo, J. A. Cano, J. De la Horra, J. Martín, D. Ríos-Insúa, and B. Betrò. An Overview of Robust Bayesian Analysis. *Test*, Vol. 3, No. 1, 1994, pp. 5–124.
17. Google Map. *South East Queensland*. https://www.google.com.au/maps/place/Queensland.
18. Tavassoli, A., A. Alsger, M. Hickman, and M. Mesbah. How Close the Models Are to the Reality? Comparison of Transit

Origin-Destination Estimates with Automatic Fare Collection Data. *Proc., 38th Australasian Transport Research Forum (ATRF)*, Melbourne, Victoria, Australia, 2016.
19. Alsger, A. A., M. Mesbah, L. Ferreira, and H. Safi. Use of Smart Card Fare Data to Estimate Public Transport Origin–Destination Matrix. *Transportation Research Record: Journal of the Transportation Research Board*, 2015. 2535: 88–96.
20. Google Inc. *General Transit Feed Specification*, 2013.
21. Oh, M.-S., and A. E. Raftery. Model-based Clustering with Dissimilarities: A Bayesian Approach. *Journal of Computational and Graphical Statistics*, Vol. 16, No. 3, 2012, pp. 559–585.
22. Fraley, C., and A. E. Raftery. How Many Clusters? Which Clustering Method? Answers via Model-based Cluster Analysis. *The Computer Journal*, Vol. 41, No. 8, 1998, pp. 578–588.
23. Fraley, C., A. E. Raftery, T. Brendan Murphy, and L. Scrucca. *Mclust Version 4 for R: Normal Mixture Modeling for Model-Based Clustering, Classification, and Density Estimation*. Technical. Report No. 597. Department of Statistics, University of Washington, USA, 2012.
24. R Core Team. *R: A Language and Environment for Statistical Computing*. R Foundation for Statistical Computing, Vienna, Austria, 2016.
25. Travel Model Validation and Reasonability Checking Manual. *The Travel Model Improvement Program*, Cambridge Systematics Inc., Massachusetts, 2010.

The Standing Committee on Public Transportation Planning and Development (AP025) peer-reviewed this paper (18-03402).

TRR
JOURNAL OF THE TRANSPORTATION RESEARCH BOARD

Article

Transportation Research Record
2018, Vol. 2672(8) 859–869

DOI: 10.1177/0361198118796356
journals.sagepub.com/home/trr
SAGE

Extended Methodology for the Estimation of a Zonal Origin-Destination Matrix: A Planning Software Application Based on Smartcard Trip Data

Sebastián Tamblay[1], Juan Carlos Muñoz[1], and Juan de Dios Ortúzar[1]

Abstract

A key planning consideration when designing transit services is the projected demand on each route, as this is a necessary input to determine frequencies and vehicle capacities. Predicting demand changes in complex networks requires a thorough modelling of the public transport system and its users' decisions. Aiming to help in this process, a new transit planning tool (STEPTRANS) is being developed, seeking to predict the reactions of public transport users to network changes, allowing better design and operation of the transport system. This tool needs a reliable origin-destination (O-D) matrix for the city of interest, which is assigned to the transit network, using state of practice route choice models. This work presents an extension of the former methodology implemented in STEPTRANS, which allows a zonal O-D matrix for a city to be estimated, given a stop-to-stop matrix obtained from smartcard trip data. The methodology and extension were applied to estimate models and O-D matrices for Santiago de Chile's integrated transit system, Transantiago. The extension focuses on three improvements: (1) addressing data limitations by using the latest large scale O-D survey available for Santiago; (2) updating the zonal system, including a new methodology for access links construction; and (3) enhancing the model formulation, incorporating trip pattern corrections and reformulating cost measures using route choice models. The inclusion of the resulting matrices in STEPTRANS will allow the evaluation of major changes to the transport system, as a zonal O-D matrix allows modelled users to choose their initial and final stops.

In recent years there has been an increasing trend for implementing information technologies, such as Automated Fare Collection (AFC) and Automatic Vehicle Location (AVL), in public transport systems. These technologies provide valuable information about trip patterns and, with adequate computational tools, this new massive data can be processed and serve as input for planning decisions.

A key planning consideration when designing transit services is the projected demand on each route, as this is a necessary input to determine frequencies and vehicle capacities. Predicting demand changes in complex networks requires a thorough modelling of the public transport system and its users' decisions.

Aiming to aid in this process, Project D10I1049-D10E1049 of the *Fondo de Fomento al Desarrollo Científico y Tecnológico* (FONDEF) is developing a public transport planning computational tool: STEPTRANS (Spanish initials for tactical-strategical planning software). This tool seeks to predict the reactions of public transport users to network changes, allowing better design and operation of the transport system (*1*).

As one of its main inputs, this tool needs a reliable origin-destination (O-D) matrix for the city of interest, which is assigned to the transit network using state of practice route choice models. The STEPTRANS base O-D matrix is the disaggregated "stop-to-stop" trip matrix developed by Munizaga and Palma (*2*) for Transantiago (and enhanced in Devillaine et al. (*3*); and Munizaga et al. (*4*)). Like many cities, Santiago de Chile has an integrated transit system, meaning that this O-D matrix includes both bus stops and metro stations (indistinctly referred to in the remainder of the paper as stops).

[1]Departamento de Ingeniería de Transporte y Logística, Pontificia Universidad Católica de Chile, Macul, Santiago, Chile

Corresponding Author:
Address correspondence to Sebastián Tamblay: sjtambla@uc.cl

However, this disaggregated stop-to-stop matrix is not really adequate for estimating the impact of projects that could change the stop where users start or end their trips, for example, extending routes, relocating bus stops, designing new routes or changing the fare structure. To study this kind of change, a zonal O-D matrix that allows the users modelled by STEPTRANS to choose their initial and final stops is needed.

To do this, Tamblay et al. (*5*) presented a methodology to estimate a zonal O-D matrix for a city, given a stop-to-stop trip matrix. The approach infers a probability for the zone of origin and destination for each trip between two stops, using a discrete choice model. Model inputs are land use and public transport information, for any period of interest in any city. The methodology was applied to estimate a model for Santiago de Chile's morning peak period.

This study presents a series of extensions to the previous methodology, focusing on three major improvements: (1) addressing data limitations; (2) updating Santiago's zonal system, including a new methodology for access links construction; and (3) enhancing the model formulation.

The remainder of this paper is structured as follows. The next section provides a literature review on smartcard O-D matrices estimation and aggregation. We then present a summary of Tamblay et al. (*5*), along with the proposed extended methodology. We follow this by introducing and describing model inputs for the case of Santiago, serving also to exemplify the data needed to apply the methodology in other cities. This is followed by a presentation of the basic zonal inference model, along with a validation analysis and the results of the proposed model extensions in Transantiago. Finally, we draw our conclusions, present future research guidelines, and provide policy applications of this study.

Background

The estimation of transit O-D matrices using smartcard data has been widely studied, aiming to exploit the increasing availability of this new automatic information. Through the application of different methodologies, O-D matrices have been estimated for many different cities (*2*–*4*, *6*–*10*). These methodologies are mostly based in the identification of boarding stops and alighting points using AFC and AVL data. In systems where passengers are forced to tap in and out in every vehicle, a full stop-to-stop trip matrix can be directly obtained by contrasting each AFC transaction timestamp with the corresponding AVL position data. In systems where passengers only tap in, then the alighting stop of every trip leg must be estimated, as has been done for Santiago and other systems were passengers are not required to tap out.

Most studies have focused on obtaining precise stop-to-stop O-D matrices, while few researchers have addressed the problem of aggregating and inferring the zonal origins and destinations of trips. Obtaining aggregated O-D matrices is important for transit modelling software as a higher degree of flexibility in their alternative routes is needed, reflecting the fact that many users are willing to walk and start their trips from other stops if better alternatives arise. Thus, proper network modelling needs to include some level of aggregation, acknowledging that trips are not really originated at stops, and stop-to-stop information only shows how users are travelling in the current network.

In cities where smartcards are associated with personal information, such as home and work address, the zonal origins and destinations of some trips can be inferred. However, in the case of Santiago and many other cities there is no such information available. For this, Amaya et al. (*11*) developed a method that estimates the residence zone of card users, but is limited to frequent users and is not enough to estimate a zonal origin and destination for every trip.

Other authors have opted to forget traditional zonal systems and instead aggregate stops to form O-D matrices (*12*–*14*). These aggregation methods have proven useful to reduce the size and ease the analysis of the O-D matrices, while also allowing for a better estimation of stop-to-stop passenger flows. However, they share the downside that the resulting O-D matrices are generally not compatible with traditional O-D survey information, census data, and administrative borders such as districts or communes.

Model Formulation

It was defined that the STEPTRANS O-D matrix had to be consistent with traditional zonal systems, to have better compatibility with other Chilean planning tools and O-D information. Thus, there was a need to develop a methodology for estimating a zonal O-D matrix based on smartcard trip information, using a traditional zonal system.

This section presents the zonal inference model formulation used in this study. First, we present the base model developed by Tamblay et al. (*5*), which assigns a probability for the zone of origin and destination for each entry in a stop-to-stop trip matrix. In other words, the model infers the access (and egress) walks for each trip between two stops, considering land use and network information of zones within walking distance. We then present the proposed extended methodology, including the enhanced model formulation, the new methodology for access links construction, and highlight the main improvements over the base model.

Base Zonal Inference Methodology

The proposed zonal inference model allows us to obtain the probability that an observed trip using stops k and l (as their boarding and alighting points, respectively), was originated in zone i and had zone j as its destination. This probability is defined as $\text{Prob}\left({}^{ij}/_{kl}\right)$.

To obtain these probabilities for each bus stop, metro station, and zone, we follow Daly (*15*) and formulate this particular gravitational model:

$$T_{ij}^{kl} = A_i B_j f_{ij}^{kl} \quad \forall(i,j)(k,l) \tag{1}$$

Subject to the constraints:

$$\sum_m \sum_n T_{mn}^{kl} = T_{kl} \quad \forall(k,l) \tag{2}$$

Where A_i and B_j are measures of the generating and attracting powers of zone i and j, linked to the population and land use information for these zones; T_{ij}^{kl} is the (unknown) number of trips made from zone i to zone j that use stops k and l as their boarding and alighting point, respectively; f_{ij}^{kl} is an inverse measure of the "cost" of choosing those stops to make the trip; T_{kl} is defined as the observed trips between public transport stops k and l, which is supposed to be known for every k and l; lastly, sets m and n are the zones of influence for stops k and l, respectively (i.e., zones within walking distance from each bus stop or metro station).

Then, by defining:

$$V_{mn/kl} = ln\left(A_m B_n f_{mn}^{kl}\right) \tag{3}$$

the latter can be represented by a disaggregate Logit model (*16*, *17*), as follows:

$$Prob\left({}^{ij}/_{kl}\right) = \frac{e^{V_{ij}/kl}}{\sum_m \sum_n e^{V_{mn/kl}}} \tag{4}$$

Different model specifications can be estimated using survey data and the Maximum Likelihood method (*18*). It is worth noting that this Logit model including size variables (A_i and B_j) is obtained without any loss of generality and imposing no restrictions on the specifications of the cost function implied (*15*).

Finally, defining T_{ij} as the trips from zone i to j, we can reconstruct each element of the desired zonal O-D matrix from the observed travel behavior between stops as shown by:

$$T_{ij} = \sum_k \sum_l Prob\left({}^{ij}/_{kl}\right) * T_{kl} \forall(i,j) \tag{5}$$

This formulation also requires to define the zones of influence (sets m and n) for each stop, and connect them with access links, defining the choice set for the Logit model. Tamblay et al. (*5*) proposed a methodology to add access links from zonal centroids to bus stops and metro stations (both for the zonal inference model and planning software). The proposed methodology encompasses two sequential tasks: (1) decide which stops are connected by access links to each zone; and (2) assign each one of them a distance value (which will be later related to the cost measure f_{ij}^{kl}).

The first is solved by connecting centroids to bus stops and metro stations that are inside the zonal boundaries, but also inside an imaginary extension of its borders defined as the influence radius.

Secondly, once defined which zones are connected to which stops, a distance value is assigned to each access link, considering land-use variables from the city blocks of each zone. The purpose was to obtain accurate values for access times as each zone is not necessarily homogenous and, for instance, the population may be concentrated in particular areas of the zone.

This way, each city block M was assigned a point in its geometrical center with two different weights linked to land use characteristics; one for its potential as a trip generator (w_M^g, linked to residential land use, i.e., housing squared meters), and the other as a trip attractor (w_M^a, linked to commercial, industrial, residential, health, educational, and offices total squared meters).

Then, distances are measured from each one of these points to each public transport stop inside the zone; finally, weighted averages of these distances are calculated using the generating and attracting weights of each city block. Thus, two distances are calculated for each stop: one corresponding to the generator link and another to the attractor link of the zone. Note that this methodology allows access time from zone i to stop k to be different from the access time from stop k to zone i, giving a better representation of heterogeneous zones (where population and trip attractors are located in different sectors).

Figure 1 shows an example for the calculation of the access link distance from zone i to stop p (d_{ip}), using a weighted average of the distances from the three city blocks that compose the zone i (d_{1p}, d_{2p}, d_{3p}).

Extended Methodology

As previously mentioned, this paper proposes three main improvements over the former work. The following section presents each of them in detail.

Addressing Data Limitations. The new methodology uses, as its estimation dataset, the latest large-scale O-D survey available for Santiago de Chile (*19*). This is a significant improvement, as the previous work used a smaller survey with lesser coverage and was not validated by the

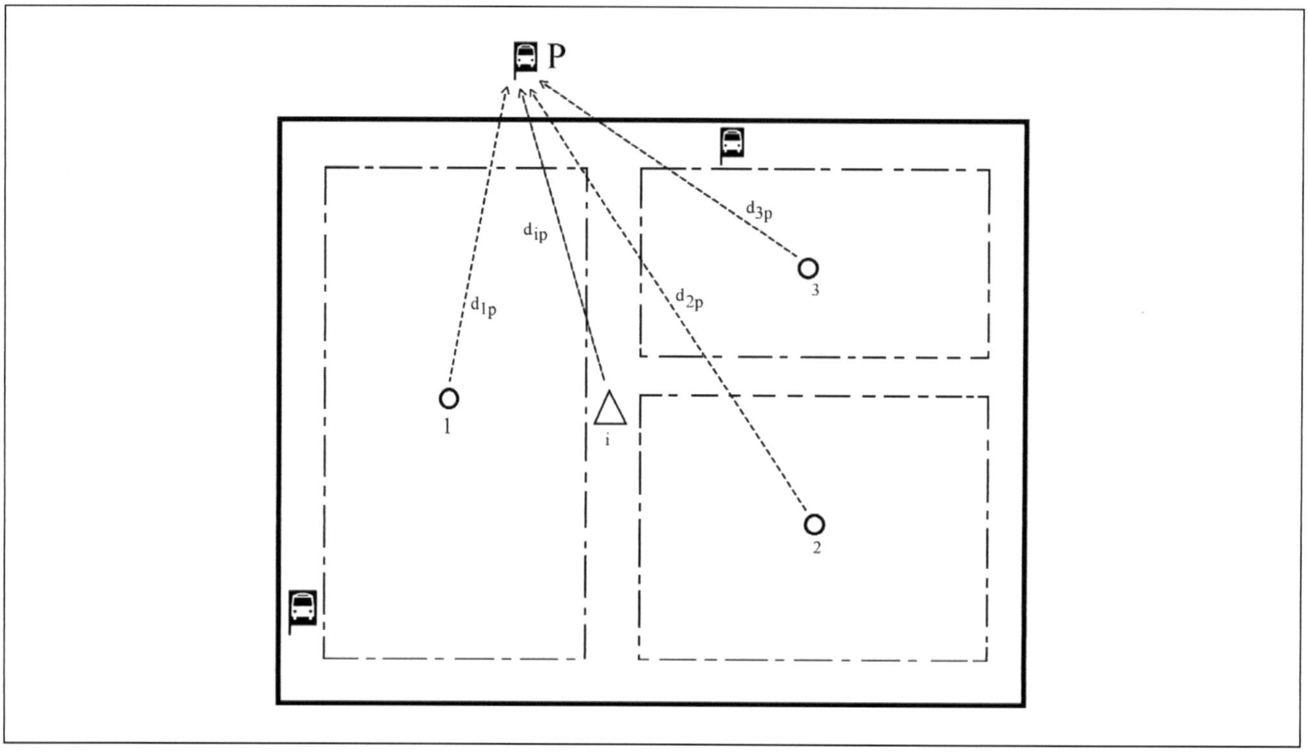

Figure 1. Access link distance calculation example.

Chilean authorities. Using the new O-D survey has the added advantage that it has data for the whole day (and not only for the morning, as the previous dataset), enabling the estimation of the model for new periods: afternoon peak and morning off-peak. Additionally, O-D surveys are periodically conducted and available for both Santiago and other large cities in the country. This way, developing a process that enables us to estimate the model directly from this kind of survey will simplify Santiago's model updates and potential STEPTRANS implementations in other cities.

Access Link Construction Methodology. The methodology proposed by Tamblay et al. (*5*) allows us to add access links to any zonal system, considering land use variables from the city blocks of each zone. However, their proposed weights (w_M^g and w_M^a) linked to land use characteristics were chosen arbitrarily and are not neccesarily accurate.

We propose an extension to their methodology that involves adding an access link distance calibration step. The objective is to calculate accurate generating and attracting weights and access distances, which are obtained from the parameters A_i and B_j of preliminary zonal inference models.

First, zonal inference models are obtained with the initial and arbitrarily chosen weights. These zonal inference models provide the relative generating and attracting powers of each land use, reflected in their parameters A_i and B_j. These relative powers are used to update the weights, which in turn leads to new models with updated parameters. This iterative procedure has proven rapidly convergent and a simple way of calculating more meaningful distances for the STEPTRANS network modelling.

Enhanced Model Formulation. In addition to the access link construction methodology, this study presents two main modelling improvements. First, using the new O-D survey as the dataset enables the estimation of a weighted Logit model, allowing the inclusion of basic sociodemographic and trip patterns corrections, thus decreasing selection bias.

Secondly, the base model's inverse cost function (f_{ij}^{kl}) was related to the access and egress links distances, from the observed stops *k* and *l* to zones *i* and *j*, respectively. However, the users' perceived cost of choosing particular stops to make a trip between two zones encompasses a process significantly more complex than a simple distance measure. This way, we included new cost measures, using route choice model utilities in STEPTRANS as an indicator of the users' perceived cost of choosing these particular stops to make a trip between those two zones.

Thus, a route and access station hierarchical Logit choice model, currently under development (with preliminary results presented in Abud, (*20*)), was included in the calibration procedure. In this choice model, users decide which route and bus stops (or metro stations) they will use to travel between two particular zones.

It is important to note that these new cost measures were calculated using the utilities of choice models that consider the access link distances outlined above. This way, the previous section is a necessary first step that not only allows us to improve our final zonal inference model, but also helps achieve a better representation of accessibility for the route choice models.

Data and Model Inputs

The essential data, needed to apply the proposed methodology to any city, can be categorized in: (1) stop-to-stop trip matrix; (2) land-use information; (3) O-D survey; and (4) zonal system.

This section presents the different sources of information that were gathered and unified to estimate and apply the zonal inference model for Santiago. Its objective is twofold: first, it presents a brief description of the data used for this application, and secondly, it helps to exemplify the kind of data needed, guiding the future application of the methodology in other cities.

Stop-to-Stop Trip Matrix

The zonal inference model works by assigning observed trips between two public transport stops to nearby zones, through a Logit model. In this study, we considered observed trips as those belonging to the STEPTRANS base stop-to-stop trip matrix (*2–4*), obtained from smartcard data for April 2013 on Santiago's network. These trips were used in T_{kl} variables defined in the previous section. Figure 2*a* presents Santiago's more than 11,000 bus stops and 100 metro stations, related to the stop-to-stop matrix input. Figure 2*b* shows Santiago's zonal system, related to the zonal O-D matrix output.

It is important to note that the way these stop-to-stop matrices were built translates into some limitations for this study. First, each trip alighting point is an estimation subject to error, as the system does not require passengers to tap out. Also, the assigned stop-to-stop matrix does not explicitly include non-integrated modes (such as bicycles, shared taxis and private vehicles) nor fare evasion, which distorts some of its trips. These limitations are extended to the zonal O-D matrix when assigned by the zonal inference model. Similarly, the stop-to-stop matrix is developed based on smartcard information and does not include trips paid with tickets. However, this restriction is less significant as tickets represented only a minor share of paid fares (being definitely removed in 2018).

Figure 2. (*a*) Transantiago's network (bus stops and metro stations in blue and red, respectively) and (*b*) The STEPTRANS zonal system.

Finally, it should be emphasized that these limitations are not inherent to the proposed zonal inference methodology. This way, if they are addressed in a new stop-to-stop matrix, a new zonal O-D matrix without these limitations can be easily obtained through the methodology proposed in this study.

Land Use Information

Generating and attracting powers of each zone (A_i and B_j, respectively) are linked to the land use information of each one of them. The latter allows the model to capture that, for example, in the morning peak, residential zones generate a larger number of trips or that business centers attract more trips.

Land-use information was included for Santiago de Chile for 2014, which contains the number of units and total squared meters dedicated to each land use classification in each zone. Some of the most relevant classifications available are: residential, commercial, educational, industrial, and health and offices buildings. This information is gathered and updated periodically by the Chilean Internal Revenue Service (*21*) for taxation and regulatory purposes, meaning that it is available to transportation planners at practically no cost.

O-D Survey

As previously mentioned, this study uses the latest large-scale O-D survey available for Santiago, as its estimation dataset for the zonal inference models of the three periods modelled. This dataset contains both socio-economic and detailed travel choice information for each of the surveyed users (*19*).

After digitation and processing, a total of 2,033 valid entries (i.e., which had clearly identified routes, origins, and destinations) remained for the morning peak period. Similarly, 1,051 valid entries remained for the afternoon peak and 1,039 for the morning off-peak period.

Zonal System

The proposed model allocates observed trips into nearby zones, following the probabilities resulting from the Logit calculations. This way, one of the key elements in determining the city's zonal system that will be used. The chosen zoning system is based in the one proposed in the SECTRA (*19*) O-D survey, but some zones were considered too large and were divided, resulting in an original zonal system with 1,171 zones (shown in Figure 2*b*).

The influence radii, necessary for the access link construction methodology, were previously defined by Tamblay et al. (*5*). They were obtained from Abud (*20*) survey data, resulting in a measure of 250 meters for bus stops and 750 meters for metro stations which represented 95% of surveyed users' access and egress walks. Each access link distance was calculated using the extended methodology proposed above, obtaining the generation and attraction weights from estimating preliminary zonal inference models. This procedure was rapidly convergent, and in only four iterations were the final access links obtained for the morning peak period.

Final Model and Zonal Matrix Estimation

This section presents model results, obtained using BIOGEME (*22*), for the three modelling periods: morning peak (trips ending between 8:00 and 9:00), afternoon peak (trips starting between 18:00 and 19:00), and morning off-peak (trips with mean time between 10:00 and 12:00).

First, the base model for each period is presented. Then, results are validated by comparing them with alternative simpler approaches. Third, the results of including the new cost measure proposed in this study are presented.

Basic Zonal Inference Models

Different specifications for the zonal inference model were estimated and compared, following the maximum likelihood procedure and a Weighted Multinomial Logit formulation. Weights were obtained from SECTRA's (*19*) O-D survey and the proposed utility function follows Tamblay et al. (*5*), with the general form:

$$V_{mn/kl} = ln(A_m) + ln(B_n) + \ln\left(f_{mn}^{kl}\right) \tag{6}$$

More specifically, the proposed functions were expressed in the following manner:

$$\begin{aligned} V_{mn/kl} = {} & \lambda_O{}^*\ln\left(\sum_P \theta_{m2_{PO}}{}^*M2_{P_{Om}}\right) \\ & + \lambda_D{}^*\ln\left(\sum_P \theta_{m2_{PD}}{}^*M2_{P_{Dn}}\right) \\ & + \ln\left(\frac{e^{(\beta_O{}^*d_{m-k})}}{(d_{m-k})^{\gamma_O}}\right) + \ln\left(\frac{e^{(\beta_D{}^*d_{l-n})}}{(d_{l-n})^{\gamma_D}}\right) \end{aligned} \tag{7}$$

Where $M2_{P_{Om}}$ and $M2_{P_{Dn}}$ are the total squared meters of land-use classification P in origin zone m or destination zone n, respectively; d_{m-k} and d_{l-n} are access distances in kilometers from origin and destination zones to the chosen initial and final stops, respectively; finally, λ, θ, β, and γ are parameters to be estimated. The first and second sum terms are related to the generating and attracting powers of zones m and n, respectively; while the last two correspond to the inverse cost measure f_{mn}^{kl},

Table 1. Zonal Inference Model for the Morning Peak Period

Associated to	Parameter	Value	*t*-test
Generation (origin)	λ_O	1	Fixed
A_i	$\theta_{m2_{CO}}$ - Commercial m^2	0	Dropped
	$\theta_{m2_{EO}}$ - Educational m^2	0	Dropped
	$\theta_{m2_{HO}}$ - Habitational m^2	1	Fixed
	$\theta_{m2_{IO}}$ - Industrial m^2	0	Dropped
	$\theta_{m2_{OO}}$ - Offices m^2	0	Dropped
	$\theta_{m2_{SO}}$ - Health m^2	0.903	1.81
Attraction (destination)	λ_D	1	Fixed
B_j	$\theta_{m2_{CD}}$ - Commercial m^2	0.348	2.31
	$\theta_{m2_{ED}}$ - Educational m^2	0.534	2.32
	$\theta_{m2_{HD}}$ - Habitational m^2	0.428	2.86
	$\theta_{m2_{ID}}$ - Industrial m^2	0.487	2.18
	$\theta_{m2_{OD}}$ - Offices m^2	0.596	2.55
	$\theta_{m2_{SD}}$ - Health m^2	1	Fixed
Cost	β_O	0	Dropped
f_{ij}^{kl}	γ_O	2.42	26.53
	β_D	−2.60	−3.75
	γ_D	1.38	3.94
Fit	Null log-likelihood	−5,519.6	
	Final log-likelihood	−4,256.0	
	Likelihood ratio test	2,527.2	
	Valid entries	2,033	

Table 2. Zonal Inference Model for the Afternoon Peak Period

Associated to	Parameter	Value	*t*-test
Generation (origin)	λ_O	1	Fixed
A_i	$\theta_{m2_{CO}}$ - Commercial m^2	0.440	2.56
	$\theta_{m2_{EO}}$ - Educational m^2	0	Dropped
	$\theta_{m2_{HO}}$ - Habitational m^2	0.330	2.90
	$\theta_{m2_{IO}}$ - Industrial m^2	0.326	1.86
	$\theta_{m2_{OO}}$ - Offices m^2	0.623	2.47
	$\theta_{m2_{SO}}$ - Health m^2	1	Fixed
Attraction (destination)	λ_D	1	Fixed
B_j	$\theta_{m2_{CD}}$ - Commercial m^2	0	Dropped
	$\theta_{m2_{ED}}$ - Educational m^2	0	Dropped
	$\theta_{m2_{HD}}$ - Habitational m^2	1	Fixed
	$\theta_{m2_{ID}}$ - Industrial m^2	0	Dropped
	$\theta_{m2_{OD}}$ - Offices m^2	0	Dropped
	$\theta_{m2_{SD}}$ - Health m^2	2.01	2.72
Cost	β_O	0	Dropped
f_{ij}^{kl}	γ_O	1.92	19.14
	β_D	−2.13	−2.48
	γ_D	1.31	2.84
Fit	Null log-likelihood	−3,070.2	
	Final log-likelihood	−2,477.6	
	Likelihood ratio test	1,185.4	
	Valid entries	1,051	

where we used a gamma function and explicitly allowed different impacts of access and egress distances on utility.

It is important to note that, for identifiability reasons, we cannot estimate a parameter for each size variable in origin and destination, so two of them are fixed to a value of one. Furthermore, both λ_O and λ_D are fixed to unitary values, to ensure that the model is independent from the zonal system used for estimation (*15*).

Tables 1, 2, and 3 shows the estimation results for this formulation, for the morning peak, afternoon peak, and

Table 3. Zonal Inference Model for the Morning Off-Peak Period

Associated to	Parameter	Value	*t*-test
Generation (origin)	λ_O	1	Fixed
A_i	$\theta_{m2_{CO}}$ - Commercial m^2	0.205	2.28
	$\theta_{m2_{EO}}$ - Educational m^2	0	Dropped
	$\theta_{m2_{HO}}$ - Habitational m^2	1	Fixed
	$\theta_{m2_{IO}}$ - Industrial m^2	0	Dropped
	$\theta_{m2_{OO}}$ - Offices m^2	0	Dropped
	$\theta_{m2_{SO}}$ - Health m^2	1.70	2.46
Attraction (destination)	λ_D	1	Fixed
B_j	$\theta_{m2_{CD}}$ - Commercial m^2	1	Fixed
	$\theta_{m2_{ED}}$ - Educational m^2	0.684	2.66
	$\theta_{m2_{HD}}$ - Habitational m^2	0.375	4.93
	$\theta_{m2_{ID}}$ - Industrial m^2	0	Dropped
	$\theta_{m2_{OD}}$ - Offices m^2	0.326	2.15
	$\theta_{m2_{SD}}$ - Health m^2	0	Dropped
Cost	β_O	−2.36	−2.30
f_{ij}^{kl}	γ_O	1.42	2.69
	β_D	0	Dropped
	γ_D	2.36	21.39
Fit	Null log-likelihood	−2,933.2	
	Final log-likelihood	−2,281.7	
	Likelihood ratio test	1,302.8	
	Valid entries	1,039	

morning off-peak, respectively. It is important to note that these results were obtained after the iterative access link construction methodology proposed above.

From results above we observe that all included parameters have the expected signs, and most of them are statistically significant. A more detailed look highlights that in the morning peak habitational and health buildings generate trips; the first one is obvious, while the latter is explained mainly by early appointments and work shift changes. In this period, all attracting variables were significant, with offices and health being the most important. On the other hand, in the afternoon peak we can observe an inverse trend, in which habitational and health buildings become the more important trip attractors. Educational land use also loses its significance, as most schools and university classes end earlier than the afternoon peak. Finally, the morning off-peak presents a similar situation to its peak counterpart; however, industries and health lose some attracting power, while commercial buildings gain. It is surprising that in this period health land use was not significant, while in other periods it was more so. A possible explanation is that the estimation dataset could have few health purpose trips, insufficient to capture their attraction effect precisely. In any event, this subject is being addressed and models are being checked for input inconsistencies.

Regarding the inverse cost function, parameters β_O and β_D dropped out (in different periods) due to their low significance and the fact that they are not fundamental (cost function maintains the expected trend without it).

Validation Analysis

Results validation was done by comparing the model's prediction to two alternative zonal assignments: equiprobable, which means assigning the trips in each stop with the same probability to each zone connected with an access link to it; and a stop's physical zone assignment, which means assigning all trips to the zone where the stop is. The model was expected to behave better than these alternatives, as the first one is a very simple solution and the second one should fail often, as zones are relatively small (less than 1 km length on average) and most zonal boundaries are defined in important streets.

The three alternative assignments (inference model, equiprobable, and stop's physical zone) were applied to each survey origin and destination stop, obtaining three predicted O-D matrices, each following one of these approaches. These predicted matrices were then compared with the survey's actual O-D matrix, and percentage absolute differences were obtained for each alternative (Table 4).

Results show that the proposed model allows for better results than applying alternative solutions. Assigning the trips to the stop's physical zone exhibits the worst results, as many users walked to stops outside them. It is also important to note that this validation was

Table 4. Percentage Absolute Differences, Proposed Model Comparison with Alternatives

	Morning peak		Afternoon peak		Morning off-peak	
	Origin	Destination	Origin	Destination	Origin	Destination
Stop's physical zone	75.8%	70.6%	77.5%	84.3%	81.8%	71.4%
Equiprobable	61.1%	69.4%	76.1%	74.5%	71.5%	76.0%
Inference model	45.1%	55.8%	62.4%	60.6%	59.7%	57.4%

Table 5. Zonal Inference Model for the Morning Peak Period Using the New Cost Function

Associated to	Parameter	Value		t-test
Generation (origin)	λ_O	1		Fixed
A_i	$\theta_{m2_{CO}}$ - Commercial m^2	0		Dropped
	$\theta_{m2_{EO}}$ - Educational m^2	0		Dropped
	$\theta_{m2_{HO}}$ - Habitational m^2	1		Fixed
	$\theta_{m2_{IO}}$ - Industrial m^2	0		Dropped
	$\theta_{m2_{OO}}$ - Offices m^2	0		Dropped
	$\theta_{m2_{SO}}$ - Health m^2	1.36		2.15
Attraction (destination)	λ_D	1		Fixed
B_j	$\theta_{m2_{CD}}$ - Commercial m^2	0.646		2.32
	$\theta_{m2_{ED}}$ - Educational m^2	0.204		1.67
	$\theta_{m2_{HD}}$ - Habitational m^2	0.175		2.56
	$\theta_{m2_{ID}}$ - Industrial m^2	0.0321		0.76*
	$\theta_{m2_{OD}}$ - Offices m^2	1.37		2.41
	$\theta_{m2_{SD}}$ - Health m^2	1		Fixed
Cost	β_P	−17.4		−10.20
f_{ij}^{kl}	γ_P	−29.8		−12.84
Fit	Null log-likelihood		−5,515.7	
	Final log-likelihood		−3,402.7	
	Likelihood ratio test		4,225.9	
	Valid entries		2,031	

constructed with the basic zonal inference models, and better fit is expected when the new cost measure is included.

Extension Results: New Cost Measure

The new cost measure, proposed above as a model extension, was studied. In this formulation, f_{ij}^{kl} variables will be related to the probability that a trip originated in zone i with its destination in zone j uses stops k and l, as their boarding and alighting points (defined as $Prob\left({}^{kl}/_{ij}\right)$), since when a route alternative is perceived as costly the probability of choosing its related stops must be low for a rational user (and vice versa for low cost routes). As previously stated, these probabilities are obtained from the STEPTRANS route and access station choice Logit model. In this model, the generalized level-of-service is composed by attributes related to the mode, different time components, transferring experience, vehicle crowding, and network topology (*1*, *20*).

This way, access and egress distances are already included in $Prob\left({}^{kl}/_{ij}\right)$ and are therefore removed from f_{ij}^{kl}, resulting in a new formulation as shown by:

$$f_{ij}^{kl} = \frac{e^{\beta_P * Prob\left({}^{kl}/_{ij}\right)}}{Prob\left({}^{kl}/_{ij}\right)^{\gamma_P}} \quad (8)$$

The morning peak model was re-estimated using this new formulation, and results are shown in Table 5. Results show a significant likelihood gain, when compared with the basic model that used a simpler cost function. Indeed, the likelihood ratio test improved from 2527.2 to 4225.9. It is important to note that two observations dropped (going from 2,033 to 2,031 valid entries), as the route choice model assigned a null probability that a trip between those zones would use the chosen stops.

We expect that the inclusion of this model in STEPTRANS will result in a better O-D matrix than the ones obtained with the simple models described above. In this regard, we expect that including these new cost measures will significantly reduce the errors presented above. We are currently working on implementing it in our planning software.

Conclusion

In conclusion, we have presented an extension for the methodology presented by Tamblay et al. (*5*), that allows us to estimate a zonal O-D matrix from an observed stop-to-stop O-D matrix. The extensions focused on three major improvements: (1) addressing data limitations; (2) updating Santiago's zonal system, including a new methodology for access links construction; and (3) enhancing the model formulation.

The extended methodology is applied to Santiago's transit system, including its morning peak, afternoon peak, and morning off-peak periods. However, the methodology and model are completely general and could be used in any public transport system with sufficient information. The application of the methodology to other cities in Chile would be particularly easy, especially where there are O-D surveys available (similar to the one used in this study for Santiago).

By including a more massive and complete data source, the latest available O-D survey for Santiago (*19*), many of the limitations of the base model were solved. However, the resulting zonal matrix is still an assignment of the stop-to-stop trip matrix and does not explicitly include non-integrated modes (such as bicycles, shared taxis and private vehicles) nor fare evasion. Correcting the distortion of evaded trips is being addressed in another study, also related to the new computational tool being developed. Likewise, we propose as future research to extend the model and apply it to non-integrated modes, allowing us to better represent a wider number of users and not only those who access their stops by walking.

Finally, we highlight the policy implications of this research. The inclusion of the resulting matrix in STEPTRANS will allow the evaluation of major changes to the transport system. Previously, transport projects that modified the stop where users started or ended their trips could not be properly evaluated, because the trip assignment stage was based on a stop-to-stop matrix. For example, it was not possible to know how many trips would start or end in a new bus stop. So, the proposed methodology and model becomes a key ingredient of the public transport planning software for Santiago, allowing the analysis of trips at a new zonal level, resulting in better and more flexible predictions.

Acknowledgments

This research was supported by FONDEF D10I1049 and FONDEF D10E1049 "Una herramienta táctico-estratégica de gestión y planificación de sistemas de transporte público urbano."

Author Contributions

The authors confirm contribution to the paper as follows: study conception and design: S. Tamblay, J. C. Muñoz, J. de D. Ortúzar; data collection: S. Tamblay; analysis and interpretation of results: S. Tamblay, J. C. Muñoz, J. de D. Ortúzar; draft manuscript preparation: S. Tamblay. All authors reviewed the results and approved the final version of the manuscript.

References

1. Raveau, S., J. C. Muñoz, C. G. Prato, A. Soto, S. Tamblay, and P. Iglesias. A Behavioural Planning Tool for Modelling Public Transport Systems. Presented at TransitData2017, Santiago, Chile, 2017.
2. Munizaga, M., and C. Palma. Estimation of a Disaggregate Multimodal Public Transport Origin–Destination Matrix from Passive Smartcard Data from Santiago, Chile. *Transportation Research Part C: Emerging Technologies*, Vol. 24, 2012, pp. 9–18.
3. Devillaine, F., M. Munizaga, and M. Trepanier. Detection of Activities of Public Transport Users by Analyzing Smart Card Data. *Transportation Research Record: Journal of the Transportation Research Board*, 2012. 2276: 48–55.
4. Munizaga, M., F. Devillaine, C. Navarrete, and D. Silva. Validating Travel Behavior Estimated from Smartcard Data. *Transportation Research Part C: Emerging Technologies*, Vol. 44, 2014, pp. 70–79.
5. Tamblay, S., P. Galilea, P. Iglesias, S. Raveau, and J. C. Muñoz. A Zonal Inference Model Based on Observed Smart-Card Transactions for Santiago de Chile. *Transportation Research Part A: Policy and Practice*, Vol. 84, 2016, pp. 44–54.
6. Barry, J., R. Newhouser, A. Rahbee, and S. Sayeda. Origin and Destination Estimation in New York City with Automated Fare System Data. *Transportation Research Record: Journal of the Transportation Research Board*, 2002. 1817: 183–187.
7. Trépanier, M., N. Tranchant, and R. Chapleau. Individual Trip Destination Estimation in a Transit Smart Card Automated Fare Collection System. *Journal of Intelligent Transportation Systems: Technology, Planning, and Operations*, Vol. 11, No. 1, 2007, pp. 1–14.
8. Gordon, J., H. Koutsopoulos, N. Wilson, and J. Attanucci. Automated Inference of Linked Transit Journeys in London Using Fare-Transaction and Vehicle Location Data. *Transportation Research Record: Journal of the Transportation Research Board*, 2013. 2343: 17–24.
9. Ma, X., Y.-J. Wu, Y. Wang, F. Chen, and J. Liu. Mining Smart Card Data for Transit Riders' Travel Patterns. *Transportation Research Part C: Emerging Technologies*, Vol. 36, 2013, pp. 1–12.
10. Alsger, A., B. Assemi, M. Mesbah, and L. Ferreira. Validating and Improving Public Transport Origin–Destination Estimation Algorithm Using Smart Card Fare Data.

Transportation Research Part C: Emerging Technologies, Vol. 68, 2016, pp. 490–506.
11. Amaya, M., R Cruzat, and M. Munizaga. Estimating the Residence Zone of Frequent Public Transport Users to Make Travel Pattern and Time Use Analysis. *Journal of Transport Geography*, Vol. 66, 2018, pp. 330–339.
12. McCord, M., R. Mishalani, and X. Hu. Grouping of Bus Stops for Aggregation of Route-Level Passenger Origin-Destination Flow Matrices. *Transportation Research Record: Journal of the Transportation Research Board*, 2012. 2277: 38–48.
13. Lee, S., M. Hickman, and D. Tong. Stop Aggregation Model: Development and Application. *Transportation Research Record: Journal of the Transportation Research Board*, 2012. 2276: 38–47.
14. Luo, D., O. Cats, and H. van Lint. Constructing Transit Origin–Destination Matrices with Spatial Clustering. *Transportation Research Record: Journal of the Transportation Research Board*, 2017. 2652: 39–49.
15. Daly, A. J. Estimating Choice Models Containing Attraction Variables. *Transportation Research Part B: Methodological*, Vol. 16, No. 1, 1982, pp. 5–15.
16. Daly, A. J. Some Developments in Transport Demand Modelling. In *Behavioural Travel Modelling* (Hensher, D. A., and P. R. Stopher, eds.), Croom Helm, London, 1979.
17. McFadden, D. Quantative Methods for Analysing Travel Behaviour of Individuals. In *Behavioural Travel Modelling* (Hensher, D. A., and P. R. Stopher, eds.), Croom Helm, London, 1979.
18. Ortúzar, J. de D., and L. G. Willumsen. *Modelling Transport*, 4th ed. John Wiley & Sons, Ltd., Chichester, 2011.
19. SECTRA. *Informe Final del estudio "Encuesta Origen Destino de Viajes 2012"*. Secretaría Interministerial de Planificación de Transporte, Santiago, 2015.
20. Abud, I. *Modelos de Elección de Paradero, Modo y Ruta para Herramientas de Planificación de Transporte Público*. Tesis de Magíster en Ciencias de la Ingeniería. Pontificia Universidad Católica de Chile, 2015.
21. Servicio de Impuestos Internos. *Roles de la Región Metropolitana de la Segunda Serie No Agrícola y Superficie Construida*. Data obtained through Chilean Transparency Law (N. 20.285). Servicio de Impuestos Internos, Santiago, 2014.
22. Bierlaire, M. BIOGEME: A Free Package for the Estimation of Discrete Choice Models. *Proc., 3rd Swiss Transportation Research Conference*, Ascona, Switzerland, 2003.

The Standing Committee on Public Transportation Planning and Development (AP025) peer-reviewed this paper (18-02920).

Article

TRR
JOURNAL OF THE TRANSPORTATION RESEARCH BOARD

Transportation Research Record
2018, Vol. 2672(8) 870–878

DOI: 10.1177/0361198118777605
journals.sagepub.com/home/trr
SAGE

The Effect of Demographic Changes on Transit Ridership Trends

Richard A. Driscoll[1], Kurt R. Lehmann[1], Steven Polzin[2], and Jodi Godfrey[2]

Abstract

Since 2013, there has been a noted overall decline in U.S. transit ridership despite continued growth in population. Transit ridership declines are not a new phenomenon, however. Since 1980, ridership has peaked and decreased seven different times. In the same time period, transit ridership per capita has decreased by nearly 15%. While economic considerations, fuel price, and changing modal choices are among the factors being evaluated in attempts to explain this decline, several unexplored causes share a strong relationship to ridership decline. Two demographic phenomena help to explain this dampening of transit usage and are the focus of this paper. First, the population is aging and there is a lower share of the population in the young age cohorts that have a higher propensity for transit use. Quantification of the impact of this phenomenon explains some of the ridership declines. Second, there has been a significant decline in the population in some of the counties with high-quality transit service and use. Simultaneously much of the population growth is occurring in counties with lower levels of transit service and use. Rapidly growing counties had half the rate of commuting to work by transit as did rapidly declining counties. In seeking to fully understand the factors that influence transit ridership trends, it is important that each of the many factors be fully understood so that policy makers and practitioners can respond and position transit accordingly.

The last 30–40 years have proved to be volatile regarding national transit ridership. According to data from APTA, the early 1980s and 1990s saw substantial periods of declining ridership, especially during the steep drop between 1990 and 1995 (Figure 1). During the housing market collapse of 2008 there was a peak in ridership followed by 2 years of decline. Ridership had also seen a decline at the beginning of the millennium and from 2014 to the present (*1*, *2*). APTA cost data, available only from 1992, also shows that national investment in bus and rail transit, including operating and capital expenses, has increased faster than population, seemingly matching the trend of continually larger shares of the population living in urban areas (Figure 2). The expected result of growing urban populations combined with generous transit investment is not only an increase in overall transit ridership, but also an increase in ridership per capita. However, this has not been the case.

Relative to 1992 levels, total investment in bus and rail transit was over 55% higher in 2014 (*1*). Although population also grew during this time frame, it only grew 25% (*3*). Therefore, one would assume that transit ridership per capita should have increased since 1992, since investment is growing faster than population (more than twice as fast). However, transit ridership per capita remained mostly stagnant from 2000 to 2014, barely increased from 1992 to 2014, and decreased from 1992 to 2016. The U.S. investment in transit in total and per capita has been increasing over the past few decades to accommodate a decreasing ridership per capita (Figures 1 and 2).

This is not to say that transit is failing in the United States. In fact, transit ridership has increased 21%, albeit through many periods of decline, since 1980 (*1*). However, it seems that transit ridership has tended to take "three steps forward, two steps back" as noted in the April 2016 *Planetizen* article "Public Transportation Ridership: Three Steps Forward, Two Steps Back?" (*4*). Why can't transit ridership see long-term, consistent growth without multi-year interruptions of decrease? Is it due to decreased costs of auto gasoline? Is it the rise of enhanced transportation network companies like Uber and Lyft? Is it because of communications-based substitutions for travel? Perhaps all of these have played a

[1]Department of Civil and Environmental Engineering, University of South Florida, Tampa, FL
[2]Center for Urban Transportation Research, University of South Florida, Tampa, FL

Corresponding Author:
Address correspondence to Richard A. Driscoll: rdriscoll@cutr.usf.edu

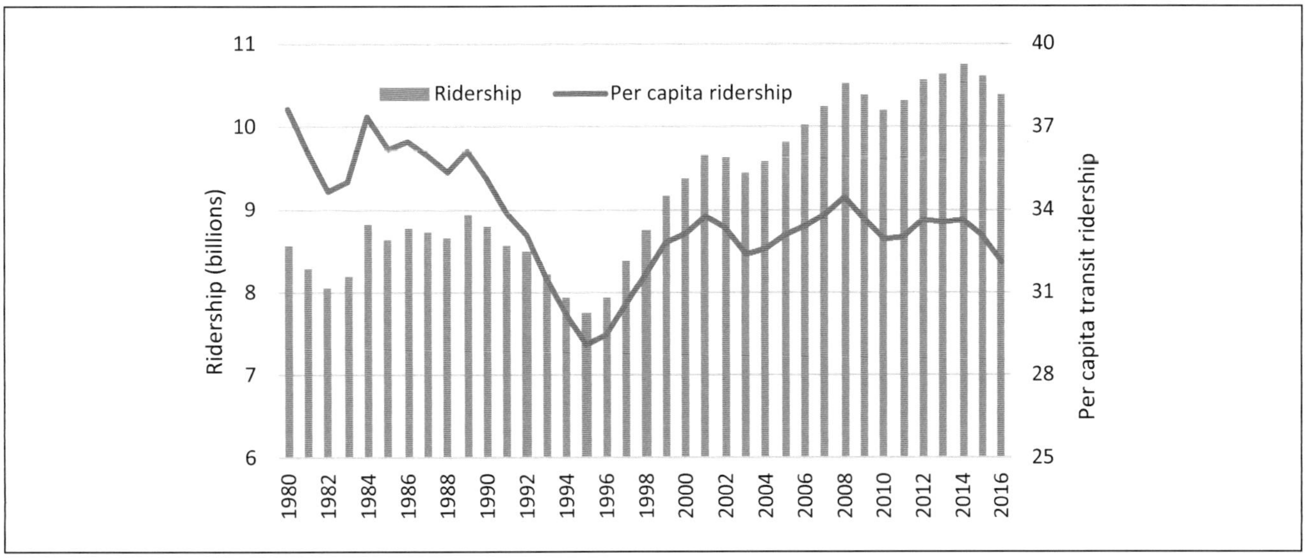

Figure 1. National transit ridership and ridership per capita, 1980–2016.

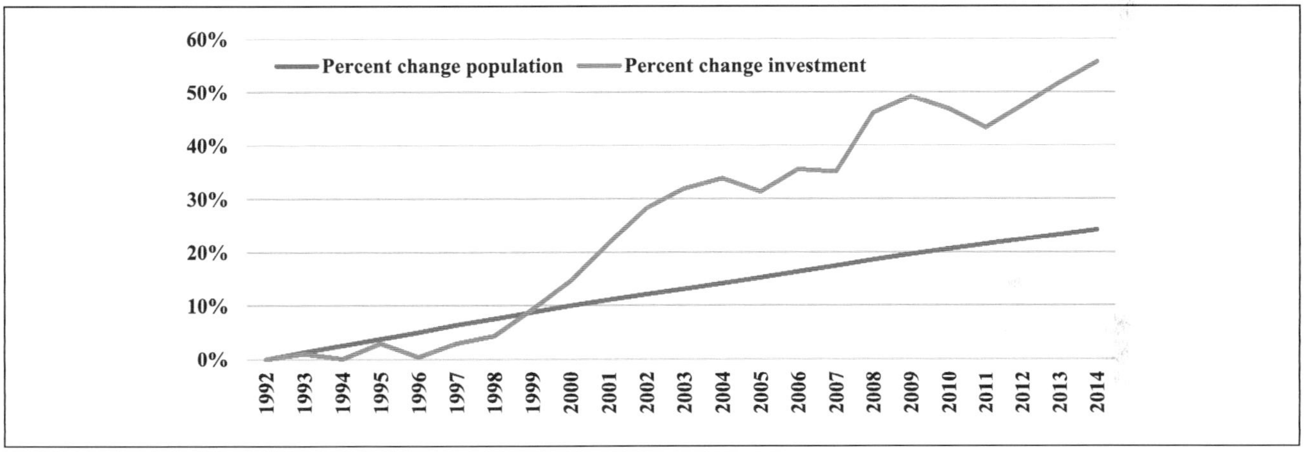

Figure 2. Population and transit investment (2014 dollars), percent changes from 1992.

minor role in the dampening of transit ridership over the last few decades. However, there are major demographic changes that have been occurring in the United States during the same period of time that may provide insight into the "unknown" reasons (*4*). These changes play a role in the uphill battle faced by the transit industry. Demographic changes include a host of factors from changes in age, ethnicity, settlement density, income levels and distribution, household composition, auto ownership, and licensure levels, and so forth. The remainder of this paper will look at two demographic considerations: the impact of age distribution on transit use and macro level aspects of settlement patterns on transit use, specifically county-level settlement trends.

Age Distribution Impacts

Background

One major factor influencing a person's travel behavior is age. People of different ages have different life priorities, work different amounts, and have different means of transportation. People's stage of life such as youth, working age, or retirement is correlated not only with how much they travel, but by what mode they travel. People aged 35–54 take more trips than other age groups, especially older cohorts, for example. Additionally, younger people tend to use transit at a higher rate than their older counterparts. This is made evident by National Household Travel Survey (NHTS) data (*5*).

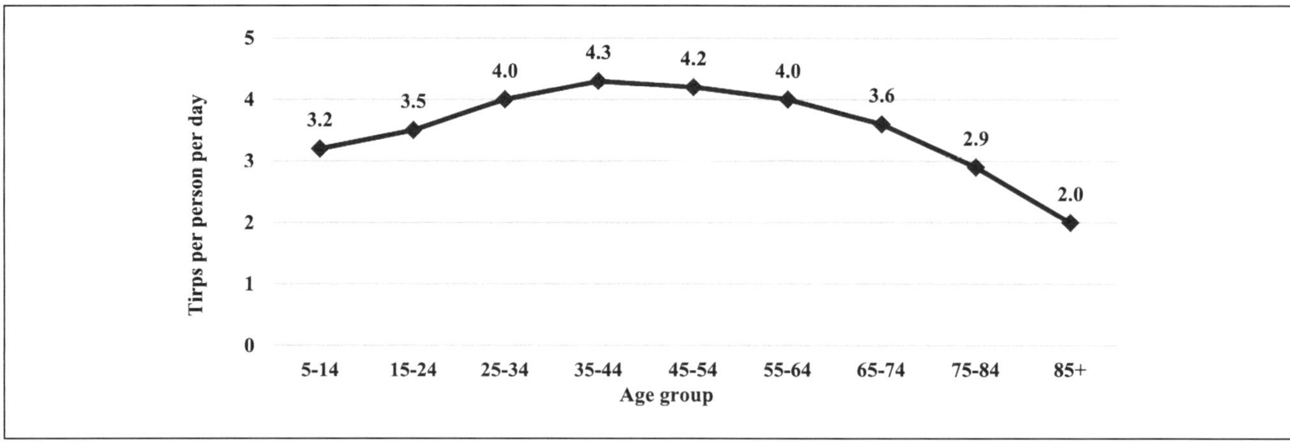

Figure 3. NHTS trips per person per day by age group, 2009.

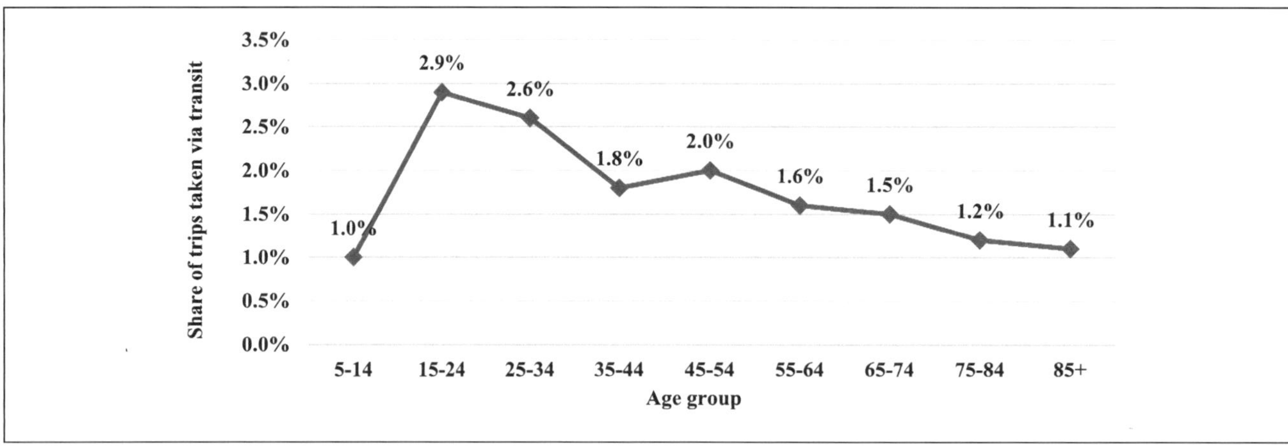

Figure 4. NHTS shares of trips via transit by age group, 2009.

Trip Rates and Mode Shares

The 2009 NHTS results, the most current data available, indicate that people between 35 and 44 years old take more trips than any other age cohort, at 4.3 trips per day (Figure 3). In comparison, people between 65 and 74 take 3.6 trips per day. Over the course of a full year, and with population in the hundreds of millions, this seemingly small difference in travel behavior results in a significant displacement of trip potential between younger-middle-aged people and the elder populous.

The 2009 NHTS results also show that people between the ages of 15 and 24 have the highest propensity toward transit use (Figure 4). This young age group takes an average of 2.9% of all their trips using transit, and the next youngest group (25–34) takes an average of 2.6%. This is contrasted with older age cohorts, whose transit trip share ranges between 1.3% and 2.0%. It makes practical sense that teenagers, college students, and young professionals are more likely to use transit than older individuals. Young people may be less able to afford private transportation, and more able to move afoot – a requirement for most transit use – than those who are older and more financially established.

Age Profile and Methods

As a consequence to the difference in travel behavior between age groups, a changing national age profile has resulted in shifts to overall travel tendencies over the last 35 years. The baby boomer population bubble shows up in a national age profile, and it has moved with time as the baby boomers have grown older (Figure 5).

In 1980, the bulge of the populous was under age 35. Although the people represented by this bulge were taking fewer trips than those over 35, nearly 3% of those trips were taken using transit. Although that share may not sound impressive, it is much higher than older age cohorts' transit share. However, by 2010, that same bulge was into the 45–54 and 55–64 age groups (*6–8*). At this point, the baby boomers were taking more trips than they were in the 1980s, but using transit much less. Could this

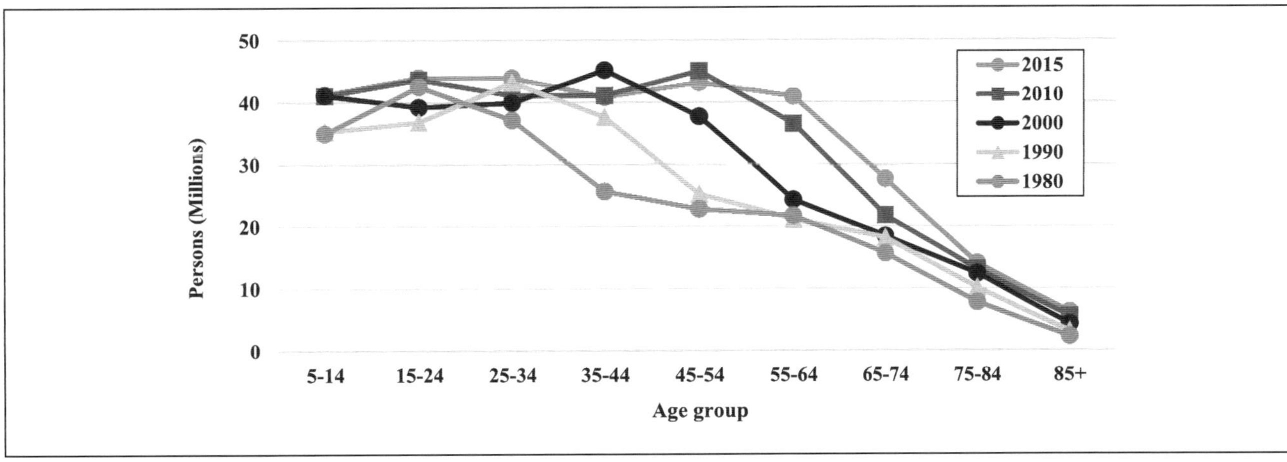

Figure 5. Select U.S. Census population age distributions, 1980–2015.

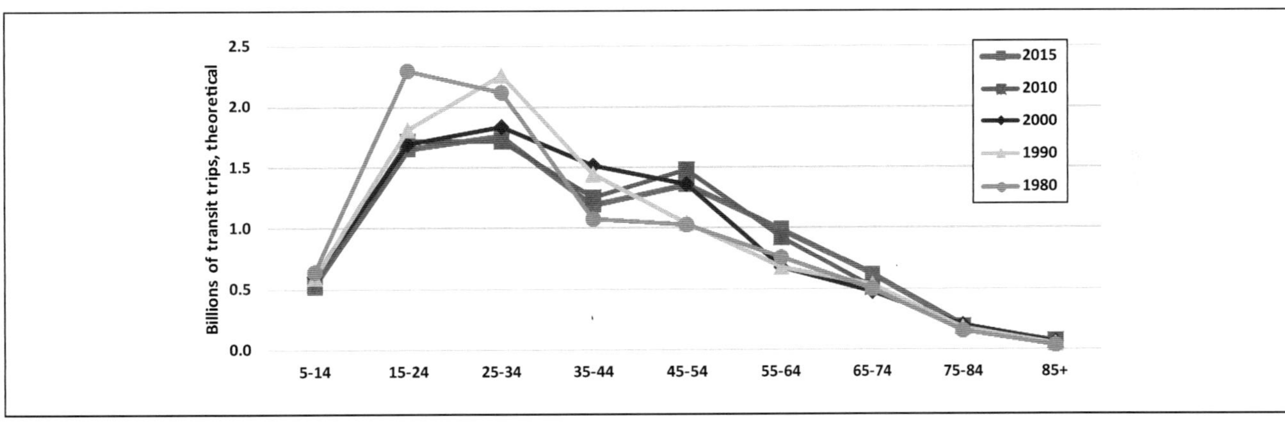

Figure 6. Theoretical transit trips by age group, adjusted for population growth (2015), NHTS 2009 trip rates and mode share.

trend in the age distribution shift lead to a shrinking propensity for the populous to use transit?

To test this theory, a model could be developed to estimate the number of theoretical transit trips that should occur under any given age distribution, using the NHTS trip rates and transit usages by age group. The robustness of this theory could be tested by applying trip rate and transit mode share data from additional NHTS survey years, 2001 and 1995 (*5*). Travel behaviors are ever changing, thus it is not certain that today's 35–44 year olds will behave identically to today's 45–54 year olds when they reach that age. However, understanding past cohort travel behaviors provides a comparative baseline. Equations 1, 2, and 3 outline the steps of the model for estimating potential transit trips (*9*):

$$\text{Total theoretical trips in year } y \text{ for age group } g = \frac{\text{total trips}}{\text{person*day}} \text{ for age group } g \text{ (NHTS)} * \frac{365 \text{ days}}{\text{year}} * \text{persons in age group } g \text{ in year } y \text{ (Census)} \quad (1)$$

$$\text{Total theoretical transit trips in year } y \text{ for age group } g = \text{Total theoretical trips in year } y \text{ for age group } g * \frac{\text{transit trips}}{\text{total trips}} \text{ for age group } g \text{ (NHTS)} \quad (2)$$

$$\text{Total theoretical unlinked transit trips in year } y\text{, adjusted for population growth (2015), for age group } g = \text{Total theoretical transit trips in year } y \text{ for age group } g * \frac{\text{total population, 2015}}{\text{total population, year } y} * \frac{1.53 \text{ unlinked trips}}{\text{linked trips}} \quad (3)$$

Implications

In theory, transit ridership per capita has had the propensity to decrease each decade since 1980 simply due to the shifting age distribution and a gradually aging populous. Thus, the age distribution effect creates the expectation of dampened overall transit ridership due to trip rate and mode share changes associated with aging. Superimposing the 1980 age distribution on the 2015

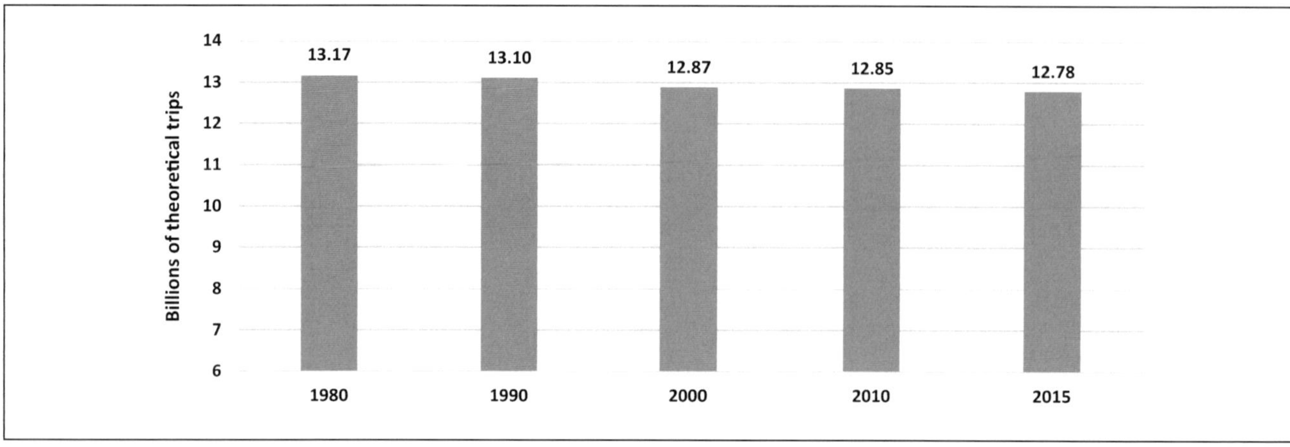

Figure 7. Total theoretical transit trips by year, adjusted for population growth (2015) and age distributions, NHTS 2009 trip rates and mode share.

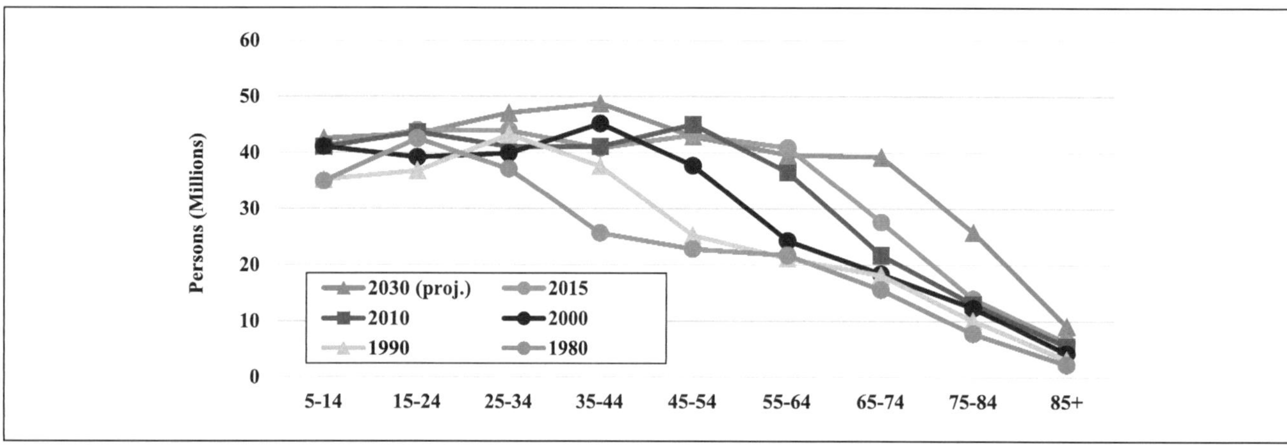

Figure 8. Select population age distributions, 1980–2015, and 2030 projection.

total population, there is a theoretical expectation for 13.17 billion unlinked transit trips. However, the same expectation using the 2015 age distribution is 12.78 billion unlinked transit trips, 3.1% fewer. The aging of the populous does not have enough power to produce decreased total transit ridership, but it does carry a suppressing effect. Therefore, the theoretical results explain part of the shrinking transit ridership per capita trend (Figures 6 and 7).

These estimations are higher than actual nationally reported transit ridership, which was 10.6 billion in 2015, for example (*2*). This could be due to sampling error on the part of the NHTS, or a disconnect between the definitions of transit as used by APTA and NHTS respondents. Perhaps not all transit use referred to in the NHTS survey responses is captured by APTA data. The results of this model using the 2009 NHTS are very similar to those using inputs from different NHTS years.

Forecasting Future Growth

In another 15 years, the bubble in the age distribution will have grown even older, and will have moved into the 65 + age group (Figure 8). As per the 2009 NHTS, people make fewer and fewer trips as they grow older, decreasing from 4.0 to 3.6 to 2.9 trips per day at 55–64, 65–74, and 75–84, respectively. Compared with 2015, U.S. Census Bureau predicts a 2030 population profile with 42% more people aged 65–74, 84% more people aged 75-84, and 48% more people aged 85 and older (*10*). The nature of travel, and thus of transit ridership, will surely evolve as the bulge in population moves into older age cohorts.

The 2015-equivalent number of transit trips expected using the 2030 age profile is 12.45 billion, an even steeper decrease than in the previous 15 years. This highlights the impending need for transit agencies to likewise evolve

Table 1. ACS Transit Commuting Share by Largest Population Increase, 2013–2016

County	Population 2016	Population change (percent change) 2013–2016	2015 commuting, share using transit
Harris County, Texas	4,589,928	243,045 (5.6%)	2.9%
Maricopa County, Arizona	4,242,997	231,778 (5.8%)	2.4%
Clark County, Nevada	2,155,664	130,568 (6.4%)	4.0%
Los Angeles County, California	10,137,915	122,479 (1.2%)	6.8%
Bexar County, Texas	1,928,680	106,624 (5.9%)	2.8%
Tarrant County, Texas	2,016,872	104,371 (5.5%)	0.6%
King County, Washington	2,149,970	104,096 (5.1%)	12.1%
San Diego County, California	3,317,749	99,330 (3.1%)	3.0%
Riverside County, California	2,387,741	96,289 (4.2%)	1.4%
Dallas County, Texas	2,574,984	95,174 (3.8%)	2.9%
Weighted average:			4.4%

their service. To keep ridership per capita from falling further, the transit industry could use its resources to appeal more to these older age cohorts, which are on the verge of growing by unprecedented volumes.

In essence, as more people have grown out of the transit-faithful age groups, the transit industry has struggled to maintain even modest ridership growth and has failed to produce growth in transit ridership per capita. Although this effect is minimal, under one-tenth of 1% per year, it is nonetheless providing downward pressure on transit use. Changes in immigration or other factors could exacerbate this trend.

Geographic Population Distribution Shift Impacts

Background

Another demographic factor influencing transit ridership derives from the fact that population growth appears to be strongest in modest and weak transit markets, whereas population declines tend to be occurring in strong transit markets. Certain areas of the country like the Northeast, Seattle, Chicago, and Los Angeles, have huge transit networks that stretch over hundreds of miles collectively, whereas other areas have far more modest transit networks. Strong transit networks are primarily a function of population density, explaining why large cities and downtown areas have such vast transit networks and are so dependent on transit for transportation.

Geographies of Interest

The American Community Survey (ACS) provides yearly data for various commuting travel and population demographics. According to the ACS, transit is the choice mode for over 18% of commuters in Cook County, Illinois, and well over half of commuters in several New York counties. The national average is 4.9% (*11*). These heavily populated cities require a transit system capable of supporting peak hour demands to relieve roadway congestion and provide for the basic travel necessities of the residents.

There are several counties across the country without a transit-intensive transportation system that are experiencing increasing population. Of the 10 counties that gained the most population between 2013 and 2016, eight had a lower than national average 2015 commuting share using transit (Table 1). Even more surprising, of the 10 counties that lost the most population in the same time period, six had a transit commute share higher than national average in 2015 (Table 2) (*11*).

The movement of individuals out of more transit-intensive counties and into lower transit-intensive counties is evident not only in the last 3 years, but also from 2000 to 2016. Population growth has been centered in and around counties that have less than a 3.0% share of commuters using transit, well below the 4.9% national average (*11*). U.S. residents are electing to live in areas traditionally characterized by lower population density. Low density reduces the ability to support access to high levels of transit, and inevitably has contributed to U.S. transit ridership per capita declines since 2000.

When examining areas where transit commuters constitute a large share of total commuters and population has decreased, it is possible to explore the reason why the per capita transit ridership in the United States is falling. Some areas of the Northeast Corridor, including Baltimore and several counties just outside of New York City, have had population decreases since 2000 thereby reducing the number of commuters utilizing transit. Similarly, the Rust Belt region has had noted population declines in the Detroit, Cleveland, and Pittsburgh areas, all of which are characterized by heavier transit use. The Chicago area has seen some of the highest population declines in the country where tens and hundreds of thousands of people have left its counties, some of which have as much as a fifth of commuters taking

Table 2. ACS Transit Commuting Share by Largest Population Decrease, 2013–2016

County	Population 2016	Population change (percent change) 2013–2016	2015 commuting, share using transit
Cook County, Illinois	5,203,499	−36,784 (−0.7%)	18.4%
Wayne County, Michigan	1,749,366	−25,257 (−1.4%)	3.2%
Cuyahoga County, Ohio	1,249,352	−13,982 (−1.1%)	4.9%
San Juan County, New Mexico	115,079	−11,439 (−9.0%)	0.3%
Allegheny County, Pennsylvania	1,225,365	−8,266 (−0.7%)	9.1%
Baltimore City, Maryland	614,664	−8,194 (−1.3%)	18.6%
Suffolk County, New York	1,492,583	−8,193 (−0.5%)	6.5%
Genesee County, Michigan	408,615	−6,938 (−1.7%)	1.2%
St. Louis City, Missouri	311,404	−6,543 (−2.1%)	9.4%
Caddo Parish, Louisiana	248,851	−6,373 (−2.5%)	1.8%
Weighted average:			11.4%

transit. This is not to say that every agency in these locations is losing ridership, but rather ridership per person in the United States is going down as a result, at least partially, of slower growth or decline in population in transit-oriented areas.

Those areas where population is increasing and transit commuters are a smaller portion of all commuters, the per capita ridership story is the same. In Texas, the area formed by the interstates connecting San Antonio, Dallas, and Houston has seen tremendous population growth since 2000, but most counties have transit commuter shares much lower than the national average. Similarly, Florida counties in the Tampa, Orlando, and Jacksonville areas constitute a large share of population growth in the state, yet very few compete with the national average share of transit commuters. The South region as a whole has had significant population growth across many counties with notable increases in and around Nashville, Atlanta, and Charlotte, North Carolina particularly along the interstate roads connecting these cities, all of which are again characterized by limited transit use. These counties may actually have some or most of their agencies increasing ridership, but since the share of commuters riding transit is lower than that of the national average, the per capita ridership decreases.

Commute Trips Method

An estimate of potential 2016 unlinked commute trips (had population grown uniformly) was compared with an estimate of actual 2016 unlinked commute trips (given actual population growth) to quantify the impact of population growth in areas with differing transit use. The comparison was performed using the following assumptions:

- constant workforce participation rates in each county's population for both 2013 and 2016,
- 2015 transit commuter share assumed for both years,
- transit commuters took two linked trips per day,
- 1.53 unlinked trips per linked trip, and
- 235 working days per year,

resulting in the development of Equation 4. The equation was then applied using 2013 and 2016 data (*9*):

$$\begin{aligned} \textit{Unlinked Commute Trips} &= \textit{Workers} \\ &* 2015\ \textit{transit commute share} * 2 * 1.53 * 235 \end{aligned} \tag{4}$$

After estimating the 2013 unlinked commute trips, a growth rate of 0.7% applied to each county to simulate uniform annual population increase for the 3 years to 2016 (*11*). The productions were then summed to obtain a national figure. This potential 2016 unlinked commute trips estimate was 5,334 million compared with the actual 2016 unlinked commute trips of 5,314 million. This indicates that transit ridership for commuting would have been 20 million greater had the population distribution remained constant.

Implications

Certainly not all transit ridership is commute based, but a significant portion is and is therefore telling of overall transit use. Potential 2016 unlinked commuter trips outperforming actual 2016 unlinked commuter trips shows that transit commute trips missed out on an additional 0.4% of commute ridership. This comes to nearly 20 million additional unlinked trips, which is synonymous with a 20 million increase in total ridership. Again, this analysis only accounts for commute trips on transit, but this shows that ridership is a function of population growth in transit-limited areas rather than a function of decreasing share of commuters using transit. Commute trips on transit constitute about 30% of all ridership, so an estimated total ridership lost to population growth outside transit-oriented areas is expected to be over 66 million from 2013 to 2016 (*12*). This represents an approximate 0.1% per year drag effect on transit ridership. While modest, it is nonetheless a component of the ridership challenge facing transit.

Conclusions and Future Work

Conclusions

National transit ridership can be difficult to predict, and it is influenced by various social, demographic, and economic characteristics. With transit investment the highest it has been in at least two decades, it is important to rationalize this magnitude of spending and utilize each dollar prudently. Since 1980, the United States has seen a per capita ridership downswing, which is concerning given the current spending levels, but provides a motivation to examine the various trends affecting ridership.

In exploring national demographics and corresponding trip rate tendencies, aging population has a downward pressure on transit ridership. As baby boomers approach retirement age they also approach an age traditionally characterized by fewer trips taken per day and a decreasing share of trips taken via transit. As this population group continues to age beyond retirement years, per capita ridership will likely continue to decrease.

The challenge for policy makers and decision makers will be to respond appropriately. Rather than reducing service as a reaction to decreasing ridership, some agencies might consider tailoring portions of their service toward addressing older cohorts' needs and preferences. One such trait might be immediate accessibility for those elderly not willing or able to walk nearly as far as those in younger age groups. Adding services that older age cohorts find useful could potentially induce higher trip rates for those cohorts by increasing mobility options.

In exploring population growth at the county level, many transit-intensive areas are not growing comparatively with transit-limited areas. Increasing population would suggest increasing ridership, and this has proven true to a point. Although total ridership has increased, ridership per capita has decreased partially as a result of where population has been growing. People are electing to reside in areas where transit commuters constitute a smaller share of the total. Population is moving to and growing in areas with less transit. Unless the transit industry better caters to the changing geographic population distribution, a downward pressure on transit ridership is expected.

Once again, decision makers must respond properly and practically. They must strategically deploy services to match changing demand. Investment in transit will influence ridership the most where population is growing the quickest. At the local level, transit agencies that adapt service swiftly to changes in settlement patterns will maintain ridership more easily than those that are slow to react to such shifts.

The April 2017 report, *Transit Ridership in the United States: The Sky is Not Falling but the Ground has Shifted*, suggests many of the explanations for transit ridership decline are evident. Beyond gasoline price decline, the article cites car ownership increase, employment growth, and parking reliability (*13*). Demographic characteristics, and other previously acknowledged factors that influence transit demand can answer some of the questions surrounding transit ridership underperformance. Demographic characteristics associated with transit dependency will remain fundamental to travel behaviors in the United States and must continually be studied to understand the trends in transit ridership.

Future Work

The purpose of this paper is primarily explorative, and sets a framework upon which to expand. Age distribution shift and geographic population shift are only two of a plethora of variables that influence transit ridership. The population age analysis relies on ACS and NHTS data, and with the release of the new NHTS in early 2018 it is now possible to expand on the existing estimations. New NHTS data will help in determining the changing ways in which people of different ages travel. The annual release of ACS data will also prove useful in expanding on the population geographic distribution analysis. The U.S. population is always growing and moving, so continuing to track this is vital in understanding where transit investment can be most effective.

An additional suggestion for future work is studying the effect of changes in the population distribution as it relates to accessibility to transit and via transit. Specifically, is new growth locating in areas with transit accessibility better or worse than the current national average? There may also be insight gained by studying the superposition of age and geographical growth to see their combined effects on transit ridership or what heterogeneity exists between the two, if any. The implications of this comparison are beyond the scope of this paper, but would prove useful in predicting future transit rider demographics and targeting transit to the needs of customers. An analysis on other demographic factors including ethnicity, income levels, household composition, and auto ownership would also help in determining the actual impact of these two theories on both nominal and per capita transit ridership.

Acknowledgments

This research was supported by the Center of Urban Transportation Research and the University of South Florida. The authors thank their supporting colleagues at the Center for Urban Transportation Research, without whom this work would not have been possible.

References

1. American Public Transportation Association. *2016 Public Transportation Fact Book Appendix A: Historical Tables.* American Public Transportation Association, Washington, D.C., 2016.
2. American Public Transportation Association. *Quarterly Public Transportation Ridership Reports, 2015.* American Public Transportation Association, Washington, D.C., 2016.
3. U.S. Census. *National Intercensal Tables, 1980-1990 and 1990-2000.* U.S. Census Bureau, Washington, D.C., 2000.
4. Polzin, S. Public Transportation Ridership: Three Steps Forward, Two Steps Back? *Planetizen*, 2016. https://www.planetizen.com/node/85595/public-transportation-ridership-three-steps-forward-two-steps-back. Accessed July 17, 2017.
5. Federal Highway Administration. National Household Travel Survey. U.S. Department of Transportation, Washington, D.C., 2009.
6. Hollmann, F. W. *U.S. Population Estimates, by Age and Sex, Race, and Hispanic Origin: 1980 to 1991.* U.S. Census Bureau, Washington, D.C., 1993.
7. Meyer, J. *Age: 2000, Census 2000 Brief.* U.S. Census Bureau, Washington, D.C., 2001.
8. Howden, L. M., and J. A. Miller. *Age and Sex Composition: 2010 Census Briefs.* U.S. Census Bureau, Washington, D.C., 2011.
9. American Public Transportation Association. *A Profile of Public Transportation Passenger Demographics and Travel Characteristics Reported in On-Board Surveys.* American Public Transportation Association, Washington, D.C., 2007.
10. U. S. Census. *2014 National Population Projections Table, 2014 to 2016.* https://www.census.gov/data/datasets/2014/demo/popproj/2014-popproj.html. Accessed July 25, 2017.
11. U.S. Census. *American Factfinder.* https://factfinder.census.gov/faces/nav/jsf/pages/index.xhtml. Accessed July 18, 2017.
12. American Association of State Highway and Transportation Officials. *Commuting in America 2013: The National Report on Commuting Patterns and Trends.* Brief 13 Trnasit Commuting. American Association of State Highway and Transportation Officials, Washington, D.C., 2015.
13. Babbitt, R. T. *Transit Ridership in the United States: The Sky Is Not Falling but the Ground Has Shifted.* Robert T. Babbitt PLLC, 2017.

The Standing Committee on Public Transportation Planning and Development (AP025) peer-reviewed this paper (18-04556).

Article

TRR
JOURNAL OF THE TRANSPORTATION RESEARCH BOARD

Transportation Research Record
2018, Vol. 2672(8) 879–889

Reprints and permissions:
sagepub.com/journalsPermissions.nav
DOI: 10.1177/0361198118790842
journals.sagepub.com/home/trr
SAGE

The Potential of Demand-Responsive Transport as a Complement to Public Transport: An Assessment Framework and an Empirical Evaluation

María J. Alonso-González[1], Theo Liu[1], Oded Cats[1], Niels Van Oort[1], and Serge Hoogendoorn[1]

Abstract

Demand-responsive transport (DRT) services (collective on-demand services, such as shared ridesourcing and microtransit) offer a collective flexible travel alternative that can potentially complement fixed transit (FT). The combination of an on-demand and line-based service holds the promise of improved mobility and increased service coverage. However, to date, it remains unknown whether DRT services deliver these much anticipated improvements. This study presents an assessment framework to evaluate the performance of DRT and related changes in accessibility, and performs an empirical analysis for a recently introduced DRT service in the Netherlands. The framework includes a performance benchmark between DRT and FT based on the computation of generalized journey times of the DRT rides and the FT alternatives, and can help identify whether DRT is used as a complement or a substitute for FT. The framework covers the spatial and temporal dimensions, and the explicit consideration of rejected trips is an integral part of the evaluation. Results suggest large accessibility improvements for DRT users, especially for some underserved origin–destination pairs.

Reducing car use is high on the agenda for transportation planners, yet the flexibility and convenience that the car provides often makes it a more attractive alternative than traditional fixed transit (FT). To compete with private car use, line-based FT could benefit from embracing new flexible services that have appeared in urban areas so as to improve and complement its services. Specifically, demand-responsive transport (DRT) services (collective on-demand services, such as shared ridesourcing and microtransit) could potentially complement FT in urban areas.

Even though DRT services are not a new innovation (they were recommended for future urban transportation in the sixties [*1*]), only recent technological advancements have enabled their real-time large-scale operation. As a result, new urban DRT services have appeared, both in the United States (e.g., Bridj, Lyftline, UberPOOL, Via) and in Europe (such as Abel in Amsterdam, Kutsuplus in Helsinki, Padam in Paris, and Radiobus di Quartiere in Milan). To better understand the utilization of these services and the role they play in relation to FT, a systematic assessment framework is proposed in this study.

Several studies have aimed at helping planners design DRT systems. These studies help estimate the required capacity for a given level of service and the resulting operating costs (*2*), or assess whether DRT should substitute FT for a given scenario (*3*, *4*). Research has also evaluated via simulation the impact that DRT services would have in real urban networks such as in Hino (Japan) (*5*), Lisbon (Portugal) (*6*), and New York City (United States) (*7*). However, little is known about how these services perform in real settings. A notable exception is Kutsuplus, "apparently the world's first fully automated, real-time demand-responsive public transport service" (*8*), which operated between 2012 and 2015 in Helsinki, and for which a final project report is available. However, despite Kutsuplus being implemented as part of the public transport system, its final report does not include an analysis on the extent to which the new

[1]Department of Transport and Planning, Delft University of Technology, Delft, The Netherlands

Corresponding Author:
Address correspondence to María J. Alonso-González: M.J.AlonsoGonzalez@tudelft.nl

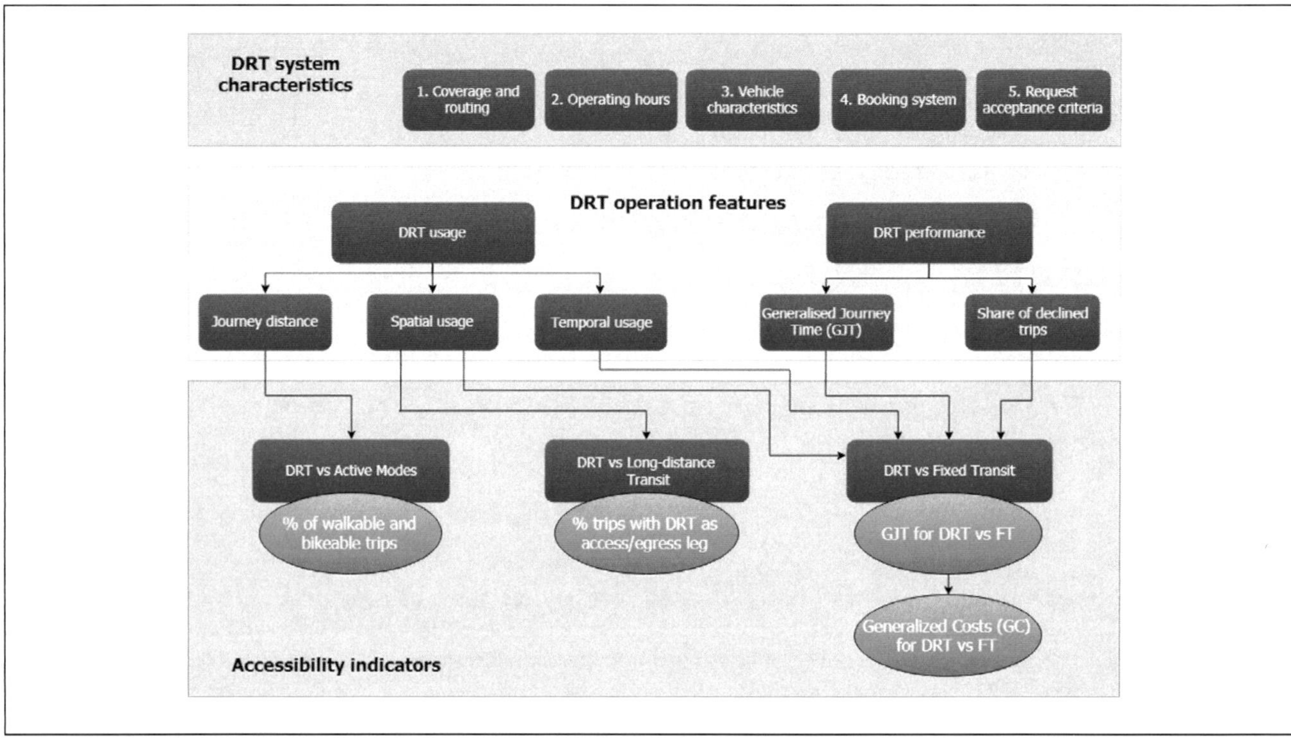

Figure 1. Assessment framework.

service improved the mobility in comparison to the already existent alternatives.

This study fills this gap by introducing a DRT assessment framework that analyzes the improvement in mobility that a DRT system adds to the transportation network based on empirically observed usage patterns. Current DRT assessment frameworks contemplate the concept of DRT at a high level without including concrete indicators (*9*), or focus on key performance indicators (KPIs) that consider DRT in isolation from other modes (*10*). In this study, the accessibility gains (i.e., the increased easiness of reaching the required or desired activities [*11*]) that DRT granted for the performed trips are examined.

The main objective of this study is to help transportation authorities in assessing the improvement in mobility that DRT users have experienced in relation to the co-existing FT alternatives. After a detailed description of the framework, it is applied to a DRT system in the Netherlands. Based on the observed results, the impacts of DRT as a complement to FT are discussed.

Methodology

The proposed DRT assessment framework covers three aspects. First, it analyzes the DRT system characteristics. Second, it examines DRT operation, involving both DRT usage and performance. Finally, based on the operational characteristics of the system, a series of accessibility indicators are examined. This framework is depicted in Figure 1 and its components are detailed in the subsequent subsections.

DRT System Characteristics

Before analyzing the DRT operation, it is necessary to understand the intrinsic characteristics of the system under consideration (upper box of Figure 1). We highlight five aspects of the DRT system:

1. Coverage and routing. Defined by the operating area and the degree of flexibility in the operation. The latter can range from a fully flexible door-to-door service, to more rigid configurations in which only partial deviations from a planned route are possible.
2. Operating hours. Equivalent to the service span considered in FT.
3. Vehicle characteristics. DRT fleets often consist of minibus vehicles, with previous research suggesting that a vehicle capacity of eight is the preferred vehicle size (*12*). Fleet size is also a key design variable of DRT systems as it will determine the different routes that can be covered simultaneously by the system.
4. Booking system. The system can allow for instant bookings in real-time, require advanced bookings or allow for both options. Traditionally, dial-a-

ride schemes relied on telephone reservations, but large-scale real-time DRT systems require Internet connection to perform the bookings in an efficient manner.

5. Request acceptance criteria. Time required to the pick-up point or vehicle availability are the often-used criteria in deciding whether a request is accepted or rejected.

DRT Operation Features

Because of its inherently dynamic interaction with demand, DRT performance is directly tied to its usage (*13*). As such, both aspects are an integral part in DRT operation (see middle box of Figure 1).

DRT Usage. DRT usage is analyzed in terms of demand patterns. Three important aspects can be highlighted: journey distance, spatial usage patterns, and temporal usage patterns. Journey distance can provide insights into the modes that would compete with DRT. For instance, short trips might substitute walking and cycling. The analysis of spatial usage reveals which origins, destinations, and routes are most frequently used. This can shed light on areas with high demand that may not be conveniently served by FT. Lastly, the temporal usage identifies peaks in the operation, both by time of the day and by day of the week. Traditionally, transport demand in general and, in particular, public transport services tend to exhibit two daily peaks following commuting patterns. Since shopping and social trips have been identified as the most recurrent trip purposes for DRT trips (*14*), different temporal usage patterns might be expected for DRT.

DRT Operation. Regarding DRT operation, the presented framework considers two aspects: the generalized journey time (GJT) and the share of declined trips. The GJT represents the perceived passenger journey time. It is calculated by multiplying the time of the different segments of the door-to-door journey by different weighting factors so as to transform them into equivalent in-vehicle time (*15*). The different segments to take into account are the walking time (t^f), the waiting time (t^w), the in-vehicle time (t^v), and the number of transfers (n^t). The GJT can be expressed as:

$$GJT = \alpha_f \cdot t^f + \alpha_w \cdot t^w + t^v + \alpha_t \cdot n^t$$

where α represents the different weighting factors. The estimation of these α has been the subject of a large number of public transport studies (e.g., *16*, *17*). It is important to note that the waiting time can be expected or unexpected, with unexpected waiting time being valued by passengers as 2 to 3 times longer than expected waiting times (*18*). As such, an accurate and up-to-date forecast of the pick-up times in DRT services is of utmost importance to lower GJT.

This study also includes the share of declined trips by the operator as a separate element in this assessment since it is an important reliability indicator of the DRT system. A systematic analysis of the declined trips can also help improve the DRT service and search for inequalities in operation (e.g., some zones may have a higher share of declined or less profitable trips and may thus suffer from more recurrent cancellations). Moreover, DRT operators could be penalized when declining trips for which no adequate FT alternative is available.

Accessibility Indicators

Following the analysis of DRT operation, this paper proposes a series of indicators that measure the change in accessibility attributed to the DRT service. These are depicted in the ovals in Figure 1.

DRT vs. Active Modes. DRT can be seen as a competitor of walking and cycling for admissible walking and biking distances. The threshold for admissible walking and cycling distances needs to be adjusted based on the DRT setting, as these values differ by region (*19*), as well as by trip purpose and population subgroup (*20*). The first accessibility indicator expresses the share of walkable and bikeable trips as follows:

$$\%walkable\ trips = \frac{number\ of\ trips\ shorter\ than\ the\ walking\ threshold}{total\ number\ of\ trips}$$

$$\%bikeable\ trips = \frac{number\ of\ trips\ shorter\ than\ the\ biking\ threshold}{total\ number\ of\ trips}$$

The smaller these shares are, the less the actual DRT usage is competing with active modes. Note that although the former is a subset of the latter, biking is not always a feasible alternative (e.g., lack of a bicycle or cycling knowledge). Including both indicators serves to, depending on the setting, better assess the actual competition of DRT with active modes.

DRT vs. Long-Distance Transit. DRT operation is often restricted to a certain area. For trips that extend beyond these boundaries, DRT can be used as a complementary mode and connect passengers with long-distance transit modes such as train, long-distance bus, or regular bus

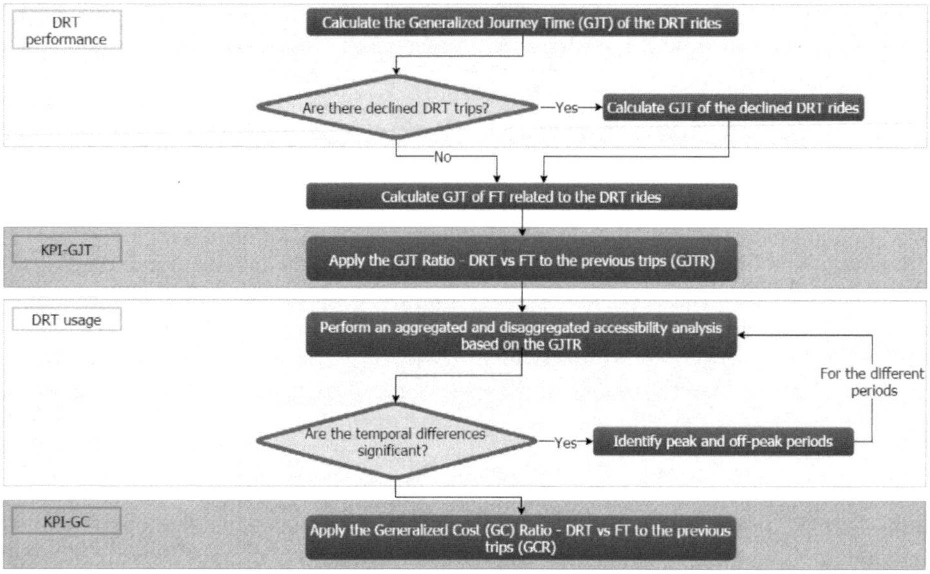

Figure 2. DRT accessibility performance benchmarks against FT.

for the case of DRT acting as a feeder. The second accessibility indicator is thus expressed as:

$$\%trips\ with\ DRT\ as\ access\ or\ egress\ leg = \frac{transit\ trips\ that\ use\ DRT\ as\ an\ access\ or\ egress\ leg}{total\ number\ of\ trips}$$

The larger this share is, the more DRT constitutes a connecting service to long-distance transit.

DRT vs. Fixed Transit. DRT can also be used as the main mode of transport, competing with FT. This is the more general scenario due to the similitudes that DRT and FT exercise. In order to address this key aspect, the assessment framework provides a more complex benchmark against FT, in which two KPIs – the GJT and the generalized costs (GC) – are calculated and compared for DRT and FT.

Following the workflow described in Figure 2, the GJT is calculated for the DRT trips and, if available, for the declined trips. The GJT of DRT is compared with the GJT that the passengers would have experienced if they had traveled using FT. Three parameters are used to calculate the GJT of the FT alternatives: booking time (as the desired start time), and origin and destination locations. Automatic vehicle location data can be used for this purpose. If this information is unavailable, scheduled information from FT journey planners can be used. The GJT ratio (GJTR) of each trip is expressed as:

$$GJTR = \frac{GJT^{DRT}}{GJT^{FT}}$$

The GJTR indicator is analyzed by considering its distribution, and performed and declined trips are assessed separately. It is important to note that the median (or the quartiles for a more detailed representation) should be used to assess the distribution rather than rely on the mean due to the asymmetry caused by the ratio calculation.

The GJTR can be further evaluated both spatially and temporally. An aggregated representation of this KPI can identify origin–destination (O-D) connections for which the DRT service has mostly improved the transit provision. A comparison between the number of performed and declined trips for the different O-D pairs can identify whether rejections are more prevalent in certain areas. Moreover, a disaggregated representation can identify connections that are underserved by conventional transit. If significant temporal differences exist (either daily or weekly), the spatial analysis can be performed for different time intervals.

The last step of the comparison between DRT and FT is based on the GC in which the GJT values are multiplied by the corresponding value of time (VoT). The monetarized value of the GC ratio (GCR) indicator is expressed, similarly to the GJTR as:

$$GCR = \frac{GC^{DRT}}{GC^{FT}}$$

Whenever the focus is set on the total monetary cost of the passenger instead of on the temporal convenience of both modes, this GC indicator can be used.

Application

Case Study Description

The assessment framework described in the previous section was applied to the DRT pilot "Breng flex," which operates in the Arnhem-Nijmegen region in the east of the Netherlands. In the following, the Breng flex service in Nijmegen is analyzed. The pilot started in December 2016 and it offers stop-to-stop connection with a total of 255 stops. The service area includes the city of Nijmegen and the neighboring municipalities of Wijchen, Berg en Dal, and Oosterhout, with a total of approximately 200,000 inhabitants. Breng flex is run by Breng, the incumbent local bus operator. With its introduction, two bus lines that linked the municipality of Wijchen, traversed the central parts of Nijmegen and terminated at the northern part of Nijmegen, were simultaneously cancelled. The Breng flex service has a fixed price of €3.50 per person and operates from 06:30 to 24:00 during the weekdays, 08:00 to 24:00 on Saturdays, and 09:00 to 24:00 on Sundays. The fleet comprises five minibuses (maximum of five seats for passengers and two wheelchairs) and four electric cars (maximum of three seats for passengers). Rides can be booked in real-time via a mobile app or a telephone number, and no advance booking option is available. Maximum waiting time is set to 20 min, rides that cannot be served within this time window will not be accepted.

The analyzed data includes the pick-up and drop-off locations for all registered ride requests – both performed and declined trips – between the 16th December 2016 and 18th May 2017. Booking times and expected in-vehicle times were available for the declined trips. Data for the performed trips also included actual in-vehicle time and waiting time. Rides with an unrealistic average travel speed – lower than 10 km/h (6.21 mph) or higher than 70 km/h (43.48 mph) – were not considered in the analysis. For declined trips, requests with the same spatial characteristics that differed by less than 5 min were included only once in the analysis to prevent multiple counts. A total of 4,719 performed trips and 130 declined trips (2.7% of all requests) were considered valid for the analysis.

Results

The average distance of the served DRT trips was 7.05 km (4.38 mi). As no sociodemographic data of the DRT users was available, the average walking and cycling distances in the Netherlands were used as thresholds for these modes, that is, 1.2 km (0.75 mi) for walking trips (*19*) and 3.6 km (2.24 mi) for cycling trips (*21*). In total, 0.1% and 16.5% of the DRT trips could have been performed on foot and by bike respectively, with the large majority of the rides covering a longer distance than the average active mode trips.

The obtained data do not allow identification of DRT trips that constitute a part of a longer transit trip. In order to have an approximation of the number of trips that could have used Breng flex as the access or egress leg, we identified the number of DRT trips that had one of the six train stations within the case study area as pick-up point (12.3% of the rides) or drop-off point (8.3%). These numbers are upper bounds and suggest that up to 20% of the rides could have been used to access or egress one of the train stations.

The other indicators compared the DRT rides to FT rides. The characteristics of the FT rides that could have substituted the performed and declined DRT rides were obtained by introducing the booking time and the start location and end locations of the DRT rides in the transit Google Maps Directions API (*22*). The Google Maps Directions API provides scheduled rather than actual transit data, hence possible deviations from plans in FT operations are not accounted for. As retrieval criterion, we selected the FT trip with "the earliest time at the destination." However, using this criterion, the option shown for 5% of the performed trips and for 13% of the declined trips (242 and 17 respectively) did not involve any transit, only walking. This is arguably an indication that the existing FT connection for these origins and destination is inadequate, or that the distance covered is rather short. To allow for a fair comparison for which FT values could be attained, these trips were not included in the subsequent comparison of DRT and FT.

For the previous trips, the GJT for the performed and declined trips were calculated (the factors used being $\alpha_f = \alpha_w = 2$ *and* $\alpha_t = 10$ [*17*]). The average waiting time of the performed trips (9.4 min) was taken as the estimated waiting time for the declined trips. The distribution of the GJTR is illustrated in Figure 3 for both served trips (*a*) and declined trips (*b*). The median for the performed trips was 0.50 (0.38 for the 25th percentile and 0.68 for the 75th percentile). In other words, for 50% of the performed trips, the FT alternative was perceived by the passengers to be over twice as long as the DRT alternative. For the declined trips, the median was 0.54 (0.45 for the 25th percentile and 0.71 for the 75th percentile). These results attest to how the DRT alternative considerably improved accessibility of the performed trips in comparison to the FT alternative.

The distribution of the GJTR is now analyzed spatially, both in an aggregated and in a disaggregated manner. Figure 4 presents the aggregated spatial analysis of

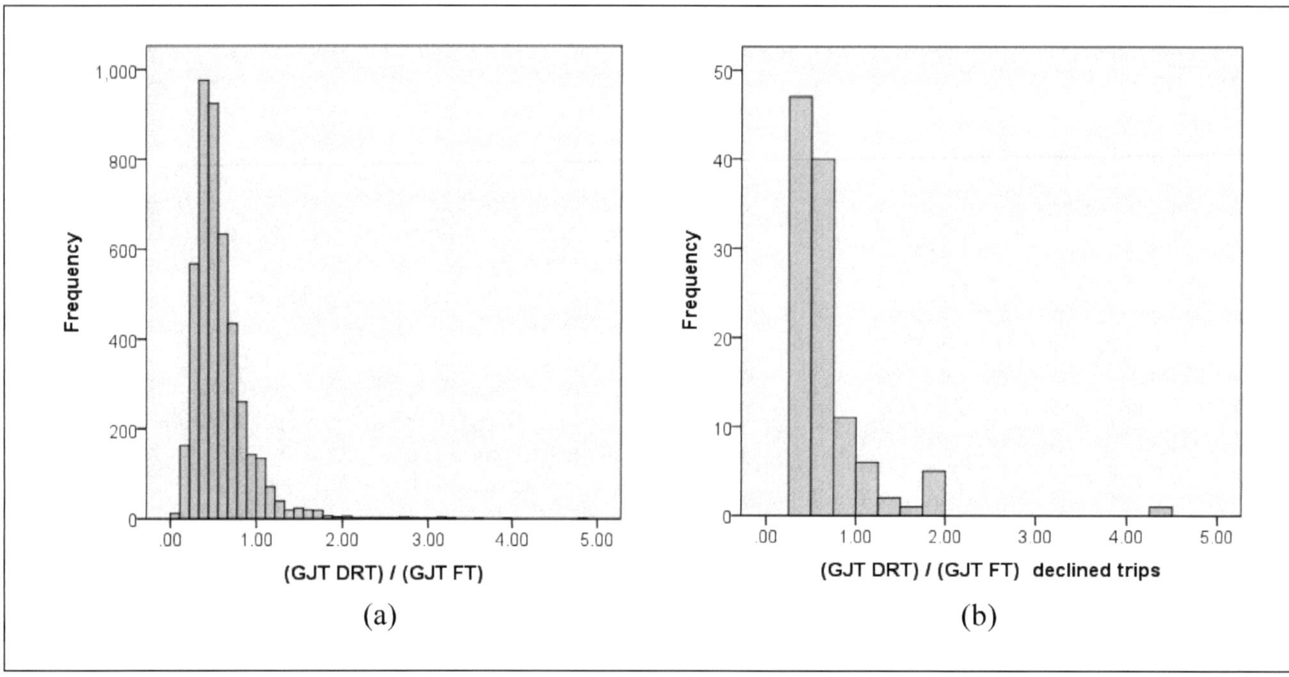

Figure 3. Distribution of the GJTR for DRT in comparison to PT (*a*) for the performed trips (N = 4,477) and (*b*) for the declined trips (N = 113).

the performed trips, with (*a*) the zonal delimitations used (in which Nijmegen has been divided into three parts by following natural water barriers), (*b*) the O-D matrix in hundreds, and (*c*) the median of the GJTR corresponding to each of the O-D pairs. The difference in the number of rides for each of the O-D pairs was partly due to the difference in the population size of the different areas. Moreover, part of the demand between Wijchen and the different areas of Nijmegen might be attributed to users of the two bus lines that were eliminated with the introduction of Breng flex. This fact could also explain why trips from Nijmegen North to Wijchen had the lowest median for the GJTR indicator (the median DRT GJT is 22% of the FT GJT value). Conversely, the O-D pair for which the increases of accessibility provided by DRT in comparison to FT were the lowest was Oosterhout–Oosterhout (GJTR = 0.69). Regarding the declined trips, the low number of occurrences (with no occurrence at all for many of the O-D pairs) did not provide a representative sample to analyze if there were statistical differences between both trip groups and were therefore not included in the shown figures. Importantly, the number of declined trips for each O-D pair correlated with the number of served trips. Thus, there were no spatial disparities in the DRT service availability.

The disaggregated spatial analysis is presented geocoded in Figure 5. The performed rides have been distributed across three graphs, (*a*), (*b*), and (*c*), according to their GJTR, whereas declined rides are illustrated in (*d*). Node size is consistent with use frequency, as well as link size and link color intensity. The direction of the trips is represented clockwise. The large majority of the trips for which FT outperformed DRT (GJTR > 1) were radial trips with an end in the center of Nijmegen, whereas recurrent links with a distinctive DRT benefit had a more tangential nature. No distinctive pattern can be observed between performed and declined trips. The large majority of the 255 DRT stops were utilized during the study period, and the stop at "Nijmegen central railway station" was one of the two most recurrently used stops.

The analysis of the evolution of the demand for Breng flex services shows a constant increase in the number of rides since the implementation of the DRT service. The number of trips in the last week quadrupled the trips in the first weeks to its introduction. Due to this large variation in demand volumes, only the data from the last 10 weeks were considered in the search for daily and weekly variations in demand levels. No statistically significant differences were found in the demand on different days of the week (p = 0.113 for one-way ANOVA) but differences in demand levels at different hours of the day were statistically significant (p = 0.001 for one-way ANOVA studying the hourly demand between 09:00 and 21:00). The hourly DRT demand did not show strong commuting peaks as in the case for FT, but smaller peaks still do exist, with the highest demand observed in the afternoon (around 15:00). A spatial representation of the peak

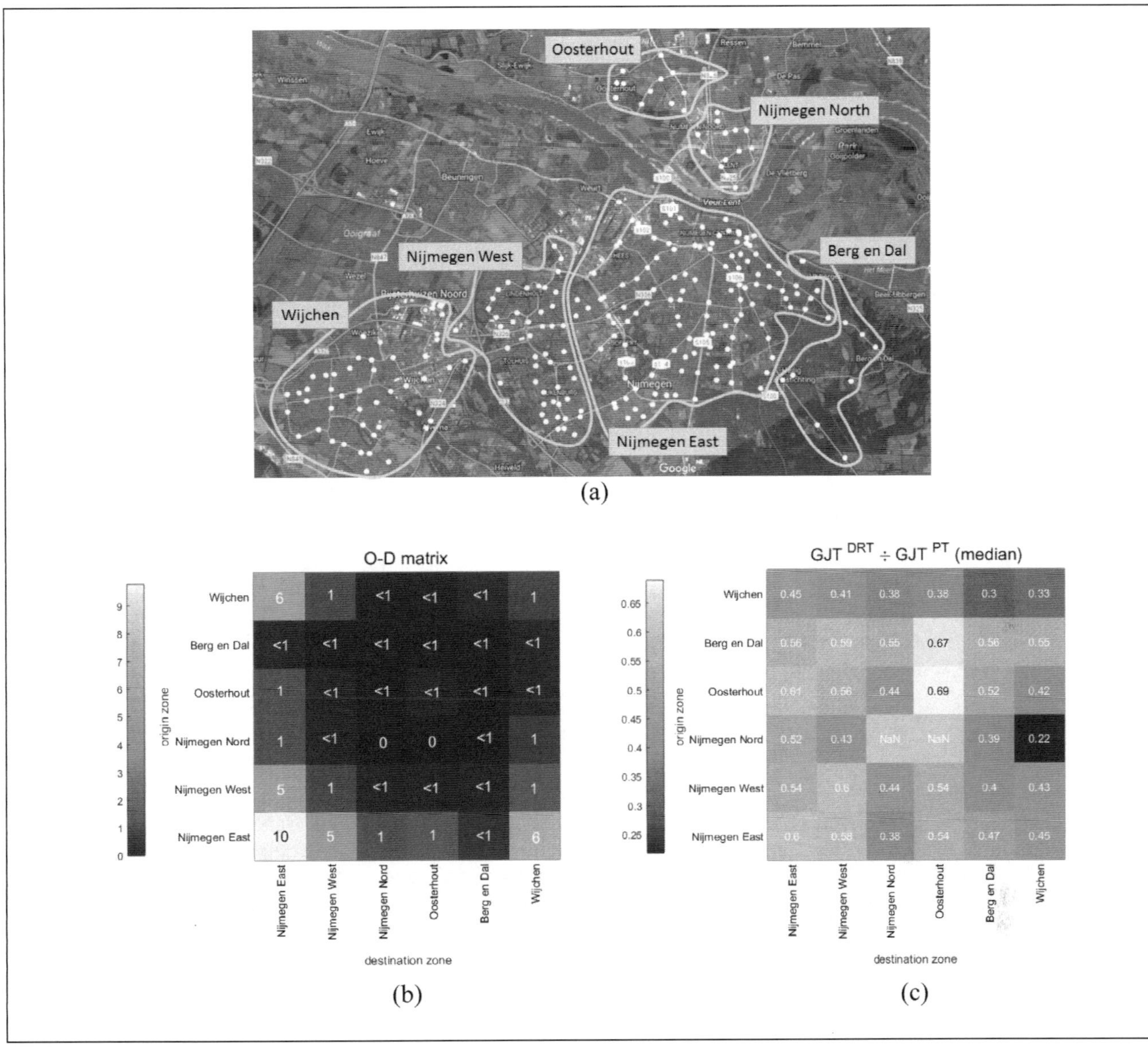

Figure 4. Aggregated spatial analysis of DRT: (*a*) zonal delimitation; (*b*) O-D matrix (in hundreds); (*c*) median of the GJTR for each O-D pair.

period could identify the highest demand areas during this timeslot. Due to space limits, this is not presented in this paper.

The last step described in the assessment framework involves the calculation of the GC of both performed and declined trips. A future modal shift from FT to DRT could arguably be expected if the GCR distribution favors DRT to a large extent. The VoT used for this study (both for DRT and FT) was the VoT calculated in Kouwenhoven et al. (*24*) for bus/tram/metro for all trip purposes in the Netherlands, which is €7.39 per hour per person in 2016 terms. The distribution of GCR is represented in Figure 6. The median for the performed trips was 0.75 (0.60 for the 25th percentile and 0.97 for the 75th percentile). For the declined trips, the median was 0.80 (0.63 for the 25th percentile and 0.99 for the 75th percentile). These results show that although DRT outperformed FT for the trips for which passengers opted for DRT, the distributions of the GC were closer to 1 than those of the GJT (Figure 3). This shift is caused by the higher fixed price for DRT than the distance-based FT fare, resulting in a smaller discrepancy in passenger disutility between the services when accounting for the price difference.

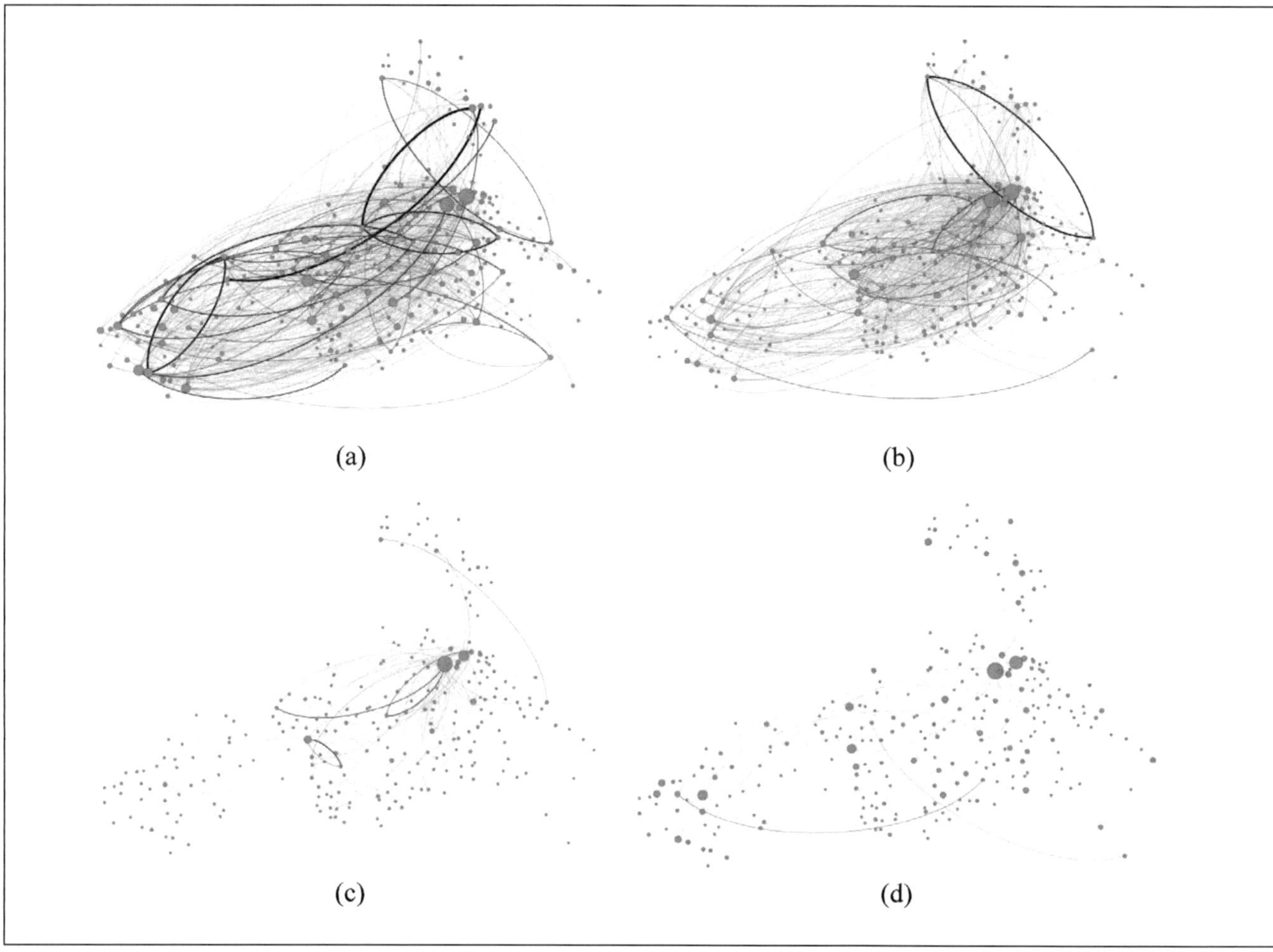

Figure 5. Disaggregated spatial analysis of DRT: (*a*) performed rides with GJTR < 0.5 (2,206 rides); (*b*) performed rides with 0.5 < GJTR < 1 (1,905 rides); (c) performed rides with GJTR > 1 (366 rides); (*d*) declined rides (113 rides). Plotted in Gephi (*23*).

Discussion

Until recently, the main reasons for introducing DRT services were to substitute FT in low demand areas or to fulfill the needs of specific sectors of the population (mainly the elderly and the disabled). The case study service was also partly introduced based on these grounds (it offers a barrier-free service, and two fixed bus lines were eliminated in conjunction with its introduction). Notwithstanding, it also offers a complementary service to FT, with the objective of improving the mobility offered in the area and to better address passengers' needs. This additional role of DRT is in line with the Mobility-as-a-Service (MaaS) paradigm, in which "mobility services [are bought] as packages based on consumers' needs instead of buying the means of transport" (*25*), and for which different mobility services complement FT.

Results of the case study show that for 50% of the performed rides, the GJT offered by the DRT alternative was 0.5 or less than the GJT offered by the available FT alternative, showing that large improvements in mobility can be attributed to DRT. Rayle et al. compared the times that performed ridesourcing trips (i.e., trips provided by transportation network companies such as Uber and Lyft) would have taken with FT (*26*). In line with the results of this study, they found that "the majority of ridesourcing trips would have taken more than twice as long if made by public transport." Both results show that, even though these services may be competitors of FT, they improve the offered mobility, with the subsequent economic and social benefits.

Despite the opportunities that DRT provides as a complement to FT, its implementation also brings a series of risks that should not be overlooked. Three main risks are identified. Firstly, for a DRT system to achieve a high degree of efficiency, a relatively large fleet is necessary, as stressed in Kutsuplus's final report (*8*). As a result, the offered DRT service could become a (possibly

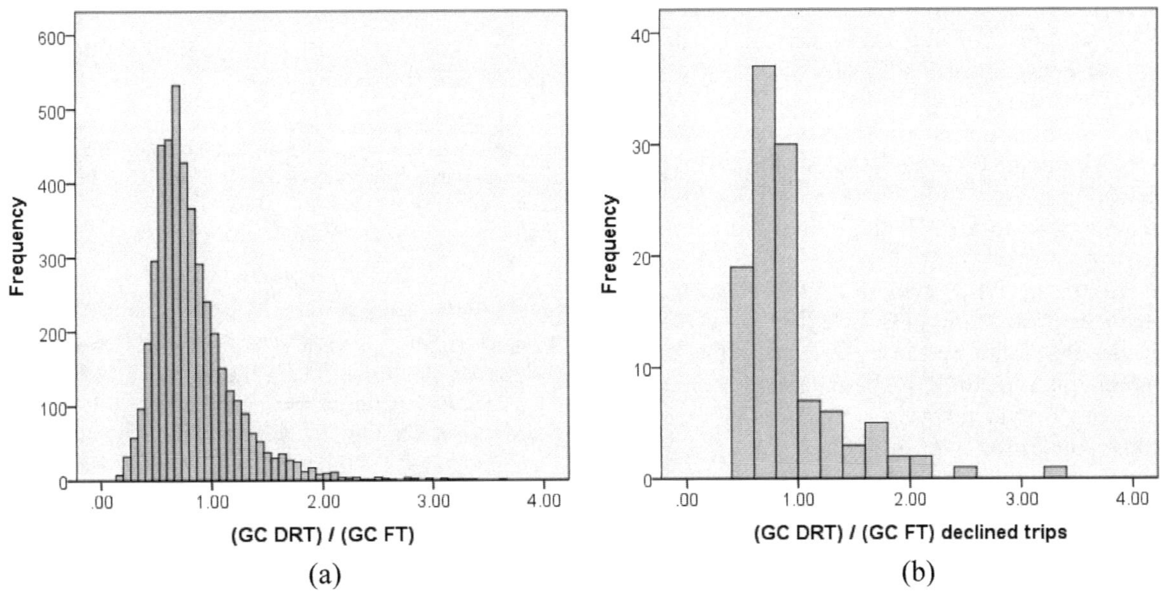

Figure 6. Distribution of the GC for DRT in comparison to PT (*a*) for the performed trips (*N* = 4,477) and (*b*) for the declined trips (*N* = 113).

heavily subsidized) individual taxi service. Secondly, if DRT provides a lower GC than FT, more users may shift from FT to DRT, with the FT revenue seeing a decrease and, in extreme cases, a possible increase in road congestion. An analysis of the incurred GC of both DRT and FT (as performed in this study) could help identify large differences in the costs that passengers incur with both services. Furthermore, FT revenue decrease could lead to the deterioration (or elimination) of the offered FT service, even though DRT may not be capable of absorbing the entire passenger demand. This second risk has been elevated by recent research, which has found that FT users are more prone to use DRT than non-FT users (*27*). Finally, DRT services in their current regulatory framework do not guarantee a certain level of service. As a result, the risk of not obtaining the desired ride exists, the importance of which increases if no alternative service (such as FT) is available. The share of declined trips could vary for different O-D pairs, with rides for less convenient/profitable locations being more likely to be declined. Thus, a spatial analysis of the declined rides should be performed together with the spatial analysis of the performed rides, as described in the suggested framework.

The above-mentioned risks can be reduced with an integrated FT–DRT approach. In an integrated FT–DRT network, the ride fee for DRT could be modified so as to obtain the desired split between the demand for DRT and FT. This price adjustment could be regulated based on different attributes. When striving for an equal urban accessibility, DRT price could be calculated as a function of FT accessibility. This would imply that the price for DRT would be higher when a competitive FT alternative exists, and lower otherwise. The accessibility indicators proposed in this study could be used to adjust the DRT price function if this approach is adopted. An alternative price adjustment approach could be based on comfort considerations as more exclusive (and less collective) mobility services provide a higher utility to the passenger. This approach, which is used in the flexible mobility on demand system presented in Atasoy et al., implies that passengers should pay a higher fee for using a premium service that is less capacity-efficient (*5*). Whatever the adjusting function is, an integrated FT–DRT approach could increase the opportunities that DRT could bring as a complement to FT and decrease the collateral risks.

Conclusion

There are a wide range of innovative transportation services that have recently appeared in urban areas. However, they are often considered in isolation, and the impacts of their interactions and implications for other modes are largely ignored (*28*). This paper presents an assessment framework for evaluating the usage of DRT. A generalized travel cost comparison was proposed as an indicator of changes in accessibility related to the introduction of DRT in relation to the FT alternatives. Moreover, the usage of DRT as the first or last leg in longer FT journeys, and the suitability of the performed rides for walking and cycling were also considered in this assessment. To the authors' knowledge, this is the first study that has measured the increase in accessibility that the implementation of DRT has granted DRT passengers, in comparison to the FT alternative. An empirical analysis was performed by applying the proposed framework to an urban DRT system in the Netherlands. Results indicate a reduction of over half of the GJTs for half the rides that were performed using the DRT service, in comparison to the FT alternatives. Results also identified areas for which this reduction was the highest, highlighting connections for which FT offers a poor alternative.

Public transport authorities could use the proposed framework for one or more of the following purposes: 1) evaluating the real performance of DRT for the different areas, including the distribution of declined rides; 2) identifying whether DRT is used as a complement of FT due to poor FT connections, or as a substitute for FT due to the higher comfort DRT provides, and 3) assessing the impacts of DRT on improving mobility.

Other than DRT, recent studies have shown that other flexible modes, such as ridesourcing and car-sharing, may also complement public transit (*29*). However, even if a series of partnerships between them and transit are already in place, data concerning their usage is necessary to enhance transit planning (*30*). The assessment framework provided in this study could be extended to analyze the usage of other flexible modes and better identify opportunities for synergies as opposed to symbiotic relations in which new services gain the most lucrative markets.

In the absence of information on individual characteristics, the shortcomings of this study pertain to the unobserved drivers of user behavior. The analysis of travel behavior under different conditions would allow estimating demand elasticities, the socioeconomic characteristics of DRT riders or differences in perceived times for DRT with respect to FT values. Future studies could include these aspects in their evaluation by incorporating travel survey data. Also, further research is needed to analyze how urban development characteristics and transit-related factors influence DRT usage, as already researched for taxi ridership (*31*).

Acknowledgments

This work was funded by NWO (The Organisation for Scientific Research from the Netherlands), as part of the SCRIPTS (Smart Cities Responsive Intelligent Public Transport Services) research project. The authors thank Breng and Arjen van Weert of HAN University for providing the data used for the case study.

Author Contributions

The authors confirm contribution to the paper as follows: study conception and design: MJA-G, OC; data collection: TL; analysis and interpretation of results: MJA-G, OC, TL; draft manuscript preparation: MJA-G, OC. All authors reviewed the results and approved the final version of the manuscript.

References

1. Cole, L. M. *Tomorrow's Transportation: New Systems For The Urban Future*, U.S. Department of Housing and Urban Development, Washington, D.C., 1968.
2. Markovic, N., M. E. Kim, and P. Schonfeld. Statistical and Machine Learning Approach for Planning Dial-a-Ride Systems. *Transportation Research Part A: Policy and Practice*, Vol. 89, 2016, pp. 41–55. https://doi.org/10.1016/j.tra.2016.05.006.
3. Kalpakcı, A., and N. Karataş Ünverdi. Integration of Paratransit Systems with Inner-City Bus Transport: The Case of Izmir. *Public Transport*, Vol. 8, No. 3, 2016, pp. 405–426. https://doi.org/10.1007/s12469-016-0136-9.
4. Li, X., and L. Quadrifoglio. Feeder Transit Services: Choosing between Fixed and Demand Responsive Policy. *Transportation Research Part C: Emerging Technologies*, Vol. 18, No. 5, 2010, pp. 770–780. https://doi.org/10.1016/j.trc.2009.05.015.
5. Atasoy, B., T. Ikeda, X. Song, and M. E. Ben-Akiva. The Concept and Impact Analysis of a Flexible Mobility on Demand System. *Transportation Research Part C: Emerging Technologies*, Vol. 56, 2015, pp. 373–392. https://doi.org/10.1016/j.trc.2015.04.009.
6. Martinez, L. M., G. H. A. Correia, and J. M. Viegas. An Agent-Based Simulation Model to Assess the Impacts of Introducing a Shared-Taxi System: An Application to Lisbon (Portugal). *Journal of Advanced Transportation*, Vol. 49, No. 3, 2015, pp. 475–495. https://doi.org/10.1002/atr.1283.
7. Alonso-Mora, J., S. Samaranayake, A. Wallar, E. Frazzoli, and D. Rus. On-Demand High-Capacity Ride-Sharing via Dynamic Trip-Vehicle Assignment. *Proceedings of the National Academy of Sciences*, Vol. 114, No. 3, 2017, pp. 462–467. https://doi.org/10.1073/pnas.1611675114.
8. Rissanen, K., Helsinki Regional Transport Authority. *Kutsuplus - Final Report*, Helsinki Regional Transport Authority (HSL) Publications, Helsinki, Finland, 2016.
9. Ferreira, L., P. Charles, and C. Tether. Evaluating Flexible Transport Solutions, Transportation Planning and Technology. *Transportation Planning and Technology*, Vol. 30, No. 2–3, 2007, pp. 249–269. https://doi.org/10.1080/03081060701395501.
10. Morse, L., M. Trompet, A. Barron, and D. J. Graham. Development of a Key Performance Indicator System to Benchmark Relative Paratransit Performance. *Transportation Research Record: Journal of the Transportation Research Board*, 2017. 2650: 1–8.

11. Yigitcanlar, T., N. Sipe, R. Evans, and M. Pitot. A GIS-Based Land Use and Public Transport Accessibility Indexing Model. *Australian Planner*, Vol. 44, No. 3, 2007, pp. 30–37. https://doi.org/10.1080/07293682.2007.9982586.
12. Khattak, A. J., and Y. Yim. Traveler Response to Innovative Personalized Demand-Responsive Transit in the San Francisco Bay Area. *Journal of Urban Planning and Development*, Vol. 130, No. 1, 2004, pp. 42–55. https://doi.org/10.1061/(ASCE)0733-9488(2004)130:1(42).
13. Lerman, S. R., M. Flusberg, W. M. Pecknold, R. E. Nestle, and N. H. M. Wilson. A Model System for Forecasting Patronage on Demand Responsive Transportation Systems. *Transportation Research Part A: General*, Vol. 14, No. 1, 1980, pp. 13–23. https://doi.org/10.1016/0191-2607(80)90090-4.
14. Jain, S., N. Ronald, R. Thompson, and S. Winter. Predicting Susceptibility to Use Demand Responsive Transport Using Demographic and Trip Characteristics of the Population. *Travel Behaviour and Society*, Vol. 6, 2017, pp. 44–56. https://doi.org/10.1016/j.tbs.2016.06.001.
15. Balcombe, R., R. Mackett, N. Paulley, J. Preston, J. Shires, H. Titheridge, M. Wardman, and P. White. The Demand for Public Transport: A Practical Guide. TRL Limited, Wokingham, Berkshire, UK, 2004.
16. Wardman, M. Public Transport Values of Time. *Transport Policy*, Vol. 11, No. 4, 2004, pp. 363–377. https://doi.org/10.1016/j.tranpol.2004.05.001.
17. Booz Allen & Hamilton (NZ) Ltd. *Valuation of Public Transport Attributes*. 2000.
18. Currie, G., and I. Wallis. Effective Ways to Grow Urban Bus Markets - a Synthesis of Evidence. *Journal of Transport Geography*, Vol. 16, No. 6, 2008, pp. 419–429. https://doi.org/10.1016/j.jtrangeo.2008.04.007.
19. International Transport Forum. *Pedestrian Safety, Urban Space and Health*. OECD Publishing, Paris, France, 2012.
20. Yang, Y., and A. V Diez-Roux. Walking Distance by Trip Purpose and Population Subgroups. *American Journal of Preventive Medicine*, Vol. 43, No. 1, 2012, pp. 11–19. https://doi.org/10.1016/j.amepre.2012.03.015.
21. KiM Netherlands Institute for Transport Policy Analysis. *Mobiliteitsbeeld 2016*. 2016.
22. Google Developers. Google Maps Directions API. *Google*. https://developers.google.com/maps/documentation/directions/. Accessed July 23, 2017.
23. Bastian, M., S. Heymann, and M. Jacomy. Gephi: An Open Source Software for Exploring and Manipulating Networks. Proceedings of the Third International ICWSM Conference, Association for the Advancement of Artificial Intelligence, 2009, pp. 361–362.
24. Kouwenhoven, M., G. C. de Jong, P. Koster, V. A. C. van den Berg, E. T. Verhoef, J. Bates, and P. M. J. Warffemius. New Values of Time and Reliability in Passenger Transport in The Netherlands. *Research in Transportation Economics*, Vol. 47, No. 1, 2014, pp. 37–49. https://doi.org/10.1016/j.retrec.2014.09.017.
25. Kamargianni, M., W. Li, M. Matyas, and A. Schäfer. A Critical Review of New Mobility Services for Urban Transport. *Transportation Research Procedia*, Vol. 14, No. 14, 2016, pp. 3294–3303. https://doi.org/10.1016/j.trpro.2016.05.277.
26. Rayle, L., D. Dai, N. Chan, R. Cervero, and S. Shaheen. Just a Better Taxi? A Survey-Based Comparison of Taxis, Transit, and Ridesourcing Services in San Francisco. *Transport Policy*, Vol. 45, 2016, pp. 168–178. https://doi.org/10.1016/j.tranpol.2015.10.004.
27. Alonso-González, M. J., N. Van Oort, O. Cats, and S. Hoogendoorn. Urban Demand Responsive Transport in the Mobility as a Service Ecosystem: Its Role and Potential Market Share. Presented at Thredbo 15 Conference, Stockholm, Sweden, 2017.
28. Ciari, F., and H. Becker. How Disruptive Can Shared Mobility Be ? A Scenario-Based Evaluation of Shared Mobility Systems Implemented at Large Scale. In *Disrupting Mobility*, Springer International Publishing, Basel, Switzerland, pp. 51–63.
29. Feigon, S., and C. Murphy. *TCRP Research Report 188: Shared Mobility and the Transformation of Public Transit*. Transportation Research Board of the National Academies, Washington, D.C., 2016.
30. New York Public Transit Association. *NYPTA White Paper on Transportation Network Company (TNC) Issues*. 2017.
31. Nam, D., K. Hyun, H. Kim, K. Ahn, and R. Jayakrishnan. Analysis of Grid Cell Based Taxi Ridership with Large Scale GPS Data. *Transportation Research Record: Journal of the Transportation Research Board*, 2016. 2544: 131–140.

The Standing Committee on Public Transportation Planning and Development (AP025) peer-reviewed this paper (18-04555).

Article

TRR
JOURNAL OF THE TRANSPORTATION RESEARCH BOARD

Transportation Research Record
2018, Vol. 2672(8) 890–900

DOI: 10.1177/0361198118782275
journals.sagepub.com/home/trr
SAGE

Water Transit and Ferry-Oriented Development in Sweden: Comparisons with System Trends in Australia

Michael Tanko[1], Matthew I. Burke[2], and Harsha Cheemakurthy[1]

Abstract

Water transit is increasingly becoming an option for cities looking to expand their public transport network and provide an alternative to land-based transport congestion. But there are challenges in realising a functional urban ferry network, especially in Sweden which faces land use policy and weather-related constraints. This paper has three related aims: i) to compare the development of water transit in Swedish and Australian cities, ii) to compare changes in patronage over time, and iii) to analyze the role of ferry-oriented development (FOD) and planning policy effects on water transit development. Our approach is exploratory and comparative, and we use the cases of Stockholm and Gothenburg in Sweden with comparison to Brisbane and Sydney in Australia. Patronage figures indicate that in both countries, water transit systems are a small part of the total public transport network, from 0.5% of total public transport trips in Stockholm to 3.7% in Brisbane. However, there has been modest growth in patronage that in relative terms has exceeded that of urban rail and bus services in all cities except Sydney. Integration of water transit with land use plans in FOD varies, with Brisbane and Sydney embarking on large-scale waterfront and transit plans focused on economic development. In Sweden integration with land use remains in early stages, with ad hoc terminal location and restrictive land use being limiting factors, especially in Stockholm. There are also conflicting strategies toward FOD in Gothenburg. Better conceptualization of land use implications for development of water transit planning is therefore suggested.

Urban water transit is undergoing a small renaissance as cities contend with increasing land-based congestion issues and promote waterfront development (*1*, *2*). In Sweden, the expansion of passenger and freight uses on inland waterways is of interest to address congestion issues, but also to facilitate a new environmentally efficient transport option to reduce urban pollution (*3*, *4*). However, there exist challenges in creating city-wide water transit systems in cities such as Stockholm. A recent International Association of Public Transport (UITP) report has shown how advances in urban water transit are generating patronage in Gothenburg, Stockholm and other water-based cities around the world (*5*). But there is limited detail on how Swedish systems are performing over time in terms of either patronage growth or generation of ferry-oriented development (FOD). By comparison, greater research effort has been placed on key Australian and U.S. urban water transit systems (*2*, *6*, *7*), which can offer insights beneficial to developing systems such as in Sweden. The focus of this paper is to outline the current scope of water transit operations, system planning and land development in Sweden, and to explore its prospects. To consider these issues and illustrate relative performance of Swedish systems, we make limited comparisons with well-established compatriot water transit networks in Brisbane and Sydney. The paper has three inter-related aims:

i. to audit current system operations in select Swedish and Australian cities, detailing operating models and system characteristics,
ii. to compare patronage performance over time from the last six years of operations in each city, and
iii. to review the experience with FOD related planning policies and prospects for the future.

The paper offers several contributions to the study of contemporary water transit. It provides a first analysis of longitudinal patronage trends in Sweden and how this compares to other examples in Australia. The paper also focuses on the significance of FOD land use policies and provides some example best practices, highlighting the subsequent implications for future water transit development in Swedish cities.

The paper is structured as follows; a review of available published literature on water transit in Sweden and Australia is presented, followed by a comparison of system characteristics and patronage. The role of FOD is then introduced and a review of current policies and examples is provided. The

[1]KTH Royal Institute of Technology, Stockholm, SE, Sweden
[2]Griffith University, Nathan Queensland, Australia

Corresponding Author:
Address correspondence to Michael Tanko: tanko@kth.se

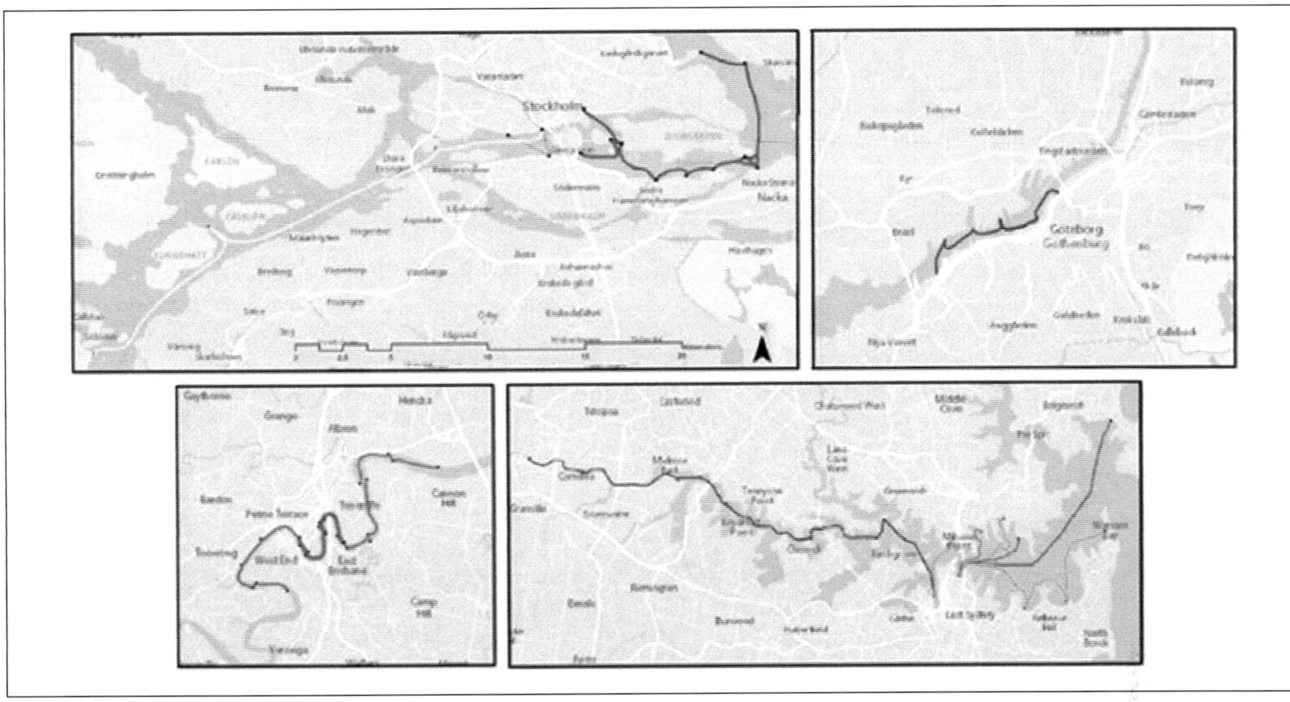

Figure 1. Clockwise from top left: Stockholm, Gothenburg, Sydney, and Brisbane.

paper concludes with a discussion of the implications of these findings for future water transit planning and FOD policy development in Sweden and suggestions for further research.

Research Context

As the two largest cities in Sweden, both Stockholm and Gothenburg possess large public transport networks incorporating bus, tram, ferry and, in Stockholm, urban rail. Both cities also feature water transport networks servicing both the inner-city region and outward toward large archipelago networks. The focus of this study is on water transit services within the inner city that are incorporated into the local public transport planning network. In Stockholm the public transport agency SL (*Storstockholms Lokaltrafik*) manages the transport provision within Stockholm County, a population of approximately 2.2 million (*8*). Boat services operated by SL include five lines: 80, 82, 84, 85 and 89. These services are contracted out to three private operators, who run the *Djurgårdsfärjan* and *Sjövägen* services, while *Waxholmbolaget* is another company who runs an inner-city service in addition to its outer archipelago services (*9*, *10*). These services share common ticketing with the wider transport network via either rechargeable smart card or individual tickets (*11*). In Gothenburg the public authority *Västtrafik* operates two commuter boat lines on the inner harbor of the Göta River. They include a linear route and a free cross-river service between Lindholmen and Stenpiren, a key transport hub where a new ferry terminal and transfer station was completed in 2016 (*12*). Services also allow transfer to other public transport modes. Brisbane and Sydney's water transport are also integrated into their public transport networks, which include bus, urban rail, ferry and, in Sydney, light rail. For a detailed overview of Australian networks see previous research on Brisbane (*2*, *13*) and Sydney (*14*).

Systems and Performance

Figure 1 illustrates the route networks of the water transit systems in Sweden (Stockholm and Gothenburg) and Australia (Brisbane and Sydney) using at-scale maps. Different colors indicate individual routes within each network. Table 1 presents an overview of the existing system characteristics. Stockholm has the larger network in Sweden which includes five routes covering 30 km, operated by a total fleet of 12 vessels (*5*). Two new routes were introduced in 2016. The first was an inner-city shuttle service connecting the southern residential and business island Södermalm with the central business district (CBD) (green line). The other was a longer suburban route west to Ekerö which has traditionally been a transport-deprived region (yellow line). As a secondary city, Gothenburg operates a smaller network mostly confined to the inner core of the city, anchored by key terminals at Stenpiren and Lindholmen, the latter of which serves as the waterfront technology hub of Gothenburg (*15*).

By comparison, Sydney has the most extensive route combination with eight separate lines. These include inner harbour services operated predominantly by high capacity vessels, and a single linear river route to the nearby city of

Table 1. Water Transit Operations in Gothenburg, Stockholm, Brisbane, and Sydney

City	Route km	Routes	Vessels	Terminals	Frequency	Network integration	Fare (USD)
Stockholm	30	5	12	23 (fixed land, floating pontoon)	12 mins peak 30 mins off peak	Yes	$5.20
Gothenburg	13	2	6	6 (fixed land, floating pontoon)	30 mins peak 60 mins off peak 7–8 mins cross-river service	Yes	$3.25 fixed single for entire route (free cross-river route, no bicycle surcharge)
Brisbane	21	5	30 (21 catamaran, 9 monohull)	25 (pontoon)	10 mins peak 15–30 mins off peak	Yes	$3.50–$4.10 zone based (free inner city "City Hopper" service)
Sydney	45 (34 km inner harbour services) (25 km Parramatta river route)	8	32	36 (fixed land and pontoon)	15–30 mins peak 30–60 mins off peak	Yes	$4.70 < 9 km $5.90 > 9 km

Sources: (*1, 2, 4, 5, 9, 12, 16, 17*).

Parramatta (blue line) operated by high speed catamarans (*16*). Sydney also possesses the longest route length of the studied systems with 70 km in total. Brisbane's network is dominated by a single linear route stopping at key inner-suburban and CBD activity centers (*2*). Brisbane has the largest fleet with 21 catamarans operating the CityCat route (blue lines) supplemented by nine monohull vessels for other shorter routes (red lines) (*17*).

Table 2 describes the fleet and contextual information for each city. Owing to the need for year-round, all-weather operations in Sweden, vessels in both Stockholm and Gothenburg are monohull and mostly powered by diesel engines, with one hybrid diesel-electric vessel in operation in Stockholm. Striking a balance of efficient year-round vessel operation is an ongoing challenge in planning water transport in Sweden due to ice conditions (*3*). Options to meet this challenge are currently being investigated by Stockholm County Council (*4*). Travel speeds are therefore generally lower in Sweden. There is also a speed limit imposed in the inner harbor of Gothenburg and Stockholm of 6 knots. By comparison, Brisbane has a speed limit in key inner reaches of 8 knots, but the majority of the CityCat route operates upwards of 20 knots.

Table 3 presents patronage comparisons of public transport networks in each city. Figure 2 shows patronage and mode share changes over the study period. As shown, Stockholm and Gothenburg mode shares are dominated by urban metro and bus services, respectively. Water transit provides much less passenger capacity and ridership in the public transport network of each city. The mode share for inner city ferries operated under the public transport in the two cities ranges from approximately 0.5–0.8% in Stockholm and Gothenburg, respectively. However, ridership is relatively strong for the system size and the lower mode share is partly due to very high patronage of rail-based and bus public transport in both cities. Relative to other modes, water transit has seen the most increase in passenger numbers over the study period, with a 49% increase in Stockholm and 78% increase in Gothenburg (*12*).

By comparison, Sydney carries by far the most passengers per year with over 15 million in 2016 (*18*). Brisbane's mode share has increased significantly from 2011 to 2016 to currently holding approximately 3.7% of Brisbane's total public transport mode share (*24*). Brisbane has only seen mode share return to 2010 levels after flooding destroyed many terminals in 2011 and required infrastructure rebuilding in the following years (*13*). Sydney is the only city of the four that experienced a decrease in ferry patronage and mode share during the study period (*18*).

Competing uses of waterways is an ongoing concern in both Gothenburg and Stockholm as they remain active ports. There is still major commercial and shipping activity which causes potential conflicts. Alternatively, much of Brisbane and Sydney's commercial port industry has moved out of the inner city to locations nearby each airport at the Port of Brisbane and Port Botany, respectively (*24*, *25*). Congestion still occurs in Sydney Harbor with cruise ships, water taxis and non-regulated ferries, and remains an ongoing issue (*26*). In Brisbane the most common conflicts arise between leisure craft and other recreational waterway users such as rowers. Weather is another factor that significantly affects operational performance of the networks in Sweden. While in Brisbane it is possible to use high speed, low wake catamarans year-round, in Sweden the need for more robust vessels with ice breaking capacity for year-round operation is a necessity.

Table 2. Water Transit Fleet and Conditions

City	Hull type	Engine type	Max speed (knots)	Speed limit	Hours	Capacity	Crew	On-board facilities	Maximum tidal range	Natural hazards
Stockholm	Monohull	Diesel engine	12	6 knots in inner harbor	6 am–11 pm	225–300	2–3	Coffee; disability access; bicycle storage	0 m	Ice, storm
Gothenburg	Monohull	Diesel engine and diesel electric	11	6 knots in inner harbor	6 am–12 am	298 (160 seated); 80 bicycle spaces; seating in/out	3	Toilets/coffee; on-board bicycle storage; disability access	0 m	Ice, storm
Brisbane	Catamaran (river route) and monohull (cross-river services)	Diesel engine	26	8 knots in inner city reach	5 am–12 am	Catamarans 149–162; Monohull 53–78; seating in/out	1–3	Coffee, snacks; disability access; limited bicycle storage	3 m	Flood, storm, fog, debris
Sydney	Monohull and catamaran (Sydney) and catamaran (Parramatta River route)	Diesel engine	12 knots inner harbor 26 knots river service	–	5 am–12 am	300–400 inner harbor; 1110 Manly route; 150–250 Parramatta route	3–6	Bicycle space, toilets, and WiFi (larger vessels, not-chartered only)	2 m	Storm, fog, flooding (Parramatta river service only)

Sources: (*1*, *2*, *4*, *5*, *13*, *16*, *17*).

Table 3. Water Transit Patronage in Gothenburg, Stockholm, Brisbane, and Sydney (000 s)

	2011	2012	2013	2014[1]	2015	2016	% change
Stockholm							
Metro	309000	322000	328000	334000	340700	347400	12.43
Bus	291000	298000	300000	307000	312500	318000	9.28
Ferry	2345	2173	2335	2461	3430	3510	49.68
LRT	45000	45700	47000	49040	52070	54488	21.08
Total	648000	668000	678100	693388	708670	723388	11.74
Metro%	47.61	48.18	48.30	48.11	47.96	47.90	0.61
Bus%	44.82	44.58	44.18	44.22	43.99	43.85	−2.21
Ferry%	0.36	0.33	0.34	0.36	0.48	0.49	33.94
LRT%	6.93	6.73	6.92	7.07	7.33	7.51	8.35
Gothenburg[2]	2011	2012	2013	2014	2015	2016	% change
Bus	177007	178289	183098	176804	178696	196475	11.00
Ferry	1419	1840	2078	2451	2380	2532	78.44
LRT	79293	80911	93002	98996	100904	104025	31.19
Total	258000	261000	278000	278000	282000	303000	17.44
Bus%	68.61	68.31	65.86	63.60	63.37	64.84	−5.49
Ferry%	0.66	0.69	0.68	0.79	0.85	0.83	51.94
LRT%	30.73	31.00	33.45	35.61	35.78	34.33	11.71
Sydney	2011	2012	2013	2014	2015	2016	% change
Metro	294457	303550	306229	315099	326448	361134	22.64
Bus	214273	219221	220151	224063	257015	290297	35.48
Ferry	14503	14768	14943	15977	14794	15410	6.25
LRT	2739	3975	4150	3889	6135	9728	255.17
Total	527983	543526	547486	561042	606407	678585	28.52
Metro%	55.77	55.85	55.93	56.16	53.83	53.22	−4.57
Bus%	40.58	40.33	40.21	39.94	42.38	42.78	5.41
Ferry%	2.75	2.72	2.73	2.85	2.44	2.27	−17.33
LRT%	0.52	0.73	0.76	0.69	1.01	1.43	176.34
Brisbane	2011	2012	2013	2014	2015	2016	% change
Metro	51600	52480	50020	49820	50420	54280	5.19
Bus	124400	120200	119770	118650	113130	119440	-3.99
Ferry	4300[4]	5200	5890	6000	6430	7000	62.79
LRT[3]	–	–	–	–	6280	7600	–
Total	178600	178300	175700	175910	176260	188360	5.46
Metro%	28.89	29.43	28.47	28.32	28.61	28.82	-0.26
Bus%	69.65	67.41	68.17	67.45	64.18	63.41	-8.96
Ferry%	2.41	2.92	3.35	3.41	3.65	3.72	54.36
LRT%	–	–	–	–	3.56	4.03	–

Sources: (*5, 10, 12, 18–23*).
[1]New linear route 85 to Nacka introduced (blue in Figure 1). Green and Yellow routes introduced in 2016 and not included in analysis.
[2]No metro rail service in Gothenburg.
[3]Light rail service introduced in 2014.
[4]Operations suspended and many terminals damaged by the January 2011 Brisbane floods (6.6 m passengers in 2010).

Ferry-Oriented Development

A key consideration of the success of water transit systems has been supportive land use around terminals and the extent of FOD. Currently, there is limited research on guidance indicating performance standards and best practice for implementing effective FOD. One of the first to link ferry development and land use was Thompson et al. (2006), who identified the potential for a network of ferry terminals to act in a transit-oriented development (TOD) function similar to other transit lines (*6*). Compared to existing TOD, they suggest that for urban water transit and FOD to be successful there should be additional focus on terminal design and proving additional facilities, such as shelter and retail arrangements. Weisbrod and Lawson also identified the potential for ferry services to be a catalyst for economic revitalization in U.S. cities, in conjunction with waterfront rejuvenation plans (*7*). They also observe that ferry passenger are often leisure users less concerned about time and may be more likely to linger around terminals and use

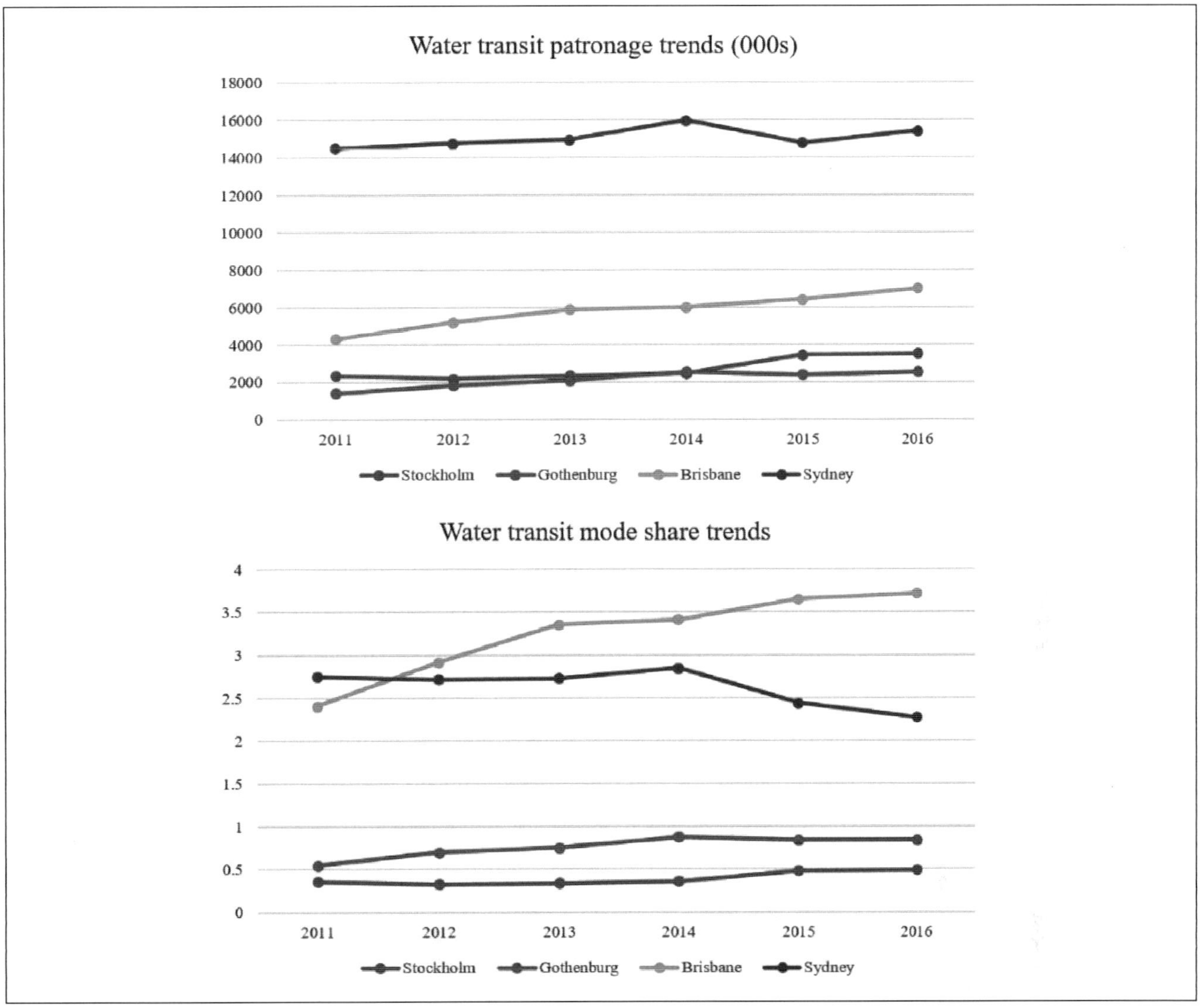

Figure 2. Water transit usage trends 2011–2016.

supporting land uses. These aspects suggest effective FOD may benefit from and creating destination that people are likely to spend more time at.

Research on economic benefits surrounding ferry terminals also suggests that they may differ from TOD and offer more development potential. Two key studies have so far explored the property value uplift effects around terminals of such systems. In Brisbane, where the system has operated since 1996, Tsai et al. (2015) found that value uplift of 4% has occurred around terminals (*27*). The New York City Economic Development Corporation, the lead organization responsible for the initial East River Ferry, found that after a three-year trial there was an 8% value uplift around ferry terminals (*28*). This result was a contributing factor in the expansion to a city-wide urban ferry system connecting all five boroughs. The development model used in New York highlights the importance of FOD success as a key factor facilitating the overall success of a water transit system.

Other models of FOD have also been increasing. In Brisbane property developers agreed to finance a new terminal in exchange for ferry service to attract potential residents. In London, similar arrangements are occurring, with discussions ongoing to service the new Battersea Power Station development. Some cities have also sought to promote the image of a 'river city' through city branding and tourism initiatives, where water transport has been one of a suite of policies coordinated with new residential and commercial land uses centered around rivers (*7*). Such results suggest there may be an increased potential for FOD, especially in conjunction with such redevelopment plans where water transport services are included early in discussions and innovative financing solutions are employed.

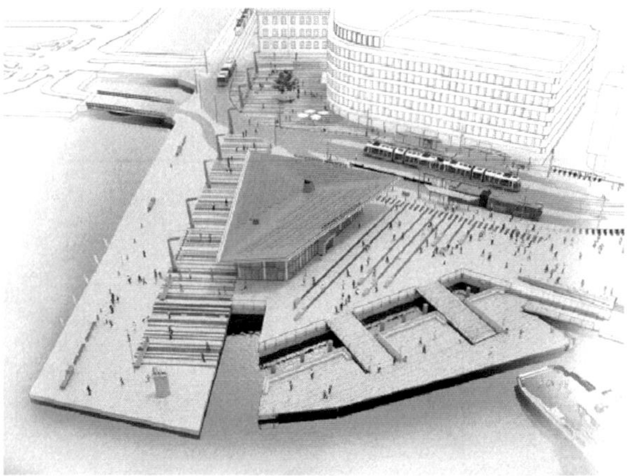

Figure 3. Stenpiren ferry terminal and transit hub in Gothenburg.

On the other hand, there have been concerns raised that developers are perhaps less willing to invest in FOD which are seen as less permanent than other TOD projects (*6*). While this may be seen as an advantage for the flexibility and adaptability of water transit systems, it may discourage investment around terminals. From current examples of successful water transit systems, best practice in terms of both economic benefits and patronage growth suggests investment in permanent fixtures is preferred. These should also be planned with the existing transport network, not only with integrated ticketing, but including uniform signage and wayfinding to and from other transit modes. In a study of Brisbane's network, it was found that 15% of all ferry journeys included another mode (*29*), therefore facilitating transfers should be a priority.

Looking at the contexts under investigation, Sweden and Australia have so far differed in the development of FOD. In Sweden, the *Gothenburg River Vision* report has set out a plan to develop the waterfront areas to provide better river access for residents and to encourage business and activity center creation along the waterfront. A key project has been the opening in 2016 of the new Stenpiren ferry terminal where it was envisioned that "an entire neighborhood will be developed with housing, businesses and meeting places" (*15*) on the FOD model, to change the perception of the river as a barrier. This has been an initiative of the municipal authority *River Bank Development*, with a new transport hub developed featuring transfer options between tram and bus services (Figure 3). There has also been collaboration with private industry in locating business in the area as part of a renewal policy agenda. Such urban strategies are increasingly common, often promoted by the urban regimes and growth coalitions that operate in most cities (*29*). Ferries are often promoted in such strategies not just for their transport function but for the imagery they provide (*1*, *6*). Gothenburg's renewal project objectives envision "*the river for everyone, not only some specific target groups, selective economic activities or some luxurious, mono-functional housing development. It is about diversity of environments and economies alongside the river*" (*30*). However, simultaneous promotion of the area as a new technology and cultural hub of the city has conflicted with this goal. A longitudinal ethnographic study of preliminary outcomes of the plan analyzing the makeup of the residential population within the redevelopment concluded that it has so far failed to achieve a more integrated and equitable population mix (*31*).

In Stockholm the waterfront development at Hammarby Sjöstad (Lake City) has been the flagship waterfront redevelopment, much discussed for its ecologically sustainable design focus. The project initially was conceived as an athlete village as part of Stockholm's unsuccessful 2012 Olympic Games bid but has subsequently been developed as a model for mixed used waterfront development. A key early consideration was transport provision. Part of this was a purpose-built pier and subsidized ferry services to connect passengers to the CBD 5 km away. However, there is currently a new train line being built nearby to connect the city to the eastern region, raising questions as to the viability of ferry services in the future.

One of the key issues impacting on FOD in Stockholm is that terminals have poor land use integration. Current piers have been located in existing available locations and not planned for future integration with waterfront development plans of the city. The new cross-river route in Södemalm exemplifies this, where there is little to signify that there is a ferry service available. This has been coupled with poor access owing to the unfavorable location which has led to criticisms of the piers and their convenience for actual commuting use (Figure 4) (*32*).

Despite these example projects, FOD is still in its infancy in Sweden. There are several factors that have contributed to this. Firstly, quite restrictive land use policies in waterfront areas exist. Regulations specify that developments must generally not infringe on public access; this creates benefits in that 80% of Stockholm's shoreline is publicly accessible (*33*). Further, additional consideration is given to the watershed when planning for urban development. Municipal authorities are also retreating in part from playing a lead role in the redevelopment process, with reduced land ownership in public hands and the cost to acquire new waterfront land for redevelopment in many cases prohibitively expensive. But urban regimes continue to look to FOD as a way to unlock new territories for redevelopment. Recently there has been interest from the Stockholm Chamber of Commerce in exploring options for development of waterfront districts for business and tourism purposes, in combination with new ferry terminal locations (*34*).

Figure 4. Stockholm route 85 pier and accessibility.

Figure 5. New Barangaroo ferry terminal, Sydney.

Australia's history with FOD has parallels. Brisbane has seen rapid redevelopment around existing and newly planned terminals as part of continuing riverfront renewals programs in different areas of the city (*2*, *13*). Similar to Gothenburg, the renewal of areas such as Tenerife and New Farm, partly on the FOD model, has seen gentrification and the displacement of low-income groups, despite early council rhetoric around equity. As noted previously property developers have paid for new ferry terminals and this has further contributed to inequalities.

Though the most recent Hamilton North-Shore development is primarily targeting higher-income populations, the state land developer involved uses profits from the site to help subsidize affordable and social housing in their other projects. As such, Brisbane has struggled to create inclusive FOD but is finding other ways to use FOD to overcome inequality. Sydney has also been actively developing new projects around water transit. An ongoing issue with the Sydney network has been capacity at single key exchange terminal at Circular Quay (*35*). A new two pier hub at Barangaroo opened in 2017 to service the newly developed precinct of commercial, hotel, residential and casino operations (Figure 5).

Discussion of Findings

Despite being a relatively minor mode of transport in Sweden there has been modest progress in development of water transit networks. Patronage is growing and in Stockholm continues to be in excess of planning authority expectations; there is also growing recognition of its potential role in future city plans (*3*). However, the conceptualization of FOD in Sweden and what it might deliver remains vague. In Gothenburg, the vision of inclusive development is proving problematic; in Stockholm interest in ferries remains more on transport functions than land development. This raises major questions for the field as to how water transit and effective FOD may be achieved. Links between government and the land development sector in ferry system development are important, as seen in Brisbane, New York and London.

But FOD should be more than just an agent of urban regimes for gentrification and eventually displacement of the poor in cities being reclaimed by the wealthy. With Sweden being in the early phases of development there is an opportunity to develop a more coherent strategy including more socially inclusive FOD.

Although we looked at Australian comparators, cities elsewhere also show ways forward. In New York, the primary goal was to stimulate waterfront precincts through a trial ferry route. Ferries have provided transit accessibility to locations that previously had effectively very little, including some socially disadvantaged communities. Stockholm is somewhat following this trail, with new suburban lines and current investigations into routes to connect inaccessible locations in the inner archipelago. The subsidising of a ferry route in Hammarby Sjöstad in conjunction with a new mixed-use development is also a positive sign for increasing feasibility of water transport in the future. This may encourage the formation of a model for linking inner archipelago islands that currently have poor accessibility to central Stockholm. While outer archipelago ferry services are often the only option in regional areas outside of the urban area, there is the possibility for introducing ferries on inner archipelagos to compete with other commuting modes such as private cars, which often travel longer on a less direct route. However, Sweden may need to loosen some land use regulations on the waterfront to encourage better FOD outcomes, whilst still maintaining through-movement on the landside and public access.

The effect of city configurations is also worth noting. Stockholm and Sydney may be considered similar in terms of the hub and spoke pattern of service to inner harbors and archipelago islands connecting to a central city hub. Brisbane and Gothenburg's structure are also similar, based around a more linear route that travels parallel to the shoreline to key activity centers, and offers cross-river connections. Locations at the end of these routes in suburban areas are being actively developed for more residential and commercial uses in both cities. Planning for water transit in either city configuration needs to be considered, with the complexity of Stockholm's archipelago network needing careful planning to design route structures that emphasize multidirectional travel, to avoid capacity issues at the central pier as has occurred in Circular Quay in Sydney.

Finally, political and institutional arrangements are worth considering. Despite government support and positive intentions, political barriers can arise in implementing water transport when land use and marine departments can result in conflicting goals, as has been the challenge in Bangkok in its efforts to expand and modernize its network (*36*). Stockholm is made up of 48 separate municipal authorities, each with the responsibility to enact a land use plan for their respective district. Therefore, while development in one key localized area was able to be achieved, this may not be the case when operating in larger context. A particular challenge is the coordination of waterfront land use policies to facilitate a uniform network across these respective boundaries.

Conclusion

This paper has looked at the development of Swedish and Australia water transit systems and provided the first comparative analysis of patronage and longitudinal performance. It has also provided a look at the importance of FOD within the Swedish political context and some examples of progress that has been made, and challenges in the future. So far Sweden has a focus on water transit development and aspirations toward inclusive FOD but has had mixed success in Gothenburg, with promotion at the same time as a technology hub being a conflicted strategy. However, the change of location for a new pier and transit hub from a previous location with little supportive land use is a positive sign. In Stockholm, a different environmental efficiency driven agenda exists. In this case the challenge is to more tightly integrate land use considerations in a clear FOD vision if success is to be achieved in expanding the water transport network.

There are limitations to the current research, though. Perhaps foremost is the lack of disaggregate analysis of water transit within the two Swedish cities. The descriptive approach used lacks power to isolate the effects of the water transit system on user behavior or land development. There remains a paucity of research in water transit systems. Our recommendations for future research to assist the Swedish systems involve several items that would assist the broader field of urban ferry planning and operations. These include:

i) Disaggregate analysis of patronage in Gothenburg and Stockholm, perhaps using fare transaction in a similar way to the previous work on Brisbane of Soltani et al. 2015 (*29*) to better understand passenger movements and inter-modal journeys within the cities.
ii) Extending the longitudinal research framework to explore such issues as network effects within water transit systems, particularly as new routes are added to the Stockholm system. Similarly, use of longitudinal pre-/post-intervention studies on the effects of new water transit piers on residential travel behavior may provide useful insights to assist future planning and policy, especially if controlled for travel attitudes and other known influences on travel behavior.
iii) The benefits that accrue in terms of property value uplift around Swedish water transit piers are not well understood. Hedonic value studies may help reveal what effects these systems are having on property values and if the results are consistent with effects seen in other cities.
iv) Work on vessel design and operations to overcome the particular problems of ice in Swedish winters that may impede high speed operations.
v) Analysis and simulation of alternative route structures to maximize patronage and efficiency, especially in Stockholm, and the differing contexts such as archipelago/island/bay configurations and their unique challenges, building on the work of Sandell (2015) in Sydney (*35*) and the recent city-wide ferry study in New York (*28*).

Acknowledgments

Michael Tanko is jointly funded as a postdoctoral researcher by the Swedish Maritime Competence Centre (Lighthouse) and KTH Royal Institute of Technology. Transport research at the Cities Research Institute, Griffith University is supported by the Queensland Department of Transport and Main Roads and the Motor Accident and Insurance Commission. Associate Professor Matthew Burke is the recipient of an Australian Research Council Future Fellowship [FT120100976].

Author Contributions

The authors confirm contribution to the paper as follows: study conception and design: Michael Tanko, Matthew Burke; data collection: Michael Tanko, Matthew Burke, Harsha Cheemakurthy; analysis and interpretation of results: Michael Tanko, Matthew

Burke; draft manuscript preparation: Michael Tanko, Matthew Burke. All authors reviewed the results and approved the final version of the manuscript.

References

1. Tanko, M., and M. I. Burke. Transport Innovations and their Effect on Cities: The Emergence of Urban Linear Ferries Worldwide. *Transportation Research Procedia*, Vol. 25, 2017, pp. 3961–3974.
2. Tanko, M., and M. I. Burke. Innovation and Transport Planning: Introducing Urban Linear Ferries in Brisbane, Australia. In *State of Australian Cities Conference 2015: Refereed Proceedings, Gold Coast: Urban Research Program at Griffith University on behalf of the Australian Cities Research Network* (P. Burton and H. Shearer, eds.), Griffith University, Brisbane, Australia, 2015.
3. Swedish Maritime Administration. *Analysis of Development Potential For Inland and Coast Islands in Sweden: Main Report*. Sjöfartsverket, Stockholm, Sweden, 2016.
4. Stenius, I., K. Garme, S. H. Kihl, and M. Burman. *Waterway 365 Sustainability and Resilience: Toward a Systems Approach*. KTH Royal Institute of Technology and Vattenbussen, Stockholm, Sweden, 2014.
5. International Association of Public Transport. *Waterborne Public Transport: Evolution, General Overview and Networks*. International Association of Public Transport (UITP), Brussels, Belgium, 2016.
6. Thompson, R., R. Burroughs, and T. Smythe. Exploring the Connections between Ferries and Urban Form: Some Considerations before Jumping on Board. *Journal of Urban Technology*, Vol. 13, No. 2, 2006, pp. 25–52. http://www.tandfonline.com/doi/abs/10.1080/10630730600872021.
7. Weisbrod, R. E., and C. T. Lawson. Ferry systems: Planning for the Revitalization of U.S. Cities. *Journal of Urban Technology*, Vol. 10, No. 2, 2003, pp. 47–68.
8. Statistiska Centralbyrån. *Population in the Country, Counties and Municipalities on 31/12/2016 and Population Change in 2016*. Stockholm, Sweden, 2016. http://www.scb.se/en/finding-statistics/statistics-by-subject-area/population/population-composition/population-statistics/pong/tables-and-graphs/yearly-statistics--municipalities-counties-and-the-whole-country/population-in-the-country-counties-and-municipalities-and-population-change/.
9. Storstockholms Lokaltrafik. *Timetables, Maps and Other Travel Information-SL–SL*. Stockholm, Sweden. 2017. http://sl.se/sv/reseinfo/.
10. Storstockholms Lokaltrafik. *Facts about SL and the County in 2014*. Stockholm County Council, Stockholm, Sweden, 2015.
11. Storstockholms Lokaltrafik. *Köpa biljett – Stockholm County Council, Stockholm, Sweden,* 2017. https://sl.se/sv/kop-biljett/.
12. Västtrafik. *Västtrafiks årsredovisning 2016* (*Annual Report 2016*). Vastra Gotaland Regionen, Gothenburg, Sweden, 2016.
13. Sipe, N., and M. Burke. Can River Ferries Deliver Smart Growth? The Experience of Brisbane's CityCats. Presented at 89th Annual Meeting of the Transportation Research Board, Washington, D.C., 2010.
14. Rissel, C., M. Crane, L. M. Wen, S. Greaves, and C. Standen. Satisfaction with Transport and Enjoyment of the Commute by Commuting Mode in Inner Sydney. *Health Promotion Journal of Australia*, Vol. 27, 2015, pp. 80–83.
15. Gothenburg County Council. *Rivercity Gothenburg Vision*. Gothenburg, Sweden, 2012.
16. Transport for NSW. *Transport for NSW Annual Report 2015-16*. New South Wales Government, Sydney, Australia, 2016.
17. Translink. *Maps | TransLink*. Queensland Government, Brisbane, Australia. 2017. https://translink.com.au/plan-your-journey/maps.
18. Transport for NSW. *All Modes Historical Patronage - Top Level Chart | Transport for NSW*. New South Wales Government, Sydney, Australia, 2016. https://www.transport.nsw.gov.au/data-and-research/passenger-travel/all-modes-patronage-historical/all-modes-historical-patronage.
19. Department of Transport and Main Roads. *Annual Report 2013-2014*. Queensland Government, Brisbane Australia, 2014.
20. Department of Transport and Main Roads. *Annual Report 2012-2013*. Queensland Government, Brisbane Australia, 2013.
21. Department of Transport and Main Roads. *Annual Report 2011-2012*. Queensland Government, Brisbane Australia, 2012.
22. Västtrafik. *Västtrafks årsredovisning 2011 (Annual Report 2011)*. Vastra Gotaland Regionen, Gothenburg, Sweden, 2015.
23. Storstockholms Lokaltrafik. *Facts about SL and the County in 2011*. Stockholm County Council, Stockholm, Sweden, 2011.
24. Department of Transport and Main Roads. *Annual Report 2014-2015*. Queensland Government, Brisbane Australia, 2015.
25. Port Authority of New South Wales. *Port Services & Facilities*. Port Authority New South Wales, 2017. https://www.portauthoritynsw.com.au/sydney-harbour/port-services-facilities/.
26. Office of Transport and Safety Investigations. *Ferry Safety Investigation Report Systemic Investigation Into Ferry Close Quarter Incidents Sydney Harbour 2010–2016*. Office of Transport and Safety Investigations, 2016. https://www.parliament.nsw.gov.au/la/papers/DBAssets/tabledpaper/webAttachments/71816/170210%20Final%200715%20170915.pdf.
27. Tsai, C. H., C. Mulley, M. Burke, and B. Yen. Exploring Property Value Effects of Ferry Terminals: Evidence from Brisbane, Australia. *Journal of Transport and Land Use*, Vol. 10, No. 1, 2017, pp. 1–19.
28. New York City Economic Development Corporation. *Citywide Ferry Study*. New York City Economic Development Corporation, 2013.
29. Soltani, A., M. Tanko, M. I. Burke, and R. Farid. Travel Patterns of Urban Linear Ferry Passengers: Analysis of Smart Card Fare Data for Brisbane, Australia. Transport Research Record: Journal of the Transportation Research Board, 2015. 2535: 79–85.
30. RiverCity Gothenburg. *International Workshop*. RiverCity Gothenburg, 2011.

31.Borggren, J., and P. Strom. On the Waterfront: Studying the Development of Residences and Workplaces at Norra Alvstranden, Gothenburg, Sweden. *Local Economy*, Vol. 29, No. 4–5, 2014, pp. 429–452.
32. Sveriges Radio. *New Stockholm Ferry Line #85 has Plenty of Room to Spare*. Sveriges Radio. 2016. Available at: http://sverigesradio.se/sida/artikel.aspx?programid=2054&artikel=6353852.
33. Stockholm Country Council. *Stockholm City Plan. Summary May 2009*. City Planning Administration, Stockholm, Sweden, 2009.
34. Stockholm Chamber of Commerce. *Stockholms Handelskammare - Kampen om kajerna*. Stockholm Chamber of Commerce, Stockholm, Sweden, 2017.
35. Sandell, R. Improving the connectivity of an urban transit ferry network through integrated regular-interval timetabling. *Proc., 37th Australasian Transport Research Forum (ATRF)*, Sydney, New South Wales, Australia, 2015.
36.Hossain, M., and P. Iamtrakul. Water Transportation in Bangkok: Past, Present, and the Future on Urban Planning. *Journal of Architectural/Planning Research and Studies*, Vol. 5, No. 2, 2007, pp. 1–24.

The Standing Committee on Ferry Transportation (AP085) peer-reviewed this paper (18-06640).

Article

TRR
JOURNAL OF THE TRANSPORTATION RESEARCH BOARD

Transportation Research Record
2018, Vol. 2672(8) 901–910

DOI: 10.1177/0361198118758671
journals.sagepub.com/home/trr
SAGE

Advanced Parking Space Management Strategy Design: An Agent-Based Simulation Optimization Approach

Cong Zhao[1,2], Siyu Li[1,2], Wei Wang[1,2], Xinghua Li[1,2], and Yuchuan Du[1]

Abstract

In this paper, we present a bi-level approach to determine optimal parking management in large parking lots with multiple user classes and different types of parking spaces. The upper level of the model aims to maximize the social benefits of the local parking system, such as the system searching time and walking time by drivers. The lower level is a simulation system constructed using the multi-agent modeling environment NetLogo. However, it is essential to understand how drivers choose their parking location, then we can describe drivers' movements and responses to different parking management strategies in the micro-simulation system. This study presents the results from a stated preference experiment on the choice behavior of drivers in Shanghai, China. The model attributes are walking distance, traveling time and uncertainties towards finding a vacant parking spot, which were selected based on a literature review. A simulated annealing algorithm is adapted to solve this bi-level simulation optimization model. The approach is applied to optimize parking management in the campus of Tongji University, China. The results show that the implementation of different duration limits is useful to adapt the time-varying demand with different user classes. The approach provides a powerful tool for strategically designing parking management, and for assessing parking policies and parking information systems.

Parking in downtown or residential areas has become a major headache for both commuters and managers in most large cities, this is mainly attributed to the imbalance between parking demand and supply (*1*). Parking management has therefore become just as important as congestion control. Previous studies have investigated various aspects of parking management and have mainly focused on parking problems in the traffic network. However, large parking lots (e.g., airport terminals or shopping centers) have different types of parking spaces (e.g., spatial location or parking charge) and multiple user classes (e.g., value of time or parking duration). Searching for an available parking spot and walking to the destination in large parking lots can often constitute an appreciable fraction of the total travel time (*2*). Therefore, advanced parking space management strategies should be proposed and implemented to improve operational efficiency. For example, parking duration affects the supply side by changing the turnover of parking spaces. Motorists also perceive this effect in their choice decision by noting that a parking search might be short if all spaces are turned over frequently (*3*). Advanced parking management strategies also include parking information, reservation and navigation, aiming to help drivers find parking spaces quickly (*4*). Yang et al. found that an appropriate combination of reserved and unreserved parking spots can temporally relieve traffic congestion at the bottleneck and hence reduce the total system cost based on an improved bottleneck model (*5*).

Currently, there are three general approaches that are used for designing and optimizing parking management: empirical, analytical, and multi-agent (MA) simulations tools. Shoup, based on the review of 16 studies of mostly American and European cities, found that the average time to search for a curbside parking space ranged between 3.5 min and 14 min, with searching flow reached 30% in business areas during rush hour (*2*). Ayala et al. found out that every year in Chicago, vehicles travelled 63 million miles to find a vacant space to park, which generates 48,000 tons of carbon dioxide into the environment (*6*). Studies like these are often collected for local projects for an area or a city, and it is therefore difficult to draw generalized conclusions from them.

[1]Key Laboratory of Road and Traffic Engineering of the Ministry of Education, Tongji University, Shanghai, China
[2]National Institute of Innovation and Development, Tongji University, Shanghai, China

Corresponding Author:
Address correspondence to Yuchuan Du: ycdu@tongji.edu.cn

In the notable analytical studies, economics, macroscopic traffic models and traffic assignment models are the major contributors to parking management. Bifulco evaluated various parking policies using a static stochastic traffic assignment model (*7*). Arnott and co-authors extended the bottleneck model to analyze the temporal-spatial equilibrium of curbside parking and have developed a series of economic analyses of downtown parking over the years (*8, 9*). Gallo et al. proposed an assignment model on urban networks to simulate parking choice, and the impact of a parking search on traffic congestion (*10*). Boyles et al. also described an equilibrium formulation for incorporating parking searches into traffic network assignment models (*11*). However, the model is based on stationary-state conditions, and cannot describe the dynamics of the system (i.e., time-varying conditions).

Models have played a limited role in the analysis of urban parking policy and practices, with some notable exceptions. Much of the modeling literature is theoretical in nature and has not been applied to real-world situations (*12*). Much of the policy-oriented work, in turn, hardly makes use of the potential offered by state-of-the-art modeling techniques (*13*). Against this background, it is proposed that an agent-based model could be used to simulate urban parking policy scenarios and analyze their impacts from a user and public policy perspective. MA simulation tools are widely used to simulate travel behavior for a large number of users. These tools allow inputs such as a non-homogeneous network and personal preferences. Bonsall and Palmer used PARKIT parking choice simulator to model car off-street parking behaviors of drivers (*14*). PARKAGENT is an agent-based model that was used to simulate the on-street parking behavior of drivers in the city of Tel Aviv (*15*). To have a proper design, understanding car drivers' movement across parking lots plays a decisive role (*16*). These studies have focused on simulating the behaviors of the whole parking process or the choice of parking facility. Vo et al. developed a multi agent-based simulation tool to demonstrate its capability of studying driver movements across parking lots, that can investigate the efficiency of the parking based on vehicle travel time and parking occupancy indicators (*17*). Among micro-simulation programs, the multi agent-based modelling simulation environment of NetLogo has shown its advantages for behavioral simulation (*4, 17*). NetLogo is an open source software and is simple but powerful, and users can create their own models (*18*). These advantages make it become a promising method for transportation researchers.

This study aims to develop a micro-simulation of drivers' movements in the parking lot to investigate the connection between micro level behaviors of individuals, and macro-level patterns coming from their interactions. It is a spatially explicit, agent-based model in a parking lot, which is based on a direct representation of every driver, and simulates the whole parking process, including driving into the parking area, searching, and exiting the parking place after a variable period of time. A bi-level simulation optimization framework for the optimal design of parking management strategies in large parking area is then presented (*19, 20*). Simulation optimization refers to determining the best values of some decision variables for a real system whose performance is evaluated by the output of a simulation model (*21*). The approach has been tested in the real case of Tongji University, China, determining the optimum management strategy and associated social cost. The article ends with a section that enumerates the most important conclusions reached.

Basic Considerations

Bi-Level Simulation Optimization System

In this paper, a bi-level simulation optimization approach for parking space level is propsed, as shown in Figure 1. The upper level is the optimization model and algorithm running with Matlab, the lower level is a multi agent simulation based on a Netlogo environment.

The multi agent-based simulation system includes four modules; spatial structure of parking area, dynamic/static parking demand, driver behavior and operational measures from the upper level. The measures for parking spaces include dynamic differential parking pricing, duration limit and identity restriction. The upper level model is detailed as follows:

$$\max_{x}/\min F\left(y, x\left(y\right)\right) \tag{1}$$

Subject to

$$g_i\left(y, x\left(y\right)\right) \leq 0,\ i = 1, \ldots, p_1 \tag{2}$$

$$h_j\left(y, x\left(y\right)\right) \leq 0,\ j = 1, \ldots, p_2 \tag{3}$$

In which:

g_i, h_j = the external constraints, respectively;
$x\left(y\right)$ = the results of parking simulation system;
y = decision variables.

Factors Influencing the Driver Behavior

Parking movement at the parking lot includes two steps, that is, route choice for parking strip, and route choice for parking space. These two steps are complex as they are influenced by many factors (*14–16, 22, 23*).

The route choice for parking strip will be influenced by three major factors.

- Travel time to the desired parking strip;
- Walking time from the parking strip to the desired destination;
- Guidance information signs (parking occupancy probability in the parking strip) and the queuing time at parking entrance.

Similar to the route choice for parking strip, the choice for parking space is also affected by many factors.

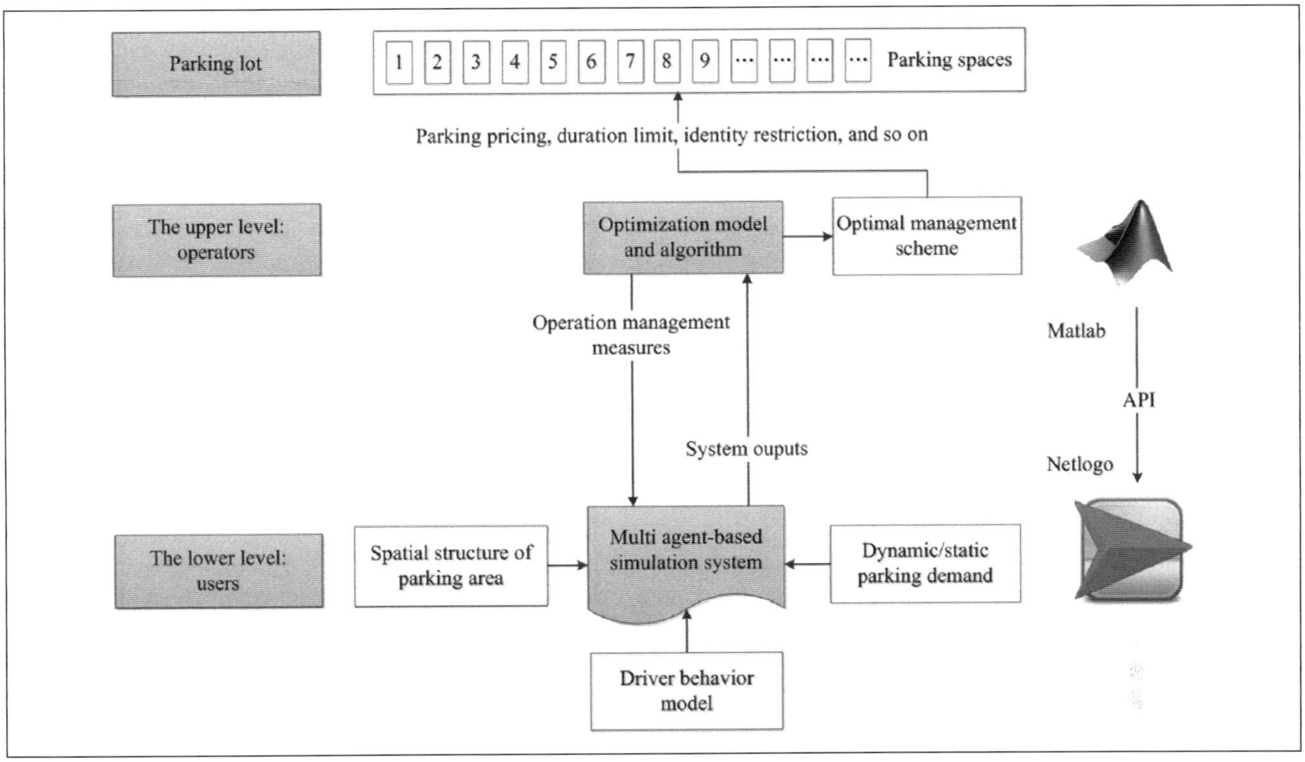

Figure 1. The framework of the bi-level simulation optimization system.

- Walking time between the location of the parking space and the final destination, driver tends to choose the parking space which is closest to the final destination;
- Parking charges or duration limit;
- Uncertain parking availability.

Model Formulation

Car Advance

The micro-simulation model works in a discrete time and space, at each time-step every vehicle can make a move, the size of which is determined by the vehicle's speed. The cars' movement is considered using car following models. The speed of a vehicle was set as no >30 km/h in the parking lots. In case the length of a parking place or the time-step are changed, all model calculations are automatically adjusted to the new values.

Follow the study of Benenson, the velocity of each car is low during a parking search (*15*). As was recorded during trips with drivers, a driver decreases his/her velocity to 20–25 km/h when starting to estimate the state of parking in the area. The speed is further reduced to 10–12 km/h when the driver starts searching parking places ahead with the aim of parking in one of them (*24*). However, to account for the interaction between parking cars, the model drivers adjust their movements in response to the car in front of them. Before advancing the parking-length interval, a driver checks whether the interval is free or not; in the latter case, the advancement is interrupted. The order in which the cars advance is established anew at random at every iteration.

Route Choice

It is assumed that the model driver possesses some knowledge of the parking lot network. When approaching a junction, the driver has to decide which direction to take to advance towards the destination. In the model, the driver's decision is governed by a logit-type formula, which is expressed as:

$$d_n = \frac{\exp\{-(\xi + \theta u_n)\}}{\sum \exp\{-(\xi + \theta u_n)\}}, \forall n \in \{1, 2, \ldots, N\}, \quad (4)$$

In which

N = the set of feasible travel directions;

u_m = the travel disutility;

ξ = the bias parameter of different travel directions;

θ = the importance of the travel disutility perception for the direction choice decision.

The travel disutility u_m is determined by three factors, travel time to the desired parking strip, walking time from

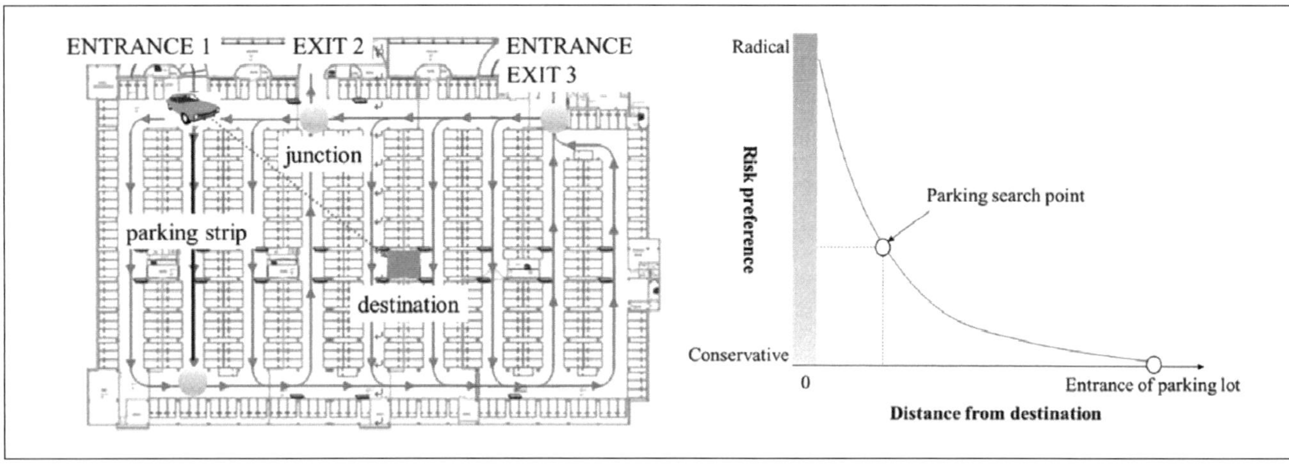

Figure 2. The relationship between the searching point and drivers' risk preference.

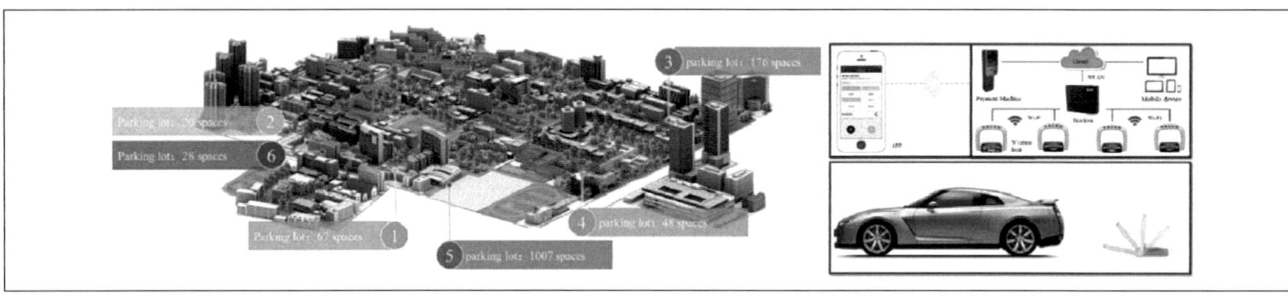

Figure 3. SiPing campus of Tongji University, Shanghai, China.

the parking strip to the desired destination and guidance information signs (parking occupancy probability in the parking strip), and the queuing time at the parking space.

Searching and Parking Behavior

The rules of agent behavior in the model depend on the drivers' risk preference when driving towards the destination from the distance d_s. The driver's behavior at this stage: decreasing speed to 25 km/h and continue driving towards the destination, then parking at the first available parking space. d_s is dependent on the drivers' risk preference, as shown in Figure 2.

Case Study

To demonstrate the capability of the developed-simulation tool and evaluate the effects of the parking management strategy on the parking efficiency, a case study on the campus of Tongji University, China, is replicated. The campus of Tongji University is located in the central area of Shanghai, where parking is difficult for both commuters and visitors, this is mainly attributed to the imbalance between parking demand and supply. The parking spaces in the campus are equipped with an intelligent parking lock, see Figure 3. It is intended that a duration limit should be set on parking spaces to increase the turnover, so the following three questions need to be answered:

1. How much duration limit do the parking spaces need?
2. How long is the limit duration?
3. Where are the locations of these duration-limit spaces?

Parking Choice Behavior Model

To understand drivers' parking location choice a series of stated preference experiments were conducted in the form of questionnaires, as shown in Figure 4. In this section, the attribute set and levels are described, where the set includes attributes accounting for the probability of acquiring a vacant parking spot upon arrival. The collected data set of 367 successful respondents is discussed in terms of descriptive statistics. This stated preference data is used to estimate various parking location choice models.

The attributes and levels included in the final design are summarized in Table 1. The final design for the choice situations was based on the priors derived by the second round of the pilot study. Again, combinations of levels and model structures were tested using the D-error as a benchmark. As

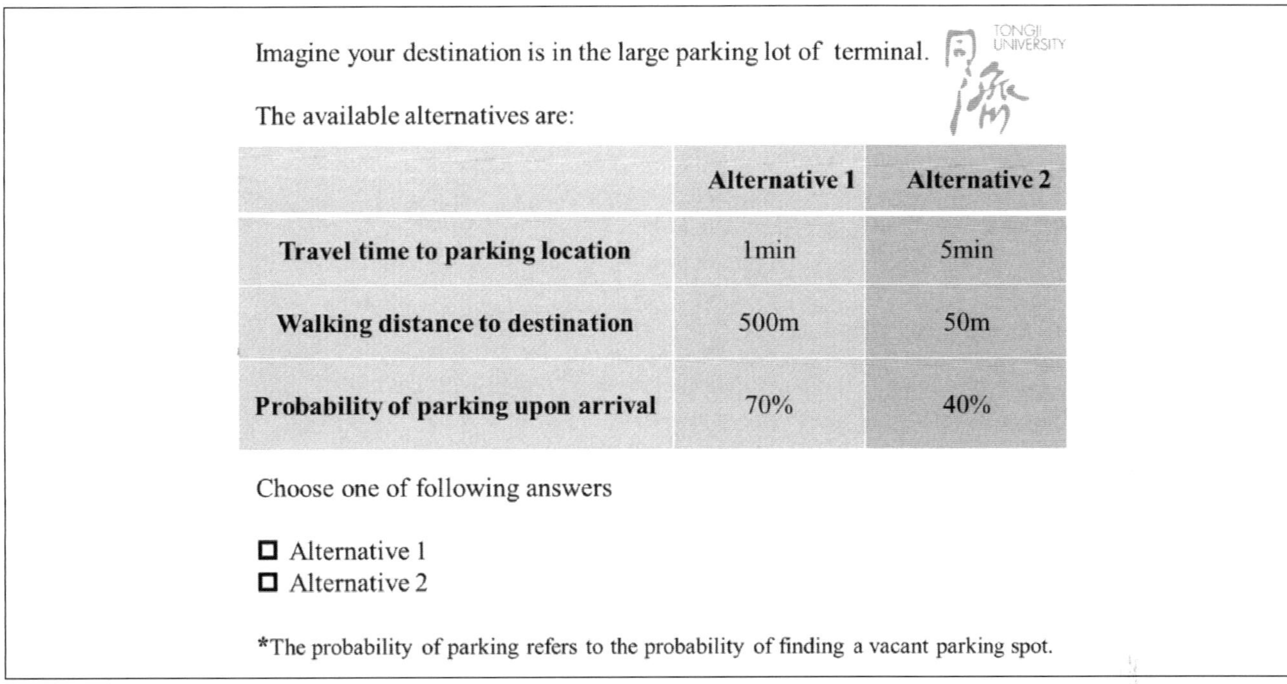

Imagine your destination is in the large parking lot of terminal.

The available alternatives are:

	Alternative 1	Alternative 2
Travel time to parking location	1min	5min
Walking distance to destination	500m	50m
Probability of parking upon arrival	70%	40%

Choose one of following answers

- ☐ Alternative 1
- ☐ Alternative 2

*The probability of parking refers to the probability of finding a vacant parking spot.

Figure 4. Screenshot of choice situation in survey.

Table 1. Survey Attributes and Values in Final Design

Attributes	Level values
Travel time to parking location (min)	1,5,10,15
Walking distance to destination (m)	50,200,500,800
Probability of parking upon arrival (%)	10,40,70,100

the priors of the model were fixed, the level values within the range of the level values used for the model calculation resulted in different models. It was found that for the probability attributes, a design of three levels would have better results in acquiring the most information (given the number of attributes and the number of choice situations). The D-error of the final design is 0.01.

Survey respondents were surveyed in the parking lots of shopping malls, airports, Universities, and so on. From 460 total respondents, 367 respondents were found to have completed the survey successfully. A binary logit model is estimated based on the stated preference experiments. Table 2 shows the results of the final model. Here all coefficients have the expected sign, where walking distance and travel time are negative and parking availability upon arrival are positive.

Statistical Models for Empirical Arriving and Departure Data

In this section, the empirical data is collected of an intelligent parking system located in Tongji University to describe parking characteristics for the demand inputs of a micro-simulation model. The intelligent parking system can record the time when users check in and check out. The data studied in this article were recorded during May 9, 2016, to June 5, 2016. This 28-day data set contains 53,305 data items.

In this article, he characteristics of the stochastic variables arriving time, parking duration, and parking volume are analyzed. The proposed distribution model, the Gaussian mixture model (GMM), is used to preserve the global characteristics of drivers while containing the information related to drivers' variability. GMM has been widely applied in the fields of data mining, machine learning and image processing. It is proven that the GMM algorithm is a mature and convenient approach to efficiently estimating the parameters. GMM can be used to yield significant clustering patterns and estimate the parameters that can characterize the behaviors of different groups of drivers based on the variables of arriving time and time duration.

$$p(x_i) = \sum_{j=1}^{M} \alpha_j N_j(x_i; \mu_j, \sigma_j) \quad (5)$$

In which

M = the number of clusters, $j = 1, \ldots, M$;

μ_j, σ_j = associated parameters that represent the mean and variance of the observation;

x_i = the probability density function of the observation;

α_j = the probability of x_i being in cluster j and $\sum_{j=1}^{M} \alpha_j = 1$;

Table 2. Estimated Discrete Choice Models

Coefficient/Constant	Estimate	Robust asymptotic standard error	Wald χ^2	Sig.
Constant β_0	−0.224	0.195	1.320	0.251
Travel time β_1	−0.0108	0.0051	4.484	0.034
Walking distance β_2	−0.0524	0.135	0.151	0.698
Probability β_3	0.0764	0.0102	56.103	0.000

The prediction accuracy of the model is: 73.1%;

$-2\left[L(0)-L(\bar{\beta})\right](\chi^2)=12.836$;

Cox & Snell R^2 = 0.278; Negelkerke R^2 = 0.337; Mcfadden = 0.241.

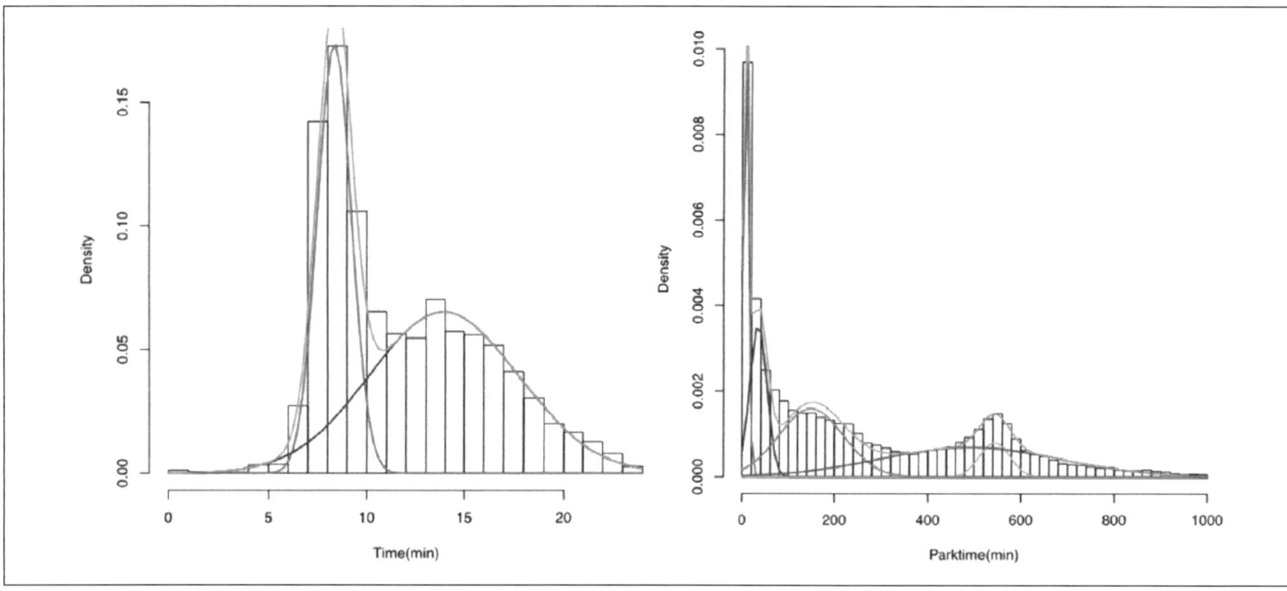

Figure 5. The estimated distribution of the commuters' arrival time and parking duration.

$N_j(x_i;\mu_j,\sigma_j)$ = the probability density function of a Gaussian distribution,

$$N_j(\mathbf{x};\mu_j,\sigma_j)=\frac{1}{\sqrt{2\pi}\sigma}\exp\left[-\frac{1}{2}\left(\frac{x-\mu_j}{\sigma_j}\right)^2\right]. \quad (6)$$

The expectation maximization (EM) algorithm was applied to estimate the probability α_j, expectations μ_j, and variance $\sigma_j, j=1,\ldots,M$, from a series of data samples. To boost estimation accuracy, an extension of the original EM algorithm, which is called the competitive expectation maximization (CEM) algorithm, is used in this article (*25*). The CEM algorithm takes advantage of some competitive mechanisms to automatically choose the number of component distributions.

Figure 5 illustrates the aforementioned two patterns of commuters, respectively. The empirical observed arriving rates are plotted by histograms. The green solid curves illustrate the estimated distribution of the GMM, and the distributions of each component Gaussian model are plotted with different colors. It can be seen that a two-component GMM is accurate enough to model their arrival patterns. Indeed, these two main components stand for the arrival peaks in the morning and afternoon, respectively. As with the parking duration, it can be seen that a five-component GMM is accurate enough to model their duration patterns.

Multi Agent Simulation Model

A multi agent-based simulation model of the large parking garage in Tongji University was constructed based on the Netlogo environment, which has double floors and 1007 parking spaces with two entrances and exits, four exit points for pedestrians are located near the entrance, as shown as Figure 6. Cars can only exit near the entrance point. Each space is equivalent to one patch in the simulation. The parking space has a square-shape with a dimension of 5.5 × 5.5

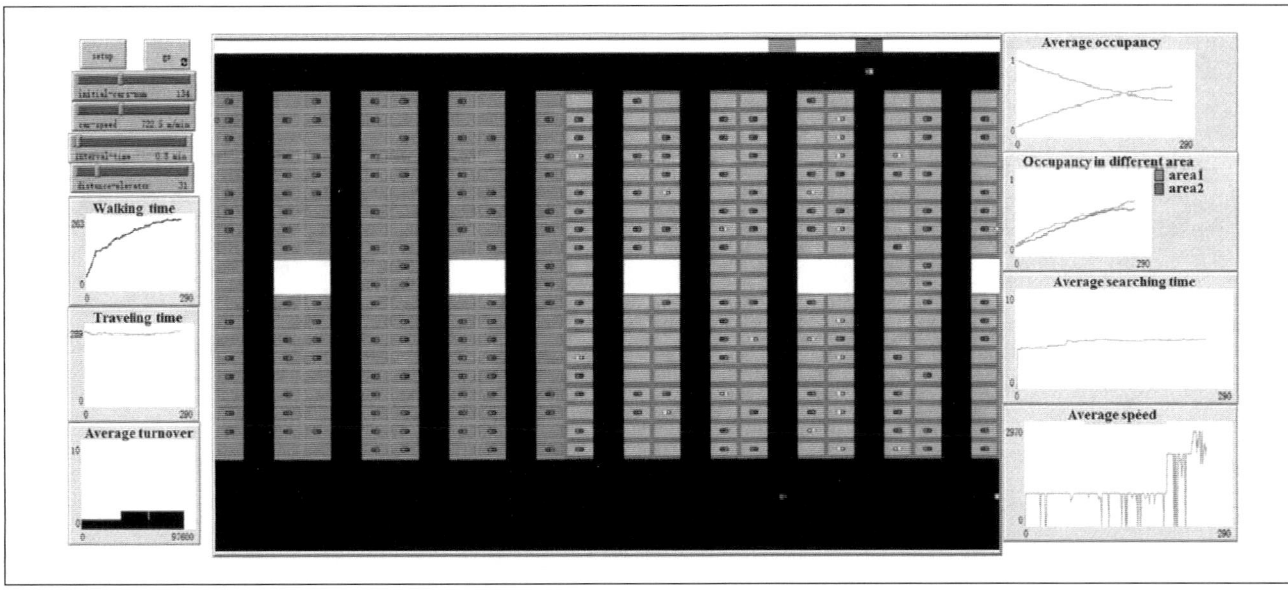

Figure 6. Simulation interface of the parking simulation.

meters which was modified from the recommendation of the Urban Plan Institute. This dimension is designed as an input button in the simulation interface.

The simulation has been modelling by measured the data of parking demand and parking choice behavior. Then a smart parking system data (average time occupancy of each parking space, average daily turnover rate) was used to verify the validity of the simulation, as shown in Figure 7. The results show that the two groups of data of the average daily turnover rate can be considered from the same distribution by a t-test, sig = 0.8870. The results of the average time occupancy show that two groups of data can be considered from the same distribution by t-test, sig = 0.7067. Therefore, it can be demonstrated that the micro parking simulation platform can effectively describe the actual parking behavior, and the platform can provide a scientific assessment tool for parking management optimization, see Table 3.

Simulation Optimization

The parking lot is divided into i zones, $i = \{1, 2, 3, \ldots, 18\}$, selecting m of them as duration limit zone, duration time is t hours, $t = \{0.5, 1, 1.5, 2, 2.5, 3, 3.5, 4, 4.5, 5, 5.5, 6\}$. The objective function of the upper model is the optimal social benefit (traveling time, searching time, and walking distance). The function is as follows:

$$Minimize\ U(T_t, T_s, T_w) = \eta_1 T_t + \eta_2 T_s + \eta_3 T_w \quad (7)$$

In which

$U(T_t, T_s, T_w)$ = objective function;

T_t = system traveling time;

T_s = system searching time;

T_w = system walking time;

η_1, η_2, η_3 = the values (RMB/h) of the traveling time, searching time, and walking time, respectively, $[\eta_1, \eta_2, \eta_3] = [40, 60, 80]$.

The solution algorithm is simulated annealing (SA). The SA algorithm in this study uses initial temperature (T_0), final temperature (T_f), temperature reduction rate (α) and the acceptance criteria for the non-improving solutions (P) as the main heuristic parameters. T_0 states the temperature at which the algorithm will start. In the beginning current temperature (T) is set to T_0. At each iteration current temperature is reduced by the ratio α. When T is below T_f the algorithm stops. When the new solution does not improve the best solution, it may be accepted with a condition depending on the value P. The steps of the proposed SA algorithm can be seen in the following.

Step 1: Set parameters T_0, T_f and α.
Step 2: Generate an initial solution (S).
Step 3: Update best solution: Best Solution = S.
Step 4: Set current temperature as the initial temperature: $T = T_0$.
Step 5: Main body of the SA:

while $T > T_f$

Create eighbor solutions: use swap, reversion and insertion with a 1/3 probability if cost of new solution < = cost of S, accept new solution: S = new solution;

else

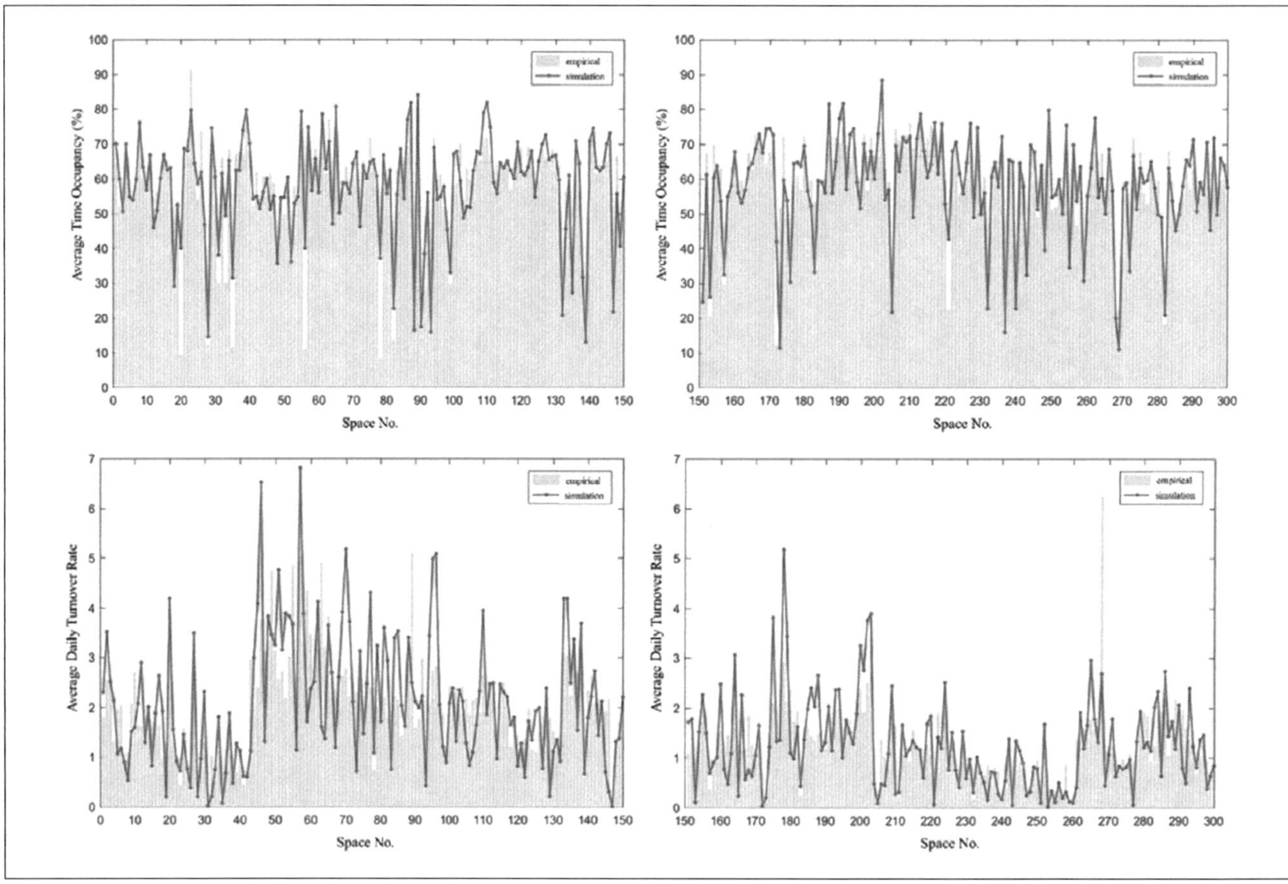

Figure 7. Comparative analysis of empirical data and simulation outputs.

Table 3. The Results of Simulation Optimization.

Total number of iterations	Convergent times	The optimal parking duration limit zone	Optimal parking limit time
80	73	{1,3,4,5,9,14}	2.5 h

Table 4. Comparison of Parameters Before and After Optimization

	Average travel time (s)	Average searching time (s)	Average walking distance (m)
Current situation	192	85	225
Parking duration limit	166	46	170

delta= (cost of new solution) – (cost of S)

P = exp(-delta/ T);

Generate a random number between 0 and 1

if random number < = P

Accept new solution: S = new solution;

if (cost of S) < = (cost of best solution) and vehicle capacities are not exceeded

Update best solution: best solution = S;

$T = \alpha * T$.

Table 4 shows that the average travel time of the parking system was reduced by 13.5%, the average searching time of the parking system reduced by 45.9%, and the average walking diatance of the parking system reduced by 13.5%. It is demonstrated that setting parking duration limit zones can effectively improve service quality and system efficiency.

Conclusion

In this study, a bi-level simulation optimization approach to design and evaluate parking management strategies in large parking lots was proposed, which can deal with multiple user classes and different types of parking spaces. The upper level of the model aims to maximize the parking system benefit, whereas the lower level is a simulation system constructed using the multi-agent modeling environment NetLogo.

In particular, amicro-simulation tool was developed that is a promising platform to simulate the characteristics of the parking layout and the drivers' behaviors. In this study, the interaction between the drivers, as well as between the drivers and the parking lot when the drivers are searching parking spaces was simulated. Therefore, this can be applied to study the factors influencing the parking behavior of drivers, and the efficiency of parking usage in the parking lot.

Our case study demonstrate that the proposed method is effective for designing and evaluating parking management strategies. The results show that the implementation of differential duration limits is useful to effectively cater to the time-varying demand with different parking needs. The approach provides a powerful tool for strategically designing parking management strategies and evaluating various parking policies.

Acknowledgments

This work was supported by a research grant (16DZ20702) from Shanghai Science and Technology Committee and a research grant from Transportation Department of Jiangxi Province. The work of the last author was supported by Program for Changjiang Scholars and Innovative Research Team in University.

References

1. Wenneman, A., K. M. N. Habib, and M. J. Roorda. Disaggregate Analysis of Relationships Between Commercial Vehicle Parking Citations, Parking Supply, and Parking Demand. *Transportation Research Record: Journal of the Transportation Research Board*, 2015. 2478: 28–34.
2. Shoup, D. C. *The High Cost of Free Parking: Updated Edition*. Routledge, Abingdon-on-Thames, 2017.
3. Sun, Y., W. Fan, and P. Schonfeld. Static Parking Choice Model with Consideration of Parking Duration. *Transportation Research Record Journal of the Transportation Research Board*, 2016. 2543: 134–142.
4. Chen, Z., Z. Xu, M. Zangui, and Y. Yin. Analysis of Advanced Management of Curbside Parking. *Transportation Research Record: Journal of the Transportation Research Board*, 2016. 2567: 57–66.
5. Yang, H., W. Liu, X. Wang, and X. Zhang. On the Morning Commute Problem with Bottleneck Congestion and Parking Space Constraints. *Transportation Research Part B Methodological*, Vol. 58, No. 4, 2013, pp. 106–118.
6. Ayala, D., O. Wolfson, B. Xu, B. DasGupta, and J. Lin. Pricing of Parking for Congestion Reduction. *Proc., 20th International Conference on Advances in Geographic Information Systems*, Redondo Beach, CA, USA,ACM, New York, 2012, pp. 43–51.
7. Bifulco, G. N. A Stochastic User Equilibrium Assignment Model for the Evaluation of Parking Policies. *European Journal of Operational Research*, Vol. 71, No. 2, 1993, pp. 269–287.
8. Arnott, R., and J. Rowse. Downtown Parking in Auto City. *Regional Science and Urban Economics*, Vol. 39, No. 1, 2009, pp. 1–14.
9. Arnott, R., and E. Inci. The Stability of Downtown Parking and Traffic Congestion. *Journal of Urban Economics*, Vol. 68, No. 3, 2010, pp. 260–276.
10. Gallo, M., L. D'Acierno, and B. Montella. A Multilayer Model to Simulate Cruising for Parking in Urban Areas. *Transport Policy*, Vol. 18, No. 5, 2011, pp. 735–744.
11. Boyles, S. D., S. Tang, and A. Unnikrishnan. Parking Search Equilibrium on a Network. *Transportation Research Part B: Methodological*, Vol. 81, 2015, pp. 390–409.
12. Inci, E. A Review of the Economics of Parking. *Economics of Transportation*, Vol. 4, No. 1, 2015, pp. 50–63.
13. Marsden, G. The Evidence Base for Parking Policies: A Review. *Transport Policy*, Vol. 13, No. 6, 2006, pp. 447–457.
14. Bonsall, P., and I. Palmer. Modelling Drivers' Car Parking Behaviour Using Data from a Travel Choice Simulator. *Transportation Research Part C: Emerging Technologies*, Vol. 12, No. 5, 2004, pp. 321–347.
15. Benenson, I., K. Martens, and S. Birfir. PARKAGENT: An Agent-Based Model of Parking in the City. *Computers, Environment and Urban Systems*, Vol. 32, No. 6, 2008, pp. 431–439.
16. Van der Waerden, P., H. Timmermans, and A. N. R. da Silva. The Influence of Personal and Trip Characteristics on Habitual Parking Behavior. *Case Studies on Transport Policy*, Vol. 3, No. 1, 2015, pp. 33–36.
17. Vo, T. T. A., P. van Der Waerden, and G. Wets. Micro-Simulation of Car Drivers' Movements at Parking Lots. *Procedia Engineering*, Vol. 142, 2016, pp. 100–107.
18. Tisue, S., and U. Wilensky. Netlogo: A Simple Environment for Modeling Complexity. Presented at International Conference on Complex Systems, Boston, USA, 2004.
19. Guo, W., Y. Zhang, M. Xu, Z. Zhang, and L. Li. Parking Spaces Repurchase Strategy Design Via Simulation Optimization. *Journal of Intelligent Transportation Systems*, Vol. 20, No. 3, 2016, pp. 255–269.
20. Chen, Z., F. He, Y. Yin, and Y. Du. Optimal Design of Autonomous Vehicle Zones in Transportation Networks. *Transportation Research Part B: Methodological*, Vol. 99, 2017, pp. 44–61.
21. Du, Y., C. Zhao, X. Zhang, and L. Sun. Microscopic Simulation Evaluation Method on Access Traffic Operation. *Simulation Modelling Practice and Theory*, Vol. 53, 2015, pp. 139–148.

22. Thompson, R. G., and A. J. Richardson. A Parking Search Model. *Transportation Research Part A: Policy and Practice*, Vol. 32, No. 3, 1998, pp. 159–170.
23. Van der Waerden, P., A. W. J. Borgers, and H. J. P. Timmermans. Travelers Micro-Behavior at Parking Lots-A Model of Parking Choice Behavior. Presented at 82nd Annual Meeting of the Transportation Research Board, Washington, D.C., 2003.
24. Carrese, S., E. Negrenti, and B. B. Belles. Simulation of the Parking Phase for Urban Traffic Emission Models. In *TRISTAN V-Triennial Symposium on Transportation Analysis*, Guadeloupe, 2004.
25. Zhang, B., C. Zhang, and X. Yi. Competitive EM Algorithm for Finite Mixture Models. *Pattern Recognition*, Vol. 37, No. 1, 2004, pp. 131–144.

The Standing Committee on Transportation Demand Management (ABE50) peer-reviewed this paper (18-05913).

The authors take sole responsibility for all views and opinions expressed in this paper. The authors declare that there is no conflict of interest regarding the publication of this paper.

Article

TRR
JOURNAL OF THE TRANSPORTATION RESEARCH BOARD

Transportation Research Record
2018, Vol. 2672(8) 911–919

DOI: 10.1177/0361198118790126
journals.sagepub.com/home/trr
SAGE

Discrete-Time Hazard Model of the Dynamic Utilization of Parking Spaces within Park-and-Ride Lots

Bibhuti Sharma[1], Mark Hickman[1], and Carlo G. Prato[1]

Abstract

Park-and-ride (P&R) offers an alternative to the exclusive use of the car when considering the door-to-door travel experience. The intended effects of good P&R services are an increase in public transport ridership, a decrease in the vehicle-kilometers traveled, and a reduction in the number of cars in the city centers. However, no consensus exists in the literature about the effectiveness of P&R, and no consideration of parking space utilization exists when searching for the optimal location of P&R lots or analyzing P&R choices. Accordingly, this study models the utilization of 7590 parking spaces in 20 P&R lots along the major arterials around Brisbane (Australia). A discrete-time logistic regression model calculates the probability that each parking space is occupied at the end of one of 60 time intervals between 4:00 a.m. and 9:00 a.m. on a weekday. The findings from the model suggest that the probability of a parking space to be occupied increases with a larger capacity of the P&R lot, a larger number of public transport services, and a lower walking time to the platforms. Moreover, the results suggest that a parking space is more likely to be occupied in P&R lots farther from the central business district (CBD) until 8.00 a.m., but it is more probable to be occupied in P&R lots closer to the CBD from 8.00 a.m. onwards. This dynamic model of parking space utilization within a P&R lot could contribute to the evaluation of future P&R utilization, the formulation of realistic optimization models, and the estimation of dynamic activity-based models.

When considering the door-to-door travel experience, park-and-ride (P&R) proposes the combination of car and public transport as an alternative to the exclusive use of the car. The rationale behind P&R is that public transport ridership will increase, the vehicle-kilometers traveled (VKT) will decrease, and the number of cars in the city centers will diminish. Consequently, traffic externalities such as congestion delays, pollutant emissions, and parking pressure will be reduced.

Extensive literature exists about the effectiveness of P&R (*1*). Consensus exists about P&R helping increase public transport ridership for rail and bus (*2–6*). However, consensus does not exist about P&R helping to decrease the VKT, as studies have found that P&R might achieve the intended goal, but examples from Europe and the U.S. show a VKT increase because of bus-based P&R (*7–10*). The effectiveness of P&R is also questioned in light of an increase in pollutant emissions that relates to the increase in VKT as well as the fact that short trips consume a very large amount of fuel (*11*, *12*), and the "cold start" and "hot soak" phenomena happen with every car trip regardless of distance (*13*, *14*).

Extensive literature exists also about the unintended effects of P&R (*1*): (i) P&R users might substitute part of their public transport trip with a car trip (*7*, *9*, *15*); (ii) P&R users do not always use public transport after parking their vehicle (*9*, *16*); (iii) P&R users often drive a detour to reach their facility (*7*). All unintended effects relate to two other streams of research focusing, respectively, on the optimal location of P&R lots and the choice of P&R lot. The first stream has proposed optimization problems in conditions of equilibrium (*17–20*) or in linear city settings (*21–24*). Most of these studies have inherent problems in the simplification of either the travel behavior representation, with travelers not affected by the provision of P&R lots (*17–19*), or the transit network, which is either devoid of common-lines (*20*) or stylized (*22–24*). The second stream has analyzed the choice of P&R as a mode being mainly affected by travel time and parking costs (*25–27*), and the choice of P&R lot as being mainly affected by travelers' characteristics (*28–30*) and lot utilization (*31*).

[1]School of Civil Engineering, The University of Queensland, St. Lucia, Brisbane, Australia

Corresponding Author:
Address correspondence to Bibhuti Sharma: b.sharma@uq.edu.au

In both the latter two streams of research, the utilization of parking spaces is rarely considered. Optimization problems implicitly share the assumption that travelers will find available parking spaces at the locations that are evaluated as optimal, while choice models implicitly consider that the possible alternatives always have available parking spaces. Intuitively, this does not appear logical, as the unavailability of parking spaces would cause travelers to either park in nearby places or continue to their destination and thus have negative effects related to the consequent increase in parking pressure and VKT (*32*, *33*). Accordingly, there exists a need for understanding the determinants of parking space utilization within a P&R lot, as also proved by recent attitudinal surveys about P&R provision showing how travelers perceive parking availability as essential for a positive attitude toward P&R (*34*, *35*). Moreover, there exists a need for considering the dynamic nature of P&R utilization, given its proven dependence on time of day (*36–39*). Understanding the dynamic utilization of P&R lots would contribute to a better evaluation of the potential for P&R infrastructure to be used, the formulation of optimization problems that would include the possibility for travelers to detour and/or extend their car trip, and the formulation of advanced activity-based models where P&R is considered as an alternative in mode choice models and activities are time-dependent in nature (*36*, *40*, *41*).

The current study analyzed P&R utilization with a dynamic model that considers the arrival at the P&R lot over sequential time intervals as well as the characteristics of the lot itself as determinants of the utilization. A discrete-time hazard model was specified to estimate the probability that a randomly selected parking space will be available in a certain time interval. A case study was selected for South East Queensland (SEQ), an area with 170 P&R lots providing 27,000 parking spaces to travelers and an average utilization rate around 80%. Interestingly, Brisbane (the capital city of Queensland) has more P&R spaces per inhabitant than the remaining large Australian cities (Adelaide, Melbourne, Perth, Sydney).

The remainder of the paper is organized as follows: the next section introduces the case study and the formulation of the model used to represent the dynamic parking availability in P&R lots; then, the model specification and estimates are presented; last, results are discussed and future research lines are drawn.

Methods

Case Study

The case study focuses on SEQ, a region with Brisbane at its core and a population of about 3 million inhabitants. In the region, there are over 170 P&R facilities that provide 27,000 parking spaces including 21,500 along the rail network. The average utilization rate, defined as the ratio between occupancy and capacity, is about 80%, and the capacity of the P&R facilities is exceeded in 35% of the facilities, although some facilities are under-utilized.

The case study is therefore suitable for the investigation of the dynamic utilization of parking spaces within the existing facilities. In the region, it is possible to observe both under- and over-utilization (*30*), and the utilization has been shown to be endogenous to the selection of the P&R lot (*31*). Most relevantly for the purpose of this study, there exists a clear dynamic in the utilization of the parking spaces, as shown in Figure 1 (*32*). While the previous study analyzed "how" the utilization changes over time (*32*), this study looks at the factors underlying the utilization of parking places in the P&R lots of the region.

Given the aim to understand the determinants of dynamic P&R utilization, a survey of P&R lots in SEQ was performed in the second half of August 2015. The selection of the period was motivated by August being a winter month where there are no school holidays affecting the commute in the region. The survey was conducted only on weekdays and focused on 20 P&R lots that covered the region around the city of Brisbane (see Figure 2). The selection of the lots was motivated by the intention to look at facilities that are more utilized according to a previous P&R study, that are along all the major radial corridors for accessing the Brisbane central business district (CBD) by either rail or busway, and that are at different distances from the Brisbane CBD. Because of our interest in higher utilized lots, this does generate some bias in our study over the full population of P&R lots.

Surveyors accessed each of the P&R lots 10 minutes prior to the arrival of the first public transport service at/near the lot, and left the P&R lot when either the lot was full or no new car arrivals were observed for 15 minutes. Surveyors stood near the entrance of each of the P&R lots and recorded the number and arrival times of the vehicles entering the lots. The registration of the vehicles was from the first four digits of the license plate: on the one hand, this allowed avoidance of double counting circulating vehicles; on the other hand, exact identification of the vehicle was avoided, thus maintaining the privacy of P&R users. As the location of the parking space has been argued to be relevant because travelers might prefer to park closer to the public transport platforms (*33*, *36*, *42*), seven zones were defined according to their walking distance in minutes from the platforms: (A) 0–2 min; (B) 2–5 min; (C) 5–8 min; (D) 8–10 min; (E) 10–12 min; (F) 12–15 min; (G) over 15 min. The surveyors logged the zone where each vehicle parked and then recorded the total number of parking spaces and the final number of occupied parking spaces at the conclusion of the survey. It should be noted that other surveys of P&R lots were

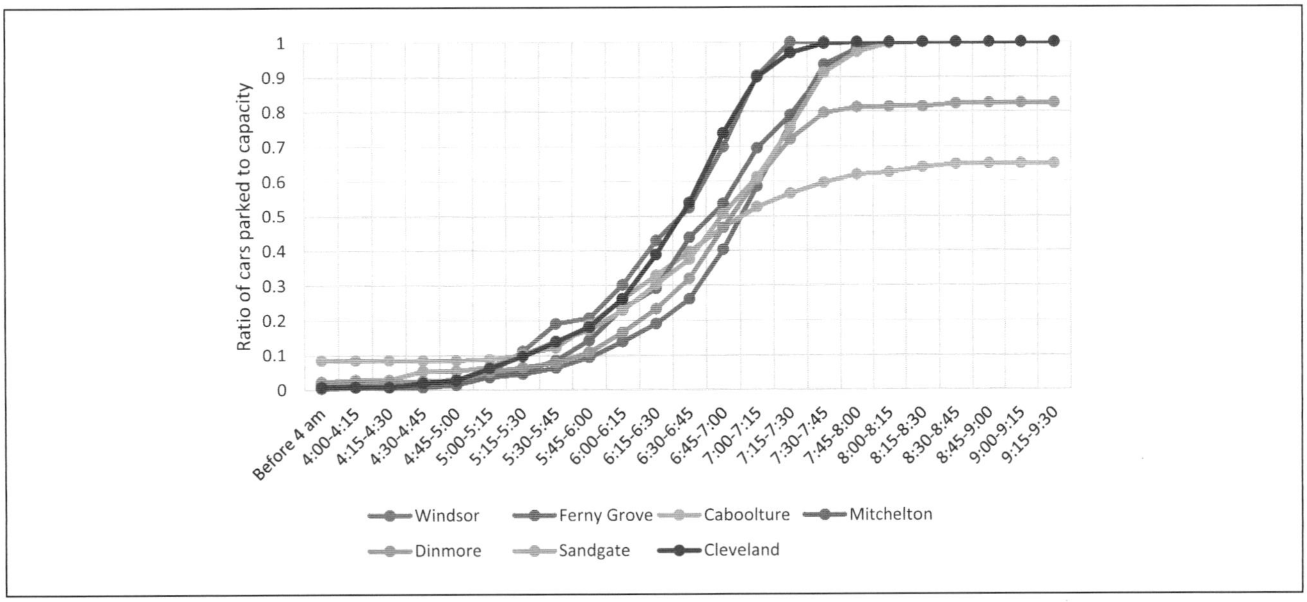

Figure 1. Utilization of P&R lots in SEQ.

performed (*5, 32, 37, 43*), but this survey has the unique features of a larger amount of P&R lots being surveyed and of the location of the parking spaces with respect to the public transport platforms being recorded.

Model

Given the aim of the study, survival analysis was applied to model the time-dependent utilization of the P&R lots. In this study, the *event* is the occupation of a parking space and corresponds to a *failure* in typical survival analysis language. All the events have an equal *risk* from the beginning of the study (i.e., all parking spaces have an equal probability of being occupied), and all events are non-repeatable (i.e., all parking spaces can be filled only once). The observation of events in this study revealed that the second condition was fulfilled, as no parking space was occupied more than once in the surveying period. Also, in this study the *risk set* consists of the parking spaces that remain eligible to be occupied in the period. Initially, all the parking spaces form the risk set and the total number of available parking spaces in the 20 P&R lots is equal to 7590.

The *start time* is 4:00 a.m. for all P&R lots, as the common start is the one that places all the population within the survey to have the same probability of experiencing the event. Then, the time is discretized in 5-minute intervals until 9:30 a.m. (i.e., 4:00–4:05 a.m., 4:05–4:10 a.m., ..., 9:25–9:30 a.m.). The *end time* is formally 9:30 a.m., although the actual time differs across parking spaces and P&R lots because either the lot is full before 9:30 a.m. or no arrivals were registered for 15 minutes. The data in this study are *censored* (*44*), with left-censored spaces corresponding to the parking spaces that were occupied before the start time, and right-censored spaces consisting of the parking spaces that were not occupied after the end time. Specifically, no left censoring was considered in this study and right censoring was explicitly counted at the end of the survey.

Given these definitions, a discrete-time hazard model was specified. In discrete-time models where time is segmented into uniform discrete intervals, exit probabilities in each time interval can be estimated by using logistic regression (*45*). In this model, an event is treated as a Bernoulli trial with two possible outcomes for each time interval: 1 if the parking space is occupied, and 0 if the parking space is censored at the end of the interval.

If a parking space i with P covariates $z_{1ij}, z_{2ij}, \ldots, z_{pij}$ has not been occupied by the end of time interval j-1, the conditional probability h_{ij} for i to be occupied in time interval j is defined as:

$$h_{ij} = P(T_i = j | T_i \geq j, Z_{1ij} = z_{1ij}, Z_{2ij} = z_{2ij}, \ldots, Z_{pij} = z_{pij}) \quad (1)$$

The probability h_{ij} can be reparametrized (*46*) and therefore can be expressed as having a logistic dependence on the time intervals and the covariates (*47*):

$$h_{ij} = \frac{1}{1 + \exp(-((\alpha_1 D_1 + \alpha_2 D_2 + \ldots + \alpha_J D_J) + (\beta_1 Z_{1ij} + \beta_2 Z_{2ij} + \ldots + \beta_P Z_{Pij})))} \quad (2)$$

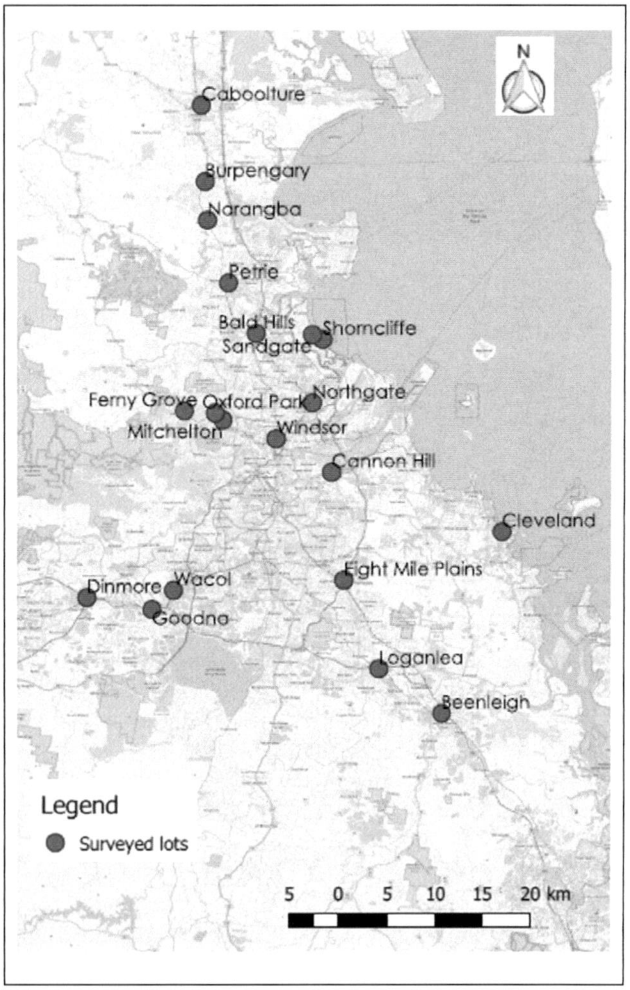

Figure 2. P&R lots surveyed.

where the dummy variables (D_1, D_2, ..., D_J) index the J time intervals, (α_1, α_2, ..., α_J) are intercept parameters to be estimated, and (β_1, β_2, ..., β_J) are slope parameters to be estimated. It should be noted that, when the values of the covariates are equal to zero, the intercept parameters (α_1, α_2, ..., α_J) capture the baseline hazard in each time interval.

A logistic transformation of both sides of Equation 2 gives the following functional form:

$$\ln\left(\frac{h_{ij}}{1-h_{ij}}\right) = (\alpha_1 D_1 + \alpha_2 D_2 + \ldots + \alpha_J D_J) + (\beta_1 Z_{1ij} + \beta_2 Z_{2ij} + \ldots + \beta_P Z_{Pij}) \quad (3)$$

The intercept and the slope parameters of the logistic discrete-time hazard model in Equations 2 and 3 are estimated by maximizing a likelihood function that is a product of two components: (i) the probability for occupied spaces; (ii) the probability for empty spaces.

The probability that an uncensored parking space i will be occupied in time interval j, given that the parking space was not occupied at the end of time interval j-1, is expressed as:

$$P(T_i = j) = P(T_i = j|T_i \geq j)P(T_i \neq j-1|T_i \geq j-1)\ldots P(T_i \neq 1|T_i \geq 1) = h_{ij}(1-h_{i(j-1)})\ldots(1-h_{i1}) = h_{ij}\prod_{k=1}^{j-1}(1-h_{ik}) \quad (4)$$

The probability that parking space i is empty at the end of the time interval is expressed as:

$$P(T_i > j) = P(T_i \neq j|T_i \geq j)P(T_i \neq j-1|T_i \geq j-1)\ldots P(T_i \neq 1|T_i \geq 1) = (1-h_{ij})(1-h_{i(j-1)})\ldots(1-h_{i1}) = \prod_{k=1}^{j}(1-h_{ik}) \quad (5)$$

Accordingly, the likelihood function is written as:

$$L = \prod_{i=1}^{N}\left(h_{ij}\prod_{k=1}^{j-1}(1-h_{ik})\right)^{\delta i}\left(\prod_{k=1}^{j}(1-h_{ik})\right)^{1-\delta i} \quad (6)$$

where N is the total number of parking spaces, $\delta_i = 1$ if parking space i is occupied and $\delta_i = 0$ if parking space i is empty.

Results

Model Specification

Given the registered vehicle arrivals between 4:00 a.m. and 9:30 a.m., and given the discretization in 5-minute intervals, the baseline hazard function captured by the parameters (α_1, α_2, ..., α_J) can be calculated. Namely, the hazard probability in any time interval can be computed as the conditional probability that a randomly selected parking space will be occupied in that time interval, given that it was not occupied in an earlier interval. Similarly, the survival probability in any time interval can be calculated as the probability that a randomly selected parking space will remain empty (i.e., survive) beyond the current time interval.

Table 1 shows the values for the baseline hazard and survival probability in the case study, while presenting every fifth time interval for brevity. The extremely small number of observations after 9:00 a.m. suggested restricting the analysis to 60 intervals between 4:00 and 9:00 a.m. The median survival is between intervals 36 and 37 (about 7:00 a.m.), and precisely it is 36.2, indicating that half of the parking spaces are occupied by about 7:00 a.m.

Figure 3 plots the estimated baseline hazard and survival probabilities to provide a visual representation of the baseline. The baseline hazard does not explain the

Table 1. Baseline Hazard and Survival Probability

Interval	Risk set	Event = 1 (occupied)	Event = 0 (empty)	Estimated hazard	Estimated survival probability
0	7590	0	7590	0.000	1.000
5	7571	9	7562	0.001	0.996
10	7476	27	7449	0.004	0.981
15	7304	55	7249	0.008	0.955
20	7023	69	6954	0.010	0.916
25	6537	127	6410	0.019	0.845
30	5760	173	5587	0.030	0.736
35	4514	330	4184	0.073	0.551
40	2928	294	2634	0.100	0.347
45	1605	217	1388	0.135	0.183
50	909	49	860	0.054	0.121
55	718	33	685	0.046	0.096
60	219	12	293	0.055	0.079

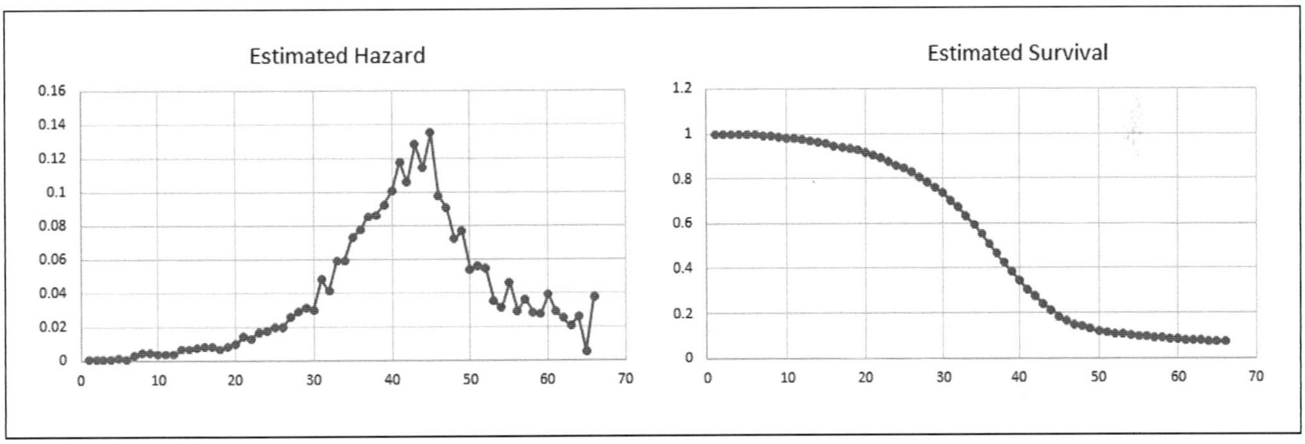

Figure 3. Plot of the estimated baseline hazard and survival probability.

factors underlying the probability of parking spaces being occupied. The specification of the model focuses on explanatory covariates that are both time independent and time dependent. After specification testing, significant time-independent covariates were the walking time within the P&R lot and the capacity of the P&R lot. The time-dependent covariates were the availability of public transport services and the distance to the Brisbane CBD as the most prominent commuting destination.

The walking time within the P&R lot was easily observed for the occupied spaces, as the surveyors registered the zone where the vehicles were parked. For the censored spaces, the walking time was assumed from the subsequent zone: for example, at the end of the survey, if all parking spaces were occupied in zone B, the next empty space was considered to be in zone C. The best model specification considered the walking time as continuous by taking the average for each zone. The capacity of the P&R lot was observed as the total number of parking spaces available at the start time, and was assumed to investigate whether P&R users have a general preference of travelers for lots of larger capacity. Accordingly, both the walking time and the capacity were defined as time-independent covariates.

The public transport service near the P&R lot was measured as the number of train and bus services within 500 m walking distance from the P&R lot in the 5-min time interval after the one when the vehicle arrived. This service measure was assumed to capture the intention of P&R users to arrive some minutes prior to the intended public transport service (*39*), and it was time dependent as the number of departing trains and buses differs for each time interval. The distance to the Brisbane CBD was measured for each P&R lot, and the best model specification considered different parameters for distance in every half hour. This distance measure was assumed to capture the fact that P&R lots might fill faster when they are close to the CBD (*48*), as well as the variation in sensitivity to distance at different times in the morning.

Table 2. Estimates of the Discrete-Time Logistic Regression Model

Variable	Estimates	Std. error	Sig.	Variable	Estimates	Std. error	Sig.
Walking time	–0.732	0.013	***	D25	–4.012	0.120	***
Capacity	0.001	0.000	***	D26	–3.996	0.120	***
Public transport services	0.125	0.007	***	D27	–3.765	0.112	***
Distance to CBD 4.00–4.30 a.m.	0.059	0.016	***	D28	–3.613	0.110	***
Distance to CBD 4.30–5.00 a.m.	0.037	0.007	***	D29	–3.549	0.108	***
Distance to CBD 5.00–5.30 a.m.	0.024	0.005	***	D30	–3.608	0.109	***
Distance to CBD 5.30–6.00 a.m.	0.028	0.004	***	D31	–3.050	0.090	***
Distance to CBD 6.00–6.30 a.m.	0.035	0.003	***	D32	–3.119	0.094	***
Distance to CBD 6.30–7.00 a.m.	0.034	0.003	***	D33	–2.755	0.087	***
Distance to CBD 7.00–7.30 a.m.	0.026	0.003	***	D34	–2.734	0.088	***
Distance to CBD 7.30–8.00 a.m.	–0.001	0.004	~	D35	–2.444	0.085	***
Distance to CBD 8.00–8.30 a.m.	–0.039	0.007	***	D36	–2.426	0.086	***
Distance to CBD 8.30–9.00 a.m.	–0.062	0.012	***	D37	–2.091	0.087	***
D1	–8.067	0.617	***	D38	–1.896	0.086	***
D2	–8.587	0.716	***	D39	–1.783	0.087	***
D3	–8.070	0.617	***	D40	–1.598	0.088	***
D4	–7.885	0.588	***	D41	–1.343	0.087	***
D5	–7.478	0.539	***	D42	–1.324	0.093	***
D6	–8.311	0.656	***	D43	–0.536	0.102	***
D7	–6.160	0.290	***	D44	–0.501	0.108	***
D8	–5.708	0.255	***	D45	–0.066	0.110	~~
D9	–5.623	0.248	***	D46	–0.356	0.126	**
D10	–5.788	0.261	***	D47	–0.206	0.134	*
D11	–5.795	0.261	***	D48	–0.337	0.149	**
D12	–5.956	0.269	***	D49	0.586	0.185	***
D13	–4.912	0.191	***	D50	0.541	0.213	**
D14	–4.962	0.193	***	D51	0.855	0.220	***
D15	–4.791	0.183	***	D52	0.986	0.229	***
D16	–4.679	0.180	***	D53	0.623	0.262	**
D17	–4.684	0.180	***	D54	0.629	0.278	**
D18	–4.918	0.190	***	D55	1.767	0.313	***
D19	–4.731	0.163	***	D56	1.457	0.352	***
D20	–4.608	0.156	***	D57	2.214	0.383	***
D21	–4.261	0.141	***	D58	2.212	0.432	***
D22	–4.326	0.141	***	D59	2.304	0.480	***
D23	–4.038	0.136	***	D60	2.926	0.441	***
D24	–4.064	0.134	***				

~ $p<1$. ~~ $p<0.8$. *$p<0.5$. **$p<0.1$. ***$p<0.01$. ****$p<0.001$. *****$p<0.0001$.

Model Estimates

Given the data and model specification, Table 2 presents the estimates for the discrete-time logistic regression model. A deviance test revealed that the baseline model had a deviance equal to 55,436.16 while the best specification model presented in the table had a deviance equal to 49,178.06, which indicates that the model is far better than the baseline one.

The estimates of the 60 parameters related to the time intervals increase progressively from the first to the last interval, suggesting that the probability of a parking space being occupied increases over time.

The estimate of the parameter related to the walking time indicates that the position of the parking space plays a role, as the further the parking space is from the public transport platforms, the lower is the likelihood of getting occupied. Considering that the parameter is −0.732 for 1-min difference in walking time, and the sample has a standard deviation of 2.01 min for the walking time. That is, any two parking spaces whose walking times are a standard deviation different would differ by the product of the two (–1.47). The exponential of this value suggests that a parking space that has 2-min more walking time than another will have about 23% less probability of being occupied with respect to that other parking space. Also, the estimate of the parameter for the capacity suggests that parking spaces within larger P&R lots are more attractive and therefore more likely to be occupied at every time interval.

The estimate of the parameter associated with the public transport services indicates that the availability of a higher number of services in the following time interval is likely to increase the probability of a parking space being occupied at each time interval. Notably, the best model specification was found for a single parameter regardless of the variation of the number of services over time. Last, the estimates of the parameters related to the distance to the CBD suggest that the probability of a parking space to be occupied increases with the distance from the CBD from 4:00 to 4:30 a.m., and so on until 8:00 a.m. However, after 8.00 a.m. the sign of the parameters suggests the opposite to be the case, namely that parking spaces closer to the CBD have a higher likelihood to be occupied.

Conclusion

This study revealed the factors underlying the probability of a parking space within a P&R lot to be occupied in a certain time interval. A discrete-time logistic regression model was applied to estimate the occupancy probability for 7590 parking spaces over 20 P&R lots that covered the major radial movements toward, and at different distances from, the Brisbane CBD.

The findings from this study indicate that capacity of the P&R lot, walking time to the platforms, number of services close to the P&R lot, and distance to the CBD are relevant factors even after controlling for the baseline probability of a parking space to be occupied over 60 time intervals between 4:00 and 9:00 a.m.

The findings about the walking time are consistent with a previous study that suggested that a later arrival time at the P&R lot implies a lower probability to get a space close to the platform, which in turn lengthens the transfer time and makes the parking space less attractive (*36*). Moreover, the variation in probability between two parking spaces separated by one standard deviation of walking time is similar to a previous value found for general parking (*42*). In other words, the relative advantage found for one parking space with respect to another within a P&R lot is comparable to the relative advantage calculated for one parking facility with respect to another.

The findings about the variation in sensitivity to distance to the CBD over time are in contrast with a previous study pointing out that parking lots in the morning peak are used in order of increasing distance from the CBD (*48*). Notably, the previous study focused on general parking. In other words, the findings from general parking cannot be generalized to the P&R context when distance to the CBD is considered. Possibly, public transport services start early at stops and stations located farther from the Brisbane CBD, and therefore the parking spaces in the P&R lots nearby are occupied earlier.

The investigation of the dynamics of parking availability at the detail of 5-min intervals suggests that there is a need for considering the dynamic nature of the occupancy of P&R lots within transport applications that consider P&R. For example, the potential use of P&R lots would be calculated more accurately by using the estimated model to forecast the occupancy of parking spaces as a function of the capacity, the public transport services, and the distance from the CBD. Similarly, the search for optimal locations of future P&R lots would be improved by relaxing the implicit assumption of availability of at least one parking space and introducing the dynamic aspect of occupation of the parking spaces. Lastly, the estimation of activity-based models that are becoming more dynamic and fine-grained in nature would certainly benefit from a detailed model able to provide an estimate of the dynamic potential occupancy of P&R lots. Further research could also consider parking fees, as all P&R lots in this case study are free of charge, but fees have been demonstrated to be relevant predictors of P&R lot choice (*25–27*).

Author Contributions

The authors confirm contribution to the paper as follows: study conception and design: Mark Hickman, Bibhuti Sharma; data collection: Mark Hickman, Bibhuti Sharma; analysis and interpretation of results: Bibhuti Sharma, Carlo Prato, Mark Hickman; draft manuscript preparation: Bibhuti Sharma, Carlo Prato, Mark Hickman. All authors reviewed the results and approved the final version of the manuscript.

References

1. Zijlstra, T., T. Vanoutrive, and A. Verhetsel. A Meta-Analysis of the Effectiveness of Park-and-Ride Facilities. *European Journal of Transport and Infrastructure Research*, Vol. 15, No. 4, 2015, pp. 597–612.
2. Kuby, M., A. Barranda, and C. Upchurch. Factors Influencing Light-Rail Station Boardings in the United States. *Transportation Research Part A: Policy and Practice*, Vol. 38, No. 3, 2004, pp. 223–247.
3. Lane, C., M. DiCarlantonio, and L. Usvyat. Sketch Models to Forecast Commuter and Light Rail Ridership: Update to TCRP Report 16. *Transportation Research Record: Journal of the Transportation Research Board*, 2006. 1986: 198–210.
4. Duncan, M. To Park or to Develop: Trade-off in Rail Transit Passenger Demand. *Journal of Planning Education and Research*, Vol. 30, No. 2, 2010, pp. 162–181.
5. Hamer, P. Analysing the Effectiveness of Park and Ride as a Generator of Public Transport Mode Shift. *Road & Transport Research*, Vol. 19, No. 1, 2010, pp. 51–61.
6. Guerra, E., and R. Cervero. Cost of a Ride: The Effects of Densities on Fixed-Guideway Transit Ridership and Costs. *Journal of the American Planning Association*, Vol. 77, No. 3, 2011, pp. 267–290.
7. Parkhurst, G. Influence of Bus-Based Park and Ride Facilities on Users' Car Traffic. *Transport Policy*, Vol. 7, No. 2, 2000, pp. 159–172.

8. Meek, S., S. Ison, and M. Enoch. Evaluating Alternative Concepts of Bus-Based Park and Ride. *Transport Policy*, Vol. 18, No. 2, 2011, pp. 456–467.
9. Mingardo, G. Transport and Environmental Effects of Rail-Based Park and Ride: Evidence from the Netherlands. *Journal of Transport Geography*, Vol. 30, 2013, pp. 7–16.
10. Duncan, M., and D. Cook. Is the Provision of Park-and-Ride Facilities at Light Rail Stations an Effective Approach to Reducing Vehicle Kilometers Traveled in a US Context? *Transportation Research Part A: Policy and Practice*, Vol. 66, No. 1, 2014, pp. 65–74.
11. Jensen, S. Driving Patterns and Emissions from Different Types of Roads. *Science of the Total Environment*, Vol. 169, 1995, pp. 123–128.
12. Chang, D., and E. Morlok. Vehicle Speed Profiles to Minimize Work and Fuel Consumption. *Journal of Transportation Engineering*, Vol. 131, No. 3, 2005, pp. 173–182.
13. Kessler, J., and W. Schroeer. Meeting Mobility and Air Quality Goals: Strategies that Work. *Transportation*, Vol. 22, No. 3, 1995, pp. 241–272.
14. de Nazelle, A., B. Morton, M. Jerrett, and D. Crawford-Brown. Short Trips: An Opportunity for Reducing Mobile-Source Emissions? *Transportation Research Part D: Transport and Environment*, Vol. 15, No. 8, 2010, pp. 451–457.
15. Parkhurst, G. Park and Ride: Could it Lead to an Increase in Car Traffic? *Transport Policy*, Vol. 2, No. 1, 1995, pp. 15–23.
16. Lindström Olsson, A. L. *Factors that Influence Choice of Travel Mode in Major Urban Areas*. Report KTH/INFRA–03/048—SE. Royal Institute of Technology, Stockholm, Sweden, 2003.
17. Horner, M. W., and S. Groves. Network Flow-Based Strategies for Identifying Rail Park-and-Ride Facility Locations. *Socioeconomic Planning Sciences*, Vol. 41, No. 3, 2007, pp. 255–268.
18. Farhan, B., and A. T. Murray. Siting Park-and-Ride Facilities Using a Multi-Objective Spatial Optimization Model. *Computers and Operation Research*, Vol. 35, No. 2, 2008, pp. 445–456.
19. Aros-Vera, F., V. Marianov, J. Mitchell, and J. P-Hub. Approach for the Optimal Park-and-Ride Facility Location Problem. *European Journal of Operations Research*, Vol. 226, No. 2, 2013, pp. 277–285.
20. Fan, W., M. B. Khan, J. Ma, and X. Jiang. Bilevel Programming Model for Locating Park-and-Ride Facilities. *Journal of Urban Planning and Development*, Vol. 140, No. 3, 2014. http://dx.doi.org/10.1061/(ASCE)UP.1943-5444.0000178.
21. Sargious, M., and N. Janarthanan. Forecasting Demand for the Park-and-Ride Mode and Determining the Optimal Location of Stations. *Canadian Journal of Civil Engineering*, Vol. 10, No. 4, 1983, pp. 695–702.
22. Wang, J. Y., H. Yang, and R. Lindsey. Modelling Park-and-Ride Service in a Linear Monocentric City. *International Journal of Urban Science*, Vol. 8, No. 1, 2004, pp. 74–89.
23. Wang, J. Y., H. Yang, and R. Lindsey. Locating and Pricing Park-and-Ride Facilities in a Linear Monocentric City with Deterministic Mode Choice. *Transportation Research Part B: Methodological*, Vol. 38, No. 8, 2004, pp. 709–731.
24. Holguin-Veras, J., W. Yushimito, F. Aros-Vera, and J. Reilly. User Rationality and Optimal Park-and-Ride Location under Potential Demand Maximization. *Transportation Research Part B: Methodological*, Vol. 46, No. 8, 2012, pp. 949–970.
25. Li, Z. C., W. Lam, S. C. Wong, D. L. Zhu, and H. J. Huang. Modeling Park-and-Ride Services in a Multimodal Transport Network with Elastic Demand. *Transportation Research Record: Journal of the Transportation Research Board*, 2007. 1994: 101–109.
26. Islam, S. T., Z. Liu, and M. Sarvi. Park-and-ride Network Design in a Bi-Modal Transport Network Optimising Network Reliability. *Proc., Australasian Transport Research Forum (ATRF)*, Sydney, New South Wales, Australia, 2015.
27. Pineda, C., C. E. Cortés, P. Jara-Moroni, and E. Moreno. Integrated Traffic-Transit Stochastic Equilibrium Model with Park-and-Ride Facilities. *Transportation Research Part C: Emerging Technologies*, Vol. 71, 2016, pp. 86–107.
28. Clayton, B., E. Ben-Elia, G. Parkhurst, and M. Ricci. Where to Park? A Behavioural Comparison of Bus Park and Ride and City Centre Car Park Usage in Bath, UK. *Journal of Transport Geography*, Vol. 36, 2014, pp. 124–133.
29. Pang, H., and A. Khani. Modeling Park-and-Ride Location Choice of Heterogeneous Commuters. *Transportation*, Vol. 45, No. 1, 2016, pp. 71–87. https://doi.org/10.1007/s11116-016-9723-5.
30. Sharma, B., M. Hickman, and N. Nassir. Park-and-Ride Lot Choice Model Using Random Utility Maximization and Random Regret Minimization. *Transportation*, 2017, pp. 1–16.
31. Sharma, B., M. Hickman, and N. Nassir. Park-and-Ride Lot Choice Model with Corrected Endogeneity. Presented at 14th World Conference on Transport Research, Shangai, China, 2016.
32. Shirgaokar, M., and E. Deakin. Study of Park-and-Ride Facilities and their Use in the San Francisco Bay Area of California. *Transportation Research Record: Journal of the Transportation Research Board*, 2005. 1927: 46–54.
33. Lai, L., and A. S. Shalaby. Cellular Automaton Simulation of Vehicle Dynamics in Park-and-Ride Facilities (No. 07-1755). Presented at 86th Annual Meeting of the Transportation Research Board, Washington, D.C., 2007.
34. He, B., W. He, and M. He. The Attitude and Preference of Traveler to the Park & Ride Facilities: A Case Study in Nanjing, China. *Procedia – Social and Behavioral Sciences*, Vol. 43, 2012, pp. 294–301.
35. Olaru, D., B. Smith, J. Xia, and T. Lin. Travellers' Attitudes towards Park-and-Ride (PnR) and Choice of PnR Station: Evidence from Perth, Western Australia. *Procedia – Social and Behavioral Sciences*, Vol. 162, 2014, pp. 101–110.
36. Tsang, F. W. K., A. S. Shalaby, and E. J. Miller. Improved Modeling of Park-and-Ride Transfer Time: Capturing the Within-Day Dynamics. *Journal of Advanced Transportation*, Vol. 39, No. 2, 2005, pp. 117–137.
37. Hamid, N. A. Utilization Patterns of Park and Ride Facilities among Kuala Lumpur Commuters. *Transportation*, Vol. 36, No. 3, 2009, pp. 295–307.

38. Chen, Z., J. C. Xia, B. Irawan, and C. Caulfied. Development of Location-Based Services for Recommending Departure Stations to Park and Ride Users. *Transportation Research Part C: Emerging Technologies*, Vol. 48, 2014, pp. 256–268.
39. Sharma, B., M. Hickman, and N. Nassir. A Study on the Utilization of Park-and-Ride Lots in South East Queensland. Presented at 38th Australasian Transport Research Forum, Melbourne, Australia, 2016.
40. Davidson, B., P. Vovsha, and J. Freedman. New Advancements in Activity-Based Models. Presented at 33rd Australasian Transport Research Forum, Adelaide, Australia, 2011.
41. Zockaie, A., M. Saberi, H. S. Mahmassani, L. Jiang, A. Frei, and T. Hou. Activity-Based Model with Dynamic Traffic Assignment and Consideration of Heterogeneous User Preferences and Reliability Valuation. *Transportation Research Record: Journal of the Transportation Research Board*, 2015. 2493: 78–87.
42. Caicedo, F., F. Robuste, and A. Lopez-Pita. Parking Management and Modeling of Car Park Patron Behavior in Underground Facilities. *Transportation Research Record: Journal of the Transportation Research Board*, 2006. 1956: 60–67.
43. Chen, C., and Y. Xie. Modeling the Safety Impacts of Driving Hours and Rest Breaks on Truck Drivers Considering Time-Dependent Covariates. *Journal of Safety Research*, Vol. 51, 2014, pp. 57–63.
44. Kleinbaum, D. G., and M. Klein. *Survival Analysis: A Self-Learning Text*. Springer Science & Business Media, New York, 2012.
45. Washington, S. P., M. G. Karlaftis, and F. L. Mannering. *Statistical and Econometric Methods for Transportation Data Analysis*. CRC Press, Oxford, 2010.
46. Cox, D. R. Regression models and life tables. *Journal of the Royal Statistical Society, Series B*, Vol. 34, No. 2, 1972, pp. 187–202.
47. Singer, J. D., and J. B. Willett. It's about time: Using discrete-time survival analysis to study duration and the timing of events. *Journal of Educational and Behavioral Statistics*, Vol. 18, No. 2, 1993, pp. 155–195.
48. Arnott, R., A. De Palma, and R. Lindsey. A Temporal and Spatial Equilibrium Analysis of Commuter Parking. *Journal of Public Economics*, Vol. 45, No. 3, 1991, pp. 301–335.

The Standing Committee on Transportation Demand Management (ABE50) peer-reviewed this paper (18-00992).

Article

TRR
JOURNAL OF THE TRANSPORTATION RESEARCH BOARD

Transportation Research Record
2018, Vol. 2672(8) 920–929

DOI: 10.1177/0361198118773893
journals.sagepub.com/home/trr
SAGE

Quantifying the Effect of On-Street Parking Information on Congestion Mitigation using a Driving Simulator

Samira Ahangari[1], Celeste Chavis[1], Mansoureh Jeihani[1], and Zohreh Rashidi Moghaddam[1]

Abstract

This study investigates travelers' reactions to different types of information, in deciding their parking choice behavior and its effect on circulation time, through a driving simulator and a stated preference (SP) survey. In the simulator-based driving experiments, we develop a 3.47 mile^2 network in the Chinatown area of Washington, D.C., with different scenarios of traffic, driving conditions, and information provision. The parking information is provided using a variable message sign (VMS) and mobile application. In all scenarios, participants can choose from three parking options with different prices and different walking distances to the destination (Verizon Garage, 11th St. Garage, and on-street parking). A sample of 76 participants with diverse socio-economic backgrounds who in total conducted 636 experiments is used. We applied a multinomial logistic regression model, linear regression, and *t*-test to analyze the collected data. We conclude that types of information and age are important determinants of drivers' parking choice and compliance behaviors. In addition, the results show that the existence of information decreases the circulation time. In addition, the parking choice behavior revealed through the driving simulator is shown to be significantly different from that stated in the survey questionnaire.

Like advanced traveler information systems (ATISs) that have been used widely to diminish congestion in urban areas, information about the availability of on-street parking is progressively significant as the additional time spent cruising for a space increases congestion. Such detailed and variable levels of parking information are becoming more freely available to the consumer with the assistance of mobile apps and sensing technologies.

When the number of vehicles travelling to urban areas increases, demand for parking spots also increases, which leads drivers to circulate in search of an optimum parking space. The time spent by drivers searching for a parking space is related to the availability of parking information. Such a search for parking occurs when an individual reaches their destination, aims to park, and drives around an area until they discover a spot that meets their particular needs for that specific trip. Therefore, a parking search consists of additional travel by drivers that occurs only upon reaching a destination (*1*). Van Ommeren et al. (*1*) estimated that because of the time spent searching for parking, commuting time increases by approximately 20%. Decreasing the time that vehicles spend searching for parking places not only alleviates the traffic congestion problem, but also the environmental burden. Studies by Waraich et al. (*2*) and Shoup (*3*) found that at certain moments in the day up to 50% of the traffic is related to cruising for parking. A noteworthy factor of the parking search is the length of time required by drivers to locate an empty parking space that efficaciously fulfils their requirements. In addition, Arnott et al. (*4*) investigated the parking search from an economic perspective and proposed an intelligent algorithm that allows drivers to choose a parking lot with the maximum probability of getting a space. Pullola et al. (*5*) modeled the availability of a parking lot using a Poisson process and proposed an algorithm that helps dynamically locate the best available parking lots closest to the driver's destination.

In addition, Caicedo (*6*) advanced a demand assignment model to estimate the benefits of manipulating information with the objective of decreasing the time and distances involved in searching for parking. The

[1]Department of Transportation and Urban Infrastructure Studies, Morgan State University, Baltimore, MD

Corresponding Author:
Address correspondence to Celeste Chavis: celeste.chavis@morgan.edu

model allows the evaluation of parking control measures and for quantification of the effect of information management. It analyzes real information to evaluate different possibilities for distributing information to influence the distance covered in reaching a parking space, the time spent searching, and the walking distance between the parking space and the destination. According to Rye et al. (*7*), if managing information about parking availability lessens the time spent circulating and seeking a free space, it could indicate a prevailing demand for a management tool. Guo et al. (*8*) provided numerous significant insights into the factors of the dissemination of travel time under the impact of on-street parking. Arnott et al. (*9*) recommended a simple model of parking congestion that focused on a driver's search for an available parking space in a spatially identical metropolis.

Tsamboulas (*10*) estimated drivers' behavior alteration from an at-present chosen parking location, based on a combination of increased/decreased walking time and higher/lower parking charges, and found a more expensive parking choice to be acceptable if it reduced walking time. Gallo et al. (*11*) developed a simulation-based parking assignment model with a hierarchical structure simulating parking location choice on a trip, including a cruising and a walking layer; studies indicated that 28–45% of traffic congestion in urban areas is because of drivers looking for a place to park. Chaniotakis et al. (*12*) analyzed drivers' choice behavior among alternative parking locations under uncertain parking availability and search times. As parking probabilities are shown to be important determining factors in drivers' parking decisions, this reaffirms that parking guidance and reservation systems could be useful in reducing the amount of traffic searching for parking. This ensures further research into drivers' acceptance of and willingness to pay for such systems.

In a similar concentration, targeted at reducing unnecessary travel time and its influence on traffic networks, Thompson et al. (*13*) developed a model to curtail parking queues and, thus, waiting time by distributing extra demand to parking facilities with spare capacity and diverting drivers from centrally located parking toward parking facilities located closer to drivers' travel origin. It is apparent from the literature that well-designed and effective parking guidance information can help advance application and management of parking resources (*6*). Moreover, guidance systems considerably improve the probability of finding an available parking space and alleviate obstruction by drivers/visitors unfamiliar with the city center. They also decrease the queues in front of parking garages and reduce total amount of vehicle-miles traveled (particularly in the city centers), average trip time, energy consumption, and air pollution (*14*).

Network traffic flow and congestion increase when vehicles search for on-street parking, which creates environmental and economic externalities. Parking search is an under-researched area, mostly addressed in recent years, although it has extensive adverse influences on individual drivers and society. To be more specific, cars searching for parking create 14% of traffic density and generate a 50% growth in congestion-related time loss (*15*). Likewise, in a review of 16 studies conducted in 11 cities in the USA, an estimated average of 30% of traffic was searching for parking, with 8.1 minutes the average search time (*3*).

In the same manner, according to Arnott et al. (*6*) the stock of cars cruising for parking increases traffic congestion in urban areas. Some researchers offered the downtown parking model that incorporated traffic congestion and saturated on-street parking from an economic perspective. As a result, in a car-dominant era, parking policies and management are essential for discouraging traffic from entering city centers, ensuring parking space availability and shifting passengers to other travel modes. Several studies investigated the effect of parking policies on traffic flow (*16*, *17*) and traffic demand management (*18*, *19*).

Improvements in technology are influencing the parking search, with search time decreasing in urban areas as technological parking solutions are introduced. For example, the SFPark scheme in San Francisco (USA) installed embedded sensors in on-street parking spaces to specify real-time parking occupancy of each space, and employed dynamically flexible parking charges according to the demand for on-street metered parking spaces (*20–22*). Websites, such as Parkopedia.co.uk, Car Parks 4U.com, and Confused.com, give drivers information in advance as to the location of parking places around a desired destination, decreasing the need for parking search upon arrival.

Some systems claim to convey real-time information of vacant parking spaces to citizens, for instance, FastPrk (www.fastprk.com). Such a dynamic technological advance in the form of mobile phone (smartphone) applications helps motorists locate, and possibly reserve and pay for, an available parking space remotely by advising drivers as to where empty spaces are located in real time and thus decreasing the requisite parking search (*23*). A 2014 scheme in Chicago (USA), ParkChicago, allowed drivers to pay for parking by mobile phone without the need to locate a meter, show a valid ticket, or return to the meter to extend parking time. Others generate mobile applications to handle parking-meter payments, as ParkMobile does in the UK (www.parkmobile.com/uk). From the first solutions for mobile payment of parking spaces such as Mobipay (*24*) to the very latest smart parking applications like Central Parking Systems, most of the research activities focused on the deployment of sensors (*25*). Mathew et al. (*26*) developed a smartphone app that reserved a parking space in an off-street lot. Drivers who had reserved a parking space beforehand decreased search time by 40%, discovered an available space during peak times, decreased overall travel time, and reduced emissions since vehicles were no longer slowing down or accelerating to find a parking space. This type of smartphone application has the potential to achieve similar impacts for on-street parking searches. Grazioli et al. (*27*) demonstrated a modular, service-oriented smart parking system, which comprises web applications for parking operators and end-users, along with mobile applications for end-users and parking controllers.

To control the District's curbside and parking properties, the District Department of Transportation (DDOT) uses an app called ParkDC that reduces the circulation time for finding parking by providing real-time information about parking availability (*28*). An advanced street parking system called PhonePark—using the GPS, accelerometer, and Bluetooth sensors on a traveler's mobile phone in conjunction with geospatial data—can automatically detect when and where the traveler parked his or her car, and when he or she left a parking slot (*29*). Likewise, a smart parking algorithm that helps drivers find a parking facility based on features such as access time, walk time, parking charge, and traffic congestion has the potential to be applied to on-street parking (*30*). Nowadays, significant efforts are being made to develop systems to help drivers on the road, providing them with diverse types of pertinent information. Soon, the request for street parking guidance services is expected to grow rapidly. Because of low power, small size, and the low cost of wireless sensor networks, these have great scope to provide an ideal solution for street parking services (*31*).

As a final point, information related to the number of obtainable parking spaces could be displayed on the main roads, streets, and intersections, or it could be distributed through the application. To quantify the impact of on-street parking information on congestion mitigation, this study seeks to link information provision on parking availability and driving circulation time using a real-world network in downtown Washington, D.C. and traffic and parking availability information provided by the District Department of Transportation.

Methodology

To study the relationship between parking information and circulation time, this study uses a driving simulator with VR-Design Studio software (formerly UC Win-Road) by Forum8 (*32*). Since stated responses in survey questionnaires may differ from participants' parking choices in the real world, combining surveys with simulator-based driving data can provide more accurate parking information models. In this study, drivers' socio-economic characteristics and attitudes were collected using online surveys before participating in the driving simulator session to ensure a diverse sample of participants.

In the driving simulator, we developed a 3.45 mi^2 network of Chinatown in Washington, D.C., including streets, signs, trees, vehicles, and buildings (Figure 1). Then, using different scenarios, we asked participants to drive from one of two origins (the intersection of 14th St. and Pennsylvania St. or the intersection of K St. and 15th St.) to the destination (Verizon Center, which since this study, has been renamed the Capital One Arena) and find parking. Participants had three parking alternatives to choose from, including Verizon Garage located at 6th St. between G St. and H St.; 11th St. Garage at F St. across 11th St. with a walking distance time of 7 minutes to the destination; and on-street parking throughout the network with a walking distance between 0 and 20 minutes.

Twelve scenarios were developed in this study to simulate different traffic conditions (a.m. peak, p.m. peak, off-peak, and weekend), trip purposes (purpose of trip requires you to arrive at a set time (time matters or not) and different levels of parking availability information (no-information, VMS, and mobile application). As shown in Table 1, Scenarios 1–4 are the base scenarios in which no information is provided to benchmark participants' circulation time in the absence of information. Scenarios 5–8 have Variable Message Signs (VMS) that provide information for available on-street parking in each intersection (Figure 2*a*). Scenarios 9–12 have information provided via mobile phone application (Figure 2*b*). It should be noted here that the scenarios were not used sequentially for a participant in the driving simulator runs; all 12 scenarios were divided into two groups (A and B) of six scenarios and each participant chose to drive between one and six scenarios.

In addition to the simulator-based driving experiments, all participants were asked to fill out three different survey questionnaires sequentially. The first questionnaire focused on participants' socio-economic characteristics including gender, age, education level, job status, driving license type, income level, and household size. The second survey, specific to the study area, elicited some attitudes toward different information types and parking choices. The third survey was conducted after the driving simulator experiment and addressed the acceptance and usefulness of parking availability information provisions.

Data Collection

To obtain an unbiased sample, participants were solicited by distributing information related to the proposed simulator-based driving experiments within the Morgan

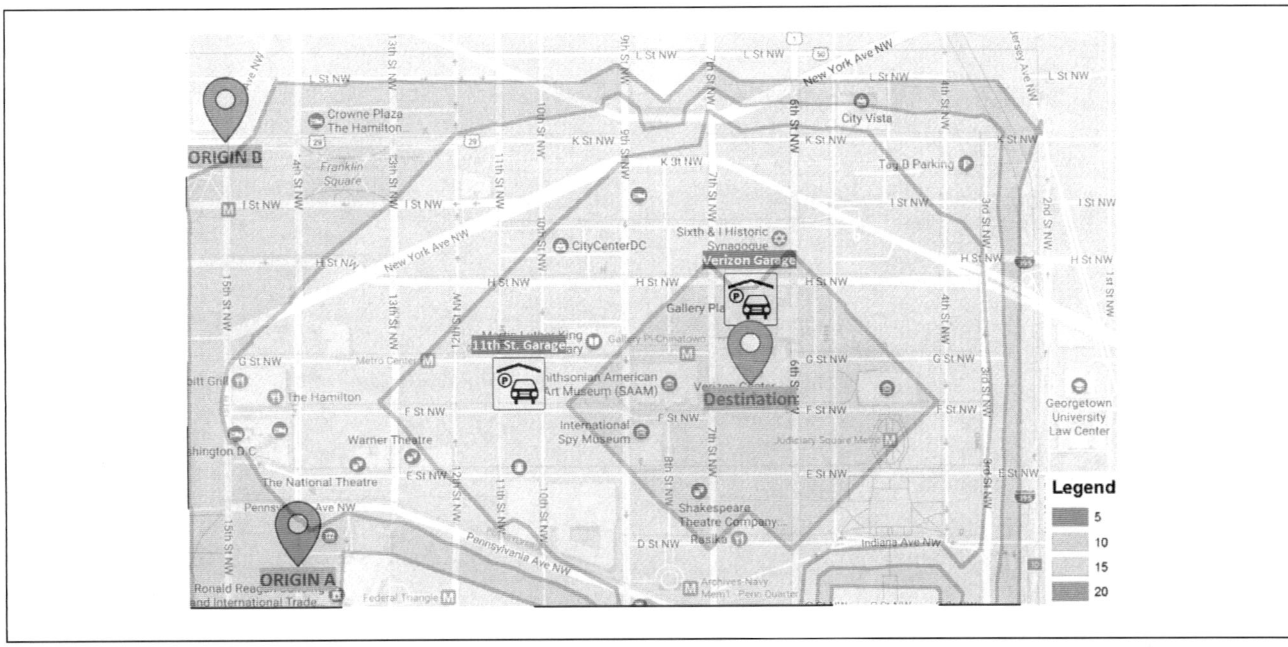

Figure 1. Study area (Washington, D.C.) with walking distance time to destination.

Table 1. Scenarios Specification

Scenario	Origin[a]	Information type	Time matters	Traffic type	Parking price: Verizon Garage	Parking price: 11th St. Garage	Parking price: On-street parking
1	K St. and 15th St.	No Information	Yes	p.m.-peak	$40	$20	$2
2	14th St. and Pennsylvania	No Information	No	a.m.-peak	$30	$20	$2
3	14th St. and Pennsylvania	No Information	No	Off-peak	$30	$20	$2
4	K St. and 15th St.	No Information	Yes	Weekend	$30	$20	$ 0
5	K St. and 15th St.	VMS	Yes	p.m.-peak	$40	$20	$2
6	14th St. and Pennsylvania	VMS	No	a.m.-peak	$30	$20	$2
7	14th St. and Pennsylvania	VMS	No	Off-peak	$30	$20	$2
8	K St. and 15th St.	VMS	Yes	Weekend	$30	$20	$ 0
9	K St. and 15th St.	App	Yes	p.m.-peak	$40	$20	$2
10	14th St. and Pennsylvania	App	No	a.m.-peak	$30	$20	$2
11	14th St. and Pennsylvania	App	No	Off-peak	$30	$20	$2
12	K St. and 15th St.	App	Yes	Weekend	$30	$20	$ 0

[a]The destination in each scenario is the Verizon Center.

State University campus and the Baltimore Metro area, in addition to posting the information on advertising websites such as Craigslist and Facebook. Seventy-six participants were recruited, who accomplished 636 successful driving experiments, with an average of 8.3 scenarios per subject. The participants were asked to fill out the first two survey questionnaires before their driving session. They filled out the third survey questionnaire immediately after completing their driving experiments in their session. The 76 participants' socioeconomic characteristics are presented in Table 2.

To simulate real-world conditions, a set of rules was developed and specified to the participants before the driving experiment. For example, for the parking alternative chosen in each scenario, a distinct amount of money (Verizon Garage: $1.50; 11th St. Garage: 75 cents; and on-street parking: 20 cents) was deducted as a parking price from their payment of $4 per scenario driven. Although this amount in reality is $40 for the Verizon garage, $20 for the 11th St. garage and $2–$3 for on-street parking (on-street parking is free on Sundays), it was reduced to the above-mentioned amounts so that participants were

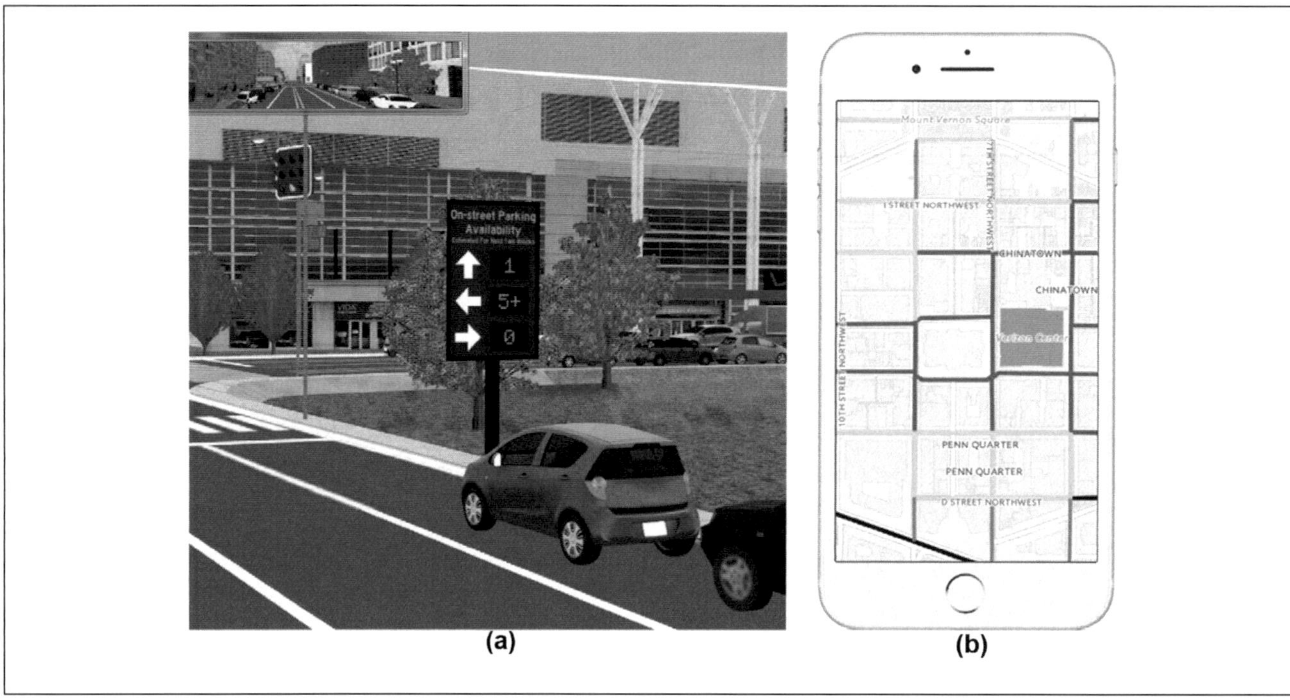

Figure 2. (*a*) Example of VMS in the driving simulator network and (*b*) Example of mobile application displaying parking conditions (Green indicates 5 + parking spots available, Yellow 2–4 spots available, and Red 1 spot available).

Table 2. Participants' Socioeconomic Characteristics

Characteristics	Percentages
Gender	
Male	57.9
Female	42.1
Age	
18–25	44.7
26–35	35.5
36–45	10.5
46–55	2.6
56–65	6.6
Education level	
High school or less	31.6
Associate degree	21.1
Bachelor degree	26.3
Post-graduate	21.1
Work status	
Unemployed	35.5
Work part-time	34.2
Work full-time	30.3
Income level	
< $20K	23.3
$20K–$30K	15
$30K–$50K	25
$50K–$75K	16.7
$75K–$100K	13.3
> $100K	6.7
Household size	
1	27
2	24.3
3	20.8
≥ 4	21.6

adequately compensated for their time. As stated earlier, some scenarios involved time matters, requiring the participant to be at the destination by a certain time. In these cases, there was a random deduction as a penalty if the subject could not reach the destination on time. The deduction varied between zero and three dollars to simulate the risk of getting to the destination late. Participants were also penalized for non-compliance with traffic signs (such as speed limit signs, running a red light), and for accidents and crashes to ensure driving realism. To make the study more realistic, after each scenario when the participants found parking, they were asked to leave the simulator and walk around the building for the duration required to reach their destination. See Figure 1 for the GIS map containing walking time buffers to the destination.

Descriptive and statistical analyses were performed using the collected data to understand travelers' circulation times and information provision. Also, a parking choice model was developed.

Results and Discussion

Attitudinal Analysis

According to the survey questionnaires, 56.45% of the participants were familiar with the study area and the other 43.55% were not. On-street parking was the stated preferred choice of 52.07% participants, while 38.82% and 9.11% indicated the 11th St. Garage and Verizon Garage as their preferred choice, respectively.

Table 3. Stated Parking in Surveys vs. Revealed Parking Choice in Driving Simulator

		Revealed			
		Verizon Garage	11th St. Garage	On-street parking	Total
Stated	Verizon Garage	0.37%	0.92%	7.82%	9.11%
	11th St. Garage	1.59%	3.91%	33.33%	38.82%
	On-street parking	2.13%	5.24%	44.70%	52.07%
	Total	4.09%	10.06%	85.85%	100%

The second questionnaire presented participants with situations similar to the designed driving simulator scenarios to analyze how their responses on surveys differ from their choice in the driving simulator.

Table 3 shows the parking choice (stated) by participants in the surveys versus the parking chosen (revealed) in the driving simulator; they are significantly different; only 48.98% of participants had identical stated and revealed parking choices. For example, 0.37% of participants chose the Verizon Garage in the survey questionnaire and chose this parking in their driving simulator experience, while 0.92% of participants stated they'd choose the Verizon Garage but chose the 11th St. Garage and 7.82% chose on-street parking instead. On the other hand, among the 4.09% of participants who chose the Verizon Garage in the driving simulator experience, 0.37% had stated they would choose it, while 1.59% had stated they would choose the 11th St. Garage and 2.13% on-street parking. Also, while 52.07% of participants stated they would choose on-street parking, only 44.70% chose it. These parking choice differences could be attributed to the difference between the real price of parking (which was provided in the survey) and the driving simulator scenarios price of parking.

t-test Analysis

To find the driving travel time, the start time and end time of each scenario for each participant were extracted from the simulator's log files; such log files include different information such as: start time, end time, speed, road, crash, and other related information. The average Driving Travel Time by types of parking information was compared for all 12 scenarios using a *t*-test. Parking information causes a significant reduction in travel time. The result of the first *t*-test that compared Driving Travel Time with No-Information to VMS-Information (Table 4) shows that the Driving Travel Time dropped from 11.25 min to 9.27 min due to the information available via VMS. The result of the second *t*-test that compared Driving Travel Time between No-Information and App-Information (Table 4) shows that the Driving Travel Time dropped from 11.26 min to 7.90 min due to the information available via App.

In a subset of scenarios (scenarios 1, 4, 5, 8, 9, and 12), participants were told that they had to arrive at their destination by a set time (i.e., time mattered). Thus, a comparison between Driving Travel Time and Types of Information was done for such scenarios. The result of the *t*-test that compared Driving Travel Time with No-Information to VMS-Information (Table 4) shows that the Driving Travel Time fell significantly from 11.21 min to 9.19 min due to the information available via VMS. The result of the *t*-test that compared Driving Travel Time between No-Information and App-Information (Table 4) shows that the Driving Travel Time dropped from 11.21 min to 7.79 min due to the information available via App. Such a decrease in Driving Travel Time can be because of the App-Information since participants pay attention to it before starting their trip, find a suitable parking space and go directly to the available space. The majority of participants do not pass by the destination before parking when they have access to App-Information. However, when there is no information or information provided via VMS, participants go to the destination and then circulate to find parking.

The comparison between Driving Travel Time and Types of Information was done for scenarios (S2, S3, S6, S7, S10, S11) in which time did not matter. The result of the *t*-test that compared Driving Travel Time between No-Information and VMS-Information (Table 4) shows that the Driving Travel Time fell significantly from 11.30 min to 9.36 min due to the information available via VMS. The result of the *t*-test that compared Driving Travel Time between No-Information and App-Information (Table 4) shows that the Driving Travel Time dropped from 11.30 min to 8.01 min due to the information available via App.

The primary purpose of this study is to investigate the impact of parking information on circulation time. For this study circulation time is defined as the time that a participant circulates looking for parking after reaching the destination (Verizon Center). As mentioned before, the Driving Travel Time was calculated with the data from simulator log files, that is, end time (time that the driver reaches the destination and finds parking) minus start time (time that the driver starts

Table 4. *t*-test Results for Driving Travel Time Comparisons

Variable	N	Mean	Std. Dev.	*t*-statistic	Sig.
Panel A. Between No-Information and VMS-Information (All Scenarios)					
Driving Travel Time when there is No Information	227	11.25	6.044	3.73	.000
Driving Travel Time when there is VMS Information	214	9.27	5. 053	3.75	.000
Panel B. Between No-Information and App-Information (All Scenarios)					
Variable	N	Mean	Std. Dev.	*t*-statistic	Sig.
Driving Travel Time when there is No Information	227	11.26	6.045	6.803	.000
Driving Travel Time when there is APP Information	195	7.90	3.569	7.058	.000
Panel C. Between No-Information and VMS-Information (Time Matters Only)					
Variable	N	Mean	Std. Dev.	*t*-statistic	Sig.
Driving Travel Time when there is No Information	112	11.21	5.721	2.850	0.005
Driving Travel Time when there is VMS Information	108	9.19	4.778	2.859	0.005
Panel D. Between No-Information and App-Information (Time Matters Only)					
Variable	N	Mean	Std. Dev.	*t*-statistic	Sig.
Driving Travel Time when there is No Information	112	11.21	5.721	5.260	.000
Driving Travel Time when there is APP Information	96	7.79	3.036	5.493	.000
Panel E. Between No-Information and VMS-Information (Time Does Not Matter Only)					
Variable	N	Mean	Std. Dev.	*t*-statistic	Sig.
Driving Travel Time when there is No Information	115	11.30	6.369	2.450	.015
Driving Travel Time when there is VMS Information	106	9.36	5.342	2.467	.014
Panel F. Between No-Information and App-Information (Time Does Not Matter Only)					
Variable	N	Mean	Std. Dev.	*t*-statistic	Sig.
Driving Travel Time when there is No Information	115	11.30	6.369	4.437	.000
Driving Travel Time when there is APP Information	99	8.01	4.032	4.582S	.000

Note: The reported significance is 2-tailed for all tests.

Table 5. *t*-test Results for Circulation Time Comparisons

Variable	N	Mean	Std. Dev.	t-statistic	Sig. (2-tailed)
Panel A. Between No-Information and VMS-Information					
Circulation time when there is No Information	227	4.37	4.628	.088	.930
Circulation time when there is VMS Information	214	4.33	5.504	.087	.930
Panel B. Between No-Information and App -Information					
Variable	N	Mean	Std. Dev.	t-statistic	Sig. (2-tailed)
Circulation time when there is No Information	227	4.38	4.628	4.902	.000
Circulation time when there is App Information	195	2.42	3.387	5.016	.000

driving), and the time when a participant reaches the destination was also extracted from the data file. So, the circulation time was calculated by deducting the time participants reach the destination from their Driving Travel Time (end time minus start time); a negative circulation time indicates the participant parked before reaching the destination which usually happened when drivers have access to information or they are familiar with the network.

The comparison between Circulation Time and Types of Information is provided in Table 5 below. The result of the *t*-test that compared Circulation Time between No-Information (scenarios 1 to 4) and VMS-Information (scenarios 5 to 8) (Table 5) shows that the Circulation Time was reduced from 4.37 to 4.33 due to the information available via VMS; that is not significant. The result of the *t*-test that compared Circulation Time between No-Information (scenarios 1 to 4) and App-Information (scenarios 9 to 12) (Table 5) shows that the Circulation Time was significantly reduced from 4.43 min to 2.42 min due to the information available via App.

Regression Analysis

A linear regression analysis was performed to identify the relationship between circulation time as a dependent variable and Income Value of the Simulator (participants received 4 dollars per scenario minus the parking fee) and types of information (No-Information and Information).

Table 6. Results for the Regression of Circulation Time on Income Value of the Simulator and Types of Information

Variable	B	Sig. (2-tailed)
Intercept	− 2.708	.000
Income value of the simulator	2.908	.000
Parking information	− .378	.000

Note: Adjusted R Square: .770.

The result (Table 6) shows that there is a positive relationship between Circulation time and Income Value of the Simulator, the more the participant circulates for parking, the more money they will receive; since there is a negative relationship between the Income Value of the Simulator, it makes sense that participants with less annual income prefer to circulate more in the simulator to earn more money. Furthermore, there is a significant negative relationship between Circulation Time and Types of Information; when participants have information related to parking space availability they will circulate less than when they do not have any information.

Parking Choice Model

A multinomial logit regression analysis was performed to identify the most appropriate independent variables to describe the probability of choosing on-street parking. Among all socioeconomic characteristics and variables related to the scenarios, only age and type of information had a significant impact on parking choice.

As discussed earlier, there are three parking choices: Verizon Garage, 11th St. Garage, and on-street parking. On-street parking was the choice of the majority of participants, since it costs less than the garages. The results of the multinomial logistic regression are illustrated in Table 7. It shows that the probabilities of choosing the Verizon Garage and 11th St. Garage are lower than that of on-street parking in all age groups of participants. Younger participants were less likely to choose to park in a garage which is reflective of their traditionally lower value of time. The time ratio variable is the ratio of total travel time (driving time + walking time) to the travel time from the origin directly to the Verizon Center. The Verizon Garage is costly but requires no walking time. An individual who spends significant time traveling is less likely to have parked in the Verizon Garage compared with parking on-street. However, the reverse is true for the 11th Street garage; a longer travel time ratio indicates that a person is more likely to park in 11th Street Garage compared with parking on-street. This result is indicative of the participant who is cost sensitive and searched for cheaper parking but was unable to find on-street parking and thus settled with parking in the cheaper, further garage.

The type of information was found to be an important factor in this model. The probability of choosing the

Table 7. Parking Choice Behavior Model

Parking choice	Variable	β	Standard error	Significance
Verizon Garage	Constant	2.38	0.888	0.01
	Age			
	18 to 25	− 2.12	0.608	0.00
	26 to 35	− 1.69	0.570	0.00
	36 to 45	− 1.27	0.692	0.07
	More than 46	Reference category		
	Types of Information			
	VMS Information	− 1.08	0.490	0.03
	App Information	− 2.35	0.771	0.00
	No Information	Reference category		
	Travel Time Ratio	− 1.41	0.383	0.00
11th St. Garage	Constant	− 0.791	0.423	0.06
	Age			
	18 to 25	− 1.62	0.419	0.00
	26 to 35	− 1.49	0.422	0.00
	36 to 45	− 1.11	0.519	0.03
	More than 46	Reference category		
	Types of Information			
	VMS Information	− 1.62	0.364	0.00
	App Information	− 1.60	0.377	0.00
	No Information	Reference category		
	Travel Time Ratio	0.264	0.0841	0.00

Note: On-Street Parking is reference category. Number of observations = 636. Adjusted *R*-square = 0.601.

Verizon Garage or 11th St. Garage compared with on-street parking is less when there is information pertinent to parking space availability via VMS or App. This could be because the information helps participants find parking that is cheaper with less circulation; thereby affecting the participants' parking choice. To test independence from irrelevant alternatives (IIA), a nested logit model was tested in which the first nest was the choice between on-street and garage parking. However, this did not improve the model.

Summary and Conclusion

This study used both stated preference (SP) survey and driving simulator techniques to evaluate the effects of different types of information related to parking space availability on parking choice and circulation behavior.

Some 56% of the participants were familiar with the study area; 52.07% of the participants stated that they preferred on-street parking, 38.82% preferred the 11th St. Garage, while only 9.11% chose the Verizon Garage.

The results illustrated that age and parking availability information affect parking choice behavior. The participants' choices were significantly different in the simulator experiment compared with the SP survey, which might be because of the existence of the real price of parking in the survey. Participants behaved differently when information existed and time mattered. Very few participants opted for the Verizon Garage, which is the most expensive garage. However, when time mattered and there was no information available, they were willing to pay the parking price and use the Verizon Garage to be on time.

The comparative analysis reveals that when no information is provided, the probability of choosing the Verizon Garage was high in the driving simulator experiments. In the presence of information, a majority of participants chose on-street parking as the information provided via app shows the parking availability in the whole network and on-street parking is the cheapest choice. Furthermore, the results show that there is a negative relationship between types of information and circulation time. We conclude that the information provision decreases the circulation time significantly. It was found that the App is more effective than VMS. Future studies may examine the relationship between price sensitivity and traffic congestion on parking choice.

Acknowledgments

This project was supported by the Mid-Atlantic Transportation Sustainability University Transportation Center (MATS-UTC), the National Transportation Center (NTC) at Morgan State University and the District Department of Transportation (DDOT).

Author Contributions

The authors confirm contribution to the paper as follows: study conception and design: SA, CC, MJ; data collection: SA, CC, MJ, ZRM; analysis and interpretation of results: SA, CC, MJ; draft manuscript preparation: SA, CC. All authors reviewed the results and approved the final version of the manuscript.

References

1. Van Ommeren, J., D. Wentink, and J. Dekkers. The Real Price of Parking Policy. *Journal of Urban Economics*, Vol. 70, No. 1, 2011, pp. 25–31.
2. Waraich, R., and K. Axhausen, Agent-based Parking Choice Model. *Transportation Research Record: Journal of the Transportation Research Board*, 2012. 2319: 39–46.
3. Shoup, D. C. Cruising for Parking. *Transport Policy*, Vol. 13, No. 6, 2006, pp. 479–486.
4. Arnott, R., and E. Inci. An Integrated Model of Downtown Parking and Traffic Congestion. *Journal of Urban Economics*, Vol. 60, No. 3, 2006, pp. 418–442.
5. Pullola, S., P. K. Atrey, and A. El Saddik, Toward an Intelligent GPS-based Vehicle Navigation System for Finding Street Parking Lots. *IEEE International Conference on Signal Processing and Communications, 2007. ICSPC 2007*, Dubai, United Arab Emirates Publisher: IEEE. 2007, pp. 1251–1254.
6. Caicedo, F. Real-time Parking Information Management to Reduce Search Time, Vehicle Displacement and Emissions. *Transportation. Research Part D: Transport and Environment*, Vol. 15, No. 4, 2010, pp. 228–234.
7. Rye, T., K. Hunton, S. Ison, and N. Kocak. The Role of Market Research and Consultation in Developing Parking Policy. *Transport Policy*, Vol. 15, No. 6, 2008, pp. 387–394.
8. Guo, L., S. Huang, J. Zhuang, and A. W. Sadek. Modeling Parking Behavior Under Uncertainty: A Static Game Theoretic Versus a Sequential Neo-additive Capacity Modeling Approach. *Networks and Spatial Economics*, Vol. 13, No. 3, 2013, pp. 327–350.
9. Arnott, R., E. Inci, and J. Rowse. Downtown Curbside Parking Capacity. *Journal of Urban Economics*, Vol. 86, 2015, pp. 83–97.
10. Tsamboulas, D. A. Parking Fare Thresholds: A Policy Tool. *Transport Policy*, Vol. 8, No. 2, 2001, pp. 115–124.
11. Gallo, M., L. D'Acierno, and B. Montella. A Multilayer Model to Simulate Cruising for Parking in Urban Areas. *Transport Policy*, Vol. 18, No. 5, 2011, pp. 735–744.
12. Chaniotakis, E., and A. J. Pel. Drivers' Parking Location Choice Under Uncertain Parking Availability and Search Times: A Stated Preference Experiment. *Transportation Research Part A: Policy and Practice*, Vol. 82, 2015, pp. 228–239.
13. Thompson, R. G., K. Takada, and S. Kobayakawa. Optimisation of Parking Guidance and Information Systems Display Configurations. *Transportation Research Part C: Emerging Technologies*, Vol. 9, No. 1, 2001, pp. 69–85.
14. Teodorović, D., and P. Lučić. Intelligent Parking Systems. *European Journal of Operational Research*, Vol. 175, No. 3, 2006, pp. 1666–1681.

15. Arnott, R., and J. Rowse. Downtown Parking in Auto City. *Regional Science and Urban Economics*, Vol. 39, No. 1, 2009, pp. 1–14.
16. Bolanowska, M., and B. Hemily. *Developing Transit-supportive Parking Policies: A Synthesis of Key Concepts from the Literature*. Report, Canadian Urban Transit Association, Canada, 2000.
17. Willson, R. Parking Policy for Transit-oriented Development: Lessons for Cities, Transit Agencies, and Developers. *Journal of Public Transportation*, Vol. 8, No. 5, 2005, p. 5.
18. Shiftan, Y., and R. Burd-Eden. Modeling Response to Parking Policy. *Transportation Research Record: Journal of the Transportation Research Board*, 2001. 1765: 27–34.
19. Petiot, R. Parking Enforcement and Travel Demand Management. *Transport Policy*, Vol. 11, No. 4, 2004, pp. 399–411.
20. Chatman, D. G., and M. Manville. Theory Versus Implementation in Congestion-priced Parking: An Evaluation of SFpark, 2011–2012. *Research in Transportation Economics*, Vol. 44, 2014, pp. 52–60.
21. Millard-Ball, A., R. R. Weinberger, and R. C. Hampshire. Is the Curb 80% Full or 20% Empty? Assessing the Impacts of San Francisco's Parking Pricing Experiment. *Transportation Research Part A: Policy and Practice*, Vol. 63, 2014, pp. 76–92.
22. Pierce, G., and D. Shoup. Getting the Prices Right: An Evaluation of Pricing Parking by Demand in San Francisco. *Journal of the American Planning Association*, Vol. 79, No. 1, 2013, pp. 67–81.
23. Nawaz, S., C. Efstratiou, and C. Mascolo. Parksense: A Smartphone Based Sensing System for On-street Parking. *Proc., 19th Annual International Conference on Mobile Computing & Networking*, Miami, Florida, USA Association for Computing Machinery (ACM). 2013, pp. 75–86.
24. Mas, I. *Competition Aspects of New Mobile Payment Networks: The Case of Mobile Payments in Spain (December 2008)*. Lydian Payments Journal, August 2010.
25. Srikanth, S. V., P. J. Pramod, K. P. Dileep, S. Tapas, and M. U. Patil. Design and Implementation of a Prototype Smart PARKing (SPARK) System Using Wireless Sensor Networks. *International Conference on Advanced Information Networking and Applications Workshops, 2009. WAINA '09*. IEEE, Bradford, UK. pp. 401–406, 2009.
26. Mathew, S. S., Y. Atif, Q. Z. Sheng, and Z. Maamar. Building Sustainable Parking Lots with the Web of Things. *Personal and Ubiquitous Computing*, Vol. 18, No. 4, 2014, pp. 895–907.
27. Grazioli, A., M. Picone, F. Zanichelli, and M. Amoretti. Collaborative Mobile Application and Advanced Services for Smart Parking. *2013 IEEE 14th International Conference on Mobile Data Management (MDM)*, IEEE, Milan, Italy. Vol. 2, pp. 39–44, 2013.
28. Dormsjo, L. A. *District Department of Transportation* (Online). Available: https://ddot.dc.gov/page/parkdc. Accessed July 26, 2017.
29. Stenneth, L., O. Wolfson, B. Xu, and S. Y. Philip. PhonePark: Street Parking Using Mobile Phones. *2012 IEEE 13th International Conference on Mobile Data Management (MDM)*, IEEE, Bengaluru, Karnataka, India. pp. 278–279, 2012.
30. Shin, J.-H., and H.-B. Jun. A Study on Smart Parking Guidance Algorithm. *Transportation Research Part C: Emerging Technologies*, Vol. 44, 2014, pp. 299–317.
31. Baronti, P., P. Pillai, V. W. Chook, S. Chessa, A. Gotta, and Y. F. Hu. Wireless Sensor Networks: A Survey on the State of the Art and the 802.15.4 and ZigBee Standards. *Computer Communications*, Vol. 30, No. 7, 2007, pp. 1655–1695.
32. Forum8. *VR-Design Studio* (Online). Available: http://www.forum8.com/vr_design_studio.htm. Accessed April 14, 2017.

The Standing Committee on Transportation Demand Management (ABE50) peer-reviewed this paper (18-03969).

Article

TRR
JOURNAL OF THE TRANSPORTATION RESEARCH BOARD

Transportation Research Record
2018, Vol. 2672(8) 930–941

DOI: 10.1177/0361198118772947
journals.sagepub.com/home/trr
SAGE

Predicting Owners' Willingness to Share Private Residential Parking Spots

Chu Zhang[1], Jun Chen[1], Zhibin Li[1], and Yuanyuan Wu[2]

Abstract

Sharing of private residential parking spots is a new pattern of parking management in China. This pattern corresponds to the booming sharing economy and is growing very fast. It can significantly improve the utilization of parking resources and relieve parking supply pressure. Based on the real data of 1-year behavioral records of owners obtained from Ding Ding Parking (DParking), an application on smart phones, as well as various field survey data, the study analyzed the influential factors and predicted owners' sharing willingness. Two Classification and Regression Trees (CART) were developed to answer questions pertaining to whether owners would share their parking spot and how long owners would share during peak periods of parking demand, respectively. The results showed good accuracy in both models and revealed that owners' self-use behavior, along with owners' private spots' physical characteristics and rental effects of the previous month, all have significant influence on owners' willingness to share. The influence of factors and their importance differ for the two models; thus, a detailed comparison is performed. The findings in this paper would be beneficial to the government's parking supply policies, as well as to third parties, so as to enhance the effective distribution of parking resources.

Private residential parking spot sharing has recently become a hot topic in China. Generally, parking demands are mainly centered around places of work in the daytime and then transfer to residential zones by early evening or night (*1*). Making private residential parking spots accessible would promote the utilization of parking resources and be more compatible with both owners' and parkers' interests (*2*). Because of the legal issues admitted by the government and the technologies developed for parking management, owners of private parking spots now have access to share spare parking resources with the public, in exchange for economic payback. Because the ownership of parking spots in residential areas tends to be scattered, transactions facilitated through an app-based peer-to-peer marketplace were adopted instead of the more traditional parking features, such as parking meters, gate machines, and so on. Some firms take on the role as a third party, so as to collect and distribute parking resources through the use of various smart phone applications, such as ROVER in the United States and DParking in China. The detailed operation steps of both owners and borrowers adopting DParking are listed in Figure 1. Thus, the new pattern of parking supply generates numerous fluctuations in shared parking studies.

Firstly, shared parking behavior includes the participation of both the owners and parkers. Current studies have already taken many factors into consideration, such as parking fees, walking distances, trip purpose, parkers' psychology, and so on (*3, 4*). However, all the studies above only considered the participation willingness of parkers, and none of the studies has explored aspects concerning the sharing willingness of owners.

Moreover, the traditional research method for willingness primarily focuses on stated preference surveys (*5, 6*); the shortcomings are obvious. The accuracy of the answers mostly depends on the quality of the questionnaires. Inevitably, vague descriptions of the preset context may tend to mislead respondents, which may result in wrong or skewed answers. Thanks to the new technologies, the revealed preference survey can now be conducted through the behavioral record documents saved in the database; thus, a real-data based study would prove to be more persuasive.

Another difference is the level of study objects. Traditional studies are based on the parking lots (*3, 7,*

[1]School of Transportation, Southeast University, Nanjing, China
[2]School of Civil and Environmental Engineering, Nanyang Technological University, Singapore

Corresponding Author:
Address correspondence to Jun Chen: chenjun@seu.edu.cn

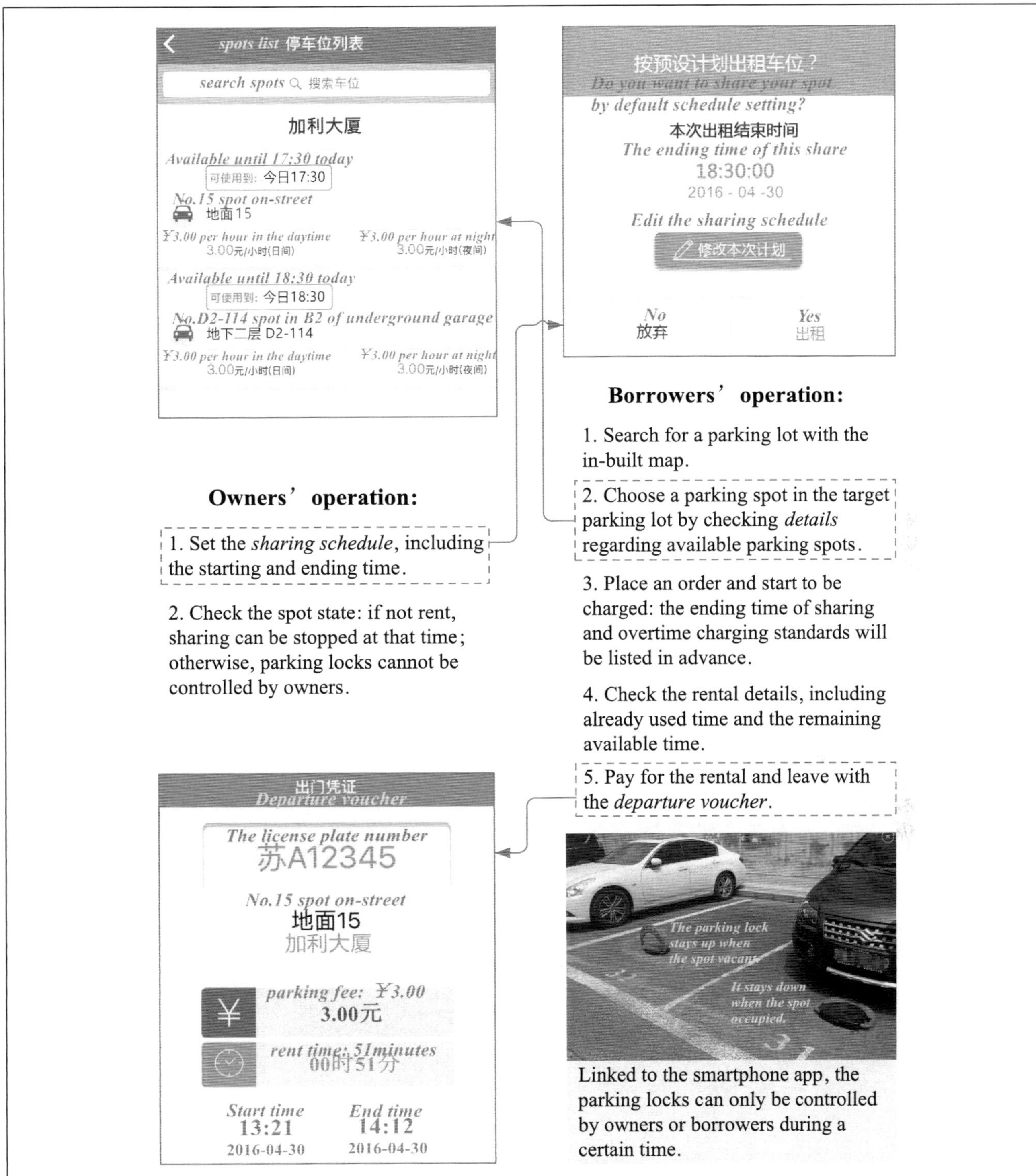

Figure 1. Operation of DParking app and parking locks.

8), whereas, in this study, every spot is assumed to be associated with its owner and to be different from others. Spot-level research is seldom up-to-date and mostly focuses on the distribution methodology (9, 10), without any adequate real-data support. Real study data encompass detailed information about both the parking spot's physical characteristics and its owner's behavioral records, perhaps for the first time.

As a result, this paper conducts a spot-level study of owners' sharing willingness driven by real-data for the first time. The primary objective of the research is to accurately investigate factors affecting the sharing willingness of private parking spots' owners. More specifically, the study presented in this paper focuses on answering the following two questions: (1) would owners decide to share or not? (2) how long would owners share their parking spots during the peak periods of parking demand?

Data

Study Area

The research was conducted in a residential building on the north side of the 2nd Ring Road of Beijing, adjacent to two office buildings and one government building. The target building consists of a two-level underground parking garage as well as a few on-street parking spots. As depicted in Figure 2, the on-street parking zone stretches for less than 200 m, and lies along the alley across from major buildings. The underground parking garage also has an entrance and an exit for automobiles separately along the parking zone. The longest distance between any spot and the furthest surrounding building is 247 m (810 ft), which can be accepted by most parkers (*11*).

The procedure for shared parking is as follows. Owners post the open-for-share information through the app, and ensure that the parking spots are available during the preset period. If the spots are not booked or rented, the owners can withdraw or modify their sharing plans at any time. Parking prices are fixed at 5 yuan (0.8 USD) per hour throughout the daytime period from 9:00 am to 19:00 pm and 1 yuan (0.2 USD) per hour for the remaining time. The automatic charge unit is per 15 min., and the charging scheme is the same on workdays and weekends.

Data Processing

Study data were obtained from DParking, an application on smart phones, and were truncated from all continuous operation records for a period of an entire year, starting at 2015/11/01 00:00:00 and ending at 2016/10/31 23:59:59. Because seasonal heterogeneity influences the travel behavior of both the owners and renters, the samples were aggregated per spot per month. Therefore, a total of 604 samples were assembled among 72 spots representing various types of owners. The statistical results revealed that the total rental time per day on workdays was 2.35 times that of weekends. Moreover, it revealed the total rental time per hour in the daytime (from 7:00 am to 19:00 pm) was 2.56 times that of night rentals. Thus, willingness to share in the daytime on workdays is crucial to meet the growing demands and promote utilization. The period from 7:00 am to 19:00 pm on workdays was defined as the peak time of parking demand in this study.

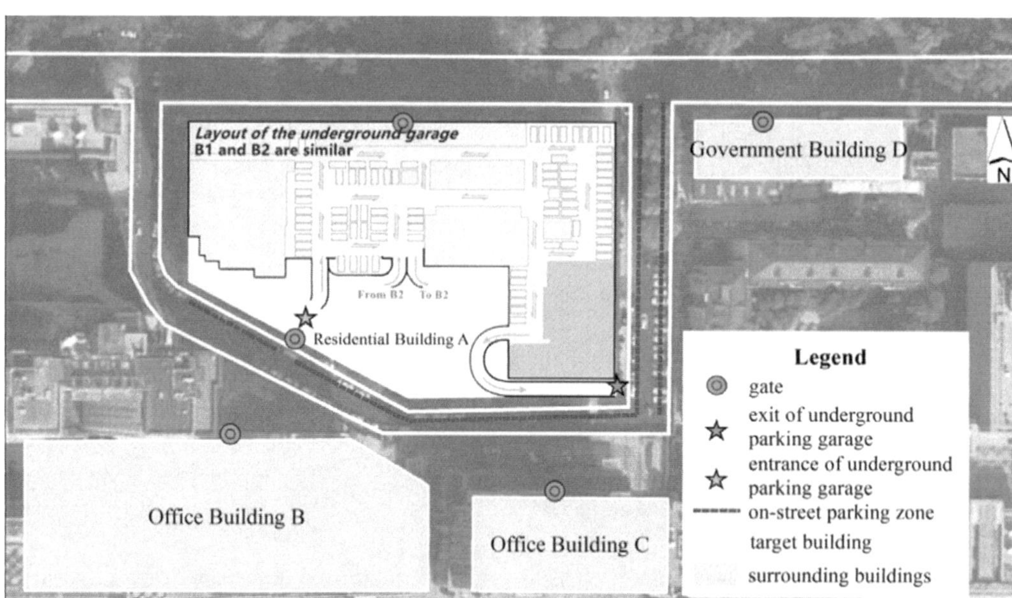

Figure 2. Layout of the study area.

Table 1. Descriptions of Variables

Variables	Descriptions	Avg	SD	Avg	SD
	Owners' sharing decisions				
osYN	Open-to-share decision	0: Unshare (227 samples)		1: Share (377 samples)	
osDtime	Average sharing time in the daytime on workdays (h)	na		5.74	3.54
	Spots' physical characteristics				
Type	0: underground; 1: on-street	0.50	0.50	0.50	0.50
Floor	1: on-street; 2: B1 floor; 3: B2 floor	1.68	0.76	1.76	0.84
distanceL	Distances from spots to the nearest surrounding building (m)	80.17	41.70	71.01	42.98
distanceA	Average distances from spots to surrounding buildings (m)	133.37	47.17	130.23	44.34
	Owners' self-use behavior				
sfTtime	Average self-use time in a day (h)	16.14	6.11	10.17	7.79
sfDtime	Average self-use time in the daytime on workdays (h)	8.06	3.13	3.94	3.50
sfDfres	Average self-use parking frequency in the daytime on workdays	2.31	0.23	1.95	0.41
	Rental effects of the previous month				
rtTtime	Average rental time in a day (h)	na		2.58	4.04
rtDtime	Average rental time in the daytime on workdays (h)			2.29	2.72
rtosDR	Ratio of rental time in sharing time in the daytime on workdays			0.52	0.29

Note: Avg = average; SD = standard deviation.

The entire study data include 377 shared samples and 227 unshared samples. Variables of the two types of samples are described separately. All 12 variables were collected and organized into four specific groups (see Table 1), including the owners' sharing decisions, the parking spots' physical characteristics, owners' self-use behavior, and the rental effects of the previous month. The owners' sharing behavior group consists of two dependent variables: osYN and osDtime; osYN is a binary variable indicating whether the sample is available for sharing or not. Additionally, osDtime, as a continuous variable only existing in the shared samples, is the average time of sharing behavior in the daytime during workdays. In the spots' physical characteristics group, Type and Floor constitute the fundamental factors, distanceL describes the minimum distances between the spot and the gate of the nearest surrounding building, and distanceA represents the average value of the distances from the spots to all the gates of the three surrounding buildings. In the owners' self-use behavior group, sfTtime is the average value of the owners' self-parking time in a day, whereas sfDtime and sfDfres depict the average time and frequency owners parked in the daytime on workdays. As for rental effects of the previous month's group, all the variables were collected from records for the previous month, the one before the statistical month, as owners' prior knowledge, which may have influenced the current month's sharing decisions. Moreover, rtTtime and rtDtime are the average rental time in a day and in the daytime on workdays, respectively. The rtosDR variable depicts the ratio of rental time in sharing time in the daytime on workdays. Like osDtime, these three variables also only exist in shared samples.

Method

Most sharing willingness studies have adopted parametric statistical approaches, such as various logit models (*2–4*). However, some researchers also built nonparametric models which can provide higher prediction accuracy like using CART (*12, 13*). CART is a typical machine learning method and has several advantages perfectly matching this study's targets. First, there are two kinds of CARTs, classification tree (CT) and regression tree (RT), distinguished by the types of target value. As in this case, one dependent variable, osYN, is dichotomous and the other, osDtime, is continuous, both CT and RT are required. Considering that the influential factors in both models are similar, adopting both CART models would benefit the comparison. Secondly, CART reflects the complex nonlinear correlation between the influencing factors and target behavior variables and, thus, produces "if–then" rules (determining the willingness according to the conditions), so as to predict owners' sharing choices. Besides, CART can graphically depict the correlation, which aids in understanding how owners make their decisions.

Although the leaf value of CT is represented by the majority and that of RT is represented by the average, the development of a CART model is the same. It generally consists of three steps: (1) growing the tree, (2) pruning the tree, and (3) selecting an optimal tree from the pruned trees. The detailed procedure relevant to this paper is as follows:

Step 1: To grow the tree is to recursively partition the target variable, so that the data in the descendent nodes are always more pure than the data in the parent node. The splitting methods are differentiated between CT and RT. For CT, node impurity based on entropy reduction was adopted, as it would be a stricter splitting criterion for a complex and chaotic set than the traditional Gini index (*14*). For RT, on the other hand, node impurity based on the least squares method was adopted.

For a given node k in CT, the node impurity $i(k)$ is calculated

$$i(k) = -\sum_{j=1}^{2} p_j \log_2 p_j \tag{1}$$

where p_j is the proportion of class j in node k.

For a given node k in RT, the node impurity $i(k)$ is calculated as

$$i(k) = \frac{1}{N}\sum_{i=1}^{N} (y_i - r_i)^2 \tag{2}$$

where N is the total number of samples in node k; and r_i represents the average value of dependent variable y_i.

Moreover, for all the possible splits in the values of the explanatory variable x, the impurity reduction on the parent node k, caused by a split s is calculated as

$$\Delta i(x, s, k) = i(k) - p_L i(k_L) - p_R i(k_R) \tag{3}$$

where k_L and k_R are the left and right child nodes of the parent node k, respectively; p_L and p_R are fractions of the records belonging to the left and right child nodes, respectively. If $\Delta i(x, s, k)>0$, this implies that the records in the descendent nodes are purer than those in the parent node. If not, then the split is not viable. The split scheme contributing to a global maximum impurity reduction constitutes the most appropriate split.

In this paper, splitting ceases when one of the following stopping rules is satisfied: (1) none of the nodes at the bottom can be split further when $\Delta i(x, s, k)$ is less than the allowed thresholds, which is 0 for CT and 0.01 for RT in this case; (2) records in the node are less than the preset minimum number, which is 4 in this case; and (3) the current tree depth has reached the preset maximum tree depth, which is 8 in this case.

Step 2: Tree pruning is used in case the tree is overfitted. The overfitted tree could result in a substantial true error rate on the new (testing) data (*14, 15*). Therefore, there is a critical need to remove some branches that do not contribute to predictive accuracy. Generally, the cost complexity measurement approach, based on the postpruning concept, is employed to prune the grown tree. The cost complexity of a tree T is then defined as the weighted sum of its complexity (i.e., number of leaf nodes) and its error, expressed as

$$R_\alpha(T) = R(T) + \alpha|\tilde{T}| \tag{4}$$

where α is the weight; $|\tilde{T}|$ is the number of leaf nodes in tree T; and $R(T)$ is the corresponding misclassification error rate, which is node impurity in this particular case.

For a given node t in the original tree T_0, tree t also has its child tree T_t. The reduction index of total loss in pruned trees is expressed as

$$g(t) = \frac{R(t) - R(T_t)}{|\tilde{T}_t - 1|} \tag{5}$$

The tree pruning process is recursively performed by determining the minimum $g(t)$ and pruning the corresponding child tree until only the root node remains. The optimal tree is in the set of pruned trees and is gained by each pruning.

Step 3: To select an optimal tree from the pruned trees is to calculate the cost complexity, based on the average values of global accuracy, or R-squared through the test data and select the one with a minimum value. In this paper, a 10-fold cross validation method was adopted for testing.

Moreover, by determining the final tree, the importance of the variables that intervene in the model can be derived from the CART model. This is the average reduction of entropy or R-squared in all the nodes. The value of the standardized importance of the independent variables reflects the impact of such predictor variables on the model (*16*).

Results

CART_1: Whether to Share or Not?

Because the unshared samples account for a small proportion of 37.6%, in order to eliminate biased results caused by unbalanced data, data preprocessing should be performed before building CART_1. Generally, the over-sampling method for expanding minority data may increase the probability of the overfitting problem compared with the undersampling method for reducing the majority data (*17*). Therefore, the random undersampling method to reduce the shared samples is applied in this study. The input data include 227 unshared samples and 227 shared samples, as the share possibilities are assumed to be equal. As for the independent variables, factors in spots' physical characteristics and owners' self-use behavior groups were included in CART_1. The dependent variable here is the categorical variable, osYN.

Based on the preset parameters (illustrated in Methods) and the 10-fold cross validation testing method, the pruned CART_1 that leads to the lowest cost complexity for the testing data was determined. Figure 3 depicts the relationship between the number of leaves and cost complexity in CART_1; cost complexity reached the bottom at 11 leaves with a value of 0.167.

Figure 4 illustrates CART_1 with 11 leaf nodes; the shaded rectangle represents one of the leaf nodes. CART_1 can be intuitively interpreted as follows: the initial split at Node 1 (the root node) is based on sfDtime. This means the most important factor affecting owners' sharing decisions is the owners' self-use of time during the daytime on workdays. CART_1 directs values of sfDtime lower than 1.90 hours to the left, forming LeafNode 1, and values larger than 1.90 hours to the right, forming Node 2. As indicated by LeafNode 1, CART_1 predicts that owners parking their own vehicles for less than 1.90 hours during the daytime on workdays will probably be willing to share. Further, CART_1 proceeds to split Node 2, in accordance with sfTtime. If sfTtime is more than 21.96 hours a day, the not-to-share decision is most likely to be made (97.2%), as is depicted in LeafNode 11. This is because the self-use parking has already occupied the whole day and the available time is too limited. When sfDtime exceeds 1.90 hours and sfTtime is less than 21.96 hours, the possibilities of owners' decisions to share or not are quite close, with values of 53.9% for not sharing and 46.1% for share.

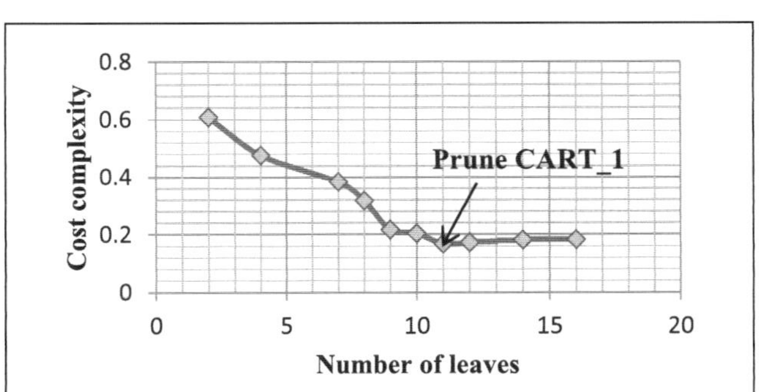

Figure 3. Procedure of selecting an optimal CART_1.

Likewise, Node 3 can be further split by the narrowed intervals of sfDtime and sfTtime, forming Node 6, LeafNode 4, Node 7, LeafNode 9, and LeafNode 10. The variety of subclasses also reflects the complexity of the decision-making process affected by owners' self-use parking time schedules. Afterwards, Node 6 directs types of less than 0.5, namely, underground spots to the left and on-street spots to the right, forming LeafNode 2 and LeafNode 3, respectively. The predicted results reveal that owners of on-street spots are more likely to share than those of underground spots. In Node 7, values of sfDfres lower than 2.02 are to the left, forming Node 8, and those above 2.02 to the right, forming LeafNode 8. The predicted result demonstrates that owners who park more than 2.02 times during the daytime on workdays are more prone to share than those who might be parking for a long period without moving. CART_1 further proceeds to split Node 8 by Floor, where values of Floor less than 2.50 form LeafNode 5, which indicate that the spots on the street and in B1 are inclined to be unshared under that specific circumstance. However, the attitudes are not clear for owners in B2, as is reflected in Node 9 (53.3% versus 46.7%). Thus, Node 9 is further split by sfDtime at the value of 8.22 hours. Values of sfDtime less than 8.22 hours tend to be unshared, whereas those exceeding 8.22 hours are more likely to share, because their parking time primarily focuses on the daytime hours, thereby leaving the nighttime available.

Lastly, a set of sharing decision rules, which can be identified by computers and adopted by managements, were extracted from 11 nodes of CART_1. The detailed rules are listed in Table 2, and it can be observed that six rules refer to share and five rules refer to unshare. The total predicting accuracy of CART_1 is 93.4%; furthermore, CART_1 performs better in predicting unshare, rather than share (97.8% versus 89.0%). To validate the results, the same dataset is adopted in a widely accepted statistical model, the binary logistic model; and the prediction accuracy is lower than that of CART_1 with similar findings.

CART_2: How Long to Share during the Peak of Parking Demand?

To predict the sharing time during the peak period of parking demand, namely, in the daytime on workdays, 377 shared samples were utilized to build CART_2. Apart from the independent variables employed in CART_1, the rental effects of the previous month's variables were also adopted in this model. Moreover, the dependent variable here is the continuous variable osDtime.

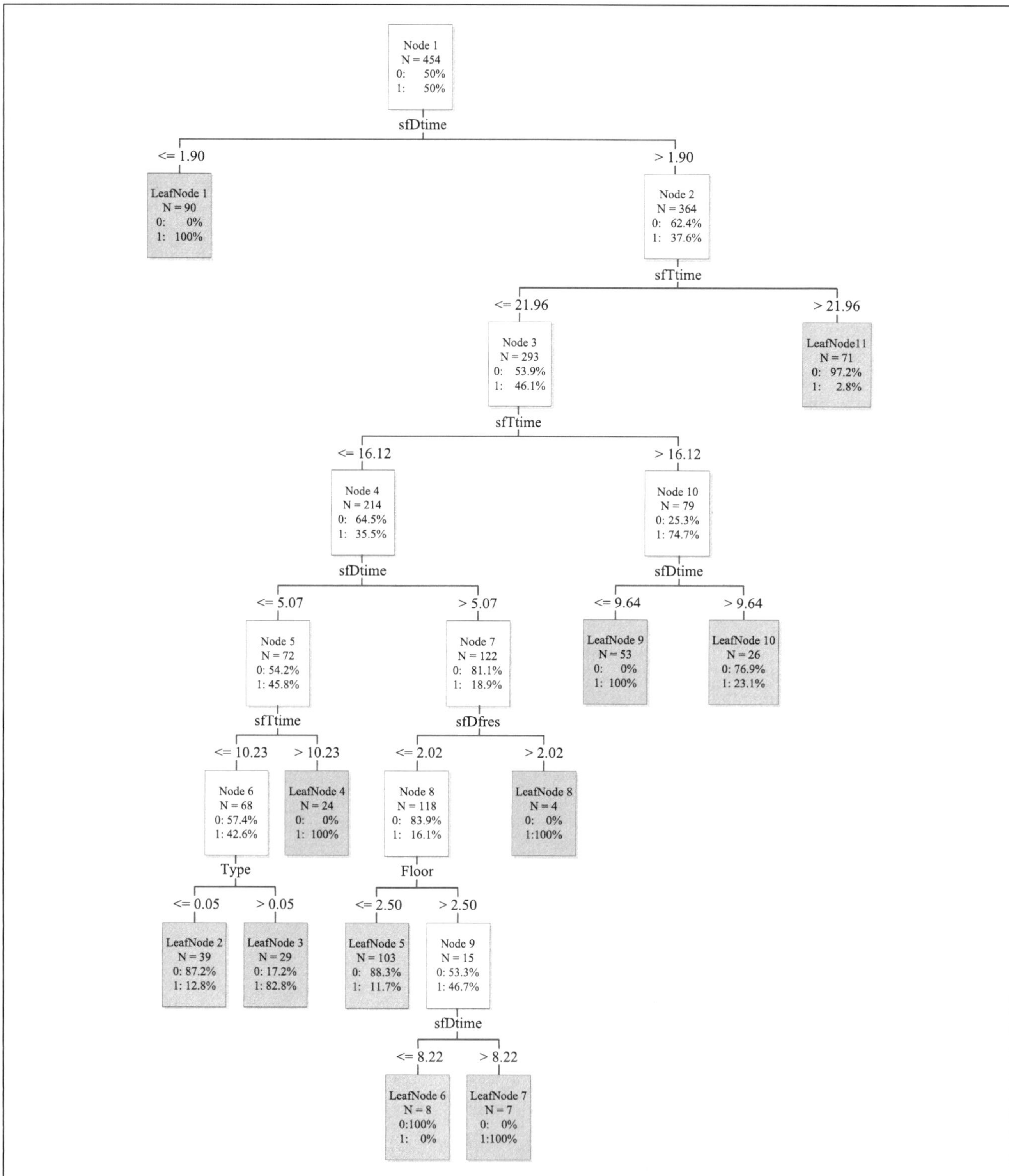

Figure 4. Graphic expression of CART_1.

Table 2. Production rules in CART_1

No.	Antecedent	Predicted osYN	Accuracy
1	sfDtime<=1.90;	share	100%
2	1.90<sfDtime<=5.07; sfTtime<=10.23; Type<=0.05;	unshare	87.2%
3	1.90<sfDtime<=5.07; sfTtime<=10.23; 0.05<Type;	share	82.8%
4	1.90<sfDtime<=5.07; 10.23<sfTtime<=16.12;	share	100%
5	5.07<sfDtime; sfTtime<=16.12; sfDfres<=2.02; Floor=<2.50;	unshare	88.3%
6	5.07<sfDtime<=8.22; sfTtime<=16.12; sfDfres<=2.02; 2.50<Floor;	unshare	100%
7	8.22<sfDtime; sfTtime<=16.12; sfDfres<=2.02; 2.50<Floor;	share	100%
8	5.07<sfDtime; sfTtime<=16.12; 2.02<sfDfres;	share	100%
9	1.90<sfDtime<=9.64; 16.12<sfTtime<=23.18;	share	100%
10	9.64<sfDtime; 16.12<sfTtime<=21.96;	unshare	76.9%
11	1.90<sfDtime; 21.96<sfTtime;	unshare	97.2%

Note: No. = number.

Based on the same preset parameters (maximum leaf numbers and maximum tree depth) and testing methodology as in CART_1 as well as the impurity reduction thresholds for RT (illustrated in Method), a pruned CART_2 that leads to the lowest cost complexity for the testing data was determined. Figure 5 depicts the relationship between the number of leaves and cost complexity in CART_2. The cost complexity reached the bottom at 15 leaves, with a value of 0.296.

CART_2 has 15 leaf nodes, where the shaded rectangles represent different levels of a willingness to share. As the total daytime period consists of 12 hours, those between 8 and 12 hours are classified as high-level in the red rectangles; those between 4 and 8 hours are classified as mid-level in the blue rectangles, and those below 4 hours are classified as being at a low-level in the green rectangles.

As graphically presented in Figure 6, CART_2 initially split at Node 1(the root node), based on sfDtime, forming Node 2 with an average sharing time of 8.99 hours and Node 9 of 2.49 hours. The significant difference in sharing time demonstrates that self-use parking time represents a priority when owners consider their sharing willingness during peak periods of parking demand. Node 2, with values of sfDtime less than 1.82 hours, is further split by sfDfres. Values of sfDfres less than 1.57 are gathered in the left child Node 3, whereas those above 1.57 are in the right child Node 7. The difference in the average sharing time in Node 3 and Node 7 (9.58 hours versus 7.86 hours) indicates that self-use parking frequency might be beneficial, so as to distinguish the high-level from the mid-level sharing time. Afterwards, Node 3 and Node 7 are further split by the rental effect of the previous month's variables rtTtime and rtosDR, as well as spots' physical characteristics variables distanceA and distanceL, forming LeafNodes 1–8. Likewise, CART_2 further split Node 9 based on the variables rtDtime, sfDtime, rtosDR, distanceL, and sfTtime into LeafNodes 9–15, shown on the right-hand side of the tree in Figure 6. Compared with LeafNodes 1-8 on the left-hand side, values of sfDtime above 1.82 exhibit a lower sharing willingness. It is also shown that all five high-level leaf nodes are on the left, whereas all four low-level leaf nodes are on the right.

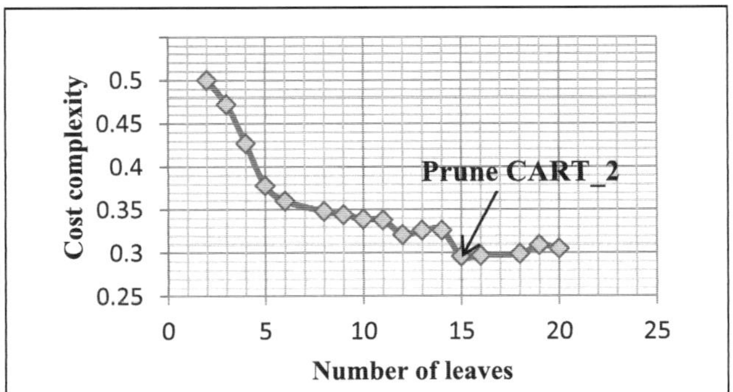

Figure 5. Procedure of selecting an optimal CART_2.

Judging from the split values and the average sharing time of the child trees, it can be inferred that a higher rental time in the previous month would boost owners' willingness to share this month, such as in Node 3, where the average sharing time is 11.11 hours for those values of rtTime more than 5.55 hours, as compared with 8.78 hours for those values of rtTime less than 5.55 hours. The same situation can be found in Node 7, Node 5, and Node 9. However, rtosDR is not an important consideration for owners, as the lower rtosDR of the previous month is related to a higher sharing willingness this month in Node 6 and Node 11. Moreover, the lower self-use parking time leads to a higher sharing time, as indicated in Node 1, Node 10, Node 12, and Node14. Another inference is that closer distances to surrounding buildings would enhance the owners' willingness to share, as in Node 4, Node 8, and Node 13.

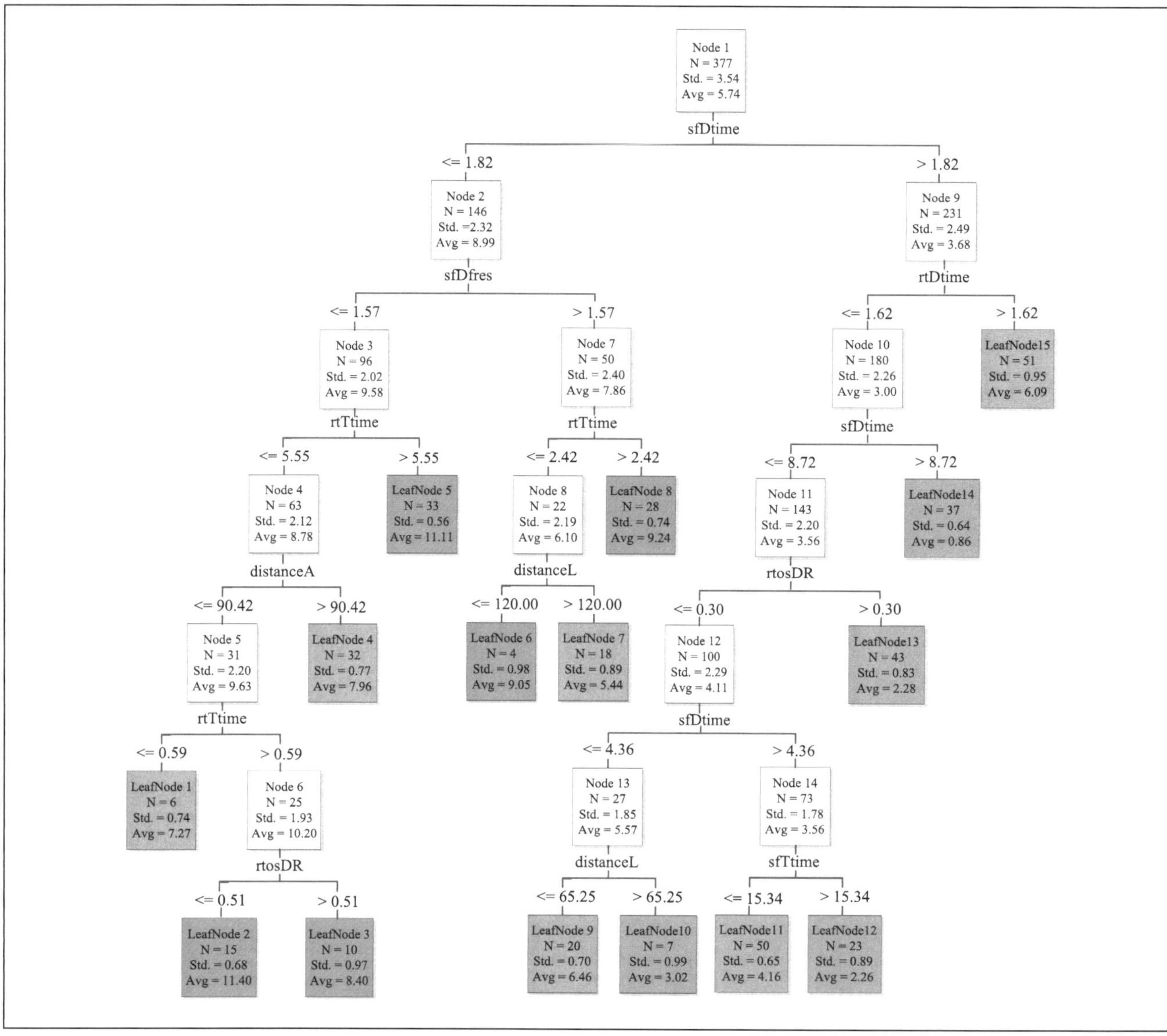

Figure 6. Graphic expression of CART_2.

Lastly, a set of sharing decision rules, which can be readily identified by computers and adopted by managements, was extracted from 15 nodes of CART_2, and the detailed rules are revealed in Table 3. It can be observed that five rules refer to a high-level sharing willingness, six rules to a mid-level and four rules to a low-level. The total predicting R-squared of CART_2 is 0.95. Therefore, CART_2 performs significantly better in high-level sharing time (R-squared = 0.96) than in a mid-level (R-squared = 0.95) or low-level (R-squared = 0.94). The same dataset is also adopted in a traditional statistical model, the linear regression model; and the R-squared is lower than that of CART_2 with no new findings of variables.

Discussions on Predictor Variables

Although the research purposes of CART_1 and CART_2 are quite diverse, considering the input variables are somewhat similar, the importance of variables (seen in Table 4) in the two models can be compared, along with several findings that are illustrated as follows.

All three variables of owners' self-use behavior, including sfDtime, sfTtime, and sfDfres, are mutually significant factors in both CART_1 and CART_2. Most especially, sfDtime wins a score of 100.00 in both models, meaning the self-use time during the peak periods of parking demand is certainly the most vital consideration in both decision-making processes. Judging from the order and importance of the variables, sfTtime and

Table 3. Production Rules in CART_2

No.	Antecedent	Predicted osDtime	SD
1	sfDtime<=1.82; sfDfres<=1.57; rtTtime<=0.59; distanceA<=90.42;	7.27	0.74
2	sfDtime<=1.82; sfDfres<=1.57; 0.59<rtTtime<=5.55; distanceA<=90.42; rtosDR<=0.51;	11.40	0.68
3	sfDtime<=1.82; sfDfres<=1.57; 0.59<rtTtime<=5.55; distanceA<=90.42; 0.51<rtosDR;	8.40	0.97
4	sfDtime<=1.82; sfDfres<=1.57; rtTtime<=5.55; 90.42<distanceA;	7.96	0.77
5	sfDtime<=1.82; sfDfres<=1.57; 5.55<rtTtime;	11.11	0.56
6	sfDtime<=1.82; 1.57<sfDfres; rtTtime<=2.42; distanceL<=120.00;	9.05	0.98
7	sfDtime<=1.82; 1.57<sfDfres; rtTtime<=2.42; 120.00<distanceL;	5.44	0.89
8	sfDtime<=1.82; 1.57<sfDfres; 2.42<rtTtime;	9.24	0.74
9	1.82<sfDtime<=4.36; rtDtime<=1.62; rtosDR<=0.39; distanceL<=65.25;	6.46	0.70
10	1.82<sfDtime<=4.36; rtDtime<=1.62; rtosDR<=0.39; 65.25<distanceL;	3.02	0.99
11	4.36<sfDtime<=8.72; rtDtime<=1.62; rtosDR<=0.39; sfTtime<=15.34;	4.16	0.65
12	4.36<sfDtime<=8.72; rtDtime<=1.62; rtosDR<=0.39; 15.34<sfTtime;	2.26	0.89
13	1.82<sfDtime<=8.72; rtDtime<=1.62; 0.39<rtosDR;	2.28	0.83
14	8.72<sfDtime; rtDtime<=1.62;	0.86	0.64
15	1.82<sfDtime; 1.62<rtDtime;	6.09	0.95

Note: No. = number; SD = standard deviation.

Table 4. Importance of Variables in Two Models

CART_1		CART_2	
Variables	Score	Variables	Score
sfDtime	100.00	sfDtime	100.00
sfTtime	63.19	rtDtime	59.32
sfDfres	18.22	rtTtime	21.55
Floor	13.45	rtosDR	18.70
Type	8.21	distanceL	16.43
		sfDfres	11.21
		sfTtime	10.94
		distanceA	5.73

sfDfres are more important in CART_1 than in CART_2. It indicates that owners may evaluate their openness-to-share decision, based primarily on their own individual parking behavior, while paying less attention to self-use parking, once they have made their decisions to share.

Moreover, the influences of the self-use variables in the two models are not quite consistent. Theoretically, the increase of self-use time would lower owners' sharing willingness. This was proved true in CART_2 (see Node 1 and Node 12). However, in CART_1, the intervals of sfDtime were split several times and the open-to-share possibility was simply not consistent with sfDtime. It indicates that the decision-making process of whether to share or not is rather complex, and the owners' willingness to share most probably resulted from the available variety of self-use parking modes.

As for spots' physical characteristics, Floor and Type are testified important to CART_1 and the other two variables, distanceL and distanceA, are essential to CART_2 (see Table 4). The disparity indicates that owners most likely value the basic construction characteristics more when considering whether to share or not, whereas in sharing practices, they may determine that the location of the spots is more important to rental effects. Therefore, their sharing willingness might be influenced more significantly by the two distance variables.

The most profound difference is the participation of variables in the rental effects of the previous month group in CART_2. All three variables ranked from second to fourth on the importance of variables list, which reveals that owners' willingness to share may be highly encouraged by the rental effects of the previous month.

Conclusions

Shared parking of private residential parking spots is an innovative policy in China. Furthermore, it can substantially improve the utilization of parking resources and profoundly relieve the pressure on parking supply, especially during the peak periods of parking demand. Based on the real data of 1-year behavioral records of owners obtained from applications on smart phones, as well as field survey data, the study presented in this paper analyzed the influential factors in owners' sharing decision mechanism. Moreover, two CART models were developed to adequately predict the willingness of spot owners to share private parking. CART_1 answers whether owners would share their spots, whereas CART_2 explains the average time owners would be willing to share during the peak periods of parking demand.

The major conclusion is that owners' self-use behavior, spots' physical characteristics and rental effects of the previous month have all been found to have a significant influence on owners' willingness to share. More specifically, it was determined that the owners always put more emphasis on their own parking demands ahead of sharing decisions. Moreover, the self-use parking modes significantly affect the owners' start-to-share decision. Once shared, the owners' sharing willingness during the peak periods of parking demand would largely rely on the previous month's rental effects, if the self-use parking demand is satisfied. Furthermore, the influential factors of spots' physical characteristics vary according to the two models, thereby indicating that the consideration is dissimilar in the two stages.

It probably is the first time that a spot-level study of owners' sharing willingness driven by real-data and modeled by both CT and RT has been conducted. The main contribution of CART_1 is to facilitate the classification of owners into several groups, based on their self-use parking habits and spots' physical factors as well as to assess the openness-to-share possibility of each group. The main contribution of CART_2 is to adequately identify the important factors affecting owners' sharing willingness during the peak periods of parking demand and, furthermore, to predict the sharing time of each owner. A combination of the two models can effectively predict the total potential of shared parking in a residential zone, which would substantially benefit the government, as relating to their parking supply policies, and also be of advantage to third parties to enhance the distribution of parking resources.

The study has the following limitations. First, the owners' willingness to share may vary according to human factors, such as age and family composition, which this study did not expound on, because of confidentiality agreements aimed at protecting the owners' privacy. Future studies will optimize the data collection methodology and acquire more demographic information by means of door-to-door canvassing in other study regions. Second, the owners' willingness to share a parking spot was not correlated with those in other residential communities; therefore, in future studies, more communities with diverse geographic land use and demographic attributes should be researched and systematically compared and evaluated.

Acknowledgments

This research is supported by the National Natural Science Foundation of China (51478111, 51508094). The authors would like to express their appreciation to the Beijing Tongyudao Technology Co., Ltd. for their data support.

Author Contributions

Study conception and design: Jun Chen, Chu Zhang; data collection: Chu Zhang, Yuanyuan Wu; analysis and interpretation of results: Zhibin Li, Chu Zhang; draft manuscript preparation: Chu Zhang. All authors reviewed the results and approved the final version of the manuscript.

References

1. Weinberger, R., M. Kodransky, J. Karlin-Resnick, A. Gauthier, Z. Gyarmati, Y. Pen, and L. Fu. *Parking Guidebook for Chinese Cities*, Handbooks, 2013.
2. Tian, G., R. Ewing, R. Weinberger, K. Shively, P. Stinger, and S. Hamidi. Trip and Parking Generation at Transit-Oriented Developments: A Case Study of Redmond Tod, Seattle Region. *Transportation*, Vol. 44, 2017, pp. 1235–1254.
3. Chaniotakis, E, and A. J. Pel. Drivers' Parking Location Choice under Uncertain Parking Availability and Search Times: A Stated Preference Experiment. *Transportation Research Part A: Policy and Practice*, Vol. 82, 2015, pp. 228–239.
4. Ibeas, A., L. Dell'Olio, M. Bordagaray, and J. D. Ortúzar. Modelling Parking Choices Considering User Heterogeneity. *Transportation Research Part A: Policy and Practice*, Vol. 70, 2014, pp. 41–49.
5. Li, Z. B., W. Wang, C. Yang, and D. R. Ragland. Bicycle Commuting Market Analysis Using Attitudinal Market Segmentation Approach. *Transportation Research Part A: Policy and Practice*, Vol. 47, 2013, pp. 56–68.
6. Efthymiou, D., C. Antoniou, and P. Waddell. Factors Affecting the Adoption of Vehicle Sharing Systems by Young Drivers. *Transport Policy*, Vol. 29, 2013, pp. 64–73.
7. Shaaban, K., and A. Pande. Classification Tree Analysis of Factors Affecting Parking Choices in Qatar. *Case Studies on Transport Policy*, Vol. 4, No. 2, 2015, pp. 88–95.
8. Rogers, J., D. Emerine, P. Haas, D. Jackson, P. Kauffmann, R. Rybeck, and R. Westrom. Estimating Parking Utilization in Multifamily Residential Buildings in Washington, D.C. *Transportation Research Record: Journal of the Transportation Research Board*, 2016. 2568: 72–82.
9. Xu, S. X., M. Cheng, X. T. Kong, H. Yang H, and G. Q. Huang GQ. Private Parking Slot Sharing. *Transportation Research Part B: Methodological*, Vol. 93, 2016, pp. 596–617.

10. Shao, C., H. Yang, Y. Zhang, and J. Ke. A Simple Reservation and Allocation Model of Shared Parking Lots. *Transportation Research Part C: Emerging Technologies*, Vol. 71, 2016, pp. 303–312.
11. Smith, M. S. *Shared Parking*. 2nd ed. Urban Land Institute, Washington, D.C., 2005.
12. Meng, Q., and J. Weng. Classification and Regression Tree Approach for Predicting Drivers' Merging Behavior in Short-Term Work Zone Merging Areas. *Journal of Transportation Engineering*, Vol. 138, 2012, pp. 1062–1070.
13. Chang, L. Y., and H. W. Wang. Analysis of Traffic Injury Severity: An Application of Non-Parametric Classification Tree Techniques. *Accident Analysis and Prevention*, Vol. 38 2006, pp. 1019–1027.
14. Breiman, L., J. H. Friedman, R. A. Olshen, and C. J. Stone. *Classification and Regression Trees*. Wadsworth Brooks/Cole, Pacific Grove, CA, 1984.
15. Weng, J., S. Xue, and X. Yan. Modeling Vehicle Merging Behavior in Work Zone Merging Areas During the Merging Implementation Period. *IEEE Transactions on Intelligent Transportation Systems*, Vol. 17, 2016, pp. 917–925.
16. Kashani, A. T., and A. S. Mohaymany. Analysis of the Traffic Injury Severity on Two Lane, Two-Way Rural Roads Based on Classification Tree Models. *Safety Science*, Vol. 49, 2011, pp. 1314–1320.
17. Lewis, D. D., and J. Catlett. Heterogeneous Uncertainty Sampling for Supervised Learning. *Machine Learning Proceedings*, 1994, pp. 148–156.

The Standing Committee on Transportation Demand Management (ABE50) peer-reviewed this paper (18-02688).

Article

TRR
JOURNAL OF THE TRANSPORTATION RESEARCH BOARD

Transportation Research Record
2018, Vol. 2672(8) 942–952

DOI: 10.1177/0361198118772948
journals.sagepub.com/home/trr
SAGE

Market Segmentation Analysis of Commuter Parking Relative to Shared Private Residential Parking Spots

Chu Zhang[1], Jun Chen[1], Zhibin Li[1], and Liangpeng Gao[1]

Abstract

Using shared private residential spots for commuter parking is a relatively new parking method in China. It promotes the utilization of residential parking resources, while reducing parking supply pressure during the day. This study adopted the market segmentation approach for analyzing this new method, based on surveying parkers' socioeconomic and attitudinal factors. The Partial Least Squares Structural Equation Modeling (PLS-SEM) was employed to identify various influential factors. The results reveal that attitudes and socioeconomic heterogeneity influence parkers' willingness and attitudes, which play a more significant role in this process. Moreover, the K-means was used to segment the market. Finally, six segments with distinct characteristics were identified, based on a willingness to utilize private residential spots for commuter parking, sensitivity to parking search time, walking distances, the economy, environmental awareness and parking payment method. Suggestions for marketing promotion for parkers of each segment are also discussed. The findings in this paper will help clarify the mechanism behind commuter-parking location choices and benefit third parties by enabling them to establish precise marketing strategies.

Commuter parking accounts for a significant percentage of parking demand in many urban metropolitan areas (*1, 2*). Compared with non-commuter parking, commuter parking generally requires extended parking time during the day near work areas. It sometimes leads to insufficient parking resources for non-commuting vehicles, which can hinder the achievement of service targets, especially when there is no public parking resource available nearby. Since it is challenging to change travel modes, transferring the parking demand spatially to other available private parking spots nearby, often referred to as shared parking, may be an alternative.

Among all the various types of land use, parking spots in residential buildings exhibit the most potential to serve as an outlet for sharing, which can provide sufficient parking resources during the day (*3*). Considering the inconvenience for the residents and extra work for the property management staff, shared commuter parking may reduce disadvantages by offering a lower frequency of in-and-out travel. Thus, this study focuses on accommodating commuter parking needs through private residential parking spots.

Current policies and technology in China make it possible to share private residential parking spots with others. Moreover, this is a market of high potential given the significant number of parking spaces (*4*); many third parties have served as middlemen between the owners of the spaces and commuters (*5*). Unlike residential parking permits in western countries, in this system the third parties play a vital role in market positioning and resource allocation. To better promote the shared parking method and attract more parkers, extensive marketing studies into the characteristics of potential parkers are required, which help the third parties with their marketing strategies.

Traditionally, the marketing segmentation approach is used in travel behavior studies to divide potential users into several submarkets with distinctive characteristics. Previous studies determined the marketing segmentation approach as helpful in improving market share and in developing strategies that best serve the various submarkets (*6, 7*). However, there have been no marketing segmentation studies of the new shared parking method.

[1]School of Transportation, Southeast University, Nanjing, China

Corresponding Author:
Address correspondence to Jun Chen: chenjun@seu.edu.cn

These would be vital for promoting the market effectively.

The primary objective of this study is to segment the market based on potential parkers' willingness to adopt the shared private residential spots for commuter parking. More specifically, this study has the following aims: (1) to identify the influential factors influencing commuting parkers' willingness to use shared private residential spots; (2) to segment the parkers into several distinct categories based on the major influential factors; (3) to discuss the characteristics of each submarket and propose suggestions for promoting the use of the new shared parking. The conclusions of this study may help identify the parkers who could be persuaded to adopt the new method and those who might be enticed out of this method or approach.

The remainder of this paper is organized as follows. The following section reviews related existing work, while Section 3 introduces the methodology for the market segmentation analysis. Sections 4 and 5 provide the model estimation results pertaining to the identification of influential factors and market segmentation, respectively. Section 6 discusses the findings and strategy options. The paper ends with concluding remarks, while Section 7 outlines future research.

Literature Review

As the proposed method of private residential shared spots for commuter parking is quite new, there is so far no research analyzing the factors that influence commuters in their choice of parking location. While aiming to discover the influential factors, related studies into parking location choices are summarized.

Generally, the major influential factors are categorized as parker-related and parking lot-related (see Figure 1). Parker-related factors are the description of parkers' trip attributes and socioeconomic characteristics adopted in parking-related studies (*8, 9*). Parking lot-related factors are utilized to consider parkers' awareness of four characteristics of parking lots: location, facility, management, and status (*10, 11*). Each of these is associated with other factors, such as location with walking distances between spots and destinations and parking searching time. As revealed in previous studies, choices of parking location represent the combined results of both parker-related and parking lot-related factors that are beneficial to segment parkers (*12*).

For different trip purposes, the decision-making mechanisms can be quite different (*13*). Current research into commuter parking mostly focuses on the travel mode choice among single-driving, carpooling-driving and transit-riding (*14*). Research concerning parking location choices for commuter purposes are infrequent; therefore, most researchers have adopted the aforementioned factors without adequate consideration of the particularity or characteristics associated with commuter parking (*15*). For example, commuter parking involves extended parking duration which significantly lowers the turnover rate, thereby jeopardizing the traffic efficiency if people park curbside. Parkers' attitudes to this problem, based on their environmental awareness, may affect the choice of location for commuter parking. Sometimes attitudinal factors are more relevant and important than instrumental variables (*16, 17*). Thus, the influential

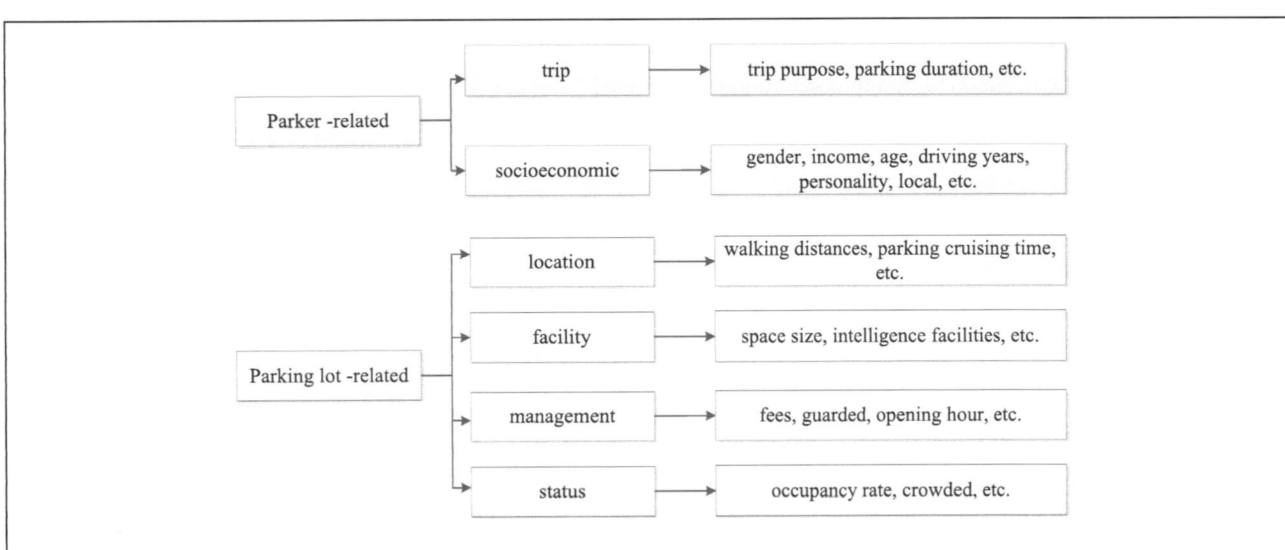

Figure 1. Factors influencing parking location choices.

factors of selecting shared private residential spots for commuter parking in this paper would cover the majority of the factors listed in Figure 1 and extended by other commuter parking-related factors.

Methodology

A comprehensive approach for the market segmentation analysis of private residential sharing spots for commuter parking is presented in this section. The data resource is first introduced, and then the main modeling methods including PLS-SEM and K-means clustering are briefly discussed.

Data Resource

The stated preference survey was conducted in 2017 in Xinjiekou, the city center of Nanjing, China. Nanjing is one of the largest cities in southeast China, as of 2016, with a population of 8.3 million and an area of 598 km^2. Thus, Xinjiekou is an example of a traditional city center with various types of land-use and one that is always short of adequate parking areas.

The respondents were carefully selected, according to several standards. Most importantly, all the respondents were car owners with experience of driving for commuting purposes. Secondly, the workplaces of the respondents were in the city center where the parking problem is quite severe; thus, respondents who were on business in Nanjing were eliminated. Moreover, various types of land-use were included to reduce the deviation caused by the respondents' working locations such as hospitals, schools, government or business buildings, hotels, shopping malls, and so forth. Considering the current commuter parking situation, both employer-paid parking (majority in governments, schools, etc.) and employee-paid parking (primarily business buildings, shopping malls, etc.) samples were collected. For those employer-paid parkers, free parking only occurs in designated places, mostly in their place of work; therefore, employees have to pay if they park elsewhere, such as when company designated parking spaces are occupied.

The survey was conducted at the same time as the promotion of an app offering a shared parking service. The respondents were informed about the advantages of private residential spots for commuter parking and received instructions for operating procedures on the app. More specifically, it was ensured that the parking fee of the proposed shared parking method would not exceed that of public parking spots. Furthermore, parkers could choose available parking spots and reserve them through smart phones, and move freely in or out of the residential buildings by displaying a parking proof provided by the app. All parking facilities would be well equipped and easily controlled by respondents, and the fee would be paid via the app, instead of in cash. Thus, all respondents were adequately familiarized with using private sharing spots for commuter parking before answering the questionnaire.

Based on previous research, the questionnaire was carefully designed to cover the aforementioned parker-related and parking lot-related variables and was extended, based on the research target of this paper. It included several parts: (1) respondents' general socioeconomic variables such as gender, age, education, income, and years of driving; (2) respondents' specified socioeconomic variables concerning the attributes of commuter parking, such as payment method and frequency of commuter parking; and (3) attitudinal questions adopted to measure the respondents' sensitivity to parking lot-related characteristics and their willingness to use private sharing spots for commuter parking.

The attitudinal questions were grouped into seven categories, each of which consisted of three or four statements. The respondents were asked to provide their level of agreement or disagreement for each statement by using a five-point scale, with "1" expressing strong disagreement and "5" expressing strong agreement.

Successive analysis was conducted on the 317 valid records remaining after various samples with missing key

Table 1. Statistics of Socioeconomic Variables

Socioeconomic variables	Description	Percentage
Gender		
Male	0	54%
Female	1	46%
Age		
Below 30	1	40%
30 to 49	2	43%
50 and higher	3	17%
Education		
Below high school	1	8%
High school	2	39%
College and higher	3	53%
Annual individual income (¥)		
Below 100 thousand	1	27%
100 to 199 thousand	2	59%
200 thousand and higher	3	14%
Driving years		
Below 3 years	1	30%
3 to 9 years	2	57%
10 years and higher	3	13%
Usual payment method		
By employer	0	38%
By employee	1	62%
Weekly frequency of commuter parking		
Below 3 times	1	23%
3 to 4 times	2	36%
5 times and higher	3	41%

Table 2. Descriptive Statistics of the Attitudinal Questions

Category	Question	Content	Mean	Std.
Sensitivity to economic conditions	Q1	I always choose the cheapest parking lot even if further away from my destination.	3.21	0.96
	Q2	If parking is free, I don't mind spending more time searching for a vacant spot.	3.64	0.87
	Q3	If the parking fee for commuting increases 10%, I will drive less.	3.44	1.37
Sensitivity to parking searching time	Q4	I prefer to reserve a parking spot before I have arrived if possible.	4.12	0.71
	Q5	I have to come early to find a spot for commuter parking.	2.66	1.42
	Q6	It usually costs me a lot of time finding a spot for commuter parking.	3.18	0.99
Sensitivity to walking distances	Q7	If I can't find an available spot nearby, I might park illegally.	2.17	1.01
	Q8	I can't tolerate a walking distance of more than 200 m even in good weather.	2.63	1.42
	Q9	I would drop driving for commuting if I had to walk more than 10 minutes after parking.	3.52	1.20
Desire for convenience	Q10	I think paying parking fees via smart phones saves a lot of trouble.	3.29	1.69
	Q11	I prefer to use a parking guidance system rather than asking directions.	3.15	1.03
	Q12	I don't like to ask for permission when entering and leaving a private parking space.	2.88	1.43
Environmental awareness	Q13	I think driving is less environmentally friendly than other modes.	3.25	0.87
	Q14	I think shared parking can make parking easier.	3.11	0.92
	Q15	I think shared parking would promote the utilization of parking resources and benefit society.	3.27	0.89
Desire for better parking environments	Q16	I prefer a less crowded parking lot.	3.69	0.90
	Q17	I prefer well-designed parking lots with bigger parking spaces.	3.72	1.15
	Q18	I prefer to use indoor navigation facilities when finding a spot.	3.54	1.29
	Q19	I prefer guarded and well-managed parking lots.	3.21	0.85
Willingness to utilize private sharing spots for commuter parking	Q20	I think using private sharing spots for commuter parking would promote my parking efficiency.	3.27	1.45
	Q21	I think commuter parking should not take up limited parking resources in work places.	2.89	1.52
	Q22	If private spots in residential buildings were available for commuter parking, I would try them.	3.51	0.82

information (such as socioeconomic variables) or exhibiting problems pertaining to logic in the attitudinal questions had been eliminated. The reliability test was performed using SPSS 21 with a Cronbach's alpha value of 0.749 (more than 0.70), proving good internal consistency. A summary of statistics from the valid samples is depicted in Table 1, while the questions and descriptive statistics are presented in Table 2.

Market Segmentation Approach

The market segmentation approach has been applied in other studies (*18, 19*). The major steps in this approach consist of PLS-SEM (to identify the factors influencing parkers' willingness to use private residential spots for commuter parking) and K-means (to segment the parkers based on significant influential factors gained from the last step).

PLS-SEM. Generally, SEM is a modeling protocol widely utilized in empirical research to investigate relationships among observed variables and unobserved latent variables. The objective for employing the SEM in this study is to estimate the coefficients of latent socioeconomic and attitudinal factors, as functions of the original description and statement variables, and identify the correlation between them.

SEM is formed with measurement models and a structural model. Measurement models depict the relationship between observed and latent variables, while the structural model depicts the relationship among latent variables. There are two types of measurement models, reflective and formative, both needed in this study. The reflective measurement models, denote the effects of an underlying latent construct, as the statements reflect parkers' latent attitudes (*20*). However, the formative measurement model measures the causes of the construct, rather than its effects (*21*). The collected socioeconomic variables reveal inconsistent correlations, despite capturing the same concept that fits this model. The expression methods (arrow direction) are also different for the two models (see in Figure 2).

PLS-SEM is one kind of SEM model which adopts the partial least squares (PLS) algorithm for parameter

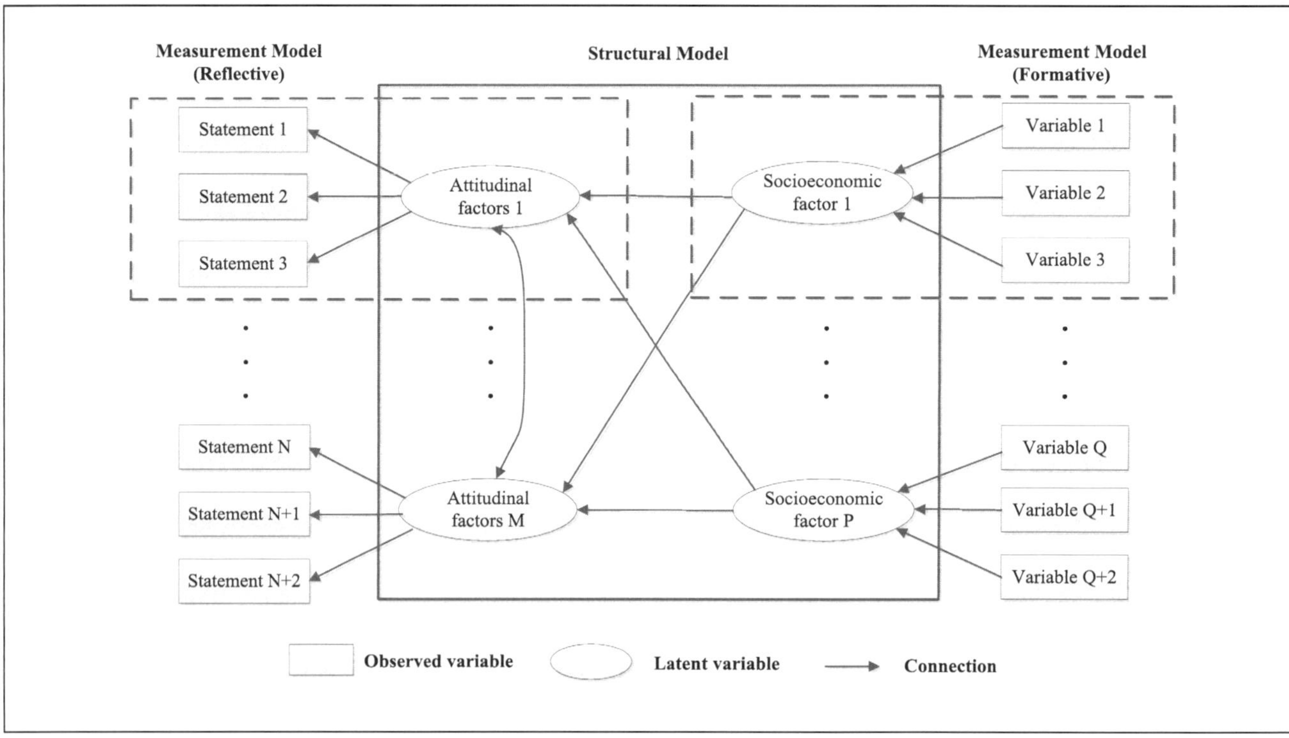

Figure 2. Illustrations of SEM models.

estimation. This model captures both the reflective measurement and formative measurement concurrently, which means parkers' socioeconomic and attitudinal characteristics can be computed and compared simultaneously.

K-means. K-means clustering is a statistical methodology that has been widely used to determine a set of subgroups by several pre-specified dividing factors (*22*). The goal of K-means clustering in this study is to identify specific groups of homogenous parkers by minimizing within-group distances, while maximizing between-group distances according to the selected factors. Moreover, based on a pre-specified number of clusters, the solution is obtained when all samples are allocated.

An optimal value of K is decided by the silhouette coefficient, which is the comprehensive judgment of clustering effects combined with cohesion and separation. Theoretically, the silhouette coefficient is a value between -1 and 1, with the larger number being preferred. However, in reality, the clusters' size could neither be too large nor too small. Too large a size usually means clusters are too heterogeneous and cannot be statistically analyzed further, while too small a size may indicate clusters fail to distinguish parkers with distinct attitudes and do not subdivide parkers well. Moreover, each cluster should be reasonably and clearly comprehended.

Identification of Influential Factors using Pls-Sem

The PLS-SEM structure is designed as follows. Attitudinal statements are categorized into seven groups as described in Table 2. The only dependent latent variable in this structure, the willingness to utilize private sharing spots for commuter parking, is one of the attitude latent variables and is assumed to be affected by other attitude latent variables. Moreover, parkers' socioeconomic latent variable, consisting of all observed socioeconomic factors, is assumed to influence all attitude latent variables (see in Figure 3).

In this model, the values of the average variance extracted (AVE) demonstrate the existence of a convergent validity by presenting values higher than the reference values of 0.5 (values of all latent variables ranging from 0.759 to 0.813). The results also prove the existence of a discriminant validity of the factors, evaluated by comparing the values of AVE with the highest square of the correlation among the factors ($r^2 = 0.603$), since all AVE values were higher than 0.603. From the values of the composite reliability (CR), we assessed the existence of a good internal consistency since the values in all cases were higher than 0.7 (values of all latent variables ranging from 0.925 to 0.963). The overall model demonstrates a good fit with NFI = 0.89.

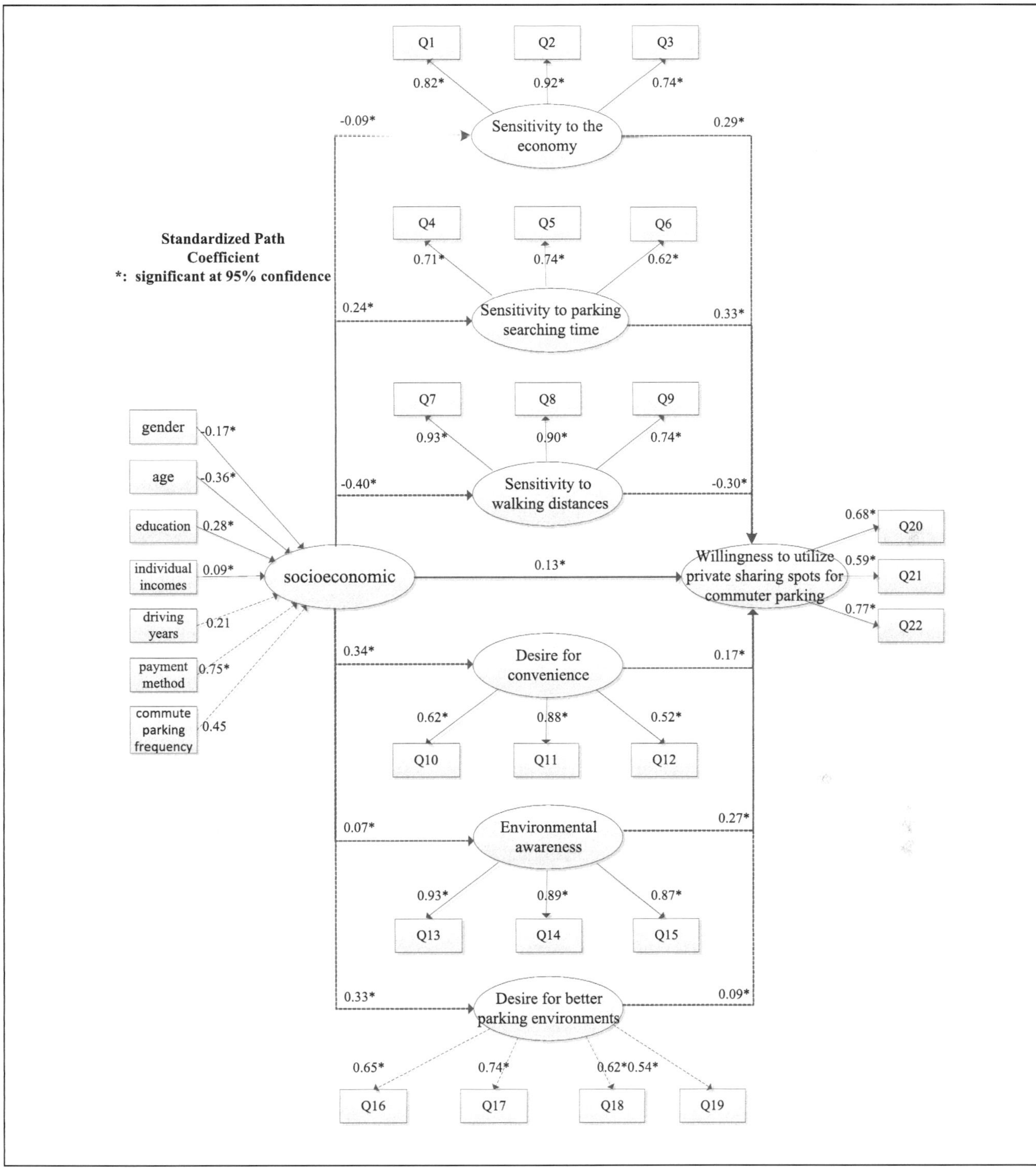

Figure 3. Estimation results of PLS-SEM model.

As illustrated in Figure 3, the numbers on the arrows represent parameter estimates of standardized loading factors. Apart from driving years and commuter parking frequency, other variables illustrate an effective connection with the target latent variable as the factor loadings are significant at a 95% confidence level. Their indirect and direct effects on the willingness to utilize private sharing spots for commuter parking are listed in Table 3.

Ordered by the total effects, parkers' sensitivity to parking searching time was revealed as the most

Table 3. Direct and Indirect Effects on the Target Latent Variable

Variables	Variable category	Direct effects	Indirect effects	Total effects
Sensitivity to parking searching time	Attitudes	0.33	0	0.33
Payment method	Socioeconomic	0	0.31	0.31
Sensitivity to walking distances	Attitudes	−0.30	0	−0.30
Sensitivity to the economy	Attitudes	0.29	0	0.29
Environmental awareness	Attitudes	0.27	0	0.27
Desire for convenience	Attitudes	0.17	0	0.17
Age	Socioeconomic	0	−0.15	−0.15
Education	Socioeconomic	0	0.11	0.11
Desire for a better parking environments	Attitudes	0.09	0	0.09
Gender	Socioeconomic	0	−0.07	−0.07
Individual incomes	Socioeconomic	0	0.04	0.04

significant factor influencing their willingness, with a total effect value of 0.33. Thus, parkers who worry about more parking searching time are more prone to select private residential spots for commuter parking, while those providing reservation services may encourage this new method. The second significant influential factor was the payment method, also the most important variable among the socioeconomic group, having a total effect of 0.31 on the target willingness variable. The positive value confirms employee-paid parking is a more popular type of shared parking than employer-paid parking. The third and fourth most significant variables reflect parkers' sensitivity to walking distances and economic conditions. These effects values indicate that private residential spots are more suitable for those concerned about frugality more than walking distances. The fifth variable was parkers' environmental awareness, which had a positive effect on willingness. The remaining variables, including parkers' attitude variables, for example, a desire for convenience and a better parking environment, as well as socioeconomic variables of age, education, gender and individual incomes, had effects of less than 0.20 on the target latent variables. Therefore, general conclusions can be inferred from the signs effecting values. Parkers would increase their willingness, based on parking convenience and a better parking environment, although these aspects are less attractive to parkers than other attitude latent variables. Besides, young, highly educated, male parkers with high individual incomes are more likely to utilize private residential spots for commuter parking, as they may have fewer security concerns than females and a higher acceptance for new technology and concepts.

Segmentation of the Market using K-Means

Preliminary experiments were conducted to determine the cluster factors used for the market segmentation. As too many factors would increase the complexity of clustering and lower the practicability, the selection of cluster factors was based on the major influential factors in the PLS-SEM model. The top five influential factors in Table 3 exhibit similar total effects and are obviously bigger than the others. Among them, the payment method is the only socioeconomic variable with a binary value, which would reduce the accuracy of the clustering results. Thus, the samples were divided into two groups based on the payment method. Samples in each group were clustered by the target variable and four attitude factors, including the willingness to use private sharing spots for commuter parking, sensitivity to parking searching time, sensitivity to walking distances, sensitivity to the economy, and environmental awareness. The average value of the statements in each attitude factor group was calculated for the subsequent clustering analysis.

The cluster number, mostly decided by the average silhouette coefficients, should also consider ease for analysis in the next step and ease of operation (number of clusters should not be too large). Thus, the average silhouette coefficients in response to the cluster numbers

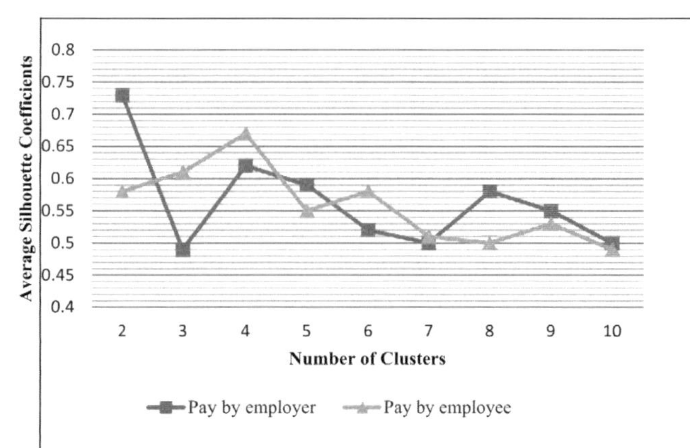

Figure 4. Determination of cluster numbers.

Table 4. Segmentation Result of Private Residential Spots for Commuter Parking

		Pay by employer		Pay by employee			
Attitudinal variables	Level	S1 (N = 47)	S2 (N = 74)	S3 (N = 19)	S4 (N = 79)	S5 (N = 65)	S6 (N = 33)
Willingness to use private sharing spots for commuter parking	High					√(3.95)	√(4.11)
	Moderate		√(2.74)		√(3.07)		
	Low	√(1.87)		√(1.61)			
Sensitivity to parking searching time	High			√(4.41)		√(3.87)	
	Moderate	√(2.82)			√(3.31)		√(3.07)
	Low		√(2.36)				
Sensitivity to walking distances	High			√(3.85)			
	Moderate	√(3.11)	√(2.76)				√(2.99)
	Low				√(1.94)	√(1.73)	
Sensitivity to economic situations	High		√(3.93)		√(3.71)	√(3.68)	
	Moderate	√(3.42)					√(2.89)
	Low			√(1.37)			
Environmental awareness	High						√(4.05)
	Moderate	√(3.39)	√(3.01)		√(3.34)	√(3.12)	
	Low			√(2.11)			

Note: √ indicates the interval the cluster center belongs to; values in parentheses indicate a cluster center in one dimension of the attitudinal variables; high level: >3.5; moderate level: 2.5–3.5; low level: <2.5.

ranging from 2 to 10 were calculated (see Figure 4). For payment by employer and by employee groups, the average silhouette coefficients reached the highest point when the cluster number was 2 and 4, respectively.

The segmentation results are listed in Table 4, while the spider charts which help to intuitively compare the six segments, are presented in Figure 5. The attributes of the cluster factors within each segment are summarized as follows:

(1) Segment 1 (S1) is a group, paid by the employer, with a low willingness. They have a moderate sensitivity to parking searching time, walking distances and economy, as well as environmental awareness.
(2) Segment 2 (S2) is a paid-by-employer group with a moderate willingness. They exhibit a low sensitivity to parking searching time, and a moderate sensitivity to walking distances and environmental awareness. They have a high sensitivity to the economy.
(3) Segment 3 (S3) is a paid-by-employee group with a low willingness. They have a high sensitivity to parking searching time and walking distances, but a low sensitivity to the economy as well as environmental awareness.
(4) Segment 4 (S4) represents a paid-by-employee group with a moderate willingness. They exhibit a moderate sensitivity to parking searching time as well as environmental awareness, but a low sensitivity to walking distances. However, they have a high sensitivity to economic frugality.
(5) Segment 5 (S5) constitutes a group, paid by employee, with a high willingness. They have a high sensitivity to parking searching time and the economy; however, they display a low sensitivity to walking distances. They have a moderate awareness of the environment.
(6) Segment 6 (S6) is a paid-by-employee group with a high willingness, and a moderate sensitivity to parking searching time, walking distances and the economy. They also exhibit a high level of environmental awareness.

Discussions on Marketing Promotion Suggestions

The analysis above reveals most of the socioeconomic variables, including gender, age, education, individual income, and payment method, as well as all the attitude variables toward parking searching time, walking distances, economy, convenience, parking environment, and environmental awareness are all significantly related to parkers' willingness to utilize private residential spots for commuter parking. Detailed suggestions based on the findings are discussed below.

Suggestions Based on Factors

For attitude variables, except for a sensitivity to walking distances, which is negatively related to the willingness to adopt the shared parking method, other attitudes toward willingness are all positively correlated. To increase the

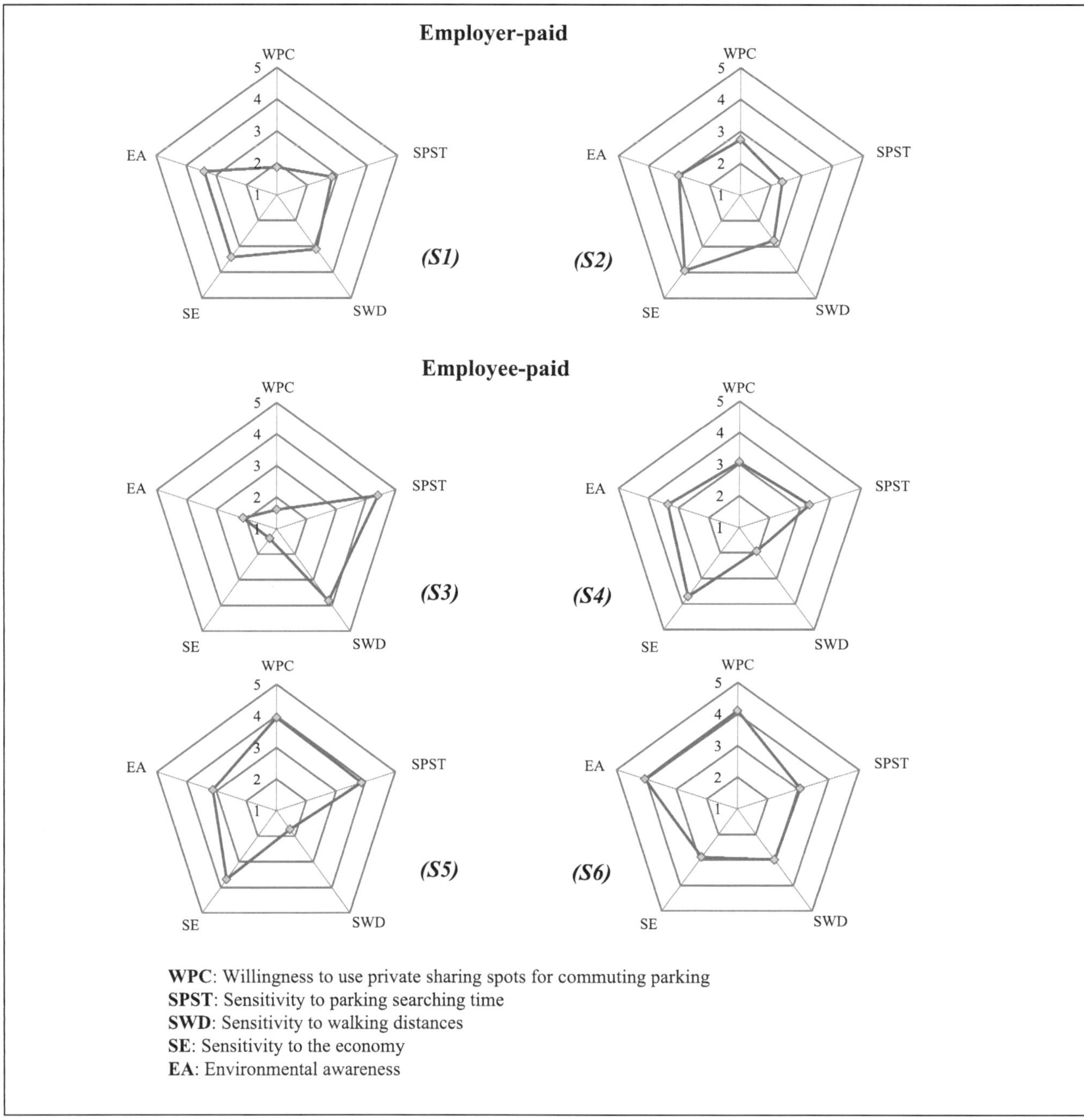

Figure 5. Spider charts for segmentation results.

attraction of this shared parking method, numerous measures might be used and it would, therefore, be beneficial to provide a better parking environment, simplify the process, and propagandize environmental protection.

Socioeconomic variables, even though they are easier to operate in practice, demonstrate a lower effect on willingness than attitude factors, which might be the results of parkers' heterogeneity. Among all the socioeconomic variables, payment method represents the only selected cluster variable of socioeconomic characteristics. Generally speaking, parkers paying for themselves have a higher willingness to adopt this method of shared parking than those used to free parking. Therefore, canceling free commuter parking is strongly preferred. Besides, different recommendation strategies would be preferred for parkers with different socioeconomic characteristics. For example, those with cheaper or closer spots should be prioritized for females or the elderly. Moreover, those further away or costlier, but with better facilities or management, should be prioritized for parkers with a higher education or income level.

Suggestions Based on Segments

The accuracy of market positioning can be enhanced through the comprehensive segmentation of parkers based on both socioeconomic and attitude variables. According to the segmentation results above, the characteristics of each segment are concluded and suggestions are proposed as follows:

Parkers in S1 and S3 both exhibit a low willingness to use private residential parking spots for commuter parking. Those in S3 would mostly remain unchanged, since their high sensitivity to walking distances exposes the weakness of this method, while their low sensitivity to the economy and low environmental awareness represent the advantages. However, those in S1 may have the opportunity to promote their willingness if the parking is no longer free.

Parkers in S2 and S4 both exhibit a moderate willingness to use private residential parking spots for commuter parking. Those in S2 have a higher willingness than those of the same payment method in S1, because their sensitivity to the economy is higher. Thus, the economic strategy along with cancelling the free parking would effectively promote the willingness of S2 parkers. Since those in S4 have a low sensitivity to walking distances and a high sensitivity to the economy, they greatly conform to the characteristics of this type of shared parking. In the results, the parkers of S4 would potentially have a higher willingness if encouraged by an optimization in the economy, parking searching time and environmental protection awareness aspects.

Parkers in S5 and S6 both exhibit a high willingness to use private residential parking spots for commuter parking. Those in S5 are quite similar to those in S4 with the only difference being the higher sensitivity to searching time. Therefore, a reservation service should be emphasized to retain their high willingness. Those in S6 do not have specific requests for walking distances, parking searching time or the economy, since they value environmental protection more. They would thus be firm supporters of this shared parking method.

Conclusions

Using private residential spots for commuter parking is a new shared parking method in China, which can promote the utilization of residential parking resources and reduce parking supply pressure during the day. This study adopted the market segmentation approach to analyze this new shared parking method based on surveying parkers' socioeconomic and attitudinal variables. The PLS-SEM was employed to identify the influential factors, while the K-means was utilized to segment the market. Suggestions for parkers of each segment were discussed.

The results demonstrate that both attitudes and socioeconomic heterogeneity influence parkers' willingness, while attitudes play a more important role in this process. Specifically, parkers' sensitivity to parking searching time, walking distances, the economy and their environmental awareness, along with their commuter parking payment method are relatively more significant and influential factors. Employee-paid parkers are more likely to adopt the new shared parking method than those employer-paid.

Parkers in S3 would be firm non-supporters, as opposed to parkers in S6. Parkers in S4 and S5 are quite similar; both fit the shared parking manner of thought based on their sensitivity to walking distances and the economy. Parkers in S1 and S2 would only change their parking places and accept the shared parking method if free commuter parking was no longer available.

The findings in this paper will clarify the mechanism behind location choices and benefit third parties to determine more accurate marketing strategies. The findings can also assist governments of other countries facing parking supply pressure to design shared parking policies.

Since the selection of socioeconomic and attitudinal factors utilized for market segmentation is predicated on the researchers' subjectivity, future studies could test other types of socioeconomic and attitudinal factors in the segmenting procedure to determine if more homogenous and reasonable market segments can be generated. Moreover, with the promotion of apps, public surveys based on real data in Nanjing will be available and, in conjunction with the stated surveys, promote the accuracy of market positioning. The authors recommend that future studies focus on these issues.

Acknowledgments

This research is supported by the National Natural Science Foundation of China (51478111, 51508094). The authors would like to express our appreciation to the senior students from the Transportation School of Southeast University and staff from Beijing Tongyudao Technology Co., Ltd. for their assistance in the survey, as well as the anonymous reviewers for donating their time and expertise in reviewing our paper and providing their professional and constructive commentary.

Author Contributions

Study conception and design: Jun Chen, Chu Zhang; data collection: Chu Zhang, Liangpeng Gao; analysis and interpretation of results: Zhibin Li, Chu Zhang; draft manuscript preparation: Chu Zhang. All authors reviewed the results and approved the final version of the manuscript.

References

1. Cervero, R. Mixed Land-Uses and Commuting: Evidence from the American Housing Survey. *Transportation Research, Part A*, Vol. 30, 1996, pp. 361–377.
2. Weinberger, R. R., M. Seaman, and C. Johnson. Residential Off-Street Parking Impacts on Car Ownership, Vehicle Miles Traveled, and Related Carbon Emissions: New York City Case Study. *Transportation Research Record: Journal of the Transportation Research Board*, 2009. 2118: 24–30.
3. Rogers, J., D. Emerine, P. Haas, D. Jackson, P. Kauffmann, R. Rybeck, and R. Westrom. Estimating Parking Utilization in Multifamily Residential Buildings in Washington, D.C. *Transportation Research Record: Journal of the Transportation Research Board*, 2016. 2568: 72–82.
4. Weinberger, R., M. Kodransky, J. Karlin-Resnick, A. Gauthier, Z. Gyarmati, Y. Pen, and L. Fu. *Parking Guidebook for Chinese Cities*. Handbooks, 2013.
5. Hamari, J., M. Sjöklint, and A. Ukkonen. The Sharing Economy: Why People Participate in Collaborative Consumption. *Journal of the Association for Information Science & Technology*, Vol. 67, 2016, pp. 2047–2059.
6. Elmore, Y. R. *TCRP Report 36: A Handbook: Using Market Segmentation to Increase Transit Ridership*. Transportation Research Board of the National Academies, Washington, D.C., 1998.
7. de Oña, J., R. de Oña, and G. López. Transit Service Quality Analysis Using Cluster Analysis and Decision Trees: A Step Forward to Personalized Marketing in Public Transportation. *Transportation*, Vol. 43, 2015, pp. 725–747.
8. Guo, L.Y., S. Huang, J. Zhuang, and A. Sadek. Modeling Parking Behavior under Uncertainty: A Static Game Theoretic Versus a Sequential Neo-Additive Capacity Modeling Approach. *Networks & Spatial Economics*, Vol. 13, 2013, pp. 327–350.
9. Ibeas, A., L. dell'Olio, M. Bordagaray, and J. de Dios Ortúzar. Modelling Parking Choices Considering User Heterogeneity. *Transportation Research Part a-Policy and Practice*, Vol. 70, 2014, pp. 41–49.
10. Chaniotakis, E, and A. J. Pel Drivers' Parking Location Choice under Uncertain Parking Availability and Search Times: A Stated Preference Experiment. *Transportation Research Part A-Policy and Practice*, Vol. 82, 2015, pp. 228–239.
11. Chen, Z., Y. Yin, F. He, and J. Lin. Parking Reservation for Managing Downtown Curbside Parking. *Transportation Research Record: Journal of the Transportation Research Board*, 2015. 2498: 12–18.
12. Shaaban, K, and A. Pande. Classification Tree Analysis of Factors Affecting Parking Choices in Qatar. *Case Studies on Transport Policy*, 2015. doi:10.1016/j.cstp.2015.11.002.
13. Marsden, G. The Evidence Base for Parking Policies—a Review. *Transport Policy*, Vol. 13, 2006, pp. 447–457.
14. Proulx, F. R., B. Cavagnolo, and M. Torres-Montoya. The Impact of Parking Pricing and Transit Fares on Mode Choice at the University of California, Berkeley. *Transportation Research Record: Journal of the Transportation Research Board*, 2014. 2469: 41–48.
15. Hess, D. Effect of Free Parking on Commuter Mode Choice: Evidence from Travel Diary Data. *Transportation Research Record: Journal of the Transportation Research Board*, 2001. 1753: 35–42.
16. Anable, J. 'Complacent Car Addicts' or 'Aspiring Environmentalists'? Identifying Travel Behaviour Segments Using Attitude Theory. *Transpot Policy*, Vol. 12, 2005, pp. 65–78.
17. Bahamonde-Birke, F. J, and T. Hanappi. The Potential of Electromobility in Austria: Evidence from Hybrid Choice Models under the Presence of Unreported Information. *Transportation Research Part A-Policy and Practice*, Vol. 83, 2016, pp. 30–41.
18. Li, Z., W. Wang, C. Yang, and D. R. Ragland. Bicycle Commuting Market Analysis Using Attitudinal Market Segmentation Approach. *Transportation Research Part a-Policy and Practice*, Vol. 47, 2013, pp. 56–68.
19. Badoe, D. A., and E. J. Miller. An Automated Segmentation Procedure for Studying Variations in Mode Choice Behavior. *Journal of Advanced Transportation*, Vol. 32, 1998, pp. 190–215.
20. Bollen, K. A., and Lennox, R. Conventional Wisdom on Measurement: A Structural Equation Perspective. *Psychological Bulletin*, Vol. 10, 1991, pp. 305–314.
21. Curtis, R., and E. Jackson. Multiple Indicators in Survey Research. *American Journal of Sociology*, Vol. 68, 1962, pp. 195–204.
22. Washington, S. P., M. G. Karlaftis, and F. L. Mannering. *Statistical and Econometric Methods for Transportation Data Analysis*, 2nd ed. CRC Press, Routledge, 2010

The Standing Committee on Transportation Demand Management (ABE50) peer-reviewed this paper (18-02700).

Sage Publications USA
sagepub.com

ISBN 978-1-5108-8708-4
9 781510 887084